USEFUL PHYSICAL QUANTITIES

Quantity	Nominal Value
Acceleration of gravity	9.8 m/s^2
Mass of earth	$5.98 \times 10^{24} \text{ kg}$
Mean radius of earth	$6.37 \times 10^6 \text{ m}$
Mass of moon	$7.35 \times 10^{22} \text{ kg}$
Mean earth–moon distance	$3.84 \times 10^8 \text{ m}$
Mass of sun	$1.99 \times 10^{30} \text{ kg}$
Mean earth–sun distance ($\equiv$1 AU)	$1.50 \times 10^{11} \text{ m}$
Sidereal day	23 h 56 min 4 s
Sidereal month	27.3 day
Speed of sound in air (20°C)	343 m/s
Latent heat of fusion of water	$3.33 \times 10^5 \text{ J/kg}$ ($\simeq$80 kcal/kg)
Latent heat of vaporization of water	$2.26 \times 10^6 \text{ J/kg}$ ($\simeq$540 kcal/kg)

Note: More precise values are given in Appendix 4.

THE GREEK ALPHABET

Letter	Symbol		Pronunciation	Letter	Symbol		Pronunciation
alpha	A	α	**al**-fa	nu	N	ν	nyoo or noo
beta	B	β	**bay**-ta	xi	Ξ	ξ	ksee
gamma	Γ	γ	**gam**-ma	omicron	O	o	**oh**-mi-cron
delta	Δ	δ	**del**-ta	pi	Π	π	pie
epsilon	E	ϵ	**ep**-si-lon	rho	P	ρ	roe
zeta	Z	ζ	**zay**-ta	sigma	Σ	σ	**sig**-ma
eta	H	η	**ay**-ta	tau	T	τ	**tah**-oo
theta	Θ	θ	**thay**-ta	upsilon	Υ	υ	**you**-psi-lon
iota	I	ι	eye-**o**-ta	phi	Φ	ϕ	phee or phy
kappa	K	κ	**kap**-pa	chi	X	χ	kye
lambda	Λ	λ	**lam**-da	psi	Ψ	ψ	psee or psy
mu	M	μ	myoo	omega	Ω	ω	oh-**may**-ga

Notes: (1) In the symbol column, the uppercase letter is given first, and then the lowercase letter. Among the lowercase letters, don't confuse ξ (xi) with ζ (zeta). (2) The pronunciations are those most common in American English.

PHYSICS

Physics in Harmony with Music
The richly varied sounds of the violin arise from the vibration of its strings. The study of vibrating strings by Pythagoras in the 4th century B.C.E. was among the very earliest quantitative analyses of physical systems. The superposition of the violin on a background of musical notation, the language in which the composer and the violinist communicate, is analogous to the intimate connection between physics and its language, mathematics. As a musical score supports the development and expression of beautiful and stimulating ideas by the violinist, so does mathematics support a like development of beautiful and stimulating ideas by the physicist. In their creative potential, music and physics are one.

PHYSICS

for Scientists and Engineers

Lawrence S. Lerner
California State University
Long Beach

Jones and Bartlett Publishers
Sudbury, Massachusetts

BOSTON ▪ LONDON ▪ SINGAPORE

Editorial, Sales, and Customer Service Offices
Jones and Bartlett Publishers
40 Tall Pine Drive
Sudbury, MA 01776
1-800-832-0034
508-443-5000
info@jbpub.com
http://www.jbpub.com

Jones and Bartlett Publishers International
Barb House, Barb Mews
London W6 7PA
UK

Library of Congress Cataloging-in-Publication Data

Lerner, Lawrence S., 1934–
    Physics for scientists and engineers / Lawrence S. Lerner.
      p.   cm.
    Includes index.
    ISBN 0-86720-479-6 (hard)
    1. Physics.   I. Title.
QC21.2.L47   1995
530—dc20   95-42899
                                         CIP

Photo and illustration credits appear on pages A27–A29, which constitute a continuation of the copyright page.

Printed in the United States of America
99 98 97   10 9 8 7 6 5 4 3 2

Chief Executive Officer: Clayton E. Jones
Chief Operating Officer: Donald W. Jones, Jr.
Vice President, Production and Manufacturing: Paula Carroll
Vice President and Editor-in-Chief: David P. Geggis
Director of Sales and Marketing: Rob McCarry
Marketing Manager: Anne T. King
Acquisitions Editor: David E. Phanco
Associate Editor: Laura Maier
Editorial Assistant: Deborah L. Haffner
Manuscript Editor: Patricia Zimmerman
Production Administrator: Mary Sanger
Senior Manufacturing Buyer: Dana L. Cerrito
Design/Layout: Deborah Schneck
Editorial Production Service: Lifland et al., Bookmakers
Illustrations: Network Graphics, JAK Graphics Ltd.,
    Tech-Graphics
Photo Research: Kristi Heffron, Mary Sanger
Cartoons: Sidney Harris
Typesetting/Separations: University Graphics, Inc.
Cover Design: Mimi Ahmed
Printing and Binding: R. R. Donnelley & Sons Company
Cover Printing: Henry N. Sawyer Co., Inc.
Cover Photograph: © Comstock Inc., 1995.

Physics for Scientists and Engineers	ISBN 0-86720-479-6
Modern Physics for Scientists and Engineers	ISBN 0-86720-487-7
Physics for Scientists and Engineers, Volume 1	ISBN 0-86720-491-5
Physics for Scientists and Engineers, Volume 2	ISBN 0-7637-0460-1

To the memory of Isidor Lerner

physicist, teacher, father

About the Author

Lawrence S. Lerner has been a member of the Department of Physics and Astronomy at California State University, Long Beach since 1969. There he has taught physics, history of science, and many interdisciplinary courses. Earlier, he held research positions at Hughes Aircraft, Hewlett-Packard, and Lockheed Aircraft Research Laboratories, working mainly in condensed-matter physics. A native of New York City, he attended Stuyvesant High School. After the eleventh grade, he entered the University of Chicago, where he earned his bachelor's, master's, and doctoral degrees.

Lerner has been honored with his university's Distinguished Teaching Award and three other awards of merit. He has been active for many years in efforts to encourage more young women to enter the sciences and engineering. He has served as Director of Cal State's General Honors Program and as Founding President of the campus Phi Beta Kappa chapter and is a member of the National Faculty for the Humanities, Arts, and Sciences.

Lerner's current research centers on history of science and science education. He is the author of more than a hundred scholarly papers, two textbooks, and a translation of a work by Giordano Bruno. Lerner and his wife, Dr. Narcinda R. Lerner, a research chemist, live in Woodside, California with their two Newfoundland dogs, and they attend the opera whenever they can.

Brief Contents

Contents

Preface   xxi

PART 1   Mechanics of Particles in One Dimension   16

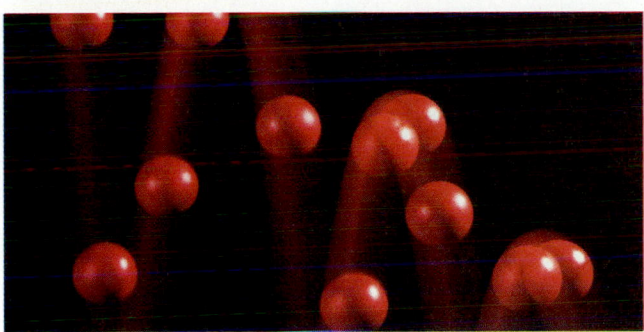

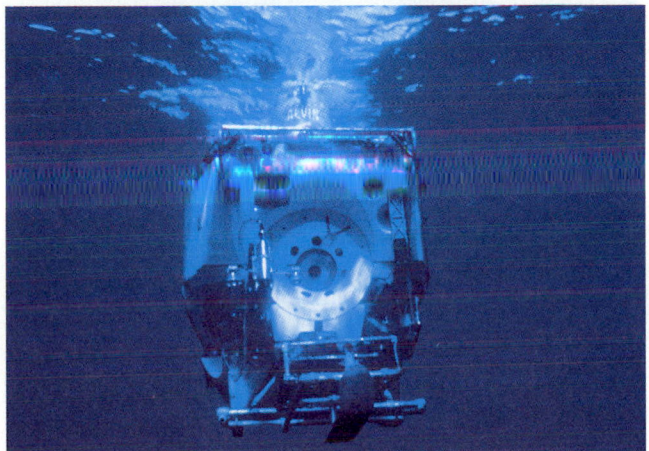

17 Temperature and Heat Flow 445

18 Kinetic Theory of Heat 477

19 Thermodynamics I: The First Law 507

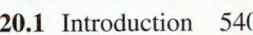
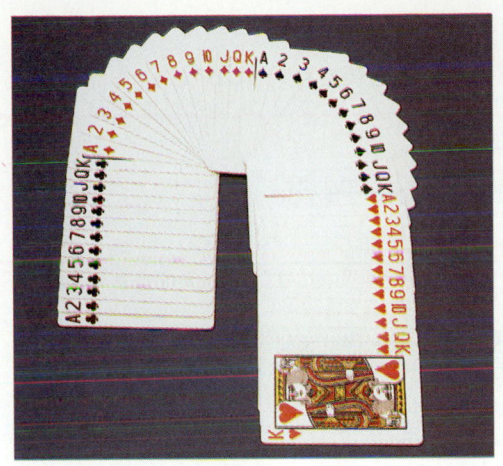

PART VII   The Mechanics of Wave Motion   564

27 Electromotive Force and Electric Current 725

28 Magnetism I: The Magnetic Force 761

29 Magnetism II: Magnetic Field of an Electric Current 793

30 Faraday's Law 825

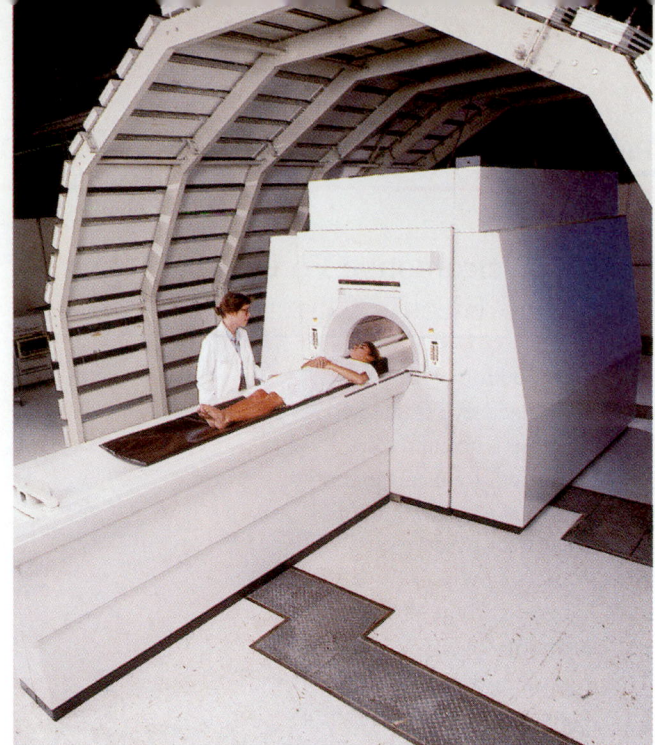

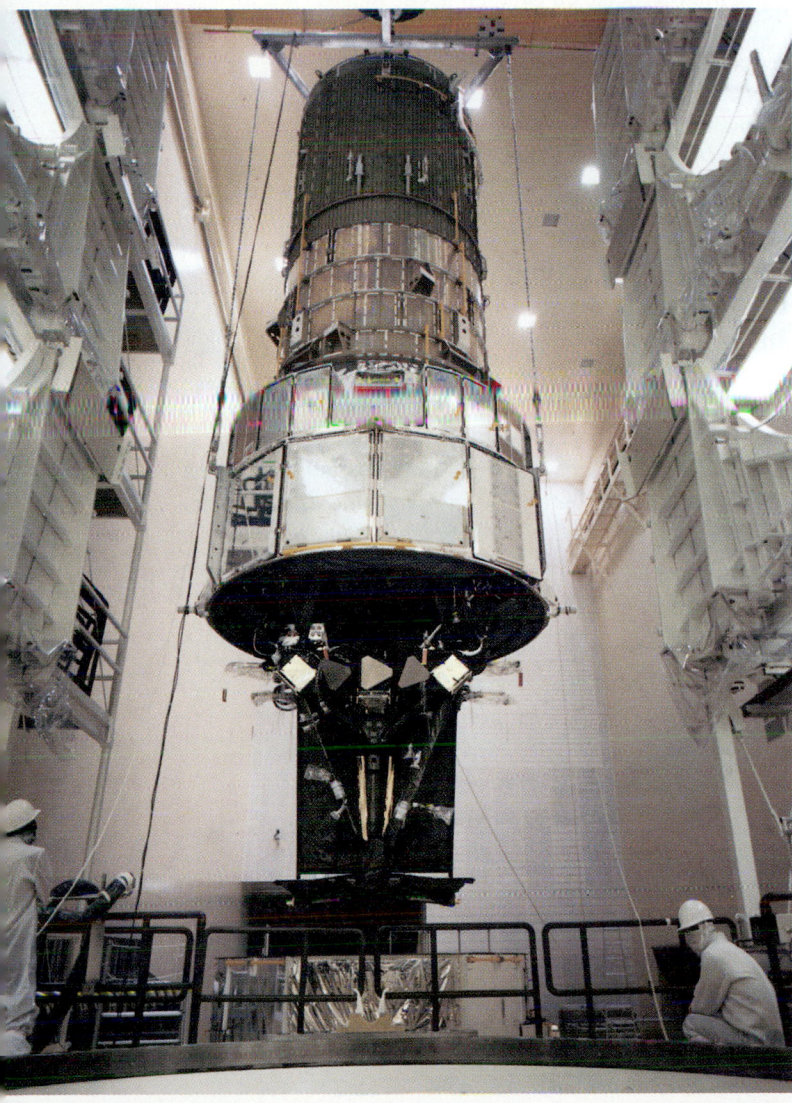

38 Relativistic Kinematics 1043

39 Relativistic Dynamics 1069

Publisher's Note: To accommodate the wide variation in the placement of relativity in physics courses, Chapters 38 and 39 on Relativistic Kinematics and Relativistic Dynamics are included in both *Physics for Scientists and Engineers* and *Modern Physics for Scientists and Engineers.* The following chapters comprise the content of *Modern Physics for Scientists and Engineers:*

Chapter 38 Relativistic Kinematics
Chapter 39 Relativistic Dynamics
Chapter 40 The Quantum of Light
Chapter 41 The Quantized Atom
Chapter 42 Wavelike Properties of Matter
Chapter 43 Quantum Systems I: Nuclei
Chapter 44 Quantum Systems II:
Many-Atom Systems
Chapter 45 Ultimate Matter?

Preface

Physics for Scientists and Engineers is intended to serve the student who is taking the standard two- or three-semester (or three- to five-quarter) introductory course for science and engineering majors who have completed (or are concurrently taking) an introductory course in calculus. This text is also available in a combined package with *Modern Physics for Scientists and Engineers* for those students who require a solid grounding in the areas beyond classical physics.

Physics is beautiful, exciting, satisfying, and very useful. Every physicist knows these things and feels them deeply; the question is how best to convey the passion and the excitement—as well as a new way of thinking, a complex logical structure, and quite a lot of information—to the wide spectrum of students whose professional goals presuppose a good grounding in physics.

Everyone who has taught calculus-level physics has considered this challenge. Having myself been committed to the profession of physics and the art of physics teaching for quite a long time, I considered how my experience and successes could be amplified beyond the relatively narrow bounds of my own classroom and office. This book is the result, and I present it in the hope that it conveys some of the delight I had in writing it.

What do students need to help them learn physics most effectively and to a depth that will sustain their further studies? Of the many such needs, the one that I have placed at the forefront is *friendly, painstaking, careful explanation.* I decided at the outset that the terse, formal style of professional scientific writing was not appropriate. The rigorous arguments of good science must be couched in terms appropriate and comfortable to beginning students, most of whom do not plan to be professional physicists and many of whom approach the course with some trepidation. Such students are not engaged by the abstract formalism of traditional style and, as a result, find studying traditional texts a discouraging experience. A literate but colloquial, friendly style is a better alternative.

The second need of students is epitomized in the maxim "practice makes perfect." Indeed, most successful students find that three-quarters or more of their physics study time is spent in working problems, either individually or in groups. Good problems provide both the opportunity and the motivation for practice. But, though problem solving is a vital part of the learning process, it is far too easy for students to overlook the forest for the trees. Too much emphasis on developing problem-solving skills can easily overwhelm the real goal, which is to build a solid and growing conceptual understanding and physical intuition.

Third, I am keenly aware of the amount of time the student spends studying alone. Moreover, many students are reluctant to ask questions in class or even in recitation sections, and the textbook becomes a resource more essential than it ought ideally to be. The textbook writer must "be with the student," writing in such a way that students can read and understand largely on their own.

End-of-Chapter Problems

Given the overall goals I had set for this book, I did not follow the easy path of building the end-of-chapter problem sets by mining the extensive resources of existing textbooks. Each chapter contains an extensive selection of problems, all of which—even the simple "confidence builder" ones—should require the student to think before reaching for the nearest equation. I do not presume to claim that my problems are completely original, but they do offer a fresh alternative to those found in existing texts. I believe that they effectively serve their role, which is to achieve the pedagogical goals stated earlier and thus to help the student achieve his or her own goal—that is, to learn physics well.

A few comments on how the problem sets are organized:

- There are more than 500 Queries, qualitative or semiquantitative in nature, which serve to enhance the student's awareness of his or her gain in understanding while challenging the student to attain still higher levels.

- There are more than 2500 quantitative Problems arranged in three groups. Group A comprises single-step problems designed to help build the student's confidence. The middle-level Group B problems require and help to build a broader understanding of the subject matter. Group C problems are more challenging.

- Series of two or more problems distributed throughout the three groups attack an interesting subject at increasing levels of sophistication. These problems are identified by a common title followed by a Roman numeral.

- Most Queries and Problems are keyed to the relevant text section by an Arabic number given parenthetically immediately after the title. More general exercises are denoted by (*G*).

Because problems are the tool by which students reinforce their skills, a *Student's Study Guide and Solutions Manual* with a problem-solving emphasis has been developed to accompany the text. Full solutions with detailed explanations are provided for 25% of the end-of-chapter problems, and Hints and Strategies are offered for solving both Queries and Problems.

Applications

Most students of university physics are not going to be physicists but plan to follow careers in engineering, chemistry, geology, and other technical or scientific fields. These students have a right to a clear explanation of why they are required to take physics. Discussions of various applications in the text are helpful, but conveying a sense of reality demands student participation. Hence, many exercises are aimed at just such participation; they deal with real (rather than artificial) applications of physical principles to many fields. A few examples:

- How did Chicago politics collide with the dream of a really high-speed link to the airport? See Problem 2.70.
- How can we infer the density of the earth's core? See Problem 11.65.
- What did Poiseuille, a physiologist, infer about blood flow in capillaries? See Problem 16.33.
- How does the output of a gasoline engine depend on the compression ratio? See Problem 19.58.
- How do electromagnetic brakes work and what are their advantages? See Query 30.5.
- Can on-site solar power meet the electrical needs of the average family? See Problem 34.42.
- Do color TV sets really generate dangerous X rays? See Problem 41.15.

Mathematical Language

Most students who enroll in the first semester or quarter of the physics course have already taken some calculus; some are taking their first calculus course concurrently. But even those students who have had some background in differentiation and integration often find their first attempts to apply their knowledge to physics problems a bewildering experience. Students have great trouble appreciating new physical concepts if they are struggling with the infrastructure of mathematical technique. With this in mind, I have tried to introduce the calculus as gradually as possible and have provided a brief review or introduction to differentiation and integration when the operations are first used in Chapters 2 and 8, respectively.

Thinking Like a Physicist

Because they spend so much time working problems and preparing for exams that require solving similar problems, students can easily fall into the serious misconception already noted in passing—that physics is nothing more than the set of techniques required to do the end-of-chapter problems. Although deep insight into the nature of physics takes a long time to develop, both the classroom instructor and the textbook author can help the student to step back once in a while and see what he or she has been doing. The "Thinking Like a Physicist" pieces distributed throughout the text are one mechanism for doing this. Some of the pieces are simple; others are intended as launching points for class discussion.

Physics as a Human Endeavor

Finally, physicists are people, and a judicious amount of discussion of some of them is appropriate: Rumford, the social reformer who was contemptuous of the poor people whose lot he improved; Young, the physician who made monumental contributions to Egyptology, physiology, and many fields of physics but could never develop a satisfactory bedside manner; Morley, who wanted to be a minister but became a chemist because he was such a bad preacher; Planck, who agonized for years over the quantum hypothesis we take for granted today; Blodgett, who blazed a trail for other women in industrial physics and gave us the anti-reflection coating; Curie, who received an honor she richly deserved only in 1995, six decades after her death. I have tried to give students a glimpse at those mentioned here, as well as many others, in the hope that the students will look further into, and take part in, the very human joys and sorrows of physics.

For, to return to my initial assertion, physics is stimulating, challenging, beautiful, and pleasurable, as well as broadly useful and professionally necessary. Learning this is up to the student, but the instructor and the textbook author can help immeasurably by making the most of the wondrous opportunities to combine the science of physics with the art of teaching. I offer this book in that spirit.

Supplements to the Text

Instructor's Solutions Manual, compiled by Dean Lee of Harvard University and a team of graduate students from Harvard, MIT, Boston College, and Boston University, contains complete solutions to all end-of-chapter problems. All solutions have been thoroughly checked by the author and a group of additional reviewers to ensure accuracy and consistency with text notation. The *Instructor's Solutions Manual* is also available on disk.

Student's Study Guide and Solutions Manual, prepared by Paul French of the State University of New York at Oneonta, summarizes important physics concepts, addresses common student misconceptions, identifies strategies for problem solving, and includes numerous worked examples as well as detailed step-by-step solutions to approximately half of the odd-numbered end-of-chapter problems from the text.

Overhead Transparencies and a *Test Bank* are available to qualified adopters. The set of full-color transparencies consists of key illustrations from the text. The complete test bank includes fill-in-the-blank, true/false, matching, multiple choice, and conceptual essay questions.

Experimental Research Notebook, made of recycled carbonless paper, is ideal for compiling laboratory data; students can submit the original and keep a copy. The labs use a quadrille graph design, and the three-hole punched pages are numbered consecutively, making the manual easy to use. The notebook is available in 100 duplicate sets of labs, to meet individual course needs.

Student Package Flexibility

Lerner's *Physics for Scientists and Engineers* is available in several student package selections. To allow for complete course flexibility, the *Standard Edition* is hardbound and includes thirty-seven chapters of classical physics and two separate chapters on relativity for a total of thirty-nine chapters. Also available is a paperbound *Volume 1* containing chapters 1–22 and a paperbound *Volume 2* containing chapters 23–45. *Modern Physics* is a paperbound volume that includes chapters 38 and 39 on relativity and chapters 40–45 with material on modern physics.

Acknowledgments

I take seriously a maxim that I often quote to my students: Anyone can write; the real task of the serious writer is in the rewriting. Many skilled physicists with long teaching experience played an essential role in the transformation of a first-draft manuscript into this book through careful and creative reviews of all or part of the work in various stages of its development. Their comments have contributed much to the clarity and accuracy of the text, and I thank them sincerely:

Donald Abernathy, *DeVry Institute of Technology*
Harold D. Bale, professor emeritus, *University of North Dakota*
John Paul Barach, *Vanderbilt University*
Robert Bauman, professor emeritus, *University of Alabama*
Colston Chandler, *University of New Mexico*
Roger W. Clapp, *University of South Florida*
Sumner P. Davis, professor emeritus, *University of California, Berkeley*
Mildred S. Dresselhaus, *Massachusetts Institute of Technology*
Lowell Eliason, *California State University, Long Beach*
Raymond Enzweiler, *Northern Kentucky University*
Donald R. Franceschetti, *Memphis State University*
Anthony P. French, *Massachusetts Institute of Technology*
Paul French, *State University of New York at Oneonta*

Edward F. Gibson, *California State University, Sacramento*
Thomas P. Greenslade, *Kenyon College*
Patrick Hammill, *San Jose State University*
Edwin Kashy, *Michigan State University*
Isidor Lerner (deceased), *University of Illinois, Chicago Circle*
Robert H. Lieberman, *Cornell University*
Ralph Llewellyn, *Central Florida University*
Bo Lou, *Ferris State University*
Oscar Lumpkin, *University of California, San Diego*
David Markowitz, *University of Connecticut*
Charles E. McFarland, *University of Missouri, Rolla*
Jack Munsee, *California State University, Long Beach*
Austin Napier, *Tufts University*
Lawrence Pinsky, *University of Houston*
Wendell H. Potter, *University of California, Davis*
C. W. Price, *Millersville University*
John P. Ralston, *University of Kansas*
Sema'an I. Salem, *California State University, Long Beach*
Scott Shepard, *Baylor University*
Christopher M. Sorensen, *Kansas State University*
Julien C. Sprott, *University of Wisconsin–Madison*
Edwin F. Taylor, *Massachusetts Institute of Technology*
George Williams, *University of Utah*
Arthur M. Yelon, *Ecole Polytechnique, Montréal*

Professors Eliason, Munsee, and Salem were particularly helpful in testing parts of the manuscript in their classes.

Narcinda R. Lerner patiently and expertly turned messy mathematical functions into neat computer-generated graphs throughout the text.

A special note of thanks to Anthony P. French of The Massachusetts Institute of Technology, who offered encouragement as well as many constructive comments in the course of his review of the manuscript and who later scrutinized the entire text for accuracy, reading every page of proof. He truly merits the epithet "lynx eyed" that Galileo took so to heart.

Grateful as I am for all of these contributions to accuracy and clarity, I wish to emphasize that any remaining errors or obscurities are solely my responsibility. If you find errors, please let me know; they will be corrected. I warmly welcome any responses to the text from instructors and students alike, as well as any suggestions for future editions.

My very special thanks are due to Sidney Harris, who contributed cartoons that offer his unique combination of a whimsical world view and insight into the nature of the physical principles that his cartoons highlight.

I also thank Dean Lee of Harvard University for the careful preparation of solutions and answers to problems. He and a team of graduate students and faculty meticulously worked all of the problems in the text.

Writing a book begins as a solitary activity but expands rapidly into a team effort as the book approaches production and publication. It is impossible to overestimate the dedication, talents, and effort applied to this work by the experts who turn a draft manuscript and scribbly sketches into a finished book. Among the many members of this team, I am particularly grateful to those who interacted with me most directly and intensely: William Bennetta, Paula Carroll, Dave Geggis, Deborah Haffner, Kristi Heffron, Anne King, Sally Lifland, Gail Magin, Laura Maier, Irene Nunes, David Phanco, Mary Sanger, Deborah Schneck, Madge Schworer, and Patricia Zimmerman. Finally, I owe much to the physics publishing experience of Don Jones, Sr., and to Art Bartlett and Paul Prindle for their faith in this work.

PHYSICS

Introduction

You can't tell the players without a program!

—BASEBALL PROGRAM HAWKER

What Is Physics?

What do you see? That depends not only on what you are looking *at*, but also on what you are looking *for*. If you adopt the most fundamental way of looking at things, you see matter interacting with other matter wherever you turn your attention. Billiard balls collide with other billiard balls, electrons with atoms, galaxies with galaxies. Turbines spin, electric light filaments glow, television sets receive signals from distant transmitters, spacecraft follow specific orbits, and your heart pumps blood through your body. As you explore more and more widely over the great theater of the universe, you find variety that seems to have no limit. But, beginning about four centuries ago, investigators using the methods of **physics** have achieved a long series of successes—many of them spectacular—in interpreting an ever-increasing wealth of observed phenomena in terms of a tightly organized, relatively small set of basic principles. These principles are inferred more or less directly from the phenomena themselves. This two-way process—interpretation in terms of principles and inference from phenomena—is a hallmark of physics. It has been widely applied to other sciences as well.

Skillfully applying this two-way process of observing and interpreting the world brings plenty of satisfaction, but physics also has great practical utility. Beginning with physical principles, we are able to predict what *will* happen under certain controlled circumstances—often with great accuracy. This is one of the reasons why physics is invaluable to the other sciences and to engineering.

Left: The subject of physics is the entire universe, of which this photograph shows but a small part.

Measurement and the International System of Units

Like all the sciences, physics is ultimately based on observation. Yet physics entails a lot more than observation. Much thinking is needed to figure out which observations, of the many one might make, are relevant to the matter at hand. Still more thinking is required to figure out how to assemble the relevant observations into a coherent picture by constructing a logical framework. Such a framework, called a **theory**, is intended to make sense of a range of phenomena. Theory construction demands a large share of the physicist's time and effort, but it enables the physicist to account for past observations and to decide how new ones should be made.

Nearly all physical observations are quantitative; they require *measurement*. Many more kinds of measurements are required in science and engineering than in everyday life. Every measurement is a comparison of a quantity with a **standard quantity**—that is, an agreed-upon quantity of the same kind. To measure a length, for example, you adopt as your standard a convenient measuring rod, whose length you use as the **unit** of length. You count the number of times that the rod fits into the length to be measured. This number gives the length in terms of the chosen unit.

The choice of the standard is arbitrary. However, several criteria must be met if a standard is to be as useful as possible.

1. **Stability**: There should be good reason to believe that the standard does not vary with time. If this criterion is satisfied, measurements made at different times, using the same standard, can be meaningfully compared.

2. **Reproducibility**: The standard should be accurately reproducible so that copies, ideally identical with the standard itself, can be used elsewhere. If this criterion is satisfied, measurements made at different places can be compared.

3. **Acceptability**: The standard should be universally accepted so as to eliminate clumsy and possibly inaccurate comparisons among measurements made with separate standards.

4. **Accessibility**: As nearly as possible, the standard should be readily accessible to everyone who needs to use it.

5. **Precision**: It should be possible to measure the standard itself with a precision at least as great as the precision with which any comparable measurement can be made. For example, if your standard of length were the distance between two blurry marks on a cheap plastic ruler, you could not use it to calibrate a micrometer that can distinguish between two blocks whose lengths are 5.0002 cm and 5.0003 cm. Similarly, it would make little sense to use a cheap clock for comparing the duration of one day (noon to noon) with the duration of the next day.

6. **Security**: The standard should be as safe as possible from—and preferably immune from—all possible causes of damage. (We will soon see how modern standards meet this criterion.)

When a standard meets these criteria as nearly as possible, it can be taken by mutual agreement as the **primary standard**. The primary standard can then be used to produce **secondary standards** calibrated in terms of the primary, and so on. A hierarchy is thus established. Secondary and lower-level standards may be constructed quite differently from the primary standard. Your ruler, for example, differs greatly from the primary standard of length, which we will consider shortly. Although much less accurate, the ruler costs much less and is much easier to use for everyday purposes.

Different secondary standards are used to measure quantities of the same kind but of substantially different magnitude. For example, a ruler is suitable for measuring the length of this page, but other means must be used to measure the diameters of atoms or the distances to neighboring galaxies. But however they may differ technically, all length-measuring devices are related by a chain of comparisons with the primary standard (at least in principle).

Development of a primary standard, together with the measuring equipment needed to make periodic comparisons between it and secondary standards, is expensive and

This surveying team measures distances with a laser rangefinder. The time required for a light pulse to travel to a reflector and back is a precise measure of the rangefinder–reflector distance.

time consuming. Fortunately, we don't need a primary standard for every quantity we want to measure. We can often define the necessary unit of measurement in terms of some other unit or units for which a primary standard exists. For instance, we often need to measure volume, such as when we measure sugar into coffee by teaspoons. But we do not need a primary standard of volume. Instead, the unit of volume is *defined* in terms of the standard of length. How would you do this?

All the units required for measurement can be defined in terms of a small number of **base units**. In current practice, the number of base units is seven. They are tabulated on the inside front cover of this book. Two of them, the units of length and time, are discussed in this chapter.

A set of such base units, together with the rules required to express all other units in terms of them, constitutes a *system* of units. The system in general use throughout the world is called (not surprisingly) the **International System of Units**. The short form **SI** (from the French *Système International*) is used more commonly than the full name. SI is not only brief but, unlike the full name, is the same in all languages.

The formal foundations of SI were laid in 1875, when seventeen nations signed a treaty called the Convention of the Meter. Under the treaty's provisions, the system of measurement and the standards became the concern of the International Bureau of Weights and Measures, located on international territory in Sèvres, a suburb of Paris. The Bureau is responsible to the General Conference of Weights and Measures, a body of experts from many nations that meets periodically to consider policy matters. International law authorizes the General Conference to modify the system whenever necessary or desirable. SI represents the current level of refinement and standardization of the **metric system**, a system of measurement that originated some 200 years ago. With a few exceptions, SI units are used throughout this book.

The metric system (which simply means "system of measurement") arose from a study undertaken by a committee of experts in France at the time of the French Revolution. Their aim was to devise a system of units that was logical, self-consistent, easy to use in calculations, and based on the best possible set of primary standards. After a period of introduction and modification that coincided with the chaos and confusion of the French Revolution, the system was made compulsory in France in 1799. As Napoleon conquered Europe, he brought the metric system with him, and it replaced confusing sets of local units. The rationality and simplicity of the metric system enabled it to prevail despite early public resistance. Scientists now use it exclusively, and it is the everyday system used in almost the entire world. The major exceptions are Great Britain and Canada, where the transition to everyday use of SI is well under way, and the United States, where the transition is much more sluggish. In the United States, engineers still often use the so-called U.S. Customary system of units.

> **U.S. Customary System**
> This system is also called the English system, but that name is neither descriptive nor satisfactory. There are in fact several "English" systems, in which the same word is used for different units.

The Unit of Time

We measure a time interval by comparing it with a unit of the same kind—a unit of time. The unit of time must be defined in terms of some physical system that behaves in a repetitive way. If we have reason to believe that the repetition is regular (we will shortly see exactly what this means), we can use the time interval between repetitions, called the **period**, to define the time unit. Many such repetitive phenomena have been (and are) used to measure time with greater or lesser convenience and accuracy. Among such phenomena are the coming of springtime, the new moon, the annual flooding of the Nile, the pulse beat, the swing of a pendulum through its low point, and the passage of the sun through its daily high point in the sky at the instant we call local noon.

But how can we tell whether the time intervals between repetitions are truly regular? The answer lies in intercomparison. For example, we can repeatedly compare two pendulums of the same length with each other and with someone's pulse beat. When we do this, we find that the pendulums agree with each other much better than either agrees with the pulse beat, and we begin to suspect that the pulse is not regular. We soon also find that, as measured by one of the pendulums, the pulse rate depends on the physical exertion of the owner of the pulse. We conclude that the pulse rate depends on at least

FIGURE 1.1 NIST-7, the seventh-generation atomic clock at the National Institute of Standards and Technology. Accurate within one part in 10^{14}, it is considered currently to be the world's most accurate clock. It has been used to keep standard time since 1993.

The Second (s)

The spelling of the unit names varies from language to language. The symbols do not. Because they are symbols, not abbreviations, no period is placed after them.

The Meter Bar

The first meter bar was constructed in 1795. Its length was intended to correspond to one ten-millionth of what was then the most reliable value for the distance from the North Pole to the equator, along the meridian of longitude passing through Paris. That terrestrial definition of the meter conveyed a sense of transnational unity, but the pole-to-equator distance could not (and cannot) be measured nearly as precisely as the distance between the lines on the meter bar.

---

one factor that we cannot easily control and is therefore not a good candidate for an accurate time standard.

The length of the solar day (noon to noon) seems a better candidate. Indeed, for many years, the base unit of time, the second, was defined in terms of the **mean solar day**, which is the length of the day averaged over one year. Specifically, the second was defined to be $\frac{1}{60} \times \frac{1}{60} \times \frac{1}{24}$, or $\frac{1}{86\,400}$, of a mean solar day.

This definition was superseded in 1967. By then, it had become possible to measure the periods of atomic phenomena far more precisely than the periods of larger systems, and to use such phenomena in establishing standards. A standard based on an atomic phenomenon is called an **atomic standard**. The atomic standard of time relies on the fact that an atom emits a specific kind of electromagnetic ("light") wave when the arrangements of its electrons undergo a specific change called an atomic transition.* Like all repetitive waves, electromagnetic waves are periodic. Since 1967, the second has been defined in terms of a particular atomic transition, in which an outer electron of a cesium-133 atom "flips" its orientation relative to the atom's nucleus. This flip causes the atom to emit a wave that has a very sharply defined period. The device used to measure the period is called an *atomic clock* (Figure 1.1). The clock contains electronic components that both stimulate and detect the repeated flipping in the many cesium atoms within the clock. The **second** is defined as the time required for 9 192 631 770 periods of the microwaves that stimulate these transitions. Like all SI units, the name *second* has an international standard symbol, which is **s**.

The numerical value in this definition of the second was so chosen as to make the new standard compatible with the old one. The new standard is, however, about 1000 times as precise as the old one. Just as fluctuations in a human pulse rate can be measured by comparing the pulse with the swing of a pendulum, fluctuations in the earth's rotation rate (which determines the length of the day) can be measured in terms of the period of the microwaves produced in an atomic clock.

As a result of the redefinition, the day is no longer exactly 86 400 s long. This is awkward for astronomers and others who continue to use the "mean solar second," defined as $\frac{1}{86\,400}$ day. To keep the two systems compatible, an extra "leap second" is added to the mean solar day every few years as needed, by international agreement.

Atomic standards have advantages other than precision over arbitrarily constructed standards. Because all atoms of a given kind are identical, there is no need to construct and maintain a standard in a central laboratory. We need not worry about the possible destruction of the standard, and we need not transport secondary standards to it for checking. Every properly equipped laboratory has equal access to the standard quantity.

The Unit of Length

The SI unit of length is the **meter** (symbol **m**). The name is derived from the Greek word *metron*, meaning "measure." From 1889 until 1960, the meter was defined as the distance, measured under specified conditions, between two fine lines scribed on a bar designed and maintained with exquisite care. The bar is still kept at the International Bureau of Weights and Measures in Sèvres.

A meter bar is an example of an **artificial standard**—that is, a standard built to serve its specific purpose. In 1960, the bar was superseded by an atomic standard. Like the standard of time, this length standard was based on the electromagnetic waves—in this case, light waves—emitted by a specific atomic transition. The light wave chosen has a very sharply defined *wavelength*, which could be measured with much greater precision than the distance between the two lines on a meter bar.

However, the 1960 standard of length was not as precise as the standard of time. In 1983, the 1960 length standard was superseded by a new standard that explicitly takes

---

*Electromagnetic waves are treated in Chapter 34 of this book, and atomic transitions are described in Chapter 41. For now, you needn't worry about the exact meaning of these terms.

advantage of the precision of the time standard. The meter is now defined to be the distance that light travels, through vacuum, in $\frac{1}{299\ 792\ 458}$ s.

Three important considerations underlie this definition. First, the speed of light is now *defined* to be *precisely* 299 792 458 meters per second. Should more precise measurements be made of the speed of light, the effect would be to change the length of the meter slightly. Second, length and time can now be measured with comparable precision. Third, and most important of all, the speed of light in vacuum is precisely the same for all observers. This is a foundation of the *special theory of relativity*, which is the subject of Chapters 38 and 39. So fundamental is this theory, and so strong is the confidence placed in it by scientists, that the definition of the meter can be soundly based on the constancy and universality of the speed of light in vacuum.

Like other local systems, the U.S. Customary system originally had its own primary standard of length, the **yard**. But, in 1959, the yard was redefined in terms of the meter:

$$1 \text{ yard (yd)} \equiv 0.9144 \text{ m} \quad \textit{exactly.}$$

Powers of Ten

The quantities that scientists must measure range enormously in magnitude. Consider length: At one end of the length scale, atomic nuclei are about a thousand-million-millionth of a meter (0.000 000 000 000 001 m) across; at the other end, the diameter of the universe is something like a hundred million million million million meters (100 000 000 000 000 000 000 000 000 m).

Two devices are used to make such quantities manageable. The first is the system of **scientific notation**. In this system, a quantity is represented by a number of convenient size multiplied by a power of ten. Here are two examples:

1. The average distance from the sun to the earth is 149 600 000 000 m. This is much more conveniently written in the form 1.496×10^{11} m. (You can check the equivalence of the two forms by beginning with the number 1.496 and moving the decimal point 11 places to the right.)

2. The time required for light to travel a distance of 1.0000 m is 0.000 000 003 335 7 s. This is better written in the form 3.3357×10^{-9} s. (You should check this equivalence as well.)

The second helpful device makes use of a set of prefixes. Beginning with any SI unit, you can construct a unit of more convenient size for your purpose by using the proper prefix. You are doubtless familiar with one such unit: the millimeter (symbol mm). The prefix *milli-* means a thousandth. A millimeter is equal to 10^{-3} m. Table 1.1 lists all the standard prefixes with their standard symbols.

Notice that, when units are formed by the use of prefixes, the relations among such units are always powers of ten: 1 kilometer = 10^3 meters; 1 nanometer = 10^{-9} meter; and so on.

Some general rules apply to the use of these prefixes:

1. Use of the prefixes centi-, deci-, deka-, and hecto-, which do not represent powers of ten with exponents divisible by 3, is discouraged for scientific purposes. For example, write 3×10^{-2} s or 30 ms but not 3 cs. Write 400 m or 0.4 km but not 4 hm. The main exception to this rule, due to strongly established custom, is the centimeter (cm).

2. Never stack prefixes. For 10^{-9} m, write 1 nm but not 1 mμm.

3. All prefixes are accented on the first syllable. Say **kil'**-o-meter, never kil-**o'**-meter; and **mi'**-cro-meter (meaning 10^{-6} m), not mi-**cro'**-meter (meaning the measuring instrument).

The convenience and utility of the powers-of-ten notation are evident in Tables 1.2 and 1.3. Table 1.2 gives approximate values, in seconds, of some time intervals that are important in the sciences. Table 1.3 gives approximate values, in meters, of some important distances. In each table, note the vast range from the smallest to the largest values. It would be awkward indeed to express these values—let alone manipulate them—without using scientific notation.

TABLE 1.1 SI Prefixes

Factor	Prefix	Symbol
10^{-24}	yocto	y
10^{-21}	zepto	z
10^{-18}	atto	a
10^{-15}	femto	f
10^{-12}	pico	p
10^{-9}	nano	n
10^{-6}	micro	μ (Greek mu)
10^{-3}	milli	m
10^{-2}	centi	c
10^{-1}	deci	d
10^{0}	—	—
10^{1}	deka	da
10^{2}	hecto	h
10^{3}	kilo	k
10^{6}	mega	M
10^{9}	giga	G
10^{12}	tera	T
10^{15}	peta	P
10^{18}	exa	E
10^{21}	zetta	Z
10^{24}	yotta	Y

TABLE 1.2 Selected Time Intervals

	Interval (s)
Planck time[a]	$\simeq 10^{-43}$
Shortest lifetime for an object to which the term "particle" can be applied meaningfully	$\simeq 10^{-23}$
Time required for fission of a uranium nucleus struck by a neutron	$\simeq 10^{-22}$
Lifetime of a neutral pion	8.7×10^{-17}
Time required for one "vibration" of a visible light wave	$1.3 - 2.3 \times 10^{-15}$
Typical time required for one vibration of an atom in a solid	10^{-13}
Lifetime of a muon	2.2×10^{-6}
Time required for one vibration of an audible sound source	$5 \times 10^{-5} - 5 \times 10^{-2}$
Time between human heartbeats (adult, resting)	8×10^{-1}
One day	8.6×10^{4}
One year	3.15×10^{7}
Typical human lifetime	2.4×10^{9}
Age of the Lascaux cave paintings	5×10^{11}
Age of earliest *Homo sapiens*	5×10^{12}
Age of the earth	1.3×10^{17}
Age of the universe	5×10^{17}
Lifetime of the proton	$> 10^{39}$

[a]When we trace the history of the universe, the Planck time is the earliest "age" after the "Big Bang" at which we can begin to apply the laws of physics as they are currently known.

TABLE 1.3 Selected Lengths

	Length (m)
Diameter of an atomic nucleus	$\simeq 10^{-15}$
Diameter of an atom	$\simeq 10^{-10}$
Size of a typical virus	1×10^{-8}
Wavelength of visible light	$4 - 7 \times 10^{-7}$
Diameter of a typical cell	7×10^{-6}
Thickness of a sheet of paper	1×10^{-4}
Height of an average woman (United States)	1.7×10^{0}
Height of the tallest building (Sears Tower, Chicago)	4.4×10^{2}
Diameter of the earth	1.3×10^{7}
Mean distance from earth to sun	1.5×10^{11}
Mean diameter of Pluto's orbit	1.2×10^{13}
One light year (distance light travels in 1 y)	9.5×10^{15}
Diameter of our galaxy (the Milky Way)	$\simeq 10^{21}$
Distance to nearest large galaxy (M31)	2×10^{22}
Distance to the farthest known quasar	2×10^{26}

Numbers and Physical Quantities

The scientist is interested in numbers not for their own sake but for their usefulness in representing physical quantities. Physical quantities are not mere abstractions; rather, they are the results of specific measurements, and the measurements usually yield numbers. The physical meaning of these numbers depends on the accuracy of the measurements and on the units in which they are expressed. We now consider unambiguous ways of expressing both.

Significant Figures

There is always a limit to the accuracy with which a measurement can be made. In presenting the result of the measurement, it is important to specify not only the value that the measurement gives, but also the values it does *not* give. Suppose, for example, that you measure the length of a sheet of paper with an ordinary ruler (Figure 1.2). You place the zero mark of the ruler next to one end of the sheet, and you read the ruler mark at the other end. The smallest divisions on the ruler are tenths of centimeters, and you find that the length you measure is a little greater than 27.9 cm. You are quite sure, however, that the end of the sheet lies closer to the 27.9 cm mark than to the 28.0 cm mark. That is, all three digits in the number 27.9 are reliably known to you, and the length of the sheet is 27.9 cm; it is *not* 28.0 cm, and it is *not* 27.8 cm either. Reliably known figures are **significant figures**.

FIGURE 1.2 Measuring the length of a sheet of paper with a ruler graduated in centimeters and subdivided into tenths of a centimeter. The sheet of paper is rolled into a cylinder by bringing edges A and B together and taping.

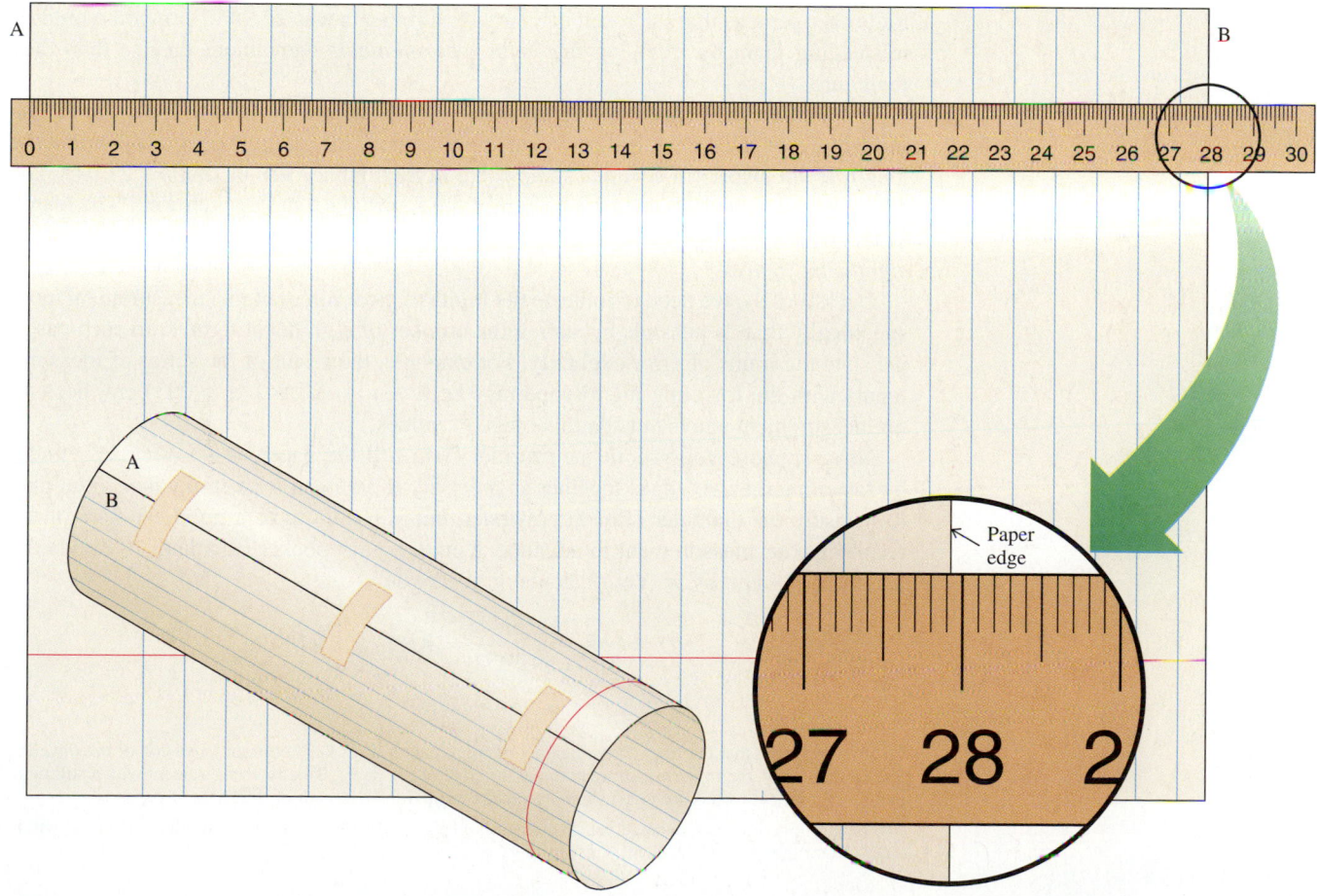

But you decide to try to measure more closely. After some study, you find that in your mind's eye you can subdivide each tenth-of-a-centimeter division of the ruler into tenths—that is, into hundredths of a centimeter—fairly consistently. You test this ability by remeasuring the length of the paper several times at different locations across its width. You find that, when you estimate the length of the paper to a hundredth of a centimeter, you most often obtain the value 27.94 cm. Occasionally you get 27.95 cm, and sometimes you get 27.93 cm, but you never obtain a measurement outside this range. You can now say that the length of the paper sheet is probably 27.94 cm and that it almost certainly lies somewhere between 27.93 cm and 27.95 cm. You are quite sure that the length is not 27.90 cm or 27.98 cm. Although you are not completely certain of the value of the last digit in your best value, 27.94 cm, it nevertheless has a real meaning. So you can now state the length of the sheet of paper to four significant figures.

But why stop there? You decide to measure the length of the sheet of paper with still greater accuracy. You know you cannot do better with the ruler, so you borrow a traveling microscope. This is a microscope mounted on a moveable carriage driven by a micrometer screw: If you move the microscope's cross hairs from one end of the paper to the other, you can read the length of the paper from the micrometer scale. The smallest divisions on the scale are 10 μm (0.001 cm) apart, and you find that you can estimate to one-tenth of a division, or 1 μm. Repeated measurements give you the following values:

$$27.9432 \text{ cm} \qquad 27.9386 \text{ cm} \qquad 27.9406 \text{ cm} \qquad 27.9443 \text{ cm}$$
$$27.9397 \text{ cm} \qquad 27.9422 \text{ cm} \qquad 27.9410 \text{ cm}.$$

Inspection of these values shows that the last (sixth) digit is certainly not significant. It seems that any digit at all can show up in the last place, so the last digit is not reliably known. The last place thus contributes no meaning to the measurement. Indeed, if you told other persons that the result of your measurements was 27.9432 cm, you would be misleading them by implying that your measurements were more precise than they really are.

What about the next-to-last (fifth) digit? It fluctuates widely, so it too is not a significant figure. However, it is not completely without meaning. It does tell you that the length of the paper (in cm) lies somewhere in the high 27.93s or the low 27.94s. The fifth digit thus confirms the significance of the fourth digit in the assertion "The length of the paper is 27.94 cm," and we can say that we know the length of the paper *to four significant figures.**

Occasionally, we need to indicate the limits of error inherent in a measurement more specifically than is possible by stating the number of significant figures. In such cases, we state the limits of error explicitly. For example, the result of the series of measurements with the traveling microscope may be given as (27.941 $\pm$ 0.003) cm, because no measurement strays outside this range of values.[†]

Now, suppose you go one step further. You roll the paper into a circular cylinder by fastening the two edges together as precisely as you can with sticky tape. You plan to measure the diameter D of the cylinder, but you can make a good estimate of the result of your measurement in advance. You use your pocket calculator to divide the length, 27.94 cm, by π. Your calculator gives you

$$D = \frac{27.94 \text{ cm}}{3.141\,592\,654} = 8.893\,578\,219 \text{ cm}.$$

---

*It is fair to guess that the limit on the number of significant figures lies not in the precision of the traveling microscope but in the imperfect straightness of the edge of the paper. That is, measurement with a still more precise instrument would not give us a more precise value for the length of the sheet of paper.

[†]To obtain this statement, average the seven measurements and note that all the individual values lie within 0.003 cm of the average.

This result is misleading. The calculator is capable of computing arithmetical results to a large number of digits. But it has no way of deciding which digits are significant. *You must decide where to cut off the number the calculator gives you so that you neither throw away significant information nor misrepresent meaningless digits as significant.*

There are mathematical rules for finding the number of significant figures in the result of a calculation, depending on the number of significant figures in each input value. We consider here an intuitive approach based directly on the meaning of the term "significant figures." In your calculations, the number 27.94 cm implies "*not 27.95 cm.*" So you repeat the calculation of D with this "nearest wrong" value to obtain

$$\frac{27.95 \text{ cm}}{3.141\,592\,654} = 8.896\,761\,318 \text{ cm.}$$

This "nearest wrong" result differs from the correct result only in the fourth and subsequent digits. Thus, the fourth digit is significant, whereas the subsequent ones are not. Rounding off the correct result to four significant figures gives the predicted value of the diameter of the paper cylinder:

$$D = 8.894 \text{ cm.}$$

The value of D thus obtained is the best that you can expect when you actually measure the diameter of the cylinder. But the measured value may well have fewer significant figures. One reason is that the cylinder is not likely to be exactly circular. That is, the *theory* on the basis of which you made your prediction of D is an imperfect description of the actual situation. Can you think of other reasons?

The following five conventions provide a clear way of expressing the number of significant figures.

1. When a number contains a decimal point—as the value 27.94 cm does—the number of significant figures is the number of digits presented. A significant figure to the right of the decimal point can be a zero; for example, the number 3.9800 has five significant figures. (The "nearest wrong" value is 3.9801, not 3.99.)

2. If a number has a value less than 0.1, the zeroes between the decimal point and the first nonzero digit are *not* significant figures. They merely establish the magnitude of the number. For example, the number 0.006 209 contains four significant figures; the number 0.006 2090 contains five.

3. Some ambiguity exists in dealing with large numbers. In the number 85 300 000, the zeroes are essential to the magnitude of the number and must appear. But are any or all of them significant figures? Most often, the writer of such a number does not intend that the zeroes be interpreted as significant figures. That is, the writer usually wants the reader to understand that the number has three significant figures. But there is no sure way of telling. It is better to present the number in scientific notation. Written as 8.53×10^7, the number clearly has three significant figures; written as 8.530×10^7, it clearly has four.

4. Some numbers are *exact* because their values have been set by definition. One example you have already seen is the U.S. Customary yard, defined to be exactly 0.9144 m. Another exact number is the constant $\frac{1}{2}$ (or 0.5) in the formula that gives the area of a triangle:

$$\text{area} = \tfrac{1}{2} \times \text{base} \times \text{altitude.}$$

The concept of significant figures does not apply to exact numbers. If you wish, however, you may think of such a number as being followed by an infinite string of zeroes, all of which are significant.

5. Sometimes a calculation contains a number representing a measurement of considerably greater precision than any other measurement appearing in the same calculation. In such cases, the value of the precise measurement does not determine the number of significant figures in the result. The most precise number may be considered exact for purposes of calculation. If this is understood, the number is often quoted in "round figures," as in Example 1.1.

EXAMPLE **1.1**

The inside diameter of precision-bore glass tubing is uniform within 3×10^{-4} cm. A sample of 3-cm precision-bore tubing is cut to a length of 6.6 cm. Find the volume within the tube in cubic centimeters.

SOLUTION: Even though the diameter of the tubing is quoted as 3 cm, the manufacturer's specifications tell you that its value is (3.0000 ± 0.0003) cm. The diameter is thus known to better than four significant figures. The length, however, is quoted to only two significant figures. You can safely argue that the un-

certainty in the length of the tube will govern the uncertainty in the volume that it contains. Because the volume of a cylinder is given by $\pi \times$ (length) $\times$ (diameter/2)^2, you have

$$\text{volume} = \pi \times 6.6 \text{ cm} \times \left(\frac{3 \text{ cm}}{2}\right)^2 = 47 \text{ cm}^3$$

to two significant figures.

If you use a pocket calculator to make this computation, why is it unnecessary to quote the number of significant figures to which π is taken?

Order of Magnitude

When making physical calculations, we often need only a very general idea of the value of a quantity. This is commonly the case when a preliminary calculation or a feasibility estimate must be made. It may suffice to estimate the **order of magnitude** of a quantity, which is the power of ten nearest to the actual value. It is often much easier to make or obtain such estimates than to obtain more precise values.

EXAMPLE **1.2**

A typical animal cell has a diameter of 7 μm. What is the order of magnitude of the number of cells in your body?

SOLUTION: There are four steps in making this estimate. First, make the crude assumption that your body is completely packed with cells, all of the same size. Second, estimate the volume of a cell, assuming that it is roughly cubic. Third, estimate the volume of your body. Fourth, divide the body volume by the cell volume to find the number of cells that your body contains.

You can carry out the actual calculation in two ways. The rougher way is to round off all values to the nearest order of magnitude at the outset. In this spirit, call the cell diameter 10 μm, or 10^{-5} m. The cell volume is then $V_c = (10^{-5} \text{ m})^3 = 10^{-15}$ m^3. Next, estimate the body volume V_b in like manner. Make the approximation that your body is a rectangular block having height of order of magnitude 1 m, and width and thickness of order of magnitude 0.1 m. (All these figures are underestimates, but the next-larger orders of magnitude would be gross overestimates.) You have $V_b = 1 \text{ m} \times 10^{-1} \text{ m} \times 10^{-1}$ m $= 10^{-2}$ m^3. Finally, divide to find the number of cells N:

$$N = \frac{V_b}{V_c} = \frac{10^{-2} \text{ m}^3}{10^{-15} \text{ m}^3} = 10^{13}.$$

The more precise method takes only a little more effort. Here, you retain one significant figure when estimating and two

(to minimize round-off errors) when calculating. The cell volume then becomes $V_c = (7 \times 10^{-6} \text{ m})^3 = 3.4 \times 10^{-16}$ m^3. (This is about one-third the value of the cruder estimate.) Again, approximate your body by a rectangular block, but use more accurate figures for height, width, and thickness. Measure yourself with a ruler, remembering that you need only one significant figure. You will obtain something like $V_b = 2 \text{ m} \times 0.4 \text{ m} \times 0.2 \text{ m} = 0.16$ m^3. (This is about 20 times the value of the cruder estimate and is likely much more accurate.) Again, dividing to find the number of cells, you will have

$$N = \frac{V_b}{V_c} \simeq 10^{14}.$$

The sign $\simeq$ means "is approximately equal to." By rounding off to the nearest order of magnitude in the last step, we acknowledge the crudeness of the calculation. Why is it sometimes worthwhile to retain significant figures in the calculation that are to be discarded at the end?

The second calculation gives a result one order of magnitude greater than the first. The first refinement of a very crude calculation results in a correction of just a factor of 10 or so. This suggests that further refinements (which could readily be made) will result in still smaller revisions of the value of N. Thus, you already have a pretty good estimate of the number of cells in your body.

Units and Conversions

A physical quantity must nearly always be expressed by means of a *unit* as well as a *magnitude*. To say that the speed of an automobile is 20, for instance, is useless unless you also specify the units in which the magnitude 20 gives the speed: meters per second or kilometers per hour or miles per hour. In making calculations, moreover, it is essential that all the physical quantities be expressed in *consistent* units. If you want to know the total amount of fuel consumed by an automobile during a long trip, you cannot simply add one purchase made in liters to another made in gallons. Consistency is guaranteed if all quantities are expressed in SI units.

Quantities not given in the desired units must be converted. Even when conversion factors are given, however, there is room for error. In a particular calculation, do you have to multiply or divide by the conversion factor? The method developed in Examples 1.3 and 1.4 minimizes the likelihood of error.

EXAMPLE 1.3

Express the time interval 25.3 minutes in seconds.

SOLUTION: You know how to express 1 min in units of seconds; you have

$$1 \text{ min} = 60 \text{ s}. \tag{1.1}$$

This expression has all the properties of an algebraic equation. Specifically, the equals sign asserts that two quantities expressed in different terms are equal to each other. You can manipulate this equation just as you can any algebraic equation. In particular, you can divide both sides by the quantity 1 min. When you do this, you obtain

$$\frac{1 \text{ min}}{1 \text{ min}} = \frac{60 \text{ s}}{1 \text{ min}}. \tag{1.2}$$

One way of interpreting this result is to say that the unit "min" in the numerator of the fraction on the left side of Equation 1.1 cancels with the identical unit in the denominator. We can generalize and say that *identical units can be canceled in fractions in the same way as identical numbers.*

Equation 1.2 tells you that the fraction (60 s)/(1 min) is another way of expressing the quantity 1:

$$1 = \frac{60 \text{ s}}{1 \text{ min}}. \tag{1.3}$$

Hence, you can multiply the quantity 25.3 min, given in the statement of the example, by this fraction without changing its value:

$$25.3 \text{ min} = 25.3 \text{ min} \times \frac{60 \text{ s}}{1 \text{ min}} = \frac{25.3 \text{ min}}{1 \text{ min}} \times 60 \text{ s}.$$

The manipulation has thus made it possible to cancel the unit "min" from the right side of the equation. You have

$$25.3 \text{ min} = \frac{25.3 \text{ min}}{1 \text{ min}} \times 60 \text{ s} = 25.3 \times 60 \text{ s}$$

$$= 1.52 \times 10^3 \text{ s}.$$

Consider Equation 1.3 again. It states that the pure number 1 on the left side is equal to a fraction whose numerator is given in seconds and whose denominator is given in minutes. How can this be correct? The answer lies in the concept of **dimensionality** of units. Both minutes and seconds are units of time; they have the same **dimension**. If you were making a graph to describe the way that a system changes with time, you could plot both minutes and seconds on the same axis, provided you were careful about the scale factor. This point is illustrated in Figure 1.3. Regardless of the particular unit used, time represents one of the dimensions of the two-dimensional space defined by

FIGURE 1.3 (*a*) A thermometer is immersed in a beaker containing cold water. At time $t = 0$, the Bunsen burner is ignited. (*b*) Plot of the water temperature, as read by the thermometer, as a function of time. The time scale can be read either in seconds (upper scale) or in minutes (lower scale).

(a)

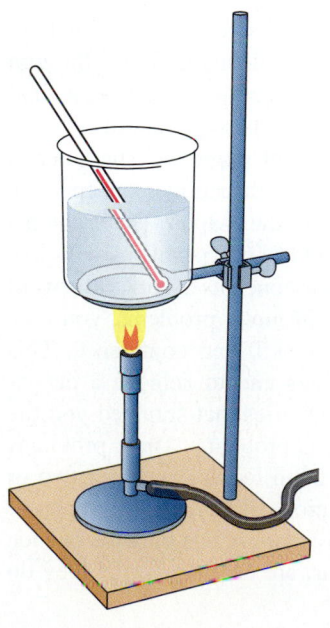

(b)

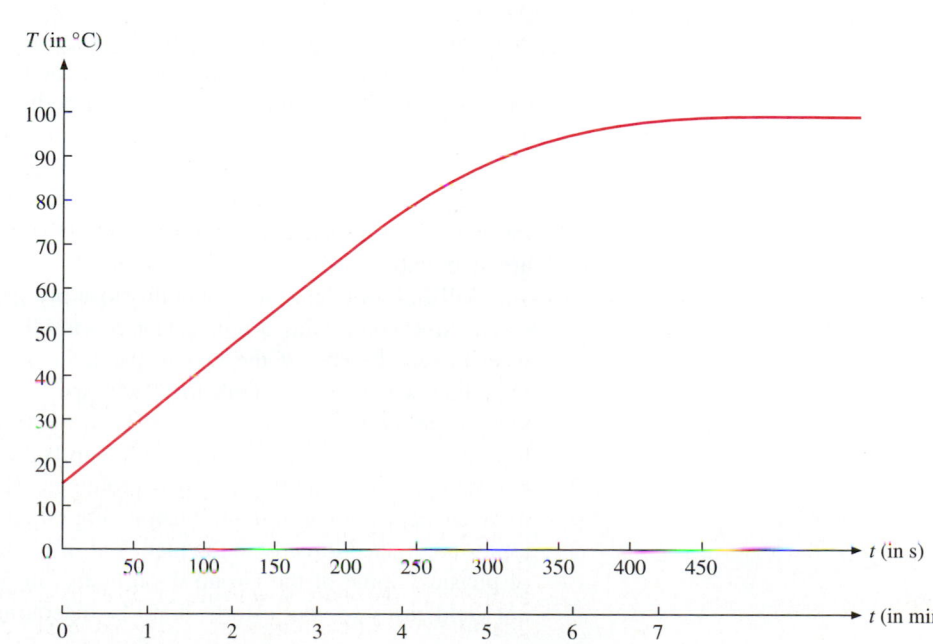

the axes of the graph. (In Figure 1.3, the other dimension happens to be temperature.) Dividing one quantity by another having the same dimensions, as was done in Equation 1.2, yields a **dimensionless quantity**, as a pure number is called in this context. In Equation 1.2, the magnitude of that quantity happens to be 1.

EXAMPLE 1.4

An automobile travels at a speed of 25.0 mi/h (miles per hour). Find the speed in m/s (meters per second). Use the following equivalencies: 1 mi = 5280 ft (feet); 1 ft = $\frac{1}{3}$ yd; 1 yd = 0.9144 m; 1 h = 60 min; 1 min = 60 s.

SOLUTION: This example is more complicated than Example 1.3 in two ways. First, two units must be converted: miles into meters, and hours into seconds. Second, each conversion requires more than one step. To separate the various operations required, express the given speed in the form

$$25.0 \, \frac{\text{mi}}{\text{h}} = 25.0 \, (\text{mi}) \left(\frac{1}{\text{h}} \right).$$

You can now deal with each term in parentheses as in Example 1.3. You multiply each unwanted unit by a series of fractions having the value 1, as follows:

$$25.0 \, (\text{mi}) \left(\frac{1}{\text{h}} \right) = 25.0 \left(\text{mi} \times \frac{5280 \, \text{ft}}{1 \, \text{mi}} \times \frac{\frac{1}{3} \, \text{yd}}{1 \, \text{ft}} \times \frac{0.9144 \, \text{m}}{1 \, \text{yd}} \right.$$
$$\left. \times \left(\frac{1}{\text{h}} \times \frac{1 \, \text{h}}{60 \, \text{min}} \times \frac{1 \, \text{min}}{60 \, \text{s}} \right). \right.$$

The fractions are chosen so that the denominator of each is the

same as the numerator of the preceding one, or vice versa. You can thus cancel the unwanted units by striking them out as shown. What is left is

$$25.0 \, \text{mi/h} = 25.0 \times 5280 \times \tfrac{1}{3} \times 0.9144 \times \tfrac{1}{60} \times \tfrac{1}{60} \, \text{m/s}$$
$$= 11.2 \, \text{m/s}.$$

What if you make a wrong choice for one or more of the fractions equal to 1? Suppose, for instance, you choose the fraction (1 mi)/(5280 ft) instead of (5280 ft)/(1 mi). In that case, you simply do not obtain the desired cancellation. Instead, you obtain

$$25.0 \, \text{mi/h} = 4.01 \times 10^{-7} \, (\text{mi}^2 \cdot \text{m})/(\text{ft}^2 \cdot \text{s}).$$

This result is true but not useful. The immediate tipoff that the answer is not what you want: Units that you do not wish to retain in the result—miles and feet—appear as squares instead of canceling. The method almost checks itself.

One word of caution: The method works only if you are careful to carry the units along at every step of the calculation. Never try to cut corners by writing 25.0 when you mean 25.0 mi/h, in the expectation of fixing things up at the end of the calculation. There are too many opportunities for error!

SECTION 1.4   How to Study Physics

As you read and study further, you will become more and more skillful at describing and predicting events yourself, and you will find increasing pleasure in your ability to do so. Starting with relatively simple situations, you will learn how to "think like a physicist," describing a certain range of observations in a general way. A worked example will show how this general description can be applied to a particular situation. You should follow these examples with pencil, paper, and calculator: Work through each step for yourself and check your intermediate results as you go, as well as the final answer. And, when you review, be sure to work through the examples carefully.

The exercises at the end of each chapter serve a variety of purposes. The Queries are qualitative; they will help you to see how well you have grasped the concepts just introduced, perhaps indicating that you need to reread parts of the chapter. The Problems are quantitative and hence more specific. Working out the Group A problems will give you skill and confidence in your ability to apply specific new concepts in straightforward cases. After completing a representative sampling of the Group A problems, you may wish to reread parts of the text in the light of your new skill and confidence. This rereading will help you both to clarify specific difficulties and to achieve a deeper understanding of the subject matter. Then return to any Queries that stumped you the first time around, and proceed to the Group B and Group C problems. These problems are more general than the Group A problems. Many of them require applying two or more concepts. Some of them illustrate important applications of physical principles to topics not treated directly in the main text or to subjects of importance in fields outside of physics. Some of the Group C problems, in particular, are challenging. If they do not yield easily to your efforts, don't be discouraged.

THINKING LIKE A PHYSICIST

Everyone knows that physics is a science. Physics is also a *craft*. That is, the working physicist approaches a problem from the point of view of a skilled craftsman. Like any craftsman, the physicist combines an overall point of view— a kind of philosophy of approach—with a "tool bag" of specific techniques. Whether or not you plan specifically to become a physicist, you need to acquire at least the basic elements of both the philosophy and the tool bag. That is, you need to learn how to think like a physicist. By doing so, you will steadily increase your ability to solve physics problems and—even more important—you will come to see physics as much more than the particular set of techniques required to solve the problems at the ends of the chapters.

Learning the craft of physics is very much up to you. Nevertheless, helpful hints along the way can be useful as you "serve your apprenticeship." You will see such hints in most chapters; they share the title "Thinking like a physicist." These hints are often of a general nature, inviting you to look beyond the particular subject matter of the main text to the broader context in which that subject matter acquires deeper meaning. Don't study these hints for the next exam; think about them as you eat pizza or get ready for bed.

Nearly all physics students find that the largest part of their study time is spent in problem solving. Careful reading of the text is indispensable, but insight into the general principles usually comes with persistent working of problems. Insight does not always dawn with a sudden, great light. You can easily satisfy yourself as to the truth of the statement "I have obtained the correct solution to Problem 9.6." It is harder to evaluate the truth of the statement "I understand the principle of conservation of linear momentum," which is the main subject of Chapter 9. But the principle will become more and more meaningful as you work the chapter's problems.

It takes time to grasp fully the important ideas of physics. Cramming at the last moment is less useful in physics than in areas where it is profitable to memorize large amounts of material, perhaps only temporarily. (Memorization is probably less useful in physics than in any other field, with the possible exception of pure mathematics.) Work the examples and problems as you read the text, using a schedule that you find convenient and that is reasonably uniform in pace. Don't gloss over something that you don't understand and go on to the next topic. Physics has a very tight structure; more often than not, each topic depends directly on what precedes it. So be persistent and thorough.

As with all subjects of study, you will begin modestly. But, as you proceed and your horizons expand, you will begin to perceive a grand order and get the hang of a general methodology. Best of all, you will enjoy more and more of the great beauty that, more than anything else, attracts physicists to their subject. It is this order and beauty that makes physics one of the great arenas of the human mind.

QUERIES

The number in parentheses after a query or problem number refers to the section of the chapter (for example, Section 1.2) on which the query or problem is based. The letter *G* denotes a more general query or problem that addresses subject matter treated in more than one section.

1.1 *(2) Long deliberations.* Suppose that a new standard of length is adopted by international agreement. **(a)** Does the length of a given object change? **(b)** If not, explain precisely what it is that does change. **(c)** In view of your answers to parts **a** and **b**, why did the General Conference of Weights and Measures redefine the meter to be the distance traveled by light through vacuum in $1/299\ 792\ 458$ s, instead of some round number, such as $1/300\ 000\ 000$ s?

1.2 *(2) Carpe diem.* The early Romans divided the day into 12 hours, beginning at sunrise and ending at sunset. Comment on the disadvantages of this method for scientific measurement.

1.3 *(G) A footling diversion.* Measure the length of your room, using the length of your foot as a unit. Justify the number of significant figures that you use in your result.

1.4 (G) *Ruler of all you survey.* Express the following lengths in meters and in centimeters: your height; the distance between the middle fingertip of your outstretched right arm and the tip of your nose as you look straight forward; the length of your right forearm from elbow to middle fingertip; the span of your right hand from the tip of the thumb to the tip of the little finger; your foot; your normal walking pace; the distance between the centers of the pupils of your eyes. Justify the number of significant figures that you give in each result. (The second through fourth distances given are the historical bases of the units called the yard, the cubit, and the span.)

PROBLEMS

GROUP A

1.1 (2) *How long, brethren?* Estimate the following in meters: (**a**) the length of an automobile, (**b**) the wingspan of a house-fly, (**c**) the height of an elephant, (**d**) the height of a very tall man, (**e**) the height of a newborn child, (**f**) the width of the street on which you live, (**g**) the distance from New York to London, (**h**) the thickness of a sheet of paper.

1.2 (2) *How "long," brethren?* Estimate the following in seconds: (**a**) the length of your life to date, (**b**) the time elapsed since the building of the Great Pyramid (2900 B.C.E.), (**c**) the time it takes to run one block, (**d**) the time until your next birthday, (**e**) the life-span of a dog, (**f**) the time it takes to dial a telephone number and receive a response, (**g**) the time elapsed since Galileo's death (9 January 1642).

1.3 (2) *How small?* Express the approximate diameter of an atomic nucleus, given in Section 1.2, in scientific notation.

1.4 (2) *How big?* Express the approximate diameter of the universe, given in Section 1.2, in scientific notation.

1.5 (2) *How many smalls in a big? I.* Using the approximate diameter of an atomic nucleus (Section 1.2) as a unit, what is the approximate diameter of the universe?

1.6 (2) *How many smalls in a big? II.* What is the ratio of the longest time in Table 1.2 to the shortest?

1.7 (2) *How many smalls in a big? III.* What is the ratio of the greatest distance in Table 1.3 to the smallest?

1.8 (2) *A power of tens.* How many micrometers (μm) are there in a terameter (Tm)?

1.9 (2) *Nice little place, . . . wasn't it?* The speed of light is close to 3×10^8 m/s. Approximately how long does it take a light signal to cross an atom of diameter 1×10^{-10} m? Express your results in seconds and in attoseconds.

1.10 (2) *Astronomical distance.* A light year (ly) is defined to be the distance that a light signal travels in one year. Express 1 ly in meters.

1.11 (2) *. . . and a star to steer her by.* A nautical mile is the length of 1 minute of arc ($\frac{1}{60}$ degree) on the earth's surface at the equator. Assuming the earth to be spherical, express 1 nautical mile in kilometers.

1.12 (3) *Square deal.* A rectangular piece of sheet metal 0.30 m long is sawed from a long strip whose width is 5.00 cm. Find the area of the piece, expressing your result to the proper number of significant figures.

1.13 (3) *I didn't know I had it in me!* To the nearest order of magnitude, compare the number of cells in your body (see Example 1.2) with the number of people on earth.

1.14 (3) *Changing gears.* Express the speed 60 mi/h (**a**) in km/h, (**b**) in m/s.

GROUP B

1.15 (2) *Passin' through Paree.* The meridian passing at mean sea level from the North Pole through Paris to the equator is called the Paris quadrant. A modern measurement gives its length to be 10 002 288.3 m. (**a**) Express the "terrestrial meter," defined as $\frac{1}{10\,000\,000}$ of the length of the Paris quadrant, in terms of the modern meter. (**b**) Expressed in parts per million, what is the degree of precision of a length measurement for which the two definitions of the meter give different results?

1.16 (2) *Tramp, tramp, tramp.* The word "mile" is derived from the Latin *milia passuum*, meaning 1000 paces. The ancient Romans carefully trained their soldiers to use a standard pace (by which they meant a pair of steps, one with each foot), and they measured distances in 1000-pace miles. Measure your pace at a brisk walk. Then, using your own height and a plausible argument as the basis for your calculation, estimate a Roman soldier's pace on the assumption that the average

soldier was about 1.65 m tall. Finally, estimate the length of a Roman mile in meters. Compare your result with the U.S. Customary statute mile, whose length is 1609.344 m.

1.17 (2) *Old times there are not forgotten.* Express the speed 60 mi/h in ft/s. When you are obliged to make calculations in U.S. Customary units, this is a handy equivalence to remember.

1.18 (2) *Keep a-inchin' along.* Given the definitions 1 yd $\equiv$ 0.9144 m and 1 in. $\equiv \frac{1}{36}$ yd, show that 1 in. = 2.54 cm *exactly*. (The sign $\equiv$ means "identical with.")

1.19 (3) *Figures are significant.* A friend observes that in Problem 1.18 the result is expressed in three digits, though in the information provided the yard is expressed in four digits. He argues that, because the inch is shorter than the yard, the proportional error in determining the end points of the measurement is greater, and therefore fewer significant figures are justified. A second friend disagrees. She argues that the result should be given to three significant figures. The yard is expressed in terms of the meter to four significant figures, and the inch is expressed in terms of the yard to two. She claims the number of significant figures in the result should be the average of 4 and 2, or 3. Who is right? If neither, give the correct answer and the reason for it.

1.20 *(G) Easy as pi.* The number of days in the so-called Gregorian year is the number of days per year averaged over the 400-year cycle of the Gregorian calendar (which is the calendar in general worldwide use). This cycle contains 97 leap years of 366 days each and 303 common years of 365 days each. (The rule is that a year is a common year if its number is not divisible by 4 or not divisible by 400 or both; otherwise it is a leap year.) **(a)** Find the number of days in a Gregorian year. **(b)** Find the number of seconds in a Gregorian year. **(c)** It is very convenient to remember the approximation

$$1 \text{ Gregorian year} \simeq \pi \times 10^7 \text{ s}.$$

To how many significant figures is this approximation correct?

1.21 *(3) The right way to square a wrong.* A circular disk of sheet metal has a measured diameter of 0.962 m. **(a)** Using the standard formula for the area of a circle, find the area of the disk, expressing your result to the proper number of significant figures. **(b)** Find the "nearest wrong" value of the area by repeating the calculation with the diameter 0.963 m. **(c)** Express the uncertainty in the area as a proportion of the area by taking the ratio

$$\frac{\text{``nearest wrong'' area} - \text{area}}{\text{area}}.$$

(d) Compare this result with the uncertainty in the diameter as a proportion of the diameter, found by taking the ratio

$$\frac{\text{``nearest wrong'' diameter} - \text{diameter}}{\text{diameter}}.$$

(e) How can you account for the difference in the results of parts **c** and **d**?

1.22 *(3) To the nth degree.* A metal sphere has a measured diameter of 0.962 m. **(a)** Using the standard formula for the volume of a sphere, find its volume, expressing your result to the proper number of significant figures. **(b–e)** Carry out the steps of Problem 1.21, parts **b–e**, with respect to the volume of the sphere. **(f)** Suppose a measured quantity q appears raised to the nth power (that is, as q^n) in the calculation of another quantity p. Compare the proportional uncertainty in p with that of the value of q itself.

1.23 *(3) What a difference a year makes!* The parents of a little girl measure her height at every birthday. She stands against a wall, and her mother holds a ruler against the top of her head, keeping it as level as possible. Her father then makes a pencil mark where the underside of the ruler touches the wall. On two successive birthdays, the recorded heights are 1.220 m and 1.271 m. **(a)** Express the girl's growth in the course of the year to the proper number of significant figures. **(b)** Compare the number of significant figures in the result with the number in the individual measurements. Comment on the difference. **(c)** Make the (optimistic) assumption that the distance between the two pencil marks can be measured within 0.0001 m. Does it follow that the girl's growth could thus be measured more precisely than in part **a**? Explain your answer.

1.24 *(3) Big, empty place.* The distance from the sun to the nearest star is about 4×10^{16} m. Assume, in the absence of more detailed information, that this is a typical spacing between stars. (The assumption is open to criticism, but never mind; you have to begin somewhere.) Assume also that the universe is spherical, having a diameter of order of magnitude 10^{26} m. Make a crude estimate of the number of stars in the universe. Then present an argument showing why this estimate is likely to be too large or too small.

1.25 *(3) Packing 'em in.* A typical atom has a diameter of 2×10^{-10} m. Estimate the number of atoms in the earth.

Part *I*

Mechanics of Particles in One Dimension

Mechanics is the study of the *motion* of bodies and of the *forces* that influence this motion. Mechanics lies at the heart of physics. Historically, it was the first branch of physics developed in quantitative detail; its first flowering dates from the publication of Isaac Newton's monumental work, the *Principia*, in 1687. Mechanics is one of the logical foundation stones of many other areas of physics, and it has essential applications in every area of physics. Moreover, mechanics has countless practical applications of its own. Not surprisingly, then, a systematic study of physics starts with mechanics.

We begin our study of mechanics with a commonplace observation. If you push on something, it will often move. Some things can be moved easily, others with more difficulty; some things move quickly, others more slowly. But these are vague statements. We wish to know: *Exactly what is the connection between force and motion?* Part I of this book deals with answering this question in a systematic, quantitative way. The answer must be based on careful definitions of the concepts *force* and *motion* and on unambiguous methods of describing them.

In Part I, and throughout most of this book, we restrict our study to objects whose speed is small compared with the speed of light (3×10^8 m/s). We concentrate chiefly on objects whose size is in the "middle range." (Roughly speaking, "middle range" refers to objects considerably larger than atoms but considerably smaller than large stars.) The vast domain thus carved out for study is that of *classical mechanics*.

Until about 1900, physicists took it for granted that the laws of classical mechanics (then simply called mechanics) could be applied to all objects regardless of their speed or size. This view, now proved wrong, was based on the universal—and indispensable—human tendency to extend generalizations from familiar areas, where they are amply justified, into unfamiliar areas. Today, we understand quite a lot about the domains of the very small, the very large, and the very fast-moving, even though classical mechanics fails to describe these domains satisfactorily. The behavior of the very small must be studied in terms of the laws of *quantum mechanics*; this is the subject of Part XII. The behavior of the very large and the very

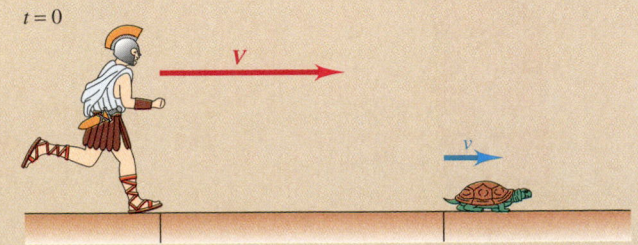

fast-moving must be studied in terms of the laws of *relativistic mechanics*, which is the subject of Part XI. A combination of the two, called *relativistic quantum mechanics*, is required to understand the behavior of very small, very fast-moving objects. Nevertheless, classical mechanics remains the key to physics and the logical starting point for a systematic study of physics.

Our first task is to develop methods for describing motion without regard to the forces associated with that motion. This subject is called **kinematics**, from the Greek word *kinematos*, meaning "motion." The simplest kind of motion is motion restricted to a straight line—that is, *one-dimensional* motion.

The kinematics of one-dimensional motion is the main subject of Chapter 2.

Our second task is to introduce and refine the concept of force. Having done so, we study the laws of nature that connect force and motion. The study of these laws is called **dynamics**, from the Greek work *dynamis*, meaning "strength." The dynamics of one-dimensional motion is the main subject of Chapter 3.

Most motion is not restricted to one dimension, and many important and interesting phenomena do not take place in a one-dimensional context. Nevertheless, the one-dimensional world gives enough scope to develop the fundamentals of mechanics. You will thus have the opportunity to achieve some mastery of these fundamentals before tackling the additional complexities of the multidimensional world—and savoring its riches—in Part II.

17

Kinematics of One-Dimensional Motion

Left: Starting from rest, the drag racer tries to cover a quarter-mile (402 m) in the least possible time. For a given time, what is the mean acceleration required? How fast does the racer cross the finish line?

What! will the line stretch out to the crack of doom?

— SHAKESPEARE, *Macbeth*

Introduction: Translation and Rotation

All real objects extend over a certain region of space. For this reason, they are called **extended bodies**, or **bodies** for short. When you throw a body—a stone, for instance—all of its parts do not move in exactly the same way. The stone spins as it flies through the air, and its parts follow different and possibly complicated paths. Nevertheless, the stone as a whole appears to follow a fairly simple path. A dramatic example of extended-body motion is shown in Figure 2.1.

The first step in studying such motion is to simplify it by breaking it into two kinds of motion. As the body as a whole moves along its fairly simple path, the parts of the body spin about a certain point in the body. The motion of the body as a whole from one place to another is called **translation**, from the Latin word meaning "carried across." The spinning motion is called **rotation**, from the Latin word meaning "wheel."

Can we separate the motion of the baton in Figure 2.1 into two parts—translation of the whole baton and rotation of the baton about a certain point—and consider each part by itself? Happily, the answer is Yes. We shall begin by considering translational motion only.

This restriction to translational motion does not divorce us from the real world. Some bodies do not rotate at all as they translate. An example is a rowboat crossing a lake. Other bodies may rotate, but it suffices to consider just their translational motion. For example, a golfer is interested in getting the ball into the cup and is concerned only

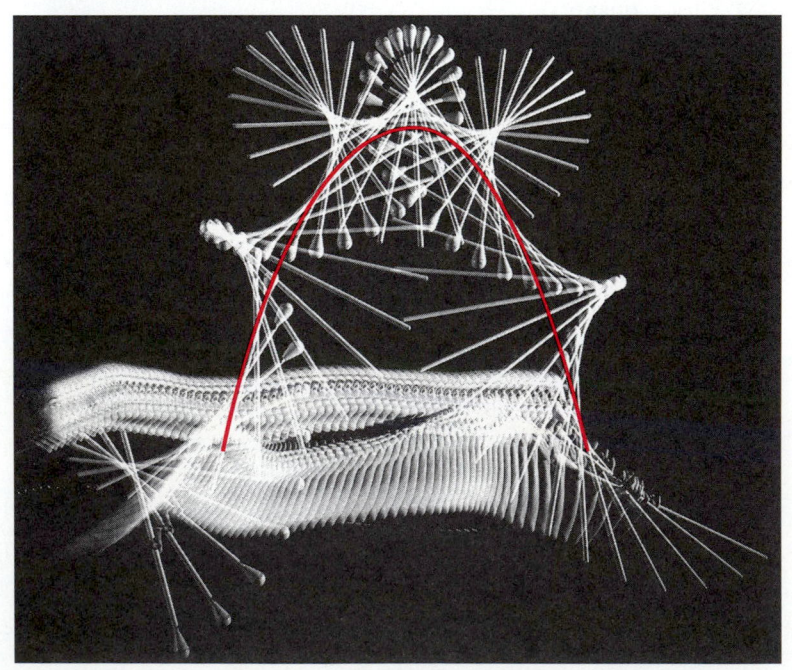

FIGURE 2.1 The flight of a baton, shown in a stroboscopic (multiple-exposure) photograph. The baton twirler throws the baton at left and catches it at right. (The lower, regular set of images is the twirler herself.) Clearly, different points of this moving extended body follow different paths. The solid curve traces the motion of a special point in the baton, called the center of mass, which follows a relatively simple path. If you trace the paths of other parts of the baton (for example, the head or the tip), you will find them much more complicated.

indirectly with how the ball spins as it flies through the air. When a body does not rotate (or when its rotation is not of interest), every part of it may be thought of as moving in the same way. We may therefore imagine that the body is just a single point. The resulting abstract object is called a **particle**. Because we want to begin our study with the simplest type of translational motion, we consider in this chapter only particles whose motion is restricted to a straight line—that is, particles whose motion is **one-dimensional**.

Position and Displacement

For everyday purposes, we can sometimes express location adequately in general terms. We can say, for example, "Savannah is in the Deep South." But that is quite different from the quantitative statement "Savannah is 1070 km due south of Cleveland." Both statements are useful in their own ways, and both are illustrated in Figure 2.2.

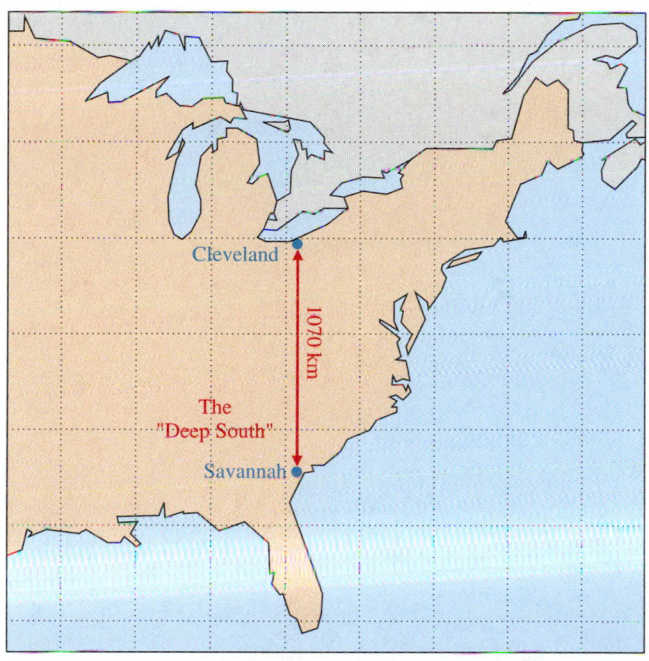

FIGURE 2.2 Two methods—one qualitative and one quantitative—for locating Savannah.

To describe the motion of a particle, we must express its position quantitatively. We must specify three things: a *reference point*, a *distance*, and a *direction*. In Figure 2.2, the reference point is Cleveland, the distance is 1070 km, and the direction is due south. We carry out the same operations in a more general way in Figure 2.3, in which we locate a point P that lies on a straight line. The reference point is the arbitrarily chosen origin O of the one-dimensional coordinate system. The distance between O and P is $|x_1|$.* (To be meaningful, the quantity $|x_1|$ must possess both a numerical value— magnitude or "size"—and a unit of length; for example, $|x_1| = 7.5$ m.) The direction from O to P also must be specified. In one dimension, we can do this in a very simple way, by attaching a sign to the quantity $|x_1|$. Most commonly, a + sign specifies the direction to the right of O and a − sign specifies the direction to the left. This is the convention used in Figure 2.3.

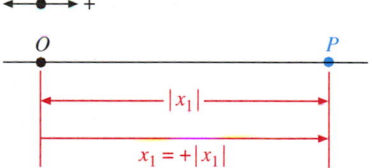

FIGURE 2.3 Locating a point P. The reference point is the origin O. The distance between points O and P is represented by the magnitude or absolute value $|x_1|$. In one dimension, the two possible choices of direction can be represented by + and − signs. The direction from O to P has been chosen as the positive direction. Consequently, the position of P with respect to O is represented by the quantity $x_1 = +|x_1|$, which incorporates a magnitude and a direction.

---

*The vertical bars around the quantity x_1 signify "absolute value of." The quantity $|x_1|$ is a positive number regardless of the sign of x_1. For example, $|+3| = 3$; $|-5| = 5$.

Taken together, the **direction** (expressed by the sign $+$ or $-$) and the **magnitude** $|x_1|$ constitute the **one-dimensional vector** x_1. For our purposes, *a vector is any quantity having both a magnitude and a direction*. The vector x_1 uniquely specifies the position of P with respect to O.

EXAMPLE 2.1

Find the distance $|x|$ between points O and P in Figure 2.4a, and find the position x of P with respect to O.

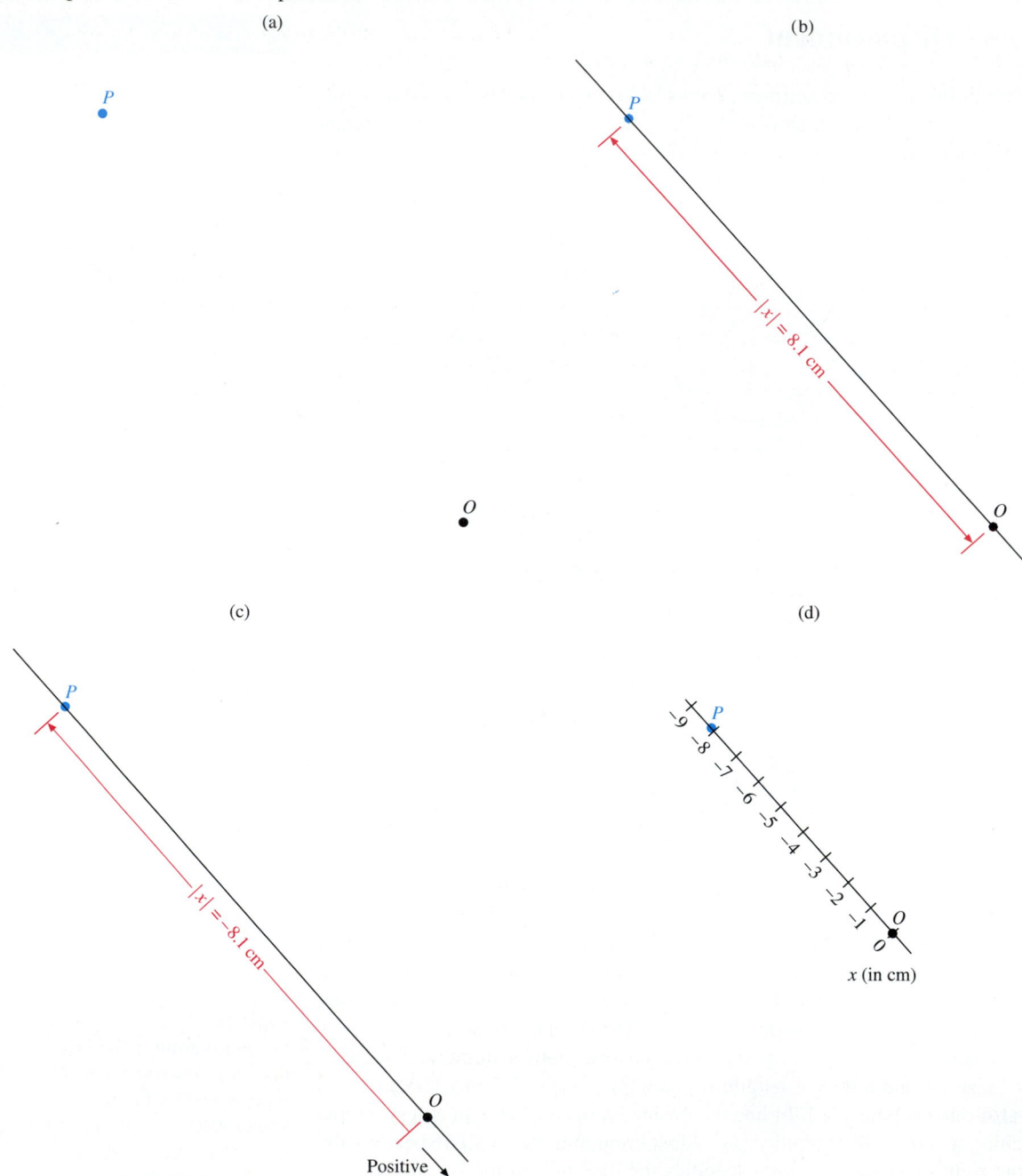

FIGURE 2.4 (*a*) The distance $|x|$ between points P and O and the position x of P with respect to O must be determined. (*b*) A coordinate axis is established on which both O and P lie, and the value of $|x|$ is measured, using a ruler graduated in cm. (*c*) The positive direction is chosen, and the position x is determined. (*d*) The process described in the first three parts of this figure is represented in a scaled graph. Scale, units, and direction are specified in the graph, and the value of the vector x, representing the position of P relative to O, can be read directly.

SOLUTION: Because O is specified as the reference point, you may as well take it as the origin of the coordinate system. Because there are no points of interest other than O and P, you are free to choose the straight line passing through O and P as the coordinate axis, as shown in Figure 2.4b.

In order to specify the distance $|x|$ between O and P, you must choose a unit of length. Simple inspection gives the order of magnitude of the distance, and the centimeter appears to be a convenient unit. Using a centimeter rule, you measure the distance and express it to two significant figures as the magnitude

$$|x| = 8.1 \text{ cm.}$$

To determine the distance—a magnitude—it is unnecessary to specify direction along the coordinate axis. In order to specify the *position* of P, however, you must adopt a sign convention for direction. Suppose you take the direction generally toward the right along the line as positive and the opposite direction as

negative, as in Figure 2.4c. Point P then lies in the negative direction from O, and the position of P is expressed by the vector

$$x = -|x| = -8.1 \text{ cm.}$$

The position of P with respect to O was established by direct measurement. Often, it is desirable to represent distances (or even, as in Figure 1.2, other quantities) by *scaling* them. (For instance, a road map might be drawn at a scale in which 1 cm on the map is equivalent to 1 km on the terrain it represents.) In Figure 2.4d, a scale has been chosen; it is represented by the division of the axis into segments, each of which represents 1 cm, even though the segments are not in fact 1 cm long. When a distance is scaled, as in Figure 2.4d, three things must be specified: (1) the quantity, (2) the unit, and (3) the scale. The first of these is specified by the label "x," the second by the notation "(in cm)," and the third by the ticks and numbers on the axis.

Because our goal here is to describe motion, we must consider *changes* of position, called **displacements**. Figure 2.5 represents a particle initially located at P_1. This position with respect to the origin O is represented by x_1. At some later time, the particle has moved to P_2. This new position with respect to the *same* origin is represented by x_2. The displacement is found by subtracting the vector x_1 from the vector x_2. This subtraction is represented graphically in the figure. The displacement is represented by the symbol Δx, defined as

$$\Delta x \equiv x_2 - x_1. \tag{2.1}$$

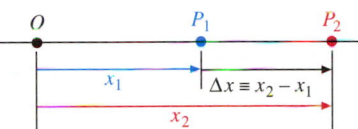

FIGURE 2.5 Displacement Δx of a particle whose initial and final positions are x_1 and x_2, respectively.

Because the displacement Δx, like the positions x_2 and x_1, has both a magnitude and a sign, it too is a one-dimensional vector. The quantity Δx is spoken "delta x." The symbol Δ (Greek delta) is widely used to represent change, and so Δx means "the change in x." It does *not* mean "delta times x."

EXAMPLE 2.2

Find the displacement of the particle in Figure 2.6, which moves from P_1 to P_2.

SOLUTION: Reading directly from the graph, you find that the initial position is given (to two significant figures) by $x_1 = 3.9$ m. The final position is given by $x_2 = 2.2$ m. Using Equation 2.1, you have

$$\Delta x = x_2 - x_1 = 2.2 \text{ m} - 3.9 \text{ m,}$$

or $$\Delta x = -1.7 \text{ m.}$$

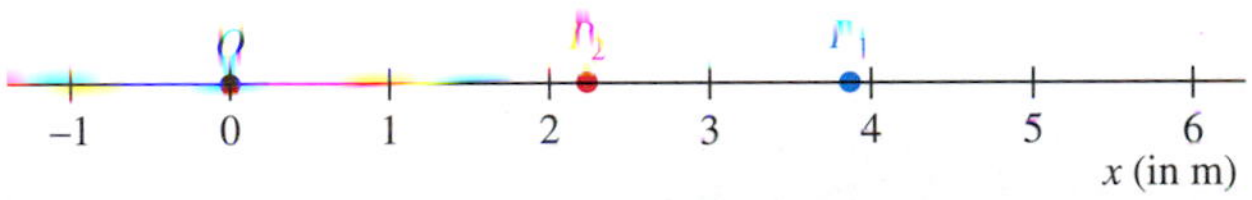

FIGURE 2.6

The displacement is negative even though both initial and final positions are represented by positive quantities. The particle has moved leftward, and because this direction was chosen as negative when the graph was drawn, the value of Δx is negative.

EXAMPLE 2.3

The displacement considered in Example 2.2 is studied by another observer, who chooses the origin O' of his coordinate system differently (Figure 2.7). Find the displacement $\Delta x'$ of the particle as this second observer sees it.

SOLUTION: You find values for x_1', the initial position of the particle at P_1, and x_2', the final position of the particle at P_2, from the new point of view. These values are different from the values x_1 and x_2 measured by observer O: $x_1' = -1.1$ m and

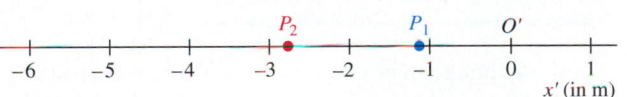

FIGURE 2.7

$x_2' = -2.8$ m. Again using Equation 2.1, you obtain

$$\Delta x' = x_2' - x_1' = -2.8 \text{ m} - (-1.1 \text{ m})$$
$$= -2.8 \text{ m} + 1.1 \text{ m} = -1.7 \text{ m.}$$

The results of Examples 2.2 and 2.3 show that

$$\Delta x' = \Delta x.$$

Unlike the vectors x_1, x_2, x_1', and x_2', each of which describes a position with respect to a particular origin, the displacement vector Δx represents a *change* in position. This change is independent of the choice of origin; in the examples, you saw that $\Delta x' = \Delta x$. Put in geometric terms, the displacement from P_1 to P_2 is the difference $\overline{OP_2} - \overline{OP_1}$ between the directed lengths—the "arrows"—from the origin to the final and initial positions. By comparing Figures 2.6 and 2.7, you can see that the difference does not depend on the choice of origin.

Because the displacement vector is independent of the choice of origin, it is called a **free vector**. In contrast, a position vector depends on the choice of origin. Such a vector is called a **bound vector** because its tail end is bound to the origin. Most (but not all) vectors of interest in physics are free vectors. Consequently, when the term "vector" is used, it almost always means "free vector."

A vector is conveniently represented by an arrow, as we have already done in Figure 2.5. The length of the arrow represents the magnitude of the displacement, and the direction of the arrow represents the direction of the displacement. Displacement is typical of physical quantities represented by free vectors. Indeed, *vector* is the Latin word meaning "carrier"; you can think of the vector as "carrying" a particle through the displacement that the vector describes.

Vectors and Scalars

Before proceeding further in our study of physics, we consider the mathematical properties of vectors in more detail. We first make a distinction between vectors and another class of quantities called *scalars*.

Vectors: A vector is a quantity that has both a magnitude and a direction. Position and displacement are two examples of vectors, and you will meet many other kinds of vectors as you continue your study of physics.

Scalars: For our purposes, a **scalar** is a quantity that has magnitude but not direction. The distance between two points is a scalar because only a magnitude (and not a direction) is needed to specify it. In Example 2.1, for example, you found that the distance $|x|$ between points O and P was 8.1 cm. It does not matter whether you measure from O to P or from P to O; the distance is the same either way.

The magnitude of a vector is one example of a scalar quantity. Many other physical quantities may be represented as scalars as well. Among them are time, volume, and temperature.

Some scalars—such as distance and volume—are always positive. It makes no sense to say that the distance between New York and Chicago is -1400 km or that a certain bottle has a capacity of -10 liters. Other scalars—temperature is one example—can be either positive or negative; on a winter day, the temperature can be 18°C indoors and -10°C outdoors. But no spatial direction is implied by the sign; when we write "-10°C," we mean only that -10°C is colder than 0°C.

Vectors for Mathematicians

The mathematician's approach to defining vectors and scalars is different from the physical approach that we have taken. To a mathematician, a vector is an ordered n-tuplet of numbers $(x_1, x_2, \ldots x_n)$, and a scalar is merely the special case of a vector in one dimension (x_1). The mathematical definitions of vector and scalar have the advantage of greater generality, and they avoid some subtle inconsistencies in the physical definitions used here. However, the physical definition of a vector has the advantage of directness; it is physically meaningful to say that any quantity having magnitude and direction is a vector quantity.

THINKING LIKE A PHYSICIST

Physicists often adopt concepts from the world of mathematics because these concepts prove useful for expressing and manipulating physical quantities. The vector is a typical mathematical concept, with properties of interest to mathematicians. Physicists have adopted vectors because they afford a convenient and powerful way of manipulating many physical quantities, of which displacement is only one. As you continue to study physics, you will learn much more about both the mathematical properties of vectors and the physical properties of the quantities they represent.

The interaction of physics and mathematics is a two-way street. Entire fields of mathematics have sprung from the solution of a physical problem. The calculus is one example.

As long as we restrict our consideration of vectors to one dimension, there is little difference between the *mathematical* rules for treating vectors and those for treating scalars. However, the *physical* meanings of vector and scalar quantities are quite different. When we treat vector quantities in two or three dimensions in Chapter 4, the different *mathematical* properties of vectors and scalars will become apparent and will demand special attention. For the present, bear in mind that *the magnitude of a vector is a positive scalar*. For example, if a displacement vector is given as -10 cm, its magnitude (the length of the vector) is the scalar $|-10 \text{ cm}| = 10$ cm.

Mean Velocity

We always observe the position of a particle at a certain time. Suppose that, in the course of measuring the displacement of a particle as in Section 2.2, we keep a record not only of the positions P_1 and P_2 but also of the times t_1 and t_2 at which the positions are observed. The displacement $\Delta x = x_2 - x_1$ occurs over a **time interval** Δt, defined to be

$$\Delta t \equiv t_2 - t_1. \tag{2.2}$$

The quantity Δt is a scalar. The sign of Δt is normally positive because t_2 normally represents a time later than t_1.

The **mean velocity** $\langle v \rangle$ of our particle in its journey from P_1 to P_2 is defined as the displacement divided by the corresponding time interval. This definition is conveniently expressed in the mathematical form

$$\langle v \rangle \equiv \frac{\Delta x}{\Delta t} = \frac{x_2 - x_1}{t_2 - t_1}. \tag{2.3}$$

The brackets $\langle \ \rangle$ denote a mean (average) quantity.* In SI, the units of $\langle v \rangle$ are m/s (spoken as "meters per second"). Because Δt is positive, the sign of $\langle v \rangle$ is the same as that of the displacement Δx with which it is associated.

EXAMPLE 2.4

In the system studied in Examples 2.2 and 2.3, the observers start a stopwatch before determining the position of the particle at P_1 and its later position at P_2. The first determination is made by both observers just as the watch reads $t_1 = 37$ s. The second determination is made when the watch reads $t_2 = 48$ s. Find the mean velocity.

SOLUTION: Regardless of which observer's data you use, the displacement of the particle has the value $\Delta x = -1.7$ m. According to Equation 2.2, the corresponding time interval is $\Delta t = 48 \text{ s} - 37 \text{ s} = 11$ s. Using Equation 2.3, you obtain the mean velocity

$$\langle v \rangle = \frac{-1.7 \text{ m}}{11 \text{ s}} = -0.15 \text{ m/s}.$$

Graphical Interpretation of Mean Velocity

When we measure the position of a particle at two different times, we may summarize the results as two pairs of numbers, (t_1, x_1) and (t_2, x_2). Figure 2.8 represents these number pairs graphically. Each point on the graph corresponds uniquely to one pair of numbers. Also shown are the time interval $\Delta t = t_2 - t_1$, represented graphically by the horizontal line segment, and the displacement $\Delta x = x_2 - x_1$, represented graphically by the directed vertical line segment (whose direction is from x_1 to x_2).

How does the graph express the mean velocity $\langle v \rangle \equiv \Delta x/\Delta t$? The mean velocity is the *slope* of the red line that joins the points (t_1, x_1) and (t_2, x_2). As a particle

---

*In an alternative notation, mean values are denoted by a bar, thus: $\bar{v}$. The two notations mean the same thing; $\bar{v} \equiv \langle v \rangle$.

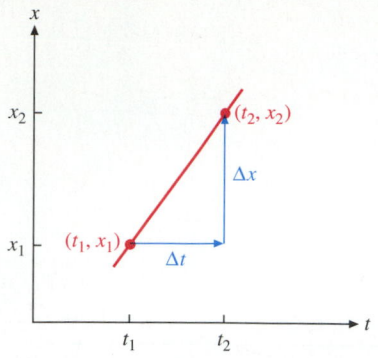

FIGURE 2.8 A two-dimensional graph, with t and x axes, is used to depict the one-dimensional motion of a particle along the x axis. Every point on the plane of the graph is uniquely represented by some pair of numbers (t, x) that denotes a possible position x of the particle at time t. The points (t_1, x_1) and (t_1, x_2) represent the positions x_1 and x_2 of the particle at two times t_1 and t_2, respectively. The slope $\Delta x/\Delta t$ of the red line represents the mean velocity $\langle v \rangle$ of the particle in the time interval beginning at t_1 and ending at t_2.

Time Is a Scalar

Why do we not refer to Δt as a *directed* line segment? Because there is only one possibility for the direction of flow of time and thus no need to specify a "direction." That is why time, unlike displacement, is a scalar quantity. (If we force Δt to have a negative value by taking $\Delta t = t_1 - t_2$, we know that we are arbitrarily "taking time backward." But, in doing this, we force a sign change in the value of Δx as well, and the sign of $\Delta x/\Delta t$ remains unchanged.)

continues to move, the mean velocity—and so the slope—may change. But let us first consider **constant velocity**. Figure 2.9 shows the result of a series of $n + 1$ measurements of a particle's position, $x_0, x_1, x_2, x_3, \ldots, x_j, \ldots, x_n$, made at evenly separated times $t_0, t_1, t_2, t_3, \ldots, t_j, \ldots, t_n$.* For convenience, the initial time t_0 has been set at zero, as it would be if you started a stopwatch at the beginning of a series of measurements.

Any pair of points can be used to calculate a mean velocity. In Figure 2.9, the result will be the same no matter which pair is chosen because every point (t_j, x_j) lies on the line whose slope is $\langle v \rangle$. In such a case, we say that *the velocity has the constant value* v. That is, we have $\langle v \rangle = v$; the mean value of any constant quantity is the quantity itself.

What if we had made an additional measurement at some time t' between two instants at which we actually made measurements? We have no sure way of saying what the corresponding position x' would have been. However, experience strongly suggests that motion is *continuous*. As long as the measurements are made close enough together, we can safely guess that any intermediate measurements would yield still more points on the same smooth, straight line. This straight line can be represented by an equation that expresses the position x at *any* time as a function of the time t. The line intercepts the x axis at x_0 and has slope v, and the equation of the line is thus

$$x = x_0 + vt \quad \text{for } v = \text{constant and } t_0 = 0. \qquad (2.4)$$

If the initial position has the arbitrary value x_i, Equation 2.4 can be written

$$x = x_i + vt \quad \text{for } v = \text{constant and } t_0 = 0. \qquad (2.5)$$

The conditions are written explicitly at the right of the equation as a reminder that *the equation will be valid only if the conditions are met.*

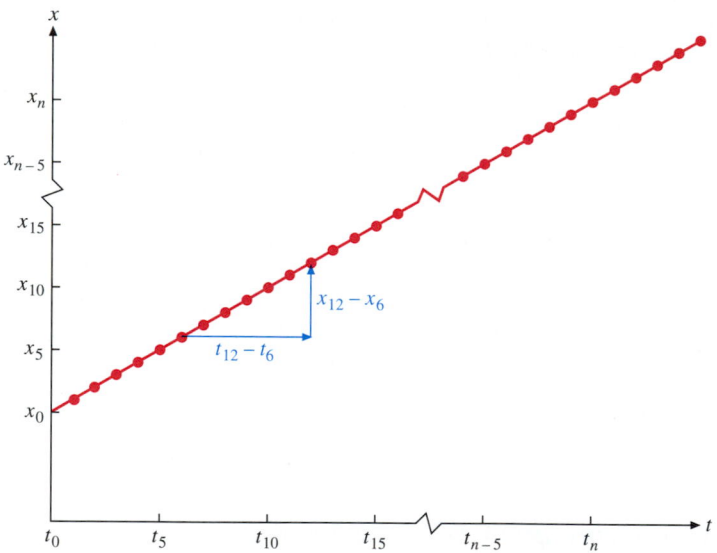

FIGURE 2.9 Position measurements, made at equal time intervals, of a particle moving with constant velocity. The velocity v may be found by measuring the slope of the line joining any pair of points (t_j, x_j). Because all the points lie on the same line, it makes no difference which pair is chosen. In the construction shown, the points (t_6, x_6) and (t_{12}, x_{12}) are used.

*In this notation, j (called the *dummy index*) represents an arbitrary integer in the range $0 \leq j \leq n$.

EXAMPLE **2.5**

The famous *shinkanzen*, or "bullet train," travels at a steady 190 km/h along the line from Tokyo via Nagoya to Osaka, a total distance of 495 km. The train covers the distance from Nagoya to Osaka in 49.0 min—that is, 0.817 h. How far is Nagoya from Tokyo? Neglect the short time that the train spends speeding up and slowing down.

SOLUTION: Choose Tokyo, the starting point of the line, as the origin ($x = 0$) of the one-dimensional coordinate system, and take the direction from Tokyo toward Nagoya and Osaka as positive. Then make a rough sketch of x as a function of t (like that of Figure 2.10) to clarify the situation. The position of the train at Nagoya, at time $t_0 = 0$, has the value x_0 that you wish to determine. When the train is at Osaka at time $t = 0.817$ h, its position has the corresponding value $x = 495$ km. You thus have values for all the quantities in Equation 2.4 except the desired quantity x_0, and you can solve the equation for x_0:

$$x_0 = x - vt = 495 \text{ km} - 190 \text{ km/h} \times 0.817 \text{ h} = 340 \text{ km}.$$

The distance from Tokyo to Nagoya is 340 km.

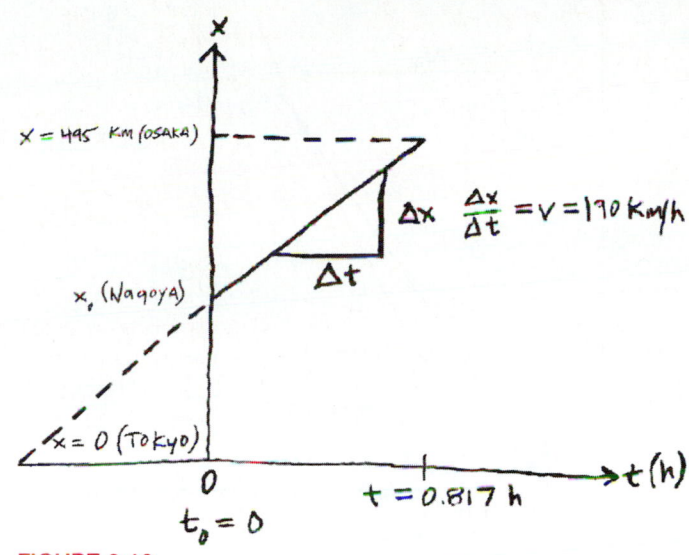

FIGURE 2.10

Instantaneous Velocity and Derivatives

Mean velocity is defined in terms of measurements of two positions, made at two distinct instants separated by a known time interval. But we do not normally experience velocity in this way. When you are driving a car, it takes just one reading of the speedometer to determine how fast you are going. Indeed, with a little experience, you can make a pretty good guess of your speed without even looking at the speedometer. Such experiences lend intuitive validation to the notion of **instantaneous velocity**—a velocity associated with a *single* instant of time. Instantaneous velocity and mean velocity may appear at first glance to be conceptually different, but they are connected in a way that we now make explicit. The connection will provide a basis for dealing with velocities that are *not* constant.

Figure 2.11 is a graph of position as a function of time for a particle whose velocity is not constant. We are interested in finding the instantaneous velocity of the particle at P—that is, at $(t, x) = (0.5 \text{ s}, 1.225 \text{ m})$. (Because our calculation is purely mathematical and contains no quantities obtained from physical measurements, all numerical values are exact.) Let us begin with the definition of mean velocity developed in Section 2.3 and show that it is consistent with our intuitive notion of instantaneous velocity.

Consider first the time interval PA shown in Figure 2.11a. This interval begins at $t = 0.5$ s and ends at $t = 2.5$ s, and thus $\Delta t = 2$ s. The mean velocity over the interval, $\langle v \rangle_{PA}$, can be found by reading the corresponding value of Δx from the graph and employing Equation 2.3, $\langle v \rangle = \Delta x / \Delta t$. We obtain

$$\langle v \rangle_{PA} = \frac{30.6 \text{ m} - 1.2 \text{ m}}{2.5 \text{ s} - 0.5 \text{ s}} = 14.7 \text{ m/s}.$$

Next, consider the shorter time interval PB. It begins at the same initial time as PA but is only half as long. As you can see, the slope $\langle v \rangle_{PB}$ is less than the slope $\langle v \rangle_{PA}$. If you carry out a calculation just like that in the preceding paragraph, you will find that it yields $\langle v \rangle_{PB} = 9.8$ m/s.

The time interval PC is half as long as PB. Again, the mean velocity is smaller; you will find that $\langle v \rangle_{PC} = 7.4$ m/s. The time interval PC is shown again on an expanded

(a)

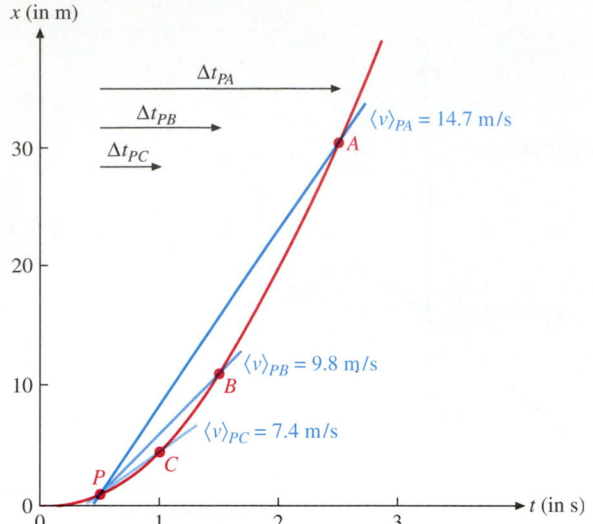

(b)

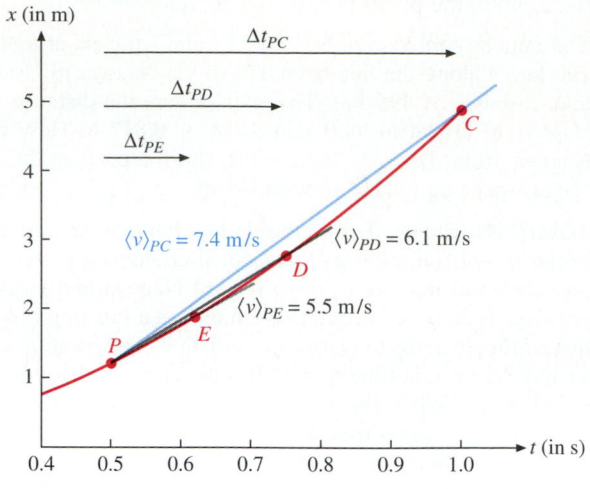

FIGURE 2.11 Depiction of the limiting process used to find the instantaneous velocity of a particle at $(t, x) = (0.5 \text{ s}, 1.225 \cdot \text{m})$. (*a*) The graph point (t, x) is denoted by the symbol P. The curve depicts the position of the particle as a function of time. The slopes of the straight lines PA, PB, and PC are the values of the mean velocity of the particle over various time intervals beginning at $t = 0.5$ s. (*b*) The part of the curve in the vicinity of point P, drawn on an expanded scale.

scale in Figure 2.11b. When the interval is cut in half twice more, the results for the mean velocity are $\langle v \rangle_{PD} = 6.1$ m/s and $\langle v \rangle_{PE} = 5.5$ m/s.

Summarizing these results, we have the sequence of mean velocities 14.7 m/s, 9.8 m/s, 7.4 m/s, 6.1 m/s, and 5.5 m/s. As the time interval gets smaller and smaller, the mean velocity diminishes. The figures suggest, however, that the value of $\langle v \rangle$ converges to a nonzero limiting value, somewhat less than 5.5 m/s, as the end point is brought as close as you wish to P. (To see this, take the differences between all the pairs of successive values of $\langle v \rangle$ that we have calculated, and observe how these differences diminish.) Inspection of the graph suggests the same thing.

In Figure 2.11, each of the straight lines PA, PB, and so forth, cuts the curve describing the actual motion of the particle in two places. Each of the straight lines is called a **chord** of the curve. As the two points at which the chord and the curve intersect are brought closer and closer together, it becomes harder and harder to distinguish between the chord and the arc of the curve that it cuts. For example, in Figure 2.11b, the curve in the region PE is hard to distinguish from a straight line. Strictly speaking, we cannot assign a fixed value to the slope of the curve in this region. But nowhere in the region is the curve very different from the chord, whose slope *does* have a fixed value.

Let us make a leap of the imagination and ask what happens when the end point of the time interval is brought *infinitesimally* close to the initial point P so that it "merges" with P; that is, the time interval is so short that it cannot be distinguished from a single instant. The straight line is no longer a chord but a **tangent**, which touches the curve at *one* place. The slope of the tangent is a velocity, but this velocity applies to a single instant and not to a time interval. Thus the velocity is no longer a mean velocity $\langle v \rangle$, but an instantaneous velocity v.

We could in principle determine the value of v by the graphical method—that is, by redrawing graphs like those of Figure 2.11 with more and more expanded scales until we obtained a result having the desired accuracy. But such a method would be impossibly laborious. Fortunately, there is a quick and fairly simple way of evaluating the slope of the tangent to a curve at any desired point, provided that the curve can be expressed as a mathematical function. The method is called *differentiation*. Still more fortunately, the position-vs.-time curves that describe the most important types of particle motion can be expressed as simple mathematical functions.

Differentiation

Differentiation is developed in detail in every introductory calculus text. If you have already studied calculus, you may wish to skip this subsection or use it as a quick review. But if you are studying calculus concurrently with physics and have not yet dealt with differentiation, the bare-bones sketch presented here will tide you over until you have had the opportunity to study differentiation more thoroughly.

The position-vs.-time graphs of Figure 2.11 depict the motion of a freely falling particle. The position of the particle at various times is represented quite well by the function

$$x = bt^2, \tag{2.6}$$

where b is a constant whose value is close to 4.90 m/s^2 in SI units (the reasons underlying this statement are the main topic of Section 2.7). The quantity x is the vertical displacement of the particle from the point where it was released at time $t = 0$, with the downward direction taken positive.

Figure 2.12 is a graph of Equation 2.6. It differs from Figure 2.11 only in that it does not depict specific numerical values. We wish to evaluate the instantaneous velocity v at an arbitrarily chosen time t_i, when the position of the particle at P is given by $x_i = bt_i^2$. To do this, we proceed exactly as in Figure 2.11 and the accompanying text but in more general terms. Just as before, the mean velocity over a finite time interval such as PA, beginning at $t = t_i$ and ending at $t = t_f$, is defined by Equation 2.3:

$$\langle v \rangle = \frac{\Delta x}{\Delta t}. \tag{2.7}$$

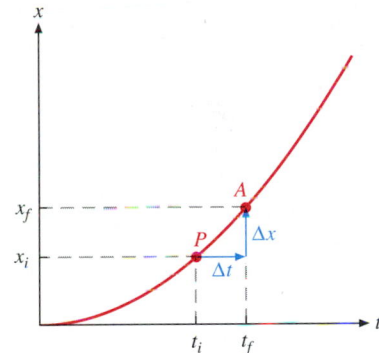

FIGURE 2.12 Generalization of Figure 2.11.

Our method depends explicitly on the possibility of expressing x as a function of t. We have just such a functional relation in Equation 2.6, and we exploit it to express the numerator of Equation 2.7 in terms of t. To begin with, we use the definition $\Delta x \equiv x_f - x_i$. Then, we substitute $x_f = bt_f^2$ and $x_i = bt_i^2$ to obtain

$$\Delta x = bt_f^2 - bt_i^2. \tag{2.8}$$

Next, we imagine that point A of Figure 2.12 slides along the curve toward point P. When this happens, the value of t_f changes. The sliding process will be easier to interpret if we eliminate t_f from Equation 2.8 by making the substitution $t_f = t_i + \Delta t$. We then have

$$\Delta x = b(t_i + \Delta t)^2 - bt_i^2.$$

The quadratic term $(t_i + \Delta t)^2$ can be expanded; this gives

$$\Delta x = b[t_i^2 + 2t_i \Delta t + (\Delta t)^2] - bt_i^2$$
$$= bt_i^2 + 2bt_i \Delta t + b(\Delta t)^2 - bt_i^2.$$

The first and last terms on the right side cancel. What remains is

$$\Delta x = 2bt_i \Delta t + b(\Delta t)^2.$$

Now that we have obtained a value for Δx in terms of t_i and Δt, we can go back and substitute it into Equation 2.7:

$$\langle v \rangle = \frac{\Delta x}{\Delta t} = \frac{2bt_i \Delta t + b(\Delta t)^2}{\Delta t},$$

or

$$\langle v \rangle = 2bt_i + b \Delta t. \tag{2.9}$$

What happens to the value of $\langle v \rangle$ as point A in Figure 2.12 slides closer and closer to point P? The sliding process is called **taking the limit** of the quantity $\langle v \rangle$ as the time interval Δt approaches zero. The concise notation that describes the process is

$$v = \lim_{\Delta t \to 0} \langle v \rangle. \tag{2.10}$$

That is, *the instantaneous velocity v is equal to the limiting value of the mean velocity $\langle v \rangle$ as the time interval Δt over which $\langle v \rangle$ is evaluated approaches zero.*

We can substitute the value of $\langle v \rangle$ given by Equation 2.9 into Equation 2.10 to obtain

$$v = \lim_{\Delta t \to 0} (2bt_i + b\,\Delta t).$$

Consider what happens to the quantity $2bt_i + b\,\Delta t$ as we carry out the limiting process. The term $2bt_i$ does not contain Δt and thus does not depend at all on the value of Δt. The term $b\,\Delta t$ is just Δt multiplied by a constant. As Δt approaches zero, so does $b\,\Delta t$:

$$v = \lim_{\Delta t \to 0} (2bt_i + b\,\Delta t) = 2bt_i. \tag{2.11}$$

We have thus made an *exact* determination of the value of the instantaneous velocity v at time t_i when $x = bt_i^2$. This result is more general than it looks at first. We made our calculation for the particular time t_i. We set no restrictions on choosing t_i, however, so t_i can be *any* time. We are therefore justified in dropping the subscript and writing

$$v = 2bt \quad \text{for } x = bt^2. \tag{2.12}$$

Our process of taking the limit can be written symbolically as

$$v = \lim_{\Delta t \to 0} \frac{\Delta x}{\Delta t}. \tag{2.13}$$

Even though the denominator of the fraction $\Delta x / \Delta t$ is made to approach zero, the fraction does not "blow up"—that is, does not increase without limit. This is because the numerator also becomes small, so the fraction remains finite. There is a special notation, and a special name, for the fraction $\Delta x / \Delta t$ after the limiting process has been carried out. We define

$$\frac{dx}{dt} \equiv \lim_{\Delta t \to 0} \frac{\Delta x}{\Delta t}. \tag{2.14}$$

The quantity dx/dt is called the **derivative of x with respect to t**. It is often spoken as "*dx* by *dt*" or "*dx dt*" but never as "*dx* over *dt*" or "*dx* divided by *dt*." You should consult an introductory calculus text for the subtle but important reasons for this distinction.

From Equations 2.13 and 2.14, it follows that

$$v = \frac{dx}{dt}. \tag{2.15}$$

That is, *instantaneous velocity is the derivative of position with respect to time.* Except when confusion might arise, the qualifier "instantaneous" is conventionally dropped and we speak simply of the velocity.

EXAMPLE 2.6

A particle moves in such a way that its position is described by the function $x = kt$, where $k = 27.4$ m/s. Using the limiting process described in the text above, find the velocity of the particle at time $t = 31.2$ s.

SOLUTION: You begin with Equation 2.7 in the form $\langle v \rangle = (x_f - x_i)/\Delta t$. Because in the present case $x = kt$, you have $x_f = kt_f$ and $x_i = kt_i$. You substitute these values into the expres-

sion for $\langle v \rangle$ to obtain

$$\langle v \rangle = \frac{kt_f - kt_i}{\Delta t} = \frac{k(t_f - t_i)}{\Delta t}.$$

But the quantity $(t_f - t_i)$ is just the quantity Δt, so $\langle v \rangle = k(\Delta t / \Delta t)$, or

$$\langle v \rangle = k. \tag{2.16}$$

To find the instantaneous velocity v, you must take the limit of the value of $\langle v \rangle$ as Δt approaches zero. In the present case, the task is a trivial one because the right side of Equation 2.16 does not contain Δt at all. Consequently, the value of the right side does not change with time; you have

$$v = \lim_{\Delta t \to 0} \langle v \rangle = \lim_{\Delta t \to 0} k = k;$$

that is,

$$v = \frac{dx}{dt} = k \quad \text{for } x = kt. \tag{2.17}$$

The velocity is constant. In this case—the simplest possible—the (instantaneous) velocity at $t = 31.2$ s is the same as the velocity at any other time—namely, $v = k = 27.4$ m/s. Both the constant-velocity case and the quadratic case $x = bt^2$, for which $v = 2bt$ (Equation 2.12), are found very frequently in physical situations.

You can confirm the constancy of v by plotting the function $x = kt$, as shown in Figure 2.13. The position of the particle is a linear function of time, and the curve $x(t)$ has a constant slope $\Delta x/\Delta t$ whose value does not depend on the interval Δt over which you measure it. One possible measurement interval Δt is shown in blue in Figure 2.13, together with the corresponding value of Δx.

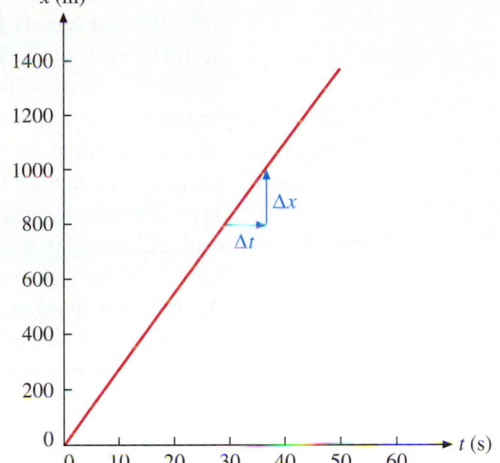

FIGURE 2.13

The process of obtaining the derivative of a function is called **differentiation**. If you have not already done so in your study of calculus, you will soon learn a complete set of rules for differentiation of functions. For the present purpose, we need two of these rules:

1. Consider any function of the form $x = qt^n$, where q and n are arbitrary constants. The rule for finding the derivative dx/dt is

$$\frac{dx}{dt} = nqt^{n-1} \quad \text{for } x = qt^n. \tag{2.18}$$

In particular, any constant c can be written $c = ct^0$. We therefore have

$$\frac{dc}{dt} = 0 \quad \text{for } c = \text{constant}. \tag{2.19}$$

Equation 2.19 states that c does not change as t changes; that is what is meant by "constant."

2. Let the function have the somewhat more complicated form

$$x = qt^n + pt^m$$

where q, p, n, and m are constants. The rule for finding the derivative is

$$\frac{dx}{dt} = nqt^{n-1} + mpt^{m-1}. \tag{2.20}$$

That is, the derivative of a sum of terms is just the sum of the derivatives of the individual terms. (This rule applies even if the terms are not powers of the variable t but other functions of t.)

We will not prove these rules, which are derived in all introductory calculus texts. However, you can see that the two results given in Equations 2.11 and 2.17 are special cases of Equation 2.18.

Velocity and Speed

In the one-dimensional world that we are considering at present, every velocity is characterized by a magnitude and a sign. In the definition

$$v = \frac{dx}{dt} = \lim_{\Delta t \to 0} \frac{\Delta x}{\Delta t},$$

both Δx (or dx) and Δt (or dt) possess a magnitude and a sign, and thus their ratios must also. Velocity is a vector quantity and the sign of the vector denotes the direction of motion.

The magnitude $|v|$ of the velocity is a scalar, called the **speed**. Do not confuse velocity and speed. Two automobiles traveling in opposite directions on the same road at 100 km/h have the same speed, but their velocities have opposite signs. The distinction is equally important for mean velocity $\langle v \rangle$ and mean speed $\langle |v| \rangle$. Suppose, for instance, that a 50-m swimming race is held in a 25-m pool, and the winner requires 45 s to traverse the pool and return to his starting point. His mean velocity is zero, because his change in position over the entire race is zero. But his mean speed is not zero. The mean speed is calculated by dividing the total distance covered by the elapsed time. The scalar distance—a quantity having magnitude but not direction—is represented by the positive number 2×25 m $= 50$ m. The elapsed time is represented by the positive number 45 s. Thus the mean speed of the swimmer is $\langle |v| \rangle = 50$ m/45 s $= 1.1$ m/s. How will his instantaneous velocity and instantaneous speed differ from this value at $t = 0.2$ s after the starting gun? at $t = 15$ s? at $t = 35$ s?

SECTION 2.5

Acceleration

To describe the motion of a particle when its velocity is not constant, we need a quantitative measure of the change of velocity with time. This measure is called *acceleration* (from the Latin word meaning ''to hasten''). Acceleration as a qualitative experience is familiar to anyone who has stood in a bus as it leaves a bus stop or comes to a sudden stop in a traffic jam. As we develop a quantitative description of acceleration, you will find that the mathematical ideas have a familiar ring.

Figure 2.14 is a plot of the instantaneous velocity v of a particle versus time t. The velocity increases rapidly at first, then less rapidly, and then more rapidly again. (The velocity of an automobile as the automatic transmission shifts gears might look like this.) At some initial time t_i the velocity is v_i; at a later time t_f the velocity is v_f. We define the **mean acceleration** $\langle a \rangle$ to be the change of velocity divided by the corresponding time interval:

$$\langle a \rangle \equiv \frac{v_f - v_i}{t_f - t_i} = \frac{\Delta v}{\Delta t}. \qquad (2.21)$$

If you compare this definition with Equation 2.3, $\langle v \rangle \equiv \Delta x / \Delta t$, you will see that mean acceleration bears the same relation to change in velocity that mean velocity bears to change in position. Specifically, the mean acceleration is the slope of the chord cutting the velocity-vs.-time curve at t_i and t_f, just as the mean velocity is the slope of the chord cutting the corresponding position-vs.-time curve.

By inspecting Equation 2.21, you can see that the dimensions of acceleration are [velocity]/[time]—that is, $\{$[length]/[time]$\}$/[time]. However, this notation is awkward and potentially confusing. Knowing that dimensions obey the same rules of multiplication and division that numbers obey, we write

$$\frac{[\text{length}]/[\text{time}]}{[\text{time}]} = [\text{length}]/[\text{time}]^2.$$

In SI, the corresponding unit is m/s^2 (spoken ''meters per second squared''); this is the SI unit of acceleration. Like displacement and velocity, acceleration is a vector quantity having both magnitude and direction. The definition of mean acceleration is applied in Example 2.7.

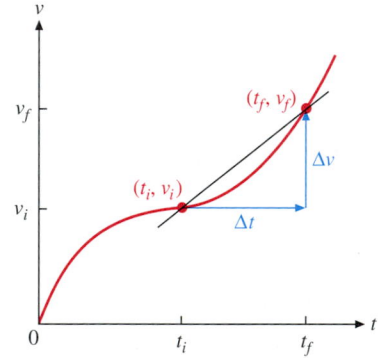

FIGURE 2.14 Plot of the velocity v of a particle as a function of time t. The slope $\Delta v / \Delta t$ of the chord shown represents the mean acceleration $\langle a \rangle$ of the particle in the time interval beginning at t_i and ending at t_f.

EXAMPLE 2.7

An automobile can accelerate from 0 to 80 km/h in 9.2 s; it can accelerate from 0 to 100 km/h in 12.3 s. Using SI units, find the mean accelerations $\langle a \rangle_1$ in the speed range 0 to 80 km/h, $\langle a \rangle_2$ in the range 0 to 100 km/h, and $\langle a \rangle_3$ in the range 80 km/h to 100 km/h.

SOLUTION: First express the given speeds in the SI unit m/s:

$$80 \ \frac{km}{h} = 80 \ km \times \frac{10^3 \ m}{1 \ km} \times \frac{1}{h} \times \frac{1 \ h}{3600 \ s} = \frac{80 \times 10^3}{3600} \ \frac{m}{s}$$

$$= 22 \ m/s.$$

You could repeat the calculation to express 100 km/h in m/s. But it is easier to note that 100 km/h = $\frac{5}{4} \times$ 80 km/h. Using the result just obtained, you have

$$100 \ km/h = \tfrac{5}{4} \times 22 \ m/s = 28 \ m/s.$$

You can now find $\langle a \rangle_1$, the mean acceleration during the first 9.2 s, by using Equation 2.21:

$$\langle a \rangle_1 = \frac{22 \ m/s - 0 \ m/s}{9.2 \ s - 0 \ s} = 2.4 \ m/s^2.$$

You find that, in the first 12.3 s (the time required for the automobile to reach 100 km/h from rest), the mean acceleration is

$$\langle a \rangle_2 = \frac{28 \ m/s - 0 \ m/s}{12.3 \ s - 0 \ s} = 2.3 \ m/s^2.$$

Finally, consider the acceleration in the speed range 80 km/h to 100 km/h, called the high-speed acceleration. The corresponding time interval begins at $t_i = 9.2$ s and ends at $t_f = 12.3$ s. So the mean acceleration $\langle a \rangle_3$ is

$$\langle a \rangle_3 = \frac{28 \ m/s - 22 \ m/s}{12.3 \ s - 9.2 \ s} = 1.9 \ m/s^2.$$

The time required to reach 80 km/h (50 mi/h) from a standing start is often quoted as a criterion of performance in automobile advertising. The time required to reach 100 km/h (60 mi/h) is quoted less often because it is longer and the mean acceleration is somewhat smaller. The high-speed acceleration $\langle a \rangle_3$ is an important measure of the car's ability to pass other cars. It is always considerably smaller than $\langle a \rangle_1$ and $\langle a \rangle_2$.

In Example 2.7, the performance of the automobile is quoted in terms of speeds rather than velocities. We follow this practice because the speed is increasing in every case, and we take for granted that the automobile is moving in the positive direction. In general, however, it is important to remember that the sign of an acceleration depends on both the magnitudes and the signs of the initial and final velocities. Consider, for example, a particle that is moving in the negative direction. Suppose that its initial velocity is $v_1 = -10$ m/s, and 5 s later its velocity is $v_f = -20$ m/s; that is, the particle is "speeding up":

$$|v_f| - |v_i| = 20 \ m/s - 10 \ m/s = 10 \ m/s.$$

Nevertheless, the velocity is decreasing (becoming more negative):

$$v_f - v_i = -20 \ m/s - (-10 \ m/s) = -10 \ m/s.$$

Thus, the mean acceleration is negative, even though the speed is increasing:

$$\langle a \rangle = \frac{v_f - v_i}{\Delta t} = \frac{-10 \ m/s}{5 \ s} = -2 \ m/s^2.$$

Instantaneous Acceleration

The **instantaneous acceleration** a of a particle is defined as the limiting value of its mean acceleration as the time interval Δt is made smaller and smaller:

$$a \equiv \lim_{\Delta t \to 0} \frac{\Delta v}{\Delta t}. \tag{2.22a}$$

Using the derivative notation developed in Section 2.4, we can write this definition in the form

$$a \equiv \frac{dv}{dt}. \tag{2.22b}$$

That is, *instantaneous acceleration is the derivative of the velocity with respect to time.*

Earning an F in kinematics.

Instantaneous acceleration is much more important in physics than mean acceleration, so we usually drop the qualifier "instantaneous." Except in the rare cases in which a possibility of confusion exists, **acceleration** implies instantaneous acceleration.

The acceleration of a particle is related to its displacement through the velocity. To see this, we can substitute Equation 2.15, $v = dx/dt$, into Equation 2.22b to obtain

$$a = \frac{d}{dt}\left(\frac{dx}{dt}\right). \qquad (2.23a)$$

That is, the acceleration is the derivative of the derivative of the displacement, where both derivatives are taken with respect to time. Taking two derivatives successively with respect to the same variable is called taking the **second derivative**. Equation 2.23a is usually written in the following shorthand notation:

$$a = \frac{d^2x}{dt^2}. \qquad (2.23b)$$

By convention, dt^2 implies $(dt)^2$. The quantity d^2x/dt^2 is usually spoken "d-squared x by dt-squared." Although the notation of Equation 2.23b is universal, some caution is required in interpreting it. It does *not* imply that some quantity "d^2" is to be multiplied by x and then divided by "d times t^2." Rather, the notation implies that the derivative of x is to be taken twice in succession with respect to t.

Figure 2.15 illustrates the relations among acceleration, velocity, and displacement. It schematically traces the motion of a subway train as it travels from one station to the next. Starting from rest, the train accelerates with positive acceleration—speeds up—until it reaches the speed limit. The train then travels at constant velocity (that is, $a = 0$) until it nears the second station. Then it accelerates with negative acceleration—

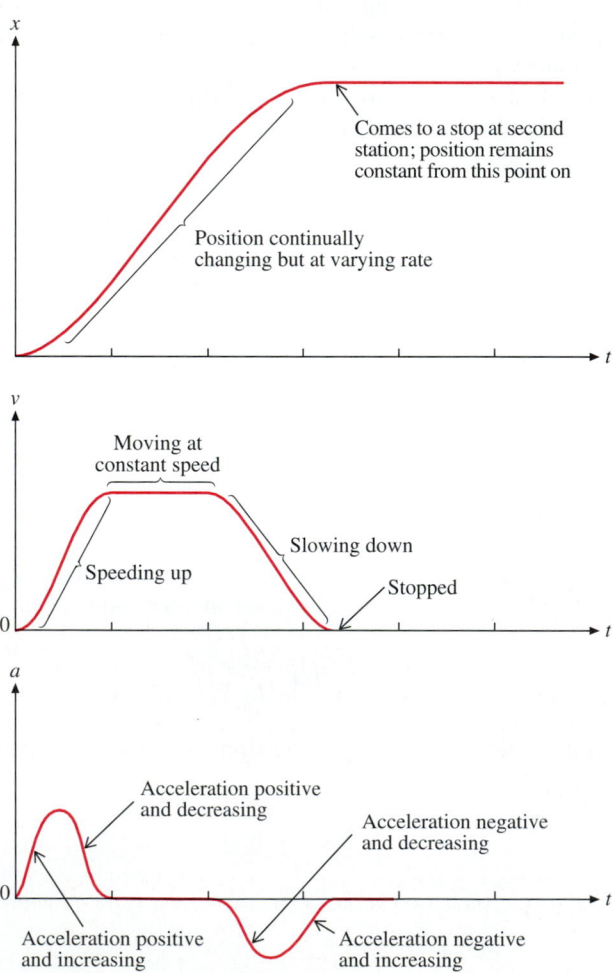

FIGURE 2.15 Schematic plot of the displacement, velocity, and acceleration of a subway train as it moves from one station to the next.

slows down—until it comes to rest in the station. The acceleration itself does not begin and end instantaneously but varies in a smooth way. Because acceleration is the rate of change of velocity, an acceleration of large magnitude is associated with a rapid change in velocity.

EXAMPLE 2.8

A falling rock moves so that its position is described by the function $x = bt^2$. Find its acceleration a.

SOLUTION: If you substitute the value $x = bt^2$ into Equation 2.23b, you obtain

$$a = \frac{d^2}{dt^2}(bt^2).$$

To make explicit what is meant by this expression, rewrite it in the equivalent form of Equation 2.23a,

$$a = \frac{d}{dt}\left[\frac{d}{dt}(bt^2)\right]. \tag{2.24}$$

The first task in evaluating this expression is to carry out the operation inside the brackets—that is, to take the derivative of the function bt^2 with respect to time. To use the general rule of

Equation 2.18, $dx/dt = nqt^{n-1}$, you equate b to q and set $n = 2$. This gives you

$$\frac{d}{dt}(bt^2) = 2bt^{2-1} = 2bt. \tag{2.25}$$

(We obtained the same result in Section 2.4.) Substituting into Equation 2.24, you have

$$a = \frac{d}{dt}(2bt). \tag{2.26}$$

Now carry out the process of differentiation a second time. You can use Equation 2.18 again. This time, you set q equal to the constant $2b$ and note that $t \equiv t^1$. Because $t^{1-1} = t^0 = 1$, you obtain

$$a = 1(2b)t^0 = 2b. \tag{2.27}$$

Let us summarize the argument of Example 2.8 as follows: *If* the position of a particle is described by a quadratic function $x = bt^2$, *then* the velocity is described by the linear function $v = 2bt$ and the acceleration has the constant value $a = 2b$. The special case of quadratic displacement, linear velocity, and constant acceleration is important because it is common in nature. In particular, freely falling bodies move in this way, as you will see in Section 2.7. Figure 2.16 illustrates the relations among acceleration, velocity, and displacement in this special case.

FIGURE 2.16 Schematic plot of acceleration, velocity, and displacement of a body whose acceleration is constant.

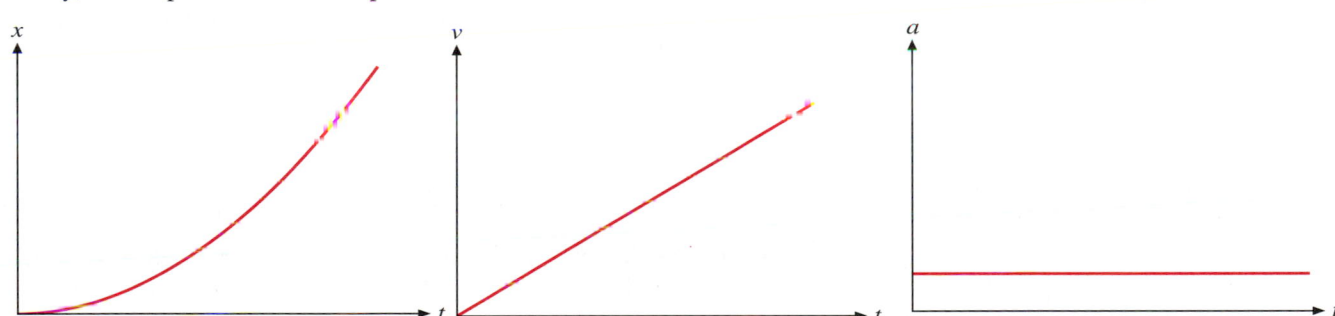

Constant Acceleration

Suppose a particle moves with constant acceleration a. How do its velocity v and position x change in time? To answer this question, we use the argument developed in Section 2.5 as a basis for working backward from acceleration to velocity to position.

According to the "if-then" argument at the end of Section 2.5, constant acceleration is a consequence of a velocity that varies linearly in time according to the equation

$$v = 2bt.$$

Because, according to Equation 2.27, $b = \frac{1}{2}a$, we can write this expression for v in the more physically meaningful form

$$v = at. \tag{2.28}$$

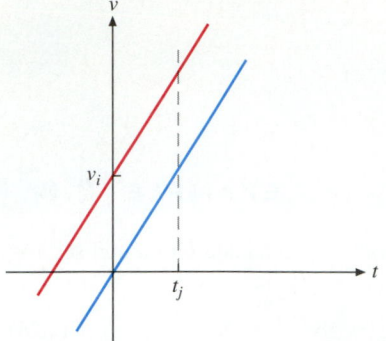

FIGURE 2.17 Illustration of the need to generalize Equation 2.28 if it is to represent the velocity of any possible particle moving with constant acceleration a. Both the blue and the red line depict motion with the same acceleration a but with different velocities at any particular time t_j. The blue line represents the special case $v = 0$ at $t = 0$. The red line represents the case $v = v_i$ at $t = 0$.

This form for the velocity is in turn a consequence of a position that varies in time according to the equation $x = bt^2$, or

$$x = \tfrac{1}{2}at^2. \tag{2.29}$$

But "if-then" statements are not automatically reversible. We cannot say that *if* the acceleration is constant, *then* the velocity will vary according to Equation 2.28 and the position will vary according to Equation 2.29. Although these equations are consistent with constant acceleration, they are not the only consistent equations possible. This point is most easily seen graphically. In Figure 2.17, the blue line passing through the origin is the graphical expression of Equation 2.28, $v = at$. The blue line does indeed have the constant slope a. But the red line, which is parallel to the $v = at$ line, also has slope a. It intersects the v axis at some value v_i; it is the graphical expression of the equation

$$v = v_i + at. \tag{2.30}$$

The physical significance of this equation can be seen by setting $t = 0$; then $v = v_i$. That is, the particle whose velocity is described by Equation 2.30 is not necessarily at rest at the initial time $t = 0$ but may have an initial velocity v_i. The value of v_i can be positive, negative, or zero. If $v_i = 0$, the particle *is* initially at rest and Equation 2.30 reduces to Equation 2.28.

So Equation 2.30 does *not* follow from Equation 2.29, $x = \tfrac{1}{2}at^2$, because Equation 2.29 is not general enough. We have already seen that differentiation leads to Equation 2.28, $v = at$. We need an equation for x having an extra term that, when differentiated with respect to time, yields the term v_i on the right side of Equation 2.30.

Let us see if adding the term $v_i t$—the distance the particle would move in time t at its initial velocity v_i, in the absence of acceleration—does the job. That is, let us generalize Equation 2.29 to the form

$$x = v_i t + \tfrac{1}{2}at^2. \tag{2.31}$$

Differentiating this equation according to the rules given in Section 2.4 yields $v = v_i + at$, which is just what we want. But Equation 2.31 still lacks complete generality. Again, we make this evident by setting $t = 0$; this yields $x = 0$. That is, Equation 2.31 describes the motion of a particle whose initial position is $x = 0$. In general, however, the initial position of the particle has some arbitrary value x_i, which may be positive, negative, or zero. We therefore add the term x_i to the right side of Equation 2.31 to obtain

$$x = x_i + v_i t + \tfrac{1}{2}at^2 \quad \text{for } a = \text{constant.} \tag{2.32a}$$

This can also be written

$$\Delta x = x - x_i = v_i t + \tfrac{1}{2}at^2 \quad \text{for } a = \text{constant.} \tag{2.32b}$$

Let us summarize the physical meaning of the three terms on the right side of Equation 2.32a. The term x_i denotes the position of the particle at time $t = 0$. The quantity v_i in the second term denotes the velocity of the particle at $t = 0$. If there is no acceleration—that is, if $a = 0$—the third term is equal to zero, the equation reduces to $x = x_i + v_i t$, and the velocity remains constant; compare with Equation 2.5.

The third term, $\tfrac{1}{2}at^2$, represents the additional displacement that occurs because the velocity is changing with time. Let us consider further the reasons for its mathematical form. By definition, $a = dv/dt$, and, because we are considering the case of constant a, we can just as well write $a = \Delta v/\Delta t$, where Δt is any time interval at all. In particular, take Δt to be the entire time interval of interest, from 0 to t. This gives

$$\frac{1}{2}at^2 = \frac{1}{2}\frac{v - v_i}{t - 0}t^2 = \frac{1}{2}(v - v_i)t \quad \text{for } a = \text{constant.} \tag{2.33}$$

On inserting this value of $\tfrac{1}{2}at^2$ into Equation 2.32b, we obtain

$$\Delta x = v_i t + \tfrac{1}{2}vt - \tfrac{1}{2}v_i t$$

or $\qquad\qquad\qquad \Delta x = \dfrac{v_i + v}{2}t \quad \text{for } a = \text{constant.} \tag{2.34}$

The term $(v_i + v)/2$ is the arithmetical average of the initial velocity v_i and the velocity v at time t. If the particle had been moving at the constant velocity $(v_i + v)/2$ during the entire time interval, it would have experienced the same displacement $x - x_i$ as it does when it accelerates with constant a from initial velocity v_i to final velocity v. The connection between the actual velocity of the accelerated particle as a function of time and the constant velocity $(v_i + v)/2$ is shown graphically in Figure 2.18. Because the acceleration is constant, the average of the initial and final velocities, $(v_i + v)/2$, is equal to the mean velocity of the particle in the time interval of interest. That is, Equation 2.34 can be written in the equivalent form

$$\Delta x = \langle v \rangle t. \tag{2.35}$$

Compare this form with Equation 2.3, $\langle v \rangle = \Delta x / \Delta t$, which is the definition of mean velocity. Equation 2.34 is valid only for constant acceleration, but there are no restrictions on the validity of Equation 2.35.

We have derived a set of general equations that describe the position and velocity of a particle as a function of time when the acceleration is constant. As a review of the mathematical connection between them, we begin with the position equation, Equation 2.32a, and differentiate to obtain the corresponding velocity equation. We have

$$x = x_i + v_i t + \tfrac{1}{2}at^2 \quad \text{for } a = \text{constant}; \tag{2.36a}$$

$$v = \frac{dx}{dt} = v_i + at \quad \text{for } a = \text{constant}. \tag{2.36b}$$

As a final check, differentiate again. Note that Equation 2.36b leads—*only* if a is constant—to

$$a \equiv \frac{dv}{dt} = \frac{d^2x}{dt^2} = a. \tag{2.36c}$$

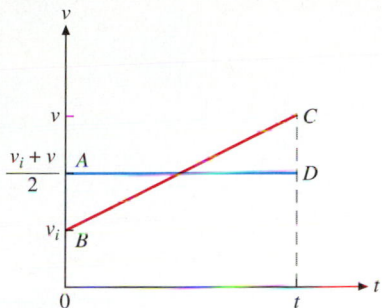

FIGURE 2.18 The velocity of a particle is described as a function of time by the line BC, whose slope is the constant acceleration a. The horizontal line AD represents a constant velocity $(v_i + v)/2$. A particle moving at this velocity will experience the same displacement in the time interval from 0 to t as one whose velocity is described by BC.

EXAMPLE 2.9

A motorist traveling at 90.0 km/h (25.0 m/s) approaches a speed zone where the limit is 40.0 km/h (11.1 m/s). She brakes moderately and reaches the speed limit in 7.3 s. Find **(a)** the acceleration of the car and **(b)** the distance it travels while slowing down. Assume that the acceleration is constant.

SOLUTION:
(a) Find the acceleration of the car.
Because a is constant, you can write $a \equiv dv/dt = \Delta v / \Delta t$. Using the numerical values given, you have

$$a = \frac{v_f - v_i}{\Delta t} = \frac{11.1 \text{ m/s} - 25.0 \text{ m/s}}{7.3 \text{ s}} = -1.9 \text{ m/s}^2.$$

The negative value of the acceleration means that the car is slowing down while moving in the direction assumed positive. (How is this assignment of the positive direction, which you normally make without thinking about it, based on the given data?)

(b) Find the distance the car travels while slowing down.
You can now use Equation 2.32b, $\Delta x = v_i t + \tfrac{1}{2}at^2$, to find the distance Δx traveled. You have

$$\Delta x = 25.0 \text{ m/s} \times 7.3 \text{ s} + \tfrac{1}{2} \times (-1.9 \text{ m/s}^2) \times (7.3 \text{ s})^2$$

$$= 183 \text{ m} - 51 \text{ m} = 130 \text{ m},$$

rounded to two significant figures.

The Relation between Speed and Distance for Constant Acceleration

Equation 2.34 expresses the displacement Δx in terms of the time t. Equation 2.36b expresses the velocity v in terms of t. The two equations can be combined to eliminate t. The result is an expression for the speed $|v|$ in terms of the position x; it is valid, as are Equations 2.34 and 2.36b, for $a = $ constant. We first solve Equation 2.36b for t, obtaining

$$t = \frac{v - v_i}{a}.$$

Substituting this value into Equation 2.34, we have

$$\Delta x = \frac{v_i + v}{2} \frac{v - v_i}{a} = \frac{(v + v_i)(v - v_i)}{2a}.$$

We expand the numerator, multiply through by $2a$, and exchange the right and left sides of the equation to obtain

$$v^2 - v_i^2 = 2a\,\Delta x \quad \text{for } a = \text{constant.} \tag{2.37}$$

In the special case where $v_i = 0$ and $x_i = 0$, this equation simplifies to the form

$$v^2 = 2ax \quad \text{for } v_i = 0,\ x_i = 0,\ a = \text{constant.} \tag{2.38}$$

Equations 2.37 and 2.38 contain the velocities v and v_i only as squares. Because the square of a number is always positive, regardless of the sign of the number, only the speeds $|v|$ and $|v_i|$, and not the velocities v and v_i, are relevant to the equation. When the direction of motion is significant, it must be recovered from the context of the problem. But often, as in Example 2.10, the direction of motion presents no difficulty.

EXAMPLE 2.10

A commercial airplane must reach a speed of 85 m/s in order to take off. With its engines set at takeoff power, the plane accelerates at an approximately constant 3.0 m/s^2. What is the minimum length of the runway required?

SOLUTION: Because the plane starts at rest and because you are free to choose the origin at the beginning of the takeoff run, you can use Equation 2.38, $v^2 = 2ax$. The quantity x then cor-

responds to the minimum runway length. Solving for x, you have

$$x = \frac{v^2}{2a} = \frac{(85 \text{ m/s})^2}{2 \times 3.0 \text{ m/s}^2} = 1200 \text{ m.}$$

Needless to say, runways intended for such airplanes are always considerably longer than 1200 m.

Summary of Kinematic Relations

In Sections 2.5 and 2.6, we studied a number of **kinematic quantities**—quantities useful for describing the motion of a particle. Among these quantities are the displacement Δx, the time t, the (instantaneous) velocity v, and the (instantaneous) acceleration a.

We have derived various equations that relate these quantities in the special case where a is constant. These equations are summarized in Table 2.1.

TABLE 2.1 Some Kinematic Relations for One-Dimensional Motion with Constant Acceleration

Equation	Application	Equation Number
$v = v_i + at$	Relates velocity to time, given v_i and a	2.30
$\Delta x = v_i t + \frac{1}{2}at^2$	Relates displacement to time, given v_i and a	2.32b
$\Delta x = \frac{1}{2}(v_i + v)t$	Relates displacement to time, given v_i and v	2.34
$v^2 - v_i^2 = 2a\,\Delta x$	Relates speed to displacement, given $|v_i|$ and a	2.37

Note: In all cases, $t_i = 0$.

SECTION 2.7   Freely Falling Bodies

How do bodies fall? This was a central unsolved problem of physics from classical Greek times until the seventeenth century. The following three examples suggest that the problem is not a simple one:

- Feathers and coins fall through air very differently.
- A block of oak will sink slowly through oil but will float in water.
- Stones of the same composition but of different sizes and shapes will not fall in the same way through water.

FIGURE 2.19 Stroboscopic photograph of two steel balls released (almost) simultaneously from rest. The time interval between exposures is a constant 0.02 s. The increasing distance between successive images makes it evident that the balls are accelerating. The motions of the two balls are nearly identical despite their different sizes. (The slight difference in motion is due to an imperfection in the release mechanism; the smaller ball was released approximately one flash interval—roughly 0.02 s—later than the larger one.) If the time elapsed between two successive images is known, the mean velocity over that interval can be deduced. This mean velocity is equal to the instantaneous velocity at the instant halfway through the time interval, as Figure 2.16 shows. From any two instantaneous velocities determined in this way, the mean acceleration for the time interval between the two corresponding instants can be found. If the same acceleration is found for all pairs of instants, the acceleration must be constant.

(From Haber-Schaim: *PSSC Physics.* Copyright 1991 by Kendall/Hunt Publishing Company. Used with permission.)

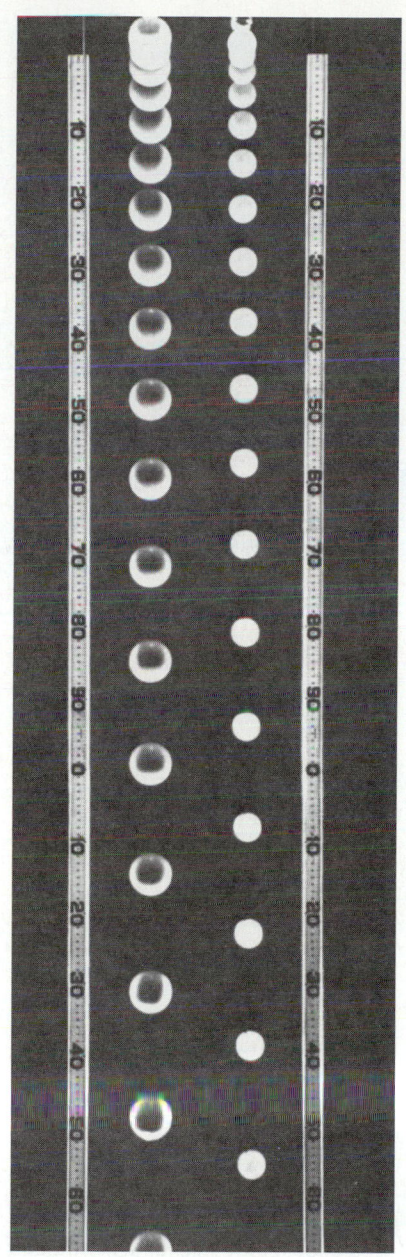

The great Italian natural philosopher Galileo Galilei (1564–1642) was the first to see that the key to the problem of falling bodies lay in identifying the simplest case, called free fall, and focusing attention on it. **Free fall** is vertical fall in the absence of any significant frictional resistance from the surrounding medium and in the absence of any significant buoyancy due to the medium. Ideally, we would like to study falling bodies in vacuum. But with a reasonable choice of body—a lead ball, for example, rather than a sheet of paper—air exerts so little resistance and so little buoyancy that the presence of air does not complicate our observations.

Freely falling bodies move rapidly, and high-speed photography is one way to study rapidly moving bodies. Figure 2.19 is a multiple-flash photograph of two metal balls released simultaneously from rest. Measurements obtained from such photographs confirm the results that Galileo obtained in a less direct way:

1. The free-fall motion of a body does not depend on the body's size, shape, or composition.
2. If the path of the fall is short in comparison with the radius of the earth—a condition not too hard to realize in the laboratory—the free-fall motion of a body is one of constant acceleration.

It follows from these observations that all bodies falling freely at the same place on earth have the same acceleration. The acceleration is called the **acceleration of gravity** and is universally denoted by the symbol g. For most purposes in this book, we will use the value

$$|g| = 9.80 \text{ m/s}^2 \qquad (2.39)$$

for its magnitude. The values of $|g|$ encountered over the earth's surface do not vary from this value by more than 0.3%—that is, by 3 parts in 1000. At any particular location, the value of $|g|$ depends on the latitude, the elevation, and the local geology. Indeed, measurements of tiny variations in $|g|$ constitute an important tool of the geologist in determining structure and locating minerals.

EXAMPLE 2.11

A stone is thrown vertically upward from a window located 20.0 m above the ground with initial velocity $v_i = 15.5$ m/s. **(a)** Find the velocity of the stone after 1 s and after 2 s. **(b)** How long does it take for the stone to reach the ground? **(c)** How high does the stone go? **(d)** At what height is the speed of the stone $|v| = 5.00$ m/s? In solving, neglect air resistance.

SOLUTION:

(a) Find the velocity of the stone after 1 s and after 2 s.

To begin with, you must establish an origin and a positive direction for the coordinate system. One possible choice is to take the origin at ground level and the upward direction as positive (Figure 2.20). You are considering a freely falling body, whose acceleration has the constant value g. To find the velocity v, you use Equation 2.30, $v = v_i + at$, which expresses v as a function of time when the acceleration is constant. You write

the equation in the form

$$v = v_i + gt.$$

In inserting numerical values into this equation, you must remember that the choice of the upward direction as positive means that g has the *negative* value

$$g = -9.80 \text{ m/s}^2.$$

(This negative value is physically plausible when you remember that the effect of gravity on the upward-thrown stone is to reduce its velocity.) You now have, for $t = 1$ s,

$$v_1 = 15.5 \text{ m/s} - 9.80 \text{ m/s}^2 \times 1 \text{ s} = 5.70 \text{ m/s}.$$

Similarly, for $t = 2$ s, you have

$$v_2 = 15.5 \text{ m/s} - 9.80 \text{ m/s}^2 \times 2 \text{ s} = -4.10 \text{ m/s}.$$

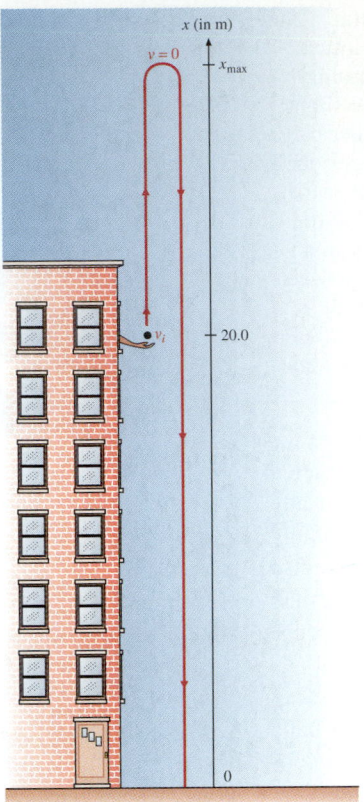

FIGURE 2.20

From these two results, it appears that the stone must reach its maximum height and begin to fall at some time between 1 s and 2 s after being thrown.

(b) How long does it take for the stone to reach the ground?

Because you know the position x of the stone when it reaches the ground, the initial position x_i, the initial velocity v_i, and the acceleration g, you can use Equation 2.34a to find the time of flight. You have

$$x = x_i + v_i t + \tfrac{1}{2} g t^2.$$

All the quantities in this equation are known except t. The equation is quadratic in t, and so you cast it into the standard form for solving a quadratic:

$$\tfrac{1}{2} g t^2 + v_i t + (x_i - x) = 0.$$

Applying the standard solution for a quadratic equation, you get

$$t = \frac{-v_i \pm \sqrt{v_i^2 - 2g(x_i - x)}}{g}.$$

You can now insert the numerical values:

$$t = \frac{-15.5 \text{ m/s} \pm \sqrt{(15.5 \text{ m/s})^2 - 2(-9.80 \text{ m/s}^2)(20.0 \text{ m} - 0)}}{-9.80 \text{ m/s}^2}$$

$$= 1.58 \text{ s} \mp 2.57 \text{ s}.$$

The solutions are thus

$$t = -0.99 \text{ s} \quad \text{and} \quad t = +4.15 \text{ s}.$$

Of the two roots of the quadratic equation, the positive root is the one that answers the question at hand: The stone strikes the ground 4.15 s after it is thrown upward from the window. Although the negative root is not the answer to the problem, it does have a hypothetical meaning. If the stone had been thrown upward from the ground in such a way as to reach the window on its way up at $t = 0$ and with velocity 15.5 m/s, it would have had to be thrown at $t = -0.99$ s.

(c) How high does the stone go?

Note that the stone attains its maximum height just at the instant when its velocity is $v = 0$. (If the velocity of the stone is positive, it is still rising; if the velocity is negative, it has already started to descend.) This observation, together with the data furnished, gives you all the information you need to solve Equation 2.37 for the desired height x:

$$v^2 - v_i^2 = 2a \, \Delta x = 2g(x - x_i).$$

You solve for x:

$$x = x_i + \frac{v^2 - v_i^2}{2g} \tag{2.40}$$

$$= 20.0 \text{ m} + \frac{0 - (15.5 \text{ m/s})^2}{2(-9.80 \text{ m/s}^2)},$$

or

$$x = 20.0 \text{ m} + 12.3 \text{ m} = 32.3 \text{ m}.$$

The result states that the stone reaches a maximum height 32.3 m above the ground (or 12.3 m above the window).

(d) At what height is the speed of the stone $|v| = 5.00$ m/s?

You again use Equation 2.40. In this case, numerical substitution gives

$$x = 20.0 \text{ m} + \frac{(5.00 \text{ m/s})^2 - (15.5 \text{ m/s})^2}{2(-9.80 \text{ m/s}^2)}$$

$$= 20.0 \text{ m} + 11.0 \text{ m} = 31.0 \text{ m}.$$

The stone passes through $x = 31.0$ m twice, once on its way up and again on its way down. Recall that Equations 2.37 and 2.40 concern only speeds and not velocities. The speed passes through the value $|v| = 5.00$ m/s twice. The first time, the velocity is $v = 5.00$ m/s; the second time, it is $v = -5.00$ m/s. How would you calculate the corresponding times?

THINKING LIKE A PHYSICIST

In this chapter, we began by developing the rules of kinematics, with emphasis on the case of constant acceleration. Then we considered the physical phenomenon of free fall and were delighted to find that the rules just developed were exactly what was needed to deal with the real physical situation.

Historically speaking, physics almost never develops so neatly. The problem of free fall intrigued serious thinkers for centuries before Galileo showed how useful it is to consider free fall as a case of motion with constant acceleration. There is almost always a close, often messy, interplay between experimental observations (for example, of freely falling bodies) and the work of developing a system of analysis required to interpret the observations (for example, the kinematic equations). The neat, logical presentation that is the hallmark of a good textbook is a far cry from the path, with its backtracks and blind alleys, that led to the original discovery.

When you are frustrated because your thought processes are not as direct and logical as the presentation in the text, take comfort in the fact that your process of discovery is more akin to Galileo's than to the development in the text!

a	acceleration	$_f$ (subscript)	final value of a varying quantity
g	acceleration of gravity	$_j$ (subscript)	the jth value of a varying quantity in a series of measurements labeled $0, 1, 2, \ldots,$ $j, \ldots, n$
t	time		
Δt	time interval		
v	velocity	$\dfrac{d}{dt}$	derivative with respect to t (of a quantity that must be specified, as in dx/dt)
Δv	change in velocity		
$|v|$	speed	$\dfrac{d^2}{dt^2}$	second derivative with respect to t (of a quantity that must be specified, as in d^2x/dt^2)
x	position		
Δx	displacement (change in position)		
b, c, m, n, p, q	constants	$|\ \ |$	absolute value of a quantity
$_i$ (subscript)	initial value of a varying quantity	$\langle\ \rangle$	mean value of a quantity over a specified interval

Summing Up

The position of a particle must be expressed with reference to a particular coordinate system. Once the system has been established, the position is given in terms of the distance and direction of the particle from the origin. Any physical quantity that possesses both magnitude and direction is a **vector quantity**. If all positions of interest lie along a line passing through the origin, a position can be specified by a one-dimensional vector; in one dimension, a sign suffices to specify the direction of a vector.

A change in position of a particle, $\Delta x = x_f - x_i$, is called the **displacement** of the particle. Displacement is a vector quantity.

Any physical quantity that has magnitude but not direction is a **scalar** quantity. The magnitude of a vector is a positive scalar.

If the positions x_i and x_f are measured at times t_i and t_f, respectively, the **mean velocity** $\langle v \rangle$ over the time interval $\Delta t = t_f - t_i$ is defined by Equation 2.3,

$$\langle v \rangle \equiv \frac{\Delta x}{\Delta t} = \frac{x_f - x_i}{t_f - t_i}.$$

Velocity is a vector quantity because it possesses magnitude and direction. If the value of $\langle v \rangle$ is the same over all time intervals of interest, the velocity is constant and is denoted by the symbol v. In this case, the position of the particle is given by Equation 2.5,

$$x = x_i + vt \quad \text{for } v = \text{constant}, \ t_i = 0.$$

The **instantaneous velocity** is defined as the limiting value of the mean velocity as Δt is reduced to an infinitesimal value. This limiting process is expressed by Equations 2.13, 2.14, and 2.15, which are summarized as follows:

$$v = \frac{dx}{dt} = \lim_{\Delta t \to 0} \frac{\Delta x}{\Delta t}$$

The magnitude $|v|$ of any velocity v is called **speed**.

Mean acceleration is defined by Equation 2.21,

$$\langle a \rangle \equiv \frac{\Delta v}{\Delta t}.$$

The **instantaneous acceleration** is defined as the limiting value of the mean acceleration as Δt is reduced to an infinitesimal value, as expressed in Equation 2.22:

$$a = \frac{dv}{dt} = \lim_{\Delta t \to 0} \frac{\Delta v}{\Delta t}.$$

At some initial time chosen to be $t = 0$, the position of a particle is x_i and its velocity is v_i. If—and only if—the acceleration a of the particle is constant, its position x and velocity v at any time t are given by Equations 2.36a and 2.36b:

$$x = x_i + v_i t + \tfrac{1}{2}at^2 \quad \text{for } a = \text{constant}$$

and

$$v = v_i + at \quad \text{for } a = \text{constant}.$$

Also, the relation between $|v|$ and displacement Δx is given by Equation 2.37,

$$v^2 - v_i^2 = 2a\,\Delta x \quad \text{for } a = \text{constant.}$$

In the absence of friction and buoyant effects, all bodies falling freely near the surface of the earth experience the same constant acceleration, whose magnitude is close to $|g| = 9.80 \text{ m/s}^2$.

KEY TERMS

Section 2.1 Translation and Rotation

extended body, particle ▪ translation, rotation ▪ one-dimensional motion ▪ displacement

Section 2.2 Position and Displacement

scalar ▪ one-dimensional vector ▪ free vector, bound vector

Sections 2.3 and 2.4 Velocity and Derivatives

constant velocity, mean velocity ▪ instantaneous velocity ▪ speed ▪ derivative

Sections 2.5 and 2.6 Acceleration

second derivative ▪ constant acceleration, mean acceleration, instantaneous acceleration

Section 2.7 Freely Falling Bodies

free fall ▪ acceleration of gravity

Queries and Problems for Chapter 2

QUERIES

2.1 *(1) Abstract with care!* In which of the following situations is it possible to consider the earth as a particle? **(a)** In studying the revolution of the earth around the sun. **(b)** In studying the way in which ocean tides are caused by the gravitational attraction of the moon and the sun. **(c)** In using slight irregularities in the orbits of artificial satellites to study the shape of the earth. **(d)** In measuring the distance from the earth to the moon.

2.2 *(3) Staying up late.* Galileo tried to measure the speed of light, without success, using the method illustrated below. He and a friend took two shuttered lanterns and practiced until the friend could open his shutter as quickly as possible when he saw Galileo opening the other shutter. On a clear night, they located themselves on the tops of two hills a few kilometers apart. After he opened his own shutter, Galileo was unable to detect any delay in seeing his friend's light, beyond the delay he had observed when they were practicing close together. Assuming that human reaction times are about 0.2 s, how far apart would they have had to be to detect even a 10% change in the delay time? The speed of light is now known to be 3×10^8 m/s.

2.3 *(3) Not so simple!* In making a journey, you cover half the distance at 30 mi/h and the other half of the distance at 60 mi/h. Explain why your average speed is not 45 mi/h.

2.4 *(4) Zeno's paradox.* The champion runner Achilles, whose speed is V, chases a tortoise whose speed has the much

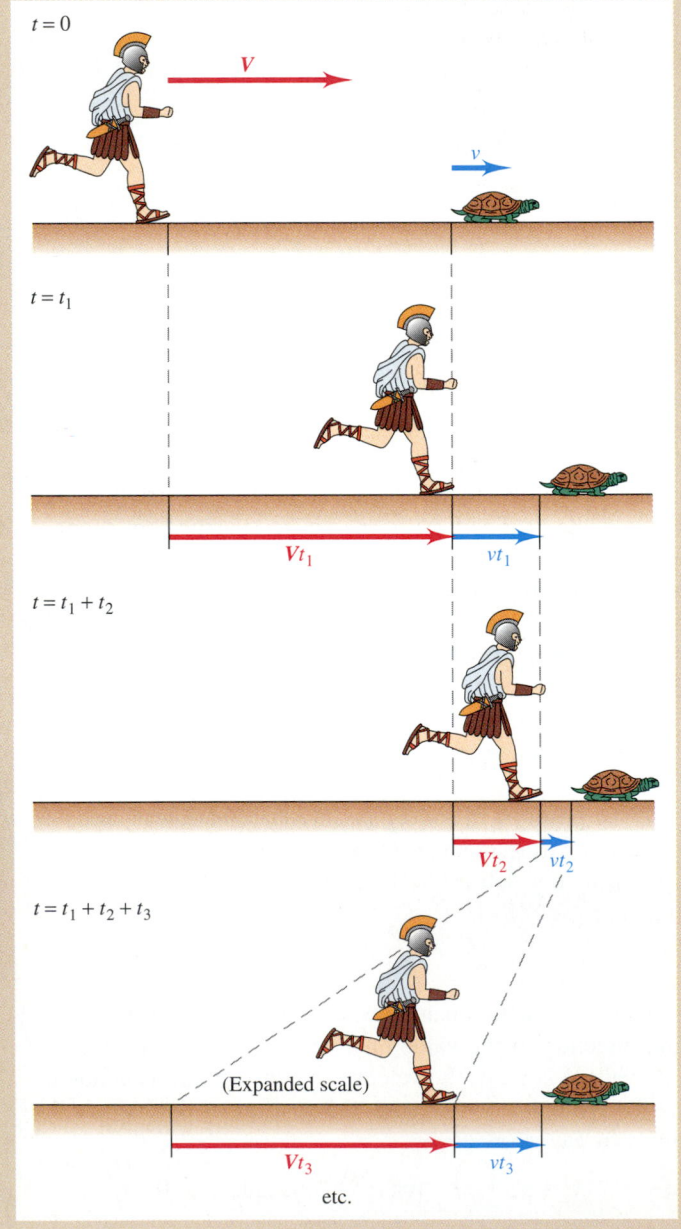

smaller value v. Zeno claims that Achilles can never catch the tortoise. His argument is illustrated in the sequence of "snapshots" shown on the preceding page. Suppose, Zeno argues, it takes Achilles a time t_1 to cover the distance to the tortoise's starting point. In the meantime, the tortoise has moved a distance vt_1. It takes Achilles an additional time $t_2 = vt_1/V$ to cover this distance. But, when he has done so, the tortoise has moved a further distance $vt_2 = v^2t_1/V$, and so on. Thus, no matter how many of these distances Achilles has covered, he must still cover another one and can therefore never catch the tortoise. Sketch a graph showing the positions of Achilles and of the tortoise as functions of time, and use the graph to demonstrate the fallacy of Zeno's argument.

2.5 *(5) It all averages out, I.* Measurements on a moving particle show that its mean velocity $\langle v \rangle$ is equal to its instantaneous velocity v at every point. What can you say about its acceleration?

2.6 *(5) It all averages out, II.* Measurements on a moving particle show that its mean acceleration $\langle a \rangle$ is equal to its instantaneous acceleration a at every point. What can you say about its velocity?

2.7 *(5) Stark realism, I.* The illustration below is a plot of position as a function of time for a hypothetical particle. Is it likely that the plot describes a real particle? Why or why not?

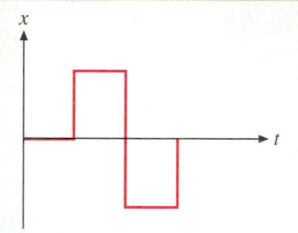

2.8 *(5) Stark realism, II.* The illustration below is a plot of velocity as a function of time for a hypothetical particle. Is it likely that the plot describes a real particle? Why or why not?

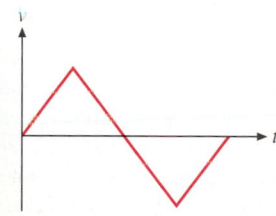

2.9 *(5) Sketchy but adequate, I.* The illustration below is a qualitative plot of displacement as a function of time for a hypothetical particle. Segments AB, CD, and EF are straight; the other segments are curved. **(a)** Make a qualitative plot of velocity as a function of time for this particle. **(b)** Make a qualitative plot of acceleration as a function of time for the particle.

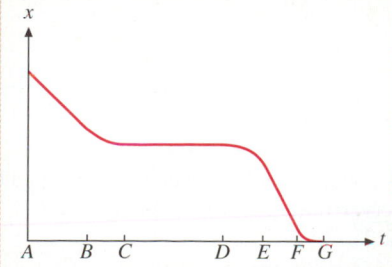

2.10 *(5) Sketchy but adequate, II.* The illustration below is a qualitative plot of velocity as a function of time for a hypothetical particle. Segments BC, DE, and GH are straight; the other segments are curved. **(a)** Make a qualitative plot of displacement as a function of time for this particle, assuming that $x = 0$ at $t = 0$. **(b)** Make a qualitative plot of acceleration as a function of time for the particle.

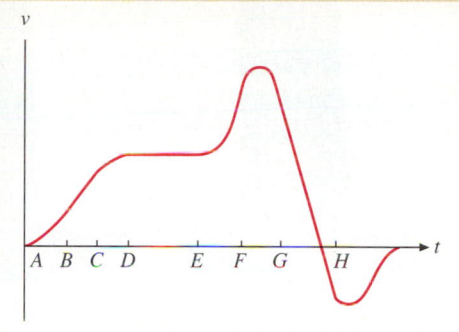

2.11 *(5) Copping a plea.* Policemen love to make their reports sound official by writing such phrases as "Subject was proceeding at an excessively high rate of speed" rather than simply stating, "Subject was proceeding at an excessive speed." If you were the defendant "subject" in a court trial based on such a report, how would you persuade the judge to dismiss the case on the grounds that there was no record of your breaking any speed limit?

2.12 *(G) There's running, and then there's running . . .* Runners in short sprints, such as the 100-yard dash, start in a crouching position with their feet against starting blocks, as in the first photograph below. Long-distance runners, such as the marathon runners shown in the second photograph, do not take such measures. Explain the difference.

2.13 *(G) Jerky motion.* A body moves in such a way that its acceleration is not constant but varies in time according to the function $a = Jt$, where J is a constant. At $t = 0$, the body is at rest at $x_i = 0$. How will the velocity v vary with time?

2.14 *(7) Logic and the Leaning Tower.* In Galileo's day, most people believed that two bodies of the same shape and composition, but of different sizes, would fall at different rates. The bigger body, they believed, would fall faster than the smaller, the rates perhaps being proportional to their weights (drawing **a** on the next page). Galileo objected to this view. He argued that the two bodies would fall at very nearly the same rate, as in drawing **b**. To support his view, he suggested the following imaginary experiment. Drop the larger body and observe its time of fall through a given height. Drop the smaller body and observe its time of fall through the same height. Tie the two together with a light string, and observe the time of fall through the same height, as in drawing **c**. Fill in the argument, and show that the assumption of different rates of fall leads to a logical contradiction.

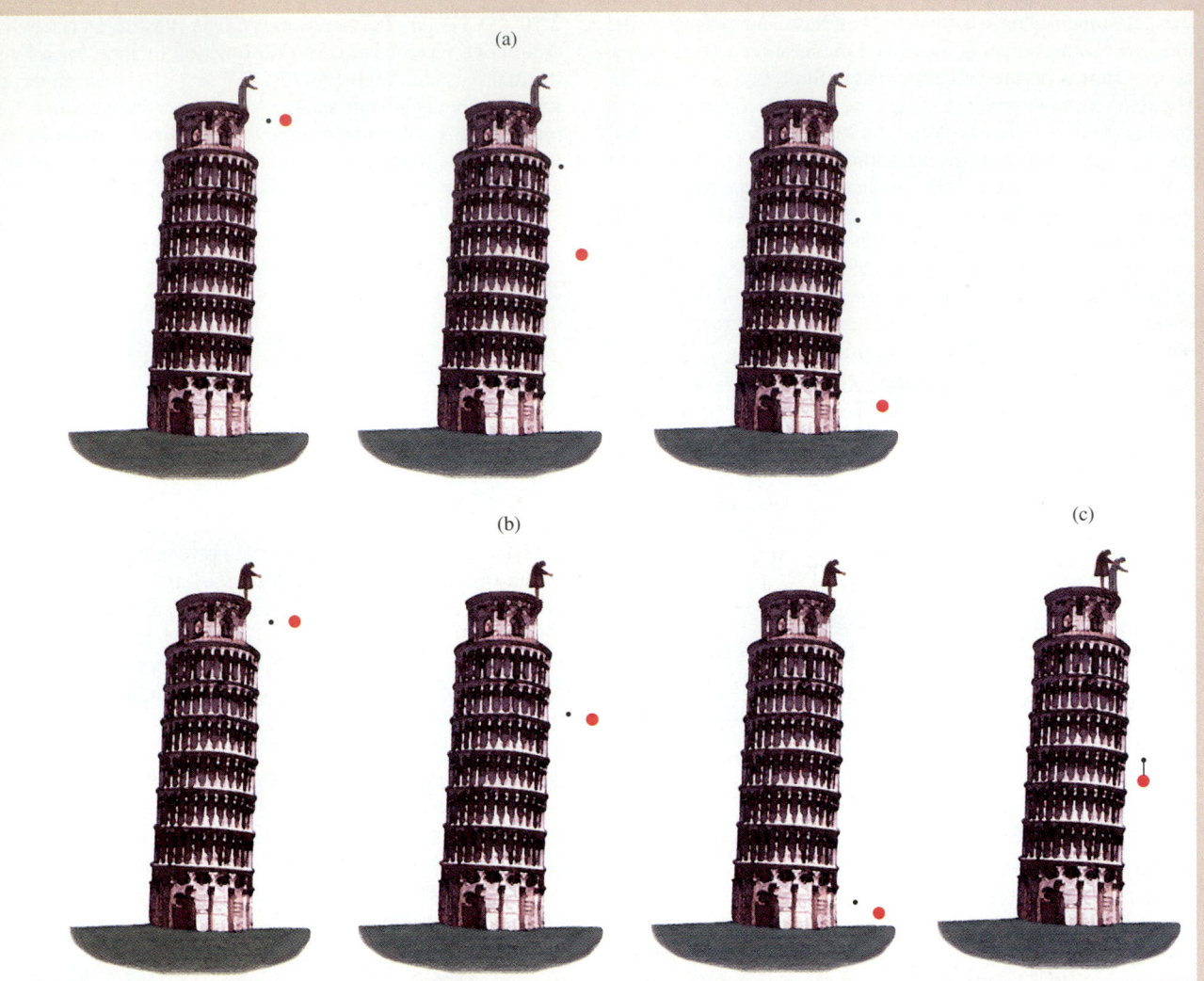

(a)

(b)

(c)

2.15 *(7) Come on, Galileo, be practical for once!* Some of Galileo's critics argued that his approach to falling bodies was useless because it was artificial and unrealistic. After all, they argued, for most falling bodies friction or buoyancy or both are important effects. For example, pebbles falling through water clearly move with essentially constant velocity and do not accelerate. It is therefore a waste of time, and perhaps a misleading exercise, to study the relatively unusual and possibly anomalous case in which the body accelerates. Besides, it is precisely in this case that the motion is so rapid that accurate measurements are impossible. Therefore, it makes more sense to concentrate on the common, easily observed situation in which there is substantial resistance or buoyancy or both and worry about the case in which they are negligible later. Criticize this argument.

2.16 *(G) Driver's-license exam.* An often-stated rule of thumb is that a good driver should maintain a minimum distance of approximately one car length between his or her own car and the car ahead for every 15 km/h of speed. Normal human response time (say, from the time that you see the brake light of the car ahead until you can step on your brake) is about 0.2 s. Make a reasonable estimate of a ''car length'' and use the estimate to comment on the wisdom of the rule. Will it make any difference if the road is wet or icy? Will it make any difference if the road is icy and the driver ahead has tire chains but you do not?

PROBLEMS

GROUP A

2.1 *(2) Anyone can watch, I.* Repeat the calculation of Example 2.2 from the point of view of an observer located at $x = -0.9$ m who measures the initial and final positions of the particle, P_1 and P_2, with respect to his own position. Show that this observer finds the same displacement as that found in Example 2.2.

2.2 *(2) Anyone can watch, II.* The illustration below shows the same coordinate system as that of Figure 2.6, except that an observer is located at an arbitrary point whose coordinate is X. He observes the displacement of the same particle studied by the observers at the two origins O and O' considered in Examples 2.2 and 2.3, and he obtains the value $\Delta x''$. Show that he finds the same value of the displacement as they do; that is, show that $\Delta x'' = x_2 - x_1$.

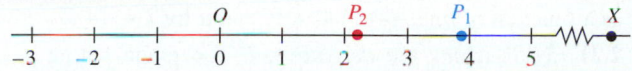

2.3 *(2) Anyone can watch, III.* Still another observer sits at the origin of Figure 2.6 but chooses her coordinate system with the leftward direction as positive; see the illustration below. Show that this observer finds the same value for the displacement of the particle as do the observers in Examples 2.2 and 2.3 and in Problem 2.1.

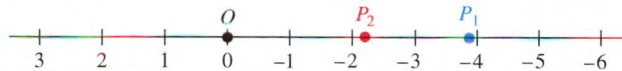

2.4 *(2) Fast action.* Traffic safety experts always advise, "Don't overdrive your headlights." That is, when you drive at night, you should not drive so fast that you could not avoid an obstacle visible at the extreme range of your headlights. Suppose that your headlights give reliable illumination out to a distance of 40 m in front of your car. How much time do you have to react and take evasive action—if necessary, to come to a stop—if you are driving at 100 km/h?

2.5 *(2) Flying shuttle, I.* An airplane files repeatedly between San Francisco and Los Angeles according to the following schedule:

Lv SFO	Ar LAX	Lv LAX	Ar SFO
7:00 A.M.	8:15 A.M.	9:00 A.M.	10:15 A.M.
11:00 A.M.	12:15 P.M.	1:00 P.M.	2:15 P.M.

(a) The distance between San Francisco and Los Angeles is 520 km. Setting your origin at San Francisco, sketch a graph of the distance covered by the airplane as a function of time. Neglect the short times required to accelerate between rest and cruising speed (which is essentially constant). **(b)** Using the same time axis, sketch a graph of the displacement of the airplane as a function of time.

2.6 *(3) Career decision.* Baseball pitchers can throw fastballs at speeds in excess of 40 m/s. The distance from the pitcher to the batter is 18 m. How much time does the batter have in which to decide whether (and how) to swing at the ball?

2.7 *(3) Tortoise and hare.* The world record for the 100-m dash is 9.93 s. For the marathon (42 195 m), the record is 2 h 9 min. **(a)** Calculate the mean speed of the runner in each case. **(b)** In which case is the runner's mean speed likely to be closer to his maximum speed? Explain your answer.

2.8 *(4) Flying shuttle, II.* For the airplane of Problem 2.5, sketch a graph of **(a)** the velocity and **(b)** the speed as a function of time. Neglect the short times required to accelerate between rest and cruising speed.

2.9 *(3) Long distance, I.* The speed of radio waves through air is essentially equal to the speed of light in vacuum,

3.00×10^8 m/s. Formerly, transatlantic phone conversations were transmitted by radio from a transmitter in New Jersey to a receiver in Ireland, a distance of 4800 km. (The transmission in the opposite direction was between stations located elsewhere but about the same distance apart.) Find the time required for a signal to traverse the Atlantic Ocean, assuming that the path follows the surface fairly closely, as shown below. If you were to ask a question over this phone connection, would you be likely to notice a delay in receiving the answer?

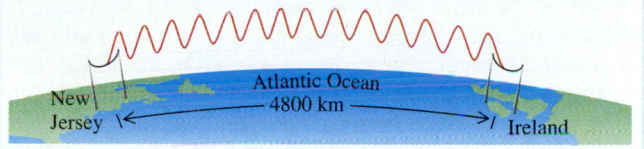

2.10 *(3) Long distance, II.* Most of the transatlantic telephone traffic today is carried over satellite links. The signal is transmitted from the ground to a satellite over the mid-Atlantic at an altitude of 3.58×10^4 km and is then retransmitted to a receiver on the other side of the ocean, as illustrated below. Find the time required for a signal to cover this path at the speed of light, assuming that the straight-line distance between the ground stations is 4700 km. If you were to ask a question over this phone connection, would you be likely to notice a delay in receiving the answer? Compare your result with that of Problem 2.9.

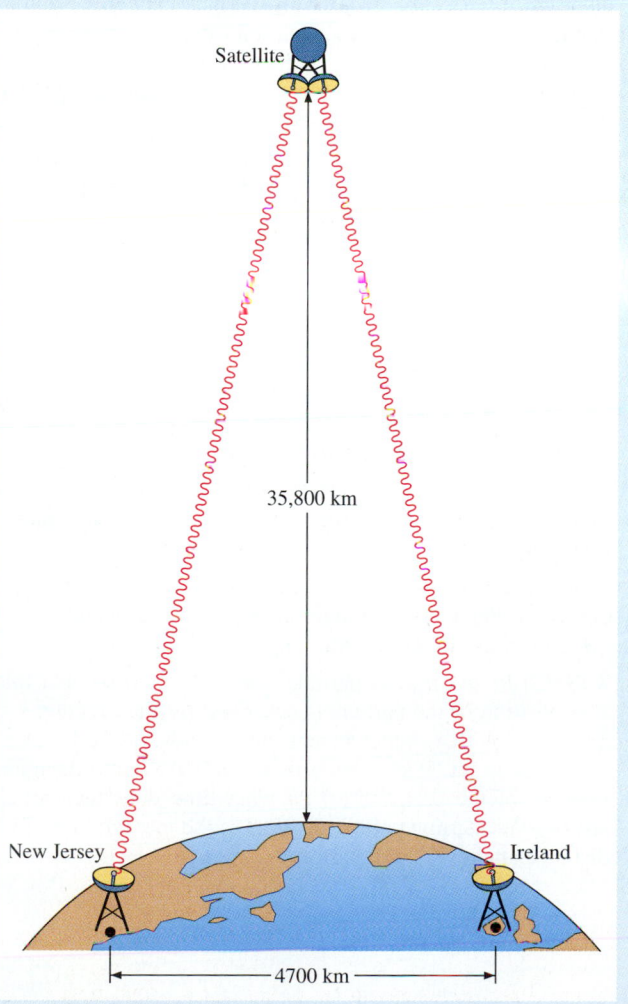

2.11 *(3) Time trap.* On many toll roads, the driver is given a toll ticket on entering and surrenders it when paying the toll on leaving. The ticket is stamped with the entrance time and place. The speed limit on such a road is 90 km/h. A driver actually drives at 120 km/h, covering a total distance of 490 km. However, he makes a 45-minute stop for lunch on the road. Will the toll-taker be able to determine that the driver has been speeding?

2.12 *(3) Gandy dancing.* The speed of sound in air is 0.340×10^3 m/s. In steel rails, the speed of sound is 5.20×10^3 m/s. While walking along a little-used railroad, you hear the sound of sledge hammers used by a working crew some distance away. When you put your ear to the rail, you observe a 2.3-s delay between the sound arriving at one ear through the rail and the sound arriving at the other ear through the air. How far away is the working crew?

2.13 *(4) I mean the average is . . .* A car is driven from A to B at a steady speed of 100 km/h. It returns at a steady speed of 50 km/h. Not counting the time spent at B, what is the mean speed for the entire journey?

2.14 *(4) More than half the fun in less than half the time.* A bicyclist climbs a hill at 10 km/h and then descends, returning to her starting point. Her mean speed for the trip is 16 km/h. What is her downhill speed?

2.15 *(4) Scout's pace, I.* Boy Scouts in a hurry sometimes alternate 100 paces at a walk with 100 paces at a trot. Suppose that the length of a walking pace (one step with each foot) is 110 cm and the length of a trotting pace is 145 cm. Suppose also that walking speed is 1.41 m/s and trotting speed is 2.57 m/s. What is the mean speed of the scouts?

2.16 *(4) Signs of the times.* A particle experiences each of the following velocity changes over a 5-s time interval. In each case, calculate the mean acceleration.
- **(a)** $v_i = 5.6$ m/s, $v_f = 12.9$ m/s
- **(b)** $v_i = 5.6$ m/s, $v_f = -12.9$ m/s
- **(c)** $v_i = -5.6$ m/s, $v_f = -12.9$ m/s
- **(d)** $v_i = -5.6$ m/s, $v_f = 12.9$ m/s
- **(e)** $v_i = 12.9$ m/s, $v_f = 5.6$ m/s
- **(f)** $v_i = 12.9$ m/s, $v_f = -5.6$ m/s
- **(g)** $v_i = -12.9$ m/s, $v_f = -5.6$ m/s
- **(h)** $v_i = -12.9$ m/s, $v_f = 5.6$ m/s

2.17 *(4) Slow down!* A driver slows her automobile from 20 m/s to 10 m/s. In the time that this takes, the auto moves 80 m. What is the value of the mean acceleration?

2.18 *(4) Taking off.* If the takeoff speed of a jet plane is 80 m/s, what is the minimum value of the acceleration required if the runway is 2 km long?

2.19 *(5) Round trip.* A particle moves along a straight line. The position of the particle is described by the function $x = 5t - 20t^2$, with x in meters and t in seconds. **(a)** Is the acceleration constant? If so, what is its value? **(b)** What is the initial velocity of the particle? **(c)** At what time does the particle return to its starting point? **(d)** What is the maximum positive distance attained by the particle?

2.20 *(5) Getting it together, I.* The position of a particle is described by the function

$$x = (25 \text{ units})t + (12 \text{ units})t^2,$$

where x is expressed in meters and t in seconds. **(a)** What are the proper units for the factor 25 in the term $(25 \text{ units})t$? **(b)** What are the proper units for the factor 12 in the term $(12 \text{ units})t^2$? **(c)** Express the velocity v of the particle as a function of time. **(d)** Express the acceleration a of the particle as a function of time. **(e)** Find x, v, and a for $t = 14$ s.

2.21 *(5) Sketching the derivative, I.* The graph below is a schematic plot of the position of a particle as a function of time. **(a)** Make a corresponding sketch of the velocity as a function of time. **(b)** Make a corresponding plot of the acceleration as a function of time.

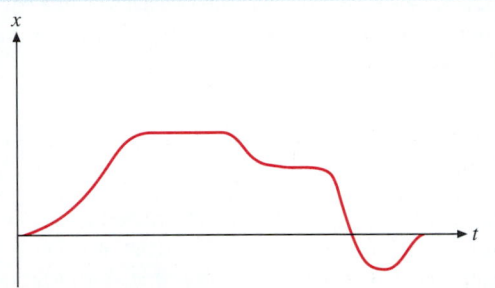

2.22 *(5) Sketching the derivative, II.* The graph below is a schematic plot of the velocity of a particle as a function of time. Make a corresponding sketch of the acceleration of the particle as a function of time.

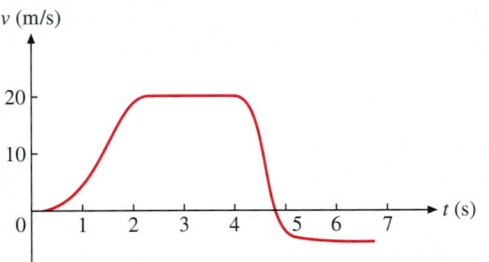

2.23 *(5) Looking for constants in a changing world.* The position of a particle is expressed by each of the following functions of time: **(a)** $x = qt$; **(b)** $x = pt^2 + qt$; **(c)** $x = pt^3$; **(d)** $x = pt^2 + qt + r$; **(e)** $x = p\sqrt{t}$. The quantities p, q, and r are constants having the proper dimensions. In which of these cases is v constant? In which is a constant?

2.24 *(6) Gentle acceleration, I.* A train accelerates at a constant 0.75 m/s^2, starting from rest. **(a)** How long does the train take to reach 50 km/h? **(b)** What distance does the train cover in reaching 50 km/h?

2.25 *(6) Gentle acceleration, II.* A train accelerates at a constant 0.68 m/s^2, starting from an initial speed of 50 km/h. **(a)** How long does the train take to reach 85 km/h? **(b)** What distance does the train cover during the acceleration from 50 km/h to 85 km/h?

2.26 *(6) Rolling down.* When a ball rolls down an incline that is not too gentle, its acceleration is approximately constant. Such a ball starts from rest and covers 1 m in the first second. **(a)** What is its acceleration? **(b)** What is its velocity 2 s after the start? **(c)** How far does the ball go in the first 3 s? **(d)** How far does the ball go during the fourth second?

2.27 *(6) Local service.* A commuting train travels between two stations located 1.8 km apart. The train accelerates for half the distance and decelerates for the remaining half of the distance. Assume that $|a| = 1.2$ m/s^2 for both acceleration and deceleration. **(a)** What is the maximum speed achieved by the train? **(b)** How long does the trip between stations take?

2.28 *(7) Watch out for tourists! I.* A physics student drops a lead weight from the top of the Leaning Tower of Pisa. How fast is the weight moving at **(a)** $t = 0.10$ s? **(b)** $t = 2.16$ s?

2.29 *(7) Watch out for tourists! II.* How far has the weight of Problem 2.28 descended at **(a)** $t = 0.10$ s? **(b)** $t = 2.16$ s?

2.30 *(7) Watch out for tourists! III.* The height of the Leaning Tower of Pisa is 54.6 m. **(a)** How fast is the lead weight of Problems 2.28 and 2.29 moving when it reaches the ground? **(b)** How long does it take for the weight to reach the ground?

2.31 *(7) Downer.* The speed of a freely falling ball increases from 20.0 m/s to 30.0 m/s. How far does it fall during this process?

2.32 *(7) An odd result, I.* A steel ball is released from rest at the top of a high tower. **(a)** Find the displacements of the ball, Δx_1, Δx_2, Δx_3, and Δx_4, during the first, second, third, and fourth seconds. **(b)** Find the ratios $\Delta x_1/\Delta x_1$, $\Delta x_2/\Delta x_1$, $\Delta x_3/\Delta x_1$, and $\Delta x_4/\Delta x_1$ of these displacements.

2.33 *(7) Getting the ol' arm in shape.* A man on a roof throws a baseball upward with an initial speed of 25 m/s. **(a)** How high above the roof does the baseball rise? **(b)** When the ball falls, it clears the edge of the roof and strikes the ground 40 m below. What is the total time of flight?

2.34 *(7) Ups and downs, I.* The stone of Example 2.11 is thrown upward from the ground at $t = -0.99$ s instead of from the window at $t = 0$. For all positive values of t, its history is identical with that described in the example. What is its initial speed?

2.35 *(7) Ups and downs, II.* **(a)** For the stone described in Example 2.11, find the times after the stone is thrown from the window at which the speed is $|v| = 5.00$ m/s. **(b)** Find the time after the stone is thrown from the window at which the speed is $|v| = 21.3$ m/s. Why is there only one time instead of two, as in part **a**?

2.36 *(7) Cytherean lightness.* The acceleration of gravity at the surface of Venus is 8.87 m/s^2. When an object is dropped from rest, how long does it take to fall 10.0 m?

2.37 *(7) Simple experiment.* An astronaut, having landed on Mars, climbs to the top of the spacecraft and drops a stone to the ground from a height of 10.6 m. She observes that it takes 2.4 s for the stone to reach the ground. What is the acceleration of gravity on the surface of Mars?

2.38 *(7) Golf course of champions.* If you can throw a golf ball upward so that it rises 10 m, how high could you throw it on the moon, where the acceleration of gravity has one-sixth its value on earth? Neglect such "trivialities" as air resistance on the earth and the impediment of the space suit you would have to wear on the moon.

GROUP B

2.39 *(3) Outdoor life.* You are standing on a hill, watching a distant woodcutter split a log. He strikes the log with his axe regularly every 4.0 s. Simultaneously with every stroke, you hear the sound of the axe striking the wood. The speed of sound is 340 m/s. **(a)** How far away is the woodsman? **(b)** Is your answer the only one possible?

2.40 *(3) Turn again, Whittington!* A plane carries 400 liters of fuel, which it consumes at the rate of 80 liters per hour. On a training exercise, the plane flies directly downwind from an airport and then returns. Because of the wind, which is steady, the plane's outward speed is 200 km/h, and its return speed is 140 km/h. What is the maximum distance from the airport at which the plane must turn back?

2.41 *(3) Flying squad.* State police forces sometimes use small airplanes to detect speeders on long, remote stretches of road. Special marks are painted on the road surface, and the pilot of the airplane measures the time required for a car to pass from one mark to the next. On observing a speeder, the pilot radios the information to a patrol car officer on the road, who stops the car and issues the speeding ticket. On a certain highway, the speed limit is 105 km/h. However, the police allow a driver's margin of error of 15 km/h and will not stop a speeder going less than 120 km/h. **(a)** If the marks are 400 m apart, which is the maximum transit time for which a driver will get a speeding ticket? **(b)** The pilot cannot measure the transit time to better than ± 0.4 s. Someone proposes that the airplane could be used more efficiently if the marks were repainted closer together, thus reducing the observation time for any one car. But this increases the error in the speed measurement, thus reducing the margin of error allowed the driver. What is the minimum distance between marks for which the error in observing speeders traveling at 120 km/h will not exceed 5 km/h?

2.42 *(3) Equal times.* Suppose that you travel a distance d, taking a total time t to do so. If you travel at speed v_1 for half the total time and at speed v_2 for half the total time, show that your mean speed $\langle v \rangle$ for the trip is given by the expression

$$\langle v \rangle = \frac{v_1 + v_2}{2}.$$

2.43 *(3) Equal distances.* Suppose that you travel a distance d. If you travel at speed v_1 for half the total distance and at speed v_2 for the other half of the distance, show that your mean speed $\langle v \rangle$ for the trip is given by the expression

$$\frac{2}{\langle v \rangle} = \frac{1}{v_1} + \frac{1}{v_2}.$$

2.44 *(G) And that inverted Bowl we call the Sky ...* One method of detecting a minute effect is through its accumulation over a long period of time. The earth's period of rotation (the length of the day) increases by 12×10^{-9} hour per year, averaged to smooth out short-term fluctuations. If the present length of the day is used to calculate the date and time of an astronomical event that was observed 2000 years ago, how great a discrepancy will there be between the calculated and observed times of the event?

2.45 *(4) Scout's pace, II.* Consider again a group of Boy Scouts in a hurry, but consider them more generally than in Problem 2.15. Suppose that the length of a walking pace is l_w and the length of a trotting pace is $l_p = bl_w$, where b is a constant greater than 1. Suppose also that walking speed is v_w and trotting speed is $v_t = cv_w$, where c is a constant greater than 1. **(a)** What is the mean speed $\langle v \rangle$ of the scouts, if they alternate n walking paces with m trotting paces? **(b)** Check your result by simplifying it for the three special cases $m = 0$, $n = 0$, and $b = c = 1$. Then show that the simplified expressions agree with what you would expect in these three special cases.

2.46 *(5) How goes it?* Is the acceleration of a particle constant if its velocity is given **(a)** by $v = cx$? **(b)** by $v^2 = hx$? The quantities c and h are constants having the proper dimensions.

2.47 *(5) The farther the faster.* The motion of a particle is described by the function $bv = c\sqrt{x}$, where b and c are constants. **(a)** Is the acceleration constant? **(b)** If so, what is its value?

2.48 *(5) Getting it together, II.* The motion of a particle is described by the function

$$x = pt - qt^2.$$

(a) If x is to be expressed in meters and t in seconds, what are the proper units for the constant factors p and q? **(b)** At what time does the particle achieve its maximum positive displacement? **(c)** What is the position of the particle at time $t = p/q$? **(d)** Through what total distance has the particle moved at time $t = p/q$? **(e)** What is the velocity of the particle at $t = 0$ and at $t = p/q$? **(f)** What is the acceleration a of the particle? **(g)** Can the position of the particle be properly described by an expression of the form

$$x = \frac{v^2 - v_i^2}{2a} ?$$

Why or why not?

2.49 *(G) Slippery fingers.* A mechanic is cleaning a steel ball bearing whose diameter is 0.800 cm. She accidentally drops it. It falls a distance of 1.20 m into a bucket of thick grease, where it comes to rest just buried under the surface of the grease. If the resistance of the air is negligible, find the mean acceleration **(a)** from the moment the ball is dropped until it hits the grease; **(b)** from the moment it hits the grease until it comes to rest; **(c)** over the entire incident.

2.50 *(G) Wild goose chase.* At time $t = 0$, you drop a ball from the top of a high tower. At time $t = \Delta t$, you drop another ball. **(a)** Neglect air resistance, and find the difference $v_1 - v_2$ between the speeds of the two balls at any time after the second ball is released (but before the first ball hits the ground). Show that $v_1 - v_2$ is independent of time. **(b)** Taking the downward direction as positive, show that the distance between the balls can be written

$$\Delta x \equiv x_1 - x_2 = gt\,\Delta t + \tfrac{1}{2}g(\Delta t)^2$$

at any time after the second ball is released (but before the first ball hits the ground). **(c)** Show that for $t \gg \Delta t$, the distance between the balls increases linearly with time as long as both balls keep falling.

2.51 *(7) Juggling practice.* A juggling student is learning how to keep three balls in the air at a time. The fastest he can throw them is one every 0.50 s. **(a)** With what initial speed must he throw the balls? **(b)** Can he practice while standing in his dormitory room, whose ceiling is 2.4 m from the floor?

2.52 *(7) On its way down.* One second after a ball is dropped from the roof of a building, it passes a ledge 30 m above the ground. With what speed does it strike the ground?

2.53 *(7) An odd result, II.* The following theorem, which we owe to Galileo, is a generalization of Problem 2.32. A body is released from rest at $t_0 = 0$, $x_0 = 0$ and moves with constant acceleration. Its position is measured at the equally spaced times $t_1 = \Delta t$, $t_2 = 2\,\Delta t$, $t_3 = 3\,\Delta t$, and so forth, as in Figure 2.19. Prove that the displacements $\Delta x_j \equiv x_j - x_{j-1}$, which occur in the successive time intervals Δt, are in the ratio

$$\Delta x_1 : \Delta x_2 : \Delta x_3 : \ldots = 1 : 3 : 5 : \ldots.$$

That is, prove that the successive displacements $x_1 - x_0$, $x_2 - x_1$, $x_3 - x_2$, ... are in the ratio of the odd numbers.

2.54 *(7) An odd result, III.* Using the result of Problem 2.53, measure several of the successive displacements in Figure 2.19 and show that the acceleration of the falling balls is in fact constant.

2.55 *(7) An odd result, IV.* The scales shown in Figure 2.19 are in centimeters. Find the time interval between the light flashes that produced the multiple-exposure photograph.

2.56 *(G) Make a wish!* A honeymooner drops a penny down a deep well. The splash is heard 3.0 s later. How far down is the water surface? Take the speed of sound in air to be 340 m/s.

2.57 *(G) On tap.* The volume of water flowing past any cross section of a stream per unit time is given by Av, where A is the cross-sectional area of the stream and v the speed of the stream at that section. If water is neither added to nor removed from the stream along the way, the product Av must be the same everywhere along the stream. The speed of a stream of water is 4.0 m/s where it leaves the faucet. What is the ratio of the cross-sectional area A_1 of the stream where it leaves the faucet to the cross-sectional area A_2 located 1.0 m below the faucet?

2.58 *(G) Speed trap.* An automobile moving steadily at 130 km/h passes a hidden highway patrol car, which immediately takes off after it. Assume that the patrol car accelerates at a constant 2.0 m/s^2. **(a)** How long will it take to catch up with the speeder? **(b)** What distance will the patrol car have covered when this happens?

2.59 *(G) Making herself welcome.* At a picnic at Cliffside Park, a baby falls over the cliff. Fortunately, Wonder Woman is a guest at the picnic. She dives off the edge 2.0 s after the start of the baby's fall. If the cliff is 100 m high, what must Wonder Woman's initial velocity be if she is to catch the baby just before he reaches the ground? Assume that both Wonder Woman and the baby experience acceleration of magnitude $|g|$ once they leave the cliff top.

2.60 *(G) Safe to pass?* You are driving on a two-lane road, stuck behind a truck traveling at a constant speed of 70.0 km/h. You wish to pass the truck, whose length is 12.2 m. The length of your car is 3.8 m. To pass safely, you need to pull out into the other lane of the road when the front of your car is no less than 7.0 m behind the truck and pull in when the rear of your car is no less than 7.0 m ahead of the truck. Suppose that your car can accelerate at a constant 2.0 m/s^2. **(a)** What is the minimum distance you need to pass? **(b)** How fast will you be going when you pull in ahead of the truck? **(c)** Suppose that cars coming the other way are going no faster than 100 km/h. For what minimum distance ahead must the road be clear of oncoming cars before you begin to pass?

2.61 *(G) Tortoise and hare.* A car waiting at a traffic light starts to move with a constant acceleration of 2.0 m/s^2 when the light turns green. Two seconds later, a truck moving in the same direction passes the light at a constant speed of 25 m/s and soon passes the car. Soon the car passes the truck. **(a)** How long after the car starts do the two overtakings occur? **(b)** How far from the light does each take place?

GROUP C

2.62 *(G) Fishing for lunch.* A vacationer gets into his outboard motorboat and leaves a dock on a river bank for a day's fishing. Just as he turns upstream, he hears a splash but pays no attention and continues at normal cruising speed. Twenty minutes later, he realizes that his watertight lunch box is gone. He turns downstream, with the motor still set at cruising speed. He is relieved to sight the lunchbox floating down the river and retrieves it at a point 1.0 km below the dock. **(a)** How long after turning around does he pick up his lunch? **(b)** What is the speed of the river current?

2.63 *(G) Ride & tether.* This little-known sport is a cross-country team race, usually through rough, mountainous terrain. Each team consists of two people and one horse. The clock is started when a team crosses the starting line with one of the people—say, Ann—riding the horse and the other—say, Bob—walking. The rider, of course, leaves the walker behind. At any time, Ann may tether the horse and continue on foot. When Bob reaches the horse he mounts, riding past Ann and continuing any distance he wishes. Bob then tethers the horse and continues on foot. When Ann reaches the horse, she mounts and rides past Bob. This alternation of walking and riding continues until all members of the team—both people and the horse—have crossed the finish line, at which time the clock is stopped. **(a)** Suppose that both people have the same mean walking speed $\langle v \rangle$ and that the horse can carry either one with mean speed $c\langle v \rangle$, where c is a constant greater than 1. What fraction of the total race distance should each of the people ride in order to minimize the total race time? Assume that the race distance is sufficiently long that Ann and Bob can alternate on the horse many times. **(b)** Suppose now, more realistically, that Ann and Bob do not have equal walking speeds. Specifically, let Ann's and Bob's mean walking speeds be $\langle v_A \rangle$ and $\langle v_B \rangle = b\langle v_A \rangle$, where b is a constant different from 1. The horse carries either person at the same mean speed $c\langle v_A \rangle$, which is faster than either can walk. What fraction of the total race distance should each person ride in order to minimize the total race time? **(c)** Now consider a still more realistic model of the actual race. Bob's legs are longer than Ann's, and he can walk faster than she can. But Ann is lighter than Bob, and the horse can carry her faster than it can carry Bob. Define an appropriate constant to describe the ratio of riding speeds, and again find the fraction of the total race distance that each person should ride in order to minimize the total race time.

2.64 *(G) Seeds of integral calculus.* In the figure below, the horizontal axis represents time and the vertical axis velocity. **(a)** If these quantities are expressed in SI, what are the proper units for expressing the area of the rectangle *EADF*? **(b)** What are the proper units for expressing *any* area on the graph?

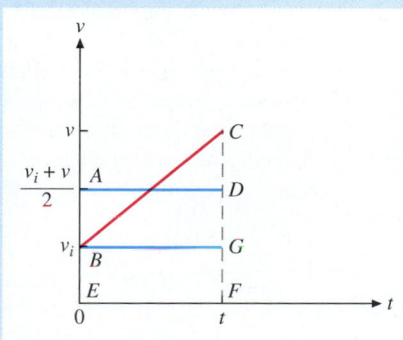

(c) What is the area of rectangle *EADF*? Compare your result with the right side of Equation 2.33. What does this area represent physically? **(d)** Use the formula for the area of a triangle to find the area of the triangle *BCG*. Compare your result with the right side of Equation 2.32. **(e)** Express the altitude of triangle *BCG* in terms of the acceleration *a*. Use this result to write the area, and compare it with the left side of Equation 2.32. **(f)** The area of the trapezoid *EBCF* can be written as the sum of the areas of the rectangle *EBGF* and the triangle *BCG*. Write the area in this form, and compare your results with Equation 2.31b. **(g)** Compare the areas *EBCF* and *EADF* directly. (Note: The relation arrived at in part **g** was first pointed out in the late Middle Ages by scholars at Merton College, Oxford. Consequently, it is sometimes called the *Merton Rule*.)

2.65 *(G) Slippery slope.* Suppose that the acceleration of a particle is not constant. Rather, the acceleration increases gradually from a small initial value at *t* = 0 to a larger value at a later time *t*. **(a)** Assume that the initial and final velocities are the same as those in the illustration for Problem 2.64, and sketch the curve that must replace the straight line *BC*. **(b)** In light of the results of Problem 2.64, what does the area bounded by *EB*, your curve, *CF*, and *FE* represent? **(c)** How does the enclosed area compare with that of the rectangle *EADF*? **(d)** Sketch a horizontal line on your graph that represents a mean velocity $\langle v \rangle$ such that $\Delta x = \langle v \rangle t$. **(e)** Why is $\langle v \rangle$ not equal to $(v_i + v)/2$?

2.66 *(7) Look out below!* A ball is dropped from the roof of a building. An observer looking out a window sees the ball fall just outside. If the window is 1.15 m tall and the ball is in view for 0.162 s, how far below the roof is the bottom of the window?

2.67 *(7) Fun with rockets.* A typical toy rocket engine burns for 1.20 s, during which the rocket accelerates with $a = 58.0 \text{ m/s}^2$. It then "coasts" to a maximum altitude and falls to the ground. Neglecting air resistance, find **(a)** the maximum altitude and **(b)** the time of flight.

2.68 *(7) Measuring g.* The diagram in the next column shows a method for measuring the acceleration of gravity precisely. A small ball is situated in an evacuated chamber (not shown) to eliminate the influence of air resistance. The ball is projected upward by a "gun." The ball passes electronic "gates" 1 and 2 as it rises, and again as it falls. Each gate is connected to a separate timer. The first passage of the ball through each gate starts the corresponding timer, and the second passage through the same gate stops the timer. The time intervals Δt_1 and Δt_2 are thus measured. The vertical distance between the two gates is *d*. Show that the acceleration of gravity is given by the expression

$$g = \frac{8d}{(\Delta t_1)^2 - (\Delta t_2)^2}.$$

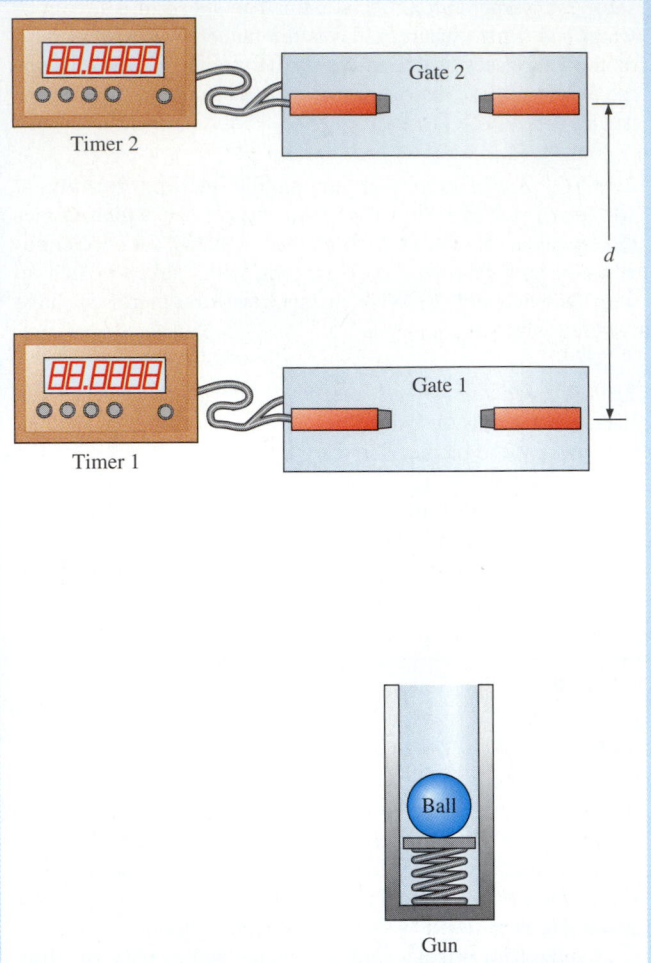

2.69 *(G) Leaning Tower of Pisa.* There is no evidence that Galileo ever in fact performed the legendary experiment in which he is supposed to have simultaneously dropped a cannonball and a musket ball from the top of the Leaning Tower of Pisa, demonstrating to an assembly of dumb-struck professors that the balls hit the ground at the same time. (See Query 2.12, which concerns a similar experiment done in imagination.) But in one of his books Galileo does discuss such an experiment and points out (correctly) that the musket ball would be "a hand's breadth" behind the cannonball when the latter reached the ground. **(a)** For the sake of argument, assume that the cannonball experiences negligible air resistance throughout its fall and accelerates downward with the standard value of *g*. Assume also that, in spite of the small air resistance that makes the musket ball reach the ground slightly later than the cannonball, its acceleration *a* is also constant. Find the ratio *a*/*g*. Because the height of the Leaning Tower is 54.6 m, let this be the release height of the two balls. Also, let a "hand's breadth" be 10 cm. **(b)** Is it plausible to assume that objects of similar shape and composition fall at rates proportional to their weights?

2.70 *(G) Speed and politics.* The following problem is modeled on at least one historical incident. You are designing a rapid-transit rail system. The rails will run down the center of an automobile expressway. You can easily design the trains and the track bed to allow for a maximum speed well in excess of the automobile speed limit of 90 km/h, and you believe that the sight of the trains speeding past them will divert motorists to the trains, thus reducing congestion, smog, and petroleum consumption. However, with passengers standing in the trains (inevitable in rush hour in spite of enthusiasts' predictions to the contrary), you cannot safely allow accelerations with magnitudes in excess of 1.5 m/s^2. Every station stop requires 45 s to load and unload passengers. **(a)** For what minimum distance between stations will the train just keep up with the automobiles? **(b)** If stations were placed this distance apart, what would be the maximum speed of the train? **(c)** Suppose you must limit maximum train speed to 130 km/h for practical and safety reasons. What is the minimum distance between stations for which the train will reach this speed? **(d)** After construction has started, every city council member whose district lies along the line insists that his or her district have a station to serve its residents. No city council district is more than 1.6 km wide. Will the rail system succeed in luring many motorists from their automobiles? **(e)** The stations do in fact end up about 1.6 km apart. Find the mean speed of the trains, including stops.

Newton's Laws of Motion

3

LOOKING AHEAD

— What is the connection between force and motion?

— Newton's first law of motion tells how a body moves when no force acts on it.

— Newton's first law is valid for observers situated in inertial reference frames.

— Mass is the measure of the inertia of a body.

— Newton's second law of motion tells how a body, observed from an inertial frame, moves when a force acts on it.

— A body in free fall is a body acted on only by the gravitational force.

— The weight of a body is the force required to support it against the gravitational force.

— When one body exerts a force on a second body, the second body exerts an equal and opposite force on the first; this is Newton's third law.

— Atwood's machine is a simple device that clearly illustrates the operation of Newton's second and third laws of motion.

Left: How much harder than the losing team must the winning team pull to drag the losers into the muddy ditch?

> *Force without wisdom falls of its own weight.*
>
> —HORACE

Inertia and Newton's First Law

We are now ready to return to the basic question that launched our study of mechanics: What is the connection between force and motion? Before we can answer this question completely, we will have to define *force* in a precise way. For the time being, think of a force as simply what you exert on a weight when lifting it or what a truck tractor exerts on a trailer to make it move.

To begin with, what happens to a body when *no* force is acting on it? Figure 3.1 shows a block sitting on a tabletop. For the moment, let us ignore any forces that might be acting on the block in the vertical direction, because the block can move only horizontally. No force is acting on the block in the horizontal direction, and it is not moving.

But *must* a body be at rest if no force is acting on it? Figure 3.2 shows that the answer is No. The disk in the figure is made of solid carbon dioxide, called dry ice. Carbon dioxide exists in its solid form only when its temperature is below $-78.5°C$. When it contacts relatively warm surroundings, dry ice *sublimes* (changes into its gaseous form). This happens continually on the lower surface of the disk. Consequently, the disk does not touch the table; it "floats" on a thin cushion of carbon dioxide gas. The gas flows away around the edges of the disk and is renewed by further sublimation. As a result, the frictional forces between the disk and the table are negligible, and we need consider only the forces that we deliberately apply to the disk. This simplifies the observations and analysis considerably.

If the disk is pushed and released—that is, if a force is briefly applied to the disk and then removed—the disk continues to move. This is shown in Figure 3.2a, a long-exposure photograph. After being pushed to the right and released, the disk continues to move to the right. Figure 3.2b is a multiflash photograph of the same process. This

$v = 0$
$a = 0$

FIGURE 3.1 A block sitting undisturbed and at rest on a tabletop. No force is acting on the block horizontally; its velocity is $v = 0$, and its acceleration is $a = 0$.

FIGURE 3.2 A dry-ice disk receives an initial push but subsequently has no forces acting on it horizontally. (*a*) Long-exposure photograph of the motion of the disk. (*b*) Multiflash photograph of the same process.

(a)

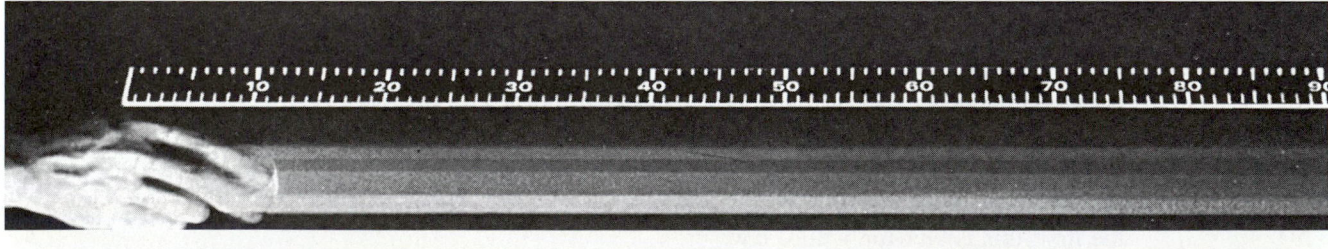

(b)

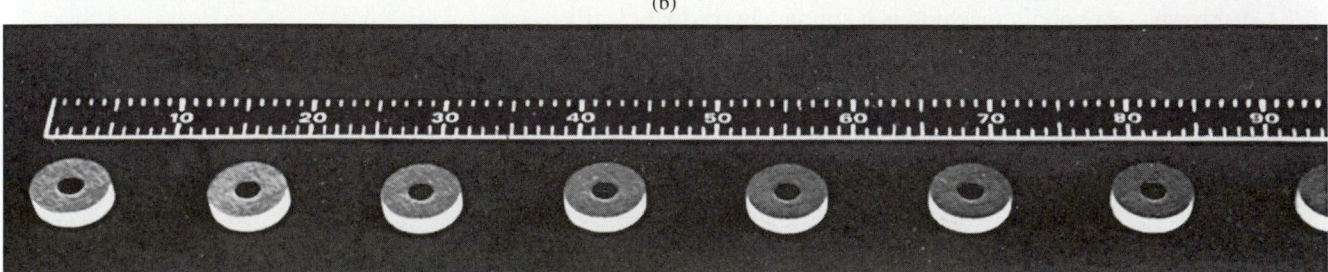

photograph shows that the disk moves equal distances in equal times; that is, *with no horizontal forces acting on the disk, it moves with constant velocity*. This observation is just one example of an experimental result so general and so important that it is given the status of a law of nature:

> *Unless there is a net external force acting on a body, its velocity (both magnitude and direction) remains unchanged.*

This statement has two names, used interchangeably. It is called the **law of inertia,** from a Latin word meaning "sluggishness." It is also called **Newton's first law of motion.** The law includes the case in which the body is at rest ($v = 0$); that is, the law applies just as well to bodies at rest as to bodies in motion. Indeed, the particular velocity of the body is not relevant to the law of inertia. The crucial point is that, in the absence of a net external force, the velocity does not change—*the acceleration is zero*. Thus, a body initially at rest remains at rest, and a body in motion continues to move at the same velocity, unless acted upon by a net external force.

What is meant by the word "net" in the statement of the law of inertia? A body may be acted on simultaneously by several forces. Consider, for example, two teams of children trying to push a large medicine ball in opposite directions. Each team exerts a force that acts on the ball from outside the ball—an *external force*. The ball does not move at all if the two teams are equally matched. Your intuition probably tells you that the forces exerted by the two teams "cancel" each other; we say in this case that the *net external force* is zero.

Now consider a moving body. If you wish to push a chair across a level floor at constant velocity, you must push with a constant force. Does this mean that moving a body with constant velocity requires application of a constant net external force? In other words, does this observation contradict the law of inertia? No, because your push is not the *only* external force exerted on the chair. As the chair slides along, the floor also exerts a force on it. This force (which happens to be a frictional force) opposes the force that you apply in such a way that the *net* force acting on the chair is zero. When bodies move in everyday situations, friction is always present to some degree. That is the main reason why the law of inertia is not obvious. (We began by considering the dry-ice disk so as to eliminate complications such as forces like friction that we cannot control directly.)

In addition to external forces, *internal forces* act within the body. When you push on the chair, it moves as a whole and does not collapse under the action of the force you apply. This is because internal forces act to keep the chair rigid. But such internal forces are not relevant to the law of inertia.

Standing on the Shoulders of a Giant

Although the law of inertia was first clearly enunciated by Galileo, the English natural philosopher Isaac Newton (1642–1727) incorporated the law into the solid logical basis on which he founded the science of mechanics in his great 1687 work, commonly called *Principia*.

EXAMPLE 3.1

A water skier is towed by a motorboat at a constant velocity of magnitude 15 km/h. The boat speeds up, and after a short interval the skier is towed at a new constant velocity of magnitude 20 km/h. What is the net force on the skier when she is moving at 15 km/h? at 20 km/h?

SOLUTION: Two major forces act on the water skier (and her skis). As shown in Figure 3.3, one of these forces is that exerted on her hands by the towrope. The other is the resistance of the water (and to a lesser extent the air). When the skier is moving in a straight line at a constant 15 km/h, her velocity is constant. According to Newton's first law, the net force on the skier (and the skis) *must* be zero. Indeed, the law itself is the basis for asserting that the force exerted by the water and the air on the skier is exactly equal in magnitude, and opposite in direction, to that exerted on her by the towrope.

When the skier is moving at 20 km/h, the force exerted on her by the towrope is greater in magnitude than that at 15 km/h.

But so is the resistive force. Again, the net force on the skier must be zero because her velocity is constant.

FIGURE 3.3

Force exerted on skier by towrope

Force exerted on skis by water and skier by air

Inertial Frames of Reference

The motion of a body has meaning only with respect to some reference point. Stationed at that reference point is an observer, real or imaginary, who measures displacements and elapsed times. On the basis of these observations, velocities can be calculated. Thus the observer is in a position to check the validity of Newton's first law of motion.

But will *any* observer conclude that Newton's first law is valid? Or must the observer meet certain requirements? Suppose that a dry-ice disk at rest sits next to observer O on a horizontal surface. She can easily verify that its velocity is not changing, because $v = 0$ at all times. Because the friction between the disk and the surface is negligible, the observer knows that the horizontal force on the disk is zero. Thus O can assert that Newton's first law is valid for the disk.

Next suppose that a second observer, O', rides past O on a golf cart. Observer O can consider O' an object of observation, and she can measure his position at several times. She can use the results of the measurement to calculate values of his velocity and acceleration as she observes them.

As O' moves past O, O', too, makes position and time observations of the dry-ice disk. Will he agree with O that Newton's first law is valid for the disk? We now show that there is no single yes or no answer; rather, the answer depends on the motion of O' as seen by O.

Consider two specific possibilities. Suppose first that O finds O' to be moving past her at a constant velocity V, as shown in Figure 3.4. From the point of view of O', O and the dry-ice disk in front of her are moving past him at a constant velocity $v' = -V$.* He can take the word of O that no net force is acting on the disk (and in principle he can check her results before and after his trip past her). Thus he, too, can conclude that Newton's first law is valid because v' remains constant in the absence of a net external force acting on the disk.

As a second possibility, suppose that O' is speeding up as he passes O. Measuring his position at several times, O finds that O' is moving with constant acceleration A, as shown in Figure 3.5. From the point of view of O', O and the disk are moving past with constant acceleration $a' = -A$. For O', then, Newton's first law is not valid because the disk is accelerating although no net external force is acting on it.

We must therefore place a limit on the law: *Newton's first law of motion is valid only for nonaccelerating observers.* The coordinate systems attached to nonaccelerating observers are called **inertial frames of reference**, or **inertial frames** for short. The coordinate systems attached to accelerating observers are called **noninertial frames of reference**, or **noninertial frames** for short. In Figure 3.4, both O and O' are situated in inertial frames. In Figure 3.5, O is situated in an inertial frame, but O' is situated in a noninertial frame.

FIGURE 3.4 Two observers, O and O', observe the motion of a dry-ice disk. From the point of view of O, the disk is at rest ($v = 0$) and O' moves with constant velocity V. From the point of view of O', both the disk and O move with constant velocity $-V$. Newton's first law is valid for both observers.

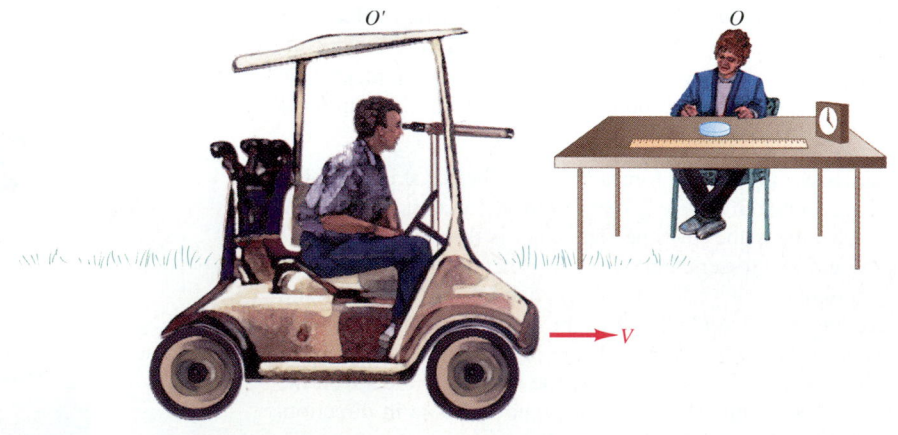

---

*It is conventional to label all quantities measured or calculated by O' with primes. Thus O' observes the velocity v'.

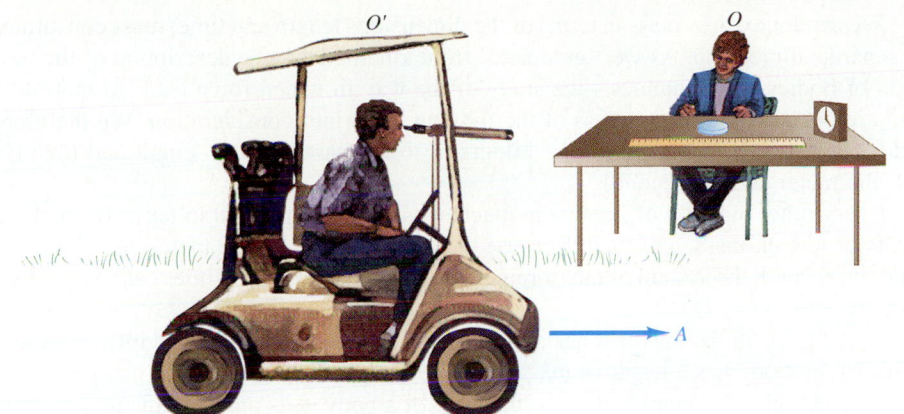

This question now arises: With respect to *what* is an observer accelerating or not accelerating? For the purpose of classical mechanics, it is satisfactory to answer: With respect to the fixed stars. Admittedly, the meaning of this statement is not absolutely clear. So far as we know, no star in the universe is fixed with respect to anything else. For practical purposes, however, it is often an adequate approximation to measure the acceleration of an observer with respect to another observer who, like O, is fixed on the surface of the earth. The translation and rotation of the earth are accelerated motions, but these accelerations are negligible compared with the accelerations in many laboratory-scale systems. When this approximation is not satisfactory (as in the study of the motions of the planets or of ocean currents), we can imagine an observer who is fixed with respect to the sun. For still larger-scale purposes, we can imagine an observer fixed with respect to the center of the galaxy, and so on. In this way, we can always imagine an observer whose frame of reference is a sufficiently close approximation to the ideal one that does not accelerate "with respect to the fixed stars."

In this book, we will only rarely consider things from the point of view of an observer in a noninertial frame. Unless they explicitly state otherwise, all discussions will pertain to inertial frames of reference.

Mass and the Unit of Mass

Let us now look more closely at the inertia, or "sluggishness," of a body. The law of inertia, as we have seen, states that a body's velocity remains constant unless a net external force acts on the body. In other words, no body will change its velocity spontaneously. Changing a body's velocity is possible, but this is easier to do for some bodies than for others. As a general rule, a bigger body is harder to speed up or slow down than a smaller one. That is why big cars generally have larger engines and stronger brakes than small cars.

But exactly what do we mean by "bigger"? We are talking about the *quantity of matter* present in a body, called its **mass** (often denoted by the symbol m). To make quantitative statements, we must define a unit of mass so that the mass of any body can be expressed in terms of that unit. Once we have done this, the mass of a body will be a quantitative measure of the body's inertia.

The original metric unit of mass was the gram, defined as the mass of 1 cm^3 of pure water. Like that of the meter, however (Section 1.2), this definition did not lead to results that were reproducible with sufficient accuracy. Since 1889, the standard has been a cylinder made of a platinum-iridium alloy kept at the Bureau of Weights and Measures in Sèvres. By definition, the **unit of mass** is the mass of that cylinder and is called the **kilogram** (symbol **kg**). Unlike the meter and the second, the kilogram has not yet been redefined in terms of an atomic standard because a sufficiently precise method for doing so does not yet exist.

Fine Points of Nomenclature

The word *gram* is derived from the Greek "gramma," the name of a commercial standard used in the Hellenistic world and later adopted widely through the Eastern Roman Empire.

Of all the SI units, the kilogram is the only one that has a multiplier prefix (kilo-) as a part of its name. This irregularity is due to the historical accident that the gram, defined in 1792, was replaced by the kilogram as the base unit of mass seven years later. Now that the kilogram is the base unit, the **gram** (symbol **g**) is defined simply as $1\text{ g} \equiv 10^{-3}\text{ kg}$.

We cannot express mass in terms of the dimensions length and time; mass constitutes a separate dimension. As we "graduate" from kinematics—the description of the motion of bodies—to dynamics—the study of the way in which force leads to motion—we are obliged to take the mass of the moving body into consideration. We therefore add a base unit of mass (in SI, the kilogram) to the base units of length and time (in SI, the meter and the second).

Every other quantity of interest in mechanics can be expressed in terms of the base units of length, mass, and time. Because the SI base units are the meter, the kilogram, and the second, the system of measurement based on them is sometimes called the MKS system.

Every body in the universe has a mass. Mass is a positive scalar quantity—positive because no body has a negative mass and scalar because the concept of mass involves magnitude only. For many purposes, the mass of a body is its only significant property. In such cases, we often refer to the body as "a mass." For example, we speak of "a ten-kilogram mass" when we mean "a body having a mass of ten kilograms." You will encounter this usage very frequently, both in this book and elsewhere. As long as you do not forget that mass is a *property* of the body and not the body itself, you will not get into difficulties.

Newton's Second Law and Force

We are now ready to consider what happens to a body when a nonzero net force acts on it. That is, we are ready to make a precise, though limited, answer to the question with which we began Part I: What is the connection between force and motion?

According to Newton's first law, a body does not accelerate in the absence of a net applied force. Let us make the plausible guess that, when a net force is applied to the body, it *does* accelerate. What is the quantitative connection between acceleration a, mass m, and force F? We will establish this connection by means of a set of experiments.

Figure 3.6 shows the first of these experiments in an idealized way. (We will discuss actual experiments in Section 3.6 and elsewhere.) A 1-kg block rests on a frictionless surface. Attached to it are a string and a spring scale that can be used to apply a horizontal force to the block and simultaneously to measure the applied force. (This applied force is the net external force applied to the block because no other horizontal forces are acting on the block.) We have not yet defined a unit of force, and so we cannot calibrate the scale. We can nevertheless use the scale to apply a *constant* force to the block by pulling so that the scale indicator, having moved from its undisturbed position to a new position, stays there.

With constant applied force, we find that the acceleration of the block is constant. (We could measure the acceleration by means of stroboscopic photography, as outlined in the legend to Figure 2.19.) By trial and error, we find the particular constant force that results in an acceleration $a_1 = 1$ m/s^2. We mark the spring scale at the indicator position that produced this result.

Next we place a second, identical 1-kg mass on top of the first one, so that the

FIGURE 3.6 (*a*) A force F_1 is applied to a 1-kg block resting on a frictionless surface. The force can be kept constant by keeping the scale indicator steadily at the same mark. When the force is kept constant, the acceleration of the block is always found to be constant. By trial and error, the scale indicator position corresponding to acceleration $a_1 = 1$ m/s^2 is found. (*b*) The experiment in part *a* is repeated. This time, a second 1-kg block is laid on top of the first one, for a total mass of 2 kg. Applying the same constant force F_1 as before, as verified by the position of the scale indicator, results in a constant acceleration $a_2 = \frac{1}{2}$ m/s^2.

(a)

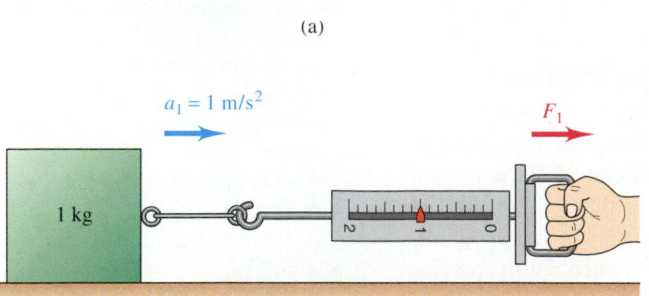

(b)

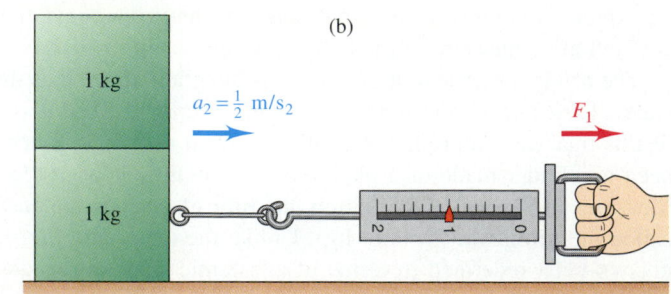

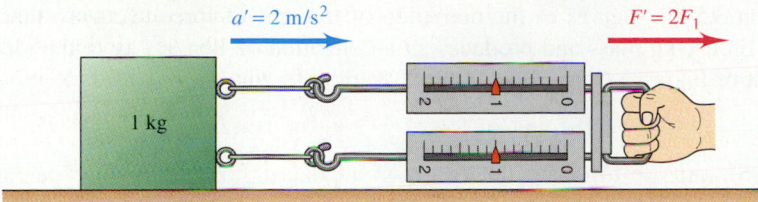

$a' = 2 \text{ m/s}^2$

$F' = 2F_1$

1 kg

FIGURE 3.7 The experiment of Figure 3.6a is repeated with two identical scales attached to the 1-kg mass. A force is applied such that each of the scale indicators is at the F_1 mark established in the preceding experiment. This means that the total force applied to the mass is now doubled to the value $F' = 2F_1$. The acceleration is again constant and has the magnitude $a' = 2 \text{ m/s}^2$.

combined mass of the blocks is 2 kg. We repeat the experiment, applying the same force as before. We find that the acceleration has the constant value $a_2 = \frac{1}{2} \text{ m/s}^2$. Repetition with a combined mass of 3 kg results in an acceleration $a_3 = \frac{1}{3} \text{ m/s}^2$, and so on. In general, *when the force is kept fixed, the acceleration is inversely proportional to the mass being accelerated*:

$$a \propto \frac{1}{m} \quad \text{for } F \text{ fixed.} \tag{3.1}$$

We must now determine the relation between acceleration and force for fixed mass. To do this, we double the force that resulted in a 1 m/s² acceleration of the 1-kg mass and measure the new acceleration of the same mass.* Measurement shows that the acceleration of the block is 2 m/s². Repetition with a force three times the original one will yield an acceleration of magnitude 3 m/s², and so on. In general, *when the mass is kept fixed, the acceleration is directly proportional to the force being applied*:

$$a \propto F \quad \text{for } m \text{ fixed.} \tag{3.2}$$

Combining Proportionalities 3.1 and 3.2, we have the more general proportionality

$$a \propto \frac{F}{m}. \tag{3.3}$$

This expression specifies the dependence of a on both F and m. Thus the expression is not subject to the restrictions on Proportionalities 3.1 and 3.2, and both the force F and the mass m can be treated as variables.

Newton's Second Law and the Unit of Force

Let us multiply Proportionality 3.3 through by m and rewrite it in the form

$$F \propto ma \tag{3.4}$$

Because units for both m and a have been defined, there is no need to define an independent unit for force F. Because the unit of force is as yet undefined, we can reexpress Proportionality 3.4 as an equation by inserting any proportionality constant we desire. We choose for that constant the dimensionless number 1, the simplest possible choice:

$$F = 1 \ ma,$$

or

$$F = ma. \tag{3.5}$$

This equation expresses **Newton's second law of motion**:

> *The net external force exerted on a body is equal to the product of its mass and its acceleration.*[†]

Because the mass m is a positive scalar, the direction of the net external force F exerted on a body must be the same as that of the acceleration a. Both the magnitude and the direction of a force must be specified to describe a force completely, and therefore *force is a vector quantity*.

---

*Without a defined scale of force, how do we know when the force is doubled? Here is one way: We use a second spring scale identical to the first. We attach the two scales side by side to the 1-kg mass, as shown in Figure 3.7. We pull the system so that each of the scales indicates that a constant force F_1 is being applied to the mass. The total force applied to the mass must be $2F_1$ (which is double the original force).

[†]Like Newton's first law, this law is valid only for observers situated in inertial frames of reference.

Equation 3.5 also serves as the definition of the unit of force. Suppose that a net force acts on a 1-kg mass and produces an acceleration of 1 m/s^2, as in the idealized experiment of Figure 3.6. It follows from Equation 3.5 that

$$1 \text{ [SI unit of force]} = 1 \text{ kg} \times 1 \text{ m/s}^2.$$

Thus the SI unit of force is the kg · m/s^2 (spoken "kilogram-meter per second squared"). This unit turns up very frequently in physics. It is therefore given the special name **newton** (symbol **N**). By definition,

$$1 \text{ N} \equiv 1 \text{ kg} \cdot \text{m/s}^2. \tag{3.6}$$

A unit so expressed in terms of SI base units is called a **derived unit.** The units of velocity and acceleration are likewise derived units. You will encounter others as well.

There is a standard set of rules for naming derived units. With a few exceptions, each unit is named after a scientist who pioneered in the area of physics to which the unit is relevant. (Currently, there are thirteen such names.) The name of the unit is always written with a lowercase initial letter (as in newton, not Newton). Sometimes the name is truncated for the sake of brevity (for example, the name Faraday yields farad, the derived unit of electrical capacitance introduced in Chapter 26). If the scientist's name includes an accent, it is dropped in the name of the unit (for example, the name Ampère yields ampere, the base unit of electric current introduced in Chapter 27). The symbol for the unit is the capitalized first letter of the unit's name; for example, the name Watt yields the unit watt (symbol W), the derived unit of power introduced in Chapter 7. An exception occurs when the initial letter has already been used as the symbol for another unit. In this case a second, lowercase letter is added to avoid duplication; for example, the name Weber yields weber (symbol Wb), the derived unit of magnetic flux introduced in Chapter 28.

Example 3.2 is a simple and direct application of Newton's second law of motion.

How Do You Say It?

The general rule for pronouncing unit names is a flexible one: If you are using the native language of the scientist after whom a unit is named, you should pronounce it as he or she did. If you are using a different language, you should pronounce the name in a manner convenient to the language that you are using.

EXAMPLE 3.2

A dry-ice disk having a mass of 0.254 kg is pulled along a smooth horizontal tabletop by a light thread. Analysis of a multiple-flash photograph indicates that the disk has a constant acceleration of magnitude 7.52 m/s^2. What assumption must you make if you are to argue that the force exerted on the disk by the thread is constant? If the force is constant, what are its direction and magnitude?

SOLUTION: Consider Equation 3.5, $F = ma$. Given the observation that the acceleration a is constant, the force F will be constant only if the mass m also is constant. Because the mass of the dry-ice disk is continually diminishing as solid carbon dioxide sublimes into the gaseous form and ceases to be part of the disk, m can never be strictly constant. However, the mass

of the disk is known only to three significant figures—say, roughly, to within the mass of a penny. As long as the observation is made over a time interval short enough so that less than a "penny's mass" of dry ice sublimes, the mass of the disk can be taken as constant. Under these circumstances, the constancy of the observed acceleration ensures that the force is indeed constant within three significant figures.

Because the mass m is a positive scalar, the direction of the force is the same as that of the acceleration. To find the magnitude of the force, you use Equation 3.5 to obtain

$$F = 0.254 \text{ kg} \times 7.52 \text{ m/s}^2 = 1.91 \text{ kg} \cdot \text{m/s}^2$$
$$= 1.91 \text{ N}.$$

Newton's First and Second Laws: The Logical Connection between Them

Newton's first law asserts that a body tends to move with constant velocity. Newton's second law provides a quantitative measure of the force that will produce a given acceleration of the mass.

At first glance, it may seem that Newton's first law is not an independent law of nature at all, but merely the special case of Newton's second law, $F = ma$, in which the net external force is zero. If $F = 0$, we have $a = F/m = 0$ and therefore $v =$ constant, as the first law states. Is this all there is to Newton's first law? No, because this interpretation of Newton's first law is based on an incomplete statement of the law.

Remember that *Newton's first law is valid only for inertial observers.* If you review the argument at the beginning of this section, you will see that this key statement is the foundation on which the justification of Newton's second law rests. Thus, *Newton's first law logically precedes the second law*; you cannot even begin to consider what happens to a body when a nonzero net external force acts on it (as seen by an inertial observer) until you have the first law as a basis for further argument.

Free Fall and Weight

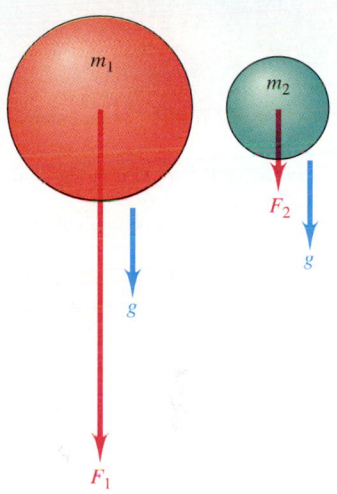

Let us reconsider the freely falling body—a body falling in the absence of buoyant and frictional effects—in the light of Newton's second law of motion. A body falling freely near the surface of the earth moves with constant acceleration g (Section 2.7).*

A freely falling body of mass m_1 *must* have a net force acting on it. How do we know this? According to Newton's second law, $F = ma$, any acceleration of a body must be due to an applied net force. If the acceleration of the body has the particular value g, the applied net force F_1 must have the value

$$F_1 = m_1 g \qquad (3.7)$$

(Figure 3.8). The force F_1 is called the **force of gravity** or the **gravitational force due to the earth**. Every mass in the universe exerts an attractive force, called the gravitational force, on every other mass. For any body near the earth's surface, the earth itself is by far the largest mass in the vicinity, and its gravitational attraction greatly exceeds all others exerted on the body.

Consider now a second body whose mass is m_2 (Figure 3.8). It, too, falls freely with the same acceleration g. The force of gravity F_2 exerted on it must be

$$F_2 = m_2 g. \qquad (3.8)$$

To eliminate the common quantity g from these equations, we divide Equation 3.7 by Equation 3.8. This yields

$$\frac{F_1}{F_2} = \frac{m_1}{m_2}. \qquad (3.9)$$

This relation is true for arbitrary values of m_1 and m_2. It follows that *the gravitational force exerted on a body by the earth is directly proportional to the mass of the body.*

FIGURE 3.8 Two bodies having different masses m_1 and m_2 fall freely with the same acceleration g. It follows from Newton's second law that the gravitational forces acting on them must be in direct proportion to their masses.

Weight

You can prevent a body from falling by hanging it from a string, as in Figure 3.9. But if the body is hanging at rest, Newton's first and second laws both tell you that the net force F exerted on the body must be zero; $F = ma = 0$. How can this be? If the mass of the body in Figure 3.9 is m, the force of gravity is $F_{grav} = mg$. This force surely does not disappear when you hang the body from a string. But now a second force F_s is exerted on the body—by the string. The net force F exerted on the body must be the sum of the two force vectors:

$$F = F_{grav} + F_s.$$

Because $F = 0$, we have $F_{grav} + F_s = 0$, or $mg + F_s = 0$. Only thus can the body be at rest, as it obviously is. Solving this last equation for F_s gives

$$F_s = -mg. \qquad (3.10)$$

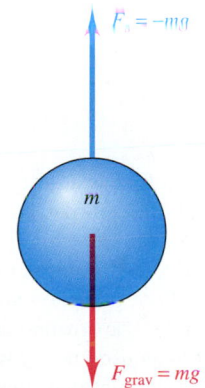

FIGURE 3.9 A body of mass m is prevented from falling by a force exerted by a string. This force is directed opposite the force of gravity and has the same magnitude, so the net external force exerted on the body is zero. In this free-body diagram, the string is not shown. It is replaced by its effect on the body, which is the upward force F_s. Such a diagram of a body shows only all of the external forces acting on it.

---

*Having defined force as the product of mass and acceleration, we can now assert that the "effects" of buoyancy and friction are in fact forces, because their presence influences the acceleration of a falling body. The reason for restricting our study to cases in which the forces of buoyancy and friction are negligible is now clear. By doing so, we remove forces that we are not yet prepared to measure.

The connection between mass and weight is a useful one.

The negative sign indicates that F_s is directed opposite to the force of gravity—that is, upward.

In the preceding paragraph, we did not specify whether the upward or the downward direction is taken to be positive. To show that it makes no difference, go through the analysis with $g = -|g|$ (upward direction positive) and then with $g = +|g|$ (downward direction positive).

The magnitude $m|g|$ of the force required to support a body is called its **weight**, often represented by the symbol W. That is,

$$W = m|g|. \tag{3.11}$$

Weight is a positive scalar quantity because it is the magnitude of the vector quantity, force. Like force, weight is expressed (in SI) in newtons.

People commonly confuse weight and mass, owing in part to the misconception that weight is expressed in kilograms, a unit of mass. When you buy goods—meat or tomatoes, for instance—you are really interested in the quantity of matter that you are getting for your money. That is, you are interested in the *mass* of your purchase. But it is almost always easiest to determine the mass of an object by weighing it—that is, by finding out how much force is required to support it against gravity. The reason that this method works is evident in Equation 3.11: The weight W is directly proportional to the mass m (as long as g is constant). The faces of market scales display not the direct result of the measurement, which is the weight W in newtons, but the indirect result $m = W/|g|$, which is the mass in kilograms. It would be easy enough to paste a new face on the scale that gave the weight in newtons, but the weight, strictly speaking, is not usually what you want to know when you weigh something.

To summarize: You often determine the mass of a body (in kg) by measuring its weight (in N). The process works because the value of g does not vary enough to be of concern in everyday transactions. It does *not* follow that the weight of a body is given in kilograms.

Almost everyone has seen pictures of astronauts floating around the cabins of their spacecraft (Figure 3.10). It is quite accurate to say that the astronauts are **weightless** because the external force required to support them (that is, to keep them from falling to the floor of the spacecraft) is zero. It does *not* follow that they are massless. Indeed, the astronaut's mass is the same as it was before the spacecraft was launched, and that mass is still evident in his or her inertia. The effects of this inertia can be seen in motion pictures, in which the astronauts must repeatedly start and stop themselves by pushing on the walls of the spacecraft as they maneuver from place to place.

FIGURE 3.10 Because Jerry L. Ross is weightless, his fellow astronaut Linda M. Godwin is actually exaggerating the effort required to keep him off the floor of Space Shuttle Atlantis. In the background are some of the restraints used to allow the astronauts to sleep without floating about and bumping into things.

EXAMPLE **3.3**

You find yourself a little ahead of your budget, so you go out to buy a lovely porterhouse steak to share with a friend. The butcher tells you that the steak "weighs" 1.21 kg. You are in a pedantic mood. How do you correct the butcher? What, in fact, is the weight of the steak?

SOLUTION: If you are larger than the butcher, you can point out to him that, although he has told you that 1.21 kg is the steak's *weight*, that quantity is really the steak's *mass*. To find its *weight*, you use Equation 3.11, $W = m|g|$, together with the standard value $|g| = 9.80$ m/s^2 given by Equation 2.39, to obtain

$$W = 1.21 \text{ kg} \times 9.80 \text{ m/s}^2 = 11.9 \text{ N}.$$

In inferring the masses of objects from their weights under everyday circumstances, it is safe to ignore variations in the value of $|g|$ from place to place. We have done so in Example 3.3. But the weight of an object does depend on the value of g, as Example 3.4 makes evident.

EXAMPLE **3.4**

The butcher of Example 3.3 tells you what he thinks of you, and you walk out of the store without buying the steak. The customer who later buys it is an astronaut. He packs it in ice and takes it on a voyage to the moon, planning to celebrate the moon landing with his crewmates. What is the weight of the steak when they have landed on the moon? The magnitude of the acceleration of gravity on the surface of the moon is $|g_m| = 1.62$ m/s^2.

SOLUTION: The mass of the steak, $m = 1.21$ kg, does not depend on its location. When the steak is unpacked on the moon, its weight is

$$W = m|g_m| = 1.21 \text{ kg} \times 1.62 \text{ m/s}^2 = 1.96 \text{ N};$$

this is about one-sixth of the weight of the steak on the surface of the earth. Nevertheless, the steak will nourish the astronauts just as much as if its weight had the terrestrial value 11.9 N, because the quantity of matter in the steak is unaffected by the voyage.

Vertical Acceleration

Suppose we want not merely to keep a body from falling, but to accelerate it upward. We need, then, an upward force whose magnitude is greater than the body's weight $m|g|$. This force can be calculated by using Newton's second law, as illustrated in Example 3.5.

EXAMPLE **3.5**

Figure 3.11 shows a body of mass 10.0 kg hanging from a spring scale, which in turn is suspended from the ceiling of an elevator car. **(a)** Find the force F_{s0} applied to the body by the scale when the elevator is at rest, and find the reading of the scale. **(b)** The elevator starts upward. The elevator accelerates with $a_1 = 1.30$ m/s^2, the upward direction taken as positive. Find the reading of the scale. **(c)** The elevator comes to rest. As it slows, its acceleration is $a_2 = -1.30$ m/s^2. (Note that a_2 is negative even though v is positive.) Find the reading of the scale.

FIGURE 3.11

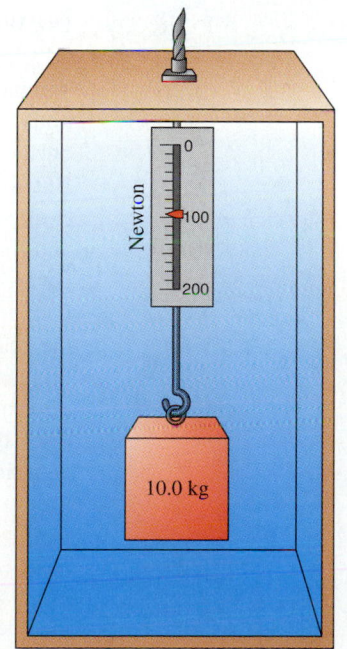

SOLUTION:
(a) What is the force F_{s0} applied to the body by the scale when the elevator is at rest? What is the scale reading?

With the elevator at rest, the scale reading is simply the weight W of the body—that is, the magnitude of the force F_{s0} exerted by the scale hook on the body. The value of W can be found by direct application of Equation 3.11. However, to prepare for parts **b** and **c** of the example, write Newton's second law for the body. As always, the proper force to insert into the equation $F = ma$ is the net force exerted on the body. Because $a = 0$, you have

$$F = F_{grav} + F_{s0} = mg + F_{s0} = 0.$$

You thus have

$$F_{s0} = -mg.$$

Before inserting numerical values, remember that the upward direction has been chosen positive. Thus $g = -9.80$ m/s^2, and

$$F_{s0} = -10.0 \text{ kg} \times (-9.80 \text{ m/s}^2) = 98.0 \text{ N}.$$

The positive value of F_{s0} follows from the choice of positive direction. The spring scale reads the weight, which is the positive scalar $W = m|g| = 98.0$ N.

(b) The elevator accelerates upward with $a_1 = 1.30$ m/s^2. What is the scale reading?

With the elevator accelerating, the quantity ma in Newton's second law is no longer zero. Therefore, the net force F is not zero. Using F_s for the force exerted on the body by the scale as the elevator accelerates, you have for the net force

$$F = F_{grav} + F_s = ma,$$

or

$$mg + F_s = ma.$$

Solving for F_s yields $F_s = ma - mg$, or

$$F_s = m(a - g). \tag{3.12}$$

The value of acceleration applicable here is $a = a_1$, so the spring force has the corresponding value $F_{s1} = m(a_1 - g)$. The numerical result is

$$F_{s1} = 10.0 \text{ kg} \times [1.30 \text{ m/s}^2 - (-9.80 \text{ m/s}^2)] = 111 \text{ N},$$

which is also the reading of the scale. The force applied by the scale must now not merely support the body against gravity, but accelerate it as well.

(c) The elevator slows with acceleration $a_2 = -1.30$ m/s^2. Find the reading of the scale.

Although the velocity vector v of the elevator is still directed upward, v is decreasing. Thus the acceleration vector a_2 is directed downward. Just as in part **b**, the force exerted by the scale is given by Equation 3.12, $F_s = m(a - g)$. But now the applicable value of acceleration is $a = a_2$, so the spring force is $F_{s2} = m(a_2 - g)$. The numerical result is

$$F_{s2} = 10.0 \text{ kg} \times [-1.30 \text{ m/s}^2 - (-9.80 \text{ m/s}^2)] = 85.0 \text{ N}.$$

Because of the negative (downward) acceleration of the elevator, the force that the scale must apply to the body is decreased. The body is now accelerating downward. The force required to produce this acceleration is supplied by the force of gravity. However, the force required is $ma_2 = 10.0$ kg $\times$ $(-1.30$ m/s$^2) = -13.0$ N. The total force of gravity is $mg = 10.0$ kg $\times (-9.80$ m/s$^2) = -98.0$ N. The difference between these two values is supplied by the remaining upward force exerted by the spring scale. If the connection between the scale and the body were cut, what would be the net force F exerted on the body? What would its acceleration be?

Many persons experience a queasy feeling in elevators as they accelerate downward. This feeling can be largely explained in terms of Example 3.5. The organs of the digestive system are suspended within the abdomen by a number of membranes, and the downward acceleration results in a reduced tension in these membranes, just as it does in the string that holds up the spring scale of Example 3.5. This reduction of tension is experienced by many persons as a discomfort centered in the abdomen. Upward acceleration, which results in increased tension in the suspensory membranes, does not seem to have so strong an effect.

Newton's Third Law of Motion

We have spent some time considering how a spring scale exerts a force on a massive body. This force may keep the body from falling or may make it accelerate, depending on the circumstances. The indicator on the scale tells us the magnitude of the force. But the very fact that the scale pointer indicates a force implies that force is being exerted not only *by* the scale but *on* the scale as well. Why else would the spring stretch and the indicator move from its zero position? Figure 3.12 shows that *two* forces take part in the interaction between the spring scale and the body. One force, $F_{scale \rightarrow body}$, is exerted *by* the scale *on* the body. The other, $F_{body \rightarrow scale}$, is exerted *by* the body *on* the scale. It is easy to confuse the two forces because they have equal magnitudes. But the confusion is eliminated when you bear in mind two facts: (1) *their directions are opposite*, and (2) *they act on different bodies*. The two forces are thus different forces even though they exist at the same time. Such a pair of forces is called an **action-reaction pair**.

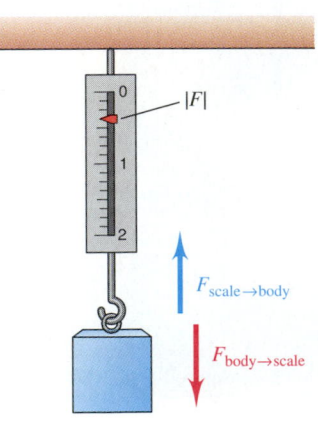

FIGURE 3.12 The force exerted by the scale on the body, $F_{scale \rightarrow body}$, and the force exerted by the body on the scale, $F_{body \rightarrow scale}$, constitute an *action-reaction pair*. Newton's third law requires that every mechanical force be a member of such a pair and that it be equal in magnitude and opposite in direction to the other member.

(a) 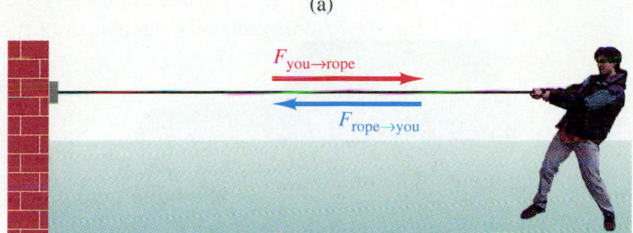 (b)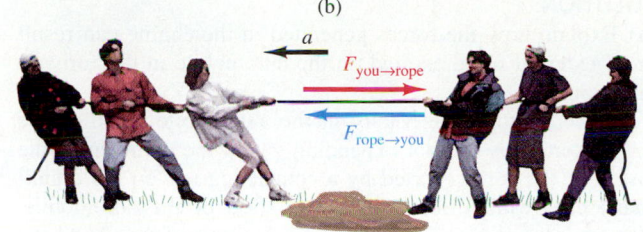

FIGURE 3.13 (*a*) You pull on a rope that is firmly attached to a wall. When you exert a force $F_{\text{you}\rightarrow\text{rope}}$ on the rope, the rope exerts a force $F_{\text{rope}\rightarrow\text{you}}$ on you. The two forces are equal in magnitude but opposite in direction. (*b*) You are on the losing side in a game of tug-of-war. The superior efforts of the winning side are propelling you toward the mudhole with acceleration *a*. The presence of an acceleration, which distinguishes this situation from the one shown in part *a*, has no effect on the applicability of Newton's third law. The forces $F_{\text{rope}\rightarrow\text{you}}$ and $F_{\text{you}\rightarrow\text{rope}}$ may be different from the ones in part *a*, but they are still equal and opposite to each other.

Action-reaction pairs are a matter of common experience. When you pull on a rope, it pulls back—a point quickly noted if you pull hard. This is true regardless of whether the pull results in an acceleration or not; see Figure 3.13.

In the *Principia*, Newton states this class of commonplace observations as a general law and follows it with examples. Here is **Newton's third law of motion** in his own words; modern terms are supplied in brackets where necessary:

> *To every action [force] there is always opposed an equal reaction [force]: or, the mutual actions of two bodies upon each other are always equal [in magnitude] and directed to contrary parts [in opposite directions].*

Newton's third law makes everyday observation quantitative; it specifies that *the two forces of an action-reaction pair are always equal in magnitude and opposite in direction*. The law also makes clear that *each of the two forces of such a pair acts on a different body*.

For concreteness, consider any two bodies, 1 and 2, that *interact*. When body 1 exerts a force $F_{1\rightarrow2}$ *on* body 2, body 2 must exert an equal and opposite force $F_{2\rightarrow1}$ *on* body 1: $F_{1\rightarrow2} = -F_{2\rightarrow1}$

When we wish to use Newton's second law, $F = ma$, to analyze the motion of a body, we must make sure that we are considering all the forces acting *on* that body and no other forces. In Example 3.5 and Figure 3.11, we were interested in the motion of the 10.0-kg mass. We therefore carefully restricted our attention to the two forces acting on it—namely, the force of gravity acting on it because of the presence of the earth and the force F_{s0}, F_{s1}, or F_{s2} exerted on it by the hook of the spring scale. In Figure 3.12, the latter force is called $F_{\text{scale}\rightarrow\text{body}}$ to make clear which object is acting on which. As Figure 3.12 shows, there is also a force $F_{\text{body}\rightarrow\text{scale}} = -F_{\text{scale}\rightarrow\text{body}}$ exerted by the body on the scale. It would not be correct to include this force in an analysis of the motion of the body, because it does not act on the body. If we wished to use Newton's second law to analyze the motion of the scale (rather than that of the body), we would indeed have to include $F_{\text{body}\rightarrow\text{scale}}$ among the forces considered. In that case, we would not include $F_{\text{scale}\rightarrow\text{body}}$.

Straight from the Horse's Mouth

To his statement of the third law, Newton adds the following examples and commentary:

Whatever draws or presses another is as much drawn or pressed by that other. If you press a stone with your finger, the finger is also pressed by the stone. If a horse draws a stone tied to a rope, the horse (if I may say so) will be equally drawn back toward the stone; for the distended rope, by the same endeavor to relax or unbend [unstretch] itself, will draw the horse as much towards the stone as it does the stone towards the horse, and will obstruct the progress of the one [the horse] as much as it advances that of the other [the stone].

EXAMPLE 3.6

The continued motion of an automobile depends ultimately on forces generated in its engine. Frictional forces, such as the force of air resistance, tend to bring the automobile to rest. **(a)** Explain how the forces generated in the engine can result in an external force exerted on the automobile in the forward direction. **(b)** Make a sketch of the external forces exerted on the automobile, and show how a net force can result in a forward acceleration. **(c)** Make a sketch of the situation in which the automobile is moving on a level road at constant velocity.

SOLUTION:

(a) Explain how the forces generated in the engine can result in an external force exerted on the automobile in the forward direction.

No force acting internally in the automobile (such as the force exerted by the hot expanding gas in the cylinder on the piston or the force exerted by a connecting rod on the crankshaft) can result directly in acceleration of the automobile. However, the force originating in the pressure of the gas on the piston is *transmitted* through the drive train to the driving wheels. (By ''transmitted'' we mean that each part in the train exerts a force on the next. Each such force is a member of an action-reaction pair, but the reaction forces need not concern us here because they are not acting on a body whose motion we wish to study.) Ultimately, a force is exerted by the driving wheels on the road surface. But this force, $F_{\text{wheels}\rightarrow\text{road}}$, cannot be directly responsible for the acceleration of the automobile, because it is not a force exerted *on* the automobile. It is the reaction force $F_{\text{road}\rightarrow\text{wheels}}$ $(= -F_{\text{wheels}\rightarrow\text{road}})$ that in fact drives the car.

(b) Make a sketch of the external forces exerted on the automobile, and show how a net force can result in a forward acceleration.

If the automobile is accelerating in the forward direction, the sum F of the driving force $F_{\text{road}\rightarrow\text{wheels}}$ and the frictional forces lumped together as F_f must be a forward-directed force. Taking the forward direction as positive, you have

$$ma = F = F_{\text{road}\rightarrow\text{wheels}} + F_f > 0, \qquad (3.13)$$

where m is the mass of the automobile and a its acceleration. You can most easily show this in a sketch like that of Figure 3.14a.

(c) As the velocity of the automobile increases, the frictional forces (notably that of air resistance) also increase in magnitude. If the engine is operated in such a way that $F_{\text{wheels}\rightarrow\text{road}}$ remains constant, the net external force exerted on the automobile as it speeds up, given by Equation 3.13, becomes smaller and

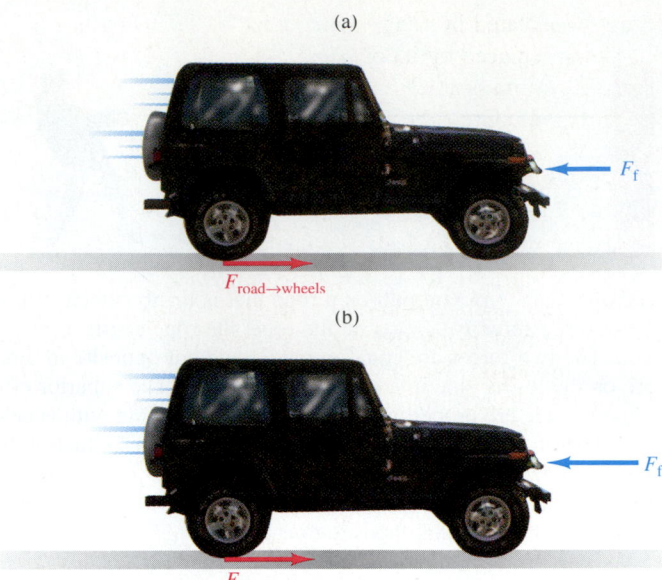

(a)

(b)

FIGURE 3.14 (a) The force $F_{\text{road}\rightarrow\text{wheels}}$, which is the reaction force to the force $F_{\text{wheels}\rightarrow\text{road}}$ provided by the engine, tends to propel the automobile forward. It is opposed by the lumped frictional forces F_f. Because the sum of the two forces is directed forward, however, the automobile accelerates. (b) The automobile is now going sufficiently faster that F_f has the same magnitude as the oppositely directed force $F_{\text{road}\rightarrow\text{wheels}}$. The net external force acting on the automobile is now zero, and its velocity is constant.

smaller, approaching zero; that is, $F_{\text{wheels}\rightarrow\text{road}} + F_f \rightarrow 0$. When $F = 0$, the velocity no longer increases but remains constant in accordance with Newton's first law. You can show this in a sketch like that of Figure 3.14b.

SECTION 3.6

Atwood's Machine and the Free-Body Diagram

Figure 3.15 depicts a device called **Atwood's machine**. Atwood's machine gives us a direct way of demonstrating Newton's laws of motion in the laboratory. In addition, the analysis of the motion of Atwood's machine serves as a prototype for almost all analyses of mechanical systems consisting of more than one body.

The machine consists of two masses, m_1 and m_2, connected by a string that passes over a pulley. We assume that the mass of the pulley is negligible, that there is no friction in the system, and that the string is massless and both perfectly flexible and perfectly inextensible (that is, incapable of being stretched).

Atwood's machine has a complication that we have not considered before: It comprises two massive bodies. Although the bodies do not move in the same way—one moves up when the other moves down—a definite linkage exists between their motions, so they cannot be considered in isolation from each other. Nevertheless, we can simplify the analysis by ''taking the system apart'' in a conceptual way. We can consider each of the two bodies separately *if* we consider all the external forces acting on each body in applying Newton's laws to it. We begin by depicting these forces in the diagrams of Figure 3.16. The difference between Figure 3.15 and Figure 3.16 is that the two bodies

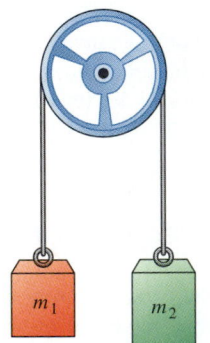

FIGURE 3.15 Atwood's machine.

are now isolated in imagination. That is, the string has disappeared in Figure 3.16. It has been replaced by its physical function, which is to exert force T_1 on mass m_1 and force T_2 on mass m_2.*

A diagram like either of those in Figure 3.16 is called a **free-body diagram** because it shows a body isolated (or freed) from all other parts of the system. It shows all the forces that act on the body. By "all the forces," we mean both the forces exerted on the body from outside the entire system (in this case, the gravitational force) and the forces exerted on the body by other parts of the system (in this case, the force due to the string tension). In the free-body diagram, we have separated the original system— the Atwood machine—into two simpler subsystems.

We are now ready to apply Newton's laws to the two subsystems of Figure 3.16a and Figure 3.16b. We take the upward direction as positive, so g has a negative value. Newton's second law applied to m_1 relates the net force to the acceleration a_1 as follows:

$$T_1 + m_1g = m_1a_1. \tag{3.14}$$

For m_2, the net force is similarly related to the acceleration a_2:

$$T_2 + m_2g = m_2a_2. \tag{3.15}$$

These two equations are called the **equations of motion** of the Atwood machine. We cannot yet solve them because they contain four unknowns: two string tension forces T_1 and T_2 and two accelerations a_1 and a_2. But we have two more pieces of information, called **constraints** on the equations of motion, from which we can construct two more equations—enough to solve for four unknowns.

The first constraint arises from the inextensibility of the string. Whatever displacement Δx_1 is experienced by m_1, the other mass m_2 must experience a displacement $\Delta x_2 = -\Delta x_1$. Consequently, the accelerations of the two masses must be equal in magnitude and opposite in direction:

$$a_2 = -a_1. \tag{3.16}$$

The second constraint arises from the way in which a massless string (or a string of negligible mass) transmits force along its length. Specifically, the tension in the string must be everywhere the same, and thus the forces exerted on the two masses by the ends of the string are equal in magnitude. Because the string bends around the pulley, both of these forces are directed upward, so

$$T_2 = T_1. \tag{3.17}$$

We now use the constraints expressed in Equations 3.16 and 3.17 to eliminate T_2 and a_2 from Equation 3.15. This yields

$$T_1 + m_2g = -m_2a_1. \tag{3.18}$$

This equation and Equation 3.14 are a pair of simultaneous equations in two unknowns, T_1 and a_1. To make this evident, we write Equation 3.18 directly under Equation 3.14:

$$T_1 + m_1g = m_1a_1,$$
$$T_1 + m_2g = -m_2a_1.$$

We subtract the second equation from the first to eliminate T_1:

$$m_1g - m_2g = m_1a_1 + m_2a_1,$$

or
$$(m_1 - m_2)g = (m_1 + m_2)a_1.$$

The acceleration a_1 of the mass m_1 on the left side of the machine is thus

$$a_1 = \frac{m_1 - m_2}{m_1 + m_2}g. \tag{3.19}$$

The value of a_2 can be found immediately by using $a_2 = -a_1$:

$$a_2 = \frac{m_2 - m_1}{m_1 + m_2}g. \tag{3.20}$$

*We use the symbol T as a reminder that the magnitude of the forces is the tension in the string.

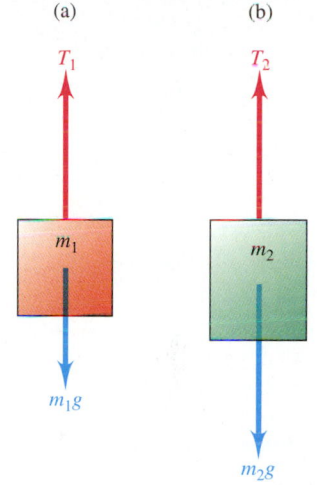

FIGURE 3.16 Free-body diagrams (a) for mass m_1 and (b) for mass m_2 of the Atwood machine of Figure 3.15. As in Figure 3.9, the physical connection to each of the masses (the string) is replaced by the forces it exerts.

The free-body diagram is a sort of pictorial accounting device, which we will use repeatedly. Our aim is to apply Newton's second law, $F = ma$, to each of the two bodies that make up the Atwood machine. In order to apply the law to a body, however, we must determine the net external force F acting on that body. This determination requires us to take into account all of the forces acting on the body. We must be careful to leave no forces out, to count none more than once, and to include no irrelevant forces (such as reaction forces exerted by the body on its surroundings). We must also be careful to keep track of the directions of the individual forces. The free-body diagram is an excellent way to accomplish these tasks.

Consider the physical meaning of these two equations. If $m_1 > m_2$, the fraction on the right side of Equation 3.19 will be positive, and the sign of a_1 will be the same as that of g. This agrees with the observation that under these circumstances m_1 will accelerate downward. The sign of the fraction in Equation 3.20 will be negative, and the sign of a_2 will therefore be opposite that of g. And indeed, m_2 will accelerate upward under these circumstances. You should satisfy yourself that the equations also agree with your predictions of the motion of the machine in the contrary case, when $m_2 > m_1$.

If $m_1 = m_2$, the acceleration of the system will be zero. If the system is initially at rest, it "balances" and remains at rest. But if it is set into motion by a push, it will remain in motion at constant speed, in agreement with Newton's first law.

Now consider the value of the fraction $(m_1 - m_2)/(m_1 + m_2)$ in Equation 3.19. Because m_1 and m_2 are positive scalars, the value of the fraction must always lie between -1 and $+1$ regardless of the values of the two masses. Therefore, the magnitude of the acceleration of the system is always less than $|g|$. In effect, Atwood's machine is a way to study falling bodies in slow motion, especially if $|m_1 - m_2|$ is small compared with $m_1 + m_2$.

How does Atwood's machine "slow down" the motion of falling bodies? Let us rewrite Equation 3.19 in a form that demonstrates that it is simply a special case of Newton's second law:

$$\underbrace{(m_1 - m_2)g}_{F} = \underbrace{(m_1 + m_2)}_{M}\underbrace{a_1}_{a}. \tag{3.21}$$

The left side of the equation gives the net force F driving the system—that is, the entire Atwood machine. The right side is the total mass of the system multiplied by its acceleration. On the left side of the equation, we see the masses in their **gravitational role**—they experience forces on account of the presence of the earth. Owing to the presence of the pulley, the gravitational forces on m_1 and m_2 work against each other and the net driving force is $(m_1 - m_2)g$. The magnitude of this quantity is less than that of either of the quantities m_1g and m_2g.

On the right side of the equation, the masses are seen in their **inertial role**—they resist any change in their velocity. When the Atwood machine accelerates, the net driving force must accelerate the sum of the masses—that is, the total inertial mass of the system $m_1 + m_2$. The quantity $m_1 + m_2$ is greater than either m_1 or m_2.

In summary, a net force smaller than the total gravitational force acting on the two bodies separately drives a system whose mass is greater than that of either of the two bodies. As a result, the acceleration of the system is smaller than that of freely falling bodies.

Unexpected Coincidence

In what we have said thus far, the inertial role of mass and the gravitational role of mass appear to be quite distinct. However, when the mass of a body is measured separately by inertial and gravitational methods, the results always turn out the same (within the limits of experimental error). Such experiments have been carried out with exquisite precision. There is no reason to expect this result in advance, and it is a surprising one. The surprise was not explained until Einstein, in his great 1915 reformulation of physics called general relativity, showed that the identity of the two values was no accident.

EXAMPLE 3.7

The Atwood machine shown in Figure 3.17 is initially in balance. On each end of the string hangs a pan with a number of small masses totaling 2.500 kg. The upper mass is 0.90 m higher than the lower one. A 5-g mass is removed from the lower pan and placed on the upper pan, and the system is released from rest. **(a)** How long after release will the two masses pass each other? **(b)** How fast will they be going at that moment?

SOLUTION:

(a) How long after release will the two masses pass each other?

The masses pass each other when each has moved through a distance $|\Delta x| = (0.90 \text{ m})/2 = 0.45 \text{ m}$. Equations 3.19 and 3.20 give the accelerations of the masses and show that these accelerations are constant. You can therefore use Equation 2.36a, the kinematic equation that relates displacement to elapsed time when acceleration is constant. The system starts at rest, and Equation 2.36a therefore simplifies to

$$\Delta x = \tfrac{1}{2}at^2.$$

From this equation, the desired time is

$$t = \sqrt{2 \left| \frac{\Delta x}{a} \right|}. \tag{3.22}$$

You need only the magnitude of the acceleration, and thus either Equation 3.19 or Equation 3.20 will serve. If you use the latter and call the common value of the magnitude of the accelerations $|a| = |a_2|$, you have

$$a = \frac{(2.500 + 0.005) \text{ kg} - (2.500 - 0.005) \text{ kg}}{2.495 \text{ kg} + 2.505 \text{ kg}} \times 9.80 \text{ m/s}^2$$

$$= 0.020 \text{ m/s}^2.$$

Inserting the numerical values into Equation 3.22 yields

$$t = \left(\frac{2 \times 0.45 \text{ m}}{0.020 \text{ m/s}^2} \right)^{1/2} = 6.7 \text{ s}.$$

(b) How fast will the masses be going at the instant of passage?

You now have enough information to find, in either of two ways, the speed of the two masses as they pass each other. One way is to use Equation 2.36b, which simplifies in the present case to the form $|v| = |a|t$, so

$$|v| = 0.020 \text{ m/s}^2 \times 6.7 \text{ s} = 0.13 \text{ m/s}.$$

You can obtain the same result beginning with Equation 2.38, $v^2 = 2ax$. What value will you have to use for x?

FIGURE 3.17

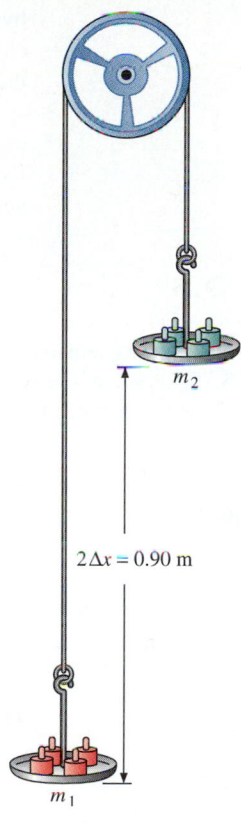

$2\Delta x = 0.90 \text{ m}$

m_2

m_1

The String Tension

We now return to the equations of motion for Atwood's machine and solve for the string tension. What follows is one of several ways to do this. Because we have $T_1 = T_2$, we drop the subscript and call the tension T. Also, we use Equation 3.16, $a_2 = -a_1$, to eliminate a_2. Equations 3.14 and 3.15 then become

$$T + m_1g = m_1a_1 \qquad \text{and} \qquad T + m_2g = -m_2a_1.$$

We wish to solve this pair of simultaneous equations for T, which requires that we eliminate the other unknown quantity, a_1. To do this, we multiply the first of these equations through by m_2 and the second by m_1, and then we isolate the terms in T on the left sides to obtain

$$m_2T = m_1m_2(a_1 - g)$$

and

$$m_1T = m_1m_2(-a_1 - g).$$

The sum of these two equations is

$$(m_1 + m_2)T = m_1m_2(-2g).$$

The solution for T is

$$T = -m_1g \, \frac{2m_2}{m_1 + m_2}. \tag{3.23a}$$

This can equally well be written

$$T = -m_2g \, \frac{2m_1}{m_1 + m_2}. \tag{3.23b}$$

The negative signs indicate that the force exerted by the string on each of the masses is upward, opposite to the downward gravitational force $m_1 g$ or $m_2 g$.

Consider, first, the physical meaning of Equation 3.23a. For the sake of argument, let m_2 be greater than m_1. The magnitude of the force required just to support m_1 without accelerating it is its weight $m_1|g|$. But the magnitude of T exceeds this value by the fraction $2m_2/(m_1 + m_2)$; and, because we assumed $m_1 < m_2$, the value of this expression must be greater than unity. Thus the force suffices not merely to support the mass m_1 but to accelerate it upward. (Compare with part **b** of Example 3.5.)

Now consider what Equation 3.23b tells us about the motion of m_2. Again, let $m_2 > m_1$. The force required to support m_2 is $-m_2 g$. But T is less than this by the factor $2m_1/(m_1 + m_2)$, whose value lies between 0 and 1. Therefore, the mass m_2 must accelerate downward. (Compare with part **c** of Example 3.5.)

EXAMPLE 3.8

Find the magnitude of the string tension T for the Atwood machine of Example 3.7 **(a)** when the system is in balance; **(b)** when a 500-g mass is transferred from m_1 to m_2.

SOLUTION:
(a) When the system is in balance, m_1 is equal to m_2 and the fraction in either Equation 3.23a or Equation 3.23b is unity. In this case

$$|T| = m_1|g| = m_2|g|,$$

or $\quad |T| = 2.500 \text{ kg} \times 9.80 \text{ m/s}^2 = 24.5 \text{ N}.$

(b) After the transfer of the 500-g mass, you have $m_1 = 2.000$ kg and $m_2 = 3.000$ kg. Using Equation 3.23a, you find the magnitude of the tension to be

$$|T| = 2.000 \text{ kg} \times 9.80 \text{ m/s}^2 \times \frac{2 \times 3.000 \text{ kg}}{5.000 \text{ kg}} = 23.5 \text{ N}.$$

The string tension is thus smaller than that in the balanced case (part **a**). What would the tension be if you transferred all of the mass to one side of the Atwood machine?

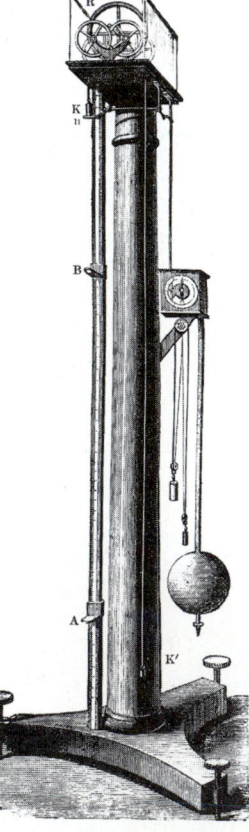

Historical Significance of Atwood's Machine

Like Newton a century earlier, George Atwood (1746–1807) was a member of Trinity College, Cambridge. Atwood described his machine in his 1784 book, *A Treatise on the Rectilinear Motion and Rotation of Bodies*. A nineteenth-century version of the Atwood machine is shown in Figure 3.18.

The Atwood machine was the first laboratory apparatus that provided direct experimental verification of Newton's laws of motion. Given the values of m_1 and m_2, you can use Newton's laws to predict the acceleration, the string tension, the positions of the masses as a function of time—in fact, every mechanical quantity relevant to the machine. These predictions can be verified in the laboratory, often with good precision. Atwood-machine experiments are commonly included in the laboratory curriculum of physics courses for just such purposes.

Do not suppose, however, that the world waited for more than a century for experimental verification of Newton's laws before accepting them as the basis of mechanics. Science hardly ever progresses in so seemingly direct a fashion. On the contrary, Newton devised his laws for the analysis of pressing problems, the most dramatic of which were astronomical in nature. (We will consider some of them in Chapter 14.) By Atwood's time, tremendous progress had been made in many fields by applying Newton's laws. Many predictions based on these laws had been verified with exquisite precision and there were no significant failures, though much remained to be explored. Newton's laws had assumed such a central role in physical theory that no serious worker doubted their validity. Even today, they are as valid within their proper realm as ever they were.

Atwood's machine was conceived from the beginning as a laboratory demonstration and teaching tool. Probably no laboratory apparatus used today has undergone so little change over so long a period of time.

FIGURE 3.18 An 1878 Atwood machine with a pre-electronic timing system. The masses K and K' differ slightly on account of the small additional mass resting on K. They are connected by the string that runs over the low-friction pulley at the top. When the masses are released the clock starts. At B, the extra mass is removed from K, and the system thereafter moves at constant speed until it strikes platform A. Measurement of the travel times from release to B and from B to A can be used to find the final speed of the system and thus its acceleration from the release point to B.

a, A	acceleration		t	time
F	net external force exerted on a body		T	force exerted on a body due to tension in attached string
$F_{subscript}$	force exerted on a body by an agent denoted by the subscript		v, V	velocity
$F_{a \to b}$	force exerted *by* body a *on* body b		W	weight
g	acceleration of gravity		Δx	displacement
m, M	mass		$\propto$	is proportional to

According to **Newton's first law of motion**, also called the **law of inertia**, the velocity of a body remains unchanged unless the body is acted upon by a net external force. This law is valid only when measurements are made by an observer situated in an **inertial frame of reference**. Indeed, measurements of the motion of a body free of applied forces can be used by an observer to test whether his frame of reference is inertial or noninertial.

The velocity of a body can be changed by applying a net external force to it. The quantitative measure of a body's inertia—its resistance to velocity change—is its **mass**. The mass of a body is determined by comparing it (usually indirectly) with the mass of the standard kilogram.

The relation among the mass m of a body, the net external force F applied to it, and its acceleration a is given by Equation 3.5, **Newton's second law of motion**:

$$F = ma,$$

The unit of force is defined by this equation. A force of **1 newton (N)** is that net external force required to make a 1-kg mass accelerate at 1 m/s^2. Newton's second law is valid only for observers in inertial frames.

From the fact that all bodies in free fall near the surface of the earth have the same acceleration g, it follows that the force exerted on a body by gravity is directly proportional to its mass; $F_{grav} = mg$. This **gravitational role** of mass is experimentally distinct from the **inertial role** of mass, which appears in Newton's second law, but the mass of a body always has the same value in both aspects. The distinction between the gravitational and inertial aspects of mass is made particularly evident in the analysis of the **Atwood machine**; according to Equation 3.21,

$$\underbrace{(m_1 - m_2)g}_{F} = \underbrace{(m_1 + m_2)}_{M}\underbrace{a_1}_{a}.$$

The **weight** of a body is defined to be the magnitude of the upward force required to support the body; $W = m|g|$. The weight, expressed in newtons, is always proportional to the mass, expressed in kilograms.

When a body is accelerated upward or downward, the net force $F = ma$ applied to it must be different from the force $-mg$ required to support it at rest. The force F_s applied by the support can be found by inserting the sum of the forces applied to the body into Newton's second law. Its value is given by Equation 3.12, $F_s = m(a - g)$.

Every mechanical force is a member of an **action-reaction pair**—a pair of forces of equal magnitude and opposite direction. This is **Newton's third law of motion**. Each of such a pair of forces, however, is exerted *by* a different body and acts *on* a different body. Members of the same action-reaction pair never appear together in the sum required to find the net force used in Newton's second law.

We summarize all three of **Newton's laws of motion** as follows:

1. *The law of inertia*: Unless there is a net external force acting on a body, the body's velocity remains unchanged, as seen by an inertial observer.

2. *The connection between force and motion*: The net external force exerted on a body is equal to the product of the mass of the body and its acceleration: $F = ma$.

3. *The law of action and reaction*: When body 1 exerts a force $F_{1 \to 2}$ on body 2, body 2 exerts an equal and opposite force $F_{2 \to 1}$ on body 1; $F_{1 \to 2} = -F_{2 \to 1}$. The two forces are called an **action-reaction pair**. The forces constituting an action-reaction pair are equal in magnitude but opposite in direction, and *each of the two forces acts on a different body*.

The analysis of the motion of **Atwood's machine** requires the application of all three of Newton's laws. The first law establishes that the system will not accelerate when it is balanced. The third law is implicit in the balance of the system and, more generally, is required to establish the connection $T_1 = T_2$ between the forces exerted by the string on each of the two masses. The second law is used, together with the analytical device called the **free-body diagram**, to write a separate **equation of motion** for each of the two masses. The system of two equations of motion and two **constraint equations** is solved to find the accelerations of the two masses, given by Equations 3.19 and 3.20, and the string tension, given by either Equation 3.23a or Equation 3.23b.

KEY TERMS

Section 3.1 Newton's First Law of Motion
inertia ▪ inertial and noninertial frames of reference

Section 3.2 Mass
SI unit of mass, the kilogram (kg)

Section 3.3 Force
Newton's second law of motion ▪ SI unit of force, the newton (N) ▪ derived unit

Section 3.4 Weight
weightlessness ▪ gravitational force

Section 3.5 Newton's Third Law of Motion
action-reaction pair

Section 3.6 Atwood's Machine
free-body diagram ▪ constraint ▪ equation of motion

Queries and Problems for Chapter 3

QUERIES

3.1 *(1) Easy catch.* The apparatus shown below is often used in lecture demonstrations. The small car runs on rails with very little friction. A steel ball sits on a compressed spring in the bottom of the funnel. As the car coasts along, the trigger underneath the funnel strikes a projection on the track and releases the spring. The ball is shot upward as the car continues to coast along. When the ball falls, it returns to the funnel. Explain.

3.2 *(1) Cruisin'.* When you are driving along a straight highway at a constant 100 km/h, what is the net force exerted on the car?

3.3 *(1) Safety first.* Before a truck driver starts off with a heavy load (such as a load of steel bars), he or she is extremely careful to check that the load is securely tied down to the truck bed. Why is this important?

3.4 *(1) Hard lesson.* A foolish passenger rides in a car without a seat belt. In a head-on collision, the passenger flies through the windshield. Is there a force that throws him in this way? If so, what is its origin? If not, account for the passenger's mishap.

3.5 *(1) Kid stuff.* A child plays in the aisle of a train traveling through a flat plain. She places a marble on the floor. When she lets it go, it rolls toward the rear of the train. Is the child in an inertial frame? Explain.

3.6 *(2) Pitfalls.* Consider the definition of 1 g as the mass of 1 cm³ of pure water. Give as many reasons as you can why this definition is not a good one.

3.7 *(2) Getting the feel of it.* For each of the following bodies, express the approximate weight in newtons and the approximate mass in kilograms: yourself; this book; an automobile; an egg; a large truck; a soccer ball; an elephant.

3.8 *(3) Not too fast!* Highway safety campaigns often use the slogan "Speed kills." Comment on this slogan.

3.9 *(3) Homage to scholasticism.* "What happens when an irresistible force acts on an immovable object?" Comment on this question in the light of Newton's second law.

3.10 *(3) Slipping through your fingers.* You have no doubt had the experience of squeezing on a small piece of ice with two fingers and having the ice shoot out from between them, sideways. Does this experience conflict with the inference, immediately following Equation 3.5, that the direction of the net force exerted on a body must be the same at that of the acceleration?

3.11 *(3) Other times, other ways, I.* In one of the English systems of units, the base units are as follows: the unit of mass is the *pound* (lb), the unit of length is the *foot* (ft), and the unit of time is the *second* (s). In this system, the unit of force is called the *poundal*, sometimes abbreviated pdl. Express the poundal in terms of the three base units.

3.12 *(3) Other times, other ways, II.* In another English system of units, the base units are not units of mass, length, and time. Instead, they are units of force (the *pound*), length (the *foot*), and time (the *second*). In this system, the unit of mass is called the *slug*. **(a)** Express the dimensions of mass as a combination of the dimensions [force], [length], and [time]. **(b)** Express the slug in terms of the three base units.

3.13 *(3) At the gallop.* In the nineteenth century, there was debate as to whether all four feet of a galloping horse are ever off the ground simultaneously. The pioneering Welsh-American photographer Eadweard Muybridge (1830–1904) showed that they are, by using a series of cameras to make a rapid sequence of photographs; see below. When the horse is off the ground, it cannot exert any force on the ground. The force-free interval must therefore reduce the value of the propulsive force, averaged over one cycle of the gallop. Why does the off-the-ground interval nevertheless contribute positively to the efficiency of the gallop?

George Eastman House

3.14 *(4) Alice falls down the rabbit-hole.* The following quotation is taken from *Alice in Wonderland*: ''Either the well was very deep, or she fell very slowly, for she had plenty of time as she went down to look about her, and to wonder what was going to happen next. . . . She looked at the sides of the well, and noticed that they were filled with cupboards and bookshelves. . . . She took down a jar from one of the shelves as she passed: it was labeled 'ORANGE MARMALADE,' but . . . it was empty: she did not like to drop the jar, for fear of killing somebody underneath, so managed to put it into one of the cupboards as she fell past it.'' Comment on Alice's experience.

3.15 *(4) Stop if you can.* You are in an elevator that is rising when the emergency brake engages. For what minimum magnitude and direction of acceleration will your feet leave the floor?

3.16 *(4) Keeping cool in a crisis.* An unfortunate passenger is inside an elevator when the cable breaks, all of the safety systems fail, and the car falls freely. Both the passenger and the car descend with acceleration g. The passenger (if he is cool enough) can argue that he is weightless in the sense that no force is required to keep him from falling to the floor of the elevator car. An observer on the ground will argue that the passenger is not weightless, because in the absence of any suspending force he is indeed falling with acceleration g. Are both right? If so, what is the reason for the difference in their points of view?

3.17 *(4) Newton and mass.* The first sentence of Newton's *Principia* gives his definition of mass: ''*The quantity of matter is the measure of the same, arising from its density and bulk conjointly.*'' That is, the mass of a body is the product of its density and its volume. Criticize this definition of mass.

3.18 *(5) Newton's horse.* Newton discusses a horse pulling a stone attached to a rope. Consider the entire system, comprising the horse, the rope, the stone, and the earth. **(a)** Sketch a free-body diagram of each component of the system. **(b)** Identify all the action-reaction pairs of forces in the system.

3.19 *(5) Horatio Alger was wrong!* Sketch a free-body diagram to show why you cannot pull yourself up by your own bootstraps.

3.20 *(5) The pair is there.* A body of mass m falls freely toward the earth. That is, the only force acting on the body is the force of gravity. According to Newton's third law, what are the magnitude and direction of the other force of the action-reaction pair? On what does that force act? Why do we not normally notice the effect of the reaction force?

3.21 *(5) A balanced view.* The lower balance illustrated below

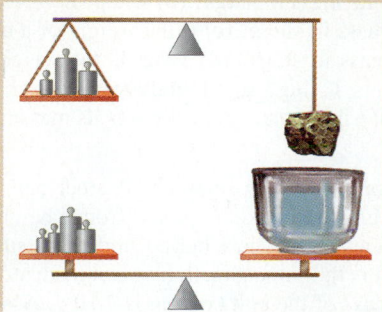

is initially balanced, with a jar of water on the right pan. The upper balance also is balanced, with a stone hanging from the right end. The upper balance is lowered so that the stone is immersed in the water but does not touch the jar. Buoyancy effects (which you need not consider in detail) disturb the equilibrium of both balances. Just enough mass is removed from the upper balance pan to restore the equilibrium of the upper balance. This mass is placed on the lower balance pan. Is the lower balance now tilted down to the left, tilted down to the right, or in equilibrium? Explain.

3.22 *(6) Nature exercises her full privileges.* Suppose that the mass of every body had different values in its inertial and gravitational aspects. Call these values m_I and m_G, respectively, and assume that, in general, m_I/m_G has different values for any two bodies. Describe the results of Galileo's ''Leaning Tower of Pisa'' thought experiment in this case. (See Problem 2.69 for a description of the experiment.)

3.23 *(6) Within a constant.* A student claims to have carried out a series of experiments like that suggested in Query 3.22. She claims to have discovered that $m_I \neq m_G$. However, she has found that the inertial value of mass for any body is always proportional to the gravitational value; that is, $m_I = km_G$, where k is a universal constant whose value is the same for all bodies. Show that this statement is equivalent to $m_I = m_G$, as far as its consequences for the laws of physics are concerned.

3.24 *(G) Hey-ho and up she rises.* Sketch a free-body diagram to show why you can pull yourself up using the apparatus shown here.

3.25 *(G) Was Newton inconsistent?* Criticize the following argument: Newton's first law and Newton's third law cannot both be true. A horse pulls on a rope whose other end is attached to a heavy sledge. If Newton's third law is true, the forward force exerted by the horse on the rope and the backward force exerted by the sledge on the rope are equal and opposite. The net force is thus zero, and Newton's first law predicts that the horse-rope-sledge system cannot move if it is initially at rest. If, however, Newton's first law is true, the fact that the sledge can be set into motion proves that the forces are not equal in magnitude, and thus Newton's third law is false.

3.26 *(G) Not too loose.* Many automobile seat belts can be adjusted so that they are a little slack and do not press uncomfortably on the wearer. However, it is wise not to allow more than the minimum amount of slack required for comfort. Explain why the seat belt should not be too loose.

3.27 *(G) Lift-off.* When a rocket designed to lift objects into orbit leaves the ground, more than 80% of its mass is the mass of the fuel. The thrust (the propulsive force exerted on the rocket) remains nearly constant as the fuel burns. Describe qualitatively the motion of the rocket up to the point at which the fuel is exhausted. Neglect air resistance and the variation of the acceleration of gravity with altitude.

3.28 *(G) More bounce to the ounce.* When you first step onto a bathroom scale, the dial reading increases to a maximum value exceeding your weight and then decreases. Why?

PROBLEMS

GROUP A

3.1 *(3) Airy descent.* A parachutist descends through the air at a constant speed $v = 9$ m/s. Her mass, including the parachute and other equipment, is 90 kg. What are the magnitude and direction of the force of air resistance acting on the parachute and the woman?

3.2 *(3) Speedup, I.* What net applied force is required to give a 5-kg mass an acceleration of magnitude 3 m/s^2?

3.3 *(3) Speedup, II.* A body experiences an acceleration of 2.85 m/s^2. If the net applied force is 12.2 N, what is the mass of the body?

3.4 *(3) Speedup, III.* A 2750-N force is applied to a body, whose mass is 6875 kg. What is the acceleration of the body?

3.5 *(3) Forcing a change, I.* A particle of mass 8.5 kg moves along a straight line. Its position is described by the function $x = 5t - 20t^2$, with x in meters and t in seconds. (This is the same function that describes the motion of the particle of Problem 2.19.) Is the force constant? If so, what is its value?

3.6 *(3) Forcing a change, II.* A particle of mass 8.5 kg moves along a straight line. Its position is described by the function $x = 5t - 20t^2 + 2t^3$, with x in meters and t in seconds. What is the value of the force at $t = 2.3$ s?

3.7 *(3) Speedup, IV.* A force of magnitude 2.65 N acts on a 4.00-kg mass for 3.00 s. What is the final speed if the mass starts from rest?

3.8 *(3) Braking force.* A car of mass 1.77 tonnes (1770 kg) moves at speed 18.5 m/s. The car is braked to a stop over a distance of 52.3 m. Find the braking force, assuming that the force is constant.

3.9 *(3) Vroom!* A sports car accelerates from rest to 100 km/h in 6.6 s. If its mass is 1725 kg (including the driver), what force must be exerted by the ground on the drive wheels? Assume (not too realistically) that the force is constant.

3.10 *(3) Performance penalty.* The sports car of Problem 3.9 now carries an additional passenger whose mass is 87 kg. How long does it take for the car to accelerate from rest to 100 km/h?

3.11 *(3) How hard?* What constant force will give a 0.40-kg mass a speed of 30 m/s in 6.0 s? The mass starts from rest.

3.12 *(3) Home run.* A baseball is pitched to a batter with a speed of 50 m/s. The batter hits it, and the ball reverses its direction, leaving the bat with a speed of 70 m/s. If the ball remains in contact with the bat for 0.025 s, what is the average value of the force exerted by the bat on the ball? The mass of the baseball is 150 g.

3.13 *(3) How long?* How long must a 100-N force act on a body of mass 20.0 kg, initially at rest, to give it a speed of 40.0 m/s?

3.14 *(3) How far?* A force of magnitude 6.40 N acts on a 5.00-kg mass. If the mass is initially at rest, how far has it moved at the end of 2.00 s?

3.15 *(3) Covering ground.* A body of mass 6.8 kg lies at rest on a frictionless surface. A constant horizontal force of magnitude 2.9 N is applied. **(a)** How long does it take for the body to attain a speed of 4.5 m/s? **(b)** How far does the body move in this time?

3.16 *(3) Still taking off.* A jet plane takes off at 80 m/s after a 2-km run. (This is the same airplane discussed in Problem 2.18.) If the mass of the plane is 115 tonnes ($\equiv 115 \times 10^3$ kg), what is the net force required? Assume that the net force is constant.

3.17 *(4) Weighty matters.* What is the weight of **(a)** a 5-kg bag of flour? **(b)** a 10-tonne boxcar?

3.18 *(4) Growing boy.* A child standing on a platform scale reads the dial and finds that his mass is 27.4 kg. What force does he exert on the platform of the scale?

3.19 *(4) Moon-light.* On the moon's surface, the acceleration of a freely falling body has magnitude $|g_m| = 1.62$ m/s^2. The mass of an astronaut is 115 kg (including life-support equipment). **(a)** What is the weight of the astronaut with equipment? **(b)** What would be the weight of the astronaut with equipment on earth?

3.20 *(4) Slugging it out.* In the U.S. Customary system of units, the unit of mass is the *slug*, defined as that mass which experiences an acceleration of magnitude 1 foot per second per second (ft/s^2) when a force of magnitude equal to a weight of 1 pound (lb) acts on it. When a mass falls freely, its acceleration has magnitude $|g| = 32.2$ ft/s^2. **(a)** If a body weighs 1 lb, what is its mass in slugs? **(b)** If a person weighs 130 lb, what is her mass in slugs? **(c)** If the weight of a body is H lb, what is its mass in slugs? **(d)** What is the weight of a body whose mass is 1 slug? **(e)** A body weighing 1 lb on earth weighs about $\frac{1}{6}$ lb on the moon. What is its mass on the moon, in slugs?

3.21 *(4) Slippery fingers revisited.* A steel ball of diameter 0.800 cm falls into a bucket of grease from a height of 1.20 m. The ball comes to rest just buried under the surface of the grease. (This is the same situation you considered in Problem 2.49.) The mass of the ball bearing is 2.10 g. Assume that air

---

resistance is negligible. **(a)** Find the mean net force exerted on the ball while it decelerates through the grease. (If the net force exerted on the ball were constant throughout the deceleration process—which is not—it would be equal to the mean net force.) **(b)** Find the mean force exerted by the grease.

3.22 *(4) Hard bounce.* A ball of mass 0.10 kg falls from a height of 1.0 m onto a microswitch that activates an electronic timer. The ball rebounds from the switch to a height of 0.50 m. The timer indicates that the ball remained in contact with the switch for 0.0020 s. **(a)** With what speed did the ball reach the switch? **(b)** With what speed did it leave the switch? **(c)** What was the mean acceleration of the ball during its contact with the switch? **(d)** What mean force did the switch exert on the ball? **(e)** What is the ratio of this force to the weight of the ball?

3.23 *(5) Sluggish.* When you drop a 1-kg mass, it falls toward the earth with acceleration g. **(a)** According to Newton's third law, there must be a reaction force paired with the gravitational force exerted on the 1-kg mass. On what body is it exerted? What are its magnitude and direction? **(b)** The mass of the earth is 6×10^{24} kg. Find the acceleration of the earth toward the 1-kg mass as the 1-kg mass falls toward the earth. **(c)** If the 1-kg mass falls from a height of 10 m, how far does the earth move before it and the 1-kg mass strike each other?

3.24 *(6) Ups and downs.* In an Atwood machine, the two masses have the values $m_1 = 4.30$ kg and $m_2 = 3.70$ kg. Find the acceleration of each mass, and find the string tension.

3.25 *(6) Displacement and acceleration.* The Atwood machine constraint equation $a_2 = -a_1$ depends on the fact that $\Delta x_2 = -\Delta x_1$. Beginning with the latter equation, derive the former.

3.26 *(6) Another way.* Use another method to solve for the speed of the Atwood machine in part **b** of Example 3.8. Hint: You know $|a|$ and $|\Delta x|$.

3.27 *(3) Pull to actuate.* In the illustration in the next column, a block of mass m_1 rests on a frictionless table. A light string passes over a massless, frictionless pulley. A constant force $F_s = 19.6$ N is applied downward on the end of the string. If $m_1 = 5.00$ kg, find the acceleration of the block.

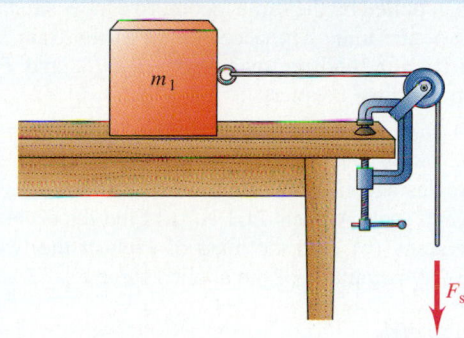

3.28 *(6) Half-Atwood.* **(a)** Sketch a free-body diagram for each of the two bodies in the frictionless device shown below. **(b)** Find the acceleration of the device. **(c)** Find the tension in the string. **(d)** Let $m_1 = 5.00$ kg and $m_2 = 2.00$ kg. Evaluate the acceleration and the tension. **(e)** Compare the results of part **d** with those of Problem 3.27. Explain the differences.

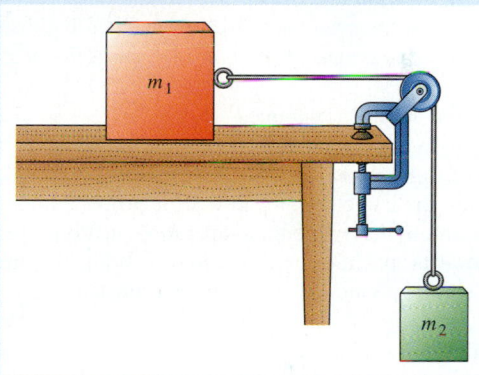

3.29 *(6) Grave matter.* Mass m_1 of an Atwood machine rests on a tabletop. The bottom of mass m_2 (which is greater than m_1) hangs at a height 1.00 m above the tabletop. The system is released, and m_2 strikes the tabletop 1.43 s later. If $m_1 = 4.50$ kg and $m_2 = 5.50$ kg, find the value of g.

GROUP B

3.30 *(3) Celerity.* It is possible to define a measure of the ease with which a body can be accelerated. To do this, write Newton's second law in the form $a = cF$. The quantity c is called the *celerity*. The bigger the value of c for a body, the more rapidly that body is accelerated by a given force. **(a)** What is the relation between celerity and mass? **(b)** Two bodies have celerities c_1 and c_2, respectively. If they are fastened together, what is the value of c, the celerity for the combined body?

3.31 *(3) Pulling out of the station, I.* A locomotive of mass 15 tonnes (15×10^3 kg) pulls a train of three cars, each of mass 8 tonnes. The train is accelerating on a level track with $a = 0.3$ m/s². **(a)** Sketch a free-body diagram for the train as a whole, another one for the locomotive, and one for each car taken separately. Neglect friction. **(b)** Find the magnitudes of the forces at each of the three couplers.

3.32 *(3) Pulling out of the station, II.* In part *a* of the drawing

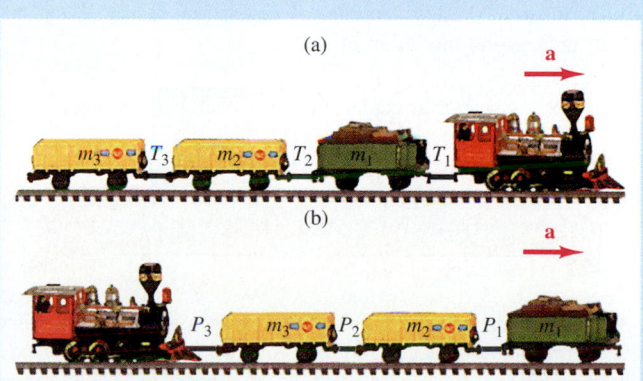

above, a locomotive pulls three cars whose masses are m_1, m_2, and m_3, respectively. The train has acceleration **a**. Neglecting friction, find the magnitudes T_1, T_2, and T_3 of the forces in the three couplers.

3.33 (3) *Pulling out of the station, III.* In part *b* of the drawing in Problem 3.32, the same train shown in part *a* is pushed rather than pulled by the same locomotive, now situated at the rear end of the train. The acceleration of the train is **a**. Neglecting friction, find the magnitudes P_1, P_2, and P_3 of the forces in the three couplers.

3.34 (3) *The scales don't lie.* A block of mass $m_1 = 12.5$ kg is pulled along a frictionless table by the pair of identical spring scales shown in the illustration below. Scale B reads 20.7 N, and scale A reads 21.1 N. (a) Find the acceleration *a* of the system. (b) Find the mass of each of the two scales. (c) Find the magnitude of the applied force F_s.

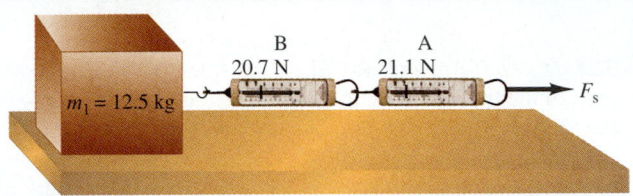

3.35 (3) *Why catapult?* Naval aircraft that take off from the decks of aircraft carriers are always launched with catapults. To see why this is necessary, consider the following typical figures:
Mass of the aircraft: 10 tonnes
Minimum takeoff speed: 85 m/s ($\approx$ 300 km/h)
Length of flight deck: 100 m
(a) During the launching, the aircraft is propelled both by the catapult and by its own engine operating at full power. What is the total propulsive force required to bring the aircraft to takeoff speed? Assume that the force is constant. (b) A typical jet engine designed for this purpose develops a maximum thrust of 10^5 N. What fraction of the total propulsive force is supplied by the engine? (c) How long a runway does the same aircraft need to take off unassisted?

3.36 (3) *Not a weighty matter, I.* Here is a way to measure an unknown mass m_2 without weighing it. A spring of negligible mass is squeezed between m_2 and a known mass $m_1 = 1$ kg. The system is laid on a frictionless table and released, as shown below. At a certain moment, the acceleration of m_1 is $a_1 = -2.20$ m/s^2 and the acceleration of m_2 is $a_2 = 1.35$ m/s^2. Find the value of m_2.

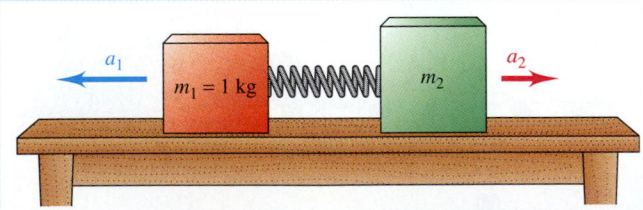

3.37 (4) *Point of view.* A passenger is in a glass-walled elevator moving with acceleration *A*. She drops a steel ball. What is its acceleration *a* as she observes it? What is its acceleration *a'* as seen by an observer standing outside the elevator?

3.38 (4) *Linking it together.* Each of the five links of the chain illustrated in the next column has mass *m*. (a) Draw a free-body diagram for the entire chain regarded as a single body. (b) Draw a free-body diagram for the top link, for the bottom link, and for the middle link. (c) Find the value of the supporting force F_c when the chain hangs at rest. (d) Now let

F_c have an arbitrary value. Find the acceleration *a* of the chain. (e) What is the net force on each link?

3.39 (4) *Accelerometer.* A mass hangs from a spring scale in an elevator car at rest. The scale is calibrated to read in kilograms, in the everyday way. The scale reading is 10.0 kg. The elevator starts, and the scale reads 11.8 kg. (a) Is the elevator going up or down? (b) What is the acceleration of the elevator?

3.40 (4) *Bosun's chair.* In the illustration for Query 3.24, the mass of the chair is 15 kg and that of the passenger 60 kg. Find the magnitude and direction of the force that the passenger must exert on the rope in order to rise at constant speed.

3.41 (5) *Lubber's chair?* A practical-minded friend argues that the arrangement in the illustration below is superior to that illustrated in Query 3.24 because the large pulley is more accessible for servicing and can be easily removed and kept out of the weather when not in use. Assuming the masses given in Problem 3.40, what force must the passenger exert on the rope to rise at constant speed?

3.42 (G) *Skydiver.* Skydivers do not experience free fall because they are subject to a substantial force due to air resistance, in addition to the gravitational force. Taking the downward direction as positive, the force of air resistance is described approximately by the relation $F_r = -cv^2$. In this relation, *c* is a constant: Its value is proportional to the frontal area presented by the skydiver to the air stream, and its value also depends on the skydiver's shape and orientation with respect to the air stream. Suppose that a skydiver of mass *m* falls in a fixed posture, so as to keep *c* constant. (a) Express the units of *c* in terms of the SI base units. (b) Show that the skydiver will attain a *terminal speed* v_f, given by the relation

$$v_f = \sqrt{\frac{mg}{c}},$$

and then will undergo no further acceleration. **(c)** For a 70-kg skydiver falling face downward in a spread-eagle posture, a typical value of v_f is 50 m/s. Estimate the value of c.

3.43 *(G) Dynamic balance.* In the system illustrated below, you can prevent m_2 from descending, thereby pulling m_3 to the right along the top of m_1, by pulling on the string attached to m_1 with force F. **(a)** Sketch free-body diagrams for the three masses. **(b)** Assuming that all surfaces are frictionless, find F.

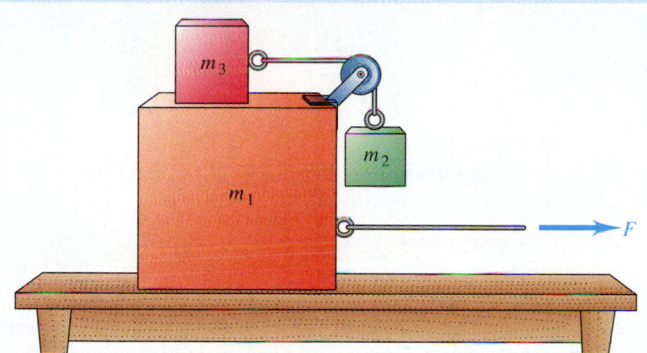

3.44 *(G) It makes a difference how you do it.* A heavy iron block is suspended by a thread, as shown below. A similar thread is tied to the bottom of the block. If a slowly increasing force P is applied by pulling on the lower thread, the upper thread breaks. But, if the lower thread is given a sudden jerk, the lower thread breaks. **(a)** Write the equation relating the forces acting on the block to the acceleration of the block. **(b)** How does this equation explain which thread breaks under what condition? (Hint: Assume that the threads are slightly extensible, so that you can imagine the block accelerating.)

3.45 *(G) Counterweight.* Elevators are always designed with counterweights, as shown in the diagram in the next column. The mass of the counterweight is approximately equal to that of the elevator car. **(a)** Suppose that the elevator car and counterweight each have mass 1000 kg. If the system is to have maximum acceleration of magnitude 1.1 m/s², what force must the motor be capable of applying to the cable? (Neglect the mass of the passengers.) **(b)** A beginning designer decides to reduce the inertia of the system, hoping thus to reduce the size of the motor required. He proposes to remove the

counterweight from the system. What force must the motor now be capable of applying to the cable to maintain the same performance characteristics?

3.46 *(G) Winter fun.* Twin children on ice skates play on a frozen pond. One has a rope tied around her waist. The other holds the free end of the rope, a distance L away from the first. The second child begins to pull on the rope. **(a)** Describe the motion of the children. **(b)** Where do they meet? **(c)** When they meet, how much rope has passed through the second child's hands?

3.47 *(6) Shifting masses, I.* Let the total suspended mass in an Atwood machine be $M = m_1 + m_2$. The masses on the two sides of the pulley are initially equal. A mass Δm is then removed from the left side of the machine and placed on the right side. **(a)** Rewrite Equation 3.19 to give the acceleration a_1 in terms of Δm and M. **(b)** Sketch a graph of a_1/g as a function of Δm in the range $-M/2 \leq \Delta m \leq M/2$. What is the significance of negative values of Δm?

3.48 *(6) Shifting masses, II.* **(a)** An Atwood machine is in balance, with $m_1 = m_2 = M/2$. Express the string tension T in terms of M. **(b)** A mass Δm is removed from the left side of the machine and placed in the right side. Derive an expression for T in terms of M and Δm. Write the expression as a product, one factor being the result of part **a** and the other a dimensionless factor containing the fraction $\Delta m/M$. **(c)** Sketch a graph of T as a function of $\Delta m/M$ in the range $-M/2 \leq \Delta m \leq M/2$. What kind of curve have you sketched?

3.49 *(6) Bound to fall.* An Atwood machine is at rest with mass m_2 located a distance $2\Delta x$ above mass m_1, as in Example 3.7. Suppose that $m_2 > m_1$. After the system is released, the two masses pass each other at time t, given by Equation 3.22. Write t as a product of the free-fall descent time through the same distance and a dimensionless factor. Interpret your result physically.

3.50 *(6) Weight-loss program.* An Atwood machine hangs from a spring scale, as shown below. When the masses are in motion, what is the scale reading? Assume that all other parts of the machine have negligible mass.

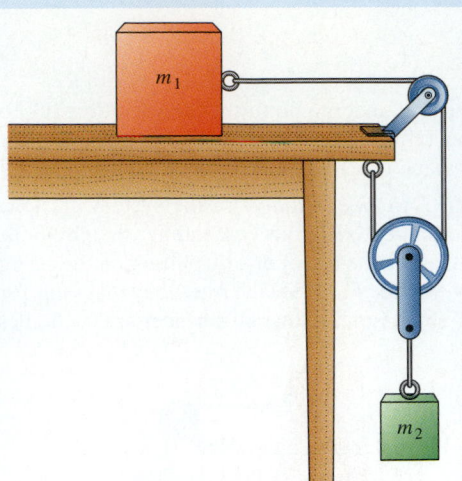

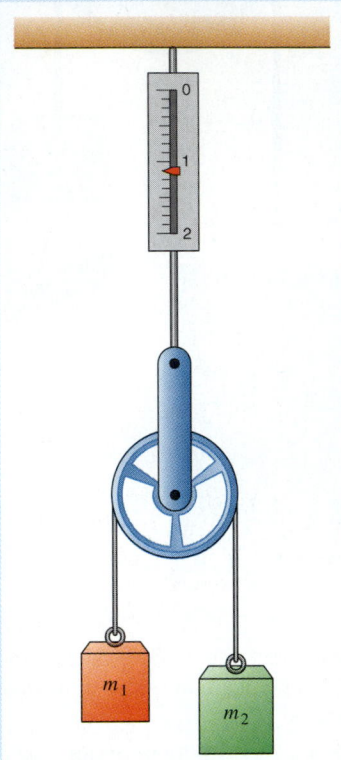

3.51 *(G) Not a weighty matter, II.* In performing the experiment described in Problem 3.36, it is much easier to measure the displacement of each of the two masses than to measure their accelerations. The accelerations a_1 and a_2 are *not* constant. Nevertheless, over any time interval Δt there exist mean accelerations $\langle a_1 \rangle$ and $\langle a_2 \rangle$. Using this fact, and remembering that $m_1 = 1$ kg, express the value m_2 of the unknown mass in terms of the displacements Δx_1 and Δx_2 of the two masses, measured over the same time interval.

3.52 *(G) Enough rope, I.* Consider the apparatus illustrated below. **(a)** Find the accelerations a_1 of m_1 and a_2 of m_2. **(b)** Find the tension T in the string. **(c)** Let $m_1 = 5.00$ kg and $m_2 = 0.800$ kg. Obtain values for a_1, a_2, and T. Assume that

the table is frictionless and that both pulleys have negligible mass and friction.

3.53 *(G) Enough rope, II.* Consider the apparatus shown below. **(a)** Find the accelerations a_1 of m_1 and a_2 of m_2. **(b)** Find the tension T in the string. **(c)** Let $m_1 = 8.00$ kg and $m_2 = 4.50$ kg. Evaluate a_1, a_2, and T. Neglect the mass and friction of the pulleys.

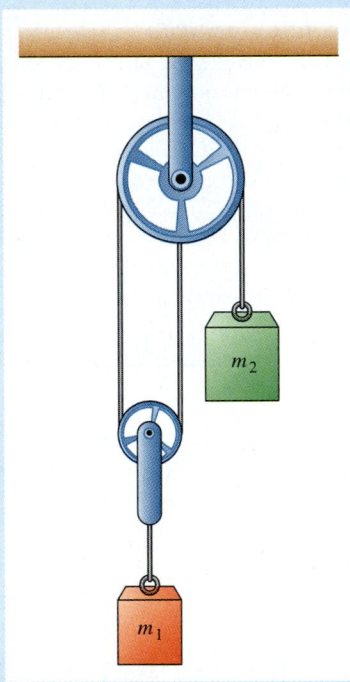

GROUP C

3.54 *(3) Flat out.* Note: To work parts **d** through **g** of this problem, you should be familiar with the technique of finding the extremal value of a function by setting its derivative equal to zero. **(a)** Once an automobile is in high gear, the driving force F_1 applied by the wheels to the road increases with increasing speed, reaches a peak value, and then decreases. The function $F_1(v)$ is complicated and depends on the design of the engine as well as the drive-train gear ratios. However, to first approximation, its magnitude can be written in the form $|F_1(v)| = p + qv - rv^2$, where p, q, and r are positive constants. Sketch a graph of $|F_1(v)|$. **(b)** The frictional resistance

that tends to slow the automobile also depends on speed. To first approximation, its magnitude can be written in the form $|F_2(v)| = h + kv^2$, where h and k are positive constants. Superimpose a sketch of $|F_2(v)|$ on the graph of part **a**. **(c)** On the same graph, sketch the net force $F(v)$ acting on the automobile. **(d)** Find the top speed v_{max} of the automobile on a level roadway. **(e)** Find the maximum acceleration a_{max}. **(f)** At what speed is the acceleration a maximum? **(g)** The performance of the automobile can be improved by changing the values of the constants h, k, p, q, and r. Suggest the kinds of design changes that will affect each constant.

3.55 (G) *Galilean transformation equations.* Observers O and O' repeat the experiment shown in Figure 3.4. This time, however, the dry-ice disk is moving from the point of view of O, as shown in the drawing below. She makes repeated observations of its position, which she calls $x_1, x_2, \ldots, x_j, \ldots$, at times $t_1, t_2, \ldots, t_j, \ldots$. Observer O' makes his own observations simultaneously and gives his results the corresponding names $x'_1, x'_2, \ldots, x'_j, \ldots$.

(a) Derive an expression for the position x'_j seen by O' in terms of x_j, t_j, and V, the velocity of O' seen by O. This equation is called the *Galilean position transformation*, a term first used by Einstein. **(b)** Using the result of part **a**, derive an expression for the instantaneous velocity v'_j of the disk, as determined by O'. Express v'_j in terms of V and the velocity v_j that O determines on the basis of her position and time measurements. This relation is called the *Galilean velocity transformation*. **(c)** Using the result of part **b**, derive an expression for the instantaneous acceleration a'_j of the disk as determined by O', in terms of the acceleration a_j determined by O. **(d)** Why is Newton's second law valid for all observers fixed in inertial frames, regardless of their velocities V relative to one another? (Note: This statement is true only in the classical domain, where $|V|$ is very much less than the speed of light.) **(e)** Suppose that O observes O' to be moving with nonzero acceleration A, as in Figure 3.5. Are the Galilean transformation equations correct? Is Newton's second law valid for O'? Why or why not?

3.56 (G) *Dynamic unbalance.* Suppose that the system shown in Problem 3.43 is initially at rest; a catch (not shown) holds m_3 in place on the upper surface of m_1. Show that when the catch is released, the initial accelerations of m_3 and m_2, as seen by an observer on the ground, are

$$a_3 = \frac{m_2}{m_2 + m_3}\, g \quad \text{and} \quad a_1 = \frac{m_2 m_3}{(m_1 + m_2)(m_2 + m_3)}\, g.$$

3.57 (G) *Think before you catch, Wonder Woman!* In Problem 2.59, an unfortunate baby falls off a cliff and Wonder Woman zooms down to catch him. As she zooms, she thinks about the physics of the catch. If the baby is to avoid injury, Wonder Woman cannot apply a force to him in excess of 3 times his body weight. **(a)** At what minimum height above the ground must she catch the baby? **(b)** If she descends with constant acceleration (of magnitude greater than $|g|$) to the

point where she catches the baby, what propulsive force must she exert, in units of her own weight? (The means by which Wonder Woman and her superhuman colleagues exert their propulsive forces seems not to be specified in the literature.)

3.58 (6) *Diff'rent strokes.* The two masses of an Atwood machine rest on a tabletop. A small upward force F_p on the pulley keeps the string taut. Suppose that $m_2 > m_1$, and consider what happens as F_p is increased. Show that **(a)** if $|F_p| < 2m_1|g|$, the system remains at rest; **(b)** if $2m_1|g| < |F_p| < 2m_2|g|$, the pulley and m_1 move upward but m_2 stays on the table; **(c)** if $|F_p| > 2m_2|g|$, the whole system leaves the table. In each case, find T, a_1, a_2, and A, the acceleration of the pulley. Assume that the pulley itself has negligible mass.

3.59 (G) *Elevating experience.* An Atwood machine with masses m_1 and m_2 hangs from the ceiling of an elevator car. The elevator moves with acceleration A. Find a_1, the acceleration of m_1 as seen by an observer in the car.

3.60 (G) *Two-faced machine.* The machine shown below will operate as a unit only as long as the string connecting m_2 and m_3 remains taut. When this string goes slack, the machine degenerates into two independent Atwood machines. For example, the system acts as a unit if $m_4 > m_3$ and $m_1 > m_2$, but it acts as two independent systems if $m_3 > m_4$ and $m_2 > m_1$. **(a)** State the general conditions under which the system acts as a unit. **(b)** Find the magnitude of the acceleration of the system under these circumstances.

Part *II*

Mechanics of Particles in Two and Three Dimensions

The world in which we live has three spatial dimensions, not just one. Thus far, we have considered only one-dimensional processes—processes in which all forces act along a straight line and all motion takes place along that line. In this restricted world, we developed and applied the basic laws of mechanics in the simplest possible context. Some systems that are not really one-dimensional, such as the Atwood machine, could be reduced conceptually to one dimension.

Now it is time to expand our horizons. Nature does not display all its riches in one dimension. Most physical processes are two- or three-dimensional in an essential way. That is, they are not reducible to one dimension and *must* be studied in the two- or three-dimensional context in which they take place. Most of the rest of our study of physics will concern multidimensional systems and phenomena. In particular, Part II of this book extends the ideas developed in Part I into a richer world. The price that we pay for added richness is added complication, but we cannot expect nature to be simple just to accommodate us.

Happily, we will need no new laws of mechanics to guide us in this exploration. But we must express the laws that we already have in a multidimensional way. The first order of business is to develop the necessary mathematical language. This language is called *vector algebra*. Though somewhat more complicated than the algebra we used to study one-dimensional mechanics, vector algebra is a straightforward and consistent extension of algebraic and geometrical ideas with which you are already familiar.

Chapter 4 begins by developing vector algebra and then applies this new tool to the study of forces. Vector algebra makes possible the study of the common situation in which

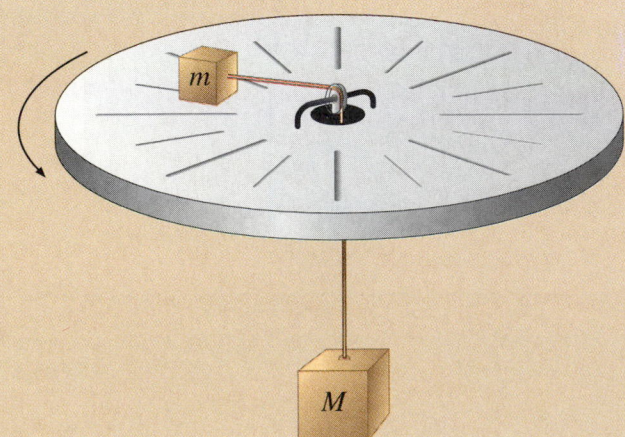

several forces, acting in different directions, are applied simultaneously to the same particle. In Chapter 4, we limit ourselves to the simplest of such cases, in which the forces do not change in time, and the resulting motion of the particle is along a straight line. A skier coming straight down a hill of uniform slope and snow coverage experiences such motion.

In Chapter 5, we study the motion of particles that follow curved paths of various sorts. A thrown ball exhibits one such type of motion, and a stone whirled on the end of a string exhibits another type.

Chapter 6 deals with *oscillatory motion*, a frequently seen type of motion that takes place when the net force applied to a particle varies with time in a repetitive fashion. A child's swing is one of the most familiar examples of such motion.

When you finish Part II, you will have acquired a fair amount of experience in analyzing particle motion by applying Newton's three laws—what we have called the *dynamical approach*. In Part III, we will consider a complementary, equally important approach based on the conservation principles of mechanics. Throughout the rest of this book — and indeed throughout your career as a scientist or engineer—you will learn to draw on both approaches as you expand your mastery of physics.

Vectors and Resolution of Forces

L O O K I N G   A H E A D

—— In extending Newton's laws to the rich two- and three-dimensional worlds, we need to exploit the properties of vectors more fully.

—— Multidimensional vectors have mathematical properties all their own, similar to but a little more complicated than those of scalars.

—— The algebraic and geometric operations needed to manipulate multidimensional vectors include vector addition and subtraction.

—— A vector can be multiplied or divided by a scalar.

—— Displacement, velocity, acceleration, and force are vector quantities.

—— A body moving on an inclined plane is subjected to several forces having various directions; we use vector addition to find the net force so that we can apply Newton's second law.

—— Contact friction is a commonly encountered force.

Left: When the cross-current is strong, the pilot must exercise great skill in manipulating the speed and direction of the ferryboat if it is not to hit the pilings as it eases into the slip.

> *Pressed by hounds on every side,*
> *the fox ran off in all directions.*
>
> —TRADITIONAL

Introduction

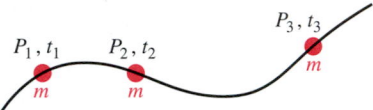

FIGURE 4.1 A particle of mass m moves along an arbitrary path. To analyze the motion of the particle, we must have a way to describe its position, displacement, velocity, and acceleration, as well as the net force acting on it, as functions of time.

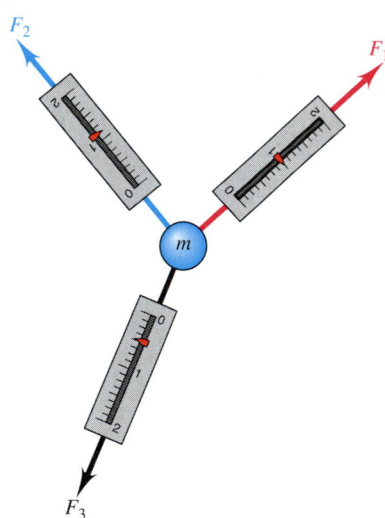

In the multidimensional world, forces can act in any direction and bodies can move in any direction. Before we can apply Newton's laws of motion to the study of such forces and motions, we must extend our mathematical language so that it can be used to describe vector quantities in any direction.

Two situations show us what we will need. In Figure 4.1, a particle of mass m moves along an arbitrary but specified path lying in the plane of the page. At time t_1 it is located at P_1. At some later time t_2 it is at P_2, and at a still later time t_3 it is at P_3. To describe the motion of the particle, we need to specify its position, displacement, velocity, and acceleration as functions of time. Only then can we deduce the net force acting on it.

Figure 4.2 shows a different but related problem. A particle of mass m lies on a frictionless surface in the plane of the page. Three strings are tied to it, and the spring scales indicate that the tensions in the strings are F_1, F_2, and F_3. The three forces do not lie along the same line. How can we ''add'' them and thus find the net force acting on the particle? We must find this net force before we can apply Newton's second law and find the acceleration of the particle. Only then can we describe its motion.

Position, displacement, velocity, acceleration, force—all these quantities can be described precisely by means of the mathematical device called the *vector*. We have already considered vector quantities in one dimension. But to deal with vectors in two or three dimensions, we must introduce some new mathematical ideas. In the interest of simplicity, we begin with vectors in two dimensions. However, generalization to three dimensions will be straightforward. The displacement vector will be our prototype for all vectors.

FIGURE 4.2 A number of forces act simultaneously on a particle. To analyze the motion of the particle, we must have a way to calculate the net force acting on it at any instant.

The Displacement Vector: Geometric Description

A particle is located at O, as shown in Figure 4.3. The particle experiences a displacement, and at some later time it is located at point P. We depict the displacement by means of the arrow labeled $\vec{A}$, as shown in the figure. Just as in one dimension, the arrow represents both a *magnitude* (the distance between O and P, which can be represented by a positive scalar) and a *direction* (the direction from O to P).

The arrow conveys more information regarding direction than is needed in one dimension, where the two possible directions are distinguished by the use of + and − signs. We will soon develop algebraic means for describing direction, but let us first explore the geometry of the situation more completely.

In two or three dimensions, just as in one dimension, *a vector is an entity that possesses magnitude and direction*. Because $\vec{A}$ represents a displacement in the two-dimensional space that is the plane of the page, it is specifically a two-dimensional displacement vector.

The symbol conventionally used to denote a vector in handwriting is an arrow over the letter, as in Figure 4.3, but this notation is inconvenient in typesetting. The universal convention in printed matter is to denote vectors by **boldface type**, thus: **A**. The two conventions mean exactly the same thing:

$$\vec{A} \equiv \mathbf{A}, \quad \vec{B} \equiv \mathbf{B}, \quad \text{and so on.}$$

There was no need in Figure 4.3 to specify any particular coordinate system before using the vector **A** to describe the displacement $O \to P$. This is because the only two quantities needed to specify a displacement—the magnitude and direction specified by the vector—do not depend on the choice of a coordinate system. This property of vectors holds in two and three dimensions just as in one dimension (Section 2.2).

Figure 4.4 illustrates this point. All the vectors shown are the *same* vector **A**. We can interpret this statement in two equivalent ways. Because a displacement does not depend on the choice of the coordinate frame from which it is viewed, we may interpret the figure as showing the same displacement vector **A** as it is seen by four observers located at the origins of four different coordinate frames. The second interpretation is often directly useful in manipulating vectors. If an operation on the vector does not disturb its magnitude or its direction, the operation cannot change the vector because a vector has no intrinsic properties other than magnitude and direction. Consequently, the four "arrows" in Figure 4.4 represent the *same* vector moved parallel to itself.

As long as a vector describes a quantity—such as a displacement—whose magnitude and direction do not depend on the choice of a particular coordinate system, we are free to move the vector parallel to itself. Most vectors encountered in the study of physics are of this sort and are called *free vectors*. As already noted in Chapter 2, the term "vector" implies a free vector unless otherwise specified. Occasionally, we will encounter vectors to which this freedom does not apply. Such vectors are called *bound vectors*.

Geometric Addition of Vectors

When a displacement is specified, as by the vector **A** in Figure 4.5, no restriction is made as to the particular path along which the displaced particle has passed from its initial to its final location. Among the infinite number of paths possible, one is the straight line from O to P, along **A** itself. But another possible path consists of the straight line from O to Q, represented by vector **B**, followed by the straight line from Q to P, represented by vector **C**. The direct path from O to P and the indirect path through Q amount to the same displacement. We state this point in the algebraic form

$$\mathbf{A} = \mathbf{B} + \mathbf{C}. \tag{4.1}$$

Equation 4.1 and the corresponding diagram of Figure 4.5 define **vector addition**. Graphically, the sum of the two vectors **B** and **C** is found by moving **C** until its tail coincides with the head of **B** and then drawing the vector **A** from the tail of **B** to the head of **C**. The result of this process is itself a vector (**A** in Figure 4.5) having a specific magnitude and direction. In concord with Equation 4.1, **A** is called the **sum** of **B** and **C**. It is also often called the **resultant** of **B** and **C**. Because the vectors being added (here **B** and **C**) and their resultant (here **A**) form a polygon, this method of vector addition is often called the "polygon method."

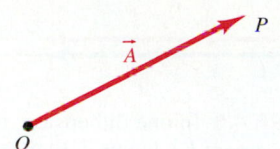

FIGURE 4.3 A particle located initially at O experiences a displacement to a final location P. The displacement is described by the arrow labeled $\vec{A}$, called a **vector**.

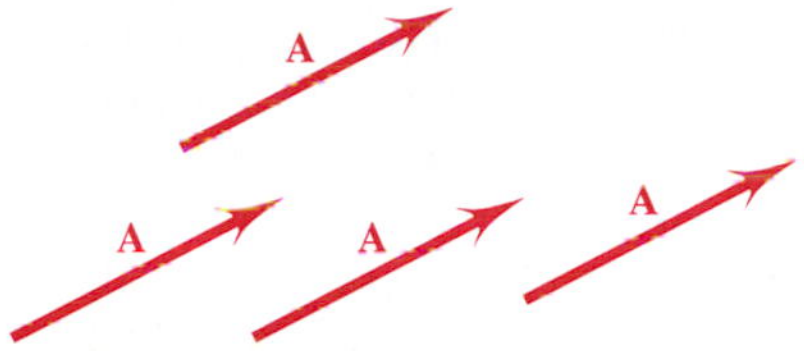

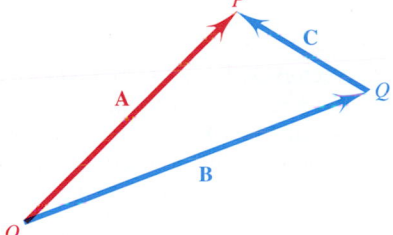

FIGURE 4.4 All the vectors shown are the same vector **A**, because all four have the same magnitude and the same direction. It makes no difference whether we imagine several observers, displaced from one another, seeing the same vector from their respective observation points, or think of the vector as being displaced parallel to itself.

FIGURE 4.5 A particle may be displaced from O to P over an infinite number of possible paths. One such path, shown in red, lies along the straight line $\overline{OP}$. Another path, shown in blue, follows the straight line $\overline{OQ}$, then the straight line $\overline{QP}$. Although the two paths are different, they are equivalent in the sense that they result in the same displacement of the particle. This equivalence is stated in the vector equation **A** = **B** + **C**.

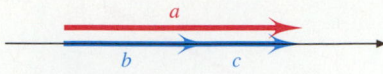

FIGURE 4.6 In one dimension, the displacement b (shown in blue) followed by the displacement c (also shown in blue) is equivalent to the displacement a (shown in red and displaced slightly for illustrative purposes). We denote this equivalence by writing the simple algebraic sum $a = b + c$. The two-dimensional situation of Figure 4.5 is analogous but not identical. The displacement **B** followed by the displacement **C** is equivalent to the displacement **A**, and this equivalence is denoted by writing **A** = **B** + **C**. The magnitude and direction of **A** can be found from the magnitudes and directions of **B** and **C**, but the process is more complicated than algebraic addition of two signed numbers. As Figure 4.5 shows, the magnitude of **A** is *not* simply the sum of the magnitudes of **B** and **C**. The direction of **A** lies somewhere ''in between'' those of **B** and **C**, and we have not yet developed a means for expressing that direction other than by actually drawing **A**.

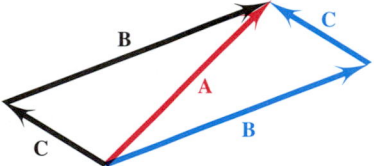

FIGURE 4.7 Proof of Equation 4.2. The vectors **A**, **B**, and **C**, reproduced from Figure 4.5, illustrate the summation rule **A** = **B** + **C**. Blue vector **C** is moved parallel to itself until its tail coincides with that of **A**; the displaced vector is shown in black. Then blue vector **B** is moved until its tail coincides with the head of **C**; this is also shown in black. The blue and black pairs of vectors **B** and **C** constitute the sides of a parallelogram of which **A** is the diagonal. We have **B** + **C** = **A** = **C** + **B**, which proves the commutative law for vector addition.

If you compare Figure 4.5 with Figure 4.6 and its legend, you will see the strong analogy between vector addition and the familiar process of algebraic addition.

Vector addition obeys the commutative law just as ordinary addition does. That is, the order of addition makes no difference, and we have

$$\mathbf{B} + \mathbf{C} = \mathbf{C} + \mathbf{B}, \tag{4.2}$$

just as $b + c = c + b$ for ordinary addition. The proof of this statement is given in Figure 4.7.

Figure 4.8 illustrates the generalization of the rule for adding two vectors to a rule for adding any number of vectors. The second through last vectors to be added are moved parallel to themselves as necessary to place each so that its tail coincides with the head of the preceding one, just as for two vectors. The resultant is then found by drawing the vector **Z** from the tail of the first vector (**A** in the figure) to the head of the last one (**E** in the figure). This operation is represented algebraically by the vector sum

$$\mathbf{Z} = \mathbf{A} + \mathbf{B} + \mathbf{C} + \mathbf{D} + \mathbf{E}. \tag{4.3}$$

Vector addition obeys the mathematical rule called the *associative law*, just as ordinary addition does. That is, we have

$$(\mathbf{A} + \mathbf{B}) + \mathbf{C} = \mathbf{A} + (\mathbf{B} + \mathbf{C}), \tag{4.4}$$

just as $(a + b) + c = a + (b + c)$ for ordinary addition. The proof of this statement is given in Figure 4.9.

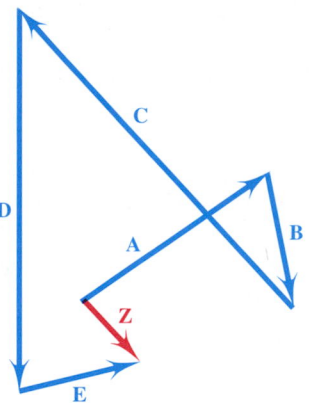

FIGURE 4.8 The rule for adding any number of vectors: Place them sequentially, tail to head. The sum vector, or resultant vector, is **Z**.

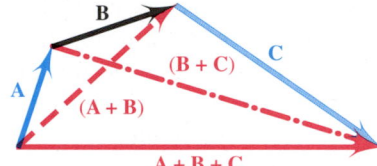

FIGURE 4.9 Proof of Equation 4.4. The partial sum (**A** + **B**) is shown as a dashed vector, and the partial sum (**B** + **C**) is shown as a dash-dot vector. The complete sum **A** + **B** + **C** is shown in solid red. It is evident from the figure that (**A** + **B**) + **C** = **A** + **B** + **C** = **A** + (**B** + **C**), which proves the associative law for vector addition.

The Null Vector, the Negative of a Vector, and Vector Subtraction

Given any vector **A** that describes a displacement from O to P, we can always draw a vector $-\mathbf{A}$ that describes the opposite displacement from P to O, as Figure 4.10 shows. The operation of adding these two vectors can be represented algebraically by the sum

$$\mathbf{A} + (-\mathbf{A}) = \mathbf{0}. \tag{4.5}$$

The resultant of the two vectors, which represents no displacement at all, is called the **null,** or **zero, vector, 0**. Because the null vector has no particular direction, it is often represented by the scalar 0, which is the convention that we will use in this book.

Just as in ordinary algebra, the vector $-\mathbf{A}$ is called the **negative of A**. The vector $-\mathbf{A}$ can be formed from **A** by simple reversal—by moving the head of the arrow representing **A** to the tail end of the arrow. This fact leads directly to the idea of **vector subtraction**. Suppose we wish to form the sum $\mathbf{B} + (-\mathbf{C})$. One way to do this is shown in Figure 4.11a. Given the vector **C**, we first form $-\mathbf{C}$ by reversal of **C**. Then we add

−C to B by moving −C parallel to itself until its tail is in contact with the head of **B**. Finally, we draw the resultant vector **B** + (−**C**) from the tail of **B** to the head of −**C**.

In Figure 4.11b, we omit the intermediate step. Instead of forming −**C**, we simply move **C** until its *head* is in contact with the head of **B**. We then draw the resultant vector from the tail of **B** to the tail of **C**. We call the operation subtraction, and its result the **vector difference B − C**. The result of this procedure is the same as that described in the preceding paragraph. That is, we can write

$$\mathbf{B} - \mathbf{C} = \mathbf{B} + (-\mathbf{C}), \tag{4.6}$$

just as in ordinary algebra.

(a)

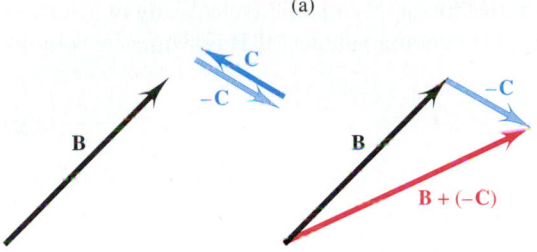

(b)

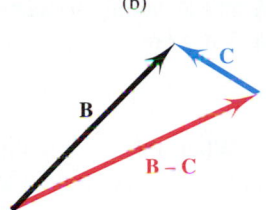

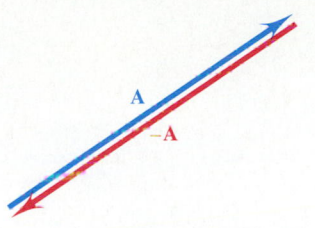

FIGURE 4.10 Construction of −**A**, the negative of **A**. The two vectors **A** and −**A** are shown slightly displaced from each other for clarity.

FIGURE 4.11 (*a*) Construction of −**C** from **C**, followed by formation of the vector sum **B** + (−**C**). (*b*) Formation of the vector difference **B** − **C**. Here, the vectors are placed head to head, and the resultant is drawn from the tail of **B** to the tail of **C**.

EXAMPLE 4.1

Show that **C** − **B** = −(**B** − **C**).

SOLUTION: The difference **C** − **B** can be formed by adding −**B** to **C**. Carry out the subtraction process just described, with the roles of **B** and **C** reversed. That is, move **B** until its head is in contact with the head of **C**. Then draw the resultant vector from the tail of **C** to the tail of **B**, as shown in Figure 4.12. In vector subtraction, the two vectors are always placed head to head. However, the direction of the resultant vector must be drawn *from* the tail of the subtrahend (here **C**) *to* the tail of the minuend (here **B**). For comparison, the difference **B** − **C** ob-

FIGURE 4.12 Illustration showing that **C** − **B** = −(**B** − **C**).

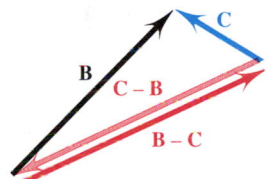

tained in Figure 4.11b is reproduced here in a darker red. The comparison shows that **B** − **C** is indeed the negative of **C** − **B**; that is,

$$\mathbf{C} - \mathbf{B} = -(\mathbf{B} - \mathbf{C}).$$

Magnitude and Direction of a Vector and the Unit Vector

It is often useful to express the magnitude and direction of a vector separately. Figure 4.13 shows an arbitrary two-dimensional vector **A**. Let us continue to assume that **A** is a displacement vector, and so its magnitude is properly expressed in terms of a unit of length. To measure the magnitude—that is, the length—of **A**, we lay a meter stick alongside it, as the figure shows. With the meter stick in place, we can read the length $|f - i|$ off the meter stick exactly as in the one-dimensional case discussed in Chapter 2. We write the result of this operation as $|\mathbf{A}|$. This use of the absolute value signs, $|\ |$, is analogous to their use for one-dimensional vectors, where $|x|$ is a positive scalar denoting the magnitude (absolute value) of x. Because $|\mathbf{A}|$ is *not* a vector but a positive scalar, the abbreviated notation A is often used to mean exactly the same thing:

$$|\mathbf{A}| \equiv A.$$

Take care, however, in interpreting this notation. A symbol written in *italics* usually denotes a quantity that can have either positive or negative value. But the magnitude of a vector (such as A) is *always a positive scalar*, as we have already seen for the one-dimensional case in Section 2.2.

We now turn to the direction of the arbitrary vector **A**. We define a special vector $\hat{\mathbf{A}}$ having the same direction as **A**. $\hat{\mathbf{A}}$ is called the **unit vector in the A direction**. More informally, $\hat{\mathbf{A}}$ is called "A-hat." Its magnitude is 1, as implied in the term "unit vector."

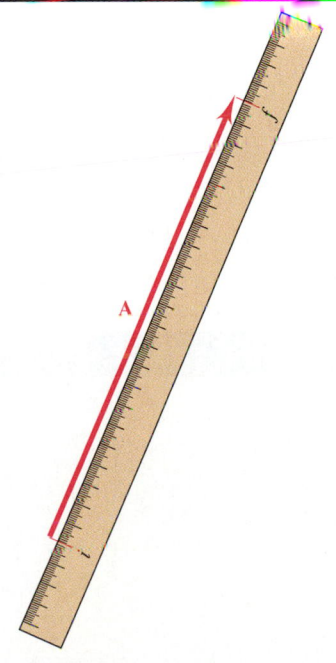

FIGURE 4.13 The magnitude of the displacement vector **A** is found by laying the meter stick along the vector and measuring its length $|f - i|$.

We can write any vector $\mathbf{A}$ in terms of its magnitude $|\mathbf{A}|$ $(=A)$ and its direction $\hat{\mathbf{A}}$:

$$\mathbf{A} \equiv |\mathbf{A}|\hat{\mathbf{A}} \quad \text{or, equivalently,} \quad \mathbf{A} \equiv A\hat{\mathbf{A}}. \tag{4.7a,b}$$

By definition, $\hat{\mathbf{A}}$ is dimensionless because direction is not a quantity that needs dimensions to have physical meaning. The dimensions of $\mathbf{A}$ are a part of the magnitude A. If, for example, $\mathbf{A} = (10 \text{ m northeastward})$, then $A = (10 \text{ m})$ and $\hat{\mathbf{A}} = (1 \text{ northeastward})$.

Multiplication and Division of a Vector by a Scalar

The idea of multiplication of a vector by a scalar is immediately suggested by Equation 4.7b, $\mathbf{A} \equiv A\hat{\mathbf{A}}$. A straightforward generalization of this equation relates any two vectors $\mathbf{A}$ and $\mathbf{B}$ having the same direction but different magnitudes. If $\mathbf{B}$ is c times as long as $\mathbf{A}$, we write

$$\mathbf{B} = c\mathbf{A}. \tag{4.8}$$

When a vector is multiplied by a *negative* scalar, the result is a vector having opposite direction. In particular, let $c = -1$, and consider the product

$$\mathbf{B} = c\mathbf{A} = (-1)\mathbf{A}.$$

Just as in ordinary algebra, we write this result in the shorter form

$$\mathbf{B} = -\mathbf{A}.$$

Now suppose that c is a scalar having any negative value. We have

$$\mathbf{B} = c\mathbf{A} = -|c|\mathbf{A} = |c|(-\mathbf{A}) = |c|A(-\hat{\mathbf{A}}).$$

That is, $\mathbf{B}$ is a vector of magnitude $|c|A$ and direction opposite that of $\mathbf{A}$.

Division of a vector by a scalar is easy to define. Every scalar c has a reciprocal $d = 1/c$. With this in mind, we define

$$\mathbf{B} = \frac{\mathbf{A}}{d} \equiv c\mathbf{A}, \quad \text{where} \quad c = \frac{1}{d}. \tag{4.9}$$

That is, *division of a vector by a scalar is identical to multiplication of the vector by the reciprocal of the scalar.*

When a vector $\mathbf{A}$ is multiplied or divided by a positive scalar, the result is a vector having the same direction as $\mathbf{A}$ but (except in the trivial case where the value of the scalar is 1) a different magnitude. When a vector $\mathbf{A}$ is multiplied or divided by a negative scalar, the result is a vector whose direction is opposite that of $\mathbf{A}$ and whose magnitude is (in general) different from that of $\mathbf{A}$.

The Displacement Vector: Algebraic Description

Vectors can be described and manipulated algebraically as well as geometrically. We can translate freely back and forth between the geometric language of Section 4.2 and the algebraic language to be introduced in this section by means of analytic geometry. Free translation not only affords us deeper insight into the mathematical properties of the vector, but also adds greatly to the power of the vector as a tool for physical analysis.

We begin to translate our geometric understanding of vectors into the language of algebra with Figure 4.14. We draw a cartesian (x-y) coordinate system in a plane that contains the arbitrary vector $\mathbf{A}$. Next, we draw two special vectors of which $\mathbf{A}$ is the sum. The first of these, $\mathbf{A}_x$, is parallel to the x axis and is drawn so that its tail coincides with the tail of $\mathbf{A}$. The second, $\mathbf{A}_y$, is parallel to the y axis and is drawn so that its head coincides with the head of **A**. The lengths of **A**$_x$ and **A**$_y$ are such that the head of the former coincides with the tail of the latter. As you can see from the figure, we indeed have

$$\mathbf{A} = \mathbf{A}_x + \mathbf{A}_y. \qquad (4.10)$$

The coordinate system itself defines the two directions that we have chosen for **A**$_x$ and **A**$_y$. In vector notation, we give the name $\hat{\mathbf{x}}$ to the positive x direction and the name $\hat{\mathbf{y}}$ to the positive y direction. Vectors such as $\hat{\mathbf{x}}$ and $\hat{\mathbf{y}}$ are called **basis vectors**; they define the basic directions in the coordinate space that they describe. As Figure 4.14 and its legend show, we can write **A**$_x$ and **A**$_y$ in terms of their own magnitudes and the basis vectors:

$$\mathbf{A}_x = |x_f - x_i|\hat{\mathbf{x}} \quad \text{and} \quad \mathbf{A}_y = |y_f - y_i|\hat{\mathbf{y}}. \qquad (4.11a,b)$$

Equations 4.11a and 4.11b are not valid, however, for vectors whose directions do not lie in the first quadrant—that is, do not lie somewhere between $\hat{\mathbf{x}}$ and $\hat{\mathbf{y}}$. You can verify this by studying the vector **B** in Figure 4.15. The figure and its legend demonstrate, however, that the following pair of equations *always* works:

$$\mathbf{A}_x = (x_f - x_i)\hat{\mathbf{x}} \quad \text{and} \quad \mathbf{A}_y = (y_f - y_i)\hat{\mathbf{y}}.$$

For brevity, we define

$$A_x \equiv x_f - x_i \quad \text{and} \quad A_y \equiv y_f - y_i. \qquad (4.12)$$

In terms of A_x and A_y, we can rewrite Equations 4.11a and 4.11b in the more convenient form

$$\mathbf{A}_x = A_x\hat{\mathbf{x}} \quad \text{and} \quad \mathbf{A}_y = A_y\hat{\mathbf{y}}. \qquad (4.13a,b)$$

Bear in mind that A_x and A_y are scalars, each of which may be either positive or negative. Hence, they *do not in general* represent the magnitudes of the corresponding vectors **A**$_x$ and **A**$_y$, though their absolute values $|A_x|$ and $|A_y|$ always do. Although the vectors of Equation 4.13 are no longer represented purely and simply as products of a magnitude and a direction, the representation is very convenient because it automatically takes care of signs and directions.

Now let us return to the arbitrary vector **A** of Figure 4.14 and Equation 4.10. Substituting Equation 4.13 into Equation 4.10, we obtain

$$\mathbf{A} = A_x\hat{\mathbf{x}} + A_y\hat{\mathbf{y}}. \qquad (4.14)$$

The scalars A_x and A_y are called the **components** of **A**. The corresponding vectors **A**$_x$ and **A**$_y$ are called the **component vectors**; they should not be confused with the components themselves.

After the coordinate system has been established, the directions $\hat{\mathbf{x}}$ and $\hat{\mathbf{y}}$ are automatically specified. Often, it is convenient not to write them explicitly. Instead, we write Equation 4.14 in the equivalent form

$$\mathbf{A} = (A_x, A_y), \qquad (4.15)$$

and **A** is then said to be written in **component form**. The order of the components is very important. You can satisfy yourself by inspecting Figure 4.14 that (A_y, A_x) would *not* be the same as (A_x, A_y). For this reason, (A_x, A_y) is called an **ordered pair**.

Figure 4.16 and its legend show that the components of a vector do not vary when the vector is moved parallel to itself—just as the vector itself does not vary when it is moved parallel to itself.

Because we can readily manipulate the components algebraically, the component form is a useful way of representing vectors. It complements the more graphic (but less easily manipulable) geometric methods developed in Section 4.2.

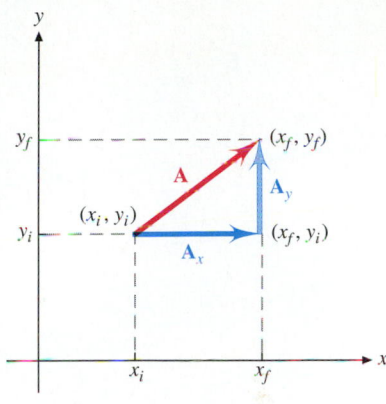

FIGURE 4.14 **A** shown as the sum of two vectors **A**$_x$ and **A**$_y$, each parallel to one of the axes of an x-y coordinate system. The magnitudes and directions of the two vectors can be read directly from the graph: $\mathbf{A}_x = A_x\hat{\mathbf{A}}_x = |x_f - x_i|\hat{\mathbf{x}}$, and $\mathbf{A}_y = A_y\hat{\mathbf{A}}_y = |y_f - y_i|\hat{\mathbf{y}}$.

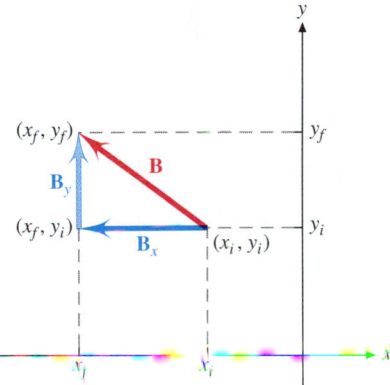

FIGURE 4.15 **B** shown as the sum of two vectors **B**$_x$ and **B**$_y$. Vector **B**$_y$ is identical with **A**$_y$ in Figure 4.14, and we can write $\mathbf{B}_y = |y_f - y_i|\hat{\mathbf{y}} = (y_f - y_i)\hat{\mathbf{y}}$ (because $y_f - y_i$ is positive). However, $\mathbf{B}_x = |x_f - x_i|(-\hat{\mathbf{x}})$. But, because $x_f - x_i$ is negative, we have $x_f - x_i = -|x_f - x_i|$, and again $\mathbf{B}_x = (x_f - x_i)\hat{\mathbf{x}}$.

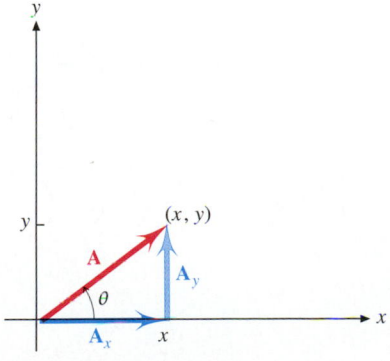

FIGURE 4.16 The vector **A** of Figure 4.14 has been moved parallel to itself so that its tail lies at $(x, y) = (0, 0)$. This process is equivalent to the one in which the coordinate system is translated so that its origin coincides with the tail of the vector.

Let us use the components (A_x, A_y) of the vector $\mathbf{A}$ to describe its magnitude and direction. In Figure 4.16, $\mathbf{A}_x$ and $\mathbf{A}_y$ are the sides of the right triangle whose hypotenuse is $\mathbf{A}$. We can thus use Pythagoras's theorem to find the length of the hypotenuse, which is the magnitude A of $\mathbf{A}$:

$$A = +\sqrt{A_x{}^2 + A_y{}^2}. \qquad (4.16)$$

We take the positive square root because it is the one having geometric meaning; A is the magnitude of a vector, which cannot be negative.

The direction $\hat{\mathbf{A}}$ is found as follows. We take $\hat{\mathbf{x}}$ to be the reference direction and use the angle θ between $\hat{\mathbf{x}}$ and $\mathbf{A}$ to specify the direction $\hat{\mathbf{A}}$, as shown in Figure 4.16. The definition of the tangent function immediately gives

$$\tan \theta = \frac{A_y}{A_x} \quad \text{or} \quad \theta = \tan^{-1} \frac{A_y}{A_x}. \qquad (4.17a,b)$$

EXAMPLE 4.2

Find the components, the magnitude, and the direction of the vector $\mathbf{A}$ in Figure 4.17.

SOLUTION: The tail of the vector lies at $(-3\text{ m}, 2\text{ m})$, and its head lies at $(-5\text{ m}, 6\text{ m})$. Taking the values to be exact, you have

$$A_x = -5\text{ m} - (-3\text{ m}) = -2\text{ m}.$$

Similarly, you have

$$A_y = 6\text{ m} - 2\text{ m} = 4\text{ m}.$$

If you prefer, you can draw a new set of axes like that shown in gray in the figure and obtain $x' = -2$ m, $y' = 4$ m. Either way, you can write $\mathbf{A}$ in component form as

$$\mathbf{A} = (-2\text{ m}, 4\text{ m}).$$

You then use Equation 4.16 to obtain

$$A = \sqrt{(-2\text{ m})^2 + (4\text{ m})^2} = \sqrt{4\text{ m}^2 + 16\text{ m}^2} = 4.47\text{ m}.$$

Equation 4.17b yields

$$\theta = \tan^{-1} \frac{4\text{ m}}{-2\text{ m}} = 116.6°.$$

[Your calculator will give you the other solution, which is the supplement of the angle desired: $\tan^{-1} (4\text{ m})/(-2\text{ m}) =$

FIGURE 4.17

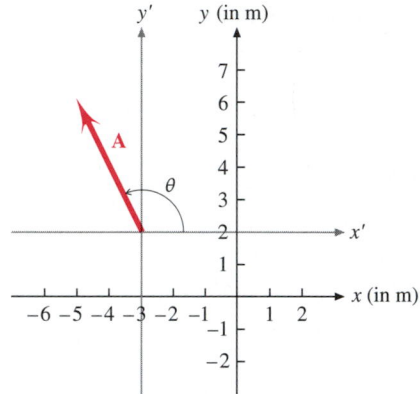

$-63.4°$. This is because the calculator is programmed to give the solution for which $|\theta|$ is smallest. But you can find the correct solution by direct inspection of Figure 4.17.]

You should satisfy yourself that the calculation carried out in this example would work out in exactly the same way if the vector were moved parallel to itself into any quadrant of the coordinate system, or even if it were moved so that it crossed one or both coordinate axes.

Polar Representation of Vectors

Equations 4.16, 4.17a, and 4.17b provide a way of finding the magnitude and direction of a vector when its x and y components are known. That is, the equations provide a way of transforming from the cartesian algebraic representation of a vector back to the geometric form in which we first encountered it. But the equations also express that magnitude and direction in the form of the algebraic quantities A and θ. These two quantities may be regarded as a pair of components just as A_x and A_y are. Specifically, we can represent the vector in the form

$$\mathbf{A} = (A, \theta). \qquad (4.18)$$

This notation is called the **polar representation** of the vector because it bears a close kinship to the polar coordinate system of describing the points in a plane. Similarly, the components (A, θ) are called **polar components**. There is an important difference between the polar and cartesian components. Although the x and y components of a vector are scalars and can have either positive or negative values, the **radial component**

A is always positive, because it represents the magnitude of the vector **A**. The **tangential component** θ is normally given as a positive angle, measured counterclockwise from the **polar axis**, which is conventionally made to correspond to the $\hat{x}$ direction.

We have used Equations 4.16, 4.17a, and 4.17b to transform the cartesian representation of a vector, $\mathbf{A} = (x, y)$, into the polar representation, $\mathbf{A} = (A, \theta)$. But we can transform the polar representation into the cartesian representation just as readily. Recall the definitions of the cosine and sine functions; we have

$$A_x = A \cos \theta \quad \text{and} \quad A_y = A \sin \theta. \qquad \textbf{(4.19a,b)}$$

The use of this polar-to-cartesian transformation is illustrated in Example 4.3.

EXAMPLE 4.3

The vector **A** in Figure 4.18 has magnitude 8.65 m. Its direction is 244°, measured from the $\hat{x}$ direction. Find its *x* and *y* components, and sketch the vector with its components.

FIGURE 4.18

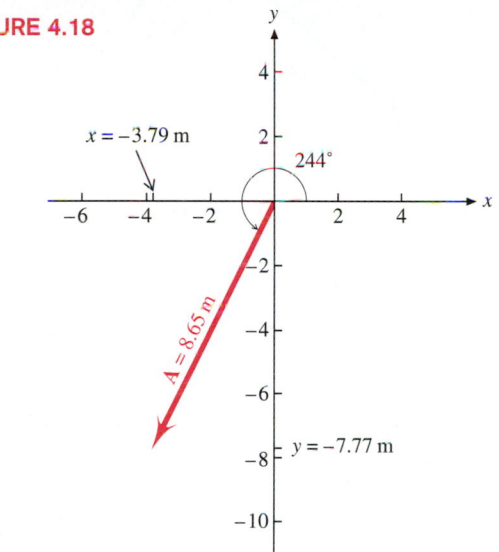

SOLUTION: Using Equation 4.19a, you obtain

$$A_x = 8.65 \text{ m} \times \cos 244° = -3.79 \text{ m};$$

the minus sign signifies that the vector is directed generally to the left, as you would expect for a vector for which θ lies in the third quadrant.

Equation 4.19b gives you

$$A_y = 8.65 \text{ m} \times \sin 244° = -7.77 \text{ m}.$$

Here, the minus sign indicates that the vector is directed generally downward, again as you would expect for a value of θ in the third quadrant. Figure 4.18 is a sketch of the results.

Addition and Subtraction of Vectors: Algebraic Method

Figure 4.19 shows three vectors, $\mathbf{A} \equiv (A_x, A_y)$, $\mathbf{B} \equiv (B_x, B_y)$, and $\mathbf{C} \equiv (C_x, C_y)$. It is evident from their geometric arrangement that

$$\mathbf{A} = \mathbf{B} + \mathbf{C}. \qquad \textbf{(4.20a)}$$

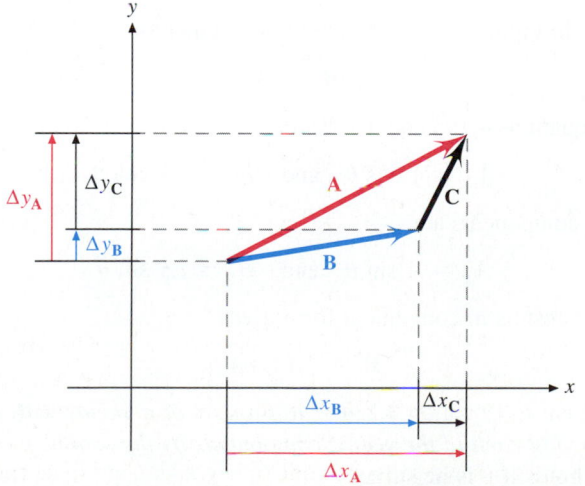

FIGURE 4.19 Construction to show that the sum of two vectors, **B** and **C**, may be found by adding their *x* and *y* components separately. Because the components are signed quantities, they may be depicted as vectors in the one-dimensional spaces respectively defined by the *x* axis and the *y* axis.

In the same figure, the three vectors are represented by their component vectors on the x and y axes, respectively. The components satisfy the relations

$$A_x = B_x + C_x \quad \text{and} \quad A_y = B_y + C_y.$$

Hence, the vector sum of Equation 4.20a can be expressed in the equivalent component form

$$\mathbf{A} = (A_x, A_y) = (B_x + C_x, B_y + C_y). \tag{4.20b}$$

The sum of two vectors may be found by algebraic addition of their corresponding components. Figure 4.20 shows that the rule applies without regard to the signs of the components.

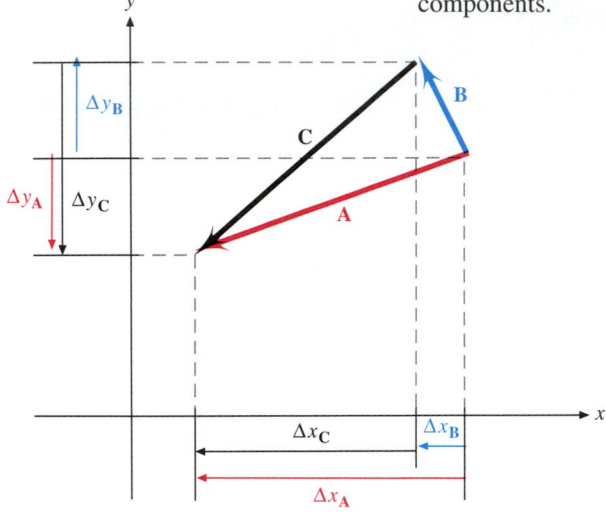

FIGURE 4.20 Construction to show that Equation 4.21b for addition of vectors component by component is valid regardless of the signs of the components.

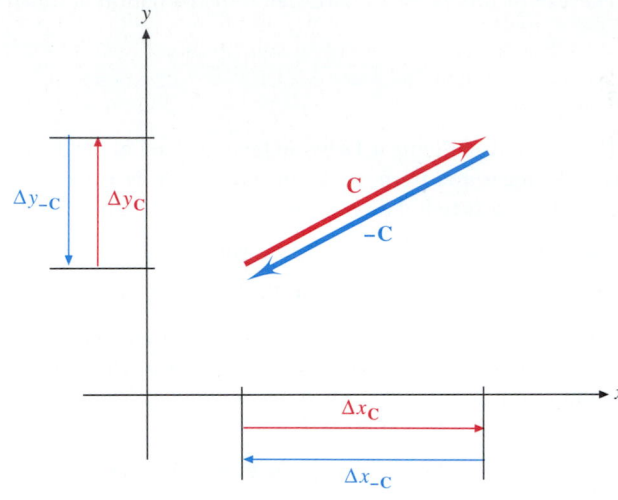

FIGURE 4.21 Construction to show that $-\mathbf{C} = (-C_x, -C_y)$.

Figure 4.21 shows how to express the negative of a vector in component form. If $\mathbf{C} = (C_x, C_y)$, we have

$$-\mathbf{C} = -(C_x, C_y) = (-C_x, -C_y). \tag{4.21}$$

The expression for vector subtraction in terms of components logically follows:

$$\mathbf{B} - \mathbf{C} = \mathbf{B} + (-\mathbf{C}) = (B_x - C_x, B_y - C_y). \tag{4.22}$$

The difference of two vectors may be found by algebraic subtraction of their corresponding components.

Multiplication of a Vector by a Scalar: Algebraic Method

The two vectors in Figure 4.22 are related by the equation

$$\mathbf{B} = c\mathbf{A}. \tag{4.23a}$$

According to Equations 4.19a and 4.19b, the x components of the two vectors are

$$A_x = A \cos \theta \quad \text{and} \quad B_x = cA \cos \theta.$$

Similarly, the y components are

$$A_y = A \sin \theta \quad \text{and} \quad B_y = cA \sin \theta.$$

Expressing these results in component form gives

$$\mathbf{B} = (cA_x, cA_y), \tag{4.23b}$$

which is equivalent to Equation 4.23a. *The product of a vector with a scalar may be found by multiplying each of the vector components by the scalar.* Can you prove that the same result holds if c is negative?

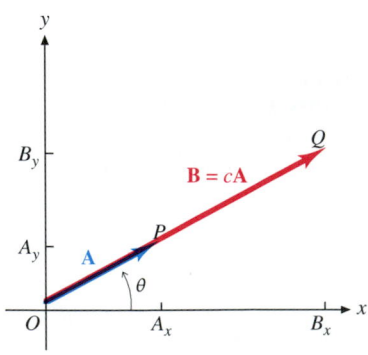

FIGURE 4.22 Construction to show that $\mathbf{B} = c\mathbf{A} = (cA_x, cA_y)$.

EXAMPLE 4.4

A hiker walks 1.5 km in the direcion 24° north of east. He then turns and walks 6.2 km in the direction 54° west of north. Using the methods of vector algebra, find how far he must walk, and in what direction, to return to his starting point. Assume there are no obstructions to walking in a straight path.

SOLUTION: Even when working with algebraic methods, it always pays to begin the analysis of a problem with a little sketch. So begin by drawing a picture roughly to scale, like that in Figure 4.23. It is convenient to choose the x axis so that $\hat{\mathbf{x}} =$ eastward, though there are other valid choices. Call the vectors describing the first and second legs of the trip **A** and **B**, respectively. The vector drawn from the head of **B** to the origin represents the leg the hiker takes back to his starting point; call it **C**.

From the sketch, you see immediately that

$$\mathbf{A} + \mathbf{B} + \mathbf{C} = 0.$$

You solve this equation for **C**, obtaining

$$\mathbf{C} = -(\mathbf{A} + \mathbf{B}). \tag{4.24a}$$

You will need this equation in component form, so you write

$$(C_x, C_y) = (-A_x - B_x, -A_y - B_y). \tag{4.24b}$$

The sketch also helps you determine the proper values of the angles needed to evaluate the components of **A** and **B**. The statement of the problem gives $\theta_A = 24°$ directly, because angles are measured from the positive x direction. The angle θ_B is given with respect to northward, which corresponds to $\theta = 90°$ in Figure 4.23. Hence, you must write $\theta_B = 90° + 54° = 144°$.

You are now ready to express **A** and **B** in component form. Using Equations 4.19a and 4.19b, you obtain

$$\mathbf{A} = (A_x, A_y) = (1.5\ \text{km} \times \cos 24°, 1.5\ \text{km} \times \sin 24°)$$
$$= (1.37\ \text{km}, 0.61\ \text{km}),$$

and

$$\mathbf{B} = (B_x, B_y) = (6.2\ \text{km} \times \cos 144°, 6.2\ \text{km} \times \sin 144°)$$
$$= (-5.02\ \text{km}, 3.64\ \text{km}).$$

You can now evaluate the components of **C**, using Equation 4.24b:

FIGURE 4.23

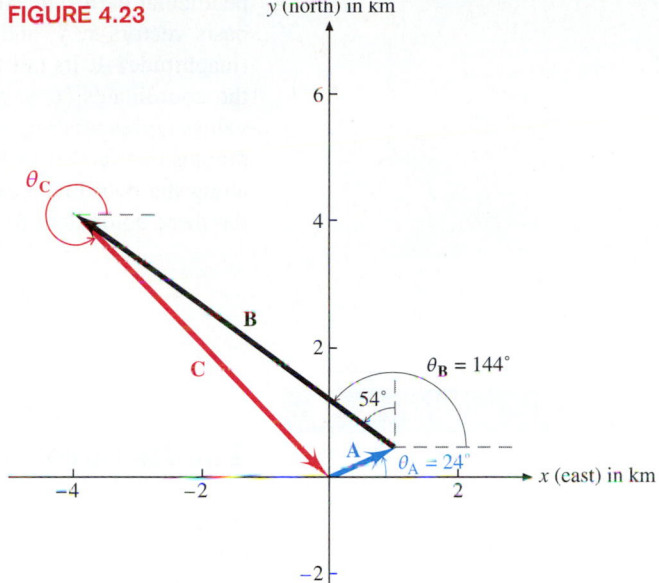

$$(C_x, C_y) = [-1.37\ \text{km} - (-5.02\ \text{km}), -0.61\ \text{km} - 3.64\ \text{km}]$$
$$= (3.65\ \text{km}, -4.25\ \text{km}).$$

The last step is to find the magnitude and direction of **C**. Using Equation 4.16, you obtain the magnitude

$$C = \sqrt{(3.65\ \text{km})^2 + (-4.25\ \text{km})^2} = 5.60\ \text{km}.$$

This is the distance the hiker must cover on his return leg. Equation 4.17b gives you the direction

$$\theta_C = \tan^{-1} \frac{-4.25\ \text{km}}{3.65\ \text{km}} = -49.3°.$$

As shown in Figure 4.23, the negative sign indicates an angle measured *clockwise* from the $\hat{\mathbf{x}}$ direction. That is, the direction the hiker must take to return to his starting point is 49.3° south of east. (As in Example 4.2, the inverse tangent also has the supplement of $-49.3°$ for a solution. Here, again, the figure makes clear which solution is the correct one.)

Vectors in Three Dimensions

It is always possible to depict any two free vectors in some plane. But we must often deal with more complex situations that require the description of one or more vectors in three-dimensional space. Figure 4.24 shows such a vector.

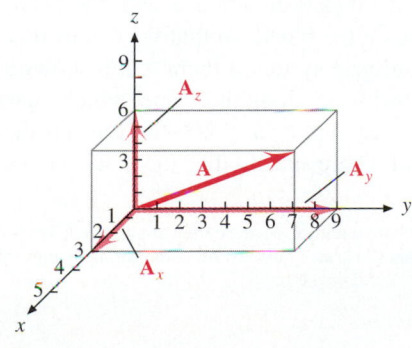

FIGURE 4.24 The vector (3, 9, 6) in a three-dimensional space defined by the coordinate system (x, y, z).

It is not difficult to extend the rules that we have developed for two-dimensional vectors into the three-dimensional space of the figure. We require three mutually perpendicular cartesian axes, labeled x, y, and z. Corresponding to these axes are the three basis vectors $\hat{\mathbf{x}}$, $\hat{\mathbf{y}}$, and $\hat{\mathbf{z}}$. The vector $\mathbf{A}$ is represented by the directed line of length (magnitude) A. Its tail lies at the origin $(0, 0, 0)$, and its head at the point specified by the coordinates (x, y, z). The components of the vector (now three in number) have values corresponding to these coordinates. (See Figure 4.16 for the corresponding two-dimensional case.) In Figure 4.24, $\mathbf{A}$ is represented as the sum of three vectors lying along the coordinate axes, $\mathbf{A}_x = x\hat{\mathbf{x}}$, $\mathbf{A}_y = y\hat{\mathbf{y}}$, and $\mathbf{A}_z = z\hat{\mathbf{z}}$. Thus we can write $\mathbf{A}$ in the three equivalent forms

$$\mathbf{A} = (A_x, A_y, A_z) \tag{4.25a}$$
$$= \mathbf{A}_x + \mathbf{A}_y + \mathbf{A}_z \tag{4.25b}$$
$$= A_x\hat{\mathbf{x}} + A_y\hat{\mathbf{y}} + A_z\hat{\mathbf{z}}. \tag{4.25c}$$

The Force and Acceleration Vectors. The Inclined Plane

We have developed the basic ideas of vector mathematics mainly in terms of the physical quantity displacement. But many other physical quantities have magnitude and direction and are therefore conveniently described in vector language. Among them are force and acceleration. We now study force and acceleration in two and three dimensions.

To specify a force, we must give its magnitude and direction. In two dimensions, a vector $\mathbf{F}$ (that is, an ordered pair of scalars) is required for the purpose. Acceleration $\mathbf{a}$ is likewise a vector quantity having magnitude and direction. Its direction $\hat{\mathbf{a}}$ is the same as the direction $\hat{\mathbf{F}}$ of the (net external) force that produces it; that is, $\hat{\mathbf{a}} = \hat{\mathbf{F}}$. The magnitude a is related to the magnitude F of the force by the now-familiar relation $F = ma$. Combining magnitudes and directions, we write **Newton's second law of motion in vector form**:

$$\mathbf{F} = m\mathbf{a}. \tag{4.26a}$$

Equation 4.26a can also be written in the alternative forms

$$F_x\hat{\mathbf{x}} + F_y\hat{\mathbf{y}} = m(a_x\hat{\mathbf{x}} + a_y\hat{\mathbf{y}}) \tag{4.26b}$$

and

$$(F_x, F_y) = m(a_x, a_y). \tag{4.26c}$$

A body is often subject to several forces acting on it simultaneously in various directions. (An example of such a body is a man walking three untrained puppies on leashes.) Before we can use one of Equations 4.26a, 4.26b, or 4.26c to find the acceleration of the body, we must have a way to evaluate the *net* force acting on it. Just as in one dimension, this requires two steps:

1. Identify all the forces acting *on* the body, taking care not to confuse them with the third-law reaction forces exerted *by* the body on its surroundings.

2. Find the sum—here, the *vector* sum—of the forces.

As you have already seen in one dimension, the first step is best carried out using the free-body diagram method. We now extend that method to two-dimensional situations.

Consider the system illustrated in Figure 4.25a. A block of mass m slides down a frictionless plane. The plane makes an angle α with the horizontal. We would like to find the net force $\mathbf{F}$ acting on the block, so that we can evaluate its acceleration $\mathbf{a}$.

First, we choose a coordinate system. Figure 4.25b shows a convenient choice. We take the positive x axis downward along the plane, which guarantees that the direction of the acceleration of the block will be $\hat{\mathbf{a}} = \hat{\mathbf{x}}$.* The y axis is then normal to the inclined plane. The term **"normal"** here means that the y axis is perpendicular to *any* line

---

*It would make no sense to compare $\mathbf{a}$ with $\mathbf{x}$, because they are different kinds of quantities. But the equation $\hat{\mathbf{a}} = \hat{\mathbf{x}}$ compares *directions*. This illustrates how useful dimensionless unit vectors are; see the discussion following Equation 4.7.

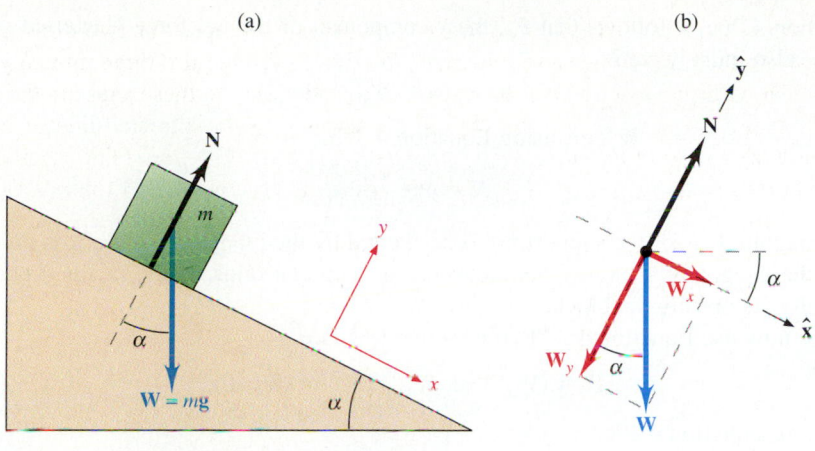

FIGURE 4.25 (*a*) A block slides down a frictionless inclined plane. (*b*) Free-body diagram corresponding to part *a*, showing the block as an isolated particle, together with vectors representing all the external forces acting on it. The component vectors $\mathbf{W}_x$ and $\mathbf{W}_y$ of $\mathbf{W}$ are shown in color. *Caution*: The two component vectors may be used to *replace* $\mathbf{W}$; they do not act *simultaneously* with $\mathbf{W}$.

drawn on the surface of the inclined plane, and not just to the *x* axis, which is the intersection of that plane with the plane of the page.

There are two forces acting on the block. The first is the gravitational force exerted by the earth. This force is represented by the vector $\mathbf{W}$, whose magnitude is the weight $W = mg$. We call the downward direction $\hat{\mathbf{g}}$, and write $\mathbf{W} = m\mathbf{g}$, in accordance with the vector form of Newton's second law. (Note that $\hat{\mathbf{g}} \neq \hat{\mathbf{y}}$.)

The second force acting on the block is the force exerted by the surface of the inclined plane. Because the plane is frictionless, this force *must* act in the direction $\hat{\mathbf{y}}$, normal to the plane. It is therefore called the **normal force** and assigned the symbol $\mathbf{N}$; in component form, $\mathbf{N} = (0, N)$.*

As always in the construction of free-body diagrams, the aim is to depict the body under study as a particle isolated from all other bodies, though subjected to all the forces that act on it. This is done in Figure 4.25b, which summarizes in graphic form the arguments made in the two preceding paragraphs.

The magnitude and direction of the net force $\mathbf{F}$ can now be calculated. Applying step 2 of the general procedure, we take the vector sum

$$\mathbf{F} = \mathbf{W} + \mathbf{N}. \tag{4.27}$$

In making vector calculations, we must at times pay strict attention to the signs of the angles and to the quadrants in which they lie. But the direction of the net force and the acceleration for a sliding block are self-evident. We may treat all angles as acute and positive and keep track of the directions separately from the magnitudes. Using this convention, we find the *x* and *y* components of $\mathbf{W}$. As Figure 4.25b shows, the angle between $\mathbf{W}$ and the *x* axis is $90° - \alpha$. Using $90° - \alpha$ for θ in Equations 4.19a and 4.19b, we obtain for the magnitudes of the components the values

$$|W_x| = W \cos(90° - \alpha) = W \sin \alpha = mg \sin \alpha \tag{4.28a}$$

and
$$|W_y| = W \sin(90° - \alpha) = W \cos \alpha = mg \cos \alpha. \tag{4.28b}$$

You can see from either part of Figure 4.25 that $\mathbf{W}$ lies in the fourth quadrant of the coordinate system; thus, its *x* component is positive and its *y* component negative. Using this information, we have

$$W_x = mg \sin \alpha \quad \text{and} \quad W_y = -mg \cos \alpha. \tag{4.29a, b}$$

The next task is to evaluate $\mathbf{N}$. We use the physical fact that the block always remains in contact with the plane, neither rising above it nor falling through it. That is, the *y* component of the block's acceleration, a_y, must be zero. From Newton's second law,

---

*Why must the force exerted by the surface on the block be normal to the plane? Suppose instead that the force were tilted slightly backward (toward the upper end of the plane) with respect to $\hat{\mathbf{y}}$. It would then have a component in the direction $-\hat{\mathbf{x}}$, upward along the plane. Such a force would tend to retard the acceleration of the block down the incline. The retarding force would constitute a frictional force, which contradicts the assumption that the plane is frictionless.

Equation 4.26c, it follows that F_y, the y component of the net force $\mathbf{F}$ exerted on the block, also must be zero:

$$F_y = N + W_y = 0.$$

This gives us $N = -W_y$, or, using Equation 4.29b,

$$N = mg \cos \alpha. \qquad (4.30)$$

The magnitude N of the supporting force exerted by the plane on the block is thus less than the weight $W = mg$ by the factor $\cos \alpha$. You can think of the inclined plane as partially supporting the block.

We now use Equation 4.27 to obtain the net force

$$\mathbf{F} = (W_x + 0, W_y + N) = (W_x, 0), \qquad (4.31)$$

or, using Equation 4.29a,

$$\mathbf{F} = (mg \sin \alpha, 0). \qquad (4.32a)$$

As expected, $\hat{\mathbf{F}} = \hat{\mathbf{x}}$; $\mathbf{F}$ is directed downward along the surface of the inclined plane. Thus, once the net force has been evaluated by two-dimensional methods, the rest of the problem reduces to one dimension. This becomes more evident if we write Equation 4.32a in the equivalent vector form

$$\mathbf{F} = mg \sin \alpha \, \hat{\mathbf{x}}. \qquad (4.32b)$$

Better yet, we express the same result in the scalar form

$$F = mg \sin \alpha, \qquad (4.32c)$$

where $\hat{\mathbf{F}} = \hat{\mathbf{x}}$ is understood. The last two forms of Equation 4.32 are often used for numerical calculation in problems of this type.

Now that we know the net force $\mathbf{F}$ acting on the block, we can use Equation 4.26a to evaluate the acceleration $\mathbf{a}$. We already know that the direction of $\mathbf{a}$ is downward along the plane in the positive x direction; $\hat{\mathbf{a}} = \hat{\mathbf{F}} = \hat{\mathbf{x}}$. Using Equation 4.32c, we find

$$a = \frac{F}{m} = \frac{mg \sin \alpha}{m},$$

or

$$a = g \sin \alpha. \qquad (4.33)$$

Compared with the free-fall value g, the acceleration is reduced by the factor $\sin \alpha$.

EXAMPLE 4.5

In Figure 4.26, the system of Figure 4.25a is shown again. Now, however, the mass of the block has the specific value 2 kg, and the inclined plane makes an angle of 27° with the horizontal. Assume that these values are precise. **(a)** Find the force exerted by the inclined plane on the block and the net force $\mathbf{F}$ acting on the block. **(b)** Find the acceleration $\mathbf{a}$ of the block.

SOLUTION:

(a) Find the force exerted by the inclined plane on the block and the net force $\mathbf{F}$ acting on the block.

The force exerted by the plane on the block is the normal force $\mathbf{N} = N\hat{\mathbf{y}}$, where N is given by Equation 4.30. Using $m = 2$ kg and $\alpha = 27°$, you have

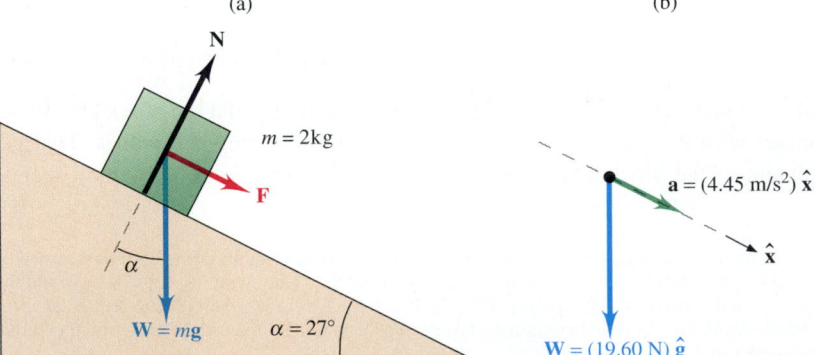

FIGURE 4.26 The system is that of Figure 4.25a, with the following given numerical values: $m = 2$ kg, $\alpha = 27°$.

$$N = 2 \text{ kg} \times 9.80 \text{ m/s}^2 \times \cos 27° = 17.5 \text{ N}.$$

(As you would expect, this is less than the weight $W = mg = 2 \text{ kg} \times 9.80 \text{ m/s}^2 = 19.6 \text{ N}$.) The net force acting on the block is $\mathbf{F} = F\hat{\mathbf{x}}$, where F is given by Equation 4.32c. Again, inserting the numerical values given, you obtain

$$F = 2 \text{ kg} \times 9.80 \text{ m/s}^2 \times \sin 27° = 8.90 \text{ N}.$$

One way to check these numerical results is to note that F and N are the magnitudes of the x and y components of the weight W. According to Equation 4.16, you must therefore have $W = \sqrt{F^2 + N^2}$. So you insert the numerical values just obtained and find that

$$\sqrt{(8.90 \text{ N})^2 + (17.5 \text{ N})^2} = 19.6 \text{ N},$$

which is indeed the value of W. The vector $\mathbf{W}$ is shown in Figure 4.25b.

(b) Find the acceleration $\mathbf{a}$ of the block.

Having determined F, you use Newton's second law in the form $a = F/m$ to evaluate a:

$$a = \frac{8.90 \text{ N}}{2 \text{ kg}} = 4.45 \text{ m/s}^2.$$

You combine this magnitude with the direction $\hat{\mathbf{a}} = \hat{\mathbf{x}}$ to obtain the complete answer $\mathbf{a} = a\hat{\mathbf{a}} = a\hat{\mathbf{x}} = 4.45 \text{ m/s}^2 \, \hat{\mathbf{x}}$.

In Example 4.6, the kinematic equations derived in Chapter 2 help us find the velocity and displacement of the block as functions of time.

EXAMPLE 4.6

The block of Example 4.5 is projected upward along the frictionless 27° inclined plane with initial speed 5.00 m/s. (a) How long does it take for the block to reach its highest point on the incline? (b) What is the position of this point, measured from the block's starting point? (c) Where is the block when its speed is 8.50 m/s?

SOLUTION: Continue to use the coordinate system of Example 4.5, with the positive x direction chosen downward along the slope. Because you are going to specify positions, you must now choose an explicit origin. The simplest choice is to fix $x = 0$ at the point from which the block is projected, at the initial time $t = 0$.

(a) At what value of t does the block reach its highest point on the incline?

As in Example 2.11, you can find the rise time by noting that the block attains its maximum height when its speed is $v = 0$. Beginning with Equation 2.36b, $v = v_i + at$, you can write

$$t = \frac{v - v_i}{a} = -\frac{v_i}{a},$$

From Example 4.5, you have $a = 4.45 \text{ m/s}^2$. The initial speed is given as $|v_i| = 5.00 \text{ m/s}$. Because the block is projected upward along the plane, in the negative direction, you have $v_i = -5.00 \text{ m/s}$. Inserting these values into the equation, you obtain

$$t = -\frac{-5.00 \text{ m/s}}{4.45 \text{ m/s}^2} = 1.12 \text{ s}.$$

(b) Where is the highest point?

One way to locate the highest point reached is to begin with Equation 2.36a, $x = x_i + v_i t + \frac{1}{2}at^2$. This equation relates x to t when the initial position and velocity, and the (constant) acceleration, are known. The initial position was chosen as $x_i = 0$. Because the time t found in part a is the time when the block reaches its highest point, you can solve immediately:

$$x = 0 - 5.00 \text{ m/s} \times 1.12 \text{ s} + \frac{1}{2} \times 4.45 \text{ m/s}^2 \times (1.12 \text{ s})^2$$
$$= -2.81 \text{ m}.$$

The block moves 2.81 m upward (and toward the left) along the plane before coming to rest at its highest point.

(c) Where is the block when its speed is 8.50 m/s?

The most direct way to find the position of the block, given its speed at that moment and its acceleration, is to use Equation 2.37, $v^2 - v_i^2 = 2a \Delta x$. Solving for Δx, you find

$$\Delta x = \frac{v^2 - v_i^2}{2a}.$$

Because $\Delta x \equiv x - x_i = x$ and $v_i = -5.00 \text{ m/s}$ in the present case, you can write

$$x = \frac{(8.50 \text{ m/s})^2 - (-5.00 \text{ m/s})^2}{2 \times 4.45 \text{ m/s}^2} = 5.31 \text{ m}.$$

That is, the block will be 5.31 m down the slope from its starting point when its speed is $|v| = 8.50 \text{ m/s}$. Its velocity is then $v = +8.50 \text{ m/s}$. Can you explain the meaning of the solution $v = -8.50 \text{ m/s}$, for which the speed is the same?

If the block in Example 4.6 is released from rest at the origin, its position depends on time in a particularly simple way. Specifically, Equation 2.36a, used in part b of the example, becomes

$$x = \frac{1}{2}at^2.$$

The dependence of position on time is quadratic, just as for freely falling bodies. However, the magnitude of the acceleration is less than the free-fall value g and can be changed by changing the incline angle α. This is the essential principle underlying the inclined-plane experiments of Galileo, discussed in Section 2.7.

Circumventing a Difficulty

Galileo used a bronze ball rolling on a polished hardwood track, not a sliding block. Galileo did not know how to analyze rolling motion quantitatively. But he did not need to, because the rolling-ball experiment still yields $x \propto t^2$ and thus substantiates his main point: A quadratic dependence of x on t implies motion with constant acceleration.

Why bother to study frictionless systems, when most systems encountered in everyday experiences are subject to considerable friction? Neglect of friction is a good example of the idealization with which physicists begin to study a class of systems. Often the idealization lays bare important principles that make the abstraction from reality well worthwhile. Neglect of friction, for example, leads to a key insight: The connection between force and motion lies in the proportionality of net force to acceleration. This insight then leads to Newton's laws of motion.

The conceptual framework of Newton's laws now turns out to be just what is needed for a fruitful study of the broad class of systems with significant friction—the systems in which the problem of connecting force and motion arose in the first place but could not be solved. Thus, the idealization proves valuable beyond the fairly small class of familiar systems in which friction is in fact negligible. But we will generalize cautiously, considering only a special kind of friction that is relatively easy to treat, beginning in Section 4.5.

SECTION 4.5 | Friction

Whenever an attempt is made to slide a body along a surface with which the body is in contact, a force is encountered that tends to prevent or retard the sliding. This familiar force is called **contact friction**. The amount of friction between two surfaces varies over a wide range and depends on many factors. The chemical composition of the surfaces, their cleanliness, and the presence or absence of a thin film of intervening fluid such as oil or water all affect the friction. The detailed treatment of contact friction is very complex; it is a specialty of mechanical engineering with strong components of physics, chemistry, and metallurgy. Here, we treat the simplest case in an approximate manner.

Figure 4.27a shows a block lying on a smooth (but not frictionless) level surface. As the corresponding free-body diagram shows, the only forces acting on the block are

FIGURE 4.27 Setting a block into motion against a frictional resistance. (*a*) The block at rest and its free-body diagram. (*b*) A force **P** applied parallel to the surface is opposed by a static frictional force $\mathbf{\Phi}_s = -\mathbf{P}$, and the block remains at rest. (*c*) With increasing **P**, the block just begins to move. The frictional force is now $\mathbf{\Phi}_k$; $\Phi_k < P$. It is also true for most pairs of surfaces in contact that $\Phi_k < \Phi_s$. (*d*) In order to keep the block moving at constant speed, P is reduced; now $P = \Phi_k$. (*e*) Schematic plot of the magnitude Φ of the frictional force **Φ** as the magnitude P of the applied force **P** is gradually increased. When P reaches the critical value $P^* = \Phi_{s\,max}$, the frictional force abruptly decreases to the value Φ_k, which remains constant as P is increased further.

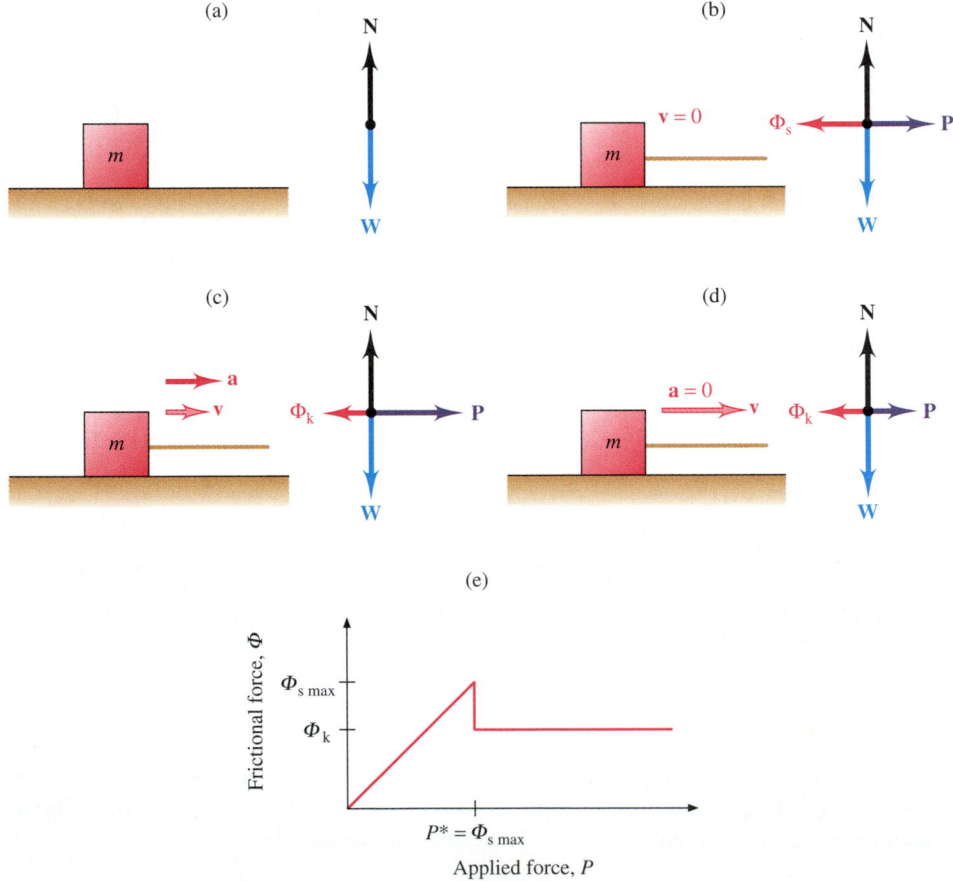

the gravitational attraction of the earth, **W**, and the opposing normal force **N** of equal magnitude, exerted by the table. We know that $\mathbf{N} = -\mathbf{W}$ because the block is motionless and the net force $\mathbf{F} = \mathbf{N} + \mathbf{W}$ on the block must therefore be zero. In Figure 4.27b, a propulsive force **P**, directed parallel to the surface, is applied to the block by means of the string. As **P** is increased gradually from an initial zero magnitude, the block at first does not move. Again, we know from the absence of motion that there must be an equal and opposite force exerted on the block, as shown in the corresponding free-body diagram. This force, Φ_s, is the **static frictional force**. When **P** reaches some critical value, the block "breaks loose" and begins to move, as shown in Figure 4.27c. At that instant, we say that the propulsive force has overcome the static frictional force. That is, the bond between the surfaces, whatever its nature may be, can no longer supply a force whose magnitude is as great as P.

If we wish to keep the block in motion at some small constant speed, we must continue to pull with a force equal and opposite to the frictional force, as shown in Figure 4.27d. For most (but not all) pairs of surfaces, however, the magnitude of the **kinetic frictional force**, Φ_k, is less than the maximum magnitude attained by Φ_s at the instant before motion begins. Recall your experience in pushing heavy objects; once you have got such an object into motion, you don't have to push so hard to keep it moving slowly, as Figure 4.27c suggests.

How big are these frictional forces? There is no general answer. But for reasonably smooth, unlubricated surfaces moving at fairly small relative speeds, the following *approximate* statements concerning the magnitudes are usually valid:

The inclined plane with a little friction.

1. The kinetic frictional force Φ_k is independent of the speed. (This is emphatically *not* true for other kinds of frictional forces, such as wind resistance.)

2. Both Φ_k and the maximum (breakaway) value of Φ_s are directly proportional to the magnitude N of the normal force exerted by each surface on the other. To express these two proportionalities mathematically, we define two **coefficients of friction** μ_k and μ_s and write

$$\Phi_k = \mu_k N \qquad \text{and} \qquad \Phi_{s\,max} = \mu_s N. \qquad \textbf{(4.34a,b)}$$

The quantity μ_k is called the **coefficient of kinetic friction**, and μ_s is called the **coefficient of static friction**. The lowercase Greek letter μ (mu) is used almost universally to represent the coefficients of friction; do not confuse μ with the letter u. Because μ is a constant of proportionality relating the magnitudes of two forces, it is a dimensionless quantity.

The coefficients of friction must be determined experimentally for every pair of surfaces, and the coefficients are very sensitive to how clean the surfaces are. Consequently, values of μ_k and μ_s should not usually be taken seriously to more than one or perhaps two significant figures. Some typical values are given in Table 4.1.

3. Other things being equal, the frictional forces are approximately independent of the total areas of the surfaces in contact, provided that the areas are not so small that one of the surfaces tends to dig into the other. Suppose, for example, we measure the force required to drag a sheet of brass along a steel surface. Then we fold the sheet in half and measure the force again. The two measurements will yield approximately the same result.

TABLE 4.1 Coefficients of Friction for Selected Pairs of Surfaces

Composition of Surfaces	Kinetic Coefficient μ_k	Static Coefficient μ_s
Ice on ice, 0°C	0.02	
Ski wax on ice, 0°C		0.04
Metal on Teflon (PTFE)	0.04	0.04
Steel on steel	0.5	0.8
Steel on aluminum	0.5	0.6
Aluminum on aluminum	1.4	1.9
Glass on glass	0.4	0.9
Oak on oak	0.4	0.6
Rubber on concrete	0.8	0.8–4

The use of the coefficients of friction to evaluate the frictional force acting on a body is illustrated in Examples 4.7 and 4.8.

EXAMPLE 4.7

A block of mass $m = 12$ kg lies on a table, just as in Figure 4.27a. The coefficients of friction between the bottom surface of the block and the top surface of the table are $\mu_s = 0.35$ and $\mu_k = 0.25$. **(a)** Find the magnitude P_1 of the horizontally directed propulsive force required to set the block into motion. **(b)** Find the magnitude P_2 of the propulsive force required to keep it in motion at a low constant speed.

SOLUTION: The magnitude of the normal force **N** exerted on the block by the table is simply the weight W of the block. Calculating to two significant figures, you find

$$W = mg = 12 \text{ kg} \times 9.8 \text{ m/s}^2 = 120 \text{ N}.$$

(a) What is the magnitude P_1 of the force required to set the block into motion?

To set the block into motion, you must just exceed the maximum static frictional force $\Phi_{s\,max}$. According to Equation 4.34b, this requires the application of a force of magnitude

$$P_1 = \mu_s W = 0.35 \times 120 \text{ N} = 42 \text{ N}.$$

(b) What is the magnitude P_2 of the force required to keep the block in motion once it is started?

Once the block is moving, the force required is given by Equation 4.34a:

$$P_2 = \mu_k W = 0.25 \times 120 \text{ N} = 30 \text{ N}.$$

In Example 4.8, we calculate the frictional force for a block on an inclined plane and consider its effect on the motion of the block.

EXAMPLE 4.8

In Example 4.5, we considered a block of mass $m = 2$ kg lying on a frictionless plane inclined at 27° to the horizontal. Figure 4.28a shows this system again, but suppose now that $\mu_s = 0.4$ and $\mu_k = 0.3$. **(a)** If the block is initially at rest, will it begin to move spontaneously? **(b)** Once the block has begun to move (either spontaneously or because it is given an initial push downward along the plane), what will its acceleration be?

SOLUTION:
(a) Will the block begin to move spontaneously?

Begin by drawing a free-body diagram of the system, like that of Figure 4.28b. As in Example 4.5, the magnitude of the normal force is less than the weight of the block. You must take this into account in using Equations 4.34a and 4.34b to calculate the kinetic and maximum static frictional forces. As in Example 4.5, you have $N = 17.5$ N. Inserting this value into Equation 4.34b gives you

$$\Phi_{s\,max} = \mu_s N = 0.4 \times 17.5 \text{ N} = 7 \text{ N}$$

to one significant figure. To see if the block will move spontaneously, you must compare this force with the available propulsive force P. In the present case, the propulsive force is the x component of the weight, $P = W_x$. As you saw in Example 4.5, its value is 8.90 N. There is thus a net force $F =$

$(8.90 − 7)$ N $= 2$ N available to set the block into motion, and it *will* move spontaneously.

(b) What is the acceleration **a** of the block, once it is in motion?

To find **a**, you need to know the net force **F** acting on the block. In Example 4.5, **F** was just $W_x\hat{x}$. Here, however, the frictional force Φ_k, whose direction is $-\hat{x}$, must also be taken into account. You use Equation 4.34a to find its magnitude:

$$\Phi_k = \mu_k N = 0.3 \times 17.5 \text{ N} = 5 \text{ N}$$

to one significant figure. So you have for the net force

$$\mathbf{F} = \mathbf{P} + \mathbf{\Phi}_k = W_x\hat{x} - \Phi_k\hat{x}$$
$$= (8.90 \text{ N} - 5 \text{ N})\hat{x} = 4 \text{ N } \hat{x}.$$

You can write this in the equivalent scalar form $F = 4$ N.
You now use Equation 4.33a to find

$$a = \frac{F}{m} = \frac{4 \text{ N}}{2 \text{ kg}} = 2 \text{ m/s}^2.$$

This is considerably less than the value $a = 4.45$ m/s² obtained in Example 4.5; friction has substantially diminished the net force F.

(a)

(b)

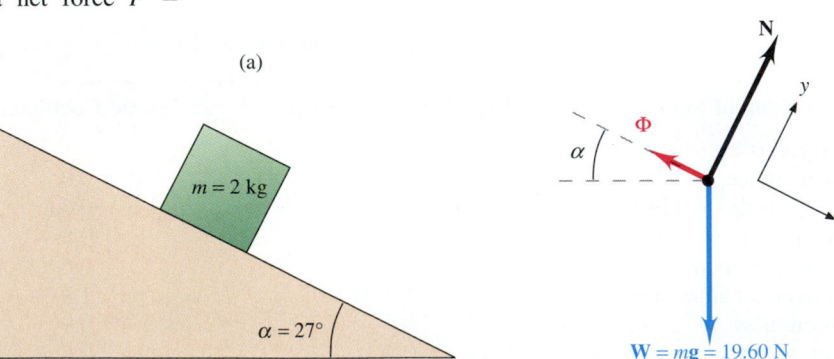

FIGURE 4.28 (a) The system.
(b) The corresponding free-body diagram.

The Critical Angles

As you know from experience, a block or similar nonrolling object will sit still on an inclined plane only up to a certain angle of incline. When that angle is reached, the block starts to slide; the propulsive force just exceeds the maximum static frictional force. Let us see how the two forces depend on the incline angle α.

THE PROPULSIVE FORCE   For a block on an inclined plane, the propulsive force is supplied by $\mathbf{W}_x$, the component vector of the gravitational force $\mathbf{W}$ along the surface of the plane. The magnitude of that vector is given by Equation 4.29a, $|\mathbf{W}_x| = mg \sin \alpha$.

THE MAXIMUM STATIC FRICTIONAL FORCE   This force is given in general by Equation 4.34b, $\Phi_{s\,max} = \mu_s N$. For a block on an inclined plane, the magnitude of the normal force N is equal to $|\mathbf{W}_y|$. Using Equation 4.34b, we thus have

$$\Phi_{s\,max} = \mu_s |\mathbf{W}_y| = \mu_s mg \cos \alpha.$$

As the incline angle α is increased, the magnitude of the propulsive force increases from its initial zero value as $\sin \alpha$, while $\Phi_{s\,max}$ decreases as $\cos \alpha$. The block will begin to move when these magnitudes are equal. This occurs at the particular angle α_s for which

$$mg \sin \alpha_s = \mu_s mg \cos \alpha_s. \tag{4.35}$$

Because $\sin \alpha / \cos \alpha = \tan \alpha$, this equation simplifies to

$$\boxed{\mu_s = \tan \alpha_s} \quad \text{or} \quad \boxed{\alpha_s = \tan^{-1} \mu_s.} \tag{4.36a,b}$$

The angle α_s is called the **critical angle for static friction**. You can readily determine the coefficient of static friction for two bodies in contact by laying the smaller one on top of the larger and tilting the larger until the smaller just begins to slide on it. Try it with a ruler or an eraser and a notebook.

One may define a **critical angle for kinetic friction**, though it is of less practical interest than the critical angle for static friction. The angle, α_k, is defined as the incline angle at which a block slides down a plane at constant speed. Then the kinetic frictional force given by Equation 4.34a, $\Phi_k = \mu_k N$, is equal and opposite to the propulsive force. Using an argument just like the one leading to Equation 4.36, we obtain the analogous expression

$$\mu_k = \tan \alpha_k \quad \text{or} \quad \alpha_k = \tan^{-1} \mu_k. \tag{4.37a,b}$$

Variants of the Atwood Machine

The system shown in Figure 4.29 is a hybrid. It is an Atwood machine in which one of the freely hanging masses has been replaced by the system we have just been studying—a block lying on an inclined plane.

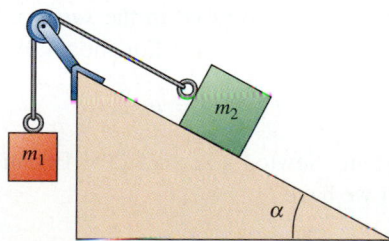

FIGURE 4.29 A variant of the Atwood machine, in which m_2 lies on a plane having incline angle α. We assume that the pulley and the string have negligible mass and that the pulley is frictionless. For the time being, we neglect the friction between m_2 and the inclined plane.

THINKING LIKE A PHYSICIST

When we study an unfamiliar system, it always pays to see if it bears any resemblance to a familiar system or systems. The system under study here combines two such systems— the Atwood machine and the block on an inclined plane. Recognizing this fact greatly reduces the amount of new analysis required.

We will use this hybrid system mainly as an exercise in extending the techniques of vector mechanics to increasingly complicated cases. But the system is a prototype for elements frequently found in real mechanical devices.

The Frictionless System

We begin, as always, with a free-body diagram of the system. It is best to consider the masses m_1 and m_2 as two separate subsystems, shown in the two parts of Figure 4.30. We are free to choose two separate systems of coordinates. For subsystem 1, we take the positive x_1 direction vertically upward. For subsystem 2, we take the positive x_2 direction downward and to the right, along the plane.

As the diagram shows, two forces act on m_1, just as in the simple Atwood machine discussed in Section 3.6. They are the gravitational force $\mathbf{W}_1 = m_1\mathbf{g}$ and the string force $\mathbf{T}_1$. The net force $\mathbf{F}_1$ acting on m_1 must be the vector sum of these forces, and Newton's second law can be written

$$m_1\mathbf{a}_1 = \mathbf{F}_1 = \mathbf{W}_1 + \mathbf{T}_1. \tag{4.38a}$$

Three forces act on m_2. Two of them we have already encountered in the simple case of the block on the frictionless inclined plane. The first is the gravitational force $\mathbf{W}_2 = m_2\mathbf{g}$ exerted by the earth. The second is the normal force $\mathbf{N}$ exerted by the surface of the plane. The third (new) force, called $\mathbf{T}_2$, is that exerted by the string. Because the string is parallel to the plane, the direction of $\mathbf{T}_2$ is upward along the plane (and thus in the negative x_2 direction) at an angle α with the horizontal. The net force $\mathbf{F}_2$ acting on m_2 must be the vector sum of $\mathbf{W}_2$, $\mathbf{N}$, and $\mathbf{T}_2$, and Newton's second law can be written

$$m_2\mathbf{a}_2 = \mathbf{F}_2 = \mathbf{W}_2 + \mathbf{N} + \mathbf{T}_2. \tag{4.38b}$$

We know in advance that the motion of each of the two subsystems will be one-dimensional. Our first task is to evaluate $\mathbf{F}_1$ and $\mathbf{F}_2$, using the algebra of vectors. Once that is done, we can manage the rest of the problem using scalar methods.

We evaluate $\mathbf{F}_1$ first. If you inspect Figure 4.30a, you will see that

$$\mathbf{W}_1 \equiv m_1 g\, \hat{\mathbf{g}} = -m_1 g\, \hat{\mathbf{x}}_1$$

and

$$\mathbf{T}_1 \equiv T\hat{\mathbf{T}}_1 = T\hat{\mathbf{x}}_1.$$

In writing these expressions, we take account of the directions of the vectors $\mathbf{W}_1$ and $\mathbf{T}_1$: $\hat{\mathbf{W}}_1 = \hat{\mathbf{g}} = -\hat{\mathbf{x}}_1$ and $\hat{\mathbf{T}} = \hat{\mathbf{x}}_1$. As is always true for the magnitudes of vectors, g and T are positive scalars. We write $|T_1|$ simply as T, and not as T_1, because it is the string tension—that is, the magnitude of the string force. As in the simple Atwood machine, the forces $\mathbf{T}_1$ and $\mathbf{T}_2$ shown in Figure 4.30b have the same magnitude, and a subscript is unnecessary.

Now that $\mathbf{W}_1$ and $\mathbf{T}$ have been expressed in the *same* one-dimensional coordinate system x_1, we can write the resultant force of Equation 4.38a in scalar form as

$$F_1 = T - m_1 g. \tag{4.39}$$

In a one-dimensional system, Newton's second law, Equation 4.38a, reduces to the scalar form $ma = F_1$, and we have

$$m_1 a = T - m_1 g. \tag{4.40}$$

As in the simple Atwood machine, the acceleration $\mathbf{a}_1$ and the acceleration $\mathbf{a}_2$ shown in Figure 4.30b have the same magnitude a, and we need not use a subscript.

Evaluating $\mathbf{F}_2$ is just a little more complicated. We know that the result will be of the form $\mathbf{F}_2 = F_2\hat{\mathbf{x}}_2$ (as in the simple inclined-plane case). Hence, we need evaluate only the x_2 components of the three vectors:

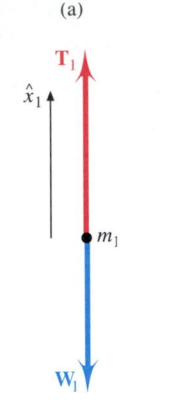

(a)

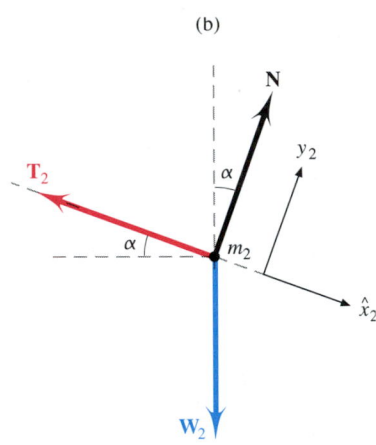

(b)

FIGURE 4.30 Free-body diagrams for the subsystems of Figure 4.29: (a) for the freely hanging mass m_1; (b) for the mass m_2 on the frictionless inclined plane. The force vectors are shown in color. The coordinate axes chosen for each subsystem are shown in black; the dashed gray lines in part *b* denote the horizontal and vertical.

$\mathbf{T}_2$: You can see from Figure 4.30b that

$$\mathbf{T}_2 \equiv T_2\hat{\mathbf{T}}_2 = -T\hat{\mathbf{x}}_2.$$

Here again, we write T for the string tension (the magnitude of the string force $\mathbf{T}_2$), just as we did for $\mathbf{T}_1$ earlier. We also take account of the direction $\hat{\mathbf{T}}_2 = -\hat{\mathbf{x}}_2$.

$\mathbf{N}$: Because $\mathbf{N}$ is directed along the y_2 axis, normal to the inclined plane, its x_2 component is zero, just as in the simple inclined-plane case.

$\mathbf{W}_2$: We evaluate W_{x_2}, the x_2 component of $\mathbf{W}_2$, just as in the simple inclined-plane case. In the present case, W_{x_2} is no longer the net propulsive force acting on m_2. But its magnitude and direction are still those given by the right side of Equation 4.32c:

$$W_{x_2} = m_2g \sin\alpha,$$

with α taken to be an acute, positive angle.

Putting the three results together, we write the net force of Equation 4.38b in scalar form as

$$F_2 = -T + 0 + m_2g \sin\alpha. \tag{4.41}$$

Again, we use Newton's second law in the one-dimensional form $ma_2 = F_2$. This gives us

$$m_2a = -T + m_2g \sin\alpha. \tag{4.42}$$

Equations 4.40 and 4.42 are a pair of simultaneous equations, which we write together for clarity:

$$m_1a = \quad T - m_1g$$

and
$$m_2a = -T + m_2g \sin\alpha.$$

Adding these equations yields

$$(m_1 + m_2)a = (m_2 \sin\alpha - m_1)g. \tag{4.43}$$

We can now solve for the magnitude of the acceleration:

$$a = \frac{m_2 \sin\alpha - m_1}{m_1 + m_2}g. \tag{4.44}$$

Let us compare this equation with the corresponding equation for the simple Atwood machine, Equation 3.20. Rewritten to keep notation consistent, that equation is

$$a = \frac{m_2 - m_1}{m_1 + m_2}g.$$

The difference between this equation and Equation 4.44 lies in the numerator of the fraction. Because in the present case m_2 is partially supported by the inclined plane, its contribution to the propulsive force is reduced by the factor $\sin\alpha$. However, m_2 continues to make a full contribution to the inertia of the system, as evidenced by the unchanged form of the denominator, $m_1 + m_2$.

EXAMPLE 4.9

The frictionless system of Figure 4.29 is exactly balanced. If m_1 = 8.2 kg and m_2 = 18.6 kg, what is the value of the angle α?

SOLUTION: If the system is balanced, the acceleration a must be zero. According to Equation 4.44, a equals zero only if the numerator of the fraction on the right side is zero. You have

$$m_2 \sin\alpha - m_1 = 0, \tag{4.45a}$$

or
$$\alpha = \sin^{-1}\frac{m_1}{m_2}. \tag{4.45b}$$

Insertion of the given numerical values into this equation gives you

$$\alpha = \sin^{-1}\frac{8.2 \text{ kg}}{18.6 \text{ kg}} = 26°.$$

a, b, etc.	scalars	$\mathbf{P}$	net propulsive force acting on a body
$\mathbf{A}, \mathbf{B}$, etc.	vectors	$\mathbf{T}, T$	string force, string tension
$|\mathbf{A}|, A$	magnitude of $\mathbf{A}$	$\mathbf{W}$	force on a body due to gravitational attraction of the earth
$|A|$	magnitude (absolute value) of A		
$\hat{\mathbf{A}}$	unit vector in the direction of $\mathbf{A}$	W_x, W_y	components of $\mathbf{W}$ along an inclined plane and normal to the plane
(A_x, A_y)	component form of $\mathbf{A}$		
$\mathbf{a}, a$	acceleration of a body, magnitude of the acceleration	α	incline angle of a plane
		α_s	critical angle for static friction
$\mathbf{F}$	net force acting on a body	θ	angle between the x axis and some vector, measured counterclockwise from $\hat{\mathbf{x}}$
$\mathbf{g}, g$	acceleration of gravity, magnitude of the acceleration of gravity		
		μ_k, μ_s	coefficients of kinetic friction, static friction
m	mass of a body	$\mathbf{\Phi}_k, \mathbf{\Phi}_s$	kinetic frictional force, static frictional force
N	normal force exerted on a block by the inclined plane on which it rests	Φ_k, Φ_s	magnitude of $\mathbf{\Phi}_k, \mathbf{\Phi}_s$
		$\Phi_{s\,max}$	maximum (breakaway) value of Φ_s

Summing Up

Displacement of a particle can be depicted by an arrow drawn from its initial to its final location. This arrow corresponds to the mathematical idea of the **vector**, an entity that possesses *magnitude* and *direction*. Vectors can be used to describe any physical quantity that has these two properties.

Many rules for mathematical manipulation of vectors are analogous to (but not identical with) rules for manipulation of scalars. Specifically, two successive displacements $\mathbf{B}$ and $\mathbf{C}$ can be represented by the **vector sum** of Equation 4.1,

$$\mathbf{A} = \mathbf{B} + \mathbf{C}.$$

Vector addition obeys the commutative law, Equation 4.2, and the associative law, Equation 4.4.

To represent the magnitude and the direction of a vector separately, we use Equation 4.7,

$$\mathbf{A} \equiv |\mathbf{A}|\hat{\mathbf{A}} \quad \text{or} \quad \mathbf{A} \equiv A\hat{\mathbf{A}},$$

where $|\mathbf{A}|$ or A is a positive scalar representing the magnitude of $\mathbf{A}$, including the appropriate units, and $\hat{\mathbf{A}}$ is the **unit vector**, having dimensionless magnitude 1, which represents the direction of $\mathbf{A}$.

Vector manipulations can be represented algebraically as well as geometrically. Any vector $\mathbf{A}$ can be represented as the sum $\mathbf{A} = \mathbf{A}_x + \mathbf{A}_y$, where the directions of the cartesian **component vectors** correspond to the **basis vectors** $\hat{\mathbf{x}}$ and $\hat{\mathbf{y}}$. We write the component vectors in the form of Equation 4.13, $\mathbf{A}_x = A_x\hat{\mathbf{x}}$ and $\mathbf{A}_y = A_y\hat{\mathbf{y}}$. The scalars A_x and A_y are the **components** of $\mathbf{A}$. A vector can be represented in the **component form** given by the **ordered pair** of Equation 4.15, $\mathbf{A} = (A_x, A_y)$.

The components can be determined algebraically from the magnitude and direction of $\mathbf{A}$. We represent the magnitude and direction of $\mathbf{A}$ in the **polar form** of Equation 4.18,

$\mathbf{A} = (A, \theta)$. Equations 4.19a and 4.19b give the components:

$$A_x = A \cos \theta \quad \text{and} \quad A_y = A \sin \theta.$$

We can proceed in the opposite direction as well by means of Equations 4.16 and 4.17:

$$A = \sqrt{A_x{}^2 + A_y{}^2} \quad \text{and} \quad \theta = \tan^{-1}\frac{A_y}{A_x}.$$

For every vector $\mathbf{A}$ there is a **negative vector** $-\mathbf{A}$ such that $\mathbf{A} + (-\mathbf{A}) = 0$. We define the **vector difference** represented by Equation 4.6,

$$\mathbf{B} - \mathbf{C} = \mathbf{B} + (-\mathbf{C}).$$

Multiplication of a vector $\mathbf{A}$ by a scalar c yields a vector having the direction $\hat{\mathbf{A}}$ and the magnitude $|c|A$. Multiplication is represented by the notation of Equation 4.8, $\mathbf{B} = c\mathbf{A}$. **Division** of a vector by a scalar is defined by Equation 4.9, $\mathbf{B} = \mathbf{A}/d \equiv c\mathbf{A}$, where $c = 1/d$.

When a block slides down an inclined plane, the motion is one-dimensional. Nevertheless, vector methods must be used to determine the magnitude of $\mathbf{F}$, because $\mathbf{F}$ is the sum of forces acting in nonparallel directions.

In the frictionless case, two such forces act. One is the gravitational force $\mathbf{W}$ acting vertically downward, and the other is the **normal force** $\mathbf{N} = (0, N)$ exerted by the surface of the plane. We choose a cartesian coordinate system with the x axis directed downward along the plane and the y axis upward along the normal to the plane. The components of $\mathbf{W}$ are given by Equations 4.29a and 4.29b,

$$W_x = mg \sin \alpha \quad \text{and} \quad W_y = -mg \cos \alpha.$$

Knowing that any motion must be along the plane, we can deduce Equation 4.30,

$$N = mg \cos \alpha.$$

The net force is then given by any form of Equation 4.32; because $\hat{\mathbf{F}} = \hat{\mathbf{x}}$, we can write the result in the scalar form of Equation 4.32c, $F = mg \sin \alpha$. The magnitude of the acceleration is given by Equation 4.33, $a = g \sin \alpha$.

When a propulsive force is exerted on a body that is in contact with some surface, a retarding force of **contact friction** is encountered. For a pair of reasonably smooth surfaces, this frictional force is often proportional to the normal force. The proportionality is expressed by Equations 4.34a and 4.34b, which give the frictional forces Φ in terms of the **coefficients of friction** μ:

$$\Phi_k = \mu_k N \quad \text{and} \quad \Phi_{s\,max} = \mu_s N.$$

The direction of the frictional force, whether static or kinetic, is always opposite that of the propulsive force. The magnitude is found by calculating N and using a tabulated value of μ_k or μ_s, whichever is appropriate.

QUERIES

4.1 *(2) All coming to naught.* Find all pairs of vectors in the figure below whose sum is the zero vector.

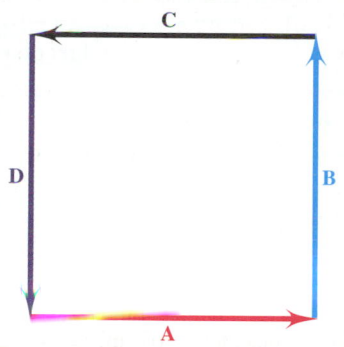

4.2 *(2) No difference at all?* Find all pairs of vectors in the figure of Query 4.1 whose difference is the zero vector.

4.3 *(2) The result(ant) is what counts, I.* In the figure of Query 4.1, find all the other pairs of vectors whose sum or difference is equal to $\mathbf{A} + \mathbf{B}$.

4.4 *(2) The result(ant) is what counts, II.* In the figure of Query 4.1, find all the other pairs of vectors whose sum or difference is equal to $\mathbf{A} - \mathbf{B}$.

4.5 *(3) Taking things apart.* A vector $\mathbf{A}$ has component vectors $\mathbf{A}_x$ and $\mathbf{A}_y$. What are the magnitude and direction of the component vector $\mathbf{A}_x = -3.5\hat{\mathbf{x}}$? What is the x component of $\mathbf{A}$?

4.6 *(3) Keeping things simple.* In cartesian coordinates, the x and y components of a vector may be either positive or negative. Explain why the r component of a vector expressed in polar form (r, θ) cannot be negative. (Hint: If r were allowed to be either positive or negative, what ambiguity would arise in the expression of a vector in polar coordinates? Why does this ambiguity not arise in cartesian coordinates?)

4.7 *(4) Vectors galore.* In this chapter, the properties of vectors have been derived mainly in terms of displacement. Why can force be treated equally well as a vector? Can you think of at least three other quantities that have vectorial properties?

4.8 *(4) Variety is the spice of life.* Explain the difference between a scalar quantity and a vector quantity. Give at least two examples of each type of quantity.

4.9 *(4) Inclined not to push too hard.* A block of mass m rests on an inclined plane. The plane is not frictionless. But, if left to itself, the block will slide down the plane. If you wish to prevent it from doing so with a minimal exertion of force, in what direction must you push? What force would you have to exert if you pulled vertically upward on the block?

4.10 *(5) ABSolutely!* Which coefficient of friction between rubber and concrete, μ_k or μ_s, is significant when an automobile is braked to a stop in a normal way? Which coefficient becomes significant in a skid? Explain completely why it is good driving practice to release the brake momentarily when a skid begins. Automatic antiskid mechanisms have long been in use on airplanes and have now become common on automobiles, where they are called antilock braking systems (ABS). An automatic device senses the beginning of a skid, releases the brake for a very short period of time, reapplies it until skidding begins again, and so on. Why is the shortest stop achieved when the tire almost, but not quite, skids on the pavement?

4.11 *(6) Signs of the times.* The conditions $T = T_1 = T_2$ and $a = a_1 = a_2$ for the string tensions and accelerations of the subsystems of the hybrid Atwood machine express the constraints on the system. In the analysis of the simple Atwood machine in Section 3.6, a similar role is played by the conditions $T_2 = T_1$, and $a_2 = -a_1$. How do you account for the sign differences?

PROBLEMS

GROUP A

4.1 *(2) Square deal, I.* The figure of Query 4.1 shows four vectors of equal magnitude 1 m. Redraw the figure to a convenient scale on graph paper, and find the magnitude and direction of each of the following vector sums: **(a) A + B, (b) A + C, (c) A + D, (d) B + C, (e) A + B + C + D**.

4.2 *(2) Square deal, II.* Redraw the figure of Query 4.1 to a convenient scale on graph paper, and find the magnitude and direction of the following vector differences: **(a) A − B, (b) A − C, (c) A − D, (d) B − C, (e) A + B − C − D**.

4.3 *(2) A special angle.* Three vectors are related by the equations **A = B + C** and $A = B = C$. Find **(a)** the angle between **A** and **B**, **(b)** the angle between **A** and **C**.

4.4 *(2) Eternal triangle.* The figure below depicts three vectors of equal magnitude, whose directions are 120° apart. **(a)** Show that, by taking the sums and differences of all possible pairs of these vectors and their negatives, you can produce resultant vectors having all directions at 30° intervals from **Â**. **(b)** Do all these resultant vectors have the same magnitude? If not, how many different magnitudes are there?

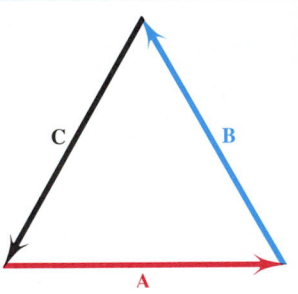

4.5 *(2) Geometric addition and subtraction* . . . Vector **A** is directed 30° north of east and has length 25 m. Vector **B** is directed 75° west of north and has length 75 m. Construct a diagram to a convenient scale, and find the magnitude and direction of **(a) A + B, (b) A − B**.

4.6 *(2) . . . and multiplication.* Given the two vectors described in Problem 4.5, construct a diagram and find the magnitude and direction of **(a) 2A + 3B, (b) 3A − 4B**.

4.7 *(3) Law & order.* On a sheet of graph paper, draw a convenient pair of coordinate axes (x, y). Draw a vector **A** from the origin to (3, 4). Then draw a vector **B** from the origin to (4, 3). **(a)** Is $A = B$? **(b)** Is **A = B**? Save your graph for Problem 4.8.

4.8 *(3) Magnitude and direction.* **(a)** Using algebraic methods, find the magnitude A and direction **Â** of the vector **A** = (3, 4). Express the direction as the angle θ from the x axis to the vector. **(b)** Find B and **B̂** for **B** = (4, 3). **(c)** If you have worked Problem 4.7, measure the magnitudes and directions directly from the graph that you drew, and compare the results with those obtained in working this problem.

4.9 *(3) Vector in a coordinate system.* The tail of a vector lies at (1.4 m, −6.3 m), and its head lies at (−10.9 m, 5.0 m). Find the components, magnitude, and direction of the vector.

4.10 *(3) Algebraic addition, I.* Find the magnitude and direction of the vector **C** = (7, 3) + (5, 8).

4.11 *(3) Algebraic subtraction.* **(a)** Find the magnitude and direction of the vector **D** = (7, 3) − (5, 8). **(b)** Find the magnitude and direction of the vector **E** = (5, 8) − (7, 3). **(c)** Can you find a simple relation between **D** and **E**?

4.12 *(3) Algebraic addition, II.* Find the magnitude and direction of the vector that is the sum of the following vectors, given in polar coordinates: (6 m, 25°), (10 m, 165°), (4 m, 320°), (8 m, 210°), (12 m, 70°).

4.13 *(3) Component vectors.* Given two component vectors $A_x = 27.7$ km **x̂** and $A_y = 13.9$ km **ŷ**, express the vector **A** in component form in cartesian coordinates and in polar coordinates.

4.14 *(3) Extending a multiplication proof.* Prove that Equation 4.23b, **B** = (cA_x, cA_y), holds for negative values of c.

4.15 *(3) Friendly meeting.* Alice and Bill happen to meet on a street corner. After a little chat, they go their separate ways. Alice walks eastward along a level street, while Bill goes northward up a steep hill. When Alice has reached the next corner along her street, 200 m away, Bill has also reached the next corner along his street, 150 m away as measured along the street and 30 m vertically upward. Find the straight-line distance between Alice and Bill at that moment.

4.16 *(4) Adding forces.* Two forces act on a body. They are directed perpendicular to one another, and one of them has magnitude 300 N. The resultant force is of magnitude 500 N.

(a) What is the magnitude of the other force? (b) What is the angle between the resultant and the 300-N force?

4.17 *(4) Change of pace.* A light airplane having a cruising speed of 200 km/h flies due north through still air. A strong wind begins to blow from the northeast with speed 100 km/h. The airplane continues to fly with its nose headed due north. What is the velocity of the airplane with respect to the ground?

4.18 *(4) Holding things up.* A block of mass 8.95 kg rests on a frictionless plane inclined at an angle of 42.0° from the horizontal. A string is attached to the block. You wish to keep the block from moving by pulling on the string in an upward direction along the plane. What force must you exert?

4.19 *(4) Slippery slope.* The photograph below is of an air track. The "glider" moves with negligible friction on a thin cushion of air, which is pumped through a series of tiny holes in the track. When the track is tilted, it becomes an essentially frictionless inclined plane. Find the incline angle for which the glider moves with acceleration (a) $0.01g$; (b) $0.1g$. (c) For each of these cases, find the time required for the glider to cover the entire 3-m length of the air track from a standing start.

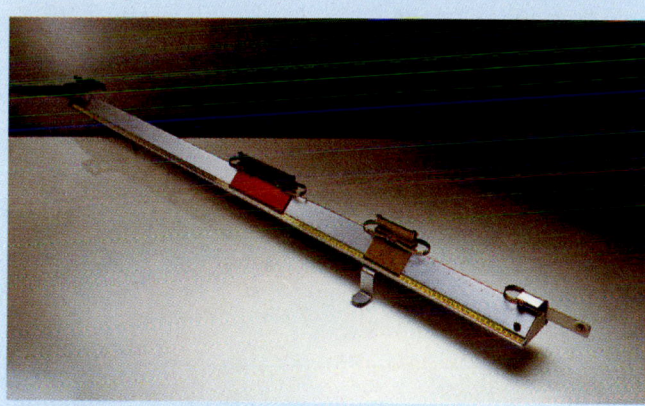

4.20 *(5) A rough measurement.* A block of mass 20 kg lies on a horizontal surface. You attach a rope to a ring on the block, and you tie the other end of the rope to a spring scale. The scale shows that it takes a horizontal pull of 90 N to set the block into motion and 65 N to keep it moving at constant speed. What are the coefficients of friction?

4.21 *(5) Moving day, I.* You want to push a 100-kg sofa across the floor. The coefficient of static friction is 0.4, and the coefficient of kinetic friction is 0.3. (a) How hard do you have to push (horizontally) to get the sofa moving? (b) How hard do you have to push to keep it moving?

4.22 *(5) Moving day, II.* In pushing a heavy box across the floor, you apply a horizontal force just sufficient to get the box moving. As the box starts to move, you continue to apply the same force. (a) Show that the acceleration of the box, once it gets started, is

$$a = (\mu_s - \mu_k)g.$$

(b) Let $\mu_s = 0.4$ and $\mu_k = 0.3$. Calculate the acceleration of the box. (c) Think of your experience in pushing heavy boxes. Is the result of part b consistent with your experience? If not, point out two assumptions made in this problem that are very likely unrealistic.

4.23 *(5) Bellywhopping.* Children who live in snowy but level regions like to "bellywhop" on their sleds. The child runs as fast as possible, carrying the sled, and then falls down on the sled and coasts as far as possible. If a child can run 3 m/s, how far will the sled go? Assume that the acceleration of the sled is constant, and assume a coefficient of friction $\mu_k = 0.04$ for steel sled runners on well-compacted, cold snow.

4.24 *(5) Better bearings.* Two flywheels are identical in every way, except that one has a bearing in which the unlubricated steel hub turns on a steel shaft, whereas the other has a bearing in which both hub and shaft are Teflon lined. If both wheels are started spinning in exactly the same way, how much longer will the one with the Teflon-lined bearing continue to turn?

4.25 *(5) Engine vs. brakes.* In most automobiles, driving force is applied to two wheels only. Assume to first approximation that these two wheels support half the weight of the car, and let the coefficient of static friction between the tires and the road be μ_s. (a) What is a_{max}, the maximum acceleration possible without skidding? (b) Using the approximate value $\mu_s = 1$ given in Table 4.1, calculate a_{max}. (Assume that the engine is powerful enough so that it is not the limiting factor—an assumption that is correct only for "muscle cars.") (c) Automobiles have brakes on all four wheels. If the automobile can reach speed v in a distance d, what is the minimum stopping distance d' from that speed?

4.26 *(5) Panic stop.* For every automobile sold in the United States, the manufacturer is required by law to specify the minimum distance in which the automobile can be stopped safely from 100 km/h on a dry road surface. This distance is typically 60 m. Refer to Table 4.1, and decide whether the limit is imposed mainly by the friction between the tires and the road or by the performance of the brake system.

4.27 *(5) Rough passage.* A block lies on a plane of incline angle 35°. The coefficient of kinetic friction between the block and the plane is 0.3. What is the acceleration of the block? (Assume that $\alpha_s < 35°$.)

4.28 *(5) Not too rough.* A block slides spontaneously down a plane of incline angle 35° when it is released from rest, with acceleration $a = 1.2$ m/s². (a) Set an upper limit on the value of the coefficient of static friction. (b) Find the coefficient of kinetic friction.

4.29 *(5) Holding things down.* A block on a plane whose incline angle θ is steeper than the critical angle $\alpha = \tan^{-1} \mu_s$ can be held at rest by pressing on it in a direction normal to the plane. Show that the force necessary is given by

$$mg \left[\frac{\sin \theta}{\mu_s} - \cos \theta \right].$$

4.30 *(6) Avoiding damage claims.* A delivery truck has nearly completed its rounds, and there is just one package in the cargo compartment. The truck is moving at 100 km/h. If the coefficient of static friction between the package and the floor is 0.4, what is the minimum distance in which the truck can come to a stop without making the package slide?

4.31 *(6) Extra burden.* In the figure on the next page, the coefficient of static friction between m_2 and the table is μ_s. When it is placed on top of m_2, the mass m_3 is just sufficient

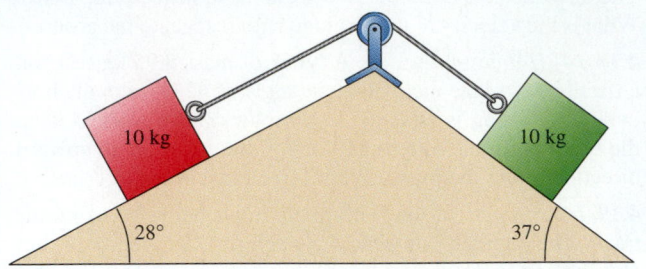

4.33 *(6) A slant and a twist.* The masses in the system of Problem 4.32 are interchanged. What is the acceleration now?

4.34 *(6) There's more than one angle, I.* The system diagrammed below has negligible friction. Find the acceleration of the system.

to keep the system from moving. **(a)** Express μ_s in terms of the masses m_1, m_2, and m_3. **(b)** Find μ_s if $m_1 = 7.0$ kg, $m_2 = 8.2$ kg, and $m_3 = 6.4$ kg.

4.32 *(6) A new slant on Atwood.* When the system diagrammed below is released, what is the acceleration? Neglect friction.

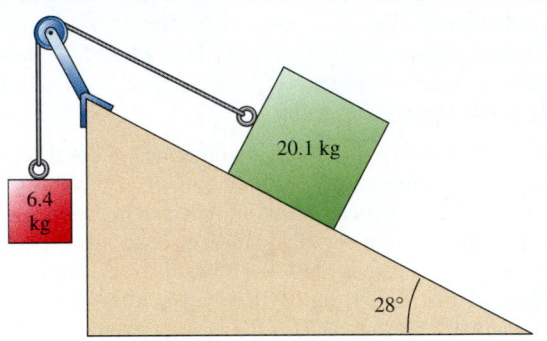

4.35 *(G) We deliver!* Beginning at the depot, a parcel-service truck driver travels 30 km north and delivers a package. Then she drives 25 km northeast and delivers another package. Then she drives 18 km southeast and delivers a third package. **(a)** When the driver has delivered her third package, what is her straight-line distance from the depot? **(b)** Find the direction from the driver's location to the depot. Express your result with respect to a reference frame in which the eastward direction is 0° and the northward direction is 90°. (This is the normal "mathematician's choice.") **(c)** Reexpress the result of part **b** with respect to a reference frame in which the northward direction is 0° and the eastward direction is 90°. (This is the conventional "navigator's choice.")

GROUP B

4.36 *(G) Treasure hunt.* You are a competitor in a treasure hunt game. The clues bring you and your team to a large tree. There you locate the final clue, which says, "The treasure is in a cave 2200 paces to the east and 2900 paces to the south of the tree." A rival team has already found the last clue, but they are still in sight; you see them quickly pacing out the distance to the east. In order to beat them, you must find the correct straight-line path from the tree to the treasure. **(a)** Find the required distance and direction. **(b)** Assuming that your team and your competitors walk at a fairly brisk 2 paces per second, is it likely that you can beat them to the treasure?

4.37 *(3) A little navigation.* Toronto is 320 km from Detroit, in a direction 26.3° north of east. Washington is 625 km from Detroit, in a direction 39.3° south of east. Neglecting the curvature of the earth, find the distance and direction from Washington to Toronto.

4.38 *(3) Find the missing component.* A vector has magnitude 36.5 m. Its x component is 25.4 m. **(a)** Find the y component of the vector. **(b)** Find the direction of the vector.

4.39 *(3) Grades as percentages.* The grade, or steepness, of a hill is often given as a percentage. To say that a grade is γ percent means that $\Delta y / \Delta l = \gamma\%$, where Δy is the vertical

ascent and Δl the distance along the hill surface. For grades of 10%, 20%, 30%, and 40%, find **(a)** the angle the slope makes with the vertical, **(b)** the horizontal distance Δx traveled during a vertical ascent of 100 m.

4.40 *(4) . . . climb nearly to the stars.* **(a)** Show that the force required to pull a cable car of mass m up a hill of grade γ percent (see Problem 4.39), neglecting friction, is $F = \frac{1}{100}\gamma\, mg$, where g is the acceleration of gravity. **(b)** The mass of a San Francisco cable car is 9.0 tonnes (9.0×10^3 kg). When it is jammed with tourists (nearly always the case), its mass is increased by about 1.5 tonnes. Neglecting friction, find the force that must be supplied by the cable to pull the car up the steepest grade on the line, which is about 20%.

4.41 *(4) Push comes to shove.* Two mechanics push a disabled automobile into their garage for repair. One pushes against the left front doorframe while steering the auto straight ahead. He exerts a force of 320 N at an angle of 30° with the forward direction. The other mechanic pushes on the right rear fender, exerting a force of 435 N at an angle of 10° with the forward direction. What is the effective forward force on the car? Do the sidewise components of the two mechanics' pushes have to cancel one another? Why or why not?

4.42 *(4) Aerial tramway.* The device shown below is used to transport materials over a canyon. A cable is stretched between two pylons, and the load is carried by a basket that hangs from a pulley riding on the cable. The basket is pulled across the canyon by means of the light lines shown. When the basket is just passing the center of the canyon, carrying a 510 kg load, the cable assumes approximately the form of two straight lines that make an angle of 160° with one another. Find the tension in the cable.

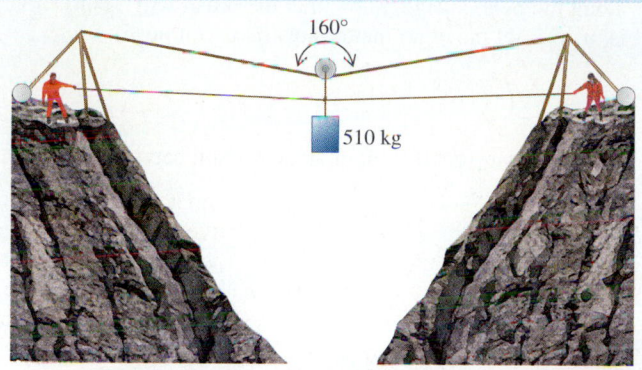

160°

510 kg

4.43 *(4) Three-sided tug of war.* Three ropes are tied together at a single point. The ropes are under tensions of 100 N, 200 N, and 250 N. If the system does not move, what are the angles between the ropes?

4.44 *(5) Double your pleasure.* In the illustration below, block 1 has mass m_1. The coefficient of static friction between block 1 and the inclined plane is μ_{1s}. Block 2 has mass m_2, and the coefficient of static friction between block 2 and the inclined plane is $\mu_{2s} < \mu_{1s}$. **(a)** Beginning at $\alpha = 0$, the incline angle is gradually increased. Show that the critical angle α_s at which the blocks will start to slide down the plane is

$$\alpha_s = \tan^{-1} \frac{\mu_{1s}m_1 + \mu_{2s}m_2}{m_1 + m_2}.$$

(b) What would happen if μ_{2s} were greater than μ_{1s}?

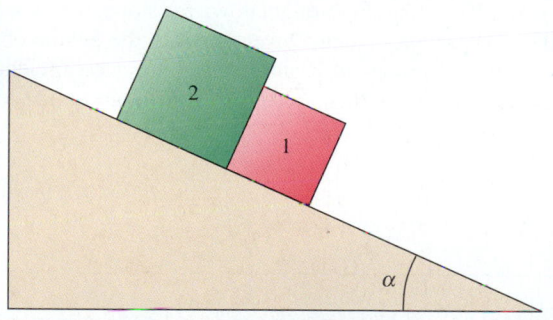

2

1

α

4.45 *(5) A new angle.* If left to itself, the block in the drawing in the next column slides down the ramp in spite of frictional resistance. **(a)** What is the maximum possible value of the static coefficient of friction μ_s? **(b)** The man is wearing boots that enable him to stand and walk on the ramp without slipping. Find the magnitude T_{1s} of the force that he must exert on the rope, at the angle β shown, to prevent the block from moving. **(c)** Find the magnitude T_{2s} of the force that he must

β

m

α

exert to start the block moving up the ramp. **(d)** Once he has started the block moving, find the magnitude T_k of the force that he must exert to keep it moving at constant speed.

4.46 *(5) Lumberyard, I.* You have a stack of identical oak planks. If you wish to pull the top n planks off the pile, how hard do you have to pull?

4.47 *(5) Lumberyard, II.* Again, you have a stack of identical oak planks. This time you wish to pull just the nth plank from the top out of the pile. How hard do you have to pull? (Assume that you have an assistant to hold the other planks in place.)

4.48 *(6) There's more than one angle, II.* Let's generalize Problem 4.34. Let the two masses be m_1 and m_2 (not necessarily the same), and let the two angles be α_1 and α_2. Find the acceleration a of the system diagrammed in Problem 4.34 in terms of m_1, m_2, α_1, and α_2. Neglect friction.

4.49 *(6) Flatwood machine.* In the drawing below, all three bodies have the same mass m. The coefficient of kinetic friction between the table and each of the two bodies on it is μ_k. **(a)** Show that the acceleration of the system is

$$a = \frac{1 - 2\mu_k}{3} g.$$

(b) Find the tension T in the string that joins the two bodies on the table.

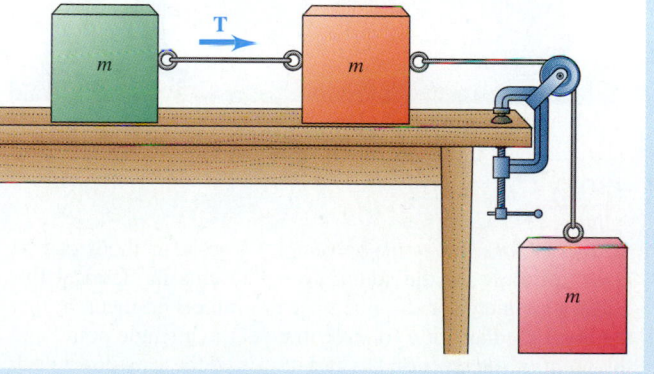

m T m

m

4.50 *(G) In the groove, I.* A block of wood slides down the right-angle channel shown on the next page. The channel is

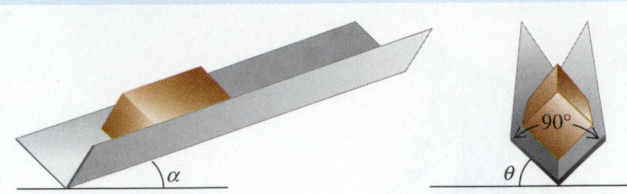

inclined at an angle α with respect to the horizontal. The angle θ is 45°; that is, the channel, seen end-on, is oriented symmetrically with respect to the vertical. If the coefficient of kinetic friction between the block and the channel is μ_k, find the acceleration of the block.

4.51 *(G) In the groove, II.* (Note: To work this problem, you should be familiar with the technique for finding extremal values of a function by setting its derivative to zero.) In the system illustrated in Problem 4.50, the angle θ has an arbitrary value. **(a)** Express the acceleration of the block in terms of the angles α and θ. **(b)** For what value of θ is the frictional force greatest and the acceleration smallest?

4.52 *(G) Ups and downs.* A body is at rest at the bottom of a plane of incline angle α, which is greater than the critical angle α_s. You give the body a shove, launching it up the plane. It reaches its maximum height in time t_1. It then slides back down; the time of descent is $t_2 = ct_1$, where c is a constant greater than 1. Show that the coefficient of kinetic friction is

$$\mu_k = \frac{c^2 - 1}{c^2 + 1} \tan \alpha.$$

4.53 *(G) Think before you pull, I.* (Note: To work this problem, you should be familiar with the technique for finding extremal values of a function by setting its derivative to zero.) The block in the illustration below rests on a level surface.

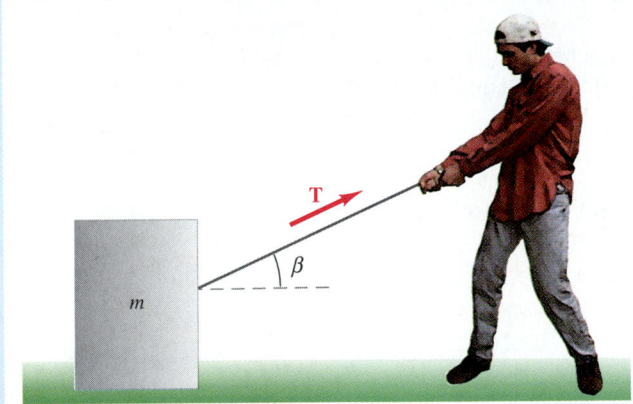

The static coefficient of friction is μ_s. The man shown in the figure wishes to start the block moving by pulling on the rope. **(a)** Find the value of the angle β that minimizes the force T he must apply. **(b)** Find the value of T in terms of μ_s, the mass m of the block, and g.

4.54 *(G) Dirty work.* It's chore time, and you are scrubbing a high wall with a brush mounted on a long pole, as shown in the illustration below. As you move the brush up and down, both the angle α and the push Π (capital pi) that you exert along the pole change. Show that the brush will "jam" (that is, it will not move no matter how hard you push) if

$$\alpha > \tan^{-1} \frac{1}{\mu_s},$$

where μ_s is the coefficient of static friction between the brush and the wall.

4.55 *(G) Shuffleboard.* A flat disk is given an initial push and slides a distance x along a level surface before coming to a stop. If the initial speed of the disk is v, what is the coefficient of friction μ_k?

4.56 *(G) Learning to ski.* A novice skier takes a short rope tow to the top of a beginners' slope, a distance along the surface of 50 m. She skis straight down the slope, whose angle is 10°, and continues along a level stretch at the bottom of the slope until she comes to a stop. Assuming a coefficient of kinetic friction $\mu_k = 0.03$, how far does she go on the level stretch?

GROUP C

4.57 *(2) Plane geometry by vectors.* Vector methods can be powerful tools for the solution of theorems in classical Euclidean geometry. You remember the important theorem that the three medians of a triangle intersect in a single point, and the point of intersection lies two-thirds of the way down each median from its vertex to the midpoint of the opposite side of the triangle. In the figure shown on the next page, *ABC* is an arbitrary triangle. $\overline{BM}$ is one of its medians, which terminates at M, the midpoint of the side $\overline{AC}$. H is the point such that $\overline{BH} = 2\overline{HM}$. O is an arbitrary point, joined to the vertices of the triangle by the vectors **a**, **b**, and **c**, and to H by the vector **e**. Show that $\mathbf{e} = \frac{1}{3}(\mathbf{a} + \mathbf{b} + \mathbf{c})$ and that the theorem is thus proved.

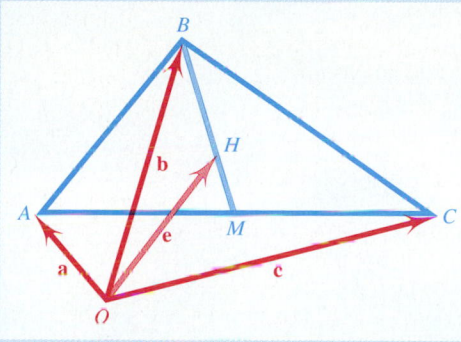

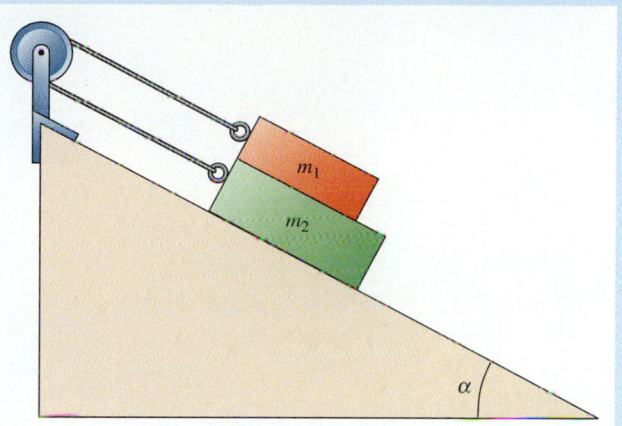

4.58 *(2) Solid geometry by vectors.* The methods of Problem 4.57 can be extended to three dimensions. In the figure below, *ABCD* is an arbitrary tetrahedron. Its faces are triangles, and *M* is the intersection point of the medians of face *ABC*. *H* is the point such that $\overline{DH} = 3\overline{HM}$. *O* is an arbitrary point, joined to the vertices of the tetrahedron by the vectors **a**, **b**, **c**, and **d**, and to *H* by **e**. Prove that $\mathbf{e} = \frac{1}{4}(\mathbf{a} + \mathbf{b} + \mathbf{c} + \mathbf{d})$ and that therefore the four lines drawn from each vertex of a tetrahedron to the intersection of the medians of the opposite side all meet in a common point, located three-quarters of the way down each line from its vertex.

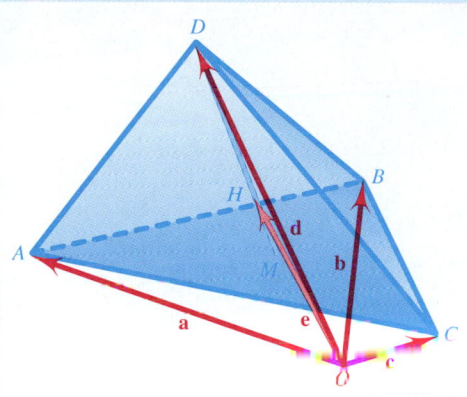

4.59 *(5) Making spaghetti.* You pour a quantity of flour of volume *V* onto a board, where it forms a conical pile. The angle that the sides of the pile make with the horizontal is called the **angle of repose**. **(a)** Express the height *h* of the pile in terms of *V* and μ_s, the coefficient of static friction of flour on flour. **(b)** The height of a pile containing a cup of flour (about 225 cm³) is 5.5 cm. Estimate μ_s.

4.60 *(G) Being very critical.* What is the critical angle for the system shown in the next column—that is, the angle α_c at which the system begins to move? Assume that $m_2 > m_1$ and that the coefficients of friction for the two pairs of surfaces in contact are the same.

4.61 *(G) Think before you pull, II.* (Note: To work this problem, you should be familiar with the technique for finding extremal values of a function by setting its derivative to zero.) The block in the figure below remains at rest if undisturbed. The man shown in the figure wishes to start the block moving down the ramp by pulling on the rope. **(a)** Given the ramp angle α and the static coefficient of friction μ_s, find the value of the angle β that minimizes the force *T* that he must apply. **(b)** Find the value of *T* in terms of α, μ_s, *m*, and *g*. **(c)** Compare your results with those for the simpler case of Problem 4.53.

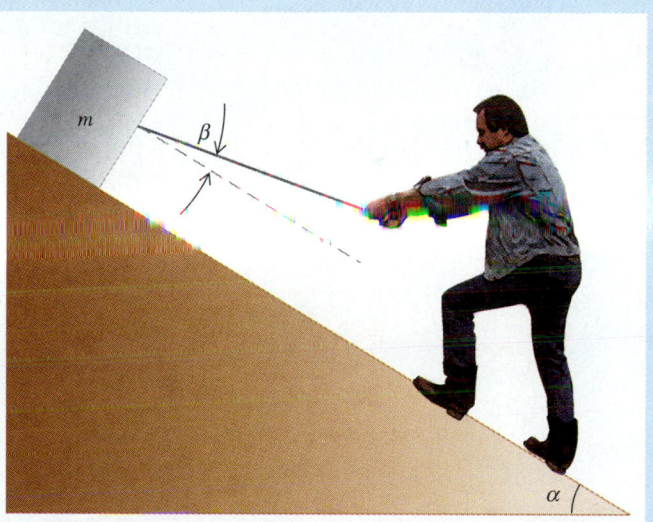

Curvilinear Motion

5

────── We apply Newton's laws to bodies whose motion does not lie along a straight line.

────── Velocity and acceleration are vectors that are derivatives of other vectors.

────── We can describe the path of a projectile by separating the motion into horizontal and vertical components, each of which we already know how to analyze.

────── In uniform circular motion, a body experiences acceleration without change of speed.

────── Earth satellites, conical pendulums, and the banking of curves are important examples of curvilinear motion.

Left: Why does the stream of water have the shape it has?

SECTION 5.1 Introduction

When the direction of the net force acting on a body is not the same as the direction in which the body is already moving, the body moves not in a straight line but along some sort of curved path. Such motion is called **curvilinear motion**. We develop the vector mathematics needed to deal with curvilinear motion in Section 5.2. In this chapter, we focus mainly on two elementary cases of curvilinear motion: *projectile motion* and *uniform circular motion*.

A projectile is a moving body such as a thrown rock, a batted baseball, or a fired bullet. A projectile is set into motion in an arbitrary initial direction—horizontal, vertical, or somewhere in between—and subsequently "coasts" under the action of gravity. In Section 5.3, our aim is to describe the path followed by the projectile.

In uniform circular motion, a particle travels in a circular path at constant speed. In the well-known biblical account in which David uses a leather sling to whirl a stone around his head before releasing it in the direction of Goliath's head, the motion approximates uniform circular motion. So do the motions of an automobile rounding a curve, the earth in its orbit around the sun, and a nuclear particle whirling around the "racetrack" of a high-energy accelerator. In all these cases, a force exerted on the particle keeps it moving in a circle instead of along the inertial straight line that it would follow if left to itself. We analyze uniform circular motion in Section 5.4 and consider some important applications in Section 5.5.

SECTION 5.2 Velocity and Acceleration as Vector Derivatives

Velocity and acceleration are defined as time derivatives of displacement, which is a vector quantity. What does it mean to take the time derivative of a vector quantity? Consider a particle moving along the arbitrary two- or three-dimensional path shown in Figure 5.1a.* At some instant, the particle is located at P. During a subsequent small time interval Δt, the particle experiences a small displacement $\Delta \mathbf{s}$. We define the **mean velocity** $\langle \mathbf{v} \rangle$ to be

$$\langle \mathbf{v} \rangle \equiv \frac{\Delta \mathbf{s}}{\Delta t}. \qquad (5.1)$$

---

*For simplicity in drawing, the path in Figure 5.1 is shown as two-dimensional, lying in a plane. But the entire argument that follows applies equally well to a three-dimensional path that does not lie in a plane.

This equation is analogous to the definition of velocity in one dimension, $\langle v \rangle \equiv \Delta x / \Delta t$. What is new here is that $\Delta \mathbf{s}$ can represent a displacement in any direction at all, and must therefore be expressed as a vector. Consequently, $\langle \mathbf{v} \rangle$ also must be a vector because it is defined in Equation 5.1 as a vector divided by the scalar Δt. (Division of a vector by a scalar is defined in Equation 4.9.)

We now define the **instantaneous velocity v**. In analogy to the one-dimensional case, we take the limit of the right side of Equation 5.1 as the time interval approaches zero:

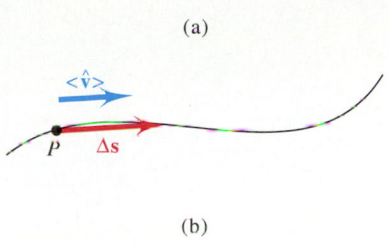

(a)

$$\mathbf{v} \equiv \frac{d\mathbf{s}}{dt} \equiv \lim_{\Delta t \to 0} \frac{\Delta \mathbf{s}}{\Delta t}. \tag{5.2}$$

In the limiting process, the chord $\Delta \mathbf{s}$ of Figure 5.1a becomes the tangent $d\mathbf{s}$ to the particle path at P, as shown in Figure 5.1b. We can write $d\mathbf{s}$ in terms of its magnitude and direction:

$$d\mathbf{s} = ds\,\hat{\mathbf{s}}. \tag{5.3}$$

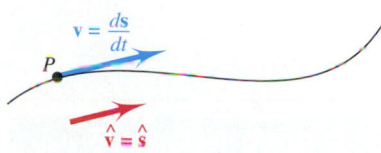

(b)

(Any vector can be written as the product of a magnitude and a unit vector representing direction. The infinitesimal vector $d\mathbf{s}$ is the product of the infinitesimal magnitude ds and the unit vector $\hat{\mathbf{s}}$. A unit vector has magnitude 1 and is thus not infinitesimal.)

We write **v** also in terms of its magnitude and direction:

$$\mathbf{v} = v\hat{\mathbf{v}}. \tag{5.4}$$

The direction of the velocity is the direction of the infinitesimal displacement:

$$\hat{\mathbf{v}} = \hat{\mathbf{s}}. \tag{5.5}$$

For the magnitude of the velocity we have

$$v = \frac{ds}{dt}, \tag{5.6}$$

just as in the one-dimensional case. And just as in one dimension, v is the **speed**.*

FIGURE 5.1 (*a*) In a time Δt, a particle experiences a displacement $\Delta \mathbf{s}$, beginning at P; $\Delta \mathbf{s}$ is a chord of the curved path of the particle. The mean vector velocity $\langle \mathbf{v} \rangle$ is defined by Equation 5.1. (*b*) As $\Delta t \to 0$, the chord becomes the tangent to the curve at P. The instantaneous vector velocity is defined by Equation 5.2. The direction $\hat{\mathbf{v}}$ is the direction $\hat{\mathbf{s}}$ of the infinitesimal displacement $d\mathbf{s}$ experienced by the particle during the infinitesimal time interval dt.

Vector Acceleration

We next define the vector acceleration. In Figure 5.2, the vector $\mathbf{v}_i$ represents the velocity of a particle at some initial time t_i. At a slightly later time t_f, the velocity is $\mathbf{v}_f$. Over the time interval Δt, the change in velocity is $\Delta \mathbf{v}$. In general $\mathbf{v}_i$, $\Delta \mathbf{v}$, and $\mathbf{v}_f$ all have different directions, as Figure 5.2 shows. We define the **mean acceleration** $\langle \mathbf{a} \rangle$ over the time interval Δt to be

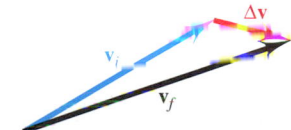

$$\langle \mathbf{a} \rangle \equiv \frac{\Delta \mathbf{v}}{\Delta t}, \tag{5.7}$$

in analogy to the one-dimensional definition $\langle a \rangle \equiv \Delta v / \Delta t$.

We now define the **instantaneous acceleration a**. Taking the limit of the right side of Equation 5.7 as Δt approaches zero, we have

$$\mathbf{a} \equiv \frac{d\mathbf{v}}{dt} \equiv \lim_{\Delta t \to 0} \frac{\Delta \mathbf{v}}{\Delta t}. \tag{5.8}$$

The vector **a** can be expressed in terms of its magnitude and direction in the form

$$\mathbf{a} = a\hat{\mathbf{a}}. \tag{5.9}$$

In general, $\hat{\mathbf{a}}$ is not the same direction as either $\hat{\mathbf{v}}_i$ or $\hat{\mathbf{v}}_f$. Rather, $\hat{\mathbf{a}}$ is the direction of the infinitesimal change $d\mathbf{v}$, the limiting value of $\Delta \mathbf{v}$ as $\Delta t \to 0$. You can see that this is so by using the first identity of Equation 5.8 to write

$$d\mathbf{v} = \mathbf{a}\,dt. \tag{5.10}$$

FIGURE 5.2 Over a small finite time interval Δt, the velocity of a particle changes from $\mathbf{v}_i$ to $\mathbf{v}_f$. In the case shown, the velocity has both increased and changed in direction. The red vector represents the small finite change $\Delta \mathbf{v} \equiv \mathbf{v}_f - \mathbf{v}_i$.

---

*We need not write $|v| = |ds/dt|$. Because the scalar ds represents a displacement in the direction $\hat{\mathbf{s}}$, it can never be negative.

The vector $d\mathbf{v}$ is expressed as the product of the vector $\mathbf{a}$ with the scalar dt; hence, $\mathbf{a}$ and $d\mathbf{v}$ must have the same direction.

Like all vectors, $\mathbf{v}$ and $\mathbf{a}$ can be written in terms of their component vectors or their components. In two-dimensional space, for example, we can write

$$\mathbf{v} = \mathbf{v}_x + \mathbf{v}_y = v_x\hat{\mathbf{x}} + v_y\hat{\mathbf{y}} = (v_x, v_y) \tag{5.11}$$

and

$$\mathbf{a} = \mathbf{a}_x + \mathbf{a}_y = a_x\hat{\mathbf{x}} + a_y\hat{\mathbf{y}} = (a_x, a_y). \tag{5.12}$$

We make direct use of the component forms in Section 5.3.

Projectile Motion

Nearly every child spends a good deal of time throwing and catching balls and thus learns to predict at least the end points of the **trajectory** (that is, the path) of such projectiles. Nevertheless, everyday experience does not seem to be a good teacher when it comes to analyzing this commonplace phenomenon. Despite experience to the contrary, many persons have a strong impression that the projectile starts out moving at least roughly in a straight line in the direction in which it was originally launched, then turns sharply downward as its initial motion "runs out," and finally falls more or less straight downward. Even experts held this view in the past (Figure 5.3). The facts are quite different, as Figure 5.4 shows. The actual trajectory of a projectile varies in curvature, but the curvature changes quite smoothly, right from the first instant. The trajectory has no straight segments, and the curve appears symmetrical, unlike those shown in Figure 5.3.

We now develop a mathematical description of the trajectory. Figure 5.5a depicts a projectile (here, a baseball) at some instant during its flight. We choose the coordinate system shown, with its origin at the beginning of the trajectory. To obtain a clear picture of the forces acting on the projectile, we use a free-body diagram (Figure 5.5b). For simplicity, we neglect air resistance—the force exerted on the projectile by the air. As is always true for an unsupported body near the earth's surface, the acceleration is

$$\mathbf{a} = \frac{\mathbf{F}}{m} = \frac{m\mathbf{g}}{m} = \mathbf{g}.$$

In our coordinate system, chosen with $\hat{\mathbf{y}}$ upward, this acceleration is

$$\mathbf{a} = -g\hat{\mathbf{y}}.$$

In Figure 5.5c, the velocity $\mathbf{v}$ has been replaced by its component vectors $\mathbf{v}_x$ and $\mathbf{v}_y$. This replacement points the way to a considerable simplification. Because $\mathbf{a}$ has no component in the x direction, $\mathbf{v}_x$ remains unchanged throughout the flight of the projectile. At the same time, $\mathbf{v}_y$ changes uniformly, in a way that we can readily describe using the kinematic ideas developed in Chapter 2.

To put it another way, the simultaneous motions of the projectile in the x and y directions are *separable*, and each can be treated according to the rules of one-dimensional kinematics. Our immediate purpose is to express x and y, the position coordinates of the projectile, as functions of time. The two-dimensional trajectory problem thus reduces to a pair of one-dimensional problems. One entails constant-velocity motion in the x direction, and the other constant-acceleration motion in the y direction.

For the constant-velocity motion in the x direction, we use Equation 2.5. This equation takes the form $x = x_i + v_x t$. Our choice of coordinate system gives us $x_i = 0$, and thus we have the simplified relation

$$x = v_x t. \tag{5.13}$$

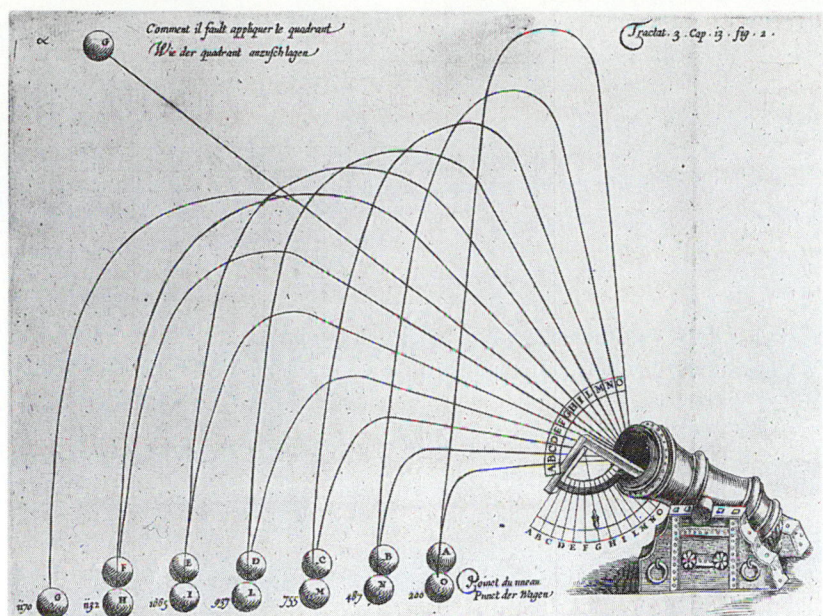

FIGURE 5.3 Representation of cannonball trajectories from a seventeenth-century practical treatise on gunnery, showing how to use a quadrant with plumb bob to determine the elevation angle of the cannon for desired ranges. The author has one important point correct: namely, there are two elevation angles for each range less than the maximum. Surprisingly, he shows that the maximum range is achieved with a 45° elevation angle; this is true, but only in the absence of air resistance. The author could never have arrived at this result purely on the basis of observation of cannonball flights, in which the air resistance is *not* negligible. Note especially that the author clings, in spite of his experience, to the notion that the trajectory consists of a straight segment at the cannon elevation angle, a sharp downward curve, and a final straight, nearly vertical segment.

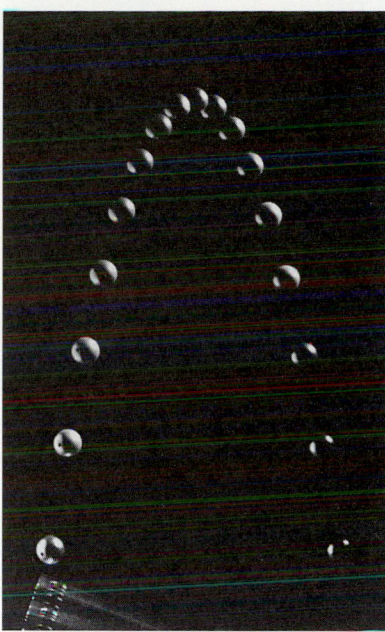

FIGURE 5.4 Multiflash photograph of a ball in flight. The trajectory is quite different from those suggested in Figure 5.3. Here, the curve has no straight segments, its curvature varies smoothly, and it is symmetrical about its midpoint. (If the initial and final parts of the trajectory appear to you to be straight at first glance, satisfy yourself that they are curved by tilting the book and looking at the photograph nearly edge on.) The photograph is reduced to one-tenth actual size, and the flash interval is 0.043 s.

For the constant-acceleration motion in the y direction, we use Equation 2.36a. This equation takes the form $y = y_i + v_{y_i}t + \frac{1}{2}at^2$, where v_{y_i} is the y component of the initial velocity $\mathbf{v}_i$. But in our coordinate system we have $y_i = 0$ and $a = a_y = -g$. Hence, the equation becomes

$$y = v_{y_i}t - \frac{1}{2}gt^2. \tag{5.14}$$

Equations 5.13 and 5.14 accomplish our immediate goal—expressing the position coordinates of the projectile as functions of time. They thus constitute the mathematical description of the trajectory that we have been seeking. But the description is not yet in a convenient form because we do not usually know v_x and v_{y_i} directly. We must express v_x and v_{y_i} in terms of the quantities that we usually do know, which are the

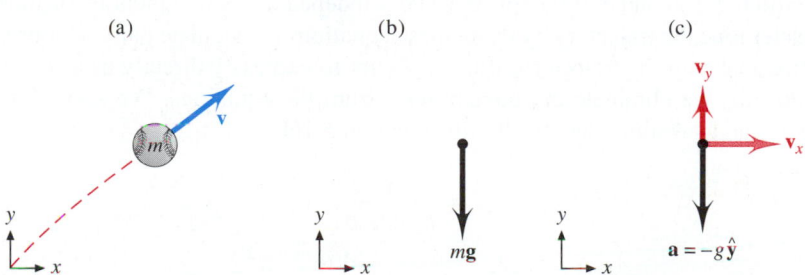

(a)   (b)   (c)

FIGURE 5.5 (a) A projectile (here shown as a baseball) at some instant during its flight. Its velocity $\mathbf{v}$ is shown; the ball is still rising. (b) Corresponding free-body diagram. The only force acting on the ball (neglecting air resistance) is the gravitational attraction of the earth. (c) The acceleration vector is shown in black; it has the fixed magnitude and direction $\mathbf{a} = -g\hat{\mathbf{y}}$ always. The component vectors shown in red, $\mathbf{v}_x$ and $\mathbf{v}_y$, have been substituted for $\mathbf{v}$. The key to further analysis, carried out in the text, is that the acceleration changes $\mathbf{v}_y$ just as if there were no horizontal motion, while $\mathbf{v}_x$ remains constant.

Physicists are not the only people who attempt to solve new problems by reducing them to one or more simpler problems whose solutions they already know. But this strategy is probably used more systematically and more consciously by physicists than by practitioners in other areas of knowledge. We have used the strategy before (as when we reduced the hybrid inclined-plane Atwood machine to a one-dimensional system in Section 4.6), and you will see it again and again.

Precise, reliable real-time calculation of projectile motion can be worth millions of dollars a year.

magnitude and direction of $\mathbf{v}_i$, the launching velocity of the projectile. Figure 5.6 shows that v_x and v_{y_i} are simply the horizontal and vertical components of the initial velocity and are therefore given by

$$v_x = v_i \cos \theta \quad \text{and} \quad v_{y_i} = v_i \sin \theta. \qquad \textbf{(5.15a, b)}$$

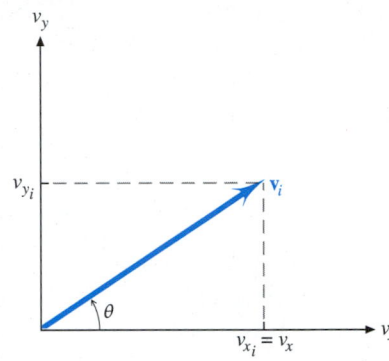

FIGURE 5.6 A projectile is launched with initial speed v_i at an angle θ above the horizontal. The v_x axis has the direction $\hat{\mathbf{x}}$; the v_y axis has the direction $\hat{\mathbf{y}}$. The desired components are given by Equations 5.15a and 5.15b.

Equation 5.15a tells us that the projectile moves in the direction $\hat{\mathbf{x}}$ at a constant speed, a speed that is equal to the component of the initial velocity $\mathbf{v}_i$ in that direction. Equation 5.15b tells us that the projectile begins to move in the direction $\hat{\mathbf{y}}$ with a speed equal to the component of the initial velocity in that direction. By substituting the values of v_x and v_{y_i} into Equations 5.13 and 5.14, we obtain

$$x = v_i t \cos \theta \qquad \textbf{(5.16a)}$$

and

$$y = v_i t \sin \theta - \tfrac{1}{2} g t^2. \qquad \textbf{(5.16b)}$$

"Parameter" Really Means Something

Two equations that express two different dependent variables in terms of a single independent variable are called **parametric equations**, and the independent variable is called the parameter. In current speech, the term "parameter" is often used to mean "limit," "condition," "variable," or even simply "quantity." The intended meaning of the term "parameter" is preserved only in precise technical discussions.

In Equation 5.16a, the quantity $v_i \cos \theta$ is a constant; thus x increases linearly with t. The first term on the right side of Equation 5.16b is likewise the product of a constant and t, and by itself would yield a linear dependence of y on t. However, the second term on the right incorporates the effect of the downward gravitational force and expresses a downward displacement that is quadratic in t. Indeed, the motion in the y direction is exactly the same as the motion of a stone projected vertically, as in Example 2.11. Figure 5.7 confirms this assertion.

Equations 5.16a and 5.16b express x and y independently as functions of time. We could determine the trajectory by using these equations to calculate pairs of coordinates (x, y) that make up the trajectory. But it is better to express y directly as a function of x. To do this, we eliminate the **parameter** t from the equations. We solve Equation 5.16a for t and substitute the result into Equation 5.16b:

$$t = \frac{x}{v_i \cos \theta},$$

and thus

$$y = v_i \frac{x}{v_i \cos \theta} \sin \theta - \frac{1}{2} g \frac{x^2}{(v_i \cos \theta)^2}.$$

This simplifies to

$$y = x \tan \theta - \frac{g}{2(v_i \cos \theta)^2} x^2. \qquad (5.17)$$

You can see that y is a quadratic function of x. Like all curves that describe quadratic functions, *the trajectory is a parabola.* Figure 5.8 shows several trajectories plotted by using Equation 5.17, with v_i fixed at the value 75 m/s—about the speed of a well-driven golf ball. Look carefully at the photographs of Figures 5.4 and 5.7 and you will see a strong resemblance to the plots of Figure 5.8.*

In launching projectiles, the child playing catch and the naval gunner face the same practical problem. The problem is to determine the **range** R— that is, the distance at which the projectile will land on level ground. Mathematically, the problem can be phrased as follows: At launch, the position coordinates of the projectile are $x = 0$, $y = 0$. What will be the value of x, the distance of the projectile from the launch point, when its height y is again zero? To answer this question, we set y equal to zero in Equation 5.17:

$$\frac{g}{2(v_i \cos \theta)^2} x^2 - x \tan \theta = 0.$$

One of the two roots of this quadratic equation is the trivial one $x = 0$; we guaranteed this result by setting the launching position at (0, 0). Putting this solution aside, we factor x out of the equation, leaving

$$\frac{g}{2(v_i \cos \theta)^2} x - \tan \theta = 0.$$

We solve this expression for x to obtain the other, more interesting root of the original quadratic equation. This special value of x is the range R:

$$R = \frac{2v_i^2}{g} \cos^2 \theta \tan \theta.$$

Because $\tan \theta = \sin \theta / \cos \theta$, this simplifies to

$$R = \frac{v_i^2}{g} 2 \sin \theta \cos \theta. \qquad (5.18)$$

Finally, we use the trigonometric identity $2 \sin \theta \cos \theta = \sin 2\theta$ to obtain

$$R = \frac{v_i^2}{g} \sin 2\theta. \qquad (5.19)$$

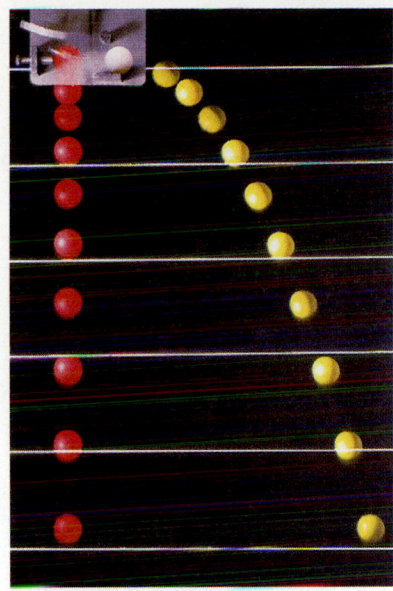

FIGURE 5.7 Multiple-exposure photograph of two balls released simultaneously from the same height. The red ball is released gently and falls vertically. The yellow ball is projected horizontally with a launch speed $v_i = v_x = 0.58$ m/s. The vertical positions of the two are always nearly the same. This observation shows that the vertical motion of the yellow ball is completely unaffected by its horizontal motion. The grid spacing is 6.35 cm ($2\frac{1}{2}$ in.), and the time interval between the light flashes that expose the photographic film is $\frac{1}{30}$ s.

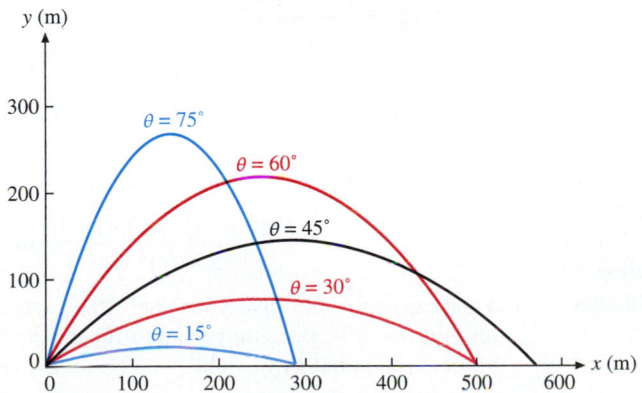

FIGURE 5.8 Trajectories plotted by using Equation 5.17. For all cases, $v_i = 75$ m/s.

*You cannot tell that a curve is really a parabola just by looking at it, but the paths in both photographs certainly *resemble* parabolas, despite the fact that they must be somewhat distorted by the effects of air resistance. A quantitative demonstration, by means of measurements on Figure 5.4, is the subject of Problem 5.3.

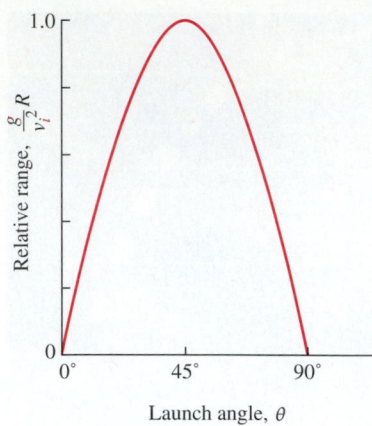

FIGURE 5.9 Plot of relative range, $R/R_{max} = Rg/v_i^2$, as a function of launch angle θ. The maximum range corresponds to $\theta = 45°$; for every other value of R, there are two complementary angles.

For any launch angle θ, the range R depends on the *square* of the launch speed. Therefore, in the absence of substantial air resistance, a relatively small increase in initial speed leads to a relatively large increase in range.

Equation 5.19 shows that, for a fixed launch speed, the range of the projectile depends on the sine of *twice* the launch angle. Figure 5.9 illustrates this dependence. (In the figure, range is plotted as the dimensionless ratio R/R_{max}, which is independent of the particular value of R_{max}.) The maximum range is attained when the sine function has its maximum value, $\sin 2\theta = 1$. That is, $2\theta = 90°$, which corresponds to the launch angle $\theta = 45°$. The value R_{max} of the maximum range is

$$R_{max} = \frac{v_i^2}{g}, \quad \theta = 45°; \quad \text{air resistance negligible.} \quad (5.20)$$

Any lesser range can be attained by using either of two angles, as Figure 5.8 suggests. Figure 5.9 shows that there are two angles between 0° and 90° for which $\sin 2\theta$ has the same value. (The exception is $\theta = 45°$.) These two angles are *complementary*; that is, their sum is 90°. One of the angles produces a low trajectory and the other a high trajectory, as is evident in Figure 5.8. In the low-trajectory case, v_x is relatively large and v_{y_i} is small. Consequently, the projectile does not rise very high and spends relatively little time in flight. However, it covers horizontal distance rapidly. In the high-trajectory case, v_x is relatively small and v_{y_i} is large. Consequently, the time of flight is long and the same horizontal distance is covered at a more leisurely pace.

In baseball, the high trajectory is called a pop fly. It is easy to catch because the fielder has plenty of time to get into position. The low trajectory, called a line drive, is much harder to catch because the ball reaches the outfield much more quickly. In military applications, the low trajectory is preferable from the point of view of accuracy. The short time of flight reduces the effects of air resistance and other less predictable influences such as local air currents, and it also makes detection and interception more difficult. The high trajectory is used with mortars, where the aim is to get the shell to its target over intervening barriers.

How high does a projectile rise? The maximum height Y depends only on v_{y_i}; the relation between the two quantities is given by Equation 2.37. If we modify the notation to suit the present case, this equation becomes $0 - v_{y_i}^2 = 2(-g)Y$ (see Example 2.11c), so

$$Y = \frac{v_{y_i}^2}{2g}. \quad (5.21a)$$

The maximum height can also be expressed in terms of the launch speed and angle, v_i and θ. Use of Equation 5.15b yields

$$Y = \frac{v_y^2 \sin^2 \theta}{2g}. \quad (5.21b)$$

EXAMPLE 5.1

A good fielder can throw a baseball with an initial speed of 35 m/s. Katchum Ketchum of the Mudville Mudhens wishes to throw to another player 75 m away. **(a)** What two launch angles θ_1 and θ_2 can he use? **(b)** How high will the ball go in each case? **(c)** What will be the time of flight in each case? Neglect air resistance.

SOLUTION:

(a) What two launch angles θ_1 and θ_2 can Katchum use?

The problem statement gives the desired range R and the value of v_i; Equation 5.19 gives R in terms of v_i and θ. You can solve this equation for 2θ, obtaining

$$2\theta = \sin^{-1} \frac{gR}{v_i^2}.$$

When you insert the numerical values, you obtain

$$2\theta = \sin^{-1} \frac{(9.80 \text{ m/s}^2) \times 75 \text{ m}}{(35 \text{ m/s})^2} = \sin^{-1} 0.60.$$

Your calculator will give you $2\theta = 36.9°$, which leads to the result $\theta_1 = 18.4°$. Because you know R and v_i to two significant figures, you round θ_1 to yield

$$\theta_1 = 18°$$

for the first possible launch angle. (However, it is not a bad idea to use the unrounded value in carrying out further calculations, to avoid possible rounding errors.) To find the second possible launch angle θ_2, remember that the calculator automatically computes first- or fourth-quadrant angles (angles between 90°

and $-90°$) for inverse trigonometric functions. There is also an angle $2\theta_2$ in the second quadrant that is equal to $\sin^{-1} 0.60$. This value of $2\theta_2$ is the supplement of $2\theta_1$, and θ_2 is therefore the complement of θ_1. You thus have the second launch angle

$$\theta_2 = 90° - \theta_1 = 72°.$$

(b) How high will the ball go for each of the two launch angles?

To find the maximum heights, you use Equation 5.21b. For the low trajectory, the equation gives

$$Y_1 = \frac{(35 \text{ m/s})^2 \times \sin^2 18.4°}{2 \times 9.80 \text{ m/s}^2} = 6.2 \text{ m}.$$

For the high trajectory, the equation gives

$$Y_2 = \frac{(35 \text{ m/s})^2 \times \sin^2 71.6°}{2 \times 9.80 \text{ m/s}^2} = 56 \text{ m}.$$

(c) What will be the time of flight for each of the two trajectories?

Each time of flight depends only on the initial vertical velocity component. You can determine these components by using the technique of Example 2.11b, which deals with the time of flight of a stone thrown vertically upward. (Here, again, the separability of the trajectory problem into horizontal- and ver-

tical-motion parts is useful.) In the notation of the present problem, Equation 2.36a becomes

$$y = y_i + v_{y_i}t - \tfrac{1}{2}gt^2.$$

The ball begins and ends its flight at $y = 0$. Also, you have $v_{y_i} = v_i \sin \theta$. The equation thus becomes

$$0 = 0 + v_i t \sin \theta - \tfrac{1}{2}gt^2.$$

One of the roots is $t = 0$; this root represents the trivial fact that the ball's flight begins at $t = 0$. Factoring out this root, you have the significant root

$$t = \frac{2v_i \sin \theta}{g}.$$

For the low trajectory, this gives you

$$t_1 = \frac{2 \times 35 \text{ m/s} \times \sin 18.4°}{9.80 \text{ m/s}^2} = 2.3 \text{ s}.$$

For the high trajectory, you find

$$t_2 = \frac{2 \times 35 \text{ m/s} \times \sin 71.6°}{9.80 \text{ m/s}^2} = 6.8 \text{ s}.$$

The player catching the ball has nearly three times as long to get into position for the high throw as for the low throw.

In analyzing projectile motion, we have found it possible to separate the motion into two parts: horizontal motion at constant velocity and vertical motion with constant acceleration. Although the problem is most naturally set up in terms of vector mathematics, the analysis proceeds largely in terms of two linked one-dimensional calculations in which vector methods are not necessary. In Section 5.4, we consider a somewhat more complicated case: uniform circular motion, in which vector methods must be used more extensively.

Uniform Circular Motion and Centripetal Force

In many and varied processes, both natural and man-made, a body moves in a circle at constant speed. We now analyze such **uniform circular motion**. To begin with, we note that the body's velocity $\mathbf{v}$ changes only because its *direction* $\hat{\mathbf{v}}$ is changing, while the *speed* v remains constant. That is, the velocity vector $\mathbf{v}$ changes its direction but not its magnitude. This point is illustrated in Figure 5.10, in which the change in velocity $\Delta \mathbf{v}$ is found by vector subtraction.

A velocity change always requires an acceleration. To evaluate this acceleration, consider Figure 5.10. If the displacement from P to Q requires a time Δt, we can write, as usual,

$$\langle \mathbf{a} \rangle = \frac{\Delta \mathbf{v}}{\Delta t}. \qquad (5.22)$$

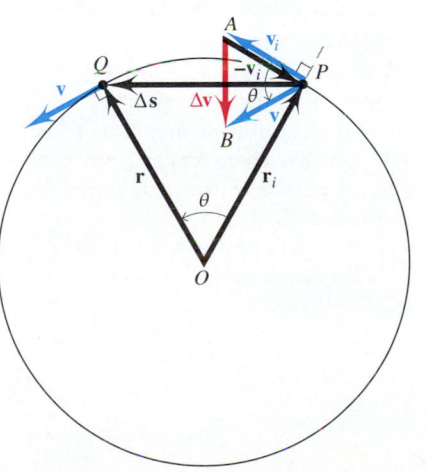

FIGURE 5.10 A body moves uniformly along a circular path of radius r about the center O. At time $t = 0$, the body is located at P; at a later time t, it has moved to Q. The velocity has changed from $\mathbf{v}_i = v\hat{\mathbf{v}}_i$ to $\mathbf{v} = v\hat{\mathbf{v}}$. Even though the speed v—the magnitude of $\mathbf{v}$—has not changed, a velocity change $\Delta \mathbf{v} = \mathbf{v} - \mathbf{v}_i$ results from the change in the direction of $\mathbf{v}$. The isosceles triangles BPA and QOP are similar, so $\langle a \rangle \equiv \Delta v/\Delta t = (v/r)(\Delta s/\Delta t) = (v/r)\langle v \rangle$. In the limit $\Delta t \to 0$, the direction of Δv approaches $-\hat{\mathbf{r}}$, while $\langle v \rangle \to v$. It follows that $\mathbf{a} = -(v^2/r)\hat{\mathbf{r}}$.

As usual, we find the instantaneous acceleration **a** by taking the limit of Equation 5.22 as Δt approaches zero. Here, however, we must pay special attention to the limiting direction of **a** as well as to its limiting magnitude.

First, consider the magnitude a. The displacement during the time Δt is the vector $\Delta \mathbf{s}$, which is the directed chord of the circle from P to Q. The vector $\mathbf{r}_i$, called the **radius vector**, is drawn to P from the center O; the radius vector **r** is similarly drawn to Q.

The vectors **v** and $\mathbf{v}_i$, and their vector difference $\Delta \mathbf{v} = \mathbf{v} - \mathbf{v}_i$, are the sides of the triangle BPA. The instantaneous velocity vectors $\mathbf{v}_i$ and **v** are tangent to the circle at P and Q. Because the tangent to a circle at any point is perpendicular to the radius vector to that point, we have $\mathbf{v}_i \perp \mathbf{r}_i$ and $\mathbf{v} \perp \mathbf{r}$. The angles QOP and BPA are equal because their corresponding sides are mutually perpendicular. Hence, the isosceles triangles QOP and BPA are similar.* The ratios of corresponding sides of similar triangles are equal, and so we have

$$\frac{\Delta v}{v} = \frac{\Delta s}{r}. \tag{5.23}$$

Because we are seeking an acceleration, we want to introduce the quantity $\langle a \rangle = \Delta v / \Delta t$ into this equation. To do so, we multiply both sides of Equation 5.23 by $v/\Delta t$ to obtain

$$\frac{\Delta v}{\Delta t} = \frac{v}{r} \frac{\Delta s}{\Delta t}.$$

We now take the limits of the two sides of this equation as $\Delta t \to 0$:

$$\lim_{\Delta t \to 0} \frac{\Delta v}{\Delta t} = \frac{v}{r} \lim_{\Delta t \to 0} \frac{\Delta s}{\Delta t}.$$

The limit on the left side has the value a, while the limit on the right has the value v. So we have $a = (v/r)v$, or

$$a = \frac{v^2}{r}. \tag{5.24}$$

This is the magnitude of the acceleration experienced by a body as it moves in a circle of radius r at constant speed v.

Next, we consider the *direction* of this acceleration. In Figure 5.10, $\Delta \mathbf{v}$ is perpendicular to $\Delta \mathbf{s}$. (This follows from the mutual perpendicularity of the other corresponding sides of the two triangles.) What happens when we make the time interval Δt smaller without limit—that is, we let $\Delta t \to 0$? The point Q comes closer and closer to P, and the chord $\Delta \mathbf{s}$ becomes the infinitesimal tangent $d\mathbf{s}$ to the circle at P. In this limiting case, the vector $\Delta \mathbf{v}$ becomes the infinitesimal vector $d\mathbf{v}$. The vector $d\mathbf{v}$ must lie along the radius OP because the perpendicular to $d\mathbf{s}$, which is tangent to the circle, is the radius OP. You can see from Figure 5.10 that $d\mathbf{v}$ is directed *inward* along the radius, opposite the direction of $\mathbf{r}$.[†] Finally, remember that $\mathbf{a} \equiv d\mathbf{v}/dt$; the direction of the acceleration is therefore the same as that of $d\mathbf{v}$. We conclude that $\hat{\mathbf{a}} = -\hat{\mathbf{r}}$.

We combine the conclusions of the two preceding paragraphs to express the vector **a**. For a particle moving in a circle of radius r at constant speed v, we have

$$\mathbf{a} = -\frac{v^2}{r} \hat{\mathbf{r}}. \tag{5.25}$$

This acceleration, experienced by any body moving in a curvilinear path, is called the **centripetal acceleration**. The name was coined by Newton from Latin roots meaning "center seeking."

Like any acceleration, centripetal acceleration of a body must be the consequence of a force acting on the body, in accordance with Newton's second law. For a body of

An Idea Whose Time Had Come

Newton probably derived Equation 5.25 about 1665 but first used the word "centripetal" in print in 1687. An analysis in the spirit of the one given here was first published in 1673 by the great Dutch physicist Christian Huygens (1629–1695), to whom the credit for the derivation is therefore due.

---

*Here is one of several ways to prove the similarity: Because $v_i = v$ and $r_i = r$, both triangles are isosceles. Two isosceles triangles with equal vertex angles must have equal base angles. Two triangles with equal corresponding angles are similar.

[†]In the limit $\Delta t \to 0$, $\hat{\mathbf{r}}$ converges toward $\hat{\mathbf{r}}_i$ and becomes the same direction.

mass m, we find the force by inserting the value of $\mathbf{a}$ from Equation 5.25 into Newton's second law in the vectorial form of Equation 4.26a, $\mathbf{F} = m\mathbf{a}$:

$$\mathbf{F} = -\frac{mv^2}{r}\,\hat{\mathbf{r}}. \tag{5.26}$$

The magnitude of the force is

$$F = \frac{mv^2}{r}. \tag{5.27}$$

Any force that results in a centripetal acceleration is called a **centripetal force**. When an automobile rounds a highway curve, the road surface exerts on the tires a centripetal force, directed toward the curve's center. This force accelerates the car inward as it rounds the curve. The gravitational force exerted on the earth by the sun is likewise a centripetal force; in its absence, the earth would move in a straight line rather than in its approximately circular orbit.

EXAMPLE 5.2

(a) Find the centripetal acceleration of the earth as it moves around the sun, using the approximation that the orbit is circular. (b) What is the gravitational centripetal force exerted by the sun on the earth?

SOLUTION:
(a) What is the centripetal acceleration of the earth as it moves around the sun?

When you apply Equation 5.26 to the sun-earth system, the unit vector $\hat{\mathbf{r}}$ represents the direction from the sun to the earth at any instant. The mean distance from the sun to the earth is given in the table of physical quantities inside the front cover as $r = 1.50 \times 10^{11}$ m. You can use this value, together with the fact that the **orbital period** of the earth—the time that it requires to circle the sun—is 1 year, to calculate the speed v as follows. The circumference of a circular orbit is

$$C = 2\pi r.$$

Using τ (the Greek letter tau) to denote the orbital period, you write

$$v = \frac{C}{\tau} = \frac{2\pi r}{\tau}. \tag{5.28}$$

You now use Equation 5.25 to find

$$\mathbf{a} = -\frac{4\pi^2 r^2}{\tau^2 r}\,\hat{\mathbf{r}} = -\frac{4\pi^2 r}{\tau^2}\,\hat{\mathbf{r}}. \tag{5.29}$$

You are now ready to insert numerical values. Table 1.2 gives $1\ \text{y} = 3.15 \times 10^7$ s. You thus have

$$\mathbf{a} = -\frac{4\pi^2 \times 1.50 \times 10^{11}\ \text{m}}{(3.15 \times 10^7\ \text{s})^2}$$
$$= -5.97 \times 10^{-3}\ \text{m/s}^2\ \hat{\mathbf{r}}.$$

(b) What is the gravitational centripetal force exerted by the sun on the earth?

You could use Equation 5.27 to calculate the centripetal force $\mathbf{F}$. Because you already have a numerical value for the centripetal acceleration $\mathbf{a}$, however, it is easier to use Newton's second law. From the table of physical quantities inside the front cover, the mass of the earth is $m = 5.98 \times 10^{24}$ kg, and thus

$$\mathbf{F} = m\mathbf{a} = 5.98 \times 10^{24}\ \text{kg} \times (-5.97 \times 10^{-3}\ \text{m/s}^2\ \hat{\mathbf{r}})$$
$$= -3.57 \times 10^{22}\ \text{N}\ \hat{\mathbf{r}}.$$

As always, the direction of the net force acting on a body (here, the earth) is parallel to the direction of its acceleration. As you can see from the numerical values for $\mathbf{a}$ and $\mathbf{F}$, the earth's inertial mass is so large by everyday standards that even a very modest acceleration—of magnitude less than 0.1% of the familiar acceleration of gravity g—requires the exertion of a huge force.

Applications of Uniform Circular Motion

In this section, we consider three applications of the concepts of centripetal acceleration and centripetal force developed in Section 5.4. They are the minimum-orbit earth satellite, the conical pendulum, and the banking of highway or railroad curves.

The Minimum-Orbit Earth Satellite

On 4 October 1957, *Sputnik I*, the first artificial earth satellite, was launched into orbit by the USSR. Since then, thousands of objects have been launched into orbits of many shapes and heights above the earth.

The earth's atmosphere sets a lower limit on the altitude of any satellite, because with decreasing altitude comes rapidly increasing atmospheric density and consequently rapidly increasing frictional resistance. In the altitude range between 160 and 200 km above the earth, the resistance becomes so great that it limits the satellite's life to just a few orbits. The early satellites were short-lived because the capabilities of the available booster rockets limited the satellites to relatively low orbits. *Sputnik I*, for example, stayed aloft for just a few weeks even though its initial, somewhat noncircular orbit took it only briefly as low as 160 km each time it circled the earth.

Because atmospheric friction sets a lower limit on the altitude of satellites, a satellite in a circular orbit at an altitude of roughly 200 km is called **a minimum-orbit earth satellite**. Such satellites are of great practical use when long life in orbit is not required, for they are relatively economical to launch.

What is the orbital period τ of a minimum-orbit earth satellite? The space shuttles launched by NASA, the U.S. space agency, usually follow orbits slightly higher than minimum; these orbits are called low orbits, and they have altitudes ranging from 240 km to 360 km. To be specific, let us consider the very nearly circular orbit of the third flight of the space shuttle *Columbia*, 22–30 March 1982. In this orbit, *Columbia*'s altitude varied between 241 km and 242 km. (The orbit was thus circular within about 1 part in 10^4.)

Let us use what we know of uniform circular motion to *predict* the orbital period of a satellite in a perfectly circular orbit of altitude 242 km and then compare the prediction with the actual period for *Columbia*. The altitude h of the orbit is only a small fraction of the earth's radius R_e ($= 6.367 \times 10^6$ m). Taking $h = 242$ km, we have

$$\frac{h}{R_e} = \frac{242 \text{ km}}{6367 \text{ km}} = 3.80 \times 10^{-2}, \tag{5.30}$$

or less than 4%. Because h/R_e is so small, the period τ for *Columbia* will be only slightly different from the period for a minimum-orbit satellite. The magnitude part of Equation 5.29, $a = 4\pi^2 r/\tau^2$, expresses a relation among the period τ, the orbit radius r, and the magnitude a of the centripetal acceleration of a body in uniform circular motion. We solve for the period, obtaining

$$\tau = 2\pi \sqrt{\frac{r}{a}}. \tag{5.31}$$

We know the orbit radius; it is just $r = R_e + h$, the sum of the earth's radius and the altitude. To find τ, we need to know the acceleration a. In view of the smallness of h/R_e, a is not very different from g, the value of the acceleration of gravity at the earth's surface. But we can do better than this approximation with only a little effort. Specifically, we can find the correction $\Delta a = g + a$ that arises because the satellite is a distance h above the surface of the earth. As you will see in detail in Chapter 14, the acceleration of a body due to the gravitational force exerted on it by the earth (or any other spherical mass) decreases as $1/r^2$; that is,

$$a \propto \frac{1}{r^2}.$$

When we differentiate both sides of this proportionality, we obtain

$$da \propto -\frac{2\,dr}{r^3};$$

this proportionality relates the infinitesimal change in acceleration da to the infinitesimal change dr from which it arises. We now have two proportionalities. We can obtain an equation by dividing the second of the two by the first; this gives us

$$\frac{da}{a} = -\frac{2\,dr}{r}. \tag{5.32}$$

Even though Δa is finite, we will not be far wrong if we use Δa to approximate the infinitesimal da and the corresponding Δr to approximate the infinitesimal dr:

$$\frac{\Delta a}{a} = -\frac{2\,\Delta r}{r}. \tag{5.33}$$

To apply the approximation, we note that the quantity $\Delta r/r$ on the right side of Equation 5.33 is just the quantity h/R_e in Equation 5.30. We thus have

$$\frac{\Delta a}{a} = \frac{\Delta a}{g} = -2 \times 3.80 \times 10^{-2} = -7.60 \times 10^{-2}.$$

Hence, the value of the acceleration of gravity at the altitude of the minimum-orbit satellite is

$$a = g\left(1 + \frac{\Delta a}{g}\right) = 9.80 \text{ m/s}^2 \times (1 - 7.60 \times 10^{-2}) = 9.06 \text{ m/s}^2.$$

We can now insert numerical values into Equation 5.31:

$$\tau = 2\pi \times \sqrt{\frac{6367 \times 10^3 \text{ m} + 242 \times 10^3 \text{ m}}{9.06 \text{ m/s}^2}} = 5.37 \times 10^3 \text{ s}.$$

This is equivalent to

$$\tau = 89.4 \text{ min}.$$

The published orbit period for the *Columbia* mission is given as "90 min ($\pm$)." This deliberately imprecise value is quoted because the orbit period varied slightly as maneuvers were carried out from time to time to accomplish the shuttle's various tasks. Nevertheless, the agreement with our calculated value of 89.4 min is excellent.

THINKING LIKE A PHYSICIST

The finite approximation method used here is worth careful study because it has wide application. Very often, we know the value of some quantity at a fixed position and wish to find the value of the same quantity at another position not far away. The proportionality $a \propto 1/r^2$ is inverse-quadratic—not the most complicated functional relation you can think of, but not the simplest, either. We can approximate the necessary correction to the acceleration of gravity by the linear relation of Equation 5.33. Complicated functions can often be approximated in this way—always bear-

ing in mind, however, that a price must be paid in loss of precision. The ease of manipulating linear relations mathematically often makes finite linear approximations valuable for "first cut" estimates. If an estimate is accurate enough for the purposes at hand, no further calculation is needed. If the estimate is not sufficiently accurate, it still furnishes a guideline for further work. (If you are familiar with the Taylor series from your study of calculus, you can see that the Taylor series is an elaboration of the approximation process used here.)

The Conical Pendulum

Children often play with one form or another of the simple device pictured in Figure 5.11a. Figure 5.11b shows a popular variant used by rodeo entertainers. The mass m, suspended by the string, revolves in a horizontal circle whose center is at A. As the mass revolves, the string, which makes a constant angle θ with the vertical, describes a cone whose vertex is the fixed point O. The string-mass system is therefore called a **conical pendulum**. In the absence of friction, the system would continue indefinitely without change once started. A real system that approximates this idealization can be kept in motion by slight movement of the string at O.

At any instant the pendulum lies in a particular vertical plane that also includes OA (Figure 5.11c). The mass experiences the two external forces shown in the free-body diagram of Figure 5.11d. One of these forces, $\mathbf{T}$, is exerted by the string. The other is the gravitational force $m\mathbf{g}$ due to the earth. To find the net force $\mathbf{F}$ acting on the mass, we need to evaluate the vector sum $\mathbf{F} = \mathbf{T} + m\mathbf{g}$. As in the case of the block sliding down the inclined plane (Chapter 4), we have information that simplifies the task. We know that the mass neither rises nor falls as it whirls in a horizontal circle. Hence, the

(a)　　　　　　　　(b)　　　　　　　(c)　　　　　　　(d)

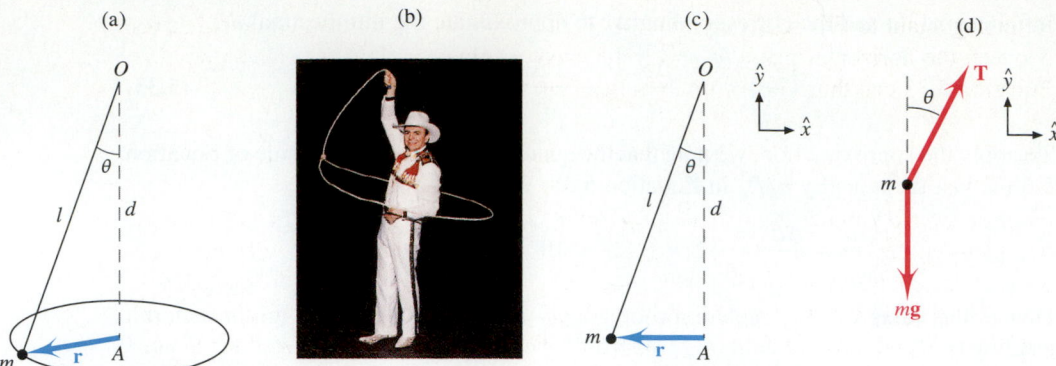

FIGURE 5.11 (*a*) A conical pendulum. (*b*) Cowboy Mike (Wooldridge) dramatically illustrates the conical pendulum. The point mass *m* is replaced by the (approximately) horizontal loop of rope. (*c*) Diagram of the vertical plane containing the string and mass at some instant. (*d*) Free-body diagram of the mass shown in part *c*.

vertical component F_y of the net force is equal to zero. Choosing $\hat{\mathbf{y}}$ in the upward vertical direction, we write

$$T \cos \theta - mg = 0.$$

Only the string tension **T** has a horizontal component. This component, $T \sin \theta$, is the centripetal force that makes the mass move in a horizontal circle. We therefore use Newton's second law in the special form of Equation 5.27 to write

$$T \sin \theta = \frac{mv^2}{r}.$$

We now have two equations that can be combined to eliminate *T*. To do this, we rewrite the first in the form $T \cos \theta = mg$. We then divide it into the second equation and obtain

$$\frac{T \sin \theta}{T \cos \theta} = \frac{mv^2}{rmg},$$

or

$$\tan \theta = \frac{v^2}{rg}. \tag{5.34}$$

This expression is not very convenient because it contains three related variables: θ, *r*, and *v*. However, we can reexpress Equation 5.34 in a more transparent form in which only two variables appear. We begin by eliminating the term v^2 from the equation. Equation 5.28, $v = 2\pi r/\tau$, relates the speed *v* of a body in uniform circular motion to its orbit radius *r* and orbit period τ. Squaring both sides of the equation, we obtain $v^2 = 4\pi^2 r^2/\tau^2$. We substitute this value for v^2 into Equation 5.34 to obtain

$$\tan \theta = \frac{4\pi^2 r}{\tau^2 g}.$$

It may seem that we have succeeded only in substituting the variable τ for the variable *v*; we still have the three variables, θ, *r*, and τ. However, the next step is to solve for τ:

$$\tau = 2\pi \sqrt{\frac{r}{g \tan \theta}}.$$

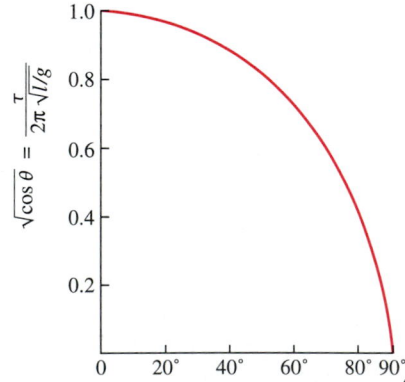

FIGURE 5.12 Plot of $\sqrt{\cos \theta}$ versus θ, showing how the period τ changes with angle θ for the conical pendulum of Figure 5.11. The period is expressed as the dimensionless ratio $\tau/(2\pi\sqrt{l/g})$, which is another way of writing $\sqrt{\cos \theta}$, as you can see by examining Equation 5.35.

We can now eliminate the variable *r*. To do this, note from Figure 5.11c that $r = l \sin \theta$. We substitute this value into the preceding expression to obtain

$$\tau = 2\pi \sqrt{\frac{l \sin \theta}{g \tan \theta}},$$

or

$$\tau = 2\pi \sqrt{\frac{l}{g} \cos \theta}. \tag{5.35}$$

126 ■ Curvilinear Motion

Referring again to Figure 5.11c, note that $l \cos \theta = d$, where d is the vertical distance between the horizontal plane in which the mass revolves and the suspension point O. Equation 5.35 can thus be written in the compact form

$$\tau = 2\pi \sqrt{\frac{d}{g}}. \qquad (5.36)$$

Equation 5.35 shows that the period τ of the conical pendulum depends on the angle θ. However, the dependence goes as $\sqrt{\cos \theta}$. As Figure 5.12 shows, τ depends only weakly on θ if θ is not too large (for example, varying θ from 0° to 30° changes τ by only 7%). Satisfy yourself on this point by direct experiment: Make a conical pendulum out of string and a convenient mass, and time the period with a wristwatch for a range of choices of θ. Can θ ever be equal to 90°? Why or why not?

THINKING LIKE A PHYSICIST

Figure 5.12 is another example of a dimensionless plot. We divide the quantity of interest (here, the period τ) by a constant quantity having the same dimensions (here, $2\pi\sqrt{l/g}$). The constant quantity that we choose is one that *characterizes* the system—a quantity that describes the important fixed characteristics of the system (here, the string length and the acceleration of gravity).

A dimensionless plot is valid for all systems of the same kind (here, all conical pendulums) and not just for a particular pendulum having a certain length. Thus, a dimensionless plot can often serve to bring out a general property of a class of systems. Refer back to Figure 5.9, in which a dimensionless plot is used to describe the dependence of the range of a projectile on the elevation angle at which the projectile is launched.

The Banking of Curves

As we have already noted, the centripetal force that accelerates an automobile toward the center of the curve that it is rounding is exerted on the tires by the road. If the road is level, the magnitude of the centripetal force required to keep the car on the road depends on the speed of the car and the radius r of the curve: $F = mv^2/r$ (Figure 5.13a). If F does not exceed the maximum static frictional force $\mu_s mg$ between the road surface and the tire, the automobile will not skid as it rounds the curve.

If the road is intended for low-speed turns (for instance, at street corners), a level road surface is quite satisfactory. A level surface is not satisfactory, however, for curves intended to be negotiated at high speeds. In such cases, there is danger of skidding on a level surface, especially if rain or snow reduces the coefficient of friction μ_s between the tires and the road. Even if skidding does not occur, there is stress on vehicles and road surface, with the possibility of passenger discomfort and of damage due to shifting loads. These difficulties can be eliminated (or at least ameliorated) by **banking** the road—that is, by raising its outer edge so that the road surface lies at an angle θ with respect to the horizontal, as shown in Figure 5.13c.

The basic idea behind the banking of curves is simple. Because it both supports the car and supplies the force that accelerates it inward toward the center of the curve, the road must exert on the tires a force **F** that has both a vertical and a horizontal component. The vertical component is the familiar force that supports the vehicle against gravitation; the magnitude of this component is mg. The horizontal component provides the centripetal force, whose magnitude is mv^2/r. As shown in Figure 5.13b, **F** is inclined inward (toward the center of the curve) and thus makes some angle θ with the vertical. If we tilt the road by the same angle θ so that its surface is normal to **F**, there will be no possibility of skidding. What is more, the force that must be applied by the cargo bed to goods or by the seats to passengers inside the vehicle is also a normal force. Neither goods nor passengers will tend to slide when the force applied to them is normal to the support on which they rest. For people, forces applied approximately along the spine are much easier to accommodate than are transverse forces.

FIGURE 5.13 (*a*) Top view of an automobile of mass *m* rounding a curve. The instantaneous velocity **v** is tangent to the circular curve segment and thus perpendicular to the radius vector **r**. (*b*) The same system viewed from behind the automobile. The $\hat{\mathbf{x}}$ direction is chosen to coincide with the direction $\hat{\mathbf{r}}$ from the center of the curve to the instantaneous position of the automobile. The $\hat{\mathbf{y}}$ direction is chosen upward, perpendicular to the $\hat{\mathbf{x}}$ direction. This is usually the most convenient choice for circular motion; the same choice was made in the analysis of the conical pendulum (Figure 5.11). (Don't confuse circular motion with motion on an inclined plane, Section 4.4, where the axes are chosen parallel and perpendicular to the plane.) The force vectors, shown lumped together at one tire only for clarity, are in fact distributed among all four tires. The vertical component vector $-m\mathbf{g}$ and the horizontal component vector $-(mv^2/r)\hat{\mathbf{r}}$, shown in black, can be replaced by their vector sum **F**, shown in blue. If $mv^2/r > \mu_s mg$, the automobile will skid. (*c*) The road surface is inclined at an angle θ with the horizontal and is thus normal to **F**. There is now no component of **F** parallel to the surface. (*d*) Free-body diagram of the forces exerted on the automobile.

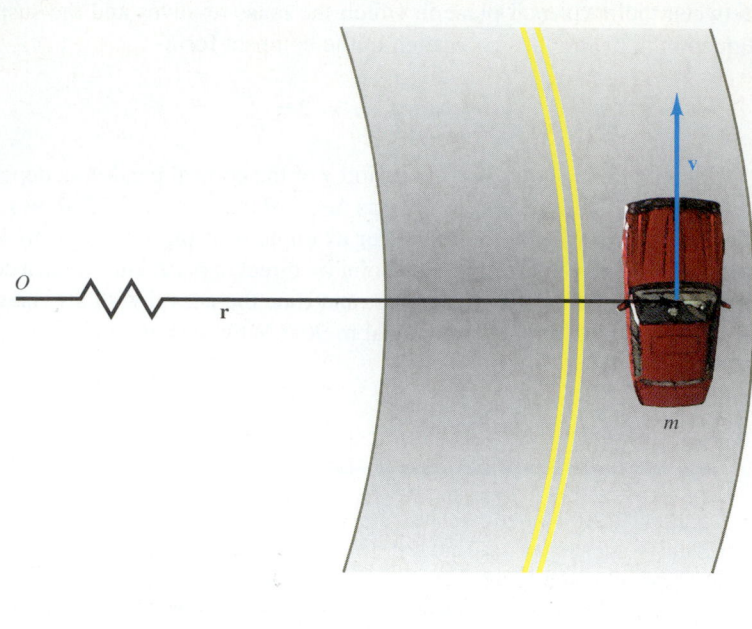
(a)

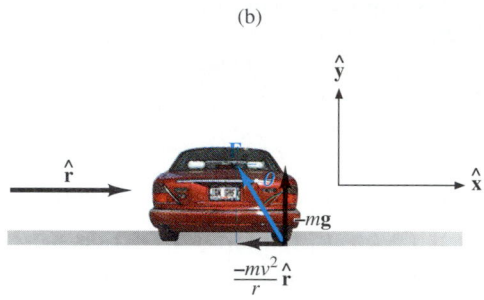
(b)

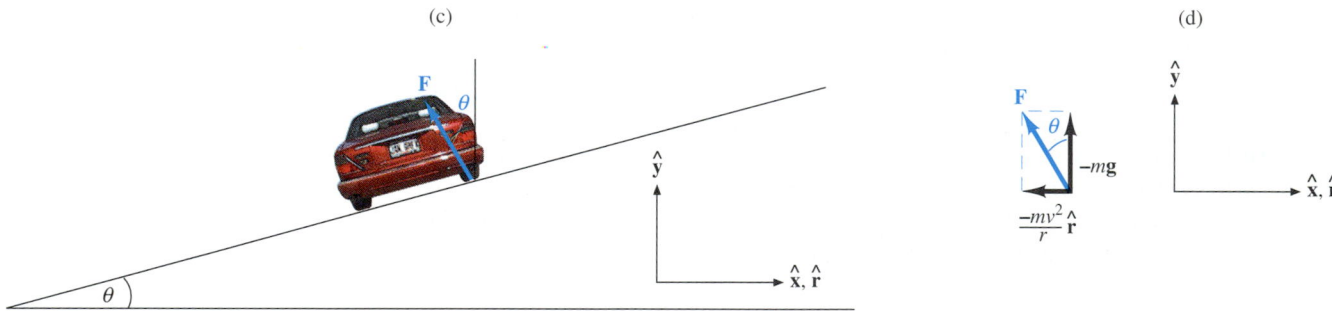
(c)   (d)

We can find the desired angle between the road surface and the horizontal, called the **angle of bank** θ, by looking at Figure 5.13d. We have

$$\tan \theta = \frac{mv^2/r}{mg} = \frac{v^2}{rg}. \tag{5.37}$$

According to this equation, the tangent of the required angle of bank is proportional to the square of the desired speed and inversely proportional to the radius of the curve. Thus, high speeds and sharp curves (that is, curves having a small radius of curvature) require large bank angles. Note that Equation 5.37 is identical with Equation 5.34, which pertains to the motion of the conical pendulum. This identity is no accident; can you explain it?

In practice, the angle of bank is influenced by additional considerations. The main one is that vehicles must be able to negotiate the curve safely and comfortably even if they are going much slower than the design speed. In particular, a vehicle must be in no danger of tipping over or sliding toward the inside of the curve, even if it comes to a stop on the curve.

EXAMPLE 5.3

(a) What is the magnitude of the force exerted by the road on an automobile of mass m as it rounds an ideally banked curve? **(b)** A commercial airline finds that most passengers feel uncomfortable if, due to the centripetal force exerted on them as the airplane makes a turn and they move along with it, they feel an increase in body weight of more than 30% above their normal weight—that is, if $F \geq 1.3mg$, where $F \equiv m|(\mathbf{g} + \mathbf{a}_c)|$. What is the maximum permissible angle of bank?

SOLUTION:

(a) What is the magnitude of the force exerted by the road on an automobile of mass m as it rounds an ideally banked curve?

By referring to Figure 5.13d, you can see that $\mathbf{F}$ is the hypotenuse of a right triangle whose sides are $-m\mathbf{g}$ and $-(mv^2/r)\hat{\mathbf{r}}$. You therefore calculate the magnitude F by using Pythagoras's theorem:

$$F = \sqrt{\left(\frac{mv^2}{r}\right)^2 + (mg)^2} = m\sqrt{\frac{v^4}{r^2} + g^2}. \quad (5.38)$$

(b) What is the angle of bank at which airline passengers feel their weight to increase by 30% above normal?

The maximum allowable angle of bank is attained when $F/mg = 1.30$. Using the value of F given by Equation 5.38,

you obtain

$$\frac{m\sqrt{\frac{v^4}{r^2} + g^2}}{mg} = \sqrt{\frac{v^4}{r^2g^2} + 1} = \sqrt{1 + \left(\frac{v^2}{rg}\right)^2} = 1.30.$$

Next, you use Equation 5.37 to substitute $\tan^2 \theta$ for $(v^2/rg)^2$. This gives you

$$\sqrt{1 + \tan^2 \theta} = 1.30.$$

The trigonometric identity $1 + \tan^2 \theta = \sec^2 \theta$ gives you

$$\sec \theta = 1.30.$$

If your calculator is smart enough, you can solve this directly by evaluating $\theta = \sec^{-1} 1.30$. Otherwise, you use the fact that $\sec \theta = 1/(\cos \theta)$ and write $\theta = \cos^{-1}(1/1.30)$. Either way, you obtain

$$\theta = 39.7°.$$

This result is unrealistically precise from the point of view of a pilot who is banking a plane, because aircraft bank-and-turn indicators read at best to the nearest degree. You therefore round the result; you have

$$\theta = 40°.$$

The Centrifugal Force*

Uniform circular motion depends on the presence of a centripetal force, directed inward toward the center of the circle. How do we reconcile this fact with the everyday impression that passengers have of being thrown *outward* as the vehicle rounds a curve?

The reconciliation depends on the very important point that *Newton's laws of motion apply only in inertial frames of reference* (Sections 3.1 and 3.2). Passengers, like the vehicles in which they are riding, are experiencing acceleration (in this case, centripetal acceleration) and are therefore *not* in an inertial frame. If we are riding in a bus rounding a curve, an observer standing on the street and watching us will corroborate this. But, when we are sitting in the bus seats, it is easy to believe that we *are* situated in an inertial frame. That is, we take the self-centered view that we are at "rest" in the physical sense, with the outside world moving past us. We therefore infer that we ought to be relaxed in our seats. When, nevertheless, we feel the vehicle seats dragging our own seats along and requiring us to pull our heads and shoulders uncomfortably inward, we instinctively take an alternative point of view. We stubbornly assert that we *are* in an inertial frame. Because we cannot "wish away" the centripetal force (a force that implies an acceleration and thus contradicts our assertion), we *invent* a force $\mathbf{F}'$ equal and opposite to the centripetal force $\mathbf{F}$. This force, $\mathbf{F}' = +(mv^2/r)\hat{\mathbf{r}}$, allows us to believe that the net force acting on us is $\mathbf{F} + \mathbf{F}' = 0$, as must be true in an inertial frame. $\mathbf{F}'$ is called the **centrifugal force**, from Latin roots meaning "center fleeing."

In sum, we have two alternative ways of accounting for our everyday experience. The first and more fundamental view is that of the inertial observer on the street. She sees that the road exerts an inward force on the bus tires and makes it follow a circular path. The bus seats do the same to us, and we must use our back and neck muscles to

---

*This subsection may be omitted without loss of continuity.

make ourselves sufficiently rigid to follow along. The second view is that of us passengers on the bus. Insisting as we do that we are in an inertial frame when sitting at rest, we invent a centrifugal force that, when added to the centripetal force, allows us to analyze events in our frame as though the frame were inertial. That is, adding in the "extra" force allows us to use Newton's laws to study events in a *noninertial frame*— a frame that is not inertial. Such an extra force is called a **d'Alembert force** [after Jean le Rond d'Alembert (1717–1783), the French physicist who first considered such forces in detail] or, more commonly, a **fictitious force**. Another familiar fictitious force is the force that seems to throw you forward onto your face when you are standing in a bus that suddenly stops. The word "fictitious" must be taken in its technical sense. Although the force is fictitious in the sense of "invented," it is certainly not fanciful, as anyone who has ridden on a badly driven bus can attest!

Calling a fictitious force into existence is not merely a way of evading the reality that the observer's reference frame is noninertial. As we have already noted in passing in the preceding paragraph, assuming the existence of the fictitious force makes it possible for such an observer to apply Newton's laws of motion. To analyze the motion of any body seen from such a frame, the observer adds $\mathbf{F}'$ in with all the other forces acting on the body. An example is described briefly in Figure 5.14 and its legend. We will not pursue this point further here, but the idea is developed further in several problems at the end of this chapter.

FIGURE 5.14 Centrifugal force. A passenger releases a ball on the floor of a bus as the bus is going around a circular curve. (*a*) From the point of view of the passengers, the ball (of mass m) accelerates outward along $\hat{\mathbf{r}}$; the passengers attribute this acceleration to the centrifugal force $\mathbf{F}'$ and argue that $\mathbf{a}' = \mathbf{F}'/m = (v^2/r)\,\hat{\mathbf{r}}$. (*b*) An observer standing by the roadside and looking into the bus sees, rather, that the ball is obeying the law of inertia, continuing at constant velocity along the tangent to the curve at the point of its release. She does not attribute acceleration along $\hat{\mathbf{r}}$ to the free ball; rather, she attributes acceleration along $-\hat{\mathbf{r}}$ to the bus and the passengers who are attached to it.

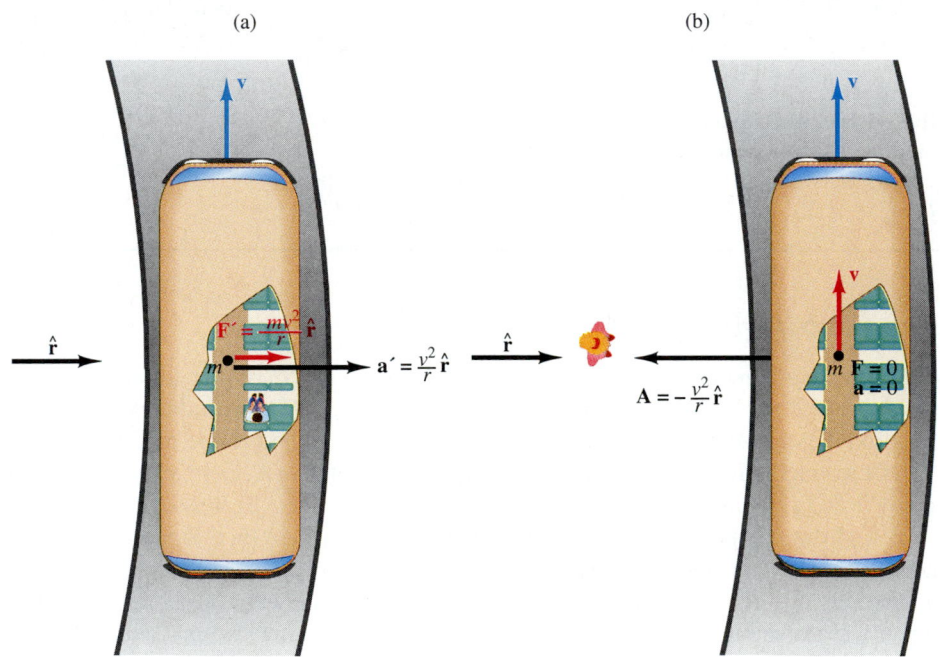

(a) (b)

Curvilinear motion is the motion that results when the direction of the net force acting on a body is not the same as, nor opposite to, the direction in which the body is moving. Two elementary cases of curvilinear motion are **projectile motion** and **uniform circular motion**.

The definitions of the kinematic vector quantities $d\mathbf{s}$ (or $\Delta\mathbf{s}$), $\mathbf{v}$, and $\mathbf{a}$ are completely analogous to those of the corresponding one-dimensional quantities ds (or Δs), v, and a. The definitions are given by Equations 5.2 and 5.8:

$$\mathbf{v} \equiv \frac{d\mathbf{s}}{dt} \equiv \lim_{\Delta t \to 0} \frac{\Delta\mathbf{s}}{\Delta t} \quad \text{and} \quad \mathbf{a} \equiv \frac{d\mathbf{v}}{dt} \equiv \lim_{\Delta t \to 0} \frac{\Delta\mathbf{v}}{\Delta t}.$$

The connection of kinematics with dynamics is through Newton's second law, which must be used in the vector form $\mathbf{F} = m\mathbf{a}$ (Equation 4.26a).

For a projectile moving without air resistance, the constant force $m\mathbf{g}$ is the only one acting. It is therefore possible to separate the motion into horizontal and vertical parts. Combining the two parts to eliminate the time dependence yields Equation 5.17,

$$y = x \tan\theta - \frac{g}{2(v_i \cos\theta)^2} x^2,$$

which describes the **trajectory** of the projectile.

If a body is to move in a circular path at constant speed, it must experience an acceleration of constant magnitude whose direction is always toward the center of the circle. This **centripetal acceleration** is given by Equation 5.25,

$$\mathbf{a} = -\frac{v^2}{r}\,\hat{\mathbf{r}}.$$

According to Newton's second law, centripetal acceleration must be the result of the application of a **centripetal force** (Equation 5.26),

$$\mathbf{F} = -\frac{mv^2}{r}\,\hat{\mathbf{r}}.$$

KEY TERMS

Section 5.1 Introduction
curvilinear motion

Section 5.2 Velocity and Acceleration
mean velocity, instantaneous velocity ▪ mean acceleration, instantaneous acceleration

Section 5.3 Projectile Motion
trajectory ▪ range

Section 5.4 Uniform Circular Motion and Centripetal Force
radius vector ▪ centripetal acceleration ▪ centripetal force

QUERIES

5.1 *(3) I got it! I got it!* A projectile is launched with initial speed v_i at an angle θ with the horizontal. At the same instant, an observer O' starts off at the launch point, moving along the (level) ground at constant speed $v' = v_i \cos\theta$ in the same direction as the projectile. Describe the motion of the projectile as seen by O'.

5.2 *(3) ... nor all your piety nor wit ...* Immediately after dropping the atomic bomb on Hiroshima in 1945, the pilot of the airplane *Enola Gay* made a sharp turn. (This maneuver is not normally used in dropping conventional chemical explosive bombs.) Explain the purpose of the maneuver.

5.3 *(3) Green grow the lilacs.* In using a garden hose to water a shrub at a distance of 3 m, you hold the nozzle at an angle of 30° above the horizontal. At what other angle could you hold the nozzle to water the same shrub? What are the advantages and disadvantages of each of the angles?

5.4 *(5) Petit prix.* Banked curves on roads or racetracks are not always desirable. What would be the effect on an auto race of a well-banked track?

5.5 *(5) Long road to nowhere.* Some high-speed automobile test tracks are designed with curves banked like that shown in part *a* opposite. Part *b* shows a vertical cross section through

(a)

(b)

an idealized test track of the kind shown in part *a*. The angle of bank is zero at the inside of the curve and gradually increases to a large angle (ideally, 90°) on the outside. Explain the advantages of such a design. If you were driving a test car around the track, how would you make optimal use of the curve design?

5.6 *(5) Move over, Col. Sanders!* Much of the design effort expended on sports cars is aimed at maximizing the speed at which they can be driven on winding roads. Specifically, they are expected to round turns safely at speeds much greater than the design speeds of the curves. Why are such cars almost always equipped with bucket seats—that is, seats having forward-projecting wings along the sides of their backs, and sometimes along the sides of their seats as well?

5.7 *(5) Level flight.* A pilot makes an airplane climb by pulling its nose up. This increases the angle at which the wings meet the oncoming airstream (called the angle of attack), resulting in an increase in the lifting force exerted by the air on the plane. Increasing the angle also increases the aerodynamic drag and thus slows the airplane down. Consider now an airplane in straight, level flight. The pilot wishes to change direction while keeping speed and altitude constant. Why must the pilot bank the airplane as it turns? Why must the pilot pull the nose up and increase the engine power as the plane enters the turn?

5.8 *(5) The colonel's lady and Rosie O'Grady.* Why are Equation 5.34, pertaining to the vertex angle of a conical pendulum, and Equation 5.37, pertaining to the angle of bank of a road, identical? Hint: Imagine a conical pendulum made by suspending a toy automobile from a string, as shown opposite. As the pendulum whirls, imagine a circular track brought up under the wheels of the toy, which supports the toy and relieves the tension in the string. At what angle to the horizontal must the track be banked to meet the wheels and thus properly replace the string?

5.9 *(5) Tricky maneuvers.* When the pilot of an airplane ex-

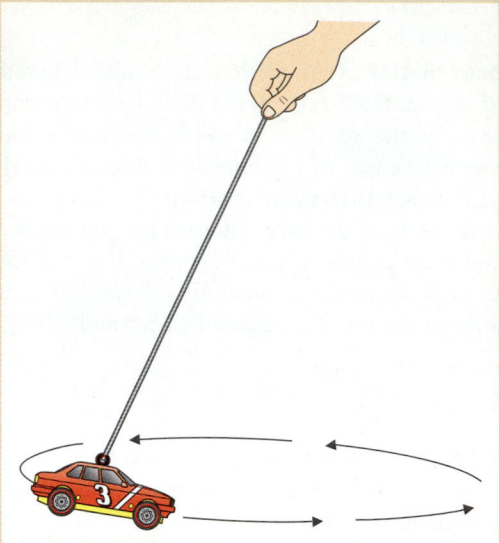

periences centripetal acceleration in the usual seated position, the blood in the pilot's body has an inertial tendency not to follow the body. If the general direction of the acceleration is from feet toward head, the blood may pool toward the feet, and the brain may be deprived of blood to the extent that the pilot "blacks out." Is this dangerous condition more likely to occur in the aerobatic maneuver called the "outside loop," in which the plane executes a vertical circle with the pilot's head directed outward, or in the "inside loop," where the pilot's head is directed inward?

PROBLEMS

GROUP A

5.1 *(2) Accelerometer.* An automobile salesman tells you that a certain used car can go from 0 to 100 km/h in 8 s flat. You know that the car is of a model well known for its pickup, but you have doubts because of its age. So you use a grease pencil to mark a vertical line on the passenger-side window and tell the salesman to make the car accelerate as fast as it can as you sit in the passenger seat. You have a string with a small mass tied at one end. As the car accelerates, you use your thumb to hold the free end of the string against the upper end of the grease-pencil line. If the acceleration is constant, what angle should the string make with the grease pencil line?

5.2 *(2) Moving right along.* The position of a particle is described by the function

$$\mathbf{r}(t) = (At^3 + Bt^2 + C)\hat{\mathbf{x}} + (Dt^2 + Et)\hat{\mathbf{y}}.$$

In this expression, A, B, C, D, and E are constants. **(a)** Find the velocity $\mathbf{v}$ as a function of time. **(b)** Find the acceleration $\mathbf{a}$ as a function of time.

Let $A = 1 \text{ m/s}^3$, $B = 5 \text{ m/s}^2$, and $D = 6 \text{ m/s}^2$. **(c)** Find the magnitude of the acceleration of the particle at $t = 3$ s. **(d)** Find the direction of the acceleration of the particle at $t = 3$ s.

5.3 *(3) Appeal to experiment.* By measurement on Figure 5.4,

show that y is indeed proportional to x^2 and the curve is thus parabolic. Using the scale factors given in the figure legend, evaluate g.

5.4 *(3) Putting the shot, I.* A champion shot putter can throw the shot about 22 m. Needless to say, the shot putter does not release the shot from ground level. But as a first approximation assume that he does, and calculate how fast he throws the shot in order to achieve the 22-m range. (More complete treatments are suggested in Problems 5.33 and 5.55.)

5.5 *(3) A miss is as good as a mile.* A hunter fires a rifle bullet at a deer but misses. If the rifle barrel is level, the hunter's shoulder is 1.50 m above the ground, and the "muzzle velocity" (initial speed) of the bullet is 600 m/s, how far away does the bullet strike the ground?

5.6 *(3) Look out below!* You throw a stone horizontally out a window, and it strikes the ground 0.65 s later at a distance of 5.3 m from the wall in which the window is located. **(a)** How high above the ground did you release the stone? **(b)** How fast did you throw the stone? **(c)** What is the speed of the stone when it hits the ground? **(d)** At what angle with the horizontal is the stone moving just before it hits the ground?

5.7 *(3) Galileo never pitched softball.* As you stand on top of a tower of height 25.3 m, you throw a stone horizontally with speed 16.8 m/s. **(a)** How long does it take for the stone to hit the ground? **(b)** At what horizontal distance x from the base of the tower does the stone strike the ground? **(c)** What is the speed of the stone just before it strikes the ground? **(d)** Just before the stone strikes the ground, what is the angle between the direction of motion of the stone and the ground?

5.8 *(3) Flying time.* You can throw a stone with initial speed v_i. Express the time of flight t of the stone as a function of the launch angle θ.

5.9 *(3) Hole in one?* A golfer drives a ball with initial speed 60 m/s at an angle of 35° above the horizontal. **(a)** Find the position of the ball with respect to the golfer 1.7 s after it leaves the club. **(b)** Find the magnitude and direction of the velocity of the ball at the same instant.

5.10 *(3) Way over par.* A less skillful golfer drives a ball with initial speed 32 m/s at an angle of 64° above the horizontal. **(a)** Find the position of the ball with respect to the golfer 3.1 s after it leaves the club. **(b)** Find the magnitude and direction of the velocity of the ball at the same instant.

5.11 *(3) War games.* A mortar is aimed at a target 500 m away. The shell leaves the mortar with $v = 300$ m/s. **(a)** Calculate the two possible angles of elevation that will yield a hit on the target. **(b)** Find the difference between the times of flight for the two trajectories.

5.12 *(3) Looking good.* A shot putter from the University of Washington in Seattle competes in the Pan American Games, held in Mexico City. At home in Seattle, where the acceleration of gravity is 9.809 m/s^2, his best effort yields 20.97 m. How far can he expect to put the shot in Mexico City, where the acceleration of gravity is 9.778 m/s^2?

5.13 *(3) So high, so far.* The maximum range of a certain gun is R_{max}. Show that if the gun is aimed straight up, it will shoot a bullet to a height $\frac{1}{2}R_{max}$.

5.14 *(3) How far to the top? I.* Derive the value of Y given by Equation 5.21b by using the fact that $y = Y$ corresponds to $x = \frac{1}{2}R$.

5.15 *(3) Playing catch.* You throw a ball with initial speed 10 m/s, at an angle 40° above the horizontal. **(a)** How high does the ball go? **(b)** How far will the ball be from you when it returns to the height of your arm? **(c)** How long does the ball take to travel this far?

5.16 *(3) Yer out!* A baseball shortstop catches a pop fly at a distance of 35 m from the batter. If the ball remained in flight for 4.3 s, find **(a)** the maximum height that the ball attained; **(b)** the horizontal and vertical components of its initial velocity; **(c)** its speed as it left the bat; **(d)** the angle with respect to the horizontal at which it left the bat; **(e)** its velocity when it is caught.

5.17 *(4) Round and round.* A particle moves at constant speed in a circle of radius 40 m. If it completes one revolution in 8.0 s, find **(a)** its speed, **(b)** the magnitude of its acceleration.

5.18 *(4) Small, whirled.* A crude but useful way of picturing a hydrogen atom was originally proposed in 1913 by the Danish physicist Niels Bohr. An electron of mass 9.1×10^{-31} kg follows a circular orbit around a much more massive proton, which we may consider to be stationary. The radius of the

orbit is 0.53×10^{-10} m, and the speed of the electron is 2.2×10^6 m/s. Find **(a)** the acceleration of the electron, **(b)** the force that must act on the electron to produce this acceleration.

5.19 *(4) Jovial.* Calculate the acceleration of Jupiter as it circles the sun. Take the radius of Jupiter's orbit to be 7.78×10^{11} m and its period to be 11.9 years. Compare your result with the acceleration of the earth about the sun, calculated in Example 5.2.

5.20 *(4) Mercurial.* Calculate the acceleration of Mercury as it circles the sun. Take the radius of Mercury's orbit to be 5.80×10^{10} m and its period to be 88.0 days. Compare your result with the accelerations of Jupiter and the earth about the sun, calculated respectively in Problem 5.19 and Example 5.2.

5.21 *(4) Fun at the amusement park, I.* The photograph below shows a popular amusement park ride. Riders walk into the vertical cylinder through a door (not shown), which is then closed. The riders stand with their backs against the wall. The cylinder starts to rotate on its axis; when it reaches full speed, the floor is lowered, leaving the riders ''glued'' to the wall by friction. **(a)** Express the minimum value of the coefficient of static friction μ_s between the riders' clothing and the wall in terms of the radius r of the cylinder, the speed v of the wall, and any constants that you need. **(b)** If $r = 2.5$ m and the minimum expected value of μ_s is 0.1, find the maximum allowable value of the rotation period τ.

5.22 *(4) Tense situation.* In the illustration below, a ball of mass m is attached to a rotating shaft by means of two stiff,

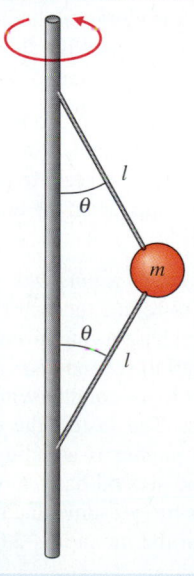

light rods, each of length l. The system rotates with period τ. **(a)** Neglect the forces on the rods that arise from the weight of the ball. Show that the tension T in each rod is given by the relation

$$T = \frac{2\pi^2 ml}{\tau^2},$$

a value independent of the angle θ that each rod makes with the shaft. **(b)** Take $m = 157$ g and $l = 23.5$ cm. The ball rotates around the shaft 865 times per minute. Find T. **(c)** For the numerical values given in part **b**, is it a reasonable approximation to neglect the weight of the ball? One way to answer this question is to compare the centripetal acceleration of the ball with g.

5.23 (5) *Space-shuttle orbit.* It is noted near the beginning of Section 5.5 that the orbit of the space shuttle *Columbia* in its third flight was circular to about 1 part in 10^4. Using the information given in the text, show that this is correct.

5.24 (5) *Conical pendulum.* A conical pendulum is made with a string 1.00 m long. **(a)** For what angle θ will the period be 1.90 s? **(b)** Find the corresponding values of r and v. **(c)** By how much must θ be changed in order to decrease the period by 5%?

5.25 (5) *Unbanked curve.* An automobile moving with speed v enters an unbanked curve of radius r. If the automobile is not to skid, what is the minimum value of the coefficient of friction μ_s between the tires and the road surface? Compare this result with Equation 5.37.

5.26 (5) *Don't spill the coffee!* An airplane flies at 120 m/s. If it is to make a turn of radius 2.8 km, at what angle should it bank?

5.27 (5) *Backtracking.* The airplane in part **b** of Example 5.3 flies at 255 m/s. If it reverses direction using the sharpest permissible turn, what is the distance between the parallel straight paths that it follows before and after the turn?

5.28 (G) *Looping the loop, I.* A stunt flier performs a vertical circular loop of radius 350 m. If she is firmly in her seat (that is, not hanging from her seat belt) when the plane is upside down at the top of the loop, what is the minimum speed of the plane at that point?

5.29 (G) *Looping the loop, II.* The stunt flier performs another vertical circular loop of radius 350 m, this time at higher speed than in Problem 5.28. When the plane is upside down at the top of the loop, the seat exerts a force on her equal to half her normal weight. What is the speed of the plane at that point?

GROUP B

5.30 (3) *Comparing height and distance.* Show that, for a projectile that follows the maximum-range parabolic trajectory for given v_i, the relation $Y = \frac{1}{4}R$ holds. Here, Y is the maximum height attained by the projectile, and R is the range.

5.31 (3) *How far to the top? II.* Show that the distance x to the point on the ground directly below the top of a parabolic trajectory is $\frac{1}{2}R$. Do this by evaluating dy/dx and setting it equal to zero.

5.32 (3) *Vantage point.* A golfer can drive a ball 200 m to its first bounce. If the golfer stands on the edge of a cliff of height 30 m and drives the ball with the same velocity $\mathbf{v}_i$, how far will the ball travel to its first bounce on the level plain below?

5.33 (3) *Putting the shot, II.* Repeat the calculation of Problem 5.4, taking into consideration the height—about 2.1 m—at which the shot putter releases the shot. Compare your result with that you obtained in Problem 5.4.

5.34 (3) *Off the end.* Starting from rest, a particle slides 1.50 m down a frictionless plane of incline angle 35°. It leaves the end of the plane and falls freely to the floor 1.35 m below. At what horizontal distance from the end of the inclined plane does it hit the floor?

5.35 (3) *The risky art of stealing bases.* In a baseball game, the catcher is about to return the ball to the pitcher when he sees the runner on first base trying to steal second. The catcher needs to throw the ball to second base, located 39 m away, as quickly as possible if the second baseman is to have a chance to tag the runner out. The fastest the catcher can throw the ball is 35 m/s. The pitcher is standing exactly midway between the catcher and second base. Does the pitcher need to duck as the ball goes by? Assume that the top of the pitcher's head, as he stands on the mound, is 2.0 m above the general level of the field.

5.36 (3) *Trading speed for height.* When a ball reaches its maximum height, its speed is one-fourth the speed with which it was thrown. What was the launch angle?

5.37 (3) *Evening things out.* For what launch angle $\theta_=$ will the maximum height reached by a particle be equal to its range?

5.38 (3) *Another angle on things, I.* You throw a ball at launch angle θ. As the ball flies through the air, you follow it with your eyes. When it reaches its maximum height Y, your eyes are directed at an angle ϕ with respect to the horizontal. **(a)** Show that

$$\tan \phi = \tfrac{1}{2} \tan \theta.$$

Perhaps it surprises you that this result does not depend on the launch speed of the ball. **(b)** Find ϕ if you throw the ball so as to maximize its range R.

5.39 (3) *Another angle on things, II.* (This is a generalization of Problem 5.38.) You throw a ball at launch angle θ. As the ball flies through the air, you follow it with your eyes. Express the angle ϕ above the horizontal, at which you see the ball, as a function of t, the time elapsed since you threw the ball.

5.40 (3) *Another angle on things, III.* Show that the angle γ between the velocity vector $\mathbf{v}$ of a projectile and the horizontal is given by the expression

$$\tan \gamma = \tan \theta - \frac{gt}{v_i \cos \theta}.$$

5.41 (4) *Fun at the amusement park, II.* Suppose some of the patrons of the ride described in Problem 5.21 secretly spray Teflon lubricant on the walls of the cylinder as they ride, thus creating trouble for subsequent riders. The park manager decides to make the ride trouble proof by modifying the cylinder to a cone segment, as shown in the illustration below. (a) Suppose that the riders are located on the part of the cone whose distance from the axis is approximately r_0. For what rotation period τ is the ride independent of the coefficient of friction between the riders' clothing and the wall? (b) Why is the modification from a cylinder to a cone segment not a complete solution to the problem of slippery walls? That is, why is some friction between the riders' clothing and the wall still necessary for stability?

5.42 *(4) Atwood spins!* A mass m rests on a turntable, as shown in the diagram below. The coefficient of static friction between the mass and the table surface is μ_s. Another mass, M, is attached to a string that passes through a hole at the center of the turntable and is then attached to m. What are the smallest and largest periods of revolution of the turntable, τ_{min} and τ_{max}, for which m remains fixed at a distance r from the center of the turntable?

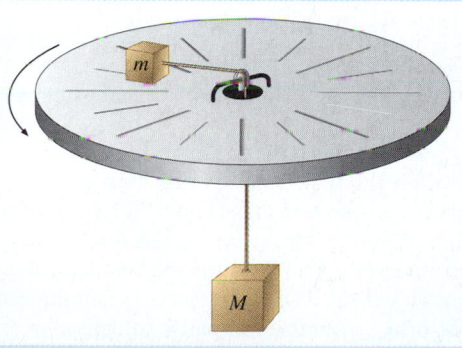

5.43 *(4) Equation 5.33 without approximation.* As noted in Section 5.5, the gravitational acceleration a above the surface of the earth depends on the distance r from the center of the earth according to the proportionality $a \propto 1/r^2$. (a) Define a proportionality constant C, and reexpress the relation between a and $\mathbf{r}$ as an equation instead of a proportionality. (b) Express the value of C in terms of R_e, the radius of the earth, and g, the acceleration of gravity at the surface of the earth. (c) Find an exact relation between a and r. (d) Express the result of part **c** in terms of g and the ratio h/R_e. (e) Take the same

numerical values for g and h/R_e used in Section 5.5, and recalculate the nominal orbit period of *Columbia*. What is the percentage error introduced by the approximation used in the text?

5.44 *(4) Bulging at the waist.* The measured acceleration of gravity at the equator is less than that at the poles. The rotation of the earth is responsible for this in two ways. The first arises from the equatorial bulge induced by the rotation: the equatorial radius of the earth is $r_e = 6.378 \times 10^6$ m, whereas the polar radius is $r_p = 6.357 \times 10^6$ m. Use the approximation of Equation 5.33 to find the percentage by which g_e, the value of the acceleration of gravity measured at the equator, differs from the value g_p, measured at the North Pole, on account of this effect.

5.45 *(5) Real and fictitious, I.* There are two ways in which the rotation of the earth results in a smaller value for the acceleration of gravity measured at the equator than that measured at the poles. The first is the subject of Problem 5.44. The second is the subject of this problem. An observer measuring the acceleration of gravity at the equator must correct for the fact that he is not an inertial observer. Specifically, he must add his own centripetal acceleration $\mathbf{a}_r$ to the value $\mathbf{g}'$ that he measures. The resulting sum is the value $\mathbf{g}$ that would be measured by an inertial observer—an imaginary nonrotating observer who could somehow hang above the earth's surface as it spun underneath her. (a) The radius of the earth at the equator is 6.38×10^6 m. Find the percentage by which g' differs from g. (b) If the earth rotated so fast that objects at the equator were weightless, how long would the day be? (c) What is the overall difference between the polar and equatorial values of the acceleration of gravity due to the causes considered in this problem and in Problem 5.44? You may wish to compare the results of your calculations with the standard values: $g_e = 9.780$ m/s^2, $g_p = 9.832$ m/s^2.

5.46 *(5) Checking an assertion.* In the discussion of the minimum-orbit earth satellite, we used the assumption that its acceleration due to the earth's gravitational attraction is proportional to $1/r^2$. Show that this statement is in agreement with the following data for the synchronous communications satellites that revolve about the earth with a period equal to the time that the earth takes to rotate once on its axis (such a satellite always remains above the same meridian of longitude on the earth as the earth rotates): orbit altitude, 3.58×10^7 m; orbit period, 23 h 56 min. (This is the time it takes for the earth to make one rotation relative to the fixed stars. The familiar 24-hour day is the time from noon to noon—the time that it takes for the earth to make one rotation relative to the sun. The extra 4 min are due to the fact that the earth is revolving about the sun. Can you see why?)

5.47 *(5) Checking clear out to the moon.* Repeat the check of Problem 5.46, using the moon instead of a synchronous satellite. The necessary data for the moon are as follows: orbit period, 27.3 days; earth–moon distance (center to center), 3.84×10^8 m. The calculation of the moon's centripetal acceleration was of great importance to Newton in establishing the law of gravitation, as we shall see in Chapter 14.

5.48 *(5) Tension in a conical pendulum.* A conical pendulum consists of a mass m at the end of a string of length l. If the period of the pendulum is τ, what is the tension T in the string?

5.49 (G) *Party trick.* You can swing a pail of water in a vertical circle without getting wet if you swing it fast enough. **(a)** Assuming that the distance from your shoulder to the surface of the water in the pail is 1.0 m, what is the minimum speed at which the water surface must be moving if you are to stay dry? **(b)** Using the result of part **a**, set an upper limit on the time you are allowed to swing the pail around once. **(c)** Why is this upper limit a safe one? That is, if you meet the criterion of part **a**, why will the time per revolution in fact be less than that calculated in part **b**?

5.50 (G) *Training for the big one.* It is fair to assume that David practiced extensively with his sling before engaging in his famous bout with Goliath. Suppose that his sling was 1.3 m long and that he swung it over his head in a horizontal circle 1.8 m above the ground. When David missed his practice target, his stone would hit the ground at a distance of 41 m. What was the centripetal acceleration of the stone in the sling just before launching?

5.51 (G) *Vectorial approach to the trajectory problem.* Let the vector from the launch point to the instantaneous position of a projected particle be $\mathbf{r}(t)$. **(a)** Express $\mathbf{r}(t)$ as a sum of two terms, one that includes the unit vector $\hat{\mathbf{x}}$ and the other the unit vector $\hat{\mathbf{y}}$. (Hint: Your expression will contain the quantities v_i, θ, and g.) **(b)** For a given instant t, find the magnitude and direction of $\mathbf{r}$. Compare your result with that of Problem 5.39. **(c)** By differentiating the result of part **a**, find the velocity $\mathbf{v}(t)$ of the particle. **(d)** For a given instant t, find the magnitude and direction of $\mathbf{v}$. Compare your result with that of Problem 5.40. **(e)** By differentiating again, show that $\mathbf{a} = -g\hat{\mathbf{y}}$.

5.52 (G) *Centripetal acceleration by differentiation.* If a particle moves in a circular orbit of radius R with period τ, its position $\mathbf{r}(t)$ can be described by the vector equation

$$\mathbf{r}(t) = R \cos \frac{2\pi}{\tau} t \, \hat{\mathbf{x}} + R \sin \frac{2\pi}{\tau} t \, \hat{\mathbf{y}}.$$

(a) Draw a sketch to show that the equation correctly describes circular motion. **(b)** By differentiation, find the velocity $\mathbf{v}(t)$. Make a sketch to show that $\mathbf{v} \perp \mathbf{r}$ at any time t. **(c)** Find the acceleration $\mathbf{a}(t)$. Compare $\mathbf{a}(t)$ with $\mathbf{r}(t)$, and make a sketch to show that $\hat{\mathbf{a}} = -\hat{\mathbf{r}}$ at any time. **(d)** By comparing $\mathbf{a}(t)$ with $\mathbf{r}(t)$, show that $a = v^2/R$, the same result obtained by a geometric argument in Section 5.4. Write the complete vectorial expression for $\mathbf{a}(t)$ in terms of $\mathbf{r}(t)$.

GROUP C

5.53 (3) *Don't try this yourself at home!* An artillery crew demonstrates its skill by firing a shell at angle θ_1 and then lowering the gun barrel and firing a second shell at a smaller angle θ_2 in such a way that the two shells collide in midair. **(a)** Show that, neglecting air resistance, the time elapsed between the two firings must be

$$\Delta t = \frac{2v_i}{g} \frac{\sin (\theta_1 - \theta_2)}{\cos \theta_1 + \cos \theta_2}.$$

(b) Let $v_i = 255$ m/s, $\theta_1 = 64.1°$, and $\theta_2 = 47.3°$. Find Δt.

5.54 (3) *Throwing uphill.* A hiker is climbing a hill of constant slope ϕ. She picks up a nice round stone and throws it uphill with speed v_i and launch angle θ measured with respect to the horizontal. **(a)** Show that the stone hits the ground at a distance R up the hill given by

$$R = \frac{2v_i^2}{g} (\sin \theta - \tan \phi \cos \theta) \frac{\cos \theta}{\cos \phi}.$$

(b) Show that the range is maximized when

$$\theta = \theta_{max} = 45° + \frac{\phi}{2}.$$

(c) Show that the maximum range is

$$R_{max} = \frac{v_i^2}{g} \frac{1}{1 + \sin \phi}.$$

5.55 (3) *Putting the shot, III.* Champion shot putters do not launch the shot at $\theta = 45°$ in order to maximize the range. It can be shown that the effect of air resistance changes the optimum value of θ by only about 0.1° and decreases the range by only about 15 cm in a championship-level 22-m throw. Nevertheless, films show that the very best shot putters use a launch angle of about 43°, whereas skilled athletes of less than star quality maximize their performances by using even smaller launch angles of about 38°. The optimum angle is always less than 45° because the shot is launched not from ground level but from a height of about $h = 2.1$ m. **(a)** Set up equations for the x and y components of motion, with initial conditions $x_i = 0$, $y_i = h$ and final conditions $x_f = R$, $y_f = 0$. Show that, for an athlete who can throw the shot at speed v, R is maximized for

$$\sin^2 \theta_{max} = \frac{1}{2} \frac{1}{1 + gh/v^2}.$$

(b) By substituting this value back into the equation for the y component of the motion, eliminate v and show that

$$R_{max} = h \tan 2\theta_{max}.$$

Compare this result with the information given for the best shot putters and the less expert ones. **(c)** Using the equation derived in part **b**, find θ_{max} for a 22-m throw and compare it with the observed value. What is the best launch angle for a mediocre shot putter who can throw only half this far? **(d)** Use the value of θ_{max} obtained in part **c** to determine the launch speed v achieved by a championship-class shot putter. Compare this with the results obtained for the simplest possible approximation of Problem 5.4 and the more sophisticated approximation of Problem 5.33. The comparisons will give you some insight into the very small differences between the performances of competition winners and those of their close runners-up. **(e)** Show that the initial and final velocity vectors are perpendicular to one another. [For further information, see D. B. Lichtenberg and J. G. Wills, *Amer. J. Phys.* **46**, 546 (1978) and S. K. Bose, *Amer. J. Phys.* **51**, 458 (1983).]

5.56 (G) *Not on the level.* A bowl of water is located with its center at the center of a turntable. The turntable is made to rotate at constant speed for some time, until all of the water is in uniform circular motion. Prove that the water surface assumes the form of a paraboloid of revolution—in other words, that the cross-section cut through the surface by any vertical plane that includes the axis of rotation is a parabola.

5.57 (G) *Real and fictitious, II.* As you have seen in Problem 5.45, an object on or near the earth's surface weighs less than it would weigh if the earth were not rotating. **(a)** Show that, assuming a perfectly spherical earth, the weight decrease is proportional to $\cos^2 \lambda$, where λ is the latitude at which the measurement is made. **(b)** At what latitude is the variation of g with latitude the greatest?

6

Oscillatory Motion

——— Simple harmonic motion is the simplest and most important form of oscillatory motion.

——— The kinematics of simple harmonic motion is the simplest case in which acceleration is not constant.

——— There is a connection between simple harmonic motion and uniform circular motion.

——— Hooke's law, the simplest possible nonconstant relation between force and displacement, often governs the motion of mechanical systems.

——— The harmonic-oscillator equation is the result of applying Newton's laws to Hooke's-law systems.

——— The harmonic-oscillator equation has simple solutions.

——— The simple pendulum is an approximate application of simple harmonic motion.

Left: More than 200 years old, this clock is driven by weights but owes its remarkable accuracy to the small-angle oscillation of the pendulum, which is approximately simple harmonic.

In Mathematicks *he was greater*
Then Tycho Brahe *or* Erra Pater*:*
For he by Geometrick *scale*
Could take the size of Pots of Ale*;*
Resolve by Sines and Tangents straight,
If Bread *or* Butter *wanted weight;*
And wisely tell what hour o'th' day
The Clock doth strike, by Algebra.

—SAMUEL BUTLER, *Hudibras*

Introduction

The clock mentioned in the epigraph above doubtless had a pendulum. It thus belonged to the vast class of artificial and natural systems whose behavior is *oscillatory*—that is, systems in which a back-and-forth motion is repeated many times or perhaps without limit. Some other oscillating systems are the quartz crystals that control most modern clocks and watches, the atoms in solids, the pistons in automobile engines, the swings in playgrounds, and your heart.

When a body oscillates—moves back and forth over the same path again and again—it reverses its direction of motion at each end of the path. Each time it reverses, it slows down as it approaches the end of the path, comes instantaneously to rest, and then proceeds in the opposite direction, speeding up as it goes. Thus, the oscillating body experiences acceleration. Indeed, the displacement, the velocity, and the acceleration all vary in an oscillatory fashion. And, because the acceleration is oscillatory, Newton's second law requires that the body must be subject to a force that also is oscillatory. The kinematics and dynamics of oscillation are the main subjects of this chapter.

We will direct most of our attention to the simplest and most important form of oscillatory motion, called *simple harmonic motion*. In this special type of oscillatory motion, displacement, velocity, acceleration, and force all vary in a way that can be described by either the sine or the cosine function, collectively called **sinusoids**. Any oscillatory motion that cannot be described so simply is called **anharmonic oscillation**.

Limiting our study to simple harmonic motion is not as restrictive as it might seem, for two reasons. The first reason is that many natural systems move in ways that are quite accurately simple harmonic. (The quartz crystals in watches are of this class.) The second reason is that it is mathematically possible to decompose any anharmonic oscillation into a sum of simple harmonic motions. (The motion of an automobile piston can be described in this way.) We will not consider anharmonic motion further in this chapter.

The Kinematics of Simple Harmonic Motion

As we have done before, we begin with kinematics—motion without regard to the forces that produce it. Consider a body that moves back and forth along the *x* axis (Figure 6.1) in such a way that its position *x* varies with time *t* according to the relation

$$x = A \cos 2\pi \frac{t}{\tau}. \tag{6.1}$$

Such a body is said to be in **simple harmonic motion**, also called **harmonic oscillation**; the body is often called a **harmonic oscillator**. The quantity τ is a constant. Because the argument* of a sinusoid must be a dimensionless quantity, t/τ must be dimensionless. Hence, τ must have the dimensions of time. We will consider the physical significance of τ shortly.

The quantity A also is a constant, called the **amplitude**. It has the dimensions of length and must be expressed in the same units as x. As time passes and t increases, the argument of the cosine, $(2\pi/\tau)t = $ constant $\times\, t$, also increases. As this happens, the value of the cosine varies repeatedly between the limits 1 and -1. At any instant when the cosine attains one of these extremal values, Equation 6.1 yields $x = \pm A$. Consequently, the amplitude A represents the greatest distance from the origin that the body attains in either direction as it oscillates back and forth. Figure 6.2a is a graph of x as a function of time; that is, it is a plot of Equation 6.1. As you can see, the motion has a well-defined *period* τ; this is the time required for the particle to go through an entire cycle of motion. The value of τ can be measured along the t axis, beginning at any point on the x-versus-t curve and ending at the nearest point at which the curve begins to repeat itself. [Caution: The period is *not* in general the time interval between two neighboring values of t at which x has the same value, because the particle passes through every x (except $x = A$, $-A$) *twice* per cycle. Satisfy yourself as to this matter by studying Figure 6.1.]

We now prove that the quantity τ in Equation 6.1 is identical with the period τ shown in Figure 6.2a. In the figure, a time interval τ is marked off beginning at the arbitrary time $t = t_1$ and ending at $t = t_1 + \tau$. This must represent one full cycle of the cosine in Equation 6.1, and thus the argument of the cosine must increase by 2π over the time

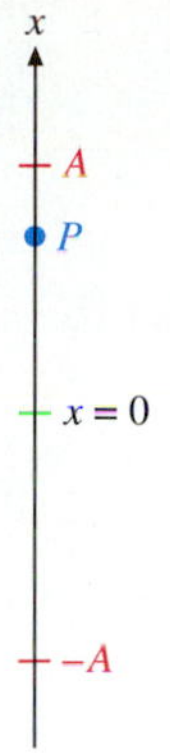

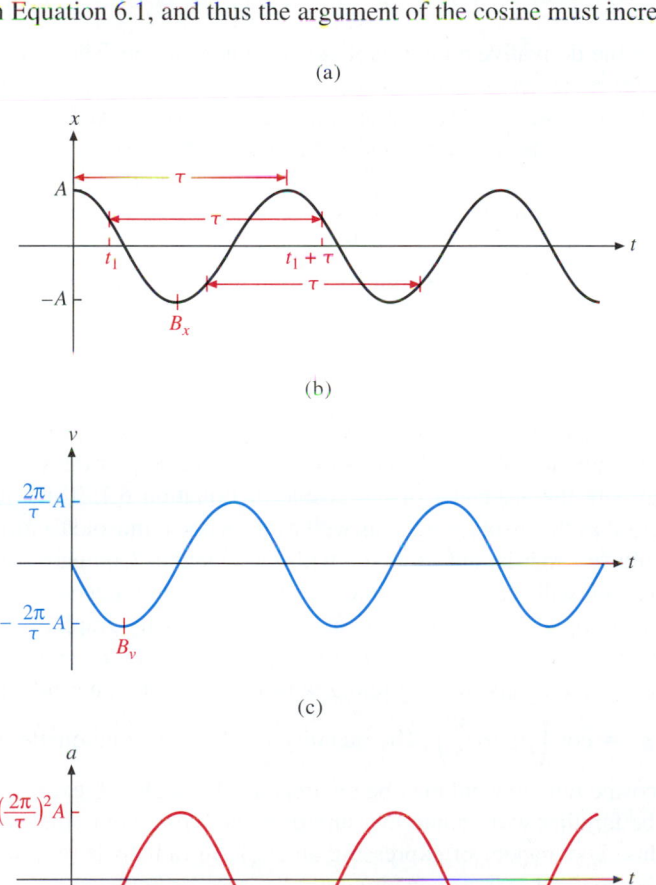

(a)

(b)

(c)

FIGURE 6.1 The particle P in one-dimensional simple harmonic motion along the x axis. The motion is described by Equation 6.1. The limits of the particle's motion, A and $-A$, are determined by the amplitude A. At the particular instant shown, the displacement x of the particle is positive.

Physics in Harmony with Music

When the string of a musical instrument vibrates in the simplest possible way, the motion of any point on the string is described by Equation 6.1. This is the basis for the name "simple harmonic motion." The study of vibrating strings goes back at least as far as Pythagoras (c. 560–c. 480 B.C.E.).

FIGURE 6.2 (*a*) Graph of the displacement x of the particle of Figure 6.1 as a function of time. Because this is a graph of Equation 6.1, it is a cosine curve with amplitude A and period τ. These two quantities completely determine a cosine curve and thus completely specify the motion of the particle. (*b*) Velocity v of the same particle as a function of t, plotted from Equation 6.5. The velocity is sinusoidal with the same period as that of the displacement, but the phase is one quarter-cycle ahead of the displacement. (*c*) Acceleration a of the same particle, plotted from Equation 6.7. Again, the acceleration is sinusoidal with the same period, but the phase is one quarter-cycle ahead of the velocity and one half-cycle ahead of the displacement.

---

*The *argument* of any function is the quantity for which the function is evaluated; for example, the argument of $\sin z$ is z.

interval. Does the argument in fact do this? Yes, because we have

$$2\pi\left(\frac{t_1 + \tau}{\tau}\right) - 2\pi\frac{t_1}{\tau} = 2\pi\frac{\tau}{\tau} = 2\pi.$$

To summarize, when the value of t in Equation 6.1 increases by an amount τ, the value of t/τ increases by 1. As a result, the argument of the cosine increases by 2π, and the cosine goes through one cycle.

The Velocity and the Acceleration

We can find the velocity v of the oscillator as a function of time by using the fundamental definition of velocity expressed in Equation 2.15,

$$v = \frac{dx}{dt}. \tag{6.2}$$

Inserting the value of x given by Equation 6.1, we have

$$v = \frac{d}{dt}\left(A\,\cos\frac{2\pi}{\tau}t\right).$$

Because the amplitude A is a constant, we can rewrite this equation in the slightly simpler form

$$v = A\frac{d}{dt}\left(\cos\frac{2\pi}{\tau}t\right). \tag{6.3}$$

We must now evaluate the derivative on the right side of this equation. The general result, which you probably remember from your study of calculus, is presented here for convenience. We will soon need to differentiate the sine function as well, and so that result also is presented. Let z be a variable and p a constant. We have

$$\frac{d(\cos pz)}{dz} = -p\,\sin pz \quad \text{and} \quad \frac{d(\sin pz)}{dz} = p\,\cos pz. \tag{6.4a,b}$$

Remembering that $2\pi/\tau$ is a constant, we apply Equation 6.4a to Equation 6.3 and obtain

$$v = -\frac{2\pi}{\tau}A\,\sin\frac{2\pi}{\tau}t. \tag{6.5}$$

This function is plotted in Figure 6.2b. The velocity, like the displacement plotted in Figure 6.2a, is a sinusoidal function of time. What is more, the argument of the sine in Equation 6.5 is identical with the argument of the cosine in Equation 6.1. Thus, the velocity has the same period as the displacement, as well as the same (sinusoidal) form. But v and x are **out of phase**—that is, out of step—with one another. Compare parts a and b of Figure 6.2, and you will see that the velocity is one-quarter of a cycle *ahead* of the displacement. (For example, the minimum point B_v on the velocity graph is one-quarter of a cycle before the minimum point B_x on the displacement graph.) You can reach the same conclusion algebraically by beginning with Equations 6.1 and 6.5 and using the relation $-\sin pz = \cos\left(pz + \frac{\pi}{2}\right)$. The quantity $\frac{\pi}{2}$ represents one-quarter of a cycle ($\frac{1}{4} \times 2\pi$) of the cosine function and may be regarded as the angle $90°$ expressed in radians. You should be familiar with radian measure of angles from your studies in trigonometry and calculus. The manner of expressing an angle in radians is reviewed in Appendix 1. Remember that 2π radians constitute a full circle, or $360°$.

We next find an expression for the acceleration of the simple harmonic oscillator by differentiating again. We begin with the definition of acceleration expressed in Equation 2.22b,

$$a \equiv \frac{dv}{dt}. \tag{6.6}$$

Combining this definition with Equation 6.5, we obtain

$$a = \frac{d}{dt}\left(-\frac{2\pi A}{\tau}\sin\frac{2\pi}{\tau}t\right) = -\frac{2\pi A}{\tau}\frac{d}{dt}\left(\sin\frac{2\pi}{t}t\right).$$

Applying Equation 6.4b then gives us

$$a = -\frac{4\pi^2}{\tau^2}A\cos\frac{2\pi}{\tau}t. \tag{6.7}$$

This function is plotted in Figure 6.2c. The acceleration is a sinusoidal function of time having the same period as the velocity and the acceleration, because all three sinusoids have the same argument. If you inspect the three parts of Figure 6.2, you will see that the acceleration is one-quarter of a cycle ahead of the velocity and one-half of a cycle ahead of the displacement. (Compare, for example, the time coordinates of the points marked B_a, B_v, and B_x on the three graphs.) You can reach the same conclusion algebraically by beginning with Equations 6.7, 6.5, and 6.1 and using the relations $-\cos pz = -\sin(pz + \pi/2) = \cos(pz + \pi)$.

Simple Harmonic Motion and Uniform Circular Motion: The Reference Circle

There is an intimate connection between simple harmonic motion and the uniform circular motion that concerned us in Chapter 5. We now explore that connection; understanding it will lead to deeper insight into both types of motion. Figure 6.3 shows a particle P moving in a circle of radius A with uniform speed V. We call this circle the **reference circle**. We denote the instantaneous position of the particle by means of the radius vector $\mathbf{A}$, which rotates about the center of the circle as time passes. We give the name **phasor** to this rotating vector. (The reason for the name will become evident shortly.) The initial conditions are chosen so that P lies on the positive x axis at the zero of time. After a time t, the particle has moved along the arc of the circle a distance Vt. Simultaneously, the angle between the x axis and $\mathbf{A}$ has increased from zero to a value θ.

FIGURE 6.3 A particle P moves at uniform speed V in a circular path of radius A, called the reference circle. The center of the circle lies at the origin of the coordinate system; the positive x direction has been chosen upward, toward the top of the page. The zero of time is chosen at an instant when the particle lies on the positive x axis at $x = A$. The system is shown at some later time t when the radius vector $\mathbf{A}$ from the origin to the particle makes an angle θ with the x axis. The position $x(t)$ of the particle P' on the x axis is given by the x component of OP. One way to visualize this is to imagine a screen S parallel to the x axis and perpendicular to the plane of the page, located to the right of the reference circle. A bright light far to the left casts a shadow of P on the screen. The motion of this shadow is identical with that of P'. The diagram shows that the velocity v of P' is the x component of $\mathbf{V}$; $v = -V\sin\theta$.

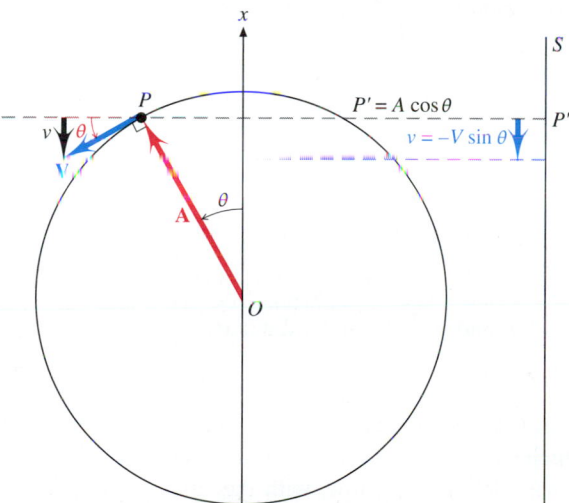

What is the value of this **angular displacement** θ as a function of t? Because the particle moves uniformly, θ must increase uniformly. Thus we can write

$$\theta = \omega t; \tag{6.8}$$

ω is a constant, called the **angular velocity**. The rotational relation expressed by this equation,

angular displacement = angular velocity × time,

is analogous to the translational relation

displacement = velocity × time.

We can evaluate θ at any time if we can express ω in terms of known quantities. To do this, we use our knowledge of the time required for the particle to complete one full circle, so that $\theta = 2\pi$ radians. As you saw in Example 5.2, this time is the *period* τ.

For $t = \tau$, Equation 6.8 becomes

$$2\pi = \omega\tau, \quad \omega\tau \text{ expressed in radians.} \tag{6.9}$$

We can find the value of τ by dividing the distance $2\pi A$ around the circle by the speed V:*

$$\tau = \frac{2\pi A}{V}. \tag{6.10}$$

Substituting Equation 6.10 into Equation (6.9) and solving for ω yields

$$\omega = \frac{V}{A}; \quad \omega \text{ expressed in radians per second.} \tag{6.11}$$

Because the radian is a dimensionless quantity, the unit radians per second (rad/s) is not really different from the unit s^{-1}. Nevertheless, using the term ''rad/s'' often helps us to remember that we are dealing with angular velocity.

Equation 6.9 can be used to express ω in terms of the period τ. Dividing through by τ, we obtain

$$\omega = \frac{2\pi}{\tau}; \quad \omega \text{ expressed in rad/s.} \tag{6.12}$$

With these remarks about uniform circular motion in mind, let us return to oscillatory motion. Specifically, as the phasor **A** in Figure 6.3 rotates uniformly around the reference circle, consider the behavior of its x component. At any particular moment, that component is $x = A \cos \theta$. By substituting in the value $\theta = \omega t$ from Equation 6.8, we can describe x as a function of time:

$$x = A \cos \omega t. \tag{6.13}$$

But this is the same as the defining equation for simple harmonic motion, Equation 6.1, which is $x = A \cos 2\pi t/\tau$. To see this, you need only substitute the value of ω given by Equation 6.12 into Equation 6.13. That is, *when a particle is in uniform circular motion, the x component of its position executes simple harmonic motion with the same period.* To take another point of view, imagine a light source far to the left of Figure 6.3, shining on the particle P as it revolves around the center of the circle and casting a moving shadow on a screen S parallel to the x axis. The ''shadow particle'' P' executes simple harmonic motion.

Next, let us turn to the velocity of P'. Refer again to Figure 6.3, and note that the velocity vector **V** of P, being tangent to the circle, is always perpendicular to the phasor **A**. Let us give the name v to the x component of **V**. The figure shows that

$$v = -|\mathbf{V}| \sin \theta.$$

We use Equations 6.8 and 6.11 to make the substitutions $\theta = \omega t$ and $V = \omega A$. This gives us

$$v = -\omega A \sin \omega t. \tag{6.14}$$

By using Equation 6.12, $\omega = 2\pi/\tau$, you can see that Equation 6.14 is the same as Equation 6.5, which expresses the velocity of a harmonic oscillator. Thus, the velocity of the ''shadow particle'' P' conforms with the rules of simple harmonic motion, as well as its displacement.

Let us now consider the acceleration of P'. As you saw in Section 5.4, the uniform circular motion of the particle P results in a centripetal acceleration

$$-\frac{V^2}{A} \hat{\mathbf{A}}.$$

The direction of this acceleration, opposite that of the phasor **A**, is denoted by the minus sign. The acceleration a of P' is the x component of the acceleration of P:

$$a = -\frac{V^2}{A} \cos \theta.$$

---

*Compare with Equation 5.28.

Again using $\theta = \omega t$ and using Equation 6.11 to substitute $V^2 = \omega^2 A^2$, we obtain a in the form

$$a = -\omega^2 A \cos \omega t. \qquad (6.15)$$

Remember that $\omega = 2\pi/\tau$ (Equation 6.12). Inserting this value of ω into Equation 6.15, we have $a = -(4\pi^2/\tau^2)(\cos 2\pi t/\tau)$. This is identical with the value of a given by Equation 6.7, which expresses the acceleration of a harmonic oscillator. Thus, the acceleration of P' conforms with the rules of simple harmonic motion, just as its velocity and displacement do. The reference circle approach shows, in a clear geometric manner, the important fact that the acceleration a of a harmonic oscillator is equal to a negative constant $(-\omega^2)$ times the displacement x ($= A \cos \omega t$):

$$a = -\omega^2 x. \qquad (6.16)$$

The acceleration and the displacement are always in opposite directions, and the acceleration is always proportional to the displacement, the proportionality constant being the square of the angular frequency, ω^2. We will make use of this remarkably simple relation between displacement and acceleration when we consider the equation of motion of a harmonic oscillator from a dynamical point of view in Section 6.4.

Frequency, Angular Frequency, and Period

In the preceding subsection, we defined the angular velocity of the particle P of Figure 6.3 by means of Equation 6.12, $\omega = 2\pi/\tau$. However, the quantity ω also enters into the description of the motion of the corresponding harmonic oscillator P', as Equations 6.13, 6.14, and 6.15 show. The physical meaning of ω is indirect for the harmonic oscillator P' because, unlike P, P' is not revolving around anything at ω radians per second. A quantity that has more direct physical meaning for an oscillator is the number of oscillations that P' completes per second. This quantity is called the **frequency**; it is conventionally represented by the symbol ν. (This is the lowercase Greek letter nu. Take great care not to write it in such a way that it can be confused with the letter v. Velocity v and frequency ν are often found in the same mathematical expressions.) The frequency ν is directly related to the period τ. Because the period is the amount of time required to complete one oscillation and the frequency is the number of oscillations completed in unit time, the relation is

$$\nu = \frac{1}{\tau}. \qquad (6.17)$$

You can see from this equation that the SI unit of frequency is (1/second), or s^{-1}. However, the quantity ν is used so often in the sciences and engineering that its unit is given its own name, the **hertz**. We define

$$1 \text{ hertz (Hz)} \equiv 1 \text{ s}^{-1}. \qquad (6.18)$$

The unit is named after the German physicist Heinrich Hertz (1857–1894). Hertz was the first to demonstrate experimentally the generation, transmission, and reception of electromagnetic waves, a process that required painstaking frequency measurement. Hertz's work is discussed in more detail in Chapter 34. In older work, especially in English, the hertz is called the cycle per second, often abbreviated cps. But this terminology is now obsolete.

The connection between frequency ν and angular velocity ω follows from a comparison of Equations 6.17 and 6.12. The comparison shows that

$$\omega = 2\pi\nu. \qquad (6.19)$$

When it applies to a harmonic oscillator rather than a revolving particle, ω is usually called the **angular frequency** rather than the angular velocity. According to Equation 6.19, it is technically correct to specify ω in Hz, the same unit used for ν. But it is not customary to do so. Even when ω does not refer to a physical rotation, it is the common practice to specify ω in the unit rad/s.

Stretching SI

When the name *hertz* was adopted internationally in the 1960s, after having been used for many years in Germany and elsewhere in Europe, there was some resistance in the United States, especially among electrical engineers accustomed to the use of cps. One such person ingeniously suggested that the unit could appropriately be renamed the steinmetz, after the German-American electrical engineer Charles Proteus Steinmetz (1865–1923), a pioneer in the theory and technology underlying the commercial use of alternating-current electricity. According to the suggestion, the unit would then be denoted by Steinmetz's initials!

The harmonic oscillator equations can be written in terms of the frequency ν as well as the period τ (as in Equations 6.1, 6.5, and 6.7) or the angular frequency ω (as in Equations 6.13, 6.14, and 6.15). Before we do so, however, we make a further generalization that applies regardless of whether ν, τ, or ω is used. Underlying the equations for x, v, and a is a special assumption about the initial conditions: that at the zero of time, $t = 0$, the oscillator was located at $x = A$ and was instantaneously at rest, with $v = 0$. This is often a convenient choice. But it is not a necessary one, and in some cases it is a choice that cannot be made. If you refer to Figure 6.2, you will see that choosing some other set of initial conditions amounts to shifting all three sinusoids by the same amount along the positive (or negative) t axis. The algebraic equivalent of this shift is the addition of a negative (or a positive) constant to the argument of each sinusoidal function. This constant, called the **phase constant**, is conventionally denoted by the Greek letter δ (delta). The phase constant is an angle because it is part of the argument of a sinusoidal function. Usually, δ is most conveniently expressed in radians. Using the phase constant, we can express the motion of a simple harmonic oscillator quite generally. We have the kinematic expressions

$$x = A \cos\left(2\pi \frac{t}{\tau} + \delta\right) = A \cos(\omega t + \delta) = A \cos(2\pi\nu t + \delta), \tag{6.20}$$

$$v = \frac{dx}{dt} = -\frac{2\pi}{\tau} A \sin\left(2\pi \frac{t}{\tau} + \delta\right) = -\omega A \sin(\omega t + \delta) = -2\pi\nu A \sin(2\pi\nu t + \delta), \tag{6.21}$$

and $$a = \frac{d^2x}{dt^2} = -\frac{4\pi^2}{\tau^2} A \cos\left(2\pi \frac{t}{\tau} + \delta\right) = -\omega^2 A \cos(\omega t + \delta) = -4\pi^2\nu^2 A \cos(2\pi\nu + \delta). \tag{6.22}$$

If δ is positive, the sinusoidal curves of Figure 6.2 are shifted to the left; if δ is negative, they are shifted to the right.

The phase constant δ can also be interpreted in terms of the reference circle, as shown in Figure 6.4. We have already described the angular displacement of the phasor **A** in the form of Equation 6.8, $\theta = \omega t$. However, this describes the special case in which the phasor lies along the positive x axis at the zero of time. We can generalize the description so that it applies to any conditions at all by rewriting the angular displacement as

$$\theta = \omega t + \delta. \tag{6.23}$$

When $t = 0$, we have $\theta = \delta$. You should satisfy yourself that the meaning of δ in Equation 6.23 corresponds to the meaning of the phase constant δ in Equations 6.20 through 6.22.

When the variable $\omega t + \delta$ appears in an equation describing a harmonic oscillator, as in Equations 6.20 through 6.22, it is called the **instantaneous phase** of the oscillator. We have already noted that a uniformly rotating radius vector such as **A** in Figures 6.3 and 6.4, which is used in the analysis of simple harmonic motion, is called a *phasor*. The name calls attention to the fact that the angular displacement θ of **A** corresponds to the instantaneous phase $\omega t + \delta$ of the associated harmonic oscillator. We will use phasors again in developing the theory of alternating-current circuits in Chapter 33 and the theory of diffraction of light in Chapter 36.

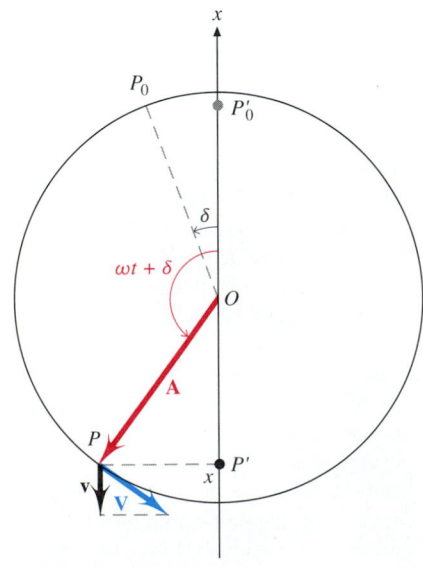

FIGURE 6.4 Generalized geometric view of the connection between a particle P' in simple harmonic oscillation and the phasor **A** at an arbitrary instant t. The magnitude of **A** is the constant A, and the instantaneous phase of **A** is $\omega t + \delta$. The instantaneous velocity of the point P at the head of **A** is given by the vector **V**. The instantaneous position of P' is given by the component vector **x** of **A**. Its velocity, given by **v**, is the vector component of **V** on the x axis. Also shown is the way in which the phase constant δ specifies the *initial* positions (at $t = 0$) P_0 and P_0'.

EXAMPLE **6.1**

The motion of an oscillator is described by Equation 6.20 in the form

$$x = A \cos (\omega t + \delta),$$

with $A = 8.52 \times 10^{-2}$ m, $\omega = 29.6$ rad/s, and $\delta = \dfrac{\pi}{3}$ rad. The initial conditions are depicted in Figure 6.5. **(a)** Find the frequency ν and the period τ of the oscillator. **(b)** Find the displacement, the velocity, and the acceleration of the oscillator at $t = 0$. **(c)** Find x, v, and a at $t = 5.50 \times 10^{-2}$ s.

Caution: When preparing to make calculations involving angles measured in radians, always make sure that your calculator is set to radian mode.

SOLUTION:
(a) Find the frequency ν and the period τ.
From Equation 6.19, you have $\nu = \omega/2\pi$. So the frequency is

$$\nu = \frac{29.6 \text{ rad/s}}{2\pi} = 4.71 \text{ Hz}.$$

To find the period, you use Equation 6.12, $\tau = 2\pi/\omega$. This gives you

$$\tau = \frac{2\pi}{29.6 \text{ rad/s}} = 0.212 \text{ s}.$$

Alternatively, you can use Equation 6.17 to write $\tau = 1/\nu = 1/(4.71 \text{ Hz}) = 0.212$ s.

(b) Find x, v, and a at $t = 0$.
At $t = 0$, the displacement equation becomes $x = A \cos \delta$. Using the numerical values given, you have for the displacement

$$x = 8.52 \times 10^{-2} \text{ m} \times \cos \frac{\pi}{3} = 4.26 \times 10^{-2} \text{ m}.$$

This displacement is one-half the amplitude, because

$$\cos \frac{\pi}{3} = \frac{1}{2}.$$

For the velocity, you use Equation 6.21 in its "ω form." Again setting $t = 0$, you have $v = -\omega A \sin \delta$. The numerical values give you

$$v = -29.6 \text{ rad/s} \times 8.52 \times 10^{-2} \text{ m} \times \sin \frac{\pi}{3} = -2.18 \text{ m/s}.$$

The minus sign indicates that the particle is moving in the negative direction. Because its displacement is positive, the particle is moving away from the point of maximum positive displacement, A, and toward $x = 0$.

You can use Equation 6.22 to write the acceleration at $t = 0$ as $a = -\omega^2 A \cos \delta$. This gives you

$$a = -(29.6 \text{ rad/s})^2 \times 8.52 \times 10^{-2} \text{ m} \times \cos \frac{\pi}{3}$$

$$= -37.3 \text{ m/s}^2.$$

The acceleration is in the negative direction, whereas the displacement is positive. As always for a harmonic oscillator, the acceleration and displacement are in opposite directions (Equation 6.16). Because the acceleration has the same sign as the velocity, the particle is speeding up as it moves toward $x = 0$.

(c) Find x, v, and a at $t = 5.50 \times 10^{-2}$ s.
When $t = 5.50 \times 10^{-2}$ s, you have, from Equation 6.20,

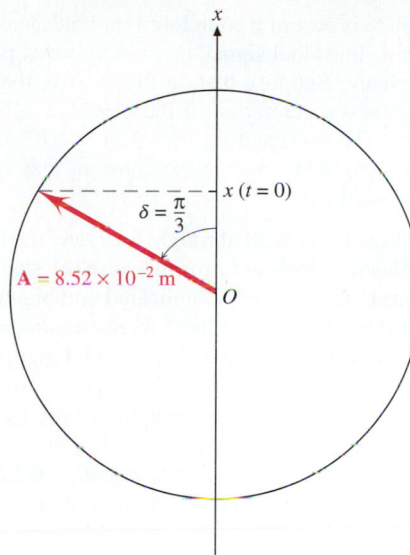

FIGURE 6.5 The phase constant δ, the amplitude A, and the initial position $x(t = 0)$ of the oscillator are shown.

$$x = 8.52 \times 10^{-2} \text{ m}$$

$$\times \cos \left(29.6 \text{ rad/s} \times 5.50 \times 10^{-2} \text{ s} + \frac{\pi}{3} \right) \quad (6.24)$$

$$= 8.52 \times 10^{-2} \text{ m} \times (-0.893),$$

or $x = -7.61 \times 10^{-2}$ m.

This negative displacement implies that the oscillator has moved from its initial positive displacement past the point $x = 0$. Mathematically, this transition from positive to negative displacement is expressed by a change in sign of the cosine. Explicitly evaluating the instantaneous phase (the quantity given in parentheses in Equation 6.24) gives 2.68 radians, which lies between $\pi/2$ rad (≈ 1.57 rad) and π rad (≈ 3.14 rad). Thus, the instantaneous phase lies in the second quadrant, and the value of the cosine is negative.

You can find the velocity from Equation 6.21, which gives

$$v = -29.6 \text{ rad/s} \times 8.52 \times 10^{-2} \text{ m}$$

$$\times \sin \left(29.6 \text{ rad/s} \times 5.50 \times 10^{-2} \text{ s} + \frac{\pi}{3} \right)$$

$$= -1.13 \text{ m/s}.$$

The oscillator is still moving in the negative direction, as the sign of v indicates.

Equation 6.22 yields the acceleration. Inserting the proper numerical values, you have

$$a = -(29.6 \text{ rad/s})^2 \times 8.52 \times 10^{-2} \text{ m}$$

$$\times \cos \left(29.6 \text{ rad/s} \times 5.50 \times 10^{-2} \text{ s} + \frac{\pi}{3} \right)$$

$$= 66.7 \text{ m/s}^2.$$

Here again, the direction of the acceleration is opposite that of the displacement. The direction of a also is opposite that of v; thus the particle is slowing down.

EXAMPLE 6.2

A flat metal plate is part of a sound-system loudspeaker. When it is driven by a sinusoidal signal, its motion is that of a simple harmonic oscillator. Suppose that, at time $t = 0$, the displacement, velocity, and acceleration of the plate are $x_0 = 6.56 \times 10^{-5}$ m, $v_0 = 1.11$ m/s, and $a_0 = -3.71 \times 10^3$ m/s^2. Find the frequency, amplitude, and phase constant that specify the motion of the oscillator.

SOLUTION: Equations 6.20 through 6.22 are most conveniently manipulated in their ω forms, because the single symbol ω takes the place of the more complicated combinations $2\pi/\tau$ or $2\pi\nu$ in the other forms. We have three simultaneous equations to be solved for ω, A, and δ. As a first step in solving these equations, note that Equation 6.20 for the displacement and Equation 6.22 for the acceleration differ only by the factor $-\omega^2$. This is consistent with the general rule given by Equation 6.16, $a = -\omega^2 x$. Indeed, if you divide Equation 6.22 by Equation 6.20, you have $a/x = -\omega^2$. In the special case $t = 0$, this becomes

$$-\omega^2 = \frac{a_0}{x_0}. \qquad (6.25)$$

You can solve this equation for the angular frequency ω, obtaining

$$\omega = \sqrt{-\frac{a_0}{x_0}}. \qquad (6.26)$$

The negative sign within the square root guarantees that ω will be a real number.

You can now insert the numerical values of the initial acceleration and displacement into Equation 6.26:

$$\omega = \sqrt{-\frac{-3.71 \times 10^3 \text{ m/s}^2}{6.56 \times 10^{-5} \text{ m}}} = 7.52 \times 10^3 \text{ rad/s}.$$

To express this result as a frequency ν, you divide by 2π in accordance with Equation 6.19:

$$\nu = 1200 \text{ Hz}.$$

The next task is to solve for the phase constant δ. The easiest way to do this is to take a hint from part **b** of Example 6.1 and write Equations 6.21 and 6.20 for the special case $t = 0$. This simplifies the argument of the two sinusoids, $\sin(\omega t + \delta)$ and $\cos(\omega t + \delta)$, leaving

$$v_0 = -\omega A \sin \delta \qquad (6.27a)$$

and

$$x_0 = A \cos \delta. \qquad (6.27b)$$

This is a pair of simultaneous equations in which the unknowns are δ and A. To eliminate A and thus solve for δ, you can divide the first of these equations by the second:

$$\frac{v_0}{x_0} = -\omega \tan \delta.$$

Solving for δ, you obtain

$$\delta = \tan^{-1}\left(-\frac{v_0}{\omega x_0}\right). \qquad (6.28)$$

All the numerical quantities on the right side of this equation are known, so you find

$$\delta = \tan^{-1}\left(-\frac{1.11 \text{ m/s}}{7.52 \times 10^3 \text{ rad/s} \times 6.56 \times 10^{-5} \text{ m}}\right)$$
$$= \tan^{-1}(-2.25).$$

If you use your calculator to evaluate $\tan^{-1}(-2.25)$, it will give you the result as the negative angle -1.15 rad. This is equivalent to the positive angle $(-1.15 + 2\pi)$ rad $= 5.13$ rad ($\approx 294°$), which lies in the fourth quadrant. But the tangent function is negative in the second quadrant as well, and there is an angle in that quadrant whose tangent has the same value. The general rule is $\tan \theta = \tan(\theta \pm \pi)$, so the other angle is $(5.13 - \pi)$ rad $= 1.99$ rad ($\approx 114°$). Which of these values is the correct phase constant δ? To find the answer, look back at Equations 6.27a and 6.27b. Both v_0 and x_0 are positive. According to the equations, this can be true only if $\sin \delta$ is negative and $\cos \delta$ is positive. This is true in the fourth quadrant, but not in the second, so you have

$$\delta = 5.13 \text{ rad}.$$

Finally, you evaluate the amplitude A. Because you already know ω and δ, any of Equations 6.20 through 6.22 can be used for this purpose by setting $t = 0$. Equation 6.27b is easiest to use because it has the simplest form of all. Solving that equation for A, you obtain

$$A = \frac{x_0}{\cos \delta} = \frac{6.56 \times 10^{-5} \text{ m}}{\cos 5.13} = 1.62 \times 10^{-4} \text{ m}.$$

One way to summarize the results of this example is to write the displacement equation, $x = A \cos(\omega t + \delta)$, in the special form

$$x = 1.62 \times 10^{-4} \text{ m}$$
$$\times \cos[(7.52 \times 10^3 \text{ rad/s})t + 5.13 \text{ rad}].$$

Equilibrium and Hooke's Law

We now proceed to the *dynamics* of oscillatory motion. A body oscillates because it is under the influence of a force—not an arbitrary force, but one that satisfies certain conditions. On the basis of everyday observations of things that oscillate, we can make a loose statement of these conditions: Whenever the body is located away from a fixed point, called the **equilibrium point**, the force that it experiences is directed *toward* the equilibrium point. This implies that the direction of the force must change whenever the body passes through the equilibrium point. When the body is *at* the equilibrium point, therefore, the force must be zero. Figure 6.6a illustrates such a case.

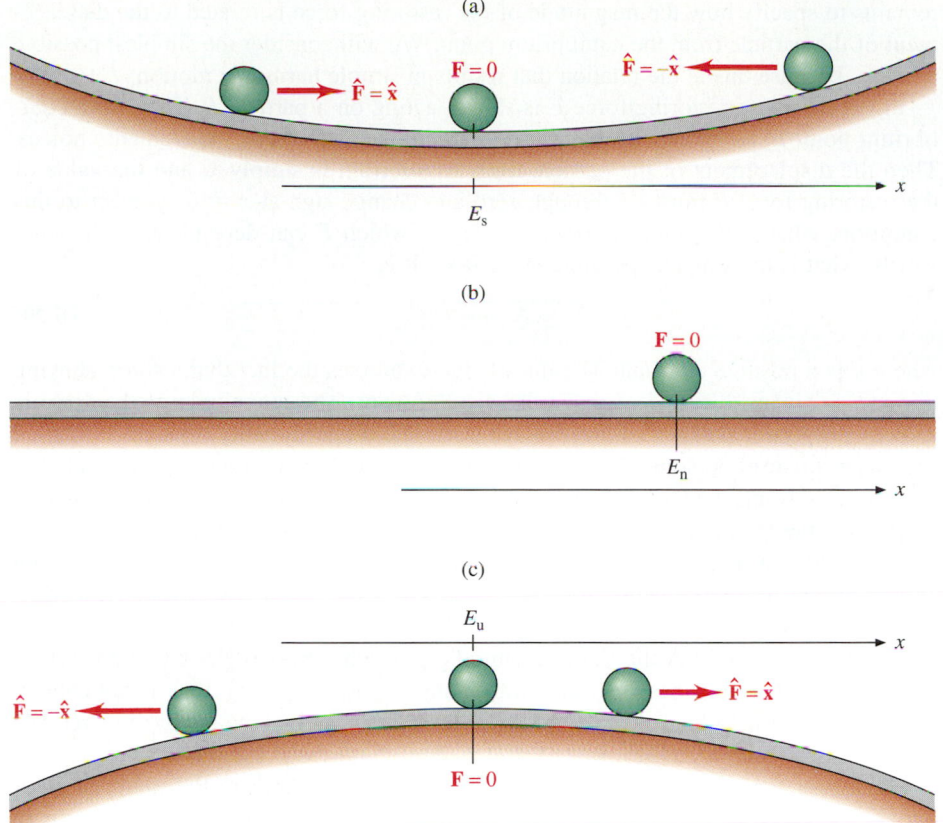

(a)

$\hat{\mathbf{F}} = \hat{\mathbf{x}}$ $\mathbf{F} = 0$ $\hat{\mathbf{F}} = -\hat{\mathbf{x}}$

E_{s} x

(b)

$\mathbf{F} = 0$

E_{n} x

(c)

E_{u} x

$\hat{\mathbf{F}} = -\hat{\mathbf{x}}$ $\hat{\mathbf{F}} = \hat{\mathbf{x}}$

$\mathbf{F} = 0$

FIGURE 6.6 Types of equilibrium. A ball rides on a track. In each case, no force acts on the ball when it is at the equilibrium point E. (*a*) The ball at E_{s} is in stable equilibrium. Should it move away from E_{s}, a force acts so as to tend to move it back to E_{s}. (*b*) Neutral equilibrium. There is no force tending to return the ball to E_{n} if it moves away from E_{n}. (*c*) Unstable equilibrium. If the ball moves away from E_{u}, a force acts so as to tend to move it still farther away.

When the body is at rest at the equilibrium point E_{s}, there is nothing in the system tending to make it move away from that point. However, the equilibrium is stronger than this negative statement implies by itself. Suppose that the body *does* move ever so slightly away from E_{s}, for whatever reason. (To give one possibility, there might be a slight vibration in the system of Figure 6.6a.) A force will immediately come into play, tending to propel the body back toward the equilibrium point. This condition is called **stable equilibrium**, and the force is called a **restoring force**.

Next, consider the system shown in Figure 6.6b. Here again, there is nothing in the system tending to make the body move away from the equilibrium point E_{n} if it is initially at rest there. However, should it move away from that point, there is nothing to make it return, either. This weaker form of equilibrium is called **neutral equilibrium**.

Finally, consider the system of Figure 6.6c. When the body is at rest at the equilibrium point E_{u}, there is again nothing in the system tending to make it move away. But, in this case, should it move away from E_{u} ever so slightly, a force will immediately come into play that tends to propel it still farther away from E_{u}. This case is called **unstable equilibrium**.

To summarize, all three types of equilibrium share the property that a body at the equilibrium point experiences no force. Each of the three types is characterized by the direction of the force that comes into play when the body is elsewhere, as follows:

Having faith in Hooke's law.

In $\begin{bmatrix} stable \\ neutral \\ unstable \end{bmatrix}$ equilibrium, any displacement of the body from the equilibrium point results in a force whose direction is $\begin{bmatrix} opposite\ that\ of\ the\ displacement \\ null \\ the\ same\ as\ that\ of\ the\ displacement \end{bmatrix}$.

In this chapter, we are concerned only with systems having a point of stable equilibrium, because oscillation occurs only in such systems. (Can you explain why?) Further, we will restrict our study to systems in which there is only one particle and the motion of that particle is one-dimensional. Even with these restrictions, however, it

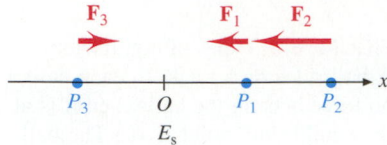

FIGURE 6.7 A particle can move along the x axis, subject to a force F described by the force law $F = -kx$. The form of this law implies the existence of a stable equilibrium point E_s at the origin. The particle is shown at three different possible positions, P_1, P_2, and P_3. In each case, the magnitude and direction of F are depicted schematically.

Robert Hooke

In 1678, Hooke published a detailed account of the behavior of springs, in which he clearly enunciated the proportionality of Equation 6.29. His study of springs, begun about 1660, led among other things to his invention of the hairspring and balance wheel escapement, which made accurate pocket watches feasible; it remained the basic timekeeping mechanism of all practical watches until the mid-1960s. Hooke was a pioneer in the design and use of microscopes, and he made many speculations concerning light, heat, and gravitation that came to fruition in the hands of later investigators. He was the first person to use the word "cell" in its biological sense, as well as one of the first to insist that fossil plants and animals provide a historical record of earth far longer than the few thousand years of the biblical account.

remains to specify how the magnitude of the restoring force is related to the displacement of the particle from the equilibrium point. We will consider the simplest possible relation, because this is the relation that results in simple harmonic motion.

In Figure 6.7, a restoring force F is shown acting on a particle P. The stable equilibrium point E_s is located at the origin (almost always the most convenient choice). Then the displacement of the particle from equilibrium is simply x, and the value of the restoring force F must go through zero and change sign at $x = 0$. Subject to this condition, what is the simplest possible way in which F can depend on x? In other words, what is the simplest possible **force law**? It is

$$F = -kx, \tag{6.29}$$

where k is a positive constant. The minus sign expresses the fact that a force obeying this law is always directed opposite the displacement. The magnitude of the force is directly proportional to that of the displacement. The proportionality constant k is called the **force constant** or, sometimes, the **spring constant**. The latter name arises from the fact that a body attached to an ordinary spring experiences a force described by Equation 6.29 when the spring is stretched by an amount x from its *relaxed* (unstretched) condition. A large value of k implies that a large force must be applied to produce a given stretch x; that is, a large value of k corresponds to a *stiff* spring. (Conversely, a small value of k corresponds to a *weak* spring.) The force law expressed by Equation 6.29 is called **Hooke's law**, after the English natural philosopher Robert Hooke (1635–1703). Hooke's law also describes the behavior of many important systems that do not contain springs in the usual sense of the word. Most solid objects obey the law when stretched or compressed, provided they are not stretched or compressed too far.

No matter what system we consider, there is *always* a limit to the range of x for which Hooke's law is valid. Even the "springiest" springs will fail to conform to Hooke's law if they are stretched too far—that is, if a critical value of x is exceeded. If they are stretched still more, they will eventually distort permanently. Figure 6.8 shows the force exerted *by* a typical coil spring with open coils (that is, coils that do not touch their neighbors when the spring is relaxed). This force is exerted *on* the body that is stretching or compressing the spring.

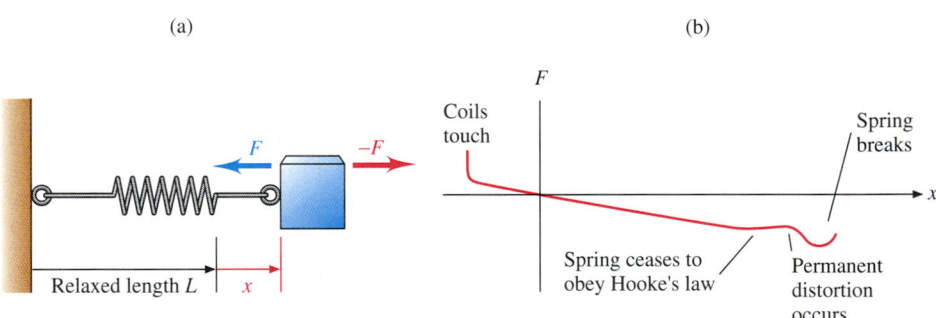

FIGURE 6.8 General properties of the force law for (*a*) a typical open coil spring of relaxed length L, such as a bedspring or an automobile suspension spring. (*b*) The quantity F shown on the ordinate axis is the force exerted *by* the spring *on* the body at its right end; it is that body which is supplying the force $-F$ that stretches the spring. The linear part of the curve, which passes through the origin, represents the "Hooke's law" region of the spring's behavior. Normally, a system is designed so that the spring cannot be stretched or compressed beyond this region. If this precaution is not taken, the spring will not fulfill its function well and will likely suffer permanent damage. If the spring is compressed too far, its neighboring coils touch and it ceases to be "springy" on the scale of magnitude for which it is designed. (Will it exhibit Hooke's-law behavior on any scale at all?) If the spring is stretched too far, it first becomes "weaker"; that is, the restoring force it provides to the body at its end no longer increases in proportion to the displacement. Further stretching results in permanent distortion, so that the spring will no longer return to its original shape. Ultimately, the spring will break.

Measuring the Force Constant

For a particular spring, the value of the force constant k can be determined by direct measurement. One simple way of doing this is shown in Figure 6.9a. The relaxed length of the spring is L. A weight W_1 is now suspended from the spring, which stretches. When the system is again at rest, the force exerted by the spring on the weight has magnitude $|F_1| = W_1$, in accordance with Newton's third law. The length $L + x_1$ of the extended spring is measured. The weight is then increased to W_2 and the new length $L + x_2$ measured. By repeating this process with several weights, one can acquire the data needed to plot a graph like that of Figure 6.9b. If the data points lie on a straight line, the spring obeys Hooke's law within the range of observation. This follows from the fact that $|F| = |-kx|$ is the equation of a straight line of slope k. In terms of any two measurements i and j, the slope of the line is the force constant:

$$k = \left| \frac{\Delta F}{\Delta x} \right| = \frac{\Delta W}{\Delta x} = \frac{W_j - W_i}{(L + x_j) - (L + x_i)} = \frac{W_j - W_i}{x_j - x_i}.$$

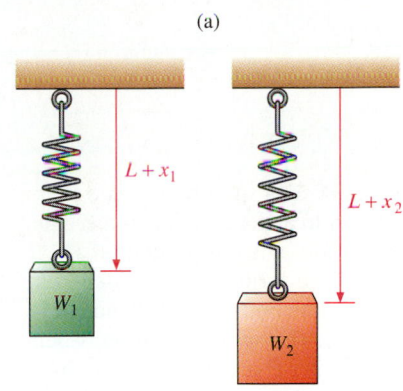
(a)

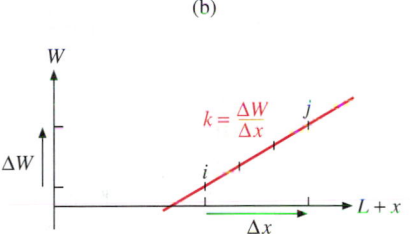
(b)

FIGURE 6.9 (a) Determining the force constant k of a coil spring. With weights $W_1, W_2, \ldots$ hanging from the spring, the corresponding lengths $L + x_1, L + x_2, \ldots$ are measured. (b) The data are plotted. The linear dependence of length on weight indicates that the spring is conforming to Hooke's law. By using two points on the curve, such as i and j, the slope $\Delta W / \Delta x$ is determined. This slope is the force constant k. Why is the slope positive, in contrast to the negative slope of the main part of the curve of Figure 6.8b? What sign convention would you use if you were measuring k for a spring designed to be compressed rather than stretched? for a spring that could be either stretched or compressed?

EXAMPLE 6.3

In the laboratory, you measure the length of a spring as a function of the weight you hang on it. You tabulate the experimental data as in Table 6.1. (a) Specify the region in which the spring obeys Hooke's law. (b) Find the force constant k.

TABLE 6.1 Data for Example 6.3

Weight W (N)	Spring Length $L + x$ (cm)
0	12.1
10.0	15.9
15.0	17.7
20.0	19.6
25.0	21.5
30.0	23.4
35.0	25.2
40.0	28.2
45.0	32.0

SOLUTION:

(a) Specify the region in which the spring obeys Hooke's law.

You begin by plotting a graph of W versus $L + x$, like that of Figure 6.10. (Although the measurements of spring length are most conveniently made in centimeters, as shown in Table 6.1, plotting them in meters allows a direct computation of the

FIGURE 6.10

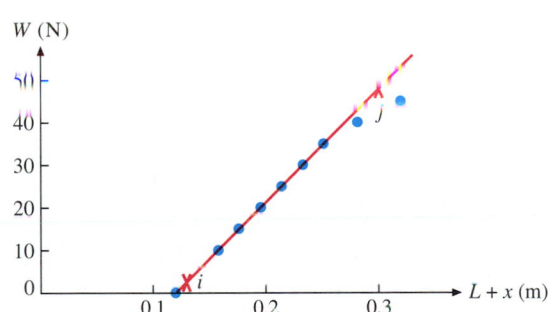

slope in the SI unit N/m.) The graph clearly shows that the linear portion of the curve ends at about 35 N. (It is almost always desirable to plot graphs as you go along when making experimental measurements. By doing so in a case like this, you greatly reduce the likelihood of overloading the spring and ruining it.)

(b) Find the force constant k.

You draw the best straight line through the data points on the linear portion of the curve. Once this is done, you can determine the slope using any two points on the line. Using, for instance, the two points marked i and j, you have

$$k = \frac{\Delta W}{\Delta x} = \frac{47.8 \text{ N} - 2.0 \text{ N}}{0.300 \text{ m} - 0.130 \text{ m}} = 269 \text{ N/m}.$$

The Dynamics of Simple Harmonic Motion

We now show that a net Hooke's-law force acting on a body sets the body into simple harmonic motion. Consider the air-track system of Figure 6.11. On account of its displacement x from the equilibrium point E_s, the body (here, the glider) is subject to a restoring force $F = -kx$. No other force acts on the body in a horizontal direction, and F is thus the net force driving it. As always in making a dynamical analysis of the motion of a system, we use Newton's second law, $F = ma$, to relate the driving force to the motion that arises from it. Using the Hooke's-law value for F, we have*

$$-kx = ma. \tag{6.30}$$

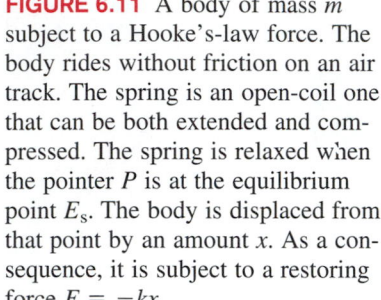

FIGURE 6.11 A body of mass m subject to a Hooke's-law force. The body rides without friction on an air track. The spring is an open-coil one that can be both extended and compressed. The spring is relaxed when the pointer P is at the equilibrium point E_s. The body is displaced from that point by an amount x. As a consequence, it is subject to a restoring force $F = -kx$.

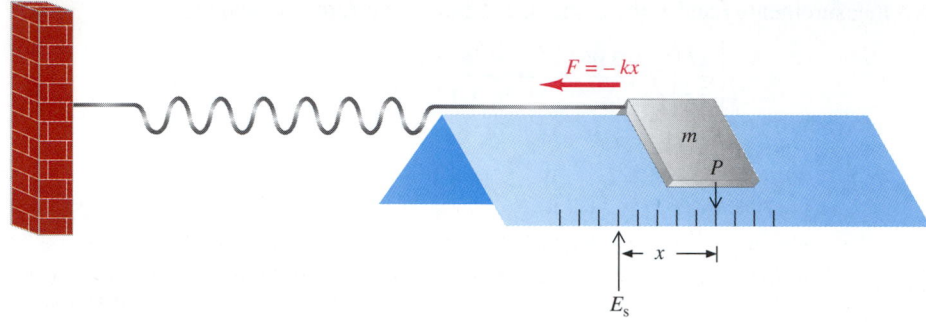

We solve this equation for the instantaneous acceleration a, obtaining

$$a = -\frac{k}{m} x. \tag{6.31}$$

For any given system, the quantity k/m is a constant. A stiff spring has a large force constant k, leading to a large acceleration for a given displacement. A large mass leads to a small acceleration because of its large inertia. The minus sign tells us that the acceleration, like the restoring force that produces it, is always directed opposite the displacement.

In solving Equation 6.31, we cannot use the kinematic equations for constant acceleration, developed in Chapter 2, because here the acceleration is *not* constant. Rather, a is directly proportional to x, and so we must find a new approach. As a first step, we remember that the acceleration is the second derivative of the displacement with respect to time; that is, $a \equiv d^2x/dt^2$. Substituting this definition into Equation 6.31, we obtain

$$\frac{d^2x}{dt^2} = -\frac{k}{m} x. \tag{6.32}$$

This equation connects the dependent variable x with the independent variable t. The displacement x appears in the equation directly and also in its second derivative. Such an equation is called a **second-order differential equation**. Although you may not yet have encountered such an equation in your study of mathematics, don't worry. You will see the equation yield to the most elementary and universal technique of equation solving—guessing at the answer and then checking to see if the guess was correct!

It is important to make an educated guess and not a random one, and we can get some hints by studying Equation 6.32. It tells us that x and its second derivative differ only by a negative constant factor $-k/m$. We have already noted the same thing about the harmonic oscillator equations. We have Equation 6.20,

$$x = A \cos (\omega t + \delta),$$

and Equation 6.22, $\qquad \dfrac{d^2x}{dt^2} = -\omega^2 A \cos (\omega t + \delta).$

---

*In writing this equation, we have assumed that the mass of the spring is negligible compared with the mass of the body. Otherwise, the inertia of the spring would figure into the equation of motion. This would introduce a complication because different parts of the spring are displaced by different amounts as the spring stretches.

This looks very promising. Let us guess that the solution to Equation 6.32 *is x =* *A* cos ($\omega t + \delta$) and insert these values into Equation 6.32 to test our guess. We obtain

$$-\omega^2 A \cos(\omega t + \delta) = \left(-\frac{k}{m}\right)[A \cos(\omega t + \delta)].$$

The minus signs, the amplitudes *A*, and the cosine terms cancel out, leaving

$$\omega^2 = \frac{k}{m}, \qquad \textbf{(6.33)}$$

or

$$\omega = \sqrt{\frac{k}{m}}. \qquad \textbf{(6.34a)}$$

What does Equation 6.34a tell us? It says that the guess that we made for *x* is indeed correct—*if and only if the angular frequency is* $\omega = \sqrt{k/m}$. To put it another way, the angular frequency of the system is determined by the force constant and the mass. And we have established that *a body subjected to a Hooke's-law force oscillates in simple harmonic motion*. Such a body therefore obeys the kinematic rules derived in Section 6.2, with the additional dynamical proviso of Equation 6.34a. For this reason, Equation 6.32, which is the equation of motion of a body subjected to a Hooke's-law force, is called the **harmonic-oscillator equation**. It is one of the most frequently encountered equations in mechanics.

Equation 6.34a can be written in equivalent forms in terms of the frequency ν and the period τ.

$$\nu = \frac{1}{2\pi}\sqrt{\frac{k}{m}} \quad \text{and} \quad \tau = 2\pi\sqrt{\frac{m}{k}}. \qquad \textbf{(6.34b,c)}$$

Equations 6.34a, 6.34b, and 6.34c do appear to conform to the intuitive notions that most persons develop on thinking about the mass-spring system. To be specific, let us discuss these qualitative notions in terms of Equation 6.34c. The mass *m* in the system functions inertially and thus seems to tend to "hold the system back." As far as this idea goes, it is certainly in concord with Equation 6.34c, which tells us that the period τ increases as the mass increases. That is, the larger the mass, the longer the time the system takes to complete one oscillation cycle. On the other hand, the spring "drives" the system, supplying the force needed to keep the mass in continuous acceleration, as required by Equation 6.31, $a = -(k/m)x$. Thus, it seems plausible that an increase in the stiffness of the spring—that is, an increase in *k*—corresponds to a decrease in the period. What is not so intuitively evident is that *the period of a simple harmonic oscillator does not depend on the oscillation amplitude A of the system*. This is evidenced by the fact that *A* appears nowhere in the equivalent forms of Equation 6.34.* We say that the system is **isochronous** (from the Greek roots meaning "same time") or that it exhibits **isochrony** (pronounced i-**sok**′-ro-ny).

EXAMPLE 6.4

Consider again the system of Example 6.3. Find its oscillation frequency ν when the spring is loaded with a body having weight 10 N and when it is loaded with a body having weight 30 N. Assume that the mass of the spring is negligible.

SOLUTION: First, you must find the net force exerted on the hanging body. If (and only if) the net force is a Hooke's-law

force, you can apply the equations that we have derived for simple harmonic motion. Let the mass of the body have the arbitrary value *m*. Figure 6.12a shows the spring hanging freely; its length is *L*. The coordinate system is set up with the positive direction downward and with the origin at the lower end of the spring, a distance *L* below the support. If the body is then attached and lowered gently until it comes to rest, the system

---

*It may seem surprising that no additional time is required to traverse a larger distance. But a larger *A* is associated with a larger mean force $\langle|F|\rangle$, and hence with a larger mean acceleration $\langle|a|\rangle$, and therefore with a larger mean speed $\langle|v|\rangle$. If you work Problem 6.32, you will see explicitly that increases in *A* and $\langle|v|\rangle$ compensate exactly.

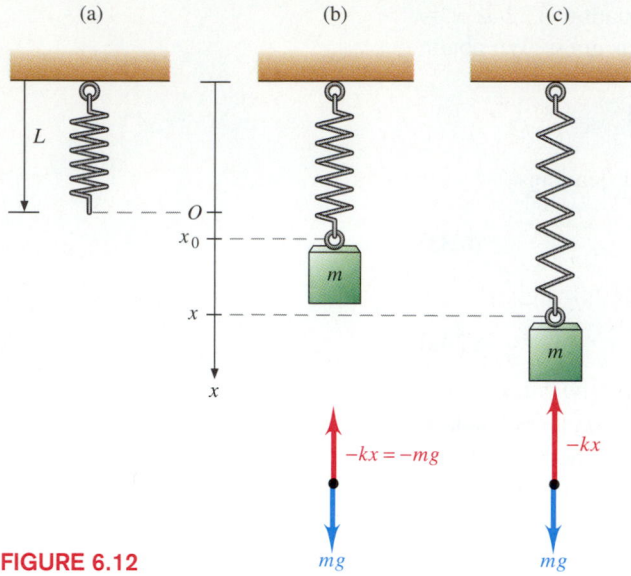

(a)   (b)   (c)

FIGURE 6.12

appears as in Figure 6.12b. The body is hanging in equilibrium; the corresponding free-body diagram is shown below the system. If the downward direction is taken as positive, the downward gravitational force exerted on the body is $F_g = mg$. If the body is to be in equilibrium, the net force acting on it must be zero. This can be true only if the spring exerts an upward force $F_s = -mg$; then we have the net force

$$F = F_g + F_s = mg - mg = 0.$$

We have assumed that the spring conforms to Hooke's law, $F = -kx$. In order to exert the force $F_s = -mg$, therefore, the spring must be stretched from its relaxed length L by an amount x_0, whose value is such that $-mg = -kx_0$, or

$$mg = kx_0. \qquad (6.35)$$

When the body is at the equilibrium point, the stretch of the spring is

$$x_0 = \frac{mg}{k}. \qquad (6.36)$$

The length $L + x_0$ is the equilibrium length of the spring with the body hanging from it.

We now prove that, if the body is set into motion, it oscillates with simple harmonic motion about the equilibrium point. Suppose that the body is pulled to the arbitrary position x and released. Figure 6.12c depicts the system just after release. (The figure shows the body located below the equilibrium position; that is, $x > x_0$. But the analysis that follows is just as valid when the body is above x_0.) The free-body diagram is once again shown below the system. The net force on the body is no longer zero; its value is

$$F = mg - kx.$$

Substituting in the value of mg given by Equation 6.35, we have $F = kx_0 - kx$, or

$$F = -k(x - x_0). \qquad (6.37)$$

This is Hooke's law again; the only change is that the net force on the body is now zero at $x = x_0$ rather than at $x = 0$. But x_0 is the equilibrium point, as Figure 6.12b shows.

You have already seen that a body experiencing a net force that conforms to Hooke's law undergoes simple harmonic motion. The hanging body satisfies this condition.

You can now proceed to solve for the desired oscillation frequencies. The first step is to determine the appropriate mass. For the 10-N weight, you have

$$m_1 = \frac{W_1}{g} = \frac{10 \text{ N}}{9.80 \text{ m/s}^2} = 1.02 \text{ kg}.$$

The corresponding frequency is given by Equation 6.34b. The force constant evaluated in Example 6.3 is $k = 269$ N/m, so you have

$$\nu_1 = \frac{1}{2\pi} \sqrt{\frac{269 \text{ N/m}}{1.02 \text{ kg}}} = 2.59 \text{ Hz}.$$

You can calculate ν_2, the frequency of the system with the 30-N weight, in the same manner. However, it is easier to note that because $m_2 = 3m_1$, you have $\nu_2 = \nu_1/\sqrt{3}$, or

$$\nu_2 = 1.50 \text{ Hz}.$$

The Force in the Mass-Spring System

In solving the equation of motion for the mass-spring system, we eliminated the force at the first step, obtaining Equation 6.30, $-kx = ma$. But it is easy to show that the force F behaves in simple harmonic manner, just as the kinematic quantities x, v, and a do. Now that we have established the simple harmonic behavior of the system, we can use Equation 6.33 to rewrite the acceleration given by Equation 6.22 in the form

$$a = \frac{d^2x}{dt^2} = -\frac{k}{m} A \cos(\omega t + \delta). \qquad (6.38)$$

We then use Newton's second law to find

$$F = ma = -kA \cos(\omega t + \delta). \qquad (6.39)$$

By comparing these two equations, you can see that the acceleration and the net force are always in phase, as they must be if they are to conform to the linear relationship $F = ma$. Equation 6.39 also shows that, as the cosine function varies between its extreme values 1 and -1, the force varies between the limits $-kA$ and kA.

Simple Harmonic Oscillators, Isochrony, and Clocks*

Isochrony—having a period independent of amplitude—is perhaps the most important property of harmonic oscillators. Because they are isochronous, small harmonic oscillating systems can be used to control the rate of clocks.

Primitive clocks do not contain oscillating control systems. Instead, they depend on some physical process that takes place at an approximately constant rate. A candle burns down, and the uppermost remaining mark on its side indicates the time. Or water drips out of a container through a small hole, and the level of the remaining water indicates the time. Or a weight on the end of a long rope unwinds the rope from a pulley at a rate controlled by the friction in the system, and a clock hand is attached to the pulley shaft. However, such clocks are never very accurate because the rate at which the essential process occurs is subject to change as conditions change. The slightest draft, for instance, influences the burning rate of the candle. Small temperature changes alter the flow rate of the water clock and the friction in the weight-driven clock.

Compare those primitive systems with accurate clocks, such as mechanical wristwatches. In a wristwatch, there is a complicated train of gears. At one end, the train is driven by the *mainspring*, and at the other end are the hands on the dial. But the watch does not depend on the natural unwinding rate of the mainspring to measure the passage of time. Instead, the unwinding rate is controlled by a small isochronous harmonic oscillator system. In mechanical watches, this oscillating system consists of a balance wheel and hairspring, as shown in Figure 6.13. (The balance wheel and hairspring are the rotational analogue of the mass-spring system, whose motion is translational. We

(a)

hairspring

balance wheel

FIGURE 6.13 (*a*) General view of the timing mechanism of a mechanical watch. The balance wheel, attached to the hairspring, oscillates by rotating back and forth on the low-friction ruby bearing. The escapement mechanism, which controls the rate at which the mainspring unwinds and thus the rate at which the watch hands move, also drives the balance wheel. The escapement consists of the balance staff (mostly not visible) and the escapement wheel, seen more clearly in the enlarged picture of part *b*. (*b*) The balance staff is forked at one end, and the fork ends *FF'* are made of ruby for low friction. Linked to the balance wheel by a pin mechanism, the balance staff and its forked ends oscillate back and forth as the balance wheel oscillates. The right tine *F'* of the fork is seen holding one of the hook-shaped teeth *E'* of the escapement wheel, which is spring loaded by a train of gears beginning at the mainspring. The right tine will move away from the tooth (arrow), allowing the tooth to escape and move to the right. The left tine *F* will move simultaneously (arrow), catching tooth *E''* of the escapement wheel. The small blow given *F* by *E''* is transmitted to the balance wheel and helps to keep the wheel in motion. The ticks you hear as the watch runs are produced by the collisions of the escapement wheel teeth with the tines of the escapement lever.

(b)

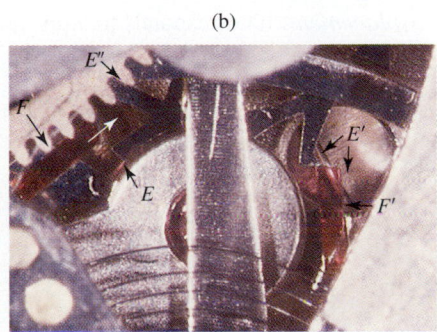

---

*This subsection may be omitted without loss of continuity.

will consider the analogy in quantitative detail in Chapter 12.) The oscillating system is linked to the main system through a small lever called the *balance staff*. At one end, the balance staff is linked to the balance wheel by small pins on the wheel, and therefore oscillates at the same frequency. The other end of the balance staff is forked, and the tines catch the teeth of a gear called the *escapement wheel*, which is part of the main-spring-driven main system. This keeps the main system from unwinding freely. With each oscillation of the balance staff, however, the escapement wheel is allowed to slip by one tooth. Thus, the main system advances by one small step with each highly uniform oscillation period of the balance-wheel system, and the position of the clock hands is controlled by the oscillator.

There is friction in even the most exquisitely designed balance-wheel system. What keeps the oscillation from dying down in a short time? Just after the balance staff releases a tooth of the escapement wheel, it catches the next tooth. Because the es-capement wheel is spring loaded, it gives a tiny impulse to the balance staff as it is caught. The impulse is transmitted to the balance wheel, giving it a small driving kick.

We can now understand why the isochrony of the oscillator is significant. When the mainspring is tightly wound, the impulses transmitted to the balance wheel are relatively large and the wheel oscillates with large amplitude. When the mainspring has partially run down, the impulses are smaller and so is the balance-wheel amplitude. But, because the system is isochronous, the amplitude change does not affect the running rate of the watch.

Other clocks use other mechanisms, both for the main drive and for the oscillating control system. In the old-fashioned pendulum clock, the place of the mainspring is taken by a heavy weight and that of the balance wheel by the approximately isochronous pendulum, which controls the escapement wheel. (Pendulums are discussed in Section 6.5.) In the modern quartz watch, the main drive system is an electronic counter. The control system is a tiny quartz crystal whose dimensions vary slightly, in an oscillatory fashion, at a stable frequency determined by the crystal's size, shape, and mechanical properties. The oscillations of the crystal are driven electrically, by means of an elec-tromechanical property called piezoelectricity. Each oscillation allows the counter to advance by one count and thus controls the reading on the watch face. Quartz watches operating in this manner are much more accurate and much cheaper than mechanical watches. The details of their operation are much more complicated than those pertaining to mechanical watches, but the underlying principle of control is exactly the same.

The Simple Pendulum

An ideal **simple pendulum** is shown in Figure 6.14a. The point mass m, called the **bob**, is suspended beneath a frictionless, rigidly anchored pivot by an inextensible, massless cord or rod of length l. To set the system into motion, the bob is drawn to one side from its vertical equilibrium position and then released. The figure depicts a moment when the cord makes an angle θ with the vertical. We call θ the *angular displacement*. It corresponds to a distance s measured along the circular path traversed

FIGURE 6.14 (*a*) A simple pendu-lum. A bob of mass m hangs at the lower end of a massless cord or rod of length l, attached at its upper end to a rigidly anchored pivot. The ideal system is frictionless. A real pendu-lum will approximate this ideal sys-tem adequately if the bob has dimen-sions small compared with l, the cord or rod has mass small compared with m, and the friction is not too great. (*b*) Free-body diagram for the simple pendulum.

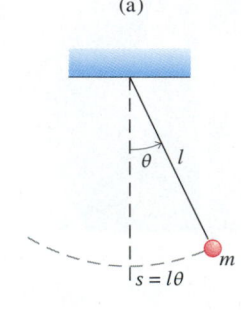

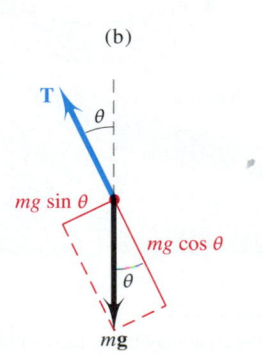

by the bob. With θ expressed in radians, the relation between them is

$$s = l\theta. \tag{6.40}$$

The free-body diagram of the pendulum is shown in Figure 6.14b. A downward gravitational force $m\mathbf{g}$ acts on the bob, as does a generally upward force T exerted by the cord. The free-body diagram is very much like that for a block resting on a frictionless inclined plane, which we considered at length in Chapter 4.

Just as in the inclined-plane case, we resolve the gravitational force into components directed opposite to $\hat{\mathbf{T}}$ (along the radius of the circular path) and perpendicular to $\hat{\mathbf{T}}$ (along the tangent to the path). These components are shown in red in Figure 6.14b. In view of the fixed length l of the cord, we need not be concerned with forces along $\hat{\mathbf{T}}$ because there is no motion along $\hat{\mathbf{T}}$. The direction of the driving force acting on the bob is thus along the tangent. The force can be written in the scalar form

$$F = -mg \sin \theta. \tag{6.41}$$

The negative sign is used to account for the fact that the sign of F is always opposite that of θ, and thus opposite that of $\sin \theta$ as well. That is, F is a restoring force. Using Equation 6.40, we can rewrite the force as

$$F = -mg \sin \frac{s}{l}. \tag{6.42}$$

But is the pendulum's oscillation a simple harmonic oscillation? The answer is No, because the restoring force does not obey Hooke's law. Equation 6.42 shows that F is not directly proportional to the displacement s.

Nevertheless, the motion of a simple pendulum is *approximately* simple harmonic, *provided that the amplitude θ_0 of the oscillation is small*. In this special case, which is often met quite adequately in practical systems, we can make the approximation

$$\sin \theta \simeq \theta \quad \text{for } \theta \text{ small and expressed in radians.} \tag{6.43}$$

Just what does "small" mean here? That depends on the accuracy required of the approximation. For $\theta = 1$ rad (about $57°$, a rather large angle for a simple pendulum), the error is $(\theta - \sin \theta)/\sin \theta = 19\%$. But, for $\theta = 0.2$ rad (about $13°$), the error is only 0.7%. For $\theta = 0.1$ rad (about $6°$), the error is 0.2%. You can readily calculate the error for any other angle you desire.

With this approximation, Equation 6.42 becomes

$$F = -\frac{mg}{l} s = (\text{constant}) \times s,$$

as required for simple harmonic motion.

Because the simple pendulum is a harmonic oscillator for small oscillations, we know that it will be isochronous. That is, its period will be independent of the oscillation amplitude, provided the amplitude does not become too large. Our main aim, as when we studied the mass-spring system treated in Section 6.4, is to obtain the angular frequency ω, the frequency ν, and the period τ. We proceed just as we did in that section. We use the value of F just found to write Newton's second law, $ma = -(mg/l)s$. We then cancel m from both sides of this equation and substitute $a = d^2s/dt^2$ to write the acceleration along the tangent to the path as

$$\frac{d^2s}{dt^2} = -\frac{g}{l} s. \tag{6.44}$$

This equation has exactly the same mathematical form as Equation 6.32,

$$\frac{d^2x}{dt^2} = -\frac{k}{m} x,$$

which we solved in Section 6.4 to determine ω, ν, and τ. So solving it would be repetitious. All we need do is compare the two. Comparison shows that the ratio g/l plays the role for the simple pendulum that is played by the ratio k/m for the mass-spring system. We can therefore immediately substitute the former for the latter in Equations 6.34a, 6.34b, and 6.34c and obtain

$$\omega = \sqrt{\frac{g}{l}}, \qquad \nu = \frac{1}{2\pi}\sqrt{\frac{g}{l}}, \quad \text{and} \quad \tau = 2\pi\sqrt{\frac{l}{g}}. \qquad \textbf{(6.45a,b,c)}$$

Can you show explicitly that the quantities g/l and l/g have the proper dimensions to satisfy these equations?

Galileo was the first to point out the near-isochrony of the simple pendulum and to make use of it, though others had probably been aware of it much earlier. It was Galileo's first important scientific discovery, made in 1583 when he was a nineteen-year-old medical student at the University of Pisa. Tradition has it that he made his observations in the Cathedral of Pisa, where the side aisles are illuminated by numerous heavy bronze lamps hanging from the very high ceiling by long chains. The lamps are just above head level and are easy to set swinging. Galileo presumably timed the oscillations as they slowly died down and noted that the period remained constant. He was in the habit of using his own pulse as a rough-and-ready timer, but he must soon have realized that the pendulum was steadier than his pulse. With this observation in mind, Galileo later turned the process around, inventing a simple medical instrument called the *pulsometer*. This consists of a small bob attached to a string having marks along its length. The physician takes the patient's pulse with one hand, swinging the pendulum with the other and letting the string slip slowly through his fingers until the periods of pendulum and pulse coincide. The mark on the string at the physician's fingers is then a measure of the pulse rate.

EXAMPLE 6.5

(a) Find the length of the simple pendulum whose period is 2 s. Because each one-way swing of such a pendulum takes one second, the pendulum is traditionally (if somewhat confusingly) called the **seconds pendulum**. **(b)** What would the period τ_m of this pendulum be on the surface of the moon, where the acceleration of gravity is $\frac{1}{6}g$?

SOLUTION:
(a) Find the length of the simple pendulum whose period is 2 s.
  You solve Equation 6.45c for l, obtaining

$$l = \frac{\tau^2 g}{4\pi^2}.$$

The numerical value of l is

$$l = \frac{(2\ \text{s})^2 \times 9.80\ \text{m/s}^2}{4\pi^2} = 0.993\ \text{m}.$$

Because this length happens to be close to 1 m, the eighteenth-century planners of the metric system considered the possibility of defining the unit of length to be the length of the seconds pendulum. Such a definition would have eliminated the need for designing, building, and maintaining a separate standard of length. However, variations of g from place to place, the difficulty of defining the length of a real pendulum whose bob is not an ideal point mass, and other complications rendered the proposal impracticable, and it was abandoned.

(b) What would the period τ_m of this pendulum be on the surface of the moon, where the acceleration of gravity is $\frac{1}{6}g$?
  There are two ways to make this calculation. The direct way is to use the value of l just calculated using Equation 6.45c and to substitute the value $\frac{1}{6}g$ for g. However, it is easier to note that τ is inversely proportional to $\sqrt{g}$. Reducing the acceleration of gravity sixfold thus increases the pendulum period by a factor $\sqrt{6}$. You have

$$\tau_m = \sqrt{6} \times 1\ \text{s} = 2.45\ \text{s}.$$

Symbols Used in Chapter 6

a	acceleration		l	length of a pendulum
A	amplitude of an oscillator, magnitude of a phasor		L	relaxed length of a spring
$\mathbf{A}$	phasor (constant-magnitude vector rotating uniformly on reference circle)		m, M	mass
			s	displacement along an arc
g	acceleration of gravity		$\mathbf{T}$	force exerted by the cord of a pendulum on the bob
k	force (or spring) constant		v	velocity of an oscillator

V	velocity of a particle in rotation	ν	frequency of an oscillator
W	weight	τ	period of an oscillator
δ	phase constant	ϕ	constant phase angle
θ, θ_0	angle of displacement, maximum angle of displacement	ω	angular frequency of an oscillator, angular velocity of a particle in rotation

The simplest type of oscillatory motion is one-dimensional **simple harmonic motion**, or **harmonic oscillation**. Equations 6.20, 6.21, and 6.22 respectively describe the displacement, velocity, and acceleration of a **harmonic oscillator** in the most general way:

$$x = A \cos \left(2\pi \frac{t}{\tau} + \delta \right)$$
$$= A \cos (\omega t + \delta) = A \cos (2\pi \nu t + \delta),$$
$$v = \frac{dx}{dt} = -\frac{2\pi}{\tau} A \sin \left(2\pi \frac{t}{\tau} + \delta \right)$$
$$= -\omega A \sin (\omega t + \delta) = -2\pi \nu A \sin (2\pi \nu t + \delta),$$
$$a = \frac{d^2 x}{dt^2} = -\frac{4\pi^2}{\tau^2} A \cos \left(2\pi \frac{t}{\tau} + \delta \right)$$
$$= -\omega^2 A \cos (\omega t + \delta) = -4\pi^2 \nu^2 A \cos (2\pi \nu + \delta).$$

The last two of these equations may be derived directly from the first by means of differentiation with respect to time. Alternatively, the three equations may be derived by considering the behavior of the x component—the "shadow"—of a **phasor A** rotating uniformly around a **reference circle** of radius A with **angular velocity** ω, so that its angular displacement is given by Equation 6.23,

$$\theta = \omega t + \delta.$$

Crucial properties of a harmonic oscillator are its **angular frequency** ω, its **period** τ, and its **frequency** ν. They are related by Equations 6.12, 6.17, and 6.19,

$$\omega = \frac{2\pi}{\tau}, \quad \nu = \frac{1}{\tau}, \quad \text{and} \quad \omega = 2\pi \nu.$$

A body is said to be in **stable equilibrium** if it always experiences a **restoring force** tending to return it to a position E_s, called the **equilibrium point**. If, moreover, the magnitude of the force is proportional to the displacement of the body from E_s, the system is said to obey **Hooke's law** (Equation 6.29),

$$F = -kx.$$

Many physical systems conform to Hooke's law, provided that the displacement x is not too large. The **force constant** k is a measure of the stiffness of the spring.

A system that conforms to Hooke's law, such as the mass-spring system, executes simple harmonic motion. The motion is described by Equation 6.32,

$$\frac{d^2 x}{dt^2} = -\frac{k}{m} x,$$

called the **harmonic-oscillator equation**. The angular frequency, frequency, and period of the mass-spring system are given by Equations 6.34a, 6.34b, and 6.34c:

$$\omega = \sqrt{\frac{k}{m}}, \quad \nu = \frac{1}{2\pi} \sqrt{\frac{k}{m}}, \quad \text{and} \quad \tau = 2\pi \sqrt{\frac{m}{k}}.$$

Harmonic oscillators are **isochronous**.

Other systems obey equations of motion whose mathematical form is the same as that of Equation 6.32. An important example is the **simple pendulum**, provided that its amplitude of oscillation is small enough to allow the approximation $\sin \theta = \theta$. *In this case only*, the equation of motion for the pendulum is the harmonic-oscillator form of Equation 6.44,

$$\frac{d^2 s}{dt^2} = -\frac{g}{l} s,$$

The angular frequency, frequency, and period of the pendulum are given by Equations 6.45a, 6.45b, and 6.45c:

$$\omega = \sqrt{\frac{g}{l}}, \quad \nu = \frac{1}{2\pi} \sqrt{\frac{g}{l}}, \quad \text{and} \quad \tau = 2\pi \sqrt{\frac{l}{g}}.$$

KEY TERMS

Section 6.2 The Kinematics of Simple Harmonic Motion
harmonic oscillation (simple harmonic motion) ▪ harmonic oscillator ▪ amplitude ▪ period, frequency, angular frequency ▪ angular displacement, angular velocity, angular acceleration ▪ reference circle ▪ phasor

Section 6.3 Equilibrium and Hooke's Law
equilibrium point ▪ stable equilibrium, neutral equilibrium, unstable equilibrium ▪ restoring force ▪ force law ▪ Hooke's law ▪ force constant (spring constant)

Section 6.4 The Dynamics of Simple Harmonic Motion
harmonic oscillator equation ▪ isochrony

QUERIES

6.1 *(2) Cranky.* The drawing below shows the way in which an automobile piston and crankshaft are linked by means of a connecting rod. As the crankshaft turns at a constant rate, the piston oscillates with the same period. Explain why the oscillatory motion of the piston is *not* simple harmonic.

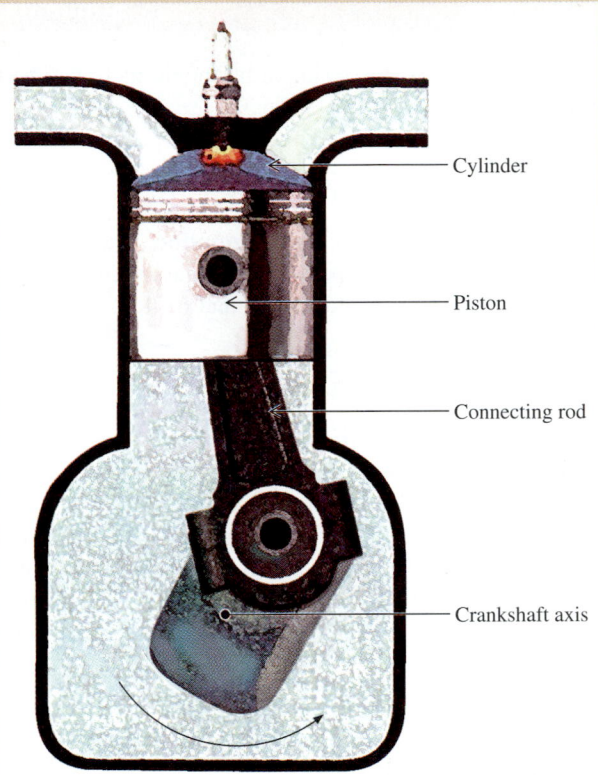

Cylinder

Piston

Connecting rod

Crankshaft axis

6.2 *(2) Circular reasoning?* Show explicitly that 1 rad ≈ 57.3°.

6.3 *(3) Asinine behavior.* Many ancient and medieval philosophers commented on the following ''paradox.'' A hungry donkey stands exactly midway between two equally delectable, equally large bundles of oats. Being equally attracted to each, he cannot decide which way to go, so he starves to death in the presence of plenty. Comment on this paradox in the light of the discussion of equilibrium in Section 6.3.

6.4 *(3) Easy as falling off a horse.* Think about equilibrium in a situation where motion must be considered in more than one dimension. What type of equilibrium is possessed by a small ball if it is located in a teacup with a rounded bottom? in a horizontal trough of cylindrical cross section like a rain gutter? on the outside surface of a spherical dome? on a saddle-shaped surface? In general, if different types of equilibrium obtain in different directions, what type of overall equilibrium exists?

6.5 *(4) Tilt!* The system of Figure 6.11 is set into motion, and its oscillation frequency is observed. Next, the entire appara-

tus is turned 90° clockwise so that the mass hangs vertically from the spring. **(a)** When the system is at rest, how far from the mark E_s will the pointer P be? **(b)** When the system is set into vertical oscillation, will the frequency be different from that observed for horizontal motion? Why or why not?

6.6 *(4) Mass, length, and time.* You need to find the mass of a brass block. You have a spring whose force constant you know to be k. Unfortunately, you don't have a ruler handy, so you can't simply hang the block from the spring and measure the amount of stretch. However, your quartz wristwatch can be used as a stopwatch. Describe a method for determining the desired mass m.

6.7 *(5) Bought on the morn that the old man was born.* Accurate pendulum clocks (such as grandfather clocks) are always driven by weights rather than by a mainspring. Why is this important for accuracy?

6.8 *(G) Family resemblance.* The two mass-spring systems in the illustration below are constructed of identical masses and springs. Both systems are set into oscillation, and their periods τ_a and τ_b are measured. (Assume that the lower spring in system b is always taut.) Is there any difference between τ_a and τ_b? Explain.

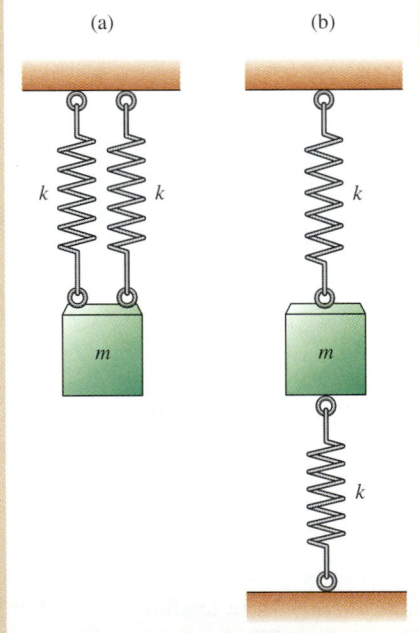

(a)   (b)

6.9 *(G) Getting jumpy.* A horizontal metal plate executes simple harmonic motion in the vertical direction. Under what circumstances will a small metal object lying on the plate begin to rattle?

6.10 *(G) Subtle connection.* Compare the small-oscillation period of a simple pendulum with that of a conical pendulum for which the cone half-angle θ is small; see Equation 5.35. In what sense is the conical pendulum a harmonic oscillator?

6.11 *(G) Conical and simple pendulums.* The drawing shown is a top view of a conical pendulum in motion. Two light sources and two screens are set up as shown, and the pendulum casts two shadows. Make an argument showing that the two shadows execute simple harmonic motion. What is their relative phase? Why is the period of a conical pendulum the same as that of a simple pendulum of the same length, provided the conical pendulum angle is small?

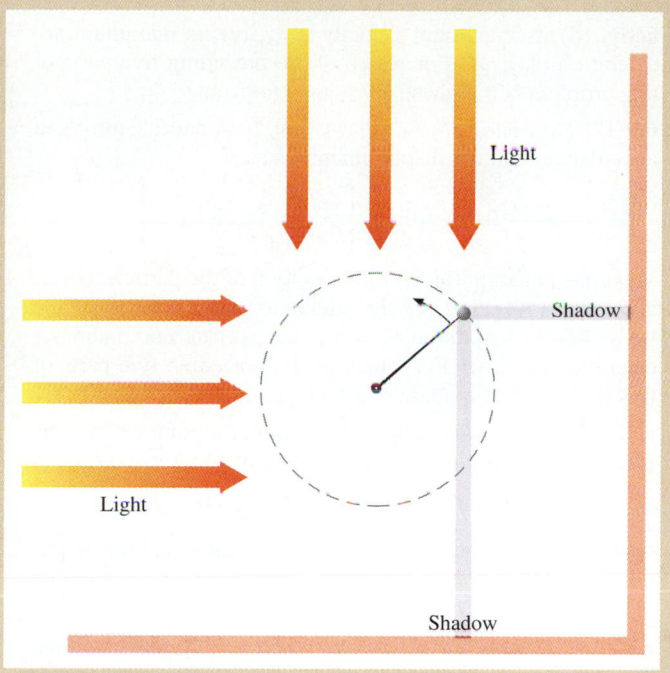

PROBLEMS

GROUP A

6.1 *(2) Functioning in harmony.* A particle undergoes harmonic oscillation with amplitude 10 cm. Beginning at a distance $x = 5$ cm from the equilibrium position, it completes 125 full oscillations in 1 min. Write the equation that describes x as a function of time.

6.2 *(2) Sweet harmony.* A moderately loud sound of frequency 3.5 kHz (=3500 Hz) results in an oscillation of the human eardrum over a range of approximately 1×10^{-6} m. Write an expression for the displacement of the eardrum as a function of time.

6.3 *(2) Little green men need astronomy, too.* A resident of the planet Pluto, looking toward the sun, sees the orbit plane of the earth almost edge-on. On the basis of his astronomical observations, which agree well with the corresponding quantities listed in Tables 1.2 and 1.3, he writes an expression describing the apparent position of the earth as a function of time. What is this expression? Ignore the motion of Pluto itself, which circles the sun only once every 247 years.

6.4 *(2) Getting the picture.* The displacement of a particle is described by the expression

$$x = A \cos \left(\omega t - \frac{\pi}{4} \right).$$

Let $A = 0.1$ m and $\omega = 20\pi$ rad/s. Make sketches of the displacement, the velocity, and the acceleration as functions of time. Draw each one below the preceding one, as in Figure 6.2. Your sketch should be sufficiently precise to clearly show the period, amplitude, phase constant, and phase difference between x and v, v and a, x and a.

6.5 *(2) It's all in there!* The displacement of a body is given by the expression

$$x = (0.3 \text{ m}) \cos [(200 \text{ rad/s}) \cdot t + 2 \text{ rad}].$$

Find the amplitude A, the frequency ν, the period τ, and the instantaneous phase δ of the body at $t = 0$.

6.6 *(2) One thing leads to another.* The amplitude of a harmonic oscillator is 0.2 m, and its period is 1.5 s. What are the maximum values of **(a)** the velocity and **(b)** the acceleration?

6.7 *(2) Other functions, other motions.* The displacement of a body is described by the function

$$x(t) = \beta_0 + \beta_1 t + \beta_2 t^2 + \beta_3 t^3,$$

where the β's are constants. **(a)** Find an expression for the velocity, and show that the velocity is independent of β_0. What is the physical meaning of the quantity β_0? **(b)** Find an expression for the acceleration, and show that the acceleration is independent of β_0 and β_1. What is the physical meaning of the quantity β_1? **(c)** The motion described by the function $x(t)$ is not oscillatory. What special mathematical property is possessed by functions that *do* describe oscillatory motion?

6.8 *(2) Extracting all the information, I.* A particle moves in accordance with the displacement equation

$$x = (0.06 \text{ m}) \cos \left[(24\pi \text{ rad/s}) \cdot t + \frac{\pi}{4} \text{ rad} \right].$$

Write the equation for **(a)** the velocity v of the particle, **(b)** its acceleration a. Find **(c)** the oscillation amplitude A of the body, **(d)** its maximum velocity v_{max}, **(e)** its maximum acceleration a_{max}. **(f)** For which of the preceding five parts of this problem is the phase constant significant?

6.9 *(2) Extracting all the information, II.* A particle moves in accordance with the displacement equation

$$x = (0.047 \text{ m}) \sin \left[\frac{2\pi t}{0.024 \text{ s}} - 4.7 \text{ rad} \right].$$

Write the equation for **(a)** the velocity v of the particle, **(b)** its acceleration a. Find **(c)** the oscillation amplitude A of the body, **(d)** its maximum velocity v_{max}, **(e)** its maximum acceleration a_{max}. **(f)** For which of the preceding five parts of this problem is the phase constant significant?

6.10 *(2) Noisemaker.* The acceleration of a point on the surface of a vibrating metal plate is described by the expression

$$a = (7.0 \times 10^2 \text{ m/s}^2) \cos \left[(5.3 \times 10^3 \text{ rad/s}) \cdot t \right].$$

(a) Find the angular frequency, the frequency, and the amplitude of the oscillation. **(b)** Write an expression for the displacement of the point as a function of time.

6.11 *(3) Forcing an extension.* When a 5-kg mass is hung from a spring, it stretches 17.0 cm. Find the value of the force constant k.

6.12 *(3) Shopping day.* When you weigh a sack of potatoes on a supermarket scale, the scale pan descends 5.73 cm. **(a)** If the sack is marked "5 kg," what is the force constant of the spring in the scale? Assume for the moment that the sack is correctly marked. **(b)** Contrary to the assumption in part **a**, sacks of potatoes rarely (if ever) contain the mass marked on the package. This is because the sacks are filled with potatoes one by one (usually by automatic machinery) until the total weight exceeds $5g$ N. If a typical sack contains 22 potatoes, estimate the variation in the length of the spring in the supermarket scale as you weigh several sacks. Assume that all the potatoes are about the same size.

6.13 *(3) Pushmi-pullyu.* The relaxed length of a spring is 25 cm, and its force constant is 120 N/m. **(a)** How much force must you exert on each end of the spring to compress it to one-half its relaxed length? **(b)** How much force must you exert on each end of the spring to double its relaxed length?

6.14 *(3) Springs in series.* Two springs of force constants k_1 and k_2 are connected end to end to make a long spring. **(a)** Show that the force constant k_s of the combined springs is given by the expression $\frac{1}{k_s} = \frac{1}{k_1} + \frac{1}{k_2}$. **(b)** Are the combined springs stiffer or weaker than the individual springs? **(c)** Let $k_1 = 245$ N/m and $k_2 = 320$ N/m. Find k_s.

6.15 *(3) Springs in parallel.* Two springs of force constants k_1 and k_2 are connected side by side, as shown in the drawing.

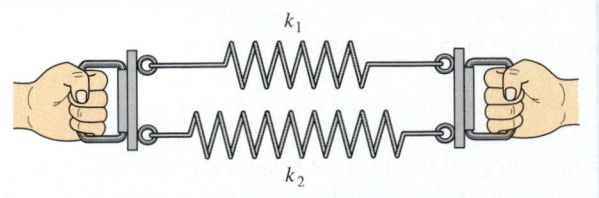

(a) Show that the force constant k_p of the combined springs is $k_p = k_1 + k_2$. **(b)** Are the combined springs stiffer or weaker than the individual springs? **(c)** Find k_p for the two springs considered in part **c** of Problem 6.14.

6.16 *(3) Working backwards, I.* You are asked to design a spring for use in a spring scale. The case has already been designed. The scale on the case is marked in such a way that the distance from the zero mark to the 25-N mark is 8.6 cm. Specify the force constant of the spring.

6.17 *(3) Weighty matters.* When you step on a bathroom scale, the platform descends 2.6 mm. The scale reads 68 kg. What is the force constant of the spring in the scale?

6.18 *(4) Most and least.* A spring hangs from a hook. A mass $m = 1$ kg, hanging from the lower end of the spring, executes harmonic oscillations with angular frequency $\omega = 25$ rad/s and amplitude $A = 1.6$ cm. If the spring has negligible mass, find the maximum and minimum values of the force that the spring exerts on the hook.

6.19 *(4) Dimensional analysis.* Show that, if the force constant k and the mass m are expressed in the proper units, the angular frequency of Equation 6.34a is expressed in the proper units.

6.20 *(4) Extracting all the information, III.* As in Problem 6.8, a particle's motion is described by the equation

$$x = (0.06 \text{ m}) \cos \left[(24\pi \text{ rad/s}) \cdot t + \frac{\pi}{4} \text{ rad} \right].$$

Suppose that the particle has mass $m = 0.2$ kg. Find **(a)** the force $F(t)$ acting on the particle, **(b)** the maximum value F_{max} of that force.

6.21 *(4) Ideal oscillator.* The mass of a mass-spring system rests on a frictionless table, as shown below. Let $m = 2.5$ kg and $k = 95$ N/m. The mass is pulled 12 cm away from the equilibrium point and released from rest. Find **(a)** the oscillation period, **(b)** the amplitude, **(c)** the phase constant, **(d)** the maximum speed of the mass, **(e)** the maximum acceleration of the mass.

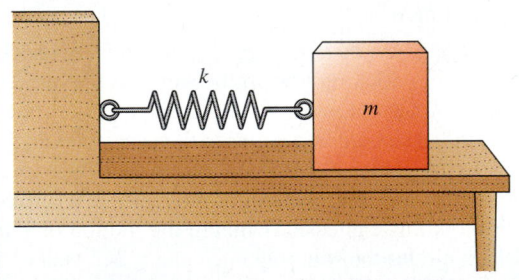

6.22 *(4) Double time, I.* Two masses are hung from identical springs and set into oscillation. If the period of system 1 is twice the period of system 2, find the ratio m_1/m_2 of the corresponding masses.

6.23 *(4) Double time, II.* Identical masses are hung from two springs and set into oscillation. If the period of system 1 is twice the period of system 2, find the ratio k_1/k_2 of the force constants of the two springs.

6.24 *(4) Making a watch serve as a ruler.* In Query 6.6, a method is suggested for determining the mass of an object using a spring of known force constant and a stopwatch. Using this method, determine the mass of a brass block that oscillates with period $\tau = 0.374$ s when hanging from a spring of force constant $k = 215$ N/m.

6.25 (4) *Space and time.* When you hang an object from a spring, the spring stretches 6.51 cm. If you set the mass-spring system into oscillation, what will the frequency ν be?

6.26 (5) *Disharmony.* On semilogarithmic paper or by using a semilogarithmic computer plotting routine, plot the "error ratio" $(\theta - \sin \theta)/\theta$ as a function of θ in the range 10^{-3} rad $\leq \theta \leq 1$ rad. How accurate is the approximation $\sin \theta \simeq \theta$ for $\theta = 10°$? 5°? 0.1°? 0.01°?

6.27 (5) *Swinging.* What is the period of a simple pendulum of length 2.00 m?

6.28 (5) *Pulsometer.* You want to make a model of Galileo's pulsometer. If it is to be useful in timing the pulses of a variety of patients, its length must be adjustable so as to make one-way swings (half cycles) correspond to pulses at rates varying

from 50/min to 140/min. At what distances from the bob should you make marks on the string to read pulses at 10/min intervals?

6.29 (5) *Starting a pendulum three ways.* **(a)** At time $t = 0$, a pendulum of length $l = 0.80$ m is released from an initial angle $\theta_0 = 3.0°$ with respect to the vertical. Write an equation that gives the subsequent displacement θ as a function of time. **(b)** The same pendulum is initially set into motion from its equilibrium position by giving it a push with $v_0 = -0.22$ m/s. Again find θ as a function of time. **(c)** The pendulum is set into motion yet again, this time by displacing it to $\theta_0 = 3.0°$ *and* giving it a push with $v_0 = -0.22$ m/s. Find θ as a function of time.

6.30 (G) *Grandpa going up! I.* A grandfather clock with pendulum of length l is located in an elevator car. **(a)** What is the frequency ν_{up} of the pendulum when the car has a constant upward acceleration $\mathbf{a}$? **(b)** What is the frequency ν_{down} when the acceleration has the equal but opposite value $-\mathbf{a}$? Assume that $a < g$, so that the pendulum string always remains taut.

GROUP B

6.31 (2) *Predicting the future.* A particle executes simple harmonic motion about $x = 0$, with angular frequency $\omega = 4$ rad/s. At a time $t = t_1$, its displacement is 25 cm and its velocity is 100 cm/s. What are its displacement and velocity at $t_2 = t_1 + 2.4$ s?

6.32 (2) *The farther, the faster.* (A knowledge of integration is required to do this problem.) As noted in the text, the isochrony of harmonic oscillators is not intuitively obvious. After all, one might argue that the oscillator must cover a distance $4A$ in the course of each oscillation; as that distance increases with increasing amplitude, it should take longer to cover it. To show definitively that this is *not* so, find the mean speed $\langle |v| \rangle$ over one quarter-cycle—say, from $t = 0$ to $t = \tau/4$. This can be done by evaluating

$$\langle |v| \rangle \equiv \frac{\int_0^{\tau/4} |v|\, dt}{\int_0^{\tau/4} dt}.$$

Show that $\langle |v| \rangle$ is directly proportional to A, so any increase in distance is exactly compensated by an increase in mean speed.

6.33 (2) *Shaking the rattle.* The top of the vibrating table illustrated below can be made to oscillate harmonically in the vertical direction. Both the amplitude and the angular fre-

quency of the oscillation are adjustable. A block lies on top of the table. As the angular frequency is increased with the amplitude held constant, the block is seen to leave the table during a part of each cycle. **(a)** At what part of the oscillation cycle is this effect first seen? **(b)** Show that the block begins to leave the table when $\omega = \sqrt{g/A}$.

6.34 (2) *What a difference a phase makes!* A particle moves in the x-y plane. Its motion is described by the equations

$$x = (10 \text{ m}) \cos \omega t \quad \text{and} \quad y = (10 \text{ m}) \cos \left(\omega t + \frac{\pi}{3} \right).$$

Make a plot of the path of the particle.

6.35 (2) *Lissajous figures.* (A computer will save a lot of work in this problem.) A particle oscillates simultaneously in the x and y directions according to the equations

$$x = \cos n\omega t \quad \text{and} \quad y = \sin m\omega t.$$

(The amplitudes are equal and are taken equal to 1 for convenience.) Make plots of the path followed for the particle when **(a)** $n = 1$, $m = 2$; **(b)** $n = 1$, $m = 3$; **(c)** $n = 2$, $m = 3$. These beautiful curves are members of the large class first described in 1815 by the American navigator, mathematician, and astronomer Nathaniel Bowditch (1773–1838). However, they are called *Lissajous figures*, after the French physicist Jules Antoine Lissajous (pronounced Lee-sah-**zhoo'**) (1822–1880), who produced them in his 1855 study of sound vibrations. Lissajous reflected a light beam successively from two small mirrors fixed to two tuning forks vibrating at right angles to one another and displayed the resulting oscillating light beam on a screen. Other Lissajous figures can be produced by changing the ratio $n:m$ and also by changing the relative phases of the x and y components.

6.36 (3) *Cutting up, I.* You have a long coil spring whose force constant is k. You cut it into two pieces, one twice as long as the other. What are the force constants of the resulting springs?

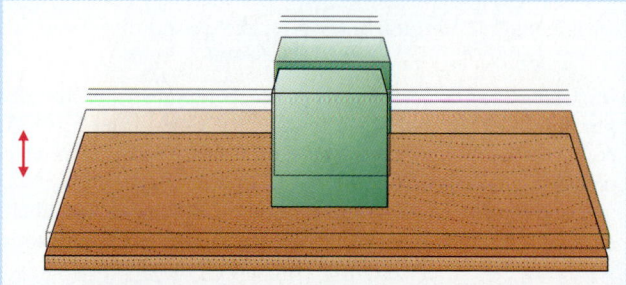

6.37 *(3) Cutting up, II.* You have a long coil spring whose force constant is k. You cut it into three equal pieces, which you hook together as shown below. What is the force constant k' of the combination? (Hint: See Problems 6.14 and 6.15.)

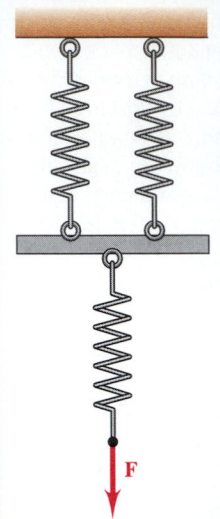

6.38 *(4) Shifting the equilibrium.* In the laboratory, you hang a pan with weights on it from a spring. When you set the system into vertical oscillation, its period is 0.48 s. You add another weight to the pan, and the new period is 0.66 s. How far has the equilibrium point shifted?

6.39 *(4) Letting go.* You suspend a spring of force constant k from a hook. You attach a mass m to the free end but hold it so that the spring is completely relaxed. You call the position of the mass $y = 0$. You release the mass suddenly, and the system begins to oscillate harmonically. **(a)** How far below the release point will the equilibrium point of the oscillation be located? **(b)** Will the mass return to the release point in the course of its oscillation? If so, at what point on the oscillation cycle will the release point be located? **(c)** How far below the release point will the lowest point of the oscillation be located? **(d)** What is the displacement y of the mass as a function of time? **(e)** What are the maximum and minimum values of the force exerted by the spring on the mass?

6.40 *(4) Hatchet job.* A mass m is suspended from a spring. The oscillation period of the system is measured and found to be τ_1. Then the spring is cut in half, the halves are hung side by side from a support, and the two half-springs are used to suspend the mass m. The oscillation period of this new system is found to be τ_2. Find the ratio τ_2/τ_1.

6.41 *(4) Pulled in both directions, I.* Find the period of oscillation of the system illustrated below. The mass is m, and

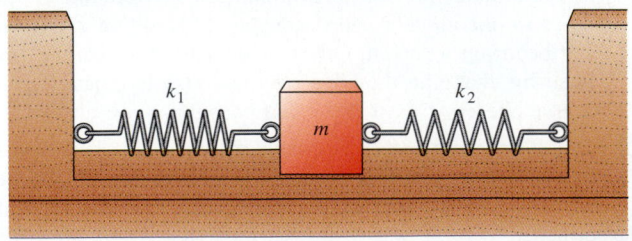

the springs have force constants k_1 and k_2. Assume that the system is frictionless and the mass of the springs is negligible. [Hints: (1) Refer to Query 6.8. (2) Are the two springs acting "in series" as in Problem 6.14 or "in parallel" as in Problem 6.15?]

6.42 *(4) Pulled in both directions, II.* **(a)** A spring of negligible mass and of force constant k is fixed in a vertical position between two supports. A small mass m is tied to the spring at a point whose distance from the upper end is a fraction ϕ of the distance between the supports; see the illustration below. Find the oscillation period τ of the system. **(b)** Let $k = 13$ N/m, $m = 0.025$ kg, and $\phi = \frac{1}{3}$, and find the value of τ.

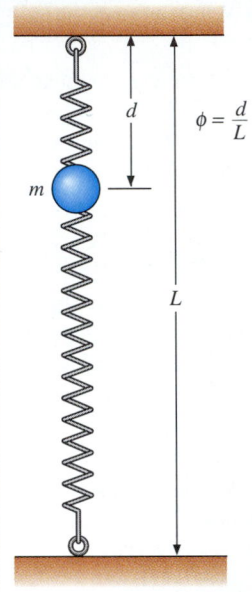

6.43 *(4) Surprisingly simple.* When a mass m is hung from a spring, it stretches the spring by an amount Δ. What is the length of the simple pendulum that has the same oscillation frequency as this mass-spring system?

6.44 *(4) A result for all seasons.* The mass of an ideal mass-spring system lies on a level frictionless surface. Suppose that you displace the mass from its equilibrium position $x = 0$ to an initial position x_0. Then you release the mass with a shove, giving it an initial velocity v_0. Beginning with Equations 6.20 and 6.21, **(a)** show that the oscillation amplitude of the system is

$$A = \sqrt{x_0^2 + \frac{m}{k} v_0^2} \; ;$$

(b) show that the phase constant is

$$\delta = \tan^{-1}\left(\sqrt{\frac{m}{k}} \frac{v_0}{x_0} \right).$$

(c) Using the results of parts **a** and **b**, express $x(t)$, the displacement of the mass as a function of time, in terms of the constants m, k, x_0, and v_0. **(d)** Express $v(t)$ and $a(t)$ in terms of the same constants.

Note that the results of parts **c** and **d** can be used to evaluate any mechanical quantity relevant to the system at any time, whatever the *initial conditions* x_0 and v_0 may be.

6.45 *(5) Good guess.* Solve Equation 6.44 by a process analogous to that used in the text to solve the mass-spring equation, Equation 6.32. That is, guess at s, evaluate d^2s/dt^2, and show that the equation is satisfied. Then show that Equations 6.45a, 6.45b, and 6.45c impose a condition on the solution that must be satisfied.

6.46 *(5) Good excuse for a Caribbean trip.* Newton predicted that the rotation of the earth on its axis results in a bulge, with the radius at the equator greater than that at the poles. This, he argued, results in a value of the acceleration of gravity that is smaller near the equator than in more northern latitudes. Toward the end of the seventeenth century, this prediction was tested experimentally. A pendulum clock that kept accurate time in London was carefully packed, sent to British Guiana (now Guyana), and set up again. **(a)** Show that for small pendulum oscillations, the relative change in pendulum period is given by

$$\frac{d\tau}{\tau} = -\frac{1}{2}\frac{dg}{g}.$$

(b) If the acceleration of gravity is $g_L = 9.81$ m/s^2 in London and $g_G = 9.78$ m/s^2 in Guyana, by how much per day does the clock fall behind in Guyana? If the clock is allowed to run for a month, what will its error be?

6.47 *(5) Interrupted pendulum.* A pendulum has a period $\tau = 2.0$ s when it swings freely. The pendulum is hung as shown in the drawing below, so that only one-fourth of its total length is free to swing to the left of the obstacle. It is displaced to position A and released. How long does it take to swing to the extreme displacement B and return to A? Assume that the displacement angle is always small.

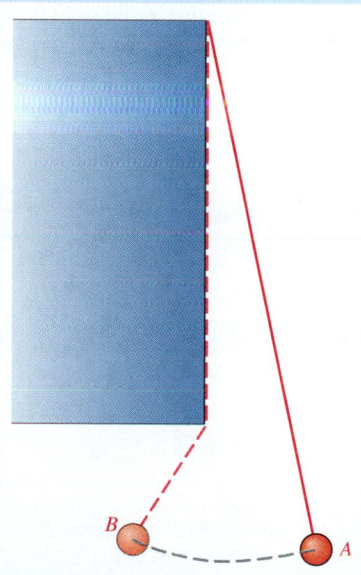

6.48 *(G) What can you do in an elevator besides stare? I.* A mass-spring system of mass m and force constant k hangs from the ceiling of a stationary elevator. With the mass at rest, an observer in the car sets up a vertical scale so that its zero point, $y = 0$, coincides with the position of the mass. The elevator starts upward with a constant acceleration a. Taking the upward direction as positive, the observer monitors the position y of the mass as a function of time. Show that y is subsequently described by the equation

$$y = \frac{a}{\omega^2}(\cos \omega t - 1) \quad \text{where } \omega \equiv \sqrt{\frac{k}{m}}.$$

[Hints: (1) Consider the way in which the equilibrium position of the mass shifts when the elevator starts, and review the method by which you solved Problem 6.39. (2) As always, you should begin the analysis of this problem by drawing a free-body diagram. However, you can simplify the diagram—and the subsequent analysis—by remembering that the net force acting on the mass before the elevator starts is zero. Thus it suffices to consider only the *additional* forces that come into play when the elevator starts.]

6.49 *(G) Sitting tight.* In the illustration below, the large mass M rides on an essentially frictionless air track. It oscillates under the influence of the attached spring with a frequency $\nu = 3$ Hz. A smaller mass m is now placed on top of M. The static coefficient of friction between the two is $\mu_s = 0.6$. If there is to be no slipping, what is the maximum value A_{max} that the amplitude can have?

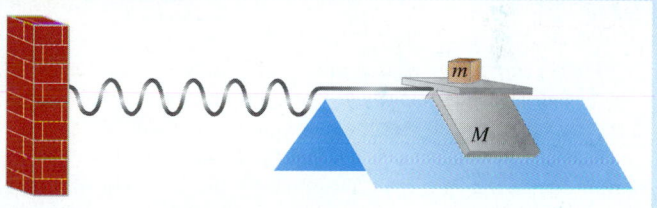

6.50 *(G) Look, ma, no springs!* The U-tube shown below is partly filled with water, as shown. Without moving the tube, you disturb the water level momentarily (say, by blowing into one end of the tube). Newton showed that the water level subsequently exhibits simple harmonic motion. Show that he was right.

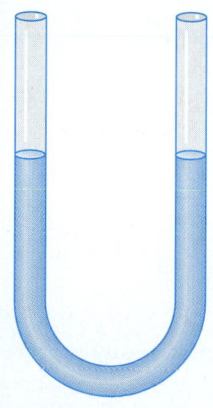

6.51 *(G) Twang!* In the apparatus diagrammed (next page), the hanging mass M keeps the string taut and stretched over the wooden wedge B. A particle of mass $m \ll M$ is tied to the string midway between A and B. The mass of the string itself is negligible. The particle is pulled a small distance

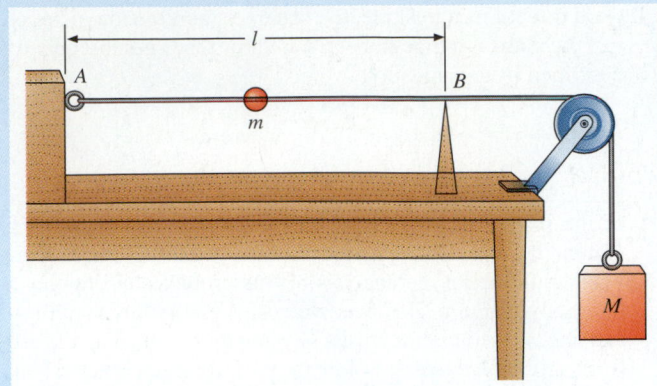

search for every possible influence that can change the period of such a pendulum, varying everything you can think of. Among other things, you experiment with changes in the mass of the bob, the material of which it is made, and its shape and volume (always, however, keeping its largest dimension small in comparison with the length of the string). You vary the length of the string. You make measurements while riding in an accelerating elevator, so as to vary the effective acceleration of gravity. When you have run out of ideas for things to vary, you find that the period depends *only* on the effective value of g and the string length l. Using dimensional analysis, show that the period of a simple pendulum in small oscillation *must* be given by the expression

$$\tau = \text{constant} \times \sqrt{\frac{l}{g}}.$$

upward and released. Prove that the string vibrates with period $\tau = \pi\sqrt{ml/Mg}$, where l is the distance between A and B.

6.52 *(G) Appeal to experiment.* Make believe, for the moment, that you have not read this chapter and know nothing about harmonic oscillations or simple pendulums. In this state of feigned ignorance, you decide to investigate the properties of simple pendulums undergoing small oscillations. You

6.53 *(G) Dull in-flight movie.* A passenger in the airplane of Example 5.3 (part **b**) passes the time by experimenting with a pendulum. As the airplane makes a turn with a banking angle of 40°, she observes a period of 1.0 s. **(a)** What is the length of the pendulum? **(b)** What is the period of the pendulum when the airplane resumes straight-line flight?

GROUP C

6.54 *(2) Four observations give it all.* Suppose that the displacement x_1 and velocity v_1 of a harmonic oscillator are known at time t_1 and that the corresponding quantities x_2 and v_2 are known at time t_2. Show that the angular frequency and amplitude of the oscillator are given by the expressions

$$\omega = \sqrt{-\frac{v_2^2 - v_1^2}{x_2^2 - x_1^2}} \quad \text{and} \quad A = \sqrt{\frac{v_2^2 x_1^2 - v_1^2 x_2^2}{v_2^2 - v_1^2}}.$$

6.55 *(3) Working backwards, II.* Two springs having unknown force constants k_1 and k_2 are attached end to end, and the combined force constant k_s is measured. The two springs are then attached side by side, and the combined force constant k_p is measured. Express k_1 and k_2 in terms of k_s and k_p.

6.56 *(G) What can you do in an elevator besides stare? II.* Consider again the elevator-mass-spring system of Problem 6.48. This time, however, the elevator starts upward with an acceleration $a = \alpha t$, with α a constant so that a increases linearly with time. **(a)** What is the net force F acting on the mass as a function of t? (See Hint 2 of Problem 6.48.) **(b)** Remembering that Newton's second law can be written in the form $F = m\,d^2y/dt^2$, write the equation of motion of the mass in a form analogous to Equation 6.32. **(c)** "Guess" that the solution to this equation is

$$y = \frac{\alpha}{\omega^3}(\sin \omega t - \omega t) \quad \text{where } \omega \equiv \sqrt{\frac{k}{m}}.$$

By substituting this quantity and its second derivative with respect to time into the equation, show that the "guess" is correct one. **(d)** Discuss this solution, showing that the motion it describes is one of simple harmonic oscillation of constant amplitude, centered about an equilibrium position that moves downward with constant speed with respect to the elevator ceiling. What are the values of the amplitude and the speed?

6.57 *(G) In a whirl.* In the apparatus illustrated below, the cylindrical mass m slides without friction on the thin rod. The combined force constant of the two springs is k. **(a)** When the system is not rotating, what is the oscillation frequency ν of the mass? **(b)** When the system rotates about its vertical axis at angular velocity Ω, where does the equilibrium point of the mass-spring system lie? **(c)** For what range of values of Ω is the equilibrium stable? For what value of Ω is the equilibrium neutral? For what range of values of Ω is the equilibrium unstable? **(d)** Assume that the value of Ω lies within the range of stable equilibrium. With what frequency ν will the mass oscillate back and forth along the rod?

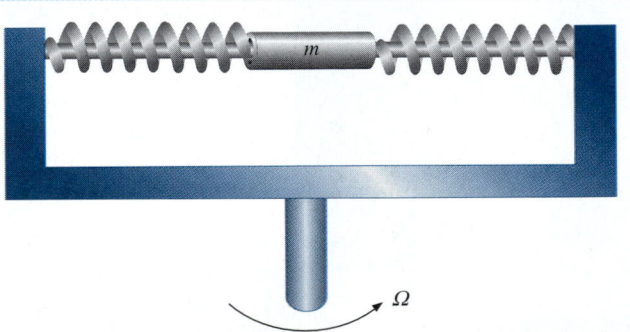

6.58 *(G) Bimodal oscillator.* The figure below shows two identical masses m connected to three springs, each having

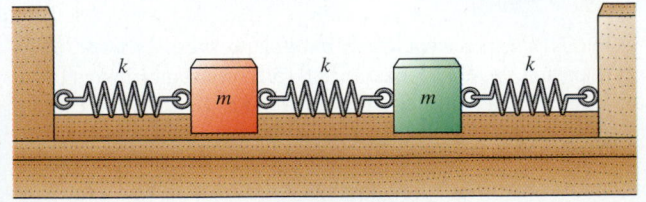

force constant k. All three springs can be either stretched or compressed. All are initially in their relaxed condition. The surface on which the masses rest is frictionless. Both masses are moved to the right by the same amount d, and they are then released simultaneously. (a) By how much is the center spring stretched? (b) What are the periods of oscillation of the masses? (c) This time, the left mass is moved a distance d to the left and the right mass is moved the same distance to the right. What is the net force acting on each mass? (d) The two masses are released simultaneously. What is the period of each?

6.59 *(G) Grandpa going up! II.* The elevator of Problem 6.30 accelerates upward with constant acceleration **a** through a distance h, beginning from rest at $t = 0$. It then reverses very quickly and heads downward with constant acceleration $-$**a**. The pendulum clock in the elevator reads correctly when the upward acceleration begins. Show that it again reads correctly at time

$$ t = \sqrt{\frac{2h}{a}} \; \frac{\sqrt{\dfrac{g+a}{g}} - \sqrt{\dfrac{g-a}{g}}}{1 - \sqrt{\dfrac{g-a}{g}}}. $$

6.60 *(G) Isochronous oscillator.* (A knowledge of integration is required to solve this problem.) A bead slides without friction on a wire bent into the form of a cycloid that lies in a vertical plane, as in the figure opposite. The cycloid is conveniently described by the pair of parametric equations

$$ x = b(\beta + \sin \beta) \quad \text{and} \quad y = b(1 - \cos \beta). $$

We are interested only in the part of the cycloid shown in the figure, which corresponds to $-\pi \le \beta \le \pi$. The parts of this problem lead to the conclusion that the oscillatory motion of the bead is *exactly* isochronous; that is, the period is independent of the amplitude. (a) Show that at a point on the cycloid corresponding to any value of β, the slope is given by $\sin \beta / (1 + \cos \beta)$. (b) Show that the restoring force acting on the bead when it is located at that point is $mg \sin (\beta/2)$. (c) Note that an infinitesimal displacement ds along the wire can be written

$$ ds = \sqrt{(dx)^2 + (dy)^2}. $$

By integrating this expression, find the distance along the wire from the equilibrium point to the bead, when its position is described by β. (d) Using the results of parts **b** and **c**, write the equation of motion of the bead. Show that it is the harmonic oscillator equation, and find the oscillation period τ of the bead.

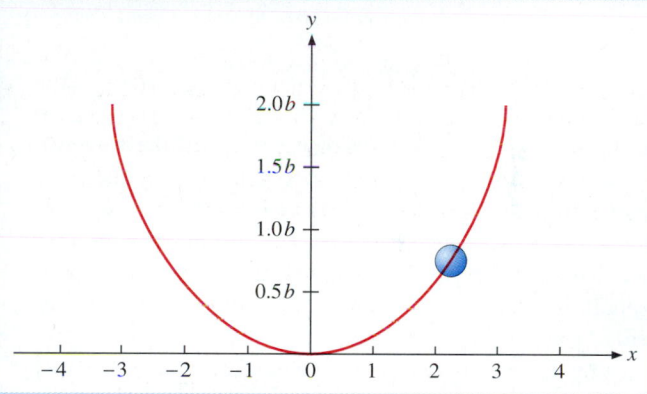

Part **III**

Conservation Principles in Mechanics

Physics is always a study of *process*—of the way systems change. Up to this point, we have described changes in physical systems *dynamically*. That is, we have followed and described the systems as the changes take place. This is what we do when we trace the trajectory of a baseball, for instance, or express the position of a harmonic oscillator as a function of time.

In physics, as in many other fields, there is another way to look at change. It might loosely be called the before-and-after method, because we look at the system at two different times and compare the appearance of the system at the later time with the appearance at the earlier time. If we know how to make this comparison for a particular system under particular circumstances, we can even *predict* the end state of the system from a knowledge of the beginning state. In taking this approach, we always apply some kind of *conservation principle*. We use some property of the system that *does not* change with time to predict what will happen to properties that *do* change.

Some conservation principles are simple and straightforward. To give a familiar example, a bank teller always has to show that the amount of money in the till at the end of the day is equal to the amount in the till at the beginning of the day. The bank is less interested in how the money is distributed among various coins, bills, checks, deposit vouchers, or money orders. And the bank is not at all interested in tracing the details of how the assortment of metal and paper items in the till changed from moment to moment in the course

of the day—as long as the money comes out right at closing time. Everyone understands that the somewhat abstract concept "money" is of more concern than its more concrete embodiment in coins, bills, and so forth. What the bank wants conserved is not, say, the total number of ten-dollar bills, but the total amount of money. "Conservation of money" is the conservation principle under which the teller must operate.

Children learn the abstraction called money early. You can sometimes induce a three-year-old to trade his dime for your nickel because the nickel is bigger. But five-year-olds will rarely if ever accept this trade because they already understand that the dime represents more of the abstract stuff "money," even though the nickel is larger.

Some conservation principles are more complex than conservation of money, and we often apply them without really analyzing what we are doing. For example, when you meet an old friend whom you have not seen for years, you probably recognize him instantly in spite of the substantial changes in his appearance. Although you are aware of the changes, you immediately and unconsciously search out certain facial characteristics that have been conserved. Your recognizing your friend does not depend on a knowledge of the day-to-day appearance of his face through the years.

In all such cases, we deal with change by identifying those features of the system under study that *do not* change—in other words, those that are *conserved*. For this reason, the before-and-after method is formally called

the *conservational approach*. The use of this approach in physics is much like its use in the everyday contexts exemplified above. In physics, however, we must be careful to describe and use conservation principles in a very precise way.

You will encounter a number of conservation principles in this book. In mechanics, we will be concerned with three of them:

1. Conservation of energy

2. Conservation of linear momentum

3. Conservation of angular momentum

The first two are the central subject matter of Part III. Conservation of angular momentum, which is intimately connected with the subject of rotation, is considered in Part IV.

A fourth principle, conservation of mass, is familiar to you if you have studied chemistry. You will see in Part XI that it merges with the energy-conservation principle into a single, very powerful conservation principle in the framework of the theory of relativity. A fifth principle, conservation of electric charge, is discussed in Part VIII. Still other conservation principles are important in the study of fundamental (subatomic) particles and in the closely related subject of cosmogony, the study of the origin of the universe. We will discuss some of these principles in Part XII.

The conservational approach in physics turns out to have a very important bonus feature that gives it significance beyond its convenience in analyzing the behavior of certain classes of systems. On the basis of experience, physicists have often come to believe that a certain quantity should be conserved under certain circumstances. But it has sometimes turned out, in the light of more critical investigation, that the quantity is *not* conserved. Such surprises have repeatedly led to powerful generalizations of conservation principles, together with unification of two or more branches of physics whose close relation was not formerly recognized. You will see in Part VI, for example, that the failure of energy to be conserved when frictional effects are present was an important clue to the discovery that heat is a manifestation of motion. Because this method of discovery has been so fruitful over the years, it is used today in a deliberate fashion: In studying physical processes that are not as yet understood in terms of a general theory, we assume that some quantity is conserved, work out the consequences of the assumption, and test the consequences experimentally. Historically, experiments that confirmed or invalidated assumed conservation principles have led to great advances in understanding how the natural world works.

The conservational approach in physics does not supplant the dynamical approach we have already considered. Rather, the two approaches complement one another. A thorough understanding of both is essential to the study of physics.

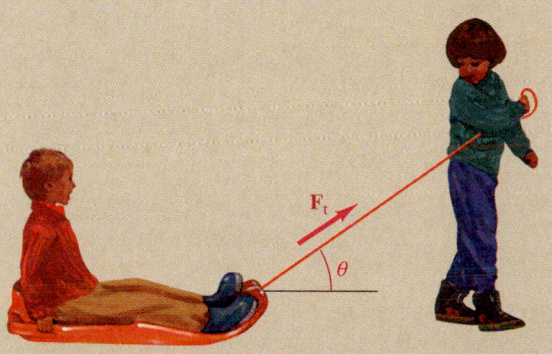

Work and Energy

L O O K I N G   A H E A D

— Work is the scalar product of two vectors, the force **F** acting on a body and the distance d**s** through which the force acts.

— When work is done by a varying force or over a curvilinear path, it is evaluated by means of the work integral.

— Work and kinetic energy are related by means of the work-energy theorem.

— Power is work done per unit time.

Left: Exactly what do we mean when we say it takes work to raise the weights?

All Nature seems at work.

—Samuel Taylor Coleridge, "Work without Hope"

A little work, a little play,
To keep us going—and so, good day!

—George du Maurier, *Trilby*

Introduction

Figure 7.1 shows before-and-after views of several familiar systems. In part *a*, each system is at rest, and in part *b*, each system has been displaced over some distance by the application of a force. As a result, something significant has happened to each one. The glider on the air track is now moving with speed *v*. The ball, which was lying on the floor, now falls and rises, gaining and losing speed as it bounces. The pendulum swings back and forth, also falling and rising and gaining and losing speed as it does so. The mass-spring system is a little different from the other three because there is no motion even after the force has acted to stretch the spring. But, if the stretched spring is attached to a suitable mechanical system, it can make that system move.

Something has been done to each of these systems to make it "active." That something is called *work*, which is the subject of Section 7.2. The property that makes each system active is called *energy*, the main subject of the rest of this chapter and of Chapter 8.

(a) (b)

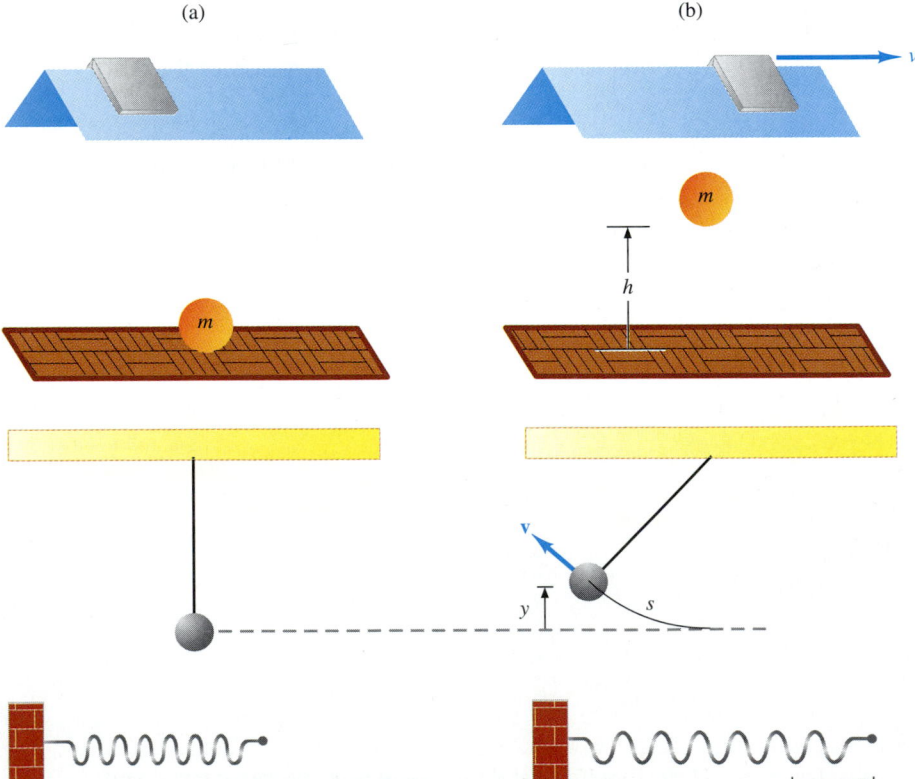

FIGURE 7.1 (*a*) Four systems in an "inert" state. The air-track glider, ball, and pendulum are at rest, and the spring is relaxed. (*b*) The same systems after work has been done on them. The glider is in motion with speed *v*. As it bounces, the ball gains and loses height and simultaneously loses and gains speed. The same is true of the pendulum, though its motion is not vertical. The spring, now stretched, possesses the potential to do work on some other mechanical system, setting the latter into motion.

Work. The Scalar Product of Two Vectors

If you apply a force to the initially stationary air-track glider of Figure 7.1, you will set it into motion. How fast the glider goes depends on four things: its mass, the magnitude and the direction of the force you apply, and the distance over which you apply the force. Let us assume for the moment that (1) you keep the magnitude of the force constant as long as you apply it and (2) you direct the force always along the air track, parallel to the motion of the glider. If you wish to give the glider a certain final speed v, what force should you apply and over what distance should you apply it? You probably know from everyday experience (for example, from riding a multispeed bicycle) that you can make a trade-off between force and distance. That is, a large force applied over a small distance will result in the same value of v as will a smaller force applied over a greater distance.

We refine and quantify these familiar ideas by defining the *physical* quantity "work" in a precise way. If a constant force F directed parallel to the motion of a body is applied as the body moves over a distance x, the **work** W done *on* the body is

$$W \equiv Fx \quad \text{for } F \text{ constant and directed parallel to } x. \tag{7.1}$$

It is evident from this definition that the same amount of work can be done by a large force applied over a small distance or a small force applied over a large distance. Let us verify that the end result of the work done on the air-track glider will be that the glider moves at the same speed either way. If the glider has mass m, its acceleration will be $a = F/m$. This constant acceleration, taking place over the distance x, will result according to Equation 2.38 in a final speed $v = \sqrt{2ax}$, or

$$v = \sqrt{\frac{2Fx}{m}} = \sqrt{\frac{2W}{m}}. \tag{7.2}$$

Because the final speed depends only on the product Fx and not on F or x individually, our assertion that the final speed depends only on the product Fx—the work done on the body—is proved.

It follows from Equation 7.1 that the SI unit of work is the newton·meter (N·m). Because this unit is encountered frequently, it has been given its own name—**joule (J)**. This name honors the English physicist James Prescott Joule (1818–1889), who devoted much of his career to experiments that placed the concepts of work and energy on a firm foundation. Thus we have

$$1 \text{ J} \equiv 1 \text{ N·m}. \tag{7.3}$$

Pronunciation Note

The pronunciations "jowell" and "jool" are both encountered in the English-speaking world. But Joule himself pronounced his name in the first of these ways, and most dictionaries give that as the only or primary pronunciation. In the town of Stone, Staffordshire, where Joule's family brewery produced Joule's Ales and Beers until the late 1970s, the name is still pronounced "jowell."

EXAMPLE 7.1

The essentially frictionless air-track glider of Figure 7.1 is initially at rest. You apply to the glider a 5.0-N force directed parallel to the track. **(a)** If you apply the force over a 10-cm distance, how much work do you do on the glider? **(b)** If you repeat the experiment starting again from rest, through what distance must you apply a 1.5-N force to achieve the same final speed?

SOLUTION:
(a) If you apply a 5-N force over a 10-cm distance, how much work do you do on the glider?

Because the force is constant and parallel to the direction of motion, you can use Equation 7.1. You have

$$W = 5.0 \text{ N} \times 0.10 \text{ m} = 0.5 \text{ J}.$$

(b) If you repeat the experiment starting again from rest, through what distance must you apply a 1.5-N force to achieve

the same final speed as in part **a**?

If the final speed is to be the same, the work done must be the same. Calling the force and application distance of the first trial F_1 and x_1, and the corresponding quantities for the second trial F_2 and x_2, you have

$$W = F_1 x_1 = F_2 x_2.$$

You solve the second equality for x_2, obtaining

$$x_2 = \frac{F_1 x_1}{F_2},$$

or

$$x_2 = \frac{5.0 \text{ N} \times 0.10 \text{ m}}{1.5 \text{ N}} = 0.33 \text{ m} = 33 \text{ cm}.$$

How could you arrive at the same answer by using the result of part **a**?

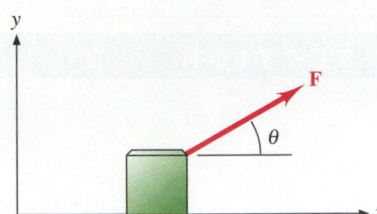

FIGURE 7.2 The block is pulled along the surface by the constant force **F**. The direction $\hat{\mathbf{F}}$ makes an angle θ with the surface, but the vertical component of **F** is not large enough to lift the block off the surface.

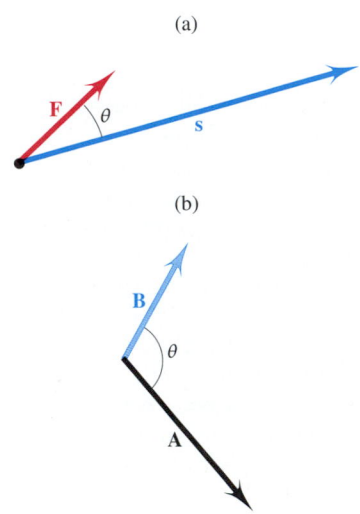

FIGURE 7.3 (*a*) A force **F** is applied to a body at an angle θ to the body's line of motion over a straight-line displacement **s**. The scalar product is defined to be $\mathbf{F} \cdot \mathbf{s} \equiv Fs \cos \theta$. (*b*) Generalization of the definition of the scalar product for any two vectors **A** and **B**.

The Scalar (Dot) Product

It often happens that a moving body is acted on by a force whose direction is not parallel to the direction of motion of the body. Figure 7.2 shows such a situation. The block is supported by a surface along which we choose the *x* axis. Applied to the block is the constant force **F**, directed at an angle θ with respect to the *x* axis. If the perpendicular component of **F** is not large enough to lift the block off the surface, the block will be pulled along the surface. But only the component parallel to the surface can propel the block along the *x* axis. It follows that the work done in moving the block through a distance *x* along the surface is the product of the parallel component $F_{\parallel}$ and the distance. That is, the work done is

$$W = F_{\parallel} x, \tag{7.4a}$$

or, because $F_{\parallel} = F \cos \theta$,

$$W = F \cos \theta \, x. \tag{7.4b}$$

The specific example of Figure 7.2 is easily generalized. In Figure 7.3a, a body moving along a straight-line path denoted by the vector **s** is acted upon by the constant force **F**. We give the name θ to the angle between **s** and **F**. (You will soon see that the sense of the angle—whether it is measured from $\hat{\mathbf{s}}$ to $\hat{\mathbf{F}}$ or from $\hat{\mathbf{F}}$ to $\hat{\mathbf{s}}$—is immaterial.) The work done by the force **F** on the body is

$$W = |\mathbf{F}| \cos \theta \, |\mathbf{s}|, \tag{7.5a}$$

or

$$W = Fs \cos \theta. \tag{7.5b}$$

You can see from these equations that the *scalar* quantity work, *W*, is found by carrying out a mathematical operation on two **vectors**, force **F** and displacement **s**. This operation is used frequently in physics, not only in calculating work but in many other contexts as well. It is therefore represented by a special notation that allows us to write Equations 7.5a and 7.5b in the equivalent but more compact form

$$\boxed{W = \mathbf{F} \cdot \mathbf{s}.} \tag{7.5c}$$

The notation of this equation can be applied quite generally to any two vectors **A** and **B** having an angle θ between them, as shown in Figure 7.3b. We define

$$\boxed{\mathbf{A} \cdot \mathbf{B} \equiv |\mathbf{A}| \cos \theta \, |\mathbf{B}| = AB \cos \theta.} \tag{7.6}$$

This mathematical operation is formally called the **scalar product** of two vectors because the result of the operation is a scalar. (In Equation 7.5c, for example, the result is the scalar *W*.) The scalar product is much more commonly (and informally) called the **dot product** because it is denoted by the boldface dot. Even though the result of the operation is a scalar, the vectorial properties of **A** and **B** are fully exploited in the operation. Their magnitudes are required to evaluate *AB*, and their directions are required to evaluate $\cos \theta$. Note in particular that

$$\mathbf{A} \cdot \mathbf{B} = AB \qquad \text{for } \theta = 0, \tag{7.7a}$$

$$\mathbf{A} \cdot \mathbf{B} = 0 \qquad \text{for } \theta = \pm\frac{\pi}{2}, \tag{7.7b}$$

and

$$\mathbf{A} \cdot \mathbf{B} = -AB \qquad \text{for } \theta = \pi. \tag{7.7c}$$

The dot product is commutative; that is,

$$\mathbf{A} \cdot \mathbf{B} = \mathbf{B} \cdot \mathbf{A}. \tag{7.8}$$

To prove this commutativity, consider the quantity $AB \cos \theta$ in Equation 7.6. Both *A* and *B* are scalars, and so you have $AB = BA$ because scalars obey the commutative law. And because the cosine is an even function—that is, $\cos \theta = \cos (-\theta)$—the sense of the angle between **A** and **B** is immaterial, as was asserted without proof in the discussion immediately preceding Equation 7.5a. You can therefore write

$$\mathbf{A} \cdot \mathbf{B} = AB \cos \theta = BA \cos \theta = \mathbf{B} \cdot \mathbf{A}.$$

The dot product also obeys the distributive law:

$$\mathbf{A} \cdot (\mathbf{B} + \mathbf{C}) = \mathbf{A} \cdot \mathbf{B} + \mathbf{A} \cdot \mathbf{C}. \qquad (7.9)$$

The proof of this statement is the subject of a problem at the end of this chapter.

EXAMPLE 7.2

A child pulls a friend on a sled behind her, as shown in Figure 7.4a. The towrope makes an angle of 35° with the horizontal.

(a) If the towing force F_t that the child applies to the rope is 42 N, find the work W_t that she does in pulling the sled a distance of 100 m.

(b) If the sled moves at constant speed, how do you account for the absence of acceleration?

(c) Over the 100-m distance, how much work is done on the sled by each of the forces acting on it? What is the total work done on the sled by all of them?

SOLUTION:

(a) If the towing force F_t that the child applies to the rope is 42 N, find the work W_t that she does in pulling the sled a distance of 100 m.

Because you are given the magnitude of the force F_t, the magnitude of the displacement x, and the angle between F_t and x, it is most convenient to use Equation 7.5b. Taking the direction of motion of the sled as the positive s direction, you have

$$W_t = F_t s \cos \theta = 42 \text{ N} \times 100 \text{ m} \times \cos 35°,$$

or $\qquad W_t = 3.4 \times 10^3 \text{ J}.$

(b) If the sled moves at constant speed, how do you account for the absence of acceleration?

The absence of acceleration implies that the net force $\mathbf{F}_{\text{net}}$ acting on the sled is zero. There must be a backward frictional force $\mathbf{F}_f$ exerted on the sled runners by the snow. Figure 7.4b is a free-body diagram of all the forces acting on the sled, whose mass (the passenger included) is m.

(c) Over the 100-m distance, how much work is done on the sled by each of the forces acting on it? What is the total work done on the sled by all of them?

You already know the work done on the sled by the towing force $\mathbf{F}_t$; this is the result of part **a**. A second force acting on the sled is the downward gravitational force $m\mathbf{g}$. As is evident from Figure 7.4b, this force is perpendicular to $\mathbf{s}$; any of Equations 7.5a, 7.5b, and 7.5c or Equation 7.7b leads to the conclusion that this force does no work on the sled because the cosine of the angle between $m\mathbf{g}$ and the direction of motion is zero; that is,

$$W_{m\mathbf{g}} = m\mathbf{g} \cdot \mathbf{s} = mg(\cos 90°)s = 0.$$

The same is true of the upward normal supporting force $\mathbf{N}$ exerted on the sled runners by the snow; that is, $W_{\mathbf{N}} = 0$.

There remains the frictional force $\mathbf{F}_f$. Because the acceleration of the sled is zero, this force must be equal and opposite to the horizontal component of the towing force $\mathbf{F}_t$. Thus, you

FIGURE 7.4

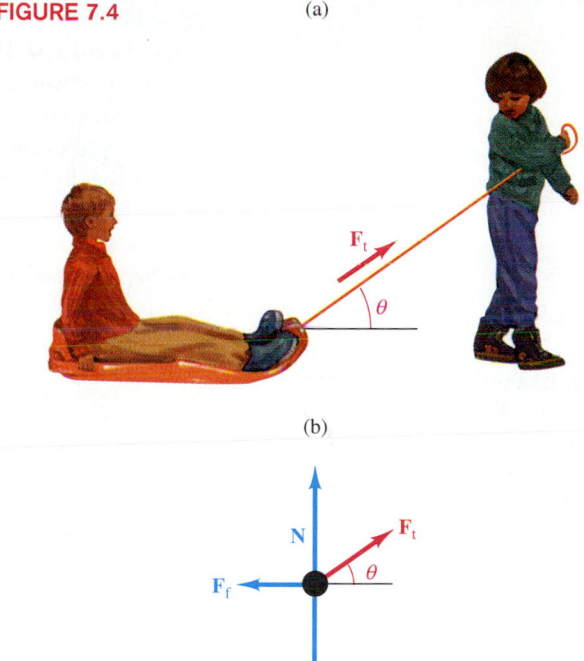
(a)

(b)

have $\mathbf{F}_f = -(F_t \cos \theta \, \hat{\mathbf{s}})$. The work done by this force is

$$W_f = \mathbf{F}_f \cdot \mathbf{s} = -F_t \cos \theta \, \hat{\mathbf{s}} \cdot \mathbf{s}.$$

The dot product is $\hat{\mathbf{s}} \cdot \mathbf{s} = 1 \cos 0° \, s = s$. Hence, you have

$$W_f = -F_t s \cos \theta = -W_t = -3.4 \times 10^3 \text{ J}.$$

You could obtain the same result by using Equation 7.7c. The negative value of this work signifies that the frictional force is directed so that it tends to bring the sled to a stop. In this example, the force is directed opposite to the motion. But any force having a component in the negative $\hat{\mathbf{s}}$ direction will result in work having a negative sign.

Adding up the work done by all the forces yields the **net work**:

$$W = W_t + W_{m\mathbf{g}} + W_{\mathbf{N}} + W_f$$
$$= W_t + 0 + 0 - W_t = 0.$$

Alternatively, you can write the net work as

$$W = \mathbf{F}_{\text{net}} \cdot \mathbf{s} = \mathbf{0} \cdot \mathbf{s} = 0.$$

Thus no *net* work is done on the sled, even though two of the individual forces do work.

Example 7.2 raises a point worth considering further. How can we say that no net work is done on the sled when the child pulling it is exerting a force on it through a distance? The answer is implicit in parts **b** and **c** of the example. The child does indeed do work $W_t = \mathbf{F}_t \cdot \mathbf{s}$ on the sled. This work is obvious to us because we have all been in a position like that of the child. At the same time that the child is doing work, however, the frictional force exerted by the snow on the sled runners is doing work on the sled; we have $W_f = \mathbf{F}_f \cdot \mathbf{s} = -\mathbf{F}_t \cdot \mathbf{s} = -W_t$. This frictional work is less obvious to us than W_t because we tend to be self-centered; we don't spontaneously think about work done by forces beyond our direct control. But the frictional work is just as real as the work done by the child.

The sign of the frictional work W_f is negative. In the example, we reached this conclusion by a dynamical argument. We reasoned that the absence of acceleration implied that the net force acting on the sled is zero, and we accounted for this by setting $\mathbf{F}_f = -F_t \cos \theta \, \hat{\mathbf{s}}$. The negative value of W_f followed directly.

Now, however, let us look at the work done by the frictional force $\mathbf{F}_f$ from a different point of view—in terms of the definition of work expressed by Equation 7.5a, $W = |\mathbf{F}| \cos \theta \, |\mathbf{s}|$. For the frictional work, we have $W_f = |\mathbf{F}_f| \cos \phi \, |\mathbf{s}|$, where ϕ is the angle between the direction of motion of the sled and the backward direction; see Figure 7.5a. According to the definition of the magnitude $|\mathbf{A}|$ of any vector $\mathbf{A}$, both $|\mathbf{F}_f|$ and $|\mathbf{s}|$ must have positive values. The angle ϕ is 180°, and therefore $\cos \phi$ is negative; thus the work W_f is negative.

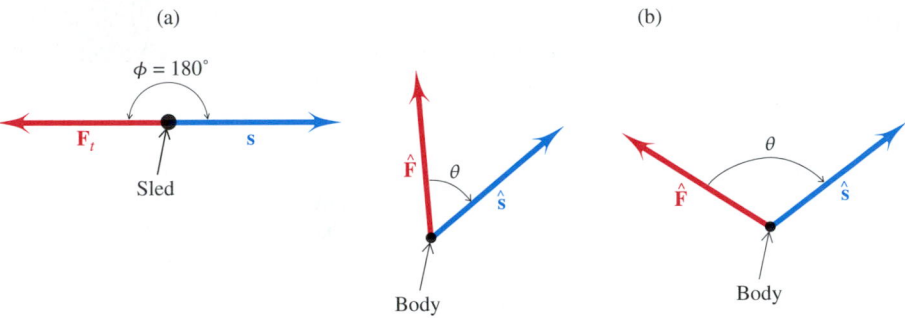

(a)   (b)

FIGURE 7.5 (*a*) The angle $\phi = 180°$ between the displacement **s** of the sled of Example 7.2 and the frictional force $\mathbf{F}_f$ exerted on the sled runners by the snow. (*b*) In general, there is an angle θ between the direction $\hat{\mathbf{F}}$ of any force **F** applied to a body and the direction of motion $\hat{\mathbf{s}}$ of the body. In the first sketch, θ is acute and the work done on the body by the force **F** is positive. In the second sketch, θ is obtuse and the work done on the body by the force **F** is negative.

In general, the sign of the work done on a body by a force depends only on the sign of the cosine of the angle θ between the direction $\hat{\mathbf{F}}$ of the force and the direction $\hat{\mathbf{s}}$ of motion of the body; see Figure 7.5b. (Note again that the sense of the angle is immaterial, and we need consider only angles in the range $0° \le \theta \le 180°$.) If $\theta < 90°$—that is, if the angle is acute—then $\cos \theta > 0$ and the work done by the force on the body is positive. If $\theta > 90°$—that is, if the angle is obtuse—then $\cos \theta < 0$ and the work is negative.

In Example 7.2, the *net* work done on the sled is zero. The net work done on a body need not be zero; we will consider such cases at length, beginning in Section 7.4. We must evaluate the net work done on the body in order to analyze the motion of the body in terms of the work done on it, just as we must evaluate the net force exerted on a body in order to make a dynamical analysis of its motion. It is often easier to evaluate the net work, because work is a scalar quantity and the necessary sum is an ordinary algebraic sum. This mathematical simplicity can be a great advantage in studying the motion of systems.

To summarize, there are two ways to evaluate the net work done on a body:

1. Evaluate the net force $\mathbf{F}_{net}$ acting on the body by taking the vector sum $\mathbf{F}_{net} = \mathbf{F}_1 + \mathbf{F}_2 + \cdots + \mathbf{F}_n$ over all the forces acting on the body. Then evaluate the net work by taking the scalar product $W = \mathbf{F}_{net} \cdot \mathbf{s}$.

2. Evaluate the work done on the body by each force individually, by calculating the scalar products $W_1 = \mathbf{F}_1 \cdot \mathbf{s}$, $W_2 = \mathbf{F}_2 \cdot \mathbf{s}$, ..., $W_n = \mathbf{F}_n \cdot \mathbf{s}$. Then evaluate the net work W by taking the scalar (algebraic) sum $W = W_1 + W_2 + \cdots + W_n$.

Both of these ways are useful.

Less strength is needed to ascend the ramp than the stairs, but the distance is greater.

Work Done by a Varying Force. Integration

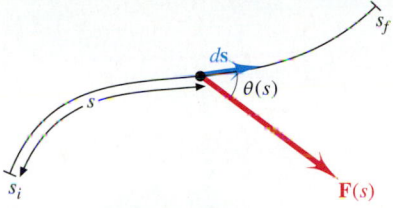

Before turning to analysis of the motion of bodies in terms of the work done on the bodies, we make one further generalization of the concept of work. In Figure 7.6, a body moves along an arbitrary path that begins at s_i and ends at s_f. As the body moves, it is acted on by a force whose magnitude and direction both vary in an arbitrary way. The position s of any point on the path is its distance from s_i, measured along the path. We can no longer define the work by means of Equation 7.5c, $W = \mathbf{F} \cdot \mathbf{s}$, because $\mathbf{F}$ is not constant and no vector $\mathbf{s}$ is defined. Nevertheless, the force is acting over a distance and thus doing work on the body, and we must find a way of expressing that work.

We begin by dividing the path into an infinite number of infinitesimal straight segments $d\mathbf{s}$. The path as a whole does not have vectorial properties, because it cannot be described in terms of a particular magnitude and a particular direction. But each segment does have vectorial properties; although its length ds is infinitesimal, each segment has a definite direction $\hat{d\mathbf{s}}$:

$$d\mathbf{s} = ds\ \hat{d\mathbf{s}}. \tag{7.10}$$

The unit vector is written $\hat{d\mathbf{s}}$ because it denotes the direction of $d\mathbf{s}$. But $\hat{d\mathbf{s}}$ is *not* infinitesimal. Like all unit vectors, it has the dimensionless magnitude 1.

As the body moves along the path, the force $\mathbf{F}$ varies. Suppose that the value of $\mathbf{F}$ at any point depends only on the body's position s, so $\mathbf{F}$ is a function of s: $\mathbf{F} = \mathbf{F}(s)$. The work done on the body by the force as the body moves through the segment $d\mathbf{s}$ located at s is infinitesimal. We call it $dW(s)$, and it has the value

$$dW(s) = \mathbf{F}(s) \cdot d\mathbf{s}. \tag{7.11a}$$

This is just the finite equation $W = \mathbf{F} \cdot \mathbf{s}$ (Equation 7.5c), rewritten so that it applies to the present infinitesimal case. Using Equation 7.6, $\mathbf{A} \cdot \mathbf{B} = |\mathbf{A}| \cos \theta\, |\mathbf{B}|$, we can write $dW(s)$ in the equivalent form

$$dW(s) = F(s) \cos \theta\,(s)\, ds. \tag{7.11b}$$

And, because $F(s) \cos \theta\,(s)$ is just $F_\parallel\,(s)$, the component of $\mathbf{F}$ parallel to $d\mathbf{s}$ at the point $\mathbf{s}$, we can also write a third equivalent form

$$dW(s) = F_\parallel\,(s)\, ds. \tag{7.11c}$$

The work W done by the force as the body traverses the entire path is found by adding up the bits of work dW done over all of the infinite number of infinitesimal segments. As you have learned in your study of calculus, this is done by carrying out the mathematical process of **integration**. The process is denoted by the **definite integral**

$$W = \int_{s_i}^{s_f} dW. \tag{7.12}$$

We can substitute for dW any of the equivalent values given by Equations 7.11a, 7.11b, and 7.11c. This gives the equivalent integrals

$$W = \int_{s_i}^{s_f} \mathbf{F}(s) \cdot d\mathbf{s}, \tag{7.13a}$$

$$W = \int_{s_i}^{s_f} F(s) \cos \theta\,(s)\, ds, \tag{7.13b}$$

and

$$W = \int_{s_i}^{s_f} F_\parallel\,(s)\, ds. \tag{7.13c}$$

Equation 7.13, in any of its three forms, constitutes the most general definition of work. Evaluating the integral in any of the forms requires an explicit knowledge of the function $\mathbf{F}(s)$ or, equivalently, $F(s) \cos \theta\,(s) = F_\parallel\,(s)$. Once this function is known, the

FIGURE 7.6 As a body moves along a path, it is subjected to a variable force $\mathbf{F}(s)$. Over the infinitesimal segment $d\mathbf{s}$ located at the position s, the force does infinitesimal work $dW = \mathbf{F}(s) \cdot d\mathbf{s} = F(s) \cos \theta(s)\, ds$.

TABLE 7.1 Integrals of Selected Functions

$$\int_a^b x^n \, dx = \frac{x^{n+1}}{n+1}\bigg|_a^b = \frac{b^{n+1}}{n+1} - \frac{a^{n+1}}{n+1} \qquad \text{for } n \neq -1$$

$$\int_a^b \frac{1}{x} \, dx = \ln x \bigg|_a^b = \ln \frac{b}{a} \qquad \text{for } n = -1$$

$$\int_a^b \sin c\theta \, d\theta = -\frac{1}{c} \cos c\theta \bigg|_a^b = \frac{1}{c}(\cos ca - \cos cb)$$

$$\int_a^b \cos c\theta \, d\theta = \frac{1}{c} \sin c\theta \bigg|_a^b = \frac{1}{c}(\sin cb - \sin ca)$$

a, b, c, and n are constants; x and θ are variables.

For any function f:

$$f(x)\bigg|_a^b \equiv f(b) - f(a)$$

$$\int_a^b cf(x) \, dx = c \int_a^b f(x) \, dx$$

evaluation is a matter of applying the art of integration, to which a considerable part of every introductory calculus course is devoted. The derivation of Equations 7.12 and 7.13 is sketched in Appendix 2.

The integrals of many functions are tabulated in calculus textbooks and more complete standard reference works. Table 7.1 lists the values of a few integrals that we will need in the next few chapters.

EXAMPLE 7.3

A coil spring is initially in its relaxed state, with length L. One end is fixed. You grab the free end and pull it through a distance A. Figure 7.7 shows the system as it appears part way through this process. If the force constant is k, how much work do you do on the end of the spring?

SOLUTION: The path traveled by the end of the spring as you apply the force $\mathbf{F}_t$ is a straight line whose direction you can designate $\hat{\mathbf{x}}$. Because the spring obeys Hooke's law, $\mathbf{F}_t$ is certainly not constant over the distance x. However, you can use Equation 7.13a to calculate the work you must do, because you can express $\mathbf{F}_t$ as a function of x. The force $\mathbf{F}_t$ is the Newton's-third-law reaction force to the Hooke's-law force $-kx\,\hat{\mathbf{x}}$. So you have $\mathbf{F}_t(x) = kx\,\hat{\mathbf{x}}$.

This force is always parallel to the infinitesimal displacement $d\mathbf{x}$. According to Equation 7.7a you have

$$\mathbf{F}_t \cdot d\mathbf{x} = F_t \, dx = kx \, dx.$$

Equation 7.13a thus becomes

$$W_t = \int_0^A kx \, dx.$$

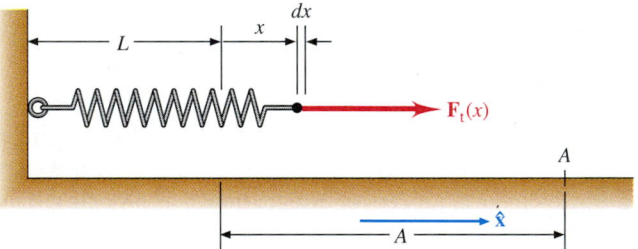

FIGURE 7.7 The spring is shown at an instant when it is stretched part way to its final extension A.

Because k is a constant, you can carry out the integration in the following steps, referring to Table 7.1 if necessary:

$$W_t = k \int_0^A x \, dx = k \frac{x^2}{2}\bigg|_0^A = \frac{1}{2}kA^2 - 0,$$

or $\qquad W_t = \frac{1}{2}kA^2.$ \hfill **(7.14a)**

Equation 7.14a shows that the work required to stretch a spring increases quadratically with the amount of stretch. This faster-than-linear increase is due to the fact that the more the spring is stretched the harder it is to stretch it further. This point is made evident in Example 7.4.

EXAMPLE **7.4**

(a) Having stretched the spring of Example 7.3 through a distance A, you now stretch it an additional distance A so that the total stretch is $2A$. How much additional work must you do to stretch the spring the additional distance?

(b) Now you gradually reduce the force that you are applying and thus allow the spring to return slowly through the distance $2A$ to its relaxed position. As the end of the spring returns to its relaxed position, how much work do you do on it?

SOLUTION:
(a) How much work must you do to stretch the spring from A to $2A$?

As in Example 7.3, you find the work by integrating $kx\,dx$. Now, however, the limits are A and $2A$ instead of 0 and A:

$$W'_t = k \int_A^{2A} x\,dx = k \left.\frac{x^2}{2}\right|_A^{2A} = \frac{1}{2}k(2A)^2 - \frac{1}{2}kA^2 = \frac{3}{2}kA^2.$$

$$(7.14b)$$

This is three times the work W_t required to stretch the spring

through the equal distance A beginning at the relaxed position.

(b) Now you allow the spring to return slowly to its relaxed position. How much work do you do on the spring in this process?

Again, you integrate $kx\,dx$. Now the limits are $2A$ and 0:

$$W''_t = k \int_{2A}^0 x\,dx = k \left.\frac{x^2}{2}\right|_{2A}^0 = 0 - \frac{1}{2}k(2A)^2 = -2kA^2.$$

$$(7.14c)$$

The work is negative because the direction of motion is opposite the direction in which you apply the force to the spring.

Add Equations 7.14a, 7.14b, and 7.14c together to find the total work that you do on the spring in the complete cycle of pulling it from rest to an extension $2A$ and then allowing it to return to the relaxed state. You have

$$W_t + W'_t + W''_t = \tfrac{1}{2}kA^2 + \tfrac{3}{2}kA^2 - 2kA^2 = 0;$$

the net work that you do over the cycle is zero.

Work and Kinetic Energy

The air-track glider of Figure 7.1 and Example 7.1 is shown again in Figure 7.8. The work done on this body by the constant force F sets it into motion; in this simple case, F is equal to the *net* force acting on the body. Equation 7.2, $v = \sqrt{2W/m}$, expresses the relation between the net work done on the body and the speed that it attains, provided that its initial speed is zero. If we solve that equation for W, we obtain

$$W = \tfrac{1}{2}mv^2. \tag{7.15}$$

The quantity on the right side of this equation is called the **kinetic energy**, for which we will use the symbol K. That is, we *define*

$$K \equiv \tfrac{1}{2}mv^2. \tag{7.16}$$

Kinetic energy (meaning literally the energy of motion) is a measure of the degree to which a system has the ability to do work on some other system. For example, suppose a spring is mounted on the end of the air track, as shown in Figure 7.8. As the glider approaches the end of the air track, the force ceases to act on it and it coasts. When the glider hits the spring, it compresses the spring. As you saw in Example 7.3, distorting a spring from its relaxed state requires work. A more massive or faster-moving glider will compress the spring more than a less massive or slower-moving one will. Equation 7.16 bears this out; it shows that the kinetic energy of a body depends on both its mass and its speed. Unlike work, kinetic energy is always a positive quantity,

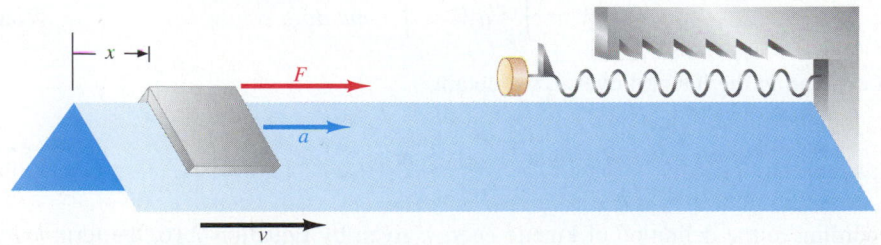

FIGURE 7.8 The air-track glider of mass m is acted on by a force F parallel to its direction of motion. Its instantaneous velocity is v. The force ceases to act, and the glider coasts until it hits the spring at the end of the track.

because m is a positive scalar and v^2 cannot be negative. Like W, K is a scalar and is measured in joules (see Problem 7.21).

We now generalize the argument that led to Equation 7.15. That is, we find the connection between the net work done on a body and its kinetic energy under the most general conditions possible:

1. the body follows an arbitrary path,
2. the net force $\mathbf{F}(s)$ varies in magnitude and direction as a function of s, and
3. the initial speed of the body has any value v_i.

We begin with Equation 7.11c, $dW(s) = F_\parallel(s)\, ds$, which expresses the work done by a force $\mathbf{F}(s)$ over an infinitesimal path segment $d\mathbf{s}$. Because we have asserted that $\mathbf{F}(s)$ is the *net* force acting on the body, we can use Newton's second law to write $F_\parallel = ma_\parallel$, where $a_\parallel$ is the component of the acceleration of the body tangential to the path. At any point s along the path, we thus have

$$dW = ma_\parallel\, ds. \tag{7.17}$$

Next, note that the tangential component $a_\parallel$ of an acceleration $\mathbf{a}$ results in a change only of the speed v, whereas the radial component $a_\perp$ results in a change only of the direction $\hat{\mathbf{v}}$. With this in mind, we can write

$$a_\parallel = \frac{dv}{dt},$$

where dv/dt is the rate of change of the body's speed. So we can write Equation 7.17 in the form

$$dW = m\frac{dv}{dt}\, ds. \tag{7.18}$$

The use of the term dv/dt in this expression implies that v is a function of t. But it is also true that v is a function of s, because the speed of the body varies in a specific way as it moves along the path under the influence of the net force $\mathbf{F}(s)$. We can therefore write

$$\frac{dv}{dt} = \frac{dv}{dt}\frac{ds}{ds} = \frac{dv}{ds}\frac{ds}{dt}.$$

And, because ds/dt is the speed v, we have

$$\frac{dv}{dt} = v\frac{dv}{ds}.$$

Inserting this value into Equation 7.18 gives

$$dW = mv\frac{dv}{ds}\, ds = mv\, dv. \tag{7.19}$$

We are now ready to evaluate the total work W done on the body as it passes over the path between s_i and s_f. To do this, we integrate dW, as in Equation 7.12. However, we must use appropriate limits of integration because dW is expressed as a function of v, not s. Fortunately, this is straightforward. When the body is located at the initial point s_i, it has some initial speed. Let us call this speed v_i. Similarly, let the final speed be v_f at the final point s_f. Equation 7.12 can then be written

$$W = \int_{s_i}^{s_f} dW = \int_{v_i}^{v_f} mv\, dv. \tag{7.20}$$

On evaluating the integral over v, we obtain

$$W = \tfrac{1}{2}mv^2\Big|_{v_i}^{v_f} = \tfrac{1}{2}mv_f^2 - \tfrac{1}{2}mv_i^2. \tag{7.21}$$

According to the definition of kinetic energy given by Equation 7.16, the term $\tfrac{1}{2}mv_f^2$

is the final kinetic energy K_f, and the term $\frac{1}{2}mv_i^2$ is the initial kinetic energy K_i. We can thus write Equation 7.21 in the form

$$W = K_f - K_i. \tag{7.22a}$$

We define the *change* in kinetic energy to be $\Delta K \equiv K_f - K_i$. Equation 7.22a then becomes

$$W = \Delta K. \tag{7.22b}$$

The change in kinetic energy of a body is equal to the net work done on it—that is, the work done on it by the net force acting on it. This statement is called the **work-energy theorem**. Because its derivation depends on the use of Newton's second law, its validity is restricted to inertial frames. We will generalize the work-energy theorem in Chapter 8.

Although the kinetic energy of a body can never be negative, the change ΔK can be either positive or negative, depending on the sign of the net work W done on the body. (If $W = 0$, there is no change in the kinetic energy and $\Delta K = 0$.)

EXAMPLE 7.5

A 1500-kg automobile is traveling at 100 km/h. The driver sees a slow truck ahead and slows down to 60 km/h. Find the initial and final energy of the automobile and the work done on it to slow it down.

SOLUTION: From the definition of kinetic energy (Equation 7.16) you have initial and final kinetic energies $K_i = \frac{1}{2}mv_i^2$ and $K_f = \frac{1}{2}mv_f^2$. Using the given numerical values and converting the speed into SI units, you have

$$K_i = \frac{1}{2} \times 1500 \text{ kg} \times \left(100 \text{ km/h} \times \frac{10^3 \text{ m/km}}{3600 \text{ s/h}}\right)^2$$

$$= 5.79 \times 10^5 \text{ J}$$

and $K_f = \dfrac{1}{2} \times 1500 \text{ kg} \times \left(60 \text{ km/h} \times \dfrac{10^3 \text{ m/km}}{3600 \text{ s/h}}\right)^2$

$$= 2.08 \times 10^5 \text{ J}.$$

According to the work-energy theorem, the work done is equal to the change in kinetic energy. You use Equation 7.22a to find

$$W = 2.08 \times 10^5 \text{ J} - 5.79 \times 10^5 \text{ J} = -3.71 \times 10^5 \text{ J}.$$

The negative value of W is consistent with the fact that the automobile slows down. What is the major source of the force that does the work on the automobile? What is the force's direction relative to the direction of motion of the auto?

THINKING LIKE A PHYSICIST

Our initial definition of kinetic energy (Equation 7.16) was a very narrow one because the argument leading to it contained three restrictions. We evaluated the work done on a body *initially at rest* (restriction 1) by a *constant* net force (restriction 2) directed *parallel to the motion of the body* (restriction 3). This evaluation led to Equation 7.15, $W = \frac{1}{2}mv^2$. We then defined the kinetic energy $K = \frac{1}{2}mv^2$, equating K to the work done on the body; that is, $W = K$. This would have been an empty definition had we not been guided by the idea that the work done on a body in putting the body into motion is recoverable as work that can be done by the body on some other system as the body comes to rest.

When we think about this idea, we are easily persuaded that the concept of kinetic energy must extend beyond the highly restricted argument leading to Equation 7.16. This is the motivation for pursuing the more intricate argument that leads to the much more general statement we call the work-energy theorem.

Power

A given amount of work W can be done quickly or gradually. The *rate* at which work is done on a system is often significant. We call this rate **power** P, defined to be

$$P \equiv \frac{dW}{dt}. \tag{7.23}$$

Because P is defined on the basis of the work dW done as the system passes over an infinitesimal path segment ds, P represents *instantaneous* power. Power may or may not be constant. But, even if P is not constant, it is possible to define the **mean power** $\langle P \rangle$ over a specified time interval $\Delta t = t_f - t_i$. If the work done during that time interval is W, we have

$$\langle P \rangle \equiv \frac{W}{\Delta t} \quad \text{over the time interval specified.*} \tag{7.24}$$

You can see by inspecting Equation 7.23 that the SI unit of power is the joule/second (J/s). This unit is given its own name, the **watt (W)**, so that

$$1 \text{ W} \equiv 1 \text{ J/s}. \tag{7.25}$$

The watt is named in honor of the Scottish engineer, inventor, and manufacturer James Watt (1736–1819). Watt made revolutionary improvements in the steam engine in the late eighteenth century and thereby made an enormous contribution to the Industrial Revolution. The steam engine remained the dominant source of motive power for more than a century; its direct descendants, including the gasoline and diesel engines, fill essential roles in the modern world.

Watt was the first person to define a unit of power. The customers who bought the steam engines built by his firm needed to know their capabilities. To express the output power of his engines in familiar terms, Watt devised the **horsepower**. The unit, now obsolete but still encountered, was intended to represent roughly the amount of power that a representative horse, working steadily, could produce. In modern terms, the *horsepower* (hp) is defined to be

$$1 \text{ hp} \equiv 746 \text{ W}. \tag{7.26}$$

We mention in passing two non-SI units of work or energy whose definitions are based on the unit of power. The **kilowatt-hour (kWh)**, a unit used mainly in the electric power industry, is defined as the amount of work done over a one-hour period by a system whose power output is a constant 1 kW ($= 10^3$ W). You should be able to show that

$$1 \text{ kWh} = 3.6 \times 10^6 \text{ J}. \tag{7.27}$$

Another nonstandard unit of work or energy is used in the electric power industry for long-range and large-scale planning purposes. This unit, the **quad**, is equal to one quadrillion British thermal units (1.055×10^{18} J). Its convenience lies in its magnitude; in each of the major forms—coal, petroleum, natural gas, hydro, and nuclear—the world use of energy is of order of magnitude 10 quads per year.

It is often convenient to express power in a way closely related to that given by Equation 7.23. We begin with Equation 7.11a, which we write in the slightly simplified form $dW = \mathbf{F} \cdot d\mathbf{s}$. Using Equation 7.23, we have

$$P = \frac{dW}{dt} = \mathbf{F} \cdot \frac{d\mathbf{s}}{dt}.$$

The term $d\mathbf{s}/dt$ is just the velocity $\mathbf{v}$. We thus have

$$P = \mathbf{F} \cdot \mathbf{v}. \tag{7.28}$$

Conservative Ratings Can Be Good Business

In reality, only very strong horses of the largest breeds are capable of working at a rate of 1 hp over an extended period. The average horse can maintain an output closer to $\frac{2}{3}$ hp. For comparison, the two earliest commercial models of Watt's steam engines were rated at 8 hp and 16 hp.

It used to be common to rate automobile engines in horsepower. The rating proved deceptive in practice, however. Manufacturers would often arrive at the value for an engine on the basis of tests in which the essential auxiliary systems of the engine—the water and oil pumps, the electric generator, and the cooling fan among them—were driven by an external power source and not by the engine itself. Some governments still tax automobiles on the basis of a horsepower rating arrived at yet more arbitrarily—for example, in terms of a formula based on piston displacement. Even without these intentional confusions, the power output of a gasoline engine depends very much on its speed, the load under which it is operating, the external temperature, and many other factors. Consequently, the horsepower rating of an engine tells the buyer relatively little about its on-the-road performance.

EXAMPLE 7.6

(a) If the automobile of Example 7.5 completes its deceleration in 9.0 s, find the mean braking power. (Include in this braking power not only the effect of the brakes, but also all other influences that tend to slow the automobile down.)

(b) As noted in Problem 4.26, a typical automobile can be made to stop from 100 km/h in a minimum distance of 60 m if the pavement is dry. Assuming that this specification applies to the automobile of part **a**, find the maximum braking power of

*A rigorous mathematical justification of this equation is the subject of Problem 7.46.

the automobile and compare this value with the power calculated in part **a**.

SOLUTION:

(a) Find the mean braking power required to bring a 1500-kg automobile to a stop from 100 km/h in 9.0 s.

You saw in Example 7.5 that the work W done over the braking interval is equal to the change in kinetic energy of the automobile, in accordance with the work-energy theorem. Calculation yielded $W = -3.71 \times 10^5$ J. You now use Equation 7.24 to write the mean power as

$$\langle P \rangle = \frac{W}{\Delta t} = \frac{-3.71 \times 10^5 \text{ J}}{9.0 \text{ s}} = -4.1 \times 10^4 \text{ W}.$$

That is, the applied braking power has magnitude 41 kW. The negative sign indicates that the automobile is slowing down. However, the sign is often not specified because it is usually obvious whether power is being applied to a system to drive it or to retard it.

(b) Find the maximum braking power of the automobile and compare this value with the power calculated in part **a**.

Here, it is best to use Equation 7.28. In the present case, with **F** parallel to **v**, this equation simplifies to $P = Fv$. You do not know the braking force F, but you can use Newton's second law to express it in terms of the corresponding acceleration a. This gives you

$$P = mav \qquad (7.29)$$

for the instantaneous power. You are not given a, but you do know the minimum braking distance s. Because the stop in question is to be made over minimum distance, the acceleration must have a maximum value. This value is most likely dictated mainly by the maximum frictional force that can be applied by the road to the tires, which does not depend strongly on the speed. In any case, it makes a reasonable first approximation to assume that the acceleration is constant. If this is so, the acceleration is related to the initial speed v_i and the braking distance s by the familiar expression $v_i^2 = 2as$, or

$$a = \frac{v_i^2}{2s}.$$

When you insert this value of a into Equation 7.29, you obtain

$$P = \frac{mv_i^2}{2s} v$$

for the instantaneous power. The quantity $mv_i^2/2s$ is a constant specified by the conditions of the problem. The quantity v is the instantaneous speed. The equation thus shows that the instantaneous power is directly proportional to the speed and has its maximum value P_{max} at the beginning of the braking process, when $v = v_i$. It follows that

$$P_{max} = \frac{mv_i^3}{2s}.$$

You are now ready to find a numerical value for P_{max}. Before you do so, however, it is convenient to convert the initial speed into SI units. You have

$$100 \text{ km/h} \times \frac{10^3 \text{ m/km}}{3600 \text{ s/h}} = 27.8 \text{ m/s}.$$

When you use this value and the other given quantities, you obtain

$$P_{max} = \frac{1500 \text{ kg} \times (27.8 \text{ m/s})^3}{2 \times 60 \text{ m}} = 2.7 \times 10^5 \text{ W} = 270 \text{ kW}.$$

This is more than six times as great as the mean power calculated for the moderate deceleration of part **a**.

---

a	acceleration
$\mathbf{A} \cdot \mathbf{B}$	scalar (dot) product of the vectors **A** and **B**
$\mathbf{F}_{net}$	net force exerted on a body
$F_{\parallel}$	component of a force **F** parallel to the displacement of the body on which the force acts
k	force constant of a spring
K	kinetic energy of a body
ΔK	change in kinetic energy
P	power
$\langle P \rangle$	mean power
s_i, s_f	initial and final points on a path
s	distance from s_i measured along a specified path

$d\mathbf{s}$	vector representing an infinitesimal displacement along the path
$\hat{d\mathbf{s}}$	unit vector in the direction of $d\mathbf{s}$
$\mathbf{s}$	vector representing displacement of a body along a straight-line path
v	speed
$\mathbf{v}$	velocity
W	work; net work done by the net force $\mathbf{F}_{net}$
W_j	work done by the force $\mathbf{F}_j$
dW	infinitesimal work done on a body as it moves through $d\mathbf{s}$
x	one-dimensional displacement

Summing Up

Suppose a force **F** acts on a body as the body moves over an infinitesimal path segment $d\mathbf{s}$. If the force is a function of the location s of the body along the path—that is, if $\mathbf{F} = \mathbf{F}(s)$—then Equation 7.11a defines the work done *by* the force *on* the body: $dW(s) = \mathbf{F}(s) \cdot d\mathbf{s}$. In this equation, the

scalar product, or **dot product**, of the two vectors is taken according to the general definition of Equation 7.6,

$$\mathbf{A} \cdot \mathbf{B} \equiv |\mathbf{A}| \cos \theta |\mathbf{B}| = AB \cos \theta.$$

The work W done by the force on the body as it traverses

the entire path from s_i to s_f can be found by summing the work dW done over the infinite number of infinitesimal segments $d\mathbf{s}$ that make up the path. The result is expressed in the integrals of Equations 7.12 and 7.13a,

$$W = \int_{s_i}^{s_f} dW = \int_{s_i}^{s_f} \mathbf{F}(s) \cdot d\mathbf{s}.$$

A special but important case is the one in which the force is constant over a straight-line path. If the straight-line displacement is expressed by the vector $\mathbf{s}$, the work done by the force on the body is given by Equation 7.5c,

$$W = \mathbf{F} \cdot \mathbf{s}.$$

The **net work** done on a body as it traverses a specified path is the work done by the *net* force acting on it.

When a body of mass m has a speed v, its **kinetic energy** (energy of motion) is defined by Equation 7.16,

$$K \equiv \tfrac{1}{2}mv^2.$$

The **work-energy theorem**, Equations 7.22a and 7.22b, states that the change in the **kinetic energy** K of a body is equal to the net work W done on the body over the specified path:

$$W = K_f - K_i = \Delta K.$$

Suppose that work dW is done on a body as it traverses

a certain infinitesimal path segment $d\mathbf{s}$ and that the time required for this traversal is dt. **Power** is defined by Equation 7.23 to be the time derivative of work—that is, the work done per unit time:

$$P \equiv \frac{dW}{dt}.$$

Power can be expressed alternatively in the form given by Equation 7.28,

$$P = \mathbf{F} \cdot \mathbf{v}.$$

Both of these equations express the instantaneous power, which may or may not be constant in time. Even if the power is not constant, Equation 7.24 can be used to express the **mean power** over a specified time interval Δt:

$$\langle P \rangle = \frac{W}{\Delta t}.$$

KEY TERMS

Section 7.2 Work. The Scalar Product of Two Vectors
work, net work ▪ scalar product (dot product)

Section 7.4 Work and Kinetic Energy
kinetic energy ▪ work-energy theorem

Section 7.5 Power
mean power

QUERIES

7.1 *(2) Waiting for Goliath.* A child swings a stone tied to a cord in a horizontal circle above his head. How much work is done on the stone by the centripetal force exerted by the string? Assume that air resistance is negligible.

7.2 *(2) An inclination toward reality.* Friction is always present in real systems like the one described in Query 7.1. Consequently, it is necessary to supply a continuous driving force to do work on the stone. How is this force applied to the string? How is it transmitted to the stone? How can you tell by watching the child that he is doing work on the stone?

7.3 *(2) Holding up the wall.* If you push hard against a heavy object for a long time without being able to move it, you get tired. You may well say that you have done a lot of work in the everyday sense of the word. But have you done work on the object in the precise physical sense of the word?

7.4 *(3) Getting along together.* Show that the dot product $\mathbf{A} \cdot \mathbf{B}$ can be written equally well as $A_{\parallel}B$ or as $B_{\parallel}A$, where $A_{\parallel}$ is the component of $\mathbf{A}$ along $\mathbf{B}$ and $B_{\parallel}$ is the component of $\mathbf{B}$ along $\mathbf{A}$.

7.5 *(3) One way or another.* In the middle term in each of the three expressions given in Table 7.1, the quantity to the left of the limit specifier $\Big|_a^b$

is a function called the **indefinite integral** of the original function. By differentiating each indefinite integral in Table 7.1, show that the result is the original function. That is, show that integration and differentiation are *inverse operations*. (The general proof of this statement is called the *fundamental theorem of calculus.*)

7.6 *(4) Driver's ed.* How does the minimum stopping distance of an automobile depend on its speed? Consider only the distance over which the brakes are actually applied, and not the driver reaction time—the time required to decide to stop and to apply the brake.

7.7 *(4) Doubling up.* A body initially at rest is acted on by a constant force. At time t, its speed is v and its kinetic energy is K; at time $2t$, its speed is $2v$ but its kinetic energy is $4K$. Account for this difference.

7.8 *(4) Keeping fit.* Estimate your kinetic energy when you are running as fast as you can.

7.9 *(4) Old & new.* A very large ocean liner has a mass of 80 000 tonnes and a cruising speed of 40 km/h. A large passenger airplane has a mass of 150 tonnes and a cruising speed of 1000 km/h. How do their kinetic energies compare at cruising speed?

7.10 *(4) Hard labor, little profit?* When you drag a heavy object over a long distance, you do a substantial amount of work. What is the kinetic energy of the object when you have finished? Does your answer conflict with the work-energy theorem? Explain.

7.11 *(4) Did you say fixed Newton or fig newton?* Is the kinetic energy of a body the same as seen by all observers who are fixed in inertial reference frames?

7.12 *(5) More power to you!* When a vehicle moves through a fluid (an airplane through air or a submarine through water), the drag force exerted by the fluid is roughly proportional to the square of the vehicle's speed. If the speed of such a vehicle is to be doubled, what increase of engine power is required?

PROBLEMS

GROUP A

7.1 *(2) Huff & puff.* A heavy chest of drawers has a mass of 150 kg. If the coefficient of kinetic friction between the feet of the chest and the floor is 0.8, how much work must you do to push it across a room that is 5 m wide?

7.2 *(2) Push comes to shove, I.* In Problem 4.41, two mechanics push a disabled automobile along a level driveway into their garage for repair. One pushes against the left doorframe while steering the automobile straight ahead. He exerts a force $\mathbf{F}_1$ of magnitude 320 N at an angle of 30° with the forward direction. The other mechanic pushes on the right rear fender, exerting a force $\mathbf{F}_2$ of magnitude 435 N at an angle of 10° with the forward direction. If they push the car a distance of 50 m, how much work does each do? How much work do they do together?

7.3 *(2) Push comes to shove, II.* **(a)** Find the magnitude and direction of the combined force $\mathbf{F}_1 + \mathbf{F}_2$ exerted by the two mechanics in Problem 7.2. **(b)** How much work do the two mechanics do together in pushing the car 50 m? Does your result agree with the one you obtained for Problem 7.2?

7.4 *(2) Working at the scalar product.* The angle between two vectors is 35°. Their magnitudes are $|\mathbf{F}| = 150$ N and $|\mathbf{s}| = 35$ m. Find their dot product.

7.5 *(2) It's a drag, I.* Suppose that, in Figure 7.2, the force is 32 N and the angle θ is 25°. Find the work done by the force in pulling the block a distance of 30 m.

7.6 *(2) It's a drag, II.* If the speed of the block in Problem 7.5 remains constant as it is pulled the 30-m distance, what are **(a)** the magnitude of the frictional force acting on it, **(b)** the work done by the frictional force, and **(c)** the net work done on it?

7.7 *(2) Scaling the heights.* **(a)** A 25.7-kg block hangs from a rope. How much work is required to raise the block through a vertical distance of 3.42 m? **(b)** Suppose instead that you drag the block up a frictionless plane of incline angle 15°. How far do you have to drag the block along the plane to achieve the same vertical height as in part **a**? How much work is required to do this?

7.8 *(2) Work, work, work.* You drag a 12-kg block up a frictionless plane of incline angle 28°, exerting a force parallel to the plane to do so. **(a)** If the speed of the block is constant, what is the magnitude of the force? **(b)** How much work have you done when you have dragged the block 7.5 m? **(c)** How much work does the gravitational force do on the block over the same distance?

7.9 *(2) Getting an angle on things.* Two vectors have magnitudes $A = 8$ and $B = 2$. Their dot product is $\mathbf{A} \cdot \mathbf{B} = 6$. Find the angle between the two vectors.

7.10 *(2) Is Dotty a square?* **(a)** Find the dot product of a vector with itself; that is, evaluate $\mathbf{A} \cdot \mathbf{A}$. **(b)** Let **x**, **y**, and **z** be mutually perpendicular vectors. Prove that $\mathbf{x} \cdot \mathbf{y} = \mathbf{y} \cdot \mathbf{z} = \mathbf{z} \cdot \mathbf{x} = 0$.

7.11 *(2) Unity and diversity.* Consider the mutually perpendicular unit vectors $\hat{\mathbf{x}}$, $\hat{\mathbf{y}}$, and $\hat{\mathbf{z}}$. Evaluate the scalar products **(a)** $\hat{\mathbf{x}} \cdot \hat{\mathbf{x}}$, $\hat{\mathbf{y}} \cdot \hat{\mathbf{y}}$, and $\hat{\mathbf{z}} \cdot \hat{\mathbf{z}}$; **(b)** $\hat{\mathbf{x}} \cdot \hat{\mathbf{y}}$, $\hat{\mathbf{y}} \cdot \hat{\mathbf{z}}$, and $\hat{\mathbf{z}} \cdot \hat{\mathbf{x}}$.

7.12 *(2) Algebraic method for the scalar product.* Consider two vectors $\mathbf{A} = (A_x, A_y, A_z)$ and $\mathbf{B} = (B_x, B_y, B_z)$. **(a)** Show that their dot product can be written

$$\mathbf{A} \cdot \mathbf{B} = A_x B_x + A_y B_y + A_z B_z;$$

that is, the dot product of two vectors is the sum of the products of their corresponding components.

7.13 *(2) Algebraic manipulation.* Consider two vectors $\mathbf{A} = 2\hat{\mathbf{x}} + 2\hat{\mathbf{y}}$ and $\mathbf{B} = 3\hat{\mathbf{x}} - \hat{\mathbf{y}}$. Find **(a)** the magnitudes A and B of the two vectors; **(b)** the angle θ between them. (Hint: Use the results of Problems 7.10 and 7.12.)

7.14 *(3) Stretching things a bit, I.* You have a spring with force constant 450 N/m. How much work do you do in **(a)** stretching the spring from its relaxed position to an extension of 10 cm, **(b)** stretching it an additional 10 cm, **(c)** allowing it to return from an extension of 20 cm to its relaxed position?

7.15 *(3) The force isn't constant but the force constant is.* In order to stretch a spring by 0.325 m, you must do 13.0 J of work on it. Find the force constant of the spring.

7.16 *(3) Obedience to the law.* When you pull on the end of a relaxed spring and stretch it by 3.1 cm, you do 1560 J of work on it. When you stretch it an additional 5.3 cm, you do 9240 J more work. Does the spring conform to Hooke's law over the entire range of stretching?

7.17 *(4) Watch your speed!* For a given vehicle, what is the approximate ratio of the safe stopping distance from 100 km/h to the safe stopping distance from 75 km/h?

7.18 *(4) Free fall.* A mass m is released from rest and allowed to fall. If air resistance is negligible, what is its kinetic energy when it has fallen through a distance y?

7.19 *(4) Overcoming friction and then some.* You push a loaded wheelbarrow of mass 130 kg, exerting a horizontal force of 120 N as you do so. As the wheelbarrow moves, the frictional force acting on it is 85 N. **(a)** If you begin from rest, how far do you push the wheelbarrow before reaching a moderate walking speed of 1.3 m/s? **(b)** Over this distance, how much work do you do on the wheelbarrow? **(c)** How much work does the frictional force do on the wheelbarrow? **(d)** What is the final kinetic energy of the wheelbarrow?

7.20 *(4) Speedup.* A block of mass 1 kg is accelerated from an initial speed of 2 m/s to a final speed of 6 m/s as it is dragged a distance of 10 m over a level floor. If the coefficient of kinetic friction between the block and the floor is 0.2, how much work must be done by the applied force?

7.21 *(4) Cheek by joule.* By direct substitution of dimensions in the definition $K = \frac{1}{2}mv^2$, show that kinetic energy is properly expressed in joules.

7.22 *(4) Still taking off, I.* Consider again the jet plane of Problems 2.18 and 3.16, which takes off at 80 m/s after a 2-km run. If the mass of the plane is 115 tonnes, what is its kinetic energy at takeoff?

7.23 *(4) Energetic activity, I.* A water skier has a mass of 47 kg. When she is towed by a boat at a speed of 30 km/h, what is her kinetic energy?

7.24 *(4) Terra firma?* Using the values given in the table titled Useful Physical Quantities (inside front cover), calculate the kinetic energy of the earth as seen by an observer located on the sun.

7.25 *(4) Stopping the train.* A railway car of mass 20 tonnes is moving at 54 km/h when the brakes are applied. **(a)** How much work must the brakes do in bringing the car to a stop? **(b)** If the brakes exert a force of 6000 N, what is the stopping distance?

7.26 *(5) Still taking off, II.* Assume that the acceleration of the airplane of Problem 7.22 is essentially constant. What is the mean power delivered by the engines to the plane during its takeoff run?

7.27 *(5) Energetic activity, II.* The force required to tow the water skier of Problem 7.23 is 85 N. What is the power delivered by the boat through the towrope to the skier?

7.28 *(5) Still taking off, III.* For the airplane of Problem 7.22, what is the instantaneous power delivered by the engines to the plane at the moment before lift-off?

7.29 *(5) . . . climb nearly to the stars.* The San Francisco cable car discussed in Problem 4.40 moves at a constant speed of 4.0 m/s. If the loaded mass of the cable car is 10.5 tonnes, how much power must the cable supply to pull the car up a 20% grade (sin θ = 0.2)? Neglect friction (not a bad approximation in this case).

7.30 *(5) Motive power, I.* As noted in Problem 3.55, the frictional force tending to slow an automobile down depends on its speed according to the approximate relation $|F_f(v)| = h + kv^2$, where h and k are positive constants. **(a)** How much power must be delivered to the ground by the wheels to keep the car going at a constant speed v? **(b)** For a medium-size automobile, typical values of the constants are $h = 250$ N and $k = 0.65$ N·s^2/m^2. How much power is required when the car is moving on a level road at 15 m/s? at 30 m/s?

7.31 *(5) Motive power, II.* The mass of the car of Problem 7.30 is 1600 kg. If it begins to climb a hill of incline angle 3°, what total power is required to keep it going at 30 m/s?

7.32 *(G) Motive power, III.* If the car of Problems 7.30 and 7.31 coasts down a hill at a constant speed of 30 m/s without engine power, how steep is the hill?

GROUP B

7.33 *(2) Smooth ride.* A train leaves a station at $t = 0$ and moves with constant acceleration. Express the total net work done on the train as a function of t.

7.34 *(2) It's a drag, III.* The coefficient of kinetic friction between a block and the surface on which it lies is μ_k. You drag the block across the surface, using a force **F** whose direction makes the angle θ with the surface. **(a)** If the force is such that the speed of the block is constant, express the magnitude F of the force in terms of the angle θ and the constants μ_k, m, and g. **(b)** Show that the work done in moving the block a distance D is

$$W = \frac{\mu_k mgD}{1 + \mu_k \tan \theta}.$$

(c) Evaluate W for the extreme cases $\theta = 0°$ and $\theta = 90°$. What is the necessary value of F for each of these cases? Describe what happens in each case, and give a physical explanation for the values of F that you obtained.

7.35 *(3) Stretching things a bit, II.* Two springs having force constants k_1 and k_2 are linked end to end. How much work is required to stretch the pair of springs by a total amount L?

7.36 *(3) Unorthodox spring.* A spring is designed in such a way that it does not obey Hooke's law. Rather, the force required to hold it stretched by an amount x is given by $F = ax + bx^2$. How much work is required to stretch it **(a)** a distance x, beginning at the relaxed position $x = 0$, and **(b)** from $x = x_1$ to $x = x_2$?

7.37 *(3) Integrated workplace.* At a certain instant, a simple pendulum of length l and mass m makes an angle θ with respect to the vertical. **(a)** What is the net force acting on the pendulum bob? **(b)** How much work, dW, does the net force do on the pendulum bob as the angle changes by the infinitesimal amount $d\theta$? **(c)** Using Equation 7.20 in the slightly modified form

$$W = \int_{\theta_0}^{0} dW,$$

find the work done by the net force on the bob as the pendulum descends from an initial angle θ_0 to the vertical position. **(d)** Let h be the initial vertical height of the pendulum bob above its lowest position. Express the result of part **c** in terms of h.

7.38 *(4) . . . off to Buffalo.* In the game of shuffleboard, a wooden disk of mass m is pushed so that it slides across a waxed wooden floor. **(a)** If the initial speed of the disk is v, how much work must be done on it by the frictional force exerted by the floor to bring it to a stop? **(b)** If it comes to a stop in a distance x, what is the coefficient of kinetic friction between the disk and the floor?

7.39 *(4) A time for forceful action.* A body of mass m is initially at rest. If a constant net force $\mathbf{F}$ acts on it for a time t, show that the final kinetic energy of the body is $K = F^2 t^2 / 2m$.

7.40 *(5) Dynamometer.* The device shown below, called a *dynamometer*, can be used to measure engine output power.

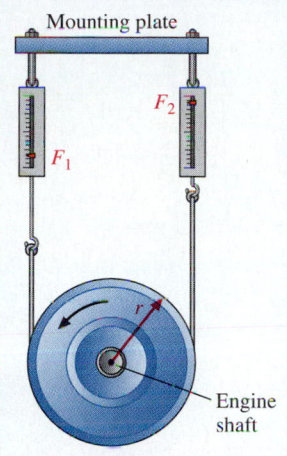

Mounting plate

F_2

F_1

r

Engine
shaft

The engine shaft is attached to the pulley, which has radius r. The belt wrapped around the pulley slips as the pulley turns. The two spring scales maintain tension in the belt. The tension, and thus the engine load, can be adjusted by raising or lowering the mounting plate to which the scales are attached. When the pulley is rotating at N revolutions per second, what is the power output P of the engine in terms of r, N, and the scale readings F_1 and F_2?

7.41 *(5) Powering a change.* Problems 2.19 and 3.5 describe a particle that moves along a straight line in such a way that its position is described by the function $x = 5t - 20t^2$, with x in meters and t in seconds. The particle's mass is 8.5 kg. **(a)** What is the kinetic energy of the particle at $t_1 = 0.1$ s? at $t_2 = 2.1$ s? **(b)** What is the mean power exerted on the particle during the time interval $t_2 - t_1$? **(c)** What is the instantaneous power at t_1? at t_2? [Hint: One way (but not the only one) of solving this problem involves the use of Equation 7.28. If you do use this method, pay careful attention to the sign of $\mathbf{F} \cdot \mathbf{v}$.]

7.42 *(G) Looking for constants in a changing world.* The position of a particle is expressed by each of the following functions of time: **(a)** $x = pt^{3/2} + q$; **(b)** $x = rt$; **(c)** $x = u\sqrt{t}$. The quantities p, q, r, and u are constants having the proper dimensions. In which of these cases is K constant? In which is P constant?

GROUP C

7.43 *(2) The function of work.* A force is given by the function $\mathbf{F} = 2x^2 \hat{\mathbf{x}} + 5y\hat{\mathbf{y}}$. Find the work done by the force as it moves a body from $(0, 0)$ to $(1, 1)$ along each of the three paths shown below: **(a)** the path $(0, 0) \to (1, 0) \to (1, 1)$; **(b)** the straight-line path $y = x$; **(c)** the parabolic path $y = x^2$.

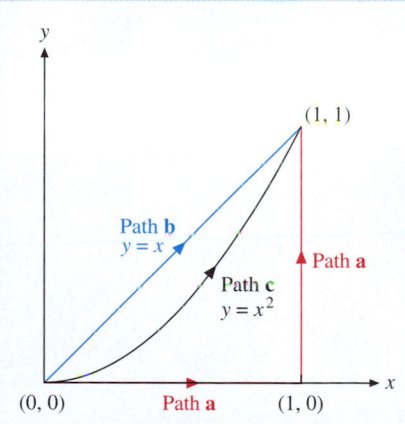

y

$(1, 1)$

Path b
$y = x$

Path a

Path c
$y = x^2$

$(0, 0)$

Path a

$(1, 0)$

x

7.44 *(4) Kinetic energy of a harmonic oscillator.* The displacement of the mass in a mass-spring system is $x = A \cos \omega t$. Show that the kinetic energy of the mass is given as a function of time by the expression $K = \frac{1}{2} k A^2 \sin^2 \omega t$, where k is the force constant of the spring.

7.45 *(4) Going round in circles.* A particle moves along a circle of radius R in such a way that its kinetic energy is given by the expression $K = cs^2$, where c is a constant and s is the distance covered along the circle. What is the magnitude of the net force acting on the particle as a function of s?

7.46 *(5) Derivation of Equation 7.24.* **(a)** Assume that the power exerted on a body is known as a function of time; that

is, $P = P(t)$. Show that the work W done on the body over the time interval $\Delta t = t_f - t_i$ is

$$W = \int_{t_i}^{t_f} P(t)\, dt.$$

(b) Suppose that you have used this equation to evaluate the work W done over a known time interval Δt. You can always find a constant that, when multiplied by Δt, will give you the known value W. What must the dimensions of that constant be? Why are you justified in calling it the mean power $\langle P \rangle$? Now show that Equation 7.24 is valid. **(c)** Generalize the argument of parts **a** and **b** to prove the **mean value theorem** of calculus. In somewhat restricted form, this theorem is as follows: Given any function $f(z)$ of an arbitrary independent variable z and given any initial and final values z_i and z_f of z, there exists a constant called the *mean value* $\langle f \rangle$ such that

$$\int_{z_i}^{z_f} f(z)\, dz = \langle f \rangle \int_{z_i}^{z_f} dz = \langle f \rangle (z_f - z_i).$$

Take one more simple step, and generalize the result of part **b** so as to *define* the mean value $\langle f \rangle$ of the arbitrary function $f(z)$ as the ratio of two integrals. Is this definition consistent with the elementary idea of an average?

7.47 *(5) Power hitter.* A batter hits a baseball of mass m. It leaves the bat with speed v_i at an angle α above the horizontal. **(a)** Express the instantaneous power exerted on the ball by the force of gravity as a function of time. **(b)** The ball is caught by an outfielder just as it returns to the height above the ground at which the batter hit it. What is the mean power $\langle P \rangle$ exerted by the gravitational force over the flight of the ball? Neglect air resistance.

Conservative Systems and Potential Energy

Left: As it bounces up and down, the ball seems to trade height for speed and vice versa, again and again. As speed and height continually change, is there something that doesn't change at all?

The more things change, the more they remain the same.

—FRENCH PROVERB

Introduction

In the Introduction to Chapter 7, we considered the possibility of making an inert system become ''active'' by doing work on it and thus giving it energy. Figure 8.1, which is almost identical with Figure 7.1, shows several systems before and after work has been done on them. Of these systems, however, only the air-track glider can be analyzed completely in terms of the work done on it and the kinetic energy that it acquires.

The three other systems are more complicated. The ball is always acted on by a downward gravitational force of magnitude mg; in order to raise the ball, a minimum upward force of magnitude mg must be applied. To raise the ball through a vertical distance h, that applied force must do an amount of work

$$W = mgh \tag{8.1}$$

on the ball. But the ball is not in motion after it has been raised and thus has no kinetic energy. (The work-energy theorem is not violated, because the *net* work done on the ball is zero. Can you see why $W_{net} = 0$, even though W in Equation 8.1 is not zero?) Although the ball has no kinetic energy, you know from experience that *because it has*

FIGURE 8.1 Four systems shown (*a*) in a state in which their energy is zero and (*b*) in a state in which their energy is not zero. Except for the air track, all of these systems acquire energy that is not kinetic energy because the energy is not manifested in motion.

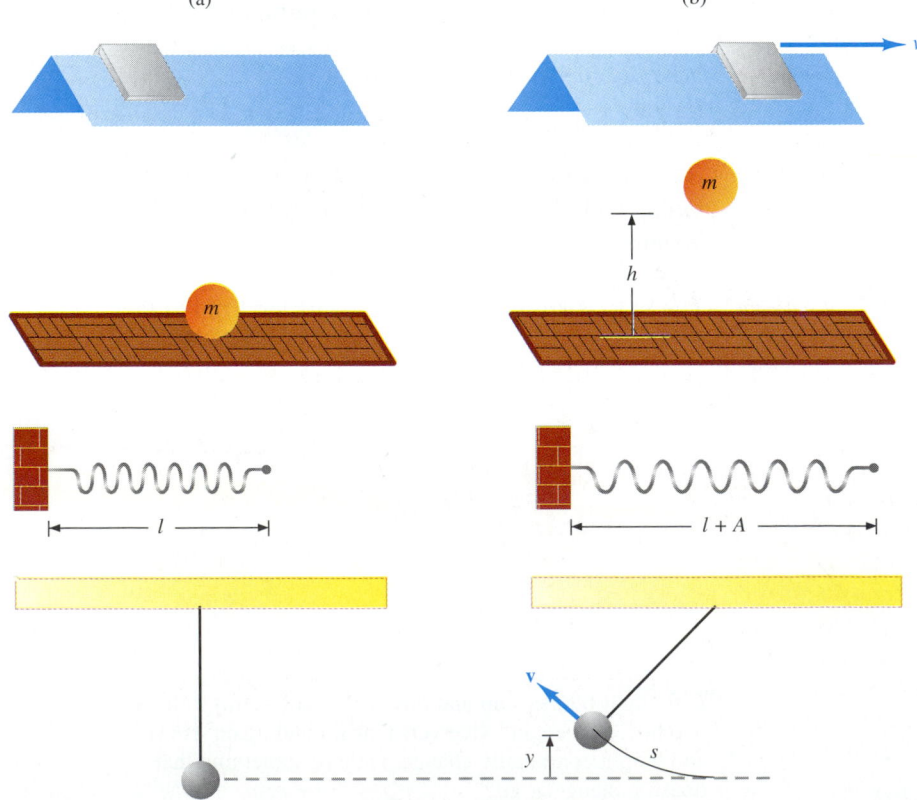

been raised, the ball has the possibility of acquiring kinetic energy. That possibility can be realized simply by allowing the ball to fall.

Similar considerations apply to the mass-spring system. In order to displace the mass a distance A from its equilibrium position, a force must be applied to the mass, and thus also to the spring. That force must do an amount of work $W = \frac{1}{2}kA^2$ on the spring, as you saw in Example 7.3. If the system is then released, the spring will return to its relaxed state, doing work on the mass and conferring kinetic energy on it in the process.

Lastly, consider the behavior of the pendulum in the absence of friction. Once the pendulum has been raised and released, it repeatedly gains and loses kinetic energy as it swings. As it descends, it speeds up. Its kinetic energy is greatest when it is at the bottom of its swing because it moves fastest at that point. As the pendulum rises again, it loses kinetic energy and comes instantaneously to rest at the high point of its arc before reversing and speeding up again. The pendulum appears to be repeatedly "trading off" its kinetic energy for some *other* form of energy and then trading this other form of energy back for kinetic energy again. The other form of energy is called *potential energy*. In this case, the trade-off takes place in such a way that the sum of the kinetic energy and the potential energy is constant over time; that is, the sum of the energies is *conserved*. The sum of the kinetic and potential energies of a system is called its *total mechanical energy*. In this chapter, we will develop means of analyzing systems in which total mechanical energy is conserved, as well as systems in which it is not.

In Section 8.2, we define potential energy in a rigorous way. In particular, we find that potential energy can be associated only with forces belonging to the special category called *conservative forces*, and we consider the properties of such forces. In Section 8.3, we study the motion of systems possessing both kinetic and potential energy, and, in Section 8.4, we extend the work-energy theorem to make it useful for studying such systems.

Potential Energy

The qualitative discussion of Section 8.1 suggests that a system can possess energy in a form other than kinetic. Moreover, it seems possible to transform that nonkinetic energy into kinetic energy, at least under certain circumstances. Let us follow that transformation quantitatively in two systems already discussed qualitatively, the raised ball and the mass-spring system.

The Raised Ball Falls

When the raised ball is released, it falls with acceleration g. When it is about to strike the ground, its speed (neglecting air resistance) is given by Equation 2.38, which becomes $v_f^2 = 2gh$ in the present notation. We can immediately evaluate the corresponding kinetic energy by using the definition $K \equiv \frac{1}{2}mv^2$; we have

$$K_f = \tfrac{1}{2}mv_f^2 = \tfrac{1}{2}m(2gh) = mgh. \tag{8.2}$$

Compare this kinetic energy with Equation 8.1, $W = mgh$, which gives the amount of work needed to raise the ball to a height h above the floor in the first place. The two values are the same! On this basis, we make a plausible assertion, which we will make rigorous after we have explored it enough to get an intuitive feeling for its meaning. Here is our assertion: In the interim between the performance of work on the ball and the appearance of kinetic energy as the ball speeds up during its fall, the energy is already present but exists in a different, nonkinetic form. This **potential energy** is conventionally denoted by the symbol U. We assert that the increase ΔK in the kinetic energy of the ball as it falls through the height h is associated with an equal decrease $-\Delta U$ in its potential energy; that is,

$$\Delta K = -\Delta U.$$

The initial kinetic energy of the ball is $K_i = 0$. We therefore have $\Delta K = K_f - K_i = K_f$, or

$$-\Delta U = U_i - U_f = K_f = mgh.$$

According to this equation, the ball possesses potential energy by virtue of its *position*. The greater the initial height h, the greater the value of $-\Delta U$ and the greater the kinetic energy of the ball just before it hits the floor.

The Mass-Spring System

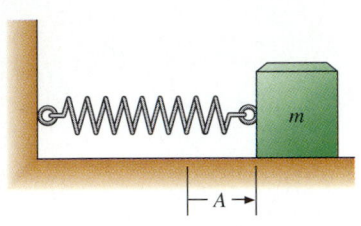

FIGURE 8.2 An idealized mass-spring system. The spring, of force constant k, has negligible mass. The mass m slides without friction on the supporting surface. The mass is initially displaced by an amount A from its equilibrium position, stretching the spring by the same amount. The system is then released.

Next, consider the mass-spring system shown in Figure 8.2. To bring this system to the stretched state shown, with the mass displaced a distance A from its equilibrium position, a force F must be applied to overcome the resistance of the spring to stretching. As you saw in Example 7.3, this force does work $W = \frac{1}{2}kA^2$. When the system is released, the spring contracts toward its relaxed state, doing work on the mass and giving it kinetic energy in accord with the work-energy theorem. Just as we calculated the kinetic energy of the falling ball just before it hits the ground, let us calculate the kinetic energy of the system at the instant when the spring has returned to its relaxed state.

When the spring is distended by an arbitrary amount x, the force it exerts on the mass is given by Hooke's law, $F = -kx$. The equation of motion for the mass is found by applying Newton's second law, which gives $-kx = ma$, or

$$a = -\frac{k}{m}x. \tag{8.3}$$

This is just Equation 6.31 rederived.

Using the definition $a \equiv dv/dt$, we have

$$dv = a\,dt.$$

We would like to integrate this equation to find the velocity $v = \int dv = \int a\,dt$ and thus the kinetic energy $K = \frac{1}{2}mv^2$. But we cannot integrate directly, because we know the acceleration as a function of x, not t. To deal with this difficulty, we write the series of equalities

$$dv = a\,dt = a\,dt\,\frac{dx}{dx} = a\,\frac{dt}{dx}\,dx.$$

Because $dt/dx = 1/v$, we have $dv = (a/v)\,dx$, or

$$v\,dv = a\,dx.$$

Next, we use Equation 8.3 to express a as a function of x. We obtain

$$v\,dv = -\frac{k}{m}\,x\,dx.$$

Both sides of this equation can be integrated directly. What are the limits of integration? The physical process begins when the system is released from the position $x = A$, where $v = 0$. We are interested in the final situation, where $x = 0$. We call the corresponding value of velocity v_f, and we then have

$$\int_0^{v_f} v\,dv = -\frac{k}{m}\int_A^0 x\,dx. \tag{8.4}$$

Carrying out the integrations (see Table 7.1) yields

$$\frac{v^2}{2}\Big|_0^{v_f} = -\frac{k}{m}\frac{x^2}{2}\Big|_A^0 ,$$

or

$$\frac{v_f^2}{2} - 0 = 0 - \frac{-kA^2}{2m}.$$

The term $v_f^2/2$ is just the final kinetic energy $K_f = \frac{1}{2}mv_f^2$ divided by m. When we multiply the equation through by m, we thus obtain

$$\tfrac{1}{2}mv_f^2 = \tfrac{1}{2}kA^2. \tag{8.5}$$

The term on the left is K_f, the kinetic energy of the mass just as the spring returns to its relaxed state, corresponding to $x = 0$. The term on the right is the work done in stretching the spring in the first place. Remember our argument that the process of raising the ball confers energy on it in the form of potential energy, and that the energy later takes the form of kinetic energy as the ball falls and speeds up. Here, we make a similar argument: The process of stretching the spring confers energy on the mass-spring system, which is "stored" in the spring in the form of potential energy. Whereas the raised ball possessed potential energy by virtue of its position, the spring possesses potential energy by virtue of its *configuration*—specifically, its distortion from the relaxed state. Just as the potential energy of the ball depends only on its height above the floor, the potential energy of the spring depends only on the amount by which it is stretched. Again using the assertion $-\Delta U = \Delta K$ (which we have not yet proved), we have

$$-\Delta U = U_i - U_f = K_f - K_i = K_f = \tfrac{1}{2}kA^2.$$

EXAMPLE 8.1

In Figure 8.3, the applied force $\mathbf{F}_a$ directed parallel to the plane of incline angle α barely suffices to pull the block of mass m up the plane. Find the work done by $\mathbf{F}_a$ as the block moves through the displacement $\mathbf{L}$ **(a)** if friction is negligible and **(b)** if the coefficient of kinetic friction between the block and the plane is μ_k.

SOLUTION:
(a) How much work is done by $\mathbf{F}_a$ over the displacement $\mathbf{L}$ if friction is negligible?

The work is

$$W = \mathbf{F}_a \cdot \mathbf{L} = F_aL. \tag{8.6}$$

You can now proceed as in Section 4.4 to determine what the magnitude of $\mathbf{F}_a$ must be if it is to pull the block up the plane at constant speed. To do so in the absence of friction requires that

$$F_a = mg\sin\alpha. \tag{8.7}$$

The term on the right is the magnitude of the gravitational force component along the plane. Inserting this value of F_a into Equation 8.6 gives you

$$W = mgL\sin\alpha.$$

But $L\sin\alpha$ is just the vertical height h through which the block is raised. Thus you can write the work done by $\mathbf{F}_a$ in the form

$$W = mgh. \tag{8.8}$$

This is the same amount of work required to raise the block vertically through the same height. Using the inclined plane reduces the force necessary by the factor $\sin\alpha$, from mg to $mg\sin\alpha$. The ratio $(mg)/(mg\sin\alpha) = 1/\sin\alpha$ is called the **mechanical advantage** of the inclined plane. But there is a price to pay: Although the force is reduced, you must exert it over the increased distance $L = h/\sin\alpha$. The overall result is that *the work done in overcoming gravity is the same either way*. Nevertheless, devices such as the inclined plane, which change the magnitude of the force required to do work, are useful over a wide range of applications. They are called **simple**

FIGURE 8.3

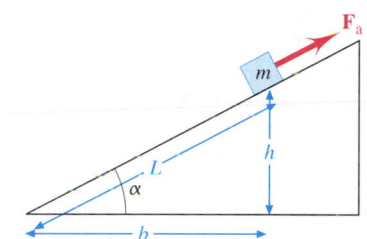

machines. A number of them are considered in problems at the end of this chapter.

In this frictionless case, all the work done by the applied force $\mathbf{F}_a$ is done in overcoming gravity and leads to an increase in the potential energy of the block. As with the raised ball, the potential energy of the block can be converted into kinetic energy simply by releasing the block and letting it slide back down the plane.

(b) How much work is done by $\mathbf{F}_a$ over the displacement $\mathbf{L}$ if the coefficient of kinetic friction is μ_k?

With friction present, the force applied to raise the block along the plane must overcome not only gravity but friction as well. As you learned in Section 4.5, the magnitude of the kinetic frictional force for a block moving on an inclined plane is $\mu_k mg\cos\alpha$. To take this force into consideration in the present case, you must modify Equation 8.7; the magnitude of the applied force needed to keep the block moving at constant speed is now

$$F_a = mg\sin\alpha + \mu_k mg\cos\alpha$$
$$= mg(\sin\alpha + \mu_k\cos\alpha).$$

As in part **a**, the work is $W = F_aL$; here, you have

$$W = mgL(\sin\alpha + \mu_k\cos\alpha).$$

As before, you can substitute $L\sin\alpha = h$. But you also have $L\cos\alpha = b$, which is the horizontal distance traversed by the block. Thus, you can write the work done by $\mathbf{F}_a$ in the form

$$W = mgh + \mu_k mgb. \tag{8.9}$$

The first term on the right side of this equation is equal in magnitude to the work done on the block by the gravitational force; it is the work done in overcoming gravity. As in part **a**, it is the potential energy gained by the block. The second term is equal in magnitude to the work done on the block by the frictional force; it is the additional work done in overcoming friction. It does *not* appear as potential energy. Unlike the potential energy, which can be converted into kinetic energy under the proper conditions, the work done in overcoming friction is gone for good, so far as the mechanical behavior of the system is concerned.

Conservative Forces

In part **b** of Example 8.1, the work W done on the block by the applied force $\mathbf{F}_a$ is divided into two parts. One part, the work mgh done in overcoming the gravitational force exerted on the block, results in a change in the potential energy of the system. The other part, the work $\mu_k mgb$ done in overcoming the frictional force exerted on the block, does nothing to change the potential energy of the system. We divide the work into these two parts because there is an important distinction between forces typefied by the gravitational force and forces typefied by the frictional force. Forces of the first kind are called *conservative forces*. Forces of the second kind are called *nonconservative* (or *dissipative*) *forces*.

There is something special about the way in which a conservative force does work on a system. We *define* a **conservative force** in terms of that special way:

Let a force do work on a system as the system moves along an arbitrary closed path—a path that begins and ends at the same point, as shown in Figure 8.4a. If and only if the total work done around the path is zero, the force is conservative.

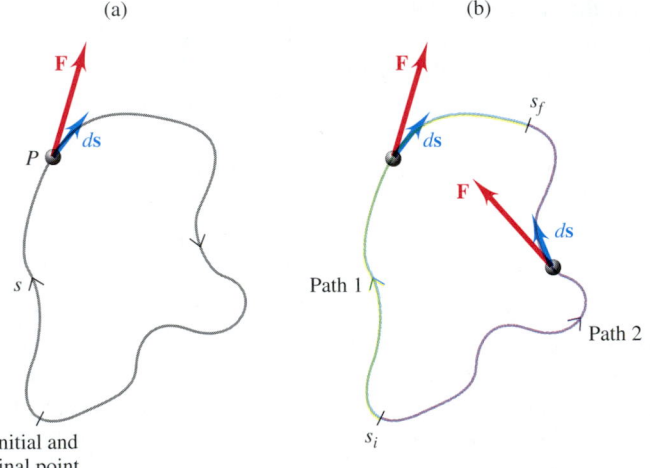

(a)   (b)

FIGURE 8.4 (*a*) The force **F** does work on the particle P as it moves around a closed path. The force **F** is shown acting on P as it moves through a representative path element $d\mathbf{s}$. (The value of **F** may depend on the location of $d\mathbf{s}$.) The force **F** is *conservative* if and only if the work that it does on P around the path is zero. (*b*) The force **F** does work on P as P moves from an initial point s_i to a final point s_f along two different paths of the infinitely many it might take. The force **F** is shown acting on P as it moves through a representative path element $d\mathbf{s}$ on each of the paths. The force **F** is conservative if and only if the work that it does on P is the same for both paths—that is, if W is independent of the path taken from s_i to s_f.

This definition can be expressed in the mathematical form

$$W = \int_{\substack{\text{closed} \\ \text{path}}} \mathbf{F} \cdot d\mathbf{s} = 0 \quad \text{for a conservative force.} \qquad (8.10)$$

Another useful way to state this definition is logically equivalent to the one above:

Let a force do work W on a system as the system moves along an arbitrary path from an initial point s_i to a final point s_f. Two such paths are shown in Figure 8.4b. If and only if W is independent of the path taken, the force is conservative.

This definition can be expressed in the mathematical form

$$W_{\text{path 1}} = W_{\text{path 2}} \quad \text{for a conservative force,}$$

where

$$W_{\text{path 1}} = \int_{\substack{s_i \\ \text{path 1}}}^{s_f} \mathbf{F} \cdot d\mathbf{s} \quad \text{and} \quad W_{\text{path 2}} = \int_{\substack{s_i \\ \text{path 2}}}^{s_f} \mathbf{F} \cdot d\mathbf{s}. \qquad (8.11)$$

The proof that the two definitions of a conservative force are equivalent is as follows. If you follow path 1 in Figure 8.4b from s_i to s_f and then follow path 2 backward from s_f to s_i, you have completed the closed path of Figure 8.4a. That is, the work integral around the closed path can be written as the sum of two parts:

$$\int_{\substack{\text{closed} \\ \text{path}}} \mathbf{F} \cdot d\mathbf{s} = \int_{\substack{s_i \\ \text{path 1}}}^{s_f} \mathbf{F} \cdot d\mathbf{s} + \int_{\substack{s_f \\ \text{path 2}}}^{s_i} \mathbf{F} \cdot d\mathbf{s}.$$

Comparing the left side of this equation with the first definition, Equation 8.10, we see that the value of the left side is zero for a conservative force. The second integral on the right side of the same equation is just the negative of the second integral in Equation 8.11. (This is an example of a general rule: The sign of an integral is reversed when the limits are interchanged.) Thus, for a conservative force you have

$$0 = \int_{\substack{s_i \\ \text{path 1}}}^{s_f} \mathbf{F} \cdot d\mathbf{s} - \int_{\substack{s_i \\ \text{path 2}}}^{s_f} \mathbf{F} \cdot d\mathbf{s},$$

or

$$\int_{\substack{s_i \\ \text{path 1}}}^{s_f} \mathbf{F} \cdot d\mathbf{s} = \int_{\substack{s_i \\ \text{path 2}}}^{s_f} \mathbf{F} \cdot d\mathbf{s}.$$

This can be written $W_{\text{path 1}} = W_{\text{path 2}}$, which is Equation 8.11. The equivalence of the two definitions is thus proved.

Because the work done by a conservative force is independent of the path taken, it depends only on the positions of the initial and final points. That is, we need not specify the path at all when evaluating the work integral

$$W = \int_{s_i}^{s_f} \mathbf{F} \cdot d\mathbf{s}.$$

What is more, *the work done by a conservative force cannot depend on any other characteristics of the process,* such as the particular time when the work is done or the speed of the system as it moves from s_i to s_f. If such a dependence did exist, we could do work on the system as it moved from s_i to s_f at a certain time (or speed) and then do work on the body as it moved from s_f back to s_i at some other time (or speed) and thus obtain a nonzero value for the round-trip work, in violation of Equation 8.10.

Another property of conservative forces can be deduced from the definitions stated earlier, though we will not give a rigorous proof here. *A conservative force $\mathbf{F}$ can be a function of position only.** It may not be a function of time, or speed, or any other characteristic of the process. You can make this statement plausible by considering what would happen if $\mathbf{F}$ were a function of, say, time. Then repeat the argument of the preceding paragraph.

We are now in a position to define a **nonconservative (dissipative) force**. A force is nonconservative if it does not fulfill the requirements of a conservative force. That is, *a force is nonconservative if work done by it depends on anything other than the initial and final positions of the system on which the force does the work.* One example of work done by a nonconservative force is the work done on a block of mass m by a frictional force $\mu_k mg$ as the block is dragged along a level floor. In this case, the amount of work done depends on the length of the path taken from the initial to the final location of the block. Another example is the force exerted on an automobile by air resistance. This force depends roughly on the square of the speed. Even if the car follows the same path in making two trips, the work done by air resistance will be different if the trips are made at different speeds.

---

*The converse of this statement—a force that is a function of position only is a conservative force—is true in one-dimensional space but not in general. A force in two- or three-dimensional space can be a function of position only and yet not be conservative. For the purposes of our study of mechanics, however, the point need not concern us.

EXAMPLE 8.2

Using the block and the frictionless inclined plane discussed in part **a** of Example 8.1, show that the gravitational force is conservative. Then use the same system with friction added, discussed in part **b** of Example 8.1, to show that the frictional force is nonconservative.

SOLUTION:

(a) Show that the gravitational force is conservative.

Use the definition of Equation 8.10, and show that the work done by the gravitational force around a closed path is zero. A good closed path to choose is the triangular path ABC shown in Figure 8.5a. We can use the result of part **a** of Example 8.1 to calculate the work done on the block as it moves along segment AB of this path, and it is easy to calculate the work done in moving the block along each of the other two segments as well. To follow path ABC, you use the string attached to the block to pull it up the plane from A to B, just as in Example 8.1. Then you move the block along BC by letting it hang from the string. Finally, you use the string to lower the block from C back to A.

The path consists of three straight segments. Call the unit vectors denoting the directions of motion along these segments $\hat{s}_1$, $\hat{s}_2$, and $\hat{s}_3$. Give the name ϕ to the angle between $\hat{s}_1$ and the downward direction $\hat{g}$ along which the gravitational force always acts. As the block rises along AB, the work W_1 done on it by the gravitational force is given by the dot product of the force with the displacement, $m\mathbf{g} \cdot L\hat{s}_1 = mgL \cos \phi$. Because $\phi = \alpha + 90°$ (why?), you have $\cos \phi = -\sin \alpha$, and the work can be written

$$W_1 = -mgL \sin \alpha = -mgh.$$

Along the second segment BC, which is horizontal, the gravitational force is perpendicular to the direction of motion, and the work is

$$W_2 = m\mathbf{g} \cdot b\hat{s}_2 = 0.$$

On the third segment CA, the gravitational force lies along the direction of motion, and the work is

$$W_3 = m\mathbf{g} \cdot h\hat{s}_3 = mgh.$$

You can now add to find the total work W. You have

$$W = W_1 + W_2 + W_3 = -mgh + 0 + mgh = 0.$$

This satisfies the condition of Equation 8.10 and demonstrates that the gravitational force is conservative. Can you reach the same conclusion by comparing the nonzero work done in going from A to C (1) along the direct path $A \rightarrow C$ and (2) along the path $A \rightarrow B \rightarrow C$?

(b) Show that the frictional force is nonconservative.

You could use the same path as that of part **a**, but it is instructive to consider instead the closed path $A \rightarrow B \rightarrow A$ shown in Figure 8.5b, which constitutes a simple round trip up and down the incline, with legs $L\hat{s}_1$ and $L(-\hat{s}_1)$.

Whichever way the block moves, the frictional force is always directed backward. Consequently, its value along the first leg is $\mu_k mg \cos \alpha (-\hat{s}_1)$. The work done on the block by the frictional force along the path $L\hat{s}_1$ is

$$W_1 = \mu_k mg \cos \alpha (-\hat{s}_1) \cdot L\hat{s}_1 = -\mu_k mgL \cos \alpha.$$

Along the second leg, the frictional force—again acting backward—is $\mu_k mg \cos \alpha (\hat{s}_1)$. The work done on the block by the frictional force along the path $-L\hat{s}_1$ is

$$W_2 = \mu_k mg \cos \alpha (\hat{s}_1) \cdot (-L\hat{s}_1) = -\mu_k mgL \cos \alpha.$$

The total frictional work along the closed path is

$$W = W_1 + W_2 = -2\mu_k mgL \cos \alpha.$$

This result is not equal to zero, and therefore the frictional force is not conservative. Consequently, it does not act to change the potential energy of the block.

FIGURE 8.5

(a)

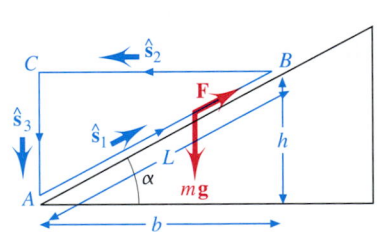

(b)

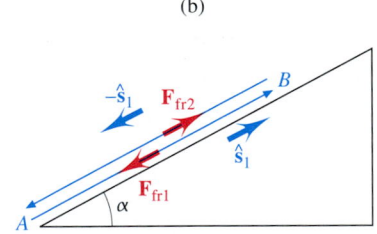

Formal Definition of Potential Energy

We are now prepared to define potential energy rigorously in terms of the work done by conservative forces. Consider a system that moves from an initial position $\mathbf{r}_i$ to a final position $\mathbf{r}_f$ along any path whatsoever, while it is acted on by a conservative force $\mathbf{F}(\mathbf{r})$. We define a **potential energy function** $U(\mathbf{r})$ whose change dU over an infinitesimal displacement $d\mathbf{r}$ is the negative of the work done by the force over $d\mathbf{r}$:*

$$dU = -dW. \tag{8.12}$$

---

*The reason for the minus sign in this definition will be discussed shortly, in Example 8.3.

But $dW = \mathbf{F}(\mathbf{r}) \cdot d\mathbf{r}$; compare with Equation 7.11a. Hence, we can write

$$dU = -\mathbf{F}(\mathbf{r}) \cdot d\mathbf{r}. \quad (8.13)$$

As the system moves from $\mathbf{r}_i$ to $\mathbf{r}_f$, the function $U(\mathbf{r})$ changes by a finite amount. We can evaluate that amount by integrating both sides of Equation 8.13 to obtain

$$\int_{\mathbf{r}_i}^{\mathbf{r}_f} dU = \int_{\mathbf{r}_i}^{\mathbf{r}_f} -\mathbf{F}(\mathbf{r}) \cdot d\mathbf{r} \quad \text{for } \mathbf{F} \text{ conservative.} \quad (8.14)$$

The right side of this equation is $-W$, the negative of the work W done by the conservative force between $\mathbf{r}_i$ and $\mathbf{r}_f$. The left side can be integrated immediately to obtain

$$U(\mathbf{r}_f) - U(\mathbf{r}_i) \equiv \Delta U.$$

Thus, Equation 8.14 becomes

$$\Delta U = -W \quad \text{for } \mathbf{F} \text{ conservative;} \quad (8.15)$$

that is, *the change in potential energy of a system is equal to the negative of the work done on it by a conservative force.* Consistent with the definition of a conservative force, the potential energy is a function of position only.

The change in potential energy is defined in terms of the potential energy function $U(\mathbf{r})$. But only the change ΔU appears in Equation 8.15, which provides the connection between potential energy and the physical quantity work. And indeed, only *differences* in potential energy have physical significance. Such a difference is not affected if we define a new potential energy function $U'(\mathbf{r})$ by adding an arbitrary constant C to $U(\mathbf{r})$. That is, if

$$U'(\mathbf{r}) = U(\mathbf{r}) + C,$$

we still have the same ΔU:

$$\begin{aligned} \Delta U &= U'(\mathbf{r}_f) - U'(\mathbf{r}_i) \\ &= U(\mathbf{r}_f) + C - [U(\mathbf{r}_i) + C] \\ &= U(\mathbf{r}_f) - U(\mathbf{r}_i). \end{aligned}$$

Although U and U' are different mathematical functions, the difference is not physically significant. Why, then, do we decide to use one function rather than another in studying a given system? We do so because it is often useful to fix the value of the potential energy at a convenient **reference point** $\mathbf{r}_0$ and measure all potential energy differences from that point. Most commonly, the value of the **reference-point energy** $U(\mathbf{r}_0)$ is fixed at zero; that is, $U(\mathbf{r}_0) = 0$. In that case, we can use Equations 8.14 and 8.15 to write

$$\Delta U = \int_{\mathbf{r}_0}^{\mathbf{r}} dU = U(\mathbf{r}) - U(\mathbf{r}_0) = U(\mathbf{r}) - 0 \quad \text{for } U(\mathbf{r}_0) = 0. \quad (8.16)$$

Once this is done, it makes physical sense to speak of *the* potential energy $U(\mathbf{r})$ at a particular position $\mathbf{r}$. When you do this, you must be consistent and express *all* potential energies with respect to the *same* reference point.

EXAMPLE 8.3

You raise a ball of mass $m = 100$ g from the floor to a height $h = 1.50$ m above the floor. The ball is at rest at both the beginning and the end of the process. **(a)** How much work W_a do you do on the ball by means of the force you apply? **(b)** How much work W_g is done on it by the conservative gravitational force? **(c)** What is the change ΔU in the potential energy of the ball? **(d)** What is the potential energy function $U(y)$ if the origin $y = 0$ is chosen at floor level and the reference

point energy $U(y) = 0$ is fixed at the origin? **(e)** What is the potential energy $U(h)$ of the ball with respect to this reference point? **(f)** If the floor is located at a height $H = 4.75$ m above the ground and the reference point energy $U'(y) = 0$ is fixed at ground level, what is the new potential energy function $U'(y)$? What is the potential energy of the ball when it is on the floor? when it is a distance h above the floor? What is the potential energy difference between these two points?

SOLUTION:

(a) How much work W_a do you do on the ball by means of the force you apply?

To raise the ball slowly against gravity, you must apply an upward force of magnitude mg as you move it upward. (If you applied a constant force of greater magnitude, the ball would not be at rest when it reached its final height.) The direction of the applied force is parallel to the path of the ball. Thus, in raising the ball through a distance Δy, you must do an amount of work $W_a = mg\,\Delta y$. If $\Delta y = h - 0 = h$, the work is

$$W_a = mgh$$
$$= 0.100 \text{ kg} \times 9.80 \text{ m/s}^2 \times 1.50 \text{ m} = 1.47 \text{ J}.$$

(b) How much work W_g is done on the ball by the conservative gravitational force?

As you raise the ball, the downward gravitational force of magnitude mg is directed opposite the direction of motion. Thus the gravitational force does work

$$W_g = -mg\,\Delta y = -mgh$$
$$= -1.47 \text{ J}.$$

(Note that the net work $W_a + W_g$ done on the ball is zero. That is why, consistent with the work-energy theorem, it has negligible kinetic energy.)

(c) According to Equation 8.12, the potential energy change of the ball is the negative of the work done by the conservative gravitational force. Thus, you have

$$\Delta U = -W_g = mgh,$$
or $$\Delta U = 1.47 \text{ J}.$$

In the light of this calculation, you can see why potential energy is defined in such a way that there is a negative sign in Equations 8.12 and 8.15. The negative sign makes the definition consistent with our intuitive ideas about work and with the qualitative ideas developed in Section 8.1. When you raise the ball, you do positive work on it. You therefore expect that its potential energy—its ability to do work on some external system—will increase. And indeed you do 1.47 J of work on the ball to increase its potential energy by 1.47 J. However, the work that you do on the ball is *not* work done by a conservative force. A conservative force must be a function of position only. But you are free to apply a force whose magnitude varies in any way you like as you raise the ball, provided only that the mass comes to rest at height h. If you wish, you can apply a force whose magnitude is very slightly and very briefly greater than mg initially, then equal to mg over most of the path, and finally very slightly and very briefly less than mg. You thus raise the ball at a slow, uniform speed. If you prefer, you can give the ball a very rapid initial acceleration by applying a large force, and later decelerate it to rest at height h with a large downward force. The two processes involve forces that vary with position in very different ways. Thus, the applied force needed to raise the ball cannot be specified as a function of position. It therefore cannot be conservative and is not directly associated with the potential energy gain of the ball. The role of the applied force is indirect; by overcoming gravitation, it makes possible the upward displacement of the ball. During this displacement, the *negative* work done by the conservative gravitational force *increases* the potential energy of the ball.

To take another point of view, consider what will happen if

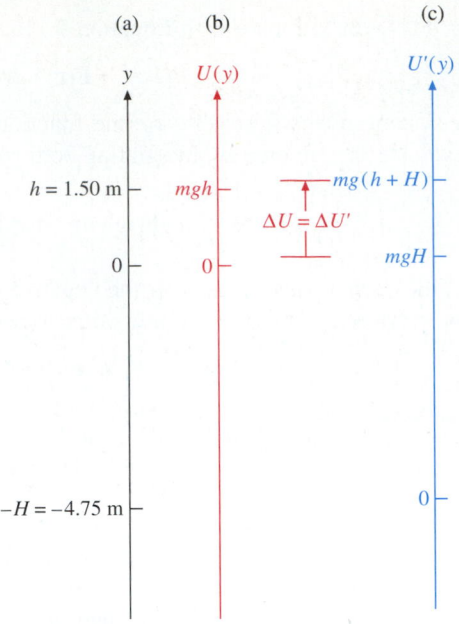

FIGURE 8.6 (*a*) Coordinate system chosen with $y = 0$. (*b*) Potential energy $U(y)$ with respect to the reference point $U(0) = 0$. (*c*) Potential energy $U'(y)$ with respect to the reference point $U'(-H) = 0$. The potential energy differences ΔU and $\Delta U'$ are the same.

you release the ball after it is raised. As the ball falls under the influence of the gravitational force alone, that force will do positive work because it and the motion have the same direction. And indeed, the kinetic energy of the ball will increase in accord with the work-energy theorem.

(d) What is the potential energy function $U(y)$ if the origin $y = 0$ is chosen at floor level and the reference point energy $U(y) = 0$ is fixed at the origin, as shown in Figure 8.6a?

If you call the floor level $y = 0$ and set $U(0) = 0$, you have from Equation 8.16

$$U(y) = \Delta U = mgy.$$

This result is shown in graphic form in Figure 8.6b.

(e) What is the potential energy $U(h)$ of the ball with respect to this reference point?

The ball is raised to $y = h = 1.50$ m. So you have $U(h) = mgh = 1.47$ J.

(f) If the floor is located at a height $H = 4.75$ m above the ground and the reference point energy $U'(y) = 0$ is fixed at ground level ($y = -4.75$ m), what is the new potential energy function $U'(y)$? What is the potential energy of the ball when it is on the floor? when it is a distance h above the floor? What is the potential energy difference between these two points?

Continuing to call the floor level $y = 0$, you now fix the reference level of the new potential energy function U' at $y = -H$, so $U'(-H) = 0$ (Figure 8.6c). Using the argument that follows Equation 8.15, you have

$$U'(y) = U(y) + \text{constant.}$$

In particular, at ground level $y = -H$ this equation becomes

$$U'(-H) = U(-H) + \text{constant.}$$

If you fix $U'(-H) = 0$, you obtain

$$0 = U(-H) + \text{constant} = -mgH + \text{constant.}$$

This enables you to evaluate the constant:

$$\text{constant} = mgH.$$

You can now express U' in terms of U for any y; you have

$$U'(y) = U(y) + mgH = mgy + mgH = mg(y + H).$$

In terms of U', the potential energy of the ball at floor level is

$$U'(0) = mg(0 + H) = mgH$$
$$= 0.100 \text{ kg} \times 9.80 \text{ m/s}^2 \times 4.75 \text{ m} = 4.66 \text{ J.}$$

At height h it is

$$U'(h) = mg(h + H)$$
$$= 0.100 \text{ kg} \times 9.80 \text{ m/s}^2 \times (4.75 \text{ m} + 1.50 \text{ m})$$
$$= 6.13 \text{ J.}$$

These values are not the same as the corresponding values of U. But their difference is

$$\Delta U' = 6.13 \text{ J} - 4.66 \text{ J} = 1.47 \text{ J,}$$

exactly the result you found for ΔU in part **c**. The identity of the two results is clear in Figure 8.6.

As you would expect, the potential energy *difference* is independent of the choice of reference level; $\Delta U' = \Delta U$. Will it make any difference if you redefine the coordinate system by choosing the ground level rather than the floor level as $y = 0$?

THINKING LIKE A PHYSICIST

Look back at the path we have traveled in developing the concept of potential energy. At the beginning of Section 8.2, we looked at the behavior of two familiar systems, a raised ball and a mass-spring system, in a mostly qualitative way. In particular, we noted that, after work done on either system by an external force "activates" the system, some time may pass before the "activation" becomes evident in the form of kinetic energy—the energy of motion of the system. This delay distinguishes the two systems from an air-track glider, whose speed changes immediately when work is done on it.

To account for the energy during the time interval, we surmised that the energy was present but "hidden" in a non-kinetic form that we called potential energy. We played with this idea and found that we could make plausible calculations involving the potential energy of the two systems. That is, we found potential energy to be a useful physical concept.

It thus appeared worthwhile to explore the concept of potential energy more deeply. We did so in a series of rather abstract mathematical derivations concerning conservative forces and potential energy functions. If you have a mathematical turn of mind, you may have found these derivations interesting from a purely mathematical point of view. But we would not be justified in taking the effort to carry the derivations out if they did not have physical implications. Without the earlier nonrigorous survey, moreover, the derivations would be hard to follow.

The payoff for this exercise in mathematical formalism is immediate. We now return to our study of physical systems in the light of the concept of potential energy. Potential energy is well defined, and its restriction to conservative forces is clear. In Section 8.3, you will see that the new concept makes possible a powerful new way of analyzing the motion of mechanical systems.

Conservation of Mechanical Energy

You saw at the beginning of Section 8.2 that it is possible to convert potential energy into kinetic energy. We now study the conversion process more generally in the light of the detailed picture of potential energy just developed. As examples, let us again use the falling ball and the mass-spring system.

The Falling Ball Revisited

At the beginning of Section 8.2, we compared the kinetic energy change ΔK of a ball just before it completes a free fall through a height h with the negative of its potential energy change $-\Delta U$ during the fall, and we found them to have the same value mgh. Now consider the ball at an instant during the fall when it is at an arbitrary height y above the floor, as shown in Figure 8.7. If we take the floor ($y = 0$) as the reference level $U(0) = 0$, the potential energy of the ball at this height is $U = mgy$. Evidently, this value is smaller than its initial potential energy at the instant of release, which was $U_i = mgh$.

Because the ball gains speed as it falls, the loss of potential energy is accompanied

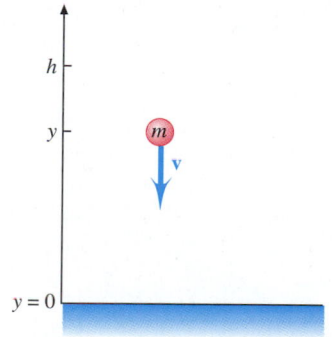

FIGURE 8.7 A ball of mass m is lifted to a height h above the floor. It is released and falls freely. When it passes through the height y, its speed is v.

by a gain of kinetic energy. What is this gain? The ball has fallen a distance $h - y$ at constant acceleration g, so we have $v^2 = 2g(h - y)$ from Equation 2.38. Hence, the kinetic energy, whose initial value was zero, has increased to

$$K = \tfrac{1}{2}mv^2 = \tfrac{1}{2}m\,2g(h - y)$$
$$= mgh - mgy = U_i - U(y).$$

That is, the gain in kinetic energy is equal to the loss in potential energy.

With this in mind, we define the **total mechanical energy** E of a system to be the sum of its kinetic and potential energies at any instant:

$$E \equiv K + U. \qquad\qquad (8.17)$$

There are other kinds of energy besides the quantity defined here, and the symbol E is used for all of them. When there is no danger of confusion, we will often shorten the name ''total mechanical energy'' to **total energy**, **mechanical energy**, or just **energy**, as appropriate.

For the falling ball at the moment when its height is y, we have

$$E = (mgh - mgy) + mgy = mgh.$$

This value is independent of the position of the ball, and it remains constant for the duration of the fall. We say that *the total mechanical energy of the system is conserved*. This is the case because during the fall only conservative forces (here, the force of gravity) do work on the ball. A system in which only conservative forces do work is called a **conservative system**.

We now generalize and prove that, *in a system in which only conservative forces do work*, total mechanical energy is conserved. This statement has profound consequences, especially when it is expanded to include other kinds of energy, as we will do later in this book. It is called the **principle of conservation of mechanical energy**, and its proof is as follows.

According to the work-energy theorem, Equation 7.22b, the net work done on a system is equal to the change in its kinetic energy:

$$W = \Delta K.$$

Now suppose that the system is conservative, so only conservative forces do work. Then Equation 8.15 applies, and we have

$$W = -\Delta U.$$

Combining these two equations gives

$$\Delta K = -\Delta U; \qquad\qquad (8.18a)$$

that is, *the increase in kinetic energy equals the decrease in potential energy*. This proves the plausible assertion that we made at the beginning of Section 8.2. We can write Equation 8.18a in the equivalent form

$$\Delta K + \Delta U = \Delta(K + U) = 0. \qquad\qquad (8.18b)$$

But the change in $K + U$ is just the change in the total mechanical energy E, and we have

$$\Delta E = 0. \qquad\qquad (8.18c)$$

It follows that

$$E = \text{constant} \quad \text{in a conservative system}; \qquad\qquad (8.19)$$

that is to say, total mechanical energy is conserved, and the proof is complete.

Because the value of U is determined only if a reference point is chosen, the value of $E = K + U$ depends on that choice. In using Equation 8.19, you must be certain that $U(\mathbf{r}_0)$ is properly specified. However, Equations 8.18a, 8.18b, and 8.18c are valid even if no reference point is specified, because they express changes only.

EXAMPLE **8.4**

An ideal Atwood machine is started from rest with its masses at the same level, as shown in Figure 8.8a. Use the principle of conservation of total mechanical energy to find **(a)** the speed of either of the masses as a function of its position and **(b)** the acceleration of either of the masses.

SOLUTION:

(a) Find the speed of either of the masses as a function of its position.

Choose the initial height of the masses as $y = 0$, as shown in Figure 8.8, and fix $U(0) = 0$ there. Then the initial total energy of the system is $E_i = U_i + K_i = 0 + 0 = 0$.

A little while after the machine is released, it will have the configuration shown in Figure 8.8b, with the smaller mass m_1 at y and the larger mass m_2 at $-y$. The potential energy of the system is then

$$U_f = m_1 gy + m_2 g(-y) = (m_1 - m_2)gy.$$

Both masses will be moving with the same speed v, and so the kinetic energy of the system is

$$K_f = \tfrac{1}{2}m_1 v^2 + \tfrac{1}{2}m_2 v^2 = \tfrac{1}{2}(m_1 + m_2)v^2.$$

The final total energy of the system is $E_f = U_f + K_f$ and, according to Equation 8.19, E_f must be equal to E_i. Thus, you have

$$(m_1 - m_2)gy + \tfrac{1}{2}(m_1 + m_2)v^2 = 0.$$

Solving for v^2 gives you

$$v^2 = \frac{m_2 - m_1}{m_2 + m_1} 2gy,$$

so the speed is

$$v = \sqrt{\frac{m_2 - m_1}{m_2 + m_1} 2gy}.$$

(b) Find the acceleration of either of the masses.

You know from Section 3.6 that the acceleration of an Atwood machine is constant. Also, the Atwood machine of this example is started from rest. The kinematic relation $v^2 = 2ay$ therefore applies to the motion of either mass. The value of v^2

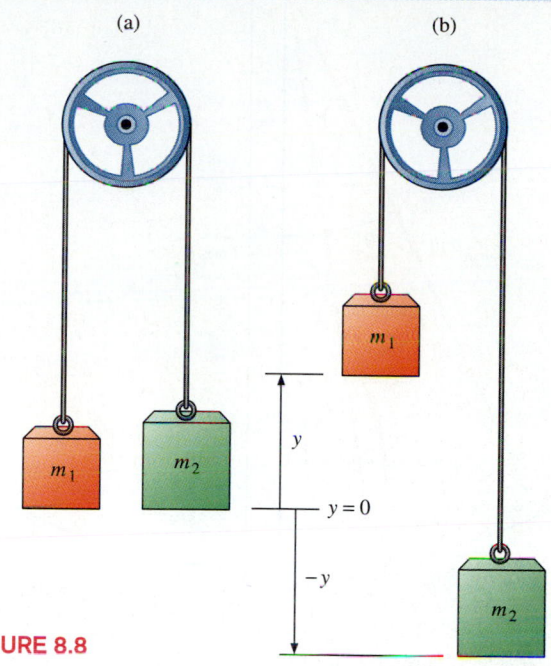

(a) (b)

FIGURE 8.8

obtained in part **a** gives

$$2ay = \frac{m_2 - m_1}{m_2 + m_1} 2gy,$$

so

$$a = \frac{m_2 - m_1}{m_2 + m_1} g.$$

You should compare this result with Equations 3.19 and 3.20, obtained by means of a more elaborate dynamical analysis.

The energy-conservation approach can yield the magnitudes of the velocity and the acceleration only, and not their directions. This is because they are obtained by calculating the kinetic energy K, which is always positive and which is directly related to v^2, the invariably positive square of the velocity. Signs of the velocity and the acceleration must be found from the context of the problem. As in the present example, this is usually not difficult to do.

For the Atwood machine, the energy-conservation principle provides a somewhat simpler way of deriving v and a than the straightforward dynamical approach of Section 3.6. There are problems for which a dynamical solution is much more difficult than a conservational one, however, as Example 8.5 illustrates.

EXAMPLE **8.5**

The bob of a simple pendulum is pulled to one side until the pendulum makes an angle θ_0 with the vertical and then released, as shown in Figure 8.9. **(a)** If the length of the pendulum is l, find the speed v of the bob and the angular speed ω of the pendulum when the pendulum makes an arbitrary angle $\theta < \theta_0$ with the vertical. Neglect air resistance. **(b)** Let $\theta_0 = 40°$ and $l = 85$ cm. What is the speed of the bob when $\theta = 12°$?

SOLUTION:

(a) Find the speed v of the bob and the angular speed ω of the pendulum when the pendulum makes an arbitrary angle $\theta < \theta_0$ with the vertical.

Here a direct calculation of the speed using Newton's second law would be complicated by the fact that the tangential component of the gravitational force changes continually with θ.

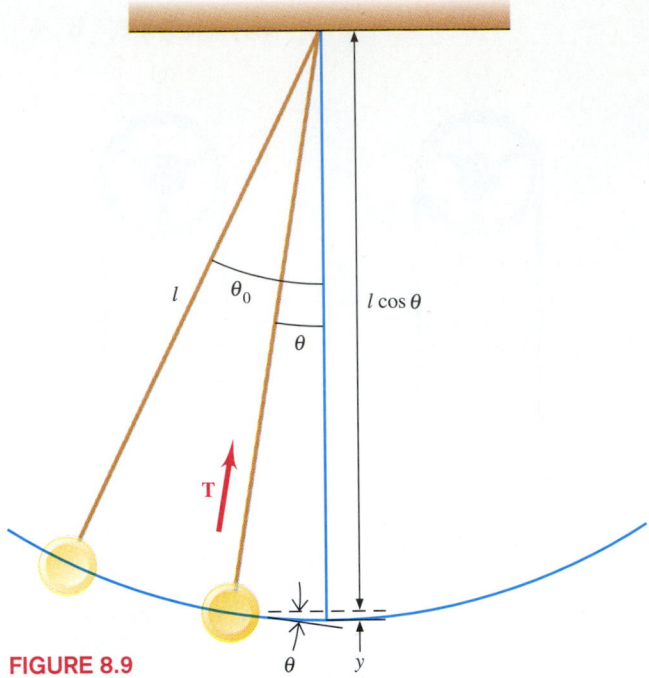

FIGURE 8.9

like that in Figure 8.9, which gives you $y = l(1 - \cos \theta)$. It follows that the potential energy can be written in the form

$$U = mgl(1 - \cos \theta).$$

Next, you note that the pendulum is released from rest at the angle θ_0. Its kinetic energy at that instant is zero, and so the total mechanical energy is equal to the potential energy. You thus have

$$E = U(\theta_0) = mgl(1 - \cos \theta_0).$$

This expression represents the potential energy at the instant of release, but it also represents the total energy at all other instants because the system is conservative.

To find the speed of the bob for an arbitrary value of θ, you evaluate the kinetic energy. From Equation 8.17 you have

$$K = E - U$$
$$= mgl(1 - \cos \theta_0) - mgl(1 - \cos \theta)$$
$$= mgl(\cos \theta - \cos \theta_0).$$

Because $K \equiv \frac{1}{2}mv^2$, the speed is given by

$$v = \sqrt{\frac{2K}{m}}, \qquad (8.20)$$

which leads to the result

$$v = \sqrt{2gl(\cos \theta - \cos \theta_0)}. \qquad (8.21)$$

This result is not restricted to small oscillations.

To find the angular speed ω, you use the relation between ω and v given by Equation 6.11. In the present notation, this relation is

$$\omega = \frac{v}{l}.$$

Inserting the value of v from Equation 8.21, you have

$$\omega = \sqrt{\frac{2g}{l}} (\cos \theta - \cos \theta_0). \qquad (8.22)$$

(b) Let $\theta_0 = 40°$ and $l = 85$ cm. What is the speed of the bob when $\theta = 12°$?

Using Equation 8.21, you obtain

$$v = \sqrt{2 \times 9.8 \text{ m/s}^2 \times 0.85 \text{ m} \times (\cos 12° - \cos 40°)}$$
$$= 1.9 \text{ m/s}.$$

Because the change in gravitational potential energy of the bob depends only on the change in its height, the speed of the bob must be the same as that for a body that has fallen vertically through the same height. You should verify this.

But because the potential energy change depends only on the change in vertical position, a calculation based on conservation of total mechanical energy is much simpler.

In an analysis of the motion of the simple pendulum, it is usual to choose the bob as the system, as was done in Section 6.5. Before proceeding, however, you must make sure that the system thus chosen is a conservative one. There are two forces acting on the bob. The first is the gravitational force, which we have already established to be conservative. The other is the force **T** exerted by the string. It does not matter whether this force is conservative or not because *it does no work on the bob*. The string force is always perpendicular to the direction of motion of the bob, so $\mathbf{T} \cdot d\mathbf{s}$ is always zero. Such a force is called a **workless constraint**. Because no other forces come into play (neglecting friction), the system is conservative.

With these preliminaries out of the way, you can begin the calculation. It is best to begin by fixing the reference point of potential energy and the value of U there. One convenient choice is to set $y = 0$ at the bottom of the swing and to fix $U(0) = 0$ there. The potential energy of the bob is then $U = mgy$. Because you need to express the result of your calculation as a function of the pendulum angle θ, you make a construction

The Mass-Spring System Revisited

As a further application of the principle of conservation of mechanical energy, consider again the mass-spring system of Figure 8.2. You saw in the derivation of Equation 8.5 that the work $W_a = \frac{1}{2}kA^2$ done by an applied force F_a in stretching the spring when the mass is displaced through a distance A is equal to the kinetic energy $\frac{1}{2}mv^2$ that the spring gives to the mass in returning to its relaxed state. We can now understand this relation in terms of energy conservation. We can also generalize it to determine the potential and kinetic energy of the system for any displacement of the mass.

First, let us evaluate the potential energy U. According to Hooke's law, the spring force is $F_s = -kx$. As the end of the spring attached to the mass is pulled through the infinitesimal amount dx, the spring force does work

$$dW_s = -kx \, dx.$$

Using Equation 8.12, $dU = -dW$, we have

$$dU = kx\,dx.$$

For a spring, it usually makes sense to choose the reference point $U(x) = 0$ at $x = 0$. So we use Equation 8.16 to write

$$U(x) = \int_0^x dU = \int_0^x kx\,dx = \tfrac{1}{2}kx^2. \tag{8.23}$$

The corresponding work done by the spring force is $W_s = -U = -\tfrac{1}{2}kx^2$. Because W_s is a function of x only, we have verification that F_s is a conservative force, and the calculation of U is thus justified.

In the special case in which the spring is fully extended, $x = A$, we have $W_s = -\tfrac{1}{2}kA^2$. The work W_s is the negative of the work W_a done by the force F_a required to stretch the spring; see Example 7.3. This is not surprising because F_s and F_a are an action-reaction pair, and both act through the same distance.

Suppose now that we stretch the spring by displacing the mass from its equilibrium position at $x = 0$ to $x = A$ and then release the mass-spring system and allow it to oscillate with amplitude A. What is the kinetic energy of the system for an arbitrary displacement x? We begin with the definition in Equation 8.17, $E \equiv K + U$, and exploit the fact that at the instant of release the energy of the system is entirely potential. We thus have

$$E = 0 + \tfrac{1}{2}kA^2.$$

Even though K and U will both subsequently change, the value of E will remain fixed in the absence of friction. This is because the mass-spring system is a conservative system.

When the displacement of the system is x, the potential energy is $\tfrac{1}{2}kx^2$ and the kinetic energy has some (as yet unknown) value K. So we have

$$K + \tfrac{1}{2}kx^2 = E = \tfrac{1}{2}kA^2.$$

Solving for K gives

$$K = \tfrac{1}{2}k(A^2 - x^2). \tag{8.24}$$

The corresponding speed is

$$v = \sqrt{\frac{2K}{m}} = \sqrt{\frac{k}{m}(A^2 - x^2)}, \tag{8.25a}$$

Because $\sqrt{k/m}$ is equal to ω, the angular frequency of the mass-spring oscillator (Equation 6.34a), this equation can be conveniently written

$$v = \omega\sqrt{A^2 - x^2}. \tag{8.25b}$$

Either form of Equation 8.25 shows how the speed increases as the displacement decreases, and vice versa. Figure 8.10 and its legend give a geometric picture of the relation between speed and displacement for a simple harmonic oscillator in the light of the energy conservation principle.

In the special case where $x = 0$, Equation 8.24 becomes

$$K_{\text{max}} = \tfrac{1}{2}kA^2.$$

That is, the kinetic energy reaches a maximum when the potential energy diminishes to zero; all of the potential energy has been converted into kinetic energy. The result agrees with that obtained by more laborious methods in the discussion leading to Equation 8.5.

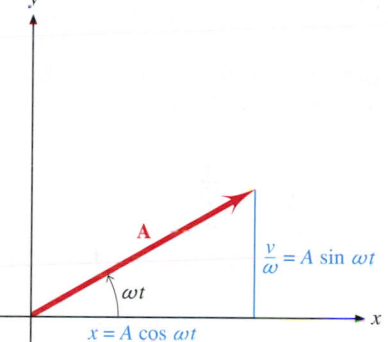

FIGURE 8.10 Phasor picture of the relation between displacement and speed in a mass-spring oscillator, as seen from the point of view of conservation of total mechanical energy. The system oscillates with angular frequency $\omega = \sqrt{k/m}$. Its motion is described in terms of the phasor **A**, which rotates with angular velocity ω, so its angular displacement is ωt. Rearranging Equation 8.25b gives $A = \sqrt{x^2 + v^2/\omega^2}$, which has the form of Pythagoras's theorem, as the illustration shows. The displacement of the mass is given by the x component of **A**, which is $A\cos\omega t$. The y component of **A** is $-v/\omega$, the velocity divided by the constant factor $-\omega$. As **A** rotates, the displacement varies sinusoidally between A and $-A$; the velocity, $v = -\omega A\sin\omega t$, varies between A/ω and $-A/\omega$ at a phase angle 90° behind x.

The Generalized Work-Energy Theorem

Consider a system that may have a number of different parts, such as the mass-spring system just discussed. As the system moves, a number of forces may do work on it at the same time. Some of these forces, like the spring force, are conservative. Others,

The fun is in converting potential energy to kinetic energy; acquiring potential energy comes first.

like the force applied to stretch the spring in the first place or any frictional forces that may be present, are nonconservative. What happens to the energy of the system as all these forces do work on it?

We begin to answer this question by reconsidering the work-energy theorem set forth in Equation 7.22b,

$$W = \Delta K. \tag{8.26}$$

In this equation, W signifies the *net* work done on the system. This quantity can be found by evaluating the work done by each of the forces and taking the algebraic sum. Before doing so, however, let us divide the forces into two classes, conservative and nonconservative. We can sum them separately and thus find the net work W_c done by the conservative forces and the net work W_{nc} done by the nonconservative forces. The overall net work W is then the sum of the two:

$$W = W_c + W_{nc}.$$

When we combine this equation with Equation 8.26, we obtain

$$W_c + W_{nc} = \Delta K. \tag{8.27}$$

Next, remember that the special property of work done by conservative forces is that it results in a change in the potential energy of the system; according to Equation 8.15, we can write $W_c = -\Delta U$. Substituting this value into Equation 8.27 gives us $-\Delta U + W_{nc} = \Delta K$, or

$$W_{nc} = \Delta K + \Delta U. \tag{8.28}$$

That is, *the net work done on the system by the nonconservative forces acting on it is equal to the sum of the changes in the kinetic and potential energies of the system.* But the sum of the changes in the two kinds of energy is just the change in the sum of the two, which means that it is the change in the total mechanical energy:

$$\Delta K + \Delta U = \Delta(K + U) = \Delta E.$$

Equation 8.28 thus becomes

$$W_{nc} = \Delta E; \tag{8.29}$$

the net work done on a system by the nonconservative forces acting on it is equal to the change in its total mechanical energy. Because this statement relates the change in *total* energy to work done by nonconservative forces, we call it the **generalized work-energy theorem**. Examples 8.6 and 8.7 suggest some of the possibilities of Equation 8.29.

EXAMPLE 8.6

A block is projected upward from the bottom of a plane of incline angle α with initial speed v_i. The block reaches a certain maximum vertical height and then slides back down the plane. (a) If its final speed at reaching the bottom again is $v_f = \frac{1}{2}v_i$, what is the value of the coefficient of kinetic friction μ_k? (b) What can you say about the value of the static coefficient μ_s? (c) Let the incline angle be $\alpha = 35°$, and find the value of μ_k.

SOLUTION:

(a) If the block returns to the bottom with half its initial speed, find μ_k.

Let the reference level for potential energy be at the bottom of the plane, and set $U = 0$ at this point. The block then begins with a total energy equal to its initial kinetic energy and ends with a total energy equal to its final kinetic energy. So you have

$$E_i = \tfrac{1}{2}mv_i^2$$

and

$$E_f = \frac{1}{2}mv_f^2 = \frac{1}{2}m\left(\frac{v_i}{2}\right)^2 = \frac{1}{8}mv_i^2.$$

The change in energy over the process is thus

$$\Delta E = E_f - E_i = -\tfrac{3}{8}mv_i^2 = -\tfrac{3}{4}E_i. \tag{8.30}$$

The negative sign means that energy is lost by the system during the process. According to Equation 8.29, this change in energy is due to the nonconservative work done on the block. In the present case, that work is done by the frictional force. As you saw in part **b** of Example 8.2, the work done over the round trip is

$$W_{nc} = -2\mu_k mgL \cos \alpha, \tag{8.31}$$

where L is the one-way length of the path of the block along the plane. The negative sign is consistent with that in Equation 8.30. But L remains to be determined.

To determine L, note that the frictional work done is proportional to the distance traveled along the plane and does not depend on the direction. Consequently, half of the total frictional work is done on the upward trip and the other half on the downward trip. It follows from Equation 8.29 that half the change in total energy occurs during each half of the trip. Call the total energy of the system at the moment when the block is at the top of its path E_{top}. You can then write

$$E_{top} - E_i = \tfrac{1}{2}\Delta E = -\tfrac{3}{8}E_i,$$

or

$$E_{top} = \tfrac{5}{8}E_i.$$

Why is this equation useful? Because you already know that $E_i = \tfrac{1}{2}mv_i^2$, and so you can now determine E_{top} in terms of the maximum height h reached by the block. You do this by noting that at the top of its path the block is instantaneously at rest and its total energy is entirely potential. Thus you have

$$E_{top} = mgh.$$

Combining these two equations and solving for h gives you

$$h = \frac{5}{16}\frac{v_i^2}{g}. \tag{8.32}$$

Now you can get a value for L. Because $L = h/\sin\alpha$, that value is

$$L = \frac{5v_i^2}{16g\sin\alpha}.$$

By inserting this value of L into Equation 8.31, you can express W_{nc} entirely in terms of known quantities. You have

$$W_{nc} = -2\mu_k mg\,\frac{5v_i^2}{16g\sin\alpha}\cos\alpha = -\frac{5\mu_k mv_i^2}{8\tan\alpha}.$$

The last step is again to use Equation 8.29, $W_{nc} = \Delta E$, and thus equate the value of W_{nc} just obtained with the value of ΔE given by Equation 8.30. This gives you

$$-\frac{3}{8}mv_i^2 = -\frac{5\mu_k mv_i^2}{8\tan\alpha}.$$

You solve for μ_k and obtain

$$\mu_k = \tfrac{3}{5}\tan\alpha. \tag{8.33}$$

(b) What can you say about the value of the static friction coefficient μ_s?

According to the description of the motion of the system, the block does not "get stuck" at the top but instead slides back down. This means that the incline is steep enough so that the static frictional force acting on the block at the instant when it is at rest at the top of its path is not sufficient to hold it at rest. Consequently, α must be greater than the critical angle α_s for static friction given by Equation 4.36a, $\mu_s = \tan\alpha_s$. You thus have

$$\mu_s < \tan\alpha,$$

or

$$\mu_s < \tfrac{5}{3}\mu_k.$$

(c) Let the incline angle be $\alpha = 35°$, and find the value of μ_k. Using Equation 8.33, you obtain

$$\mu_k = \tfrac{3}{5}\tan 35° = 0.42.$$

It follows also that

$$\mu_s < 0.70.$$

Once the ratio of the final speed of the block to its initial speed is given, the value of μ_k depends only on the incline angle α. The maximum height h can be determined from Equation 8.32 if v_i is given. It does not depend explicitly on the value of μ_k. However, the ratio v_f/v_i (which does depend on α) is built into that equation. For some ratio other than $\tfrac{1}{2}$, the fraction in Equation 8.32 will be different from $\tfrac{5}{16}$.

EXAMPLE 8.7

A grandfather clock has a pendulum whose period is $\tau = 2$ s. Substantially all of the 5-kg mass of the pendulum is concentrated in the bob. In normal operation, the amplitude of the swing is $\theta_0 = 0.12$ rad (about 7°). If the driving mechanism is interrupted, the pendulum swing dies down to half this amplitude in $N = 30$ swings. Estimate how much power the pendulum requires for normal operation.

SOLUTION: You saw in Example 8.5 that the total mechanical energy of a simple pendulum of amplitude θ_0 is

$$E_{\theta_0} = mgl(1 - \cos\theta_0),$$

where m is the mass of the pendulum and l its length. If the amplitude is allowed to diminish to $\theta_0/2$, its energy will diminish correspondingly to

$$E_{\theta_0/2} = mgl\left(1 - \cos\frac{\theta_0}{2}\right).$$

The change in energy is the negative quantity

$$\Delta E = E_{\theta_0/2} - E_{\theta_0} = mgl\left(\cos\theta_0 - \cos\frac{\theta_0}{2}\right).$$

This loss of energy must be made up by the driving mechanism if the clock is to continue to operate with a constant pendulum amplitude θ_0. That is, the driving mechanism must do positive work $W = -\Delta E$ on the pendulum over the time during which ΔE is lost to friction. The time t can be written as the product of the pendulum period τ and the number of swings N; that is, $t = N\tau$. If you make the crude assumption that the energy loss is linear in time, the mean power (Equation 7.24) delivered to the pendulum must be

$$\langle P\rangle \equiv \frac{W}{t} = -\frac{\Delta E}{N\tau} = \frac{mgl}{N\tau}\left(\cos\frac{\theta_0}{2} - \cos\theta_0\right).$$

According to Equation 6.45c, the period of a simple pendulum of length l is

$$\tau = 2\pi\sqrt{\frac{l}{g}}.$$

You can solve this equation for l and then use the result to eliminate l from the equation for $\langle P\rangle$. You have $l = \tau^2 g/4\pi^2$, and thus

$$\langle P\rangle = \frac{mg^2\tau}{4\pi^2 N}\left(\cos\frac{\theta_0}{2} - \cos\theta_0\right). \tag{8.34}$$

You are now ready to insert numerical values into Equation 8.34:

$$\langle P \rangle = \frac{5 \text{ kg} \times (9.8 \text{ m/s}^2)^2 \times 2 \text{ s}}{4\pi^2 \times 30} \times \left[\cos \frac{0.12 \text{ rad}}{2} - \cos (0.12 \text{ rad}) \right]$$

$$= 4 \times 10^{-3} \text{ W}.$$

Symbols Used in Chapter 8

A	maximum distance through which a spring is stretched or compressed; amplitude of a harmonic oscillator	$\mathbf{s}$	displacement along straight path
		$\hat{\mathbf{s}}$	direction of motion of body on which work is being done
b	horizontal distance through which a body is moved	U	potential energy
		W	work
E	total mechanical energy	W_c	net work done by conservative forces acting on a system
$\mathbf{F}_a$	force applied to a system to do work on it		
F_s	Hooke's-law force exerted by a spring	W_{nc}	net work done by nonconservative forces acting on a system
h	vertical distance through which a body is moved		
		W_a	work done by the applied force $\mathbf{F}_a$
i, f (subscript)	initial, final values of a quantity	α	incline angle of a plane
k	force constant of a spring	α_s	critical angle for static friction
K	kinetic energy	θ	instantaneous angular displacement of a pendulum
l	length of a simple pendulum		
$L, \mathbf{L}$	path length along an inclined plane, displacement along the plane	θ_0	angular amplitude (maximum angular displacement) of a pendulum
$\langle P \rangle$	mean power	μ_k, μ_s	kinetic and static coefficients of friction
$\mathbf{r}$	position vector	τ	period of a pendulum
s	length of path over which work is done	ω	angular speed; angular frequency

Summing Up

When the position, or configuration, of a system is changed, there may be a change in its **potential energy**. This will be the case if one or more conservative forces do work on the system during the process. A **conservative force** is any force that satisfies Equation 8.10,

$$W = \int_{\substack{\text{closed} \\ \text{path}}} \mathbf{F} \cdot d\mathbf{s} = 0,$$

or its logical equivalent, Equation 8.11,

$$W_{\text{path 1}} = \int_{\substack{s_i \\ \text{path 1}}}^{s_f} \mathbf{F} \cdot d\mathbf{s} = W_{\text{path 2}} = \int_{\substack{s_i \\ \text{path 2}}}^{s_f} \mathbf{F} \cdot d\mathbf{s}.$$

Conservative forces can be functions of position only; $\mathbf{F} = \mathbf{F}(\mathbf{r})$. A **nonconservative**, or **dissipative**, **force** is one that does not satisfy the criteria for a conservative force.

Suppose that a conservative force $\mathbf{F}(\mathbf{r})$ does work dW on a system as the system moves through a displacement $d\mathbf{r}$. The **potential energy function** $U(\mathbf{r})$ is one of a family of functions that satisfy Equation 8.12,

$$dU = -dW.$$

It follows that, when the system moves over any path connecting the points $\mathbf{r}_i$ and $\mathbf{r}_f$, it experiences a potential energy change that, according to Equation 8.15, is

$$U(\mathbf{r}_f) - U(\mathbf{r}_i) \equiv \Delta U = -W.$$

For any system on which a conservative force acts, this relation is satisfied by an infinite number of potential energy functions. All of these functions, however, differ only by a constant. Thus, all of them will yield the same value of ΔU between the same $\mathbf{r}_i$ and $\mathbf{r}_f$. By fixing a **reference point** $\mathbf{r}_0$ at which $U(\mathbf{r}_0) = 0$, we single out a particular potential energy function. It is then possible to speak of *the* potential energy $U(\mathbf{r})$ of the system, as specified in Equation 8.16,

$$\Delta U = \int_{\mathbf{r}_0}^{\mathbf{r}} dU = U(\mathbf{r}) - U(\mathbf{r}_0)$$

$$= U(\mathbf{r}) - 0 \quad \text{for } U(\mathbf{r}_0) = 0.$$

We define the **total mechanical energy** of a system as

the sum of its kinetic and potential energies, according to Equation 8.17,

$$E \equiv K + U.$$

A system is called a **conservative system** if work is done on it by conservative forces only. (This is the case either if no nonconservative forces are present or if such forces are present but do no work. Forces of the latter sort are called **workless constraints**.) In a conservative system, E does not change with time. This statement is called the **principle of conservation of mechanical energy**; it can be stated in any of the ways given by Equations 8.18a, 8.18b, 8.18c and 8.19: For a conservative system,

$$\Delta K = -\Delta U; \quad \Delta K + \Delta U = \Delta(K + U) = 0;$$
$$\Delta E = 0; \quad E = \text{constant}.$$

If a system is not conservative, the forces doing work on it can be sorted into two classes, conservative and noncon-servative. According to the **generalized work-energy theorem**, Equation 8.29, the work W_{nc} done on the system by the nonconservative forces is equal to the change in the total mechanical energy of the system:

$$W_{nc} = \Delta E.$$

KEY TERMS

Section 8.2 Potential Energy
potential energy ▪ conservative force, nonconservative force (dissipative force) ▪ function of position only

Section 8.3 Conservation of Mechanical Energy
total mechanical energy ▪ conservative system ▪ principle of conservation of mechanical energy ▪ workless constraint

Section 8.4 The Generalized Work-Energy Theorem
generalized work-energy theorem

QUERIES

8.1 *(2) Something missing?* In part **b** of Example 8.1, you saw that the work required to pull a block up an inclined plane with friction present can be expressed in the form of Equation 8.9, $W = mgh + \mu_k mgb$, in which neither the distance L nor the incline angle α appears. How do you explain this, in view of the fact that work done against friction is directly proportional to the distance traversed?

8.2 *(2) Distinction without a difference.* We have made a distinction between the positional potential energy of a system, exemplified by the gravitational energy mgy, and the configurational potential energy of a system, exemplified by the spring energy $\frac{1}{2}kx^2$. Show that the distinction depends entirely on the way in which the system boundaries are chosen. First, reconsider the system consisting of a ball acted on from outside the system by the external gravitational force. Redefine the boundaries so that the system includes the earth as well as the ball, so that the gravitational force is internal to the system. Is the potential energy of the redefined system positional or configurational? How does the redefinition affect the value of the potential energy? [Assume that the same choice of $U(y) = 0$ is used for both systems.] Second, reconsider the potential energy of a mass-spring system—in which the spring force is internal—as the potential energy of a smaller system. This smaller system consists of the mass alone, and the spring force is external to the system. Discuss the same two questions for this case.

8.3 *(2) Liberal force?* Suppose you grab the mass of a mass-spring system and pull on it, stretching the spring. Your friend, watching you, argues as follows. The force exerted by the spring on the mass is a conservative force. The force you exert on the mass is the third-law reaction force to the spring force and is therefore always equal and opposite to the spring force. Thus, the force you exert must be a conservative force as well. Is your friend right or wrong? Explain your answer.

8.4 *(2) Abnormal?* In calculating the work done on the block of Example 8.1, we did not consider the normal force **N** exerted on the block by the inclined plane. Why is this omission justified? What kind of force is **N**?

8.5 *(2) Power play.* You can estimate your maximum short-time power output by running up a flight of stairs as fast as you can. What measurements must you make? How do you carry out the calculation? Most healthy young persons turn out roughly 500 W. What is your output?

8.6 *(3) Psychoceramics.* Every so often, someone invents the perpetual-motion machine shown below. An endless belt is stretched around two low-friction pulleys. Spaced uniformly along the belt are metal containers like opened tin cans. The

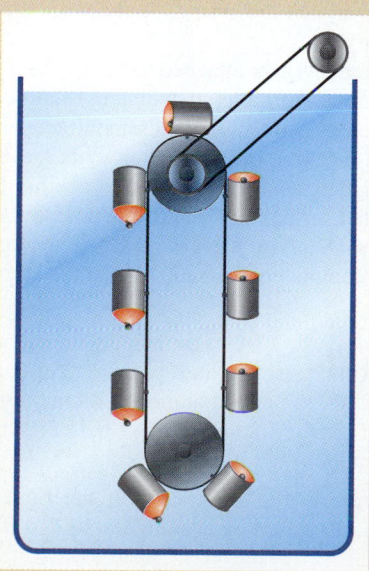

missing end of each can is covered by a flexible rubber sheet in the middle of which is glued a lead weight. The cans are filled with air, and the whole apparatus is submerged in water. The auxiliary pulley and belt deliver the output power to some useful machine. The inventor argues that the volume of the cans on the left side of the system is larger than that of the equal number of cans on the right, owing to the distention of the rubber sheets by the lead weights. The left side is therefore more buoyant than the right, and the whole system revolves clockwise. Analyze the system in the light of the generalized work-energy theorem. A detailed dynamical analysis is much more complicated. If you took the trouble to carry it out, do you think the result would be more pleasing to the inventor?

Comment on your answer in view of the logical connections between the dynamical and the energy-conservation approaches to mechanics.

8.7 *(3) Not even wrong!* Inventors of perpetual-motion machines often complain bitterly that physicists and engineers are unwilling to study their inventions in detail. Consequently, they claim, they are denied the services of the experts who could discover the "technical reasons" why—as seems always to be the case—the machines don't quite work. (The system considered in Query 8.6 is a case in point.) Why are physicists and engineers reluctant to spend the time to study such inventions?

PROBLEMS

GROUP A

8.1 *(2) Pocket summary, I.* A 12-kg mass is lifted from the floor to a point 1.7 m above it. The process is slow, so the mass never has appreciable kinetic energy. **(a)** During the lifting, how much work is done on the mass by nonconservative forces? **(b)** How much work is done on it by conservative forces? **(c)** What is the net work done on it? **(d)** What is the change in its potential energy?

8.2 *(2) Pocket summary, II.* The mass in Problem 8.1 is allowed to fall. **(a)** During the fall, is the system consisting of the mass conservative?

Just before the mass hits the floor, **(b)** how much work has been done on it during its descent? by what force or forces? **(c)** what is its kinetic energy?

8.3 *(2) Springing to life.* One end of a spring is fixed to a ring screwed into a tabletop, and the free end is stretched horizontally by 30 cm. A force of 160 N is required to hold it. Assume that the spring conforms to Hooke's law. **(a)** How much work is required to stretch the spring? Is the work conservative or nonconservative?

A 5.5-kg mass is now attached to the free end of the spring. The mass rests on the tabletop, which is frictionless. The system is released. When the spring has contracted halfway back to its relaxed position, **(b)** how much work has been done by the spring on the mass? **(c)** how much work has been done by the mass on the spring? **(d)** what is the potential energy of the mass-spring system? **(e)** what is its kinetic energy? **(f)** When the spring is completely relaxed, what is the kinetic energy of the system?

8.4 *(2) Plane facts, I.* A ramp has an incline angle of 25°. How much work must you do if you are to drag a 50-kg mass a distance of 100 m upward along the ramp? Assume that the coefficient of kinetic friction is $\mu_k = 0.35$ and that you apply the force parallel to the plane.

8.5 *(2) Wedge.* In the illustration in the right column, the heavy object A is raised by pushing the wedge B horizontally.

Neglecting friction, what is the mechanical advantage of the system?

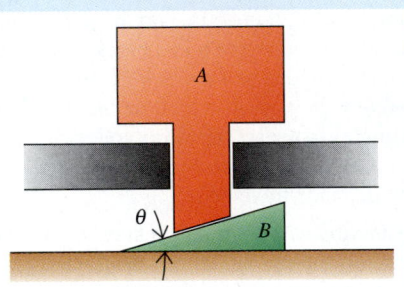

8.6 *(2) Fun in the lab.* In a standard student experiment, a spring gun launches a ball of mass m horizontally. The gun is on a table a distance h above the floor. The ball is loaded into the gun by compressing the spring through a distance d. When the gun is "fired," the ball hits the ground a horizontal distance x away from the gun. **(a)** Express the spring constant k in terms of x, m, h, d, and any necessary constants. **(b)** Find k if $x = 2.36$ m, $m = 55.2$ g, $h = 83.8$ cm, and $d = 5.32$ cm.

8.7 *(2) Takeoff.* An airplane takes off and climbs to its cruising altitude. When it reaches an altitude of 5 km, its speed is 720 km/h. What is the ratio of the work done to increase its potential energy to that done to increase its kinetic energy?

8.8 *(3) Double your fun.* Suppose you want to double the horizontal range of a spring gun like the one described in Problem 8.6. By what factor must you increase d, the distance through which the spring is compressed?

8.9 *(3) Improving on Galileo?* A scientifically minded but reckless tourist throws a stone of mass m horizontally with speed v_0 from the top of the Leaning Tower of Pisa. Taking $U = 0$ at ground level, find the kinetic energy, the potential energy, and the total energy of the stone in flight as a function of time. The Leaning Tower is 54.6 m high. If $v_0 = 15$ m/s and $m = 250$ g, evaluate U, K, and E at $t = 2$ s.

8.10 *(3) Mixed flavors.* At a certain time t_1, a particle has kinetic energy 50 J and potential energy 50 J. At a later time t_2, the particle is at rest with potential energy 60 J. Over the time interval from t_1 to t_2, how much work is done on the particle by **(a)** conservative forces? **(b)** nonconservative forces?

8.11 *(3) Speedy calculation.* You throw a ball vertically upward, and it leaves your hand at a height h above the ground with speed v_0. **(a)** What is the total mechanical energy E of the ball at the moment of release? **(b)** What is the kinetic energy of the ball at a later time when its height above the ground is x? Express your result in terms of v_0, h, and x. **(c)** How does the answer to part **b** imply that there is a maximum height H above which the ball will not rise? Using the answer to part **b**, find H. **(d)** Although h is measured from ground level in parts **a** and **b**, show that the answer to part **b** does not depend on the reference level from which you measure h. If $h = 1.8$ m and $v_0 = 7.4$ m/s, find **(e)** H and **(f)** v when $x = 0.86$ m.

8.12 *(3) The stiffer the faster.* A mass oscillates on the end of a spring with amplitude 12 cm. If the kinetic energy of the mass reaches a maximum value of 1 J, what is the force constant of the spring?

8.13 *(3) Halfway measures.* You stretch the spring of a certain mass-spring system 14.0 cm from its relaxed state; this requires 21.2 J of work. You then release the system. What is its kinetic energy when the stretch of the spring is 7.0 cm? (Hint: There's a straightforward way to do this problem. There's also an easy shortcut that requires almost no calculation.)

8.14 *(3) Division of energy, I.* Find the ratio K/U for a mass-spring system oscillating with amplitude A when its displacement from equilibrium is **(a)** $\frac{1}{2}A$, **(b)** $\frac{1}{4}A$, **(c)** $-\frac{1}{3}A$.

8.15 *(3) Division of energy, II.* The motion of a mass-spring oscillator is described by the equation $x = A \cos 2\pi t/\tau$. Find the ratio K/U for **(a)** $t = \frac{1}{12}\tau$, **(b)** $t = \frac{1}{8}\tau$, **(c)** $t = \frac{1}{6}\tau$.

8.16 *(3) Deconstruction business.* A wrecking ball is a device consisting of a massive steel sphere suspended from a cable whose upper end is attached to a crane; see the photograph below. By moving the crane, the operator makes the ball swing away from the vertical. When the ball swings back, it collides with the structure to be demolished. **(a)** If the cable

length is 14.5 m and the ball swings so that the cable makes a maximum angle of 36° with respect to the vertical, what is the speed of the ball when its hits a wall if the cable angle is then 6°? **(b)** If the mass of the ball is 1 tonne (1000 kg), what is the kinetic energy of the ball when it strikes the wall?

8.17 *(3) Slide, I.* A block of mass $m = 1$ kg slides down an inclined plane 2 m high and 4 m long (measured along the incline). The coefficient of kinetic friction is $\mu_k = 0.07$. **(a)** What is the kinetic energy of the block when it reaches the bottom of the plane? **(b)** The block continues to slide over a level plane with the same value of μ_k. How far does it go before coming to rest?

8.18 *(3) Slide, II.* A block slides a distance L down an inclined plane and then an equal distance along a horizontal surface having the same coefficient of kinetic friction μ_k. If the plane angle is 12°, what is the value of μ_k?

8.19 *(3) Slide, III.* A block of mass 3 kg slides down a plane of incline angle $\alpha = 30°$, starting at a height of 0.5 m. If its speed at the bottom is 2.55 m/s, **(a)** what is the value of μ_k? **(b)** how much frictional work is done?

8.20 *(3) Bumper tag.* The block in the illustration below slides down the frictionless track and strikes the sticky bumper attached to the spring. The spring is compressed a distance A, and the mass-spring system subsequently oscillates with period τ. Show that τ is given by the expression

$$\tau = \pi A \sqrt{\frac{2}{gh}}.$$

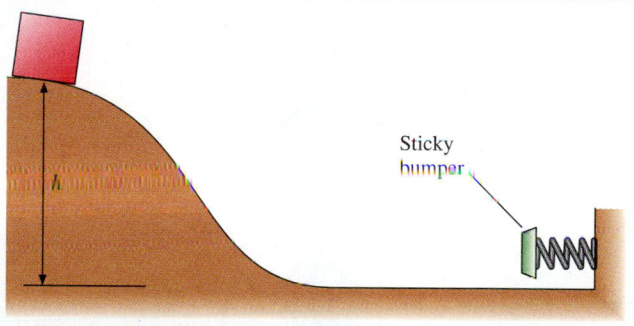

Sticky bumper

8.21 *(3) Plane facts, II.* When you have accomplished the task described in Problem 8.4, you let the block go. How fast will it be going when it reaches the bottom?

8.22 *(3) Aye, there's the rub! I.* Consider a mass-spring system just like that in Problem 8.3, except that now the coefficient of kinetic friction between mass and tabletop is $\mu_k = 0.1$. What is the kinetic energy of the system at the instant when the spring is relaxed for the first time?

8.23 *(3) Dying down.* A mass-spring oscillator consists of a mass m hanging from a spring of force constant k. It is set into oscillation with amplitude A but, owing to friction, the amplitude diminishes gradually. When a time T has passed, the amplitude is nine-tenths its original value. **(a)** If the system is to be driven with constant amplitude A, approximately how much power $\langle P \rangle$ must be delivered to the system by the driving mechanism? **(b)** Find $\langle P \rangle$ if $A = 18$ cm, $k = 12$ N/m, $m = 0.25$ kg, and $T = 4.0$ min.

8.24 *(3) Lever.* The drawing below shows the familiar simple machine called the *lever*. Imagine that the lever is initially horizontal and is then raised through a small angle δ. Use the generalized work-energy theorem to find the mechanical advantage, defined to be mg/F.

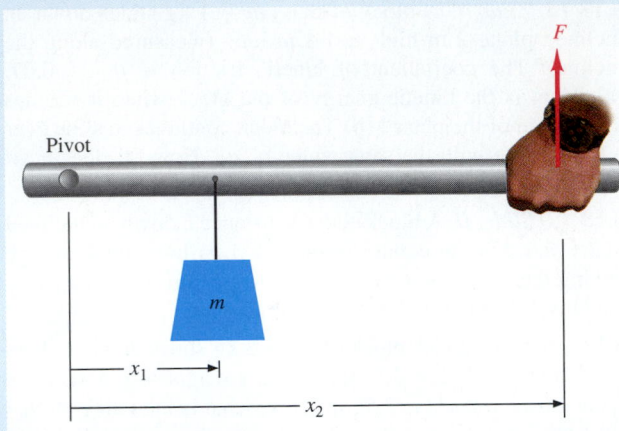

8.25 *(3) Block and tackle.* The simple machine shown below has been used for millennia to raise heavy weights. The combination of ropes and pulleys is called a *block and tackle*. Use the generalized work-energy theorem to show that the mechanical advantage of the system is $mg/F = N$, where N is the number of rope segments that support the weight, shown to be 3 in the drawing. Neglect friction (*not* a good approximation for real systems).

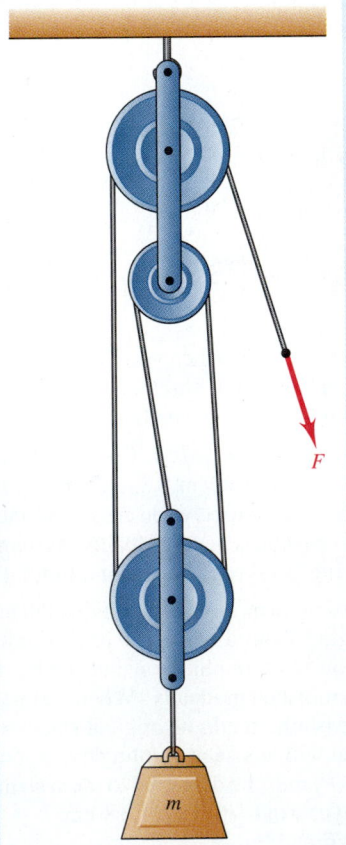

8.26 *(3) Half-Atwood.* In the illustration below, the mass of block A is 50 kg and that of block B is 20 kg. The system is released from rest. If the horizontal surface is frictionless, how fast are the blocks moving when block B has descended 7 m?

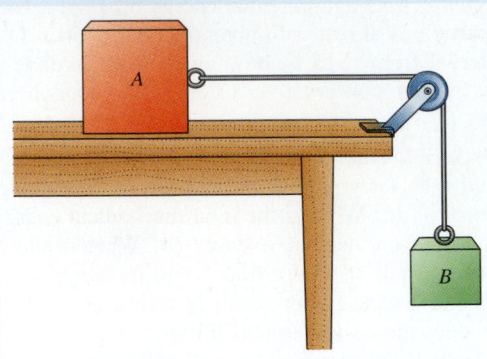

8.27 *(3) Wand.* A rod of negligible mass and length l is pivoted at its center, as shown below. A small mass m is fixed to its left end, and a mass $2m$ is fixed to the right end. The system is released. **(a)** What is the speed v of the two masses when the rod is vertical? **(b)** What is the angular speed ω of the system at that instant?

8.28 *(3) Bump and bounce.* The Atwood machine of Example 8.4 is released from rest, as described in the example. A tabletop is located a distance h below the initial level of the masses. After the larger mass strikes the table and comes abruptly to rest, how much higher does the smaller mass go?

8.29 *(3) Half-Atwood with friction.* In the figure below, the coefficient of kinetic friction between m_1 and the tabletop is μ_k. The system starts from rest and descends a distance y.

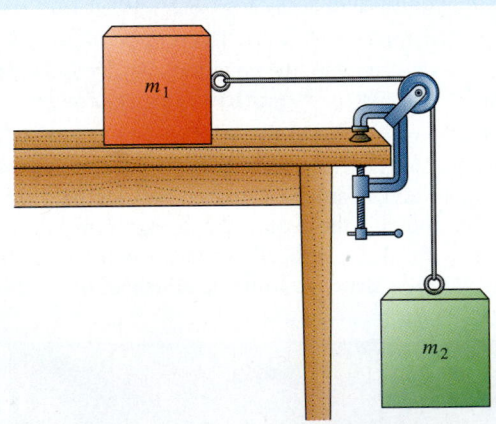

(a) Use the generalized work-energy theorem to show that the speed of the system is then

$$v = \sqrt{\frac{2(m_2 - \mu_k m_1)gy}{m_1 + m_2}}.$$

(b) Use the result of part **a** to find the acceleration of the system.

8.30 *(3) Jumping jack, I.* A ball of mass 10 g is placed on the upper end of a vertical spring inside a guiding tube, as shown below. The spring is then compressed 1 cm and released. If the force constant is 600 N/m, what is the greatest upward velocity attained by the ball? How high does it go above its position at rest on the compressed spring? Neglect the initial compression of the spring due to the weight of the ball.

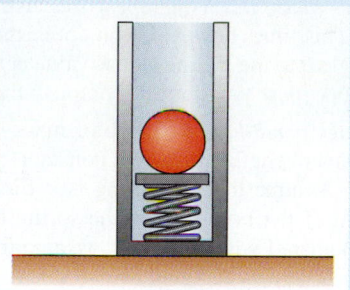

8.31 *(3) Batting practice, I.* You hit a baseball of mass m with initial speed v_i and launch angle θ. For the instant when the ball reaches the top of its trajectory, find (neglecting air resistance) **(a)** its potential energy and **(b)** its kinetic energy. **(c)** For what value of θ are the potential energy and the kinetic energy equal? **(d)** Show that energy is conserved; that is, show that the total energy of the ball at the top of the trajectory is equal to its energy just after you hit it.

8.32 *(3) Batting practice, II.* You hit a baseball with initial speed v_i. What is its kinetic energy when it reaches height y?

GROUP B

8.33 *(2) Hanging question.* A mass m hangs from a spring of force constant k_1. The spring in turn hangs at rest from another spring of force constant k_2. What is the ratio U_2/U_1 of the potential energy stored in the two springs?

8.34 *(2) Pyramid builder.* You are designing a ramp to be used in raising large blocks of mass M to a height h. You are free to choose the ramp angle α. Show that, if the coefficient of kinetic friction between the blocks and the ramp is μ_k, the work required for each block is $W = Mgh(1 + \mu_k \cot \alpha)$.

8.35 *(2) Rolling hills.* Equation 8.9, $W = mgh + \mu_k mgb$, gives the work done in dragging a block up a hill of constant steepness. Most hills do not have uniform steepness from bottom to top, though. Show that Equation 8.9 is valid anyway, as long as the force is always applied parallel to the surface and the coefficient of kinetic friction is the same everywhere.

8.36 *(2) Constant and conservative.* As a particle moves around the rectangular path of the drawing below, it experiences a force given by the expression $\mathbf{F} = 5\hat{\mathbf{x}} - 3\hat{\mathbf{y}}$. By integrating the work done on the particle by this force as the particle makes a complete circuit of the path, show that the force is conservative. (This exemplifies a theorem not proved in this chapter: Any force that is constant is conservative.)

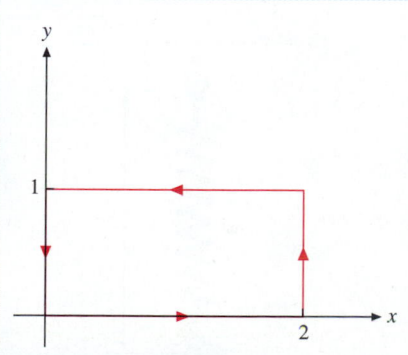

8.37 *(2) Working backwards.* In this chapter, we have dis-

cussed means of calculating the potential energy when the force is known. But the reverse process is at least as useful. **(a)** Begin with the one-dimensional form of Equation 8.14, which can be written

$$\int_{x_i}^{x_f} dU = \int_{x_i}^{x_f} -F \, dx,$$

and show that, if you know the potential energy $U(x)$, you can express the associated conservative force in the form

$$F = -\frac{dU}{dx}. \tag{1}$$

(b) Show that Equation 1 is independent of the choice of the reference point for the potential energy. **(c)** Show that Equation 1 gives the correct result for the conservative forces associated with (i) the potential energy of a raised mass and (ii) the potential energy of a stretched spring.

8.38 *(3) Equilibrium.* The potential energy of a system is $U(x)$. **(a)** Show that, if there is a value x_0 for which $dU/dx = 0$, x_0 is an equilibrium point for the system. [Hint: Sketch a graph of $U(x)$ versus x for a spring of relaxed length L, with the origin at the fixed end of the spring.] **(b)** Show that, if $d^2U/dx^2 > 0$ at x_0, the equilibrium is stable. What is the physical meaning of $d^2U/dx^2 = 0$? of $d^2U/dx^2 < 0$? **(c)** Check your results by considering a spring of relaxed length L, for which $U(x) = \frac{1}{2}k(x - L)^2$.

8.39 *(3) Engineering for amusement, I.* You are designing a loop-the-loop ride for an amusement park, shown schematically (next page). The car starts at a height h with negligible speed and coasts along the track, going around the inside of the vertical circle of radius r, whose bottom is at ground level. In your initial design, you neglect friction. **(a)** What is the minimum value of h for which the car will stay on the track at the top of the circle? **(b)** When the car makes an angle θ with the vertical, as shown in the illustration, what is the magnitude of the normal force exerted on the car by the track?

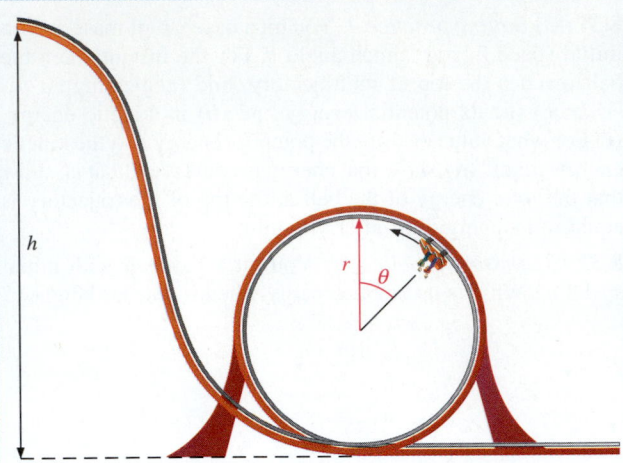

8.40 *(3) Engineering for amusement, II.* At a later stage of the design project described in Problem 8.39, you build a scale model of the ride. To test it, you place a small body of mass *m* on a miniature electronic scale and put the body-scale assembly into the model car. You then experiment with various starting heights *h*. In one experiment, you note that when the car reaches the top of the loop the scale reads the ordinary weight *mg* of the body. What will the scale read when the car reaches the bottom of the loop? Are you likely to use this combination of conditions in the actual ride?

8.41 *(3) Chain drive.* The illustration below shows a uniform chain of mass *m* and length *L* hanging over the edge of a frictionless table. **(a)** If you pull the end of the chain horizontally, how much work must you do to get the entire chain onto the table? **(b)** Suppose you stop pulling on the chain when just a little bit of it still hangs over the edge. When you let go, the chain slips back over the edge. When the last bit of chain leaves the table, how fast is the chain moving?

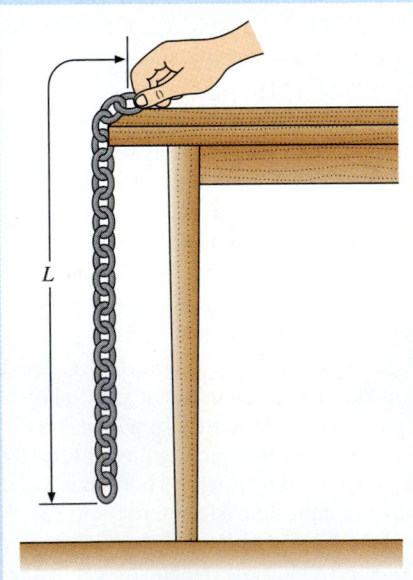

8.42 *(3) Look out, Goliath! I.* A child whirls a stone tied to the end of a string of length *l* in a vertical circle. **(a)** What is the value of v_t, the speed of the stone at the top of the circle,

if it is just sufficient to keep the string taut? **(b)** What is the corresponding speed v_b of the stone at the bottom of the circle?

8.43 *(3) Look out, Goliath! II.* The stone of Problem 8.42 is now whirling with any v_t greater than that required to keep the string taut. How much greater than the string tension at the top of the circle is the tension at the bottom of the circle?

8.44 *(3) Plane facts, III.* You repeat the task described in Problem 8.4. This time, however, you apply the force to the block at an angle β to the incline whose value is $\beta = \tan^{-1} \mu_k$. How much work must you do to accomplish the task?

8.45 *(3) Springy pendulum, I.* A bob of mass *m* hangs from a spring of relaxed length *l* and force constant *k* (part *a* of the figure below). Assume that the spring is sufficiently stiff so that the weight of the bob extends the spring by an amount quite small compared with *l*. That is, assume that $mg/k \ll l$.

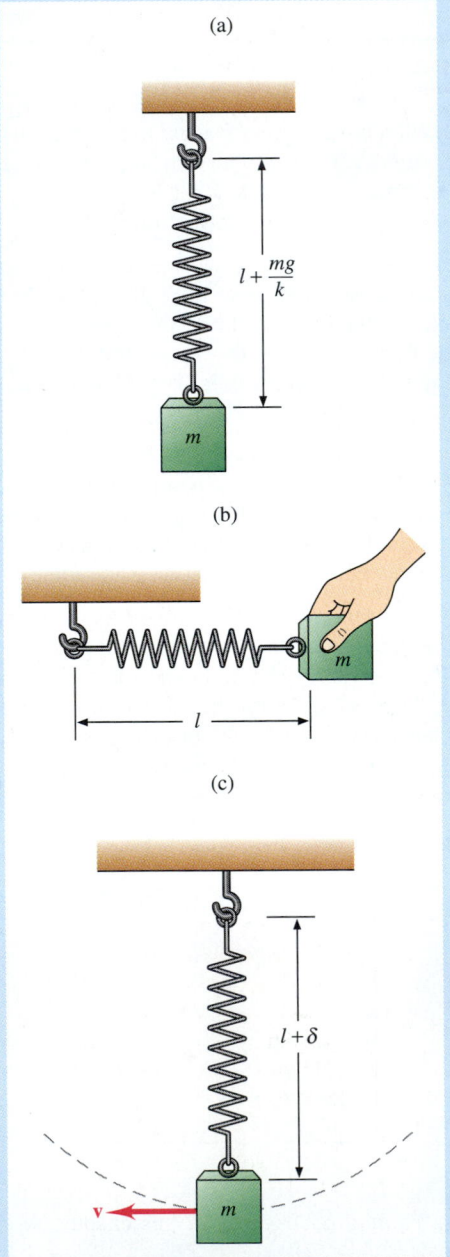

The bob is raised so that the pendulum is in the horizontal position with the spring relaxed (part *b* of the figure) and is then released. **(a)** When the pendulum passes through the vertical position (part *c* of the figure), what is the extension δ of the spring? **(b)** With $m = 0.5$ kg, $l = 0.10$ m, and $k = 1000$ N/m, what is the value of δ?

8.46 *(3) Tense moment.* A circus acrobat of mass *m* swings from a trapeze of length *l*. He starts the swing at a platform whose location is such that the trapeze rope makes an initial angle θ_0 with the vertical. Find the tension in the trapeze rope as a function of the angle θ that it makes with the vertical.

8.47 *(3) Galileo's interrupted pendulum.* Although Galileo lived long before the introduction of the concept of energy, his simple device shown below indicates that he had a strong intuitive understanding of the transformation of energy between its kinetic and potential forms. A pendulum having a string of length *l* is released from the horizontal position shown. A peg *P* can be placed at any height *y* above the bottom of the pendulum's swing. **(a)** For what range of *y* will the string wind up on the peg, remaining taut through the entire process? **(b)** For what range of *y* will the pendulum snag itself on the peg, with the bob looping over the peg at the end of a slack string? **(c)** For what range of *y* will the pendulum swing back and forth? Will the string be taut through the entire swing?

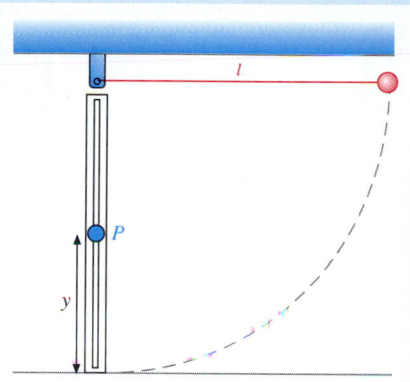

8.48 *(3) Off the wall.* A small ice cube rests at the top of a large, smooth hemispherical dome. It starts to slide downward with negligible friction. At what angle θ_c from the top will it leave the dome? Can you explain qualitatively why the result is independent of the radius of the dome?

8.49 *(3) Find the equation.* A mass-spring oscillator has total mechanical energy $E = 6 \times 10^{-5}$ J. The spring is fully extended at time $t = 2$ s; at that instant, the force exerted by the spring is $F_s = 2.2 \times 10^{-3}$ N. The period is $\tau = 3$ s. Write *x* as a function of *t*.

8.50 *(3) Springy revolution.* In the illustration in the next column, a small steel block of mass *m* rests on a frictionless air table. A string leads from the block through a small hole at the center of the table and then to a very flexible spring of force constant *k* and negligible mass, whose other end is anchored. When the block is at rest at the center hole of the table, the spring is relaxed. By stretching the spring, the experimenter can move the block to any part of the table. The block is pulled away from the center hole through a distance *r*, stretching the spring. The block is then projected in such a

way that it moves in a circular orbit of radius *r* around the center hole. **(a)** Find the relation between the kinetic energy *K* and the potential energy *U* of the system and the total energy *E* of the system. Take $U(r) = 0$ at $r = 0$. **(b)** Find the speed *v* and the angular speed Ω of the block in its circular orbit. **(c)** As the block revolves in a circle with angular speed Ω, the experimenter quickly places a magnet on the table at a point radially outward from the instantaneous position of the block and then removes it. The resulting short-duration magnetic force pulls the block slightly outward without otherwise disturbing it. The block subsequently oscillates in the radial direction with angular frequency ω while continuing its revolution around the center hole. Show that $\omega = \Omega$. **(d)** On the basis of the answer to part **c**, what is the name of the curve traced by the block as it simultaneously oscillates in and out and revolves round and round?

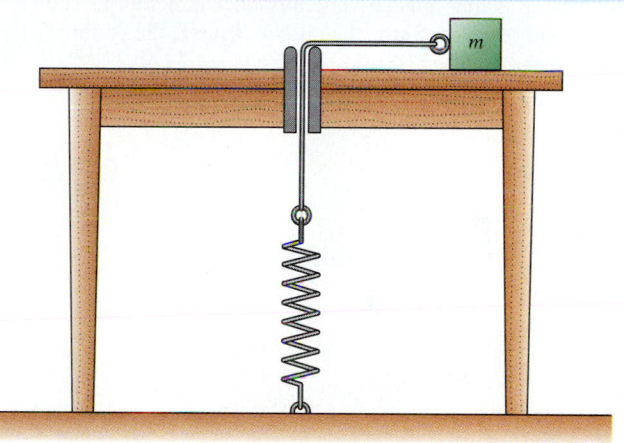

8.51 *(3) Weighted revolution.* The apparatus illustrated below is similar to that described in Problem 8.50. The block on the table is set into motion in a circular orbit of radius *r* around the center hole, as in Problem 8.50. Here, however, the string links the block of mass *m* on the table to an equal mass *m* hanging from its lower end rather than to a spring. **(a)** Find the speed *v* and the angular speed Ω of the block in its circular orbit. **(b)** Find *K*, *U*, and *E* as functions of *r*. Take $U(r) = 0$ at $r = 0$.

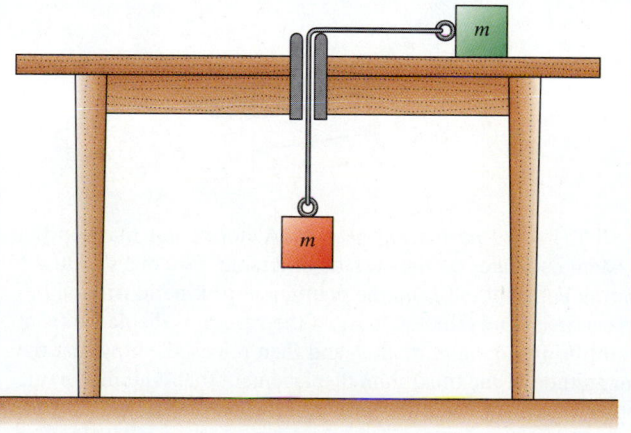

8.52 *(3) Jumping jack, II.* Consider the system illustrated in Problem 8.30 in a general way. When you place the ball on the spring, the weight of the ball compresses the spring a distance mg/k. Then you press down on the system, compressing the spring a further distance l. Finally, you release the system and the ball rises to a height D above its position on the compressed spring. **(a)** Show that

$$D = \frac{k}{2mg}\left(l + \frac{mg}{k}\right)^2.$$

(b) Suppose that the mass of the ball is small and the spring is stiff, so $mg/k \ll l$. Find the value of D. If you have worked Problem 8.30, compare the two results.

8.53 *(3) Jumping jack, III.* Two small blocks of equal mass m are joined together by a spring of force constant k and negligible mass. Starting at the relaxed configuration of the system, the two blocks are squeezed together by an amount δ. The system is then placed vertically on a tabletop, as shown below, and released. Show that, if $\delta > 3mg/k$, the system will jump off the table.

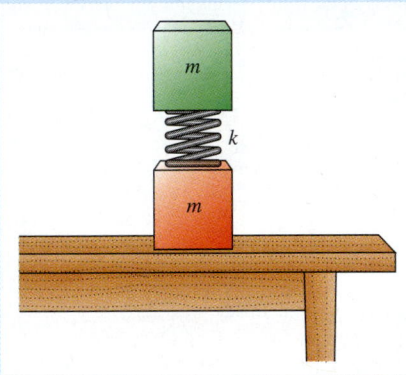

8.54 *(3) Shuttle.* The track illustrated below consists of two frictionless circular arcs of height h joined by a flat section of length h. The coefficient of friction μ_k of the flat section is small but not zero. If a small block is released from the upper left corner of the track, as shown, how many one-way trips will it make across the flat section before coming to rest?

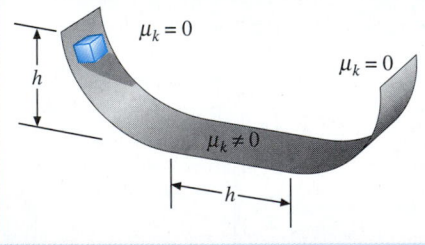

8.55 *(3) Aye, there's the rub! II.* A horizontal mass-spring system oscillates on the surface of a table. The mass is m, the spring constant is k, and the coefficient of kinetic friction between mass and tabletop is μ_k. If the system is displaced from equilibrium by an amount A and then released, show that the magnitude of the maximum displacement that it attains on the other side of the equilibrium point is

$$x_{\max} = A - \frac{2\mu_k mg}{k}.$$

8.56 *(3) Potential energy of a harmonic oscillator.* The displacement of the mass in a mass-spring system is $x = A \cos \omega t$. **(a)** Find the potential energy of the system as a function of time. **(b)** The kinetic energy is given as a function of time by the expression $K = \frac{1}{2}kA^2 \sin^2 \omega t$; see Problem 7.44. Show that the system satisfies the principle of conservation of mechanical energy.

8.57 *(3) Loading the scale, I.* A block of mass m is attached to a spring scale, as in part a of the illustration below. The block is held at $\mathbf{y} = 0$, where the spring is relaxed and the scale reads zero. The block is then released. The maximum scale reading corresponds to the lowest position reached by the block, at $\mathbf{y} = -Y\hat{\mathbf{y}}$ (where the direction of its motion reverses); see part b of the illustration. **(a)** What is the change ΔU_g in the gravitational potential energy of the mass-spring system? **(b)** What is the change ΔU_s in the potential energy of the spring? **(c)** What is the overall change ΔU in the potential energy of the system? **(d)** Use the energy conservation principle to show that, if friction is negligible, then $Y = 2mg/k$ and the maximum scale reading is $2mg$, or twice the weight of the block. **(e)** Describe qualitatively the subsequent motion of the system.

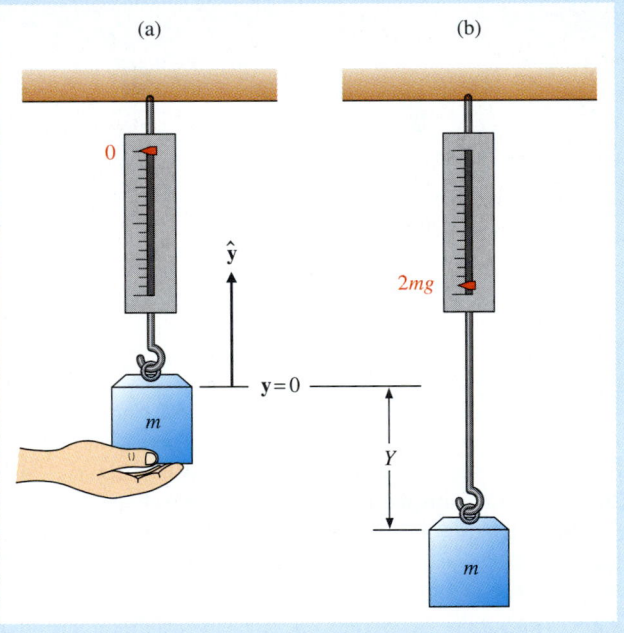

8.58 *(3) Loading the scale, II.* **(a)** Consider the scale of Problem 8.57 at the moment when the scale indicator moves downward past the mark mg. What is the change in the potential energy of the system from its initial value? **(b)** What is the kinetic energy of the block at that moment? **(c)** Compare the result of part **b** with that obtained for the horizontal mass-spring oscillator in Equation 8.5.

8.59 *(3) Differential pulley.* The system illustrated on the next page, called the *differential pulley*, serves the same purpose as the block and tackle (Problem 8.25). However, it is possible to obtain any mechanical advantage desired without the need for a messy network of rope. The upper part of the system consists of two pulleys of radius r_1 and r_2. They are joined rigidly together and rotate as a unit on the same axle. An endless chain is threaded in two loops around the upper pulleys; one loop passes through the lower pulley as shown. In the grooves of the upper pulleys are teeth that engage the chain links and prevent slipping. The weight is raised or lowered by pulling on the free loop of chain. Use the generalized work-energy theorem to show that, neglecting friction, the mechanical advantage of the system is

$$\frac{mg}{F} = \frac{1}{1 - \dfrac{r_1}{r_2}}.$$

This can be quite large if r_1 is only slightly less than r_2.

GROUP C

8.60 *(2) A radical force?* A force is represented by the function $\mathbf{F}(x, y) = cxy^2(\hat{\mathbf{x}} + \hat{\mathbf{y}})$, where c is a constant of dimensions [force]/[length]^3. Show that, although $\mathbf{F}$ is a function of position only, it is not conservative. To do so, evaluate the work done by the force around the square path $(0, 0) \to (1, 0) \to (1, 1) \to (0, 1) \to (0, 0)$. (Hint: In integrating along a path parallel to the x axis, remember that y has a constant value; in integrating along a path parallel to the y axis, remember that x has a constant value.)

8.61 *(3) Pyramid building made a little easier.* The system shown below is just like that of Figure 8.3 and Example 8.1, except that here the force $\mathbf{F}_a$ makes an angle β with the plane. **(a)** If the coefficient of friction is μ_k, show that the work required to raise the block to a height h is

$$W = mg\,\frac{h + \mu_k b}{1 + \mu_k \tan \beta}.$$

subject to the condition $\beta < 90° - \alpha$. **(b)** Show that pulling at an angle, as shown in the figure below, reduces the work required to do the job.

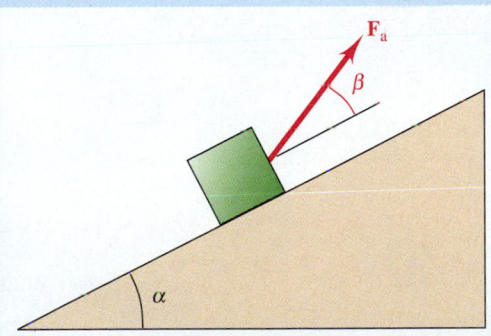

8.62 *(3) Springy pendulum, II.* As in Problem 8.45, a pendulum is made of a bob of mass m suspended from a spring of length l and force constant k. The pendulum is raised to the horizontal position (with the spring relaxed) and released. **(a)** This time, do not assume that $mg/kl \ll 1$, but only that the spring is not so weak that the pendulum's path when

the spring is vertical deviates appreciably from horizontal. Show that the extension of the spring is then

$$\delta = \frac{3mg}{4k} - \frac{l}{4}\left[1 - \sqrt{1 + 18\frac{mg}{kl} + \left(\frac{3mg}{kl}\right)^2}\,\right].$$

(b) Why are you justified in discarding the other quadratic solution, which leads to a positive sign immediately in front of the square root in the preceding equation? **(c)** Now suppose that the spring is stiff enough that the approximation $mg/kl \ll 1$ is applicable. Expand the square root and show that you obtain the simple solution of Problem 8.45. (Hint: Remember that for $\alpha \ll 1$, you can write $\sqrt{1 + \alpha} \simeq 1 + \frac{1}{2}\alpha$.) **(d)** Use the numerical values given in Problem 8.45 to find the exact value of δ. Compare your answer with the approximate value you obtained in Problem 8.45.

8.63 *(3) Look out, Goliath! III.* As in Problem 8.42, a stone tied to the end of a string of length l whirls in a vertical circle. Again the speed v_t at the top of the circle is just sufficient to keep the string taut. Compare the period of revolution T with **(a)** the time t required for the stone to fall straight down from the top to the bottom of the circle, beginning with a downward vertical velocity of magnitude v_t, and **(b)** the half-period (one-way swing time) $\tau/2$ of the stone and string oscillating as a small-amplitude pendulum. Hint: You will need to use the value of the definite integral

$$\int_0^{2\pi} \frac{d\theta}{\sqrt{3 - 2\cos\theta}} = 4.04.$$

8.64 *(3) Aye, there's the rub! III.* As in Problem 8.55, a mass-spring system oscillates with the mass resting on a table. The coefficient of friction is μ_k. **(a)** Call the initial oscillation amplitude A_0, and show that after N full cycles the amplitude has diminished to $A_N = A_0 - 4N\mu_k mg/k$. **(b)** Suppose that the energy loss per cycle, ΔE, is small compared with the initial total energy E_0, so the system oscillates many times before coming to rest. Find the energy loss ΔE_N during the Nth cycle and the corresponding total energy E_N, and show that $\Delta E_N/E_N$ is a constant independent of N, both for N fairly small and for N near its limiting value.

9

Linear Momentum

Left: The rocket isn't "pushing against the ground." Why is it accelerating?

That momentum . . . nothing has been able to resist.

—EDMUND BURKE, *Reflections on the Revolution in France*

SECTION 9.1 Introduction

In this chapter, we study two important physical quantities: *impulse* and *linear momentum*, defined in Section 9.2. The relation between impulse and linear momentum is analogous to that between work and energy. Closely connected to the relation between impulse and linear momentum is the *conservation of linear momentum*, which is the subject of Section 9.3. The conservation of linear momentum, like the conservation of energy, is a powerful analytical tool as well as a foundation stone in the edifice of physics.

Mechanical systems in which linear momentum is conserved constitute a large and varied class. We will use the conservation of linear momentum to study *collisions*—events in which two bodies exert forces on one another over an interval of time. Collisions are very common events, and they have enormous diversity. The interaction between billiard balls is a collision. So is the encounter of a space probe with the planet Jupiter as the probe swings by on its way to another planet farther out; there is a collision without actual contact because Jupiter and the space probe exert gravitational forces on each other. At the other end of the scale of size, experiments in which elementary particles collide are the most important means used to study the fundamental building blocks of matter.

In this chapter, we develop the basic concepts of collision theory in a general way but concentrate on applications in one dimension. In Chapter 10, we will be able to consider collisions from the standpoint of both linear momentum and total mechanical energy. From that standpoint, we will consider two-dimensional collisions in the important special case in which energy is conserved.

Linear momentum conservation is the principle underlying rocket propulsion. This important application is the subject of Section 9.4. Closely connected with linear momentum conservation and collision theory is the concept of *center of mass*, which is the subject of Section 9.5.

In Section 9.6, we reformulate Newton's laws of motion from the point of view of linear momentum conservation. This reformulation puts the definitions of force and mass on a sounder logical basis than that developed in Chapter 3.

SECTION 9.2 Impulse and Linear Momentum

We began our study of work and energy in Chapter 7 by defining the work done by a force $\mathbf{F}(s)$ on a body or a system of bodies. That definition may be expressed most generally in the form of Equation 7.13a:

$$W \equiv \int_{s_i}^{s_f} \mathbf{F}(s) \cdot d\mathbf{s}.$$

That is, work is an integral of force over space. We found that, under circumstances that can be precisely specified, work results in a change in the total mechanical energy of the body or system. The connection between work and energy—the generalized work-energy theorem—has proved to be a powerful tool for studying the motion of systems. But, in order to use this tool, we must express the force **F** as a function of position s.

Sometimes, however, a force exerted on a body is expressed as a function of *time* rather than position. In analogy to Equation 7.13a, we define the **impulse J** imparted to the body by the force $\mathbf{F}(t)$ over the time interval beginning at t_i and ending at t_f:

$$\mathbf{J} \equiv \int_{t_i}^{t_f} \mathbf{F}(t)\, dt. \tag{9.1}$$

Thus, impulse is an integral of force over time. The unit of impulse is N · s; it has no special name. The scientific meaning of *impulse* is closely related to the commonplace meaning; impulse is what you feel when you catch a hard-hit baseball.

There is an important mathematical difference between impulse and work. Work is a scalar quantity because it depends on the dot product of two vectors (Section 7.2). Impulse is a vector quantity because it depends on the product of a vector and a scalar.

By using Newton's second law in the form $\mathbf{F}(t) = m\mathbf{a}(t)$, we can write Equation 9.1 as

$$\mathbf{J} = \int_{t_i}^{t_f} m\mathbf{a}(t)\, dt.$$

Evaluating the integral yields

$$\mathbf{J} = m(\mathbf{v}_f - \mathbf{v}_i), \tag{9.2a}$$

or

$$\mathbf{J} = m\,\Delta\mathbf{v}. \tag{9.2b}$$

The quantity $m\,\Delta\mathbf{v}$ is not unfamiliar to you. When you catch a hard-hit baseball, your arms recoil, acquiring a change of velocity $\Delta\mathbf{v}$; you must exert some muscular effort to bring them to rest again. We define the **linear momentum p** (often shortened to **momentum**) to be

$$\mathbf{p} \equiv m\mathbf{v}. \tag{9.3}$$

We use this definition to rewrite Equation 9.2a as

$$\mathbf{J} = \mathbf{p}_f - \mathbf{p}_i, \tag{9.4a}$$

or

$$\mathbf{J} = \Delta\mathbf{p}. \tag{9.4b}$$

In these equations, $\mathbf{p}_f$ is the momentum of the system at time t_f and $\mathbf{p}_i$ is the momentum at time t_i; $\Delta\mathbf{p}$ is thus the change in momentum of the system over the time interval from t_i to t_f. Either of the two equations expresses the fact that *the impulse imparted to a system is equal to the change in the linear momentum of the system*. This statement is called the **impulse-momentum theorem**. Because its derivation is analogous to that of the work-energy theorem, you will not be surprised that the connection the impulse-momentum theorem makes between impulse and momentum is analogous to the connection between work and energy made by the work-energy theorem.

You can see from the definition $\mathbf{p} \equiv m\mathbf{v}$ that the unit of momentum is the product of the unit of mass and the unit of velocity. The SI unit of momentum is thus kg·m/s, which has no special name. From Equation 9.4b it follows that the units of impulse and momentum must be identical. Can you show that this is so?

Strictly speaking, the impulse-momentum theorem is valid only if the force $\mathbf{F}(t)$ in Equation 9.1, which defines impulse, is the *net* force acting on the body. Consider what happens, however, when a baseball bat hits a pitched ball. During the brief period of contact between bat and ball, the force exerted on the ball by the bat is very much greater than the force $m\mathbf{g}$ exerted on the ball by gravity. Thus, *ignoring $m\mathbf{g}$ in calculating the impulse imparted to the ball during its brief interaction with the bat yields an excellent approximation*. This statement is true even though $m\mathbf{g}$ is the major force exerted on the ball during all but a tiny part of its flight from pitcher to batter to

When Must We Specify "Linear"?

Linear momentum is always given its full name in contexts where there is a possibility of confusion between it and a different, equally important physical quantity called angular momentum (Chapter 13). Otherwise, the short name *momentum* is usually used for linear momentum. The name *angular momentum* is never shortened.

FIGURE 9.1 High-speed photograph
of a bat hitting a baseball, showing
the deformation due to the collision.

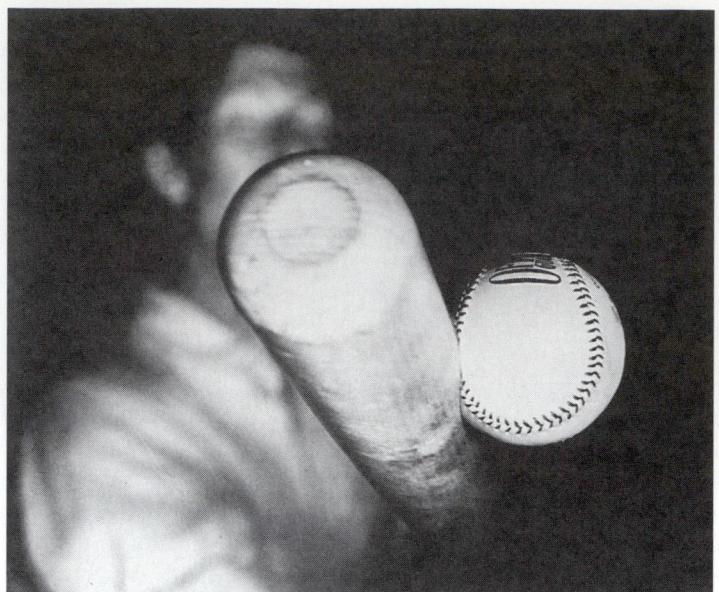

outfielder. The interaction between ball and bat is called a *collision*. For some collisions,
it is possible to consider all the forces involved, but often, as is the case for the bat and
ball, there is no need to do so.

In the most familiar type of collision, two solid bodies come into direct contact with
one another, as the bat does with the ball or as happens when two billiard balls collide.
In such a **contact collision**, the force exerted on each body by the other results in
deformation of the two bodies (Figure 9.1). Most (but not all) solids are more or less
elastic. That is, they resist deformation to some degree and tend to return to their
undisturbed shapes. Each of the two bodies in a contact collision exerts a force on the
other. The direction of these forces is such that the speed of each body relative to the
other diminishes and becomes zero at the instant of maximum deformation. If the
collision is elastic, the deformation subsequently decreases, and each body is accelerated
away from the other. Finally, the contact is terminated and the collision is over. The
magnitude of the contact force exerted by each body on the other is shown schematically
in Figure 9.2a. Figure 9.2b shows the effect on the velocity of one of the bodies of the
contact force exerted on it by the other body.

Colliding bodies may interact without actually touching each other, as when a space
probe makes a swing around Jupiter en route from the earth to Saturn; the interaction
forces are gravitational (Figure 9.3). For the space probe on its way to Saturn, the
Jupiter encounter—the interval during which Jupiter exerts a gravitational force on the
probe sufficient to exert a strong influence on its trajectory—lasts for a few days of a
voyage that takes years. In calculating the trajectory of the probe during the encounter,
we can neglect the gravitational force exerted on the probe by the sun (at least to first

(a)

Schematic shape of one colliding body

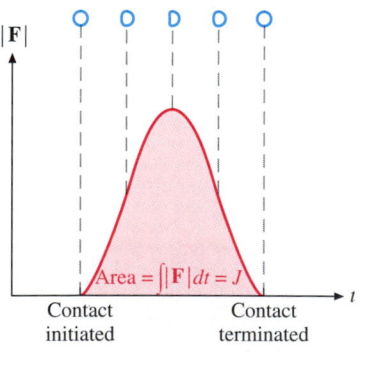

Area $= \int |\mathbf{F}| dt = J$

Contact
initiated

Contact
terminated

(b)

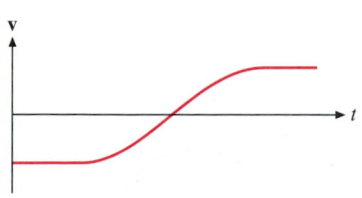

FIGURE 9.2 (*a*) Schematic plot as a function of time of the magnitude of
the force exerted by each body on the other in a two-body contact collision. The area under the curve is the magnitude of the impulse imparted
by each body to the other, according to Equation 9.1. (*b*) Schematic plot
as a function of time of the velocity of one of the colliding bodies. The
slope of this curve is the acceleration of the body; the instant of greatest
acceleration is the instant at which the greatest force is applied by the
other body.

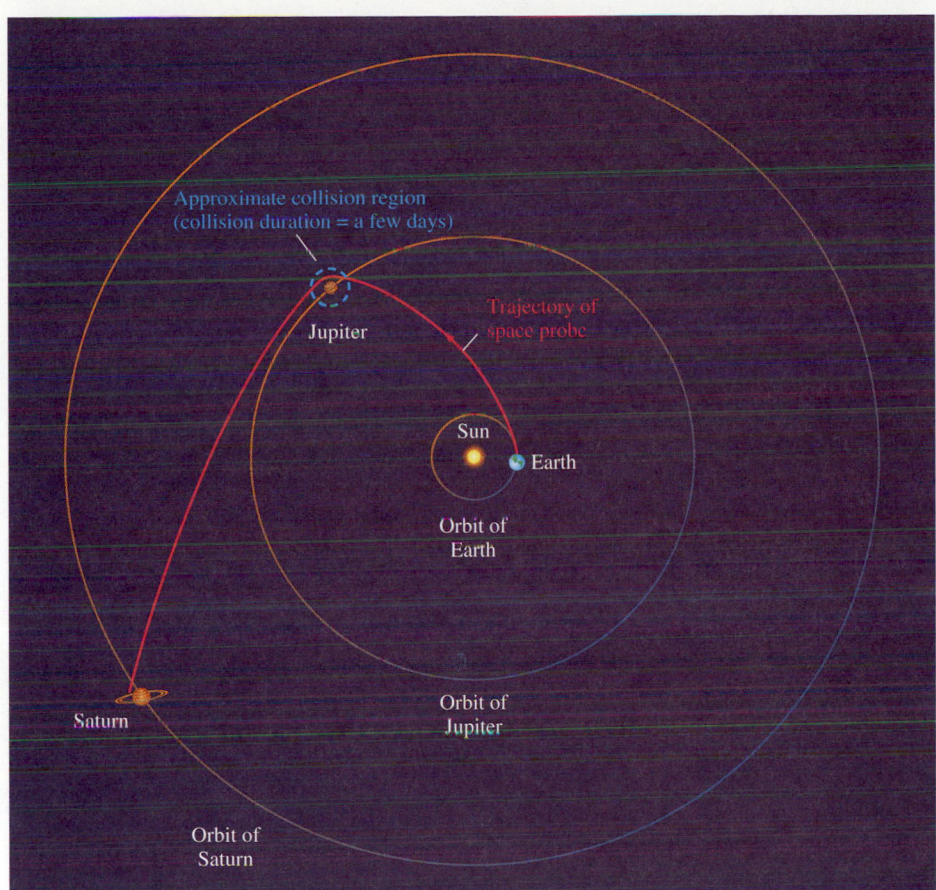

Approximate collision region
(collision duration = a few days)

Jupiter

Trajectory of
space probe

Sun

Earth

Orbit of
Earth

Orbit of
Jupiter

Saturn

Orbit of
Saturn

Grand Tour

From 1977 to 1989, the spacecraft *Voyager II* carried out a spectacular series of collisions in successive encounters with the planets Jupiter, Saturn, Uranus, and Neptune, as well as with several of their moons, before leaving the solar system and heading into interstellar space. At each encounter, the spacecraft relayed back to earth more information about the surfaces of the planets, their moons, and their rings than had been accumulated in all the previous observations.

approximation) even though it is the only significant force during most of the rest of the voyage. The encounter between the probe and Jupiter is of short duration and can be regarded as a collision.

EXAMPLE 9.1

In a pinball machine (Figure 9.4), the ball rolls down the inclined baseboard, accelerating as it goes. At an instant when its speed is 0.85 m/s, it hits a flipper. At the same instant, the player actuates the flipper and the ball moves off with the direction of its velocity reversed and a speed of 9.2 m/s. The mass of the ball is 57 g. **(a)** What is the momentum of the ball just before it hits the flipper? **(b)** What is its momentum just after it leaves the flipper? **(c)** What is the impulse imparted to the ball by the flipper?

FIGURE 9.4

SOLUTION:

(a) What is the momentum of the ball just before it hits the flipper?

Because all motion takes place along a line, you may as well treat the problem as one-dimensional. In one dimension, Equation 9.3 takes the form $p \equiv mv$. Take the initial direction of motion, downward along the baseboard of the machine, to be the negative direction. You use Equation 9.3 to find the initial momentum of the ball just before its collision with the flipper:

$$p_i = mv_i = 5.7 \times 10^{-2} \text{ kg} \times (-0.85 \text{ m/s})$$
$$= -4.9 \times 10^{-2} \text{ kg·m/s}.$$

The negative direction of the momentum corresponds to the negative direction of the velocity to which it corresponds.

(b) What is the momentum of the ball just after it leaves the flipper?

You write

$$p_f = mv_f = 5.7 \times 10^{-2} \text{ kg} \times 9.2 \text{ m/s} = 52 \times 10^{-2} \text{ kg·m/s}.$$

As a result of the collision, the direction of the ball has been reversed and the magnitude of its momentum considerably increased.

Section 9.2   Impulse and Linear Momentum ■ **221**

(c) What is the impulse imparted to the ball?

The change in momentum is the result of the impulse imparted to the ball by the flipper. You have no way of evaluating the force $\mathbf{F}(t)$ exerted by the flipper on the ball, and you therefore cannot use Equation 9.1 to evaluate the impulse. This is a very common situation in contact collisions. However, you can use Equation 9.4a, in the one-dimensional form $J = p_f - p_i$, to evaluate the impulse:

$$J = 52 \times 10^{-2} \text{ kg·m/s} - (-4.9 \text{ kg·m/s}) = 57 \text{ N·s}.$$

The collision of Example 9.1 satisfies the requirement that we have established for a ''short time'' collision. Before and after the collision, the ball experiences a net force downward along the inclined baseboard. That is why it accelerates downward along the baseboard. Over a sufficient time, the downward force will impart a considerable impulse to the ball. This force does not cease to operate during the brief collision of the ball with the flipper. But it is no longer the *net* force exerted on the ball. During most of the collision interval, indeed, the downward force is negligibly small compared with the quite large force exerted by the flipper. That is why we can neglect all forces other than that exerted by the flipper when we calculate the change in the momentum of the ball during the collision.

Conservation of Linear Momentum

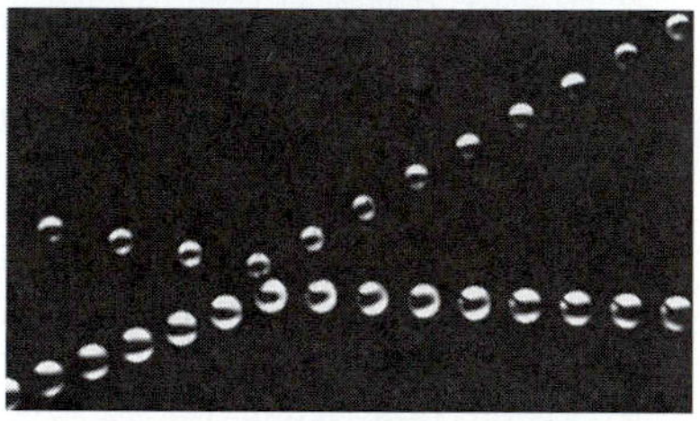

When two bodies collide, each imparts an impulse to the other. We continue our study of impulse and momentum by considering (1) the effect of the collision on each of the two bodies and (2) the effect of the collision on the system consisting of both bodies.

For simplicity, we assume that each body is subject to negligible net force both before and after the collision. This situation is exemplified by the system shown in the stroboscopic photograph of Figure 9.5. Two hard balls suspended from very long strings come in contact with one another. The collision is so brief that it occurs in the $\frac{1}{30}$-s interval between successive light flashes, and so we do not quite see the balls in contact. During the very short time over which the balls are in contact and exerting forces on one another, the forces vary rapidly in an unknown manner. Even though we do not know the forces explicitly as functions of time, we can still use Equation 9.1 to express the impulses $\mathbf{J}_1$ and $\mathbf{J}_2$ imparted to each of the bodies. Let us call the force exerted by body 2 on body 1 $\mathbf{F}_{2\to1}(t)$ and the force exerted by body 1 on body 2 $\mathbf{F}_{1\to2}(t)$. We then write

$$\mathbf{J}_1 = \int_{t_i}^{t_f} \mathbf{F}_{2\to1}(t)\, dt \quad \text{and} \quad \mathbf{J}_2 = \int_{t_i}^{t_f} \mathbf{F}_{1\to2}(t)\, dt. \tag{9.5a,b}$$

At any instant, the two forces are equal and opposite in accordance with Newton's third law. Thus we have

$$\mathbf{F}_{1\to2}(t) = -\mathbf{F}_{2\to1}(t).$$

Substituting this value for $\mathbf{F}_{1\to2}(t)$ into Equation 9.5b gives us

$$\mathbf{J}_2 = -\int_{t_i}^{t_f} \mathbf{F}_{2\to1}(t)\, dt.$$

We then compare this last equation with Equation 9.5a to obtain

$$\mathbf{J}_2 = -\mathbf{J}_1; \tag{9.6}$$

that is, *the impulse imparted to each body in a two-body collision is equal and opposite to the impulse imparted to the other.* Equation 9.6 expresses Newton's third law in the form of a time integral.

As a result of the collision, each of the two bodies experiences a change of momentum. To find the relation between these changes, we apply the impulse-momentum

FIGURE 9.5 A collision between two balls that approach one another from the left of the photograph and rebound toward the right. The time interval between flashes is $\frac{1}{30}$ s.

Notation Convention

Quantities pertaining to particular parts of a system are labeled with subscripts such as $_1$ and $_2$; quantities pertaining to the entire system do not carry such labels.

theorem in the form of Equation 9.4b, $\mathbf{J} = \Delta\mathbf{p}$, to each body. This gives us

$$\mathbf{J}_1 = \Delta\mathbf{p}_1 \quad \text{and} \quad \mathbf{J}_2 = \Delta\mathbf{p}_2. \tag{9.7a,b}$$

It follows immediately from Equations 9.6 and 9.7 that

$$\Delta\mathbf{p}_2 = -\Delta\mathbf{p}_1; \tag{9.8}$$

that is, *as a result of the collision, each body experiences a momentum change equal to and opposite the momentum change experienced by the other body.*

Finally, consider the change in linear momentum experienced by the entire system consisting of the two bodies. The quantities $\Delta\mathbf{p}_1$ and $\Delta\mathbf{p}_2$ are vectors, and the momentum change $\Delta\mathbf{p}$ of the entire system is the vector sum

$$\Delta\mathbf{p} = \Delta\mathbf{p}_1 + \Delta\mathbf{p}_2. \tag{9.9}$$

When we combine this equation with Equation 9.8, we obtain $\Delta\mathbf{p} = \Delta\mathbf{p}_1 - \Delta\mathbf{p}_1$, or

$$\Delta\mathbf{p} = 0; \tag{9.10a}$$

that is, *the linear momentum of a two-body system is unchanged by a collision taking place within it.* This is one statement of the **principle of conservation of linear momentum**, often called the **momentum-conservation principle** for short. Our present discussion is limited to two-body systems, but we will remove this limitation in Section 9.5.

It is sometimes useful to rewrite Equation 9.10a in terms of the initial (precollision) momentum of the system $\mathbf{p}_i$ and the final (postcollision) momentum $\mathbf{p}_f$. We have $\Delta\mathbf{p} = \mathbf{p}_f - \mathbf{p}_i$; Equation 9.10a immediately gives us

$$\mathbf{p}_f = \mathbf{p}_i. \tag{9.10b}$$

The momentum conservation principle is a very general one. Indeed, we know of no instances in which it is violated, although the definition of momentum must be generalized beyond $\mathbf{p} \equiv m\mathbf{v}$ in considering systems that do not conform to the rules of Newtonian mechanics (for example, the systems considered in Chapter 39).

The momentum-conservation principle is independent of the principle of conservation of total mechanical energy. Different criteria govern the applicability of the two principles to a given system:

- If mechanical energy is to be conserved in a system, all forces (both internal and external) acting on it must either be conservative (Section 8.2) or be workless constraints (Section 8.3).

- If linear momentum is to be conserved in a system, no net **external force** may act on it; that is, no force may act on the system from outside the system. We say that such a system is **isolated**. But **internal forces**—forces exerted by one part of the system on another part—have no bearing on whether the linear momentum of the system is conserved.

Whether a force is internal or external to a system depends on the way in which the system is chosen. Consider, for example, the system of Figure 9.6a, consisting of the masses m_1 and m_2 and the string that connects them. The force $\mathbf{T}_1$ exerted by the string on m_1 is an internal force, as is the force $\mathbf{T}_2$ exerted by the string on m_2. The force $\mathbf{F}$ is an external force, as are the gravitational forces $m_1\mathbf{g}$ and $m_2\mathbf{g}$ acting on the two bodies. (Don't be misled by the fact that the arrow representing $m_1\mathbf{g}$ is drawn so that it doesn't cross the system boundary. This is merely because force vectors are conventionally drawn with their tails fixed to the bodies on which they act. The significant point is that the force $m_1\mathbf{g}$ is exerted on m_1—and thus on the system as a whole—by the earth, which lies outside the system boundary.)

Now consider the system of Figure 9.6b, consisting of m_1 only. The forces $\mathbf{F}$ and $m_1\mathbf{g}$ are external forces, as they are in Figure 9.6a. Now, however, $\mathbf{T}_1$ is an external force as well because it is exerted on the system by the string, which lies outside the system boundary. We need not consider the forces $\mathbf{T}_2$ and $m_2\mathbf{g}$ at all because they are not exerted on the system.

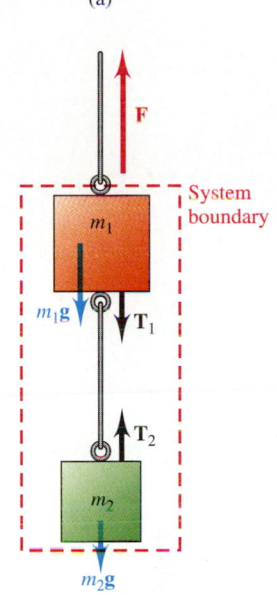

(a)

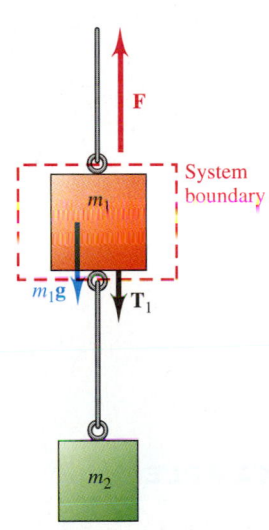

(b)

FIGURE 9.6 (*a*) The system is chosen to include masses m_1 and m_2 and the string connecting them. The forces $\mathbf{F}$, $m_1\mathbf{g}$, and $m_2\mathbf{g}$ are external forces; $\mathbf{T}_1$ and $\mathbf{T}_2$ are internal forces. (*b*) The system is chosen to include m_1 only. Here again, $\mathbf{F}$ and $m_1\mathbf{g}$ are external forces; now $\mathbf{T}_1$ also is an external force. The forces $\mathbf{T}_2$ and $m_2\mathbf{g}$ are not drawn; they are irrelevant because they do not act on the system or any part of it.

Collisions in One Dimension

As a first application of the principle of conservation of linear momentum, consider a collision between two bodies having masses m_1 and m_2. They move along the x axis and collide in such a way that the force exerted by each on the other lies exactly along the x axis. That is what we mean by a **head-on collision**; neither body ever experiences a force having a y or z component, and neither accelerates in any direction not along the x axis. We can therefore carry out the analysis in one dimension, using one-dimensional vector quantities. As you will see, one-dimensional collisions are considerably simpler to analyze than multidimensional collisions.

As is usually the case in collisions, we cannot evaluate the forces $F_{1 \to 2}$ and $F_{2 \to 1}$ exerted on each body by the other at any moment during the collision. Thus we cannot use Newton's second law to answer the question: How does the collision affect the motion of either body? However, we can arrive at the answer by using the momentum-conservation principle because the system comprising the two bodies is isolated. That is, either no net external force is acting on the system or else (like the gravitational force acting on a baseball while it is in contact with the bat) any such force is negligible for the duration of the collision. Because momentum is conserved, Equation 9.8 is valid. Expressed in one-dimensional vector notation, the equation takes the form

$$\Delta p_2 = -\Delta p_1.$$

The change in momentum of each body is the difference between its final momentum and its initial momentum. We can write

$$\Delta p_1 \equiv p_{1f} - p_{1i} = m_1 v_{1f} - m_1 v_{1i} = m_1(v_{1f} - v_{1i})$$

and

$$\Delta p_2 \equiv p_{2f} - p_{2i} = m_2 v_{2f} - m_2 v_{2i} = m_2(v_{2f} - v_{2i}).$$

Equating Δp_2 to $-\Delta p_1$ gives us

$$m_2(v_{2f} - v_{2i}) = -m_1(v_{1f} - v_{1i}). \tag{9.11a}$$

The term in parentheses on the left side of this equation is the change in velocity of body 2: $(v_{2f} - v_{2i}) = \Delta v_2$. Similarly, the term in parentheses on the right side of the equation is the change in velocity of body 1: $(v_{1f} - v_{1i}) = \Delta v_1$. We can thus express Equation 9.11a in the equivalent forms

$$m_2 \, \Delta v_2 = -m_1 \, \Delta v_1 \tag{9.11b}$$

and

$$\frac{\Delta v_2}{\Delta v_1} = -\frac{m_1}{m_2}. \tag{9.11c}$$

Equation 9.11c makes it evident that the changes in velocity of the two bodies are opposite in direction and have magnitudes in inverse ratio to their masses. If you played marbles as a child, your experience probably agrees in a general way with this statement.

EXAMPLE 9.2

In the physics lab, you do a collision experiment with two gliders on an air track, as shown in Figure 9.7. The first glider has mass $m = 0.70$ kg; by timing its passage over a known distance, you find that its initial velocity is $v_{1i} = 2.5$ m/s. The second glider has mass $m_2 = 0.30$ kg; you find its initial velocity to be $v_{2i} = -1.5$ m/s. After the collision, you find that the final

(a)

(b)

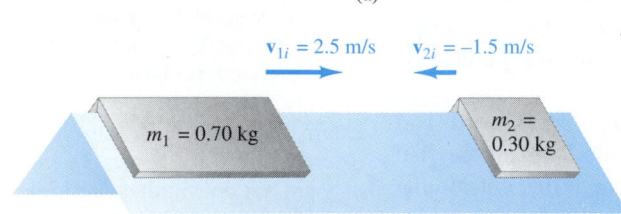

FIGURE 9.7 (*a*) The system before the collision.
(*b*) The system after the collision.

velocity of the first glider is $v_{1f} = 0.10$ m/s. Before using your direct observations of distance and time to calculate v_{2f}, the final velocity of the second glider, you decide to predict what it will be on the basis of the values of the three velocities you have already calculated. What value do you expect to find?

SOLUTION: You can solve Equation 9.11a for v_{2f}:

$$v_{2f} = -\frac{m_1}{m_2}(v_{1f} - v_{1i}) + v_{2i}$$

$$= -\frac{0.70 \text{ kg}}{0.30 \text{ kg}} \times (0.10 \text{ m/s} - 2.5 \text{ m/s}) + (-1.5 \text{ m/s}),$$

or

$$v_{2f} = 4.1 \text{ m/s}.$$

As a check on your calculation, you can compare the initial momentum of the two-body system with its final momentum and show that the two are indeed equal, as Equation 9.10b requires:

$$p_i = m_1 v_{1i} + m_2 v_{2i}$$

$$= 0.70 \text{ kg} \times 2.5 \text{ m/s} + 0.30 \text{ kg} \times (-1.5 \text{ m/s})$$

$$= 1.3 \text{ kg·m/s}.$$

$$p_f = m_1 v_{1f} + m_2 v_{2f}$$

$$= 0.70 \text{ kg} \times 0.10 \text{ m/s} + 0.30 \text{ kg} \times 4.1 \text{ m/s}$$

$$= 1.3 \text{ kg·m/s}.$$

This result is in agreement with the condition $p_f = p_i$, as it must be.

In this chapter we have been concerned mainly with collisions at the everyday scale of size. In Chapter 18, however, you will see that collisions on the molecular scale underlie the macroscopic sensation of heat, as well as the rich store of related *thermal phenomena*.

Avoiding a Pitfall: Momentum Conservation and Energy Conservation Are Not the Same Thing

As we have already noted, the conditions under which momentum conservation applies are different from the conditions under which the principle of conservation of mechanical energy is applicable (Section 8.3). It is worth reiterating the difference because many students tend at first to confuse the two conservation principles—which are quite distinct. In Table 9.1, the necessary conditions for the two principles are summarized. Make sure that you know the differences; more important, make sure that you understand the reasons that underlie the differences.

TABLE 9.1 Criteria for Applicability of Conservation Principles

Conservation Principle	Restrictions on Internal Forces	Restrictions on External Forces
Total mechanical energy (Section 8.3)	Must be conservative or must be workless constraints	Must be conservative or must be workless constraints
Linear momentum	No restrictions	Net force must be zero

Reaction Propulsion and Rockets

Most of the familiar vehicles of everyday life propel themselves by pushing on some external thing, which pushes back in accordance with Newton's third law. Wheeled vehicles push on road surfaces or tracks, boat propellers push on water, and propeller-driven airplanes push on the surrounding air. In the rocket engine, however, we have a system that operates most efficiently where there is no surrounding medium at all—in other words, where there is nothing to push against. To understand how rockets work, we must analyze the process of propulsion in a less intuitive but more general way, in terms of momentum conservation in an isolated system consisting of the vehicle itself and its expelled gases.

Rocket Propulsion

The propulsion produced by a rocket engine can be analyzed by applying the momentum-conservation principle. We suppose for simplicity that the vehicle is in deep space, far from the earth or any other body large enough to exert a significant gravitational force. For the purposes of the analysis, however, we assume that the vehicle is watched by an *inertial observer* standing on a small asteroid conveniently located nearby, as shown in Figure 9.8.

The engine expels a stream of hot gas at a constant speed v_{ex} *relative to the rocket nozzle*. This quantity v_{ex} is called the **exhaust speed**. (The term ''exhaust velocity'' is often used but is misleading because v_{ex} expresses a magnitude only. It is thus a speed, not a velocity.) Because gas is continuously expelled, the vehicle's mass changes at a rate dm/dt. (Note that dm/dt is a negative quantity because the vehicle mass is decreasing.) Both v_{ex} and dm/dt are constants whose values are determined by the design characteristics of the engine.

We now consider the momentum of the system at the beginning and the end of a time interval dt. The system we choose consists of the vehicle and all the fuel not yet expelled at a certain time t. At time t, represented in Figure 9.8a, the mass of the vehicle is m and its velocity relative to the observer on the asteroid is v. The momentum of the vehicle at t is thus

$$mv.$$

At a time dt later, depicted in Figure 9.8b, the mass of the vehicle has decreased and has the new value $m + (dm/dt)\, dt = m + dm$. The velocity of the vehicle relative to the observer has increased to the new value $v + dv$. The momentum of the vehicle at $t + dt$ is thus

$$(m + dm)(v + dv).$$

If we are to apply the momentum-conservation principle, we must consider the *same* system at time $t + dt$ as we did at time t. During the interval dt, however, the rocket expels a quantity of gas, and we must also take the momentum of this gas into account in order to evaluate the total system momentum at $t + dt$. As the mass of the vehicle changes at the rate dm/dt, the mass of the gas that has been expelled changes at the rate $-dm/dt$. Thus the mass of the gas expelled during the time interval dt is $(-dm/dt)\, dt = -dm$. The velocity of the gas relative to the observer on the asteroid is $v - v_{ex}$.

(a)   (b)

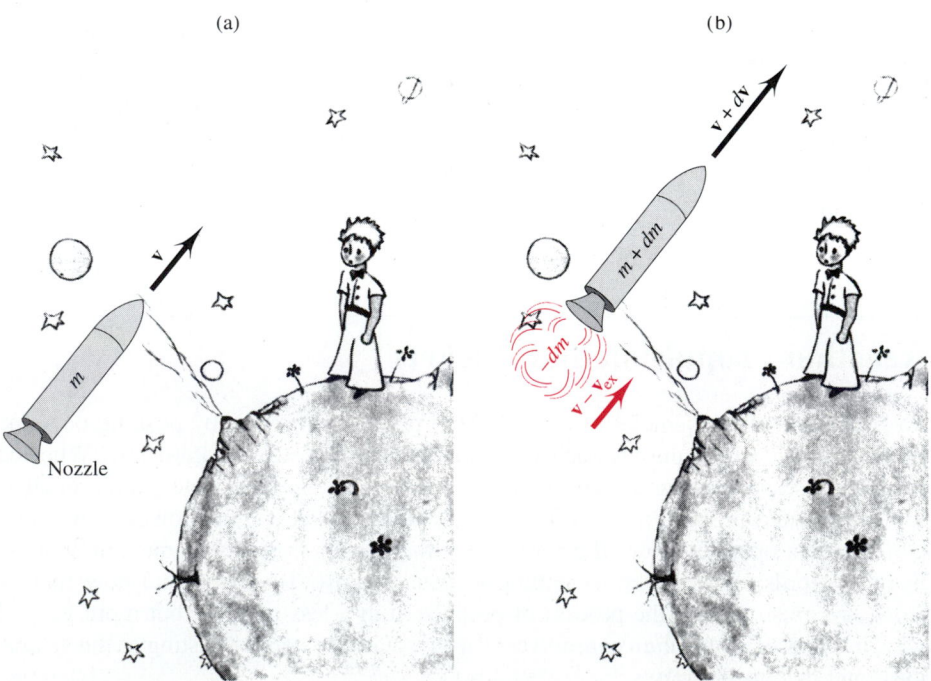

FIGURE 9.8 An inertial observer on a small asteroid watches a space vehicle pass by with its engine operating. (*a*) At time t, the mass of the vehicle is m and its velocity as seen by the observer is v; its momentum is mv. (*b*) At time $t + dt$, the mass of the vehicle is $m + dm$ and its velocity as seen by the observer is $v + dv$. The mass of the gas expelled during the infinitesimal time interval dt is $-dm$, and its velocity as seen by the observer is $v - v_{ex}$. The momentum of the system is $(m + dm)(v + dv) - dm(v - v_{ex})$. Note that, although the exhaust gas always moves backward with respect to the rocket nozzle, its velocity relative to the observer may be negative or positive depending on whether $|v| < v_{ex}$ or $|v| > v_{ex}$.

The momentum of the gas at $t + dt$ is therefore the product

$$-dm(v - v_{ex}).$$

The momentum of the entire system at $t + dt$ is the sum

$$(m + dm)(v + dv) - dm(v - v_{ex}).$$

Because the rocket and its expelled gas constitute an isolated system—that is, one on which no external forces act—we can use the momentum-conservation principle to equate the final momentum at $t + dt$ and the initial momentum at t. This gives us

$$(m + dm)(v + dv) - dm(v - v_{ex}) = mv.$$

Expanding the left side of this equation, we obtain

$$mv + m\, dv + dm\, v + dm\, dv - dm\, v + dm\, v_{ex} = mv.$$

This simplifies to

$$m\, dv + dm\, dv + v_{ex}\, dm = 0.$$

The quantity $dm\, dv$ is the product of two infinitesimals and so may be neglected. We can thus collect all the terms involving mass on one side of the equation and all the terms involving velocity on the other. This gives

$$\frac{1}{v_{ex}}\, dv = -\frac{dm}{m}. \tag{9.12}$$

Equation 9.12 relates the change in vehicle mass to the change in vehicle velocity over an infinitesimal time interval dt. In order to find the relation over a finite time interval, we must integrate both sides of the equation between the limits of that time interval. As time elapses from an initial t_i to a final t_f, the velocity of the vehicle changes from v_i to v_f. In the same time interval, the mass of the vehicle changes from m_i to m_f. So we have

$$\frac{1}{v_{ex}} \int_{v_i}^{v_f} dv = -\int_{m_i}^{m_f} \frac{dm}{m}. \tag{9.13}$$

(Because v_{ex} is a constant, it can be taken outside the integral.) Carrying out the integrations (Table 7.1) gives

$$\frac{1}{v_{ex}} v \Big|_{v_i}^{v_f} = -\ln m \Big|_{m_i}^{m_f}.$$

Evaluating the left side of this equation yields

$$\frac{1}{v_{ex}}(v_f - v_i) = \frac{\Delta v}{v_{ex}}.$$

The right side yields $-\ln m_f + \ln m_i$, which is more conveniently expressed in the form

$$\ln \frac{m_i}{m_f}.$$

Equating the left and right sides and solving for Δv gives the result

$$\Delta v = v_{ex} \ln \frac{m_i}{m_f}. \tag{9.14}$$

The exhaust speed v_{ex} is a positive number, and the logarithm is positive because $m_i > m_f$. Thus, as you would expect, Δv is positive. Equation 9.14 is sometimes called the **Tsiolkovskii formula**.

Equation 9.14 shows that it is important in rocket design to maximize the exhaust speed because the desired speed increase Δv is directly proportional to v_{ex}. The quantity m_f/m_i is called the **gross payload ratio**; it is the ratio of the vehicle mass when the fuel is exhausted to the fully fueled vehicle mass. The reciprocal of the gross payload ratio is the argument of the natural logarithm in Equation 9.14. Therefore, to obtain the

A Rocket-Propulsion Pioneer

The Tsiolkovskii formula was first derived in 1897 by the Russian engineer Konstantin Eduardovich Tsiolkovskii (1857–1935). Debarred by his deafness from entering school as a child, Tsiolkovskii educated himself and went on to a distinguished career. He is best remembered as a pioneer in modern rocket propulsion technology.

largest possible velocity increase in burning all the fuel, the gross payload ratio must be minimized; that is, a lot of fuel must be burned. Practical considerations place a lower limit of somewhere between 0.2 and 0.1 on the gross payload ratio. That is, not more than from 80% to 90% of the launch mass of the vehicle can be fuel because it is simply not possible to reduce the mass of the engine, the fuel tanks, and the overall structure of the vehicle without limit. (This difficulty can be circumvented to a degree by using *multistage rockets*, in which relatively small but complete rocket-driven vehicles ride piggyback on larger ones. This approach is the subject of Problem 9.58.)

Thrust

So far, we have taken the system under consideration to be the rocket-propelled vehicle together with the gas it expels. This system is an isolated one, and no external forces act on it. Now, we make a subtle shift in point of view and consider the rocket-propelled vehicle alone as the system under consideration. We take this new point of view when we wish to account for the acceleration of the vehicle by means of an equivalent force acting on it. We begin with Equation 9.12, which we rearrange in the form

$$-v_{ex}\, dm = m\, dv.$$

Dividing through by dt gives

$$-v_{ex}\frac{dm}{dt} = m\frac{dv}{dt}.$$

We have defined dm/dt as a negative quantity (p. 226); we can therefore rewrite the left side of this equation as $v_{ex}\,|dm/dt|$. The quantity on the right side of the equation has the form of a mass times an acceleration; we can use Newton's second law to substitute an equivalent force of magnitude F for the right side of the equation. This gives

$$v_{ex}\left|\frac{dm}{dt}\right| = F. \tag{9.15}$$

The quantity F is called the **thrust** of the rocket. You can think of the thrust as the force that would produce the acceleration dv/dt if it were exerted on the rocket by an imaginary towline. As the equation shows, the thrust is proportional to both the exhaust speed and the rate at which fuel is burned. If a vehicle is to be launched from the earth, the thrust must exceed the fully loaded weight, $m_i g$. Most manned space vehicle designs provide for a value of F that is three or four times the weight. For unmanned vehicles, still larger factors are used.

Specific Impulse

We can multiply Equation 9.15 through by $|dt/dm|$ to obtain

$$v_{ex} = \frac{F\, dt}{dm}. \tag{9.16}$$

The quantity $F\, dt$ is the impulse imparted to the vehicle during the time interval dt. Thus the right side of this equation is the impulse per unit mass dm of fuel expelled through the rocket nozzle and is called the **specific impulse**. According to the impulse-momentum theorem, specific impulse represents the momentum added to the vehicle per unit mass of fuel consumed. To optimize the design of a rocket engine, specific impulse must be maximized so as to get the most out of the fuel. Equation 9.16 shows that the specific impulse is simply equal to the exhaust speed of the expelled gas, which must likewise be maximized.

Good engine design comes close to the theoretical limit on v_{ex} set by the chemical energy per unit mass stored in the fuel. The energy per unit mass is maximized in turn by the use of fuel mixtures having low molecular masses and large reaction energies per molecule, two properties that fortunately tend to go together. In principle, the best possible mixture would be hydrogen and fluorine, which react to produce hydrogen fluoride. But fluorine and hydrogen fluoride are both so extremely toxic and corrosive that the combination is never used. The most widely used fuel mixture for large space boosters is hydrogen-oxygen, whose specific impulse ($v_{ex} \simeq 5100$ m/s) is only slightly less than that of hydrogen-fluorine.

Specific impulses very much greater than 5100 m/s can be attained with nonchemical propulsion systems. In one system that has been tested extensively, called an ion propulsion engine, the propellant gas (usually lithium or sodium vapor) is ionized and then accelerated by an electric field. Exhaust speeds more than 10^2 times as great as those possible with chemical fuels have been achieved. The possibility of achieving very high vehicle speeds with small amounts of propellant makes ion propulsion an attractive prospect for voyages much longer than those achieved to date, such as those required for interstellar missions.

EXAMPLE 9.3

An interplanetary probe is launched from a space station. The specific impulse of the propulsion system is 3000 N·s/kg. The fuel load constitutes 80% of the launch mass, and the vehicle proper an additional 16%, leaving 4% for the net payload—the equipment whose delivery to an outer planet is the purpose of the mission. **(a)** Relative to the space station, what will be the final speed of the probe? **(b)** What final speed can be achieved if the mass of the net payload is reduced by 25%?

SOLUTION:

(a) Find the final speed of the probe relative to the space station.

According to Equation 9.16, the exhaust speed v_{ex} is equal to the specific impulse, and so you have $v_{ex} = 3000$ m/s. The gross payload (vehicle plus net payload) is 20% of the launch mass. You set $v_i = 0$ because the probe is initially at rest with respect to the space station. Equation 9.14 then gives you

$$v_f = \Delta v = 3000 \text{ m/s} \times \ln \frac{1}{0.20} = 4830 \text{ m/s}.$$

(b) What final speed can be achieved if the mass of the net payload is reduced by 25%?

In the initial configuration of the interplanetary probe, the gross payload m_f constitutes 20% of the launch mass, and the net payload constitutes 4% of the launch mass. The gross payload is five times the net payload. If the net payload is reduced by 25%, the gross payload is thus reduced by only 5%—that is, one-fifth of 25%. In the new configuration, the gross payload is therefore 95% × 20% = 19% of the launch mass—a rather small reduction. The launch mass m_i, which is 25 times as great as the net payload in the initial configuration, is now reduced by one twenty-fifth of 25%—that is, by only 1%. The new launch mass is thus 99% of the launch mass in part **a**. Therefore Equation 9.14 yields

$$v_f = 3000 \text{ m/s} \times \ln \frac{0.99}{0.19} = 4950 \text{ m/s}.$$

The increase in the final speed is about 2.6% at a cost of one-quarter of the net payload. You can see why vehicle mass reduction is such a crucial aspect of rocket design.*

The Center of Mass

For every system of particles, a point called the *center of mass* can be defined. Because an extended body can be considered a continuous collection of point particles, a center of mass can be defined for extended bodies as well. The first part of this section concerns the definition of the center of mass, and the second part considers some of the mechanical properties of the center of mass that make it an important concept in mechanics. In particular, you will see the role played by the center of mass in applying the principle of momentum conservation and the laws of motion to systems that cannot be considered particles.

You probably have an intuitive concept of center of mass as a sort of "midpoint" or "balance point" of a body. For a simple, uniform body—say, a meter stick—the center of mass lies at the middle—that is, at the 50-cm mark of the stick. We now define the center of mass formally and devise a way to determine its position precisely.

---

*In James Michener's novel *Space*, the need to reduce mass provides a crucial turning point in the plot.

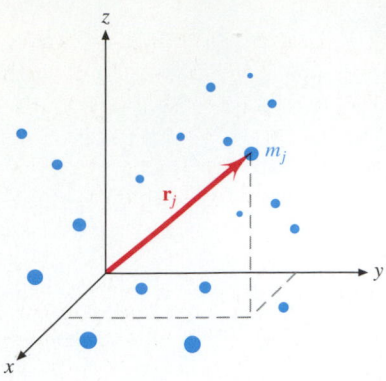

FIGURE 9.9 A system of n particles. The jth particle has mass m_j. With respect to the chosen coordinate system, its location is specified by the radius vector $\mathbf{r}_j = (x_j, y_j, z_j)$.

Figure 9.9 depicts a system of n particles with masses $m_1, m_2, m_3, \ldots, m_j, \ldots, m_n$. A coordinate system is chosen, and the location of each particle is specified by means of a position vector. For the jth particle, which is typical, this vector is $\mathbf{r}_j = (x_j, y_j, z_j)$. The location of the **center of mass** of the system is specified by the vector $\mathbf{R} = (X, Y, Z)$ defined by the equation

$$\mathbf{R} \equiv \frac{m_1\mathbf{r}_1 + m_2\mathbf{r}_2 + m_3\mathbf{r}_3 + \cdots + m_j\mathbf{r}_j + \cdots + m_n\mathbf{r}_n}{m_1 + m_2 + m_3 + \cdots + m_j + \cdots + m_n}. \tag{9.17a}$$

This equation expresses $\mathbf{R}$ as the mean position of the mass in the system. You can probably guess why any quantity evaluated the way $\mathbf{R}$ is evaluated here is called a weighted average. The denominator of the fraction on the right side of the equation is the sum of all the masses in the system—that is, the total mass M. The numerator is more compactly represented in summation notation. The equation then takes the form

$$\mathbf{R} \equiv \frac{1}{M} \sum_{j=1}^{n} m_j\mathbf{r}_j. \tag{9.17b}$$

The two forms of Equation 9.17 involve vector summation. It is usually most convenient to carry out this summation component by component. In component form, Equation 9.17b becomes

$$X = \frac{1}{M} \sum_j m_j x_j, \quad Y = \frac{1}{M} \sum_j m_j y_j, \quad Z = \frac{1}{M} \sum_j m_j z_j, \tag{9.18a,b,c}$$

where it is understood that the sums are to be taken over all the particles in the system.

EXAMPLE 9.4

Find the position of the center of mass of the three-particle system shown in Figure 9.10. The particles lie at the vertices of an isosceles triangle of base 8 cm and altitude 6 cm.

SOLUTION: In calculating the center of mass of a system, you are free to locate the origin and orient the coordinate axes in any way you wish. However, a wise choice of coordinates often makes the calculation both easier and clearer in meaning. In Figure 9.10, the origin has been chosen at the center of the base of the triangle, with the x direction along the base. In this example, it is not necessary to specify a z axis because the three-particle system lies in a plane, and so the solution of Equation 9.18c is trivial: $Z = 0$.

The total mass of the system is $M = 4m$. Taking the particles from left to right, in order of increasing x, Equation 9.18a gives you

$$X = \frac{1}{4m} [m(-4 \text{ cm}) + 2m(0) + m(4 \text{ cm})]$$

$$= 0.$$

Taking the particles from bottom to top, in order of increasing

FIGURE 9.10

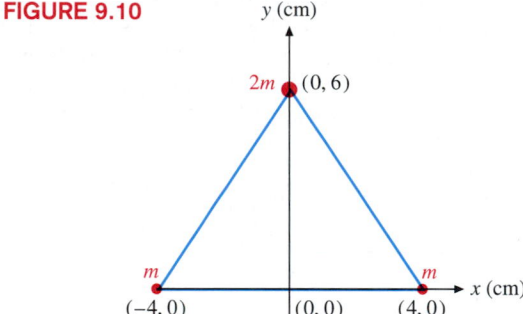

y, Equation 9.18b gives you

$$Y = \frac{1}{4m} [m(0) + m(0) + 2m(6 \text{ cm})]$$

$$= \tfrac{1}{4} \times 12 \text{ cm} = 3 \text{ cm}.$$

Thus the center of mass of the three-particle system is located at $\mathbf{R} = (0 \text{ cm}, 3 \text{ cm})$. That is, the center of mass lies along the altitude, halfway from the base to the upper vertex.

If you recalculate the center of mass of the system in Example 9.4 using a different set of coordinates, you will obtain different coordinates for both the positions of the masses and the center of mass. However, *the position of the center of mass is fixed with respect to the system itself. The center of mass is a property of the body independent of the body's orientation or its surroundings.* If you are not convinced of this point, recalculate the example with the origin at, say, the lower left-hand particle.

The altitude of the isosceles triangle in Example 9.4 is a *symmetry axis* of the system because the masses are distributed symmetrically about it. If a system has such a symmetry axis, the center of mass always lies along it. This property of symmetry is most easily seen when, as in the example, a coordinate axis is chosen to coincide with the symmetry axis. Then you can see immediately that terms like $m_j x_j$ in Equation 9.18a cancel in pairs (as do the first and last terms in the equation for X in Example 9.4). If there are two or more symmetry axes, the center of mass must lie at their intersection. Would the altitude of the triangle in Figure 9.10 still be a symmetry axis if the mass of the lower right-hand particle were doubled without changing anything else?

Center of Mass of an Extended Body

An extended body may be regarded as a continuous collection of point particles. In Figure 9.11a, a body of arbitrary shape is divided into small elements. The typical jth element has mass Δm_j, and the position of the center of mass is given by Equation 9.17b. In the present case, that equation becomes an approximation:

$$\mathbf{R} \simeq \frac{1}{M} \sum_j \Delta m_j \, \mathbf{r}_j.$$

The reason why the equation becomes an approximation is that the vector $\mathbf{r}_j$ cannot precisely express the position of the entire element Δm_j because this element has finite size and so its position cannot be precisely represented by a single point. When we take the limit $\Delta m \to 0$, however, and increase the number of elements correspondingly, each element of mass Δm_j shrinks down to an infinitesimal element of mass dm. Now, not only is the mass of the element infinitesimal, but so is its extent in space (Figure 9.11b). Thus the position vector $\mathbf{r}$ becomes a precise indicator of the element's position. In the limit, the center of mass of the extended body is expressed as the integral

$$\mathbf{R} = \frac{1}{M} \int_{\text{body}} \mathbf{r} \, dm. \tag{9.19}$$

Like the finite sum of Equation 9.17, this vector integral can be written component by component:

$$X = \frac{1}{M} \int_{\text{body}} x \, dm, \quad Y = \frac{1}{M} \int_{\text{body}} y \, dm, \quad Z = \frac{1}{M} \int_{\text{body}} z \, dm. \tag{9.20a,b,c}$$

In order to evaluate the integral of Equation 9.19, the integrand must be expressed in terms of a single variable. The same is true of Equations 9.20a, 9.20b, and 9.20c. Often, this can be done quite directly, as in Example 9.5.

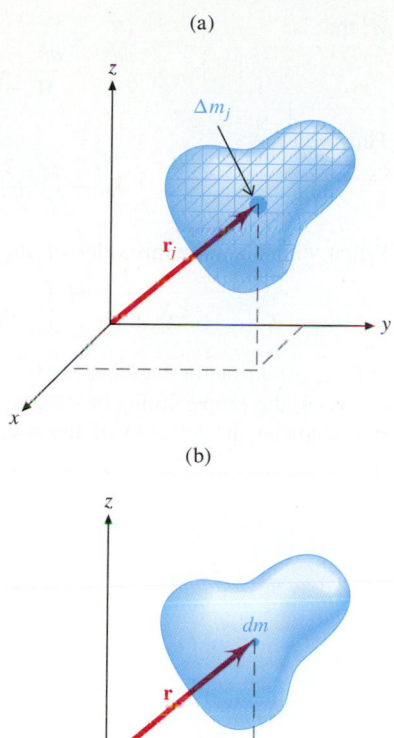

(a)

(b)

FIGURE 9.11 (*a*) An extended body is divided into elements. The typical jth element has mass Δm_j. Its position is described approximately by the vector $\mathbf{r}_j$. (*b*) Here, the division of the body into elements has been extended to the limit in which there are infinitely many infinitesimal elements each having mass dm. The position of any such element is exactly described by the corresponding vector $\mathbf{r}$.

EXAMPLE   9.5

Figure 9.12 shows a uniform rod of mass M and length L. Find the position of the center of mass.

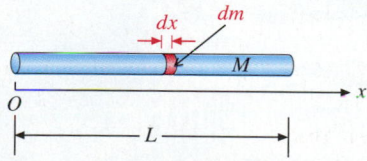

FIGURE 9.12 The uniform rod has mass M and length L. The x axis is chosen along the rod, with the origin at the left end of the rod. The rod is divided into elements of length dx, one of which is shown.

SOLUTION: The fact that the rod has a point of symmetry at its center suggests that the center of mass lies there. (See the discussion following Example 9.4.) If you choose the x axis along the rod, you can use Equation 9.20a to show that the suggestion is correct. To evaluate the integral in that equation, you must express the integrand, $x \, dm$, in terms of a single variable. The most convenient way to do this is to express dm in terms of dx. Divide the rod into infinitesimal elements of length dx, one of which is shown in Figure 9.12. How can you express dx in terms of dm? Just note that the rod is uniform. Consequently, the element's length dx is the same fraction of the entire length L as the element's mass dm is of the entire mass

M; that is,

$$\frac{dx}{L} = \frac{dm}{M}.$$

Thus you have

$$dm = \frac{M}{L}\, dx.$$

When you substitute this value of dm into the integral, you get

$$X = \frac{1}{M}\frac{M}{L}\int_{\text{rod}} x\, dx.$$

(The constant factor M/L has been moved out of the integral.) What are the proper limits of integration? Suppose you choose the origin at the left end of the rod. Then the rod begins at

$x = 0$ and ends at $x = L$, so the equation becomes

$$X = \frac{1}{L}\int_0^L x\, dx$$

$$= \frac{1}{L}\frac{x^2}{2}\Big|_0^L = \frac{1}{L}\frac{L^2}{2},$$

which yields the result

$$X = \frac{L}{2}.$$

This agrees with the expected result: the center of mass is at the midpoint of the rod. What would have happened if you had chosen the midpoint as the origin in the first place?

Motion of the Center of Mass

You have seen Figure 9.13 before; it is the twirling-baton photograph near the beginning of Chapter 2. At that point, the photograph was used to emphasize the fact that the overall motion of an extended body can be quite complicated. In looking at the photograph again here, note that a curve has been superimposed on it. That curve connects subsequent positions of the center of mass of the baton as the baton flies through the air. The curve is quite simple; it is the parabolic trajectory that would be followed by a *particle* having the same mass as the baton if the particle were thrown with the same impulse as the baton. The center of mass appears to act like an isolated particle; so far as the *translational* motion of the baton is concerned, the baton can be replaced by a particle located at its center of mass, having the same mass as the baton. We often used this property of the center of mass implicitly when we treated extended bodies as if they were particles. We now rigorously justify a statement we have used up to now without thinking much about it: As far as translational motion is concerned, a body of

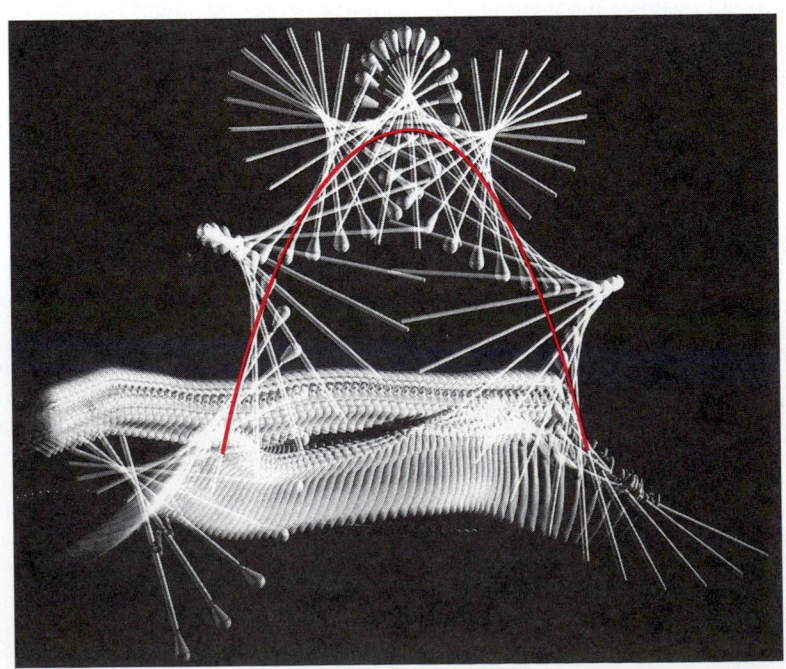

FIGURE 9.13 Multiflash photograph showing the flight of a baton. The successive positions of the center of mass of the baton are joined by the superimposed curve. (Parade batons like this one are weighted at one end so that the center of mass is not at the midpoint.) Although the motion of the baton is complicated, its center of mass moves in a parabolic trajectory, just like that of a particle. Thus the *translational* motion of the baton as a whole can be considered that of a particle located at the center of mass.

mass M can be represented by a particle of mass M located at the center of mass of the body. (We study rotational motion of extended bodies in Chapters 11 through 13.)

Consider a system whose total mass M remains unchanged with time. The system could be a large, irregular body such as an iceberg breaking off the edge of a glacier and falling into the sea, or it could be a charge of buckshot flying through the air. Because the position $\mathbf{R}$ of the center of mass of any system has been defined (Equation 9.17b or Equation 9.19), we can define the *velocity* $\mathbf{V}$ *of the center of mass* as well. We do this by taking the time derivative of Equation 9.17b. This gives us

$$\mathbf{V} = \frac{d\mathbf{R}}{dt} = \frac{1}{M} \sum_j \frac{d}{dt} m_j \mathbf{r}_j.$$

Because each m_j is constant, the time derivative of each term yields

$$m_j \frac{d\mathbf{r}_j}{dt} = m_j \mathbf{v}_j = \mathbf{p}_j.$$

The equation for $\mathbf{V}$ thus becomes

$$\mathbf{V} = \frac{1}{M} \sum_j \mathbf{p}_j. \tag{9.21}$$

Now, $\Sigma_j \mathbf{p}_j$ is just the sum of the momenta of all the particles making up the system. Hence it is the total momentum $\mathbf{P}$ of the system. Using this fact and multiplying through by M gives us

$$M\mathbf{V} = \mathbf{P}. \tag{9.22}$$

That is, *the total momentum of a system is equal to the product of the mass of the system and the velocity of the center of mass.* As far as momentum is concerned, we can replace the entire system (no matter how complicated it may be) with a single particle of the same mass located at the center of mass.

Next, we consider the *acceleration* $\mathbf{A}$ *of the center of mass* of the system. We take the time derivative of Equation 9.21 and obtain

$$\mathbf{A} = \frac{d\mathbf{V}}{dt} = \frac{1}{M} \sum_j \frac{d\mathbf{p}_j}{dt}.$$

The time derivative of each term in the sum can be written

$$\frac{d\mathbf{p}_j}{dt} = m_j \frac{d\mathbf{v}_j}{dt} = m_j \mathbf{a}_j.$$

According to Newton's second law, $m_j \mathbf{a}_j$ is equal to $\mathbf{F}_j$, which is the net force acting on the jth particle. So we have for the acceleration of the center of mass

$$\mathbf{A} = \frac{1}{M} \sum_j \mathbf{F}_j. \tag{9.23}$$

Now, each particle j may have many forces acting on it, and thus each net force $\mathbf{F}_j$ may be the sum of many individual forces. Some of them are *internal* forces. The cohesive forces that keep a kicked football from flying into bits are internal forces. We call the vector sum of the internal forces acting on the jth particle $\mathbf{F}_{j\,\text{int}}$.

The remainder of the forces acting on the system are *external* forces. The gravitational forces acting on the parts of the baton of Figure 9.13 and making the baton follow a parabolic path are of this kind. We call the vector sum of the external forces acting on the jth particle $\mathbf{F}_{j\,\text{ext}}$. Thus the net force acting on the jth particle is given by the vector sum

$$\mathbf{F}_j = \mathbf{F}_{j\,\text{int}} + \mathbf{F}_{j\,\text{ext}}.$$

Substituting this sum into Equation 9.23 and multiplying through by M, we obtain

$$M\mathbf{A} = \sum_j \mathbf{F}_{j\,\text{int}} + \sum_j \mathbf{F}_{j\,\text{ext}}. \tag{9.24}$$

We now make a delicate argument concerning the internal forces. Consider any two particles, j and k, that belong to the system. If j exerts a force on k, Newton's third law requires that k exert an equal and opposite force on j. Thus, when we carry out the first summation in Equation 9.25, adding up *all* the internal forces acting on *all* the particles that make up the system, each force will cancel the other member of its action-reaction pair. Consequently, the first term on the right side of the equation is equal to zero.

What remains is the sum of all the external forces. Because this sum is just the net external force $\mathbf{F}_{ext}$ acting on the system, Equation 9.24 becomes

$$MA = \mathbf{F}_{ext}. \tag{9.25}$$

The external forces acting on a system may be applied to different parts of the system, which may or may not be bound to one another by internal forces. Nevertheless, *the vector sum* $\mathbf{F}_{ext}$ *is equal to the product of the total mass M of the system and the acceleration A of its center of mass.* That is, as far as translational motion is concerned, the entire system may be reduced to a point particle whose mass is equal to the total mass of the system and whose location is at the center of mass of the system. Equation 9.25 has the same form as Newton's second law for a particle. This identity of form is the justification for applying Newton's second law to extended bodies.

When we combine Equation 9.25 with Equation 9.22, we obtain

$$\mathbf{F}_{ext} = MA = M\frac{d\mathbf{V}}{dt} = \frac{d(M\mathbf{V})}{dt},$$

or

$$\mathbf{F}_{ext} = \frac{d\mathbf{P}}{dt}. \tag{9.26}$$

Consider what happens when the net external force acting on the system is zero. Then $d\mathbf{P}/dt = 0$, which means that the momentum of the system is constant:

$$\Delta\mathbf{P} = 0 \quad \text{for an isolated system.} \tag{9.27}$$

This is the principle of momentum conservation for an isolated system. Compare Equation 9.27 with Equation 9.10a, $\Delta\mathbf{p} = 0$, which states the principle for isolated two-body systems. We have already considered momentum conservation in detail for two-body systems. *Equation 9.27 extends the principle and its consequences to all isolated systems, no matter how complicated their internal interactions may be.*

Logically equivalent to this statement is the following one: *In the absence of a net external force, the momentum of the center of mass of a system is constant*; that is,

$$\mathbf{P} = M\mathbf{V} = \text{constant} \quad \text{for an isolated system.} \tag{9.28}$$

Because the system is isolated, M is constant. It follows that

$$\mathbf{V} = \text{constant.} \tag{9.29}$$

The velocity of the center of mass of an isolated system remains constant.

We have thus extended the principle of conservation of linear momentum to systems of any complexity whatever. We have proved, moreover, that the motion of the center of mass of any system can be analyzed by using a generalization of Newton's laws of motion. These ideas are applied in Example 9.6.

The exploding skyrocket exemplifies the constancy of the center-of-mass velocity of an isolated system. Can you see how?

EXAMPLE 9.6

You are a member of a student rocket club that builds an experimental two-stage solid-fuel rocket. To test it, you and the other club members go out to the country and launch the rocket eastward. The first stage works well: After a short burn, the fuel is exhausted and the rocket coasts in a parabolic trajectory. An explosive bolt is supposed to separate the second stage from the first just as the rocket reaches the top of its trajectory at a horizontal distance $x_t = 550$ m downrange (east) from the launching point. (We use the subscript $_t$ to denote "top.") Un-

fortunately, the small explosion ignites some residual fuel in the first stage. The second stage never ignites. Instead, the rocket breaks into three fragments, which fall toward the ground as shown in Figure 9.14. After a hunt, you find one fragment in an open field at a location $x_1 = x_t = 550$ m east and $y_1 = 120$ m north of the launch point. You find the second fragment in the same field at a location $x_2 = x_t = 550$ m east and $y_2 = 65$ m south of the launch point. The third fragment has fallen into woods and cannot be found. You note, however, that the

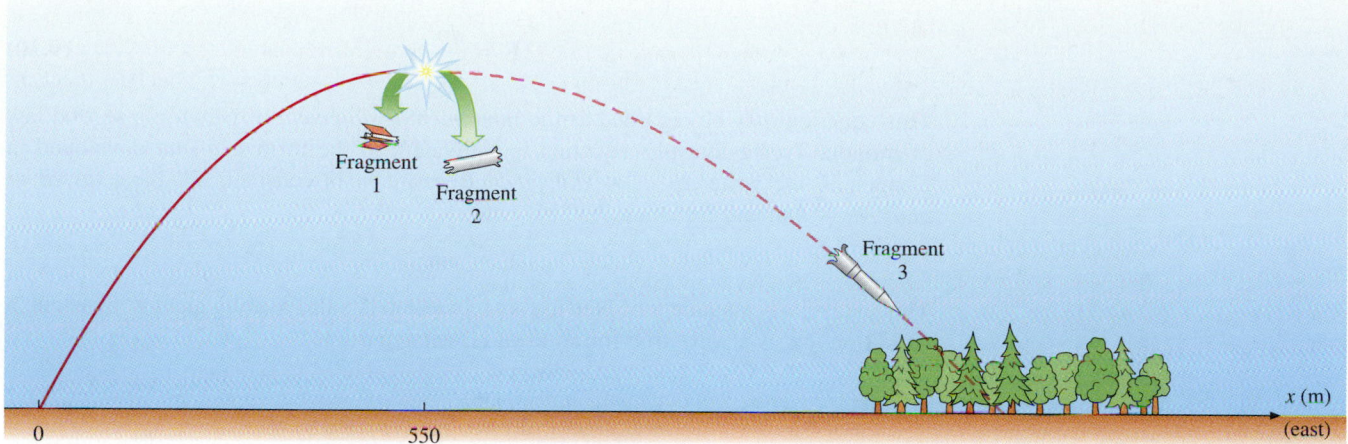

FIGURE 9.14 The rocket breaks up into three pieces at the top of its parabolic trajectory. The path followed by the center of mass is not affected by the explosion, which produces only forces internal to the system.

first two fragments are of approximately equal mass m. From a knowledge of the rocket structure, you deduce that the lost fragment has approximate mass $2m$. Moreover, a quick review of the videotape of the test launch suggests that all three fragments fell in such a way that they probably struck the ground at about the same time. Where do you advise your fellow club members to concentrate their search for the missing fragment?

SOLUTION: Because you are only trying to define a general search location, you neglect that part of the mass of the system consisting of spent gases and any possible small fragments. To the extent that this approximation is valid, you can assert that the three major fragments constitute the system of interest. Consequently, all the forces generated by the explosion are internal to the system. Furthermore, you neglect air resistance. Then the only significant external force acting on the system is the downward gravitational force $M\mathbf{g}$. Finally, the shortness of the first-stage burn allows you to approximate the actual trajectory with a parabolic one. That is, you neglect the distance traveled by the rocket under power and concern yourself only with the free-fall path followed by the rocket system after engine shutdown.

For purposes of analyzing the translational motion of the system, you can replace the system by its center of mass. According to Equation 9.25, the acceleration of the center of mass is

$$\mathbf{A} = \frac{M\mathbf{g}}{M} = \mathbf{g}.$$

That is, the acceleration of the center of mass is identical with the acceleration of the rocket before the explosion but after burnout of the first stage. It follows that the trajectory of the

center of mass of the system consisting of the three fragments is a continuation of the trajectory of the rocket.

From this argument, you can conclude that the center of mass "strikes the earth" at the same place where the coasting rocket would have done so—at a distance $X = 2x_t = 1100$ m east of the launch point. The system is isolated until the instant of impact—an instant common to all three fragments—when strong contact forces come into play. Hence the center-of-mass picture of the system remains valid until impact. (If the fragments had not struck the ground at the same time, the picture would be valid only until the first fragment struck.)

To find x_3, the x coordinate of the missing rocket fragment, you apply Equation 9.18a. This gives you

$$X = \frac{mx_1 + mx_2 + 2mx_3}{4m}.$$

Using the known values of X, x_1, and x_2, you solve for x_3 to obtain

$$x_3 = 3x_t = 1650 \text{ m}.$$

You find y_3 in a similar manner. Equation 9.18b gives you

$$Y = \frac{my_1 + my_2 + 2my_3}{4m}.$$

Because $Y = 0$, this becomes

$$y_3 = 0 - \tfrac{1}{2}y_1 - \tfrac{1}{2}y_2 = -\tfrac{1}{2}(120 \text{ m} - 65 \text{ m})$$
$$= -28 \text{ m}.$$

You therefore begin to look for the lost fragment at a location 1650 m east and 28 m south of the launch point.

Now that we have considered momentum conservation in systems of unlimited complexity, let us return to the simplest case. This will allow us to check that our new, general way of looking at motion from the viewpoint of momentum conservation is entirely consistent with something with which you are by now familiar—Newton's second law for a particle. Consider what happens when a system consists of just a single particle of mass m and momentum $\mathbf{p}$. For such a system, all forces are external (in other words, the phrase "internal force" has no meaning for an ideal particle). If the net force acting on the particle is **F**, we can write Equation 9.26 in the form

$$\mathbf{F} = \frac{d\mathbf{p}}{dt}. \tag{9.30}$$

This equation may be regarded as the *fundamental statement* of Newton's second law of motion. To see that this equation is equivalent to the form $\mathbf{F} = m\mathbf{a}$ developed in Chapter 3, we write $d\mathbf{p}/dt = d(m\mathbf{v})/dt$. Because m is constant, we have $d\mathbf{p}/dt = m\,d\mathbf{v}/dt = m\mathbf{a}$, and Equation 9.30 becomes

$$\mathbf{F} = m\mathbf{a}.$$

This elementary statement of Newton's second law is valid as long as m = constant, a condition that is always true for an ideal classical particle.

SECTION 9.6

Newton's Second Law Reconsidered*

By now, you probably feel quite familiar with Newton's second law, which you have used in a variety of contexts. You probably feel comfortable with the use of the quantities force and mass, which are central to the law and always appear in its applications. There is, however, a conceptual weakness in the way that Newton's second law was developed in Chapter 3. We can now deal with that weakness, using the law of conservation of linear momentum.

If you scan through Sections 3.2 and 3.3, you will find that the development follows a course like this:

1. The concept "force" is invoked without rigorous definition, by refinement of the everyday meaning of the term.

2. Mass is defined as "quantity of matter." The unit of mass is then defined by assigning 1 unit of mass (1 kg in SI) to an agreed-upon but otherwise arbitrary object. This approach is quantitatively satisfactory, but the definition of mass is not so much a definition as a reexpression of the term "quantity of matter," which is itself undefined.

3. Even though force is not rigorously defined, a method of applying a constant force to a body is made plausible by means of a thought experiment involving a spring scale. A method is then devised for applying any integral multiple of that constant force to a body. The thought experiment involves measurement of acceleration, which is well defined in terms of the primitive quantities displacement and time. The result of the thought experiment is expressed in the proportionality $F \propto ma$. The unit of force is established by defining 1 N $\equiv$ 1 kg·m/s^2. The proportionality then becomes the equation $F = ma$, which is Newton's second law in one dimension. The thought experiment thus makes possible a transition from the use of the word "force" in an undefined but intuitive way to a use that is well defined in terms of the quantities mass and acceleration.

The difficulty in this approach lies in the weakness of the definition of mass. Because the definition of force is made in terms of mass, the difficulty propagates and involves Newton's third law as well. This logical weakness certainly does not cast doubt on the validity of using Newton's laws to analyze classical mechanical systems, however. You have already seen far too much consistency in the conceptual structure we have developed, and had far too much success in achieving results using that structure, to allow room for such doubt. Nevertheless, we wish to base classical mechanics on a sound logical foundation. An operational procedure for doing this was first set forth in detail by the Austrian physicist, physiologist, and philosopher Ernst Mach (1838–1916). The discussion that follows runs along substantially the same lines as Mach's.[†]

---

*This section may be omitted without loss of continuity.

[†]For more details, see E. Mach, *The Science of Mechanics* (LaSalle, Ill.: Open Court Publishing, 1960). The discussion paraphrased here begins on page 264.

We begin by pretending we have never encountered the concept of mass. Even though we do not know the meaning of mass, we can nevertheless carry out a series of thought experiments involving collisions between pairs of bodies. Although we will perform the experiments in imagination only, there is no part of them that you could not carry out in a laboratory, using an air track. For the sake of concreteness, suppose we *are* doing the experiments on an air track, using a large collection of gliders as the colliding bodies. Using various pairs of gliders, we measure their initial and final velocities in many collisions. At first we choose pairs at random. But soon we find that for *certain pairs* of gliders—gliders j and k, for instance—the change in velocity of glider j is the negative of the change in velocity of glider k:

$$\Delta v_j = -\Delta v_k. \qquad (9.31)$$

We say that *any two gliders that satisfy this condition have equal mass.*

But we want to be able to say more than this about the quantity to which we have now given the name mass, and so we undertake another experiment. We label one of the gliders of an equal-mass pair 0. Using this glider, we carry out a collision experiment with any other glider, which we label 1. In general, Equation 9.31 will not be satisfied. Nevertheless, Δv_0 can always be related to Δv_1 by means of some constant α: The result of the collision experiment using gliders 0 and 1 can be expressed as

$$\Delta v_0 = -\alpha \, \Delta v_1. \qquad (9.32a)$$

Carrying out a collision experiment with glider 0 and still another glider, which we label 2, we express the result in terms of the constant β:

$$\Delta v_0 = -\beta \, \Delta v_2. \qquad (9.32b)$$

We can continue to repeat this process with glider 0 and any other glider. The results of these experiments can be expressed in terms of equations that are identical in form with Equations 9.32a and 9.32b but contain the constants γ, δ, ϵ, ...

The next step is to carry out a collision experiment with any two of the gliders just used, other than glider 0. Suppose we do this with gliders 1 and 2. We find that

$$\alpha \, \Delta v_1 = -\beta \, \Delta v_2. \qquad (9.33)$$

Any other pair of gliders will collide in a way described by a similar equation. That is, *no new constants need be defined to describe the results.* We infer that each glider (or, in general, any body at all) possesses a property whose value can be used to describe the outcome of a collision with *any other* glider (or body). That property is the **mass** of the body.

To express mass quantitatively, we choose a body arbitrarily and define its mass to be the **unit mass**. Let us choose glider 0 and call its mass the unit mass m_0. We can then rewrite Equations 9.32a and 9.32b in the form

$$m_0 \, \Delta v_0 = -m_1 \, \Delta v_1, \quad m_0 \, \Delta v_0 = -m_2 \, \Delta v_2, \quad \ldots .$$

In these equations, the masses m_j are defined in terms of the unit mass:

$$m_1 \equiv \alpha m_0, \quad m_2 \equiv \beta m_0, \quad \ldots .$$

In terms of these masses m_1 and m_2, Equation 9.33 can be written

$$m_1 \, \Delta v_1 = -m_2 \, \Delta v_2. \qquad (9.34)$$

For *any* two bodies j and k, the corresponding relation is

$$m_j \, \Delta v_j = -m_k \, \Delta v_k. \qquad (9.35)$$

The advantage of this formulation over Equation 9.33 is that this formulation makes clear the fact that mass m is a property inherent in every body in the universe. Moreover, the mass of a body does not depend on which other body it happens to interact with.

Equation 9.35, together with the argument leading up to it, constitutes a rigorous operational definition of mass. The definition takes for granted only two points: the meaning of displacement and the meaning of time, both of which are implicit in making the necessary velocity measurements. The equation also expresses the observed result of a vast family of collision experiments. Rewritten in the form

$$m_j \, \Delta v_j + m_k \, \Delta v_k = 0,$$

Equation 9.35 constitutes a statement of the principle of conservation of linear momentum for an isolated two-body system. In effect, the definition of mass rests on this principle.

Newton's Second Law and the Definition of Force

Once mass has been defined, it is not difficult to take the next step and define force. Let us reconsider the air-track experiment described by Equation 9.35 and concentrate on body k,* whose momentum changes by the amount $\Delta p_k = m_k \, \Delta v_k$. We argue that this change is produced by an impulse J acting on body k. That is, we *define* the impulse acting on any body during a collision as

$$J = \Delta p \equiv m \, \Delta v = \int_{v_i}^{v_f} m \, dv = \int_{t_i}^{t_f} m \, \frac{dv}{dt} \, dt.$$

Taking the time derivative of the extreme left and right terms, we obtain

$$\frac{dJ}{dt} = m \, \frac{dv}{dt} = ma. \tag{9.36}$$

Finally, we define the **force** F to be the time derivative of the impulse:

$$F \equiv \frac{dJ}{dt}.$$

Equation 9.36 then assumes the familiar form of Newton's second law,

$$F = ma. \tag{9.37}$$

Although we have made the argument in a one-dimensional context for the sake of simplicity, it can be generalized to two or three dimensions without introducing anything fundamentally new.

---

*In making this shift in point of view, we must redefine the system under consideration. From a two-body system in which all the forces are internal and momentum is conserved, we go to a one-body system subject to external force—that is, to a system in which momentum is not conserved. Compare this shift with the one in the argument leading to a definition of rocket thrust in Section 9.4.

The logical flow of the argument of this section is as follows:

1. A tacit agreement as to the meaning of displacement and time, taken together with a thought experiment whose results tacitly imply the principle of conservation of linear momentum,

leads to

2. a definition of equality of mass of two bodies. Taking the unit of mass to be the mass of an arbitrary body and then obtaining experimental results in conformity with momentum conservation

leads to

3. a definition of mass (whose value for any particular body A does not depend on the mass of the body B with which A collides) and a convenient way of expressing the linear momentum of any body. The definition of mass, together with a review of the thought experiment from a shifted point of view in which the system is redefined,

leads to

4. a definition of impulse,

which leads to

5. a definition of force and Newton's second law.

Summing Up

The **impulse J** imparted to a body by a force $\mathbf{F}(t)$ is defined by Equation 9.1,

$$\mathbf{J} \equiv \int_{t_i}^{t_f} \mathbf{F}(t)\, dt.$$

The **linear momentum p** is defined by means of Equation 9.3,

$$\mathbf{p} \equiv m\mathbf{v}.$$

This leads to Equation 9.4b, the **impulse-momentum theorem**,

$$\mathbf{J} = \Delta\mathbf{p}.$$

This theorem is strictly valid only when the impulse imparted by $\mathbf{F}(t)$ is the *net* force acting on the body. It often happens, however, that the impulse is imparted over a short time during which $\mathbf{F}(t)$ is by far the largest force present,

and so any other forces present can be neglected. Such an interaction is called a **collision**.

When two bodies collide, it follows from Newton's third law that the impulses imparted by each to the other are equal and opposite. As a consequence, the total momentum change of the two-body system is zero. This is expressed in Equation 9.10a, the **principle of conservation of linear momentum**,

$$\Delta\mathbf{p} = 0.$$

The operation of a rocket engine can be understood by applying the momentum conservation principle to the system comprising the rocket-propelled vehicle and the gases expelled by the rocket during a time interval dt. The equation of motion of this system is Equation 9.12,

$$\frac{1}{v_{ex}}\, dv = -\frac{dm}{m}.$$

The velocity increment of the vehicle over a period during which its total mass diminishes from m_i to m_f is expressed in Equation 9.14,

$$\Delta v = v_{ex} \ln \frac{m_i}{m_f}.$$

Because the vehicle accelerates with acceleration a, it is convenient to think of the equivalent force $F = ma$ that acts on it. In order to take this viewpoint, the system must be redefined to include only the vehicle and not the expelled gas. The force then becomes an external force called the **thrust** F, given by Equation 9.15,

$$F = v_{ex} \left| \frac{dm}{dt} \right|.$$

The location of the **center of mass** of a system of particles is defined by Equation 9.17b,

$$\mathbf{R} \equiv \frac{1}{M} \sum_{j=1}^{n} m_j \mathbf{r}_j.$$

If the system is an extended body, the center of mass is more conveniently defined in the integral form of Equation 9.19,

$$\mathbf{R} = \frac{1}{M} \int_{body} \mathbf{r} \, dm.$$

The translational motion of an isolated system can be described in terms of the motion of its center of mass. The total momentum $\mathbf{P}$ of the system is given by Equation 9.22,

$$\mathbf{P} = M\mathbf{V},$$

where $\mathbf{V}$ is the velocity of the center of mass. The equation of motion for the center of mass is Equation 9.25 or Equation 9.26,

$$\mathbf{F}_{ext} = M\mathbf{A} \quad \text{or} \quad \mathbf{F}_{ext} = \frac{d\mathbf{P}}{dt}.$$

In these equations, $\mathbf{F}_{ext}$ is the net external force acting on the system. Subject to this definition of $\mathbf{F}_{ext}$, the equations have a form identical with that of Newton's second law for particles. This identity of form justifies the treatment of the translational motion of extended systems of total mass M as though the systems were particles of mass M situated at the center of mass of the system. The momentum conservation principle is thus extended to systems comprising any number of particles.

The basic definitions of mass and force are essential to mechanics. These definitions can be placed on a firm operational basis through a series of thought experiments whose results depend on the momentum conservation principle. The thought experiments lead to a definition of the mass of a body. Force is then defined by means of a redefinition of the system, as was done in defining the thrust of a rocket.

KEY TERMS

Section 9.2 Impulse and Linear Momentum
impulse ▪ linear momentum ▪ impulse-momentum theorem ▪ collision

Section 9.3 Conservation of Linear Momentum
principle of conservation of linear momentum ▪ external force, internal force, isolated system ▪ head-on collision

Section 9.4 Reaction Propulsion and Rockets
exhaust speed ▪ gross payload ratio ▪ equivalent force, thrust ▪ specific impulse

Sections 9.5 and 9.6 The Center of Mass and Newton's Second Law Reconsidered
center of mass ▪ thought experiment ▪ mass, unit mass ▪ force

Queries and Problems for Chapter 9

QUERIES

9.1 *(2) Pinball pleasures.* The argument of Example 9.1 depends on the tacit assumption that the gravitational force component that drives the ball down the sloping baseboard is much smaller than the force exerted on the ball by the flipper. In view of the fact that we do not know the value of the flipper force at any moment, how can the assumption be justified?

9.2 *(2) Diverse behavior.* A rubber ball and a lump of clay of equal mass strike a wall with equal speed. The ball bounces off; the clay sticks. Which collision imparts more impulse to the wall?

9.3 *(2) Slippery fingers, again!* An automobile mechanic drops a ball bearing of diameter 1 cm. It falls from a height of 30 cm into a container of grease where it comes to rest just buried under the surface of the grease. What is the total impulse imparted to the ball bearing?

9.4 *(3) One law violation leads to another.* Imagine a universe in which Newton's third law is not true. Show that the momentum of an isolated two-particle system is not conserved.

9.5 *(3) Braking the law?* A cannon is mounted on a railroad flatcar. If the brakes are not set when the cannon is fired, the car recoils, moving along the track. If the brakes are set, the car stays in place. Does this mean that setting the brakes invalidates the principle of momentum conservation? Explain.

9.6 *(3) The dangers of pearl diving.* Why is a chinaware plate more likely to chip when you hit it against another chinaware plate of equal mass than when you hit it at the same speed against a wooden platter of equal mass?

9.7 *(3) Stopping is hard to do.* An airplane touches down on an airport runway and comes to a stop. Is its momentum conserved? Can you describe a system whose momentum *is* conserved in the process?

9.8 *(3) Point of view.* Consider two head-on collisions. In the first, a bowling ball moving with speed v strikes a stationary Ping-Pong ball. In the second, a Ping-Pong ball moving with the same speed v strikes a stationary bowling ball. Show that the two collisions look exactly the same from the point of view of an observer riding on the bowling ball.

9.9 *(4) Speed limit?* Your friend makes the following argument in favor of using ion propulsion instead of chemical propulsion for interstellar spaceship voyages. The distance to the nearest star, Alpha Centauri, is about 10^{16} m. The exhaust speed of the best chemical propellants is about 5×10^3 m/s. A rocket moving at this speed will take a clearly impractical 60 000 years to get to Alpha Centauri. However, ion rockets can produce exhaust speeds roughly 10^3 times as great as those of chemical rockets, thus reducing the length of the voyage a thousandfold, to a time comparable to a human lifetime. This is why ion rockets should be used for interstellar travel. Criticize this argument.

9.10 *(4) New dimensions?* Specific impulse, defined by Equation 9.16, is usually given in units of N·s/kg. Show that this is equivalent to m/s, as the equation requires.

9.11 *(4) Next stop, Alpha Centauri!* Because of its large specific impulse, a rocket operating on the ion propulsion principle is a prime candidate for propelling a vehicle on a very long voyage, such as a voyage to one of the nearest stars. The thrust of an ion rocket is very small compared with that of even a modest chemical rocket. Why doesn't the smallness of the thrust matter?

9.12 *(5) No calculation needed.* You wish to find the center of mass of a flat, thin, but irregularly shaped object. One method, illustrated at the top of the next column, is as follows. Use a string to hang the object from any point on its periphery. From the same point, hang a plumb bob (a small weight attached to a string). Draw a line along the plumb bob. Then repeat the process, using any other point on the periphery of the object. The center of mass lies at the intersection of the two lines. Explain why this process works.

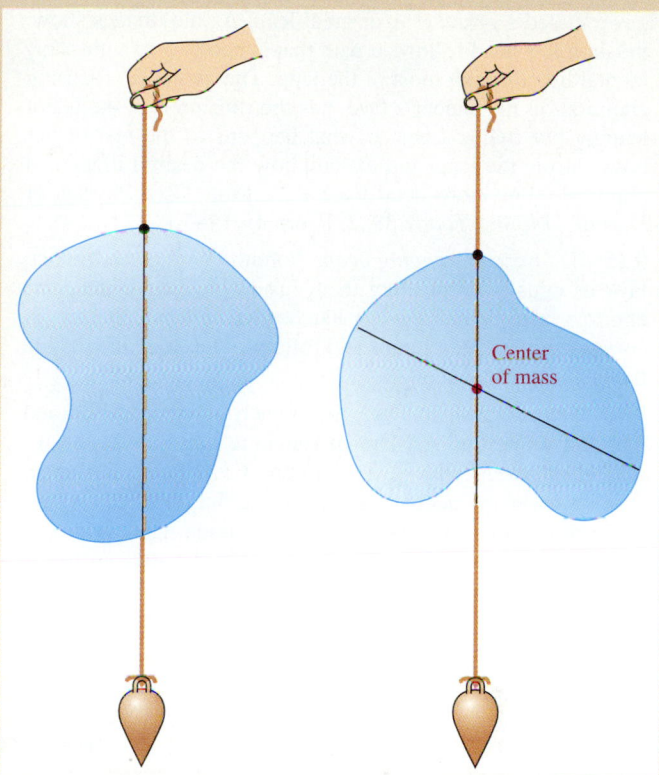

Center of mass

9.13 *(5) Bending the rules just a bit?* In championship high jumping and pole vaulting, the winner usually clears the bar at a height only half a centimeter or so greater than the best height of the second placer. In view of this extremely close competition, a high jumper or pole vaulter must use every trick in the book. In particular, he must clear the bar in such a way that his center of mass actually passes under the bar. Explain how this is done.

9.14 *(5) Ballet magic.* Anyone can broad jump and have his or her center of mass follow a parabolic trajectory, as does the center of mass of the baton in Figure 9.13. This is what we expect to see when we observe broad jumpers. The intent of the ballet dancer, however, is seemingly to defy the laws of physics and thus to bring magic to the stage. The figure below shows a dancer performing the classical ballet man-

euver called a *grand jeté*, or great leap. As the images show, the dancer's head follows a path that is nearly a straight, horizontal line through most of the leap. The audience, focusing attention on the dancer's face, has the illusion that she is not leaping but flying. Look at what happens to the rest of her body during the leap, and explain how the desired illusion is obtained. (For more detail, see K. Laws, ''The Physics of Dance,'' *Physics Today* **38**, 2, February 1985.)

9.15 (5) *Mexican jumping bean.* A moth common in Mexico lays its eggs by depositing them inside growing beans, one egg to a bean. When the egg hatches into a larva, the apparently inanimate bean jumps and rolls about as the larva inside the bean twitches. Explain.

9.16 (5) *Polymath.* In 1897, the French playwright Edmond Rostand wrote his best-known romantic play, about the remarkable seventeenth-century figure Cyrano de Bergerac. Among his many accomplishments, the fictionalized Cyrano devised a number of ways to go to the moon. Here is one of the ways: ''Seated on an iron plate, [I] hurl a magnet in the air—the iron follows—I catch the magnet—throw again—and so proceed indefinitely.'' Criticize Cyrano's plan.

9.17 (5) *For champion billiard players only.* Suppose three bodies collide simultaneously. Show that the momentum of the three-body system is conserved in the collision.

9.18 (5) *sss-BOOM-aaah!* A skyrocket is designed to explode into many luminous particles, making a beautiful display. Why do the particles appear as an expanding but coherent spherical system rather than a random collection of particles?

9.19 (G) *A real expert practices every chance he gets.* A juggler stands quietly on a bathroom scale, holding a number of balls. If he starts to juggle the balls so that some balls are always in the air, does the average reading of the scale change? Explain.

9.20 (G) *Major leagues.* A professional baseball player hits a fast-ball pitch and knocks it into the stands for a home run. What impulse does he impart to the ball? Make reasonable estimates of the necessary quantities.

PROBLEMS

GROUP A

9.1 (2) *Dropping the ball.* A 5-kg ball is released from rest and falls freely. **(a)** What are the magnitude and direction of the impulse imparted to the ball over the 0.5-s interval following its release? **(b)** What is the momentum of the ball at the end of that interval? **(c)** What is the final velocity of the ball?

9.2 (2) *Forcing the issue, I.* The illustration below is a plot of the magnitude of the force acting on a 3-kg body as a function of time. **(a)** Find the magnitude of the impulse imparted to the body. **(b)** If the body is initially at rest and no other forces are acting on it, what is its speed at $t = 9$ s? **(c)** What constant force, acting over the same time interval, would accelerate the body to the same final speed?

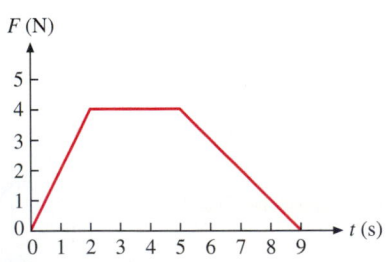

9.3 (2) *Forcing the issue, II.* A body of mass 2.5 kg is moving in the positive x direction with speed 12 m/s. A constant force **F** acts on it for 3 s, after which the body is moving with the same speed but in the positive y direction. Find the magnitude and direction of **F**.

9.4 (2) *Strenuous game, I.* In a handball game, the ball hits the wall head on at a speed of 20 m/s and rebounds at a speed of 14 m/s. If the mass of the ball is 130 g, find **(a)** the impulse imparted to the wall by the ball; **(b)** the change in momentum of the ball.

9.5 (2) *Magnitude and direction.* A particle of mass 85 kg moves with velocity $\mathbf{v} = (3.2\hat{\mathbf{x}} + 8.1\hat{\mathbf{y}})$ m/s. Find **(a)** the x and y components of the momentum and **(b)** the magnitude and direction of the momentum **p**.

9.6 (2) *Momentum vs. energy.* A 2000-kg automobile moves with speed 30 m/s. **(a)** What is the magnitude of its momentum? **(b)** At what speed will a 20 000-kg truck have the same momentum? **(c)** Under these conditions, will the car and the truck have equal kinetic energies? If not, what is the ratio of the energies?

9.7 (2) *Thinking big.* The earth's mass is 5.97×10^{24} kg, and the radius of its orbit about the sun is 1.49×10^{11} m. **(a)** Find the magnitude of the earth's momentum at any instant, as seen by an imaginary observer on the sun. **(b)** Find the change in the earth's momentum over a six-month interval. **(c)** Find the change in the earth's momentum over a three-month interval.

9.8 (3) *Freight yard routine.* An empty freight car of mass 10 tonnes coasts along a straight track with speed 1.0 m/s. It couples with a similar car, initially at rest. After coupling, the two cars move as a unit with speed 0.5 m/s. **(a)** Show that momentum is conserved in the collision. **(b)** Show that kinetic energy is *not* conserved.

9.9 *(3) Recoil.* A rifle of mass 4.6 kg fires a 7.1-g bullet with a muzzle speed of 630 m/s. What is the initial speed of the rifle butt as it pushes against the shoulder of the person holding it? Neglect the momentum of the gases that leave the muzzle of the rifle.

9.10 *(4) Boom!* A large gun is mounted on a railway flatcar; the total mass is 1.5×10^4 kg. The gun is aimed at an angle 30° above the horizontal, and the car is initially stationary. After the gun is fired, how fast does the flatcar move? Take the mass of the shell to be 130 kg and its muzzle speed to be 950 m/s. Why is the impulse imparted to the flatcar by the gun not equal to the final momentum of the car?

9.11 *(4) Festivities.* Every Fourth of July, a small Pennsylvania town fires an antique cannon first used in the American Revolution. Its mass is 1.2×10^3 kg, and that of the cannonball is 5.0 kg. The ball takes 0.011 s to travel the 1.5-m length of the barrel. If the muzzle is elevated to 20° above the horizontal, find **(a)** the initial recoil speed of the cannon and **(b)** the mean vertical force exerted by the cannon on the ground.

9.12 *(4) Fun on the water, I.* A children's summer camp organizes a variety relay race. One of the children is to receive the baton from a runner on his team, run down a pier, jump into a canoe, and paddle across a lake. The canoe is initially at rest at the end of the pier. The child runs at 8.0 km/h as he jumps into the canoe. If his mass is 35 kg and that of the canoe is 45 kg, how fast is the canoe going as it leaves the pier?

9.13 *(4) Fun on the water, II.* In the race described in Problem 9.12, the opposing team decides to get an advantage. As one child on this team runs down the pier at 8.0 km/h carrying the baton, another child gets the canoe going by paddling the length of the pier, achieving a maximum speed of 5.3 km/h. The running child catches up and jumps into the canoe at the end of the pier. If each child has mass 35 kg, how fast is the canoe going as it leaves the pier?

9.14 *(4) Practice makes perfect.* Figure skaters Alice (mass 46 kg) and Bob (mass 68 kg) are just beginning to practice a new routine. They skate toward each other, each with speed 4.3 m/s. They intend to link arms and spin, but they miscalculate and collide head on. If they hold on to each other, what is their final velocity? Which one experiences the greater acceleration?

9.15 *(4) Shoot 'em up, I.* A machine gun fires 600 bullets per minute with muzzle speed 500 m/s. Each bullet has mass 7 g. What is the equivalent mean recoil force?

9.16 *(4) Tough job.* A fire hose delivers water at 300 liters per minute. The water leaves the nozzle at a speed of 25 m/s. What is the magnitude of the force that must be exerted on the hose by the firefighters who hold it? The mass of a liter of water is 1 kg.

9.17 *(4) Space mission, I.* A rocket of mass 1.4×10^5 kg is launched from a space station in outer space, where there are no appreciable gravitational forces. **(a)** If the exhaust speed is 4800 m/s, what is the speed of the rocket relative to the space station when 3.3×10^4 kg of fuel has been burned? **(b)** If the burning of 3.3×10^4 kg of fuel requires 620 s, what is the force exerted on the vehicle?

9.18 *(4) Space shuttle, I.* The fuel supply for the main engines of the NASA space shuttle consists of 102×10^3 kg of liquid hydrogen and 609×10^3 kg of liquid oxygen. The engines burn for a total of 8.5 min, during which they generate an approximately constant thrust of 5.0×10^6 N. Find the specific impulse achieved and compare it with the theoretical value quoted in the text.

9.19 *(5) Fourth of July.* A skyrocket is flying at 10 m/s when it breaks into two parts. One part, whose mass is 40% of the whole, continues to move in the same direction but with its speed increased to 25 m/s. Find the speed of the other part.

9.20 *(5) Center of mass.* Find the coordinates of the center of mass of the three-particle system illustrated below. Compare your result with that of Example 9.4. Why does the center of mass of the present system lie away from the altitude of the triangle?

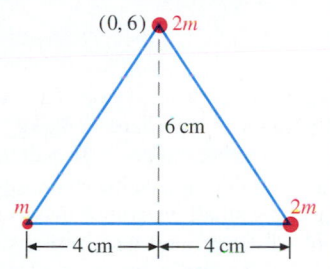

9.21 *(5) Earth and moon.* The distance from the center of the earth to the center of the moon is close to 60 times the radius of the earth. The mass of the moon is $\frac{1}{81}$ that of the earth. Show that the center of mass of the earth-moon system lies within the earth.

9.22 *(5) Center of mass of a right triangle.* The figure below shows a uniform plate in the form of a right triangle of mass M, with base b and altitude h. **(a)** Demonstrate that the mass dm of the infinitesimal segment shown satisfies the ratio

$$\frac{dm}{M} = \frac{y\,dx}{\frac{1}{2}bh}.$$

(b) Show that the x coordinate of the center of mass is $X = \frac{2}{3}b$. **(c)** *Without any further calculation,* show that the y coordinate of the center of mass is $Y = \frac{1}{3}h$.

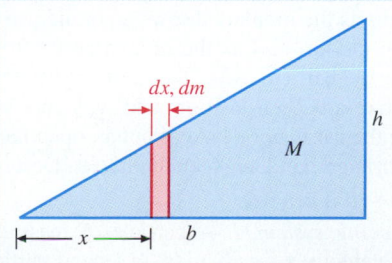

9.23 *(5) Center of mass of an isosceles triangle.* A uniform plate is in the form of an isosceles triangle. Show that the center of mass lies on the altitude, two-thirds of the way from the vertex to the base, and that this result is independent of the altitude and the base.

9.24 *(5) Center of mass of an arbitrary triangle.* Show that the center of mass of a uniform triangular plate lies at the intersection of its medians. (This intersection is called the **centroid** of the triangle.)

9.25 *(5) Center of mass of a trapezoid, I: brute force.* Find the center of mass of the isosceles trapezoidal plate shown below by direct application of Equation 9.20.

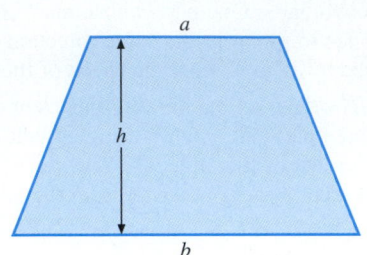

9.26 *(5) Never stand up in a small boat! I.* On a quiet lake, a fisherman sits in the stern of his canoe. The mass of the man is 85 kg, and the mass of the canoe is 35 kg. The fisherman walks to the bow of the boat to get fresh bait. **(a)** If the canoe is 5.2 m long, how far does the canoe move relative to a point on shore? Neglect the small difference between the length of the canoe and the distance over which the fisherman moves. **(b)** A canoe is close to symmetrical about its center, so its center of mass is near the center of the canoe. Would the result of part **a** be different for an asymmetrical rowboat, whose center of mass is closer to the stern than to the bow?

9.27 *(5) Never stand up in a small boat! II.* A child of mass 35 kg stands up in the stern of a rowboat. She walks from the stern to the bow, a distance of 4.8 m. If the mass of the rowboat is 95 kg, how far does the rowboat move with respect to the shore? How far does the child move with respect to the shore? Neglect any motion of the water.

9.28 *(5) Monkey business.* A railroad boxcar rests on frictionless wheels. The mass of the car is 40 tonnes, and its length is 20 m. Inside the car at one end is a large crate of oranges, each of mass 0.3 kg. A monkey trapped inside the car relieves his boredom by flinging the oranges to the opposite end of the car where they hit the wall and fall to the floor. The monkey throws each orange with speed 15 m/s. **(a)** Describe the motion of the car as the monkey throws an orange, as the orange flies through the air, and as the orange comes to rest at the other end of the car. **(b)** After the monkey has thrown 1000 oranges, what are the position and velocity of the car? **(c)** Suppose the car is not a boxcar but an open flatcar, and so the thrown oranges land between the tracks. Describe the motion of the car in this case.

9.29 *(5) Moving system, I.* A particle of mass $m_1 = 3$ kg moves with velocity $\mathbf{v}_1 = 8\hat{\mathbf{x}}$ m/s. A second particle of mass $m_2 = 7$ kg moves with velocity $\mathbf{v}_2 = 4\hat{\mathbf{y}}$ m/s. Find **(a)** the velocity of the center of mass of this two-particle system and **(b)** the total momentum of the system.

9.30 *(5) Moving system, II.* A particle of mass $m_1 = 4.5$ kg moves with speed $v_1 = 6.8$ m/s in a direction 28° north of east. A second particle of mass $m_2 = 9.3$ kg moves with speed $v_2 = 2.1$ m/s in a direction 56° west of north. Find **(a)** the velocity of the center of mass of this two-particle system and **(b)** the total momentum of the system.

9.31 *(5) Moving system, III.* A particle of mass $m_1 = 2.3$ kg moves with velocity $\mathbf{v}_1 = (7.2\hat{\mathbf{x}} + 3.0\hat{\mathbf{y}} - 8.6\hat{\mathbf{z}})$ m/s. A second particle of mass $m_2 = 6.4$ kg moves with velocity $\mathbf{v}_2 = (3.6\hat{\mathbf{x}} - 1.2\hat{\mathbf{y}} + 4.5\hat{\mathbf{z}})$ m/s. A third particle of mass $m_3 = 1.3$ kg moves with velocity $\mathbf{v}_3 = (-4.4\hat{\mathbf{x}} - 1.8\hat{\mathbf{y}} + 6.1\hat{\mathbf{z}})$ m/s. Find **(a)** the velocity of the center of mass of this three-particle system and **(b)** the total momentum of the system.

9.32 *(5) Oops!* You are standing in a shop, holding a handsome vase of mass m that has attracted your attention, when you let it slip and it falls to the floor. It breaks into three pieces, two of mass $\frac{1}{4}m$ and one of mass $\frac{1}{2}m$. If the two pieces of equal mass slide along the floor with equal initial speed v, but at right angles to one another, what is the initial velocity of the third piece?

9.33 *(6) Mass according to Mach.* You perform a collision experiment with two gliders on an air track. The first glider has mass $m_1 = 0.662$ kg. Its initial velocity is $v_{1i} = 1.85$ m/s, and its final velocity is $v_{1f} = -1.32$ m/s. The initial velocity of the second glider is $v_{2i} = -2.17$ m/s, and its final velocity is $v_{2f} = 3.22$ m/s. What is its mass m_2?

9.34 *(G) Letting go.* In the drawing below, a spring of negligible mass is compressed between two blocks of mass $m_1 = 2.23$ kg and $m_2 = 6.47$ kg. The blocks rest on a frictionless surface. The string that holds the two blocks together is burned, allowing the spring to expand and push the blocks apart. **(a)** If the velocity of block 1 is $\mathbf{v}_1 = -3.55\hat{\mathbf{x}}$ m/s, find the velocity $\mathbf{v}_2$ of block 2. **(b)** Find the total kinetic energy of the system. **(c)** If the spring is initially compressed 5.21 cm from its relaxed length, find the force constant of the spring. Assume that total mechanical energy is conserved as the spring expands.

9.35 *(G) Radioactive decay.* The mass of the nucleus of a radium atom is 3.77×10^{-25} kg. At some unpredictable instant, the nucleus undergoes radioactive decay by ejecting an alpha particle whose mass is 6.68×10^{-27} kg. The kinetic energy of the alpha particle is 7.26×10^{-16} J. Assuming that the mass of the system does not change appreciably during the decay process, find **(a)** the momentum of the alpha particle, **(b)** the speed of the remaining nucleus (called the daughter nucleus), and **(c)** the kinetic energy of the daughter nucleus.

9.36 (2) *Building momentum.* A body of mass m is dropped from a height h. **(a)** What is the impulse imparted to the body by the gravitational force? **(b)** What is the momentum of the body just before it hits the ground? Neglect air resistance.

9.37 (2) *Impulsive action.* You throw a ball upward with initial speed 12 m/s. **(a)** What is the momentum of the ball when it reaches its maximum height? **(b)** Use the impulse-momentum theorem to find how long the ball takes to reach its maximum height.

9.38 (2) *Building momentum?* **(a)** Show that the impulse imparted to an automobile when it hits a wall at 90 km/h is about the same as it would be if the auto fell to the ground from the top of a 10-story building. Take the height of a story to be about 3 m. **(b)** What is the equivalent height if, instead of a wall, the car hits an identical car going 90 km/h in the opposite direction?

9.39 (3) *Forcing the issue, III.* For the 3-kg body subjected to the force depicted in Problem 9.2, find both **(a)** the momentum and **(b)** the speed as functions of time.

9.40 (3) *Forcing the issue, IV.* A body of mass m, initially at rest, is subjected to a force $F = F_0 \sin^2 t$. Find both **(a)** its momentum and **(b)** its speed as functions of time during the interval $0 \leq t \leq 2\pi$. (Hint: $\int \sin^2 t\, dt = \frac{1}{2}t - \frac{1}{4}\sin 2t$.)

9.41 (3) *Forcing the issue, V.* The displacement of a body attached to a spring is described by the equation $x(t) = A \cos \omega t$, where $\omega^2 = k/m$ is the ratio of the force constant to the mass. **(a)** Using Hooke's law to determine the spring force $F(t)$, find the impulse $J(t)$ imparted to the mass as a function of time. **(b)** Beginning with the equation for $x(t)$, find the velocity $v(t)$ and thus determine the momentum $p(t)$ of the body. **(c)** Compare the results of parts **a** and **b**.

9.42 (3) *Strenuous game, II.* Equation 9.1 defines impulse as the integral over time of a force $\mathbf{F}(t)$. Most commonly, the function $\mathbf{F}(t)$ is not known. **(a)** Use the mean value theorem of calculus to show that you can nevertheless express the impulse in terms of a *mean force* $\langle \mathbf{F} \rangle$ by writing

$$\mathbf{J} = \langle \mathbf{F} \rangle \, \Delta t,$$

where Δt is the duration of the impulse. (You can find a statement of the mean value theorem in part **c** of Problem 7.46 or in your calculus text.) **(b)** Use the result of part **a** to find the mean force exerted by the wall on the handball of Problem 9.4 if the ball is in contact with the wall for 2.2 ms.

9.43 (3) *Splat!* A ball of clay has mass m and diameter D. It is dropped onto a concrete slab from a height $h \gg D$. The height h is just sufficient so that the ball is flattened into a thin pancake. Show that the mean force exerted on the ball during impact is approximately mgh/D. (Hint: In order to calculate the collision time, you must make a reasonable estimate of the mean speed of the ball during the collision.)

9.44 (4) *Winter wonderland.* An ice skater whose mass is 70 kg and who is initially at rest throws a 1.2-kg piece of ice horizontally with speed 15 m/s. If the coefficient of kinetic friction between the skates and the ice is 0.03, how far will the skater move backward before coming to rest again?

9.45 (4) *Ballistics laboratory.* A bullet of mass 12.0 g embeds itself in a wooden block of mass 1.50 kg that lies initially at rest on a frictionless table. If the final speed of the block is 5.48 m/s, what was the speed of the bullet?

9.46 (4) *Shoot 'em up, II.* Show that the equivalent mean recoil force $\langle F \rangle$ exerted on a machine gun is directly proportional to the frequency ν with which it fires bullets.

9.47 (4) *Making cookies.* A baker pours sugar from a height h into a bowl on a kitchen scale. The first sugar reaches the bowl at $t = 0$. If he pours at a rate of σ kg/s, show that the scale reading (set to zero when the bowl is empty) is $\sigma(gt + \sqrt{2gh})$. What happens to the scale reading when he stops pouring?

9.48 (4) *Space shuttle, II.* When a significant gravitational force is exerted on a rocket as it is launched, that force must be taken into account in analyzing the motion of the rocket. **(a)** Assuming that g remains essentially constant over the period of rocket burn, show that the equation of motion becomes

$$ma = v_{ex} \left| \frac{dm}{dt} \right| - mg.$$

(b) Show that, if the fuel is burned over a time t, the final speed of the rocket is

$$v_f = v_{ex} \ln \frac{m_i}{m_f} - gt.$$

Compare this result with Equation 9.14, which was derived for the zero-gravity case. Why is it important in the present case to burn the fuel as fast as possible? **(c)** The launch mass of the NASA space shuttle is 2.0×10^6 kg. The combined thrust of the two solid-fuel booster engines and the three hydrogen-oxygen main engines is 3.1×10^7 N. What is the acceleration of the vehicle at lift-off? **(d)** The solid-fuel boosters exhaust their fuel in 150 s and are jettisoned. Just before the booster fuel is exhausted, the mass of the vehicle has diminished to 0.8×10^6 kg. What is the acceleration at this point? The altitude of the vehicle at this moment is about 55 km, and you can assume to an adequate approximation that the value of g is not too much smaller than the sea-level value.

9.49 (4) *Steering jets.* A spaceship of mass M is moving through deep space at constant velocity $\mathbf{v}_0$. In order to change its direction, a steering jet is fired at right angles to the axis of the ship. The exhaust speed is v_{ex}. When a fuel mass m has been expended, **(a)** through what angle has the spaceship turned? **(b)** what is its speed?

9.50 (4) *Space mission, II.* A space vehicle is launched from a space station far from the earth. The initial mass of the vehicle is m_0; during the fuel burn, its mass changes at the rate dm/dt. **(a)** Find the speed of the vehicle relative to the space station at an arbitrary time t during the burn period. **(b)** Consider again the vehicle of Problem 9.17. How fast is the vehicle going at $t = 620$ s? at $t = 1240$ s? **(c)** Show that, at time t, the distance r traveled from the space station is

$$r = v_{ex}t - v_{ex}\left(t + \frac{m_0}{dm/dt}\right) \ln\left(1 + \frac{dm/dt}{m_0}t\right).$$

(d) How far has the vehicle of Problem 9.17 traveled at $t = 620$ s? at $t = 1240$ s?

9.51 *(5) Center of mass of a semicircular rod.* A uniform rod is bent into a semicircle of diameter $2R$, as shown below. Find the center of mass.

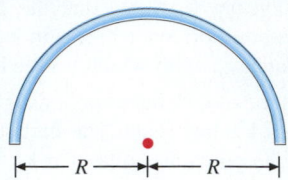

9.52 *(5) Center of mass of a semicircular plate.* Find the center of mass of a plate of uniform thickness that has the form of a semicircle of diameter $2R$. (Hint: Divide the plate into a set of very many nested semicircular rods having a common center, and use the result of Problem 9.51.)

9.53 *(5) Center of mass of a cone.* Find the center of mass of a right circular cone of altitude h.

9.54 *(5) Center of mass of a bracket: a bit of trickery.* An L-shaped bracket is formed by making a right-angle bend in a uniform strip of steel. The two arms of the bracket have lengths A and B. Find the center of mass of the bracket. (Hint: Where are the centers of mass of the two arms taken separately?)

9.55 *(5) Center of mass of a trapezoid, II: more trickery.* Refer to the illustration in Problem 9.25 and note that the isosceles trapezoid can be thought of as an isosceles triangle from which a piece, also an isosceles triangle, has been removed.

Substitute for the "original triangle" a properly located particle of the same mass. Then substitute for the missing piece a properly located particle whose mass is the *negative* of the mass of the missing piece. Finally, locate the center of mass of the two particles. Compare your result with that obtained by direct calculation in Problem 9.25.

9.56 *(5) Center of mess.* Using the method of Problem 9.55, find the center of mass of the object shown below. The two circles (the disk and the hole) are tangent.

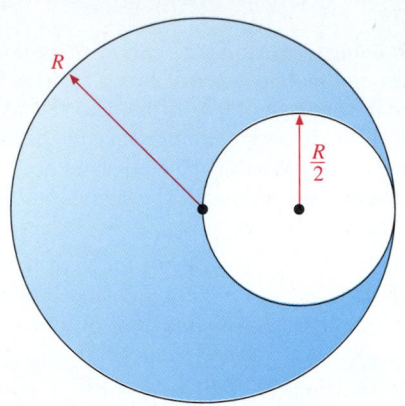

9.57 *(5) Skyrocket.* A skyrocket explodes just as it reaches the top of its trajectory at height h above the ground. The display, consisting of many luminous particles, spreads out as an expanding, brilliant sphere. The bottom of the sphere just touches the ground when its radius is R. With what speed were the luminous particles ejected by the explosion? (Hint: Draw a sketch before you start to calculate.)

GROUP C

9.58 *(4) Multistage rocket.* It is possible to improve on the performance of a rocket-propelled vehicle for a given gross payload ratio by using a multistage arrangement. In this arrangement, a large vehicle, called the first stage, has a smaller complete vehicle, called the second stage, as its net payload. In principle, there may be as many more stages as desired, but the practical limit appears to be three or four. As each stage exhausts its fuel, its remaining structure (engine, tanks, and so forth) is jettisoned, and then the next stage fires. Consider a system with n stages. For simplicity, assume that the gross payload ratio m_f/m_i is the same for all stages. Show that, when all n stages have been fired, the total speed increment of the final stage is

$$\Delta v = n v_{\text{ex}} \ln \frac{m_i}{m_f}.$$

Compare this result with the Tsiolkovskii formula for a single-stage rocket (Equation 9.14).

9.59 *(5) Pappus's theorem, I.* The illustration opposite shows an arbitrary plane figure whose area is A. It lies in the xy plane, but the x axis does not pass through it. You can generate a solid figure by rotating the plane figure about the x axis. Show that the volume V of the solid thus generated is

$$V = 2\pi AY,$$

where Y is the y coordinate of the center of mass of the plane figure. This proof is called **Pappus's theorem**, after the fourth-century Alexandrian mathematician who first devised it.

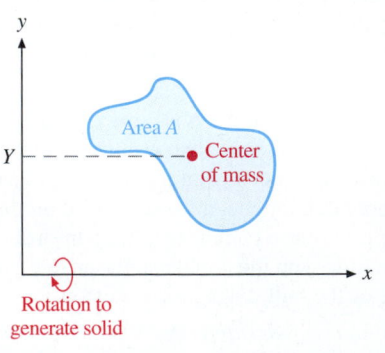

9.60 *(5) Pappus's theorem, II.* The illustration (p. 247) shows a circular torus (a donut) of major radius R and minor radius a. Use Pappus's theorem to show that its volume is $2\pi^2 a^2 R$.

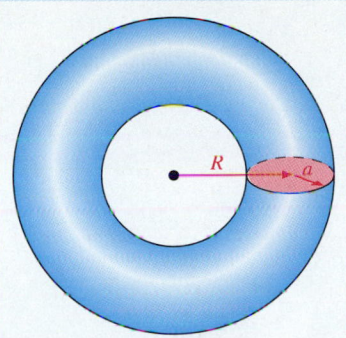

9.61 *(5) Pappus in reverse.* Work backward, using Pappus's theorem, and find the center of mass of a semicircular plate.

9.62 *(G) Leaky car.* A railroad gondola car is filled with sand. The total mass is m. Beginning at rest at $t = 0$, it is accelerated by a locomotive pulling with constant force F. There is a leak in the bottom of the car, through which sand escapes at a rate σ kg/s. Neglecting friction, find the speed of the car as a function of time.

9.63 *(G) Sierra snowstorm.* A diesel locomotive pushes a powerful rotary snow plow at a speed of 5.2 m/s. The plow throws snow to the side (perpendicular to its direction of motion) at a rate of 2.4×10^3 kg/s. **(a)** What force is the locomotive exerting on the plow, over and above that required to push it in the absence of snow? **(b)** In practice, the plume of snow is ejected not exactly to the side but at a backward angle with respect to the motion of the plow. Why is this advantageous?

9.64 *(G) Rope trick.* A heavy rope, whose mass per unit length is λ kg/m, lies coiled on a table. You wish to pull the rope upward at a constant speed v. Show that you must exert a force that varies with the amount of rope l already lifted off the table according to the rule $F = \lambda(gl + v^2)$.

9.65 *(G) Conveyor belt.* In a concrete plant, sand is fed to a mixer by means of a conveyor belt. The sand drops onto the belt from a hopper at a controlled rate σ kg/s; see the illustration below. **(a)** Neglecting friction in the belt mechanism, what force F is required to drive the belt at the constant speed v? **(b)** Does the proportionality $F \propto v$ (rather than $F \propto a$) constitute a violation of Newton's second law? Why not? (Hint: Begin with Equation 9.30 and note that, for the system comprising the belt and the sand on it, m is *not* constant.)

9.66 *(G) The hard weigh.* The device shown at the top of the next column is situated on a sensitive electronic scale. The upper compartment is filled with fine sand, and the lower compartment is empty. The small hole in the bottom of the upper compartment is closed. At time $t = 0$, the hole is opened without disturbing the scale reading. The sand flows through the hole at a rate σ kg/s. Describe what happens to the scale reading up to the time when all the sand is at rest in the bottom compartment.

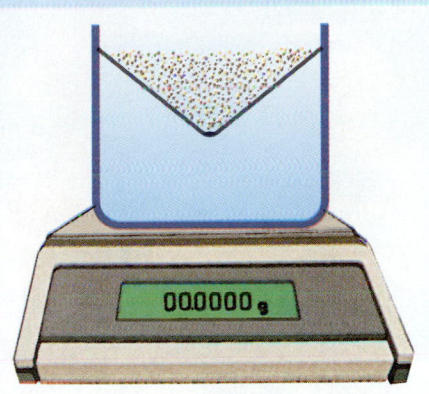

9.67 *(G) Reduced mass.* The figure below shows two masses, m_1 and m_2, connected by a spring of relaxed length L and force constant k. The system is located on a frictionless table. You pull the two masses apart to a separation $L + D$ and release the system, thus setting it into oscillation. You are careful to release the two masses simultaneously. **(a)** Describe the motion of the center of mass of the system. **(b)** Choose a coordinate system with its origin at the center of mass. This is called the *center-of-mass coordinate system* or the *barycentric system*. Find the equilibrium distance of each mass from the origin of this system. **(c)** From the point of view of an observer at the center of mass, what is the effective length of the spring to which m_1 is attached? the spring to which m_2 is attached? **(d)** From the point of view of an observer at the center of mass, what is the effective force constant of the spring to which m_1 is attached? the spring to which m_2 is attached? **(e)** What is the angular frequency ω of the oscillation of m_1? of m_2? Express the result(s) in terms of the quantity μ, defined by the relation

$$\frac{1}{\mu} \equiv \frac{1}{m_1} + \frac{1}{m_2}.$$

The quantity μ is called the *reduced mass* of the system. **(f)** If you now hold m_1 fixed in place, what will be the angular frequency ω_2 with which m_2 oscillates? If you hold m_2 fixed, what will be the angular frequency ω_1 with which m_1 oscillates? Express the result(s) of part **e** in terms of ω_1 and ω_2.

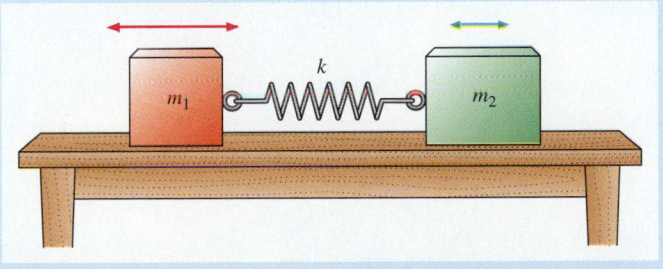

10

Linear Momentum and Mechanical Energy

LOOKING AHEAD

——— Momentum and mechanical energy are distinct quantities that are conserved under different circumstances.

——— In totally inelastic collisions, momentum is conserved but mechanical energy is not.

——— The ballistic pendulum is analyzed as a two-part process: momentum is conserved in the first part but energy is not; energy is conserved in the second part but momentum is not.

——— Elastic collision in one dimension is the simplest example of a process in which momentum and mechanical energy are both conserved.

——— In two-dimensional collisions, momentum conservation may be applied component by component.

Left: The momentum of the white cue ball is transferred in a complicated way to the other fifteen balls. The rapid deceleration of the cue ball is evident, as is the rapid acceleration of some of the other balls. Much of the mechanical energy is transferred from the cue ball to the others. What kind of measurements can be used to determine how much is lost?

Like to a double cherry, seeming parted
But yet an union in partition;
Two lovely berries molded on one stem.

—SHAKESPEARE, *A Midsummer-Night's Dream*

It is always good
When a man has two irons in the fire.

—BEAUMONT AND FLETCHER, *The Faithful Friends*

SECTION 10.1   Introduction

Linear momentum and mechanical energy are distinct quantities, but there are intriguing parallels between them. Each lies at the heart of a profound and powerful conservation principle. In the three preceding chapters, we explored the two quantities separately, and we exploited the corresponding conservation principles separately. We now use the two concepts together to study a variety of mechanical systems.

SECTION 10.2   Momentum and Energy

When a collision takes place within an isolated system, linear momentum is always conserved. *Sometimes,* mechanical energy is conserved as well. A collision in which mechanical energy is conserved is called an **elastic collision**. A collision in which mechanical energy is not conserved is called an **inelastic collision**. If we know what happens to the energy of a system as a collision occurs within it, we have information in addition to what momentum conservation tells us. In this section, we take advantage of the additional information.

Totally Inelastic Collisions in One Dimension

We consider first the case of a head-on (one-dimensional) collision in which the two colliding bodies stick together, as when a ball strikes a lump of clay. This special type of inelastic collision is called **totally inelastic**. The 1994 collision of comet Shoemaker-Levy with Jupiter was a spectacular example of such a collision.

The analysis of totally inelastic collisions is simplified by the fact that the final velocities of the two bodies are identical;

$$v_{1f} = v_{2f} \equiv v_f.$$

Indeed, we take the common final velocity of the colliding bodies as the defining property of a totally inelastic collision. We could use this equality to simplify Equation 9.11a, $m_2(v_{2f} - v_{2i}) = -m_1(v_{1f} - v_{1i})$, which applies to all one-dimensional collisions. But it is easier to begin with Equation 9.10b,

$$\mathbf{p}_f = \mathbf{p}_i,$$

in its one-dimensional form

$$p_f = p_i.$$ (10.1)

When we insert the appropriate masses and velocities, this equation gives us

$$(m_1 + m_2)v_f = m_1 v_{1i} + m_2 v_{2i}.$$

Solving for v_f, we have

$$v_f = \frac{m_1 v_{1i} + m_2 v_{2i}}{m_1 + m_2}.$$ (10.2)

Consider an important special case: a totally inelastic collision in which one of the bodies (say, m_2) is initially at rest. In this case, Equation 10.2 becomes

$$v_f = \frac{m_1}{m_1 + m_2} v_{1i} \quad \text{for } v_{2i} = 0.$$ (10.3a)

If you rewrite this equation in the form

$$\frac{v_f}{v_{1i}} = \frac{m_1}{m_1 + m_2},$$ (10.3b)

you can see that the ratio of the final velocity of the combined bodies to the initial velocity of body 1 is equal to the ratio of m_1 to the total mass of the system. This agrees with everyday experience: When you roll a small lump of clay at a stationary bowling ball, the clay-ball combination hardly moves at all. But when you roll a bowling ball at a small lump of clay, the ball picks up the clay and rolls on without much decrease in speed.

We now show that mechanical energy is not conserved in a totally inelastic collision. In particular, we ask: What proportion of the initial energy E_i of the system is lost in the case described by Equation 10.3? Because m_2 is initially at rest, the initial energy of the system is just the kinetic energy of m_1:

$$E_i = \tfrac{1}{2} m_1 v_{1i}^2.$$

The final energy of the system is the kinetic energy

$$E_f = \tfrac{1}{2}(m_1 + m_2)v_f^2.$$

When we substitute the value of v_f given by Equation 10.3a, this equation becomes

$$E_f = \frac{1}{2} \frac{m_1^2}{m_1 + m_2} v_{1i}^2.$$

We now have the information necessary to find the proportional energy loss

$$\left| \frac{\Delta E}{E_i} \right| = \frac{E_i - E_f}{E_i}.$$

Into this expression, we insert the values of E_i and E_f just obtained. This yields

$$\left| \frac{\Delta E}{E_i} \right| = \frac{\dfrac{1}{2} m_1 v_{1i}^2 - \dfrac{1}{2} \dfrac{m_1^2}{m_1 + m_2} v_{1i}^2}{\dfrac{1}{2} m_1 v_{1i}^2},$$

which simplifies to

$$\left| \frac{\Delta E}{E_i} \right| = 1 - \frac{m_1}{m_1 + m_2} \quad \text{for } v_{2i} = 0,\ v_{1f} = v_{2f} \text{ (totally inelastic collision).}$$ (10.4)

According to this equation, the proportional energy loss depends only on the ratio of the mass of the initially moving body to the total mass of the system. (Compare with Equation 10.3b.) Does Equation 10.4 make sense physically? To see that it does, let us explore two extreme cases. First, consider what happens if $m_1 \ll m_2$. This case is exemplified by a bullet burying itself in a bale of cotton. We expect the initial kinetic energy of the system to dissipate completely. And, indeed, Equation 10.4 yields $|\Delta E/E_i| \simeq 1$; the loss of energy is essentially equal to the initial energy. Next, consider what happens if $m_1 \gg m_2$. This case is exemplified by a rolling boulder picking up a small piece of mud. We expect the boulder to be affected very little. And, indeed, Equation 10.4 yields $|\Delta E/E_i| \simeq 0$.

EXAMPLE 10.1

In a railroad switching yard, a freight car is made to roll at 2.3 m/s and couples with a stationary train made up of seven cars identical with the rolling car. **(a)** What is the final speed of the eight-car combination, and what is the proportional energy loss in the coupling process? **(b)** What would be the final speed and the proportional energy loss if the seven-car train were made to roll into a stationary single car at the same initial speed of 2.3 m/s?

SOLUTION:
(a) Find v_f and $|\Delta E/E_i|$ when the moving single car couples with the stationary seven-car train.

According to Equations 10.3a and 10.4, both v_f and $|\Delta E/E_i|$ depend on the mass ratio $m_1/(m_1 + m_2)$. Call the mass of one freight car M. Because the cars are identical, the mass of the seven-car train is $7M$, and the total mass of the eight cars coupled together is $8M$. The ratio of the initially moving mass to the total mass of the system is thus

$$\frac{m_1}{m_1 + m_2} = \frac{M}{8M} = \frac{1}{8}.$$

Inserting this value of the mass ratio into Equation 10.3a, together with the value $v_i = 2.3$ m/s, you obtain

$$v_f = \tfrac{1}{8} \times 2.3 \text{ m/s} = 0.29 \text{ m/s}.$$

Inserting the value of the mass ratio into Equation 10.4, you

have

$$\left| \frac{\Delta E}{E_i} \right| = 1 - \frac{1}{8} = \frac{7}{8} = 0.88.$$

Most of the initial kinetic energy—seven-eighths—is lost, and the final speed is only one-eighth the initial speed.

(b) Find v_f and $|\Delta E/E_i|$ when the moving seven-car train couples with the stationary single car.

In this case, the initial moving mass is $7M$; the total mass of the system is still $8M$. You now have the mass ratio

$$\frac{m_1}{m_1 + m_2} = \frac{7M}{8M} = \frac{7}{8}.$$

Repeating the calculations of part **a** with this mass ratio, you obtain

$$v_f = \frac{7}{8} \times 1.3 \text{ m/s} = 2.0 \text{ m/s} \quad \text{and} \quad \left| \frac{\Delta E}{E_i} \right| = 1 - \frac{7}{8} = \frac{1}{8} = 0.13.$$

Only one-eighth of the initial kinetic energy is lost, and the final speed is seven-eights as great as the initial speed. The results of the calculations are consistent with the qualitative remarks made earlier; the collision of part **a** is something like a small clay ball rolled into a bowling ball, whereas the collision of part **b** is something like a rolling bowling ball picking up a small piece of clay.

The Ballistic Pendulum

A classical application of the principles developed in the preceding subsection is the **ballistic pendulum**. This device, shown in Figure 10.1, has been used for centuries to determine the speed of bullets and other similar projectiles whose speed is too great for easy direct measurement.

The operation of the ballistic pendulum is analyzed in two stages. During the very brief first stage, the bullet collides totally inelastically with the block, which is initially at rest. As you learned in the preceding subsection, mechanical energy *is not* conserved during this process. However, the momentum of the bullet-block system *is* conserved, and we can equate the initial and final momenta according to Equation 10.1. In the notation of Figure 10.1, this gives

$$mv = (M + m)V, \tag{10.5}$$

where V is the instantaneous speed of the bullet and block just after the collision.

As a consequence of the collision, the block swings, rising through a maximum vertical height h before swinging back. During this second stage, momentum is *not* conserved because the system is acted on by the nonzero external force that is the vector sum of the gravitational force and the string forces. However, total mechanical energy *is* conserved because, once the collision is over, all the forces acting on the block conform to the criteria given in Table 9.1: The gravitational force $(M + m)\mathbf{g}$ acting on

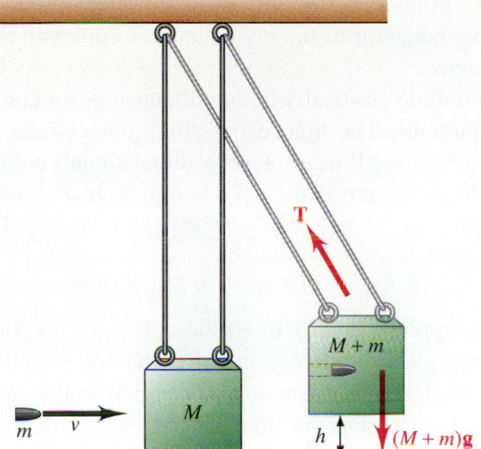

FIGURE 10.1 A ballistic pendulum. The bullet, having mass m and initial speed v, buries itself in the wooden block of mass M. The block subsequently swings, rising through a maximum vertical height h.

the system is conservative, and the force $\mathbf{T}$ exerted by the strings is a workless constraint because the strings do not stretch. We can therefore write

$$\tfrac{1}{2}(M + m)V^2 = (M + m)gh.$$

We solve this equation for V, obtaining the familiar result

$$V = \sqrt{2gh}.$$

Substituting this result into Equation 10.5 and solving for the bullet speed v, we obtain

$$v = \frac{M + m}{m}\sqrt{2gh}. \qquad \textbf{(10.6)}$$

The ballistic pendulum is useful because the block can easily be made much more massive than the bullet. The fraction on the right side of Equation 10.6 is quite large. Consequently, a high bullet speed corresponds to a modest rise h, which can be measured conveniently.

EXAMPLE 10.2

A bullet of mass $m = 15.8$ g is fired from a .38 caliber pistol into the block of a ballistic pendulum. The block has mass $M = 5.68$ kg. The bullet buries itself in the block, and the bullet-block combination rises through a vertical distance $h = 7.78$ cm. **(a)** What is the speed of the bullet just before it hits the block? **(b)** What proportion of the bullet's mechanical energy is dissipated as it buries itself in the block?

SOLUTION:
(a) What is the speed of the bullet just before it hits the block?
   You use Equation 10.6 to find the bullet speed:

$$v = \frac{5.68\ \text{kg} + 0.0158\ \text{kg}}{0.0158\ \text{kg}}$$
$$\times \sqrt{2 \times 9.80\ \text{m/s}^2 \times 7.78 \times 10^{-2}\ \text{m}}$$
$$= 445\ \text{m/s}.$$

This is roughly one and a half times the speed of sound, a typical value for a bullet just after it leaves a pistol.

(b) What proportion of the bullet's kinetic energy is dissipated? According to Equation 10.4, you have

$$\left|\frac{\Delta E}{E_i}\right| = 1 - \frac{0.0158\ \text{kg}}{5.68\ \text{kg} + 0.0158\ \text{kg}} = 0.997.$$

That is, 99.7% of the bullet's kinetic energy is dissipated in the collision. The remaining 0.3% of the initial energy is sufficient to raise the fairly massive block through a vertical distance that is readily measurable.

Elastic Collisions in One Dimension

We now consider collisions in which no mechanical energy is lost. Such *elastic collisions* are common on the microscopic scale, as when two gas molecules collide. Ideal elasticity is rare on the macroscopic scale, but it is often well approximated. This is the case when a steel ball bounces off a polished granite block or when two billiard balls collide. The surprising behavior of the toy called the Superball depends on its approximation to ideal behavior.

When two bodies collide elastically in one dimension, we can describe the collision by means of two equations. The first, expressing conservation of linear momentum, applies without distinction to all head-on (one-dimensional) collisions. The initial momentum of the two-body system is $p_i = p_{1i} + p_{2i} = m_1 v_{1i} + m_2 v_{2i}$. The final momentum of the system is $p_f = p_{1f} + p_{2f} = m_1 v_{1f} + m_2 v_{2f}$. Equating the final and initial momenta, we have

$$m_1 v_{1f} + m_2 v_{2f} = m_1 v_{1i} + m_2 v_{2i}. \tag{10.7}$$

The second equation, applicable *only* to elastic collisions, expresses the conservation of total mechanical energy. Both before and after the collision, the mechanical energy of the two-body system is entirely kinetic and is equal to the sum of the kinetic energies of the two bodies. Equating the final kinetic energy of the system to its initial kinetic energy, we have $K_{1f} + K_{2f} = K_{1i} + K_{2i}$, or

$$\tfrac{1}{2} m_1 v_{1f}^2 + \tfrac{1}{2} m_2 v_{2f}^2 = \tfrac{1}{2} m_1 v_{1i}^2 + \tfrac{1}{2} m_2 v_{2i}^2. \tag{10.8}$$

Equations 10.7 and 10.8 constitute a pair of simultaneous equations, so they can be solved for any two unknown quantities if all other quantities are known. For example, if the masses and the initial velocities are known, it is possible to solve for the final velocities.

A simple criterion characterizes all elastic collisions: *In an elastic collision, the speed of either body with respect to the other—their* **relative speed**—*is unchanged*. We prove this criterion here for one-dimensional collisions only. We begin by rearranging the momentum-conservation equation, Equation 10.7, gathering the terms involving m_1 on one side and the terms involving m_2 on the other. We obtain

$$m_1(v_{1f} - v_{1i}) = -m_2(v_{2f} - v_{2i}). \tag{10.9}$$

We do the same with the energy-conservation equation, Equation 10.8:

$$m_1(v_{1f}^2 - v_{1i}^2) = -m_2(v_{2f}^2 - v_{2i}^2).$$

The velocity terms in this equation factor to yield

$$m_1(v_{1f} - v_{1i})(v_{1f} + v_{1i}) = -m_2(v_{2f} - v_{2i})(v_{2f} + v_{2i}). \tag{10.10}$$

We now divide Equation 10.10 by Equation 10.9 and obtain

$$v_{1f} + v_{1i} = v_{2f} + v_{2i}.$$

Finally, we gather the final (f) terms on one side of the equation and the initial (i) terms on the other. This gives

$$v_{1f} - v_{2f} = -(v_{1i} - v_{2i}) \quad \text{for } \Delta E = 0. \tag{10.11}$$

That is, *the velocity of each body relative to the other is reversed by the collision*. This result is valid only for head-on collisions.

To express the relative *speed*—the speed of each body as seen by an observer located on the other—we take the absolute value of both sides. The result is

$$|v_{1f} - v_{2f}| = |v_{1i} - v_{2i}| \quad \text{for } \Delta E = 0. \tag{10.12}$$

This proves the assertion that the relative speed of the bodies is unchanged by the collision. The proof of this statement for collisions that are not one-dimensional is the subject of Problems 10.43 and 10.44.

The constancy of relative speed expressed in Equation 10.12 is exploited when a space probe is sent to the outer parts of the solar system. The probe is aimed so that it passes very close to Jupiter, approaching from "behind" with respect to Jupiter's motion around the sun, as shown in Figure 9.3. As a result of the gravitational interaction between Jupiter and the probe, the probe's velocity is changed in both magnitude and direction. As noted in Section 9.2, the encounter may be regarded as a collision. Because the only force involved in the collision is the conservative gravitational force, the collision is elastic. Jupiter's mass is so large that its speed relative to the sun (≈ 13 km/s) is unaffected by the collision. Therefore, the fact that the speed of the probe relative to Jupiter remains unchanged implies that the speed of the probe relative to the sun must increase. To get a rough idea of what happens, suppose that the probe approached Jupiter directly from behind and departed directly ahead (which is not the case). In order to satisfy Equation 10.12, the speed of the probe relative to the sun would have to increase by about twice Jupiter's speed relative to the sun, or by about 26 km/s. In practice, smaller but substantial increases in speed (and thus kinetic energy) relative to the sun are achieved at no cost in fuel. In a typical mission, the Jupiter encounter increases the speed of the probe relative to the sun from 17 km/s to 28 km/s; its kinetic energy is nearly tripled.

Elastic Collision in the Case $v_{2i} = 0$

It very often happens that one of the bodies involved in a collision is initially at rest. We label this body 2 and thus set $v_{2i} = 0$. In this case, the momentum-conservation condition of Equation 10.7 becomes

$$m_1 v_{1f} + m_2 v_{2f} = m_1 v_{1i}, \tag{10.13}$$

and the elastic-collision criterion of Equation 10.11 becomes

$$v_{2f} - v_{1f} = v_{1i}.$$

We can use this equation to eliminate either v_{2f} or v_{1f} from Equation 10.13. Eliminating either of these final velocities leads to a solution for the other. The results are

$$v_{1f} = \frac{m_1 - m_2}{m_1 + m_2} v_{1i} \quad \text{and} \quad v_{2f} = \frac{2m_1}{m_1 + m_2} v_{1i} \quad \text{for } v_{2i} = 0. \tag{10.14a,b}$$

Two subcases are of special interest. In the first, a moving body strikes a much more massive stationary one, as when a Ping-Pong ball strikes a bowling ball; $m_2 \gg m_1$. Equations 10.14a and 10.14b then take the form

$$v_{1f} \simeq -v_{1i} \quad \text{and} \quad v_{2f} \simeq 0.$$

That is, the light body bounces off the heavy one with nearly unchanged speed while the heavy one hardly moves.

In the second subcase, a moving body strikes a much less massive stationary one, as when a bowling ball strikes a Ping-Pong ball; $m_2 \ll m_1$. Equations 10.14a and 10.14b take the form

$$v_{1f} \simeq v_{1i} \quad \text{and} \quad v_{2f} \simeq 2v_{1i}.$$

That is, the motion of the heavy body is hardly affected by the collision while the light body moves ahead of the heavy one with a speed twice as great as the speed of the heavy one.

To an observer located on the bowling ball, the motion of the Ping-Pong ball looks exactly the same in the two subcases just discussed. Can you explain this?

EXAMPLE 10.3

A neutron is moving with velocity v_{ni}. It makes an elastic head-on collision with a carbon atom, whose mass is 11.9 times as great. The carbon atom is initially at rest. **(a)** What are the velocities v_{nf} and v_{Cf} of the neutron and the carbon atom, respectively, after the collision? **(b)** In terms of the initial kinetic energy of the neutron (which is the total initial kinetic energy K of the system), what are the kinetic energies of the two particles after the collision?

SOLUTION:
(a) What are the velocities v_{nf} and v_{Cf} of the neutron and the carbon atom, respectively, after the collision?

Because the collision is elastic and the carbon atom is initially at rest, you can use Equation 10.14a to find the final velocity of the neutron. If you call the mass of the neutron m_n, the equation gives you $v_{nf} = \dfrac{m_n(1 - 11.9)}{m_n(1 + 11.9)} v_{ni} = -0.845 v_{ni}$.

The negative sign indicates that the neutron rebounds.

Equation 10.14b gives you the final velocity of the carbon atom:

$$v_{Cf} = \frac{2m_n}{m_n(1 + 11.9)} v_{ni} = 0.155 v_{ni}.$$

(b) In terms of the initial kinetic energy of the neutron (which is the total initial kinetic energy K of the system), what are the kinetic energies of the two particles after the collision?

Because you already know the final velocity of the neutron, the easiest way to calculate its kinetic energy is to use the definition of kinetic energy. You write

$$K_{nf} = \tfrac{1}{2}m_n v_{nf}^2 = \tfrac{1}{2}m_n(-0.845 v_{ni})^2 = \tfrac{1}{2}m_n v_{ni}^2(0.845)^2$$

$$= (0.845)^2 K$$

$$= 0.714K.$$

To calculate the kinetic energy of the carbon atom most eas-

ily, remember that energy is conserved in the collision and that the system consists of the neutron and the carbon atom only. Thus K_{Cf} is the difference between the unchanged total energy of the system, K, and the final energy of the neutron, $K_{nf} = 0.715K$. You therefore subtract to find

$$K_{Cf} = K(1 - 0.714) = 0.286K.$$

As a check, you can use the result for v_{Cf} from part **a** to write

$$K_{Cf} = \tfrac{1}{2}m_C v_{Cf}^2 = \tfrac{1}{2}m_C(0.155 v_{ni})^2.$$

To express the quantity on the right in terms of K, you multiply it by m_n/m_n. This gives you

$$K_{Cf} = \frac{1}{2}\, m_n v_{ni}^2\, \frac{m_C}{m_n}\, (0.155)^2 = K \times 11.9 \times (0.155)^2$$

$$= 0.286K.$$

This agrees with the result obtained by subtraction.

Equal Masses in One-Dimensional Elastic Collision

Another important class of head-on elastic collision is that in which the two colliding bodies have equal masses—that is, when $m_1 = m_2$. In this case, the momentum-conservation condition, Equation 10.7, simplifies to

$$v_{1f} + v_{2f} = v_{1i} + v_{2i} \quad \text{for } m_1 = m_2.$$

Because we are considering elastic collisions, Equation 10.11 applies as well:

$$v_{1f} - v_{2f} = -(v_{1i} - v_{2i}) \quad \text{for } \Delta E = 0.$$

By first adding the second of these equations to the first and then subtracting the second from the first, we obtain

$$v_{1f} = v_{2i} \quad \text{and} \quad v_{2f} = v_{1i} \quad \text{for } m_1 = m_2. \tag{10.15a,b}$$

That is, *two bodies of equal mass exchange velocities in an elastic collision.*

In the special case in which body 2 is initially at rest, we have

$$v_{1f} = 0 \quad \text{and} \quad v_{2f} = v_{1i} \quad \text{for } m_1 = m_2 \text{ and } v_{2i} = 0. \tag{10.16a,b}$$

That is, the momentum of the system, initially possessed entirely by body 1, is completely transferred to body 2 in the collision. Billiard and pool players are familiar with this phenomenon. You can demonstrate it even more conveniently on any reasonably smooth tabletop, using two pennies.

Elasticity Nomenclature

The terminology used in describing collisions is unfortunately not standardized. Collisions that are neither elastic nor totally inelastic are sometimes called *partially elastic* or *partially inelastic*. When these terms are used, elastic collisions are often called *totally elastic collisions*. We will not use these terms. In this book, *elastic* is reserved to describe collisions for which $\Delta E = 0$ and $\epsilon = 1$; all other collisions are inelastic.

Inelastic Collisions

Most collisions in common experience are neither elastic nor totally inelastic, but lie somewhere in between. To characterize such collisions conveniently, we take note of the fact that the relative speed of the colliding bodies is unchanged in elastic head-on collisions (Equation 10.12) and is reduced to zero in totally inelastic collisions. In a collision that is neither elastic nor totally inelastic, the relative speed of the colliding bodies is reduced to some fraction ϵ (Greek epsilon) of its original value, but not to zero. The quantity ϵ is called the **coefficient of restitution**:

$$\epsilon \equiv \left| \frac{v_{1f} - v_{2f}}{v_{1i} - v_{2i}} \right|; \quad 0 \le \epsilon \le 1. \tag{10.17}$$

A totally inelastic collision is thus an inelastic collision in the extreme case $\epsilon = 0$.

As long as the bodies are not permanently deformed in a collision, ϵ is usually independent of the relative speed of the collision, or nearly so. The coefficient of restitution must be evaluated separately for every pair of colliding objects. When various objects are bounced off a very hard surface, however, the measured values of ϵ may be regarded as belonging to the objects themselves.

EXAMPLE 10.4

When a golf ball is dropped from a height $h_0 = 1.00$ m onto a smooth flagstone, as in Figure 10.2, it rebounds to a height $h_1 = 68$ cm. **(a)** Find the value of the coefficient of restitution. **(b)** To what height h_2 will the ball rise on the next bounce?

SOLUTION:
(a) Evaluate ϵ.
  The speed v_i of the ball just before it hits the ground, when

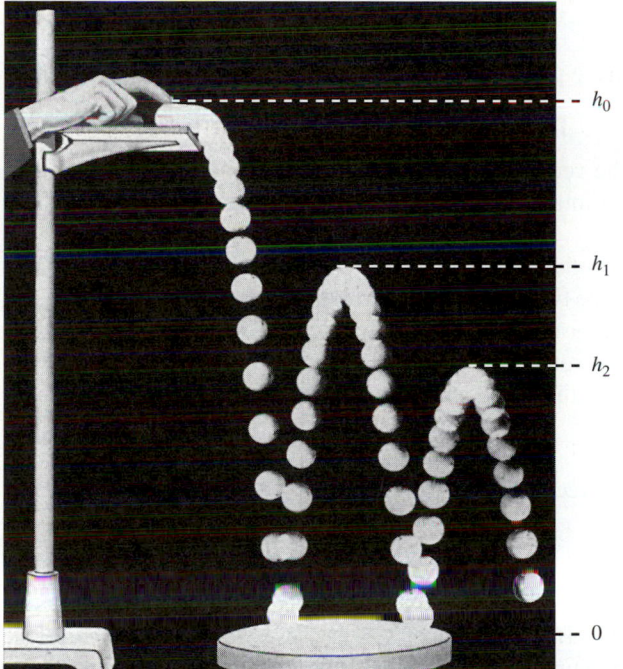

FIGURE 10.2 One way of measuring the coefficient of restitution.

it has fallen from rest at height h_0, is

$$v_i = \sqrt{2gh_0}.$$

Similarly, its speed v_f just after it leaves the ground is

$$v_f = \sqrt{2gh_1}$$

because it rises to height h_1 on the first rebound. The speed of the ball is, in both cases, equal to the relative speed of the colliding objects because the flagstone with which it collides remains immobile. Thus you can use Equation 10.17 to obtain the general result

$$\epsilon = \frac{v_f}{v_i} = \sqrt{\frac{h_1}{h_0}}. \qquad (10.18)$$

This is often a convenient way of measuring coefficients of restitution.
  Using the numerical values given for the golf ball, you have

$$\epsilon = \sqrt{\frac{0.68 \text{ m}}{1.00 \text{ m}}} = 0.83.$$

(b) To what height h_2 will the ball rise on the next bounce?
  Because the coefficient of restitution is approximately independent of the impact speed, you can use Equation 10.18 again to predict the second rebound height h_2. Now the initial height is the first rebound height h_1, and you have

$$\epsilon = \sqrt{\frac{h_2}{h_1}},$$

so
$$h_2 = h_1 \epsilon^2$$
$$= 0.68 \text{ m} \times (0.83)^2 = 0.46 \text{ m}.$$

As a check, you should satisfy yourself by direct measurement on Figure 10.2 that the coefficient of restitution is the same for both bounces.

Collisions in Two Dimensions

Most commonly, the motion of two colliding bodies is not restricted to one dimension. Billiard balls usually do not collide head on; nor do atoms. And the collision between the spacecraft and Jupiter shown in Figure 9.3 is clearly two-dimensional. The underlying principle—conservation of momentum—is the same as for the one-dimensional case. In the interest of simplicity, the discussion that follows is restricted to the case in which body 2 is initially at rest.
  Figure 10.3 shows body 1 moving toward a collision with body 2, which is at rest. The collision is not head on. The "miss distance" b shown in the figure is called the **impact parameter**. The coordinate system is chosen with the x axis along $\hat{\mathbf{p}}_{1i}$. Because

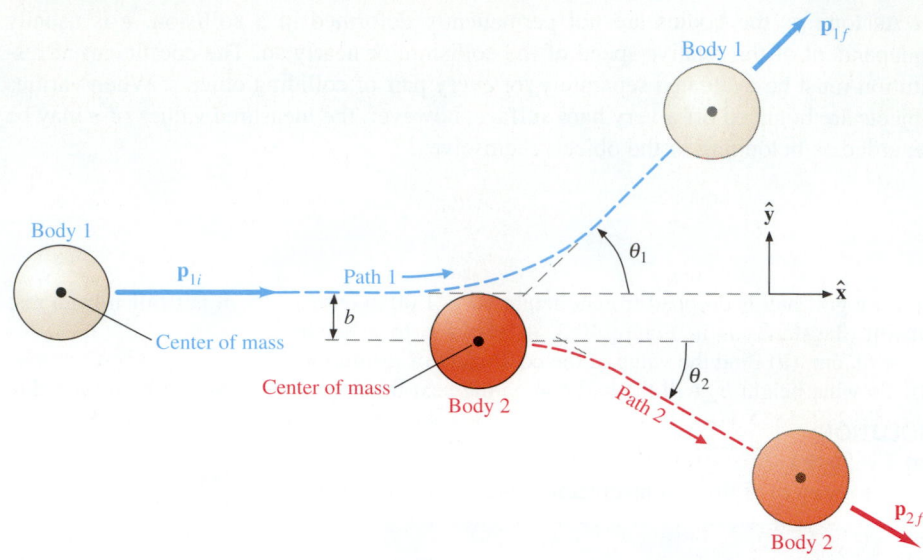

FIGURE 10.3 A collision in two dimensions. The initial momentum of body 1 is $\mathbf{p}_{1i}$, and body 2 is initially at rest at the origin of the coordinate system. The x axis is chosen along the direction $\hat{\mathbf{p}}_{1i}$. The "miss distance" b is the perpendicular distance from the center of mass of body 2 to the straight-line extension of the initial path of the center of mass of body 1. This distance is called the *impact parameter*. If a contact collision like that between billiard balls is to occur, the impact parameter must be less than the sum of the radii of the two balls, and path 1 has a sharp corner at the point of contact. But, for interactions involving long-range forces, such as the gravitational, electric, and magnetic forces, there is no such restriction on the impact parameter, and the curvature of the paths may remain significant over a substantial distance. The latter situation is the one shown in the figure; the sharp corner is merely the limiting case in which the force exerted by each body on the other varies very rapidly with the distance between their centers.

body 2 is initially at rest, we have $\mathbf{p}_{2i} = 0$ and the total initial momentum of the system is

$$\mathbf{p}_i = p_{1i}\hat{\mathbf{x}} + 0(\hat{\mathbf{y}}).$$

That is, the x component of the vector describing the total momentum of the system is equal to the magnitude of the initial momentum of body 1; the y component of the vector is zero.

When the collision is over—that is, when the force exerted by each body on the other has become negligible—the two bodies move off with constant velocities at angles θ_1 and θ_2, as shown. These angles are called the **scattering angles**. The total momentum of the system is now the vector sum $\mathbf{p}_{1f} + \mathbf{p}_{2f}$. If the two-body system is isolated, its total momentum is conserved in the collision. We express this fact in the equations

$$\mathbf{p}_{1f} + \mathbf{p}_{2f} = \mathbf{p}_{1i}; \quad \mathbf{p}_{2i} = 0. \tag{10.19}$$

As usual, we can evaluate this vector sum component by component. The sum of the x components is

$$p_{1f} \cos \theta_1 + p_{2f} \cos \theta_2 = p_{1i}. \tag{10.20a}$$

Because of the way in which we have chosen the coordinate system, the y component of the initial momentum is zero. Because momentum is conserved in the collision, the y component of the final momentum is likewise zero. So the sum of the y components is

$$p_{1f} \sin \theta_1 - p_{2f} \sin \theta_2 = 0. \tag{10.20b}$$

(The negative sign arises from the way we have chosen the sense of θ_2 in Figure 10.3.) Equations 10.20a and 10.20b are equivalent to Eq. 10.19, but they make explicit the fact that *momentum is conserved component by component*. They are a pair of simultaneous equations that can be solved for any two unknown quantities if the three other quantities are known.

Elastic Collisions in Two Dimensions

If a collision is known to be elastic, we can add more information to that provided by Equation 10.19. Because the initial kinetic energy of body 2 is zero, the energy-conservation equation, Equation 10.8, simplifies to the form

$$m_1 v_{1f}^2 + m_2 v_{2f}^2 = m_1 v_{1i}^2. \tag{10.21}$$

It is convenient to rewrite this equation in terms of momentum. To do this, we need to express terms of the form mv^2 in terms of the vector $\mathbf{p}$. Note that we can use Equation 7.7a, $\mathbf{A} \cdot \mathbf{B} = AB$ for $\theta = 0$, to express the dot product $\mathbf{p} \cdot \mathbf{p}$ as

$$\mathbf{p} \cdot \mathbf{p} = p^2. \tag{10.22a}$$

Because $p^2 = (mv)^2$, we have

$$\mathbf{p} \cdot \mathbf{p} = (mv)^2 = m^2 v^2. \qquad (10.22b)$$

If you compare Equation 10.22b with Equation 10.21, you will see that each of the terms in Equation 10.21 is of the form p^2/m. Specifically, the equation can be written

$$\frac{p_{1f}^2}{m_1} + \frac{p_{2f}^2}{m_2} = \frac{p_{1i}^2}{m_1} \quad \text{for elastic collision with } \mathbf{p}_{2i} = 0. \qquad (10.23)$$

This equation, taken together with Equations 10.20a and 10.20b, makes a set of three simultaneous equations that can be solved for any three unknown quantities if the other quantities are known.

Collisions provide most of the fun in bumper-car rides. Momentum is conserved; mechanical energy is not.

Equal Masses in Two-Dimensional Elastic Collision

Let us specialize still further to the case in which the two masses in elastic collision are equal; $m_1 = m_2$. In this case, the simplified energy-conservation equation, Equation 10.23, simplifies still further to the form

$$p_{1f}^2 + p_{2f}^2 = p_{1i}^2 \quad \text{for elastic collision with } \mathbf{p}_{2i} = 0 \text{ and } m_1 = m_2. \qquad (10.24)$$

We want to combine the information provided by this equation with that provided by the momentum-conservation condition, Equation 10.19. To do so, we square both sides of Equation 10.19. Because it is a vector equation, we carry out the squaring process by taking the dot product of each side with itself, thus:

$$(\mathbf{p}_{1f} + \mathbf{p}_{2f}) \cdot (\mathbf{p}_{1f} + \mathbf{p}_{2f}) = \mathbf{p}_{1i} \cdot \mathbf{p}_{1i}.$$

Carrying out the scalar multiplication term by term gives

$$\mathbf{p}_{1f} \cdot \mathbf{p}_{1f} + \mathbf{p}_{2f} \cdot \mathbf{p}_{2f} + 2\mathbf{p}_{1f} \cdot \mathbf{p}_{2f} = \mathbf{p}_{1i} \cdot \mathbf{p}_{1i}.$$

We use Equation 10.22a to rewrite this equation in the simplified form

$$p_{1f}^2 + p_{2f}^2 + 2\mathbf{p}_{1f} \cdot \mathbf{p}_{2f} = p_{1i}^2.$$

Remember that this equation expresses the fact that momentum is conserved in the collision. From this equation, we subtract Equation 10.24, which expresses the fact that energy is conserved in the collision as well. The result is

$$2\mathbf{p}_{1f} \cdot \mathbf{p}_{2f} = 0. \qquad (10.25)$$

There are three possible ways for Equation 10.25 to be true:

1. $\mathbf{p}_{2f} = 0$. This is the trivial case in which body 1 completely misses body 2, and there is no collision at all.

2. $\mathbf{p}_{1f} = 0$. This is the head-on collision already discussed; see Equations 10.16a and 10.16b.

3. The angle between $\mathbf{p}_{1f}$ and $\mathbf{p}_{2f}$, which we call $(\mathbf{p}_{1f}, \mathbf{p}_{2f})$, is 90°. That is, the sum of the angles θ_1 and θ_2 in Figure 10.3 must be $\theta_1 + \theta_2 = (\mathbf{p}_{1f}, \mathbf{p}_{2f}) = 90°$. To see that this is so, use the definition of the dot product in the form $\mathbf{A} \cdot \mathbf{B} = AB \cos (\mathbf{A}, \mathbf{B})$ to rewrite Equation 10.25:

$$2p_{1f}p_{2f} \cos (\mathbf{p}_{1f}, \mathbf{p}_{2f}) = 0.$$

If the value of the cosine is to be zero, its argument must be 90°. The converse also is true: Suppose a body collides with a stationary body of equal mass. *If the paths of the two bodies after collision are mutually perpendicular, the collision is elastic.*

EXAMPLE 10.5

In an experiment, a van de Graaff accelerator produces a beam of aluminum ions (atoms from which an electron has been removed) that strike a thin aluminum foil target. The ions in the beam have a speed of 4.71×10^6 m/s. One of them collides with an aluminum atom in the foil, ionizes it, and knocks this newly formed ion out of the foil. A detector is located so that the line between it and the target makes an angle $\theta_1 = 25.0°$ with respect to the beam (Figure 10.4). This detector registers

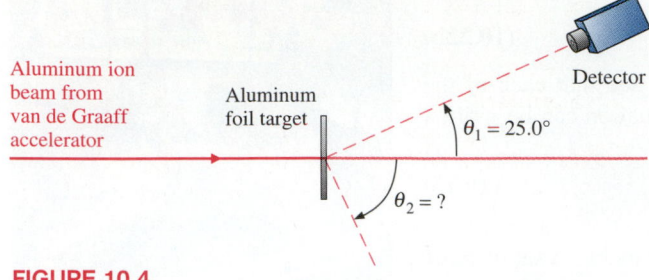

Aluminum ion beam from van de Graaff accelerator

Aluminum foil target

Detector

$\theta_1 = 25.0°$

$\theta_2 = ?$

FIGURE 10.4

the arrival of an aluminum ion. It is known from other experiments that the collision is elastic. Find **(a)** the scattering angle θ_2 at which the other ion in the collision is scattered and **(b)** the speeds of the two ions immediately after the collision. Neglect any interactions with other atoms in the foil.

SOLUTION:

(a) Find θ_2.

Because the collision is known to be elastic, you can use condition 3 in the preceding list to find

$$25.0° + \theta_2 = 90°,$$

so

$$\theta_2 = 65.0°.$$

(b) Find the speeds of the two ions after the collision.

Using the definition $p = mv$ and the fact that the masses of the two atoms are equal, you can write Equations 10.20a and 10.20b in the form

$$v_{1i} = v_{1f} \cos \theta_1 + v_{2f} \cos \theta_2 \quad \text{(10.26a)}$$

and

$$0 = v_{1f} \sin \theta_1 - v_{2f} \sin \theta_2. \quad \text{(10.26b)}$$

Next, you solve this pair of equations for v_{1f} by eliminating

v_{2f}. To do this, multiply the second equation through by $\cos \theta_2 / \sin \theta_2$ and add the resulting equation to the first equation. When you do this, you obtain

$$v_{1i} = v_{1f} \left(\cos \theta_1 + \frac{\sin \theta_1 \cos \theta_2}{\sin \theta_2} \right).$$

Solving for v_{1f} and inserting the numerical values gives you

$$v_{1f} = \frac{4.71 \times 10^6 \text{ m/s}}{\cos 25.0° + \dfrac{\sin 25.0° \times \cos 65.0°}{\sin 65.0°}}$$

$$= 4.27 \times 10^6 \text{ m/s}$$

for the ion collected by the detector.

You can now insert this value into Equation 10.26b to obtain v_{2f}. You have

$$v_{2f} = v_{1f} \frac{\sin \theta_1}{\sin \theta_2} = 4.27 \times 10^6 \text{ m/s} \times \frac{\sin 25.0°}{\sin 65.0°}$$

$$= 1.99 \times 10^6 \text{ m/s}$$

for the other ion in the collision.

You can check to see if this result conforms to the energy-conservation condition given by Equation 10.24. First, cast that equation into a convenient form by dividing it through by m^2 and taking the square root of both sides. This gives you

$$\sqrt{v_{1f}^2 + v_{2f}^2} = v_{1i}.$$

Inserting the values just obtained for the final speeds into the left side of this equation, you obtain

$$\sqrt{(4.27 \times 10^6 \text{ m/s})^2 + (1.99 \times 10^6 \text{ m/s})^2}$$
$$= 4.71 \times 10^6 \text{ m/s};$$

this agrees with the given value of v_{1i}.

Symbols Used in Chapter 10

b	impact parameter	m, M	mass
E	energy	$\mathbf{p}, p$	momentum
g	magnitude of the acceleration of gravity	v, V	velocity in one dimension, speed
h	vertical height to which a ballistic pendulum rises	$\mathbf{v}$	velocity
h_j	bounce height of a ball on the jth bounce	ϵ	coefficient of restitution
K	kinetic energy	θ_1, θ_2	scattering angle of body 1, 2

Summing Up

In Chapter 9, you learned that the momentum-conservation principle can be made to yield important information about the behavior of an isolated system in which a collision occurs. In Chapter 10, you have learned that further information can be obtained if the initial and final mechanical energies of the system are known as well.

Collisions are classified according to the fate of the initial mechanical energy of the colliding bodies:

- A collision is **elastic** if E is conserved—that is, if $\Delta E = 0$.

- A collision is **inelastic** if E is not conserved—that is, if $\Delta E \neq 0$.

- An inelastic collision is **totally inelastic** when the two colliding bodies stick together—that is, if $\mathbf{v}_{1f} = \mathbf{v}_{2f}$.

In a head-on (one-dimensional) totally inelastic collision, the common final velocity v_f of the colliding bodies is given by Equation 10.2,

$$v_f = \frac{m_1 v_{1i} + m_2 v_{2i}}{m_1 + m_2}.$$

If body 2 is initially at rest, this result simplifies to Equation 10.3a,

$$v_f = \frac{m_1}{m_1 + m_2} v_{1i} \quad \text{for } v_{2i} = 0.$$

The proportional energy loss in such a collision is given by Equation 10.4, for $\Delta E = 0$,

$$\left| \frac{\Delta E}{E_i} \right| = 1 - \frac{m_1}{m_1 + m_2} \quad \text{for } v_{2i} = 0, v_{1f} = v_{2f}.$$

In a head-on (one-dimensional) elastic collision, the velocity of each colliding body relative to the other is reversed by the collision. This fact is expressed by Equation 10.11,

$$v_{1f} - v_{2f} = -(v_{1i} - v_{2i}) \quad \text{for } \Delta E = 0.$$

Because speed is the magnitude of velocity, it follows that the *speed* of each body relative to the other is unchanged by the collision. This fact is expressed by Equation 10.12,

$$|v_{1f} - v_{2f}| = |v_{1i} - v_{2i}|.$$

If body 2 is initially at rest, a stationary observer can use Equations 10.14a and 10.14b to find the velocities of the two bodies after a head-on elastic collision:

$$v_{1f} = \frac{m_1 - m_2}{m_1 + m_2} v_{1i} \quad \text{and} \quad v_{2f} = \frac{2m_1}{m_1 + m_2} v_{1i} \quad \text{for } v_{2i} = 0.$$

If the colliding bodies have equal masses, a head-on elastic collision has the effect of exchanging their velocities according to Equations 10.15a and 10.15b:

$$v_{1f} = v_{2i} \quad \text{and} \quad v_{2f} = v_{1i} \quad \text{for } m_1 = m_2.$$

An *inelastic collision* is characterized by the ratio of the final relative speed of the colliding bodies to their initial relative speed. This ratio, called the **coefficient of restitution**, is given by Equation 10.17,

$$\epsilon \equiv \left| \frac{v_{1f} - v_{2f}}{v_{1i} - v_{2i}} \right|; \quad 0 \le \epsilon \le 1.$$

The upper limiting value $\epsilon = 1$ implies an elastic collision, in agreement with Equation 10.12. The limiting value $\epsilon = 0$ implies a *totally* inelastic collision.

For collisions that are not head on, we define the **impact parameter** b to be the perpendicular distance by which the center of mass of body 1 misses the center of mass of body 2. After the collision, the two bodies move off in directions described by the **scattering angles** θ_1 and θ_2. We can always choose the $\hat{\mathbf{x}}$ direction along the initial path of body 1. This ensures that the total momentum of the system, both before and after the collision, is expressed by a vector whose y component is zero. Equations 10.20a and 10.20b follow:

$$p_{1f} \cos \theta_1 + p_{2f} \cos \theta_2 = p_{1i}$$

and $\quad p_{1f} \sin \theta_1 - p_{2f} \sin \theta_2 = 0 \quad \text{for } \mathbf{p}_{2i} = 0.$

If the collision described by these equations is elastic, further information can be obtained from Equation 10.23, which expresses the fact that mechanical energy is conserved:

$$\frac{p_{1f}^2}{m_1} + \frac{p_{2f}^2}{m_2} = \frac{p_{1i}^2}{m_1} \quad \text{for } \Delta E = 0 \text{ and } \mathbf{p}_{2i} = 0.$$

In a collision between two bodies of equal mass, with $v_{2i} = 0$, that is neither head on nor a complete miss, the simultaneous application of momentum- and energy-conservation conditions imposes a strong requirement on the outcome: The angle between the paths of the two bodies must be 90°.

KEY TERMS

Section 10.2 Momentum and Energy
elastic collision, inelastic collision, totally inelastic collision ▪ ballistic pendulum ▪ relative speed ▪ coefficient of restitution

Section 10.3 Collisions in Two Dimensions
impact parameter ▪ scattering angle

QUERIES

10.1 *(2) What's conserved? What isn't?* In which of the following cases is linear momentum conserved? In which is total mechanical energy conserved? In answering, carefully define the system to which your answer applies. **(a)** Two billiard balls collide. **(b)** A bullet buries itself in a block of wood. **(c)** A ball falls freely. **(d)** A block slides down a frictionless inclined plane. **(e)** A batter hits a baseball. **(f)** The baseball flies through the air. (Neglect friction.) **(g)** A puck moves along a level air table. **(h)** A pendulum swings. **(i)** A mass oscillates up and down under the combined action of a spring and the gravitational force. **(j)** A mass-spring system suspended from a hook oscillates under the action of the gravitational force.

10.2 *(2) Same difference? I.* Two bodies of different mass have the same kinetic energy. Which one has the greater momentum?

10.3 *(2) Same difference? II.* Which is more dangerous, a bullet having the same kinetic energy as a rolling bowling ball or a bullet having the same momentum as the bowling ball? Explain.

10.4 (2) *Benign neglect.* In analyzing the ballistic pendulum, why can you neglect the action of the external gravitational force during the collision stage?

10.5 (2) *Youthful indiscretion.* Benjamin Thompson (of whom you will read much more in Chapter 17) claimed to have invented the double ballistic pendulum, illustrated below, about 1778. In fact, Thompson's work with this device largely repeated work that had been published in 1742 by the English gunsmith and engineer Benjamin Robins. Thompson was subjected to much criticism on this account, but he nevertheless procured election to the Royal Society in 1781, at age 27, on the basis of his claim. The trigger pull required to fire the gun is made very light so that the act of pulling the trigger does not itself set the gun swinging significantly. The bullet—a lead ball—is fired into the large wooden block. The device can be used to measure the degree to which the gun's performance is degraded if there is a loose fit between the ball and the gun bore, allowing some of the gas produced by the explosion of the gunpowder to blow out of the gun without propelling the ball. Explain.

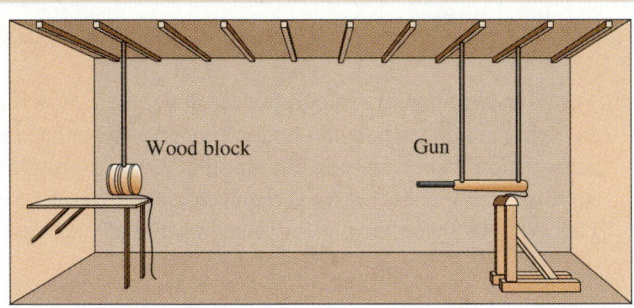

10.6 (2) *Same difference? III.* A body falls freely from initial height h. As it falls, its total mechanical energy remains constant but its momentum increases. Explain.

10.7 (3) *Basic relation.* Show that the kinetic energy K of a body of mass m can be written in terms of its momentum **p** in the form

$$K = \frac{p^2}{2m}.$$

Using this result, show that Equations 10.23 and 10.24 are direct expressions of the conservation of mechanical energy in a system whose potential energy does not change.

10.8 (3) *Moderation in all things.* In a nuclear reactor, fission of atomic nuclei results in the emission of neutrons of mass m_n. These neutrons then collide with other nuclei, resulting in their fission. This is the famous nuclear *chain reaction*. The neutrons emerge from the fissioned nuclei with large kinetic energies. In reactors of the most common type, the effectiveness of neutrons in producing further nuclear fissions is increased by slowing the neutrons down. This is done by introducing a material into the reactor called a *moderator*, with whose nuclei the neutrons can make repeated elastic collisions. Explain why carbon (nuclear mass $\approx 12m_n$) makes a better moderator than lead (nuclear mass $\approx 207m_n$).

10.9 (3) *Double collision.* An alpha particle (mass $4m_p$) makes an elastic collision with a proton (mass m_p), moves a short distance, and makes an elastic collision with another proton. What is the direction of motion of the second proton relative to the direction of motion of the first?

PROBLEMS

GROUP A

10.1 (2) *Other things being equal.* When a body collides totally inelastically with a body of equal mass that is initially at rest, what fraction of the initial kinetic energy of the system is lost?

10.2 (2) *Coupling.* A loaded freight car of mass 22 tonnes moves at 1.8 m/s toward an empty freight car (mass 10 tonnes) that is at rest. The two cars couple and move together. **(a)** Find the final speed of the two cars. **(b)** What fraction of the initial kinetic energy is lost?

10.3 (2) *Sticking together, I.* A 5-kg mass collides totally inelastically with a 2-kg mass, which is initially at rest. After the collision, the energy of the system is 5 J. What was the kinetic energy of the 5-kg mass before the collision?

10.4 (2) *Sticking together, II.* Two bodies collide totally inelastically. Before the collision, the velocity of body 1 is $v_{1i} = 2$ m/s and the velocity of body 2 is $v_{2i} = -4$ m/s. The common final velocity of the bodies is $v_f = 1$ m/s. What is the ratio K_1/K_2 of the initial kinetic energies of the two bodies?

10.5 (2) *Pop!* Two blocks of mass m_1 and m_2 are in contact. A small explosive charge located between them is detonated, and the blocks move off in opposite directions. As they do so, **(a)** what is the relation between the momenta p_1 and p_2 of the two blocks? **(b)** what is the relation between their kinetic energies?

10.6 (2) *Diff'rent strokes.* A 3-kg mass moving with velocity 4 m/s overtakes and collides head on with a 5-kg mass whose velocity is 2 m/s. Find the velocities of the masses **(a)** if the collision is totally inelastic; **(b)** if the collision is elastic.

10.7 (2) *Stopping 'em dead.* A body of mass m_1 moves with initial velocity v_1. It makes a head-on, totally inelastic collision with a body of mass m_2. If both bodies come to rest in the collision, what is the initial velocity v_2?

10.8 (2) *Ballistic pendulum, I.* A bullet of mass 11.6 g is fired into the block of a ballistic pendulum. The block, whose mass is 10.0 kg, rises through a vertical distance of 3.64 cm. **(a)** What was the speed of the bullet? **(b)** What fraction of the initial kinetic energy of the bullet is lost in the collision?

10.9 (2) *Ballistic pendulum, II.* A bullet having mass 8.6 g and speed 550 m/s is fired into the 6.3-kg block of a ballistic pendulum. If the strings of the pendulum are 1.2 m in length, what is the maximum angle that the strings make with the vertical?

10.10 (2) *Energy transfer.* A body of mass m makes an elastic head-on collision with a body of equal mass initially at rest. What fraction of the initial kinetic energy of the first body is transferred to the second body in the collision?

10.11 (2) *Double pendulum, I.* Two pendulums of equal length have identical bobs made of sticky clay. The left-hand bob is displaced (part *a* of the drawing below) and then released. It strikes the other bob and sticks, and the two continue to swing together (part *b* of the drawing). Find the maximum height h_f to which they rise in terms of the initial height h_i of the single pendulum.

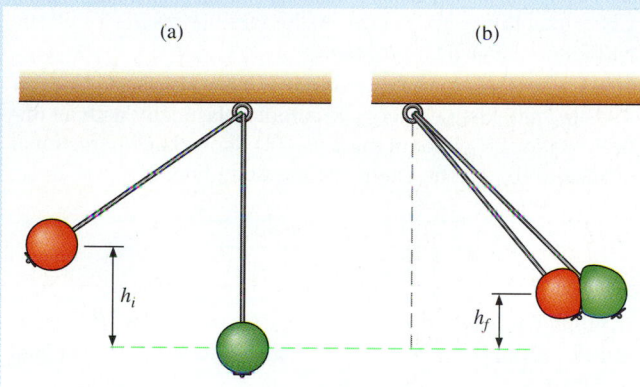

(a)   (b)

10.12 (2) *Double pendulum, II.* Two pendulums of equal length have identical steel bobs of equal mass. The left-hand bob is displaced (part *a* of the drawing in Problem 10.11) and then released. It collides elastically with the other bob. Describe the subsequent motion of the system, giving the maximum height to which the bobs rise.

10.13 (2) *Slide and collide.* In the diagram below, a block of mass 12 kg lies at a vertical height 7.3 m on a frictionless inclined plane. It slides down the plane and collides elastically with a second block whose mass is 6.8 kg. Find the final velocities of the two blocks.

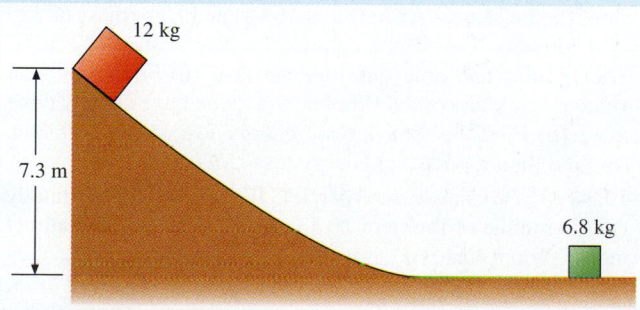

12 kg

7.3 m

6.8 kg

10.14 (2) *Compression, I.* In the illustration below, a metal cup filled with clay, having total mass m_2, is mounted on a spring having force constant k. A steel ball of mass m_1 strikes the clay and sticks. (a) If the initial speed of the ball is v_{1i}, find the distance A through which the spring is compressed. Assume that the table is frictionless. (b) If $v_i = 3.5$ m/s, $m_1 = 125$ g, $m_2 = 287$ g, and $k = 472$ N/m, find A.

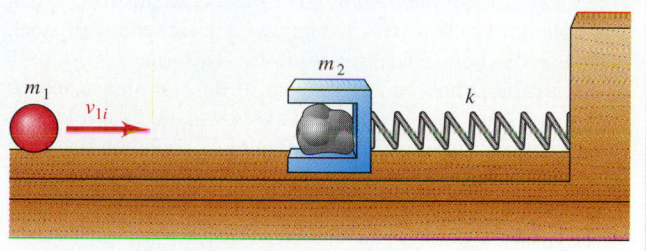

m_1 v_{1i} m_2 k

10.15 (2) *Compression, II.* The clay-filled metal cup of Problem 10.14 is replaced with a hard metal plate of the same mass. The steel ball strikes the plate with speed v_{1i} and bounces off elastically. (a) Find the distance A through which the spring is compressed. (b) If $v_i = 3.5$ m/s, $m_1 = 125$ g, $m_2 = 287$ g, and $k = 472$ N/m, find A.

10.16 (2) *Bouncing ball, I.* A ball is dropped from a height of 2.00 m and rebounds to 1.62 m. Find the coefficient of restitution for the ball and the floor.

10.17 (2) *Bouncing ball, II.* A ball is dropped from a height of 5.50 m. The coefficient of restitution for the ball and the floor is 0.76. Find the height of the first two bounces.

10.18 (2) *Bouncing ball, III.* A ball is dropped from a height of 2.55 m. On the second bounce, it rises to 1.33 m. Find the coefficient of restitution.

10.19 (3) *Elastic?* An air puck of mass 55 g is moving at velocity $\mathbf{v} = 8.2$ m/s in the positive x direction when it collides with a stationary air puck of equal mass. If the two pucks move off in directions that make angles 38° and 33° with the x direction, is the collision elastic?

10.20 (3) *Billiards 101.* A billiard player shoots a ball with initial speed 7.0 m/s. It strikes a stationary billiard ball and moves off at an angle of 34° to its original path. Assume that the collision is essentially elastic, and neglect any effects having to do with the spin of the ball. Find the speed and direction of motion of the second ball.

10.21 (3) *A midsummer-night's entertainment.* An air-table puck collides with a stationary puck of the same mass. The two pucks move off at scattering angles $\theta_1 = 52°$ and $\theta_2 = 16°$. If the initial momentum of the first puck has magnitude $p_{1i} = 0.84$ kg·m/s, what are the final momenta of the two pucks? Is the collision elastic? If not, what fraction of the initial energy of the system is lost in the collision?

10.22 (2) *Ballistic pendulum, III.* A bullet of mass 53.5 g is shot at the block of a ballistic pendulum whose mass is 10.3 kg. However, the block is too small, and the bullet passes through the block. If the initial speed of the bullet is 425 m/s and the block rises a vertical distance of 18.5 cm, how fast is the bullet going when it leaves the block?

10.23 (2) *Double pendulum, III.* Consider again the system illustrated in Problem 10.11. The two bobs are made of steel, and collisions between them are elastic. This time, let the mass of the left-hand bob be $2m$ and that of the right-hand bob be m. The left-hand bob is displaced as shown and then released. (a) How high, and in which direction, does each bob rise after the first collision? (b) Where will the second collision between the bobs take place? (c) What will the velocity of each bob be just before the second collision? (d) How high, and in which direction, does each bob rise after the second collision?

10.24 (2) *Timing the bounces, I.* You drop a ball from a height h_0 and use a stopwatch to time the interval Δt between the first impact with the floor and the second. (a) Express the coefficient of restitution ϵ in terms of h_0, Δt, and necessary constants. (b) If the release height is 2.0 m and the interval is 1.2 s, find ϵ.

10.25 (2) *Timing the bounces, II.* (a) Express the ratio $\Delta t_{j+1}/\Delta t_j$ of the time intervals between successive bounces of a ball in terms of the coefficient of restitution ϵ. Neglect air friction. (b) Use this result to show that, if the ball is initially released from height h_0, the quantity

$$\sqrt{\frac{2h_0}{g}}\,\frac{1+\epsilon}{1-\epsilon}$$

represents an upper limit on the total time from the release of the ball until it stops bouncing. (The fact that the ball may subsequently roll along the floor is irrelevant.) Hint: Use the sum of the geometric series

$$1 + a + a^2 + a^3 + \cdots = \sum_{j=0}^{\infty} a^j = \frac{1}{1-a}.$$

(c) Explain qualitatively why the ball will not continue to bounce for quite this length of time.

10.26 (2) *Elastic and totally inelastic.* In an air-track experiment, you make glider 1 collide elastically with glider 2, which is initially at rest. You find the final velocity of glider 2 to be v_{2f}. Then you repeat the experiment using the same initial conditions. This time, however, you put a drop of instant glue on the bumper of glider 2 so that the collision is totally inelastic. Express the common final velocity v_f of the two gliders in terms of v_{2f}, the result obtained in the first experiment. Note that the result is independent of the masses of the two bodies.

10.27 (2) *Superelastic collision.* An athlete performs on a trampoline. She descends from a height $h_1 = 2.0$ m and rebounds to a height $h_2 = 2.3$ m. (a) What is the value of the coefficient of restitution ϵ? (b) In what way is the collision between the athlete and the trampoline different from the collisions discussed in Section 10.2? (c) A collision of this sort is called *superelastic*. Define a superelastic collision in terms of the value of ϵ.

10.28 (2) *Inelastic collision, I.* A head-on collision takes place between body 1, whose mass and velocity are 6.4 kg

and 26.4 m/s, and body 2, whose mass and velocity are 4.7 kg and 0. If the coefficient of restitution is $\epsilon = 0.53$, what are (a) the final velocities and (b) the final kinetic energies of the two bodies?

10.29 (2) *Narrowing the gap.* A body strikes an initially stationary body of equal mass. Suppose that the collision is inelastic, so Equation 10.25 no longer holds. Replace Equation 10.24 with an inequality, and show that the angle between the paths of the two bodies after collision must be less than 90°.

10.30 (2) *Inelastic collision, II.* Show that, when a body makes a head-on collision with an initially stationary body of equal mass, the final velocities bear the relation

$$v_{1f} = \frac{1-\epsilon}{1+\epsilon}\,v_{2f},$$

where ϵ is the coefficient of restitution for the pair of bodies. Check that this result reduces to the familiar results for elastic and totally inelastic collisions.

10.31 (2) *Elastic energy transfer, I.* An air-track glider of mass m_1 and kinetic energy K collides elastically with an initially stationary glider of mass m_2. (a) Show that the fractional change in the kinetic energy of the first glider is

$$\left|\frac{\Delta K}{K}\right| = \frac{4m_1 m_2}{(m_1 + m_2)^2}.$$

(b) Find the mass ratio $\alpha \equiv m_1/m_2$ for which $\Delta K/K$ is maximized. Compare your result with Equations 10.16a and 10.16b.

10.32 (2) *Elastic energy transfer, II.* An air-table puck of mass m_1 has kinetic energy K. It collides elastically with an initially stationary puck of mass m and rebounds in a direction perpendicular to its original path. (a) Show that the fractional change in the kinetic energy of the first puck is

$$\left|\frac{\Delta K}{K}\right| = \frac{2m_1}{m_1 + m_2}.$$

(b) For what range of values of the mass ratio $\alpha \equiv m_1/m_2$ is such a 90° rebound possible?

10.33 (2) *Practice makes perfect, I.* The collision described in Problem 9.14 is a one-dimensional, totally inelastic collision: Figure skaters Alice (mass 46 kg) and Bob (mass 68 kg) skate directly toward one another, each with speed 4.3 m/s. They collide and hold onto one another. (a) Find their final velocity. (If you worked Problem 9.14, you have already done this.) (b) Find the total kinetic energy lost in the collision. (c) Find the proportional energy loss $|\Delta E/E_i|$.

10.34 (3) *Practice makes perfect, II.* Alice and Bob finally get the routine of Problem 10.33 down pat and try something more complicated. As shown in the diagram (p. 265), they skate at right angles to one another, again at speed 4.3 m/s. They meet at O and hold onto one another. Taking Bob's original direction of motion as the positive x direction, find (a) the magnitude and direction of their final velocity, (b) the total kinetic energy lost in the collision, and (c) the proportional energy loss $|\Delta E/E_i|$. Assume that Alice and Bob turn their skates in the direction in which they tend to move after the collision, so friction with the ice is negligible.

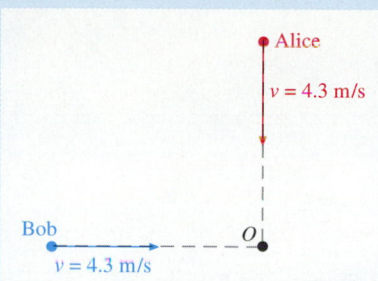

Alice
$v = 4.3$ m/s

Bob
$v = 4.3$ m/s

O

10.35 *(3) Practice makes perfect, III.* Alice and Bob repeat the maneuver of Problem 10.34, except that their initial paths now subtend an angle of 135°. Find **(a)** the magnitude and direction of their final velocity, **(b)** the total kinetic energy lost in the collision, and **(c)** the proportional energy loss $\Delta E/E_i$.

10.36 *(3) Law of reflection.* A particle of mass m collides elastically with a smooth wall. Consider the wall to be a stationary body of mass $M \gg m$. Show that, if the path of the particle as it approaches the wall makes an angle θ with the normal to the wall, (1) the particle must rebound in the plane defined by the path of approach and the normal and (2) the rebound path lies at an equal angle θ on the other side of the normal. (This is often called the *law of reflection*. It is stated succinctly by saying that *the angle of reflection is equal to the angle of incidence.*)

10.37 *(3) Momentum conservation and center of mass.* An isolated system consists of two particles whose initial momenta are $\mathbf{p}_{1i}$ and $\mathbf{p}_{2i}$. The particles collide. In the light of the principle of momentum conservation, what is the effect of the collision on the motion of the center of mass of the system? Compare your result with that obtained in Section 9.5.

10.38 *(G) Compression, III.* Consider again the apparatus illustrated in Problem 10.14. Suppose the ball strikes the clay-filled cup at $t = 0$, when the cup is at rest at $x = 0$. Assuming that negligible time is required for the ball to come to a stop in the clay, write an expression $x(t)$ for the position of the cup.

10.39 *(G) Compression, IV.* Consider again the apparatus of Problem 10.15. The clay-filled cup shown in Problem 10.14 has been replaced by a metal plate. Suppose the ball strikes the plate at $t = 0$, when the plate is at rest at $x = 0$. Write an expression $x(t)$ for the position of the plate.

10.40 *(G) Fatal accident.* A drunken driver crashes his car into a parked car whose brakes are set. The two cars move off together, the locked wheels of the parked car leaving distinct skid marks. The investigating police officer finds that the skid marks are 12.0 m long. If the mass of the moving car is 1450 kg and the mass of the parked car is 1880 kg, how fast was the driver going before the collision? Assume that the driver never applied the brakes of his car, and take the coefficient of friction between the locked wheels of the parked car and the road to be 1.0.

GROUP C

10.41 *(2) Symmetry arguments save calculation!* You can derive Equations 10.16a and 10.16b without any calculation at all by means of a thought experiment and a symmetry argument. Suppose that body 2 lies initially at rest on a frictionless table. Suppose also that a canopy is situated over the general area where body 2 is located so that it cannot be seen and its exact position is not known. Body 1, which is identical with body 2 in every way, is set into motion toward the canopy with velocity v_1 and disappears under it. A short time later, a single body emerges from under the opposite edge of the canopy. Show that there are only two possible ways of accounting for the emergence of one puck, that Equations 10.16a and 10.16b represent one of the ways, and that the experiment cannot distinguish between the possibilities.

10.42 *(2) Energy loss in elastic collisions.* Consider a series of inelastic collisions having identical initial conditions but differing values of ϵ. Because mechanical energy is lost in inelastic collisions, your intuition has probably already prompted you to the view that the largest possible loss of energy takes place in a totally inelastic collision; $\epsilon = 0$. **(a)** For the special case of a collision in which body 2 is initially stationary, show that

$$\left| \frac{\Delta K}{K} \right| = m_2 \left(\frac{1 - \epsilon^2}{m_1 + m_2} \right).$$

(b) Show that this quantity increases as ϵ decreases, with the largest possible value of $\Delta K/K$ corresponding to $\epsilon = 0$.

10.43 *(3) Maintain speed, I.* Show that Equation 10.12, $|v_{1f} - v_{2f}| = |v_{1i} - v_{2i}|$, is valid for *any* elastic collision between two bodies. To do this, begin by considering the consequences of energy conservation for such a collision from the point of view of an observer located on body 1 (for whom v_1 is always zero) and write a relation between the initial and final speeds of body 2. Then note that the *relative* velocity of two bodies is independent of the velocity of the observer and rewrite the speed relation for the same collision from the point of view of an arbitrary observer.

10.44 *(3) Maintain speed, II.* The result of Problem 10.43 can be obtained by means of a different argument as well. Begin by setting up an x-y coordinate system so that the y axis lies along the line joining the centers of the two colliding bodies at the moment of collision. Then answer the following questions: **(a)** What is the direction of the impulse imparted by each particle to the other? What is the effect of these impulses on the velocities of the particles? **(b)** What happens to the component of the velocity of each particle in the direction perpendicular to the impulse? **(c)** What happens to the speed of each particle?

10.45 *(3) Maximum deviation.* A body of mass m_1 collides elastically with a stationary body of lesser mass m_2. Show that the maximum scattering angle $\theta_{1 \text{ max}}$ is $\theta_{1 \text{ max}} = \sin^{-1} \dfrac{m_2}{m_1}$.

10.46 *(3) Billiards 201.* A billiard ball moving with kinetic energy K along the x axis collides elastically with an identical, stationary billiard ball. At the moment of impact, the line between the centers of the balls makes an angle α with the x axis. Neglecting spin, show that at the moment of maximum deformation of the two balls, the potential energy of deformation is $\frac{1}{2} K \cos^2 \alpha$.

Part IV

Rotational Mechanics

At the beginning of our study of mechanics, we divided motion into two kinds: translation and rotation. We then set the subject of rotation aside and considered pure translation at length.

We now pick up the thread laid aside and develop the mechanics of rotation. We will be able to proceed much more quickly than we did in studying translational mechanics because many of the ideas developed in the context of translation are adaptable to rotation, either directly or by close analogy. In addition, the skills and insights you have developed in learning about translational mechanics will be continually useful.

We begin our study of rotational mechanics in Chapter 11, focusing our attention on the simplest case: A rigid body rotates about a fixed axis, as a flywheel does. We then extend some of our findings to the slightly more complicated case in which a rigid body rotates about an axis that is not fixed but moves without changing its orientation, as the axle of a wheel rolling along a straight path does. In Chapter 12, we apply the laws of rotational mechanics developed in Chapter 11 to a variety of cases.

In Chapter 13, we define the fundamental quantity angular momentum. This is done partly by analogy with *linear* momentum and partly by extension of the ideas developed in Chapters 11 and 12. Having considered angular momentum in the fixed-axis case, we next develop the vector meth-

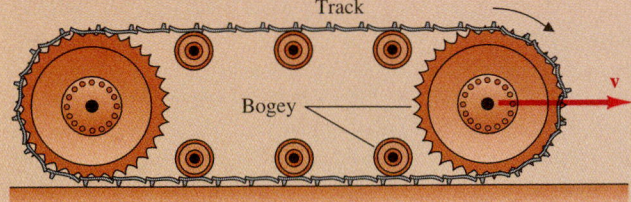

Track

Bogey

v

ods needed to extend the subject into a three-dimensional context. In the light of this extension, you will find answers to such questions as these: Why does a spinning top not fall over? How does the top behave? How can we justify what we all do intuitively—namely, separate the motion of rigid bodies into translational and rotational parts?

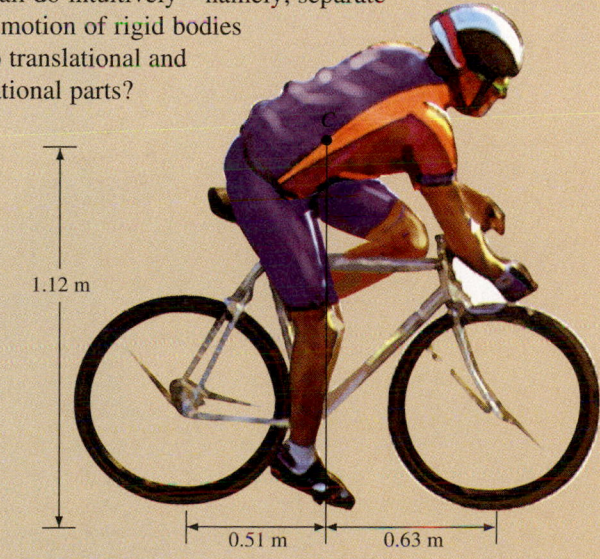

1.12 m

0.51 m   0.63 m

It is neither possible nor necessary within the scope of this book to develop rotational mechanics in a completely general form. Nevertheless, the subject matter of Parts II through IV constitutes a system of mechanics—called classical or Newtonian mechanics—that is complete in concept if not in all details. Parts II through IV thus accomplish the first task that we set ourselves at the beginning of the book: to have a solid basis for the study of the motion of an enormous variety of physical systems.

Rotation about a Fixed Axis

Left: A good player has considerable control over the relation between a Yo-Yo's translational and rotational motions—and lots of fun, too!

'Round and 'round and 'round she goes;
Where she stops, no one knows.

—Carnival Barker

Introduction

In Section 2.1, we distinguished between two ways in which an extended body can move. We called these motions *translation* and *rotation*. Up to this point, we have concentrated on translation, the kind of motion in which the entire body changes its location while its orientation remains unchanged. In rotation, the orientation of the body changes. Even without a formal definition of "orientation," your intuition probably suggests that any motion whatsoever of a body can be considered to be translation, rotation, or a combination of translation and rotation. The way to separate general motion into translational and rotational parts will be developed in Chapter 12 and extended in Chapter 13.

We begin by considering the simplest type of rotation, which is subject to two restrictions. First, we focus attention on rigid bodies. A **rigid body** is a body within which the distance between any two points does not change, no matter how the body moves. (Many real objects can be considered rigid for practical purposes, even though no perfectly rigid body exists.) Second, we restrict ourselves to *rotation about a fixed axis*. The hands of a clock, the blade of a kitchen blender, and a knife sharpener's grinding wheel provide familiar examples of rotation about a fixed axis. In contrast, a wobbling, spinning top is an example of rotation where there is no fixed axis. When a bicycle moves in a straight line, its wheels are not rotating about fixed axes. However, we will see that some of the concepts of fixed-axis rotation can be applied to cases in which the axis moves without changing its orientation.

You will soon see that there is a remarkable parallelism between the mathematical description of fixed-axis rotation of a rigid body and the mathematical description of one-dimensional translational motion. We will take the greatest possible advantage of this parallelism. Analogies will emerge at almost every step, and they will enable you to apply directly the physical insight you have already gained in dealing with translational motion.

Kinematics of Rotation about a Fixed Axis

As we did in studying translational mechanics, we first devise a way of describing rotation about a fixed axis without regard to the forces that produce the rotation. That is, we develop the *kinematics* of fixed-axis rotation. You are already familiar with a related subject—the kinematics of a particle in uniform circular motion. We developed this subject in Section 5.4 for its own sake, and in more detail in Section 6.2 because of its great usefulness in studying the important kind of translational motion called

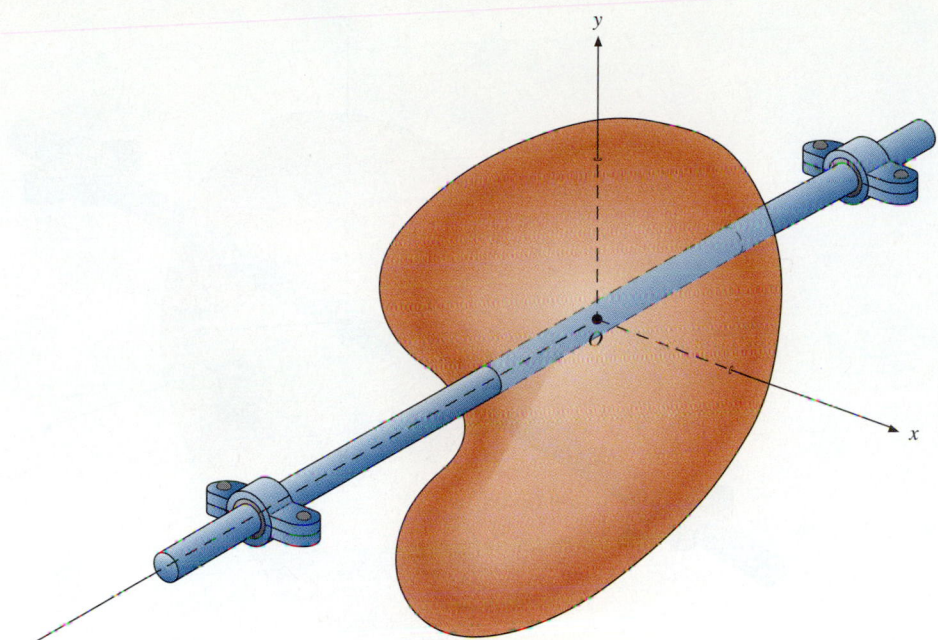

simple harmonic motion. Much of what we do here will amount to generalizing the ideas developed earlier for uniform circular motion and changing our point of view so as to stress the rotation of a body as a whole, rather than the translation of its parts along circular paths.

Figure 11.1 is a view of an arbitrarily shaped rigid body mounted on a shaft. The shaft is firmly mounted in fixed bearings. The body is free to turn about the centerline of the shaft, which is called the **axis of rotation**, but it is not free to move in any other way. The body turns with respect to a fixed cartesian coordinate system, defined as shown in Figure 11.1. The z axis is conventionally chosen so that it lies along the axis of rotation.

Because the body is rigid, it must rotate about the axis as a unit. If any part of the body rotates through a certain angle about the axis, every other part must rotate through the same angle. (For example, if any part of the body has passed through one complete circle about the axis, every other part must have done so as well.) We therefore need describe the rotation of just one point in the body—any point not on the axis will do—in order to describe the rotation of the whole body. The task of describing the rotation of the rigid body is thus reduced to the task of describing the circular motion of a particle.

Angular Displacement

Figure 11.2 on the next page depicts the point P we have chosen to represent the rotation of the entire rigid body of Figure 11.1. At time $t = 0$, P lies on the x axis at a distance r from the origin O. At a time dt later, P has moved through the infinitesimal circular arc shown, whose length is ds, and it has rotated through the infinitesimal angle $d\theta$. The angle $d\theta$ is the **angular displacement** of P; it is the infinitesimal analogue of the angular displacement θ defined in Section 6.2. Like θ, $d\theta$ must be expressed in radians. (See Appendix 2 for a discussion of radian measure.)

The importance of $d\theta$ to the study of rigid-body rotation is this: Not all points in the body lie at a distance r from the origin, and not all points move through arcs of length ds. But all points in the body experience the same angular displacement $d\theta$ in the same time interval of duration dt. It is therefore most convenient to describe the rotation of a rigid body in terms of angular displacement.

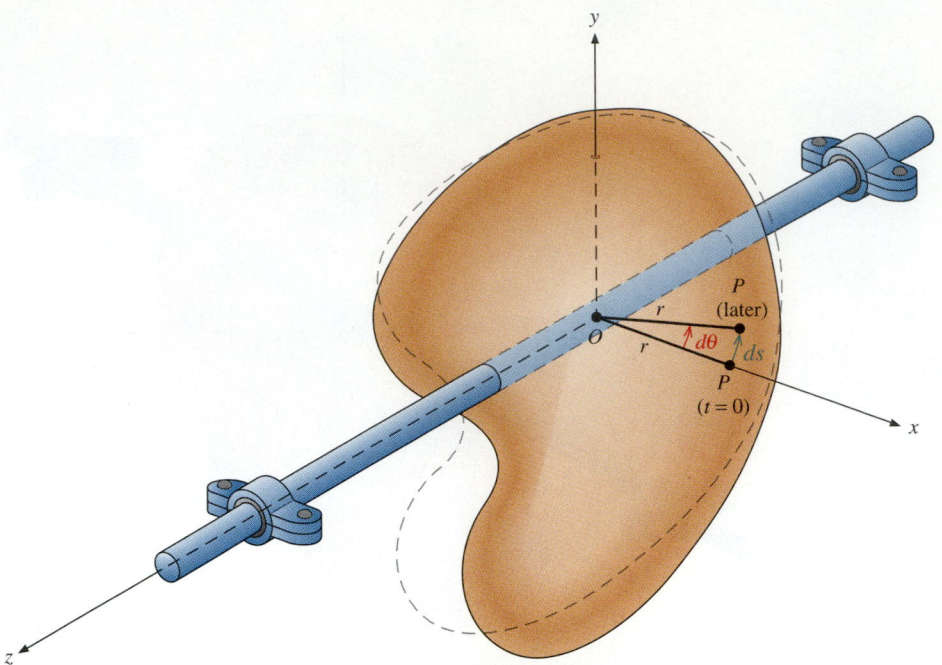

FIGURE 11.2 A single point located in a rigid body can be used to represent the rotation of the body about a fixed axis. The plane of rotation (the *x-y* plane) coincides with the plane of the page, and the axis of rotation (the *z* axis) is normal to the page, passing through it at the origin *O*. Point *P* moves through a circular arc whose length *ds* depends on the particular point chosen. But all points in the body experience the same angular displacement *dθ*.

Sign Convention for Sense of Rotation

The body can rotate about a given axis in only two ways, called **senses**—counterclockwise or clockwise. All possible (infinitesimal) angular displacements can be described by assigning a positive or negative sign to the value of *dθ*. Under these circumstances, the angular displacement is a signed number analogous to the translational displacement *dx* of a particle in one dimension. To be consistent with angle conventions we have already adopted, we call counterclockwise rotations positive and clockwise rotations negative. Thus, in the situation shown in Figure 11.2, *dθ* has a positive value. Although it has magnitude and sign, *dθ* is not a vector—not even a one-dimensional vector—because the sign represents not a *direction* (say, north or up) but a *sense*. Nevertheless, because *dθ* can be described completely by a magnitude and a sign for fixed-axis rotation, it has *mathematical* properties like those of the one-dimensional displacement vector *dx*.

If we had chosen the initial position of point *P* not on the *x* axis but elsewhere, as in Figure 11.3, it would have been necessary to specify an **angular position** *θ* (and specifically an initial angular position *θ_i*) for *P*. Angles *θ* are commonly (but not invariably) measured from the positive *x* direction, as shown in Figure 11.3.

Angular displacements need not be infinitesimal. In Figure 11.3, *P* has undergone a finite angular displacement $\Delta\theta \equiv \theta - \theta_i$.

Angular Velocity

In one-dimensional translational kinematics, we defined the velocity *v* to be the displacement per unit time: $v \equiv dx/dt$. Here, analogously, we define the **angular velocity** *ω* to be the angular displacement per unit time:

$$\omega \equiv \frac{d\theta}{dt}. \tag{11.1}$$

The sign of *ω* is consistent with that of the angular displacement. If the rotation is counterclockwise, as in Figure 11.2, *dθ* is positive and *ω* is positive. If the rotation is clockwise, *dθ* is negative and so is *ω*. When *dθ* is expressed in radians, Equation 11.1 dictates that *ω* must be expressed in rad/s; this is the SI unit for angular velocity. Because it is defined to be the ratio of arc length to radius vector length, the radian is a dimensionless quantity. Thus the unit rad/s is equivalent to the unit s^{-1}.

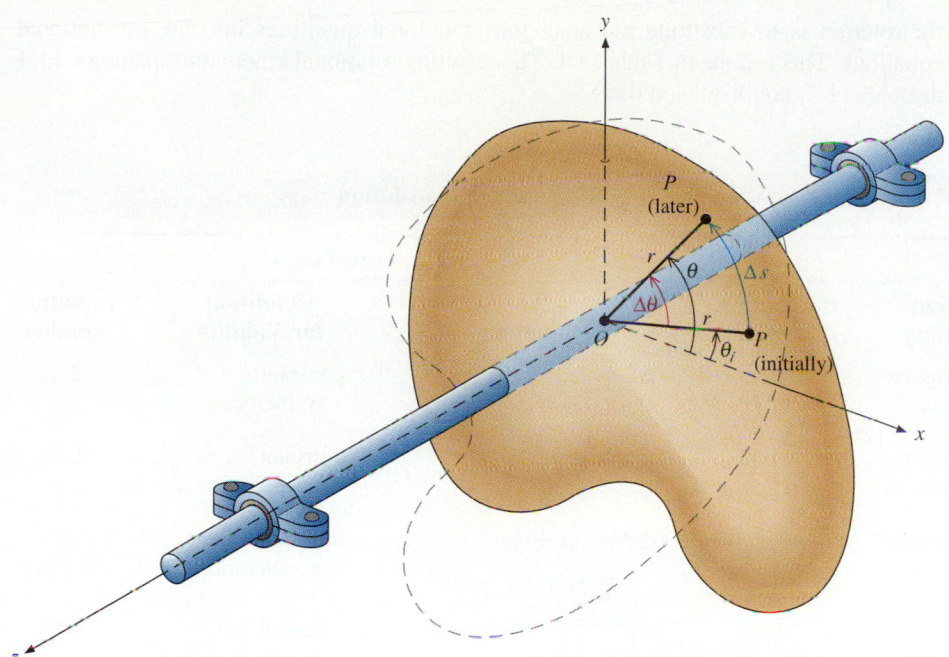

FIGURE 11.3 When a point is not initially located on the x axis ($\theta_i \neq 0$), the angular displacement $\Delta\theta$ is the difference between the final and initial angular positions of P; $\Delta\theta \equiv \theta - \theta_i$. The angle $\Delta\theta$ is the ratio of the circular arc length Δs to the radius r. Thus $\Delta\theta$ is the finite analogue to the infinitesimal angular displacement $d\theta$ of Figure 11.2. But the arc length Δs is not the finite analogue to the infinitesimal translational ds. Can you explain why?

In calculations in which rotation speeds are given in other units, such as rev/min or deg/s, it is often necessary—and always safe—to convert into rad/s to assure that the calculation is made in consistent units. (Remember that 1 rev = 2π rad; 1 rad = $360°/2\pi \approx 57.3°$.)

Angular Acceleration

In one-dimensional translational kinematics, we defined the acceleration a to be the change in velocity per unit time: $a \equiv dv/dt \equiv d^2x/dt^2$. Here, we again proceed by analogy and define the **angular acceleration** α to be the change in angular velocity per unit time:

$$\alpha \equiv \frac{d\omega}{dt} = \frac{d^2\theta}{dt^2}. \tag{11.2}$$

The SI unit for angular acceleration can be inferred from Equation 11.2; it is rad/s^2, or (equivalently) s^{-2}. Satisfy yourself as to the proper sign for α in the following cases: a speeding up of counterclockwise rotation; a slowing down of counterclockwise rotation; a speeding up of clockwise rotation; and a slowing down of clockwise rotation.

The Kinematic Equations for Rotational Motion

In Chapter 2, we paid considerable attention to the derivation of equations relating the quantities displacement Δx, velocity v, acceleration a, and time t. Because the case of constant velocity and the case of constant acceleration are so important in dynamics, we paid special attention to them. Fortunately, it is not necessary to carry out the analogous processes here for the corresponding rotational quantities: angular displacement $\Delta\theta$, angular velocity ω, angular acceleration α, and time t. We need only note that the derivation of the translational kinematic equations depended entirely on the fundamental defining relations

$$a \equiv \frac{dv}{dt} = \frac{d^2x}{dt^2}. \tag{11.3}$$

Because the rotational quantities are related in a manner mathematically identical with that given by Equation 11.3, all we need do to obtain the desired kinematic relations for rotation is to substitute the analogous rotational quantities into the translational equations. This is done in Table 11.1. The resulting rotational kinematic equations, 11.4 through 11.7, are displayed there.

TABLE 11.1 Kinematic Equations for Fixed-Axis Rotation and One-Dimensional Translation

Rotation			Translation		
Equation	**Condition for Validity**	**Equation Number**	**Equation**	**Condition for Validity**	**Equation Number**
$\Delta\theta = \omega t$; $\theta = \theta_i + \omega t$	constant angular velocity ω	11.4a 11.4b	$\Delta x = vt$; $x = x_i + vt$	constant velocity v	2.4 2.5
$\Delta\omega = \alpha t$; $\omega = \omega_i + \alpha t$	constant angular acceleration α	11.5a 11.5b	$\Delta v = at$; $v = v_i + at$	constant acceleration a	2.30
$\Delta\theta = \omega_i t + \frac{1}{2}\alpha t^2$; $\theta = \theta_i + \omega_i t + \frac{1}{2}\alpha t^2$	constant angular acceleration α	11.6a 11.6b	$\Delta x = v_i t + \frac{1}{2}at^2$; $x = x_i + v_i t + \frac{1}{2}at^2$	constant acceleration a	2.32b 2.32a
$\omega^2 - \omega_i{}^2 = 2\alpha\,\Delta\theta$	constant angular acceleration α	11.7	$v^2 - v_i{}^2 = 2a\,\Delta x$	constant acceleration a	2.37

EXAMPLE 11.1

An automobile wheel accelerates from rest with constant angular acceleration $\alpha = 5.80 \text{ rad/s}^2$. When 9.50 s has elapsed, what are **(a)** the angular velocity and **(b)** the total angular displacement of the wheel?

SOLUTION:
(a) Find the angular velocity of the wheel at $t = 9.50$ s.

From the viewpoint of an observer standing on the road, the rotation of the wheel is not fixed-axis rotation because the axle participates in the translational motion of the automobile. However, you can view the rotation of the wheel as fixed-axis rotation if you imagine yourself observing the wheel from a perch on the axle. Equation 11.5b, in Table 11.1, expresses angular velocity in terms of (constant) angular acceleration and time;

because the wheel starts from rest, you set $\omega_i = 0$. This gives you

$$\omega = \omega_i + \alpha t = 0 + 5.80 \text{ rad/s}^2 \times 9.50 \text{ s}$$
$$= 55.1 \text{ rad/s}.$$

(b) Find the total angular displacement of the wheel.

Equation 11.6a, in Table 11.1, expresses angular displacement in terms of acceleration and time. Again you use the fact that $\omega_i = 0$ to obtain

$$\Delta\theta = \omega_i t + \tfrac{1}{2}\alpha t^2 = 0 + \tfrac{1}{2} \times 5.80 \text{ rad/s}^2 \times (9.50 \text{ s})^2$$
$$= 262 \text{ rad}.$$

How many revolutions has the wheel turned?
Can you obtain the result to part **b** by using Equation 11.7?

Relation between Rotational and Translational Quantities

There is a physical as well as a mathematical connection between the fixed-axis rotational quantities $\Delta\theta$, ω, and α and the corresponding translational quantities. Consider an object at P in Figure 11.3 rotating along the circular arc Δs. At any instant, the translational speed of this object is $v = ds/dt$. (Note that this quantity is a speed and not a velocity.) We wish to relate v to the angular speed ω. To do so, we write the relation between a circular arc of length s and the angle θ it subtends in the form

$$s = r\theta. \tag{11.8a}$$

This equation arises directly from the definition of θ in radians. Other forms we will find useful are the incremental form

$$\Delta s = r\,\Delta\theta \tag{11.8b}$$

and the infinitesimal form

$$ds = r\, d\theta. \tag{11.8c}$$

In the infinitesimal case, the length of the circular arc is indistinguishable from the length of the straight chord that transects it.

By differentiating both sides of Equation 11.8a with respect to time, we obtain $ds/dt = d(r\theta)/dt$. Because r is constant for fixed-axis rotation of a rigid body, the equation becomes

$$\frac{ds}{dt} = r\,\frac{d\theta}{dt}.$$

The quantity ds/dt is the translational speed v, and $d\theta/dt$ is the angular speed ω. We thus obtain the relation

$$v = r\omega \tag{11.9}$$

between the translational speed v of a point at a distance r from the axis and the angular speed ω of the rotating system. (Note that the equation does *not* specify the sense of the rotation.)

We now repeat the process and differentiate both sides of Equation 11.9 with respect to time. Again remembering that r is a constant, we obtain

$$\frac{dv}{dt} = r\,\frac{d\omega}{dt}, \tag{11.10}$$

or

$$a_t = r\alpha.$$

We use the subscript "t" as a reminder that a_t is the *tangential* acceleration—the translational acceleration of any point a distance r from the axis in the direction along the tangent to the circle along which the point moves. There is a radial (centripetal) acceleration $a_r = v^2/r$ as well, but we are not considering it here. Equation 11.10 relates the magnitude of the tangential acceleration to the magnitude α of the angular acceleration of the rotating system.

EXAMPLE 11.2

The automobile wheel of Example 11.1 has a radius of 33 cm. **(a)** Neglecting the flattening of the tire where it makes contact with the ground, find the translational speed of the automobile when 9.50 s has elapsed. **(b)** How far does the automobile travel during the 9.50-s interval?

SOLUTION:
(a) Find the speed of the automobile at $t = 9.50$ s.

From the point of view of the observer on the axle, the ground moves by as the tread (outer surface) of the tire pushes on it. If the tire does not slip as it rolls along the ground, the speed of the automobile must be equal to the speed v of a point on the tread of the tire. You already know the angular speed ω of the tire at $t = 9.50$ s from part **a** of Example 11.1. To find v, you use Equation 11.9, which relates v to ω. You insert the value $\omega = 55.1$ rad/s to find the surface speed

$$v = r\omega = 0.33 \text{ m} \times 55.1 \text{ rad/s} = 18 \text{ m/s}.$$

Because it is customary to express automobile speeds in km/h, you write

$$v = 18 \text{ m/s} \times \frac{10^{-3} \text{ km}}{1 \text{ m}} \times \frac{3600 \text{ s}}{1 \text{ h}} = 65 \text{ km/h}.$$

(b) How far does the automobile travel during the 9.50-s interval?

For the observer on the axle, the ground moves a distance equal to the path length Δs covered by a point on the tire tread during the same time. You can use Equation 11.8b, together with the result from part **b** of Example 11.1, to write

$$\Delta s = 0.33 \text{ m} \times 262 \text{ rad} = 87 \text{ m}.$$

Can you think of another way to obtain the same result?

You can use Equations 11.8, 11.9, and 11.10 to derive Equations 11.4, 11.5, 11.6, and 11.7 from the corresponding translational equations given in Table 11.1. One of these derivations is shown in Example 11.3. Note, however, that the derivation is valid for magnitudes only, and not for directions.

EXAMPLE 11.3

Derive Equation 11.4a, $\Delta\theta = \omega t$, from the corresponding translational equation, $\Delta x = vt$.

SOLUTION: First, you note that the translational equation expresses a displacement Δx in one dimension. However, you can use it to express a displacement along a curved path provided you consider magnitudes only, and not directions. As usual in such cases, emphasize this point by using Δs to denote distance along a curved path. The translational equation then becomes

$$\Delta s = vt,$$

where v denotes the constant speed along the path. For a circular arc such as that in Figure 11.3, Equation 11.8b is valid, and you can make the substitution $\Delta s = r\,\Delta\theta$ into the preceding equation. Equation 11.9 also is valid, and you can make the substitution $v = r\omega$. This gives

$$r\,\Delta\theta = r\omega t.$$

Dividing through by r yields

$$\Delta\theta = \omega t,$$

which is Equation 11.4a.

Dynamics of Rotation about a Fixed Axis

Since the translational-rotational analogy has been so helpful in developing rotational kinematics, let us extend it to rotational *dynamics*—that is, to the study of the equations that relate motion to its causes. In order to write such equations of motion for rotation, we will have to define rotational quantities analogous to the translational dynamical quantities force F and mass m. We will do so in very much the same way as we derived the kinematic quantities θ, ω, and α by analogy to the corresponding translational quantities x, v, and a_t.

Rotational Analogue of Newton's First Law for Fixed-Axis Rotation

Everyone has had many opportunities to observe a wheel spinning by itself on good bearings. It is only a slight extension of that experience to suppose that a wheel completely isolated from outside influence except for the constraint imposed by its fixed axis—that is, a wheel having perfect friction-free bearings and not subject to air resistance—will continue to spin indefinitely with constant angular velocity ω. Thus the rotational analogue to Newton's first law, $v = $ constant for an isolated body, is the statement

$$\omega = \text{constant} \quad \begin{array}{l} \text{for a rotating rigid body} \\ \text{constrained by a fixed axis but otherwise isolated.} \end{array} \qquad \textbf{(11.11)}$$

Torque

Rather than considering Equation 11.11 in detail, we will do better to generalize our inquiry and find the rule governing angular acceleration when ω is not constant. We can certainly accelerate a wheel by applying a force. But you know from experience that the same force applied to different parts of the wheel has different effects.

The simple rotatory device called the **lever**, shown in Figure 11.4, can be a source of insight into the effect of a force on a wheel. Note that you can regard a lever as a slice cut from a wheel so as to include the axis. The axis of a lever is usually called its **fulcrum**. The weight mg is raised by exerting a just-adequate force F. Because the weight is raised very slowly, the kinetic energy of the system is always negligible. Suppose that the lever rotates through the infinitesimal angle $d\theta$. Because the force F is applied at a distance r_2 from the fulcrum that is different from the distance r_1 at which the weight mg is attached, the force is applied through a distance $ds_2 = r_2\,d\theta$, whereas the weight is raised through a distance $ds_1 = r_1\,d\theta$. In this process, the work done on the lever by the applied force F is $F\,ds_2 = (r_2\,d\theta)F$. The work done on the lever by the string attached to the weight is $(-mg)\,ds_1 = (r_1\,d\theta)(-mg)$. The minus sign accounts for the fact that the string applies a force to the lever in the direction

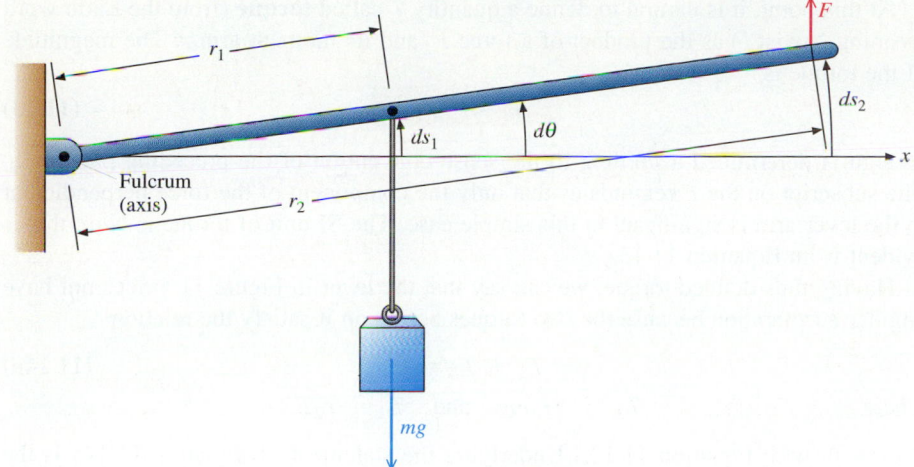

FIGURE 11.4 A weight mg hangs from a lever at a distance r_1 from the fulcrum. The weight is balanced by a vertical force F, applied at a distance r_2 from the fulcrum. The angle $d\theta$, exaggerated for clarity, is actually small, so the lever is nearly horizontal.

opposite to the direction of motion. Because the lever does not gain kinetic energy in the process, the net work done on it is zero, in accordance with the work-energy theorem:

$$(r_1\, d\theta)(-mg) + (r_2\, d\theta)F = 0.$$

It may seem trivial to rewrite this equation in the form

$$(-r_1 mg + r_2 F)\, d\theta = 0,$$

But, because $d\theta$ is not zero (though it may be arbitrarily small), the left side of the equation can be equal to zero only if

$$r_1 mg = r_2 F, \tag{11.12}$$

so far as magnitudes are concerned. The distances r_1 and r_2 are called the **lever arms** or **moment arms** of the corresponding forces. (The terms *moment arm* and *lever arm* mean exactly the same thing, but it is customary to refer to the lever arm only when discussing levers; moment arm is used more generally.)

We now see that *the product of a force and its moment arm, and not the force alone, is of significance in rotation.* To put it another way, the *point of application* of the force is significant here. The force itself, being a vector, is not changed if it is moved without changing its direction. But the effect it has on rotation *is* changed. For this reason, a force vector that tends to produce rotation belongs to the class called *bound vectors*; see Section 4.2.

What of the directions of the forces acting on a lever? A force applied parallel to the lever arm has no effect at all on the rotation of the lever. If the direction of the force makes an arbitrary angle with the lever arm, the parallel component is likewise ineffective and only the perpendicular component has a tendency to rotate the lever. Evidently, a force of a given magnitude applied to a lever is most effective when the direction of the force is perpendicular to the lever arm. In this section, we will consider *only* forces (or their components) that are *perpendicular* to their lever arms.

It is not just the direction of the force that is of significance in rotation. In Figure 11.4, an upward force balances a downward one. But if the fulcrum were located between the points of application of the two forces, rather than on the same side of both, a downward force would be needed to balance a downward force. Thus, *we are concerned with the sense in which the force tends to produce rotation* and not merely with the direction of the force. In Figure 11.4, for example, the product $r_2 F$ of the applied force and its lever arm tends to produce counterclockwise rotation. We therefore assign a positive sign to this product, in accordance with the convention established in Section 11.2. And, because the product $r_1 mg$ of the weight and its lever arm tends to produce clockwise rotation, we assign a negative sign to it. Note that exactly the same argument would hold true if the lever were in a vertical position and the forces were horizontally directed.

At this point, it is natural to define a quantity T called **torque** (from the Latin word meaning "twist") as the product of a force $F_\perp$ and its moment arm r. The magnitude of the torque is

$$|T| \equiv |rF_\perp|; \qquad (11.13)$$

its sign is determined according to the sense convention of the preceding paragraph. The subscript on the F reminds us that only the component of the force perpendicular to the lever arm is significant in this simple case. The SI unit of torque is N·m; this is evident from Equation 11.13.

Having thus defined torque, we can say that the lever in Figure 11.4 does not have angular acceleration because the two torques acting on it satisfy the relation

$$T_1 + T_2 = 0, \qquad (11.14a)$$

where $$T_1 = -r_1 mg \quad \text{and} \quad T_2 = r_2 F.$$

(Compare with Equation 11.12.) Underlying the statement of Equation 11.14a is the stipulation that the moment arms of both torques are measured from the fulcrum, which is the fixed axis of rotation. We therefore speak of these torques as *torques about the axis*. In this chapter, we will consider only such torques.

Newton's Second Law for Rotation of a Particle about a Fixed Axis

If there are N torques acting on a body, we assert that

$$\sum_{j=1}^{N} T_j = 0 \quad \text{if and only if} \quad \omega = \text{constant}. \qquad (11.14b)$$

This is the rotational analogue of Newton's second law for translation in one dimension, in the special case in which the N forces applied to a body add up to zero:

$$\sum_{j=1}^{N} F_j = 0 \quad \text{if and only if} \quad v = \text{constant}.$$

Equation 11.14b makes a more general statement than does Equation 11.11 in that Equation 11.14b allows for the presence of torques acting on a rotating body, provided that their *sum* is zero. A familiar example of such a situation is a gear wheel in the train of gears in a mechanical watch. Both neighboring gears in the train exert torques on the gear in question, but its angular velocity is constant because the torques are equal and opposite and so add to zero.

The next step in developing an equation of motion for rotation is a small one. So far, we have focused on the condition for balance of the lever. If we want to raise the weight in Figure 11.4, we must increase the applied force at least briefly so that the net torque is at least briefly greater than zero. The lever will then accelerate from rest ($\omega = 0$) to some positive angular velocity about its fulcrum. It is plausible to assert that the angular acceleration will be proportional to the net applied torque:

$$T \propto \alpha \qquad (11.15)$$

in analogy to Equation 3.2, $F \propto a$.

What is the proportionality constant that relates the rotational quantities T and α, analogously to the way that mass relates F and a in the translational form of Newton's second law, $F = ma$? To answer this question, let us consider a special system in which the rules of translational and rotational mechanics can be applied simultaneously. Because we already know the rules for translation, we can deduce the rules for rotation by demanding that the two sets of rules lead to consistent results.

Our special system, illustrated in Figure 11.5, consists of a particle of mass m attached to the end of a massless string of length r, whose other end can rotate freely about a pivot. The entire system lies on a level, frictionless table, so we need not be concerned with gravitational forces.

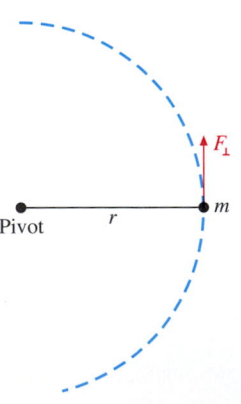

FIGURE 11.5 Diagram for inferring the rotational form of Newton's second law, $\tau = I\alpha$, from the translational form $F = ma$. The system, seen from above, rests on a frictionless table not shown. The circular path of the particle is shown as a dashed line. The applied force $F_\perp$ is always perpendicular to the string and thus tangent to the path.

If a force $F_\perp$ is applied to the particle, always in a direction perpendicular to the string and in the horizontal plane, the particle will experience an acceleration a_t in that direction, tangential to the circular path of the particle. The magnitude of the acceleration is given by the scalar relation

$$a_t = \frac{F_\perp}{m}.$$

(11.16)

This follows immediately from Newton's second law. Any force applied by the string can produce only centripetal acceleration. Such a force cannot affect the result just stated because a string can exert a force only along its length.

Multiplying Equation 11.16 through by r makes it possible to reexpress the equation in terms of the torque T:

$$ra_t = \frac{rF_\perp}{m} = \frac{T}{m}.$$

(11.17)

To complete the transition to rotational language, we need to reexpress Equation 11.17 in terms of an angular acceleration α. To do so, we note that a_t is given by the kinematic relation of Equation 11.10, $a_t = r\alpha$. We can thus substitute $r\alpha$ for a_t in Equation 11.17 to obtain

$$r^2\alpha = \frac{T}{m},$$

or

$$T = (mr^2)\alpha.$$

(11.18)

We have found the desired proportionality constant between torque and angular acceleration because the quantity mr^2 is indeed a constant—thus our assertion that $T \propto \alpha$ is verified for the special case of a rotating system consisting of a single particle. We define the **moment of inertia** I of the particle about the specified axis to be the proportionality constant:

$$I \equiv mr^2.$$

(11.19)

The SI unit of I is kg·m^2; this follows immediately from the definition. The moment of inertia of a system is a measure of the **rotational inertia** of the system (the effort required to impart a given angular acceleration to it), just as the mass of a system is a measure of the translational inertia of the system (the effort required to impart a given translational acceleration to it).

We can now write

$$T = I\alpha \quad \text{for rotation of a particle about a fixed axis.}$$

(11.20)

This equation is a restricted form of **Newton's second law for rotation**. The equation is necessarily analogous to the one-dimensional translational form $F = ma$ because it was deduced from that form.

Newton's Second Law for Rotation of a Rigid Body about a Fixed Axis

We must now generalize the argument leading to Equation 11.20 so that it applies to a rigid body rotating about a fixed axis. The generalization has a happy result: Equation 11.20 is just as valid for a rigid body as it is for a particle. However, we must generalize the definition of moment of inertia (Equation 11.19) so that it has meaning for an extended body.*

Figure 11.6a shows a collection of independent particles, each of which is attached to a massless rigid rod and can rotate in the plane of the page about the common axis A. A typical particle has mass Δm_j and is located at a distance r_j from the axis. The

---

*This generalization may be omitted without loss of continuity. The reader who wishes to do so should skip to Equation 11.25 and take its validity for granted, together with the general definition of moment of inertia, Equation 11.24.

(a)

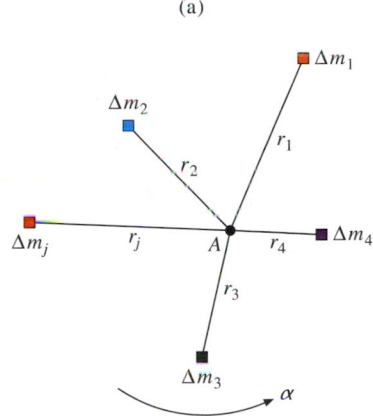

(b)

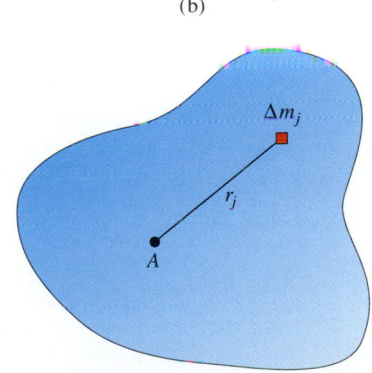

FIGURE 11.6 (a) A collection of independent particles rotating in the plane of the page about a common fixed axis A. A typical particle has mass Δm_j and follows a circular path at a distance r_j from the axis. All particles have the same angular acceleration α. (b) A rigid body rotating about the fixed axis A is subdivided in imagination into many mass elements. The typical element j has mass Δm_j and lies at a distance r_j from the axis.

particle is acted on by a torque T_j. According to Equation 11.18, the particle experiences an angular acceleration α_j whose magnitude is

$$\alpha_j = \frac{T_j}{r_j^2 \, \Delta m_j}.$$

The particles may have different masses, and the torques acting on them may be different as well. But let us suppose for the sake of argument that all the angular accelerations are the same. In that case, we give all of them the common name α:

$$\alpha_1 = \alpha_2 = \cdots = \alpha_j = \cdots \equiv \alpha.$$

It follows that

$$\frac{T_1}{r_1^2 \, \Delta m_1} = \frac{T_2}{r_2^2 \, \Delta m_2} = \cdots = \frac{T_j}{r_j^2 \, \Delta m_j} = \cdots, \qquad \textbf{(11.21)}$$

even though the numerators and denominators of the fractions are not all the same.

At this point we use a theorem of arithmetic, which you can readily verify. Given a set of equal fractions,

$$\frac{\nu_1}{\delta_1} = \frac{\nu_2}{\delta_2} = \cdots = \frac{\nu_j}{\delta_j} = \cdots,$$

the sum of the numerators divided by the sum of the denominators is equal to the individual fractions:

$$\frac{\displaystyle\sum_j \nu_j}{\displaystyle\sum_j \delta_j} = \frac{\nu_j}{\delta_j}. \qquad \textbf{(11.22)}$$

When we apply this theorem to Equation 11.21, we obtain

$$\frac{\displaystyle\sum_j T_j}{\displaystyle\sum_j r_j^2 \, \Delta m_j} = \frac{T_j}{r_j^2 \, \Delta m_j}.$$

The sum in the numerator on the left side of this equation is the net external torque T exerted on the system. (The distinction between external and internal torques is identical with the distinction between external and internal forces; see Section 9.3.) The fraction on the right side is the angular acceleration α. On multiplying both sides of the equation by the denominator of the left side, we obtain

$$T = \left(\sum_j r_j^2 \, \Delta m_j \right) \alpha. \qquad \textbf{(11.23)}$$

We now *define* the moment of inertia of the system to be the term in parentheses in this equation:

$$I \equiv \sum_j r_j^2 \, \Delta m_j. \qquad \textbf{(11.24)}$$

This is a natural definition; it states that the moment of inertia of the entire system is the sum of the moments of inertia $r_j^2 \, \Delta m_j$ of the elements of which it consists. With this definition, Equation 11.23 assumes the simple form

$$T = I\alpha \quad \text{for rotation of an extended body about a fixed axis.} \qquad \textbf{(11.25)}$$

Equation 11.25 is identical in form with Equation 11.20, even though Equation 11.25 is more general because it is not restricted to a single particle, as Equation 11.20 is.

We are now ready for the final step in generalizing Newton's second law for rotation: We extend the foregoing argument so that we can apply it to a rigid body. So far, we have considered a system consisting of a collection of independent particles. We may certainly regard a rigid body as consisting of a collection of adjacent mass elements, as we did, for example, in defining the center of mass in Section 9.5. This viewpoint is illustrated in Figure 11.6b, in which a typical mass element Δm_j is shown located at a distance r_j from the axis of rotation. When the rigid body is thus considered, Equation

11.24—which defines the moment of inertia—is applicable without further ado. (We will see in the next subsection, and in Section 11.5, how to use this equation in an actual calculation.)

But the particles that the rigid body comprises are not independent of one another. In considering the system of Figure 11.6a, we had to make the special assumption that all the particles experienced the same angular acceleration α. For the mass elements of a rigid body, this special assumption becomes an automatic condition because any difference in the angular acceleration of two elements would result in a change of the distance between them, in contradiction to the definition of a rigid body. The uniform angular acceleration of the rigid body is brought about by internal forces exerted by the elements of the body on one another so as to prevent distortion of the body. In Section 9.5, we separated the forces exerted on an element into internal and external forces and showed that the sum of all internal forces is zero. A similar argument here leads to the conclusion that no net torque can result from internal forces. The net torque exerted on the body is therefore the external torque T. We conclude that the equation $T = I\alpha$ applies to a rigid body as well as to a collection of independent particles having equal angular accelerations. Using the definition of I given by Equation 11.24, we can write Newton's second law for rotation with its accompanying condition:

$$T = I\alpha \quad \text{for rotation of a rigid body about a fixed axis.} \qquad (11.26)$$

Although this equation has exactly the same form as Equation 11.20, it is much more generally applicable.

The identity of the forms of Equations 11.20 and 11.26 has a translational analogy. Equation 9.25, $\mathbf{F}_{\text{ext}} = M\mathbf{A}$, describes the motion of the center of mass of an extended system of mass M. Its form is identical with that of Newton's second law for a particle, $\mathbf{F} = m\mathbf{a}$.

Moment of Inertia of an Extended Body

By comparing Equation 11.26 with the corresponding translational equation, $\mathbf{F}_{\text{ext}} = M\mathbf{A}$, you can immediately see that moment of inertia is the rotational analogue of mass that we have been seeking. Moment of inertia is somewhat more complicated to deal with computationally than is mass in the translational case because, in general, not all the mass elements in a rotating object are at the same distance r from the axis of rotation. But let us begin our study of the moment of inertia of an extended body by considering the simplest case—that in which essentially all the mass *is* at the same distance from the axis. An idealization of such a body is a hoop of radius R, constrained to rotate about an axis normal to its own plane and passing through its center (Figure 11.7). All of the mass of the hoop is located at the same distance from the center. (A bicycle wheel, because the tire and rim are thin and much more massive than the spokes, is a fair approximation to this ideal hoop.) To find the moment of inertia of the entire hoop about this axis, we divide the hoop into small mass elements Δm_j, one of which is shown in Figure 11.7. All of these elements lie at the same distance R from the axis; that is, $r_j = R$. Equation 11.24 thus assumes the simple form

$$I = \sum_j R^2 \, \Delta m_j = R^2 \sum_j \Delta m_j.$$

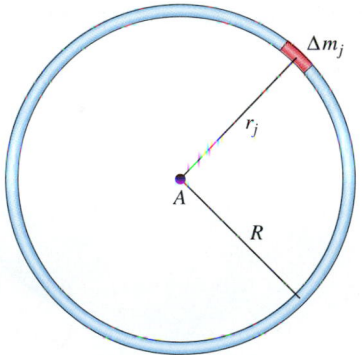

FIGURE 11.7 In a hoop, all mass elements Δm_j lie at the same distance $r_j = R$ from the axis of rotation.

The sum is thus reduced to a sum over all the mass elements. But this sum is just the mass M of the entire wheel:

$$\sum_j \Delta m_j = M.$$

Consequently, the entire wheel has moment of inertia

$$I = MR^2. \qquad (11.27)$$

The close resemblance of this result to Equation 11.19, which gives the moment of inertia of a particle to be $I = mr^2$, will probably not surprise you.

Unlike the mass of a body, which is the same no matter how the body is constrained, the moment of inertia of a body depends explicitly on the location of the axis of rotation because this location determines the value of r_j appropriate to each mass element Δm_j. The hoop, for example, will have a different moment of inertia if it is made to rotate about an axis passing through a point on its rim. However, Equation 11.25 is valid for *any* location of the axis of rotation, as long as the appropriate value of I is used.

EXAMPLE 11.4

A bicycle wheel has mass $M = 1.2$ kg and radius $R = 33$ cm. Find the angular acceleration α and the tangential acceleration a_t when a constant tangential force $F = 4.0$ N is exerted by the drive chain on the rim of the driving sprocket, whose radius is $r = 5.0$ cm.

SOLUTION: First draw the system to make the geometry clear, as in Figure 11.8. The driving torque is

$$T = rF = 0.050 \text{ m} \times 4.0 \text{ N} = 0.20 \text{ N·m}.$$

If you assume that essentially all the mass of the system is concentrated at the periphery of the wheel—that is, that the spokes and driving sprocket have negligible mass—the moment of inertia of the system is

$$I = MR^2 = 1.2 \text{ kg} \times (0.33 \text{ m})^2 = 0.13 \text{ kg·m}^2.$$

You can now apply the rotational form of Newton's second law to obtain the angular acceleration

$$\alpha = \frac{T}{I} = \frac{0.20 \text{ N·m}}{0.13 \text{ kg·m}^2} = 1.5 \frac{N}{kg·m} = 1.5 \frac{kg·m/s^2}{m·kg}$$

$$= 1.5 \text{ s}^{-2}.$$

FIGURE 11.8

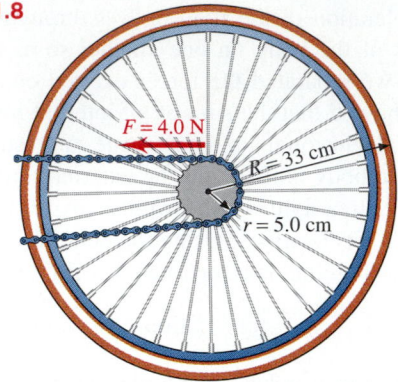

$F = 4.0$ N
$R = 33$ cm
$r = 5.0$ cm

Now you use Equation 11.10 to obtain the tangential acceleration of a point on the periphery of the wheel. You have

$$a_t = \alpha R = 1.5 \text{ s}^{-2} \times 0.33 \text{ m} = 0.51 \text{ m/s}^2.$$

There is also a radial component of acceleration, the centripetal acceleration $a_r = v^2/R = \omega^2 R$. This component is not constant; it increases as the angular speed of the wheel increases.

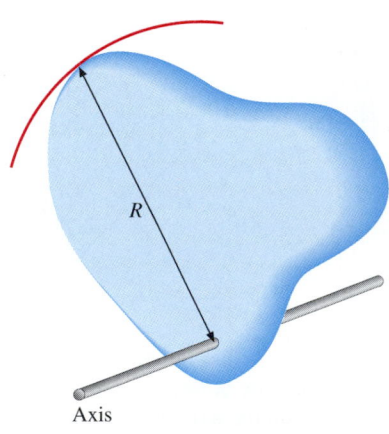

FIGURE 11.9 A body of arbitrary size and shape mounted on an arbitrary fixed axis, showing the maximum radius R of Equation 11.28.

The moment of inertia of a rigid body of a given mass about a given axis depends on the size and shape of the body. For any object other than a hoop rotating about a normal axis through its center, some of the mass must be located at a distance from the axis that is less than the maximum radius R of the object. Therefore the moment of inertia

$$I = \sum_j r_j^2 \Delta m_j$$

must be less than MR^2, which is the moment of inertia of a hoop of the same mass and radius. Indeed, because the contribution of any mass element Δm_j depends on the *square* of its distance from the axis of rotation, the moment of inertia of a body of arbitrary shape can be much less than MR^2. In any case, however, we can write

$$I = cMR^2, \tag{11.28}$$

where M is the total mass of the body, R is its maximum radius with respect to the axis of rotation (Figure 11.9), and c, the "shape factor," is some fraction whose value lies between 0 and 1.

Calculation of Moments of Inertia

We now develop a method for calculating the moment of inertia for a rotating system more complicated than the hoop and the special axis treated in Section 11.3. The method amounts to explicit evaluation of the constant c in Equation 11.28, $I = cMR^2$.

Figure 11.10 represents an arbitrarily shaped rigid body of mass M attached to the fixed axis shown. In imagination, we divide the body into an infinite number of infinitesimal mass elements dm. One of these elements is shown; it lies a distance r from the axis. The moment of inertia dI of this element about the axis is

$$dI = r^2\, dm. \tag{11.29}$$

As usual, we can in principle find the moment of inertia of the entire body by integrating Equation 11.29 over the infinite number of infinitesimal elements that make up the body. This process is represented by the equation

$$I = \int_{\substack{\text{entire}\\\text{body}}} dI = \int_{\substack{\text{entire}\\\text{body}}} r^2\, dm, \tag{11.30}$$

which corresponds to the finite process of summation represented by Equation 11.24.

Although Equation 11.30 is quite generally valid, it is possible to carry out the integration analytically only for bodies whose shape can be expressed analytically. Fortunately, a very large proportion of the rigid bodies encountered in common experience have this property, at least approximately. Example 11.5 applies Equation 11.30 to a cylinder mounted on a shaft that coincides with its symmetry axis.

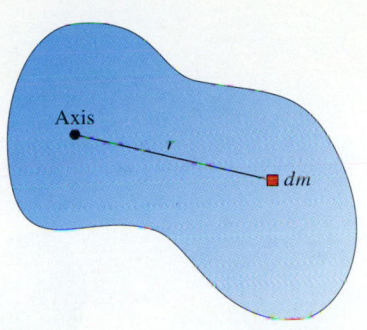

FIGURE 11.10 A rigid body can be divided into infinitesimal mass elements dm.

EXAMPLE 11.5

A cylindrical grindstone has radius $R = 23$ cm, thickness $Z = 6.0$ cm, and mass $M = 26$ kg. Find its moment of inertia.

SOLUTION: You begin by drawing a sketch of the grindstone, as in Figure 11.11. In order to apply Equation 11.29, you must divide the grindstone into infinitesimal elements. In doing so, you take advantage of its cylindrical symmetry. Specifically, all parts of a thin, ring-shaped element of radius r, width dr, and thickness Z lie at the same distance from the axis. Thus the moment of inertia dI of one such element is related to its mass dm by Equation 11.29, $dI = r^2\, dm$. To use this equation, you must find the mass dm. Similarly to Example 9.5, you express dm in terms of the volume dV of the element. It is reasonable to assume that the grindstone is uniform. Hence the fraction of the total mass contained in the element is equal to the fraction of the total volume contained in the element:

$$\frac{dm}{M} = \frac{dV}{V},$$

where V is the volume of the entire grindstone. You thus have $dm = M\, dV/V$.

Next you express dV in terms of the dimensions of the ring-shaped element: $dV = 2\pi r\, dr\, Z$. Substituting these values into the expression for dm that you have just derived gives you

$$dm = M\,\frac{2\pi r\, dr\, Z}{\pi R^2 Z} = \frac{2M}{R^2}\, r\, dr.$$

This is the value you need to find dI, the moment of inertia of the ring-shaped element:

$$dI = r^2\, dm = \frac{2M}{R^2}\, r^3\, dr.$$

The moment of inertia of the entire grindstone can now be

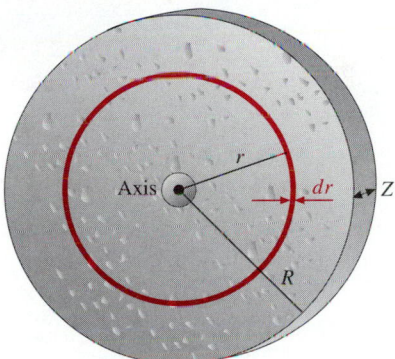

FIGURE 11.11 Calculating the moment of inertia of a cylindrical grindstone. In subdividing the disk into mass elements, it is important to exploit the symmetry of the system.

found by using Equation 11.30. You calculate

$$I = \int_{\substack{\text{grind-}\\\text{stone}}} dI = \frac{2M}{R^2} \int_0^R r^3\, dr$$

$$= \frac{2M}{R^2}\, \frac{r^4}{4}\Bigg|_0^R = \frac{2M}{R^2}\, \frac{R^4}{4},$$

or

$$I = \tfrac{1}{2}MR^2. \tag{11.31}$$

Note that the thickness Z of the grindstone does not appear in this result. Can you explain why?

You can now find the numerical value of the moment of inertia of the grindstone:

$$I = \tfrac{1}{2} \times 26\ \text{kg} \times (0.23\ \text{m})^2 = 0.69\ \text{kg·m}^2.$$

Equation 11.31 shows that for a cylindrical disk the value of the constant c in Equation 11.28, $I = cMR^2$, is $c = \tfrac{1}{2}$. To put it another way, a given mass distributed uniformly over a cylindrical disk contributes to the moment of inertia about the axis half as effectively as the mass would if it were concentrated at the rim of the same disk.

Table 11.2 gives the value of c, and of the moment of inertia, for a number of frequently encountered geometric forms about the axes shown. The method of Example 11.5 can be used to find the result in each case.

TABLE 11.2 Moment of Inertia of Uniform Bodies of Mass M Having Various Shapes

	Shape	Shape Factor c	Moment of Inertia I
	Thin cylindrical shell about its symmetry axis	1	MR^2
	Solid cylinder about its symmetry axis	$\frac{1}{2}$	$\frac{1}{2}MR^2$
	Thick-walled hollow cylinder about its symmetry axis		$\frac{1}{2}M(R_1^2 + R_2^2)$
	Slender rod (any uniform cross-sectional shape) about perpendicular axis through center	$\frac{1}{12}$	$\frac{1}{12}ML^2$
	Slender rod (any uniform cross-sectional shape) about perpendicular axis through end	$\frac{1}{3}$	$\frac{1}{3}ML^2$
	Rectangular block about perpendicular axis through center of face		$\frac{1}{12}M(a^2 + b^2)$
	Thin spherical shell about diameter	$\frac{2}{3}$	$\frac{2}{3}MR^2$
	Solid sphere about diameter	$\frac{2}{5}$	$\frac{2}{5}MR^2$

The Parallel-Axis Theorem

In most of the cases illustrated in Table 11.2, the moment of inertia is given about an axis of symmetry. As Example 11.5 suggests, it is usually easiest to calculate I about such an axis. Because the calculations depend on the fact that the body is in each case uniform, the axis of symmetry in each case must pass through the center of mass. Fortunately, there is a simple relation between a body's moment of inertia I_{cm} about any axis that passes through the center of mass and the body's moment of inertia $I_{\parallel}$ about any other axis parallel to the first axis. If the distance between the two axes is D and the mass of the body is M, the relation between the two moments of inertia is

$$I_{\parallel} = I_{cm} + MD^2. \tag{11.32}$$

Thus, if I_{cm} is known, $I_{\parallel}$ can easily be calculated. This relation is called the **parallel-axis theorem**, or **Steiner's theorem**, after the Swiss mathematician who first derived it. The proof follows.

Figure 11.12 is a view of a body of mass M and arbitrary shape, seen along an axis of rotation A that passes through its center of mass. The z axis is chosen to lie along this axis (normal to the plane of the page), so that the axis is described by the coordinates $(0, 0, z)$. A second axis A' lies parallel to A and is separated from A by a distance D. We choose the positive x direction to be that from A to A'. The axis A' is then described by the coordinates $(D, 0, z)$.

The typical mass element Δm_j has coordinates (x_j, y_j, z_j). It lies a distance r_j from A and a distance r'_j from A'. Its contribution to the moment of inertia $I_{\parallel}$ of the body about A' is $r'^2_j \Delta m_j$. According to the law of cosines (Appendix 3), r'^2_j can be written

$$\begin{aligned} r'^2_j &= r^2_j + D^2 - 2Dr_j \cos\theta \\ &= r^2_j + D^2 - 2Dx_j. \end{aligned}$$

Summation over all contributions $r'^2_j \Delta m_j$ thus yields for $I_{\parallel}$ the expression

$$I_{\parallel} = \sum_j r'^2_j \Delta m_j = \sum_j r^2_j \Delta m_j + D^2 \sum_j \Delta m_j - 2D \sum_j x_j \Delta m_j. \tag{11.33}$$

The quantity D has been passed through the summation signs because it is a constant and does not depend on j. According to Equation 11.24, the first sum to the right of the second equals sign in Equation 11.33 is equal to I_{cm}. The second sum yields the total mass M of the body. The third sum is the product of M with the x coordinate of the center of mass. This follows immediately from Equation 9.18a, which gives that coordinate as

$$\frac{1}{M} \sum_j x_j \Delta m_j.$$

But we have chosen the z axis, $(0, 0, z)$, to lie along the rotation axis A, which passes through the center of mass. Hence, we have $\sum_j x_j \Delta m_j = 0$. The equation for $I_{\parallel}$ reduces to

$$I_{\parallel} = I_{cm} + MD^2,$$

which is what we set out to prove.

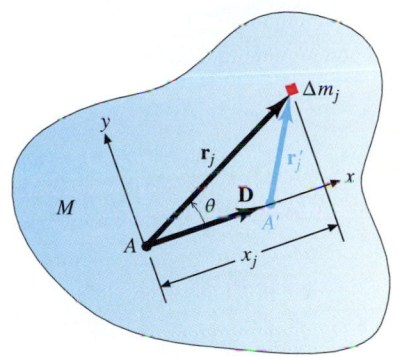

FIGURE 11.12 The parallel-axis theorem. The axis of rotation A is normal to the plane of the page.

EXAMPLE 11.6

Find the moment of inertia of a uniform solid sphere of mass M and radius R about an axis tangent to its surface.

SOLUTION: For any axis tangent to a sphere, there exists a parallel axis along a diameter. This parallel axis passes through the center of mass of the sphere. The distance between the two axes is R, and Equation 11.32 thus takes the form

$$I_{tangent} = I_{cm} + MR^2.$$

According to Table 11.2, $I_{cm} = \frac{2}{5}MR^2$ for a solid sphere. So you have

$$I_{tangent} = \frac{2}{5}MR^2 + MR^2 = \frac{7}{5}MR^2.$$

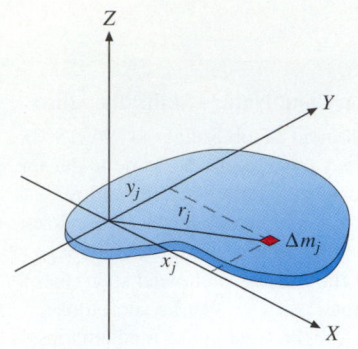

FIGURE 11.13 The perpendicular-axis theorem.

The Perpendicular-Axis Theorem

Figure 11.13 depicts a uniform body that is thin and flat but is otherwise of arbitrary shape. (It may even have holes in it.) Any two perpendicular axes of rotation X and Y lie in the plane of the body, and Z is an axis of rotation that passes through the intersection of X and Y and is perpendicular to them. (In Figure 11.13, the intersection of the axes lies within the body, but this is not necessary.) The moments of inertia about these three axes—respectively, I_X, I_Y, and I_Z—bear the relation

$$I_Z = I_X + I_Y. \tag{11.34}$$

This relation is called the **perpendicular-axis theorem**.

The proof is as follows. Consider a mass element such as Δm_j in Figure 11.13. Its distance from the Z axis is r_j, and its contribution to I_Z is thus $r_j^2 \, \Delta m_j$. Summation yields the total moment of inertia about that axis,

$$I_Z = \sum_j r_j^2 \, \Delta m_j.$$

The distance of the same mass element from the X axis is y_j, and so a similar argument yields the total moment of inertia about the X axis,

$$I_X = \sum_j y_j^2 \, \Delta m_j.$$

In like manner, we find

$$I_Y = \sum_j x_j^2 \, \Delta m_j.$$

Adding I_X and I_Y gives

$$I_X + I_Y = \sum_j (y_j^2 + x_j^2) \, \Delta m_j = \sum_j r_j^2 \, \Delta m_j = I_Z,$$

and the perpendicular-axis theorem is proved.

EXAMPLE 11.7

Find the moment of inertia of a hoop of mass M and radius R about a rotation axis along one of its diameters.

SOLUTION: The moment of inertia I_Z of the hoop about an axis through its center and normal to its plane is given by Equation 11.27, $I_Z = MR^2$. According to the perpendicular-axis theorem, this moment of inertia must be equal to the sum $I_X + I_Y$ of the

moments of inertia about two perpendicular axes, each of which is a diameter of the hoop. Because the hoop is symmetrical, its moment of inertia must be the same about all diameters, so $I_X = I_Y$. It follows that $I_X + I_Y = 2I_X = I_Z = MR^2$, and the desired result is

$$I_X = \tfrac{1}{2}MR^2.$$

Rotational Kinetic Energy, Work, and Power

In Section 11.3, we developed rotational quantities analogous to the dynamical quantities of translational mechanics. We now extend the analogy to include kinetic energy.

Consider again the situation of Figure 11.5, in which a particle of mass m is constrained by a string of length r to rotate about a frictionless pivot. Let the angular velocity of the particle be ω. From Equation 11.9, we know that the tangential speed of the particle is $v = r\omega$. Thus its kinetic energy must be

$$K = \tfrac{1}{2}mv^2 = \tfrac{1}{2}mr^2\omega^2.$$

Because mr^2 is the moment of inertia I of the point mass about the pivot (Equation 11.19), we can express the kinetic energy of the particle in the rotational form

$$K = \tfrac{1}{2}I\omega^2. \tag{11.35}$$

The kinetic energy for an extended rigid body of mass M can be found by subdividing the body into infinitesimal mass elements dm. According to Equation 11.29, such a mass element has a moment of inertia $dI = r^2\,dm$. The kinetic energy of such a mass element can be expressed in its rotational form by rewriting Equation 11.35 in the differential form

$$dK = \tfrac{1}{2}\omega^2\,dI.$$

The kinetic energy for the entire body is then found by integrating the contributions dK of all the mass elements of which it is made up:

$$K = \int_{\substack{\text{entire}\\ \text{body}}} dK = \int_{\substack{\text{entire}\\ \text{body}}} \tfrac{1}{2}\omega^2\,dI.$$

But the angular velocity ω is the same for all parts of a rotating rigid body. Hence the constant $\tfrac{1}{2}\omega^2$ can be passed through the integral sign. We obtain

$$K = \tfrac{1}{2}\omega^2 \int_{\substack{\text{entire}\\ \text{body}}} dI,$$

or

$$K = \tfrac{1}{2}I\omega^2. \tag{11.36}$$

Compare with the translational form $K = \tfrac{1}{2}Mv^2$. Note again that I plays a role in rotation analogous to the role of M in translational motion.

EXAMPLE 11.8

The system shown in Figure 11.14 consists of a mass m_1 hanging from a string wrapped around a drum. The drum has moment of inertia I and radius R. The system is released from rest. **(a)** Neglecting friction, find the speed of m_1 when it has descended a distance h. **(b)** Suppose that the drum is a solid cylinder and that its mass is $M = m_1$. Find v.

SOLUTION:
(a) Find the speed of m_1 when it has descended a distance h.

In the absence of friction, the system comprising the mass and the drum is conservative, and the total mechanical energy change during the process is zero: $\Delta E = \Delta K + \Delta U = 0$ (Equations 8.18b and 8.18c). Because the system is initially at rest, its initial kinetic energy is zero, and you have $K = -\Delta U$ at any time. At the instant when m_1 has descended a distance h, the kinetic energy of the system must be

$$K = -(-m_1gh) = m_1gh.$$

This kinetic energy is divided into two parts. The first is the translational kinetic energy K_t of m_1, which is moving with instantaneous speed v. It has the value $K_t = \tfrac{1}{2}m_1v^2$. The second is the rotational kinetic energy K_r of the drum, which is rotating with instantaneous angular speed ω. It has the value $K_r = \tfrac{1}{2}I\omega^2$. Together, K_t and K_r constitute the kinetic energy K of the entire system. So you have

$$m_1gh = K_t + K_r = \tfrac{1}{2}m_1v^2 + \tfrac{1}{2}I\omega^2.$$

Just as in Example 11.2, v and ω are connected by Equation 11.9, $v = r\omega$. Setting $r = R$ and using this relation to eliminate ω from the energy equation, you obtain

$$\frac{1}{2}mv^2 + \frac{1}{2}I\frac{v^2}{R^2} = m_1gh.$$

Even if the moment of inertia of the drum is not specified in terms of its mass M, you can always use Equation 11.28 to write

FIGURE 11.14

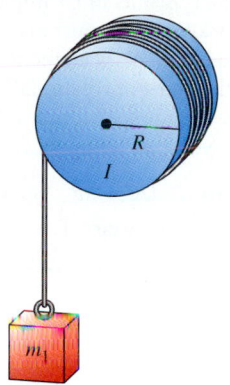

$I = cMR^2$, where c is the shape factor. This substitution gives you

$$\frac{1}{2}m_1v^2 + \frac{1}{2}cMR^2\frac{v^2}{R^2} = m_1gh,$$

or

$$v = \sqrt{2\,\frac{m_1}{m_1 + cM}\,gh}. \tag{11.37}$$

Compare this result with that for a freely falling body, $v = \sqrt{2gh}$. The speed of m_1 in the present case is smaller than the speed of the freely falling body by the factor $\sqrt{m_1/(m_1 + cM)}$ because some of the kinetic energy is manifested in the rotation of the drum rather than in the translation of the falling mass. How does the acceleration of m_1 compare with that of a freely falling body? How does the present system compare with the Atwood machine of Examples 3.8 and 8.4?

(b) Find v if the drum is a solid cylinder of mass $M = m_1$.

According to Table 11.2, the shape factor for a solid cylinder is $c = \tfrac{1}{2}$. Equation 11.37 thus simplifies to the form

$$v = \sqrt{\tfrac{4}{3}gh}.$$

What is the acceleration of m_1?

Work Done on a Rotating System

The energy of a rotating system can be changed by doing work on it, just as with translating systems. In Example 11.8, for instance, you could choose to regard the falling mass and the rotating drum as two separate systems linked by the string. From this point of view, the drum accelerates because work is being done on it by the string. (What is the ultimate source of this work?) In Example 11.9, we develop a convenient rotational form for describing work.

EXAMPLE 11.9

A bicycle sprocket wheel having moment of inertia I is accelerated from rest to a final angular velocity ω by means of a constant tangential force F applied to its teeth by a drive chain. Derive the rotational expression for the work done on the wheel. Be guided by analogy to the translational equation $W = F\,\Delta s$, which expresses the work done on a body by a force of constant magnitude acting along the direction of motion.

SOLUTION: You begin by drawing a sketch of the system, as in Figure 11.15. Call the radius of the sprocket wheel R. Now consider what happens as the chain pulls on the wheel. When the chain is pulled a distance Δs, it does an amount of work $W = F\,\Delta s$ on the wheel. Just as in Examples 11.2 and 11.8, the periphery of the sprocket wheel must also move a distance Δs. The corresponding angular displacement is $\Delta\theta = \Delta s/R$ (see Equation 11.8b). In the spirit of rotational mechanics, you want to express the work equation in terms of $\Delta\theta$ rather than Δs. Because $\Delta s = R\,\Delta\theta$, you can write

$$W = F(R\,\Delta\theta) = (RF)\,\Delta\theta \quad \text{for constant } F$$

or $\qquad W = T\,\Delta\theta \quad$ for constant T. **(11.38)**

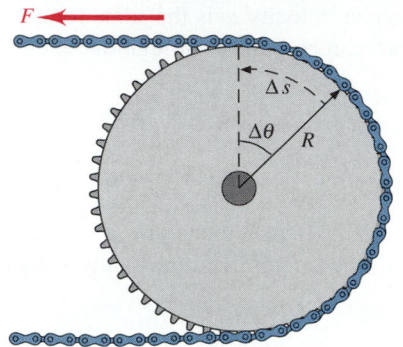

FIGURE 11.15 The usefulness of expressing work in terms of torque in rotating systems.

You have thus expressed the work done on the wheel in terms of rotational quantities only: the torque T and the angle $\Delta\theta$ through which the wheel turns under the action of the torque.

Equation 11.38 can readily be generalized. We rewrite the equation so that it applies to an infinitesimal rotation $d\theta$:

$$dW = T\,d\theta. \tag{11.39}$$

The restriction to constant torque no longer applies because even a variable torque will have a definite value over the infinitesimal angular displacement $d\theta$.

If torque is applied to a body over a finite angular displacement beginning at θ_i and ending at θ_f, the work done by the torque is expressed by the integral

$$W = \int_{\theta_i}^{\theta_f} T\,d\theta. \tag{11.40}$$

This expression simplifies to Equation 11.38 *only* if T is constant and can therefore be taken outside the integral.

The work-energy theorem, Equation 7.22b, applies to rotational as well as translational motion. Suppose a torque T acts on a rigid body whose moment of inertia is I. We use Equation 11.26, $T = I\alpha$, to write Equation 11.40 in the form

$$W = \int_{\theta_i}^{\theta_f} I\alpha\,d\theta.$$

We now manipulate the factor $\alpha \, d\theta$ in the integrand. We have

$$\alpha \, d\theta = \frac{d\omega}{dt} \, d\theta = \frac{d\theta}{dt} \, d\omega = \omega \, d\omega.$$

We can thus write Equation 11.40 in a form that can be integrated:

$$W = \int_{\omega_i}^{\omega_f} I\omega \, d\omega = \tfrac{1}{2}I\omega^2 \Big|_{\omega_i}^{\omega_f},$$

which yields

$$W = \tfrac{1}{2}I(\omega_f^2 - \omega_i^2) = K_f - K_i = \Delta K. \qquad \textbf{(11.41)}$$

This is the work-energy theorem in rotational form. Compare with Equations 7.22a and 7.22b.

Power

Power also can be expressed in terms appropriate to rotating systems. We begin with Equation 11.39, $dW = T \, d\theta$. Using the definition of power given by Equation 7.23, $P \equiv dW/dt$, we have

$$P = T\frac{d\theta}{dt} = T\omega. \qquad \textbf{(11.42)}$$

As the engine makes each revolution, its output torque varies greatly. Because of the flywheel's large moment of inertia, the corresponding angular acceleration of the system is very modest and the engine runs smoothly.

Compare with the translational form given by Equation 7.28, $P = \mathbf{F} \cdot \mathbf{v}$. In the special case where $\mathbf{F}$ and $\mathbf{v}$ are parallel, this expression becomes $P = Fv$. The analogy to Equation 11.42 is striking.

EXAMPLE 11.10

A 70-kg man rides a bicycle as hard as he can. Estimate the power he can apply to the bicycle. Take the crank length (the distance from the pedal to the axis of the driving sprocket) to be 17 cm.

SOLUTION: Think about what you do when you ride a bicycle hard. You exert your full weight first on one pedal and then on the other. (If you are a really good rider, you can substantially increase the force exerted on the pedal by pulling upward on the handlebars as you push down on the pedals, but neglect this additional force in your estimate.)

The maximum torque is applied to the pedal when the crank is horizontal, so the force exerted by the foot is perpendicular to the moment arm, and $F = F_\perp$. For the 70-kg rider, this maximum torque is

$$T = rmg = 0.17 \text{ m} \times 70 \text{ kg} \times 9.8 \text{ m/s}^2 = 120 \text{ N·m}.$$

The mean torque, averaged over a full rotation of the driving sprocket, cannot be this great because the rider's weight is applied perpendicular to the moment arm only briefly each time the pedals make a half-turn. For the purposes of this estimate, make a plausible approximation and take the mean torque to be one-half the maximum torque; $\langle T \rangle = 60$ N·m.

To calculate the power, you must estimate the angular speed of the driving sprocket. A bicycle rider who is not an expert can manage to turn the driving sprocket through one revolution in about 0.8 s without reducing the force applied to the pedals.

(A professional can do a good deal better!) Assume that the bicycle is a good one with a multispeed gear shift, so the rider can choose a gear that gives the driving sprocket the optimal angular speed regardless of the translational speed of the bicycle. Using the estimate 0.8 s, you have for the power, averaged over the rotation of the crank,

$$\langle P \rangle = \langle T \rangle \omega = 60 \text{ N·m} \times \frac{1}{0.8 \text{ s/rev}} \times 2\pi \text{ rad/rev}$$

$$= 500 \text{ W};$$

that is, about two-thirds of a horsepower.

As a rough check, you can recalculate the result in purely translational terms. Think again about what you do when you ride a bike hard. You put your full weight on a pedal until it is at the bottom of its travel. Then you transfer your weight to the other pedal, which is at the top of its travel. In doing so, you raise most of your weight—but not all of it—through a vertical distance equal to twice the crank radius. You do this twice during each rotation of the driving sprocket. The rider of this example, doing the same, raises himself through 2×17 cm in $\tfrac{1}{2} \times 0.8$ s. You use Equation 7.24:

$$\langle P \rangle = \frac{W}{\Delta t} = \frac{mgh}{\Delta t} = \frac{70 \text{ kg} \times 9.8 \text{ m/s}^2 \times 0.34 \text{ m}}{0.4 \text{ s}} = 600 \text{ W}.$$

Given the crudeness of the assumptions made, the two results are in satisfactory agreement.

In this chapter, we have been guided largely by analogy. The analogy between one-dimensional translational mechanics and fixed-axis rotational mechanics is doubly appealing. It has physical appeal because it conforms to our physical intuition; it has mathematical appeal because the mathematics is nearly identical in the two cases and because it is so easy to "translate" from the translational form to the rotational form and back by means of such relations as $v = r\omega$.

Analogy is perhaps the most powerful device the physicist can use to mobilize what he or she knows in penetrating into unknown territory. But analogy is dangerous precisely because it is so appealing to use whole chunks of our knowledge of a familiar world in explaining an unknown world. We know in advance that the unknown world is different from the known world; that is why it is unknown! When

Niels Bohr explored the structure of the atom in 1913, for example, he imagined that an atom looked like a tiny solar system with electron "planets" orbiting about a nucleus "sun." Of course, Bohr knew that an atom is *not* a solar system; the analogy is only the first step toward understanding, not the final theory. The final theory of the atom is much broader than, and quite different in many ways from, the picture Bohr first set forth.

We must thus always be careful about the limits of analogy. We must validate the analogy cautiously, by showing that it provides useful solutions to real problems, as we did in Example 11.10 and elsewhere in this chapter. And ultimately, as every physicist knows, we must establish a rigorous basis for our inquiry into the new world. With respect to rotation, these two tasks constitute the main thrust of the next two chapters.

Symbols Used in Chapter 11

a_t, a_r	tangential acceleration, radial (centripetal) acceleration
c	ratio of the moment of inertia of a system to that of a hoop of equal mass and maximum radius
D	distance between two parallel axes, one of which passes through the center of mass of a body
$F_\perp$	force exerted perpendicular to a radius vector
I	moment of inertia
K	kinetic energy
K_r, K_t	rotational, translational kinetic energy
m	mass of a particle
M	mass of an extended body
$\Delta m_j, dm$	mass of a finite or infinitesimal element of an extended body
r, r_j	distance from an axis of rotation to a point
R	radius of an extended body; distance from rotation axis to most distant part of an extended body

$s, \Delta s, ds$	distance, finite or infinitesimal displacement measured along an arc
T	torque about a fixed axis
ΔU	change in potential energy of a system
V, dV	volume of a rigid body, of an infinitesimal element of a rigid body
W	work
X, Y, Z	rotation axes
α	angular acceleration
θ	angular position
$\Delta\theta, d\theta$	finite or infinitesimal angular displacement
ν_j, δ_j	numerator and denominator of the jth fraction in a set of equal fractions
ω	angular velocity or angular speed
$\Delta\omega, d\omega$	finite or infinitesimal change in angular velocity

Summing Up

The rotation of a rigid body about a fixed axis is most conveniently described in terms of the kinematic quantities angular displacement $d\theta$ or $\Delta\theta$, angular velocity ω, and angular acceleration α. These rotational quantities are related to the corresponding translational quantities by Equations 11.8c, 11.9, and 11.10:

$$ds = r\,d\theta,$$
$$v = r\omega,$$
and
$$a_t = r\alpha.$$

The rotational quantities are related to one another by Equation 11.2,

$$\alpha \equiv \frac{d\omega}{dt} = \frac{d^2\theta}{dt^2},$$

which is completely analogous to the equations relating the corresponding translational quantities.

The analogy can be extended to provide a complete set of kinematic equations for fixed-axis rotation. These equations are listed in Table 11.1, together with their translational analogues.

Torque is the analogue in rotation of force in translation.

In the context of fixed-axis rotation, torque is defined by means of Equation 11.13,

$$|T| \equiv |rF_\perp|.$$

Moment of inertia, the analogue in rotation of inertial mass in translation, is defined for a point mass by Equation 11.19,

$$I \equiv mr^2.$$

A rotational form of Newton's second law that is valid for fixed-axis rotation is given by Equation 11.20,

$$T = I\alpha.$$

This equation bears a close resemblance to Newton's second law for translation in one dimension, $F = ma$.

For an extended rigid body, the moment of inertia is always less than that of an ideal hoop of the same mass and of radius equal to the maximum extent of the body from the axis. This is expressed in Equation 11.28,

$$I = cMR^2;$$

the shape factor c is never greater than 1.

The moment of inertia for any body about a specified axis can be calculated by dividing the body into mass elements dm and summing their contributions dI, according to Equation 11.30,

$$I = \int_{\substack{\text{entire} \\ \text{body}}} dI = \int_{\substack{\text{entire} \\ \text{body}}} r^2 \, dm.$$

If we know the value of I_{cm}—a body's moment of inertia about an axis passing through the body's center of mass—the value of I about any parallel axis is given by Equation 11.32, the parallel-axis theorem:

$$I_\parallel = I_{cm} + MD^2.$$

Given any two mutually perpendicular axes lying in the plane of a uniform, thin, flat body, the moments of inertia about the two axes are related to that about a third, mutually perpendicular axis by Equation 11.34, the perpendicular-axis theorem:

$$I_Z = I_X + I_Y.$$

The rotational kinetic energy of a body is given by Equation 11.35,

$$K = \tfrac{1}{2}I\omega^2.$$

The work done on a body by a torque is given by Equation 11.39,

$$dW = T \, d\theta.$$

Taken together with Equations 11.2 and 11.13, these equations make possible the analysis of the motion of systems containing rotating as well as translating elements.

A rotational form of the definition of power follows directly from Equation 11.39 and is given by Equation 11.42:

$$P = T\frac{d\theta}{dt} = T\omega.$$

KEY TERMS

Section 11.2 Kinematics of Rotation about a Fixed Axis
rigid body ▪ axis of rotation ▪ sense of rotation ▪ angular displacement, angular position, angular velocity, angular acceleration

Section 11.3 Dynamics of Rotation about a Fixed Axis
lever, fulcrum, moment arm ▪ torque ▪ moment of inertia ▪ rotational inertia ▪ Newton's second law for rotation

Section 11.4 Calculation of Moments of Inertia
parallel-axis theorem ▪ perpendicular-axis theorem

QUERIES

11.1 *(2) Be sensible!* In describing a wheel to a friend, you tell him that it is rotating counterclockwise. If you are to be sure that he understands your description, what more must you specify?

11.2 *(2) Innocent game.* A child plays with a Yo-Yo, letting it go so that it descends and spins as the string unwinds. When the string is fully unwound, the Yo-Yo continues to spin and rises again to the child's hand as the string winds up. For two successive up-and-down cycles, make a rough sketch of the angular velocity and the angular acceleration of the Yo-Yo about its own axis as functions of its distance below the child's hand.

11.3 *(3) Why plumbing pays better than physics.* A plumber seeking to loosen a rusty pipe joint with a wrench will sometimes slip a piece of pipe over the end of the wrench handle and pull on the pipe, using it as an extension of the handle. Explain.

11.4 *(3) Without a net.* Why does a tightrope walker carry a long pole? Why is the pole sometimes weighted at the ends?

11.5 *(3) Balancing act.* It is very difficult to balance a pencil oriented vertically with its point on the end of your finger, but it is quite easy to balance a broomstick in the same way. Explain.

11.6 *(3) Dry lab.* Devise a simple experimental method for measuring the moment of inertia of an object about some specified axis.

11.7 *(3) Are there no limits?* Can the constant c in Equation 11.28, $I = cMR^2$, ever be greater than 1? If not, explain. If so, make a sketch of a system for which $c > 1$.

11.8 *(3) Inertial guidance.* Explain the function of the flywheel in automobile engines. If you were designing a 4-cylinder engine and an 8-cylinder engine of equal power, which one (if either) should have the bigger flywheel?

11.9 *(4) Perpendicular-axis theorem?* You can draw three mutually perpendicular axes through the midpoints of opposite faces of a cubic block. Symmetry requires that the moments of inertia I_X, I_Y, and I_Z about these axes be equal. Does this requirement conflict with the perpendicular-axis theorem? Explain.

11.10 *(5) Everywhere you turn.* Describe as many machines as you can in which rotational kinetic energy is converted into translational kinetic energy or vice versa.

PROBLEMS

GROUP A

11.1 *(2) Unwinding.* A weight hangs from a string wrapped around a drum. When the weight has descended through a distance of 5.0 m, the drum has made 4.6 revolutions. Find the radius of the drum.

11.2 *(2) Pull to turn.* In the system shown below, the pulley has radius $R = 12.0$ cm. **(a)** Find the angular displacement of the pulley in radians when point P on the rope is raised 3.50 m. **(b)** How far does the weight rise in the same process? Does the difference between this result and the displacement of point P have any bearing on the answer to part **a**?

11.3 *(2) Astronomical matters.* **(a)** What is the angular speed of the earth about the sun? **(b)** of the moon about the earth?

11.4 *(2) Given a little latitude.* What is the angular speed about the earth's axis of **(a)** a point on the equator? **(b)** a point at latitude 40°?

11.5 *(2) Straightening the curves.* A pulley of radius 24.0 cm rotates about a fixed axis through an angle $\Delta\theta = \pi/3$ rad. **(a)** What is the length of the path through which a point on the rim of the pulley passes? **(b)** What is the magnitude of the translational displacement experienced by the same point? **(c)** Express the difference between the answers to parts **a** and **b** as a percentage of the answer to part **a**. Explain the difference. **(d)** The pulley rotates through an angle $\Delta\theta = 1.75 \times 10^{-2}$ rad, or approximately 1°. Repeat the calculations of parts **a** and **b**, and again express the difference as a percentage of the former. Compare with the result of part **c**.

11.6 *(2) A fast turn-on.* When a grinding wheel is turned on, it takes 12.0 s to reach its full speed of 3600 rev/min. **(a)** Calculate the angular acceleration of the wheel, assuming that it is constant. **(b)** How many revolutions does the wheel make as it speeds up?

11.7 *(2) Faster and faster.* A wheel initially at rest experiences a constant angular acceleration. It makes 1 revolution in the first second. How long does the wheel take to complete 2 revolutions, starting from rest?

11.8 *(2) Cool it.* A two-speed fan is operating at its high speed, 40 rad/s. When it is switched to "low," it accelerates to its new speed, 20 rad/s, at an approximately constant angular acceleration $\alpha = -10$ rad/s. Express the angular displacement of the fan during the acceleration period in radians and in revolutions.

11.9 *(2) Doing it another way.* Calculate the angular displacement of the wheel in Example 11.1, using Equation 11.7.

11.10 *(2) Keep on trucking.* In order to monitor the durability of their tires, long-distance trucking operators install on each wheel hub a device called a "hubdometer," which counts the number of revolutions made by the wheel. A trucker finds that a tire needs to be replaced when it has made 20 million revolutions. **(a)** If the tire diameter is 1.1 m, express the service life of the tire in kilometers. **(b)** If the truck travels at an average speed of 80 km/h, find the operating lifetime of the tire. **(c)** Use the result of part **b** to find the average angular speed of the wheel.

11.11 *(2) Still unwinding.* The weight of Problem 11.1 requires 4.0 s to descend from rest through a distance of 5.0 m. Find the angular acceleration and the final angular velocity of the drum.

11.12 *(3) Holding steady.* In the lever system of Figure 11.4, the distances r_1 and r_2 are 0.45 m and 1.23 m, respectively. The hanging weight has mass $m = 11.5$ kg. **(a)** Find the torque exerted about the fulcrum by the weight. **(b)** What must the magnitude of the force **F** be if the system is to remain horizontal? **(c)** Find the answers to parts **a** and **b** if the fulcrum and the point of application of the externally applied force **F** are interchanged.

11.13 *(3) Tinkertoy.* The square structure shown below consists of four point masses connected by rods of negligible mass. Find the moment of inertia of the structure about the following axes: **(a)** axis *A*, passing through the center of the structure and normal to its plane; **(b)** axis *B*, passing through one of the point masses and normal to the plane of the structure; **(c)** axis *CC'*, passing through two adjacent point masses; and **(d)** axis *DD'*, along the diagonal of the structure.

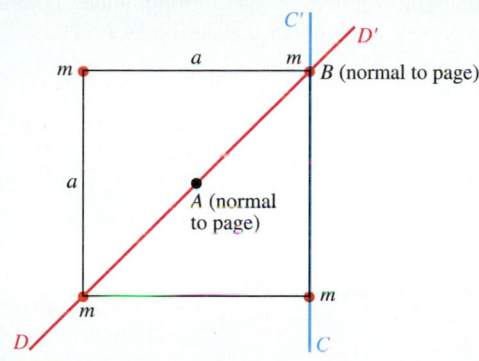

11.14 *(4) Small shift.* The moment of inertia of a uniform sphere about an axis through its center is given in Table 11.2. Show that an axis about which the moment of inertia is twice as great passes through the sphere.

11.15 *(4) Just one moment.* Calculate the moment of inertia of a uniform thin rod of length *L* about an axis perpendicular to the length of the rod and passing through its center. Check your result with Table 11.2.

11.16 *(4) Around the end.* Use the parallel-axis theorem to calculate the moment of inertia of the rod of Problem 11.15 about an axis perpendicular to the rod and passing through one end. Check your result with Table 11.2.

11.17 *(4) Rotating cube.* Illustrated below is a uniform cubic block of mass *M* and side *a*. Find the moment of inertia about **(a)** axis *A*, passing through the midpoints of opposite faces; **(b)** axis *B*, which bisects one of the cube faces; and **(c)** axis *C*, along one of the cube edges.

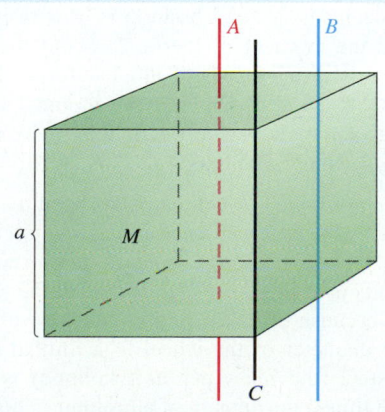

11.18 *(4) Changing the pivot, I.* The wooden form shown at the top of the next column is uniform in thickness but

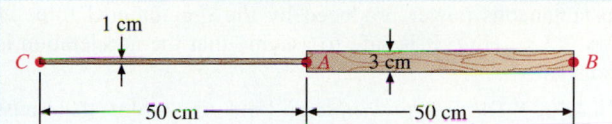

not in width. The mass of the form is 54.0 g. Find the moment of inertia about an axis normal to the page and passing through the form at **(a)** point *A*, **(b)** point *B*, and **(c)** point *C*.

11.19 *(4) Macaroni factory.* A thick-walled tube has inner radius R_1 and outer radius R_2. Derive the value of the moment of inertia about the symmetry axis given in Table 11.2.

11.20 *(4) Thinning down.* Show that the expression for the moment of inertia of a hollow cylinder about its symmetry axis reduces to the expression for a thin cylindrical shell as $R_1 \rightarrow R_2$.

11.21 *(4) Something wrong?* The mass of the earth is 5.99×10^{24} kg, and its radius is 6.27×10^6 m. Its moment of inertia about its rotation axis is 8.04×10^{37} kg·m². What can you infer about the distribution of mass through the earth?

11.22 *(4) Two hard ways to spin a hoop.* Find the moment of inertia of a hoop of mass *M* and radius *R* about an axis that is **(a)** in the plane of the hoop and tangent to the hoop and **(b)** normal to the plane of the hoop and tangent to the hoop.

11.23 *(5) Over the top.* A uniform thin rod of length 0.83 m hangs from a frictionless pivot at its upper end. You hit the lower end of the rod with a small bat. How fast must you make the lower end of the rod move if it is to swing up and over the pivot?

11.24 *(5) Snare drum.* A 2.8-kg mass is attached to a string that is wound around a solid cylindrical drum of mass 8.3 kg. The mass is released and allowed to descend. Neglecting friction, what is the angular acceleration of the drum?

11.25 *(5) Whirling descent.* The uniform cylinder illustrated below is released at *t* = 0 and begins to descend as the strings unwind. If its mass is 8.3 kg and its radius is 1.4 cm, find **(a)** the linear and angular speeds of the cylinder when it has descended 1 m, **(b)** the linear and angular accelerations of the cylinder, **(c)** the tension in each of the two strings, and **(d)** the

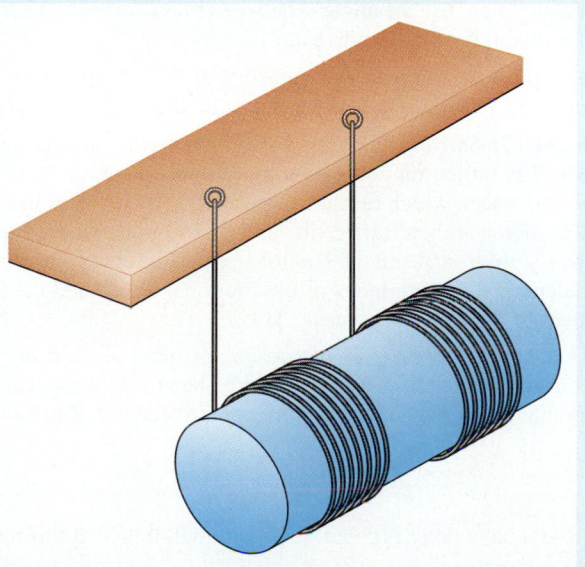

instantaneous power produced by the gravitational force at $t = 2.3$ s. (Hint: It is safe to assume that the acceleration is constant.)

11.26 *(5) Turning to business.* Find the torque that must be exerted by a motor on a drive shaft that rotates at 2000 rev/min and transmits 30 kW of power.

11.27 *(5) Shifting gears.* The maximum torque that can be exerted by a gasoline or diesel engine depends on the angular speed ω of the engine and peaks at a value of ω that is fixed for any particular engine. (This point is raised in Problem 3.54.) Use this fact to explain the function of the transmission in an automobile.

11.28 *(G) Starting up.* A cylindrical flywheel, initially at rest, has mass 11.0 kg and radius 50.0 cm. A tangential force $F_\perp = 400$ N is applied to the rim for 15.0 s. **(a)** Find the angular displacement of the wheel during this time. **(b)** What are the final values of the angular velocity and the kinetic energy? **(c)** Was the power input to the system constant during the process? Explain.

11.29 *(G) Tim-berr!* A uniform rod of length L is pivoted at one end, as shown opposite. The gravitational force exerted on a body may be regarded as acting on it at its center of mass. (This plausible statement is proved rigorously in Section 12.2.) **(a)** Calculate the torque about the pivot exerted by the rod's weight as a function of the angle θ. **(b)** Find the angular acceleration of the rod as a function of θ. **(c)** Show that, when $\theta = \sin^{-1} \sqrt{\frac{2}{3}}$, the vertical component of the translational acceleration of the free end of the rod is equal to g. **(d)** Unwanted tall brick chimneys are often demolished by setting off a small explosive charge at the base. As a chimney topples, it breaks at some point along its length. Thus the debris falls in an area whose radius is less than the original height of the chimney. Use the result of part **c** to explain why. (Hint: Although brick-and-mortar construction is very strong under compressive forces, it is very weak under tensile forces.)

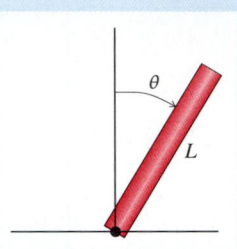

GROUP B

11.30 *(2) A new angle on centripetal acceleration.* A mass m attached to the end of a string of length r whirls in a horizontal circle around a pivot. Express the centripetal acceleration of the mass in terms of its angular velocity ω.

11.31 *(2) A new angle on centripetal force.* Find the tension in the string of Problem 11.30 in terms of ω.

11.32 *(2) All kinds of acceleration, I.* Beginning at rest, a flywheel "spins up" with constant angular acceleration α. At an arbitrary time t, find **(a)** the tangential acceleration a_t of a point on the wheel a distance r from the axis, **(b)** the radial acceleration a_r of the same point, and **(c)** the magnitude and direction of the total acceleration $\mathbf{a}$ of the point. To express the direction, use the angle ϕ between the radius vector $\hat{\mathbf{r}}$ from the axis to the point and vector $\mathbf{a}$.

11.33 *(2) All kinds of acceleration, II.* For the point on the wheel of Problem 11.32, express a_r in terms of a_t, r, and t.

11.34 *(3) Starting and stopping.* The motor of a grinding wheel is turned on and exerts a constant torque of 100 N·m on the wheel, which reaches a speed of 1500 rev/min in 25 s. The motor is then turned off, and in 250 s the wheel comes to a stop as a result of friction (which is always present). Calculate the magnitudes of the frictional torque and the moment of inertia of the wheel.

11.35 *(3) Dumbbell.* Two particles of mass m_1 and m_2 are connected by a massless rigid rod of length R. Show that the moment of inertia of the system about its center of mass is

$$I = \frac{m_1 m_2}{m_1 + m_2} R^2.$$

(If you have done Problem 9.67, you will note that this result can be written $I = \mu R^2$, where μ is the reduced mass of the system.)

11.36 *(3) Spinning wheel.* A disk-shaped wheel of mass 5 kg and radius 20 cm is mounted on an axis passing through its center. The angular speed of the wheel varies in time according to the expression $\omega = 7$ rad/s $+ (8 \text{ rad/s}^2)t$. **(a)** Is the applied torque constant, or does it vary with time? **(b)** Find the torque.

11.37 *(3) Brakes.* A flywheel having moment of inertia 58.2 kg·m² rotates with initial angular speed 34.8 rad/s. If the wheel comes to a stop in 21.5 s, what torque acts on it?

11.38 *(4) Bending a wire loop.* A thin wire of length L is joined end to end and formed into a circular loop. The moment of inertia of the loop about an axis normal to its plane and through its center is I_0. **(a)** The loop is now reshaped into a square. Find the moment of inertia I_1 about the same axis. **(b)** The wire is now squashed into the form of a doubled straight wire of length $L/2$. Find the moment of inertia I_2 about an axis perpendicular to the doubled wire and through its center, $L/4$ from either end.

11.39 *(4) Retooling.* Your company has been making a disk-shaped steel flywheel for an engine manufacturer. The manufacturer wants you to change to an aluminum wheel having the same moment of inertia. The thickness of the wheel cannot be changed because of space limitations. However, you can increase the diameter of the wheel to a limited degree. The density of steel (the mass per unit volume) is 2.91 times (nearly three times) the density of aluminum. Show that, nevertheless, you need increase the diameter of the wheel by only 31%.

11.40 *(4) Rectangular plate.* A thin, uniform rectangular plate has length a and width b. Calculate the moment of inertia about an axis **(a)** in the plane of the plate, midway between the long sides; **(b)** in the plane of the plate, midway between the short sides; and **(c)** normal to the plane of the plate and passing through its center. Check your result for part **c** with that given in Table 11.2. (Hint: How can you do part **c** without integrating?)

11.41 *(4) Well, more or less.* Two balls each have radius 5 cm and mass 1 kg. They are connected by a rigid rod of negligible mass and length 40 cm, so the distance between the centers of the balls is 50 cm. The system is fixed to an axis perpendicular to the rod and passing through the midpoint of the rod. **(a)** Calculate the approximate moment of inertia of the system by replacing each ball by a particle of the same mass located at the center of mass of the ball. **(b)** Calculate the exact value of the moment of inertia of the system. **(c)** What is the percentage error that arises from the approximation of part **a**?

11.42 *(4) Onion rings.* The moment of inertia of a thin spherical shell about a diameter, given in Table 11.2, can be derived as follows. Divide the shell into "onion ring" elements as shown below. Show that the ratio of the mass dm of any element to the total mass m of the shell is $dm/m = \frac{1}{2}\sin\theta\,d\theta$. Next, write an expression for the contribution ΔI of this element to the moment of inertia of the entire shell. Finally, integrate between the proper limits to find I for the entire shell.

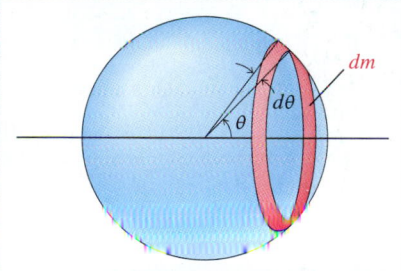

11.43 *(4) Shell game.* Use the result of Problem 11.42 to find the moment of inertia of a uniform solid sphere of mass M and radius R about a diameter, and compare the result with that given in Table 11.2. To carry out the calculation, consider the following steps. **(a)** Subdivide the sphere into concentric shells of infinitesimal thickness dr. What is the volume of a typical shell of radius r? **(b)** What is the mass dM of this typical shell, in terms of the total sphere mass M? **(c)** What is the moment of inertia dI of the shell? **(d)** Integrate between the proper limits to find the moment of inertia I of the entire sphere.

11.44 *(4) Changing the pivot, II.* For the wooden form illustrated in Problem 11.18, find the distance from the small end to the point about which the moment of inertia is a minimum.

11.45 *(4) Missing part.* A circular hole of radius $R/2$ is cut out of a circular disk of radius R, as shown at the top of the next column. Find the moment of inertia of what remains of the disk **(a)** about an axis normal to the page and passing through O, the center of the disk; and **(b)** about an axis normal to the page and passing through P, the center of the hole.

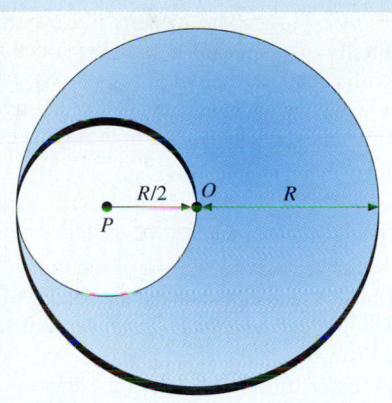

11.46 *(4) Fat cylinder.* Consider a cylinder of radius r, length l, and mass M, pivoted about the axis shown below. **(a)** Find the moment of inertia. **(b)** As a check on the result of part **a**, show that it reduces to values given in Table 11.2 for $l \to 0$ and for $r \to 0$. **(c)** Show that, for fixed mass, the moment of inertia of a cylinder about the axis shown has an extreme value if the ratio of length to radius is $l/r = \sqrt{\frac{3}{2}}$. **(d)** Is the extreme value of part **c** a maximum or a minimum? **(e)** What is the extreme value of the moment of inertia?

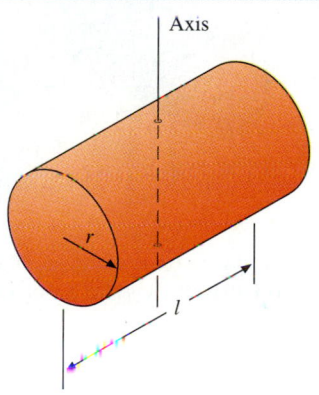

11.47 *(5) Round and round.* Compare the kinetic energy of the earth's rotation about its own axis with the kinetic energy of its revolution about the sun. Use $I = 8.04 \times 10^{37}$ kg·m² for the moment of inertia of the earth about its axis.

11.48 *(5) Winding down.* Suppose that, in the system illustrated below, $M = 30.0$ kg, $m = 2.50$ kg, $R = 5.00$ cm, and $r = 0.250$ cm. The system is initially at rest. The friction in the bearings is negligible, as is the mass of the shaft. Find **(a)** the speed of m when it has descended a distance of 1.75 m and **(b)** the acceleration of m.

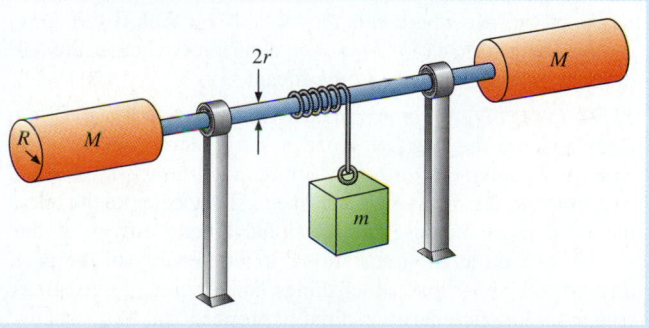

11.49 *(5) Winding up.* The system illustrated in Problem 11.48 is initially at rest, with *m* hanging a considerable distance below the shaft. The shaft is spun up by a strong torque and rapidly achieves an angular speed of 50 rev/min. If the system subsequently turns freely, how high will the mass *m* rise before it comes to rest?

11.50 *(5) Applied kinematics.* Find $|a|$, the magnitude of the acceleration of the mass m_1, for the system of Example 11.8.

11.51 *(5) Runaway rope, I.* A rope of total mass *m* and length *l* is wound in a single layer around a cylinder of mass *M* and radius *r*, as illustrated below. The cylinder axis has negligible friction. A little bit of the rope hangs down, and its weight starts the cylinder turning and the rope unwinding. Find the angular speed of the cylinder when a length *x* of the rope has unwound. (Hint: When a length *x* has unwound, what is the mass of the unwound length and where is the center of mass of that length?)

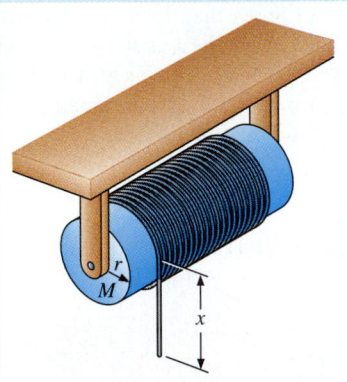

11.52 *(5) Runaway rope, II.* For the system illustrated in Problem 11.51, show that the angular acceleration of the cylinder is

$$\alpha = \frac{mgx}{lr(\frac{1}{2}M + m)}.$$

11.53 *(5) Correction factor?* Working as an engineer for an automobile company, you are assigned to design the brakes for a new model. You must make sure that the brakes have the capacity to convert the kinetic energy of the car into heat energy within the distance specified for bringing the car to a stop from any given speed *v*. The translational kinetic energy of the car is $K_t = \frac{1}{2}Mv^2$, where *M* is the mass of the car. In addition, however, the car possesses rotational kinetic energy K_r because of the rotation of the wheels. You need to know whether you must design additional capacity into the brakes to allow for K_r. **(a)** Show that the greatest possible value of the ratio of the rotational to the translational kinetic energy is $K_r/K_t = 4m/M$, where *m* is the mass of a wheel. **(b)** Making reasonable estimates for *M* and *m*, determine whether the additional braking capacity is significant.

11.54 *(5) Is bigger better?* In designing a racing bicycle, you want to keep the mass as small as possible consistent with structural soundness, in order to minimize the work required to accelerate the bike to a given speed. To accelerate the bike, the rider must increase the rotational kinetic energy of the wheels as well as the translational kinetic energy of the bike as a whole. Show that, other things being equal, the required rotational kinetic energy is directly proportional to the radius

of the wheels; in the absence of other considerations, the smallest possible wheels would be the best.

11.55 *(5) Atwood machine with a real pulley, I.* The diagram below represents a real Atwood machine. The pulley, of radius *R* and mass *M*, has a nonzero moment of inertia *I*. (However, we still neglect friction.) **(a)** Extend the method used in Problem 11.50 to find $|a|$, the magnitude of the acceleration of either hanging mass. Express your result in terms of *I* and *R*, and also in terms of the product *cM*. Check your result by showing that it reduces to the result for an ideal Atwood machine when $I = 0$. **(b)** Let the pulley be in the form of a cylindrical disk of mass $M = m_1 + m_2$. Find $|a|$. **(c)** Comment on the physical meaning of the product *cM*.

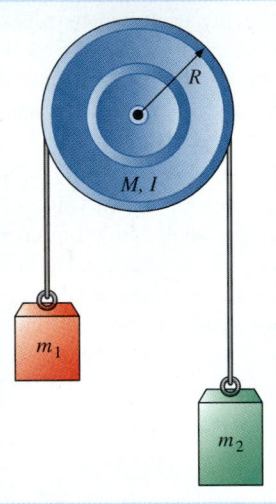

11.56 *(5) Half-Atwood again.* In the illustration below, the pulley has the form of a uniform disk. The friction in the pulley is negligible, but the coefficient of kinetic friction between the table and the block resting on it is $\mu_k = 0.10$. Let $m_1 = 1.2$ kg, $m_2 = 7.1$ kg, and $m_3 = 9.6$ kg. Find the acceleration of the blocks.

11.57 *(5) Tick-tock.* **(a)** The balance wheel of a mechanical watch (that is, the oscillating wheel that controls the rate at which the watch runs) is connected to a fine spiral spring called the hairspring; see Figure 6.13. The spring exerts a torque on the wheel whose magnitude is proportional to the wheel's angular position θ measured from a fixed reference direction $\theta = 0$; that is, $T = -\kappa\theta$, where κ is a constant. If the wheel is at rest with $\theta = 0$, how much work is required to turn it to some angular position θ? **(b)** In a certain mechanical wristwatch, $\kappa = 2.07 \times 10^{-7}$ N·m/rad. Find the work required to turn the wheel through an angle of 1.20 rad from its equilibrium position.

11.58 *(G) Overcoming friction.* The large disk-shaped fly-wheel illustrated below is made to spin by the small wheel that contacts it at its rim. The small wheel applies a constant force $F = 1060$ N. Friction in the bearing exerts a retarding torque $T_f = 48.5$ N·m on the large wheel. If the angular acceleration of the wheel is 92.3 rad/s^2, what is its mass?

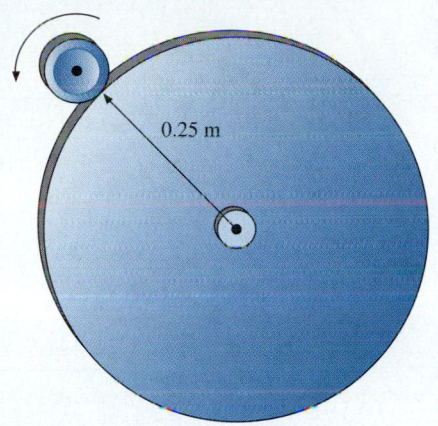

0.25 m

11.59 *(G) Adding fractions.* Prove the theorem stated in Equation 11.22: If the members of a set of fractions are equal, the sum of their numerators divided by the sum of their denominators is equal to any of the fractions.

11.60 *(G) How fast?* Show that the angular velocity of the rod illustrated in Problem 11.29, expressed as a function of the angle θ, is

$$\omega = \sqrt{\frac{3g}{L}(1 - \cos\theta)}.$$

Use your result for part **a** of Problem 11.29. You will find the following substitution useful:

$$\alpha\,d\theta = \frac{d\omega}{dt}\,d\theta = \frac{d\theta}{dt}\,d\omega = \omega\,d\omega.$$

11.61 *(G) How fast? faster.* You can solve Problem 11.60 without integrating by using energy conservation, as follows. **(a)** What is the change in the gravitational potential energy of the system when the rod has fallen from its initial vertical position, $\theta_i = 0$, to some arbitrary angular position θ? **(b)** What is the rotational kinetic energy of the rod at this moment? **(c)** What is the angular velocity ω?

11.62 *(G) Skidding to a stop.* A cylindrical roller of radius R rolls along a tabletop until it hits a wall, as shown below.

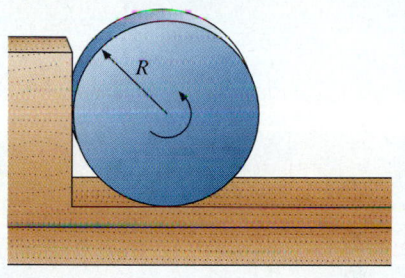

When the roller hits the wall, it starts to skid against both the tabletop and the wall. The coefficient of kinetic friction between the roller and both surfaces is μ_k. **(a)** Find the angular acceleration of the roller. **(b)** Let the angular speed of the roller just after it contacts the wall be ω_0. Show that the number of revolutions the wheel makes in coming to a stop is

$$N = \frac{\omega_0^2 R(1 + \mu_k^2)}{8\pi\mu_k g(1 + \mu_k)}.$$

GROUP C

11.63 *(4) Moment of inertia of a cone.* Find the moment of inertia of a right circular cone of mass M and base radius R about its symmetry axis. (Hint: The technique is similar to that of Problem 11.42.)

11.64 *(G) Atwood machine with a real pulley, II.* Derive the equations of motion for the Atwood machine illustrated in Problem 11.55 in a straightforward dynamical way. Specifically, find the accelerations of the two hanging masses, the angular acceleration of the pulley, and the string tensions F_1 and F_2 on *both* sides of the pulley. Neglect friction.

11.65 *(G) Armchair geology.* Problem 11.21 suggests that the density of the earth is not uniform but, instead, increases with increasing depth. Now, let's carry out a crude quantitative argument that gives surprisingly accurate results for the density of the matter at the earth's center and in its crust. Use the following values in your calculations; they are obtained from experimental and astronomical measurements:

- mass of earth: $M = 5.99 \times 10^{24}$ kg
- radius of earth (assumed spherical): $R = 6.37 \times 10^6$ m

- moment of inertia of earth about its rotation axis: $I = 8.04 \times 10^{37}$ kg·m^2

(a) If the earth's density ρ is not uniform, the simplest remaining possibility is a linear variation, $\rho = \rho_0 - \gamma r$. Here, ρ is the density of the matter at a distance r from the center, ρ_0 is the density at the center, and γ is a positive constant. Express the mass M of the earth in terms of ρ_0, γ, and R. **(b)** Using an argument similar to that of Problem 11.43, express the earth's moment of inertia in terms of ρ_0, γ, and R. **(c)** Use the results of parts **a** and **b** to solve for ρ_0 and γ, and find numerical values for these quantities. [The best currently available estimate gives $\rho_0 \simeq (12$ to $18) \times 10^3$ kg/m^3.] **(d)** Using the results of part **c**, find ρ_R, the mean density of matter in the outermost layer of the earth. Compare this result with the density of basalt, the most important component of this layer, which is approximately 2.9×10^3 kg/m^3. **(e)** Give two reasons why the assumption used in this problem—that the density of the earth's matter increases linearly with increasing depth—cannot be correct in spite of the remarkably good comparison with known values that it yields.

Applications of Rotational Mechanics

—— A stationary body may have several forces and torques acting on it. Statics is the study of these forces and torques.

—— The torsion pendulum is the rotational analogue of the mass-spring system. What are the laws that govern the motion of the torsion pendulum?

—— When an extended body is suspended from a pivot, it swings. How does the oscillation of this physical pendulum compare with that of the simple pendulum?

—— When a body translates, and also rotates so that the direction of the rotation axis does not change, the laws governing the rotational motion are the same as those that apply to fixed-axis rotation.

—— For such a body, the kinetic energy can be separated into translational and rotational parts.

—— What are the laws governing the motion of a body that rolls without slipping?

Left: Racing requires continual adjustment of the relation between the torque exerted by the bicyclist on the pedals and the force exerted by the rear wheel on the ground.

> *One good turn deserves another.*
>
> —PETRONIUS, *Satyricon*

Introduction

Rotational mechanics, like translational mechanics, has application to a vast variety of physical situations. In this chapter, we consider some of these applications.

We begin, in Section 12.2, with the application that is simplest in principle. This application is *statics*: We study bodies that neither rotate nor translate, even though external forces act on them, and we discover a set of conditions that govern the forces acting on a static system. Knowing these conditions is essential to the design of mechanical structures and is a major concern of mechanical engineering.

In the next two sections, we expand the study of oscillating systems that we began in Chapter 6. Section 12.3 treats the *torsion pendulum*, whose oscillation is the rotational analogue of the translational oscillation of the mass-spring system. Section 12.4 deals with the *physical pendulum*. This name is given to any pendulum that is not *simple*—in other words, any pendulum that does not consist of an ideal point mass attached to a massless string.

In Section 12.5, we take another step in broadening our theory of mechanics by considering an important special class of systems in which translation and rotation take place simultaneously. In such a system, the rotation axis moves but its orientation does not change. This happens, for example, when a cylinder rolls on a flat surface.

Statics

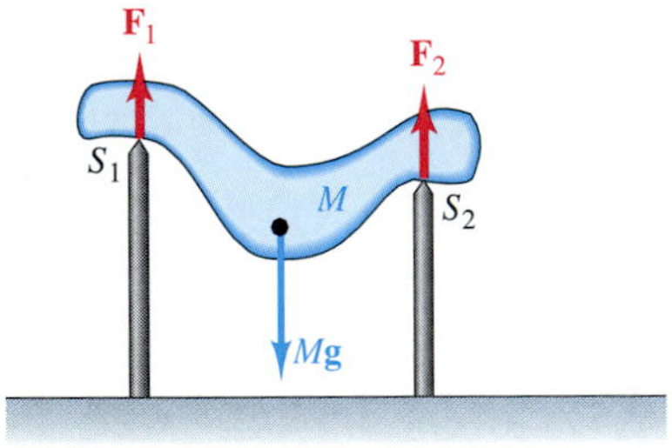

FIGURE 12.1 An irregularly shaped plate of mass M is supported by two columns at points S_1 and S_2. The point contacts are frictionless; consequently, we know that the supporting forces $\mathbf{F}_1$ and $\mathbf{F}_2$ are directed vertically.

In Figure 12.1, two columns support the irregularly shaped plate of mass M by making frictionless point contact with it at S_1 and S_2. The plate is motionless, neither translating nor rotating. Let us consider the absence of translation first. In the absence of translation, the acceleration $\mathbf{A}$ of the center of mass of the plate must be zero, and the translational form of Newton's second law for an extended body (Equation 9.26) then requires that the net external force $\mathbf{F}_{\text{ext}}$ acting on the body be zero as well:

$$\mathbf{F}_{\text{ext}} = M\mathbf{A} = 0. \tag{12.1}$$

Because $\mathbf{F}_{\text{ext}}$ is the sum of all the individual external forces $\mathbf{F}_j$ acting on the plate, we can write the **static condition for forces** as

$$\sum_j \mathbf{F}_j = 0. \tag{12.2}$$

Because the plate is not rotating, the net external torque T acting on it must be zero as well. Thus Equation 11.26, Newton's second law for rotation, is satisfied:

$$T = I\alpha = 0 \quad \text{for rotation about a fixed axis.} \tag{12.3}$$

But what fixed axis are we to choose for evaluating T? For the static case, *any axis will do* because, whichever axis you choose, every part of the body has zero angular acceleration about it. However, once an axis is chosen, consistency requires that all torques must be calculated with reference to it. (As you will see, a clever choice of axis can sometimes reduce the labor of calculation.)

Because the net torque about any axis is the sum of the individual torques T_j applied by external forces, the **static condition for torques** may be written

$$\sum_j T_j = 0 \quad \text{all } T_j \text{ referred to the same axis.} \quad (12.4)$$

Equations 12.3 and 12.4, taken together, summarize the conditions that must be satisfied if a body is to remain stationary when external forces and torques act on it.

The Center of Gravity

In using Equation 12.2 to analyze the system of Figure 12.1, we calculate the net force **F** by adding the forces $\mathbf{F}_1$, $\mathbf{F}_2$, and $M\mathbf{g}$ vectorially. To use Equation 12.4, we must know the point of application of each force so that we can measure its moment arm about the chosen axis. The points of application of $\mathbf{F}_1$ and $\mathbf{F}_2$ are simply S_1 and S_2, where the supporting columns make contact with the plate. But the gravitational force is distributed over the entire plate. Can we substitute for this distributed force a gravitational force applied at a single point? We can, after we prove that

1. the gravitational force on an extended body of mass M acts as if it were applied to a *particle* of the same mass M located at a point called the **center of gravity** of the body; and

2. if **g** has the same value over the entire body (the usual case), the center of gravity coincides with the center of mass.

Here is the proof of these two statements. Consider the arbitrary body shown in Figure 12.2. We choose a coordinate system with the x and z axes horizontal and the y axis vertically upward. We choose a rotation axis A that coincides with the z axis (whose positive direction is out of the page). In imagination, we divide the body into vertical strips (parallel to the y-z plane). We choose the widths of the strips so that each strip has the same infinitesimal mass dm. The strip shown lies at the particular coordinate x. The gravitational force exerted on it is $-dm\, g\hat{\mathbf{y}}$, perpendicular to the x axis. We find the magnitude of the infinitesimal torque dT on the strip due to gravity by applying Equation 11.13, $|T| \equiv |rF_{\perp}|$:

$$|dT| = |x\, dm\, g|.$$

Next, note that the weight of the particular strip shown will tend to make the body rotate clockwise around the axis A. Hence, we must assign a negative sign to the torque in accordance with the convention established in Section 11.3. So we have

$$dT = -gx\, dm. \quad (12.5)$$

This sign convention applies as well to strips dm located to the left of the origin. There the value of x will be negative, and each dT will be positive, consistent with the fact that the weight of any strip to the left of the origin tends to make the body rotate counterclockwise around the axis A.

Because Equation 12.5 applies to the entire body, we can find the net gravitational torque T exerted on the body about A by integrating over the body:

$$T = \int_{\text{body}} dT = \int_{\text{body}} (-gx\, dm).$$

If, as is almost always the case, the value of g is everywhere the same over the body, the net torque takes the simpler form

$$T = -g \int_{\text{body}} x\, dm. \quad (12.6)$$

We now use this expression for the torque exerted on the body by gravitation to find X_{cg}, the x coordinate of the point in the body at which a downward force of magnitude Mg exerts the same torque as the actual, distributed gravitational force. That is, we wish to find the particular value $x = X_{\text{cg}}$ for which $T = -MgX_{\text{cg}}$. (The Mean Value Theorem of calculus assures us that such a value of x exists.) We insert this value of T into

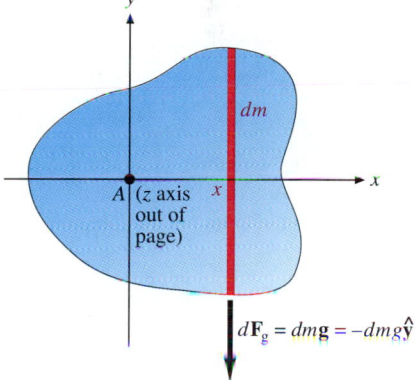

FIGURE 12.2 Determination of X_{cg}, the x coordinate of the center of gravity of an arbitrary body. A coordinate system is chosen as shown, with the x axis horizontal, the z axis (also horizontal) directed out of the page, and the y axis vertical with the positive direction upward. The rotation axis is chosen to coincide with the z axis. The body is subdivided into vertical strips of mass dm; the one shown has the particular coordinate x. The gravitational force $d\mathbf{F}_g$ acting on the strip also is shown.

Equation 12.6 to find

$$-MgX_{cg} = -g \int_{body} x \, dm,$$

which yields

$$X_{cg} = \frac{1}{M} \int_{body} x \, dm. \qquad (12.7)$$

But the right side of this equation is identical with that of Equation 9.22a, which gives X_{cm}, the x coordinate of the center of mass. So we have

$$X_{cg} = X_{cm}. \qquad (12.8a)$$

We could repeat the procedure to find the y and z coordinates of the center of gravity, but we can write down the results without taking the trouble:

$$Y_{cg} = Y_{cm} \quad \text{and} \quad Z_{cg} = Z_{cm}. \qquad (12.8b,c)$$

This proves the theorem. Because X, Y, and Z are components of a vector $\mathbf{R}$ from the origin to the point of interest, the theorem may also be written in the equivalent vectorial form

$$\mathbf{R}_{cg} = \mathbf{R}_{cm}. \qquad (12.9)$$

In the next subsection, we will find the center of gravity useful in determining the forces acting on a stationary body.

Application of the Static Conditions

We now apply the static conditions, Equations 12.2 and 12.4, to finding the forces acting on a body at rest. As a first application, let us return to the system of Figure 12.1, which is reproduced in Figure 12.3. Because the system consists of only a plate and two supports, determining the forces is a two-dimensional problem. We can therefore write Equation 12.2, the static condition for forces, as a pair of scalar equations involving the components of the $\mathbf{F}_j$. Taken together with Equation 12.4, the static condition for torques, these force equations constitute a set of simultaneous equations.

In Figure 12.3, we superpose an x-y coordinate system on the mechanical system of Figure 12.1. We locate the origin at S_1, and we locate the imaginary rotation axis there as well, normal to the plane of the page. Let the x coordinate of the center of gravity be X_{cg} and that of the support point S_2 be x_2. Because we have assumed that the contacts at S_1 and S_2 are frictionless, it follows that the support forces $\mathbf{F}_1$ and $\mathbf{F}_2$ are vertical; the forces are of the form

$$\mathbf{F}_1 = (0, F_1) \quad \text{and} \quad \mathbf{F}_2 = (0, F_2).$$

Because the gravitational force also is vertical, equating the sum of the horizontal force components gives only the trivial result $0 + 0 + 0 = 0$. In the vertical direction, Equation 12.2 gives

$$\sum_j F_{yj} = F_1 + F_2 - Mg = 0. \qquad (12.10)$$

Equation 12.4 gives

$$T_1 + T_2 + T_{cg} = 0 \cdot F_1 + x_2 F_2 - X_{cg} Mg = 0. \qquad (12.11)$$

Three things should be noted about this equation. First, because the x axis is horizontal and all the forces are vertical, each force is perpendicular to its moment arm; this means that, in each case, $F = F_\perp$. Second, choosing the rotation axis at A automatically gives the value zero to T_1. Third, the signs of the two nonzero torques are chosen to be consistent with their senses.

Equations 12.10 and 12.11 are a pair of simultaneous equations, which we now solve for the forces F_1 and F_2 that support the body. We first eliminate F_2 from the two equations. The torque equation, 12.11, gives

$$F_2 = Mg \frac{X_{cg}}{x_2}.$$

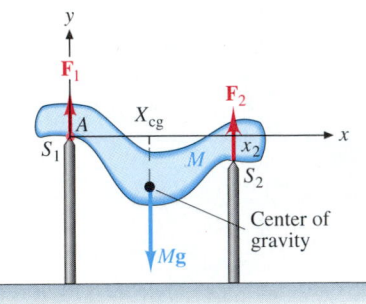

FIGURE 12.3 The system of Figure 12.1 with a coordinate system and a rotation axis A drawn in. The axis directions are chosen as in Figure 12.2. The support point S_1 is at the origin. The x coordinate of the center of gravity is X_{cg}, and that of S_2 is x_2.

Inserting this value for F_2 into the force equation, 12.10, we obtain

$$F_1 = Mg\left(1 - \frac{X_{cg}}{x_2}\right).$$

We have thus determined the support forces. As you may have expected, the way in which the total force $F_1 + F_2$ required to support the body is distributed between the two points S_1 and S_2 depends on the horizontal distance of each point from the center of gravity. Can you see how these two distances are made manifest in the equations for F_1 and F_2?

The following series of three examples develops the technique for solving statics problems by application to successively more complicated situations.

EXAMPLE 12.1

In Figure 12.4, a meter stick of mass $M = 125$ g is supported by a knife edge. A weight of mass $m_1 = 255$ g hangs from the meter stick at the 30.0-cm mark, and a weight of mass $m_2 = 263$ g hangs at the 80.0-cm mark. Where must the knife edge be located if the system is to balance?

SOLUTION: This system resembles the one of Figure 12.3, in that all the forces are vertical when the system is in balance. You can determine the support force F_s exerted by the knife edge immediately by inspection of Figure 12.4, followed by a little addition. However, by making a clever choice of rotation axis you can solve the problem without even calculating F_s! Specifically, choose the axis *at* the knife edge. This is the location you want to find; call it x. Choose the origin at the zero mark of the meter stick (its left end). Then, when you determine x, its value will be the meter stick reading at the proper location of the knife edge.

With the origin chosen at the zero mark, the positions of the weights are $x_1 = 30.0$ cm and $x_2 = 80.0$ cm. You know that the center of gravity of the uniform meter stick must lie at its midpoint, so you have $X_{cg} = 50.0$ cm for the stick. All the forces applied to the stick are perpendicular to the stick, and you write the static condition for torques as

$$\sum_j T_j = (x - x_1)(m_1 g) - (X_{cg} - x)(Mg) - (x_2 - x)(m_2 g)$$

$$= 0.$$

In choosing the signs of the three terms in the equation, you must make assumptions as to the senses of the torques, even though you don't know in advance whether the knife edge should be to the left of X_{cg}, as shown, or to the right of it. Assume that the first torque is counterclockwise and the other two are clockwise. This is consistent with the way the figure is drawn, with the knife edge located between the weights and to the left of the stick's center of gravity. Whether the assumption

FIGURE 12.4

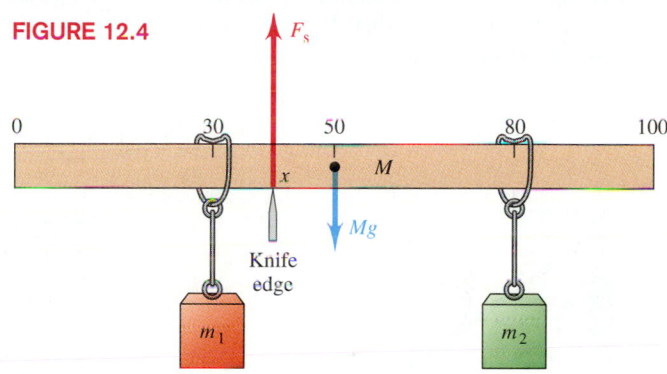

is right or wrong does not matter. Even if the assumption is wrong (if, for instance, it turns out that the system balances with the knife edge on the right side of X_{cg}), the signs of the difference terms in parentheses will automatically take care of matters. You should satisfy yourself on this point.

The acceleration of gravity g cancels out of the equation. Solving for x gives you

$$x = \frac{x_1 m_1 + X_{cg}M + x_2 m_2}{m_1 + M + m_2}. \qquad (12.12)$$

Inserting the numerical values, you obtain

$$x =$$

$$\frac{30.0 \text{ cm} \times 255 \text{ g} + 50.0 \text{ cm} \times 125 \text{ g} + 80.0 \text{ cm} \times 263 \text{ g}}{255 \text{ g} + 125 \text{ g} + 263 \text{ g}}$$

$$= 54.3 \text{ cm}.$$

Thus, balance requires that the knife edge be located at the 54.3-cm mark. This does indeed contradict the assumption made in drawing Figure 12.4.

In working this example, there was no need to convert into SI units. Refer back to Equation 12.12 and see why this is so.

EXAMPLE 12.2

Figure 12.5a shows a structure that carries the power lines for a two-track light-rail transit line. The mass of the horizontal beam is $M = 80.0$ kg, and its length is $L = 6.00$ m. The beam is symmetrical about its midpoint and is pivoted at the vertical support post. The guy wire that supports the outer end of the

beam has negligible mass and makes an angle $\theta = 35°$ with the horizontal. The system is designed to support maximum downward loads $F_1 = F_2 = 3000$ N from each of the two power lines. The lines are located at distances $x_1 = 2.13$ m and $x_2 = 5.63$ m from the vertical post. What is the minimum breaking strength B required of the guy wire? What is the minimum compression C the horizontal beam must withstand without buckling? What are the magnitude and direction of the force $\mathbf{P}$ exerted on the left end of the beam by the pivot?

SOLUTION: It is always a good idea to clarify a system on which a complicated set of forces is acting by drawing a free-body diagram. Figure 12.5b shows the forces acting on the horizontal beam. You set up the problem by using the static conditions under which the system will just hold—that is, under which the cable is on the verge of breaking. The horizontal components of all the forces exerted on the beam must add to zero; so must the vertical components. There are only two forces that have horizontal components: $\mathbf{P}$, exerted by the pivot, and the maximum force $\mathbf{B}$ exerted by the guy wire. The direction of $\mathbf{P}$ is unknown, but call the angle between $\mathbf{P}$ and the horizontal ϕ, as the figure shows. (Note that we have no reason to believe that $\phi = 0$ or that $\phi = \theta$.) You can write the horizontal-component equation

$$P \cos \phi - B \cos \theta = 0.$$

(a)

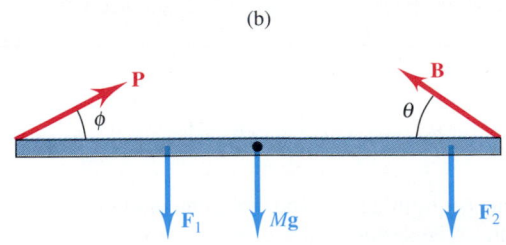

Guy wire

6.00 m

$x_2 = 5.63$ m

$\mathbf{P}$

ϕ

$\theta = 35°$

$P \cos \phi$

$x_1 = 2.13$ m

$\mathbf{F}_1$

Mg

$\mathbf{F}_2$

MARKET STREET
8

(b)

$\mathbf{P}$

ϕ

$\mathbf{B}$

θ

$\mathbf{F}_1$

Mg

$\mathbf{F}_2$

FIGURE 12.5

Adding from left to right, you have for the vertical-component equation

$$P \sin \phi - F_1 - Mg - F_2 + B \sin \theta = 0.$$

Finally, the torques exerted on the beam must add to zero. You don't know the direction of $\mathbf{P}$, and hence you cannot easily calculate the torque exerted by the pivot. You therefore sweep the matter under the rug for the time being by taking the axis at the pivot. Adding from left to right gives you the torque equation

$$0 - x_1 F_1 - \frac{L}{2} Mg - x_2 F_2 + LB \sin \theta = 0.$$

It is possible to solve these equations in symbolic form for the desired quantities B, C, and P. But the solutions will not be applicable beyond this specific example, and so nothing is lost by taking the more convenient path of inserting numerical values at the point in the solution where the symbolic equations become complicated. You can proceed in a number of ways; one way is as follows. Because the axis chosen coincides with the pivot, the torque equation has only one unknown, B. Solving gives

$$B = \frac{x_1 F_1 + \frac{1}{2} L M g + x_2 F_2}{L \sin \theta}$$

$$= \frac{2.13 \text{ m} \times 3000 \text{ N} + \frac{1}{2} \times 6.00 \text{ m} \times 80.0 \text{ kg} \times 9.8 \text{ m/s}^2 + 5.63 \text{ m} \times 3000 \text{ N}}{6.00 \text{ m} \times \sin 35°},$$

which yields

$$B = 7450 \text{ N}.$$

This is the minimum force the guy wire must bear without breaking—the minimum breaking strength of the wire.

Next, note that the horizontal-component equation contains only B and the unknown $P \cos \phi$. Now, $P \cos \phi$—the component of $\mathbf{P}$ along the beam—is the compressive force C the beam must withstand. With B known, you can solve for C immediately, obtaining

$$C = P \cos \phi = B \cos \theta = 7450 \text{ N} \times \cos 35°$$

$$= 6100 \text{ N}.$$

The vertical-component equation contains only one remaining unknown term, $P \sin \phi$. Because $F_1 = F_2$, solving gives you

$$P \sin \phi = 2F_1 + Mg - B \sin \theta$$
$$= 2 \times 3000 \text{ N} + 80.0 \text{ kg} \times 9.8 \text{ m/s}^2$$
$$- 7450 \text{ N} \times \sin 35°,$$

or $$P \sin \phi = 2510 \text{ N}.$$

Now that you have the horizontal and vertical components of the pivot force $\mathbf{P}$, you can find its magnitude and direction. The magnitude is

$$P = \sqrt{(P \cos \phi)^2 + (P \sin \phi)^2} = \sqrt{(6100 \text{ N})^2 + (2510 \text{ N})^2}$$

$$= 6600 \text{ N}.$$

The direction relative to the horizontal is given by the angle

$$\phi = \tan^{-1} \frac{P \sin \phi}{P \cos \phi} = \frac{2510 \text{ N}}{6100 \text{ N}}$$

$$= 22.4°.$$

EXAMPLE **12.3**

A ladder rests on the floor of a room, leaning against the wall. If the coefficient of static friction between ladder and floor is μ_f and that between ladder and wall is μ_w, what is the minimum angle θ that the ladder can make with the floor if the ladder is not to slip? Assume that the mass of the ladder is distributed uniformly.

SOLUTION: You begin by making a sketch, as in Figure 12.6a, and a free-body diagram, as in Figure 12.6b. The weight of the ladder is Mg. Because the ladder is uniform, the center of gravity is at its midpoint, a distance $L/2$ from either end. The normal force exerted on the ladder by the floor is directed vertically upward and has magnitude N_f. The corresponding maximum static frictional force exerted by the floor is directed horizontally toward the wall and has magnitude $\Phi_f = \mu_f N_f$. This maximum force is crucial because you want to know when the ladder will just not slip. The normal force exerted on the ladder by the wall is directed horizontally away from the wall and has magnitude N_w. The corresponding maximum static frictional force is directed vertically upward and has magnitude $\Phi_w = \mu_w N_w$.

The static condition for forces yields the horizontal-component and vertical-component equations

$$\mu_f N_f - N_w = 0 \quad \text{and} \quad N_f - Mg + \mu_w N_w = 0.$$

Taking the rotation axis at the point where the ladder touches the floor is a reasonable choice. When you do so, taking the forces from left to right gives you the torque equation

$$-\frac{L}{2} Mg \cos \theta + L N_w \sin \theta + L \mu_w N_w \cos \theta = 0.$$

The second term involves the component of the normal force perpendicular to the ladder at the point where it touches the wall, and the third term involves the component of the frictional force perpendicular to the ladder at the same point. These three equations have three unknowns (N_f, N_w, and θ) and can be solved in a straightforward way. You should do this for practice.

This is a case in which making a clever choice of axis can reduce your calculational labor considerably. The best choice to make here is one that leads to a zero value for as many individual torques as possible.

First, note that the torque produced by the gravitational force $M\mathbf{g}$ on the ladder is zero about an axis chosen *anywhere* along the vertical line passing through the midpoint, because the moment arm is zero for any such choice. We say that a force always produces a zero torque along its **line of action**—the straight line on which the force vector lies.

Next, use the components N_f and Φ_f of the force exerted on the ladder by the floor to draw the resultant force $\mathbf{F}$. Call the angle between $\mathbf{F}$ and the floor ξ (lowercase Greek xi). Extend the line of action of $\mathbf{F}$ until it crosses the line of action of $M\mathbf{g}$, as shown in Figure 12.6a. The intersection A of the two lines of action is a point about which neither of the forces produces a torque. Thus, A looks like a good place for the intersection of the axis of rotation with the plane of the page.

There is a bonus in choosing an axis of rotation through A. The only other force exerted on the ladder is that exerted by the wall. You can find this force, $\mathbf{W}$, by using its components N_w and Φ_w to construct it. Call the angle between $\mathbf{W}$ and the horizontal η (lowercase Greek eta). The line of action of $\mathbf{W}$

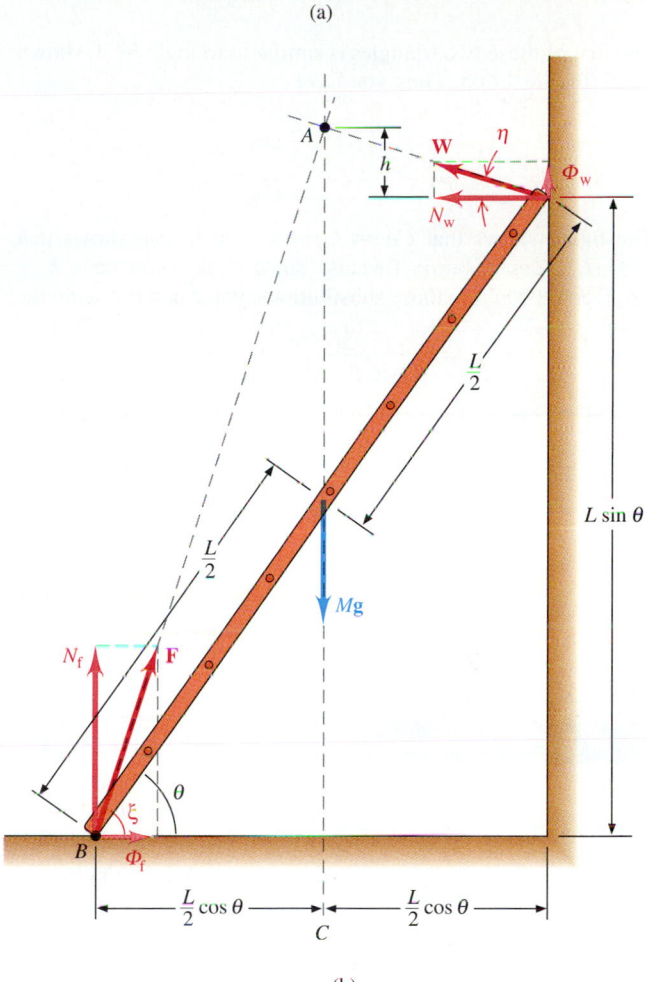

(a)

(b)

FIGURE 12.6

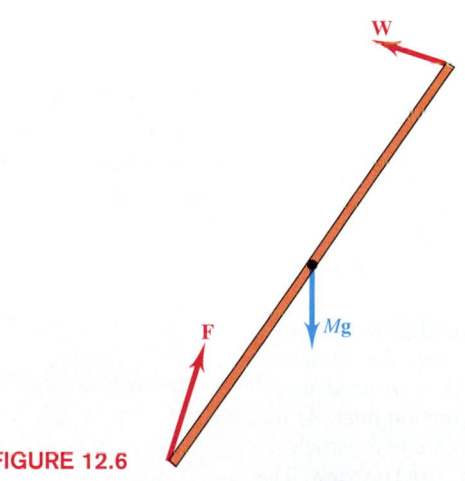

must *also* pass through A. Otherwise $\mathbf{W}$ would produce a nonzero torque, and the net torque on the ladder would not satisfy the static condition.

Now consider the triangle formed by the force vector $\mathbf{F}$ and its components. The base angle of the triangle is ξ, and you have

$$\cot \xi = \frac{\Phi_f}{N_f} = \mu_f.$$

For the corresponding triangle formed by **W** and its components, you have

$$\tan \eta = \frac{\Phi_w}{N_w} = \mu_w.$$

The first of these two triangles is similar to triangle BCA, shown in the Figure 12.6a. Thus you have

$$\mu_f = \cot \xi = \frac{\frac{L}{2} \cos \theta}{\overline{CA}}.$$

The figure shows that $\overline{CA} = L \sin \theta + h$. It also shows that $h = (L/2 \cos \theta) \tan \eta$. Because $\tan \eta = \mu_w$, you have $h = \frac{1}{2}\mu_w L \cos \theta$. Making these substitutions, you can write μ_f in the form

$$\mu_f = \frac{\frac{1}{2}L \cos \theta}{L \sin \theta + \frac{1}{2}\mu_w L \cos \theta}.$$

Cancel the L's and multiply through by twice the denominator of the right side to obtain

$$2\mu_f \sin \theta + \mu_w\mu_f \cos \theta = \cos \theta.$$

Divide through by $\cos \theta$ to obtain

$$2\mu_f \tan \theta + \mu_w\mu_f = 1.$$

The solution for θ, the angle at which the ladder just begins to slide, is

$$\theta = \tan^{-1}\frac{1 - \mu_w\mu_f}{2\mu_f}.$$

You should show that the three simultaneous equations derived near the beginning of this example yield the same result.

Example 12.3 illustrates an important theorem of statics: *If a system subject to three nonparallel external forces is in static equilibrium, the three lines of action of the forces must pass through a common point.* Forces whose lines of action pass through a common point are called **concurrent forces**. Although we will not prove this theorem formally, the proof is implicit in the paragraph of the example that begins "There is a bonus . . . ."

The Torsion Pendulum

We now apply the principles of rotational mechanics to systems that *do* rotate, in contrast to the static systems of Section 12.2. We begin with the **torsion pendulum**, shown in idealized form in Figure 12.7. (The word *torsion* comes from a Latin root meaning "twisting.") As you will soon see, the torsion pendulum is the rotational analogue of the mass-spring system. The vertical wire, called a **torsion fiber**, supports a horizontal member, consisting of a rigid, massless beam of length $2R$ with a particle of mass $\frac{1}{2}M$ at each end. The moment of inertia of the horizontal member about the torsion fiber is $I = MR^2$. The moment of inertia of the torsion fiber is negligible.

The equilibrium position of the system is the one in which the torsion fiber is relaxed—that is, not twisted. If the system is twisted through an angle θ from equilibrium, the fiber will exert a **restoring torque** on the beam. The sense of this torque will always be opposite that of the angular displacement because the fiber tends to untwist itself.

It is very often true that the magnitude of the restoring torque is proportional to the angular displacement. In this case, we can write the relation between T and θ in the form

$$T = -\kappa\theta. \tag{12.13}$$

FIGURE 12.7 An ideal torsion pendulum. (*a*) Side view. The massless beam of length $2R$ is suspended at its midpoint by the torsion fiber. At the ends of the beam are two particles, each of mass $\frac{1}{2}M$. (*b*) Top view. The system shown at an instant when the beam is displaced through an angle θ from its equilibrium position. The tendency of the torsion fiber to untwist results in a restoring torque T.

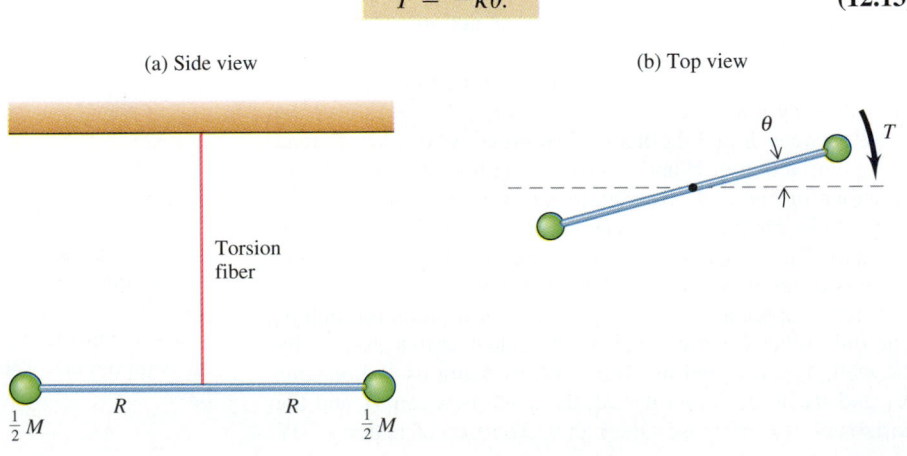

(a) Side view

(b) Top view

Torsion fiber

R R

$\frac{1}{2}M$ $\frac{1}{2}M$

This equation is the rotational analogue of the translational equation (6.29), $F = -kx$. The analogy is so close that we call Equation 12.13 the **rotational form of Hooke's law**. When there is no danger of confusion between them, both equations are simply called Hooke's law. The proportionality constant κ (lowercase Greek kappa) in the rotational form is called the **torsion constant** or **torque constant**. It is analogous to the force constant k in the translational form; κ is a measure of how hard it is to twist a torsion fiber, just as k is a measure of how hard it is to stretch a spring. The SI unit of κ is N·m/rad.

Just as the inertia of the mass in a mass-spring system makes the mass overshoot the equilibrium position at which the spring is relaxed, the moment of inertia of the torsion pendulum makes the horizontal member overshoot the equilibrium position at which the torsion fiber is relaxed. Thus twisted again, the torsion fiber exerts a torque on the horizontal member, accelerating it to rest and making it rotate back toward equilibrium again. The result is rotational oscillation. Each complete (two-way) rotational oscillation of the pendulum is analogous to a complete (two-way) oscillation of the mass-spring system.

The motion of the torsion pendulum is analyzed mathematically in exactly the same way as the mass-spring system was analyzed in Section 6.4. We begin with Equation 11.32, $T = I\alpha$, which we rewrite in the form $T = I\, d^2\theta/dt^2$. Combining this rotational form of Newton's second law with Equation 12.13, the rotational form of Hooke's law, yields

$$\frac{d^2\theta}{dt^2} = -\frac{\kappa}{I}\,\theta. \tag{12.14}$$

This is the equation of motion for a simple harmonic oscillator. It is mathematically identical with Equation 6.32,

$$\frac{d^2x}{dt^2} = -\frac{k}{m}\,x.$$

Consequently, the solutions of Equation 12.14 are identical in mathematical form with the solutions of Equation 6.32 and can be written down directly by making the proper translation-to-rotation substitutions. The angular frequency* Ω of the translational simple harmonic oscillator is given by Equation 6.34a, $\Omega = \sqrt{k/m}$. Making the substitutions $k \rightarrow \kappa$ and $m \rightarrow I$, we obtain

$$\Omega = \sqrt{\frac{\kappa}{I}} \tag{12.15a}$$

for the angular frequency of a rotational simple harmonic oscillator. This result can be expressed equally well in terms of the frequency ν and the period τ. Analogous to Equations 6.34b and 6.34c, we have

$$\nu = \frac{1}{2\pi}\sqrt{\frac{\kappa}{I}} \quad \text{and} \quad \tau = 2\pi\sqrt{\frac{I}{\kappa}}. \tag{12.15b,c}$$

The quantities Ω, ν, and τ are independent of the amplitude of oscillation, just like the corresponding quantities for translational harmonic oscillation.

The torsion pendulum has many and varied applications in physics and engineering. The great French military engineer and physicist Charles Augustin de Coulomb (1736–1806) was the first to recognize the extremely broad possibilities and to exploit some of them. We will touch on several important applications in Chapters 14, 15, 16, and 23. Here, however, let us look at the torsion pendulum in its most familiar form, as the **balance wheel**, the time-regulating element in a mechanical watch. We have already considered this application qualitatively in Section 6.4. Example 12.4 considers the matter quantitatively.

---

*We use an uppercase omega (Ω) for angular frequency instead of the lowercase omega (ω) used in Chapter 6 because now we are using ω to denote instantaneous angular velocity, a different quantity that is also important in rotating systems.

EXAMPLE 12.4

The oscillating balance wheel of Figure 12.8 is mounted in very low friction jeweled bearings. The restoring torque is supplied by the fine spiral-wound *hairspring* visible in the photograph. A typical balance wheel has a mass of 22 mg and a radius of 4.2 mm. Almost all mechanical watches are designed to tick five times a second. Because a watch ticks twice for each oscillation of the balance wheel, the oscillation frequency of the wheel is very close to 2.5 Hz. Assuming to first approximation that all of the mass of the wheel is concentrated in its rim, estimate the torsion constant of the hairspring.

SOLUTION: From Equation 12.15b, you have $\kappa = 4\pi^2\nu^2 I$. To the accuracy of the assumption concerning the mass distribution, you can write $I = MR^2$ (Equation 11.27). It follows that

$$\kappa = 4\pi^2\nu^2 MR^2$$
$$= 4\pi^2 \times (2.5 \text{ s}^{-1})^2 \times 22 \times 10^{-6} \text{ kg} \times (4.2 \times 10^{-3} \text{ m})^2,$$

or

$$\kappa = 9.6 \times 10^{-8} \text{ N·m/rad}.$$

FIGURE 12.8 Photograph of the balance wheel and hairspring of a mechanical watch. For more details, see Figure 6.13.

SECTION 12.4

The Physical Pendulum

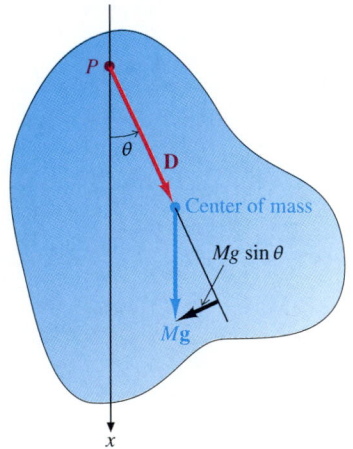

FIGURE 12.9 A physical pendulum. The arbitrarily shaped body of mass M can swing about the pivot P. The vector from the pivot to the center of mass is **D**. At the instant shown, **D** makes an angle θ with the downward vertical, which is chosen to be the reference direction $\theta = 0$.

In Chapter 6, we studied the motion of the simple pendulum—a point mass attached to a massless string. Real pendulums are never so simple, because point masses and massless strings are idealizations. We now use rotational mechanics to consider how any pivoted body oscillates about its pivot.

In Figure 12.9, a body of mass M and arbitrary shape hangs from a pivot placed at an arbitrary point P (other than its center of mass). Such a system is called a **physical pendulum**. Let us see how this system oscillates.

When the center of mass of the body is directly below the pivot, the system is in equilibrium. As you can see from Figure 12.9, when the center of mass is in any other position, the gravitational force exerts a torque that tends to rotate the system back toward equilibrium.

The vector **D** in Figure 12.9 is drawn from the pivot to the center of mass of the body. At the instant depicted in the figure, **D** makes an angle θ with the vertical; we take the downward vertical direction as $\theta = 0$. The gravitational force Mg may be regarded as acting on the center of mass, as shown. The component of Mg perpendicular to **D** is $Mg \sin \theta$. Because the only other external force acting on the body is the force exerted by the pivot, the net torque about the pivot is

$$T = -D\,Mg \sin \theta. \qquad (12.16)$$

As we did in analyzing the simple pendulum, let us consider the important special case in which θ is small enough that $\sin \theta \simeq \theta$. Equation 12.16 then simplifies to

$$T = -DMg\,\theta \quad \text{for } \sin \theta \simeq \theta. \qquad (12.17)$$

The quantity DMg is a constant. Thus, Equation 12.17 is of the form of Hooke's law for rotation, $T = -\kappa\theta$, with

$$\kappa = DMg. \qquad (12.18)$$

It follows immediately from the argument of Section 12.3 that the physical pendulum executes simple harmonic oscillations, in accordance with Equation 12.14. The angular frequency Ω of the oscillation is found by substituting Equation 12.18 into Equation

12.15a. This gives

$$\Omega = \sqrt{\frac{DMg}{I}} \quad \text{for } \sin \theta \simeq \theta. \tag{12.19a}$$

The corresponding values of the frequency ν and the period τ are

$$\nu = \frac{1}{2\pi} \sqrt{\frac{DMg}{I}} \quad \text{and} \quad \tau = 2\pi \sqrt{\frac{I}{DMg}} \quad \text{for } \sin \theta \simeq \theta. \tag{12.19b,c}$$

Equation 12.19b provides an often-convenient way of measuring the moment of inertia of a body about an axis not located at the center of mass. The mass of the body is measured, and the position of its center of mass determined. Once the center of mass is located, the distance D between the center of mass and the pivot can be measured. Finally, the body is made to oscillate as a pendulum, and the oscillation frequency ν is measured. It follows from Equation 12.19b that the moment of inertia is

$$I = \frac{DMg}{4\pi^2 \nu^2}. \tag{12.20}$$

Center of Oscillation and Radius of Gyration

Every physical pendulum is characterized by its mass M, its moment of inertia I about the pivot, and the distance D from the pivot to the center of mass. These quantities determine the angular frequency Ω of the pendulum. For every physical pendulum, there is an **equivalent simple pendulum** whose angular frequency is the same as that of the physical pendulum for small oscillations. The angular frequency of the equivalent simple pendulum is given by Equation 6.45a, $\Omega = \sqrt{g/l}$. So we have, from Equation 12.19a,

$$\sqrt{\frac{DMg}{I}} = \sqrt{\frac{g}{l}}.$$

We solve this equation for the length of the equivalent simple pendulum:

$$l = \frac{I}{MD}. \tag{12.21}$$

The point at $\mathbf{l} \equiv l\hat{\mathbf{D}}$, which is a distance l from the pivot in the direction $\hat{\mathbf{D}}$, is called the **center of oscillation** of the physical pendulum. A physical pendulum oscillates as if all of its mass were concentrated at the center of oscillation, $\mathbf{l}$, just as an extended body translates as if all of its mass were concentrated at the center of mass, $\mathbf{D}$ (Section 9.5). You can see from Equation 12.21 that the two points $\mathbf{l}$ and $\mathbf{D}$ are not in general the same.

Related to the center-of-mass distance D and the center-of-oscillation distance l is a third distance called the **radius of gyration** γ, defined to be

$$\gamma \equiv \sqrt{\frac{I}{M}}. \tag{12.22}$$

What is the physical significance of the radius of gyration? Let us use Equation 12.22 to express the moment of inertia in terms of γ; we have

$$I = M\gamma^2. \tag{12.23}$$

That is, the moment of inertia of the body about the specified axis is equal to the moment of inertia about the same axis of a particle of equal mass M at a distance γ from the axis. Remember that the moment of inertia of a body about a specified axis may be expressed by Equation 11.28, $I = cMR^2$. In this equation, R is the distance from the axis to the most distant part of the body and c is the shape factor—some constant whose value, between 0 and 1, depends on the distribution of mass in the body. By comparing Equations 11.28 and 12.23, we see that the radius of gyration is related to the shape factor c and the maximum radius R by the equation

$$\gamma^2 = cR^2. \tag{12.24}$$

Thus γ is always less than R; that is, the radius of gyration of a body always lies within the extreme bounds of the body. (In the limiting case of a hoop of radius R rotating about its center, we have $\gamma = R$.)

The parallel-axis theorem, Equation 11.32, assumes a particularly transparent form when expressed in terms of the radius of gyration. Beginning with the theorem in its simple form, $I_{\parallel} = I_{\mathrm{cm}} + MD^2$, we use Equation 12.23 to obtain

$$\gamma_{\parallel}^{\,2} = \gamma_{\mathrm{cm}}^{\,2} + D^2. \tag{12.25}$$

The appeal of this form of the parallel-axis theorem lies in the fact that all three terms in the equation are the squares of lengths, a fact that emphasizes the essentially geometric nature of the moment of inertia; the mass, a dynamical quantity, is not present.

By combining Equations 12.22 and 12.21 to eliminate I/M, we obtain

$$\gamma = \sqrt{lD}. \tag{12.26}$$

That is, the radius of gyration γ about a given axis is the geometric mean of the center-of-oscillation distance l and the center-of-mass distance D.

EXAMPLE 12.5

A uniform disk of mass M and radius r hangs from a pivot at its rim, as shown in Figure 12.10. **(a)** Find the moment of inertia I; the value of the shape factor c in Equation 12.24; the position D of the center of mass and the position l of the center of oscillation (both with respect to the pivot); the radius of gyration γ; and the period of oscillation τ. **(b)** If the disk is now hung from a new pivot located at the center of oscillation, find the new center of oscillation distance l' and the new period τ'.

SOLUTION:

(a) Find I, c, D, l, γ, and τ.

To find the moment of inertia, you use the value given in Table 11.2 for a uniform cylindrical disk about its center, together with the parallel-axis theorem, Equation 11.32. This gives you

$$I = \tfrac{1}{2}Mr^2 + Mr^2 = \tfrac{3}{2}Mr^2.$$

The value of c depends on the value of R, the distance from the pivot to the most distant part of the body. For the disk suspended from its rim, R is the diameter $2r$. In terms of R, the moment of inertia is $I = \tfrac{3}{2}M(R/2)^2 = \tfrac{3}{8}MR^2$, so the constant c has the value $c = \tfrac{3}{8}$.

By symmetry, the center of mass lies at the center of the disk. Thus, the center-of-mass distance is the edge-to-center distance; $D = r$.

You can use Equation 12.21 to obtain the center-of-oscillation distance

$$l = \frac{\tfrac{3}{2}Mr^2}{Mr} = \tfrac{3}{2}r.$$

Thus the center of oscillation is located halfway between the center of the disk and the point on the rim opposite the pivot, as shown in Figure 12.10.

You can now use Equation 12.26 to find the radius of gyration. You have

$$\gamma = \sqrt{lD} = \sqrt{(\tfrac{3}{2}r)r} = \sqrt{\tfrac{3}{2}}\,r.$$

This distance, approximately $1.22r$, is marked off (by red dots) along the vertical diameter of the disk in Figure 12.10.

You obtain the value of the oscillation period by using Equa-

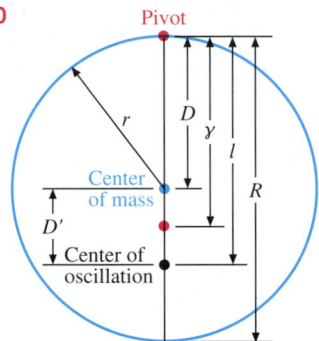

FIGURE 12.10

tion 12.19c:

$$\tau = 2\pi \sqrt{\frac{I}{DMg}} = 2\pi \sqrt{\frac{\tfrac{3}{2}Mr^2}{rMg}} = 2\pi \sqrt{\frac{3}{2}\frac{r}{g}}.$$

Another way of obtaining the same result is to remember that l is the length of the equivalent simple pendulum, whose period is $2\pi\sqrt{l/g}$. Using the value $l = \tfrac{3}{2}r$ obtained earlier gives you again $\tau = 2\pi\sqrt{\tfrac{3}{2}r/g}$.

(b) The disk is rehung from a pivot at l. Find the new center of oscillation l' and the new period τ'.

The new pivot is halfway from the center of the disk to its rim. Thus the center of mass is located a distance $D' = \tfrac{1}{2}r$ away from the pivot. In order to determine l', you need to obtain the new moment of inertia. Again applying the parallel-axis theorem, you have

$$I' = \tfrac{1}{2}Mr^2 + M(\tfrac{1}{2}r)^2 = \tfrac{3}{4}Mr^2.$$

Again applying Equation 12.21, you obtain the new center of oscillation distance

$$l' = \frac{I'}{MD'} = \frac{\tfrac{3}{4}Mr^2}{M(\tfrac{1}{2}r)} = \frac{3}{2}\,r.$$

Thus you have $l' = l$.

Because the oscillation period depends only on the value of l, it follows immediately that

$$\tau' = \tau = 2\pi \sqrt{\frac{3}{2}\frac{r}{g}}.$$

The results $l' = l$ and $\tau' = \tau$ in Example 12.5 are not accidental. *When a physical pendulum is repivoted at its old center of oscillation, the old pivot becomes the new center of oscillation.* The two are called **conjugate points**. The distance l between conjugate points on a body hung as a physical pendulum is the length of the *same* equivalent simple pendulum for both pivots; the periods are therefore identical for the two pivots and (by definition) for the equivalent simple pendulum. [The Dutch physicist Christian Huygens (1629–1695), sometimes considered the inventor of the pendulum clock, was the first to point this out.] The proof of this statement is based on a direct generalization of the reasoning of Example 12.5; it is the subject of Problem 12.41. The concept of conjugate points underlies the operation of a device for precise measurement of the acceleration of gravity g called the *Kater pendulum*; this is the subject of Problem 12.57.

THINKING LIKE A PHYSICIST

Analogy has carried us a long way. In Chapter 11, we framed analogies, based on one-dimensional translational motion, that led us into the study of fixed-axis rotation. In this chapter, we have used further analogy, based on our study of translational oscillation in Chapter 6, to study rotational oscillation. The analogies have enabled us to understand such diverse subjects as static structures, torsional oscillators, and physical pendulums. Equally important, the success of the arguments based on the analogies gives us confidence that something logically firmer than mere analogy underlies the success of the arguments. We continue to press the analogical picture to its limits. Those limits are broad, but they are limits nevertheless.

Combined Translational and Rotational Motion

The translational kinetic energy of a nonrotating body of mass M is $K_t = \frac{1}{2}Mv^2$, where v is the speed of the body. The rotational kinetic energy of a body of moment of inertia I about a fixed axis is $K_r = \frac{1}{2}I\omega^2$, where ω is the angular speed of the body about the axis. Often, a body translates and rotates simultaneously, as a ball usually does as it flies through the air (Figure 12.11).

An important (and familiar) special case of a body translating and rotating simultaneously is that of **rolling motion**. A bowling ball, an automobile tire, and even an irregular boulder rolling down a mountainside exemplify rolling motion. The motion of a rolling body is evidently more complicated than the motion of a body in simple translation or in fixed-axis rotation, and three questions are raised in preparing to analyze the motion:

1. Is it possible to separate the total kinetic energy of the body into translational and rotational parts in a unique way, and thus to simplify the analysis of the motion? (We have already proceeded in this way—but without justification—in Chapter 11.)

2. In calculating K_t, we must use a velocity $\mathbf{v}$. Is the value $\mathbf{v}_{cm}$—the special value of $\mathbf{v}$ for the center of mass—the proper one to use in the expression for K_t?

3. Is the axis A through the center of mass the proper one about which to take I and ω?

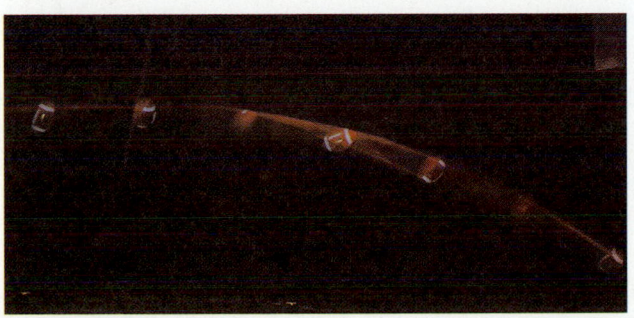

FIGURE 12.11 This badly thrown football is simultaneously translating and rotating.

As your intuition probably suggests, the answer to all three questions is Yes. This important point may be stated as a theorem: *The kinetic energy K of a rigid body is equal to the sum of* (1) *the translational kinetic energy K_t of a particle of equal mass located at the center of mass of the body and* (2) *the rotational kinetic energy K_r of the body about an axis that passes through the center of mass and maintains a fixed orientation as the center of mass moves.* That is,

$$K = K_t + K_r = \tfrac{1}{2}Mv_{cm}^2 + \tfrac{1}{2}I_{cm}\omega^2 \quad \text{for an axis having fixed orientation.} \quad \textbf{(12.27)}$$

In this equation, both I_{cm} and ω are taken *about the axis specified.*

Proof of the Theorem $K = K_t + K_r$*

In Figure 12.12, a body of arbitrary shape is translating and rotating simultaneously. We have singled out a typical mass element Δm_j of the body. An observer O is located at the origin of the inertial system xyz. From her point of view, the position of Δm_j is given at a certain time t by the vector $\mathbf{r}_j$, and the position of the center of mass of the body is given by the vector $\mathbf{R}_{cm}$. Another observer O' is located at the center of mass at A and is moving with it. He sets up his own coordinate system $x'y'z'$, whose axes are parallel to xyz. [This primed coordinate system is *not* necessarily inertial (Section 3.1).] From his point of view, the position of Δm_j is given at the same time t by the vector $\mathbf{r}_j'$. The diagram shows that $\mathbf{r}_j$ and $\mathbf{r}_j'$ bear the relation

$$\mathbf{r}_j = \mathbf{R}_{cm} + \mathbf{r}_j'.$$

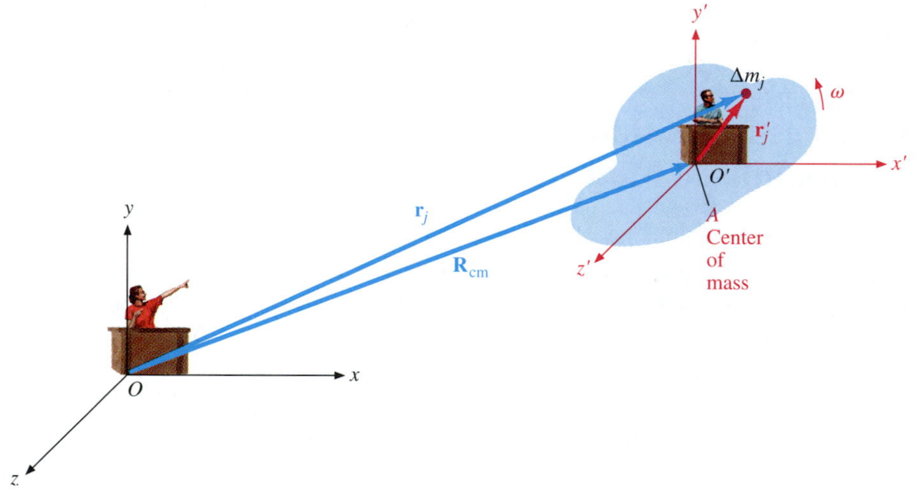

FIGURE 12.12 Observer O in the inertial xyz coordinate system sees a rigid body in simultaneous translation and rotation. The instantaneous position of the center of mass of the body is $\mathbf{R}_{cm}$; that of the mass element Δm_j is $\mathbf{r}_j$. A rotation axis A passes through the center of mass. A second observer O' located at A makes observations with respect to the coordinate system $x'y'z'$, whose axes are parallel to those of the unprimed system. For him, Δm_j is located at $\mathbf{r}_j'$.

Differentiating this relation among positions with respect to time gives the corresponding relation among the instantaneous velocities,

$$\mathbf{v}_j = \mathbf{v}_{cm} + \mathbf{v}_j'. \quad \textbf{(12.28)}$$

Observer O calculates the kinetic energy ΔK_j of the mass element as seen from her point of view. She makes no attempt to distinguish between translational and rotational motion, but simply writes

$$\Delta K_j = \tfrac{1}{2}\Delta m_j\, \mathbf{v}_j^2.$$

We can make a partial translation into the language of the primed coordinate system of O' by substituting the value of $\mathbf{v}_j$ from Equation 12.28. This gives

$$\begin{aligned}
\Delta K_j &= \tfrac{1}{2}\Delta m_j(\mathbf{v}_{cm} + \mathbf{v}_j')^2 = \tfrac{1}{2}\Delta m_j(\mathbf{v}_{cm} + \mathbf{v}_j')\cdot(\mathbf{v}_{cm} + \mathbf{v}_j') \\
&= \tfrac{1}{2}\Delta m_j(\mathbf{v}_{cm}\cdot\mathbf{v}_{cm} + \mathbf{v}_j'\cdot\mathbf{v}_j' + 2\mathbf{v}_{cm}\cdot\mathbf{v}_j') \\
&= \tfrac{1}{2}\Delta m_j(v_{cm}^2 + v_j'^2 + 2\mathbf{v}_{cm}\cdot\mathbf{v}_j').
\end{aligned}$$

---

*The proof may be omitted without loss of continuity. The theorem is applied in Example 12.6.

To find the total kinetic energy K of the body, we must sum over all elements j:

$$K = \sum_j \Delta K = \tfrac{1}{2} \sum_j \Delta m_j \, v_{cm}{}^2 + \tfrac{1}{2} \sum_j \Delta m_j \, v_j'^2 + \sum_j \Delta m_j \, \mathbf{v}_{cm} \cdot \mathbf{v}_j'. \quad \textbf{(12.29)}$$

Let us consider the three terms to the right of the second equals sign term by term, beginning with the third term. The quantity $\mathbf{v}_{cm}$ does not depend on j and can thus be taken out of the sum. So the third term becomes

$$\mathbf{v}_{cm} \cdot \sum_j \Delta m_j \, \mathbf{v}_j'.$$

But we have $\mathbf{v}_j' = d\mathbf{r}_j'/dt$. The mass Δm_j is constant in time, and so we can write the third term as

$$\mathbf{v}_{cm} \cdot \frac{d}{dt}\left(\sum_j \Delta m_j \, \mathbf{r}_j' \right).$$

What is the quantity in parentheses? According to the definition of the center of mass (Equation 9.17b), the quantity is M times the position of the center of mass in the center-of-mass (primed) coordinate system. Hence its value is zero, and so is the value of the third term on the right in Equation 12.29.

The first term on the right in Equation 12.29 is easy to evaluate. The quantity $v_{cm}{}^2$ does not depend on j, and so the term becomes

$$\tfrac{1}{2}\left(\sum_j \Delta m_j \right) v_{cm}{}^2 = \tfrac{1}{2} M v_{cm}{}^2.$$

Translational and rotational motion combined. If he applies the brake too heavily, the rider can find himself rotating over the front wheel, with painful results.

To evaluate the second term on the right in Equation 12.29, we use Equation 11.9, which has the general form $v = r\omega$. In the present case the equation becomes $v_j' = r_j'\omega$, where ω is the angular velocity about the axis and is common to all parts of the rigid body. The second term can thus be written

$$\tfrac{1}{2} \sum_j \Delta m_j \, r_j{}^2 \, \omega^2 = \tfrac{1}{2}\left(\sum_j \Delta m_j \, r_j{}^2 \right) \omega^2.$$

The term in parentheses is the moment of inertia I_{cm} about the specified axis passing through the center of mass. So the second term on the right in Equation 12.29 boils down to $\tfrac{1}{2} I_{cm}\omega^2$.

Adding together the three terms on the right side of Equation 12.29 yields the result

$$K = \tfrac{1}{2} M v_{cm}{}^2 + \tfrac{1}{2} I_{cm}\omega^2 + 0 = K_t + K_r,$$

which is Equation 12.27. The theorem is thus proved.

THINKING LIKE A PHYSICIST

Like all sciences, physics is firmly based on logic, and most physicists admire rigorous logic. The theorem $K = K_t + K_r$ (Equation 12.27) can be admired for its own beauty, but physicists expect more than logical rigor and elegance. The theorem gives us a basis for moving on to cases of rotation more complicated than fixed-axis rotation, beginning with our study of rolling motion. We will need still further generalization to deal with rotation in general, and this generalization will lead us still further beyond the analogy with which we started in Chapter 11.

EXAMPLE 12.6

A tennis ball rolls without slipping down an inclined plane. If it starts from rest, how fast is it going when it has descended a vertical distance h? Neglect frictional energy losses.

SOLUTION: In Chapter 8, you learned how to apply the generalized work-energy theorem, $W_{nc} = \Delta U + \Delta K$ (Equation 8.28), to the analysis of a body sliding down an inclined plane.

By taking advantage of the theorem $K = K_t + K_r$, you can do the same thing with a rolling body. In the absence of dissipative friction, the nonconservative work is $W_{nc} = 0$. Letting the mass of the tennis ball be M, you write the generalized work-energy theorem in the form

$$\Delta U + \Delta K = -Mgh + \tfrac{1}{2} M v_{cm}{}^2 + \tfrac{1}{2} I_{cm}\omega^2 = 0.$$

Let the radius of the ball be R. Because a tennis ball approximates a hollow spherical shell, you can use the value $I = \frac{2}{3}MR^2$ given in Table 11.2. So the conservation equation becomes

$$Mgh = \frac{1}{2}Mv_{cm}^2 + \frac{1}{2} \times \frac{2}{3}MR^2\omega^2.$$

The fact that the ball is rolling without slipping imposes a relation between the angular and translational speeds. As you saw in Example 11.2, the relation is $R\omega = v_{cm}$. You use this relation to eliminate $R^2\omega^2$ from the conservation equation and obtain

$$2gh = v_{cm}^2 + \frac{2}{3}v_{cm}^2.$$

The translational speed of the ball is thus

$$v_{cm} = \sqrt{\tfrac{6}{5}gh}.$$

The next-to-last equation indicates that the kinetic energy is divided up in the proportion $K_t = \frac{3}{2}K_r$; in other words, three-fifths of the total kinetic energy is in translational form and the remaining two-fifths is in rotational form. This proportion depends only on the moment of inertia of the rolling body—that is, on the way its mass is distributed.

Dynamics of Rolling

In the absence of friction, no body would roll downhill. Rather, it would simply slide. But, in rolling motion without slipping, the part of the body in instantaneous contact with the surface is constrained to be motionless by the force of static friction $\Phi \le \mu_s N$. This is made evident in Figure 12.13, which shows a uniform, round body rolling down an inclined plane.

In the preceding subsection, you learned that the kinetic energy of a body can be separated into translational motion of its center of mass and rotational motion about its center of mass. In view of this, you will probably find it plausible that a dynamical analysis—that is, an analysis involving forces, accelerations, torques, and angular accelerations—can be carried out in the same spirit. We will here assume that this statement is not only plausible but true, deferring its proof to Section 13.6.

As always in dynamics, we seek a connection between acceleration and force. In order to find the acceleration a_{cm} of the center of mass of the rolling body we must use Newton's second law for translational motion, $\Sigma\mathbf{F} = M\mathbf{a}_{cm}$, with the vector sum taken over all external forces applied to the body. As Figure 12.13 shows, this law takes the form

$$Mg \sin\theta - \Phi = Ma_{cm}. \qquad \textbf{(12.30)}$$

The fact that the body is not slipping as it rolls tells us only the maximum possible value of the static frictional force, and not its actual value Φ. However, we can obtain further information about Φ by considering the rotation of the body. When we take the axis of rotation at A, through the center of mass, Newton's second law for rotation takes the form $\Sigma T = I_{cm}\alpha$. Here, I_{cm} is the moment of inertia about A, and the sum is to be taken over all external torques applied to the body. If the distance from the center of mass to the point of contact is R, the law takes the form

$$R\Phi = I_{cm}\alpha. \qquad \textbf{(12.31)}$$

Equations 12.30 and 12.31 are a pair of equations having three unknowns: Φ, a_{cm}, and α. We now wish to relate α to a_{cm}, which will enable us to eliminate α from Equation 12.31. We will then have a pair of simultaneous equations in two unknowns, which we can solve for Φ and a_{cm}. The relation we need is intuitively "obvious" but a little tricky to prove. Imagine yourself located at A, moving with the center of mass of the body. You observe the point on the surface of the body that is in contact with the plane at O. The angular acceleration of the point is α, and the magnitude of its translational acceleration is given by Equation 11.10, which is $a = R\alpha$ for a body of radius R. Because the body is rolling without slipping, the point O on the plane, with which the body is in contact, also has translational acceleration of magnitude a as seen from A.

Now, imagine your friend to be located at O. For her, O is motionless and you are accelerating; the magnitude of your acceleration is a_{cm}. Because you are located at the center of mass of the body and are moving with it, the magnitude of the acceleration a_{cm} of A, as seen from O, is equal to the magnitude of the acceleration $a = R\alpha$ of O,

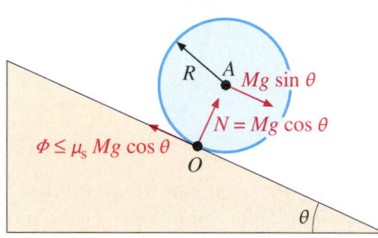

FIGURE 12.13 A round body of mass M and radius R rolls without slipping down a plane of incline angle θ. The component of the gravitational force along the incline is $Mg \sin\theta$. The normal force exerted on the body by the plane is $N = Mg \cos\theta$. The frictional force Φ is exerted tangential to the body. We do not know the magnitude of the force directly; all we can say is that it cannot exceed the value $\mu_s N$, or the body would slip. Whatever the magnitude of Φ, it exerts a torque about axis A of magnitude $R\Phi$.

as seen from A; that is, $a_{cm} = a$, and therefore

$$a_{cm} = r\alpha. \qquad (12.32)$$

Thus Equation 12.31 becomes $R\Phi = I_{cm}a_{cm}/R$. We solve for the frictional force in terms of the linear acceleration:

$$\Phi = \frac{I_{cm}a_{cm}}{R^2}.$$

Now that we have found the frictional force, we know all of the external forces acting on the body. We substitute the value of Φ into Equation 12.30 to obtain

$$Mg \sin \theta - \frac{I_{cm}a_{cm}}{R^2} = Ma_{cm}.$$

All of the quantities in this equation are known except a_{cm}, and the equation is readily solved for the translational acceleration of the rolling body:

$$a_{cm} = \frac{g \sin \theta}{1 + \dfrac{I_{cm}}{MR^2}}. \qquad (12.33)$$

This result may look a little messy. But it can be greatly simplified, and its physical meaning made quite clear, by expressing I_{cm} in the form of Equation 11.28, $I = cMR^2$:

$$a_{cm} = \frac{g \sin \theta}{1 + c}. \qquad (12.34)$$

The numerator of the fraction is the (uniform) acceleration experienced by a body sliding without friction down a plane with the same incline angle θ; see Section 4.4. The denominator, whose value always lies between 1 and 2, tells us by what factor the translational acceleration is reduced for the rolling body. This reduction is due to the fact that some of the original gravitational potential energy is converted into *rotational* kinetic energy as the body descends. Because the *total* kinetic energy is equal to the change in potential energy over the height through which the body descends, rotational kinetic energy can be present only at the expense of translational kinetic energy. When the body has descended a distance x along the plane, we have

$$K_t = \tfrac{1}{2}Mv^2 = \tfrac{1}{2}M(2a_{cm}x) = Ma_{cm}x.$$

Thus we have $a_{cm} = K_t/Mx$, and any reduction of K_t results in a reduction of a_{cm} by the same factor. That factor is $1 + c$, the denominator of Equation 12.34.

It may surprise you that the result expressed in Equation 12.34 has nothing to do with the size of the body, but depends only on its form. For what form of rolling body will the acceleration be smallest?

> **An Explanation That Had to Wait**
>
> The observation that Equation 12.34 depends only on shape was first established clearly by Galileo before 1610, on a purely experimental basis. But a theoretical understanding had to await the development of Newtonian dynamics in the latter half of the same century.

EXAMPLE 12.7

A billiard ball and a tennis ball roll down the same inclined plane. What is the ratio t_b/t_t of the times required by the two balls to reach the bottom?

SOLUTION: Because the centers of mass of both balls traverse the same path and both experience uniform (though different) accelerations, you have the kinematic relation (Equation 2.29)

$$\tfrac{1}{2}a_b t_b^2 = \tfrac{1}{2}a_t t_t^2.$$

Solving for the ratio of the times gives you

$$\frac{t_b}{t_t} = \sqrt{\frac{a_t}{a_b}}.$$

The values of the accelerations are given by Equation 12.34. When you insert them, you obtain

$$\frac{t_b}{t_t} = \sqrt{\frac{1 + c_b}{1 + c_t}}.$$

From Table 11.2, you find $c_b = \tfrac{2}{5}$ for the solid billiard ball and $c_t = \tfrac{2}{3}$ for the hollow tennis ball. So the ratio of the times is

$$\frac{t_b}{t_t} = \sqrt{\frac{1 + \tfrac{2}{5}}{1 + \tfrac{2}{3}}} = \sqrt{\frac{21}{25}} = 0.92.$$

The billiard ball takes about 92% as long as the tennis ball to cover the same distance.

Instantaneous-Axis Approach to Rolling Motion*

There is a way of viewing rolling motion as purely rotational, which is often convenient. As a body rolls along a surface, the point in contact with the surface is instantaneously at rest. For that instant, we can think of the whole body as rotating about this fixed point, shown as the **instantaneous axis** O in Figure 12.14. (At a later time, the instantaneous axis is located elsewhere, because then a different point on the periphery of the body is in contact with the surface.) The analysis of rolling thus reduces to analysis of fixed-axis rotation, and the only equation of motion to be considered is the rotational equation $T = I\alpha$. The cost of this simplification is having to regard the axis as constantly changing. But you will see that this introduces less difficulty than you might think.

As a first step in analyzing rotation about the instantaneous axis, we prove that *the angular velocity ω of the body about the instantaneous axis is the same as its angular velocity ω' about its center of mass.* Suppose that the body, of radius R, is rolling with speed v_{cm}. This is the translational speed of the center of mass from the point of view of a stationary observer at O. Because the center of mass is a distance R from the instantaneous axis, we can write $v_{cm} = R\omega$ (Figure 12.14a).

From the point of view of an observer at A, the center of mass is at rest and the surface is moving with speed v_{cm} (Figure 12.14b). Because the part of the body in contact with the surface does not slip, the observer at A must see that part of the body as having the same speed as the surface; hence $R\omega' = v_{cm}$. Comparing the two values of v_{cm}, we have $R\omega = R\omega'$, so $\omega = \omega'$. This equality has been derived for the two special points O and A. But, because the body is rigid, the angular speed ω of A about O must be the same as the angular speed of any other part of the body about O. For the same reason, ω' is the angular speed of any part of the body about A, and the theorem is proved.

If we can find the angular velocity about the instantaneous axis, then we immediately know the angular velocity about the center of mass. From the latter, we can deduce the translational motion of the body. The method is illustrated in Example 12.8.

(a)

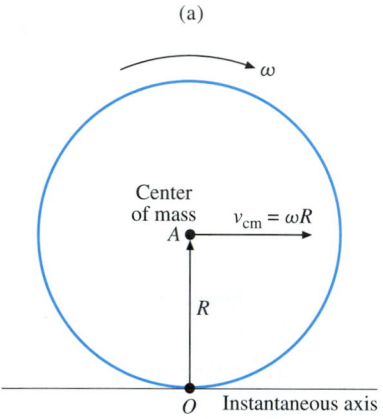

(b)

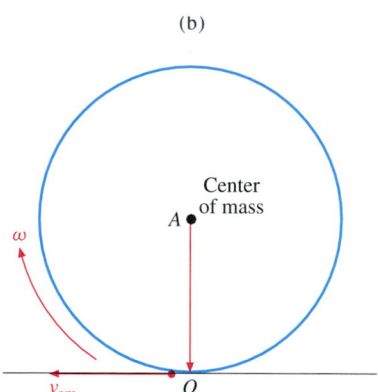

FIGURE 12.14 Comparison of (*a*) rolling motion from the instantaneous-axis point of view with (*b*) rolling motion from the center-of-mass–axis point of view. In part *a*, the center of mass rotates about O and its translational speed is $v_{cm} = \omega R$. In part *b*, the point at O rotates about A and its translational speed is $v_{cm} = \omega' R$; the translational motion of the center of mass itself with respect to the surface is not shown. Vectors in black refer to the stationary coordinate system whose origin is at O. Vectors in red refer to the moving coordinate system whose origin is at the center of mass at A.

EXAMPLE 12.8

A billiard ball rolls down a plane of incline angle θ. Use the instantaneous-axis approach to find the translational acceleration of the ball.

SOLUTION: You fix the instantaneous axis of the ball at O, its point of contact with the plane, as shown in Figure 12.15. With this choice of axis, the only torque acting on the ball is that exerted by the component of the gravitational force parallel to the plane. If you call the radius of the ball R, the equation of motion $T = I\alpha$ takes the form

$$RMg \sin \theta = I\alpha.$$

Here you must be careful. The moment of inertia that you use

FIGURE 12.15

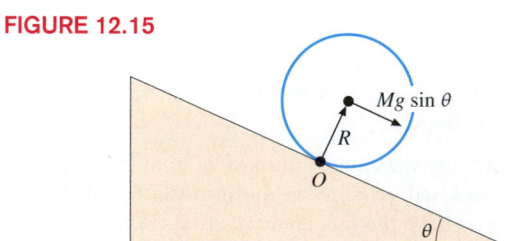

in this equation must be I_O, *the moment of inertia about the instantaneous axis O.* So you use the center-of-mass value from Table 11.2 together with the parallel-axis theorem (Section

---

*This subsection may be omitted without loss of continuity.

11.4) to find

$$I_O = \tfrac{2}{5}MR^2 + MR^2 = \tfrac{7}{5}MR^2.$$

The equation of motion thus becomes

$$RMg \sin \theta = \tfrac{7}{5}MR^2\alpha.$$

Solving this equation for the angular acceleration gives you

$$\alpha = \frac{5}{7}\frac{g}{R} \sin \theta.$$

Because the angular velocity about the instantaneous axis is the same as that about the center of mass, the same must be true of their time derivatives, the corresponding angular accelerations. So the value of α you have just found is also the proper value for rotation about the center of mass. You can use this value to find the translational acceleration $a_{cm} = R\alpha$; this yields

$$a_{cm} = \tfrac{5}{7}g \sin \theta.$$

Compare this result with the one you obtain when you use Equation 12.34. The derivation of Equation 12.34 depended on a translational equation of motion as well as a rotational equation of motion. However, the moment of inertia used in the latter was the center-of-mass value.

A, a, a	acceleration	
a_{cm}	magnitude of the acceleration of the center of mass of a body	
c	ratio of the moment of inertia of a system to that of a hoop of equal mass and maximum radius	
D	vector from axis to center of mass	
F	force	
$F_\perp$	component of **F** perpendicular to moment arm	
I	moment of inertia (axis must be specified)	
k	force constant	
K	kinetic energy	
K_r, K_t	kinetic energy of rotation, translation	
l, l	vector from axis to center of oscillation, length of equivalent simple pendulum	
m	mass of a particle	
dm, Δm	mass element of a body	
M	mass of a body	
N	normal component of force exerted on a body by a surface	
$\mathbf{r}_j$, $\mathbf{v}_j$	radius vector, velocity vector of element j in a stationary reference frame	

$\mathbf{r}'_j$, $\mathbf{v}'_j$	radius vector, velocity vector of element j in a moving reference frame	
R	distance from axis to most distant part of a body	
$\mathbf{R}_{cg}$, $\mathbf{R}_{cm}$	position vector of the center of gravity, center of mass	
T	torque (axis must be specified)	
ΔU	change in potential energy	
$\mathbf{v}_{cm}$	velocity of the center of mass of a body	
X_{cg}, Y_{cg}, Z_{cg}	coordinates of the center of gravity	
X_{cm}, Y_{cm}, Z_{cm}	coordinates of the center of mass	
α	angular acceleration	
γ	radius of gyration	
ξ, η, ϕ	direction angle of a force with respect to a reference direction	
κ	torsion constant	
μ	coefficient of friction	
ν, τ	frequency, period of an oscillator	
Φ	frictional force	
ω	angular velocity, angular speed	
Ω	angular frequency of an oscillator	

If a rigid body is neither to translate nor to rotate, the external forces acting on it must satisfy the **static condition for forces**, Equation 12.2, and the **static condition for torques**, Equation 12.4:

$$\sum_j \mathbf{F}_j = 0 \quad \text{and}$$

$$\sum_j T_j = 0 \quad \text{all } T_j \text{ referred to the same axis.}$$

In finding the torque exerted by the gravitational force on the body, the mass may be regarded as concentrated at the center of gravity, which usually coincides with the center of mass.

The **torsion pendulum** is the rotational analogue of the mass-spring system. It comprises a body of moment of inertia I subject to a restoring torque exerted by a **torsion fiber**. If the pendulum is to undergo simple harmonic oscillation, the torsion fiber must obey Equation 12.13, the **rotational form of Hooke's law**:

$$T = -\kappa\theta.$$

The negative sign indicates that the torque is a **restoring** torque, whose magnitude is determined by the **torsion constant** κ. The angular frequency, frequency, and period of oscillation of a torsion pendulum subject to Hooke's law are independent of amplitude and are given by Equations 12.15a, 12.15b, and 12.15c:

$$\Omega = \sqrt{\frac{\kappa}{I}}, \quad \nu = \frac{1}{2\pi}\sqrt{\frac{\kappa}{I}}, \quad \text{and} \quad \tau = 2\pi\sqrt{\frac{I}{\kappa}}.$$

The **physical pendulum** is a rigid body having moment of inertia I about a pivot located a distance D from the center of mass of the body. Its angular frequency, frequency, and period are given by Equations 12.19a, 12.19b, and 12.19c. These equations are obtained from Equations 12.15a, 12.15b, and 12.15c by making the substitution $\kappa = DMg$.

Just as the simple pendulum in small oscillations approximates a simple harmonic oscillator, so does the physical pendulum in small oscillations. The physical pendulum oscillates as if it were a simple pendulum with its mass concentrated at the **center of oscillation l**. The length l of this **equivalent simple pendulum** is given by Equation 12.21,

$$l = \frac{I}{MD}.$$

The center of oscillation and the pivot point are **conjugate points**; the periods of oscillation about them are the same.

The **radius of gyration** γ of a body of mass M about an axis of rotation is defined by Equation 12.22, $\gamma \equiv \sqrt{I/M}$. This radius is the distance from the axis at which a particle of the same mass M would have the same moment of inertia. The radius of gyration is related to l and D by Equation 12.26, $\gamma = \sqrt{lD}$.

Provided the orientation of the rotation axis remains unchanged, the total kinetic energy K of a rigid body of mass M may be divided into translational and rotational parts according to Equation 12.27,

$$K = K_t + K_r = \tfrac{1}{2}Mv_{cm}^2 + \tfrac{1}{2}I_{cm}\omega^2.$$

That is, the translational kinetic energy may be regarded as that of a particle of the same mass M located at the center of mass, and the rotational kinetic energy may be taken about an axis through the center of mass, as though this axis were fixed.

The translational and rotational forms of Newton's second law may be used together to treat **rolling motion**. In particular, the acceleration of a body rolling down an inclined plane is given by Equation 12.34,

$$a_{cm} = \frac{g\,\sin\theta}{1 + c}.$$

An alternative approach to analyzing rolling motion is to consider the purely rotational motion of the body about its **instantaneous axis**. The single equation of motion needed is then $T = I\alpha$, where the torque and the moment of inertia are to be taken about the instantaneous axis.

KEY TERMS

Section 12.2 Statics
static condition for forces ▪ static condition for torques ▪ center of gravity

Section 12.3 The Torsion Pendulum
torsion pendulum ▪ torsion fiber ▪ restoring torque ▪ rotational form of Hooke's law, torque constant

Section 12.4 The Physical Pendulum
physical pendulum ▪ equivalent simple pendulum ▪ center of oscillation ▪ radius of gyration ▪ conjugate points

Section 12.5 Combined Translational and Rotational Motion
rolling motion ▪ instantaneous axis

QUERIES

12.1 *(2) A new angle.* The illustration below is a top view of

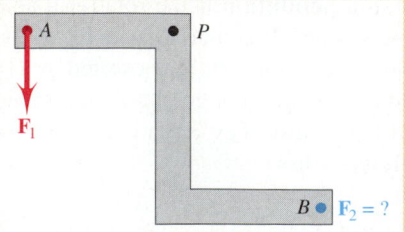

an angle iron pivoted at point P. The three segments of the iron are of equal length. A force $\mathbf{F}_1$ is applied at A as shown. If a force $\mathbf{F}_2$ of minimum magnitude is to be applied at B to balance $\mathbf{F}_1$, what should its direction be? With no more calculation than you can do in your head, express the magnitude F_2 in terms of F_1.

12.2 *(2) Is easy hard?* When you drag a heavy easy chair backward across a rug, why is it often necessary to pull on it at a point below the top of the chair back?

12.3 *(2) Nosedive.* When the driver of a moving automobile applies the brakes hard, the front end "dives" as the front springs compress. Draw a sketch and explain.

12.4 *(2) One up on your friend.* When two persons carry a heavy object (such as a piece of furniture) up a stairway, why does the person on the downstairs end usually carry a greater load?

12.5 *(2) Inclined to be safe.* New ladders always come with safety instructions. One of these instructions specifies the maximum safe horizontal distance between the bottom of the ladder and the wall against which the ladder is leaning. What is the reason for this instruction?

12.6 *(2) One way or another.* A road maintenance worker wants to move a loaded wheelbarrow from the roadway over a curb onto the sidewalk. If a board is laid down to make a crude ramp, he pushes the wheelbarrow up the ramp. But, if there is no ramp, he turns around and pulls the wheelbarrow over the curb. Explain.

12.7 *(3) Mechanical precision.* The balance wheels of high-quality mechanical watches are usually made of a special alloy called Invar, which expands very little with increasing temperature. The hairsprings are made of another special alloy, Elinvar, whose elastic properties vary very little with temperature changes. Why is this done?

12.8 *(3) Marital bliss.* The drawing below depicts a wedding ring suspended from a knife edge. The ring can be made to oscillate in two ways. It can oscillate in its own plane, normal to the knife edge, as in part *a*, or it can oscillate normal to its own plane, as in part *b*. For small oscillations, will the two periods be the same or different?

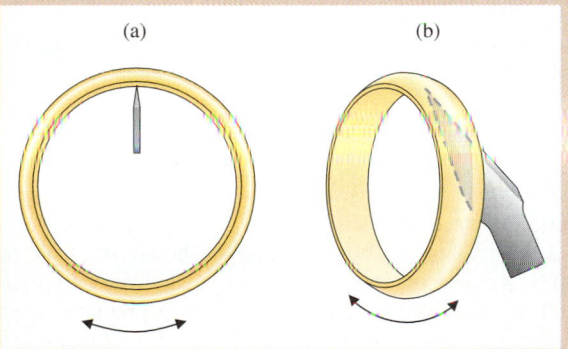

(a)   (b)

12.9 *(3) A new twist.* A torsion pendulum has period τ. An astronaut takes the pendulum to the moon, where the acceleration of gravity has one-sixth its value on earth. What is the period of the torsion pendulum on the moon?

12.10 *(4) Poor pulsometer.* You hold a uniform meter stick at the 45-cm mark by squeezing it between the thumb and index finger of your right hand. With your left hand, you start the stick swinging. Then you relax your grip slightly, so that the stick slips between your fingers at it swings. What happens to the period of oscillation as the stick slips slowly from the 45-cm mark to the 0-cm mark?

12.11 *(5) If it were smart, could it roll up?* In the illustration at the top of the next column, a dumbbell-shaped body rolls down an inclined plane. Explain why the acceleration is

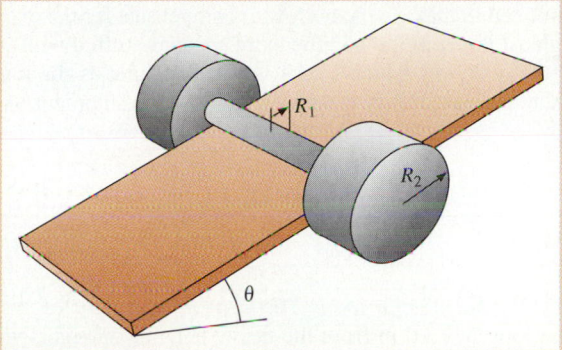

properly described by Equation 12.33 but not by Equation 12.34. What are the precise limitations on the applicability of the latter equation?

12.12 *(5) Experts ain't always right!* The illustration below is adapted from an article on bouncing balls in *Scientific American*. It depicts the bounces of an initially spinless ball that bounces elastically and is so rough that it does not slip while in contact with the surface. What is wrong with the picture?

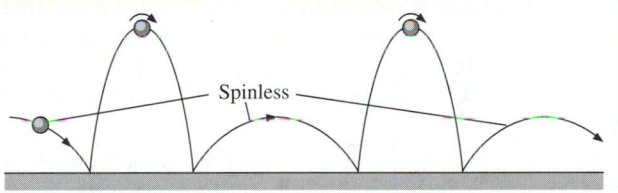

Spinless

12.13 *(5) Easter egg roll.* Suppose you want to extend the treatment of rolling motion in Section 12.5 to cover objects of rounded but noncircular cross section (for example, a cylinder of egg-shaped cross section). What additional torque must you take into consideration? Describe qualitatively the effect of this torque on the motion.

12.14 *(G) Skidding paradox.* When an automobile is braked too suddenly on an icy surface, it may "spin out." That is, the automobile may start to rotate out of control about a vertical axis. **(a)** Sketch a top view of an automobile, and describe the conditions under which it will become unstable as it is braked. **(b)** Suppose that the front-wheel brakes have failed, so only the rear-wheel brakes work. Is spin-out likely to occur? Assume for simplicity that the center of mass of the automobile is midway between the front and rear axles. Assume also that under icy conditions the maximum possible deceleration is so small that the normal force on front and rear wheels remains almost the same. (Hint: The coefficient of static friction is usually larger than the coefficient of kinetic friction.) **(c)** Now suppose that the rear-wheel brakes have failed, so only the front-wheel brakes work. Is spin-out likely to occur? [Note: This "paradoxical" result is discussed at length in W. G. Unruh, Instability in Automobile Braking, *Am. J. Phys.* **52** (10), 903–909 (1984).]

12.15 *(G) Your terrorist is my freedom fighter.* Terrorists hatch a scheme to smuggle slugs of ^{235}U, the fissionable isotope of uranium, into a country where they plan to fabricate a bomb. Each slug is hidden in a hollow in the base of a heavy brass ashtray intended for export. The ashtrays are roughly dish shaped. Because uranium is very dense, sufficient brass is removed in making the hollow to compensate for the excess weight of the uranium and the extra space is stuffed with cotton; the doctored ashtrays thus weigh the same as the much more numerous ordinary ashtrays in the same shipment. As a customs officer, you must find the doctored ashtrays quickly, at the site, and with no equipment beyond what you can scrounge from the local maintenance shop. Devise a simple method.

PROBLEMS

GROUP A

12.1 *(2) Sharing the load.* The center of gravity of a log 7.0 m long lies 3.0 m from the heavy end. Two loggers wish to carry the log. If one carries it at a point 0.4 m from the heavy end, where should the other carry it to even the load?

12.2 *(2) Hanging in the balance, I.* A meter stick is to be balanced on a knife edge located at the 30.0-cm mark. If a 10.0-kg mass hangs from the stick at the 0-cm mark, what mass must be hung from the 100-cm mark to make the system balance? What are the magnitude and direction of the force exerted on the meter stick by the knife edge? Neglect the mass of the meter stick.

12.3 *(2) Hanging in the balance, II.* Suppose the meter stick in Problem 12.2 has mass 250 g. Without detailed calculating, find where a 500-g mass could be hung to compensate for the mass of the meter stick and thus maintain the balance already found.

12.4 *(2) Hang it all!* A body of irregular shape hangs from two vertical cables located 2.50 m apart. The tension in one cable is 3580 N, and the tension in the other is 4730 N. **(a)** What is the mass of the body? **(b)** What is the horizontal distance from the first cable to the center of gravity of the body?

12.5 *(2) Cable and strut, I.* For the system illustrated below, find **(a)** the tension in the cable and **(b)** the magnitude and direction of the force exerted on the strut by the wall. Neglect the mass of the strut.

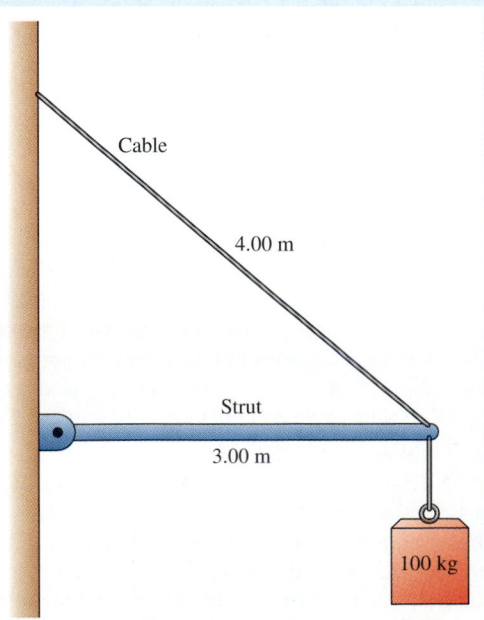

12.6 *(2) Cable and strut, II.* The mass of the strut in the figure below is 100 kg. If the vertical load supported at the end is 4000 N, find **(a)** the tension in the horizontal cable and **(b)** the magnitude and direction of the force exerted on the strut by the ground.

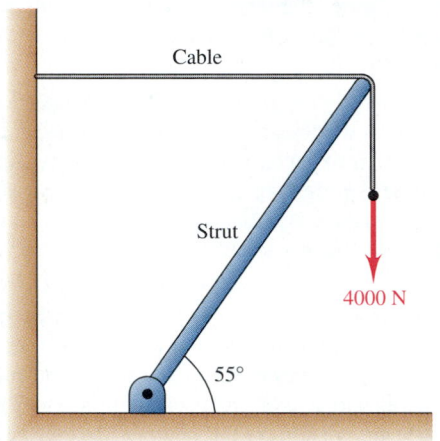

12.7 *(2) Everything hinges on zero net torque.* A typical interior house door is 2.75 m high and 0.70 m wide. The hinges are mounted 0.24 m from the top and bottom of one side. If the door is uniform and its mass is 25.0 kg, find the horizontal components of the force on each hinge.

12.8 *(2) Peculiar weighing device.* With the mass m removed from the system illustrated below, the spring scale is set to read zero. When the mass is hung as shown, the scale reads 800 N. Assume that the spring in the scale is stiff enough not to allow significant variation in the angle θ, and find the value of m.

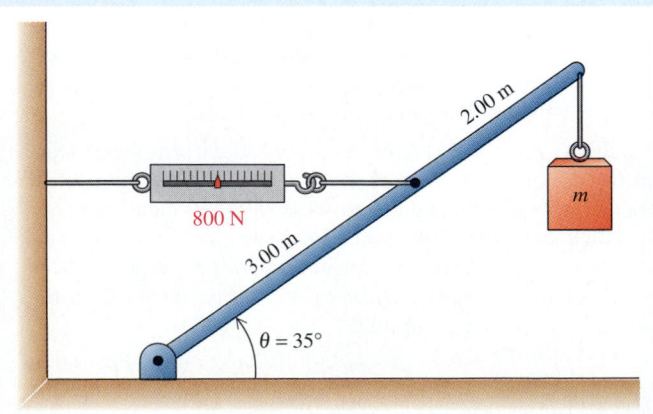

12.9 *(2) Slipping ladder, I.* A uniform ladder of mass 20 kg is 8.00 m long. Its upper end leans against a wall, where the friction is negligible. The coefficient of static friction between ladder and ground is 0.45. How far from the wall can the lower end be placed before the ladder starts to slip of its own accord?

12.10 *(2) Steelyard blues.* The illustration below depicts the simple—and very old—weighing device called the *steelyard*.

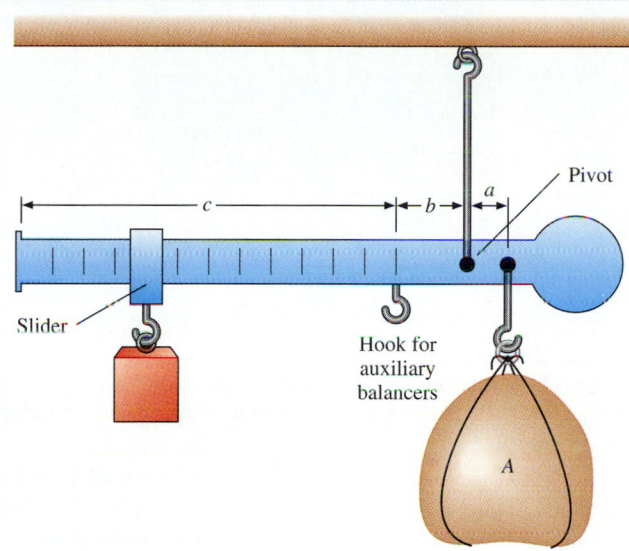

(The word is not connected with the modern meaning of "steel," but comes from a Middle Low German word meaning "sample court," a place in medieval London where Hanseatic cloth merchants did business and where such a public scale was kept.) The object to be weighed is hung from the hook at the right of the pivot, and the slider at the left is moved along the range *C* until the arm balances. If necessary, auxiliary balancers are hung from the hook at the left of the pivot. With no load attached, the system is in balance when the slider is at the right end of its travel. When the slider is at the left end of its travel, it balances a load 40 times its own weight. **(a)** What is the relation among the distances *a*, *b*, and *c* in the figure? **(b)** A set of auxiliary balancers is provided so that the steelyard can be used to weigh any object up to a maximum weight 15 times the capacity with the slider alone. Specify the weights of a minimum set of such auxiliary balancers. Express your answers in terms of the weight *W* of the slider.

12.11 *(3) Twisting in time, I.* A uniform circular disk of mass 1.207 kg and radius 7.500 cm is made into a torsion pendulum by suspending it at its center from a wire so that the disk lies in a horizontal plane. If the torsion constant of the wire is 7.639×10^{-2} N·m/rad, find the oscillation period of the pendulum.

12.12 *(3) Twisting in time, II.* The disk of Problem 12.11 is rehung by attaching the end of the same wire to its rim so that the disk hangs in a vertical plane. What is the oscillation period of this new torsion pendulum?

12.13 *(3) Having a ball.* A torsion pendulum is made by welding a steel wire to a steel ball of mass 64.1 g and radius 1.25 cm at a point on its surface. If the period of the pendulum is 1.17 s, find the torsion constant of the wire.

12.14 *(3) After the ball.* The steel ball of Problem 12.13 is replaced with a cylinder having the same mass and the same diameter. The wire is attached to the cylinder at the center point of one of its ends. **(a)** Find the period of the new pendulum. **(b)** If a series of bodies, all having the same mass and radius but different shapes, are hung successively from the same torsion fiber, which shape will result in the smallest period? the greatest period?

12.15 *(3) Changing times.* You suspend a stiff rod of negligible mass at its center from a thin wire. You make a torsion pendulum by attaching two equal masses at equal distances 1.55 cm from the center. The frequency of oscillation is 0.408 Hz. If you wish to change the frequency to 0.204 Hz, to what new (equal) distances from the center of the rod must you move the masses?

12.16 *(3) Pocket watch.* A designer decides to replace the balance wheel of the watch of Example 12.4 with a wheel that will make the watch tick once a second instead of five times a second. No other changes are to be made. If the new wheel is to have the same shape and thickness as the old one, what must the radius of the new wheel be?

12.17 *(4) Swing the ring.* A segment of heavy-walled steel pipe is suspended from a knife edge, as shown below. **(a)** Express the oscillation period τ in terms of the inner and outer radii R_1 and R_2. **(b)** If the inner diameter is 25.0 cm and the wall thickness is 2.5 cm, what is the value of τ?

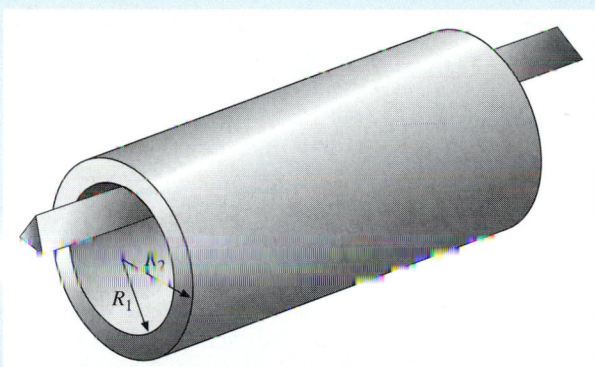

12.18 *(4) Stick-tock, I.* A meter stick is suspended from a pivot at its zero mark. **(a)** Where is the center of oscillation? **(b)** What is the radius of gyration? **(c)** What is the period of oscillation?

12.19 *(4) Stick-tock, II.* The meter stick of Problem 12.18 is now suspended from a pivot at its *x*-cm mark. **(a)** Where is the center of oscillation? **(b)** What is the radius of gyration? **(c)** Show that the period of oscillation is

$$\tau = 2\pi \sqrt{\frac{z}{g} \left(1 + \frac{1}{12z^2} \right)},$$

where we define the dimensionless quantity z as $z \equiv |(50 - x)/100|$ and g can thus be expressed in SI. **(d)** By taking z very large, show that the result of part **c** is consistent with the period of the meter stick hung as a pendulum from a very long string attached at the 0-cm mark—that is, for $z \gg 1$ m.

12.20 *(4) Grandfather had no equal.* The diagram on the next page shows the pendulum of a grandfather clock. The pen-

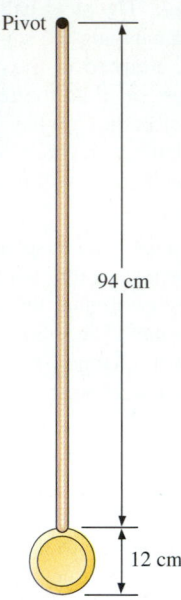

Pivot

94 cm

12 cm

sliding it to the desired point in the narrow slot and clamping it there. The disk is made to oscillate, as shown by the red arrow. Find the oscillation frequency ν as a function of r.

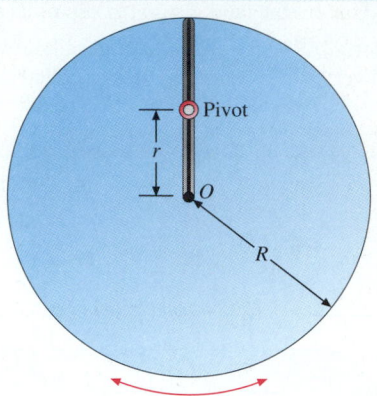

Pivot

r

O

R

dulum consists of a brass disk of diameter 12.0 cm and mass 2.04 kg attached to a hardwood rod of length 94.0 cm and mass 0.042 kg. **(a)** Find the center of mass of the pendulum. **(b)** Find the length of the equivalent simple pendulum. **(c)** If you assume for simplicity that the actual pendulum is a simple pendulum whose mass is concentrated at the center of mass, what error do you make in the period of the pendulum?

12.21 *(4) Diskulum.* The uniform disk illustrated at the top of the next column hangs from an adjustable pivot. The pivot can be moved to any distance r from the center of the disk by

12.22 *(5) Any shape you like.* Find the acceleration of **(a)** a hoop, **(b)** a solid cylinder, **(c)** a thin-walled hollow ball, and **(d)** a solid ball, all of which roll without slipping down a plane of incline angle θ.

12.23 *(5) Roll 'em, I.* A spherical ball of mass 0.45 kg rolls without slipping down a plane of incline angle $\theta = 32°$. Find **(a)** the component of the gravitational force parallel to the plane and **(b)** the magnitude of the frictional force exerted on the ball by the plane.

12.24 *(5) Roll 'em, II.* What is the solution to Problem 12.23 if the ball is replaced by a thick-walled hollow cylinder having the same mass, and the inner diameter of the cylinder is one-half its outer diameter?

GROUP B

12.25 *(2) Wedged in.* A uniform sphere of mass M rests on two flat surfaces, as shown below. If the surfaces make angles ξ and η with the horizontal, find the force exerted on the sphere by each surface.

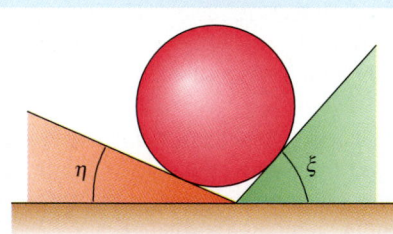

η

ξ

12.26 *(2) Round wheel and square curb, I.* A wheel has mass M and radius R. You wish to pull it over a curb of height h by exerting a horizontal force F on its axle. **(a)** Find the required magnitude of F. **(b)** Make a rough sketch of F as a function of the ratio $\rho \equiv R/h$.

12.27 *(2) Rearranging the furniture, I.* A rectangular dining-room table is 2.36 m long and 0.92 m high. It has a leg at each corner. The table is pulled along the floor at constant speed by a horizontal force applied to the tabletop at the center

of one end. If the coefficient of kinetic friction between table legs and floor is 0.35, find the vertical and horizontal components of the forces exerted by the floor on each leg in terms of mg, the weight of the table.

12.28 *(2) Rearranging the furniture, II.* The table of Problem 12.27 is moved into another room whose floor is covered with a rug. Continued attempts to pull it as described in that problem result in the lifting of the rear legs off the floor. Find the minimum value of the coefficient of friction between legs and rug.

12.29 *(2) Slipping ladder, II.* An 85-kg man starts to climb the ladder of Problem 12.9. How far can he climb before it starts to slip?

12.30 *(2) Slipping ladder, III.* A uniform ladder rests against a vertical wall. The ladder is stable only if the angle between it and the floor is greater than 48°. If the coefficient of static friction between the ladder and the floor is 0.4, find the corresponding coefficient between ladder and wall.

12.31 *(2) Slipping ladder, IV.* Obtain the result of Example 12.3 by direct solution of the three simultaneous equations derived there.

12.32 *(2) The glass is steady.* A cylindrical drinking glass has height h, radius r, and mass m. Assume that the mass of the bottom of the glass is negligible so that substantially all of the mass is in the walls. By pouring sand into the glass, you can vary the vertical position Y of the center of gravity of the system. **(a)** If you tip the glass, it is stable—that is, it returns to its original position when released—up to a certain tilt angle θ, measured from the vertical. Beyond that angle, the glass falls over. Express θ in terms of Y and r. Assume that the sand does not shift as the glass is tipped. **(b)** When the glass is filled to the top with sand, the mass of the sand is $M = \mu m$. What is the mass of the sand in the glass when it is filled to a height $y < h$? **(c)** Find Y for the glass-sand system as a function of y. **(d)** Show that the system has maximum stability—that is, θ has its maximum value—when the glass is filled to the level

$$y = \frac{\sqrt{1 + \mu} - 1}{\mu} h.$$

(e) Find y/h for $\mu = 4$.

12.33 *(3) Harmonic-oscillator solution.* Show explicitly that a solution of Equation 12.14 is

$$\theta = \theta_0 \cos\left(2\pi \frac{t}{\tau} + \delta\right),$$

where θ_0 is the oscillation amplitude of the torsion pendulum. What are the angular velocity ω and the angular acceleration α of the torsion pendulum as functions of time? (Hint: Refer to Equations 6.20 through 6.22.)

12.34 *(3) Moment of inertia "scale."* The torsion pendulum device shown below can be used to measure the moment of inertia I of a body about any desired axis. First, the device is set into torsional oscillations with the pan empty, and the period τ_0 is determined. Then the body whose moment of inertia is to be measured is placed on the pan so that the desired axis coincides with the center of the pan, and the new period τ_1 is measured. Express I in terms of τ_0 and τ_1.

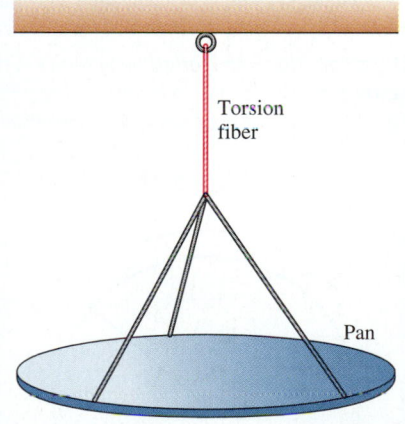

Torsion fiber

Pan

12.35 *(3) Another kind of torsion pendulum.* In the figure at the top of the next column, a uniform disk of mass M and radius R is mounted on a fixed horizontal axle passing through its center. The two springs attached at a point on the rim of the disk have a combined force constant k. Find the frequency of oscillation of the system.

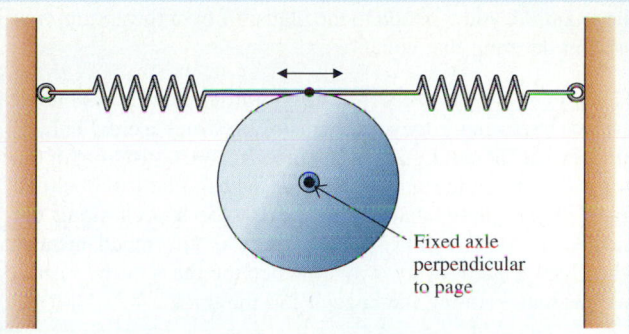

Fixed axle perpendicular to page

12.36 *(4) Minimizing the period, I.* If a rigid body is made into a physical pendulum by swinging it from a pivot, the period of oscillation will depend on the location of the pivot. Prove that the *shortest* period is obtained if the pivot is located so that its distance from the center of mass is equal to the radius of gyration about the center of mass.

12.37 *(4) Minimizing the period, II.* **(a)** Where should you place the pivot on a uniform meter stick in order to minimize its oscillation period? **(b)** What will the minimum period be?

12.38 *(4) The power of negative thinking.* A uniform disk of radius R is mounted on a horizontal axis normal to the disk and passing through its center. A circular hole of radius $R/2$ is cut in the disk, centered halfway between the center of the disk and its rim. Find the period of the resulting physical pendulum.

12.39 *(4) Swinging plate, I.* A uniform rectangular plate of sides X and Y is hung from a pivot at the middle of one of the sides of length X. **(a)** If X is fixed and Y varied, for what ratio X/Y will the period of oscillation be a minimum? **(b)** What will the period be?

12.40 *(4) Swinging plate, II.* Consider again the plate of Problem 12.39. For what ratio X/Y will the period of oscillation be the same as that of a simple pendulum of length Y?

12.41 *(4) Conjugate points.* Prove the theorem stated immediately after Example 12.5: If a physical pendulum having mass M and moment of inertia I about its pivot is repivoted at its center of oscillation, the original pivot becomes the new center of oscillation and the oscillation period is unchanged.

12.42 *(5) Off the wall.* A small, uniform solid ball rests at the top of a large hemispherical dome. It is released and starts to roll downward without slipping. At what angle θ from the top will it leave the dome? Compare your result with that of Problem 8.48.

12.43 *(5) An energetic approach.* Derive Equation 12.33, which gives the acceleration of a body rolling without slipping down an inclined plane, by starting with Equation 12.27 and using conservation of mechanical energy. (Hint: Consider the change in potential energy of the body as it rolls a distance l along a plane of incline angle θ. Use the no-slipping condition to write the final kinetic energy in terms of translational quantities only. Then use the fact that the acceleration is constant to relate the distance traversed to the acceleration.)

12.44 *(5) Be dynamic!* Analyze the problem of Example 12.6 using Equation 12.34. By obtaining the same result as that of the example, you will add to the plausibility of the assumption used in deriving that equation.

12.45 (5) *Caterpillar tractor.* The figure below shows one of the two track-and-bogey mechanisms on which a crawler tractor moves. The total mass of the mechanism, consisting of the bogeys (the driving and supporting wheels) and the beltlike track, is M. Show that, when the tractor moves along the ground at speed v, the kinetic energy of the mechanism is Mv^2, independent of the way in which the mass of the system is distributed among the bogeys and the track.

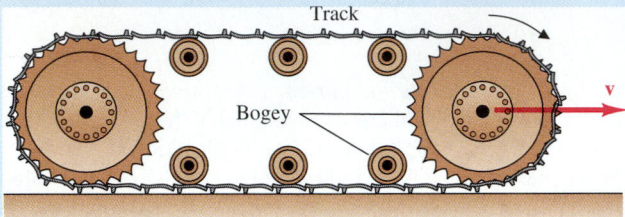

12.46 (5) *Pulling the rug out from under.* A bowling ball rests on a carpet. A child grabs the end of the carpet and pulls it out from under the ball with acceleration a. What is the acceleration of the ball if it does not slip?

12.47 (5) *No spool like an old spool, I.* A Yo-Yo lies on a rough table, as shown below. As you probably know from childhood experience, you can make the Yo-Yo roll without slipping by pulling on the string gently. **(a)** Which way will it roll for each line of action shown for the applied force? (The line of action of $\mathbf{F}_3$ passes through the point of contact O.)

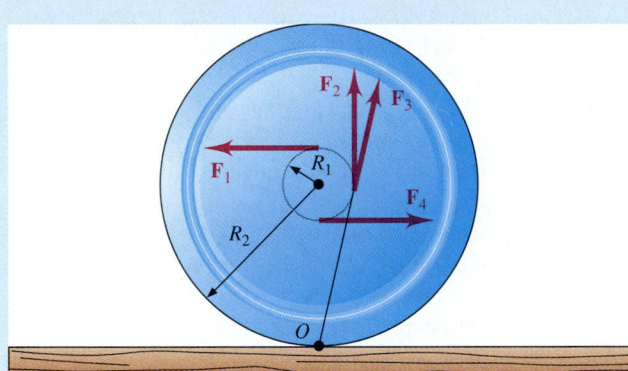

(b) Show that if $\mathbf{F}_1$ and $\mathbf{F}_4$ have equal magnitudes, the ratio of the corresponding accelerations they produce is

$$\frac{a_1}{a_4} = -\frac{R_2 + R_1}{R_2 - R_1}.$$

(c) If $\mathbf{F}_2$ has the same magnitude as $\mathbf{F}_1$, what is the ratio a_1/a_2? **(d)** If the instantaneous speed of the pulled end of the string is v_0, what is the velocity of the center of mass of the Yo-Yo when the force has the directions $\hat{\mathbf{F}}_1$, $\hat{\mathbf{F}}_2$, and $\hat{\mathbf{F}}_4$? [Hints: Take the (unknown) moment of inertia to be $I = cMR_2^2$. In order to avoid sign confusion, take the positive direction to the left so that a positive acceleration will correspond to a positive sense of angular acceleration.]

12.48 (5) *Rolling dumbbell, I.* Suppose that essentially all the mass of the dumbbell illustrated in Query 12.11 is concentrated in the end pieces. Derive an expression for the acceleration a as the dumbbell rolls without slipping down the inclined plane. What quantity replaces the constant c in Equation 12.34?

12.49 (5) *Something has to slip!* In the illustration below, the coefficient of kinetic friction between the uniform cylindrical roller and the inclined plane is μ_k. The string is wound around the roller. The frictional force between roller and plane is small enough so that the roller slips, the string unwinds, and the roller descends. **(a)** Find the force F exerted by the string on the roller. **(b)** Show that the acceleration is

$$a = \tfrac{2}{3}g(\sin\theta - 2\mu_k\cos\theta).$$

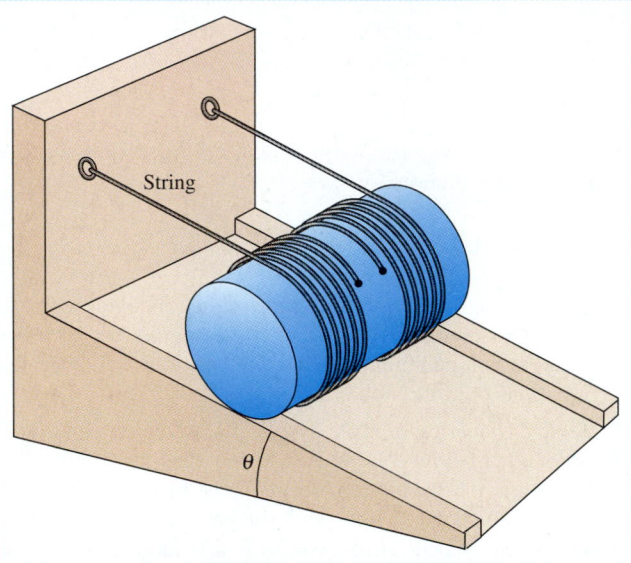

(Hint: You can solve this problem by using either dynamical methods or the generalized work-energy theorem. If you use the latter, think *carefully* about the frictional work done as the roller descends a distance L along the plane.)

12.50 (5) *Different axes, same angular velocity.* This problem concerns another proof of the theorem illustrated in Figure 12.14. The diagram below represents a wheel that rolls with-

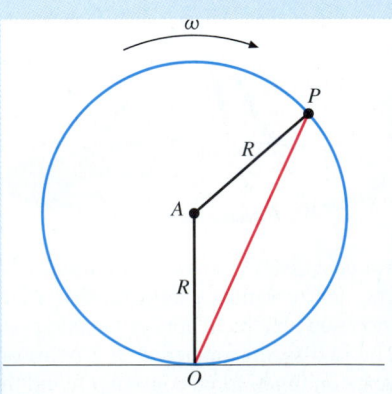

out slipping. Its angular velocity about the axis A passing through its center is ω. Show that the angular velocity about the instantaneous axis O, which passes through the point of contact of the wheel with the surface, has the same value ω. (Hint: Consider the instantaneous linear velocity of the point P, as seen by observers at O and A, and the velocity of A as seen from O. Remember that the direction of motion of a point, as seen by an observer located at any axis, is perpendicular to the radius vector from that axis to that point.)

12.51 *(5) Something's slipping!* A symmetrical round body rolls down a broad inclined plane. Show that, if the rolling is to take place without slipping, the incline angle θ must be less than the *critical angle for rolling without slipping*:

$$\theta < \tan^{-1}\left(\frac{1+c}{c}\,\mu_s\right),$$

where μ_s is the coefficient of static friction between the body and the plane. Compare this result with the critical angle for sliding motion, $\tan^{-1}\mu_s$, given by Equation 4.36b.

12.52 *(G) Bifilar pendulum.* A uniform rod is suspended at its ends by two threads of equal length h, as shown opposite.

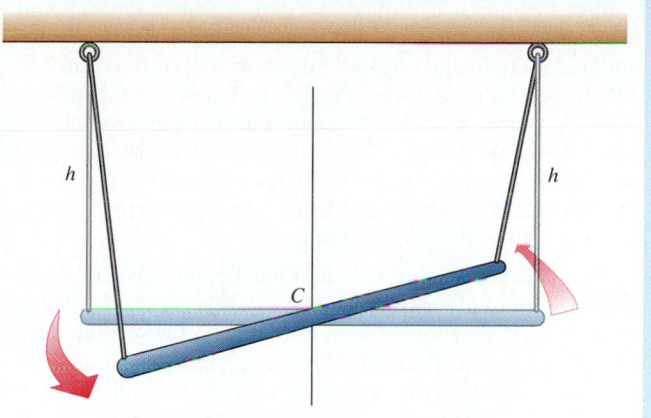

The system is displaced from equilibrium by rotating it through a small angle about a vertical axis passing through the midpoint C of the rod. What is the angular frequency of oscillation of the system? This system is called a *bifilar* (two-thread) *pendulum*.

GROUP C

12.53 *(2) Round wheel and square curb, II.* As in Problem 12.26, you wish to pull a wheel of mass M and radius R over a curb of height h by pulling on its axle. This time, however, you pull at an angle θ with respect to the horizontal. **(a)** Show that the force required has magnitude

$$F = Mg\,\frac{\sqrt{2\rho - 1}}{(\rho - 1)\cos\theta + \sqrt{2\rho - 1}\,\sin\theta},$$

where $\rho \equiv R/h$. **(b)** Show by appropriate differentiation that this force is minimized if it is applied perpendicular to the radius that joins the center of the wheel to the point of contact with the curb. Your result should agree with the one you obtain by a qualitative argument beginning with the elementary definition of torque. **(c)** Show that this minimum force has magnitude

$$F_{min} = Mg\,\frac{\sqrt{2\rho - 1}}{\rho}.$$

(d) Check the results of parts **a** and **c** by considering the extreme cases $\rho = 1$ and $\rho = \infty$. What happens if $\rho < 1$? What happens if $\theta < 0$?

12.54 *(2) Rearranging the furniture, III.* You are dragging a table along the floor at constant speed. The legs are located at the corners of the table. **(a)** Find the angle θ with the horizontal at which you should pull in order to minimize the force required. **(b)** Compare the result with that of Problem 4.53, and thus show that the result is independent of the placement of the table legs with respect to the corners of the table.

12.55 *(2) Bucking bicycle.* A bicycle and rider are illustrated in the next column. The point C indicates the center of mass of the bicycle and rider taken together. **(a)** When the bicycle is rolling freely, what are the normal forces N_f and N_r exerted by the ground on the front and rear wheels? **(b)** The rider stops

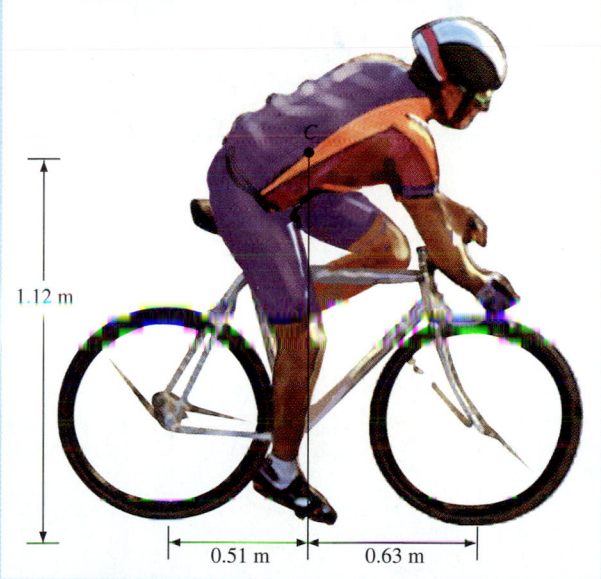

the bicycle by jamming on the front wheel brake only, so the front wheel skids. The coefficient of friction between tire and road is $\mu_k = 0.6$. Find the new normal forces N_f' and N_r', as well as the acceleration a' of the bicycle. (Hint: What torque, in addition to those you considered in part **a**, must you take into account now?) **(c)** Later, the rider stops the bicycle by jamming on the rear wheel brake only, so the rear wheel skids. Find the normal forces N_f'' and N_r'', as well as the acceleration a'' of the bicycle now. **(d)** Which of the two means of stopping the bicycle is faster? Which is safer?

12.56 *(4) Quantitative wedding ring.* Find the period(s) of oscillation for the wedding ring illustrated in Query 12.8. Its inner diameter is 2.20 cm, its outer diameter is 2.90 cm, and a cross-sectional cut through the ring is semicircular. [Hint: You *could* calculate the moment of inertia of a torus (''doughnut'') of semicircular cross section about its axis of symmetry. Before you take the time to do so, calculate the moment of inertia for a thick-walled cylinder with the same radius and for a thin hoop whose radius is the average of the inner and outer radii. Compare these moments of inertia and see if a substantial error is introduced by using the approximation that the ring is nearly enough a thin hoop.]

12.57 (4) *Kater's pendulum.* In principle, the acceleration of gravity g can be measured by timing the period of a simple pendulum. In practice, however, the method is severely limited because no real pendulum has a bob that is a point mass. It is thus impossible to locate the center of oscillation with high precision. This difficulty is circumvented by the use of the ingenious pendulum (below) devised in 1817 by the British military engineer, geodesist, and metrologist Henry Kater (1777–1835). Under ideal conditions, the Kater pendulum can be used to determine g to a precision of order one part in 10^7; for almost a century and a half, it was the best method available.

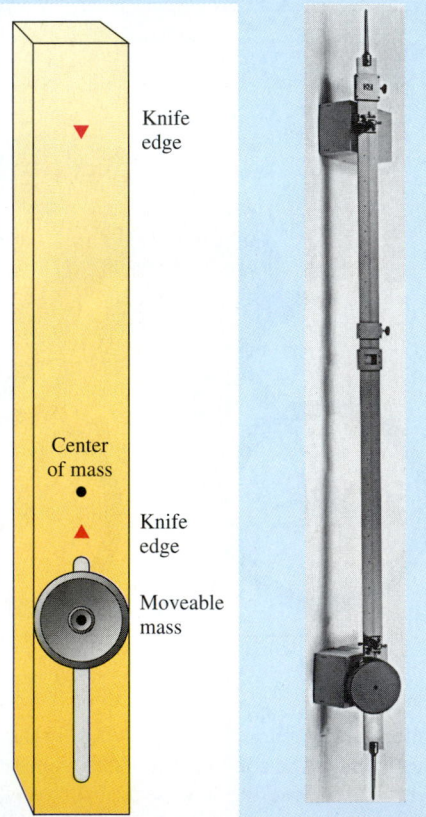

Knife edge

Center of mass

Knife edge

Moveable mass

The pendulum is deliberately made asymmetrical by addition of the large moveable mass, whose position can be adjusted with a micrometer screw (not shown). By trial and error, the adjustment is used to make the oscillation periods τ_1 and τ_2 about the two knife edges as close as possible to one another. The knife edges then lie approximately—but not exactly—at conjugate points. The center of mass of the entire system is shown for this particular adjustment of the moveable mass. Because of the asymmetry of the pendulum, the two knife edges lie at quite different distances D_1 and D_2 from the center of mass. **(a)** Let the radius of gyration about the center of mass be γ_0. Use the parallel-axis theorem, Equation 12.25, to show that the corresponding radii of gyration about the two knife edges satisfy the relations

$$\gamma_1{}^2 = \gamma_0{}^2 + D_1{}^2 \quad \text{and} \quad \gamma_2{}^2 = \gamma_0{}^2 + D_2{}^2.$$

(b) Using the result of part **a**, express $\tau_1{}^2/4\pi^2$ in terms of $\gamma_0{}^2$, $D_1{}^2$, and g, and $\tau_2{}^2/4\pi^2$ in a corresponding form involving $D_2{}^2$. **(c)** Now show that

$$\frac{g}{4\pi^2} \frac{\tau_1{}^2 + \tau_2{}^2}{D_1 + D_2} = \frac{\gamma_0{}^2}{D_1 D_2} + 1$$

and

$$\frac{g}{4\pi^2} \frac{\tau_1{}^2 - \tau_2{}^2}{D_1 - D_2} = 1 - \frac{\gamma_0{}^2}{D_1 D_2}.$$

(d) Using this result, obtain the expression

$$\frac{8\pi^2}{g} = \frac{\tau_1{}^2 + \tau_2{}^2}{D_1 + D_2} - \frac{\tau_1{}^2 - \tau_2{}^2}{D_1 - D_2}.$$

The periods τ_1 and τ_2 can be measured very precisely by counting the swings of the pendulum over a carefully measured interval of many hours. The distance $D_1 + D_2$ between the knife edges can be measured very precisely by interferometric methods. But the denominator of the second term on the right can be obtained with at best much poorer precision; the necessary measurements depend on balancing the pendulum on a knife edge to locate the center of mass. Why does this relatively low precision not degrade the precision of the value of g? In particular, why is it so important that the pendulum be asymmetrical?

12.58 (5) *Any thickness you like.* A thick-walled cylinder rolls without slipping down a plane of incline angle θ. Show that the acceleration is

$$a = \frac{2g \sin \theta}{\sigma^2 - 2\sigma + 4},$$

where σ is the ratio of the wall thickness to the outer radius.

12.59 (5) *No spool like an old spool, II.* The system shown in Problem 12.47 rolls without slipping when the string is pulled. Express the acceleration a of the Yo-Yo in terms of the force $\mathbf{F}$, the coefficient c, and the radii R_1 and R_2 for an arbitrary angle θ between the positive (leftward) direction and the line of action of $\mathbf{F}$. As a check, find the value of a for the special case $\mathbf{F} = \mathbf{F}_3$ shown in the figure.

12.60 (5) *No spool like an old spool, III.* Consider again the system illustrated in Problem 12.47, subject to the force $\mathbf{F}_1$. Show that, if the system is constructed so that $R_1/R_2 = c$, the Yo-Yo will roll without slipping even on a frictionless surface.

12.61 (5) *No spool like an old spool, IV.* By pulling at the correct angle θ, you can make the Yo-Yo illustrated in Problem 12.47 roll on a frictionless surface without slipping, even if the Yo-Yo is not constructed to satisfy the condition of Problem 12.60. **(a)** Find the required value of θ. **(b)** Does such a value of θ exist for every possible spool? If not, what is the condition for which it does exist? Express your result in terms of R_1, R, and the radius of gyration γ of the spool about its axis. What is the physical meaning of this condition?

12.62 *(5) Rolling dumbbell, II.* If the dumbbell of Problem 12.48 is to roll without slipping, what condition must replace that of Problem 12.51, which applies to bodies rolling on their outermost surfaces?

12.63 *(G) Skidding start.* A massive disk of radius R is mounted on a light axle. With the axle held horizontal, the disk is made to spin with angular speed ω_0. It is then lowered gently onto a level table, and released as soon as its rim makes contact with the tabletop. The disk begins to roll along the table, skidding and picking up speed as it goes. **(a)** Show that, as the disk skids, its translational and angular accelerations bear the relation

$$a = \tfrac{1}{2}R\alpha.$$

(b) Show that the disk will continue to skid for a time

$$t = \frac{R\omega_0}{3\mu_k g},$$

where μ_k is the coefficient of kinetic friction between the rim of the disk and the tabletop. **(c)** Show that, when the disk stops skidding and begins to roll without slipping, its speed is

$$v = \tfrac{1}{3}R\omega_0.$$

(d) Show that two-thirds of the initial kinetic energy is lost in the skidding process. **(e)** Suppose that the spinning body is a sphere of the same radius. When the skidding stops, will v be greater or less than that for the disk? What if the body is a hoop?

12.64 *(G) Atwood's oscillator?* A pulley in the form of a uniform disk of mass M and radius R is mounted on a hori- zontal axle, as shown below. A compact body of mass m is mounted on the rim of the pulley. A weight pan hangs from a string wound around the pulley. When a certain amount of weight is piled on the pan, the system will be in equilibrium when the radius vector from the axle to m makes an angle ϕ with the vertical. Show that, when the system is displaced by a small amount from this equilibrium, it executes simple har- monic motion with angular frequency

$$\Omega = \sqrt{\frac{2mg\cos\phi}{[M + 2m(1 + \sin\phi)]R}}.$$

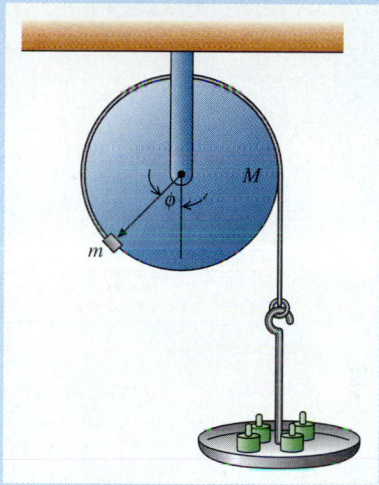

Angular Momentum

Left: The gyroscope doesn't really defy gravity!

When true simplicity is gain'd,
To bow and to bend we shan't be asham'd,
To turn turn will be our delight
'Til by turning turning we come round right.

—Traditional Shaker Song, "Simple Gifts"

Introduction

In this chapter, the last of three dealing with the mechanics of rotation, we develop the concept of *angular momentum*. Angular momentum is the most fundamental rotational quantity, in the same sense that linear momentum is the most fundamental translational quantity. That is, rotational dynamics can be based on the fundamental *principle of conservation of angular momentum*. Building a logical framework for rotational dynamics on this basis is analogous to building translational dynamics on the principle of conservation of linear momentum, as we did toward the end of Section 9.5 and in Section 9.6.

We can make connections between linear momentum and angular momentum, as we do in Section 13.2 and subsequent sections. You should bear in mind, however, that linear momentum and angular momentum are distinct, and they are equally fundamental. It is possible to be misled into thinking that angular momentum is nothing more than linear momentum considered from a new point of view, but this is emphatically not true.

In Section 13.2, we develop the concept of angular momentum by analogy with the corresponding translational quantity, linear momentum, as we have done successfully with other rotational quantities. We restrict our initial consideration to fixed-axis rotation. In Sections 13.3 and 13.4, we go on to develop the mathematical apparatus needed to treat rotational mechanics in a more general way. That apparatus centers on the mathematical operation called the *vector product*. Although mechanical rotation will be our first application of the vector product, this mathematical operation is a powerful tool for treating *all* physical processes that are inherently three-dimensional. We will use the vector product extensively in treating electromagnetism.

In Sections 13.5 and 13.6, we use the vector product to generalize many of the ideas considered in Chapters 11 and 12 and thus to develop a completely general set of laws of Newtonian mechanics. Section 13.7 treats the phenomenon called *precession*.

Angular Momentum in Fixed-Axis Rotation

In Figure 13.1, a body rotates about the frictionless fixed axis A with angular velocity ω. The moment of inertia of the body about A is I. Remembering that I is the rotational analogue of the mass M and ω is the fixed-axis rotational analogue of the translational velocity $\mathbf{v}$, let us define the **angular momentum L about the axis** A:

$$L \equiv I\omega \quad \text{for rotation about a fixed axis,} \qquad (13.1)$$

analogous to the definition of linear momentum, $\mathbf{p} \equiv m\mathbf{v}$, given by Equation 9.3.

From Equation 13.1, you can readily derive the SI unit of angular momentum, kg·m²/s. The unit has no special name.

Continuing to argue by analogy, we know that linear momentum is conserved in a system on which no net external force acts. Now, the rotational analogue of force is the torque T. Is it true that angular momentum is conserved in a system on which no net external torque acts? We have already noted (Section 11.3) that a rotating rigid body constrained by a (frictionless) fixed axis, such as the body of Figure 13.1, rotates with constant angular velocity if it is isolated—that is, if no external torque acts on it. This statement, which is represented mathematically by Equation 11.11, is the rotational analogue of Newton's first law. Because I is constant, Equation 13.1 ensures that L will be constant if ω is constant. Thus angular momentum is certainly conserved in an isolated system rotating about a fixed axis.

What happens when the body of Figure 13.1 is not isolated, but rather is acted on by an external torque T? For the sake of analogy, consider Newton's second law in the form of Equation 9.30,

$$\mathbf{F} = \frac{d\mathbf{p}}{dt};$$

(13.2)

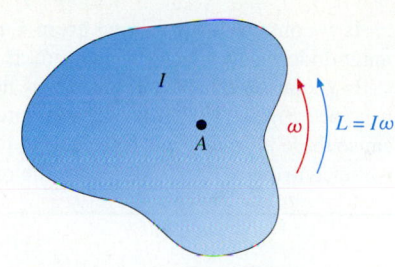

FIGURE 13.1 A rigid body rotates at angular velocity ω about the friction-less fixed axis A, which is normal to the page. The moment of inertia of the body about A is I. The angular momentum of the body is defined to be $L \equiv I\omega$.

the rate of change of linear momentum is equal to the net external force. Substituting the analogous rotational quantities—torque T for force $\mathbf{F}$, and angular momentum L for linear momentum $\mathbf{p}$—gives us

$$T = \frac{dL}{dt};$$

(13.3)

the rate of change of angular momentum is equal to the net external torque. Is this consistent with what you have already learned? To see that it is, remember how we showed in Section 9.5 that Equation 13.2 is equivalent to the familiar form $\mathbf{F} = m\mathbf{a}$:

$$\mathbf{F} = \frac{d\mathbf{p}}{dt} = \frac{d(m\mathbf{v})}{dt} = m\frac{d\mathbf{v}}{dt} = m\mathbf{a}.$$

We can do just the same thing with Equation 13.3. We have

$$T = \frac{dL}{dt} = \frac{d(I\omega)}{dt} = I\frac{d\omega}{dt} = I\alpha.$$

Eliminating the three intermediate steps leaves the familiar form of Newton's second law for fixed-axis rotation, $T = I\alpha$, which is Equation 11.20. In the absence of a net external torque, this equation reduces to $\alpha = 0$, which implies $\omega = $ constant and $L = $ constant.

EXAMPLE **13.1**

A particle of mass m is fixed to the end of a massless rod of length R and rotates in a horizontal plane with constant angular velocity ω about axis A. At the instant depicted in Figure 13.2, the velocity of the particle is $\mathbf{v}$. Express the angular momentum L of the particle about A in terms of the magnitude p of its linear momentum. Is L conserved? Is p conserved? Is $\mathbf{p}$ conserved?

SOLUTION: According to Equation 11.19, the moment of inertia of the particle about A is $I = mR^2$. You can therefore use the definition of angular momentum, Equation 13.1, to write

$$L = mR^2\omega.$$

(13.4)

Next, you regroup the factors thus:

$$L = mR\,R\omega.$$

But $R\omega$ is just v, the *speed* of the particle. So you have

$$L = mRv.$$

(13.5a)

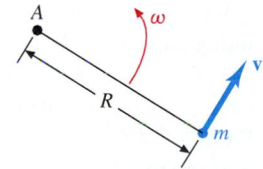

FIGURE 13.2 A particle of mass m at the end of a massless rod of length R rotates at constant angular velocity ω about axis A. The instantaneous velocity of the particle is $\mathbf{v}$.

The product mv is the magnitude p of the linear momentum. Thus Equation 13.5a can be put in the form

$$L = Rp.$$

(13.5b)

Is angular momentum conserved? Because ω is constant, $L = I\omega$ is evidently conserved. This is what you would expect for an isolated system according to the argument following Equation 13.1. (Because R is constant, the speed $v = R\omega$ is constant as well.)

Is p conserved? It follows from Equation 13.5b that p, the magnitude of the linear momentum, is constant.

Is **p** conserved? No, because **v** is not constant even though v is; see Figure 13.2. Linear momentum is not conserved because there is an external force acting on the particle. That force is the centripetal force exerted by the rod, in whose absence the particle would not move in a circle. But the *angular momentum* of the particle about the axis A is conserved nevertheless; the torque about A exerted by the centripetal force is always zero because the line of action of the centripetal force always passes through A.

The centripetal force acting on the particle of Example 13.1 belongs to the very important class called central forces. A **central force** is one whose line of action always passes through the same fixed point no matter where the body on which the force acts is located. The fixed point is called the **force center**. (In Example 13.1, the force center is located at the intersection of the axis A with the rod.) Central forces are special because their action does not affect the angular momentum of the system on which they act.

We have dealt in Example 13.1 with a special case in which the path followed by the body is a circle whose center coincides with the force center. However, you should not be misled by the words "central" and "center." A body moving under the influence of a central force does not necessarily move in a circle. For example, you will see in Chapter 14 that the earth moves in an elliptical orbit under the influence of the gravitational force exerted by the sun. The gravitational force is a central force whose center is at the sun. However, the orbit is an ellipse—not a circle—and the sun lies not at the center of the ellipse but at one of its foci.

Angular momentum is *always* conserved in an isolated system. Example 13.2 exploits this principle in a case in which energy is not conserved and the moment of inertia of the system is not constant.

EXAMPLE 13.2

In a spectacular maneuver of figure skating, a skater extends his arms and starts himself spinning on the point of one blade by pushing hard on the ice with the other blade. While spinning, he pulls his arms inward, crossing them in front of his chest, and his angular speed increases dramatically.

With his arms extended, the skater's moment of inertia about the vertical axis passing through the skate point is $I_i = 3.6$ kg·m^2. With his arms pulled in as far as possible, his moment of inertia about the same axis is $I_f = 1.1$ kg·m^2. **(a)** If his initial angular speed is $\omega_i = 0.85$ rev/s, find the skater's final angular speed. **(b)** Find the increase in his rotational kinetic energy during the maneuver. What is the source of the additional energy?

SOLUTION:
(a) Find the skater's final angular speed.

The reaction force to the skater's initial push on the ice sets him into rotation by exerting a torque on him. If we neglect the small frictional torque exerted on the skate point by the ice, there are no external torques acting on the skater once the push is completed. Thus the skater is an isolated system and his angular momentum must be conserved. You can therefore write

$$L_i = L_f,$$

or

$$I_i\omega_i = I_f\omega_f.$$

You find the skater's final angular speed by solving for ω_f:

$$\omega_f = \frac{I_i}{I_f}\,\omega_i = \frac{3.6 \text{ kg·m}^2}{1.1 \text{ kg·m}^2}\,\omega_i = 3.3\omega_i.$$

That is, his angular speed increases more than threefold as a consequence of the reduction of his moment of inertia by the same factor. You have

$$\omega_f = 3.3 \times 0.85 \text{ rev/s} = 2.8 \text{ rev/s}.$$

The speedup is dramatic in spite of the fact that the skater's arms account for only a fairly small part (typically about 12%) of his total mass. But the radius of gyration of the extended arms is considerably greater than that of the rest of the body. Because it is the square of the radius of gyration that enters into the moment of inertia ($I = m\gamma^2$), the proportion I_i/I_f is large, as the preceding calculation confirms.

(b) Find the increase in the skater's rotational kinetic energy during the maneuver. What is the source of the additional energy?

In calculating the initial and final rotational kinetic energies, you must express the ω's in radians per second. The skater's initial kinetic energy (Equation 11.35) is

$$K_i = \tfrac{1}{2}I_i\omega_i^2$$
$$= \tfrac{1}{2} \times 3.6 \text{ kg·m}^2 \times (2\pi \text{ rad/rev} \times 0.85 \text{ rev/s})^2$$
$$= 51 \text{ J}.$$

His final kinetic energy is

$$K_f = \tfrac{1}{2}I_f\omega_f{}^2$$
$$= \tfrac{1}{2} \times 1.1 \text{ kg·m}^2 \times (2\pi \text{ rad/rev} \times 2.8 \text{ rev/s})^2$$
$$= 170 \text{ J}.$$

Taking the difference, you find that the skater's kinetic energy increases by the amount

$$\Delta K = K_f - K_i = 120 \text{ J}.$$

Where does the additional energy come from? Work is done by the centripetal force that the skater exerts on his arms to pull them in. The potential energy of the system does not change. If we neglect friction, no other work is done on the system. According to the work-energy theorem, the work done by the skater (which happens to be internal work) must appear in the form of rotational kinetic energy. Which conservation principle dictates that the work not appear in the form of translational kinetic energy?

Angular Impulse

We now consider the behavior of a rotating system when it is not isolated—that is, when an external torque is exerted on it and angular momentum is *not* conserved. Again we proceed by analogy, this time to Equation 9.1, which defined translational impulse **J** in terms of the force **F**(t) applied to a system over a time interval beginning at t_i and ending at t_f:

$$\mathbf{J} \equiv \int_{t_i}^{t_f} \mathbf{F}(t)\, dt.$$

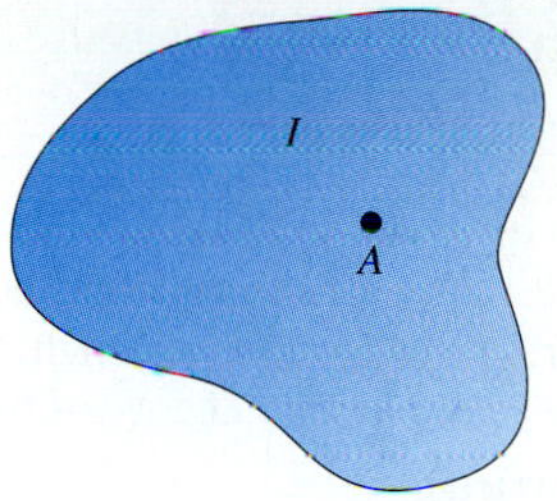

Suppose that a torque $T(t)$ is applied to the system of Figure 13.3 over a time interval beginning at t_i and ending at t_f. We define the **angular impulse** G:

$$G \equiv \int_{t_i}^{t_f} T(t)\, dt. \tag{13.6}$$

FIGURE 13.3 A system consists of a rigid body constrained by a fixed axis A, which is normal to the page.

The SI unit of angular impulse follows directly from this definition; it is N·m·s.

Translational impulse is of interest because it is related to the change of momentum of the system to which it is imparted. According to the impulse-momentum theorem (Equation 9.4b), we have $\mathbf{J} = \Delta\mathbf{p}$. We are aiming at an analogous result for rotation. With this aim in mind, we rewrite Equation 13.3 in the form

$$T(t)\, dt = dL(t)$$

and substitute this value into Equation 13.6 to obtain

$$G = \int_{t_i}^{t_f} T(t)\, dt = \int_{t_i}^{t_f} dL(t).$$

The expression following the second equals sign integrates immediately, and we have the result

$$G = L_f - L_i, \tag{13.7a}$$

or

$$G = \Delta L; \tag{13.7b}$$

the angular impulse imparted to a system is equal to the change in its angular momentum. This is the time-integral form of Newton's second law for fixed-axis rotation. It is the rotational analogue that we sought to the impulse-momentum theorem, $\mathbf{J} = \Delta\mathbf{p}$.

By using Equation 13.1, $L = I\omega$, we can express Equation 13.7a in the alternative form

$$G = I_f\omega_f - I_i\omega_i. \tag{13.7c}$$

EXAMPLE 13.3

Two identical flywheels, each having moment of inertia I, are mounted on shafts having a common axis, as shown in Figure 13.4. The shafts are mounted on bearings having negligible friction. A clutch of negligible mass, mounted between the shaft ends, can be engaged so as to connect the flywheels. With the clutch disengaged, flywheel 1 has angular velocity ω_i and flywheel 2 is at rest. The clutch is engaged, slipping briefly until the two shafts have a common angular velocity ω_f. Find **(a)** the

FIGURE 13.4

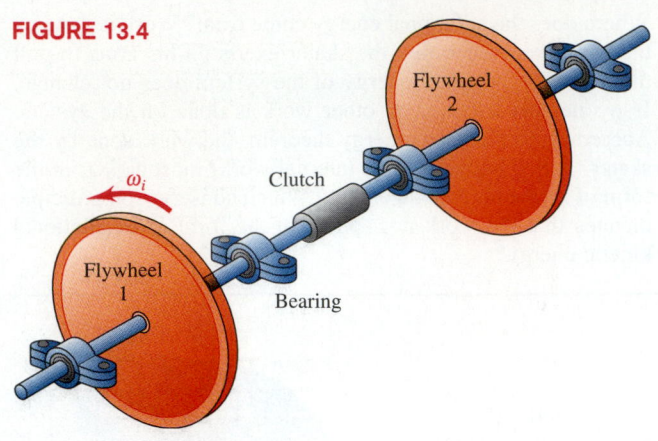

Flywheel 2

Clutch

ω_i

Flywheel 1

Bearing

value of ω_f in terms of ω_i, **(b)** the angular impulse $G_{c\to 2}$ imparted by the clutch to shaft 2, **(c)** the angular impulse $G_{c\to 1}$ imparted by the clutch to shaft 1, and **(d)** the ratio of the final to the initial kinetic energy of the system.

SOLUTION:

(a) Express ω_f in terms of ω_i.

Because the flywheels have a common axis, it is possible to determine the angular momentum L of the entire system simply by adding the angular momenta L_1 and L_2 of the two flywheels:

$$L = L_1 + L_2.$$

If the friction in the bearings is negligible, the system is isolated. The clutch is mounted on the shafts and is not otherwise connected mechanically to the outside world. Hence, when the clutch is engaged, the torques that it exerts on the two shafts are *internal* to the system and cannot change its angular momentum. You can therefore use the angular-momentum conservation principle to equate the initial and final angular momenta of the system: $L_i = L_f$. This equality gives you

$$I\omega_i + 0 = 2I\,\omega_f.$$

The solution for ω_f is

$$\omega_f = \frac{\omega_i}{2}.$$

(b) Find $G_{c\to 2}$.

The change in angular momentum of flywheel 2 is

$$\Delta L_2 = I(\omega_f - 0) = \tfrac{1}{2}I\omega_i. \tag{13.8}$$

According to Equation 13.7, this change must be equal to the angular impulse that produced it. So you have

$$G_{c\to 2} = \tfrac{1}{2}I\omega_i. \tag{13.9}$$

(c) Find $G_{c\to 1}$.

You proceed as in part **b** to find the change in angular momentum of flywheel 1:

$$\Delta L_1 = I(\omega_f - \omega_i) = -\tfrac{1}{2}I\omega_i. \tag{13.10}$$

The minus sign implies that the angular momentum of flywheel 1 has decreased. Engagement of the clutch has resulted in the transfer of half the angular momentum of the system from flywheel 1 to flywheel 2. Again using Equation 13.7, you obtain

$$G_{c\to 1} = -\tfrac{1}{2}I\omega_i. \tag{13.11}$$

The angular impulses are equal and opposite.

(d) Find K_f/K_i.

The initial energy of the system is $K_i = \tfrac{1}{2}I\omega_i^2$. The final kinetic energy is

$$K_f = \tfrac{1}{2}(2I)\omega_f^2 = I(\tfrac{1}{2}\omega_i)^2 = \tfrac{1}{4}I\omega_i^2.$$

Hence you have

$$K_f/K_i = 1/2.$$

That is, one-half of the initial energy of the system is dissipated by the friction in the clutch. This result does not depend on the rate at which the clutch is applied; compare it with the result of Problem 12.63.

The process described in this example is the rotational analogue of a totally inelastic collision between two bodies of equal mass. Indeed, the action of the clutch in linking the two wheels together is analogous to the action of the magnets or latches that can be used to make two air-track gliders stick together in a totally inelastic collision. Are you surprised that the energy loss in this example is 50%? If you are, compare the result with Equation 10.4, in the special case $m_1 = m_2$.

Newton's Third Law for Rotation

In Example 13.3, Equations 13.9 and 13.11 illustrate a significant point that you may have taken for granted because of the close analogy with the translational situation. *When one body imparts an angular impulse to a second body, the second imparts an equal and opposite angular impulse to the first.* We can express this relation by means of the equation

$$G_{1\to 2} = -G_{2\to 1}. \tag{13.12}$$

(The fact that the clutch of Example 13.3 acts as an intermediary is not important.)

Equation 13.12 can be derived from the principle of conservation of angular momentum. Consider an isolated system consisting of subsystems 1 and 2, like the two flywheels of Example 13.3. The total angular momentum L of the system cannot change, though its parts L_1 and L_2 may change. So we have

$$0 = \Delta L = \Delta L_1 + \Delta L_2. \tag{13.13}$$

This immediately yields

$$\Delta L_2 = -\Delta L_1. \tag{13.14}$$

But applying Equation 13.7 to the present situation gives $\Delta L_2 = G_{1\rightarrow 2}$ and $\Delta L_1 = G_{2\rightarrow 1}$. Substituting these values into Equation 13.14, we obtain Equation 13.12. This completes the derivation.

The principle enunciated by Equation 13.12 may be expressed in terms of torque. To do this, use Equation 13.6 to write

$$\int_{t_i}^{t_f} T_{1\rightarrow 2}(t)\, dt = -\int_{t_i}^{t_f} T_{2\rightarrow 1}(t)\, dt. \tag{13.15a}$$

Next, differentiate both sides with respect to time. This operation undoes the integration. At any instant t, we have

$$T_{1\rightarrow 2} = -T_{2\rightarrow 1}; \tag{13.15b}$$

when one body exerts a torque on a second body, the second body exerts an equal and opposite torque on the first. This result is called **Newton's third law for rotation**. It is analogous to Newton's third law for translation, $\mathbf{F}_{1\rightarrow 2} = -\mathbf{F}_{2\rightarrow 1}$.

The Direction of Rotational Vector Quantities*

We now lay the groundwork for extending our study of rotation beyond fixed-axis cases. Our approach is guided by analogy to what we did in Chapter 4, where we extended our study of translational mechanics from one dimension to two and three dimensions.

What is needed? In one-dimensional space, there are only two possible directions, and so the two possible signs of an algebraic quantity suffice to describe translational motion. Sometimes, no further information is needed. For example, in studying collisions between two gliders on a level air track, it is immaterial whether the air track is aligned north–south or east–west. But, even for the one-dimensional case, an observer in three-dimensional space must often have additional information concerning the orientation of the coordinate axis. For example, the observer must know the incline angle of a plane to calculate the acceleration of a block sliding on it.

Analogously, in fixed-axis rotation, the two possible signs of an algebraic quantity suffice to distinguish between the two possible senses of rotation. As long as the orientation of the axis is either immaterial or obvious, no further information is needed. That is to say, we are able to treat fixed-axis rotation in a one-dimensional mathematical context, even though the physical phenomenon of rotation can never take place in a one-dimensional space.

But suppose we want to study the system of Figure 13.5. Two tops spin on a glass table. At the instant shown, the axes of the tops are mutually perpendicular. What is the total angular momentum of the two-top system? The senses of rotation alone cannot provide a basis for an answer. Indeed, the senses are ambiguous. Observer A, looking vertically downward, might maintain—at least loosely—that the sense of rotation of top 1 is negative (clockwise) and that of top 2 positive (counterclockwise). But observer B, looking vertically upward from beneath the table, will maintain the opposite. For her, top 1 is rotating counterclockwise while top 2 is rotating clockwise.

---

*Study of Sections 13.3 and 13.4 may be deferred until the cross product is needed in Chapter 27. The remaining material in this chapter (Sections 13.5 through 13.7) may be omitted without loss of continuity, provided that the proof in Section 14.3 of Kepler's second law (which depends on material in Sections 13.3 through 13.5) is skipped.

FIGURE 13.5 Two tops spin on a
glass table. The sense of rotation of
each top is indicated by a curved arrow. At the instant shown, the axes
of the tops are mutually perpendicular. Observer *A*, looking straight
down, calls the sense of rotation of
top 1 negative and that of top 2 positive. Observer *B*, looking straight up
from beneath the table, asserts the
opposite. Observer *C*, looking along
the axis of top 1, can make no statement as to the sense of rotation of
top 2.

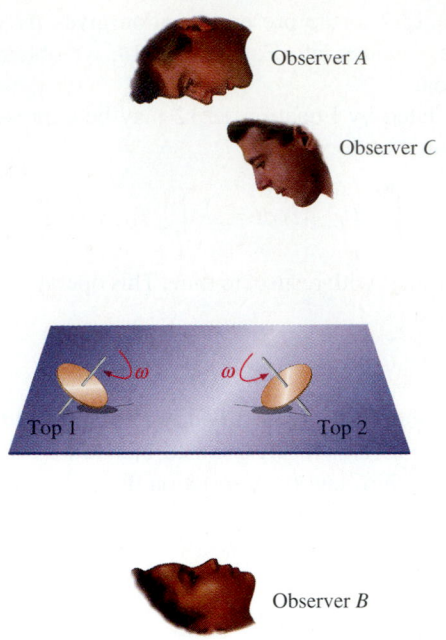

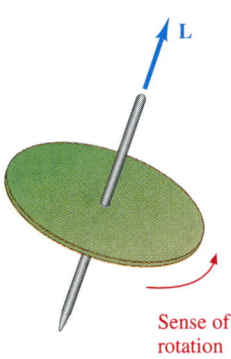

FIGURE 13.6 The sense of rotation
of the top is indicated by the curved
red arrow. The magnitude and orientation of the angular momentum are
denoted unambiguously by the vector **L** drawn along the rotation axis.
At this point in our development of
rotation vectors, however, the direction shown for **L** is only one of two
possibilities.

For observer *C*, looking along the axis of top 1, the situation is still worse. He can
agree with *A* that top 1 is rotating clockwise. But the axis of top 2 is perpendicular to
his line of vision, and he cannot meaningfully assign either the term "clockwise" or
the term "counterclockwise" to its rotation.

Just as we did in multidimensional translational mechanics, we use the more powerful directional properties of vectors to extend our treatment of rotational mechanics.
To describe the angular momentum of a top, for example, we use the magnitude of a
vector to describe the magnitude of the angular momentum and the direction of the
vector to describe the orientation of the rotation axis. That is, the angular momentum
is directed along the rotation axis, as shown in Figure 13.6.

It may seem strange to you at first that a motional quantity like angular momentum
is denoted by a vector whose direction is the only one along which motion is *not*
occurring. But that is precisely why the axial direction is so useful. As noted in Section
11.2, the parts of a rotating rigid body move in all possible directions normal to the
axis. Hence a vector in the plane of rotation *cannot* be used to specify the rotation. But
the parts of the body all share a common rotation about a common axis. Once the
orientation of the axis is specified, the only information still to be given is the sense of
rotation about the axis, out of the two possible. We now describe a universal convention
for expressing that sense unambiguously for all observers, in terms of the direction of
the vector.

The Right-Hand Rule

In Figure 13.7a, a set of cartesian axes has been embedded in the spinning top of Figure
13.6. The *z* axis has been chosen to coincide with the rotation axis, and the *x* and *y*
axes turn with the top. The three axes have been specified so that they constitute a
right-handed coordinate system. The reason for the name is made evident in Figure
13.7b, which shows a right hand. The coordinate system is embedded in the hand in
such a way that the sense of rotation from $\hat{x}$ to $\hat{y}$ is the sense in which the fingers point,
and the thumb points along $\hat{z}$.

There is an equivalent way of visualizing a right-handed coordinate system, one that
some persons find easier to remember. In Figure 13.7c, a coordinate system is embedded
in a right-handed (standard) wood screw. The *x* and *y* axes lie in the head, and the *z*
axis lies along the shank with $\hat{z}$ directed from the head toward the point. Turning the

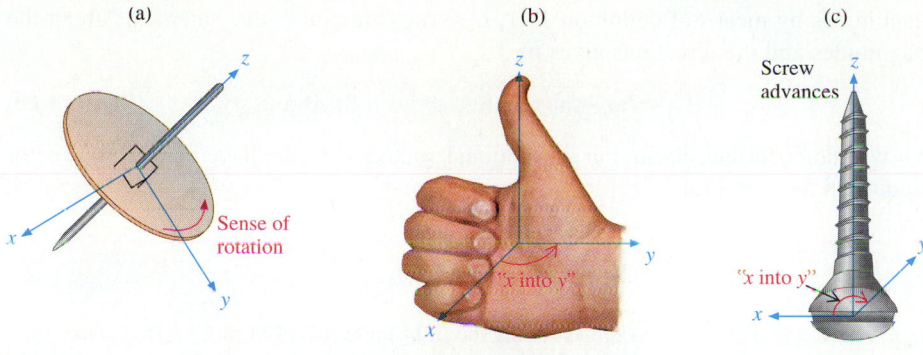

FIGURE 13.7 (*a*) Establishing a right-handed coordinate system in the top. (*b*) The right-hand rule for ordering the three axes. The fingers curl along the sense of rotation from $\hat{\mathbf{x}}$ toward $\hat{\mathbf{y}}$. (*c*) Using a right-handed screw for the same purpose. Rotation of the screw head in the sense from $\hat{\mathbf{x}}$ toward $\hat{\mathbf{y}}$ makes the screw advance "into the wood," in the $\hat{\mathbf{z}}$ direction.

screw so that $\hat{\mathbf{x}}$ rotates toward $\hat{\mathbf{y}}$ results in the advance of the screw into the wood, in the positive z direction. (Note the connection with the fingers-and-thumb picture just described. If you could grab the screw and turn it in the sense of your fingers, it would advance in the direction of your thumb.)

Imagine that you are driving the screw into a wooden board. Normally the screwdriver and the screw lie between you and the board. You turn the screwdriver to drive the screw away from you, into the wood. Under these circumstances, it is reasonable to speak of driving the screw by turning it clockwise. But sometimes it is necessary to drive a screw into the back side of a board, where restricted space makes it impossible for you to get behind it. Then you stand in front of the board with your arms around its sides; the board is between you and the screw. In order to drive the screw into the wood—now toward you—you must turn it counterclockwise!

Unlike the "clockwise-counterclockwise" convention, the right-hand rule is the same for all observers, no matter what their location or orientation. Turning the screw in the sense from $\hat{\mathbf{x}}$ toward $\hat{\mathbf{y}}$ *always* makes the screw advance into the wood, in the $\hat{\mathbf{z}}$ direction.

The direction of a rotation vector, such as **L** in Figure 13.8, is chosen in accord with the **right-hand rule**. If the sense of rotation is that from $\hat{\mathbf{x}}$ toward $\hat{\mathbf{y}}$, as in Figure 13.8a, **L** points in the positive z direction. That is, $\hat{\mathbf{L}} = \hat{\mathbf{z}}$. If the sense of rotation is from $\hat{\mathbf{y}}$ toward $\hat{\mathbf{x}}$, as in Figure 13.8b, **L** points in the negative z direction. That is, $\hat{\mathbf{L}} = -\hat{\mathbf{z}}$. You can verify these statements by orienting your right hand with the thumb along the positive z axis, as shown in Figure 13.7b.

Although we have described the right-hand rule in terms of the angular momentum vector **L**, the same convention applies to the angular velocity vector $\boldsymbol{\omega}$, the angular acceleration vector $\boldsymbol{\alpha}$, and the torque vector **T**, among other quantities. In both parts of Figure 13.8, for example, the direction of $\boldsymbol{\omega}$ is the same as that of **L**; $\hat{\boldsymbol{\omega}} = \hat{\mathbf{L}}$. Indeed, we can immediately write a relation between **L** and $\boldsymbol{\omega}$. We have already related their

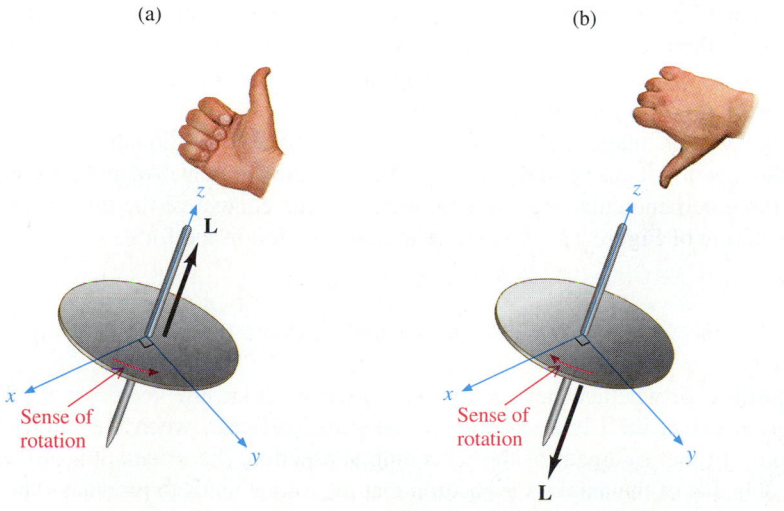

FIGURE 13.8 The direction of **L** established uniquely by the right-hand rule. (*a*) The sense of rotation is from $\hat{\mathbf{x}}$ toward $\hat{\mathbf{y}}$. Consequently, $\hat{\mathbf{L}} = \hat{\mathbf{z}}$. (*b*) The sense of rotation is that from $\hat{\mathbf{y}}$ toward $\hat{\mathbf{x}}$. Consequently, $\hat{\mathbf{L}} = -\hat{\mathbf{z}}$.

magnitudes by means of definition 13.1, $L \equiv I\omega$. Combining the statements about the magnitudes and the directions gives us

$$\boxed{L \equiv I\omega \quad \text{for rotation about a fixed axis.}} \tag{13.16}$$

We will defer further discussion of rotational vectors until we have defined the vector product in Section 13.4.

EXAMPLE 13.4

What is the direction of the angular velocity vector that describes the rotation of the earth about its own axis?

SOLUTION: The earth rotates "from west to east." That is, the sense of rotation is such as to move New York toward Barcelona, Melbourne toward Auckland, and Shanghai toward Los Angeles. Using the right-hand rule of Figure 13.7b, you see that, if your fingers curl from west to east, your thumb points north. So the angular velocity vector points northward along the earth's axis. Alternatively, if you imagine a right-handed screw fixed in the earth and turning with the earth, it advances northward.

The Vector (Cross) Product

(a)

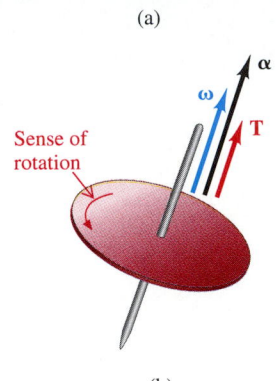

Sense of rotation

(b)

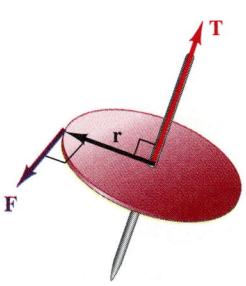

FIGURE 13.9 (a) A top rotates with increasing angular velocity ω. The direction of ω does not change. The angular acceleration $\alpha = d\omega/dt$ must thus have the same direction as ω. The torque responsible for α, $T = I\alpha$, must also have the same direction. (b) A simple way of producing the torque T of part a. The force F is applied perpendicular to r, which is perpendicular to the top axis $\hat{T}$. These three essential quantities cannot be represented in fewer than three dimensions.

Rotation is an inherently three-dimensional process. Let us explore this point, using the top shown in Figure 13.9 as an example. The top is initially at rest and is then accelerated to some final angular velocity ω without change of the orientation of its axis. Because $\hat{\omega}$ does not change, the angular acceleration vector, $\alpha = d\omega/dt$, must have the same direction as ω. The direction of ω and α is that of the axis, as shown in Figure 13.9a.

The angular acceleration is the result of an applied torque. The torque also must be represented by a vector because it is proportional to the vector α. The proportionality constant is the moment of inertia I, a positive scalar:

$$T = I\alpha. \tag{13.17}$$

This is Newton's second law for rotation expressed in vectorial form. It is evident from the equation that the direction of T must be the same as that of α, as Figure 13.9 shows.

The torque is the result of the application of a force F to the top at a point whose location is specified by the position vector r. A given torque T can result from many different combinations of r and F, but Figure 13.9b shows a particularly simple one. The point of application of the force is on the periphery of the top and is specified by a position vector r, which is perpendicular to the axial direction $\hat{T}$. The force F is tangent to the periphery and perpendicular to r.

The inherent three-dimensionality of the simple dynamical process of accelerating a top is now evident. The vectors r, F, and T are mutually perpendicular and cannot be represented in fewer than three dimensions. (Even in a more complicated case in which these three vectors are not mutually perpendicular, they still span three-dimensional space.) We need a quantitative way of expressing the relation among the three quantities, both as to magnitude and as to direction.

We dealt with magnitude in Section 11.3. According to Equation 11.13, $|T| = |rF_\perp|$, the magnitude of the torque is the product of the moment arm and the component of the force perpendicular to the moment arm. In the context of the three-dimensional representation of Figure 13.10, we rewrite this equation in the form

$$|T| = r \sin \phi \, F, \tag{13.18}$$

where ϕ is the (smaller) angle between $\hat{r}$ and $\hat{F}$. (Note that $\sin \phi$ is always positive because $0° \le \phi \le 180°$.)

Regardless of whether they are mutually perpendicular, the vectors r and F define a plane except in the trivial case where they are collinear, when the torque is zero. According to the argument of the preceding subsection, the orientation of the torque vector T is that of the axis of the rotation that the torque tends to produce—the normal

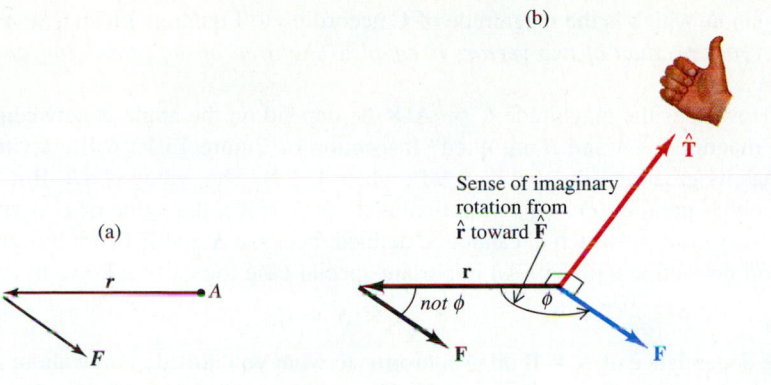

(a)

(b)

Sense of imaginary
rotation from
$\hat{\mathbf{r}}$ toward $\hat{\mathbf{F}}$

$\hat{\mathbf{T}}$

$\mathbf{r}$

not ϕ

ϕ

$\mathbf{F}$

$\mathbf{F}$

r

$\bullet A$

$\mathbf{F}$

to the plane defined by **r** and **F**. This is made evident in Figure 13.10b. The remaining ambiguity of direction (in which of two possible ways does **T** point?) is resolved by means of the right-hand rule. (As Figure 13.10b shows, **r** and **F** do not have to be mutually perpendicular in order to imagine a rotation from $\hat{\mathbf{r}}$ toward $\hat{\mathbf{F}}$ through the smaller angle between them.) We call a normal whose direction is assigned in accordance with the right-hand rule a **right-handed normal**.

The operations described in the two preceding paragraphs establish the magnitude and direction of the vector **T** in terms of the vectors **r** and **F**. Taken together, the operations make up a *vector operation* called the **vector product**. The vector product is represented by the notation

$$\mathbf{T} \equiv \mathbf{r} \times \mathbf{F}. \qquad (13.19)$$

Because the boldface multiplication sign × is universally used in modern notation to represent the vector product, it is familiarly (and most often) called the **cross product**. "Vector product" and "cross product" mean exactly the same thing. In speaking of **r** × **F**, we say "**r** cross **F**."

To summarize, *the magnitude of the vector (or cross) product* **r** × **F** *is* $|\mathbf{r} \times \mathbf{F}| = r \sin \phi \, F$, *and its direction* $\hat{\mathbf{T}}$ *is that of the right-handed normal to the plane defined by* **r** *and* **F**.

A cross product can be defined for *any* two vectors **A** and **B**, as shown in Figure 13.11. In general, it is necessary for clarity to move **A** and **B** until they lie tail to tail and thus define a plane, as shown in Figure 13.11b. This can always be done, because moving a vector without disturbing its magnitude and direction does not change the vector. The vector **C** shown in Figure 13.11b is the cross product

$$\mathbf{C} = \mathbf{A} \times \mathbf{B}. \qquad (13.20)$$

By definition, the magnitude of **C** is

$$C = |\mathbf{C}| = A \sin \phi \, B; \qquad (13.21)$$

the direction $\hat{\mathbf{C}}$ is that of the right-handed normal to the plane defined by **A** and **B**.

For unit vectors $\hat{\mathbf{A}}$ and $\hat{\mathbf{B}}$, the cross product is

$$\hat{\mathbf{A}} \times \hat{\mathbf{B}} = \mathbf{C} = \hat{\mathbf{C}} \sin \phi. \qquad (13.22)$$

Because each of the two unit vectors has magnitude 1, the magnitude of their cross product is 1 (sin ϕ) 1 = sin ϕ. The direction is determined just as for nonunit vectors.

Properties of the Cross Product

The magnitude of the cross product can be visualized in a simple geometric way, shown in Figure 13.12. Using **A** and **B** as sides, we construct a parallelogram. If we consider **A** to be the base of the parallelogram, its altitude is $B \sin \phi$. If instead we consider **B** to be the base, the altitude is $A \sin \phi$. Either way, the area of the parallelogram is

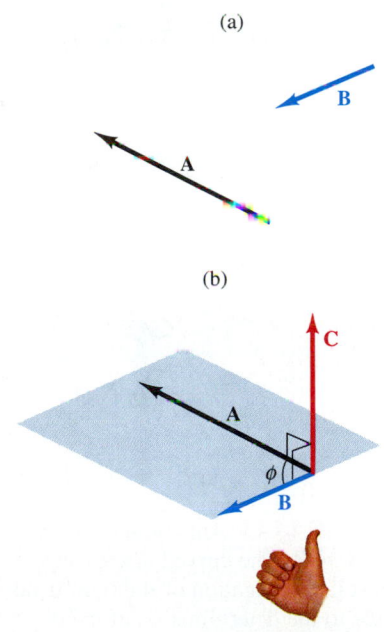

(a)

B

A

(b)

C

A

ϕ

B

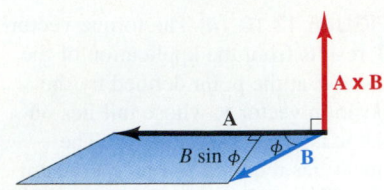

FIGURE 13.12 Construction to show that $|\mathbf{A} \times \mathbf{B}|$ is equal to the area of the parallelogram determined by $\mathbf{A}$ and $\mathbf{B}$.

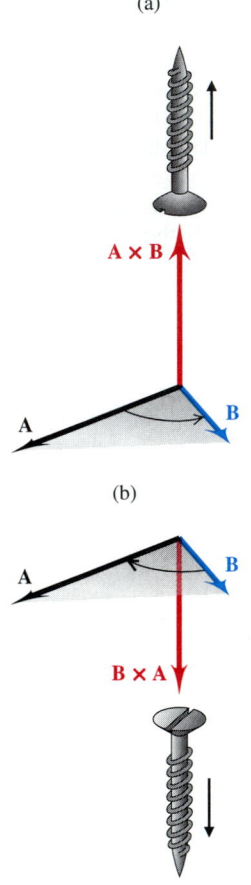

(a)

(b)

FIGURE 13.13 Anticommutativity of $\mathbf{A} \times \mathbf{B}$. The curved arrows suggest the application of the right-hand rule to the determination of the direction of (a) $\mathbf{A} \times \mathbf{B}$ and (b) $\mathbf{B} \times \mathbf{A}$. The screws are shown as aids.

$AB \sin \phi$, which is the magnitude of $\mathbf{C}$ according to Equation 13.21. *The magnitude of the cross product of two vectors is equal to the area of the parallelogram defined by them.*

How does the magnitude $C = |\mathbf{A} \times \mathbf{B}|$ depend on the angle ϕ between $\mathbf{A}$ and $\mathbf{B}$ if the magnitudes A and B are fixed? Inspection of Figure 13.12 will convince you that C has its greatest value for $\phi = 90°$, when $\mathbf{A} \perp \mathbf{B}$. That value is AB. But, if $\mathbf{A}$ and $\mathbf{B}$ are either parallel ($\phi = 0°$) or antiparallel ($\phi = 180°$), the value of C is zero. (This is the very case for which $\hat{\mathbf{C}}$ cannot be defined, because $\mathbf{A}$ and $\mathbf{B}$ lie on the same line and so do not define a plane.) An important special case logically follows immediately:

$$\mathbf{A} \times \mathbf{A} = 0. \tag{13.23}$$

The dependence of $\mathbf{A} \times \mathbf{B}$ on ϕ conforms to what you already know about the physical properties of the torque $\mathbf{T} = \mathbf{r} \times \mathbf{F}$. The torque is greatest when the force is perpendicular to the moment arm, whose direction is $\hat{\mathbf{r}}$. When the force lies along the moment arm, the torque is zero.

Figure 13.13 illustrates a remarkable feature of the cross product—its *anticommutativity*. Figure 13.13a shows the cross product $\mathbf{A} \times \mathbf{B}$. Figure 13.13b shows the cross product $\mathbf{B} \times \mathbf{A}$. The magnitudes are the same, both being equal to the area of the parallelogram whose sides are $\mathbf{A}$ and $\mathbf{B}$. But *the directions are opposite*; that is,

$$\mathbf{B} \times \mathbf{A} = -\mathbf{A} \times \mathbf{B}. \tag{13.24}$$

The cross product is said to be **anticommutative** because this property is contrary to the more familiar property of scalars, $ab = ba$, called the *commutative law*.

We state here, without proof, some other important properties of the cross product:

Distributivity: $\qquad\qquad \mathbf{A} \times (\mathbf{B} + \mathbf{C}) = \mathbf{A} \times \mathbf{B} + \mathbf{A} \times \mathbf{C}. \tag{13.25}$

Nonassociativity: $\qquad\qquad (\mathbf{A} \times \mathbf{B}) \times \mathbf{C} \neq \mathbf{A} \times (\mathbf{B} \times \mathbf{C}). \tag{13.26}$

Chain rule for differentiation with respect to a variable q: $\qquad \dfrac{d}{dq}(\mathbf{A} \times \mathbf{B}) = \dfrac{d\mathbf{A}}{dq} \times \mathbf{B} + \mathbf{A} \times \dfrac{d\mathbf{B}}{dq}. \tag{13.27}$

(In further manipulations of such a derivative, beware of the anticommutativity of the cross products.)

Algebraic Representation of the Vector Product

As we have done for other vector operations, we now represent $\mathbf{A} \times \mathbf{B}$ in algebraic form. We begin by writing $\mathbf{A}$ and $\mathbf{B}$ as sums of their component vectors:

$$\mathbf{A} \times \mathbf{B} = (A_x\hat{\mathbf{x}} + A_y\hat{\mathbf{y}} + A_z\hat{\mathbf{z}}) \times (B_x\hat{\mathbf{x}} + B_y\hat{\mathbf{y}} + B_z\hat{\mathbf{z}}).$$

Next, we multiply out component by component, using the distributive law (Equation 13.25) and the commutative law for scalars. This tedious but straightforward task yields

$$
\begin{aligned}
\mathbf{A} \times \mathbf{B} = \quad & A_xB_x\,\hat{\mathbf{x}} \times \hat{\mathbf{x}} + A_xB_y\,\hat{\mathbf{x}} \times \hat{\mathbf{y}} + A_xB_z\,\hat{\mathbf{x}} \times \hat{\mathbf{z}} \\
+ \ & A_yB_x\,\hat{\mathbf{y}} \times \hat{\mathbf{x}} + A_yB_y\,\hat{\mathbf{y}} \times \hat{\mathbf{y}} + A_yB_z\,\hat{\mathbf{y}} \times \hat{\mathbf{z}} \\
+ \ & A_zB_x\,\hat{\mathbf{z}} \times \hat{\mathbf{x}} + A_zB_y\,\hat{\mathbf{z}} \times \hat{\mathbf{y}} + A_zB_z\,\hat{\mathbf{z}} \times \hat{\mathbf{z}}.
\end{aligned} \tag{13.28}
$$

This sum contains all nine possible cross products of pairs of the unit vectors $\hat{\mathbf{x}}$, $\hat{\mathbf{y}}$, and $\hat{\mathbf{z}}$. Of these, three are cross products of a unit vector with itself and are equal to zero, according to Equation 13.23. The values of the other six can be found by referring to Figure 13.14. Three of the six constitute the **cyclic rule**

$$\hat{\mathbf{x}} \times \hat{\mathbf{y}} = \hat{\mathbf{z}}, \quad \hat{\mathbf{y}} \times \hat{\mathbf{z}} = \hat{\mathbf{x}}, \quad \hat{\mathbf{z}} \times \hat{\mathbf{x}} = \hat{\mathbf{y}}. \tag{13.29a,b,c}$$

The other three can be found by combining the cyclic rule with the anticommutative law (Equation 13.24); for example, $\hat{\mathbf{y}} \times \hat{\mathbf{x}} = -\hat{\mathbf{z}}$.

When we apply these results to Equation 13.28, we obtain

$$\mathbf{A} \times \mathbf{B} = \hat{\mathbf{x}}(A_yB_z - A_zB_y) + \hat{\mathbf{y}}(A_zB_x - A_xB_z) + \hat{\mathbf{z}}(A_xB_y - A_yB_x). \tag{13.30a}$$

You should not waste time trying to memorize this relation. You will soon learn it as you use it, because its cyclic properties are so remarkable. However, it is simpler to visualize and easier to use if it is expressed as a determinant;

$$\mathbf{A} \times \mathbf{B} = \begin{vmatrix} \hat{\mathbf{x}} & \hat{\mathbf{y}} & \hat{\mathbf{z}} \\ A_x & A_y & A_z \\ B_x & B_y & B_z \end{vmatrix}.$$

(13.30b)

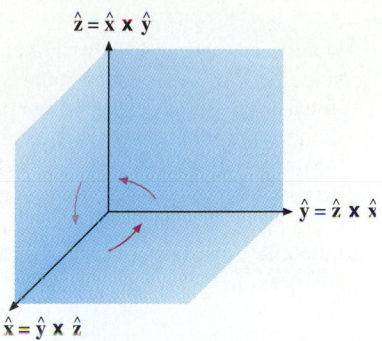

FIGURE 13.14 The cyclic rule for cross products of unit vectors. The three curved arrows suggest the rotations leading to the results $\hat{\mathbf{x}} \times \hat{\mathbf{y}} = \hat{\mathbf{z}}$, $\hat{\mathbf{y}} \times \hat{\mathbf{z}} = \hat{\mathbf{x}}$, and $\hat{\mathbf{z}} \times \hat{\mathbf{x}} = \hat{\mathbf{y}}$.

EXAMPLE 13.5

(a) Find the magnitude and direction of the torque exerted on a rigid body about the origin (0, 0, 0) of a cartesian coordinate system embedded in the body when a force $\mathbf{F} = (2, 3, 5)$ N is applied at the point $\mathbf{r} = (4, 1, -3)$ m, as shown in Figure 13.15.
(b) What is the angle ϕ between $\mathbf{r}$ and $\mathbf{F}$?

SOLUTION:
(a) Find $\mathbf{T}$.

Equation 13.19 gives the torque as $\mathbf{T} = \mathbf{r} \times \mathbf{F}$. If you wish to use the determinant form of Equation 13.30 (13.30b) to evaluate the cross product, you set $\mathbf{r} = \mathbf{A}$ and $\mathbf{F} = \mathbf{B}$, and you substitute in the values of the components of $\mathbf{r}$ and $\mathbf{F}$ to obtain

$$\mathbf{T} = \begin{vmatrix} \hat{\mathbf{x}} & \hat{\mathbf{y}} & \hat{\mathbf{z}} \\ 4 \text{ m} & 1 \text{ m} & -3 \text{ m} \\ 2 \text{ N} & 3 \text{ N} & 5 \text{ N} \end{vmatrix}.$$

By evaluating this determinant or by using Equation 13.30a, you obtain

$$\begin{aligned} \mathbf{T} = \quad & \hat{\mathbf{x}}[1 \text{ m} \times 5 \text{ N} - (-3 \text{ m}) \times 3 \text{ N}] \\ + \; & \hat{\mathbf{y}}[(-3 \text{ m}) \times 2 \text{ N} - 4 \text{ m} \times 5 \text{ N}] \\ + \; & \hat{\mathbf{z}}[4 \text{ m} \times 3 \text{ N} - 1 \text{ m} \times 2 \text{ N}], \end{aligned}$$

which boils down to

$$\mathbf{T} = (14\hat{\mathbf{x}} - 26\hat{\mathbf{y}} + 10\hat{\mathbf{z}}) \text{ m·N} = (14, -26, 10) \text{ N·m}.$$

This torque is shown in Figure 13.15.

(b) What is the angle ϕ between $\mathbf{r}$ and $\mathbf{F}$?

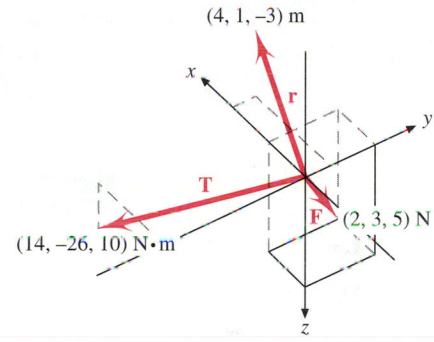

FIGURE 13.15 The scale of $\mathbf{r}$ is 1:1. The scale of $\mathbf{F}$ is 1 N:1 m, and the scale of $\mathbf{T}$ is 4 N·m:1 m. Because of the perspective of the drawing, the scales are not uniform, but apparent directions are preserved.

Because you have evaluated $\mathbf{T}$, it is convenient to use Equation 13.18, $T = r \sin \phi F$, to find ϕ. Using Pythagoras's theorem to evaluate the magnitudes of the three vectors, you have

$$\begin{aligned} \phi &= \sin^{-1} \frac{T}{rF} \\ &= \sin^{-1} \sqrt{\frac{14^2 + (-26)^2 + 10^2}{[4^2 + 1^2 + (-3)^2](2^2 + 3^2 + 5^2)}} \\ &= 83°. \end{aligned}$$

THINKING LIKE A PHYSICIST

We have defined the cross product of two arbitrary vectors by generalizing the relation between a torque $\mathbf{T}$, on the one hand, and the moment arm $\mathbf{r}$ and force $\mathbf{F}$ that give rise to it, on the other. In this chapter, you will see the cross product applied to other rotational quantities as well. The inherent mathematical three-dimensionality of the cross product makes it "natural" for the study of many other inherently three-dimensional physical phenomena. You will see this especially in connection with the study of electromagnetism, beginning in Chapter 27.

Perhaps it seems peculiar to see the term "natural" used in connection with the cross product. If you are making first acquaintance with the cross product and find it very artificial and complicated, you are not alone. What connection can there be between the screw—an artificial object on which our explanation of the cross product is partially based—and such natural things as rotating planets or galactic electromagnetic fields, whose properties are described in terms of the cross product? Surely, the wood screw did not serve as a model for the universe! Nevertheless, the behavior of the

Vector Representation of Rotational Quantities

(a)

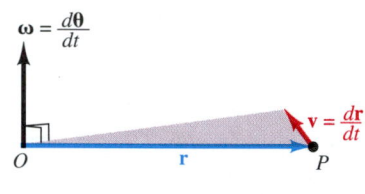

(b)

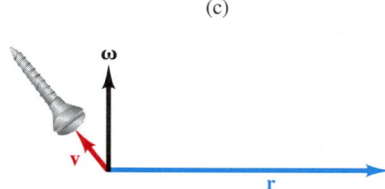

(c)

FIGURE 13.16 (a) The position vector r locates the point P with respect to O. When r rotates through an infinitesimal angle $d\boldsymbol{\theta}$, P moves through the infinitesimal displacement $d\mathbf{r}$ during the same time dt. (Remember that the direction of $d\boldsymbol{\theta}$ is the right-handed normal to the plane defined by r and $d\mathbf{r}$. The magnitude $d\theta$ is shown.) (b) The angular and linear velocities bear the relation $\mathbf{v} = \boldsymbol{\omega} \times \mathbf{r}$. If P is part of a rigid body, all parts of the body will have the same $\boldsymbol{\omega}$ but not the same v. (c) The vectors of part b have been redrawn so that their tails coincide, making it easy to apply the right-hand rule and verify the directional relation $\hat{\mathbf{v}} = \hat{\boldsymbol{\omega}} \times \hat{\mathbf{r}}$.

We now apply the cross product to defining important rotational quantities other than the torque vector **T**. We begin by deriving the relation between the angular velocity vector $\boldsymbol{\omega}$ and the linear velocity vector **v**. In Figure 13.16a, the initial location of the point P with respect to an arbitrary point O is specified by the vector **r**. During the infinitesimal time interval dt, **r** rotates through an infinitesimal angle of magnitude $d\theta$ about O. That is, **r** has rotated to a new position shown as **r′**. During the time dt, P experiences an infinitesimal linear displacement of magnitude $|\mathbf{r}' - \mathbf{r}| = |d\mathbf{r}|$. Because $|d\mathbf{r}| = r\,d\theta$, this displacement satisfies the scalar equation

$$\frac{|dr|}{dt} = \frac{d\theta}{dt}\, r. \tag{13.31a}$$

Using the definitions $|dr|/dt \equiv v$ and $d\theta/dt \equiv \omega$, we can rewrite Equation 13.31a in the form

$$v = \omega r. \tag{13.31b}$$

We wish to generalize Equations 13.31a and 13.31b to vector equations that express directions as well as magnitudes.

We begin with the vector $d\boldsymbol{\theta}$, whose magnitude is $d\theta$ and whose direction is the right-handed normal to the plane defined by **r** and $d\mathbf{r}$ (Figure 13.16a). We define the angular velocity $\boldsymbol{\omega}$ by analogy to the definition of the linear velocity $\mathbf{v} \equiv d\mathbf{r}/dt$. We have

$$\boldsymbol{\omega} \equiv \frac{d\boldsymbol{\theta}}{dt}. \tag{13.32}$$

The magnitude of $\boldsymbol{\omega}$ is the angular speed ω. Because dt is a scalar quantity, the direction of $\boldsymbol{\omega}$ is the same as the direction of $d\boldsymbol{\theta}$, just as the direction of **v** is the same as the direction of $d\mathbf{r}$ (Figure 13.16b).

By applying the right-hand rule to the three mutually perpendicular vectors in Figure 13.16c—**v**, $\boldsymbol{\omega}$, and **r**—you can verify that their directions satisfy the relation

$$\hat{\mathbf{v}} = \hat{\boldsymbol{\omega}} \times \hat{\mathbf{r}}. \tag{13.33}$$

We multiply each side of this equation with the corresponding side of Equation 13.31b to obtain $v\hat{\mathbf{v}} = \omega r(\hat{\boldsymbol{\omega}} \times \hat{\mathbf{r}})$. Because every vector is equal to the product of its magnitude and its direction, we have

$$\mathbf{v} = \boldsymbol{\omega} \times \mathbf{r}. \tag{13.34}$$

Angular Acceleration

We now derive the relation between the linear and angular acceleration vectors, beginning with Equation 13.34. We use the chain rule (Equation 13.27) to differentiate Equation 13.34 with respect to time. We obtain

$$\frac{d\mathbf{v}}{dt} = \frac{d\boldsymbol{\omega}}{dt} \times \mathbf{r} + \boldsymbol{\omega} \times \frac{d\mathbf{r}}{dt}. \tag{13.35}$$

The quantity $d\mathbf{v}/dt$ is the translational acceleration $\mathbf{a}$. The factor $d\boldsymbol{\omega}/dt$ is the vectorial generalization of the scalar we have called the angular acceleration, $\alpha = d\omega/dt$. So it is natural to define $d\boldsymbol{\omega}/dt$ as the angular acceleration $\boldsymbol{\alpha}$:

$$\boldsymbol{\alpha} \equiv \frac{d\boldsymbol{\omega}}{dt}. \tag{13.36}$$

The last term in Equation 13.35 contains the factor $d\mathbf{r}/dt$, which is just the translational velocity $\mathbf{v}$. We can thus write Equation 13.34 in the form

$$\mathbf{a} = \boldsymbol{\alpha} \times \mathbf{r} + \boldsymbol{\omega} \times \mathbf{v}. \tag{13.37a}$$

Figure 13.17a shows the instantaneous vector $\boldsymbol{\omega}$ in the context of a top experiencing angular acceleration $\boldsymbol{\alpha}$ under the influence of an imposed torque $\mathbf{T}$. The radius vector $\mathbf{r}$ and the velocity $\mathbf{v}$ are shown at the same instant for a point P on the periphery of the top. In Figure 13.17b, the vectors $\boldsymbol{\alpha}$, $\mathbf{r}$, and $\mathbf{v}$ have been moved so that their tails coincide. This makes it easy to use the right-hand rule and see that the direction of the vector $\boldsymbol{\alpha} \times \mathbf{r}$ is $\hat{\mathbf{v}}$, the tangent to the circle in which P is moving. Because $\boldsymbol{\alpha}$ and $\mathbf{r}$ are mutually perpendicular, the magnitude of the vector is $|\boldsymbol{\alpha} \times \mathbf{r}| = \alpha r$. Thus the first term on the right in Equation 13.37a represents the **tangential acceleration**

$$\mathbf{a}_t = \alpha r \hat{\mathbf{v}},$$

which has to do with the change in the speed v of P.

Using the right-hand rule again, you will see that the direction of the vector $\boldsymbol{\omega} \times \mathbf{v}$ is $-\hat{\mathbf{r}}$, the inward radial direction, as shown in Figure 13.17c. (The vector $\boldsymbol{\omega}$ has been drawn again with its tail at P, to make the application of the right-hand rule easier.) Because $\boldsymbol{\omega}$ and $\mathbf{v}$ are mutually perpendicular, the magnitude of their cross product is

$$|\boldsymbol{\omega} \times \mathbf{v}| = \omega v = \frac{v}{r} v = \frac{v^2}{r}.$$

Thus the second term on the right side of Equation 13.37a represents the **centripetal acceleration**, or **radial acceleration**,

$$\mathbf{a}_r = -\frac{v^2}{r} \hat{\mathbf{r}},$$

which has to do with the change in the direction of motion of P. The minus sign accounts for the fact that $\hat{\mathbf{a}}_r = -\hat{\mathbf{r}}$; $\mathbf{r}$ is directed outward from the axis, whereas $\mathbf{a}_r$ is directed inward toward the axis.

FIGURE 13.17 (a) A top experiences angular acceleration $\boldsymbol{\alpha}$ under the influence of the torque $\mathbf{T}$. The angular velocity $\boldsymbol{\omega}$ is shown at a certain instant, as are the radius vector $\mathbf{r}$ to point P and the velocity $\mathbf{v}$ of P. The tangential acceleration $\mathbf{a}_t$ and the radial acceleration $\mathbf{a}_r$ are evaluated by use of Equations 13.37a and 13.37b. (b) Showing that $\boldsymbol{\alpha} \times \hat{\mathbf{r}} = \hat{\mathbf{v}}$. For clarity, $\mathbf{v}$ has been moved to make its tail coincide with the tails of the other two vectors. (c) Showing that $\hat{\boldsymbol{\omega}} \times \hat{\mathbf{v}} = -\hat{\mathbf{r}}$. (d) The top, as seen by an observer at S in part a. The acceleration $\mathbf{a}$ of P is the sum of $\mathbf{a}_t$ and $\mathbf{a}_r$.

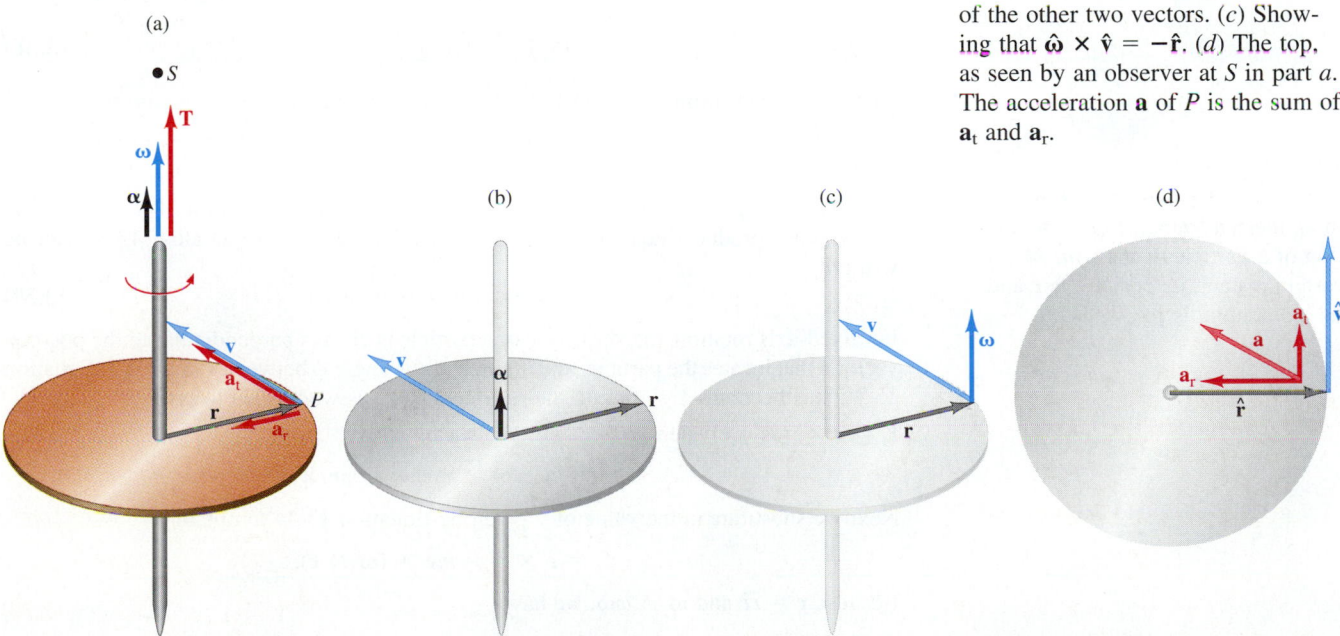

As shown in Figure 13.17d, the acceleration **a** is the vector sum of the tangential and radial accelerations:

$$\mathbf{a} = \mathbf{a}_t + \mathbf{a}_r, \quad \text{where } \mathbf{a}_t \equiv \boldsymbol{\alpha} \times \mathbf{r} \text{ and } \mathbf{a}_r \equiv \boldsymbol{\omega} \times \mathbf{v}. \tag{13.37b}$$

Although we have derived Equations 13.37a and 13.37b in the context of circular motion, they are completely general and are valid even when **r** is not constant and when the vectors involved in the cross products are not mutually perpendicular.

EXAMPLE 13.6

A top is spun up from rest to an angular speed $\omega = 21$ rad/s by accelerating it uniformly over a time interval $\Delta t = 0.25$ s. Find the magnitude and direction of the linear acceleration of a point on the top located 3.5 cm from the axis at the instant just before the spin-up is completed.

SOLUTION: In order to use Equation 13.37b, you must find the magnitude of the angular acceleration $\boldsymbol{\alpha}$. Because the acceleration is uniform, you have

$$\alpha = \frac{\Delta\omega}{\Delta t} = \frac{(21 - 0) \text{ rad/s}}{0.25 \text{ s}} = 84 \text{ rad/s}^2.$$

Because $\boldsymbol{\alpha}$ and **r** are mutually perpendicular, the tangential acceleration has magnitude

$$a_t = |\boldsymbol{\alpha} \times \mathbf{r}| = \alpha r = 84 \text{ rad/s}^2 \times 0.035 \text{ m} = 2.9 \text{ m/s}^2.$$

And, because $\boldsymbol{\omega}$ and **v** are mutually perpendicular, the radial acceleration has magnitude

$$a_r = |\boldsymbol{\omega} \times \mathbf{v}| = \omega v = \omega(\omega r) = \omega^2 r$$
$$= (21 \text{ rad/s})^2 \times 0.035 \text{ m} = 15 \text{ m/s}^2.$$

The net linear acceleration is

$$\mathbf{a} = \mathbf{a}_t + \mathbf{a}_r = 2.9 \text{ m/s}^2 \, \hat{\mathbf{v}} - 15 \text{ m/s}^2 \, \hat{\mathbf{r}};$$

the minus sign in the second term accounts for the relation $\hat{\mathbf{a}}_r = -\hat{\mathbf{r}}$. The magnitude of **a** is

$$a = \sqrt{(2.9 \text{ m/s}^2)^2 + (15 \text{ m/s}^2)^2} = 16 \text{ m/s}^2$$

to two significant figures. The direction of the net acceleration lies at a small angle "forward" (toward $\hat{\mathbf{v}}$) from $-\hat{\mathbf{r}}$. The angle that **a** makes with $-\hat{\mathbf{r}}$ is

$$\phi = \tan^{-1} \frac{2.9}{15} = 11°.$$

Newton's Second Law for Rotation

We have justified Newton's second law for rotation, in the scalar form $T = I\alpha$, largely by analogy with the translational case. We now generalize to a vectorial formulation and provide a more rigorous justification for the law at the same time.

In Figure 13.18, an inertial observer at O watches the motion of a particle of mass m. At a certain instant, the particle is located at **r** and has linear momentum **p**. According to Equation 13.5a, which is restricted to fixed-axis rotation, the magnitude of the angular momentum is $L = rmv$. We now *define* the angular momentum of the particle to be

$$\mathbf{L} \equiv \mathbf{r} \times \mathbf{p}. \tag{13.38}$$

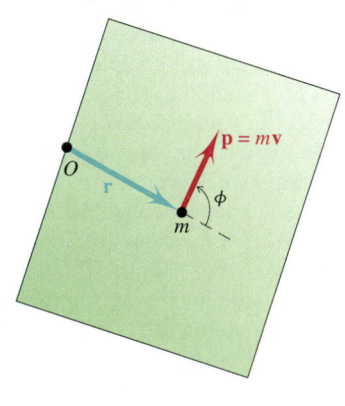

FIGURE 13.18 Observer O, who is in an inertial frame, studies the motion of a particle of mass m. At a certain instant, its position is **r** and its momentum is **p**.

To justify this definition, we show that (1) the restricted definition $L = rmv$ is a special case of Equation 13.38, and (2) Equation 13.38 is equivalent to definition 13.16, $\mathbf{L} \equiv I\boldsymbol{\omega}$:

1. $L = rmv$ is a special case of $\mathbf{L} = \mathbf{r} \times \mathbf{p}$. According to the definition of the magnitude of the cross product (Equation 13.21), the magnitude part of Equation 13.38 can be written

$$L = r \sin \phi \, p = rmv \sin \phi. \tag{13.39}$$

For fixed-axis rotation, the velocity **v** of a particle is always perpendicular to the position vector **r** that locates the particle, and the sine of the angle ϕ between **v** and **r** is 1. Equation 13.39 thus becomes $L = rmv$ in the restricted case, as we wished to prove.

2. $\mathbf{L} = \mathbf{r} \times \mathbf{p}$ is equivalent to $\mathbf{L} = I\boldsymbol{\omega}$. We write

$$\mathbf{L} = \mathbf{r} \times \mathbf{p} = \mathbf{r} \times m\mathbf{v} = m\mathbf{r} \times \mathbf{v}.$$

Next we substitute in the value of **v** given by Equation 13.34 to obtain

$$\mathbf{L} = \mathbf{r} \times \mathbf{p} = m\mathbf{r} \times (\boldsymbol{\omega} \times \mathbf{r}).$$

Because $\mathbf{r} = r\hat{\mathbf{r}}$ and $\boldsymbol{\omega} = \omega\hat{\boldsymbol{\omega}}$, we have

$$\mathbf{L} = \mathbf{r} \times \mathbf{p} = mr^2\omega[\hat{\mathbf{r}} \times (\hat{\boldsymbol{\omega}} \times \hat{\mathbf{r}})].$$

The quantity in brackets is a unit vector that gives the direction of **L**; that is, the quantity $[\hat{\mathbf{r}} \times (\hat{\boldsymbol{\omega}} \times \hat{\mathbf{r}})] = \hat{\mathbf{L}}$. To evaluate this quantity, we use the right-hand rule twice. The cross product is nonassociative, and so we must evaluate $(\hat{\boldsymbol{\omega}} \times \hat{\mathbf{r}})$ first. Because $\boldsymbol{\omega} \perp \mathbf{r}$, the result will be a unit vector; call it $\hat{\mathbf{u}}$. As you can see from Figure 13.19, $\hat{\mathbf{u}}$ lies in the plane of the particle's motion, perpendicular to **r** and in the forward direction. The remaining cross product, $\hat{\mathbf{r}} \times \hat{\mathbf{u}}$, yields the unit vector $\hat{\boldsymbol{\omega}}$, so the overall result is

$$\mathbf{r} \times \mathbf{p} = mr^2\omega\hat{\boldsymbol{\omega}} = I\boldsymbol{\omega}.$$

This proves our assertion that Equation 13.38 is equivalent to Equation 13.16; $\mathbf{L} = \mathbf{r} \times \mathbf{p} = I\boldsymbol{\omega}$.

We now use Equation 13.38 to obtain the rotational form of Newton's second law. Taking the time derivative of both sides of the equation, we have

$$\frac{d\mathbf{L}}{dt} = \frac{d}{dt}(\mathbf{r} \times \mathbf{p})$$

$$= \frac{d\mathbf{r}}{dt} \times \mathbf{p} + \mathbf{r} \times \frac{d\mathbf{p}}{dt}. \tag{13.40}$$

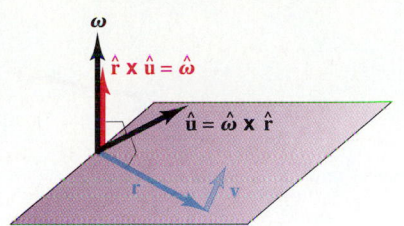

FIGURE 13.19 Showing that $\hat{\mathbf{r}} \times (\hat{\boldsymbol{\omega}} \times \hat{\mathbf{r}}) = \hat{\boldsymbol{\omega}}$.

The first term on the right side of this equation is zero. To see this, we recall that, for any vector **A**, $\mathbf{A} \times \mathbf{A} = 0$ (Equation 13.23). Now we can write

$$\frac{d\mathbf{r}}{dt} \times \mathbf{p} = \mathbf{v} \times m\mathbf{v} = m\mathbf{v} \times \mathbf{v} = 0.$$

Finally, we use Equation 13.2 to make the substitution $d\mathbf{p}/dt = \mathbf{F}$, and Equation 13.40 becomes

$$\frac{d\mathbf{L}}{dt} = \mathbf{r} \times \mathbf{F},$$

or

$$\frac{d\mathbf{L}}{dt} = \mathbf{T}. \tag{13.41}$$

Consider what happens in the special case of rotation of a rigid body about a fixed axis. In this case I is constant. If we differentiate Equation 13.16, $\mathbf{L} = I\boldsymbol{\omega}$, with respect to time, we have

$$\frac{d\mathbf{L}}{dt} = \frac{d}{dt}I\boldsymbol{\omega} = I\frac{d\boldsymbol{\omega}}{dt} = I\boldsymbol{\alpha}.$$

Combining this with Equation 13.41 gives the result

$$\mathbf{T} = I\boldsymbol{\alpha}, \tag{13.42}$$

which is the familiar form of Newton's second law for rotation. We arrived at the same result less rigorously in the brief argument leading to Equation 13.17. Note that Equation 13.42 is less general than Equation 13.41; the former depends on the proviso $I = $ constant. In complete analogy, Newton's second law for translation is less general in the form $\mathbf{F} = m\mathbf{a}$ than in the form $\mathbf{F} = d\mathbf{p}/dt$ because the former depends on the proviso $m = $ constant.

For fixed-axis rotation, the generalization from a particle to a rigid body is simple. The moment of inertia is no longer mr^2, as it is in the argument leading to Equation 13.38. Nevertheless, for any rigid body of mass M rotating about a specified axis, there is always a radius of gyration γ such that $I = M\gamma^2$, and Equation 13.42 still holds.

Rotation and Translation

In Chapters 11 and 12, we took for granted the idea that the motion of a body can be separated unambiguously into translational and rotational parts. Although this idea is easy to accept as "common sense," we now consider it in a precise way. Our aim is to show rigorously that it is true, subject to specified conditions.

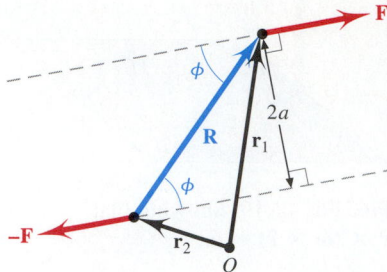

FIGURE 13.20 A couple, consisting of the equal and opposite pair of forces **F** and −**F**, whose lines of action are separated by the perpendicular distance 2*a*. The torques exerted by the two forces are calculated about the arbitrary point *O*. The points of application of **F** and −**F** are specified respectively by the position vectors **r**₁ and **r**₂.

Couples

You learned in Section 9.5 that study of the translational motion of a system of mass *M* under the influence of a net external force can be simplified. Specifically, the motion can be reduced to that of a particle of the same mass located at the center of mass of the system and subject to the same net external force. This result is summarized in Equation 9.25, $M\mathbf{A} = \mathbf{F}_{ext}$. In particular, if the net external force is zero, the system will move inertially, with translational acceleration $\mathbf{A} = 0$.

It does not follow, however, that the system will experience zero angular acceleration **α**. Consider a system subject to a pair of forces such as the pair shown in Figure 13.20. The two forces, **F** and −**F**, are equal and opposite. Their lines of action are separated by a perpendicular distance 2*a*. Such a force pair is called a **couple**. A familiar example of a couple is the pair of forces that act on the ends of a magnetic compass needle when it is not aligned in the north–south direction.

Let us calculate the net torque **T** exerted by the two forces of a couple about an arbitrary point *O*. To make the diagram of Figure 13.20 as simple as possible, we have shown *O* lying in the plane defined by **F** and −**F**, but this is not necessary to the argument that follows.

The point of application of **F** is located with respect to *O* by the position vector **r**₁. The vector **r**₂ serves the same purpose for −**F**. The net torque is

$$\mathbf{T} = \mathbf{T}_1 + \mathbf{T}_2 = \mathbf{r}_1 \times \mathbf{F} + \mathbf{r}_2 \times (-\mathbf{F}).$$

The distributive law for cross products and the commutativity of the scalar −1 allow us to write this equation in the form

$$\mathbf{T} = (\mathbf{r}_1 - \mathbf{r}_2) \times \mathbf{F}.$$

But $\mathbf{r}_1 - \mathbf{r}_2$ is the vector **R** shown in Figure 13.20, and the net torque exerted by the couple is therefore

$$\mathbf{T} = \mathbf{R} \times \mathbf{F}. \tag{13.43a}$$

The vector **R** is independent of the location of the point *O* about which we chose to calculate **T**. That is to say, *the torque exerted by a couple is the same about all centers of rotation*. The magnitude *T* of the torque is $R \sin \phi\, F$, where ϕ is the angle shown in Figure 13.20. But, because $R \sin \phi = 2a$, we have

$$T = 2aF; \tag{13.43b}$$

the magnitude of the torque exerted by a couple is the product of the distance between the lines of action of the two forces and the magnitude of either force. It follows that *the magnitude of the torque exerted by the couple is independent of the points of application of the forces*. Finally, the lines of action of the two forces define the plane in which the vector **R** must lie as well. The unit vector $\hat{\mathbf{T}}$ is normal to that plane. If you imagine the couple as turning the head of a right-handed screw, $\hat{\mathbf{T}}$ is the direction in which the screw advances.

Motion of a Rigid Body Subject to an Arbitrary External Force

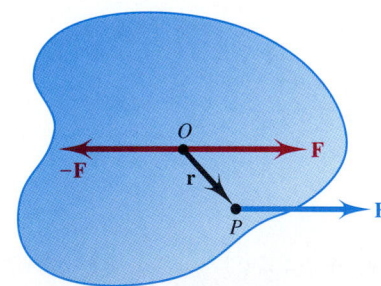

FIGURE 13.21 A rigid body, whose center of mass is at *O*, is subjected to a force **F** applied at *P*, whose position with respect to *O* is represented by **r**. Two equal and opposite forces **F** and −**F**, shown in red, are applied at *O*. The force **F** at *P* and −**F** constitute a couple, which produces rotation only. The force **F** at *O* produces translation only.

We are now ready to address the main topic of this section. We prove that the motion of a rigid body is uniquely separable into translational and rotational parts. Suppose that the body of Figure 13.21 is subjected to the external force **F** applied at a point *P*, whose position with respect to *O*, the center of mass, is **r**. Imagine that the body is also subjected to equal and opposite forces **F** and −**F** applied at the center of mass. This pair of equal, opposite forces will not affect the motion of the body because their vector sum is zero.

Consider now the ''original'' force **F** applied at *P* and the force −**F** applied at *O*. They constitute a couple that exerts a torque of magnitude 2*aF*. They will produce rotation of the body but not translation. The remaining force **F**, applied to the center of mass, will produce translation but not rotation. In this way, *any external force applied to a rigid body can be resolved into a couple and a force applied to the center of mass.* Because this statement is true for any single external force applied to a rigid body, it is true for any number of forces applied simultaneously. The resulting torque vectors can be added to find the net torque, and thus to determine the rotational part of the motion. The resulting forces applied to the center of mass can be added vectorially to find the net force, and thus to determine the translational part of the motion. These verbal statements can be summarized in the following equations:

$$\mathbf{F} = \sum_j \mathbf{F}_j = M\mathbf{a} \quad \text{for translation of the center of mass,} \tag{13.44a}$$

and

$$\mathbf{T} = \sum_j \mathbf{T}_j = I\boldsymbol{\alpha} \quad \text{for rotation about the center of mass.} \tag{13.44b}$$

These equations may be regarded as a complete statement of Newton's second law. As always for Newton's second law, they are valid only for inertial observers.

Precession

Almost everyone has played with tops as a child. When you did, you probably noticed that the motion of the top consists of more than mere spinning about its axis. The axis of the top is usually inclined with respect to the vertical. When the top is spinning fairly fast, the most noticeable aspect of its motion (aside from the spinning itself) is a slow turning of the rotation axis about the vertical, suggested in Figure 13.22. This turning motion is called **precession of the axis** or simply **precession**. When precession takes place, the axis is fixed neither in position nor in orientation, and the analysis must be carried out in vectorial terms.

Before considering the precession of a top, let us consider the similar but somewhat simpler device shown in Figure 13.23a, which is commonly used in lecture demonstrations. A bicycle wheel is modified by adding weights to its rim so as to increase its moment of inertia. The wheel is mounted on low-friction bearings on a shaft that has two convenient handles.

When the wheel is not spinning, its behavior holds no surprises. If you hold it at both ends of the axle (as shown) and twist, the device simply turns whichever way your

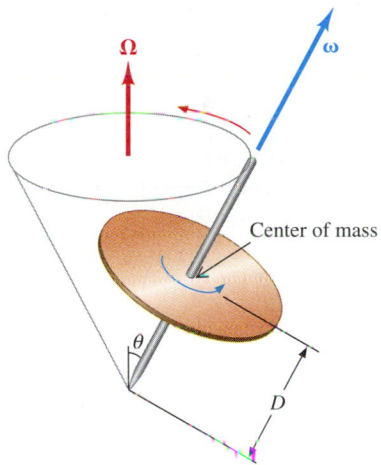

FIGURE 13.22 A toy top spins with a fairly large angular velocity $\boldsymbol{\omega}$. Most commonly, the axis is not vertical but is inclined at an angle θ. The axis slowly revolves about the vertical, describing a cone with vertex angle 2θ. This slow revolution, called precession, takes place with angular velocity $\boldsymbol{\Omega}$.

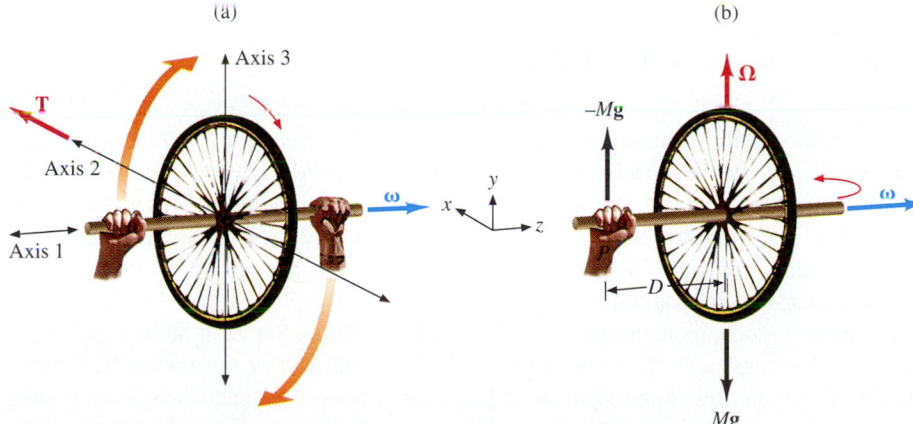

FIGURE 13.23 (*a*) A bicycle wheel, weighted at the rim to increase its moment of inertia, is mounted on an easily held axle. The wheel spins with angular velocity $\boldsymbol{\omega}$ about axis 1. If you apply a couple about axis 2 with your hands (in the sense shown by the arrows), the system does not rotate about that axis. Instead, it rotates about axis 3. (*b*) If one hand is completely removed, the system does not topple but precesses about the vertical with angular velocity $\boldsymbol{\Omega}$, supported (by the other hand) at the single off-center point P.

hands go. For example, if you twist the device about axis 2 (as suggested by the arrows above and below the two hands in Figure 13.23a), the axle rotates from a horizontal to a vertical orientation. But when the wheel is *spinning* about axis 1 (the axle), the device behaves in a very different, surprising way. If you hold the device by the handles and try to twist it about an axis perpendicular to the axle, the axle will not follow your hands. For example, if you try to twist the device about axis 2 (Figure 13.23a), it tends to twist out of your hands; instead of turning about axis 2, it turns about axis 3, which is perpendicular to axes 1 and 2. If you remove your right hand, as in Figure 13.23b, the released handle does not fall toward the floor. Instead, the device seemingly defies gravity as the axle sweeps out a horizontal circle, rotating about the support provided by your left hand. In the analysis that follows, you will see that gravity is not in fact defied and that, in behaving as it does, the system is obeying Newton's second law for rotation.

Because the wheel is spinning about the axle with angular velocity $\boldsymbol{\omega}$, the system has angular momentum $\mathbf{L} = I\boldsymbol{\omega}$, where I is its moment of inertia about the axle. When the system is supported by both hands, it is at rest as far as translation of the center of mass is concerned, although the wheel is spinning. The gravitational force $M\mathbf{g}$ acts at the center of mass, and each hand supports half the weight by exerting an upward force $-M\mathbf{g}/2$ on one of the handles. The net external force and the net external torque acting on the system are both zero.

Now suppose you take away your right hand. Consider the system immediately afterward, at the instant shown in Figure 13.23b. The downward gravitational force is unchanged, but the remaining hand, located a distance D from the center of mass, now supports the entire weight of the system by exerting an upward force $-M\mathbf{g}$. The two forces $M\mathbf{g}$ and $-M\mathbf{g}$ constitute a couple. According to Equation 13.43b, the magnitude of the couple is MgD. If the coordinate axes are chosen as shown in Figure 13.23b, a screw, oriented with its head in the y-z plane and turned by the couple, would advance in the positive x direction (into the page). Thus the torque exerted by the couple can be written

$$\mathbf{T} = MgD\hat{\mathbf{x}}.$$

Alternatively, you can evaluate $\mathbf{T}$ by summing the separate torques exerted about the center of mass by the upward and downward forces. The support force is $-M\mathbf{g} = Mg\hat{\mathbf{y}}$, and so you obtain

$$-D\hat{\mathbf{z}} \times Mg\hat{\mathbf{y}} + 0 = -MgD\hat{\mathbf{z}} \times \hat{\mathbf{y}} = MgD\hat{\mathbf{x}},$$

which is the same result.

The application of this torque to the system results in a change of its angular momentum according to Equation 13.41, $\mathbf{T} = d\mathbf{L}/dt$. Over the time interval dt, the change of angular momentum is $d\mathbf{L} = \mathbf{T}\,dt$, or

$$d\mathbf{L} = MgD\,dt\,\hat{\mathbf{x}}. \tag{13.45}$$

But the initial angular momentum is $\mathbf{L} = I\omega\hat{\mathbf{z}}$. That is, *the change in angular momentum is perpendicular to the initial angular momentum*. Because $d\mathbf{L}$ is perpendicular to $\mathbf{L}$, the effect of $d\mathbf{L}$ is to change the *direction* of $\mathbf{L}$ but not its *magnitude*. (This is analogous to the case of centripetal acceleration in which an infinitesimal velocity change $d\mathbf{v}$ affects the direction but not the magnitude of the velocity vector $\mathbf{v}$ to which $d\mathbf{v}$ is perpendicular; see Section 5.4.)

The new angular momentum, $\mathbf{L}' = \mathbf{L} + d\mathbf{L}$, is the vector sum shown in Figure 13.24. As $\mathbf{L}$ changes its direction, so does the moment arm of the torque $\mathbf{T}$. Consequently, $\mathbf{T}$ remains always perpendicular to $\mathbf{L}$, and so does $d\mathbf{L}$. Hence the system rotates in the z-x (horizontal) plane—about an axis in the vertical $\hat{\mathbf{y}}$ direction—at some angular velocity $\boldsymbol{\Omega}$. This rotation is called precession.

Gravity is not defied because the gravitational force is one member of a couple. The other member is the support force exerted on the axle. The couple exerts a torque on the system, whose angular momentum therefore changes. But the initial angular momentum $\mathbf{L}$ and the change $d\mathbf{L}$ are both in the horizontal plane. Thus the new angular momentum $\mathbf{L}'$ also is in the horizontal plane. The rotation of the angular momentum vector in the horizontal plane corresponds to the precession of the axle in that plane.

We now evaluate the magnitude and direction of $\boldsymbol{\Omega}$. As you can see from Figure

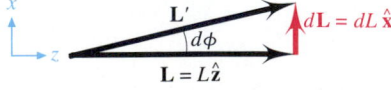

FIGURE 13.24 Vector diagram showing the change $d\mathbf{L}$ in angular momentum of the system of Figure 13.23b over the time interval dt. The view is downward from above the system; $\hat{\mathbf{L}}$ coincides with the axle of the bicycle wheel. The precession angle is $d\phi = dL/L$.

13.24, the axle has turned during the time interval dt through an angle $d\phi$ given by the ratio

$$d\phi = \frac{dL}{L}.$$

For the numerator dL of this ratio, we use the value $MgD\,dt$ given by Equation 13.45. For the denominator, we can simply write $L = I\omega$. This gives

$$d\phi = \frac{MgD}{I\omega}\,dt.$$

Finally, we divide through by dt to find the magnitude of the angular velocity, $\Omega = d\phi/dt$. We obtain

$$\Omega = \frac{MgD}{I\omega}. \qquad\qquad \textbf{(13.46a)}$$

Note that MgD is the magnitude T of the torque applied to the system by the couple consisting of the gravitational force $M\mathbf{g}$ and the support force $-M\mathbf{g}$ (Figure 13.23b). We can thus write Ω in the form

$$\Omega = \frac{T}{I\omega}. \qquad\qquad \textbf{(13.46b)}$$

The direction of $\boldsymbol\Omega$ is found by applying the right-hand rule. In the case shown in Figure 13.23b, the axle is rotating from the $\hat{\mathbf{z}}$ direction toward the $\hat{\mathbf{x}}$ direction, and $\hat{\boldsymbol\Omega}$ is directed vertically upward, in the $\hat{\mathbf{y}}$ direction; see Equation 13.29c. What would happen if the wheel were rotating in the sense opposite that shown?

The quantity $\boldsymbol\Omega$ is called the **precession velocity**. (The more precise term *angular velocity of precession* is not generally used.)

Let us review the physical significance of the factors in Equation 13.46a. Increasing any of the quantities in the numerator increases the torque, and thus the value of Ω. The same effect can be achieved by hanging an extra weight from the free end of the axle. Increasing either of the terms in the denominator increases the angular momentum, and thus decreases the ratio $\Delta L/L$, which is proportional to Ω. Note in particular that, as the system slows down (as a real system does owing to friction), the precession velocity increases. You have probably noted this effect in playing with tops.

In the foregoing analysis of precession, we ignored the fact that the system has an additional angular momentum $\boldsymbol\Lambda$ (uppercase Greek lambda) due to the precession itself. The precession may be regarded as a process in which the center of mass of the system describes a circular orbit of radius D about the support point P in Figure 13.23b, and $\boldsymbol\Lambda$ is sometimes called the *orbital angular momentum* of the system. (It follows that $\boldsymbol\Lambda = MD^2\Omega\hat{\mathbf{y}}$.) Strictly speaking, the angular momentum that is changed by the applied torque is not the "spin angular momentum" $\mathbf{L}$, but the sum $\mathbf{L} + \boldsymbol\Lambda$, which is the total angular momentum of the system. This introduces complications that become significant when Λ is not small compared with L. In Example 13.7, you will see that the neglect of $\boldsymbol\Lambda$ is not serious if the system spins with a reasonably large value of ω.

EXAMPLE **13.7**

A weighted bicycle wheel of the type just discussed has radius 33.0 cm. Its mass, which is mostly concentrated at the rim, is 13.4 kg. **(a)** If the wheel is made to spin at 200 rev/min and is supported at a point on the axle a distance 20.0 cm from the hub, what is the precession velocity? **(b)** Find the magnitude and direction of the total angular momentum $\mathbf{L} + \boldsymbol\Lambda$, and compare it with $\mathbf{L}$.

SOLUTION:
(a) Find the precession velocity.

The moment of inertia of the wheel about the axis is

$$I = MR^2 = 13.4\text{ kg} \times (0.330\text{ m})^2 = 1.46\text{ kg·m}^2.$$

The angular speed is 200 rev/min = 20.9 rad/s, and so the spin angular momentum has magnitude

$$L = I\omega = 1.46\text{ kg·m}^2 \times 20.9\text{ rad/s} = 30.5\text{ kg·m}^2/\text{s}.$$

You can use these values in Equation 13.46a to obtain

$$\Omega = \frac{13.4\text{ kg} \times 9.80\text{ m/s}^2 \times 0.200\text{ m}}{30.5\text{ kg·m}^2/\text{s}} = 0.861\text{ rad/s}.$$

At this angular velocity, the system will make one precessional rotation every 7.30 s.

(b) Find $\mathbf{L} + \mathbf{\Lambda}$, and compare it with $\mathbf{L}$.

You could use the perpendicular-axis theorem, followed by the parallel-axis theorem, to find the moment of inertia of the wheel about a vertical axis through the support point. But, because the mass of the wheel is concentrated in its rim, all parts of which are equidistant from the support point, it is easier to use Pythagoras's theorem to find the square, $r^2 = D^2 + R^2$, of that distance. You can then write the magnitude of the angular momentum as

$$\Lambda = Mr^2\Omega$$
$$= 13.4 \text{ kg} \times [(0.330 \text{ m})^2 + (0.200 \text{ m})^2] \times 0.861 \text{ rad/s}$$
$$= 1.72 \text{ kg·m}^2/\text{s}.$$

Because $\hat{\boldsymbol{\omega}}$ is horizontal and $\hat{\boldsymbol{\Omega}}$ is vertical, you use Pythagoras's theorem again to find the magnitude $|\mathbf{L} + \mathbf{\Lambda}|$ of their sum. You

have

$$|\mathbf{L} + \mathbf{\Lambda}| = \sqrt{(30.5 \text{ kg·m}^2/\text{s})^2 + (1.72 \text{ kg·m}^2/\text{s})^2}$$
$$= 30.5 \text{ kg·m}^2/\text{s}.$$

That is, the precessional angular momentum is completely negligible within three significant figures, as far as magnitude is concerned.

Next, consider the angle ϕ that the total angular momentum $\mathbf{L} + \mathbf{\Lambda}$ makes with the horizontal. You have

$$\theta = \tan^{-1} \frac{1.72}{30.5} = 3.23°.$$

Because the torque vector is perpendicular to $\mathbf{L} + \mathbf{\Lambda}$ as well as to $\mathbf{L}$, the precession calculation is not affected by the assumption $\mathbf{L} \simeq \mathbf{L} + \mathbf{\Lambda}$.

The Top

It is not difficult to generalize from the bicycle wheel of the preceding subsection, whose axis is horizontal, to a top like that of Figure 13.22, whose axis makes an angle θ with the vertical. The torque that produces precession is still exerted by a couple consisting of the gravitational force $M\mathbf{g}$ and the support force $-M\mathbf{g}$. The latter force is now exerted on the lower tip of the top by the surface on which it spins. The only change from the bicycle wheel case is that now the torque has magnitude

$$T = MgD \sin \theta,$$

as shown in the vector diagram of Figure 13.25. The direction $\hat{\mathbf{T}} = \hat{\mathbf{x}}$ is unchanged. Thus the change in angular velocity over the time interval dt is now

$$d\mathbf{L} = MgD \sin \theta \, dt \, \hat{\mathbf{x}}.$$

FIGURE 13.25 Vector diagram showing the change $d\mathbf{L}$ in angular momentum of the top of Figure 13.22 over the time interval dt. The coordinate system is the same as that chosen in Figures 13.23 and 13.24. The precession angle is now $d\phi = dL/L \sin \theta$.

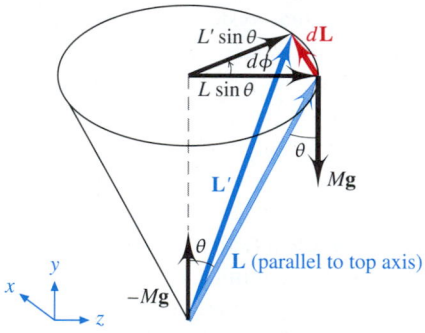

Again, we wish to find the angle $d\phi$ through which the top precesses during the interval dt. As Figure 13.25 shows, however, the angle is not dL/L, as it was in the case of the bicycle wheel, because the angular momentum vector does not trace out a circle of radius L. Instead, the radius of the circle is $L \sin \theta$. We thus have

$$d\phi = \frac{dL}{L \sin \theta} = \frac{MgD \sin \theta}{I\omega \sin \theta}.$$

We thus obtain the precession velocity

$$\Omega = \frac{d\phi}{dt} = \frac{MgD}{I\omega},$$

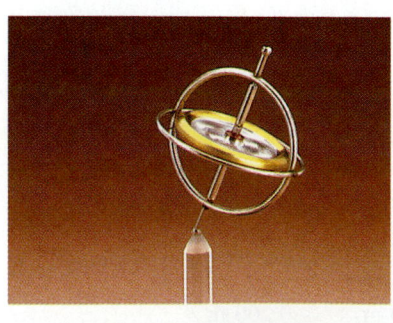

Even when you understand why a spinning top precesses instead of falling over, the sense of wonder persists.

which is the same as the result for precession in the horizontal plane, given by Equation 13.46a. The gravitational torque is reduced, in the present case, by the fact that the force is not perpendicular to the moment arm. But the horizontal component of the spin angular momentum, which is the only part of the angular momentum affected by the torque, is reduced proportionately. Thus the final result is not changed.

$\mathbf{a}, a$	linear acceleration	m	mass of a particle
$\mathbf{A} \times \mathbf{B}$	vector (cross) product of $\mathbf{A}$ and $\mathbf{B}$	M	mass of a system or a rigid body
$\mathbf{a_r}, \mathbf{a_t}$	radial, tangential acceleration	$\mathbf{p}, p$	linear momentum
D	distance from the pivot of a pendulum to its center of mass	$\mathbf{v}, v$	linear velocity, speed
$\mathbf{F}, F$	force	$\mathbf{T}, T$	torque
$\mathbf{G}, G$	angular impulse	$\boldsymbol{\alpha}, \alpha$	angular acceleration
I	moment of inertia	γ	radius of gyration
$\mathbf{J}$	linear impulse	$\boldsymbol{\Lambda}, \Lambda$	angular momentum due to precession
K	kinetic energy	ϕ	angle between two vectors
$\mathbf{L}, L$	angular momentum	$\boldsymbol{\omega}, \omega$	angular velocity, angular speed
		$\boldsymbol{\Omega}, \Omega$	precession velocity

Angular momentum L is the rotational analogue of linear momentum **p**. Angular momentum is conserved in an **isolated system**—a system subjected to zero net external torque. This definition of an isolated system differs from the definition appropriate to translational motion. Angular momentum may be conserved when linear momentum is not, and vice versa.

For a particle rotating about a fixed axis at a distance R from that axis, the angular momentum is given in scalar form by Equation 13.4,

$$L = mR^2\omega.$$

In the same restricted case, Equation 13.5b gives the relation $L = Rp$. For a rigid body rotating about a fixed axis, we generalize Equation 13.4 and define the angular momentum as in Equation 13.1, $L \equiv I\omega$.

The angular momentum of a system may be changed by imparting to it an **angular impulse G**. The scalar angular impulse G, defined by Equation 13.6, may be used in the analysis of fixed-axis rotation. The change in L is given by Equation 13.7,

$$G = \Delta L.$$

This relation is analogous to, but distinct from, the translational relation $\mathbf{J} = \Delta\mathbf{p}$.

When two subsystems constituting an isolated system interact, they impart angular impulses to each other and angular momentum is transferred from one to the other. The total angular momentum of the system is conserved. The impulses obey the condition given by Equation 13.12,

$$G_{1\rightarrow2} = -G_{2\rightarrow1}.$$

Equivalently, the torques exerted by each subsystem on the other must be equal and opposite at any instant, according to Equation 13.15b,

$$T_{1\rightarrow2} = -T_{2\rightarrow1}.$$

Either of these equations may be regarded as **Newton's third law for rotation**. Although they are restricted to fixed-axis rotation in this scalar form, they are generally valid in vector form.

Vector quantities are required to describe rotation in a general way. The magnitude of such rotational quantities as angular momentum **L**, angular velocity **ω**, and angular acceleration **α** is defined just as for fixed-axis rotation, where scalars suffice. For rigid-body rotation, these vectors are parallel to the axis of rotation, in the direction defined by the **right-hand rule**.

The right-hand rule is used in defining the direction $\hat{\mathbf{C}}$ of the **vector product**, or **cross product**, of two vectors, written as in Equation 13.20,

$$\mathbf{C} = \mathbf{A} \times \mathbf{B}.$$

The magnitude of the cross product is

$$C = |\mathbf{A} \times \mathbf{B}| = A \sin \phi \, B,$$

where ϕ is the (smaller) angle between $\hat{\mathbf{A}}$ and $\hat{\mathbf{B}}$. Some of the important rules for manipulation of cross products are given in Equations 13.22 through 13.30.

Consider a particle of mass m located at **r** with respect to an observer and moving with linear velocity **v**. The relation among **v**, **r**, and the angular velocity **ω** is given by Equation 13.34,

$$\mathbf{v} = \boldsymbol{\omega} \times \mathbf{r}.$$

The acceleration of the particle is given by Equation 13.37b:

$$\mathbf{a} = \mathbf{a_t} + \mathbf{a_r},$$

where $\mathbf{a_t} \equiv \boldsymbol{\alpha} \times \mathbf{r}$ and $\mathbf{a_r} \equiv \boldsymbol{\omega} \times \mathbf{v}$. The vector angular momentum of the particle is defined by Equation 13.38 or Equation 13.16:

$$\mathbf{L} \equiv \mathbf{r} \times \mathbf{p} \quad \text{or} \quad \mathbf{L} \equiv I\boldsymbol{\omega}.$$

The motion of a rigid body can be separated into two parts. The first part is translation of the center of mass, described by the equation of motion (Equation 13.44a),

$$\mathbf{F} = \sum_j \mathbf{F}_j = M\mathbf{a}.$$

The second part is rotation of the body about the center of mass, described by the equation of motion (Equation 13.44b),

$$\mathbf{T} = \sum_j \mathbf{T}_j = I\boldsymbol{\alpha}.$$

Together, these equations constitute a complete statement of Newton's second law of motion. The separation into parts depends on the possibility of resolving each force acting on a body into a force acting at the center of mass, which results in translation only, and a **couple**, which results in rotation only. The torque exerted by a couple is given by Equation 13.43a, $\mathbf{T} = \mathbf{R} \times \mathbf{F}$. The magnitude of the couple is given by Equation 13.43b, $T = 2aF$.

When a rotating system such as a top is subjected to an external torque, it precesses. The direction of the precession velocity $\boldsymbol{\Omega}$ is the right-handed normal to the plane defined by $\mathbf{L}$ and $\mathbf{T}$, and the magnitude of $\boldsymbol{\Omega}$ is given by Equation 13.46b,

$$\Omega = \frac{T}{I\omega}.$$

Table 13.1 lists some vector relations important to the study of rotation, together with their translational analogues.

KEY TERMS

Section 13.2 Angular Momentum in Fixed-Axis Rotation
angular momentum ▪ angular impulse ▪ Newton's third law for rotation ▪ central force, force center

Section 13.3 The Direction of Rotational Vector Quantities
right-handed coordinate system ▪ right-hand rule

Section 13.4 The Vector (Cross) Product
right-handed normal ▪ anticommutativity of cross product

Sections 13.5 and 13.6 Vector Representation of Rotational Quantities; Rotation and Translation
tangential acceleration ▪ centripetal (radial) acceleration ▪ couple

Section 13.7 Precession
precession of the axis ▪ precession velocity

TABLE 13.1 Vector Equations for Rotation and Translation

Rotational Equation	Reference	Connecting Equations	Reference	Translational Equation	Reference
$\mathbf{L} = I\boldsymbol{\omega}$	Eq. 13.16	$\mathbf{L} = \mathbf{r} \times \mathbf{p}$ $\mathbf{v} = \boldsymbol{\omega} \times \mathbf{r}$	Eq. 13.38 Eq. 13.34	$\mathbf{p} = M\mathbf{v}$	Eq. 9.3
$\mathbf{G} = \Delta\mathbf{L}$	Eq. 13.7b (scalar form)	$\mathbf{G} = \mathbf{r} \times \mathbf{J}$		$\mathbf{J} = \Delta\mathbf{p}$	Eq. 9.4b
$\mathbf{T} = \dfrac{d\mathbf{G}}{dt} = \dfrac{d\mathbf{L}}{dt}$	Eq. 13.41	$\mathbf{T} = \mathbf{r} \times \mathbf{F}$	Eq. 13.19	$\mathbf{F} = \dfrac{d\mathbf{J}}{dt} = \dfrac{d\mathbf{p}}{dt}$	Eqs. 9.1, 9.30, 13.2
$\mathbf{T} = I\boldsymbol{\alpha}$	Eq. 13.42	$\mathbf{a} = \boldsymbol{\alpha} \times \mathbf{r}$ $+ \boldsymbol{\omega} \times \mathbf{v}$	Eq. 13.37a	$\mathbf{F} = M\mathbf{a}$	

Queries and Problems for Chapter 13

QUERIES

13.1 *(2) Expanding frontiers.* Example 13.1 concerns a particle attached to a massless rod. In the analysis, the particle by itself was taken to be the system. Suppose instead that you regard the system as comprising the particle, the rod (rigid and massless) to which the particle is attached, and the axis to which the rod is attached. What are the net external force and torque acting on this new system? What quantities are conserved?

13.2 *(2) Form follows function.* Because it has the mathematical form $G = \int T \, dt$, angular impulse must have the units of torque $\times$ time, expressed in SI as N·m·s. Angular mo-

mentum, having the mathematical form $L = I\omega$, must have the units of mr^2/t, expressed in SI as kg·m²/s. According to Equation 13.7, the two sets of units must be identical. Show that this is so.

13.3 *(2) Chopping logic.* The helicopter in the photograph, like most helicopters, has a single main rotor mounted on a vertical axis. The main rotor provides lift, as well as horizontal propulsion when the entire helicopter is tilted in the desired direction. In addition to the main rotor, there is a small rotor mounted on one side of the tail, with the rotor axis horizontal and perpendicular to the fuselage of the helicopter. Explain

the purpose of this small rotor. How can you design a helicopter that does not need such a rotor?

13.4 *(2) Can the center hold?* Which of the following forces are central forces? **(a)** The gravitational force

$$\mathbf{F}_{grav} = -\frac{Gm_1m_2}{r^2}\,\hat{\mathbf{r}}; \quad G,\ m_1,\ \text{and}\ m_2\ \text{constants.}$$

(b) The electric force

$$\mathbf{F}_e = \frac{1}{4\pi\epsilon_0}\frac{q_1q_2}{r^2}\,\hat{\mathbf{r}}; \quad \epsilon_0,\ q_1,\ q_2\ \text{constants.}$$

(c) The magnetic force

$$\mathbf{F}_m = q\mathbf{v} \times \mathscr{B},$$

where q is a constant, $\mathbf{v}$ is the velocity of the particle on which the force is to be calculated, and the "magnetic induction" $\mathscr{B}$ is a vector whose magnitude and direction may vary with position $\mathbf{r}$. **(d)** The spring force

$$\mathbf{F}_s = -k\mathbf{r}; \quad k\ \text{a constant.}$$

13.5 *(3) Turnabout is fair play.* Suppose you interchange the x and y axes in Figure 13.7b. Show that you can construct this coordinate system by following the instructions in the text but using your *left* hand.

13.6 *(3) Three-fifths of a five-finger exercise?* You can use the thumb, index finger, and middle finger of your right hand to construct a right-handed coordinate system by holding them as shown below. (This is the only way you can comfortably

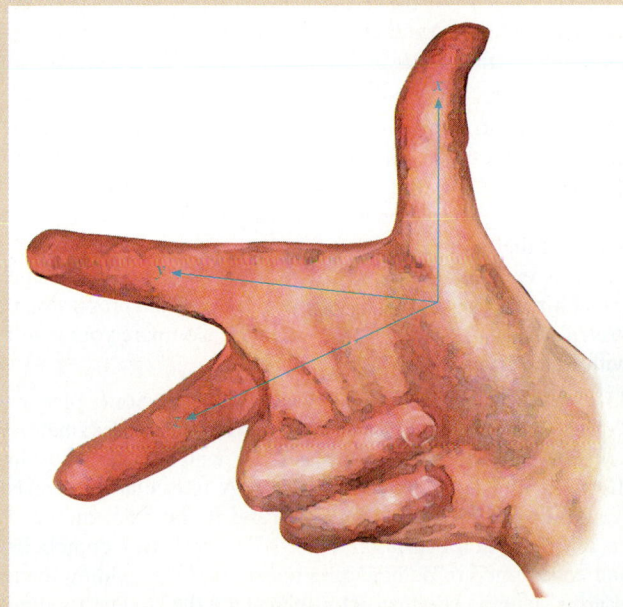

extend those three fingers in mutually perpendicular directions.) Label the thumb x, the index finger y, and the middle finger z. By appropriate rotations of your wrist, show that the cyclic rule (Equations 13.29a, 13.29b, and 13.29c) is satisfied.

13.7 *(4) Crossfire.* If $\mathbf{C} = \mathbf{A} \times \mathbf{B}$, it is always true that $\hat{\mathbf{C}}$ is normal to the plane defined by $\mathbf{A}$ and $\mathbf{B}$. Then why in general is $\hat{\mathbf{C}} \neq \hat{\mathbf{A}} \times \hat{\mathbf{B}}$? Under what *special* circumstances is it true that $\hat{\mathbf{C}} = \hat{\mathbf{A}} \times \hat{\mathbf{B}}$?

13.8 *(5) Just rollin' along.* You watch a ball roll down a plane that is inclined downward toward the right. What is the direction of the angular momentum of the ball?

13.9 *(5) Once a month.* Consider (1) the angular velocity $\boldsymbol{\omega}_o$ and angular momentum $\mathbf{L}_o$ of the moon's revolution in its orbit around the earth and (2) the angular velocity $\boldsymbol{\omega}_s$ and angular momentum $\mathbf{L}_s$ of the moon's rotation about its own axis. From the point of view of an observer moving with the center of mass of the earth-moon system, which of the following statements are true? **(a)** $\boldsymbol{\omega}_s = \boldsymbol{\omega}_o$. **(b)** $\boldsymbol{\omega}_s = -\boldsymbol{\omega}_o$. **(c)** $\mathbf{L}_o = \mathbf{L}_s$. **(d)** $\mathbf{L}_o > \mathbf{L}_s$.

13.10 *(5) Revolutionary thinking.* Which is greater, the angular momentum of the moon's monthly rotation about the earth or the angular momentum of the moon's annual rotation about the sun? (The moon–sun distance is roughly 400 times the moon–earth distance.)

13.11 *(5) Spin control.* In a common lecture demonstration, a student sits on a high stool whose seat and footrest are mounted on a low-friction bearing. The instructor hands the student a spinning bicycle wheel weighted like that described in Section 13.7. The axis of the wheel is vertical, and the wheel is spinning counterclockwise as seen from above. The student now inverts the wheel, turning the axle so that it is again vertical with the wheel spinning clockwise. What happens to the student? Explain.

13.12 *(6) Single or couple?* You can start a flywheel spinning by applying a force to its rim. Does this contradict the argument at the beginning of Section 13.6, according to which a system can be made to rotate without translation only if a couple is applied to it, rather than a single force? Explain.

13.13 *(7) Constant **a**, constant $\boldsymbol{\Omega}$?* A top precesses because the couple comprising the gravitational force and the support force exerts a constant torque on it. Given Newton's second law, $\mathbf{T} = I\boldsymbol{\alpha}$, why doesn't the precession speed up?

13.14 *(7) Two-wheeler.* Why does a moving bicycle stay up even if the rider leans to one side? What happens to this stability as the bicycle slows to a stop?

13.15 *(7) Flying lesson.* Single-engine airplanes are designed so that the propeller turns clockwise as seen by the pilot in the cabin behind it. When you turn a single-engine airplane in one direction, the nose tends to rise, and, when you turn the airplane in the other direction, the nose tends to drop; you must use the controls to compensate for this tendency. Explain what causes the tendency. In particular, does the nose tend to rise or fall when you turn left?

13.16 *(G) The wild Australian coriolis.* Consider a northern-hemisphere ocean current such as the Gulf Stream. As any small volume of water moves northward on the ocean surface, its distance from the earth's axis decreases. Explain why northern-hemisphere currents tend to rotate in a clockwise sense, as seen from above. (Southern-hemisphere currents tend to rotate counterclockwise for essentially the same reason. However, the argument suggested here can only partially explain the phenomenon. A complete analysis leads to the definition of a fictitious force, related to the centrifugal force and called the *Coriolis force*.)

13.17 *(G) Planetary embryology.* Astronomers are in general agreement that the solar system had its origin in a lens-shaped cloud of gas and small particles; its diameter was 10^3 times

that of the present solar system or greater, and it rotated slowly about its symmetry axis. Roughly 5×10^9 years ago, most of the cloud collapsed and condensed to form the sun, but some was left behind to condense into the much smaller planets and their satellites. Explain how this view is consistent with the observation that, with very few exceptions, the angular momentum vectors of revolution and rotation of all of the bodies in the solar system point in the same general northward direction as that of the earth's rotation.

PROBLEMS

GROUP A

13.1 *(2) Spinning wheel, I.* A flywheel of mass 155 kg has the form of a disk of diameter 37.5 cm. Find its angular momentum when it is rotating at 2550 rev/min.

13.2 *(2) Spinning wheel, II.* A brake is applied to the wheel of Problem 13.1, and the angular speed of the wheel is reduced to 1630 rev/min. What angular impulse is imparted to the wheel by the brake?

13.3 *(2) Angular momentum without angles?* A particle of mass 2.3 kg moves in the positive y direction, at speed 10.3 m/s, at a distance $+1.8$ m from the y axis. Find the angular momentum of the particle **(a)** about the origin; **(b)** about the point $(4.1, -0.3)$ m.

13.4 *(2) A rolling ball loses no fuzz.* A tennis ball of mass 41 g and diameter 6.4 cm rolls without slipping down an inclined plane, starting from rest. When the ball has descended a vertical distance of 25 cm, what is its angular momentum about the axis of rotation through its center?

13.5 *(2) A rolling ball grows no hair, either.* A billiard ball of mass 420 g and diameter 5.1 cm rolls without slipping down an inclined plane, starting from rest. When the ball has descended a vertical distance of 25 cm, what is its angular momentum about the axis of rotation through its center?

13.6 *(2) Angular momentum and kinetic energy, I.* You have seen in Query 10.7 that the translational kinetic energy K_t and linear momentum $\mathbf{p}$ of a body of mass m bear the relation $K = p^2/2m$. Find the analogous relation between rotational kinetic energy K_r and angular momentum L.

13.7 *(2) Angular momentum and kinetic energy, II.* A small flywheel rotates with angular speed 3000 rev/min. Its kinetic energy is 6000 J. Find its angular momentum.

13.8 *(2) Earth and moon, I.* The distance from the earth to the moon is about 60 times the radius of the earth. The mass of the moon is about $\frac{1}{81}$ that of the earth. The angular speed of the earth's rotation about its own axis is about 27 times as great as that of the moon about the earth. Make the crude assumptions that the earth is a uniform sphere and that the moon revolves about the center of the earth, and neglect the rotation of the moon about its own axis. What fraction of the total angular momentum of the earth-moon system is due to the moon's motion?

13.9 *(2) Putting the shoe on the other foot.* The mass of Ju-

piter is about $\frac{1}{1000}$ that of the sun, and the distance from the sun to Jupiter is about 1000 times the radius of the sun. Jupiter revolves around the sun about once every 12 years; the sun rotates on its own axis about once every 30 days. **(a)** Show that Jupiter carries a considerably larger proportion of the angular momentum of the solar system than the sun does. **(b)** Does Jupiter or the sun have greater kinetic energy of rotation about the center of mass of the solar system (which lies close to the center of the sun)?

13.10 *(2) Child-powered carousel.* Many playgrounds have small carousels. To first approximation, such a carousel consists of a disk-shaped turntable, of mass 45 kg and radius 1.2 m, mounted on a vertical shaft. A child of mass 30 kg takes a running start along a path tangent to the rim of carousel and jumps on. **(a)** If the carousel is initially at rest and the child's running speed is 4.5 m/s, what is the angular speed of the carousel after the child jumps on? **(b)** What is the fractional loss $\Delta K/K$ in the kinetic energy of the system?

13.11 *(2) Merry-go-round catch.* A merry-go-round makes a complete circle once every 10 s. A child riding on a dragon located 3.0 m from the axis throws a ball of mass 0.25 kg radially outward. Another child, riding on a sea serpent located 6.5 m from the axis, catches the ball. What angular impulse must the second child impart to the ball?

13.12 *(2) Linking flywheels.* Consider a two-flywheel system like that of Example 13.3, in which the moment of inertia of flywheel 1 is I and that of flywheel 2 is nI. **(a)** If flywheel 1 has initial angular velocity ω_i and flywheel 2 is initially at rest, find the final angular velocity ω_f of the system after the clutch is engaged. **(b)** Find the ratio K_f/K_i of the final and initial kinetic energies of the system. **(c)** Find the proportional loss of energy, $|\Delta E/E_i| = (E_i - E_f)/E_i$. Compare your result with Equation 10.4.

13.13 *(2) The three-dumbbell demonstration.* Some physics instructors have a taste for practical jokes. In the days before lawsuits became stylish, they found a welcome occasion in the following lecture demonstration. A muscular student is seated on a high stool whose seat and footrest are mounted on a low-friction bearing. He is given two heavy dumbbells and encouraged to demonstrate his strength by holding them at arms' length. The instructor then starts the student rotating and asks him to pull the dumbbells quickly in toward his chest.

If the dumbbells are heavy enough, the sudden increase in angular speed is often sufficient to make the student fall off the stool. The traditional name of this exercise—"the three-dumbbell demonstration"—is self-explanatory as well as smug, though you may draw your own conclusion concerning whom it refers to. Suppose that the student's moment of inertia, not including the dumbbells, is 4.2 kg·m² with arms extended and 1.3 kg·m² with arms pulled in. Let the mass of each dumbbell be 10 kg. When the student's arms are extended, the dumbbells are held at a distance 0.86 m from the axis of the stool. Pulled in and held across his chest, the dumbbells lie at a negligible distance from the axis. **(a)** If the instructor starts the rotation at an initial angular speed of 1.2 rev/s, what is the final angular speed? Neglect the moment of inertia of the stool. **(b)** Explain qualitatively why the student falls off the stool.

13.14 *(2) Puzzling gadget, I.* The gadget illustrated below

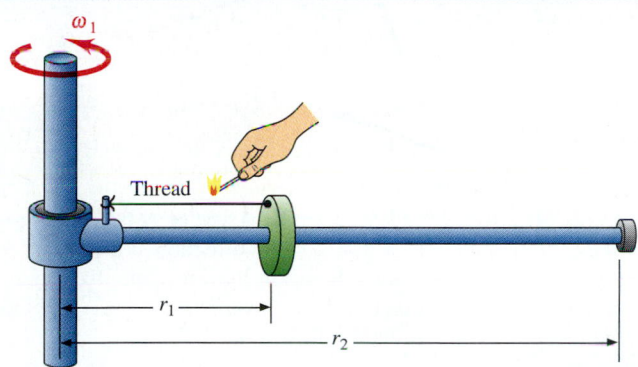

consists of a rod of length $r_2 = 0.80$ m and negligible mass on which a small cylindrical slug of mass 0.750 kg is free to slide without friction. One end of the rod is fixed to a vertical shaft on which the rod can rotate with negligible friction. At the other end is a stop that prevents the slug from slipping off. The slug is initially held at a distance $r_1 = 0.23$ m by a thread. The system is set spinning with angular speed $\omega_1 = 125$ rev/min. As it spins, the thread is burned and the slug slides out to the end of the rod. **(a)** Find the final angular speed ω_2 of the system. **(b)** Find the initial and final kinetic energy of the system.

13.15 *(3) Directional signals, I.* If $\hat{\mathbf{A}} = -\hat{\mathbf{z}}$ and $\mathbf{B} = -\hat{\mathbf{y}}$, find the direction of **(a)** $\mathbf{A} \times \mathbf{B}$; **(b)** $\mathbf{B} \times \mathbf{A}$.

13.16 *(3) Directional signals, II.* Vectors **A** and **B** lie in the

x-y plane. Find the direction of $\mathbf{A} \times \mathbf{B}$ if **(a)** **A** lies in the first quadrant and **B** in the second quadrant; **(b)** **A** lies in the first quadrant and **B** in the fourth quadrant.

13.17 *(4) Cross-product practice, I.* Find $\mathbf{A} \times \mathbf{B}$ if **(a)** $\mathbf{A} = \hat{\mathbf{x}}$ and $\mathbf{B} = \hat{\mathbf{x}} + \hat{\mathbf{y}}$; **(b)** $\mathbf{A} = \hat{\mathbf{x}}$ and $\mathbf{B} = \hat{\mathbf{x}} + \hat{\mathbf{z}}$; **(c)** $\mathbf{A} = \hat{\mathbf{x}} + \hat{\mathbf{y}}$ and $\mathbf{B} = \hat{\mathbf{y}} + \hat{\mathbf{z}}$.

13.18 *(4) Getting an angle on things.* If $\mathbf{A} = 5\hat{\mathbf{x}} - 3\hat{\mathbf{y}}$ and $\mathbf{B} = 2\hat{\mathbf{x}} + \hat{\mathbf{y}}$, find **(a)** $\mathbf{A} \times \mathbf{B}$; **(b)** the angle between **A** and **B**. (Hint: $|\mathbf{A} \times \mathbf{B}| = A \sin \phi\, B$.)

13.19 *(4) Cross-product practice, II.* Find $\mathbf{A} \times \mathbf{B}$ if **(a)** $\mathbf{A} = 5\mathbf{y}$ and $\mathbf{B} = 2\hat{\mathbf{x}} + 2\hat{\mathbf{y}}$; **(b)** $\mathbf{A} = 4\hat{\mathbf{x}} - \hat{\mathbf{y}}$ and $\mathbf{B} = 3\hat{\mathbf{z}}$; **(c)** $\mathbf{A} = (1, 0, 3)$ and $\mathbf{B} = (0, -2, 0)$; **(d)** $\mathbf{A} = (1, 2, 3)$ and $\mathbf{B} = (3, 2, 1)$.

13.20 *(4) Cross-product practice, III.* For each of the four pairs of vectors in Problem 13.19, find the angle between **A** and **B**.

13.21 *(4) Calculating torque, I.* A force $\mathbf{F} = (7, 3, 0)$ N acts on a particle located at $\mathbf{r} = (2, -1, 0)$ m. Find the torque about the origin.

13.22 *(4) Calculating torque, II.* A force $\mathbf{F} = (6, -4, 1)$ N is applied to a point with coordinates $(-0.2, 1.6, -0.5)$ m. Find the torque exerted about the origin.

13.23 *(4) Very special case.* Two vectors **A** and **B** satisfy the equation

$$\mathbf{A} \cdot \mathbf{B} = |\mathbf{A} \times \mathbf{B}|.$$

What can you say about **A** and **B**?

13.24 *(4) Dotting every* **A** *and crossing every* **B**. Show that

$$\mathbf{A} \cdot \mathbf{B} \times \mathbf{A} = 0.$$

13.25 *(5) Acting on impulse.* The equation $\mathbf{G} = \mathbf{r} \times \mathbf{J}$ in Table 13.1 is not derived explicitly in the text. Give a derivation.

13.26 *(5) A new twist.* The equation $\mathbf{T} = d\mathbf{G}/dt$ in Table 13.1 is not derived explicitly in the text. Give a derivation.

13.27 *(7) Hanging a left.* A 10-kg mass is hung from the free end of the weighted bicycle wheel–axle system of Example 13.7, at a distance 40.0 cm from the support point. Find the precession velocity Ω.

13.28 *(7) Childhood revisited.* The rotor (wheel) of a toy gyroscope has radius 3 cm and mass 150 g. The rotor is mounted at the midpoint of the axis, which is 7 cm long. If the gyroscope precesses through one complete turn every 5 s, estimate the angular speed of the rotor.

GROUP B

13.29 *(2) Less than meets the eye.* The electron has been subjected to a wide variety of experimental measurements. Some of the results are approximately as follows:

- Mass: $m_e = 9 \times 10^{-31}$ kg.
- Intrinsic angular momentum: $L = 5 \times 10^{-35}$ kg·m²/s.
- Effective radius: $r_{eff} \leq 5 \times 10^{-15}$ m.

It is tempting to visualize the angular momentum by imagining the electron to be a little spherical ball of radius r_e that spins about an axis through its center, as (for example) the earth does. But this picture cannot be correct. To see one rea-

son why, find the necessary linear speed of a point on the equator of the electron "ball" if the values given above are to be satisfied. Compare this speed with the speed of light, $c = 3 \times 10^8$ m/s. One of the fundamental facts of nature is that c is an upper limit on the speed at which matter can move.

13.30 *(2) Too warm.* If the earth were to warm up sufficiently to melt the polar ice caps, the general sea level would rise by about 60 m. Take 6.4×10^6 m for the radius of the earth and 8.0×10^{37} kg·m² for its moment of inertia, and show that the length of the day would increase by about 1 s. (The mass of 1 m³ of water is 10^3 kg.)

13.31 (2) *Puzzling gadget, II.* Consider the gadget of Problem 13.14 in a more general way. Let the rod and the slug each have mass m. Call the initial distance from the central shaft to the slug r_1, and call the length of the rod r_2. **(a)** If the system rotates with initial kinetic energy K_1, find its kinetic energy K_2 after the thread is burned but just before the slug hits the stop. Why are you justified in conserving kinetic energy? **(b)** Just before the slug hits the stop, what is its radial speed (the component of its velocity along the rod)? What is its tangential speed (the component of its velocity perpendicular to the rod)? **(c)** Find the kinetic energy K_2' after the slug hits the stop.

13.32 (2) *Earth and moon, II.* The calculation of Problem 13.8 is based on the crude assumption that the moon revolves about the center of the earth. In fact, both earth and moon revolve about their common center of mass. In addition, the moon rotates once on its own axis each time it circles the earth. (That is why we always see the same side of the moon.) **(a)** Where is the center of mass of the earth-moon system?

Find the fraction of the total angular momentum of the earth-moon system due to **(b)** the orbital motion of the earth; **(c)** the rotation of the earth about its axis; **(d)** the orbital motion of the moon; **(e)** the rotation of the moon about its axis. Use the ratios given in Problem 13.8, and take the radius of the moon to be 0.27 that of the earth. Assume both moon and earth to be uniform spheres (of different densities).

13.33 (2) *The tie that binds.* The disk illustrated below, whose mass is M and whose radius is R, rotates without friction on a hollow vertical shaft. A small body of mass m can slide frictionlessly along a radial guide. The small body is initially at the periphery of the disk. A string attached to the body passes along the guide and into and through the hollow shaft. The system has initial angular speed ω_0. The string is pulled (see black arrow) until the body lies a negligible distance from the shaft. **(a)** Find the final angular speed ω_f of the disk. **(b)** How much work is done in pulling the string?

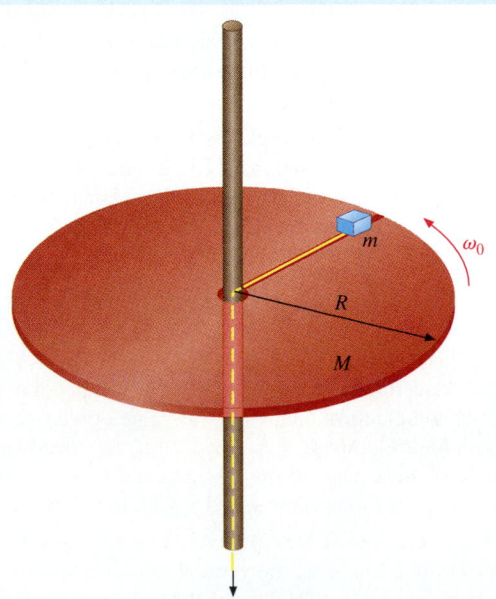

13.34 (2) *Momentous occasion, I.* A solid sphere and a solid cylinder having the same mass and radius roll the same dis-

tance down the same inclined plane. Both start from rest and both roll without slipping. Show that the angular momenta of the two about their centers of mass are in the ratio

$$\frac{L_{sph}}{L_{cyl}} = 0.83.$$

13.35 (4) *Drawing parallels.* In the illustration below, the vectors **A** and **B** define a parallelogram. The line of **A** and the side of the parallelogram opposite **A** are extended indefinitely. Show that *any* vector **B′** drawn so that its tail lies on the line of **A** and its head on the parallel line satisfies the relation

$$\mathbf{A} \times \mathbf{B} = \mathbf{A} \times \mathbf{B'}.$$

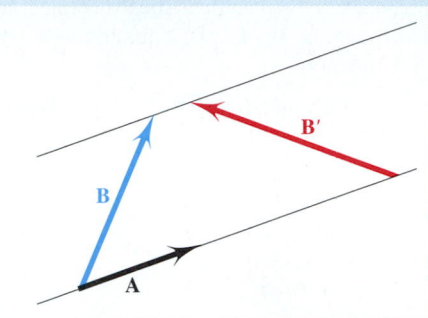

13.36 (4) *Take care when crossing.* A vector **A** lies in the x-y plane. Its magnitude is 3.4, and its direction is 36° from $\hat{\mathbf{x}}$ toward $\hat{\mathbf{y}}$. A second vector **B** lies in the y-z plane. Its magnitude is 2.7, and its direction is 67° from $-\hat{\mathbf{y}}$ toward $-\hat{\mathbf{z}}$. Find the magnitude and direction of $\mathbf{A} \times \mathbf{B}$.

13.37 (4) *Volume operation.* The double operation represented by $\mathbf{A} \cdot \mathbf{B} \times \mathbf{C}$ is called the *scalar triple product*. **(a)** Why is it unnecessary to use parentheses to denote the fact that the cross product must be carried out before the dot product? **(b)** Show that $\mathbf{A} \cdot \mathbf{B} \times \mathbf{C}$ is the volume of the parallelepiped whose edges are defined by the vectors **A**, **B**, and **C**. **(c)** Show that $\mathbf{A} \cdot \mathbf{B} \times \mathbf{C} = \mathbf{A} \times \mathbf{B} \cdot \mathbf{C}$.

13.38 (4) *Scalar triple product.* Show that the scalar triple product can be written in the determinant form

$$\mathbf{A} \cdot \mathbf{B} \times \mathbf{C} = \begin{vmatrix} A_x & A_y & A_z \\ B_x & B_y & B_z \\ C_x & C_y & C_z \end{vmatrix}.$$

13.39 (5) *Angular momentum of a free particle.* A particle of mass m moves freely with linear momentum **p**. At a certain instant, it is located at **r** with respect to a point O, as shown.

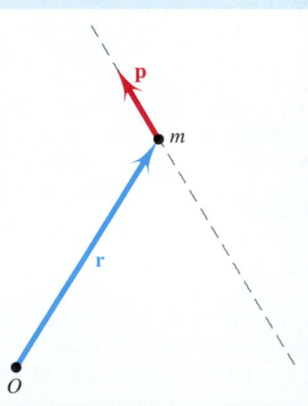

(a) At this instant, what is the angular momentum **L** of the particle with respect to O? (b) Does **L** change with time? Why or why not?

13.40 (5) *Central force.* A central force acts on a body as the body moves along an arbitrary path. Prove that the torque exerted on the body by the force is always zero and that the central force therefore cannot change the angular momentum of the body.

13.41 (5) *Modeling an ice skater.* A particle of mass m lies on a frictionless table, as illustrated below. The particle is tied to a string that passes through a small hole in the table. As long as the other end of the string is held fixed, the mass must move in a circle centered on the hole. Suppose that the distance from the hole to the mass is r_i and that the mass moves with angular velocity ω_i. (a) What are the angular momentum L_i and the kinetic energy K_i of the system? (b) The lower end of the string is now pulled slowly downward, reducing the radius of the circle in which the mass moves. At an instant when the radius is r, a force $F(r)$ is required to hold the string. Is $F(r)$ a central force? (c) The string is pulled until r attains a final value r_f. What is the final value L_f of the angular momentum of the system? What is the final value ω_f of the angular velocity? What is the final value K_f of the kinetic energy? (d) Evaluate $F(r)$ explicitly, and find how much work W must be done to reduce the radius from r_i to r_f. (e) Use the values of K_i and W to find K_f. Does your result agree with that of part **c**?

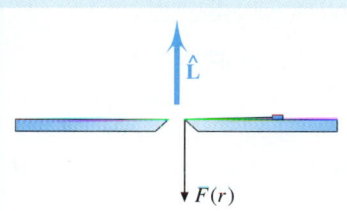

13.42 (5) *Determining the angular momentum.* At a certain instant, a particle is located at **r** with respect to an observer O and is moving with velocity **v**. Show that the angular momentum of the particle with respect to the observer can be written as the determinant

$$\mathbf{L} = \begin{vmatrix} \hat{\mathbf{x}} & \hat{\mathbf{y}} & \hat{\mathbf{z}} \\ r_x & r_y & r_z \\ v_x & v_y & v_z \end{vmatrix}.$$

13.43 (6) *Conserve everything!* A dumbbell consists of two small bodies, each of mass $m/2$, fastened to the ends of a light rod of length l. The dumbbell lies on a frictionless table, as shown below. Another small body, of mass m, approaches the

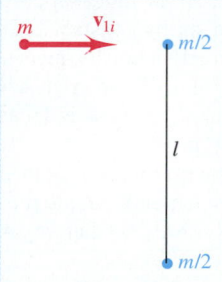

dumbbell with velocity $\mathbf{v}_{1i}$ in a direction perpendicular to the length of the dumbbell. It collides elastically with one of the ends of the dumbbell. Show that after the collision (a) the velocity of the free body is $\mathbf{v}_{1f} = \frac{1}{3}\mathbf{v}_{1i}$; (b) the center of mass of the dumbbell has velocity $\mathbf{v}_{cm} = \frac{2}{3}\mathbf{v}_{1i}$; (c) the angular speed of the dumbbell about its center of mass is $\omega = 4v_i/3l$.

13.44 (6) *Equivalent couples.* The illustration below shows a couple $\mathbf{T}_1$ consisting of the forces $\mathbf{F}_1$ and $-\mathbf{F}_1$ applied along lines of action separated by $2a_1$. A second couple $\mathbf{T}_2$ consists of the forces $\mathbf{F}_2$ and $-\mathbf{F}_2$ separated by $2a_2$. The relation $2a_2F_2 = 2a_1F_1$ is satisfied. Use a construction in the spirit of Figure 13.20 to show that $\mathbf{T}_2 = \mathbf{T}_1$. (Hints: A couple is not affected by moving the force vectors along their own lines of action. Applying a pair of equal and opposite forces along the diagonal shown in red does not change anything.)

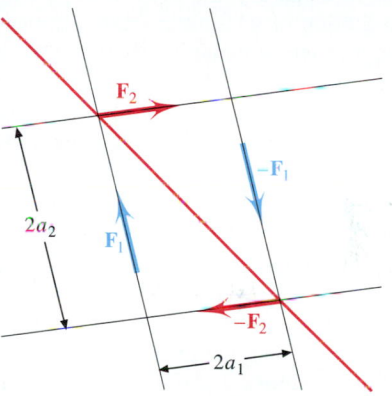

13.45 (6) *Different axes, same angular velocity.* Use the vector relation $\mathbf{v} = \boldsymbol{\omega} \times \mathbf{r}$ to prove that, when a wheel rolls without slipping, its angular velocity about the instantaneous axis is the same as that about the center; see the figure below. Compare with the proof illustrated by Figure 12.14 and the proof that is the subject of Problem 12.50.

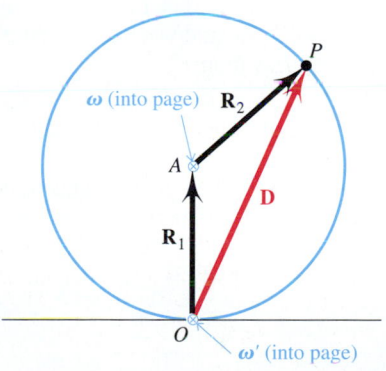

13.46 (6) *Momentous occasion, II.* A solid sphere and a solid cylinder having the same mass and radius roll the same distance down the same inclined plane. Both start from rest and both roll without slipping. Show that the angular momenta of the two, taken about a fixed point on the surface of the plane, are in the ratio

$$\frac{L'_{sph}}{L'_{cyl}} = 0.97.$$

(Hint: See Problem 13.34.)

13.47 *(6) Momentous occasion, III.* A body having uniform radius and shape factor c rolls without slipping down an inclined plane, starting from rest. **(a)** Show that, at any instant, the ratio of the angular momentum L' taken about a fixed point on the surface of the plane to the angular momentum L taken about the center of mass of the body is

$$\frac{L'}{L} = 1 + \frac{1}{c}.$$

(b) What is the shape of the body for which this ratio is greatest—that is, for which the spin angular momentum of the body makes the smallest proportional contribution to L'? Explain your result in a qualitative way.

13.48 *(6) Separating angular momentum into spin and orbital parts.* An observer O at rest watches a body of mass M move past her, spinning as it goes. A second observer O' moves with the center of mass of the body. Compare their observations of the motion of an arbitrary mass element Δm_j of the body. By appropriate summation, show that $\mathbf{L}_t$, the total angular momentum of the body as seen by O, is equal to the sum of the angular momentum $\boldsymbol{\Lambda}$ of the center of mass about O (which we have called the orbital angular momentum) and the angular momentum $\mathbf{L}$ of the body about its center of mass (which we have called the spin angular momentum). (Hint: See Figure 12.12.)

13.49 *(6) Quick route to Equation 12.34.* Starting at rest, a round body of radius R slides down a frictionless plane of incline angle θ without rolling. **(a)** Using the initial point of contact of the body with the plane as a reference point, find the angular momentum L_s of the sliding body as a function of time. **(b)** The same body is now made to roll without slipping down a rough plane having the same incline angle. If the moment of inertia of the body about the axis through its center is $I = cMR^2$, find the total angular momentum L_r of the rolling body as a function of time about the same reference point as in part **a**, and show that $L_r = L_s$. **(c)** Explain how this result can be regarded as a special case of the general result of Problem 13.48. **(d)** Use the fact that $L_r = L_s$ to show that the acceleration of a body rolling down an inclined plane without slipping is given by Equation 12.34,

$$a_{cm} = \frac{g \sin \theta}{1 + c}.$$

13.50 *(6) Rolling with slipping.* In general, a round body may slip as it rolls down an inclined plane. Generalize the argument of Problem 13.49 to show that, if at any instant the speed of the center of mass of the body is v, its angular speed will be

$$\omega = \frac{v_s - v}{cR},$$

where v_s is the speed the body would have had at the same instant if it had slid frictionlessly (without rolling) down the same plane.

13.51 *(6) Linear momentum and angular momentum.* Observer O makes measurements on a system of particles (which may or may not constitute a rigid body) and finds the system to have linear momentum $\mathbf{p}$ and angular momentum $\mathbf{L}$. Observer O' is located at $\mathbf{R}$ with respect to O. **(a)** Express the

angular momentum $\mathbf{L}'$ that O' measures in terms of $\mathbf{L}$, $\mathbf{R}$, and $\mathbf{p}$. **(b)** Under what circumstances will $\mathbf{L}$ and $\mathbf{L}'$ be equal? That is, what is the condition for which the angular momentum of the system is independent of the position of the observer?

13.52 *(6) Backspin.* A child throws a hoop so that it strikes the ground with backspin, as shown below. Suppose that the angular momentum of the hoop—referred to O, the point of initial contact with the ground—is zero. **(a)** With O as the reference point, what is the net torque at any instant as the hoop skids along the ground? Neglect all friction other than the skidding friction between hoop and ground. **(b)** What is the angular momentum of the hoop at any time? **(c)** When the skidding stops, what is the final speed v_f of the center of mass of the hoop?

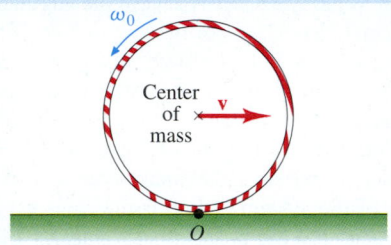

13.53 *(7) Orbital angular momentum of a top.* For the top of Figures 13.22 and 13.25, show that the angular momentum resulting from the precession of the center of mass about the vertical is given by the expression

$$\boldsymbol{\Lambda} = \frac{M^2 g D^3 \sin^2 \theta}{L} \hat{\mathbf{y}},$$

where L is the magnitude of the spin angular momentum of the top (its angular momentum about its axis), provided that $\Lambda \ll L$.

13.54 *(7) Spherical top.* A top is made by attaching a small pointed tip to the surface of a solid spherical ball. Express the precession velocity for such a top in terms of its spin angular speed ω, its radius R, and necessary constants.

13.55 *(7) Conical top.* A traditional toy top, approximately conical in form, is made to spin about its vertex. Show that its precession velocity is

$$\Omega = \frac{1}{\omega} \frac{5g}{2h \tan^2 \phi},$$

where h is the height of the top and ϕ is the half-angle of the cone vertex. (Hint: You will save time if you use the results of Problems 9.53 and 11.63.)

13.56 *(G) In & out, round & round.* In the illustration at the top of the next page, an air table supports a mass-spring system with mass m and spring constant k. The mass is connected to the spring not directly, but by means of a string passing through a small frictionless hole in the center of the table. The spring is relaxed when the mass is located at the hole. You stretch the spring by pulling the mass outward to a distance r_1 and release the system by shoving the mass along the table surface in a direction perpendicular to the spring with an initial speed v_1. If you choose v_1 so that $kr_1 = mv_1^2/r_1$, the mass

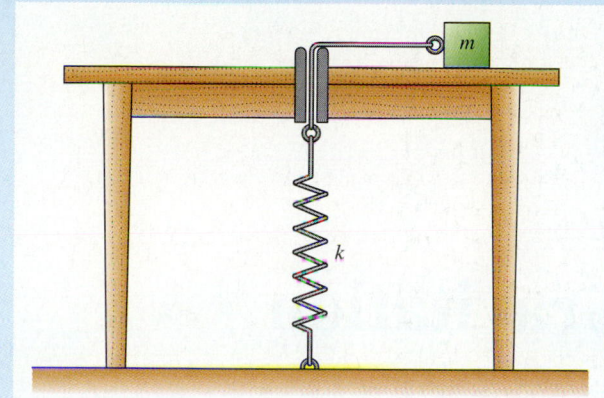

will revolve around the hole in a circular path. But suppose v_1 is smaller than this value. As a result, the system whirls around the hole as the mass oscillates toward and away from the hole. Neglecting the effect of gravity on the spring, show that, after the system is set into motion, the minimum distance between the mass and the hole is

$$r_2 = v_1 \sqrt{\frac{m}{k}}.$$

(Hint: What quantities are conserved?)

GROUP C

13.57 *(6) No-skid bounce.* A toy marketed under the name Superball experiences a nearly elastic collision when bounced against a hard surface. In addition, the coefficient of static friction between ball and surface is so great that there is no skidding even if the ball is spinning rapidly. The ball is uniform and solid. Suppose you throw it toward the right against a sidewalk with speed v_0 at an angle θ, as shown in the diagram below. If you give it a strong counterclockwise spin having the proper value ω_0, the ball will rebound vertically. **(a)** Assuming that the bounce is elastic, show that the proper value is

$$\omega_0 = \frac{3}{4} \frac{v_0 \cos \theta}{R},$$

where R is the radius of the ball. [Hint: See Query 12.12. You will need to set up a pair of simultaneous equations containing the unknown post-bounce quantities ω and v. One of the equations arises from energy conservation. To set up the other, consider the way in which the normal and frictional forces operate to convert kinetic energy back and forth between its translational and rotational forms. In particular, remember that you can write the initial translational kinetic energy as $\frac{1}{2}mv_0^2(\cos^2 \theta + \sin^2 \theta)$.] **(b)** What are the sense and magnitude of the final angular velocity of the Superball? **(c)** Show that conservation of total mechanical energy in an elastic collision requires that the final speed be $v = v_0 \sin \theta$.

mounted between two bearings. Two rods of length y and negligible mass are mounted on the shaft perpendicular to its length and a distance $2z$ apart. At the end of each rod is a particle of mass m. The shaft is **statically balanced**. That is, when it is at rest, it shows no tendency to rotate to any particular equilibrium position. However, it is **dynamically unbalanced**, as the following discussion shows. **(a)** Find the center of mass of the system. **(b)** When the shaft rotates with angular velocity ω, show that the magnitude of the angular momentum about the center of mass may be written

$$L = 2m\sqrt{y^2 + z^2}\, y\omega.$$

(c) Show that, at any instant, the direction of **L** is perpendicular to the line joining the two particles and lies in the plane defined by that line and the shaft. **(d)** Show that the component vector of angular momentum parallel to the shaft, $\mathbf{L}_z$, is constant but that the component vector perpendicular to the shaft, $\mathbf{L}_\perp$, varies with time. **(e)** Find the magnitude and direction of the torque exerted by the shaft on its bearings. **(f)** Find the magnitude and direction of the force exerted on either bearing at the instant depicted in the figure. **(g)** The dynamical unbalance implied in the answers to parts **e** and **f** is almost always undesirable because it leads to vibration and excessive bearing wear. How can you balance the shaft dynamically? That is, how can you add masses to the system in such a way that the rotating shaft will exert no torque on its bearings?

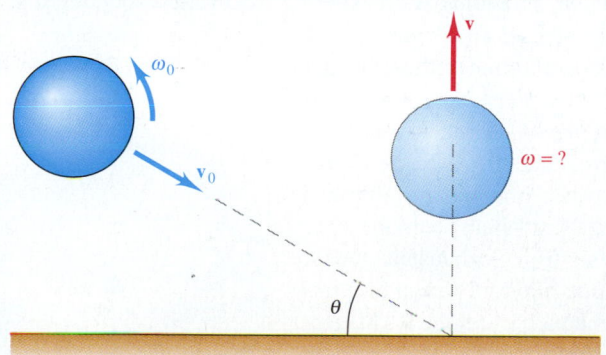

13.58 *(G) Dynamically unbalanced shaft.* The figure opposite is a simplified and idealized representation of a situation often encountered in engineering design. (Indeed, you may have encountered the situation in the context of balancing automobile wheels.) A shaft of length D and negligible mass is

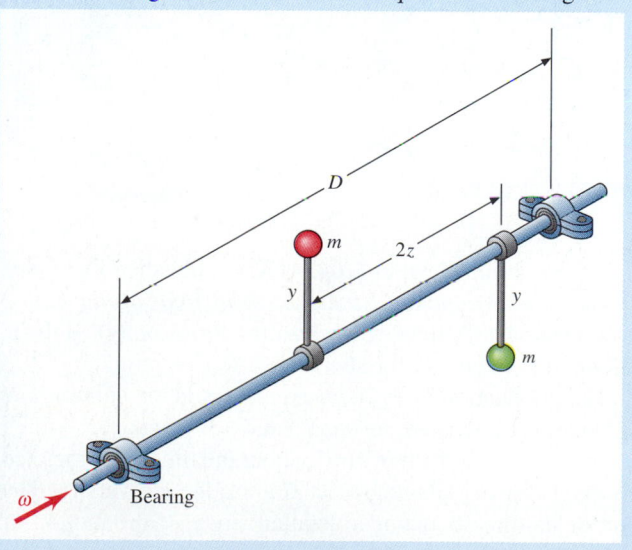

Part V

Universal Gravitation

In 1687, Isaac Newton (1642–1727) published his major work, *Philosophiae Naturalis Principia Mathematica* (Mathematical Principles of Natural Philosophy), universally called *Principia* for short.

The publication of *Principia* represented the culmination of Newton's intensive, sustained study of mechanics, which had begun to bear fruit in 1665, about the time he completed his Bachelor of Arts degree at Cambridge University. The first of its three books is a detailed study of the motion of bodies under the influence of a variety of forces, with emphasis on central forces. Without considering the origins of the forces, Newton uses the three laws named after him to deduce the paths of the bodies. This is a major aspect of what we call Newtonian, or classical, mechanics, though there is much more to Newtonian mechanics, as you have already seen. Other matters are discussed in Book II of *Principia*, including the design of isochronous pendulums, the compression of gases, and the motion of bodies through fluids. Very much more still remained to be worked out by others in subsequent centuries, and the possibilities of Newtonian mechanics are far from exhausted today. But Newton's great and immediate triumph was his rigorous demonstration in Book III that the universe is held together, on the global and astronomical scale, by a universal gravitational force that every body in the universe exerts on every other body.

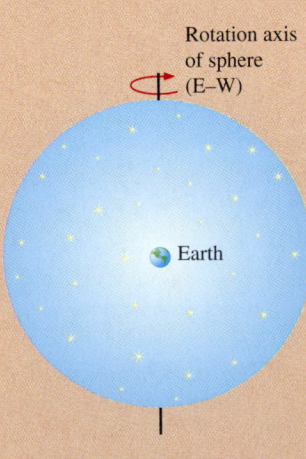

Rotation axis
of sphere
(E–W)

Earth

We will now follow Newton's lead: in Chapter 14, we will apply what we have developed of mechanics in the earlier part of this book to a study of the effects of the gravitational force. Newton fully understood the significance of what he had done, and he also had expectations for the future that turned out to be prophetic. He says in his preface:

... I offer this work as the mathematical principles of philosophy [science], for the whole [task] of philosophy seems to consist in this—from the phenomena of motions to investigate the forces of nature, and then from these forces to demonstrate . . . other phenomena. . . . [B]y the propositions mathematically demonstrated, . . . I derive from [astronomical observations] the forces of gravity with which bodies tend to the sun and the several planets. Then from these forces, by other propositions which are also mathematical, I deduce the motions of the planets, the comets, the moon, and the sea [the tides]. I wish we could derive the rest of the phenomena of Nature by the same kind of reasoning from mechanical principles, for

I . . . suspect that they may all depend upon certain forces by which the particles of bodies, by some causes hitherto unknown, are either mutually impelled towards one another, and cohere in regular figures, or are repelled and recede from one another. These forces being unknown, philosophers have hitherto attempted the search of Nature in vain; but I hope the principles here laid down will afford some light either to this or some truer method of philosophy.

In his *Dialogue* of 1632, Galileo argued persuasively but in mainly qualitative terms that the same laws govern the motions of the heavens and the motions of bodies on earth. Newton's arguments are not only rigorous but quantitative: They can be used to make very precise predictions that can be confirmed by observation or experiment. In this sense, Newton's work represents the fulfillment of Galileo's program. If modern physical science was conceived in Galileo's *Dialogue*, it was Newton's *Principia* that gave it birth.

Gravitation

How do the planets move? The solution of this problem was seriously pursued for more than two millennia.

Isaac Newton based a dynamic description of planetary motion on his three laws of motion and his law of universal gravitation. Newton thus completed what Galileo had begun: He showed that the laws of physics extend to the astronomical realm.

The gravitational force exerted by one particle on another is proportional to the masses of the bodies and inversely proportional to the square of the distance between them.

How can we measure the proportionality constant in the law of gravitation (that is, the universal gravitational constant)?

What is a gravitational field? What is the potential energy of a body in a gravitational field?

Left: Composite photograph of the fragments of Comet P/Shoemaker-Levy 9 heading for collision with Jupiter, taken by the Hubble Space Telescope on 17–18 May 1994. Newton's second law of motion, together with his law of gravitation, made it possible to predict the details of the impact well before the event. The breaking of the comet into fragments is likewise a consequence of the law of gravitation, as is the motion of Io, the moon seen to the left of and a little below its own shadow.

> [T]here is a power of gravity pertaining to all bodies, proportional to the several quantities of matter which they contain. . . . The force of gravity towards the several equal particles of any body is inversely as the square of the distance between their centres.
>
> —ISAAC NEWTON, *Principia*

Introduction

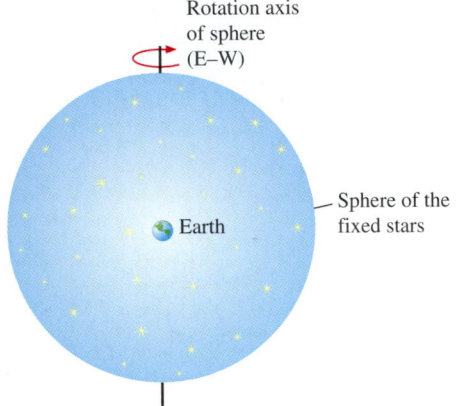

FIGURE 14.1 Model accounting for the grossest features of the observed motion of the heavens. The stars appear to be fixed to a large spherical shell concentric with the earth. The spherical shell appears to rotate once a day about an axis passing through the poles of the earth. The sense of rotation is from east to west, as shown. Which part of the stellar display an observer sees depends on his location on the earth. It also depends on the location of the sun, because the stars can be seen only when the sun is below the observer's horizon. The horizon is located by drawing the radius through the observer's location and then constructing the plane of the great circle normal to that radius.

If you watch the heavens night after night, as people have done since prehistoric times, a paradox emerges from the beauty of the continuously changing display. On the one hand, there is great simplicity; the myriad "fixed" stars appear to move in majestic unity, as though they were all attached to the inside of a giant sphere that rotates daily on an axis passing through the poles of the earth (Figure 14.1). On the other hand, there is bewildering complexity; the sun, moon, and planets not only share in the general motion of the fixed stars, but move relative to the stars. Although some of these paths are very complicated, there are tantalizing regularities to them. A diligent observer is easily persuaded that it should be possible to predict the motions with great accuracy on the basis of a relatively simple model.

Efforts to do this go as far back as human records go. Some of the efforts have been aimed at a kinematic description, with the limited goal of predicting the angular position of (say) a planet at any given time. Other efforts have aimed at a dynamic description, intended not only to predict the observed angular positions, but to understand the basic "machinery" underlying the observed motions.

But it turned out not to be an easy task to frame even a satisfactory kinematic model of the planets, the sun, and the moon—let alone a dynamic one. To the task the astronomers of the Hellenistic world turned their great talents in geometry. The long-term efforts of many were summarized and crowned in the second century C.E. by the work of the Alexandrian Claudius Ptolemy (c. 100–170). Ptolemy published a comprehensive work consisting of models, tables, and calculational methods and known by the name *Almagest*. This name, a Latinized form of the Arabized Greek one given it by the medieval Arabs who preserved it and copied it, means simply "The Greatest."

The Ptolemaic system is **geocentric**; it takes the very commonsensical view that the earth lies at the center of the universe, all of which revolves about the earth. The complicated motion of the planets is explained by means of a complex (and flexible) system of "wheels within wheels." Some astronomers believed that the Ptolemaic system, or some modification of it, was a description of the universe as it actually is. More astronomers probably took the view that the system was merely a mathematical artifice, useful for calculating and predicting angular positions, and that the "real" state of affairs was quite different—perhaps totally inaccessible to the human mind.

The Ptolemaic system was elaborated in a variety of ways for many centuries. But, in spite of all efforts, the system became less and less satisfactory. As observations of astronomical positions accumulated, making them fit the predictions of the Ptolemaic

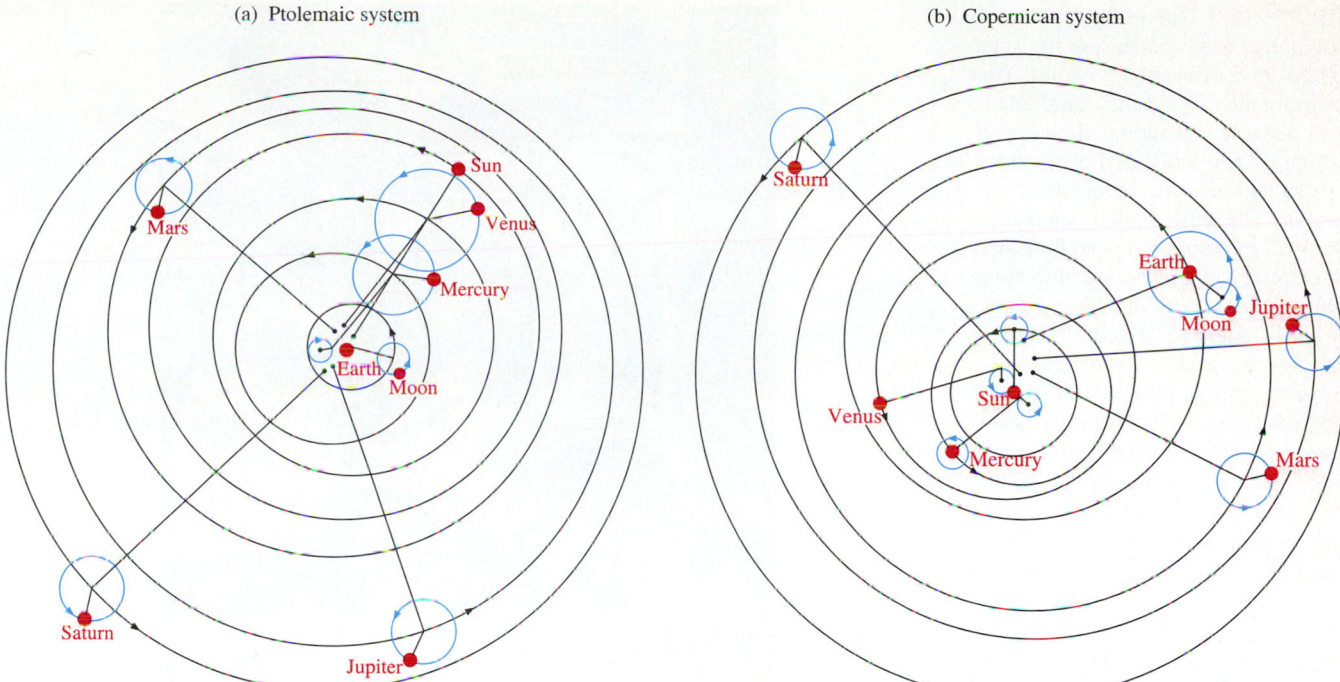

(a) Ptolemaic system   (b) Copernican system

FIGURE 14.2 Comparison of Copernicus's universe with a contemporary Ptolemaic version. All curves are circles; the large, black ones are called deferents, and the smaller, blue ones are called epicycles. The planets move around the epicycles, whose centers simultaneously move around the deferents. The center of each deferent, called its eccentric, is indicated by the inner end of its radius. The Ptolemaic model also includes a point called the equant for each deferent; the angular velocity is constant not about the eccentric but about the equant. In both pictures, some of the eccentrics themselves revolve on small deferents. In neither picture is it possible to fit the observations with orbits having a common center.

model required increasingly complicated and implausible adjustments of the model. The tables of the *Almagest* became less and less helpful. For this reason, King Alfonso the Wise of Spain (1221–1284) commissioned new astronomical tables; their intricacy is said to have prompted him to remark: "Had I been present at the creation, I would have suggested something simpler."

With this state of affairs, there was bound to be a reawakening of interest in alternatives to the Ptolemaic system, and it was in this context that a revolutionary step was taken by the Polish Prussian church administrator and amateur astronomer usually known by the Latin form of his name, Nicolaus Copernicus (1473–1543). Copernicus revived the long-neglected **heliocentric** view, putting the sun rather than the earth at the center of the universe. His motives for doing so were partly mystical and religious. But he also strove for a simplified kinematic explanation of the motions of the solar system. In this he was not completely successful. The Ptolemaic and Copernican models are compared in Figure 14.2. A quick comparison will suggest to you that the simplification is not dramatic. Moreover, there is little to choose between the two models on the grounds of the precision of their positional calculations.

The need for better agreement between theory and observation was made more pressing by the work of the Danish astronomer Tycho Brahe (1546–1601). In the twenty-year period from 1576 to 1596, Tycho made meticulous observations on nearly every clear night. For the purpose, he devised the best angle-measuring sights ever used up to the time; one of them is shown in Figure 14.3.

The most important contribution of Tycho Brahe was made indirectly, through the use of his detailed observational data by the German astronomer Johannes Kepler (1571–1630). Profoundly mystical by nature, Kepler was obsessed with the idea that the universe could be interpreted in the language of mathematics. Using the Copernican system as a basis, he spent many years in laborious trial-and-error calculations aimed at showing that the observational data could be accounted for on a simple geometric basis. In 1600, he was hired by Tycho Brahe as an assistant. The relation between the two men was stormy and unproductive, but Kepler came into possession of Tycho's data after the latter's death in 1601. These data were Kepler's gold mine. He knew that the largest angular error did not exceed 2 arc minutes, and he would accept no model of planetary motion that did not agree with the observations within this limit.

FIGURE 14.3 One of Tycho Brahe's quadrants, used to measure the angular position of heavenly bodies. This instrument is called the mural quadrant because it is mounted on a wall, which is seen indistinctly behind the clocks and beneath the quadrant proper. The angle scale is marked on the wall in degrees. Tycho is shown at extreme right, at F, sighting along the line defined by the sliding sight E on the angle scale and a fiducial (reference) mark at the central axis of the quadrant, located on the wall opposite at upper left. One assistant uses a candle to read the angle scale and the clocks, while another records data at the table at the lower left of the picture. In the inset within the quadrant is shown a larger portrait of Tycho, as well as various instruments and activities at his observatory, Uraniborg. The Latin inscription gives the date of the engraving as 1587 and Tycho's age as 40. Perhaps the function of the dog was to keep Tycho's feet warm on cold nights!

FIGURE 14.4 Kepler's first law. Every ellipse may be defined in terms of two points called its **foci** f_1 and f_2. For every point on the circumference of the ellipse, the sum of the distances to the foci is the same; $r + s =$ constant. The sun lies at one focus of the elliptical orbit of every planet. The planet is most distant from the sun at its **aphelion**. It is closest to the sun at its **perihelion**. Aphelion and perihelion lie at the ends of the **major** (longest) **axis** of the ellipse. The midpoint of the major axis is the center O of the ellipse. It lies midway between the foci f_1 and f_2, a distance λ from each. The **minor** (shortest) **axis** of the ellipse also passes through the center and is perpendicular to the major axis. The mean sun–planet distance $\langle r \rangle$ is called the **mean radius**. It is equal to the **semimajor axis**; $\langle r \rangle = \rho$. The eccentricity (deviation from circularity) of the planetary orbit is much exaggerated in this drawing. The orbits of the planets (except Pluto's) are in fact so close to circular that you cannot detect the deviation from circularity by simply looking at an accurate plot of such an orbit.

Kepler's Laws

After herculean trial-and-error efforts, Kepler discovered the three laws of planetary motion on which his reputation rests. The first two he published in 1609; the third, in 1619. For Kepler, these laws were purely kinematic. That is to say, their validity rested entirely on the beautiful precision with which they agreed with Tycho's excellent observations. **Kepler's laws** are as follows:

1. The orbits of the planets (including the earth) are ellipses, with the sun at one focus (Figure 14.4).

2. The radius vector from the sun to a planet sweeps out equal areas in equal times. We say that the **areal velocity** of the planet is constant (Figure 14.5).

3. Consider any two planets having periods τ_1 and τ_2 and mean radii (mean distances from the sun) ρ_1 and ρ_2. (In Figure 14.4, ρ is shown for one planet.) The ratio of the squares of the periods is equal to the ratio of the cubes of the mean radii:

$$\frac{\tau_1^{\,2}}{\tau_2^{\,2}} = \frac{\rho_1^{\,3}}{\rho_2^{\,3}}. \qquad \textbf{(14.1)}$$

Although they are purely empirical, Kepler's laws brought quantitative certainty to the Copernican system. As framed by Copernicus, the system yielded planetary position

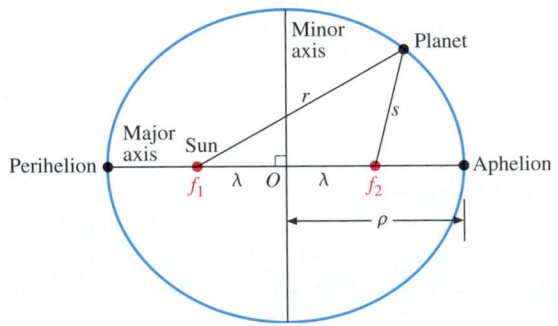

predictions somewhat more simply than the Ptolemaic system, but no more accurately. But Kepler's laws yield precise predictions *that have meaning only in terms of the Copernican system.* Moreover, Kepler's laws are more universal than he himself could have expected. When Galileo used his telescope in 1610 to discover the moons of Jupiter, he found to his delight that they resembled a Copernican solar system in miniature. When precise measurements were later made, it turned out that this miniature system obeys Kepler's laws just as well as the solar system itself.

In the political and religious turmoil of the late sixteenth and early seventeenth centuries, it was possible to read nonscientific messages into theories of the universe. Some thinkers did exactly this, often with turbulent consequences. In 1600, the eccentric Neapolitan philosopher-priest Giordano Bruno (b. 1548?) was burned at the stake in Rome for building what was officially construed as a heretical religious-political vision on the basis of the Copernican system. In 1634, Galileo was convicted by the Roman Inquisition of "vehement suspicion of heresy" for "holding as true the false doctrine ... that the Sun is the center of the world and immovable and that the Earth moves ..." He was forced to "abjure, curse, and detest ... [these] errors and heresies ..."* and condemned to lifelong house arrest, his books burned and forbidden. Nevertheless, by the time of Galileo's death just a few years later in 1642, there was no serious thinker who doubted the correctness of the Copernican picture of the universe, as modified by Kepler and enriched by the telescopic discoveries of Galileo and others.

Still, a dynamical picture was lacking. Attempts had been made to explain the motions of the planets. Both Kepler and the English physician-physicist William Gilbert (1544–1603) had suggested that the sun exerted a magnetic force on the planets. Galileo had considered the possibility that the law of inertia dictated circular rather than straight-line motion. But it was left to Newton to formulate a satisfactory solution. That solution and its consequences are the subject of the remainder of this chapter.

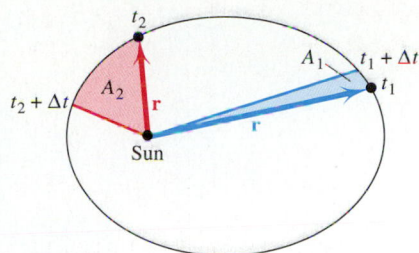

FIGURE 14.5 Kepler's second law. The planet is shown at two different locations in its orbit at times t_1 and t_2. In equal time intervals Δt, the **radius vector r** from the sun to the planet sweeps out areas A_1 and A_2. The law states that $A_1 = A_2$. That is, the **areal velocity** dA/dt is constant.

Newton's Law of Universal Gravitation

Newton's bold step was to assert that the force responsible for the motion of the planets in their observed orbits was the same force that makes a ripe apple fall—the **gravitational force**. Put more broadly, the gravitational force is **universal**. That is, *every bit of matter in the universe exerts a gravitational force on every other bit of matter.* To become more than an assertion, this statement must be tested against observation. But first it must be cast into a quantitative form.

Direction of the Force

Consider two bodies, 1 and 2, that exert gravitational forces on each other. Let us focus our attention on the force $\mathbf{F}_{1\to2}$ that body 1 exerts on body 2. If the particular gravitational force with which we are most familiar—the force exerted by the earth on the apple—is typical, the force is attractive and central.† So we can write

$$\hat{\mathbf{F}}_{1\to2} = -\hat{\mathbf{r}}_{1\to2}, \tag{14.2a}$$

where $\hat{\mathbf{r}}_{1\to2}$ is the unit vector from body 1 to body 2.

Mass Dependence of the Force

How does $\mathbf{F}_{1\to2}$ depend on the masses m_1 and m_2 of the two bodies? We can draw an inference from our knowledge of the special case in which body 1 is the earth and body

---

*Translation quoted from G. de Santillana, *The Crime of Galileo* (Chicago: University of Chicago Press, 1955), pp. 306–310.

†Remember that the force exerted on a body is a central force if the line of action of the force always passes through the same point, called the force center. See Section 13.2.

2 is a falling body near the surface of the earth. We already know from falling-body experiments that the gravitational force exerted by the earth is proportional to the mass of the body on which it acts; we have $F_{1 \rightarrow 2} \propto m_2$. That is why the gravitational acceleration g is the same for all bodies. But the force must also be proportional to the mass m_1 of the body that is exerting the force on body 2; $F_{1 \rightarrow 2} \propto m_1$. This follows from Newton's third law and the fact that the two bodies may be regarded as an isolated system. So we have for the mass dependence of the gravitational force

$$F_{1 \rightarrow 2} \propto m_1 m_2. \tag{14.2b}$$

Distance Dependence of the Force

How does the force depend on the distance between bodies 1 and 2? It seems plausible that the force will become weaker as the bodies are separated from each other. Newton asserted specifically that the force would depend inversely on the *square* of the distance:

$$F_{1 \rightarrow 2} \propto \frac{1}{r^2}. \tag{14.2c}$$

(Note: We need not specify a subscript for r, because $\mathbf{r}_{1 \rightarrow 2}{}^2 = \mathbf{r}_{2 \rightarrow 1}{}^2$.) We will soon make an argument to justify the plausibility of this choice. For the moment, we simply assert that it leads to verifiable results and that no other choice does so. A direct check is afforded by the argument of Problems 5.46 and 5.47, where the assumption of an inverse-square law is tested against the orbital periods of a synchronous artificial satellite and of the moon.

Newton's Law of Gravitation

Putting together the three relations 14.2a, 14.2b, and 14.2c, we obtain the proportionality

$$\mathbf{F}_{1 \rightarrow 2} \propto -\frac{m_1 m_2}{r^2} \hat{\mathbf{r}}_{1 \rightarrow 2}. \tag{14.3a}$$

As usual, we prefer to deal with equations rather than proportionalities. So we define a proportionality constant G, called the **universal gravitational constant**, and write

$$\mathbf{F}_{1 \rightarrow 2} = -G \frac{m_1 m_2}{r^2} \hat{\mathbf{r}}_{1 \rightarrow 2}. \tag{14.3b}$$

This equation is called **Newton's law of universal gravitation**. The value of the universal gravitational constant is determined by experiment. The modern value is

$$G = (6.672\ 59 \pm 0.000\ 85) \times 10^{-11} \ \text{N·m}^2/\text{kg}^2. \tag{14.4}$$

The value of G is the most poorly known of all the important constants of nature because the gravitational force is so weak. Only because we are so close to the earth, with its enormous mass, do we feel a substantial gravitational force.

EXAMPLE 14.1

Express the acceleration of gravity $\mathbf{g}$ in terms of Newton's law of gravitation.

SOLUTION: A small body of mass m near the surface of the earth experiences a gravitational force $\mathbf{F} = m\mathbf{g}$ as a result of the presence of the earth, whose mass is M. (The quantity m here corresponds to m_2 in Equation 14.3b, and M corresponds to m_1.) The direction of $\mathbf{F}$ is certainly downward toward the center of the earth; that is, $\hat{\mathbf{F}} = -\hat{\mathbf{r}}$, where $\mathbf{r}$ is the vector from the center

of the earth to the body. So you can equate $\mathbf{F}$ to the gravitational force $\mathbf{F}_{1 \rightarrow 2}$ in Equation 14.3b and obtain

$$m\mathbf{g} = -G \frac{Mm}{r^2} \hat{\mathbf{r}}.$$

Canceling m from both sides of the equation, you obtain

$$\mathbf{g} = -\frac{GM}{r^2} \hat{\mathbf{r}}. \tag{14.5}$$

Gravitational Attraction of a Spherical Body

Just what is the meaning of the quantity r in Equation 14.5? According to the argument that leads to Newton's law of gravitation, the net attractive force exerted on the small body by the earth is made up of contributions of all the mass elements that constitute the earth. But the distances of those mass elements from the small body vary widely, from essentially zero to the diameter of the earth, 1.28×10^7 m. Your intuition probably tells you that the proper distance to use is the distance from the center of the earth. This intuition is indeed correct; Newton proved by integration that *the gravitational force exerted by a solid sphere of mass M on a body outside the sphere is equal to the force that would be exerted on the body by a particle of mass M located at the center of the sphere.* This result is true even if the sphere is not of uniform density (as the earth is not), provided that each shell making up the sphere is uniform. So the proper value of r to be inserted into Equation 14.5 is the radius of the earth.

We will not repeat Newton's proof of this theorem here. It is not difficult, but it is fairly lengthy. A proof given in 1839 by the German mathematician Carl Friedrich Gauss requires essentially no calculation and is a source of much deeper insight into the meaning of the inverse-square law. *Gauss's law*, as the central result of this proof is called, has very important applications in the theory of electricity, and we examine it in detail in Chapter 24. The spirit of Gauss's law is outlined toward the end of Section 14.3, as well as in several problems at the end of this chapter.

EXAMPLE **14.2**

The rocks found at or near the surface of the earth have an average density ρ of approximately 3×10^3 kg/m³. Make the crude assumption that the density of the earth is uniform, and obtain a first approximation to the value of the universal gravitational constant G. Is this approximation likely to yield a value that is too large or too small?

SOLUTION: If the earth were a uniform sphere of radius r, its mass would be

$$M = \rho \times \text{volume} = \tfrac{4}{3}\pi\rho r^3.$$

According to Equation 14.5, the magnitude of the acceleration of gravity is $g = MG/r^2$. Substituting the estimate for M into this equation, you have

$$g = \frac{\tfrac{4}{3}\pi\rho r^3 G}{r^2} = \tfrac{4}{3}\pi\rho r G.$$

So you find that

$$G = \frac{3}{4\pi}\frac{g}{\rho r} \tag{14.6}$$

$$= \frac{3}{4\pi} \times \frac{9.8 \text{ m/s}^2}{3 \times 10^3 \text{ kg/m}^2 \times 6.4 \times 10^6 \text{ m}}$$

$$= 1 \times 10^{-10} \text{ N·m}^2/\text{kg}^2.$$

This value is certain to be too large. The reason is that the density of the matter making up the earth increases with depth; see Problems 11.21 and 11.65. The uniform-density approximation underestimates the mass of the earth. It therefore overestimates the strength of the gravitational force, of which G is the measure. This is evident in Equation 14.6, which shows that G is inversely proportional to the mean density ρ of the earth. The underestimate of ρ used here thus leads to an overestimate of G. Nevertheless, the approximation is not bad for a first try. It differs from the modern value given in Equation 14.4 by less than a factor of two.

The Apple, the Moon, and the Solar System

Now that we have expressed the law of gravitation in the quantitative form of Equation 14.3b, $\mathbf{F}_{1\to2} = -G(m_1 m_2/r^2)\hat{\mathbf{r}}_{1\to2}$, let us follow Newton in testing it. It has been known since the time of the ancient Greeks that the moon orbits the earth in an approximately circular orbit whose radius R is close to 60 times the radius r of the earth. Newton asserted that the centripetal acceleration **a** of the moon was a result of the gravitational force exerted on it by the earth. If this assertion is correct, then **a** must be equal to the gravitational force given by Equation 14.3b divided by the mass m of the moon, in accordance with Newton's second law of motion. The magnitude of **a** is thus

$$a = \frac{F_{\text{earth} \to \text{moon}}}{m} = \frac{GM}{R^2},$$

where M is the mass of the earth. Let us compare a with the acceleration of an apple falling off a tree in Newton's mother's garden, **g**, which is a gravitational acceleration by definition. Using Equation 14.5, we have the ratio

$$\frac{a}{g} = \frac{\dfrac{GM}{R^2}}{\dfrac{GM}{r^2}} = \left(\frac{r}{R}\right)^2.$$

Note how the quantity G, whose value Newton did not know with any precision, drops out of the equation. The ratio r/R is 1/60, and hence

$$a = g \times \left(\frac{1}{60}\right)^2 = \frac{9.80 \text{ m/s}^2}{3600} = 2.72 \times 10^{-3} \text{ m/s}^2.$$

Now let us compare this value of a with the actual centripetal acceleration a_c of the moon. This quantity can be obtained purely kinematically, using the sidereal period of the moon ($\tau = 27.3$ days) and the mean radius of its nearly circular orbit ($R = 3.84 \times 10^8$ m) as suggested in Problem 5.47. We have

$$a_c = \omega^2 R = \frac{4\pi^2}{\tau^2} R,$$

where ω is the angular speed of the moon about the earth. Inserting the numerical values, we obtain

$$a_c = 4\pi^2 \times \frac{3.84 \times 10^8 \text{ m}}{(27.3 \text{ days} \times 8.64 \times 10^4 \text{ s/day})^2}$$
$$= 2.72 \times 10^{-3} \text{ m/s}^2.$$

Compare this with the acceleration due to the gravitational force. As Newton put it, they "answer pretty nearly."

Kepler's Laws and Newton's Laws

Having successfully shown that Newton's laws of gravitation and of motion are valid at least as far away from us as the moon, we next extend the laws beyond the "sphere of the moon" and show that they account for the motions of the planets around the sun. This is a big step in generalization because we will, for the first time, make calculations based on the assumption that a body other than the earth—the sun—exerts a gravitational force on other bodies. We will begin by deriving Kepler's third law from Newton's laws for the special case of circular orbits.* For a circular orbit, the mean orbit radius ρ for each planet is just the radius of the circle.† Then we will show that Kepler's second law is a direct consequence of the law of conservation of angular momentum and of the fact that the sun exerts a central force on the planets.

Sidereal and Synodic Periods

The sidereal period is the time required for the moon to complete its orbit, returning to the same position with respect to the stars in the background. The more familiar synodic period from full moon to full moon, about 29.5 days, is longer because the moon must "catch up" with the motion of the earth around the sun.

Newton's Nostalgic Look Back

About a year before his death, Newton wrote, ". . . And the same year [1666] I began to think of gravity extending to yᵉ orb of the Moon & . . . from Kepler's [third law I deduced the inverse-square property of the gravitational force] . . . & thereby compared the force requisite to keep the Moon in her Orb with the force of gravity at the surface of the earth, and found them answer pretty nearly. . . . For in those days I was in the prime of my age for invention & minded Mathematicks & Philosophy more then at any time since."

---

*The circular orbit is a good first approximation for most of the planets.

†We use ρ in this chapter to represent two quite different quantities: orbit radius and mean density of the earth. Although this double use has obvious undesirable aspects, use of ρ for both quantities is so strongly established that it is best to stay with convention. Fortunately, the two quantities represented by ρ do not appear in the same equations, and you are not likely to confuse them.

Kepler's Third Law for Circular Orbits

Consider a planet of mass m having a circular orbit of radius ρ around the sun, whose mass is M. The centripetal force required to keep the planet in that orbit is $m\omega^2\rho$, where ω is the (constant) angular speed of the planet around the sun. Because that force is provided by the gravitational interaction, we write

$$m\omega^2\rho = G\frac{Mm}{\rho^2}. \tag{14.7}$$

We cancel out the planetary mass m and multiply through by ρ^2 to obtain

$$\omega^2\rho^3 = GM.$$

Finally, we reexpress this equation in terms of the orbit period τ instead of the angular speed ω. Because ω is constant for a circular orbit, we have $\omega = 2\pi/\tau$. Making the substitution $\omega^2 = 4\pi^2/\tau^2$ and dividing through by $4\pi^2$ gives us

$$\frac{\rho^3}{\tau^2} = \frac{GM}{4\pi^2}. \tag{14.8}$$

The right side of this equation is a constant that has the same value for all planets (or, for that matter, all satellites) in orbit about the same **primary**, whose mass is M. Equation 14.8 is thus a general form of Kepler's third law. To cast the equation into Kepler's form (Equation 14.1), consider any two planets, 1 and 2, in circular orbits about the sun. Using Equation 14.8 twice, we obtain

$$\frac{\rho_1^3}{\tau_1^2} = \frac{\rho_2^3}{\tau_2^2}.$$

A little rearrangement immediately gives

$$\frac{\tau_1^2}{\tau_2^2} = \frac{\rho_1^3}{\rho_2^3},$$

which is Equation 14.1.

As Kepler originally conceived it, his third law was a (very successful) attempt to fit planetary observations with a simple empirical rule. But the fact that Kepler's third law can be deduced from Newton's laws of gravitation and of motion makes Kepler's law a crucial test for Newton's laws. This test is one passed with flying colors. Table 14.1 exhibits the best modern values of the mean radius and period for the nine known

> **Some Astronomical Terminology**
> The **primary** is the body around which one or more **secondary** bodies orbit, under the influence of the central gravitational force exerted on them by the primary. For the planets, the primary is the sun; the moon and many artificial satellites are secondaries whose primary is the earth.

TABLE 14.1 Comparison of Planetary Mean Orbital Radii and Periods with Kepler's Third Law

Planet	Mean Orbit Radius ρ (AU[a])	Period τ (y[b])	ρ^3/τ^2 (AU^3/y^2)
Mercury	0.387 44	0.240 899	1.002 2
Venus	0.722 81	0.615 185	0.997 84
Earth	1	1	1
Mars	1.523 3	1.880 82	0.999 22
Jupiter	5.202 5	11.861 3	1.000 8
Saturn	9.540 7	29.456 8	1.000 9
Uranus	19.190	84.008 1	1.001 3
Neptune	30.086	164.784	1.002 9
Pluto	39.507	248.35	0.999 76

[a]Planetary orbit radii are usually expressed as multiples of the mean orbit radius of the earth, which is called the **astronomical unit** (AU). This is because until quite recently the astronomical unit itself was known with much less precision than ratios of the mean radii.
[b]Planetary periods are conveniently expressed as multiples of that of the earth, which is 1 year (y) by definition. The periods here are all the sidereal periods that would be seen by an observer looking into the solar system from outside.

planets. The last column in the table lists the values of ρ^3/τ^2 calculated from these data. In the units used, (astronomical units)^3/(year)^2, the value of ρ^3/τ^2 for the earth is 1 by definition. Deviations from this value for the other planets measure the degree to which the theory fails to fit the data. Nearly all of the deviation is due to **perturbations**, as gravitational forces exerted by the planets on one another are called. Jupiter, which possesses more than 70% of the mass of the solar system not in the sun, has by far the greatest effect. Detailed computation of perturbations is complicated because interplanetary distances are constantly changing. But such calculations have been made, and perturbations, together with other still smaller effects, account for the entire deviation from unity of the values in the last column of Table 14.1.

Kepler's Second Law

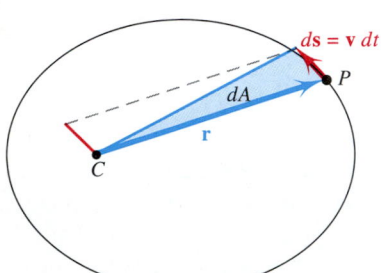

FIGURE 14.6 Diagram for deriving Kepler's second law from the law of conservation of angular momentum. The planet P moves with velocity $\mathbf{v}$ under the influence of a central force. The force center is at C. Over the time interval dt, the planet moves through the distance $|\mathbf{v}\,dt| = |d\mathbf{s}|$, and the radius vector $\mathbf{r}$ sweeps out an area $dA = \frac{1}{2}|\mathbf{r} \times \mathbf{v}|\,dt$.

Figure 14.6 shows a planet P moving along its orbit under the influence of a central force. The position of the planet with respect to the force center C is described by the radius vector $\mathbf{r}$. The instantaneous velocity of the planet, tangent to the orbit, is $\mathbf{v}$. In accordance with Equation 13.38, the angular momentum of the planet with respect to the force center is

$$\mathbf{L} = \mathbf{r} \times \mathbf{p} = m\mathbf{r} \times \mathbf{v}.$$

Over an infinitesimal time interval dt, the planet moves a distance $\mathbf{v}\,dt = d\mathbf{s}$. So we have

$$\frac{1}{m}\,\mathbf{L}\,dt = \mathbf{r} \times \mathbf{v}\,dt = \mathbf{r} \times d\mathbf{s}.$$

The magnitude $|\mathbf{r} \times d\mathbf{s}|$ is the area of the parallelogram shown in Figure 14.6; this is a property of the cross product (Section 13.4). The triangle constructed by drawing the diagonal of the parallelogram has one-half the area of the parallelogram. But the area of the triangle is the area dA swept out by the planet in the time dt. So we have

$$dA = \frac{1}{2}\,|\mathbf{r} \times d\mathbf{s}| = \frac{L}{2m}\,dt. \qquad \textbf{(14.9a)}$$

As you learned in Section 13.2, the angular momentum L of the planet must be constant because the only force acting on the planet is a central force. The mass also is constant, and so the factor $L/2m$ on the right side of the equation is constant. It follows that dA, the area swept out by the radius vector, is proportional to the time dt, or

$$\frac{dA}{dt} = \frac{L}{2m} = \text{constant.} \qquad \textbf{(14.9b)}$$

This proves Kepler's second law. In the proof, we have not used the fact that the force acting on the planet is an inverse-square force, but only that it is a central force. Thus Kepler's second law is valid for *all* central forces.

Kepler's First Law

In the form in which Kepler originally enunciated it, his first law states that each planetary orbit is an ellipse with the sun at one focus. This law is presupposed in the general statement of the third law. Only if the orbits *are* ellipses is it possible to speak of comparing orbits in terms of their semimajor axes, as the third law does.

However, Newton used the inverse-square nature of his law of gravitation, together with the laws of mechanics, to derive a more general form of Kepler's first law. The proof is too complicated mathematically to consider in this book. But the generalized result, which is valid for all inverse-square forces, is easily stated:

1. The orbit of a body moving under the influence of an inverse-square force is always a **conic section**—a circle, an ellipse, a parabola, or (one branch of) a hyperbola; see Figure 14.7. This is true whether the force is attractive (as the gravitational force is) or repulsive (as the inverse-square electric force can be).

2. If the force is repulsive, the orbit is always a hyperbola. If the force is attractive, the form of the orbit depends on the total mechanical energy of the body.

3. If the orbit is a circle, the center of force lies at its center. If the orbit is a parabola, the center of force lies at its focus. If the orbit is an ellipse or a hyperbola, the center of force lies at one of the two foci.

Why an Inverse-Square Law?*

The success of Kepler's laws in describing the motion of the planets around the sun, and of satellites around their primary planets, amply justifies the assumption that the law of gravitation is an inverse-square law. But there is another way of looking at the matter that is simpler and at the same time more subtle. Can it be that the inverse-square nature of the gravitational force—or (for that matter) of the electric force you will encounter in Chapter 23—is a direct consequence of the three-dimensionality of space?

In Figure 14.8, two concentric spheres of radii r_1 and r_2 are drawn centered on a body of mass M (say, the sun). Planets 1 and 2 are located at points on the surfaces of the respective spheres. For the sake of simplicity, we let both planets have the same mass m. Each planet experiences a gravitational force due to the presence of the sun. Why? We cannot answer this question in a completely satisfactory way. But we can imagine an "influence" that has the sun as its source and that "spreads out" through all space in all directions from the sun. Because there is nothing special about any particular direction, we argue that this influence, whatever it may be, spreads out uniformly in all directions. When we make this argument, we say that space is **isotropic**—the same in all directions.

We can represent this influence, as in Figure 14.8, by an array of N lines spreading out in all directions from the sun. The number of lines we use is arbitrary, but our assumption of isotropy demands that they spread out with the same density in all directions, as shown. The lines extend to infinity, representing the fact that the gravitational force is universal. As we move away from the sun, the lines "thin out." Nevertheless, the number of lines penetrating each sphere is the same; this is true as well for any other sphere you might draw around the sun.

We argue that the weakening of the gravitational force with distance is represented by the thinning out of the lines. We go one step further and assert that the number of

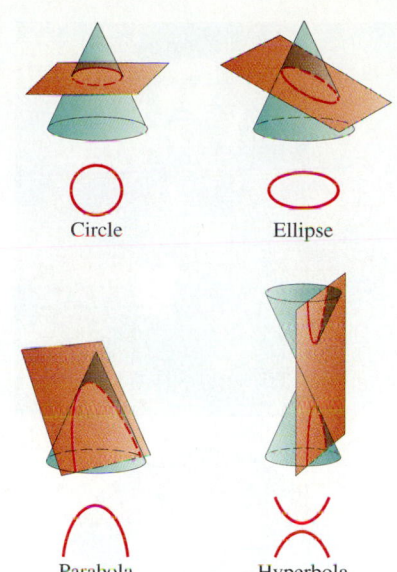

Circle   Ellipse

Parabola   Hyperbola

FIGURE 14.7 The conic sections, which are the only possible orbits for a body under the influence of an inverse-square force. When the plane cutting a right circular cone is oriented normal to the cone axis, the section is a circle. When the plane is oriented parallel to the side of the cone, the section is a parabola. At intermediate orientations, the section is an ellipse whose eccentricity depends on the plane orientation. When the plane is oriented still further away from the normal to the cone axis than the parallel to the cone side, it cuts the cone twice and produces the two branches of a hyperbola.

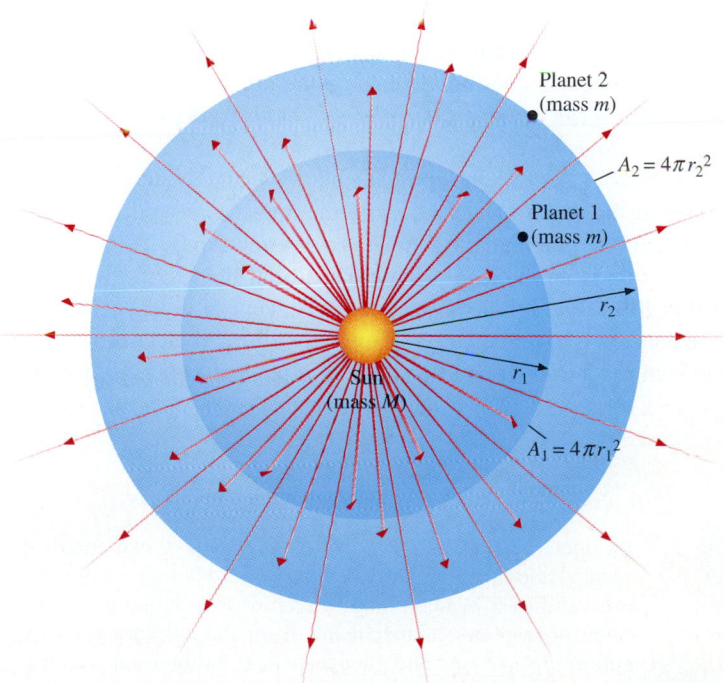

Planet 2
(mass m)

$A_2 = 4\pi r_2^2$

Planet 1
(mass m)

r_2

r_1

Sun
(mass M)

$A_1 = 4\pi r_1^2$

FIGURE 14.8 Diagram illustrating the connection between the inverse-square nature of the gravitational force and the three-dimensionality of space.

---

*This subsection may be omitted without loss of continuity.

Johannes Kepler (top) and Isaac Newton (bottom). Kepler worked out the kinematics of the solar system; more than half a century later, Newton worked out the dynamics.

lines penetrating sphere 1 per unit area is a measure of the gravitational force F_1 exerted on planet 1 at the surface of sphere 1 and that the number of lines penetrating sphere 2 per unit area is a measure of the gravitational force F_2 exerted on planet 2 at the surface of sphere 2. That is, we can write the proportionalities

$$F_1 \propto \frac{N}{A_1} \quad \text{and} \quad F_2 \propto \frac{N}{A_2},$$

where A_1 and A_2 are the surface areas of the two spheres. Because the lines are distributed uniformly in all directions, the number of lines per unit area is the same all over a sphere and is equal to the total number of lines divided by the total surface area of the sphere. So we can write

$$F_1 \propto \frac{N}{4\pi r_1^2} \quad \text{and} \quad F_2 \propto \frac{N}{4\pi r_2^2}.$$

We can turn these two proportionalities into a single equation by dividing the first by the second. This gives

$$\frac{F_1}{F_2} = \frac{\dfrac{N}{4\pi r_1^2}}{\dfrac{N}{4\pi r_2^2}} = \frac{r_2^2}{r_1^2}. \tag{14.10}$$

(Now you can see why it does not matter how we choose N!) Because r_1 and r_2 are chosen arbitrarily, this equation states that the gravitational force due to a body such as the sun is inversely proportional to the square of the distance.

The entire argument leading to Equation 14.10 hinges on the purely geometric fact that the surface area of a sphere is proportional to the square of its radius. And why this proportionality? Because a sphere is a three-dimensional body. For comparison, the circumference of a circle—the two-dimensional equivalent of a sphere—is proportional to the first power of its radius. Can you make an argument to show that the gravitational force in an imaginary two-dimensional universe would be proportional to r^{-1}?

We now justify the assertion, made in Section 14.2, that the gravitational force exerted by a spherical body of mass M on a particle located outside itself is not changed if the spherical body is replaced by a point mass M located at its center. In Figure 14.8, the lines representing the gravitational influence lie along radii of the spherical sun. This is a consequence of the symmetry of a sphere about its center. If the sun is made to contract about its center (without changing its mass), nothing is changed. The pattern of lines penetrating each of the outer spheres (on which planets 1 and 2 are located in Figure 14.8) is exactly the same and does not depend on the volume of the sun. Indeed, *the same is true for sphere 1 if the sun is expanded (without changing its mass) so that it just fills the space inside the sphere.*

If you substitute the earth for the sun, the apple for planet 1, and the moon for planet 2, the argument applies to the case of the moon and the apple. Can you see why the difference in mass between the moon and the apple is not important?

The "influence" just discussed is called the **gravitational field**, and the lines are called **field lines** or **flux lines**. We will not consider the concept of field further in this chapter. However, the concept is extremely important in the study of electromagnetism, and we will return to it in much more detail in Chapter 24 and subsequent chapters.

THINKING LIKE A PHYSICIST

That "thinking like a physicist" is a special way of thinking was a new idea when Galileo and Newton fused the observational astronomy of Tycho and Kepler with the new laws of physics. Galileo and Newton postulated laws of physics (sometimes demonstrating their validity with laboratory ex-

periments, as Galileo did by rolling balls down inclined planes) and then demonstrated that the *same* laws of physics are valid over a much larger domain. In extending the domain of the laws of mechanics from the laboratory to the sun-moon system, and thence to the entire solar system, these pioneering physicists founded our understanding of the mechanics of the solar system but did much more as well. They forged an intellectual tool that has been used by scientists ever since. At least one commentator has said that Galileo's foundation of a scientific methodology was by far the most important of his many contributions to knowledge.

When we are confident that certain physical laws are valid in a familiar domain, we boldly attempt to apply them to a broader, unfamiliar domain, and we are often successful, as Galileo and Newton were. So powerful is this methodology that we can easily become overconfident in it, coming to believe that the domain of validity of a set of physical laws is limitless. Indeed, the domain of classical mechanics, the science founded on the astronomical analyses that are the main subject of this chapter, is so wide that its boundaries remained beyond view for more than two centuries. Nevertheless, we must be prepared to encounter limits to the domain over which any set of laws is applicable. The limits of classical mechanics were discovered about the turn of the twentieth century by Planck, Einstein, and others.

One consequence of these limits is implicit in our justification in this subsection of the inverse-square nature of the gravitational force on a geometric basis. You may be delighted to find in Chapter 24 that the argument works just as well for the electric force; the domain of the justification is thus extended. But today fundamental forces are known that are *not* inverse-square forces; these are the forces that determine the structure of the atomic nucleus. Classical mechanics, which implicitly incorporates classical three-dimensional geometry, does not apply to objects that are very small (as atoms are) or very large (as the universe as a whole is) or that move with great speed (as certain energetic atomic particles do). We did not develop the laws of classical mechanics in these domains; and while we may hope to emulate Galileo and Newton in extending the laws into an unfamiliar region, we must be prepared to find that our arguments fail there. Nevertheless, classical mechanics is the indispensable starting point for discovery of new, more broadly applicable laws. And those new laws are consistent with the old ones in the domain where the old ones are valid—the domain of the middle-sized and not-too-fast.

When we embark to discover new worlds, we necessarily do so in the light of what we know of the old world. If we are fortunate enough to discover a new world, we may justly be delighted in its similarities to the old. But we must learn to take delight in and learn to understand what is new and different, too. As a bonus, the new insights that our new, broader view opens to us will deepen our understanding of what is already familiar. Discovering the failure of a set of laws—the boundaries of a physical domain—is not a disappointment but a welcome challenge.

Determination of the Universal Constant *G*

We made a crude estimate of the value of *G* in Example 14.2, but it is important to obtain a better value. We need a measurement based on a better knowledge of the two test masses between which the gravitational force is measured. As already noted, the difficulty to be overcome is the weakness of the gravitational interaction. In Example 14.2, we circumvented the difficulty by using a very large mass (that of the earth as a whole) but at the cost of not knowing that mass very well. Choice of a smaller, more readily measurable mass means that a very small force will have to be measured.

In the 1780s, the English clergyman and astronomer John Michell (1724?–1793) attempted to make just such a measurement in the laboratory, using laboratory-size bodies whose masses could be measured with considerable precision. Michell proposed to measure the very small force by using a torsion balance (Section 12.3), which he invented about 1768. (Coulomb generally gets credit for his independent reinvention of the torsion balance, in recognition of his thorough analysis of its action and of the very diverse uses to which he put it.)

Michell died before he had achieved significant results with his apparatus. It came eventually into the hands of the rich eccentric Henry Cavendish (1731–1810). Cavendish rebuilt and greatly improved the apparatus, identifying many sources of error and protecting against them. We treat Cavendish's experiment in some detail not only because of its importance, but also because it furnishes clear examples of the thoughtful planning, meticulous measurement, and thorough analysis that are nearly always required in experimental science. The apparatus in its final form is shown in Figure 14.9. The entire system was set up in a basement, to minimize vibration and temperature fluctuations.

The torsion balance is so sensitive that it takes a very long time to come to rest. The equilibrium position is closely estimated by observing the extremes of several succes-

Discovery Its Own Reward

Cavendish was a recluse with a passion for the physical sciences and mathematics. He first achieved distinction in 1766 by isolating the element hydrogen, though he attributed a different identity to it in the context of then-current chemical theory. Nevertheless, he showed in 1784 that hydrogen was a constituent of water. A perfectionist, he declined to publish the results of many of his researches, which are consequently credited to other, later workers.

FIGURE 14.9 Side and top views of Cavendish's torsion balance, taken from his 1798 paper. Two relatively small lead balls, each of mass 729.77 g, hang from the ends of the light (about 170 g) but strong structure consisting of the 1.86-m-long wooden rod *hmh* and the silver guy wire *hgh*. The structure hangs from the torsion fiber *lg*, made of a silvered copper wire about 1 m long. Having large moment of inertia and a weak torsion fiber, the torsion pendulum has a very long period—about 14 min for the torsion fiber used in the final measurements. Two large bodies, *WW*, hang from a separate structure. Each has an effective mass $M = 154.6$ kg. As shown in the top view, the bodies can be swung so as to attract the small balls on the torsion balance from either side, the effective center-to-center distance being 0.220 m. The balance is protected from drafts and convection currents arising from the heat of the observer's body by a wooden case with small observing windows. All necessary adjustments are made remotely by means of systems of pulleys, rods, and gears. The motion of the torsion balance is observed by means of small ivory scales fixed on the case near *AA*, just below the ends of the beams. On the beam ends themselves are vernier scales which swing past the fixed scales. A displacement of 0.2 mm is easily measured. The scales are illuminated by the lanterns *L* and read by means of the small telescopes below the lanterns.

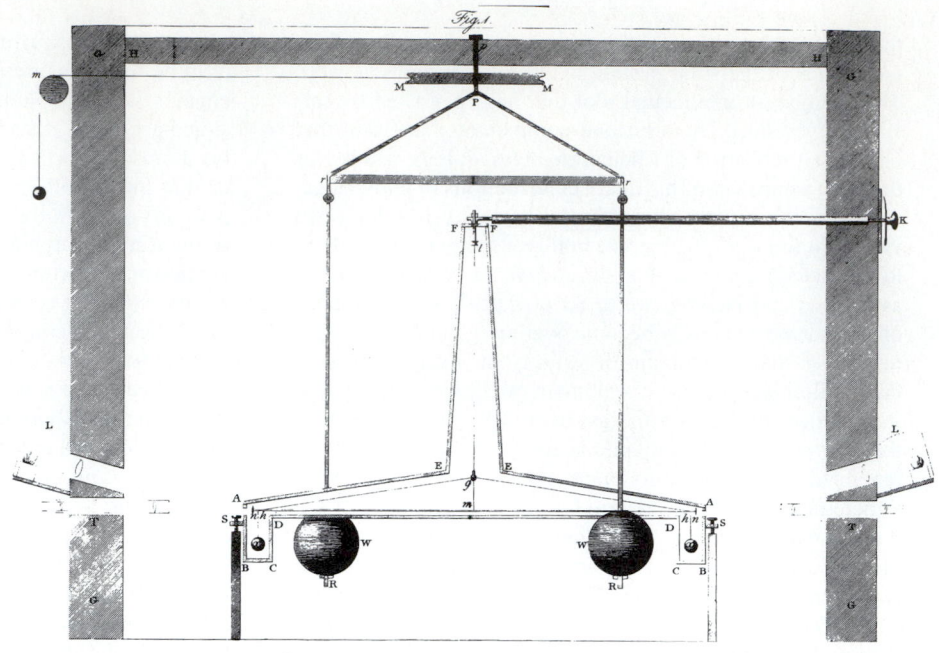

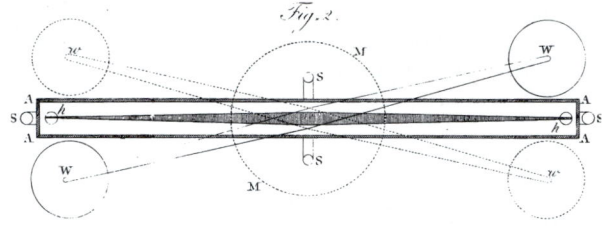

sive swings. (This takes much patience because each one-way swing takes about 7 minutes.) After the equilibrium position has been established, with the large bodies *WW* at lower left and upper right, they are swung around to lower right and upper left, and the process is repeated. The displacement δ between the two equilibrium positions of the end of the balance arm is the essential product of the measurement.

The experimental data are used to find the value of *G* on the basis of the following analysis. According to Equation 12.15c, the period of the torsion balance, considered as a torsion pendulum, is

$$\tau = 2\pi \sqrt{\frac{I}{\kappa}},$$

where *I* is the moment of inertia of the system and κ is the torsion constant of the suspension wire. To first approximation, we consider only the moment of inertia of the small spheres *hh*, each having mass *m* and hanging a distance *r* from the rotation axis. We neglect the moment of inertia of the wooden rod from which the spheres hang. This approximation gives us the period

$$\tau = 2\pi \sqrt{\frac{2mr^2}{\kappa}}. \tag{14.11}$$

Within the accuracy of our approximation, the torque *T* exerted on the torsion balance is the result of the gravitational force exerted on each small hanging sphere by the adjacent large body *W*, whose mass is *M*. Because the spheres are swinging, their distances from the large bodies vary. But let us call the mean distance *D* and argue (again approximately) that the force exerted on each small sphere has magnitude $F = GMm/D^2$. The magnitude of the total torque on the torsion balance is thus

$$|T| = 2Fr = 2\frac{GMm}{D^2}\, r.$$

When the large bodies are swung from one of the positions shown in the top view in Figure 14.9 to the other, the direction of the force on the torsion balance is reversed. The *change* in torque is of magnitude

$$\Delta T = 2|T| = 4\frac{GMm}{D^2} r.$$

As a result of this change, the equilibrium position of the balance shifts through an angle $\Delta\theta$. We use the rotational form of Hooke's law (Equation 12.13) to express the relation between the magnitudes ΔT and $\Delta\theta$:

$$\frac{4GMmr}{D^2} = \kappa\,\Delta\theta.$$

Solving for κ, we obtain

$$\kappa = \frac{4GMmr}{D^2\,\Delta\theta}.$$

We insert this value of κ into Equation 14.11; this yields

$$\tau = 2\pi\sqrt{\frac{D^2 r\,\Delta\theta}{2GM}}.$$

(The mass m of the small spheres cancels out.) We now solve this equation for G. We have

$$G = \frac{2\pi^2 D^2 r\,\Delta\theta}{M\tau^2}.$$

Finally, note that the quantity $r\,\Delta\theta$ is equal to the displacement δ of the equilibrium position of the end of the torsion balance arm, which is what the observer reads on the scales. So we have the result

$$G = \frac{2\pi^2 D^2 \delta}{M\tau^2}. \tag{14.12}$$

All the quantities on the right side of the equation are known, and G can be evaluated.

EXAMPLE 14.3

In a typical experimental run, Cavendish obtained the following data. (The values have been translated into modern units.)

- Oscillation period: $\tau = 852$ s.
- Shift of equilibrium position: $\delta = 7.45$ mm.

Using the approximation implicit in Equation 14.12, find the value of G. The necessary instrument characteristics are quoted in the legend for Figure 14.9.

SOLUTION: Equation 14.12 gives

$$G = \frac{2\pi^2 \times (0.220\ \text{m})^2 \times 7.45 \times 10^{-3}\ \text{m}}{154.6\ \text{kg} \times (852\ \text{s})^2}$$

$$= 6.34 \times 10^{-11}\ \text{N·m}^2/\text{kg}^2.$$

The value obtained in Example 14.3 must be corrected to allow for effects we have neglected in the approximation leading to Equation 14.12. In particular, taking into account the moment of inertia of the balance arm leads to a 3.5% increase in the value of G calculated in Example 14.3. Other minor corrections are required as well. When Cavendish had made all these corrections, he expressed his results in terms of the mean density of the earth. Averaged and reexpressed in terms of the gravitational constant, his results correspond in modern terms to the value $G = 6.67 \times 10^{-11}\ \text{N·m}^2/\text{kg}^2$. It has taken nearly two centuries to obtain three more decimal places.

Mass of the Earth, the Sun, and the Planets

Given the value of the universal gravitational constant G, the mass of any body that possesses a satellite can be readily determined by using Kepler's third law. We solve

Equation 14.8, $\rho^3/\tau^2 = GM/4\pi^2$, for M and obtain

$$M = \frac{4\pi^2\rho^3}{G\tau^2}.$$ (14.13)

Knowledge of the mean orbit radius ρ of the satellite and its period τ immediately yields a value of M.

EXAMPLE 14.4

(a) The mean radius of the moon's orbit around the earth is 3.84403×10^8 m, and the sidereal period of the moon is 2.36055×10^6 s. Find the mass of the earth. (b) The mean radius of the earth's orbit around the sun is 1.4957×10^{11} m, and the sidereal period of the earth is 3.1558150×10^7 s. Find the mass of the sun.

SOLUTION:

(a) Find the mass of the earth.

Using the data for the moon in Equation 14.13, you obtain for the mass of the earth

$$M_{earth} = \frac{4\pi^2 \times (3.84403 \times 10^8 \text{ m})^3}{6.6732 \times 10^{-11} \text{ N·m}^2/\text{kg}^2 \times (2.36055 \times 10^6 \text{ s})^2}$$

$$= 6.0306 \times 10^{24} \text{ kg}.$$

(b) Find the mass of the sun.

In like manner, you use the data for the earth in Equation 14.13 to obtain the mass of the sun. You have

$$M_{sun} = \frac{4\pi^2 \times (1.4957 \times 10^{11} \text{ m})^3}{6.6732 \times 10^{-11} \text{ N·m}^2/\text{kg}^2 \times (3.1558150 \times 10^7 \text{ s})^2}$$

$$= 1.9876 \times 10^{30} \text{ kg}.$$

As you can see from these figures, the sun is *much* more massive than the earth; the mass ratio is more than 300 000 to 1. For both earth and sun, the limit of precision on the determination of M is set by the precision with which we know G.

EXAMPLE 14.5

Use the result of part **a** of Example 14.4 to determine the acceleration of gravity g at the surface of the earth. Take the earth's radius to be $r = 6.367 \times 10^6$ m. Compare the result with the standard value $g_s = 9.80$ m/s^2, and account for the difference.

SOLUTION: Because the earth is nearly spherical, you can equate its gravitational attraction for a body at its surface with that of a particle of mass M_{earth} located at the center of the earth, a distance r from the surface. Thus you can use the magnitude part of Equation 14.5 to write

$$g = \frac{GM_{earth}}{r^2}$$

$$= \frac{6.673 \times 10^{-11} \text{ N·m}^2/\text{kg}^2 \times 6.031 \times 10^{24} \text{ kg}}{(6.367 \times 10^6 \text{ m})^2}$$

$$= 9.93 \text{ m/s}^2.$$

This value is about 1% larger than the standard value g_s. A part of the difference is due to the imperfect sphericity of the earth. The rest of the difference is due mainly to the reduction of the gravitational acceleration by the outward centrifugal acceleration that an observer on the surface of the earth experiences on account of the earth's rotation about its axis. The centrifugal acceleration is associated with the fictitious centrifugal force discussed in Section 5.5. See also Problem 5.45.

Gravitational Potential Energy

The great age of the solar system—about 5 billion years—immediately suggests that its total mechanical energy E is conserved or very nearly conserved, because the planets still circle the sun, and the satellites their planets. Certainly E is conserved over such relatively short times as tens of thousands of years—times that are still sufficiently long to assure that even the outermost planets make many revolutions around the sun. As far as the gravitational force is concerned, we are not surprised that E is conserved. The gravitational force exerted on the planet by the sun, $\mathbf{F} = -G(Mm/r^2)\hat{\mathbf{r}}$, is a function of the planet's position **r** only and is thus conservative; see Section 8.2. The same is true of the much smaller gravitational forces exerted by the various members of the solar system on one another. If any dissipative forces are present, they are very small indeed. Hence conservation of energy becomes an important tool in studying planetary orbits. Because artificial satellites obey the same laws of gravitation and mechanics, energy conservation is essential to a treatment of their orbits as well.

A planet's distance from the sun is not constant as the planet moves along its orbit, but varies between a maximum distance at a point called **aphelion** and a minimum distance at a point called **perihelion** (Figure 14.4). The planet moves ''uphill'' toward aphelion (slowing down as it does so) and ''downhill'' toward perihelion (speeding up as it does so). There is a continual interchange of energy between its kinetic and potential forms. We can find the change ΔU in a planet's potential energy as its position **r** with respect to the sun changes from $\mathbf{r}_i$ to $\mathbf{r}_f$ by using Equation 8.14,

$$\Delta U = \int_{\mathbf{r}_i}^{\mathbf{r}_f} dU = \int_{\mathbf{r}_i}^{\mathbf{r}_f} -\mathbf{F(r)} \cdot d\mathbf{r}; \qquad (14.14)$$

F(r) and $d\mathbf{r}$ are shown in Figure 14.10. Into this general equation we insert the gravitational force exerted on the planet by the sun. We then have

$$\Delta U = +GMm \int_{\mathbf{r}_i}^{\mathbf{r}_f} \frac{1}{r^2}\, \hat{\mathbf{r}} \cdot d\mathbf{r}.$$

The fact that the gravitational force is a central force simplifies matters considerably, because the dot product $\hat{\mathbf{r}} \cdot d\mathbf{r}$ is just the scalar dr—the change in the planet's distance from the sun—regardless of the direction of motion of the planet. So the equation becomes

$$\Delta U = GMm \int_{r_i}^{r_f} \frac{dr}{r^2}. \qquad (14.15)$$

We have changed the limits of integration from vector quantities to scalar quantities because only their magnitudes will matter in evaluating the limits of the scalar integral. That is, the change ΔU in potential energy depends only on the change in the distance r of the planet from the sun and not on its particular initial and final positions $\mathbf{r}_i$ and $\mathbf{r}_f$. (As always in a calculation of potential energy, ΔU is independent of the path the planet takes from its initial to its final position; see Section 8.2.)

We now evaluate the integral in Equation 14.15. We obtain

$$\Delta U = GMm \left(-\frac{1}{r} \right) \Bigg|_{r_i}^{r_f},$$

which yields the result

$$\Delta U = GMm \left(\frac{1}{r_i} - \frac{1}{r_f} \right). \qquad (14.16)$$

It is usually most convenient to choose the zero of potential energy at $r = \infty$. That is, the planet's potential energy is taken to be zero when the planet is so far from the sun that the gravitational influence of the sun vanishes. When we make this choice, we are choosing $r_i = \infty$ in Equation 14.16. Once the choice is made, we can speak of *the* gravitational potential energy U of the planet when it is a distance $r_f = r$ from the sun. We have

$$U = GMm \left(\frac{1}{\infty} - \frac{1}{r} \right) = GMm \left(0 - \frac{1}{r} \right),$$

or

$$U = -\frac{GMm}{r}. \qquad (14.17)$$

The fact that the potential energy is always negative with this choice of zero presents no difficulty. The gravitational potential energy of the planet does increase with increasing distance from the sun, as you would expect. The dependence of U on r is

Not Mysterious, Just Greek

The terms **aphelion** and **perihelion** derive from Greek roots meaning ''away from the sun'' and ''in the neighborhood of the sun,'' respectively. For satellites of the earth, the corresponding terms are **apogee** and **perigee**, and there are corresponding (though less used) terms for other primary bodies as well.

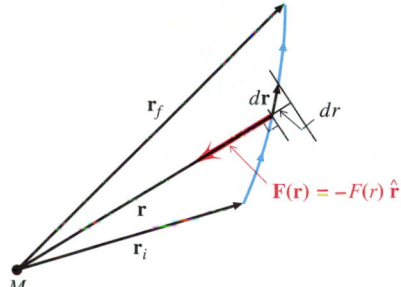

FIGURE 14.10 Calculating the change ΔU in the gravitational potential energy of a planet as it moves along its orbit from $\mathbf{r}_i$ to $\mathbf{r}_f$. At the arbitrary intermediate point **r**, the change in potential energy dU over the infinitesimal path element $d\mathbf{r}$ is $-\mathbf{F(r)} \cdot d\mathbf{r} = F(r)\, dr$, where dr is the radial component of $d\mathbf{r}$.

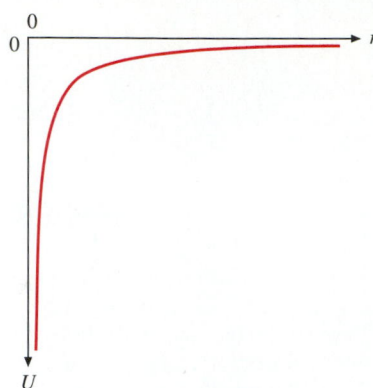

FIGURE 14.11 Schematic plot of the gravitational potential energy U of a planet as a function of its distance r from the sun.

shown graphically in Figure 14.11. For small values of r, U increases rapidly; for large values of r, U goes asymptotically to zero.

For the sake of concreteness, we have spoken of a planet under the gravitational influence of the sun. But Equations 14.16 and 14.17 apply equally to any body k, whose mass is m, under the gravitational influence of another body j, whose mass is M, provided that body j is

1. a particle; or
2. a sphere made up of uniform shells, in which case r is measured from the center of body j; or
3. so distant from body k that its irregularity of form is insignificant—the value of r does not depend significantly on where in body k the origin is chosen.

If body j does not satisfy one of these conditions, the force on body k is not a central force. Then Equations 14.16 and 14.17 are not applicable because they are derived from Equation 14.15, which is not valid for noncentral forces.

EXAMPLE 14.6

Find the gravitational potential energy of a 100-kg artificial satellite when it is resting on its booster rocket, ready to be launched, and when it is in a geosynchronous orbit of radius $r_s = 42\ 180$ km.* Take the radius of the earth to be $r_e = 6367$ km.

SOLUTION: You must apply Equation 14.17 twice. You can do this directly, but you can save some work by referring to Equation 14.5 and noting that at the surface of the earth you have $g = GM/r_e^2$. Comparing this with Equation 14.17 gives

$$U_e = -\frac{GMm}{r_e} = -gmr_e \qquad (14.18)$$

for the potential energy of the satellite before launch. Inserting the numerical values, you obtain

$$U_e = -9.80 \text{ m/s}^2 \times 100 \text{ kg} \times 6.367 \times 10^6 \text{ m}$$
$$= -6.24 \times 10^9 \text{ J}.$$

The potential energy at a distance r_s from the center of the earth is $U_s = -GMm/r_s$. Dividing this equation by Equation 14.18 and solving for U_s, you have

$$U_s = U_e \frac{r_e}{r_s}$$

$$= -6.24 \times 10^9 \text{ J} \times \frac{6.367 \times 10^6 \text{ m}}{42.18 \times 10^6 \text{ m}}$$

$$= -0.942 \times 10^9 \text{ J}.$$

The magnitude $|U_s|$ is about 15% of the magnitude $|U_e|$. By the time the satellite is raised by the synchronous orbit, it has about 85% of the potential energy required to remove it to infinity—that is, to remove it entirely from the earth's gravitational influence.

Escape Speed

Example 14.6 suggests that, if a body is given enough mechanical energy to rise to an infinite distance above the earth (or, practically speaking, a very large distance), it will never return. It is no longer bound to the earth by gravitation, but is free to move independent of any gravitational force exerted on it by the earth. In principle, the necessary energy could be conferred on the body by giving it enough kinetic energy at the surface of the earth so that it "coasts" to infinity. If we neglect air resistance, the coasting process is conservative. So we have for the required initial kinetic energy

$$K_i = \Delta U,$$
$$\tfrac{1}{2}mv_i^2 = U_\infty - U_e = 0 - U_e,$$

or

$$\tfrac{1}{2}mv_i^2 = \frac{GMm}{r_e}.$$

*A geosynchronous orbit is one for which the satellite period is 1 day, so the satellite always remains over the same meridian of longitude. Geosynchronous satellites whose orbit planes coincide with the equatorial plane of the earth are widely used for telecommunications and broadcasting because the satellite dishes—ground antennas that transmit to and receive from them—need not be continuously re-aimed. See Problems 5.46, 14.21, and 14.29.

By solving for v_i, we can evaluate the initial speed that will enable a body to escape from the earth. We have

$$v_i = \sqrt{\frac{2GM}{r_e}}. \qquad (14.19a)$$

It is often convenient to use Equation 14.18 to recast this equation into the form

$$v_i = \sqrt{2gr_e}. \qquad (14.19b)$$

The quantity v_i is the **escape speed**, or, as it is commonly if imprecisely called, the **escape velocity**. A similar speed can be calculated for the escape of a body from the gravitational influence of any astronomical object if the body is located at or above the surface at a distance r from the center of the object.

EXAMPLE 14.7

Calculate the escape speed for a body launched from the surface of the earth, neglecting air resistance.

SOLUTION: Inserting the necessary numerical values into Equation 14.19b, you have

$$v_i = \sqrt{2 \times 9.80 \text{ m/s}^2 \times 6.367 \times 10^6 \text{ m}} = 11.2 \text{ km/s}.$$

This value should not be taken too seriously for launches from the surface of the earth. Air resistance is not negligible, and kinetic energy is not conferred on the satellite instantaneously by the rocket booster. But there are practical cases for which the escape speed may have direct meaning. One example is the launch of a lunar return module from the surface of the moon.

Total Mechanical Energy in a Circular Orbit

To stay in orbit, a body must have kinetic as well as potential energy. For any orbit, the total energy E is just the sum of the kinetic and potential energies:

$$E = K + U.$$

For a circular orbit, the gravitational force is just equal to the centripetal force required to keep the body in that orbit:

$$\frac{GMm}{r^2} = \frac{mv^2}{r}.$$

We multiply through by r to obtain

$$\frac{GMm}{r} = mv^2.$$

The left side of this equation is just $-U$; the right side is $2K$. So we have

$$K = -\tfrac{1}{2}U \quad \text{for circular orbits.} \qquad (14.20)$$

The total mechanical energy is thus

$$E = \tfrac{1}{2}U \quad \text{for circular orbits.} \qquad (14.21)$$

EXAMPLE 14.8

Compare the energy E_m required to launch a spacecraft from the moon's surface into a minimum orbit about the moon with the energy E_{esc} required to send it on a mission from the moon to a distant planet.

SOLUTION: Because the moon has no atmosphere, a minimum orbit is one that skims the surface. Thus the additional gravitational potential energy required for the launch is negligible. The total energy required is just the necessary kinetic energy. So you have

$$E_m = K = -\tfrac{1}{2}U_m,$$

where U_m is the potential energy of the spacecraft on the surface of the moon. To escape from the moon, the spacecraft must be removed to a great distance. This requires a change in its potential energy:

$$E_{esc} = \Delta U = U_\infty - U_m = -U_m.$$

Comparing E_m with E_{esc}, you see that the escape launch requires just twice as much energy as the minimum-orbit launch.

Gravitational Mass and Inertial Mass*

You are by now quite familiar with Newton's law of gravitation and Newton's second law of motion:

$$\mathbf{F} = m\left(-\frac{GM}{r^2}\,\hat{\mathbf{r}}\right) \quad \text{and} \quad \mathbf{F} = m\mathbf{a}.$$

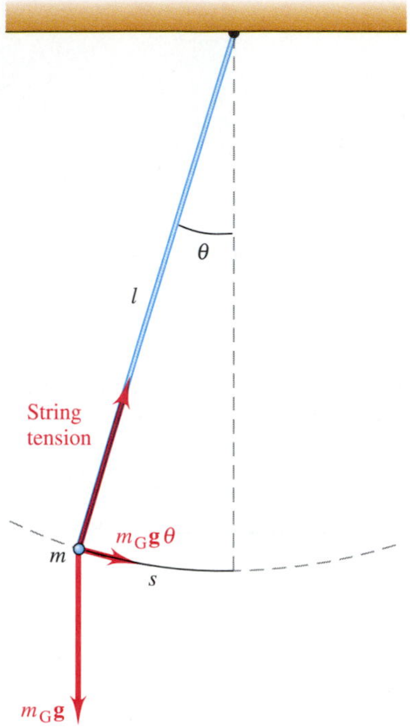

Although the mass m appears in both equations, its role in the first is quite different from that in the second. In the law of gravitation, m is an expression of how hard a certain body is pulled by some other body at a given distance. We call this the **gravitational manifestation** of mass. In the law of motion, m is an expression of the degree to which the body resists acceleration under the influence of a given external force. We call this the **inertial manifestation** of mass.

The remarkable thing is that the same quantity, m, seems to fulfill these two apparently unrelated roles. How do we know this? Newton was the first to raise the question. Let us follow his argument by reanalyzing the motion of a simple pendulum making small oscillations, which we first considered in Section 6.5. Figure 14.12 shows a simple pendulum of length l with a bob having mass m, at an instant when the bob is deflected from the vertical through a small arc of length s. There is a gravitational force acting on the bob. This force arises from the gravitational manifestation of the mass of the bob. So let us be specific and write the force as $m_G\mathbf{g}$. As you learned in Section 6.5, the string tension must also be taken into consideration in calculating the net force acting on the bob. For small deflections, the net force may be expressed in the scalar form

$$F = -m_G g\,\frac{s}{l}.$$

As we have often done before, we use Newton's second law of motion to equate this force with the product of the mass and the acceleration. But in Newton's second law we are concerned with the inertial manifestation of the mass of the bob. So let us be specific and write

$$F = m_I a = m_I \frac{d^2 s}{dt^2}.$$

Combining the two equations gives the equation of motion

$$\frac{d^2 s}{dt^2} = -\frac{m_G}{m_I}\frac{g}{l}\,s. \tag{14.22}$$

This is the same as Equation 6.44 except that the right side includes the factor m_G/m_I. This factor did not appear in Equation 6.44 because we did not distinguish between m_G and m_I.

In spite of the new feature, Equation 14.22 is still a harmonic-oscillator equation because it still has the general form

$$\frac{d^2 s}{dt^2} = -\omega^2 s.$$

The angular frequency ω is found by comparing the general form with the equation of motion; we have

$$\omega = \sqrt{\frac{m_G}{m_I}}\,\sqrt{\frac{g}{l}}. \tag{14.23}$$

This differs from the solution given by Equation 6.45a only in the term under the first radical sign.

The ratio m_G/m_I is in some degree arbitrary. The inertial mass is defined in terms of a Mach experiment of the kind discussed in Section 9.6. But the numerical value of any inertial mass depends on the choice of the standard inertial mass, which is essentially arbitrary. Similarly, but independently, the gravitational masses of bodies are

FIGURE 14.12 A simple pendulum of length l has a bob of mass m. The pendulum is deflected from the vertical through the arc length s. The gravitational force acting on the bob is a consequence of the gravitational manifestation m_G of the mass and has the value $m_G\mathbf{g}$. For small deflection, the magnitude of the net force acting on the bob is $m_G g\theta = m_G g s/l$.

---

*This section may be omitted without loss of continuity.

determined by using a device such as a beam balance to compare them with an arbitrarily chosen standard gravitational mass.

We can remove a degree of arbitrariness by choosing the standard inertial mass and the standard gravitational mass to be the *same* object—say, the platinum-iridium standard kilogram in Paris. If we were to make a pendulum out of the standard kilogram, it would obey Equation 14.23. The ratio m_G/m_I would be 1 by definition, and Equation 14.23 would simplify to the familiar form $\omega = \sqrt{g/l}$.

But if we take seriously the possibility that m_G and m_I are distinct quantities, we have no reason to expect that the ratio m_G/m_I will be the same for some other substance as it is for the platinum-iridium alloy of the standard kilogram. If the ratio depended on the substance, there would be a directly observable consequence: Simple pendulums made with bobs of various substances would have angular frequencies ω different from one another.

Newton tested this possibility by comparing the periods of pendulums identical in every way except in the material of which the bobs were made. He found that the gravitational and inertial masses of bodies do not differ from one another by as much as 0.1%. This result has since been refined to an extraordinary degree. In a long series of experiments beginning in the late 1880s and continuing almost until his death, the Hungarian physicist Loránd, Baron Eötvös* (1848–1919) used a torsion balance to detect any difference between the ratio m_G/m_I for a bob made of platinum and the corresponding ratio for bobs made of several other substances. He could not detect any such difference as great as 5 parts in 10^9. Because of the great elegance of the experimental design and the meticulous care with which it was carried out over a period of years, all experiments intended to detect a difference between gravitational and inertial mass are called Eötvös experiments. More modern Eötvös experiments using quite different techniques have shown that m_G/m_I cannot differ from 1 by more than 1 part in 10^{12}.

THINKING LIKE A PHYSICIST

When two apparently independent quantities such as m_G and m_I differ by so little, we may well suspect that they are in fact the same quantity seen in different aspects. Within the framework of classical physics, there is no satisfactory way to deal with this suspicion. We must simply be satisfied to accept the experimental results. But the suspicion was a major motivation for Einstein when he formulated the radically new view of the universe called the general theory of relativity. The assertion that the suspicion is true—that m_G and m_I are not merely *exactly equal* numerically but are the *same physical entity* seen two different ways—is called the **principle of equivalence**. This principle is the basis for the general theory of relativity.

Modern physicists often express this kind of suspicion in terms of what are only half-jokingly called the "totalitarian principles of physics":

1. Nature cannot do anything that is forbidden by physical laws.

2. Nature *must* do everything that is *not* forbidden by physical laws.

The first of these "principles" simply expresses the basic confidence in the laws of nature without which it is impossible to do science. The second "principle" does not have so obvious a basis. Taken together, the "principles" certainly do not have the force of physical law. Nevertheless, they have often proved helpful to scientists confronted with a research puzzle.

a	acceleration	D	mean distance between large and small masses in the Cavendish balance
A	surface area of an imaginary sphere centered on the sun	E	total mechanical energy
dA	area swept out by the radius vector to a planet in time dt	$\mathbf{F}_{1\to2}$	gravitational force exerted by body 1 on body 2
		$\mathbf{g}, g$	acceleration of gravity at the surface of the earth

*Often known by the German form of his name, Roland von Eötvös. The name is pronounced (very roughly) as uht-vush.

G	universal gravitational constant	T	torque
I	moment of inertia	U	potential energy
K	kinetic energy	$\mathbf{v}, v$	linear velocity, speed
l	length of a simple pendulum	δ	displacement of the equilibrium position of the end of the arm of the Cavendish balance
$\mathbf{L}, L$	angular momentum		
m, M	mass	$\Delta\theta$	angular shift of the equilibrium position for the Cavendish balance
m_G, m_I	gravitational, inertial manifestations of mass		
N	number of field lines	κ	torsion constant
$\mathbf{p}$	linear momentum	ρ	(1) mean density of a body, (2) mean orbit radius
$\mathbf{r}_{1\to2}$	radius vector from body 1 to body 2	τ	period
s	arc length of a pendulum swing	ω	orbital angular velocity or speed
$d\mathbf{s}$	infinitesimal displacement vector		

Summing Up

Kepler's three empirical laws, stated in Section 14.1, account very accurately for the observed motions of the planets. **Kepler's laws** can be not only derived but also generalized on the basis of Newton's three laws of motion, taken together with Newton's **law of universal gravitation**, Equation 14.3b:

$$\mathbf{F}_{1\to2} = -G\frac{m_1 m_2}{r^2}\,\hat{\mathbf{r}}_{1\to2}.$$

The fact that this can be done provides a spectacular demonstration of the broad applicability of classical mechanics, as well as an indispensable tool for astronomy.

Newton's generalizations of Kepler's laws are stated in Section 14.3. **Kepler's third law** for any two bodies orbiting the same primary is expressed by Equation (14.1),

$$\frac{\tau_1^{\,2}}{\tau_2^{\,2}} = \frac{\rho_1^{\,3}}{\rho_2^{\,3}}.$$

More fundamentally, the mean orbital radius and period of any body in a circular or elliptical orbit about its primary are given by Equation 14.8,

$$\frac{\rho^3}{\tau^2} = \frac{GM}{4\pi^2}.$$

This law is proved for circular orbits in Section 14.3. **Kepler's second law** (Equation 14.9b) is the law of constant areal velocity,

$$\frac{dA}{dt} = \frac{L}{2m} = \text{constant.}$$

In Section 14.3, this law is shown to be a consequence of angular-momentum conservation for any body moving under the influence of a central force. **Kepler's first law** requires a conic-section orbit for any body moving under the influence of an inverse-square force.

The inverse-square nature of the gravitational force exerted by a particle—or by a spherical body at a point outside itself—is justified at the end of Section 14.3. The justification is intimately connected with the three-dimensional character of space.

Because of the weakness of the gravitational interaction, it is very difficult to determine the value of the **universal gravitational constant** G. The method of determination is described in Section 14.4; the best modern value is given by Equation 14.4,

$$G = (6.672\ 59 \pm 0.000\ 85) \times 10^{-11}\ \text{N·m}^2/\text{kg}^2.$$

The gravitational force is conservative. Hence it is possible to define the **change in gravitational potential energy** ΔU of a body as it moves under the influence of a central gravitational force by means of Equation 14.16,

$$\Delta U = GMm\left(\frac{1}{r_i} - \frac{1}{r_f}\right).$$

The reference point for which $U = 0$ is usually chosen at $r = \infty$. The **gravitational potential energy** U is then given by Equation 14.17,

$$U = -\frac{GMm}{r}.$$

We use conservation of total mechanical energy to define the **escape speed** v_i from the surface of a planet (or other astronomical body) of mass M and radius r. If the planet is the earth and atmospheric resistance is neglected, the escape speed is given by Equation 14.19a or Equation 14.19b,

$$v_i = \sqrt{\frac{2GM}{r_e}} = \sqrt{2gr_e},$$

where r_e is the radius of the earth. Similar expressions can be written for any other astronomical body.

If a body moves in a circular gravitational orbit, its kinetic energy E and its total mechanical energy E are related to its gravitational potential energy by Equations 14.20 and 14.21,

$$K = -\tfrac{1}{2}U \quad \text{and} \quad E = \tfrac{1}{2}U.$$

These are not bad approximations for most of the planets, whose orbits are ellipses of small eccentricity.

The mass of a body takes two very different roles in determining its behavior. In gravitational interactions, it is the **gravitational manifestation** expressed in the quantity m_G that is significant. In relating force and acceleration by means of Newton's second law, it is the **inertial manifestation** expressed in the quantity m_I that is significant. Experiments show that $m_G = m_I$ for all bodies to a high degree of precision. From the point of view of classical mechanics, this is an intriguing coincidence. From the point of view of the general theory of relativity, the experimental identity is a consequence of a deep identity between the two apparently distinct roles played by mass.

KEY TERMS

Section 14.1 Introduction
geocentric (Ptolemaic) and heliocentric (Copernican) systems ▪ Kepler's laws ▪ areal velocity

Section 14.2 Newton's Law of Universal Gravitation
universal gravitational force

Section 14.3 The Apple, the Moon, and the Solar System
primary and secondary bodies ▪ isotropy of space ▪ gravitational field

Sections 4.4 and 4.5 Gravitational Potential Energy, Gravitational Mass, and Inertial Mass
escape speed ▪ gravitational manifestation, inertial manifestation of mass ▪ principle of equivalence

QUERIES

14.1 *(3) Easier on the bounce?* Until recently, the ratios of the mean orbit radii of the planets were known much more precisely than their absolute values. Explain why it is easier to obtain the ratios than the absolute distances. About 1960, it became possible to measure the earth–Venus distance quite precisely at a given instant by reflecting radar signals off the surface of Venus. Describe in principle how such a measurement can be used to provide a precise value for the astronomical unit (Table 14.1) and thus precise absolute values for the mean radii.

14.2 *(3) Ah, Cytherea, bright yet mysterious!* Until the advent of interplanetary space probes, the mass of Venus, our closest planetary neighbor, was known with far less precision than that of Saturn, which is much more distant. Explain.

14.3 *(3) Kepler hires on as a NASA consultant.* A satellite is launched from Cape Canaveral (latitude 28°N). Is it possible to put the satellite into an orbit that circles the earth at constant latitude 28°N?

14.4 *(3) Star wars.* During the Cold War, the United States had an elaborate early-warning radar network intended to detect any swarm of ICBMs that might be launched from the USSR, which lay roughly to the west (or to the east, depending on which way you want to look) of the United States. Why were most of the radar stations located in high northern latitudes?

14.5 *(3) Spy satellites.* **(a)** Suppose you launch a satellite from a site at latitude λ. Explain why it is easier to put the satellite into an orbit whose plane makes an angle greater than λ with the plane of the earth's equator than to put the satellite into an orbit for which the angle is less than λ. **(b)** The former USSR launched most of its satellites from Baikonur (in Kazakhstan), whose latitude is 47°N. The United States launched (and still launches) most of its military satellites from Vandenberg Air Force Base (in California), whose latitude is 35°N. What are the relative advantages or disadvantages of the two sites for launching spy satellites intended mainly to observe the other country?

14.6 *(4) Heavy responsibility.* In the analysis of the Cavendish experiment, the mass of the lead spheres that hang from the ends of the balance cancels out. Thus the value of G does not depend directly on m; see Equation 14.13. Why is it nevertheless important to maximize m?

14.7 *(5) Infinite wisdom.* According to Equation 14.17, the value of the gravitational potential energy U of a body "blows up" ($U \rightarrow \infty$) when $r \rightarrow 0$. Why does this mathematical blow up not present any physical difficulty?

14.8 *(5) Finding the energy to do it right.* Equation 14.17, $U = -GMm/r$, gives the potential energy of a secondary of mass m in the presence of a primary of much larger mass M. Strictly speaking, is it proper to attribute the potential energy to the secondary? (Hint: Suppose a system consists of two bodies of comparable masses m_1 and m_2 separated by a distance r. What is the potential energy of body 2 due to the presence of body 1? What is the potential energy of body 1 due to the presence of body 2? Is the energy of the entire system $-GMm/r$ or $-2GMm/r$?)

14.9 *(G) A in literature, C in physics.* In his 1865 novel, *From the Earth to the Moon*, Jules Verne made an (unnecessarily) elaborate calculation to find the point along the way at which the gravitational forces exerted on the space vehicle by the earth and the moon are equal. At that point in the voyage, the crew, their dog, and all the furniture in the vehicle rose gently off the floor and came to rest on the ceiling (thereafter the floor). Explain why this account is in error. In particular, why are all the objects in the spacecraft weightless throughout the journey, in spite of the gravitational forces exerted on them by the earth and the moon? (Hint: Review the definition of weight given in Section 3.4.)

PROBLEMS

GROUP A

14.1 (2) *But does it really work?* Show that Equation 14.5 yields a plausible value for the acceleration of gravity, using the values of the necessary constants given in this chapter and on the inside front cover.

14.2 (2) *Light-headed.* The mass of the moon is 7.35×10^{22} kg, and its radius is 1.74×10^6 m. Find the acceleration of gravity at the surface of the moon.

14.3 (2) *Weak interaction.* Two 500 000-tonne supertankers traveling through a strait pass within 1 km of each other. Assuming to first approximation that they interact gravitationally like particles, find the magnitude of the gravitational force exerted by each on the other.

14.4 (2) *Straight shot.* A spacecraft travels (unrealistically) from the earth to the moon along a straight-line path. At what point along the path are the gravitational attractions of the earth and the moon on the spacecraft equal and opposite? Take the earth–moon distance to be 3.86×10^8 m and the mass of the moon to be $\frac{1}{81}$ that of the earth.

14.5 (2) *Overshot.* A spacecraft is launched from the far side of the moon and travels away from the earth along the extension of the line connecting the earth and the moon. At what point along the path are the gravitational attractions of the earth and the moon on the spacecraft equal?

14.6 (2) *Mars? Gee!* The mass of Mars is 6.42×10^{23} kg, and its radius is 3.39×10^6 m. Find the acceleration of gravity at the surface.

14.7 (2) *Born under the sign of Dr. Jones?* If, as astrologers maintain, the position of the planets at the instant of a child's birth profoundly affects the course of the child's life, the influence must be due to the gravitational force exerted by the planets on the child because no other interactions of even remotely comparable magnitude exist between the planets and the child. **(a)** Let us consider Saturn—a planet reputed to have much influence on mood and disposition—under the most favorable circumstances. The mass of Saturn is 5.7×10^{26} kg. Its distance of closest approach to the earth is 1.2×10^{12} m.

What is the magnitude of the gravitational force exerted by Saturn on a newborn baby of mass 3.2 kg? **(b)** What is the magnitude of the gravitational force exerted on the same baby by the obstetrician, whose mass is 64 kg and whose center of mass is located 0.45 m from that of the baby at the instant of birth? **(c)** What is the ratio of the two forces you have calculated? Is it allowable to discount the influence of the obstetrician in astrological calculations?

14.8 (2) *High up, I.* At what altitude above the surface of the earth is the acceleration of gravity equal to 1 m/s²?

14.9 (2) *High up, II.* At what altitude above the surface of a planet of radius R is the acceleration of gravity equal to one-half its surface value?

14.10 (3) *Successful job-hunting strategy.* In 1610, Galileo, a Tuscan working in Venice, used his telescope to discover four of the satellites of Jupiter. Desiring to find employment back home, Galileo named the satellites the Medicean Stars in honor of the Grand Duke of Tuscany, Cosimo II de'Medici— a gentle flattery that could not have hurt, because Cosimo subsequently hired him. The name didn't stick, but Table 14.2 gives the mean orbit radii and periods of the satellites. Show that the satellites obey Kepler's third law (Equation 14.1).

14.11 (3) *Small calculation, big masses.* Use the data of Table 14.2 to find the masses of Mars, Jupiter, and Saturn.

14.12 (3) *Newtonian dodge.* Newton had no numerical value for G. Nevertheless, he calculated that the mass of the sun is 3.3×10^5 times that of the earth. Use the data of Example 14.4 to show how he did this.

14.13 (3) *Long, slow sweep.* The aphelion and perihelion distances of Pluto from the sun are respectively 7.375×10^{12} m and 4.443×10^{12} m. Find the period of Pluto.

14.14 (3) *Waiting for 2063.* Halley's comet last passed through perihelion in 1986 and is expected again in 2063. The perihelion distance of the comet is 0.59 AU. What is its aphelion distance?

TABLE 14.2 **Orbital Data for Some Satellites of Mars, Jupiter, and Saturn**

Primary	Satellite	Mean Orbit Radius ρ (m)	Period τ (s)
Mars	Phobos	9.408×10^6	
	Deimos	2.3457×10^7	$1.090\ 77 \times 10^5$
Jupiter	Io	4.219×10^8	$1.528\ 59 \times 10^5$
	Europa	6.712×10^8	$3.068\ 24 \times 10^5$
	Ganymede	1.071×10^9	$6.181\ 75 \times 10^5$
	Callisto	1.883×10^9	$1.441\ 93 \times 10^6$
Saturn	Titan	1.227×10^9	$1.379\ 12 \times 10^6$

14.15 (3) *Quick service.* The comet with the shortest known period is Encke's comet. Its semimajor axis is 3.3×10^{11} m. Find the period.

14.16 (3) *Russian moonlet.* One of the Vostok series of Russian spacecraft was put into an orbit whose maximum altitude was 244 km. The orbital period was 88.1 min. Find the minimum altitude.

14.17 (3) *A moon named Fear.* (a) Using the data given in Table 14.2, find the period of Phobos, the inner satellite of Mars. (b) The martian "day" (the time it takes for Mars to rotate on its axis once) is 1.03 earth days. Which way does Phobos appear to move across the martian heavens? How long is a Phobos "month" as seen by an observer on Mars? (Hint: As seen from above the north pole of Mars, the sense of Phobus's revolution about Mars and that of Mars's rotation about its own axis are both counterclockwise.)

14.18 (3) *Duck! Here it comes again!* The mass of Mars is 0.107 that of the earth. Its radius is 3.39×10^6 m. What is the speed of an artificial satellite in a minimum orbit about Mars? The atmosphere of Mars is so thin that the satellite can circle Mars at a very low altitude.

14.19 (3) *Time zones.* New York City, Peoria (Illinois), Laramie (Wyoming), and Susanville (California) are located at roughly the same latitude. Their longitudes are close to 15° apart; this longitude difference corresponds to a one-hour time difference, and the cities are one time zone apart from their neighbors. (a) Is it possible to put a satellite into a polar orbit—an orbit passing over both poles—such that the satellite passes over each city at noon local time? If so, what is the radius of the orbit? (b) Is it possible to put the satellite into a polar orbit such that it passes over New York City and Laramie at noon local time? If so, what is the radius of the orbit?

14.20 (3) *Thinking big.* The solar system revolves about the center of mass of our galaxy with period 2.5×10^8 years. The distance to the center of mass is about 2.2×10^{20} m. What is the mass of that part of the galaxy within the orbit of the solar system? (You will see in Chapter 24 why the mass outside the orbit has no effect on the orbit.)

14.21 (3) *Synchronous satellite, I.* An earth satellite has a circular orbit that lies in the equatorial plane of the earth. Find the orbit radius for which the satellite will appear to hover indefinitely above a fixed point on the earth's equator. Take the length of the sidereal day as 23 h 56 min, or 8.616×10^4 s.

14.22 (3) *Meteor shower.* Meteoroids are rocks of various sizes occurring in swarms, some of which probably have their origin in the partial or complete breakup of a comet. Such meteoroids share the cometary orbit, which is usually a very eccentric (elongated) ellipse with the sun at one focus. When the earth crosses this orbit, the result is a meteor shower as the rocks pass through the atmosphere and are heated until they incandesce. For the purposes of this problem, there is little error in assuming that the meteoroid orbit is a parabola rather than an ellipse. (a) Expressed in terms of v_e, the orbital speed of the earth, what is the speed of the meteoroids as they cross the earth's orbit? (b) The meteoroids can collide with the earth at any angle, from head-on to overtaking from the rear. Find the maximum and minimum possible speeds of meteoroids relative to the earth. (c) Calculate the maximum kinetic energy per unit mass of meteoroids, as seen from the earth. (d) The energy required to vaporize most solids is of order 10^6 J/kg. Why do only the largest meteoroids reach the surface of the earth, where they are called meteors?

14.23 (3) *Spinning like crazy, I.* What would the length of the day be if the earth rotated so rapidly on its axis that objects at the equator were weightless? Neglect the distortions in the shape of the earth that would occur under these circumstances. (A situation of this sort has been fundamental to the plots of a number of science-fiction novels.)

14.24 (4) *Spinning like crazy, II.* Show that a planet of mean density ρ will disintegrate if the length of its day is less than the critical value $\tau = \sqrt{3\pi/G\rho}$.

14.25 (5) *Nuclear physics.* Under proper circumstances, a deuteron (a nucleus of deuterium, or heavy hydrogen) can be broken apart into a proton and a neutron. (a) Making the crude assumption that the proton and neutron inside the deuteron are separated by the sum of their radii as independent particles, or 2×10^{-15} m, determine how much energy would be required to separate the deuteron into two independent particles if the force binding them together were gravitational. (b) The energy actually required is 3.57×10^{-13} J. Express the gravitational energy as a fraction of the actual energy to get an idea of the difference in strength between the so-called strong nuclear interaction that binds deuterons and other nuclei and the gravitational interaction.

14.26 (5) *Strong-arm geology.* You are exploring the asteroids, having heard tales of an asteroid made of nearly pure gold covered by a thin surface layer of ordinary dirt. You have just landed on one whose diameter you have measured to be 25 km. You pick a small rock off the surface and throw it as hard as you can. You find that it does not return. By throwing a few more rocks, you find that this happens only if you throw as hard as you can. (a) If the tales are true, what will the escape speed be? The density of gold is 19.3×10^3 kg/m³. (b) Is the asteroid really made mainly of gold?

14.27 (5) *Small potatoes.* Two spheres, each of mass 1000 kg, are located with their centers 2.5 m apart. What is the gravitational potential energy of the system?

14.28 (5) *By the light of the silvery earth.* Using the values of the mass and radius of the moon given in Problem 14.2, find the energy required to launch a 100-kg spacecraft from the surface of the moon into (a) a minimum orbit and (b) a minimum escape orbit. Assume that the rocket exhausts its fuel very quickly, so the energy required to accelerate the fuel supply is negligible.

14.29 (5) *Synchronous satellite, II.* Suppose that the satellite of Problem 14.21 is launched into the same orbit, but in such a way that it revolves about the earth in a westward direction. (Of course, it won't be a synchronous satellite in that case!) What is the proportional increase in the work that must be performed on the satellite if it is to be put into a westward rather than an eastward orbit, launched from a site on the equator? Assume that the rocket goes straight up and then turns parallel to the earth's surface to give the satellite the necessary tangential speed.

14.30 (5) *Moving out fast.* The radius of the sun is 6.96×10^8 m. Using the value of its mass found in Example

14.4, find (a) the escape speed for a body at the surface of the sun and (b) the acceleration of gravity at the surface of the sun.

14.31 (5) *Astronaut primer.* The radius of the moon is 1740 km. The escape speed for a body at its surface is 2.37×10^3 m/s. Find (a) the mass of the moon and (b) the acceleration of gravity at its surface.

14.32 (5) *Minimum-orbit satellite, I.* The escape speed from the surface of Titan, the largest of the satellites of Saturn, is 816 m/s. What is the speed of an artificial satellite in a minimum orbit about Titan?

14.33 (5) *Minimum-orbit satellite, II.* An artificial satellite is placed in a minimum orbit about a planet of mass M and radius r. What is the period of the satellite?

14.34 (5) *The "satellite paradox."* An artificial satellite is in an orbit so low that there is a small but significant amount of air friction, and the satellite gradually descends to earth over many revolutions. Show that the kinetic energy of the satellite actually *increases* in the process, and it moves faster and faster.

14.35 (5) *Gone forever.* A spacecraft is in an orbit around the sun that coincides with the orbit of the earth. What additional speed must be given it if it is to leave the solar system completely?

14.36 (5) *A little extra goes a long way.* Express the escape speed from a planet in terms of the speed of a spacecraft in a minimum orbit about the planet. Neglect the effect of the small distance between the planet's surface and the minimum orbit.

14.37 (5) *All the way out.* What speed must be given a spacecraft launched from the earth if it is to leave the solar system? (Hint: Don't forget that the spacecraft is launched from a moving platform.)

14.38 (5) *Energy plot.* A spacecraft of mass 1000 kg is in a circular orbit of radius r around an asteroid of mass 1.50×10^{19} kg. On the same set of axes, make rough but quantitative plots of (a) the potential energy, (b) the kinetic energy, and (c) the total mechanical energy of the spacecraft, all as functions of r.

14.39 (G) *Mercury's orbit at a glance.* As seen from the earth, the angle between Mercury and the sun reaches a maximum value (called the maximum elongation) of about 28°. Use this information to estimate the aphelion distance of Mercury in astronomical units.

GROUP B

14.40 (2) *Going up!* (a) Beginning with the magnitude part of Equation 14.5, show that the rate dg/dr at which the acceleration of gravity varies with r is

$$\frac{dg}{dr} = -\frac{2GM}{r_e^3} \quad \text{for } r > r_e,$$

where r_e is the radius of the earth. (b) You use a gravimeter (a device that measures the acceleration of gravity) to determine g at a certain location. Then you raise the gravimeter to a height $h = \Delta r$ and measure g again. What is the proportional change $\Delta g/g$? (c) If the acceleration of gravity is 9.80 m/s² at a certain sea-level point, what is its value at an altitude of 8900 m above that point? (This is approximately the height of Mt. Everest.) (d) Measurement of very small variations in g is an important technique in geophysical research and mineral prospecting. Geologists often use portable gravimeters so sensitive that they are capable of detecting a difference in g between the top of a table and the floor on which it rests. What precision is required of such a gravimeter?

14.41 (2) *It's high tide twice a day.* You "learned" as a child that the tides are due mainly to the gravitational attraction of the moon. But it's a little more complicated than that. True, the moon attracts the ocean water nearest it, but it attracts the rest of the earth as well. What counts is the way the force *varies* with distance from the moon. Take the center of the moon as the origin, and let the distance from the center of the moon to the center of the earth be R (see part a of the figure opposite). Consider a particle located somewhere in the earth along the moon-earth axis, a distance $R + r$ from the origin. (a) Let a be the gravitational force per unit mass exerted on the particle by the moon, and calculate da/dr, the rate of variation of a with respect to r. (b) Find a_c, the force per unit

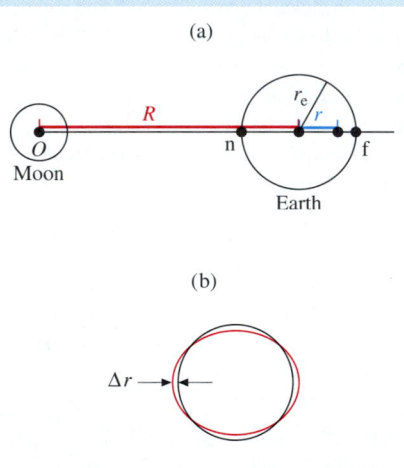

mass exerted on a particle located at the center of the earth. (c) Find a_n, the force per unit mass at n, the point on the earth nearest the moon, and a_f, the force per unit mass at f, the point on the earth farthest from the moon. (d) Define $\Delta a_1 \equiv a_n - a_c$ and $\Delta a_2 \equiv a_c - a_f$. Show that

$$\Delta a_1 = \Delta a_2 = -\frac{2Gm}{(R + r_e)^3} r_e,$$

where m is the mass of the moon and r_e is the radius of the earth. (e) Explain why there are approximately two high tides a day. (f) Imagine that the earth is completely covered by water. The water must flow until its surface has a constant value of net "acceleration of gravity" due to both the earth and the moon. As shown in part b of the figure, the ocean surface deviates from the shape it would have in the absence of the moon (shown in black) to a shape (shown in red) whose surface is shifted by a maximum distance Δr. Use the result of part **a** of Problem 14.40, and show that

$$\Delta r = \frac{m}{M}\left(\frac{r_e}{R + r_e}\right)^3 r_e.$$

(g) The height difference between high and low tides in mid-ocean (where the large distorting effect of coastlines is absent) is $\frac{3}{2}\Delta r$. [The factor $\frac{3}{2}$ arises from the fact that at the high-tide point (part *b* of the figure) the lunar and terrestrial gravitational forces are in line, whereas at the low-tide point they act at right angles.] Remembering that the mass of the moon is $\frac{1}{81}$ that of the earth and its distance from the earth is 60 times the earth's radius, find the height of midocean tides. The result is quite close to what is observed at such midocean locations as Hawaii.

14.42 *(3) Welcome to the club!* The orbital period of Halley's comet is 76.2 years. Its orbit, unlike the orbits of the planets, is so highly eccentric that the aphelion distance is very much greater than the perihelion distance. Estimate the aphelion distance. Refer to Table 14.1, and determine whether the aphelion of Halley's comet lies well within, far beyond, or at about the same distance from the sun as the orbits of the outermost planets—Neptune and Pluto. (Note: Pluto's orbit is rather eccentric, to the extent that it spends a fair proportion of its time inside the orbit of Neptune. But the orbit is not nearly as eccentric as that of Halley's comet.)

14.43 *(3) The little dog laughed.* A well-known nursery rhyme runs

> Hey diddle, diddle,
>
> The cat and the fiddle,
>
> The cow jumped over the moon . . .

Pictures in nursery-rhyme books often show the cow in a pasture with the moon above. **(a)** If, however, "jumping over the moon" means leaving the earth, circling the moon, and returning to the earth at about the same place from which she left, the cow cannot see the moon from the place where she begins her jump. Explain. (Hint: See Problem 9.21.) **(b)** The radius of the earth is $\frac{1}{60}$ the earth–moon distance, the radius of the moon is $\frac{1}{220}$ the earth–moon distance, and the mass of the moon is $\frac{1}{81}$ that of the earth. Show that the minimum time required for the cow's jump is 0.37 sidereal month, or 10 days.

14.44 *(3) Scenic tour.* A satellite is in a circular polar orbit—that is, an orbit whose plane contains the axis of the earth—at altitude 350 km. If the satellite passes directly over New York City (latitude 41°N, longitude 75°W), over what state does the satellite pass one revolution later, when it reaches the same latitude? (You will need to consult a map of the United States.)

14.45 *(G) Plane facts.* **(a)** Use the principle of conservation of angular momentum to show that, when a particle moves under the influence of a central force, its orbit must lie in a plane. **(b)** If a satellite is launched by propelling it upward and eastward from a site at latitude λ, what will be the inclination angle ι between the plane of the orbit and the plane of the earth's equator? See the drawing at the top of the next column. **(c)** If an orbit of greater inclination angle is desired, what must be done? **(d)** Why is it not possible to achieve

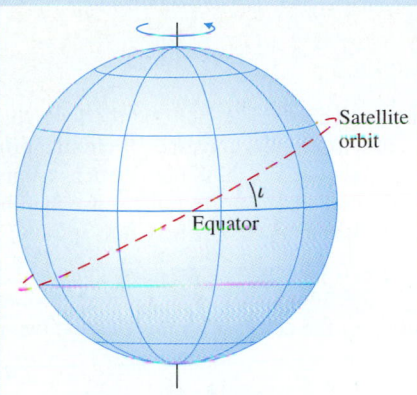

a smaller inclination angle without post-launch maneuvering of the space vehicle? What happens, for example, if the launch is directed in a southeastward direction from a site in the northern hemisphere?

14.46 *(3) Synchronous satellite, III.* An earth satellite is launched in a circular orbit whose radius is the same as that of the synchronous satellite of Problem 14.21, except that the plane of the orbit is inclined to the equatorial plane at an angle ι, as in Problem 14.45. Show that an observer on the earth will see the satellite oscillate in a north-south direction, and find the period and amplitude of the oscillation.

(Note: From a launching site not on the equator, it is harder to achieve the equatorial orbit than an inclined orbit. But the ground antennas required to receive signals from a synchronous satellite in an equatorial orbit are much simpler and cheaper than those required for a satellite in an inclined orbit because they do not need continuous adjustment to follow the satellite. Consequently, communications satellites are usually placed in equatorial orbits in spite of the extra expense. But the very special equatorial orbit required is now getting crowded to the point where international regulation may soon be needed.)

14.47 *(3) Fancy meeting you again!* **(a)** Two artificial satellites are in circular earth orbits in the same plane. The orbit radius of one satellite is r; the other satellite is in a slightly larger orbit of radius $r + \Delta r$. What is the time interval t between successive closest approaches of the satellites? **(b)** Suppose that $r = 11 \times 10^3$ km and $\Delta r = 25$ km. Find the value of t.

14.48 *(3) Peculiar universe.* Imagine a universe in which the law of gravitation has the inverse-first-power form

$$\mathbf{F} \propto -\frac{Mm}{r}\,\hat{\mathbf{r}}.$$

(a) Show that the speed of a planet in a circular orbit is independent of the orbit radius. **(b)** Show that in such a universe Kepler's third law would take the form

$$\frac{r_1}{r_2} = \frac{\tau_1}{\tau_2}.$$

14.49 *(3) Not the best of all possible worlds.* Imagine a universe in which the law of gravitation has the inverse-cube form

$$\mathbf{F} = -G' \frac{Mm}{r^3} \hat{\mathbf{r}},$$

where G' is a universal constant. Find the form that Kepler's third law would take in this universe for circular orbits. (Note: Noncircular orbits are not possible in such a universe, and even circular orbits are unstable in the sense that the slightest perturbation makes them "blow up" into spirals.)

14.50 (4) *Doing it Cavendish's way.* Cavendish expressed the result of his experiment not in terms of G but in terms of the mean density ρ of the earth. (a) Show that the two are related by the expression

$$\rho = \frac{3}{4\pi} \frac{g}{GR},$$

where R is the radius of the earth and g is the acceleration of gravity corrected to remove the effect of the earth's rotation. (b) Find the mean density ρ and compare it with the value, based on the density of surface rocks, used in Example 14.2.

14.51 (4) *Swinging on the moon.* The period of a pendulum oscillating on the surface of the moon is 2.46 times as great as its period on the surface of the earth. Given that the mass of the moon is $\frac{1}{81}$ that of the earth, find the radius of the moon.

14.52 (4) *Magic transformation.* Suppose that the length scale of the solar system were suddenly changed by a factor ξ. This change is to include the radii of all the bodies in the system as well as the distances between them. The mean densities of the bodies, however, remain unaffected. Show that all the orbital periods would remain unchanged.

14.53 (5) *Gravitational potential energy of a three-particle system.* A system consists of three particles. (a) Show that the gravitational potential energy of the system can be written

$$U = -G\left(\frac{m_1 m_2}{r_{12}} + \frac{m_2 m_3}{r_{23}} + \frac{m_3 m_1}{r_{31}}\right),$$

where r_{jk} is the distance between body j and body k. (b) Is it possible to speak of the potential energy of one particular particle in this system? See Query 14.6.

14.54 (5) *Moon shuttle.* A space station is in permanent circular orbit around the moon, 2000 km above the moon's sur-

face. A shuttle leaves the space station for the moon. It is launched tangent to the station orbit and fires its rocket engines briefly in such a way that no further engine power is needed until it is ready to decelerate for touchdown. In answering the following questions, neglect the rotation of the moon about its axis. The moon has negligible atmosphere. Its radius is 1738 km. (a) Where on the moon will the shuttle land, relative to the point on the moon's surface directly below the space station at the instant of launching? (b) What is the value of the mean radius of the shuttle orbit? (c) How long does the trip from space station to lunar surface take? (d) In which direction is the shuttle launched from the space station? (e) What is the speed of the shuttle relative to the moon, immediately after the initial firing of its engines? (f) What is its speed relative to the space station? (g) What is its speed when it lands?

14.55 (5) *Tektites.* Small, glassy buttonlike objects called tektites are found strewn over areas in various locations over the earth. (Tektite fields have been found in Australia, the Americas, Antarctica, and the ocean bed.) Tektites are manifestly foreign to the geology of the localities where they are found, and their composition is distinct from those of other meteorites. One view of the origin of tektites is that they were ejected from the moon by the impact of meteorites on the lunar surface. At what minimum speed must a body be ejected from the lunar surface if it is to reach the earth? (Hint: A minimum-speed tektite must have negligible speed when it passes through a certain point on its moon-earth trajectory.)

14.56 (G) *Wheels within wheels.* (a) Using the masses found in Example 14.4, find the ratio of the gravitational forces exerted on the moon by the earth and the sun. The distance from the sun to the moon is about 390 times that from the earth to the moon. (b) Explain qualitatively why it is possible to treat the earth-moon system as isolated when calculating the motion of the moon about the earth, the result of part **a** notwithstanding.

14.57 (G) *Minimum-orbit satellite, III.* A planet has mean density ρ. Show that the period τ of a minimum-orbit satellite about the planet is independent of the mass M and radius r of the planet and that it depends only on $1/\sqrt{\rho}$.

GROUP C

14.58 (3) *Flatlanders.* Imagine a universe that has only two space dimensions. The "flatlanders" who inhabit this universe perceive it as a vast, planar space. Use the symmetry argument of Section 14.3 to show that the gravitational force in such a universe would be proportional to $1/r$.

14.59 (3) *Double star.* A quite large proportion of all stars are members of double star systems. Consider such a system in which two stars of roughly comparable mass orbit about their common center of mass with period τ. Consider a pair of such stars separated by a distance R and having total mass M. Show that Kepler's third law takes the remarkably simple form

$$\frac{R^3}{\tau^2} = \frac{GM}{4\pi^2}.$$

Most double star systems are too distant from us for a direct determination of R. However, the period can often be determined from the periodic variation in the Doppler shift in the spectra of the stars (Chapter 22). Spectroscopic measurement often leads to a good estimate of the masses as well, and the separation can thus be found.

14.60 (4) *G by the method of Bouguer and Maskelyne.* You plan to repeat a measurement of G that gave the best value available before Cavendish's work. You will hang a plumb line near the base of a mountain. Using the deviation of the plumb line from the astronomically determined vertical and an estimate of the density of the mountain, you will determine the mean density ρ_e of the earth and thus estimate the value of G. You are thinking of using an approximately conical mountain whose height above the surrounding plain is 4900 m. Its base, at the level of the plain, is a circle of diameter 17 km. Assume the mountain to be uniform with a density $\rho_m = 3.0 \times 10^3$ kg/m^3. If you hang a plumb line at the base of the mountain, what order of magnitude of deviation of the line from the vertical must you be prepared to measure? In making this estimate, ignore problems that arise from the fact that the presence of the mountain distorts the crust on which it rests. (This distortion is due to what geologists call the *principle of isostasy*, according to which a mountain must float on the underlying rock much as an iceberg floats on water.) (Hint: See Problem 9.53. Assume that the gravitational attraction of a conical body is the same as if its entire mass were concentrated at its center of mass.) This method was first used by Pierre Bouguer (1698–1758) in the course of a geodetic expedition lasting from 1735 to 1744. He used Mt. Chimborazo, a large volcano situated in an otherwise flat plain in Ecuador. Nevil Maskelyne (1732–1811) improved on this work between 1774 and 1776, using Mt. Schiehallion in Scotland. He obtained the value

$$G = 8.1 \times 10^{-11} \text{ N·m}^2/\text{kg}^2.$$

14.61 *(4) It doesn't quite average out.* Equation 14.12 expresses the value of G obtained in the Cavendish experiment in terms of D, the equilibrium center-to-center distance between the small lead balls at the ends of the balance arm and the large spheres that attract them. But the balance beam is in fact oscillating, its ends moving through a distance δ. Because the gravitational force varies inversely as the square of the center-to-center distance, it is not quite correct to use D as the mean distance. **(a)** Use the definition of the mean value of a function $f(x)$,

$$\langle f \rangle = \frac{\displaystyle\int_a^b f(x)\,dx}{\displaystyle\int_a^b dx},$$

to show that the mean force exerted on one of the balance balls over one half-oscillation is

$$\langle F \rangle = \frac{GMm}{D^2} \frac{1}{1 - \dfrac{\delta^2}{4D^2}},$$

so the value of G obtained from the experiment is

$$G = \frac{2\pi^2 D^2 \delta}{M\tau^2} \left(1 - \frac{\delta^2}{4D^2}\right).$$

That is, the correction factor is $1 - (\delta^2/4D^2)$. (Hint: See Problem 7.46.) **(b)** Using the data given in Example 14.3 and in the legend for Figure 14.9, show that the correction is about 0.03% and is thus not significant in the context of Cavendish's experiment.

14.62 *(G) Gravitational field.* The **gravitational field vector** $\mathbf{\Gamma}(\mathbf{r})$ is defined to be the gravitational force acting on a body of mass m located at $\mathbf{r}$, divided by m:

$$\mathbf{\Gamma}(\mathbf{r}) \equiv \frac{1}{m}\,\mathbf{F}(\mathbf{r}).$$

(a) What is the value of $\mathbf{\Gamma}(\mathbf{r})$ in the vicinity of a particle of

mass M? Assume that no other masses of comparable magnitude are present. **(b)** What is the value of $\mathbf{\Gamma}$ at the surface of the earth? **(c)** In considering the basis of the inverse-square law in Section 14.3, we argued that the number of field lines originating at a body can be chosen arbitrarily, because we were interested mainly in their pattern of symmetry. However, there are occasions when it is convenient to specify N quantitatively. Beginning with the proportionality $F \propto N/A$ suggested in the text, we *define* the gravitational field $\mathbf{\Gamma}$ at any location to be

$$\mathbf{\Gamma}(\mathbf{r}) \equiv -\frac{N}{A}\,\hat{\mathbf{r}};$$

the gravitational field is equal in magnitude to the number of field lines per unit area penetrating a small imaginary area normal to the lines at the location of interest, and it is directed toward the force center. (This is not the most general definition possible, but it will suffice for the present purpose.) Express N in terms of M, G, and necessary constants. **(d)** Define the **gravitational potential**

$$V \equiv -\int \mathbf{\Gamma}(\mathbf{r}) \cdot d\mathbf{r}.$$

Express V in terms of the gravitational potential energy of a body located in the field, and show that V is to $\mathbf{\Gamma}$ as U is to $\mathbf{F}$.

14.63 *(G) Gravitational field inside a hollow sphere without calculation.* Using a symmetry argument like that used when we considered the basis for the inverse-square law in Section 14.3, draw the gravitational field lines emanating from a hollow spherical shell of mass M. What is the force exerted on a small mass m located *inside* the shell? [Hints: (1) You already know that the pattern of field lines outside the sphere must be the same as if they originated at an equal mass located at the center of the sphere. (2) You know that field lines emanate from mass elements; they cannot begin at an empty point.]

14.64 *(G) Gravitational field inside a uniform solid sphere.* Imagine a small hole drilled along a diameter of a uniform solid sphere of radius R. **(a)** Using the result of Problem 14.63, show that the gravitational field inside the hole, at a distance r from the center of the sphere, is proportional to r. **(b)** Find the force exerted on a particle of mass m located in the hole at r, and show that such a particle will experience simple harmonic motion. **(c)** What is the period of the simple harmonic motion? **(d)** Suppose that a hole could be drilled through the earth along a diameter. If the hole were evacuated, how long would it take for a stone dropped into it at one end to reach the other end? Make the crude assumption that the density of the earth does not vary with depth. **(e)** Now suppose that a straight tunnel is drilled through the earth in any direction— say, from New York to London. A rail car is designed to roll with negligible friction through the tunnel, which is evacuated. Show that the motion of the car is simple harmonic and that the half-period is the same as that found in part **d**. Roughly speaking, how would travel time through such a tunnel compare with New York to London service by SST? (This system has often been proposed, but no one has figured out how to dig the tunnel!) In calculating the travel time from New York to London, why is the nonuniform density of the earth relatively unimportant?

Part VI

The Physics of
Many-Particle Systems

The physical universe is not restricted to rigid bodies. The springs in your bed, the coffee you drank for breakfast, and the air in the last breath you took are just three examples of nonrigid bodies.

In the following six chapters, we take the next step in expanding our study to include systems of increasing complexity and richness. We began this book with a study of particles in ones and twos. Fortunately, we found that extended bodies can be considered particles for many (but not all) purposes. We then proceeded to the next level of complexity and found ways of describing motions in which rigid bodies cannot be reduced to particles.

We enter a still richer world in Chapter 15, with a description of the macroscopic behavior of bodies—both solid and fluid—when they aredeformed. (A body that deforms violates thedefinition of a rigid body.) In Chapter 16, we study the macroscopic behavior of moving fluids. As we progress, we find an increasing need to think about what is happening at the molecular level if we are to attain a satisfactory understanding of the gross behavior of a deformable body. At the molecular level, the body appears as a collection of a vast number of particles. This is the reason for the title of Part VI.

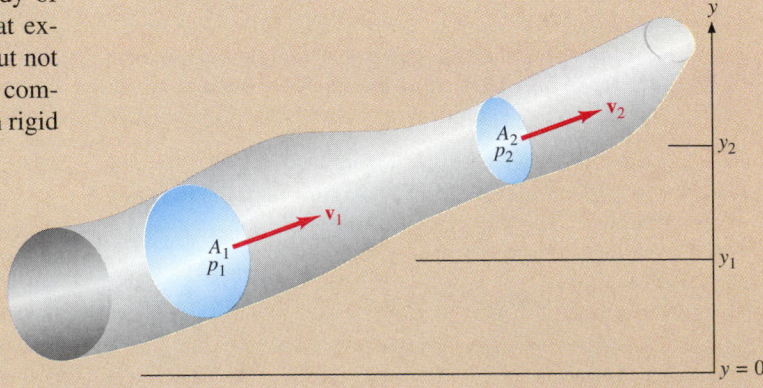

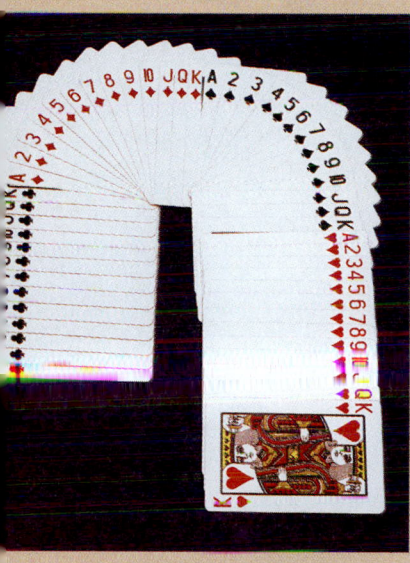

In Chapter 17, we embark on a study of the *thermal behavior* of bodies—that is, their behavior when their temperature is varied and heat is exchanged among them. The subject at first seems almost completely unconnected with anything that has come before. But in Chapter 18 we come to see how phenomena that we perceive as thermal on the macroscopic level are manifestations of mechanical behavior on the microscopic level. Although we start by merely inferring what is happening on the microscopic level, we find increasingly direct ways of carrying out microscopic observations. Indeed, we achieve a quite general understanding, in principle, of the essentially mechanical nature of thermal phenomena. And we build a fairly extensive (though far from complete) collection of tools for applying the principles to important cases.

In Chapters 19 and 20, we return to macroscopic systems with the insight afforded by microscopic studies. We consider *thermodynamic systems*—in particular, the important class of such systems called *heat engines*. Chapter 19 mainly concerns the *first law of thermodynamics*. Grounded on insight into the microscopic world, this law extends the principle of energy conservation—a principle of mechanics—to *heat*, a quantity that on first encounter seems unrelated to mechanics. Chapter 20 mainly concerns the *second law of thermodynamics*. Initially, this law appears simply as a precise statement of "commonsense" restrictions on the performance of heat engines, over and above the requirement that they conform to the principle of energy conservation. At the end of Chapter 20, we look briefly into the essentially *statistical* nature of the second law. We see that its significance arises out of the fact that we are studying systems containing many, many particles.

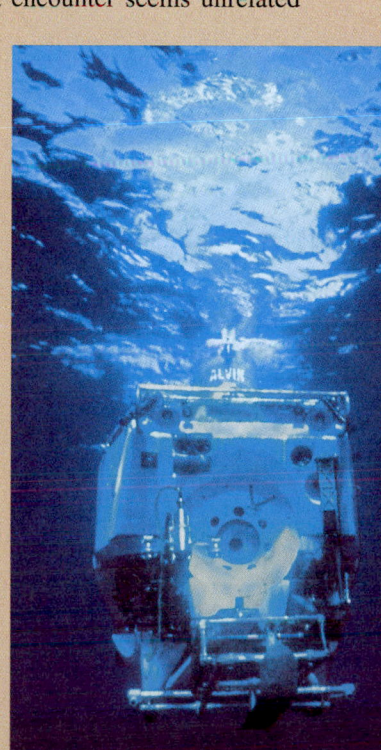

Statics of Continuous Media

When a solid body is subjected to external forces in such a way that it does not move, it distorts.

The strain of a rod—the change in length per unit length—depends on the stress—the force per unit cross-sectional area—and on the material of which the rod is made.

The relation between stress and strain of a body is expressed in terms of a set of constants called elastic moduli, whose values depend on the material of which the body is made.

A fluid, unlike a solid, does not ''spring back'' when a shear stress is released. However, fluids do resist the isotropic stress called pressure.

The pressure in a fluid depends on density and depth.

A body partly or completely immersed in a fluid experiences a buoyant force equal to the weight of the fluid displaced.

If the temperature is held constant, the pressure of a fixed quantity of confined gas is inversely proportional to its volume.

Left: The deep submersible research vehicle *Alvin* just beginning a descent. *Alvin* is designed to withstand enormous hydrostatic pressures.

*By convention there is color, . . . sweetness, . . .
bitterness, but in reality there are atoms and space.*

—Democritus (c. 460–c. 400 b.c.e.)

*[A] theory of matter is a policy rather than a creed;
its object is to connect . . . apparently diverse
phenomena, and above all to suggest, stimulate and
direct experiment.*

—J. J. Thomson (1907)

SECTION 15.1 Introduction

All real bodies are deformable. Solid bodies can be stretched, compressed, and bent, as you know from much and varied experience. Bodies of fluid are more easily deformed than solid bodies, as you have often observed on lowering your solid self into a bathtub filled with fluid water.

In this chapter, we study the static properties of a body—by which we mean the properties exhibited by a body when it is subjected to pairs of forces that are equal and opposite, so the body distorts but does not move as a whole. In principle, static properties can be traced to the forces exerted on one another by the enormous number of atoms that make up the body. When we take this microscopic view, we are considering the body as a system of many particles to which we can apply the laws of mechanics. The microscopic approach thus extends the range of application of these laws, which we have applied until now only to particles and to bodies that can be considered rigid.

But we can understand much about the behavior of extended bodies on the basis of an approach that is mainly or entirely *macro*scopic. We consider a body, such as a bar of metal, to be a **continuous medium**. That is, we ignore its microscopic atomic structure and think of it as homogeneous (smooth rather than grainy). In the interest of simplicity, we restrict our study to **isotropic bodies**—that is, bodies whose properties are the same in all directions.

THINKING LIKE A PHYSICIST

Our view in this chapter is *phenomenological*; that is, we are satisfied for the time being to develop a consistent description of the behavior of deformable bodies and do not attempt to derive this description from their underlying structure. Although it does not give a complete picture of a system, a phenomenological theory enables us to make important practical predictions (for example, the maximum safe tension in a bridge cable). Equally important, today's good phenomenological theory of a system is a source of essential insights into the form that tomorrow's more fundamental description of the system must take. For instance, Kepler's phenomenological laws of planetary motion con-

tributed mightily to Newton's fundamental insights into the laws of motion and gravitation that underlie the motions of the planets.

Moreover, experience shows that tomorrow's "fundamental" theory is not the last word on the subject; it is often the next day's phenomenological jumping-off point for a new and deeper, more fruitful theory. For example, discrepancies between the observed motion of Mercury and the detailed theory of planetary motion based on Newton's insights furnished an important starting point for Einstein's still more general theory of gravitation.

The first part of the chapter deals with solids. We begin in Section 15.2 with a one-dimensional study of what happens to a solid rod when forces are applied along its length. But a rod becomes thinner when it is stretched; to deal with this phenomenon, we take a three-dimensional view in Section 15.3. In Section 15.4 we study *shear*, the effect of applying a couple—a pair of equal and opposite forces having different lines of action—to a solid body that is not free to rotate.

We then turn to a study of fluids. In Section 15.5, we consider the way forces are transmitted through fluids. Section 15.6 treats the floating and sinking of solids immersed in fluids. In Section 15.7, we concentrate on the mechanical properties of the easily compressible fluids called *gases*.

Uniaxial Stress and Strain

Figure 15.1 shows a rod of length l and uniform cross-sectional area A. Two oppositely directed forces of equal magnitude F are applied to the ends of the rod. The forces in Figure 15.1a are directed away from each other and are called **tensile** forces because they place the rod under tension. The forces in Figure 15.1b are **compressive** forces because they are directed toward each other and place the rod under compression.

The application of such forces has the following consequences:

1. If the magnitude F of the applied forces is not too large, the change in length $|\Delta l|$ of the rod is proportional to F:

$$|\Delta l| \propto F \quad \text{or} \quad F = k|\Delta l|. \tag{15.1a,b}$$

These relations are forms of Hooke's law. For a given test sample, the proportionality constant k (the force constant) is the same for tensile and compressive forces.*

2. For rods identical except for length and subjected to identical forces, the change in length is proportional to the length l:

$$|\Delta l| \propto l. \tag{15.2}$$

3. For rods identical except for cross-sectional area A and subjected to identical forces, the change in length is inversely proportional to A:

$$|\Delta l| \propto \frac{1}{A}, \tag{15.3}$$

These assertions are readily verifiable by straightforward measurement. But proportionalities 15.2 and 15.3 can also be deduced from Hooke's law, by means outlined in Queries 15.1 and 15.2.

We can combine proportionalities 15.1a, 15.2, and 15.3 to obtain

$$|\Delta l| \propto \frac{Fl}{A}.$$

We divide both sides of this relation by l so as to express the fractional change in length $|\Delta l|/l$ in terms of the applied force per unit area:

$$\frac{|\Delta l|}{l} \propto \frac{F}{A}. \tag{15.4}$$

Tensile forces produce an increase in l, and so Δl is positive; compressive forces produce a decrease in l, and so Δl is negative. With this in mind, we define the *fractional change in length* $\Delta l/l$ to be the **strain** ϵ (Greek epsilon):

$$\epsilon \equiv \frac{\Delta l}{l}; \quad \epsilon > 0 \text{ for tensile forces, } \epsilon < 0 \text{ for compressive forces.} \tag{15.5}$$

Beyond Hooke's Law

If a rod is deformed beyond its **elastic limit**, the simple proportionality of Hooke's law fails. If the rod is subjected to further deformation, the **yield limit** is reached; the rod is distorted permanently and does not return to its original shape when the force is removed. With still further deformation, the rod reaches its **ultimate strength** and breaks.

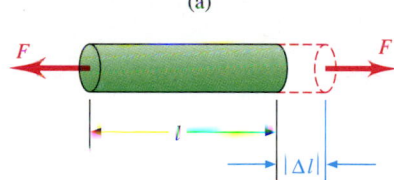

FIGURE 15.1 A rod of length l and cross-sectional area A is subjected to uniaxial (*a*) tensile stress and (*b*) compressive stress by a pair of forces of magnitude F. In each case, the rod experiences a change in length of magnitude $|\Delta l|$.

---

*But it is not always practical to subject the same test sample to tension and compression. For example, it is easy to stretch a fine wire but much more difficult to compress it.

Strain is a dimensionless quantity.

Now consider the force per unit area, which appears on the right side of proportionality 15.4. For consistency with Equation 15.5, we must assign a positive sign to F/A if the forces are tensile and a negative sign if the forces are compressive. With this in mind, we define the **uniaxial stress** σ:

$$\sigma \equiv \frac{F}{A}; \quad \sigma > 0 \text{ for tensile forces}, \ \sigma < 0 \text{ for compressive forces.} \qquad (15.6)$$

Uniaxial stress is always applied to a body by two equal and opposite forces having the same line of action but different points of application. It is evident from the definition that the SI unit of stress is N/m^2. This unit is given the name **pascal** (Pa):

$$1 \text{ Pa} \equiv 1 \text{ N/m}^2. \qquad (15.7)$$

Young's Modulus

Proportionality 15.4 is conveniently expressed in terms of the two quantities we have just defined: *The strain of a body is proportional to the uniaxial stress applied to the body.* Because this statement is based on Hooke's law, it is limited for every body to the range of stress within which Hooke's law is obeyed. As usual, we prefer to express the statement as an equation. We introduce the proportionality constant Y and write

$$\sigma = Y\epsilon. \qquad (15.8a)$$

The quantity Y is called **Young's modulus**, after Thomas Young (1773–1829), whose contribution to the development of the energy-conservation principle was noted in Section 8.3. Because ϵ is dimensionless, the SI unit of Young's modulus is the pascal.

Young's modulus is a measure of the "stiffness" of a solid material. Rewriting Equation 15.8a in the form

$$Y = \frac{\sigma}{\epsilon}, \qquad (15.8b)$$

we see that this "stiffness" is expressed quantitatively as the applied stress divided by the uniaxial strain that results from application of the stress. Because Young's modulus depends only on the material of which a body is made and is independent of the shape or size of the body, Equation 15.8a may be regarded as a generalization of Hooke's law. Table 15.1 lists typical values of Young's modulus for a variety of solid materials. Note

The Pascal

The unit of pressure is named after the French philosopher, mathematician, and physicist Blaise Pascal (1623–1662) in recognition of contributions discussed in Section 15.5, and of his prediction of a decrease in atmospheric pressure with increasing altitude. At his suggestion, Pascal's brother-in-law organized a mountain-climbing picnic, with a sufficient number of servants to tote a barometer as well as an elegant lunch. At predetermined times, the barometer was set up and read. A servant who had been left at home read another barometer at the same times for comparison. Pascal's predictions were amply confirmed.

TABLE 15.1 Typical Values of Young's Modulus Y, Shear Modulus G, and Poisson's Ratio ν

Material	Y (10^{10} Pa)	G (10^{10} Pa)	ν
Aluminum	7	2.5	0.36
Brass	10	3.8	0.37
Copper	12.5	4.6	0.37
Ceramic (typical)	1		
Glass	5	2	0.2
Gold	8.1	2.9	0.42
Iron (cast)	15	6.0	0.27
Steel	21	8.3	0.28
Lead	1.5	0.54	0.43
Magnesium	4.2	1.6	0.31
Plastic (typical)	0.4	0.1	0.4
Silver	7.6	2.7	0.38
Tin	5	2	0.34
Tungsten	36	13	0.35
Tungsten carbide	53	22	0.22

Note: The values of these elastic constants for a particular sample depend to a considerable degree on the thermal, mechanical, and chemical details of its preparation and prior history. Consequently, the term "typical" in the title of the table should be taken seriously.

The designers of this bridge knew quite a lot about the elastic properties of the materials used to build it.

that nearly all the values quoted lie within a relatively limited range of less than two orders of magnitude; the stiffest solid is about a hundred times as stiff as the least stiff.

EXAMPLE 15.1

Consider a rod of length l and cross-sectional area A that obeys Hooke's law. Show that the material of which the rod is made conforms to Equation 15.8, and express Young's modulus in terms of the Hooke's-law force constant k of the rod.

SOLUTION: Because the rod obeys Hooke's law, you can express the rod's response to a pair of applied forces by means of Equation 15.1b, $F = k|\Delta l|$. This equation relates the magnitude of the forces to the change in length, and you wish to rewrite the relation in terms of the stress F/A and the strain $\Delta l/l$. To do this, divide the equation through by A, and multiply the right side by l/l. You then have

$$\frac{F}{A} = \frac{kl}{A}\frac{\Delta l}{l},$$

or

$$\sigma = \frac{kl}{A}\epsilon.$$

(Why can you drop the absolute value signs from Δl?) The coefficient kl/A is a constant for any given rod. The equation is identical with Equation 15.8a, $\sigma = Y\epsilon$, if

$$Y = k\frac{l}{A}. \tag{15.9a}$$

This equation is used in determining the value of Young's modulus for a material. A bar made of the material, of a known length and cross-sectional area, is clamped between the jaws of a *strength-testing machine*. The machine applies a controllable force to the bar, and the change in length is measured as a function of the applied force. The slope dF/dl of the plot of F versus l yields the force constant k; Y is then found by use of Equation 15.9a.

Mechanical designers often find it useful to work in the other direction as well. If you rewrite Equation 15.9a in the form

$$k = Y\frac{A}{l}, \tag{15.9b}$$

you can predict the force constant k of a bar of given dimensions made of a specified material whose Young's modulus is known. This is important information in predicting the behavior of structures under the forces they are designed to withstand.

EXAMPLE 15.2

Working on a bridge project, you are responsible for designing the rectangular steel frame shown in Figure 15.2a. The sides are to be of lengths $a = 1.000$ m and $b = 0.650$ m. Each long side will be rigidly attached to a large structural assembly, but the short sides will be free. The design specifications require that the frame not distort from rectangular form by more than

FIGURE 15.2

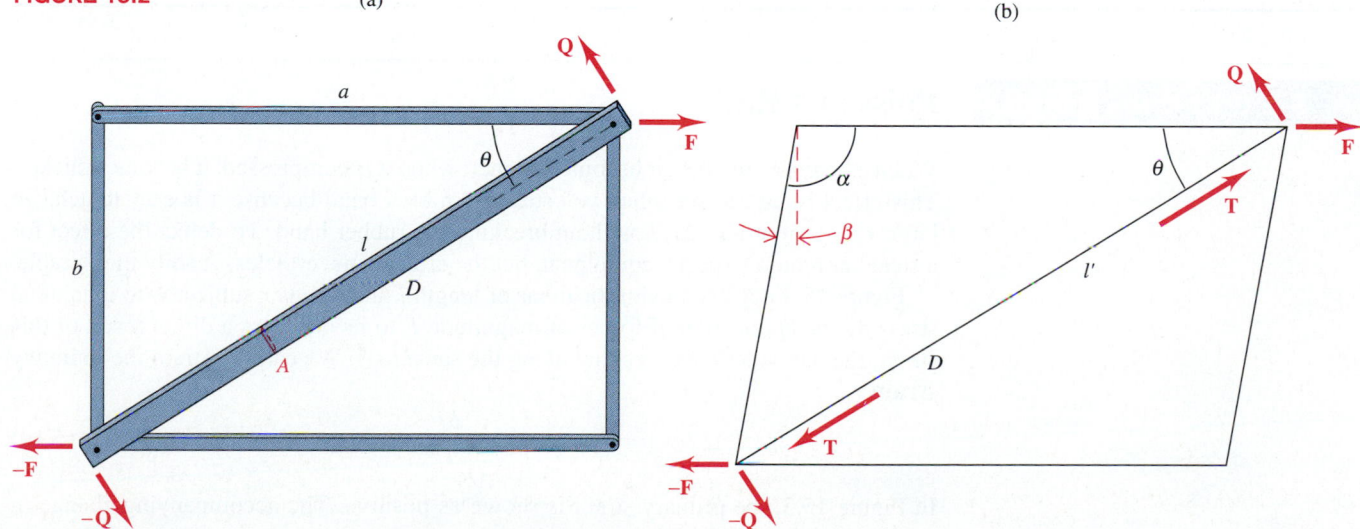

(a) (b)

1° when subjected to the couple **F** and −**F** shown, with $F = 8 \times 10^5$ N. What cross-sectional area A must you specify for the diagonal member D, whose length is l? Assume that member D bears all the load applied by the couple.

SOLUTION: The forces applied to the frame constitute a couple and are thus not uniaxial. To reduce the problem to a study of the uniaxial stress exerted on D, you must first determine the tensile force T to which D will be subjected. Besides the forces F constituting the couple, the structure must be subjected to other forces constituting a couple of opposite sense. (In the absence of such an additional couple, the structure would rotate.) A possible couple is that exerted by the forces **Q** and −**Q** shown in Figure 15.2. Referring to the diagram, you see that D sustains a tensile force given by the expression

$$T = \frac{F}{\cos \theta} = F \frac{l}{a}.$$

Next, you must find the change in length Δl experienced by D under the load T. You draw a sketch of the structure in its distorted configuration, as in Figure 15.2b. The distorted length of D is $l' = l + \Delta l$. The angle β by which the corners are distorted from right angles is much exaggerated for clarity. The corner angle α is simply $90° + \beta$. Using the law of cosines to express l'^2, you have

$$l'^2 = a^2 + b^2 - 2ab \cos \alpha.$$

(It is plausible to assume that the lengths a and b do not change appreciably.) Because the square of the original length of D is $l^2 = a^2 + b^2$, you have

$$l'^2 - l^2 = -2ab \cos \alpha.$$

But $\cos \alpha = -\sin \beta$, and so you have

$$l'^2 - l^2 = 2ab \sin \beta \simeq 2ab\beta. \qquad (15.10)$$

The approximation $\sin \beta \simeq \beta$ is permissible because the distortion angle β is small. (In using this approximation, you must remember to express β in radians.)

Now you are ready to find Δl. The left side of Equation 15.10 factors to give $(l' - l)(l' + l)$. The first factor is the desired Δl.

The second factor is essentially $2l$ because the change in length of the member is small. So the equation becomes

$$\Delta l = \frac{2ab\beta}{2l} = \frac{ab\beta}{l}. \qquad (15.11)$$

You can now relate Δl to the applied force. From Hooke's law (Equation 15.1b), you have

$$\Delta l = \frac{T}{k} = \frac{Fl}{a}\frac{1}{k}.$$

You use Equation 15.9b to express the unknown force constant in terms of the known values of Young's modulus, the length l of the diagonal member D, and the desired design value of the cross-sectional area A:

$$\Delta l = \frac{Fl}{a}\frac{l}{YA} = \frac{Fl^2}{YAa}. \qquad (15.12)$$

Combining Equations 15.11 and 15.12 gives

$$\frac{ab\beta}{l} = \frac{Fl^2}{YAa},$$

which you can immediately solve for A:

$$A = \frac{Fl^3}{Ya^2b\beta} = \frac{F(a^2 + b^2)^{3/2}}{Ya^2b\beta}.$$

You can express this result a little more neatly in the form

$$A = \frac{1}{\beta}\frac{Fa}{Yb}\left(1 + \frac{b^2}{a^2}\right)^{3/2}. \qquad (15.13)$$

Note that the required cross-sectional area is inversely proportional to the allowable distortion angle (for β small). It is also inversely proportional to Young's modulus—the stiffer the material, the less thick the diagonal member D need be.

You can now insert the numerical values, referring to Table 15.1 for the value of Young's modulus for steel. Remembering that $\beta = 1° = 1.75 \times 10^{-2}$ rad, you have

$$A = \frac{1}{1.75 \times 10^{-2}}\frac{8 \times 10^5 \text{ N} \times 1.000 \text{ m}}{21 \times 10^{10} \text{ Pa} \times 0.650 \text{ m}}[1 + (0.650)^2]^{3/2}$$

$$= 5.7 \times 10^{-4} \text{ m}^2.$$

This is the cross-sectional area of a square rod with sides 2.4 cm.

Poisson's Ratio

When a body is stretched, it becomes thinner; when it is compressed, it becomes thicker. This effect is easy to see when you stretch a rubber band because it is easy to achieve large values of strain, $\Delta l/l$, without breaking the rubber band. To detect the effect for a steel bar requires special equipment, but the effect is nevertheless readily measurable.

Figure 15.3 depicts a cylindrical bar of length l and radius r subjected to a uniaxial stress by the application of forces of magnitude F to its ends. As a direct result of this stress, the bar experiences a strain along the same axis. We call the strain the **primary strain** ϵ_p:

$$\epsilon_p = \frac{\Delta l}{l}. \qquad (15.14)$$

In Figure 15.3, the primary strain is shown as positive. The accompanying change in

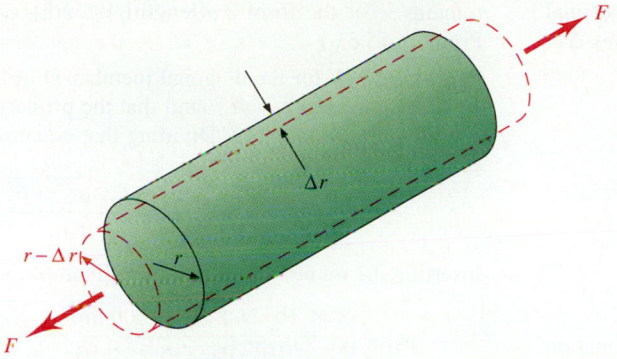

FIGURE 15.3 A bar has length l and circular cross section with radius r. It is subjected to a uniaxial stress $\sigma = F/A$ and experiences the primary strain $\epsilon_p = \Delta l/l = \sigma/Y$. The cross-sectional area A of the bar changes. This change can be expressed in terms of the induced strain $\epsilon_i = \Delta r/r$. The sign of the induced strain is always opposite that of the primary strain.

cross-sectional area (in this case, a reduction) may be thought of in terms of a proportional change $\Delta r/r$ in the radius of the cross section. *This proportional change is itself a strain.* Because it is a consequence of—is induced by—the primary strain, it is called an **induced strain** ϵ_i. That is, we define

$$\epsilon_i \equiv \frac{\Delta r}{r}. \tag{15.15}$$

Because of the way we define a cross section, the induced strain is in the plane normal to the primary strain. In an isotropic substance, there is no reason to expect different induced strains in different directions lying in this plane. And indeed the induced strains are found to be the same in all these directions. That is, the cross section of the bar remains circular when the bar is stressed along its axis.

A body subjected to a uniaxial strain always experiences a change in volume. As you would expect, a positive strain (stretch) increases the volume; a negative strain (compression) reduces it. The change in cross-sectional area compensates partly (but not completely) for the change in length. Thus the volume change of the bar is less than you would expect on the basis of the change in length of the bar alone. The degree to which the compensation takes place depends on the material. It is expressed by **Poisson's ratio,***

$$\nu \equiv -\frac{\epsilon_i}{\epsilon_p} = \frac{\Delta r/r}{\Delta l/l}. \tag{15.16}$$

Because the induced and primary strains are always of opposite sign, the minus sign is introduced into the definition to make Poisson's ratio a positive quantity. Poisson's ratio is given for a number of materials in Table 15.1. The values are largest for soft materials such as plastics and smallest for hard materials such as glass and tungsten carbide. But Poisson's ratio is never very different from $\frac{1}{3}$.

EXAMPLE 15.3

(a) A rod under uniaxial stress experiences a primary strain ϵ_p. Express the proportional change $\Delta V/V$ in the rod volume in terms of ϵ_p and Poisson's ratio ν. **(b)** What is the proportional change in volume of the diagonal member D of the structure considered in Example 15.2?

SOLUTION:
(a) Express $\Delta V/V$ in terms of ϵ_p and ν.

Consider again a cylindrical rod. (It will turn out that the result is independent of this assumption, but having a particular shape in mind makes the argument more concrete.) The volume of the rod is $V = \pi r^2 l$, where l is its length and r its radius.

You can apply the product rule of differentiation to this relation and express an infinitesimal volume change dV:

$$dV = 2\pi lr \, dr + \pi r^2 \, dl.$$

The proportional change in volume is then

$$\begin{aligned}
\frac{dV}{V} &= \frac{2\pi lr \, dr + \pi r^2 \, dl}{\pi r^2 l} \\
&= 2\frac{dr}{r} + \frac{dl}{l}.
\end{aligned}$$

---

*After the French mathematician and physicist Siméon-Denis Poisson (1781–1840).

This result, though strictly true only for infinitesimal changes dV, applies approximately for small finite changes ΔV. You can therefore write

$$\frac{\Delta V}{V} = 2\frac{\Delta r}{r} + \frac{\Delta l}{l}.$$

But $\Delta l/l$ is the primary strain ϵ_p, whereas $\Delta r/r$ is the induced strain ϵ_i. So you have

$$\frac{\Delta V}{V} = 2\epsilon_i + \epsilon_p = \epsilon_p\left(1 + 2\frac{\epsilon_i}{\epsilon_p}\right).$$

Finally, you use the definition of Poisson's ratio (Equation 15.16) to rewrite this equation in the form

$$\frac{\Delta V}{V} = \epsilon_p(1 - 2\nu) \quad \text{for uniaxial stress.} \quad \textbf{(15.17)}$$

Although we have been considering a rod of circular cross section, the expression we have derived contains only dimensionless quantities. The expression is independent of the dimensions of the rod and is quite generally true for any body made of isotropic material and subjected to uniaxial stress. Why does the result turn out this way? Because the area of any cross section, whatever its shape, is expressed in the dimensional form [constant][length]^2. When dV is found by differentiation and the ratio dV/V is taken, the constant factor cancels out and what

remains is of the form $2\,d[\text{length}]/[\text{length}]$, or $2\epsilon_i$. (See also Problem 15.31.)

(b) Find $\Delta V/V$ for the diagonal member D of Example 15.2.

In Example 15.2, you found that the primary strain for D is given by Equation 15.12. Dividing that equation through by l, you have

$$\epsilon_p = \frac{\Delta l}{l} = \frac{Fl}{YAa} = \frac{F\sqrt{a^2 + b^2}}{YAa}.$$

Inserting the numerical values from Example 15.2, you obtain

$$\epsilon_p = \frac{8 \times 10^5\,\text{N} \times \sqrt{(1.000\,\text{m})^2 + (0.650\,\text{m})^2}}{21 \times 10^{10}\,\text{Pa} \times 5.7 \times 10^{-4}\,\text{m} \times 1.00\,\text{m}}$$

$$= 7.97 \times 10^{-3},$$

where the last figure is probably not significant but is carried over for the sake of further calculation. The strain is less than 1% and thus satisfies the requirement of Equation 15.17 that $\Delta V/V$ be small.

You refer to Table 15.1 for the value of Poisson's ratio for steel and obtain the desired result

$$\frac{\Delta V}{V} = (7.97 \times 10^{-3})(1 - 2 \times 0.28) = 3.5 \times 10^{-3}.$$

That is, the diagonal member D of Example 15.2 experiences a volume increase of about 0.3%.

Shear Stress and Strain

As you saw in Example 15.2, stress can be applied to a body by a couple as well as by a pair of equal and opposite forces directed along the same line of action. In that example, however, it was possible to reduce the problem to one of uniaxial stress of the diagonal member D.

Such a reduction is *not* possible in general. Consider the situation shown in Figure 15.4. The block, made of isotropic material, has top and bottom surfaces of area A. Any plane parallel to the top and bottom surfaces that cuts the block defines a cross section whose area also is A. The upper and lower surfaces are securely fastened to two rigid plates. A couple is applied by pulling on the plates as shown. [As in Example 15.2, the fact that the block is at rest implies that an equal and opposite couple (not shown) is present as well.]

The forces $\mathbf{F}$ and $-\mathbf{F}$ are *parallel* to the cross section. This contrasts with uniaxial stress, in which case the forces are *normal* to the cross-sectional surface of interest. As a result of the way in which the forces are applied, the block is subjected to a **shear stress** σ, defined by the equation

$$\sigma \equiv \frac{F}{A}. \quad \textbf{(15.18)}$$

FIGURE 15.4 Definition of shear stress and shear strain. The block is made of uniform isotropic material. It has uniform cross-sectional area A and height h. Two forces of magnitude F, applied along the top and bottom surfaces as shown, constitute a couple. The upper surface is displaced relative to the lower surface by a small distance Δl, and the corner angles change by a small amount $\beta \approx \Delta l/h$.

The term "shear stress" is used because the stress is of the type applied by the blades of a pair of shears to the object it is cutting. It is evident from the definition that the SI

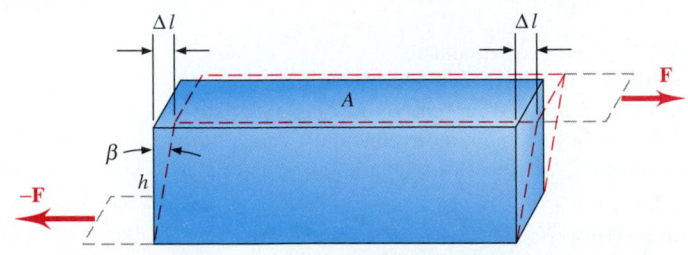

unit of shear stress is the pascal. Equation 15.18 is *mathematically* identical with Equation 15.6, which defines uniaxial stress. However, the different orientation of the forces relative to the cross section makes an important *physical* difference between the two cases. The physical difference does not show up in the equations because of the scalar nature of the quantities they define. Unless there is a possibility of confusion, the same symbol σ is used for both kinds of stress.

As a result of the shear stress applied to the block of Figure 15.4, the block distorts. The corner angle changes by an amount β from the unstressed situation. This corresponds to a displacement of the upper surface through a distance Δl with respect to the lower surface. As we did in considering the rectangular structure of Example 15.2, we use the small-angle approximation and express the angle β in radians as the ratio

$$\beta = \frac{\Delta l}{h}. \tag{15.19}$$

The quantity β is called the **shear strain**. Like the uniaxial strain ϵ defined by Equation 15.5, β is the ratio of a change in length to a length and is therefore a dimensionless quantity.

Shear Modulus

In complete analogy to what we did in the uniaxial case, we define the **shear modulus** G to be the ratio of stress to strain:

$$G \equiv \frac{\sigma}{\beta}. \tag{15.20}$$

The SI unit of the shear modulus is the pascal. Typical values of G are listed in Table 15.1. For most solid materials, the value of the shear modulus is about one-third that of Young's modulus.

EXAMPLE 15.4

In Example 15.2, you designed a structure that would distort through 1° under the application of forces of magnitude $F = 8 \times 10^5$ N along its upper and lower members. The necessary thickness of the steel diagonal member turned out to be 2.4 cm. If you substitute a solid steel plate of the same thickness for the structure, what is the magnitude of the forces F' that would need to be applied to produce the same distortion?

SOLUTION: The required shear stress is found by using Equation 15.20 to write $\sigma = \beta G$. From Equation 15.18, you have $F' = \sigma A$. So you can combine the two equations to obtain

$$F' = \beta G A.$$

In the present case, the desired area A is that of the top edge of the plate. (Or, what amounts to the same thing, it is the area of the bottom edge or of any horizontal cross section through the plate.) If you call the thickness t, you have $A = at$, and you can write the force as

$$
\begin{aligned}
F' &= \beta G a t \\
&= 1.75 \times 10^{-2} \text{ rad} \times 8.3 \times 10^{10} \text{ Pa} \times 1 \text{ m} \times 0.024 \text{ m} \\
&= 3.5 \times 10^7 \text{ N}.
\end{aligned}
$$

By comparing this value of F' with the corresponding value of F applied to the structure of Example 15.2, you can see that the solid plate is about 44 times as "stiff" as the structure. But it is heavier as well; this point is the subject of Problem 15.43.

Fluids. Pressure and Bulk Modulus

SECTION 15.5

We define a body of matter as **fluid** if it has no *static* resistance to shear stress. This property of fluids contrasts with the behavior of solids, which do have static resistance to shear stress. In our particular corner of the universe, the most familiar fluid states are the *liquid* and *gaseous* states.

The most familiar fluid of all is water, and you can use a bowl of water to satisfy yourself that the definition of a fluid, in terms of its lack of static resistance, makes sense. Holding your hands palms together, dip them into the water. Then separate them

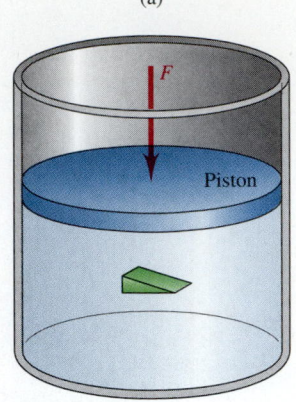

(a)

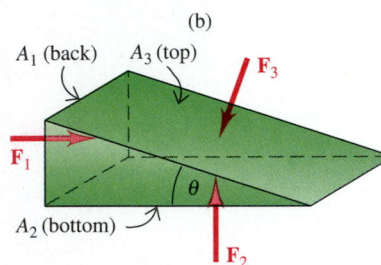

(b)

A_1 (back) A_3 (top) **F_3**

F_1

θ

A_2 (bottom)

F_2

FIGURE 15.5 Diagram for deriving Pascal's principle of stress isotropy in fluids. (*a*) A wedge-shaped element with wedge angle θ is part of a body of fluid. The fluid, confined in a cylinder, is subjected to a compressive stress by the force F acting on the piston. (*b*) Enlarged picture of the wedge, showing the normal forces F_1, F_2, and F_3 acting respectively on surfaces having areas A_1, A_2, and A_3.

slowly by sliding the palms along each other. You will feel resistance from the water— *but only as long as your hands are moving*. Once they are separated and at rest again, the water will show no tendency at all to push them back together.

As this simple experiment shows, a fluid reacts to shear stress not statically but *dynamically*, by *flowing*. Conversely, *when a fluid is at rest, there can be no shear stress present*. We then say that the fluid is in **hydrostatic equilibrium**. The absence of shear stress in hydrostatic equilibrium has a very important consequence for stresses in fluids, which we now derive.

Figure 15.5a shows a fluid contained at rest in a cylinder. A piston at the top is used to exert a force on the fluid, which is thus subjected to a compressive stress. Now, consider the wedge-shaped element of fluid, with wedge angle θ. This element is shown enlarged in Figure 15.5b. Because the element is at rest, we can conclude that there are no shear stresses acting on it. It follows that *all forces exerted on the surfaces of the wedge must be normal to the surfaces*.

Let us consider in particular the forces $\mathbf{F}_1$, $\mathbf{F}_2$, and $\mathbf{F}_3$ exerted respectively on the surfaces whose areas are A_1, A_2, and A_3. Because the element is at rest, the horizontal components of these forces must add to zero and so must the vertical components. So we have

$$F_1 = F_3 \sin \theta \quad \text{and} \quad F_2 = F_3 \cos \theta.$$

Now, consider the areas on which these forces act. You can see from inspection of Figure 15.5b that they bear the relations

$$A_1 = A_3 \sin \theta \quad \text{and} \quad A_2 = A_3 \cos \theta.$$

Using these four equations, we can compare the compressive stresses acting on the three surfaces of the wedge. We have

$$\frac{F_1}{A_1} = \frac{F_2}{A_2} = \frac{F_3}{A_3},$$

or

$$\sigma_1 = \sigma_2 = \sigma_3.$$

The stresses are equal. But the wedge angle θ is completely arbitrary. Hence the compressive stress acting on *any* surface in a fluid is independent of the orientation of the surface. We usually speak of such an isotropic stress in terms of the closely related quantity *pressure*. We define the **pressure** p to be the force per unit area exerted on any surface (real or imaginary) at a given location:

$$p \equiv \frac{F}{A} \quad \text{for } p \text{ isotropic.} \tag{15.21a}$$

Because pressure is exerted uniformly in all directions, we can think of pressure as the three-dimensional analogue of uniaxial stress.

It is sometimes more convenient to define pressure in a derivative form so that it applies to the infinitesimal force applied to an infinitesimal area element:

$$p \equiv \frac{dF}{dA} \quad \text{for } p \text{ isotropic.} \tag{15.21b}$$

This form of the definition allows us to speak without ambiguity of "the pressure at a point." Using either form of the definition, you can see that the SI unit of pressure is the pascal. This is not surprising because the definition of pressure is similar to the definitions of uniaxial and shear stress. However, a different sign convention is used. For uniaxial stress, it was logical to assign a negative sign to compressive stresses. But fluids in hydrostatic equilibrium are nearly always under compression; the walls of the container that holds a fluid exert compressive stress on the fluid. So we assign a *positive* value to the pressure when a fluid is under compression; that is, $p = -\sigma$.

From the defining property of a fluid—its complete lack of static resistance to shear stress—we have derived the conclusion that dF/dA is isotropic in a fluid. Although it is not a fundamental proposition of physics, this conclusion is traditionally dignified by the name **Pascal's principle**, or **Pascal's law**, in recognition of its importance.

Pascal's principle is often stated as follows: *A change Δp in the pressure at any point in a body of fluid is accompanied by an equal change at all other points in the body of fluid.* We will justify this form of the principle shortly.

EXAMPLE 15.5

The device in Figure 15.6 is called a hydraulic lift. The small and large cylinders, each with a close-fitting piston, are connected by a pipe of any convenient shape and length. The system is filled with a light oil called hydraulic fluid. The radii of the pistons are r_1 and r_2, respectively. The system is initially at rest, with piston 1 held in place by the stops. An automobile of mass M rests on piston 2. The masses of the two pistons are small compared with M. **(a)** What is the magnitude of the external force F_1 that must be exerted on piston 1 in order to begin to raise the automobile? **(b)** If the mass of the automobile is 1800 kg and the pistons have radii $r_1 = 2.5$ cm and $r_2 = 10$ cm, find the necessary value of F_1 and the pressure in the fluid when the automobile starts to rise.

FIGURE 15.6

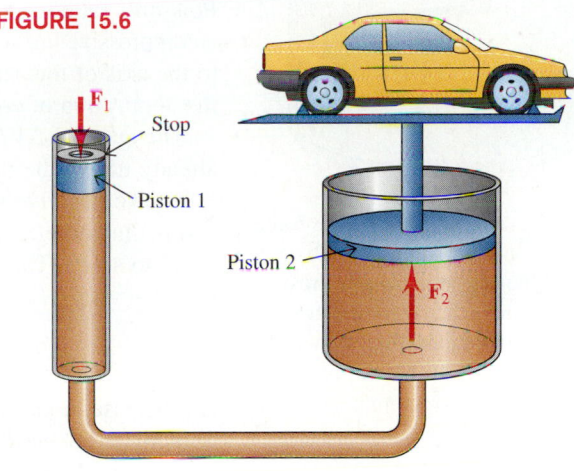

SOLUTION:
(a) What force F_1 is required to raise the automobile?
   Call the areas of the pistons A_1 and A_2. Exerting a force on piston 1 will increase the pressure $p = F_1/A_1$ that this piston exerts on the hydraulic fluid. According to Pascal's principle, that increase in pressure will be the same throughout the system and, in particular, at the surface where the hydraulic fluid comes into contact with piston 2. The force F_2 exerted by the fluid on piston 2 is

$$F_2 = pA_2.$$

If F_2 just exceeds the weight Mg of the automobile, piston 2 will rise. So you have $Mg = pA_2$, and the necessary pressure is

$$p = \frac{Mg}{A_2}. \tag{15.22}$$

To exert this pressure, you must press on piston 1 with a force

$$F_1 = pA_1 = Mg\frac{A_1}{A_2}.$$

Because the pistons are circular, the area ratio in this equation is $\pi r_1^2/\pi r_2^2$, and the equation becomes

$$F_1 = Mg\left(\frac{r_1}{r_2}\right)^2. \tag{15.23}$$

From this equation, you can see that a hydraulic press can be a very effective force multiplier. (But you don't get something

for nothing; see Query 15.10 and Problem 15.11.) It is easy to make the difference between r_1 and r_2 large, and the force multiplication ratio F_1/Mg depends on the square of the ratio of the radii.

(b) Find the numerical value of F_1.
   You insert the known values of M, r_1, and r_2 into Equation 15.23 to obtain the force

$$F_1 = (1.8 \times 10^3 \text{ kg} \times 9.8 \text{ m/s}^2)\left(\frac{2.5 \text{ cm}}{10 \text{ cm}}\right)^2$$

$$= 1.1 \times 10^3 \text{ N}.$$

Equation 15.22 gives you the pressure in the hydraulic fluid. You have

$$p = \frac{Mg}{A_2} = \frac{F_2}{A_2} = \frac{1.8 \times 10^3 \text{ kg} \times 9.8 \text{ m/s}^2}{\pi \times (0.10 \text{ m})^2}$$

$$= 5.6 \times 10^5 \text{ Pa}.$$

(The ratio F_1/A_1 will give you the same result.) As you will soon see, this is about five times the pressure exerted by the earth's atmosphere at sea level.

Bulk Modulus

In Figure 15.7, a solid block of isotropic material having length l, width w, and thickness t is placed in a fluid-filled cylinder. The block is then subjected to a pressure increase Δp by means of a force exerted on the piston. This isotropic stress results in an isotropic strain—that is, a proportional change $\Delta V/V$ in the volume of the block.

   How can we calculate the change in volume? The isotropic pressure is equivalent to three simultaneous uniaxial stresses acting along the length, width, and thickness axes of the block. Each of these uniaxial stresses produces a uniaxial strain. The proportional volume change resulting from any one of these uniaxial strains is given by Equation

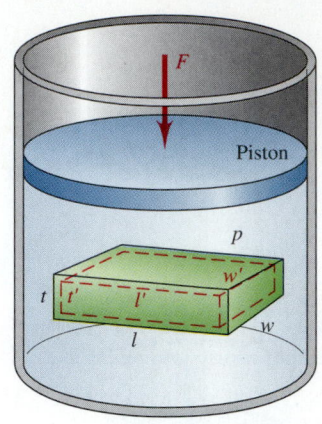

FIGURE 15.7 By means of the force F exerted on the piston, the submerged block is subjected to pressure p. Because the block experiences reductions in its length, width, and thickness, its volume is reduced as well.

15.17; it is $\epsilon_p(1 - 2\nu)$. (Because the block is isotropic, the *proportional* changes in the three directions are the same even though l, w, and t are not equal.) The total proportional volume change of the block is the sum of the three uniaxial effects and is thus

$$\frac{\Delta V}{V} = 3\epsilon_p(1 - 2\nu) \quad \text{for isotropic stress.} \tag{15.24}$$

The factor $(1 - 2\nu)$ implies that the change in volume is less than it would be if Poisson's ratio were zero. You can see this physically as follows. A body subjected to a compressive uniaxial stress gets shorter and fatter. The increase in size perpendicular to the axis of this stress partly counteracts the decrease in size resulting directly from the application of compressive stress along other axes.

The quantity $\Delta V/V$ is the three-dimensional analogue to the uniaxial strain $\Delta l/l$. As already noted, the pressure p is the three-dimensional analogue to the uniaxial stress $-\sigma$. (The negative sign arises from the opposite sign conventions used in defining p and σ.) In complete analogy to what we did in the uniaxial case, we define the ratio of stress to strain. This ratio is called the **bulk modulus** B:

$$B \equiv -\frac{\Delta p}{\Delta V/V} = -V\frac{\Delta p}{\Delta V}. \tag{15.25a}$$

An increase in pressure always results in a decrease in volume, and so the minus sign is included in the definition to make B a positive quantity. The SI unit of the bulk modulus is the pascal. Because we sometimes find it convenient to work in terms of derivatives, we can also define the bulk modulus in the equivalent form

$$B \equiv -V\frac{dp}{dV}. \tag{15.25b}$$

In experimental measurements, the common procedure is to vary the pressure and determine the resulting change in volume rather than vice versa. When this is done, it is convenient to use the reciprocal of the bulk modulus, called the **compressibility** κ:

$$\kappa \equiv \frac{1}{B} = -\frac{1}{V}\frac{dV}{dp}. \tag{15.26}$$

We have now defined four quantities having to do with the elastic properties of materials. They are Young's modulus Y, the shear modulus G, the bulk modulus B, and Poisson's ratio ν. The four are not independent of one another, as Example 15.6 shows.

EXAMPLE 15.6

Express the bulk modulus in terms of Young's modulus and Poisson's ratio.

SOLUTION: You begin with the definition of B given by Equation 15.25a, rewriting it to express B in terms of the small but finite strain $\Delta V/V$ that results from the application of a small but finite pressure p to a sample like that shown in Figure 15.7. If the pressure on the sample is initially zero, you have $\Delta p = p$, and

$$B = -\frac{p}{\Delta V/V}.$$

As noted in the discussion leading to Equations 15.24 and

15.25, applying the pressure p to the sample is equivalent to the simultaneous application of identical uniaxial compressive stresses $-\sigma$ to all sides of the sample. So you can substitute $\sigma = -p$. You use Equation 15.24 to express the resulting three-dimensional strain dV/V:

$$B = \frac{\sigma}{\epsilon_p}\frac{1}{3(1 - 2\nu)}.$$

But the ratio σ/ϵ_p is Young's modulus Y. The equation thus becomes

$$B = \frac{Y}{3(1 - 2\nu)}. \tag{15.27}$$

Equation 15.27 sets a limit on possible values of Poisson's ratio. Because the moduli B and Y are always positive, ν can never be greater than $\frac{1}{2}$. A body for which $\nu = \frac{1}{2}$ must be perfectly incompressible—a property no real body possesses.

Pressure and Depth in Incompressible Fluids

Even in the absence of an external force exerted on a fluid (for example, by a piston), the weight of the fluid itself gives rise to pressure within it. We now develop a quantitative relation between pressure and depth for the special case of an incompressible fluid. As you will soon see, an incompressible fluid is an idealization well approximated by most liquids.

The container in Figure 15.8 holds a fluid at rest. Imagine a horizontal surface of area A located at an arbitrary depth y below the top of the fluid. The column of fluid lying above this surface has mass m and exerts a force $m\mathbf{g}$ on the surface. The column of fluid lying below the imaginary surface must exert a reaction force $-m\mathbf{g}$ on the same surface in order to support the column of fluid lying above it; otherwise the fluid would not be at rest.

The weight mg of the supported column can be expressed in terms of its volume Ay and its density ρ. The **density** is defined to be

$$\rho \equiv \frac{\text{mass}}{\text{volume}}. \tag{15.28}$$

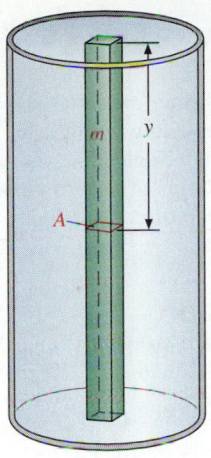

FIGURE 15.8 To maintain hydrostatic equilibrium, the column of fluid below the area A must support the weight of the column above it. If the fluid is incompressible ($\rho = $ constant), the weight of the upper column is $mg = \rho Ayg$. The pressure at depth y is thus $p(y) = \rho gy$.

The volume V occupied by a given mass depends on the pressure exerted on it by its surroundings. According to Equation 15.25a, a change Δp in pressure results in a change in volume $\Delta V = (-V/B) \Delta p$. Because V and B are always positive, an increase in pressure always results in a decrease in volume. Equation 15.28 shows that this decrease in volume leads to an increase in density. To calculate the force exerted by a column of fluid, we need to know its mass. If the density of the fluid is not constant, we cannot *in general* calculate the mass of the column by multiplying its volume by a constant ρ. However, the familiar fluids called liquids are a special case; a substantial increase in pressure produces only a small—often negligible—reduction in volume. For many practical purposes, we can consider liquids to be incompressible. This is the most striking property that distinguishes liquids from gases, the other familiar class of fluids. The density of an incompressible fluid is independent of the pressure to which it is subjected.

Thus, if the fluid in the column above A in Figure 15.8 is incompressible, the density of the fluid is the same everywhere. We can therefore use Equation 15.28 to calculate the mass of the column. We have $m = \rho Ay$, and the corresponding weight is

$$mg = \rho Ayg. \tag{15.29}$$

The stress exerted by the column on the area A is simply this weight divided by A. But, because we are considering a fluid, Pascal's principle guarantees that the stress is isotropic and can be represented by a pressure $p = mg/A$. We thus have

$$p = \rho gy \quad \text{for an incompressible fluid.} \tag{15.30}$$

The pressure at any point in an incompressible fluid is directly proportional to the depth of the fluid at that point. That is why the term "pressure head" (or simply "head") is often used to denote pressure due to the weight of a column of liquid, as described by Equation 15.30.

Suppose that a liquid is contained in a cylinder closed by a piston that exerts pressure. The total pressure at any point in the liquid is the sum of the externally applied pressure due to the piston and the internal pressure due to the weight of the liquid above the point. Now you can see why Pascal's principle is best stated in terms of pressure changes. The pressures at two points in the liquid at different depths will be different. But any *change* in the pressure—whether due to a change in the piston force or due to a depth change resulting from addition or removal of liquid—will be the same at both places. In most practical applications, either the external or the internal pressure is negligible.

EXAMPLE 15.7

(a) Find the pressure at the deepest point in the ocean, 10 915 m below the surface. The density of seawater is 1.02×10^3 kg/m^3. Neglect the variation of ρ and g with depth. **(b)** The bulk modulus of water is 2.06×10^9 Pa. To what extent is the assumption of constant density, used in part **a**, in error?

SOLUTION:

(a) What is the pressure 10 915 m below the ocean surface?

From Equation 15.30, $p = \rho gy$, you have

$$p = 1.02 \times 10^3 \text{ kg/m}^3 \times 9.80 \text{ m/s}^2 \times 10.915 \times 10^3 \text{ m}$$
$$= 1.09 \times 10^8 \text{ Pa}.$$

This is roughly 1000 times the pressure of the atmosphere at sea level.

(b) Find the error resulting from the assumption that the density of seawater is constant from the ocean surface to the bottom.

From Equation 15.25a, you have

$$\frac{\Delta V}{V} = -\frac{\Delta p}{B}. \qquad (15.31)$$

Assume that the change in density is not large. (Finding out if this is so is the aim of this part of the example.) If your assumption is correct, you can approximate the finite change $\Delta\rho$ by treating it like the infinitesimal quantity $d\rho$. So you use the definition of ρ given in Equation 15.28 to write

$$\Delta\rho = \Delta\left(\frac{m}{V}\right) = -\frac{m}{V^2}\Delta V.$$

(In differentiating, you treat m as a constant because you are seeking the change in density of a fixed mass of water as its volume is changed.) The proportional change in density is thus

$$\frac{\Delta\rho}{\rho} = -\frac{m}{V^2}\Delta V \frac{V}{m} = -\frac{\Delta V}{V}. \qquad (15.32)$$

Combining this result with that of Equation 15.31, you have

$$\frac{\Delta\rho}{\rho} = \frac{\Delta p}{B}$$
$$= \frac{1.09 \times 10^8 \text{ Pa}}{2.06 \times 10^9 \text{ Pa}} \approx 5 \times 10^{-2}.$$

So the proportional increase in the density of the seawater is something like 5% from surface to bottom. This result is not exact because the increase in density itself leads to an increase in the pressure at the bottom, which in turn implies a further increase in the density of the water at the bottom. But the approximation is close enough because it sets a limit of *less than* 5% on the error in the pressure calculation of part **a**—something like $2\frac{1}{2}$%. Can you see why the error is closer to $2\frac{1}{2}$% than to 5%?

The calculation of Example 15.7 shows why it is so often justifiable to treat liquids as incompressible fluids.

The Barometer

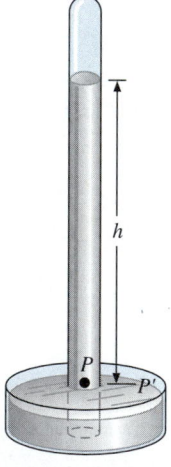

FIGURE 15.9 The mercury barometer. To set up the barometer, the tube, which is about 1 m long, is completely filled with mercury. With the open end temporarily plugged, the tube is inverted and the open end placed in the open dish of mercury. When the end is unplugged, mercury flows out of the tube until the mercury level in the tube attains a stable height h above the mercury level in the dish.

Figure 15.9 is a schematic representation of the **barometer**, a simple device used to measure atmospheric pressure. The operation of the barometer is explained on the basis of Pascal's principle taken together with Equation 15.30. The barometer is set up as described in the legend for Figure 15.9. In principle, any liquid can be used to fill the barometer, but mercury is the substance of choice because its great density makes possible an instrument of reasonable size.

Why is the mercury level at a height h (about 76 cm at sea level) above the level in the dish? Consider the points marked P and P' in Figure 15.9. Both points are at the level of the mercury surface in the dish, though P is inside the tube. According to Equation 15.30, there can be no difference in the pressures at the two points. Now, the pressure at P is evidently exerted by the column of mercury directly above it and has the value ρgh, where ρ is the density of mercury. The pressure at P' cannot be due to the weight of a column of mercury because there is no mercury above this point. The only fluid above P' is air, and the necessary pressure can be exerted only by the column of air above P'. So the pressure of the air at P' is also equal to ρgh. Any change in atmospheric pressure will be reflected in a change in the height h of the mercury column. The column height can be measured with precision, and hence the barometer is a good pressure gauge. As you have long known, the small fluctuations in h (usually ranging over less than 4 cm, or about 5%, at any fixed location) are widely used to provide data for analyzing weather conditions.

When the plugged end of the filled barometer tube is submerged in the mercury in the dish and the plug removed, the mercury level in the tube is greater than h. Consequently, the pressure at P is temporarily greater than the pressure at P'. Under these circumstances, mercury must flow out of the tube until the pressures equalize and there is thus no net force to drive mercury from P toward P'. What is left above the mercury in the tube? Essentially nothing. So the pressure at P is due entirely to the weight of the mercury column and has the value ρgh.

It is often convenient to express pressures as multiples of the standard atmospheric pressure, rather than in pascal. We define the non-SI unit **atmosphere**:

$$1 \text{ atmosphere (atm)} \equiv 1.013\ 25 \times 10^5 \text{ Pa.} \qquad \textbf{(15.33a)}$$

Other related nonstandard units for pressure are sometimes encountered. Among them is the **torr**, or **Torr** [also called the millimeter of mercury (mmHg)], defined to be the pressure that will support a mercury column 1 mm high:

$$1 \text{ torr} \equiv 133.322 \text{ Pa.} \qquad \textbf{(15.33b)}$$

Also used occasionally is the **bar**:

$$1 \text{ bar} \equiv 10^5 \text{ Pa.} \qquad \textbf{(15.33c)}$$

The millibar (mbar) is still used in U.S. meteorological practice, but most other countries use the kilopascal (kPa); 1 mbar = 0.1 kPa.

EXAMPLE 15.8

Observation of the small variations in atmospheric pressure at fixed locations is a fundamental tool for weather prediction. *Standard* atmospheric pressure at sea level is somewhat arbitrarily fixed to be that pressure which supports a mercury column exactly 760 mm high when the temperature of the mercury is 0°C. The density of mercury at 0°C is 13.5955×10^3 kg/m^3. The standard value of the acceleration of gravity is taken to be $9.806\ 36$ m/s^2. Express the standard atmospheric pressure in pascal.

SOLUTION: Using Equation 15.30, you obtain

$$p = \rho gh = 13.5955 \times 10^3 \text{ kg/m}^3 \times 9.806\ 36 \text{ m/s}^2 \times 0.76 \text{ m}$$
$$= 1.013\ 25 \times 10^5 \text{ Pa.}$$

Archimedes' Principle

You have certainly experienced the change in water level that occurs when you get into a bathtub. The level change is a consequence of the *displacement* of water by your body.* The space occupied by your body is no longer available to the water originally located there. If you fill the bathtub to the rim before you get in, the volume of the water that overflows will be equal to the volume of the part of you that is submerged.

You have also experienced the **buoyant force**—the upward force resulting from the submersion of your body in the water—that contributes so much to making the bathtub a pleasant and relaxing place. Although you probably do not float, you feel when you are immersed in water as if you weigh much less than your normal weight.

These two phenomena are connected by **Archimedes' principle**: *A body partially or completely immersed in a fluid experiences a buoyant (upward) force equal to the weight of the fluid displaced by the body.* The proof is as follows. In Figure 15.10, an arbitrarily shaped body of volume V is placed in a container filled to the top with a fluid of density ρ_f. The body is shown completely immersed, but complete immersion is not essential to the proof. To begin with, imagine the situation before the body was immersed. The region now occupied by the body was filled with fluid, whose weight

---

*The meaning of the word ''displacement'' in this context is related to but different from the meaning intended when we speak of the displacement of a particle from one position to another. What we mean here is that the water moves out of the space preempted by your body.

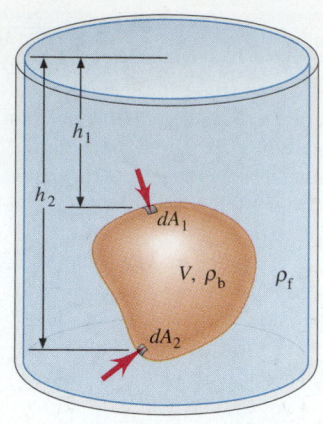

FIGURE 15.10 Archimedes' principle. A body of volume V and density ρ_b is immersed in a fluid of density ρ_f. The fluid that originally occupied the region now occupied by the body has been displaced. The buoyant force exerted by the neighboring fluid on the body is equal to the force exerted by the neighboring fluid on the same region before the body was present. Hydrostatic equilibrium requires that the latter force be equal to the weight of the displaced fluid. The buoyant force is the vector sum of the elemental normal forces of magnitude $p\,dA$ exerted on the body by the neighboring fluid. Two such elements are shown.

was $\rho_f Vg$. Because the fluid as a whole was in hydrostatic equilibrium, the net upward force on the fluid in that region was equal to the weight of the fluid occupying that region.

What is the source of that net upward force? At every point on the boundary of the region, the neighboring fluid exerts pressure on the surface of the region. Owing to this pressure, a normal force of magnitude $p\,dA$ is exerted on every boundary element of area dA. We know that the horizontal components of these elemental forces cancel out. Otherwise the fluid that originally occupied the region would have moved horizontally and would not have been in hydrostatic equilibrium. But the vertical components do not cancel because the pressure increases with depth. The generally upward forces on the lower surfaces of the region are therefore greater than the generally downward forces on the upper surfaces.

Now consider what happens when the body has displaced the fluid. The pressure at every point on the surface of the body is unchanged from the value at the same location when the body was not present. This is because the pressure at any point depends only on the depth of that point below the fluid surface. Hence the net force exerted by the surrounding fluid on the body is exactly the same as that exerted on the region before the body was present. But we know the latter to be $\rho_f Vg$, the weight of the displaced fluid. Hence this must also be the buoyant force exerted on the body. Archimedes' principle is thus proved.

Why do some bodies float and others sink? Consider what happens as the body in Figure 15.10, originally held above the surface, is lowered slowly into the fluid. As the body enters the fluid and displaces more and more of it, the buoyant force increases. But, *if the density of the fluid is greater than the density ρ_b of the body*, the buoyant force will become equal to the weight of the body before the entire body is immersed. When that equality is achieved, the net force on the body—the vector sum of the downward gravitational force mg and the upward buoyant force—will be zero, and the body will float. If the density of the body is greater than that of the fluid, the equality will not be achieved even when the body is entirely immersed; if the body is not supported, it will sink to the bottom. Nevertheless, the net downward force on the body will be less than its weight; specifically, the magnitude of the force will be

$$F = \text{weight} - \text{buoyant force} = (\rho_b - \rho_f)Vg. \tag{15.34}$$

EXAMPLE 15.9

According to legend, a traveling salesman offered to sell a solid gold crown to the tyrant of Syracuse (in Sicily) at a high price. The tyrant had reason to suspect that the crown was really made of a much cheaper alloy of gold with silver and perhaps some copper. He asked Archimedes to determine the truth of the matter. Deeply intrigued with the problem, and having pondered it without success for some time, Archimedes decided to relax in a hot bath. As he got into the tub and had the familiar experiences described at the beginning of this section, the solution flashed into his head. He is supposed to have run stark naked through the streets of Syracuse shouting, "Eureka!" ("I have found it!") How did Archimedes solve the problem?

SOLUTION: Archimedes attached the crown to a simple scale and weighed it twice: once in air (in the usual way) and once immersed in a container of water. He then repeated the process with a sample of what he knew to be pure gold.

To see how the analysis proceeds on the basis of these measurements, call the weight of the crown in air W_c, its weight in water W_c', its volume V, and the density of the material of which

it is made ρ_c. According to Equation 15.34, the weight of the crown in water is

$$W_c' = (\rho_c - \rho_w)Vg,$$

where ρ_w is the density of water. The weight of the crown in air is $W_c = \rho_c Vg$. Dividing the first weight by the second, you have

$$\frac{W_c'}{W_c} = 1 - \frac{\rho_w}{\rho_c}.$$

The volume of the crown drops out of this expression and so need not be determined. The ratio of the weights in water and in air depends only on the densities of water and the crown. If you know the densities of water and gold, you can substitute these values on the right side of the equation and compare the resulting value of the right side with the experimentally determined ratio on the left side. Archimedes did not have tables of density, and so he compared the weight ratio with that he obtained for the sample known to be pure gold. According to the legend, the ratios were different and Archimedes saved the tyrant from being swindled. We hope he was suitably rewarded!

Boyle's Law

The most striking property of the class of fluids called gases is the ease with which they can be compressed. Like liquids, gases have zero shear modulus, and bodies of gas therefore have no definite shape. Unlike liquids, however, gases have no definite volume either. The volume of a body of gas of given mass and composition depends on the pressure to which it is subjected.

Because the indefinite volume of a body of gas is one of its most confusing aspects, an indispensable step in bringing order out of confusion is that of expressing gas volume as a function of other quantities. If a body of gas has a fixed mass, its volume is strongly influenced by changes in pressure and temperature. In this chapter, we will consider pressure only, deferring a consideration of the influence of temperature to Chapter 17. In order to exclude temperature effects, we will consider experiments in which the initial and final temperatures of the gas are the same. That is, we will investigate the properties of gases under constant-temperature, or **isothermal**, conditions. Such experiments were first carried out in a precise way in 1662 by the Anglo-Irish physicist and chemist Robert Boyle (1627–1691) on the basis of strong suggestions contained in the work of earlier investigators.

Boyle's simple apparatus is shown schematically in Figure 15.11. The J-shaped tube is closed at one end. The closed arm is calibrated so that the volume V contained between the closed end and any point in the arm can be read directly from the scale. Mercury poured into the tube traps air in the closed end. The amount (mass) of air trapped can be varied by tilting the tube back and forth, allowing bubbles of air to pass into or out of the closed region. This is done until the level of mercury is the same in both arms of the tube, as shown in Figure 15.11a. The pressure p of the trapped air must then be the same as the pressure p_0 of the surrounding atmosphere. The mercury level in the closed end also yields the corresponding volume of air V_0.

Next, additional mercury is poured into the open end of the tube. The tube is held vertically so that additional air can neither enter nor escape from the trapped volume in the closed end. The added mercury changes the liquid levels in both arms of the tube, but the two levels are no longer equal. Why is this so? Because the rise of the mercury level in the closed arm means that the trapped air is *compressed*—confined in a reduced volume. Compression requires an increase in the pressure exerted on the trapped air, and thus (according to Pascal's principle) in the pressure of the trapped air,

Gases and Chaos

Gases are much harder to "grab" than liquids, in more than the literal sense of the word. The Flemish natural philosopher Jan Baptista van Helmont (1577 or 1580–1644) was among the first to have a clear idea that the concept "gas" was broader than the concept "air"—that there are gases other than air having distinct physical and chemical properties. The formlessness of gases led van Helmont to coin the word *gas* in 1632 from the Greek word "chaos." Van Helmont wrote, "This vapor I called *Gas*, not far different from the Chaos of the ancients." In Flemish, the letter *g* is pronounced like the Greek chi (χ), which is the first letter of the word "chaos" ($\chi\acute{\alpha}o\sigma$.) The sound does not exist in English, but is usually rendered "ch" or "kh."

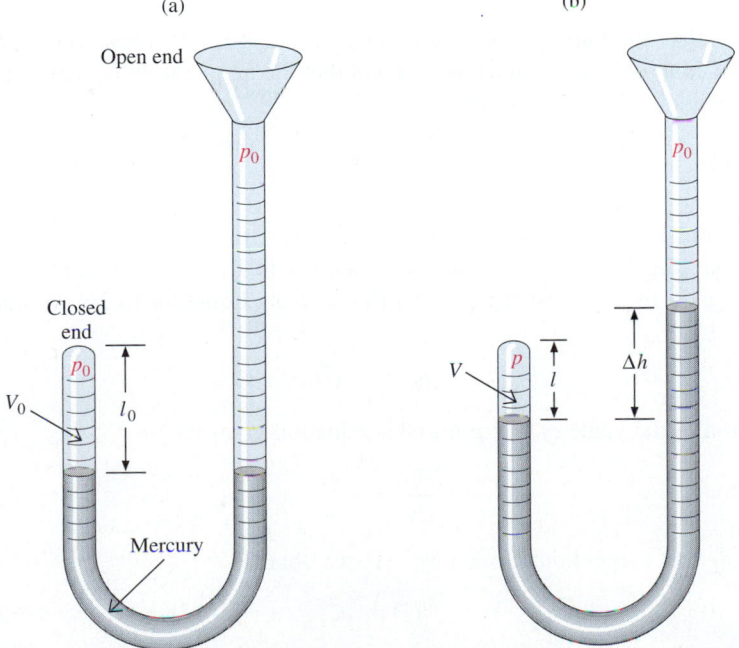

(a)   (b)

FIGURE 15.11 Boyle's law apparatus. (*a*) Mercury is poured into the tube, and the amount of air trapped in the closed arm is varied (by tilting the apparatus back and forth) until the mercury level is the same in the two arms. The pressure of the trapped air is then equal to the atmospheric pressure; $p = p_0$. On the basis of previous calibration, the volume $V = V_0$ of the trapped air is read from the scale on the tube. (*b*) More mercury is added to vary p and V. The new pressure p is found from the difference Δh in the mercury levels in the two arms. The new volume V is read from the scale.

The necessary additional pressure is provided by the extra height Δh of the mercury column in the open end of the tube:

$$p = p_0 + \rho g\, \Delta h,$$

where ρ is the density of mercury. The value of p_0 can be determined by reading a barometer, and the pressure p of the trapped air can thus be determined. The mercury level in the closed end also yields the corresponding volume V of the trapped air. More mercury can be poured into the open end, yielding a series of readings of pressure versus volume.

The entire experiment can be repeated with different amounts of trapped air or with other gases. In each experiment, the amount of trapped gas and the temperature are held constant. Consider any two measurements, j and k, among a series of measurements made of the pressures and corresponding volumes in such an experiment. Over a wide range of pressures, the measurements conform quite accurately to the relation

$$\frac{p_j}{p_k} = \frac{V_k}{V_j}. \tag{15.35a}$$

This relation can be expressed more generally as the proportionality

$$p \propto \frac{1}{V} \quad \text{for mass and temperature constant,} \tag{15.35b}$$

or by the corresponding equations

$$pV = \text{constant} \quad \text{or} \quad p_j V_j = p_k V_k \quad \text{for mass and temperature constant.} \tag{15.35c,d}$$

In any of these four forms, the relation between pressure and volume is called **Boyle's law**. Boyle compared the process of compressing the trapped gas to that of compressing a spring, and he spoke of the effect as the "spring of air." Boyle's law is just one part of a more general relation among the quantities pressure, volume, mass, and temperature. We will consider this matter further in Chapter 18.

There are limits to the conditions under which a gas obeys Boyle's law. At temperatures near or above room temperature and pressures not too much higher than atmospheric pressure, however, most familiar gases—including air and its component elements oxygen and nitrogen—obey Boyle's law quite well.

Bulk Modulus of a Gas

Let us calculate the bulk modulus of a gas that obeys Boyle's law. Suppose that we have confined a particular amount of gas and that the temperature is kept constant. We rewrite Equation 15.35b in the form

$$p = \frac{c}{V}, \tag{15.36}$$

where the constant of Equation 15.35b happens to have the particular value c. According to Equation 15.25b, the bulk modulus is defined to be $B \equiv -V\,(dp/dV)$. We can evaluate dp/dV for the present case by differentiating Equation 15.36; we find

$$\frac{dp}{dV} = -\frac{c}{V^2}.$$

Substitution of the value $c/V = p$ into this equation gives us

$$\frac{dp}{dV} = -\frac{p}{V}.$$

Multiplying this expression through by $-V$, we obtain

$$B = p. \tag{15.37}$$

For a gas that obeys Boyle's law, the bulk modulus is equal to the pressure. In order to emphasize one of the essential conditions for the validity of Boyle's law, we give the name **isothermal bulk modulus** to the quantity B given by Equation 15.37.

The striking difference between the effort required to compress gases on the one hand and that required to compress liquids or solids on the other is made evident by comparing bulk moduli. At a pressure of 1 atm (of order 10^5 Pa), the bulk modulus of gases is of order 10^5 Pa. But the bulk moduli of liquids are of order 10^9 Pa, and those of solids are of order 10^{10} Pa—four or five orders of magnitude greater. Thus, beginning at atmospheric pressure and doubling the pressure results in a halving of the volume of a sample of gas. The same pressure change results in a volume change of order 1 part in 10^4 for a liquid sample and 1 part in 10^5 for a solid sample. That is why we can consider liquids and solids as incompressible for many purposes.

A	cross-sectional area of a body		Y	Young's modulus
B	bulk modulus		β	shear strain
G	shear modulus		ϵ	strain
k	force constant		ϵ_p, ϵ_i	primary strain, induced strain
l	length of a body		κ	compressibility
p	pressure		ν	Poisson's ratio
T	tensile force		ρ	density
V	volume		σ	stress

Summing Up

Equal and opposite forces applied to a body along the same axis are called **tensile** or **compressive** forces, depending on whether they tend to stretch or squeeze the body. The force per unit cross-sectional area is called the **uniaxial stress** σ and is defined by Equation 15.6,

$$\sigma \equiv \frac{F}{A}.$$

This stress results in a proportional change in the length of the body along the axis. This proportional change is called the **strain** ϵ and is defined by Equation 15.5,

$$\epsilon \equiv \frac{\Delta l}{l}.$$

By definition, tensile stress and strain are positive; compressive stress and strain are negative.

If the body is isotropic and obeys Hooke's law, the strain is proportional to the stress. This relation is expressed by Equation 15.8a,

$$\sigma = Y\epsilon.$$

The proportionality constant Y is called **Young's modulus**. For a bar of uniform cross section, Y is related to the Hooke's-law force constant k by Equation 15.9b,

$$k = Y\frac{A}{l}.$$

When a body is stressed uniaxially, the **primary strain**

ϵ_p is not the only consequence. There is also **induced strain** ϵ_i in all directions perpendicular to the stress axis. The sign of ϵ_i is always opposite that of ϵ_p, so there is partial compensation for the volume change due to the primary strain. **Poisson's ratio** is defined by Equation 15.16,

$$\nu \equiv -\frac{\epsilon_i}{\epsilon_p}.$$

When a couple is applied to a body not free to rotate, the body is subjected to a **shear stress**. The result is a distortion expressed by Equation 15.19 as the **shear strain** $\beta = \Delta l/h$. The **shear modulus** G is defined according to Equation 15.20 as the ratio

$$G \equiv \frac{\sigma}{\beta}.$$

Matter is said to be in a **fluid** state if it cannot sustain a static shear stress. Within a body of fluid or at its surface, the stress at any point must be isotropic. This statement is called **Pascal's principle**. The negative of the isotropic stress $p = -\sigma$ is called the **pressure**. It is defined by either of Equations 15.21a and 15.21b:

$$p \equiv \frac{F}{A} \quad \text{or} \quad p \equiv \frac{dF}{dA} \quad \text{for } p \text{ isotropic.}$$

When a fluid experiences a change Δp in the pressure to which it is subjected, there is a proportional change $\Delta V/V$ in its volume. The **bulk modulus** B is defined by either of

Equations 15.25a and 15.25b,

$$B \equiv -V\frac{\Delta p}{\Delta v} \quad \text{or} \quad B \equiv -V\frac{dp}{dV}.$$

Equation 15.26 defines the reciprocal of the bulk modulus, called the **compressibility** κ:

$$\kappa \equiv \frac{1}{B} = -\frac{1}{V}\frac{dV}{dp}.$$

In an incompressible fluid ($\kappa = 0$), the pressure is directly proportional to the depth, as given by Equation 15.30,

$$p = \rho g y.$$

A body immersed in a fluid experiences a buoyant force equal to the weight of the fluid displaced by the body. This result is called **Archimedes' principle**. The body will float if its density is less than that of the displaced fluid.

A **gas** is a fluid that is readily compressible. At fixed temperature, the volume of a gas is inversely proportional to its pressure over a wide range of pressure. This proportionality, Equation 15.35a, is called **Boyle's law**. For any gas that conforms to Boyle's law, the bulk modulus is equal to the pressure; $B = p$ (Equation 15.37).

Queries and Problems for Chapter 15

QUERIES

15.1 *(2) Everything in proportion, I.* A rod of uniform cross section is subjected to tension by a pair of equal and opposite forces of magnitude F. Make an argument to show that the change in length Δl is proportional to the relaxed length l. (Hint: What happens when two or more identical springs are linked end to end, as in Problem 6.14?)

15.2 *(2) Everything in proportion, II.* A rod of length l and uniform cross section is subjected to tension by a pair of equal and opposite forces of magnitude F. Make an argument to show that the change in length Δl is inversely proportional to the cross-sectional area A. (Hint: What happens when two or more identical springs are placed side by side, as in Problem 6.15?)

15.3 *(3) Direction counts!* A uniaxial stress is applied to a bar made of an isotropic material. As a result of this stress, the bar experiences a primary strain ϵ_p along the stress axis and a smaller induced strain ϵ_i along any transverse direction. How can the two strains be different when the material is isotropic?

15.4 *(5) Everything normal here.* As you know from experience, a free liquid surface at rest is always normal to the local vertical. Deduce this fact from the definition of a fluid as a body of matter that has no static resistance to shear stress.

15.5 *(5) Just a touch.* Imagine a body made of a hypothetical material having a negative bulk modulus. Describe its behavior when a very small stress is applied to it.

15.6 *(5) Benign neglect.* In deriving Equation 15.21a, we neglected the weight of the wedge of fluid. Justify this neglect.

15.7 *(5) Singling out three.* In deriving Equation 15.24, we regarded isotropic pressure as equivalent to three uniaxial stresses acting along the length, width, and thickness axes of a solid block. In view of the fact that the pressure is the same in all directions, why isn't it necessary to consider stresses acting in all other directions as well?

15.8 *(5) More of the same.* The bottoms of the two vessels illustrated below have equal areas, and the vessels are filled to the same depth with the same liquid. The pressures at the bottoms of the two vessels are the same, and the total forces exerted by the liquid on the two bottoms are the same. Yet A contains more liquid than B. Explain the apparent contradiction.

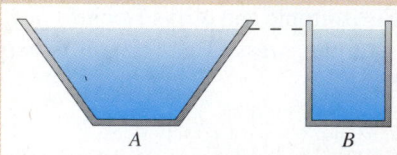

15.9 *(5) The properties of nothing.* To prove that the role of the torricellian vacuum in the behavior of a mercury barometer is completely passive, Boyle enclosed a barometer in a large glass vessel from which he could pump the air. Describe the behavior of the barometer (a) as the air is slowly pumped out of such a vessel and (b) as air is allowed to flow slowly back into the vessel. How do the observations bear out Boyle's assertion that the torricellian vacuum is not somehow "pulling" on the mercury column?

15.10 *(5) No free lunch.* As you saw in Example 15.5, a hydraulic lift can be used as a force multiplier. But show that the force multiplication ratio F_2/F_1 is obtained at the cost of piston travel and that the work done *by* piston 2 is not greater than the work done *on* piston 1.

15.11 *(6) Moderation in all things.* A block is made of solid material whose density is exactly the same as that of water. **(a)** What happens if you lower the block to the surface of a container of water and release it? **(b)** What happens if you submerge the block in the water and release it?

15.12 *(6) Getting a lift.* Why does a helium-filled balloon rise through the air?

15.13 *(7) Escape.* Crew members of submarines are provided with escape breathing devices that allow them to survive during the period of many minutes needed to float to the surface, should the submarine be disabled when submerged. The instructions for use of the device place great emphasis on the need to exhale, despite the instinct to hold one's breath when submerged. Explain the importance of exhaling (aside from the fact that you can't breathe in if you don't breathe out).

15.14 *(G) Stretch or twist?* In designing a coil spring, are you mainly concerned with Young's modulus or the shear modulus of the material you plan to use? Explain.

15.15 *(5) Identity crisis.* If you strike a block of wax with a sharp blow, thus applying a large stress suddenly, the block will shatter. If, however, you apply the same stress gradually and continue to apply it for some time, the block of wax will flow—that is, distort permanently. Is the wax a solid? a liquid? Explain.

15.16 *(G) Oops!* The Boeing 707, the first widely used commercial passenger jet airplane, went into service in 1960. Early models experienced difficulty when the high heels worn by (then almost exclusively female) flight attendants dented and even punctured the thin sheet metal of the cabin floor, which had to be redesigned with a considerable weight penalty. Why did the flight attendants' high heels do damage when the predominantly male builders, most of them much heavier than the flight attendants, had not encountered this difficulty during the many hours they spent inside the cabin?

15.17 *(G) Which way?* A child stands in a crowded bus. He has a helium-filled balloon, the string of which he ties to a convenient handle. The bus begins to move, the standing passengers lurching toward the back of the bus. What happens to the balloon? (Hint: Draw a free-body diagram for the balloon.)

15.18 *(G) High rise.* Balloons used for high-altitude research are initially filled with helium to a fairly small fraction of their total capacity. The helium forms a bubble in the upper part of the balloon (a closed-end sleeve made of thin plastic), and the lower part of the balloon is simply collapsed. Explain what happens as the balloon rises. Will the balloon be more or less buoyant as it rises to higher altitudes? Neglect temperature changes.

15.19 *(G) Cartesian diver.* Pictured below is a simple device, invented by the French mathematician and philosopher René Descartes, that you may have played with as a child. To set up the device, you fill a test tube with water and submerge it in a large bottle of water in the inverted position shown. Using a straw, you blow air into the tube, displacing some of the water, until the tube is barely buoyant—that is, until it barely rises when released. Then you insert a stopper into the neck of the bottle. By pressing on the stopper with your hand, you can make the test tube sink; when you remove your hand, the tube rises again. By proper manipulation, you can move the tube to any depth you want. Explain how the device works.

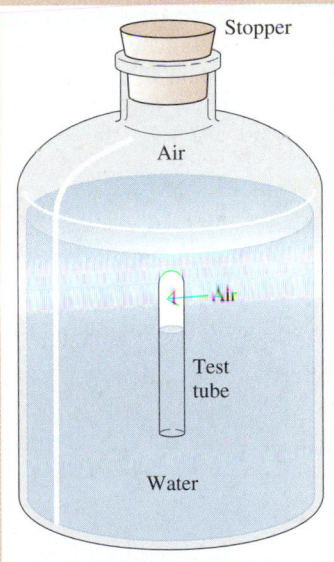

PROBLEMS

GROUP A

15.1 *(2) Stretching things a bit.* A brass rod has length 1.67 m, width 2.0 cm, and thickness 1.5 cm. It is subjected to a tension of 5.6 tonnes along its length. Find **(a)** the tensile stress σ, **(b)** the resulting primary strain ϵ_p, and **(c)** the change Δl in length.

15.2 *(2) Brute force.* How much force, applied to the ends of a cylindrical steel rod of length 10 m and diameter 2.5 cm, is required to stretch the rod by 1.5 cm?

15.3 *(2) Bars in series, I.* A brass bar and an aluminum bar of equal length and cross-sectional area are linked end to end. Stress is applied to the ends of the resulting bar. If the strain of the aluminum bar is 1.0×10^{-4}, what is the strain of the brass bar?

15.4 *(3) Stretch and reduce.* **(a)** What is the change in diameter of the steel rod of Problem 15.2? **(b)** What is the change in volume?

15.5 *(3) Thinning down, too.* For the rod of Problem 15.1, find the change Δw in width and the change Δt in thickness.

15.6 *(3) And yet expanding.* Find the volume change ΔV of the rod discussed in Problems 15.1 and 15.5.

15.7 *(4) Twisted steel.* Forces of equal magnitude are applied to a square steel plate, as shown below. If the plate is 14 cm wide and 0.55 cm thick, what must the magnitude of the forces be to produce a shear strain of 3.5×10^{-3}?

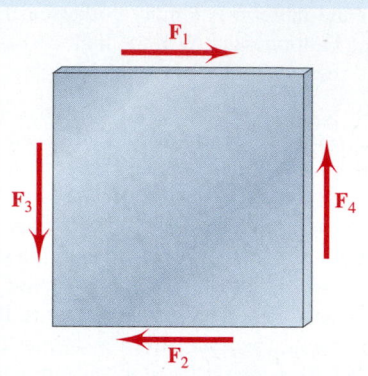

15.8 *(4) Skewed.* Illustrated below is a rectangular plate of copper of thickness 1.5 cm. **(a)** A couple is applied to the top and bottom edges of the plate, and the corner angles change by 0.62°. What is the magnitude F_1 of the forces that make up the couple? **(b)** A couple is applied to the left and right edges of the plate, and again the corner angles change by 0.62°. What is the magnitude F_2 of the forces? **(c)** Are the couples equal to each other?

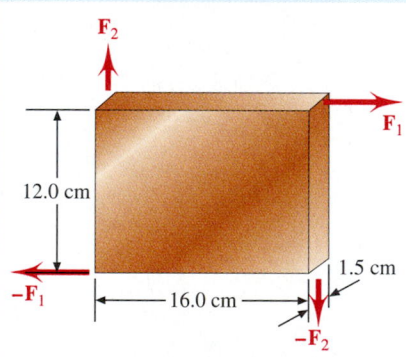

15.9 *(5) Reducing the bulk.* The bulk modulus of magnesium is 3.7×10^{10} Pa. What pressure is required to reduce the volume of a block of magnesium by 0.2%?

15.10 *(5) Steel in a steely grip.* When a pressure of 10^4 atm is exerted on a sample of steel, its volume decreases by 0.68%. Find the bulk modulus of this particular type of steel.

15.11 *(5) Grease rack.* Consider the hydraulic automobile lift of Example 15.5. **(a)** If the automobile resting on piston 2 is to be raised 10 cm, how far must piston 1 be moved? **(b)** Neglecting friction, calculate the work that must be done on piston 2 and the work that must be done on piston 1 in the process of part **a**, and show that energy is conserved.

15.12 *(5) Diving practice.* How far below the surface of a lake is the total pressure 2 atm?

15.13 *(5) Water barometer.* If a barometer is filled with water instead of mercury, how high will the water column be when the atmospheric pressure is 1.013×10^5 Pa? [The German bureaucrat and natural philosopher Otto von Guericke (1602–1686) set up a water barometer in such a way that the tube passed through the roof of his house at a height about equal to the average height of the water column. A life-size wooden doll floating at the surface rose through the roof to "enjoy" good weather and retreated under the roof in bad weather.]

15.14 *(5) Hanging barometer.* A mercury barometer hangs from a wire whose other end is fastened to the ceiling of a railroad car. The height of the mercury column is 76.0 cm. The train approaches a station and slows down with constant acceleration of magnitude $a = 1.20$ m/s². What is the length of the mercury column in the barometer tube as the train decelerates?

15.15 *(5) And all the king's horses . . .* In 1654, von Guericke built a pair of hollow bronze hemispheres of radius 25 cm. The two hemispheres could be fitted together edge to edge to form an airtight sphere. Through a valve built into one of the hemispheres, von Guericke pumped most of the air out of the sphere. In a famous demonstration, illustrated below, he hitched a team of eight horses to each hemisphere. The horses could barely pull the hemispheres apart. With what force could each horse pull? (The hemispheres are named after the town of Magdeburg, about 100 km west of Berlin, where von Guericke lived.)

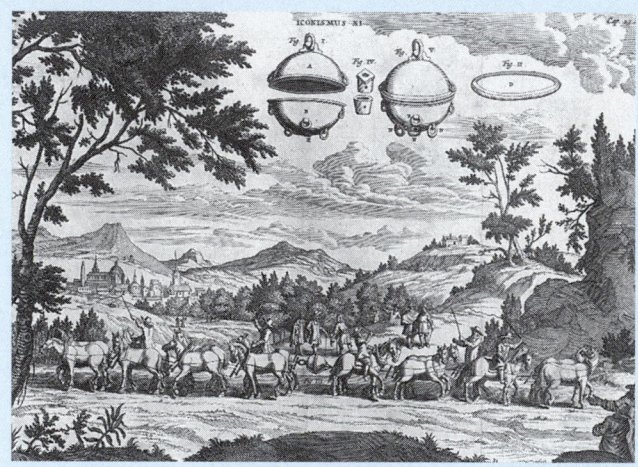

15.16 *(5) Uniform atmosphere.* Suppose that air were an incompressible fluid so that the density of the atmosphere was everywhere equal to its sea-level value $\rho_0 = 1.223$ kg/m³. How high above sea level would the top of such an atmosphere be for the sea-level pressure to have its actual value, $p_0 = 1.013 \times 10^5$ Pa? (Note that the tops of many mountains would be above such a uniform atmosphere.)

15.17 *(5) Tight seal.* A cylinder 5 cm high is oriented with its flat end surfaces horizontal. Each of the end surfaces has area 3 cm². The edge of the bottom surface is fitted with a flexible O-ring that can be used to make a watertight fit against the bottom of a tank. The tank is filled to a height of 7 cm. Because water cannot flow between the tank and the cylinder, the latter remains at the bottom. **(a)** Find the magnitude and direction of the force exerted by the water on the cylinder.

(b) The cylinder is lifted slightly to break the watertight seal. What are the magnitude and direction of the force exerted by the water on the cylinder now?

15.18 *(6) A wet weigh.* The chunk of metal shown below has mass 0.450 kg, and its volume is 100 cm^3. The mass of the jar and water together (not including the chunk of metal) is 1.00 kg. Find the reading (in N) on each scale.

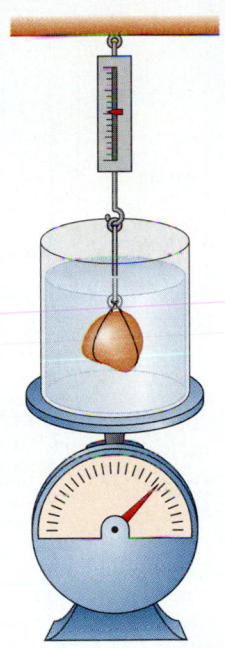

15.19 *(6) Sinking.* A block of aluminum weighs 98.0 N in air and 61.7 N in water. What is the density of aluminum?

15.20 *(6) Floating.* A cube having sides 10.0 cm long is made of wood whose density is 0.73 × 10^3 kg/m^3. **(a)** When the cube is placed in a bucket of water, how far above the water level is the top of the cube? **(b)** Coins are stacked on the top of the cube until it is flush with the water surface. What is the total weight of the coins?

15.21 *(6) Helping Archimedes out, I.* The density of gold is 19.3 × 10^3 kg/m^3. **(a)** Suppose the weight in air of the crown offered to the tyrant of Syracuse was 53.2 N. What would its weight in water be if it were genuine? **(b)** Archimedes is supposed to have successfully carried out his task (outlined in Example 15.9). If this legend is true, were the scales available to him accurate to better than ±5%?

15.22 *(6) Ballooning.* The density of helium is 0.139 that of air at the same temperature and pressure. If a balloon is to have a gross mass of 1.0 tonne, what is the minimum volume of helium required to lift it? Take the density of air to be 1.2 kg/m^3.

15.23 *(6) Almost as efficient, much safer.* Besides being fairly cheap, hydrogen is the least-dense gas; from that point of view, it is the substance of choice for filling lighter-than-air balloons. Unfortunately, it is extremely flammable, and there have been terrible disasters resulting from its use. The density of helium gas is twice that of hydrogen, and air is about 14 times as dense as hydrogen. Show that the gross lift of a helium-filled balloon—that is, the upward force the helium exerts on the balloon envelope and everything hanging from it— is 92% of the lift of an identical balloon filled with hydrogen.

15.24 *(7) Inflation.* The volume of air contained in an automobile tire is approximately 24 liters (1 liter = 10^3 cm^3). **(a)** If the pressure inside the tire is 3.3 atm, how many liters of air at atmospheric pressure must enter the pump when you fill the tire? (Hint: Remember that the tire contains air at 1 atm pressure before you begin to fill it.) **(b)** Why is it possible to solve this problem without converting everything into SI units?

15.25 *(7) Boyle lore.* A Boyle's-law apparatus like that of Figure 15.11 is filled with mercury as shown in part *a* of the figure. Mercury is then poured into the open end until $\Delta h = 3.4$ cm. **(a)** Find the change in the height of the mercury level in the closed end of the tube due to the addition of mercury in the open end. Take $p_0 = 1$ atm and $l_0 = 12.5$ cm. **(b)** Why don't you need to know the diameter of the tube to solve this problem?

GROUP B

15.26 *(2) Bars in series, II.* Two bars of equal length and cross section are joined end to end. They are made of materials having Young's moduli Y_1 and Y_2, respectively. Show that, when the resulting compound bar is subjected to stress, its overall length changes as if it were a uniform bar whose Young's modulus Y satisfies the relation

$$\frac{1}{Y} = \frac{1}{2}\left(\frac{1}{Y_1} + \frac{1}{Y_2}\right).$$

(Hint: See Problem 15.3.)

15.27 *(2) Sagging under the weight.* A steel cable 4 cm in diameter is stretched horizontally between two supports 4 m apart. The tension in the cable is not large, but it is sufficient so that the cable is nearly straight. (Note that no tension, however large, can make a horizontal cable perfectly straight.) A 1-tonne mass is now suspended at the middle of the cable. How far does the cable sag? That is, how far is the middle of

the cable below its initial location? (Hint: What is the stress in the cable, and how does it depend on the strain?)

15.28 *(2) Stored energy, I.* When a rod is subjected to uniaxial stress and experiences strain, it acquires potential energy in much the same way as a spring does when stretched. Show that the potential energy U stored per unit volume of the rod is

$$\frac{U}{V} = \frac{1}{2}Y\epsilon^2.$$

15.29 *(2) Stored energy, II.* A rod of length $2R$ and cross-sectional area A is made of a material having density ρ and Young's modulus Y. In the figure on p. 418, the rod is shown mounted on a shaft through its center and perpendicular to its length. The shaft rotates with angular speed ω. **(a)** Show that at a point a distance r from the axis, the stress is

$$\sigma = \sigma(r) = \tfrac{1}{2}\rho\omega^2(R^2 - r^2).$$

(b) Because the stress varies along the rod, the strain will vary as well. Consider the infinitesimal slab of the rod shown in the figure, whose unstressed thickness is dr. Find the strain

$$\epsilon = \epsilon(r) = \frac{\Delta(dr)}{dr}$$

experienced by the slab when the system is rotating. (c) Show that the proportional change in length of the rod as a whole is

$$\frac{\Delta r}{2R} = \frac{\rho\omega^2 R^2}{3Y}.$$

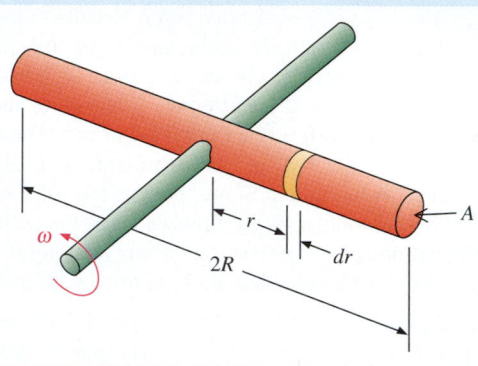

(d) Using the results of part b and of Problem 15.28, show that the total elastic potential energy stored in the rod is

$$U = \frac{2\rho^2\omega^4 AR^5}{15Y}.$$

(e) Suppose the rod is made of steel of density $\rho = 7.86 \times 10^3$ kg/m³, is 40 cm long from end to end, and has cross-sectional area 1.0 cm². If the maximum allowable strain anywhere in the rod is $\epsilon_{max} = 0.01$, what is the maximum allowable angular speed ω_{max}? Express your results in rev/min as well as in rad/s. (f) Find the elastic energy stored in the rod at this speed, and show that this energy is very small compared with the rotational kinetic energy of the rod.

15.30 *(2) Springtime.* A mass m hangs from the end of a metal wire of length l and radius r. The mass is pulled down a little way, stretching the wire, and is then released. (a) Assuming that the mass of the wire is much smaller than m, express the period of oscillation of the system in terms of l, r, and Y, the value of Young's modulus for the metal. (b) Find the period for a 10-kg mass that hangs from an aluminum wire of diameter 1.0 mm and length 7.5 m.

15.31 *(3) Squaring accounts.* A rectangular metal block has length l, width w, and thickness t. The value of Poisson's ratio for the metal is ν. The block is stressed along its long axis and experiences a primary strain ϵ_p. Using an analysis similar to that of part a of Example 15.3, show explicitly that the relative change in volume $\Delta V/V$ is the same as that given by Equation 15.17 for a cylindrical rod.

15.32 *(5) Tire store.* A mechanic raises an automobile of mass 1500 kg off the floor on a grease rack and installs new tires. She fills the tires with air to a pressure 2.34×10^5 Pa above atmospheric pressure and then lowers the automobile back to the ground. The tires have radius 32.4 cm and width 16.5 cm. By what amount do the tires flatten as they touch the ground

and carry the weight of the automobile? Make the approximation that the tires are disk shaped and do not bulge at the sides.

15.33 *(5) Density measurements.* A U-tube is partially filled with water. A quantity of oil (which does not mix with water) is poured into one arm of the tube. The system comes to rest as shown below. Find the density of the oil.

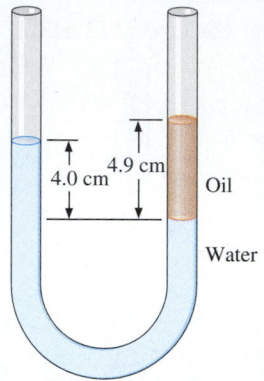

15.34 *(5) Center of pressure.* A dam of width w holds back a reservoir. At the dam, the depth of the reservoir is h. (a) Show that the water exerts on the dam a horizontal force whose magnitude is

$$F = \tfrac{1}{2}\rho gwh^2.$$

(Hint: The water pressure is a function of the depth; you have to integrate.) (b) This force exerts a torque T on the dam. Take the torque axis along the bottom edge of the dam, parallel to its width, and show that

$$T = \tfrac{1}{6}\rho gwh^3.$$

(c) By comparing the results of parts a and b, show that the vertically distributed force exerted by the water on the dam can be replaced by a single horizontal force applied at a height $\tfrac{1}{3}h$ above the bottom of the dam. This point is called the *center of pressure*.

15.35 *(5) Dam site.* A farmer wants to dam a small stream to make a pond for watering his cattle. The easiest way he can think of to do it is as follows. He builds an open rectangular box with walls 3 m high. The box is 5 m wide and long enough so that when it is placed across the stream the flow of water is completely blocked. He places the box in the stream and fills it with stones. The mean density of the fill (including water that leaks into the box) is 2×10^3 kg/m³. As the pond fills, the water level rises until water spills over the top of the dam. Will the dam overturn, or is it stable?

15.36 *(5) Spinning gracefully.* You place a soup bowl on a turntable so that the center of the bowl is located on the turntable axis. You fill the bowl partially with water and set the turntable spinning with angular speed ω. Show that the equilibrium shape of the water surface is a paraboloid of revolution (the surface traced when a parabola is rotated about its symmetry axis). This result is the basis of spin casting, a method sometimes used to manufacture paraboloidal reflectors for such applications as headlight reflectors. A refined version of the method has been used to cast mirrors for very large astronomical telescopes of diameters from 1.8 m to 8 m. [See *Science* **234**, 1495 (19 December 1986).]

15.37 (5) *Sloppy way to make a barometer.* A worker at an instrument factory is assembling a barometer. He takes a tube 1 m long having one closed end and fills it with mercury. However, he carelessly fails to fill it to the top and leaves an air-filled space 0.85 cm long between the mercury level and the open end of the tube. He then plugs the open end, inverts the tube, places the open end in a dish of mercury, and removes the plug. Where will the mercury level be in the tube? Assume that the (correct) barometric pressure is 760 torr.

15.38 (6) *Helping Archimedes out, II.* The crown salesman of Problem 15.21 could have made quite a lot of money by selling as pure gold a crown that contained as little as 10% silver. The density of silver is 10.5×10^3 kg/m^3, and the density of gold-silver alloys can be predicted fairly closely from the proportions of gold and silver. What is the smallest weight difference that Archimedes' scale had to detect to be sure of detecting 10% silver in the 53.2-N crown?

15.39 (6) *Archimedes in two flavors.* A container is partly filled with carbon tetrachloride (CCl$_4$) of density 1.6×10^3 kg/m^3. Water (which does not mix with CCl$_4$) is poured on top. A block of plastic of density 1.3×10^3 kg/m^3 is dropped into the container. The water is deep enough so that the plastic is entirely submerged. What fraction of the volume of plastic block is submerged in CCl$_4$?

15.40 (6) *Isostasy.* The part of the earth's *crust*—its outermost layer—consisting of the continents is made mainly of granite and similar rocks, of mean density 2.65×10^3 kg/m^3. As shown below, the continents float on the earth's *mantle*, the much thicker layer that lies just below the crust. The major component of the mantle is basalt, with density 2.85×10^3 kg/m^3. (Under the stresses present at appreciable depth below the surface, the basalt behaves like a fluid; see Problem 15.48.) The mean height of the continents above sea level is 0.75 km, and the mean depth of the oceans is 4.25 km. **(a)** Why is it reasonable to argue that the thickness of the earth's crust under the oceans is small compared with that of the continental crust, as shown in the figure? **(b)** Making the assumption of part **a** and taking into account the buoyant effect of both the basalt and the ocean water, show that the continents extend about 45 km below the level of the ocean floor. (This calculation agrees fairly well with observations made by seismic methods.)

The statement that the continents float in approximate hydrostatic equilibrium on the mantle is called the *principle of isostasy.* (See Problem 14.60.)

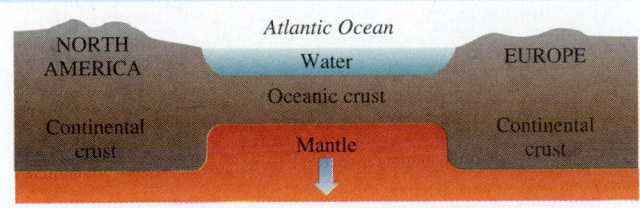

15.41 (6) *Flotation paradox.* It is possible to float an object in a body of water whose total mass is less than the mass of the object. Pictured in the next column is a glass cylinder of cross-sectional area 60 cm^2, into which 500 cm^3 of water (of

mass 500 g) has been poured. A hollow cylinder is made of plastic. Its mass is 1000 g, its cross-sectional area is 50 cm^2, and its height is 30 cm. The plastic cylinder is lowered into the water. Show that the plastic cylinder will float when its lower 20 cm is immersed in water and that this flotation will occur before the plastic cylinder touches bottom.

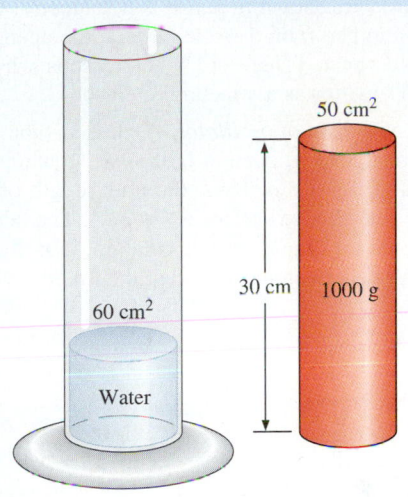

15.42 (7) *Getting bigger.* A bubble of gas of radius R rises from the bottom of a lake. Just before the bubble breaks through the surface, its radius has doubled. Assuming that the bubble is big enough so that the effect of surface tension is negligible, how deep is the lake?

15.43 (G) *Even trade.* The structure illustrated in part *a* below is similar to that considered in Examples 15.2, 15.3, and 15.4. The present structure is simpler in that it is square, with sides a. The diagonal member D is made of steel with density ρ and has a square cross section with edges t and area A ($=t^2$). The other members are made of the same steel but have much lighter construction. **(a)** Beginning with Equation 15.13, find the force F required to distort the structure through an angle β. **(b)** Assuming that member D contains most of the mass m of the structure, find F/m. This ratio may be regarded as a figure of merit for the structure because one usually wishes to design a structure combining the greatest possible strength with the smallest possible mass. **(c)** The solid square plate illustrated in part *b* is made of the same steel and has the same overall dimensions as the structure in part *a*, with sides a and thickness t. Using the result of Example 15.4, find the figure

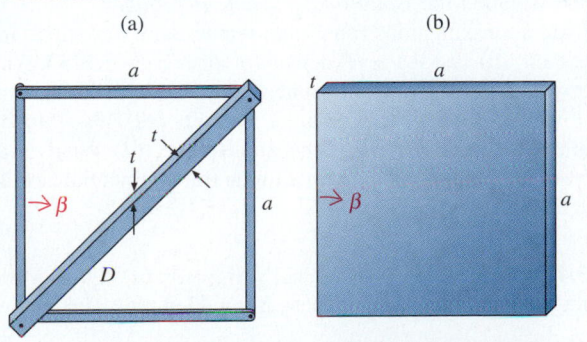

of merit F'/m' for the plate. **(d)** Compare the results of parts **b** and **c** by taking the ratio

$$\frac{F/m}{F'/m'}.$$

Show that the ratio is not very different from 1 and thus that the open structure shown in part *a* is not superior to the solid plate shown in part *b* on the sole basis of the open structure's resistance to shear. (There are other reasons why the open structure is superior as a structural element.)

15.44 *(G) New kind of oscillator.* A glass U-tube of uniform cross-sectional area A, open at both ends, contains a volume V of liquid of density ρ. Thus, the total length of the liquid column from surface to surface is $l = V/A$. The tube is briefly jiggled to set the liquid into oscillation. The figure below shows the system at an instant when the liquid surface on the

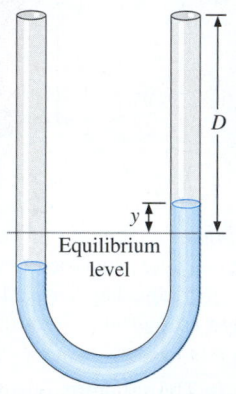

Equilibrium level

right lies a distance y from its equilibrium position. **(a)** Neglecting friction, show that the liquid experiences a restoring force of the form $F = -ky$ and that y therefore varies in simple harmonic fashion. Express the force constant k in terms of ρ. **(b)** Show that the period of oscillation τ_0 depends not on ρ but only on l and is thus independent of the particular liquid used. This problem was first proposed and solved by Newton.

15.45 *(G) The "spring of air."* One end of the tube of Problem 15.44 is now closed. The system is in equilibrium with the two surfaces at the same level and the trapped air at atmospheric pressure p_0. The distance from the equilibrium level to the closed end of the tube is D. **(a)** Show that, when the liquid column is made to oscillate, the oscillation is *not* simple harmonic because (assuming that the temperature remains constant)

$$\frac{dF}{dy} = k - \frac{p_0 AD}{(D-y)^2} \neq \text{constant}.$$

Thus, Boyle's use of the term "spring" must not be taken to imply that the trapped gas acts like a Hooke's-law spring. **(b)** Consider *small* oscillations of the liquid column, for which $y \ll D$. Using the approximation $(1-\delta)^n \simeq 1 - n\delta$, valid for $\delta \ll 1$, show that the force constant k obtained in Problem 15.44 is increased by an approximate additive constant $p_0 A/D$ and that the oscillation is therefore approximately simple harmonic. Show also that, owing to the presence of the trapped air, the period is decreased from τ_0 to

$$\tau_0\left(1 - \frac{p_0}{4\rho g D}\right).$$

GROUP C

15.46 *(7) Barometric equation for an isothermal atmosphere.* To first approximation, assume that the temperature of the atmosphere is the same everywhere and that its height is not so great that variations in the acceleration of gravity are significant. **(a)** If p_0 and ρ_0 are respectively the pressure and the density of the air at sea level, show that the pressure p varies with altitude h according to the equation

$$p = p_0 e^{-(\rho_0/p_0)gh}.$$

This is called the *isothermal barometric equation.* More realistic approximations must take temperature variations into account. **(b)** The sea-level density of air is $\rho_0 = 1.223 \text{ kg/m}^3$. At what altitude is the atmospheric pressure equal to 0.5 atm? At what altitude is the pressure equal to p_0/e? This altitude is called the *scale height* of the atmosphere. **(d)** What is the pressure at 10 km, a common altitude for commercial airplane flights?

15.47 *(4) Coulomb as a materials engineer.* **(a)** A thin-walled tube has radius r, wall thickness Δr, and length l. The material

of which the tube is made has shear modulus G. Show that the torque T required to twist the tube through an angle θ is

$$T = \frac{2\pi r^3 \, \Delta r \, G}{l}\,\theta.$$

[Hints: (1) In imagination, slit the tube along its length and open it into a flat sheet, as on p. 421. What couple, applied to the top and bottom edges of the sheet, will produce a shear strain equivalent to that experienced by the twisted tube? (2) Don't confuse θ with β.] **(b)** Beginning with the result of part **a**, find the torque required to twist a solid wire of radius R, length l, and shear modulus G through an angle θ. The fact that $T \propto R^4$ makes possible the building of torsion pendulums with a wide range of properties. **(c)** The wire of part **b** is used as the torsion fiber of a torsion pendulum with moment of inertia I. Express the shear modulus in terms of I, l, R, and the measured period τ of the pendulum. This convenient method of measuring G requires only a small sample of the material of interest. The method was Coulomb's first important application of the torsion pendulum. The analysis sketched in this problem is originally due to him.

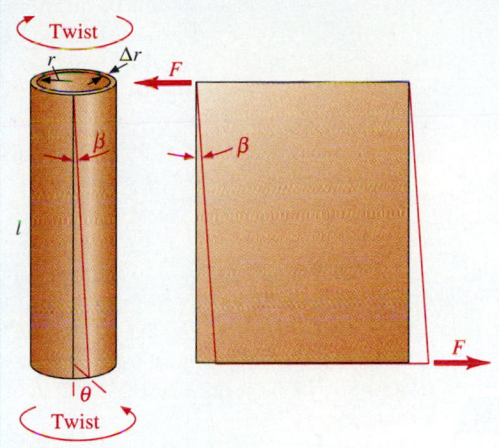

15.48 *(5) The asthenosphere.* If a solid body is subjected to large stress but is confined so that it cannot break, it will begin to flow like a syrupy fluid when the strain exceeds about 0.02. Because the pressure inside the earth increases with depth, there is a depth at which the solid rock, which we normally think of as rigid, flows readily in response to any disturbance in equilibrium. The region where this is the case is called the *asthenosphere* (from a Greek word meaning "weak"). Because its physical properties differ appreciably from those of the solid uppermost layer of the earth called the crust, the asthenosphere marks the boundary between the crust and the layer below, called the mantle. Take the density of rock near the surface of the earth to be 3×10^3 kg/m^3 and the bulk modulus to be 5×10^{10} Pa. Neglecting the variation of g with depth, estimate the depth below the surface at which the asthenosphere begins and thus estimate the thickness of the earth's crust.

15.49 *(5) Strange planet.* An interstellar explorer discovers a remarkable planet made entirely of a uniform incompressible fluid. The radius of the planet is R, and the acceleration of gravity at its surface is g. **(a)** What is the pressure at the center of the planet? [Hints: (1) Use the result of part **a** of Problem 14.64. (2) Remember the essential result of Problems 14.62 and 14.63: The acceleration of gravity $a(r)$ within the planet, at a distance r from its center, is found by considering the mass contained within the sphere of radius r.] **(b)** The mass of the earth is $M = 5.99 \times 10^{24}$ kg, and its radius (assuming it to be spherical) is $R = 6.37 \times 10^6$ m. Make the *very* crude assumption that the density of the earth is uniform throughout, and find the pressure at its center.

15.50 *(5) The shape of the earth.* Because of its rotation, the earth is not precisely spherical but bulges at the equator. Newton devised the following approach to approximating the difference ΔR between the polar and equatorial radii of the earth. Assume the earth to be homogeneous. Imagine two tunnels

dug to the center of the earth, one beginning at the north pole and the other at some point on the equator. **(a)** At any point in the polar tunnel, the acceleration of gravity is simply $a(r)$, found in Problem 15.49. But in the equatorial tunnel, the value $a(r)$ must be reduced by a term owing to the rotation of the earth in order to obtain the effective value $a'(r)$ of the acceleration. Find $a'(r)$. **(b)** Imagine that the tunnels meet at the center of the earth and that both are filled to the surface with water. Show that

$$\int_0^{R+\Delta R} [a(r) - a'(r)]\, dr = \int_R^{R'} a(r)\, dr \simeq g\,\Delta R,$$

where g is the surface acceleration of gravity at the pole, R is the polar radius, and R' is the equatorial radius. **(c)** How much longer must the equatorial tunnel be than the polar tunnel? **(d)** Evaluate the result of part **c** numerically. Compare your result with the measured difference, which is 14.4 km, or 0.226%.

15.51 *(5) Familiar planet.* Even though the material of which the earth is composed is "incompressible" under the conditions usually encountered on its surface, the pressures within the earth are so large that density varies substantially with depth. This is one important reason why the approximation in part **b** of Problem 15.49 is not a very good one. As an improved approximation, assume a linear variation of the form

$$\rho(r) = \rho_0 - \gamma r,$$

where ρ_0 is the density of the matter at the center of the earth, γ is the rate of change of density, and $\rho(r)$ is the density at a distance r from the center of the earth. In Problem 11.65, this approximation was used, together with the known values of the earth's mass M and moment of inertia I, to determine the values of the constants ρ_0 and γ. To do the present problem, you need not have done Problem 11.65, but you do need to know the results it yields; they are $\rho_0 = 14.1 \times 10^3$ kg/m^3 and $\gamma = 1.80 \times 10^{-3}$ kg/m^4. **(a)** Express the acceleration of gravity $a(r)$ in terms of ρ_0, γ, r, and the universal gravitational constant G. **(b)** Show that the pressure at the center of the earth can be written

$$p_0 = \frac{4\pi G}{3}\left(\frac{1}{2}\rho_0^2 R^2 - \frac{7}{12}\rho_0\gamma R^3 + \frac{3}{16}\gamma^2 R^4\right).$$

(c) Using the numerical values given, find p_0. Compare with the result obtained using the cruder approximation of Problem 15.49, and show that the improved approximation yields a value about twice as great. Your result should be within about 7% of the best available value, $p_0 = 3.62 \times 10^{11}$ Pa $\pm 0.5\%$. [See R. Snyder, "Two-Density Model of the Earth," *Am. J. Phys.* **54**, 511 (1986); H. Jeffreys, *The Earth*, 6th ed. (Cambridge University Press, 1976).] **(d)** Why would you expect that the value of p_0 obtained in this problem is still an underestimate?

Dynamics of Fluids

- The flow of a fluid through a pipe is measured by the volume per unit time passing through any cross section of the pipe. This quantity is called the volume flux.

- For an incompressible fluid in the steady state, the volume flux must be the same in all parts of the pipe; this is the basis for the continuity equation.

- The flow of an ideal, frictionless fluid is described by a work-energy relation called Bernoulli's equation, which relates changes in pressure, height, and speed of the fluid.

- Any real fluid exhibits internal friction called viscous drag, which resists any force that tends to make the fluid move.

- In laminar flow, the simplest form of viscous flow, we regard the fluid as consisting of a set of thin layers, each slipping with respect to its neighbors.

Left: It takes energy to make waves.

It is not possible to step twice into the same river.

—Heraclitus (c. 540–c. 480 B.C.)

SECTION 16.1

Introduction

Fluids *flow*; this is their most striking characteristic. In this chapter, we are concerned with the dynamics of fluid flow, called **hydrodynamics**. The field of hydrodynamics is extensive and varied, and there are many important cases in which the application of the basic principles can be quite complicated. But many of the fundamental ideas can be approached in a straightforward way, as we do in this chapter.

We begin by considering fluid flow under ideal circumstances—those in which there is no friction and for which we can invoke conservation of mechanical energy. In Section 16.2, we set up an avenue of approach useful for all studies of fluid flow. In particular, we set up the *continuity equation*. In Section 16.3, we derive the master equation that describes ideal fluid flow, called *Bernoulli's equation*, and consider a number of important applications. In Section 16.4, we study the phenomenon of *fluid friction*, in which the motion of one part of a body of fluid relative to another part leads to a dissipation of energy. Section 16.4 is concerned with the simplest case of fluid flow, that of *laminar flow*, which we treat in a quantitative way.

SECTION 16.2

The Steady State and the Continuity Equation

In the interest of simplicity, we begin our study of fluid flow by considering the behavior of an **ideal fluid**. Such a fluid satisfies the following conditions:

1. The fluid is *incompressible*. As you learned in Chapter 15, this is very often a good approximation for liquids. But results we obtain by using this assumption are applicable to gases only in systems where the pressure variations are small enough that the density variations from point to point are negligible. To give just one example, it is often acceptable to make this approximation in the analysis of air flow past airplane wings.

2. The fluid is **nonviscous**. By this we mean that the fluid moves past the walls of a pipe (or other solid structure with which it comes into contact) without any friction and that one part of the fluid can slide past another without friction. The only fluid for which this is strictly true is liquid helium at temperatures below about $-271°C$. But it is not a bad approximation for many liquids under conditions that are often realized.

Figure 16.1 shows a pipe whose direction and cross-sectional shape and area vary in an arbitrary way. An ideal fluid flows through the pipe in the sense shown. When we start up the flow in the system (by turning on a pump, say), the flow is *unsteady*. By "unsteady" we mean that any quantity you might measure at a given point in the system, such as fluid velocity or pressure, varies with time. But we concentrate here on the **steady state**, in which the velocity of the fluid passing any given point, such as P in Figure 16.1, remains constant. This does not mean that the velocity of a given element of the fluid remains constant as the element flows along. An element must speed up as it passes through a narrow part of the pipe, where the element has to "make room" for other elements behind it, and slow down where the pipe widens and there is more room. (Otherwise the fluid would have to be compressed to fit into the confined space and therefore could not be incompressible.) And the direction of motion of the element must change as the pipe twists and turns.

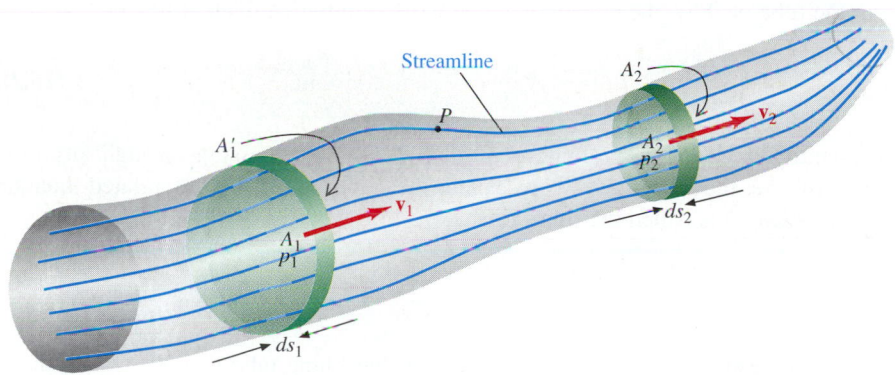

Streamline

FIGURE 16.1 A pipe of arbitrarily varying diameter and direction carries a steady-state stream of ideal fluid. Shown are a number of streamlines. At any point on a streamline, such as P, the fluid velocity is always the same. The fluid mass entering the region between A_1 and A_2 during time dt occupies the volume element $A_1 \, ds_1$ at the end of the time interval. This mass must be equal to the mass occupying the corresponding volume $A_2 \, ds_2$. The equality implies the continuity equation, $A_1 v_1 = A_2 v_2$.

In addition to focusing our attention on steady-state flow, we assume that the flow is *irrotational*. That is, there are no eddies, or swirls, in the flow; if you imagine a small solid flake carried along by the fluid, the flake does not rotate as it moves.

We represent the path of an element of fluid by a curve called a **streamline**. A number of streamlines are shown in Figure 16.1. As the element passes any point on the streamline, the direction of the velocity of the element lies along the tangent to the streamline at that point.

As the element moves along its streamline, it is followed by a stream of other similar elements. In the steady state, the streamlines are fixed, and the velocity of every element as it passes a given point on the streamline is identical with the velocity of every other element that has passed (or will pass) the same point.

A bundle of streamlines is called a **tube of flow**. All the fluid passing through a tube of flow must penetrate any cross section that cuts the tube of flow. Two such cross sections, with areas A_1 and A_2, are shown in Figure 16.1. No fluid can enter or leave a tube of flow through its sides because all fluid moves along streamlines and any streamline must be either within or without the tube of flow. In particular, all of the fluid passing through the pipe of Figure 16.1 constitutes a tube of flow. (But this large tube of flow may be divided into a number of thinner tubes of flow as well.)

We now express mathematically the following simple statement: Because matter is neither created nor destroyed in the region of the tube of flow between A_1 and A_2, the mass of the fluid within the region must remain constant in time. Therefore, if a mass m of fluid enters the region at A_1, an equal mass must leave at A_2. Speaking loosely but familiarly, *what comes in must go out*.

To express this statement quantitatively, consider what happens to the elements of fluid that lie on the cross section A_1 at a certain instant. At a time dt later, these elements have moved a distance ds_1 along their respective streamlines and lie on the surface A_1'. The thin cylinder bounded by A_1 and A_1' is filled with fluid that has crossed A_1 during the time interval dt. The mass of this fluid is

$$\rho A_1 \, ds_1,$$

where ρ is the density of the fluid. But if the mass within the region bounded by A_1 and A_2 is to remain constant, an equal mass of fluid must leave the region by crossing A_2 during the same time dt. This second mass occupies the thin cylinder bounded by A_2 and A_2', and so we have the equality

$$\rho A_2 \, ds_2 = \rho A_1 \, ds_1. \tag{16.1}$$

Next, note that $ds_1 = v_1 \, dt$, where v_1 is the speed of the fluid as it passes through A_1. For fluid passing through A_2, we have the similar relation $ds_2 = v_2 \, dt$. When we substitute these values into Equation 16.1 and cancel common factors from the two sides of the equation, we obtain the result

$$A_2 v_2 = A_1 v_1. \tag{16.2a}$$

This is the **continuity equation**. Because the cross sections can be chosen anywhere along the tube of flow, the equation can be written in the equivalent form

$$Av = A\frac{ds}{dt} = \frac{dV}{dt} = \text{constant}. \tag{16.2b}$$

The quantity dV/dt is the volume of fluid per unit time passing through any cross section of the tube of flow. It is called the **volume flux**. The related quantity $\rho \, dV/dt = dm/dt$ is called the **mass flux**.

The Concept of Flux

You will see the concept of flux applied to many physical quantities other than volume and mass, in such diverse areas as acoustics, electromagnetism, and optics. In many of these applications, we consider a flux of something more abstract than a fluid. Hence, the importance of understanding the context in the present, concrete sense extends beyond the realm of hydrodynamics. *Flux* is simply the Latin word for flow.

EXAMPLE 16.1

A bottling machine at a winery fills one 0.750-liter bottle every 0.950 s. If the diameter of the filling tube is 1.00 cm, find the speed of the wine stream.

SOLUTION: From Equation 16.2b you have

$$v = \frac{1}{A}\frac{dV}{dt}.$$

Because the radius of the filling tube is 5.00×10^{-3} m, the numerical values give

$$v = \frac{1}{\pi \times (5.00 \times 10^{-3} \text{ m})^2} \frac{0.750 \times 10^{-3} \text{ m}^3}{0.950 \text{ s}} = 10.1 \text{ m/s}.$$

Bernoulli's Equation

We now apply the generalized work-energy theorem to fluid flow and thus derive an equation that relates the height, pressure, and speed of a fluid as it moves from place to place. Figure 16.2 again shows the pipe of Figure 16.1, whose direction, shape, and

FIGURE 16.2 Diagram for deriving Bernoulli's equation. The pressure at A_1 is p_1, and the pressure at A_2 is p_2.

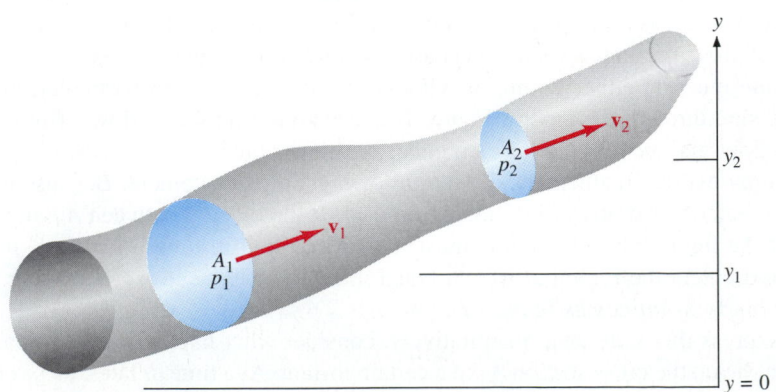

cross-sectional area vary in an arbitrary way. The pipe runs up and down with respect to a reference level $y = 0$; consequently, the fluid pressure inside the pipe varies from place to place. Because there is a pressure p_1 at A_1, the fluid immediately to the left of A_1 exerts a force $F_1 = p_1 A_1$ on the fluid at that surface. Any fluid entering the region bounded by A_1 and A_2 must push away some of the fluid already there. This pushing requires that work be done on the fluid within the region.

According to Equation 7.28, the rate at which work is done—in other words, the power—may be expressed as $P = \mathbf{F} \cdot \mathbf{v}$. (In what follows, be careful not to confuse power P with pressure p.) The fluid speed at A_1 is v_1, and the streamlines are (by definition) normal to the cross section. Thus work is done at A_1 at the rate $P_1 = \mathbf{F}_1 \cdot \mathbf{v}_1 = F_1 v_1$. When we express the force as the product of pressure and area, the power input to the fluid within the region is

$$P_1 = p_1 A_1 v_1 = p_1 \frac{dV}{dt}. \tag{16.3a}$$

That is, the power is the product of the pressure with the volume flux. We need not specify a subscript for dV/dt because this quantity is constant throughout the pipe, in accordance with the continuity equation.

Similarly, fluid leaving the bounded region at A_2 must do work in pushing the fluid just to the right of A_2 out of the way. The power output from the region is

$$-P_2 = -F_2 v_2 = -p_2 A_2 v_2 = -p_2 \frac{dV}{dt}. \tag{16.3b}$$

The minus sign expresses the fact that this power represents a flow of energy out of the region between A_1 and A_2. The *net power input* to the region is

$$P = P_1 - P_2 = (p_1 - p_2) \frac{dV}{dt}. \tag{16.4}$$

Unless the pressure is the same at A_1 and A_2, the power input expressed by Equation 16.4 is not zero.

The power input to the region between A_1 and A_2 that we have just calculated implies a flow of energy into the region. The energy flow is a result of mechanical work done either by fluid as it pushes its way into the region or on fluid as it is pushed out of the region. But this is not the only way in which the mechanical energy contained in the region changes. Fluid flowing into the region has potential energy by virtue of its height y and kinetic energy by virtue of its speed v. Consequently, the mechanical energy of the fluid is added to the mechanical energy of the region as fluid enters the region. Similarly, fluid flowing out of the region carries mechanical energy with it and thus removes mechanical energy from the region. To complete our calculation of the mechanical energy flowing into the region, we calculate the net flows of potential and kinetic energy into the region separately, beginning with the potential energy.

During a time interval dt, a mass of fluid dm enters the region between A_1 and A_2 at height y_1 and an equal mass leaves the region at height y_2. Hence the potential energy carried by the fluid into the region is $dm\, gy_1$, and the potential energy carried by the fluid out of the region is $-(dm\, gy_2)$. On the basis of these two quantities, we can write the net rate of flow of potential energy into the region, dU/dt, in the form

$$\frac{dU}{dt} = \frac{dm\, gy_1}{dt} - \frac{dm\, gy_2}{dt},$$

or

$$\frac{dU}{dt} = \rho g (y_1 - y_2) \frac{dV}{dt}. \tag{16.5}$$

For the particular region shown in Figure 16.2, dU/dt is negative because the exit is higher than the entrance; $y_1 < y_2$. That is, in this particular case, fluid flows out of the region with greater potential energy than it had when entering the region, implying a net flow of energy out of the region.

Let us now turn to the kinetic energy of the fluid. During the time interval dt, the mass of fluid dm enters the region between A_1 and A_2 with speed v_1 and an equal mass leaves the region with speed v_2. Hence the net rate of flow of kinetic energy into the region is

$$\frac{dK}{dt} = \frac{\frac{1}{2} dm\, v_1^2}{dt} - \frac{\frac{1}{2} dm\, v_2^2}{dt},$$

or

$$\frac{dK}{dt} = \frac{1}{2} \frac{dm}{dt} (v_1^2 - v_2^2).$$

Because $dm/dt = \rho\, dV/dt$, we can rewrite this equation in the form

$$\frac{dK}{dt} = \frac{1}{2}\, \rho(v_1^2 - v_2^2)\, \frac{dV}{dt}. \tag{16.6}$$

For the particular region shown in Figure 16.2, dK/dt is negative because the pipe is wider at A_1 than at A_2, and hence $v_1 < v_2$. That is, in this particular case, fluid flows out of the region with greater kinetic energy than it had when entering the region, implying a net flow of energy out of the region.

We are now ready to put things together and evaluate the total energy flow into the region between A_1 and A_2. In the steady state, the total mechanical energy of the fluid within the region must remain constant, even though fluid is continuously flowing into and out of the region. In other words, *the net rate of flow of energy into the region must be zero*:

$$P + \frac{dU}{dt} + \frac{dK}{dt} = 0.$$

Using Equations 16.4, 16.5, and 16.6, we have

$$(p_1 - p_2)\frac{dV}{dt} + \rho g(y_1 - y_2)\frac{dV}{dt} + \frac{1}{2}\rho(v_1^2 - v_2^2)\frac{dV}{dt} = 0.$$

Canceling and rearranging, we obtain

$$p_1 + \rho g y_1 + \tfrac{1}{2}\rho v_1^2 = p_2 + \rho g y_2 + \tfrac{1}{2}\rho v_2^2. \tag{16.7a}$$

Because the subscripts 1 and 2 refer to arbitrarily chosen locations in the same tube of flow, we can also write this equation in the form

$$p + \rho g y + \tfrac{1}{2}\rho v^2 = \text{constant}. \tag{16.7b}$$

A third way of expressing this result is in terms of the *changes* $\Delta p \equiv p_2 - p_1$, $\Delta y \equiv y_2 - y_1$, and $\Delta(v^2) \equiv v_2^2 - v_1^2$, where the subscripts 1 and 2 refer to any two points in the tube of flow:

$$\Delta p + \rho g\,\Delta y + \tfrac{1}{2}\rho\,\Delta(v^2) = 0. \tag{16.7c}$$

In any of these forms, Equation 16.7 is called **Bernoulli's equation**, after the Swiss physician, mathematician, and physicist Daniel Bernoulli (1700–1782), who first published it in 1738.

Although it enables us to assert that the energy of a region is constant, the steady-state condition is *not* an energy-conservation condition. For example, every bit of fluid that enters the particular region shown in Figure 16.2 leaves the region at a higher level and at a greater speed and thus with increased energy. That is why a pump is required to maintain the steady state.

Situations are often encountered in which one of the terms in Equation 16.7c is either equal to zero or negligibly small. In the remainder of this section, we consider the simplifications of Bernoulli's equation that result in the three cases $\Delta(v^2) = 0$, $\Delta p = 0$, and $\Delta y = 0$.

There is something new, and somewhat subtle, in the analysis leading to Bernoulli's equation. Up to this point in our study of mechanics, we have always been interested in the energy of a body or of a system. Here, however, we have in effect evaluated the "energy of a region." Strictly speaking, a region does not possess mechanical energy of itself; what we are really talking about here is the energy of the matter contained in the region. It is thanks to the steady-state condition that we are able to think of this energy as "belonging" to the region, even though there is a constant turnover in the matter within the region as fluid leaves at one end and is replaced by fluid entering at the other end. This viewpoint is often useful in studying flow, because we are usually interested in the flow through a system and not in the particular fluid contained in a particular region of the system at a particular time.

EXAMPLE 16.2

This example is concerned with Bernoulli's equation in the special case where the term $\Delta(v^2)$ in Equation 16.7c is equal to zero. The large vertical pipe of Figure 16.3 is kept filled by the water source shown at the top, even though water constantly flows out of the pipe through the spout at the bottom. The diameter of the pipe is uniform. Find the difference in pressure Δp between points S_2, at a depth d below the top of the pipe, and S_1, at the top of the pipe.

FIGURE 16.3

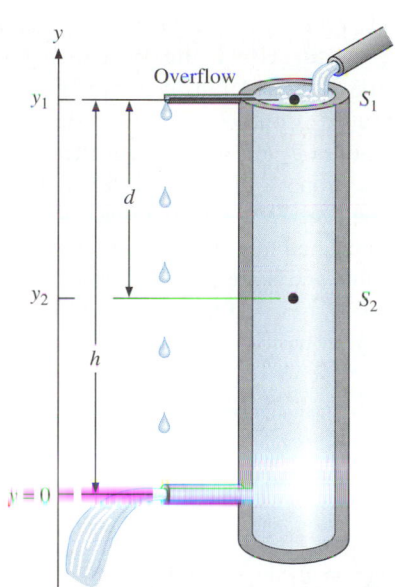

SOLUTION: You cannot simply apply Equation 15.30, $p = \rho g y$, because water is flowing through the pipe and the system is thus not in hydrostatic equilibrium. Instead, you use Bernoulli's equation in the form of Equation 16.7c. In the present case, the uniform diameter of the pipe yields an important simplification: According to the continuity equation (Equation 16.2a), the speed of the water must be the same at both points. Thus Equation 16.7c assumes the form

$$\Delta p = -\rho g \, \Delta y.$$

Next, note that $\Delta y = y_2 - y_1 = -d$; the water is flowing downward. So you have

$$\Delta p = \rho g d \quad \text{for } \Delta(v^2) = 0. \tag{16.8}$$

For uniform flow speed, Bernoulli's equation thus yields the same result as the hydrostatic equation (Equation 15.30), which applies only when $v = 0$. Equation 16.8 may be regarded as a generalization of Equation 15.30.

EXAMPLE 16.3

In this example, we consider the special case of Bernoulli's equation in which $\Delta p = 0$. Consider again the system of Figure 16.3, which we studied in Example 16.2. What is the speed of the jet of water emerging from the spout, a distance $h = 3.5$ m below the water surface?

SOLUTION: The water in the jet has emerged from the spout and is not confined. It is therefore at atmospheric pressure, as is the water at the surface, at S_1. (You need not worry about the very small atmospheric pressure difference over the height difference h.) Thus you can simplify Equation 16.7c to the form

$$\rho g \, \Delta y + \tfrac{1}{2}\rho \, \Delta(v^2) = 0.$$

On solving for $\Delta(v^2)$, you obtain

$$\Delta(v^2) = -2g \, \Delta y. \tag{16.9a}$$

In the present case, the continuity equation assures you that the speed v_1 of the water at the top of the vertical pipe is negligibly small in comparison with the speed v_2 of the water leaving the spout, because the spout is much narrower than the pipe. So you take $\Delta(v^2) = v_2{}^2 - 0 = v_2{}^2$. Also, you have $-\Delta y = -(0 - y_1) = h$. So you can solve for v_2; you have

$$v_2 = \sqrt{2gh} \quad \text{for } \Delta p = 0. \tag{16.9b}$$

For $h = 3.5$ m, the stream speed is

$$v_2 = \sqrt{2 \times 9.8 \text{ m/s} \times 3.5 \text{ m}} = 8.3 \text{ m/s}.$$

The remarkable thing about Equation 16.9b is that the speed (though not necessarily the direction) of the emerging water stream is the same as that of the water that falls past the same

level from the overflow at the top of the system. That is, the water emerging from the spout acquires the same kinetic energy as it is propelled through the pressure drop $\Delta p = \rho g h$ as it would acquire in falling freely through the distance h. This result is called **Torricelli's theorem**.

The Venturi Effect

We now consider Bernoulli's equation in the third special case, $\Delta y = 0$. This is probably the most interesting of the three special cases, and it has some surprising results.

In Figure 16.4, ideal fluid flows through a level pipe from the cross section A_1 to the smaller cross section A_2. The continuity equation assures us that the fluid speed will increase. What is the corresponding pressure change? Because $\Delta y = 0$, solving Equation 16.7c for Δp gives

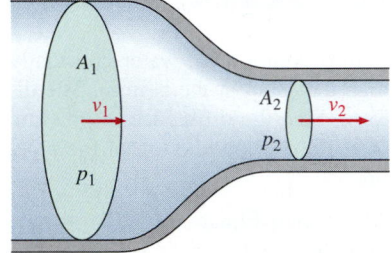

or

$$\Delta p = -\tfrac{1}{2}\rho\, \Delta(v^2), \tag{16.10a}$$

$$p_2 - p_1 = -\tfrac{1}{2}\rho(v_2{}^2 - v_1{}^2). \tag{16.10b}$$

According to either of these equations, *an increase in fluid speed implies a decrease in pressure*. This statement is called the **Venturi effect**, after the Italian physicist Giovanni Battista Venturi (1746–1822), who first discussed it in detail in 1797. The Venturi effect may seem to contradict intuition because you know that the faster a body moves, the greater the impulse it exerts on another body on impact. But there is no impact here. Any such impact would imply the exertion of a backward-directed force by the other body on the fluid. On the contrary, the acceleration of the fluid as it passes from A_1 to A_2 in Figure 16.4 implies that it is subjected to a net *forward* force in the intervening region. Hence intuition is not applicable and there is no contradiction.

We can express the pressure change in terms of the cross-sectional areas A_1 and A_2. We use Equation 16.2a, $A_1 v_1 = A_2 v_2$, to write

$$v_2{}^2 = v_1{}^2 \frac{A_1{}^2}{A_2{}^2}.$$

When we use this value of $v_2{}^2$, the quantity in parentheses on the right side of Equation 16.10b becomes

$$v_2{}^2 - v_1{}^2 = v_1{}^2\left(\frac{A_1{}^2}{A_2{}^2} - 1\right).$$

The pressure change given by Equation 16.10b can thus be written

$$p_2 - p_1 = -\frac{1}{2}\,\rho v_1{}^2\left(\frac{A_1{}^2}{A_2{}^2} - 1\right). \tag{16.11}$$

FIGURE 16.4 The Venturi effect. Fluid flows through a level pipe whose cross-sectional area changes from A_1 to A_2; there is a resulting change in the speed of the fluid from v_1 to v_2 and a corresponding pressure change from p_1 to p_2.

EXAMPLE 16.4

Water flows through the horizontal pipe shown in Figure 16.5. The diameter of the pipe is 3.00 cm at Q_1 and 1.00 cm at Q_2. The U-tube connecting Q_1 and Q_2 is partially filled with mercury; the upper parts of the U-tube are filled with water. Call the density of water ρ_w. The density of mercury is $\rho_{Hg} = 13.6\rho_w$. If the speed of the water flowing past Q_1 is $v_1 = 0.562$ m/s, what is the height difference h between the mercury levels in the two arms of the U-tube?

SOLUTION: First, you must find the pressure difference Δp between Q_1 and Q_2. With r_1 the radius at Q_1 and r_2 the radius at Q_2, you use Equation 16.11 to find the pressure difference

from Q_2 to Q_1:

$$p_1 - p_2 = \frac{1}{2}\,\rho_w v_1{}^2\left(\frac{r_1{}^4}{r_2{}^4} - 1\right).$$

Because the ratio of the radii is the same as the ratio of the corresponding diameters, you can use the latter in the numerical calculation:

$$p_1 - p_2 = \tfrac{1}{2} \times 1.00 \times 10^3 \text{ kg/m}^3$$
$$\times (0.562 \text{ m/s})^2[(3.00)^4 - 1]$$
$$= 1.26 \times 10^4 \text{ Pa}.$$

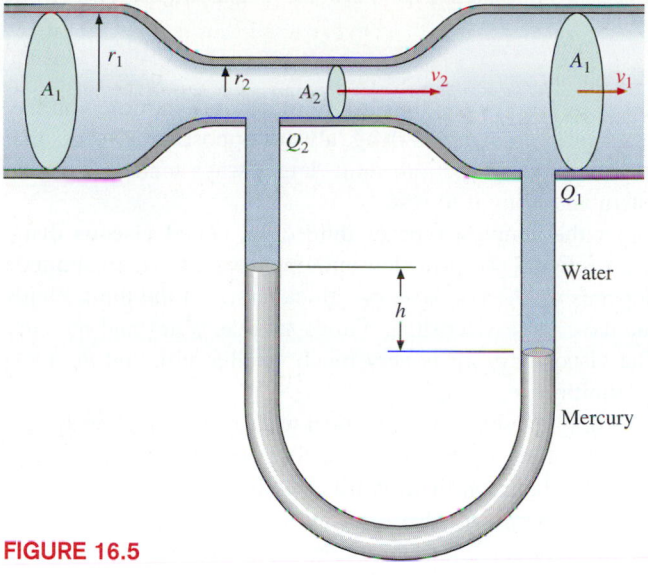

FIGURE 16.5

This pressure difference is of order 0.1 atmosphere.

Now you are ready to find the height difference h in the U-tube. The pressure difference is balanced by a rise in the mercury surface in the left branch of the U-tube and a corresponding fall in the surface in the right branch, which produces

a compensating pressure difference $\rho_{Hg}gh$. But the lowering of the mercury column on the right results in a lengthening of the water column above it, with a corresponding shortening of the water column on the left. The additional water column on the right offsets the pressure difference produced by the additional mercury column on the left by an amount $\rho_w gh$, and the total pressure difference is

$$p_1 - p_2 = \rho_{Hg}gh - \rho_w gh = (\rho_{Hg} - \rho_w)gh.$$

Solving for h, you obtain

$$h = \frac{p_1 - p_2}{g(\rho_{Hg} - \rho_w)} = \frac{p_1 - p_2}{g(13.6 - 1)\rho_w}$$

$$= \frac{1.26 \times 10^4 \text{ Pa}}{9.80 \text{ m/s}^2 \times 12.6 \times 1.00 \times 10^3 \text{ kg/m}^3} = 0.102 \text{ m}$$

$$= 10.2 \text{ cm}.$$

The threefold increase in pipe diameter from Q_2 to Q_1 implies a ninefold increase in cross-sectional area, and thus a ninefold increase in the quite modest initial speed v_1 of the water stream. The resulting shift in the mercury column is substantial.

Suppose that the U-tube opening at Q_1 were placed in the wide part of the main tube upstream of Q_2 rather than downstream of Q_2 as in Figure 16.5. Given that the fluid flow is approximately ideal, would the change make any difference in the value of h?

The Venturi effect is the basis of many parlor tricks, in which a light object is held in stable equilibrium by a blast of air that, it seems, should blow it away. Figure 16.6 shows one variation. By blowing into the top end of the hole in the spool, you can keep a small sheet of paper suspended in contact with the lower end of the spool. Why does this work? It is true that the air inside the spool is at a pressure very slightly higher than atmospheric because you are pushing the air through the spool with your lungs. But what is far more significant is the high speed of the air as it passes through the narrow horizontal space between the flat bottom end of the spool and the sheet of paper. The resulting Venturi effect leads to a pressure difference between the lower and upper surfaces of the paper, and the resulting upward force holds the paper in place.

There are much more important applications of the Venturi effect. Among them are the perfume atomizer, the automotive carburetor (Query 16.5), and various speed indicators used on boats and aircraft (Problems 16.8, 16.24, and 16.26).

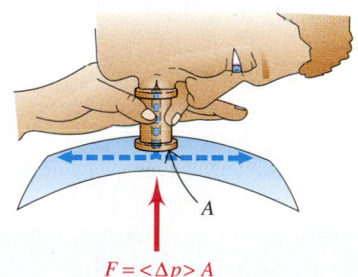

$$F = \langle \Delta p \rangle A$$

FIGURE 16.6 A parlor trick demonstrating the Venturi effect. Air passes through the small space between the paper and the bottom of the spool, of area A, at high speed. Consequently, the air pressure is reduced below p_{atm}. As a result, there is a pressure difference between the lower and upper sides of the paper. The value of the pressure difference, averaged over the bottom surface of the spool, is $\langle \Delta p \rangle$. The resulting net upward force, $F = \langle \Delta p \rangle A$, can easily exceed the weight of the paper. (A small straight pin, not shown, is usually stuck through the paper and into the hole in the spool, to keep the paper from slipping sideways.)

> **Plausible Isn't the Same as True**
> High-school science texts often attribute the lift of an airplane wing to the Venturi effect. According to the standard argument, the asymmetric shape of the wing forces air to flow faster over the upper surface than over the lower, resulting in a pressure drop above the wing and thus an upward force on the wing. You can readily see that this is incorrect: Even a flat plate produces lift when inclined with respect to the airstream; thus shape is not the basic factor underlying lift. Besides, many airplanes can fly upside down! In any case, the flow of air over a wing is turbulent and thus not irrotational, and the ideal-fluid picture developed in this chapter cannot be applied. For a correct nonmathematical account of aerodynamic lift, see Peter P. Wegener, "The Science of Flight," *American Scientist* **74**, 268 (1986). A quantitative discussion whose mathematical level you will find accessible is Klaus Weltner, "A Comparison of Explanations of the Aerodynamic Lifting Force," *Am. J. Phys.* **55**(1), 50 (1987).

Laminar Flow

When you move your hand through water and thus set the water into motion, you feel frictional resistance. This resistance, often called **drag**, always opposes any forces that tend to make a fluid move. Like sliding friction, fluid drag always tends to remove kinetic energy from the system and bring it to rest.

In this section, we consider the simplest type of fluid drag, called **viscous drag**. Viscous drag is the dominant frictional effect in slow-moving systems. The magnitude of the viscous drag force depends on the *viscosity*, or "thickness," of the fluid. Fluids such as molasses and grease have large viscosities. Fluids such as water and mercury have smaller viscosities. The viscosity of air is very much smaller still, and the ideal fluid has zero viscosity by definition.

We can understand viscous drag in a phenomenological way by studying the system shown in Figure 16.7. Two very large parallel plates, each of surface area A, are separated by a distance D. The space between them is filled with a fluid. Plate 1 is stationary; an external mechanism (not shown) drives plate 2 with constant velocity $\mathbf{v}_0$ relative to plate 1. In Figure 16.7, the direction of motion is the x direction.

If the fluid has nonzero viscosity, the moving plate drags it along. How fast does the fluid move? Measurement shows that, *if v_0 is not too large, the motion of the fluid is everywhere in the x direction*. The velocity of the fluid varies linearly with distance y from the stationary plate according to the simple rule

$$\mathbf{v}(y) = v_0 \, \frac{y}{D} \, \hat{\mathbf{x}}. \tag{16.12}$$

That is, the fluid immediately adjacent to plate 1 is essentially at rest; the fluid immediately adjacent to plate 2 moves with velocity $\mathbf{v}_0$ (and is thus at rest with respect to plate 2); and the fluid velocity varies smoothly and uniformly between the two plates, being everywhere parallel to the velocity of plate 2. In view of the smooth variation of velocity, it is useful to think of the fluid as being divided into an infinite number of parallel layers of infinitesimal thickness dy. Each of these layers is called a **lamina** (plural **laminae**). This comparatively simple type of flow, in which each lamina slips past its neighbors with relative speed dv, is called **laminar flow**.

A force $\mathbf{F}$ must be applied to plate 2, as shown in Figure 16.7, to maintain the constant velocity $\mathbf{v}_0$ in the presence of the viscous drag exerted on the plate by the fluid. At the same time, a force $-\mathbf{F}$ must be applied to plate 1 to keep it in place. The two forces constitute a couple; thus there is a shear stress $\sigma = F/A$ applied to the fluid between the plates. But a fluid can respond to a shear stress only by moving. Consequently, plate 2 moves through an ever-increasing distance $X = v_0 t$ relative to plate 1. That is, the constant shear stress σ results in a shear strain $\beta = X/D$ that increases without limit.

This situation contrasts with that of solids, in which the stress is directly proportional to the strain and the proportionality constant is the shear modulus G defined by Equation 15.20, $G \equiv \sigma/\beta$. But, in the present case of laminar flow, the stress is directly proportional to the *time derivative of the strain*. We call the proportionality constant η (Greek

FIGURE 16.7 Idealization of laminar flow. Two large parallel plates each have area A. The space between them, of width D, is filled with fluid. Under the action of the constant force F, plate 2 moves in the x direction with constant velocity $\mathbf{v}_0$ relative to plate 1. Three adjacent laminae of fluid, a, b, and c, are shown, each of width dy.

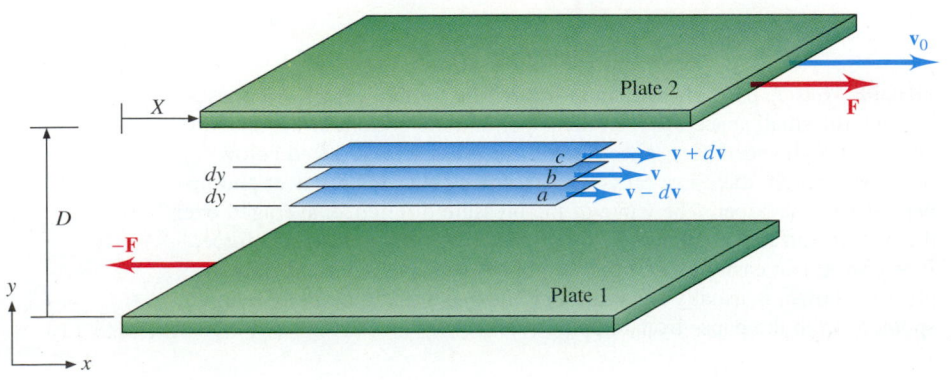

eta) and define

$$\eta \equiv \frac{\sigma}{d\beta/dt} = \frac{\sigma D}{dX/dt},$$

or

$$\eta = \frac{\sigma D}{v_0}.$$ (16.13)

The quantity η is a property of the fluid and is called the **viscosity** (or sometimes the **coefficient of viscosity**). It is the quantitative measure of the "thickness" of the fluid that we have been seeking. Because the SI unit of stress is the pascal, it follows from Equation 16.13 that the SI unit of viscosity is the pascal-second (Pa·s).* The viscosities of a variety of fluids are listed in Table 16.1.

Equation 16.13 properly describes the behavior of fluids only up to some critical value of v_0 at which the flow ceases to be laminar. The critical value of v_0 depends on the fluid, the temperature and pressure, and the shape and size of the system (for example, the shape and size of a solid body moving through a fluid). Above that value, other mechanisms of fluid drag, usually lumped under the term *turbulence*, become dominant. Turbulent flow (which we do not consider in this book) dissipates mechanical energy more effectively than laminar flow.

How is viscosity measured? One device that is often useful, especially for liquids, is the *rotating-cylinder viscometer* shown schematically in Figure 16.8. The outer cylinder, which has a bottom so as to contain the liquid, is mounted on a variable-speed turntable. An inner cylinder hangs from a torsion balance calibrated so that its scale directly reads the torque exerted on the inner cylinder. The two cylinders are coaxial. The difference between their radii, $r_2 - r_1$, is much smaller than either radius. Thus the annular (ring-shaped) space between the two cylinders is a reasonable approximation of the planar space between the two parallel plates of Figure 16.7.

The space between the cylinders is filled to a depth h with the fluid whose viscosity is to be measured. The outer cylinder is made to rotate, and it pulls the fluid along with it. As a result of the viscous drag of the fluid, the inner cylinder experiences a torque T. Application of Equation 16.13 then yields the value of the viscosity η; the details are worked out in Example 16.5.

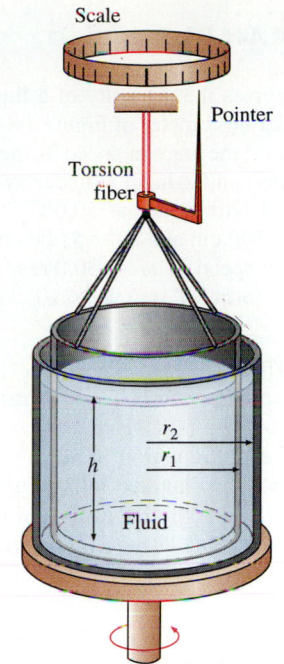

FIGURE 16.8 Rotating-cylinder viscometer.

TABLE 16.1 Viscosity of Selected Fluids at Atmospheric Pressure

	Material	Temperature (°C)	η (Pa·s)
Liquids	Acetone	20	0.327×10^{-3}
	Ethyl alcohol	20	1.200×10^{-3}
	Glycerin	0	12.11
		20	1.490
	Helium	−272	0
	Ice (under glacier conditions)	0 (approx.)	1×10^{13}
	Mercury	20	1.554×10^{-3}
	Olive oil	20	8.4×10^{-2}
	Pitch	15	1×10^{9}
	Sugar (sucrose, molten)	125	190
	Tin	300	1.73×10^{-3}
	Water	0	1.787×10^{-3}
		20	1.002×10^{-3}
		100	0.282×10^{-3}
	Wax (typical)	10	5×10^{5}
Gases	Air	0	1.708×10^{-5}
		20	1.819×10^{-5}
	Helium	−257	0.27×10^{-5}
		20	1.941×10^{-5}
	Water (steam)	100	1.255×10^{-5}

*In older literature, the non-SI unit *poise* (P) is often encountered. Conversion into SI is simple; 1 Pa·s = 10 P.

EXAMPLE 16.5

(a) Express the viscosity of a liquid contained in the rotating-cylinder viscometer of Figure 16.8 in terms of the cylinder radii r_1 and r_2, the angular speed ω, the liquid depth h, and the torque T exerted on the inner cylinder by the liquid. (b) The viscometer is filled with water at 50.0°C. The inner and outer radii are $r_1 = 4.800$ cm and $r_2 = 5.000$ cm. The depth h is 11.2 cm, the angular speed is $\omega = 30.0$ rev/min, and the torsion balance reads a torque $T = 6.96 \times 10^{-5}$ N·m. Find the viscosity of the water.

SOLUTION:

(a) Express the viscosity in terms of quantities that can be measured by using the viscometer.

To apply Equation 16.13, $\eta = \sigma D/v_0$, you must find (1) the shear stress σ applied to the fluid at the two surfaces where it contacts the cylinder walls and (2) the linear speed v_0 of the outer wall. The distance D between the plates is just the radius difference; $D = r_2 - r_1$.

You know both the torque $T = Fr_1$ exerted by the fluid on the inner cylinder and the wetted area A of the cylinder. So you have all the information you need to calculate the shear stress exerted by the fluid on the inner cylinder (or by the inner cylinder on the fluid):

$$\sigma = \frac{F}{A} = \frac{T/r_1}{2\pi r_1 h} = \frac{T}{2\pi r_1^2 h}.$$

Next, you use the known value of the angular speed of the outer cylinder to calculate the linear speed v_0:

$$v_0 = \omega r_2.$$

So Equation 16.13 becomes

$$\eta = \frac{T(r_2 - r_1)}{2\pi \omega r_1^2 r_2 h}. \tag{16.14}$$

(b) From the data given, calculate the viscosity of water.

The given angular speed, 30.0 rev/min, is equivalent to π rad/s. Inserting the other given values as well into Equation 16.14, you find that the viscosity of water at 50.0°C is

$$\eta = \frac{6.96 \times 10^{-5} \text{ N·m} \times 2.00 \times 10^{-3} \text{ m}}{2\pi \times \pi \text{ rad/s} \times (4.800 \times 10^{-2} \text{ m})^2 \times 5.00 \times 10^{-2} \text{ m} \times 11.2 \times 10^{-2} \text{ m}}$$

$$= 5.47 \times 10^{-4} \text{ Pa·s}.$$

Not surprisingly, this value lies between the values given in Table 16.1 for water at 20°C and 100°C.

A Microscopic View of Viscosity*

Why are fluids viscous? A detailed answer is quite complicated, but the following qualitative discussion is close to correct for gases. Consider two adjacent laminae, such as those labeled b and c in Figure 16.7. The velocity of the fluid in a lamina is equal to the mean velocity of the molecules contained in that lamina. Thus the mean velocity of the molecules in lamina b is $\mathbf{v}$, while the molecules in lamina c have the slightly greater mean velocity $\mathbf{v} + d\mathbf{v}$. As you will see in detail in Chapter 18, each molecule possesses a random velocity whose magnitude is usually quite a lot larger than that of the mean velocity. Consequently, molecules are continually migrating in large numbers between the two laminae. *On the average,* molecules passing from lamina c to lamina b will be moving too fast for their "new" lamina by an amount $d\mathbf{v}$ and will slow down as a result of collisions with the molecules in lamina b. The result is a transfer of momentum from faster-moving laminae to their neighboring slower-moving laminae and thus eventually to plate 1. Because the original source of this transferred momentum is plate 2, the overall result is a transfer of momentum from plate 2 to plate 1. If no external forces were applied, this momentum transfer would soon diminish to zero the speed of plate 2 with respect to plate 1. The externally applied couple required to maintain constant $\mathbf{v}$ consists of the forces required to overcome viscous drag.

The reduction in the velocity of the molecules in the direction of laminar flow is due to the fact that their rebound directions after collision tend to be random. This randomization process, discussed in detail in Chapter 18, amounts to an increase in the thermal energy of the fluid at the expense of its macroscopic kinetic energy. That is, the process is dissipative, or "frictional."

In liquids there is an additional, stronger interaction between molecules in adjacent laminae, owing to the intermolecular forces that distinguish liquids from gases. The result is qualitatively the same as that in a gas (though quantitatively greater): There is a transfer of momentum from faster-moving to slower-moving laminae, resulting in a viscous drag.

---

*This subsection may be omitted without loss of continuity.

Stokes's Law

Up to this point, our consideration of laminar flow has been restricted to the simple planar geometry of Figure 16.7. The slightly more complicated geometry of the rotating-cylinder viscometer of Figure 16.8 was attacked by assuming that the circular geometry approximated the planar case. When still more complicated situations are encountered, the mathematical analysis quickly becomes more involved. Small bodies moving slowly through fluids constitute a very important class of such situations because we often need to know the speed dependence of the force required to move a body through a given fluid.

Fortunately, it is possible to understand the essential aspects of such motion without detailed mathematical analysis. Figure 16.9 depicts the streamlines for fluid flowing at a low speed past a body of arbitrary but reasonably simple shape. (In practice, we are usually concerned with bodies moving through a fluid at rest. But, because it is only the relative velocity that counts, it does not matter which way we look at the situation.) Far away from the body, all of the fluid moves at the same velocity, called the **freestream velocity v***. As an element of fluid passes near the body, its velocity changes from the freestream value in a complicated way that depends on the location of the element. In particular, a fluid element that makes contact with the body must come essentially to rest because it becomes part of the lamina immediately adjacent to the stationary body. Fluid elements passing the body at progressively greater distances experience progressively smaller changes in velocity. The fluid thus experiences a dynamic shear, just as in the simpler situation of Figure 16.7. In Figure 16.9, which depicts low-speed flow, each region of fluid bounded by a pair of adjacent streamlines may be regarded as a lamina whose velocity is only slightly different from those of the adjacent laminae. Viscous drag between adjacent laminae tends to equalize their velocities, and all of the fluid must therefore resume the freestream velocity after it is well past the body. Once this has happened, there is no relative motion within the fluid and the viscous drag effects cease in that part of the fluid.

Equation 16.13 was derived for a special, simple geometry and cannot be applied to the present case. But the flow is laminar in the present case, and so the equation describing the force required to move the body through the fluid must be *dimensionally* consistent with Equation 16.13. To see what the desired equation must look like, we use Equation 16.13 to write the dimensional relation

$$[\eta] = [\text{stress}] \frac{[\text{length}]}{[\text{velocity}]} = \frac{[\text{force}]}{[\text{length}]^2} \frac{[\text{length}]}{[\text{velocity}]}.$$

Solving for [force], we have

$$[\text{force}] = [\eta][\text{length}][\text{velocity}]. \tag{16.15}$$

We can use this dimensional equation to write a relation between the force F required to move the body through the fluid (or to move the fluid past the body) and the quantities on the right side of the equation. To do this, we first note that Equation 16.15 tells us that the two sides are dimensionally equal but tells us nothing about quantitative equality. Thus, in substituting quantities for dimensions, we must allow for the presence of a dimensionless constant, which we will call C. Second, we deal with the following question: What specific force, length, and velocity are appropriate to the situation shown in Figure 16.9? The external force F applied to the body (the gravitational force exerted on a raindrop, say) is the only possible source of the work that must be done on the system in overcoming the fluid resistance. The velocity must be that of the body relative to distant fluid: the freestream velocity **v***.

What is the desired length? The exact answer is not obvious. But the body is not highly irregular, and the radius r shown in Figure 16.9 must be of the right order of magnitude. In any case, there are no other conceivably meaningful lengths in Figure 16.9. Any discrepancy between r and the "real" length we need can be included in the constant C. When we have done all these things, we can write the scalar equation

$$F = C\eta r v^*. \tag{16.16}$$

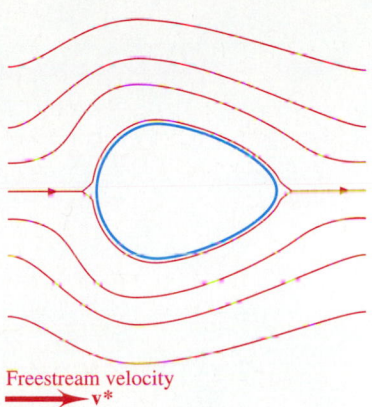

Freestream velocity
v*

FIGURE 16.9 Streamlines showing laminar flow past a solid body of arbitrary but reasonably regular shape. The fluid between any two immediately adjacent streamlines constitutes a lamina whose velocity differs only slightly from that of the adjacent lamina.

We do not know the proper value of the constant C, which will in any case depend on the specific shape of the body. But it is fair to guess that its value will not be different from 1 by many orders of magnitude.

Evaluation of C for a body of any given shape must usually be carried out by numerical analysis, often very lengthy and difficult. But there is one case for which the value of C can be found analytically: where the body is spherical. Even that calculation is quite complicated, and we will not carry it out here but merely state the result $C = 6\pi$.* We thus have

$$F = 6\pi\eta r v^* \quad \text{for laminar flow past a spherical body.} \quad \textbf{(16.17)}$$

This result is called **Stokes's law** after the British mathematical physicist Sir George C. Stokes (1819–1903), who first derived it in 1845.

THINKING LIKE A PHYSICIST

The argument used to derive Equation 16.16 is called a **dimensional argument**. Such arguments are often useful for their simplicity, their generality, and the insight they afford into complex situations. Dimensional arguments have their limitations; we have already noted, for example, the detailed analysis required to find the factor $C = 6\pi$ in Stokes's law, and a factor of 6π (roughly, 20) cannot be ignored when it comes to making quantitative calculations. Nevertheless, Equation 16.16 summarizes nearly all the physics entailed in laminar drag forces and does so in a way independent of such details as the shape of the body.

Physicists are widely reputed to be very persnickety about precision, but precision is only part of the story. An important secret of success in physics is always to make calculations neither more nor less precise than the situation requires. Insufficient precision yields results of little or no use; excessive precision can obscure physical insight. Knowing *how* precise to be in a particular situation is a skill that comes with experience.

EXAMPLE 16.6

Atmospheric water condensing in a cloud first forms tiny droplets. The cloud will produce rain—that is, water drops will fall out of the cloud into the clear space below—only if the droplets grow (by further condensation and collision) to a size at which their rate of fall is appreciable. Calculate the terminal speed of a raindrop whose diameter is 0.1 mm, and use your result to explain why rain never contains drops whose diameter is smaller than about 0.1 mm. Make the assumption that the flow of air around a falling drop of this size is laminar. Assume also that the drop is spherical, and take the temperature to be 20°C.

SOLUTION: Equations 16.16 and 16.17 show that, under conditions of laminar flow, the force exerted on a body by viscous drag is directly proportional to the speed of the body relative to the fluid. If a body of weight mg is falling through a quiet fluid, the body will accelerate until the magnitude of the drag force is equal to mg. At this point, the body has attained its **terminal speed**. Because the terminal speed is equivalent to the freestream speed of the fluid relative to the body, you can readily find the terminal speed for a spherical body by using Equation 16.17. You have

$$mg = 6\pi\eta r v^*,$$

or

$$v^* = \frac{mg}{6\pi\eta r}.$$

Calling the density of the body ρ_b, you can write the mass of the spherical raindrop in the form $m = \frac{4}{3}\pi\rho_b r^3$. So the terminal speed is

$$v^* = \frac{2\rho_b r^2 g}{9\eta}. \quad \textbf{(16.18)}$$

You can now use the value of η given for air in Table 16.1 to calculate v^*. You obtain

$$v^* = \frac{2 \times 1 \times 10^3 \text{ kg/m}^3 \times (0.05 \times 10^{-3} \text{ m})^2 \times 9.80 \text{ m/s}^2}{9 \times 1.82 \times 10^{-5} \text{ Pa·s}}$$

$$= 0.3 \text{ m/s}.$$

This is the speed of an extremely slow walk—about 1 km/h. Note in Equation 16.18 that the terminal speed depends on the square of the raindrop radius. Thus the speed of raindrops increases quite rapidly as their size increases, as long as the air flow around them remains laminar.

In fact, laminar flow ceases to apply to raindrops of diameter greater than about 0.2 mm. This is fortunate, because v^* is proportional to r^2. If a large raindrop of diameter 4 mm experienced laminar flow, its terminal speed would be $(4/0.2)^2 \times 0.3 \text{ m/s} = 120 \text{ m/s}$—fast enough to be lethal!

Drops of diameter appreciably smaller than 0.1 mm will fall so slowly that their motion will be dominated by air currents. Even if such a drop escapes from the cloud into the air below, it will evaporate long before it reaches the ground.

---

*This value certainly justifies our guess that C is not different from 1 by many orders of magnitude. We can guess further that C does not differ from 1 (or 6π) by many orders of magnitude for shapes other than spherical.

A	area	U	potential energy
D	distance between two plates, characteristic length of a system	v	speed, velocity
		v_0	relative speed of two viscometer plates
h	depth in a fluid	$\mathbf{v}^*, v^*$	freestream velocity, speed
K	kinetic energy	V	volume
p	pressure	β	shear strain
P	power	η	coefficient of viscosity
r	radius	ρ	density
t	time	ρ_b, ρ_f	density of a solid body, of a fluid
T	torque		

An **ideal fluid** is **incompressible** and **nonviscous**. As an ideal fluid flows through a system, every element of the fluid follows a smooth path called a **streamline**. A bundle of streamlines is called a **tube of flow**. Fluid can enter or leave a segment of a tube of flow only through the ends (1 and 2) of the tube. As a result, the flow is described by the **continuity equation**, Equation 16.2a or 16.2b:

$$A_2 v_2 = A_1 v_1 \quad \text{or} \quad \frac{dV}{dt} = \text{constant}.$$

The quantity dV/dt is called the **volume flux**.

The energy content of a bounded region in a tube of flow is affected in the following three ways:

1. Work must be done to push fluid into the region through end 1. This work implies a power input P_1 to the region, given by Equation 16.3a. Work must be done to push fluid out of the region through end 2. This work implies a power output $-P_2$ from the region, given by Equation 16.3b. The net power input P to the region is given by Equation 16.4.

2. Fluid entering the region has gravitational potential energy determined by the height of end 1. Energy is thus carried into the region. Fluid leaving the region has gravitational potential energy determined by the height of end 2. Energy is thus carried out of the region. The net potential energy input rate dU/dt to the region is given by Equation 16.5.

3. Fluid entering the region has kinetic energy determined by the entry speed v_1. Energy is thus carried into the region. Fluid leaving the region has kinetic energy determined by the exit speed v_2. Energy is thus carried out of the region. The net kinetic energy input rate dK/dt is given by Equation 16.6.

The steady-state condition requires that the energy content of the region remain constant. Thus the three effects just enumerated must add to zero. This condition leads to the three forms of **Bernoulli's equation** (Equation 16.7):

$$p_1 + \rho g y_1 + \tfrac{1}{2}\rho v_1{}^2 = p_2 + \rho g y_2 + \tfrac{1}{2}\rho v_2{}^2,$$
$$p + \rho g y + \tfrac{1}{2}\rho v^2 = \text{constant},$$

and $\Delta p + \rho g \,\Delta y + \tfrac{1}{2}\rho \,\Delta(v^2) = 0.$

Bernoulli's equation has three simple special cases:

1. Equation 16.8 holds for any two points in a system where v is the same:
$$\Delta p = \rho g d.$$

2. Equation 16.9a holds for any two points in a system where p is the same. In particular, if $v_1 = 0$ and $\Delta y = h$, the equation simplifies still further to Equation 16.9b, **Torricelli's theorem**,
$$v_2 = \sqrt{2gh}.$$

3. Equation 16.10b holds for any two points in a system at the same level:
$$p_2 - p_1 = -\tfrac{1}{2}\rho(v_2{}^2 - v_1{}^2).$$

This expresses the **Venturi effect**.

Under conditions of **laminar flow**, the shear stress experienced by the fluid is directly proportional to the time derivative of the strain. The proportionality constant is called the **viscosity**, defined by Equation 16.13:

$$\eta \equiv \frac{\sigma D}{v_0}.$$

When a spherical body is moved through a fluid in such a way that the flow past the body is laminar, the viscous drag force is given by Equation 16.17, **Stokes's law**:

$$F = 6\pi\eta r v^*,$$

where v^* is the **freestream speed**.

KEY TERMS

Section 16.2 The Steady State and the Continuity Equation
ideal fluid, incompressibility, nonviscosity ▪ steady state ▪ streamline, tube of flow ▪ continuity equation, volume flux, mass flux

Section 16.3 Bernoulli's Equation
power input to a region, energy of a region ▪ Bernoulli's equation ▪ Torricelli's theorem ▪ Venturi effect

Section 16.4 Laminar Flow
viscous drag ▪ lamina, laminar flow ▪ viscosity ▪ freestream velocity ▪ Stokes's law ▪ terminal speed

QUERIES

16.1 *(2) Weekend chores.* You can use the nozzle of a garden hose to vary the cross-sectional area of the opening. When you reduce the area, the speed of the emerging water increases. How do you account for this? As the speed increases, does the volume flux increase as well?

16.2 *(2) Curves ahead.* An ideal fluid flows through a pipe, shown below, whose cross section is a narrow rectangle. The bend is circular. A few streamlines are shown in the straight parts of the pipe, where the streamlines may be regarded as uniformly distributed. What is the density of streamlines at *a* compared with the density at *b*? At which of the two points is the flow speed greater? At which is the pressure greater?

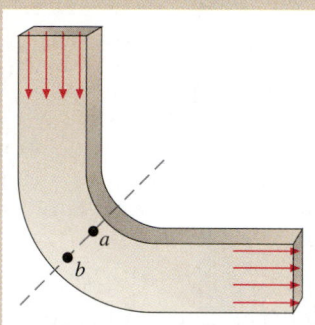

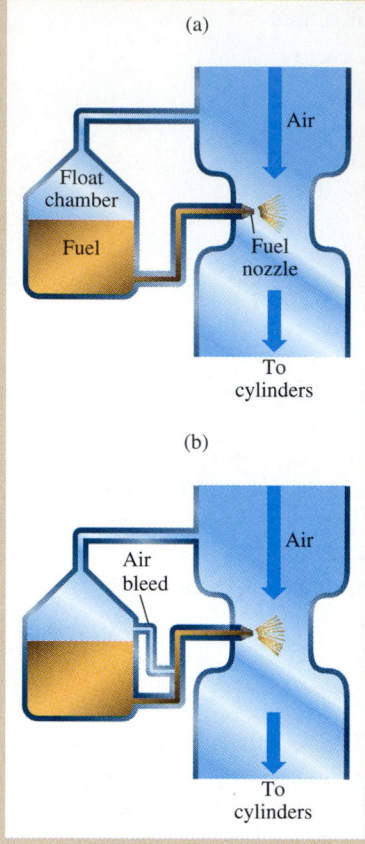

16.3 *(2) Open up your Golden Gate . . .* The entrance to San Francisco Bay is roughly 1.5 km wide and roughly 100 m deep at its deepest point. The tidal current reaches a maximum speed of about 2 m/s. Using these values and making plausible assumptions, estimate the amount of water that flows out of the bay during one ebb cycle (the period of approximately 6 hours beginning midway between high tide and low tide and ending midway between low tide and the next high tide).

16.4 *(3) Topsy-turvy.* Suppose the steady-state flow through the pipe segment of Figure 16.2 were downward instead of upward. Would energy be conserved?

16.5 *(3) Carburetor.* In many automobile engines, fuel is sprayed into the incoming stream of air in the carburetor. It is very important to keep the fuel-to-air mass ratio constant (typically, about 1:15) over a wide range of operating conditions. In particular, this constancy must be maintained as the speed of the engine varies and the volume flux of incoming air varies in direct proportion. **(a)** A primitive carburetor is sketched in part *a* of the figure opposite. Fuel is fed to the nozzle from a small constant-level reservoir called the float chamber. Explain the operation of the carburetor. Assuming that the rate of fuel flow is directly proportional to the pressure difference that drives it, how does the fuel flow rate vary with engine speed? **(b)** A more realistic arrangement is shown in part *b*. A tube called the *air bleed* is added to the system shown in part *a*. How does the fuel flow rate vary with engine speed in this arrangement? Discuss how, by proper adjustment of such physical features as the tube diameters, the fuel flow rate can be made to vary approximately linearly with engine speed.

16.6 *(4) Stoking the fires of industry.* As it comes out of the cow, milk consists of a suspension of small globules of a fatty fluid called cream in a denser, nonfatty fluid called skim milk. If left alone for an hour or so, the milk will separate, the cream rising to the top. In dairies, the process is greatly speeded up by putting the milk into a rapidly spinning container called a cream separator. Explain how the separator works.

16.7 *(G) Confluence.* Consider two rivers that flow together in fairly flat country, as the Ohio and the Mississippi do, or the White Nile and the Blue Nile. The river downstream from the confluence can accommodate the flow of both tributaries by being wider than either tributary, by being deeper than either, by flowing faster than either, or by any combination of the three. Which of the three possibilities is likely to be the dominant one? Explain.

16.8 *(G) Listen carefully!* Turn on a hot-water faucet that has not been used for a while so that the initial flow is of cold water. Let the water run until it becomes hot, and listen to the sound made by the water striking the sink. How do you account for the change in the sound as the water warms up?

16.9 *(G) Raindrops falling on my head.* The raindrops that fall out of a cloud vary in size. Give two reasons why there is a smaller range in the sizes of the drops that reach the ground.

PROBLEMS

GROUP A

16.1 (2) *Making soda pop.* Carbon dioxide gas of density 7.5 kg/m³ flows through a pipe of diameter 2.0 cm. During a time interval of 20 min, 0.55 kg of gas flows out of the pipe. Find the flow speed of the gas.

16.2 (2) *Dishwashing diversion.* A vertical stream of water issues from the end of a kitchen faucet of cross-sectional area A_0. The speed of the stream at the faucet is v_0. The stream descends some distance H before it breaks up into drops. Show that, at a distance $h < H$ below the faucet, the cross-sectional area A of the stream is

$$A = \frac{A_0 v_0}{\sqrt{v_0^2 + 2gh}}.$$

16.3 (3) *Keeping it full, I.* A tank has a hole in its bottom. If you must pour water into the tank at a rate 0.36 liter per second to maintain the depth of the water in the tank at a constant 9.6 cm, what is the area of the hole? (Remember that 1 liter = 1×10^{-3} m³.)

16.4 (3) *Keeping it full, II.* A container of height 85 cm has a circular hole of radius 1.2 mm in the bottom. At what rate must you pour water into the container to keep it full?

16.5 (3) *Downpour.* An open-topped cylindrical tank of diameter 1.20 m has a hole of diameter 1.0 cm in its bottom. The tank is initially filled with water and then allowed to drain. How fast is the water level dropping when the depth is 0.32 m?

16.6 (3) *Watering the lawn.* A stream of water flows out of a garden hose with speed 25 m/s. What is the pressure inside the hose, just upstream from the nozzle?

16.7 (3) *Instant fountain.* The bent pipe shown below is closed at the upper end except for a small hole. The hole is flush with the water surface. The horizontal part of the pipe lies just below the surface of a stream whose flow speed is 2.4 m/s. **(a)** How fast does the water in the jet emerge from the small hole? **(b)** How high does the jet rise? **(c)** Compare the speed of the jet when it returns to the stream surface with the flow speed of the stream. Neglect friction.

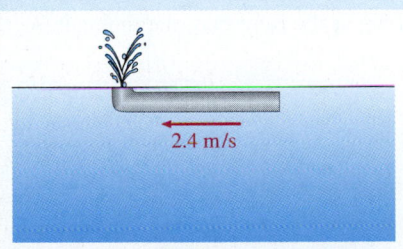

2.4 m/s

16.8 (3) *Crude measurement.* In the system shown at the top of the next column, the liquid-level difference h between the two vertical tubes is 10.0 cm. What is the flow speed of the liquid in the main tube? (The bent tube is called a *Pitot tube*; see Problem 16.26.)

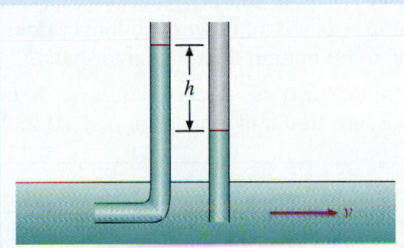

16.9 (3) *On the level, I.* The large tank illustrated below is filled to a depth h with an ideal fluid. A horizontal pipe leads from the bottom of the tank to a drain. A vertical standpipe is attached to the horizontal pipe. **(a)** If the drain opening is plugged, what is the height of the fluid column in the standpipe? **(b)** Now the drain opening is unplugged, and the fluid starts to drain. What is the new height of the fluid in the standpipe?

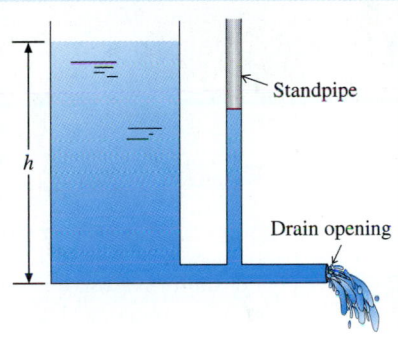

Standpipe

Drain opening

16.10 (3) *On the level, II.* The system illustrated below is similar to that of Problem 16.9, except that here the open end of the drain is constricted so that the cross-sectional area of the tube at A is one-half the cross-sectional area at B. When the ideal fluid from the tank is draining, what is the height of the fluid in the standpipe?

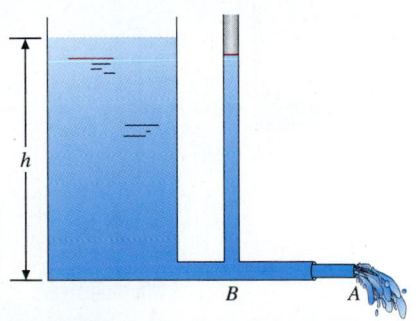

B   A

16.11 (4) *Terminal speed, I.* You drop a small copper ball of density 8.92×10^3 kg/m³ and diameter 1.00 mm into a container of glycerin whose temperature is 20°C. Find the terminal speed of the ball.

16.12 *(4) Terminal speed, II.* You repeat the experiment of Problem 16.11, this time using a mixture of copper balls of diameters 1.00 mm and 3.00 mm. The balls reach their terminal speeds over quite short descent distances. If the depth of the glycerin is 0.570 m, how much longer does it take the smaller balls to hit bottom than the larger balls?

16.13 *(4) Another way to measure viscosity.* A tall cylinder contains a viscous liquid of density $\rho_f = 1.20 \times 10^3 \text{ kg/m}^3$.

A steel ball bearing of diameter 2.00 mm is dropped into the liquid and quickly reaches its terminal speed. There are two marks on the cylinder, 20.0 cm apart. The ball takes 3.52 s to fall from the upper mark to the lower one. Assuming the flow around the ball bearing to be laminar, find the viscosity of the liquid. The density of the steel is $7.82 \times 10^3 \text{ kg/m}^3$. (Hint: Don't forget to use Archimedes' principle to account for buoyancy.)

GROUP B

16.14 *(3) Primitive lawn sprinkler.* Diagrammed below is a top view of a cylindrical can mounted on a turntable. The can is filled with water. At a depth h below the water surface are two horizontal tubes of length l and cross-sectional area A, with right-angle bends at their ends. Show that, as the water jets emerge from the tubes, there is a torque T exerted on the system given by the expression

$$T = 4\rho g h (r + l) A,$$

where ρ is the density of the water.

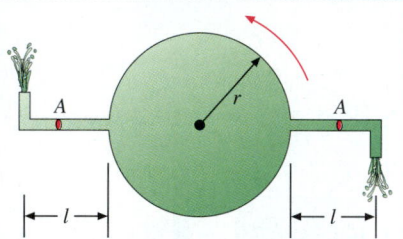

16.15 *(3) Siphon.* A pipe of uniform cross section is used to siphon an ideal fluid of density ρ from one tank to another, as shown below. The lower end of the pipe is above the liquid level in the receiving tank. If the atmospheric pressure is p_a, find **(a)** the speed of the fluid as it passes through the pipe; **(b)** the pressure of the fluid at points 1, 2, 3, 4, and 5. (The value of p_a is essentially the same at both ends of the pipe.)

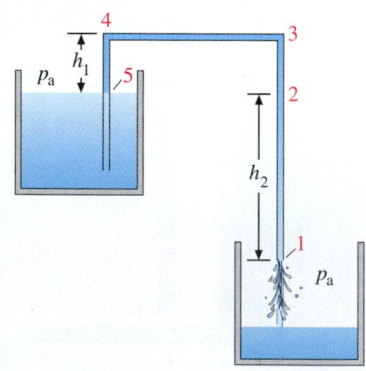

16.16 *(3) Playing with water, I.* You take a large cylindrical can and drill a vertical series of small holes in its side at various distances above its bottom. You place the can on a level sidewalk and fill it with water to a level H above the sidewalk.

As water flows out of the can through the holes, you replenish the water supply so that the water level remains at H. **(a)** Consider two holes, one a distance h below the water surface and the other an equal distance h above the sidewalk. Show that the jets of water emerging from the two holes hit the sidewalk at the same point. **(b)** For what value of h will the range R of the jet—the horizontal distance from the hole to the point where the water hits the sidewalk—be a maximum?

16.17 *(3) Delivery rate.* A water tank, open on top and initially full, drains through a hole of area A in its bottom. Express the rate of flow out of the tank—the volume flux dV/dt—in terms of the (varying) depth h of water in the tank.

16.18 *(3) Bathtub drain, I.* Assume to first approximation that a bathtub has vertical walls and a flat bottom, so the horizontal cross-sectional area A_1 is the same at all levels. At the bottom of the tub is a drain whose cross-sectional area is A_2. The tub is initially filled with water to a depth H. **(a)** Assuming the water to be an ideal fluid, show that the time required to empty the tub is

$$t = \frac{A_1}{A_2} \sqrt{\frac{2H}{g}}.$$

(b) Show that it takes about 2.4 times as long for the last half of the water to drain as for the first half.

16.19 *(3) Bathtub drain, II.* If the bathtub of Problem 16.18 is kept full as water flows out of it, what time t' is required for a volume of water equal to the capacity of the tub to flow out through the drain? Express your result in terms of t, the time found in part **a** of Problem 16.18.

16.20 *(3) Solder joint.* In the drawing below, a conical tube is soldered into the side of a tank. At the left end of the tube, its radius is A_1; at the right end, the radius is A_2. Show that,

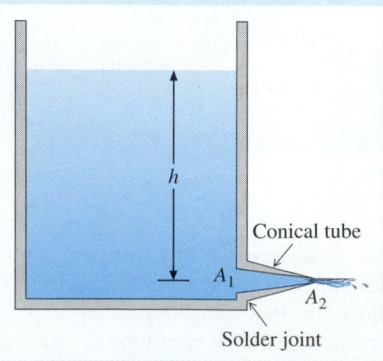

if the depth of the water in the tank is h, the solder joint must be able to withstand a force

$$F = \rho g h \frac{(A_1 - A_2)^2}{A_1}$$

if it is not to fail.

16.21 *(3) Syringe.* The cylinder below initially contains a volume V of ideal fluid of density ρ. The small orifice at the end of the cylinder has cross-sectional area A. If you exert a constant force on the plunger, how much work must you do to empty the cylinder in time t?

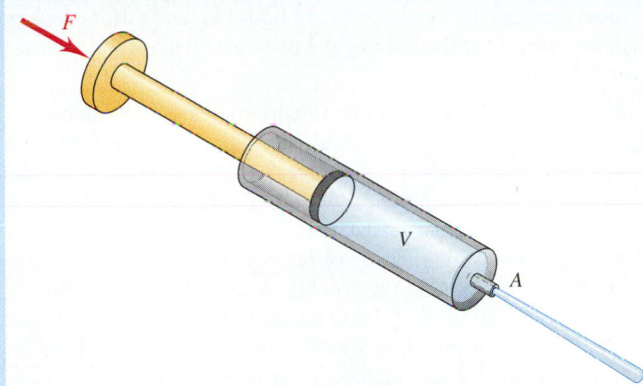

16.22 *(3) Mariotte bottle.* The simple device shown below was invented by the French physicist Abbé Edme Mariotte (d. 1684). As water flows out of the outlet at O, air bubbles into the bottle through the inlet at I and replaces the water. The bottle has the remarkable and useful property that the flow rate at O remains constant as the water level inside the bottle falls, as long as the water level remains above I. **(a)** What is the pressure p_I at I? **(b)** Find the water flow speed v at the outlet. **(c)** Find the pressure of the air above the water surface inside the bottle as a function of the (varying) depth h of I below the water surface.

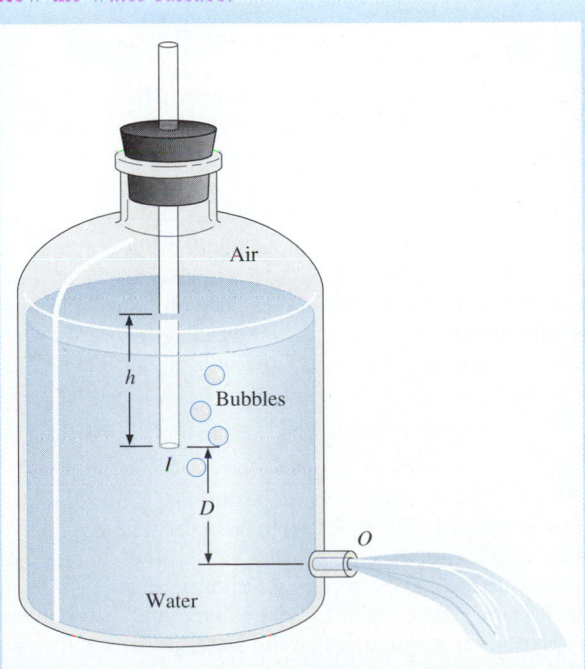

16.23 *(3) Vena contracta.* Sketched below is a thin-walled tank having a small circular hole of area A_1 in its side. As you can see from the sketch, the cross-sectional area of the water jet emerging from the hole does not have constant area A_1 because this would require the streamlines to have sharp kinks in them where they pass through the hole. Instead, the cross-sectional area of the jet diminishes for some distance after the jet leaves the hole. The area attains a minimum value A_2 where the streamlines are essentially parallel to one another. This location is called the *vena contracta* (Latin for "narrow conduit"). A detailed analysis is very difficult, even for the simplest case of a circular hole in a thin plate. But, for this case, the result is $A_2 = \pi/(2 + \pi) \approx 0.611 A_1$. If the hole is circular with radius 1.20 cm and the depth of the hole below the water surface is 2.5 m, what is the volume flux dV/dt through the hole?

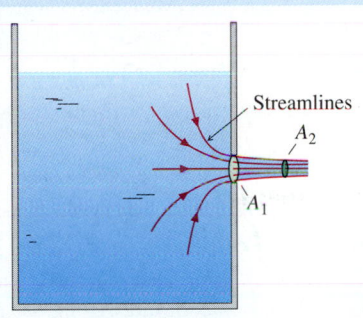

16.24 *(3) Venturi flowmeter.* The device shown below can be used as a flowmeter. The horizontal main pipe has a cross-sectional area A_1 except at the constriction, where the area is A_2. When an ideal fluid flows through the main pipe, there is a difference h in the fluid levels in the two small vertical standpipes. **(a)** Show that the flow speed v_1 at any point in the main pipe except at the constriction is

$$v_1 = \frac{\sqrt{2gh}}{\sqrt{\dfrac{A_1^2}{A_2^2} - 1}} = \frac{A_2}{\sqrt{A_1^2 - A_2^2}} \sqrt{2gh}.$$

(b) Show that the volume flux in the main pipe is

$$\frac{dV}{dt} = \frac{A_1 A_2}{\sqrt{A_1^2 - A_2^2}} \sqrt{2gh}.$$

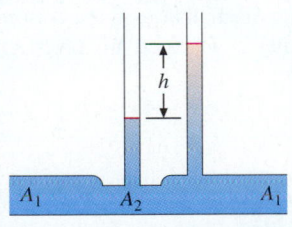

16.25 *(3) Pressure and potential energy.* A volume dV of fluid is subjected to a pressure p. Show explicitly that the product $p \, dV$ represents a potential energy dU. (Hint: Suppose that the volume lies at depth h. If you imagine that it was pushed from the surface through the surrounding fluid to that position, how much work was required?)

16.26 *(3) Pitot-Prandtl tube.* The device illustrated below is called the *Pitot-Prandtl tube*, after the French hydraulic engineer Henri Pitot (1695–1771) and the German aerodynamic engineer Ludwig Prandtl (1875–1953). It is used as a speed indicator for boats and, with some modification, as an airspeed indicator for aircraft. The freestream speed of the fluid past the device is v^*. The pressure gauge measures the pressure difference Δp between points a and b. If the outer casing of the device is properly shaped, the speed v_a of the fluid past point a is essentially equal to v^*. At point b, called a *stagnation point*, the speed of the fluid is essentially zero. Find the pressure difference Δp. Express the freestream speed v^* in terms of Δp and the density ρ_f of the fluid. (See Problem 16.8.)

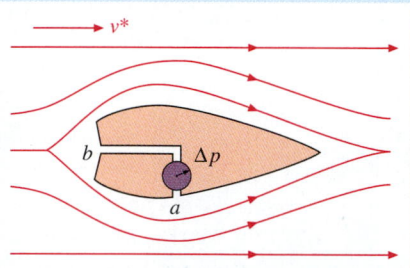

16.27 *(4) Bernoulli's equation with viscosity.* Pictured below is a tank filled with a viscous liquid. Along the drain pipe, at

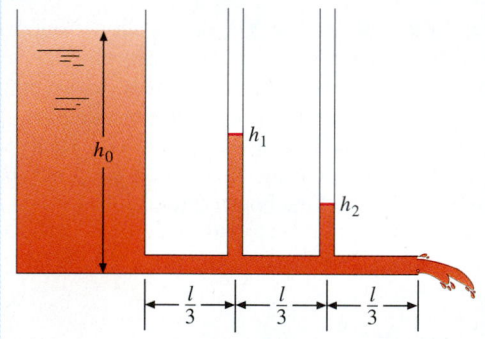

distances one-third and two-thirds of its length, are standpipes. The depth of the liquid in the tank is $h_0 = 0.70$ m. The liquid levels in the two standpipes are $h_1 = 0.40$ m and $h_2 = 0.20$ m. If the density of the liquid is 0.85×10^3 kg/m³, find the flow speed of the liquid in the drain pipe. (Hint: See Problem 16.9.)

16.28 *(4) Fast and easy.* A simple commercial viscometer frequently used when great precision is not needed consists of a glass tube into which is poured the liquid whose viscosity is to be measured. The tube is not quite filled, so a small air bubble remains in the tube when it is closed. The tube is inverted, and the bubble rises. A stopwatch is used to time the passage of the bubble between two marks on the side of the tube. Express the viscosity of the liquid in terms of its density ρ, the radius r of the bubble, the distance l between the marks, and the time t.

16.29 *(4) Disk brake?* In the illustration below, a thin disk of radius R is located in an oil-filled container in the form of a squat cylinder. The clearance between the disk faces and the upper and lower container walls is D. The disk is driven at angular speed ω by a shaft that passes into the container through an oil-tight bearing of negligible friction. The viscosity of the oil is η, and the conditions are such that the flow is laminar. **(a)** Show that, ignoring end effects due to the fluid between the periphery of the disk and the vertical walls of the container, the torque required to drive the disk is

$$T = \frac{\pi \eta \omega R^4}{D}.$$

(b) Let $R = 10$ cm, $\omega = 600$ rev/min, $\eta = 8.5 \times 10^{-2}$ Pa·s, and $D = 1.0$ mm. Find the power required to drive the disk.

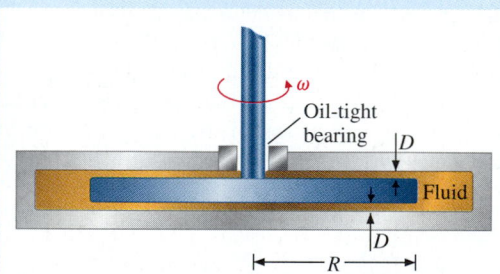

GROUP C

16.30 *(3) Spin dry.* Shown below is a test tube filled with water and closed with a loose-fitting stopper that allows air to pass freely. There is a small hole in the bottom of the test tube, at a. The test tube is fastened to a vertical shaft that ro-

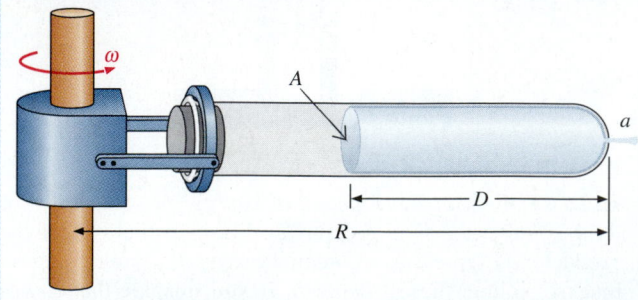

tates with angular speed ω. The distance from the center of the shaft to the small hole is R. As the tube spins, water sprays out of the hole. Show that the speed with which the water flows through the hole depends on the length D of the water column remaining in the test tube according to the relation

$$v = \omega D \sqrt{\frac{2R}{D} - 1}.$$

16.31 *(3) Playing with water, II.* Consider again the system of Problem 16.16. This time, you have a small nozzle that you can fit tightly into any hole in the side of the can. You are free to adjust the angle θ between the nozzle and the horizontal at will. Define the range R to be the horizontal distance from the spout to the point at which the water jet strikes the sidewalk. **(a)** Show that

$$(1 + \tan^2 \theta)R^2 = \frac{2v^2}{g}(R \tan \theta + h), \tag{1}$$

where $v = \sqrt{2g(H - h)}$ is the speed at which the jet emerges from the spout. **(b)** For a given value of h, there is some value of θ—call it θ'—that gives the maximum range R'. Show that this value is given by

$$\tan \theta' = \frac{v^2}{gR'} = \frac{2(H - h)}{R'}. \tag{2}$$

(Hint: You could obtain this result by solving Equation 1 for R, evaluating $dR/d\theta$, setting the result equal to zero, and solving for $\tan \theta$. But this will quickly tangle you up in a messy calculation. A better way is as follows. First, take $d/d\theta$ of both sides of Equation 1, remembering that R is itself a function of θ. Second, in the resulting equation, set $dR/d\theta = 0$. Third, remembering that the value of R corresponding to this condition is $R = R'$, evaluate $\tan \theta'$.) **(c)** Using Equations 1 and 2, show that, for this optimum angle θ', the range has the value

$$R' = 2\sqrt{H(H - h)}. \tag{3}$$

(d) Finally, use Equations 2 and 3 to evaluate the optimum angle explicitly, and show that

$$\tan \theta' = \sqrt{1 - \frac{h}{H}}. \tag{4}$$

(e) For what values of h and θ will R have its maximum value? What is that value? Compare your result with that for a thrown ball. [Note: Implicit in the derivation of Equations 2, 3, and 4 is the assumption that the cross-sectional area of the can is very much greater than that of the nozzle. But the results do not depend on that assumption. For details, see I. R. Lapidus, *Am. J. Phys.* **53** (1985), pp. 1141 and 1182.]

16.32 *(3) Radial flow.* The liquid-filled tank illustrated below has a drain of radius R_1 at the bottom. The drain is covered with a large circular plate of radius R_2, with $R_2 \gg R_1$. The space between the plate and the bottom of the tank is quite narrow, so the drain does not fill with liquid. Show that the pressure in the space between the plate and the bottom of the tank is

$$p = p_{\text{atm}} + \rho g h \left(1 - \frac{R_1^2}{r^2}\right),$$

where r is the distance from the axis of the drain ($R_1 < r < R_2$), h is the depth of the liquid in the tank, and ρ the density of the liquid.

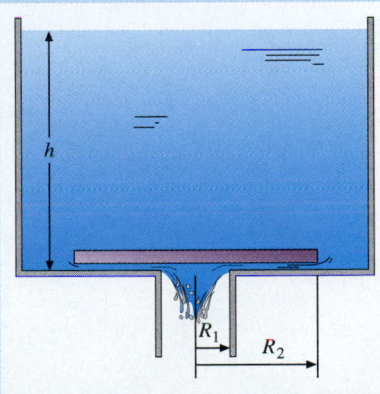

16.33 *(4) Poiseuille's law.* A fluid of viscosity η is driven through a cylindrical pipe of radius R and length l by a pressure difference Δp between the pipe ends. The flow is laminar. The geometry dictates that the laminae are coaxial cylindrical shells. A typical shell has radius r, thickness dr, and speed v. The outermost shell, in contact with the inner surface of the pipe, is essentially at rest. The fluid speed v increases with decreasing r; you could determine v as a function of r if you knew dv/dr. To find dv/dr, refer to Equation 16.13, which applies to the planar geometry of Figure 16.7. You can immediately rearrange that equation into the form $\sigma = \eta v/D$. In the planar case, the rate of change of fluid speed with distance from the stationary plate is linear; $dv/dy = v/D$. So you can just as well write

$$\sigma = \eta \frac{dv}{dy}.$$

In the present case of cylindrical geometry, v does *not* vary linearly with r. But you can still write the corresponding relation in the derivative form $\sigma = \eta(dv/dr)$. Now consider a cylindrical "plug" of fluid of radius $r < R$, flowing through the pipe. This plug is driven through the pipe by the net force exerted on its ends by the pressure difference Δp. The driving force is

$$F_{\text{driving}} = \Delta p(\text{end area}) = \pi r^2 \Delta p.$$

In the steady state, the net force on the plug must be zero. So the driving force must be exactly balanced by the viscous drag force exerted on the plug by the slower-moving fluid immediately outside it. The magnitude of the drag force is given by the product of the stress σ and the surface area of the plug:

$$F_{\text{drag}} = \sigma(\text{surface area}) = \sigma 2\pi r l.$$

(a) Show that the desired quantity dv/dr can be written

$$\frac{dv}{dr} = -\frac{\Delta p}{2\eta l} r. \tag{1}$$

(b) By integrating Equation 1 from R to r, determine v as a function of r. **(c)** Show that the volume flux through the pipe is

$$\frac{dV}{dt} = \frac{\pi}{8\eta} \frac{\Delta p}{l} R^4. \tag{2}$$

It is no surprise that dV/dt depends linearly on the pressure difference per unit length, $\Delta p/l$ (as long as the flow remains laminar). But the R^4 dependence may surprise you. Equation 2 is called *Poiseuille's law*, after the French physiologist Jean Louis Marie Poiseuille (1799–1869), who first found the relation empirically in 1841. (Very roughly, the name is pronounced Pwa-**zuh**'-yeh.) **(d)** The flow of blood in capillaries (small vessels) is essentially laminar. In many animals (including human beings), involuntary muscles can make small variations in the diameter of the capillaries near the surface of the body. Explain the utility of this function.

16.34 *(4) Glacial speed.* A certain glacier consists of a mass of ice 100 m thick, lying on a mountainside inclined at an angle of 25° with the horizontal. Estimate the flow speed of the glacier. [Hint: The shear stress is greatest near the bottom of the glacier. So to first approximation you can assume (1) that nearly all the flow takes place in the bottom (say) 10% of the glacier and (2) that the stress is uniform in that layer. (This is only one of several mechanisms responsible for the flow of glaciers.)]

Temperature and Heat Flow

— Temperature is measured by using a thermometer and an established temperature scale.

— The constant-pressure gas thermometer depends on the observation that, over a wide range of temperature and pressure, most gases closely approximate an ideal gas, whose volume depends on temperature in a simple way.

— The gas thermometer serves as the basis for the kelvin temperature scale.

— Heat flow is a form of energy transfer.

— Heat flow into a body is often manifested in a temperature increase; the heat flow per unit temperature change is called the heat capacity of the body.

— When two bodies are in thermal contact, heat always flows from the warmer one to the cooler one until they achieve equal temperature at thermal equilibrium.

— Heat flow into or out of a body can also result in a phase change, during which the temperature remains constant.

Left: There is a connection between mechanical energy and heat energy.

I hunted out the secret spring of fire, . . . which when revealed became the teacher of each craft to men, a great resource.

—AESCHYLUS, *Prometheus Bound*

Having still fresh in my memory the accidents I had so often met with in eating hot apple-pies, I was very impatient . . . to see if apples . . . really possess a greater power of retaining Heat than [water].

—BENJAMIN THOMPSON, COUNT RUMFORD

Introduction

This is the first of four chapters dealing with *temperature* and *heat*. These two concepts are so closely related to each other that the distinction between them was not clearly understood until the eighteenth century. The earlier state of confusion is still evident in our language today. We ''heat'' an object to increase its temperature, and we say the object is ''hot'' when we mean that its temperature is high.

The study of **thermal phenomena**—phenomena having to do with temperature changes and heat exchanges—has been regarded as a part of physics for centuries. But what have temperature and heat to do with the central subject of physics, which is mechanics? In particular, what do they have to do with the mechanics of many-body systems, the overall subject of Part VI? The answer to this question will begin to emerge in Section 17.7, and the next three chapters deal largely with the important consequences of the answer. As we have done with other topics, we begin with a phenomenological survey of the thermal properties of matter. Later, we use the phenomenological picture that emerges as a basis for a mechanical theory of thermal phenomena.

Sections 17.2 and 17.3 are concerned with *thermometry*, the measurement of temperature. The most familiar kind of thermometer depends on the phenomenon of *thermal expansion*—the dimensional change of a body that results from change in its temperature. We take thermal expansion for granted in Section 17.2, but then in Sections 17.3 and 17.4 we lift ourselves by our bootstraps and consider thermal expansion in more detail. In Section 17.5, we define *heat flow* in a phenomenological way and devise methods of measuring heat flow from one body to another. Section 17.6 deals with the well-known heat-related phenomena called *phase changes*, such as melting and boiling. Finally, in Section 17.7, we quantify the familiar phenomenon of frictional heating and see how this quantification leads toward the connection between thermal physics and mechanics.

Thermometers and Thermometry

People have been measuring temperature in a rough way since time immemorial. A mother feels her child's forehead for the fever that can mean illness. A cook throws a few drops of water into his frying pan and estimates the pan's temperature from the behavior of the drops. A blacksmith watches the color of red-hot iron with a practiced eye to judge when it is ready to hammer.

To study thermal phenomena quantitatively, we must go beyond these crude, qualitative methods of measuring temperature. We need a **thermometer**—a reasonably rugged device capable of making reproducible, quantitative temperature measurements. Such measurements, in turn, require a *temperature scale*.

The thermometer with which you are probably most familiar is the mercury-in-glass thermometer (Figure 17.1). Its operation depends on the empirical observation that the volume occupied by a fixed mass of mercury varies slightly but reproducibly with temperature. The mercury expands—its volume increases—when it is warmed, and it contracts—its volume decreases—when it is cooled. Many other physical quantities also vary with temperature, and any such quantity can be chosen as a basis for the operation of a thermometer (though some are more practical than others). Among the physical quantities used in this way are the electrical resistance of a metal wire; the color of a hot, glowing body; and the pressure of a fixed mass of gas confined in a fixed volume. Any such physical quantity is called a **thermometric parameter**.

Temperature Scales

Everyone knows that a pot of water boils more vigorously as the size of the flame under it is increased. It is easy—but wrong—to infer that the temperature of briskly boiling water is higher than that of water just at the boil. In 1724, Fahrenheit, using superior thermometers he had made himself, showed that the temperature at which a pure liquid boils *does not* depend on the rate at which it boils but *does* depend on the pressure of the surrounding atmosphere. A similar statement applies to the melting temperature of pure solids (or, what amounts to the same thing, the freezing temperature of pure liquids).

On the basis of this information, Fahrenheit devised a **two-point temperature scale**, which was soon improved upon by others. Any such two-point scale is constructed by calibrating a standard thermometer according to the following rules:

1. Choose a system in which a certain phenomenon is believed to occur always at the same temperature—for example, the melting of pure ice when the barometric pressure is 1 atmosphere. Immerse the standard thermometer in the system and mark the thermometer reading. This reading is the *first fixed point*.

2. Choose a system in which a second phenomenon is believed to occur always at the same temperature—for example, the boiling of pure water when the barometric pressure is 1 atm. Immerse the standard thermometer in the system and mark the thermometer reading. This reading is the *second fixed point*.

3. Divide the interval between the two fixed-point marks on the thermometer into a convenient number of equal parts. Each of these parts is called a **degree** (°).

Once the fixed-point phenomena have been chosen, three arbitrary choices remain to be made: the numerical values to be assigned to the two fixed points and the number of degrees between them. By 1750, the following conventions had been widely adopted, and they remained the basis for the standard temperature scale until 1954:

1. The temperature at which pure ice melts at 1 atm (the **ice point**) is assigned the value 0°.

2. The temperature at which pure water boils at 1 atm (the **steam point**) is assigned the value 100°.

3. Implicit in the two preceding conventions is the fact that there are 100 degrees between the fixed points. Any scale for which this is the case is called a **centigrade** (Latin for "hundred degree") scale.

The particular centigrade scale thus defined was called the **Celsius scale**, after the Swedish astronomer Anders Celsius (1701–1744), who pioneered a temperature scale similar to the modern one. By convention, we express a temperature measured on this scale in **degrees Celsius** (°C), but we speak of a temperature *difference* or a temperature *change* in **Celsius degrees** (C°).

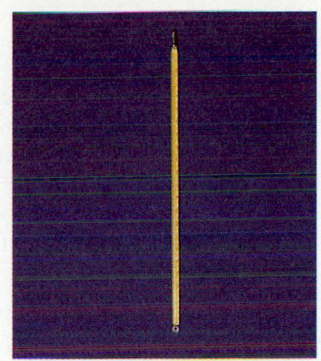

FIGURE 17.1 The mercury-in-glass thermometer. The thick-walled glass tube with its small, uniform inside diameter is called a **capillary tube**. The capillary tube terminates in a relatively large, thin-walled bulb, which contains most of the volume of the thermometer. Mercury is added to the system until the bulb and part of the capillary are filled. The air is removed by means of a vacuum pump, and the upper end is sealed. Like most materials, mercury expands when its temperature is increased. The only free region is the upper part of the capillary tube, and so the mercury level rises in the tube. Because the inside diameter of the tube is so small, a small increase in mercury volume results in a significant change in mercury level, and it is this change that is read on the scale. (The glass expands, too, but the volume of glass varies considerably more slowly as a function of temperature than does the volume of mercury.)

Good Thermometers: New Technology Spurs Scientific Progress

In 1724, the Danzig-born physicist and instrument maker Daniel Fahrenheit (1686–1736) set up a famous instrument shop in Amsterdam, where he made many thermometers not very different in appearance from today's mercury-in-glass instruments. Fahrenheit's thermometers were stable and reliably comparable. Their wide distribution throughout Europe in the 1720s contributed much to the acceleration of research into thermal phenomena.

EXAMPLE 17.1

(a) On the Fahrenheit scale, the ice point is 32°F and the steam point is 212°F. A Fahrenheit thermometer reads f. What is the corresponding reading t of a Celsius thermometer at the same location? **(b)** Normal human body temperature on the Celsius scale is conventionally assigned the value 37.0°C. What is the corresponding value on the Fahrenheit scale?

SOLUTION:

(a) Express the Celsius temperature t in terms of the corresponding Fahrenheit temperature f.

The relation between the Fahrenheit and Celsius scales is linear, as Figure 17.2 shows. It is evident from the figure that you can relate temperatures on the two scales by equating the ratios

$$\frac{f - 32°F}{212°F - 32°F} = \frac{t - 0°C}{100°C - 0°C}. \tag{17.1}$$

Solving for f, you obtain

$$f = \frac{180\ F°}{100\ C°}\, t + 32°F,$$

which simplifies to

$$f = \frac{9\ F°}{5\ C°}\, t + 32°F. \tag{17.2a}$$

This equation also gives the relation between the "sizes" of the degrees on the two scales. You have

9 Fahrenheit degrees = 5 Celsius degrees. **(17.2b)**

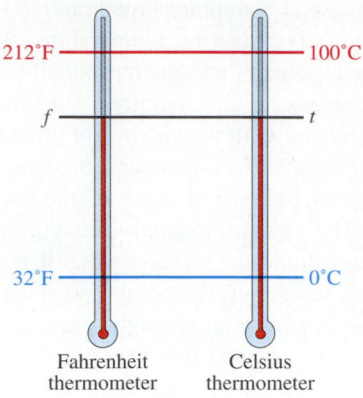

FIGURE 17.2 Comparison of the Fahrenheit and Celsius scales.

(b) Express normal human body temperature on the Fahrenheit scale.

You can use Equation 17.2a to find the value f_b of conventional human body temperature on the Fahrenheit scale. You have

$$f_b = \frac{9\ F°}{5\ C°} \times 37.0°C + 32°F = 98.6°F.$$

The Triple Point of Water

The two-point scale has a practical shortcoming: When a thermometer is calibrated at the fixed points, the pressure must be controlled with great care, especially at the boiling point because the boiling temperature of liquids depends on pressure in a fairly sensitive way. As a consequence, the precision of the calibration is limited.

We can circumvent this difficulty for one of the fixed points by taking advantage of an important physical phenomenon. Water exists in the forms ice, liquid water, and steam, called **phases**. As is true for any such substance, the three phases of pure water can exist together in equilibrium *only* at a single combination of temperature and pressure, called the **triple point**. Because the pressure is unique, the experimenter need only look into the apparatus and see if all three phases are in fact present and that no phase is growing rapidly at the expense of the others. If these conditions are met, the pressure and temperature are automatically fixed. For water, the triple-point pressure is 6.11×10^2 Pa, or about 6×10^{-3} atm. On the Celsius scale used until 1954, the triple-point temperature is close to 0.01°C. (The qualification "close to" is due mainly to the experimental difficulties in determining the ice point.) Because the triple point of water is a more reproducible fixed point than the ice point, the temperature scale was redefined in 1954, using the triple point of water as a fixed point. Specifically, the triple-point temperature t_{tp} is *defined* to be

$$t_{tp} \equiv 0.01°C. \tag{17.3}$$

The ice point is then no longer *exactly* 0°C. But the choice of the definition of Equation 17.3 assures us that temperatures measured on the new scale will agree closely with temperatures measured on the old scale. For most purposes, the difference is small enough to be negligible.

The Constant-Pressure Gas Thermometer and Charles's Law

In general, thermometric parameters are not linear functions of one another. Suppose, for example, that we have a mercury-in-glass thermometer and an electrical resistance thermometer that have been carefully calibrated so that both read 0°C when immersed in an ice-water bath and 100°C when immersed in a boiling-water bath. If both thermometers are then immersed in a bath and the mercury-in-glass thermometer reads 50°C, the resistance thermometer will read a slightly different temperature; if the bath temperature is adjusted until the resistance thermometer reads 50°C, the mercury-in-glass thermometer will read a slightly different temperature.

If we can find a class of thermometers that do not disagree as the two thermometers just described do, we have some justification for using one of the class as a basis for redefining the temperature scale. This section is concerned with the successful search for such thermometers and the far-reaching consequences of the search.

In considering Boyle's law in Section 15.7, we saw the importance of keeping the temperature of a body of gas constant when investigating the way that pressure depends on volume. Here we turn the tables: We keep the pressure of a body of gas constant and investigate the way that volume depends on temperature. The device we use to carry out this investigation is called a **constant-pressure gas thermometer** (Figure 17.3).

FIGURE 17.3 Constant-pressure gas thermometer. The horizontal glass tube is connected at its right end to a chamber filled with gas, whose pressure may be adjusted by the experimenter. Gas is trapped in the bulb-tube system by a drop of mercury. A scale behind the tube is calibrated to read the volume V of the gas trapped in the thermometer as a function of the position of the mercury drop. The pressure p of the trapped gas is always the same as that in the pressure-control chamber, which is held constant during a series of measurements. The temperature of the trapped gas is the same as that of the bath, which can be varied at will by means of heating or refrigeration units (not shown) or both.

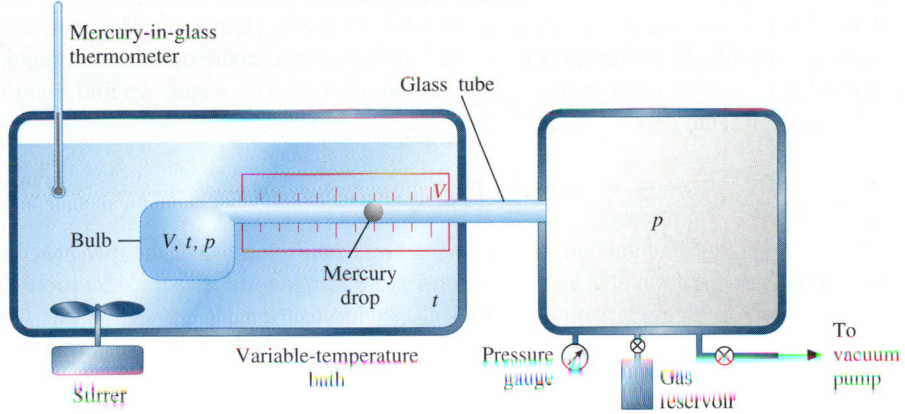

We measure the volume V of the gas trapped in the bulb of this thermometer as a function of the Celsius temperature t indicated by the adjacent mercury-in-glass thermometer. In view of the precision with which the triple-point temperature can be attained, let us use as a reference value the volume V_{tp} at $t_{tp} = 0.01°C$. We then make measurements at other Celsius temperatures and plot the volume ratio V/V_{tp} as a function of temperature (Figure 17.4). The first striking observation that emerges from such measurements is that the plot of V/V_{tp} versus t is nearly (but not exactly) a straight line [as was confirmed by observations begun about 1787 by the French physicist Jacques A. C. Charles (1746–1823) and refined by many other subsequent observers]. Ignoring the small deviation from linearity for the moment, we say that the slope of the line has the *constant* value

$$\beta \equiv \frac{d(V/V_{tp})}{dt} = \frac{1}{V_{tp}}\frac{dV}{dt}. \qquad (17.4)$$

The slope β—the fractional change in volume per unit change in temperature at constant pressure—is called the **volume coefficient of thermal expansion at constant pressure** or simply the **coefficient of thermal expansion** when there is no likelihood of confusion. Because dV/V_{tp} is a dimensionless quantity, the unit of β is $(C°)^{-1}$ when the temperature is expressed in Celsius degrees. The coefficient of thermal expansion is

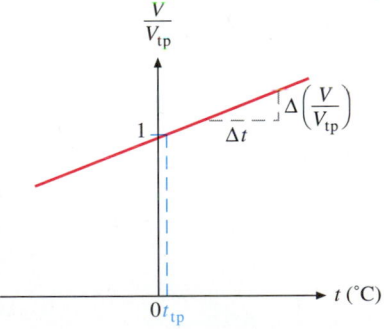

FIGURE 17.4 Plot of the volume ratio V/V_{tp} of the gas trapped in the bulb of a constant-pressure gas thermometer as a function of the Celsius temperature t. At $t = 0$, the volume is V_{tp} and the ratio is 1. The slope of the straight line is $\beta \equiv \Delta(V/V_{tp})/\Delta t = (1/V_{tp})\,dV/dt$.

analogous to the isothermal compressibility κ defined in Equation 15.26. The compressibility, which can be written $\kappa = |(1/V)\,dV/dp|$, is the fractional change in volume per unit change in pressure at constant temperature.

All gases **condense**—either liquefy or solidify—if the temperature is lowered far enough. If β is to remain constant, we must take care that the temperature of the constant-pressure gas thermometer is well above the temperature at which condensation takes place for the particular gas in use. We must also take care that the pressure of the gas is not too great in the temperature range in which it is used, for reasons described in detail in Section 18.2. But if these conditions are observed, *the coefficient of thermal expansion β is very nearly the same for all gases.*

The Ideal-Gas Thermometer and the Thermodynamic Temperature Scale

The similar thermal behavior of all gases suggests that, in defining a temperature scale, it is better to compare two or more constant-pressure gas thermometers (containing different gases and possibly having different pressures) with one another, rather than comparing a constant-pressure gas thermometer with a mercury-in-glass or other type of thermometer, as we did in the preceding discussion. We find that the agreement between two gas thermometers is better than the agreement between two thermometers of arbitrary types. Moreover, the small discrepancies between two constant-pressure gas thermometers become even smaller as the gas pressures are made smaller. Finally, observation shows that the behavior of constant-pressure gas thermometers filled with helium is nearly independent of the pressure of the helium. Hence we adopt the low-pressure helium-filled gas thermometer as the standard and compare other thermometers with it. Making careful measurements with this standard thermometer, we find that the coefficient of thermal expansion is

$$\beta = \frac{1}{273.16}\,(\text{C}°)^{-1} \quad \text{for helium at low pressure.} \tag{17.5}$$

The logical and experimental considerations leading to the adoption of the low-pressure helium-filled gas thermometer as the standard are analogous to the considerations in Section 1.2, where a comparison of various clocks led to the adoption of the one best suited as a standard. The use of a gas thermometer as the standard was first suggested in 1802 by the French physicist and chemist Joseph Louis Gay-Lussac (1778–1850) and independently by the British chemist John Dalton (1766–1844). Both made pioneering attempts at evaluating the slope of the curve of Figure 17.4, and both obtained the value $1/267$°C. Not bad for a first try!

The linear relation between V/V_{tp} and t displayed in Figure 17.4 raises an intriguing question. As the temperature of the gas trapped in the constant-pressure thermometer decreases, its volume becomes smaller—at least, until the gas condenses. But suppose we had an **ideal-gas thermometer**, containing a gas that *never* condenses no matter how low the temperature. At what temperature t_0 would the volume of such a thermometer go to zero?

To answer this question, we take advantage of the constancy of β to rewrite Equation 17.4 in terms of finite differences:

$$\beta = \frac{1}{V_{\text{tp}}}\frac{\Delta V}{\Delta t}. \tag{17.6}$$

Let us solve this equation for $\Delta t = t_0 - t_{\text{tp}}$. We obtain

$$\Delta t = \frac{1}{\beta}\frac{\Delta V}{V_{\text{tp}}} = \frac{1}{\beta}\frac{V - V_{\text{tp}}}{V_{\text{tp}}} = \frac{1}{\beta}\left(\frac{V}{V_{\text{tp}}} - 1\right). \tag{17.7}$$

Setting $V = 0$ and using the value $\beta = 1/273.16\,(\text{C}°)^{-1}$ (Equation 17.5), we have

$$\Delta t = t_0 - t_{\text{tp}} = -\frac{1}{\beta} = -273.16\ \text{C}°.$$

We use the value $t_{tp} = 0.01°C$ (Equation 17.3) and find

$$t_0 = -\frac{1}{\beta} + t_{tp} = (-273.16 + 0.01)°C,$$

or

$$t_0 = -273.15°C. \qquad (17.8)$$

THINKING LIKE A PHYSICIST

Does an ideal-gas thermometer make physical sense? Consider helium, which at low pressure condenses at about $t = -272°C$. We can use Equation 17.7 to find the corresponding volume ratio:

$$\frac{V}{V_{tp}} = \beta\,\Delta t + 1 = -\frac{272}{273.16} + 1 \approx 0.004.$$

That is, the helium in a low-pressure constant-pressure thermometer will not condense until the gas volume is only about 0.4% of the volume at the triple point. The step from a helium-filled thermometer to an ideal-gas thermometer is not great!

The approach of the constant-pressure gas thermometer to zero volume is just one of many phenomena which indicate that $t_0 = -273.15°C$ represents the **absolute zero** of temperature. Thus, although we have derived the value of t_0 on the restricted basis of gas thermometer measurements, its significance will turn out to be a universal one.

In view of this universality, we define the **thermodynamic temperature scale**. The unit of temperature in this scale is the basic SI unit of temperature; it is called the **kelvin** (**K**). Temperatures on the thermodynamic scale are denoted by the uppercase symbol T and are designated by the unit K:

$$T \text{ (in K)} = t \text{ (in °C)} + 273.15 \text{ K}, \qquad (17.9)$$

where the lowercase t represents the Celsius temperature. On the thermodynamic scale, the ice point is close to 273.15 K, the triple point of water is *exactly*

$$T_{tp} \equiv 273.16 \text{ K}, \qquad (17.10)$$

and the steam point is close to 373.15 K.* The sum in Equation 17.9 is correct in spite of the different units in which the two terms are expressed because the "size" of the kelvin is the same as that of the Celsius degree:

$$1 \text{ Celsius degree (C°)} \equiv 1 \text{ kelvin (K)}. \qquad (17.11)$$

The linear relation between V/V_{tp} and temperature can be expressed very simply when the temperatures are expressed in kelvin. Figure 17.5 is just Figure 17.4 replotted using the Kelvin temperature scale. The plot of V/V_{tp} as a function of T passes through the origin, and we have

$$\frac{V}{V_{tp}} = \beta T \quad \text{for } p = \text{constant.} \qquad (17.12)$$

The value of β is expressed in the unit $(C°)^{-1}$ in Equation 17.5. According to identity 17.11, β can be expressed immediately in K^{-1}:

$$\beta = \frac{1}{273.15} \text{ K}^{-1} = \frac{1}{T_{tp}} \quad \text{for an ideal gas.} \qquad (17.13)$$

Thus Equation 17.12 can be written

$$\frac{V}{V_{tp}} = \frac{T}{T_{tp}} \quad \text{for } p = \text{constant.} \qquad (17.14)$$

---

*Note that the symbol ° is not used to denote temperature differences or temperatures expressed in kelvin. Thus the notational distinction between C° (expressing temperature difference) and °C (expressing a temperature) on the Celsius scale is unnecessary on the thermodynamic scale.

Lord Kelvin

The British physicist William Thomson, Lord Kelvin (1824–1907) was one of the scientific giants of the nineteenth century, both in theory and in experiment. In addition to doing pioneering work in thermal physics and refrigeration engineering, he was an innovator in electromagnetic theory and measurement, contributed in an essential way to the technology of the first transatlantic telegraph cables and to the development of commercial electricity, and did significant work in other fields, notably fluid dynamics.

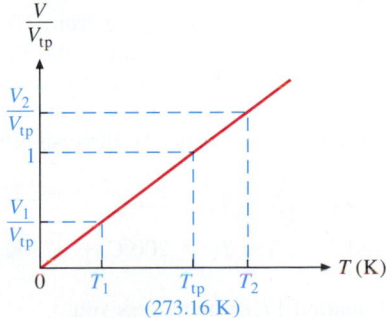

FIGURE 17.5 Figure 17.4 replotted using the thermodynamic temperature scale. The slope of the curve has the constant value $\beta = (1/V_{tp})\,dV/dT = 1/273.16 \text{ K}^{-1}$.

Let us now express Equation 17.14 in a more general form. Consider the particular volume V_1 corresponding to temperature T_1 and the particular volume V_2 corresponding to T_2 in Figure 17.5. From inspection of the figure, you can obtain the relation

$$\frac{V_1}{V_2} = \frac{T_1}{T_2} \quad \text{for } p = \text{constant;} \tag{17.15}$$

at constant pressure, the volume of a body of gas varies proportionally with the thermodynamic temperature. This relation is called **Charles's law** or **Gay-Lussac's law**. The name applies equally to Equation 17.15, Equation 17.14, Equation 17.12, or any other statement that expresses the proportionality.

Henceforth we take the thermodynamic scale to be the fundamental scale of temperature. We therefore *redefine* the Celsius scale—which is *not* the SI temperature scale—in terms of the thermodynamic scale. The Celsius temperature t is defined to be

$$t \text{ (in °C)} \equiv T - 273.15 \text{ K.} \tag{17.16}$$

Because the definition of the thermodynamic scale depends on only one arbitrary fixed point—the triple point of water—it is a **one-point scale**. The Celsius scale as redefined in terms of the thermodynamic scale by means of Equation 17.16 also is a one-point scale.

To summarize, we have arrived at the one-point thermodynamic temperature scale based on the highly reproducible triple point of water, taken together with a new insight into the absolute zero of temperature. The thermodynamic temperature scale has been adopted as the SI scale of temperature. The familiar (and practically useful) Celsius scale has been redefined in terms of the thermodynamic scale and is related to it by an additive constant; when conversion is necessary, it can be carried out simply.

EXAMPLE 17.2

(a) The device shown in Figure 17.6 consists of a leakproof piston that can move in and out of a cylinder. The system is initially at temperature $t_1 = 120°C$ and contains steam at pressure $p_1 = 1.00$ atm. It is heated until its temperature is $t_2 = 200°C$. During the process, the piston is allowed to move in such a way that the pressure remains constant. What is the ratio V_2/V_1 of the new volume of the cylinder to the initial volume? **(b)** The piston is now pushed back into the cylinder until the volume is again V_1. During this process, the temperature is kept constant. What is the final pressure p_3 inside the cylinder?

SOLUTION:

(a) If the system is heated from 100°C to 200°C at constant pressure, what is the volume ratio V_1/V_2?

To use Charles's law in the form of Equation 17.15, you must first use Equation 17.9 to express the temperatures on the thermodynamic scale. To three significant figures, you have

$$T_1 = 120°C + 273 \text{ K} = 393 \text{ K}$$

and $\qquad T_2 = 200°C + 273 \text{ K} = 473 \text{ K.}$

Equation 17.15 then gives you

$$\frac{V_2}{V_1} = \frac{T_2}{T_1} = \frac{473 \text{ K}}{293 \text{ K}} = 1.61.$$

That is, the cylinder has expanded by 61%.

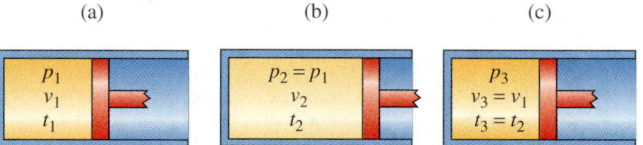

FIGURE 17.6 Example 17.2. (*a*) The initial situation. (*b*) The situation after the volume of the system has increased because of a "Charles's law" temperature increase at constant pressure. (*c*) The final situation, after the pressure of the system has increased because of a "Boyle's law" decrease in volume at constant temperature.

(b) The gas in the cylinder is now compressed back to its initial volume while the temperature is held constant. Find the final pressure p_3.

Because the temperature is held constant as the gas in the cylinder is recompressed, you can use Boyle's law in the form of Equation 15.36c, $pV = $ constant, to find the final pressure p_3. You have $p_2V_2 = p_3V_3$. Given that $p_2 = p_1$ and $V_3 = V_1$, you can write

$$p_1V_2 = p_3V_1.$$

Solving for p_3, you have

$$p_3 = p_1 \frac{V_2}{V_1} \tag{17.17}$$

$$= 1.00 \text{ atm} \times 1.61 = 1.61 \text{ atm.}$$

The ratio of final pressure to initial pressure is $p_3/p_1 = 1.61$; when the system is returned to its original volume, its pressure has increased by 61%. It is no accident that the proportional change is the same in parts **a** and **b**. We will return to this point in Section 18.2.

Thermal Expansion

SECTION 17.4

Volume change with temperature, called **thermal expansion**, is a common property of matter generally. We have already considered how the phenomenon of thermal expansion is turned to good use in the mercury-in-glass thermometer. Here, we consider thermal expansion in a quantitative way. Our approach will be entirely phenomenological; although we will not be able to obtain much insight into the complicated interatomic interactions that underlie thermal expansion, we will be able to describe the behavior of bodies subjected to temperature change over a fairly wide range of circumstances. And, as always, a phenomenological picture is an essential first step toward a deeper understanding.

Volume Expansion of Liquids and Solids

When pressure is held constant, the volume of a liquid or solid body changes much less with changing temperature than does the volume of a body of gas. You are already familiar with the analogous statement that applies at constant temperature: The volume of a liquid or solid body changes much less with changing pressure than does the volume of a body of gas.

The general theory of volume expansion for liquids and solids is much more complicated than that for gases. But in many practical cases an empirical approach is adequate. Suppose that a body is initially at temperature T_i and has volume V_i. The temperature is changed to some other value T_f, not too different from T_i. Because the volume of a liquid or solid body changes only gradually with changing temperature, we approximate the volume change by means of a linear relation:

$$\frac{V_f}{V_i} = 1 + \beta(T_f - T_i). \tag{17.18}$$

As for gases, β is the volume coefficient of thermal expansion. Its value must be determined by direct measurement on the substance of interest in the proper temperature range.

Because the volume change is usually small, Equation 17.18 is conveniently re-expressed in terms of the fractional volume change $\Delta V/V = (V_f - V_i)/V$. Subtracting 1 from both sides of Equation 17.18, we obtain $V_f/V_i - 1 = \beta(T_i - T_f)$, or

$$\frac{V_f - V_i}{V_i} = \beta(T_f - T_i). \tag{17.19a}$$

We can rewrite this equation in terms of the temperature change ΔT and the corresponding volume change ΔV:

$$\frac{\Delta V}{V} = \beta \, \Delta T. \tag{17.19b}$$

Why is it usually not necessary to specify that the value of V in the denominator on the left side is V_i and not V_f?

The volume coefficient of expansion β is tabulated for various liquids and solids in Table 17.1. As the table suggests, β is generally greater for liquids than for solids. Note also that nearly all substances expand with increasing temperature. Water is exceptional; it expands with *decreasing* temperature in the limited temperature range between 0°C and about 4°C. This phenomenon is related to a more familiar one: Unlike most other

TABLE 17.1 Volume Coefficients of Thermal Expansion β

Material	Temperature (°C)	β [(C°)$^{-1}$ or K^{-1}]
Liquids		
Benzene	20	1.237×10^{-3}
Bromine	20	1.132×10^{-3}
Carbon tetrachloride	20	1.236×10^{-3}
Mercury	20	1.8175×10^{-4}
Water	2	-0.331×10^{-4}
	3.98	0
	20	2.06×10^{-4}
	60	5.14×10^{-4}
	90	6.73×10^{-4}
Solids		
Carbon (diamond)	40	3.54×10^{-6}
Glass (typical)	0 to 100	2.6×10^{-5}
Ice	−20 to −1	1.13×10^{-4}
Iron	0 to 100	3.55×10^{-5}
Rock salt (NaCl)	55	1.21×10^{-4}
Silver	0 to 100	5.83×10^{-5}
Sulfur	30	2.23×10^{-4}

liquids, water expands when it freezes. There are a few other substances that have this peculiar property; the best known are the metals bismuth and antimony. If a substance expands on freezing, the solid is less dense than the corresponding liquid and will float on a pool of the liquid. Such materials are called *icelike*.

EXAMPLE 17.3

You wish to engrave a scale on a new, unmarked mercury-in-glass thermometer. The thermometer contains 4.20×10^{-8} m^3 of mercury, most of it in the bulb. The diameter of the capillary tube is 6.65×10^{-5} m. You immerse the thermometer in a mixture of water and ice and mark the mercury level. Then you immerse the thermometer in boiling water and mark the new mercury level. How far apart are these two marks? How far apart must you make the one-degree marks? Neglect both the thermal expansion of the glass and the gradual variation in the coefficient of expansion of mercury over the temperature range of interest.

SOLUTION: You use Equation 17.19b to write the change ΔV in the volume of the mercury due to the temperature change ΔT; you have

$$\Delta V = V\beta \, \Delta T.$$

The mercury is free to expand only into the cylindrical capillary tube. So you can express ΔV in terms of the capillary diameter d and the mercury level change Δl:

$$\Delta V = \pi \left(\frac{d}{2}\right)^2 \Delta l.$$

Combining these two equations, you obtain

$$\Delta l = \frac{4V\beta \, \Delta T}{\pi d^2}.$$

The initial temperature is 0°C and the final temperature is 100°C, so you have $\Delta T = 100$ K. Using the value of β given for mercury in Table 17.1, you obtain

$$\Delta l = \frac{4 \times 4.20 \times 10^{-8} \text{ m}^3 \times 1.818 \times 10^{-4} \text{ K}^{-1} \times 100 \text{ K}}{\pi \times (6.65 \times 10^{-5} \text{ m})^2}$$

$$= 0.220 \text{ m}.$$

A distance of 22 cm between the boiling-point and freezing-point readings is typical of thermometers intended for use in this range. Each degree spans $\frac{1}{100}$ of this distance, so the degree marks are 0.22 cm apart. It would not be difficult to estimate temperatures within 0.1°C with such a thermometer, provided it was carefully calibrated.

Linear Expansion of Solids

With solid bodies, it is often more useful to consider the change in length (or width or height) rather than the change in volume. Can we assume that the fractional change in length, $\Delta l/l$, is a linear function of temperature? We assumed linearity for the proportional change in volume in deriving Equation 17.19b, $\Delta V/V = \beta\,\Delta T$, and justified the assumption on the basis of the smallness of $\Delta V/V$. We use the same approach here in considering changes in length. Consider a rod of length l made of a uniform solid. We write the empirical relation

$$\frac{\Delta l}{l} = \alpha\,\Delta T. \tag{17.20}$$

The proportionality factor α is called the **linear coefficient of thermal expansion**. Because it is the one most frequently quoted for solids, this coefficient is often called simply the *coefficient of expansion* when there is no likelihood of confusion.

When we solve Equation 17.20 for α, we obtain

$$\alpha = \frac{1}{l}\frac{\Delta l}{\Delta T}. \tag{17.21}$$

This equation may be regarded as the definition of the coefficient of linear expansion. It follows directly from Equation 17.21 that the SI unit for α is K^{-1}. Values of α are listed in Table 17.2. These values validate the assumption that the proportional change $\Delta l/l$ is small unless the temperature change is very large. Metals for the most part have larger coefficients of expansion than nonmetals.

For most isotropic solids, the relation between the two coefficients of expansion α and β is given fairly accurately by the expression

$$\beta = 3\alpha. \tag{17.22}$$

The derivation of this relation is similar to the derivation of Equation 15.27, according to which the relation between Young's modulus Y and the bulk modulus B is $Y = 3B(1 - 2\nu)$. But Equation 17.22 is simpler to derive because there is no thermal analogy to Poisson's ratio ν; see Problem 17.57.

Large structures must have expansion joints to allow for temperature changes.

TABLE 17.2 Linear Coefficients of Thermal Expansion α

Material	Temperature (°C)	$\alpha\ [(C°)^{-1}$ or $K^{-1}]$
Aluminum	27	2.32×10^{-5}
Brass (typical)	25 to 100	1.8×10^{-5}
Carbon (diamond)	27	1.1×10^{-6}
Carbon (graphite)	27	8.8×10^{-6}
Copper	27	1.68×10^{-5}
Ice	-10	5.58×10^{-5}
Invar (64% Fe, 36% Ni)	0 to 100	1×10^{-6}
Iron	27	1.17×10^{-5}
Steel (typical)	20 to 100	1.2×10^{-5}
Sulfur	27	6.1×10^{-5}

EXAMPLE 17.4

(a) Until about 1950, the usual practice in laying steel railroad rails was to leave a small gap between rail sections to allow for thermal expansion. Suppose that rails are laid on a day when the temperature is 15°C. If the maximum temperature expected is 50°C and the length of each rail section is 10 m, how big a gap should be left? (b) The disadvantage of rail gaps is that they make for a rough ride, which contributes to the wear of both trains and track. Modern practice is to use continuous rail, assembled on site by welding the rail sections together. Continuous rails must be fastened to the ties strongly enough so that

the rails will not bow or buckle on hot days. Assuming again that the rails are laid when the temperature is 15°C and that the maximum temperature expected is 50°C, what is the maximum compressive stress $|\sigma|$ to be expected in the rails? If the cross-sectional area of the rail is a typical 50 cm², what is the corresponding compressive force?

SOLUTION:

(a) How big a gap should be left between rail sections to allow for thermal expansion?

From Table 17.2, you take the value of the linear coefficient of expansion for steel, $\alpha = 1.2 \times 10^{-5}$ K^{-1}. Because the maximum temperature increase expected is $\Delta T = |50°C - 15°C| = 35$ K, you use Equation 17.20 to write

$$\frac{\Delta l}{l} = 1.2 \times 10^{-5} \text{ K}^{-1} \times 35 \text{ K} = 4.2 \times 10^{-4}.$$

That is to say, the proportional change in the length of the rails is 0.042%. The change in length of each 10-m rail section is l times this proportion, or

$$\Delta l = 10 \text{ m} \times 4.2 \times 10^{-4} = 4.2 \times 10^{-3} \text{ m} = 4.2 \text{ mm}.$$

The gap between adjacent rail sections must be at least this large.

(b) If the rails are welded together, find the maximum compressive stress and compressive force.

Once the rails are welded together, their overall length can-not change. Thus the expansion must be opposed by a compressive stress. The strain ϵ produced by this stress must be just equal to the proportional expansion $\Delta l/l$ that would result in its absence; $|\epsilon| = |\Delta l/l|$. But the strain is related to the stress by Equation 15.8a, $\sigma = Y\epsilon$, where Y is Young's modulus. So you have

$$|\sigma| = Y\frac{\Delta l}{l}.$$

Using the value of Y given for steel in Table 15.1, you obtain

$$|\sigma| = 21 \times 10^{10} \text{ Pa} \times 4.2 \times 10^{-4} = 8.8 \times 10^{7} \text{ Pa}.$$

The corresponding force is

$$F = \sigma A = 8.8 \times 10^{7} \text{ Pa} \times 5.0 \times 10^{-3} \text{ m}^2 = 4.4 \times 10^{5} \text{ N}.$$

This is roughly equal to the weight of a 50-tonne mass. Although the proportional expansion $\Delta l/l$ that must be counteracted is small, steel is a very stiff substance, and the necessary compressive force F is therefore quite large.

Heat Flow

On a given stove, a pot full of cold water takes much longer to bring to the boil than a similar pot holding only a small amount of water. Although change of temperature certainly plays a role here, the quantity of central interest is the quite different one we call *heat flow*. One of the great scientific advances of the mid-nineteenth century was the verification of a view that has its roots in antiquity: *Heat flow is a form of energy transfer.* You will see ample verification of this statement in this and the next three chapters. At the outset of our study, however, we will consider heat flow in a purely phenomenological way. Indeed, much of the subject matter of this section does not depend explicitly on the fact that heat flow is a form of energy transfer. Nevertheless, you should not let this important reality stray far from your mind.

A Phenomenological View of Heat Flow. Specific Heat Capacity

To raise the temperature of a given sample of matter by an amount ΔT, we must provide for a certain heat transfer Q into the sample. The value of Q depends not only on ΔT, but also on the mass m of the sample and on a quantity called the **specific heat capacity** c, which is a property of the particular substance of which the sample is made. The empirical equation that relates Q to m, c, and ΔT for any amount of any substance is

$$Q = mc\,\Delta T. \qquad \textbf{(17.23)}$$

Water is taken to be the standard substance in this primitive relation, just as it is in the primitive definition of temperature scales. Specifically, the numerical value of c is taken to be 1 for water, and the amount of heat transfer that increases the temperature of 1 kg of liquid water by 1 K is called 1 **kilocalorie (kcal)**.* From Equation 17.23 you can see that, if Q is expressed in kilocalories and m in kilograms, the unit of c is kcal/kg·K. For water, c is 1 kcal/kg·K. Values of c are given in Table 17.3. The values are quoted for specified temperatures because c is itself a function of temperature. However, c usually varies only gradually with temperature and can be taken as a constant if the temperature change ΔT in Equation 17.23 is not too large.

---

*The kilocalorie is *not* an SI unit, but it is often convenient nevertheless. We will soon redefine the kilocalorie more precisely in terms of SI units.

TABLE 17.3 **Specific Heat Capacity _c_ of Selected Substances**

Material	Temperature (°C)	c (kcal/kg·K)	c (J/kg·K)
Water	15	1 (by definition)	4.1858×10^3
Ice	−4.9	0.493 2	2.065×10^3
Steam (1 atm)	110	0.481	2.01×10^3
Aluminum	25	0.215 4	0.9017×10^3
Ammonia (liquid)	20	1.125	4.709×10^3
Bromine (liquid)	25	0.113 1	0.4735×10^3
(solid)	−23	0.088 59	0.3709×10^3
Carbon (diamond)	25	0.122	0.509×10^3
(graphite)	25	0.170	0.711×10^3
Carbon dioxide (dry ice)	−225	0.124	
Copper	25	0.092 0	0.385×10^3
Ethanol	25		2.43×10^3
Gold	25	0.030 8	0.129×10^3
Iron	25	0.107	0.449×10^3
Lead	25	0.030 9	0.129×10^3
Lithium	25	0.852	3.56×10^3
Mercury	25	0.033 4	0.140×10^3
Sodium chloride	0	0.204	0.854×10^3

If the temperature of a body increases as heat transfer takes place between the body and its surroundings, ΔT is positive and Q is positive. We then say that _heat flows into the body_. If the temperature of the body decreases as heat is transferred, ΔT is negative and Q is negative. We then say that _heat flows out of the body_.

EXAMPLE 17.5

(a) A lead ball of mass 150 g is heated from 0.0°C to 100.0°C. How much heat flows into the ball? **(b)** If the same amount of heat flows out of an equal mass of water whose temperature is initially 50.0°C, what will be its final temperature t_f?

SOLUTION:

(a) How much heat flows into the ball?

According to Table 17.3, the specific heat capacity of lead is $c_{Pb} = 0.0309$ kcal/kg·K. The temperature change of the lead ball is $\Delta T_{Pb} = 100.0$ K. Using Equation 17.23, you obtain

$$Q = 0.150 \text{ kg} \times 0.0309 \text{ kcal/kg·K} \times 100.0 \text{ K}$$
$$= 0.464 \text{ kcal.}$$

(b) If the same amount of heat flows out of an equal mass of water whose temperature is initially 50.0°C, what will be its final temperature t_f?

For the equal mass of water, you can use Equation 17.23 to determine the temperature change ΔT_w and thus t_f. Using the symbol c_w to represent the specific heat capacity of water, you have

$$\Delta T_w = \frac{Q}{m c_w} = \frac{0.464 \text{ kcal}}{0.150 \text{ kg} \times 1 \text{ kcal/kg·K}} = 3.09 \text{ K.}$$

This temperature change is much smaller than the 100 K change you found in part **a** for lead. Indeed, you can show that the ratio of the two temperature changes is given by

$$\frac{\Delta T_{Pb}}{\Delta T_w} = \frac{c_w}{c_{Pb}};$$

for the same heat flow, the temperature changes of the two substances are in inverse ratio of their specific heat capacities. Having evaluated ΔT_w, you can now find t_f. You have

$$t_f = t_i - \Delta T_w = 50.0°\text{C} - 3.09 \text{ K} = 46.9°\text{C.}$$

Heat Flow and Thermal Equilibrium

How can we predict, in a particular case, which way heat will flow when two bodies are placed in **thermal contact**—that is, a contact that makes heat flow possible? The answer is simple—and consistent with common experience. Suppose that two bodies 1 and 2 having different initial temperatures T_{1i} and T_{2i} are placed in thermal contact. Suppose also that $T_{1i} > T_{2i}$. If there is no outside intervention, the warmer body (body 1) will cool down and the cooler body (body 2) will warm up until both bodies achieve a common temperature T_f somewhere between the two initial temperatures; that is,

$$T_{1i} > T_f > T_{2i}. \tag{17.24}$$

No further temperature change will take place. We then say that the two bodies are in **thermal equilibrium** with each other, or that the system comprising the bodies is in a state of thermal equilibrium.

Equally important, the reverse process *never* takes place. You have never seen a case in which, in the absence of outside intervention, the temperature of the warmer body increases while that of the cooler body decreases. Neither has anyone else! To put it another way, a system that is in thermal equilibrium will remain in thermal equilibrium indefinitely if there is no outside intervention. We will consider this deceptively simple point in detail in Section 20.3.

A body that is not in thermal equilibrium with its surroundings, but whose temperature does not change with time, is said to be **thermally isolated** from the surroundings. According to Equation 17.23, the fact that $\Delta T = 0$ implies that $Q = 0$; heat cannot flow either into or out of the body. Complete thermal isolation of a body is usually not possible. However, thermal isolation can often be well approximated by surrounding the body with a material that does not conduct heat well. Such a material is called a **thermal insulator**. Goose down is a well-known thermal insulator. A material that does conduct heat well, as do most metals, is called a **thermal conductor**.

Total Heat Capacity

For a particular sample of mass m and specific heat capacity c, how much heat must flow into the sample if its temperature is to rise by 1 K? From Equation 17.23, $Q = mc\,\Delta T$, we have

$$\frac{Q}{\Delta T} = mc.$$

The product mc is a fixed quantity for a particular body, and we give it the name C. We can then write

$$C = mc = \frac{Q}{\Delta T}. \tag{17.25}$$

The quantity C is called the **total heat capacity** of the body. If Q is expressed in kilocalories, C is expressed in kcal/K.

Thermometry and the Zeroth Law of Thermodynamics

When you wish to measure the temperature of a body, you place a thermometer in thermal contact with it. This is a process we usually take for granted, as we have done so far in this chapter. But the process deserves a deeper look because it is fundamental to the physics of thermal phenomena.

The thermometer is itself a body. Before the thermometer and the other body are placed in thermal contact, they are usually not at the same temperature. If they are not, the temperatures of both change until they come to the same temperature and are thus in thermal equilibrium with each other. Because the thermometer usually has a much smaller total heat capacity than the body whose temperature it is used to measure, the temperature of the thermometer usually changes much more than the temperature of the other body. You see this change in the familiar motion of the end of the mercury column when you dip an ordinary thermometer into a container of liquid.

What makes temperature measurement useful is the fact that the same thermometer can be used to compare the temperatures of two or more other bodies. For example, you can dip a thermometer into one cup of water, wait until the thermometer is in thermal equilibrium with the water, and read the temperature. Then, by repeating the

Beware of Loose Terminology

Both the specific heat capacity c of a substance and the total heat capacity C of a body are sometimes called simply "heat capacity." In this book, we will always distinguish carefully between C and c.

procedure with another cup of water, you can tell whether or not the two cups of water are at the same temperature. That is, you can tell whether or not the two cups of water are in thermal equilibrium with each other even though they are not in thermal contact with each other and no heat can flow from one to the other.

The fact that you can make this comparison between two bodies not in thermal contact with one another is stated formally as a basic law of nature, called the **zeroth law of thermodynamics**:

Consider three bodies, *a*, *b*, and *c*. Suppose that *a* and *c* are in thermal equilibrium with one another. Suppose also that *b* and *c* are in thermal equilibrium with one another. Then *a* and *b* are in thermal equilibrium with one another.

Because two bodies in thermal equilibrium have equal temperatures, it follows that, if $T_a = T_c$ and $T_b = T_c$, then $T_a = T_b$. (The law is *not* merely the mathematical statement that two quantities equal to a third are equal to each other.) The law makes thermometry possible. In that sense, the zeroth law logically precedes the other laws of thermodynamics, which we will study in the following three chapters. Because the zeroth law is so easily taken for granted, the importance of stating it explicitly in laying a logical foundation for the physics of thermal phenomena was not perceived until after the names "first law of thermodynamics" and "second law of thermodynamics" had become firmly established—thus the peculiar but apt name "zeroth law of thermodynamics."

Phase Changes and Latent Heat

We have already noted that many substances can exist in several *phases*; water, for instance, can exist as ice, liquid water, or steam. If the pressure is fixed, the *phase change* or *phase transition* between two phases—a familiar example is melting—takes place at a sharply defined temperature.

As you know, however, the process is not abrupt. The device of Figure 17.7a can be used to show what happens if you begin with a block of ice of mass *m*, whose temperature is well below the melting point, and measure its temperature repeatedly as you make heat flow into it at a constant rate. If you measure the temperature inside the device as a function of time, you will obtain a graph like that of Figure 17.7b. At first, the temperature of the ice rises at a constant rate that depends on the total heat capacity $C_{ice} = mc_{ice}$; no liquid water appears until the temperature reaches 0°C. Then the ice begins to melt, but the temperature remains constant at 0°C until all the ice is gone. Then—and only then—the temperature again begins to rise. The rate of temperature increase depends on the total heat capacity $C_{water} = mc_{water}$ and remains constant until the temperature reaches 100°C. Then the water begins to boil. Again the temperature remains constant even though the rate of heat flow into the system remains constant.

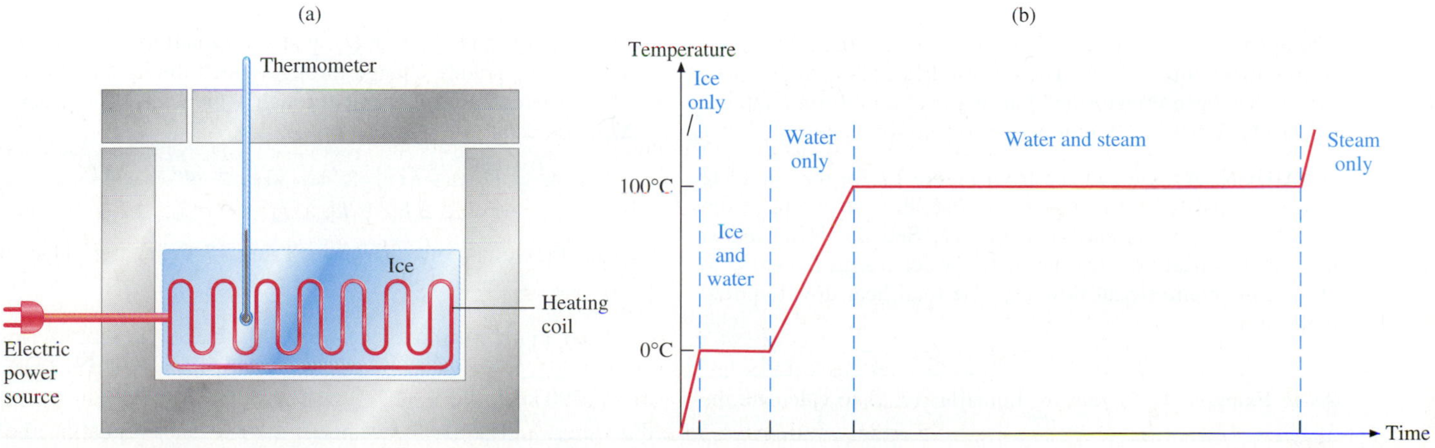

FIGURE 17.7 (*a*) An insulated container holds a block of ice whose initial temperature is well below the melting point. An electric heater, whose coils are distributed throughout the block, is turned on, and there is a constant flow of heat into the ice. The process continues as the ice melts and as the resulting water warms and then boils. A small hole in the container allows the escape of steam, so the pressure inside is always 1 atm. The experiment is discontinued shortly after all the water has boiled. The temperature is monitored as a function of time by means of the thermometer. (*b*) Idealized plot of temperature as a function of time. Time is shown on an arbitrary but linear scale.

(a)

(b)

When the last liquid water has disappeared and only steam remains, the temperature begins to rise above 100°C.

That temperature remains constant during the phase change implies that *the phase change is a result of heat flow into the system*, just as the temperature changes are. Because ice cannot exist above 0°C at atmospheric pressure, the phase change must go to completion before further temperature change can take place. The heat transfer required to change the phase of a unit mass of a given substance is called **latent heat**— that is, hidden heat—because the heat transfer process does not reveal itself in a temperature change. The latent heat L depends on the substance, the particular phase change, and the pressure. The total heat flow Q required to change the phase of a body of mass m is

$$Q = mL. \qquad (17.26)$$

If Q is expressed in kilocalories, the units of L are kcal/kg. Latent heats for selected substances undergoing specified phase changes are given in Table 17.4. As you can see, the latent heat of boiling is usually considerably greater than the latent heat of melting for the same substance.

TABLE 17.4 **Latent Heat L of Selected Substances**

Material	Phase Change and Temperature (°C)	L (kcal/kg)	L (J/kg)
Water	Melting, 0	79.71	3.335×10^5
	Boiling, 100	539.4	2.257×10^6
Aluminum	Melting, 660.4	94.5	3.95×10^5
Ammonia	Boiling, −33.4	327.7	1.371×10^6
Carbon dioxide ("dry ice")	Sublimation,[a] −78.5	137.0	5.732×10^5
Carbon tetrachloride	Melting, −22.9	3.8	1.6×10^4
	Boiling, 76.8	46.6	1.95×10^5
Copper	Melting, 1083.4	48.9	2.05×10^4
Ethanol (ethyl alcohol)	Melting, −114.4	26.05	1.090×10^5
Gold	Melting, 1064.43	15.0	6.28×10^4
Helium	Boiling, −268.935 (4.215 K)	5.0	2.1×10^4
Lead	Melting, 327.3	5.9	2.47×10^4
Mercury	Melting, −38.87	2.78	1.16×10^4
	Boiling, 356.72	70.44	2.947×10^4
Sodium chloride (rock salt)	Melting, 800	124	5.19×10^5
	Boiling, 1457	691	2.89×10^6

[a]At atmospheric pressure, carbon dioxide undergoes a solid-to-gas phase change (or vice versa). Such a change is called **sublimation**.

EXAMPLE 17.6

The apparatus of Figure 17.7a contains 7.5 kg of ice at a uniform temperature −12°C. The electric heater is operated until the ice has been transformed into water at a uniform temperature 19°C. What is the total heat flow into the system?

SOLUTION: As you can see from Figure 17.7b, the complete process consists of three stages. First, the ice warms from −12°C to 0°C; this requires heat flow Q_1. Second, the ice melts; this requires heat flow Q_2. Third, the water warms from 0°C to 19°C; this requires heat flow Q_3. The total heat flow required is the sum

$$Q = Q_1 + Q_2 + Q_3.$$

As in Example 17.5, you use Equation 17.23 to calculate the quantities Q_1 and Q_3. For Q_2, you use Equation 17.26. If you call the temperature change through which the ice passes ΔT_1 and the temperature change through which the water passes ΔT_3, you can write

$$Q = mc_{\text{ice}}\, \Delta T_1 + mL_{\text{ice}\to\text{water}} + mc_{\text{water}}\, \Delta T_3$$
$$= m(c_{\text{ice}}\, \Delta T_1 + L_{\text{ice}\to\text{water}} + c_{\text{water}}\, \Delta T_3).$$

Using the values of c given in Table 17.3 and the value of $L_{\text{ice}\to\text{water}}$ given in Table 17.4, you obtain

$$Q = (7.5 \text{ kg})\left(0.49\, \frac{\text{kcal}}{\text{kg·K}} \times 12 \text{ K} + 80\, \frac{\text{kcal}}{\text{kg}} + 1\, \frac{\text{kcal}}{\text{kg·K}} \times 19 \text{ K}\right)$$
$$= 790 \text{ kcal}.$$

In Example 17.6, most of the heat input to the system goes into the melting process rather than the two temperature change processes. Without going into detail, we can make a plausible guess as to the reason. In melting, a substance undergoes a radical change in the way that neighboring molecules are bound to one another; much energy must be provided (or removed) to effect such a change. In a temperature change, on the other hand, a substance undergoes a gradual change in the intermolecular distance, as suggested by the small magnitude of the thermal expansion coefficient of solids and liquids. Compared with melting, temperature change is a process requiring relatively little energy.

The Mechanical Equivalent of Heat

So far, we have dealt with heat in a purely phenomenological way. We have developed a method for measuring heat flow in terms of an empirical unit (the kilocalorie) that bears no apparent relation to the SI units defined earlier. We now consider experimental evidence that leads us to a fundamental understanding: *Heat flow into a body is a process in which energy is transferred to the body from its surroundings.*

The insights we need arise naturally from our familiarity with *frictional heating*, the heating that takes place when you rub your hands together vigorously on a cold day or pull a nail out of a board. Work is done on the nail as it slides with considerable friction over the surface of the wood in which it is embedded. Seen at the microscopic level, the resulting overall distortion of the surface of the nail is due to a displacement of its surface atoms relative to their neighbors farther inside. This relative motion is quickly transmitted to all the atoms of the nail by the forces that bind them together. The result is a complicated, disorderly oscillation of the atoms; the macroscopic work done on the nail as it moves has increased the energy of the atoms.

On the macroscopic scale, we cannot see the microscopic motion of the atoms because it does not contribute to motion of the nail as a whole. But what we do "see" is macroscopic evidence of the motion: heating of the nail. The view that thermal phenomena, such as the heating of the nail, are the macroscopic manifestation of the energy of disordered, microscopic motion is called the **kinetic theory of heat**. When we wish to speak in macroscopic terms of how the frictional work done on it affects the nail, we say that the work has increased the **internal energy** of the nail, of which the energy of disordered, microscopic motion is a part. (Part of the internal energy of a system may be in other forms, but this need not concern us. As you will see, we are interested only in *changes* of the internal energy.)

The scientific community accepted the kinetic theory of heat slowly, over a period of many years ending about 1850. The realization that heat flow is an energy transfer process led to a great broadening of the concept of conservation of energy. The power of this broadened concept will become evident in the next three chapters.

Various processes, including work against friction, can result in the *evolution of heat* in a system. When we say that **heat is evolved** in a system, we mean that a thermal phenomenon such as temperature increase or melting takes place in the system as a result of some process other than flow of heat into the system. The amount of heat evolved, Q, can be measured in terms of the observed thermal phenomenon—for example, by using Equation 17.23, $Q = mc \, \Delta T$.

The fact that processes in which work is done against friction are accompanied by the evolution of heat does not, by itself, prove that heat flow is a form of energy transfer.* But we can provide a very solid basis for the assertion by carrying out an experimental program to demonstrate the following four points:

1. When frictional work is done (by rubbing two bodies together, say), heat is evolved as long as the frictional work continues.

Caloric: Plausibility Isn't Enough

Long before definitive evidence existed to support the idea that heat flow was a manifestation of microscopic processes, the view was accepted at least tentatively by many natural philosophers. Among them were Huygens, Newton, Hooke, and Daniel Bernoulli. Others, however, took the quite different view that heat flow was the flow of a material fluid called *caloric*. This view infers the existence of stuff called "heat" from observations of a process called *heat flow*. The inference, though plausible, is now known to be erroneous, and the caloric theory is therefore incorrect. Just because we give the name "X flow" to a process that we observe, we cannot conclude that X exists; if we make this conclusion without further evidence, we are fooling ourselves with a word game. Nevertheless, we still use the term "heat flow" as though something were flowing. And we still use the term "heat capacity," which evokes an image of the amount of caloric a body can hold.

---

*To give a simple analogy, you can always condense water out of a match flame by bringing the flame near a cold surface. It does not follow that water is a form of flame.

2. The total heat evolved, Q, is directly proportional to the total frictional work done, W:

$$W = J Q. \tag{17.27}$$

The proportionality constant J in this equation is called the **mechanical equivalent of heat**. In the units we are using, W is measured in joules and Q in kilocalories. The corresponding unit for J is thus J/kcal.

3. The value of J does not depend on how the frictional work is done or on the rate at which it is done.

4. Other dissipative processes, not involving contact friction between two solid surfaces (such as the stirring of a liquid or the passage of electric current through a heater wire), result in the evolution of heat with the *same* proportionality constant J.

By showing that these four points hold true, we will demonstrate not only that work done on a body is equivalent to heat flowing into the body, but also that the heat evolved in the body as a result of work done on it is independent both of the way in which the work is done and of the rate at which it is done. The main subject of the next three subsections is the carrying out of this four-point program.

The Rumford Cannon-Boring Experiment

The first point—that frictional work can evolve heat indefinitely—was demonstrated in a dramatic manner in the 1790s by the American physicist Benjamin Thompson, Count Rumford (1753–1814). In the course of overseeing the boring of brass castings to make cannons, Rumford had been impressed by the large amount of heat evolved in the process. According to the widely accepted caloric theory, the metal shavings cut by the boring tool could not hold as much caloric fluid as the solid casting had held, just as a sponge cut into thin shavings will hold very little water. Thus the caloric "leaked out" during the boring and manifested itself as heat flow. To discredit this view, Rumford ran the boring machine with a completely dull tool. In 45 minutes the tool, driven by two horses, wore less than 6 g of brass off a 50-kg casting (in modern units). But the temperature of the casting nevertheless rose by almost 40 K. Rumford asked, "Is it possible that the very considerable quantity of Heat . . . could have been produced by so inconsiderable a quantity of metallic dust? and this merely in consequence of *a change* of its capacity for Heat?"

Count Rumford: Spy, Scoundrel, Physicist, Engineer, Social Reformer

Born in colonial Massachusetts, Benjamin Thompson married a rich widow at the age of 19 and soon became a Loyalist in the American Revolution. Abandoning his wife and infant daughter, he worked as a spy (probably as a double agent) and in 1776 fled to England, where he advanced his career through a combination of brilliance, intrigue, and corruption. Returning briefly to America as a military officer, he earned the enduring hatred of the farmers of Long Island through his high-handedness and callousness. He spent most of the rest of his life in Bavaria (where he was made a Count of the Holy Roman Empire in spite of the fact that he was a Protestant), England, and France. He made numerous advances in the physics of heat and light, almost all of which (like the cannon-boring experiment) stemmed directly from practical investigations. His inquiry into the insulating power of fabrics intended for military uniforms led him to elucidate the phenomenon of convection and to ascertain that water attains its maximum density at 3.9°C. In the course of his search for better heating fuels and lighting sources, he invented the combustion calorimeter and the Rumford photometer. Modern cooking pots, stoves, the drip coffee pot, and central steam heat are based on some of his inventions. In spite of his contempt for common people, he was a great social reformer; among other things, he devised elaborate systems for the mass feeding and employment of the poor, founded the first free school for poor children and the first public park in Europe (the English Garden in Munich), and pioneered the use of the potato as a staple food in Bavaria. He founded the Royal Institution in London (later the home of Humphry Davy and Michael Faraday) and endowed the Rumford Premiums, pioneer prizes for scientific achievement. In 1804, Rumford married the brilliant, rich, and powerful widow of the great chemist Antoine Lavoisier. Their incompatibility led to a separation, and he died a semirecluse near Paris. For an entertaining short biography of Rumford, see Sanborn C. Brown, *Count Rumford: Physicist Extraordinary* (Garden City, N.Y.: Doubleday, 1962).

In further experiments, Rumford showed that water placed in a jacket surrounding the boring machine could be made to boil indefinitely. He concluded:

... the Heat generated by friction ... appeared evidently to be *inexhaustible*. ... [A]nything which any *insulated* body ... can continue to furnish *without limitation* cannot possibly be a *material substance*; and it appears to me to be ... impossible to form any distinct idea of anything capable of being excited ... in [this] manner ... except it be MOTION.

Rumford often demonstrated this experiment for visitors, and it never failed to impress them. He wrote, "It would be difficult to describe the surprise and astonishment expressed in the countenances of the bystanders, on seeing so large a quantity of cold water heated, and actually made to boil, without any fire."

THINKING LIKE A PHYSICIST

The kinetic and caloric theories of heat could both be used to account for many (if not all) of the thermal properties of matter as they were known toward the end of the eighteenth century. Confronted with these two distinct and conflicting theories, Rumford conceived and carried out a *crucial experiment*—an experiment whose results could be reconciled with one of the theories but not the other. A crucial experiment that results in a clear choice between two theories, both of which plausibly account for a wide range of phenomena, is a rarity and is not usually accepted as the last word until a wider range of experimental evidence is available. After all, both of the theories could be wrong in the light of further evidence—or the experimental results could be wrong! Nevertheless, a crucial experiment is at the very least a rallying point for physicists, who are encouraged by its results to undertake further experiments.

Joule's Friction Calorimeter

Rumford's demonstrations provoked a growing interest in the kinetic theory of heat and particularly in accurately determining the value of the mechanical equivalent of heat J defined by Equation 17.27. This interest culminated in the life work of Joule. In a long series of meticulous experiments covering more than three decades beginning in 1840, Joule employed various forms of energy to evolve heat using every method at his disposal. In each case, he carefully measured the work done and the heat evolved. Of his many experiments, the one illustrated in Figure 17.8 is the most famous. This experiment—essentially a refinement of the one in which Rumford made water boil in the jacket surrounding the cannon casting—makes possible a reliable quantitative measurement of J. Joule's experiment thus goes beyond Rumford's demonstration of the first point of the experimental program and proceeds to the second: The heat Q evolved is directly proportional to the frictional work W done.

An experimental run with Joule's device, called a friction calorimeter, begins with a measurement of the temperature inside the calorimeter. The weights are allowed to descend, driving the calorimeter paddles and thus doing work against the fluid friction in the system. When the weights have reached bottom, a clutch mechanism is used to disengage them from the calorimeter, and the windlass crank is used to raise them. The clutch is reengaged, and the weights are allowed to descend again. This process is repeated, typically about twenty times. Finally, the temperature inside the calorimeter is measured again. Small corrections must be made to allow for the flow of heat through the walls of the calorimeter as its temperature rises above that of its surroundings, the residual friction of the bearings, the small but nonzero kinetic energy of the weights as they hit bottom, and the sound energy emitted by the system as it runs; we will not consider them here. A typical experimental run is considered in Example 17.7.

Joule: The Fruits of Meticulousness and Precision
James Prescott Joule (1818–1889) was born into a family of prosperous brewers. Like Cavendish and Young before him, Joule was among the brilliant amateurs who dominated British physics from the death of Newton until the mid-nineteenth century.

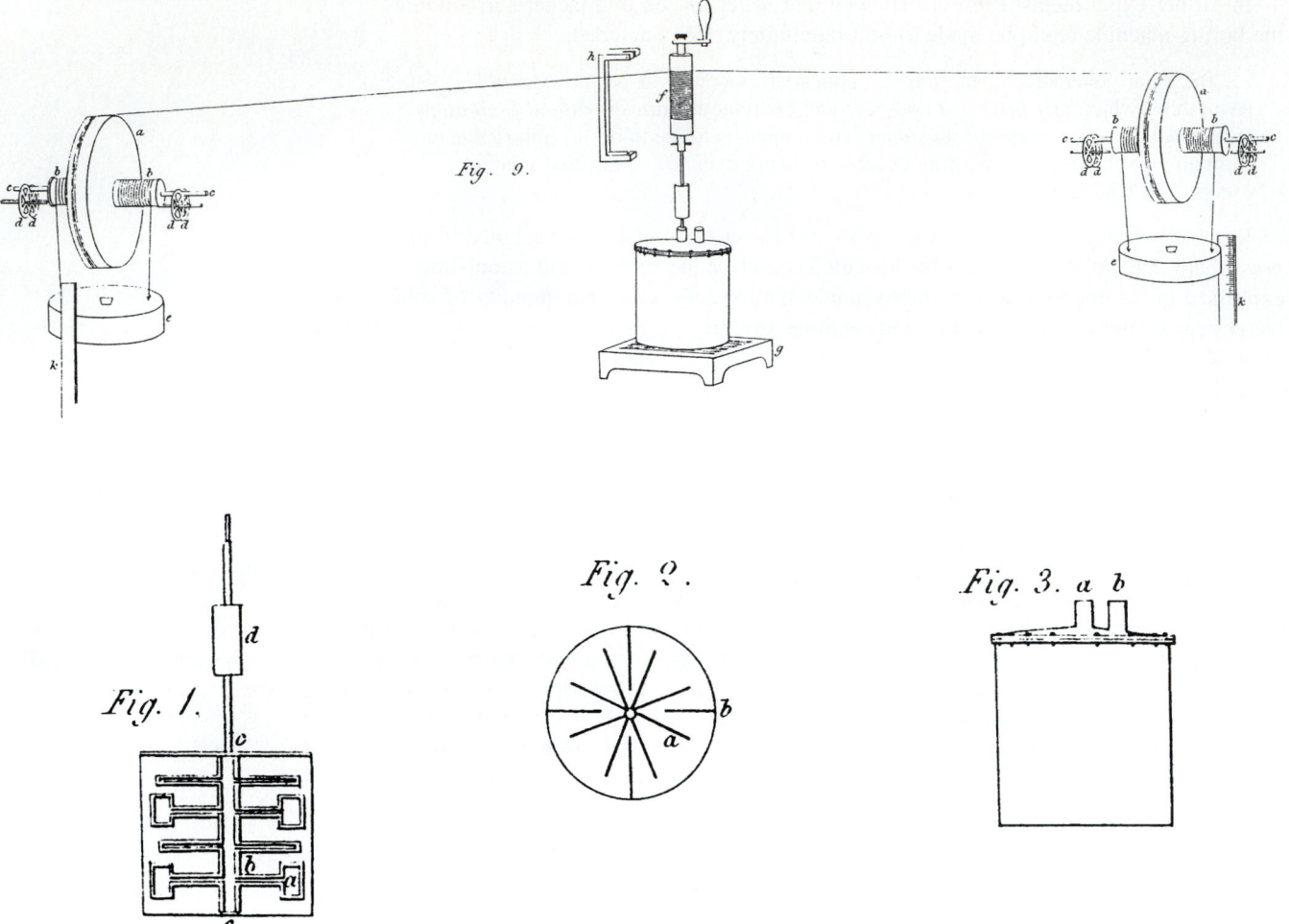

FIGURE 17.8 Joule's friction calorimeter, taken from his account of the experiment. The calorimeter proper is labeled Fig. 9; Fig. 1 is a cutaway view of the same apparatus, and Fig. 2 is a top view. In Fig. 1, the paddles, *a*, are made to rotate by the drive shaft, *dc*. The container is filled with water, and the temperature is measured by means of a very sensitive mercury-in-glass thermometer inserted through bushing *b* in Fig. 3. The drive shaft is driven by the windlass, *f*, shown in Fig. 9. The windlass in turn is driven by the cords attached to the twin pulley systems shown at left and right. (For clarity, the twin systems are shown rotated about vertical axes from their actual orientation, in which the shafts are normal to the plane of the page.) The primary motive force is furnished by heavy weights, *e*, whose descent distance is measured by means of scales, *k*. When the weights descend, their gravitational potential energy *U* is converted into kinetic energy of motion of the system. This kinetic energy is in turn converted into heat by means of the fluid friction that arises as the paddles are driven through the water in the calorimeter. Good bearings are used throughout to minimize friction everywhere else in the system.

EXAMPLE 17.7

Joule obtained the following data from one of his experimental runs. (All quantities are converted into modern units.)

- Mass m of driving weights e: 26.32 kg
- Total vertical descent h of weights: 31.85 m
- Total heat capacity C of calorimeter: 6.316 kcal/K
- Temperature rise ΔT inside calorimeter: 0.316 K

- Acceleration of gravity g at Joule's laboratory in Manchester: 9.814 m/s²

Using these data, calculate the value of J.

SOLUTION: From Equation 17.27, you have

$$J = \frac{W}{Q}. \tag{17.28}$$

As the paddles do frictional work on the water, the internal energy of the calorimeter is increased; the driving weights descend and their gravitational potential energy decreases. Thus the total frictional work W is equal to the total decrease in gravitational potential energy $-\Delta U$ of the weights (neglecting the small losses elsewhere in the system). You have

$$W = -\Delta U = mgh = 26.32 \text{ kg} \times 9.814 \text{ m/s}^2 \times 31.85 \text{ m}$$
$$= 8227 \text{ J}.$$

Because the total heat capacity C of the calorimeter is given, you can use Equation 17.25, $C = Q/\Delta T$, to find the evolved heat Q resulting from the frictional process. You write

$$Q = C \Delta T = 6.316 \text{ kcal/K} \times 0.316 \text{ K}$$
$$= 1.996 \text{ kcal}.$$

Inserting these numerical values for Q and W into Equation 17.28 and rounding off to three significant figures, you obtain the result

$$J = \frac{8227 \text{ J}}{1.996 \text{ kcal}} = 4120 \text{ J/kcal}.$$

This value lies within 2% of the best modern measurements.

Completing the Four-Point Program

We now turn to the last two points of our four-point program aimed at demonstrating that heat flow is a form of energy transfer. To demonstrate point 3—the value of J does not depend on how the frictional work is done or on the rate at which it is done—Joule repeated the experiment illustrated in Figure 17.8, replacing the water-filled calorimeter with a mercury-filled calorimeter. He then used a "dry" calorimeter, in which the frictional action is between solid disks. Within experimental error, the same value of J is found in all cases.

Joule demonstrated point 4—other dissipative processes result in the evolution of heat with the *same* proportionality constant J—by using forms of energy other than mechanical energy to increase the internal energy of the calorimeter. Most notably, he carried out experiments in which electrical energy was used in this fashion. We will discuss such experiments in Section 27.6.

We conclude that a particular thermal phenomenon in a given system—for example, a particular temperature change in a given calorimeter—can be produced *either* by a given heat transfer Q to the system *or* by transfer of energy $W = JQ$ to the system that results in evolution of heat Q in the system. The only plausible interpretation of this conclusion is that heat transfer is simply a form of energy transfer.

Redefining the Kilocalorie

The four-point program for demonstrating that heat flow is a form of energy transfer has been amply fulfilled by the work of Rumford, Joule, and many others. Much work has been devoted to the precise determination of the mechanical equivalent of heat J. The best modern value is

$$J = 4186 \text{ J/kcal}. \tag{17.29}$$

As you have seen, the kilocalorie is a convenient unit for heat measurement in processes in which the only thing that happens is heat flow from one body to another and there is no conversion of other kinds of energy into internal energy or vice versa. But there are many important processes in which such conversion does take place; among them are heat engine cycles and chemical reactions.

Once we are confident that heat flow is a form of energy transfer, it makes no sense *fundamentally* to measure heat flow in any units other than energy units. Mixing energy units and heat units in the same equation requires the use of the conversion factor J, and this is a nuisance. More seriously, the value of J will change repeatedly as new, more precise measurements become available.

In order to avoid these difficulties, the kilocalorie has been *redefined* in terms of the joule:

$$1 \text{ kcal} \equiv 4186 \text{ J}. \tag{17.30}$$

That is, 1 kcal is no longer the amount of heat flow required to increase the temperature of 1 kg of water by 1 K. Not surprisingly, the value in Equation 17.30 was chosen so that the new definition corresponds closely to the old one.

Having recognized that heat flow is a form of energy transfer, we can apply the principle of energy conservation in a much broader way. Until now, we have been able to consider conservation only of mechanical energy. And, paradoxically, mechanical energy is conserved only in conservative systems. We can now extend the conservation principle to systems in which friction is present. We assert that in any isolated system

$$\Delta K + \Delta U + Q = 0; \qquad (17.31)$$

the sum of the change in (macroscopic) kinetic energy, the change in potential energy, and the heat evolved in the system is zero. Example 17.8 involves an application of this generalized form of the principle of energy conservation.

EXAMPLE 17.8

A lead rod of mass $M = 0.72$ kg is propelled into a long, narrow tube containing water, as shown in Figure 17.9. The temperature of the rod and that of the water are equal. The speed v of the rod as it strikes the water surface is 15 m/s. The mass m of the water is 0.039 kg, and its initial depth h is 0.50 m. The lead rod, which fits fairly closely in the tube, slows down as it enters the water, sinking slowly to the bottom as the water flows past it through the narrow space between rod and tube wall. **(a)** How much heat Q is evolved in the process? **(b)** Neglecting the heat capacity of the tube, find the temperature rise ΔT of the system.

FIGURE 17.9

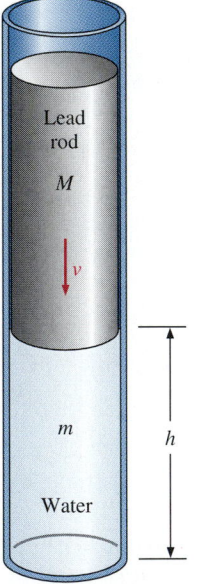

Lead rod
M

v

m

h

Water

SOLUTION:
(a) Find Q.

You begin by calculating the change ΔK in the kinetic energy of the system. Because only the lead rod is moving initially, you have $K_i = \frac{1}{2}Mv^2$. And, because the entire system ends up at rest (once the spraying and splashing of the water have ceased), you have $K_f = 0$. Thus the kinetic energy change is

$$\Delta K = 0 - \tfrac{1}{2} \times 0.72 \text{ kg} \times (15 \text{ m/s})^2 = -81 \text{ J}.$$

Next, you calculate the change ΔU in the gravitational po-

tential energy of the system from the initial instant when the lower end of the rod strikes the water surface to the final instant when the rod end comes to rest at the bottom of the tube. This change consists of two parts. The first part, ΔU_{Pb}, is a decrease resulting from the descent of the center of mass of the rod. The second part, ΔU_w, is an increase resulting from the ascent of the center of mass of the water as it is displaced by the lead rod.

To calculate ΔU_{Pb}, you note that the center of mass of the lead rod descends a distance equal to the initial depth of the water in the tube. So you have

$$\Delta U_{Pb} = Mg(-h) = 0.72 \text{ kg} \times 9.8 \text{ m/s}^2 \times -0.50 \text{ m}$$
$$= -3.5 \text{ J}.$$

It is more complicated to make a precise calculation of the change in position of the center of mass of the water because the rod does not quite fill the tube. Thus the final configuration of the water is an annulus from the bottom of the tube to the top of the rod and a cylinder above the rod. But note that the total mass of the water is less than 5% of that of the lead rod, and that ΔU_{Pb} is less than 5% of ΔK. Thus ΔU_w is less than 5% × 5% = 0.25% of ΔU_{Pb}; you can justify neglecting ΔU_w entirely, and you set $\Delta U \simeq \Delta U_{Pb}$.

You are now ready to calculate Q. Using Equation 17.31, you have

$$Q = -(\Delta K + \Delta U)$$
$$= 85 \text{ J}.$$

(b) Find the temperature rise of the system.

To calculate ΔT you use Equation 17.23, $Q = mc\,\Delta T$. Both the lead rod and the water must be warmed. But, because the initial and final temperatures of both are the same, one value of ΔT applies to the entire system, and so you can write

$$Q = Mc_{Pb}\,\Delta T + mc_w\,\Delta T = (Mc_{Pb} + mc_w)\,\Delta T,$$

where c_{Pb} is the specific heat capacity of lead and c_w that of water. You solve for ΔT, obtaining

$$\Delta T = \frac{Q}{Mc_{Pb} + mc_w}.$$

Because your solution for part **a** expresses Q in joules, you must use values of specific heat capacity expressed in J/kg·K (Table

17.3). You have

$$\Delta T = \frac{85 \text{ J}}{0.72 \text{ kg} \times 0.13 \times 10^3 \text{ J/kg·K} + 0.039 \text{ kg} \times 4.2 \times 10^3 \text{ J/kg·K}}$$
$$= 0.33 \text{ K}.$$

As in the Joule experiment of Example 17.7, the temperature rise is not large. From the values in Table 17.3, you can see that the specific heat capacity of most substances is within an order of magnitude of 1000 J/kg·K. That is, it takes quite a lot of mechanical energy to produce a modest temperature rise. This point is also implicit in the equivalency 1 kcal = 4186 J.

Summing Up

Any physical property that varies with temperature is a **thermometric parameter**. A particular thermometer, based on any thermometric parameter, can be used to construct a **two-point temperature scale**. Until 1954, the **Celsius scale** was a two-point scale, defined by calling the freezing temperature 0°C and the boiling temperature 100°C.

In the interest of improved precision, the **triple point of water** is chosen as a fixed point instead of the freezing point. When a low-pressure gas thermometer is used at temperatures well above the condensation temperature of the gas, every one-degree change of temperature at constant pressure results in a fractional change in the volume of the trapped gas that is independent of the temperature and is nearly the same for all such thermometers. For the standard low-pressure helium thermometer, the fraction is (Equations 17.4 and 17.5)

$$\beta \equiv \frac{1}{V_{tp}} \frac{\Delta V}{\Delta t} = \frac{1}{273.16} \text{ (C°)}^{-1}.$$

Using this thermometer, we can measure any temperature t by comparing the corresponding volume V to the volume V_{tp} at the triple-point temperature t_{tp}. According to Equation 17.7, we have

$$t - t_{tp} = \frac{1}{\beta} \left(\frac{V}{V_{tp}} - 1 \right).$$

If the gas in the thermometer did not condense, its volume would go to zero at the Celsius temperature given by Equation 17.8:

$$t_0 = -\frac{1}{\beta} + t_{tp} = -273.15°C.$$

No lower temperature has physical meaning in terms of the gas thermometer, t_0 is called the **absolute zero of temperature**.

On the **thermodynamic temperature scale**, the absolute zero of temperature is called 0 K. The scale is then completely defined by fixing the triple point of water according to Equation 17.10,

$$T_{tp} \equiv 273.16 \text{ K}.$$

The Celsius scale is *redefined* on the basis of the thermodynamic scale according to Equations 17.16 and 17.11,

$$t \equiv T - 273.15 \text{ K}; \quad 1 \text{ Celsius degree (C°)} \equiv 1 \text{ kelvin (K)}.$$

Both the Celsius scale and the thermodynamic scale are **one-point scales**.

To a good approximation, the volume of most materials varies in a linear fashion with temperature, in accordance with Equation 17.19b,

$$\frac{\Delta V}{V} = \beta \Delta T;$$

β is called the **volume coefficient of thermal expansion**.

For solids, it is often convenient to consider the proportional change in length $\Delta l/l$ with temperature, rather than $\Delta V/V$. Equation 17.20,

$$\frac{\Delta l}{l} = \alpha \, \Delta T,$$

is often a good approximation; α is called the **linear coefficient of thermal expansion**.

When two bodies having different initial temperatures are placed in **thermal contact**, heat flows from the warmer body to the cooler one until their temperatures attain the same value T_f. The value of T_f always lies between the two initial temperatures (Inequality 17.24). The amount of heat Q that flows into a body is proportional to the temperature change ΔT it experiences, according to Equation 17.23:

$$Q = mc \, \Delta T.$$

We define the unit of Q, the **kilocalorie**, in a tentative phenomenological way: The amount of heat required to raise the temperature of 1 kg of water by 1 K is 1 kcal. It follows that the **specific heat capacity** c for water has the value 1 kcal/kg·K. For other substances, c must be determined experimentally. The proportionality constant $C = mc$ (Equation 17.25) is called the **total heat capacity** of the body. The sign of Q is the same as the sign of ΔT. We say that heat flows *into* a body when the temperature of the body increases and $Q > 0$; heat flows *out of* a body when the temperature of the body decreases and $Q < 0$.

There is a discontinuous change in c when the body experiences a phase change. While a body is undergoing a phase change, its temperature remains constant. Nevertheless, the process requires a flow of heat into or out of the body. The amount of heat Q required is proportional to the mass of the body according to Equation 17.26,

$$Q = mL;$$

the constant L, called the **latent heat**, is the heat required to effect the phase change per unit mass of the body.

When work W is done in a frictional or other dissipative process, heat Q is evolved. The two quantities are always proportional to each other according to Equation 17.27,

$$W = JQ;$$

the proportionality constant J is called the **mechanical equivalent of heat**. The value of J is completely independent of the form of energy involved in the performance of the work and of all of the details of the process. This experimental observation is powerful evidence in support of the assertion that *heat flow is a form of energy transfer*. On the basis of this assertion, we redefine the kilocalorie in terms of the fundamental unit of energy, the joule, by means of Equation 17.30,

$$1 \text{ kcal} \equiv 4186 \text{ J}.$$

We expand the principle of conservation of energy to include heat energy. The expanded form of the principle, expressed in Equation 17.31,

$$\Delta K + \Delta U + Q = 0,$$

is much more widely applicable than the form restricted to the potential and kinetic forms of macroscopic mechanical energy.

KEY TERMS

Section 17.2 Thermometers and Thermometry
thermometer, thermometric parameter ▪ two-point temperature scale, ice point, steam point ▪ centigrade scale ▪ Celsius scale ▪ triple-point temperature and pressure

Section 17.3 The Constant-Pressure Gas Thermometer
coefficient of thermal expansion ▪ ideal gas thermometer ▪ thermodynamic (one-point) temperature scale ▪ Charles's or Gay-Lussac's law

Section 17.4 Thermal Expansion
linear coefficient of thermal expansion

Section 17.5 Heat Flow
specific heat capacity, total heat capacity ▪ thermal contact ▪ thermal equilibrium ▪ thermal insulator, thermal conductor ▪ zeroth law of thermodynamics

Section 17.6 Phase Changes and Latent Heat
latent heat

Section 17.7 The Mechanical Equivalent of Heat
kinetic theory of heat ▪ internal energy ▪ evolution of heat ▪ mechanical equivalent of heat

QUERIES

17.1 *(2) Don't miss the point!* One step in redefining the Celsius scale from a two-point scale to a one-point scale involves setting $t_{tp} = 0.01°C$. How does this change affect the size of the degree? Is the one-point Celsius scale a centigrade scale exactly? approximately?

17.2 *(2) Reverse swedish?* In devising his temperature scale, Celsius assigned the temperature 100° to the ice point and 0° to the steam point. Does this assignment result in a valid two-point scale? Could the scale be modified to a one-point scale? Explain.

17.3 *(3) Galileo's thermoscope.* Pictured below is a crude thermometer invented by Galileo, called a *thermoscope*. The narrow vertical tube is partially filled with a column of colored liquid a few centimeters long. The temperature of a body is measured by placing it in contact with the bulb at the top of the tube. Explain the operation of the thermoscope. To what errors is it liable if the connection to the outside air (at the right of the flask) is open? if it is sealed?

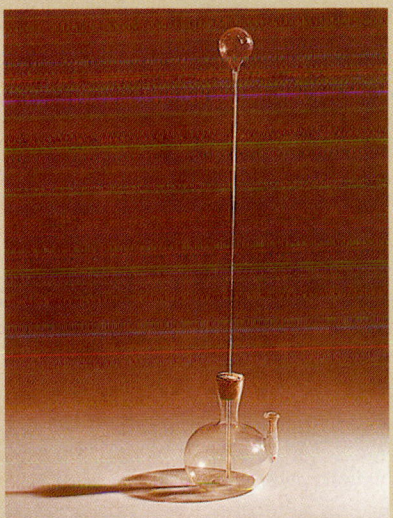

17.4 *(3) Sign of the times.* The volume coefficient of expansion is defined by Equation 17.4, $\beta \equiv (1/V_{tp})(dV/dt)$. The compressibility is defined by Equation 15.26, $\kappa \equiv -(1/V)(dV/dp)$. Explain why one of these definitions contains a minus sign and the other does not.

17.5 *(3) Hootenanny.* One verse of a folk song goes

> It froze clear down to China,
> It froze to the heavens above;
> At a thousand degrees below zero
> It froze my logger love.

Comment.

17.6 *(4) Everything else being equal . . .* If all materials had the same expansion coefficient, could you build a liquid-in-glass thermometer?

17.7 *(4) Quick dip.* When you suddenly immerse a mercury-in-glass thermometer in a container of boiling water, the mercury level falls momentarily before it begins to rise. Explain.

17.8 *(4) Design criteria.* Why is mercury a good choice for the liquid in a general-purpose thermometer?

17.9 *(4) Neatness counts.* Atoms in crystalline solids are arranged in a very orderly fashion. On the basis of this fact, explain why most substances contract on freezing.

17.10 *(4) Skating on thin ice.* What would happen to lakes in winter if ice were denser than water, as is the case for most solids and their corresponding liquids? Describe the resulting freezing process, the transfer of heat between lake and air, and the melting process in the spring; compare with what actually happens.

17.11 *(4) Type-casting off Broadway.* Despite many technological advances, the highest-quality printing is still done by the old-fashioned letterpress method. Type is made by casting a low-melting metal alloy in a precision mold. In the printing process, the typeface is wetted with ink and pressed against the paper. Each piece of type must be a faithful replica of the mold if the printed matter is to be legible and attractive. Explain why the alloy used, called type metal, must be icelike.

17.12 *(4) Keep your eye on the donut, not the hole.* A metal plate has a circular hole in it. When the plate is heated uniformly, what happens to the diameter of the hole? Explain.

17.13 *(4) Ruling the bends.* Depicted below is a *bimetallic strip*, made of thin sheets of copper and iron bonded together. Explain how the strip can be used as a thermometer.

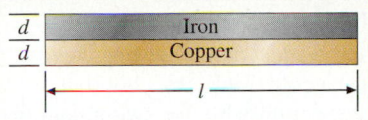

17.14 *(5) Sommelier practice.* A room-temperature bottle of white wine, placed in a 5°C refrigerator, requires several hours to cool to the best temperature for drinking (about 10°C). If the bottle is placed in a −18°C freezer, cooling to 10°C requires about half an hour. If the bottle is immersed in a bucket of ice cubes, the same cooling requires only about 15 minutes, even though the ice is not as cold as the freezer; the process is hastened even further by adding a little water to the ice. Explain.

17.15 *(6) Espresso bar.* The tasty Italian milk-and-coffee drink called *cappuccino* is made by heating the initially cold milk and rendering it foamy by bubbling steam through it from the *caffè espresso* machine. In view of the fact that steam condenses in the milk during the process, explain why it is possible to make the milk quite hot without making it unpleasantly watery at the same time.

17.16 *(6) Why boilers are licensed.* Burns produced by the exposure of skin to steam are often more severe than burns produced by boiling water. Explain.

17.17 *(6) Cash flow.* Accountants describe the activity of a company in terms of net cash flow, defined as the difference between the total money taken in by the company per unit time and the total money spent by the company per unit time. What happens to a company that has positive cash flow? negative cash flow? If the cash flow is positive, can you conclude that piles of money are accumulating in the company office? Does measurement of cash flow imply the existence of cash? Does measurement of heat flow imply the existence of heat?

17.18 *(7) No free lunch.* In the Joule experiment, macroscopic kinetic energy is converted into the internal energy of the calorimeter. What is the source of the kinetic energy?

17.19 *(7) Keep fit, clothes fit.* Exercise is an important part of a weight-reduction program because it tends to keep the dieter's mind off food, moderates appetite and feelings of hunger, improves muscle tone to counteract the sagging of skin and connective tissue that often accompanies fat loss, and often improves the dieter's general psychological outlook. But the direct contribution of exercise to weight loss, through ''burning off fat,'' is minor. Explain.

17.20 *(7) Getting down to earth.* When a spacecraft reenters the atmosphere, its mechanical energy is dissipated largely through frictional heating. Refer to Chapter 14, and estimate the energy per unit mass that must be dissipated in bringing the spacecraft to rest on the ground. Compare this quantity with typical latent heats of melting and boiling listed in Table 17.4, and see why it's not so simple to bring the spacecraft down in one uncooked piece! (In single-use reentry vehicles, the frictional heat is usually dissipated by means of a complex combination of mechanisms collectively called *ablation*, in which a plastic nose cone decomposes chemically and emits gas that forms an insulating layer, chars, glows, and radiates.)

17.21 *(G) Philosophical reflection.* The English philosopher Francis Bacon asserted that a container of warm water put in a cold place will freeze more quickly than an identical container of cool water in the same place. You have probably heard this assertion cited as one of those "strange but true facts of nature." Is there any objective basis for it?

PROBLEMS

GROUP A

17.1 *(2) Getting the feel of it.* Give approximate Celsius temperatures for **(a)** a cold drink; **(b)** a hot (but drinkable) cup of coffee; **(c)** water comfortable for swimming; **(d)** bath water; **(e)** an oven set for baking a cake; **(f)** air comfortable for taking off your winter coat on a windless evening.

17.2 *(2) Réaumur scale.* On French farmhouses and barns, you occasionally find thermometers using the Réaumur scale, a two-point scale on which the ice and steam points are fixed respectively at 0°Ré and 80°Ré. **(a)** Is the Réaumur scale a centigrade scale? **(b)** Derive an equation for converting Celsius temperatures into Réaumur temperatures. **(c)** Express normal human body temperature (37°C) on the Réaumur scale.

17.3 *(2) Different scale, same reading.* A Celsius thermometer and a Fahrenheit thermometer read the same. What is the temperature?

17.4 *(3) Seems hotter!* On a hot summer day, the temperature is 35°C. Express the temperature in kelvin.

17.5 *(3) Spreading out.* At the ice point, the volume of the nitrogen contained in a constant-pressure gas thermometer is 35.62 cm³. What will the volume be at the steam point?

17.6 *(3) Much warmer.* Hydrogen boils at 20.3 K. If 12.2 cm³ of hydrogen gas is warmed from the boiling point to room temperature (20°C) at constant pressure, what is the final volume of the gas?

17.7 *(3) Seasonal change.* Compare the density of the air on a hot summer day, when the temperature is 35°C, with the density on a cold winter day, when the temperature is −15°C. Assume that the barometric pressure is the same on both days and that the composition of the air is the same.

17.8 *(3) Rankine temperature scale.* The Rankine scale is defined by choosing the size of the degree to be the same as that of the Fahrenheit degree and choosing the zero point (0°R) to coincide with the absolute zero of temperature. **(a)** Express the Rankine degree in terms of the Celsius degree. **(b)** Find the value of absolute zero on the Fahrenheit scale. **(c)** Express the ice point, the steam point, and the triple point of water on the Rankine scale.

17.9 *(4) The whole truth?* The density of water at 4°C is 1.00×10^3 kg/m³. The density of ice at 0°C is 0.9168×10^3 kg/m³. Using information from Table 17.1, determine whether the expansion of water as it is cooled below 4°C accounts for this entire difference. If it does not, what percentage of the difference is thus accounted for?

17.10 *(4) Volumetric correction.* A chemist's volumetric flask is made with a long, narrow cylindrical neck on which an engraved mark denotes the capacity calibrated (usually) at 15°C. If a 1-liter volumetric flask is immersed in a boiling water bath and slowly filled to the mark with oil, what volume of oil is contained in the flask?

17.11 *(4) Graduated cylinder.* A graduated Teflon cylinder is filled to the 100-cm³ mark with water whose temperature is 80°C. If the water is chilled to 10°C and poured quickly into an identical cylinder whose temperature is 80°C, what will be the volume reading on the graduated scale before the cylinder cools down appreciably? (Water does not wet Teflon, so you need not worry about water remaining in the first cylinder.)

17.12 *(4) Real mercury-in-glass thermometer.* In a real mercury-in-glass thermometer, the glass envelope expands as well as the mercury contained in it. In making the calculation of Example 17.3, we neglected the expansion of the glass envelope. Correct for the expansion of the envelope, using the volume expansion coefficients β given for mercury and glass in Table 17.1. Specifically, show that the actual change in length $\Delta l'$ of the mercury column is about 86% of the value Δl calculated in Example 17.3.

17.13 *(4) Stretching things a little.* You have an aluminum bar whose length is 1.000 m when the temperature is 15°C. At what temperature will its length be 1.001 m? 0.999 m?

17.14 *(4) Feelin' groovy.* The photograph below shows the Queensboro Bridge in New York City. As in all cantilever bridges, the central span between the two main towers consists

of two structurally independent sections, each of which hangs from one of the towers. The total central span length is 360 m. To allow for expansion and contraction where the two sections join, it is necessary to provide an *expansion joint* which allows relative movement. How much relative movement must be allowed for in designing for temperature extremes of −30°C and 50°C? (The high extreme takes into consideration the effects of direct heating by the sun.) The bridge is made of steel.

17.15 (4) *Coefficient of expansion of nothing?* A square copper plate 10 cm on a side has a circular hole drilled through it. At 10°C, the diameter of the hole is 2.50 cm. What is the diameter of the hole when the plate is heated to 250°C? (Hint: See Query 17.12.)

17.16 (4) *Shrink fit.* Sometimes it is necessary to make a band fit very tightly around a circular disk. A classical example is the fitting of iron tires on wooden wagon wheels. One way of doing this is by *shrink fitting.* The band is designed so that its inner diameter is smaller than the outer diameter of the disk on which it is to be fitted. The band is heated, slid over the disk, and then allowed to cool. (Alternatively, the disk can be chilled below room temperature, the band slid on, and the disk allowed to warm up.) Suppose a wheelwright wishes to fit an iron tire onto a wheel whose diameter is 1.850 m. If he uses a tire whose inner diameter at room temperature is 1.845 m, to what temperature must he heat it before fitting it onto the wheel?

17.17 (5) *Calorimetry, I.* If you pour 65 g of ethanol at 3.0°C into 98 g of water at 72°C, what is the final temperature of the mixture?

17.18 (5) *Calorimetry, II.* You wish to determine the specific heat capacity of magnesium. You obtain a sample and determine that its mass is 36.8 g. You prepare a calorimeter (an insulated container) containing 232 g of water at 15.2°C. You heat the magnesium sample in boiling water and then drop it into the calorimeter. The final temperature of the system is 19.3°C. What is the value of c for magnesium? Neglect the heat capacity of the calorimeter container.

17.19 (5) *Calorimetry, III.* You would like to use the technique of Problem 17.18 to verify the published value of the specific heat capacity of sodium. But sodium reacts violently with water, and so you cannot use water in the calorimeter. Instead, you fill the calorimeter with 135 g of trichloroethylene, whose specific heat capacity is 0.223 kcal/kg·K. The initial temperature of the calorimeter is 14.7°C. You use a separate trichloroethylene bath to warm a 7.55-g sample of sodium to 42.5°C. If the published value of c for sodium is 0.293 kcal/kg·K, what final temperature should you expect in the calorimeter?

17.20 (5) *Calorimetry, IV.* A 500-g lump of copper is dropped into a calorimeter containing 1.00 kg of water at 7.50°C. The final temperature of the system is 21.22°C. What was the original temperature of the copper?

17.21 (5) *Calorimetry, V.* You can get an estimate of the temperature of the flame in a kitchen gas stove as follows. Partially fill a metal measuring cup with water, and measure the temperature. Choose a medium-size steel nail, and measure its mass. Using a pair of pliers, hold the nail as completely in the flame as possible until the nail is thoroughly heated. Then drop the nail into the water, and measure the final temperature of the system. Suppose that in such an experiment the mass of the water is 125 g, the mass of the nail is 8.5 g, the initial temperature of the water is 8.0°C, and the final temperature of the system is 14.1°C. What was the initial temperature of the nail? Why is this likely to be an underestimate of the temperature of the flame?

17.22 (6) *Letting off steam.* If you begin with a 14-kg block of ice whose temperature is 260 K, how much heat must you transfer to it to boil it all away?

17.23 (6) *Do it in a hood!* If you begin with solid carbon tetrachloride at its melting point, how much heat must you transfer to it to boil it all away? Take the specific heat capacity of carbon tetrachloride to be 0.206 kcal/kg·K.

17.24 (7) *Missing information, I.* In Table 17.3, one entry for carbon dioxide is missing. Calculate the missing value.

17.25 (7) *Missing information, II.* In Table 17.3, one of the entries for ethanol is missing. Calculate the missing value.

17.26 (7) *Keeping warm, and then some.* (a) An adult man engaged in quiet activity in surroundings at a comfortable temperature evolves heat at a rate of about 100 W when awake and at about half that rate when asleep. Approximately how many "dietitian's calories" of food must he eat per day to maintain constant weight, assuming an 8-hour sleep period? (The dietitian's calorie is equal to the kilocalorie.) (b) A trained athlete engaged in an endurance sport such as bicycle road racing or triathlon may well eat 6000 kcal of food a day. Taking the human system as a whole, from digestive tract to muscle, about 10% of the input energy, over and above the maintenance energy discussed in part **a**, is converted into muscular output. How much mechanical work does such an athlete perform in a day? Assuming that the major output takes place over 10 hours, what is the average muscular power output of the athlete? (c) An input ranging from 6000 to 7000 kcal is close to the maximum amount of food a human digestive system can process per day. Why do English Channel swimmers lose weight, even on this maximum intake?

17.27 (7) *Scientific honeymoon.* Joule and his new wife honeymooned in Switzerland. High waterfalls are a spectacular feature of the Swiss Alps, especially in summer when the falls are fed by rapidly melting ice. Joule had brought along some sensitive thermometers, and he showed by direct measurement that the water in the pool at the bottom of a waterfall is a little warmer than the water at the top, where the fall begins. (a) What is the mechanism by which kinetic energy is converted into heat? (b) On the same scale, make schematic sketches of the gravitational potential energy, the kinetic energy, the heat energy, and the total energy of the water as a function of distance fallen from the top. (Assume that the water falls vertically, with little obstruction from top to bottom.) (c) What is the top-to-bottom temperature difference for a waterfall 250 m high? (Note: You might think that the modest temperature rise would be completely masked by cooling of the water due to evaporation. But the copious spray that such waterfalls produce assures saturation of the atmosphere with water vapor for some distance around the waterfall. As a result, the water that reaches the bottom pool has suffered negligible evaporation.)

17.28 *(3) Thermometer conversion.* The volume of gas V_i contained in a constant-pressure gas thermometer, whose pressure is p_0, is measured at the ice point. The surroundings of the thermometer are then warmed to the steam point, where the volume is V_s. With the system kept at this temperature, the pressure in the thermometer is increased to the new value p_1, at which the thermometer volume is again equal to V_i. **(a)** Show that $p_1/p_0 = V_s/V_i = T_s/T_i$. **(b)** What would happen to the pressure in the thermometer if volume rather than pressure were held constant during warming? Show that the thermometer is thus converted into a *constant-volume* gas thermometer. Make a schematic sketch of a constant-volume gas thermometer, and explain how you would go about measuring temperature with it.

17.29 *(3) Another form of Charles's law.* To generalize the result of Problem 17.28, note that Equation 17.15 states Charles's law in the form $V_1/V_2 = T_1/T_2$ for p constant, where T_1 and T_2 are any two temperatures on the thermodynamic scale. Show that this relation, taken together with Boyle's law, implies $p_1/p_2 = T_1/T_2$ for V constant. This result is also called Charles's law or Gay-Lussac's law.

17.30 *(3) Unusual thermometer.* The area of the upper piston illustrated below is A_1, and that of the lower piston is A_2. Both pistons are exposed to the outside air, whose pressure is p_0. The pistons, of combined mass m, are connected by a light, inextensible string, and they can slide through their tubes without friction. **(a)** What is the pressure of the gas between the pistons? **(b)** Find $\Delta h/\Delta T$, the height h through which the pistons rise per unit temperature increase.

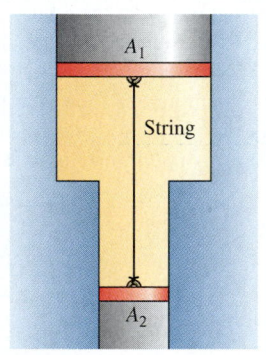

17.31 *(4) Comparison rule.* Two rulers made of different materials have the same length l at temperature t_0. At temperature t_1, ruler B is longer than ruler A by Δl. Express the coefficient of expansion α_B of the material of which ruler B is made in terms of the coefficient α_A of the material of ruler A.

17.32 *(4) Spreading out, I.* A square piece is cut from a copper sheet. When the piece is heated from 20°C to 150°C, find the percentage increase in **(a)** the volume, **(b)** the thickness, and **(c)** the area of one of the faces.

17.33 *(4) Spreading out, II.* A cubic block of volume V is made of a material whose linear coefficient of thermal expansion is α. If the temperature of the block is increased by Δt, show that the increase ΔA in the surface area is

$$\Delta A = 12\alpha V^{2/3} \Delta t.$$

17.34 *(4) Spreading out, III.* **(a)** Repeat Problem 17.33 for a sphere of volume V. **(b)** Compare the results for the two problems, and show that the difference between the two results lies entirely in a constant factor.

17.35 *(4) Spreading it thinner.* As the temperature of a body changes, its density ρ changes as well. Show that the rate of change is $d\rho/dt = -\beta\rho$. Account for the presence of the minus sign in this result.

17.36 *(4) Temperature correction of a mercury barometer, I.* As you saw in Example 15.8, standard atmospheric pressure is defined as the pressure required to support a mercury column exactly 760 mm high, when the temperature is 0°C. **(a)** Usually, a barometer is read at some different temperature. Explain why a correction to the mercury column height reading must be made. **(b)** If the barometer reading is h and the Celsius temperature is t, show that the barometric pressure corresponds to the corrected column height $h' = h(1 - \beta t)$, where β is the volume coefficient of expansion of mercury and the expansion of the glass barometer tube is neglected.

17.37 *(4) Everybody likes to slow down on a warm day.* The rate of a pendulum clock depends on the period of the pendulum, which depends in turn on the length of the pendulum. But if the temperature changes, the length of the pendulum also changes and the clock rate changes. **(a)** Consider a typical clock pendulum, consisting of a heavy brass bob hung at the end of a long, thin steel rod. Such a pendulum is a good approximation to a simple pendulum. Suppose the clock runs correctly at some temperature t_0 with a pendulum period τ_0. Show that when the temperature differs from t_0 by an amount Δt, the proportional error in the rate of the clock is approximately

$$\frac{\Delta\tau}{\tau_0} = \tfrac{1}{2}\alpha\,\Delta t,$$

where α is the linear coefficient of expansion of steel. **(b)** Show that the clock will be in error by about $\tfrac{1}{2}$ s per day for each one-degree variation of temperature above or below t_0. **(c)** What would the corresponding error be if the pendulum were made of Invar instead of steel?

17.38 *(4) Taken as a whole.* A bar consists of two sections of different materials, having coefficients of expansion α_1 and α_2. The length of section 1 is a fraction f_1 and the length of section 2 is a fraction f_2 of the total length of the bar. **(a)** Show that the bar, taken as a whole, has a coefficient of expansion $\alpha = \alpha_1 f_1 + \alpha_2 f_2$. **(b)** Using aluminum and brass, you want to make up a 1.00-m bar having $\alpha = 2.0 \times 10^{-5}\ \text{K}^{-1}$. How long should the two sections be?

17.39 *(4) Beg to differ.* At 0°C, the length of an aluminum rod is 1.00 m and that of a copper rod is 1.38 m. **(a)** What is the difference between the lengths of the two rods when the temperature is 100°C? **(b)** Show that the difference in length between any two rods is independent of temperature if their lengths are chosen so that $l_2/l_1 = \alpha_1/\alpha_2$.

17.40 *(4) Temperature-compensated pendulum, I.* The *gridiron pendulum* illustrated on p. 473 is designed to compensate for the error described in Problem 17.37. The heavy weight, containing most of the mass of the pendulum, is suspended at its midline. It is connected to the long, relatively light steel pendulum rod by means of the suspension shown in the drawing. The vertical rods making up the suspension are made alternately of aluminum and steel. The length of the pendulum, from the pivot to the midline of the heavy weight, is 1.100 m. (a) What should the length h of the suspension bars be if the period of the pendulum is to be independent of temperature? (Hint: See part **b** of Problem 17.39.) (b) Generalize the pendulum illustrated below. Let the pivot-to-midline distance be l, let the linear coefficients of expansion of the two metals be α_1 and α_2, and suppose that there are $2n$ vertical suspension rods on each side of the main pendulum rod. Find the proper value of h/l.

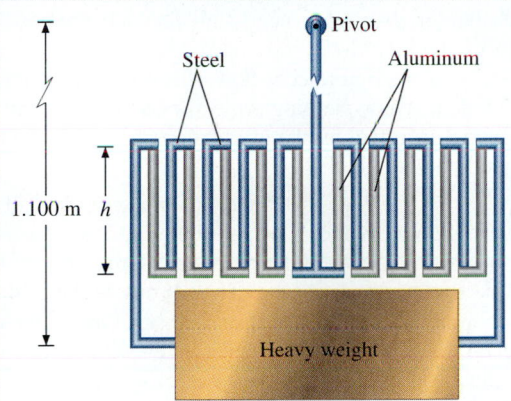

17.41 (4) *The center cannot hold.* A long, thin metal bar of length l is clamped rigidly at its ends at temperature t_0. When the temperature t is increased, the expanding bar will bow out, as shown below. If the bowing is not too large, a fair first approximation to the shape of the bar is two equal straight segments in the form of a wide V. (a) What is the arch δ of the bow as a function of t? (This is the distance between the corner of the V and the straight line that represents the form of the bar at t_0.) (b) Suppose the bar is made of aluminum and its length is 1.000 m at 15°C. What will be the value of δ at 95°C?

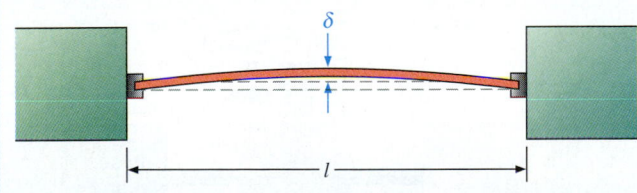

17.42 (4) *Two ways at once.* A rectangular plate is made of a material having a linear coefficient of expansion α. At temperature t_0, the sides of the plate have length a and b so its area is $A = ab$. Show that, when the temperature is changed to a new value t, the plate experiences a proportional change in area $\Delta A/A = 2\alpha$. What approximation do you have to make in deriving this result?

17.43 (5) *All bent out of shape.* A rectangular copper plate of length x and width y is clamped snugly at two opposite corners, as shown in the next column. The temperature of the

plate (but not of the clamps) is raised by Δt. (a) Find the angle γ. (b) Find the shear stress σ in the plate. (c) If the plate is three times as long as it is wide, what temperature difference will give $\gamma = 89.9°$? What is the corresponding value of σ?

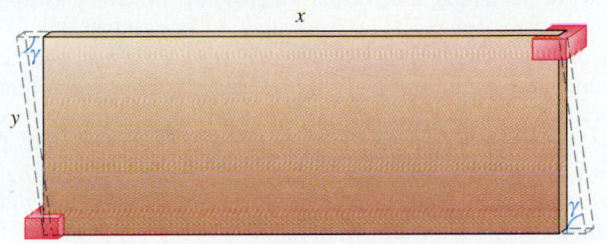

17.44 (5) *Newton's law of cooling.* When the temperature of a body is higher than that of its surroundings, the body cools down. Newton found by experiment that, if the temperature difference ΔT is not too large, the rate of cooling is proportional to the temperature difference: $d(\Delta T)/d\tau = -\kappa \, \Delta T$. In this equation τ represents time and κ is an empirical constant whose value depends on the size and shape of the cooling body, the nature of its thermal contact with its surroundings, and other factors. Suppose that, at some initial time τ_0, the temperature difference is ΔT_0. (a) Show that the temperature difference at any later time τ is $\Delta T = \Delta T_0 e^{-\kappa(\tau-\tau_0)}$. (b) Sketch a graph of $\Delta T/T_0$ versus time.

17.45 (5) *Homespun calorimetry.* In colonial America, a drink called flip was popular on cold winter nights. Cider at room temperature, often spiced and mixed with whiskey, was poured into a mug, and a heavy iron poker heated in the fireplace was then plunged into the drink to make it piping hot. Make plausible estimates of the initial and final temperatures of the drink and of the diameters of the mug and the poker. Use this information to estimate the initial temperature of the poker. Take the density of iron to be $8 \times 10^3 \ kg/m^3$, and neglect the heat lost to the atmosphere by the production of steam when the poker is plunged into the liquid.

17.46 (5) *Combustion calorimeter.* The figure below is a schematic drawing of a *combustion calorimeter*. This device is used to measure the heat evolved when fuel or other combustible material is burned. The heat evolved per unit mass, called the *heat of combustion K*, is a figure of merit that can be used for comparison of fuels. The value of K for a given

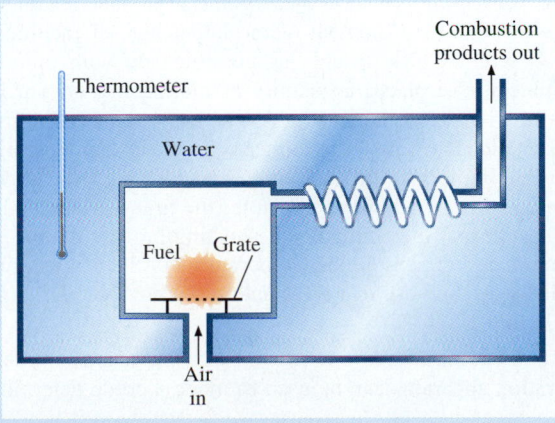

fuel is also helpful in estimating the efficiency of a particular furnace in which the fuel is used.

The combustion calorimeter amounts to a stove with a long chimney, all surrounded by a water jacket. Owing to the length of the chimney, the gaseous combustion products are essentially at the temperature of the water by the time they escape into the atmosphere. The temperature difference between the incoming air and the outgoing gas is negligible because the flame temperature (the initial temperature of the combustion products) is very much greater than the temperature of either the water or the incoming air. (a) The heat of combustion of a typical crude oil (petroleum) sample is 1.08×10^4 kcal/kg. What temperature increase ΔT will you observe when you burn a 350-g sample of the oil in a combustion calorimeter if the water jacket contains 236 kg of water? Neglect the total heat capacity of the rest of the calorimeter. (b) A 1.50-kg sample of anthracite coal is burned in the same combustion calorimeter. The temperature rise of the water is $\Delta T = 41.1$ K. Find the value of K for the coal sample.

17.47 *(5) Cold facts.* At very low temperatures, the specific heat capacity of crystalline solids varies rapidly with temperature, according to the relation $c = aT^3$, where a is a constant. Find the heat flow Q required to warm such a crystal, of mass m, from T_i to T_f.

17.48 *(6) If you've got lemons ...* On arriving home from school on a hot day, a boy makes lemonade. The temperature of the lemonade, made with tap water, is a not-very-chilly 14°C, and so the boy adds ice. He drops an ice cube of mass 25 g and temperature −8°C into a glass containing 250 g of the lemonade. (a) Does all the ice melt? (b) What is the final temperature of the lemonade? (c) If the final temperature is above 0°C, how much more ice at −8°C would just suffice to bring the temperature of the lemonade to 0°C? Neglect the heat capacity of the glass and heat flow into or out of the system.

17.49 *(6) For the big kids.* One way to make iced coffee is to brew the coffee and then pour it over ice. Cookbooks always recommend that the coffee brewed for iced coffee be stronger than the two tablespoons of coffee per cup of water that is best for hot coffee. Suppose the freshly brewed coffee is at 88°C and the ice is at the melting point. If all the ice is melted when the temperature of the mixture is 5°C, how many tablespoons of coffee should be used per cup of hot water? (Neglect the additional strengthening of the recipe required because taste buds are less sensitive at low temperatures.)

17.50 *(6) Crude but clever measurement.* A student wants to make a rough measurement of the latent heat of melting of ice, L, with as little trouble as possible and with minimal equipment. She places a quantity of melting ice on a paper towel to remove as much liquid water as possible. She then drops the ice into a metal container and sets the container over a Bunsen burner. At the same time, she starts a timer. Stirring vigorously as the ice melts, she notes the time τ when the last ice disappears. She continues the heating process and uses a thermometer to measure the temperature t of the water at time 2τ. How can she use this information to determine L? (Hint: See Figure 17.7b.)

17.51 *(7) Joule in the classroom.* A simple classroom demonstration apparatus can be used to make a crude determina-

tion of the mechanical equivalent of heat J. A cardboard mailing tube of length l is capped at both ends. The tube contains about a handful of lead shot. The system is allowed to come to room temperature t_0. Then the tube is repeatedly inverted so that the shot fall from end to end. A count is kept of the number of inversions n. Finally, the temperature increase $\Delta t = t - t_0$ of the shot is measured. (a) Express the value of J in terms of l, n, Δt, and the necessary constants. (b) In a typical experiment, the tube is 1.5 m long and is inverted 25 times. If the temperature increase is 2.5 C°, find the value of J. Why is this experiment likely to yield too-large values of J?

17.52 *(7) Rifle range.* A lead rifle bullet whose muzzle speed is 500 m/s is fired at a nearby target made of hardened steel. Will the bullet melt on impact? [Hint: Because hardened steel is much harder than lead, nearly all the deformation takes place in the bullet. Also, the impact takes place over such a short time that negligible heat flows from the bullet into the steel. (Modern armor-piercing projectiles depend for much of their destructive effect on the production at impact of a high-speed jet of liquid metal.)]

17.53 *(7) Prony brake.* The device shown schematically below, called a Prony brake, is often used to measure the power output of an engine as a function of the angular speed of its shaft. The essential component of the Prony brake is the assembly comprising the pair of friction disks. One of the disks is attached to the engine shaft and rotates with it. The other disk is attached to a torque-measuring device, which holds the disk stationary. The frictional torque exerted by each disk on the other can be controlled by a hydraulic or spring-loaded device (not shown) that varies the force with which the disks are pressed together. Considerable heat is evolved, and so the disk assembly is water cooled. In making a measurement, one varies the disk pressure until the desired angular speed ω of the engine is achieved and then measures the torque τ. (a) If $\omega = 160$ rad/s and $\tau = 340$ N·m, find the output power of the engine being tested. (b) Cooling water flows through the system at 45.0 liters per minute. What is the difference in temperature between the output and input streams?

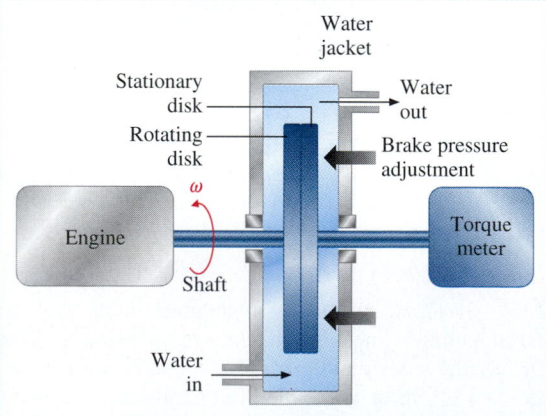

17.54 *(G) Heat of solution.* When 1.0×10^{-2} kg of sodium bromide (NaBr) at 25°C dissolves in 1.00 kg of water at the same temperature, the temperature of the solution is greater than the initial temperature by 0.014 C°. (a) How much heat is evolved in the process of solution? Express your result in kilocalories and in joules. **(b)** What is the source of the energy that manifests itself in the temperature increase? **(c)** The quantity of NaBr contains approximately 5.8×10^{22} atoms. How much energy is transferred to the solution by each atom in the NaBr as it dissolves? (Don't bother to make a distinction between sodium and bromine atoms, but quote an average value.) **(d)** Suppose you raise the temperature of the solution to the boiling point. How much additional heat must you transfer to the solution to boil away the water and end up with solid NaBr again? **(e)** Many substances (such as NaCl) *reduce* the temperature of water when they dissolve. Compare the heat required to boil off the water from such a solution with the heat of vaporization of pure water.

GROUP C

17.55 *(4) Temperature correction of a mercury barometer, II.* **(a)** Refer to the result of Problem 17.36. On a very hot day, when the temperature is 40°C, what is the proportional correction $|h' - h|/h$ that must be applied to the observed column height h of a mercury barometer? **(b)** Weather services usually quote barometric pressures with a precision of about 1 part in 10^3. Refer to the result of Problem 17.12, and determine whether such a measurement requires further correction for the expansion of the glass barometer tube, at least on very hot days. If it does, is the volume coefficient β or the linear coefficient α the appropriate quantity to use in correcting? (Hint: Note carefully the physical difference between the present problem and Problem 17.12.) Write the formula for the twice-corrected value h'' of the mercury column height in terms of h, t, and the appropriate expansion coefficients.

17.56 *(4) Temperature-compensated pendulum, II.* One way of making a pendulum whose period is independent of the temperature was considered in Problem 17.40. The figure below depicts another type of temperature-compensated pendulum. The pendulum bob consists of a set of tubes partially filled with a considerable mass of mercury and suspended from a long, relatively light steel rod. To a good approximation, the center of gyration of the pendulum lies at the center of mass of the mercury. The length of the pendulum, from the pivot to the bottom of the mercury tubes, is 1.100 m. What should be the depth of the mercury if the period of the pendulum is to be independent of temperature?

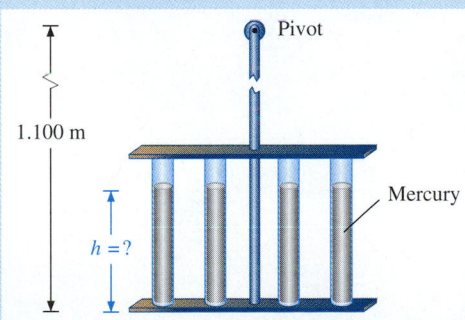

17.57 *(4) Mind your alphas and betas.* Derive Equation 17.22, $\beta = 3\alpha$, for the relation between the volume and linear coefficients of thermal expansion for an isotropic solid. (Hint: Review the argument leading to Equation 15.27. Also see Problem 17.42.)

17.58 *(4) Unshrinkable.* An aluminum wire and a steel wire of equal diameter are welded together end to end. With the room temperature at 30°C, the welded wire is clamped between two sturdy vises so that its free length is 1.000 m. The weld is exactly in the middle. The wire is under a small tension so that it does not sag. The room temperature is then lowered to 0°C. Assume that the distance between the vises does not change. The necessary values of Young's modulus are given in Table 15.1. **(a)** Find the tension in the wire. **(b)** Where is the weld located with respect to the vise holding the aluminum end of the wire?

17.59 *(4) Bimetallic thermostat.* The bimetallic strip shown in part *a* of the figure below is made by bonding thin, straight leaves of copper and iron together at temperature t_0. The thickness of each leaf is d, and the length of each is l_0 at t_0. As the temperature increases above t_0, the compound strip bends as shown in part *b* of the figure. At some temperature t_1, the two electrical contacts touch and an electrical control system is activated. The arrangement can be used to control the temperature of the air surrounding the bimetallic strip by turning on and off a device such as an oven or an air conditioner. **(a)** Show that the bimetallic strip bends into a circular arc and that, at temperature t, the angle θ between the tangents to the two ends is

$$\theta = \frac{l_0}{d} (\alpha_{Cu} - \alpha_{Fe})(t - t_0),$$

where α_{Cu} and α_{Fe} are the linear coefficients of expansion of copper and iron. [Hint: When the bimetallic strip is bent, each leaf is under tensile stress near its convex surface and under compressive stress near its concave surface. You will want to consider the behavior of the unstressed median (middle) plane of each of the two leaves.] **(b)** If the distance between the two electrical contacts is D_0 when the temperature is t_0, find the temperature t_1 at which the contacts touch. **(c)** Let $t_0 = 15$°C, $l_0 = 3.50$ cm, $d = 0.125$ mm, and the desired turn-on temperature $t_1 = 25$°C. How should you set the gap D_0 between the electrical contacts at t_0?

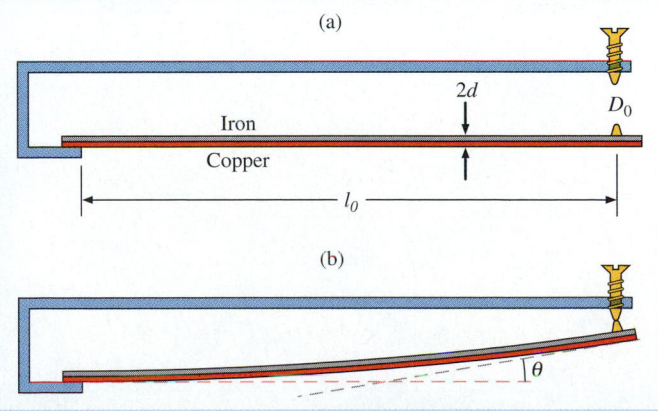

Kinetic Theory of Heat

L O O K I N G   A H E A D

——— Under a wide range of conditions, the pressure, volume, mass, and temperature of a real gas are described with good accuracy by an equation called the ideal-gas law.

——— The kinetic model of an ideal gas, based on a small number of assumptions, can be used to derive the ideal-gas law.

——— By a simple extension of the kinetic theory, we can understand why the molar specific heat of many solids has a universal value that can be expressed in terms of a constant derived from experiments on gases.

——— The speed of gas molecules, the pressure of the atmosphere as a function of altitude, and many other observations can be understood on the basis of the kinetic theory.

Left: At the 8847-m altitude of Mt. Everest, the atmosphere is pretty thin. The kinetic theory of gases can be used to predict the barometric pressure quite accurately.

*Motion . . . is the genus of which heat is a species . . .
[it is not merely that] heat generates motion or that
motion generates heat (though both are true in
certain cases) but that Heat itself, its essence and
quiddity, is Motion and nothing else . . . not uniformly
of the whole body together, but in the smaller parts
of it.*

—FRANCIS BACON, *Novum Organum* (1620)

SECTION 18.1 | Introduction

In this chapter, we accomplish the task begun in Section 17.7: We develop a logical framework for understanding thermal phenomena in purely mechanical terms. This framework is called the *kinetic theory of heat*. Based on just a few fairly simple ideas, the kinetic theory leads us to insight into a wide range of thermal phenomena. Moreover, the theory is open ended, and so it can be extended far beyond the limited realm in which we develop it.

We begin in Sections 18.1 and 18.2 by developing a dynamical theory of the thermal behavior of gases because gases have the simplest structure of all the states of matter. The rest of the chapter deals with applications and extensions of the theory. In Section 18.3, we use the theory to predict the heat capacity of gases and compare the result with experiment. In Section 18.4, we extend the kinetic theory to solids and show that the theory predicts their heat capacities as well. In Section 18.5, we consider the microscopic properties of a gas in more detail and discover the way in which the energy of the gas as a whole is distributed among its molecules.

You have already seen that heat flow is a macroscopic manifestation of microscopic motion. Specifically, heat flow is a form of energy transfer. One particularly persuasive piece of evidence in this direction is the Joule experiment, discussed in Section 17.7. To see what heat energy "looks like," however, we must do our looking on the microscopic scale.

THINKING LIKE A PHYSICIST: THE IDEA OF A MODEL

We begin our deeper look into the nature of heat energy by developing a **model**. That is, we sketch a simple picture of what a gas "looks like" on the microscopic scale. At the outset, such a model is purely a figment of the imagination—though we are not likely to make much progress unless the figment is based on serious thought! We then apply the principles of Newtonian mechanics to the model and *deduce* its thermal properties. If these deduced thermal properties correspond to the results of actual macroscopic measurements,

we begin to place confidence in the model. Specifically, we come to accept the model as a fruitful *representation* of the actual system. But we must never deceive ourselves that the model *is* the system.

A minimum requirement for a valid kinetic model of a gas is that it must make predictions that agree with the observed basic macroscopic properties of gases. These basic properties must certainly include Boyle's law and Charles's law.

The Kinetic Model
and the Ideal-Gas Law

Our primitive model is intended to give us a bare-bones picture of a gas, which we call an *ideal gas*. The properties of an ideal gas are simpler than those of any real gas. But the properties of real gases closely approach those of the ideal gas under a wide range of conditions. In due time, we will relax the simplifying assumptions to some extent.

The Model

Our kinetic model is based on four assumptions. The first two apply to all gases, ideal or otherwise:

1. A gas consists of a very large number of molecules.
2. The molecules of the gas are in continual motion.

The third and fourth assumptions apply to the ideal gas only:

3. The molecules of an ideal gas are identical particles—geometric points with equal mass m.
4. An ideal-gas molecule exerts a force on another body only when the two are in contact.

Assumption 1 will not surprise you, because we have already made fruitful arguments on the basis that everyday-sized samples of ordinary matter are composed of vast numbers of molecules. The motion of these molecules, assumption 2, is supported by a wide variety of experimental evidence. The very word *kinetic* implies molecular motion, and you will soon see its central importance to the theory.

Assumption 3 may seem strange at first. But consider the fact that a gas is much less dense than the liquid and solid states of the same substance. When water boils at atmospheric pressure, for instance, the resulting steam occupies almost 1700 times the volume occupied by the liquid. If we assume that the molecules in liquids and solids are pretty much in contact with one another, the distance between molecules in a gas is on the average many molecular diameters. That is, the molecules are far apart. In shrinking the molecules to zero size, we are making this point in a simplified, if somewhat exaggerated, way.

Assumption 4 is related to assumption 3. Taken together, they imply that the forces acting between molecules fall off rapidly with distance. We know that the intermolecular forces in liquids and solids have appreciable magnitude; that is why liquids and solids do not fall apart spontaneously. The fact that liquids boil abruptly (without passing through some kind of gradual transition) suggests that the intermolecular bonds break abruptly. That is, the forces rapidly become negligible with increasing distance when that distance reaches a critical value not much larger than the center-to-center distance between two molecules in actual contact. Once the intermolecular distances exceed this critical value, the molecules of the resulting gas are substantially independent. As you will see in the next subsection, this greatly simplifies our analysis by enabling us to consider molecules one at a time.

A Single Ideal-Gas Molecule in a Box

The molecules of an ideal gas interact only when they are in contact with one another. But, because they are geometric points, they *never* make contact with one another. Thus every molecule in an ideal gas acts independently of every other molecule. We exploit this total independence in our model-building strategy: First, we study the behavior of a *single* molecule; then, we predict the properties of the gas as a whole by arguing that it consists of a large number of independent single molecules.

Consider a single molecule confined in a cubic box of sides l, as shown in Figure 18.1. We choose the coordinate axes along the edges of the box, as shown in the figure. The molecule, of mass m, has an arbitrary velocity **v**. As it moves, it collides repeatedly

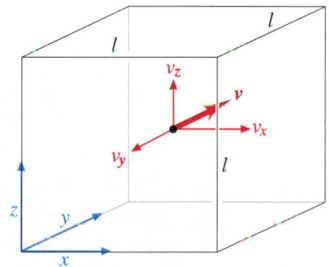

FIGURE 18.1 A cubic box with sides l contains a single molecule of an ideal gas. The molecule, which has zero size, has mass m and velocity **v**.

with the walls of the box. We assume that the collisions are elastic, so the magnitude $|\mathbf{v}|$ does not change on collision. We assume further that the collisions are **specular**: The molecule bounces off the walls the way that light is reflected from a mirror (the Latin word for mirror is *speculum*). For example, when the molecule strikes one of the two walls normal to the x axis, the x component of its velocity, v_x, is reversed while the other components, v_y and v_z, remain unchanged. (This assumption is not essential to the analysis that follows, but it simplifies the analysis considerably.)

When the molecule strikes, say, the right-hand wall, the x component of the velocity of the molecule changes from v_x to $-v_x$ while the y and z components of the velocity are unchanged. We do not know which wall the molecule will strike next. But, regardless of the order of its collisions with the various walls, we do know that its transit time from the right-hand wall to the left-hand wall is l/v_x. (We know this because any intervening collisions with other walls will not affect v_x.) The collision with the left-hand wall will again reverse the x component of the velocity of the molecule, and it will again head toward the right-hand wall with speed v_x. Thus the time between successive collisions with the right-hand wall is

$$\Delta t = \frac{2l}{v_x}. \tag{18.1}$$

In the course of each collision with the right-hand wall, the molecule experiences a change in momentum equal to the difference between its after-collision and before-collision momenta:

$$\Delta(m\mathbf{v}) = (-mv_x) - (mv_x) = -2mv_x. \tag{18.2}$$

According to Newton's second law, the force exerted on the molecule by the right-hand wall is given by the time derivative of the momentum: $\mathbf{F} = d(m\mathbf{v})/dt$. Because we do not know the (very short) time required for the collision, we cannot evaluate the force. But we can evaluate the *mean* force exerted on the molecule by the wall over many collisions. The mean force is

$$\langle \mathbf{F} \rangle = \frac{\Delta(m\mathbf{v})}{\Delta t},$$

where the numerator is the momentum change, given by Equation 18.2, and the denominator is the time elapsed between collisions, given by Equation 18.1. So we have

$$\langle F \rangle = \frac{-2mv_x}{2l/v_x} = -\frac{mv_x^2}{l}.$$

According to Newton's third law, the mean force exerted on the wall by the molecule is $-\langle F \rangle$. The area of the wall is l^2, and so the mean force per unit area—the "pressure"—exerted on the wall by the molecule is

$$\text{``pressure''} = -\frac{\langle F \rangle}{l^2} = \frac{mv_x^2}{l^3}.$$

But l^3 is the volume V of the box. So we have

$$\text{``pressure''} = \frac{mv_x^2}{V}. \tag{18.3}$$

Many Ideal-Gas Molecules in a Box

You may think it artificial to derive a pressure from a series of microscopic impulses exerted on a wall by a single molecule—and it is. That is why we have placed the word "pressure" in quotes. But in the next step we will make the idea realistic. Having considered a single molecule, we now fill the box with a very large number N of molecules. Each molecule collides with the wall repeatedly, so the wall experiences a constant bombardment of molecules all over its surface. Under these circumstances, the idea of pressure becomes quite natural.

At this point, we follow the strategy outlined at the beginning of the preceding subsection and invoke the independence of the molecules. According to assumption 3 of the ideal-gas model, the molecules are geometric points. Having zero size, they present no target, and they *never* collide with one another, but only with the walls. Each molecule, acting entirely independently of the others, exerts a "pressure" given by Equation 18.3 on the right-hand wall of the box. In general, each molecule has a different velocity component v_x. But we are interested in the total pressure p exerted by all the molecules on the right-hand wall. That pressure is found by adding the individual contributions of all N molecules:

$$p = \sum_{j=1}^{N} \frac{m v_{x_j}^2}{V}.$$

We have assumed that all the molecules have the same mass, and so we can take m as well as V outside the summation:

$$p = \frac{m}{V} \sum_{j=1}^{N} v_{x_j}^2. \tag{18.4}$$

We do not know the value of v_x^2 for any of the molecules—let alone for all N of them. But whatever the values are, there must be some *mean* value

$$\langle v_x^2 \rangle = \frac{\sum_{j=1}^{N} v_{x_j}^2}{N}.$$

Thus the sum in Equation 18.4 is equal to $N\langle v_x^2 \rangle$. Making this substitution gives

$$p = \frac{Nm\langle v_x^2 \rangle}{V}. \tag{18.5}$$

We want to express this result in terms of the mean energy $\langle \epsilon \rangle$ of the gas molecules. To do this, we note first that the energy of a molecule is entirely kinetic.* Thus a molecule has energy

$$\epsilon = \tfrac{1}{2}mv^2 = \tfrac{1}{2}m(v_x^2 + v_y^2 + v_z^2). \tag{18.6}$$

The mean energy of the molecules is

$$\langle \epsilon \rangle = \tfrac{1}{2}m(\langle v_x^2 + v_y^2 + v_z^2 \rangle).$$

That is, $\langle \epsilon \rangle$ is $m/2$ times the mean value of the sum of three quantities. We now use a mathematical theorem without proof:[†] The mean value of a sum of terms is equal to the sum of the mean values of the terms. We thus have

$$\langle \epsilon \rangle = \tfrac{1}{2}m(\langle v_x^2 \rangle + \langle v_y^2 \rangle + \langle v_z^2 \rangle). \tag{18.7}$$

To simplify this equation, we invoke a symmetry argument. We have chosen the x, y, and z directions arbitrarily in Figure 18.1. Although the values of v_x^2, v_y^2, and v_z^2 are very unlikely to be equal for any one molecule, there is no reason to believe that the *mean* values of these quantities, taken over all the molecules, will be different from one another. So we have

$$\langle v_x^2 \rangle = \langle v_y^2 \rangle = \langle v_z^2 \rangle. \tag{18.8}$$

It follows immediately that

$$\langle \epsilon \rangle = \tfrac{3}{2}m\langle v_x^2 \rangle. \tag{18.9}$$

---

*In most cases we can neglect the gravitational potential energy of the molecules. The box is of modest size, and the change in potential energy from top to bottom is negligible.

[†]The proof is the subject of Problem 18.36.

Thus we can rewrite Equation 18.5 in the form

$$p = \frac{2}{3}\frac{N\langle\epsilon\rangle}{V}.$$

(18.10a)

And, because $N\langle\epsilon\rangle$ is the *total* energy E of all the molecules in the box, we have also

$$p = \frac{2}{3}\frac{E}{V}.$$

(18.10b)

Finally, we invoke symmetry again to argue that our calculation of the pressure exerted by the gas molecules on the right-hand wall of the box applies equally well to all the other walls. Equation 18.10 thus gives *the* pressure of the gas in the box. This result is consistent with Pascal's principle, which states that the pressure of a fluid at a given location must be the same in all directions.

Comparing the Model with the Real World

We obtained Equation 18.10a by applying principles of mechanics to the ideal-gas model. Let us see how well the equation corresponds to direct observations of the properties of gases. Multiplying Equation 18.10a through by V gives

$$pV = \tfrac{2}{3}N\langle\epsilon\rangle.$$

(18.11)

The right side of this equation will be constant if N and $\langle\epsilon\rangle$ are constant. The condition $N = $ constant implies that a fixed number of molecules—and therefore a fixed mass of gas—are trapped in the box. If $\langle\epsilon\rangle$ is constant as well, we have

$$pV = \text{constant.}$$

This is just Boyle's law (Equation 15.35c). But Boyle's law is valid only if the mass and the temperature of the gas are held constant. We have just argued that the mass is constant because N is constant. We thus reach a conclusion that depends on an important condition:

a. **The ideal-gas model yields Boyle's law**

   if

b. **the temperature T of a gas—and only the temperature—is related to the mean energy $\langle\epsilon\rangle$ of its molecules.**

The *conclusion* (**a**) is a check on the correspondence of our model to physical reality. The *condition* (**b**) yields a new insight, which we pursue further.

Equation 18.10a can be solved for V to yield

$$V = \frac{2}{3}\frac{N\langle\epsilon\rangle}{p}.$$

(18.12)

Let us compare this model-derived equation with the empirical Charles's law. Equation 17.14 gives Charles's law in the form $V/V_{tp} = T/T_{tp}$, for quantity of gas and p constant. For our present purpose, it is more convenient to note that the triple-point values V_{tp} and T_{tp} are fixed and to rewrite the law as

$$V = \text{constant} \times T \quad \text{for quantity of gas and } p \text{ constant.}$$

(18.13)

Eliminating V from Equations 18.12 and 18.13 and solving for $\langle\epsilon\rangle$, we obtain

$$\langle\epsilon\rangle = \text{constant} \times \frac{3p}{2N} T \quad \text{for quantity of gas and } p \text{ constant.}$$

(18.14)

We must retain the conditions given for Equation 18.14 because they are necessary for the validity of Charles's law, on which the validity of the equation depends. But the conditions imply that the quantities p and N, which entered Equation 18.14 via model-derived Equation 18.12, are constant. Thus all the quantities except T on the right side of Equation 18.14 are constants. Let us lump them all together into a single constant, which we call $\frac{3}{2}k$. (The reason for retaining the fraction $\frac{3}{2}$ will become apparent in Section 18.5.) We then have

$$\langle \epsilon \rangle = \tfrac{3}{2}kT. \tag{18.15}$$

This equation makes condition **b** more specific:

b′. The mean energy of the molecules of an ideal gas is directly proportional to the thermodynamic temperature.

The proportionality constant k is a universal constant called **Boltzmann's constant**. It relates molecular energy to temperature because it is a measure of the energy possessed by the "average" molecule per kelvin of temperature. The currently accepted value of Boltzmann's constant is

$$k = 1.380\ 658 \times 10^{-23} \text{ J/K}. \tag{18.16a}$$

For many purposes, the value may be expressed with adequate precision as

$$k = 1.381 \times 10^{-23} \text{ J/K}. \tag{18.16b}$$

By turning Equation 18.15 around and expressing T in terms of $\langle \epsilon \rangle$, we can achieve a new, more fundamental definition of temperature. We have

$$T = \frac{2\langle \epsilon \rangle}{3k}. \tag{18.17}$$

This equation may be regarded as a *definition of temperature in terms of purely mechanical quantities*. Thus we no longer need consider temperature as a quantity outside the realm of mechanics.

The Ideal-Gas Law

We now combine Equation 18.15, $\langle \epsilon \rangle = \frac{3}{2}kT$, with Equation 18.11, $pV = \frac{2}{3}N\langle \epsilon \rangle$.

$$pV = NkT. \tag{18.18}$$

This equation is called the **ideal-gas law**. Any equation that relates the pressure, volume, temperature, and quantity of a substance is called an **equation of state**, and so Equation 18.18 is also called the **equation of state of an ideal gas**.

Boyle's law and Charles's law are special cases of the ideal-gas law. If N and T are held constant, Equation 18.18 becomes $pV = $ constant, which is Boyle's law. If N and p are held constant, Equation 18.18 becomes $V/T = $ constant, which is a form of Charles's law.

A remarkable property of the ideal-gas law is the way the quantity of gas appears in it. The factor N represents the *number* of molecules present and is independent of their mass m. That is, Equation 18.18 applies to all gases that approximate the ideal gas, regardless of their chemical composition. Indeed, if any gas approximating an ideal gas is contained in a volume V at pressure p and temperature T, the number of molecules is determined:

$$N = \frac{pV}{kT}. \tag{18.19}$$

Hence *equal volumes of different gases at the same temperature and pressure contain equal numbers of molecules*. This statement is known as **Avogadro's law**, after the Italian physicist and chemist Amedeo Avogadro (1776–1856). (Avogadro was the first person to state clearly that the molecules of a gas are on the average far apart compared with their diameters; this underlies assumption 3 of our ideal-gas model.)

Ludwig Boltzmann

Max Planck was the first to name the constant k in honor of the Austrian physicist Ludwig Boltzmann (1844–1906). Boltzmann was one of the pioneers of kinetic theory and its profound generalization, statistical mechanics. Boltzmann's explicit use of the assumption that matter is composed of atoms (or molecules) exposed him to incessant criticism from his colleagues at the University of Vienna, notably Ernst Mach. Mach's objection to the atomic theory was based on the view—not unreasonable at the time—that no one was ever likely to "see" an atom and that it was unwise to base scientific theory on entities undetectable in principle. Depressed, Boltzmann committed suicide only a year before the controversy was settled resoundingly in his favor by a rising tide of direct experimental evidence of the existence of atoms. Some of Boltzmann's contributions are discussed in Section 18.5 and in Chapter 20.

EXAMPLE 18.1

Chemists refer to "standard conditions of temperature and pressure" (often abbreviated STP). By convention, STP means that the pressure is 1 atm (1.013×10^5 Pa) and the temperature is 0°C (273.15 K).

Find the number of molecules of oxygen (O_2) present in a 1-m^3 container at STP.

SOLUTION: Using Equation 18.19, you have for the number of molecules

$$N = \frac{1.013 \times 10^5 \text{ Pa} \times 1 \text{ m}^3}{1.38 \times 10^{-23} \text{ J/K} \times 273.15 \text{ K}}$$
$$= 2.69 \times 10^{25}.$$

Underlying this calculation is the tacit assumption that oxygen behaves like an ideal gas under these conditions. This assumption will be justified in Example 18.3 and in Section 18.3.

Equation 18.18 contains a mixture of macroscopic quantities (p, V, T) and microscopic quantities (N, k). In performing experiments on gases, however, we can most easily measure macroscopic quantities, such as the total mass $M = Nm$. It is therefore useful to recast Equation 18.18 in entirely macroscopic terms. We build the necessary bridge between the macroscopic and the microscopic by expressing the molecular mass m in both microscopic and macroscopic terms. By definition, 1 **unified mass unit (u)** is one-twelfth of the mass of one carbon-12 (^{12}C) atom.* The mass of a molecule of a substance, *expressed in unified mass units*, is called the **molecular mass** μ of that substance.† To avoid confusion, we will use the term *mass of the molecule* to signify the mass m expressed in kilograms.

Having defined a new unit of mass—the unified mass unit—we need to convert masses expressed in kilograms (kg) into unified mass units (u), and vice versa. Let us call the conversion factor A; we have

$$1 \text{ kg} = A \text{ u} \quad \text{or} \quad 1 \text{ u} = \frac{1}{A} \text{ kg}. \qquad \textbf{(18.20a,b)}$$

The values of A and its reciprocal, found experimentally, are

$$A = 6.022\,136\,7 \times 10^{26} \text{ u/kg} \quad \text{and} \quad \frac{1}{A} = 1.660\,540\,2 \times 10^{-27} \text{ kg/u}.$$
$$\textbf{(18.21a,b)}$$

We will usually use the rounded values

$$A = 6.022 \times 10^{26} \text{ u/kg} \quad \text{and} \quad \frac{1}{A} = 1.661 \times 10^{-27} \text{ kg/u}. \quad \textbf{(18.21c,d)}$$

The straightforward relation between the two units of mass has a remarkable consequence. Given any particular atom (or molecule), we can express its mass m in two equivalent ways:

$$m = \mu \text{ u} = \frac{\mu}{A} \text{ kg}.$$

Next, suppose we collect A of these atoms. The mass of the collection, M_0, can likewise be expressed in two equivalent ways:

$$M_0 = A\mu \text{ u} = \mu \text{ kg}.$$

Whatever the numerical value of μ may be, the mass M_0 of A atoms has the same value in the macroscopic unit, kg, as the mass m of one atom has in the microscopic unit, u. For example, one atom of ^{12}C has mass 12 u (by definition), and A atoms of ^{12}C have mass 12 kg. Similarly, one molecule of chlorine gas, Cl_2, has mass 70.9 u; A molecules of Cl_2 have mass 70.9 kg.

The constant A, whose value is given by Equations 18.21a and 18.21c, is called **Avogadro's number**. The huge value of Avogadro's number reflects the fact that atoms and molecules are very small on the everyday scale of size. A lot of molecules are needed to make up a mass whose order of magnitude is 1 kilogram!

Taking Physical Liberties

For reasons of simplicity, we deviate slightly in this book from SI practice. In SI, Avogadro's number is defined in terms of the gram-mole (mol or g mol), which is the quantity of a substance whose mass in *grams* is equal to the atomic mass (or molecular mass). In SI, the value of A is $6.022 \times 10^{23} \text{ mol}^{-1}$, a number you may remember from your study of chemistry. The SI value is convenient for chemists, who measure out gram quantities in the laboratory much more commonly than kilogram quantities. But the kilomole and the value of Avogadro's number in kmol^{-1} are more convenient for the physicist.

---

*1 u is approximately, but not exactly, equal to the mass of the smallest atom, hydrogen (^{1}H). Carbon-12 is used for the standard rather than ^{1}H because it lends itself to more precise measurements.

†The misnomer ''molecular weight'' was universally used until recently, and its use is still fairly common. Don't be misled; we are interested in the mass, not the weight.

If a substance is pure, it consists of a large number of *identical* atoms, molecules, or other **elementary entities** (basic building blocks). If a sample of a substance contains Avogadro's number of elementary entities, it contains one **kilomole (kmol)** of the substance. The number n of kilomoles in a sample of a substance is a macroscopic measure of the amount of the substance present. The macroscopic measure n is related to the microscopic measure N, the number of molecules (or other elementary entities) present, by the equation

$$n = \frac{N}{A}. \qquad (18.22)$$

This equation is the bridge we set out to build between the macroscopic and microscopic worlds.

We can rewrite Equation 18.22 in the form $A = N/n$. From this form, it is clear that the units of A can be written as molecules per kilomole. Note, however, that "molecules" is nothing more than an abbreviated way of saying "number of molecules"; that is, "molecules" is a dimensionless number. Consequently, the common practice is to express A in the unit kmol^{-1}, rather than the equivalent unit u/kg.

EXAMPLE **18.2**

For the system of Example 18.1, express the amount of O_2 gas present in kilomoles and in kilograms. The atomic mass of oxygen (O) is 15.9994 u.

SOLUTION: You know the system contains 2.69×10^{25} molecules, and so you use Equation (18.22) to write

$$n = \frac{2.69 \times 10^{25} \text{ molecules}}{6.022 \times 10^{26} \text{ molecules/kmol}} = 4.47 \times 10^{-2} \text{ kmol}.$$

The mass of the oxygen gas is $M = mN$. You express the mass m of the molecule in terms of the molecular mass μ, which gives you $M = \mu N/A$. To determine μ, you note that there are two atoms to a molecule, so

$$\mu = 2 \times 15.9994 \text{ u} = 32.0 \text{ u}$$

to three significant figures. You therefore have

$$M = mN = (1.661 \times 10^{-27} \text{ kg/u} \times 32.0 \text{ u}) \times 2.69 \times 10^{25}$$
$$= 1.43 \text{ kg}.$$

We can now rewrite the ideal-gas law in entirely macroscopic terms. We use Equation 18.22 to replace the microscopic quantity N in Equation 18.18, $pV = NkT$, with the macroscopic quantity n. This gives

$$pV = n\,Ak\,T.$$

The product Ak combines the microscopic quantity k, which relates the energy of molecules to their temperature, with the very large quantity A, which counts molecules on the macroscopic level. We call the product the **universal gas constant** R:

$$R \equiv Ak. \qquad (18.23a)$$

R is a universal constant because A and k are universal constants. Its value is

$$R = 6.022\,136\,7 \times 10^{26} \times 1.380\,658 \times 10^{-23} \text{ J/kmol·K},$$

so
$$R = 8.314\,510 \times 10^3 \text{ J/kmol·K.} \qquad (18.23b)$$

We will usually use the rounded value

$$R = 8.314 \times 10^3 \text{ J/kmol·K.} \qquad (18.23c)$$

Note that the value of R is neither very large nor very small.

Rewriting the ideal-gas law in terms of R, we have

$$pV = nRT. \qquad (18.24)$$

In this form, the ideal-gas law is expressed in terms of macroscopic quantities only.

EXAMPLE 18.3

(a) What is the volume occupied by 1 kmol of an ideal gas under standard conditions of temperature and pressure? (b) What is the mean distance between molecules? (c) The density of liquid oxygen is 1.149×10^3 kg/m³. Assuming that the molecules of a liquid are essentially in contact with one another, use this information to calculate roughly the diameter of an O_2 molecule, and compare with the result of part **b**.

SOLUTION:

(a) What is the volume occupied by 1 kmol of an ideal gas at STP?

You solve Equation 18.24 for V, obtaining

$$V = \frac{nRT}{p}.$$

At STP, you have for 1 kmol

$$V = \frac{1 \times 8.314 \times 10^3 \text{ J/K} \times 273.15 \text{ K}}{1.013 \times 10^5 \text{ Pa}} = 22.42 \text{ m}^3.$$

(b) Find the mean intermolecular distance.

Imagine that you can divide the box of volume V into A tiny cubes, each of which contains one molecule on the average. The volume of each such cube is

$$\frac{V}{A} = \frac{22.42 \text{ m}^3}{6.022 \times 10^{26}} = 3.72 \times 10^{-26} \text{ m}^3.$$

The mean distance Λ between molecules is the length of the side of one such cube:

$$\Lambda = (3.72 \times 10^{-26} \text{ m}^3)^{1/3} = 3.34 \times 10^{-9} \text{ m}.$$

(c) Given the density of liquid oxygen, roughly calculate the diameter of an O_2 molecule.

From Example 18.2, you know that the molecular mass of an O_2 molecule is $\mu = 32.0$ u, so the mass of 1 kmol is $M_0 = 32.0$ kg. One cubic meter of liquid O_2 thus contains

$$n = \frac{M}{M_0} = \frac{1.149 \times 10^3 \text{ kg}}{32.0 \text{ kg}} = 35.9 \text{ kmol}.$$

The number of molecules in 1 m³ is this number multiplied by Avogadro's number, or

$$N = nA = 35.9 \text{ kmol} \times 6.022 \times 10^{26} \text{ kmol}^{-1}$$
$$= 2.16 \times 10^{28}.$$

Because the oxygen molecules share the 1 m³ volume approximately equally, each molecule in the liquid occupies a volume approximately equal to

$$\frac{1 \text{ m}^3}{N} = 4.63 \times 10^{-29} \text{ m}^3.$$

Oxygen molecules are not cubic in shape. But, for a rough comparison with the intermolecular distance Λ in the gas, you will be close enough if you calculate an "average" diameter d equal to the cube root of this volume. You have

$$d = (4.63 \times 10^{-29})^{1/3} = 3.6 \times 10^{-10} \text{ m}.$$

The value of Λ is roughly 10 times this diameter. So the oxygen gas molecules are indeed far apart at STP, and assumption 3 of the ideal-gas model is a reasonable approximation for oxygen (and for many other gases as well) under these conditions.

Molar Heat Capacity and Equipartition

In our continuing search for experimental tests of the ideal-gas model, we now use the model to predict the specific heat capacity of an ideal gas and compare the prediction with experiment. Equation 17.25 defines the total heat capacity C of any substance in an empirical way. For a body made of a given substance, the total heat capacity is Q, the quantity of heat flow into (or out of) the body, divided by the temperature change ΔT:

$$C = \frac{Q}{\Delta T}. \tag{18.25}$$

The specific heat capacity c of a substance is the total heat capacity of a sample of the substance divided by the mass M of the sample—that is, the heat capacity per unit mass:

$$c = \frac{C}{M} = \frac{1}{M} \frac{Q}{\Delta T}. \tag{18.26a}$$

When considering ideal gases, whose behavior conforms to the law $pV = nRT$, we find it more convenient to deal with quantities of matter measured in kilomoles (the unit of n) rather than in kilograms (the unit of M). With this in mind, we define the **molar heat capacity** c' to be the total heat capacity of a sample of the substance divided by the number of kilomoles n contained in the sample—that is, the heat capacity per kilomole:

$$c' = \frac{C}{n} = \frac{1}{n} \frac{Q}{\Delta T}. \tag{18.26b}$$

Because heat flow is a form of energy transfer, heat flow into (or out of) a body must result in a change in the total internal energy E of the body. Suppose that *no process other than heat flow* is operating to change the energy of the body. Then the entire change in E is due to the heat flow, and we have $Q = \Delta E$. In particular, we must suppose that no mechanical work is being done on the gas. One way to do mechanical work on a body of gas is to compress it. So we require that the volume occupied by the gas remain constant during the heat-flow process. Then we can write Equation 18.26b in the form

$$c'_v = \frac{1}{n} \frac{\Delta E}{\Delta T}, \qquad (18.27)$$

The subscript "v" indicates that the specific heat capacity c'_v applies to a process taking place at constant volume. Therefore, c'_v is called the **molar heat capacity at constant volume**.

Heat capacity is not in general independent of temperature. But even if c'_v is a function of T, we can rewrite Equation 18.27 in the differential form

$$c'_v = \frac{1}{n} \frac{dE}{dT}. \qquad (18.28)$$

We are now ready to use the results of the kinetic theory to *predict* the value of the molar heat capacity at constant volume for an ideal gas. According to Equation 18.15, the mean energy of the molecules is $\langle \epsilon \rangle = \frac{3}{2}kT$. If a gas sample contains N molecules, the total energy of the sample is

$$E = N\langle \epsilon \rangle = \tfrac{3}{2}NkT. \qquad (18.29)$$

We differentiate this equation in order to obtain the value of dE/dT that we need to insert into Equation 18.28. We have $dE = \frac{3}{2}Nk\, dT$, or

$$\frac{dE}{dT} = \frac{3}{2} Nk. \qquad (18.30a)$$

And, because $Nk = nAk = nR$, we can also write

$$\frac{dE}{dT} = \frac{3}{2} nR. \qquad (18.30b)$$

Substitution into Equation 18.28 gives

$$c'_v = \tfrac{3}{2}R = 1.25 \times 10^4 \text{ J/K} \qquad (18.31)$$

for the molar heat capacity of an ideal gas at constant volume.

We now compare this result with experimental values obtained for the molar heat capacities of real gases. A selection of such values is given in Table 18.1. In each case, the value is quoted for a temperature well above the temperature at which the gas condenses into a liquid or a solid. (We can be sure that a gas near its condensation temperature will *not* act like an ideal gas because an ideal gas does not condense.)

For monatomic gases, the agreement between kinetic theory and experiment is dramatic. In nearly every case, the measured value lies within 1% of $\frac{3}{2}R$, and the disagreement in the worst case is less than 2%.

For diatomic gases, the agreement between our theory and experiment is not good. We should not be too surprised at this because a diatomic molecule resembles a point mass less well than a monatomic molecule does. Indeed, if we picture a monatomic molecule as a tiny spherical ball, we are led to think of a diatomic molecule as something like a dumbbell—that is, as two linked balls. But in spite of the disagreement between the measured values and the simple theory, we can draw two significant inferences from the values in Table 18.1. The first is that the molar heat capacities of diatomic gases are all nearly the same. The second is that the values lie close to $\frac{5}{2}R$. In the next subsection, we will modify the kinetic theory in the light of these inferences. In doing so, we will broaden the concept of an ideal gas to embrace "realistic" molecules that are not point masses.

TABLE 18.1 Molar Heat Capacity at Constant Volume for Typical Gases

Gas	Temperature (K)	c_v' (J/kmol·K)	c_v'/R
Monatomic Gases			
Helium, He	300	1.25×10^4	1.50
Argon, Ar	300	1.25×10^4	1.50
Sodium, Na	1100	1.26×10^4	1.51
Potassium, K	1200	1.27×10^4	1.52
Cadmium, Cd	1200	1.25×10^4	1.50
Mercury, Hg	700	1.25×10^4	1.50
Zinc, Zn	1500	1.25×10^4	1.50
Diatomic Gases			
Hydrogen, H_2	300	2.05×10^4	2.46
Nitrogen, N_2	300	2.08×10^4	2.50
Oxygen, O_2	300	2.10×10^4	2.53
Carbon monoxide, CO	300	2.10×10^4	2.53
Hydrogen chloride, HCl	300	2.10×10^4	2.53
Polyatomic Gases			
Water, H_2O	700	2.84×10^4	3.42
Carbon dioxide, CO_2	300	2.81×10^4	3.38
Ammonia, NH_3	300	2.85×10^4	3.43
Methane, CH_4	300	2.71×10^4	3.26

Equipartition of Energy

Why is the molar heat capacity of a diatomic gas equal to $\frac{5}{2}R$? The key to answering this question lies in reviewing the monatomic case, for which

$$c_v' = \tfrac{3}{2}R = A \times 3 \times \tfrac{1}{2}k.$$

Let us write Equation 18.7 in the form

$$\langle \epsilon \rangle = \tfrac{1}{2}m\langle v_x^2 \rangle + \tfrac{1}{2}m\langle v_y^2 \rangle + \tfrac{1}{2}m\langle v_z^2 \rangle, \tag{18.32}$$

or

$$\langle \epsilon \rangle = \langle \epsilon_x \rangle + \langle \epsilon_y \rangle + \langle \epsilon_z \rangle. \tag{18.33}$$

Each of the three terms on the right side of this equation is the mean energy of the molecules that is associated with their motion along one of the three principal directions. The three terms, which just suffice for a complete description of the energy of the molecules, are called **degrees of freedom**. We argued on the basis of symmetry that these three mean energies must be equal to one another; see Equation 18.8. Because $\langle \epsilon \rangle = \tfrac{3}{2}kT$, it follows that *a mean energy $\tfrac{1}{2}kT$ is associated with each degree of freedom.* This equal division of mean energy among the degrees of freedom is called **equipartition**.

The ideal monatomic gas we have just considered exemplifies a very general statement, called the **equipartition theorem**. We can state the equipartition theorem as follows. Suppose that

1. a system consists of many identical bodies;
2. the energy ϵ of each body can be written as the sum of f terms, each of which has the quadratic form $\epsilon_j = a_j \chi_j^2$, where the a_j are constants and the χ_j are variables. [For example, $\epsilon_x = (m/2)v_x^2$.]

Then the mean energy associated with each term f is $\langle \epsilon_j \rangle = \tfrac{1}{2}kT$, and the mean

*Keep in mind that this argument applies to *mean* energies. There is no restriction on the values of ϵ_x, ϵ_y, or ϵ_z for any particular molecule.

energy of the bodies is

$$\langle \epsilon \rangle = \sum_{j=1}^{f} \langle \epsilon_j \rangle = \frac{f}{2} kT.$$

We have already justified the equipartition theorem for an ideal monatomic gas by showing that the theorem accounts for the observed value of the molar heat capacity, $\frac{3}{2}R$. Rather than prove the theorem rigorously, we show that it accounts quantitatively for the molar heat capacity of a diatomic gas and qualitatively for the molar heat capacity of gases consisting of still more complex molecules.

Figure 18.2 is the simplest possible picture of a diatomic molecule; there are two particles, each of mass $m/2$, joined by a rigid rod of negligible mass. Because this molecule is not a point particle, it can rotate as well as translate. The translation of the center of mass of the diatomic molecule is described by the velocity components v_x, v_y, and v_z, just as in the monatomic case. But we must supplement this information with a description of the rotation of the molecule about its center of mass; see Section 13.6, especially Equations 13.44a and 13.44b. To accomplish this, we fix a set of cartesian axes ξ, η, ζ in the molecule, as described in Figure 18.2 and its legend. We can then describe the rotation about the center of mass in terms of the angular velocity components ω_ξ and ω_η about the ξ and η axes, as well as the corresponding moments of inertia I_ξ and I_η. We need not be concerned with rotation about the ζ axis because the moment of inertia about that axis is $I_\zeta = 0$. This is because the two atoms constituting the molecule are themselves particles, even though the molecule is not.

The rotational kinetic energy ϵ_r of the molecule is

$$\epsilon_r = \tfrac{1}{2}I_\xi \omega_\xi^2 + \tfrac{1}{2}I_\eta \omega_\eta^2. \tag{18.34}$$

There is no potential energy to consider, and so the total energy of the molecule is found by adding ϵ_r to the three translational terms that give the energy for a monatomic ideal gas. When we do this, we obtain

$$\epsilon = \tfrac{1}{2}mv_x^2 + \tfrac{1}{2}mv_y^2 + \tfrac{1}{2}mv_z^2 + \tfrac{1}{2}I_\xi \omega_\xi^2 + \tfrac{1}{2}I_\eta \omega_\eta^2. \tag{18.35}$$

The mean energy of the molecules is the sum of the mean values for the five terms:

$$\langle \epsilon \rangle = \tfrac{1}{2}m\langle v_x^2 \rangle + \tfrac{1}{2}m\langle v_y^2 \rangle + \tfrac{1}{2}m\langle v_z^2 \rangle + \tfrac{1}{2}I_\xi \langle \omega_\xi^2 \rangle + \tfrac{1}{2}I_\eta \langle \omega_\eta^2 \rangle. \tag{18.36}$$

The diatomic molecule has five degrees of freedom, compared with the three degrees of freedom of the monatomic molecule. We now assert that *equipartition applies to the rotational degrees of freedom as well as the translational degrees of freedom.* We will justify this assertion in the following subsection. Because the mean energy of each degree of freedom is $\frac{1}{2}kT$, the mean energy of the diatomic molecule is

$$\langle \epsilon \rangle = \tfrac{5}{2}kT. \tag{18.37}$$

The total energy of N molecules of gas is $E = N\langle \epsilon \rangle$, so we have

$$E = \tfrac{5}{2}NkT = \tfrac{5}{2}nRT. \tag{18.38}$$

The molar heat capacity is

$$c_v' = \frac{1}{n}\frac{dE}{dT} = \frac{5}{2}R. \tag{18.39}$$

This result agrees quite well with the values listed for diatomic gases in Table 18.1.

For gases whose molecules have more than two atoms, the molar heat capacity is greater than $\frac{5}{2}R$ because there are more than five degrees of freedom. In such a molecule, the atoms do not usually lie in a straight line, and we cannot picture the molecule as a dumbbell. We must therefore consider the rotational energy about all three of the axes ξ, η, and ζ. As you will see in Section 18.4, it may also be necessary to consider additional degrees of freedom introduced by oscillation of the atoms with respect to the center of mass—that is, to take into account the nonrigidity of the bonds joining the atoms. The increase of molar heat capacity with increasing molecular complexity is evident in Table 18.1.

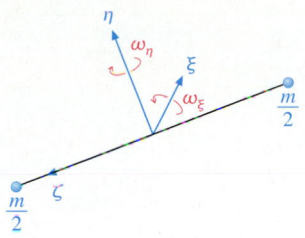

FIGURE 18.2 The simplest model of a diatomic molecule. Two point particles, each of mass $m/2$, are linked by a rigid, massless rod. To describe the rotation of this molecule, we choose the center of mass as the origin of a set of cartesian axes ξ, η, ζ, with the ζ axis along the rod. The rotation of the rod can be expressed in terms of the angular velocity components ω_ξ and ω_η. The moment of inertia of the molecule about the ξ axis is I_ξ, and the moment of inertia about the η axis is I_η. The moment of inertia about the ζ axis is $I_\zeta = 0$ because the two atoms making up the molecule are particles.

Molecules of Finite Size. Energy Density*

In developing the ideal-gas law, $pV = nRT$, we assumed that the gas molecules were points. Such molecules can collide with the walls of the container but not with one another. We exploited this point in calculating the mean force exerted by a molecule on a wall. In particular, we found that the molecule hit the wall at uniform intervals $\Delta t = 2l/v_x$. But, in fact, under ordinary conditions a molecule collides with other molecules much more frequently than it collides with a wall. Remarkably, this great disparity between our ideal gas and a real gas does not affect the pressure, as long as the molecules interact with one another only during the instant of contact.[†] To understand qualitatively why the disparity does not affect the gas pressure, note that each such collision results in a transfer of momentum. The momentum is transferred from molecule to molecule until, inevitably, some molecule strikes the wall and exchanges momentum with it. What really matters is the transfer of momentum between the gas as a whole and the walls. Collisions between molecules do not affect the total momentum of the gas; they merely serve to redistribute it from molecule to molecule. A detailed dynamical treatment verifies this qualitative argument.

But we can completely avoid a discussion of the details of the collisions and still show quantitatively that molecule-molecule collisions do not affect the pressure exerted by the gas on the container walls. We begin with Equation 18.12, $V = \frac{2}{3}N\langle\epsilon\rangle/p$, and solve for the pressure:

$$p = \frac{2}{3}\frac{N\langle\epsilon\rangle}{V}.$$

Because the total energy of the gas is $E = N\langle\epsilon\rangle$, we can write

$$p = \frac{2}{3}\frac{E}{V}. \tag{18.40}$$

The macroscopic quantity E/V is the energy of the gas per unit volume; it is called the **energy density**. The completely macroscopic Equation 18.40 states the remarkable fact that the pressure exerted by the gas on the walls—a phenomenon having to do with the behavior of the gas at its boundaries—depends only on the energy density, which is a property of the gas as a whole. But the macroscopic energy density does not depend on the microscopic details of how the molecules collide with one another. Consequently, the pressure p must be independent of the details of the intermolecular collisions. Thus the ability of the ideal-gas model to predict the ideal-gas law does not depend on the unrealistic assumption that the molecules are point masses.

The Dulong-Petit Law

We now use the ideas of kinetic theory to consider crystalline solids. Specifically, can we predict the molar heat capacity of such a solid by applying the kinetic theory to the simplest possible model of a solid?

In 1819, the French chemist Pierre Louis Dulong (1785–1838) and the physicist Alexis-Thérèse Petit (1791–1820) noted that, though the specific heat capacities c of crystalline solids vary considerably, their molar heat capacities c' nearly all cluster about the value $3R$. This statement is called the **Dulong-Petit law**. It is illustrated in Table 18.2. The values of the specific heat capacity decrease with increasing atomic mass. But the molar heat capacities all lie within 10% of the value $3R$, most of them well within that limit.

Figure 18.3 depicts the simplest possible model of a crystalline solid. We know that the atoms in such solids are arranged in regular order. Here, we represent the atoms as

---

*This subsection may be omitted without loss of continuity.

[†]But see Problem 18.60 for the consequences of the fact that the molecules occupy a part of the total volume of the container.

Substance	Atomic mass (u)	c (J/kg·K)	c'/R
Lithium	6.941	35.6×10^2	2.97
Aluminum	26.98	9.00×10^2	2.92
Copper	63.55	3.87×10^2	2.96
Iodine	126.90	2.10×10^2	3.20
Lead	207.21	1.28×10^2	3.18
Uranium	238.03	1.16×10^2	3.31

(a) (b)

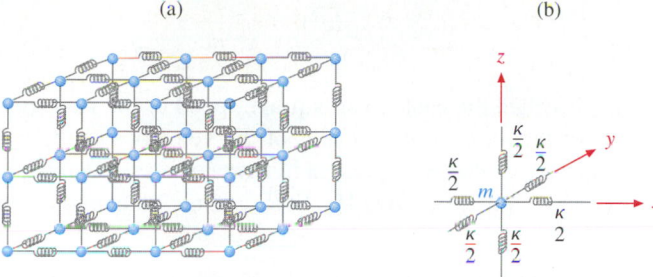

FIGURE 18.3 The simplest model of a crystalline solid. (*a*) The atoms of the solid are particles of mass m arrayed in a cubic structure. Each particle is connected to the six atoms nearest it by springs of force constant $\kappa/2$. These springs represent the bonds that give the solid its rigidity. (*b*) Detail showing a single particle and the springs to which it is attached.

identical particles of mass m, stacked in a cubic array. The fact that the solid has mechanical strength assures us that the atoms are bound together. We represent these bonds by an array of identical springs of force constant $\kappa/2$. (In this chapter and in Chapters 19 and 20, we reserve the symbol k for Boltzmann's constant.) Each atom is bound to its six nearest neighbors by six such springs, two along each principal direction. Displacing an atom a distance x from its equilibrium position along the x axis results in a restoring force $F_x = -\kappa x$, as do displacements along the y and z directions in a similar fashion.

As in a gas, the atoms of the solid are in constant disordered motion. The energy of that motion is the internal energy of the solid. But the atoms in the solid, unlike the atoms of a gas, are not completely free to move. Rather, they oscillate about their equilibrium positions. At any instant, an atom is displaced from its equilibrium position by a distance $\mathbf{r} = (x, y, z)$ and is moving with velocity $\mathbf{v} = (v_x, v_y, v_z)$. Its total energy ϵ is the sum of the kinetic energy of its motion and the potential energy of the springs to which it is attached. That sum is

$$\epsilon = \tfrac{1}{2}mv_x^2 + \tfrac{1}{2}mv_y^2 + \tfrac{1}{2}mv_z^2 + \tfrac{1}{2}\kappa x^2 + \tfrac{1}{2}\kappa y^2 + \tfrac{1}{2}\kappa z^2. \tag{18.41}$$

There are *six* degrees of freedom. According to the equipartition theorem, the mean energy of the atoms is $\tfrac{1}{2}kT$ per degree of freedom, so we have

$$\langle \epsilon \rangle = 6 \times \tfrac{1}{2}kT = 3kT,$$

and the total energy of a crystal containing N atoms is $E = 3NkT = 3nRT$.

To find the molar heat capacity, we use Equation 18.28 to obtain

$$c' = \frac{1}{n}\frac{dE}{dT} = 3R. \tag{18.42}$$

This is the Dulong-Petit law. The fact that we can so successfully extend a theory of gases to deal with solids suggests that the theory has a remarkable capacity for generalization.

Counting Degrees of Freedom
Unfortunately, there are two conflicting conventions for counting degrees of freedom. We adopt the convention of counting each term as a degree of freedom, whether it is a position term such as x or a velocity term such as v_x. The other convention counts terms in pairs; that is, the terms in x and v_x pertain to the same degree of freedom. If you adopt the latter convention, you must count differently in the argument leading to Equation 18.42. But the result turns out the same. When reading other books, make sure you know which convention is being used.

The Speed of Gas Molecules

Let us look again on the microscopic scale at an ideal gas. The mean energy of the molecules is related to the temperature by Equation 18.15,

$$\langle \epsilon \rangle = \tfrac{3}{2}kT.$$

If we consider only the translational kinetic energy, we can write

$$\tfrac{1}{2}m\langle v^2 \rangle = \tfrac{3}{2}kT.$$

This equation applies to polyatomic as well as monatomic molecules because the energy of the additional degrees of freedom of polyatomic molecules plays no part in translational motion. Solving for $\langle v^2 \rangle$, we obtain

$$\langle v^2 \rangle = \frac{3kT}{m}. \qquad \text{(18.43)}$$

Taking the square root of both sides of this equation gives

$$v_{\text{rms}} \equiv \sqrt{\langle v^2 \rangle} = \sqrt{\frac{3kT}{m}}. \qquad \text{(18.44a)}$$

The quantity v_{rms} is called the **root-mean-square speed** or the **rms speed** for short. It is the square root of the mean of the squares of the velocities of the individual molecules.* The rms speed can also be expressed in terms of macroscopic quantities. To do this, note that $kT/m = AkT/Am = RT/M_0$. So Equation 18.44 can be written

$$v_{\text{rms}} = \sqrt{\frac{3RT}{M_0}}. \qquad \text{(18.44b)}$$

EXAMPLE 18.4

Find the rms speed at room temperature (300 K) for H_2 and O_2 molecules. Take the atomic mass of oxygen to be 16.0 u and that of hydrogen to be 1.01 u.

SOLUTION: By using Equation 18.44b instead of Equation 18.44a, you can save yourself the trouble of converting the atomic masses into kilograms. Because the oxygen molecule is diatomic, you have a molecular mass of 32 u, and 1 kmol of O_2 has mass 32 kg. This gives you

$$v_{\text{rms}}(O_2) = \sqrt{\frac{3 \times 8.31 \times 10^3 \text{ J/kmol·K} \times 300 \text{ K}}{32 \text{ kg/kmol}}}$$

$$= 483 \text{ m/s}.$$

You could repeat the calculation to find the result for hydrogen, but there is an easier way. Equation 18.44b tells you that, at a given temperature, the rms speeds of gases are related inversely as the square root of the ratio of their masses per kilomole. Thus you can write

$$v_{\text{rms}}(H_2) = v_{\text{rms}}(O_2) \times \sqrt{\frac{M_0(O_2)}{M_0(H_2)}}$$

$$= 483 \text{ m/s} \times \sqrt{\frac{32.0}{2.02}} = 1920 \text{ m/s}.$$

The room-temperature rms speeds of gas molecules are considerable—of order 1 km/s.

Because the density of air decreases with increasing altitude, climbers use pure oxygen to maintain an acceptable partial pressure of oxygen in their lungs.

The Barometric Equation[†]

Nearly everyone knows that the pressure of the atmosphere diminishes with increasing altitude. The point was first demonstrated quantitatively by an experiment suggested in 1646 by Pascal (see sidebar, Section 15.5).

The underlying physical reason for the variation of pressure with altitude is most easily seen if we work downward from the top of the atmosphere. Each element of atmosphere is subjected to the pressure generated by the weight of the atmosphere above it. This is true of all fluids, but the compressibility of gases introduces another factor. With increasing pressure, the density of the gas increases as well. Thus each volume element of air is denser than those above it and contributes more to the incremental increase of pressure as we go downward. In what follows, we will develop this idea

---

*Note that in general $\langle v^2 \rangle \neq \langle v \rangle^2$. To find $\langle v^2 \rangle$, you square the individual quantities and then average them. To find $\langle v \rangle^2$, you average the quantities and then square the result. The inequality will be immediately evident if you carry out both processes for the set of numbers $-2, -1, 1, 2$. Also try the set $2, 3$.

[†]The remainder of this chapter may be omitted without loss of continuity.

quantitatively. Specifically, we will derive an expression for the pressure of the atmosphere as a function of height above sea level.

The density of the atmosphere is affected by both temperature and pressure, and both vary with altitude. For reasons of simplicity we will consider, somewhat unrealistically, an *isothermal atmosphere*—that is, an atmosphere whose temperature is the same from top to bottom.

EXAMPLE 18.5

To see how the assumption of an isothermal atmosphere simplifies the analysis of pressure as a function of altitude, show that the density ρ of the atmosphere is directly proportional to the pressure p and inversely proportional to the temperature T.

SOLUTION: You can write the ideal-gas law (Equation 18.18) in the form

$$\frac{N}{V} = \frac{p}{kT}. \qquad (18.45)$$

The quantity N/V is the number of molecules of gas per unit volume. Now, the density ρ of the gas is its mass per unit volume; $\rho = M/V$. Because $M = Nm$, you can multiply Equation 18.45 through by m to obtain

$$\rho = \frac{Nm}{V} = \frac{m}{k}\frac{p}{T}. \qquad (18.46)$$

For a particular gas, m/k is a constant. Thus the density of the atmosphere depends just as sensitively on the temperature as it does on the pressure.

Figure 18.4 shows an imaginary box containing an element of atmosphere at an arbitrary altitude h above sea level. The box has volume dV. The area of its top and bottom surfaces is a, and its height is dh. The pressure at the top surface of the box is p, and the pressure at the bottom surface has the slightly greater value $p + dp$. The pressure increment dp is due to the weight of the gas contained in the box. If the gas has mass dM, its weight is $g\,dM$, and we can write

$$dp = g\,\frac{dM}{a} = g\,\frac{\rho\,dV}{a}.$$

We can substitute $dV/a = |dh|$. So we have

$$dp = -\rho g\,dh. \qquad (18.47)$$

The minus sign is needed when we express dp in terms of dh because h increases in the upward direction while p increases in the downward direction.

Into Equation 18.47 we substitute the value of ρ given by Equation 18.46. This gives

$$dp = -\frac{mg}{kT}\,p\,dh.$$

Dividing through by p yields

$$\frac{dp}{p} = -\frac{mg}{kT}\,dh. \qquad (18.48)$$

We can integrate both sides of this equation, provided that mg/kT is a constant. We have already assumed that T is constant, and m will be constant if the atmosphere has uniform composition throughout. It remains to assume that the acceleration of gravity g does not vary appreciably over the height of the atmosphere. Because the atmosphere is a thin layer (something like 30 km) compared with the radius of the earth (6400 km), the error introduced by this assumption is negligible compared with that introduced by assuming an isothermal atmosphere.

Integration of Equation 18.48 (Table 7.1) yields

$$\ln p(h) = -\frac{mg}{kT}\,h + C. \qquad (18.49)$$

We write the pressure as $p(h)$ to make explicit the fact that pressure is a function of

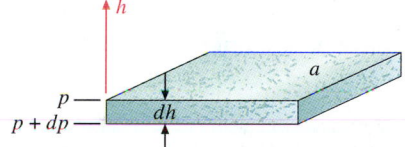

FIGURE 18.4 Atmosphere at an arbitrary altitude h above sea level is contained in an imaginary box of height dh and area a. The pressure at the top surface is p, and that at the bottom surface is $p + dp$.

altitude. We must evaluate the integration constant C. To do this, we note that, at sea level ($h = 0$), the atmospheric pressure p must have its sea-level value $p_0 = 1.013 \times 10^5$ Pa. So, setting $h = 0$ in Equation 18.49, we have

$$\ln p_0 = 0 + C \quad \text{or} \quad C = \ln p_0.$$

Because $\ln p(h) - \ln p_0 = \ln [p(h)/p_0]$, Equation 18.49 becomes

$$\ln \frac{p(h)}{p_0} = -\frac{mg}{kT} h. \tag{18.50}$$

Taking the antilogarithm of both sides of this equation gives

$$\frac{p(h)}{p_0} = e^{-(mg/kT)h}, \tag{18.51a}$$

which can also be expressed in the form

$$p(h) = p_0 e^{-(mg/kT)h}. \tag{18.51b}$$

Either of Equations 18.51a and 18.51b is called the **barometric equation** or, more precisely, the **isothermal barometric equation**. According to the barometric equation, the pressure of the atmosphere tails off asymptotically to zero with increasing altitude h. And, according to Equation 18.46, the density ρ of the atmosphere is proportional to the pressure. Thus the density also tails off asymptotically to zero, and there is no precisely definable "top" to the isothermal atmosphere. Figure 18.5 is a plot of Equation 18.51a.

FIGURE 18.5 Plot of atmospheric pressure versus altitude for an atmosphere composed of air (mean molecular mass 29 u), assuming a uniform temperature $T = 273$ K.

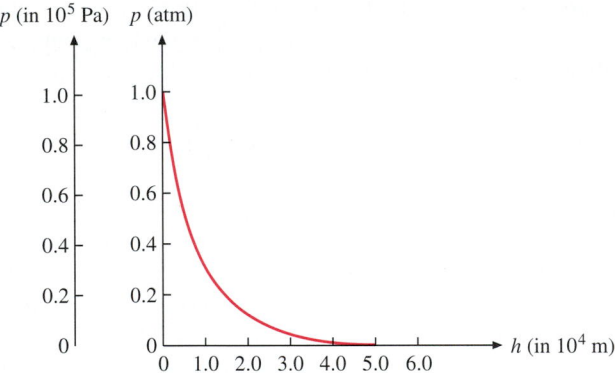

EXAMPLE 18.6

(a) What is the pressure of the atmosphere at the top of Mt. Everest, whose height is 8847 m? Take the mean molecular mass of air to be 29 u, and assume a constant temperature $T = 273$ K. **(b)** At what altitude does the atmospheric pressure have one-half its sea-level value?

SOLUTION:

(a) Find the atmospheric pressure at 8847 m.

To express the pressure at the top of Mt. Everest in atmospheres, use the barometric equation in the form of Equation 18.51a. To use this equation, you need to find the mean mass of air molecules in kilograms. You have

$$m_{air} = 29 \text{ u} \times 1.66 \times 10^{-27} \text{ kg/u} = 4.81 \times 10^{-26} \text{ kg}.$$

So Equation 18.51a becomes

$$\frac{p}{p_0} = \exp\left(-\frac{4.81 \times 10^{-26} \text{ kg} \times 9.80 \text{ m/s}^2}{1.38 \times 10^{-23} \text{ J/K} \times 273 \text{ K}} \times 8847 \text{ m}\right)$$

$$= 0.330.$$

Because $p_0 \equiv 1$ atm, you have

$$p = 0.330 \text{ atm}.$$

You can readily express this result in pascal. Multiply p/p_0 by p_0 in Pa to obtain

$$p = 0.330 \times 1.013 \times 10^5 \text{ Pa} = 3.35 \times 10^4 \text{ Pa}.$$

(b) Find the altitude at which $p = \frac{1}{2}p_0$.

The simplest place to begin is Equation 18.50. You solve for h, obtaining

$$h = -\frac{kT}{mg} \ln \frac{p}{p_0}.$$

Because you wish $p/p_0 = \frac{1}{2}$, you have

$$h = -\frac{1.38 \times 10^{-23} \text{ J/K} \times 273 \text{ K}}{4.81 \times 10^{-26} \text{ kg} \times 9.80 \text{ m/s}^2} \times \ln \frac{1}{2} = 5540 \text{ m}.$$

There are many mountains higher than this in the Himalayas and the Andes, as well as a few elsewhere.

The Maxwell-Boltzmann Velocity Distribution

At the beginning of this section, we derived Equation 18.44a, which expresses the root-mean-square speed of the molecules of a gas in terms of the temperature and molecular mass of the gas. Some molecules are moving faster than v_{rms}, and some are moving slower. Although the value of v_{rms} gives us some information as to the speed of the molecules, we would like to have *complete* information on the subject. That is, we would like to know the form of the function $\nu(\mathbf{v})$ that answers the following question: For a gas of molecular mass m, whose temperature is T, what proportion ν of the molecules of a gas have any given velocity $\mathbf{v}$?

We begin to answer this question by looking closely at Equation 18.51b,

$$p(h) = p_0 e^{-mgh/kT}.$$

For our present purposes, we want to rewrite this equation in terms of the number of molecules per unit volume, to which we assign the symbol ρ'. According to Equation 18.46, the mass density is $\rho = Nm/V = mp/kT$. So ρ' is just equal to ρ/m, or

$$\rho' \equiv \frac{N}{V} = \frac{p}{kT}. \tag{18.52}$$

In terms of ρ', we can rewrite the isothermal barometric equation in the form

$$\rho'(h) = \text{constant} \times e^{-mgh/kT}. \tag{18.53a}$$

Next we note that the term mgh in the numerator of the exponent is the gravitational potential energy $U(h)$ of a molecule whose altitude is h. Let us emphasize this point by writing Equation 18.53a in the form

$$\rho'(h) = \text{constant} \times e^{-U(h)/kT}. \tag{18.53b}$$

We now make a plausible argument, which we will later verify: The number of molecules per unit volume as a function of h depends on their potential energy according to Equation 18.53b. But gravitational potential energy $U(h)$, a function of height, is freely convertible into kinetic energy $K(\mathbf{v})$, a function of velocity. So, in analogy to Equation 18.53b, we should be able to write

$$\nu(\mathbf{v}) = C e^{-K(\mathbf{v})/kT}. \tag{18.54}$$

Here $\nu(\mathbf{v})$ is the number of molecules of gas per unit volume that have the particular velocity $\mathbf{v} = (v_x, v_y, v_z)$. Let us write this expression out explicitly in terms of the velocity. We have

$$\nu(v_x, v_y, v_z) = C e^{-m(v_x^2 + v_y^2 + v_z^2)/2kT}, \tag{18.55a}$$

or

$$\nu(\mathbf{v}) = C e^{-mv_x^2/2kT} e^{-mv_y^2/2kT} e^{-mv_z^2/2kT}. \tag{18.55b}$$

Equation 18.55b is the **Maxwell-Boltzmann velocity distribution**. It is plotted as a function of v_x in Figure 18.6, with v_y and v_z held constant. Similar curves can be plotted

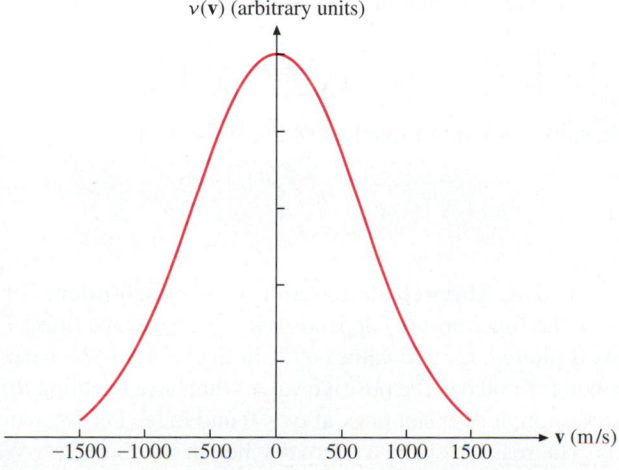

$\nu(\mathbf{v})$ (arbitrary units)

$-1500 \quad -1000 \quad -500 \quad 0 \quad 500 \quad 1000 \quad 1500$ $\mathbf{v}$ (m/s)

FIGURE 18.6 The Maxwell-Boltzmann velocity distribution (Equation 18.55b).

as functions of v_y and of v_z. You may be familiar with this bell-shaped curve; it is a special case of the more general *Gaussian distribution*. The symmetry of the curve arises from the fact that ν depends on $e^{-v_x^2}$. The symmetry guarantees that the mean value of v_x is $\langle v_x \rangle = 0$. The value of ν attains a maximum at $v_x = 0$; thus, this is also the *most probable value* of v_x: $v_{x\ mp} = 0$. The number of molecules having velocity v_x diminishes rapidly as $|v_x|$ increases.

The Maxwell-Boltzmann Speed Distribution

The *speed distribution* is of much more interest than the velocity distribution of Equations 18.55a and 18.55b. Those equations give ν as a function of some combination of v_x, v_y, and v_z. But there are many other combinations of these three variables that give the *same* value of the kinetic energy K and thus the same speed $v = \sqrt{2K/m}$. How can we determine the speed distribution function?

Any particular molecule has velocity $\mathbf{v} = (v_x, v_y, v_z)$. This particular value of $\mathbf{v}$ can be represented as a point in a cartesian coordinate system whose axes are v_x, v_y, and v_z; see Figure 18.7. Such a three-dimensional space is called a **velocity space**.

Every point $\mathbf{v}$ lies on the surface of some sphere whose center is the origin. Consider the set of points of which this spherical surface is composed. Each point represents a different value of the vector $\mathbf{v}$—that is, a different combination of components (v_x, v_y, v_z). But the equation of the spherical surface is $v^2 = v_x^2 + v_y^2 + v_z^2 = $ constant. Thus all molecules whose velocities are represented by points on the surface have the same *speed v*. (The molecules must also have equal kinetic energy $K = \frac{1}{2}mv^2$.) The area of the spherical surface is $4\pi v^2$.

The speed distribution $\nu(v)$ will be different from the velocity distribution $\nu(\mathbf{v})$. The reason is that the number of molecules having speed v is related to the number of molecules having velocity $\mathbf{v}$ by the weighting factor $4\pi v^2$, which is a measure of the number of different $\mathbf{v}$ corresponding to the same v. Using Equation 18.55a, we have

$$\nu(v) = 4\pi v^2 \nu(\mathbf{v}) = C(4\pi v^2)e^{-mv^2/2kT}. \tag{18.56}$$

It remains to evaluate the constant C. To do this, we remember that $\nu(v)$ is the number of molecules per unit volume having speed v, while $\rho' = N/V$ is the total number of molecules per unit volume. Therefore, if we sum ν over all possible values of v, we must obtain ρ':

$$\int_0^\infty \nu(v)\, dv = \rho'.$$

Inserting the explicit value of ν from Equation 18.56, we have

$$4\pi C \int_0^\infty v^2 e^{-mv^2/2kT}\, dv = \rho'. \tag{18.57}$$

We define $a \equiv m/2kT$, whereupon the integrand becomes $v^2 e^{-av^2}$. The resulting definite integral is a standard one tabulated in most mathematical handbooks. Its value is

$$\int_0^\infty v^2 e^{-av^2}\, dv = \frac{1}{4}\sqrt{\frac{\pi}{a^3}} = \frac{\sqrt{\pi}}{4}\left(\frac{2kT}{m}\right)^{3/2}.$$

Substituting this value back into Equation 18.56, we obtain

$$\nu(v) = 4\pi\rho'\left(\frac{m}{2\pi kT}\right)^{3/2} v^2 e^{-mv^2/2kT}. \tag{18.58}$$

This equation is called the **Maxwell-Boltzmann speed distribution**. For any particular molecular mass m, the function $\nu(v)$ depends only on the temperature T. The function for hydrogen gas is plotted, for two values of T, in Figure 18.8. Because ν is a function of speed, its domain is limited to the positive values that have meaning for speed. Unlike the velocity distribution, it does not peak at $v = 0$ and indeed is not symmetrical about any vertical axis. The reason for the asymmetry lies in the factor v^2. When v is small

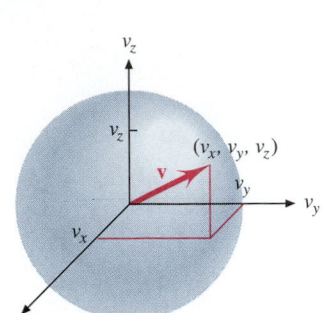

FIGURE 18.7 The velocity $\mathbf{v}$ of any particular molecule can be represented as a point in a three-dimensional velocity space defined by the cartesian axes v_x, v_y, and v_z. Any such point lies on the surface of some sphere centered at the origin. Although all points on that surface represent different values of $\mathbf{v}$, they all have the same value of $\mathbf{v}^2 = v_x^2 + v_y^2 + v_z^2$. Thus, they all have the same speed $v = \sqrt{\mathbf{v}^2}$. And all molecules represented by these points have the same kinetic energy $K = \frac{1}{2}mv^2$.

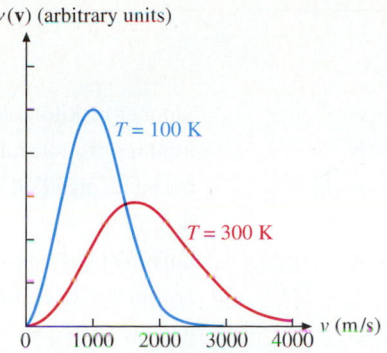

FIGURE 18.8 The Maxwell-Boltzmann speed distribution for hydrogen gas at 300 K and 100 K. The mass of the H_2 molecule is taken to be 3.35×10^{-27} kg.

(that is, $mv^2/2kT \ll 1$), the exponential factor has a value close to 1 and varies slowly, so the v^2 term dominates. Thus the curve departs from the origin in roughly parabolic fashion. With further increase in v, however, the exponential factor begins to diminish in value at a significant rate and the curve ceases to rise parabolically. Ultimately, the exponential factor dominates; the curve reaches a maximum and begins to fall. For large values of v (that is, $mv^2/2kT \gg 1$), the curve approaches zero asymptotically in an essentially exponential fashion.

Three different "average" speeds are associated with the Maxwell-Boltzmann speed distribution. The first is the root-mean-square speed $v_{rms} = \sqrt{3kT/m}$. The **most probable speed** v_{mp} is the value of v for which $\nu(v)$ has its maximum value. It can be shown that

$$v_{mp} = \sqrt{\frac{2kT}{m}} = 0.817v_{rms}; \qquad (18.59)$$

see Problem 18.54. The **mean speed** $\langle v \rangle$ is the subject of Problem 18.55. It can be shown that

$$\langle v \rangle = \sqrt{\frac{8kT}{\pi m}} = 0.922v_{rms}. \qquad (18.60)$$

The difference in the values of these averages is due to the fact that the distribution curve is asymmetrical.

How well does the Maxwell-Boltzmann distribution function correspond to experimental results? Although there are many indirect lines of evidence that give confidence in the correctness of the distribution function, it is not easy to make direct measurements of molecular speeds. Nevertheless, the results of an experiment carried out by Miller and Kusch in 1955 are convincing. These results are displayed in Figure 18.9. The details of the experiment are the subject of Problem 18.70.

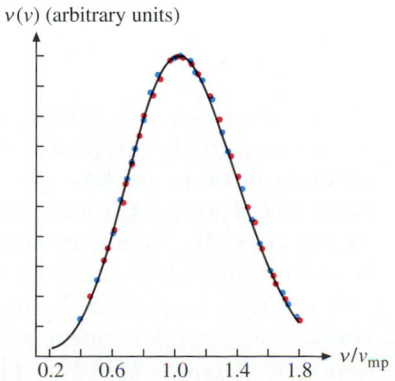

FIGURE 18.9 Correspondence between the experimental results of Miller and Kusch for thallium vapor (red and blue dots) and the Maxwell-Boltzmann distribution function. The red dots represent observations made at 944 K; the blue dots represent observations at 870 K. To make it possible to plot both sets of observations on the same curve, $\nu(v)$ is plotted for each temperature as a function of v/v_{mp}, where $v_{mp} = \sqrt{2kT/m}$ is the most probable speed for that temperature. (Derivation of the value of v_{mp} is the subject of Problem 18.54.)

a	area	n	number of kilomoles in a sample of matter
A	Avogadro's number	N	number of molecules or atoms in a sample
c	specific heat capacity	p, p_0	pressure, standard atmospheric pressure
c', c_v'	molar heat capacity, molar heat capacity at constant volume	Q	heat flow
C	total heat capacity	R	universal gas constant
E	total energy of a body of gas	T	absolute temperature
f	number of degrees of freedom	U	potential energy
$\mathbf{F}$	force exerted on a molecule when it collides with a wall	$\mathbf{v}, v$	velocity, speed
$\langle \mathbf{F} \rangle$	mean force exerted on a molecule over successive collisions with the same wall	$v_{rms}, v_{mp}, \langle v \rangle$	root-mean-square speed, most probable speed, mean speed
g	acceleration of gravity	V	volume
h	altitude	$\epsilon, \langle \epsilon \rangle$	energy of a molecule (or atom), mean energy
I_ξ, I_η, I_ζ	moment of inertia of a molecule about axes fixed in the molecule	κ	force constant
		Λ	side of a cube containing one molecule on average
k	Boltzmann's constant	μ	molecular (or atomic) mass in u
K	kinetic energy	$\nu(\mathbf{v}), \nu(v)$	number of molecules per unit volume having velocity $\mathbf{v}$, speed v
l	length		
m	mass of a molecule in kg	ρ	density (mass per unit volume)
M, M_0	mass of a sample of gas, of 1 kmol of a gas	ρ'	number of molecules per unit volume
		χ	an arbitrary variable

Summing Up

In order to pursue the idea that heat is a form of energy, we devise a **kinetic model** of an **ideal gas**. By applying the laws of mechanics to the model, we derive Equations 18.10a and 18.10b,

$$p = \frac{2}{3} \frac{N\langle \epsilon \rangle}{V} = \frac{2}{3} \frac{E}{V},$$

which relate the pressure and volume of an ideal gas to the energy of its molecules. From this result follow Boyle's law, Charles's law, and the **ideal-gas law**. The ideal-gas law is given in mixed macroscopic and microscopic terms by Equation 18.18,

$$pV = NkT,$$

and in purely macroscopic terms by Equation 18.24,

$$pV = nRT.$$

The value of **Boltzmann's constant** k, a universal constant, is given by Equation 18.16. The value of the **universal gas constant** R depends on the following chain of argument:

1. The **atomic mass unit** (u) is defined as $\frac{1}{12}$ the mass of a ^{12}C atom—that is, approximately the mass of a ^1H atom. The unified mass unit is related to the kilogram by any of Equations 18.20 and 18.21.

2. The mass μ of a molecule (or atom) expressed in u is called its **molecular mass** (or atomic mass).

3. A quantity of homogeneous matter whose mass in kilograms (kg) is equal to its molecular (or atomic) mass in unified mass units (u) constitutes 1 **kilomole** (kmol) of substance.

4. It follows that 1 kmol of any substance contains a number of molecules (or other elementary entities) given by the universal constant A, called **Avogadro's number**. The constant A is the reciprocal of the number of atomic mass units in a kilogram; the value of A is given by Equation 18.22.

5. The **universal gas constant** R is defined by Equation 18.23a, $R \equiv Ak$. Its value is given by Equations 18.23b and 18.23c.

From the ideal-gas law and the equivalence of heat and energy, we derive the **molar heat capacity at constant volume**, given by Equation 18.31,

$$c_v' = \tfrac{3}{2}R.$$

This result agrees very well with the experimental values for monatomic gases. For polyatomic gases, we invoke the **equipartition theorem** and take into account the energy associated with **degrees of freedom** other than the three translational ones. The agreement with experimental values is good for many diatomic gases.

When the equipartition theorem is applied to a simple kinetic model of solids having six degrees of freedom, the theory yields Equation 18.42, which predicts a molar heat capacity $c' = 3R$. This agrees well with the experimentally based rule called the **Dulong-Petit law**.

In an isothermal atmosphere of uniform composition, the pressure varies with altitude according to Equation 18.51b, the **isothermal barometric equation**:

$$p(h) = p_0 e^{-(mg/kT)h}.$$

Arguing by analogy from the relation between $p(h)$ and the potential energy $U(h)$ to the relation between $\nu(\mathbf{v})$ and the kinetic energy $K(\mathbf{v})$, we obtain the **Maxwell-Boltzmann velocity distribution** given by either of Equations 18.55a and 18.55b. We then devise a way to "count" the number of values of $\mathbf{v}$ corresponding to the speed v. From this, we derive Equation 18.58, the **Maxwell-Boltzmann speed distribution**:

$$\nu(v) = 4\pi\rho'\left(\frac{m}{2\pi kT}\right)^{3/2} v^2 e^{-mv^2/2kT}.$$

KEY TERMS

Section 18.2 The Kinetic Model and the Ideal-Gas Law
specular reflection ▪ Boltzmann's constant, universal gas constant ▪ ideal-gas law ▪ equation of state ▪ molecular mass ▪ Avogadro's number

Section 18.3 Molecular Heat Capacity and Equilibrium
molar heat capacity ▪ degrees of freedom ▪ equipartition theorem ▪ energy density

Section 18.5 The Speed of Gas Molecules
root-mean-square speed, most probable speed, mean speed ▪ isothermal barometric equation ▪ Maxwell-Boltzmann velocity distribution, speed distribution ▪ velocity space

QUERIES

18.1 *(2) Changing sides.* Suppose you were to repeat the derivation of Equation 18.10a, $p = \frac{2}{3}N\langle\epsilon\rangle/V$, for a gas contained in a box having unequal sides l, w, t. Would you get a different result? Explain.

18.2 *(2) Spreading out.* When water boils at atmospheric pressure, the volume occupied by the resulting steam is about 1700 times the volume occupied by the water. Is this observation consistent with the result of Example 18.3?

18.3 *(2) Doing things in a big way.* Outline an experiment, in which all measurements are macroscopic, for determining the value of the universal gas constant R.

18.4 *(3) Coming together, I.* Consider a gas whose molecules have finite size. Suppose we mix a sample of gas having high temperature T_h with a sample having low temperature T_c. Corresponding to the initial temperatures are mean molecular energies $\langle\epsilon\rangle_h$ and $\langle\epsilon\rangle_c$, with the energies of individual molecules in each of the two sets distributed about the mean values in some way. Describe qualitatively how thermal equilibrium is achieved after mixing, so that all the molecules can be characterized by a single temperature T and a single mean energy $\langle\epsilon\rangle$? What principle of mechanics underlies the equilibration process?

18.5 *(3) Equipartition in action.* Imagine several dumbbell molecules like that of Figure 18.2. One happens to have a small speed as it approaches a wall of the container, but it has a large rotational speed. Another has a large translational speed but is spinning slowly. When the molecules strike the wall, what happens, on the average, to their translational and rotational kinetic energies? (Hint: See Problems 13.52 and 13.58.) Explain how your description of the collisions accords with the equipartition theorem.

18.6 *(5) Coming together, II.* You have equal volumes of hydrogen gas and oxygen gas at the same temperature and pressure. As Example 18.4 shows, the values of v_{rms} for the two gases differ considerably. Yet, when you mix them, there is no need for them to attain thermal equilibrium because they are already at the same temperature. Explain.

18.7 *(5) Weather wise.* Meteorologists often speak of the "pressure altitude." A given pressure altitude designates an imaginary surface or contour in the atmosphere on which the barometric pressure is everywhere the same. Meteorological balloons usually drift along at a constant pressure altitude. What information can you get by monitoring the actual altitude of such a balloon as it drifts?

18.8 *(5) The squeeze is on.* A container of gas has temperature T. The gas is compressed isothermally. What happens to the root-mean-square speed?

18.9 *(5) Big, bigger, biggest.* Explain qualitatively why v_{rms} is greater than $\langle v \rangle$ and why both are greater than v_{mp}.

18.10 *(5) Long gone.* The escape speed from the surface of the moon is 2380 m/s. According to Example 18.4, the rms speed of O_2 molecules at 300 K is 483 m/s; the speed of N_2 molecules is not very different. Why doesn't the moon have an atmosphere?

18.11 *(5) Sound reasoning.* As Example 18.4 shows, the rms speeds of gas molecules at room temperature are considerable. For comparison, the room-temperature speed of sound in oxygen is 332 m/s; that for hydrogen is 1350 m/s. Give a qualitative reason why the speed of sound in a gas has very roughly the same value as v_{rms}.

18.12 *(G) Sharp eyes.* Describe two different experimental methods for determining that the molecules of iodine vapor are diatomic.

PROBLEMS

GROUP A

18.1 *(2) Mean molecular energy, I.* What is the mean energy of the molecules in a monatomic gas when the temperature is 300 K? 1000 K?

18.2 *(2) Mean molecular energy, II.* What is the mean energy of the molecules in a diatomic gas when the temperature is 300 K? 1000 K?

18.3 *(2) How much?* Nitrogen is contained in a cylinder of volume $1.2 \times 10^{-2} \text{ m}^3$. When the temperature is 290 K, the pressure is 8.10×10^6 Pa. What is the mass of the gas contained? The mass of a nitrogen molecule is 4.65×10^{-26} kg.

18.4 *(2) How hot?* A container of volume $8.20 \times 10^{-4} \text{ m}^3$ contains 2.13 g of oxygen gas. The pressure is 2.36 atm. What is the temperature of the gas?

18.5 *(2) Storage.* A cylinder of volume 0.2 m^3 contains oxygen at pressure 10^7 Pa. The temperature is 0°C. Find the volume that would be occupied by the gas at STP.

18.6 *(2) No safety factor?* A 25-liter steel tank can withstand a maximum pressure of 150 atm without failing. If the temperature of the tank never exceeds 40°C, what is the maximum mass of H_2 it can contain? Neglect the deviation of the pressurized hydrogen from ideal-gas behavior.

18.7 *(2) Ullage.* When full, a gas cylinder contains 10 kg of gas at 100 atm pressure. Gas is removed from the cylinder until the pressure has dropped to 25 atm. What is the mass of the removed gas?

18.8 *(2) In my little room?* If your bedroom has dimensions 5.50 m × 4.25 m × 2.50 m, find the number of air molecules it contains when the pressure is 1 atm and the temperature is 22°C.

18.9 *(2) Not too close! I.* At 1 atm pressure and 100°C, a given mass of steam occupies 1672 times the volume occupied by an equal mass of water. What is the approximate ratio of the mean distance between the water molecules in steam to the diameter of the water molecules?

18.10 *(2) Another way.* **(a)** Show that the number of kilomoles of a substance is given by the ratio $n = M/\mu$. **(b)** Use this equation to solve Example 18.2.

18.11 *(2) Underpressure.* How many kilomoles of gas are contained in a tank of volume 15 m³ if the pressure inside the tank is 9.8×10^4 Pa and the temperature is 295 K?

18.12 *(2) Overpressure, I.* A strong tank is half filled with liquid water and sealed. When the temperature is raised to 400°C, all the water turns to steam. Find **(a)** the pressure and **(b)** the density of the steam at this temperature. Neglect the deviation of the steam from ideal-gas behavior.

18.13 *(2) Mixup.* Air consists mainly of N_2 and O_2, mixed in the ratio 4 : 1. The atomic mass of N is 14 u, and that of O is 16 u. Find the mean molecular mass of air.

18.14 *(2) Don't sniff! I.* At constant temperature and pressure, ammonia is formed in the following gaseous chemical reaction:

1 volume nitrogen (N_2) + 3 volumes hydrogen (H_2) →
2 volumes ammonia.

What is the chemical formula of ammonia?

18.15 *(2) Don't sniff! II.* A container whose volume is $2.5 \times 10^{-2} \text{ m}^3$ holds 65.0 g of gaseous oxide of sulfur under standard conditions of temperature and pressure. The atomic mass of sulfur is 32 u. What is the chemical formula of the gas?

18.16 *(2) Heavy subject.* A lecture hall has a floor area of 100 m². The ceiling height is 4.50 m. What is the mass of the air contained in the hall when the temperature is 290 K and the barometric pressure is 1.008×10^5 Pa? Take the molecular mass of air to be 29 u.

18.17 *(2) Correction, please!* It is not usually possible to carry out gaseous chemical reactions at standard conditions of temperature and pressure. But you can always correct. Suppose you have carried out a reaction and the resulting gaseous product fills the container, whose volume is $1.000 \times 10^{-3} \text{ m}^3$. The temperature is 325.0 K, and the pressure is 3.362×10^5 Pa. **(a)** How many kilomoles of product do you have? **(b)** What would the volume of the product be at STP?

18.18 *(2) Volume business.* **(a)** How much internal energy is present in 1 kmol of helium at standard conditions of temperature and pressure? Take the molecular mass to be $\mu = 4.003$ u. **(b)** If you could extract all the energy and convert it into electrical energy, how long could you operate a 100-W light bulb?

18.19 *(3) Fast gas.* Oxygen gas is contained in a cylinder whose volume is 0.0235 m³. The total translational kinetic energy of the molecules in the gas is 5.03×10^3 J, and the rms speed of the molecules is 1.97×10^3 m/s. **(a)** Find the mass of the oxygen. **(b)** What is the pressure inside the cylinder? **(c)** What is the total rotational kinetic energy of the molecules of the gas?

18.20 *(3) Heat capacity ratio.* Show that for any substance, $c'/c = \mu$.

18.21 *(3) That much?* **(a)** What is the internal energy of 1 kmol of H_2 gas at STP? of O_2 gas? **(b)** Find the ratio of the internal energy of 1 kmol of H_2 gas to that of 1 kmol of Ar gas.

18.22 *(3) Weighing molecules by heating them.* The specific heat capacity of a diatomic gas at constant volume is 82.7 J/kg·K. What is the mass of 1 kmol of the gas?

18.23 *(3) A little of each.* What is the molar heat capacity at constant volume of a mixture of gas containing 3 kmol of argon and 2 kmol of nitrogen?

18.24 *(3) Chemical change.* Oxygen and hydrogen are mixed in the correct proportions to make steam. **(a)** What is the ratio of the molar heat capacity at constant volume of the resulting steam to that of the initial mixture? **(b)** What is the ratio of the final and initial heat capacities at constant volume?

18.25 *(3) The burden of freedom.* Estimate the molar heat capacity for a gas whose molecules have **(a)** seven degrees of freedom, **(b)** twelve degrees of freedom.

18.26 (3) *The ideal of freedom.* Show that the molar heat capacity of an ideal gas having f degrees of freedom is $fR/2$.

18.27 (4) *Light and heavy.* What is the mass of a piece of aluminum that has the same total heat capacity as 1 kg of lead?

18.28 (5) *All in proportion.* The molecular mass of nitrogen gas (N_2) is 28.01 u. Find the rms speed at room temperature.

18.29 (5) *Getting thin.* The highest altitude at which human beings live continuously without descending to lower altitudes is about 4200 m above sea level. What is the barometric pressure at that altitude, assuming a uniform atmospheric temperature $T = 280$ K?

18.30 (5) *Bottom dead center.* The lowest place on the surface of the earth is the Dead Sea, which lies 397 m below sea level. Find the barometric pressure at the Dead Sea when the sea-level pressure has the standard value p_0. Use $T = 310$ K.

18.31 (5) *Pressure difference.* For passenger comfort, commercial airplanes are usually pressurized to an "equivalent altitude" of about 2100 m. When an airplane is flying at 9000 m (a typical cruising altitude), what is the pressure dif-

ference between the inside and outside of the cabin? Assume $t = -20°C$.

18.32 (5) *Gradual change.* Show that, at altitudes not too far above sea level, the pressure of the atmosphere decreases by about 1% for every 100 m of altitude.

18.33 (5) *Giant among pygmies.* A typical dust particle has mass 10^{-10} kg. If such particles are suspended in air at 300 K, what is their root-mean-square speed? Compare your result with that found for O_2 molecules in Example 18.4.

18.34 (5) *rms minus mp.* At what temperature is the difference between v_{rms} and v_{mp} equal to 75 m/s for nitrogen gas?

18.35 (5) *Popularity contest.* What is the ratio of the number of molecules in a gas having speed v_{rms} to the number having speed v_{mp}? That is, what is the value of

$$\frac{v(v_{rms})}{v(v_{mp})}?$$

Does this ratio depend on the particular gas? on the temperature?

GROUP B

18.36 (2) *Proving a fine point.* Beginning with the definition of a mean value,

$$\langle a \rangle \equiv \frac{1}{N} \sum_{j=1}^{N} a_j,$$

prove that $\langle a^2 + b^2 \rangle = \langle a^2 \rangle + \langle b^2 \rangle$.

18.37 (2) *Good vacuum.* With care, it is possible to obtain a pressure of order 10^{-14} atm in a container at room temperature (288 K). (a) What is the number of gas molecules per unit volume remaining in the container? (b) If the residual gas is mainly H_2, what is its total mass?

18.38 (2) *Collector's item.* You go scuba diving, carrying along a 1-liter screw-top jar to collect seaweed from the sea bottom. The jar is filled with air; when you put the cover on, the temperature was 24°C and the pressure was 1.01×10^5 Pa. When you reach the bottom at a depth of 11.2 m, you unscrew the lid while holding the jar mouth down. How much water flows into the jar if the water temperature is 12°C? Take the density of sea water to be 1.03×10^3 kg/m³.

18.39 (2) *Outgassing.* In order to produce a good vacuum inside a container, it is usually necessary to *outgas* the container. This is done by warming the walls of the container during the pumping-out process. This warming "boils off" the gas molecules bound to the walls by *adsorption* (see Problems 18.63 and 18.64) and allows them to be pumped out. In a typical case, a spherical glass bulb of radius 10 cm has on its surface a layer of adsorbed gas one molecule thick. The radius of the molecules is about 3×10^{-10} m. To see the importance of outgassing, suppose that the container were completely evacuated and then the gas molecules were boiled off the surface but not pumped out of the container. What would be the resulting pressure inside the container?

18.40 (2) *Dalton's law.* A heavy-walled vessel of volume 1.00 m³ contains 1.40 kg of oxygen and 0.90 kg of water. The

vessel is sealed and heated. When the temperature is 500°C, all the water has turned to steam. What is the pressure inside the vessel? (Hint: The total pressure is the sum of the *partial pressures*—that is, of the pressures of the oxygen and the steam considered separately. This fact is called *Dalton's law*.)

18.41 (2) *Overpressure, II.* A pressure vessel of volume 1 m³ contains 0.8 kg of water ($\mu = 18$ u) and 1.6 kg of oxygen. What is the pressure in the vessel when the temperature is raised to 500°C? Assume that all the water has turned to steam, and neglect the deviation of the gas mixture in the vessel from ideal-gas behavior.

18.42 (2) *Overpressure, III.* In a room where $T = 300$ K and $p = 1$ atm, you pour 2.5 g of liquid diethyl ether ($\mu = 74$) into an air-filled 2-liter container and quickly seal it. The highly volatile ether evaporates. What is the pressure inside the container?

18.43 (2) *Dissociation.* If you heat a sealed container of CO_2 gas, the pressure will increase. However, the pressure does not increase in accordance with Charles's law because some of the CO_2 molecules dissociate according to the reaction

$$2\ CO_2 \rightarrow 2\ CO + O_2.$$

If at a certain temperature the pressure is 25% higher than you would expect on the basis of Charles's law, what proportion of the CO_2 molecules in the container have dissociated? Neglect deviation from ideal-gas behavior.

18.44 (2) *Dumas bulb.* Following up on ideas of the Italian chemist Stanislao Cannizzaro (1826–1910), the French chemist Jean-Baptiste Dumas (1800–1884) devised a method for measuring the molecular masses of volatile liquids. A glass bulb with a thin neck is evacuated and sealed at the tip, and its mass m_e is measured by weighing. Then the seal is opened and a small amount of the liquid of interest is put into the bulb. The bulb is placed in a bath of boiling water until the bulb is filled with vapor (all the air is expelled) and there is no excess liquid. With the bulb still in the bath, the tip is sealed again. The bulb is removed from the bath, and its mass, now m_v, is measured. Finally, the bulb is filled with pure (cold) water and its new mass m_w determined. (The bits of glass broken off the tip when the bulb is unsealed are carefully saved and weighed as appropriate.) **(a)** Express the molecular mass μ of the liquid in terms of m_e, m_v, and m_w. (The variation of the density of water with temperature is too small to matter; set $\rho_w = 1 \times 10^3$ kg/m^3.) **(b)** In an experiment to determine the molecular mass of carbon tetrachloride (CCl_4), the following values are obtained: $m_e = 67.95$ g; $m_v = 72.41$ g; and $m_w = 972.68$ g. Find the value of μ. (Assume that the barometric pressure is exactly 1 atm.)

18.45 (2) *Not too close! II.* Using the result of Problem 18.9, find the approximate diameter of a water molecule.

18.46 (3) *Equipartition and wall collisions.* Imagine an O_2 molecule as a dumbbell of length 2.9×10^{-10} m, at whose ends are particles of atomic mass 16.0 u each. **(a)** Assuming equipartition, find the rms angular speed ω_{rms} of the molecule when $T = 273$ K. **(b)** Find the corresponding rms tangential speed of the atoms about the center of mass. Compare this result with the rms translational speed v_{rms}. Will collisions of O_2 molecules with the container walls be an efficient means of redistributing energy between the rotational and translational degrees of freedom?

18.47 (3) *Heat and pressure, I.* **(a)** A diatomic gas in a sealed container of volume V has pressure p_0. How much heat transfer Q into the gas is required to double its pressure? **(b)** Find the value of Q if the gas is O_2, the container volume is 2.0×10^{-3} m^3, and the initial pressure is 1.0 atm.

18.48 (4) *Compounding the problem, I.* One kilomole of solid potassium (K) reacts with 1 kmol of solid iodine (I) to produce 1 kmol of potassium iodide (KI). When everything has cooled down (the reaction evolves quite a lot of heat), is the total heat capacity of the solid KI equal to 3R, 6R, or 1.5R? Explain.

18.49 (4) *Compounding the problem, II.* Two kilomoles of solid potassium reacts with 1 kmol of gaseous chlorine (Cl_2) to produce 2 kmol of solid potassium chloride, according to the reaction

$$2 \text{ K} + Cl_2 \rightarrow 2 \text{ KCl}.$$

What is the total heat capacity of the product?

18.50 (5) *On the average.* A box contains 30 steel balls, each of mass m. One ball has speed v, two have speed $2v$, five have speed $3v$, ten have speed $4v$, seven have speed $5v$, three have speed $6v$, and two have speed $7v$. **(a)** Find v_{rms}, v_{mp}, and $\langle v \rangle$. **(b)** If the volume of the box is V, find the total energy, the energy density, and the pressure exerted by the balls on the walls of the box. (Neglect gravity.)

18.51 (5) *Fraction of molecules within a speed range.* When the temperature is 0°C, what fraction of the molecules in a container of oxygen gas have speeds in the range 100 m/s $< v < 110$ m/s? (Hint: In a speed range this narrow, you are justified in using a constant value of v; say, $v = 105$ m/s.)

18.52 (5) *Centrifuge.* A *centrifuge* is a device in which a tube of fluid can be rotated around an axis at high angular speed.

This problem suggests the utility of the centrifuge. **(a)** As the centrifuge rotates with angular speed ω, the radial acceleration at a distance r from the axis is $g_{eff} = r\omega^2$. Express the corresponding potential energy $U(r)$ of a small particle of mass m located at r. **(b)** When the centrifuge has run long enough for the system to reach equilibrium, show that the particle concentration $\nu(r)$ at a distance r from the axis is

$$\nu(r) = \nu_0 e^{mr^2\omega^2/2kT},$$

where ν_0 is the concentration at (or, practically speaking, near) the axis. **(c)** Suppose that the tube, of length 5.00 cm, contains a dilute solution of protein molecules in water. The tube is located so that one end is near the axis of rotation and the other end is near the periphery of the centrifuge. The molecular mass of the protein is 3.22×10^4 u, and its density is 1.23×10^3 kg/m^3. If the centrifuge rotates at 12,000 rev/min and the temperature is 290 K, find the ratio ν/ν_0 for $r = 5.00$ cm. (Hint: To allow for the buoyancy of the water in which the protein is dissolved, you must multiply m by a correction factor; see part **a** of Problem 18.66.)

18.53 (5) *Characteristic height.* Imagine an atmosphere having the same sea-level pressure as ours but having uniform density equal to the sea-level density of our atmosphere. **(a)** What is the height of such a uniform atmosphere? This height is called the *characteristic height* of our atmosphere. **(b)** What is the pressure of our atmosphere at an altitude equal to the characteristic height, assuming isothermal conditions? **(c)** What is the pressure at twice the characteristic height?

18.54 (5) *Most probably.* Show that Equation 18.59 correctly gives the most probable speed v_{mp} for the molecules of a gas. (Hint: Begin with Equation 18.58, and set $dv/dv = 0$.)

18.55 (5) *Mean speed.* Show that Equation 18.60 correctly gives the mean speed $\langle v \rangle$ for the molecules of a gas. Hint: Use the general definition of a mean value in terms of the ratio of two definite integrals. You will need to use the result

$$\int_0^\infty v^3 e^{-av^2} \, dv = \frac{1}{2a^2}.$$

18.56 (5) *Log-log plot.* Suppose that you make a plot on log-log paper of v_{rms} versus T for an ideal gas. (Alternatively, you could plot $\ln v_{rms}$ versus $\ln T$ on ordinary graph paper.) Show that the plot is a straight line having slope $\frac{1}{2}$.

18.57 (G) *Heat and pressure, II.* A sealed container holds 10.0 g of oxygen gas at pressure p. The temperature is 280 K. **(a)** If you heat the gas until the rms speed of the molecules is doubled, what will the final temperature be? **(b)** How much heat transfer into the gas is required? **(c)** What will the final pressure be?

18.58 (G) *Putting on the pressure.* **(a)** Show that, for an ideal gas,

$$p = \tfrac{1}{3}\rho\langle v^2 \rangle.$$

(b) Suppose that the density of a gas in a container is 6.05×10^{-2} kg/m^3 and $v_{rms} = 512$ m/s. Find the pressure of the gas. **(c)** If the gas is oxygen, what is the temperature?

18.59 (G) *Tying up the package.* A 10-liter flask contains O_2 gas at $p = 1$ atm and $T = 290$ K. Find **(a)** the mass of the gas, **(b)** the internal energy, **(c)** the number of molecules present,

(d) the mean kinetic energy of the molecules, (e) their mean *translational* kinetic energy, (f) their mean speed $\langle v \rangle$, and (g) their rms speed.

18.60 *(G) Clausius equation of state.* Because the molecules of a gas occupy space, less than the entire volume V of a container is available to a particular molecule. Correction for this fact requires modification of the ideal-gas law. (a) If the molecules are not particles, the simplest remaining possibility is that they are hard spheres of radius a. Consider a pair of particular molecules j and k that are about to collide with one another, and show that molecule k, by its presence, excludes molecule j from a spherical volume $\frac{32}{3}\pi a^3$. (b) If a container contains N molecules, show that the total volume unavailable to the molecules is $b = \frac{16}{3}\pi N a^3$. (c) Show that the equation of state, corrected for the unavailable volume, is

$$p(V - b) = NkT; \quad \text{that is,} \quad p\left(V - \tfrac{16}{3}\pi N a^3\right) = NkT.$$

This equation is called the *Clausius equation of state*, after the Prussian-Polish physicist Rudolf Clausius (1822–1888). Clausius was one of the leaders in the development of thermodynamics and a pioneer in statistical mechanics. (d) Using the data given in Example 18.3, find the error introduced by neglecting the term b in calculating N for O_2 at STP in Example 18.1. (e) For O_2 at 0°C, at what pressure does use of the uncorrected ideal-gas law introduce an error of 1%? Under these conditions, what is the mean intermolecular distance, expressed as a multiple of a? (f) How well do oxygen and gases of similar molecular diameter, such as nitrogen, conform to the ideal-gas law at STP?

18.61 *(G) Collision time and mean free path.* As a spherical gas molecule of radius a moves, it sweeps out a cylindrical path. If the molecule makes collisions with other molecules, the path will be a long "crooked cylinder," changing direction at every collision. (a) Consider the straight line traced by the center of a molecule j between collisions. Show that a collision with another molecule k will occur if the distance between the center of k and the line, called the *impact parameter*, is less than $2a$. (b) Show that collisions between molecule j and other molecules will occur just as frequently if you imagine molecule j to be a disk of radius $2a$, moving normal to its surface, and all the other molecules to be points. What is the area σ of the disk, called its *collision cross section*? (c) Show that the mean time between collisions of molecule j with other molecules, called the *collision time* or the *mean free time*, is

$$\tau = \frac{V}{N}\frac{1}{\sigma \langle v \rangle},$$

where V is the volume of the container, N is the number of molecules present, and $\langle v \rangle$ is the mean speed given by Equation 18.60. (d) Show that the mean distance between collisions, called the *mean free path*, is

$$\lambda = \frac{V}{N}\frac{1}{\sigma}.$$

(e) Find the collision time and the mean free path for the oxygen molecules in the system discussed in Examples 18.1 and 18.3. (f) In a typical television tube, the electrons travel about 40 cm from the electron gun to the face of the tube where the picture is formed. If not more than 1 electron in 1000 is to collide with a molecule of the residual gas in the tube, what is the order of magnitude of the maximum allowable pressure inside the tube? Assume that the molecules of the residual gas all have about the same diameter as that of oxygen. (For this order-of-magnitude calculation, ignore the difference in radius between the electrons and the gas molecules.)

18.62 *(G) Hotfoot.* One kilomole of a monatomic ideal gas is confined in a container. The temperature is 20°C. How much heat transfer Q to the gas is needed to increase the rms speed of its molecules by 1%?

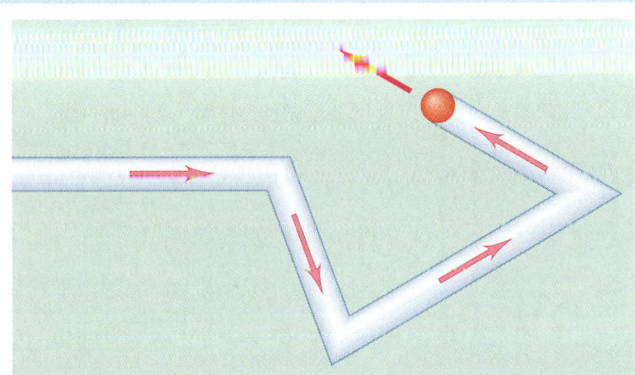

GROUP C

18.63 *(2) Adatoms.* In the phenomenon called *adsorption*, atoms of a gas strike a very clean solid surface and are bound to the surface. In some cases, the adsorbed atoms, called *adatoms*, constitute a single atomic layer called a *monolayer*. While they are bound to the surface, the adatoms can often migrate freely along the surface, interacting with each other like the molecules of a gas. Make an argument analogous to that in Section 18.2, and show that the mean energy of the molecules is

$$\langle \epsilon \rangle = kT.$$

18.64 *(G) Nothing ever stays clean!* It is notoriously difficult to do experiments requiring absolutely clean surfaces. To see why, suppose you have a device that can cleave a crystal inside a vacuum chamber and thus expose a perfectly uncontaminated surface. Any gas atom that strikes the surface will be adsorbed and stick to it, thus covering an approximate area πa^2, where a is the radius of the gas molecule. If the temperature is 300 K, at what pressure do you have about 1 minute to work before the surface is entirely covered with gas molecules? Assume that the gas is O_2, whose molecular diameter is about 2.9×10^{-10} m. For this calculation, it is adequate to assume that all the gas molecules have the same speed v_{rms}.

18.65 (G) *Brownian motion.* In 1827, the Scottish botanist Robert Brown (1773–1858) first observed what we today call the *Brownian motion*. Fluids called *colloids* contain tiny particles suspended in a liquid medium. The colloidal particles have diameters ranging from 1 nm to 1 μm; the largest of these particles can be seen through an ordinary microscope. The particles are in what Brown called ''perpetual motion.'' (The Brownian motion is particularly easy to see under low microscope magnification in cigarette smoke.) Consider a particle of diameter 1.0 μm and density 1.1×10^3 kg/m³, suspended in a colloid whose temperature is 300 K. **(a)** Why is the particle in motion? **(b)** What is its rms speed? **(c)** How many degrees of freedom does it have? **(d)** What is its total energy? **(e)** To see why the Brownian motion is not observed for particles large enough to be seen by the naked eye, show that v_{rms} is inversely proportional to the $\frac{3}{2}$ power of the particle diameter.

18.66 (G) *Brown, Boltzmann, Avogadro, Perrin & Co.* The French physicist J. B. Perrin (1870–1942) used the Brownian motion in colloids (see Problem 18.65) in two ingenious ways to determine the values of Boltzmann's constant and Avogadro's number. One of his methods, carried out in 1909, is sketched here. Consider a colloid whose particles have uniform radius 2.12×10^{-7} m. The particles, whose density is 1.254×10^3 kg/m³, are suspended in water, whose density is 0.998×10^3 kg/m³ at the temperature of the experiment. A tiny vial of the colloid, about 0.01 cm deep, is placed on the stage of a high-power microscope. The temperature of the vial is a uniform 20°C. Because of the optical properties of a microscope, the observer can focus only on a thin horizontal layer of the drop at one time. (In Perrin's experiments, the thickness of this observable layer was about 1 μm.) By turning the focus knob of the microscope, one can look at such thin layers distributed vertically through the vial and count the number of particles in the field of view for each layer. The microscope is calibrated so that its vertical displacement can be measured with high precision. In one measurement, Perrin found that the particle count diminished by a factor 2.1 when he raised the microscope a distance 30.0 μm. **(a)** Show that, owing to buoyancy effects, the effective weight of a colloidal particle suspended in water is not mg, but $mg\,\Delta\rho/\rho$, where ρ is the density of the particle and $\Delta\rho$ is the difference in density between the particle and water. Rewrite the isothermal barometric equation, taking this modification into account. **(b)** Find the value of Boltzmann's constant. Using the value of the universal gas constant R, which can be obtained from macroscopic measurements, evaluate Avogadro's number. This experiment makes no sense at all outside the framework of a molecular view of matter. Consequently, the successful performance and interpretation of the experiment was regarded at the time as definitive proof that matter is indeed made up of atoms and molecules. Perrin won the Nobel Prize

in physics in 1926 for this and other experimental demonstrations of the atomic nature of matter.

18.67 (5) *Hydrogen-rich atmosphere.* There is very little hydrogen (H_2) in the atmosphere. At sea level, it makes up 1 part in 10^4 of the mixture we call air. In contrast, nitrogen constitutes close to 80% of the sea-level atmosphere. Use the isothermal-atmosphere model to show that at high enough altitudes the (very rarefied) atmosphere consists largely of hydrogen. To do this, calculate the altitude at which the concentration of H_2 ($\mu = 2.00$) is equal to that of N_2 ($\mu = 28.0$). Let $T = 275$ K.

18.68 (5) *Maxwell-Boltzmann energy distribution.* For a monatomic ideal gas, the energy of any molecule has the purely kinetic form $\epsilon = \frac{1}{2}mv^2$. Show that the function describing the number of molecules having any energy ϵ is

$$\nu(\epsilon) = 2\pi\rho'\left(\frac{1}{\pi kT}\right)^{3/2}\sqrt{\epsilon}\,e^{-\epsilon/kT}.$$

This is the *Maxwell-Boltzmann energy distribution.* The term $e^{-\epsilon/kT}$ is called the *Boltzmann factor.*

Hint: $\qquad \dfrac{d\nu(\epsilon)}{d\nu(v)} = \dfrac{d\epsilon}{dv} = mv.$

18.69 (5) *More general way to a familiar result.* Use the result of Problem 18.68 to find the mean energy $\langle\epsilon\rangle$ of N molecules, and show that

$$\langle\epsilon\rangle = \tfrac{3}{2}kT,$$

which agrees with the result obtained in a less general and less rigorous way in the argument leading to Equation 18.15. [Hint: Remember that $\langle\epsilon\rangle = (1/N)\int\epsilon\nu(\epsilon)\,d\epsilon$, with the integral taken over all possible values of energy.]

18.70 (5) *Confirming the Maxwell-Boltzmann speed distribution.* In 1955, Miller and Kusch carried out an experiment to verify the form of $\nu(v)$ given by Equation 18.58. The apparatus is shown here schematically. The entire system is contained in a vacuum chamber (not shown). The oven, whose

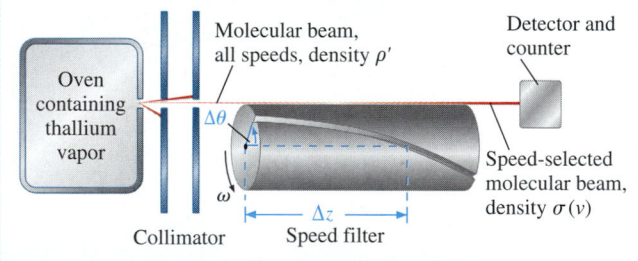

temperature is carefully regulated, contains a quantity of the metal thallium, which is slightly volatile in the temperature range about 900 K at which the experiment is performed. Thus the oven is filled with a low-pressure ideal gas of thallium molecules. Some molecules pass through the small hole in the oven. Most of these are stopped by inelastic collision with the *collimator*, consisting of two plates with small holes in them. Only those molecules that happen to be moving in the direction shown can pass through the holes. The resulting unidirectional stream of molecules is called a *molecular beam*. The molecular beam enters the rotating-drum *speed filter*, which can be used to select out of the beam just those molecules having speed v. The filter consists of a cylindrical drum with one or more helical grooves cut in its surface (like screw threads). The *pitch* of the groove is $\Pi \equiv \Delta\theta/\Delta z$; this is the angle through which the groove "turns" per unit length of the drum. The cylinder rotates with adjustable angular speed ω. **(a)** Express the speed of the molecules in terms of Π and ω. **(b)** What is the most probable speed, v_{mp}, for thallium molecules ($\mu = 204.4$ u) when the beam temperature is 944 K? **(c)** Let $\Pi = 0.400$ rad/m. For what value of ω will molecules with speed v_{mp} pass through the filter?

With the drum rotating at angular speed ω, the number of molecules passing through the speed filter and striking the detector per unit time, $\sigma(v)$, is measured. With the use of these data, values of ν can be plotted for values of v over the entire range of interest by varying ω. **(d)** Suppose that, in the beam emerging from the collimator, the density of molecules having speed v is $\rho'(v)$. Show that when molecules of speed v are passing through the speed filter, the number per unit time striking the detector conforms to the proportionality

$$\sigma(v) \propto v\rho'(v).$$

(e) Show that the speed distribution of the gas molecules in the oven is therefore given by the proportionality

$$\nu(v) \propto \frac{1}{v}\,\sigma(v).$$

This is the quantity plotted in Figure 18.9.

Thermodynamics I
The First Law

A process in which heat flows into (or out of) a system and work is done on (or by) the system is called a thermodynamic process. The operation of a heat engine exemplifies such a process.

We divide all of the energy flow into a system into two categories: heat flow into the system and work done on the system. According to the first law of thermodynamics, the sum of these two quantities is the change in the internal energy of the system.

At any instant, a thermodynamic system can be described by the values of quantities called its state variables. In a thermodynamic process, the values of the state variables change.

The first law of thermodynamics can be expressed in terms of the three state variables pressure p, volume V, and temperature T, together with a fourth state variable called the entropy S.

As a heat engine operates, the thermodynamic process can be described by a point moving along a closed curve on the p-V diagram.

The Carnot cycle is an ideal cycle in which all the heat flow into the engine occurs at one fixed temperature and all the heat flow out of the engine occurs at a second fixed temperature.

Left: Robert Stephenson's Locomotive "Rocket," the first practical steam locomotive. A heat engine is any device that converts heat energy (microscopic mechanical energy) into macroscopic mechanical energy.

SECTION 19.1   Introduction

In Section 17.5, we studied processes such as friction in which mechanical work done on a body increases the heat energy of the body. Loosely speaking, such processes *convert mechanical work into heat*; energy is transformed from one form into another. In this chapter and the next, we consider more general processes, in which heat flows into (or out of) a system at the same time that mechanical work is being done on (or by) the system. Such processes are called **thermodynamic processes**. We will particularly aim at an understanding of **heat engines**. A heat engine is a device especially designed to accept a flow of heat and to do work on some external system. Again loosely speaking, a heat engine *converts heat into mechanical work*.

In this chapter, we consider work and heat flow on equal footing. Central to this consideration is the *first law of thermodynamics*, which is a special form of the law of conservation of energy and which formalizes some ideas we have already considered.

Thermodynamic Processes

Figure 19.1 depicts just one of many possible *thermodynamic systems*. This particular system consists of three parts: a cylinder; a leakproof piston, which, with its piston rod, can move without friction in and out of the cylinder; and a *working fluid*, which fills the cylinder. The boundary of the system is shown, and everything outside this boundary is called the **outside world** or the **surroundings**.* The **working fluid** that fills the cylinder is usually a gas but, under certain circumstances, might be a liquid or even a solid. In this chapter, however, we concentrate on systems in which the working fluid is an ideal gas. In addition, we consider only **closed systems**, by which we mean systems in which the amount of working fluid (specified by *n*, the number of kilomoles) remains constant.

The First Law of Thermodynamics

A thermodynamic system can interact with the outside world in many ways. For thermodynamic purposes, we divide all such interactions into two classes, both typefied in the system of Figure 19.1:

1. *Heat flow* into or out of the system can be driven by a temperature difference across the boundary, as indicated by the two-headed arrow labeled *Q*.

2. *Work* comprises all other interactions in which there is an exchange of energy between the system and its surroundings. In Figure 19.1, mechanical work can be done on or by the system through motion of the piston rod, which is connected to some external mechanism (not shown). This work is denoted by the two-headed arrow labeled *W*. In other systems, work can take other forms (such as electrical work). But all forms are lumped together as work.

In general, both classes of interactions—heat flow and work—between the system and the outside world take place simultaneously.

At any instant, the system possesses a certain *internal energy E*. This internal energy comprises not only the thermal energy of the working fluid, but any other forms of energy that may be present, such as the gravitational potential energy of the working fluid. But we are interested mainly in *changes* ΔE in the internal energy. The only way that the internal energy of the system can change is if energy flows into or out of the system. Because all such flows of energy belong to either class 1 (heat flow) or class 2 (all other kinds of energy transfer), we can write

$$\Delta E = Q + W. \qquad (19.1)$$

This "balance sheet" equation is called the **first law of thermodynamics**. In using the equation, we adopt the following sign convention: *Q and W are positive if they tend to increase E and negative if they tend to decrease E*. That is, heat flowing *into* the system (say, from a hot plate with which it is in thermal contact) and work done *on* the system (say, by pushing down on the piston rod against the pressure of the working fluid) are positive quantities. Heat flowing *out of* the system (say, to a cold plate with which it is in thermal contact) and work done *by* the system (say, by expansion of the working fluid against an opposing force exerted on the piston rod from outside) are negative quantities.

A **thermodynamic process** is one in which there is an exchange of heat and work between the system and the outside world. All thermodynamic processes must obey Equation 19.1. To analyze a process, we need to be able to specify the state of the system at any instant by means of such quantities as pressure *p*, volume *V*, and temperature *T*, collectively called **state variables**. These quantities are related to one another by the **equation of state** of the working fluid. As you learned in Section 18.2, the ideal-gas law, $pV = nRT$, is an example of an equation of state.

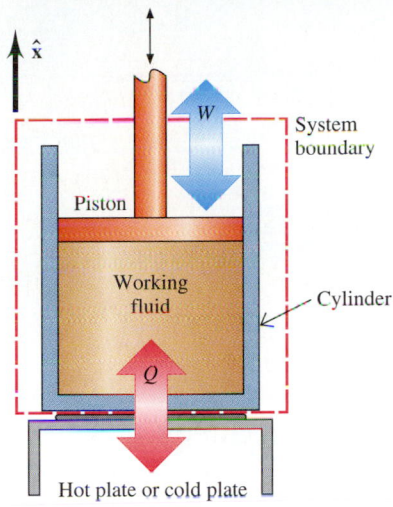

FIGURE 19.1 A representative thermodynamic system, consisting of a working fluid contained in a cylinder with a leakproof, frictionless piston. The dashed red rectangle depicts the boundary between the system and its surroundings. The system can exchange heat *Q* with its surroundings. Work *W* is done on or by the system when the piston moves and the working fluid consequently is compressed or expands.

The Sign of Work

Unfortunately, there is no universal sign convention for work. Some engineers and scientists prefer to call work done by the system positive, on the ground that that is what engines are designed to do. When reading other literature on thermodynamics, make sure you know which convention is being used.

---

*We choose the boundary to suit our purposes, just as we did in defining mechanical systems and drawing their free-body diagrams. As in the mechanical case, we must stick to a boundary once we have chosen it.

It is always possible to specify the state of a system throughout a thermodynamic process if the process is *quasistatic*. A **quasistatic process** is one in which the system changes slowly enough that each succeeding state through which it passes is essentially in equilibrium. Many processes of practical importance are quasistatic or nearly so. The operation of a heat engine of the kind shown in Figure 19.1 will usually be quasistatic if the speed of the piston does not exceed the speed of sound. This condition is quite well approximated in nearly all steam engines. For gasoline engines, the approximation is still good enough to be very useful.

If a process is quasistatic, we can apply the first law of thermodynamics to the change that takes place between two states that differ infinitesimally. (In the system of Figure 19.1, for example, the infinitesimal difference might be caused by the motion of the piston through the infinitesimal distance dx.)

For an infinitesimal change, we can write Equation 19.1 in the differential form

$$dE = đQ + đW. \tag{19.2}$$

The symbol $đ$ denotes an *inexact differential*. The quantities $đQ$ and $đW$ are infinitesimal, just as dE is. But, unlike the *exact differential dE*, they are *not* differentials of some function. To see that they are not, note that the internal energy of a body can be written as a function of temperature. [For n kmol of a monatomic ideal gas, you have $E(T) = \frac{3}{2}nRT$.] But you cannot write a function $Q(T)$, because that would imply "the amount of heat contained in a body at temperature T." This is meaningless; as we saw in Chapter 18, *heat* is not defined as a physical quantity; *heat flow* signifies a flow of energy across a system boundary in a certain form. Likewise, a function $W(T)$ would have the meaningless implication "the amount of work contained in a body." It is only the *sum* of heat and work in Equation 19.2 that specifies a definite change dE in the internal energy of the system.

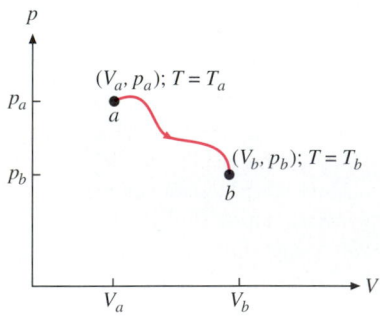

FIGURE 19.2 A p-V diagram, showing a hypothetical quasistatic process in which the system passes in a continuous manner from an initial state p_a, V_a, T_a to a final state p_b, V_b, T_b.

The p-V Diagram

As a thermodynamic process takes place, the quantities p, V, and T change in general. If the process is quasistatic, each quantity varies in a continuous fashion, passing gradually from an initial state a specified by the values p_a, V_a, and T_a to a final state b specified by the values p_b, V_b, and T_b. Any one of these variables can be considered a function of any other, and the functional relation between them can be depicted as a curve on a two-dimensional graph. A very common choice—though by no means the only useful one—is the **p-V diagram**, in which the pressure p is plotted as a function of the system volume V. Such a diagram is shown for a hypothetical quasistatic process in Figure 19.2. Starting from the initial point (V_a, p_a), the system passes through a continuous series of intermediate points (V, p) until the process ends at the final point (V_b, p_b).

The curve shown is not the only possible path from (V_a, p_a) to (V_b, p_b). Indeed, the particular path followed by the system depends on the nature of the process and on the equation of state of the working fluid. We now consider a number of special processes and the corresponding special paths in detail. These special processes are sources of insight into thermodynamic processes in general and lay the groundwork for the study of heat engines.

In thermodynamic processes, some quantity is often held constant. Boyle's law, for example, governs the relation between the pressure and the volume of a given quantity of an ideal gas under *isothermal* conditions, in which T does not change. Other types of thermodynamic processes are defined according to which quantity is held constant. The types are summarized in Table 19.1.

TABLE 19.1 Types of Thermodynamic Processes in Closed Systems (n = constant)

Process Name	Fixed Quantity	Varying Quantities	Must $Q = 0$?
Isothermal	Temperature T	p, V	No
Isobaric	Pressure p	V, T	No
Isometric	Volume V	p, T	No
Adiabatic	$Q = 0$	p, V, T	Yes

Isothermal Processes

Figure 19.3 depicts the same system shown in Figure 19.1. Suppose that the system is compressed isothermally at temperature T from an initial volume V_a to a final (smaller) volume V_b. This could be done by immersing the system in a bath having the desired temperature T and then pushing the piston inward slowly enough that the temperature of the working fluid never deviates appreciably from T. If the working fluid is an ideal gas, how much work must be done on the system to carry out the compression as the pressure of the system changes from its initial value p_a to its final value p_b? That is, how much work $W_{a \to b}$ is required to take the system from state a, characterized by p_a, V_a, T, to state b, characterized by p_b, V_b, T?

Let us choose the outward direction of piston motion to be the positive x direction. (This choice is convenient because an increase in x corresponds to an increase in V.) If the area of the piston is A, the force required to drive the piston inward against the pressure exerted by the working fluid is of magnitude $F = pA$. The work done in moving the piston through an inward displacement $-dx$ is

$$d\!\!\!/W = -pA \, dx.$$

But $A \, dx$ is just the change dV in the volume of the cylinder. So we can write

$$d\!\!\!/W = -p \, dV. \tag{19.3}$$

The minus sign is necessary because work being done *on* the system ($d\!\!\!/W > 0$) always involves a negative change in the volume ($dV < 0$).

To find the total work $W_{a \to b}$ done on the system, we integrate Equation 19.3 between the limits V_a and V_b:

$$W_{a \to b} = \int_a^b d\!\!\!/W = -\int_{V_a}^{V_b} p \, dV. \tag{19.4}$$

To evaluate the integral, we must express p as a function of V. We can do this if we know the equation of state of the working fluid. In the present case, because we have assumed that the working fluid is an ideal gas, the proper equation of state is the ideal-gas law, $pV = nRT$. So we write

$$p = \frac{nRT}{V}. \tag{19.5}$$

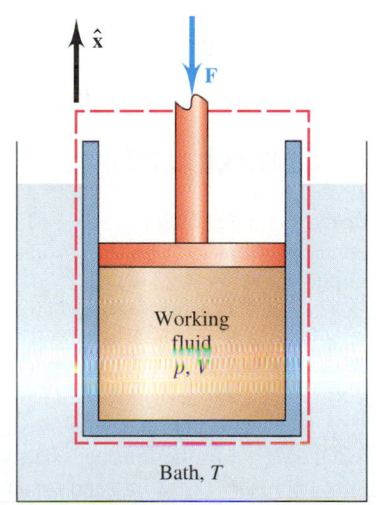

FIGURE 19.3 The system of Figure 19.1 undergoing an isothermal process. The bath maintains the system at a fixed temperature T as p and V vary. The dashed red box represents the system boundary.

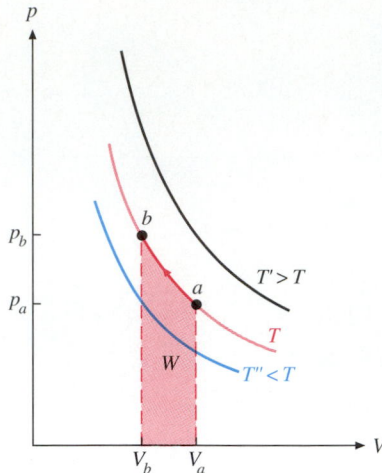

p_b

p_a

b

a

$T' > T$

T

W

$T'' < T$

V_b V_a

V

FIGURE 19.4 The p-V diagram for the isothermal process of Figure 19.3. Also shown are two other isotherms, one for $T' > T$ (in black) and the other for $T'' < T$ (in blue).

We substitute this value of p into Equation 19.4 to obtain

$$W_{a \to b} = -\int_{V_a}^{V_b} \frac{nRT}{V} \, dV.$$

Now we use the condition that the process is isothermal. Because T is a constant, we can write the integral in the form

$$W_{a \to b} = -nRT \int_{V_a}^{V_b} \frac{dV}{V} = -nRT \ln \frac{V_b}{V_a},$$

or

$$W_{a \to b} = nRT \ln \frac{V_a}{V_b} \quad \text{for an ideal gas in an isothermal process.} \quad (19.6)$$

Let us see what this isothermal ideal-gas process looks like on a p-V diagram. Equation 19.5 expresses p as a function of V. Because nRT is constant, p is inversely proportional to V, and the curve describing the thermodynamic process is the segment $a \to b$ of a hyperbola, shown in solid red in Figure 19.4. The entire curve is called an **isotherm** because any process described by a segment of the curve is an isothermal process. Other processes described by segments of any other hyperbola that could be drawn on the graph also are isotherms, but the temperatures of the processes differ from isotherm to isotherm.

The work integral of Equation 19.4, evaluated explicitly in Equation 19.6, is the integral of the function $p(V)$ between the limits specified. Thus *the work W done on the system during any thermodynamic process is given by the area under the curve on the p-V diagram*, shaded red in Figure 19.4.

EXAMPLE 19.1

For the system shown in Figure 19.3, suppose that the cylinder contains helium gas and that, in its initial state a, the cylinder volume is $V_a = 4.23 \times 10^{-3} \text{ m}^3$ and the pressure is $p_a = 1.19 \times 10^5$ Pa (a little more than 1 atm). The temperature of the system is $T = 280$ K. **(a)** What will be the pressure p_b in state b, when the volume of the cylinder is reduced to $V_b = 0.581 \times 10^{-3} \text{ m}^3$? **(b)** How much work $W_{a \to b}$ is required to carry out the compression quasistatically? **(c)** What is the change $\Delta E_{a \to b}$ in the internal energy of the helium? **(d)** What is the heat flow $Q_{a \to b}$ into the system during the process? Assume that helium approximates an ideal gas throughout the process.

SOLUTION:

(a) Find p_b.

You can use Boyle's law in the simple form of Equation 15.36c, $pV = \text{constant}$. This gives you

$$p_b = p_a \frac{V_a}{V_b}$$

$$= 1.19 \times 10^5 \text{ Pa} \times \frac{4.23 \times 10^{-3} \text{ m}^3}{0.581 \times 10^{-3} \text{ m}^3}$$

$$= 8.66 \times 10^5 \text{ Pa}.$$

(b) How much work $W_{a \to b}$ is required?

To find the work done on the helium in the process, you need to use Equation 19.6. You do not know the value of n, the number of kilomoles of helium present in the cylinder. However, you can write the ideal-gas law in the general form $n = pV/RT$. In particular, this law applies to the initial state a, and

so you can write

$$n = \frac{p_a V_a}{RT}.$$

Substituting this value of n into Equation 19.6, you obtain

$$W_{a \to b} = p_a V_a \ln \frac{V_a}{V_b}$$

$$= 1.19 \times 10^5 \text{ Pa} \times 4.23 \times 10^{-3} \text{ m}^3$$

$$\times \ln \frac{4.23 \times 10^{-3} \text{ Pa}}{0.581 \times 10^{-3} \text{ m}^3} = 999 \text{ J}.$$

Note that this result does not depend on the temperature.

(c) Find $\Delta E_{a \to b}$, the change in the internal energy of the helium.

For an ideal gas, the internal energy E depends *only* on the temperature, the number of molecules present, and the number of degrees of freedom of the molecules; specifically, $E = (f/2)NkT$. (You can verify this by referring to Equations 18.29 and 18.38.) Here, a fixed quantity of a monatomic gas is confined in the system, and so N and f are constant; that is, $\Delta N = 0$ and $\Delta f = 0$. And because the process is isothermal—$\Delta T = 0$—you know without calculation that

$$\Delta E_{a \to b} = 0 \quad \text{for an ideal gas in an isothermal process.} \quad (19.7)$$

(d) Find $Q_{a \to b}$.

Because the internal energy of the system has not changed and because mechanical work $W_{a \to b} = 999$ J has been done on the system, the first law of thermodynamics states that an equal amount of energy has flowed out of the system in the form of heat. According to Equation 19.1, $\Delta E = Q + W$, you have for the heat flow into the system

or

$$0 = Q_{a \to b} + W_{a \to b} = Q_{a \to b} + 999 \text{ J},$$

$$Q_{a \to b} = -999 \text{ J}.$$

Why is this value negative?

Isobaric and Isometric Processes

The path shown in Figure 19.4 is not the only possible one between states a and b, as we have already noted. Let us use an alternative path to return the system to state a, as shown in Figure 19.5. The final state b of the isothermal process of Figure 19.4 now becomes the initial state for the new process, which is described by the path $b \to c \to a$. The first part of this path, $b \to c$, takes place at constant pressure p_b. This part of the process is therefore isobaric (Table 19.1). The entire line of which $b \to c$ is a segment is called an **isobar**. The second part of the path, $c \to a$, takes place at constant volume $V_c = V_a$. This part of the process is therefore isometric. The entire line of which $c \to a$ is a segment is called an **isometric**. (The alternative name, *isochore*, will not be used in this book.)

One way to ensure that the system passes along the path $b \to c$, without changing its pressure, is to place a weight mg on the piston. Assuming the weight of the piston to be negligible, the pressure must always have the value $p \equiv F/A = mg/A$, regardless of the volume of the cylinder. But, with the pressure fixed, how can we make the volume change? The only way to do this is to change the temperature. Indeed, note that point c lies on the isotherm for the temperature $T' > T$. So we must remove the cylinder of Figure 19.3 from the bath at T and warm it with some external heat source, as shown in Figure 19.6a, until its temperature reaches T'.* In this isobaric process, the pressure has the constant value p_b. Thus the work done on the system as it expands from V_b to V_a is

$$W_{b \to c} = -p_b \int_{V_b}^{V_c} dV = p_b(V_b - V_c). \tag{19.8}$$

Because $V_b < V_a$, the value of W is negative, which means that the system does work

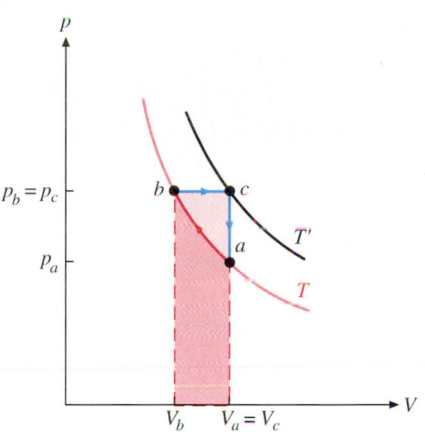

FIGURE 19.5 A p-V diagram representing the return of the system of Figure 19.1 from state b to state a by the path $b \to c \to a$. The path $b \to c$ is isobaric; the path $c \to a$ is isometric. The isotherm T' on which c lies is shown.

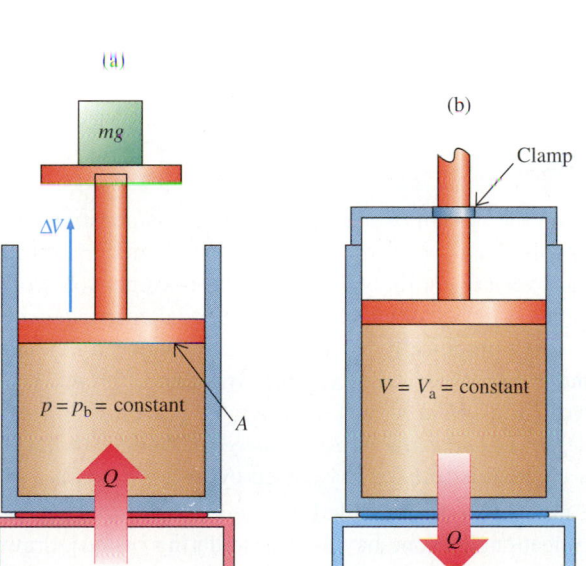

FIGURE 19.6 (a) The system of Figure 19.1 undergoing the isobaric process $b \to c$. The fixed weight mg on the piston ensures a constant internal pressure p_b while heat flows into the system from the hot plate and the piston moves outward, increasing the cylinder volume from V_b to V_c ($= V_a$). (b) The system undergoing the isometric process $c \to a$. The piston is locked in place, ensuring constant volume V_a, while heat flows out of the system to the cold plate and the pressure falls from p_b to p_a. All parts of the system except the bottom surface are surrounded by insulation (not shown), so heat flows only through the bottom.

---

*The temperature change inside the system must take place slowly enough to satisfy the quasistatic condition. In principle, we could immerse the system in a series of baths, each a little warmer than the preceding one, until the desired temperature is achieved. But in actual systems the heat flow is usually sufficiently slow that we need make no special effort to satisfy the condition.

on its surroundings. Specifically, the work is done in raising the weight mg as the piston is forced outward.

Now let us turn to the isometric process described by the path $c \rightarrow a$. We can ensure a constant volume by locking the piston in place, as shown in Figure 19.6b. To reduce the pressure from p_b to p_a, we must reduce the temperature from T' to its original value T. In Figure 19.6b, this temperature reduction is achieved by placing the system on a cold plate. No work is done in the process because $V = V_a$ throughout, and so we have from Equation 19.4

$$W_{c \rightarrow a} = -\int_{V_c = V_a}^{V_a} p \, dV = 0. \tag{19.9}$$

Adding the work done over the two segments, we find the total work done along the path $b \rightarrow c \rightarrow a$:

$$W_{b \rightarrow c \rightarrow a} = W_{b \rightarrow c} + W_{c \rightarrow a} = p_b(V_b - V_a) + 0 = p_b(V_b - V_a). \tag{19.10}$$

The magnitude of this work, $|W_{b \rightarrow c \rightarrow a}|$, is equal to the area of the rectangle under the path $b \rightarrow c$. It is not equal to the work $W_{a \rightarrow b}$ that was done in taking the system from a to b along the isotherm T, given by Equation 19.6:

$$W_{a \rightarrow b} = nRT \ln \frac{V_a}{V_b}.$$

To make a more direct comparison between $W_{b \rightarrow c \rightarrow a}$ and $W_{a \rightarrow b}$, we use the ideal-gas law to make the substitution $nRT = p_b V_b$ in Equation 19.6 and obtain

$$W_{a \rightarrow b} = p_b V_b \ln \frac{V_a}{V_b}. \tag{19.11}$$

This is clearly different from $W_{b \rightarrow c \rightarrow a}$, given by Equation 19.10. *The work done on a system as it passes from one state to another depends on the path.*

The Heat-Engine Cycle

SECTION 19.3

It is the dependence of work on the path that makes heat engines possible. Consider what happens to our system as we take it along the closed path $a \rightarrow b \rightarrow c \rightarrow a$ in Figure 19.5. Such a closed path is called a **heat-engine cycle**. The total work done on the system is

$$W_{\text{cyc}} = W_{a \rightarrow b} + W_{b \rightarrow c} + W_{c \rightarrow a} = p_b V_b \ln \frac{V_a}{V_b} + p_b(V_b - V_a) + 0. \tag{19.12}$$

The work done on the system over a complete cycle is negative because $V_a > V_b$. (A rigorous proof of this assertion is the subject of Problem 19.33.) The negative value of W_{cyc} means that over the entire cycle the system does net work of magnitude $|W_{\text{cyc}}|$ *on* its surroundings. A device of which this is true is called a *heat engine*. The magnitude $|W_{\text{cyc}}|$ is represented graphically by the difference between the area under path $b \rightarrow c$ and the area under path $a \rightarrow b$ in Figure 19.5. But the difference is just the area inside the closed curve. *The work done by a heat engine per cycle is thus given by the area within the cycle, as it is depicted in a p-V curve.* Although we have derived this result for a particular cycle, it is true for any closed cycle depicted on a p-V curve.

James Watt invented a device called the *engine indicator*, which, when attached to a reciprocating heat engine (one having an oscillating piston), draws its p-V curve. A late-nineteenth-century version of the device is shown in Figure 19.7. The p-V curve, called an *indicator diagram*, is shown for a typical steam engine in Figure 19.7c. The output work of the engine per cycle is proportional to the area enclosed by the curve; the engine power is equal to the output work per cycle divided by the cycle time.

When we obtain work from a heat engine, where does the energy come from? If the system of Figures 19.3 and 19.6 passes through one entire cycle, beginning and ending in the state a shown in Figure 19.5, its initial internal energy is equal to its final internal

Steam-Engine Diagnostics

The details of the engine-indicator curve can be used to diagnose design errors and inefficient operating conditions and thus to improve engine performance. Note, for example, the pressure oscillation at the upper-left corner of Figure 19.7c. This oscillation is the result of non-quasistatic passage of steam into the cylinder when the inlet valve opens suddenly.

(a)   (b)   (c)

FIGURE 19.7 (a) An engine indicator. In this cutaway view, a cord is wound around the bottom of the pivoted drum D and is led through the pulley assembly Q, R to a linkage that connects with the piston rod of the engine to be tested. The drum is mounted on the vertical pivot K, and the cord pulls against the drum spring D.S. so that the drum oscillates in synchrony with the engine piston. A sheet of paper is wrapped around the drum and held by the clips CL. The spring-loaded piston P in the small cylinder to the left is connected by tubing to the engine cylinder head and thus acts as a pressure gauge. Linked to P is the lever L, whose right end P.P. rises and falls with rising and falling cylinder pressure. When the pen at P.P. is pressed against drum D, as shown in part b, it writes on the rotating paper. When the paper is unwrapped, the vertical axis is the p axis. Because the horizontal axis measures the displacement of the engine piston, it also measures the engine cylinder volume V. The engine cycle is thus displayed on a p-V diagram. (b) The engine indicator in use. The operator presses the pen against the drum at P during one or more engine cycles. (c) An indicator diagram for a typical steam engine.

energy because E depends only on the state. Thus the energy change is $\Delta E = 0$ over the cycle. Combining this condition with the first law of thermodynamics, we have $\Delta E = W + Q = 0$ over the cycle. So we have

$$-W_{a \to b \to c \to a} = Q_{a \to b \to c \to a} \quad \text{for a heat-engine cycle.} \quad \textbf{(19.13a)}$$

This equation can be written in the brief and more general form

$$-W_{\text{cyc}} = Q_{\text{cyc}}, \quad \textbf{(19.13b)}$$

where Q_{cyc} is the net heat input to the engine over the cycle.

EXAMPLE 19.2

The system discussed in Example 19.1 is taken through the cycle $a \to b \to c \to a$ shown in Figure 19.5. Find the values of (a) $W_{b \to c}$; (b) the work W_{cyc} done over one cycle; (c) the temperature change $\Delta T = T' - T$ of the system in the process $b \to c$; (d) $Q_{c \to a}$; (e) $Q_{b \to c}$.

SOLUTION:
(a) Find $W_{b \to c}$.

To find the work done on the system over the path $b \to c$, you use Equation 19.8:

$$W_{b \to c} = p_b(V_b - V_a)$$
$$= 8.66 \times 10^5 \text{ Pa} \times (0.581 - 4.23) \times 10^{-3} \text{ m}^3$$
$$= -3160 \text{ J.}$$

The negative value means that the system does work on its surroundings in the process $b \to c$.

(b) Find W_{cyc}.

You found the value $W_{a \to b} = 999$ J in Example 19.1. According to Equation 19.9, $W_{c \to a} = 0$, and so you have for the total work

$$W_{cyc} = W_{a \to b} + W_{b \to c} = 999 \text{ J} - 3160 \text{ J} = -2160 \text{ J}.$$

Note that the magnitude of the total cyclical work is less than the magnitude of $W_{b \to c}$, because $W_{c \to a} = 0$ and $W_{a \to b} > 0$. Bringing the system from a back to b requires compressing the working fluid—that is, doing work *on* the system. All heat-engine cycles have positive-W parts as well as negative-W parts; the engine is useful because $W_{cyc} < 0$. The part of an engine cycle over which $W < 0$ is often called the *power stroke*.

(c) Find the temperature change in the process $b \to c$.

Because the process $b \to c$ is isobaric, you can use Charles's law in the form of Equation 17.15, $V_1/V_2 = T_1/T_2$, to write

$$\frac{T_c}{T_b} = \frac{T'}{T} = \frac{V_c}{V_b}.$$

You solve for T' and use the fact $V_c = V_a$ to obtain

$$T' = T \frac{V_c}{V_b} = T \frac{V_a}{V_b}$$

$$= 280 \text{ K} \times \frac{4.23 \times 10^{-3} \text{ m}^3}{0.581 \times 10^{-3} \text{ m}^3} = 2040 \text{ K}.$$

Finally, you subtract $T = 280$ K from T':

$$\Delta T = T' - T = 2040 \text{ K} - 280 \text{ K} = 1760 \text{ K}.$$

The isobaric expansion from b to c requires a considerable temperature increase.

(d) Find $Q_{c \to a}$, the heat flow into the system during the process $c \to a$.

You can express Q in terms of the quantity of matter present, the specific heat capacity, and the temperature change (Section 17.5). In the present case, the quantity of matter is expressed in moles and the process is isometric. Consequently, the specific heat capacity you need is the molar heat capacity at constant volume, c_v'. You have

$$Q_{c \to a} = nc_v'(T - T').$$

The difference $T - T'$ is the negative of the quantity ΔT calculated in part **c**. (That is, the temperature decreases as the pressure decreases during the isometric process $c \to a$.) So you have

$$Q_{c \to a} = nc_v'(-\Delta T).$$

Because helium is a monatomic gas, you use the value $c_v' = \frac{3}{2}R$ (Equation 18.31). You could solve numerically for n, but it is easier to make the substitution $n = p_a V_a / RT$. This gives you

$$Q_{c \to a} = \frac{3}{2} p_a V_a \frac{-\Delta T}{T}$$

$$= 1.5 \times 1.19 \times 10^5 \text{ Pa} \times 4.23 \times 10^{-3} \text{ m}^3 \times \frac{-1760 \text{ K}}{280 \text{ K}}$$

$$= -4750 \text{ J}.$$

The negative value of $Q_{c \to a}$ means that heat flows *out of* the system during the process $c \to a$. Consider the meaning of this result in the context of the first law of thermodynamics. The process is isobaric, and therefore no work is done (Equation 19.9). Yet the temperature decreases as the system passes from c, on the isobar T', to a, on the isobar $T < T'$ (Figure 19.5). The internal energy of an ideal gas is proportional to the temperature, and thus E must decrease in the process; that is, $\Delta E < 0$. Consequently, you have $\Delta E = Q + W = Q + 0 = Q < 0$. When no work is done, a decrease in the system temperature implies a flow of heat out of the system and a decrease in the internal energy of the system.

(e) Find $Q_{b \to c}$, the heat flow into the system during the process $b \to c$.

You use Equation 19.13 to write

$$-W_{cyc} = Q_{a \to b} + Q_{b \to c} + Q_{c \to a}.$$

So you have

$$Q_{b \to c} = -W_{cyc} - Q_{a \to b} - Q_{c \to a}.$$

All the quantities on the right have been calculated in this example and in Example 19.1, and you solve to find

$$Q_{b \to c} = 2160 \text{ J} + 999 \text{ J} + 4750 \text{ J} = 7910 \text{ J}.$$

Note that $Q_{b \to c}$ is greater than Q_{cyc}. This is because heat flows out of the system during the process $c \to a$; as you have seen, $Q_{c \to a}$ is negative.

Heat Capacity at Constant Pressure. The Heat Capacity Ratio γ

Consider again a system like that of Figure 19.6 being heated isobarically. An arbitrary working fluid—not necessarily an ideal gas—is contained within the system, which is heated quasistatically over a temperature range ΔT. When heat flows into the system and the temperature rises, the internal energy E of the working fluid increases. But the piston rises as the working fluid expands, and so the system does *positive* work $-(-p \, \Delta V) = p \, \Delta V$ *on* its surroundings. This work is done at the expense of the internal energy of the working fluid; thus E increases *less* than it would under conditions of constant volume. Because T depends on E, a given heat input Q to the system under conditions of constant pressure results in a smaller temperature increase than the same heat input under conditions of constant volume. Conversely, a given temperature increase ΔT requires a greater input of heat at constant pressure than is required at constant volume. So the heat capacity of a given body depends on which of the two conditions obtains. We now consider the two heat capacities in detail.

A heat capacity always relates a heat flow Q to a temperature change ΔT. We have already discussed the molar heat capacity at constant volume, c_v', for ideal gases (Section 18.3). Here we express c_v' in purely empirical terms for an arbitrary substance. For a system of fixed volume containing n kmol of any working fluid, such as the system of Figure 19.6b, we write

$$Q_v = nc_v' \, \Delta T \quad \text{for a process in which } V = \text{constant.} \tag{19.14}$$

We use the notation Q_v to remind us that this is the heat flow into the system required to raise the temperature through ΔT under constant-volume conditions. With the piston unlocked and the pressure rather than the volume held constant, as in Figure 19.6a, the **molar heat capacity at constant pressure** c_p' of the working fluid is

$$Q_p = nc_p' \, \Delta T \quad \text{for a process in which } p = \text{constant.} \tag{19.15}$$

In the constant-volume case, no work is done; $W = 0$. The first law of thermodynamics thus assumes the simple form $\Delta E = Q$, and Equation 19.14 becomes

$$\Delta E = Q_v = nc_v' \, \Delta T \quad \text{for } V = \text{constant.} \tag{19.16}$$

In the constant-pressure case, however, we have $W = -p \, \Delta V$ for the work done on the system. This (negative) work tends to reduce the internal energy of the working fluid. As a result, the heat flow into the system required for the same temperature change ΔT now has a value Q_p, which is greater than Q_v. The first law of thermodynamics takes the form $\Delta E = Q_p - p \, \Delta V$, or

$$Q_p = \Delta E + p \, \Delta V \quad \text{for } p = \text{constant.}$$

Substituting the value of Q_p given by Equation 19.15 and the value of ΔE given by Equation 19.16, we have

$$nc_v' \, \Delta T + p \, \Delta V = nc_p' \, \Delta T.$$

When we solve for c_p', we have

$$c_p' = c_v' + \frac{p}{n} \frac{\Delta V}{\Delta T}. \tag{19.17}$$

Most substances expand when heated; that is, $\Delta V / \Delta T$ is usually greater than zero. So for most substances we have $c_p' > c_v'$. There are a few exceptions, such as water in the temperature range from 0°C to 4°C.

If we know the volume coefficient of expansion $\beta = (1/V)(\Delta V/\Delta T)$ for a substance, we can calculate the difference $c_p' - c_v'$ for that substance. Because β is small for solids and liquids (Table 17.1), $c_p' - c_v'$ is usually so small that it can be neglected. That is why we usually quote a single value c for solids and liquids, as we did in Section 17.5.

The situation is quite different for gases, whose volume coefficients of expansion are close to $\beta = 1/273.16 \text{ K}^{-1}$. This is an order of magnitude greater than the coefficients of typical liquids and two orders of magnitude greater than those of typical solids. If we know the equation of state, we can calculate $\Delta V/\Delta T$ and thus find an explicit expression for c_p'. For an ideal gas, the equation of state is $pV = nRT$, and so in an isobaric process we have $p \, \Delta V = nR \, \Delta T$, or

$$\frac{\Delta V}{\Delta T} = \frac{nR}{p}.$$

Substituting this value into Equation 19.17, we obtain

$$c_p' = c_v' + R. \tag{19.18}$$

For a monatomic ideal gas $c_v' = \frac{3}{2}R$, and so we have

$$c_p' = \frac{5}{2}R \quad \text{for a monatomic ideal gas.} \tag{19.19}$$

More generally, for an ideal gas having f degrees of freedom, the molar heat capacity at constant volume is

$$c_v' = \frac{f}{2} R. \qquad \textbf{(19.20a)}$$

The corresponding molar heat capacity at constant pressure is

$$c_p' = \frac{f + 2}{2} R \quad \text{for an ideal gas.} \qquad \textbf{(19.20b)}$$

We define the **heat capacity ratio** γ:

$$\gamma \equiv \frac{c_p'}{c_v'}. \qquad \textbf{(19.21)}$$

Using Equations 19.20a and 19.20b, we obtain the very neat result

$$\gamma = \frac{f + 2}{f} \quad \text{for an ideal gas.} \qquad \textbf{(19.22)}$$

Experimental values of c_v', c_p', and γ are listed for various gases in Table 19.2, along with f for each gas and the value of γ calculated by using Equation 19.22. The agreement between experiment and theory is generally quite good.

TABLE 19.2 Molar Heat Capacities of Gases and Their Ratios

Gas	c_v' (J/kmol·K)	c_p' (J/kmol·K)	γ	f	γ (Eq. 19.21)
Monatomic Gases					
Helium, He	1.25×10^4	2.08×10^4	1.66	3	1.667 $(= \frac{5}{3})$
Sodium, Na	1.26×10^4	2.12×10^4	1.68	3	1.667
Mercury, Hg	1.25×10^4	2.09×10^4	1.67	3	1.667
Diatomic Gases					
Hydrogen, H_2	2.05×10^4	2.87×10^4	1.40	5	1.400 $(= \frac{7}{5})$
Nitrogen, N_2	2.08×10^4	2.92×10^4	1.404	5	1.400
Hydrogen chloride, HCl	2.10×10^4	2.96×10^4	1.41	5	1.400
Polyatomic Gases					
Water, H_2O	2.84×10^4	3.72×10^4	1.31	6	1.333 $(= \frac{8}{6})$
Ammonia, NH_3	2.85×10^4	3.73×10^4	1.310	6	1.333
Methane, CH_4	2.71×10^4	3.55×10^4	1.31	6	1.333

Adiabatic Processes

We now turn to the fourth important type of thermodynamic process listed in Table 19.1—the adiabatic process. During an **adiabatic process**, no heat flows into or out of the system.* In principle, an adiabatic process can be carried out by surrounding the system with a perfect thermal insulator. In practice, however, the adiabatic condition is most often achieved through the speed of the process. When the volume of a gasoline-engine cylinder increases during the power stroke, for example, there is not enough time for appreciable heat to pass through the cylinder walls before the stroke is completed, and thus the process is essentially adiabatic. In the analysis that follows, we will restrict our attention to systems containing ideal gases.

*The word *adiabatic* derives from a Greek root meaning "impassable," implying that the system boundaries are impassable to heat.

Figure 19.8 shows a cylinder-piston system containing an ideal gas and surrounded by ideal insulation. The system undergoes an adiabatic expansion from a state b, characterized by p_b, V_b, T_b, to a state c, characterized by p_c, V_c, T_c. What happens as the cylinder volume increases by an infinitesimal amount dV? The pressure changes by some corresponding amount dp, and the temperature by some corresponding amount dT. The correspondence among the changes is found by differentiating the ideal-gas law, $pV = nRT$, to obtain $d(pV) = nR\,dT$. The chain rule for differentiation yields

$$p\,dV + V\,dp = nR\,dT. \tag{19.23}$$

The Adiabatic Pressure-Volume Relation

By eliminating the temperature from Equation 19.23, we can express the relation between the two remaining variables, pressure and volume. (This relation will enable us to make a p-V plot of an adiabatic process.) In the special case of an ideal gas, we can exploit the fact that the internal energy is a function of temperature only. If the gas has f degrees of freedom, the functional relation is

$$E = \frac{f}{2}\,nRT. \tag{19.24}$$

(Compare with Equations 18.29 and 18.38, which express E for the special cases $f = 3$ and $f = 5$.) Differentiating yields

$$dE = \frac{f}{2}\,nR\,dT. \tag{19.25}$$

We now introduce the condition that the process is adiabatic, and thus $đQ = 0$. The infinitesimal form of the first law of thermodynamics (Equation 19.2) simplifies to

$$dE = đW = -p\,dV. \tag{19.26}$$

Equating the two expressions for dE given by Equations 19.26 and 19.25, we have

$$-p\,dV = \frac{f}{2}\,nR\,dT.$$

We solve for dT, obtaining

$$dT = -\frac{2p\,dV}{fnR}.$$

This is the expression we need to eliminate dT from Equation 19.23. The result is

$$p\,dV + V\,dp = -nR\,\frac{2p\,dV}{fnR},$$

or

$$p\,dV + V\,dp = -\frac{2}{f}\,p\,dV.$$

We gather terms in $p\,dV$, which gives us

$$\left(1 + \frac{2}{f}\right)p\,dV = -V\,dp. \tag{19.27}$$

We now use Equation 19.22, $\gamma = (f + 2)/f$, to simplify the term in parentheses:

$$1 + \frac{2}{f} = \frac{f + 2}{f} = \gamma.$$

With this simplification, Equation 19.27 becomes

$$\gamma p\,dV = -V\,dp.$$

We divide this equation through by pV to separate terms in V from terms in p and get the result

$$\gamma\,\frac{dV}{V} = -\frac{dp}{p}. \tag{19.28}$$

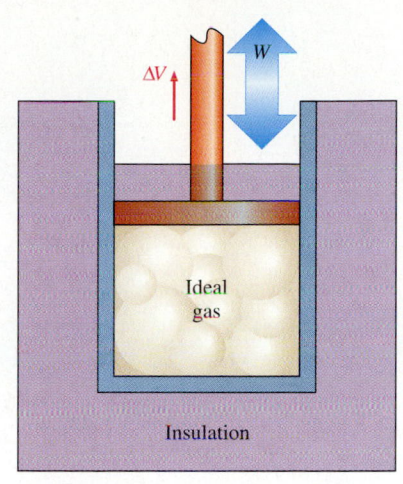

FIGURE 19.8 A cylinder-piston system containing an ideal gas is surrounded by a perfect insulator, so $Q = 0$ in any process. As the volume of the cylinder changes quasi-statically by an amount $V = A\,dx$ and the piston does work $đW$, the internal energy changes by an amount dE.

This differential equation expresses the desired relation between an infinitesimal change dp and the corresponding change dV. But we are interested in the finite process $b{\to}c$. So we write the equation in integral form, choosing the limits of the two integrals to correspond properly:

$$\gamma \int_{V_b}^{V_c} \frac{dV}{V} = -\int_{p_b}^{p_c} \frac{dp}{p}.$$

Carrying out the integration, we obtain

$$\gamma \ln \frac{V_c}{V_b} = \ln \frac{p_b}{p_c}.$$

We can eliminate the logarithms by exponentiating both sides of the equation:

$$\left(\frac{V_c}{V_b}\right)^{\gamma} = \frac{p_b}{p_c},$$

or $p_b V_b{}^{\gamma} = p_c V_c{}^{\gamma}$ for ideal gases in adiabatic processes.   **(19.29a)**

Another way of expressing this result is

$pV^{\gamma} = \text{constant}$ for ideal gases in adiabatic processes.   **(19.29b)**

Compare this result with Boyle's law, $pV = \text{constant}$, which describes the behavior of ideal gases in isothermal processes.

Figure 19.9 shows two curves, $pV^{\gamma} = \text{constant}$, plotted on a p-V diagram. Such curves are called **adiabatics**. The adiabatics plotted here are for a monatomic ideal gas, with $\gamma = \frac{5}{3}$. Shown for comparison are two isotherms. Because γ is always greater than 1, the adiabatics are steeper than the isotherms. That is, the magnitude of the slope of the adiabatic that passes through any particular point (V, p) is always greater than the magnitude of the slope of the isotherm through the same point.

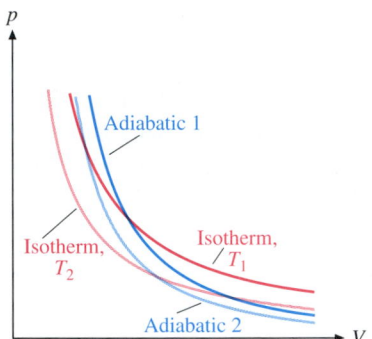

FIGURE 19.9 A p-V plot of two adiabatics, $pV^{\gamma} = \text{constant}$, for a monatomic ideal gas, $\gamma = \frac{5}{3}$. Each adiabatic corresponds to a particular choice of the constant. Two isotherms are shown for comparison.

The Adiabatic Temperature-Volume and Temperature-Pressure Relations

Relations describing the adiabatic behavior of an ideal gas in terms of V and T, and in terms of p and T, are readily derived from either of Equations 19.29a and 19.29b. We need only use the ideal-gas law, $pV = nRT$, to eliminate p or V from these equations. The details are left for Problems 19.19 and 19.20; here we merely state the results:

$$T_b V_b{}^{\gamma-1} = T_c V_c{}^{\gamma-1} \quad \text{or} \quad TV^{\gamma-1} = \text{constant}; \qquad \textbf{(19.30a,b)}$$

$$p_b{}^{(1-\gamma)/\gamma} T_b = p_c{}^{(1-\gamma)/\gamma} T_c \quad \text{or} \quad p^{(1-\gamma)/\gamma} T = \text{constant}. \qquad \textbf{(19.31a,b)}$$

Compare these results with Charles's law in the form $V/T = \text{constant}$, which describes the behavior of ideal gases in isobaric processes, and in the form $p/T = \text{constant}$, which describes the behavior of ideal gases in isometric processes.

The Carnot Cycle

Figure 19.10 depicts a heat-engine cycle consisting of an isothermal expansion $a{\to}b$, an adiabatic expansion $b{\to}c$, an isothermal compression $c{\to}d$, and an adiabatic compression $d{\to}a$. Such a cycle can be achieved in principle by combining processes we have already considered, as shown in Figure 19.11. The cycle has the particularly simple feature that all the heat flow Q_{in} *into* the engine takes place at the fixed temperature T_{hi} and all the heat flow $-Q_{\text{out}}$ *out of* the engine takes place at the fixed temperature T_{lo}. This special cycle is called the **Carnot cycle**, in honor of its inventor, the French physicist Nicolas L. Sadi Carnot (1796–1832).

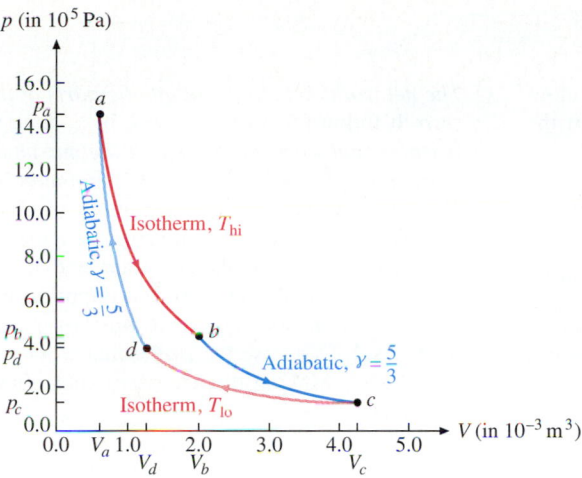

FIGURE 19.10 A Carnot engine cycle, consisting of two adiabatics and two isotherms. The cycle is analyzed in Examples 19.3 and 19.4.

Unsatisfied with the limited efficiency of steam engines, 19th-century engineers devised various internal-combustion-engine cycles, among them the Otto cycle. A modern version is shown here. Though the efficiency of an Otto engine is less than the ideal Carnot efficiency (see Problem 20.33), the value of T_{hi} is much greater than that for the steam engine, and the Otto engine has proliferated in the hundreds of millions.

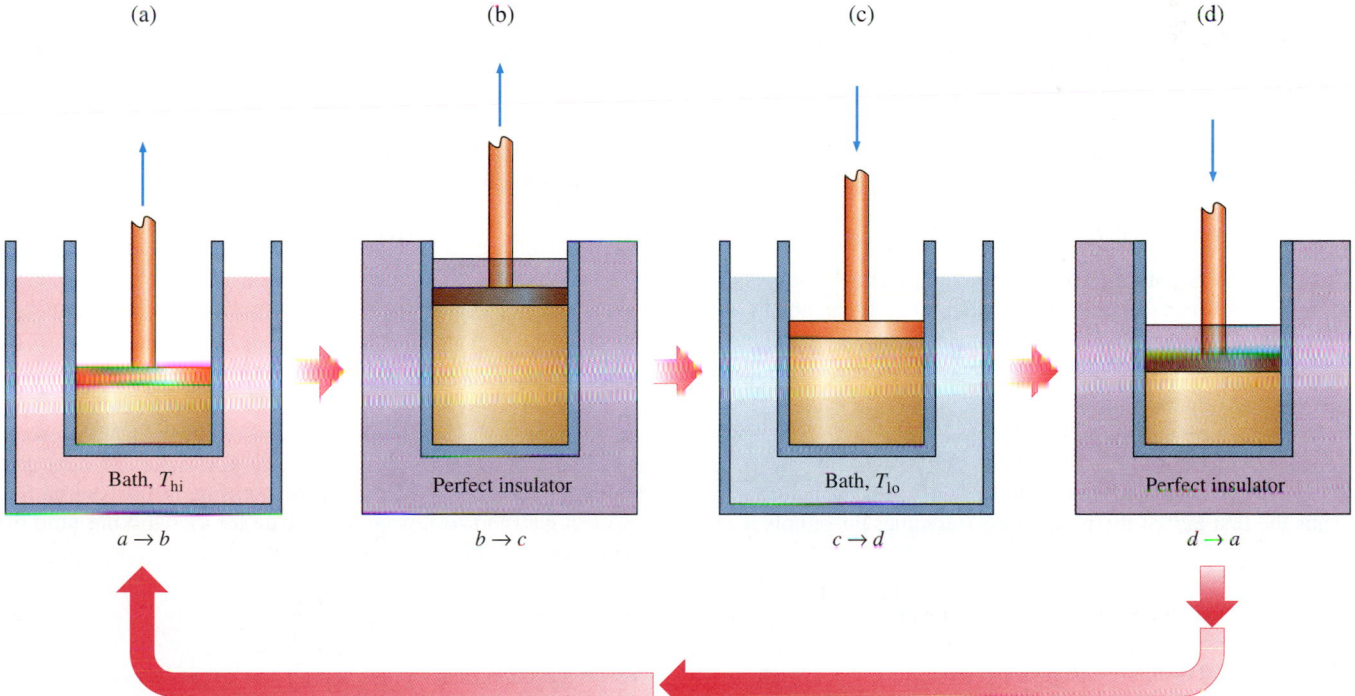

FIGURE 19.11 Idealized method of realizing a Carnot cycle. (*a*) The system is immersed in a bath at temperature T_{hi} and is allowed to expand quasistatically, doing work on an external mechanism as heat flows into the system from the bath. (*b*) The system is enclosed in a perfect insulator. The quasistatic expansion continues—now adiabatically ($Q = 0$)—until the temperature has decreased to T_{lo}. (*c*) The system is immersed in a bath at temperature T_{lo} and compressed quasistatically (by an external source of work) as heat flows out of it into the bath. (*d*) The system is again enclosed in a perfect insulator. The quasistatic compression continues, but now adiabatically, until the temperature of the system has increased back to T_{hi}. The cycle may be repeated indefinitely, as suggested by the arrow from part *d* to part *a*.

EXAMPLE 19.3

Suppose that the working fluid of a Carnot engine is a monatomic ideal gas. **(a)** Show that the cylinder volumes in the states a, b, c, and d (Figure 19.10) bear the relation

$$\frac{V_a}{V_d} = \frac{V_b}{V_c}. \tag{19.32}$$

(b) Find an expression for the work W_{cyc} done in the Carnot cycle. **(c)** As the piston moves through its cycle, the volume of the cylinder varies from its minimum value $V_a = 0.581 \times 10^{-3}$ m³ to its maximum value $V_c = 4.23 \times 10^{-3}$ m³. The working-fluid pressures corresponding to these volumes are $p_a = 14.44 \times 10^5$ Pa and $p_c = 1.19 \times 10^5$ Pa. The temperature along the isotherm $a \rightarrow b$ is $T_{hi} = 500$ K, and the temperature along the isotherm $c \rightarrow d$ is $T_{lo} = 300$ K. How much work is done in the cycle?

SOLUTION:

(a) Show that Carnot-cycle cylinder volumes conform to Equation 19.32 if the working fluid is an ideal gas.

For the temperatures at the ends of the two isothermal processes, $a \rightarrow b$ and $c \rightarrow d$, you have

$$T_a = T_b \quad \text{and} \quad T_c = T_d. \tag{19.33}$$

For the temperatures at the ends of the two adiabatic processes, $b \rightarrow c$ and $d \rightarrow a$, you use Equation 19.30a to write

$$T_a V_a^{\gamma-1} = T_d V_d^{\gamma-1} \quad \text{and} \quad T_b V_b^{\gamma-1} = T_c V_c^{\gamma-1}.$$

You can rearrange these equations to obtain

$$\left(\frac{V_a}{V_d}\right)^{\gamma-1} = \frac{T_{lo}}{T_{hi}} \quad \text{and} \quad \left(\frac{V_b}{V_c}\right)^{\gamma-1} = \frac{T_{lo}}{T_{hi}}.$$

According to Equation 19.33, the two temperature ratios on the right sides of these equations are equal. So you can equate the left sides of the two equations and take the $1/(\gamma - 1)$ power of both sides to obtain Equation 19.32.

(b) Find an expression for W_{cyc} over the Carnot cycle.

You have already explored the work done in an isothermal process in Example 19.1, so look first at the two adiabatic parts of the cycle, $b \rightarrow c$ and $d \rightarrow a$. On each of these paths, $Q = 0$. Thus the first law of thermodynamics assumes the simple form

$$W = \Delta E.$$

In an ideal gas, the internal energy depends only on the temperature. For a monatomic ideal gas in particular, you can write $E = \frac{3}{2}nRT$ and thus, for the process $b \rightarrow c$,

$$W_{b \rightarrow c} = \Delta E = \tfrac{3}{2}nR(T_c - T_b).$$

Similarly, you have for the process $d \rightarrow a$

$$W_{d \rightarrow a} = \tfrac{3}{2}nR(T_a - T_d).$$

The total work done over the adiabatic parts of the cycle is the sum

$$W_{adiabatic} = W_{b \rightarrow c} + W_{d \rightarrow a} = \tfrac{3}{2}nR(T_c - T_b + T_a - T_d).$$

But you can take advantage of the isothermal parts of the cycle, $a \rightarrow b$ and $c \rightarrow d$, to write $T_b = T_a$ and $T_c = T_d$. This gives you

$$W_{adiabatic} = \tfrac{3}{2}nR(T_d - T_a + T_a - T_d) = 0.$$

The net work done in the adiabatic parts of the Carnot cycle is zero. It follows that *all the work W_{cyc} done in a Carnot cycle is isothermal work, performed at the maximum system temperature $T_a = T_b \equiv T_{hi}$ and at the minimum temperature $T_c = T_d \equiv T_{lo}$.* As noted at the beginning of this example, all the heat flow into and out of a Carnot engine is isothermal heat flow. These two facts taken together lend the Carnot cycle much of its fundamental simplicity and importance.

Now you must evaluate the work over the isothermal parts of the cycle. You use the isothermal work equation (Equation 19.6) twice. Making the necessary substitutions of subscripts, you have

$$W_{a \rightarrow b} = nRT_{hi} \ln \frac{V_a}{V_b} \quad \text{and} \quad W_{c \rightarrow d} = nRT_{lo} \ln \frac{V_c}{V_d}.$$

The sum of these two quantities gives W_{cyc}, because $W_{adiabatic} = 0$. Adding gives you

$$W_{cyc} = W_{isothermal} = W_{a \rightarrow b} + W_{c \rightarrow d}$$
$$= nRT_{hi} \ln \frac{V_a}{V_b} + nRT_{lo} \ln \frac{V_c}{V_d}.$$

To eliminate V_c/V_d from this equation, you use Equation 19.32 to make the substitution $V_c/V_d = V_b/V_a$:

$$W_{cyc} = nRT_{hi} \ln \frac{V_a}{V_b} + nRT_{lo} \ln \frac{V_b}{V_a}.$$

To simplify, remember that $\ln(x/y) = -\ln(y/x)$ and rewrite the first term on the right to obtain

$$W_{cyc} = -nRT_{hi} \ln \frac{V_b}{V_a} + nRT_{lo} \ln \frac{V_b}{V_a},$$

or

$$W_{cyc} = -nR(T_{hi} - T_{lo}) \ln \frac{V_b}{V_a}. \tag{19.34}$$

(c) Evaluate numerically the work done over the cycle.

The temperatures you need to know to use Equation 19.34 are given, as is the volume V_a. You must still evaluate the product nR and the volume V_b. To find nR, you write the ideal-gas law in the form $nR = pV/T$. This expression is always true for an ideal gas, and so it is certainly true for the working fluid of the engine when it passes through state a:

$$nR = \frac{p_a V_a}{T_{hi}}.$$

You have numerical values for all the quantities on the right side of this equation, and thus you will be able to evaluate nR.

The easiest way to find V_b is to use Equation 19.30a on the adiabatic $b \rightarrow c$:

$$V_b = \left(\frac{T_{lo}}{T_{hi}}\right)^{1/(\gamma-1)} V_c.$$

By inserting the values of nR and V_b into Equation 19.34 and doing some algebraic manipulation, you can obtain

$$W_{cyc} = p_a V_a \left(1 - \frac{T_{lo}}{T_{hi}}\right)\left(\frac{1}{\gamma - 1} \ln \frac{T_{lo}}{T_{hi}} + \ln \frac{V_c}{V_a}\right).$$

Now you are ready to substitute the known numerical values:

$$W_{cyc} = 14.44 \times 10^5 \text{ Pa} \times 0.581 \times 10^{-3} \text{ m}^3$$

$$\times \left(1 - \frac{300 \text{ K}}{500 \text{ K}} \right)$$

$$\times \left(\frac{1}{\frac{5}{3} - 1} \ln \frac{300 \text{ K}}{500 \text{ K}} + \ln \frac{4.23 \times 10^{-3} \text{ m}^3}{0.58 \times 10^{-3} \text{ m}^3} \right)$$

$$= 410 \text{ J}.$$

Entropy: A First Look

In thermodynamic processes, we study the way in which such quantities as T, p, and V change. We have named these quantities *state variables* because they have specific values for any thermodynamic state. We have considered the four important special types of thermodynamic processes listed in Table 19.1: isothermal, isobaric, isometric, and adiabatic. Each of the first three of these types is characterized by one state variable—T, p, or V—that does not change in the course of the process. In contrast, we have characterized adiabatic processes by requiring $Q = 0$. Because zero is a constant, it may seem at first glance that $Q = 0$ is the same kind of characterization as the other three. Not so! As you have already seen, *a thermodynamic system does not possess a certain amount of heat, and Q is not a state variable*. The quantity Q represents the amount of energy that passes into or out of the system *in a particular form*. So specifying $Q = 0$ is *not* the same sort of thing as specifying $T = $ constant, $p = $ constant, or $V = $ constant.

How can we put the characterization of an adiabatic process on the same footing as the characterization of the other three types? We must invent a state variable—call it S—that does not change in adiabatic processes. We can then characterize an adiabatic process by the property $S = $ constant, or $\Delta S = 0$. What is more, we can then legitimately talk about a derivative of S, such as dS/dT, which we cannot do for $đQ$.

What does the desired quantity look like? First of all, we know that it must remain constant when no heat flows into or out of the system; that is, $dS = 0$ when $đQ = 0$. And it must change—$dS \neq 0$—when $đQ \neq 0$. These conditions on dS will be true if dS is defined as the product of $đQ$ and some other quantity.

But what should the "other quantity" be? We can get some guidance from the first law of thermodynamics in the form of Equation 19.2,

$$dE = đQ + đW.$$

This equation has the disadvantage that it contains inexact differentials. Equation 19.3, $đW = -p\,dV$, furnishes a partial remedy because we replace one of the inexact differentials, obtaining

$$dE = đQ - p\,dV. \tag{19.35}$$

Now we would like to make a substitution for $đQ$ analogous to the one made for $đW$ so that Equation 19.35 will contain only exact differentials and so that the two terms on the right will be of similar mathematical form. What will the desired term look like? We want it to be the product of dS and some other quantity. Now, $đQ$ pertains to thermal processes, in which the temperature changes. So the desired term should contain a factor T. Let us try

$$đQ = T\,dS. \tag{19.36}$$

In terms of this choice, the first law of thermodynamics assumes the symmetrical form

$$dE = T\,dS - p\,dV. \tag{19.37}$$

This equation contains no inexact differentials, and it expresses energy changes in terms of the three familiar quantities p, V, and T, together with the new quantity S, which we have set on an equal footing with the other three. The implication is that these four quantities just suffice to describe all quasistatic thermodynamic processes of the kind we have studied in this chapter.

The quantity S is called the **entropy** of the system. The name, invented about 1835 by Rudolf Clausius (Problem 18.60), is derived from a Greek root meaning "transformation."[*] The SI unit of entropy is J/K, as is evident from inspection of Equation 19.37.

Entropy Change in Adiabatic and Isothermal Processes

Let us test our new quantity, the entropy S, to see if it satisfies the conditions for which it was designed. First of all, is dS indeed an exact differential? To see that it is, simply solve Equation 19.37 for dS:

$$dS = \frac{1}{T}\,dE + \frac{p}{T}\,dV.$$

Because both dE and dV are exact differentials and because p and T are well-defined thermodynamic quantities, dS must be an exact differential.

Next, is $\Delta S = 0$ in an adiabatic process? Because $T \neq 0$ in Equation 19.36, dS must be equal to zero whenever $đQ = 0$. But we have $đQ = 0$ by definition throughout any adiabatic process, and so indeed $dS = 0$ everywhere in such a process. Hence, for an entire quasistatic process such as $b \rightarrow c$ in Example 19.3, we have

$$T\,\Delta S = \int_b^c T\,dS = \int_b^c đQ = Q = 0 \quad \text{for an adiabatic quasistatic process.}$$

$$\textbf{(19.38a)}$$

Because $T\,\Delta S = 0$ and T is not equal to zero, we have

$$\Delta S = 0 \quad \text{or} \quad S = \text{constant} \quad \text{for an adiabatic quasistatic process.} \quad \textbf{(19.38b,c)}$$

In view of Equation 19.38c, adiabatic quasistatic processes are sometimes called **isentropic** processes.

Finally, does the entropy of the system change—is $\Delta S \neq 0$—when a process is not adiabatic? Let us look, for example, at the isothermal process $a \rightarrow b$ in Example 19.3. As you saw in part **d** of Example 19.1, $\Delta E = 0$ in an isothermal process; consequently, the first law of thermodynamics simplifies to $Q = -W$. The heat input to the system is

$$Q_{a \rightarrow b} = nRT_{\text{hi}} \ln \frac{V_b}{V_a} = p_a V_a \ln \frac{V_b}{V_a}. \quad \textbf{(19.39)}$$

But we also have, integrating Equation 19.36,

$$Q_{a \rightarrow b} = \int_a^b đQ = \int_a^b T\,dS. \quad \textbf{(19.40)}$$

Because T has the constant value T_{hi}, it can be taken outside the integral, and we write

$$Q_{a \rightarrow b} = T_{\text{hi}} \int_a^b dS = T_{\text{hi}}(S_b - S_a) = T_{\text{hi}}\,\Delta S_{a \rightarrow b} \quad \text{for an isothermal process.}$$

$$\textbf{(19.41)}$$

---

[*]The reason for this name will become apparent in Chapter 20; it turns out that entropy is a measure of the energy in the system that has been "transformed" so as to be unavailable for conversion into mechanical work.

Combining this equation with Equation 19.39 gives

$$\Delta S_{a \to b} = \frac{p_a V_a}{T_{\text{hi}}} \ln \frac{V_b}{V_a} \quad \text{for an isothermal process.} \quad (19.42)$$

The entropy change $\Delta S_{a \to b}$ of the system will be positive if $V_b > V_a$—that is, if the system is allowed to expand. Under these circumstances, work is done by the system and heat flows into it. In the opposite case, $\Delta S_{a \to b}$ will be negative if $V_b < V_a$—that is, if the system is compressed. Under these circumstances, work is done on the system and heat flows out of it.

The thermodynamic quantity S thus serves the purpose for which it was devised. In particular, it can be used to place adiabatic processes on the same footing as the other three quasistatic thermodynamic processes we have considered. This point is illustrated in Table 19.3, which restates the information of Table 19.1 in the improved fashion that the concept of entropy makes possible.

TABLE 19.3 Types of Thermodynamic Processes in Closed Systems (n = constant)

Process Name	Fixed Quantity	Varying Quantities
Isothermal	Temperature T	p, V, S
Isobaric	Pressure p	V, T, S
Isometric	Volume V	p, T, S
Adiabatic (isentropic)	Entropy S	p, V, T

EXAMPLE 19.4

(a) Evaluate the entropy change of the system of Example 19.3 as it passes through the isothermal expansion $a \to b$. (b) Find the entropy change of the system as it passes through the Carnot cycle once.

SOLUTION:

(a) Evaluate $\Delta S_{a \to b}$.

Inserting the values from Example 19.3 into Equation 19.42, you have

$$\Delta S_{a \to b} =$$

$$\frac{14.44 \times 10^5 \text{ Pa} \times 0.581 \times 10^{-3} \text{ m}^3}{500 \text{ K}} \times \ln \frac{1.966 \times 10^{-3} \text{ m}^3}{0.581 \times 10^{-3} \text{ m}^3}$$

$$= 2.05 \text{ J/K}.$$

(b) Evaluate ΔS_{cyc} for the Carnot cycle.

You know that the entropy change is zero in the adiabatic processes $b \to c$ and $d \to a$. You could derive the equation for the entropy change in the isothermal process just as we did for the process $a \to b$ in the argument leading to Equation 19.42. But

you can obtain the same result more easily by merely substituting into that equation the subscripts proper to the process $c \to d$:

$$\Delta S_{c \to d} = \frac{p_c V_c}{T_{\text{lo}}} \ln \frac{V_d}{V_c}.$$

The total entropy change in the cycle is the sum $\Delta S_{a \to b} + \Delta S_{c \to d}$. To simplify this sum, you use the ideal-gas law to write $p_c V_c / T_{\text{lo}} = p_a V_a / T_{\text{hi}}$:

$$\Delta S_{\text{cyc}} = \frac{p_a V_a}{T_{\text{hi}}} \left(\ln \frac{V_b}{V_a} + \ln \frac{V_d}{V_c} \right).$$

When you use Equation 19.32 to write $V_d / V_c = V_a / V_b$, you find that the sum in parentheses becomes $\ln 1 = 0$. So you have for the entire cycle

$$\Delta S_{\text{cyc}} = 0.$$

The entropy change in a Carnot cycle is zero. This is one of the most important properties of the Carnot cycle; we will explore the implications in Chapter 20.

Entropy Change in Isobaric and Isometric Processes

In isobaric and isometric processes, the temperature is not constant. Thus we cannot calculate the entropy change ΔS by using the procedure that led to Equation 19.42 because that procedure included the step $\int T \, dS = T \int dS$. Instead, we return to the basic relation given by Equation 19.36, $dQ = T \, dS$, which we can write in the form

$$dS = \frac{dQ}{T}. \quad (19.43)$$

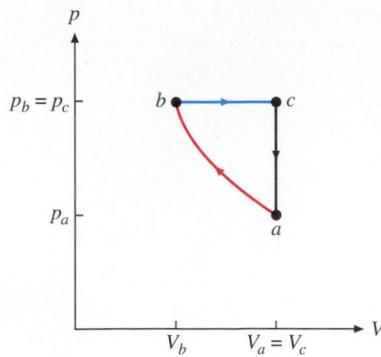

FIGURE 19.12 The cycle of Figure 19.5; $a \rightarrow b$ is an isotherm, $b \rightarrow c$ is an isobar, and $c \rightarrow a$ is an isometric.

For a thermodynamic process $j \rightarrow k$, the total entropy change can be written as the integral

$$\Delta S_{j \rightarrow k} = \int_j^k \frac{dQ}{T}.$$

(19.44)

We cannot integrate this equation directly because dQ is an inexact differential and it makes no sense in general to speak of T as a function of Q.

However, it is possible to express dQ in terms of exact differentials in special cases. Consider the isobaric process $b \rightarrow c$ of Figure 19.5, which is reproduced here as Figure 19.12. For such a process, we can use Equation 19.15 in differential form to write

$$dQ = nc_p' \, dT \quad \text{for } p = \text{constant.}$$

(19.45)

We substitute this value of dQ into the right side of Equation 19.43 to obtain

$$\Delta S_{b \rightarrow c} = \int_{T_b}^{T_c} nc_p' \frac{dT}{T}.$$

Because nc_p' is a constant, we can integrate immediately. The result is

$$\Delta S_{b \rightarrow c} = nc_p' \ln \frac{T_c}{T_b} \quad \text{for } p = \text{constant.}$$

(19.46)

To find the corresponding result for an isometric process—process $c \rightarrow a$ in Figure 19.12, for instance—we repeat the argument while using Equation 19.14 in the differential form

$$dQ = nc_v' \, dT \quad \text{for } V = \text{constant.}$$

(19.47)

The result is

$$\Delta S_{c \rightarrow a} = nc_v' \ln \frac{T_a}{T_c} \quad \text{for } V = \text{constant.}$$

(19.48)

Equations 19.46 and 19.48 are expressed in terms of the temperatures of the initial and final states. But they can also be expressed in terms of the corresponding pressures and volumes if the equation of state of the working fluid is known. See Problems 19.21 and 19.22 for the case in which the working fluid is an ideal gas.

The T-S Diagram

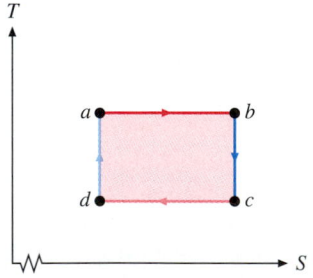

FIGURE 19.13 A *T-S* diagram for a Carnot cycle. The horizontal lines are isotherms, and the vertical lines are adiabatics.

A very useful complement to the *p-V* diagram of a heat engine is its **T-S diagram**. The *T-S* diagram for the Carnot cycle is especially simple, as Figure 19.13 shows. The two isotherms $a \rightarrow b$ and $c \rightarrow d$ are represented by horizontal line segments, $T = $ constant; the two adiabatics $b \rightarrow c$ and $d \rightarrow a$ are represented by vertical line segments, $S = $ constant. Thus the *T-S* curve of the Carnot cycle is a rectangle of area $\Delta T \, \Delta S$. *This is true for any working fluid*, which makes the *T-S* diagram more general than the ideal-gas *p-V* diagram of Figure 19.10.

The area $|\Delta T \, \Delta S|$ represents the net heat input $|Q_{cyc}|$ to the engine over the cycle, just as the area $|\int p \, dV|$ in *V-p* space represents the net work output $|W_{cyc}|$ from the engine over the cycle. But, for the Carnot cycle, $|\Delta T \, \Delta S|$ is much easier to evaluate graphically because of the simple form of the curve. Now, remember that $\Delta E_{cyc} = 0$ over the engine cycle. It follows that

$$\boxed{|Q_{cyc}| = \Delta T \, \Delta S = |W_{cyc}|;}$$

(19.49)

the area $|\Delta T \, \Delta S|$ *in S-T space represents the magnitude of the work output* $|W_{cyc}|$ *as well as the magnitude of the heat input* $|Q_{cyc}|$. Thus we can use the simple *T-S* curve to evaluate $|W_{cyc}|$ in lieu of the more complicated *p-V* curve. Even for an engine cycle whose *T-S* curve is not as simple as the Carnot curve, the *T-S* plot is frequently useful in analysis. We will use *T-S* diagrams often in Chapter 20.

EXAMPLE **19.5**

Plot a *T-S* curve for the Carnot engine considered in Examples 19.3 and 19.4, and show that the work output found by calculating $|W_{cyc}| = |Q_{cyc}| = |\Delta T \, \Delta S|$ is the same as that obtained directly from the *p-V* plot.

SOLUTION: The temperatures are $T_a = T_b = T_{hi} = 500$ K and $T_c = T_d = T_{lo} = 300$ K, so you have

$$|\Delta T| = 200 \text{ K.}$$

In part **a** of Example 19.4, you found the entropy change ΔS over the isotherm $a{\to}b$ in Figure 19.10. In part **b** of the same example, you found the entropy change over the entire cycle to be $\Delta S_{cyc} = 0$. It follows that the magnitudes of the entropy changes over the two isotherms are equal:

$$|\Delta S_{a{\to}b}| = |\Delta S_{c{\to}d}| = 2.05 \text{ J/K.}$$

Now, in no place is *the* entropy of a state calculated. But this does not matter because only entropy *differences* are significant in thermodynamic calculations. A kindred situation arises in calculating potential energies, where the choice of a reference value is arbitrary. Boltzmann showed, using the methods of statistical mechanics, that the absolute entropy S of a body does have a unique meaning (see Section 20.7). So you make a plot like that in Figure 19.14, in which the T axis has an absolute scale with a definite zero point, but the zero point on the S axis is arbitrary. This is represented in Figure 19.14 by the break in

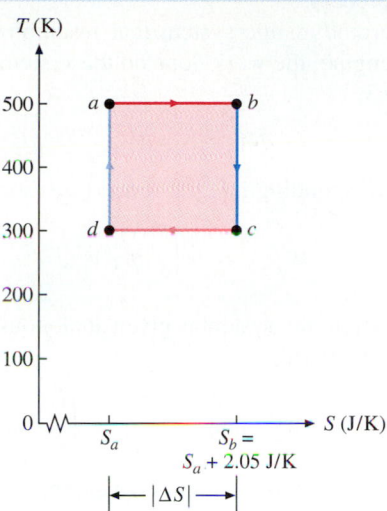

FIGURE 19.14 *T-S* diagram. The states a, b, c, and d are identical with the corresponding states shown in Figure 19.10.

the S axis. The area $|\Delta T \, \Delta S|$ is the magnitude $|Q_{cyc}|$ of the net heat input.

You can now readily calculate

$$|W_{cyc}| = |\Delta T \, \Delta S| = 200 \text{ K} \times 2.05 \text{ J/K} = 410 \text{ J.}$$

This result agrees with the result of Example 19.3.

Symbols Used in Chapter 19

a, b, c, d	thermodynamic states		p	pressure
A	area		Q, dQ	heat flow
c_v', c_p'	molar heat capacity at constant volume, constant pressure		R	universal gas constant
			S	entropy
cyc (subscript)	denotes the value of a quantity over a cyclic process		T	absolute temperature
			V	volume
E	internal energy		W, dW	work
f	number of degrees of freedom		β	volume coefficient of expansion
$j{\to}k$ (subscript)	denotes a thermodynamic process beginning at state j and ending at state k		γ	heat capacity ratio
n	quantity of working fluid, in kmol			

Summing Up

All energy exchanges between a **thermodynamic system** and its **surroundings** are divided into *heat* and *work*. Conservation of energy requires that any change ΔE or dE in the **internal energy** E of the system be the sum of the heat and the work. This requirement, stated mathematically in Equations 19.1 and 19.2, is called the **first law of thermo-**

dynamics:

$$\Delta E = Q + W \quad \text{or} \quad dE = dQ + dW.$$

Suppose that a **thermodynamic process** begins at state a and ends at state b. The way in which the energy change ΔE is divided into heat Q and work W depends on the path—

that is, the way in which the **state variables** p, V, T, and S vary in the process.

For a thermodynamic system that resembles a reciprocating heat engine, the work done on the system is given by Equation 19.3,

$$d W = -p\,dV,$$

or by the corresponding integral form (Equation 19.4),

$$W_{a\to b} = -\int_{V_a}^{V_b} p\,dV.$$

The heat flow into the system is given analogously by Equations 19.36 and 19.40,

$$d Q = T\,dS \quad \text{and} \quad Q_{a\to b} = \int_a^b T\,dS.$$

The first of these equations defines the **entropy** of the system in its differential form dS. Entropy can also be defined in the corresponding integral form (Equation 19.44),

$$\Delta S_{j\to k} = \int_j^k \frac{d Q}{T}.$$

For a thermodynamic system, the sum of Equations 19.3 and 19.36 expresses an infinitesimal energy change dE in terms of the state quantities. This sum is given by Equation 19.37,

$$dE = T\,dS - p\,dV.$$

For any substance, the **molar heat capacity at constant pressure**, c_p', is defined by Equation 19.15,

$$Q = nc_p'\,\Delta T,$$

and is related to the molar heat capacity at constant volume by Equation 19.17,

$$c_p' = c_v' + \frac{p}{n}\frac{\Delta V}{\Delta T}.$$

For the special case of an ideal gas, this relation can be written in the form of Equation 19.18 or that of Equation 19.20b,

$$c_p' = c_v' + R = \frac{f+2}{2}R.$$

The **heat capacity ratio** γ is defined by Equation 19.21,

$$\gamma \equiv \frac{c_p'}{c_v'}.$$

For an ideal gas, γ has the value given by Equation 19.22,

$$\gamma = \frac{f+2}{f}.$$

A thermodynamic process is conveniently represented on a ***p-V* diagram**. On such a diagram, an **isobar**—the path followed by a system undergoing an **isobaric process**—is represented by a horizontal line, and an **isometric**—the path followed by a system undergoing an **isometric process**—by a vertical line. For the special case in which the **working fluid** is an ideal gas, an **isotherm**—the path followed by a system undergoing an **isothermal process**—is represented by a hyperbola whose equation (Equation 19.5) is

$$p = nRT\frac{1}{V}.$$

An ideal-gas **adiabatic** is the steeper curve representing Equation 19.29b,

$$pV^\gamma = \text{constant}.$$

Equations 19.29a, 19.30, and 19.31 give alternative forms of the adiabatic ideal-gas relation.

A **heat-engine cycle** is a thermodynamic process in which a system passes along a closed path. The net internal energy change along such a path must be zero; thus the first law of thermodynamics assumes the form of Equation 19.13b,

$$-W_{\text{cyc}} = Q_{\text{cyc}}.$$

The **Carnot cycle** is a special heat-engine cycle bounded by two isotherms and two adiabatics. Because $Q = 0$ in an adiabatic process, all the heat input to a Carnot engine occurs at a single temperature T_{hi}, and all the heat output from the engine occurs at a single temperature T_{lo}. The net work done in the adiabatic parts of the cycle is zero, so all the net work done in the cycle takes place on the isotherms. Another important property of the Carnot cycle is that the net entropy change in the cycle, ΔS, is equal to zero.

The Carnot cycle is especially simple to analyze on a ***T-S* diagram**, where the cycle takes the form of a rectangle. The heat flow per cycle and the work per cycle are related by Equation 19.49,

$$|Q_{\text{cyc}}| = |\Delta T\,\Delta S| = |W_{\text{cyc}}|.$$

KEY TERMS

Sections 19.2 and 19.3 Thermodynamic Processes
thermodynamic system ▪ closed system ▪ first law of thermodynamics ▪ state variable, equation of state ▪ quasistatic process ▪ *p-V* diagram, isobar, isometric, isotherm ▪ heat-engine cycle

Section 19.4 Heat Capacity
molar heat capacity at constant pressure c_p' ▪ heat capacity ratio γ

Sections 19.5 and 19.6 Adiabatic Processes, Entropy
adiabatic process ▪ Carnot cycle ▪ *T-S* diagram

QUERIES

19.1 *(2) Exacting argument.* Considering that đW is an inexact differential, why is it permissible to write an equation such as $W = \int đW$?

19.2 *(2) Making a b-line?* In the figure below, two paths are shown from state a to state b. The paths are symmetrical with respect to the isobar from a to b. **(a)** Which is greater, Q along path 1 or Q along path 2? **(b)** Which is greater, W along path 1 or W along path 2?

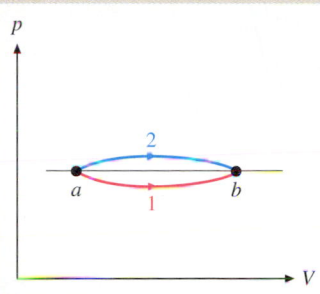

19.3 *(2) Going to c.* In the figure below, two paths are shown from state a to state c. The paths are symmetrical with respect to the isometric from a to c. **(a)** Which is greater, Q along path 3 or Q along path 4? **(b)** Which is greater, W along path 3 or W along path 4?

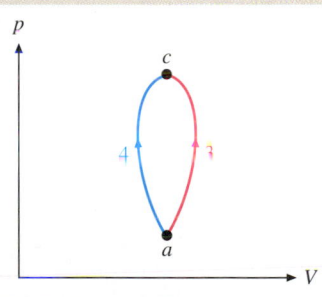

19.4 *(2) Same difference?* Why is it unnecessary to specify that an isothermal process is quasistatic?

19.5 *(2) State of affairs.* For any sample of matter, it is possible—at least in principle—to write an equation of state of the form $T = f(p,V)$. (For an ideal gas, the function has the explicit form $T = pV/nR$, but f is more complicated for other substances.) Explain why, given a sample of any substance, every point on a p-V diagram automatically fixes the temperature of the sample.

19.6 *(3) Contrariwise.* All the heat-engine cycles considered in this chapter have clockwise paths. What would happen if the same paths were traversed in the counterclockwise sense?

19.7 *(4) Where it's at.* As a practical matter, only a small fraction of the heat capacity of a heat engine resides in the working fluid. Why, then, is it possible (at least to first ap-

proximation) to ignore the heat capacity of the parts of the engine other than the working fluid?

19.8 *(4) Nothing to worry about.* Why is $c'_p - c'_v$ equal to zero for an incompressible fluid?

19.9 *(4) The medium is the message.* A quantity of helium undergoes a thermodynamic process in which its molar heat capacity has the constant value $\frac{5}{2}R$. Which kind of process is it?

19.10 *(4) Keeping cool under pressure?* Why does the latent heat of boiling of a liquid depend on the atmospheric pressure?

19.11 *(5) Expert's slope.* Give a qualitative, physical explanation for the fact that an adiabatic curve on a p-V diagram is steeper than an isotherm on the same diagram. (Hint: See Figure 19.9 and Equation 19.29b.)

19.12 *(5) There are adiabatics, and then there are adiabatics.* In the figure below, one of the curves is an adiabatic for helium and the other an adiabatic for carbon dioxide. Which is which? Explain your answer.

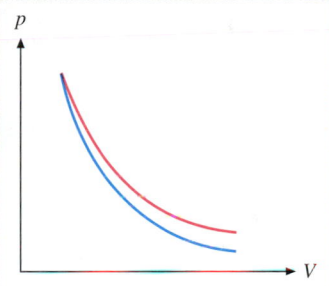

19.13 *(6) Slow and easy.* Why are Equations 19.38a and 19.38b valid only for quasistatic processes?

19.14 *(6) Similarity in diversity.* Why are the SI units of entropy the same as those of total heat capacity?

19.15 *(6) Simplicity out of complexity.* On a p-V diagram, the shape of the Carnot cycle depends on the equation of state of the working fluid. But on a T-S diagram, the Carnot cycle is a rectangle for any working fluid. Explain.

19.16 *(6) The road taken.* A quantity of ideal gas is made to pass from state a to state b by two different processes. In process 1, an isometric is followed by an isobar; in process 2, an isobar is followed by an isometric. **(a)** In which process is $|W|$ greater? **(b)** In which process is $|Q|$ greater? **(c)** In which process is $|\Delta S|$ greater? Explain your answers.

19.17 *(G) What's the difference?* Beginning in the state specified by p_a and T_a, an ideal gas is made to expand (1) isothermally and (2) adiabatically. In both cases, the final volume is V_b. **(a)** In which process is the final pressure greater? **(b)** In which process is $|W|$ greater?

19.18 *(G) Keep on target!* Sketch a p-V diagram for the thermodynamic process that takes place when a bullet is fired from a gun. Begin at the moment when the explosive propellant charge is detonated, and follow the process through the rapid heating and pressurization of the chamber behind the bullet, the change that occurs as the bullet moves down the barrel, and what takes place when the bullet leaves the barrel.

19.19 *(G) Gamut.* Each of the four curves in the figure below represents the same sample of gas going through one of the processes listed in Table 19.3. Which curve describes which process?

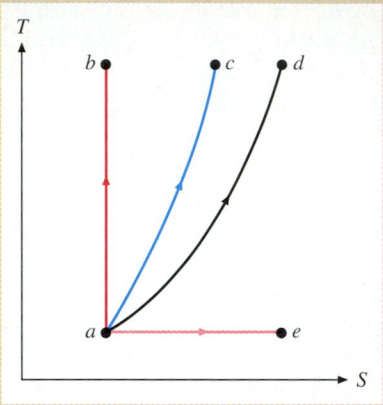

19.20 *(G) Another view.* In the figure below, three processes are described by straight lines of different slopes on a graph of $|W|$ versus $|\Delta T|$. One of the processes is isobaric; the other two are adiabatic, with argon (Ar) the working fluid for one and nitrogen (N_2) the working fluid for the other. **(a)** Assign each straight line to the proper process. **(b)** What would an isotherm look like on this plot? an isometric?

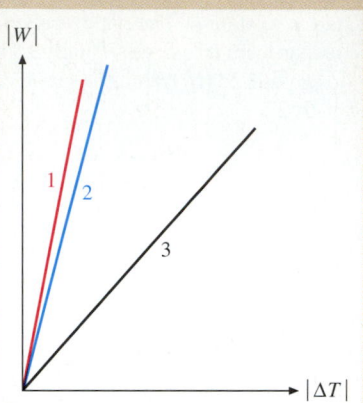

PROBLEMS

GROUP A

19.1 *(2) Be explicit!* For the system of Example 19.1, find the numerical value of n.

19.2 *(2) Reverse.* Equation 19.6 expresses the work done on a system containing an ideal gas in an isothermal process $a{\rightarrow}b$. What is the work done on the system in the isothermal process $b{\rightarrow}a$?

19.3 *(2) Letting go.* How much work is done by 1 kmol of an ideal gas in an isothermal process in which the pressure of the gas changes from 3 atm to 1.5 atm at room temperature?

19.4 *(2) Isotherm.* A cylinder-piston system containing an ideal gas expands isothermally from initial volume 0.865×10^{-3} m^3 to final volume 4.92×10^{-3} m^3. **(a)** If the initial pressure is 7.94×10^5 Pa, which is the final pressure? **(b)** What is the change ΔE in the internal energy of the working fluid? **(c)** What is the value of the work W done on the system? **(d)** What is the heat flow Q into the system?

19.5 *(2) Beginner's slope.* An isotherm of an ideal gas on a p-V diagram passes through the point (V_0, p_0). Show that the slope of the isotherm at that point is

$$\frac{dp}{dV} = -\frac{p_0}{V_0}.$$

19.6 *(3) Square deal.* A hypothetical heat engine passes through the cycle shown in the next column. **(a)** Over

which part of the cycle is work done by the system on its surroundings? What is the magnitude $|W_{out}|$ of that work? **(b)** Over which part of the cycle is work done on the system by its surroundings? What is the magnitude $|W_{in}|$ of that work? **(c)** What is the net heat input Q_{cyc} to the system over the cycle?

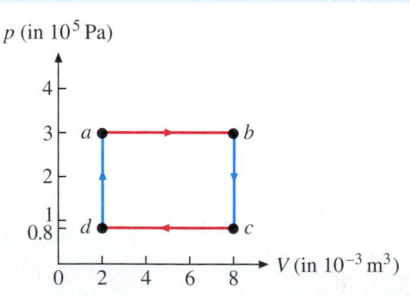

19.7 *(3) Triangular passage, I.* A hypothetical heat engine passes through the cycle $a{\rightarrow}c{\rightarrow}d{\rightarrow}a$ shown in blue on the next page. **(a)** Over which part of the cycle is work done by the system on its surroundings? What is the magnitude $|W_{out}|$ of that work? **(b)** Over which part of the cycle is work done on the system by its surroundings? What is the magnitude $|W_{in}|$ of that work? **(c)** What is the net heat input Q_{cyc} to the system over the cycle?

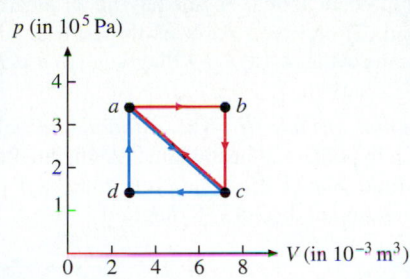

p (in 10⁵ Pa)

V (in 10⁻³ m³)

19.8 *(3) Triangular passage, II.* A hypothetical heat engine passes through the cycle $a{\to}b{\to}c{\to}a$ shown in red in Problem 19.7. **(a)** Over which part of the cycle is work done by the system on its surroundings? What is the magnitude $|W_{out}|$ of that work? **(b)** Over which part of the cycle is work done on the system by its surroundings? What is the magnitude $|W_{in}|$ of that work? **(c)** What is the net heat input Q_{cyc} to the system over the cycle? **(d)** Compare the result of part **c** with the result of part **c** of Problem 19.7. Explain why the two results are the same or why they are different.

19.9 *(3) Point of view, I.* The *V-T* diagram below is for the cycle of a hypothetical heat engine having an ideal gas for its working fluid. Sketch the same cycle on **(a)** a *p-V* diagram and **(b)** a *p-T* diagram.

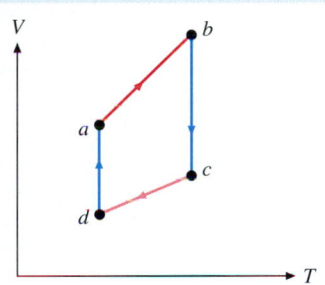

19.10 *(4) Small difference.* Using information from Table 17.1, find $c_p' - c_v'$ for mercury and for iron, and confirm that it is not usually necessary to make the distinction between the two heat capacities for liquids and solids.

19.11 *(4) Soaking up heat.* What is the molar heat capacity at constant pressure of helium gas (He)? of oxygen gas (O_2)?

19.12 *(4) Unfolding the ratio.* For a certain ideal gas, $\gamma = 1.4$. Find the values of c_v' and c_p'.

19.13 *(5) Pressure relief.* An ideal gas undergoes an adiabatic compression. Its final temperature is twice its initial temperature. What is the ratio of the final pressure to the initial pressure if the gas is **(a)** monatomic? **(b)** diatomic?

19.14 *(5) The squeeze is on.* A cylinder contains 2.55 kmol of oxygen at 1.00 atm and 20°C. If the gas is compressed adiabatically to 8.67 atm, what are the final volume and temperature?

19.15 *(5) The easy way.* Without calculation, find Q_{cyc}, the heat input over one cycle, for the Carnot engine considered in Example 19.3.

19.16 *(5) Filling in the gaps.* In Example 19.3, find the values of n, p_b, p_d, V_b, and V_d.

19.17 *(5) Dieseling.* The *compression ratio r* of an internal combustion engine is defined as the ratio of the maximum cylinder volume to the minimum cylinder volume—that is, the ratio of the volumes with the piston all the way out and all the way in. The compression ratio of a typical Diesel engine is $r = 15$. If air is admitted to the cylinder at temperature 300 K and then compressed adiabatically, what is the final temperature in the cylinder? Assume that air is a diatomic ideal gas.

19.18 *(5) Secrets of the molecule.* An unknown (but pure) gas is allowed to expand adiabatically and quasistatically until its volume has doubled. The temperature of the gas falls from 300 K to 227 K. Is the gas monatomic, diatomic, or polyatomic?

19.19 *(5) Temperature and volume, adiabatically.* Beginning with Equation 19.29a and following the method sketched in the text, derive Equation 19.30a.

19.20 *(5) Pressure and temperature, adiabatically.* Beginning with Equation 19.29a and following the method sketched in the text, derive Equation 19.31a.

19.21 *(6) Eliminating the temperature, I.* For an isobaric process involving an ideal gas, express the entropy change ΔS in terms of n, c_p', and the initial and final pressures and volumes only.

19.22 *(6) Eliminating the temperature, II.* For an isometric process involving an ideal gas, express the entropy change ΔS in terms of n, c_v', and the initial and final pressures and volumes only.

19.23 *(6) Entropy of boiling.* What is the entropy change when 1 kg of water at 100°C and 1 atm pressure changes to steam?

19.24 *(6) Entropy of freezing.* What is the entropy change when 1 kg of water at 0°C and 1 atm pressure changes to ice?

19.25 *(6) Entropy of mixing, I.* For 1 kg of liquid water at 100°C mixed with 1 kg of liquid water at 0°C, find **(a)** the final temperature of the mixture, **(b)** the entropy change of the hot water as it cools, and **(c)** the entropy change of the cold water as it warms. **(d)** What is the entropy change of the entire system during the mixing process? You will find necessary information in Table 17.3.

19.26 *(G) Eclectic heat-engine cycle.* A hypothetical heat engine using an ideal gas as a working fluid passes through the cycle shown below. Path $a{\to}b$ is isobaric, $b{\to}c$ is adiabatic,

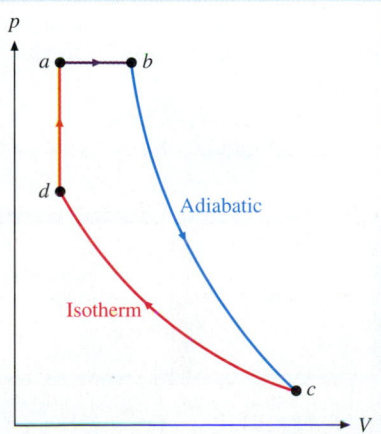

$c \rightarrow d$ is isothermal, and $d \rightarrow a$ is isometric. Fill in the table with the notations <0, 0, or >0 to indicate the signs of the seven quantities along each path.

process	W	Q	ΔE	Δp	ΔV	ΔT	ΔS
$a \rightarrow b$							
$b \rightarrow c$							
$c \rightarrow d$							
$d \rightarrow a$							

19.27 *(G) Point of view, II.* The p-V diagram below is for the

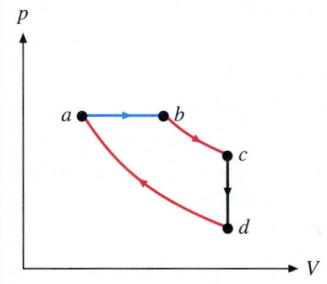

cycle of a hypothetical heat engine having an ideal gas for its working fluid. The curved parts of the path are isotherms. Sketch the same cycle on **(a)** a p-T diagram, **(b)** a V-T diagram, and **(c)** a T-S diagram.

19.28 *(G) Point of view, III.* The p-T diagram below is for the cycle of a hypothetical heat engine having an ideal gas for its working fluid. Sketch the same cycle on **(a)** a p-V diagram, **(b)** a V-T diagram, and **(c)** a T-S diagram.

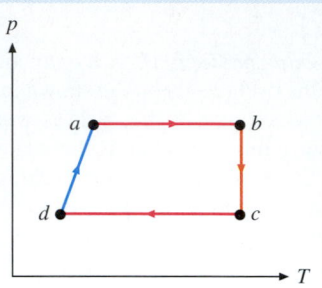

GROUP B

19.29 *(2) Thermodynamics of boiling.* A cylinder fitted with a frictionless piston of negligible mass is filled with 1 kg of water whose temperature is just at the boiling point. The outer surface of the piston is in contact with the atmosphere, assuring that the pressure inside the cylinder is always 1 atm. The cylinder is heated until all the water has just boiled, yielding steam at the boiling point. The cylinder volume is then 1.67 m³. **(a)** How much work must be done on the atmosphere as the water expands in turning to steam? (Hint: Why can you neglect the volume of the liquid water?) **(b)** How much heat energy must flow into the cylinder in the boiling process? **(c)** What is the change of internal energy as the water turns into steam? **(d)** What is the change of the total translational kinetic energy of the molecules? **(e)** What is the change of the potential energy of the molecules? **(f)** Suppose that 1 kg of water is boiled in open air, instead of in a cylinder. Would this make any difference in the answers to parts **a** through **e**? Explain your answer.

19.30 *(2) Easy work, I.* In the diagram below, the piston is locked in place at the middle of the cylindrical chamber. To

the left of the piston is n kmol of an ideal gas at pressure p_0. The space to the right of the piston is evacuated. The cylinder walls conduct heat readily, and the temperature of the gas is always equal to the external temperature T. The piston is now allowed to move slowly until it reaches the right end of the chamber. **(a)** What is the change ΔE of the internal energy of the gas? (Hint: Refer to the discussion of the energy of an ideal gas in Section 18.2.) **(b)** Find the work W done on the gas. **(c)** What is the heat flow into the gas through the cylinder walls?

19.31 *(2) Joule free expansion.* In the figure below, a thermally insulated container is divided into two equal volumes by a partition made of a thin sheet of a material such as Saran Wrap. To the left of the partition is confined n kmol of an

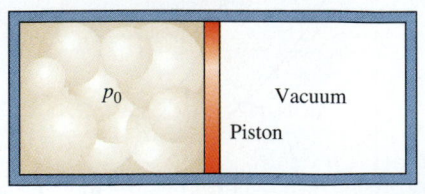

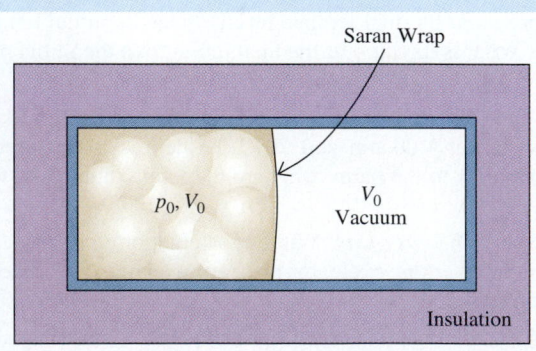

ideal gas at pressure p; the space to the right is evacuated. The partition is punctured, and the gas suddenly expands to fill the entire chamber. When Joule performed an experiment of this type, he found no temperature change. **(a)** Is the expansion quasistatic? Can you describe the process by means of a curve on a p-V diagram? Can you describe the initial and final states of the system by means of points on a p-V diagram? **(b)** Find W, Q, and ΔE. **(c)** Is Joule's result consistent with the discussion of the energy of an ideal gas in Section 18.2? Explain your answer. **(d)** If the experiment were repeated with a substance other than an ideal gas, would ΔT be zero? Explain your answer. (Hint: See Problem 19.30.)

19.32 *(2) Under pressure.* A cylinder-piston system contains n kmol of an ideal gas in a volume V_1. The piston is moved slowly, compressing the gas isothermally into a smaller volume V_2. **(a)** How much work is done on the gas? **(b)** What is the change ΔE of the internal energy of the gas? **(c)** How much heat flows into the gas?

19.33 *(3) Checking out the math.* In the discussion following Equation 19.12, we assert that W_{cyc} is negative for the heat-engine cycle of Figure 19.5. Prove the inequality

$$p_b V_b - p_b V_a + p_b V_b \ln \frac{V_a}{V_b} < 0 \quad \text{if } V_a > V_b.$$

19.34 *(3) The crooked made straight.* The heat-engine cycle shown below passes through the same points a, b, and c as the cycle of Figure 19.5. Now, however, the isotherm is replaced by a straight-line path $a{\to}b$. **(a)** What is the work W_{cyc} done on the system over one cycle? Compare your result with that obtained in Example 19.2. Would you expect the two results to be the same? **(b)** What is the heat input to the system, Q_{cyc}, over one cycle?

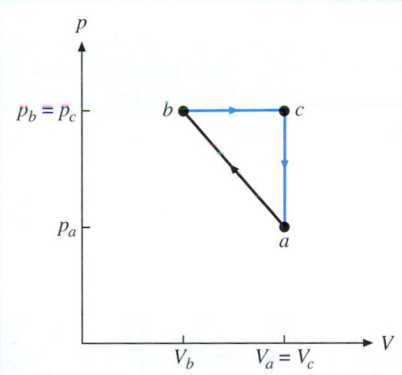

19.35 *(4) Circular reasoning?* A hypothetical heat engine goes through the cycle shown at the top of the next column. **(a)** Find W_{cyc}. **(b)** Does your result depend on the fact that the choice of scale is such that the path is circular? (Hint: Suppose the vertical scale were expanded by a factor of two. How would you calculate W_{cyc}?)

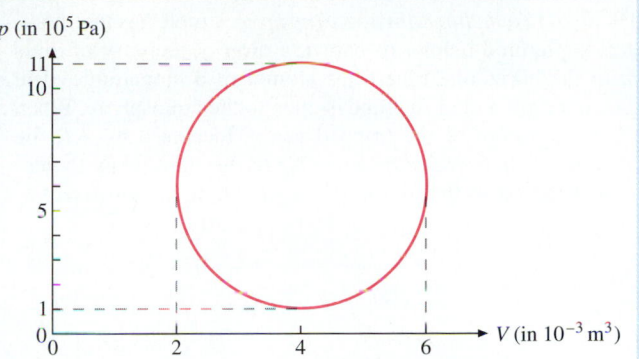

19.36 *(4) Same heat, different temperature change.* A cylinder-piston system contains a certain molar quantity of a monatomic ideal gas, and a second, identical cylinder-piston system contains the same molar quantity of a diatomic ideal gas. Both systems are at the same pressure p and temperature T_0. The same amount of heat Q is made to flow into each system, and each expands quasistatically against a constant external load—that is, at constant pressure. **(a)** If the temperature of the monatomic system increases by $T_1 - T_0 = 10$ K, what is the temperature increase $T_2 - T_0$ of the diatomic system? **(b)** What is the difference $W_2 - W_1$ between the work done on the diatomic system and the work done on the monatomic system?

19.37 *(4) Peculiar gas.* The internal energy of a hypothetical (nonideal) gas is given by the expression

$$E = a \ln \frac{T}{T_0} + b \ln \frac{p}{p_0},$$

where $a = 3.00 \times 10^3$ J, $b = 7.00 \times 10^3$ J, $T_0 = 200$ K, and $p_0 = 1.00 \times 10^4$ Pa. The gas is heated and expands isobarically. The pressure has the constant value $p = 10.0 \times 10^4$ Pa, and the volume change is $\Delta V = 5.00 \times 10^{-4}$ m^3. If the process requires a heat input $Q = 500$ J, what is the temperature change if the process begins at T_0?

19.38 *(4) Push comes to shove.* The two cylinder-piston systems illustrated below are linked. Cylinder 1 is filled with a certain molar quantity of a monatomic ideal gas, and cylinder 2 is filled with an equal molar quantity of a diatomic ideal gas. The entire apparatus is situated inside an oven whose temperature is T_a. The cylinder volumes have the same initial value V_0. When the oven temperature is raised to T_b, what is the volume change ΔV of cylinder 1? (Hint: Be careful; this one's just a little tricky!)

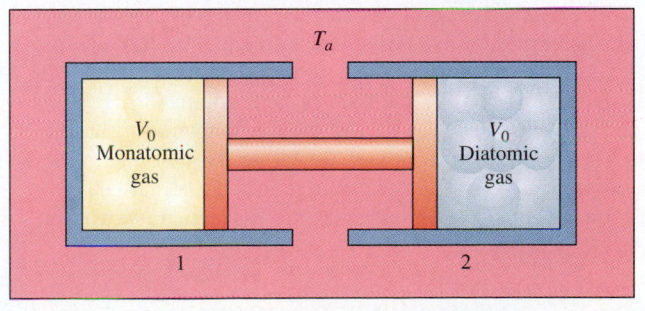

19.39 *(4) Something always turns up.* A simple gas thermometer is pictured below. In part *a*, a drop of mercury of mass *m* in the horizontal tube traps *n* kmol of a monatomic ideal gas; the right end of the tube is open to the atmosphere. When the temperature of the trapped gas is increased by ΔT, the mercury drop moves a distance Δx to the right. Now the device is turned so that the tube is vertical, as shown in part *b*.

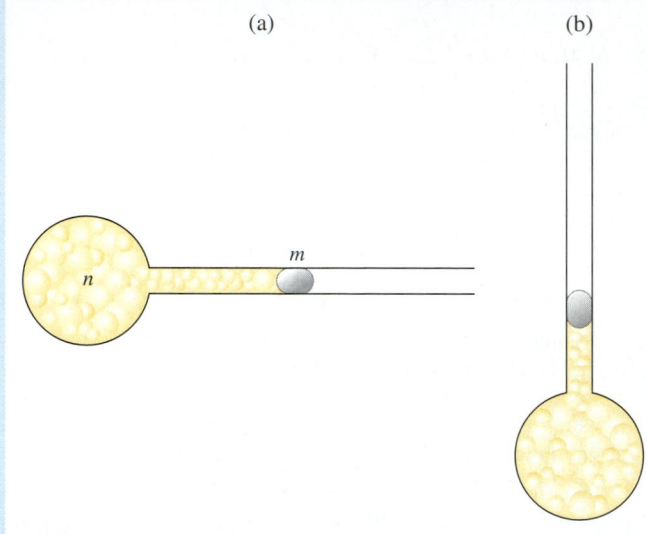

To move the mercury drop upward through the same Δx, the temperature of the trapped gas must be increased by $\Delta T'$. Show that

$$\Delta T' = \Delta T + \frac{2mg}{5nR}\,\Delta x.$$

Note that this result is independent of the atmospheric pressure, the trapped volume, and the diameter of the tube.

19.40 *(5) Easy work, II.* The system of Problem 19.30 is wrapped in ideal insulation, so that no heat energy can flow in or out, and the system is allowed to expand just as in Problem 19.30. **(a)** Find the work done on the gas in the process. **(b)** What is the change ΔE of the internal energy of the gas? **(c)** If the gas has *f* degrees of freedom, show that its temperature changes according to the relation

$$\Delta T = \frac{p_0 V_0}{nR}\,(2^{1-\gamma} - 1).$$

(d) The expansion is carried out with equal molar quantities of (1) a monatomic ideal gas and (2) a diatomic ideal gas, using the same initial pressure p_0 and volume V_0. Find the ratio $\Delta T_{\text{monatomic}}/\Delta T_{\text{diatomic}}$ of the two temperature changes.

19.41 *(5) Unexpected bonus.* **(a)** Using the first law of thermodynamics, show that the change in the internal energy of an ideal gas undergoing an adiabatic process can be written

$$\Delta E = W = nc'_v\,\Delta T.$$

(Hint: An adiabatic process for which the working fluid is an ideal gas is certainly not isometric. To justify the use of Equation 19.16 in this process, think about how the internal energy of an ideal gas depends on the volume.) **(b)** Is the result of part **a** true for *any* working fluid whose molar heat capacity at constant volume is c'_v?

19.42 *(5) Crude measurement of* γ. A cylinder-piston system like that of Figure 19.6 contains an ideal gas for which γ is to be determined. The weight of the piston is such that the gas pressure is 1 atm when the temperature is 273 K. The area of the piston is 1.0×10^{-3} m², and the distance from the cylinder head to the piston is 0.250 m. A 20-kg mass is placed on top of the piston, and the piston descends until its distance from the bottom of the cylinder is 0.116 m. The process is fast enough so that it can be considered adiabatic. **(a)** Find the value of γ for the gas. **(b)** Why is this not a very accurate way to determine γ?

19.43 *(5) Intermediate slope.* A cylinder-piston system has volume V_0 and contains an ideal gas at pressure p_0. The system is compressed or expanded adiabatically. Show that the curve on a *p-V* diagram that describes the process has slope

$$\frac{dp}{dV} = -\gamma p_0 V_0^{\gamma} V^{-\gamma-1}.$$

19.44 *(6) Unusual substance.* When a sample of a certain substance is heated, the entropy increases linearly with temperature: $\Delta S = a\,\Delta T$, where *a* is a constant. Express the total heat capacity *C* of the sample as a function of temperature.

19.45 *(6) Entropy of mixing, II.* A quantity of a certain substance is at temperature T_1. An equal quantity of the same substance is at a lower temperature T_2. The two are brought into thermal contact. Find **(a)** the final temperature of the mixture, **(b)** the entropy change of the warmer sample as it cools, and **(c)** the entropy change of the cooler sample as it warms. **(d)** What is the entropy change of the entire system during the mixing process?

19.46 *(6) All's well that ends.* In the figure below, a sample of ideal gas passes from state *a* to state *b* by the direct process $a \rightarrow b$. Alternatively, the sample passes from *a* to *b* by the process $a \rightarrow c \rightarrow d \rightarrow b$. Show that the change of entropy of the sample is the same both ways.

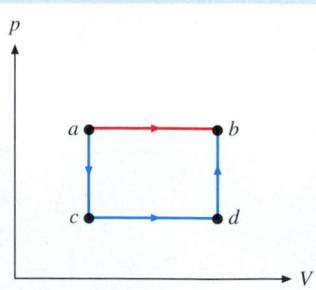

19.47 *(6) No end in sight.* The rectangular path $a \rightarrow b \rightarrow d \rightarrow c \rightarrow a$ illustrated in Problem 19.6 represents the cycle of a hypothetical engine. Find ΔS_{cyc} for the engine cycle.

19.48 *(6) Feed-pump thermodynamics.* The steam boilers used to drive turbines operate at quite high pressures. It is therefore necessary to force the feed water into the boiler by pressurizing it with a *feed pump*. The water is not heated significantly in its passage through the pump and, because liquid water is negligibly compressible, the volume of the water does not change either. Sketch the adiabatic path describing the compression of the feed water **(a)** on a *p-V* diagram and **(b)** on a *T-S* diagram.

19.49 *(6) Carnot engine, I.* The *p-V* diagram below is of a monatomic ideal-gas Carnot cycle. **(a)** Complete the diagram

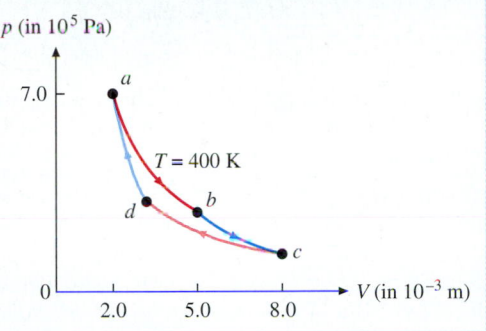

by calculating p_b, p_c, p_d, and V_d. **(b)** Find the work W along each of the four paths. **(c)** Find the net work W_{cyc}. **(d)** Find the heat input Q along each path. **(e)** Calculate the entropy change ΔS along each path. **(f)** Sketch the *T-S* diagram corresponding to your completed *p-V* diagram.

19.50 *(6) Carnot engine, II.* A Carnot engine delivers 10 kW of power at 150 rev/min. If the engine operates between the temperatures $T_{hi} = 450$ K and $T_{lo} = 275$ K, sketch the *T-S* diagram of the engine cycle.

19.51 *(G) Isotherm for boiling water.* A small quantity of liquid water whose temperature is just below the boiling point is put into a constant-load cylinder-piston system like that of Figure 19.6. Heat is made to flow into the system, and the water gradually turns to steam. Sketch the isothermal path followed by the system **(a)** on a *p-V* diagram and **(b)** on a *T-S* diagram.

19.52 *(G) Three ways.* In three separate processes, the temperature of a quantity of ideal gas is changed by the same amount ΔT. One process is isobaric, one is isometric, and one is adiabatic. For each process, express the increase in internal energy, ΔE, in terms of n, R, γ, and ΔT only.

GROUP C

19.53 *(5) Advanced slope, I.* An isotherm and an adiabatic for an ideal gas pass through the same point on a *p-V* diagram. At the common point of the two curves, find the ratio of their slopes,

$$\frac{(dp/dV)_{\text{adiabatic}}}{(dp/dV)_{\text{isothermal}}}.$$

19.54 *(5) Advanced slope, II.* Two cylinder-piston systems contain a monatomic ideal gas and a diatomic ideal gas, respectively. The initial volumes V_0 are equal, as are the initial pressures p_0. As the systems are compressed or expanded, their states will lie on two adiabatics that cross at (V_0, p_0). Show that, at this crossing point, the slopes of the two adiabatics have the ratio

$$\frac{(dp/dV)_{\text{monatomic}}}{(dp/dV)_{\text{diatomic}}} = \frac{25}{21}.$$

19.55 *(5) Adiabatic atmosphere.* The barometric equation (Equation 18.51b),

$$p(h) = p_0 e^{-(mg/kT)h},$$

gives the pressure of an isothermal atmosphere as a function of altitude. But the assumption that the atmosphere is isothermal is not very realistic. Now we will be more realistic. Imagine a mass of air that rises—perhaps because it has been slightly warmed. As this air mass rises into a region of lower pressure, it expands. In expanding, it does work against the surrounding air. Because the air mass does not exchange heat with its surroundings very quickly, the process is essentially adiabatic: As the air mass does work, its internal energy decreases and thus its temperature decreases as well. **(a)** Show

that the pressure is given as a function of altitude h by the expression

$$p = p_0 \left(1 - \frac{\gamma - 1}{\gamma} \frac{\rho_0}{p_0} gh \right)^{\gamma/(\gamma - 1)},$$

where ρ_0 is the sea-level density and p_0 is the sea-level pressure. (Hint: Use Equation 18.47 to express ρ as a function of V, and then apply the adiabatic equation for an ideal gas.) **(b)** Using the result of part a, show that the adiabatic atmosphere has a finite height

$$H = \frac{\gamma}{\gamma - 1} \frac{p_0}{\rho_0 g}.$$

(c) Find the value of H, assuming $\gamma = \frac{7}{5}$ for air (a fairly good approximation because air is nearly all nitrogen and oxygen), $p_0 = 1.013 \times 10^5$ Pa, and $\rho_0 = 1.29$ kg/m³. **(d)** Show that the density ρ is given as a function of altitude by the expression

$$\rho = \rho_0 \left(1 - \frac{h}{H} \right)^{1/(\gamma - 1)}.$$

(e) Show that the temperature of the adiabatic atmosphere is given as a function of altitude by the expression

$$T = \frac{\mu p_0}{k \rho_0} \left(1 - \frac{h}{H} \right),$$

where $\mu = 29$ u is the mean molecular mass of air and k is Boltzmann's constant. What is the temperature at $h = H$? **(f)** Calculate the *temperature gradient, dT/dh*, and show that it is about -1 K per 100 m.

19.56 *(5) Rüchardt's method (1929).* Depicted below is a simple apparatus that can be used to determine the value of γ for a gas in a particularly elegant way. A large bottle is fitted at its neck with a one-hole stopper through which passes a precision-bore glass tube. A steel ball bearing of mass m, very precisely spherical with radius r, just fits inside the tube. The bottle is filled with the gas for which γ is to be evaluated.

Sidearm

Replacement gas in

When the ball is in the tube, the gas leaking through the narrow space between the ball and the inside wall of the tube forms a gas cushion that keeps the ball from touching the wall. Thus there is very little friction between the ball and the tube wall. The leakage of gas past the ball and out of the system is compensated for by means of a very slow flow of gas into the bottle through the side arm. If the ball were lowered gently into the tube, it would compress the gas to an equilibrium pressure p_0, at which the gas just supports the ball. Give the name V_0 to the volume of gas trapped in the bottle and tube by the ball under these circumstances.

In the actual experiment, the ball is dropped into the tube and oscillates about its equilibrium position with frequency ν, which is measured. The aim of this problem is to show that

$$\gamma = \frac{4\nu^2 m V_0}{p_0 r^4}. \tag{1}$$

(a) The frequency is typically of order 1 Hz. So quickly and so slightly does the temperature of the trapped gas rise and fall as it is compressed and expands that the process is quite accurately adiabatic. Show that the force exerted by the trapped gas on the ball is

$$F = \frac{p_0 V_0^\gamma}{V^\gamma} \, \pi r^2.$$

(b) To prove that the oscillation of the ball is simple harmonic, you must show that F depends in linear fashion on the displacement y of the ball from its equilibrium position. To begin with, show that

$$\frac{dF}{dy} = \pi r^2 \frac{dF}{dV} = (\pi r^2)^2 (p_0 V_0^\gamma)\left(-\frac{\gamma}{V^{\gamma+1}}\right).$$

(c) Because the bottle is very big, the proportional change in total volume V as the ball oscillates in the tube is small. So in the equation of part **b**, you can make the approximation $V \simeq V_0$. Show that this gives you a force constant

$$\kappa \equiv -\frac{dF}{dy} = \frac{(\pi r^2)^2 p_0 \gamma}{V_0}.$$

(d) Use the standard relation between the mass m, the force constant κ, and the frequency ν of a harmonic oscillator to obtain Equation 1.

19.57 *(5) Adiabatic spring.* In Problem 15.45, you studied the oscillation of a column of liquid in a glass U-tube closed at one end. As the liquid oscillates, the trapped air is compressed as the liquid level rises in the closed arm and expands as the liquid level falls. But in Problem 15.45 isothermal conditions were assumed. This is unrealistic for the fairly rapid oscillations to be expected, and an accurate result requires the use of adiabatic conditions. **(a)** With the two liquid surfaces at the same level, the trapped air is at atmospheric pressure p_0. Show that, when the liquid column is made to oscillate, the oscillation is not simple harmonic because

$$\frac{dF}{dy} = -2\rho g A - \frac{\gamma p_0 A D^\gamma}{(D-y)^{\gamma+1}} \neq \text{constant}.$$

In this equation, y is the displacement from equilibrium position of the liquid surface in the closed arm, ρ is the density of the liquid, g is the acceleration of gravity, A is the cross-sectional area of the tube, and D is the distance from the equilibrium level to the closed tube end. Compare this result with the result of the isothermal calculation in part **a** of Problem 15.45,

$$\frac{dF}{dy} = -2\rho g A - \frac{p_0 A D}{(D-y)^2}.$$

(b) Consider *small* oscillations of the liquid column, for which $y \ll D$. Show that dF/dy has an approximately constant value and that the oscillation is thus approximately simple harmonic. If τ_0 is the oscillation period of the same liquid column in a U-tube open at both ends (Problem 15.44), show that the period for small oscillations in the closed tube is

$$\tau_{\text{adiabatic}} = \tau_0\left(1 - \frac{\gamma p_0}{4\rho g D}\right).$$

Compare this result with the isothermal result

$$\tau_{\text{isothermal}} = \tau_0\left(1 - \frac{p_0}{4\rho g D}\right).$$

(c) Suppose that the liquid is mercury ($\rho = 13.6 \times 10^3$ kg/m^3) and that $D = 2.0$ m. Calculate τ_0, $\tau_{\text{isothermal}}$, and $\tau_{\text{adiabatic}}$. Take γ for air to be 1.40. How important are the corrections?

19.58 *(5) Otto cycle.* The *p-V* diagram below is of the *Otto cycle*, which is an idealization of the operation of the common gasoline engine. Paths $a{\rightarrow}b$ and $c{\rightarrow}d$ are adiabatics; paths $b{\rightarrow}c$ and $a{\rightarrow}d$ are isometrics. The process $c{\rightarrow}d$ corresponds to the compression stroke of the piston, during which the gasoline-air mixture is compressed. The process $d{\rightarrow}a$ corresponds to the burning initiated by the spark plug, which is so rapid that (to first approximation) the piston does not move. [This idealization is not too realistic, because (1) the rapid burning is not quasistatic and (2) the value of n is changed by the chemical reactions taking place.] The process $a{\rightarrow}b$ is the power stroke, during which the expanding hot gas pushes the piston out of the cylinder, driving the crankshaft. The process $b{\rightarrow}c$ represents the rapid decompression of the gas in the cylinder when the exhaust valve is opened (somewhat unrealistically, because n, the quantity of gas in the cylinder, does not remain constant as the exhaust gas rushes out). **(a)** Show that $T_b/T_a = T_c/T_d$. **(b)** Use the result in part **a** of Problem 19.41 to show that

$$W_{\text{cyc}} = -nc_v'[(T_a - T_d) - (T_b - T_c)].$$

(c) Define the compression ratio $r \equiv V_b/V_a$, and show that

$$W_{\text{cyc}} = -nc_v'(T_a - T_d)(1 - r^{1-\gamma}).$$

(d) Show that this result can be written in the form

$$W_{\text{cyc}} = \frac{1}{\gamma - 1}\left(\frac{p_d}{p_a} - 1\right) p_a V_a(1 - r^{1-\gamma}).$$

19.59 *(5) Rankine cycle.* The *p-V* diagram below is of the ideal-gas *Rankine cycle*, which is an idealization of the operation of steam turbine systems. Paths $a{\rightarrow}b$ and $c{\rightarrow}d$ are isobars; paths $b{\rightarrow}c$ and $d{\rightarrow}a$ are adiabatics. Assume that the working fluid is an ideal gas. (This is not realistic for steam turbine systems, because the steam is condensed into water for part of the cycle.) **(a)** Show that $V_b/V_a = V_c/V_d$, just as in the Carnot cycle. **(b)** Show that $T_b/T_a = T_c/T_d$, just as in the Otto cycle of Problem 19.58. **(c)** Calculate W_{cyc}, and show that it is given by the expression

$$W_{\text{cyc}} = \frac{\gamma}{\gamma - 1}\left(\frac{V_a}{V_b} - 1\right)(p_b V_b - p_c V_c).$$

(d) Show that the result of part **c** can be written in the simpler form

$$W_{\text{cyc}} = nc_p'\left(\frac{V_a}{V_b} - 1\right)(T_b - T_c).$$

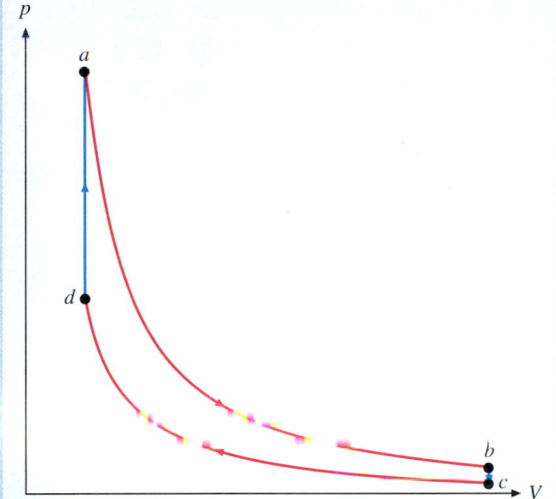

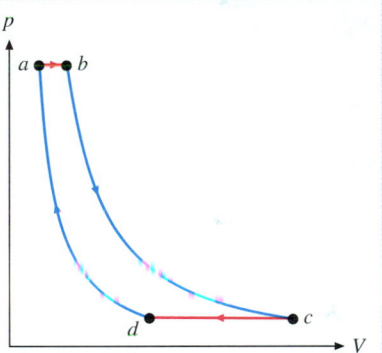

Thermodynamics II
The Second Law

— The thermodynamic efficiency of a heat engine is defined as (work output per cycle)/(heat input per cycle).

— For a Carnot engine, this efficiency is completely determined by the input and output temperatures.

— According to the second law of thermodynamics, heat cannot flow spontaneously from a cooler body to a warmer one.

— A consequence of the second law is Carnot's theorem: No heat engine can be more efficient than a Carnot engine operating between the same two heat reservoirs.

— A closed system that is not in thermodynamic equilibrium evolves toward equilibrium because the equilibrium state is the state of greatest randomness and hence of highest probability.

— Entropy is a measure of the randomness of the state of a system.

Left: Can you obtain this sequence by shuffling?

> *"Take some more tea,"* the March Hare said to
>   Alice, very earnestly.
> *"I've had nothing yet,"* Alice replied in an offended
>   tone: *"so I can't take more."*
> *"You mean you can't take* less," *said the Hatter:*
>   *"it's very easy to take* more *than nothing."*
>
> —Lewis Carroll, *Alice's Adventures in Wonderland*

> *I love to gamble so much, I'd be happy to break even.*
> *But I never seem to do as good as that.*
>
> —Newspaper interview with a gambler

SECTION 20.1

Introduction

The first law of thermodynamics places restrictions on what a heat engine can do. These restrictions all amount to a prohibition against creating energy from nothing. For example, a heat engine cannot produce mechanical work $-W$ without taking in *at least* an equal amount of heat Q. A hypothetical engine that violates this rule is called a **perpetual-motion machine of the first kind** because it violates the first law. Needless to say, no one has ever invented such a machine and no one ever will, though some people keep trying.

But it is not hard to conceive of a heat engine that, though it conforms to the first law, still violates our commonsense notions of what can and cannot be done. There is plenty of energy in the air around us; you can easily calculate that the thermal kinetic energy of the molecules in a cubic meter of air at 0°C totals more than 250 000 J. There would be no violation of the first law in using a heat engine to extract this energy and convert it into useful mechanical work. Why can't we design a jet engine that admits air at the front, extracts some of the heat energy from the air, and converts this energy into work by forcefully ejecting the same air out the tailpipe at a lower temperature, thus propelling the airplane? Such an engine would consume no fuel at all!

No one has ever done this. In spite of the plentiful supply of heat energy in our surroundings, we still have to pay for fuel to run heat engines. You may well suspect that there lurks (in the shadows up to this point) another law of thermodynamics that forbids the kind of engine just described. There is such a law, called the *second law of thermodynamics*. In this chapter, we consider this law in the context of the heat engines discussed in Chapter 19. You will see how the law guarantees that no one ever will build the wonderful fuelless jet engine. But the law has much wider applications as well, and we touch on some of them in Section 20.8.

As an introduction to the second law of thermodynamics, we will determine the efficiency of the Carnot engine in Section 20.2. Section 20.3 is concerned with the second law proper. In Section 20.4, we will study *refrigerators*, as heat engines run in reverse are called. In Section 20.5, we derive a simple and beautiful theorem, due to Carnot, which shows that no heat engine can be more efficient than a Carnot engine. Section 20.6 introduces the third law of thermodynamics.

Natural Laws and Court Injunctions

By the 1880s, the U.S. Patent Office was wasting so much time examining patent applications for perpetual-motion machines that it has required a *working* model with every such application ever since. Nevertheless, eager inventors have sometimes obtained court orders forcing the government to spend time and money reviewing their inventions. In 1985–86, for instance, the National Bureau of Standards (now the National Institute of Standards and Technology) spent $75,000 to show that a "perpetual-motion machine" submitted in 1981 had an optimal efficiency of only 67%; similar commercial devices routinely have efficiencies above 90% (but not above 100%!). See "Report of Tests on Joseph Newman's Device," U.S. Department of Commerce, National Bureau of Standards, NBSIR 866-3405, June 1986; also *Science* **233**, 154 (1986).

The concept of entropy, introduced in Section 19.6, acquires further significance in the context of the second law of thermodynamics, and we consider this significance in detail throughout the chapter. In Section 20.7, we discuss the implications of the second law and the meaning of entropy in the broader context of *statistical mechanics*.

Efficiency

In almost every human activity, one is concerned about return on investment. Factories are concerned with productivity, investors with interest rates and bond yields, and automobile drivers with gas mileage. All such quantities are special cases of **efficiency**, defined broadly as

$$\text{efficiency} \equiv \frac{\text{how much you get}}{\text{how much you pay}}. \tag{20.1}$$

We continue our study of thermodynamics by applying this general concept to the heat engine. We define "how much you get" to be the net work output $-W_{\text{cyc}}$ of the engine per cycle. And we define "how much you pay" to be the heat input Q_{in} over the cycle. The ratio of these two quantities is called the **thermodynamic efficiency** η (lowercase Greek eta):

$$\eta \equiv \frac{-W_{\text{cyc}}}{Q_{\text{in}}}. \tag{20.2}$$

The negative sign is a convenience; it makes η a positive quantity. Thermodynamic efficiency is a dimensionless number because it is the ratio of two quantities having the dimensions of energy. Efficiency is often expressed as a percentage.

The overall (measured) efficiency of a real engine is always less than the thermodynamic efficiency—often substantially less. The discrepancy between overall and thermodynamic efficiencies is due to frictional dissipation of some of the output work, loss of some of the input heat through imperfect insulation, and the need to drive such auxiliary devices as fuel pumps, cooling fans, and lubrication systems—necessary subsystems that divert output work from the "business end" of the engine. Much engineering time is spent on minimizing these losses. For a real engine, thermodynamic efficiency is an unattainable goal but an indispensable design criterion.

EXAMPLE 20.1

Figure 20.1 (like Figure 19.5) shows the heat-engine cycle studied in Examples 19.1 and 19.2. The cycle consists of an isotherm $a \rightarrow b$, an isobar $b \rightarrow c$, and an isometric $c \rightarrow a$. Find the thermodynamic efficiency of the engine. The results found in Examples 19.1 and 19.2 and relevant here are

$$Q_{a \rightarrow b} = -999 \text{ J}; \quad Q_{b \rightarrow c} = 7910 \text{ J};$$
$$Q_{c \rightarrow a} = -4750 \text{ J}; \quad W_{\text{cyc}} = -2160 \text{ J}.$$

SOLUTION: Because W_{cyc} is given, all you need calculate is the heat input Q_{in}. To do this, you inspect the three values of Q given for the three parts of the cycle and note that two of them are negative. That is, they denote heat flowing *out of* the engine and so do not contribute to Q_{in}. All the heat input therefore takes place over path $b \rightarrow c$, and you have

$$Q_{\text{in}} = Q_{b \rightarrow c} = 7910 \text{ J}.$$

So Equation 20.2 gives

$$\eta = \frac{-(-2160 \text{ J})}{7910 \text{ J}} = 0.273, \quad \text{or} \quad \eta = 27.3\%.$$

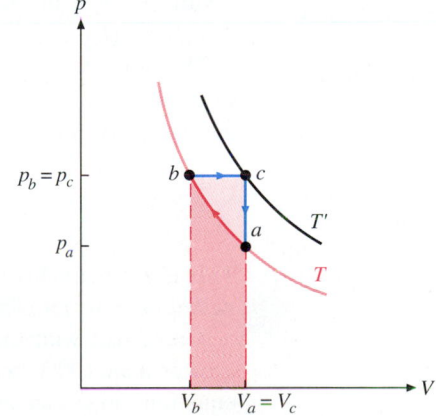

FIGURE 20.1

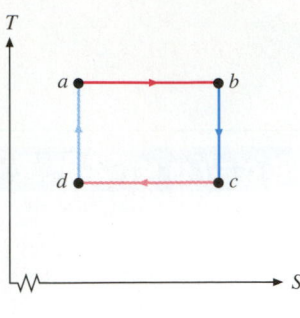

(a)

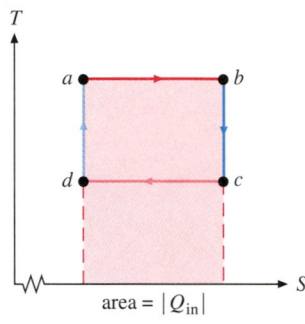

(b)

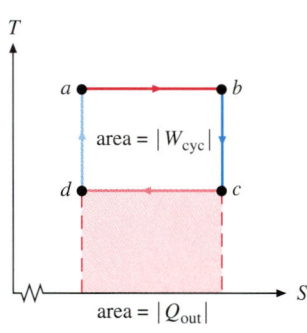

(c)

FIGURE 20.2 (*a*) *T-S* diagram of a Carnot cycle. The horizontal paths are isotherms, and the vertical paths are adiabatics. (*b*) The magnitude of the heat input $|Q_{in}|$ is the area under path $a \rightarrow b$. (*c*) The magnitude of the heat output $|Q_{out}|$ is the area under path $c \rightarrow d$. The magnitude of the cyclic work $|W_{cyc}|$ is the area of the rectangle enclosed by the cycle and is the difference between $|Q_{in}|$ and $|Q_{out}|$.

Carnot Efficiency

The efficiency of a Carnot cycle can be expressed in a particularly simple way. Figure 20.2 (like Figure 19.13) is a *T-S* diagram of the Carnot cycle. Heat flow takes place between the system and its surroundings only along the isotherm $a \rightarrow b$ at temperature T_{hi} and along the isotherm $c \rightarrow d$ at temperature T_{lo}. The magnitude $|Q_{a \rightarrow b}|$ of the heat flow along $a \rightarrow b$ is represented by the area under $a \rightarrow b$, and $Q_{a \rightarrow b}$ is given by Equation 19.41,

$$Q_{a \rightarrow b} = T_{hi} \Delta S_{a \rightarrow b}. \tag{20.3a}$$

Similarly, the magnitude $|Q_{c \rightarrow d}|$ of the heat flow along $c \rightarrow d$ is represented by the area under $c \rightarrow d$, and

$$Q_{c \rightarrow d} = T_{lo} \Delta S_{c \rightarrow d}. \tag{20.3b}$$

By inspecting the figure, you can see that $\Delta S_{a \rightarrow b}$ is positive and

$$\Delta S_{c \rightarrow d} = -\Delta S_{a \rightarrow b} \tag{20.4}$$

is negative. So $Q_{a \rightarrow b}$ represents **input heat**, and $Q_{c \rightarrow d}$ represents **output heat**, also called **waste heat**. The heat input to the system per cycle is

$$Q_{in} = Q_{a \rightarrow b} = T_{hi} \Delta S_{a \rightarrow b}. \tag{20.5a}$$

This is the quantity that appears in the denominator of Equation 20.2, the definition of η.

Similarly, the output heat is

$$Q_{out} = Q_{c \rightarrow d} = T_{lo} \Delta S_{c \rightarrow d} = -T_{lo} \Delta S_{a \rightarrow b}. \tag{20.5b}$$

Consistent with our convention that heat flow out of the system is represented by a negative quantity, Q_{out} has a negative value. In Example 20.1, the contributions to Q_{in} were identified by their positive values. Similarly, you can readily identify all the contributions to Q_{out} over any heat-engine cycle simply by looking for negative quantities. The net heat flow over the cycle is the sum

$$Q_{cyc} = Q_{in} + Q_{out} = (T_{hi} - T_{lo}) \Delta S_{a \rightarrow b}. \tag{20.5c}$$

Now we can determine the cyclic work W_{cyc}, which appears in the desired expression for thermodynamic efficiency. According to Equation 19.13b, $-W_{cyc} = Q_{cyc}$ around a heat-engine cycle. So we can immediately write*

$$W_{cyc} = -(T_{hi} - T_{lo}) \Delta S_{a \rightarrow b}. \tag{20.6}$$

As you learned from the discussion related to Figure 19.13, the magnitude of Q_{cyc} is vividly evident in the *T-S* diagram of the engine cycle, shown here as Figure 20.2; $|W_{cyc}| = |Q_{cyc}|$ is the area of the rectangle bounded by the cycle.

We now have all the information we need to find the thermodynamic efficiency η^c of a Carnot engine. Into Equation 20.2, $\eta = -W_{cyc}/Q_{in}$, we insert the values of W_{cyc} and Q_{cyc} given by Equations 20.6 and 20.5a. We thus obtain

$$\eta^c = \frac{T_{hi} - T_{lo}}{T_{hi}}. \tag{20.7}$$

This is a remarkably simple result. *The thermodynamic efficiency of a Carnot engine, called the* **Carnot efficiency**, *depends only on the input and output temperatures.*

Consistent with the first law, Equation 20.7 does not allow for Carnot efficiency greater than 100% because absolute temperatures are always positive. Moreover, the equation shows that a Carnot engine could have 100% efficiency—that is, could convert *all* the input heat into work—only if its output temperature were absolute zero. Not only is this impracticable, but it is strictly impossible for reasons discussed in Section 20.6.

---

*Compare Equation 20.6 with Equation 19.34, which gives W_{cyc} in terms of n, R, and the initial and final temperatures and volumes. Because it does not depend on the assumption that the working fluid is an ideal gas, Equation 20.6 is more general than Equation 19.34.

EXAMPLE 20.2

The best modern steam-turbine electrical generating plants use input steam at about 850 K and exhaust the steam to a cooling condenser where the temperature is about 300 K. Find the efficiency of a Carnot engine operating between these temperatures.

SOLUTION: Equation 20.7 gives you

$$\eta^{c} = \frac{850 \text{ K} - 300 \text{ K}}{850 \text{ K}} = 0.65,$$

or about 65%. This efficiency is not attainable by a real steam-turbine system. However, some steam plants have achieved efficiencies in excess of 50%.

Our fundamental definition of thermodynamic efficiency (Equation 20.2) expresses η in terms of input heat and output work. In our derivation of Equation 20.7, we eliminated the term $-W_{cyc}$ from the expression for efficiency by using the first law in the form $-W_{cyc} = Q_{cyc}$. We now express η explicitly in terms of heat alone, in a general way that is applicable to any heat engine.

Whatever the details of an engine cycle, all heat flow can be assigned to either Q_{in} or Q_{out}. According to the first law of thermodynamics, the net internal energy change in a cycle is zero, and so we have

$$Q_{in} + Q_{out} + W_{cyc} = 0. \qquad (20.8)$$

Rearranging, we write

$$-W_{cyc} = Q_{in} + Q_{out} \quad \text{and} \quad Q_{in} = -(W_{cyc} + Q_{out}). \qquad (20.9a,b)$$

The efficiency of the engine, according to Equation 20.2, is

$$\eta = \frac{-W_{cyc}}{Q_{in}} = \frac{Q_{in} + Q_{out}}{Q_{in}}. \qquad (20.10)$$

This may be regarded as a general definition of thermodynamic efficiency. Here it is especially evident that η is always less than 1, because Q_{out} is a negative quantity.

Equation 20.9a repeats algebraically the message of the *T-S* diagram: *A heat engine takes in heat energy Q_{in}, converts part of it into output work W_{oyo}, and rejects the remainder as waste heat Q_{out}.* We framed the discussion leading to this statement in terms of the Carnot cycle for concreteness, but it is valid for any heat-engine cycle. The statement is expressed in pictorial terms by the *pipeline diagram* of Figure 20.3. The Carnot engine is conventionally represented by the large circle. During each cycle, the engine accepts heat Q_{in} from a **heat reservoir**. A heat reservoir is a body whose total heat capacity is sufficiently large that the temperature of the reservoir does not vary during the heat transfer under study. Such a reservoir, when it furnishes heat to the engine, is called a **heat source**. The amount of heat input to the engine per cycle is represented by the width of the pipe, as the channel joining the heat source to the engine is called. The part of Q_{in} converted into output work W_{cyc} is represented by the width of the horizontal pipe. The remainder—the rejected waste heat Q_{out}—is represented by the width of the lower vertical pipe, which conducts the heat to a second heat reservoir. Such a reservoir, which accepts heat from the engine, is called a **heat sink**.

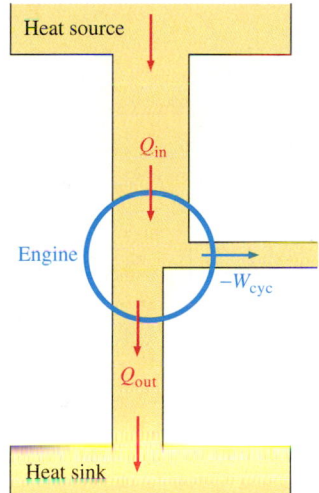

FIGURE 20.3 Pipeline diagram of a heat engine. During each engine cycle, the heat source supplies heat Q_{in} to the engine. The engine converts part of this input heat into work $-W_{cyc}$ and supplies the work to the outside world (perhaps by means of a rotating shaft). The engine rejects the rest of the heat, Q_{out}, to the heat sink. The width of each pipe conventionally represents the amount of energy involved.

The Second Law of Thermodynamics

Figure 20.4a shows a generalized process in which heat flows from a warm heat reservoir at T_{hi} to a cooler one at the lower temperature T_{lo}. This process is completely spontaneous; it is often a nuisance that can be minimized but not eliminated. (For example, the flame in a water heater must turn on from time to time even when no water is drawn because heat inevitably flows from the hot water through the imperfect

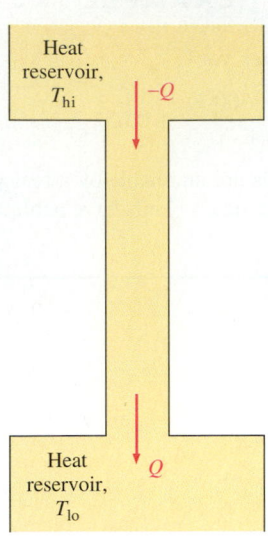

(a)

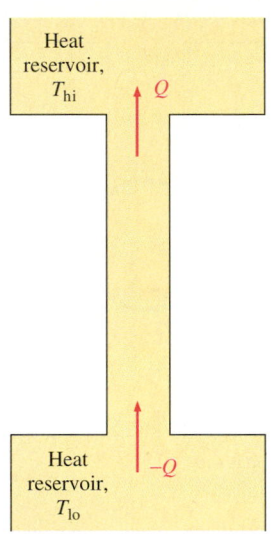

(b)

FIGURE 20.4 (*a*) A commonplace process. (*b*) A forbidden process.

insulation to the cool surrounding air.) But no one has ever seen the reverse process, shown in Figure 20.4b. These two complementary observations embrace such a wide variety of experience that we summarize them in the form of a law of nature:

No process is possible in which the *only* event is the transfer of heat from a cooler body to a warmer one.

This is called the **Clausius statement of the second law of thermodynamics**.

The Second Law and Entropy

Entropy has an intimate connection with the second law. We begin our study of this connection by calculating the entropy change of the two-reservoir system in the commonplace process of Figure 20.4a. Then we do the same for the impossible process of Figure 20.4b.

To find the net entropy change ΔS_{sys} of the system of Figure 20.4a, we add the entropy changes ΔS_{hi} and ΔS_{lo} that take place in the two heat reservoirs as a certain amount of heat $|Q|$ flows from the warmer one to the colder. The heat flowing out of the warm reservoir is $-Q$, the minus sign accounting for the fact that the flow tends to decrease the internal energy of the reservoir. Because the temperature of this reservoir has the constant value T_{hi}, Equation 19.44, $\Delta S = \int (dQ/T)$, simplifies to the form

$$\Delta S_{hi} = \frac{-Q}{T_{hi}}. \tag{20.11a}$$

The same amount of heat flows into the cool reservoir; here the proper sign is positive because the flow tends to increase the internal energy of the reservoir. So we have

$$\Delta S_{lo} = \frac{Q}{T_{lo}}. \tag{20.11b}$$

The entropy change for the system is

$$\Delta S_{sys} = \Delta S_{hi} + \Delta S_{lo} = Q\left(\frac{1}{T_{lo}} - \frac{1}{T_{hi}}\right). \tag{20.12}$$

Because $T_{lo} < T_{hi}$, *the entropy of the system increases in this commonplace process, in which heat flows spontaneously from a warmer part of a system to a cooler part.* By a **spontaneous process**, we mean one that takes place without intervention from outside the system. If a system is isolated, all processes taking place within it are spontaneous.

Now let us turn to the impossible process of Figure 20.4b. Here the heat flow out of the cooler reservoir is $-Q$ and the heat flow into the warmer reservoir is Q. The corresponding entropy changes are

$$\Delta S_{lo} = \frac{-Q}{T_{lo}} \quad \text{and} \quad \Delta S_{hi} = \frac{Q}{T_{hi}}. \tag{20.13a,b}$$

The entropy change for the system is

$$\Delta S_{sys} = Q\left(\frac{1}{T_{hi}} - \frac{1}{T_{lo}}\right). \tag{20.14}$$

Because $T_{hi} > T_{lo}$, *the entropy of the system decreases in this impossible process, in which heat flows spontaneously from a cooler part of a system to a warmer part.*

In either process, the larger the temperature difference $T_{hi} - T_{lo}$, the larger $|\Delta S_{sys}|$. Indeed, let us consider the special case in which the temperature difference is infinitesimal:

$$T_{hi} = T_{lo} + dT.$$

What happens as heat flows from the (infinitesimally) warmer reservoir to the cooler one in this case? According to Equation 20.12, the net entropy change is

$$\Delta S_{sys} = Q\left(\frac{1}{T_{lo}} - \frac{1}{T_{lo} + dT}\right).$$

We now show that this entropy change is vanishingly small. We begin by reducing the fractions to a common denominator:

$$\Delta S_{\text{sys}} = Q\,\frac{T_{\text{lo}} + dT - T_{\text{lo}}}{T_{\text{lo}}(T_{\text{lo}} + dT)} = \frac{Q}{T_{\text{lo}}}\left(\frac{dT}{T_{\text{lo}} + dT}\right).$$

The denominator of the fraction in parentheses is the sum of a finite quantity and an infinitesimal quantity. Therefore, the value of the fraction is not significantly affected if we neglect the term dT in the denominator, writing $dT/(T_{\text{lo}} + dT) = dT/T_{\text{lo}}$. We then have

$$\Delta S_{\text{sys}} = \frac{Q}{T_{\text{lo}}^2}\,dT.$$

The quantity on the right is infinitesimal because it is the product of a finite factor and the infinitesimal dT. The finite factor can be as large as we like because there is no reason to limit the amount of heat Q that flows in the process under discussion. Thus, *any* amount of heat can flow from one reservoir to the other with only an infinitesimal— and thus negligible—entropy change. So we have

$$\Delta S_{\text{sys}} = 0 \quad \text{for finite heat flow over a temperature difference } dT. \quad \text{(20.15)}$$

We have reached the conclusion embodied in Equation 20.15 by considering the process of Figure 20.4a. But, if you repeat the calculation for the process of Figure 20.4b, you will obtain the same result. *Given two heat reservoirs having an infinitesimal temperature difference dT, any amount of heat can flow from one to the other with negligible entropy change ΔS_{sys}.* We will use this result in the next subsection.

Reversibility

Processes in which heat flows across infinitesimal temperature differences are quasistatic. As an example, consider an isolated system in which a gas-filled cylinder is immersed in a constant-temperature bath (Figure 20.5). An infinitesimal expansion dV of the cylinder produces an infinitesimal decrease in the temperature of the gas inside it. The temperature decrease results in turn in a flow of heat into the cylinder from the bath. The entropy of the cylinder increases, but the entropy of the bath decreases by

(a)

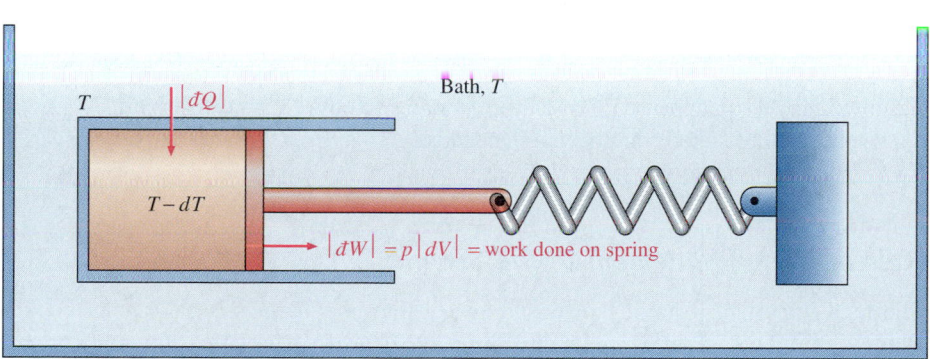

(b)

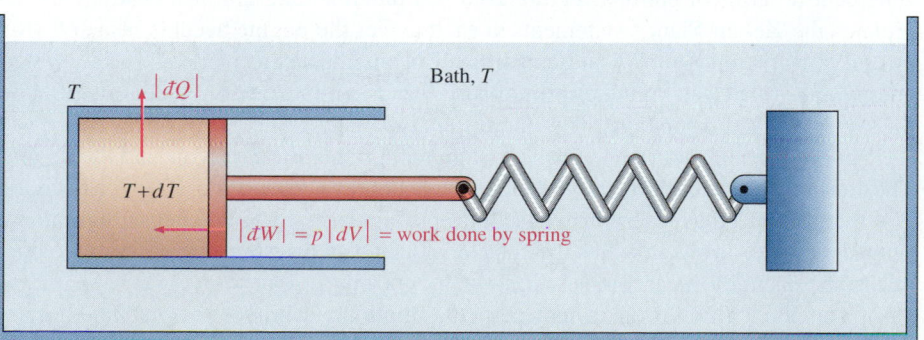

FIGURE 20.5 A reversible process. A system consists of a cylinder, a piston, and a bath of temperature T. (a) The piston moves to the right, and the cylinder volume increases by $|dV|$. The temperature of the working fluid falls to $T - dT$, and heat $|Q|$ flows into the cylinder from the bath. Work $|dW| = p|dV|$ is done by the piston on the spring. There is no entropy change for the system. (b) The spring pushes the piston to the left, and the cylinder volume decreases by $|dV|$. Work $|dW| = p|dV|$ is done by the spring on the piston. The temperature of the working fluid rises to $T + dT$, and heat $|Q|$ flows from the cylinder into the bath. Again, there is no overall entropy change.

the same amount. Thus the entropy of the cylinder-plus-bath system remains constant. If the cylinder is then compressed infinitesimally by $-dV$, the reverse process takes place and again the entropy of the system does not change. In sum, the system can be restored to its original state, with no net change in p, V, T, or S, by reimposition of the original *constraint*—in this case, restoration of the piston to its original position. For this reason, a process that can be run in either direction without any entropy change is called a **reversible process**. If a reversible process is to take place, it must still obey the first law of thermodynamics. The work $-|dW|$ done by the expanding cylinder (on another part of the system, represented by the spring) must be stored as potential energy so that work $|dW|$ is available to recompress the cylinder in the reverse process.

Any process in which heat flows across a *finite* temperature difference is *irreversible*. The system cannot be restored to its original state without outside intervention. The entropy of the system has increased, and there is no way to decrease it spontaneously. For example, consider a container divided into two parts by a removable partition. One part of the container is filled with hot water and the other part with cold water. If we remove the partition, the two bodies of water will mix and attain some intermediate temperature. If we then restore the original constraint by replacing the partition, we will not return the system to its original state.

We summarize the relation between entropy and the second law of thermodynamics as follows:

1. When heat is transferred from a warmer body to a cooler body over a finite temperature difference, the entropy change ΔS_{sys} of the system is positive; that is, the entropy S_{sys} *increases*. Such processes may take place spontaneously.

2. When heat is transferred from a cooler body to a warmer body over a finite temperature difference, the entropy change ΔS_{sys} of the system is negative; that is, the entropy S_{sys} *decreases*. Such processes may not take place spontaneously; there must be action on the system from outside.

3. When heat is transferred over an infinitesimal temperature difference dT, the entropy change ΔS_{sys} of the system is zero; that is, the entropy S_{sys} *does not change*. Such processes are quasistatic and reversible.

With these results in mind, we may restate the **second law of thermodynamics in terms of entropy change**: *When an isolated system undergoes a thermodynamic process, the entropy of the system **must** either increase or remain the same.* The entropy of a part of the system may decrease, but only if the entropy of other parts increases by at least as much. Such an entropy decrease takes place in the cooler reservoir in Figure 20.4a, at the expense of an entropy increase in the warmer reservoir.

The law can be stated equally well in negative terms: *No process is possible in which the entropy of an isolated system decreases.*

Kelvin-Planck Statement of the Second Law

The final important form of the second law of thermodynamics is called the **Kelvin-Planck statement**: *No thermodynamic process is possible in which the only event is the conversion of heat into work.* This statement of the law is readily derived from the statement in terms of entropy. Figure 20.6 is a pipeline diagram of a heat engine that violates the Kelvin-Planck statement. In each cycle, the engine accepts heat $|Q|$ from a reservoir at temperature T and converts all of this heat into work $|W|$. The flow of energy out of the system in the form of work has no influence on the entropy of either the system or its surroundings. Thus the system is isolated as far as entropy is concerned.

What is the entropy change of the system comprising the reservoir and the engine? The entropy change of the reservoir due to the heat flow $-Q$ is $\Delta S = -Q/T$, a decrease. But there is no other reservoir whose entropy increases by a compensating amount. Thus the process is one in which the entropy of an isolated system decreases in violation of the statement of the second law in terms of entropy.

The engine of Figure 20.6 is not a perpetual-motion machine of the first kind because it does not violate the first law of thermodynamics. But it does violate the second law

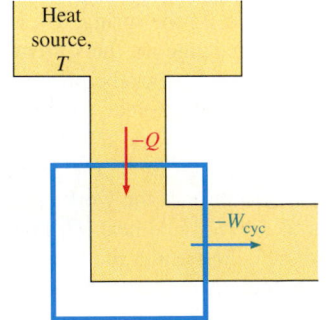

FIGURE 20.6 A perpetual-motion machine of the second kind. A heat engine accepts heat energy from a source and converts all of it into work in violation of the Kelvin-Planck statement of the second law.

and is therefore called a **perpetual-motion machine of the second kind**. The jet engine of Section 20.1, which extracts heat from the air flowing through it and converts it into mechanical thrust, is a perpetual-motion machine of the second kind. You can now see why this implausible device is indeed impossible.

Reversible Engines and Refrigerators

All the heat-engine cycles we have considered so far are described by *clockwise* passage around a path in *p-V* space or *T-S* space. In each case, this passage involves a net conversion of heat into work. Under what circumstances can we make such an engine go in the opposite sense so that its cycle is described by a closed *counterclockwise* path?

Such a path reversal can be achieved if all the thermodynamic processes making up the cycle are *reversible*. A finite process is reversible if it consists exclusively of a series of reversible infinitesimal processes. For example, the infinitesimal expansion of Figure 20.5a can be followed by similar ones until the cylinder has expanded by a finite amount and the spring has been compressed by a finite amount. The system can then be restored to its original state by a series of infinitesimal processes like that of Figure 20.5b.

The cycle $a \rightarrow b \rightarrow c \rightarrow d \rightarrow a$ in Figure 20.2 is reversible because each infinitesimal step is reversible. In the system comprising the Carnot engine *and* the two reservoirs, the entropy does not change as the engine operates. The system is *not* isolated, however, because it is linked to an outside world through the engine shaft. But the shaft conveys mechanical work only and therefore cannot effect a change $dS = đQ/T$ on the system.

Nevertheless, it is useful to consider entropy changes in a *subsystem*: the engine proper. This subsystem is not isolated from its heat reservoir and certainly does experience entropy changes. (But the entropy change of the subsystem over a complete cycle must be zero, as the *T-S* diagram shows.) We can run the subsystem through the process described by the counterclockwise path $a \rightarrow d \rightarrow c \rightarrow b \rightarrow a$ shown in Figure 20.7. This process consists of the adiabatic expansion $a \rightarrow d$ followed by the isothermal compression $d \rightarrow c$, the adiabatic compression $c \rightarrow b$, and the isothermal expansion $b \rightarrow a$. The work W' and the heat flow Q' along each path segment are the negatives of the corresponding quantities W and Q in the clockwise process of Figure 20.2. For example, compare $Q'_{b \rightarrow a}$ with $Q_{a \rightarrow b}$, given by Equation 20.3a:

$$Q'_{b \rightarrow a} = T_{hi} \Delta S_{b \rightarrow a} = T_{hi}(-\Delta S_{a \rightarrow b}) = -Q_{a \rightarrow b}.$$

The same statement applies to the complete cycle:

$$W'_{cyc} = -W_{cyc} \quad \text{and} \quad Q'_{cyc} = -Q_{cyc}. \quad \textbf{(20.16a,b)}$$

What we are doing now is performing work *on* the subsystem—turning the shaft. The low-temperature reservoir is now the heat source, and the high-temperature reservoir is the heat sink. That is, the input heat Q'_{in} and the output heat Q'_{out} are related to the corresponding quantities for the clockwise cycle by the equations

$$Q'_{in} = -Q_{out} \quad \text{and} \quad Q'_{out} = -Q_{in}. \quad \textbf{(20.17a,b)}$$

This point is illustrated in the pipeline diagram of Figure 20.8, which looks exactly like that of Figure 20.3, except that the arrows are reversed and the labels changed. An engine that can be run in reverse in this fashion is called a **reversible engine**, a concept originated by Carnot.

Refrigerators and Heat Pumps

When a heat engine is run in reverse, it is called either a **refrigerator** or a **heat pump**, depending on the application. The device accepts mechanical work, takes heat in from a low-temperature reservoir, and rejects heat to a high-temperature reservoir.

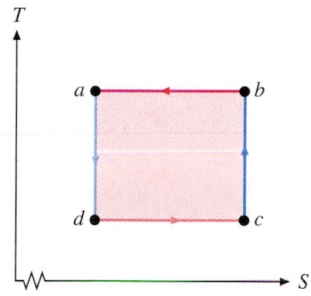

FIGURE 20.7 A Carnot cycle run in reverse. Quasistatic processes are generally reversible; the cycle shown here is simply the quasistatic process of Figure 20.2 reversed.

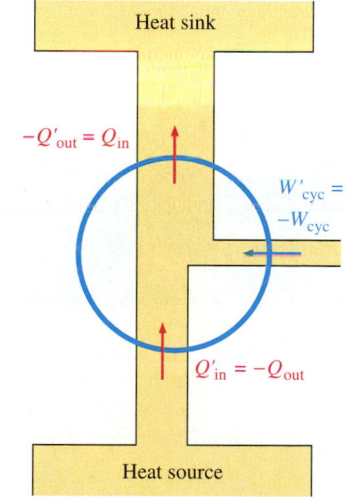

FIGURE 20.8 Pipeline diagram of a heat engine run in reverse. Compare with Figure 20.3.

The first law requires that

$$Q'_{in} + Q'_{out} + W'_{cyc} = 0;$$ (20.18)

compare with Equation 20.8.

The refrigerator (or heat pump) does not violate the Clausius statement of the second law of thermodynamics. Heat is indeed transferred from a cooler body to a warmer one, but that is not the *only* event taking place. At the same time that heat is being transferred, work is being done on the engine (through the shaft) from outside the system.

The purpose of a refrigerator is to keep the low-temperature heat source colder than its surroundings by removing heat from it and rejecting heat to the warmer surroundings. What we are *paying for* is the work W'_{cyc} needed to run the refrigerator. What we are *getting* is Q'_{in}. So, in the same spirit in which we defined the thermodynamic efficiency of a heat engine, we now define the **refrigerator coefficient of performance** ω_r:

$$\omega_r \equiv \frac{Q'_{in}}{W'_{cyc}}.$$ (20.19)

Solving Equation 20.18 for W'_{cyc} and substituting, we have

$$\omega_r = \frac{Q'_{in}}{-(Q'_{in} + Q'_{out})}.$$ (20.20a)

Remembering that $Q'_{in} > 0$ and $Q'_{out} < 0$, we can write ω_r in the equivalent form

$$\omega_r = \frac{|Q'_{in}|}{|Q'_{out}| - |Q'_{in}|}.$$ (20.20b)

The value of ω_r is always greater than 1. That is, the heat energy "pumped out" of the cold chamber is of greater magnitude than the mechanical energy used to pump it.

The purpose of a heat pump is to keep a heat reservoir warmer than its surroundings by removing heat from the cold surroundings and "pumping the heat" into the warm reservoir. Such devices are used for home heating in mild climates; the house is the warm reservoir (the heat sink) and the outdoors the cold reservoir (the heat source). An attractive feature of a heat pump is that the same machinery used to warm the house in winter can cool the house in summer. Automatic valves reverse the piping connections from the two heat reservoirs to the heat pump so that in summer the house becomes the cold reservoir (the heat source) and the outdoors the warm reservoir (the heat sink). The heat pump is thus converted into a refrigerator.

The **heat pump coefficient of performance** is defined to be

$$\omega_{hp} \equiv \frac{-Q'_{out}}{W'_{cyc}}.$$ (20.21)

By means of an argument similar to the one that led to Equation 20.20b, we can obtain

$$\omega_{hp} = \frac{|Q'_{out}|}{|Q'_{out}| - |Q'_{in}|}.$$ (20.22)

These unobtrusive air-conditioner housings contain heat engines run in reverse.

Carnot's Theorem

We have established that the efficiency of a Carnot engine depends only on the source and sink temperatures, according to Equation 20.7. We now prove the important **Carnot's theorem**: *No heat engine operating between two arbitrary heat reservoirs can have an efficiency greater than that of a Carnot engine operating between the same two reservoirs.*

By Carnot's time, steam engines had been widely used for almost half a century, over which time they had improved dramatically in reliability and efficiency. Nearly all of the improvement had been based on practical observation and not grounded in physical theory; indeed, a key motivation for developing a science of thermodynamics was the need to improve the efficiency of steam engines at a time when empirically based improvements seemed to have reached a point of diminishing returns. Many persons must have asked, What are the limits, if any, on the amount of work that can be performed by a steam engine using a given amount of fuel?—a question with important practical as well as theoretical implications.

Carnot's theorem is the central result of his only published work, a small book titled *Reflections on the motive power of fire and of the means suitable for developing it* (1824). In this work, Carnot framed his exposition in terms of the caloric theory of heat. But he had doubts as to the validity of the caloric theory and purposely made his arguments in such general terms that the results did not depend on any specific assumptions involving caloric. Carnot's book was almost entirely ignored at the time of its publication, and his work became known only after his untimely death

at the age of 36 in 1832. In 1834, the French civil engineer B. P. E. Clapeyron (pronounced Clah-pay-**rohn'**) (1799–1864) rediscovered Carnot's work and restated his arguments in a transparent—but partly erroneous—form. It was through Clapeyron's account that William Thomson (later Lord Kelvin) learned of Carnot's work. Thomson then recast Carnot's work in terms of the kinetic theory of heat. (According to notes found after his death, Carnot had come to accept the kinetic theory fully by 1830.) Thomson's integration of Carnot's work into the fabric of kinetic theory was crucial in inducing many scientists to accept the kinetic theory.

This little history exemplifies several recurrent features of physical thought:

1. A sufficiently general argument can circumvent specific uncertainties.

2. Two or more new scientific ideas that can be worked into a broader context often strengthen one another and together can be a source of considerably deeper insight than any of the ideas alone.

3. Science does not always progress smoothly or logically, even though it is described in textbooks as if it did.

Carnot's theorem is proved by contradiction of its negative. We assume that a "super-engine"—one whose efficiency is greater than η^c—*is* possible. Imagine that we have a particular Carnot engine. We design a super-engine so that its output work per cycle is the same as that of the Carnot engine. In Figure 20.9, the equality of the output work per cycle for the two engines is represented by the equal thickness of the output pipelines.

Thermodynamic efficiency is defined by Equation 20.2, $\eta = -W_{\text{cyc}}/Q_{\text{in}}$. For the present case, where all the heat flow into either engine takes place at the fixed temperature T_{hi}, we can write $Q_{\text{in}} = Q_{\text{hi}}$. For either engine, then, we have $\eta = -W_{\text{cyc}}/Q_{\text{hi}}$. Because the super-engine is the more efficient, it must take in less heat in performing the same work; that is, $|Q^s_{\text{hi}}| < |Q^c_{\text{hi}}|$. This inequality is represented in Figure 20.9 by the narrower pipeline supplying heat to the super-engine.

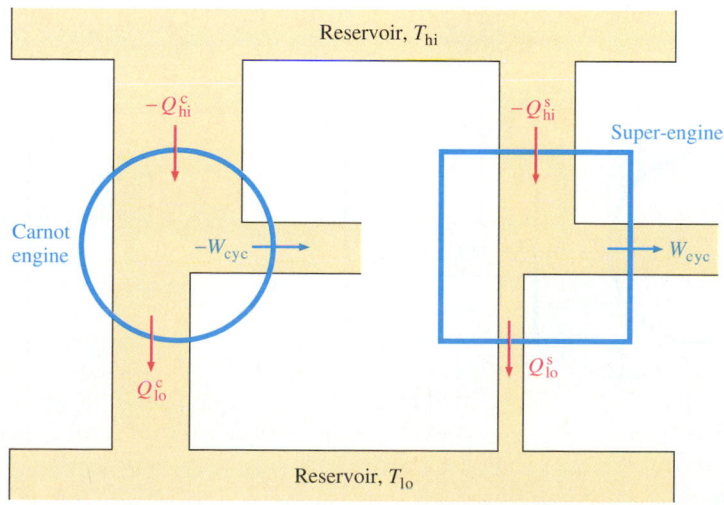

FIGURE 20.9 A Carnot engine and a super-engine whose thermodynamic efficiency is greater than that of the Carnot engine. The two engines are designed to produce the same output work per cycle. The Carnot engine is conventionally represented by a circle and the super-engine, because it is a non-Carnot engine, by a square.

Now we are ready for the cleverest part of the proof. Since the Carnot engine is reversible, we run it in reverse as a heat pump, using the super-engine to drive it, as shown in Figure 20.10a. What is the net result of all the processes? The super-engine does work just sufficient to run the Carnot heat pump, and the system as a whole neither does work on nor has work done on it by the outside world. Heat $|Q^s_{hi}|$ flows out of the warm reservoir into the super-engine, which rejects heat $|Q^s_{lo}|$ to the cool reservoir. The first law of thermodynamics requires that

$$|Q^s_{hi}| - |Q^s_{lo}| = |W_{cyc}|.$$

Simultaneously, the Carnot heat pump accepts heat $|Q^c_{lo}|$ from the cool reservoir and rejects heat $|Q^c_{hi}|$ to the warm reservoir. The first law requires that

$$|Q^c_{hi}| - |Q^c_{lo}| = |W_{cyc}|.$$

Eliminating $|W_{cyc}|$ from these two equations and rearranging, we obtain

$$|Q^c_{hi}| - |Q^s_{hi}| = -(|Q^s_{lo}| - |Q^c_{lo}|). \tag{20.23}$$

But the left side of this equation is the net heat flow Q_{hi} into the warm reservoir, and the right side is the net heat flow $-Q_{lo}$ out of the cool reservoir. This net result is shown in Figure 20.10b.

The result is consistent with the first law; all the energy flowing out of the cool reservoir ends up in the warm reservoir, and energy is neither created nor destroyed. But Equation 20.23 violates the Clausius statement of the second law (as well as everyday experience). As Figure 20.10b shows, the equation predicts that heat will flow from a cool reservoir to a warm one with nothing else happening. We conclude that an engine more efficient than a Carnot engine when both operate between the same reservoirs is impossible, and Carnot's theorem is proved.

There is a corollary to Carnot's theorem: *No reversible engine can be less efficient than a Carnot engine when both operate between the same reservoirs.* To prove this, assume that the non-Carnot engine represented by the square in Figure 20.10a—no longer a super-engine—is reversible and *less* efficient than the Carnot engine. Use the Carnot engine to run the square engine in reverse as a heat pump, and repeat the proof of Carnot's theorem.

We have thus proved that, when operated between the same reservoirs,

no engine, reversible or irreversible, can be more efficient than a Carnot engine,

and that

no reversible engine can be less efficient than a Carnot engine.

It follows that, when operating between the same reservoirs,

all reversible engines have the same efficiency as a Carnot engine.

FIGURE 20.10 (*a*) The super-engine is used to run the Carnot engine in reverse as a heat pump. The work output of the super-engine just suffices to run the Carnot heat pump. (*b*) The net result of all the processes operating in part *a*. Heat is transferred from the cooler reservoir to the warmer one, and *nothing else happens*.

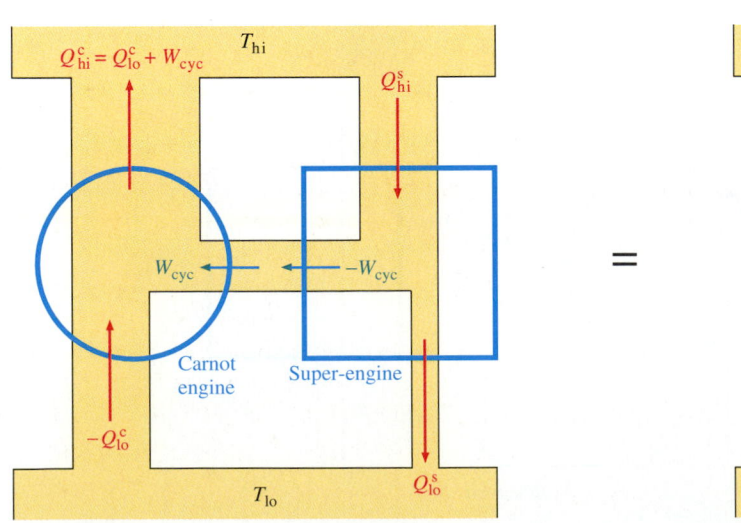

The Third Law of Thermodynamics

We have considered heat and thermodynamics on the basis of three laws, called the zeroth, first, and second laws of thermodynamics. We need one more law to found thermodynamics on a completely rigorous basis.

The **third law of thermodynamics** was first stated explicitly in 1910 by the German physical chemist Walther Nernst (1864–1941) and is sometimes called Nernst's law. We touched on this law briefly in relation to Carnot's theorem. It can be stated in two forms, which are logically equivalent:

1. A refrigeration process can in principle reduce the temperature of a heat source as close to absolute zero as desired but can never attain absolute zero.

2. As the temperature of a system is lowered toward absolute zero, its entropy tends asymptotically toward a fixed value.

Entropy and Randomness*

The concept of entropy (Section 19.6) has ramifications far beyond the realm of heat engines. As a conclusion to our discussion of the physics of thermal phenomena, therefore, we briefly consider entropy in this broader context.

In Section 20.3, we considered some commonplace irreversible processes, such as the flow of heat from a warm reservoir to a cold one. We called their reverse processes "impossible" because no one has ever seen one of them happen. To emphasize the fact that such pairs of processes are of many kinds, we cite a few here:

1. A warm body and a cool body are placed in thermal contact. The warm body cools and the cool body warms until the two are at the same temperature.

1′. Two bodies in thermal contact have the same temperature. One warms up as the other cools down.

2. A drop of red dye is deposited carefully, without stirring, on the surface of a quantity of quiet water. The dye diffuses into the water until the entire system is a uniform pale pink.

2′. The dye in a quantity of pale pink water concentrates in one place until all the water is colorless except for one intensely red drop.

3. The air in a leaky automobile tire escapes through a tiny hole until the pressure inside the tire is equal to the pressure of the surrounding atmosphere.

3′. The air pressure inside a leaky tire increases as air passes into the tire through a tiny hole.

4. An unused deck of playing cards is in standard order, with suits in the order spades, hearts, diamonds, clubs and each suit ordered ace through king. A thorough shuffle leaves them in an order that, though definite, is unpredictable.

4′. A deck of cards in a definite but arbitrary order is shuffled. After the shuffle, the deck is in standard order.

We have not considered a method for calculating the entropy of every one of these systems. Nevertheless, we know that entropy increases in all of the unprimed processes and decreases in all of the primed processes. As you have already seen, all the primed processes are possible if the systems are *not* isolated. A refrigerator can accomplish process 1′, a still can accomplish process 2′, a pump can accomplish process 3′, and a child with nothing better to do can accomplish the result of process 4′. In all these cases, it is possible to decrease the entropy of a system by doing work on the system, using energy from a source outside the system. But the total entropy of the universe increases (or at least remains the same) as the mechanism that lowers local entropy simultaneously increases the entropy of the surroundings.

---

*This section may be omitted without loss of continuity.

Underlying such entropy changes is a statistical reason. It is most easily explained in terms of processes 4 and 4′. You will probably agree that, when playing cards are shuffled, every possible final particular order of the cards, called a **microstate**, is just as likely as every other microstate. This statement is called the **postulate of equal *a priori* probability**. Why is the standard order never (well, hardly ever!) the result of a shuffle? Because the standard order is only one possibility among so very many—something like 1 possibility in 8.1×10^{67}, a number about 10^{41} times as large as Avogadro's number.

As the number of members of a system—cards in a deck or molecules in a thermodynamic system—increases, the number of possible microstates increases very much faster. The number of microstates for a deck of just 52 cards is the extremely large number just quoted, 8.1×10^{67}. The number of molecules in a thermodynamic system—typically, Avogadro's number, or something like 10^{27}—is enormously greater than the number of cards in a deck. Consequently, the number of possible microstates—say, the different ways of arranging one kilomole of molecules in a box of volume 22.4 m^3—is something like $10^{10^{28}}$, an incomprehensibly large number.

In order to understand what happens statistically in a thermodynamic process, we need to count the number of microstates. There are ways of counting for a system of any size. But we can avoid mathematical complications by considering the essential ideas of statistical thermodynamics in a system so simple that we can count the microstates on our fingers. Specifically, consider a coin-tossing game involving four identical coins. You win if you throw two heads and two tails, and you lose otherwise. What odds should you have?

You do not care which two coins come up heads and which two come up tails. (Indeed, if the coins are truly identical you cannot tell which is which after tossing them.) We say that you are interested only in the **macrostate** "two heads"; any microstate that belongs to this macrostate will do.

Because all microstates are equally probable, the probability of a macrostate is proportional to the number of microstates that belong to it. This point is illustrated in Table 20.1 for the four-coin toss. You have three chances in eight that you will throw two heads and two tails; we say that the *normalized probability* of that macrostate is $\frac{3}{8}$. So you will lose the game over the long run if the odds you are offered are less than 8 for 3.

As the number of members of the system increases, the relative likelihood of the most probable macrostate increases very rapidly. For an ordinary thermodynamic system, the likelihood is overwhelming that the macrostate you observe is the most probable one (or, as you will soon see, one very much like it). Now you can see why, for instance, the pale pink water does not transform itself into colorless water with a drop of red in it. A microstate in which the red dye is concentrated in one small region is

TABLE 20.1 Macrostates and Microstates for a Four-Coin Toss

Macrostate	Microstates	Number of Microstates	Normalized Probability
4 heads	HHHH	1	$\frac{1}{16}$
3 heads	HHHT,HHTH,HTHH,THHH	4	$\frac{4}{16} = \frac{1}{4}$
2 heads	HHTT,HTHT,THHT,THTH,TTHH,HTTH	6	$\frac{6}{16} = \frac{3}{8}$
1 head	TTTH,TTHT,THTT,HTTT	4	$\frac{4}{16} = \frac{1}{4}$
0 heads	TTTT	1	$\frac{1}{16}$
	Total number of microstates:	16	

just as probable as one with the dye distributed throughout the water. But there are so very many more microstates in which the red dye molecules are distributed than microstates in which they are concentrated that the latter never happens for all practical purposes. Analogous statements can be made about the other pairs of processes with which we began this section.

A thermodynamic state of a system is what we have here more precisely called a macrostate. Boltzmann was the first to express the entropy of a macrostate as a function of the number of microstates w belonging to it. The relation, which we do not derive here, is

$$S = k \ln w, \qquad (20.24)$$

where k is Boltzmann's constant. This relation was considered so important that it became the inscription on his tombstone (shown in slightly different notation in Figure 20.11).

FIGURE 20.11 Ludwig Boltzmann's gravestone, located in the Central Cemetery of Vienna. The formula $S = k \log W$ expresses the absolute value of the entropy in terms of the vast number W of microstates accessible to the system. Boltzmann's constant k also appears in the formula. (Here, log implies $\log_e$.)

EXAMPLE 20.3

You have a tray on which four coins rest, all heads up. You jiggle the tray vigorously, disturbing the coins so that they repeatedly bounce up from the tray, flip, and land on the tray again. Find the entropy of the system in the initial state and during the jiggling process.

SOLUTION: The tray is initially in one specific microstate, and so $w = 1$. The initial entropy of the system is then

$$S_i = k \ln 1 = 0.$$

While the tray is jiggling, you can see the coins as they lie briefly on the tray before bouncing off again. If you make a list of your observations, the entries in your list will vary among the 16 microstates available to the system according to Table 20.1. You thus have

$$S = k \ln 16 = 2.8k;$$

the entropy has increased.

In Example 20.3, adding energy to the system by jiggling the tray increases the number of microstates available to the system and thus makes possible an increase of its entropy. What happens to the coins when you stop jiggling? They come to rest "frozen" in a single microstate. You do not know which coin is which, and thus you know only the macrostate to which the observed microstate belongs. Only in the four-heads and four-tails macrostates do you know the microstate as well as the macrostate (because those macrostates have only one microstate each). You have 2 chances out of 16 that one of these macrostates will turn up. However, you have 14 chances out of 16 that the final entropy will be greater than the initial value $S_i = 0$. If, for example, the coins end up in the most probable macrostate—two heads and two tails—the final entropy will be

$$S_f = k \ln 6 = 1.8k.$$

In this four-coin system, with the initial conditions given, S_f is greater than S_i only 14 times out of 16, which is most of the time but certainly not always. That is, *the second law of thermodynamics is a statistical—not an absolute—statement.*

But, for a system with very many members, the tendency toward the most probable macrostate is much stronger than it is for the small system we have been studying. You can see this in Table 20.2, which is similar to Table 20.1 but deals with a 100-coin toss.

Note that the macrostates close to 50 heads (for example, 49 heads or 47 heads) are only moderately less probable than 50 heads. But the probability of a macrostate falls off very rapidly beyond the near-50 region. Even though 100 is much smaller than the number of molecules in a typical thermodynamic system, the probability of a distribution of heads and tails differing by even as much as 10% from a 50-50 split—that is, a 55-45 distribution—is about 1%. It would take about two billion throws to get an 80-20 distribution.

Strictly speaking, then, it is incorrect to say that the entropy of a typical thermodynamic system *never* decreases. But suppose the probability of such a decrease is a typical 1 in 10^{80}. The age of the universe is of order 10^{20} milliseconds. If you were to observe the system every millisecond for all this time, the odds against your ever seeing an entropy decrease would still be 10^{60} to 1. So you are safe in maintaining for any conceivable practical purpose that the entropy of the system never decreases.

TABLE 20.2 Macrostates and Microstates for a 100-Coin Toss

Macrostate (Number of Heads n)	Number of Microstates Belonging to the Macrostate	Normalized Probability $P(n)$ of the Macrostate[a]	$P(n)/P(50)$[b]
0	1	7.9×10^{-31}	9.9×10^{-30}
1	100	7.8×10^{-29}	9.9×10^{-28}
2	4.9×10^3	3.9×10^{-27}	4.9×10^{-26}
3	1.6×10^5	1.3×10^{-25}	1.6×10^{-24}
5	7.5×10^7	5.9×10^{-23}	7.5×10^{-22}
10	1.7×10^{13}	1.4×10^{-17}	1.7×10^{-16}
20	5.4×10^{20}	4.2×10^{-10}	5.3×10^{-9}
30	2.9×10^{25}	2.3×10^{-5}	2.9×10^{-4}
35	1.1×10^{27}	8.6×10^{-4}	1.1×10^{-2}
40	1.4×10^{28}	1.1×10^{-2}	0.14
45	6.1×10^{28}	4.8×10^{-2}	0.61
47	8.4×10^{28}	6.7×10^{-2}	0.84
48	9.3×10^{28}	7.4×10^{-2}	0.92
49	9.9×10^{28}	7.8×10^{-2}	0.98
50	1.0×10^{29}	8.0×10^{-2}	1

Total number of microstates: $2^{100} \approx 1.3 \times 10^{30}$

[a]Normalized probability is the number of microstates belonging to a macrostate (column 2) divided by the total number of microstates (2^{100}).
[b]$P(50)$ is the probability of the most probable macrostate (50 heads and 50 tails).

a, b, c, d	thermodynamic states		S	entropy
c_v, c_v'	specific heat capacity (per unit mass), molar heat capacity		T	thermodynamic temperature
^{c} (superscript)	labels quantity referring to a Carnot engine		w	number of microstates accessible to a system
k	Boltzmann's constant		$W, d W$	work
m	mass		η, η^c	thermodynamic efficiency, Carnot efficiency
n	number of moles		ω_r, ω_{hp}	coefficient of performance of a refrigerator, of a heat pump
$Q, d Q$	heat flow			
^{s} (superscript)	labels quantity referring to a "super-engine"			

The behavior of every thermodynamic system must conform to four **laws of thermodynamics**:

- The **zeroth law** (Section 17.5):

  Any two bodies in thermal equilibrium with a third body are in thermal equilibrium with each other.

- The **first law** (Equation 19.1):

  $$\Delta E = Q + W.$$

- The **second law** may be expressed in three logically equivalent forms:

  The Clausius statement: *No thermodynamic process is possible in which the only event is the transfer of heat from a cooler body to a warmer one.*
  The Kelvin-Planck statement: *No thermodynamic process is possible in which the only event is the conversion of heat into work.*
  The statement in terms of entropy: *No thermodynamic process is possible in which the entropy of an isolated system decreases.*

  In **reversible** processes, the combined entropy of a system and its surroundings does not change. In **irreversible** processes, the entropy increases.

- The **third law** (Section 20.7) may be expressed in two logically equivalent forms:

  1: *A refrigeration process can in principle reduce the temperature of a heat source as close to absolute zero as desired but can never attain absolute zero.*
  2: *As the temperature of a system is lowered toward absolute zero, its entropy tends asymptotically toward a fixed value.*

The **thermodynamic efficiency** of any heat engine is given by Equations 20.2 and 20.10,

$$\eta \equiv \frac{-W_{cyc}}{Q_{in}} = \frac{Q_{in} + Q_{out}}{Q_{in}}.$$

An engine can be run backward to function as a **refrigerator** or a **heat pump**. If the engine is **reversible**, the heat flow Q' and the work W' are related to the "forward" values Q and W by Equations 20.16 and 20.17. Corresponding to the engine efficiency η are the **coefficients of performance** given by Equations 20.19, 20.20b, 20.21, and 20.22:

$$\omega_r \equiv \frac{Q'_{in}}{W'_{cyc}} = \frac{|Q'_{in}|}{|Q'_{out}| - |Q'_{in}|},$$

and

$$\omega_{hp} \equiv \frac{-Q'_{out}}{W'_{cyc}} = \frac{|Q'_{out}|}{|Q'_{out}| - |Q'_{in}|}.$$

For the special case of a Carnot engine operating between a heat source at T_{hi} and a heat sink at T_{lo}, Equation 20.7 gives the **Carnot efficiency**

$$\eta^c = \frac{T_{hi} - T_{lo}}{T_{hi}}.$$

Carnot's theorem states:

No heat engine can be more efficient than a Carnot engine operating between the same two reservoirs.

There are two corollaries:

No reversible engine can be less efficient than a Carnot engine operating between the same two reservoirs.

and therefore

All reversible engines operating between the same two reservoirs are equally efficient.

We never see a thermodynamic process, involving a system of ordinary size, that violates the second law of thermodynamics. Such a violation may be regarded as an evolution of the system from a more orderly to a less orderly state *in the absence of interaction between the system and its surroundings*. Seen from a statistical point of view, such a process does not in fact violate any laws of physics. The process is simply so improbable that, for all conceivable practical purposes, it will never happen.

Queries and Problems for Chapter 20

QUERIES

20.1 *(3) Driving Miss Daisy.* In an air gun, air rapidly injected at high pressure into the small space behind the projectile drives the projectile down the gun barrel. Neglecting the friction between projectile and barrel, explain why the process is reversible up to the point where the projectile reaches the end of the barrel and is irreversible thereafter.

20.2 *(3) Warm-up.* When you park a car in the sun with the windows closed, the temperature inside the car quickly rises to a value much higher than that of the surrounding air. Explain why this situation does not violate the second law of thermodynamics.

20.3 *(3) While you're on your way, . . .* Two heat reservoirs having a finite temperature difference are placed in thermal contact by putting a copper bar between them. The resulting heat-flow process involves an increase in total entropy and is irreversible. If, however, the same amount of heat flows out of the warmer reservoir through a reversible engine instead of a copper bar, the process involves no change in entropy and is reversible. Explain.

20.4 *(3) One thing and another.* For a real heat engine, list as many sources of irreversible behavior as you can think of.

20.5 *(4) No place like home.* Why are heat pumps less attractive for house heating in cold climates than in mild climates?

20.6 *(4) Sun belt.* Heat pumps intended for home use are available in two general types. One type, intended for places like Florida, where the water table is just below the ground surface, uses well water for the heat source in cold weather and for the heat sink in hot weather, when the system is run in reverse as an air conditioner. The other type, more common in most parts of California, where the water table is too far below the ground surface to be readily accessible, uses the outside air for the same purpose. Which system is likely to be cheaper to run?

20.7 *(5) Good old days.* Shown in the next column is a *p-V* diagram of the cycle of the simplest form of reciprocating engine. Steam is admitted to the cylinder beginning at *a*. The pressure quickly (and essentially isometrically) reaches the boiler pressure p_b at *b*. The cylinder then expands isobarically along *b→c* as steam is admitted. At *c*, called *cutoff*, the inlet valve is closed, and the cylinder continues to expand adia-

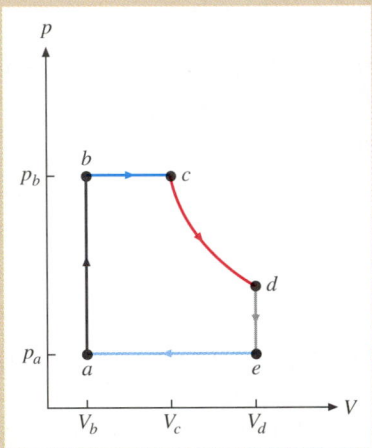

batically along *c→d* to its maximum volume V_d. The exhaust valve is then opened, and the pressure quickly falls (essentially isometrically) along *d→e*, to approximately atmospheric pressure p_a. The piston then pushes the remaining steam out of the cylinder along the isobar *e→a*. The power output of the engine is varied by adjusting the point at which cutoff takes place. Where should cutoff take place for maximum power? for maximum efficiency?

20.8 *(7) Basic poker.* In the simplest form of draw poker, each player receives five cards. The hands are then compared, and the best (least probable) hand wins. Many beginners believe that three of a kind (three cards having the same face value) is more probable than two pair (two pairs of cards, each pair having a common face value). The argument goes thus: Suppose you have one pair. You need have drawn only one more "correct" card to get three of a kind, whereas you must have drawn two more "correct" cards to get two pair. But, as every poker player knows, three of a kind beats two pair. Criticize the argument, and show why it is incorrect.

20.9 *(7) Law mosaic.* So-called creation scientists often argue that biological evolution is a scientific impossibility because it violates the second law of thermodynamics. The argument goes like this: Complex organisms are more elaborately organized than simple ones, in the same sense that a deck of 52 playing cards in standard order is more elaborately organized than a set of four coins all lying heads up. Therefore, the evolution of complex organisms from simpler ones involves a decrease in entropy, which violates the second law. Criticize this argument.

20.10 *(7) Maxwell's demon.* In his landmark textbook, *Theory of Heat*, James Clerk Maxwell proposed a diabolical method for violating the second law. Gas is contained in a box divided into two parts by a partition with a door in it. The temperatures of the two parts of the box are initially the same. A tiny chap, called Maxwell's demon, stands by the door and watches molecules as they approach in their random motion. When a molecule moving with greater than average speed approaches from the left, the demon opens the door and lets it into the right part of the box. When a molecule moving with less than average speed approaches from the right, the demon opens the door and lets it into the left part of the box. After a while, the temperature of the right part is greater than that of the left part. **(a)** Why does this process violate the second law? **(b)** Wait! The demon must see the approaching molecules so as to decide whether to open the door. Explain qualitatively how the demon's need to see can restore the second law to universal validity. [For a quantitative discussion, see L. Brillouin, *Science and Information Theory*, 2d ed. (New York: Academic Press, 1962), chapter 13.]

20.11 *(7) Sporting chance.* Many of the features of the thermal behavior of matter arise from the statistical behavior of the vast numbers of molecules (or other elementary entities) that make up ordinary bodies. A body as a whole behaves in a predictable manner as a consequence of the random behavior of the individual molecules. For analogous reasons, the apparently unrelated world of gambling games also is governed by the laws of statistics. (Indeed, much of the early modern development of statistics arose from the interest of gamblers.) Because of the common roots of thermodynamics and gambling games in statistical laws, the laws of thermodynamics are often expressed in gamblers' language, thus:

0. You won't be luckier than any other player.

1. You can't win more than what is in the pot.

2. The best you can do in the long run is break even.

3. If you want to leave the game, you can't cash in all of your chips.

Explain the connections between these "gamblers' laws" and the laws of thermodynamics as they are stated formally.

20.12 *(7) The gambler's fallacy.* In a certain game played with two dice, a player betting on 12 wins 20 times the amount bet if 12 turns up. The probability of 12 turning up on any one throw is only $\frac{1}{36}$, and so the bet appears to be a losing proposition. However, a player notes that 12 has not turned up in the last 30 throws. She reasons that 12 is coming up; that is, 12 should turn up in one of the next 6 throws. Therefore, the 20-to-1 bet is very attractive. Criticize this argument. What basic principle of probability does it violate?

20.13 *(7) "They" owe me.* If you throw a penny ten times, the most likely outcome is that you get 5 heads and 5 tails. Suppose, however, that you get 6 heads and 4 tails. It follows that "they" owe you 2 tails. Indeed, if you keep throwing the penny, you will soon get two more tails; "they" are quite punctilious about settling their debts. Analyze this argument. Is it consistent with the principles of *a priori* and *a posteriori* probability?

20.14 *(G) Straight arrow.* You watch a movie in which pool balls, initially at rest and scattered over the table, begin to move, pick up speed, collide in complicated ways, and (except for the cue ball) come to rest in a regular triangular array as the cue ball moves rapidly toward the cue. You conclude that the film has been run backward. **(a)** Are there any laws of mechanics (such as Newton's laws of motion) on which you can base that conclusion? Justify your conclusion on **(b)** a macroscopic thermodynamic argument and **(c)** a microscopic statistical argument.

PROBLEMS

GROUP A

20.1 *(2) Steady work, I.* A steam engine takes in 85 kcal of heat energy from the boiler in each cycle and rejects 78 kcal to the condenser. Neglecting other losses, determine **(a)** how much work the engine does per cycle; **(b)** the thermodynamic efficiency of the engine.

20.2 *(2) (De)generation.* In an electric-generating plant, 92% of the heat of combustion of the fuel is transferred to the steam in the boiler. The turbine converts 52% of the input heat energy into mechanical energy, which drives the generator. The generator converts 98% of the mechanical energy into electrical energy. **(a)** What is the thermodynamic efficiency of the turbine? **(b)** What is the overall efficiency of the plant?

20.3 *(2) Oceans of power.* The temperature of the surface water of the Caribbean Sea is about 28°C. At depths below about 200 m, the temperature is close to 4°C. **(a)** If you could design a seaside generating plant in which a Carnot engine used the surface water as a heat source and the deep water for a heat sink, how many tonnes of cold water per hour would you have to pump to the surface to generate 1 MW of electric power? Assume a design in which the cooling-water temperature increases by 6 C° in the condenser. **(b)** What are likely to be the main design problems in designing a practical power plant of this type? (Don't be too pessimistic; pilot plants show some promise!)

20.4 *(2) Steady work, II.* A steam engine produces 7.5 kW of output power when it operates at 725 rev/min. The engine is double acting—that is, it gives two power strokes per revolution of the shaft. If the efficiency of the engine is 10%, **(a)** how much heat energy must be provided per hour? **(b)** how much heat energy is rejected to the condenser per hour? Express your results in joules and in kilocalories.

20.5 *(2) Squaring the cycle, I.* A hypothetical engine operates with a cycle consisting of two isobars and two isometrics. The working fluid consists of 1 kmol of monatomic ideal gas. If the minimum and maximum volumes are $V_{min} = 25$ m^3 and $V_{max} = 50$ m^3 and the minimum and maximum pressures are $p_{min} = 1$ atm and $p_{max} = 2$ atm, what is the efficiency of the engine? (Hint: Sketch the cycle before you begin to calculate.)

20.6 *(2) Squaring the cycle, II.* **(a)** Find the maximum and minimum temperatures in the cycle described in Problem 20.5. **(b)** What would be the efficiency η^c of a Carnot engine operating between these two temperatures? Compare η^c with the value of η obtained in Problem 20.5.

20.7 *(2) What comes in must go out.* In each cycle, a Carnot engine accepts 2500 J of heat energy from a heat source at $T_{hi} = 450$ K and rejects heat to a sink at temperature $T_{lo} = 320$ K. **(a)** How much work is done per cycle? **(b)** How much heat is rejected to the sink per cycle?

20.8 *(2) Cooling power.* A typical nuclear electric power plant has a rated output of 800 MW and an overall efficiency of 33%. If the plant is to be cooled by water whose temperature is not to rise more than 15 K in the cooling process, at what rate must cooling water be pumped through the plant?

20.9 *(2) Lumberjack.* A strong man in top condition, working hard for a 10-hour day, can sustain a power output of about 100 W. (Much higher outputs can be produced for shorter periods.) He eats about 7000 kcal of food a day, which is about 5000 kcal over the quantity required to sustain him in a resting state. What is the efficiency of the process by which he converts food into work? [Compare with Problem 17.26. For more details concerning the power output and efficiency of human beings and other animals, see G. B. Benedek and F. M. H. Villars, *Physics* (Reading, Mass.: Addison-Wesley, 1973), Section 5D.]

20.10 *(3) Steam power.* A typical reciprocating steam engine, rated at 20 kW, consumes 8.2 kg of coal per hour. The heat of combustion of the coal is 3.2×10^7 J/kg. The temperature of the flame is 1500°C. The boiler temperature is 200°C, and the condenser temperature is 40°C. **(a)** What is the overall efficiency η of the engine? **(b)** Heat transfer from the very hot flame to the much cooler steam in the boiler is a highly irreversible process. Calculate the entropy change per hour due to this process. **(c)** What would be the efficiency η^c of a Carnot engine operating with the same boiler and condenser? **(d)** If a Carnot engine could be operated directly between the flame and the heat sink, what would its efficiency η'^c be? **(e)** Compare the three efficiencies. Where does the major loss occur?

20.11 *(4) Versatile gadget.* A Carnot engine runs with efficiency $\eta^c = 34\%$. The heat-sink temperature is 20°C. **(a)** What is the temperature of the heat source? **(b)** What is the coefficient of performance of the engine run in reverse as a heat pump? **(c)** What is the coefficient of performance of the engine run in reverse as a refrigerator?

20.12 *(4) Filling in the gaps.* Using the method suggested in the text, derive Equation 20.22.

20.13 *(4) High performance, I.* Show that, if an engine of efficiency η is run in reverse as a heat pump, its coefficient of performance is

$$\omega_{hp} = \frac{1}{\eta}.$$

20.14 *(4) High performance, II.* Find the relation between the heat-pump and refrigerator coefficients of performance of a given device.

20.15 *(4) High performance, III.* Show that, if an engine of efficiency η is run in reverse as a refrigerator, its coefficient of performance is

$$\omega_r = \frac{1}{\eta} - 1.$$

20.16 *(4) Housewarming.* A heat pump is to be used to keep the temperature of a house at 21°C when the outside temperature is 5°C. Heat leaks from the house to the outside at a rate of 15 kW. **(a)** What would be the coefficient of performance of a Carnot heat pump operating under these conditions? **(b)** What power input would the Carnot heat pump require? Compare your result with the 15 kW required to heat the house directly—say, with an electric heater. **(c)** Real heat pumps are considerably less efficient than Carnot heat pumps. A typical coefficient of performance is 3. What power input would a real heat pump require to maintain the desired temperature in the house?

20.17 *(4) Carnot heat pump.* Show that the coefficient of performance of a Carnot heat pump is

$$\omega^c_{hp} = \frac{T_{hi}}{T_{hi} - T_{lo}}.$$

20.18 *(4) Carnot refrigerator.* Express the coefficient of performance ω^c_r of a Carnot refrigerator in terms of the source and sink temperatures.

20.19 *(4) Does the light really go out when you close the door?* A typical home freezer is required to maintain a steady internal temperature of -18°C. When the outside (room) temperature is 21°C, the 250-W motor runs about 15% of the time. **(a)** What would be the coefficient of performance of a Carnot refrigerator operating under these conditions? **(b)** The coefficient of performance of ordinary household freezers is usually about 3.5. What is the rate of heat leakage into the freezer through the walls?

20.20 *(4) Ice cubes and hot tea.* A Carnot refrigerator removes heat from a quantity of water at 0°C and rejects it to another quantity of water at 100°C. How much cold water is frozen when 1 kg of hot water is turned into steam?

20.21 *(4) Cool comfort.* The *cooling capacity* of an air conditioner is the rate at which it can remove heat energy from the space it is cooling at specified outside and inside temperatures. A typical cooling capacity for a window air conditioner is 1800 W when the outside temperature is 32°C and the desired room temperature is 22°C. **(a)** If the air conditioner were a Carnot device, what would its refrigerator coefficient of performance be? (Hint: See Problem 20.18.) **(b)** What output power would be required of the motor that drives the air conditioner? **(c)** Real air conditioners of 1800-W cooling capacity have motors that draw 600 W from the power line. Assuming that the efficiency of the motor is a typical 85%, what is ω_r for the air conditioner? **(d)** Express ω_r as a fraction of ω^c_r.

20.22 *(7) Fast shuffle.* The number of different ways in which a shuffled deck of 52 playing cards can be ordered is

$$52 \times 51 \times 50 \times \cdots \times 3 \times 2 \times 1 \equiv 52!.$$

(This number is called "52 factorial.") (a) Derive this result. To do so, imagine that you are blindfolded. The cards are scattered on a table; you are picking them up one by one and making a neat stack of them. (b) The value of 52! is approximately 8.07×10^{67}. If you could shuffle the cards and determine their order once a second (that's pretty fast!), how many years would it take before you would likely come up with the deck in standard order? (c) Compare the result of part b with the age of the universe, approximately 2×10^{10} years. If a little green man has been shuffling once a second all that time, what are the odds against his having come up with the standard order by now?

20.23 (7) *Shuffling up some entropy.* Using the value of 52! given in Problem 20.22, calculate the entropy change resulting from shuffling a brand new deck of cards. Express your result (a) as a multiple of Boltzmann's constant k; (b) as a numerical value. (c) Note that the value found in part b is many orders of magnitude smaller than the entropy change $\Delta S = 2.05$ J/K calculated for an ordinary process in an ordinary thermodynamic system in Example 19.4 or the values of similar magnitude calculated in Problems 19.23 through 19.25. How do you explain this large discrepancy?

GROUP B

20.24 (2) *Economy car.* The gas consumption of a small automobile in steady travel at 75 km/h is 22 km/liter. The overall efficiency of the car in converting the chemical energy of the gasoline into mechanical work is 11%. The heat of combustion of gasoline is 8.52 kcal/liter. What is the magnitude of the force exerted by the wheels on the road?

20.25 (2) *Carnot knowledge.* Beginning with Equation 19.34, $W_{cyc} = -nR(T_{hi} - T_{lo}) \ln (V_b/V_a)$, find the efficiency of a Carnot engine that employs an ideal gas for its working fluid. Show that the result is the same as that derived more generally and expressed by Equation 20.7.

20.26 (2) *Taking the heat.* A Carnot engine takes in heat $|Q_{hi}|$ and rejects heat $|Q_{lo}|$. Express the ratio $|Q_{hi}/Q_{lo}|$ in terms of the efficiency η^c of the engine.

20.27 (2) *Circular reasoning?* Find the thermodynamic efficiency of the engine cycle illustrated here. (Hint: See Problem 19.35.)

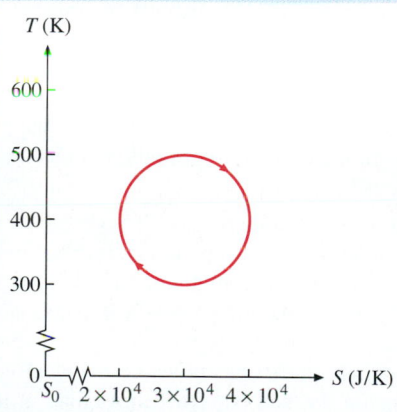

20.28 (3) *Joule free expansion.* The following experiment, which is also the subject of Problem 19.31, was first performed by Joule. A gas-tight container is divided into two equal parts, each of volume V_a, by a thin wall. One part contains n kmol of air at atmospheric pressure p_a, and the other is evacuated. The temperature of the air is measured carefully. The thin wall is then ruptured, and air very quickly fills the entire container. The temperature is measured again. (a) In this experiment, Joule was unable to detect any temperature change. How does this observation support the assertion that the air closely approximates an ideal gas? (b) The expansion

process is *irreversible.* Although experiment shows that the initial and final states have the same temperature, the process cannot be plotted on a p-V diagram because there are no well-defined intermediate states between the initial and final states. However, the initial and final states—call them a and b—are well defined. The state variables pertaining to them do not depend on the way in which the states were attained. With this in mind, consider the simplest possible *quasistatic, reversible* process $a{\to}b$ that can take the system from a to b. Sketch the process on a T-S diagram and on a p-V diagram. (c) Show that the entropy change ΔS of the system over any process (reversible or irreversible) that goes from a to b is

$$\Delta S = nR \ln 2.$$

(d) What is the entropy change of the surroundings over the reversible process? What is the net entropy change of the universe? (e) What is the entropy change of the surroundings over the irreversible free-expansion process? What is the net entropy change of the universe? (f) How do your results concord with the second law of thermodynamics?

20.29 (3) *Mix and stir.* A container holds 1 kmol of ideal gas at temperature T_1. An identical container holds 1 kmol of ideal gas at temperature T_2. The two containers are brought into thermal contact. Show that the entropy of the two-container system always increases unless $T_1 = T_2$. (Hint: For any two unequal numbers α and β, $\frac{1}{2}(\alpha + \beta) > \sqrt{\alpha\beta}$.)

20.30 (4) *Supercooling.* (a) Calculate the entropy change of 1 kg of water at 0°C as it freezes and is then cooled to −5°C. Use the value of the specific heat capacity of ice given in Table 17.3. (b) Pure water in a clean container can be *supercooled*— that is, cooled to a temperature below 0°C without freezing— if the cooling is carried out slowly in a vibration-free environment. Supercooling cannot be carried out indefinitely; at some temperature below the ordinary freezing point, the water freezes spontaneously and quite suddenly. Calculate the entropy change of 1 kg of water at 0°C as it is supercooled to −5°C and then freezes. (The specific heat capacity of water and its latent heat of freezing vary negligibly over this temperature range.) (c) Using the second law of thermodynamics, explain the difference between the results of parts a and b. What kind of thermodynamic process is the sudden freezing of supercooled water? (d) If the results of parts a and b were not different, how could you run a perpetual-motion machine of the second kind?

20.31 (5) *Graphical view of Carnot's theorem.* In the figure below, a Carnot cycle is shown on a *T-S* diagram. **(a)** Express the Carnot efficiency in geometric terms, as a ratio of areas.

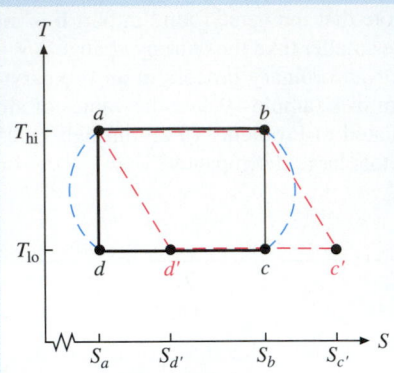

(b) One way to try to draw a more efficient cycle, given the same limiting temperatures T_{hi} and T_{lo}, is to change the shape of the cycle. First, try substituting for the rectangle a parallelogram, as shown. What happens to the efficiency? **(c)** Next, try increasing the area enclosed by the rectangle by substituting bowed sides for the straight sides. The resulting engine produces more work than the original Carnot engine, but what is its efficiency? **(d)** Consider all possible cycles lying completely between the same limiting temperatures. Which cycle "casts the smallest shadow" on the S axis per unit enclosed area?

20.32 (5) *Squaring the cycle, III.* A heat engine whose working fluid is an ideal gas operates on a cycle consisting of two isobars and two isometrics. Express the thermodynamic efficiency of the engine in terms of γ only.

20.33 (5) *Otto cycle.* Problem 19.58 concerns the Otto cycle, as the idealized gasoline-engine cycle is called. **(a)** Express Q_{in} in terms of T_a and T_d. **(b)** Using the result of part **c** of Problem 19.58 (which is stated in that problem), show that the efficiency of an Otto engine can be expressed in the simple form

$$\eta = 1 - r^{1-\gamma},$$

where r is the compression ratio V_b/V_a.

20.34 (5) *Diesel cycle.* The *p-V* diagram below shows an idealization of the cycle of the Diesel engine [named after its inventor, the German engineer Rudolf C. K. Diesel (1858–1913)]. Air is drawn into the cylinder during the intake stroke

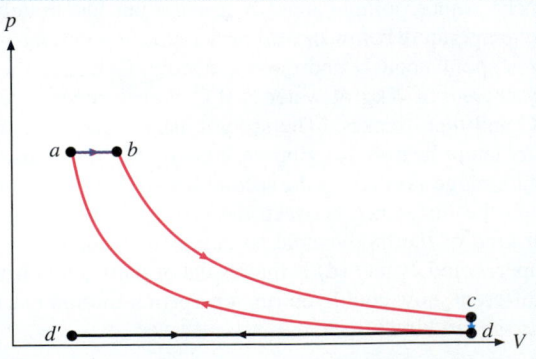

$d' \rightarrow d$. The intake valve is then closed, and the air is compressed adiabatically along $d \rightarrow a$. The compression ratio is large, and T_a is quite high. Fuel (usually a medium-weight petroleum oil, but Diesel engines are not fussy about fuel) is sprayed into the cylinder, beginning at a, by a precision high-pressure injector. The first fuel to enter the cylinder ignites immediately. In the combustion process $a \rightarrow b$, the rate of fuel injection (and thus the rate of burning) is controlled in such a way that, as the piston moves and the cylinder expands, the process is isobaric up to b, where injection terminates and combustion ceases. The remaining expansion $b \rightarrow c$ is adiabatic. When the cylinder is fully expanded at c, valves open and there is a rapid isometric pressure drop to the exhaust pressure p_d. The piston then drives the remainder of the spent gas out of the cylinder in the exhaust stroke $d \rightarrow d'$. **(a)** Show that the efficiency of the cycle is

$$\eta = 1 - \frac{1}{\gamma}\frac{T_c - T_d}{T_b - T_a}.$$

(b) As in the Otto cycle (see Problem 20.33), define the compression ratio to be $r \equiv V_d/V_a$. Also, define the *isobaric expansion ratio* $\beta \equiv V_b/V_a$. Show that the efficiency can be written

$$\eta = 1 - \frac{\beta^\gamma - 1}{\gamma r^{\gamma-1}(\beta - 1)}.$$

(c) Show qualitatively that, as in the simple steam cycle described in Query 20.7, output work and efficiency depend on β in opposite ways. (Engine design requires a careful compromise.) **(d)** For a typical Diesel engine, $r = 16$ and $\beta = 2.5$. Find the theoretical efficiency η.

20.35 (7) *But what if one lands on edge?* Construct a table like Table 20.1 for a five-coin toss. **(a)** Compare the probability of either of the two most probable outcomes with the probability of the most probable outcome for the four-coin toss. **(b)** Compare the probability of either of the two least probable outcomes with the probability of the corresponding outcomes for the four-coin toss. **(c)** What happens to the "sharpness" of the distribution of outcomes as the number of coins increases?

20.36 (7) *Don't hold your breath!* If your bedroom is of ordinary size (about $4 \times 4 \times 3$ m^3), it contains about 2 kmol of air molecules. In imagination, divide the room into two equal parts, one of which contains your bed. Because any molecule is just as likely to be in one part of the room as the other, the probability of any particular molecule being in the part that does not contain the bed is $\frac{1}{2}$. **(a)** Show that, at any instant, the probability that you will suffocate in your sleep because all the air molecules are in the other half of the room is about $2^{-1.2 \times 10^{27}}$. **(b)** The speed of an air molecule is about 300 m/s on the average at room temperature. What is the order of magnitude of the time Δt it takes for a molecule to cross from one part of the room to the other? **(c)** If the calamity described in part **a** is not taking place right now, you need not worry that it will happen for the next Δt or so. But you can never be too careful! Suppose you check the situation every Δt. How many years do you have before you are likely to succumb to the terrible fate of suffocation in your sleep? **(d)** The population of the world is of order 10^{10} persons. How long is it likely to be before *anyone* dies in his or her sleep in this terrible way?

[Note: The results of parts **c** and **d** are gross underestimates because it takes much longer than the time interval Δt to suffocate (or explode into the vacuum). Take heart!]

20.37 *(7) Gibbs's paradox.* **(a)** A partition divides a box into two equal parts. One part is filled with N_2, and the other part is filled with an equal quantity of CO. (Note that the two gases have equal molecular masses, so the quantities are equal both in mass and in number of molecules.) When you remove the partition, the molecules mix. Using the result of part **c** of Problem 20.28, show that the entropy change is $\Delta S = 2nR \ln 2$. **(b)** You repeat the experiment, beginning this time with equal quantities of N_2 filling both parts of the box. When you remove the partition and the molecules from the two parts mix, what is the entropy change of the system? **(c)** Now you replace the partition. Is the final state of the system distinguishable from the initial state? If so, how can you account for a nonzero answer to part **b**? If not, what is the essential difference between the experiments of parts **a** and **b**? **(d)** This paradox, originally proposed by the American theoretical physicist Josiah Willard Gibbs (1839–1903), was resolved in the late 1920s, when it became clear on the basis of quantum mechanics that identical particles (such as two N_2 molecules) are indistinguishable *in principle*. (You can't paint little numbers on them!) Explain how this indistinguishability resolves the paradox.

20.38 *(G) Choo-choo.* The most common form of steam locomotive is driven by a reciprocating engine having two cylinders. One cylinder drives the wheels on the left side of the locomotive through a set of driving rods, and the other drives the wheels on the right side in the same manner. In some locomotives, the two cylinders make up a *double-expansion engine*, in which the exhaust steam from the first, high-pressure cylinder is used to drive the second, low-pressure cylinder. The cylinders are usually located at the front end of the locomotive, with their piston rods horizontal. See the reproduction opposite, in which the different sizes of the two cylinders are evident. **(a)** One advantage of a two-cylinder system is the simple way in which power can be delivered to the driving wheels on the two sides of the locomotive. But that can be done just as well with two independent single-expansion cylinders. The real advantage is more subtle. During each cycle, there is significantly less irreversible heating and cooling of the cylinder walls by the steam than would be the case for an equivalent single-cylinder engine, and the efficiency is improved. Explain why this is so. (Hint: When the engine is running at normal speed, the cylinder walls maintain a nearly constant temperature on account of their large heat capacity.) **(b)** In designing a locomotive to optimize the advantage considered in part **a**, how should the total pressure drop between the boiler pressure and the final exhaust at atmospheric pressure be distributed between the two cylinders? **(c)** For smooth and efficient running, maximum traction, and minimum wheel and rail wear, the output power of the two cylinders should be equal. If the diameter of the high-pressure piston is R, what should be the diameter of the low-pressure piston? (Note: By the late nineteenth century, nearly all large stationary and ship-propulsion engines used a three-cylinder, triple-expansion system. Today, reciprocating steam engines are obsolete and rarely seen in practical service.)

low-pressure cylinder   high-pressure cylinder   M.F.Kotowski ©1995

GROUP C

20.39 *(6) Thinking about the third law.* Suppose you use a hypothetical refrigerator to cool a body from an initial temperature T_{hi} to a final temperature T_{lo} as close to absolute zero as you can get it. The total heat capacity of the body, $C(T)$, is some function of temperature. Thus, you can express the heat dQ extracted from the body as its temperature changes by dT as $dQ = C(T)\, dT$. **(a)** Find the entropy change ΔS of the body as its temperature is decreased from T_{hi} to T_{lo}. **(b)** What can you infer about the value of $C(T)$ as $T \rightarrow 0$? **(c)** The result you obtained in part **b** is true for *any* body—including the refrigerator you are using to cool the body itself. Why can't you use a refrigerator—even a hypothetical one—to attain absolute zero?

20.40 *(7) Kelvin's trouble-free separator.* In his first paper concerning the statistical behavior of gases, Kelvin solved the following problem. Under standard conditions, a 1-liter cylinder holds $5n$ molecules, consisting of approximately n molecules of oxygen and $4n$ molecules of nitrogen, where $n \simeq 5 \times 10^{21}$. **(a)** Show that the probability P of finding all of the oxygen in the left one-fifth of the cylinder, and all of the nitrogen in the right four-fifths, is

$$P = \left(\tfrac{1}{5}\right)^n \left(\tfrac{4}{5}\right)^{4n}.$$

(b) To see why people prefer to go to the trouble of building and operating elaborate machinery to separate oxygen from nitrogen, find the numerical value of P.

20.41 *(G) Improved efficiency.* An inventor comes to you, asking for financial support for the development of the heat-engine system shown at the top of the next page. He claims that its efficiency is greater than that of a simple Carnot engine operating between the hottest feasible heat source (temperature T_{hi}) and the coldest available heat sink, whose temperature T_{cw} is that of the available cooling water. Heat engine E operates between a heat source at temperature T_{hi} and a

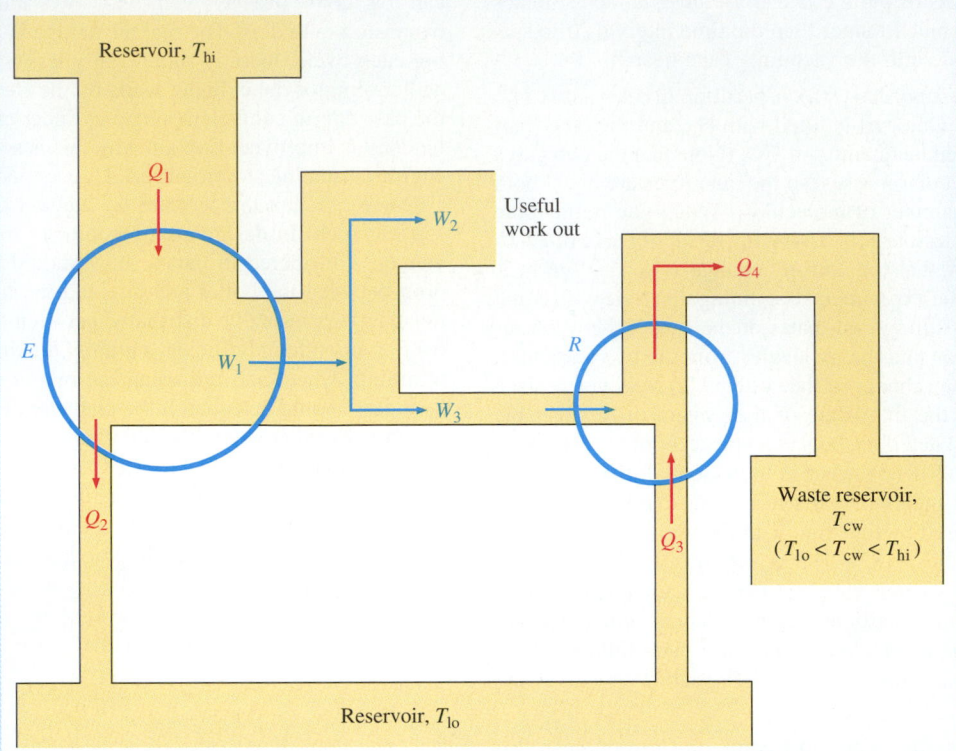

heat sink at T_{lo}. In order to maximize the efficiency of E, the inventor keeps T_{lo} well below T_{cw}. He does this by using part of the output power of the engine to drive refrigerator R, which cools the heat sink and rejects the waste heat to cooling water at T_{cw}. The rest of the output is available for useful work. Using a Carnot engine and refrigerator to set limits on what is practically possible, evaluate the utility of the inventor's project. Specifically, evaluate the efficiency of the system in terms of the three temperatures.

20.42 *(G) The Clausius-Clapeyron equation.* You probably know that, when a liquid is boiled in a container of fixed volume, the pressure rises very rapidly as the temperature increases. (This is what happens in a pressure cooker, up to the point where the safety valve begins to release steam.) The quantitative relation between pressure and boiling temperature is called the *Clausius-Clapeyron equation.* It can be derived by clever use of the Carnot cycle, as follows. Imagine a Carnot engine whose working fluid is water present simultaneously as both liquid and steam. The figure below is a p-V diagram

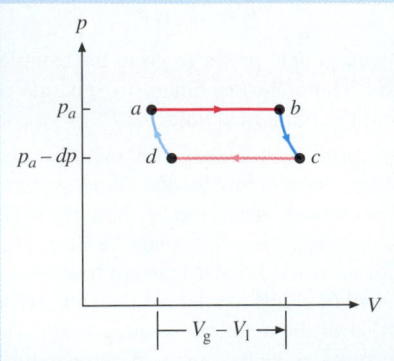

of the Carnot cycle. In the process $a \rightarrow b$, the system expands isothermally, absorbing latent heat L per unit mass as some of the liquid, of mass M, changes into steam. In boiling, the volume of this mass of liquid changes from V_l to V_g. Because it is a boiling process, $a \rightarrow b$ is isobaric at pressure p_a, as well as isothermal. Process $b \rightarrow c$ is an *infinitesimal* adiabatic expansion, in which the pressure decreases from p_a to $p_a - dp$, with corresponding changes dV in volume and dT in temperature. Process $c \rightarrow d$ is an isothermal compression, which is also isobaric as the mass M of steam changes back into water. Finally, $d \rightarrow a$ is an infinitesimal adiabatic compression back to the initial state. **(a)** What is the input heat $đQ_{in}$ per cycle? **(b)** What is the cyclic work $đW_{cyc}$? **(c)** Express the Carnot efficiency of the cycle in terms of temperature only, and show that

$$\frac{dT}{T} = \frac{V_g - V_l}{ML}\, dp \simeq \frac{V_g}{ML}\, dp.$$

(d) Make the (crude) assumption that steam approximates an ideal gas even when it is just at its boiling temperature, and derive the relation

$$\frac{dp}{p} = \frac{\mu L}{R}\frac{dT}{T^2},$$

where μ is the molecular mass of water (or of whatever other substance you may wish to boil) and $\mu L \equiv \Lambda$ is the latent heat of boiling *per kilomole.* The form $dp/dT = \mu Lp/RT^2$ of this equation is the *differential form of the Clausius-Clapeyron equation.* **(e)** Assuming that Λ does not vary significantly with temperature, integrate both sides of the equation of part **d** to obtain the desired relation between pressure and boiling temperature,

$$p = Ce^{-\Lambda/RT},$$

where C is a constant of integration. This is the *integral form of the Clausius-Clapeyron equation*. **(f)** Find the value of C for water. **(g)** Pressure cookers for household use usually operate at temperatures close to 110°C. Use the differential form of the Clausius-Clapeyron equation to find the overpressure (the pressure in excess of atmospheric pressure) needed to produce this temperature. Assume, for this modest pressure, that the standard value of the latent heat at atmospheric pressure is close to correct.

20.43 *(G) Do mountaineers like soft-boiled eggs?* Assume the validity of the isothermal barometric equation (Equation 18.51), and let the fixed temperature of the atmosphere be T_a. **(a)** Using the Clausius-Clapeyron equation (see Problem 20.42), show that at altitude h the boiling temperature $T(h)$ of water is given by the equation

$$\frac{1}{T(h)} = \frac{1}{373.15 \text{ K}} + \frac{gh}{LT_a},$$

where g is the acceleration of gravity and L is the latent heat of boiling of water (per kilogram). **(b)** What is the boiling temperature of water at an altitude of 3500 m? Take $T_a = 280$ K.

Part *VII*

The Mechanics of Wave Motion

When you pluck a guitar string near one end, the entire string is soon set into motion. More generally, when an extended body is disturbed at a certain point, the disturbance travels throughout the body. This is the phenomenon called *wave propagation*, and the disturbance is called a *mechanical wave* or just a *wave* for short. Very often, our interest centers on the disturbance, rather than on the detailed motion of the various parts of the body. In such cases, we think of the extended body as the *medium* in which the wave propagates. We concentrate on the *form* of the disturbance that carries energy and information through the medium. In doing so, we study on a formal level what your own sense of hearing does all the time. You can interpret sounds of all sorts in exquisite detail, extracting information from the form of the sound waves without reference to the motion of the medium—in this case, the air— through which the sound waves travel from their sources to your ear.

Although our interest centers on the wave, we cannot ignore the mechanical properties of the medium because those properties govern the way the wave propagates and, in particular, its speed. The study of waves, just like any other mechanical phenomenon, is ultimately based on Newton's laws.

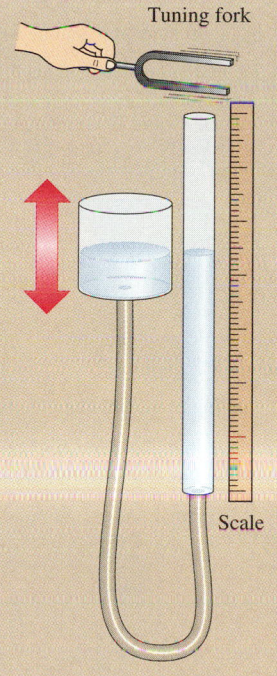

Tuning fork

Scale

In Chapter 21, we develop the kinematics and dynamics of mechanical waves in the relatively simple context of a one-dimensional medium—the vibrating string. The largest class of musical instruments, the stringed instruments, affords the most familiar examples of the vibrating string. The ends of the string have great influence on the waves in the string; we will consider these influences, called *boundary conditions*.

In Chapter 22, we apply the principles developed in the context of the vibrating string to a wider range of media. Some media have properties that have no analogues in the vibrating string, and we extend the theory of waves to deal with these new properties. In view of the importance of sound, we focus much of our attention on that particular wave phenomenon.

There are other wave phenomena—notably light—that occur in the absence of a mechanical medium. We defer consideration of such nonmechanical waves to Part X. Even though light has many properties very different from those of mechanical waves, most of the basic physical and mathematical ideas developed in the next two chapters are essential to the study of light and related wave phenomena.

Waves I

Mostly Transverse Waves

L O O K I N G   A H E A D

When a disturbance travels through a medium (as ripples travel on the surface of water), we are mainly interested in the motion of the disturbance rather than the restricted motion of any portion of the medium.

Disturbances travel through a medium at a speed that depends on the properties of the medium.

An oscillating source that is coupled to a medium results in the propagation of a periodic wave through the medium. The wave is characterized by its frequency and wavelength.

Any number of waves can pass through the same region at the same time. Often, the displacement of the medium under the influence of all the waves is the sum of the displacements produced by the waves individually; this is the superposition principle.

When a wave travels down a frictionless string, it is reflected repeatedly at the ends. Superposition of the waves traveling in opposite directions results in standing waves.

Left: The tremendous energy required to form the mountain of water that sweeps the surfer along was imparted to the water by a storm hundreds of miles away and was carried by the waves. None of the water originally disturbed by the storm is present here.

567

*Life is a wave, which in no two consecutive moments
of its existence is composed of the same particles.*

—JOHN TYNDALL, *Fragments of Science*

Introduction

A body is located at some distance from us. How can we set it into motion? One way is to throw something at it; in the collision, the distant body acquires kinetic energy. But there is another, much more common way to transmit energy across a distance—the *propagation of waves*. Imagine a quiet pond on a windless morning. A leaf floats motionless on the surface. Suddenly a fish jumps, and the water surface is disturbed. Ripples spread out in all directions over the surface, as in Figure 21.1. When we speak of ripples, we mean that the water at any particular location oscillates up and down with respect to the undisturbed water surface. Eventually, ripples reach the leaf, remote from the original disturbance. The leaf begins to oscillate up and down. The motion of the leaf is connected only *indirectly* with the original disturbance. None of the water directly moved by the fish ever touches the leaf. Indeed, nothing in the entire system moves very far from its original undisturbed position—except, perhaps, the fish.

Thus, in studying waves we are primarily concerned not with the motion of a particle or an extended body, or even with the flow of a fluid. Rather, we are concerned with the motion of a *disturbance*.

There are many phenomena in which the central action is the propagation of a disturbance away from a source. A guitarist plucks a string at one point at one instant, but the whole string (and indeed the whole guitar) vibrates for some time afterward. You clap your hands, and you can be heard for some distance in all directions. A fault deep in the earth slips; the resulting oscillations are measurable at seismographic stations all over the world.

All of these phenomena are examples of *mechanical waves*, which are produced when an initial local disturbance sets a *medium* into oscillation. For the guitar string, the medium is the string itself; the wave propagates back and forth along the string in one dimension. For the pond, the medium is the water surface, and the wave propagates over the two-dimensional surface. For the clapped hands, the medium is the air; for the

FIGURE 21.1 A set of circular ripples propagating on the surface of a pond.

earthquake, the medium is the earth itself. In these last two cases, the waves propagate through the three-dimensional media.

Not all waves are mechanical. Other kinds of waves must be understood in more abstract terms because there is no medium and thus nothing to disturb mechanically. The most familiar *nonmechanical wave phenomenon* is light, which can and does propagate through the vacuum of interstellar space. Nonmechanical waves are the subject of Chapter 31. In this chapter and the next one, we are concerned exclusively with mechanical waves. For the most part, we will study one-dimensional waves because mathematical analysis is easiest in one dimension. Fortunately, one-dimensional waves exhibit nearly all the fundamental properties of waves.

We begin in Section 21.2 by considering how a long, stretched string acts when it is plucked once—that is, when a single point on the string is pulled in a direction perpendicular to the string and released. In Section 21.3, we study the effects of a repeated disturbance, in which a device attached to the point on the string drives that point in harmonic oscillation perpendicular to the string. In both of these sections, we will be concerned only with the kinematics of the resulting wave motion—that is, with a description of how the disturbance propagates along the string.

In Section 21.4, we derive some mathematical properties of waves and then briefly study the dynamics of the vibrating string. That is, we consider not only the motions, but the forces that generate them. We took an analogous approach in Part I, where we began our study of particle motion with kinematics and then went on to study dynamics.

In Section 21.5, we consider quantitatively the way in which waves carry energy from one place to another. Section 21.6 concerns the *superposition principle*, which describes what happens when several waves pass simultaneously through the same region. In Section 21.7, we study the reflection of waves at the boundary of the propagation medium and learn how reflection leads to the phenomenon of *standing waves*.

Propagation of a Pulse on a Long, Stretched String

The formation and propagation of a *pulse* on a long string are shown in idealized fashion in Figure 21.2. The idealization is not farfetched, as you can see by comparison with the set of photographs shown in Figure 21.3. Figure 21.2a shows the undisturbed string, stretched tight so that it does not sag. The figure shows only the part of the string from the point of disturbance toward the right; for the time being, we ignore what happens to the left of the point of disturbance. In Figure 21.2b through d, the fingers holding the string snap it quickly, bringing it first up and then down and finally to rest at the initial position. The point of disturbance is called a **source**.

The disturbance distorts the string. Initially, the distortion is limited to the immediate vicinity of the source. As the source returns to its original position, the distortion moves down the string. This distortion is called a **transverse pulse**.* The pulse moves because the displaced part of the taut string pulls on the neighboring part to the right and displaces it upward; you can see this in Figure 21.2e through g, which represents the string at instants separated by equal time intervals.

The pulse moves equal distances in these equal time intervals. So we can say that the pulse propagates at a constant velocity, called the **wave velocity**. It is important to bear in mind that what is moving at the wave velocity is *not* any material object, but a *shape*. When the various elements of the string move, they move up and down. Because they do not move in the same direction as the pulse moves, they certainly do not possess the wave velocity.

---

*By *transverse*, we mean that the distortion of the string—the pulse—is perpendicular to the direction in which the pulse moves. We consider longitudinal pulses, in which the distortion lies along the direction of propagation, in Section 22.3.

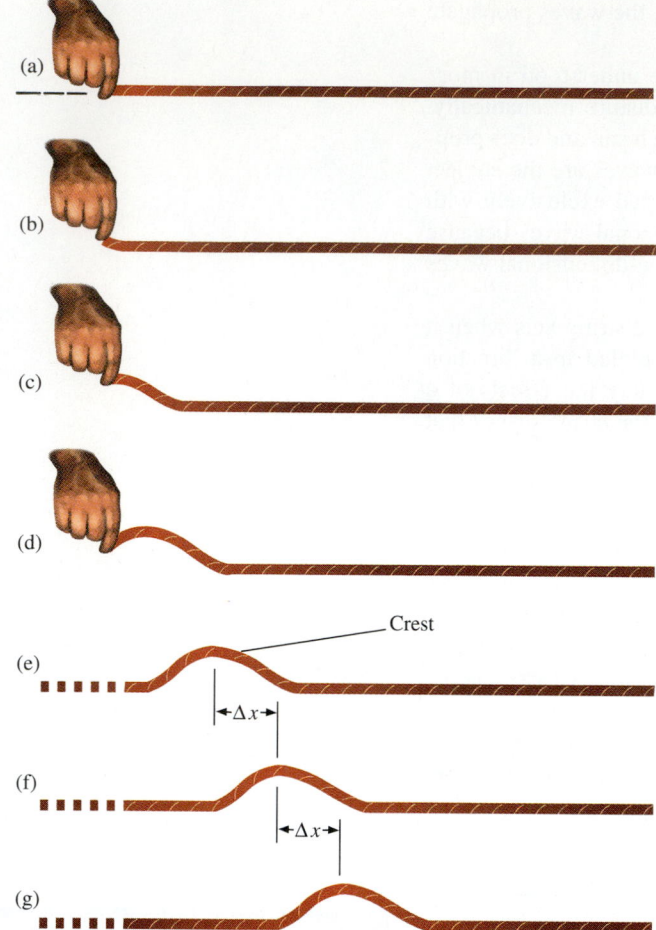

(a)

(b)

(c)

(d)

Crest

(e)

$\leftarrow\!\Delta x\!\rightarrow$

(f)

$\leftarrow\!\Delta x\!\rightarrow$

(g)

FIGURE 21.2 A single pulse-forming disturbance on a very long stretched string. The string stretches far to the left of the diagram as well (as suggested by the dashed line in part *a*), but we ignore that part of the string. The time interval between successive diagrams is constant. (*a*) The undisturbed string. (*b–d*) An external disturbance results in the formation of a pulse. (*e–g*) The pulse-forming event has ended. The pulse propagates to the right at constant speed, as elements of the string disturbed earlier return to their undisturbed positions sooner than do elements disturbed later. The maximum transverse (crosswise) displacement of the string is the pulse amplitude *A*.

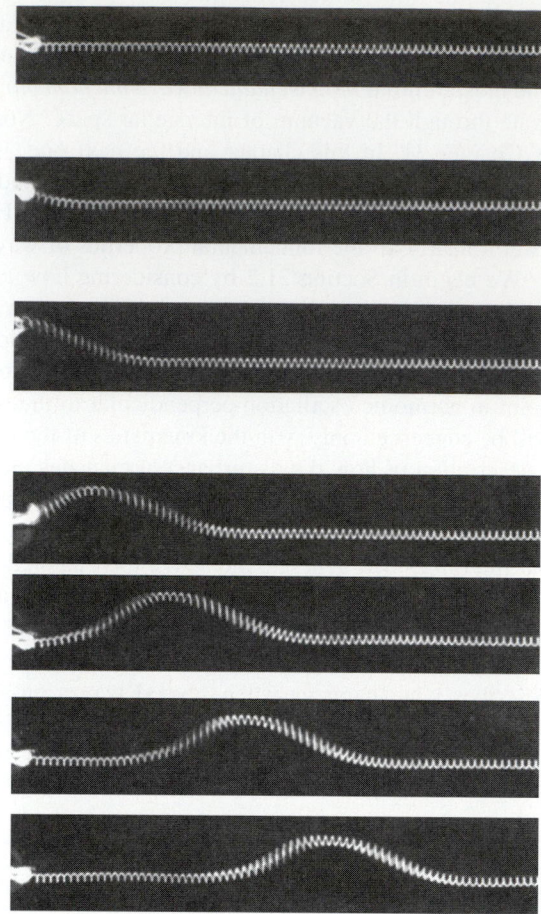

FIGURE 21.3 Photographs of the phenomenon shown in idealized form in Figure 21.2. The time interval between successive photographs is constant. A long coil spring (similar to the toy called a Slinky) is used in place of a string because it makes possible a large transverse displacement and a relatively small wave velocity. But the behavior shown is essentially the same as that of a string.

(From Haber-Schaim: *PSSC Physics.* Copyright 1991 by Kendall/Hunt Publishing Company. Used with permission.)

How do we express the wave velocity mathematically? Let us pick a particular identifiable feature on the pulse—say, the highest point on it, called the **crest**—and measure the distance Δx that this point moves in time Δt. This distance is shown in Figure 21.2. We define the wave velocity to be

$$v \equiv \frac{\Delta x}{\Delta t}. \tag{21.1a}$$

If we choose the time interval to have the infinitesimal value dt, the corresponding displacement is dx, and we write

$$v \equiv \frac{dx}{dt}. \tag{21.1b}$$

In either form, this definition is completely analogous to the definition of the velocity of a particle moving in one dimension.

The Wave Speed

What determines the speed at which the pulse moves along the string? Figure 21.4 shows a segment of a string along which a pulse is traveling in the positive x direction. The source is somewhere to the left of the segment shown. We take the positive y direction upward. At some time t, the leading edge of the pulse just reaches the previously undisturbed point on the string at x, as indicated by the kink (sharp corner) at that point in Figure 21.4a. The string to the left of x is distorted by the pulse. Upward displacement of the point at x begins at t.

(a) Time t

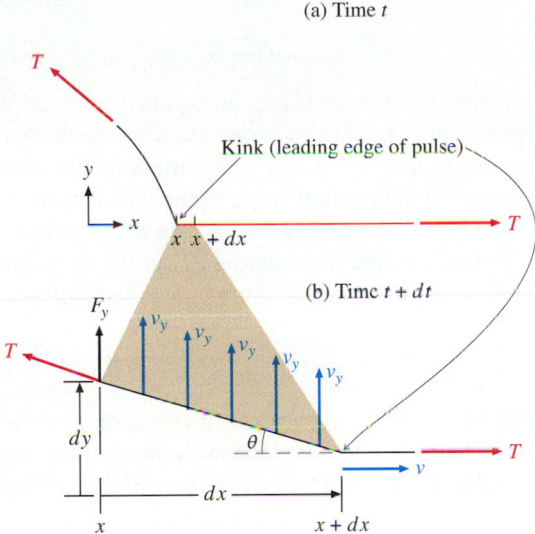

(b) Time $t + dt$

The magnification in Figure 21.4b shows an infinitesimal element of the string, located immediately to the right of x, at time $t + dt$. By that time, the leading edge of the pulse has moved from x to $x + dx$. During the time interval dt, the part of the string to the right of x begins bit by bit to follow the point at x in moving upward, as the leading edge of the pulse moves with velocity v. What is the length dx of the part of the string that has been set into motion over the time interval dt? According to Equation 21.1b, it is the product of the time with the wave speed:

$$dx = v\, dt. \qquad (21.2)$$

The string is under tension. That is to say, every segment of it experiences equal forces of magnitude T at its left and right ends. Distorting a segment of the string changes the directions of these tensile forces because such forces must always lie along the string. We know that the x components of the two forces must remain equal in magnitude because no segment of the string moves to the left or the right. Introducing the pulse into the string stretches the string slightly, and this increases the tension slightly. But, if the pulse is small and the string long, we can neglect the change in T. We restrict our consideration to such small pulses, in which the maximum transverse displacement, called the **pulse amplitude** A, is very much smaller than the length of the string. Pulses on strings often conform to this restriction quite well. For example, you can make a piano string sound quite loudly without producing an easily visible transverse displacement.

We do not know the exact shape of the string element distorted during dt. Because of its infinitesimal length, however, we introduce no error in the present analysis if we represent the element as straight. The element makes an angle θ with the x axis. The angle is fairly large in Figure 21.4, but this is for clarity only. In fact, θ is quite small; this is a consequence of the restriction to small pulses.

In representing the infinitesimal string element as straight, we imply that all parts of it are moving upward at the same speed v_y. (The displacement is smaller toward the right because points to the right begin to move later as the leading edge of the pulse travels to the right.) The point at x has been moving upward with speed v_y for the entire time interval dt, and so its transverse displacement is

$$dy = v_y \, dt. \tag{21.3}$$

(Remember that v_y and v are quite different quantities.) Using Equations 21.2 and 21.3, we can write

$$\frac{dy}{dx} = \frac{v_y \, dt}{v \, dt}.$$

You can see by inspecting Figure 21.3b that the slope of the segment is $dy/dx = \tan \theta$. It follows that

$$v_y = v \tan \theta. \tag{21.4}$$

The entire element from x to $x + dx$ is moving with speed v_y at time $t + dt$. At time t, the element was entirely at rest. Thus its momentum has been changed over the time interval dt by an amount $v_y \, dm$, where dm is the mass of the element. The impulse responsible for this change in momentum is due to the unbalanced force on the element in the y direction. As long as the leading edge of the pulse has not reached the right end of the element, the tensile force of magnitude T on the right end of the element is in the x direction and has no y component. But the tensile force of equal magnitude T on the left end has a y component

$$F_y = T \sin \theta. \tag{21.5}$$

Thus F_y is the magnitude of the net transverse force exerted on the element over the time interval dt. (Before and after this time interval, the leading edge of the pulse lies outside the element, and hence $F_y = 0$ except during dt.) The impulse imparted by F_y is thus

$$F_y \, dt = T \sin \theta \, dt.$$

We now equate the impulse to the momentum change:

$$T \sin \theta \, dt = v_y \, dm.$$

Using the value of v_y given by Equation 21.4, we can write this equation in the form

$$T \sin \theta \, dt = v \tan \theta \, dm.$$

We now exploit the fact that θ is small, so $\sin \theta \simeq \tan \theta$. This allows us to write the simplified relation

$$T \, dt = v \, dm. \tag{21.6}$$

The mass dm of the infinitesimal segment depends on the segment length dx and on the mass per unit length μ of the string:

$$dm = \mu \, dx.$$

Substituting this value into Equation 21.6 and rearranging give

$$\frac{T}{\mu} = v \frac{dx}{dt}.$$

According to Equation 21.2, the derivative dx/dt is just the wave speed v. So we have $v^2 = T/\mu$, or

$$v = \sqrt{\frac{T}{\mu}}. \tag{21.7}$$

A pulse on a stretched string travels with a speed proportional to the square root of the tension and inversely proportional to the square root of the mass per unit length of the string. This result makes sense physically. The greater the tension, the greater the unbalanced force F_y and the greater its tendency to pull the string upward and "snap the kink along." The greater the mass per unit length, the greater the inertia of the string, which tends to make it resist being snapped.

The quantities T and μ are peculiar to a stretched string. In other media, the wave speed depends on other quantities. But the argument leading to the value of v is always analogous to the one we have just carried out.

EXAMPLE 21.1

A mass $M = 13.7$ kg hangs from a uniform rope, as shown in Figure 21.5. The rope is 20.0 m long and has mass $m = 1.24$ kg. If you pluck the rope at the upper end, how long will it be before the hanging mass begins to swing?

SOLUTION: When you pluck the rope, you create a pulse, which then travels down the rope. The hanging mass will begin to swing when the pulse reaches it a time t later. You have $t = L/v$, where L is the length of the rope and v is the wave speed at which the pulse travels. Using Equation 21.7, you can write

$$t = \frac{L}{\sqrt{T/\mu}}.$$

Because the mass m of the rope is much less than M, you ignore m in calculating the tension and set $T = Mg$. The mass per unit length is $\mu = m/L$. You thus have

$$t = L\sqrt{\frac{m}{MgL}} = \sqrt{\frac{mL}{Mg}}.$$

Inserting the numerical values, you obtain

$$t = \sqrt{\frac{1.24 \text{ kg} \times 20.0 \text{ m}}{13.7 \text{ kg} \times 9.80 \text{ m/s}^2}} = 0.430 \text{ s}.$$

Although this is not a very long time, you can easily see such a pulse traveling down the rope.

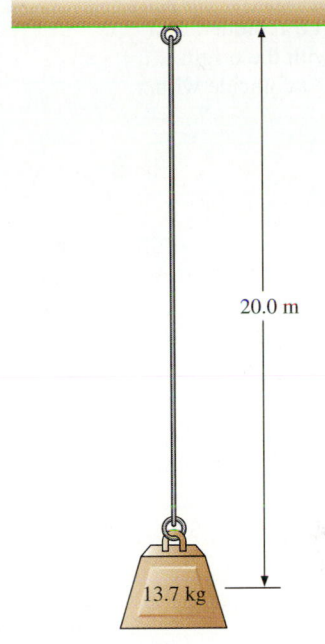

20.0 m

13.7 kg

FIGURE 21.5

Periodic Wave Trains

A source disturbance is not necessarily a single pulse. Indeed, in most cases of physical interest, a source is made to oscillate indefinitely about its equilibrium position. Under these circumstances, each pulse is followed immediately by another pulse of opposite transverse displacement. The pulses follow one another through the medium with speed v and form a **wave train**. The ripples in Figure 21.1 constitute a wave train.

We now develop mathematical language for describing the oscillation of the source, as well as the shape and motion of the wave train that propagates from the source as a result of the source oscillation. Suppose that a source located at $x = 0$ on a string is suddenly displaced from $y = 0$ to $y = A$ at time $t = 0$ and thereafter executes simple harmonic motion with period τ and amplitude A about its equilibrium position. That is, the position of the source is described by the equations*

$$y = 0 \qquad \text{for } t < 0; \qquad \text{(21.8a)}$$

$$y = A \cos 2\pi\left(\frac{t}{\tau}\right) \qquad \text{for } t \geq 0. \qquad \text{(21.8b)}$$

Figure 21.6a is a graph of the position of the source as a function of time. As always, the oscillation period (shown in red) is related to the frequency ν and angular frequency

---

*If you do not feel secure in your command of the kinematics of simple harmonic motion, this is a good time to review Section 6.2.

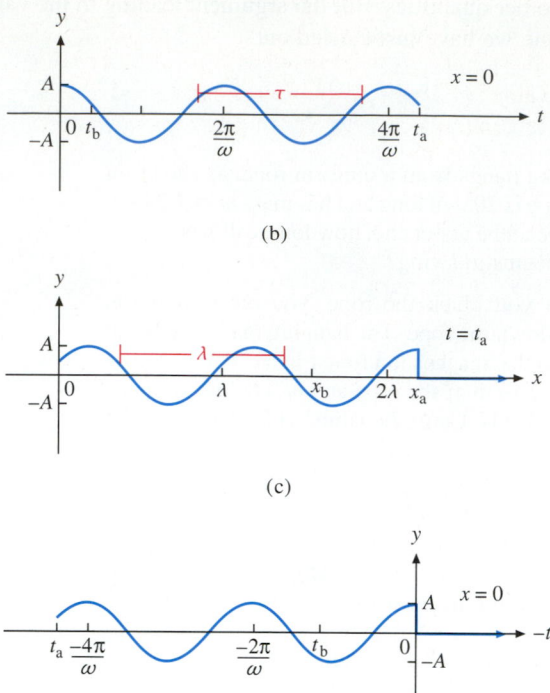

FIGURE 21.6 (a) Transverse displacement y of the source—the particular point on the string at $x = 0$—as a function of time t. The period τ is shown. (b) Transverse displacement y of *all* points $x > 0$ on the string as a function of x, at the particular time $t = t_a$. The wavelength λ is shown. (c) Part a replotted as a function of $-t$, with the origin shifted to make $-t$ coincide with x_a of part b.

ω by the equations

$$\tau = \frac{1}{\nu} = \frac{2\pi}{\omega}. \tag{21.9}$$

As a result of the source oscillation, wave trains propagate down the string in both directions with speed v. For the moment, we consider only the wave train moving to the right. Figure 21.6b shows what this wave train looks like at a particular time $t_a > 0$, just when the leading edge has reached the point x_a. (The abscissa is now an x axis, in contrast with the t axis of Figure 21.6a.) The wave train is periodic; that is, the shape repeats itself. The length occupied by a single repetition of the wave train is called the **wavelength** λ. The shape of the string is described by the equations

$$y = 0 \qquad \text{for } x > x_a; \tag{21.10a}$$

$$y = A \cos 2\pi\left(\frac{x}{\lambda}\right) \qquad \text{for } x \leq x_a. \tag{21.10b}$$

The value of λ in Equation 21.10b is yet to be determined. The location of the cutoff point—x_a at time t_a—is determined by the wave speed v. Because the source begins to oscillate at $t = 0$, the leading edge of the disturbance just reaches x_a when

$$x_a = vt_a. \tag{21.11}$$

The discontinuity at x_a is due to the initial sudden displacement of the source.

There is a remarkable connection between the two graphs of Figure 21.6a and b. The first graph is a plot of the position of a single point on the string as a function of time, and the second is a plot of the positions of all parts of the string at a particular time. In spite of the different horizontal axes, each graph looks just like the other, inverted about the y axis and shifted. That is, each graph looks like the other run backward and shifted to make the ends of the graphs coincide. This is made evident by Figure 21.6c, which is Figure 21.6a replotted as a function of $-t$ instead of t, with t_a located under the origin of Figure 21.6b.

Why this close connection between a time plot and a space plot? Remember that any disturbance originating at the source propagates down the string and that all such disturbances travel at the same speed. Thus, the behavior of the string at any point x at a given instant is identical with the behavior of the source at a certain earlier time. (By *behavior* we mean the instantaneous transverse displacement y, the velocity v_y, and the acceleration a_y of the point.) At a given instant, points successively farther from the source (at greater values of x) mimic the source at successively earlier times. To put it another way, it takes time $t = x/v$ for information about the behavior of any part of the string to reach another part of the string a distance x away. Consequently, *the behavior of the entire string at any instant is a record of the behavior of the source point over time.* Going forward in space is like going backward in time! As time passes, the shape shown in Figure 21.6b for the particular instant t_a travels down the string with speed v. What is true for the special position and time x_a and t_a is true in general:

$$x = vt. \tag{21.12}$$

The wavelength λ in Equation 21.10b serves the same purpose in the space plot of Figure 21.6b that the period τ serves in the time plot of Figure 21.6a. In the time plot, the cosine function of Equation 21.8b completes one cycle when its argument $2\pi t/\tau$ changes by 2π; thus the source completes one oscillation in time $\Delta t = \tau$. Analogously, in the space plot, the cosine function of Equation 21.10b completes one cycle when its argument $2\pi x/\lambda$ changes by 2π; thus a single wave cycle occupies a space $\Delta x = \lambda$, as shown in Figure 21.6b.

The final, quantitative step in describing the close connection between the x and $-t$ plots of Figure 21.6b and c is to find the scale factor that connects the analogous quantities λ and τ. Consider any point such as t_b in Figure 21.6c and the corresponding point x_b in Figure 21.6b. Because the points correspond and the curves have the same form, we can use Equations 21.10b and 21.8b to write

$$A \cos 2\pi \frac{x_b}{\lambda} = A \cos 2\pi \frac{t_b}{\tau}.$$

This equation is satisfied if

$$\frac{x_b}{\lambda} = \frac{t_b}{\tau}. \tag{21.13}$$

We solve for λ and use Equation 21.12 to obtain

$$\lambda = \frac{x_b \tau}{t_b} = \frac{v t_b \tau}{t_b},$$

which yields

$$\lambda = v\tau. \tag{21.14}$$

Because the wave speed v is a property of the medium, the period of the source oscillation determines the wavelength of the wave train. Conversely, given the wavelength, one can infer the period of the source oscillation.

It is useful to rewrite Equation 21.14 in a slightly different form. Because $\tau = 1/\nu$, we have $\lambda = v/\nu$. Solving for v gives

$$v = \lambda \nu. \tag{21.15}$$

For a periodic wave train, the wave speed is the product of the wavelength and the frequency. Although we have derived Equation 21.15 for the special case of a wave propagating along a one-dimensional medium, it is true for waves in multidimensional media as well.

EXAMPLE 21.2

A coloratura soprano sings her high F, whose frequency is 1397 Hz. What is the wavelength of the sound she produces as it propagates through the air? Take the speed of sound to be 347.2 m/s.

SOLUTION: From Equation 21.15 you have

$$\lambda = \frac{v}{\nu} = \frac{347.2 \text{ m/s}}{1397 \text{ Hz}},$$

so

$$\lambda = 0.2485 \text{ m.}$$

Wave Functions and the Wave Equation

Sinusoids

A **sinusoid** is any function that differs from the sine function only by a phase constant (Section 6.2). In particular, the function $\cos x = \sin (x + \pi/2)$ is a sinusoid, the phase constant being $\delta = \pi/2$.

Equations 21.10a and 21.10b afford a way to represent the form of a sinusoidal wave train on a string at a particular time. We say that these equations represent a **time-independent wave function**. Equations 21.8a and 21.8b afford a way to represent a particular point on the string (for example, the source point) as a function of time. We say that these equations represent a **space-independent wave function**. We now develop a way to represent the entire wave at any time, which we call a **complete wave function** or simply a **wave function**. In the interest of simplicity, let us suppose that the disturbance began to propagate from the source a long time ago (at a time $t_0 \ll 0$), so the wave is uniformly sinusoidal over whatever region we care to consider.

The key to generalizing the wave function lies in the relation between two time-independent wave functions representing the shape of the string at two different times. Two such functions are shown in Figure 21.7. (You can think of them as snapshots of the string made at the two times.) The black curve represents the string at $t = 0$, and the red curve represents the string at a later time t. The only difference between the two curves is that the latter one is shifted to the right by a distance vt. In other words, during the time interval t, the whole wave has traveled the distance vt to the right. (That is why we call it a *traveling wave*.) Working backward in time, you can see that the transverse displacement $y(x, t)$ of the string at any location x at any time t is the same as the transverse displacement $y(x - vt, 0)$ of the string at the location $x - vt$, nearer to the source, at the earlier time $t = 0$:

$$y(x, t) = y(x - vt, 0). \tag{21.16}$$

FIGURE 21.7 A time-independent sinusoidal wave function represented at time $t = 0$ (black) and at time t (red). In the intervening time interval, the wave has traveled a distance vt to the right. Therefore, the displacement y of point x_a at time t is the same as the displacement of point $x_a - vt$ was at time $t = 0$.

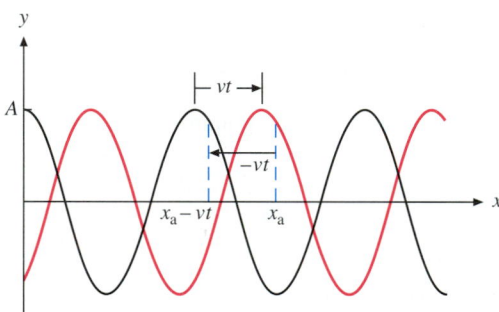

This relation is shown in Figure 21.7 for the particular point x_a.

We can evaluate $y(x - vt, 0)$. To do this, we substitute the argument $x - vt$ into the time-independent wave function given by Equation 21.10b:

$$y(x, t) = A \cos\left[\frac{2\pi}{\lambda} (x - vt)\right] \quad \text{for a wave traveling in the positive direction.} \tag{21.17a}$$

This function is not completely general for two reasons. First, it describes only waves traveling in the positive x direction. A wave traveling in the negative direction is represented by substituting $-v$ for v in Equation 21.17a:

$$y(x, t) = A \cos\left[\frac{2\pi}{\lambda} (x + vt)\right] \quad \text{for a wave traveling in the negative direction.} \tag{21.17b}$$

Second, Equations 21.17a and 21.17b require a choice of $t = 0$ such that the displacement of the source is A at that time. To allow a free choice of the instant $t = 0$, we introduce a *phase constant* δ into the argument of the cosine. The function then assumes the general form

$$y(x, t) = A \cos\left[\frac{2\pi}{\lambda}(x \mp vt) + \delta\right].$$ (21.18)

This is the *complete wave function of a one-dimensional sinusoidal wave*.

It is often convenient to write Equation 21.18 in one of several alternative forms. By using Equation 21.15, $v = \lambda\nu$, you can quickly obtain the form

$$y(x, t) = A \cos\left[2\pi\left(\frac{x}{\lambda} \mp \nu t\right) + \delta\right].$$ (21.19)

Although this form is often useful, we would still like to make explicit the symmetry between the x term and the t term in the argument of the cosine. This symmetry is evident in graphic form in the similarity between parts a and b of Figure 21.6. To make the symmetry equally evident in algebraic form, we write the argument of the cosine as

$$2\pi\frac{x}{\lambda} \mp 2\pi\nu t + \delta = 2\pi\frac{x}{\lambda} \mp \omega t + \delta.$$

In analogy to the relation $\omega = 2\pi/\tau$, we define

$$k \equiv \frac{2\pi}{\lambda}.$$ (21.20)

The quantity k is called the **wave number** because its value is the number of waves—that is, complete cycles of the oscillation—contained in 2π meters of the string (or other medium).* As you can see from Equation 21.20, the SI unit of k is m^{-1}. (Compare with the SI unit of ω, s^{-1}.) In terms of k and ω, Equation 21.19 becomes

$$y(x, t) = A \cos(kx \mp \omega t + \delta).$$ (21.21)

This equation is neat in form and makes clear the similarity of the roles played by x and t in the wave function.

In terms of the wave number and the angular frequency, the wave speed is

$$v = \frac{\omega}{k}.$$ (21.22)

The derivation of this expression from Equation 21.15 is the subject of Problem 21.8.

THINKING LIKE A PHYSICIST

We have traveled a difficult mathematical path in this section, but the difficulty is not in the mathematics (which is rather simple). Rather, the difficulty lies in expressing, in precise mathematical terms, the consequences of the deceptively simple *physical* statement made just before Equation 21.12: *The behavior of the string at any instant is a record of the behavior of the source point over time. Going forward in space is like going backward in time!* This statement is made explicit in Equation 21.21: Changing kx by a certain amount has the same effect on y as changing $-\omega t$ by the same amount.

A goodly part of learning physics is learning how to express physical ideas readily in the language of mathematics.

Although pure mathematics has its own difficulties, you will find it helpful not to confuse those difficulties with the difficulties of translating physical ideas into mathematical language. When we succeed in expressing a physical concept in clear mathematical terms, we are usually rewarded with insights into the behavior of the system described by the mathematics that are deeper than those we could achieve by trying to pursue the concept without mathematics, because with mathematics we can "talk" clearly about the concept. The consequences of the bare statement "Going forward in space is like going backward in time" are difficult to explore; you will see that expression of that statement in Equation 21.21 will take us a long way.

*Do not confuse the wave number with the Hooke's-law force constant, for which the same symbol k is used. They are quite different quantities.

The General One-Dimensional Traveling Wave Function

The shape of a wave on a string need not be sinusoidal. Indeed, the only requirement for a traveling wave is that it must travel! That is, the wave must maintain its shape as it moves away from its source. This requirement means that the transverse displacement $y(x, t)$ of the string at any point x, at any time t, must be the same as the displacement $y(x \mp vt, 0)$ at the point $x \mp vt$, at $t = 0$. We have written the wave equation for a stretched string in a variety of forms, given by Equations 21.18, 21.19, and 21.21, all of which satisfy the requirement. In all of these equations, the argument of the cosine is of the form (constant) $\times$ $(x \mp vt)$ + (another constant), and so we can write the right-hand side of any of these equations in the general mathematical form $f(x \mp vt)$. We assert (and will soon show) that, for any wave that propagates in one dimension, the wave function *must* have the specific form $f(x \mp vt)$; that is,

$$f(x, t) = f(x \mp vt). \qquad (21.23)$$

This is the general expression for a one-dimensional wave function. What it says is subtle but important. The function that describes a traveling wave cannot be just any function of x and t. It cannot, for example, have any of the forms $f = x - 2vt$, or $f = A(\cos ax + \sin bt)$, or $f = x^2/vt$.

By writing the wave function as $f(x, t)$, we express still another generalization. We recognize the fact that, in other media, a wave involves disturbances other than the transverse displacement y of a string. That is, the quantity represented by the function f is not necessarily a transverse displacement y; if we are studying sound waves, for example, f represents the pressure of the air.

The Wave Equation

In spite of the vast variety of phenomena in the domain of classical particle mechanics, all particle motion (and all rigid-body translation) is governed by the single differential equation called Newton's second law of motion. In one dimension, that equation has the form

$$F = m \frac{d^2x}{dt^2}.$$

In their equally vast variety, all waves are governed by a single differential equation called the **wave equation**. In one dimension, as we shall soon see, that equation has the form

$$\frac{\partial^2 f(x, t)}{\partial t^2} = v^2 \frac{\partial^2 f(x, t)}{\partial x^2}, \qquad (21.24a)$$

or, for short,

$$\frac{\partial^2 f}{\partial t^2} = v^2 \frac{\partial^2 f}{\partial x^2}. \qquad (21.24b)$$

The symbol ∂ may be new to you. It represents the operation of **partial differentiation**. Because f is a function of the two variables x and t, we must distinguish between its derivatives with respect to x and its derivatives with respect to t. In taking the **partial derivative** $\partial f / \partial t$, we treat x as a constant. In so doing, we are looking at the way in which a particular point on the string moves in time; that is, we are looking at the *space-independent* properties of the wave. Similarly, in taking the partial derivative $\partial f / \partial x$, we treat t as a constant. In so doing, we are looking at the shape of the wave at a particular instant; that is, we are looking at the *time-independent* properties of the wave.

The wave equation relates the space-independent and time-independent properties of the wave to one another. To be more specific, consider the vibrating string, for which the equation assumes the form

$$\frac{\partial^2 y}{\partial t^2} = v^2 \frac{\partial^2 y}{\partial x^2}. \qquad (21.25)$$

The left side of the equation represents the transverse *acceleration* of any point x on the string. The right side is the product of the square of the wave speed and the *curvature* of the string at the same point x.

We will not derive the wave equation rigorously in this book, but we now justify it and make it plausible. Refer back to the derivation of the wave speed v in Section 21.2, where we considered the unbalanced force on an infinitesimal element of string. That element, shown in Figure 21.4b, had the special property that it was displaced from its equilibrium position at all points except at its right end. It was this asymmetry that resulted in the unbalanced net force given by Equation 21.5, $F_y = T \sin \theta$. But now consider the more general element of string shown in Figure 21.8. Both ends are displaced from equilibrium. As a result, the forces exerted on the element by neighboring elements of string at both ends have y components. The magnitude of the net force on the element is the difference $|F_y| = |T_y{}^+ - T_y{}^-|$ between these y components. Because the length of the segment is fixed, the magnitude of the net force depends on how sharply the segment is curved. The greater the curvature, the bigger the difference between the two y components and the greater the net force. Now, the curvature is the second partial derivative $\partial^2 y / \partial x^2$, which appears on the right side of Equation 21.25. The curvature varies with time, as different parts of the wave pass through the element. By taking the (time-independent) partial derivative with respect to x, we consider a particular instant.

The unbalanced force F_y results in a transverse acceleration $\partial^2 y / \partial t^2$ of the string element. This quantity constitutes the left side of Equation 21.25. The greater the curvature, the greater the acceleration will be, as the equation shows. The proportionality constant between the curvature and the acceleration is the square of the wave speed.

Why is the proportionality constant v^2? Equation 21.25 certainly requires that the constant have the dimensions [length]^2/[time]^2, which are the dimensions of v^2. But let us look at the matter physically in the particular case of the string, where $v^2 = T/\mu$. The greater the tension in the string, the greater the unbalanced force $|F_y|$ will be when the string element acquires a curvature. A greater value of $|F_y|$ results in a greater transverse acceleration $|\partial^2 y / \partial t^2|$. On the other hand, an increase in μ, the mass per unit length of the string, results in increased inertia and thus a diminution in the transverse acceleration.

In other media, the value of v^2 depends on quantities other than T and μ. But v^2 always has the general form

$$v^2 = \frac{\text{force term}}{\text{inertia term}}. \qquad (21.26)$$

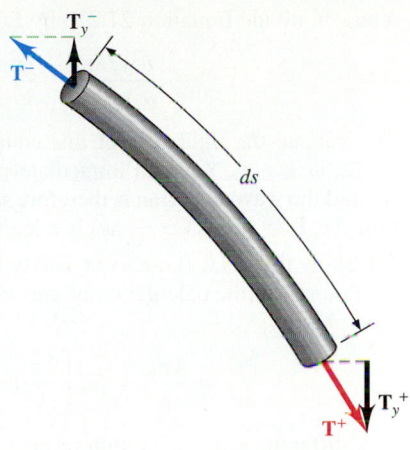

FIGURE 21.8 An infinitesimal element of length ds on a string through which a wave is propagating. The magnitude T of the tension is the same at both ends, but the directions are different owing to the curvature of the element. The net transverse force on the element has magnitude $|F_y| = |T_y{}^+ - T_y{}^-|$. Both the curvature and the angles with respect to the x direction are exaggerated for clarity. In reality, they are quite small, and so $ds \simeq dx$.

EXAMPLE 21.3

(a) Show that the function $y(x, t) = A \cos (kx - \omega t)$, which is Equation 21.21 taken with the negative sign and $\delta = 0$, satisfies the wave equation (Equation 21.25) for any values of k and ω.
(b) Show that the function

$$y'(x, t) = A \cos \left[\frac{2\pi}{\lambda} (x - 2vt) \right]$$

does *not* satisfy the wave equation.

SOLUTION:
(a) Show that $y(x, t)$ satisfies the wave equation.

The function does not itself appear in the wave equation, but two second partial derivatives, $\partial^2 y / \partial t^2$ and $\partial^2 y / \partial x^2$, do appear. You calculate each one in two steps. For the first derivative of $y(x, t)$ with respect to time, you have

$$\frac{\partial y}{\partial t} = -\omega A[-\sin (kx - \omega t)] = \omega A \sin (kx - \omega t). \quad (21.27a)$$

Note that only t is treated as a variable here, so the chain rule

results in the appearance of the factor $-\omega$.

Taking the second derivative of $y(x, t)$ with respect to time gives

$$\frac{\partial^2 y}{\partial t^2} = -\omega^2 A \cos (kx - \omega t). \qquad (21.27b)$$

Now you find the partial derivatives of $y(x, t)$ with respect to x. The first derivative is

$$\frac{\partial y}{\partial x} = -kA \sin (kx - \omega t). \qquad (21.28a)$$

Differentiating a second time, you obtain

$$\frac{\partial^2 y}{\partial x^2} = -k^2 A \cos (kx - \omega t). \qquad (21.28b)$$

To satisfy Equation 21.25, you must have

$$\frac{\partial^2 y / \partial t^2}{\partial^2 y / \partial x^2} = v^2.$$

You can divide Equation 21.27b by Equation 21.28b to obtain

$$\frac{\partial^2 y/\partial t^2}{\partial^2 y/\partial x^2} = \frac{\omega^2}{k^2}.$$

To evaluate the right side of this equation, you use Equation 21.22, $\omega/k = v$. You can immediately conclude that $\omega^2/k^2 = v^2$, and the wave equation is therefore satisfied. Hence the function $y(x, t) = A \cos (kx - \omega t)$ is a legitimate wave function.

(b) Show that $y'(x, t)$ does not satisfy the wave equation.

You repeat the calculation of part **a**, this time using $y'(x, t)$. You obtain

$$\frac{\partial y'}{\partial t} = \frac{4\pi v}{\lambda} A \sin \left[\frac{2\pi}{\lambda} (x - 2vt) \right],$$

and, differentiating again with respect to time,

$$\frac{\partial^2 y'}{\partial t^2} = -\left(\frac{4\pi v}{\lambda} \right)^2 A \cos \left[\frac{2\pi}{\lambda} (x - 2vt) \right].$$

Taking the x derivatives, you have

$$\frac{\partial y'}{\partial x} = -\frac{2\pi}{\lambda} A \sin \left[\frac{2\pi}{\lambda} (x - 2vt) \right],$$

and

$$\frac{\partial^2 y'}{\partial x^2} = -\left(\frac{2\pi}{\lambda} \right)^2 A \cos \left[\frac{2\pi}{\lambda} (x - 2vt) \right].$$

Dividing the second time derivative by the second space derivative gives

$$\frac{\partial^2 y'/\partial t^2}{\partial^2 y'/\partial x^2} = 4v^2.$$

But the right side of this equation must be equal to v^2 if the wave equation is to be satisfied. Because the value is not v^2 but $4v^2$, the function y' is *not* a legitimate wave function.

Energy in Waves

It is easy to see that waves carry energy. The medium acquires energy from the source because the source sets the medium into motion locally. As a wave train travels away from the source into a previously undisturbed part of the medium, the wave train sets that part of the medium into oscillation. Thus the wave carries energy from the source throughout the medium. This energy can subsequently be extracted from the medium, as when a leaf floating on a pond surface is set into motion by passing ripples. In this section, we study the way in which a vibrating string carries energy.

Suppose that a wave traveling down a long string is represented by the wave function $y(x, t) = A \cos (kx - \omega t)$. You saw in Example 21.3 that at time t the transverse velocity v_y of a string element located at x is

$$v_y = \frac{\partial y}{\partial t} = \omega A \sin (kx - \omega t).$$

If the length of the element is dx, its mass is $dm = \mu \, dx$. Its kinetic energy is

$$dK = \tfrac{1}{2} dm \, v_y{}^2.$$

On inserting into this expression the values of v_y and dm, we obtain for the kinetic energy of the element

$$dK = \tfrac{1}{2} \mu \, dx \, \omega^2 A^2 \sin^2 (kx - \omega t). \tag{21.29}$$

Next, let us find the potential energy of the string element. Figure 21.9 shows the element that lies between x and $x + dx$. Because the element is not horizontal, it must be slightly stretched from its undisturbed length dx. To stretch the element requires work, which is stored as elastic potential energy in the segment. Let us call the amount of stretch δ.* The work that stretches the string is done by the string tension T; thus we have

$$dU = dW = T\delta. \tag{21.30}$$

We now evaluate δ. Because the string element is infinitesimal, the stretched length ds is well approximated by the straight-line segment shown in blue in Figure 21.9.

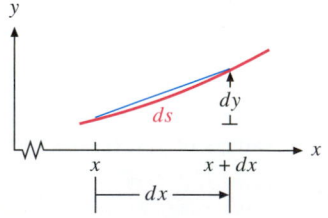

FIGURE 21.9 A string element lying between x and $x + dx$ has length ds slightly greater than dx. The length ds, shown in red, is well approximated by the chord shown in blue.

---

*The quantity δ is very small indeed, as it is the small stretch of an infinitesimal segment dx. That is why we treat T as constant in calculating the work done in stretching (Equation 21.30).

Using Pythagoras's theorem, we have $ds = \sqrt{(dx)^2 + (dy)^2}$. The amount of stretch is thus

$$\delta = ds - dx = \sqrt{(dx)^2 + (dy)^2} - dx = dx\left[\sqrt{1 + \left(\frac{dy}{dx}\right)^2} - 1\right].$$

Because δ is small, we can use the approximation $\sqrt{1 + \alpha} = 1 + \frac{1}{2}\alpha$, which gives us

$$\delta = dx\,\frac{1}{2}\left(\frac{dy}{dx}\right)^2.$$

Inserting this value into Equation 21.30, we find for the potential energy

$$dU = \frac{1}{2}T\left(\frac{dy}{dx}\right)^2 dx. \tag{21.31a}$$

The quantity dy/dx was evaluated in Example 21.3 and is given by Equation 21.28a. When we insert the square of this value into Equation 21.31a, we obtain

$$dU = \tfrac{1}{2}Tk^2A^2 \sin^2(kx - \omega t)\,dx. \tag{21.31b}$$

Let us compare this value with the value of dK given by Equation 21.29. Note that both dK and dU vary with x and t and that they depend on x and t in exactly the same way. That is, both dK and dU reach their maximum values at the same time and place, for which $\sin^2(kx - \omega t) = 1$; both have the value 0 when $\sin^2(kx - \omega t) = 0$. This behavior may surprise you because it differs from the behavior of an isolated harmonic oscillator such as a mass-spring system. As you learned in Section 8.3, the mass-spring oscillator continuously transfers its mechanical energy between kinetic and potential forms; K is maximum when U is zero, and vice versa.

The difference between the string element and the isolated harmonic oscillator is this: The string element is *not* isolated. Energy is continuously transferred between neighboring elements; that is how the wave carries energy.

We are now ready to evaluate the total mechanical energy of the string element. We have

$$dE = dK + dU = \tfrac{1}{2}A^2(\mu\omega^2 + Tk^2)\sin^2(kx - \omega t)\,dx. \tag{21.32}$$

We can express this equation more transparently by using Equations 21.7 and 21.22 to rewrite the term Tk^2. From these equations, we have $T = \mu v^2$ and $k^2 = \omega^2/v^2$, and thus $Tk^2 = \mu\omega^2$. Equation 21.32 thus takes the simpler form

$$dE = \mu\omega^2A^2 \sin^2(kx - \omega t)\,dx. \tag{21.33}$$

We define the **energy density** to be the energy per unit length dE/dx. From Equation 21.33, we have immediately

$$\frac{dE}{dx} = \mu\omega^2A^2 \sin^2(kx - \omega t). \tag{21.34}$$

The energy density of a particular string element (x fixed) varies periodically with time, attaining both its maximum value $\mu\omega^2A^2\,dx$ and its minimum value 0 twice during each period τ of the wave. This behavior is shown in Figure 21.10.

Energy Flux

As a wave travels down a stretched string, mechanical energy flows into and out of any particular element of string. This fact is evident in the periodic variation of dE/dx given by Equation 21.34; the energy is coming from somewhere and going somewhere else. It is tempting to think of the flow of energy as being analogous to the flow of energy through a pipe that carries a flowing fluid, as we did for the system considered in Sections 16.2 and 16.3. In that system, a fluid flows through a pipe and we consider the energy "belonging to" a certain fixed section of pipe. (Review in particular the two paragraphs after Equation 16.7c.) In the present case, we regard the energy as belonging to the wave and passing through the string. As in the fluid-flow case, we are

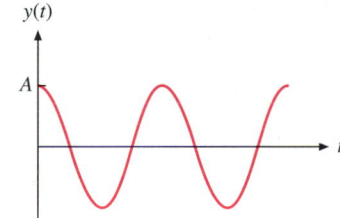

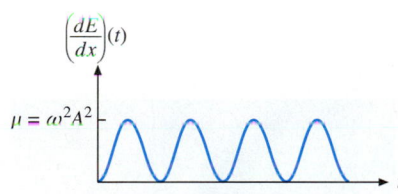

FIGURE 21.10 Comparison of y and dE/dx as functions of t for a fixed location on a string on which a sinusoidal wave is traveling to the right.

taking a liberty: Mechanical energy, strictly speaking, is a property of matter (the elements of the string) and not of form (the shape of the wave). Nevertheless, it is useful to think of a wave as carrying energy through its medium.

The energy carried by the wave per unit time past a given fixed point is called the **energy flux** Φ, defined as

$$\Phi \equiv \frac{dE}{dt}. \tag{21.35}$$

It is evident from this definition that in SI, energy flux is expressed in J/s, or W. (But note that, though energy flux has the same dimensions as power, they are physically different quantities. Power denotes work done per unit time, whereas energy flux denotes energy flowing past a specified boundary per unit time.)

We can express Φ in terms of the energy density dE/dx by observing that the wave velocity is defined to be $v = dx/dt$; hence we have $dt = dx/v$. Substituting this expression for dt into Equation 21.35, we obtain

$$\Phi = v\,\frac{dE}{dx}. \tag{21.36}$$

Using the value of dE/dx given by Equation 21.34, we obtain

$$\Phi = v\mu\omega^2 A^2 \sin^2(kx - \omega t) \quad \text{for a sinusoidal traveling wave.} \tag{21.37}$$

The energy flux varies over the wave cycle just as the energy density does. However, we can obtain a *mean value* for the energy flux by averaging Φ over one cycle. The details are the subject of Problem 21.47; here we merely state the result,

$$\langle\Phi\rangle = \tfrac{1}{2}\mu\omega^2 A^2 v; \tag{21.38}$$

the mean value is one-half the maximum value, which you can evaluate by setting $\sin^2(kx - \omega t)$ in Equation 21.37. The proportionality between Φ and A^2 applies to waves in general.

You should compare the mean energy flux with the analogous (though more concrete) quantities *volume flux* and *mass flux* defined toward the end of Section 16.2. You will see still further examples of fluxes in subsequent chapters, notably in connection with electrostatics in Chapter 24.

Waves in Multidimensional Media

In the absence of friction, a wave will travel indefinitely down an infinite string without any change in its amplitude. The situation is different, however, for waves propagating through two- or three-dimensional media. In two and three dimensions, the wave amplitude diminishes with increasing distance from the source even in the absence of friction. We now study this point quantitatively. The analysis is closely analogous to the discussion toward the end of Section 14.3, in which the inverse-square nature of the gravitational force is made plausible.

Consider first a two-dimensional wave train of ripples on a water surface, produced by a disturbance at some point. If we disturb the water at a single point, by means of a vibrator that drives the surface up and down, ripples will move outward in a circular pattern, as shown in Figure 21.11.

As the waves move outward, the circles become larger and the energy carried by each ripple is therefore spread out over an increasing circumference. As a result, the amplitude of the ripples decreases as they spread out. We can calculate the value of A as a function of r, the distance from the source, by means of the diagram shown in Figure 21.12. Consider the region situated between the imaginary boundary circles of radius r_1 and r_2. In the steady state, the energy contained in the region must remain constant. Hence the mean energy flux out of the region across the outer boundary circle at r_2 must be equal in magnitude to the mean energy flux into the region across the inner boundary circle at r_1:

$$|\langle\Phi\rangle(r_2)| = |\langle\Phi\rangle(r_1)|. \tag{21.39}$$

FIGURE 21.11 Water-surface ripples propagating outward in a circular pattern from a point source.

But let us consider the energy flux per unit length of boundary, which we call the **intensity** I. (Note that, in the definition of intensity, the length is everywhere *transverse* to the direction of wave propagation.) We have

$$I(r_1) = \frac{\langle \Phi \rangle(r_1)}{2\pi r_1} \quad \text{and} \quad I(r_2) = \frac{\langle \Phi \rangle(r_2)}{2\pi r_2}.$$

In this two-dimensional context, the SI unit of I is J/s·m, or W/m. Consequently, we have

$$\frac{I(r_2)}{I(r_1)} = \frac{r_1}{r_2} \quad \begin{array}{l}\text{for waves emanating from} \\ \text{a point source on a plane surface.}\end{array} \tag{21.40}$$

Thus the two-dimensional geometry of the situation results in an *inverse-first-power law* for wave intensity.

We now reexpress Equation 21.40 in terms of the amplitude A. Because $\langle \Phi \rangle \propto A^2$ (Equation 21.38) and $\langle \Phi \rangle = 2\pi r I$, it must also be true that $I \propto A^2$. So we have

$$\frac{A^2(r_2)}{A^2(r_1)} = \frac{I(r_2)}{I(r_1)}. \tag{21.41}$$

Taken together with Equation 21.40, this yields the relation

$$\frac{A^2(r_2)}{A^2(r_1)} = \frac{r_1}{r_2} \quad \begin{array}{l}\text{for waves emanating from} \\ \text{a point source on a plane surface,}\end{array}$$

so

$$\frac{A(r_2)}{A(r_1)} = \sqrt{\frac{r_1}{r_2}} \quad \begin{array}{l}\text{for waves emanating from} \\ \text{a point source on a plane surface.}\end{array} \tag{21.42}$$

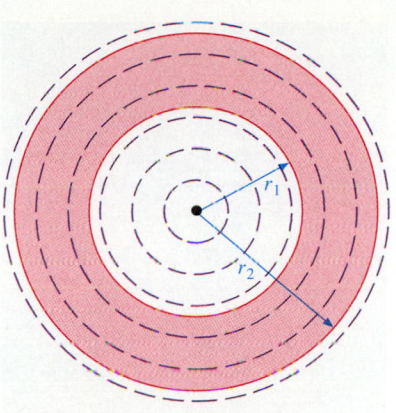

FIGURE 21.12 Diagram for finding the amplitude of waves in a two-dimensional medium as a function of distance r from the source.

We now extend our argument to three-dimensional space. Consider waves propagating away from a point source in a three-dimensional medium—for example, sound waves in air. Here the imaginary boundary surfaces are spheres instead of circles and the quantity of interest is the energy flux passing through a unit area on the surfaces of these spheres instead of the energy flux passing through a unit length on the circumferences of circles. Again, note that the boundary surfaces are everywhere *transverse* to the direction of propagation of the wave. The analysis is entirely analogous to that just carried out in two dimensions, and the details are the subject of Problem 21.51. In three dimensions, the intensity is defined to be the energy flux per unit (transverse) *area*, and the results are as follows:

$$\frac{I(r_2)}{I(r_1)} = \frac{r_1^2}{r_2^2} \quad \begin{array}{l}\text{for waves emanating from a point} \\ \text{source in three-dimensional space.}\end{array} \tag{21.43}$$

This is an *inverse-square law*. In three dimensions, where the boundary surface is two-dimensional, I is expressed in J/(s·m²), or W/m². The corresponding amplitude ratio is

$$\frac{A(r_2)}{A(r_1)} = \frac{r_1}{r_2} \quad \begin{array}{l}\text{for waves emanating from a point} \\ \text{source in three-dimensional space.}\end{array} \tag{21.44}$$

As you can see from the foregoing two- and three-dimensional arguments, the amplitude and intensity of waves emanating from a point source depend on the number of dimensions spanned by the medium that carries the waves.

The Superposition Principle

In the context of classical mechanics, waves and particles seem in many ways to have contrary properties. On the one hand, a particle is *localizable*. That is, the accuracy with which we can specify the position of a particle is limited only by the accuracy of

FIGURE 21.13 Superposition of waves. Each raindrop falling on the pond surface has produced a set of circular waves. The vertical displacement of the water surface as a function of position and time is described by a wave function obtained by adding together the wave functions that describe the individual circular wave trains.

our measurement process. On the other hand, a wave is *diffuse;* it is spread out over its medium and is not located at a particular point. Indeed, a sinusoidal wave train of indefinite length on an infinite string is "everywhere" and thus has no particular location at all.

Another familiar property of particles is their exclusiveness. Two particles cannot occupy the same position simultaneously. We now explore the contrary property of waves: *Not only can any number of waves occupy the same space simultaneously, but each passes through that space as though the others were not there.* This property is made evident in a qualitative way in Figure 21.13.

The property has an important consequence. If several waves, each described by its own wave function, are passing through a particular location, the displacement of the medium at that location at a particular time is just the sum of all the wave functions evaluated at that location and time. This is called the **superposition principle**.

For small waves on a string, we can express the superposition principle in the form

$$Y(x, t) = y_1(x, t) + y_2(x, t) + \cdots + y_n(x, t). \tag{21.45a}$$

Here the y_j are the individual wave functions, and their sum, the wave function $Y(x, t)$, describes the resultant behavior of the medium (in this case, the string) as a function of position and time. We can write Equation 21.45a more compactly as

$$Y(x, t) = \sum_{j=1}^{n} y_j(x, t). \tag{21.45b}$$

Fourier Synthesis*

Out of all the infinite number of wave forms possible, we have concentrated on sinusoidal waves. We have done this partly because of their simplicity, but there is another excellent reason for doing so. In 1822, the French mathematician Jean-Baptiste Joseph Fourier (1768–1830) proved a remarkable result, called **Fourier's theorem**, which applies to mathematical functions in general and therefore to wave functions in particular. Fourier's theorem is valid for any independent variable, but for concreteness we express the theorem in terms of the independent variable t. We state Fourier's theorem in three forms, in order of increasing generality.

FOURIER'S THEOREM I If a function $f(t)$ is periodic over the entire domain of the independent variable t ($-\infty \leq t \leq \infty$) and if the first derivative df/dt is defined for all values of t, then the function can be represented as a series having a *finite* number of terms of the form

$$f(t) = A_1 \cos(\omega t + \delta_1) + A_2 \cos(2\omega t + \delta_2)$$
$$+ A_3 \cos(3\omega t + \delta_3) + \cdots + A_n \cos(n\omega t + \delta_n), \tag{21.46a}$$

or, more compactly,

$$f(t) = \sum_{j=1}^{n} A_j \cos(j\omega t + \delta_j). \tag{21.46b}$$

The process of "constructing" a function by adding simple sinusoids is called *Fourier synthesis.* Each term in the sum is called a *Fourier component*, and the series is called a *Fourier series*. The phase constants δ_j have no special relation to one another. The frequency of the leading term, $\nu_1 = \omega/2\pi$, is called the **fundamental frequency**. It is the frequency in which the function f is periodic. The higher-frequency terms are called

---

*This subsection may be omitted without loss of continuity.

harmonics; the successive integral values of j in the terms $j\omega t$ guarantee that the frequencies of successive terms are in the ratio $1:2:3:\cdots:n$. The presence of harmonics results in alteration of the shape of the wave from a simple sinusoidal but does not affect the frequency of the wave.

FOURIER'S THEOREM II   If a function $f(t)$ is periodic over the entire domain of the independent variable t ($-\infty \le t \le \infty$) but there are values of t for which the first derivative df/dt is *not* defined (that is, the function has sharp corners), then the function can be represented as a series having an *infinite* number of terms of the form

$$f(t) = A_1 \cos(\omega t + \delta_1) + A_2 \cos(2\omega t + \delta_2) \tag{21.47a}$$
$$+ A_3 \cos(3\omega t + \delta_3) + \cdots ,$$

or
$$f(t) = \sum_{j=1}^{\infty} A_j \cos(j\omega t + \delta_j). \tag{21.47b}$$

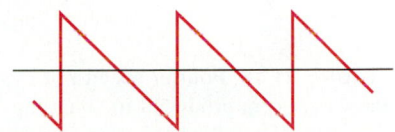

FIGURE 21.14 A sawtooth wave.

Each term is still a harmonic with a frequency that is an integral multiple of the fundamental frequency.

A function of this sort is the sawtooth wave shown in Figure 21.14. The specific Fourier series for the sawtooth wave is

$$f(t) = \sin \omega t + \tfrac{1}{2}\sin 2\omega t + \tfrac{1}{3}\sin 3\omega t + \cdots , \tag{21.48a}$$

or
$$f(t) = \sum_{j=1}^{\infty} \frac{1}{j}\sin(j\omega t). \tag{21.48b}$$

That is, all the δ_j in Equation 21.47b are set equal to $\pi/2$ [remember that $\cos(x + \pi/2) = \sin x$], and $A_j = 1/j$. It is not possible in practice to sum an infinite number of terms. But often, as in Equations 21.48a and 21.48b, the series converges fairly rapidly. The rapidity of the convergence is due to the fact that the *Fourier coefficients* A_j decrease as j increases and so higher-order terms make only a small contribution to the sum. Figure 21.15 shows how the sawtooth wave of Figure 21.14 can be built up from a series of terms like those in Equation 21.48a. For the purposes of the plot, the value of the angular frequency is chosen to be $\omega = 2\pi$ rad/s. Only one cycle is shown; that is, the plot is restricted to values of t between 0 and 1. Successive graphs display the results when the series is cut off at $j = 1, 2, 3, 5, 24,$ and 25. Comparing the last two graphs should convince you that the contribution of higher-order terms is small. You should find it plausible, as well, that the sum of the infinite series is the sawtooth wave of Figure 21.14.

FOURIER'S THEOREM III   *Any* function $f(t)$, periodic or nonperiodic, can be represented by an infinite series of sinusoidal terms. Here, however, the frequencies of the contributing terms are not a discrete series; instead, they are distributed continuously. The sum of Equation 21.46b must be replaced by an integral. We will not consider this case in detail.

It follows from Fourier's theorem that all wave forms, no matter how complicated, can be synthesized by adding sinusoids. Thus a study of sinusoidal waves leads directly to an understanding of the behavior of waves of any form.

Fourier Analysis

The problem that most often arises in physics is the inverse of Fourier synthesis. Given a wave form, we wish to find its Fourier components. The process by which this is accomplished is called *Fourier analysis*. If the phase constants δ_j are not of interest, as is often the case, the problem boils down to finding the values of the Fourier coefficients A_j. Electronic devices called Fourier analyzers or spectrum analyzers can carry out this process on any wave form fed to them. The result is most commonly presented as a *Fourier spectrum*. The Fourier spectrum is quite characteristic of its wave form f, as

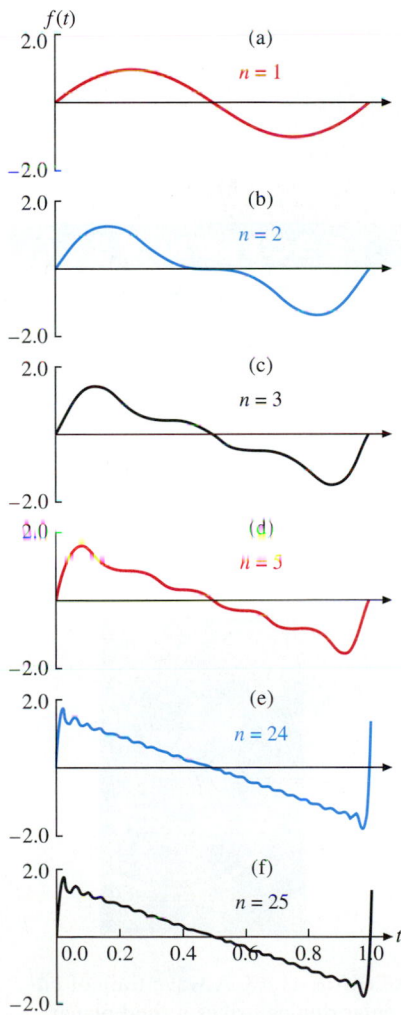

FIGURE 21.15 Graphs of one period of the Fourier series whose typical term is $j^{-1}\sin 2\pi jt$, summed over increasing numbers of terms.

is made evident in Figure 21.16, which compares Fourier spectra of wave forms produced by an ordinary violin and by a Stradivarius violin sounding the same note. Consistent with Fourier's theorem I, these spectra are *discrete spectra*, also called *line spectra*. A little calculation with the frequencies given in Figure 21.16 will convince you that only sinusoids whose frequencies are related by simple numerical ratios contribute to the wave forms $f(t)$ of the violins. If a wave form is not periodic, the corresponding Fourier spectrum is a **continuous spectrum**.

FIGURE 21.16 Fourier spectra of the wave form produced in sounding the open A string on (*a*) an ordinary violin and (*b*) a Stradivarius violin.

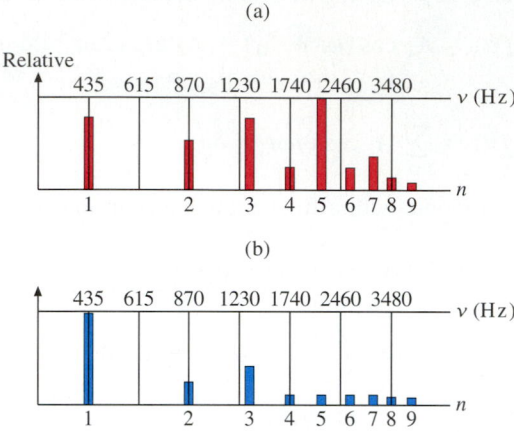

SECTION 21.7 Standing Waves

When a wave reaches the boundary of the medium through which it is propagating, it is reflected. In Figure 21.17, a vibrating point source has generated a wave train consisting of about a half-dozen ripples on a water surface. The ripples, spreading in concentric circles, strike a rigid planar barrier and are reflected.

Let us consider the relatively simple case of reflection of a pulse at the end of a string. Figure 21.18 illustrates what happens when the right end of the string is (*a*) fixed rigidly and (*b*) free. When the end is rigidly fixed, the wave is inverted when it is reflected. To see why this inversion occurs, we must consider the way in which the element at the end of the string is different from all other elements. First, consider an arbitrary element *e*. When it is displaced transversely by a pulse moving to the right, *e* drags its right-hand neighbor *e'* along. Because of the inertia of *e'*, it exerts a reaction force on *e* that accelerates *e*. But the end element has as its right-hand neighbor a body of essentially infinite mass. The reaction force exerted on the end element is therefore greater than that on an arbitrary element. As a result, the end element is snapped downward, overshooting the equilibrium position and initiating an inverted pulse.

When the right end is free, as in Figure 21.18b, there is reflection without inversion. Why? In this case, the end element differs from all other elements in that its right end is not attached to a neighboring element. When displaced by a positive pulse moving to the right, the end element is snapped upward like the free end of a whip. When it reaches the displacement *A*, at which other elements have zero transverse velocity, it still possesses upward momentum. You can see that the end element still possesses upward momentum by comparing the first and sixth photographs of Figure 21.18b. The displacement of the end of the spring in the sixth photograph is substantially greater than the amplitude of the pulse, as measured on the first photograph. (Don't confuse the end of the spring with the fixed end of the thread to which it is attached.) The extra momentum is the source of an upward force on the neighbor to the left, and a positive pulse is thus propagated toward the left.

As a result of the reflection that takes place at the end of a stretched string of finite length, waves must travel in both directions on the string. The leftward moving waves superpose with the rightward moving waves. In the remainder of this section, we will

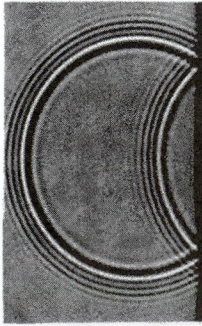

FIGURE 21.17 A wave train of circular ripples strikes a rigid planar barrier and is reflected.

(a)   (b)

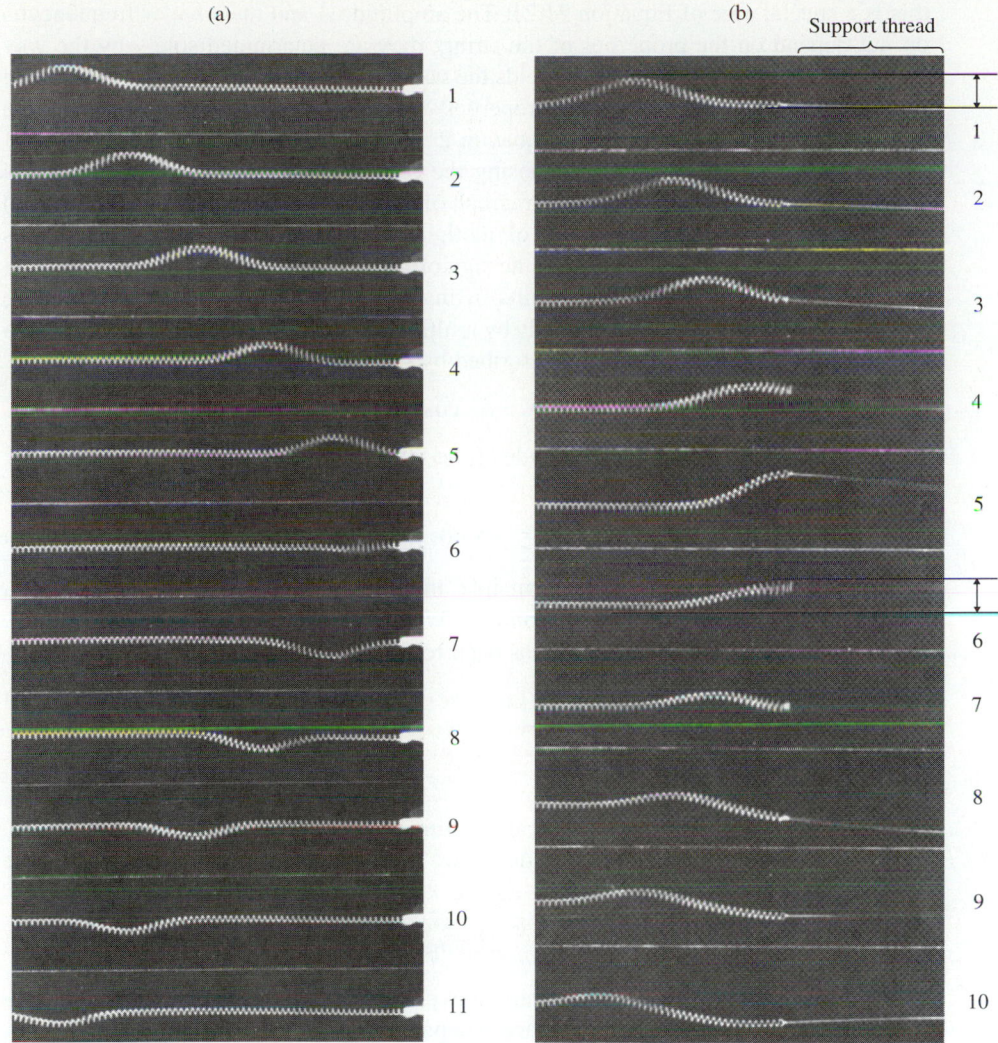

FIGURE 21.19 Reflection of a pulse traveling down a spring. (*a*) The right end of the spring is rigidly fixed, and so the pulse is inverted on reflection. (*b*) The right end of the spring is essentially free, and so the pulse is reflected without inversion. (The end of the spring is fastened to a light support thread, which provides the necessary tension. The role of the mass of the thread in the motion of the system is negligible.)

(From Haber-Schaim: *PSSC Physics.* Copyright 1991 by Kendall/Hunt Publishing Company. Used with permission.)

see how the superposition can result in *standing waves*—that is, waves that do not appear to move in either direction along the string.

Reflection of a Sinusoidal Wave Train at a Fixed End

We can readily apply the foregoing observations to a continuous wave train. We again regard the wave train as a series of pulses of alternating sign, just as we did in Section 21.3. Suppose the incident wave is described by the wave function

$$y(x,\ t) = A \cos (kx - \omega t); \qquad\qquad \textbf{(21.49a)}$$

this is a special case of Equation 21.21. The amplitude A and the angular frequency ω do not depend on the properties of the string; they are determined solely by the way the source moves. Equation 21.22 yields the value of the wave number, $k = \omega/v$. The wave number does depend on the properties of the string because it depends on the wave speed v. The minus sign in Equation 21.49a indicates the positive direction of motion. We have set $\delta = 0$ by choosing the proper initial time $t = 0$. (The initial conditions described in the second paragraph of Section 21.3 accomplish this purpose.)

Reflection reverses the direction of motion of the wave train; we represent this reversal mathematically by changing the sign of ωt in Equation 21.49a. If the right end of the string is fixed, reflection results in inversion of the wave train as well. We represent this inversion mathematically by multiplying the right side of Equation 21.49a by -1. Thus the reflected wave is described by the function

$$y'(x, t) = -A' \cos (kx + \omega t).$$

In the absence of friction, the amplitude of the reflected wave is the same as that of the incident wave; $A' = A$. So we have

$$y'(x, t) = -A \cos (kx + \omega t). \tag{21.49b}$$

According to the superposition principle, the shape of the string is described by a wave function that is the sum of y and y'. When enough time has passed so that a substantial length of the wave train has been reflected, the wave function is

$$Y(x, t) = A[\cos (kx - \omega t) - \cos (kx + \omega t)]. \tag{21.50}$$

Standing Waves on a String

Let us look more closely at the physical meaning of the total wave function $Y = y + y'$ given by Equation 21.50. In order to do so, we simplify the term in brackets, using the trigonometric identity $\cos (\alpha - \beta) - \cos (\alpha + \beta) = 2 \sin \alpha \sin \beta$. We then have

$$Y(x, t) = 2A \sin kx \sin \omega t. \tag{21.51}$$

This simplification is remarkable because it expresses Y in terms of a product of a time-independent factor, $\sin kx$, and a space-independent factor, $\sin \omega t$. Figure 21.19 is a graph of the time-independent factor $2A \sin kx$ at one of the many instants $t = (n + \frac{1}{2})\pi/\omega$ when $\sin \omega t = 1$. The wave form, which describes the shape of the vibrating

FIGURE 21.19 Plot of the wave function of Equation 21.51 at an instant when $\sin \omega t = 1$.

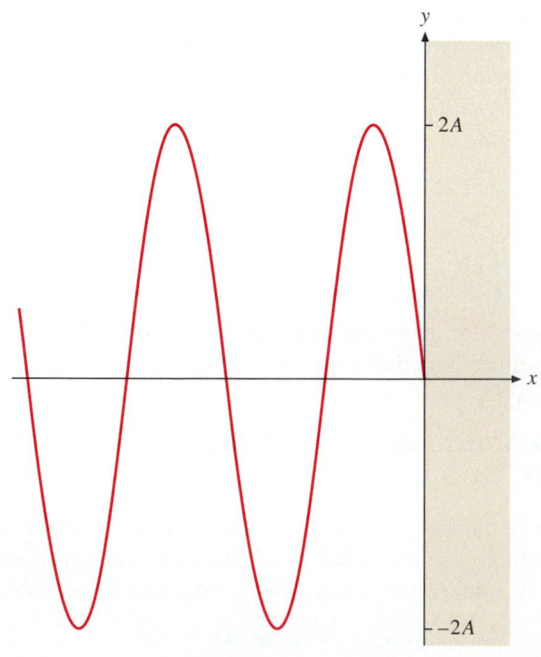

string, is a simple sine wave. However, the amplitude 2A of the wave form is *twice* the amplitude of the sinusoidal traveling waves whose superposition produces it.

What happens to the sine wave described by Equation 21.51 as *t* varies? Let us focus on an element of the string located at arbitrary *x*. This element executes harmonic motion in the *y* direction, the extremal displacements being $Y_{max} = 2A \sin kx$ and $Y_{min} = -2A \sin kx$. (One of these values is attained every time $\sin \omega t = \pm 1$.) If we look along the string, turning our attention to a succession of string elements, the values of Y_{max} and Y_{min} depend on $\sin kx$ and hence sinusoidally on the location *x*.

Now, let us consider special values of *x* such that $\sin kx = 0$. At these locations, $Y = 0$ at all times and the string element is motionless. These locations (Figure 21.20) are called **nodes** (from a Latin word meaning "knots"). Midway between nodes, where $\sin kx = \pm 1$ and therefore $Y = 2A \sin \omega t$, the string oscillates through the full wave amplitude 2A. These locations are called **antinodes**. The behavior of the string over time is shown in Figure 21.20.

As Figure 21.20 shows, the sinusoid that describes the wave grows and shrinks repeatedly as time passes, but never moves in the *x* direction; in particular, the positions of the nodes and antinodes are fixed. We call such a wave a **standing wave**. Although a standing wave is produced by the superposition of two traveling waves, of equal amplitude but moving in opposite directions, the behavior of the standing wave is very different from that of either of its component traveling waves.

The time-independent part of Equation 21.51, $2A \sin kx$, remains fixed as time passes. However, the remaining factor $\sin \omega t$ varies sinusoidally between the limits 1 and -1, changing the height of the wave. We say that the factor **modulates** the amplitude of the wave.

The energy flux of a standing wave is zero. There are two ways to see this. First, *the velocity of a standing wave is zero.* The mean energy flux given by Equation 21.38 thus has the value $\langle \Phi \rangle = \frac{1}{2}\mu\omega^2 A^2 v = 0$. Second, the values of $\langle \Phi \rangle$ for the two component traveling waves are equal and opposite, so the net energy flux is zero. This also can be seen from Equation 21.38. The values of μ, ω, and *A* are the same for both the incident and the reflected wave, but their velocities have opposite signs.

A standing wave appears on the string, with properties dramatically different from those of traveling waves, because we have imposed a deceptively simple condition on the string: we have fixed (anchored) the right end of the string, and it cannot move. Such a condition is called a **boundary condition**.

Standing waves can be beautiful!

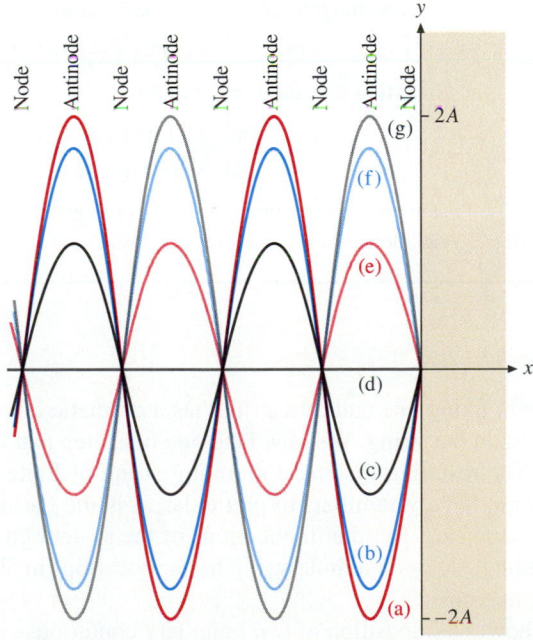

FIGURE 21.20 (*a*–*g*): Plots of the wave function of Equation 21.51 at the successive times $t = 3\pi/6\omega$, $4\pi/6\omega, 5\pi/6\omega, \ldots, 9\pi/6\omega$.

EXAMPLE 21.4

A very long, horizontal steel wire is fastened to a wall at its right end, $x = 0$. The wire has mass per unit length $\mu = 3.83 \times 10^{-4}$ kg/m and is subjected to a tension $T = 200$ N. At some point far to the left of the wall, the wire is attached to a vibrator that oscillates harmonically in the vertical direction with amplitude $A = 0.268$ cm and frequency $\nu = 120$ Hz. **(a)** When a standing wave has been established in the part of the string near the wall, find the spacing between the nodes. **(b)** Find the amplitude A_s of the standing wave. **(c)** Find the magnitude of the maximum displacement Y_{max} of the point on the string at $x = -12.5$ m. **(d)** Find the maximum transverse speed of this point. **(e)** Find the magnitude of the maximum transverse acceleration of this point.

SOLUTION:

(a) Find the distance between neighboring nodes.

From Figure 21.19 you can see that the distance d between nodes is one-half the wavelength; $d = \lambda/2$. You use Equation 21.15, $v = \lambda\nu$, and Equation 21.7, $v = \sqrt{T/\mu}$, to write

$$d = \frac{1}{2\nu}\sqrt{\frac{T}{\mu}} = \frac{1}{2 \times 120 \text{ Hz}}\sqrt{\frac{200 \text{ N}}{3.83 \times 10^{-4} \text{ kg/m}}} = 3.01 \text{ m}.$$

(b) Find A_s.

The amplitude A_s of the wave is the greatest displacement achieved by the string anywhere. According to Equation 21.51, this extreme displacement is equal to twice the amplitude of the source oscillator. So you have

$$A_s = 2A = 2 \times 2.68 \times 10^{-3} \text{ m} = 5.36 \times 10^{-3} \text{ m}.$$

(c) Find Y_{max} at $x = -12.5$ m.

According to Equation 21.51, every element of the string achieves its maximum displacement when $|\sin \omega t| = 1$. At such times, you have

$$|Y|_{max} = 2A|\sin kx|. \qquad (21.52)$$

To find the value of k in this expression, you can use Equation 21.20 to write $k = 2\pi/\lambda = \pi/d$. Taken together with the result of part **a**, this gives

$$|Y|_{max} = 5.36 \times 10^{-3} \text{ m} \times \left|\sin\frac{\pi \times (-12.5 \text{ m})}{3.01 \text{ m}}\right|$$

$$= 2.48 \times 10^{-3} \text{ m}.$$

(d) Find the maximum transverse speed of the point on the wire at $x = -12.5$ m.

To find the transverse velocity v_y of any point on the string at any time, you differentiate Equation 21.51 with respect to

time to obtain dY/dt:

$$v_y = \frac{dY}{dt} = 2A \sin kx(\omega \cos \omega t). \qquad (21.53)$$

Because you are interested only in the speed, you take the absolute values of both sides of this equation:

$$|v_y| = |2A \sin kx(\omega \cos \omega t)|.$$

This speed attains its maximum value $|v_y|_{max}$ when $|\cos \omega t| = 1$:

$$|v_y|_{max} = \omega(2A|\sin kx|).$$

You use Equation 21.52 to write this result in the condensed form

$$|v_y|_{max} = 2\pi\nu|Y|_{max}.$$

Inserting the numerical values, you obtain

$$|v_y|_{max} = 2\pi \times 120 \text{ Hz} \times 2.48 \times 10^{-3} \text{ m} = 1.87 \text{ m/s}.$$

The maximum speed attained by any element on the string depends on its location. But all elements attain their maximum speeds simultaneously, at all the instants when $|\cos \omega t| = 1$. At these same instants, $\sin \omega t = 0$. It follows that maximum speed is attained when $Y = 0$, as you would expect. The maximum speed is *out of phase* with the maximum displacement by one-quarter of a cycle, or $\pi/2$.

(e) Find $|a|_{max}$ for the point at $x = -12.5$ m.

To find the acceleration a of any point on the string at any time, you differentiate Equation 21.53 with respect to time to find d^2Y/dt^2:

$$a = \frac{d^2Y}{dt^2} = -2\omega^2 A \sin kx \cos \omega t,$$

or, using Equation 21.51,

$$a = -\omega^2 Y. \qquad (21.54)$$

Can you show that the acceleration is one-half cycle, or π rad, out of phase with the displacement? What is the phase difference between acceleration and velocity?

The magnitude of the maximum acceleration is

$$|a|_{max} = \omega^2|Y|_{max} = 4\pi^2\nu^2|Y|_{max}.$$

Inserting the numbers, you have

$$|a|_{max} = 4\pi^2 \times (120 \text{ Hz})^2 \times 2.48 \times 10^{-3} \text{ m}$$
$$= 1.41 \times 10^3 \text{ m/s}^2.$$

This is a very considerable acceleration, almost 150 times the acceleration of gravity.

Standing-Wave Modes

As we have just seen, fixing one end of a string has a dramatic effect on the nature of the wave propagating in the string. We now take one final step and fix the other end of the string as well. The resulting system, a vibrating string of finite length L stretched between two supports, is very familiar. In particular, it is the basic mechanism of all stringed musical instruments. But the implications of the system go far beyond music; the analysis of standing waves on a finite string has applications in almost every branch of physics and engineering.

Let us now see how the imposition of *two* boundary conditions—fixing the string at both ends—imposes quite severe limitations on the behavior of waves on the string. To begin with, consider an ideal, frictionless string. In the absence of friction, there is no need to drive the string continuously in order to sustain vibrations. A single initial disturbance (produced, for example, by plucking the string or striking it with a hammer) makes the string vibrate indefinitely, as the disturbance is reflected again and again (each time with a phase inversion) at the two ends. The result is a standing wave. As you saw in the preceding subsection, the energy flux for a standing wave is zero. Hence the energy initially added to the system remains in the system.

Suppose we excite the string by plucking it once, near one end. The result is a standing wave, as shown in the time-exposure photograph of Figure 21.21. Although this wave is not a pure sine wave, the wavelengths of the components are not just any wavelengths. The reason lies in the additional restrictions placed on the wave function by the second fixed boundary.

For convenience, we take the origin at the right end of the string. If the length of the string is L, the left end of the string is at $x = -L$. We express the fact that the right end of the string is fixed by means of the boundary condition

$$Y(0, t) = 0 \quad \text{for all } t.$$

The left end of the string is fixed as well, and so we also have the boundary condition

$$Y(-L, t) = 0 \quad \text{for all } t.$$

What does this additional condition imply? We already know that the boundary condition at $x = 0$ leads to the requirement that the wave be a standing wave, described by Equation 21.51. We therefore use Equation 21.51, $Y = 2A \sin kx \sin \omega t$, as a starting point and impose the new condition:

$$0 = 2A \sin (-kL) \sin \omega t \quad \text{for all } t.$$

The only way to satisfy this equation is to set $\sin kL = 0$. But $\sin kL = 0$ only if $kL = n\pi$, where n is any integer. So we have

$$k = \frac{n\pi}{L}. \tag{21.55}$$

We can substitute the definition $k \equiv 2\pi/\lambda$ into this equation to obtain

$$\frac{\lambda}{2} = \frac{L}{n} \quad \text{or} \quad 2L = n\lambda; \quad n = 1, 2, 3, \ldots \tag{21.56a,b}$$

The only standing waves permitted on a string fixed at both ends are those whose half-wavelength fits into the length of the string an integral number of times.

In a long string fixed at one end, the wavelength of the standing waves is determined by the angular frequency ω of the source, through the relation $\lambda = 2\pi v/\omega$. Because ω can have any value, so can λ.* But, with both ends of the string fixed, only those frequencies that satisfy Equation 21.56 can result in standing waves.

There are still an infinite number of ways in which the string can vibrate in standing waves, corresponding to all the values of n in Equations 21.56a and 21.56b. Each of these ways is called a **standing-wave mode**, a **harmonic mode**, or simply a **mode**.[†] There are two systems for naming modes, an unfortunate situation that can be confusing unless you are careful. In one system, we speak of the first mode, second mode, third mode, and so forth, as corresponding to $n = 1, 2, 3, \ldots$. In the other system, already mentioned briefly in the context of Fourier's theorem, the first mode is called the **fundamental mode** or simply the **fundamental**. The **first harmonic** (or, as musicians call it, the **first overtone**) corresponds to $n = 2$ and is thus the second mode. Similarly, the

FIGURE 21.21 Time-exposure photograph of a standing wave on a string fixed at both ends.

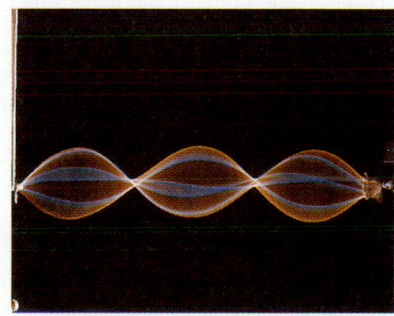

FIGURE 21.22 A string excited to vibrate in its third standing-wave mode by means of an oscillator at one end.

---

*The same point was made in terms of k following Equation 21.49a.
[†]In real strings, the modes corresponding to large n are usually damped out rapidly by friction. Nevertheless, ten or more modes are often significant in determining the tone quality of stringed instruments.

second harmonic (or second overtone) is the third mode for which $n = 3$, and so on. The third mode is the one shown in the multiple-exposure photograph of Figure 21.22. The half-wave fits into the string three times. Put otherwise, the length of the string is $\frac{3}{2}$ waves. In order to excite this pure sinusoidal wave, the left end of the string is excited by a harmonic oscillator whose frequency is carefully set to the value dictated by the string parameters.

The term *harmonic*, which you have encountered in many contexts, has its origin here. The ancient Greeks knew that harmonious musical sounds could be produced by combining tones whose wavelengths were in the ratio of small integers. That is, such tones belonged to the same harmonic series. Other combinations, they found, were often disagreeable, or *dissonant*.

EXAMPLE 21.5

Suppose the length of the string in Figure 21.22 is 1.00 m and its mass per unit length is 7.95×10^{-3} kg/m. If the tension in the string is 19.8 N, what is the frequency ν_3 necessary for exciting the standing-wave mode shown? What frequency ν_1 would excite the fundamental?

SOLUTION: From either of Equations 21.56a and 21.56b, you have $\lambda = 2L/n$. You can thus use Equation 21.15, $v = \lambda\nu$, and Equation 21.7, $v = \sqrt{T/\mu}$, to write

$$\nu = \frac{nv}{2L} = \frac{n}{2L}\sqrt{\frac{T}{\mu}}. \qquad (21.57)$$

For the mode shown in Figure 21.22, you have $n = 3$, so

$$\nu_3 = \frac{3}{2 \times 1.00 \text{ m}} \times \sqrt{\frac{19.8 \text{ N}}{7.95 \times 10^{-3} \text{ kg/m}}} = 74.9 \text{ Hz}.$$

You could use this same equation to calculate the fundamental frequency by setting $n = 1$. But it is easier simply to take the ratio of the two frequencies given by Equation 21.57 written twice; you then have

$$\frac{\nu_1}{\nu_3} = \frac{1}{3}.$$

Solving numerically, you obtain

$$\nu_1 = \tfrac{1}{3}\nu_3 = \tfrac{1}{3} \times 74.9 \text{ Hz} = 25.0 \text{ Hz}.$$

Symbols Used in Chapter 21

A	amplitude		$Y(x, t)$	total displacement due to superposition of displacements $y_j(x, t)$
$f(x, t)$	one-dimensional wave function		δ	phase constant
F	force		$\Phi, \langle\Phi\rangle$	energy flux, mean energy flux
I	intensity, defined as energy flux per unit transverse length (in two dimensions) or per unit transverse area (in three dimensions)		λ	wavelength
			μ	mass per unit length
k	wave number		ν	frequency
L	length of a string		τ	period
T	string tension		ω	angular frequency
v	wave velocity		∂	symbol denoting partial differentiation
v_y	transverse velocity			

Summing Up

When a continuous medium is briefly disturbed at one point, called a **source**, a **pulse** propagates throughout the medium. The propagation speed, called the **wave speed** v, depends on the properties of the medium. For a stretched string, the wave speed is given by Equation 21.7,

$$v = \sqrt{\frac{T}{\mu}}.$$

If the source disturbance continues over time, the resulting disturbance is called a **wave train**. If the disturbance is periodic with frequency ν, the wave train also is periodic, with wavelength λ. The frequency and the wavelength are related through the wave speed by Equation 21.15,

$$v = \lambda \nu.$$

The propagation of waves in one dimension is typefied by the vibrating string. The shape of the string—its vertical displacement y as a function of position x—varies with time, so $y = y(x, t)$. In general, the argument of a one-dimensional **wave function** $f(x, t)$ *must* have the form given by Equation 21.23,

$$f(x, t) = f(x \mp vt).$$

The simplest and most important type of one-dimensional wave function is the **sinusoidal** wave function. For a vibrating string, a sinusoidal function is described by either Equation 21.18 or Equation 21.21:

$$y(x, t) = A \cos \left[\frac{2\pi}{\lambda} (x \mp vt) + \delta \right];$$

$$y(x, t) = A \cos (kx \mp \omega t + \delta).$$

All waves are governed by the **wave equation**. Equation 21.24b is the one-dimensional form of the wave equation:

$$\frac{\partial^2 f}{\partial t^2} = v^2 \frac{\partial^2 f}{\partial x^2}.$$

When a wave of amplitude A and angular frequency ω travels along a string, every element dx of the string has energy dE, given by Equation 21.33,

$$dE = \mu \omega^2 A^2 \sin^2 (kx - \omega t) \, dx.$$

As a result, energy flows past any fixed location at a rate given by the **mean energy flux** $\langle \Phi \rangle$. According to Equation 21.38,

$$\langle \Phi \rangle = \tfrac{1}{2} \mu \omega^2 A^2 v.$$

The energy content of any wave is proportional to the square of its amplitude. When waves propagate over a planar surface from a point source, the amplitude varies with distance r from the source according to Equation 21.42,

$$\frac{A(r_2)}{A(r_1)} = \sqrt{\frac{r_1}{r_2}}.$$

When waves propagate from a point source in three-dimensional space, the amplitude varies with distance according to Equation 21.44,

$$\frac{A(r_2)}{A(r_1)} = \frac{r_1}{r_2}.$$

Waves obey the **superposition principle**. For small waves on a stretched string, the superposition principle can be expressed in the form of Equation 21.45b,

$$Y(x, t) = \sum_{j=1}^{n} y_j(x, t).$$

According to Fourier's theorem, any periodic wave form can be constructed out of sinusoids by means of linear superposition, according to Equations 21.46 or 21.47.

A wave train reaching the boundary of the medium is reflected in a way that depends on the **boundary conditions**. In a long string fixed rigidly at one end, the reflected wave has the same amplitude as the incident wave but is inverted; the phase is shifted by π rad. To the contrary, if the end of the string is free, the reflection occurs without inversion. The result of superposition of incident and reflected waves is a **standing wave**, for which $v = 0$. If the wave function of the incident wave is that of Equation 21.49a,

$$y(x, t) = A \cos (kx - \omega t),$$

the reflected wave is described by Equation 21.49b,

$$y'(x, t) = -A \cos (kx + \omega t).$$

Superposition yields the standing wave function of Equation 21.51,

$$Y(x, t) = 2A \sin kx \sin \omega t.$$

The standing wave has **nodes** at the evenly spaced points where $\sin kx = 0$ and **antinodes** at the points midway between the nodes, where $\sin kx = \pm 1$.

If a string of length L is fixed at both ends, its **standing-wave modes** are restricted to those satisfying the condition given by either of Equations 21.56a and 21.56b:

$$\frac{\lambda}{2} = \frac{L}{n} \quad \text{or} \quad 2L = n\lambda; \quad n = 1, 2, 3, \ldots .$$

KEY TERMS

Section 21.2 Propagation of a Pulse
source ▪ transverse pulse ▪ wave velocity ▪ crest ▪ amplitude

Section 21.3 Periodic Wave Trains
wave train ▪ wavelength

Section 21.4 Wave Functions and the Wave Function
time-independent wave function ▪ space-independent wave function ▪ complete wave function ▪ wave number ▪ partial differentiation ▪ partial derivative

Section 21.5 Energy in Waves
energy density ▪ energy flux ▪ intensity

Section 21.6 The Superposition Principle
Fourier's theorem ▪ fundamental frequency ▪ harmonics ▪ continuous spectrum

Section 21.7 Standing Waves
node, antinode ▪ boundary condition ▪ standing-wave mode, harmonic mode, mode ▪ fundamental, overtone

QUERIES

21.1 *(2) Getting the edge.* The kink in Figure 21.4 implies a discontinuity in $\partial y/\partial x$ at that point. What does this imply concerning the source disturbance? Is a perfectly sharp corner physically realistic? Why or why not?

21.2 *(2) Catching the wave.* Why does a surfer have to paddle vigorously toward shore to catch a ride on a wave? Once the ride has begun, what is the speed of the surfer? Why does the surfer keep moving without further propulsive effort? Why is the surfer's position with respect to the wave fairly stable?

21.3 *(3) Hi-fi.* Quality long-playing phonograph records record sounds faithfully up to a frequency of about 18 kHz. Why do the grooves on such records come to an end not too close to the center of the disk?

21.4 *(3) SP-LP-SLP.* Why do magnetic tapes recorded at high tape speeds produce higher-quality reproductions than tapes recorded at lower speeds?

21.5 *(4) Don't be fazed!* A wave on a long string is described by the time-independent function $y(x) = A \cos (kx)$. **(a)** Sketch the wave function. **(b)** Just below y, sketch $v_y(x)$. **(c)** Just below v_y, sketch the transverse acceleration $\partial v_y/\partial t$.

21.6 *(4) Making waves.* You drop a stone into a calm pond. When it has moved 10.5 m away from its source, one of the ripples is 1.3 cm high. How high will the same ripple be when it is 35.2 m from the source? Neglect friction.

21.7 *(6) Disappearing act.* In frame 6 of Figure 21.18a, the pulse seems to have disappeared. Yet it reappears in the next and subsequent frames, having been reflected and inverted. What is happening in frame 6?

21.8 *(7) Nature informing art.* In nearly all stringed instruments, the low-pitched strings are made of thicker wire than the high-pitched ones. Why would it be undesirable to use the same wire for all strings?

21.9 *(7) Is nothing forbidden?* In Equation 21.55, n can be any integer. Why is the value of n restricted to strictly positive integers (integers greater than 0) in the standing-wave condition given by Equations 21.56a and 21.56b?

21.10 *(7) Musician's choice.* You can make a guitar string sound by plucking it anywhere along its length. But the quality of the sound, which has to do with the relative amplitude of the harmonics, depends on where you pluck the string. Explain.

21.11 *(7) Sound sweetly.* It is usually desirable to excite the strings of stringed instruments in such a way as to minimize the amplitude of the seventh harmonic because this harmonic contributes a dissonant character to the sound. In the piano, this minimization is accomplished by placing each hammer so that it strikes its string one-seventh of the length from one end. Explain why this works.

21.12 *(7) Tight string at loose ends.* A heavy cord is stretched between two light threads so that the ends of the cord are essentially free to move transversely even though it is under tension. (In Figure 21.18b, a string is supported in this way at one end only.) Sketch the first three standing-wave modes on this string. Do the modes conform to Equation 21.56b?

PROBLEMS

GROUP A

21.1 *(2) Checking dimensions.* Show that Equation 21.7, $v = \sqrt{T/\mu}$, is dimensionally correct.

21.2 *(2) How fast?* A string of mass 1.54 g and length 2.06 m is subjected to a tension of 128 N. Find the wave speed.

21.3 *(2) Lose weight and run faster.* Two wires are made of the same material. The diameter of wire 2 is twice the diameter of wire 1. Find the ratio v_2/v_1 of the wave speeds in the two wires when they are under the same tension.

21.4 *(2) Hanging rope, I.* A uniform rope having a substantial length L and mass per unit length μ hangs from a rigid support. You can make a pulse travel up the rope by jiggling the bottom end. Show that the speed of the pulse is not constant but is given by $v = \sqrt{g(L - z)}$, where z is the distance of the pulse from the top of the rope.

21.5 *(2) Performance under stress.* A wire, made of material having density ρ, is subjected to uniaxial stress σ. How fast will a pulse travel down the wire?

21.6 *(2) No strain, no gain.* A wire is subjected to tension that results in strain ϵ. **(a)** Express the wave speed v in the wire in terms of ϵ and of Young's modulus Y and the density ρ of the wire material. **(b)** For steel, $Y = 2.1 \times 10^{11}$ Pa and $\rho = 7.8 \times 10^3$ kg/m^3. Find v in a steel wire stretched by 1% of its relaxed length.

21.7 *(2) Road narrows ahead.* A long wire is welded at one end to another long wire made of the same material but having only half the cross-sectional area. A pulse, traveling down the thicker wire at speed v, passes into the thinner one. What is the new speed v' of the pulse?

21.8 *(3) Traveling right along.* A 60-Hz vibrator excites traveling waves in a long string. If the wavelength is 0.83 m, what is the wave speed?

21.9 *(3) Sound reasoning.* A sound wave of frequency 470 Hz propagates through air. If the wavelength is 72 cm, what is the speed of sound?

21.10 *(3) Real deep.* Some organs have pipes that produce 16.50-Hz tones. What is the wavelength of this tone in air? Take the speed of sound to be 347.2 m/s.

21.11 *(3) Squeaky.* The wavelength of the highest note produced by a piccolo is 7.306 cm. What is the corresponding frequency when the speed of sound is 347.2 m/s?

21.12 *(3) Somewhere in between.* The frequency of the musical note middle C is 261.63 Hz. What is the wavelength of the sound wave that playing the note excites in air if the speed of sound is 347.2 m/s?

21.13 *(3) Excitement.* A long string having mass per unit length 2.75 g/m is subjected to a tension of 154 N. If the string is excited by a source whose frequency is 75.2 Hz, what is the wavelength of the waves?

21.14 *(4) Two dialects of the same language.* Derive Equation 21.22, $v = \omega/k$, from Equation 21.15.

21.15 *(4) Still another way.* Reexpress the equation $y = A \cos (kx - \omega t)$, using constants λ and τ instead of k and ω.

21.16 *(4) Can you describe it?* A sinusoidal wave travels in the negative direction on a long string. The amplitude is 2.2×10^{-2} m, the frequency is 270 Hz, and the wave speed is 230 m/s. Write the function that describes the wave.

21.17 *(4) Propaganda, I.* A source at $x = 0$ excites waves in a long, stretched string. The displacement of the source is described by the equation $y(t) = (5.0 \times 10^{-3} \text{ m}) \cos (\pi t/2)$. The speed of waves on the string is 220 m/s. **(a)** Write the wave equation $y(x, t)$ for the string. **(b)** Write the space-independent wave equation for the point $x = 530$ m. **(c)** Write the time-independent wave equation for the instant $t = 4.1$ s. Assume that the source began to oscillate a long time before $t = 0$.

21.18 *(4) Propaganda, II.* A source at $x = 0$ excites waves in a long, stretched string. The displacement of the source is described by the equation $y(t) = (1.6 \times 10^{-2} \text{ m}) \sin (500 \pi t)$. The speed of waves on the string is 170 m/s. **(a)** Write the wave equation $y(x, t)$ for the string. **(b)** Write the space-independent wave equation for the point $x = 0.75$ m. **(c)** Write the time-independent wave equation for the instant $t = 0.035$ s. **(d)** What is the transverse speed of the point $x = 0.75$ m at time $t = 0.035$ s? Assume that the source began to oscillate a long time before $t = 0$.

21.19 *(4) Propaganda, III.* A wave on a long string has amplitude $A = 5.0$ cm, wavelength λ, and period τ. If $y(0, 0) = A$, find the displacement of the point $x = \lambda/12$ at $t = \tau/6$.

21.20 *(4) Propaganda, IV.* The oscillation of a source is described by the function $y(0, t) = A \cos (2\pi t/\tau)$. At time $t = 5\tau/6$, the point at $x = 4.0$ cm has displacement $y = A/2$. What is the wavelength of the wave?

21.21 *(4) The shape of things.* A wave on a long string is described by the function $f(x, t) = A \cos (kx - \omega t)$, with $A = 0.50$ mm, $\omega = 1800 \text{ s}^{-1}$, and $k = 21 \text{ m}^{-1}$. **(a)** What is the ratio of the amplitude to the wavelength? **(b)** What is the ratio of the maximum speed of a particle on the string to the wave speed?

21.22 *(4) More than a hand-waving solution, I.* A wave on a string is described by the function $y = A \cos (kx - \omega t + \delta)$. Let $A = 2.47 \times 10^{-3}$ m, $k = 20.9 \text{ m}^{-1}$, $\omega = 157 \text{ s}^{-1}$, and

$\delta = 1.06$ rad. Calculate the wavelength, frequency, and speed of this wave.

21.23 *(4) More than a hand-waving solution, II.* For the wave of Problem 21.22, find the following quantities at $t = 108$ s at the point $x = -6.94$ m: **(a)** the transverse displacement; **(b)** the transverse velocity; **(c)** the transverse acceleration.

21.24 *(4) Empty-handed.* Show that the function $f(x, t) = f(ax^2 + bt)$ does not satisfy the wave equation.

21.25 *(5) More than a hand-waving solution, III.* **(a)** For a string tension of 150 N, find the energy density dE/dx and the mean energy flux $\langle \Phi \rangle$ carried by the wave of Problems 21.22 and 21.23. **(b)** What is the mass per unit length of the string?

21.26 *(5) Sounding off.* A 5-W source of small size produces sound waves. What is the intensity 15 m from the source?

21.27 *(5) Hark!* At a distance 5.5 m from a point source, the intensity of a sound wave is 10^{-7} W/m^2. If you can barely detect sound of intensity 10^{-12} W/m^2, how far from the source can you hear the sound?

21.28 *(5) Going with the flow.* Show that a periodic wave in a string carries a mean energy flux

$$\langle \Phi \rangle = \tfrac{1}{2}\omega^2 A^2 \sqrt{T\mu}.$$

21.29 *(6) Imposing result.* Two sinusoidal waves travel simultaneously along a string. They are represented by the functions

$$y_1 = A \cos (kx - \omega t) \quad \text{and} \quad y_2 = A \cos (kx + \omega t).$$

(a) Express the shape of the string, $Y(x, t)$, in such a way as to show that there is a standing wave on the string. **(b)** Why is your result not identical with that of Equation 21.51?

21.30 *(7) Counting nodeses.* The distance between the first and fourth nodes of a standing wave is 30 cm. What is the wavelength?

21.31 *(7) In place.* A stretched string of length 1.20 m vibrates in its nth mode. The points on the string (*not* antinodes) that oscillate with $y_{max} = 3.5$ cm are evenly spaced, 15 cm apart. **(a)** What is the value of n? **(b)** What is the amplitude $2A$ of the standing wave?

21.32 *(7) Fundamental physics, I.* A string of length 1.34 m is stretched between two rigid supports. The tension is 660 N, and the mass per unit length is 4.81×10^{-4} kg/m. What is the fundamental frequency of the string?

21.33 *(7) Fundamental physics, II.* A string of length 0.8 m and mass 30 g is subjected to a tension of 6000 N. What is the frequency of the first harmonic?

21.34 *(7) A la mode.* A wire 7.5 m long, having mass 55 g, is subjected to a tension $T = 320$ N. What are the frequencies of the first three standing-wave modes?

21.35 *(7) Less fattening without the ice cream?* A wire 15 m long, having the same mass as the wire in Problem 21.34, is subjected to the same tension. What are the frequencies of the first three standing-wave modes?

21.36 *(7) Coming up short in tight times.* A string of length L is under tension T, and its fundamental frequency is ν. If the string is shortened by 30% (to $0.70L$) and the tension is increased by 60% (to $1.60T$), find the ratio ν'/ν of the new fundamental frequency to the original one.

21.37 *(2) Hanging rope, II.* **(a)** When the pulse traveling up the rope of Problem 21.4 reaches the top, it is reflected and travels back down. Show that the round-trip time is $t = 4\sqrt{L/g}$. **(b)** Compare the round-trip time t with the period τ of the rope swinging as a physical pendulum—that is, as a rigid body (see Equation 12.19c). Show that τ/t is an irrational number approximately equal to 1.28 and hence that the swinging of the rope as a pendulum is not an efficient way to set up standing waves in the rope.

21.38 *(2) Spinning loops, I.* An astronaut in a zero-gravity environment ties the ends of a length of string together. He throws the loop with a twist (the way you would toss a Frisbee) so that it forms a circular loop spinning with angular velocity ω. He then forms a small pulse in the spinning loop by striking it gently with a pencil point. What is the speed with which the pulse travels along the string? Show that the astronaut sees two pulses, one apparently motionless and the other traveling around the loop with angular velocity 2ω.

21.39 *(3) Platter.* A standard phonograph record rotates at 33 rev/min. If a 100-Hz note is recorded on a groove located at the outside edge of the record, 15 cm from the center, what length along the groove is occupied by one wavelength of the note?

21.40 *(4) Legitimate.* Show that any function of the form $f(x, t) = f(\alpha x + t)$ satisfies the wave equation. What is the physical meaning of the quantity α?

21.41 *(4) Illegitimate.* By direct substitution into the wave equation, show that $f(x, t) = x^2/vt$ is not a legitimate wave function.

21.42 *(4) Illegitimate?* **(a)** Is $f(x, t) = x^2 + v^2t^2$ a legitimate wave function? **(b)** Is $f(x, t) = x^2 + 2vtx + v^2t^2$?

21.43 *(4) A fresh viewpoint.* Alice observes a wave traveling on a long, stretched string and finds that the wave is described by the function $y(x, t) = A \cos(kx - \omega t)$. Her lab partner, Zebulon, observes the same wave. However, at time $t = 0$, he begins to walk parallel to the string at velocity V. Show that, when Zebulon observes the wave, he finds that it is described by the wave function

$$y'(x, t) = A \cos\left[k(x - Vt) - \left(1 - \frac{V}{v}\right)\omega t \right].$$

21.44 *(4) Wave goodbye!* In all real strings, some friction is present and the wave dies down as it moves along. This dying down is called *damping*. A damped wave on a long string can be represented by the function

$$f(x, t) = Ae^{-qx} \cos(kx - \omega t),$$

where q is a positive constant smaller than k. **(a)** Show that $f(x, t)$ satisfies the wave equation. **(b)** What is the speed of the wave? **(c)** Sketch a graph of v as a function of q^2/k^2 in the range $0 \le q^2/k^2 < 1$. [Hint: $(d/dx)\, e^{-qx} = -qe^{-qx}$.]

21.45 *(4) Polarization.* A long string stretched along the x axis is simultaneously excited by transverse simple harmonic oscillations of equal amplitude along the y axis and along the z axis. **(a)** Suppose the two oscillations are in phase; that is, they are described by the functions

$$y(0, t) = A \cos \omega t \quad \text{and} \quad z(0, t) = A \cos \omega t.$$

By itself, the first of these functions would produce a traveling wave said to be *polarized in the y direction*. That is, the string would oscillate in the x-y plane. By itself, the second function would produce a wave *polarized in the z direction*. That is, the string would oscillate in the x-z plane. Describe the space-independent motion of an element at any point x along the string when both oscillations are taking place. What does the wave look like? **(b)** Suppose that the two oscillations are out of phase by $\pi/2$; that is, they are described by the functions

$$y(0, t) = A \cos \omega t \quad \text{and} \quad z(0, t) = A \sin \omega t.$$

Describe the space-independent motion of an element at any point x along the string. What does the wave look like? (Hint: See Section 6.2.)

21.46 *(5) Just average, I.* Beginning with Equation 21.34, show that, at any particular instant, the energy contained on one wavelength of a vibrating string is $E_\lambda = \frac{1}{2}\mu\omega^2A^2$. Hint: You will need the value of the definite integral

$$\int_0^{2\pi/k} \sin^2 kx \, dx = \pi/k.$$

21.47 *(5) Just average, II.* Beginning with Equation 21.37, derive Equation 21.38. (Hint: See Problem 21.46.)

21.48 *(5) Active sun, I.* When the sun is directly overhead, the solar intensity (energy flux per unit area) at the top of the earth's atmosphere is $\langle \Phi \rangle_e = 1.4\,\text{kW/m}^2$. What is the total solar energy flux falling on the earth? Take the earth's radius to be 6.4×10^6 m.

21.49 *(5) Active sun, II.* The mean radius of the orbit of Mars is 1.5 AU (By definition, the mean orbit radius of the earth is 1 AU $\simeq 1.5 \times 10^{11}$ m.) Use the value of $\langle \Phi \rangle_e$ given in Problem 21.48 to find the solar intensity falling on Mars.

21.50 *(5) Active sun, III.* The radius of the sun is 7.0×10^8 m. **(a)** What is the intensity at the surface of the sun? **(b)** What is the total energy flux of the sun? Use the value of $\langle \Phi \rangle_e$ given in Problem 21.48.

21.51 *(5) Waves in three dimensions.* Develop in detail the argument sketched at the end of Section 21.5, and derive Equations 21.43 and 21.44, which give the intensity and amplitude for waves propagating from a point source in a three-dimensional medium. (Hint: See Problem 14.58.)

21.52 *(6) Hooke's law.* The superposition principle discussed in Section 21.6 is precisely correct only if, when the medium is disturbed, the restoring forces obey Hooke's law. To show in a crude way that the Hooke's-law assumption is valid for a stretched string, imagine that the midpoint of a string under tension T is displaced a small distance $y \ll L$, as shown here. Show that $F_y = -cy$, where c is a constant.

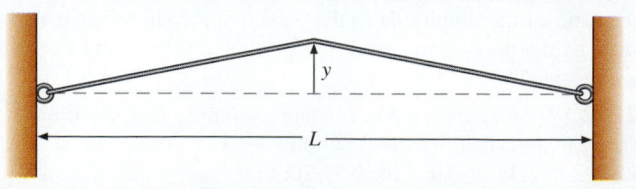

21.53 (6) *Square deal.* Using a programmable calculator or a personal computer, sum five or more terms of the series

$$y(x) = \frac{4}{\pi} \sum_{j=0}^{\infty} \frac{1}{2j+1} \sin[(2j+1)\pi x].$$

[This series amounts to setting $A_j = 1/(2j+1)$ and all $\delta = \pi/2$ in Equation 21.47b.] Evaluate $y(x)$ at 20 or more points in the interval $0 \le x \le 2$, and plot these points to show that the sum approximates a square wave.

21.54 (6) *Crosscut.* Using a programmable calculator or a personal computer, sum five or more terms of the series

$$f(t) = \frac{1}{2} - \frac{4}{\pi^2} \sum_{j=0}^{\infty} \frac{1}{(2j+1)^2} \cos[(2j+1)\pi t].$$

[This series amounts to setting $A_j = 1/(2j+1)^2$ and all $\delta = 0$ in Equation 21.47b and shifting the f axis.] Evaluate $f(t)$ at 20 or more points in the interval $0 \le t \le 2$, and plot these points to show that the sum approximates the "crosscut" sawtooth wave shown here.

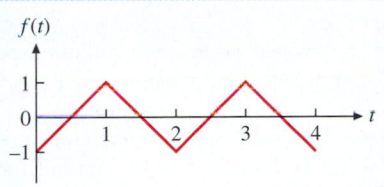

21.55 (7) *Cut and try.* A stretched wire has fundamental frequency ν. If the wire is cut to 65% of its length and the tension is increased by 70%, what is the new fundamental frequency ν'?

21.56 (7) *Piano string.* (a) Express the fundamental frequency of a wire fixed rigidly at both ends in terms of its tension, its length, its radius, and the density of the material of which it is made. (b) The density of steel is 7.86×10^3 kg/m^3. What is the tension required for a steel piano string tuned to middle C ($\nu = 261.63$ Hz)? The diameter of the string is 0.750 mm, and its length is 50.1 cm. (c) A standard piano has 243 strings. The tension varies from string to string but not by much, so all the strings are under roughly the same tension. (The wide range of pitch is achieved largely by varying the length and thickness of the strings.) What is the approximate total force that the piano frame must bear?

21.57 (7) *Being precise about it.* You cut two equal lengths L from a spool of piano wire of cross-sectional area A and attach a clamp to each end of each wire. When you apply a tension to each wire, its length increases slightly and its mass per unit length decreases slightly. Suppose you apply tension T to wire 1 and tension $2T$ to wire 2. (a) Ignoring the slight changes of length and mass per unit length, find the ratio ν_2/ν_1 of the fundamental frequencies of the two wires. (b) Now take the changes into account. Express the strain of wire 1, $\epsilon_1 = \Delta L_1/L$, in terms of Young's modulus Y and the cross-sectional area A. (c) Show that the ratio of frequencies is

$$\frac{\nu_2}{\nu_1} = \sqrt{2}\sqrt{\frac{1+\epsilon_1}{1+2\epsilon_1}}.$$

(d) Find the correction factor (the factor that multiplies $\sqrt{2}$) for $\epsilon_1 = 2\%$. (e) Is the correction significant if a tensile force of 1000 N is applied to the 50.1-cm length of piano wire described in Problem 21.56? Use $Y = 2.1 \times 10^{11}$ Pa.

21.58 (7) *Trickier than it looks.* The apparatus used for a standard laboratory experiment is shown below. You cut a string

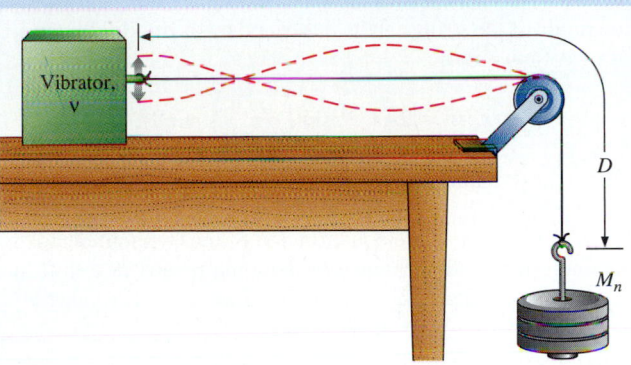

of length D and weigh it to determine its mass m. You then tie one end of the string to a 120-Hz vibrator. You pass the string over the pulley and tie the other end to a holder on which you can pile weights. (Neglect the small length of string required to tie the knots.) You adjust the hanging weight until you observe standing waves on the string; each standing-wave mode n corresponds to a weight $M_n g$. (The point of attachment to the vibrator is neither a node nor an antinode; the point of contact with the pulley is always a node.) You measure the wavelengths λ_n of several different standing-wave modes.

(a) Express λ_n in terms of M_n, m, g, D, and ν. (b) Suppose $D = 1.00$ m and $m = 1.04 \times 10^{-3}$ kg. The maximum value of M for which standing waves are observed is $M_1 = 2.97$ kg. What is the next-smaller value of M for which standing waves are observed? (c) You find that the measured values of λ deviate from those predicted by the result of part a. The deviation is largest for the lowest-n modes and typically amounts to about 2% for $n = 1$. The deviation arises from the stretching of the string, which reduces the value of μ by a different amount $\Delta\mu$ for each value of M_n. You can allow for this effect by measuring the force (Hooke's-law) constant c of the string. (This is easily done by measuring the change in length of the string when you hang weights on it.) If the uncorrected value of wavelength predicted by the result of part a is λ_n, show that the corrected value λ'_n, which agrees quite well with the value measured directly on the vibrating string, is

$$\lambda'_n = \lambda_n\left(1 + \frac{M_n g}{2cD}\right).$$

(Hint: Remember that $\Delta\mu \ll \mu$.)

21.59 (7) *They also function who only stand and wait.* Show explicitly that a sinusoidal wave on a stretched string of length L is described by the function

$$Y = 2A \sin\left(n\pi \frac{x}{L}\right) \sin \omega t.$$

21.60 (7) *Running in place.* (a) Show that the standing-wave function Y given in Problem 21.59 satisfies the wave equation. (b) What is the value of v^2 obtained in part **a**? (c) The standing wave is certainly not moving. What is the physical meaning of v?

21.61 (G) *The song is gone but the memory lingers.* When you pluck a guitar string, its vibration gradually dies down as the energy dissipates into the surrounding air. The nth vibration mode of the string can be described by the modified standing-wave equation

$$Y(x, t) = 2Ae^{-\gamma t} \sin\left(n\pi \frac{x}{L}\right) \sin \omega t,$$

where the *damping coefficient* γ satisfies the requirement $0 \le \gamma \ll 2\pi/L$. (Compare with Problem 21.44.) (a) Show that Y satisfies the wave equation. (b) Show (without detailed calculation) that the presence of damping results in a shift in the angular frequency ω of the mode away from the value ω_0 that corresponds to $\gamma = 0$. Is the shift toward higher or lower frequency?

21.62 (G) *Energy in a standing wave.* A standing wave on a string has amplitude $2A$ and angular frequency ω. The string is fixed at both ends and is vibrating in its first (fundamental) mode with wavelength λ_1. (a) Express the energy density in the string as a function of x, with $-L \le x \le 0$. (b) Show that the energy of the string is

$$E_1 = \tfrac{1}{2}\mu\omega_1^2 A^2 \lambda_1.$$

[Hints: (1) The standing-wave function of Equation 21.51 is equivalent to the sum of two traveling waves given by Equation 21.50. (2) See Problem 21.46.] (c) What is the energy of the same string when it vibrates with the same amplitude in its nth mode? (d) Find the ratio E_n/E_1. (e) Express E_1 in terms of the string mass m, and show that the result is independent of the length of the string.

GROUP C

21.63 (G) *Hanging rope, III.* Let the rope of Problems 21.4 and 21.37 have mass m. If you attach a mass M to the bottom end of the rope, the system becomes similar to that of Example 21.1 if $M \gg m$. In particular, if you start the bottom end swinging, the system may be considered to be approximately a simple pendulum. But the bottom end, because it is swinging through nearly horizontal small oscillations, may also be considered to be the source of a wave train. (a) Find the speed of the traveling wave excited in the rope by the swinging mass at its end. (b) Express the wavelength in terms of M, m, and L. (c) Using the values of M, m, and L given in Example 21.1, find the wavelength. (d) As the system swings, why does the rope remain straight, swinging slightly back and forth as a unit about the vertical (the way a rigid rod would swing), rather than assuming some curved form $f(x, t)$ and thus behaving as a medium in which a wave train propagates upward? (e) You can modify the system to make it behave more like a vibrating string and less like a swinging simple pendulum by varying the ratio M/m. Describe qualitatively what happens as you do so.

21.64 (G) *Hanging rope, IV.* Now consider quantitatively the process suggested in part **e** of Problem 21.63. In what follows, define the ratio $\rho \equiv M/m$. (a) If ρ is not very much greater than 1, you must consider the swinging system to be a physical pendulum. Show that the oscillation period of such a pendulum is

$$\tau = 2\pi \sqrt{\frac{L}{g}} \sqrt{\frac{\rho + \tfrac{1}{3}}{\rho + \tfrac{1}{2}}}.$$

(Hint: Use Equation 12.19c.) (b) Show that the round-trip travel time for a pulse generated at the bottom end of the rope is

$$t = 4\sqrt{\frac{L}{g}} (\sqrt{\rho + 1} - \sqrt{\rho}).$$

(Hint: See part **b** of Problem 21.37.) (c) It is intriguing to think (on the basis of the answers to parts **a** and **b**) that, if you adjusted ρ to the proper value, the oscillation period of the pendulum would be equal to a standing-wave period for the rope and thus the swing of the pendulum would excite standing waves in the rope. (If this situation could be achieved, the wavelength would be four times the length of the rope, with a node at the top and an antinode at the bottom.) But show that the desired condition, called a *resonance condition*, cannot be achieved. To do this, first find the value of ρ that leads to the largest possible value of t, and show that

$$t_{max} = 4\sqrt{\frac{L}{g}},$$

which is the result you obtained in part **a** of Problem 21.37. Then find the value of ρ that leads to the smallest possible value of τ, and show that

$$\tau_{min} = 2\pi \sqrt{\frac{L}{g}} \sqrt{\frac{2}{3}} \simeq 5.13 \sqrt{\frac{L}{g}}.$$

Compare these two results, and draw your conclusion. (d) Is a resonance condition possible using a standing-wave mode with larger n? Why or why not?

21.65 (G) *Funny rope.* In Problem 21.4, you saw that the wave speed v in a uniform, freely hanging rope varies with the distance z from the top. This is because the tension T varies while the mass per unit length μ remains constant. Suppose you want to design a freely hanging rope in which v is independent of z. For such a rope, μ cannot be constant, and you must specify the function $\mu(z)$. (a) Show that $dT = -g\mu \, dz$. (b) By solving $dT = -g\mu \, dz$ for μ, show that, if the rope is infinitely long, the desired form of $\mu(z)$ is

$$\mu(z) = \mu_0 e^{-(g/v^2)z},$$

where μ_0 is the mass per unit length of the rope at the top ($z = 0$). (c) Express the wave speed v in terms of μ_0 and the total mass m of the rope. (d) Express μ in terms of μ_0, m, and z only. (e) Suppose you want to cut the rope off at length L.

Show that you can still have constant wave speed on the finite rope if you hang a mass M on the end such that

$$M = me^{-\mu_0 L/m}.$$

21.66 *(G) Spinning loops, II.* Consider again the spinning loop of Problem 21.38. Suppose there is a red mark at one point on the loop. If he is dexterous enough, the astronaut can excite a periodic wave train by striking the loop repeatedly at the red mark with the pencil point, with frequency ν. **(a)** To begin with, think about the system from the point of view of an observer located at the center of the loop and rotating with it so that the red mark—and the string as a whole—appears motionless to her. To make matters simpler without losing any of the essential features of the system, suppose that the oscillations are excited by the rotating observer. By means of a light thread attached to the red mark, she makes the red mark move radially in such a way that $\Delta r = A \cos \omega t$. The amplitude A is much smaller than the radius r of the loop. As a result of the oscillation, the red mark is a source point, and two waves travel around the loop in opposite senses. Calling the source point $s = 0$ and measuring distances s along the string, write the superposed wave function $\Delta R(s, t)$. **(b)** Because the string is an endless loop, moving along the string in the direction of increasing s eventually brings you back to where you started. Use this observation to write a *periodic boundary condition* on the wave function ΔR—that is, a boundary condition that requires the wave function to repeat itself at every turn around the circumference. (Note: A similar boundary condition limits electrons to specific orbits around the nuclei in atoms.) **(c)** Show that, from the point of view of the rotating observer, the result of applying the periodic boundary condition to $\Delta R(s, t)$ is the standing-wave condition

$$\lambda = \frac{2\pi r}{n}; \quad n = 1, 2, 3, \ldots$$

and r is the radius of the loop. What is the physical meaning of this condition on λ? What is the corresponding frequency ν of the standing wave? **(d)** Show that, in the absence of friction, the nth mode can be established permanently by exciting only n oscillations. **(e)** Now return to the point of view of the astronaut. Show that what appears as a standing wave to the rotating observer appears to the astronaut as a traveling wave that rotates with angular velocity ω and thus moves with speed ωr. Show also that the astronaut could excite the same wave pattern by pressing a frictionless roller against the loop in oscillatory fashion as it rotates past him, thus making the loop oscillate with any frequency ν that satisfies the boundary condition.

Waves II

Mostly Longitudinal Waves and Sound

- Both transverse and longitudinal waves can propagate through a solid, but only longitudinal waves can propagate through a fluid.

- Sound usually means a longitudinal wave propagating through air. We characterize a sound according to its loudness (related to the sound-wave intensity), its pitch (related to the sound-wave frequency), and its quality (related to its mixture of frequencies).

- When two sound waves of slightly different frequencies are superposed, we hear the slow waxing and waning of intensity called beats.

- Standing waves can be produced in pipes; this is the principle on which many musical instruments are based.

- If either a sound source or an observer (or both) is moving with respect to the air, the observer detects a sound whose frequency is different from the source frequency. This is the Doppler effect.

Left: Damage due to the Kobe earthquake of 1995. Solids can propagate both longitudinal and transverse waves.

Sound is naught but air y-broken
And every speaking that is spoken
Loud or privy, foul or fair,
In its substance is but air.

—CHAUCER, *The House of Fame* (c. 1380)

Introduction

We now consider the propagation of waves in extended solid and fluid media. We will bring to bear nearly all the ideas developed in Chapter 21, where we studied mainly the simple, essentially one-dimensional behavior of waves on a stretched string.

We begin in Section 22.2 with a study of *shear waves* in solids. There is a close kinship between shear waves propagating through a long rod and transverse waves in a long string. Consequently, the extension from strings to rods is direct. Section 22.3 treats *longitudinal waves*, which involve disturbance of the medium back and forth along the direction of the wave motion rather than transverse to it. Sound in air is the most familiar example of a longitudinal wave phenomenon. In Section 22.4, we consider waves in fluid media and use ideas from Section 21.2 to derive the speed of sound in air. Section 22.5 concerns the physical aspects of the process by which the ear detects sound, a subject called *physical acoustics*. Section 22.6 applies the superposition principle to sound waves, and Section 22.7 deals with the important special case of standing sound waves in pipes. In Section 22.8, we study the *Doppler effect*, the set of phenomena that occur when the wave source, the observer, and/or the medium move.

SECTION 22.2

Propagation of Waves in Solids

Figure 22.1a shows a long, solid bar of uniform cross-sectional area a. In Figure 22.1b, the bar has been struck transverse to its length at the point labeled $x = 0$. Such a disturbance might be produced by striking a railroad rail with a hammer.

The disturbance moves along the bar in both directions, as shown in Figure 22.1c. The motion of the disturbance is *uniform*. That is, all points on the same cross section (at the same x) experience the same displacement y at the same time t. All such points define a two-dimensional surface that we call a **wave front**. As time passes, the wave front moves with the wave speed v characteristic of the material of which the bar is made.

The statements of the preceding paragraph hold equally well if the disturbance is periodic, as in Figure 22.2. At any time t, all points in the bar that are located on any cross-sectional plane (such as the plane at x_a) experience the same transverse displacement and lie on the same planar wave front. Each such plane is analogous to a single point on a vibrating string. We can think of the wave as a series of planar wave fronts traveling along the bar. We therefore call the wave a **plane wave**. It is possible to draw an infinite number of wave fronts, but it is usually most useful to draw a particular set of wave fronts that lie one wavelength λ apart, as shown in Figure 22.2.

602 ■ Waves II

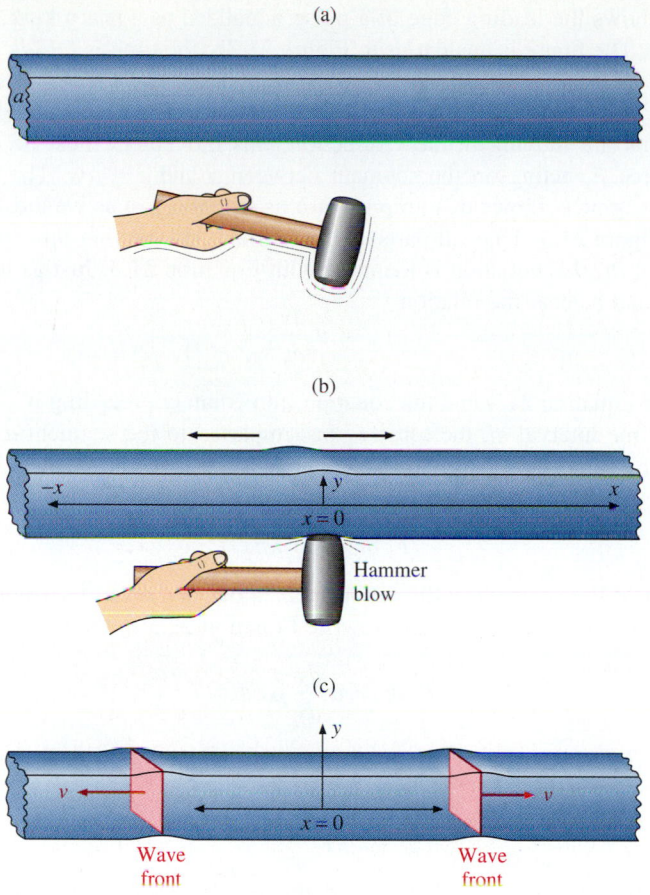

(a)

(b)

Hammer
blow

(c)

Wave
front

Wave
front

FIGURE 22.1 (*a*) A long, solid bar, undisturbed. (*b*) A transverse disturbance is produced in the bar. The displacement *y* is much exaggerated for clarity. (*c*) The disturbance travels as a pulse in both directions along the bar, with speed *v*. Two wave fronts are shown.

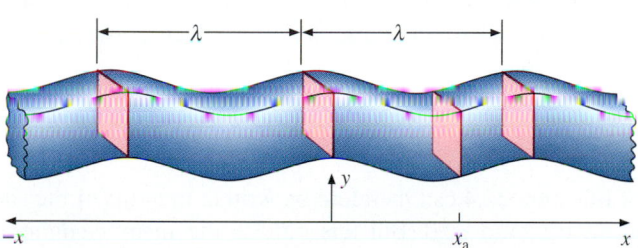

FIGURE 22.2 A periodic wave travels along a bar. The wavelength λ separates two wave fronts having the same phase. As in Figure 22.1, the transverse displacement is much exaggerated.

If the plane wave happens to be sinusoidal, it can be represented by a wave function identical with the wave function that represents a sinusoidal wave on a string:

$$y(x, t) = A \cos (kx \mp \omega t + \delta). \tag{22.1}$$

Because the argument or phase ($kx \mp \omega t + \delta$) of the cosine is the same for all points on the wave front, these points are said to be **in phase** with one another.

Speed of Transverse Waves in a Solid

As you learned in Chapter 21, a wave can propagate through a medium only if there is a restoring force that tends to resist the distortion imposed on the medium by the wave. For a stretched string, the restoring force arises from the externally imposed tension in the string. A solid bar, however, need not be under external tension to resist transverse distortion because its resistance to shear (expressed quantitatively by the shear modulus *G*) provides the necessary restoring force.

With this difference between strings and bars in mind, we can find the speed of transverse waves in a solid bar by modifying the derivation of Section 21.2 slightly.

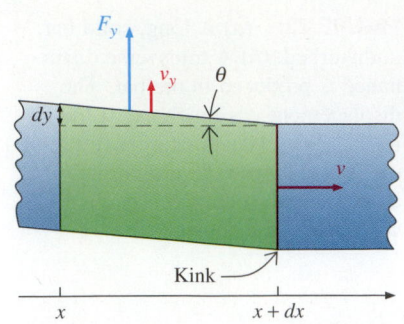

F_y

v_y θ

dy

v

Kink

x $\qquad$ $x + dx$

FIGURE 22.3 The leading edge of a disturbance, shown as a sharp kink, propagates down a bar of cross-sectional area a. A short piece of the bar is shown at time $t + dt$, when the kink has reached $x + dx$. At time t, the kink was at x, and so the propagation speed is $v = dx/dt$. All parts of the green segment, through which the kink has just passed, are moving upward with speed v_y; the cross section at x has been displaced through $dy = v_y\, dt$.

Figure 22.3 shows the leading edge of a pulse, idealized as a sharp kink, propagating through a bar. The figure is analogous to Figure 21.3b; the time is $t + dt$, and the kink has reached $x + dx$. The kink is moving to the right at the wave speed v, so $dx = v\, dt$; this equation is identical with Equation 21.2. Elements of the bar immediately to the left of the kink are moving in the y direction; this movement must be due to a net transverse force F_y acting on the segment between x and $x + dx$. The infinitesimal length of the segment allows us to represent it as straight, just as we did for the string segment of Figure 21.3. Thus all parts of the segment are moving upward with speed v_y, so $dy = v_y\, dt$; this equation is identical with Equation 21.3. Just as for the string, the speeds v and v_y bear the relation

$$v_y = v \frac{dy}{dx}; \tag{22.2}$$

compare with Equation 21.4 and the equation immediately preceding it.

Over the time interval dt, the force F_y has imparted to the segment dx an impulse $F_y\, dt$. Over the same time interval, the segment has acquired an upward momentum $v_y\, dm$. So we have, just as for the string, the impulse-momentum relation

$$F_y\, dt = v_y\, dm. \tag{22.3}$$

The mass dm of the segment is the product of its density ρ and its volume $a\, dx$. We use this relation and Equation 22.2 to rewrite Equation 22.3 in the form

$$F_y\, dt = v \frac{dy}{dx} \rho a\, dx.$$

Because $v = dx/dt$, we can divide this equation through by dt and rearrange to obtain

$$\frac{F_y}{a} \frac{1}{dy/dx} = \rho v^2. \tag{22.4}$$

The quantity F_y/a is the shear stress σ, defined by Equation 15.18 as the shear force (force applied parallel to the surface) per unit area of the surface to which it is applied. Corresponding to this shear stress is a *shear strain* β. The left surface of the bar segment is displaced through dy relative to the right surface; see Section 15.4. According to Figure 15.4 and Equation 15.20, the shear strain is defined as the ratio of the displacement to the thickness. In the present case, we have

$$\beta = \frac{dy}{dx}. \tag{22.5}$$

The left side of Equation 22.4 can therefore be written in terms of the shear stress and the shear strain as the ratio σ/β. But this ratio is the *shear modulus* G defined by Equation 15.21, $G \equiv \sigma/\beta$. Thus Equation 22.4 assumes the simple form $G = \rho v^2$. Solving for the wave speed v, we obtain

$$v = \sqrt{\frac{G}{\rho}} \quad \text{for transverse waves.} \tag{22.6}$$

Because the transmission of transverse waves through a solid has to do with the shear properties of the solid, such waves are often called **shear waves**. According to Equation 22.6, the speed of shear waves is greatest in solids that are stiff (difficult to shear) and that have low density. Note the kinship between Equation 22.6 and Equation 21.7, $v = \sqrt{T/\mu}$, which expresses the speed of transverse waves in a string in terms of the properties of the string.

EXAMPLE 22.1

(a) Find the propagation speed of shear waves in steel. Take the density of steel to be $7.83 \times 10^3\ \text{kg/m}^3$. **(b)** Diamonds are single crystals. As with all single crystals, the elastic properties of diamond are anisotropic. When shear waves propagate in a direction normal to a particular set of planes called [100] planes, their speed is 12 800 m/s. The density of diamond is $3.51 \times 10^3\ \text{kg/m}^3$. Find the value of the shear modulus of diamond for distortions along the [100] planes.

SOLUTION:

(a) Find v for shear waves in steel.

According to Table 15.1, the shear modulus of a typical steel is 8.3×10^{10} Pa. Using Equation 22.6, you find

$$v = \sqrt{\frac{8.3 \times 10^{10} \text{ Pa}}{7.83 \times 10^3 \text{ kg/m}^3}} = 3000 \text{ m/s}.$$

(b) Find G for diamond when it is sheared along the [100] planes.

You can use Equation 22.6 to write

$$G = \rho v^2 = 3.51 \times 10^3 \text{ kg/m}^3 \times (1.28 \times 10^4 \text{ m/s})^2$$
$$= 5.75 \times 10^{11} \text{ Pa}.$$

The value of v for diamond is the greatest for any solid. Why is this not surprising?

Longitudinal Waves

So far, we have considered transverse pulses and periodic waves. We now turn to pulses and waves that propagate from a source that disturbs the medium *longitudinally*—back and forth along the axis of propagation. Pulses and waves produced in this way are called **longitudinal pulses** and **longitudinal waves**. In Figure 22.4a, a long bar of a solid material is disturbed by being struck at one end with a flat-headed mallet. As a result of the blow, the element of the bar in the immediate vicinity of the end is compressed. The darker blue shading represents the (small) local increase in density of the solid. Because the solid is elastic, the compression produces restoring forces. As shown in Figure 22.4b, the tendency of the compressed element is to expand to its original length. In doing so, the element exerts a compressive force on the neighboring element immediately to the right. Continuation of this process results in the propagation of a pulse along the bar with a speed v characteristic of the material, as shown in Figure 22.4c. As the pulse passes through any part of the bar, matter is displaced first to the right and then to the left, back to its undisturbed position. This forward-and-back motion is longitudinal to the axis of propagation.

Because solids are quite rigid, a small displacement of a bit of matter relative to its neighbor is accompanied by a considerable stress. As a result, the displaced matter is restored to its undisturbed state without ever having moved very far. As you have seen, small displacement is typical of transverse waves as well.

(a)

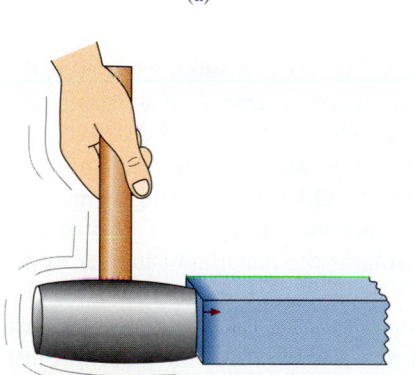

(b)

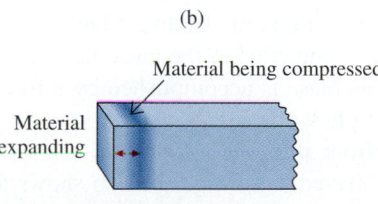

Material being compressed

Material expanding

(c)

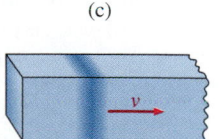

FIGURE 22.4 Excitation and propagation of a longitudinal pulse. (*a*) A mallet blow on the end of a solid bar compresses the element of the bar adjacent to the end. The darker blue shading in this element represents the small increase in density. The red arrow represents the displacement of a typical bit of matter within the element during compression. (*b*) The end element is in the process of expanding, as elastic forces restore it to its original density. The exertion of these restoring forces results in the compression of the neighboring element to the right. The red arrows represent the displacements of typical bits of matter as the end element expands and the neighboring element is compressed. (*c*) A region of compressed matter propagates through the bar with speed v.

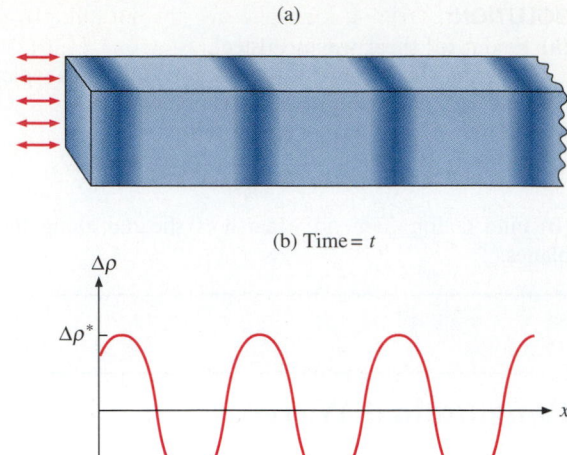

(a)

FIGURE 22.5 (*a*) A device (not shown) attached to the end of the bar drives it back and forth over a small displacement (exaggerated for clarity). As a result, a periodic longitudinal wave propagates down the bar. The periodic variation of blue shading represents the time-independent variation in density along the bar at a particular time *t*. (*b*) A more convenient and more quantitative way of representing the variation of density about the undisturbed value ρ_0, with amplitude $\Delta\rho^*$.

(b) Time = *t*

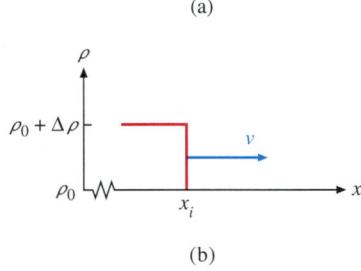

(a)

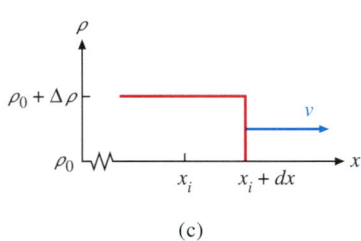

(b)

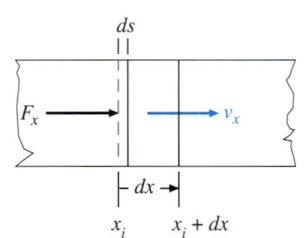

(c)

FIGURE 22.6 An idealized longitudinal pulse passes through a bar segment of undisturbed thickness dx. (*a*) Graph of $\rho(x, t)$, the density of the bar as a function of position at t, the time at which the pulse enters the segment at x_i. Because the pulse is abrupt, it results in a discontinuous increase in the density of the material from ρ_0 to $\rho_0 + \Delta\rho$ at all locations $x \le x_i$. (*b*) Graph of $\rho(x, t + dt)$, when the pulse has reached $x_i + dx$. (*c*) At time $t + dt$, the increased density $\Delta\rho$ of the entire segment may be thought of as produced by a compression that displaces the left end of the segment through a distance ds.

Now consider a source at the end of the bar that alternately pushes and pulls on the end of the bar. As a result, the density of the end element periodically rises above and falls below its undisturbed density ρ. These density variations propagate along the bar as a periodic longitudinal wave, as shown in Figure 22.5a. A useful way to represent these density variations is shown in Figure 22.5b. Note that the ordinate axis represents not displacement y but density change $\Delta\rho$. Do not be misled into thinking that the oscillatory curve represents a transverse wave.

In spite of the physical differences between longitudinal and transverse waves, the wave functions that represent them have very much the same *mathematical* form. If the longitudinal wave is sinusoidal, we can represent it by the function

$$\rho(x, t) = \rho_0 + \Delta\rho^* \cos (kx \mp \omega t + \delta). \tag{22.7}$$

In this expression, the amplitude $\Delta\rho^*$ is the magnitude of the extremal change of density from the undisturbed value ρ_0. Most commonly, $\Delta\rho^*$ is very much smaller than ρ_0.

Speed of Longitudinal Waves

To derive an expression for the speed of longitudinal waves in a solid, we need make only a few modifications to the derivation for shear waves. Figure 22.6 shows an idealized longitudinal pulse propagating through a bar whose undisturbed density is ρ_0. The pulse is abrupt; the discontinuous increase of ρ at the leading edge corresponds to the kink at the leading edge of the transverse wave. At all locations through which the leading edge of the pulse has passed, the density is increased to $\rho_0 + \Delta\rho$. This density increase is accomplished by a force F_x, which displaces the material of the bar to the right with velocity v_x. We consider a bar segment of undisturbed thickness dx, extending from x_i to $x_i + dx$. Figure 22.6a shows the situation at time t, when the pulse has just arrived at x_i. Figure 22.6b shows the situation at time $t + dt$, when the pulse has just completed its passage through the segment, to which it has imparted an impulse $F_x \, dt$. At this time, all the matter in the segment has acquired velocity v_x; its momentum is therefore $v_x \, dm$. So we have

$$F_x \, dt = v_x \, dm; \tag{22.8}$$

compare with Equation 22.3, the corresponding relation for shear waves.

We first evaluate the left side of this equation. At time $t + dt$, the material at the left end of the segment has moved to the right through the infinitesimal distance

$$ds \equiv v_x \, dt, \tag{22.9}$$

while the material at the right end has not quite begun to move; see Fig. 22.6c. Thus the segment has experienced a compressive strain (Equation 15.5) of magnitude

$$\epsilon = \frac{ds}{dx}. \tag{22.10}$$

This strain is due to the uniaxial compressive stress $\sigma = F_x/a$. For small strains, the stress and the strain are related by Young's modulus according to Equation 15.8a, $\sigma = Y\epsilon$. Equating these two expressions for σ gives

$$Y\epsilon = \frac{F_x}{a}.$$

Rearranging this equation and multiplying both sides by dt, we obtain

$$F_x\, dt = Y\epsilon a\, dt. \tag{22.11}$$

We now evaluate the right side of Equation 22.8, with a view to reexpressing $v_x\, dm$ in terms of the wave speed v instead of the displacement speed v_x. From Equation 22.9, the displacement speed is $v_x = ds/dt$. Because the pulse propagates at the wave speed $v = dx/dt$, we have

$$\frac{v_x}{v} = \frac{ds/dt}{dx/dt} = \frac{ds}{dx} = \epsilon,$$

or

$$v_x = v\epsilon.$$

And, because the mass of the element is $dm = \rho a\, dx$, the momentum of the segment at time $t + dt$ is

$$v_x\, dm = v\epsilon\rho a\, dx. \tag{22.12}$$

We now have values for both sides of Equation 22.8. Substituting Equations 22.11 and 22.12 into Equation 22.8, we obtain

$$Y\, dt = \rho v\, dx.$$

As we did for shear waves, we divide through by dt and obtain $Y/\rho = v^2$, or

$$v = \sqrt{\frac{Y}{\rho}} \quad \text{for longitudinal waves.} \tag{22.13}$$

Because the transmission of longitudinal waves through a solid has to do with the compressional properties of the solid, such waves are often called **compressional waves**. Compare Equation 22.13 with Equation 22.6 for shear waves, $v = \sqrt{G/\rho}$.

As noted in Section 15.4, the value of Young's modulus is about three times that of the shear modulus for most materials. It follows that, in most materials, the speed of compressional waves exceeds that of shear waves by the approximate factor $\sqrt{3}$.

EXAMPLE 22.2

You and a friend stand some distance apart by the side of the same little-used railroad track. While you hold your ear next to the steel rail, your friend strikes the rail with a sledgehammer. You hear two distinct sounds. If the time interval between them is 0.20 s, how far apart are you and your friend?

SOLUTION: When your friend hits the rail, the complex interaction between hammer and rail guarantees that both shear and compressional waves will be excited. Let L be the distance between you and your friend, v_s the speed of shear waves in steel, and v_c the speed of compressional waves in steel. You have for the corresponding transit times

$$t_s = \frac{L}{v_s} \quad \text{and} \quad t_c = \frac{L}{v_c}.$$

The time interval between the arrivals of the two waves at your ear is $t_s - t_c \equiv \Delta t$. You have

$$\Delta t = \frac{L}{v_s} - \frac{L}{v_c} = L\,\frac{v_c - v_s}{v_s v_c}.$$

Solving for L, you obtain

$$L = \Delta t\,\frac{v_c}{v_c/v_s - 1}. \tag{22.14}$$

You use Equation 22.13 and Table 15.1 to calculate v_c:

$$v_c = \sqrt{\frac{21 \times 10^{10}\ \text{Pa}}{7.83 \times 10^3\ \text{kg/m}^3}} = 5180\ \text{m/s}.$$

Next, you use Equations 22.13 and 22.6 to find

$$\frac{v_c}{v_s} = \sqrt{\frac{Y}{G}}$$

(22.15)

$$= \sqrt{\frac{21 \times 10^{10} \text{ Pa}}{8.3 \times 10^{10} \text{ Pa}}} = 1.59.$$

You now have all the numbers you need to calculate L; Equation 22.14 gives you

$$L = 0.2 \text{ s} \times \frac{5180 \text{ m/s}}{1.59 - 1} = 1760 \text{ m}.$$

Locating Earthquakes

The reasoning of Example 22.2 provides the basis for a method seismologists use to locate the source of an earthquake. In seismology, the shear waves are called *S waves* and the compressional (or pressure) waves are called *P waves*. The earthquake produces waves of both kinds, as does the sledgehammer in Example 22.2. The S and P waves travel at different speeds and so are detected separately at seismograph stations. The records from any three stations can be used to locate the earthquake by triangulation, as Figure 22.7 shows.

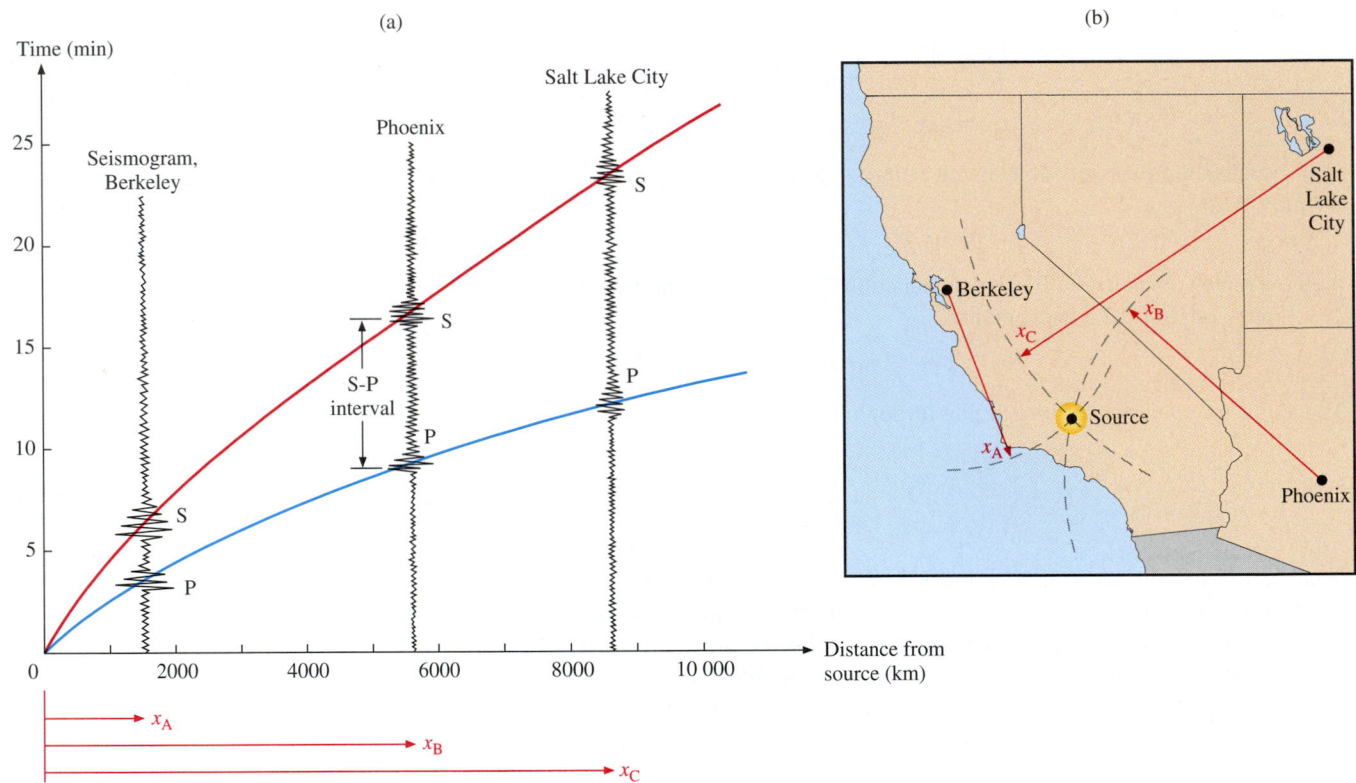

FIGURE 22.7 (*a*) Seismograph records from three stations are plotted vertically, parallel to a time axis. In each case, the S–P interval yields the distance of the station from the earthquake. (The curves labeled S and P, whose slopes are the S and P wave speeds, are not exactly straight lines because the properties of the medium vary with depth below the surface.) (*b*) Method of locating the epicenter—the surface point directly above the earthquake source—using the S–P intervals for three stations. (The symbols correspond to those in part *a*, but the numerical values do not.)

(Adapted from F. Press and R. Siever, *Earth*, 2nd ed. San Francisco: W. H. Freeman, 1978.)

| SECTION 22.4 | ## Waves in Fluids |

If they lie in the proper frequency range, both shear waves and compressional waves are perceived as sound waves, as noted in Example 22.2. Sometimes, as in Example 22.2, such waves can be perceived directly as sound without traveling through air. To give another example, you will feel as if your entire head is filled with sound if you hold a vibrating tuning fork between your teeth. The medium carrying the sound to your auditory nerve is your teeth and bones. Bone-conduction hearing aids, which are helpful in compensating for certain kinds of deafness, operate on this principle.

However, the medium that carries most of the sounds we hear is the air. Sound waves traveling through air (or through any other fluid) are always compressional waves because *fluid media cannot transmit shear waves*. Remember that fluids have no resistance to shear stress. A source that shears a fluid therefore gives rise to no restoring force, and no wave is produced.

Speed of Compressional Waves in Fluids

To calculate the speed of compressional waves in fluids, we modify the argument leading to Equation 22.13. We again use Figure 22.6, which for the purposes of the present discussion we will take to show the propagation of a compressional pulse down a fluid-filled tube. Such a pulse could be generated at the end of a fluid-filled tube by suddenly pushing a piston a short distance into the tube. Here, as in a solid bar, the result is the propagation of a pulse of increased pressure down the tube. Associated with the increased pressure is a compression of the fluid.

The derivation of the wave speed, which can be carried out just as in the solid-medium case, yields Equation 22.13,

$$v = \sqrt{\frac{Y}{\rho}}.$$

We must take care, however, to interpret the quantity Y in a way that is meaningful for a fluid. For fluids, we speak of the pressure change Δp rather than the uniaxial stress σ. But, as noted in Section 15.5, the two are related: $\Delta p = -\sigma$. In the same spirit, we can interpret the uniaxial strain ϵ in terms of the fractional volume change of the element. The undisturbed volume is $dV = a\,dx$, and the fractional change is $\Delta(dV)/dV$. So we can write Y in the form

$$Y = \frac{\sigma}{\epsilon} = \frac{-\Delta p}{\Delta(dV)/dV} = -dV\frac{\Delta p}{\Delta(dV)}.$$

Now, $-dV\,\Delta p/\Delta(dV)$ is the bulk modulus B of the fluid (Equation 15.26a). So in the present case we have $Y = B$, and the wave speed can be written

$$v = \sqrt{\frac{B}{\rho}} \quad \text{for waves propagating through fluids.} \tag{22.16}$$

Although we have been dealing here with the propagation of a single compressional pulse through a fluid, the result applies just as well to a periodic wave, which consists of alternating compression and **rarefaction** (reduction of pressure to below the undisturbed value).

Speed of Sound in Air

Equation 22.16 expresses a result published by Newton in 1687. Newton took one more step, applying the result to the most important special case: the propagation of sound in air. To do this, he used Boyle's law to set $B = p$, as we did in Equation 15.38. He then obtained the result $v = \sqrt{p/\rho}$. But Newton and others found that the value of the speed of sound thus calculated is smaller than measured values by about 18%, a discrepancy too large to be ascribed to experimental error.

It was not until 1802 that Laplace discovered the source of the discrepancy. Boyle's law applies to isothermal conditions only. But the rapid compression and rarefaction of a gas as sound waves pass through it is essentially *adiabatic*. Even at the lowest sound frequencies—approximately 20 Hz—negligible heat flows from a region warmed by compression to the neighboring regions cooled by rarefaction before the wave moves on and reverses the temperature differences.

How Do We Know That There Is Fluid in the Earth's Core?

The absence of shear waves in fluid media furnishes the primary evidence for the fluidity of the earth's core. Figure 22.7 shows the increase of the S–P interval for seismic stations located at increasing distances from the earthquake. Beyond a certain distance, however, the S wave disappears. This is because the direct wave path passes through the liquid core, which cannot transmit S waves. The part of the earth's surface where S waves are not observed from a particular earthquake is called the *shadow zone*. From observations of shadow zones, the radius of the core has been determined with considerable accuracy; the mean value is close to 3470 km, which is about half the radius of the earth. The core thus constitutes about one-eighth of the earth's volume.

Newton's expression for the speed of sound (Equation 22.16) is not wrong. But in order to obtain the correct value for the speed of sound in a gas, we must use not the isothermal (Boyle's-law) bulk modulus $B_{iso} = p$ but the **adiabatic bulk modulus** B_a. That is, we must write

$$v = \sqrt{\frac{B_a}{\rho}} \quad \text{for the speed of sound in a gas.} \tag{22.17}$$

It remains to calculate the adiabatic bulk modulus for an ideal gas. (For our present purpose, air under ordinary conditions approximates an ideal gas quite well.) We write the bulk modulus in the form of the general definition (Equation 15.26b):

$$B \equiv -V \frac{dp}{dV}. \tag{22.18}$$

We now evaluate the derivative dp/dV for an adiabatic process. From Equation 19.29a, we have

$$pV^\gamma = p_0 V_0^\gamma, \tag{22.19}$$

where p_0 and V_0 are the pressure and volume of the undisturbed gas. Dividing through by V^γ gives

$$p = p_0 V_0^\gamma \frac{1}{V^\gamma}.$$

We differentiate both sides of this equation and obtain

$$dp = -\gamma p_0 V_0^\gamma \frac{dV}{V^{\gamma+1}},$$

or

$$\frac{dp}{dV} = -\frac{\gamma p_0 V_0^\gamma}{V^{\gamma+1}}.$$

The changes in pressure and volume produced by the passage of the sound wave are quite small, and so we can make the approximations $p_0 \simeq p$ and $V_0 \simeq V$. This yields the result

$$\frac{dp}{dV} = -\frac{\gamma p}{V}.$$

Comparing this equation with Equation 22.18, we obtain the *adiabatic bulk modulus* B_a:

$$B_a = \gamma p.$$

Inserting this result into Equation 22.17 gives the desired value of v:

$$v = \sqrt{\frac{\gamma p}{\rho}}. \tag{22.20}$$

For air, the value of γ is close to 1.40. Taking into consideration the factor $\sqrt{\gamma} = \sqrt{1.40} = 1.18$, we can account for the 18% discrepancy between Newton's calculation and the measured speed of sound. The corrected result (Equation 22.20) agrees very well with measured values.

EXAMPLE 22.3

Find the speed of sound in dry air at standard conditions of temperature and pressure. For STP, take the density of dry air to be 1.293 kg/m³ and take γ to be 1.402.

SOLUTION: Using Equation 22.20, you find

$$v = \sqrt{\frac{1.402 \times 1.013 \times 10^5 \text{ Pa}}{1.293 \text{ kg/m}^3}} = 331.4 \text{ m/s}.$$

The best measured value is $v = 331.45$ m/s.

Temperature Dependence of the Speed of Sound

By applying the ideal-gas law to Equation 22.20, we can determine how the speed of sound depends on the properties of the gas. We use the ideal-gas law in the macroscopic form of Equation 18.24, $pV = nRT$. The density of the gas is $\rho = M/V$, and so the ideal-gas law can be written

$$\frac{p}{\rho} = \frac{n}{M} RT.$$

Now, M/n is the mass per kilomole M_0, which is numerically equal to the molecular mass μ of the gas. So we have

$$\frac{p}{\rho} = \frac{RT}{M_0}.$$

Inserting this quantity into Equation 22.20, we obtain the result

$$v = \sqrt{\frac{\gamma RT}{M_0}}. \qquad (22.21a)$$

The speed of sound in an ideal gas is proportional to the square root of the absolute temperature and inversely proportional to the square root of the mass per kilomole of the gas.

Compare the speed of sound given by Equation 22.21a with the root-mean-square molecular speed given for an ideal gas by Equation 18.44b,

$$v_{rms} = \sqrt{\frac{3RT}{M_0}}.$$

There is an intimate connection between the rms molecular speed and the speed of sound in an ideal gas. Specifically, we have

$$v = \sqrt{\frac{\gamma}{3}} \, v_{rms}. \qquad (22.21b)$$

For a diatomic gas like air, we have $\sqrt{\gamma/3} = 0.68$; the speed of sound is about two-thirds v_{rms}.

EXAMPLE 22.4

(a) Find the speed of sound in helium at STP. Take $M_0 = 4.00$ kg/kmol. (b) Find the speed of sound in air at atmospheric pressure and room temperature, taken to be 22°C.

SOLUTION:
(a) Find v for helium at STP.
  From Table 19.2, you have $\gamma = \frac{5}{3}$. So Equation 22.21 gives you

$$v = \sqrt{\frac{5}{3} \times \frac{8.314 \times 10^3 \text{ J/kmol·K} \times 273 \text{ K}}{4.00 \text{ kg/kmol}}} = 972 \text{ m/s}.$$

This is considerably greater than the value found for air in Example 22.3.

(b) Find v for air at 1 atm and 22°C.
  You could use Equation 22.21 to find the desired result. But you already know v_0 (the speed of sound at 0°C) from Example 22.3, and so it is easier to take the ratio

$$\frac{v}{v_0} = \sqrt{\frac{T}{T_0}},$$

where $T_0 = 273.15$ K. This gives you

$$v = 331.4 \text{ m/s} \times \sqrt{\frac{295.15 \text{ K}}{273.15 \text{ K}}} = 344.5 \text{ m/s}.$$

The Ear and Hearing

Hearing begins only when a sound wave reaches the ear canal. The sound wave must propagate through the canal, undergoing frequency-dependent absorption and reflection along the way. The wave, thus modified, strikes the eardrum, which is set into vibration in a complicated (and quite nonlinear*) manner. The vibrating eardrum in turn excites vibration in the very complex nonlinear system, part solid and part liquid, that constitutes the middle and inner ear. Within the inner ear, a complex, nonuniform array of *hair cells* converts the mechanical oscillations into electrical signals in the *auditory nerve*. These signals are then analyzed and interpreted by the auditory nerve and the brain. We will concentrate mainly on what happens up to the point at which the sound wave reaches the eardrum.

The human ear, as well as the ears of most other animals, has evolved in response to environmental pressures that place a premium on the ability to operate over a wide range of conditions. The ear must be sensitive to very faint sounds and yet must be resistant to damage from intense ones; it subjectively (but reliably) interprets the energy flux per unit area falling on it as the sensation called **loudness**. In addition, the ear must respond sensitively to, and distinguish among, sounds covering a wide range of frequencies; it subjectively (but reliably) interprets the frequency of the sound falling on it as the sensation called **pitch**. Finally, the ear must be able to make fine distinctions among sounds that are not very different from one another, such as the voices of two persons of the same sex and age or an oboe and an English horn; in effect, the ear performs a Fourier analysis of the sound and provides an output that the brain interprets as **sound quality** or (as it is called in music) **timbre**.

Energy Flux and Loudness

At the frequency at which it is most sensitive, about 3.5 kHz, the normal ear can just detect sounds whose intensity (energy flux per unit area) is about 10^{-13} W/m². Corresponding to this intensity are a pressure amplitude Δp^* of order 10^{-5} Pa and a displacement amplitude s^* of order 10^{-12} m, which is about 0.01 times the diameter of a typical atom. The human ear is sensitive indeed, but many animals have hearing that is more sensitive by two orders of magnitude. The human ear will sustain damage if it is subjected to midfrequency sound waves of intensity greater than about 10^{-1} W/m². Thus the human ear is useful over something like twelve orders of magnitude of intensity. The frequency dependence of sensitivity is shown in Figure 22.8; the outer curve, the *envelope of hearing*, is a standard based on measurements of many human subjects. The envelope for an individual person may vary significantly from the standard. Most notably, the right side of the envelope invariably moves to the left as a person ages or is exposed to excessively loud sound.

The frequency range from 300 Hz to 8 kHz is the most important for comprehension of speech. As you can see from Figure 22.8, the intensity for the faintest sounds audible over this range has an average value roughly equal to 10^{-12} W/m². This value is called the **threshold of hearing** I_0:

$$I_0 \equiv 10^{-12} \ \text{W/m}^2. \tag{22.22}$$

Many animals have more sensitive hearing than humans.

Because the ear is useful over such a wide intensity range, the sense of loudness must be highly compressed. That is, the subjective impression of loudness must vary slowly with changes in the energy flux incident on the ear. (The senses of sight and touch are likewise compressed.) How are loudness and intensity connected? Physiologists and psychologists have answered this question on the basis of extensive experiments. In a typical experiment, a subject hears a pair of sounds of the same frequency but different intensities and is asked to judge if one sound is perceptibly louder than the other. Such experimental results tell us that the response is roughly logarithmic. That is,

perceived loudness $\propto$ log (intensity).

---

*By *nonlinear*, we mean that the system does not conform to Hooke's law. See Problem 22.73.

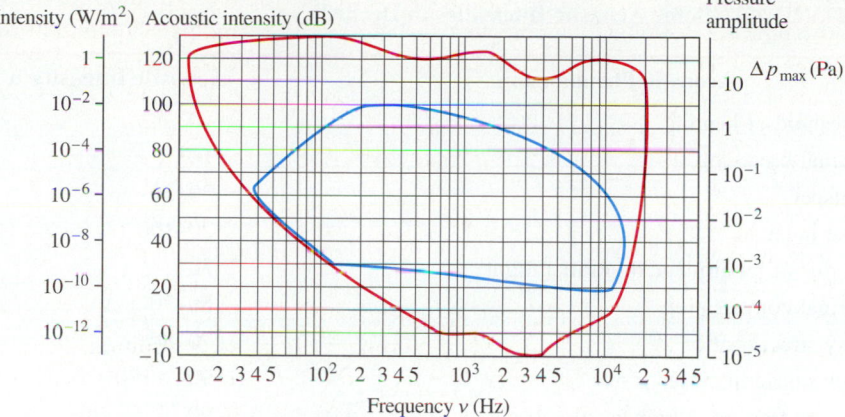

FIGURE 22.8 Envelope of sensitivity of the human ear. Sounds having frequencies to the left or right of the outer envelope, as well as sounds having intensities (or acoustic intensities) below the envelope, cannot be heard by the human ear. Sounds having intensities (or acoustic intensities) above the envelope can be heard but will cause pain or will damage the ear or both. For this reason, the upper bound of the envelope is sometimes called the **threshold of pain**. Music is for the most part confined within the inner (blue) envelope.

We take advantage of this response characteristic to define a quantitative scale of perceived loudness. If the intensity of a sound wave is I, we define the **acoustic intensity** β to be

$$\beta \equiv \log_{10} \frac{I}{I_0}, \tag{22.23}$$

where I_0 is defined by Equation 22.22.* Equation 22.23 fixes the threshold of hearing at an acoustic intensity 0. Although the quantity β is dimensionless, it is assigned a unit name, the **bel**.[†] The logarithmic relation between perceived loudness, measured in bels, and I, measured in W/m^2, is evident. The intensity of a sound must be increased tenfold in order to double the acoustic intensity β. (The use of the logarithm to the base 10 is an arbitrary convention.)

Because a 1-bel change represents a considerable change in perceived loudness, acoustic intensity is more commonly expressed in **decibels** (dB):

$$1 \text{ dB} \equiv 0.1 \text{ bel.}$$

Quantities measured in decibels are denoted by the symbol α. In terms of the decibel, Equation 22.23 becomes

$$\alpha = 10 \log_{10} \frac{I}{I_0}. \tag{22.24}$$

In the region near the center of the envelope of hearing, a change in α of 1 dB is approximately the minimum perceptible change. Thus the decibel is a unit of convenient size. The bel and the decibel are not SI units but have wide application nonetheless. Some typical acoustic intensities are listed in Table 22.1.

---

*Don't confuse intensity I—energy flux per unit area—with acoustic intensity β—a dimensionless quantity—just because the names are similar and the two quantities are related.

[†]Named after the Scottish-American phonologist, teacher of the deaf, acoustician, inventor, and artist Alexander Graham Bell (1847–1922), best known for his crucial contribution to the development of a practical telephone (1876).

TABLE 22.1 Some Acoustic Intensities in Decibels

Acoustic Phenomenon	Acoustic Intensity α (dB)
Threshold of hearing	0
Normal breathing	10
Whisper	20–30
Most music	20–100
Interior of jet airplane in normal flight	50
Normal conversation	50–70
Busy street traffic	70–80
Rock music in closed room	90–>120
Subway train (San Francisco)	90
Subway train (New York)	100
Temporary or permanent impairment of hearing on sustained exposure	110–120
Threshold of pain	120
Jet airplane taking off, from side of runway	150
Immediate serious ear damage	160

EXAMPLE 22.5

For sound of frequency 5 kHz and intensity 50 dB, what fractional increase in intensity is required for a just-perceptible increase in acoustic intensity?

SOLUTION: These conditions lie in the midrange of the envelope of hearing, so a 1-dB change is just perceptible. You can write the change as

$$1 \text{ dB} = \alpha_f - \alpha_i,$$

where α_i and α_f are the initial and final acoustic intensities. Using Equation 22.24 and calling the initial and final intensities

I_i and I_f, you have

$$1 \text{ dB} = 10 \log_{10} \frac{I_f}{I_0} - 10 \log_{10} \frac{I_i}{I_0},$$

or

$$0.1 = \log_{10} \frac{I_f}{I_i}.$$

You exponentiate both sides to obtain

$$\frac{I_f}{I_i} = 10^{0.1} = 1.26.$$

That is, you can just detect a 26% increase in intensity.

Frequency and Pitch

The subjective sensation of pitch corresponds to the frequency of the sound that excites the ear. Only sounds having a well-defined fundamental frequency or a fairly narrow frequency range have identifiable pitch. The sounds we call *noises* consist of a superposition of waves having a broad distribution of frequencies and do not have any particular pitch.

When waves of very low frequency propagate through air, we feel them as vibrations. We do not "hear" them; the ear is no more sensitive than the rest of the body to low-frequency vibrations. Depending on the intensity and the individual ear, sound becomes perceptible as such in the frequency range from 12 Hz to 25 Hz. Like the lower limit, the upper limit varies over about a factor of two from person to person. However, the upper limit changes much more with age. Most normal persons younger than 16 or so can hear sounds of frequency up to 18 kHz, and many can hear up to 20 kHz. At age 45, the limit is usually closer to 12 kHz, and it continues to diminish with increasing age. When the limit falls to about 8 kHz, difficulties in interpreting speech begin. Elderly persons often find it difficult to distinguish between consonant pairs such as *b-p* or *s-f*, whose Fourier spectra differ mainly in their high-frequency components. Loss of high-frequency sensitivity is also a result of extended listening to very loud music.

Rock musicians and fans, most of whom are young, often have frequency-response patterns characteristic of persons in their seventies.

The sensation of pitch is a roughly logarithmic function of frequency;

$$\text{pitch} \propto \log (\text{frequency}).$$

If frequencies of a series of sounds are in the ratio $1:2:4:\cdots:2^n$, we perceive their pitches as equally spaced and say that they are one **octave** apart. (The first two modes of a vibrating string are one octave apart.) The frequency range of the ear is about $1:1000$; because $\log_2 1000 \simeq 10$, this corresponds to a pitch range of about ten octaves. The standard piano keyboard covers a little more than seven octaves.

Sound Quality

Most musical instruments produce a harmonic sequence of frequencies. At least some of the harmonics have amplitudes of the same general magnitude as the fundamental, and occasionally even larger amplitudes. Nevertheless, we perceive the pitch as that of the fundamental. But the mixture of harmonics—the Fourier spectrum—is crucial in establishing the characteristic "signature" of an instrument. The Fourier spectrum also determines the quality of nonmusical sounds, such as speaking voices. The ear, the auditory nerve, and the brain constitute a remarkable Fourier analyzer; everyone of normal hearing is capable of identifying sounds by analyzing them and comparing them with a vast mental library of sounds previously heard and identified.

Superposition of Sound Waves

(a)

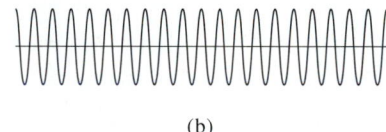

We rarely hear sounds whose wave forms are simple sinusoids. Even a single tone produced by a single musical instrument is usually quite complex. Some simplification does result, however, from the fact that *the ear is insensitive to phase differences*. To be specific, suppose that you are listening to a sound consisting of a superposition of two sound waves from separate sources. If the phase relation between the two components is changed (for example, by moving one source slightly with respect to the other), your ear will perceive no difference in sound quality. For this reason, it is often possible to consider superposition of sound waves without regard to phase.

(b)

Beats

When you listen to two sources of sound having the same amplitude but slightly different frequencies, you do not hear two separate pitches. Rather, you perceive a sound that has a single pitch but slowly waxes and wanes in loudness with a frequency called the **beat frequency**. This oscillation of loudness is called **beating**, and the loud intervals are called **beats**. Figure 22.9 shows this phenomenon graphically. Parts *a* and *b* are graphs of two sinusoidal wave functions whose frequencies differ by 10%; $\nu_2 = 1.1\nu_1$. The sum of the two functions is shown in Figure 22.9c. The wave function of the superposition is characterized by two frequencies. The higher frequency is the **sum frequency**,

(c)

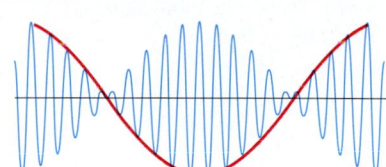

$$\nu_+ = \tfrac{1}{2}(\nu_1 + \nu_2). \tag{22.25a}$$

As you can see, the sum frequency is the average of ν_1 and ν_2. It is the frequency of the rapid oscillations in Figure 22.9c. The perceived pitch is associated with the sum frequency. The lower frequency is the **difference frequency**, ν_-, where

$$\nu_- = \tfrac{1}{2}(|\nu_2 - \nu_1|). \tag{22.25b}$$

The difference frequency is the frequency of the sinusoid shown in red in Figure 22.9c. The slow beats are heard at twice the difference frequency; the beat frequency is

FIGURE 22.9 The production of beats. (*a*) Plot of the wave function $f_1(t) = A \cos 2\pi\nu_1 t$. (*b*) Plot of the wave function $f_2(t) = A \cos 2\pi\nu_2 t$, with $\nu_2 = 1.1\nu_1$. (*c*) Superposition of the two functions. The sinusoid in blue has the sum frequency. The envelope shown in red has the difference frequency. Because the amplitude of the sum-frequency wave reaches its maximum value twice in each difference-frequency cycle, the beat frequency is $\nu_b = \nu_2 - \nu_1$.

$\nu_b = 2\nu_-$:

$$\nu_b = |\nu_2 - \nu_1|. \qquad \text{(22.25c)}$$

We now demonstrate this conclusion rigorously. Let the frequency difference be $\Delta\nu \equiv \nu_2 - \nu_1$, and suppose that both wave functions are simple sinusoids of equal amplitude. Then at a fixed location, the variations in air pressure due to each wave taken separately are

$$\Delta p_1(t) = \Delta p^* \cos 2\pi\nu_1 t \quad \text{and} \quad \Delta p_2(t) = \Delta p^* \cos 2\pi(\nu_1 + \Delta\nu)t,$$

where Δp^* is the amplitude of the pressure variation. The superposed waves are represented by the sum

$$\Delta P(t) = \Delta p^*[\cos 2\pi\nu_1 t + \cos 2\pi(\nu_1 + \Delta\nu)t].$$

Using the standard trigonometric identity for the sum of two cosine functions, we obtain

$$\Delta P(t) = 2\,\Delta p^*\{\sin\tfrac{1}{2}[2\pi\nu_1 t + 2\pi(\nu_1 + \Delta\nu)t]\cos\tfrac{1}{2}[2\pi\nu_1 t - 2\pi(\nu_1 + \Delta\nu)t]\}.$$

This simplifies to

$$\Delta P(t) = 2\,\Delta p^* \sin\,[2\pi(\nu_1 + \tfrac{1}{2}\Delta\nu)t]\cos 2\pi\,\frac{\Delta\nu}{2}\,t. \qquad \text{(22.26a)}$$

In terms of ν_1 and ν_2, this equation can be written in the equivalent forms

$$\Delta P(t) = 2\,\Delta p^* \sin 2\pi\left(\frac{\nu_1 + \nu_2}{2}\right)t \cos 2\pi\left(\frac{\nu_1 - \nu_2}{2}\right)t \qquad \text{(22.26b)}$$

and

$$\Delta P(t) = 2\,\Delta p^* \sin 2\pi\nu_+ t \cos 2\pi\nu_- t, \qquad \text{(22.26c)}$$

where ν_+ and ν_- are given by Equations 22.25a and 22.25b.

Although Equations 22.26a, 22.26b, and 22.26c are true in general, we are now interested in the case where $\Delta\nu \ll \nu_1, \nu_2$. In this case, the difference frequency often lies below the range of audibility and is not detected as a sound. Instead, it modulates the amplitude of the sound, which has the (audible) sum frequency. Why do we hear beats at *twice* the difference frequency? Our ears detect a sinusoidal wave having the audible frequency ν_+, whose amplitude has the time-dependent value $2\,\Delta p^*|\cos 2\pi\nu_+ t|$. This amplitude achieves its maximum value, $2\,\Delta p^*$, *twice* in every cycle of the cosine function. Because ν_1 and ν_2 are so close, we cannot distinguish between the pitches associated with ν_1, ν_2, and $\tfrac{1}{2}(\nu_1 + \nu_2)$.

Beats and Musical Dissonance

The human ear can detect beats as individual pulses up to about 7 Hz. At higher frequencies, the beats sound like a dissonant buzz. The beat frequency that produces the harshest dissonance depends on the pitch of the modulated sound. At low pitch, the dissonance is most evident at a beat frequency of about 15 Hz; in the midrange, at about 40 Hz; and, at high pitch, at about 250 Hz. The presence of such frequencies in a musical sound gives the impression of dissonance. Two adjacent notes on the piano will produce beat frequencies in this range and so are perceived as dissonant. But even two notes not so close together may sound dissonant if a beat frequency in the proper range results when the fundamental or harmonics of one are superposed with the fundamental or harmonics of the other.

AM Radio

Beats are an acoustic version of the phenomenon called **amplitude modulation** (AM for short), which is an important means of transmitting radio signals. In radio transmission, a radio wave having the sum frequency, typically 1 MHz, serves as what is called the carrier wave. The modulation signal, of much lower frequency, is detected as a component of an electromagnetic (radio) wave and translated electronically and electromechanically into sound.

EXAMPLE 22.6

The note called middle C has frequency $\nu_C = 523.3$ Hz, and the adjacent C-sharp has frequency $\nu_{C\sharp} = 554.4$ Hz. Find the beat frequency.

SOLUTION: From Equation 22.25c, you have

$$\nu_b = 554.4 \text{ Hz} - 523.3 \text{ Hz} = 31.1 \text{ Hz}.$$

For these midrange tones, the beat frequency will produce a strong impression of dissonance.

Standing Waves in Pipes

You learned in Chapter 21 how sounds of controlled frequency can be produced by exciting standing waves in stretched strings. Let us now consider the analogous phenomenon in which sounds of controlled frequency are produced by exciting standing waves in columns of air; this phenomenon is the basis of the classes of musical instruments called *brasses* and *woodwinds*.

Open Pipes

Figure 22.10 shows the simplest possible device for producing standing waves in a column of air. It consists of a cylindrical pipe open at both ends. Such a pipe is called an **open pipe**. (This idealized form is closely approximated by many real organ pipes.) Near one end is a source of sound, which may be any of several types. In the type shown, air at a pressure slightly greater than atmospheric is blown into the pipe past a flexible flap of wood or metal, called the **reed**. The reed oscillates, alternately opening and closing the air passage and thus admitting air to the pipe in puffs. The puffs produce the pressure variations characteristic of sound waves.

As always, the production of standing waves depends on the imposition of boundary conditions. In the open-pipe case, there must be **pressure nodes** at (or just outside) the ends of the pipe. Pressure nodes, analogous to nodes on a vibrating string, are locations where the pressure does not vary with time. The reason is simple: The pressure variations that occur in the limited confines of the pipe cannot lead to substantial pressure changes in the atmosphere at large.

Mathematically as well as physically, this boundary condition is completely analogous to the condition imposed on a string fixed at both ends, discussed in Section 21.7. All that we need to do is substitute the *pressure wave functions* $p(x, t)$ and $P(x, t)$ for the string displacement functions $y(x, t)$ and $Y(x, t)$ in Equations 21.49 through 21.51. The *standing-wave condition* is then given by Equation 21.56a or 21.56b,

$$\frac{\lambda}{2} = \frac{L}{n} \quad \text{or} \quad 2L = n\lambda; \quad n = 1, 2, 3, \ldots . \quad \text{(22.27a,b)}$$

To first approximation, L is the length of the tube.

The physical meaning of Equations 22.27a and 22.27b is made evident in Figure 22.11a,b,c, which shows the first three standing-wave modes in an open pipe. The outermost curves represent the extremal variations of pressure, $\Delta P = \pm 2 \Delta p^*$ above and below atmospheric pressure, as a function of position in the pipe. Just as with the vibrating string shown in Figure 21.19, the extremal variation is modulated by a time-dependent sinusoidal function. This modulation is represented by the curves of smaller amplitude; compare Figures 21.18 and 21.21. To express ΔP as a function of position and time, we exploit the analogy with the vibrating string and write the complete wave function of the standing wave by direct substitution into $Y(x, t) = 2A \sin kx \sin \omega t$

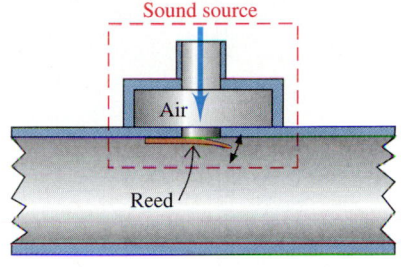

FIGURE 22.10 Idealized open pipe with reed.

FIGURE 22.11 (*a–c*) The first three standing-wave modes in an open pipe, shown in terms of pressure variations. Each sinusoid represents the pressure as a function of position at a different instant in the cycle. The symbol N_p denotes a pressure node; A_p denotes a pressure anti-node. (*a'–c'*) The first three standing-wave modes in an open pipe, shown in terms of displacement variations. Each sinusoid represents the displacement as a function of position at a different instant in the cycle. The symbol N_d denotes a displacement node; A_d denotes a displacement antinode. Displacement nodes correspond to pressure antinodes, and vice versa.

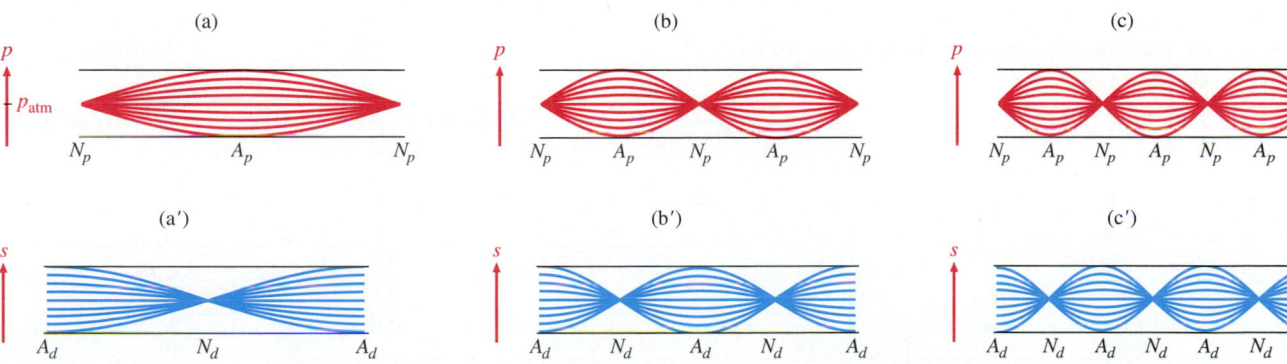

(Equation 21.51). We thus obtain an expression valid for sound waves:

$$\Delta P(x, t) = 2 \Delta p^* \sin kx \sin \omega t. \tag{22.28}$$

Using the standing-wave condition in the forms of Equations 22.27 and 21.55, $k = n\pi/L$, we can write the standing-wave function as

$$P(x, t) = 2 \Delta p^* \sin\left(\frac{n\pi}{L} x\right) \sin\left(\frac{n\pi v}{L} t\right). \tag{22.29}$$

More than one mode may be present in the sound produced by the pipe, just as more than one mode may be present in a vibrating string. The quality of the sound is determined by the harmonic content.

Pressure Nodes and Displacement Antinodes

So far, we have described sound waves in terms of their characteristic pressure variations. That is, we have explicitly or implicitly considered wave functions of the form $\Delta p(x, t)$. These pressure variations must be the result of longitudinal displacement of air in a nonuniform way. We now consider sound waves from the complementary point of view: we consider the oscillatory displacement of air that accompanies oscillatory pressure variation.

Consider what happens at a pressure antinode. You can see from inspecting Figure 22.11a,b,c that $\partial p/\partial x$ is equal to zero there at all times. That is, there is never a **pressure gradient**—a variation of pressure with location—to drive air into or out of the narrow region immediately surrounding the pressure antinode. Thus the air at the pressure antinode does not move; *the pressure antinode is a **displacement node***.

Now, consider what happens at a pressure node. You can see from Figure 22.11a,b,c that, at any instant, the pressure gradient $\partial p/\partial x$ [$= \partial(\Delta P)/\partial x$] has its extremal value at the pressure node. This extremal pressure gradient results in an extremal displacement of the air at the pressure node; *the pressure node is a **displacement antinode***. Figure 22.11a′,b′,c′ depicts the first three standing-wave modes in the same open pipe shown in Figure 22.11a,b,c, but in terms of the *displacement wave function* $\Delta S(x, t)$ rather than the pressure wave function $\Delta P(x, t)$.

Closed Pipes

A pipe closed at one end is conventionally called a **closed pipe**; see Figure 22.12. What kind of standing waves can exist in such a pipe? As always, we begin to answer the question by considering the boundary conditions. There is nothing new about the open end of the pipe; it must be the location of a pressure node. But the closed end must be the location of a displacement node because the rigid end wall allows no movement of air in its immediate vicinity. Consequently, there is a pressure antinode at the closed end.

It is possible, but unnecessary, to derive the standing-wave equation corresponding to these two boundary conditions. It is simpler to draw the first three standing-wave modes, as in Figure 22.13. You can see by inspection that the wavelengths of these modes are

$$\lambda = \tfrac{4}{1}L, \tfrac{4}{3}L, \tfrac{4}{5}L.$$

In general, we have

$$\frac{\lambda}{4} = \frac{L}{m} \quad \text{or} \quad 4L = m\lambda; \quad m = 1, 3, 5, \ldots . \tag{22.30a,b}$$

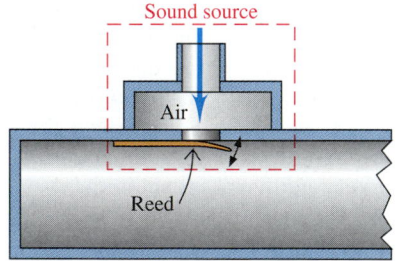

FIGURE 22.12 An idealized closed pipe with reed.

FIGURE 22.13 The first three standing-wave modes in a closed pipe, shown in terms of pressure variations.

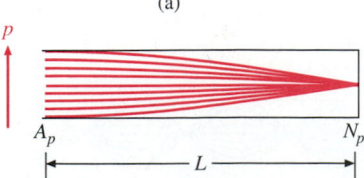

(a)

(b)

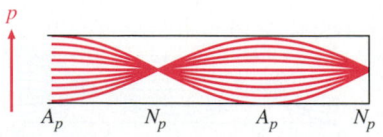

(c)

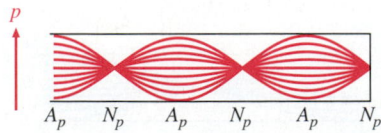

Compare this with the open-pipe condition given by Equation 22.27a or 22.27b:

$$\frac{\lambda}{2} = \frac{L}{n} \quad \text{or} \quad 2L = n\lambda; \quad n = 1, 2, 3, \ldots .$$

There are two differences:

1. The fundamental wavelength of a closed pipe is twice the fundamental wavelength of an open pipe of the same length.
2. The even-numbered modes are absent for a closed pipe.

The absence of even-numbered modes for closed pipes results in a marked difference between the timbres of musical instruments based on open and closed pipes. The difference is easiest to note in organs, which have ranks of both closed and open pipes.

EXAMPLE 22.7

A closed pipe and an open pipe have the same length. Show that no mode of the closed pipe has the same wavelength as any mode of the open pipe.

SOLUTION: Call the wavelengths of the closed and open pipes λ_c and λ_o, respectively. From Equations 22.30 and 22.27, you have

$$\lambda_c = \frac{4L}{m} \quad \text{and} \quad \lambda_o = \frac{2L}{n}.$$

If the two pipes are to have any modes with equal wavelengths, there must be some value of m and some value of n for which $\lambda_c = \lambda_o$. If this is so, you can write $4L/m = 2L/n$, or

$$m = 2n.$$

But n is an integer and m is an odd integer, and no odd integer can be twice another integer. So the two pipes have no modes of equal wavelength.

You can obtain the same result less elegantly by writing out a few terms in the series that give the wavelengths of the modes for the two pipes:

$$\frac{\lambda_c}{L} = \frac{4}{1}, \frac{4}{3}, \frac{4}{5}, \frac{4}{7}, \frac{4}{9}, \ldots$$

and

$$\frac{\lambda_o}{L} = \frac{2}{1}, \frac{2}{2}, \frac{2}{3}, \frac{2}{4}, \frac{2}{5}, \ldots .$$

The series have no terms in common.

The Doppler Effect

You have doubtless had the experience of standing at the side of a road as an automobile passes you with its horn blowing. Just as the auto passes, there is a distinct fall in the pitch you hear. This is one example of the **Doppler effect,*** an effect that applies to all waves, mechanical or otherwise.

Figure 22.14 shows the Doppler effect for surface ripples on water. The end of the vibrating rod is dipped into the water and excites ripples. The rod (the source) is moving to the right at constant speed. You can see the crowding of the ripples in the forward direction and the increase in their spacing in the backward direction.

Figure 22.15 is a drawing of the essential features of the photograph of Figure 22.14. The source is moving to the right along the x axis at speed v_s; the points 1 through 7 on the axis denote positions of the source at seven successive instants one wave period τ apart. At each instant, the source produces a wave front that propagates as a circle centered on the point. (Once produced, the motion of a wave front has nothing to do with the subsequent behavior of the source, and the wave front remains centered on the point where it was produced.) Successively smaller circles denote wave fronts produced by the source at successively later instants, when the source was successively farther to the right. All circles expand at the same speed, which is the speed v of waves in the medium. At the instant shown, wave front 1 has expanded to the radius shown from

---

*Named after the Austrian mathematician and physicist Christian Doppler (1803–1853), who first described the effect in 1845.

FIGURE 22.14 The Doppler effect for water surface ripples. The source is moving at constant speed to the right, with respect to the water, and the observer (the camera) is stationary with respect to the water.

(From Haber-Schaim: *PSSC Physics.* Copyright 1991 by Kendall/Hunt Publishing Company. Used with permission.)

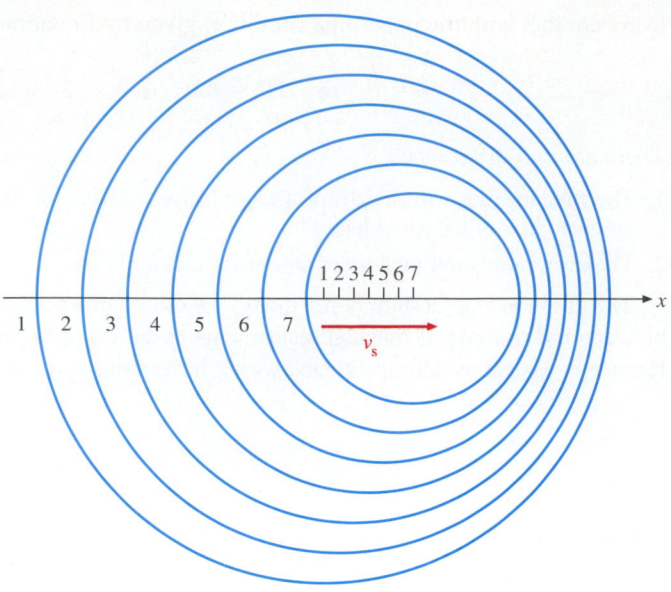

FIGURE 22.15 Successive wave fronts emitted by a moving source. Each wave front originated at the source point having the same number and propagates as an circle centered on that source point and expanding at the wave speed v.

source point 1 at its center. Wave front 2, produced one period later at source point 2, has expanded to a smaller radius from source point 2 at its center, and so on.

For the sake of simplicity, we analyze the Doppler effect in detail only for the one-dimensional case. Suppose that the source, the medium, and the observer are all moving along the same line, shown as the x axis in Figure 22.16. Let us measure all distances and speeds with respect to a fixed point $x = 0$ on that line. (In the case of sound, the source and the observer might be in different automobiles moving along the same straight road while a wind blows in the direction of the road.) We fix our coordinate system with respect to the fixed point and call the direction of motion of the wave fronts the positive x direction. In this fixed coordinate system, let the velocities of the source, the medium, and the observer be v_s, v_m, and v_o, respectively. (The signs of these quantities may be positive or negative, depending on the directions of motion. Note, however, that the wave velocity v is always positive because of the way we defined the positive direction.)

In Figure 22.16, wave front 1 was emitted at the instant $t = 0$ when the source was located at point 1. A time equal to the period τ has passed since then. The source is now located at point 2, a distance $v_s\tau$ from point 1. The wave front has traveled through the medium a distance $v\tau$. But, in the same time, the medium itself has moved a distance $v_m\tau$. So wave front 1 has moved a distance $(v + v_m)\tau$ from point 1. At time $t = \tau$,

FIGURE 22.16 The Doppler effect in one dimension. The speed of the wave front is v. The velocities of the source, the medium, and the observer are v_s, v_m, and v_o, respectively.

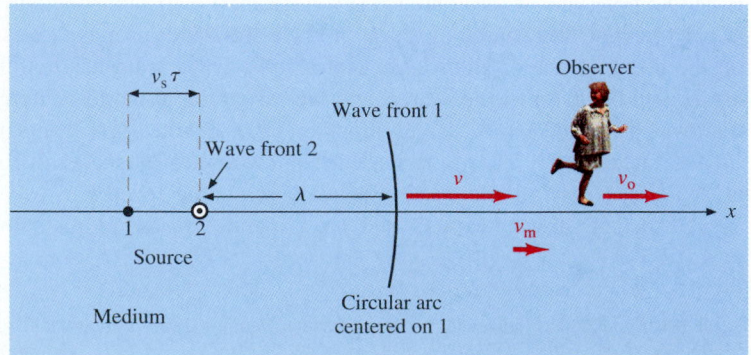

wave front 2 is emitted. (For clarity, this wave front is shown at a time slightly later than τ, when it is a small circle about the source.)

Because the two wave fronts are emitted one wave period apart, the distance between them is one wavelength λ. Along the positive x direction, this distance is

$$\lambda = (v + v_m)\tau - v_s\tau,$$

or
$$\lambda = (v + v_m - v_s)\tau. \tag{22.31}$$

Eventually, wave front 1 catches up with the observer, followed at a time τ' later by wave front 2. What is the value of τ', which is the wave period from the point of view of the observer? During the interval τ', the observer moves a distance $v_o\tau'$. Thus, to catch up with the observer, wave front 2 must travel a distance $\lambda + v_o\tau'$ past the point where wave front 1 catches up. The speed of the wave is $v + v_m$, and so we have

$$\tau' = \frac{\lambda + v_o\tau'}{v + v_m}.$$

We substitute into this equation the value of λ given by Equation 22.31 and solve for τ':

$$\tau' = \tau\frac{v + v_m - v_s}{v + v_m - v_o}. \tag{22.32}$$

We are now ready to compare the frequency ν_o heard (or otherwise detected) by the observer with the frequency ν_s of the source. Because frequency is the reciprocal of period, we have

$$\nu_s = \frac{1}{\tau} \quad\text{and}\quad \nu_o = \frac{1}{\tau'}.$$

Substitution of these values into Equation 22.32 gives us

$$\nu_o = \nu_s\frac{v + v_m - v_o}{v + v_m - v_s}. \tag{22.33a}$$

This is the *general one-dimensional Doppler formula*. The same result is often expressed in terms of the **Doppler shift**, defined to be

$$\Delta\nu = \nu_o - \nu_s = \nu_s\left(\frac{v + v_m - v_o}{v + v_m - v_s} - 1\right). \tag{22.33b}$$

Two conditions must be satisfied if this formula is to be valid. First, it is valid only if $v > v_s - v_m$; if this is not so, wave front 1 will arrive at the observer after wave front 2 rather than before it. Second, it is valid only if $v > v_o - v_m$; if this is not so, the waves will never catch up with the observer. In most familiar cases involving sound, these conditions are met because the speed of sound is usually considerably greater than that of any part of the system of interest.

If only the source is moving, Equation 22.33 assumes the simpler form

$$\nu_o = \nu_s\frac{v}{v - v_s}. \tag{22.34}$$

If only the observer is moving, Equation 22.33 simplifies to

$$\nu_o = \nu_s\frac{v - v_o}{v}. \tag{22.35}$$

EXAMPLE 22.8

A major device used by naval vessels to detect submarines is **sonar** (an acronym for **so**und **na**vigation **a**nd **r**anging). A sonar set consists of a sound source and a sensitive microphone, both placed below the waterline. The source emits short bursts of sound, and the microphone detects any sound that is reflected back. The time interval between emission and detection can be used to find the distance to the reflecting object. At the same time, the Doppler effect can be used to determine the speed of the reflecting object.

Suppose a destroyer is moving due west at 9.5 m/s with

respect to the ocean bottom. An ocean current moves due east at 1.3 m/s. If a submarine is heading due east, directly toward the destroyer, at 14.0 m/s, what is the value of the observed Doppler shift? Take the speed of sound in seawater to be 1470.0 m/s and the source frequency to be 8 kHz.

SOLUTION: There are really two Doppler shifts taking place in sequence. In the first, sound emitted by the source passes through the water and strikes the submarine, which takes the role of the observer. When it reflects the sound, the submarine acts like a source and the destroyer is the observer.

You must take care to assign the proper signs to the three velocities, using the convention that the positive x direction is that in which the sound propagates. In the first shift, you set $v_s = 9.5$ m/s, $v_m = -1.3$ m/s, and $v_o = -14.0$ m/s. So Equation 22.33a gives you

$$\nu_o = \nu_s \frac{1470.0 \text{ m/s} - 1.3 \text{ m/s} + 14.0 \text{ m/s}}{1470.0 \text{ m/s} - 1.3 \text{ m/s} - 9.5 \text{ m/s}} = 1.016\nu_s.$$

In the second shift, the sound travels in the opposite direction, which you now take to be the positive x direction. You then have $v_s' = -v_o = 14.0$ m/s, $v_m' = -v_m = 1.3$ m/s, and $v_o' = -v_s = -9.5$ m/s. Inserting the primed quantities into Equation 22.33a, you obtain

$$\nu_o' = \nu_s' \frac{1470.0 \text{ m/s} + 1.3 \text{ m/s} + 9.5 \text{ m/s}}{1470.0 \text{ m/s} + 1.3 \text{ m/s} - 14.0 \text{ m/s}} = 1.016\nu_s'.$$

Reflection by the submarine does not change the sound frequency. So you have $\nu_s' = \nu_o$. Putting the two results together gives you

$$\nu_o' = 1.016 \times (1.016\nu_s) = 8260 \text{ Hz}.$$

The overall Doppler shift is

$$\nu_o' - \nu_s = 260 \text{ Hz}.$$

This use of Doppler shifts to measure speeds is called *Doppler sonar.* The analogous use of radar (popular among traffic police) is called *Doppler radar.*

Symbols Used in Chapter 22

a	cross-sectional area
A	amplitude
B, B_{iso}, B_a	general, isothermal, adiabatic bulk modulus
F	force
G	shear modulus
I, I_0	intensity, threshold of hearing
k	wave number
L	length
M_0	mass of 1 kmol of a gas
n	positive integer, number of moles
$p, \Delta p$	pressure, pressure variation
$\Delta P(x, t)$	wave function for superposed pressure waves
$\Delta S(x, t)$	wave function for superposed displacement waves
ds	infinitesimal longitudinal displacement
v	wave velocity or speed
v_s, v_m, v_o	velocity of source, medium, observer
v_x, v_y	longitudinal, transverse velocity of medium
R	universal gas constant
T	temperature
V	volume
Y	Young's modulus
$Y(x, t)$	wave function for superposed transverse waves
α	acoustic intensity in decibels
β	shear strain, acoustic intensity in bels
γ	heat capacity ratio
δ	phase constant
ϵ	uniaxial strain
λ	wavelength
μ	molecular mass
ν	frequency
ρ	density
σ	stress
τ	period
ω	angular frequency

Summing Up

Both **transverse** and **longitudinal waves** can be transmitted through a solid medium. All points in the medium having the same phase at the same time are said to be **in phase** with one another and constitute a **wave front**. A wave may be visualized as a series of wave fronts.

The speeds of transverse (shear) waves and longitudinal (compressional) waves are given by Equations 22.6 and 22.13:

$$v_{transverse} = \sqrt{\frac{G}{\rho}} \quad \text{and} \quad v_{longitudinal} = \sqrt{\frac{Y}{\rho}}.$$

Transverse waves cannot propagate through fluids. For fluid media, the longitudinal wave speed is expressed in terms of the bulk modulus according to Equation 22.16,

$$v = \sqrt{\frac{B}{\rho}}.$$

If the fluid is a gas, it is necessary, in using this equation, to specify the *adiabatic* bulk modulus according to Equations 22.17 and 22.20,

$$v = \sqrt{\frac{B_a}{\rho}} = \sqrt{\frac{\gamma p}{\rho}}.$$

If, as is most commonly the case, the gas conforms to the ideal-gas law, Equation 22.20 can be expressed in the form of Equation 22.21a,

$$v = \sqrt{\frac{\gamma RT}{M_0}}.$$

In response to the stimulus of sound energy flux, the ear produces a sensation of **loudness**. Loudness, or **acoustic intensity**, is expressed in **decibels** by means of Equations 22.24 and 22.22,

$$\alpha = 10 \log_{10} \frac{I}{I_0}; \quad I_0 \equiv 10^{-12} \text{ W/m}^2.$$

The ear perceives **pitch** in response to the stimulus of frequency and **sound quality** (or **timbre**) in response to the Fourier spectrum of the sound.

Beats are heard when two sound waves of equal amplitude and slightly different frequency are superposed. The perceived pitch of the sound depends on the **sum frequency**, given by Equation 22.25a,

$$\nu_+ = \tfrac{1}{2}(\nu_1 + \nu_2).$$

The beats are heard at the **beat frequency** given by Equation 22.25c,

$$\nu_b = |\nu_1 - \nu_2|.$$

The formation of standing waves in air-filled pipes is analogous to the formation of standing waves in a stretched string. For an **open pipe**, the standing-wave condition is given by Equation 22.27a or 22.27b,

$$\frac{\lambda}{2} = \frac{L}{n} \quad \text{or} \quad 2L = n\lambda; \quad n = 1, 2, 3, \ldots.$$

For a **closed pipe**, the standing-wave condition is given by Equation 22.30a or 22.30b,

$$\frac{\lambda}{4} = \frac{L}{m} \quad \text{or} \quad 4L = m\lambda; \quad m = 1, 3, 5, \ldots.$$

If a source emits sound with frequency ν_s, an observer detects sound at a different frequency ν_o if the source or the observer (or both) is in motion. This is the **Doppler effect**. The two frequencies are related by Equation 22.33a,

$$\nu_o = \nu_s \frac{v + v_m - v_o}{v + v_m - v_s}.$$

The **Doppler shift**, $\Delta\nu = \nu_o - \nu_s$, is given by Equation 22.33b.

KEY TERMS

Sections 22.2 and 22.3 Propagation of Waves in Solids and Longitudinal Waves

wave front, plane wave ▪ shear wave, compressional wave ▪ longitudinal pulse, longitudinal wave

Section 22.4 Waves in Fluids

isothermal, adiabatic bulk modulus

Section 22.5 The Ear and Hearing

loudness, pitch, timbre ▪ octave ▪ threshold of hearing ▪ acoustic intensity, bel, decibel

Section 22.6 Superposition of Sound Waves

beat frequency, beat ▪ sum frequency, difference frequency

Section 22.7 Standing Waves in Pipes

pressure node, antinode ▪ displacement node, antinode

Section 22.8 The Doppler Effect

Doppler shift

QUERIES

22.1 *(4) Squeaky.* In a common lecture demonstration, a student with a deep bass voice inhales a quantity of helium gas. His voice immediately becomes high-pitched. Explain.

22.2 *(4) No matter.* The speed of waves propagating through a liquid is given by Equation 22.16. Why is it unnecessary to specify that the adiabatic bulk modulus be used?

22.3 *(4) Serious matter.* Explain qualitatively why $B_a > B_{iso}$ for gases.

22.4 *(4) Ball-park comparison.* Make a simple qualitative argument to explain why the speed of sound in a gas is not very different from the rms speed of the gas molecules.

22.5 *(4) Ups and downs.* How does the speed of sound depend on altitude in an isothermal atmosphere?

22.6 *(5) Decimal system.* The base-10 logarithm is used in defining the bel and decibel. Why is the choice of the base arbitrary and purely a matter of convenience?

22.7 *(5) Napier's choice.* Why is the base-2 logarithm used in Section 22.5 in the discussion of the connection between frequency and pitch?

22.8 *(7) Sousa's favorite.* The length of a piccolo is half that of a flute. Otherwise, the two instruments are very similar. What is the ratio of the frequency of the lowest note on a piccolo to that of the lowest note on a flute?

22.9 *(7) Leakproof.* What is the standing-wave condition for an air-filled pipe closed at *both* ends?

22.10 *(7) To what end?* Without calculation, find the standing-wave condition for a stretched string excited as shown here.

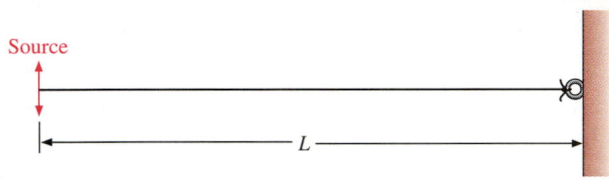

22.11 *(7) But no musician has more than ten fingers!* Why does a clarinet require more holes than a flute? (Hint: The flute is essentially an open pipe, and the clarinet a closed pipe.)

22.12 *(8) Quick change.* You and your friend stand at different distances from a railroad track. A train passes with its whistle blowing. **(a)** Do you and your friend hear the same frequency shift as the train passes? **(b)** As the train passes, you perceive a more rapid frequency shift than your friend does. Who is standing closer to the track?

22.13 *(8) Sound reasoning.* Two boats, one directly behind the other, move through a lake at the same velocity. Each boat has an underwater loudspeaker that emits beeps of frequency v, and an observer on each boat listens to the beeps emitted by the other boat. **(a)** Is the time required for a beep to travel from the boat in front to the one behind the same as the time required for a beep to travel in the opposite direction? **(b)** Are the wavelengths of the beeps traveling in the two directions the same or different? **(c)** Are the frequencies detected at the two boats the same or different? If different, which frequency is higher?

22.14 *(G) Faster than a speeding bullet.* If you have ever seen a military airplane fly overhead at high speed, you may have noted that the engine noise seems to emanate from a point a considerable distance behind the plane. Explain. How could you use your observations to make a rough estimate of the speed of the plane?

22.15 *(G) Refraction of sound.* Suppose you are outdoors with a friend on a windy day. Your friend, who is some distance upwind of you, cannot hear you when you shout. But you can hear your friend quite well. This common experience is often explained by saying that "The wind blows the sound away." But that is absurd; even a strong wind has a speed less than 5% of the speed of sound. To understand the real reason, begin with the fact that the wind near the ground has a *velocity gradient*; the speed of the wind is zero just at the ground, and it increases steadily with height. **(a)** Consider a sound pulse represented by a plane wave front initially oriented perpendicular to the ground. What happens to the wave front as it moves upwind? **(b)** Now consider a similar wave front moving downwind. What happens? How do you explain the difficulty of communicating upwind and the ease of communicating downwind? (Note: This is a special case of the phenomenon called *refraction*, which we will consider in much more detail in the context of optics in Part X.)

PROBLEMS

Notes: **(1)** In the following problems, take the speed of sound in air to be 344 m/s unless otherwise specified. **(2)** Values of the elastic moduli of various solids may be found in Table 15.1.

GROUP A

22.1 *(2) How fast?* Find the speed in aluminum **(a)** of shear waves and **(b)** of compressional waves. Take the density of aluminum to be 2.70×10^3 kg/m^3.

22.2 *(3) Shear presumption.* What is the speed of shear waves in steel?

22.3 *(3) Copper your bets when you bet your coppers!* Find the speed in copper **(a)** of compressional waves and **(b)** of shear waves. Take the density of copper to be 8.92×10^3 kg/m^3.

22.4 *(3) Lead-ing question.* The speed of compressional waves in lead is 1150 m/s. Find the speed of shear waves in lead.

22.5 *(3) Rational procedure.* Find v_c/v_s, the ratio of the speed of compressional waves to that of shear waves, in aluminum.

22.6 *(3) Significant difference.* A seismograph is located at a moderate distance D (a few kilometers) from a mine. The rock between the mine and the seismograph is fairly uniform in composition. A blast is detonated in the mine, and the seismograph registers the arrival of P and S waves at instants separated by the interval Δt. Show that

$$D = \frac{v_P v_S}{v_P - v_S} \Delta t,$$

where v_P and v_S are the speeds of P and S waves through the rock.

22.7 *(4) 200 proof.* At 20°C, the density of ethyl alcohol is 789 kg/m^3, and the speed of sound in it is 1213 m/s. Find the bulk modulus.

22.8 *(4) Fuel for thought.* At 20°C, the density of kerosene is 0.82×10^3 kg/m^3, and the speed of sound in it is 1330 m/s. What is the compressibility of kerosene?

22.9 *(4) Echo sounder, I.* Mariners often use an echo sounder to determine water depth. This device emits a burst of sound and measures the time Δt elapsed until an echo from the bottom is detected. Find Δt for the greatest depth in Lake Michigan, 281 m. Take the bulk modulus of water to be 2.03×10^9 Pa.

22.10 *(4) Echo sounder, II.* The time delay measured by an echo sounder on a ship at sea is 2.53 s. How deep is the water? Take the density of seawater to be 1.030×10^3 kg/m^3 and the compressibility to be 4.6×10^{-10} Pa^{-1}.

22.11 *(4) Light-footed.* Find the speed of sound in hydrogen gas when the temperature is 15°C.

22.12 *(4) Seasonal change.* Find the speed of sound in air on a hot day when the temperature is 35°C and on a cold day when the temperature is −15°C.

22.13 *(4) Day for rubber mouthpieces.* A high-school marching band warms up for a New Year's Day parade in a gym where the temperature is 13°C. The players tune to a standard A, 440 Hz. They begin to play and march into the street, where the temperature is a very chilly −20°C. What is the frequency of the A when it is played in the street? (The change is an easily heard half tone; on cold days, the "slide" as the band marches into the cold can be spectacular!)

22.14 *(4) Massive sound.* The density of a certain diatomic gas is 1.39 kg/m^3 at 1 atm pressure. What is the speed of sound in the gas?

22.15 *(4) Speedy calculation.* You measure the speed of sound in a diatomic gas and find $v = 332$ m/s. What is the root-mean-square speed of the gas molecules?

22.16 *(4) Energetic pursuit.* You measure the speed of sound in a sample of helium and find $v = 965$ m/s. What is the kinetic energy of the molecules in 1 kmol of the gas?

22.17 *(5) Middling loud.* The effective area of the human eardrum is about 0.5 cm^2. What is the energy flux incident on the eardrum when the acoustic intensity is 60 dB?

22.18 *(5) Doubling in brass.* When a trumpeter plays loudly, the orchestra conductor hears a sound with acoustic intensity 90 dB. What will the acoustic intensity be if two trumpeters play together?

22.19 *(5) Violin section.* A typical symphony orchestra has 14 first violinists. How much louder do they sound, playing together, than a single violinist? Explain why it is possible for a soloist playing a violin concerto to be heard above the violins in the orchestra.

22.20 *(5) Ouch!* What is the intensity of sound incident on a normal ear when the owner of the ear begins to feel pain?

22.21 *(5) Going to any lengths.* What are the wavelengths of the lowest- and highest-pitch sounds that can be heard by a young person with normal hearing? Take $\nu_{min} = 14$ Hz and $\nu_{max} = 18$ kHz.

22.22 *(6) Out of tune.* You compare a tuning fork of unknown frequency with a tuning fork whose frequency is 440 Hz. When you sound them together, you hear beats of frequency 5 Hz. **(a)** What are the two possible frequencies of the unknown fork? **(b)** You find that you can increase the beat frequency by sticking a small piece of chewing gum on each end

of the unknown fork. Which of the two answers of part **a** is the frequency of the unloaded fork?

22.23 *(6) Beating the game.* A string under tension of 145 N is in tune with a 440-Hz tuning fork. If the tension is increased by 5 N, how many beats per second are heard?

22.24 *(6) Quick change.* A tube with a cap on one end produces a tone whose fundamental frequency is 130.8 Hz. **(a)** If the cap is removed, what is the new fundamental frequency? **(b)** How long is the tube?

22.25 *(7) Down deep.* The lowest C of an organ, whose frequency is 16.35 Hz, is produced by a closed pipe. **(a)** How long is the pipe? **(b)** What is the fundamental frequency of an open pipe of the same length?

22.26 *(7) The long and short of it.* The second mode ($m = 3$) of a closed organ pipe has the same frequency as the second mode ($n = 1$) of an open organ pipe. Express the length of the open pipe in terms of that of the closed pipe.

22.27 *(7) Here, doggie!* Dogs can readily hear sounds of frequency 25 kHz, which are inaudible to human beings. How long should a closed-end "silent" dog whistle be?

22.28 *(7) Count 'em.* Standing waves are excited in a pipe of length 85.0 cm. How many modes are there below 1250 Hz if the pipe is **(a)** open and **(b)** closed at one end?

22.29 *(7) Ostentatious.* The members of a new church take great pride in their organ, whose many ranks of pipes makes a magnificent display on the walls. They point with particular pride to the rank of huge pipes behind the altar, the longest of which is 22 m long. What can you say about this pipe?

22.30 *(7) The parts fit.* The human ear canal is irregular in shape, but its acoustic behavior is very roughly that of a cylindrical tube of length 2.4 cm. What is the lowest frequency of sound that will set up standing waves in the canal? Compare this result with the frequency at which the ear is most sensitive.

22.31 *(8) Foggy.* A foghorn emits sound signals at 35 Hz. If a ship approaches the foghorn at 7 m/s, what is the frequency of the sound heard by the pilot? Take the speed of sound to be 330 m/s. Is listening to the pitch of a foghorn likely to be a practical method for judging the velocity of currents in the water through which a ship is passing?

22.32 *(8) Rolling along, I.* A source that emits sound at 1 kHz is anchored at the bottom of a river. Microphone 1 is anchored on the river bottom 100 m upstream from the source, and microphone 2 is anchored on the bottom 100 m downstream from the source. If the river current has speed 10 km/h, find the frequencies ν_1 and ν_2 detected by the microphones. Take the speed of sound in water to be 1440 m/s.

22.33 *(8) Rolling along, II.* Consider again the river-bottom sound source of Problem 22.32. This time, microphones 1 and 2 float down the river with the current. At a certain time, microphone 1 is upstream and microphone 2 is downstream from the source. Find the Doppler shifts $\Delta\nu_1$ and $\Delta\nu_2$ detected by the microphones.

22.34 *(8) Fast drop.* A 1.6-kHz sound source is attached to the underside of a sailplane. A skydiver jumps out of the sailplane. What is the frequency of the sound she hears 2.5 s after jumping? Neglect air friction and the sink rate of the sailplane.

22.35 *(8) Swing band?* A miniature loudspeaker emits a steady sound of frequency 2000 Hz. You tie it to the end of a string and swing it around your head in a horizontal circle of radius 75 cm. If the loudspeaker circles your head 2.25 times per second, what is the range $\Delta \nu$ of sound frequencies heard by your friend standing 8 m away from you?

22.36 *(G) Ultrasonic imaging.* Medical ultrasonic imaging devices use sound whose frequency is about 7 MHz to "illuminate" organs and other features within the body. As a general rule, you cannot obtain an image of any object smaller than the wavelength of the waves used to illuminate it. Taking the speed of sound in typical tissues to be 1500 m/s, determine whether the device could be used to detect a tumor of diameter 2 mm.

GROUP B

22.37 *(3) A good guess based on minimal information.* On a seismograph, you observe the P and S waves from a distant earthquake. The time interval between them is 2.0 min. Make a rough estimate of the distance between you and the earthquake site. If the waves are direct (not reflected), how far off is your estimate likely to be? (Hint: Look at Table 15.1 and note that Young's modulus does not vary over a very wide range from one solid to another; the ratio G/Y varies even less. What about the range of variation of ρ? How does v_s depend on Y and ρ?)

22.38 *(3) Keep cool by keeping busy, I.* At the time of the severe Whittier, California, earthquake of 1987, I was in my office in Long Beach. On feeling the first shock, I began to count seconds, at the same time making a quick check around to see if there was anything likely to fall on me. (There wasn't.) Approximately how much time passed before I felt the second strong shock? The distance from Whittier to Long Beach is about 22 km.

22.39 *(3) Keep cool by keeping busy, II.* At the time of the severe Loma Prieta earthquake of 1989, I was in my home in Woodside, California, talking on the phone to my friend Bill in the Marina district of San Francisco. Feeling the small initial shake, I told Bill, "We're having a little earthquake." Ten seconds later, Bill reported that he felt it, too. (Then the phone went dead.) The distance from Woodside to the Marina district is about 48 km. Using Figure 22.7a, show that the earthquake source, Woodside, and the Marina district lie approximately in a straight line.

22.40 *(3) Putting in some numbers.* Using Figure 22.7a, find the minimum and maximum speeds **(a)** of S waves and **(b)** of P waves. Express your results in m/s. **(c)** For nearly all solid materials, Y, G, and ρ all increase with increasing pressure. On the basis of Figure 22.7a and a simple geometric argument, determine whether Y and G increase more or less rapidly with increasing pressure than ρ. Neglect the change of chemical composition of the earth with depth.

22.41 *(4) Speedy measurement.* The speed of sound in argon (a monatomic gas) is 307.8 m/s at 0°C. Find the molecular mass of argon.

22.42 *(4) Espresso.* Find the speed of sound in steam at the boiling point. (The water molecule has six degrees of freedom.)

22.43 *(4) A molecular point of view.* The mean energy of the molecules in a sample of nitrogen gas (N_2) is 9.4×10^{-21} J.

Find the speed of sound in this sample. Take the mass of a nitrogen molecule to be 28 u.

22.44 *(4) Build a fire under it.* Show that the speed of sound increases by 0.6 m/s for each 1 C° increase in temperature.

22.45 *(4) Pressure wave function and displacement wave function, I.* We have considered longitudinal waves mainly in terms of pressure variations. For a sinusoidal traveling wave, you can write the *pressure wave function*

$$\Delta p(x, t) = \Delta p^* \cos (kx - \omega t). \tag{1}$$

But you can also describe the same wave in terms of the *displacement wave function*

$$s(x, t) = s^* \cos (kx - \omega t + \delta). \tag{2}$$

(a) What is the connection between the two descriptions?
(b) Show that, for any fluid,

$$-\frac{\partial s}{\partial x} = \frac{\Delta p}{B} = \frac{\Delta p}{\rho v^2}.$$

22.46 *(4) Pressure wave function and displacement wave function, II.* **(a)** By taking the partial derivative of Equation 2 with respect to x in the preceding problem and using the result of part **b**, show that the displacement and pressure amplitudes bear the relation

$$s^* = \frac{\Delta p^*}{\rho \omega v}.$$

(b) Show that the pressure and the displacement are out of phase by one-quarter of a wavelength.

22.47 *(4) Displacement wave.* A sound wave is represented by the function

$$s(x, t) = (6.2 \times 10^{-5} \text{ m}) \cos [(1800 \text{ s}^{-1})t - (5.3 \text{ m}^{-1})x].$$

(a) Find the ratio of the displacement amplitude to the wavelength of the sound. **(b)** Find the maximum speed of the displacement oscillations and the ratio of that speed to the wave speed of the sound.

22.48 *(4) Pressing problem.* Beginning with Equation 22.20, show that the speed of sound in air increases (or does not increase) with increasing pressure. Assume that the temperature is constant and that air is an ideal gas.

22.49 *(4) Upper-atmosphere physics.* It is difficult to measure the temperature of the upper atmosphere directly with a thermometer because the density of the gas is so low that one cannot be sure the thermometer has reached thermal equilibrium with the gas. The following indirect method has been used successfully. A small research rocket, which is fired straight up, liberates a series of small explosive charges at uniformly spaced altitudes. The charges are made to explode simultaneously, and the arrival of the sounds at the ground is timed. Suppose that the sound from the charge exploded at 20.5 km altitude arrives at the ground 3.38 s after the sound from the charge exploded at 19.5 km. What is the temperature at 20 km? Assume that the composition of the air does not change appreciably with altitude.

22.50 *(4) Make a wish!* The water level in a well is a distance y below ground level. You drop a small stone into the well, and you hear the splash a time t later. **(a)** Show that

$$y = \frac{v^2}{g}\left(1 - \sqrt{1 + \frac{2gt}{v}} + \frac{gt}{v}\right).$$

(b) If the elapsed time is 3.8 s, how far down is the water surface? Let the temperature be 13°C. **(c)** Show that, if you neglect the time required for the sound of the splash to travel to ground level when you calculate y, you overestimate the depth of the well by about 11%. **(d)** Show that, if the well is shallow enough, the result of part **a** simplifies to $y = \frac{1}{2}gt^2$. Hint: In expanding the square root, you will have to use the first *three* terms of the binomial series:

$$(1 \pm x)^n = 1 \pm nx \pm \tfrac{1}{2}n(n-1)x^2.$$

22.51 *(4) Speed of sound in an adiabatic atmosphere.* Using the results of Problem 19.55, express the speed of sound as a function of altitude h for an adiabatic atmosphere. (You need not have worked Problem 19.55 because the answer is given.)

22.52 *(4) Energy transport in sound waves.* Equation 21.38 expresses the energy flux for a vibrating string. Beginning with the wave functions given in Problem 22.46 and using an argument parallel to the one leading to Equation 21.38, derive the following expressions for the intensity of a sound wave:

$$I = \tfrac{1}{2}\rho\omega^2 s^{*2}v = \frac{\Delta p^{*2}}{2\rho v}.$$

Note that, though these expressions differ in detail from Equation 21.38, the intensity depends on the square of the amplitude in all cases.

22.53 *(5) Very sensitive!* **(a)** Using the result of Problem 22.52, find the pressure amplitude Δp_0 of a sound at the threshold of hearing, 10^{-12} W/m². **(b)** Find the corresponding displacement amplitude if the frequency of the sound is 3 kHz. **(c)** Find the pressure amplitude for a sound just loud enough to damage the ear.

22.54 *(6) Fine tuning.* The second-lowest G on a piano has frequency 98.00 Hz. When the note is played, the hammer strikes two strings, which should be tuned identically. A piano tuner tunes one of the strings correctly. When he plays the note, he hears beats with frequency 2.0 Hz. By what fraction must he change the tension of the improperly tuned string?

22.55 *(7) Fun with water.* The apparatus for a popular laboratory experiment is illustrated in the next column. You can adjust the water level in the long vertical tube by raising and lowering the reservoir. For the most part, the sound of the tuning fork is heard only faintly. But for certain water levels,

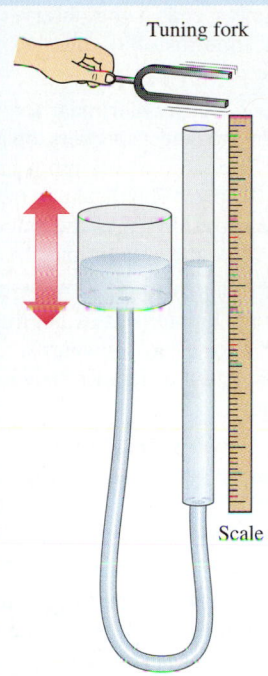

Tuning fork

Scale

the sound is quite loud. You find that, when you use a G tuning fork ($\nu = 392.0$ Hz), the sound is loudest for two water levels 43.36 cm apart. What is the speed of sound?

22.56 *(7) Pressure and displacement wave functions for standing waves.* **(a)** Write the displacement wave function $S(x, t)$ that corresponds to the value of $\Delta P(x, t)$ given by Equation 22.28. **(b)** Using an argument similar to that of Problem 22.46, show rigorously that pressure antinodes correspond to displacement nodes and vice versa; that is, show that the standing pressure and displacement waves are out of phase with one another.

22.57 *(7) Why is my dog howling? I.* A copper rod 50 cm long is clamped at its midpoint but is otherwise free. How many standing compressional wave modes are there in the frequency range 20 kHz $< \nu <$ 50 kHz? What are their frequencies? Take the density of copper to be $\rho = 8.92 \times 10^3$ kg/m³.

22.58 *(7) Why is my dog howling? II.* For the copper rod of Problem 22.57, how many standing shear wave modes are there in the frequency range 20 kHz $< \nu <$ 50 kHz? What are their frequencies?

22.59 *(8) Whistle blower, I.* As you stand by a railroad track, a train passes with its whistle blowing. As it passes, you hear a frequency shift equal to 19% of the frequency of the whistle. How fast is the train going?

22.60 *(8) Whistle blower, II.* Standing by a railroad track, you hear the whistle of an approaching train; the frequency is ν_1. After the train passes, you hear the whistle at frequency ν_2. Taking the speed of sound to be v, find **(a)** the speed of the train and **(b)** the frequency ν_0 of the whistle as the engineer hears it.

22.61 *(8) Out of tune.* A tuning fork approaches an observer at speed v_f. An identical tuning fork moves away from the observer at the same speed. If the observer hears a beat frequency ν_b, find v_f in terms of ν_b, the fork frequency ν, and the speed of sound v.

22.62 *(8) Speed cop.* A Doppler radar set, of the kind commonly used by police to detect speeders, operates at frequency 2.415×10^{10} Hz. The radio waves propagate at the speed of light, $c = 3.00 \times 10^8$ m/s. The radar can measure automobile speeds within ± 0.5 m/s. What is the smallest Doppler frequency ratio ν_o'/ν_s that the system must be able to detect?

22.63 *(8) Special Doppler shift, I.* According to Equations 22.34 and 22.35, the Doppler shift is not the same for a moving observer as it is for a moving source. But show that, for $v_s \ll v$ and $v_o \ll v$, the difference between the two effects becomes negligible.

22.64 *(8) Special Doppler shift, II.* Make a schematic plot of the Doppler frequency ratio ν_o/ν_s as a function of v_o for the special case $v_s = 0$. On the same axes, plot ν_o/ν_s as a function of v_s for the special case $v_o = 0$. Does your graph bear out the result of the Problem 22.63?

22.65 *(8) Doppler's fjord.* A tour boat in a Norwegian fjord approaches the shoreline, which is essentially a vertical cliff. When the boat's whistle is sounded, the passengers hear beating between the sound reaching them directly from the whistle and the sound reflected from the shore. **(a)** Express the beat frequency ν_b in terms of the whistle frequency ν_s, the speed v_s of the boat, and the speed of sound. **(b)** Show that, to a good approximation, the beat frequency is

$$\nu_b = 2\nu_s \frac{v_s}{v}.$$

(c) If the frequency of the whistle is 100 Hz and the beat frequency is 1 Hz, how fast is the boat going? Take $v = 330$ m/s.

22.66 *(8) Sundays on Lake Wobegon.* Two excursion boats sail in opposite directions, each with speed 8 m/s. There is a concert band on each boat. The wind is blowing from boat 1 toward boat 2 at 6 m/s. Just after the two boats have passed one another, an oboist on boat 1 sounds an A ($\nu = 440$ Hz) for her band to tune to. The musicians on boat 2 hear the oboe and also tune their instruments to it. What frequency do the passengers on boat 1 hear from boat 2? It is a warm summer day; take the speed of sound to be 347 m/s.

22.67 *(5) Acoustic intensity and pressure amplitude.* **(a)** Using the result of Problem 22.52, show that the acoustic intensity in dB can be written

$$\alpha = 20 \log_{10} \frac{\Delta p^*}{\Delta p_0},$$

where Δp_0 is the pressure amplitude of a sound wave at the threshold of hearing, evaluated in part **a** of Problem 22.53. **(b)** Express α in terms of the displacements s^* and s_0.

22.68 *(G) One-dimensional case.* Equations 21.40 and 21.43 give $I(r)$, the dependence of the intensity on the distance from a point source, for two-dimensional and three-dimensional media, respectively. Find the corresponding dependence for a uniform bar.

22.69 *(G) Clever way to measure Young's modulus.* The device shown in the next column is called Kundt's tube, after its inventor, the German physicist A. Kundt (1839–1894). A

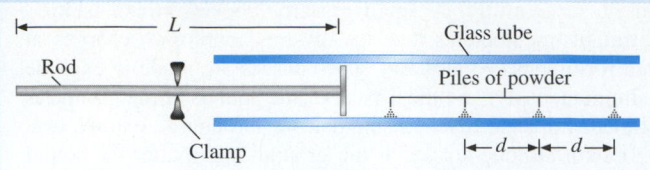

metal rod has a plastic disk of negligible mass attached to one end. The rod is clamped at its midpoint. The disk fits loosely into a glass tube. Sprinkled along the tube is a quantity of fine powder. Longitudinal standing waves can be excited in the rod by rubbing it with a rosined cloth. Because the midpoint of the rod is clamped, this point must be a displacement node; the free ends are displacement antinodes. As the rod vibrates, the longitudinal oscillation of the disk excites sound waves in the air in the glass tube. The rod assembly is moved into and out of the tube until the oscillation frequency of the disk corresponds to a standing-wave frequency of the tube. It is easy to see when this is accomplished; the powder, disturbed by the oscillating air, dances along the bottom of the tube and accumulates into regularly spaced piles located at the displacement nodes. **(a)** Express Young's modulus of the metal of which the rod is made in terms of its density ρ, the rod length L, the speed v of sound in air, and the spacing d between adjacent piles of powder. **(b)** Suppose that the rod is 2.00 m long and is made of brass, whose density is 8.92×10^3 kg/m^3. If $d = 14.5$ cm, find Young's modulus for brass.

22.70 *(G) Kundt's double tube.* Kundt used the apparatus shown below for careful measurements of the heat capacity ratios γ of various gases. Here, the apparatus is set up for measurement of carbon dioxide. For convenience, the metal rod is supported by spacers at the ends of the tubes, as shown.

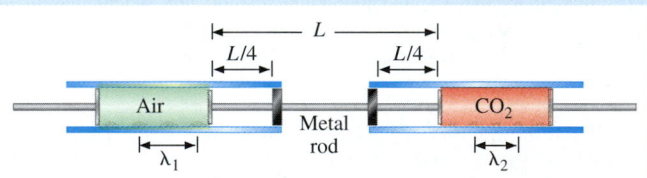

Because the support points must be nodes, the rod is excited in its second mode, but that makes no difference in what follows. The rod is rubbed until piles of powder are formed in both tubes. The wavelengths λ_{air} and λ_{CO_2} of the standing waves in the two tubes are measured. **(a)** Express the wavelength ratio $\lambda_{CO_2}/\lambda_{air}$ in terms of γ_{air}, γ_{CO_2}, and the densities ρ_{air} and ρ_{CO_2}. **(b)** Kundt found that the wavelength ratio was 0.780. Knowing that $\gamma_{air} = 1.403$ and that $\rho_{CO_2}/\rho_{air} = 1.53$, what value did he find for γ_{CO_2}?

22.71 *(G) Equal temperament.* In Western music, each octave consists of 13 notes (including the note that ends the octave and begins a new octave). We adjust the 12 pitch intervals between these steps so that we perceive them as equal. The 13 notes constitute the *chromatic scale*. Show that the ratio of the frequencies of any two adjacent notes in the chromatic scale is $1 : 2^{1/12}$ ($\approx 1 : 1.05946$). This *well-tempered*, or *equal-tempered*, system of tuning has the advantage that music written in all keys sounds equally good. J. S. Bach wrote his *Well-Tempered Clavier* (1722 and 1744), comprising short works in all possible keys, to demonstrate this advantage. When playing a clavier (a term that means any keyboard instrument), the musician cannot make small pitch adjustments, as is easily possible with the voice or the violin. For this reason, modern keyboard instruments are nearly always tuned in the well-tempered system.

GROUP C

22.72 *(4) Temperature change.* A sound source and an observer are located a distance L apart. The temperature varies in linear fashion between them, having the value T_s at the source and T_o at the observer. How long does it take for sound to travel the distance L?

22.73 *(5) Nonlinearity of the ear.* The ear deviates significantly from Hooke's law in its response to sound waves of commonly encountered acoustic intensity. A pure sinusoidal 500-Hz tone, for example, produces perceptible second-mode (1000-Hz) response at 40 dB and perceptible third-mode (1500-Hz) response at 50 dB. To explore some of the most important consequences of nonlinearity, assume the simplest possible nonlinear "model ear," whose eardrum experiences displacement s in response to pressure Δp according to the rule

$$s = a\,\Delta p + b\,\Delta p^2, \tag{1}$$

where a and b are constants. (Note that this expression becomes Hooke's law if $b \to 0$.) Next, suppose that the oncoming sound wave is purely sinusoidal; that is,

$$\Delta p(t) = \cos 2\pi\nu t, \tag{2}$$

where, for simplicity, the amplitude is taken to be $A = 1$. **(a)** Substitute Equation 2 into Equation 1 and use the appropriate trigonometric identity to express $s(t)$ in terms of sines and cosines to the first power only. Show that the ear perceives a sound having a component with frequency 2ν as well as the original frequency ν. **(b)** Suppose now that the incident wave contains two components with frequencies ν_1 and ν_2; that is,

$$\Delta p(t) = \cos 2\pi\nu_1 t + \cos 2\pi\nu_2 t. \tag{3}$$

Show that the ear will perceive not only sounds with the frequencies ν_1, ν_2, $2\nu_1$, and $2\nu_2$, but also *combination tones*: the *sum tone* of frequency $\nu_1 + \nu_2$ and the *difference tone* of frequency $|\nu_1 - \nu_2|$. When produced in the ear, these tones are called *aural harmonics*. When produced by nonlinearities in electronic systems and (especially) in loudspeakers, combination frequencies are called *intermodulation* (IM) *distortion.* The ear always produces such distortion, except in the presence of quite faint sounds, but the brain "knows" how to interpret it. The goal of the sound-reproduction engineer is to introduce as little IM distortion as possible because the listener will necessarily interpret it as an integral part of the sound incident on his or her ear. **(c)** Show that the difference tone between any two successive harmonic modes is the fundamental tone, and thus explain how it is possible to listen to recognizable music when using a cheap sound system, in which the speaker cannot reproduce the lowest frequencies present in the original music. **(d)** The ratio b/a for our model ear varies with frequency and increases as the acoustic intensity increases. Suppose you wish to use an excellent sound system to listen to a high-quality recording of a musical performance. Why is it necessary, for faithful reproduction of the original music, to adjust the volume control so that the sound level is as close as possible to the sound level at the original

performance?

22.74 *(8) Mach cone.* Repeat the construction of Figure 22.16 for the case $v_s > v$; that is, let the speed of the source exceed the wave speed. (For purposes of construction, it is convenient to choose $v_s = 2v$.) **(a)** Show that the tangent to the wave fronts is a V-shaped envelope whose apex lies at the present position of the source and whose axis is the path of the source. In three dimensions, the "V" becomes the surface of a cone, called the *Mach cone.* **(b)** Show that the half-angle at the apex of the cone is

$$\theta = \sin^{-1}\frac{v}{v_s};$$

θ is called the *Mach angle,* and the dimensionless ratio v/v_s is called the *Mach number.*

22.75 *(G) Absorption of sound by air.* No real medium transmits waves perfectly. Rather, the intensity of a plane wave is reduced by a certain fraction per meter of travel, as various "friction" mechanisms convert the energy of the wave into heat energy. **(a)** Show that the intensity of a plane wave is given as a function of distance by the equation

$$I(x) = I_0 e^{-cx},$$

where I_0 is the intensity at some specified location (typically the source) and c is the *absorption,* or *attenuation, coefficient.* **(b)** Suppose that the wave in question is a plane sound wave. Show that the acoustic intensity α changes linearly with distance; that is, $d\alpha/dx$ can be expressed as a constant having units dB/m. Show that $d\alpha/dx = -c/0.23$. (Hint: $\ln z = 2.3 \log_{10} z$.)

22.76 *(G) Attenuation of spherical waves.* **(a)** Using the result of Problem 22.75, show that the intensity of sound in air falls off with distance r from a point source according to the equation

$$I(r) = \langle\Phi\rangle_{r_0}\frac{e^{-cr}}{r^2},$$

where $\langle\Phi\rangle_{r_0}$ is the mean energy flux penetrating a spherical surface of radius r_0 centered on the source. (Hint: First consider the attenuation of the flux. Then consider the way the flux is spread over a spherical surface of radius r, and find I.) **(b)** For moist air, the attenuation coefficient for sound in the low-frequency range (where thunder is usually heard) is approximately $9 \times 10^{-4}\ \text{m}^{-1}$. Suppose that two observers watch a lightning flash. Observer O, 2 km from the flash, hears a thunderclap of acoustic intensity 100 dB. What is the acoustic intensity heard by observer O', who is 4 km from the flash? Is the diminution in acoustic intensity between the two observers due mostly to the inverse-square law or mostly to the absorption of the air? (Note: The attenuation coefficient for high frequencies is much greater than that for low-frequency sound. That explains why distant thunder is always heard as a low-pitched rumble. High-pitched thunder is heard only from very nearby lightning flashes.)

Part VIII

Electrostatics

On the scale of planets—and on any larger scale, such as the scale of galaxies—the gravitational force is the "glue" that holds the universe together. But on any smaller scale the gravitational force is weak (we feel the gravitational attraction of the earth only because the earth has very large mass and we are very close to it). Yet we, and objects about our size or smaller, do not fall apart spontaneously. Thus there must be a force, much stronger than gravitation on our own scale of size, that provides the glue that holds everyday things together.

It turns out that a single force, the *electromagnetic force*, dominates over the entire range of size from objects about as big as an atom to objects quite a bit bigger than ourselves (but not as large as a small astronomical body). It is this force, in one form or another, that keeps a lump of steel, say, from disintegrating or collapsing in on itself. The electromagnetic force is manifested in the "contact force" that we exert when we push on something and in the frictional forces on sliding bodies. The electromagnetic force is the intermolecular binding force that must be overcome when a liquid boils.

On a still smaller scale—for things about the size of atomic nuclei or smaller—other fundamental forces called the *strong* and *weak nuclear forces* come into play and dominate. We will consider these forces briefly in Part XII.

The electromagnetic force is the subject of Parts VIII and

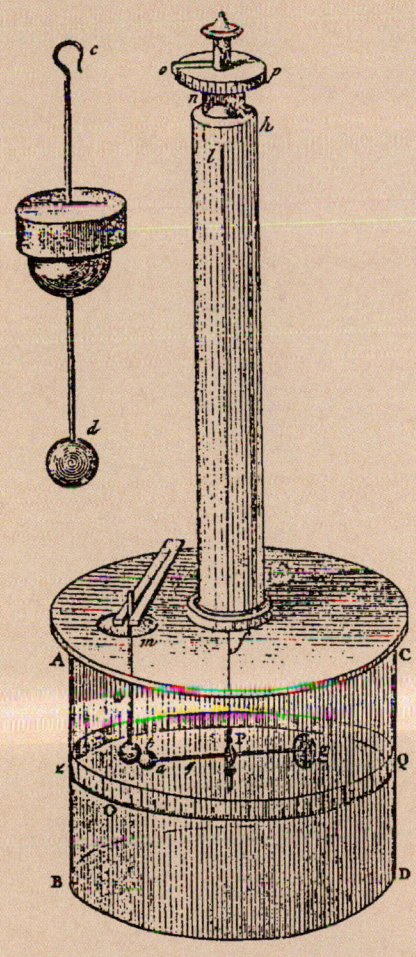

IX. In the same way that the gravitational force arises from the fundamental property of matter called *mass*, the electromagnetic force arises from an equally fundamental property of matter called *electric charge*. Part VIII deals with *electrostatics*—the study of how two or more electric charges interact when they are at rest. We begin our study of electric charge in Chapter 23 by quantifying it. Under static conditions, the electromagnetic force has an especially simple form and is called the *electric force*. Like the gravitational force, the electric force is central and conservative. These properties enable us to develop powerful methods of treating situations in which the electric force acts. In Chapter 24 we elaborate on the concept of *electric field*, and in Chapter 25 we develop the related concept of *electric potential*. In Chapter 26 we consider an important practical device called the *capacitor*.

In addition to having intrinsic interest, the study of electrostatics prepares the way for our study of electric charges in motion. The latter subject, dealt with in Part IX, is called *electromagnetism*.

The Electric Force and the Electric Field

- The force exerted by one stationary electric charge on another—the electrostatic force—is one of the fundamental forces of nature.

- Electric charge is immobile in dielectric materials but mobile in conducting materials.

- There are two kinds of electric charge, arbitrarily called positive and negative. Bodies with like charges repel each other; bodies with unlike charges attract each other.

- In processes that create and destroy electric charge, the algebraic sum of the total charge remains constant; this is the principle of charge conservation.

- The force exerted by one small charged body on another is proportional to the product of the charges on the bodies and inversely proportional to the square of the distance between them.

- At the level of atoms and above, electric charge exists only in integral multiples of a small, quantized unit denoted as e.

- In the vicinity of a source charge q, a small test charge q_t experiences a force F. The electric field at the location of q_t is defined to be F/q_t.

- At a given point, the electric field is the vector sum of the fields arising from the individual source charges in the vicinity; this is the superposition principle.

Left: Static cling can be a nuisance, but the electric force is one of the fundamental forces of nature.

Communism is Soviet government plus the electrification of the whole country.

—V. I. Lenin (1920)

I sing the body electric,
The armies of those I love engirth me . . . ,
They will not let me off till I
. . . charge them full with the charge of the soul.

—Walt Whitman, Leaves of Grass

SECTION 23.1   Introduction

The applications of electricity have made an impact far beyond the realm of physics itself. In this chapter, however, we focus on the physics of **electrostatics**, the study of the interactions between electric charges that are at rest (or at most moving very slowly). The ancient Greeks knew of the attractive force that a piece of amber* rubbed with a cloth can exert on bits of matter. Indeed, the Greek word for amber, *elektron*, is the source of the modern word *electric*. But the scientific study of electricity did not begin until 1600, when the English physician William Gilbert (1544–1603) published his pioneering work *De Magnete* (On the Magnet). On the basis of a series of carefully designed experiments, Gilbert distinguished clearly between *electric* and *magnetic* forces. By means of experiments on static systems, he showed conclusively that, in spite of intriguing similarities (to which we will return in Part X), the two forces were quite distinct.

By 1750 or so, all the basic qualitative facts concerning the electric force were known, and devices had been invented for the improved production, storage, and detection of electric charge. Section 23.2 concerns the qualitative behavior of the electric force. Further progress required a quantitative approach, of the kind that had proved so spectacularly successful in the study of mechanics. The revolutionary quantitative work of Coulomb, discussed in Section 23.3, was the foundation of the modern theory of

THINKING LIKE A PHYSICIST

We know today that, though the electric and magnetic forces are distinct, they are intimately related. That is why we speak of electromagnetism as a unified subject. However, Gilbert's distinction between the two forces in static systems was an absolute prerequisite to the quantitative study of both forces. Only on the basis of quantitative understanding of electricity and magnetism in static systems was it possible to understand their unification at a deeper level.

Electromagnetism is one of the parts of physics in which the path of learning runs more or less parallel to the path of historical development. First, we distinguish between *electrostatics* and *magnetostatics* and study the two subjects separately. Then we study the remarkable connection between electricity and magnetism that becomes clear in *moving* systems. Finally, we come to see how electricity and magnetism are two manifestations of the same basic phenomenon, electromagnetism.

---

*Amber, often used in jewelry, is the beautiful translucent yellow fossilized resin of ancient plants.

electrostatics, which is the subject of the rest of Part VIII. His results, expressed in *Coulomb's law*, establish the relation connecting the quantity of electric charge, the distance between two charges, and the force exerted on each by the other, thus making possible the definition of a unit of charge. In Section 23.4, we consider the *electric field* associated with point charges. Section 23.5 is concerned with calculating the electric field due to a continuous distribution of charges.

Electric Charge: Qualitative Effects

Objects made of many substances besides amber can be **electrically charged** by rubbing them against dissimilar objects. We can observe the charged state because charged bodies exert an electric force on other bodies—for example, they tend to attract other bits of matter. Bodies that can be readily charged by rubbing are called **dielectrics** or **insulators**.

The electric force is often attractive; this is evident to anyone who has removed clothing from a dryer and experienced "(electro)static cling." It is a little more difficult to discover that the electrostatic force can be repulsive as well as attractive. Only in 1733, more than a century after Gilbert, was electrostatic repulsion clearly demonstrated by the French physicist Charles du Fay (1698–1739).

Dielectrics

Figure 23.1 shows three idealized experiments that demonstrate the properties of the electric force. In part *a*, the silk cloth that has been used to rub the suspended glass rod is brought near the rod. The rod swings toward the cloth, implying that the cloth exerts an attractive force on the rod. In part *b*, both glass rods have been rubbed with silk cloths. The suspended rod swings away from the hand-held rod, implying that the latter exerts a repulsive force on the former. In part *c*, a glass rod that has been rubbed with a silk cloth is brought close to bits of tissue paper. The bits of paper fly up to the rod and cling to it for a time, implying that the rod exerts an attractive force on the paper.

These experimental results (and others like them) lead us to the following general rules:

1. When two bodies are rubbed together, it is not possible to create electric charge on only one of them. Either neither is charged or both are.

2. Two bodies rubbed together and thus charged exert *attractive* forces on one another (Figure 23.1a).

3. When two bodies made of the same material are charged in the same way (say, by rubbing with the same kind of cloth), each exerts a *repulsive* force on the other (Figure 23.1b).

4. The force exerted by a charged body on an uncharged body is always *attractive* (Figure 23.1c).

These and many other qualitative phenomena can be explained in a consistent way on the basis of a phenomenological theory called (for historical reasons) the **two-fluid theory**. The theory rests on a basic assumption: All matter is permeated with two kinds

FIGURE 23.1 (*a*) When a glass rod is rubbed with a silk cloth, both the rod and the cloth exhibit the electric force. When the rubbed rod is hung from a thread and the cloth is brought near it, the rod swings toward the cloth, demonstrating **electrostatic attraction**. (*b*) Two glass rods are rubbed with silk cloths. One is hung as in part *a*. When the other is brought nearby, the hanging rod swings away, demonstrating **electrostatic repulsion**. (*c*) When a charged rod is brought close to bits of tissue paper or sawdust, the bits are initially attracted to and cling to the rod. After a short time they jump off, being vigorously repelled from the rod.

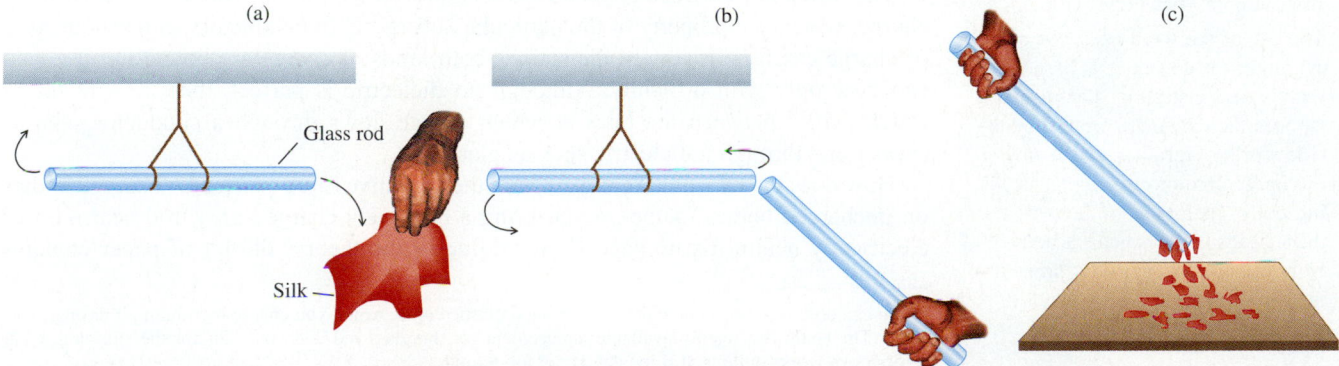

(a)   (b)   (c)

Glass rod

Silk

of electric charge (the two "fluids"). Following a convention established by Benjamin Franklin, we call the two kinds of charge **positive** and **negative**. When the two are present in a body in equal amounts, a body is electrically **neutral**; that is, it does not exhibit the behavior of charged bodies described in the preceding list. If an excess of either positive or negative charge is present, the body is **positively** or **negatively charged** and does exhibit such behavior. Which kind of charge we call positive and which negative is entirely arbitrary, but Franklin's convention has been universally adopted: When a glass rod is rubbed with a silk cloth, we say that the glass becomes *positively* charged and the silk becomes *negatively* charged.

The phenomenon of charging by rubbing is readily explained on the basis of the two-fluid theory. We argue that rubbing results in a net transfer of one kind of charge from one body to the other. This leaves one of the bodies with an excess of positive charge and the other body with an excess of negative charge. Implicit in this picture is an important point: *The charging process neither creates nor destroys charge; it only redistributes it.* This is one statement of the **principle of conservation of charge**. On the basis of the evidence that we have considered so far, the principle is largely speculative. But there is ample justification for it on the microscopic scale.

Conductors

Another crucial set of observations was made in 1729 by the English investigator Stephen Gray (d. 1736). Many substances (notably metals) *cannot* be charged by holding them in one hand and rubbing them with a cloth held in the other hand. Gray found that these substances *conduct* electric charge—that is, allow it to move freely from one part of the body to another. If one end of a long metal wire is attached to a charged dielectric, the electric force can be observed at the other end of the wire (Figure 23.2).

FIGURE 23.2 A long, stiff metal wire, which cannot be charged by the process shown in Figure 23.1a, is fastened to a glass rod. The rod is then charged by rubbing it with silk. When the free end of the wire is brought near the hanging rod, which also is charged, the hanging rod is repelled.

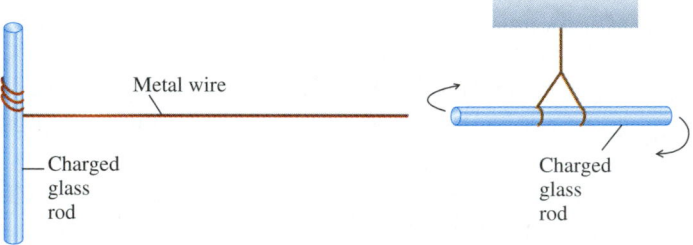

Metal wire

Charged glass rod

Charged glass rod

It is easy to see why an **electrical conductor**—a body made of a substance that conducts electric charge—cannot be charged by rubbing in the ordinary way. Suppose, for example, that you hold a lump of copper in one hand and rub it against a piece of glass held in your other hand. Electric charge that is transferred to the copper surface where it rubs against the glass can flow away through the copper and then through your body (also a conductor) to the ground, thus dispersing so widely that no electric force can be detected. Indeed, Gray showed that conductors *could* be charged by rubbing, if they were supported on good dielectrics.

The difference between dielectrics and conductors has to do with the *mobility* of the charge, which is a property of the particular substance. In conductors, at least one kind of charge can flow freely. In dielectrics, both kinds of charge are bound in place and can flow only with difficulty. Although no dielectric is perfect, there is a factor of roughly 10^{20} between the rates at which charge flows through a conductor such as copper and through a dielectric such as glass.

How does the two-fluid theory explain the attractive force exerted by charged bodies on uncharged bodies? Suppose you bring a positively charged glass rod near a bit of electrically neutral tissue paper.* According to the theory, the bit of paper contains

---

*You can do this experiment at home by using a plastic comb, which you charge by running it through your hair. (The comb is a readily available replacement for the glass rod and your hair for the silk cloth.) The experiment works quite well if the day is not too humid.

positive and negative charge in equal quantities. The net positive charge on the rod repels the positive charge in the paper and attracts the negative charge in the paper. Because the charge in the paper is slightly mobile, it flows until there is a slight excess of negative charge at the end of the paper nearest the rod and a slight excess of positive charge at the other end.

The force exerted on one charge by another depends on the distance between them. We will consider this point quantitatively in Section 23.3, but for the moment we need only make the plausible assertion that the force diminishes with distance. The magnitude of the attractive force exerted by the positive charge in the rod on the nearby negative charge in the paper is greater than the magnitude of the repulsive force exerted by the charge in the rod on the more distant positive charge on the paper. Thus there is a net attractive force on the paper. If the force is large enough, the paper will jump to the rod.

After it has remained in contact with the rod for a while, the paper jumps off quite forcefully. (You can readily convince yourself that the paper is not merely falling off the rod; if it is located on the upper surface of the rod, the paper will jump upward before falling down.) The repulsion of the paper by the rod occurs because positive charge leaks slowly from the rod onto the paper until the paper has a net positive charge, like that on the rod. The net force on the paper is then repulsive and propels the paper away from the rod.

Microscopic Basis for the Electric Force

The phenomenological macroscopic theory just described can be justified in terms of a microscopic picture of matter developed much later, beginning at the very end of the nineteenth century. Here we present the bare essentials of the microscopic picture. We will study it in much more detail in Part XII.

We have already considered (especially in Chapter 18) some of the direct evidence that matter is composed of atoms. But an atom is made up of still smaller components. Specifically, an atom contains a tiny *nucleus* whose diameter is of order 10^{-15} m. The nucleus comprises **neutrons** and **protons**, whose numbers depend on the particular type of atom. The nucleus is surrounded by a relatively large "cloud" of **electrons**. This cloud, whose diameter is of order 10^{-10} m—100 000 times that of the nucleus—makes up most of the volume of the atom. When the atom is in its normal (electrically neutral) state, the number of electrons in the cloud is equal to the number of protons in the nucleus.

Neutrons, protons, and electrons are the most stable, and therefore the best known, of the *fundamental particles*. All three possess the inherent property we have called *mass*. But, in addition, electrons and protons possess another, *equally inherent* property called **electric charge** or **charge** for short. Like mass, charge has a magnitude; we will develop a way to quantify charge in Section 23.3.

By direct experiment, we find that electrons exert repulsive electric forces on one another. The same is true for protons. Now, like particles certainly possess like charges (if they are charged at all). It follows that *particles possessing like charges repel one another*. But a proton and an electron exert attractive forces on one another. In view of the behavior of like charges, we must conclude that (1) protons and electrons possess *unlike charges* and (2) *particles possessing unlike charges attract one another*. Indeed, it is this electrical attraction that holds the atom together. The electrical neutrality of the normal atom results from its equal numbers of protons and electrons having *equal and opposite charge*.

It turns out that the sign of the charge on the proton is the same as that on Franklin's silk-rubbed glass. Consistent with this established convention, we say that *the charge of the proton is positive and the charge of the electron is negative*. An atom that loses one or more electrons acquires a net positive charge and is called a **positive ion**. An atom that gains one or more extra electrons acquires a negative charge and is called a **negative ion**.

Electrostatic Copier

In a typical copier, a metal drum is thinly coated with the semiconductor selenium. The drum is in a dark chamber, so the selenium is in the dielectric state. At the start of the process, the selenium is charged. A lens then projects onto the drum the image to be copied. Where light falls on the drum, the selenium becomes a conductor and the charge quickly leaks off via the metal drum. The drum then has charged and uncharged areas corresponding to the dark and light areas of the original image. Next the drum is pressed against a mat made of a dielectric material. Part of the charge, carrying the image, transfers to the surface of the mat. The mat is brought into contact with a toner, a mixture of carbon black and a waxy substance. The toner clings to the charged areas but falls off the uncharged areas. The mat is then pressed to the copy paper, which is heated to melt the toner. The toner soaks into the paper, forming a permanent image.

Conductors, Dielectrics, and Semiconductors

In solids, the atoms are firmly bound into position. Electrical conduction becomes possible when some of the electrons are bound only weakly to their atoms. In metals, some electrons are entirely free of individual atoms and are bound only to the metal as a whole. Such *free electrons* are responsible for the very good conductive properties of most metals. In most oxides and sulfides, on the other hand, all of the electrons are tightly bound, and these substances are good dielectrics.

In most familiar liquids, the situation is different. Water with ordinary salt (NaCl) dissolved in it is typical of *ionic solutions*. The salt is present in solution in the form of positive sodium ions and negative chlorine ions. Both types of ions are mobile and can carry electric charge through the solution, which is therefore a conductor. Here ions transport electric charge, rather than electrons, and the phenomenon is called *ionic conduction*. Whether they are electrons or ions, the microscopic particles that transport electric charge through a substance are called **charge carriers**.

The molecules of gases can be ionized as well, and ionized gas can conduct electric charge. Every electron torn loose from an atom is a charge carrier. The positive ion left behind is also a charge carrier, though its much greater mass renders it less efficient than an electron in this respect. Although it can conduct electric charge, the ionized gas remains electrically neutral as a whole. Ionized but electrically neutral fluids constitute a distinct state of matter and are called **plasmas**. The neon inside the familiar neon tube is in the plasma state; so are the luminous matter at the surface of the sun and the part of the earth's upper atmosphere called the ionosphere.

In most solids, electrons are either quite free or quite firmly bound, and most solids therefore fall squarely into the categories of conductors and dielectrics. The class of substances called **semiconductors** is an important exception. Silicon is the most familiar semiconductor. Depending on a variety of circumstances, a semiconductor can be a good dielectric, a good conductor, or something in between.

SECTION 23.3

Coulomb's Law

How can we express quantitatively the electric force **F** exerted by one charged body on another? We already know how the direction of the force depends on the signs of the two charges. It is reasonable to assume that the magnitude of the force depends on the quantity q_1 of charge on body 1 and on the quantity q_2 of charge on body 2.* We also know that the force depends on the distance between the charged bodies. We invoked such a dependence in Section 23.2 to account for the observation that a charged body can attract a neutral body.

To make the study of the distance dependence as simple as possible, consider the force exerted by *point charge* 1 on *point charge* 2. A **point charge**—a finite charge located at a geometric point—is an idealization analogous to that of a point mass (a particle). If we consider two point charges, we can precisely define the vector $\mathbf{r}_{1\to2}$ that extends from body 1 to body 2. (We approximate this ideal situation when the distance between the two charged bodies is large compared with the size of either.) At this juncture, we are tempted to draw on our experience with the gravitational force. We must do so tentatively, however, because argument by analogy must always be checked directly before it can be taken seriously. With this reservation in mind, let us take as a model Newton's law of gravitation in the form of Equation 14.3a,

$$\mathbf{F}_{1\to2} \propto -\frac{m_1 m_2}{r^2}\,\hat{\mathbf{r}}_{1\to2},$$

---

*We must be careful here. We have not yet made clear what we mean by "quantity of charge" and have thus left a gap in our argument. We must (and will) fill this gap soon.

where r is the distance between the two bodies. We argue that the electric charge q takes a role in determining electric forces analogous to the role the mass m takes in determining gravitational forces. If the electric force is likewise an inverse-square force, it should have the form

$$\mathbf{F}_{1 \to 2} \propto \frac{q_1 q_2}{r^2} \, \hat{\mathbf{r}}_{1 \to 2}. \qquad (23.1)$$

In this proportionality, the direction of $\mathbf{F}_{1 \to 2}$ takes care of itself if we attach to each q the proper sign of charge. For like charges ($++$ or $--$), we have $\hat{\mathbf{F}}_{1 \to 2} = \hat{\mathbf{r}}_{1 \to 2}$; for unlike charges ($+-$ or $-+$), $\hat{\mathbf{F}}_{1 \to 2} = -\hat{\mathbf{r}}_{1 \to 2}$.

Proportionality 23.1 is called **Coulomb's law** because in 1785 Coulomb verified that it was a correct description of the electric force by means of a series of beautiful experiments using a torsion balance. Because of the importance of Coulomb's work, the electric force is often called the **Coulomb force**.

Coulomb's Experiment*

Figure 23.3 is a picture of Coulomb's electrostatic torsion balance, taken from his 1785 paper. The entire balance is shown in Coulomb's Fig. 1. The torsion fiber f, made of fine bronze wire, hangs from the small knob at the top, shown as part no. 1 in the

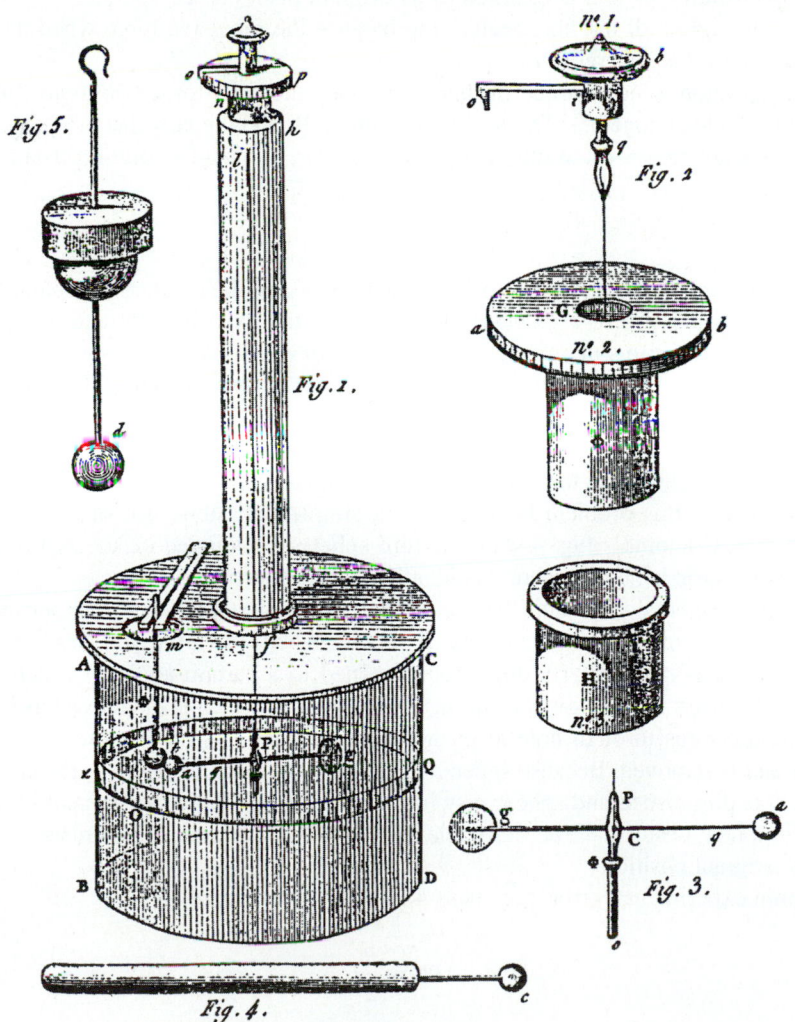

FIGURE 23.3 The torsion balance used by Coulomb to verify Equation 23.1.

___
*This subsection may be omitted without loss of continuity.

exploded detail of Fig. 2. The knob b has a pointer o, which can be rotated around the angle scale ab on the collar (part no. 2). The angle scale can be zeroed by rotating the collar with respect to part no. 3, which is the top of the tall chimney. The torsion fiber supports the beam at P, as shown in detail in Fig. 3. The beam consists of a thin rod q made of solid shellac, which Coulomb chose for its relative strength, lightness, and good dielectric properties. At one end of the beam is a sphere a made of the light pith of elder bush covered with gold leaf. At the other end is the cardboard counterbalance g. The entire torsion balance assembly is mounted on the flat cover AC, which rests on the glass cylinder BD. Marked on the periphery of the cylinder is a second angle scale OQ which can be used to read the orientation of the balance beam. This scale can be zeroed by rotating the cover AC. A V-clamp is shown holding a dielectric wand, which hangs through the hole m. At the end of the wand is a sphere b, identical with sphere a on the balance beam. Compare Coulomb's apparatus with the Cavendish balance of Figure 14.9.

The experiment has two parts. The first part concerns the dependence of the electric force on distance. Both angle scales are zeroed, and the distance r between spheres a and b is measured. The wand is removed from its clamp, and sphere b is given an electric charge. The wand is then returned to the position shown in Figure 23.3. Uncharged sphere a is attracted to charged sphere b, touches it, and immediately acquires a like charge from it. As a result, sphere a is repelled and rotates to a new position farther from b than its original position. The top knob is twisted, increasing the torque in the fiber until sphere a is returned to its original position. The pointer o now reads the torsion angle θ of the fiber required to balance the repulsive force when the separation between the spheres is r.

The top knob is now turned further, increasing the fiber torque so as to force the two spheres closer together. The new torsion angle θ' and the new distance r' between the spheres are measured. Within experimental error, the results conform to the rule

$$\frac{\theta'}{\theta} = \left(\frac{r}{r'}\right)^2. \tag{23.2}$$

The torsion angle is proportional to the torque applied to the balance beam and thus to the force holding sphere a in place. The fact that the experimental data conform to Equation 23.2 confirms the *inverse-square behavior of the electric force*.

The second part of the Coulomb experiment aims at demonstrating that the electric force is directly proportional to each of the charges q. To do this, it suffices to show that the force is proportional to either one of the charges—say, to q_1.

In doing this, we face the difficulty that we have as yet no way of quantifying q. We circumvent this problem by means of a simple but subtle use of symmetry. In Figure 23.3, Coulomb's Fig. 4 shows a third sphere, c, mounted on the end of a convenient dielectric handle. Sphere c is identical with sphere b.

The experimental procedure begins just as in the first part of the experiment. A charge is put on sphere b, and uncharged sphere a swings toward it, touches it, acquires a charge, and is repelled. The top knob is turned to a certain angle θ, which causes sphere a to return to its original position. Now sphere c, which is uncharged, is inserted into the apparatus through hole m in the cover. Sphere c is made to touch sphere b briefly and is removed. Because sphere c acquires charge from sphere b, the charge on sphere b is diminished and sphere a swings toward it. A second adjustment of the top knob, to a new position θ', partially relaxes the torsion fiber and again restores sphere a to its original position.

Within experimental error, the result is always

$$\theta' = \frac{\theta}{2}; \tag{23.3}$$

the repulsive force is halved owing to removal of some of the charge from sphere b. But how much charge was removed? In view of the essential identity of spheres b and *c, half the charge must have been removed.* So we have

$$F \propto q_1,$$

in agreement with Equation 23.1. The reasoning leading from $F \propto q_1$ to the conclusion that $F \propto q_1 q_2$ is parallel to that leading from the gravitational relation $F \propto m_1$ to Equation 14.2b, $F \propto m_1 m_2$.

The Coulomb experiment is remarkably similar in concept to the Cavendish experiment considered in Section 14.4. But the experimental difficulties that need to be overcome are quite different. In the Cavendish experiment, nearly all the difficulties arise from the weakness of the gravitational interaction, which dictates the measurement of extremely small forces. The electrostatic interaction is much stronger, and the magnitude of the forces in the Coulomb experiment does not present great measurement difficulties. Rather, the big problem is charge leakage.

The SI Unit of Charge

We are now in a position to define a unit of charge. We begin with Coulomb's law in the form of Proportionality 23.1,

$$\mathbf{F}_{1 \to 2} \propto \frac{q_1 q_2}{r^2} \, \hat{\mathbf{r}}_{1 \to 2}.$$

As we did for Newton's law of gravitation, we would like to convert this proportionality into an equation. In the analogous gravitational proportionality,

$$\mathbf{F}_{1 \to 2} \propto -\frac{m_1 m_2}{r^2} \, \hat{\mathbf{r}}_{1 \to 2},$$

the units of all the quantities are defined, and all that is necessary is to determine experimentally the numerical value of the proportionality constant G. Equation 14.3b,

$$\mathbf{F}_{1 \to 2} = -G \, \frac{m_1 m_2}{r^2} \, \hat{\mathbf{r}}_{1 \to 2},$$

then determines the units of G, which in SI must be $N \cdot m^2 / kg^2$.

To express Coulomb's law as an equation, we must deal with an additional problem. When we developed Newton's law of gravitation, we had already defined mass as one of the basic dimensions—mass, length, and time—of SI and had established a unit of mass, the kilogram. Electric charge q, which appears in Proportionality 23.1, cannot be expressed in terms of these basic dimensions. Consequently, we must expand our set of basic dimensions to include a fourth dimension, *electric charge*, for which we must then establish an SI unit. With this in mind, we proceed in two steps:

1. As a preliminary to defining it, we give the unit of electric charge a name, the **coulomb** (C). That is, the quantities q_1 and q_2 that appear in our desired equation will be expressed in coulombs. With this in mind, we use a proportionality constant to rewrite Coulomb's law as an equation:

$$\mathbf{F}_{1 \to 2} = (\text{constant}) \, \frac{q_1 q_2}{r^2} \, \hat{\mathbf{r}}_{1 \to 2}.$$

From this equation, it follows that the SI units of the proportionality constant are $N \cdot m^2 / C^2$. For reasons that will soon become clear, we write the constant in the form

$$(\text{constant}) \equiv \frac{1}{4 \pi \epsilon_0}.$$

The quantity ϵ_0 (''epsilon-zero'') that appears in this definition is called the **permittivity of free space**. (You will see the reason for this name in Section 26.4.) We can now write Coulomb's law in its most frequently used form,

$$\mathbf{F}_{1 \to 2} = \frac{1}{4 \pi \epsilon_0} \, \frac{q_1 q_2}{r^2} \, \hat{\mathbf{r}}_{1 \to 2}. \tag{23.4}$$

To be sure, writing the proportionality constant in the form $1/4\pi\epsilon_0$ is somewhat awkward. But the awkwardness is a small price to pay for later profit: The proportionality constant in other equations, derived from Coulomb's law but used much more frequently, will turn out to be 1.

2. We *define* the value of the proportionality constant:

$$\frac{1}{4\pi\epsilon_0} \equiv 10^{-7}c^2 \text{ N·m}^2/\text{C}^2, \qquad (23.5a)$$

where c is the speed of light in m/s. (The rationale for defining $1/4\pi\epsilon_0$ in terms of the speed of light lies in the fundamental connection between electric charge and *electromagnetic radiation*, an important phenomenon of which light is the most familiar example. We will consider electromagnetic radiation in Chapter 34.) Because c is defined rather than measured (Section 1.2), it is an *exact* quantity: $c \equiv 2.997\,924\,58 \times 10^8$ m/s. Thus, according to Equation 23.5a, $1/4\pi\epsilon_0$ is exact as well. For many purposes, it suffices to use the rounded value

$$\frac{1}{4\pi\epsilon_0} = 8.99 \times 10^9 \text{ N·m}^2/\text{C}^2. \qquad (23.5b)$$

It follows immediately that the value of ϵ_0, rounded to three significant figures, is

$$\epsilon_0 = 8.85 \times 10^{-12} \text{ C}^2/\text{N·m}^2. \qquad (23.6)$$

With the value of ϵ_0 thus fixed and F and r expressed in SI units, the magnitude of the coulomb is determined by Equation 23.4, as Example 23.1 shows.

EXAMPLE 23.1

One coulomb is quite a lot of charge on the everyday scale of things. To see that this is so, **(a)** find the magnitude of the force exerted on a 1-C charge by a second 1-C charge located 1 km away; **(b)** find the magnitude of two equal charges located 1 m apart if they exert forces of magnitude 1 N on one another.

SOLUTION:
(a) Find the magnitude of the force exerted on a 1-C charge by a second 1-C charge located 1 km away.

Because you are interested only in magnitudes, you write Equation 23.4 in the form

$$F = \frac{1}{4\pi\epsilon_0}\frac{q_1 q_2}{r^2}. \qquad (23.7)$$

Inserting the numerical values, you have

$$F = 8.99 \times 10^9 \text{ N·m}^2/\text{C}^2 \times \frac{(1 \text{ C})^2}{(1 \times 10^3 \text{ m})^2} = 8990 \text{ N}.$$

This is about equal to the weight of a 1-tonne mass—that is, the gravitational force exerted on such a mass by the earth, whose mass is of order 10^{29} kg.

(b) Find the magnitude of two equal charges located 1 m apart if they exert forces of magnitude 1 N on one another.

Because you have $q_1 = q_2 \equiv q$, you can rearrange Equation 23.7 to obtain

$$q = \sqrt{4\pi\epsilon_0 F}\, r = \sqrt{\frac{1}{8.99 \times 10^9 \text{ N·m}^2/\text{C}^2} \times 1 \text{ N} \times 1 \text{ m}}$$
$$= 1.05 \times 10^{-5} \text{ C} = 10.5 \text{ μC}.$$

In SI units, the numerical value of the charge required to produce a substantial force is quite small.

How Much Is a Quantum?

Quantum is a Latin word meaning "so much" or "this much." While the quantum of charge has magnitude $|e|$, protons and many other "fundamental" particles are made up of even more fundamental particles called **quarks**. Various kinds of quarks have charges $\pm\frac{1}{3}e$ and $\pm\frac{2}{3}e$. To date, free quarks have not been observed. In any case, the charges on quarks are quantized, even though the quantum is not e.

The Elementary Charge and the Principle of Charge Quantization

In Problem 26.46 we will discuss an experimental method for measuring the charge of an electron. The magnitude of that charge is universally assigned the symbol e. The best modern value is given in Appendix 4; we will usually use the rounded value

$$e = 1.60 \times 10^{-19} \text{ C}. \qquad (23.8)$$

Because the electron is negatively charged, each electron carries a charge $-e$. The charge on the proton has the exactly opposite value, $+e$. The quantity e is called the **elementary charge** or the **quantum of charge**—that is, the irreducible "building block" of electric charge, because every known free particle has charge ne, where n is an integer (positive, negative, or zero). This observation is so general that it is called the **principle of charge quantization**.

How "big" is the elementary charge e? Example 23.2 places this question in a meaningful context.

EXAMPLE **23.2**

The mass of an electron is $m_e = 9.11 \times 10^{-31}$ kg; that of a proton is $m_p = 1.67 \times 10^{-27}$ kg. Find the ratio F_e/F_g of the electric force and the gravitational force exerted by the proton on the electron (or vice versa).

SOLUTION: Because the electron and the proton have opposite charge, the Coulomb force is attractive, as the gravitational force always is. Thus the two forces act in the same direction, and you need consider only their magnitudes. From Equation 14.3b and Equation 23.4, you have

$$F_g = G\frac{m_e m_p}{r^2} \quad \text{and} \quad F_e = \frac{1}{4\pi\epsilon_0}\frac{e^2}{r^2}.$$

Taking the ratio, you have

$$\frac{F_e}{F_g} = \frac{1/4\pi\epsilon_0}{G}\frac{e^2}{m_e m_p}. \tag{23.9}$$

Note that this ratio does not depend on the distance between the electron and the proton; because both forces are inverse-square forces, their ratio is the same at all distances. When you insert the numerical values, you obtain

$$\frac{F_e}{F_g} = \frac{8.99 \times 10^9 \text{ N·m}^2/\text{C}^2}{6.67 \times 10^{-11} \text{ N·m}^2/\text{kg}^2}$$

$$\times \frac{(1.60 \times 10^{-19} \text{ C})^2}{9.11 \times 10^{-31} \text{ kg} \times 1.67 \times 10^{-27} \text{ kg}}$$

$$= 2.27 \times 10^{39}.$$

The electric force between the two particles is so much greater than the gravitational force that the latter need never be taken into consideration. Indeed, no conclusive measurement of gravitational force exerted on an electron has yet been made, though there is some evidence that an electron near the surface of the earth would fall with acceleration g if no other forces were acting on it.

The force that holds atoms together is the electric attraction between the positively charged nucleus and the electron cloud surrounding it. The force that holds atoms together in larger bodies is likewise electric, though the details are quite complicated.

EXAMPLE **23.3**

In his experiment for determining the universal gravitational constant G, Cavendish had to take great care that his results were not distorted by unwanted forces arising from small electric charges on the test masses of his torsion balance. Using the data given in the legend for Figure 14.9, find the amount of charge that, if present on each of the two swinging masses m and on each of the two large stationary masses M, would introduce a 10% error.

SOLUTION: The ratio of the forces is given by Equation 23.9. To fit the present situation, you change the notation; call the charge on each of the four bodies q. You then have the force ratio

$$\frac{F_e}{F_g} = \frac{1/4\pi\epsilon_0}{G}\frac{q^2}{Mm} = 0.1.$$

Solving for q, you obtain

$$q = \sqrt{\frac{0.1 GMm}{1/(4\pi\epsilon_0)}}$$

$$= \sqrt{\frac{0.1 \times 6.67 \times 10^{-11} \text{ N·m}^2/\text{kg}^2 \times 155 \text{ kg} \times 0.730 \text{ kg}}{8.99 \times 10^9 \text{ N·m}^2/\text{C}^2}}$$

$$= 2.9 \times 10^{-10} \text{ C}.$$

If you divide this result by e, you find that q is the magnitude of the charge on approximately 2×10^9 excess electrons—a truly small charge on the macroscopic scale, where particle counts are on the scale of Avogadro's number, 6×10^{26}. This point is pursued in more detail in Problem 23.26.

The gravitational interaction is the dominant interaction in the universe on the scale of astronomical bodies, in spite of its great weakness compared with the electrostatic interaction. Why is this? The very strength of the electrostatic interaction implies a strong tendency for macroscopic bodies to become electrically neutral. If a body has even a slight excess positive charge, it tends to attract electrons from its surroundings until it is neutral. In particular, no electrostatic corrections at all need be made in the most precise calculations of the motions of the planets.

The Electric Field: Force per Unit Charge

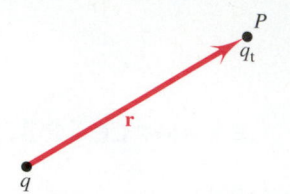

FIGURE 23.4 The location of P with respect to the point source charge q is given by the position vector $\mathbf{r}$.

So far, we have considered only the electric force exerted by one of two point charges on the other. But we frequently need to calculate the force exerted on a point charge by several other charges. We approach this problem by considering again the simplest case of two point charges.

In Figure 23.4, the point P lies in the vicinity of a charge q, which we call a **source charge** because it is the source of the force that will be exerted on any charge placed at P. The position of P with respect to q is given by the vector $\mathbf{r}$. Suppose that we place another charge at P. We call this charge the **test charge** q_t because we use it to test the influence of the source charge. By convention, *the test charge is always positive*. What force $\mathbf{F}$ will be exerted on q_t? As you know from Coulomb's law, the answer depends on the value of q_t. Given q and $\mathbf{r}$, we can predict $\mathbf{F}$ for *any* q_t. To emphasize this point, we define the **electric field** $\mathscr{E}$ to be the *force per unit test charge*:

$$\mathscr{E} \equiv \frac{\mathbf{F}}{q_t}. \qquad (23.10)$$

It follows immediately that the SI unit of electric field is N/C. According to Coulomb's law (Equation 23.4), we have

$$\mathscr{E} = \frac{1}{4\pi\epsilon_0} \frac{q}{r^2} \hat{\mathbf{r}}. \qquad (23.11)$$

Because of the way we have defined $\mathscr{E}$, its value is independent of the value of the test charge. Indeed, we will argue later that the electric field exists at P whether or not a test charge is present there. Nevertheless, in order to detect a field, we must place a test charge in it. Then the field reveals its presence by the force exerted on the test charge. From this point of view, we can write the electric force in the form

$$\mathbf{F} = q_t \mathscr{E}. \qquad (23.12)$$

The force vector is parallel or antiparallel to the field vector, depending on whether q_t is positive or negative.

Let us apply the concept of electric field to evaluating the force exerted on a test charge placed in the vicinity of a number of point source charges $q_1, q_2, \ldots, q_n$. Like gravitational forces, electric forces exerted simultaneously on the same body add vectorially. Each source charge exerts a force on q_t independent of the others. For example, the force $\mathbf{F}_1$ exerted on q_t owing to the presence of q_1 is just the same as if all the other charges were absent. This statement, which is confirmed by experiment, is called the **superposition principle**.* And $\mathbf{F}_1$ can be calculated by means of Coulomb's law (Equation 23.4):

$$\mathbf{F}_1 = \frac{1}{4\pi\epsilon_0} \frac{q_1 q_t}{r_{1\to t}^2} \hat{\mathbf{r}}_{1\to t}.$$

Thus the resultant force $\mathbf{F}$ exerted on the test charge is given by the sum

$$\mathbf{F} = \mathbf{F}_1 + \mathbf{F}_2 + \mathbf{F}_3 + \cdots + \mathbf{F}_n$$

$$= \frac{1}{4\pi\epsilon_0} \left(\frac{q_1 q_t}{r_{1\to t}^2} \hat{\mathbf{r}}_{1\to t} + \frac{q_2 q_t}{r_{2\to t}^2} \hat{\mathbf{r}}_{2\to t} + \frac{q_3 q_t}{r_{3\to t}^2} \hat{\mathbf{r}}_{3\to t} + \cdots + \frac{q_n q_t}{r_{n\to t}^2} \hat{\mathbf{r}}_{n\to t} \right).$$

The quantity q_t appears in every term, and we can factor it out to obtain the electric field $\mathscr{E}$. Using Equation 23.10, we have

---

*Note that the same term is used, with analogous meaning, in the context of wave superposition in Chapters 21 and 22.

$$\mathscr{E} = \frac{1}{4\pi\epsilon_0}\left(\frac{q_1}{r_{1\to t}^2}\hat{\mathbf{r}}_{1\to t} + \frac{q_2}{r_{2\to t}^2}\hat{\mathbf{r}}_{2\to t} + \frac{q_3}{r_{3\to t}^2}\hat{\mathbf{r}}_{3\to t} + \cdots + \frac{q_n}{r_{n\to t}^2}\hat{\mathbf{r}}_{n\to t}\right), \quad \textbf{(23.13a)}$$

or

$$\mathscr{E} = \frac{1}{4\pi\epsilon_0}\sum_{j=1}^{n}\frac{q_j}{r_{j\to t}^2}\hat{\mathbf{r}}_{j\to t}. \qquad \textbf{(23.13b)}$$

In this sum, each term is the contribution of one source charge to the electric field. That is, the individual electric field contributions $\mathscr{E}_j$ add vectorially just like the individual force contributions:

$$\mathscr{E} = \sum_{j=1}^{n}\mathscr{E}_j. \qquad \textbf{(23.14)}$$

For a given array of source charges, the electric field depends on the location at which we wish to evaluate it. If we establish a coordinate system, every point in space can be specified by a vector $\mathbf{r}$, and $\mathscr{E}$ is a function of $\mathbf{r}$:

$$\mathscr{E} = \mathscr{E}(\mathbf{r}). \qquad \textbf{(23.15)}$$

Any vector that is a function of a position vector in this way is called a **field vector**. We will consider another such vector, the *magnetic field* $\mathscr{B}$, in Chapter 28.

EXAMPLE 23.4

Two charges, $q_1 = -2.00 \times 10^{-6}$ C and $q_2 = 3.00 \times 10^{-6}$ C, are located respectively at $(0, 0)$ and $(0.3\text{ m}, 0)$, as shown in Figure 23.5a. **(a)** Find the electric field vector $\mathscr{E}$ and its magnitude and direction at point P, located at $(0, 0.6\text{ m})$. **(b)** If a charge $Q = -2.50 \times 10^{-6}$ C is placed at P, find the magnitude and direction of the electric force exerted on it.

SOLUTION:
(a) Evaluate $\mathscr{E}$ at $(0, 0.6\text{ m})$.

From Equation 23.14, you have $\mathscr{E} = \mathscr{E}_1 + \mathscr{E}_2$. As usual in problems involving vectors, you calculate the electric field component by component. The charge q_1 at $(0, 0)$ will not make a contribution to the x component of the electric field at P, so you have

$$\mathscr{E}_x = 0 + \mathscr{E}_{2x}.$$

Let $\mathbf{r}_1$ and $\mathbf{r}_2$ be the vectors from the two charges to P. You can specify the direction $\hat{\mathbf{r}}_2$ by means of the angle θ_2 shown in Figure 23.5a. Using Equation 23.13b, you write

$$\mathscr{E}_x = \frac{1}{4\pi\epsilon_0}\frac{q_2}{r_2^2}\cos\theta_2. \qquad \textbf{(23.16a)}$$

Similarly, you have for the y component of the field

$$\mathscr{E}_y = \frac{1}{4\pi\epsilon_0}\left(\frac{q_1}{r_1^2} + \frac{q_2}{r_2^2}\sin\theta_2\right). \qquad \textbf{(23.16b)}$$

You have $r_1 = 0.6$ m and thus $r_1^2 = 0.36$ m^2; you must evaluate r_2^2 and r_2. Pythagoras's theorem gives

$$r_2^2 = (0.3\text{ m})^2 + (0.6\text{ m})^2 = 0.45\text{ m}^2,$$

and thus $\quad r_2 = 0.671$ m.

Now you use the value of r_2 to evaluate $\cos\theta_2$ and $\sin\theta_2$:

$$\cos\theta_2 = \frac{-0.3}{0.671} = -0.447; \quad \sin\theta_2 = \frac{0.6}{0.671} = 0.894.$$

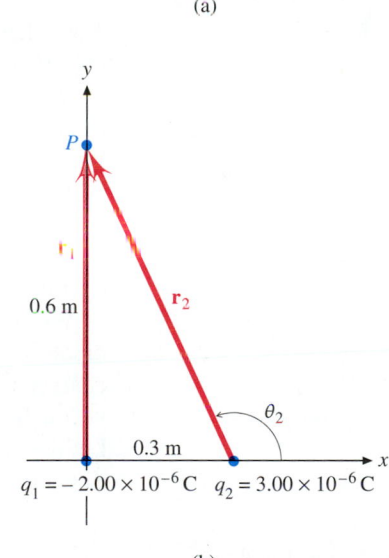

(a)

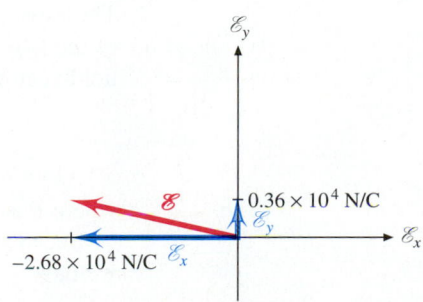

(b)

FIGURE 23.5

Inserting these values into Equations 23.16a and 23.16b, you obtain

$$\mathscr{E}_x = 8.99 \times 10^9 \text{ N·m}^2/\text{C}^2 \times \frac{3.00 \times 10^{-6} \text{ C}}{0.45 \text{ m}^2} \times (-0.447)$$

$$= -2.68 \times 10^4 \text{ N/C}$$

and $\mathscr{E}_y = 8.99 \times 10^9 \text{ N·m}^2/\text{C}^2$

$$\times \left(\frac{-2.00 \times 10^{-6} \text{ C}}{0.36 \text{ m}^2} + \frac{3.00 \times 10^{-6} \text{ C}}{0.45 \text{ m}^2} \times 0.894 \right)$$

$$= 0.36 \times 10^4 \text{ N/C}.$$

You can now write the electric field vector in component form:

$$\mathscr{E} = (-2.68 \times 10^4, \, 0.36 \times 10^4) \text{ N/C}. \quad \textbf{(23.17)}$$

It is a good idea at this point in the calculation to make a sketch of the components, as shown in Figure 23.5b.

Now you can calculate the magnitude and direction of the resultant field $\mathscr{E}$:

$$\mathscr{E} = \sqrt{\mathscr{E}_x^2 + \mathscr{E}_y^2} = 2.70 \times 10^4 \text{ N/C},$$

and

$$\phi = \tan^{-1} \frac{\mathscr{E}_y}{\mathscr{E}_x} = 172.3°.$$

(The diagram of Figure 23.5b prevents you from erroneously assigning to the inverse tangent the value of the supplementary angle $\phi = 352.3°$.)

(b) If a charge $Q = -2.50 \times 10^{-6}$ C is placed at P, find the magnitude and direction of the electric force exerted on it.

To find the force exerted by the field on the charge Q when it is placed at P, you use Equation 23.12 together with the value of $\mathscr{E}$ given by Equation 23.17. You obtain

$$\mathbf{F} = -2.50 \times 10^{-6} \text{ C} \times (-2.68 \times 10^4, \, 0.36 \times 10^4) \text{ N/C}$$

$$= (6.70 \times 10^{-2}, \, -0.91 \times 10^{-2}) \text{ N}.$$

The direction of $\mathbf{F}$ is opposite that of $\mathscr{E}$ because Q is a negative charge. The magnitude of $\mathbf{F}$ is most easily found by writing

$$F = |Q\mathscr{E}| = 2.50 \times 10^{-6} \text{ C} \times 2.70 \times 10^4 \text{ N/C}$$

$$= 6.75 \times 10^{-2} \text{ N}.$$

The direction of $\mathbf{F}$ is the supplement of ϕ:

$$172.3° + 180° = 352.3°.$$

The Electric Dipole

The calculation of the field of two point charges, developed in Example 23.4, has an important special case. An **electric dipole** consists of two equal and opposite charges, q and $-q$, separated by a distance $2b$, as shown in Figure 23.6. Electric dipoles occur in nature in a variety of situations. The hydrogen fluoride molecule (HF) is typical. When a hydrogen atom combines with a fluorine atom, the single electron of the former is strongly attracted to the latter and spends most of its time near the fluorine atom. As a result, the molecule consists of a strongly negative fluorine ion some (small) distance away from a strongly positive hydrogen ion, though the molecule is electrically neutral overall.

We now calculate the electric field at a point P lying on the *median plane* of a dipole—the plane normal to and bisecting the **dipole axis**, as the line joining the two charges is called. In Figure 23.6, the vector from the positive charge to P is $\mathbf{r}_+$ and the vector from the negative charge to P is $\mathbf{r}_-$. Both vectors have the same magnitude r. Let the contributions of the two charges to the electric field at P be $\mathscr{E}_+$ and $\mathscr{E}_-$. Using Equation 23.11 twice, we have

$$\mathscr{E}_+ = \frac{1}{4\pi\epsilon_0} \frac{q}{r^2} \hat{\mathbf{r}}_+ \quad \text{and} \quad \mathscr{E}_- = \frac{1}{4\pi\epsilon_0} \frac{q}{r^2} (-\hat{\mathbf{r}}_-). \quad \textbf{(23.18)}$$

These two field vectors are shown in Figure 23.6. You can see that the z components of the two vectors cancel, while their x components are equal (and negative). The two field vectors add to give the total field

$$\mathscr{E} = 2 \frac{1}{4\pi\epsilon_0} \frac{q}{r^2} \cos\theta \, (-\hat{\mathbf{x}}), \quad \textbf{(23.19)}$$

where θ is the angle between $\mathscr{E}_+$ and $\mathscr{E}$, as shown in Figure 23.6.

Because θ is also the base angle of the triangle defined by the two charges and P, we have

$$\cos\theta = \frac{b}{r} = \frac{b}{\sqrt{z^2 + b^2}}. \quad \textbf{(23.20)}$$

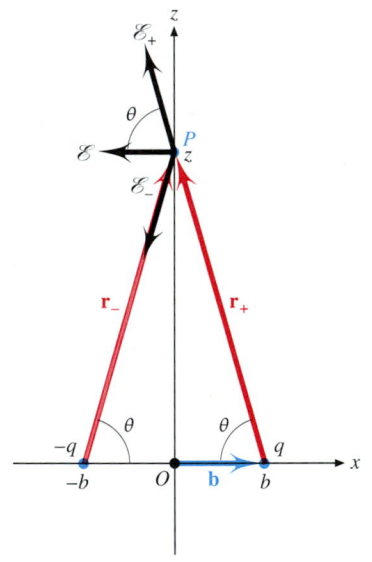

FIGURE 23.6 An electric dipole.

We substitute this value of cos θ into Equation 23.19 and also substitute $r^2 = z^2 + b^2$. This gives us

$$\mathscr{E} = -\frac{1}{4\pi\epsilon_0} \frac{1}{(z^2 + b^2)^{3/2}} 2qb\hat{\mathbf{x}} \quad \text{at a point on the median plane.} \quad (23.21)$$

The vector $2qb\hat{\mathbf{x}}$ lies in the direction from the negative charge to the positive charge. The vector is the product of $|q|$, the magnitude of either of the charges, and $2\mathbf{b}$, the vector from the negative charge to the positive charge. This quantity is called the **electric dipole moment p**, or simply the **dipole moment** when there is no likelihood of confusion:

$$\mathbf{p} \equiv 2q\mathbf{b}. \quad (23.22)$$

We have generalized by defining $\mathbf{p}$ in terms of the vector $\mathbf{b}$ rather than the unit vector $\hat{\mathbf{x}}$ because the direction $\hat{\mathbf{b}}$ has physical meaning independent of the choice of the coordinate system.

In terms of $\mathbf{p}$, Equation 23.21 takes the form

$$\mathscr{E} = -\frac{1}{4\pi\epsilon_0} \frac{\mathbf{p}}{(z^2 + b^2)^{3/2}}. \quad (23.23)$$

Note that $\mathscr{E}$ depends on the *product* $\mathbf{p}$ of the dipole charge $|q|$ and the charge separation $2\mathbf{b}$, and not independently on either.

At large distances, $z \gg b$, Equation 23.23 assumes the simplified form

$$\mathscr{E} \simeq -\frac{1}{4\pi\epsilon_0} \frac{\mathbf{p}}{z^3}. \quad (23.24)$$

That is, the dipole electric field falls off as the inverse *cube* of the distance, in contrast with the inverse-square behavior typical of a single point charge.

Electric Field of a Continuous Charge Distribution

It is often necessary to calculate the electric field due to source charges distributed continuously over a specified region. A similar problem arose in Section 14.2, where we needed to evaluate the gravitational force exerted on a particle by a nearby spherical mass. But the problem arises much more often in electrostatics.

To treat continuous charge distributions, we rewrite the sum of Equation 23.11 in integral form. We have done the same thing before in other contexts—for example, in deriving the expressions for the center of mass of an extended body (Section 9.5) and for its moment of inertia (Section 11.4).

Figure 23.7 represents a region, of arbitrary shape, through which electric charge is distributed continuously. We wish to find the electric field at P due to the presence of the charge. In imagination, we divide the region into infinitesimal elements, each of which contains charge dq. A typical element is shown in the figure. Because the element is infinitesimal, the vector $\mathbf{r}$ from it to P is well defined, and we can use Equation 23.11 to express the contribution of the element to the total electric field at P. To do so, we rewrite Equation 23.11 in differential form:

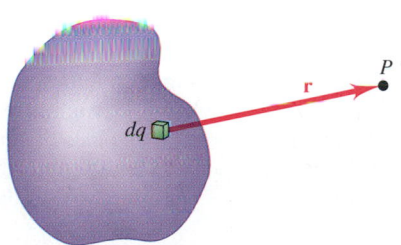

FIGURE 23.7 A uniformly charged body is divided into infinitesimal elements each having charge dq. The vector $\mathbf{r}$ locates point P with respect to the typical charge element shown.

$$d\mathscr{E} = \frac{1}{4\pi\epsilon_0} \frac{dq}{r^2} \hat{\mathbf{r}}. \quad (23.25)$$

The total electric field $\mathscr{E}$ is then the sum of all these infinitesimal contributions:

$$\mathscr{E} = \int_{\text{region}} d\mathscr{E} = \frac{1}{4\pi\epsilon_0} \int_{\text{region}} \frac{\hat{\mathbf{r}}}{r^2} dq. \quad (23.26)$$

To evaluate this integral analytically, we must express the integrand in functional form. It is often possible to do so, as Examples 23.5 and 23.6 show.

EXAMPLE 23.5

A loop of radius b is formed of thin metal wire (Figure 23.8). A quantity of charge q is placed on the loop. Find the electric field at a point P on the axis of the loop, a distance z from the plane of the loop.

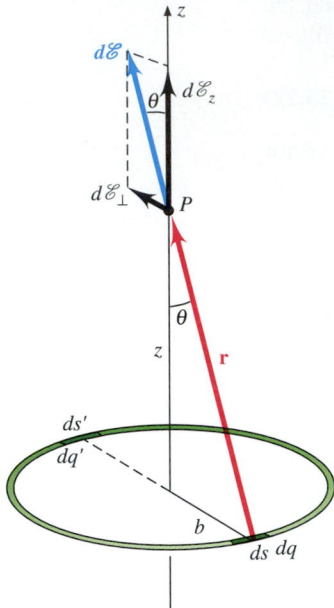

FIGURE 23.8 The field contribution $d\mathcal{E}$ lies in the plane defined by $\mathbf{r}$ and the loop axis.

SOLUTION: You begin by evaluating the quantity dq in Equation 23.26. Because metal is a conductor, the charges, repelling one another, will move freely through the wire until an equilibrium distribution is attained. (This takes a very short time.) In view of the circular symmetry of the loop, the equilibrium charge distribution must be a uniform one. That is, each infinitesimal length element ds along the loop must contain the same charge dq as every other element. Consequently, the ratio dq/q of the elemental charge to the entire charge is equal to the ratio of ds to the entire circumference of the wire:

$$\frac{dq}{q} = \frac{ds}{2\pi b}.$$

So you have

$$dq = \frac{q}{2\pi b}\, ds.$$

The vector $\mathbf{r}$ locates the point P with respect to ds. You use Equation 23.25 to write the contribution of the charge at ds to

the electric field at P as

$$d\mathcal{E} = \frac{1}{4\pi\epsilon_0}\frac{q\, ds}{2\pi b r^2}\,\hat{\mathbf{r}}. \qquad (23.27)$$

The magnitudes of the electric field contributions of all the loop elements are the same, but their directions are different. Fortunately, you can exploit the symmetry of the system to deal with this complication. Let the loop lie in the x-y plane so that the loop axis coincides with the z axis. Resolve $d\mathcal{E}$ into its components $d\mathcal{E}_z$ along the z axis and $d\mathcal{E}_\perp$ perpendicular to the axis. Next, consider what the element ds', which lies exactly opposite ds on the loop, contributes to the electric field. The component $d\mathcal{E}_\perp'$ is equal to and opposite $d\mathcal{E}_\perp$. In adding up the contributions of the elements around the loop, you can pair every ds with its opposite, and the components perpendicular to the z axis cancel pair by pair. (Compare with the cancellation of the component parallel to the z axis for the unlike charges of the electric dipole in Section 23.4.) Thus you can conclude, without actual calculation, that $\mathcal{E}_\perp = 0$.

It remains to evaluate $d\mathcal{E}_z$. Referring to the diagram, you have

$$d\mathcal{E}_z = |d\mathcal{E}| \cos\theta = d\mathcal{E}\frac{z}{r}.$$

But you have a value for $d\mathcal{E}$ in Equation 23.27. Using this value, you find

$$d\mathcal{E}_z = \frac{1}{4\pi\epsilon_0}\frac{qz\, ds}{2\pi b r^3} = \frac{1}{4\pi\epsilon_0}\frac{q}{2\pi b}\frac{z}{(z^2 + b^2)^{3/2}}\, ds.$$

This is the quantity that must be integrated around the loop. But note that z and b have the same value for all ds. So the integral has the simple form

$$\mathcal{E} = \int_{\text{loop}} d\mathcal{E}_z = \frac{1}{4\pi\epsilon_0}\frac{q}{2\pi b}\frac{z}{(z^2 + b^2)^{3/2}}\int_{\text{loop}} ds.$$

The sum of all the length elements ds around the loop is just the circumference $2\pi b$. So you have

$$\mathcal{E} = \frac{1}{4\pi\epsilon_0}\, q\,\frac{z}{(z^2 + b^2)^{3/2}}. \qquad (23.28)$$

As a check on this result, consider the extreme cases $z = 0$ and $z \gg b$. The case $z = 0$ corresponds to the situation in which P lies in the plane of the loop, so $\mathcal{E} = \mathcal{E}_\perp$. But you have already seen that $\mathcal{E}_\perp = 0$. And indeed, Equation 23.28 reduces to $\mathcal{E} = 0$ for $z = 0$.

At large distances z, it becomes difficult to tell the difference between a loop and a point. And indeed, the right side of Equation 23.28 approaches $(1/4\pi\epsilon_0)q/z^2$, the same as the electric field due to an equal point charge q located at the center of the loop.

EXAMPLE 23.6

A thin circular dielectric disk of radius R is uniformly charged; the total charge is Q. Find the electric field at a point P located on the axis of the disk at a distance z from the disk.

SOLUTION: This calculation is an extension of the calculation of Example 23.5. The situation is complicated by the fact that

the elements of charge are not equidistant from P. However, the result of Example 23.5 makes a good starting point. Consider the disk an array of thin concentric rings, each of width dx, as shown in Figure 23.9. A particular ring, whose radius is x, has area $2\pi x\, dx$ and charge dQ. The entire disk has area πR^2 and charge Q. Because the disk is uniformly charged, the ratio

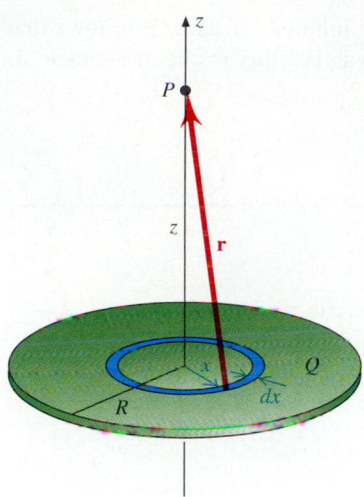

FIGURE 23.9

of the charges is the same as the ratio of the areas:

$$\frac{dQ}{Q} = \frac{2\pi x\,dx}{\pi R^2} = \frac{2}{R^2} x\,dx.$$

Thus the charge on the ring of radius x is

$$dQ = \frac{2Q}{R^2} x\,dx. \qquad (23.29)$$

The contribution of this ring to the electric field at P is given by Equation 23.28. However, you must change the notation to take account of three features of the present problem. First, the charge on the ring of infinitesimal width dx has the infinitesimal value dQ rather than the finite charge q of Equation 23.28. Sec-

ond, the radius x of the ring is a variable in the present problem, in contrast with the fixed radius b of Equation 23.28. Third, the ring makes an infinitesimal contribution $d\mathscr{E}$ to the total electric field at P. With these points in mind, you modify Equation 23.28 to read

$$d\mathscr{E} = \frac{1}{4\pi\epsilon_0}\,dQ\,\frac{z}{(z^2 + x^2)^{3/2}}.$$

Into this equation, you insert the value of dQ given by Equation 23.29 and obtain

$$d\mathscr{E} = \frac{1}{4\pi\epsilon_0}\,\frac{2Q}{R^2}\,\frac{z}{(z^2 + x^2)^{3/2}}\,x\,dx.$$

All contributions $d\mathscr{E}$ lie in the axial direction, and so you can add their magnitudes algebraically. That is, you find the magnitude of the total electric field $\mathscr{E}$ at P by integrating over the range of the variable x:

$$\mathscr{E} = \frac{Qz}{4\pi\epsilon_0 R^2} \int_0^R \frac{2x\,dx}{(z^2 + x^2)^{3/2}}. \qquad (23.30)$$

You can evaluate the necessary indefinite integral by making the substitution $u = z^2 + x^2$. It follows that $du = 2x\,dx$, and you have

$$\int \frac{2x\,dx}{(z^2 + x^2)^{3/2}} = \int \frac{du}{u^{3/2}} = -\frac{2}{\sqrt{u}} = -\frac{2}{\sqrt{z^2 + x^2}}.$$

With this result, Equation 23.30 becomes

$$\mathscr{E} = -\frac{Qz}{4\pi\epsilon_0 R^2}\,\frac{2}{\sqrt{z^2 + x^2}}\,\Bigg|_0^R.$$

Evaluation at the limits gives you the result

$$\mathscr{E} = \frac{Q}{2\pi\epsilon_0 R^2}\left(1 - \frac{z}{\sqrt{z^2 + R^2}}\right). \qquad (23.31)$$

It is sometimes useful to express Equation 23.31 in terms of the *charge per unit area* σ. Because the area of the disk is πR^2, σ is given by

$$\sigma = \frac{Q}{\pi R^2}. \qquad (23.32)$$

In terms of σ, Equation 23.31 becomes

$$\mathscr{E} = \frac{\sigma}{2\epsilon_0}\left(1 - \frac{z}{\sqrt{z^2 + R^2}}\right). \qquad (23.33)$$

An interesting special case is the limit $z \ll R$, where P is very close to the disk (or the disk is very large). In this case, the second term within the parentheses approaches zero, and

$$\mathscr{E} = \frac{\sigma}{2\epsilon_0}. \qquad (23.34)$$

That is, the electric field near the disk is independent of the distance z from the disk. Although we have derived Equation 23.34 for a circular disk, it is easy to see that it applies in the region close to any uniformly charged flat plate. Let us generalize the meaning of R so that it signifies the distance from the point on the plate directly under P to the nearest point on the boundary of the plate. If $z \ll R$, we cannot "see" the boundary from P, and it may as well be infinitely distant. It follows that the shape of the boundary is immaterial. We will rederive this result in a much simpler way in Section 24.4.

e	quantum of charge, $\simeq 1.60 \times 10^{-19}$ C	q, Q	electric charge
$\mathscr{E}, \mathscr{E}$	electric field, magnitude of electric field	ϵ_0	permittivity of free space, $\simeq 8.85 \times 10^{-12}$ C^2/N·m^2
G	gravitational constant	σ	charge per unit area
$\mathbf{p}$	electric dipole moment		

Summing Up

Electric charge is an inherent property of fundamental particles. All such particles carry charges that are either 0 or integral multiples of $\pm e$ (except for quarks, not considered further here). Particles carrying nonzero charge exert a *repulsive force* on particles of *like* charge and an *attractive force* on particles of *unlike* charge. This force is called the **electric**, or **Coulomb**, **force**.

Notable among the fundamental particles are the stable ones of which normal matter is made: the **neutron** (charge 0), the **proton** (charge e), and the **electron** (charge $-e$).

When fundamental particles are created or destroyed, the net charge does not change in the process. This is the basis of the **principle of charge conservation**.

The electrical properties of fundamental particles are reflected on the macroscopic scale. A body that contains equal numbers of electrons and protons is **electrically neutral**. A body can acquire an electric charge by means of a transfer of charged particles from another body. Each of the two bodies then possesses a net charge, which is positive or negative depending on whether protons or electrons are present in excess. A charged body exerts an electric force on other charged bodies. The force is repulsive if the (net) charges are of the same sign and attractive if the charges are of opposite sign.

Most substances can be classified as **conductors**, in which charge is mobile, or **dielectrics**, in which charge is more or less immobile. The properties of the important class of **semiconductors** lie between these extremes.

The electric force is expressed quantitatively by Equation 23.4, which is one form of **Coulomb's law**:

$$\mathbf{F}_{1 \to 2} = \frac{1}{4\pi\epsilon_0} \frac{q_1 q_2}{r^2} \hat{\mathbf{r}}_{1 \to 2}.$$

In SI, the value of the proportionality constant $1/4\pi\epsilon_0$ is fixed by Equations 23.5a and 23.5b. The value of the SI unit of charge, the **coulomb (C)**, is thereby fixed as well.

In using Coulomb's law to predict the force exerted on a test charge q_t in the presence of several other point charges

q_j, we exploit the superposition principle. It is convenient to express the force on q_t when it is located at $\mathbf{r}$ in terms of the **electric field** $\mathscr{E}(\mathbf{r})$. The electric field is defined by Equation 23.10 as the force per unit test charge,

$$\mathscr{E} = \frac{\mathbf{F}}{q_t}.$$

Like the electric force, the electric field conforms to the **superposition principle**, which can be expressed in the form of Equations 23.14 and 23.13b,

$$\mathscr{E} = \sum_{j=1}^{n} \mathscr{E}_j = \frac{1}{4\pi\epsilon_0} \sum_{j=1}^{n} \frac{q_j}{r_{j \to t}^2} \hat{\mathbf{r}}_{j \to t}.$$

For continuous charge distributions, the same statement is expressed in the integral form of Equation 23.26,

$$\mathscr{E} = \int_{\text{region}} d\mathscr{E} = \frac{1}{4\pi\epsilon_0} \int_{\text{region}} \frac{\hat{\mathbf{r}}}{r^2} \, dq.$$

An **electric dipole** consists of a pair of equal and opposite charges, q and $-q$, separated by a distance $2b$. The **electric dipole moment p** is defined by Equation 23.22, $\mathbf{p} \equiv 2q\mathbf{b}$, where $2\mathbf{b}$ is the vector from $-q$ to q.

KEY TERMS

Sections 23.1 and 23.2 Introduction and Electric Charge

electrostatics ▪ electric charge ▪ dielectric, insulator, conductor, semiconductor ▪ positive, negative, neutral ▪ principle of conservation of charge ▪ fundamental particle, neutron, proton, electron ▪ charge carrier

Section 23.3 Coulomb's Law

point charge ▪ Coulomb force ▪ permittivity of free space ▪ elementary charge, quantum of charge

Section 23.4 The Electric Field

source charge, test charge ▪ superposition principle ▪ electric field vector ▪ electric dipole, dipole moment

QUERIES

23.1 *(2) +, −, and ?* On the basis of the empirical qualitative rules governing the way in which electric charges attract or repel one another, prove that *three* kinds of electric charge cannot exist.

23.2 *(2) Thunderstorm watch.* When Franklin was experimenting with atmospheric electricity and lightning, he naturally wanted to know when a charged thundercloud was approaching. He set up the alarm shown here. Explain how it worked.

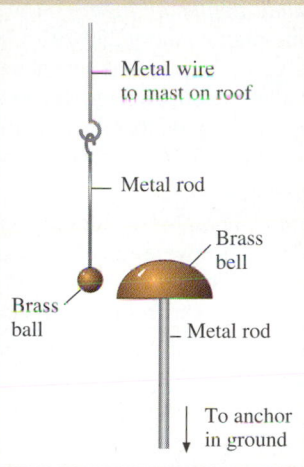

Metal wire to mast on roof

Metal rod

Brass bell

Brass ball

Metal rod

To anchor in ground

23.3 *(2) Free and yet not free.* Even though the electrons in a piece of metal are free of individual atoms, they must be bound to the metal as a whole. Make an argument supporting this statement.

23.4 *(2) Electricity made visible.* The device shown below is

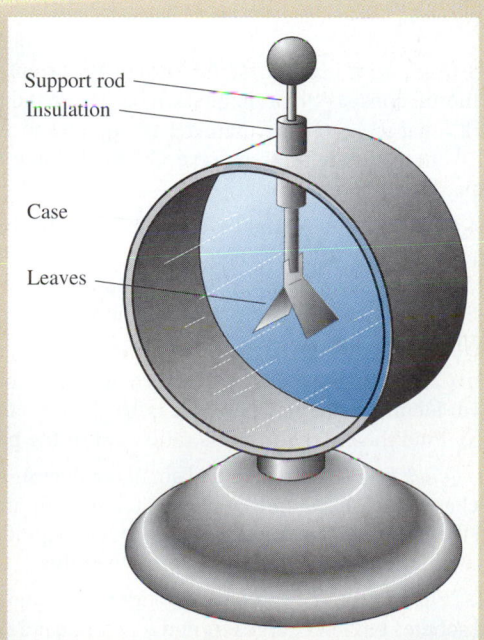

Support rod

Insulation

Case

Leaves

called a *gold-leaf electroscope.* It can be used to detect the presence of electric charge and its sign. A metal rod with a knob at its top passes through an insulating sleeve into a case having glass windows. Draped over a horizontal arm at the bottom of the metal rod is a small sheet of gold leaf, which is very thin, light, and flexible. **(a)** Describe what happens when the electroscope is charged by touching a charged wand to the top of the rod. Does the sign of the charge on the wand make any difference? **(b)** Suppose the electroscope is charged positively. What happens when a negatively charged wand is brought near (but not allowed to touch) the electroscope rod?

23.5 *(2) No charge.* You cover one end of a glass rod with a silk cap. You twist the rod back and forth inside the cap, rubbing the two together briskly. Without removing the cap, you test for charge with an electroscope but detect none. When you remove the cap and test the rod and the cap separately, they exhibit opposite charges. What conclusion can you draw from this experiment?

23.6 *(2) Inductive thinking.* The figure below illustrates the process of *charging by induction.* **(a)** You bring a charged wand near one side of the electroscope knob but do not touch it. Assume for the sake of argument that the wand is positively charged. **(b)** Holding the wand in place, you touch the opposite side of the knob momentarily with your free hand. Because your body can conduct electric charge, you thus bring the electroscope knob into brief electrical contact with the

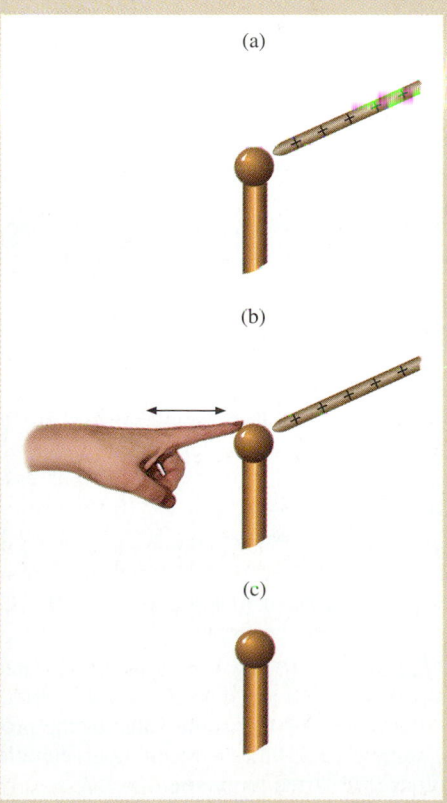

(a)

(b)

(c)

(electrically neutral) earth. **(c)** You remove the wand. At every step, sketch the distribution of excess charge on the electroscope rod and the position of the electroscope leaf. What is the sign of the final charge on the electroscope?

23.7 *(2) Detective work.* Two objects, only one of which is electrically charged, attract one another. How can you determine which object is the charged one?

23.8 *(3) Reversal.* In developing Newton's law of gravitation, we begin with a defined unit of mass and use experimental measurements to determine the value of the constant G. In developing Coulomb's law, we begin with a defined value of the constant $1/4\pi\epsilon_0$ and use this value to define the unit of charge. Discuss the reasons for proceeding in opposite directions in the two cases. Is the analogy between the two end results—Newton's law of gravitation and Coulomb's law—impaired by the difference in procedure? [Hint: Mass has two manifestations, inertial and gravitational (Section 14.6). Which manifestation is relevant in establishing SI?]

23.9 *(3) Cardboard cutout.* Why did Coulomb design the counterbalance of his torsion balance beam in the form of a disk (instead of, say, another sphere)?

23.10 *(3) Attractive idea?* Coulomb performed his torsion-balance experiment using like charges, which results in the fixed charged sphere repelling the charged sphere on the balance beam. It would be a little more difficult, but not impossible, to charge the fixed and moving spheres with unlike charges. Why could you not perform the experiment successfully in this way?

23.11 *(3) Other systems, other rules.* In SI, electric charge is regarded as having a dimensionality not reducible to mass, length, and time. But this is a requirement of the particular system of units and is not imposed by the laws of physics. In one non-SI system that used to be quite popular, called the cgs-esu system, the unit of charge is called the franklin (Fr) or (more prosaically) the electrostatic unit of charge (esu). One franklin is the amount of charge that, when present on each of two particles separated by 1 cm, results in a force of 1 dyne. (1 dyne $\equiv$ 1 g·cm/s^2 = 10^{-5} N.) Write Coulomb's law in this system. What is the value of the proportionality constant? The proportionality constant in the cgs-esu system is a dimensionless quantity *by definition*. What are the dimensions of the franklin?

23.12 *(3) Too many significant figures?* The speed of light is defined as the nine-digit quantity 299 792 458 m/s. **(a)** Show that the exact value of $1/4\pi\epsilon_0$ has seventeen digits. **(b)** Explain why there is no inconsistency in expressing $1/4\pi\epsilon_0$ as a quantity having more digits than the quantity c on which it is based. **(c)** How many digits are there in the exact value of ϵ_0?

23.13 *(3) Changing the rules.* Suppose that the force that held the solar system together were the electric force rather than the gravitational force. In what ways would the behavior of the solar system be different from what it actually is?

23.14 *(5) Getting directions.* What is the direction of the electric field along the half of the axis of a charged loop on each side of the plane of the loop if the charge is positive? negative?

PROBLEMS

GROUP A

23.1 *(3) Signs & portents, I.* Find the magnitude and direction of the force exerted on a -1-C charge when it is 10 m away from a $+1$-C charge.

23.2 *(3) Signs & portents, II.* A 23-μC charge is 45 cm away from a -74-μC charge. What are the magnitude and direction of the force exerted on the former?

23.3 *(3) Pushing it, I.* Two small spheres are positively charged and so repel one another. If the repulsive force on either has magnitude 6×10^{-2} N when the spheres are 20 cm apart, what is the force when they are 10 cm apart?

23.4 *(3) In the vicinity.* The Coulomb force acting on a point charge of 35×10^{-9} C is 2.2×10^{-2} N. If the force is due to another point charge of magnitude 45×10^{-9} C, what is the distance between the charges?

23.5 *(3) Pushing it, II.* Two protons are located 2.9×10^{-10} m apart. **(a)** Find the repulsive force exerted by each proton on the other. **(b)** Using the value of the proton mass given in Example 23.2, find the resulting acceleration of one of the protons if it is free to move.

23.6 *(3) Hark, hark, the quark!* According to current theory, the π^+ meson consists of an up quark (charge $\frac{2}{3}e$) and a down antiquark (charge $\frac{1}{3}e$). If the diameter of the meson is about 2×10^{-15} m, estimate the repulsive Coulomb force exerted by the quark on the antiquark.

23.7 *(3) Counterforce.* A small dielectric sphere having mass 0.55 kg and charge 0.13 μC hangs from a thread. Another sphere having the same charge is brought up from below until the tension in the thread is reduced to one-third its original value. What is the distance between the spheres?

23.8 *(3) Charge-to-mass ratio.* The gravitational force between two identical, charged particles is 10^{-10} times the electric force. Find the charge-to-mass ratio q/m of the particles.

23.9 *(3) Quick kiss, I.* Two identical small conducting spheres having charges $+1.1$ μC and -0.37 μC are brought into contact and then placed 20 cm apart. What are the magnitude and direction of the force exerted by each on the other?

23.10 *(3) Quick kiss, II.* You have two identical small metal-covered spheres having charges q_1 and $q_2 = -2q_1$. You place them a distance d apart and measure the force F on sphere 1. You then allow the spheres to touch and again place them a distance d apart. The force on sphere 1 is now F'. Find the ratio F'/F. (Hint: Pay attention to the signs of F and F'.)

23.11 *(4) Third party.* **(a)** Two 30×10^{-9} C point charges, A and B, lie 2.4 cm apart. Find the magnitude and direction of the force exerted on A. **(b)** A third, equal point charge C is placed so as to form an equilateral triangle with A and B. Find the magnitude and direction of the force on A.

23.12 *(4) Clear field.* At a distance of 10 m from a point charge, the electric field due to the charge is 10 N/C. What is the magnitude of the charge?

23.13 *(4) Plane facts.* A 45-μC charge is located at the point $\mathbf{r}_0 = (3$ m, 4 m$)$ in the x-y plane. Find the magnitude and direction of the electric field at $\mathbf{r} = (7$ m, -4 m$)$.

23.14 *(4) Going places.* An electron in a TV tube is subjected to an electric field of magnitude 4.8×10^4 N/C. Find the acceleration of the electron, whose mass is 9.1×10^{-31} kg.

23.15 *(4) Square deal.* Four positive charges, each of magnitude 1×10^{-10} C, lie at the corners of a square. A negative charge is placed at the center of the square. If the net force on each of the five charges is zero, find the value of the negative charge.

23.16 *(4) First step in making a map.* A source charge $q = 50$ μC is located at $\mathbf{r}_s = (2$ m, 3 m$)$. Find the electric field $\mathscr{E}$ at $\mathbf{r} = (8$ m, -5 m$)$.

23.17 *(4) Hexagons, I.* The sides of a regular hexagon are 4.0 cm long. Three positive charges q and three negative charges $-q$ are located at the vertices of the hexagon in the order $(+\;+\;+\;-\;-\;-)$. If $q = 2.5 \times 10^{-9}$ C, find the magnitude and direction of the electric field at the center of the hexagon. (Hint: Make a sketch showing the electric field vectors at the center due to each of the six charges.)

23.18 *(4) Hexagons, II.* The charge arrangement of Problem 23.17 is changed so that the order of the charges is $(+\;+\;-\;-\;+\;-)$. Find the magnitude and direction of the electric field at the center of the hexagon.

23.19 *(4) Oil drop.* If you spray oil from an atomizer, most of the tiny droplets acquire small charges. Suppose you spray the oil into a region of uniform, downward-directed vertical electric field whose magnitude you can adjust at will. Observing the mist of droplets through a small telescope, you pick out a single droplet and adjust the field until the droplet is just suspended, neither rising nor falling. **(a)** Express the net charge q on the droplet in terms of $\mathscr{E}$, the radius r of the droplet, the density ρ of the oil, and the acceleration of gravity. **(b)** Let $\rho = 0.91 \times 10^3$ kg/m^3, $r = 0.923 \times 10^{-6}$ m, and $\mathscr{E} = 6.12 \times 10^4$ N/C. How many excess electrons are present on the drop? (Note: This problem sketches a part of R. A. Millikan's famous oil-drop experiment for determining the charge on the electron, which he carried out from 1909 to 1913. But there is more to Millikan's work; it is not possible to make a direct measurement of r with adequate accuracy. The oil-drop experiment is discussed in more detail in Problem 26.46.)

23.20 *(4) In the middle of things.* Find the magnitude and direction of the electric field of a dipole at the midpoint of the dipole axis. Show that this result agrees with the result you obtain by using Equation 23.13b.

23.21 *(5) Held off.* A pendulum suspended from a point on a large vertical dielectric sheet is illustrated below. The string is dielectric. The pendulum bob, whose radius is small compared with the distance between bob and sheet, has mass 1.25 g and charge 2.48×10^{-9} C. The sheet has uniform charge per unit area 4.73×10^{-5} C/m^2. Find the angle θ between the pendulum and the vertical.

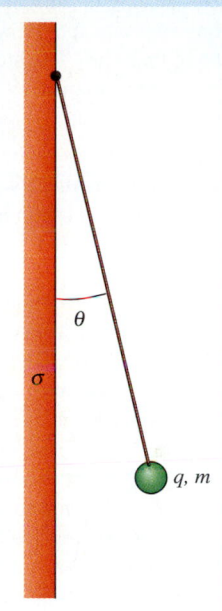

23.22 *(5) In the field.* A circular loop of diameter 35 cm carries a uniformly distributed charge $q = -2.4 \times 10^{-8}$ C. Find the magnitude and direction of the electric field at a point on the loop axis 25 cm away from the plane of the loop.

23.23 *(5) Spreading out.* A thin metal disk of radius B has a detachable rim. The rim is removed and charged, and the field $\mathscr{E}_r$ measured at a point on the axis at a distance $z = L$ from the plane of the rim. The rim is then replaced on the disk. The charge spreads uniformly over the disk and rim. The new field $\mathscr{E}_d$ is measured at $z = L$. Find the ratio $\mathscr{E}_d/\mathscr{E}_r$.

23.24 *(5) Line and plane.* A large dielectric sheet has uniform charge per unit area σ. A long, thin dielectric rod, situated parallel to the sheet and close to it, has charge per unit length λ. **(a)** Find the force per unit length exerted on the rod if $\sigma = 3.2 \times 10^{-5}$ C/m^2 and $\lambda = 2.4 \times 10^{-7}$ C/m. **(b)** If the distance between sheet and rod is decreased by one-half, what is the force per unit length?

23.25 *(5) Half measures.* A stiff metal wire is formed into a semicircle of radius b. A positive charge q is placed on the wire. **(a)** If you assume the charge distribution to be uniform, what is the direction of the electric field $\mathscr{E}$ at the center (of curvature) of the semicircle? **(b)** Show that the magnitude of $\mathscr{E}$ is $\mathscr{E} = q/(2\pi^2\epsilon_0 b^2)$.

23.26 (3) *Unwanted charge.* Consider the deviation from neutrality of the 155-kg lead masses M in Example 23.3. **(a)** The atomic mass of lead is about 207 u. How many atoms are contained in each 155-kg mass? **(b)** Each lead atom contains 82 electrons. What is the equilibrium number of electrons in the mass? **(c)** What is the fractional change in the number of electrons when the lead mass is charged as described in Example 23.3?

23.27 (3) *Imbalance.* Two 1-g copper spheres are 1 m apart. If 1% of the electrons in each sphere were removed, what would the force between them be? The atomic mass of copper is 63.5 u, and a copper atom contains 29 electrons.

23.28 (3) *Maintaining neutrality in weighty affairs.* Einstein's general theory of relativity predicts a deviation of the motion of Mercury from the orbit determined with all classical perburbations accounted for. The deviation is 43 arc seconds per century. The orbit period of Mercury is about 88 days. The mass of Mercury is 3.3×10^{23} kg, and the mass of the sun is 2.0×10^{30} kg. **(a)** To what order of magnitude of precision must the position of Mercury be observed in order to detect Einstein's predicted deviation? **(b)** Observation shows that the electrostatic force exerted by the sun on Mercury is not sufficient to introduce uncertainty as great as 4.3 arc seconds per century (10% of Einstein's predicted effect). Suppose that Mercury and the sun possess equal excess electric charge $|q|$. What is the maximum possible value of $|q|$? **(c)** Assuming that the excess charge is negative, express the result of part **b** as a number n of electrons. **(d)** The sun is composed mainly of hydrogen, whose atomic mass is 1 u and whose atoms (in their un-ionized state) consist of 1 proton and 1 electron. What is the total number N of electrons in the sun? **(e)** What is the upper limit on the fraction n/N?

23.29 (3) *Coulomb's pendulum?* Two identical small, metal-covered Styrofoam spheres of mass m are suspended from threads of equal length l. The upper ends of the threads are attached to the same support. One of the spheres is held aside, and the other is given a charge q. The uncharged sphere is then released and strikes the charged sphere. The system reaches equilibrium as shown below, with the spheres sepa-

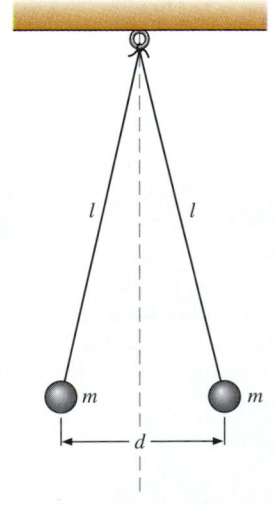

rated by the distance d. **(a)** Show that

$$q = \frac{4d\sqrt{\pi\epsilon_0 mg}}{[(2l/d)^2 - 1]^{1/4}}.$$

(b) Show that if q is small, so that $d \ll l$, you have $d \propto q^{2/3}$.

23.30 (3) *Leaky pendulum.* Suppose that, in the system of Problem 23.29, the threads are not perfectly dielectric, so the charge leaks off each sphere at a rate dq/dt. Over a period of time not too long, dq/dt is constant. As a result of the charge leakage, the distance x between the two spheres decreases with time. Show that, if $x \ll l$, the speed of either sphere with respect to the other is $v = \alpha/\sqrt{x}$, and evaluate the constant α in terms of l, m, and dq/dt.

23.31 (3) *Not too small to measure.* The hydrogen atom can be pictured in a simple, inexact but often useful way as a miniature planetary system, in which a single electron moves in a circular orbit around a much more massive proton. The ionization energy of the atom—the amount of energy required to remove the electron from its orbit to an infinite distance—is known to be 2.18×10^{-18} J. What is the diameter of a hydrogen atom? (Hint: See Equations 14.20 and 14.21.)

23.32 (4) *Electrons, baseballs . . .* An electron is moving in the x direction with velocity $\mathbf{v_0}$. It enters a region of uniform field $\mathscr{E} = \mathscr{E}\hat{\mathbf{y}}$. **(a)** Show that the trajectory of the electron is parabolic. **(b)** Suppose that the initial kinetic energy K of the electron is known. When the electron has traveled a distance x through the field, it is displaced a perpendicular distance y. Express the charge $-e$ on the electron in terms of $\mathscr{E}$, K, x, and y.

23.33 (4) *Unstable, I.* **(a)** A charge q_1 is located at $x = 0$, and a charge $q_2 = -\alpha q_1$ (α a positive constant) is located at $x = d$. **(a)** Find the value of the coordinate x at which $\mathscr{E} = 0$. **(b)** Sketch the position of this point for the case $\alpha > 1$. **(c)** Sketch the position of this point for the case $\alpha < 1$.

23.34 (4) *Unstable, II.* Two identical charges q are separated by a distance $2b$. Choose the x direction along the line joining the charges, and show that **(a)** the electric field at the midpoint between the charges is zero; **(b)** a charge of the *same* sign as q lying at the midpoint is in stable equilibrium for small displacements along the x axis but in unstable equilibrium for small displacements perpendicular to the x axis; **(c)** for a charge of sign *opposite* to q, the statements of part **b** are reversed. This problem illustrates a principle called *Earnshaw's theorem*: There can be no stable equilibrium for a body if the only forces acting on it are electric. This is one reason (but not the only one) for concluding that the electric force is not the only one acting within the atomic nucleus.

23.35 (4) *Squarely on the hypotenuse.* Three point charges are located at the vertices of an isosceles right triangle whose hypotenuse has length $2s$. The charge at the right angle is $+2q$, and the charges at the $45°$ angles are q and $-q$. Find the magnitude and direction of the field at the midpoint of the hypotenuse.

23.36 *(4) Hexagons, III.* Six equal point charges q are placed at the corners of a regular hexagon of sides a. **(a)** What is the value of $\mathscr{E}$ at the center of the hexagon? **(b)** One of the charges is removed from its corner and placed at the center. Show that the force exerted on this charge has magnitude $F = q^2/(4\pi\epsilon_0 a^2)$ and that its direction is from the center toward the empty corner. (Hint: Solving this problem is laborious if you just dive in and start to calculate. But it requires almost no calculation if you think first about symmetry and sketch a force diagram.)

23.37 *(4) Superposition of dipoles, I.* **(a)** Using the superposition principle, show that the arrangement of the three point charges in the figure below appears to a distant observer to be an electric dipole, at least insofar as the field it produces in the plane $z = \frac{1}{2}$ is concerned. **(b)** Show that $\mathbf{p} = 2q\hat{\mathbf{y}}$. (Hint: Can the superposition principle be applied to the vector $\mathbf{p}$ as it can to the vector $\mathscr{E}$?)

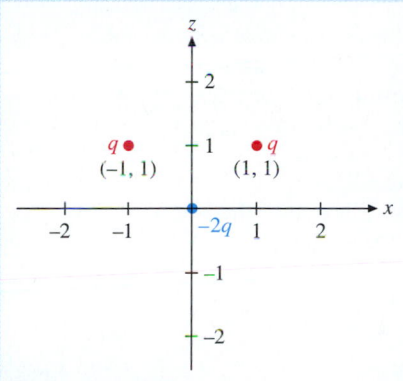

23.38 *(4) Superposition of dipoles, II.* Assume for the sake of argument that a water molecule consists of an oxygen ion having charge $-2e$ and two hydrogen ions each having charge e. The *bond angle*—the angle between the lines joining the center of the oxygen ion to the two hydrogen ions—is $105°$. The dipole moment of the water molecule is $p = 6.20 \times 10^{-30}$ C·m. **(a)** If the charges were fixed point charges, how far would either positive charge e be from the negative charge $-2e$? (Hint: Refer to Problem 23.37.) **(b)** The actual hydrogen-oxygen *bond length*—the distance from the oxygen nucleus to either hydrogen nucleus—is 9.58×10^{-11} m. On a sketch roughly to scale, show where the charges lie relative to the nuclei. Is the ionic picture of the water molecule a tenable one?

23.39 *(5) Distributed dipole.* In the figure below, two conducting wires are held together mechanically, but separated electrically, by a small plastic holder. Each wire is of length b; one has charge per unit length λ and the other charge per unit length $-\lambda$. Show that the system is an electric dipole

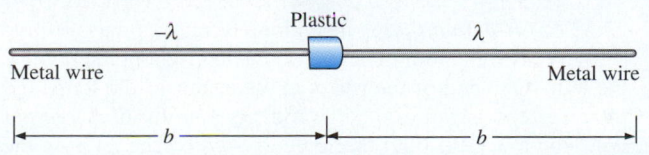

whose moment has magnitude $p = \lambda b^2$. (Hint: What is the dipole moment of a pair of elements, of length dx, located symmetrically with respect to the plastic holder?)

23.40 *(5) To the max, I.* **(a)** Find the value of z for which the electric field on the axis of a charged loop has its maximum magnitude. **(b)** Show that this magnitude is $\mathscr{E} = q/(6\sqrt{3}\pi\epsilon_0 b^2)$, where q is the total charge on the loop and b is its radius.

23.41 *(5) Out of uniform.* The figure below depicts a dielectric loop of radius b on which charge is distributed so that $\lambda = \lambda_0 \cos\theta$. **(a)** Show that the electric field at the center of the loop lies in the plane of the loop and has the direction $\theta = \pi$ and that its magnitude is $\mathscr{E} = \lambda_0/(4\epsilon_0 b)$. (Hint: Consider as a group the field contributions of the four loop elements located at θ, $-\theta$, $\theta + \pi$, and $-\theta + \pi$.) **(b)** Evaluate $\mathscr{E}$ as a function of z along the axis of the loop. **(c)** Show that, for $z \gg b$, the field along the axis looks like that of a dipole, and express the dipole moment p in terms of λ_0 and the area A of the loop.

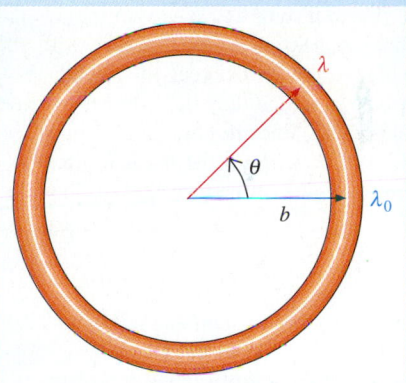

23.42 *(5) Distance blurs distinctions.* Equation 23.31 gives the electric field of a uniformly charged circular disk at a point on the disk axis at distance z from the disk. Show that, for $z \gg B$, the equation reduces to Coulomb's law; that is, from far away, the charged disk looks like a point charge.

23.43 *(5) Capacitor.* Two circular disks, each of area A, are situated parallel to one another. The distance between them is small compared with their radii. Both disks are uniformly charged; their charges per unit area are σ for one and $-\sigma$ for the other. Show that the force exerted by each on the other is $F = \sigma^2 A/2\epsilon_0$. The two disks thus charged make up an important electrical device called a *capacitor*, discussed in Chapter 26.

23.44 *(5) Tapering off.* A charged circular disk has radius 25 cm. At what distance from the disk, along its disk axis, does the value of $\mathscr{E}$ deviate by 5% from the value close to the disk?

23.45 *(5) Field of a long charged wire.* A very long wire has uniform charge per unit length λ. **(a)** Find the direction of $\mathscr{E}$ at a point P, a perpendicular distance r from the wire, for λ positive and for λ negative. **(b)** Show that

$$\mathscr{E} = \frac{\lambda}{2\pi\epsilon_0}\frac{1}{r}.$$

(Hint: You can divide the wire into infinitesimal segments dx and integrate over x. But it is a little easier to express x and dx in terms of the angle θ between the perpendicular to P and the vector to P from a particular segment dx and then to integrate over θ, as shown here.)

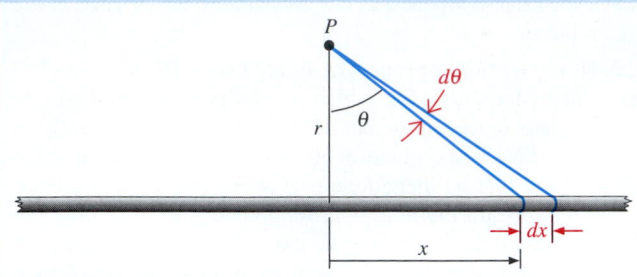

23.46 (5) *To the max, II.* Two very long, parallel dielectric threads are separated by a distance d. The threads have identical charge per unit length λ. The two threads are shown in the figure in the next column, together with their median plane—the plane midway between the threads and perpendicular to the plane in which they lie. (a) Using the result of part **b** of Problem 23.45, find the magnitude and direction of the electric field at a point P on the median plane located a distance Z from the plane of the threads. (b) Show that this field

has its maximum magnitude at $Z = d/2$ and that this magnitude is $\mathscr{E}_{max} = \lambda/\pi\epsilon_0 d$.

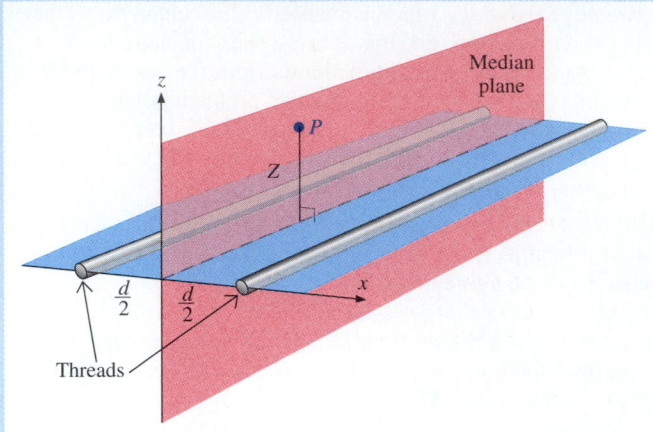

23.47 (G) *Torque on a dipole in an external field.* An electric dipole of moment $\mathbf{p} = 2q\mathbf{b}$ lies in a uniform electric field $\mathscr{E}$. (a) Suppose that the angle between $\hat{\mathbf{p}}$ and $\mathscr{E}$ is θ. Draw a sketch of the dipole, showing the directions of the forces on the two point charges q and $-q$. (b) Show that the torque exerted on the dipole by the electric field has magnitude $T = p\mathscr{E} \sin \theta$. (c) Show that the torque is given by the vector product $\mathbf{T} = \mathbf{p} \times \mathscr{E}$.

GROUP C

23.48 (3) *Flexible dimensionality.* As Query 23.11 suggests, the dimensionality of a physical quantity is not completely independent of the choice of the system of units. To see this in the more familiar context of mechanics, you can use Newton's law of gravitation to define a two-dimensional system of mechanical units as follows. (a) Write the acceleration of gravity, g, in terms of the mass and radius of the earth expressed in kilograms and meters, respectively. But let the proportionality constant G be the dimensionless number 1. What is the resulting numerical value of g? Does it look familiar? (b) What are the units of g in this system? (c) Call the unit of time in this system a "tau." Show that 1 tau $= 1 \text{ m}^{3/2}\cdot\text{kg}^{-1/2}$. (d) Express 1 tau in seconds. (e) In SI, dimensional analysis of purely mechanical laws requires the use of three base units with the irreducible dimensions mass, length, and time. In the system cooked up here, time is a derived unit and base units of only mass and length are required. Is it possible in principle to cook up a one-dimensional system of base units?

23.49 (3) *Quick kiss, III.* Two identical small, metal-covered spheres have charges q_1 and $q_2 = \beta q_1$, where β is a positive or negative constant. The force F exerted on sphere 1 is measured with the spheres a distance d apart. The spheres are allowed to touch and are again removed to a separation d. The new force F' on sphere 1 is measured. (a) Express the ratio F'/F as a function of β. (b) Make a sketch of this function. (c) Show that the function has a discontinuity at $\beta = 0$. What is the physical meaning of this discontinuity? (c) Show that the maximum value of the function for negative values of β

is $F'/F = 0$ at $\beta = -1$. What is the physical meaning of this situation? (d) Show that the minimum value of the function for positive values of β lies at $\beta = 1$.

23.50 (4) *Odd polygons.* In Problem 23.36, you considered the force on a charge displaced from one corner of a hexagon to its center. Consider now a regular polygon with an *odd* number of corners—say, a heptagon—with equal charges q lying at these corners. Here, again, move one of the charges to the center. Show that the magnitude of the force is very similar to that which you found for the hexagon and can be expressed in a general way for any regular polygon as $F = q^2/(4\pi\epsilon_0 r^2)$, where r is the radius of the circle circumscribed about the polygon. Show also that, as for the hexagon, the force is directed from the center toward the empty corner. (Hint: Even though the number of sides is odd, you can still develop a symmetry argument.)

23.51 (5) *Inverse fourth power.* A charge q is placed on a conducting ring of radius b, and a charge $-q$ is placed at the center point. (a) Evaluate the electric field on the ring axis. (b) Show that, at distances $z \gg b$, the electric field is directed toward the center of the ring and obeys an inverse-fourth-power law. (c) Show that $\mathscr{E} \to 0$ as $b \to 0$.

23.52 (5) *Hauling it in.* A metal loop of radius b has positive charge q. A long, thin dielectric rod extends along the axis of the loop. One end of the rod is at the center of the loop, and the other end is far away. The rod has a uniform charge per unit length $\lambda < 0$. Find the force exerted by the loop on the rod.

23.53 *(5) Tightening the loop.* A loop is made of a limp metal wire. When a charge q is placed on it, the loop assumes a circular shape with radius b owing to the mutual repulsion of the charge elements. The wire is thus under tension; don't attempt to calculate the tension here. What is the *increase* in the tension of the wire if a point charge q is placed at the center of the circle?

23.54 *(5) Finite rod.* A straight dielectric rod of length $2l$ having charge per unit length λ lies along the x axis, with its midpoint at $x = 0$. **(a)** Show that the magnitude of the electric field on the z axis (the perpendicular bisector of the rod) is

$$\mathscr{E} = \frac{\lambda}{2\pi\epsilon_0} \frac{1}{z} \frac{1}{\sqrt{1 + z^2/l^2}}.$$

(b) Investigate this result in the cases $z \ll l$ and $z \gg l$; to what simpler situations do these results correspond? **(c)** Show that the magnitude of the electric field at a point on the x axis outside the rod is

$$\mathscr{E}' = \frac{\lambda}{2\pi\epsilon_0} \frac{l}{x^2 - l^2}.$$

What happens when $x \gg l$?

23.55 *(5) Half of very long?* A long dielectric rod having uniform charge per unit length λ lies along the x axis. Its left end is at $x = 0$, and its right end is very far away. Find the magnitude and direction of the electric field at $(0, z)$—that is, at a point along the perpendicular to one end of the rod.

23.56 *(5) 'Round the bend.* A long wire has the form shown below. The long sections are mutually perpendicular, and the bend is circular with radius r. The charge per unit length is λ. Find the magnitude and direction of the electric field at the point O, which is the center of curvature of the bend. (Hint: The result of Problem 23.55 will be helpful.)

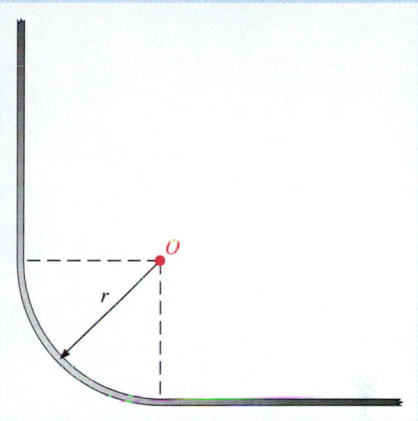

Electric Flux and Electric Field

A flux line is the path followed by a test charge that is allowed to move slowly in the electric field arising from one or more source charges.

The concept of electric flux, though abstract, is an extension of the concepts of fluid flux and of energy flux.

Flux lines always originate at a positive charge and terminate at a negative charge; they never touch or cross one another.

The electric field at any point is the flux-line density at that point—the number of flux lines per unit area normal to the flux.

Gauss's law is simply a mathematical way of saying that anything that passes out of a closed surface must have come from within it.

Gauss's law, taken together with symmetry considerations, is a powerful way of finding the electric field in the vicinity of a symmetrical charge distribution with a minimum of calculation.

Left: Animals that hunt their prey in turbid waters cannot rely on vision. Many such animals have independently evolved nervous-system organs that can generate electric fields and detect distortions of those fields produced by nearby objects. The duck-billed platypus is one of these electric predators.

> *It is in the highest degree astonishing to see what a large number of general theorems, the mathematical deduction of which requires the highest powers of mathematical analysis, . . . [Faraday] found . . . without the help of a single mathematical formula . . .*
>
> —HERMANN VON HELMHOLTZ

> *I was at first almost frightened . . . when I saw such mathematical force made to bear upon the subject, and then wondered to see that the subject stood it so well.*
>
> —MICHAEL FARADAY

SECTION 24.1 — Introduction

In Section 23.4, we defined electric field as the electric force exerted on a test charge divided by the charge q_t. In the rest of Chapter 23, we calculated electric fields as a function of position in the vicinity of a variety of source charge distributions. From this perspective, electric field appeared as little more than a simple convenience. In this chapter, we view electric field in a subtly different way. Out of this shift in viewpoint will emerge the beginnings of *field theory*—a powerful tool for studying electricity, as well as magnetism and other physical phenomena. We will find in this new approach a simple, effective way of calculating fields, called *Gauss's law*. Gauss's law will prove even more useful as a source of insight into the interaction of electric charges with one another. We by no means exhaust the possibilities of the electric field in this chapter; other implications appear throughout Part VIII and thereafter.

SECTION 24.2 — Electric Charge and Electric Flux

Given any particular array of source charges, the electric field is a function of position. In Section 23.4, we noted this point in passing. Equation 23.15 states it explicitly:

$$\mathscr{E} = \mathscr{E}(\mathbf{r}).\tag{24.1}$$

It makes sense, then, to *map* the electric field in the region surrounding the source charges. There are several ways to do this, each having its own advantages; we explore one of them in this chapter and another in the next.

Mapping with Flux Lines

A direct way of mapping the electric field is based on the simple thought experiment sketched in Figure 24.1. In imagination, we attach a known (positive) test charge q_t to a small spring scale. If it is located anywhere in an electric field $\mathscr{E}(\mathbf{r})$, the test charge will experience a force $\mathbf{F}(\mathbf{r}) = q_t \mathscr{E}(\mathbf{r})$ whose direction is the same as that of the field.

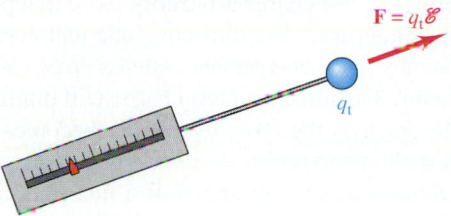

FIGURE 24.1 A thought experiment for mapping the electric field. The test charge q_t is attached to a small spring scale. The magnitude and direction of the force $\mathbf{F}$ on the test charge can be measured at any location.

Suppose we begin with the test charge at $\mathbf{r}_1$ and allow it to move *slowly* wherever the electric force tends to take it. It will follow a particular path, as shown in Figure 24.2. Such a path is called a **flux line**. If instead we begin with the charge at some other point $\mathbf{r}_2$, the test charge will in general trace a different flux line, also shown in Figure 24.2. *The electric force experienced by a test charge at any location* $\mathbf{r}$ *is always tangent to the flux line passing through* $\mathbf{r}$. It follows that *two flux lines cannot touch or cross one another.* Nor can flux lines branch. To contradict these negative statements is to imply that there are points in space where a test charge simultaneously experiences a net electric force in two directions, which is absurd.

> **One Concept, Several Names**
> Other terms are used synonymously with flux line. The original usage, still often heard, is *line of force.* This term, first used in 1629 by the Italian physicist Niccolò Cabeo (1586–1650) to denote the patterns formed by iron filings in the vicinity of a magnet, was adopted by Faraday among others. But the line itself does not exert a force on anything, and the terms **flux line** and **field line** are preferred in modern usage.

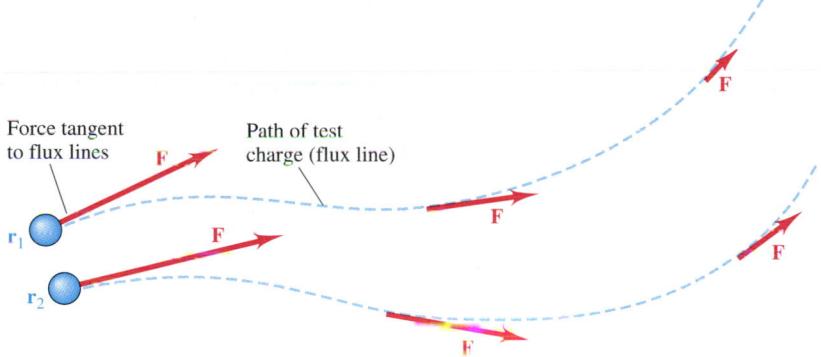

Force tangent to flux lines

Path of test charge (flux line)

FIGURE 24.2 The test charge is placed in a region where an electric field exists. If it is placed at $\mathbf{r}_1$, it tends to move in the direction of the electric force it experiences at $\mathbf{r}_1$. If the test charge is allowed to move slowly, it follows a path called a flux line. If the test charge is placed at some other point $\mathbf{r}_2$, it will in general follow a different flux line.

It is especially simple to map the flux lines in the vicinity of a single positive point source charge. We choose the coordinate system so that the source charge lies at the origin, $\mathbf{r} = 0$. Then the force on the test charge is always in the direction $\hat{\mathbf{r}}$, as shown in the map of Figure 24.3.

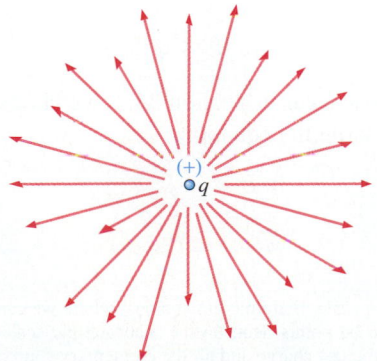

FIGURE 24.3 Map of the flux lines surrounding a positive point source charge q located at $\mathbf{r} = 0$. According to Coulomb's law, the direction of the force on a test charge located at any other point $\mathbf{r}$ is $\hat{\mathbf{F}} = \hat{\mathbf{r}}$.

We assign a *sense* to each flux line in a natural way. *The sense of the flux line is the sense in which a (positive) test charge tends to move along it.* Where, then, does the flux line begin? The answer to this question is clear in Figure 24.3; all of the flux lines begin at the source charge q. It is not difficult to extend this conclusion to more complex distributions of charge. A positively charged plate, for example, contains many point source charges. But, if we drag a test charge back toward the plate along a flux line, we can (in principle) bring the test charge arbitrarily close to a particular source charge from which the flux line emanates.* We thus conclude that *every flux line begins at a positive charge.* That is why these charges are called *source charges.*

Where do flux lines end? The force on a test charge will diminish with distance from the source charge(s). In practice, the force eventually becomes too small to measure. But, in principle, the test charge experiences *some* force at any finite distance. Thus, *a flux line cannot come to an end in empty space.* If a flux line did end in empty space, the force experienced by a test charge moving along it would go *discontinuously* from some nonzero value (perhaps small) to zero when the test charge reached the end of the line. This is not consistent with Coulomb's law.

Is the flux line endless? To answer this question, consider the flux lines associated with a *negative* charge, as shown in Figure 24.4. A (positive) test charge in the vicinity of this charge experiences an attractive force. If the test charge is allowed to move slowly under the influence of this force, it will arrive arbitrarily close to the negative charge and go no farther. Thus the flux line *ends* at the negative charge. It would be logical to call such charges ''sink charges'' in analogy to the terminology we use in thermodynamics, where heat flows from heat sources to heat sinks. But it is more usual to call charges at which flux lines end *negative source charges,* and we will follow this convention.

Consistent with the principle of charge conservation is the assertion that *every flux line begins at a positive charge and ends at a negative charge.* If the algebraic sum of the electric charge in the entire universe is zero, there are no ''orphan'' flux lines; however long it is, every flux line that begins at a positive charge will ''find'' a negative charge on which to terminate.

Let us summarize the flux-line properties we have deduced to this point:

1. Flux lines originate *only* at positive charges and terminate *only* at negative charges.
2. Flux lines cannot cross one another.

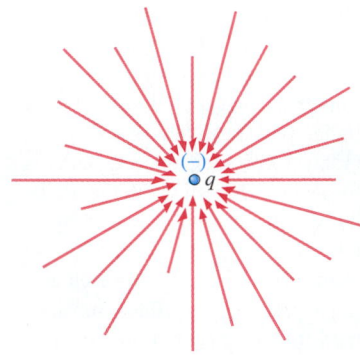

FIGURE 24.4 Map of the flux lines surrounding a negative point charge located at $\mathbf{r} = 0$. According to Coulomb's law, the direction of the force on a test charge at any other point $\mathbf{r}$ is $\hat{\mathbf{F}} = -\hat{\mathbf{r}}$.

Quantifying Flux Lines

If you draw a small spherical surface around a positive source charge, the surface comprises an infinite number of points, at any of which you could locate a (point) test charge and thus begin to trace a flux line. But a map showing an infinite number of flux lines would be of no use at all; it would be no more than a black blob. In Figures 24.3 and 24.4, we avoided this difficulty by mapping the field around a point charge by means of *representative* flux lines. We now make this idea quantitative. Your intuition probably suggests that the total **electric flux** Φ—the number of flux lines beginning or ending at a source charge—is proportional to the magnitude of the charge; that is,

$$\Phi \propto q. \tag{24.2a}$$

For reasons that will become clear in Section 24.3, we choose the proportionality constant to be $1/\epsilon_0$. We then write the equation

$$\Phi = \frac{1}{\epsilon_0} q. \tag{24.2b}$$

---

*This picture breaks down on the scale of atomic dimensions, where we cannot consider the protons, which are the actual source charges, to be points. But, on the macroscopic scale of our present discussion, it is entirely permissible to consider the test charge and all the elementary source charges to be geometric points.

The sign of q determines the sense of the flux; a positive charge implies outgoing flux lines and a negative charge implies incoming flux lines.

Flux and Field

Once electric flux has been quantified in this manner, we can establish a connection between flux and electric field. We draw an imaginary sphere of radius r centered on a point charge q, as in Figure 24.5. How much flux penetrates the surface of this sphere?

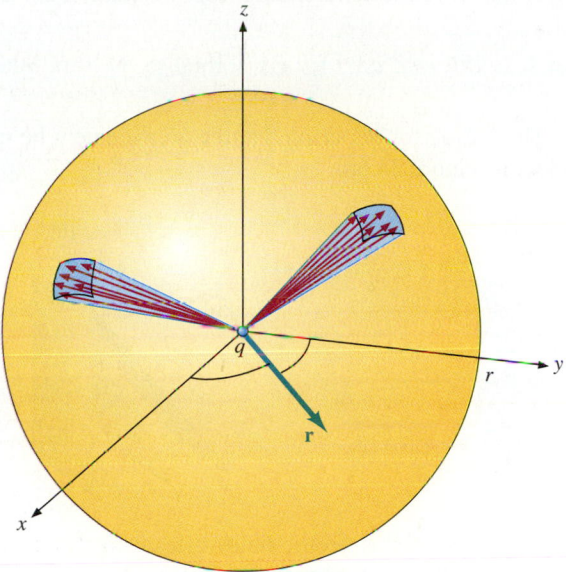

FIGURE 24.5 An imaginary sphere of radius r is drawn centered on a point source charge q. Some flux lines are shown. They must emanate from (or terminate at) q symmetrically in all directions. The value of Φ/A is thus the same at all points on the surface of the sphere.

Every flux line that emanates from (or terminates at) q must penetrate the surface because there are no other charges within the sphere. Equation 24.2b therefore gives the total flux penetrating the surface as $\Phi = q/\epsilon_0$. This result is independent of the radius r (as long as r is not increased so much that the sphere contains other charges).

Because of the symmetry of the situation, we argue that the pattern of flux lines is the same in every direction. There are not more flux lines, for example, in the $(+, +, +)$ octant of the three-dimensional coordinate system shown in Figure 24.5 than there are in the $(-, -, -)$ octant. With this in mind, we calculate Φ/A, the *flux per unit area* penetrating the sphere. The flux per unit area must be the same all over the surface, and so we simply divide the total flux by the total area, obtaining

$$\frac{\Phi}{A} = \frac{q/\epsilon_0}{4\pi r^2} = \frac{1}{4\pi\epsilon_0}\frac{q}{r^2}. \tag{24.3}$$

Compare this result with the magnitude part of Equation 23.11,

$$\mathscr{E} = \frac{1}{4\pi\epsilon_0}\frac{q}{r^2}.$$

This is the form of Coulomb's law that gives the magnitude of the electric field at a distance r from a point charge q. Because the right sides of the two equations are the same, we can write

$$\mathscr{E} = \frac{\Phi}{A}. \tag{24.4a}$$

On the basis of this relation and earlier discussion, we add a third item to our list of the properties of flux lines:

3. The electric field $\mathscr{E}$ at a given location is tangent to the local flux lines and has the direction given by the sense of the flux lines and a magnitude equal to the flux per unit area penetrating a surface normal to the flux lines.

We have obtained this result for the special, highly symmetrical case of a sphere centered on a point source charge, but it must be true in general. To see this, suppose we use a test charge to measure the magnitude and direction of the field $\mathscr{E}$ at some point P in space. We can do this without knowing anything about the configuration of the source charge or charges that give rise to the field. (Indeed, we would have to do a survey over space to establish whether the field is due to a point charge or to some more complicated array of source charges.) All that counts in establishing the magnitude of $\mathscr{E}$ at one particular location P is the value of $d\Phi/dA$ over an infinitesimal area element that includes P. We thus have the differential form of Equation 24.4a,

$$\mathscr{E} = \frac{d\Phi}{dA} \quad \text{or} \quad d\Phi = \mathscr{E}\,dA \quad \text{for } dA \text{ normal to the flux lines.} \qquad \textbf{(24.4b,c)}$$

See Figure 24.6a. The reason for requiring that the area element be normal to the flux lines is made evident in Figure 24.6b.

FIGURE 24.6 (*a*) The relation $\mathscr{E} = d\Phi/dA$ for the infinitesimal area dA surrounding the point P. The surface must be drawn normal to the flux lines. (*b*) The same amount of flux penetrates all three area elements shown. To give a unique meaning to $d\Phi/dA$, we must specify that dA is normal to the flux lines.

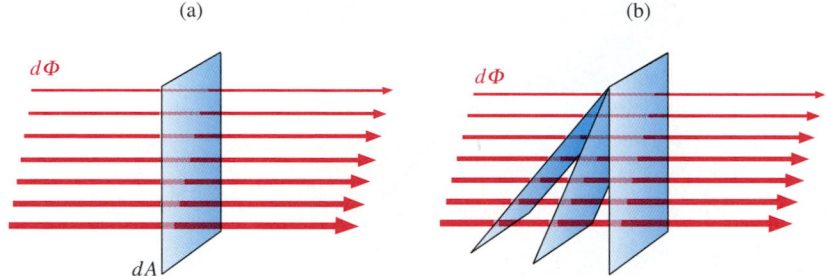

In general, the magnitude of the electric field varies from place to place. This variation requires that the flux lines crowd together and spread apart, as shown in Figure 24.7. In general, flux lines are not parallel to their neighbors. In choosing an area element dA, however, we can always satisfy the condition that it be normal to *all* the flux lines penetrating it. All we need do is choose dA small enough, in the spirit of the differential calculus. Then the deviation from parallelism of the flux lines penetrating dA becomes negligible.

It is always possible to map an electric field by using Coulomb's law and the superposition principle repeatedly at a sufficient number of points within the region of

FIGURE 24.7 If $\mathscr{E}$ varies from place to place—for example, from P_1 to P_2—the flux lines must converge or diverge and cannot be exactly parallel to their neighbors.

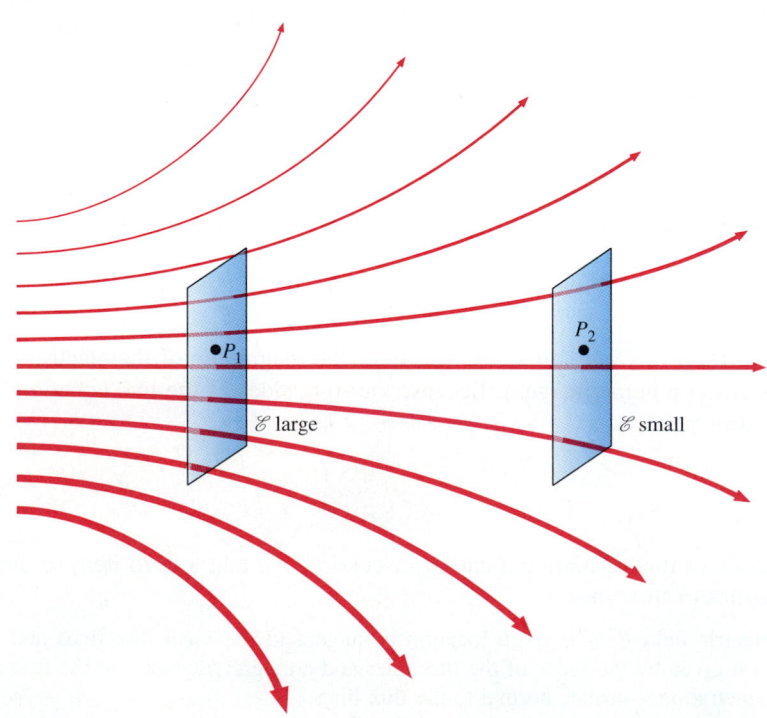

interest. But doing so can be laborious, even with a computer. Often, a qualitative flux-line map is adequate. If the array of source charges is not too complicated, the map can be sketched by using the rules we have developed for the "behavior" of flux lines, as shown in Example 24.1.

EXAMPLE 24.1

Sketch the flux-line pattern in the vicinity of an electric dipole, consisting of a pair of equal and opposite charges q and $-q$.

SOLUTION: By symmetry, the pattern will be the same in any plane containing q and $-q$, and so you can make the necessary sketch in two dimensions, as shown in Figure 24.8.

To begin with, the straight line from q to $-q$ is certainly a flux line because the forces exerted by both source charges on a test charge located on this line are in the same direction.

Next, consider the situation at locations very close to the positive charge. At such locations, the force on a test charge due to the positive charge is much greater than that due to the negative charge. You can argue that the test charge barely "sees" the negative charge. Under these circumstances, the flux lines must emanate from q almost as if $-q$ were not there at all. That is, they must emanate from q radially with uniform density, as shown. For the same reason, flux lines must terminate at $-q$ radially with uniform density, as shown.

Given both ends of the lines, you can use the symmetry of the situation to sketch their connecting parts. As the test charge moves farther and farther away from q, the repulsive force due to q becomes progressively weaker and the attractive force due to $-q$ comes into play with increasing significance. The result is that the line will begin to curve toward $-q$. The field is strongest in the region between the two source charges, where the field contributions of the two superpose constructively. This is evident in the crowding of flux lines in this region. In the regions near the top and the bottom of Figure 24.8, the two field contributions superpose destructively, the field is weak, and the lines are spread far apart. To the sides, the field falls off rapidly, as you would expect from the $1/r^3$ behavior of a dipole. This is evident in the rapid "thinning" of the lines toward the sides.

FIGURE 24.8

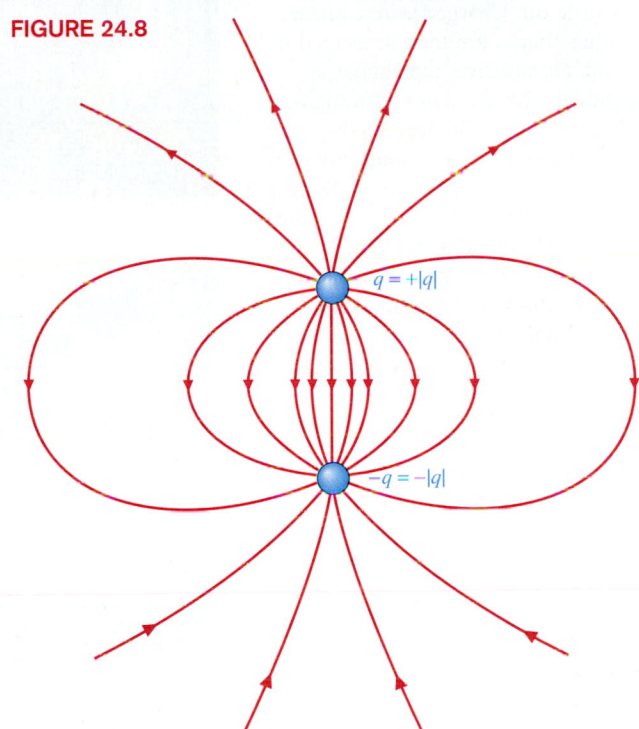

You cannot sketch all the lines completely because you will encounter some flux lines—perhaps very few—no matter how far you go away from the two source charges. This saves you from answering an embarrassing question: What happens to the line that emanates from q in the direction exactly away from $-q$?

The dipole flux-line map constructed in Example 24.1 is visualized in the photograph of Figure 24.9a (next page), which you should compare with Figure 24.8. The maps for some other charge distributions are shown as well.

What Is Electric Flux?

The flux-line picture was conceived and first used in electrical studies by the great British natural philosopher Michael Faraday (1791–1867). In part, Faraday invented flux lines to compensate for his lack of mathematical training. He never mastered the integral calculus and found difficulty in using the superposition principle to calculate fields, as we have done in Section 23.5. Much more important, however, Faraday found in the flux picture a vivid way of visualizing fields that the more formal methods did not provide. You have seen some of the fruits of the flux picture in this section. But, as the epigraphs at the head of this chapter suggest, the flux picture has profound implications that took decades to work out completely.

You may have been puzzled by the use of the word *flux*, which means "flow," in the context of electrostatics, where nothing moves. If so, you are entirely justified. We

FIGURE 24.9 Photographs visualizing the flux-line maps for (*a*) an electric dipole; (*b*) two equal charges of the same sign; (*c*) a uniformly charged flat plate; (*d*) a pair of oppositely charged parallel plates. The photographs are obtained by suspending (dielectric) grass seed in a dielectric oil. Charged wires, plates, or other shapes are then immersed in the oil. (In practice, the charge is maintained on the wires by a high-voltage source.) The tiny seeds, which are substantially longer than they are wide, tend to line up along the electric field. This is because the electric field *induces* (brings into being) a dipole moment in each seed (Section 26.4). The torque exerted on the dipole by the external electric field then swings the seeds into line with the field; see Problems 23.20 and 23.47.

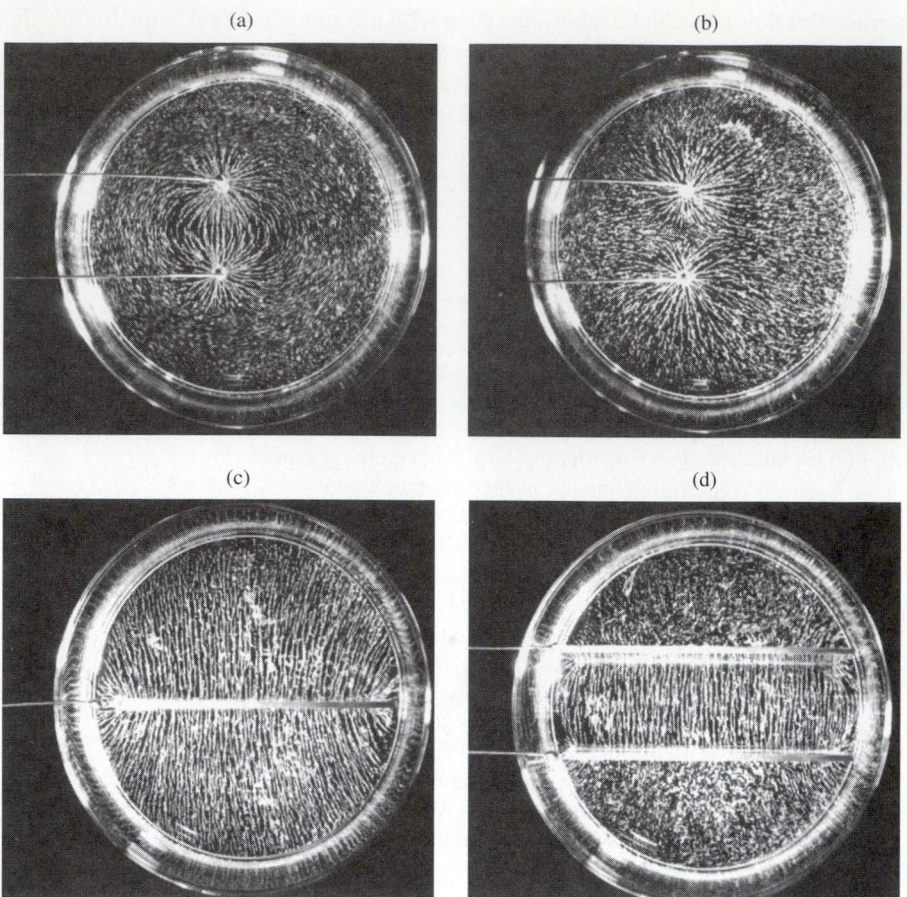

used flux in a completely literal sense in the context of fluid flow in Chapter 16, where we defined both a volume flux and a mass flux. When we defined energy flux in Chapter 21, we made the concept of flux a somewhat more abstract one. When energy flows along a vibrating string, there is nothing tangible flowing—no transport of matter along the string. Nevertheless, energy flux is both a well-defined concept and a useful one.

Here we abstract the idea of flux one step further. *Nothing is flowing in the electrostatic case.* In particular, the electric field is not in motion. Nevertheless, electric flux lines have many properties analogous to the properties of other kinds of flux. For example, electric flux lines tend to spread out into the space available to them, just like the streamlines of a flowing fluid. Like streamlines, flux lines begin at sources and end at sinks (negative sources). Their number per unit cross-sectional area—their density—has a direct meaning: electric field. You are already familiar with the analogous definition of *intensity* as energy flux per unit cross-sectional area. (Indeed, electric field is also called **electric field intensity**.)

In defining electric flux, Faraday had more directly physical ideas in mind. He asked himself, How can an electric charge exert a force on another charge across the space between them? In answering this question, he imagined space to be filled with an elastic substance (usually called *ether* by nineteenth-century physicists). This substance, he supposed, was stressed by the presence of a charge and transmitted that stress to the other charge. Faraday called this the electrotonic (electric-stress) state. (Imagine all space as being filled with an intangible Jello and the charges as two grapes embedded in the Jello. If you disturb one grape, the other will soon begin to move as well.) To Faraday, the flux lines represented lines of stress in the ether.

A fundamental understanding of this force emerged in the 1940s from the theory called *quantum electrodynamics.* We know today that no ether is required to explain the electric force. Nevertheless, the concept of electric flux continues to be useful. Simply bear in mind that the term *electric flux* implies not physical flow but only analogy to other less abstract forms of flux that *do* involve flow.

Michael Faraday was the son of a very poor blacksmith. His childhood in the teeming slums of London was one of deprivation and not infrequent hunger. Of his brief formal education, he later said, "[It] was of the most ordinary description consisting of little more than the rudiments of reading, writing and arithmetic at a common day school." Being slight of stature, Faraday was unsuited to follow his father in blacksmithing. Consequently, as he later wrote,

> I entered the shop of a bookseller and bookbinder at the age of 13, in the year 1804, remained there eight years, and during the chief part of the time bound books. Now it was in those books, in the hours after work, that I found the beginning of my philosophy. There were two that especially helped me, the "Encyclopaedia Britannica," from which I gained my first notions of electricity, and Mrs. Marcet's "Conversations on Chemistry," which gave me my foundation in that science. Do not suppose that I was a very deep thinker, or was marked as a precocious person. I was a very lively, imaginative person, and could believe in the "Arabian Nights" as easily as in the "Encyclopaedia"; but facts were important to me, and saved me. I could trust a fact, and always cross-examined an assertion. So when I questioned Mrs. Marcet's book by such little experiments as I could find means to perform, and found it true to the facts as I could understand them, I felt that I had got hold of an anchor in chemical knowledge, and clung fast to it. Thence my deep veneration for Mrs. Marcet . . .

In 1812, Faraday borrowed a shilling from his older brother and attended a series of lectures on chemistry at the Royal Institution, founded by Count Rumford (see Section 17.7) in part to further the scientific education of poor young men like Faraday. These twelve lectures, and four more a little later, constituted his entire formal scientific education. He was so entranced that he resolved to abandon a promising career as a bookbinder to devote himself completely to natural philosophy—a daring step in a day when most English natural philosophers were gentlemen of independent means. With this in mind, he bound his meticulous lecture notes into a book, which he presented to the Director of the Royal Institution, the noted chemist Sir Humphry Davy, together with a request for employment. Davy hired him as a bottle washer, general lab assistant, and (much against Faraday's wishes) valet. Faraday washed bottles for more than ten years but found time to make a series of increasingly significant contributions to electricity and chemistry. His invention of the first electric motor in 1821 brought him public notice—and promotion away from the sink. He eventually rose to succeed Davy as Director; the rest of his life was devoted to the Royal Institution, where he lived as well as worked.

Gaussian Surfaces and Gauss's Law

About 1839, the great German mathematician and physicist Carl Friedrich Gauss (1777–1855) devised a new way to express Coulomb's law, based on the flux-line picture. Figure 24.10a, like Figure 24.5, shows a point charge q around which an imaginary spherical surface of radius r has been drawn. We call this surface a **Gaussian surface**. You have already seen how Equation 24.2b, $\Phi = q/\epsilon_0$, combined with the spherical symmetry of the system, yields

$$\mathscr{E} = \frac{\Phi}{A} = \frac{1}{4\pi\epsilon_0}\frac{q}{r^2}; \qquad (24.5)$$

this is essentially Equation 24.3 and is a form of Coulomb's law.

We now take a simple but very important step: We distort the spherical surface. You can imagine the surface as being made of an ideal rubber sheet that can be stretched as much as you like but cannot be torn. The "rubber" is also perfectly impermeable to charge, so no distortion can transfer a charge from inside the surface to outside it, or vice versa. Figure 24.10b shows the result of such a distortion. The important point to note is that *the total flux Φ penetrating the surface has not changed.* What if we take the point charge and smear it out in an arbitrary way so that q is distributed over an arbitrary region lying entirely within the Gaussian surface? *The total flux Φ penetrating the surface is still the same.* Mathematicians say that the total flux Φ penetrating the

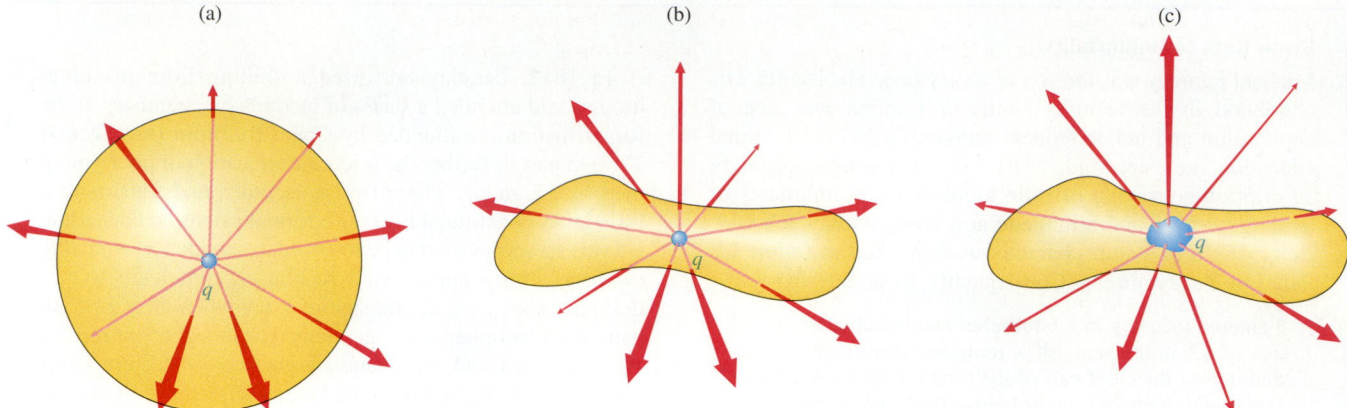

(a)   (b)   (c)

FIGURE 24.10 (*a*) All the flux lines emanating from the point charge q penetrate the surface of the imaginary sphere. The total flux is $\Phi = q/\epsilon_0$. (*b*) The imaginary spherical surface is distorted into a (closed) surface of arbitrary shape, still containing the point charge q. The total flux penetrating the surface is unchanged, though the flux per unit area is no longer uniform over the surface. (*c*) The charge q is distributed over a body of arbitrary shape that is entirely contained within the imaginary surface. The total flux penetrating the surface is still the same as in parts *a* and *b*, though the flux lines may be distributed over the surface in a different way than in part *b*.

surface is a **topological property** and not a **metric property** of the system. That is, Φ depends not on sizes and shapes but on what is called the **connectivity** of the surface—the way in which the surface separates inside from outside.

Next, consider a positive charge q' lying outside a Gaussian surface, as shown in Figure 24.11. Some of the flux lines emanating from the charge penetrate the surface in the inward sense. Because every flux line must end on a negative charge, every line that enters the region surrounded by the surface either must leave the region by penetrating the surface again in the outward sense or must terminate on some negative charge within the surface. Figure 24.11 shows a negative charge q inside, with $|q| < q'$. Flux lines a and b terminate on the inside charge; each penetrates the surface once. Flux lines c and d do not terminate on the inside charge. Line c penetrates twice—once inward and once outward. Line d penetrates four times—twice inward and twice outward. If we assign the value 1 to each outward penetration and the value -1 to each inward penetration, the sum for the four flux lines shown is

$$\underset{a}{-1} \;\; + \;\; \underset{b}{(-1)} \;\; + \;\; \underset{c}{(-1 + 1)} \;\; + \;\; \underset{d}{(-1 + 1 - 1 + 1)} \;\; = -2.$$

The entire net contribution to the nonzero sum is due to lines a and b, the ones that terminate on the inside charge. But these lines penetrate the surface and terminate on the inside charge regardless of where outside the Gaussian surface they originate. It

FIGURE 24.11 A charge q' lies outside a Gaussian surface. Some of the flux lines emanating from the charge penetrate the surface, and some of these terminate on the smaller negative charge q inside the surface. But the net number of penetrations is independent of the location of q', as long as it remains outside the surface.

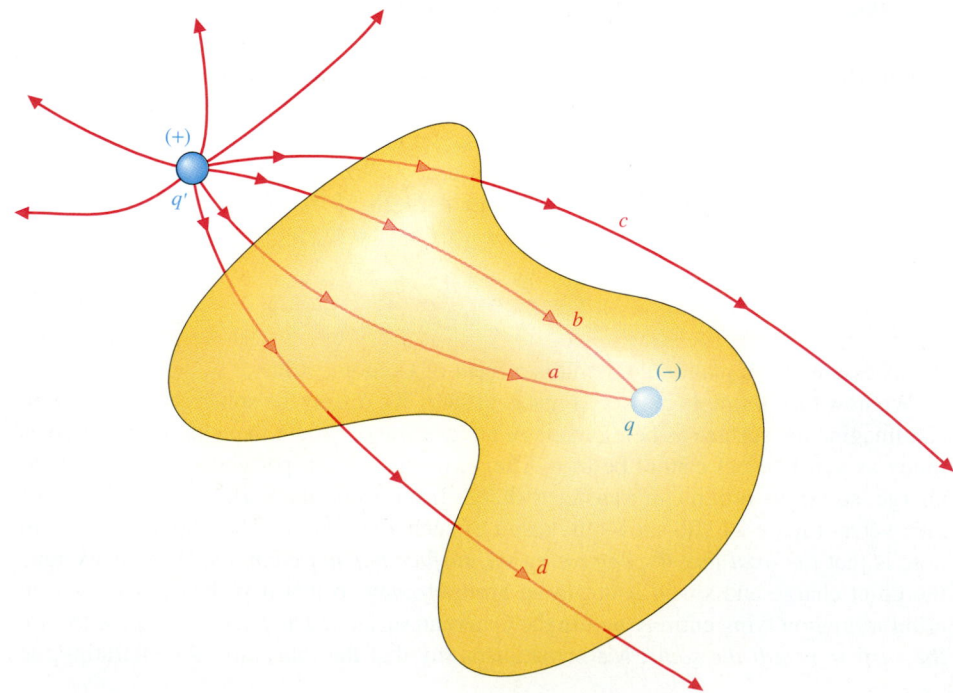

follows that *the total flux penetrating a Gaussian surface is independent of the distribution of charges outside it.* Thus, *in calculating the flux that penetrates a Gaussian surface, we need consider only the charge within it.*

Gauss's Law

Let us now connect our topological argument, which deals with the flux penetrating a Gaussian surface, with the electric field $\mathscr{E}$ at any point on the surface. We have already considered such a connection in Equation 24.4b, which gives Gauss's law in the restricted form

$$\mathscr{E} = \frac{d\Phi}{dA};$$

here dA is an infinitesimal area normal to the flux lines and $d\Phi$ is the flux penetrating that area. This equation is not applicable to a Gaussian surface of arbitrary shape because an area element on such a surface is not in general normal to the flux lines penetrating it. But we can overcome this difficulty by a straightforward generalization of Equation 24.4b.

We begin by defining a vector that describes not only the area of a surface element, but its orientation as well. Figure 24.12a shows an element of area dA. It may be a part of a larger curved surface, but if dA itself is sufficiently small it is indistinguishable from a planar surface element. We draw a vector $d\mathbf{A}$ according to the following recipe:

1. The direction $d\hat{\mathbf{A}}$ is taken normal to the surface element. Because the direction of the normal to a plane is determined by the orientation of the plane, we specify the orientation of the plane when we specify the direction of the normal.

2. The magnitude dA is equal to the magnitude of the area.

The vector $d\mathbf{A} \equiv dA\,d\hat{\mathbf{A}}$ then expresses both the area of the element and its orientation.

This recipe for defining $d\hat{\mathbf{A}}$ is incomplete because it does not distinguish between the vector $d\mathbf{A}$ shown and its negative. We eliminate this ambiguity by adding a third step to the recipe:

3. When the area element is a part of a closed surface, the outward sense determines the direction of the unit normal vector $d\hat{\mathbf{A}}$ (Figure 24.12b).

Now consider the dot product $\mathscr{E} \cdot d\mathbf{A}$. The flux lines penetrate dA in some arbitrary direction, and $\mathscr{E}$ is tangent to the flux lines (all of which are essentially parallel over the infinitesimal area element). Let θ be the angle between $\mathscr{E}$ and $d\mathbf{A}$. As Figure 24.12c shows, the same flux lines that penetrate dA also penetrate an element of area $dA' = dA \cos \theta$ that lies normal to $\mathscr{E}$. This element satisfies the condition for using Equation 24.4c to write $d\Phi = \mathscr{E}\,dA'$; we have

$$d\Phi = \mathscr{E}\,dA \cos \theta.$$

We write this in the more convenient and more general form

$$d\Phi = \mathscr{E} \cdot d\mathbf{A}. \tag{24.6}$$

The *total* flux Φ penetrating the Gaussian surface is found by integrating Equation 24.6 over the whole surface:

$$\Phi = \int_{\substack{\text{closed} \\ \text{surface}}} d\Phi = \int_{\substack{\text{closed} \\ \text{surface}}} \mathscr{E} \cdot d\mathbf{A}. \tag{24.7}$$

If we know the net charge q enclosed within the Gaussian surface, we can also use Equation 24.2b, $\Phi = q/\epsilon_0$, to evaluate the total flux. Equating the two values of Φ, we obtain **Gauss's law**,

$$\frac{q}{\epsilon_0} = \int_{\substack{\text{closed} \\ \text{surface}}} \mathscr{E} \cdot d\mathbf{A}. \tag{24.8}$$

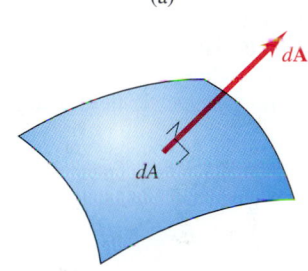

(a)

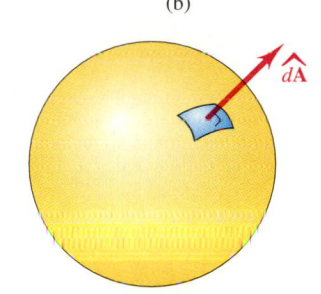

(b)

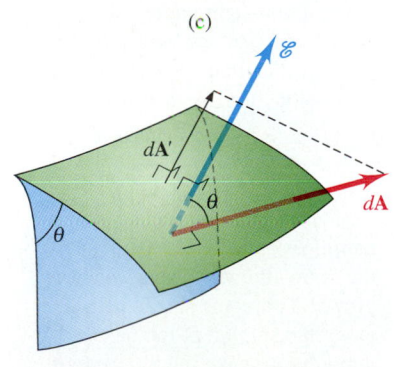

(c)

FIGURE 24.12 (*a*) The vector $d\mathbf{A}$ is normal to the area element and has magnitude dA equal to the area of the element. (*b*) When an area element is part of a closed surface, the normal vector directed outward is chosen as positive. (*c*) Verification that $\mathscr{E} \cdot d\mathbf{A} = \mathscr{E}\,dA = d\Phi$.

Electric Field inside a Uniformly Charged Spherical Shell

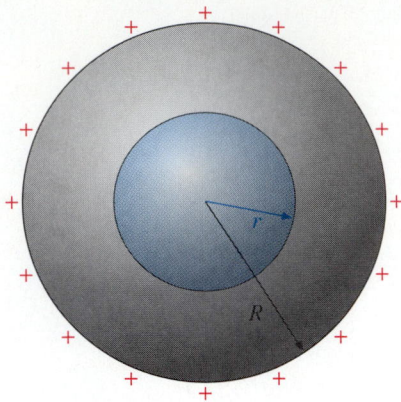

When combined with the symmetry properties of certain charge distributions, Gauss's law becomes a powerful tool for evaluating the electric field due to these charge distributions, almost without calculation. As an introduction to the use of Gauss's law in this way, we use Equation 24.6 to evaluate the electric field inside a spherical shell *on* which a charge q is uniformly distributed. Figure 24.13 shows a charged shell of radius R. To determine the value of the electric field *inside* the sphere, we draw a spherical Gaussian surface of radius $r < R$ within the shell and concentric with it.

Because there is no charge within this Gaussian surface, we know that the total flux Φ penetrating it must be zero. We use the symmetry of the system to show that the flux $d\Phi$ penetrating any element dA of the surface is likewise zero. Suppose that a flux line emerges from the enclosed region through some surface element $d\mathbf{A}$. Then it must have entered the region through some other element $d\mathbf{A}'$. But the symmetry dictates that flux must penetrate all elements in the same way. We cannot have flux lines passing inward through some elements and outward through others. Thus $d\Phi = 0$, and Equation 24.6 becomes

$$0 = \mathscr{E} \cdot d\mathbf{A}.$$

FIGURE 24.13 A charge q is distributed uniformly over a spherical shell of radius R. The spherical Gaussian surface of radius $r < R$ is used to determine the electric field inside the shell.

Because $d\mathbf{A} \neq 0$, we must have $\mathscr{E} = 0$ everywhere on the Gaussian surface. But we have placed no restriction on the radius r of the surface, except that it be smaller than the shell radius R. We conclude that *the electric field due to charge uniformly distributed over a spherical shell is zero everywhere inside the shell.*

Electric Field outside a Uniformly Charged Spherical Shell

We now use Gauss's law to evaluate the electric field at any point outside a uniformly charged spherical shell. We draw a spherical Gaussian surface of radius $r > R$, concentric with the shell, as shown in Figure 24.14. Next, we use the symmetry of the system as a guide in drawing the flux lines. Because there is charge (taken to be positive for the sake of argument) on the surface of the shell, flux lines must originate there. These lines cannot enter the shell (why not?); they must go outward. Symmetry dictates that the lines must be distributed uniformly all around the shell and must be directed radially as shown.

We now apply Equation 24.8 to the spherical Gaussian surface. The electric field $\mathscr{E}$ is parallel to the local area vector $d\mathbf{A}$ everywhere on the surface, so $\mathscr{E} \cdot d\mathbf{A} = \mathscr{E}\, dA$.

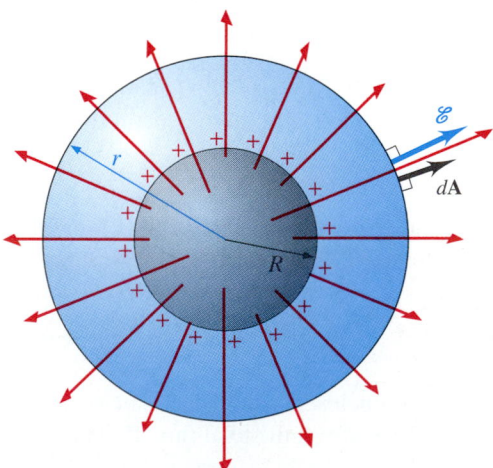

FIGURE 24.14 A charge q is distributed uniformly over a spherical shell of radius R. The spherical Gaussian surface of radius $r > R$ is used to determine the electric field at a distance r from the center of the shell.

Tonsorial Topology

Why must flux lines radiate directly outward? Suppose you argue instead that the lines emanate outward from a charged spherical shell but at a non-right angle with respect to the surface, which is the same all over the shell. A theorem of topology asserts that this is impossible. A whimsical but vivid way of stating the theorem is as follows: Suppose someone gives you a wonderful hair tonic that can grow hair even on a billiard ball. You rub the tonic all over the ball, and hair grows! You comb the hair down along the surface of the ball, attempting to make the finished job look perfectly uniform. But no matter how you do it, you will always have either one or more parts (lines of discontinuity), or two or more points of discontinuity, or some combination of parts and points. Think about it!

Also by symmetry, the magnitude $\mathscr{E}$ must be the same everywhere on the surface. The charge q on the shell lies entirely within the surface, and we have

$$\frac{q}{\epsilon_0} = \int_{\substack{\text{Gaussian}\\\text{sphere}}} \mathscr{E} \cdot d\mathbf{A} = \int_{\substack{\text{Gaussian}\\\text{sphere}}} \mathscr{E}\, dA = \mathscr{E} \int_{\substack{\text{Gaussian}\\\text{sphere}}} dA.$$

Having taken $\mathscr{E}$ outside the integral, we are left with a very simple integral. Its value is just the surface area of the sphere, $A = 4\pi r^2$. So we have $q/\epsilon_0 = 4\pi r^2 \mathscr{E}$, or

$$\mathscr{E} = \frac{1}{4\pi\epsilon_0}\frac{q}{r^2}. \tag{24.9}$$

The electric field due to charge uniformly distributed over a spherical shell is the same, everywhere outside the shell, as the field due to an equal point charge located at the center of the shell.

EXAMPLE 24.2

A charge Q is distributed uniformly throughout a dielectric sphere of radius R. Find $\mathscr{E}(r)$, the electric field as a function of distance from the center of the sphere.

SOLUTION: Because you know the expressions for the electric field of a uniformly charged shell both inside and outside the shell, you begin by dividing the solid sphere in imagination into a nest of concentric spherical shells, each of thickness dr. To evaluate the field outside the sphere, you need make no further calculations. Each shell makes a contribution to the field just as if its charge were concentrated at the center of the sphere. Thus the contribution of all the shells, taken together, is the same as that of a point charge Q at the center of the sphere. According to Equation 24.9, you have

$$\mathscr{E}(r) = \frac{Q}{4\pi\epsilon_0}\frac{1}{r^2} \quad \text{for } r > R. \tag{24.10}$$

The direction of the field at any external point $\mathbf{r}$ is $\pm\hat{\mathbf{r}}$, depending on the sign of Q.

Next, you find the field inside the sphere, at a distance $r < R$ from the center. You begin by noting that only that part of the charge located within the sphere of radius r contributes to the field at r. The rest of the charge is "outside" r—more precisely, outside the Gaussian sphere whose radius is r—and thus makes a zero contribution to the field at r.

What part q of the total charge Q lies "inside" r? Because the charge is uniformly distributed throughout the volume of the dielectric sphere, you have

$$\frac{q}{Q} = \frac{v}{V},$$

where V is the volume of the entire sphere and v is the volume of the part within the radius r. Using the standard formula for

the volume of a sphere, you have

$$q = \frac{\frac{4}{3}\pi r^3}{\frac{4}{3}\pi R^3}\, Q = \frac{r^3}{R^3}\, Q.$$

At r, the field due to the charge q is the same as that of a point charge q located at the center of the sphere. So you have

$$\mathscr{E}(r) = \frac{1}{4\pi\epsilon_0}\frac{q}{r^2} = \frac{1}{4\pi\epsilon_0}\frac{Q}{R^3}\, r. \tag{24.11}$$

That is, within the sphere the electric field increases linearly with distance from the center. Inside as well as outside the sphere, the direction of the field at any point lies along the radius through that point.

As a check, you can show that Equations 24.10 and 24.11 give the same result at the surface of the sphere, $r = R$. Figure 24.15 is a plot of $\mathscr{E}(r)$.

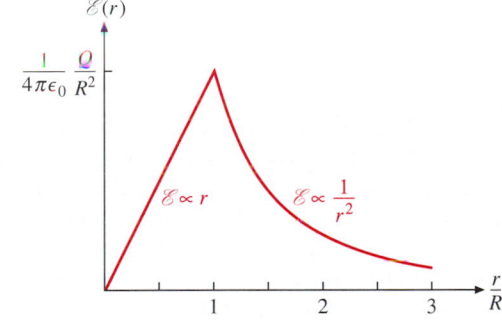

FIGURE 24.15 Plot of the electric field $\mathscr{E}(r)$ due to a charge Q distributed uniformly throughout a sphere of radius R as a function of distance r from the center of the sphere.

If you refer back to the final subsection of Section 14.3, you will see that the Gauss's-law arguments we are making here in the context of electric charges and electric fields apply equally well to gravitating masses and gravitational fields. (The gravitational field $\mathbf{\Gamma} \equiv \mathbf{F}/m_t$ is the gravitational force per unit mass exerted on a test mass m_t in the vicinity of other masses m.) In order to compare the acceleration g at the surface of the earth with the gravitational acceleration of the moon due to the earth, Newton needed to show that the gravitational attraction of a spherical body is equal to that of a point mass located at the center of the body. This point is discussed in Sections 14.2 and 14.3. With considerable difficulty, Newton devised a "brute force" proof. A similar proof can be made in the electrical case, using Coulomb's law, the electric analogue of Newton's law of gravitation. But Gauss's law makes the proof simple and direct, as you have just seen.

Applications of Gauss's Law to Systems of Lower Symmetry

By combining Gauss's law with the symmetry of a spherical charge distribution, we have been able to deduce the electric field as a function of r. In this section, we apply Gauss's law to a series of systems of progressively lower symmetry.

EXAMPLE 24.3

A long, thin, straight wire has uniform charge per unit length λ. Use Gauss's law to find the magnitude and direction of the electric field at a distance r from the wire, far from the ends of the wire. (The "brute force" approach to this problem, using Coulomb's law, is the subject of Problem 23.45.)

SOLUTION: As in the charged-sphere case, you begin by drawing the proper Gaussian surface. Such a surface has two properties: The product $\mathscr{E} \cdot d\mathbf{A}$ is readily evaluated everywhere, and the integration of Equation 24.8 is readily carried out. The trick in finding such a surface is to exploit the symmetry of the physical system. Just as the symmetry of an isolated point is spherical, the symmetry of a long, thin wire is cylindrical. So you draw as the Gaussian surface a cylinder of radius r coaxial with the wire, as shown in Figure 24.16. Because a Gaussian surface must be closed, you terminate the cylinder with flat end pieces; call the length of the cylinder l.

The symmetry also dictates the direction of the flux lines. The lines must lie in planes normal to the wire, and they must radiate uniformly in all directions in those planes, as shown in Figure 24.16. (If a flux line were not perpendicular to the wire, it would point toward one end of the wire and away from the other. But the ends are far away and so cannot influence the direction of the flux lines.) As you can see from the figure, $\mathscr{E} \cdot d\mathbf{A} = \mathscr{E}\, dA$ everywhere on the curved surface of the cylinder. Also by symmetry, the magnitude $\mathscr{E}$ is the same everywhere on the curved surface, and $\mathscr{E}$ can be taken outside the integral in applying Gauss's law.

What about the ends of the cylinder? The flux lines that lie along these end planes are everywhere perpendicular to the elemental area vectors $d\mathbf{A}$ that describe the ends (and are parallel to the cylinder's axis). So you have $\mathscr{E} \cdot d\mathbf{A} = 0$ everywhere on the ends; thus the flux lines at the ends contribute nothing to the integral of Gauss's law.

The total charge contained within the Gaussian surface is $q = \lambda l$. You can use Equation 24.8 to write

$$\frac{\lambda l}{\epsilon_0} = \mathscr{E} \int_{\substack{\text{Gaussian} \\ \text{cylinder}}} dA.$$

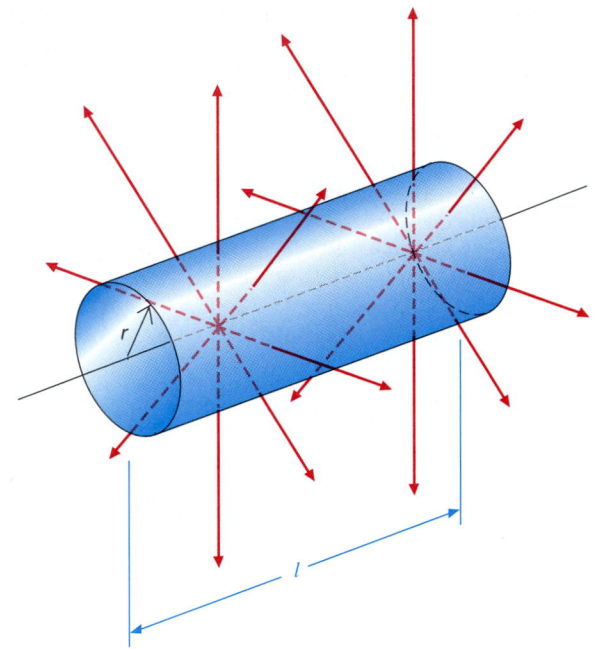

FIGURE 24.16 Radial flux lines and the cylindrical Gaussian surface.

The integral is just the surface area of the curved part of the cylinder, which is $2\pi r l$. The equation thus becomes $\lambda l/\epsilon_0 = 2\pi r l \mathscr{E}$, and you have

$$\mathscr{E} = \frac{1}{2\pi\epsilon_0} \frac{\lambda}{r}. \tag{24.12}$$

Owing to the cylindrical symmetry of the system, the field is inversely proportional to the distance r from the wire. The direction of the field at any point $\mathbf{r}$ lies along the perpendicular from $\mathbf{r}$ to the wire.

The cylindrical symmetry of Example 24.3 is a lower order of symmetry than the spherical symmetry of Example 24.2. In spherical symmetry, all directions are alike. In cylindrical symmetry, all directions in the plane normal to the symmetry axis are alike, but the direction along the symmetry axis is different. Example 24.4 explores the still lower symmetry of the plane.

EXAMPLE 24.4

A large, flat plate has uniform charge σ per unit area. Use Gauss's law to find the magnitude and direction of the electric field in the vicinity of the plate, far from the edges. (See the solution to this problem given by Equation 23.34, obtained by means of Coulomb's law.)

SOLUTION: Assume for the sake of argument that the charge is positive. Symmetry requires that the flux lines emanate uniformly from the charged plate along normals to it. The symmetry permits no distinction between the "up" and "down" directions, and so the flux line distribution must have the form shown in Figure 24.17. If you draw the Gaussian surface as shown, with all faces either parallel to or perpendicular to the plate, the flux lines will everywhere be either parallel to the faces or normal to them. (Would it make any difference if you drew the Gaussian surface as a cylindrical "pillbox" rather than a parallelepiped?) Let A be the area of the top face and thus also of the bottom face.

The area vectors describing the four side faces of the Gaussian surface are all parallel to the plate. Thus the flux lines are everywhere perpendicular to the area vectors, and you can set the product $\mathscr{E} \cdot d\mathbf{A}$ equal to zero along these side faces. (Compare with the similar argument made for the ends of the cylinder in Example 24.3.) Because the flux lines penetrate both top and bottom faces in the outward sense, you have $\mathscr{E} \cdot d\mathbf{A} = \mathscr{E} \, dA$ for area elements $d\mathbf{A}$ on both faces. The charge within the Gaussian surface is $q = \sigma A$. Thus Equation 24.8 becomes

$$\frac{\sigma A}{\epsilon_0} = \mathscr{E} \int_{\substack{\text{top and} \\ \text{bottom} \\ \text{faces}}} dA.$$

The value of the integral is $2A$. So you have $\sigma A/\epsilon_0 = 2A\mathscr{E}$, or

$$\mathscr{E} = \frac{\sigma}{2\epsilon_0}. \qquad (24.13)$$

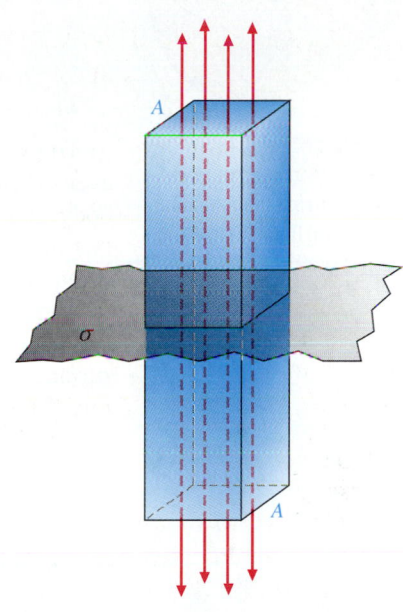

FIGURE 24.17 Flux lines normal to a plane and the corresponding Gaussian surface.

This is identical with Equation 23.34 but requires much less calculation to obtain. Owing to the planar symmetry of the system, the field is independent of the distance r from the plate. The direction of the field at any point not near the edges of the plate lies along the normal from the plate to that point. (If the plate were infinite in extent, the field would be the same everywhere outside it. For a finite plate, Equation 24.13 holds for any point whose distance from the nearest edge of the plate is much greater than its distance above the plate.)

There is an intriguing connection between the order of symmetry of a system of charges and the r dependence of its electric field. We can summarize the connection as follows. Let $\mathscr{E}(r) \propto r^{-n}$. Then the value of n depends on the symmetry thus:

Spherical or point (3-dimensional) symmetry:	$n = 2$
Cylindrical or line (2-dimensional) symmetry:	$n = 1$
Planar (1-dimensional) symmetry:	$n = 0$

Static Charge Distribution in a Conductor of Arbitrary Shape

Although we have depended heavily on the symmetry properties of various systems in applying Gauss's law, the law has powerful applications to systems of no particular symmetry. Figure 24.18 shows a body of arbitrary shape, made of a conducting material.

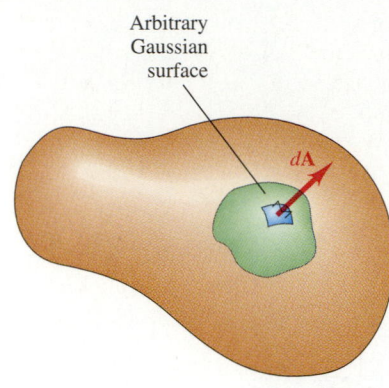

FIGURE 24.18 Electric charge in an equilibrium distribution over a conducting body of arbitrary shape. The arbitrary Gaussian surface inside the body is used to prove theorem 2.

The body may be solid or it may have one or more cavities within it. A quantity of electric charge is placed on the body, so its net charge is no longer zero. Charge flows freely through the body until an equilibrium charge distribution is achieved. We now prove three closely related theorems for this equilibrium state of a conducting body:

1. The electric field within the body is everywhere zero.

2. All the net charge lies on the outer surface of the body.

3. The electric field at a point P just outside the body is directed normal to the surface and has magnitude

$$\mathscr{E} = \frac{\sigma}{\epsilon_0}, \tag{24.14}$$

where σ is the (net) charge per unit area on the part of the surface near P.

To prove theorem 1, remember that there is always free charge within a conductor, even when it is neutral overall. If the field has a nonzero value anywhere within the body, some free charge will "see" this field and move. This contradicts the assumption that the charge is in equilibrium; indeed, the charge will move until it *is* distributed in an equilibrium fashion. When this is achieved, $\mathscr{E} = 0$ within the body.

To prove theorem 2, draw an arbitrary Gaussian surface within the body, as shown in Figure 24.18. According to theorem 1, $\mathscr{E}$ must be equal to zero everywhere on this surface. Thus the integrand $d\Phi = \mathscr{E} \cdot d\mathbf{A}$ is everywhere zero. Consequently, the *total* flux Φ penetrating the Gaussian surface must be zero. From Gauss's law, the net charge q contained within the Gaussian surface is therefore zero. This argument is independent of the size and shape of the Gaussian surface, as long as it lies entirely within the body. Thus we conclude that all the charge must lie on the surface of the body.

The proof of theorem 3 has two parts. First, suppose that a flux line leaves the surface in a nonnormal direction. Because the electric field is tangent to the flux line, there must then be a component of the field along the surface. But such a field component would lead to motion of charge along the surface, again contradicting the assumption of equilibrium.

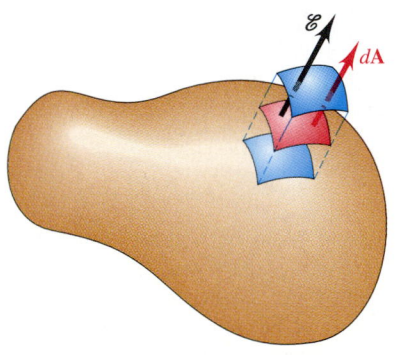

FIGURE 24.19 Construction to prove theorem 3. The Gaussian box is constructed partly inside and partly outside the conducting body. The flat ends of the cylinder are oriented parallel to the surface element $d\mathbf{A}$ of the body surface.

Now that we have established that $\mathscr{E}$ is normal to the body surface, we draw a small Gaussian "box"—a parallelepiped just like the one we used in Example 24.4 to evaluate the field near a charged flat plate. This Gaussian box is shown in Figure 24.19. In our consideration of the flat plate, we asserted on the basis of symmetry that the value of σ was uniform over the plate. We cannot do the same here. (Indeed, σ can vary considerably over the surface of the body, as we will see in Section 25.4.) But we can always draw the Gaussian box small enough to ensure that the charge dq on the enclosed body surface element $d\mathbf{A}$ is distributed essentially uniformly over the element. Then we can write $dq = \sigma\, dA$, with σ substantially constant over $d\mathbf{A}$.

According to theorem 1, we must have $\mathscr{E} = 0$ inside the body. Thus no flux penetrates the part of the Gaussian box inside the body. Outside the body, $\mathscr{E}$ is normal to the body surface and flux penetrates only the upper part of the Gaussian box. For this infinitesimal box, we write Gauss's law in the form

$$\frac{dq}{\epsilon_0} = \frac{\sigma\, dA}{\epsilon_0} = \mathscr{E} \cdot d\mathbf{A} = \mathscr{E}\, dA.$$

We thus have

$$\mathscr{E} = \frac{\sigma}{\epsilon_0},$$

which is Equation 24.14.

EXAMPLE 24.5

A uniform electric field $\mathscr{E}$ exists over a very large region of space. A flat conducting plate, oriented so that its faces are normal to the field, is inserted into the field (Figure 24.20). Find the charge per unit area on each surface, at locations not near the edges of the plate.

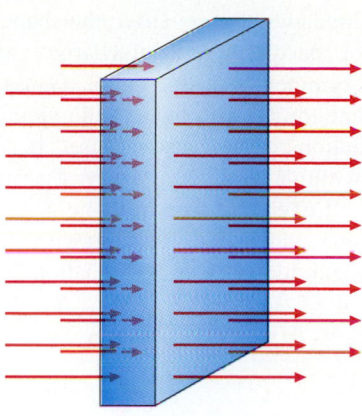

FIGURE 24.20

SOLUTION: Because the plate is a conductor, you know that the field must be zero inside it. But the presence of the finite plate cannot affect the field at indefinitely distant points. That is, the uniform field can be distorted only within the general vicinity of the plate. Thus the flux lines, which were continuous in the absence of the plate, must terminate on the left surface of the plate and resume at the right surface. This requires that there be a uniform distribution σ_- of *negative* charge on the left surface, on which the uniform flux lines that fill the left side of the space can terminate. Similarly, there must be a uniform distribution of *positive* charge σ_+ on the right surface, at which uniform flux lines that fill the right side of the space can originate. Because the electric field has the same value $\mathscr{E}$ everywhere outside the plate, you must have $|\sigma_-| = |\sigma_+| \equiv \sigma$. And from Equation 24.14 you have $\sigma = \epsilon_0 \mathscr{E}$. So the surface charges per unit area are

$$\sigma_- = -\epsilon_0 \mathscr{E} \quad \text{and} \quad \sigma_+ = \epsilon_0 \mathscr{E}.$$

Because of the symmetry of the situation, the presence of the plate does not distort the electric field outside it at all, even at points quite close to its surface. Why would this not be true for a conducting body of arbitrary shape?

THINKING LIKE A PHYSICIST

We have interwoven three conceptual threads to fabricate a powerful tool. The three threads are Faraday's flux lines, the Gaussian surface, and symmetry. At first glance, all three appear to be qualitative rather than quantitative concepts. (That is, a flux line extends continuously from its source to its termination; a closed surface separates inside from outside, whatever its shape may be; when a sphere or a long cylinder has been rotated about a symmetry axis, you cannot see any change.) Nevertheless, we have used these apparently qualitative concepts to develop a method for making quantitative field calculations as well as for achieving in-sights into the workings of the inverse-square Coulomb force (or any other inverse-square force).

As the epigraphs at the beginning of this chapter suggested, deep mathematical ideas are not necessarily associated with complicated calculation. Indeed, one of the problems in mastering the subject matter of this chapter is that you cannot surmount conceptual difficulties by burying yourself in mechanical calculation and hoping that insight will emerge if you just keep calculating long enough. That doesn't mean working problems won't help; just make sure you understand what you are doing when you calculate!

Summing Up

A **flux line** is the path taken by a test charge if it is allowed to move slowly under the influence of the electric force due to the presence of other charges. The particular flux line followed by the test charge depends on the starting point chosen. The totality of flux lines constitutes a map of the electric field.

Flux lines possess the following three important qualitative properties:

1. They emanate from positive source charges and terminate at negative source charges.

2. They cannot branch or cross one another.

3. The electric field $\mathscr{E}$ at a given location is tangent to the local flux lines; $\mathscr{E}$ has the direction given by the sense of the flux lines and a magnitude equal to the flux per unit area penetrating a surface normal to the flux lines.

The number Φ of flux lines emanating at a source charge is proportional to the charge q. The proportionality can be expressed in the form of Equation 24.2b,

$$\Phi = \frac{1}{\epsilon_0} q.$$

Equation 24.6 relates the number of flux lines $d\Phi$ that penetrate an oriented area element $d\mathbf{A}$ to the local electric field $\mathscr{E}$:

$$d\Phi = \mathscr{E} \cdot d\mathbf{A}.$$

That is, the magnitude of the electric field is equal to the flux per unit area normal to the flux lines; see Equations 24.4a, 24.4b, and 24.4c.

Gauss's law (Equation 24.8) relates the total flux Φ penetrating a closed **Gaussian surface** to the net charge q contained within the surface:

$$\frac{q}{\epsilon_0} = \int_{\substack{\text{closed} \\ \text{surface}}} \mathscr{E} \cdot d\mathbf{A}.$$

This relation holds without regard to the shape of the Gaussian surface or the distribution of charge within it. Taken together with symmetry considerations, Gauss's law can be used to evaluate the electric field in the vicinity of various charge distributions. It can also be used to show that the electric field within a conducting body of arbitrary shape is zero and that all excess charge must lie at its surface. The field just outside such a body is everywhere normal to the surface; its magnitude is given by Equation 24.14,

$$\mathscr{E} = \frac{\sigma}{\epsilon_0}.$$

KEY TERMS

Section 24.2 Electric Charge and Electric Flux
flux line, electric flux ▪ source charge ▪ electric field

Section 24.3 Gaussian Surfaces and Gauss's Law
Gaussian surface ▪ Gauss's law

Queries and Problems for Chapter 24

QUERIES

24.1 (2) *Freehand.* Sketch the flux-line pattern in the vicinity of two equal positive charges q.

24.2 (2) *L-bow.* Can a flux line have a sharp kink in it? Why or why not?

24.3 (2) *A special place.* Consider the point midway between two equal positive point charges. What is the electric field at this point? What do the flux lines look like in the immediate vicinity of the point?

24.4 (2) *To what end?* In parts b and c of Figure 24.9, the flux lines appear to terminate not at an electrode—a conducting wire, plate, or other body—but at the outer boundary of the liquid. If the electrodes seen in the photographs are positive, where is the negative electrode?

24.5 (2) *The perils of language.* In this chapter and the preceding one, we have followed common practice in using the world "field" in two related but distinct senses. One of these senses is made explicit in Equation 24.1, $\mathscr{E} = \mathscr{E}(\mathbf{r})$. The other sense is that of Equation 23.10, $\mathscr{E} \equiv \mathbf{F}/q$. Distinguish clearly between these two senses. Why is it justifiable to use the same term for both?

24.6 (2) *Tell it what to do.* Draw a flow chart for a computer program to be used in drawing a flux-line map.

24.7 (3) *Way in or far out?* What is the net flux penetrating the Gaussian surface shown in figure a? in figure b?

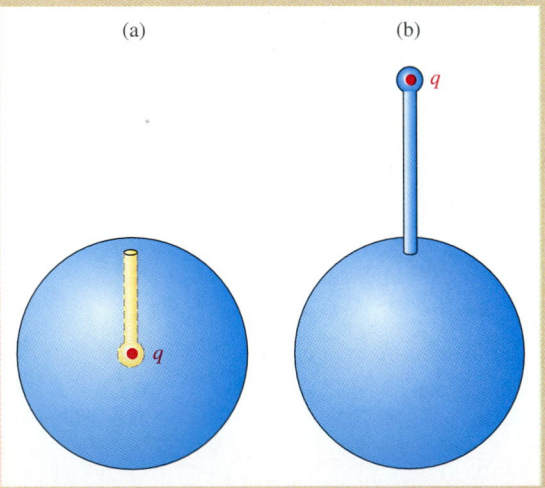

(a)   (b)

q

q

24.8 (3) *Topology lesson.* The figure on the next page shows a torus (an ideal doughnut) and two charges. Charge q lies at the center of the "cake" part of the doughnut, and charge q' lies at the center of the hole. What is the net flux penetrating the surface of the torus? (Hint: If you eat the doughnut, is it "inside" or "outside" your body when it sits in your stomach?)

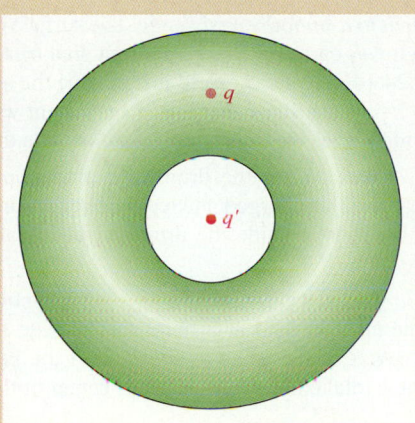

24.11 *(3) Is an ashtray a charger?* Inside a metal saucepan that rests on a dielectric pad, you place a small glass ashtray. On the ashtray, you rest a metal sphere containing charge q. You then put the cover on the saucepan and place a charge q' on the saucepan. **(a)** What is the total charge on the inside surface of the saucepan, including the cover? **(b)** What is the total charge on the outside surface?

24.12 *(3) Grave considerations.* A group of bodies has a total mass M. In imagination, surround the bodies with a closed surface. Write a form of Gauss's law that connects M with the gravitational flux Φ_G that penetrates the surface.

24.13 *(3) Field work.* If a spherical shell has a uniform charge per unit area σ, the electric field inside the shell is zero everywhere. Under what circumstances is this true for a *dielectric* shell? Is the field zero inside a dielectric shell of arbitrary shape, on which σ is constant? The field *is* zero inside a conducting shell of arbitrary shape. Is the value of σ constant over such a shell?

24.9 *(3) Cast your net.* What is the net flux penetrating a Gaussian surface drawn around an electric dipole?

24.10 *(3) Partial view.* A Gaussian sphere surrounds an electric dipole. Is the electric field zero everywhere on the sphere? If you divide the surface into hemispheres by means of a cut normal to the dipole axis and passing through its midpoint, what can you say about the flux penetrating the two hemispheres?

24.14 *(4) Paradox?* The magnitude of the electric field immediately outside a conducting body is given by Equation 24.14, $\mathscr{E} = \sigma/\epsilon_0$, where σ is the surface charge per unit area in the neighborhood of the point where $\mathscr{E}$ is to be evaluated. This value is just twice that for the field in the vicinity of a charged conducting flat plate, given by Equation 24.13. Explain.

PROBLEMS

GROUP A

24.1 *(2) SI-entific.* What is the SI unit for electric flux, Φ?

24.2 *(2) Go with the flux.* Find the flux Φ associated with an electric charge $q = -2.5$ C.

24.3 *(2) Skewed flux, I.* A circular paper disk of radius 8.7 mm is located in a uniform electric field of magnitude 3300 N/C. If the flux penetrating the disk is 0.45 N·m²/C, find the angle between the flux lines and the axis of the disk.

24.4 *(2) Skewed flux, II.* A rectangular surface of sides 1.2 cm and 1.7 cm is located in a uniform electric field of magnitude 750 N/C. The angle between the flux lines and the surface is 32°. Find the flux penetrating the surface.

24.5 *(2) Empty sphere.* A sphere of radius R is placed in a uniform external electric field $\mathscr{E}$. Find the magnitude Φ of the flux penetrating the sphere.

24.6 *(3) Ball game.* The surface of a metal sphere of radius R carries a charge per unit area σ. **(a)** What is the flux Φ emanating from the surface of the sphere? **(b)** Find the electric field $\mathscr{E}(r)$ for $r < R$ and for $r > R$. If $R = 2.3$ m and $\sigma = -4.7 \times 10^{-6}$ C/m², **(c)** evaluate Φ and **(d)** evaluate $\mathscr{E}$ at $r = 6.0$ m.

24.7 *(3) Russian dolls, I.* Two concentric hollow spheres made of thin metal have radii $r_1 = 22.3$ cm and $r_2 = 47.6$ cm.

The charges on the spheres are $q_1 = -37.4$ nC and $q_2 = 51.2$ nC. Find the magnitude of the electric field at **(a)** $r = 15.5$ cm, **(b)** $r = 36.9$ cm, and **(c)** $r = 64.8$ cm.

24.8 *(3) Russian dolls, II.* Two concentric hollow spheres made of thin metal have radii $r_1 = 32.4$ cm and $r_2 = 58.7$ cm. At a point 81.3 cm from the center of the spheres, the electric field is directed outward and has magnitude 228 N/C. At a point 46.2 cm from the center, the electric field is directed inward and has magnitude 936 N/C. Find the charges q_1 and q_2 on the spheres.

24.9 *(3) Pooling resources.* A spherical drop of mercury of radius 1.1 mm is charged so that the electric field 2.6 mm from its center is 88 N/C. If the drop merges with seven other identical drops having the same charge to form a larger spherical drop, what is the electric field **(a)** at the same distance and **(b)** at half this distance from the center of the merged drop?

24.10 *(3) Doubling up.* Two long, parallel wires are separated by a distance $2d$. The charge per unit length on each is λ. **(a)** What is the value of $\mathscr{E}$ at a point midway between the wires? **(b)** Consider the plane that passes through the midline between the wires and is perpendicular to the plane of the wires. What is the direction of $\mathscr{E}$ at a point in this plane?

(c) Sketch a graph of $\mathscr{E}$ versus z. **(d)** At what distance z from the midline does the magnitude of the electric field attain its maximum value $\mathscr{E}_{max}$? **(e)** What is the value of $\mathscr{E}_{max}$?

24.11 *(4) Imperfection.* A large steel casting contains a void of irregular shape. Within the void is a point charge $q = 3.8 \times 10^{-6}$ C. Evaluate the electric field at a point within the void at a distance 2.9 cm from the charge if the charge is distant from any part of the steel.

24.12 *(4) Cash or charge?* A U.S. quarter has a diameter of 2.4 cm. If the quarter is placed in a uniform electric field of magnitude $\mathscr{E} = 3.0 \times 10^4$ N/C and is oriented with its faces normal to the flux lines, find **(a)** the charge density σ on each face of the quarter and **(b)** the total charge on each face.

24.13 *(4) Beat to a smooth consistency.* The surface of a metal sphere of radius R carries a charge per unit area σ. According to Equation 24.14, the electric field just outside the sphere has magnitude $\mathscr{E} = \sigma/\epsilon_0$. Show that this is consistent with Equation 24.9, which gives the field at any point outside the sphere.

24.14 *(4) Copycat.* When the drum in a photocopying machine (Section 23.2) is charged and ready for exposure to light, the electric field just outside the drum surface is about 2×10^5 N/C. Find σ.

24.15 *(4) Consistency test.* Beginning with Equation 24.14, show that the electric field at a point just outside a charged conducting sphere is the same as the field at the same point due to the same total charge located at the center of the sphere.

GROUP B

24.16 *(2) Cube, I.* A point charge q lies at the center of one face of an imaginary cube. Find the total flux penetrating the other five faces of the cube.

24.17 *(2) Cube, II.* A point charge q lies at one corner of an imaginary cube. **(a)** Find the total flux penetrating the faces of the cube. **(b)** Find the flux penetrating each of the three faces that join at the corner where the charge is located. **(c)** Find the flux penetrating each of the other three faces.

24.18 *(2) Through the hoop.* An electric dipole consists of charges q and $-q$ separated by a distance $2b$. A circle of radius R, oriented normal to the dipole axis, lies with its center at the center of the dipole axis. Show that the magnitude of the flux penetrating the circle is

$$|\Phi| = \frac{q}{\epsilon_0}\left(1 - \frac{1}{\sqrt{1 + R^2/b^2}}\right).$$

24.19 *(2) Any way you slice it . . .* A dielectric sphere of radius R carries a uniform charge per unit volume ρ. If you slice through the sphere, you define a circle whose center lies a distance $a \leq R$ from the center of the sphere. Show that the magnitude of the flux penetrating this circle is

$$|\Phi| = \frac{\pi\rho}{3\epsilon_0}a(R^2 - a^2).$$

(Hint: Consider a point on the circle that lies a distance r from the center of the sphere and evaluate $\mathscr{E}$. Then, remembering that you are really interested only in the component of $\mathscr{E}$ normal to the circle, evaluate that component. Call it $\mathscr{E}_z$, and show that $\mathscr{E}_z$ has the same value everywhere on the circle.)

24.20 *(3) Improving on Newton, I.* A dielectric sphere of radius R carries a uniform charge per unit volume ρ. Find the value of $\mathscr{E}(\mathbf{r})$ at a point $\mathbf{r}$ inside the sphere.

24.21 *(3) Thick wall.* A thick-walled dielectric spherical shell has inner radius a and outer radius b. The dielectric carries a uniform charge per unit volume ρ. Find $\mathscr{E}$ as a function of distance r from the center of the sphere for $r < a$, for $a < r < b$, and for $r > b$.

24.22 *(3) Rutherford atom.* About 1909, Ernest Rutherford began to probe the structure of atoms by bombarding them with charged particles and observing the deflection of the particles (Section 41.3). In one very crude model that Rutherford used to interpret his observations, he considered the atom a point charge $+Q$ in the center of a sphere of radius R over which a charge $-Q$ is uniformly distributed. Show that the electric field at a point P inside such an atom is

$$\mathscr{E} = \frac{Q}{4\pi\epsilon_0}\left(\frac{1}{r^2} - \frac{r}{R^3}\right)\hat{\mathbf{r}},$$

where $\mathbf{r}$ is the vector from the center of the sphere to P.

24.23 *(3) Out of uniform, I.* Electric charge is distributed through a solid dielectric sphere of radius R in such a way that the charge per unit volume is directly proportional to the distance r from the center of the sphere. That is, $\rho = ar$, where a is a constant. Find the magnitude $\mathscr{E}$ of the electric field as a function of r.

24.24 *(3) Out of uniform, II.* Electric charge is distributed through a solid dielectric sphere of radius R in such a way that the charge per unit volume is inversely proportional to the distance r from the center of the sphere. That is, $\rho = b/r$, where b is a constant. Find the electric field inside the sphere and show that it is independent of r.

24.25 *(3) Out of uniform, III.* The charge per unit volume ρ in a dielectric sphere of radius R varies according to the relation

$$\rho = \rho_0\left(1 - \frac{r}{R}\right),$$

where ρ_0 is the value of ρ at the center of the sphere. **(a)** Find the magnitude $\mathscr{E}(r)$ of the electric field for $r < R$. **(b)** Find $\mathscr{E}(r)$ for $r > R$. **(c)** For what value of r is $\mathscr{E}(r)$ a maximum? What is the value of $\mathscr{E}_{max}$?

24.26 *(3) Gauss's law and Coulomb's law.* **(a)** Draw a spherical Gaussian surface around a point charge q. Show that Gauss's law yields a value of the electric field identical with the value given by Coulomb's law and that the two laws are thus logically equivalent. **(b)** Using the result of part **a**, show that violation of theorem 1 immediately preceding Equation 24.14 implies violation of Coulomb's law. **(c)** The figure (p. 679) shows a way of verifying Gauss's and Coulomb's laws experimentally. A small conducting sphere carrying charge q, suspended from a dielectric thread, is lowered into a larger hollow conducting sphere through a small hole. Where is the charge on the small sphere? on the inner and outer surfaces

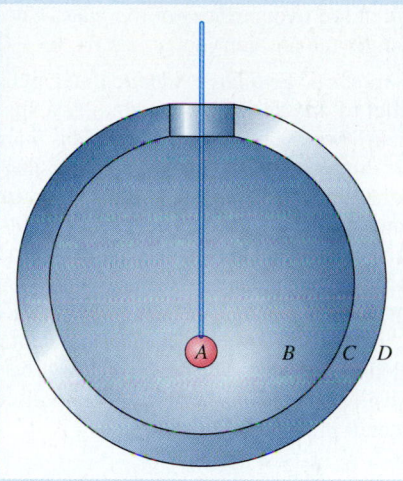

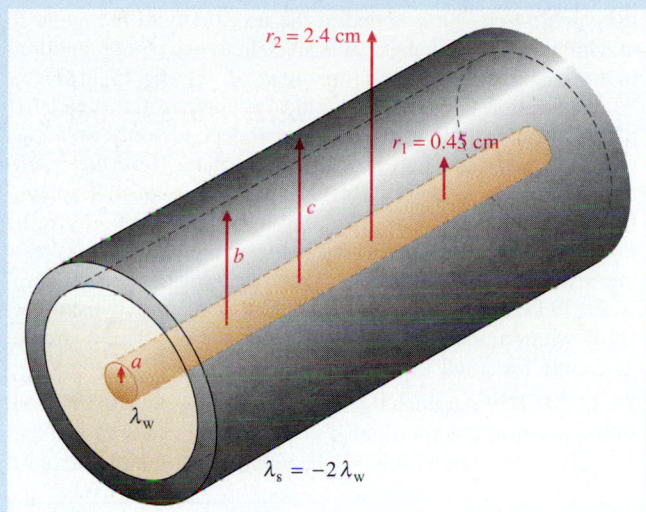

of the large sphere? What is the value of the electric field at a point in region A? region B? region C? region D? **(d)** The small sphere is allowed to touch the inner surface of the outer sphere and is then brought back to the position shown in the figure. What are the answers to the questions of part **c** now? **(e)** The small sphere is withdrawn from the large one through the hole. Do you expect any change from the situation in part **d**? The experiment of parts **c** through **e** is called the *Faraday ice-pail experiment* because Faraday performed it using a metal ice pail. Benjamin Franklin had made similar observations as early as 1755. The experiment has been repeatedly refined and Coulomb's law thus verified with increasing precision. This verification can be expressed in terms of the value of the exponent of r in Coulomb's law, which is now known not to deviate from 2 by as much as 1 part in 10^{16}.

24.27 *(3) Improving on Newton, II.* Suppose that you could drill a small hole through the earth along a diameter. Use the result of Problem 24.20 together with Gauss's law to find the time it would take for a stone dropped in at one end to reach the other end. (Because the problem is fanciful already, don't worry about the effects of air friction! Hint: Show that the motion of the stone is simple harmonic. Note: This problem is essentially identical with Problem 14.64, parts **a** through **d**.)

24.28 *(3) Shell game.* A long cylindrical shell of radius R carries a uniform surface charge σ per unit area. **(a)** Use Gauss's law to find the magnitude $\mathscr{E}(r)$ of the electric field due to this charge distribution, both for $r < R$ and for $r > R$. **(b)** Express λ, the charge per unit length of the shell, in terms of σ. **(c)** Express the result of part **a** in terms of σ, and compare with the result of Example 24.3.

24.29 *(3) Nesting instinct.* A long dielectric solid cylinder of radius R carries a uniform charge per unit volume ρ. **(a)** Find $\mathscr{E}(r)$ for $r > R$. **(b)** Find $\mathscr{E}(r)$ for $r < R$. **(c)** Express λ, the charge per unit length of the cylinder, in terms of ρ. **(d)** Compare the result of part **a** with that of Example 24.3. **(e)** Make a sketch of $\mathscr{E}(r)$ as a function of r.

24.30 *(3) Coaxial cable.* The figure at the top of the next column shows a long, straight metal wire of radius a surrounded by a metal cylindrical shell of inner radius b and outer radius c. The axes of wire and cylinder coincide. The wire carries a charge per unit length λ_w, and the shell carries a charge per unit length $\lambda_s = -2\lambda_w$. **(a)** Find the magnitude and direction of the electric field as a function of position $\mathbf{r}$, measured from the axis. Do this for $r < a$, $a < r < b$, $b < r < c$, and $r > c$. **(b)** Let $\lambda_w = 7.2 \times 10^{-7}$ C/m. Find the magnitude of the electric field at the points in the figure labeled $r_1 = 0.45$ cm and $r_2 = 2.4$ cm. **(c)** What is the charge per unit area σ on the inner surface of the shell? on the outer surface?

24.31 *(3) Hanging askew.* In the figure below, a long dielectric rod carries uniform charge per unit length λ. A pendulum consisting of a dielectric thread of length l and a bob having

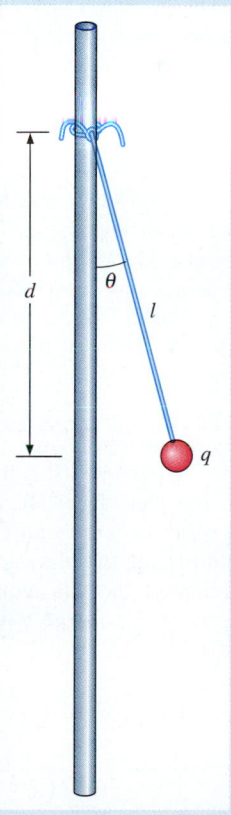

mass m and charge q is tied to the rod. **(a)** Find the angle θ at which the thread hangs. (It is all right to leave your solution in terms of trigonometric functions of θ.) **(b)** In Problem 23.21, the bob hangs in the field of a charged flat dielectric plate. In that problem, the value of θ depends only on m, q, and the charge per unit area σ on the plate. Here the result depends on m, q, and λ. But it also depends on l. Explain. **(c)** Let $l = 25$ cm, $m = 1.2$ g, and $\lambda = 1.3 \times 10^{-6}$ C/m. If the angle is $\theta = 5°$, what is the charge on the bob? **(d)** Express the result of part **b** directly in terms of θ for the special case where θ is small. **(e)** If the pendulum is disturbed from the equilibrium angle θ, it will oscillate. Is the oscillation simple harmonic for small θ? Why or why not?

24.32 *(4) Tilt!* An uncharged flat conducting plate is inserted into a uniform electric field $\mathscr{E}$. The angle between $\mathscr{E}$ and the normal to the plate is θ. Find the value of the charge per unit area for each of the two surfaces of the plate. Check to make sure that you get the answer you expect for $\theta \to 0$.

24.33 *(4) Capacitor.* Two large, identical flat metal plates are placed parallel to one another, separated by a small distance d. (By "small," we mean that d is much less than the length and width of the plates.) One plate is given a charge per unit area σ, and the other a charge per unit area $-\sigma$. **(a)** Draw the field-line pattern of the system. **(b)** Using a properly drawn Gaussian surface, find the magnitude of the electric field at an arbitrary point in the space between the plates and at an arbitrary point elsewhere. **(c)** Compare your result with Equations 24.13 and 24.14. Explain. This arrangement of a pair of plates is called a *capacitor*. The capacitor has great theoretical and practical importance in electricity; we will consider capacitors in detail in subsequent chapters.

GROUP C

24.34 *(3) Wrong atom, right size.* In 1911, Rutherford, Geiger, and Marsden demonstrated that the positive charge in an atom is concentrated in a tiny nucleus (Problem 24.28). Before that time, J. J. Thomson's "plum pudding" model was taken seriously as a realistic model for atoms. In Thomson's model, the electrons are small particles embedded in a continuous smudge of positive charge, like raisins in an English plum pudding [which usually does not contain plums but, like an American fruit cake, consists of raisins (and other bits of fruit) embedded in a dough]. Imagine that a hydrogen atom consists of a spherical positive cloud of radius R and total charge e, with a single electron (of charge $-e$ and mass $m_e \simeq 9.1 \times 10^{-31}$ kg) embedded at its center, where it is in equilibrium. If such an electron is displaced, it will vibrate about the center. **(a)** Show that the vibration is simple harmonic, with angular frequency

$$\nu = \frac{e}{2\pi} \sqrt{\frac{1}{4\pi\epsilon_0 m_e R^3}}.$$

(b) It was known in Thomson's time that an oscillating electric charge should emit light having the oscillation frequency ν (Chapter 34). It was also known that, when hydrogen emits light, the lowest frequency observed is 6.0×10^{14} Hz. What is the radius of the hydrogen atom according to this model? Compare your result with the value $R \simeq 1.5 \times 10^{-10}$ m obtained by other means.

24.35 *(4) Everything in your plate.* A large dielectric plate of thickness $2d$ carries a uniform charge per unit volume ρ. **(a)** Let the normal to the plate be the z direction, and fix $z = 0$ halfway between the plate faces. Find $\mathscr{E}(z)$ for $z < d$. **(b)** Show that your result is consistent with Equation 24.13, which gives the field outside the plate as $\mathscr{E} = \sigma/2\epsilon_0$. **(c)** A small hole is drilled through the plate along the z direction. A particle of mass m, carrying a charge q of sign opposite that

in the plate, is dropped into the hole. Show that the particle experiences simple harmonic motion, and find the oscillation frequency ν.

24.36 *(4) Gaussian excavation, I.* Consider again the charged solid dielectric sphere of Problem 24.20. **(a)** Use the result of that problem and immediately write down the value of $\mathscr{E}$ at a point A inside the sphere. **(b)** Imagine that you begin at A and remove the material (together with the charge it carries) from a spherical region. Show that the value of $\mathscr{E}$ everywhere within the region is the same as that which you found in part **a**. (Assume that you do not break through the surface of the solid sphere.)

24.37 *(4) Gaussian excavation, II.* A long dielectric cylinder carries a charge per unit volume ρ. Inside the cylinder is a cylindrical cavity whose axis is parallel to that of the cylinder but does not coincide with it. The location of the cavity axis with respect to the cylinder axis is given by the vector $\mathbf{A}$. Show that the field everywhere inside the cavity is $\mathscr{E} = \rho\mathbf{A}/2\epsilon_0$.

24.38 *(G) Divide and conquer, I.* Figure a (p. 681) shows an arbitrary body through which electric charge is distributed in an arbitrary way. The body is divided into two regions, A and B. **(a)** Show that the electric force that the charge in region A exerts on the charge in region B is equal to the force that the charge on the entire body exerts on the charge in region B. **(b)** Figure b shows a uniformly charged spherical shell of radius R, having total charge Q. Imagine the shell to be divided into two hemispheres, as shown in the figure. Using the result of part **a**, find the electric force $\mathbf{F}$ exerted on each hemisphere by the charge on the other hemisphere. Show that $\mathbf{F}$ is the same as if the charges $Q/2$ on the two hemispheres were concentrated at two points along the diameter shown in the figure, each a distance $R/2$ from the center of the sphere.

(a)

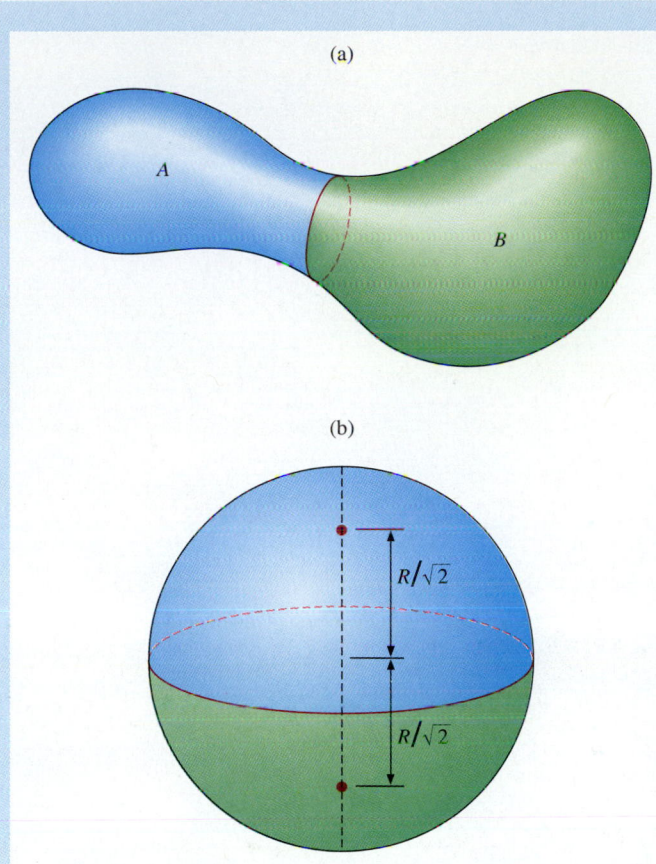

A

B

(b)

$R/\sqrt{2}$

$R/\sqrt{2}$

24.39 *(G) Divide and conquer, II.* Use the result of part **a** of Problem 24.44 to show that the force exerted on one hemisphere of a uniformly charged solid dielectric sphere by the other hemisphere is

$$\mathbf{F} = \frac{1}{4\pi\epsilon_0} \frac{3Q^2}{16R^2} \hat{\mathbf{R}},$$

where Q is the total charge on the sphere and R is its radius. (Hint: You can obtain this result by starting with the result of part **b** of Problem 24.38. But it is easier to cut the hemisphere into slices parallel to the plane dividing the two hemispheres. Then refer to Problem 24.19.) At what points x and $-x$ would two point charges, $Q/2$, exert the same forces on one another as do the hemispherical charge distributions?

24.40 *(G) The hard way.* Using Coulomb's law only, show that the electric field due to a charge q distributed uniformly through a sphere is $\mathscr{E}(r) = (1/4\pi\epsilon_0)(q/r^2)$ at a point P outside the sphere. (Hint: You will have to integrate over elements of the sphere, chosen in such a way that all parts of each element are equidistant from P. To do this, first subdivide the sphere into spherical shells of thickness dr. Then subdivide each shell into rings, each of which subtends an angle $d\theta$ as seen from the center of the sphere, and find the charge dq on each such ring. Newton used this method to prove that the gravitational field outside a uniform sphere of mass M is identical with that of a point mass M located at the center of the sphere.)

Electric Potential

As with every conservative force, a potential energy is associated with the electrostatic force.

The electric potential V is defined as the electrostatic potential energy U per unit charge of a test charge q_t.

The SI unit of electric potential is the volt.

The electric field at any point is the negative of the gradient of the electric potential at that point.

The potential at a point is the algebraic sum of the potentials arising from all the point source charges in the vicinity.

An equipotential surface—a surface on which the potential is everywhere the same—is the three-dimensional electrical analogue to the contour lines on an ordinary map.

Equipotential surfaces are everywhere normal to the flux lines that penetrate them.

Left: The potential difference built up between cloud and ground by frictional charge separation in the atmosphere can amount to hundreds of thousands of volts.

> *Potential, in electrical science, has the same relation to Electricity* [electric charge] *that Pressure, in Hydrostatics, has to Fluid, or that Temperature, in Thermodynamics, has to Heat. Electricity, Fluids, and Heat all tend to pass from one place to another, if the Potential, Pressure, or Temperature is greater in the first place than in the second.*
>
> —JAMES CLERK MAXWELL, *A Treatise on Electricity and Magnetism* (1873)

SECTION 25.1 Introduction

The electric force, like the gravitational force, is conservative. Consequently, the role of *electric potential energy U* in electrical theory is analogous to the role of gravitational potential energy in gravitational theory (Chapters 8 and 14).

When a test charge q_t is in an electric field, the value of U depends on q_t. We define the *electric potential V* to be the potential energy per unit charge, U/q_t. Like potential energy, potential is a scalar quantity. Potential is independent of q_t in the same way the electric field vector $\mathscr{E}$ is. Indeed, there is a close connection between V and $\mathscr{E}$; we will develop this connection in Section 25.2. In Section 25.3, we study the special case of an electric dipole. In Section 25.4, we extend the geometric picture of the electric field that proved so useful in Chapter 24.

SECTION 25.2 Electric Potential Energy and Electric Potential

An electric field $\mathscr{E}$ is completely specified when we give its value at every location $\mathbf{r}$. This is the meaning of Equation 23.15, $\mathscr{E} = \mathscr{E}(\mathbf{r})$. When a test charge q_t is located at some point $\mathbf{r}$ in an electric field, the electric force acting on q_t is $\mathbf{F}(\mathbf{r}) = q_t\mathscr{E}(\mathbf{r})$. Like $\mathscr{E}(\mathbf{r})$, $\mathbf{F}(\mathbf{r})$ is a function of position only. It is thus a conservative force for the same reasons that the gravitational force is conservative (Sections 8.2 and 14.5).

In Figure 25.1, a test charge initially located at $\mathbf{r}$ moves through an infinitesimal displacement $d\mathbf{r}$. Several forces may be acting on q_t at once, but we are interested here in the work dW done on the charge by the electric force alone. That work is

$$dW = \mathbf{F}(\mathbf{r}) \cdot d\mathbf{r} = q_t\mathscr{E}(\mathbf{r}) \cdot d\mathbf{r}.$$

When any conservative force $\mathbf{F}(\mathbf{r})$ acts on a body through a displacement $d\mathbf{r}$, there is a potential energy change dU equal to the negative of the work done; $dU = -dW$ (Equation 8.12). In the present case, we have

$$dU = -q_t\mathscr{E}(\mathbf{r}) \cdot d\mathbf{r}. \tag{25.1}$$

To find the potential energy change ΔU over a finite displacement $\mathbf{r}_f - \mathbf{r}_i$, we integrate Equation 25.1:

$$\Delta U = -q_t \int_{\mathbf{r}_i}^{\mathbf{r}_f} \mathscr{E}(\mathbf{r}) \cdot d\mathbf{r}. \quad (25.2)$$

This result does not depend on the path taken from $\mathbf{r}_i$ to $\mathbf{r}_f$ because the force is conservative.

We define the **electric potential difference** dV or ΔV—called the **potential difference** for short when there is no chance of confusion—to be the *electric potential energy change per unit test charge*. We can write this definition in the infinitesimal (differential) form

$$dV \equiv \frac{dU}{q_t} = -\mathscr{E}(\mathbf{r}) \cdot d\mathbf{r} \quad (25.3a)$$

or in the finite (integral) form

$$\Delta V \equiv \frac{\Delta U}{q_t} = -\int_{\mathbf{r}_i}^{\mathbf{r}_f} \mathscr{E}(\mathbf{r}) \cdot d\mathbf{r}. \quad (25.3b)$$

It is often convenient to choose a reference point $\mathbf{r}_0$ and set $U(\mathbf{r}_0) = 0$. We then speak of the potential energy $U(\mathbf{r})$ of the system, as we first did in the discussion that immediately follows Equation 8.16. We use the same device for potential difference. We set $V(\mathbf{r}) = 0$ at a convenient reference point $\mathbf{r}_0$ and define the **electric potential** (or simply the **potential**) $V(\mathbf{r})$ with respect to that point:

$$V(\mathbf{r}) = -\int_{\mathbf{r}_0}^{\mathbf{r}} \mathscr{E}(\mathbf{r}) \cdot d\mathbf{r}. \quad (25.4)$$

For a given system, $U(\mathbf{r})$ and $V(\mathbf{r})$ must be defined with respect to the same reference point. Then we can rewrite Equation 25.3b in the form

$$V(\mathbf{r}) = \frac{U(\mathbf{r})}{q_t}. \quad (25.5)$$

Potential and potential energy are closely related quantities having similar names, but they are *not* the same quantity. You should take care not to confuse them.*

It follows from the definition $dV \equiv dU/q_t$ that the SI unit of electric potential difference—and of electric potential—is the joule per coulomb, J/C. Because of its importance, this unit is given the special name **volt** (**V**):

$$1 \text{ V} \equiv 1 \text{ J/C}. \quad (25.6)$$

Take care to avoid confusion between the symbol V (italic), which represents electric potential, and the symbol V (not italic) for the unit in which that potential is expressed. Maintaining this distinction in handwriting is difficult, and you must be careful.

The name volt also leads to the informal term **voltage**, used as a synonym for potential difference. This usage is most common in practical discussions, where such phrases as "the voltage between two points in a circuit" are used. Quite often, a point in an electric circuit is **grounded**, or **earthed**. That is, the point is connected directly or indirectly to the earth so that its potential is that of the earth, called the **ground potential** and chosen as the reference potential $V = 0$. We then speak of the "voltage to ground" of any other point in the circuit.

The Relation between Potential and Field

Suppose a test charge experiences an infinitesimal displacement in the direction of the electric field vector $\mathscr{E}$. In this case, we can write Equation 25.3a in the scalar form

$$dV = -\mathscr{E}(\mathbf{r}) \, dr \quad \text{for } d\mathbf{r} \parallel \mathscr{E}(\mathbf{r}). \quad (25.7a)$$

*See part **d** of Problem 14.63 for an analogous definition of gravitational potential.

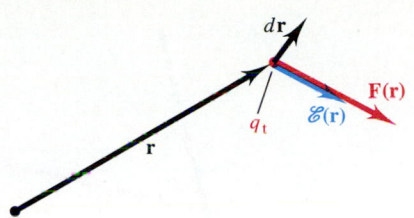

FIGURE 25.1 A test charge q_t lies in an electric field $\mathscr{E}(\mathbf{r})$ at the point $\mathbf{r}$. The test charge experiences the electric force $\mathbf{F}(\mathbf{r})$ as it moves through the infinitesimal displacement $d\mathbf{r}$ (perhaps under the influence of some other force or forces, not shown, in addition to the electric force). The work done by the electric force is $dW = \mathbf{F}(\mathbf{r}) \cdot d\mathbf{r}$, and the potential energy change is $dU = -dW = -\mathbf{F}(\mathbf{r}) \cdot d\mathbf{r}$.

Volta: A Life in Science and Politics

The name *volt* honors the Italian physicist Alessandro G. A. A. Volta (1745–1827), who is best known for his invention in 1800 of the electric battery, or "voltaic pile," discussed in Chapter 27. This invention gave rise to both the study of electric currents and the science of electrochemistry. Volta spent most of his life in Como, in northern Italy. He was an ardent supporter of Napoleon, who eventually made him a count. Volta's career benefited in proportion to Napoleon's seesaw conquest of Italy (1797–1805) and suffered eclipse with Napoleon's defeat in 1814.

The electric field at **r** then has magnitude

$$\mathscr{E}(\mathbf{r}) = -\frac{dV(\mathbf{r})}{dr}. \qquad (25.7\text{b})$$

We say that the electric field is the negative of the **gradient** (that is, the "steepness") of the electric potential—the change in V per unit displacement. The same is true in three-dimensional situations. But, in generalizing to three dimensions, we must modify Equation 25.7a. At any location **r**, we wish to express the potential difference dV across a displacement $d\mathbf{s}$ that is not necessarily parallel to $\mathscr{E}$. In Figure 25.2, the angle between $\mathscr{E}$ and $d\mathbf{s}$ is θ. We have

$$dV = -\mathscr{E}\cos\theta\, ds.$$

Now, $\mathscr{E}\cos\theta$ is just the component of $\mathscr{E}$ along $d\mathbf{s}$. Let us give this component the name $\mathscr{E}_{\hat{\mathbf{s}}}$. We can then rewrite Equation 25.7a in the generalized form

$$dV = -\mathscr{E}_{\hat{\mathbf{s}}}\, ds. \qquad (25.7\text{c})$$

Rearranging this equation to express the field component in terms of the potential difference, we obtain

$$\mathscr{E}_{\hat{\mathbf{s}}} = -\frac{dV}{ds}, \qquad (25.7\text{d})$$

which is a generalization of the one-dimensional Equation 25.7b. *The change of potential per unit displacement in a given direction $\hat{\mathbf{s}}$ is equal to the negative of the component of the electric field in that direction.*

Equations 25.7b and 25.7d lead to an alternative name for the unit of electric field, the volt per meter (V/m). Because of the importance of potential in electrical theory, the name V/m is probably used more frequently than N/C. But both have identical significance in terms of the base units meter, kilogram, second, and coulomb:

$$1 \text{ V/m} = 1 \text{ N/C}.$$

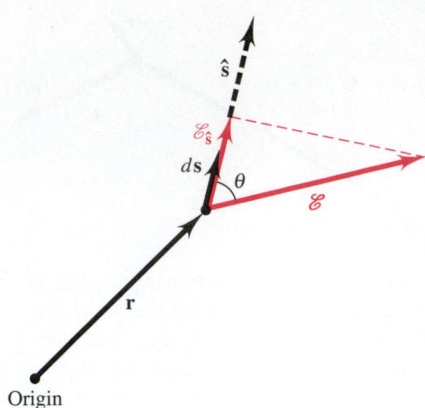

FIGURE 25.2 At a certain location **r**, the electric field is $\mathscr{E}$. The arbitrary displacement $d\mathbf{s}$ makes an angle θ with $\mathscr{E}$. The component of $\mathscr{E}$ along $d\mathbf{s}$ is $\mathscr{E}_{\hat{\mathbf{s}}} \equiv \mathscr{E}\cos\theta$.

EXAMPLE 25.1

A large flat dielectric plate has uniform charge σ per unit area. Take the reference point, at which $V = 0$, at the surface of the plate. (This choice is not the only one possible, but it is convenient.) **(a)** Find the potential V at a height h above the plate. **(b)** Let $\sigma = 2.10 \times 10^{-8}$ C/m². Find the potential at $h = 6.50$ cm. **(c)** What is the electric potential energy of a point charge $q_1 = 2.50 \times 10^{-6}$ C at $h = 6.50$ cm? **(d)** What is the electric potential energy of a point charge $q_2 = -2.50 \times 10^{-6}$ C at this location?

SOLUTION:

(a) Express V as a function of h.

According to Equation 24.13 or Equation 23.34, the electric field $\mathscr{E}$ in the vicinity of the plate has the constant magnitude

$$\mathscr{E} = \frac{\sigma}{2\epsilon_0};$$

the direction of $\mathscr{E}$ is away from the plate for $\sigma > 0$. So you can write Equation 25.4 in the simplified form

$$V(h) = -\mathscr{E}\int_0^h dr,$$

or

$$V = -\frac{\sigma h}{2\epsilon_0}. \qquad (25.8)$$

If $\sigma > 0$, the potential is negative for all values of h. (What happens when $\sigma < 0$?)

(b) Letting $\sigma = 2.10 \times 10^{-8}$ C/m², find V at $h = 6.50$ cm.

Inserting numerical values into Equation 25.8, you obtain

$$V = -\frac{2.10 \times 10^{-8} \text{ C/m}^2 \times 6.50 \times 10^{-2} \text{ m}}{2 \times 8.85 \times 10^{-12} \text{ C}^2/\text{N·m}^2} = -77.1 \text{ V}.$$

(c) Find U for a point charge $q_1 = 2.50 \times 10^{-6}$ C at $h = 6.50$ cm.

Using Equation 25.5, you can immediately write

$$U = q_1 V = 2.50 \times 10^{-6} \text{ C} \times (-77.1 \text{ V})$$
$$= -1.93 \times 10^{-4} \text{ J}.$$

Why the negative value of U? The charge on the plate is positive, and the positive charge q_1 is repelled from the plate. As the charge moves away from the plate, it either gains kinetic energy or does work on some object external to the system. The potential energy of the system is thus diminished, and $U(h) < 0$.

(d) Find U for a point charge $q_2 = 2.50 \times 10^{-6}$ C at $h = 6.50$ cm.

For the negative charge q_2, you have

$$U = q_2 \mathscr{E} = -2.50 \times 10^{-6} \text{ C} \times (-77.1 \text{ V})$$

$$= 1.93 \times 10^{-4} \text{ J}.$$

The negative charge is attracted to the positively charged plate, and external work must be done on it to move it from the plate to height h. This external work increases the potential energy of the system.

We can generalize parts **c** and **d** of Example 25.1 in two statements:

1. If a body at **r** is positively charged, the sign of its potential energy $U(\mathbf{r})$ is the *same* as the sign of the potential $V(\mathbf{r})$.

2. If the body is negatively charged, the sign of its potential energy $U(\mathbf{r})$ is *opposite* the sign of the potential $V(\mathbf{r})$.

Total Energy of a Charged Particle

When a particle having charge q and mass m moves from a point where the potential is V_i to a point where the potential is V_f, its potential energy changes by an amount

$$\Delta U = q(V_f - V_i) = q \, \Delta V;$$

see Equation 25.3b. Suppose the electric force is the only force acting on the particle. Then its total energy E must be conserved. Hence its kinetic energy K must change in such a way that $\Delta K = -\Delta U$. If the initial kinetic energy is zero—that is, if the particle is initially at rest—we can write

$$\tfrac{1}{2}mv^2 = -q \, \Delta V. \tag{25.9}$$

The circumstances under which Equation 25.9 is applicable are encountered frequently when charged elementary particles move through an electric field in an evacuated chamber. Indeed, electric fields are often used to manipulate the particles. The most familiar application of this technique is in the television picture tube. Electrons are accelerated through a substantial potential difference and thus given substantial kinetic energy. When an electron strikes the fluorescent material coating the inside of the screen end of the tube, part of the energy transferred to the material is manifested as a flash of light. A controlled and directed stream of electrons can thus "paint" a picture on the fluorescent screen.

The Electron-Volt

When an electron "falls"—or, more precisely, is accelerated—through a potential difference $|\Delta V| = 1$ V, it acquires kinetic energy—and loses potential energy—of magnitude

$$|\Delta K| = |-\Delta U| = |q \, \Delta V| = e \times 1 \text{ V}; \tag{25.10}$$

here e is the quantum of charge—the magnitude of the charge on the electron, given by Equation 23.8. We insert the value of e into this equation to obtain

$$e \times 1 \text{ V} = 1.60 \times 10^{-19} \text{ C} \times 1 \text{ V} = 1.60 \times 10^{-19} \text{ J}. \tag{25.11}$$

(A more precise value is given in Appendix 4.) This quantity makes a convenient unit of energy for the study of electrons and other microscopic charged particles because such particles are often made to pass through potential differences measured in volts. The new unit is given the name **electron-volt (eV)**:

$$1 \text{ eV} \equiv e \text{ J} \simeq 1.60 \times 10^{-19} \text{ J}. \tag{25.12}$$

Thus, for example, the energy change of an electron that has passed through a potential difference of 20 000 volts is simply 20 000 eV (or 20 keV). It is simpler to express energy directly in terms of the potential difference than to multiply the charge of the electron by the potential difference in volts and thus express the energy in joules.

Although the electron-volt is not an SI unit, it is very widely used. In conversation, the informal name "ee-vee" is most common.

EXAMPLE 25.2

In a typical television tube, the electrons are accelerated through a potential difference of magnitude 2×10^4 V before striking the screen. Find (a) the kinetic energy of an electron, in eV and in J, just before it strikes the screen and (b) the corresponding speed of the electron. The electron starts essentially at rest.

SOLUTION:

(a) Express K in eV and in J.

You can write the kinetic energy in eV without calculation. Beginning with Equation 25.10, set $q = e$ and substitute $|\Delta V| = 2 \times 10^4$ V for $|\Delta V| = 1$ V to obtain

$$K = |e \, \Delta V| = e \times 2 \times 10^4 \text{ V} = 2 \times 10^4 \times 1 \text{ eV}$$
$$= 2 \times 10^4 \text{ eV}.$$

Expressed in eV, the kinetic energy of a particle having charge $|e|$ is numerically equal to the potential difference through which it has "fallen." The convenience of the unit lies mainly in this point.

In joules, you have from Equation 25.11

$$K = 1.60 \times 10^{-19} \text{ J/eV} \times 2 \times 10^4 \text{ eV} = 3.20 \times 10^{-15} \text{ J}.$$

If you prefer, you can carry out the same calculation by using Equation 25.10 and writing

$$K = |e \, \Delta V| = 1.60 \times 10^{-19} \text{ C} \times 2 \times 10^4 \text{ V} \times 1 \frac{\text{J/C}}{\text{V}}$$
$$= 3.20 \times 10^{-15} \text{ J}.$$

The numbers are the same; you should make sure you understand why the calculation works in both sets of units.

(b) Find the speed of the electron.

Because $K = \frac{1}{2}mv^2$, the speed is

$$v = \sqrt{\frac{2K}{m}} = \sqrt{\frac{2 \times 3.20 \times 10^{-15} \text{ J}}{9.11 \times 10^{-31} \text{ kg}}} = 8.38 \times 10^7 \text{ m/s}.$$

This is more than one-quarter the speed of light. It doesn't take much energy to make an electron move fast!

Electric Potential in the Vicinity of a Point Charge

The electric field in the vicinity of a point source charge q is given by Coulomb's law in the form of Equation 23.11,

$$\mathscr{E}(\mathbf{r}) = \frac{1}{4\pi\epsilon_0} \frac{q}{r^2} \hat{\mathbf{r}}.$$

Consider an arbitrary point whose location with respect to the source charge q is given by the vector $\mathbf{r}$. We can use Equation 25.4, $V(\mathbf{r}) = -\int_{\mathbf{r}_0}^{\mathbf{r}} \mathscr{E} \cdot d\mathbf{r}$, to evaluate the potential at $\mathbf{r}$. But we must first choose a reference point r_0. When we are treating a point source charge, the most convenient choice of reference point is almost always the same one made in Section 14.5, where we evaluated the gravitational potential energy in the vicinity of a point mass. As in that case, we set $V = 0$ at $\mathbf{r}_0 = \infty$. Equation 25.4 then becomes

$$V = -\frac{q}{4\pi\epsilon_0} \int_{\infty}^{\mathbf{r}} \frac{1}{r^2} \hat{\mathbf{r}} \cdot d\mathbf{r}. \tag{25.13}$$

It does not matter what path we choose for integrating because the dot product $\hat{\mathbf{r}} \cdot d\mathbf{r}$ is always equal to the radial component dr.* Equation 25.13 simplifies to the scalar form

$$V = -\frac{q}{4\pi\epsilon_0} \int_{\infty}^{r} \frac{dr}{r^2}, \tag{25.14a}$$

which integrates to yield

$$V = \frac{1}{4\pi\epsilon_0} \frac{q}{r}. \tag{25.14b}$$

Because this potential is derived directly from Coulomb's law, it is often called the **Coulomb potential**. It is the same at all $\mathbf{r}$ lying on the same sphere centered on the source charge q. Can you see why?

---

*Proof: $\hat{\mathbf{r}} \cdot d\mathbf{r} = \hat{\mathbf{r}} \cdot (\hat{\mathbf{r}} \, dr) = (\hat{\mathbf{r}} \cdot \hat{\mathbf{r}}) \, dr = dr$.

EXAMPLE **25.3**

A metal sphere of radius $R = 25.0$ cm carries a charge $q = -5.37 \times 10^{-6}$ C. Find the potential at (a) $r = 50.0$ cm, (b) $r = 25.0$ cm, and (c) $r = 12.5$ cm.

SOLUTION:
(a) Find V at $r = 50.0$ cm.

At $r = 50.0$ cm, outside the sphere, the field is the same as that of an equal point charge at the center (Section 24.3). So to find V, you can apply Equation 25.14b directly:

$$V = (8.99 \times 10^9 \text{ N·m}^2/\text{C}^2) \times \frac{-5.37 \times 10^{-6} \text{ C}}{0.500 \text{ m}}$$

$$= -9.66 \times 10^4 \text{ V}.$$

(b) Find V at $r = 25.0$ cm.

At $r = 25.0$ cm, on the surface of the sphere, you can still treat the source charge as if it were located at the center of the sphere. Thus you could repeat the entire calculation of part **a** with the new value of r. But it is easier to note that the only change is a halving of the value of r in Equation 25.14b. So the result must be twice that of part **a**, and you can immediately write

$$V = 2 \times (-9.66 \times 10^4 \text{ V}) = -1.93 \times 10^5 \text{ V}.$$

(c) Find V at $r = 12.5$ cm.

Here you must be careful because the point lies within the sphere. Because the sphere is a conductor, all the charge lies on its surface (again, see Section 24.3). The electric field is zero everywhere inside the sphere. Thus, integrating inward from $r = R$ yields no further change. At *any* point inside the sphere the potential is

$$V = -1.93 \times 10^5 \text{ V}.$$

We generalize the conclusion of part **c** of Example 25.3 in the following statement: *The electric potential is the same at all points in any region in which $\mathscr{E} = 0$.* Such a region is called a **field-free region**. The value of the potential in a field-free region is determined by (1) the choice of the reference point $\mathbf{r}_0$ at which $V = 0$ and (2) the distribution of charge outside the region.

Superposition of Potentials

Figure 25.3 shows an arbitrary array of point charges $q_1, q_2, \ldots, q_j, \ldots, q_n$. Equation 25.14b can be used to express the potential, at some point P, due to each charge separately. Let r_j be the distance between the jth charge and P. We then have simply

$$V_j = \frac{1}{4\pi\epsilon_0} \frac{q_j}{r_j}. \qquad (25.15)$$

The potential at P due to all the charges is found by direct addition of the individual Coulomb contributions:

$$V = \frac{1}{4\pi\epsilon_0} \left(\frac{q_1}{r_1} + \frac{q_2}{r_2} + \cdots + \frac{q_j}{r_j} + \cdots + \frac{q_n}{r_n} \right) = \frac{1}{4\pi\epsilon_0} \sum_{j=1}^{n} \frac{q_j}{r_j}, \qquad (25.16)$$

or

$$V = \sum_{j=1}^{n} V_j. \qquad (25.17)$$

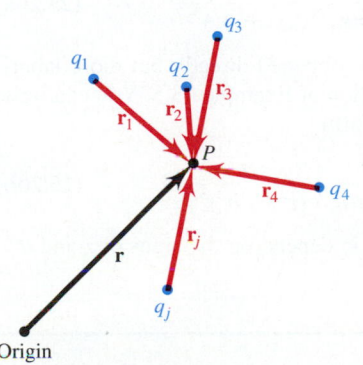

FIGURE 25.3 A point P lies in the vicinity of an arbitrary array of charges q_j. Each vector $\mathbf{r}_j$ locates P with respect to one of the charges.

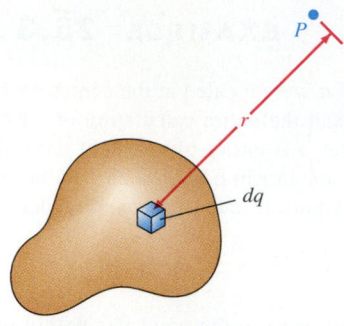

FIGURE 25.4 A continuous charge distribution is divided into infinitesimal elements dq, each lying at a particular distance r from P.

This is the **superposition principle for electric potentials**. It is particularly simple to apply because potentials are scalars and add in simple scalar fashion. Because any charge distribution can in principle be constructed from point charges, the superposition principle for potentials is quite general.

Equation 25.16 expresses the potential due to a discrete array of point charges. If the charge distribution is continuous, we express the corresponding result in integral form. In imagination, we divide the charge distribution into infinitesimal elements, as shown in Figure 25.4. Each element lies at a particular distance r from P and makes an infinitesimal contribution

$$dV = \frac{1}{4\pi\epsilon_0}\frac{dq}{r} \tag{25.18a}$$

to the potential at P. According to the superposition principle, the potential at P is

$$V = \int dV = \frac{1}{4\pi\epsilon_0}\int_{\substack{\text{charge}\\\text{distribution}}}\frac{dq}{r}. \tag{25.18b}$$

To evaluate this integral, we must know the distance r between P and each elementary charge dq. Examples 25.4 and 25.5 show how V is found.

EXAMPLE 25.4

A loop of radius b is formed of thin metal wire, as shown in Figure 25.5. A quantity of charge q is placed on the loop. Find **(a)** the electric potential at P, a point on the loop axis lying a distance z from the plane of the loop, and **(b)** the electric field at the same point. (Compare with Example 23.5.)

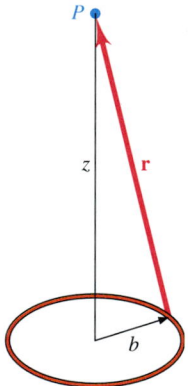

FIGURE 25.5

SOLUTION:
(a) Find V at P.

As you can see from Figure 25.5, all points on the loop are the same distance r from P, and r is given by the Pythagorean sum

$$r = \sqrt{z^2 + b^2}.$$

Because V is a scalar, it does not matter that the direction of $\mathbf{r}$ varies from point to point. The distance r is the same for all points on the ring, so you do not have to divide it into elements. Rather, you can immediately use Equation 25.14b to write

$$V = \frac{1}{4\pi\epsilon_0}\frac{q}{r} = \frac{1}{4\pi\epsilon_0}\frac{q}{\sqrt{z^2 + b^2}}. \tag{25.19}$$

As a check, consider what happens to Equation 25.19 in the extreme case $z = 0$, where P is at the center of the loop. The equation becomes $V = (1/4\pi\epsilon_0)(q/b)$. This is the Coulomb potential at a distance b from a point charge q. The fact that q is distributed around the circular loop does not affect the result because only the distance of P from the source charge is significant, and not the direction from the source charge to P. What happens when $z \gg b$?

(b) Find the electric field vector $\mathscr{E}$ at P.

You must determine both the direction and the magnitude of $\mathscr{E}$. The direction follows from the symmetry of the situation; the argument of Example 23.5 leads to the conclusion that the field $\mathscr{E}$ at any point on the z axis lies along the axis.

According to Equation 25.19, the potential along the axis is a function of z only; that is, $V(\mathbf{r}) = V(z)$. Hence you can use Equation 25.7b, $\mathscr{E}(\mathbf{r}) = -dV(\mathbf{r})/dr$, to write

$$\mathscr{E}(z) = -\frac{dV(z)}{dz}.$$

To use this relation, you differentiate both sides of Equation 25.19 with respect to z. You immediately obtain the result

$$\mathscr{E}(z) = \frac{1}{4\pi\epsilon_0}\,q\,\frac{z}{(z^2 + b^2)^{3/2}}. \tag{25.20a}$$

This is the same result you obtained directly but more laboriously in the vector calculation of Example 23.5. You can write the result in the vectorial form

$$\mathscr{E}(\mathbf{z}) = \frac{1}{4\pi\epsilon_0}\,q\,\frac{z}{(z^2 + b^2)^{3/2}}\,\hat{\mathbf{z}}. \tag{25.20b}$$

How does the direction of $\mathscr{E}$ depend on the signs of z and q?

EXAMPLE 25.5

Figure 25.6 shows a uniformly charged circular disk of radius R; the total charge is Q. Find **(a)** the electric potential V and **(b)** the electric field $\mathscr{E}$, both at a point P located on the disk axis at a distance z from the disk.

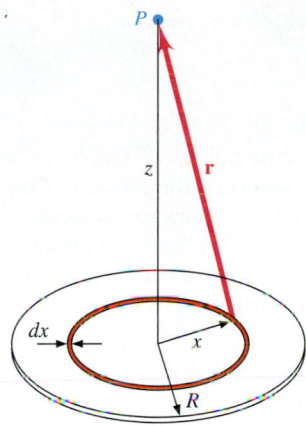

FIGURE 25.6

SOLUTION:
(a) Find V at P.

In this case, P does not lie at the same distance from all of the charge. So you divide the disk into rings of radius x and thickness dx, just as in Example 23.6. According to Equation 23.29, a typical ring carries charge

$$dQ = \frac{2Q}{R^2} x \, dx.$$

The contribution dV made by this ring to the potential at P is given by Equation 25.18a, $dV = (1/4\pi\epsilon_0)(dQ/r)$, where r is the distance from any point on the ring to P. Since $r = \sqrt{x^2 + z^2}$, you write

$$dV = \frac{1}{4\pi\epsilon_0} \frac{2Q}{R^2} \frac{x}{\sqrt{x^2 + z^2}} \, dx.$$

For any particular P, the value of z is fixed. So you can use Equation 25.18b to sum the contributions dV of all the rings that make up the disk:

$$V = \frac{1}{4\pi\epsilon_0} \frac{2Q}{R^2} \int_0^R \frac{x \, dx}{\sqrt{x^2 + z^2}}$$

$$= \frac{1}{4\pi\epsilon_0} \frac{2Q}{R^2} \sqrt{x^2 + z^2} \Big|_0^R , \qquad (25.21)$$

which yields $\quad V = \dfrac{Q}{2\pi\epsilon_0 R^2} (\sqrt{R^2 + z^2} - z)$.

(b) Find $\mathscr{E}$ at P.

As you saw in Example 23.6, the electric field at P lies along the axis of the disk. You find the magnitude of the field by using Equation 25.7b in the form $\mathscr{E}(z) = -(1/4\pi\epsilon_0)(dV/dz)$:

$$\mathscr{E}(z) = -\frac{Q}{2\pi\epsilon_0 R^2} \frac{d}{dz} (\sqrt{R^2 + z^2} - z) \qquad (25.22a)$$

$$= \frac{Q}{2\pi\epsilon_0 R^2} \left(1 - \frac{z}{\sqrt{R^2 + z^2}} \right).$$

This is the result of Equation 23.31, obtained directly but with more labor in Example 23.4. You can write the result in the vectorial form

$$\mathscr{E}(\mathbf{z}) = \frac{Q}{2\pi\epsilon_0 R^2} \left(1 - \frac{z}{\sqrt{R^2 + z^2}} \right) \hat{\mathbf{z}}. \qquad (25.22b)$$

Equipotentials and Flux Lines

Equation 25.3a expresses the relation between electric potential and electric field in the form

$$dV = -\mathscr{E}(\mathbf{r}) \cdot d\mathbf{r}.$$

This equation shows that a displacement $d\mathbf{r}$ will in general entail a change in potential. If, in particular, $d\mathbf{r}$ is parallel to $\mathscr{E}$, the change in potential per unit distance will have its maximum value. This happens when $d\mathbf{r}$ lies along the local flux line. Indeed, the flux line is made up of an infinite number of infinitesimal displacements of this kind. The flux line is thus the "*path of steepest descent.*" (Here "descent" implies decrease in V.) It is analogous to the steepest path down a mountain. There are many such steepest paths, though only one passes through any particular point.

But what if we choose $d\mathbf{r}$ perpendicular to $\mathscr{E}$? In that case, the dot product is equal to zero and there is *no* change of potential over the displacement; $dV = 0$. As Figure 25.7 shows, the directions satisfying this requirement define the plane that is normal to

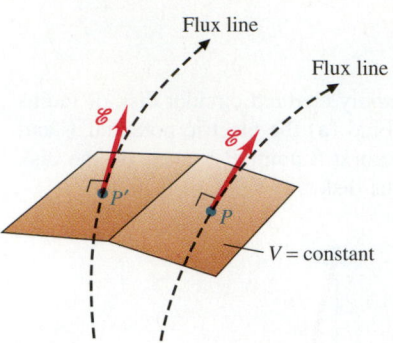

FIGURE 25.7 At the arbitrary point P, the electric field is $\mathscr{E}$. The flux line to which $\mathscr{E}$ is tangent is the "path of steepest descent" for potential. Consequently, the potential is constant on the infinitesimal surface element normal to $\mathscr{E}$. The neighboring element centered on P' also has constant potential. Because the two elements are joined along one edge, they make up a larger surface of constant potential. A finite equipotential surface is made up of many such adjoining elements.

$\mathscr{E}$ at P. The infinitesimal area element centered on $\mathscr{E}$ is a part of this plane. Because any displacement over the element leads to zero change in potential, *the potential is constant over this element.*

Adjacent to this area element and sharing a common edge with it is another infinitesimal area element (not necessarily coplanar with the first element). The same argument can be made with respect to the flux line that passes through this second element. By piecing together many such adjoining elements, we can construct a finite curved surface over which the potential is constant. This surface is called an **equipotential surface** or simply an **equipotential**.

The equipotential surface is a three-dimensional analogue of the familiar two-dimensional *contour line*, used on many maps to delineate constant elevations, and the *isobar*, used in weather maps to delineate constant barometric pressure. Indeed, arrays of equipotentials can be used as an alternative to flux-line arrays to map an electric field. Because the equipotentials are everywhere normal to the flux lines penetrating them, either map can be generated from the other.

For arbitrary charge distributions, the equipotential-surface map can be complicated, just as the flux-line map can be. However, like flux lines, equipotential surfaces obey rules:

1. Two equipotential surfaces on which the potentials are different cannot touch or cross one another.

2. Either a closed equipotential surface contains charge or else the electric field is zero everywhere in the space enclosed by the surface.

To prove the second part of rule 2, remember that, in the absence of charge within the closed surface, all flux lines within it must originate and terminate outside it. If the electric field within the surface is not zero everywhere, such flux lines must exist. Imagine a flux line that penetrates the surface inward at one point and outward at another. Because $\mathscr{E} \cdot d\mathbf{s}$ has the same sign everywhere along this line, integrating along this line from the entrance point to the exit point will yield a nonzero potential difference between the two points, which contradicts the statement that the surface is an equipotential.

In Chapter 24, we developed a technique that used Gauss's law together with symmetry considerations to deduce the electric fields of spherical, cylindrical, and planar charge distributions. A major part of the technique is to draw the Gaussian surface so that flux lines penetrate it along normals. Such Gaussian surfaces are equipotentials. Because these Gaussian surfaces can be drawn at arbitrary distances from the charge distributions, a nest of such surfaces constitutes a family of equipotentials. This point is illustrated in Figure 25.8.

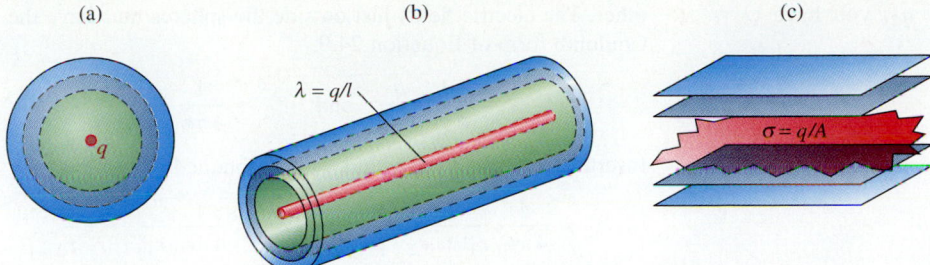

FIGURE 25.8 Nests of equipotential surfaces surrounding symmetrical charge distributions reflect the symmetry of the charge distributions. The equipotentials surround (*a*) a point charge; (*b*) a linear charge; (*c*) a planar charge. The sets of equipotentials in parts *b* and *c* are open-ended surfaces because the charge distributions themselves are open ended.

We now prove the following important theorem: *In equilibrium, the surface of a conductor of arbitrary shape is an equipotential.* Suppose there is a potential difference between any two points on the surface. Then an electric field must exist between them. This field will drive a flow of free charge until the field decreases to zero and the surface becomes an equipotential. An alternative proof can be made by beginning with the first part of theorem 3 in Section 24.4: The electric field just outside a charged conducting body is directed normal to the surface. Because the surface is everywhere normal to the field, the surface must coincide with an equipotential.

This theorem has a significant consequence: *In equilibrium, the electric field within a conducting body of arbitrary shape is everywhere zero.* We have already proved this statement in terms of flux; it is theorem 1 in Section 24.4. The proof in terms of equipotentials requires the use of rule 2 for equipotentials; this proof is the subject of Query 25.6.

EXAMPLE 25.6

Two conducting spheres, having radii R_1 and R_2, are located a considerable distance apart. They are connected by a fine conducting wire, as shown in Figure 25.9. Taken together, the two spheres contain a charge Q. (**a**) Find the charges q_1 and q_2 on the surfaces of the two spheres. (**b**) What is the magnitude of the electric field immediately outside each of the spheres?

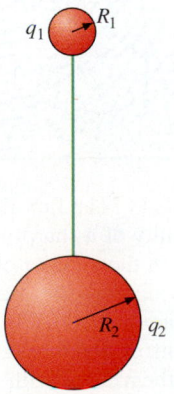

FIGURE 25.9

SOLUTION:
(**a**) Find q_1 and q_2.

Because the two surfaces are joined by a conducting wire, they constitute a single equipotential surface. (The surface of the wire also is a part of this equipotential.) Consequently, the potential V is the same at both surfaces. The charge distribution on the surface of each sphere is uniform because the two are so far apart that the field of each is negligible in the vicinity of the other. According to Equation 25.14b, you have

$$V = \frac{1}{4\pi\epsilon_0}\frac{q_1}{r_1} = \frac{1}{4\pi\epsilon_0}\frac{q_2}{r_2},$$

so

$$\frac{q_1}{q_2} = \frac{r_1}{r_2}. \qquad (25.23)$$

You can use this equation to express q_2 in terms of q_1:

$$q_2 = q_1\frac{r_2}{r_1}.$$

Because the total charge is $Q = q_1 + q_2$, you have $Q = q_1(1 + r_2/r_1)$, or

$$q_1 = \frac{Q}{1 + \dfrac{r_2}{r_1}}.$$

In similar manner, you can find

$$q_2 = \frac{Q}{1 + \dfrac{r_1}{r_2}}.$$

(b) Find $\mathscr{E}_1$ and $\mathscr{E}_2$, the magnitudes of the fields immediately outside the two spheres.

Because the spheres are far apart, the flux-line distribution just outside each sphere is not influenced by the presence of the other. The electric fields just outside the spheres thus have the Coulomb form of Equation 24.9,

$$\mathscr{E}_1 = \frac{1}{4\pi\epsilon_0}\frac{q_1}{r_1^2} \quad \text{and} \quad \mathscr{E}_2 = \frac{1}{4\pi\epsilon_0}\frac{q_2}{r_2^2}.$$

Inserting the values of q_1 and q_2 just obtained, you find that

$$\mathscr{E}_1 = \frac{1}{4\pi\epsilon_0}\frac{1}{r_1 + r_2}\frac{Q}{r_1} \quad \text{and} \quad \mathscr{E}_2 = \frac{1}{4\pi\epsilon_0}\frac{1}{r_1 + r_2}\frac{Q}{r_2}.$$

You can conveniently compare these two results by taking their ratio:

$$\frac{\mathscr{E}_1}{\mathscr{E}_2} = \frac{r_2}{r_1};$$

the electric fields at the surfaces are inversely proportional to the radii of the spheres.

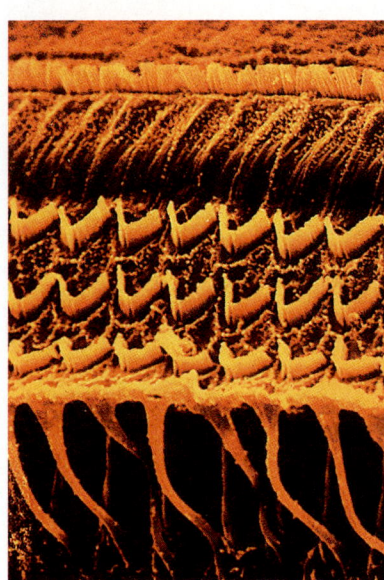

Microphoto (×3000) of the "hair cells" in the cochlea of a mammalian ear. These are the cells that transform sound energy into the electrical energy of nerve impulses. Like all nerve cells, these highly elongated cells have a modest potential difference across the cell membrane. The curvature at the tip is so large that the electric field across the membrane is near breakdown. The very slight disturbance produced by a sound wave is sufficient to produce electric breakdown and thus a nerve impulse that the brain interprets as a sound.

The final result of Example 25.6 can be extended to conducting bodies of arbitrary shape. The surface of any such body can be pieced together by using small bits of spherical surfaces of varying radii, as shown in Figure 25.10. At any point just outside the surface, the field is inversely proportional to the radius of the local spherical bit. The reciprocal of this radius is called the **curvature**; sharply curved parts of the surface have large curvatures. Thus the field is directly proportional to the curvature. Consequently, the field outside an irregular body is greatest where the curvature is greatest. The photograph of Figure 25.11 illustrates this clearly.

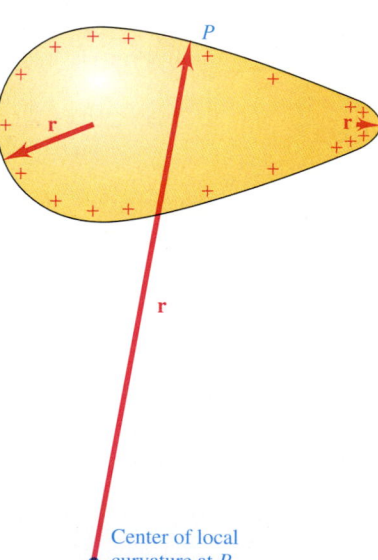

FIGURE 25.10 A body of arbitrary shape has a surface of varying curvature. The electric field at any point just outside the surface is directly proportional to the local curvature. Consequently, the surface charge must be densest in regions of high curvature.

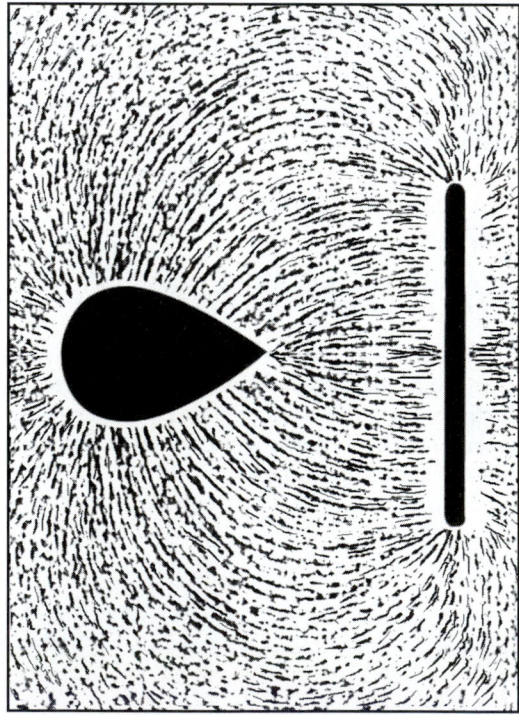

FIGURE 25.11 The flux-line pattern in the vicinity of a sharp-pointed conductor in the vicinity of a flat plate carrying opposite charge. The flux lines are made visible by suspending bits of fine thread in oil; the bits align themselves with the field. See Figure 24.9, in which grass seed is used in place of thread.

This relation between the field just outside a conducting surface and the curvature of the surface is exploited when it is desirable to produce a large electric field by means of a modest potential difference. As mentioned in the margin note on the photocopying machine near the end of Section 23.2, it is necessary to charge the selenium-coated drum onto which the image is to be projected. This is usually done by applying a potential difference between the drum and a ''brush'' made of fine metal wires with sharp ends, located a short distance away from (but not touching) the delicate drum surface. The electric field can readily be made large enough—about 3×10^6 V/m—so as to tear electrons off some of the air molecules and thus to ionize them. This results in a so-called *corona discharge*, in which the air conducts electricity, and charge can pass onto the drum.

THINKING LIKE A PHYSICIST

Both the power and the beauty of physics stem in large measure from a property of physics that is especially visible in the subject matter of this chapter: A relatively small number of basic ideas can be used to understand a wide variety of phenomena. Electric potential bears strong analogies to temperature and pressure, as Maxwell notes at the beginning of this chapter. The quantities U and V are equally applicable to electric and gravitational fields, and we have made much of this analogy. Equipotentials have much in common with contour lines, and flux lines with paths of steepest descent. As a consequence of these and other analogies, mastery of the theory of one subject area gives one a flying start in

learning another area; that is power. Understanding many aspects of the physical world in terms of a limited number of basic concepts affords insights that are broad as well as deep; that is beauty.

You can find support for these assertions in a little self-reflection. You have probably not found it easy to learn the material in Chapters 23 through 25, but you have probably found that steady application yields success. Now, try to recreate your state of mind when you began the study of mechanics—an apparently different subject—some months ago. How much progress could you have made in studying electricity if you had started right in on Chapter 23 then?

Symbols Used in Chapter 25

e	quantum of electric charge		$V, \Delta V$	electric potential, potential difference
$\mathscr{E}, \mathscr{E}$	electric field		ϵ_0	permittivity of free space
K	kinetic energy		ρ	charge per unit volume
q, Q	electric charge		σ	charge per unit area
U	electric potential energy			

Summing Up

Electric potential difference ΔV is defined by Equation 25.3b:

$$\Delta V \equiv \frac{\Delta U}{q_t} = -\int_{r_i}^{r_f} \mathscr{E}(\mathbf{r}) \cdot d\mathbf{r}.$$

That is, ΔV is the potential energy change ΔU per unit charge q_t experienced by a test charge as it is moved through an electric field $\mathscr{E}(\mathbf{r})$. By choosing a reference point $\mathbf{r}_0$ at which $V = 0$, we make it possible to speak of the **electric potential** $V(\mathbf{r})$ at any point in space. According to Equations 25.4 and 25.5,

$$V(\mathbf{r}) = \frac{U(r)}{q_t} = -\int_{\mathbf{r}_0}^{\mathbf{r}} \mathscr{E}(\mathbf{r}) \cdot d\mathbf{r}.$$

The unit of electric potential difference and electric potential is the **volt**, defined by Equation 25.6:

$$1 \text{ V} \equiv 1 \text{ J/C}.$$

Beginning with this unit, we define a convenient unit of energy called the **electron-volt (eV)** according to Equations 25.11 and 25.12,

$$1 \text{ eV} \equiv e \times 1 \text{ V} = e \text{ J}.$$

The electric field $\mathscr{E}(\mathbf{r})$ and the potential $V(\mathbf{r})$ at the same point are related by Equation 25.3a,

$$dV = -\mathscr{E}(\mathbf{r}) \cdot d\mathbf{r}.$$

When the displacement is parallel to the field, this equation assumes the scalar form $dV = -\mathcal{E}(\mathbf{r})\,dr$. The inverse relation can readily be written in this case; it is given by Equation 25.7b,

$$\mathcal{E}(\mathbf{r}) = -\frac{dV(\mathbf{r})}{dr}.$$

Like forces and electric fields, electric potentials obey the **superposition principle** expressed by Equations 25.16 and 25.17,

$$V = \sum_{j=1}^{n} V_j = \frac{1}{4\pi\epsilon_0} \sum_{j=1}^{n} \frac{q_j}{r_j}.$$

The superposition principle is expressed in integral form by Equation 25.18b,

$$V = \int dV = \frac{1}{4\pi\epsilon_0} \int_{\substack{\text{charge}\\\text{distribution}}} \frac{dq}{r}.$$

An **equipotential surface** is a surface on which the potential has the same value at all points. Equipotentials are everywhere normal to the flux lines that penetrate them. The surface of a conducting body in equilibrium is always an equipotential. It follows that the electric field just outside such a surface is largest where the curvature of the surface is greatest.

KEY TERMS

Section 25.2 Electric Potential Energy and Electric Potential
potential, potential difference, voltage ▪ potential gradient ▪ electron-volt ▪ field-free region

Section 25.4 Equipotentials and Flux Lines
equipotential surface ▪ path of steepest descent

Queries and Problems for Chapter 25

QUERIES

25.1 *(2) Remarkable difference?* What is the potential difference between the midpoint on the axis of an electric dipole and a point at infinity?

25.2 *(2) The bigger they come . . .* An α particle (the nucleus of a helium atom) has charge 2e. What is the change in the potential energy of the particle, expressed in eV, when it has fallen through a potential difference of 5 V?

25.3 *(4) . . . or to take arms against a sea of troubles . . .* Sketch the equipotentials surrounding a uniformly charged thin rod of finite length. To do this, consider (1) what the surfaces far from the wire must look like and (2) what the surfaces close to the wire must look like. Then sketch plausible transition surfaces in the middle distance. (Hint: Why is it possible to do this in a two-dimensional sketch?)

25.4 *(4) Looking beneath the surface.* Beginning with the assertion that, in equilibrium, the surface of a conducting body

of arbitrary shape is an equipotential, prove that the electric field inside the body is everywhere zero.

25.5 *(4) Packing it in.* You wish to store the greatest possible charge on a conducting body of surface area A, located in air. What is the optimal shape of the body?

25.6 *(4) Charge and change.* Consider the argument we used in Section 25.4 to prove the second part of rule 2, which pertains to a charge-free region within an equipotential surface. If there is a point charge in the region, you can still integrate along a flux line through the region from one point on the surface to another, but the argument no longer leads to the conclusion that the region is field free. Explain why.

25.7 *(G) Vindicating Maxwell.* To show that Maxwell's remarks at the beginning of this chapter are correct, write equations analogous to Equation 25.5 for thermodynamic work dW and heat flow dQ.

PROBLEMS

GROUP A

25.1 *(2) Same difference.* Show that 1 V/m = 1 N/C.

25.2 *(2) Work it in.* A point charge $q_1 = 20\ \mu C$ is located 40 cm from another point charge $q_2 = 35\ \mu C$. How much work is required to bring q_1 to a point 25 cm from q_2?

25.3 *(2) Negative thinking.* A charge $q = -5.5 \times 10^{-6}$ C is moved from a point where the potential is 35 V to a point where the potential is 52 V. What is the change in the potential energy of the system?

25.4 *(2) Electronic chicken.* Two electrons move directly toward one another. Their initial speed, when they are distant from one another, is 1.0×10^6 m/s. What is their distance of closest approach? Take the mass of each electron to be 9.1×10^{-31} kg.

25.5 *(2) Protonic chicken.* Two protons move directly toward one another with the same initial speed as that of the electrons in Problem 25.4. What is their distance of closest approach? The mass of a proton is 1.7×10^{-27} kg.

25.6 *(2) Grasping the difference.* A flat dielectric plate of area 12.5 m² carries a uniformly distributed charge $Q = -1.20 \times 10^{-7}$ C. Point P lies 3.54 cm above the surface of the plate, and point P' lies 7.92 cm above the surface. What is the potential difference $\Delta V = V - V'$ between P and P'? Neglect edge effects.

25.7 *(2) Temporary triangle.* Three equal charges, $q = 2.65 \times 10^{-6}$ C, are located far from one another. How much work must you do to bring them to the corners of an equilateral triangle of sides 8.50 cm?

25.8 *(2) In the middle.* What is the potential at the center of the triangle formed by the three charges of Problem 25.7?

25.9 *(2) Eternal triangle.* Three point charges are located as shown below. Find the potential **(a)** at point A and **(b)** at point B.

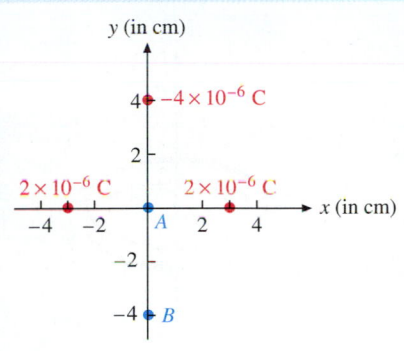

25.10 *(2) Pushing particles around.* **(a)** An electron is initially at rest. What is its speed when it has been accelerated through a potential difference of 1 V? **(b)** What is the speed of a proton accelerated in like manner?

25.11 *(2) Falling through a potential difference.* An electron lies far from a much more massive proton. Make the unrealistic assumption that there are no other charges in the vicinity. If the electron is released from rest, what is its speed when it is 2 nm from the proton?

25.12 *(2) Moving right along.* An electron accelerates from rest through a potential change $+300$ V. **(a)** What is the final kinetic energy of the electron in eV? in J? **(b)** What is the final speed of the electron? Take $m = 9.1 \times 10^{-31}$ kg.

25.13 *(2) Ionization.* The *ionization energy* of a hydrogen atom—the energy required to remove the electron far from the proton—is 13.6 eV. Express this energy in joules.

25.14 *(2) The long way.* The dielectric rod illustrated below carries a uniformly distributed charge q. Determine the potential at P, located on the rod axis a distance x from one end. Take $V = 0$ at $r_0 = \infty$.

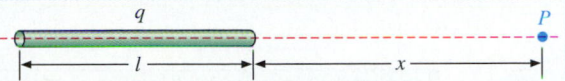

25.15 *(2) Chargers.* An isolated flat conducting plate of length l, width w, and thickness t carries charge q. **(a)** What is the value of the charge per unit area? **(b)** A similar plate carrying charge $-q$ is brought near the first plate and situated opposite it at a distance d, with faces of area lw parallel to each other. What is the value of the charge per unit area on the first plate now? Explain the change, if any. Ignore edge effects; that is, let d be much smaller then l and w.

25.16 *(2) Hole in the middle.* An annulus of inner radius R_1 and outer radius R_2 is cut out of a thin sheet of metal and given a charge Q. What is the potential at a point P, located on the axis of the annulus a distance z from the plane of the annulus?

25.17 *(2) Check at infinity.* Show that the potential of a charged disk, given by Equation 25.21, behaves as you would expect it to at large distances $z \gg R$ from the plate.

25.18 *(2) Right up close.* What is the potential difference ΔV between the surface of a disk of radius R, carrying a uniform charge per unit area σ, and a point at $r = \infty$?

25.19 *(2) Is a sketch of a field a potential landscape?* For the charged sphere of Example 25.3, sketch graphs of **(a)** $\mathscr{E}$ and **(b)** V as functions of r.

25.20 *(2) Without and within.* Just outside the surface of a charged sphere of radius 8.2 cm, the potential is 300 V. What is the potential **(a)** at a point 28.7 cm from the center of the sphere? **(b)** at a point 4.1 cm from the center?

25.21 *(2) Realizing one's potential.* The surface charge per unit area on a metal sphere of radius 1.15 cm is -2.38×10^{-7} C/m². What is the potential at a point 37.5 cm from the center of the sphere?

25.22 *(2) Getting harder.* A 1.25-μC charge is brought from a great distance to the surface of a metal sphere of radius 3.82 cm. How much work is required to repeat the process with **(a)** a second identical charge? **(b)** a third identical charge? **(c)** a fourth identical charge?

GROUP B

25.23 *(2) Double your pleasure.* Two equal charges q (of the same sign) are located on the x axis at $x = b$ and $x = -b$. **(a)** Taking $V = 0$ at infinity, find the potential at an arbitrary point on the z axis. **(b)** Show that, if $z \gg 2b$, the result simplifies to the expression for the potential due to a charge $2q$ located at the origin. **(c)** Express the electric field as a function of z. **(d)** Find the value of z for which $\mathscr{E}$ is a maximum, and evaluate $\mathscr{E}_{\max}$.

25.24 *(2) Wired.* A long, thin, straight wire carries charge per unit length λ. Find the potential difference ΔV between two points located at distances r_1 and r_2 from the wire. Why can't you choose $r = \infty$ or $r = 0$ as the reference point where $V = 0$?

25.25 *(2) Cut wire, I.* A straight wire of length l carries uniformly distributed charge q. **(a)** Set $V = 0$ at $r_0 = \infty$, and show that the potential at point P in the figure (p. 698), located

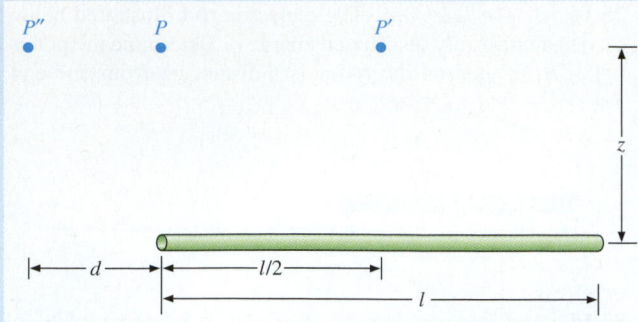

a distance z from the wire on a perpendicular passing through one end of the wire, is

$$V = \frac{1}{4\pi\epsilon_0} \frac{q}{l} \ln\left[\frac{l}{2}\left(1 + \sqrt{1 + \frac{z^2}{l^2}}\right)\right].$$

(b) Why can you choose $r_0 = \infty$ here, when you could not do so in Problem 25.24?

25.26 *(2) Tubular.* A cylindrical metal shell of radius R has a large but finite length L. It carries a charge per unit area σ. Taking $V = 0$ at $r = R$, show that the potential at a distance r from the axis of the shell is

$$V = \frac{R\sigma}{\epsilon_0} \ln\frac{R}{r} \quad \text{for } R < r \ll L$$

and $\qquad\qquad V = 0 \qquad\qquad$ for $r < R$.

25.27 *(2) Push comes to shove.* A long, straight wire has charge per unit length λ. An α particle of charge $+2e$ and mass 6.7×10^{-27} kg has speed 2.2×10^5 m/s when it is 1.2 cm from the wire. When the particle has moved to a distance 4.7 cm from the wire, its speed is 2.9×10^6 m/s. Find λ.

25.28 *(2) Uneven distribution.* Charge is distributed throughout a certain region of space in such a way that the charge per unit volume is $\rho = ax$. Show that the potential difference between points x_1 and x_2 in the region is $V = a(x_2{}^3 - x_1{}^3)/6\epsilon_0$. (Hint: In imagination, slice the region into layers of thickness dx that lie parallel to the y-z plane.)

25.29 *(2) Potential of a charged sphere, I.* A solid dielectric sphere of radius R carries a uniformly distributed charge Q. Find the potential as a function of distance r from the center. As in Example 25.3, fix V_0 at $r_0 = \infty$.

25.30 *(2) Potential of a charged sphere, II.* Charge is distributed nonuniformly throughout a solid dielectric sphere. The charge per unit volume ρ is a function of radius r only. If the potential inside the sphere is $V = ar^2 + b$ (a and b constants), show that $\rho = -6\epsilon_0 a$.

25.31 *(2) Long lines.* Shown in the next column is an end view of two very long parallel dielectric threads separated by a distance $2d$. The upper and lower threads carry charges per unit length λ and $-\lambda$. Show that the potential at the point P, whose coordinates are (r, θ), is $V = (\lambda d/2\pi\epsilon_0 r)\cos\theta$ if $r \gg 2d$.

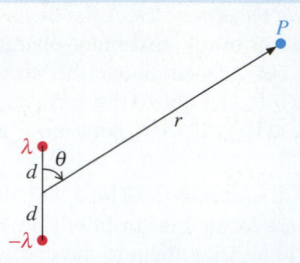

25.32 *(2) Circular reasoning.* Shown below are two coaxial wire rings of radius R, separated by a distance $2d$. The upper and lower rings carry charges q and $-q$. **(a)** Find the potential at a point on the axis located a distance z from the midpoint between the two rings. **(b)** Find the electric field at the same point.

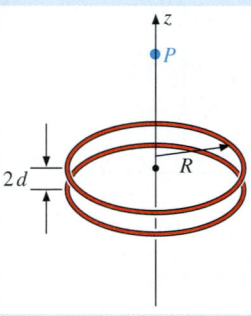

25.33 *(G) Square one.* Two charges q and two charges $-q$ are located at the corners of a square of side l. Taking $U = 0$ with the charges very far apart, find the potential energy of the system if the charges are arranged in the order

(a) $\begin{matrix} q & -q \\ -q & q \end{matrix}$; **(b)** $\begin{matrix} q & q \\ -q & -q \end{matrix}$.

25.34 *(G) Solid.* A long solid dielectric cylinder of radius R carries uniform charge per unit volume ρ. Using the reference point $r = R$ for $V = 0$, show that the potential at a distance r from the axis of the cylinder is

$$V = \frac{\rho(R^2 - r^2)}{4\epsilon_0} \quad \text{for } r < R$$

and has the value determined in Problem 25.26 for $r > R$. Check by showing that the results of the two problems agree at $r = R$. (Hint: How can you compare ρ in this problem with σ in Problem 25.26?)

25.35 *(G) Energy of a charged sphere.* A solid sphere of radius R carries charge Q, distributed uniformly. As a result, the sphere has a *self-energy*. You can see this by imagining the sphere to be assembled by bringing bits of matter of volume $d\tau$ and charge $dq = \rho\,d\tau$ up to it from infinity. Show that the self-energy is

$$U = \frac{3}{20\pi\epsilon_0}\frac{Q^2}{R}.$$

25.36 *(G) Electron radius.* On the basis of relativistic considerations, to be discussed in Chapter 39, it can be shown that, if the energy of an electron were greater than approximately

$U = m_e c^2 \simeq 5 \times 10^5$ eV (where c is the speed of light), the electron would have to have a greater mass than it actually has. Assume that all of the energy U is the self-energy considered in Problem 25.35. On this basis, set a lower limit on the radius of an electron. Compare your result with the radius of a typical atomic nucleus, $\simeq 3 \times 10^{-15}$ m. Although the nucleus contains nearly all of the mass of an atom, each electron in the atom is roughly as big as the nucleus. You should not take the meaning of this radius in the literal classical sense because the electron is surely not a little red ball!

GROUP C

25.37 *(2) Cut wire, II.* **(a)** Refer to the drawing in Problem 25.25, and find the potential at point P' located opposite the midpoint of the wire. (Hint: Don't start from scratch. Begin with the result of Problem 25.25.) **(b)** Evaluate the electric field $\mathscr{E}$ at point P'.

25.38 *(2) Cut wire, III.* Again refer to the drawing in Problem 25.25, and find the potential at point P'' located a distance z from the wire axis on a perpendicular passing through the axis a distance d from the end of the wire. (The hint here is the same as that for Problem 25.37.)

25.39 *(2) Madelung energy.* Consider a very long linear array of N charges q and N charges $-q$, spaced at equal distances R, thus:

$$\cdots \quad q \qquad -q \qquad q \qquad -q \qquad q \qquad -q \quad \cdots$$
$$\quad \leftarrow R \rightarrow \quad \leftarrow R \rightarrow \quad \leftarrow R \rightarrow \quad \leftarrow R \rightarrow \quad \leftarrow R \rightarrow$$

This configuration is important because it is the one-dimensional analogue of a cubic ionic crystal such as sodium chloride (NaCl). **(a)** Consider a charge $-q$ located somewhere in the line, far from the ends. Show that its potential energy can be written in the form

$$U = -\frac{\alpha q^2}{4\pi\epsilon_0 R},$$

where α is a constant called the *Madelung constant*. **(b)** What is the total energy (called the *Madelung energy*) of this one-dimensional "crystal"? (Remember that N is very large.)

(c) Show that the Madelung constant of the alternating one-dimensional array is

$$\alpha = 2 \ln 2.$$

Hint: $\ln(1 + x) = x - \dfrac{x^2}{2} + \dfrac{x^3}{3} - \dfrac{x^4}{4} + \cdots.$

The Madelung energy is the most important part of the energy that binds ionic crystals together. Calculation of the Madelung constant for real three-dimensional crystals often involves tricky summations and is usually carried out by use of a computer. But the principle is the same as in one dimension.

25.40 *(G) Do-∇.* **(a)** An electric potential is described in a cartesian frame by the function $V(\mathbf{r}) = V(x, y, z)$, which has arbitrary form. Show that the field at any point is

$$\mathscr{E}(x, y, z) = -\nabla V(x, y, z)$$
$$\equiv -\left(\hat{\mathbf{x}} \frac{\partial}{\partial x} + \hat{\mathbf{y}} \frac{\partial}{\partial y} + \hat{\mathbf{z}} \frac{\partial}{\partial z} \right) V(x, y, z).$$

The expression in large parentheses is called an *operator* because it has meaning only in the sense that it prescribes a series of operations on a function [here $V(x, y, z)$]. The special symbol ∇, called *del*, is used because this operator is an important one in advanced vector analysis. **(b)** Suppose that the potential is represented by the function

$$V = ax + bxy + c(z^2 - x^2),$$

where a, b, and c are constants. Use the result of part a to evaluate $\mathscr{E}(x, y, z)$.

Capacitance and Electric Field Energy

LOOKING AHEAD

A parallel-plate capacitor consists of a pair of identical conducting plates, separated by a small distance, that carry equal and opposite charges.

A capacitor can be used to store electric charge q; its capacitance C is defined to be the stored charge (of either sign) per unit potential difference V between the plates: $C \equiv q/V$.

Capacitors can be connected into networks whose equivalent capacitances can be calculated.

The capacitance of a capacitor can be increased by placing a dielectric between the plates.

The work required to place charge on the plates of a capacitor can be recovered.

The potential energy of the charged capacitor is stored in the electric field between the plates.

Left: The energy stored in the electric field of a charged capacitor is invisible but (as the discharge spark shows) not negligible.

> *Not by age but by capacity is wisdom acquired.*
>
> —PLAUTUS, *Trinummus*

Introduction

In the first part of this chapter (Sections 26.2 and 26.3), we use the concept of electric potential to examine the *capacitor*, an important device for storing electric charge and potential energy in electrical systems. In Section 26.4, we consider *polarization*, in which the imposition of an electric field on a dielectric causes a net separation of positive and negative charge. We also consider the most important practical consequence of polarization—the way the presence of a dielectric affects the properties of a capacitor. In Section 26.5, we look at the energy stored in an electric field, either in a capacitor or elsewhere.

Capacitors and Capacitance

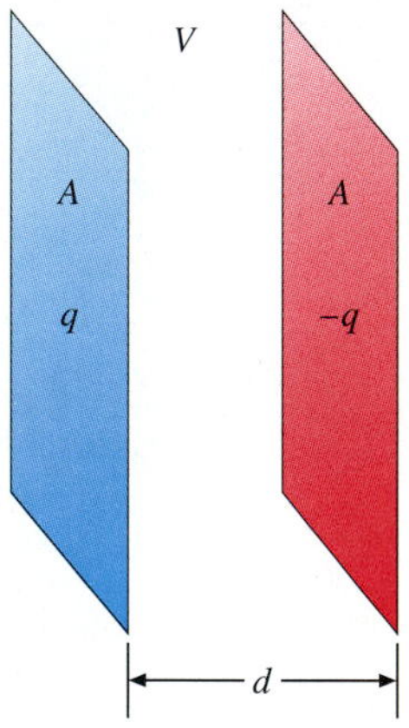

FIGURE 26.1 A parallel-plate capacitor. Two plates, each having area A, are separated by a distance d. One plate carries a charge q, and the other a charge $-q$. Both charges are uniformly distributed. The potential difference between the plates is V.

Figure 26.1 shows a device called a **parallel-plate capacitor**. It consists of two identical conducting plates, each of area A, separated by a distance d. Initially, the plates are electrically neutral. By one of several possible means (for example, by connecting the plates to the terminals of a battery), we remove charge q from one plate and place it on the other. As a result, the two plates have equal and opposite charges, $-q$ and q.

Let d be much less than the smallest linear dimension of either plate. Taking advantage of this close spacing of the plates, we concentrate on the planar symmetry that is characteristic of the greater part of the plates and ignore the *edge effects* that arise from the violation of planar symmetry near their edges.

Consider first the field $\mathscr{E}_L$ due to the left (positively charged) plate only (Figure 26.2a). According to Equation 23.34, the field has magnitude $\mathscr{E} = \sigma/2\epsilon_0 = q/2\epsilon_0 A$ and is directed normal to and away from the plate on both sides. Take the direction toward the right as the positive x direction. We then have

$$\mathscr{E}_L = -\frac{q}{2\epsilon_0 A}\,\hat{\mathbf{x}} \quad \text{in region 1} \quad \text{and} \quad \mathscr{E}_L = \frac{q}{2\epsilon_0 A}\,\hat{\mathbf{x}} \quad \text{in regions 2 and 3.}$$

Next, consider the field $\mathscr{E}_R$ due to the right (negatively charged) plate only (Figure 26.2b). We have

$$\mathscr{E}_R = \frac{q}{2\epsilon_0 A}\,\hat{\mathbf{x}} \quad \text{in regions 1 and 2} \quad \text{and} \quad \mathscr{E}_R = -\frac{q}{2\epsilon_0 A}\,\hat{\mathbf{x}} \quad \text{in region 3.}$$

We now use the superposition principle (Section 23.4) to find the field due to the pair of plates (Figure 26.2c). In regions 1 and 3, $\mathscr{E}_L$ and $\mathscr{E}_R$ have opposite directions and the fields due to the plates superpose to yield

$$\mathscr{E} = 0 \quad \text{outside the plates.} \tag{26.1a}$$

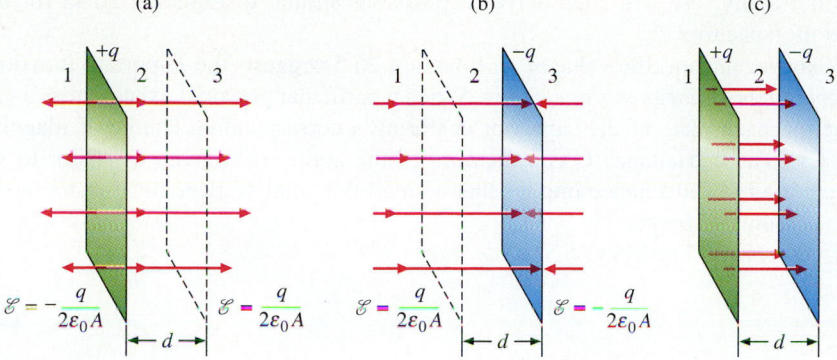

In region 2, $\mathscr{E}_L$ and $\mathscr{E}_R$ have the same direction and the fields superpose to yield

$$\mathscr{E} = \frac{q}{\epsilon_0 A}\,\hat{\mathbf{x}} \quad \text{between the plates.} \tag{26.1b}$$

In a capacitor, the potential difference between the two plates is conventionally expressed by setting $V = 0$ for the negatively charged plate. The potential of the positively charged plate is then conventionally called the "potential across the capacitor" or sometimes the "voltage across the capacitor." The potential across a parallel-plate capacitor is readily calculated. We evaluate $-\int \mathscr{E} \cdot d\mathbf{s}$ across the region between the plates (region 2), beginning at the negatively charged plate and proceeding in the $-\hat{\mathbf{x}}$ direction to the positively charged plate. Because the flux lines are straight and uniform, we have $V = (-\mathscr{E})(-d)$. Taken together with Equation 26.1b, this gives us

$$V = \frac{qd}{\epsilon_0 A} \quad \text{for a parallel-plate capacitor.} \tag{26.2}$$

Capacitance

Let us solve Equation 26.2 for q:

$$q = \frac{\epsilon_0 A}{d}\,V \quad \text{for a parallel-plate capacitor.} \tag{26.3}$$

It is evident from this equation that, for a given capacitor, the magnitude $|q|$ of the charge stored on each plate is directly proportional to the potential $|V|$ across the capacitor. The proportionality constant is called the **capacitance** C. For the special case of the parallel-plate capacitor, Equation 26.3 gives us

$$C = \frac{\epsilon_0 A}{d}, \tag{26.4a}$$

or

$$C = \epsilon_0 \Lambda. \tag{26.4b}$$

The capacitance is the product of ϵ_0 and a factor Λ that is dimensionally a length (here, $\Lambda = A/d$) whose value is determined by the geometry of the capacitor. In terms of the capacitance, we can write

$$q = CV, \tag{26.5}$$

where it is understood that q and V are taken to be the magnitudes of the charge on either plate and the potential difference between the plates—that is, q is a convenient shorthand for $|q|$ and V for $|V|$, and therefore C is always a positive quantity in accordance with Equation 26.4a. Equations 26.4b and 26.5 apply to *any* capacitor. For capacitors other than parallel-plate capacitors, however, $\Lambda \neq A/d$ and Equation 26.4a does not apply. We will soon derive expressions similar to Equation 26.4a for other kinds of capacitors.

Whatever the specific value of C, Equation 26.5 suggests the important function of a capacitor as a *charge-storage device*. When a particular potential exists across a given capacitor, each plate of the capacitor is storing a corresponding charge of magnitude $q = CV$. The capacitance C is a measure of the ability of a given capacitor to store charge; a large capacitance implies that a small potential suffices for the storage of a large amount of charge.

If you rewrite Equation 26.5 in the form

$$C = \frac{q}{V}, \tag{26.6}$$

you can see immediately that the SI unit of capacitance is the coulomb per volt (C/V). In view of its importance, the unit is given the special name **farad** (**F**), after Michael Faraday:

$$1 \text{ F} \equiv 1 \text{ C/V.} \tag{26.7}$$

EXAMPLE 26.1

A capacitor consists of two circular metal disks, each of radius 15 cm, separated by 0.5 mm. **(a)** What is the capacitance of the device? **(b)** What is the magnitude of the charge stored on each disk when the potential across the capacitor is 150 V?

SOLUTION:

(a) Evaluate C.

You use Equation 26.4a to write

$$C = \frac{8.85 \times 10^{-12} \text{ C}^2/\text{N·m}^2 \times \pi \times (0.15 \text{ m})^2}{5 \times 10^{-4} \text{ m}}$$

$$= 1.25 \times 10^{-9} \text{ F} = 1.25 \text{ nF.}$$

(b) Find q when $V = 150$ V.

Equation 26.5 gives you

$$q = 1.25 \times 10^{-9} \text{ F} \times 150 \text{ V} = 1.88 \times 10^{-7} \text{ C.}$$

The capacitor discussed in Example 26.1 is of a size very much on the everyday scale of things. But its capacitance, expressed in farads, is a quite small number. That is to say, 1 F is a very large amount of capacitance on the everyday scale of things. This follows from the now-familiar fact that 1 C is a very large amount of charge on the same scale. As a consequence, the capacitance of practical capacitors is often expressed in microfarads (μF) or picofarads (pF). However, special-purpose capacitors are available commercially up to about 1 F.

Capacitors of Other Symmetries

Any two conducting bodies carrying equal and opposite charges constitute a capacitor whose capacitance can be expressed in terms of Equation 26.6, $C = q/V$. If the bodies are irregular in shape, there is no convenient way to calculate the capacitance, and determining C depends on simultaneous measurement of q and V. But most practical capacitors have either the planar symmetry we have just considered or cylindrical or spherical symmetry. We now consider these last two cases.

The Cylindrical Capacitor

Figure 26.3 shows a **cylindrical capacitor**, consisting of two coaxial cylinders of radii R_1 and R_2 and length l, carrying charges q and $-q$, respectively. As in the parallel-plate case, we assume that the length l is substantially greater than R_2, so we can neglect edge effects.

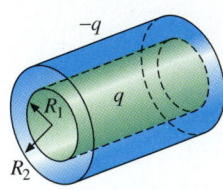

FIGURE 26.3 A cylindrical capacitor.

As in the parallel-plate case, we calculate the potential across the capacitor by integrating $-\mathcal{E} \cdot d\mathbf{s}$. The cylindrical case is a little more complicated because the field is not constant but is given by Equation 24.12,

$$\mathcal{E} = \frac{1}{2\pi\epsilon_0} \frac{\lambda}{r} = \frac{1}{2\pi\epsilon_0} \frac{q}{lr}.$$

We choose a path of integration along a radius. That is, $\mathcal{E} \cdot d\mathbf{s} = \mathcal{E} \, dr$. To be consistent with our definition of the "potential across a capacitor," we integrate inward from the negative to the positive plate:

$$V = -\frac{1}{2\pi\epsilon_0} \frac{q}{l} \int_{R_2}^{R_1} \frac{dr}{r} = -\frac{1}{2\pi\epsilon_0} \frac{q}{l} \ln r \Big|_{R_2}^{R_1},$$

so

$$V = \frac{q}{2\pi\epsilon_0 l} \ln \frac{R_2}{R_1}. \tag{26.8}$$

Using the general relation $C = q/V$ given by Equation 26.6, we obtain the capacitance

$$C = \frac{2\pi\epsilon_0 l}{\ln(R_2/R_1)}. \tag{26.9}$$

Like that for the parallel-plate capacitor, this equation has the general form $C = \epsilon_0 \Lambda$. Here the "geometry factor" is $\Lambda = 2\pi l/\ln(R_2/R_1)$.

The Spherical Capacitor

Figure 26.4 shows a **spherical capacitor**, consisting of two concentric spherical shells of radii R_1 and R_2, carrying charges q and $-q$, respectively.

The electric field between the shells is given by Equation 24.9, $\mathcal{E} = (1/4\pi\epsilon_0)(q/r^2)$. To calculate the potential difference between the shells, we again choose a path of integration inward along a radius, so $\mathcal{E} \cdot d\mathbf{s} = \mathcal{E} \, dr$. Integrating from the negative to the positive plate, we have

$$V = -\frac{q}{4\pi\epsilon_0} \int_{R_2}^{R_1} \frac{dr}{r^2} = \frac{q}{4\pi\epsilon_0} \frac{1}{r} \Big|_{R_2}^{R_1},$$

so

$$V = \frac{q}{4\pi\epsilon_0} \left(\frac{1}{R_1} - \frac{1}{R_2} \right). \tag{26.10}$$

The relation $C = q/V$ now gives us

$$C = \frac{4\pi\epsilon_0}{\dfrac{1}{R_1} - \dfrac{1}{R_2}}. \tag{26.11}$$

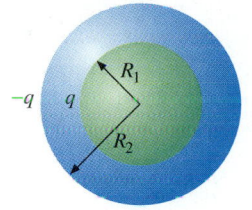

FIGURE 26.4 A spherical capacitor.

What is the value of Λ for the spherical capacitor?

Spherical capacitors are often used for precision measurements. Because they are inherently free of edge effects, the calculated capacitance is very close to the actual capacitance. It is usually necessary to make a tiny hole in the outer shell to allow the passage of a conductor for charging the inner shell. But the errors arising from this violation of spherical symmetry can be kept quite small.

Imagine that the outer shell of a spherical capacitor is allowed to expand without limit; that is, let $R_2 \rightarrow \infty$. For the parallel-plate and cylindrical capacitors, the analogous process invalidates the assumption of negligible edge effects, and the calculated values of C lose significance. But this is not so for the spherical capacitor, and we have

$$C = 4\pi\epsilon_0 R \tag{26.12}$$

for the *capacitance of an isolated spherical shell*. We drop the subscript on R because it is unnecessary in this case.

EXAMPLE 26.2

Calculate the capacitance of the earth, taking its radius to be 6.37×10^6 m.

In making this calculation, why does it not matter that the earth is solid and not a spherical shell?

SOLUTION: Using Equation 26.12, you have

$$C = 4\pi \times 8.85 \times 10^{-12} \ \text{C}^2/\text{N·m}^2 \times 6.37 \times 10^6 \ \text{m}$$
$$= 7.08 \times 10^{-4} \ \text{F}.$$

Capacitors in Parallel and in Series

In the construction of electric circuits, it is often necessary to connect two or more capacitors together into a *network*. The simplest networks—called *parallel* and *series networks*—are also the most important. In network diagrams like that of Figure 26.5, capacitors are represented by a universal symbol that suggests a pair of flat plates, each connected to a wire. This symbol is used regardless of the actual form of the capacitor.

In Figure 26.5, capacitor 1, having capacitance C_1, and capacitor 2, having capacitance C_2, are shown connected by the conducting wires that run from terminal A to the left sides of both capacitors and from the right sides of both to terminal B. We say that the two capacitors are connected **in parallel**—that is, "side by side." Consequently, any potential difference applied between the terminals at A and B is applied across both capacitors.* What is the capacitance C of the network lying between terminals A and B? That is, what single **equivalent capacitor** would have the same capacitance C as the two capacitors linked together as shown?

FIGURE 26.5 Two capacitors in parallel, with their equivalent capacitor shown in red.

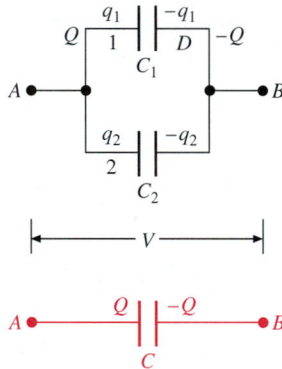

Suppose we charge this parallel network by removing positive charge Q from the system at B and putting it back into the system at A. As a result, there is a potential difference V between A and B. When we regard the entire system as a single equivalent capacitor, it must conform to the definition $q = CV$ of Equation 26.5. That is, we have

$$Q = CV. \tag{26.13}$$

The excess positive charge Q in the left half of the system is distributed in some fashion between the left-hand plates of the two capacitors so that

$$Q = q_1 + q_2. \tag{26.14}$$

Similarly, the excess negative charge $-Q$ remaining in the right half of the system is distributed so that $-Q = -q_1 - q_2$.

---

*Because the left-hand plates of the two capacitors are linked by conducting wires, they form a continuous conductor and must have the same potential. The same is true of the right-hand plates. Thus the potential differences across the two capacitors must be the same.

For the two capacitors taken individually, the relation $q = CV$ becomes

$$q_1 = C_1V \quad \text{and} \quad q_2 = C_2V,$$

where the potential difference V is the same for both. We insert these values of q_1 and q_2 and the value of Q given by Equation 26.13 into Equation 26.14:

$$C_1V + C_2V = CV.$$

The factor V cancels out, leaving us with the result

$$C = C_1 + C_2 \quad \text{for two capacitors in parallel.} \qquad \textbf{(26.15a)}$$

This argument can be extended to yield an equivalent capacitance C for any number n of capacitors in parallel, as shown in Figure 26.6:

$$C = C_1 + C_2 + \cdots + C_j + \cdots + C_n. \qquad \textbf{(26.15b)}$$

The equivalent capacitance of capacitors in parallel is the sum of their individual capacitances.

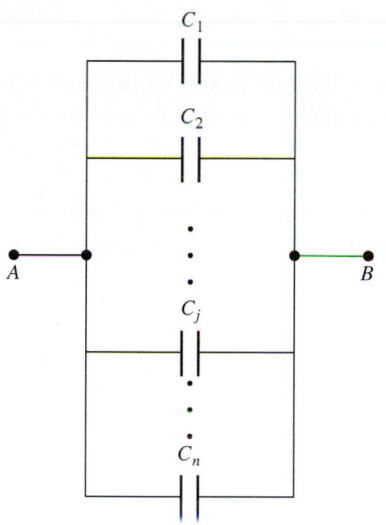

FIGURE 26.6 An arbitrary number n of capacitors in parallel.

In Figure 26.7, a capacitor of capacitance C_1 and a capacitor of capacitance C_2 are shown connected **in series**—that is, "in a string." Consequently, any potential difference V between terminals A and B must be the sum of potential differences across the two capacitors.* What is the capacitance C of the network lying between points A and B? That is, what single *equivalent capacitor* would have the same capacitance C as the two capacitors linked together?

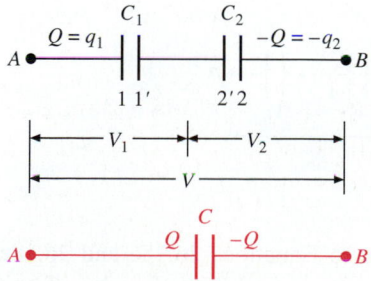

FIGURE 26.7 Two capacitors in series, with their equivalent capacitor shown in red.

---

*Because they are linked by a conducting wire, plates 1' and 2' must be at the same potential. The same is true of terminal A and plate 1, as well as of plate 2 and terminal B. Thus a potential difference between A and B must be distributed between the two gaps.

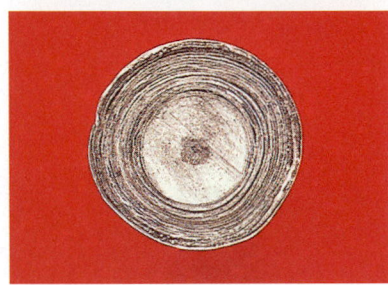

Some practical air-gap and dielectric-filled capacitors. The cutaway view shows the metal-dielectric-metal sandwich of a parallel-plate capacitor, rolled up for compactness.

Suppose that we charge this series network by removing positive charge Q from the system at B and putting it back into the system at A. Plate 1 and point A are connected by a conducting path to each other and to nothing else. The same is true of plate 2 and point B. Therefore, plates 1 and 2 must have equal and opposite charges Q and $-Q$, respectively.

The middle link of the system, consisting of plates $1'$ and $2'$ and the conductor connecting them, has no conducting path to the outside world and must remain electrically neutral overall. But the surface of this middle link is an equipotential, and the link must therefore be a field-free region. Therefore, the field originating on plate 1 and the field terminating on plate 2 must not penetrate this region. This can be true only if the positive charge Q on plate 1 induces an equal and opposite charge $-Q$ on plate $1'$ while the negative charge $-Q$ on plate 2 induces an equal and opposite charge Q on plate $2'$.

As a consequence of the charges on the plates of capacitor 1, there is a potential difference V_1 between them. Similarly, there is a potential difference V_2 between the plates of capacitor 2. We can use Equation 26.5, $q = CV$, to express these potential differences in the form

$$V_1 = \frac{Q}{C_1} \quad \text{and} \quad V_2 = \frac{Q}{C_2}. \tag{26.16a,b}$$

These two potential differences must add up to the total potential difference between A and B, since the potential difference between plates $1'$ and $2'$ is zero. So we have

$$V = V_1 + V_2. \tag{26.17}$$

For the equivalent capacitor C, we must have $V = Q/C$. We insert this value of V, together with the values of V_1 and V_2 given by Equations 26.16a and 26.16b, into Equation 26.17, to obtain

$$\frac{Q}{C} = \frac{Q}{C_1} + \frac{Q}{C_2}.$$

The factor Q cancels out, leaving us with the result

$$\frac{1}{C} = \frac{1}{C_1} + \frac{1}{C_2} \quad \text{for two capacitors in series.} \tag{26.18a}$$

This argument can be extended to yield an equivalent capacitance C for any number n of capacitors in series, as shown in Figure 26.8:

$$\frac{1}{C} = \frac{1}{C_1} + \frac{1}{C_2} + \cdots + \frac{1}{C_j} + \cdots + \frac{1}{C_n}. \tag{26.18b}$$

The reciprocal of the equivalent capacitance of capacitors in series is the sum of the reciprocals of their individual capacitances.

FIGURE 26.8 An arbitrary number n of capacitors in series.

Many persons find the derivations leading to Equations 26.15a and 26.18a tricky and even confusing on the first encounter. But the logical flow of each will be clear to you if you bear in mind the principle the two derivations have in common. In each case, there is a quantity—Q for the parallel network and V for the series network—whose value for the equivalent capacitor is just the *sum* of the values for the individual capacitors, and there is another quantity—V for the parallel network and Q for the series network—whose value is the *same* for the equivalent capacitor and for the individual capacitors. That is, we have for the parallel network

$$Q = q_1 + q_2 \quad \text{and} \quad V = V_1 = V_2.$$

And we have for the series network

$$V = V_1 + V_2 \quad \text{and} \quad Q = q_1 = q_2.$$

In each case, we use the definition $q = CV$ (or $V = q/C$) three times to yield values for the terms in the sum equation. Then we use the set of equalities to simplify the sum. The common factor cancels out, leaving the desired result.

EXAMPLE 26.3

You have three capacitors of equal capacitance $C_1 = C_2 = C_3 = C = 0.50$ µF. Find the values of equivalent capacitance you can obtain by connecting them in all possible ways.

SOLUTION:

(a) To begin with, you have the trivial case $C_a = 0.50$ µF, which you obtain simply by using any one of the three capacitors by itself.

Next, consider two of the capacitors used together.

(b) You can connect them in parallel, which gives you

$$C_b = C + C = 2C = 1.00 \text{ µF.}$$

(c) Or you can connect them in series, in which case you have

$$\frac{1}{C_c} = \frac{1}{C} + \frac{1}{C} = \frac{2}{C},$$

which gives you $C_c = \frac{1}{2}C = 0.25$ µF.

Next, use all three. There are four possibilities, as Figure 26.9 shows.

(d) The parallel connection (Figure 26.9a) gives you

$$C_d = C + C + C = 3C = 1.50 \text{ µF.}$$

(e) For the series connection (Figure 26.9b), you have

$$\frac{1}{C_e} = \frac{1}{C} + \frac{1}{C} + \frac{1}{C} = \frac{3}{C},$$

which gives you $C_e = \frac{1}{3}C = 0.17$ µF.

(f) In Figure 26.9c, you have two capacitors in parallel, the pair being in series with the third capacitor. You analyze this network by dividing it into manageable *subnetworks*. First, consider the parallel subnetwork consisting of C_2 and C_3. From part **b**, you have the equivalent capacitance $C_{23} = 2C$. When you replace C_2 and C_3 with their equivalent capacitor, you have the series network shown in the last part of Figure 26.9c. For this network you write

$$\frac{1}{C_f} = \frac{1}{C} + \frac{1}{C_{23}} = \frac{1}{C} + \frac{1}{2C} = \frac{3}{2C}.$$

You solve to obtain

$$C_f = \frac{2}{3}C = 0.33 \text{ µF.}$$

(a)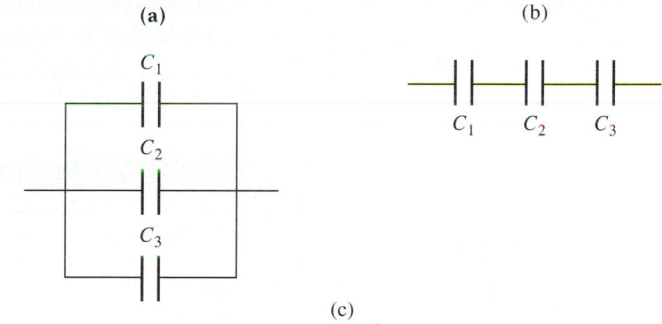

(b)

(c)

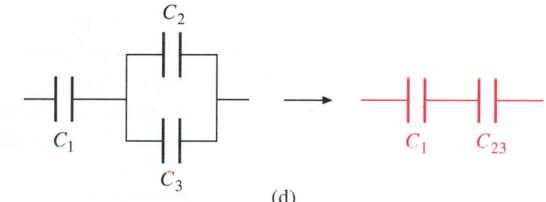

(d)

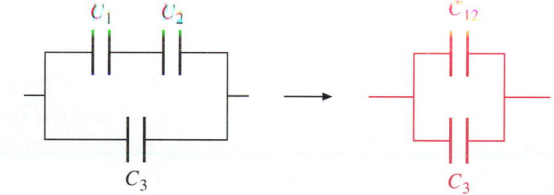

FIGURE 26.9

(g) In Figure 26.9d you have two capacitors in series, the pair being in parallel with the third capacitor. You begin with the series subnetwork consisting of C_1 and C_2. From part **c**, you have the equivalent capacitance $C_{12} = \frac{1}{2}C$. Placing C_{12} in parallel with C_3, you obtain

$$C_g = C_{12} + C_3 = \frac{3}{2}C = 0.75 \text{ µF.}$$

If you arrange these seven results in order of their capacitances, you have the possibilities $\frac{1}{3}C$, $\frac{1}{2}C$, $\frac{2}{3}C$, C, $\frac{3}{2}C$, $2C$, and $3C$. Note that the reciprocal of every possible equivalent capacitance is also a possible equivalent capacitance. Can you see why?

Dielectrics

So far, we have studied the simplest forms of capacitors, consisting of a pair of plates separated by empty (or air-filled) space. In practice, nearly all capacitors are constructed with a thin layer of a dielectric material between the plates. This dielectric layer has three functions:

1. It can serve as a structural element that rigidly bonds the two plates quite close together (thus making possible relatively large values of capacitance) while preventing them from making electrical contact.

2. The charge that can be stored on capacitor plates is limited by the **breakdown potential**—the potential difference at which a destructive spark will jump across the gap between the plates. Many dielectrics have greater **dielectric strength** than air. That is, they will sustain greater electric fields than air without breaking down. A greater breakdown potential increases the maximum allowable potential across the capacitor for a given plate separation and consequently increases the charge that can be stored on the capacitor. The dielectric strengths of various materials are listed in Table 26.1.

3. Because dielectrics are *polarizable*, the presence of a dielectric increases the capacitance of a capacitor. We now consider this matter in some detail.

TABLE 26.1 Dielectric Constant and Dielectric Strength for Selected Materials

Material	Dielectric Constant κ	Dielectric Strength (V/m)
Vacuum	1	
Air (dry)	1.0006	$1-3 \times 10^6$
Polyethylene	2.3	50×10^6
Polystyrene	2.6	25×10^6
Mylar (polyester)	3.5	$6-13 \times 10^6$
Paper (typical)	3.5	14×10^6
Mineral oil (typical)	4.5	12×10^6
Porcelain (typical)	7	4×10^6
Water	80	
Strontium titanate	230	15×10^6

We begin with a purely phenomenological discussion of the role of a dielectric in increasing the capacitance of a capacitor. Consider a parallel-plate capacitor having a gap of width d between the plates. When the gap is empty, the capacitance C_0 is given by Equation 26.4a,

$$C_0 = \epsilon_0 \frac{A}{d},$$

where A is the area of either plate. But, if a slab of a dielectric material, so shaped that it just fits, is inserted into the gap, the capacitance is observed to increase to a larger value,

$$C = \kappa C_0. \tag{26.19}$$

From this experimental point of view, κ is a purely empirical factor determined by measurement. It is called the **dielectric constant**. Its value depends on the particular dielectric material used. The dielectric constants of a variety of materials are listed in Table 26.1.

By inserting the value of C_0 for a parallel-plate capacitor into Equation 26.19, we can obtain

$$C = \kappa \epsilon_0 \frac{A}{d}. \tag{26.20}$$

The product $\kappa \epsilon_0$ for a given dielectric is called its **permittivity** ϵ; that is, we have

$$\epsilon = \kappa \epsilon_0. \tag{26.21}$$

A parallel-plate capacitor filled with a dielectric of permittivity ϵ has capacitance

$$C = \epsilon \frac{A}{d}. \qquad (26.22)$$

Similar expressions apply to dielectric-filled capacitors of other geometries. In general, the capacitance of any capacitor can be written

$$C = \epsilon\Lambda = \kappa\epsilon_0\Lambda. \qquad (26.23)$$

Here, as before, Λ depends on the geometry of the device. Although we have defined dielectric constant κ and permittivity ϵ in the context of the parallel-plate capacitor, the effect of inserting a dielectric into the gap of a capacitor is the same regardless of the geometry of the capacitor—the capacitance is increased by a factor κ. For empty space, the dielectric constant has the value 1 by definition, and $\epsilon = \epsilon_0$. That is why ϵ_0 is called the *permittivity of free space.*

EXAMPLE 26.4

(a) You connect the plates of an air-gap parallel-plate capacitor, of capacitance $C_0 = 5.3 \times 10^{-9}$ F, to the terminals of a battery having a fixed potential of $V_0 = 9.0$ V. Charge $|q_0|$ thus flows onto the capacitor plates, and the potential across the capacitor is V_0. You disconnect the battery. Then you slip a close-fitting slab of polyethylene into the gap. Find the new capacitance C and the new potential V across the capacitor. (b) You repeat the process of part **a**, except that you leave the capacitor connected to the battery as you insert the polyethylene slab. Find the final charge $|q|$ on the capacitor plates.

SOLUTION:

(a) Find the new capacitance C and the new potential V across the gap of an isolated, charged capacitor after you have slipped a dielectric slab between the plates.

The charge on the capacitor plates is

$$|q_0| = C_0V_0 = 5.3 \times 10^{-9} \text{ F} \times 9.0 \text{ V} = 4.8 \times 10^{-8} \text{ C}.$$

When you insert the polyethylene slab into the gap, you increase the capacitance. Using the value of κ given in Table 26.1, you find the new value

$$C = \kappa C_0 = 2.3 \times 5.3 \times 10^{-9} \text{ F} = 12.2 \times 10^{-9} \text{ F}.$$

Because the capacitor is electrically isolated, the charge on the plates cannot change when you insert the dielectric slab. So you have $|q| = |q_0|$. You can thus write the chain of equalities

$$CV = |q| = |q_0| = C_0V_0.$$

Solving for V in terms of C, C_0, and V_0, you obtain

$$V = \frac{C_0}{C} V_0 = \frac{V_0}{\kappa}. \qquad (26.24)$$

This gives you the numerical result

$$V = \frac{9.0 \text{ V}}{2.3} = 3.9 \text{ V}.$$

(b) Find the new charge on a capacitor connected to a battery after you have slipped a dielectric slab between the plates.

With the battery connected, the potential across the capacitor remains constant as you insert the slab; $V = V_0$. Because insertion of the slab increases the capacitance from C_0 to $C = \kappa C_0$, more charge must flow onto the plates when you insert the dielectric. This time you write the chain of equalities

$$|q| = CV = \kappa C_0 V_0 = \kappa|q_0|.$$

This gives you the numerical result

$$|q| = 2.3 \times 4.8 \times 10^{-8} \text{ C} = 1.1 \times 10^{-7} \text{ C}.$$

Both cases illustrate the increased capacitance due to the insertion of the dielectric. In the first case, the capacitor holds a constant charge at a reduced potential. In the second case, the capacitor holds an increased charge at a constant potential.

Microscopic Basis for the Dielectric Constant*

In part **a** of Example 26.4, insertion of a dielectric slab diminishes the potential across the gap. It follows that *the electric field in the gap must diminish as well.* Some of the flux lines originating at the charges on the positive plate must terminate on the dielectric slab while some of the flux lines terminating on the negative plate must originate on the dielectric slab (Figure 26.10). This can happen only if one surface of the dielectric

———————
*This subsection may be omitted without loss of continuity.

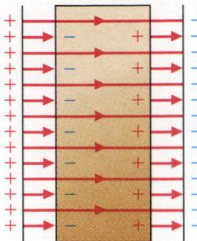

FIGURE 26.10 A capacitor containing a dielectric slab. Some flux lines originate and terminate on the dielectric. This implies the presence of surface charge.

FIGURE 26.11 (*a*) A molecule, viewed as a pair of overlapping equal and opposite charges. (*b*) An externally imposed electric field $\mathscr{E}_0$ exerts equal and opposite forces on the positive and negative charges that constitute the molecules of a dielectric. The molecules distort and acquire an *induced dipole moment*. The electric field of the induced dipole generally opposes the external field; averaged over many molecules, there is a uniform induced field $\mathscr{E}_d$. The net field $\mathscr{E}'$ within the dielectric is the vector sum $\mathscr{E}' = \mathscr{E}_0 + \mathscr{E}_d$. (*c*) The induction of dipoles does not affect the net charge contained in a small region well within the dielectric because inward-migrating charge just compensates for outward-migrating charge. A similar region containing the surface does acquire a net charge.

acquires a net excess of negative charge while the other surface acquires a net excess of positive charge. The dielectric slab contains equal amounts of positive and negative charge and is electrically neutral overall. But, when the dielectric lies within the electric field of the charged capacitor plates, it acquires an **induced dipole moment**. It is as though the "center of gravity" of the positive charge in the dielectric has shifted (very slightly) with respect to the "center of gravity" of the negative charge.

Let us examine this shift on the microscopic level. The following discussion, though rather crude, is correct in its essentials. In Figure 26.11a, we imagine a molecule as consisting of two blobs of electric charge, one positive and one negative. They are linked by a strong spring, which represents the strong electric attraction between them. (If such attraction did not exist, the substance would be a conductor, not a dielectric.) When there is no external electric field, the two blobs of charge overlap and the spring has zero length. But the external field $\mathscr{E}_0$ pulls the blobs apart to an equilibrium separation determined by the strength of the spring (Figure 26.11b).

(a) (b) (c)

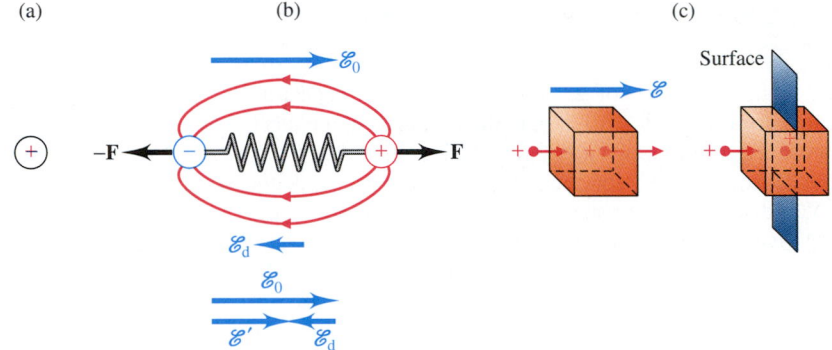

Thus distorted, the molecule becomes an electric dipole. As you can see, the sense of the dipole flux lines is generally opposite that of $\mathscr{E}_0$. Associated with this flux is a dipole electric field. For an isolated dipole, this field is a function of position with respect to the dipole. But what we see on the macroscopic scale is an average over vast numbers of molecular dipoles. On the average, there is a (macroscopically) uniform electric field $\mathscr{E}_d$, called the **depolarization field**. According to the superposition principle, the field $\mathscr{E}'$ within the dielectric is given by the vector sum

$$\mathscr{E}' = \mathscr{E}_0 + \mathscr{E}_d,$$

whose magnitude is *less than* that of $\mathscr{E}_0$. The dielectric constant is defined as the ratio of the magnitudes of the external field and the internal field:

$$\kappa \equiv \frac{\mathscr{E}_0}{\mathscr{E}'}. \tag{26.25}$$

Consider the tiny region of Figure 26.11c, which is nevertheless large enough to contain many molecules. If the region lies well within the dielectric slab, it remains electrically neutral as the imposition of the external field induces an internal dipole moment. As the molecules are distorted by the external field, positive charge leaving the region through its right-hand surface is replaced by positive charge entering it through its left-hand surface. A corresponding statement applies to the negative charge. But, at the surfaces of the dielectric slab, this compensation cannot take place. Thus one surface acquires an effective negative charge and the other an effective positive charge of equal magnitude. We say that the slab is **polarized**.

In vacuum, the dielectric constant has the value $\kappa = 1$ because there is no matter to be polarized, and thus $\mathscr{E}_d = 0$. In a conductor, $\kappa = \infty$. This follows from the fact that $\mathscr{E}'$, the denominator in Equation 26.25, must be zero within a conductor.

Some substances consist of molecules that have a dipole moment even in the absence of an external electric field. Water is a notable case; see Problem 23.38. In such substances, the macroscopic polarization arises from a somewhat different microscopic process. When there is no external electric field, the individual molecules tend to orient themselves so that their dipole moments cancel. Imposing an external field $\mathscr{E}_0$ tends to align the dipoles and thus to produce a net macroscopic polarization that reduces the internal field $\mathscr{E}'$. Materials whose molecules have permanent dipole moments tend to have quite large dielectric constants; for water, $\kappa = 80$.

Energy of the Electric Field

Work must be done to charge a capacitor. Suppose we start with the capacitor of Figure 26.12 in an uncharged state. We remove an infinitesimal element of positive charge dq from plate 2 and transport it to plate 1. The first time we do this, there is no field between the plates and the process requires no work. But, as we transport new charge, the charge already transferred repels the new charge. When the potential difference between the plates has the arbitrary value V, transport of further charge dq requires that we do work

$$dW = V\,dq$$

on the system (compare with Equations 25.3a and 25.5). Because the work is being done on the system against the conservative electric force, there is a buildup of potential energy in the system. When the charge has reached the particular value Q, the potential energy is

$$U = \int dU = \int_0^Q V\,dq.$$

We can write $V\,dq$ in terms of a single variable by using the relation $V = q/C$:

$$U = \int_0^Q \frac{1}{C}\,q\,dq.$$

Carrying out the integration, we obtain

$$U = \frac{1}{2}\frac{Q^2}{C}. \tag{26.26a}$$

Since $Q^2 = C^2V^2$, we can write the potential energy in the alternative form

$$U = \tfrac{1}{2}CV^2. \tag{26.26b}$$

Compare these results with the expression for the energy stored in a stretched spring given by Equation 8.23,

$$U = \tfrac{1}{2}kx^2,$$

where k is the force constant of the spring and x is the extension. Why do Equations 26.26a and 26.26b have the same mathematical form as the spring equation?

By comparing Equation 26.26a with Equation 8.23, you can see that the charge $|q|$ on the capacitor plates is analogous to the extension x of the spring, and the force constant k, which measures the stiffness of the spring, is analogous to $1/C$, the reciprocal of the capacitance. The analogy between capacitances and springs will be useful when we study changing electric currents in Chapters 32 and 33.

Potential energy is a property of a conservative system as a whole. But it is often useful to think of the energy as being "stored" in a particular way in the system. In the present situation, we can think of the energy as being stored in the electric field of the capacitor. Let us continue to think in terms of the parallel-plate capacitor. For this device, the capacitance is $C = \epsilon A/d$. So we can write Equation 26.26b in the form

$$U = \frac{1}{2}\frac{\epsilon A V^2}{d}.$$

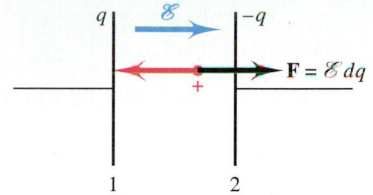

FIGURE 26.12 Charging a capacitor. Elements of charge dq are removed from plate 2 and moved to plate 1. With each transfer, the field opposing the transfer increases, requiring an increase in the amount of work required per element dq as the process continues.

Because the field in the gap of the capacitor is uniform, we have $V = \mathscr{E}d$. The energy stored in the capacitor is thus

$$U = \tfrac{1}{2}\epsilon\mathscr{E}^2 Ad. \qquad (26.27)$$

The factor Ad on the right side of this equation is just the volume between the plates, where the entire field is confined (neglecting edge effects). We divide through by this volume and define the **energy per unit volume u**, also called the **energy density**:

$$u \equiv \frac{U}{\text{volume}} = \frac{1}{2}\,\epsilon\mathscr{E}^2 = \frac{1}{2}\,\kappa\epsilon_0\mathscr{E}^2. \qquad (26.28)$$

We have defined u for the special case of the parallel-plate capacitor, where the field is uniform between the plates and zero everywhere. In this case, u has the same value throughout the capacitor gap. But note that Equation 26.26 expresses u in terms of ϵ and $\mathscr{E}$ only and contains no geometric factors at all. For this reason, it is valid for *any* electric field.

EXAMPLE 26.5

Consider again the spherical capacitor of Figure 26.4, consisting of two concentric spherical shells of radii R_1 and R_2. Assume that the space between the shells is empty. **(a)** Find the energy density u at a point r between the shells, when each shell carries a charge of magnitude $|Q|$. **(b)** Use this result to calculate the total energy in the field of the capacitor and the capacitance of the device. **(c)** Find the total energy contained in the electric field of a conducting sphere of radius R that carries charge Q.

SOLUTION:
(a) Calculate $u(r)$ between the shells.

The electric field at r is the Coulomb field $\mathscr{E} = Q/4\pi\epsilon_0 r^2$. There is no dielectric in the capacitor, and thus $\kappa = 1$ and $\epsilon = \epsilon_0$. So you use Equation 26.28 to write

$$u = \frac{1}{2}\,\epsilon_0\left(\frac{Q}{4\pi\epsilon_0 r^2}\right)^2,$$

which simplifies to

$$u = \frac{Q^2}{32\pi^2\epsilon_0 r^4}. \qquad (26.29)$$

The energy density falls off quite rapidly with r.

(b) Calculate the total electric field energy and the capacitance.

Because the energy density is a function of r, you must integrate u over the range $R_1 \le r \le R_2$ to find the total energy U. To do this, you subdivide the space between the inner and outer capacitor plates into spherical shells of thickness dr, as shown in Figure 26.13. The volume of such a shell is the product of its area and its thickness, $4\pi r^2\,dr$. Therefore, the potential energy contained within the shell is

$$dU = u\,4\pi r^2\,dr = \frac{Q^2}{8\pi\epsilon_0}\frac{dr}{r^2}.$$

The total energy contained in the capacitor is

$$U = \int dU = \frac{Q^2}{8\pi\epsilon_0}\int_{R_1}^{R_2}\frac{dr}{r^2} = \frac{Q^2}{8\pi\epsilon_0}\left.\frac{-1}{r}\right|_{R_1}^{R_2},$$

which yields the result

$$U = \frac{Q^2}{8\pi\epsilon_0}\left(\frac{1}{R_1} - \frac{1}{R_2}\right). \qquad (26.30)$$

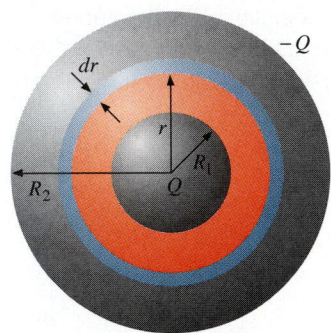

FIGURE 26.13 The energy density u is constant over a volume element of radius r and thickness dr.

Now that you know U, you can use Equation 26.26a, $U = Q^2/2C$, to evaluate the capacitance C. You have

$$C = \frac{Q^2}{2U} = \frac{4\pi\epsilon_0}{\dfrac{1}{R_1} - \dfrac{1}{R_2}}.$$

This is identical with Equation 26.11, the result obtained by more direct means in Section 26.2. However, the method sketched here is easier to use than the direct method under some circumstances.

(c) Find the total electric field energy of a charged conducting sphere.

As in Section 26.2, you consider the isolated charged sphere as the inner plate of a spherical capacitor whose other plate has infinite radius. Equation 26.30 then becomes

$$U = \frac{Q^2}{8\pi\epsilon_0 R}. \qquad (26.31)$$

The energy of the field is inversely proportional to the radius of the spherical surface on which the charge is located. This relation suggests that we cannot carry the idea of a geometric point charge too far.

A	area	V	electric potential
C	capacitance	ϵ, ϵ_0	permittivity, permittivity of free space
d	plate separation in a parallel-plate capacitor	κ	dielectric constant
e	quantum of electric charge	λ	charge per unit length
$\mathscr{E}, \mathscr{E}$	electric field	Λ	geometric factor in capacitance; see Equation 26.23
q, Q	electric charge	ρ	charge per unit volume
u	energy density (energy per unit volume)	σ	charge per unit area
U	electric potential energy		

Summing Up

A **capacitor** consists of two conducting bodies (usually called **plates**) carrying equal and opposite charges q and $-q$. The **capacitance** of a capacitor is defined by Equation 26.5,

$$C = \frac{q}{V},$$

where V is the magnitude of the potential difference between the plates. The SI unit of capacitance is the **farad** (**F**), defined by Equation 26.7,

$$1 \text{ F} \equiv 1 \text{ C/V}.$$

For any capacitor, the capacitance has the form of Equation 26.23,

$$C = \epsilon\Lambda = \kappa\epsilon_0\Lambda.$$

In this relation, the **permittivity** ϵ and (equivalently) the **dielectric constant** κ depend on the nature of the dielectric matter in the capacitor. The quantity Λ depends on the geometry (the size and shape) of the capacitor. For symmetrical arrangements of the plates, it is possible to determine the capacitance by analytical methods. In the absence of a dielectric, these values are given by Equation 26.4 for the parallel-plate capacitor, by Equation 26.9 for the cylindrical capacitor, and by Equation 26.11 for the spherical capacitor. If the capacitors are filled with dielectric, the factor ϵ (or $\kappa\epsilon_0$) must replace the factor ϵ_0 in each case.

The **equivalent capacitance** of two capacitors in parallel and in series is given respectively by Equations 26.15a and 26.18a:

$$C = C_1 + C_2 \quad \text{for two capacitors in parallel}$$

and $\quad \dfrac{1}{C} = \dfrac{1}{C_1} + \dfrac{1}{C_2} \quad$ for two capacitors in series.

The work done in charging a capacitor is stored as potential energy U. This potential energy is expressed by either Equation 26.26a or Equation 26.26b,

$$U = \frac{1}{2}\frac{Q^2}{C} \quad \text{or} \quad U = \frac{1}{2}CV^2.$$

It makes sense to regard this energy as stored in the electric field of the capacitor. We then define the **energy density** u, given by Equation 26.28:

$$u = \tfrac{1}{2}\epsilon\mathscr{E}^2 = \tfrac{1}{2}\kappa\epsilon_0\mathscr{E}^2.$$

The fact that this expression for u does not depend on the size and shape of the capacitor lends support to the view that the energy is stored in the field.

KEY TERMS

Section 26.2 Capacitors and Capacitance
capacitor ▪ capacitance ▪ farad

Section 26.3 Capacitors in Parallel and in Series
network, equivalent capacitor

Section 26.4 Dielectrics
breakdown potential, dielectric strength ▪ dielectric constant, permittivity ▪ induced dipole moment ▪ polarization

Section 26.5 Energy of the Electric Field
energy density

QUERIES

26.1 *(2) En garde!* Although the dimensions of a parallel-plate capacitor can be measured quite precisely, its capacitance generally differs from that predicted by Equation 26.4a, on account of the edge effects. **(a)** Will edge effects result in an actual capacitance greater than or less than that predicted by Equation 26.4a? **(b)** For precision measurements, a *guard-ring capacitor* is often used. Such a device is illustrated below.

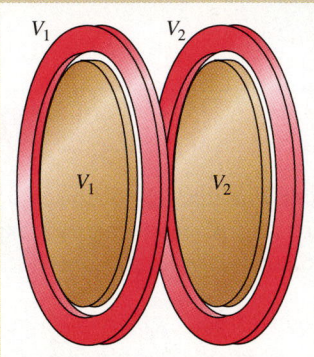

The disk-shaped plates are surrounded by guard rings that fit closely but do not touch the plates. The guard rings are charged to the same potentials V_1 and V_2 as the corresponding plates by an independent source of potential and are thus electrically isolated from the circuit of which the capacitor itself is part. Explain why this arrangement minimizes edge effects so that the capacitance of the device is close to that obtained by using Equation 26.4a.

26.2 *(2) Gauss to the rescue!* In Section 26.2, we showed that the electric field of the pair of plates that make up a parallel-plate capacitor is confined to the space between the plates (neglecting edge effects). The argument depended on the uniform, equal magnitude of the fields of the two plates taken separately, which in turn depended on the fact that $|\sigma_+| = |\sigma_-|$. This is not the case for cylindrical and spherical capacitors. Show that the field in such capacitors is nevertheless confined to the space between the plates.

26.3 *(2) Lambie-pi?* What is the value of Λ for a spherical capacitor?

26.4 *(2) Vast rondure.* When we speak of the capacitance of an isolated sphere, we imply that the sphere is one plate of a capacitor. Where is the other plate and why does its geometry not matter?

26.5 *(3) Merger.* Suppose that two parallel-plate capacitors have the same plate spacing d, though their capacitances are not necessarily equal. Make a nonmathematical argument to show that their equivalent capacitance in parallel is $C = C_1 + C_2$, by considering what happens if you join the capacitors by placing the positively charged plates in edge-to-edge contact and placing the negatively charged plates in edge-to-edge contact.

26.6 *(4) Watered down?* Why is no value given in Table 26.1 for the dielectric strength of water?

26.7 *(5) Reciprocal relation.* The force constant k of a spring is a measure of the stiffness of the spring. Why is capacitance analogous to the *reciprocal* of k?

PROBLEMS

GROUP A

26.1 *(3) Networking, I.* A 25-pF capacitor and a 37-pF capacitor are connected in parallel. Find the equivalent capacitance.

26.2 *(3) Networking, II.* The capacitors of Problem 26.1 are connected in series. What is the equivalent capacitance?

26.3 *(3) Networking, III.* Three capacitors are connected in parallel. Their capacitances are C, $2C$, and $3C$. What is the equivalent capacitance?

26.4 *(3) Networking, IV.* If the three capacitors of Problem 26.3 are connected in series, what is the equivalent capacitance?

26.5 *(3) Pairing off.* You have a 1.60-μF capacitor and a 2.80-μF capacitor. What is their equivalent capacitance if they are connected **(a)** in parallel? **(b)** in series?

26.6 *(3) Sum of the parts?* In the figure below, the equivalent capacitance of the network between terminals A and B is 3.6 μF. What is the capacitance of C_2?

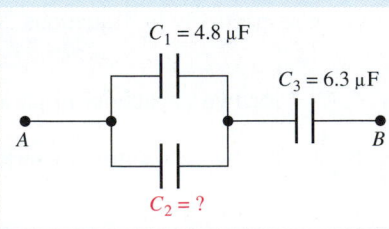

26.7 *(3) Pick and choose.* You can make contact with any pair of the three terminals X, Y, and Z illustrated (p. 717).

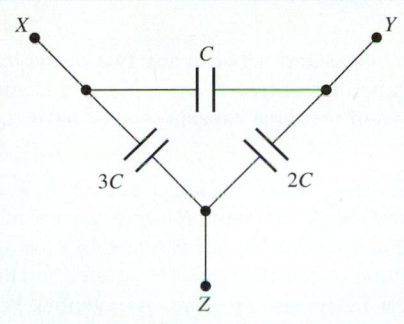

How many different equivalent capacitances can you obtain and what are their values?

26.8 *(3) More is less sometimes, . . .* Show that the capacitance of a series network consisting of any number of capacitors is always less than the capacitance of the smallest capacitor in the network.

26.9 *(3) . . . but not always.* Show that the capacitance of a parallel network consisting of any number of capacitors is always greater than the capacitance of the largest capacitor in the network.

26.10 *(4) Breaking down.* A metal sphere of radius 50 cm is submerged in a tank of mineral oil. **(a)** How much charge can you put on the sphere before the oil breaks down and allows the sphere to discharge? **(b)** Just before breakdown, what is the potential of the sphere?

26.11 *(G) Convergence of opposites.* An electron begins to move away from the negative plate of a charged parallel-plate capacitor at the same instant that a proton begins to move away from the positive plate. How far from the positive plate do they meet if the distance between the plates is 5.00 cm?

26.12 *(4) Plastic film.* A 100-pF capacitor has a polystyrene dielectric 0.032 mm thick. What is the area of the plates?

26.13 *(G) Squeezing out a derivation.* Two parallel-plate capacitors having equal plate areas A but different plate spacings d_1 and d_2 are connected in series as shown below. **(a)** What happens to the equivalent capacitance C of the system as the lead connecting plates 2 and 3 is shortened until the two plates merge into a single plate? **(b)** What happens to C if that single plate is removed from the system? **(c)** Show that

$$\frac{1}{C} = \frac{1}{C_1} + \frac{1}{C_2}.$$

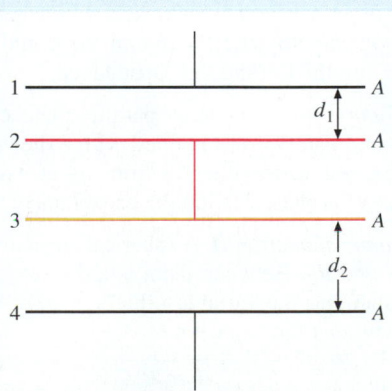

26.14 *(3) Three in one.* In the capacitor network illustrated below, all four capacitors have the same capacitance C. Find the equivalent capacitance between terminals **(a)** H and J, **(b)** H and K, and **(c)** J and K.

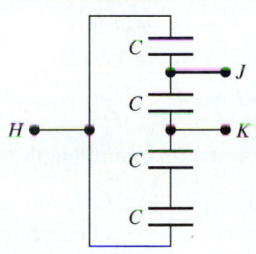

26.15 *(4) Filling in, I.* A parallel-plate capacitor has capacitance C_0 when the space between the plates is empty. Half of the space between the plates is filled with a material having dielectric constant κ_1 and the other half with a material having dielectric constant κ_2, as shown here. What is the new capacitance C?

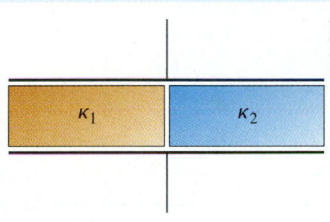

26.16 *(4) Filling in, II.* The parallel-plate capacitor shown below has plates of area A spaced a distance d apart. A metal plate of thickness $t < d$ is slipped between the plates, not making contact with either of them. What is the capacitance of the system?

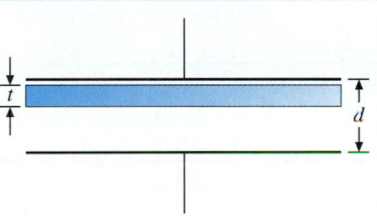

26.17 *(3) Connections.* The potential V_0 across a charged 150-pF capacitor is 285 V. The leads are connected to the leads of a 250-pF capacitor, as shown below. What is the final potential across the capacitors?

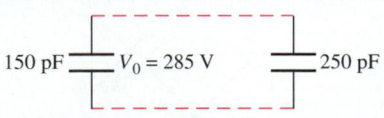

26.18 (2) *Pair of wires.* Two long parallel wires, each of radius a, are separated by a distance $b \gg a$. Show that the capacitance per unit length of the pair is

$$\frac{C}{l} = \pi\epsilon_0 \ln \frac{b}{a}.$$

26.19 (2) *Wire and plane.* A long wire of radius a lies at a constant distance $b \gg a$ from a conducting plane. Find the capacitance of the system per unit length of the wire.

26.20 (2) *Pair of spheres.* Two metal spheres of radius r are separated by a distance $R \gg r$. Find the capacitance of the system. [Hints: (1) Because $R \gg r$, you can assume that the charge on each sphere is distributed uniformly. (2) One path of integration is particularly simple.]

26.21 (2) *Club sandwich.* In the figure below, each of the four plates has area A. The separations all have the same value d. **(a)** Find the capacitance between M and N when the plates are connected as shown in part *a.* **(b)** Find the capacitance between P and R when the plates are connected as shown in part *b.*

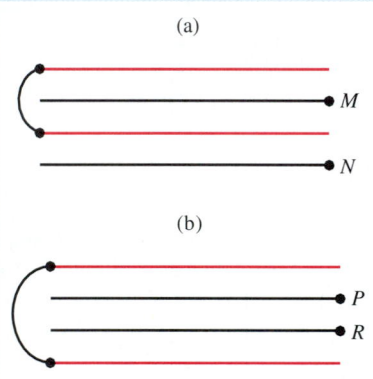

26.22 (3) *Tangle, I.* Find the capacitance between points A and B in this figure.

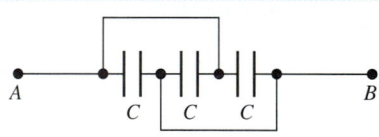

26.23 (3) *Narrowing the possibilities.* Depicted below is a triangular network of three capacitors. The equivalent capac-

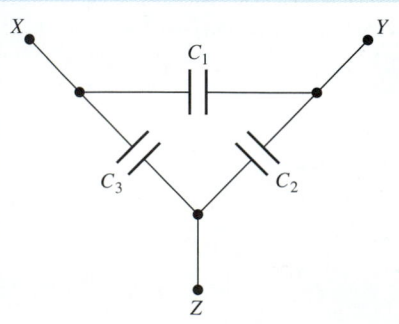

itance can be measured between any two of the terminals X, Y, and Z. Show that, if two of these equivalent capacitances are equal, two of the three capacitors must have equal capacitance.

26.24 (4) *Geiger counter.* A Geiger-counter tube consists of a tubular metal shell of radius R_2 with a wire of radius R_1 strung along its axis. The device is sealed by closing the ends, and the resulting sealed chamber is evacuated and then refilled with gas at a fairly low pressure. A potential V is applied between the shell and the wire. (The wire is positive, but that is not important for this problem.) The potential is adjusted until the field in the immediate vicinity of the wire is just a little less than the breakdown field for the gas. An energetic particle (for example, a cosmic-ray particle) that passes through the region near the wire will collide with gas molecules and ionize them, thus triggering electric breakdown. The resulting sudden flow of electric charge from wire to shell is readily observed and is used to detect the triggering particle. For a typical tube, we have $R_1 = 25$ μm, $R_2 = 1.5$ cm, and $V = 600$ V. What is the magnitude of the electric field adjacent to the wire?

26.25 (4) *Polarization.* A device having a permanent electric dipole moment is called an *electret.* One way of making an electret is to pour molten wax into a rod-shaped mold and allow it to cool and harden in the presence of a strong external electric field. As the wax hardens, the induced dipole moments of the wax molecules are "frozen" into place and persist even after the external field is removed. **(a)** Show that two electrets placed together in a bag and shaken gently will tend to adhere together with their dipole axes antiparallel to one another. **(b)** What is the dipole moment of a large collection of electrets after they have been shaken as in part **a**? **(c)** What happens to the collection of electrets if it is subjected to a large external electric field? Compare your argument with the discussion in Section 26.4.

26.26 (4) *Two in one.* The plates of a capacitor each have area A and are separated by distance d. The space between the plates is partly filled with a slab of dielectric of thickness $t < d$ and dielectric constant κ, as illustrated in Problem 26.16. What is the capacitance? (Hint: Show that the system is equivalent to two capacitors in series.)

26.27 (4) *Doubling up.* A capacitor consists of two metal plates, each of area 1.0 cm^2. Sandwiched between the plates are a sheet of polyethylene 0.015 mm thick and a sheet of Mylar 0.025 mm thick. Find the capacitance.

26.28 (4) *Funny dielectric, I.* A parallel-plate capacitor of plate area A and gap width d is filled with a dielectric whose dielectric constant varies linearly from κ_1 at plate 1 to the greater value κ_2 at plate 2. Find the capacitance.

26.29 (4) *Funny dielectric, II.* A spherical capacitor has shells of radius R_1 and R_2. Between them is a dielectric whose dielectric constant has the variable value $\kappa = a/r$, where a is a constant. Show that the capacitance is

$$C = \frac{4\pi\epsilon_0 a}{\ln\left(\dfrac{R_2}{R_1}\right)}.$$

26.30 *(4) Unequal treatment.* You connect two air-filled parallel-plate capacitors, each of capacitance C, in series across the terminals of a battery of voltage V. The electric field in the gap of each capacitor is $\mathscr{E}_0$. **(a)** If you fill the gap of one of the capacitors with a dielectric of dielectric constant κ, what is the new field $\mathscr{E}$ in that capacitor? **(b)** Show that, as you fill the gap, charge

$$|q| = \frac{CV}{2} \frac{\kappa - 1}{\kappa + 1}$$

will flow onto the plates.

26.31 *(4) Bûche de Noël.* A cylindrical capacitor consists of an inner metal shell of radius R_1 surrounded by dielectric 1 out to r and then by dielectric 2 out to R_2, which is the radius of the outer metal shell. The dielectric constants of the two dielectrics are κ_1 and κ_2, and their dielectric strengths are $\mathscr{E}_{1\,max}$ and $\mathscr{E}_{2\,max}$. When the voltage across the capacitor is slowly increased, both dielectrics break down simultaneously. What is the relation among the dielectric constants, the dielectric strengths, and the radii r and R_1?

26.32 *(4) Cloak.* A metal sphere of radius R_1 is surrounded by a concentric dielectric shell of inner and outer radii R_1 and R_2. If the dielectric constant of the shell is κ, show that the capacitance of the metal sphere is

$$C = \frac{4\pi\kappa\epsilon_0 R_1}{1 + (\kappa - 1)\dfrac{R_1}{R_2}}.$$

26.33 *(5) Forcing the issue, I.* The plates of a parallel-plate capacitor each have area A and are separated by a distance x. **(a)** When the plates carry charges q and $-q$, what is the energy stored in the capacitor? **(b)** Using the result of part **a**, find the force F exerted on each plate by the field of the other. Compare your result with that of Problem 23.43.

26.34 *(5) Forcing the issue, II.* Suppose now that the capacitor plates of Problem 26.33 are connected to the terminals of a battery so that the potential V between them remains constant (and the charge varies) as the distance between them is varied. **(a)** Find the force F exerted on each plate by the field of the other. **(b)** Why isn't the answer the same as that of Problem 26.33?

26.35 *(5) Unmaking fudge, I.* In deriving Equation 26.29, we used Equation 26.28, the derivation of which was based on a study of the parallel-plate capacitor. In order to apply Equation 26.28 to a spherical capacitor, we fudged a bit, making the correct but tricky argument that Equation 26.28 is applicable to all capacitors because it contains no geometric factors. In this problem, you will derive Equation 26.29 without reference to any other form of capacitor. **(a)** Rewrite Equation 26.11 to express the capacitance of a spherical capacitor whose plates have radius R and $R + \Delta r$. **(b)** Write the energy U of this capacitor when each plate carries a charge of magnitude $|q|$. **(c)** You cannot assume that this energy is distributed uniformly throughout the volume τ between the plates. But you can still find the *mean* energy per unit volume $\langle u \rangle$; to

do this, you divide U by τ. **(d)** Now, take the limit of $\langle u \rangle$ as $\Delta r \to 0$. Show that you get Equation 26.29.

26.36 *(5) Forceful capacitor, I.* The capacitor illustrated below has capacitance C_0 when the space between the plates is filled with air. The plates are given charges q and $-q$. A slab of dielectric constant κ, which just fits between the plates, is slipped part way between the plates. **(a)** Explain why there is a force F on the slab. **(b)** When the slab has entered the capacitor to a distance x, show that the force on it has magnitude

$$F = \frac{q^2}{2C_0} \frac{\kappa - 1}{w\left[1 + \dfrac{(\kappa - 1)x}{w}\right]^2}$$

and that its direction is inward. Neglect edge effects. (Note: The system must be in a state of stable equilibrium with the dielectric all the way into the space between the plates. But the preceding expression does not go to zero when $x = w$.

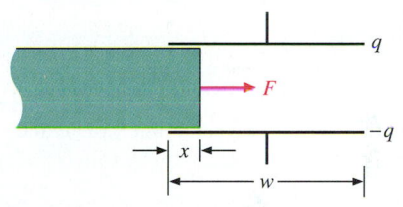

This is precisely because of the neglect of edge effects. Symmetry considerations dictate that the contribution of the edge effects to the force must dominate when the dielectric is all the way in.)

26.37 *(5) Forceful capacitor, II.* Suppose now that the capacitor of Problem 26.36 is connected to a battery that keeps the potential across the plates at a constant value V as the dielectric is slid in or out. **(a)** What is the direction of the force F on the slab? **(b)** Find the magnitude of the force and show that, unlike the force in Problem 26.36, this force has a value independent of x as long as the space between the plates is not completely filled. Neglect edge effects.

26.38 *(G) Where's the limit?* Two capacitors are connected in series. The first one has capacitance $C_1 = 1.0\ \mu\text{F}$ and breakdown voltage $V_1 = 6 \times 10^3\ \text{V}$. The second has capacitance $C_2 = 2.0\ \mu\text{F}$ and breakdown voltage $V_2 = 4 \times 10^3\ \text{V}$. **(a)** What is the maximum voltage V that can be imposed between the terminals of this series network? **(b)** Under maximum-voltage conditions, what is the charge on each capacitor?

26.39 *(g) Headstand.* The *coefficient of potential K* is defined as the reciprocal of capacitance: $K \equiv 1/C$. **(a)** Show that K is analogous to the force constant k of a spring. **(b)** Express the equivalent coefficient of potential for a series network of capacitors in terms of the coefficients of potential of the individual capacitors. **(c)** Express the equivalent coefficient of potential for a parallel network of capacitors in terms of the coefficients of potential of the individual capacitors.

26.40 *(3) Variable capacitor.* The photograph below is of a variable capacitor of the kind often used to tune radios. It

consists of a stack of moveable plates, each of area A and all connected together, between which is sandwiched a similar set of stationary plates. The capacitance of the device is changed by rotating the moveable plates and thus varying the effective area of the capacitor. Let the total number of plates be n. **(a)** Show that, when the device is set for maximum capacitance (with the plates completely interleaved), the capacitance is

$$C = (n - 1)\epsilon_0 \frac{A}{d},$$

where d is the distance between a fixed plate and an adjacent moveable plate. **(b)** Show that this result is independent of whether n is even or odd.

26.41 *(3) Tangle, II.* Find the capacitance between points A and B in this figure.

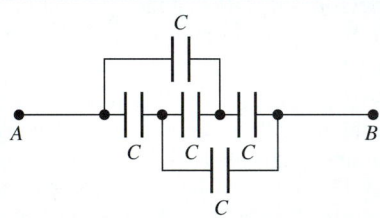

26.42 *(3) Endless string.* Find the equivalent capacitance between points A and B in the figure below. The chain consists of an infinite number of units; each unit consists of two capacitors of capacitance C. One such unit is set off by a red rectangle. (Hint: If you removed the first unit, what would be the capacitance of the remaining chain between A' and B'?)

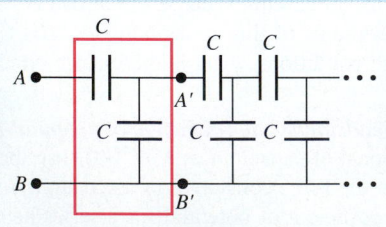

26.43 *(5) Unmaking fudge, II.* Here is a way to obtain Equation 26.28 in a manner completely independent of the geometry of the capacitor for which it is derived. **(a)** Show that, if the potential across an arbitrary capacitor is V when the charge on each plate is $|q|$, the stored energy is $U = \frac{1}{2}qV$. **(b)** In imagination, draw a closed equipotential surface that sur-

rounds one of the plates but not the other. Show that this is a useful Gaussian surface. Then use Gauss's law to express q in terms of $\mathscr{E}$ and the dielectric constant. **(c)** Draw another equipotential surface just outside the first one. Express the potential difference dV between them in terms of the field $\mathscr{E}$ (not necessarily the same everywhere on the surfaces) and the distance $d\mathbf{s}$ between them, measured along the corresponding flux line. Then show that the energy contained in the space between the surfaces is

$$dU = \tfrac{1}{2}\kappa\epsilon_0 \, \mathscr{E}^2 \, d\tau,$$

where $d\tau$ is the volume of the space between the equipotentials. Finally, use this result to obtain Equation 26.28.

26.44 *(G) Di(p)electric, I.* The capacitor depicted below consists of two horizontal plates of area A and separated by a distance d. The plates are given charges per unit area σ and $-\sigma$ and thereafter remain isolated. The charged capacitor is

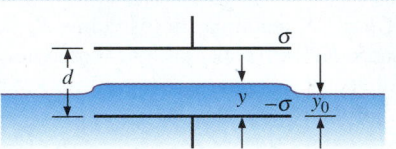

dipped into a large container of dielectric liquid of density ρ and dielectric constant κ so that the undisturbed liquid level (the level far from the capacitor) lies a distance y_0 above the lower plate. Between the plates, the liquid level has a different value, y. **(a)** Express the capacitance in terms of d, y, κ, and A. **(b)** Write the potential energy U of the system, taking into account both the electric potential energy and the gravitational potential energy of the liquid. (Hint: Take the zero of gravitational potential energy at $y = y_0$.) **(c)** Because the system is isolated, any change in the liquid level cannot result in a change in U. Show that

$$y - y_0 = \frac{\sigma^2}{2\rho g} \frac{\kappa - 1}{\epsilon_0 \kappa}.$$

26.45 *(G) Di(p)electric, II.* The capacitor shown below con-

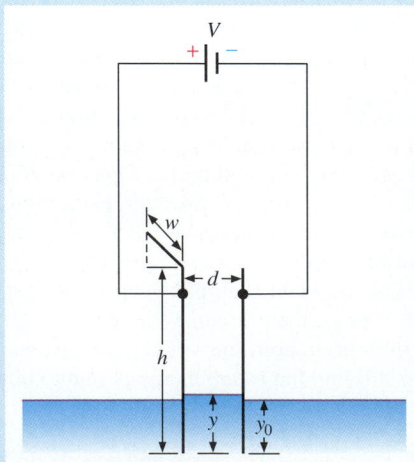

sists of two vertical plates, of height h and width w, separated by a distance d. The uncharged capacitor is dipped into the dielectric liquid described in Problem 26.44 so that the liquid level lies a distance y_0 above the bottom edges of the plates. The plates are connected to the terminals of a battery, which thereafter maintains the potential across the capacitor at a constant value V. Show that the liquid rises to a higher level y between the plates, and find $y - y_0$. (Hint: In contrast with the situation in Problem 26.44, the energy of the system consisting of the capacitor and the liquid does not remain constant. A rise in the liquid level implies an increase in the capacitance and therefore a flow of charge onto the plates in which the battery does work on the system.)

26.46 (G) *Millikan oil-drop experiment.* Between 1909 and 1912, Robert A. Millikan made the first really accurate measurements of the quantum of charge e. (Indeed, his major contribution was to demonstrate that electric charge exists *only* in integral multiples of the quantum e.) His apparatus is sketched below. The "business end" of the device is the two carefully aligned parallel plates M and N, situated a distance d apart, across which a precisely known potential V can be applied. By means of the atomizer labeled D, A, s, and r, oil can be sprayed in fine droplets, some of which drift downward through the small hole p in plate M. The droplets are side lighted by a beam of light entering the apparatus through window g on the left. Contrasted against the flat black interior of the chamber, the droplets are readily seen by the observer, who looks through a small telescope and a window (not shown) in the direction normal to the page. The X-ray tube χ on the right is turned on briefly, and the air in the chamber is ionized. Some of the droplets pick up one or a few ions or free electrons and end up with electric charge ne, where n is a fairly small positive or negative integer.

The observer concentrates on a single droplet. With V set to zero, the droplet falls at a speed $v_\downarrow$ determined by Stokes's law (Equation 16.17), $mg = 6\pi\eta r v_\downarrow$. Here η is the viscosity of the air and r is the radius of the (very closely spherical) droplet. The speed is measured by timing the passage of the droplet across a known distance. The potential across the plates is then adjusted to a value V such that the droplet moves upward with a convenient speed $v_\uparrow$, which is measured in the same way. **(a)** Express the ratio $v_\downarrow/v_\uparrow$ in terms of ne, V, d, and the weight mg of the droplet. **(b)** Using the result of part a, express the charge ne in terms of $v_\downarrow$, $v_\uparrow$, the density ρ of the oil, the droplet radius r, and necessary constants. Neglect the buoyancy of the air. **(c)** If the motion of the droplets is to obey Stokes's law, they must be so tiny that their radii cannot be measured directly with any accuracy. But show that you can express r in terms of the known quantities η, ρ, and g and the measurable quantity $v_\downarrow$. **(d)** Finally, show that the charge on the droplet is

$$ ne = \frac{4\pi}{3} \frac{d}{V} \frac{1}{\sqrt{\rho g}} \left(\frac{9\eta}{2}\right)^{3/2} \sqrt{v_\downarrow}\,(v_\uparrow + v_\downarrow). $$

Measurements on numerous droplets yield values of ne that are indeed integral multiples of the same quantity e (with n usually small), and that charge is therefore quantized. Millikan's best value, expressed in modern units, was $e = 1.592 \times 10^{-19}$ C. Compare this with the modern value (Appendix 4) obtained by other means, and note that Millikan was overoptimistic in giving four significant figures. Nevertheless, he won the Nobel Prize in 1923, in part for this work.

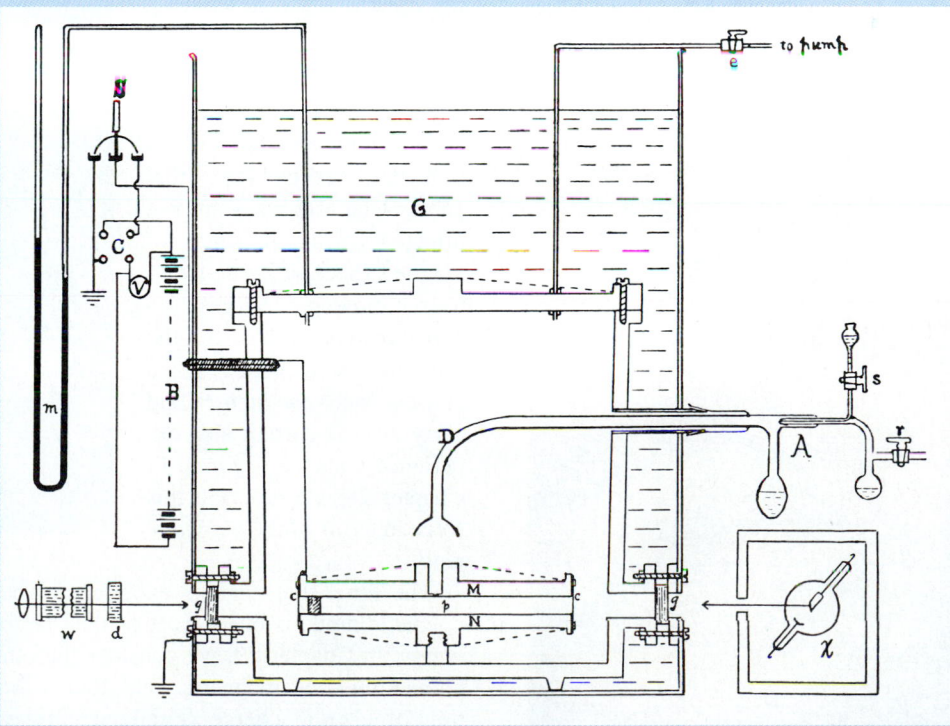

Part *IX*

Electric Charge in Motion

The next eight chapters deal with the properties that electric charges exhibit only when they are in motion. In Chapter 27, we develop quantitative methods of expressing and dealing with steady flow of electric charge through closed loops called *electric circuits*. This flow is described in terms of *electric current*. We also consider the devices, called *sources of electromotive force*, by which such steady flow can be maintained.

We have already noted Gilbert's conclusion that static electric charges are not affected by the presence of magnets (Section 23.1). However, moving charges do experience a force in the presence of magnets. As a consequence, *conductors* that carry electric current also experience a *magnetic force*. In Chapter 28, we describe this force in terms of the *magnetic field*.

The force that charges experience when they move in magnetic fields is one aspect of the *electromagnetic interaction*. In Chapter 29, we consider the other aspect of that interaction—the magnetic field that arises from the motion of electric charge. Fundamental laws analogous to

Coulomb's law and Gauss's law, respectively called the *Biot-Savart law* and *Ampère's law*, are central to this chapter.

Chapter 30 treats the last of the fundamental laws of electromagnetism: *Faraday's law*. This law describes the way in which a changing magnetic field gives rise to an electric field. Faraday's law has familiar practical applications, such as the electric generator and electric motor, as well as important scientific implications.

Taken together, Ampère's law and Faraday's law describe the electromagnetic phenomenon called *inductance*, in which a changing electric current gives rise to a changing magnetic field, which in turn gives rise to an electric field. Inductance is the subject of Chapter 31.

In Chapter 32, we digress from the mainstream of scientific development to study the behavior of electric circuits in which the current is not steady, but rather varies with time. Among such systems are *alternating-current circuits*, which we consider in some detail in Chapter 33.

Chapter 34 presents the grand summation of the laws of electromagnetism called *Maxwell's equations*. These equations not only provide a compact, symmetrical basis for understanding all electromagnetic effects, but lead to the prediction of *electromagnetic radiation*. Among the many forms of radiation are light, X rays, and all of the various classes of "radio waves."

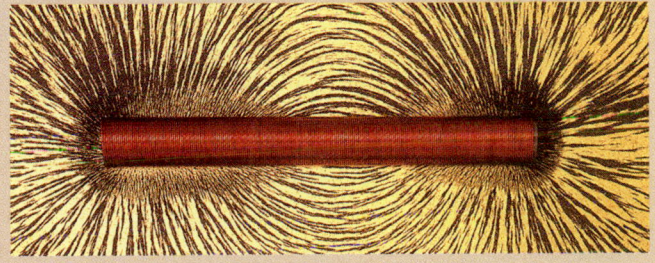

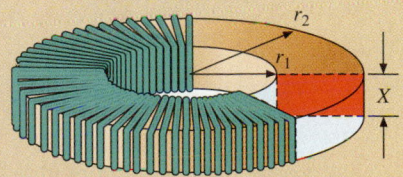

Electromotive Force and Electric Current

— A sustained potential difference can be maintained across a conducting path by means of a source of electromotive force (emf).

— Under the influence of the potential difference, an electric current will flow.

— Electric current i is the electric charge flux dq/dt.

— In many materials (such as metals), the current is proportional to the potential difference. The proportionality, called Ohm's law, can be written in the form $V = Ri$, where R is the electrical resistance of the conducting path.

— Some understanding of electrical resistance can be attained in terms of an extension of the ideal-gas model: The free electrons that carry the electric charge collide repeatedly with the ions that are the other constituents of the conductor.

— Whether or not a conductor conforms to Ohm's law, the power dissipated in it is described by Joule's law: $P = Vi$.

— Electrical resistors can be connected into networks whose equivalent resistance can be calculated.

Left: As charge leaks off this student's halo of hair into the surrounding air, the van de Graaff generator acts as a source of emf, maintaining the potential difference between her and the ground by transporting more charge on a moving belt.

> *. . . we must take the current when it serves,*
> *Or lose our ventures.*
>
> —SHAKESPEARE, *Julius Caesar*

Introduction

The most important applications of electricity are based on the sustained *flow* of electric charge from one place to another. In order to root our study of *electric current* in familiar territory, we think in terms of water driven through a system of pipes by a pump. As always, we must not push this analogy too far; nevertheless, it is a good beginning.

In Section 27.2, we consider some of the "pumps" that drive electric charge through a system of conductors. These pumps, called *sources of electromotive force*, provide a potential difference between two terminals, as the familiar electric battery does. In Section 27.3, we devise ways to describe electric current; in Section 27.4, we derive *Ohm's law*, which describes the current-carrying properties of the most familiar conductors, metal wires, in terms of a quantity called *electrical resistance*. In Section 27.5, we consider the flow of electric charge through metals from the microscopic point of view. An expression for the power required to maintain a current through a resistive element is derived in Section 27.6, and Section 27.7 treats the analysis of current through networks consisting of combinations of sources of electromotive force and resistive elements.

Electromotive Force and Its Sources

Figure 27.1 schematically shows the water analogy that will guide us in our study of electric currents. The pump produces a pressure difference $\Delta p = p_{hi} - p_{lo}$ between its outlet O and inlet I. Driven by this pressure difference, water flows through the pipes in the sense of decreasing pressure, as shown by the arrows. If its pumping capacity is sufficient, the pump will maintain the pressure difference Δp even if the valve is opened wider. Why is the pressure difference necessary to maintain the water flow in the pipe? Because the water is subject to friction, which converts the potential energy inherent in the pressure into heat energy. If the pump is turned off, the water soon ceases to circulate.

Inside the pump, water flows in the sense of increasing pressure. That is what the pump is for! An external energy source—for example, a gasoline engine—drives the pump impeller, which continually forces water from I, where the pressure is low, to O, where the pressure is high. As Bernoulli's equation makes clear (see Section 16.3), this increase of pressure is tantamount to increasing the potential energy of the water. We thus have the following energy-flow cycle:

1. External mechanical energy is converted by the pump into potential energy of water.

2. The potential energy is converted into kinetic energy of flow as the water moves through the pipes outside the pump.

3. The kinetic energy is converted into heat energy by frictional processes.

4. The process continues as long as external mechanical energy is supplied to the pump.

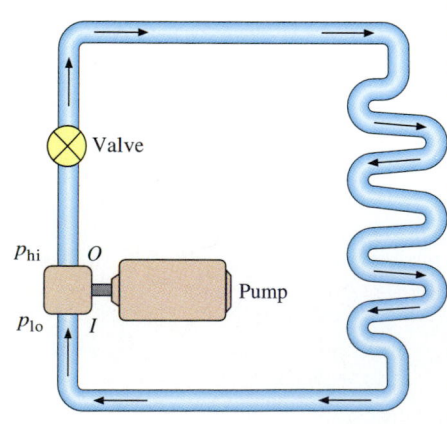

FIGURE 27.1 A pump produces a pressure difference $\Delta p = p_{hi} - p_{lo}$ between its outlet O and inlet I. This pressure difference drives the flow of water against friction through a system of pipes.

The power required to drive this process depends on the details of the system. For a given system, the power depends on the magnitude of the water flow. But, even if the valve is closed completely so that no water flows, there will be a pressure difference between the outlet and the inlet of the pump.

The process of pumping water from a low to a high pressure, and thus increasing its energy, has an electrical analogue: The potential energy of a quantity of charge is increased by transporting the charge from one location to another against an electric field. (This is what happens when a capacitor is charged.) There are many ways of carrying out this transport process. One device in which the process is particularly easy to follow is the **van de Graaff generator**, shown in Figure 27.2. By means of the corona discharge at L (Section 25.3 and Table 26.1), electric charge is "sprayed" by the sharp-pointed metal comb onto the moving belt, which thus acquires a (usually) positive charge. The potential difference required to charge the belt in this fashion is typically a few thousand volts. The charge rides up the belt. Work must be done by the motor driving the belt to force the charge on the belt into the hollow dome against the electric field due to the positive charge that the belt has previously transported to the dome. At H, the belt discharges to a second corona device. The charge immediately passes to the outside surface of the dome, in accordance with Gauss's law. (Even though there is considerable charge on the dome surface, the electric field inside it is essentially zero except at the comb, where the arriving charge on the belt produces a considerable local field.)

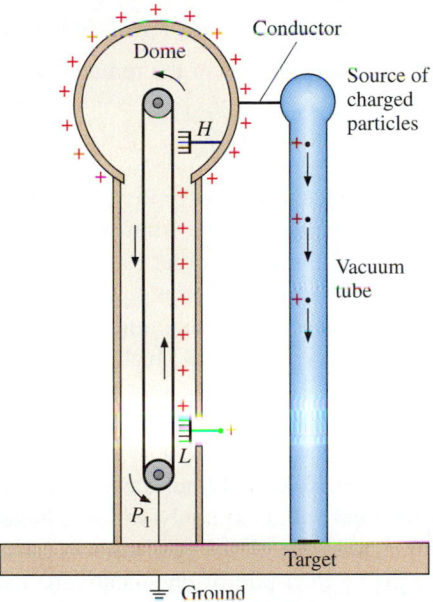

FIGURE 27.2 A van de Graaff generator. A motor (not shown) drives pulley P_1 and thus maintains the endless belt in motion. Charge "sprayed" onto the belt at L is removed at H, resulting in a large potential difference between the dome and the ground. The part of the device described up to this point is analogous to the pump of Figure 27.1. Charged particles produced near the dome are driven through the vacuum tube toward the target, which is at ground potential. The part of the device described in the preceding sentence is analogous to the pipe of Figure 27.1.

When the generator has been operating for a short time, the charge built up on the dome results in a large potential difference V between dome and ground. (Compared with V, the potential difference required to maintain the corona discharges is negligible.) Particles having charge ne are produced by means not shown at the *source*, represented as a blue sphere. The particles are accelerated through a vacuum tube toward the ground. They thus acquire kinetic energy $K = -\Delta U = |neV|$. This flow of charge is analogous to the flow of water through the pipe in Figure 27.1. In the steady state, charge is transported to the dome by the belt just as fast as it flows back to the ground through the vacuum tube.

An important general point emerges from the details of the van de Graaff generator: *An electric potential difference is produced by a system that does external work on electric charge.* The means used to do the work are not necessarily electrical. (In the van de Graaff generator, the work is done by the motor that drives the belt. But the fact that the motor is an electric motor is irrelevant to the operation of the generator.)

The external work per unit charge that must be expended to produce an increase ΔV in the *electric* potential of the charge is called the **electromotive force**. You should make a special point of noting that electromotive force is *not* a force in the scientific sense of the word. It would be better to call it "electromotive work per unit charge." But the name, which originated in the early nineteenth century when scientific terms were not used as precisely as they are today, has stuck. Luckily, "electromotive force" is a large mouthful, and it is almost always shortened in English to **emf**, spoken as e-m-f.

The external force that acts on electric charge in producing an emf can be a non-conservative force. But the electric force that opposes this external force *is* a conservative force, and the resulting electric potential difference ΔV is therefore well defined.* We express emf quantitatively in the same units as electric potential—volts—and we usually use the same symbol V (or ΔV). This does not lead to confusion because we are almost always interested only in the potential difference ΔV *produced by* the emf.

The device that produces the emf—called a **source of emf**—is usually part of an **electric circuit**, which is a closed path for circulation of charge, just as the system of Figure 27.1 is a closed path for circulation of water. Indeed, we can describe the cycle of charge circulation in terms closely analogous to those used for the water system:

1. External energy is converted by the source of emf into potential energy of electric charge.
2. The potential energy is converted into kinetic energy of flow as the charge moves through the part of the circuit external to the source of emf.
3. The kinetic energy is converted into heat by the "frictional" process that underlies the phenomenon of *electrical resistance*.
4. The process continues as charge continuously arrives at the source of emf.

We will explore all aspects of this process in the remainder of this chapter.

Sources of emf

Sources of emf operate on a wide variety of principles, but all have one property in common: They continually separate positive from negative charge. To do this, they perform work in a nonelectrical fashion in order to overcome the attractive force that tends to recombine the charges. Once the charge separation has been achieved, there is the possibility of the separated charges recombining through a path outside the source of emf. In recombining, the charges can do as much work as was done on them by the source of emf in separating them. Because the source of emf can continually replenish the separated charges, steady electrical work can be obtained. The overall result is that a nonelectrical power source is converted into a source of electric power.

The van de Graaff generator is certainly not a commonplace source of emf. Much more familiar are the **chemical cell**, commonly called a battery[†] (Figure 27.3a), and the electromagnetic device called the electric generator (Chapter 30).

The first battery, consisting of a pile of chemical cells called a *voltaic pile*, was devised by Alessandro Volta in 1800. When two dissimilar metals are immersed in an ionic solution such as a weak acid, as in Figure 27.3a, a potential difference appears between them.

To understand why the potential difference appears, consider the interaction between the zinc plate, called an **electrode**, and the solution, called the **electrolyte**. A small amount of zinc dissolves in the electrolyte in the form of positive ions. There is thus an excess of electrons in the zinc electrode, which acquires a negative charge. The dissolution-ionization of zinc soon ceases when the potential difference between the electrode and the electrolyte becomes large enough to prevent further migration of positive ions into the electrolyte (Figure 27.3b). The value of this **electrode potential** is characteristic of the metal of which the electrode is made.

A simple chemical cell. When the jar is filled with electrolyte, a potential difference appears between the dissimilar metal electrodes.

---

*In analogy, the force your arm exerts on a mass that you are raising is not conservative, but the gravitational force that opposes this force *is* conservative, and the resulting gravitational potential difference is well defined.
[†]Strictly speaking, a battery is a series of cells connected together so that the potential differences across their terminals add. But the term "battery" is often used loosely.

(a)

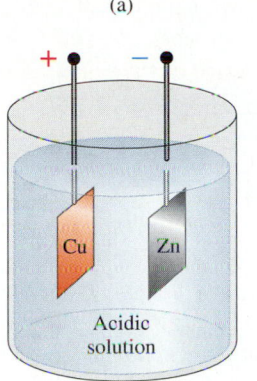

(b)

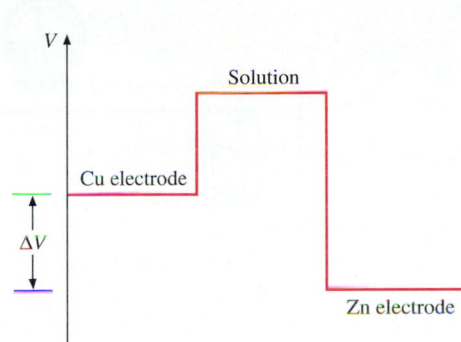

FIGURE 27.3 (*a*) A chemical cell. (*b*) Schematic plot of the electric potentials of the three parts of the cell, all taken relative to the same arbitrary reference level. ΔV is the potential difference between the electrodes.

In effect, the electrode and the electrolyte constitute the plates of a charged capacitor. If a connection is made between these capacitor plates, by an external conducting path between electrode and electrolyte, the connection allows for their "discharge." However, the chemical process that gives rise to the electrode potential—the passage of zinc ions into solution—operates continually to maintain the potential. This chemical process is the "nonelectric" source of work that drives the separation of charge typical of sources of emf.

A like process takes place between the electrolyte and the other electrode (taken to be copper in Figure 27.3a). Here, a small amount of copper goes into solution in the form of positive ions; the resulting potential difference between electrode and electrolyte is shown in Figure 27.3b. Because of its polarity, this potential difference tends to oppose the propulsion of charge through an external conductor by the zinc-electrolyte "capacitor." But the magnitude of the potential difference at the copper electrode is smaller than that at the zinc electrode, and so there is a net potential difference (ΔV in Figure 27.3b) available to drive charge through an external connector.

THINKING LIKE A PHYSICIST

Volta's work was based on the earlier work of the Italian physiologist Luigi Galvani (1737–1798), who, in 1786, was working with a dissected frog's leg hung from a copper hook attached to an iron stand. He noted that the leg twitched whenever it touched the stand. He attributed this quasi-mystically to an "animal electricity," of which the frog's leg was the source, and in five years of further effort he failed to get any deeper insight into what was happening. Volta, convinced that one should postulate novel laws of nature to explain phenomena only when all known possibilities have been exhausted, inquired whether the observations could be understood in terms of "ordinary" electricity. In a careful series of experiments, he showed that, though Galvani's observations did indeed have an electrical basis, the "source of the electricity" lay in the contact between the dissimilar metals and that the frog's leg was simply the detector. (Volta was thus the first to point the way to the intimate involvement of electric impulses in the functioning of nerves and muscles.) He followed the disproof of Galvani's animal electricity with the invention of his electrical pile. Although the invention of the voltaic pile was the most important achievement of his long career, Volta was probably also the first person to understand and use the relations $q = CV$ (Equation 26.5) and $V = U/q$ (Equation 25.5) and the first to make the volume-versus-temperature observations in gases that led to Charles's law (Section 17.3). He made important advances in the techniques of chemistry and static electricity as well.

Volta's success, where Galvani had bogged down, exemplifies the power of what philosophers of science call the conservative approach of science. Extended experience with the explanatory power of scientific theory tends to make scientists very slow to invoke brand-new hypotheses to explain new observations. Rather, they make every effort to bring new observations under the umbrella of existing theory or a modest extension of existing theory. Every success of such efforts renders the theory still more powerful, and these efforts make up almost all of what scientists do in the realm of interpretation of experimental results. Only on rare occasions is it necessary to invent brand-new theory; these occasions, on which the old theory must fail and the new theory must pass very stringent tests, are called scientific revolutions. Newton's work (Chapters 2 through 5 and 14) constituted such a revolution, and we will study another one more explicitly in Chapters 38 and 39. The conservatism of scientists underlies their reluctance to give credence to reports of such things as animal electricity, extrasensory perception, or Velikovskian cataclysms.

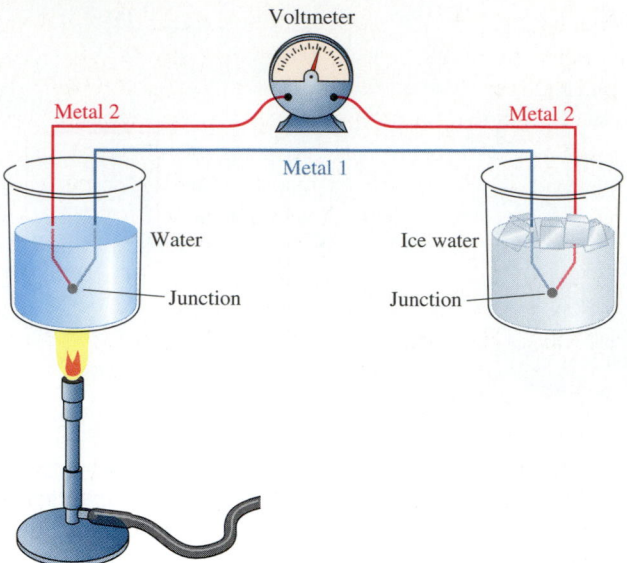

FIGURE 27.4 A thermocouple. Wires made of dissimilar metals, 1 and 2, are welded together at two junctions. The temperature difference between the two junctions drives charge from one toward the other and thus results in a potential difference between the junctions. Hence, the thermocouple is a source of emf. The potential difference is measured by means of the voltmeter.

Although the electrode potential depends on the particular metals used, it is always of order 1 V, in contrast with the potential differences in kilovolts that can be readily produced by rubbing a glass rod with a piece of fur. Strictly speaking, the glass-fur combination is a source of emf. But, in spite of the modest potential difference between the electrodes of a chemical cell, the cell is much more useful as a source of emf because the potential difference can be maintained even when substantial charge is allowed to flow from one electrode to the other through an external conductor.

In *thermoelectric devices*, the emf is produced by maintaining a temperature difference across the device. One such device, the *thermocouple*, is shown in Figure 27.4. Thermocouples are not often used today as sources of potential difference for driving electric currents because thermocouple emfs are quite small (typically of order 10 mV) and because the conversion of heat energy into electrical energy is inefficient. However, the potential difference between the terminals is a very reliable and sensitive measure of the temperature difference between the two junctions, and thermocouples are often used in thermometry.

When we design a practical source of emf, we are concerned with its efficiency, the power source required to drive it, and its suitability for the intended application. But when we study the current driven by the source through an external network, we care only about the potential difference between the terminals of the source. Under these circumstances, we depict the source of emf as the "black box" shown in Figure 27.5a— a device whose details we completely neglect. The terminal having the higher potential is labeled "+" and is called the **positive terminal**; the terminal having the lower potential is labeled "−" and is called the **negative terminal**. That is to say, $V_+ > V_-$.

The potential difference $V_+ - V_-$ is very often represented by the symbol V, rather than ΔV. When you follow this practice, you should be sure that there is no likelihood of confusion.

The standard symbol for a source of emf is shown in Figure 27.5b. By convention, the longer of the two parallel lines represents the positive terminal (even if no + sign is shown) and the shorter line represents the negative terminal.

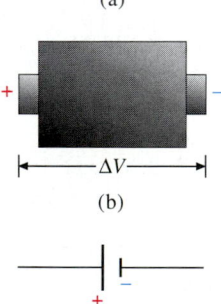

(a)

(b)

FIGURE 27.5 (*a*) A source of emf shown as a "black box." (*b*) The standard symbol for a source of emf. The longer of the two parallel lines always represents the positive terminal.

Electric Current

By definition, an electric conductor is a body containing mobile electric charge. When a conductor is used to link the two terminals of a source of emf, the potential difference between the terminals results in an electric field whose flux lines begin at the positive terminal, run through the conductor, and terminate at the negative terminal.* Under the influence of this field, the mobile charge flows. As noted in Chapter 23, the mobile charge may be positive, negative, or both, depending on the conductor. Positive charge flows in the sense of the flux lines, from the positive to the negative terminal, whereas negative charge flows in the opposite sense, from the negative to the positive terminal. But regardless of these details, the net result is a *transfer of positive charge from the positive to the negative terminal* through the external circuit.

We define the **electric current**, or simply the **current**, to be *the net amount of positive charge passing per unit time across any section through the conductor in the sense from the positive toward the negative terminal.* Two such sections, M and N, are shown in Figure 27.6. It is customary to use the symbol i for current, and we write the definition as

$$i \equiv \frac{dq}{dt}. \tag{27.1}$$

The sense of the current *outside* the source of emf is conventionally taken as the sense from the positive to the negative terminal. In the most familiar conductors, metals, the **charge carriers**—the mobile charged particles that carry the current—are electrons, which flow in the opposite sense. But the fact that electrons flow in the sense opposite that of the conventional current is of no consequence for our present purposes. It is, however, important to follow a fixed convention, and the positive-to-negative convention is universally used.

You can see from Equation 27.1 that the unit of electric current is the coulomb per second (C/s). In view of its importance, this unit is given the special name **ampere (A)**:

$$1 \text{ A} \equiv 1 \text{ C/s}. \tag{27.2}$$

In many types of conductors over wide ranges of conditions, the flow of charge quickly achieves a steady state and then remains a steady-state process. This is certainly

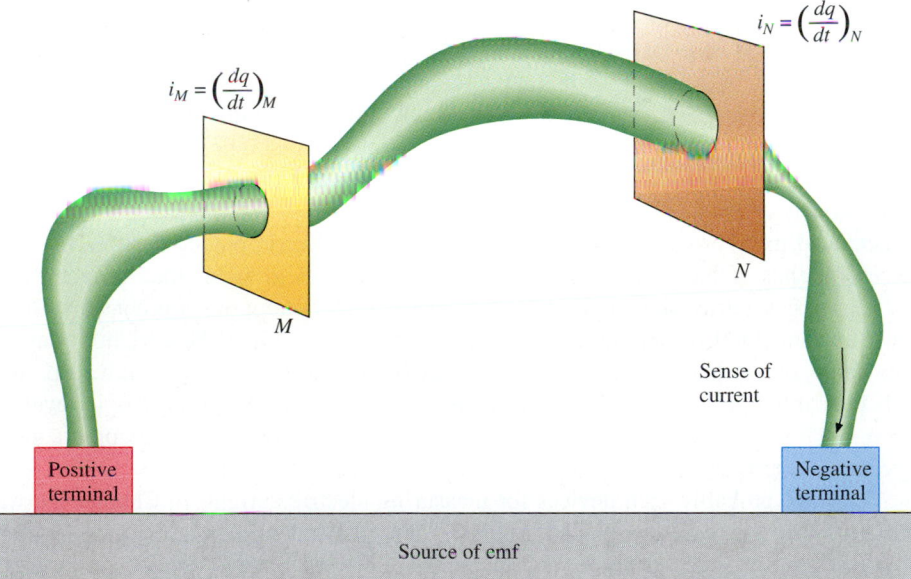

FIGURE 27.6 Electric charge flows through a conductor. Regardless of the sign(s) of the mobile charge, the net result is a flow of positive charge in the sense shown. The current through surface M is $i_M = (dq/dt)_M$; the current through N at the same time is $i_N = (dq/dt)_N$. In the steady state, $i_M = i_N$.

---

*This does not violate the rule that requires zero field within a conductor in equilibrium because the conductor is not in equilibrium under these circumstances, though it may well be in a steady state. You may wish to review the distinction between the equilibrium state and the steady state (Section 16.2).

true for the metal wires that conduct electricity in everyday applications. Under steady-state conditions, the current at any instant must be the same across all imaginary sections through the wire. In Figure 27.6, for example, $i_M = i_N$; otherwise, charge would build up at some point in the wire. Compare this argument with the argument concerning the flow of water through a pipe in Section 16.2. For water, a buildup of mass at some point in the pipe is prevented by the large bulk modulus (small compressibility) of water. The great strength of the electric (Coulomb) force serves the same function for the electric charge in a conductor.

Current Density

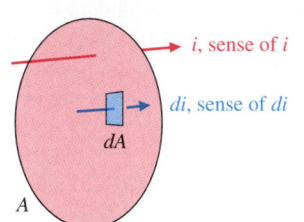

FIGURE 27.7 A conductor of cross-sectional area A carries a current i. An element di of the current passes through the area element dA. The current density at dA is defined as $j \equiv di/dA$.

Suppose that a conductor of cross-sectional area A carries a current i. A part of the current, di, passes through any element dA of the cross section, as shown in Figure 27.7. We define the **current density** j to be

$$j \equiv \frac{di}{dA}.$$

(27.3)

That is, *the current density is the current per unit cross-sectional area*. The unit of current density must be A/m^2.

If the current density is uniform throughout the cross section, we can write the simple relation

$$i = jA.$$

(27.4a)

If the current density is not uniform, we can still write the more general relation

$$i = \int_{\substack{\text{cross} \\ \text{section}}} j\, dA.$$

(27.4b)

Electrical Resistance and Ohm's Law

A free electric charge in a steady electric field will accelerate without limit. If the speed of the electrons in a metal wire increased continuously when a field was applied, they would transport charge at an increasing rate. That is, a constant field would result in a continuously increasing current dq/dt. How can we explain the fact that a steady state is achieved in a current-carrying conductor? A fixed electric potential difference between the ends of the conductor implies a steady electric field within the conductor and thus a steady electric force on the mobile carriers. If the current remains constant, there must be some kind of "friction" that opposes the electric force. This friction plays a role analogous to that of the fluid friction that limits the flow of water in a pipe. We will look more closely into the mechanism of this friction on a microscopic level in Section 27.5. Here, we develop a phenomenological view of the macroscopic consequences of the friction.

You have probably seen devices for measuring electric current. In Chapter 29, we will study the physical law, called Ampère's law, that underlies the operation of the simplest class of such devices. For the present, we take the operation of current-measuring devices for granted. Any such device is called an **ammeter** and is conventionally represented by the symbol shown in Figure 27.8.

FIGURE 27.8 The conventional symbol for an ammeter.

Armed with an ammeter and a source of emf having a known potential difference V between its terminals, we can carry out a series of experiments to determine the relation between the current i carried by a given conductor and the potential difference V that drives the current. (This series of experiments is readily carried out; devices are widely available that supply an emf V that can be varied by the turn of a knob.) If V is not too large, samples of many conducting materials exhibit the simple relation

$$i \propto V;$$

(27.5a)

the current is directly proportional to the potential difference that drives it. As usual, we prefer to express the relation between i and V as an equation, and so we write

$$V = iR. \tag{27.5b}$$

The constant R in Equation 27.5b is called the **electrical resistance** or simply the **resistance**. It is a measure of the difficulty of driving a current through a conductor; a large value of R means that a large potential difference is required to drive a modest current.

Any relation that expresses the proportionality between i and V, including Equations 27.5a and 27.5b, is called **Ohm's law**. Like Hooke's law, Ohm's law is not a universal law of nature but a convenient way of expressing a commonly encountered relation. Equation 27.5b is the form of Ohm's law encountered most frequently in practical electrical work. (There is no special reason for writing the equation $V = iR$ instead of putting the proportionality constant first and writing $V = Ri$. But $V = iR$ is established by custom.)

By inspecting Equation 27.5b, you can see that the SI unit of resistance is the volt per ampere (V/A). This unit is given the name **ohm** (Ω):

$$1\ \Omega \equiv 1\ \text{V/A}. \tag{27.6}$$

The symbol Ω (uppercase Greek omega) is used instead of O to prevent confusion with zero when one writes, for example, $R = 100\ \Omega$.

A sample whose behavior conforms to Ohm's law over some range of i and V is called **ohmic**. Ohmic behavior implies that R is a constant over the pertinent ranges of i and V. (Compare this implication with the analogous use of the term "force constant" for the quantity k in Hooke's law, $F = kx$.)

Resistivity

What determines the resistance of an ohmic sample? For concreteness, consider a set of samples in the form of wires made of the same material, each wire having a uniform cross-sectional area (Figure 27.9). Measurements of corresponding pairs of values V and i bear out what you would probably predict: R is proportional to the sample length l and inversely proportional to the cross-sectional area A:

$$R \propto \frac{l}{A}. \tag{27.7a}$$

We write this relation as an equation,

$$R = \rho \frac{l}{A}. \tag{27.7b}$$

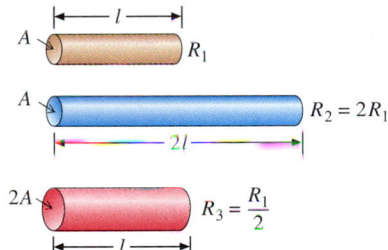

FIGURE 27.9 In any wire, resistance is directly proportional to length and inversely proportional to cross-sectional area.

The proportionality constant ρ is called the **resistivity**. By rewriting Equation 27.7b in the form $\rho = RA/l$, you can determine that the SI unit of resistivity is the ohm-meter ($\Omega \cdot$m).

The resistivity of a sample is independent of the dimensions of the sample and is thus characteristic of the material of which the sample is made. Some typical resistivities are given in Table 27.1.

Resistivity depends on temperature, sometimes in a complicated way. For many conductors (including most metals in the temperature range around 300 K), however, ρ does not change rapidly with changing temperature. For many purposes, it suffices to use a linear relation to approximate the temperature dependence:

$$\rho(t) = \rho^*[1 + \alpha(t - t^*)]. \tag{27.8}$$

In this equation, ρ^* is the resistivity of the sample at some reference temperature t^* (often 20°C), α is an empirical constant called the **temperature coefficient of resistivity**, and t is the Celsius temperature at which the resistivity has the value $\rho(t)$. Values of α for some substances are listed in Table 27.1. In the room-temperature range, α is positive for metals. Note, however, the negative value of α for carbon. This latter behavior is characteristic of the class of materials called *semiconductors*.

As you can see from Table 27.1, the range of resistivities encountered in matter is vast, extending over some 26 orders of magnitude—a larger range than is encountered in the measurement of any other physical quantity on the everyday scale of things. You can also see why it is possible to classify most materials unambiguously as either conductors or insulators. Most of the materials we think of as conductors have resistivities far less than 1 $\Omega \cdot m$, and most of the materials we think of as insulators have resistivities far greater than 1 $\Omega \cdot m$. Semiconductors span the otherwise almost empty midrange of resistivities. The resistivity of a sample of semiconductor depends in a dramatic way on the amount and type of impurities it contains, as well as on temperature. The room-temperature resistivity of silicon samples can readily be manipulated over ten or more orders of magnitude.

Superconductors have unique conducting properties. When the temperature of many metals and certain compounds is lower than a characteristic temperature called the *critical temperature* T_c, the resistivity vanishes completely; $\rho = 0$ *exactly*. This remarkable property and other physical properties with which it is linked give superconductors great commercial potential as well as intrinsic interest. The values of T_c for pure elements are quite low, the highest being about 9 K for niobium. Such temperatures are attainable only by costly techniques involving liquid helium. However, an intensive and continuing search for substances with much higher values of T_c has proceeded over several decades, and there has been much improvement. At present, superconducting magnets made of Nb_3Sn and similar compounds having $T_c \simeq 20$ K are widely used for producing high magnetic fields without large electric power costs. A variety of important medical imaging techniques depend on superconducting magnets. Recently,

TABLE 27.1 Resistivity ρ^* and Temperature Coefficient of Resistivity α for Selected Materials at 20°C

Material	Resistivity ($\Omega \cdot m$)	Temperature Coefficient α (K^{-1})
Metals		
Silver	1.59×10^{-8}	3.8×10^{-3}
Copper, hard drawn	1.77×10^{-8}	3.8×10^{-3}
Copper, annealed	1.72×10^{-8}	3.9×10^{-3}
Gold	2.44×10^{-8}	3.4×10^{-3}
Aluminum	2.82×10^{-8}	3.9×10^{-3}
Iron, 99.98% pure	10×10^{-8}	5×10^{-3}
Tungsten, drawn	5.6×10^{-8}	4.5×10^{-3}
Lead	22×10^{-8}	3.9×10^{-3}
Mercury	95.783×10^{-8}	8.9×10^{-4}
Alloys		
Brass	7×10^{-8}	2×10^{-3}
Manganin	44×10^{-8}	1×10^{-5}
Constantan	49×10^{-8}	1×10^{-5}
Nichrome	1×10^{-6}	4×10^{-4}
Nonmetals		
Carbon (typical)	5×10^{-5}	-5×10^{-4}
Water, pure	2×10^{5}	
Water, tap	10^{1}–10^{3}	
Insulators		
Glass	10^{10}–10^{14}	
Sulfur	10^{15}	
Polyethylene	10^{16}	

superconducting oxides have been devised with critical temperatures considerably greater than 70 K, making it possible to accomplish the necessary refrigeration with cheap and plentiful liquid nitrogen. Although numerous technical problems remain— some of them difficult—superconducting long-distance power transmission lines should become feasible. Superconducting power lines will avoid two major costs of the present-day high-voltage lines. First, some 10% to 20% of the power generated at remote hydroelectric stations is now lost in transmission (Section 27.6), and this lost power will be salvaged. Second, there is no need for high voltage in a superconducting cable, and so the cable can be buried in a trench. Thus there is no need for the expensive, ugly towers and rights of way of present-day lines. Many other important practical applications of superconductivity have been envisioned as well.

> **Relatively Cheap**
> Cryogenic engineers like to point out that the price of liquid nitrogen is about the same as the price of milk, while the price of liquid helium is about the same as the price of high-quality Scotch whiskey. In spite of worldwide inflation, this statement has remained true for about five decades.

EXAMPLE 27.1

B&S (American standard) 14-gauge annealed copper wire is commonly used in household wiring. Its diameter is 0.1628 cm. **(a)** Find the resistance R_{20} of a 12.0-m length of this wire at 20°C. **(b)** Find the resistance R_{-5} of the same wire at −5°C. Neglect the change in the length of the wire due to thermal contraction.

SOLUTION:
(a) What is the resistance of 20 m of 14-gauge wire at 20°C?

To use Equation 27.7b, $R = \rho l/A$, you must calculate the cross-sectional area of the wire:

$$A = \pi \times (\tfrac{1}{2} \times 1.628 \times 10^{-3} \text{ m})^2 = 2.082 \times 10^{-6} \text{ m}^2.$$

Refer to Table 27.1 for $\rho = \rho^*$ for annealed copper at 20°C. Equation 27.7b now gives you

$$R_{20} = 1.72 \times 10^{-6} \ \Omega\cdot\text{m} \times \frac{12.0 \text{ m}}{2.082 \times 10^{-6} \text{ m}^2}$$

$$= 9.91 \times 10^{-2} \ \Omega.$$

(b) What is the resistance of the same wire at −5°C?

You could use Equation 27.8 to find $\rho(-5°\text{C})$ and repeat the calculation of part **a** to obtain R_{-5}. But it is easier to note first that

$$\frac{R_{-5}}{R_{20}} = \frac{\rho(-5°\text{C})l/A}{\rho^*l/A} = \frac{\rho(-5°\text{C})}{\rho^*}.$$

From Equation 27.8, you have

$$\frac{\rho(t)}{\rho^*} = 1 + \alpha(t - t^*).$$

Combining these two equations, you obtain

$$R_{-5} = R_{20}[1 + \alpha(-5°\text{C} - 20°\text{C})].$$

Finally, you insert the necessary numerical values:

$$R_{-5} = 9.91 \times 10^{-2} \ \Omega$$

$$\times [1 + 3.9 \times 10^{-3} \text{ K}^{-1} \times (-25 \text{ K})]$$

$$= 8.94 \times 10^{-2} \ \Omega.$$

Why is it possible to retain three significant figures in the result, even though α is given to only two significant figures?

EXAMPLE 27.2

The wire of Example 27.1 is rated to carry a maximum current $i_{max} = 15$A. When it is carrying this current, what is the **voltage drop** across the wire—that is, the potential difference between the positive and negative ends? Assume that the temperature of the wire is 20°C. (This assumption introduces some inaccuracy because the wire warms up slightly when it carries its maximum rated current. But i_{max} is itself an approximate value with a substantial safety factor built in.)

SOLUTION: You use Ohm's law in the form of Equation 27.5b to write

$$V = iR = 15 \text{ A} \times 9.91 \times 10^{-2} \ \Omega = 1.5 \text{ V}.$$

Electric Field, Current Density, and Ohm's Law

In the form $V = iR$ (Equation 27.5b), Ohm's law describes the behavior of an entire conductor when a potential difference V is imposed between its ends. In this form, Ohm's law depends on two kinds of factors: geometric and intrinsic. We have already made this point clear; Equation 27.7b, $R = \rho l/A$, expresses resistance in a way that explicitly separates the length and cross-sectional area from the resistivity, an intrinsic property of the conducting material. We now wish to express Ohm's law in a way that is independent of the size and shape of the conductor and is thus applicable to any part of it. We will express Ohm's law in a way that eliminates l and A completely.

FIGURE 27.10 A small chunk
within a conductor carries current di.
The chunk has cross-sectional area
dA and thickness dl, and the current
density is j. The potential drop
across the chunk is dV.

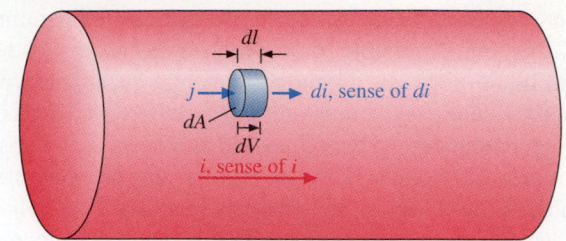

Figure 27.10 enables us to look inside a conductor carrying a current i. An infinitesimal part of this current, di, passes through the infinitesimal disk-shaped chunk of conductor we have chosen to inspect. The cross-sectional area of the chunk is dA, and its thickness (in the direction of the current) is dl. Within the chunk, the current density has the same value j everywhere. (We can always pick a chunk small enough so that this is the case.) The quantities di, j, and dA are related by Equation 27.3, $j = di/dA$.

Driving the current through the chunk is a potential difference $dV = \mathscr{E}\,dl$, where $\mathscr{E}$ is the magnitude of the electric field in the vicinity of the chunk. For this infinitesimal chunk, Ohm's law takes the form

$$dV = di\,R,$$

where R is the resistance of the chunk. Into this equation, we substitute $\mathscr{E}\,dl$ for dV, $j\,dA$ for di, and (using Equation 27.7b) $\rho\,dl/dA$ for R. This gives us

$$\mathscr{E}\,dl = j\,dA\,\rho\,\frac{dl}{dA},$$

which simplifies to yield

$$\boxed{\mathscr{E} = j\rho.} \tag{27.9}$$

This form of Ohm's law, expressed in terms of electric field and current density, is applicable to any location in a conductor, even if the conductor is irregular so $\mathscr{E}$ and j are not everywhere the same.

In Section 27.5, we will find an alternative form of Equation 27.9 useful. Specifically, we define the **conductivity** σ to be the reciprocal of the resistivity ρ:

$$\sigma \equiv \frac{1}{\rho}. \tag{27.10}$$

In SI, σ is expressed in the unit $(\Omega\cdot m)^{-1}$. In terms of σ, we can write Ohm's law in the form

$$\boxed{j = \sigma\mathscr{E}.} \tag{27.11}$$

EXAMPLE 27.3

For the 12.0-m length of 14-gauge wire considered in Examples 27.1 and 27.2, find **(a)** the magnitude of the electric field at any point in the wire, **(b)** the conductivity of the copper of which the wire is made, and **(c)** the current density in the wire when it carries 15 A.

SOLUTION:

(a) Find $\mathscr{E}$ at any point in the wire.

Because the wire is uniform in cross section, the electric field is everywhere the same. Therefore, you can write

$$|\mathscr{E}| = \frac{dV}{dl} = \frac{V}{l}.$$

From Example 27.2, you have $V = 1.5$ V. So the electric field

is

$$\mathscr{E} = \frac{1.5\ \text{V}}{12.0\ \text{m}} = 0.13\ \text{V/m}.$$

(b) Find σ.

Using the resistivity given for annealed copper in Table 27.1, you use Equation 27.10 to find

$$\sigma = \frac{1}{1.72 \times 10^{-8}\ \Omega\cdot\text{m}} = 5.81 \times 10^{7}\ (\Omega\cdot\text{m})^{-1}.$$

(c) If $i = 15$ A, find j.

You can find j in either of two ways. One is to take advantage of the uniformity of the current and write Equation 27.3 in the form $j = i/A$. The other is to use the results of parts **a** and **b** together with Equation 27.11 to write

$$j = \sigma \mathscr{E} = 5.81 \times 10^7 \ (\Omega \cdot m)^{-1} \times 0.13 \ \text{V/m}$$
$$= 7.6 \times 10^6 \ \text{A/m}^2.$$

You should convince yourself that the units in this calculation come out right.

Look carefully at the result of part **c**. A quite small electric field gives rise to a huge current density. Indeed, a copper bar of cross-sectional area 1 m^2 would carry almost 10 million amperes—that is, $\simeq 10^7$ C/s—when subjected to the same electric field. This is a vivid consequence of the small resistivity typical of metals.

A Microscopic View of Electrical Resistance

On the macroscopic scale, Ohm's law accurately describes the passage of current through many conductors under a wide range of conditions. As we have already noted, the existence of resistance implies a frictional process. But electric current involves the motion of microscopic charged particles, and, if we are to understand the mechanism of the frictional process, we must do so at the microscopic level.

For the sake of concreteness, let us consider a metal wire of uniform cross section (Figure 27.11). Suppose that a potential difference is imposed between the ends of the wire, and choose the positive direction from $+$ to $-$, consistent with the conventional sense of current i. Because the wire is uniform, the potential difference results in a uniform electric field $\mathscr{E}$ in the positive direction.

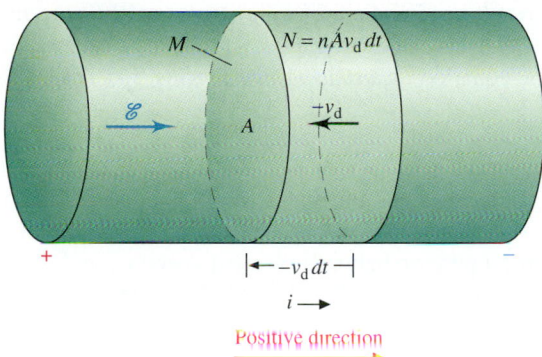

FIGURE 27.11 If the drift speed is v_d, the electron gas contained in the imaginary cylinder of area A and length $v_d \, dt$ will pass through the section M in time dt.

In metals, the charge carriers are electrons. In the presence of the electric field, these electrons are driven along the wire in the negative direction by a force $-e\mathscr{E}$. Under the combined action of this force and the frictional process, the electrons move along the wire with a constant *mean* speed v_d, called the **drift speed**. The corresponding **drift velocity** is expressed by the one-dimensional vector $\mathbf{v}_d = -v_d$.

Consider the analogous case of gas flowing in a pipe, driven by a fixed pressure difference between the ends of the pipe. On the microscopic scale, the individual gas molecules are moving with great speeds in all directions, but the net flow resulting from this random motion is zero. The macroscopic motion of the molecules, considered collectively as a gas, can be described by a flow velocity analogous to the drift velocity of the electrons. Although the magnitude of the flow velocity is usually very much less than the mean thermal speed v_{rms} of the molecules, the *same* flow velocity is imposed on *all* the molecules. Consequently, it is the flow, not the thermal motion, that we perceive on the macroscopic scale. In the spirit of this argument, we imagine that the free electrons in the metal wire make up an **electron gas**.

We bridge the gap between the macroscopic and microscopic views by making a connection between the current i, which is determined by a macroscopic measurement, and the drift velocity $-v_d$, which describes the mean motion of the microscopic electrons.

We treat the flow of the electron gas in a manner very much like the one we used in Section 16.2 to treat a fluid flowing in a pipe. Consider again the wire of uniform cross-sectional area A, shown in Figure 27.11. In a time dt, the electron gas moves through a distance $v_d \, dt$. Consequently, all of the electron gas contained in the cylinder of area A and length $v_d \, dt$ flows in the negative direction past the imaginary cut M through the wire. The volume of this cylinder is $A v_d \, dt$. Suppose that the number of free electrons per unit volume of the metal, called the **carrier concentration** or the **electron concentration**, is n. Then the total number N of electrons flowing past M in the time dt is the product of the electron concentration and the cylinder volume:

$$N = n A v_d \, dt.$$

The charge on each electron is $-e$. Thus the total charge flowing past M in the negative direction during the time interval is $-Ne = -ne A v_d \, dt$. But we are interested in the current i, which by its definition $i \equiv dq/dt$ (Equation 27.1) is related to the charge dq flowing in the *positive* direction. Because negative charge flowing in the negative direction is equivalent to positive charge flowing in the positive direction, we have

$$dq = ne A v_d \, dt.$$

Thus the current is

$$i \equiv \frac{dq}{dt} = ne A v_d. \tag{27.12}$$

This result is independent of the sign of the charge carriers. If we consider carriers of charge $+e$, their drift velocity is $\mathbf{v}_d = +v_d$. If we repeat the derivation with these changes, the result is still $i = ne A v_d$.

To express the result of Equation 27.12 in terms of the current density j, we use Equation 27.3, $j = di/dA$. As in Example 27.3, we take advantage of the uniformity of the wire to write the equation as a ratio of finite terms, $j = i/A$. We then have

$$j = ne v_d. \tag{27.13}$$

If the charge carriers are electrons, the direction of j is opposite that of the drift velocity $-v_d$, owing to the negative charge $-e$ of the electron. That is, $j = n(-e)(-v_d) = ne v_d$.

If we can establish a relation between the drift speed v_d and the electric field $\mathscr{E}$ that drives the electron gas, we can use Equation 27.13 to determine the relation between j and $\mathscr{E}$. That is, we can test our electron-gas model by seeing if it conforms to Ohm's law, just as in Chapter 17 we tested the ideal-gas model to see if it conformed to Boyle's and Charles's laws. The crucial relation between v_d and $\mathscr{E}$ depends on the friction mechanism. We now look more closely at this mechanism.

The "box" that confines the electron gas is the wire. Since the metal is neutral overall, the wire must contain a large number of immobile positive ions whose total charge is equal to and opposite that of the electrons that constitute the gas. These ions exert an attractive electric force on the electrons and thus "hold them in the wire." The ions are arranged in a regular *crystal lattice*, with an average spacing d between neighbors. (In general, the value of d lies somewhere between 10^{-10} m and 10^{-9} m.)

Suppose that the wire is in thermal equilibrium. The electrons collide frequently with the ions, exchanging energy in the collisions. As a result, the electron gas is also in thermal equilibrium. Associated with this thermal equilibrium is a velocity distribution. In the absence of an external electric field, the *mean* velocity of the electrons is zero because they are no more likely to go in one direction than in another. (See the discussion after Equation 18.55b.)

When an external electric field $\mathscr{E}$ is applied, things change. Each electron now experiences a force $\mathbf{F} = -e\mathscr{E}$. According to Newton's second law of motion, the corresponding acceleration of an electron is

$$\mathbf{a} = \frac{-e\mathscr{E}}{m}, \tag{27.14}$$

where m is the mass of the electron. Because the mean velocity is zero, we need consider only this acceleration in calculating the drift velocity of the electrons.

But Equation 27.14 applies only between collisions. In the spirit of kinetic theory, we assert that the collisions are completely *randomizing*. On the average, we argue, electrons rebound with equal probability in all directions. Although the velocity of an individual electron just after a collision can have a wide range of magnitudes and any direction, *on the average the velocity of an electron immediately after collision is zero*.

It follows that, on the average, the speed of an electron increases linearly with the time elapsed since the last collision. At time t, the mean speed is

$$\langle v \rangle = at = \frac{e\mathcal{E}t}{m}.$$

The time between collisions varies, but there must be some **mean free time**, or **collision time**, τ. On the average, the speed of an electron just before a collision is

$$\langle v \rangle = a\tau = \frac{e\mathcal{E}\tau}{m}.$$

Because the electron accelerates uniformly from rest to $\langle v \rangle$ between collisions, the electron gas must drift at a speed $\langle v \rangle / 2$. But, in recognition of the crudeness of the calculation, we drop the factor $\frac{1}{2}$ and write the approximate result

$$v_d \simeq \langle v \rangle = \frac{e\mathcal{E}\tau}{m}. \tag{27.15}$$

We now insert this value into Equation 27.13 to obtain

$$j = \frac{ne^2\tau}{m}\mathcal{E}. \tag{27.16}$$

Compare this result with Ohm's law in the form $j = \sigma\mathcal{E}$. If the factor $ne^2\tau/m$ is a constant, the electron-gas model will indeed obey Ohm's law. According to the model, the conductivity of a metal is

$$\sigma = \frac{ne^2\tau}{m}. \tag{27.17}$$

Under what circumstances will σ be constant, as required by Ohm's law? The quantities e and m are constant. The electron concentration n will change only if the free electrons are so energetic that they further ionize the ions with which they collide. That can happen in some substances but not in metals. Likewise, the collision time remains approximately constant over a wide range of conditions. Thus, Ohm's law is satisfied.

The "frictional" process that underlies electrical resistance is the randomizing collisions. The ordered energy of the drift motion of the electron gas, taken as a whole, is continually converted by the collisions into the disordered motion we call heat. It will not surprise you that conductors warm up as they carry electric current.

Our crude model of a metal, based on kinetic theory, is at least realistic enough to yield the basic proportionality $j \propto \mathcal{E}$ of Ohm's law. In Example 27.4, we push our luck and put some numbers into Equation 27.17 to see whether the model is consistent with measured values of σ.

EXAMPLE 27.4

Brass is an alloy of copper and zinc mixed in a ratio of roughly 2:1. The density of brass is 8.4×10^3 kg/m³, and the mean atomic mass of the atoms is 64 u. **(a)** Assuming that the atoms are arranged in a cubic lattice, find the distance d between neighboring atoms. **(b)** Assuming that each atom contributes one free electron to the metal (and thus becomes a singly charged positive ion), calculate the electron concentration n.

(c) Calculate the drift speed that corresponds to a current density $j = 10^7$ A/m². (This is the order of magnitude of the current density found for the wire of Example 27.3.) **(d)** Using the electrical conductivity of brass given in Table 27.1, estimate the collision time τ at 20°C (293 K). **(e)** Assuming that the electrons act like the molecules of an ideal gas, calculate the mean speed v_{rms} of the electrons at 20°C. **(f)** Using the result of parts **c** and

d, estimate the mean distance λ traveled by electrons between collisions, called the **mean free path**, and compare this distance with the distance d you found in part **a**.

SOLUTION:

(a) Find the interatomic distance d.

Because the mean atomic mass of brass is 64 u, 1 kmol of brass has mass 64 kg. Using the given value of the density, you have for the volume of 1 kmol

$$\frac{64 \text{ kg}}{8.4 \times 10^3 \text{ kg/m}^3} = 7.6 \times 10^{-3} \text{ m}^3.$$

This volume contains Avogadro's number of atoms, and so the number of atoms per unit volume, n_i, is

$$n_i = \frac{6.02 \times 10^{26}}{7.6 \times 10^{-3} \text{ m}^3} = 7.9 \times 10^{28} \text{ m}^{-3}.$$

If you imagine the atoms to be located at the centers of adjacent cubes, stacked together in a vast array (Figure 27.12), each atom occupies a cubic volume equal to

$$\frac{1}{n_i} = 1.3 \times 10^{-29} \text{ m}^3.$$

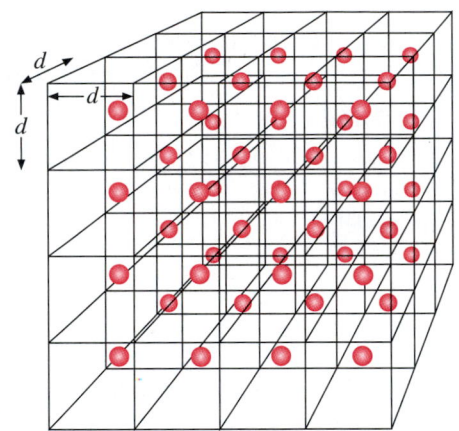

FIGURE 27.12 Atoms in a cubic lattice. Drawing a network of cubes, each centered on an atom, makes it clear that the volume per atom is $1/n_i$, the reciprocal of the number of atoms (or ions) per unit volume.

The distance between neighbors is equal to the side of the cube, which is just the cube root of this volume. So you have

$$d = (1.3 \times 10^{-29} \text{ m}^3)^{1/3} = 2.3 \times 10^{-10} \text{ m}.$$

(b) Find the electron concentration.

On the assumption that each atom contributes one free electron, you have $n = n_i$, or

$$n = 7.9 \times 10^{28} \text{ electrons/m}^3.$$

(c) If $j = 10^7 \text{ A/m}^2$, find the drift speed v_d.

You use Equation 27.13 to write

$$v_d = \frac{j}{ne}.$$

So you have

$$v_d = \frac{10^7 \text{ A/m}^2}{7.9 \times 10^{28} \text{ m}^{-3} \times 1.6 \times 10^{-19} \text{ C}}$$

$$= 7.9 \times 10^{-4} \text{ m/s}.$$

This is about 1 mm/s; the drift speed of electrons in metals is quite small. The electron concentration is so large that the electrons do not have to drift very fast to transport charge at a considerable rate.

(d) Estimate τ at 20°C.

From Equation 26.17, you have

$$\tau = \frac{\sigma m}{ne^2} = \frac{m}{\rho ne^2}.$$

You insert the numerical values to find

$$\tau = \frac{9.1 \times 10^{-31} \text{ kg}}{7 \times 10^{-8} \text{ }\Omega\text{·m} \times 7.9 \times 10^{28} \text{ m}^{-3} \times (1.6 \times 10^{-19} \text{ C})^2}$$

$$= 6.4 \times 10^{-15} \text{ s},$$

where we carry an extra significant figure for the sake of further calculation. You should satisfy yourself that the product of units, $\text{kg}/(\Omega\text{·m·m}^{-3}\text{·C}^2)$, is indeed equivalent to seconds.

(e) Find v_{rms} for the electrons at $T = 293$ K.

For an ideal gas, the root-mean-square speed is given by Equation 18.44a,

$$v_{rms} = \sqrt{\frac{3kT}{m}}.$$

So you have

$$v_{rms} = \sqrt{\frac{3 \times 1.38 \times 10^{-23} \text{ J/K} \times 293 \text{ K}}{9.1 \times 10^{-31} \text{ kg}}}$$

$$= 1.2 \times 10^5 \text{ m/s}.$$

This value is greater than the drift speed by the factor 10^8.

(f) Estimate the mean free path λ.

The value of v_{rms} is so much greater than v_d that the collision time is determined entirely by the former. The mean free path is just the product of v_{rms} and the collision time τ:

$$\lambda = v_{rms}\tau$$

$$= 1.2 \times 10^5 \text{ m/s} \times 6.4 \times 10^{-15} \text{ s} = 7.7 \times 10^{-10} \text{ m}.$$

Compare this value with the interionic spacing $d = 2.3 \times 10^{-10}$ m found in part **a**. The mean free path is about three times d, which means that an electron on the average collides with about every third ion.

The very crude calculation of Example 27.4 lends plausibility to our equally crude picture of electrons as the molecules of an ideal gas. This picture is called the **free-electron** or **Drude theory of metals** [after the German physicist Paul K. L. Drude (1863–1906), whose original development (1900, with elaborations through 1902) we have followed fairly closely].

Limitations of the Free-Electron Theory

As Example 27.4 shows, the free-electron theory gives a plausible account of the mechanism of electrical conduction in brass at room temperature. The results are about equally good for other alloys, such as Nichrome and constantan. But suppose we apply the same calculation to a pure metal, such as copper. The interionic spacing, atomic mass, and chemical valence of copper are very similar to the average values for the copper and zinc of which brass is made. But according to Table 27.1, the resistivity of copper is only one-fourth that of brass. The implication is that the mean free path of electrons in copper is about $12d$, compared with $3d$ for brass. Our crude picture of electrons colliding with ions, as though Ping-Pong balls were colliding with bowling balls, does not seem compatible with this result.

There are further difficulties with the free-electron theory. As you will see in Problem 27.34, the theory predicts that the resistivity of a metal should be proportional to the square root of the temperature. But the observed temperature dependence is expressed by Equation 27.8, which for our present purposes boils down to $\rho \propto T$, not $\rho \propto \sqrt{T}$.

We know today that the behavior of microscopic particles, such as electrons, is governed by the laws of quantum mechanics. A detailed quantum-mechanical theory of electric conduction of metals was developed in the 1930s, and the predictions of the theory agree well with a wide variety of experimental measurements.

Joule's Law

As charge flows along a conductor, the potential energy of the charge decreases. Nevertheless, the mean kinetic energy of drift of the charge carriers, $\frac{1}{2}mv_d^2$, does not increase. Thus energy is dissipated when a current passes through a conductor. This is not surprising in view of the fact that electrical resistance arises from a kind of friction. If we are interested only in the amount of energy dissipated, however, the details of the dissipation process are not important.

Suppose that a source of emf imposes a voltage V across a conductor, as shown in Figure 27.13. When an amount of charge q is transported from A to B, the work done to maintain the flow of the charge is $W = |qV|$ (Section 25.2). Because power P is defined to be the rate at which work is done, $P = dW/dt$, we can write

$$P = V\frac{dq}{dt}.$$

But dq/dt is the current i passing through the conductor. So the power can be written

$$P = Vi. \tag{27.18}$$

The power dissipated in a conductor is equal to the product of the potential difference across the conductor and the current passing through it. By "dissipated" we mean that the electrical potential energy required to drive the charge through the conductor is converted into heat energy in the conductor. (That is how electric heaters work.) Equation 27.18 is called **Joule's law.**

The equation $P = Vi$ is quite general and does not depend on the electrical properties of the conductor. *If* the conductor is ohmic, however, we can use Ohm's law to eliminate either i or V from Joule's law. We thus obtain the important special forms

$$P = V\frac{V}{R} = \frac{V^2}{R} \quad \text{and} \quad P = (iR)i = i^2R \quad \text{for ohmic conductors.} \tag{27.19a,b}$$

Equations 27.18 and 27.19 are valid even if the voltage and current vary with time, because $P \equiv Vi$ has a defined value at any instant. However, P then also varies with time. We can calculate the mean power $\langle P \rangle$ if we know the mean values of the other quantities in Equation 27.18, 27.19a, or 27.19b. To give one important example of the

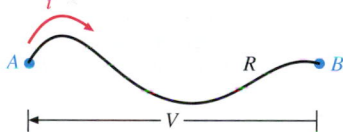

FIGURE 27.13 A conductor has resistance R between points A and B. When the potential difference across the conductor is V, the current is i.

Joule and Electrical Energy

Joule devoted a large part of his career to demonstrating the equivalence of energy in all its forms. His experimental results were pivotal in establishing the validity of the principle of energy conservation. In the 1840s, Joule carried out a series of experiments in which he immersed electrical conductors in calorimeters. He showed that a given electrical power input, Vi, always produces the same rate of temperature rise in the calorimeter as a numerically equal mechanical friction power input Fv. Today, the principle of energy conservation is so thoroughly accepted that we find Joule's result unsurprising. However, Joule's law is still important in determining the power consumption of electrical devices.

use of mean values, the voltage at which electricity is supplied commercially is not constant but varies sinusoidally with a frequency of 60 Hz or 50 Hz. However, the voltage quoted by the power company is almost always a mean value—specifically, the rms value. Example 27.5 demonstrates the use of Joule's law in connection with mean values.

EXAMPLE 27.5

An electric space heater has two settings. The high setting is 1500 W, and the low setting is 1200 W. **(a)** If the (mean) line voltage is 110 V, find the mean current and the resistance associated with the high setting. **(b)** The low setting is achieved by providing a reduced mean voltage across the heating coil. Assuming that R does not change significantly with temperature (not a very good assumption for a heating coil), find the mean values of the voltage and the current for the 1200-W setting. **(c)** In fact, the resistance of the coil does change with temperature. Will this change result in greater or lesser values of $\langle V \rangle$ and $\langle i \rangle$ than those calculated in part **b**?

SOLUTION:

(a) Find $\langle i \rangle$ and R for the 1500-W setting.

You use Equation 27.18 to write

$$\langle i \rangle = \frac{\langle P_{\text{hi}} \rangle}{\langle V \rangle} = \frac{1500 \text{ W}}{110 \text{ V}} = 13.6 \text{ A.}$$

You can use Ohm's law or either form of Equation 27.19 to find the resistance of the coil. If you choose Equation 27.19a, you write

$$R = \frac{\langle V \rangle^2}{\langle P \rangle} = \frac{(110 \text{ V})^2}{1500 \text{ W}} = 8.07 \ \Omega.$$

(b) Find $\langle V \rangle$ and $\langle i \rangle$ for the 1200-W setting, assuming R does not change when the heater is switched to the low setting.

You have from Equation 27.19a

$$\langle V \rangle = \sqrt{\langle P_{\text{lo}} \rangle R} = \sqrt{1200 \text{ W} \times 8.07 \ \Omega} = 98.4 \text{ V.}$$

Again, you can choose among several ways to calculate $\langle i \rangle$. The easiest way is to apply Ohm's law, using the mean values of the pertinent quantities:

$$\langle i \rangle = \frac{\langle V \rangle}{R} = \frac{98.4 \text{ V}}{8.07 \ \Omega} = 12.2 \text{ A.}$$

Note that a 20% reduction in power output is achieved by means of 10% reductions in both voltage and current.

(c) Does taking account of the dependence of resistance on temperature lead to greater or lesser values of $\langle V \rangle$ and $\langle i \rangle$ than those calculated in part **b**?

When the heater is switched from the high to the low setting, the temperature of the heating element decreases. If the element is made of metal, its resistance decreases as well. From Equation 27.19a, you can see that dissipating a specified amount of power in a smaller resistance requires a voltage smaller than that found in part **b**. From Equation 27.19b, you can see that the value of the current is *larger* than the value you found in part **b**.

Direct-Current Circuits

The variety of electric circuits is limitless. However, each such circuit must contain at least one source of emf and at least one other circuit element that are connected by leads to form a closed loop through which charge can flow, driven by the potential difference between the terminals of the source. A source that always drives current through the loop in the same sense is called a **direct-current source** or **dc source**. If, moreover, the voltage between the terminals does not vary, the source is called a **steady source of direct current** or simply a **steady source**. In this chapter we consider only steady sources.

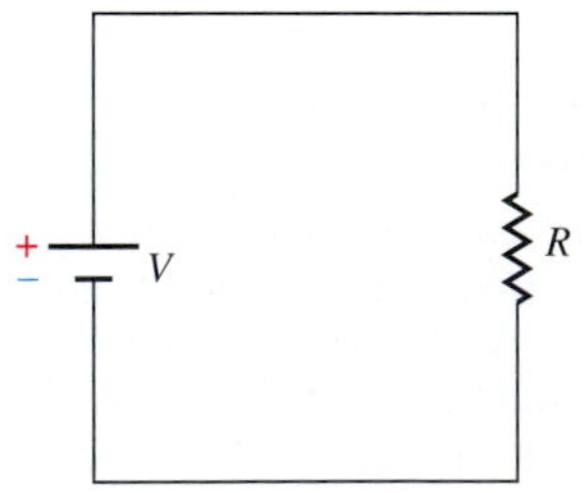

Figure 27.14 is a schematic diagram of the simplest possible circuit, consisting of two circuit elements and the leads that connect them. One element is a source of emf. The other, an ohmic circuit element, is called a **resistor**. In the diagram, all elements and leads are represented by their conventional symbols. In any circuit consisting of steady sources of emf and resistors only, the current at any point in the circuit remains constant over time. Such circuits are called **direct-current circuits** or **dc circuits**.

All real sources of emf have some **internal resistance**. That is, when current flows through the circuit—and hence through the source of emf as well—a certain amount of electrical energy is converted into heat energy within the source. When the internal resistance is comparable to other resistances in the circuit, it must be taken into account. Figure 27.15 shows how this situation is represented symbolically. But sources of emf

FIGURE 27.14 Diagram of the simplest possible circuit. A steady source of emf, at left, is connected to a resistor, at right, by wires whose resistance is very much smaller than that of the resistor and is conventionally assumed to be zero.

are usually chosen so that their internal resistance is small compared with other resistances in the circuit and thus can be ignored.

In this section, we study dc circuits of increasing complexity. In the simple circuit of Figure 27.14, the current is everywhere the same and can be calculated immediately by means of Ohm's law in the form $i = V/R$. Practical electric circuits are almost always more complicated than this. Nevertheless, it is always possible to analyze them by application of fairly simple rules.

Resistors in Series and in Parallel

Figure 27.16 represents a circuit consisting of a source of emf and two resistors connected in series. As we did for capacitors, we ask: What is the resistance R of the network lying between A and B? That is, what single **equivalent resistor** R would have the same resistance as the two resistors linked together?

Because there is only one path for electric current to follow, i must have the same value everywhere in the circuit. The potential difference between A and B is V. This potential difference must somehow be divided into two parts, V_1 and V_2, as shown, subject to the condition

$$V = V_1 + V_2.$$

Each of the two resistors obeys Ohm's law, and so we can write

$$V_1 = iR_1 \quad \text{and} \quad V_2 = iR_2.$$

Combining these three equations, we obtain

$$V = i(R_1 + R_2).$$

The equivalent resistor R, connected across the source of emf in place of R_1 and R_2, would also obey Ohm's law, so

$$V = iR.$$

Comparing this equation with the preceding one, we obtain $iR = i(R_1 + R_2)$, or

$$R = R_1 + R_2 \quad \text{for resistors in series.} \tag{27.20a}$$

This result is readily extended to a network consisting of n resistors in series:

$$R = R_1 + R_2 + \cdots + R_i + \cdots + R_n \tag{27.20b}$$

Each of the quantities $V_j = iR_j$ represents the difference in potential between the terminal at which the current enters the resistor and the terminal at which it leaves the resistor. As noted in Example 27.2, such a potential difference is conventionally called a *voltage drop* because it represents the decrease in potential experienced by a positive test charge when it moves through a circuit element. Because resistors obey Ohm's law, $V = iR$, the voltage drop across a resistor is sometimes called an **iR drop**.

In Figure 27.17, the two resistors are connected in parallel. (The terminology is identical with that used in Section 26.3 with respect to capacitors.) The voltage drop across each resistor is equal to the source voltage V. The current i, however, divides into two branches, which carry currents i_1 and i_2. As you can see from Figure 27.17, we must have

$$i = i_1 + i_2.$$

We apply Ohm's law to the two resistors, as well as to the equivalent resistance R with which we can replace them. We have

$$i_1 = \frac{V}{R_1}, \quad i_2 = \frac{V}{R_2}, \quad \text{and} \quad i = \frac{V}{R}.$$

Inserting these values into the preceding equation, we obtain $V/R = V/R_1 + V/R_2$, or

$$\frac{1}{R} = \frac{1}{R_1} + \frac{1}{R_2} \quad \text{for resistors in parallel.} \tag{27.21a}$$

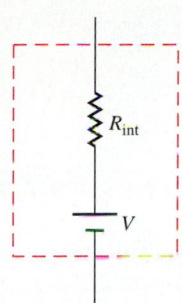

FIGURE 27.15 A real source of emf, having significant internal resistance, is conventionally represented as an ideal source coupled with a resistor. The dashed box indicates that V and R_{int} pertain to the same circuit element.

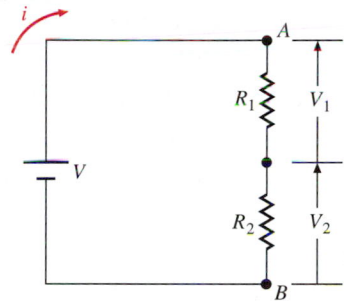

FIGURE 27.16 A circuit consisting of two resistors in series and a source of emf.

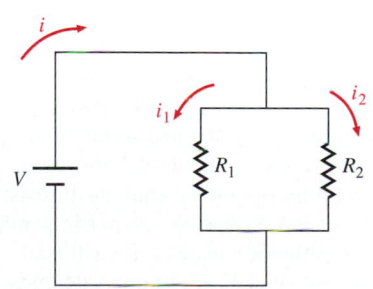

FIGURE 27.17 A circuit consisting of two resistors in parallel and a source of emf.

This result is readily extended to a network consisting of *n* resistors in series. The result is

$$\frac{1}{R} = \frac{1}{R_1} + \frac{1}{R_2} + \cdots + \frac{1}{R_j} + \cdots + \frac{1}{R_n}. \qquad \text{(27.21b)}$$

EXAMPLE 27.6

Find the equivalent resistance of the network between points *A* and *B* in Figure 27.18a.

SOLUTION: You begin by finding the equivalent resistance R_{23} of the parallel subnetwork between *C* and *B*. Using Equation 27.21a, you find

$$\frac{1}{R_{23}} = \frac{1}{56\ \Omega} + \frac{1}{97\ \Omega} = 2.8 \times 10^{-2}\ \Omega^{-1},$$

so $\qquad R_{23} = 36\ \Omega.$

The equivalent resistance R_{23} is in series with R_1, as shown in Figure 27.18b. Using Equation 27.20a, you obtain the overall equivalent resistance

$$R = R_1 + R_{23} = 85\ \Omega + 36\ \Omega = 120\ \Omega$$

to two significant figures.

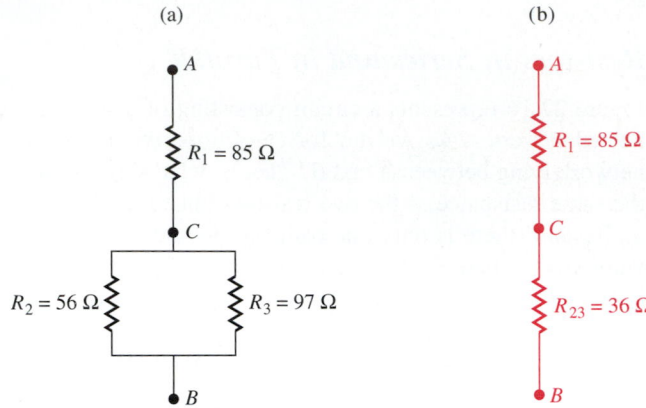

FIGURE 27.18

Kirchhoff's Rules

Not all circuits can be broken down into series and parallel subnetworks. For example, two circuits that cannot be so broken down are shown in Figure 27.19. However, it is always possible to analyze such circuits by applying a set of three definitions and two rules, devised by Kirchhoff in 1845 and 1846, when he was still a student.

Definition 1. Any circuit can be divided into a number of closed **loops**. In Figure 27.19a, one loop is the closed path *adca* (shown in yellow). A second loop is *cdbc* (in green), and a third loop is *adbca* (in blue). Paths in which steps are retraced, such as the figure-eight path *abdcbdea*, are not loops.

Definition 2. A **node** is any point where three or more conductors come together. In Figure 27.19a, the nodes are at *c* and *d*.

Definition 3. A **branch** is a path between two nodes that does not contain any other nodes. The circuit of Figure 27.19a contains three branches: *cad*, *cd*, and *cbd*.

The Node Rule. The algebraic sum of the currents entering a node is zero. For the purposes of this rule, a current is positive if it flows into the node, and negative if it flows out of the node (Figure 27.20). The node rule is just a quantitative way of saying that charge cannot accumulate at a node.

FIGURE 27.19 Two circuits not reducible to parallel and series subnetworks. (*a*) The network between *a* and *b* must have an equivalent resistance. But R_1 and R_2 are not in parallel. Although one end of each is at the potential V_a, their opposite ends are at different potentials because there is in general an *iR* drop across R_3. (*b*) With multiple sources of emf present, the concept of equivalent resistance cannot be applied.

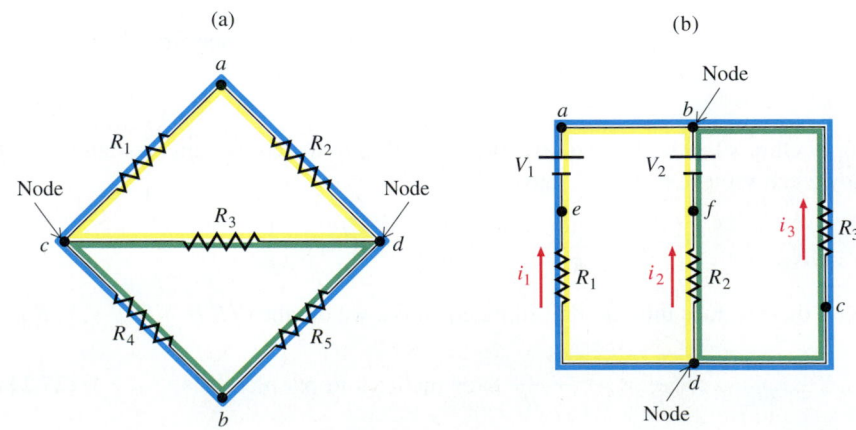

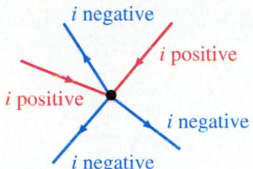

FIGURE 27.20 The sign convention for the node rule.

The Loop Rule. The algebraic sum of the potential differences across all the circuit elements in a loop is zero. If this were not so, it would not be possible to assign a unique potential to every point in the circuit (Section 8.2).

Let us use Kirchhoff's rules to analyze the passage of current through the circuit of Figure 27.19b. The first step is to assign a sense to the current in each of the branches *dcb*, *dfb*, and *deab*. The assignment is completely arbitrary. You will soon see that, if you make the wrong choice, you will merely obtain a negative value for the current. In Figure 27.19b, we have chosen all three currents i_1, i_2, and i_3 upward, as indicated by the arrows.

There are two nodes, at *b* and at *d*. We apply the node rule to write the equations

$$i_1 + i_2 + i_3 = 0 \quad \text{for node } b$$

and
$$-i_1 - i_2 - i_3 = 0 \quad \text{for node } d. \tag{27.22}$$

The second of these equations can be obtained from the first by the trivial operation of multiplying by -1. This observation illustrates a general rule: *The number of independent equations obtained by applying the node rule is one fewer than the number of nodes in the circuit.*

Next, we apply the loop rule. To do so, we must adopt a sign convention. Although several such conventions are in use, we adopt the following one because it is consistent with the signs of the potential energy changes experienced by a positive test charge as it is taken around the loop.

Sign convention a. As we pass through a source of emf from the negative to the positive terminal, the potential change is positive. For example, in passing through V_1 in Figure 27.19b in the sense of i_1, we see a potential change $+V_1$—an increase in potential. If we passed through V_1 in the opposite sense, we would see a potential change $-V_1$.

Sign convention b. As we pass through a resistor in the sense of the current, the potential change is negative. For example, in passing through R_1 in the sense of i_1, we see a potential change $-i_1R_1$—a decrease in potential.

With these rules in mind, consider the loop *deabfd*, shown in yellow in Figure 27.19b. We pass first through branch *deab* in the same sense as the current i_1 and then through branch *bfd* in the sense opposite that of the current i_2. The loop rule gives

$$-i_1R_1 + V_1 - V_2 + i_2R_2 = 0. \tag{27.23a}$$

Similarly, loop *dfbcd*, shown in green, yields the equation

$$-i_2R_2 + V_2 + i_3R_3 = 0. \tag{27.23b}$$

We could apply the loop rule again, to loop *deabcd* (shown in blue), but the resulting equation would yield no new information. *The number of independent equations obtained by applying the loop rule is one fewer than the number of loops in the circuit.*

Equations 27.22 and 27.23 constitute a set of three simultaneous equations that describe the currents in the circuit of Figure 27.19b. If V_1, V_2, R_1, R_2, and R_3 are known, the equations can be solved for the currents i_1, i_2, and i_3. However, the set of equations can be solved for any three of the eight quantities if the other five are known.

It is possible to solve the simultaneous equations in terms of the symbols and then to insert the desired numerical values at the end of the calculation. But there is little point in doing this because every circuit is different and therefore no useful generalization can be made. Besides, the algebra can get quite messy. Usually, the numbers are inserted on an as-you-go basis as the equations are solved. This procedure is illustrated in Example 27.7.

EXAMPLE 27.7

The circuit of Figure 27.21 is the same as that of Figure 27.19b, except that an ammeter has been inserted into branch *deab* so that the sense and magnitude of the current i_1 can be determined. Given the values in the figure, find the three unknown quantities i_2, i_3, and R_1.

SOLUTION: Because $i_1 = 1.6$ A, Equation 27.22 becomes

$$1.6 \text{ A} + i_2 + i_3 = 0 \quad \text{(node equation)}.$$

From Equation 27.23a, you have

$$-1.6 \text{ A} \times R_1 + 12.0 \text{ V} - 4.5 \text{ V} + 3.8 \ \Omega \times i_2 = 0$$
$$\text{(loop equation)}.$$

And Equation 27.23b gives you

$$-3.8 \ \Omega \times i_2 + 4.5 \text{ V} + 5.2 \ \Omega \times i_3 = 0 \quad \text{(loop equation)}.$$

As always in solving simultaneous equations, there are many ways to proceed. One way is to solve the first equation for i_3 and use the result to eliminate i_3 from the third equation:

$$i_3 = -1.6 \text{ A} - i_2,$$

and thus

$$-3.8 \ \Omega \times i_2 + 4.5 \text{ V}$$
$$+ 5.2 \ \Omega \times (-1.6 \text{ A}) + 5.2 \ \Omega \times (-i_2) = 0.$$

The only unknown in this equation is i_2, and you solve to obtain the first result:

$$i_2 = -0.4 \text{ A}.$$

It is now easy to substitute this value back into the node equa-

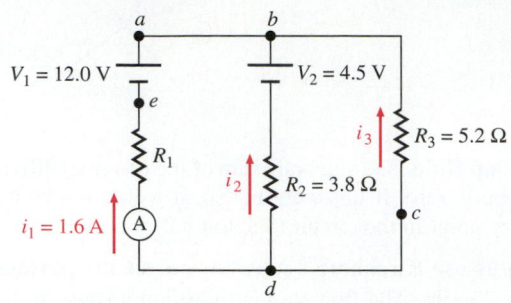

FIGURE 27.21

tion to obtain the second unknown, i_3:

$$i_3 = -1.6 \text{ A} - (-0.4 \text{ A}) = -1.2 \text{ A}.$$

The negative values of i_2 and i_3 mean that the original choice of sense for these two currents was the wrong one. That is, the corresponding arrows in Figure 27.21 should point downward. There is no easy way to predict this sense for i_2 without actually going through the calculation, but you might have guessed the downward sense of i_3 in advance. Can you see how?

You can also substitute the value of i_2 into the remaining loop equation:

$$-1.6 \text{ A} \times R_1 + 7.5 \text{ V} - 0.4 \text{ A} \times 3.8 \ \Omega = 0.$$

You now solve for the third and final unknown, R_1:

$$R_1 = 3.7 \ \Omega.$$

Symbols Used in Chapter 27

A	area		q	electric charge
e	quantum of charge		R	electrical resistance
$\mathscr{E}, \mathscr{E}$	electric field		t	time
i	electric current		t, T	temperature, absolute temperature
j	current density		v_d	drift speed
l	length		V	potential difference, voltage drop
m	mass		α	temperature coefficient of resistance
n	electron concentration		ρ	electrical resistivity
N	total number of electrons		σ	electrical conductivity
P	power		τ	collision time

Summing Up

A **source of emf** is any device that uses energy from some source (usually not electric) to "pump" charge from a lower to a higher state of potential energy. Among such devices are the van de Graaff generator, the chemical cell, the thermocouple, and the electric generator. When the terminals of

a source of emf are connected to an external network that provides a path for flow of charge—an **electric current**—the system constitutes an **electrical circuit**.

The measure of current is the amount of charge that passes a given point in a circuit per unit time, as expressed by Equation 27.1,

$$i \equiv \frac{dq}{dt}.$$

The SI unit of current is the **ampere**, defined by Equation 27.2, $1 \text{ A} \equiv 1 \text{ C/s}$.

Except in the case of superconductors, current i must be driven through a conductor by a potential difference V sustained between its ends. For many conductors under a wide variety of conditions, i is proportional to V. Such conductors are said to be **ohmic**; they conform to **Ohm's law**, which can be expressed in the form of Equation 27.5b,

$$V = iR.$$

The quantity R is a constant called the **resistance** of the conductor. The SI unit of resistance is the **ohm** (Ω), defined by Equation 27.6, $1 \text{ } \Omega \equiv 1 \text{ V/A}$. The resistance of a particular conductor depends on both its size and its shape, as well as on properties peculiar to the material of which it is made. The two kinds of dependence can be separated by writing the resistance in the form of Equation 27.7b,

$$R = \rho \frac{l}{A}.$$

The quantity ρ is called the **resistivity**. Its value depends on the particular conducting material, on its purity and mechanical history, and on the temperature.

In Equations 27.9 and 27.11, Ohm's law is written in terms of the resistivity ρ or its reciprocal, the **conductivity** σ:

$$\mathcal{E} = j\rho \quad \text{or} \quad j = \sigma\mathcal{E}.$$

Unlike Equation 27.5b, which applies only to a conductor taken as a whole, these equations can be applied locally to any point within a conductor. The quantity $\mathcal{E}$ is the electric field at that point, and j is the **current density** defined by Equation 27.3, $j \equiv di/dA$.

On the microscopic scale, a current passes through a metal when the free electrons of the metal are accelerated by the imposition of an electric field. Electrical resistance arises from the repeated collisions of the electrons with the ions that constitute the crystal lattice. These collisions randomize the motion of the electrons and thus convert the ordered energy of motion of the electron gas into heat energy, the energy of disordered motion of the individual electrons and ions. The steady-state result of the actions of the driving field and of the randomizing collisions is that the electron gas as a whole moves with the **drift speed** given by Equation 27.15,

$$v_{\text{d}} = \frac{e\mathcal{E}\tau}{m},$$

where τ is the **collision time**. The drift speed is connected to the current density by Equation 27.13,

$$j = nev_{\text{d}}.$$

Combining these equations leads to Equation 27.16, $j = (ne^2\tau/m)\mathcal{E}$. This is a form of Ohm's law *if* $\sigma = ne^2\tau/m$ is constant. This is usually the case for metals at a given temperature.

The power required to drive a current through a conductor is given by Equation 27.18, **Joule's law**:

$$P = Vi.$$

If the conductor is ohmic, this result can be written in either of the forms of Equation 27.19,

$$P = \frac{V^2}{R} \quad \text{or} \quad P = i^2R.$$

A **direct-current circuit**, or **dc circuit**, is one in which the current at any point remains constant over time. Any circuit consisting entirely of **steady sources** of emf and **resistors** is a dc circuit.

The resistance R equivalent to that of two resistors R_1 and R_2 in series is given by Equation 27.20a,

$$R = R_1 + R_2.$$

For two resistors in parallel, the equivalent resistance is given by Equation 27.21a,

$$\frac{1}{R} = \frac{1}{R_1} + \frac{1}{R_2}.$$

More complicated dc networks can be analyzed by means of **Kirchhoff's rules**, given in Section 27.7.

KEY TERMS

Section 27.2 Electromotive Force and Its Sources
emf ▪ electric circuit ▪ electrode ▪ electrode potential ▪ positive terminal, negative terminal

Section 27.3 Electric Current
charge carrier ▪ ampere ▪ current density

Section 27.4 Electrical Resistance and Ohm's Law
ammeter ▪ ohm ▪ resistivity, conductivity ▪ temperature coefficient of resistivity ▪ voltage drop

Section 27.5 A Microscopic View of Electrical Resistance
drift speed, drift velocity ▪ electron gas ▪ carrier (electron) concentration ▪ collision time (mean free time) ▪ mean free path ▪ free-electron theory

Sections 27.7 and 27.8 Joule's Law and Direct-Current Circuits
direct-current (dc) source ▪ resistor, equivalent resistor ▪ Kirchhoff's rules

QUERIES

27.1 *(2) Crucial distinction.* Explain *carefully* the difference between the emf of a battery and the potential difference across it. Which of these quantities is the voltage drop?

27.2 *(2) Electron tour.* Figure 27.3b shows the potentials at various locations in a chemical cell. Sketch a similar graph showing the electric potential energy U_e of an electron as it is moved from one part of the cell to another.

27.3 *(2) Honorable discharge?* A charged capacitor can be used (at least briefly) to make current flow through a circuit—a resistor connected between its plates, say. Is the capacitor a source of emf? Explain.

27.4 *(3) Whatever it is, it flows . . .* Compare the definition of electric current, $i \equiv dq/dt$, with Equation 16.2b ($Av = dV/dt$), which defines the volume flux of a moving fluid, and with Equation 21.35 ($\Phi \equiv dE/dt$), which defines the energy flux of a wave. In what sense is each of these quantities literally a flux? What do the three quantities have in common, and in what ways do they differ?

27.5 *(3) . . . downstream.* It is possible to define *current lines*, which describe the paths of flow of electric charge through a conductor, in a manner quite analogous to the way in which streamlines were defined for a flowing fluid in Chapter 16 and somewhat more distantly analogous to the way in which electric flux lines were defined in Chapter 24. Remember that the electric field vector $\mathscr{E}$ at any point is tangent to the electric flux line passing through that point. What vector is similarly tangent to the fluid-flow streamline? What vector is similarly tangent to an electric current line?

27.6 *(4) By dint of effort, show there is more than ensity in common!* In Section 21.5, we defined the intensity $I \equiv dE/dA$ to be the energy flux per unit area carried by a wave front. Compare the current density j with this quantity.

27.7 *(4) The world turned upside down?* The equivalent resistances of two resistors in series and in parallel are given respectively by Equations 27.20a and 27.21a,

$$R = R_1 + R_2 \quad \text{(series)} \quad \text{and} \quad \frac{1}{R} = \frac{1}{R_1} + \frac{1}{R_2} \quad \text{(parallel)}.$$

For two capacitors in series and in parallel, the corresponding relations are given respectively by Equations 26.18a and 26.15a,

$$\frac{1}{C} = \frac{1}{C_1} + \frac{1}{C_2} \quad \text{(series)} \quad \text{and} \quad C = C_1 + C_2 \quad \text{(parallel)}.$$

Explain this apparent reversal.

27.8 *(5) Well trained electrons?* You have seen that the drift speeds of electrons in metallic conductors are quite small—typically, 1 mm/s—even when the conductors carry substantial current densities. But consider an ordinary ceiling light, connected to its wall switch by wires that are several meters long. When you flip the switch, the light goes on so promptly that you cannot detect a delay. How do you reconcile these two observations?

27.9 *(6) High voltage.* Consider a long power transmission line that can carry a maximum current i. What is the advantage of increasing the potential of the line with respect to ground? (In Sweden, where there is abundant hydroelectric power in the sparsely populated north and demand in the urbanized south, dc transmission lines about 500 km long operate at 1 MV.)

27.10 *(6) A series of enlightenments.* Both a 60-W light bulb and a 100-W bulb are intended for use at 110 V. Which bulb has the higher resistance? Which bulb is brighter when the two are connected in parallel? Which is brighter when the two are connected in series?

27.11 *(7) One way or another.* In using an ammeter, you must connect it in series with the conductor through which you wish to measure i. On the other hand, if you wish to use a voltmeter—say, to measure the iR drop across a resistor—you must connect the voltmeter in parallel with the resistor. Draw a sketch of the two arrangements, and explain them.

27.12 *(7) What goes around comes around.* For a circuit having n nodes and m loops, the node rule yields $n - 1$ independent equations and the loop rule yields $m - 1$ independent equations. Why is this so? How many unknown quantities can you determine for such a circuit?

27.13 *(G) Burnout.* You may have noticed that, when an incandescent light bulb burns out, it almost always happens when you turn it on. Explain why. [Hints: (1) The burnout is usually accompanied by a bright flash. (2) The burnout occurs at a thin place in the filament. (3) The thin place was there the last time the bulb was operating, but it did not burn out then.]

PROBLEMS

GROUP A

27.1 *(3) Rush hour.* A wire carries a 1-A current. How many electrons flow past a given cross section through the wire in 1 s?

27.2 *(3) In the meantime.* The potential drop across a 55-μF capacitor is initially zero. The capacitor is charged until the potential drop is 950 V. If the charging takes 1.7 ms, what is the average value $\langle i \rangle$ of the current that flows in the capacitor leads during this time?

27.3 *(3) Current and charge.* The current through a conductor varies with time according to the relation $i = i_0 + bt$, where $i_0 = 4$ A and $b = 3$ A/s. What is the total charge that flows through any cross section of the conductor between $t = 2$ s and $t = 4$ s?

27.4 *(4) The siemens.* Ohm's law can be written $i = SV$, where the *conductance S* is the reciprocal of the resistance R. The SI unit for conductance S is called the *siemens* (S), in honor of the German-British engineer, inventor, and industrialist Karl Wilhelm von (later Sir William) Siemens (1823–1883). **(a)** Define the siemens in terms of more elementary electrical units. **(b)** What is the conductance of a 150-Ω resistor? **(c)** Express the SI unit of conductivity σ in terms of the siemens.

27.5 *(4) The international ohm.* Throughout the nineteenth century, standard resistors could be made that were much more stable than the available sources of emf V. It was still more difficult to maintain a constant current. Consequently, the use of a reliable standard resistor in a role analogous to that of the standard kilogram became attractive. In 1860, William Siemens proposed that the standard of resistance be a column of pure mercury of uniform cross-sectional area 1 mm^2 and length 1 m, held at 0°C. **(a)** What is the resistance of this column in terms of the modern SI ohm, defined as 1 V/A? **(b)** What should be the length of a mercury column of cross-sectional area 1 mm^2 if its resistance is to be 1 Ω? Be sure to use the full precision of the data given in Table 27.1. **(c)** To take full advantage of the precision with which lengths and masses could be measured in the mid-nineteenth century, which was much greater than the precision with which electrical quantities could then be measured, the *international ohm* was defined to be the resistance of a column of pure mercury, held at 0°C, whose length was 1.06300 m, whose cross-sectional area was uniform, and whose mass was 14.4521 g. The density of mercury at 0°C is 13.5955 $\times$ 10^3 kg/m^3. What is the cross-sectional area of the column?

27.6 *(4) Winding a resistor.* You have a spool of 40-gauge Nichrome wire, which you want to use to make a 500-Ω resistor. The diameter of the wire is 0.0799 mm. What is the length you need?

27.7 *(4) Warm-up exercise, I.* The resistance of a copper wire is 20 Ω when the temperature is 12°C. A current is passed through the wire, and it warms up. If the resistance of the wire is now 24.6 Ω, what is its temperature?

27.8 *(4) Warm-up exercise, II.* A resistor, consisting of a coil of iron wire, is connected to a source of emf and an ammeter. When the temperature is 0°C, the ammeter reads 20 mA. When the temperature is 50°C, what is the ammeter reading? Assume that the emf remains constant and that the ammeter has negligible resistance.

27.9 *(4) Hot work.* The tungsten filament of a light bulb is 20 cm long. At operating temperature, 2500°C, the resistance of the filament is 200 Ω. **(a)** What is its room-temperature resistance? **(b)** What is the diameter of the filament? Neglect changes in dimension due to thermal expansion.

27.10 *(4) Gently down the stream.* A metal wire of diameter 3.5 mm carries a 13-A current. Find the current density, assuming it to be uniform.

27.11 *(5) Drifting along.* If the current-carrying wire of Problem 27.10 is made of metal containing 4.5 $\times$ 10^{27} electrons/m^3, find the electron drift speed.

27.12 *(6) Overload.* A portable electric generator is designed to produce 10 A at 110 V. **(a)** When connected to an excessive load—a load whose resistance is too small—the generator produces 12 A, but the output voltage falls to 105 V. What is the effective internal resistance of the generator? **(b)** When connected to a still greater overload, the generator produces 13 A at an output voltage of 98 V. Is the generator ohmic? **(c)** Even for a nonohmic circuit element, it is possible (and sometimes useful) to define the *differential resistance* $r(V) = dV/di$. Find an approximate value of r for the generator under the conditions of part **b**. Compare this value with the result of part **a**.

27.13 *(6) Seeing the light.* Two 110-V light bulbs have resistances 240 Ω and 360 Ω, respectively. Which bulb is brighter? How many times more power does it consume than the other bulb?

27.14 *(6) Reddy kilowatt.* Electrical energy is often measured for practical purposes in the non-SI unit *kilowatt-hour* (kWh). One kWh is the amount of work done in one hour at a constant rate of 1 kW. **(a)** Express 1 kWh in joules. **(b)** If you begin with water whose temperature is 10°C, how much can you turn into steam with 5 kWh of electrical energy?

27.15 *(6) A quadrillion here, a quadrillion there . . .* Another non-SI unit sometimes used in electric power planning is the quad, defined to be 1 quadrillion (10^{15}) British thermal units (Btu), or 2.930 $\times$ 10^{11} kWh. **(a)** Express 1 quad in joules. **(b)** Modern electric generating plants typically have output capacities of order 1000 MW (10^9 W). How many such plants are required to produce 1 quad of energy per year?

27.16 *(6) Teakettle, I.* An electric teakettle has two heating elements. With the "high" element turned on, it takes 10 min to boil the water in the kettle; with the "low" element turned on, it takes 20 min. How long will it take to boil the water with both elements connected **(a)** in parallel? **(b)** in series?

27.17 *(7) A parallel way of expressing matters.* A parallel network consists of two resistors R_1 and R_2. Show that the equivalent resistance is

$$R = \frac{R_1 R_2}{R_1 + R_2}.$$

27.18 *(7) Any way you can.* You have three resistors, each having $R = 57$ Ω. **(a)** How many different equivalent resistances can you obtain, using one, two, or three of the resistors? **(b)** What are the equivalent resistances?

27.19 *(7) Net work, I.* Find the equivalent resistance between A and B of the network shown here.

27.20 *(7) Net work, II.* Find the equivalent resistance between *A* and *B* of the network shown here.

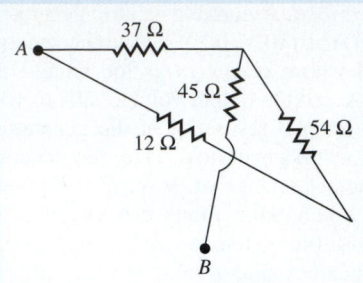

27.21 *(7) A matter of deduction, Dr. Watson!* The illustration below is of a segment of a larger circuit. If the two ammeters indicate the currents shown, what is the value of the resistor *R*?

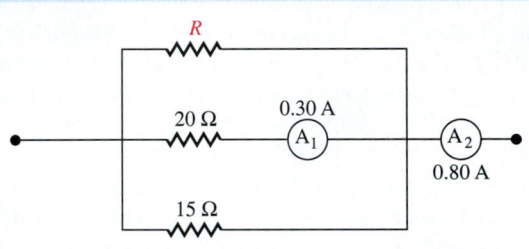

27.22 *(7) Getting it now, Dr. Watson?* Find the value of the resistor *R* in the figure below.

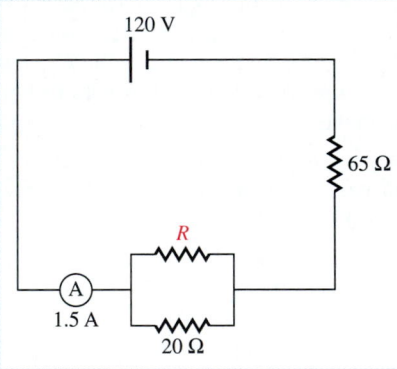

27.23 *(7) Topological games.* Show that the resistance network of figure *a* (below) is equivalent to that of figure *b*, and calculate the equivalent resistance between *A* and *D*.

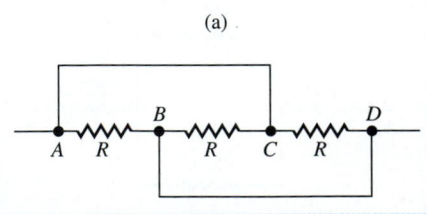

(a)

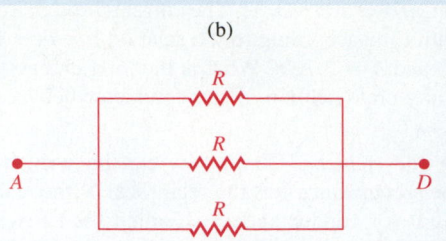

(b)

27.24 *(7) Ammeter into voltmeter.* The figure below shows how to use an ammeter as a *voltmeter*. You connect the ammeter in series with a resistor whose known resistance, R_V, is very much greater than that of the ammeter. If the ammeter reads a current *i*, what is the voltage *V* supplied by the source of emf?

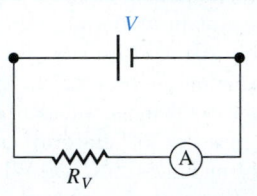

27.25 *(7) Making the adjustment.* In the circuit shown below, the variable source of emf *V* is adjusted until ammeter A_1 reads 0.18 A. Assuming that the resistance of the ammeter is negligible, **(a)** what is the value of *V*? **(b)** what is the reading of ammeter A_2?

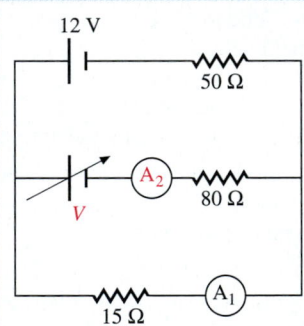

27.26 *(G) Mismatch.* An electric appliance rated for 110 V is plugged into a 120-V outlet. As a result, the output voltage of the outlet drops to 100 V. If the resistance of the wires supplying the outlet is 3 Ω, what is the power dissipated in the appliance?

27.27 *(G) Local motion.* A wire of diameter 2.3 mm carries a 4.2-A current. Connected in parallel with this wire is a second wire of the same material and of equal length but of diameter 5.1 mm. **(a)** What is the current density in the second wire? **(b)** Find the ratio v_d'/v_d of the drift speed in the second wire to that in the first.

27.28 *(G) Making do.* You have two 75-W light bulbs and one 150-W bulb, all rated for 110 V. All you have available is a 220-V line. How can you connect the bulbs so that all three will operate at their rated voltage?

27.29 *(2) Faraday's constant.* In a dilute solution of silver chloride (AgCl) in water, nearly all of the silver chloride *dissociates.* That is, it is present in the form of equal quantities of positive Ag^+ ions and negative Cl^- ions, all of which carry charges of magnitude e. As electric charge is carried by these ions through a cell containing a silver chloride solution, silver plates out on the negative electrode. How much charge must pass through the cell in order to plate out 1 gram-mole (the mass of 6.02×10^{23} atoms) of silver? This quantity of charge is called *Faraday's constant* or the *faraday* (not to be confused with the farad).

27.30 *(4) The advantages of aluminum.* Power-transmission lines are usually built with conductors of aluminum rather than copper, even though the resistivity of aluminum is considerably greater than that of copper (Table 27.1). One reason for doing so is the subject of this problem. The allowable mass of the conductors is determined by the load-bearing capacity of the transmission towers (which typically cost quite a lot more than the conductors themselves). **(a)** Consider two conductors of the same length, made of any two conducting substances 1 and 2. If the masses of the conductors are to be equal, express the resistance ratio R_1/R_2 in terms of the corresponding resistivities ρ_1 and ρ_2 and the densities δ_1 and δ_2. **(b)** Show that the resistance of an aluminum conductor is less than one-half that of a copper conductor of the same mass. Take the density of copper to be 8.92×10^3 kg/m³ and that of aluminum to be 2.70×10^3 kg/m³.

27.31 *(4) Massive resistance.* Equal masses of iron and aluminum are made into uniform wires having equal resistance. **(a)** Which wire is longer? **(b)** What is the ratio of the lengths of the wires? Take the densities to be $\delta_{Fe} = 7.8 \times 10^3$ kg/m³ and $\delta_{Al} = 2.7 \times 10^3$ kg/m³. (Hint: Don't use the same symbol for density and resistivity.)

27.32 *(4) Justification by works.* In calculating the resistance change of a wire with temperature, the common practice is to consider only the change due to the temperature coefficient of resistivity, α, and to neglect the change in dimension of the wire due to thermal expansion. **(a)** Without making any approximations, write an expression for $\Delta R/R_0$, the relative change of resistance of a uniform wire when the temperature of the wire is changed by an amount Δt. Assume that the linear approximations of Equations 27.8 and 17.20 are valid. To avoid confusion, use the symbol λ for the coefficient of linear expansion, and express your result in terms of α and λ. **(b)** Because your result is valid only to linear approximation, you will lose nothing in accuracy if you simplify the result of part **a** by discarding all terms in powers of Δt higher than the first. Write $\Delta R/R_0$ in this approximation. **(c)** Using Tables 27.1 and 17.2, show that, for nearly all metals, the error introduced by neglecting thermal expansion is only about $\frac{1}{2}$%.

27.33 *(5) Mobility.* The *mobility* of a charge carrier (for example, a free electron in a metal) is defined to be the drift speed per unit applied electric field:

$$\mu \equiv \frac{v_d}{\mathcal{E}}.$$

(a) What is the SI unit of mobility? **(b)** Express the mobility of a free electron in a metal in terms of its charge and mass and the collision time. **(c)** Express the conductivity σ in terms of μ. **(d)** Write Ohm's law in terms of the mobility. **(e)** In ionic solutions, charge is usually carried by ions of both signs, whose mobilities are, respectively, μ_+ and μ_-. Write Ohm's law for a dilute solution of silver chloride, containing N kmol/m³ each of Ag^+ and Cl^- ions.

27.34 *(5) Temperature dependence.* **(a)** Show that the free-electron theory yields the value

$$\lambda = \frac{\sigma}{ne^2} \sqrt{3kTm}$$

for the mean free path of electrons in a metal. **(b)** Show that this result, taken together with the assumptions underlying the model, yields the temperature dependence $\rho \propto \sqrt{T}$. Be explicit about the necessary assumptions.

27.35 *(6) Teakettle, II.* You are asked to design an electric teakettle that can heat 3 liters of water to boiling in 8 min. **(a)** Assuming that the initial water temperature is 10°C, what must the power of the heating coil be? **(b)** You plan to use Nichrome wire of diameter 0.42 mm to make the coil. What length of wire do you need? Use the resistivity of Nichrome at 55°C.

27.36 *(7) Cut & paste.* You have an uninsulated wire of length L and resistance R. If you cut the wire into n equal pieces and make the pieces into a bundle, what is the end-to-end resistance of the bundle?

27.37 *(7) Ring around the rosy.* A wire of resistance R is joined end to end and formed into a circular loop. Terminals A and B are located on the loop, as shown below. The resistance between the terminals is r. Express r/R as a function of the angle θ between the terminals. (Hint: The loop radius need not appear in your calculation.)

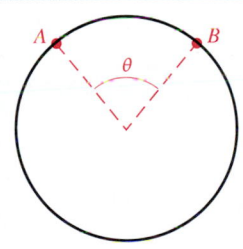

27.38 *(7) Ammeter.* In the figure below, an ammeter is connected so as to measure the current passing through a simple

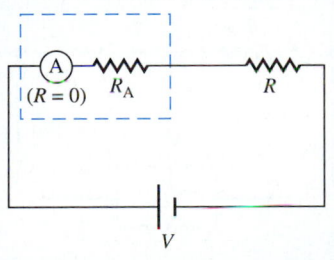

circuit. Ideally, an ammeter has zero resistance. But any real ammeter must have some internal resistance, as represented by the resistor R_A in series with an ideal ammeter. **(a)** With the ammeter removed from the circuit, the current has some (unknown) value i. When the ammeter is inserted into the circuit, what is the value i' of the current read by the meter? **(b)** Express, in terms of R and R_A, the error $(i - i')/i$ introduced into the circuit by the internal resistance R_A.

27.39 *(7) Ammeter and voltmeter.* The figure below illustrates two ways of using an ammeter and a voltmeter to measure the resistance R of an unknown resistor. If both meters are ideal—that is, if the resistance of the ammeter is zero and the resistance of the voltmeter is infinite—the two circuits give the same result. The resistance is simply the voltmeter reading divided by the ammeter reading: $R = V/i$. But a real ammeter has nonzero resistance R_A, and a real voltmeter has finite resistance R_V. **(a)** Suppose that you set up the circuit of figure a and obtain meter readings V_1 and i_1. Express the resistance R in terms of these readings and R_V. **(b)** Now you set up the circuit of figure b and obtain meter readings V_2 and i_2. Express R in terms of these readings and R_A. **(c)** Show that, if you know R_V, you can find R_A by means of the relation

$$R_A = \frac{V_2}{i_2} - \frac{V_1 R_V}{i_1 R_V - V_1}.$$

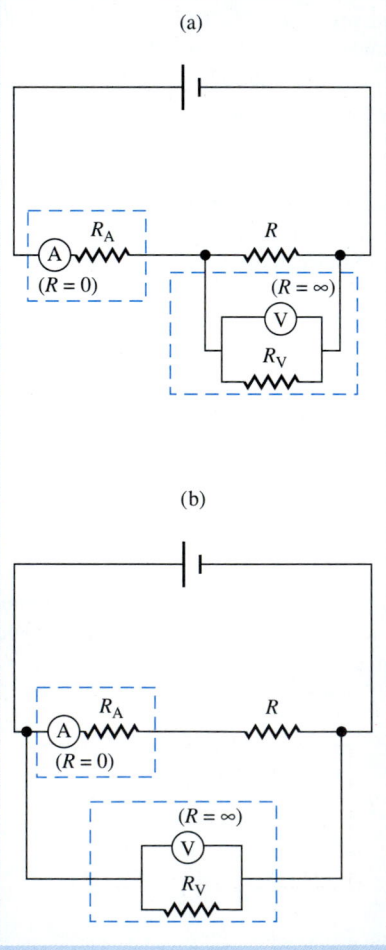

(a)

(b)

27.40 *(7) Figure of merit.* The needle of a galvanometer (an instrument like an ammeter that can be used to detect small currents) reads at full scale when the current passing through it is i_m. The *figure of merit* M of the galvanometer is defined as the reciprocal of the full-scale current:

$$M \equiv \frac{1}{i_m}.$$

The figure of merit is usually expressed in ohms per volt. **(a)** Suppose you have a galvanometer for which $i_m = 1 \times 10^{-3}$ A, so $M = 1000$ Ω/V. If the resistance of the galvanometer is 50 Ω, what is the maximum voltage it can measure? **(b)** You want to use the galvanometer as a voltmeter to measure voltages up to 5 V. What resistance must you place in series with it? **(c)** What is the value of M when the meter is set up to measure 5 V full scale? **(d)** Consider the circuit shown below. The potential difference between A and B is evidently 5 V. You connect the voltmeter of parts **b** and **c** between A and B. What will the meter reading be? What is the disadvantage of using this voltmeter in this circuit? **(e)** You have another galvanometer whose figure of merit is 10^4 Ω/V. Its internal resistance, like that of the first galvanometer, is 50 Ω. Repeat the calculations of parts **a** through **d** for this galvanometer. What is the advantage in a larger figure of merit?

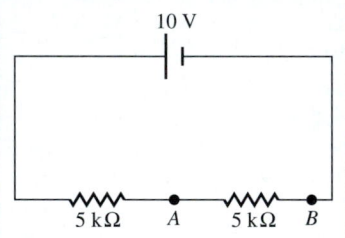

27.41 *(7) Making an ammeter.* You wish to convert the galvanometer of Problem 27.40 into an ammeter whose full-scale reading is 1A. **(a)** Using the galvanometer and a resistor R, called a *shunt*, sketch a circuit diagram that will accomplish this task. **(b)** What is the required value of the shunt resistance R?

27.42 *(7) Ohmmeter.* In the figure below, the galvanometer of Problems 27.40 and 27.41 is connected in series with a 1-V source of emf and a 950-Ω resistor. **(a)** If you place a conductor of negligible resistance between A and B, what will be the reading of the galvanometer needle? (Express as a fraction of full scale.) **(b)** For what resistance R_1 connected between A and B will the galvanometer read at half of full scale? **(c)** If you connect a 3000-Ω resistance between A and B, what will

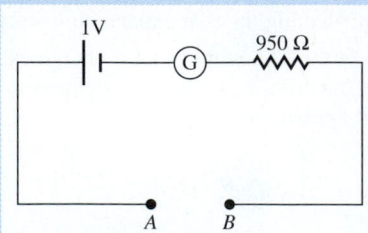

be the galvanometer reading? **(d)** Draw a diagram of the galvanometer face, calibrated to read as an ohmmeter. That is, calibrate the dial to read the resistance between A and B directly in ohms. Plot at least six values of resistance.

27.43 *(7) The standard cell dilemma.* Every laboratory has one or more standard cells. The emf of a standard cell is quite stable over time, which makes the cell useful for calibration purposes. Unfortunately, standard cells have nonnegligible internal resistance R_s. Suppose that you try to measure the emf V of a standard cell with a voltmeter whose internal resistance is R_V (Problem 27.39). **(a)** Show that your reading V' will be in error, and express the error ratio $(V - V')/V$ in terms of R_s and R_V. **(b)** You could reduce this error by substantially increasing the value of the voltmeter resistor R_V. Assuming that everything else remains the same, what difficulty do you run into then?

27.44 *(7) The potentiometer.* The figure below shows a device, called a *slide-wire potentiometer*, that can be used to measure an unknown emf, V_x, by comparing it with the emf V_s of a standard cell. This is done without drawing any current

at all from either cell, thus circumventing the difficulty that is the subject of Problem 27.43. Besides the standard cell and the unknown emf, there is a *working cell.* Although its emf V_w is not as stable as that of the standard cell over long periods, the working cell has a quite small internal resistance. (Automobile-type storage batteries are often used as working cells.) The slide wire is a long, highly uniform wire, of length L, whose resistance per unit length does not vary from point to point over the wire. A galvanometer (Problem 27.40) completes the system. **(a)** With the standard cell in the circuit, as in figure a, the sliding contact represented by the arrow is moved back and forth along the slide wire until the galvanometer reads zero. Express the iR drop across the entire slide wire in terms of V_s, L, and the distance d_s of the sliding contact from the left end of the wire. **(b)** The standard cell is now replaced by the source of unknown emf, as in figure b. The slide wire is again adjusted until the galvanometer reads zero. Express V_x in terms of V_s, d_s, and d_x. **(c)** A simplified version of the potentiometer is widely used in electronic practice to supply a controllable voltage to a circuit component. The arrangement is shown in figure c. For proper operation, what restriction must you place on the resistances r and R?

27.45 *(7) Wheatstone bridge, I.* Although the device shown below was invented by Samuel Christie, it is named after the British physicist Sir Charles Wheatstone (1802–1876), who used it extensively in investigations of Ohm's law in 1843. The bridge can be used to make quick, fairly accurate measurements of the resistance of an unknown resistor, and various modifications of the bridge have wide application in electronic circuits. A and B are known fixed resistors, and C is a variable resistor whose resistance can be read with accuracy at any setting. (The arrow through C signifies a variable device.) The galvanometer is an instrument whose indicator responds to small currents, but the scale need not be graduated to give quantitative measurements of nonzero current values. In use, C is adjusted until the galvanometer detects no current. **(a)** With $i_G = 0$, what is the potential difference between c and d? **(b)** Use the result of part **a**, together with Kirchhoff's rules, to show that $A/B = C/X$ and that the resistance of the unknown resistor is therefore

$$X = C\frac{B}{A}.$$

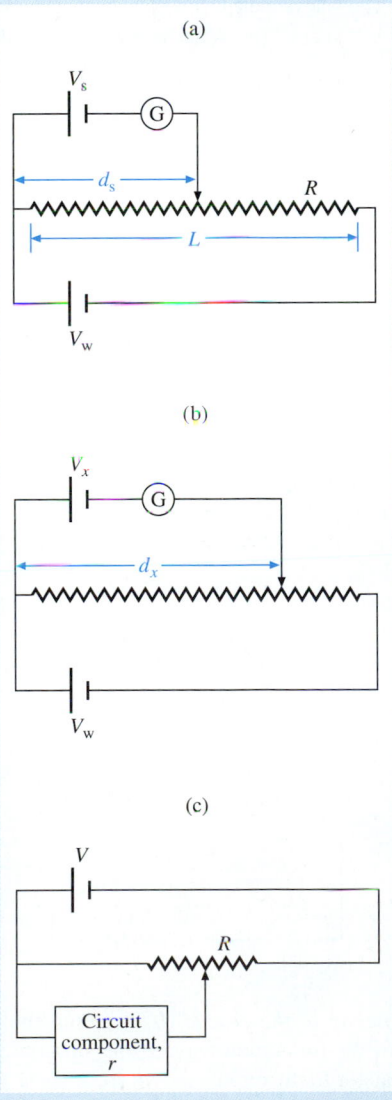

(a)

(b)

(c)

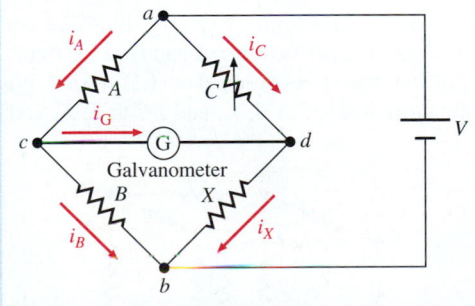

(c) What are the advantages of the Wheatstone bridge? Specifically, what quantities pertinent to complete analysis of the circuit do not affect the measured value of X?

27.46 *(7) Resistance independent of length.* **(a)** When a certain resistance R_0 is connected between C and D in figure a (below), the resistance measured between A and B also is R_0. Find the value of R_0 required to satisfy this condition. **(b)** The circuit of figure b is made up by stringing together two of the units shown in figure a. The system is terminated with the resistance R_0 found in part **a**. What is the equivalent resistance between A and B? **(c)** If n units are strung together and the system again terminated with a resistance R_0, what is the resistance between A and B? (This problem has an important application to coaxial cables used in such high-frequency applications as FM and television. You may have seen such cables specified, for example, as ''8-ohm cables.'')

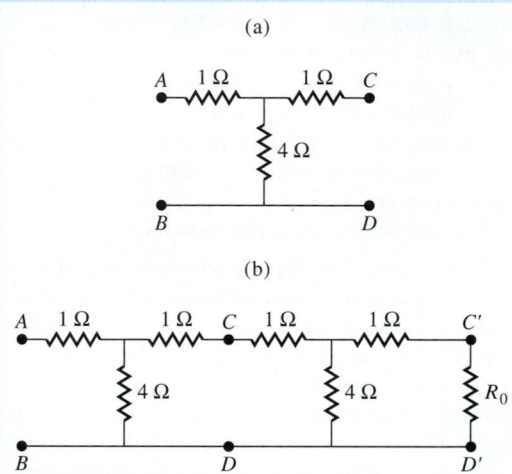

(a)

(b)

27.47 *(7) Infinite network.* The figure below shows part of an infinite string of resistors. What is the equivalent resistance between A and B? (Hint: If you cut off everything to the left of A' and B', what would be the equivalent resistance between A' and B'?)

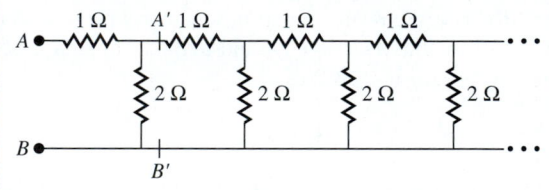

27.48 *(7) Framing a fearful symmetry.* Find the equivalent resistance of the network between A and H in the figure below. For each of the twelve resistors, $R = 1\ \Omega$. (Hint: What happens to the current at A? at B, C, and D? at E, F, and G?)

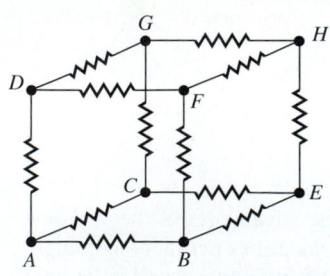

27.49 *(7) Real battery.* In the circuit shown below, the point C is grounded; this is represented by the standard symbol. The battery has emf 12 V and internal resistance 40 Ω. **(a)** Why does no current flow from C to the ground? **(b)** What is the current i through the circuit? **(c)** What is the voltage between the battery terminals A and B? **(d)** What is the potential of B with respect to the ground?

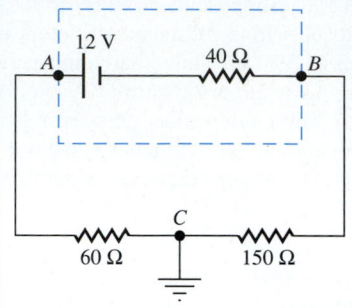

27.50 *(7) Powerful method.* **(a)** Determine the currents at A, B, and C in the figure below. **(b)** How much power is dissipated in each resistor? **(c)** What is the power output of each cell?

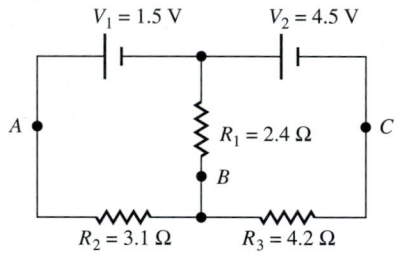

27.51 *(7) Gustav to the rescue, I.* **(a)** Find the magnitudes and senses of the currents in the three branches of the circuit shown below. **(b)** What is the power dissipated in each resistor? **(c)** What is the power output of each cell?

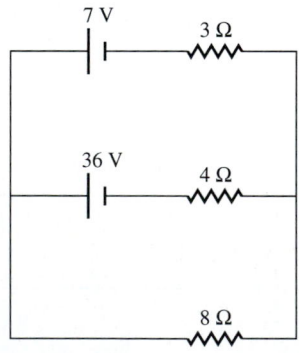

27.52 *(7) Gustav to the rescue, II.* **(a)** Find the magnitudes and senses of the two unknown currents and the value of the unknown resistor in the circuit shown (p. 755). **(b)** What is the

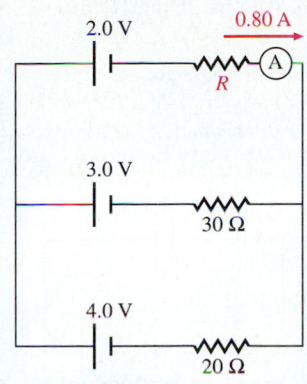

power dissipated in each resistor? **(c)** What is the power output of each cell?

27.53 *(7) Gustav to the rescue, III.* What is the charge q on the capacitor in the circuit shown below when current has been passing through the system for a long time?

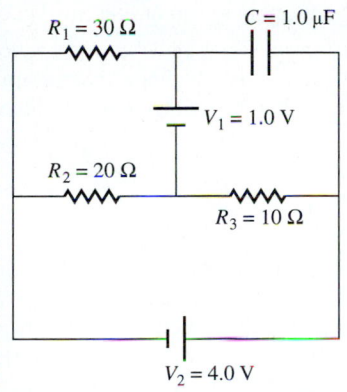

27.54 *(G) Semiconductors.* The electrical properties of semiconductors are not well described by the free-electron theory. In semiconductors that do not contain too much impurity, the number of free carriers depends on temperature T according to the approximate relation

$$n \propto e^{-E_g/2kT},$$

where T is the temperature in kelvin, k is Boltzmann's constant, and E_g is an energy, characteristic of the semiconductor, called the *band-gap energy*. **(a)** Show that the resistivity of a semiconductor varies *exponentially* with the temperature. **(b)** The resistivity of pure silicon at 0°C is 50 times the resistivity at 50°C. Use this information to estimate the band gap energy of silicon. Express your result in electron-volts.

27.55 *(G) The virtues of high voltage.* A 250-W spotlight is designed to operate at 110 V. It is connected to a 110-V wall outlet 100 m away by a pair of 14-gauge copper wires (diameter 0.1628 cm). What is the resistance **(a)** of the spotlight? **(b)** of the wires? **(c)** Find the current passing through the system. **(d)** What is the voltage across the spotlight? **(e)** How much electric power is consumed by the system? **(f)** What fraction of the total power is available to the spotlight proper?

Garden lighting systems are often designed to operate at 12 V, for safety. Suppose that the wall outlet is replaced with a 12-V power supply and the spotlight is replaced with one of the same wattage, designed to operate at 12 V. **(g)** What is the resistance of the 12-V spotlight? **(h)** If the wires are not changed, find the current passing through the system. **(i)** How much electric power is consumed by the system? **(j)** What fraction of the total power is available to the spotlight proper? (Assume, somewhat unrealistically, that the resistance of the spotlight is that obtained in part **g**.) **(k)** Is the low-voltage system practical without modification? **(l)** Suppose that you want to replace the wiring, using wires of large enough diameter so that the fraction of the total power available to the spotlight is the same as that found in part **f**. What would the diameter of the wires be? Would such wires be practical?

27.56 *(G) Subway lighting.* The older parts of the New York subway system operate on 660 V dc. Before the cars were converted to fluorescent lighting, strings of six 110-V bulbs were connected together in series to avoid the risks of using high-voltage bulbs. But if ordinary bulbs were used, the burning out of any one of the six would result in the loss of lighting from all six. To prevent such loss of lighting, special bulbs were used, as illustrated below. In the base of each bulb is a pair of spring-loaded contacts separated by a thin piece of paper. **(a)** Explain what happens when the filament burns out. **(b)** What is the voltage across each of the remaining bulbs? **(c)** Assume (somewhat unrealistically) that the resistance of the remaining filaments does not change. By what percentage does the power consumption of the string of bulbs change when one of them burns out? Is this an increase or a decrease? Is the subway car more or less brightly lit after the bulb burns out?

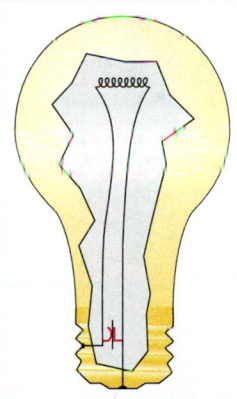

27.57 *(G) Matching resistances, I.* A circuit consisting of a source of emf V that supplies power to a load (say, an electric heater) is illustrated below. The internal resistance of the

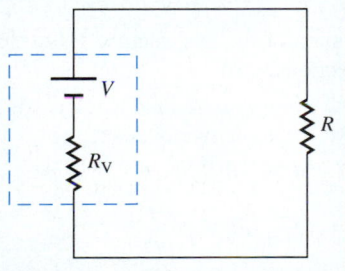

source of emf is R_V, and the resistance of the load is R. **(a)** Express the power P delivered to the load in terms of V, R_V, and R. **(b)** Show that P is maximized if $R = R_V$. This is a simple example of the important principle of *impedance matching*, to be discussed in more detail in Problem 33.40.

27.58 *(G) Freedom of choice.* You have 40 cells, each of which has emf 1.6 V and internal resistance 0.30 Ω. You want to connect them into a battery that will supply maximum current to a load whose resistance is 3.0 Ω. You can connect the cells together in series, in parallel, or in any combination of series and parallel networks. How should you connect them?

27.59 *(G) Heat resistant.* You have two conductors, made of different materials, whose resistances at 20°C are R_1* and $R_2* = kR_1*$ respectively, where k is a constant that you can measure. The temperature coefficients of resistivity of the two conductors are α_1 and α_2. **(a)** Show that, when the two conductors are connected in series, the temperature coefficient α of the entire system is

$$\alpha = \frac{\alpha_1 + k\alpha_2}{1 + k}.$$

(b) You can make a cylindrical conductor whose resistance is independent of temperature by stacking alternate layers of Nichrome and carbon, as shown below. What is the necessary

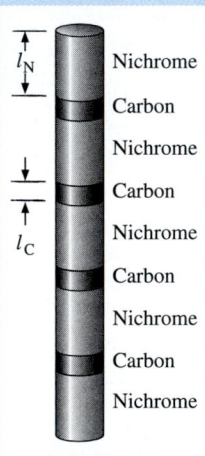

ratio l_N/l_C of the lengths of the Nichrome and carbon segments?

27.60 *(G) Battery charger.* In the figure below, a battery charger is used to charge a 12-V automobile battery. **(a)** If the internal resistance of the battery is 40 mΩ and the charging

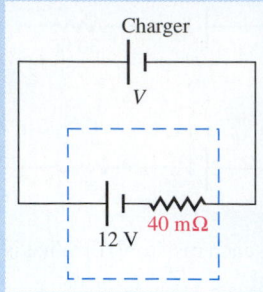

current is 30 A, what is the emf of the charger? Neglect the internal resistance of the charger. **(b)** What is the efficiency of the charging process? That is, what fraction of the energy supplied to the battery is stored as electrical energy?

27.61 *(G) Jump start.* Two batteries are connected positive terminal to positive terminal and negative terminal to negative terminal. **(a)** Show that the voltage measured between the terminals is

$$V = \frac{V_1 R_2 + V_2 R_1}{R_1 + R_2},$$

where V_1 and R_1 are the emf and internal resistance of battery 1 and V_2 and R_2 the emf and internal resistance of battery 2. **(b)** What is the power dissipated in the system?

GROUP C

27.62 *(4) Current density as a vector.* Equation 27.3, $j = di/dA$, is valid only for a cut of area dA that is oriented normal to the length of a uniform wire. But show that the equation can be generalized by defining a current density vector **j** and an area element vector $d\mathbf{A}$. Specifically, show that

$$di = \mathbf{j} \cdot d\mathbf{A}$$

for any orientation of the two vectors. (Hint: Refer to Query 27.5 and to Section 24.3.)

27.63 *(4) Not a dunce cap.* A solid metal casting has the form of a truncated cone, as shown opposite. Show that the resistance of the casting between the two faces is

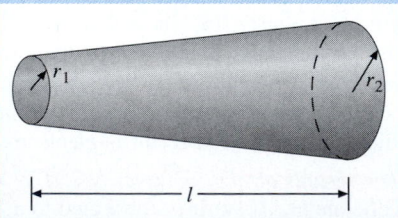

$$R = \rho \frac{l}{\pi r_1 r_2},$$

where ρ is the resistivity of the metal.

27.64 *(7) Voltmeter error.* In the figure (p. 757), a source of emf V drives a current through two equal resistors R arranged in series. **(a)** With the voltmeter not connected, what is the voltage drop V_1 across either resistor? **(b)** The voltmeter, having internal resistance R_V (Problem 27.39), is now connected across one of the resistors, as shown. Call the voltmeter reading V_1', and show that the presence of the voltmeter in the

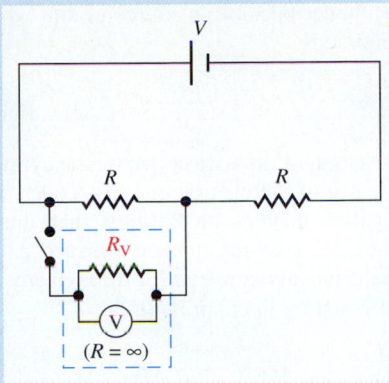

circuit introduces a *voltmeter error*:

$$\left| \frac{V_1 - V_1'}{V_1} \right| = \frac{R}{2R_V + R}.$$

27.65 *(7) Difference of opinion.* Each of the two identical voltmeters in the figure below has internal resistance R_V (Problem 27.39). The readings on the voltmeters are V and $V - \Delta V$, where ΔV is positive. **(a)** Which voltmeter reads V? **(b)** Show that

$$R_V = \left(\frac{V}{\Delta V} - 2 \right) R.$$

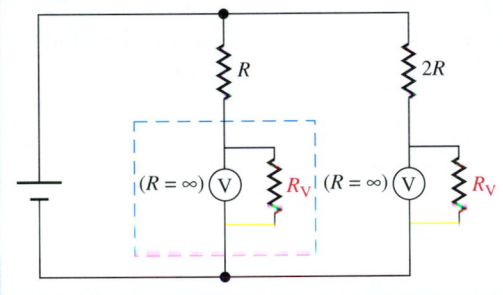

27.66 *(7) Square with diagonals.* The network illustrated below is constructed so that the resistance of every branch is R, even though the branches are not all of the same length. **(a)** Find the equivalent resistance between A and C. (Hint: Which branches can be removed without affecting the resistance of the network?) **(b)** Show that the equivalent resistance

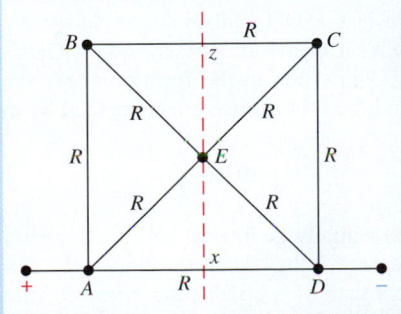

between A and D is $\frac{8}{15}R$. (Hint: Consider the line of symmetry denoted by the dashed red line.) What can you say about points x, E, and z? Draw an equivalent network.

27.67 *(7) A property of insect screens.* The figure below shows part of an infinite square grid of wires. The resistance of any segment of wire between two adjacent nodes, such as A and B, is R. By carrying out the steps of this problem, show that the resistance of the grid between A and B is $R/2$. **(a)** Imagine that the (very distant) edges of the grid are connected to a conducting frame of negligible resistance. If A is connected to the positive terminal of a source of emf and the frame is connected to the negative terminal, there is a total current i_∞. What are the currents through the branches AB, AF, AG, and AH? **(b)** Now connect the negative terminal of the source of emf to B, and connect the positive terminal to the frame. What is the total current? What are the currents through AB, CB, DB, and EB? **(c)** Show that the pattern of current in the grid of the figure, with connections at A and B, is a *superposition* of the patterns of parts **a** and **b**. In particular, how much current passes into or out of the frame? **(d)** In terms of i_∞, what are the superposed currents in AB, AF, AG, and AH? What is the total current leaving A? **(e)** With the source of emf connected to A and B, what is the actual current i_{AB} in AB in terms of V and R? What is the relation between i_∞ and i_{AB}? **(f)** What is the relation between the total current i leaving A and i_{AB}? What is the effective resistance of the network?

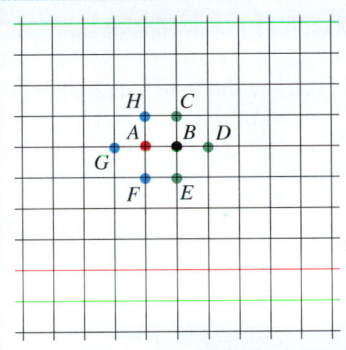

27.68 *(7) Wheatstone bridge, II.* You should solve Problem 27.45 before working this problem. In the Wheatstone bridge illustrated below, the galvanometer has internal resistance r (Problem 27.40). **(a)** Show that, if $A = B$, the current through the galvanometer has magnitude

$$i_G = V \frac{C - X}{(A + 2r)(C + X) + 2CX}.$$

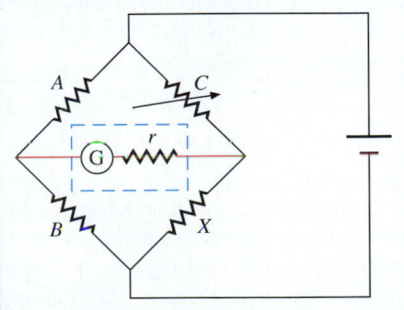

(b) Assume that the deflection of the galvanometer needle is directly proportional to i and that r is negligible compared with the other resistances in the system. Define $\delta \equiv C - X$. Show that, when δ is small (that is, the bridge is nearly balanced), the galvanometer reading is proportional to δ. Show also that the Wheatstone bridge is most sensitive for small values of X.

27.69 *(7) Matching resistances, II.* You have a source of emf consisting of n storage batteries, each of emf v and internal resistance r. You wish to use them to drive a load whose resistance is R. By connecting the batteries in series, in parallel, or in any series-parallel combination, you can vary both the output voltage and the effective internal resistance of the source of emf. Show that you will obtain maximum current through the load if the effective internal resistance is equal to the load resistance R. Is this the condition for maximum output power as well?

27.70 *(G) The Tolman-Stewart experiment.* In 1917, the American physicists Tolman and Stewart performed an experiment that gives direct support to the view that (1) the free carriers in metals are electrons and (2) these electrons really do act something like the molecules of a gas. The apparatus is shown below. An insulated metal wire is wound into a coil on a strong cylinder of radius r that can be rotated at great angular speed. A brake (not shown) can be used to bring the rotating cylinder to a stop very quickly—that is, with great (though not constant) angular acceleration. Sliding contacts make electrical contact to the ends of the wire. The contacts are connected to the terminals of a *ballistic galvanometer*, a device that can measure the total charge that flows through it during a short interval of time.

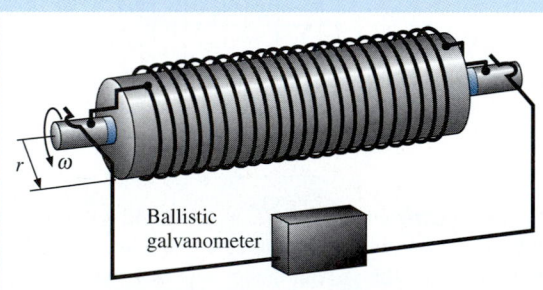

Ballistic galvanometer

(a) The cylinder, whose initial angular speed is ω, is braked to a stop. At some instant in the braking, the angular acceleration is α. Show that, if the carriers are charged "gas molecules" that are free to move within the wire, the result of the acceleration will be an electric field whose direction is along the wire and whose magnitude is

$$\mathscr{E} = \frac{m\alpha r}{q},$$

where m is the mass of a charge carrier and q is the magnitude of its charge. (It is safe to assume that the charge carriers move so quickly that a steady state is reached at every instant, even though the acceleration changes rapidly on the macroscopic scale.) **(b)** Show that on account of this internal field, the

decelerating cylinder becomes a source of emf whose instantaneous magnitude is

$$V = \frac{m\alpha rL}{q},$$

where L is the length of the wire coiled on the cylinder. **(c)** Let the total resistance of the electric circuit—coiled wire, leads, and ballistic galvanometer—be R. Show that, during the infinitesimal time interval dt, this emf drives a charge dQ through the ballistic galvanometer (or through any other point in the circuit) given by the expression

$$dQ = \frac{mrL}{qR}\,\alpha\,dt.$$

(d) Show that the total charge Q that passes through the ballistic galvanometer during the braking is

$$Q = \frac{mrL}{qR}\,\omega$$

and thus that the charge-to-mass ratio of the charge carriers—a microscopic quantity—is expressed in terms of the experimentally measured quantities by means of the equation

$$\frac{q}{m} = \frac{rL\omega}{RQ}.$$

(e) How can you determine the sign of the charge on the free carriers? The values of q/m thus obtained can be compared with the charge-to-mass ratio for free electrons in vacuum (determined by J. J. Thomson in 1897; see Chapter 28). For wires made of several metals, the results are as follows:

Metal	q/m (C/kg)
Copper	-1.60×10^{11}
Silver	-1.49×10^{11}
Aluminum	-1.54×10^{11}
Electrons in vacuum	-1.76×10^{11}

The values for all three metals are somewhat smaller than that for electrons in vacuum. We know today, on the basis of quantum-mechanical considerations, that electrons in metals behave as if they have an *effective mass* m^*, which is not the same as that of a free electron in vacuum.

27.71 *(G) Geometric connection.* Two electrodes of arbitrary shape are immersed in a fluid medium of uniform conductivity σ. The resistance between the electrodes is R. The conducting medium is then removed, and the capacitance of the two-electrode system is C. **(a)** In imagination, surround one of the electrodes with a Gaussian surface. Using Gauss's law, together with Ohm's law in the form $j = \sigma\mathscr{E}$, show that the resistance and the capacitance are connected by the relation

$$\sigma R = \frac{\epsilon_0}{C}.$$

To see the essentially geometric nature of this relation, note that both sides have the dimension [length]$^{-1}$. **(b)** A submarine cable consists of a conducting core of radius r_1 and a coaxial conducting sheath of inner radius r_2, separated by an insulator of (small) conductivity σ. (The entire cable is covered on the outside with an insulating layer, which you should neglect in this problem.) Find the leakage resistance per unit length of the cable, due to the nonzero conductivity of the insulator. **(c)** The resistivities of practical insulating materials range from 10^8 to 10^{13} $\Omega \cdot m$. For a cable 1000 km long, is insulation leakage likely to be a problem?

Magnetism I
The Magnetic Force

Left: A bar magnet is a magnetic dipole; the poles are the regions where most of the magnetic flux leaves and reenters the magnet. Iron filings magnetized by the field align themselves with flux lines, making the field pattern visible.

Magnet, n. Something acted on by magnetism.
Magnetism, n. Something acting upon a magnet.

—AMBROSE BIERCE, *The Devil's Dictionary*

Introduction

The fascinating attractive properties of magnets have been known since ancient times. Indeed, the word *magnet* comes from the ancient Greek place name Magnesia (the modern town Manisa, near Izmir in western Turkey), where the natural magnets called *lodestones* were found.

Natural and artificial *permanent magnets*, such as the compass needle, remain as useful today as they have been for centuries. However, the most fundamental aspects of magnetism are not best understood in terms of those familiar objects. Nevertheless, we begin our study of magnetism with a qualitative survey of the properties of magnets. These properties are the subject of Section 28.2.

Using the properties of magnets as a starting point, we develop the concept of *magnetic field $\mathcal{B}$* in Section 28.3. We then consider the *magnetic force* experienced by an electric charge when it moves through a magnetic field. In this force, we see for the first time the connection between electricity and magnetism. The connection turns out to be far more profound than the relatively superficial similarities between magnets and charged dielectrics would suggest. The remainder of the chapter explores some important consequences of the magnetic force.

A lodestone. This naturally magnetic rock contains a large proportion of magnetic iron oxides.

Magnets and Magnetic Poles

Figure 28.1 shows a **bar magnet**—a permanent magnet in the shape of a long, thin bar. Such magnets can be made from naturally occurring lodestones by cutting the stones into the desired shape. But much better ones are made today from carefully engineered alloys of iron, nickel, and other metals.

Any bar magnet suspended from a thread soon assumes a roughly northward-southward orientation. The end that points northward is called the *north-seeking pole* or, more commonly, the **north pole**. The southward-pointing end is called the *south-seeking* or **south pole**. This tendency of bar magnets to orient themselves along a roughly north-south line underlies the operation of all *magnetic compasses* (Figure 28.2).

Simple experimentation with magnets leads to the following general observations, which closely resemble the corresponding properties of electric charges:

1. When two magnets are brought close together, there is an *attractive* force between the north pole of one and the south pole of the other. We say that **unlike magnetic poles attract one another** (Figure 28.3a).

2. When two magnets are brought close together, there is a *repulsive* force between the two north poles and also between the two south poles. We say that **like magnetic poles**

Geographic south (approximate)

Geographic north (approximate)

FIGURE 28.1 A bar magnet suspended from a thread, showing the north and south poles.

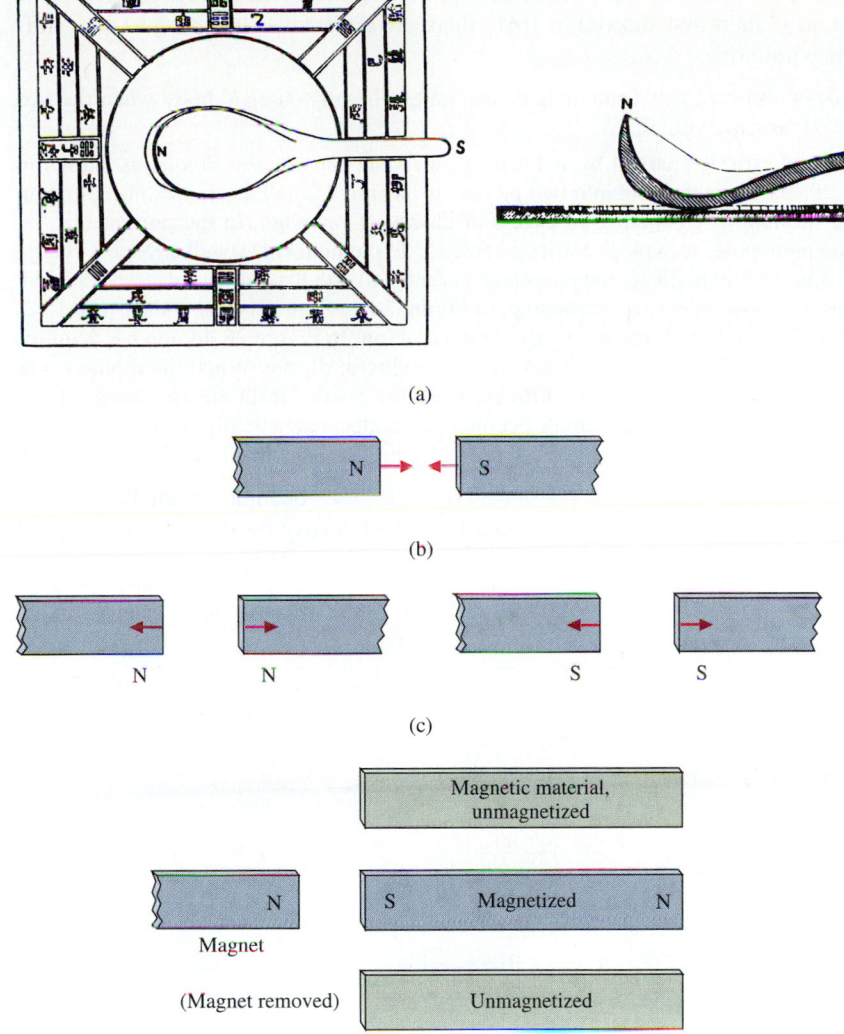

(a)

(b)

(c)

repel one another (Figure 28.3b). Magnetic repulsion was first demonstrated in 1269 by the French philosopher Pierre de Maricourt, who wrote under the name Petrus Peregrinus (Peter the Pilgrim).

3. Most objects are not obviously affected by magnets. But there is a class of materials, called **ferromagnetic materials** or **magnetic materials** for short, on which a magnet exerts an *attractive* force. The most familiar ferromagnetic materials are iron, nickel, and many of their alloys, but there are others as well.

3a. Any bar of magnetic material brought close to a permanent magnet acts like a magnet itself, with poles as shown in Figure 28.3c. The bar is then said to be **magnetized**, and it exhibits **induced magnetization**. For most magnetic materials, such as soft iron, the induced magnetization vanishes when the permanent magnet is removed. (Such materials are sometimes called ''magnetically soft.'') As Figure 28.3c shows, this behavior calls to mind the electrostatic process of charging by induction (see Query 23.6).

3b. Permanent magnets are made of a special subclass of ferromagnetic materials that are said to possess **magnetic remanence**. When a bar of such a material is magnetized by induction under proper conditions, it retains its magnetic properties even when the inducing magnet is removed. (Such materials are sometimes called ''magnetically hard.'') One way to make a permanent magnet is to heat a bar of magnetically hard material to redness and then let it cool in the presence of a strong magnet. Analogous to permanent magnets are devices called *electrets*, which can be given a permanent electric dipole moment by heating and then cooling them in a strong electric field (see Problem 26.26).

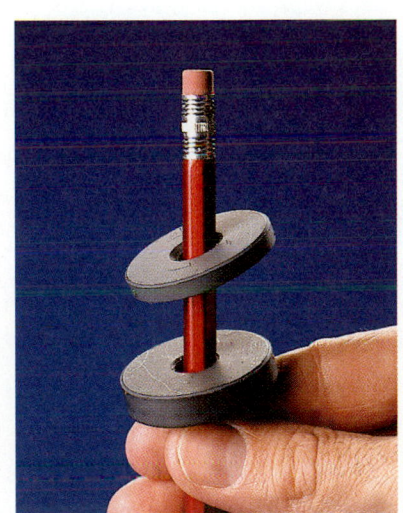

Like poles repel one another. In these magnets, the poles are located on the flat faces.

On the basis of the analogous behavior of magnet poles and electrically charged bodies, it is easy to jump to the conclusion that the magnetic and electric forces are the same. But, as Gilbert first showed in 1600, there are differences that are just as significant as the similarities:

4. No force is observed between a magnet and an electrically charged body when the two are placed near each other.

5. When a dielectric is charged by induction, it becomes an *electric dipole*, as shown in Figure 28.4a. If it is then cut into two pieces, as in Figure 28.4b, each piece becomes an **electric monopole**, containing an excess of charge of one sign. In the presence of the inducing monopole, each piece carries an electric dipole moment as well as excess charge of one sign. In Figure 28.4c, the monopole is removed and the two cut pieces separated. Each piece is now an electric monopole. In Figure 28.4d–f, the corresponding process is carried out with a bar of iron magnetized by induction. In Figure 28.4d, the bar acquires a pair of opposite poles; we say it becomes a **magnetic dipole**. When the dipole is cut (Figure 28.4e), the result is quite different from the result for an electric dipole: Each piece of the original magnetic dipole becomes a smaller magnetic dipole. But, when the inducing magnet is removed, each piece reverts to the unmagnetized state (Figure 28.4f).

When a permanent magnet is cut into two, each piece becomes a smaller magnet (Figure 28.4g,h). (The experiment was first described clearly by Peregrinus in 1269.)

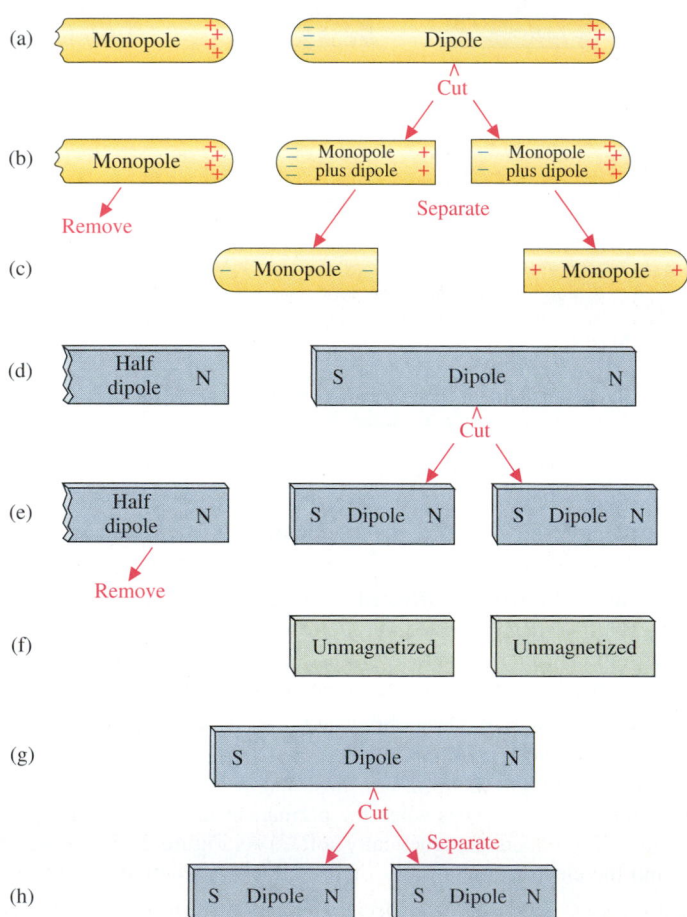

FIGURE 28.4 (*a*) An uncharged dielectric rod is brought near an electric charge and becomes an electric dipole. (*b*, *c*) When the rod is cut in two, it is made into two electric monopoles of opposite sign. (*d*) An unmagnetized iron rod is brought near a magnetic pole and becomes a magnetic dipole. (*e*) When the rod is cut in two, it is made into two smaller magnetic dipoles. (*f*) When the inducing magnet is removed, both pieces revert to the unmagnetized state. (*g*, *h*) Cutting a permanent magnet (a magnetic dipole) in two produces two smaller permanent magnets (dipoles).

On the basis of these and many other experiments, we make the generalization that *magnetic monopoles are never observed.*

Up to this point, our qualitative development has run largely parallel to that of Section 23.2, in which we considered the properties of electric charges. It may occur to you that the next logical step is to make the discussion quantitative by carrying out an experiment like Coulomb's experiment. It is not possible to isolate a magnetic pole on the end of a torsion balance, as Coulomb did with electric charge. But a reasonable approximation can be made by using two bar magnets sufficiently long so that two of their poles (one pole of each) are much closer together than the other two. This experiment was performed in 1750 by John Michell, who invented the torsion balance. In 1785, Coulomb repeated the experiment with much greater precision. The force exerted by a magnetic pole of one magnet on a pole of another does indeed conform to the inverse-square law. This result is called *Coulomb's law for magnetism.*

Unlike Coulomb's law for electric charge, the law for magnetism is something of a blind alley because we cannot express the magnetic result in terms of a quantity analogous to electric charge. When a body carries 1 C of charge, for example, it contains a certain number of excess elementary electric monopoles—say, electrons. It is possible to define a quantity called "magnetic pole strength," as we will do in Problem 28.55. But a magnet pole is only a vaguely defined region in which the magnetic activity of a magnet appears to be concentrated. The pole does not contain a certain number of elementary magnetic monopoles.

Because the approach via Coulomb's law is not profitable for magnetism, we turn to an alternative. In the next section, we consider the effects of a *magnetic field.*

Rarity Doesn't Guarantee Nonexistence

There is no guarantee that magnetic monopoles will never be found in nature. Indeed, the discovery of elementary magnetic monopoles, analogous to the electron or other electrically charged elementary particles, would remove a troubling asymmetry from nature. But intensive searches have so far been unsuccessful. At best, magnetic monopoles are so rare in our sphere of observation that we can safely develop a theory of electromagnetism on the presumption that they do not exist.

Magnetic Flux and Magnetic Field

In studying electrostatics, we made the very fruitful assumption that the space surrounding an electric charge is permeated with *electric flux.* Analogously, we assume that the space surrounding a magnet is permeated with **magnetic flux.** Although we cannot yet quantify this flux, we can use symmetry arguments to sketch the general distribution of the flux lines. Still better, we can make the flux-line pattern of a magnet visible by immersing the magnet in a transparent plastic container filled with glycerine in which miniature iron needles are suspended, as shown in Figure 28.5a. The iron needles are

(a)   (b)

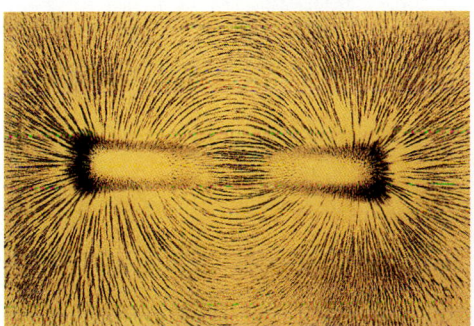
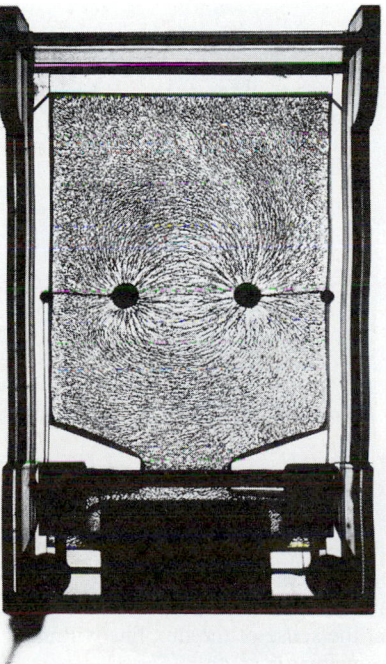

FIGURE 28.5 (*a*) Flux-line pattern of a bar magnet (a magnetic dipole). A magnet is surrounded by a transparent plastic container filled with a suspension of tiny iron needles in glycerine. The needles, acting like tiny induced magnets, orient themselves along the flux lines to produce the pattern. (*b*) Electric field of an electric dipole. Note the similarity between the fields of the magnetic and electric dipoles.

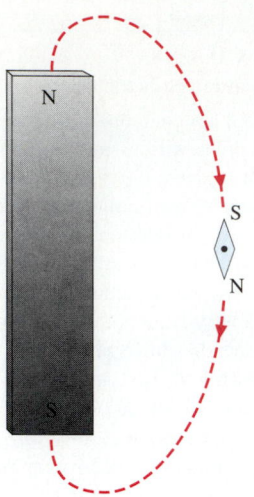

FIGURE 28.6 A tiny compass needle orients itself tangent to the flux line passing through its position in the vicinity of a bar magnet. By convention, the north pole of the compass points out the sense of the flux line. That is, the flux line emerges from the magnet near its north pole and reenters the magnet near its south pole.

in effect tiny compass needles, each aligned with the tangent to the flux line at its location. For comparison, Figure 28.5b is a photograph in which the flux lines of an electric dipole are visualized by means of small plastic fibers suspended in oil.

The magnetic and electric flux-line patterns exhibit the same symmetry, which is characteristic of any dipole. But an essential difference between the two comes to light when we inquire into the *sense* of the flux lines. For the electric dipole, the sense of the flux line passing through any point is determined by the direction of the force on a (positive) test charge. Regardless of the local direction of a flux line, its sense everywhere is the one that leads from the positive to the negative charge.

How can we determine the sense of magnetic flux lines? As already noted, a tiny compass will align itself with the tangent to the flux line at its location. By convention, *the orientation of the north pole of the compass determines the sense of the magnetic flux line*. At any point *outside* the bar magnet, the sense of the flux lines can be deduced from observations 1 and 2 in Section 28.2. The north pole of the compass is repelled by the north pole of the bar magnet, while the south pole is attracted. Hence, the sense of the flux line is the one that leads from the north to the south pole of the bar magnet (Figure 28.6).

On the basis of analogy with the electric-dipole case, it would be easy to infer that the sense of the flux lines within the bar magnet is the same as the sense outside, from north pole to south pole—to jump to the conclusion that "north" is the magnetic name for "positive" and "south" is the magnetic name for "negative." But this inference is not justified. Up to this point, we have considered no evidence that bears on the matter. We need experimental results. The necessary experiment can be done by making a small cavity in the magnet and inserting the tiny compass, as shown in Figure 28.7. When this is done, the north pole of the compass points toward the *north* pole of the magnet. That is, *within the magnet the flux runs from south pole to north pole*.

It now becomes evident that *magnetic flux lines are closed curves having no ends*, as shown in Figure 28.8. This conclusion is intimately connected with the nonexistence of magnetic monopoles. Remember that electric flux lines can originate or terminate only at electric charges. In the absence of magnetic monopoles, there are no entities on which a magnetic flux line can begin or end.

FIGURE 28.7 Investigating the sense of the flux lines within the bar magnet. A small cavity is made, and the compass needle is inserted in it. The north pole of the compass indicates that the sense of the flux line within the magnet is south to north.

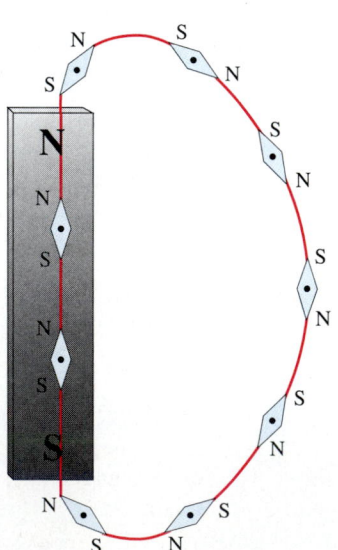

FIGURE 28.8 Tiny compass needles are placed at various locations on the same flux line, both outside and inside a bar magnet. The orientation of the north poles of the compass needles determines the sense of the flux line, which is a closed curve.

We say that the closed magnetic flux lines **circulate** through the magnet and the space surrounding it. The magnet is made of a material that has a high **permeability** for magnetic flux; that is, magnetic flux finds an easy path through it. Consequently, the flux lines are highly concentrated within the magnet and widely distributed outside it. The *poles* are merely the regions where most of the flux lines pass through the surface of the magnet, in conformity with the symmetry of the system (Figure 28.5a). For this reason, a magnetic dipole is not a true dipole, as is an electric dipole. But, as long as we consider the behavior of a bar magnet from a point of view *external* to it, the magnet behaves like a dipole, making it convenient to regard the magnet in that light.

Flux and Field

Magnetic field $\mathscr{B}$ is related to magnetic flux Φ_m just as electric field $\mathscr{E}$ is related to electric flux Φ_e.* (Hereafter, we use subscripts to distinguish between the two fluxes.) In both cases, the direction of the field is that of the tangent to the flux line at the location of interest, the positive direction being the direction consistent with the sense of the flux line. In an electric field, the direction $\mathscr{E}$ is the direction of the force exerted on a positive test charge at that location. In a magnetic field, the direction $\mathscr{B}$ is the direction in which the north pole of a compass needle points at that location.

The relation between electric flux and electric field is given by Equation 24.6,

$$d\Phi_e = \mathscr{E} \cdot d\mathbf{A}. \tag{28.1a}$$

In this equation, $d\mathbf{A}$ is a vector whose magnitude is that of the area element dA and whose direction is normal to the element (Section 24.3). In the special case where $d\mathbf{A}$ is parallel to $\mathscr{E}$, we can write the relation between electric flux and electric field in the scalar form of Equation 24.4b,

$$\mathscr{E} = \frac{d\Phi_e}{dA}. \tag{28.1b}$$

The relation between magnetic flux and magnetic field is completely analogous:

$$d\Phi_m = \mathscr{B} \cdot d\mathbf{A}. \tag{28.2a}$$

When $d\mathbf{A}$ is parallel to $\mathscr{B}$, this relation takes the scalar form

$$\mathscr{B} = \frac{d\Phi_m}{dA}. \tag{28.2b}$$

Magnetic Field and Magnetic Force[†]

Equations 28.2a and 28.2b express the relation between magnetic flux and field. We must still connect these two quantities with other physical quantities. Specifically, we must define $\mathscr{B}$ in terms of a force. In the electric-field case, we have the simple relation of Equation 23.10,

$$\mathscr{E} \equiv \frac{\mathbf{F}}{q_t}.$$

Because there are no magnetic monopoles, there is no magnetic quantity analogous to q_t, and so the analogous relation between force and magnetic field is of no physical use.

---

*The formal name for the quantity $\mathscr{B}$ we are defining here is **magnetic induction**. This name is used when it is necessary to distinguish between $\mathscr{B}$ and a related quantity called the *magnetic field intensity* $\mathscr{H}$. We will not consider the latter quantity in this book, and we follow common practice in using the informal term *magnetic field* for $\mathscr{B}$.

[†]The law governing the magnetic force is expressed in terms of the vector (cross) product. If you have not already become acquainted with the cross product in connection with angular momentum, you should study Section 13.4 before reading the rest of this and subsequent chapters.

Since we cannot proceed by analogy, we instead introduce a fundamental relation among force, magnetic field, and *electric* charge. This relation, based on experimental observation, is one of the fundamental links between electricity and magnetism to which we referred in the introduction to Part IX. As Gilbert showed in 1600, a magnetic field exerts no force on an electric charge *at rest*. Not until 1820 did it become apparent that *an electric charge does experience a force when it **moves** in a magnetic field*. This force is called the **magnetic force**. The magnetic force is more complicated than the electric force because the magnetic force is **velocity dependent**. That is, its magnitude F and direction $\hat{\mathbf{F}}$ depend on the magnitude v and direction $\hat{\mathbf{v}}$ of the velocity of the charge as well as on the magnitude $\mathscr{B}$ and direction $\hat{\mathscr{B}}$ of the field. Nevertheless, the magnetic force on an electric charge is simpler and more fundamental than the force exerted by one magnet on another.

Force and Field for $\mathbf{v} \perp \mathscr{B}$

We are now ready to discuss the magnetic force quantitatively, framing our discussion in terms of a series of experiments in which an electric charge moves through a uniform magnetic field. We summarize the results of such a series of experiments:

1. When a charge q moves *parallel* or *antiparallel* to the magnetic field $\mathscr{B}$, the charge experiences no force.
2. When the direction of motion $\hat{\mathbf{v}}$ of the charge does not coincide with $\hat{\mathscr{B}}$, a force $\mathbf{F}$ is exerted on the charge. The direction of $\mathbf{F}$ is always *perpendicular* to the plane defined by $\mathbf{v}$ and $\mathscr{B}$.
3. The magnitude of $\mathbf{F}$ depends on the angle θ between $\mathbf{v}$ and $\mathscr{B}$. Specifically, F is proportional to $\sin \theta$ and reaches a maximum when $\theta = 90°$.
4. The magnitude of $\mathbf{F}$ is proportional to the charge q.
5. The magnitude of $\mathbf{F}$ is proportional to the speed v of the moving charge.

We write a proportionality that contains all the information concerning magnitudes in these five results:

$$F \propto qv \sin \theta. \tag{28.3}$$

This proportionality gives us an opportunity to define a unit of magnetic field $\mathscr{B}$. In SI, $\mathscr{B}$ is simply the proportionality constant that converts the proportionality into an equation:

$$\mathscr{B} \equiv \frac{F}{qv \sin \theta}. \tag{28.4a}$$

The magnetic force thus has magnitude

$$F = qv\mathscr{B} \sin \theta. \tag{28.4b}$$

In the special case where $\mathbf{v} \perp \mathscr{B}$, we have $\sin \theta = 1$ and the equation simplifies to

$$F = qv\mathscr{B} \quad \text{for } \mathbf{v} \perp \mathscr{B}. \tag{28.5}$$

The relation among the directions of $\mathbf{v}$, $\mathscr{B}$, and $\mathbf{F}$ is shown for a positive charge q in Figure 28.9a and for a negative charge in Figure 28.9b.

All of the quantities on the right side of Equation 28.4b are defined, and so we can use the equation to define the SI unit of magnetic field $\mathscr{B}$. This unit is called the **tesla (T)**:

$$1 \text{ T} \equiv 1 \frac{\text{N} \cdot \text{s}}{\text{C} \cdot \text{m}}. \tag{28.6}$$

In the light of this definition, we use Equation 28.2a, $d\Phi_{\mathrm{m}} = \mathscr{B} \cdot d\mathbf{A}$, to define the corresponding unit of magnetic flux. It follows immediately from this equation that the unit is the T·m^2. This unit is given the special name **weber (Wb)**:*

$$1 \text{ Wb} \equiv 1 \text{ T·m}^2. \tag{28.7}$$

---

*After the German physicist Wilhelm E. Weber (1804–1891), who pioneered in electromagnetic measurements and made important contributions in geomagnetism.

(a)

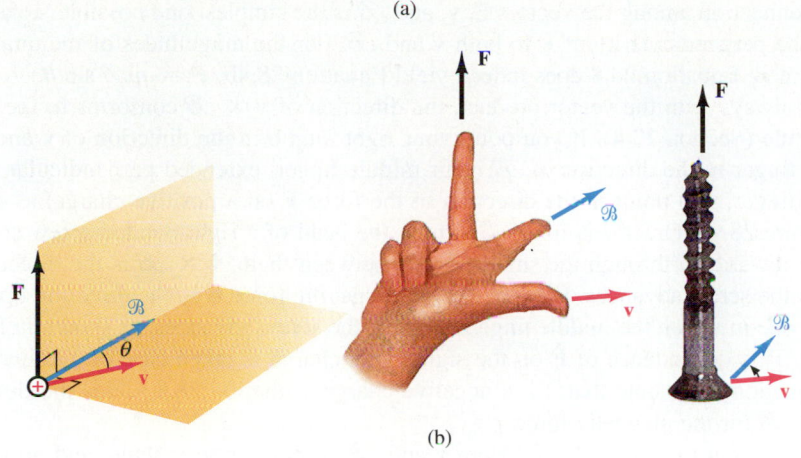

(b)

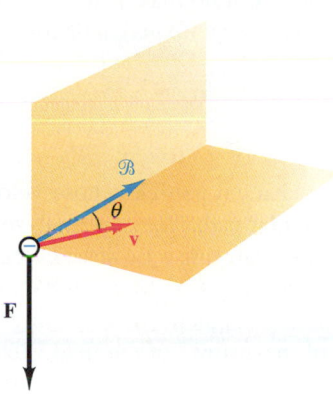

In older literature, the name *weber per square meter* (Wb/m^2) is used for the tesla. A non-SI unit, the *gauss* (G), is also often used; 1 G = 10^{-4} T.

According to Equation 28.5, a 1-C charge moving at 1 m/s in a direction perpendicular to a 1-T magnetic field will experience a force of 1 N. Table 28.1 lists some typical magnetic field magnitudes.

TABLE 28.1 Magnitudes of Some Magnetic Fields

Environment	$\mathscr{B}$ (T)
Smallest field achieved in laboratory	2×10^{-14}
Interstellar space	$< 10^{-9}$
Earth's field at surface	5×10^{-5}
Permanent magnets	10^{-2}–1
Iron-core electromagnets	up to 3
Superconducting-coil magnets	up to 20
High-current coils	up to 40
Pulsed coils (duration $\simeq$ 1 ms)	10–70
Greatest field achieved in laboratory (explosive compression of pulsed coil, duration $\simeq$ 1 μs)	10^3
Surface of pulsar	$\simeq 10^8$
Surface of atomic nucleus	$\simeq 10^{12}$

General Expression for the Magnetic Force

The dependence of the magnetic force **F** on q, **v**, and $\mathscr{B}$ is conveniently written in terms of the vector product relation

$$\mathbf{F} = q\mathbf{v} \times \mathscr{B}. \tag{28.8}$$

The connection among the vectors **F**, **v**, and $\mathscr{B}$ is the simplest one possible, consistent with the perpendicularity of **F** to both **v** and $\mathscr{B}$. For the magnitudes of the quantities concerned, Equation 28.8 does indeed yield Equation 28.4b, $F = qv\mathscr{B}\sin\theta$.

As always with the vector product, the direction of $\mathbf{v} \times \mathscr{B}$ conforms to the right-hand rule (Section 13.4). If you point your right thumb in the direction of **v** and your index finger in the direction of $\mathscr{B}$, your middle finger, extended perpendicular to the index finger, will point in the direction of the force **F** on a positive charge, as shown in Figure 28.9a. Or, if you imagine turning the head of a right-handed screw so as to turn **v** toward $\mathscr{B}$ through the smaller angle between them, $\hat{\mathbf{v}} \times \hat{\mathscr{B}}$ is the direction in which the screw advances. For a negative charge, the force is in the direction opposite to the one in which the middle finger points or the screw advances, as shown in Figure 28.9b. This dependence of $\hat{\mathbf{F}}$ on the sign of the charge is taken care of automatically by Equation 28.8; note that, for a negative charge, Equation 28.8 yields the direction $-\hat{\mathbf{v}} \times \hat{\mathscr{B}}$ for the magnetic force.

What about the case $\mathbf{v} \parallel \mathscr{B}$? Here **v** and $\mathscr{B}$ do not define a plane, and no unique direction can be established for **F**. But no matter; we have $\sin\theta = 0$, and so $\mathbf{F} = 0$; *an electrically charged particle experiences no magnetic force when moving along a magnetic flux line.*

Magnetic Orbits

In a uniform magnetic field, a charged particle whose velocity is perpendicular to the field direction will follow a circular path. We now analyze this motion.

Consider a particle of mass m, carrying a positive charge q and moving with speed v. The particle possesses kinetic energy $K = \frac{1}{2}mv^2$. In Figure 28.10a, the particle moves in the plane normal to a uniform magnetic field $\mathscr{B}$ directed into the page. (Each ✕ represents the "tail feathers of the arrow" of the field vector. In the same convention, a vector directed outward from the page is denoted by a dot representing the "head of the arrow.") Because **v** is perpendicular to $\mathscr{B}$, the magnetic force given by Equation 28.6, $\mathbf{F} = q\mathbf{v} \times \mathscr{B}$, has magnitude $qv\mathscr{B}$. The direction of the force is shown; you should check it using the right-hand rule.

FIGURE 28.10 (*a*) A particle with positive charge q moves with velocity **v** perpendicular to the uniform magnetic field $\mathscr{B}$. (See text for the significance of the ✕'s.) The magnetic force **F** is shown. The path of the particle is a circle of radius r; the position of the particle at a later time is shown in lighter colors. (*b*) The situation is the same as that in part *a*, except that here the charge is of opposite sign; $q' = -q$. As a result, $\mathbf{F}' = -\mathbf{F}$, and the particle moves in a circular orbit of opposite sense.

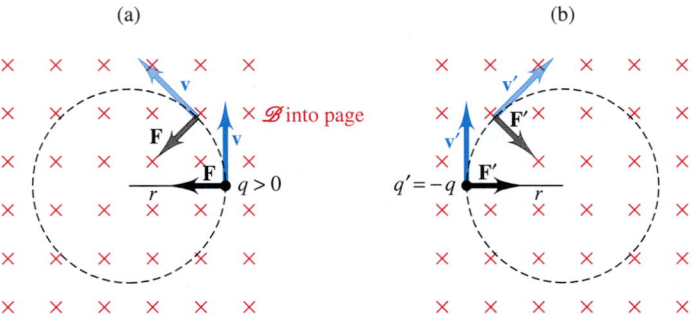

Because **F** is always perpendicular to **v**, **F** is a *centripetal force*. For such a force, we have (Section 5.4)

$$F = qv\mathscr{B} = \frac{mv^2}{r},$$

where r is the radius of the curved path in which the particle moves. Solving the equation for r yields

$$r = \frac{mv}{q\mathscr{B}}. \tag{28.9}$$

Because it changes the direction, but not the magnitude, of the velocity, the centripetal force cannot change the kinetic energy of the particle. Thus *the magnetic force can do no work on a charged particle.* The kinetic energy and the speed must therefore remain constant. Thus r remains constant and *the path of a charged particle moving in a plane normal to a uniform magnetic field is a circle whose radius is directly proportional to the momentum mv of the particle and inversely proportional to its charge and to the magnitude of the magnetic field.*

The situation in Figure 28.10b is identical with that of Figure 28.10a, except that in part *b* the sign of the charge is negative; $q' = -q < 0$. The radius of the circular orbit is still given by Equation 28.9. But the sense in which the particle travels around the circle is reversed because the direction of the magnetic force is reversed; $\mathbf{F}' = -\mathbf{F}$.

EXAMPLE 28.1

An α particle (the nucleus of a helium atom) is accelerated by a van de Graaff generator and moves through a vacuum tube with kinetic energy $K = 10$ MeV. The α particle passes between the poles of a large electromagnet where the field is of magnitude 1.85 T. If the radius of the orbit followed by the α particle in the magnetic field is 24.6 cm, find the mass of the particle. The charge on an α particle is $+2e$.

SOLUTION: To use Equation 28.9, you must know the speed of the α particle. Since you know the kinetic energy, you express the speed in terms of K, writing $v = \sqrt{2K/m}$. Substituting

this value into Equation 28.9, you obtain

$$r = \frac{\sqrt{2mK}}{q\mathcal{B}}.$$

You can now solve for m:

$$m = \frac{(rq\mathcal{B})^2}{2K}$$

$$= \frac{(2.46 \times 10^{-1} \text{ m} \times 2 \times 1.60 \times 10^{-19} \text{ C} \times 1.85 \text{ T})^2}{2 \times 10 \times 10^6 \text{ eV} \times 1.60 \times 10^{-19} \text{ J/eV}}$$

$$= 6.6 \times 10^{-27} \text{ kg}.$$

As Example 28.1 shows, the magnetic force can be used to measure the mass of a particle whose electric charge is known. The family of instruments called **mass spectrometers** is based on this possibility. J. J. Thomson first used a mass spectrometer in 1912 to show that there are two kinds of neon, now called *isotopes*, of slightly different atomic mass. Most chemical elements have two or more isotopes, whose nuclei differ in the number of neutrons they contain. This difference underlies the mass difference, which makes possible isotope separation by mass spectrometric techniques. In contrast, chemical behavior is determined almost entirely by the atomic electrons, whose number is determined by the number of protons in the nucleus. Consequently, isotope separation by chemical methods is usually not feasible.

Modern mass spectrometry has evolved into an exquisitely precise measurement technique. Mass spectrometry is combined with the quite different technique called gas chromatography in the device called the GC-MS, which has become one of the most important tools of chemical analysis.

The Cyclotron Effect

The time τ required for a charged particle to complete one revolution around its circular orbit in a uniform magnetic field is given by the simple relation

$$\tau = \frac{2\pi r}{v}.$$

From Equation 28.9, we have $r/v = m/q\mathcal{B}$, and hence

$$\tau = \frac{2\pi m}{q\mathcal{B}}. \qquad \textbf{(28.10a)}$$

For a particle of fixed mass and charge, revolving in a fixed magnetic field, this period of revolution is a constant, called the **cyclotron period**. The corresponding angular frequency,

$$\omega_c = \frac{2\pi}{\tau} = \frac{q}{m}\mathcal{B}, \qquad \textbf{(28.10b)}$$

S magnet pole

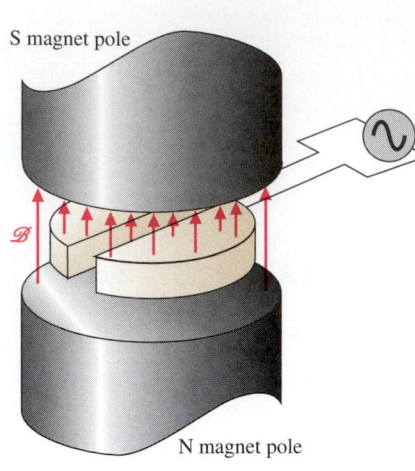

$\mathscr{B}$

N magnet pole

(b)

$\mathscr{B}$ out of page

FIGURE 28.11 (*a*) Schematic diagram of a cyclotron. (*b*) The particle orbit plane, as seen by an observer looking down on the dees along the direction $-\hat{\mathscr{B}}$. The path of a particle consists of semicircular segments within the dees, connected by segments of spirals in the space between the dees. For clarity, only a few orbits are shown and the radius change between subsequent semicircular segments is much exaggerated. Taken as a whole, the path is approximately a spiral of many turns.

is called the **cyclotron frequency**. If the speed of the particle is increased by some external means (remember that the magnetic field cannot do this because it can do no work on the particle), the orbit radius will increase proportionately and the period will not change. This isochronous property is often called the **cyclotron effect**.

The cyclotron effect is named for its first practical application, the particle accelerator called the **cyclotron**, invented in 1929 by Ernest O. Lawrence (1901–1958) to serve as a source of very energetic charged particles. By 1932, Lawrence's student M. Stanley Livingston (1905–1986) had built a cyclotron that surpassed other types of accelerators in performance, and the cyclotron in various improved forms remained the dominant type of accelerator for fundamental-particle research until the 1960s. The cyclotron still finds special applications in "medium energy" nuclear physics, in the range above 20 MeV, which is the upper limit for the more precisely controllable van de Graaff machine.

The principle of operation of the cyclotron is illustrated in Figure 28.11. The two hollow D-shaped sheet-metal structures, called "dees," are located in a vacuum chamber having the form of a squat cylinder. The vacuum chamber lies between the poles of a large electromagnet so that the flux lines are normal to the dees. Charged particles (usually protons or deuterons—the nuclei of deuterium, or "heavy hydrogen") are injected at low energy near the center of the device.

Consider a particle of charge q and mass m, which happens to be moving in a circular orbit in a plane normal to the magnetic field. The particle will move in a circular orbit of small radius. But suppose that, just as it traverses the space between the dees, an electric potential is imposed between the dees. If the resulting electric field in the space is in the right direction, the particle will accelerate. The radius of its orbit will increase because, as Equation 28.9 indicates, r is proportional to v. Now suppose that, in the time it takes for the particle to complete a semicircle, $\tau/2 = \pi m/q\mathscr{B}$, the polarity of the potential between the dees is reversed. The particle is again accelerated when it crosses the space. If the applied potential oscillates with a frequency $\nu = \omega_c/2\pi$, the particle will receive a "kick" and accelerate every time it traverses the space. It can thus acquire a large energy by means of repeated traverses through a relatively small potential difference, typically a few thousand volts. The acceleration process repeats until the radius of the orbit approaches that of the dees. When a particle achieves a sufficiently large orbit radius, it strikes a target located in its path, and the collisions of such particles with the atoms in the target can be studied. Alternatively, properly applied auxiliary electric or magnetic fields can be used to extract the particle beam from the machine for experimentation elsewhere.

An advantage of the cyclotron is its ability to produce a continuous stream of energetic particles. At any instant, particles of all energies are moving in orbits of various radii; as the most energetic particles are extracted, new ones are continually injected. The process works up to about 100 MeV. At this energy, the mass m of the particles begins to increase appreciably as a function of their energy, in accordance with the special theory of relativity (see Chapter 39). As m changes, the cyclotron frequency changes as well, drifting away from the frequency of the potential difference applied between the dees. The most energetic particles, in the outermost orbits, no longer arrive at the space between the dees at the proper time for further acceleration.

The cyclotron effect is applied in a quite different context to the study of the properties of the charge carriers in conducting solids. This application, called **cyclotron resonance**, is the subject of Problem 28.53.

SECTION 28.5

Force on a Current-Carrying Wire

When a wire carries electric current i, the charge carriers within the wire move with drift velocity $\mathbf{v}_d$. (Because of the vector nature of the magnetic force, we cannot treat drift in a one-dimensional context, as we did in Section 27.5.) If the wire is situated in a magnetic field $\mathscr{B}$, a force $\mathbf{F}$ is exerted on the wire.

Figure 28.12 shows a long, straight wire of uniform cross-sectional area A. The ends of the wire are connected to the terminals of a source of emf, and the wire carries a current i. We choose the direction $\hat{s}$ along the wire from positive to negative; this choice is consistent with the conventional sense of i in the wire. We assume for the sake of argument that the carriers have positive charge q. (You will soon see that the result does not depend on this assumption.) It follows that $\mathbf{v}_d = +v_d\hat{s}$, where v_d is the drift speed.

According to Equation 28.7, the force $\mathbf{f}$ exerted on an individual charge carrier due to the presence of the magnetic field is

$$\mathbf{f} = q\mathbf{v}_d \times \mathscr{B}.$$

The wire contains n carriers per unit volume. Because the volume of a segment of the wire of length s is As, the segment of wire contains nAs carriers. The total force $\mathbf{F}$ exerted on these carriers, which is also the force exerted on the segment of wire, is

$$\mathbf{F} = nAs\mathbf{f} = snA\,q(\mathbf{v}_d \times \mathscr{B}) = snA\,qv_d\mathscr{B}(\hat{s} \times \hat{\mathscr{B}}). \qquad (28.11)$$

We want to express the macroscopic force $\mathbf{F}$ in terms of the macroscopic current i rather than the microscopic drift velocity v_d. To do this, we use Equation 27.12 without making the restriction $q = e$:

$$i = nA\,qv_d.$$

Substituting this expression for i into Equation 28.11, we obtain for the magnetic force exerted on the wire

$$\mathbf{F} = i\mathscr{B}s\,\hat{s} \times \hat{\mathscr{B}}. \qquad (28.12a)$$

We can write this equation in a slightly simpler form in terms of the vector $\mathbf{s} \equiv s\hat{s}$, whose magnitude is the length of the wire and whose direction conforms with the sense of the current. Also, $\mathscr{B}\hat{\mathscr{B}} = \mathscr{B}$, so we have

$$\mathbf{F} = i\mathbf{s} \times \mathscr{B}. \qquad (28.12b)$$

The direction of the force is shown in Figure 28.12. Giving the name θ to the angle between $\mathscr{B}$ and $\hat{s}$, we write the magnitude F of the force in the simple form

$$F = i\mathscr{B}s\,\sin\theta. \qquad (28.12c)$$

The magnitude of the magnetic force per unit length is proportional to the current, to the magnetic field, and to the sine of the angle between the wire and the field vector. In the special case where the wire is parallel to $\mathscr{B}$, the force is zero.

What if the charge carriers are negative rather than positive? We must then make two changes in the derivation of Equations 28.12a, 28.12b, and 28.12c: For q we substitute $-q$, and for $\mathbf{v}_d = v_d\hat{s}$ we substitute $\mathbf{v}_d = -v_d\hat{s}$ (Section 27.5). The two sign changes cancel, and the result is not changed. The essential point is that the direction of the force $\mathbf{F}$ depends on the sense of the current i and not on the sign or magnitude of the carrier charge. As already noted in Section 27.5, the sense of the current is independent of the sign of the carrier charge. Thus it is not possible to tell anything about the sign of the charge carriers from the direction of the magnetic force $\mathbf{F}$.

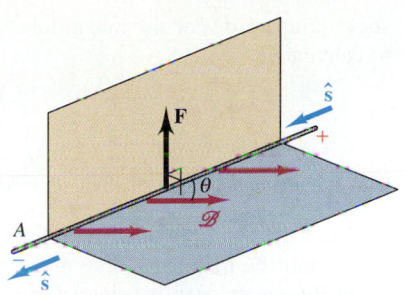

FIGURE 28.12 Diagram for determining the force on a current-carrying wire located in a uniform magnetic field $\mathscr{B}$. The angle between the wire and $\mathscr{B}$ is θ.

EXAMPLE 28.2

You have a copper wire taped along the full length of a meter stick. The wire carries a current $i = 1.5$ A. You are located near the earth's equator, where the magnetic flux lines are nearly horizontal (that is, parallel to the surface of the earth) and the field is of magnitude 1.3×10^{-4} T. Assume to first approximation that the field direction is from due south to due north, and find the magnitude and direction of the force when the meter stick is oriented so that the current runs **(a)** west to east, **(b)** northwest to southeast, and **(c)** south to north.

SOLUTION:
(a) Evaluate $\mathbf{F}$ when the meter stick is oriented west to east.

From Equation 28.12a, the direction of the force is $\hat{\mathbf{F}} = \hat{s} \times \hat{\mathscr{B}}$. Using the right-hand rule (Figure 28.9), you point your thumb eastward and your index finger northward, and you find your middle finger pointing upward. Or, if you prefer, imagine a vertically oriented right-handed screw whose head is rotated in the sense east to north, and note that the screw moves upward. Either way, you see that the force on the wire (and the meter stick) is upward. For the magnitude, you use Equation 28.12c to calculate

$$F = 1.5 \text{ A} \times 1.3 \times 10^{-4} \text{ T} \times 1 \text{ m} \times \sin 90°$$
$$= 2.0 \times 10^{-4} \text{ N}.$$

(b) Evaluate **F** when the meter stick is oriented northwest to southeast.

The current and the magnetic field still lie in the horizontal plane, and the magnetic force must therefore be vertical. But, when the meter stick is oriented in the northwest-to-southeast direction, the right-hand rule gives you a downward direction for the force. The angle between the wire and the field is 45°. You can use this value in Equation 28.12c again to calculate the force, but it is easier to use the result of part **a** to write

$$\mathbf{F} = 2.0 \times 10^{-4} \text{ N} \times \sin 45° = 1.4 \times 10^{-4} \text{ N}.$$

(c) Evaluate **F** when the meter stick is oriented south to north.

Here the current is parallel to the magnetic flux lines, so $\sin\theta = 0$ and the force is zero.

SECTION 28.6

Torque on a Current-Carrying Loop

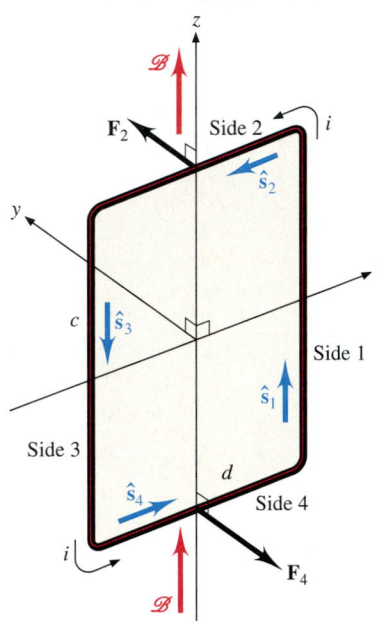

FIGURE 28.13 A rectangular loop of length c and width d carries a current i. The loop is located in the x-z plane, in a region of uniform magnetic field $\mathscr{B}$. The magnetic field is directed along $\hat{\mathbf{z}}$, in the plane of the loop.

Figure 28.13 shows a rectangular loop of sides c and d, lying in a uniform magnetic field $\mathscr{B}$. The loop carries a current i in the sense shown. (The source of emf that drives the current is not shown; it could be a small battery located somewhere in the loop.)

The figure shows a special orientation of the loop. The loop lies in the x-z plane, and $\mathscr{B}$ is parallel to the z axis. Sides 1 and 3 are parallel to the magnetic field, and so they experience no magnetic force. Sides 2 and 4 are perpendicular to the field, and so they are subjected to magnetic forces $\mathbf{F}_2$ and $\mathbf{F}_4$. According to Equation 28.12c, the magnitudes of these forces are

$$F_2 = F_4 = id\mathscr{B}.$$

According to the right-hand rule, the directions of the forces are $\hat{\mathbf{F}}_2 = \hat{\mathbf{y}}$ and $\hat{\mathbf{F}}_4 = -\hat{\mathbf{y}}$. The net force on the loop is thus zero. However, the lines of action of these two equal and opposite forces are displaced from one another, so $\mathbf{F}_2$ and $\mathbf{F}_4$ constitute a *couple* (Section 13.6). In general, the magnitude of the torque exerted by a couple is given by Equation 11.13,

$$T = |rF_\perp|.$$

In this equation, r is the distance between the points of application of the two forces, called the *moment arm*, and $F_\perp$ is the component of either force perpendicular to the moment arm. In the present case, with the forces perpendicular to the moment arm c, the torque is

$$T = cF_2 = cid\mathscr{B}.$$

Now cd is the area A of the loop, and so we can write the torque in the form

$$T = iA\mathscr{B} \quad \text{for } \mathscr{B} \text{ in the plane of the loop.} \tag{28.13}$$

The quantity iA, the product of the current and the area enclosed by the current, is called the **magnetic dipole moment** μ:

$$\mu \equiv iA. \tag{28.14}$$

You can see immediately from this definition that μ is expressed in the SI units A·m². In terms of the magnetic dipole moment, we can write the magnitude of the torque as

$$T = \mu\mathscr{B} \quad \text{for } \mathscr{B} \text{ in the plane of the loop.} \tag{28.15}$$

Now, suppose that the loop of Figure 28.13 is rotated about the x axis until it makes an angle θ with the y axis, as shown in Figure 28.14. Sides 1 and 3 are no longer parallel to $\mathscr{B}$, and so the forces on them are no longer zero. But the forces are equal and opposite, and they lie on the same line of action. Thus they have no effect on the loop as a whole. Sides 2 and 4 are still perpendicular to $\mathscr{B}$, and the forces $\mathbf{F}_2$ and $\mathbf{F}_4$ are unchanged in magnitude and direction from the situation in Figure 28.13. However, the component of $\mathbf{F}_2$ perpendicular to the moment arm is now $F_2 \sin\theta$. The resulting torque has magnitude

$$T = cF_2 \sin\theta = cid\mathscr{B} \sin\theta,$$

or

$$T = iA\mathscr{B} \sin\theta. \tag{28.16a}$$

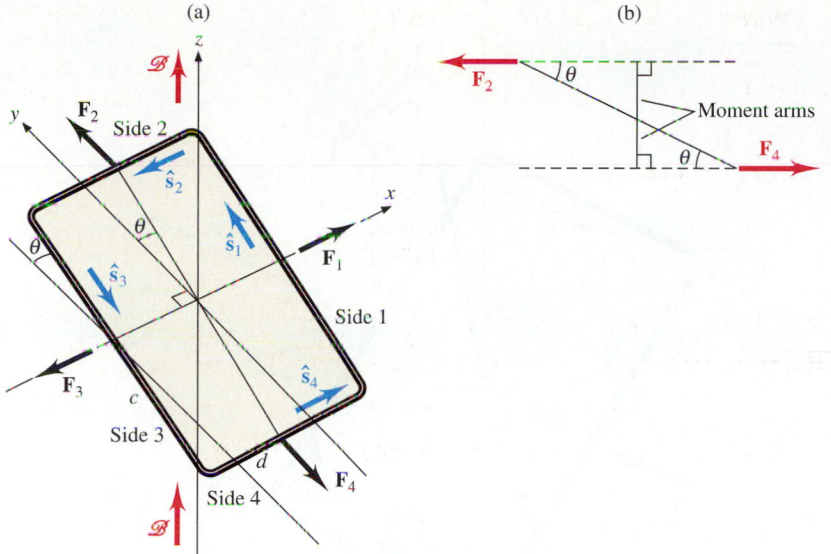

(a)   (b)

This expression for the torque is more general than Equation 28.13 and reduces to that equation in the special case $\theta = 90°$. Equation 28.16a can also be written

$$T = \mu \mathscr{B} \sin \theta. \tag{28.16b}$$

The torque tends to rotate the loop toward the plane normal to $\mathscr{B}$ and thus to reduce the value of θ. Such a rotation reduces the magnitude of the torque; when $\theta = 0$, the torque vanishes. Imagine the loop to be replaced by a bar magnet oriented normal to the plane of the loop, as shown in Figure 28.15. In the magnetic field $\mathscr{B}$, the magnet experiences a torque that tends to orient it along the flux lines. As we noted in Section 28.2, a bar magnet is a magnetic dipole. The product $\mu = iA$ is a measure of the "strength" of the equivalent magnet. Because the behavior of the current-carrying loop is equivalent to that of a bar magnet, we are justified in speaking of the *magnetic dipole moment* of the loop.

We have derived Equations 28.16a and 28.16b for a rectangular loop. But Equation 28.16a expresses the torque in terms of the area A; c and d do not appear explicitly. The shape of the loop is thus immaterial. We omit a rigorous proof of this statement.

General Expression for the Torque on a Current-Carrying Loop

We are now prepared to write a vector expression that gives both the magnitude and the direction of the torque exerted by a uniform magnetic field on a current-carrying loop. Figure 28.16 depicts the same situation as Figure 28.14. However, we now use the area vector $\mathbf{A}$ to represent both the area and the orientation of the loop. The magnitude of $\mathbf{A}$ is $A = cd$, and its direction $\hat{\mathbf{A}}$ is normal to the plane of the loop. The angle between $\mathbf{A}$ and the z axis is the same angle θ as the angle between the plane of the loop and the x-y plane. As we have done before (see the discussion after Equation 13.18), we adopt the convention of choosing $\hat{\mathbf{A}}$ along the *right-handed normal* to the loop. Imagine a right-handed screw whose head is turning in the sense of the current i around the loop; such a screw advances in the direction of $\mathbf{A}$.

In terms of the vectors $\mathbf{A}$ and $\mathscr{B}$, the torque on the loop can be expressed as the vector

$$\mathbf{T} = i\mathbf{A} \times \mathscr{B}. \tag{28.17}$$

The magnitude of $\mathbf{T}$ expressed in this equation is $T = iA\mathscr{B} \sin \theta$, which is identical with Equation 28.16a. You can confirm the direction of $\mathbf{T}$ shown in Figure 28.16 by applying the right-hand rule to the cross product $\mathbf{A} \times \mathscr{B}$.

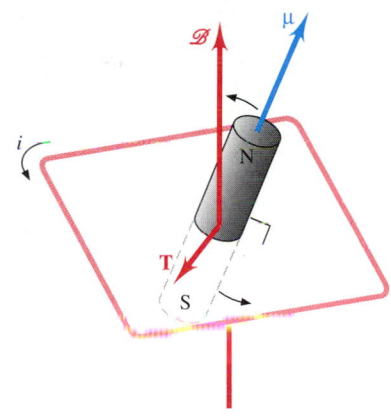

FIGURE 28.15 In a magnetic field, the torque on a current-carrying loop is equivalent to that on a bar magnet—a magnetic dipole—oriented normal to the plane of the loop.

FIGURE 28.16 The arrangement of Figure 28.15, reexpressed in terms of the area vector **A** of the loop. Vector **A** is the right-handed normal to the plane of the loop; when the head of an imaginary right-handed screw is turned in the sense of the current i circulating around the loop, the screw advances in the direction **Â**. The magnetic field $\mathscr{B}$ is directed along $\hat{z}$, and the direction of the resulting torque $\mathbf{T} = i\mathbf{A} \times \mathscr{B}$ is shown. The torque vanishes when the loop has turned until $\mathbf{A} \parallel \mathscr{B}$.

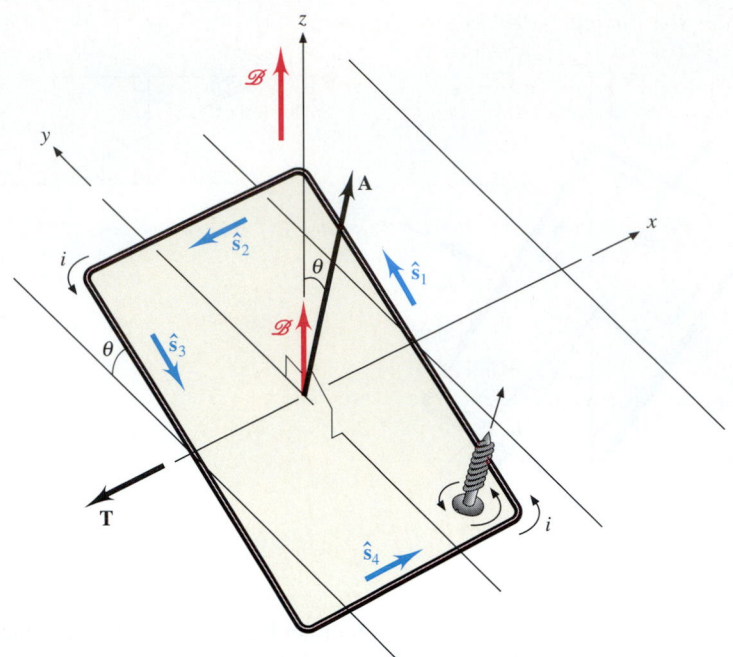

The Magnetic Moment Vector μ

In Equation 28.14, we defined the magnetic dipole moment $\mu \equiv iA$, a scalar quantity. We now define the **magnetic dipole moment vector μ**:

$$\boldsymbol{\mu} \equiv i\mathbf{A}. \tag{28.18}$$

The direction of $\boldsymbol{\mu}$ is the same as that of **A**; its magnitude is $\mu = iA$.

We can make immediate use of the definition of $\boldsymbol{\mu}$. Substituting the definition into Equation 28.17, we obtain

$$\mathbf{T} = \boldsymbol{\mu} \times \mathscr{B}. \tag{28.19}$$

EXAMPLE 28.3

Here is a way to measure the earth's magnetic field. You build a rectangular wooden frame out of a balsa-wood strip, as shown in Figure 28.17. The mass of the frame is m. You run the wire around the frame N times, as suggested by the figure; the mass of the wire is negligible compared with that of the frame. You hang the frame from a flexible string whose torsion constant is quite small. The length of the horizontal members of the frame is c, and the length of the vertical members is d. You use a small battery to run a current i through the coil. With the current on, the frame oscillates through a small angle with period τ. **(a)** What is the magnitude $\mathscr{B}$ of the local magnetic field? As in Example 28.2, assume that you are near the equator, so the magnetic flux lines run parallel to the earth's surface. **(b)** Let $m = 200$ g, $c = 120.0$ cm, $d = 50.0$ cm, $N = 200$ turns, and $i = 3.40$ A. If you find that the frame oscillates with period $\tau = 5.11$ s, what is the magnitude of the magnetic field?

SOLUTION:
(a) Evaluate $\mathscr{B}$.

You use Equation 28.16a to write $T = -Nicd\mathscr{B} \sin \theta$. The minus sign is used to signify that the frame tends to return to an equilibrium orientation normal to $\mathscr{B}$ whenever it is rotated away from that orientation in either sense. The factor N arises

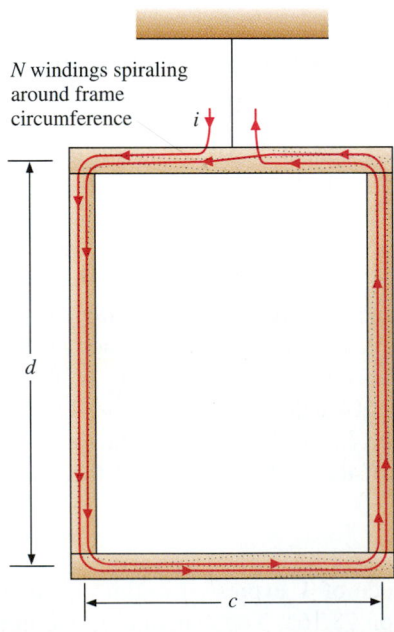

FIGURE 28.17

from the fact that Equation 28.16a gives the torque for each single turn of wire; the torque acting on the N-turn coil is simply N times that acting on each turn. As long as the oscillations are small, you can use the approximation $\sin \theta \simeq \theta$, and so the torque is

$$T = -Nicd\mathcal{B}\theta.$$

The torque is directly proportional to the angular displacement from equilibrium; this is the rotational form of Hooke's law, $T = -\kappa\theta$ (Equation 12.13). The torsion constant is

$$\kappa = Nicd\mathcal{B}.$$

Because the supporting string is quite flexible, you can neglect its additional contribution to the torsion constant.

The period τ of a rotational simple harmonic oscillator is given by Equation 12.15c,

$$\tau = 2\pi\sqrt{\frac{I}{\kappa}},$$

where I is the moment of inertia of the frame about the string axis. So you must calculate I. To do this, call the mass of each of the two horizontal members m_c and the mass of each of the two vertical members m_d. The total mass of the vertical members of the frame is then $2m_d$. Because these two members are located a distance $c/2$ from the axis, they make a contribution

$$2m_d\left(\frac{c}{2}\right)^2 = \frac{1}{2}m_d c^2$$

to the moment of inertia of the frame.

Similarly, the total mass of the horizontal members of the frame is $2m_c$. If you need to, refer to Table 11.2 to obtain the moment-of-inertia contribution

$$\tfrac{1}{12}(2m_c c^2) = \tfrac{1}{6}m_c c^2.$$

You add the two contributions to obtain the total moment of inertia of the frame:

$$I = \tfrac{1}{6}m_c c^2 + \tfrac{1}{2}m_d c^2 = \tfrac{1}{6}c^2(m_c + 3m_d).$$

Now that you know κ and I, you can write the oscillation period as

$$\tau = 2\pi\sqrt{\frac{c^2(m_c + 3m_d)}{6Nicd\mathcal{B}}}.$$

Squaring both sides of the equation and simplifying, you obtain

$$\tau^2 = \frac{2\pi^2(m_c + 3m_d)}{3Ni\mathcal{B}}\frac{c}{d}.$$

As you would expect, the period decreases as the magnetic field increases. You solve for $\mathcal{B}$:

$$\mathcal{B} = \frac{2\pi^2(m_c + 3m_d)}{3Ni\tau^2}\frac{c}{d}.$$

You are given the total mass m of the frame, rather than the partial masses m_c and m_d. To express $\mathcal{B}$ in terms of m, c, and d, you note that the total length of the members constituting the frame is $2(c + d)$; hence, the mass per unit length of the balsa-wood strip is $\lambda = m/2(c + d)$. You thus have

$$m_c = \lambda c = \frac{mc}{2(c + d)} \quad \text{and} \quad m_d = \lambda d = \frac{md}{2(c + d)}.$$

In terms of these quantities, the magnetic field becomes

$$\mathcal{B} = \frac{\pi^2 m(c + 3d)}{3Ni\tau^2/2(c + d)}\frac{c}{d}. \tag{28.20}$$

(b) Given $m = 200$ g, $c = 120.0$ cm, $d = 50.0$ cm, $N = 200$ turns, $i = 3.40$ A, and $\tau = 5.11$ s, evaluate $\mathcal{B}$.

You insert the given numbers into Equation 28.20 and solve to obtain

$$\mathcal{B} = \frac{\pi^2 \times 0.200 \text{ kg} \times [1.20 \text{ m} + (3 \times 0.500 \text{ m})]}{3 \times 200 \times 3.40 \text{ A} \times (5.11 \text{ s})^2 \times (1.20 \text{ m} + 0.500 \text{ m})}$$

$$\times \frac{1.20 \text{ m}}{0.500 \text{ m}}$$

$$= 1.41 \times 10^{-4} \text{ T}.$$

The Galvanometer and the dc Electric Motor

According to Equation 28.16a, $T = iA\mathcal{B}\sin\theta$, the torque exerted on a current-carrying loop by a magnetic field is proportional to the current through the loop. For a coil of N turns, the corresponding expression is

$$T = NiA\mathcal{B}\sin\theta. \tag{28.21}$$

We can exploit this relation to build a device for measuring current. Any device that measures current by measuring the torque on a current-carrying loop situated in a known magnetic field is called a **galvanometer**.* By means of its clever geometry, the **d'Arsonval galvanometer**,† shown in Figure 28.18, fixes the value of the factor $\sin\theta$ in Equation 28.21 at 1. The torque T is thus directly proportional to the current i, and the scale of the device is linear.

---

*The term was first used by Ampère in 1820, in honor of Luigi Galvani, whose work laid the foundations for the study of steady electric currents.

†Named after its inventor, the French physiologist and physicist Jacques Arsène d'Arsonval (1851–1940), who devised it in the 1880s.

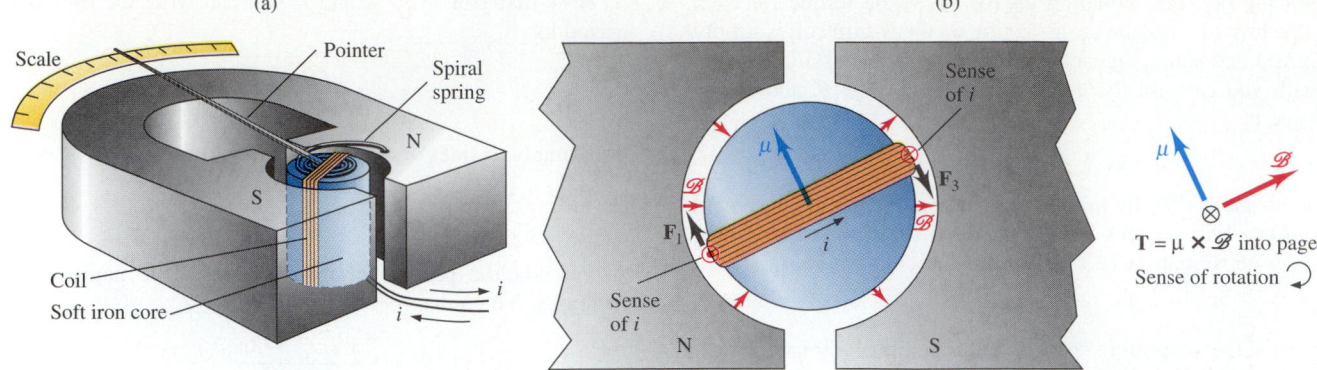

(a)

Scale · Pointer · Spiral spring · N · S · Coil · Soft iron core · i · i

(b)

Sense of i · μ · $\mathcal{B}$ · F_3 · F_1 · $\mathcal{B}$ · i · Sense of i · N · S

μ · $\mathcal{B}$ · $T = \mu \times \mathcal{B}$ into page · Sense of rotation

FIGURE 28.18 The d'Arsonval galvanometer. (*a*) General arrangement. (*b*) Top view of the moving-coil assembly, showing the way in which the magnet poles and the soft iron core are arranged to assure that the coil always lies in a radial magnetic field of constant magnitude.

Because of (1) the high permeability of both the permanent magnet and the soft iron core and (2) the low permeability of air (Section 28.3), the magnetic flux lines follow paths that minimize their length in air. Nearly all the flux lines therefore pass radially through the air gap between the magnet and the core. Within the limit of travel of the coil (typically about 100°), the magnetic torque exerted on it is independent of its position.

As Figure 28.18 shows, a current passing through the coil in the sense shown tends to turn the coil clockwise. This turning is opposed by the small spiral spring, which obeys the rotational form of Hooke's law. The coil (and the attached pointer) comes to rest when the spring and the magnetic field exert equal and opposite torques on the coil. Because all other quantities are fixed by the design of the galvanometer, the angle of rotation of the pointer is proportional to the current.

The dc Motor

As Figure 28.19 shows, the operation of the **direct-current (dc) motor** is based on the same fundamental idea as the operation of the d'Arsonval galvanometer. A coil is wound as shown around a shaft-mounted cylindrical iron core, called the **armature** or **rotor**. In the **permanent-magnet motor** shown, the armature lies in the field of a magnet very much like that of the galvanometer. In the figure, the coil is represented by a single turn for clarity; in practice, many turns are used. Current is made to pass through the coil from an external source of emf through contacts called **brushes** that slide on the two shaft-mounted segments that constitute the **split-ring commutator**. The sense of the current and the corresponding magnetic moment vector μ of the coil are shown. A torque is exerted on the coil by the external magnetic field in such a way as to align μ with the north-south direction, and the armature turns counterclockwise. Just as the coil reaches its equilibrium orientation (with the plane of the coil vertical and μ horizontal), the rotation of the commutator results in reversal of the connection

FIGURE 28.19 A simple dc motor.

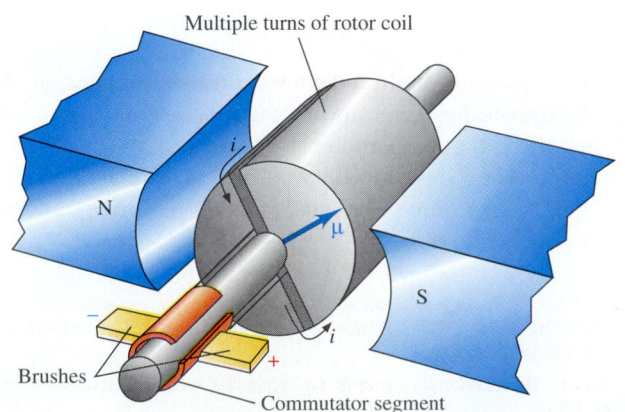

Multiple turns of rotor coil · i · N · μ · S · − · + · i · Brushes · Commutator segment

between the brushes and the commutator segments and thus reversal of the sense of the current through the coil. The magnetic moment vector is then oriented in the south-north direction, and the torque continues to turn the coil.

A motor with a single coil is not very satisfactory because its output torque varies sinusoidally once every half-revolution. For smoother operation, armatures are usually wound with several coils having different orientations.

The operation of the electric motor cannot be fully described in the context of the present chapter. We will consider the motor further in Chapter 30, in the light of another law of electromagnetism called *Faraday's law*.

The Lorentz Force

We have studied two kinds of forces that can be exerted on an electrically charged particle: the electric force $\mathbf{F}_e = q\mathscr{E}$ and the magnetic force $\mathbf{F}_m = q\mathbf{v} \times \mathscr{B}$. Taken together, they constitute *all* of the forces that can be exerted on a particle *by virtue of its electric charge q*. Their sum $\mathbf{F}$ is called the **Lorentz force***:

$$\mathbf{F} = \mathbf{F}_e + \mathbf{F}_m = q(\mathscr{E} + \mathbf{v} \times \mathscr{B}). \qquad (28.22)$$

We now consider the Lorentz force in some important special cases.

The Thomson e/m Experiment

In 1897, the British physicist Joseph John Thomson (1856–1940) measured the charge-to-mass ratio of the electron for the first time. To do this, he used the apparatus sketched in Figure 28.20. This apparatus belongs to the class of devices called **cathode-ray tubes**, of which the modern television picture tube is the most familiar type. A high vacuum is essential to the performance of the experiment because electrons must traverse the length of the tube without colliding with residual gas molecules. (In Thomson's day, achieving an adequate vacuum was one of the hardest parts of the experiment. Thomson's success was a technical triumph.)

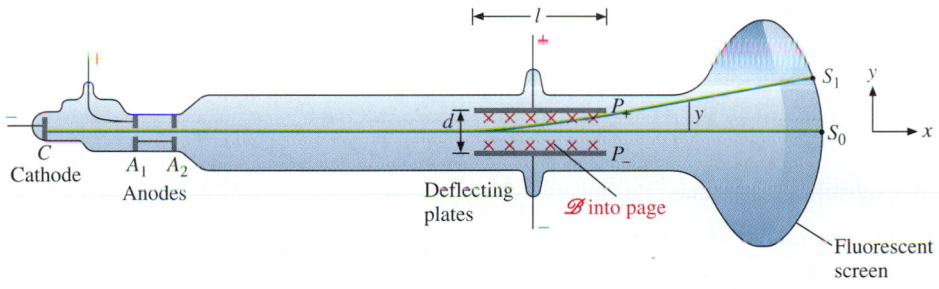

FIGURE 28.20 The cathode-ray tube used by J. J. Thomson to measure e/m for the electron. The electrons are accelerated to 5–10×10^4 eV by the electron gun at the left. They may be subjected to an electric field by means of the pair of horizontal plates, and to a magnetic field directed normal to the plane of the page.

Electrons are extracted from the **cathode** C. A large potential difference between C and the metal plate A_1—typically about 50 kV—accelerates the electrons, and they arrive at A_1 with a speed v (which we do not know). There is a small hole in A_1 and a similar hole in A_2. Only those electrons headed straight down the axis of the tube pass into the space to the right of A_2. (This thin stream of electrons was given the name "cathode ray" at a time when it was not clear that it was composed of charged particles, and television picture tubes are still called cathode-ray tubes.) If undisturbed, the electrons strike the fluorescent zinc sulfide screen at the right end of the tube at the point S_0. The impact produces a spot of light.

The parallel plates P_+ and P_- lie a distance d apart. If a potential difference V is applied between them, the resulting transverse electric field is $\mathscr{E} = V/d$. While passing

---

*The name is used in recognition of the leading role played by the Dutch theoretical physicist Hendrik Antoon Lorentz (1853–1928) in developing the modern understanding of electric and magnetic fields.

between the plates, the electrons experience a force $F = e\mathscr{E}$, and the electron stream is bent into a parabolic trajectory. The transverse acceleration is

$$a_y = \frac{F_e}{m} = \frac{e}{m}\mathscr{E}.$$

Once the electrons pass the plates, they continue in a straight line and strike the fluorescent screen at S_1. The displacement y of the electrons as they emerge from the plates can be calculated from the displacement $\overline{S_0S_1}$ of the light spot on the screen and the geometry of the tube.

If the plates P are of length l, the time required for an electron to pass through the electric field is $t = l/v$. Because the acceleration a_y is constant, we can write

$$y = \frac{1}{2}a_y t^2 = \frac{1}{2}\frac{e}{m}\frac{\mathscr{E}l^2}{v^2}. \tag{28.23}$$

If we knew the value of v, we could evaluate the ratio e/m.

The value of v is obtained by taking advantage of the velocity-dependent nature of the magnetic force. A magnetic field is imposed on the apparatus. Its direction is normal to the plane of the page, with the north pole above and the south pole below the page. By using the right-hand rule, you can convince yourself that the magnetic force exerted on negatively charged particles moving toward the right is in the $-y$ direction. That is, the magnetic force is directed opposite the electric force. Because the electric and magnetic fields required to do this are mutually perpendicular, they are called **crossed fields**.

For the sake of simplicity, let us assume that the region in which the electrons are exposed to the magnetic field coincides with the region in which they are exposed to the electric field. If the magnetic field is adjusted until the light spot moves back to its undisturbed position S_0, the Lorentz force—the total force on the electrons—must be zero:

$$\mathbf{F} = \mathbf{F}_e + \mathbf{F}_m = 0,$$

or

$$\mathbf{F}_e = -\mathbf{F}_m.$$

Considering the magnitude part of this equation only, we use the definition of the Lorentz force (Equation 28.22) to write

$$\mathscr{E} = v\mathscr{B}.$$

We thus have a value for the speed v of the electrons as they pass between the plates:

$$v = \frac{\mathscr{E}}{\mathscr{B}}.$$

We substitute this value into Equation 28.23 to obtain

$$y = \frac{1}{2}a_y t^2 = \frac{1}{2}\frac{e}{m}\frac{\mathscr{B}^2 l^2}{\mathscr{E}}.$$

Finally, we solve for e/m, obtaining

$$\frac{e}{m} = \frac{2y\mathscr{E}}{(\mathscr{B}l)^2}. \tag{28.24}$$

All the quantities on the right side of the equation can be measured and the charge-to-mass ratio thus determined.

Thomson's 1897 value, expressed in SI units, was $e/m \simeq 10^{11}$ C/kg. The best modern value is

$$\frac{e}{m} = 1.758\ 819\ 62 \times 10^{11}\ \text{C/kg}. \tag{28.25}$$

We can obtain a value of the electron mass m by combining the value of e/m with the value of e. To three significant figures we have

$$m = \frac{e}{e/m} = \frac{1.602 \times 10^{-19}\ \text{C}}{1.759 \times 10^{11}\ \text{C/kg}} = 9.11 \times 10^{-31}\ \text{kg}. \tag{28.26}$$

The Long Gestation of the Electron

Although obtaining the most precise value possible for e/m was of concern to Thomson, his primary purpose was to show that cathode rays are streams of fundamental particles of a single kind, having a particular charge and a particular mass, and thus to establish firmly the existence of such particles. At the time, many physicists—particularly those in Germany—took the view that cathode rays were not streams of particles at all, but a form of electromagnetic radiation. Thomson showed that the ratio e/m was the same for cathode rays emitted from all sorts of cathodes, and with several kinds of residual gas in the tube. He showed, moreover, that the value of e/m is at least of order 10^3 as great for the particles in cathode rays as for any ion. Thus, Thomson's relatively crude measurements were more than adequate to show that the particles were not ions.

The idea that a fundamental quantum of charge e exists originated in 1891 with the Irish physicist G. Johnstone Stoney (1826–1911), who invented the word "electron." Stoney's arguments were based on the process of electrolysis and culminated a line of thinking begun by Faraday. It took some time before all physicists were willing to associate this quantum of charge with a particle having a specific mass. Even Thomson avoided calling the particles in his cathode-ray tube electrons for many years after 1897, preferring the less specific term "corpuscles." All reasonable doubts were laid to rest by Millikan's determination of e a little more than a decade later; see Problem 26.47.

The Hall Effect

When a conductor carries a current i, the charge carriers move with a drift velocity $\mathbf{v}_d$ directed along the length of the conductor. If we place the conductor in a magnetic field, as in Figure 28.21, the moving carriers experience a magnetic force that tends to deflect them. In 1879, the American physicist Henry Rowland (1848–1901) suggested to his graduate student Edwin H. Hall (1855–1938) that he try to detect this deflection. Hall succeeded, and the deflection is called the **Hall effect**.

Let us assume that the charge carriers are all of the same kind and have positive charge $q = +e$, as in Figure 28.21a. When the sense of current through the sample is as shown, the direction of the drift velocity is the positive x direction; $\mathbf{v}_d = v_d\hat{\mathbf{x}}$. A uniform magnetic field is applied in the positive z direction. According to the right-hand rule, the magnetic force on the carriers is in the $-y$ direction:

$$\hat{\mathbf{F}}_m = \hat{\mathbf{v}}_d \times \hat{\mathscr{B}} = \hat{\mathbf{x}} \times \hat{\mathbf{z}} = -\hat{\mathbf{y}};$$

see Equation 13.29. As a result of this magnetic deflection, positive charge tends to pile up on the upper surface of the sample, leaving a deficit of positive charge—a net negative charge—on the lower surface. This process very rapidly brings into existence an *electric* field $\mathscr{E} = \mathscr{E}_y\hat{\mathbf{y}}$, called the **Hall field**. Like the Hall field, the electric force $\mathbf{F}_e = e\mathscr{E}$ exerted on the carriers is in the positive y direction. The buildup of charge on the upper and lower surfaces reaches a steady state when $\mathbf{F}_e$ and $\mathbf{F}_m$ are equal and opposite, as shown in Figure 28.21.

The Hall field is uniform throughout the sample. As a result, the potential difference between the upper and lower surfaces, called the **Hall voltage**, is

$$|V_H| = \mathscr{E}_y Y, \tag{28.27}$$

where Y is the width of the sample. Note that the upper surface is at the higher potential.

In order to evaluate $\mathscr{E}_y$, we use Equation 28.22 to write

$$\mathbf{F} = q(\mathscr{E} + \mathbf{v}_d \times \mathscr{B}) = 0.$$

In the present case, the electric and magnetic parts of the force are equal and opposite, and so we have

$$\mathscr{E}_y - v_d\mathscr{B} = 0,$$

or

$$\mathscr{E}_y = v_d\mathscr{B}. \tag{28.28}$$

We do not know the drift speed v_d, but we can use the Equation 27.12, $i = neAv_d$, to eliminate it from the equation. Remember that n is the carrier concentration; in the present case, the cross-sectional area A of the conductor is YZ. So the drift speed is

$$v_d = \frac{i}{neYZ}.$$

We can also use Equation 28.27 to replace the unknown electric field with the measurable Hall voltage. When we make these substitutions, Equation 28.28 becomes

$$\frac{V_H}{Y} = \frac{i\mathscr{B}}{neYZ}.$$

The Hall voltage thus has magnitude

$$V_H = \frac{1}{ne}\frac{i\mathscr{B}}{Z}. \tag{28.29}$$

Equation 28.29 is remarkable in two ways. First, the only sample dimension appearing in the equation is the thickness—the dimension in the direction of the magnetic field. Evidently, we can maximize the Hall voltage, and thus facilitate its measurement, by making the sample as thin as possible. Second, the Hall voltage yields a direct measure of the carrier concentration n:

$$n = \frac{i\mathscr{B}}{V_H eZ}. \tag{28.30}$$

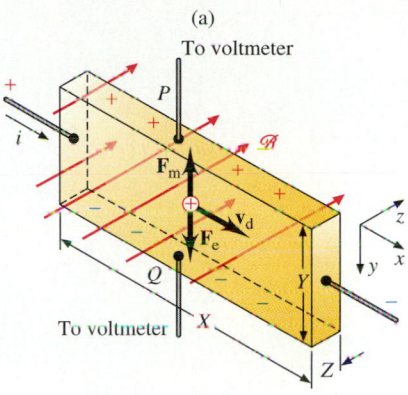

(a)

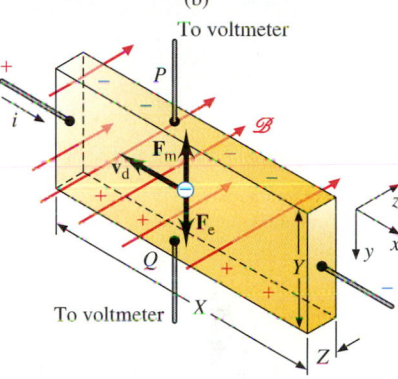

(b)

FIGURE 28.21 The Hall effect.
(*a*) The charge carriers are positive.
(*b*) The charge carriers are negative.

All the quantities on the right side of this equation are directly measurable.

Now let us consider the sign of the Hall voltage further. In Figure 28.21b, everything is the same as in Figure 28.21a, except that in part b the charge of the carriers has the *negative* value $q = -e$. With the sense of the current unchanged, the carriers now flow in the $-x$ direction; $\hat{\mathbf{v}}_d = -\hat{\mathbf{x}}$. The direction of the magnetic force $q\mathbf{v}_d \times \mathbf{\mathcal{B}}$ on the carriers is now $\hat{\mathbf{F}}_m = -(-\hat{\mathbf{x}} \times \hat{\mathbf{z}}) = -\hat{\mathbf{y}}$, *the same as for positive carriers*. Now *negative* charge tends to pile up on the upper surface of the sample, leaving excess positive charge on the lower surface. The electric field is directed upward; $\mathbf{\mathcal{E}}_y = -\hat{\mathbf{y}}$. *Changing the sign of the charge carriers changes the sign of the Hall field and thus also of the Hall voltage.* With this in mind, we define the **Hall coefficient**

$$R_H \equiv \frac{1}{nq},$$

(28.31)

where $q = e$ for positive carriers and $q = -e$ for negative carriers. This definition assures that the sign of the Hall coefficient is the same as that of the charge carriers. The SI unit for the Hall coefficient is m^3/C, as you can see from its definition. In terms of the Hall coefficient, Equation 28.29 becomes

$$V_H = |R_H| \frac{i\mathcal{B}}{Z}.$$

(28.32)

Thus the Hall effect can be used to determine both the sign and the concentration of the charge carriers. The Hall effect is one of the simplest and most powerful tools for investigating the properties of conducting solids. Although there are some exceptions (for example, beryllium, aluminum, and arsenic), the Hall coefficient of metals is in general negative; this is yet another piece of evidence tending to confirm that the charge carriers in metals are electrons. In semiconductors and some other substances, charge carriers of both signs may be present simultaneously. The analysis then becomes much more complicated; nevertheless, the Hall effect remains a very useful tool.

EXAMPLE 28.4

You have a sample of silver in the form of a rectangle of thickness 0.105 mm. When you place it in a magnetic field of magnitude $\mathcal{B} = 2.18$ T and pass a current $i = 10.0$ A through it, you measure a Hall voltage $V_H = 17.4$ μV. **(a)** Calculate the Hall coefficient. **(b)** Calculate the concentration n of the electrons that carry current in silver. **(c)** The mass density of silver is 10.5×10^3 kg/m^3, and its atomic mass is 107.9 u. Find the number of ions per unit volume, n_i. Use this result to find the "average" number of free electrons contributed to the metal per ion.

SOLUTION:

(a) Calculate the Hall coefficient R.

You rearrange Equation 28.32 in the form

$$|R_H| = \frac{V_H Z}{i\mathcal{B}}$$

$$= \frac{17.4 \times 10^{-6} \text{ V} \times 1.05 \times 10^{-4} \text{ m}}{10.0 \text{ A} \times 2.18 \text{ T}}$$

$$= 8.38 \times 10^{-11} \text{ m}^3/\text{C}.$$

You determine the sign of the Hall coefficient by comparing the sign of the measured Hall voltage with the two possibilities obtained by using the right-hand rule. The measurement will yield the result shown in Figure 28.21b, and the Hall coefficient is negative:

$$R_H = -8.38 \times 10^{-11} \text{ m}^3/\text{C}.$$

(b) Find the free electron concentration n.

You use Equation 28.31 to write

$$n = \frac{1}{R_H q} = \frac{1}{|R_H|e}$$

$$= \frac{1}{8.38 \times 10^{-11} \text{ m}^3/\text{C} \times 1.60 \times 10^{-19} \text{ C}}$$

$$= 7.46 \times 10^{28} \text{ m}^{-3}.$$

(c) Find the number of silver ions per unit volume and the average number of free electrons per ion.

You calculate n_i as you did for brass in Example 27.4. For silver, the volume of 1 kmol is

$$\frac{107.9 \text{ kg}}{10.5 \times 10^3 \text{ kg/m}^3} = 1.03 \times 10^{-2} \text{ m}^3.$$

The ion concentration is

$$n_i = \frac{6.02 \times 10^{26}}{1.03 \times 10^{-2} \text{ m}^3} = 5.86 \times 10^{28} \text{ m}^{-3}.$$

For the "average" number of free electrons per ion you have

$$\frac{n}{n_i} = \frac{7.46 \times 10^{28} \text{ electrons/m}^3}{5.86 \times 10^{28} \text{ ions/m}^3} = 1.27 \text{ electrons/ion}.$$

In chemical reactions, silver is usually monovalent and sometimes divalent. The "in between" result, 1.27 electrons/ion, suggests that the contribution of free electrons to a metal by its constituent atoms is not a simple detachment of the valence electrons from the atoms, as happens, for example, when silver is dissolved in nitric acid and a solution of Ag^+ and NO_3^- ions is formed.

A	area vector		$\hat{s}$	unit vector in an arbitrary direction
$\mathscr{B}, \mathscr{B}$	magnetic field		**T**, T	torque
$\mathscr{E}, \mathscr{E}$	electric field		$\mathbf{v}_d, v_d$	drift velocity, drift speed
e	elementary charge		κ	torsion constant
F, f	force		λ	mass per unit length
K	kinetic energy		$\boldsymbol{\mu}, \mu$	magnetic dipole moment
n	carrier concentration		τ	period
N	number of turns of a coil		Φ_e, Φ_m	electric flux, magnetic flux
q	electric charge		ω_c	cyclotron frequency

Summing Up

To an external observer, a bar magnet appears to be a **magnetic dipole**. Outside the magnet, flux lines run from a region near one end of the bar—the north pole—to a region near the other end of the bar—the south pole. Inside the magnet, however, the flux lines run from south pole to north pole; the flux lines are continuous, closed curves without beginning or end. This accords with the generalization, based on negative observation, that *magnetic monopoles do not exist*.

Magnetic field $\mathscr{B}$ is defined by analogy with electric field, according to Equation 28.2a,

$$d\Phi_m = \mathscr{B} \cdot d\mathbf{A}.$$

When an electric charge q moves with velocity **v** in a magnetic field $\mathscr{B}$, a **magnetic force F**, expressed by Equation 28.8, is exerted on the charge:

$$\mathbf{F} = q\mathbf{v} \times \mathscr{B}.$$

This equation is the basis of the definition of the unit of magnetic field, the **tesla (T)**. The magnitude of the magnetic field is given by Equation 28.4a,

$$\mathscr{B} \equiv \frac{F}{qv \sin \theta}.$$

If $F = 1$ N, $q = 1$ C, and $v = 1$ m/s, then $\mathscr{B} = 1$ T. Thus, according to Equation 28.6,

$$1 \text{ T} \equiv 1(\text{N·s})/(\text{C·m}).$$

Because the magnetic force is exerted on a charged particle in a direction perpendicular to the particle velocity **v**, the force changes the direction but not the magnitude of **v**. Hence, *a magnetic field can do no work on a charged particle*. In the absence of other forces acting on the particle, it will follow a circular orbit whose radius is given by Equation 28.9,

$$r = \frac{mv}{q\mathscr{B}};$$

the sense of rotation about the circle depends on the sign of q. Equations 28.10a and 28.10b give the period of rotation τ, called the **cyclotron period**, and the angular frequency ω_c, called the **cyclotron frequency**:

$$\tau = \frac{2\pi m}{q\mathscr{B}} \quad \text{and} \quad \omega_c = \frac{q}{m}\mathscr{B}.$$

When a current-carrying wire lies in a magnetic field, the drift velocity of the charge carriers results in a magnetic force on the wire as a whole. If the magnetic field is uniform and the wire is straight with directed length **s**, this force is given by Equation 28.12b,

$$\mathbf{F} = i\mathbf{s} \times \mathscr{B}.$$

Its magnitude is given by Equation 28.12c,

$$F = i\mathscr{B}s \sin \theta,$$

where θ is the angle between the wire and $\mathscr{B}$.

A current-carrying planar loop acts as a **magnetic dipole**. Its **magnetic dipole moment** is given by Equation 28.18,

$$\boldsymbol{\mu} \equiv i\mathbf{A},$$

where **A** is the directed area—the vector expressing the area and orientation of the loop. The value of $\boldsymbol{\mu}$ does not depend on the shape of the loop. In a uniform magnetic field, the loop experiences a torque given by Equation 28.19,

$$\mathbf{T} = \boldsymbol{\mu} \times \mathscr{B}.$$

The electric force $\mathbf{F}_e$ and the magnetic force $\mathbf{F}_m$ are the only forces that can be exerted on a particle on account of electric charge q on the particle. Their sum, expressed in Equation 28.22, is called the **Lorentz force**:

$$\mathbf{F} = \mathbf{F}_e + \mathbf{F}_m = q(\mathscr{E} + \mathbf{v} \times \mathscr{B}).$$

The Lorentz force on a moving charged particle can be set to zero by proper adjustment of the magnitudes of **crossed fields**—electric and magnetic fields oriented perpendicular to each other and to the velocity of the particle. This is the basis for the Thomson e/m experiment and for the **Hall effect**. Measurement of the Hall effect makes possible a determination of the sign and the concentration n of the charge carriers in a conductor. Equation 28.29 gives the **Hall voltage** for carriers having positive charge e:

$$V_H = \frac{1}{ne} \frac{i\mathscr{B}}{Z}.$$

For carriers having negative charge $-e$, the Hall voltage has the same magnitude but opposite polarity. The **Hall coefficient**, given by Equation 28.31, characterizes the carriers:

$$R_H \equiv \frac{1}{nq}; \quad \begin{array}{l} q = e \text{ for positive carriers;} \\ q = -e \text{ for negative carriers.} \end{array}$$

KEY TERMS

Section 28.2 Magnets and Magnetic Poles
north pole, south pole ▪ induced magnetization, magnetic remanence ▪ electric monopole, magnetic dipole

Sections 28.3 and 28.4 Magnetic Flux, Magnetic Field, and Magnetic Force
tesla, weber ▪ mass spectrometer ▪ cyclotron period, cyclotron frequency, cyclotron effect, cyclotron resonance

Section 28.6 Torque on a Current-Carrying Loop
magnetic dipole moment

Section 28.8 The Lorentz Force
cathode, cathode-ray tube ▪ crossed fields ▪ Hall effect, Hall field, Hall voltage, Hall coefficient

---

Queries and Problems for Chapter 28

QUERIES

28.1 (2) *Close embrace.* Two identical bar magnets lie side by side, north pole to south pole. Sketch the flux-line pattern inside and outside the magnets. Use arrows to denote the sense of the lines.

28.2 (2) *Well, there are only two sexes!* How could you demonstrate that a third kind of magnetic pole (other than north and south) does not exist?

28.3 (2) *Forging ahead.* You have an ingot of an alloy that can be made into a good permanent magnet. You also have tools that can be used to form the ingot into any shape you wish. How should you shape the magnet, especially its poles, to produce a uniform magnetic field over a substantial part of the space between the poles?

28.4 (4) *Steering mechanism.* "Cosmic rays" are energetic charged particles that strike the earth. Observation shows that more cosmic-ray particles strike the earth in the general vicinity of the north and south magnetic poles than near the equator. Why is this?

28.5 (4) *Wasted on polar bears?* The aurora borealis, or northern lights, often makes a brilliant nighttime display in the sky at high latitudes. A similar display in the skies of the Southern Hemisphere is called the aurora australis. The sun emits fairly energetic charged particles (collectively called the solar wind) that collide with and ionize atoms in the upper atmosphere, at altitudes of about 100 km. Light is emitted by the ions when they recombine with electrons. Why is the aurora rarely seen at low latitudes?

28.6 (4) *Footprints.* When an energetic charged particle passes through the sensitive emulsion of a photographic film, it ionizes atoms in its path. The effect on the emulsion is similar to what happens when the film is exposed to light; when the film is developed, the path of the particle shows up as a black track that can be seen under a low-power microscope. When such a film is placed in a uniform magnetic field whose direction is normal to the film, the charged-particle tracks in the plane of the film are inward spirals. Explain.

28.7 (4) *FM.* In the 1950s, a number of particle accelerators called *synchrocyclotrons* were built to produce particles at energies exceeding 100 MeV, the approximate limit for the isochronous cyclotron. In such machines, a burst of particles is injected at low energy near the center of the dees. As the particles accelerate beyond 100 MeV and their mass increases, the frequency of the accelerating electric field is varied. Is the necessary variation an increase or a decrease?

28.8 (5) *Twice as good as guessing.* Which is the north pole and which is the south pole in this picture?

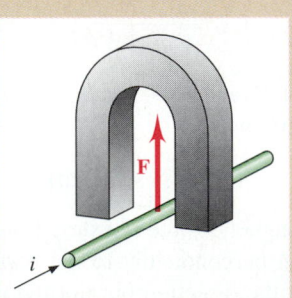

28.9 *(5) Versatile pump.* The *electromagnetic pump* shown below has no moving parts. This is a virtue when one is dealing with the corrosive, radioactive sodium-potassium alloy used as a coolant in certain types of nuclear reactors. The pump is also useful in moving blood through heart-lung machines, where even gentle mechanical agitation can cause clotting. Explain the operation of the pump.

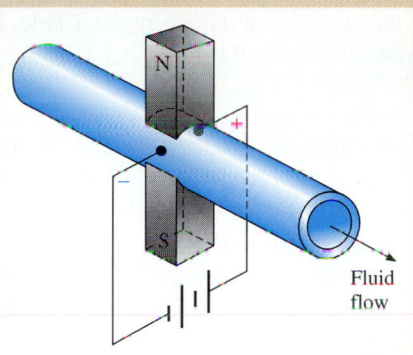

Fluid flow

28.10 *(5) All together, now!* In determining the magnetic force **F** exerted on a current-carrying wire, we simply asserted that **F** was equal to the sum of the forces **f** on the charge carriers. Using the free-electron theory, which views electrons as gas molecules confined in the wire, explain how the force is transmitted from the electrons to the wire as a whole.

28.11 *(6) Going to extremes.* The torque exerted on a current-carrying loop in a uniform magnetic field tends to rotate the loop toward an equilibrium orientation. Does this process tend to maximize or minimize the magnetic flux that penetrates the loop?

28.12 *(7) Pregnant with promise.* In 1821, Faraday invented the device shown below, which was probably the first electric motor ever built. The cup contains mercury; a bar magnet is

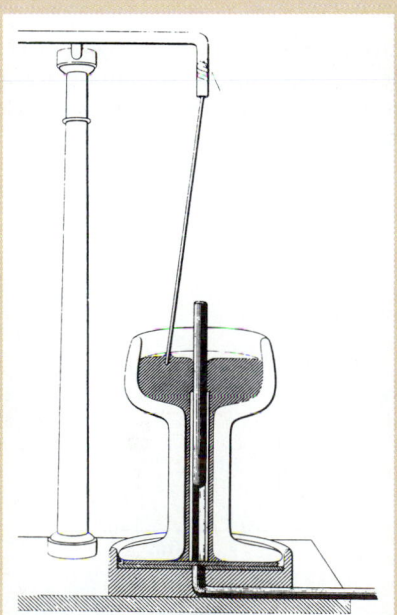

fixed along the axis of the cup. A light metal rod is suspended from the upper support in such a way that the rod can swing freely. When current passes through the upper support, the rod, and the mercury in the cup, the lower end of the rod moves in a circular path around the magnet. Explain. If the upper end of the bar magnet is a north pole and the sense of the current is downward, what is the sense of rotation of the rod?

28.13 *(7) Better things for better living.* You want to redesign a d'Arsonval galvanometer to improve its figure of merit (Problem 27.40). What can you change to accomplish this?

28.14 *(7) Another way.* In Section 28.7, the d'Arsonval galvanometer is discussed in the context of the torque exerted on a current-carrying loop in a magnetic field. Develop an equally valid qualitative discussion in terms of the force exerted on a current-carrying wire in a magnetic field.

28.15 *(8) Hall monitor.* If you measure the Hall voltage for a sample of beryllium, you will obtain the polarity shown here. What is the sign of the charge carriers in beryllium?

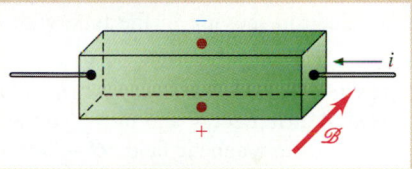

28.16 *(8) Misalignment voltage.* **(a)** In measuring the Hall effect, it is extremely important that the line joining points P and Q in Figure 28.21 be precisely parallel to the y axis. Explain. **(b)** It is not always possible to satisfy this requirement, and so an arrangement called a *phantom probe*, shown here, is sometimes used instead. Explain the operation of the phantom probe.

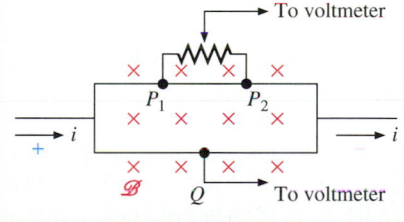

To voltmeter

To voltmeter

28.17 *(8) Magnetoresistance.* When a conducting wire is placed in a magnetic field so that the wire lies parallel to $\mathcal{B}$, the resistance of the wire is found to increase slightly over the field-free value. This phenomenon is called *longitudinal magnetoresistance*. Using the free-electron theory, explain why you would expect an increase in resistance.

GROUP A

28.1 *(3) Flux and field.* A circular loop of radius $r = 10$ cm lies in a magnetic field of magnitude 0.85 T. **(a)** If the loop is oriented so that its axis is parallel to the magnetic field vector, find the flux Φ_m that penetrates the loop. **(b)** Find Φ_m if the angle between the loop axis and the field vector is 28°.

28.2 *(3) Field and flux.* A sheet of cardboard, of area 600 cm², lies in a uniform magnetic field. The angle between the flux lines and the surface of the cardboard is 43°. Find the total flux that penetrates the cardboard.

28.3 *(4) Bend in the stream.* An α particle has charge $2e$ and mass 6.6×10^{-27} kg. A stream of α particles is accelerated through a potential difference of 1.5 MV and passes into a region where the magnetic field has the uniform magnitude 1.2 T. What is the maximum force that the field can exert on an α particle in the stream?

28.4 *(4) Hard and fast.* A vacuum tube lies in a uniform magnetic field $\mathscr{B} = (0.3\hat{x} + 0.2\hat{y})$ T. What is the magnetic force **F** on an electron moving in the tube with speed $\mathbf{v} = (3 \times 10^6)\hat{y}$ m/s?

28.5 *(4) Magnetic speedometer.* An electron is accelerated through a potential difference V and moves in the $\hat{x}$ direction into a region of uniform magnetic field $\mathscr{B} = (4.0 \times 10^{-2})\hat{y}$ T. The electron moves in a circle of radius 2.3 cm. Find **(a)** the speed v of the electron, **(b)** the kinetic energy K of the electron, and **(c)** the potential difference V. Take $e/m = 1.76 \times 10^{11}$ C/kg.

28.6 *(4) Different ways.* When it moves in a direction perpendicular to a magnetic field $\mathscr{B}$ with speed v, a proton experiences a force of magnitude F. If the proton moves at the same speed in the same magnetic field, but in a different direction, the force is $F/2$. What is the angle between the proton's velocity and the magnetic field?

28.7 *(4) Microcarousel, I.* An electron moves in a circular orbit in a magnetic field of magnitude 8.55×10^{-3} T. If the energy of the electron is 105 keV, what is the radius of the orbit?

28.8 *(4) Microcarousel, II.* An electron and a proton move in a plane normal to the flux lines of a uniform magnetic field. What is the ratio r_p/r_e of their orbit radii if they have **(a)** the same speed? **(b)** the same kinetic energy?

28.9 *(4) It's all in the timing.* A proton and an α particle move in circular orbits in a magnetic field. Find the ratio τ_α/τ_p of the periods of the two orbits. The α particle has twice the charge of the proton and four times the mass.

28.10 *(4) Isotope separation.* Potassium ions having charge e are accelerated through a potential difference. The ions then pass into a region where there is a magnetic field directed perpendicular to the motion of the ions. Potassium has two isotopes, of mass 39 u and 41 u. **(a)** Find the ratio r_{39}/r_{41} of their orbit radii. **(b)** Where in their circular orbits are the ions of the two isotopes most distant from one another?

28.11 *(4) Sizing it up.* A cyclotron can accelerate protons to an energy of 45 MeV. The magnitude of the magnetic field is 1.85 T. What is the necessary diameter of the dees?

28.12 *(5) Field investigation.* A straight wire of length 2.0 m carries a 64-A current. The wire is located in a uniform magnetic field and makes a 35° angle with the magnetic flux lines. The force on the wire is 5.5 N; find $\mathscr{B}$.

28.13 *(5) Current affairs.* A long, straight wire makes a 24° angle with the flux lines of a uniform 5.2-T field. If the magnetic force per unit length of the wire is 4.3 N/m, what current i does the wire carry?

28.14 *(5) Another turn.* The meter stick of Example 28.2 is oriented so that the current in the wire flows vertically downward. Find the magnitude and direction of the force exerted on the stick by the earth's magnetic field.

28.15 *(5) Current balance.* A current-carrying wire of length 80.0 cm is perpendicular to a uniform magnetic field of magnitude $\mathscr{B} = 1.28$ T. The wire is attached to a spring balance, which records a force 2.74 N. What is the current i in the wire?

28.16 *(5) Same difference.* In the figure below, the straight conductor and the conductor with the kink carry the same current and lie in the same magnetic field. Show that the forces on the two wires are equal.

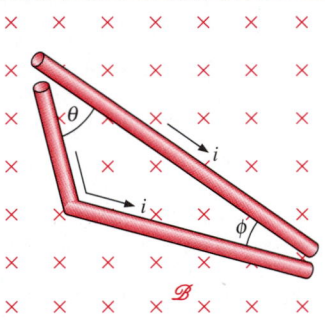

28.17 *(6) Eternal triangle, I.* The triangular conductor shown below carries a 2.5-A current. A uniform 1.25-T magnetic field is oriented as shown. Find **(a)** the magnitude and direction of the magnetic force exerted on each side of the triangle and **(b)** the torque exerted on the triangle. (Hint: Does your calculation of torque depend on your choice of axis?)

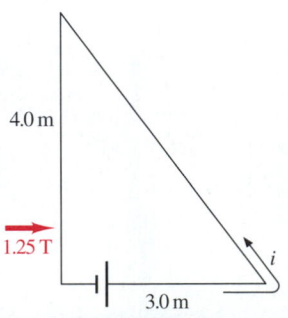

28.18 *(6) Eternal triangle, II.* The conductor of Problem 28.17 is rotated so that its plane is normal to the 1.25-T magnetic field. Find **(a)** the magnitude and direction of the magnetic force exerted on each side of the triangle and **(b)** the torque exerted on the triangle.

28.19 *(6) A matter of moment.* A circular coil of 30 turns has radius 10.0 cm. **(a)** What is the magnitude of the magnetic moment if the coil carries a current $i = 2.55$ A? **(b)** What is the maximum magnitude of the torque exerted on the coil if it is located in a magnetic field of magnitude 0.835 T?

28.20 *(6) Squaring the circle.* The coil of Problem 28.19 is bent into a square. If the current is unchanged, what is the magnetic moment of the coil now?

28.21 *(8) It could be bigger.* A proton moves in a straight line through a region where uniform electric and magnetic fields exist. Let $\mathscr{E} = 2.2 \times 10^3$ V/m and $\mathscr{B} = 0.65$ T. **(a)** What is the smallest possible speed v of the proton? **(b)** What are the conditions under which v has this smallest possible value?

28.22 *(8) Cross purposes.* An electron beam passes through the system of Figure 28.20. The length l of the plates P_+ and P_- is 5.4 cm, and the distance between them is 0.85 cm. When a potential difference $V = 300$ V is applied between the plates, the emerging electron beam is deflected through a distance $y = 0.11$ cm. **(a)** What is the kinetic energy of the electrons in the beam? **(b)** If the deflection is restored to zero by the imposition of a magnetic field on the region between the plates, what is the value of $\mathscr{B}$?

28.23 *(8) Velocity selector.* The apparatus shown below is enclosed in an evacuated chamber (not shown). Positive ions, all having the same charge, emerge from the source with a

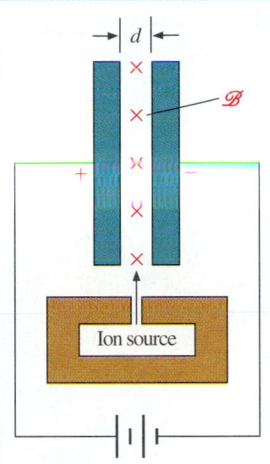

distribution of velocities. Only those ions having a particular velocity **v** can emerge at the upper end of the *velocity selector*. There is a potential difference V between the plates and a uniform magnetic field $\mathscr{B}$ directed into the page. What is the value of v for the emerging particles?

28.24 *(8) Bainbridge mass spectrometer.* The ions of Problem 28.23 emerge from the velocity selector into a region where the magnetic field is the same as that between the plates, but the electric field is zero. **(a)** Suppose that the source emits singly ionized atoms of the element thallium (Tl). Thallium has two naturally occurring isotopes, ^{203}Tl and ^{205}Tl, whose masses are, respectively, 203 u and 205 u. Sketch the trajectories of the ions of the two isotopes. (Hint: See Problem 28.10.) **(b)** When the magnetic field has turned the ions through an angle of 180°, they strike a cold plate on which they are collected. If $\mathscr{B} = 0.55$ T and the electric field in the velocity selector is $\mathscr{E} = 4.5 \times 10^4$ V/m, what is the distance separating the two isotopes at the collector? (Take 1 u = 1.66×10^{-27} kg.)

28.25 *(8) Sound choice.* Hall chose gold for his first measurements of the effect named after him because gold leaf, which is very thin, is readily available. Suppose you have a sheet of gold leaf of thickness 1.2×10^{-8} m. If you make a Hall sample and pass 15 mA through it in a magnetic field $\mathscr{B} = 0.80$ T, what Hall voltage will you measure? Use $R_H = -7.2 \times 10^{-11}$ m^3/C.

28.26 *(8) Carrier concentration, I.* **(a)** The Hall coefficient of gold is given in Problem 28.25. What is the sign of the current-carrying charge? Assuming that the magnitude of the charge on the carriers is e, what is the carrier concentration? (Take 1 u = 1.66×10^{-27} kg.) **(b)** The atomic mass of gold is 197.2 u, and its density is 19.3×10^3 kg/m^3. On the average, how many charge carriers are contributed by each atom?

28.27 *(8) Carrier concentration, II.* In measuring the Hall effect in a sample of sodium, you use a magnetic field $\mathscr{B} = 1.00$ T and a current whose density in the sample is $j = 2.00 \times 10^6$ A/m^2. You measure a Hall field $\mathscr{E}_y = 5.07 \times 10^{-4}$ V/m. What is the carrier concentration in sodium?

28.28 *(8) Deck the Hall.* A 20-A current flows through a copper plate of thickness 0.52 mm and width 10 mm. A 1.2-T field is oriented parallel to the thickness, and a Hall voltage $V_H = 3.1$ μV is measured. Find **(a)** the carrier concentration and **(b)** the drift speed of the carriers.

GROUP B

28.29 *(3) Gauss's law for magnetism.* Consider a Gaussian (closed) surface located in a region where there is a magnetic field (not necessarily uniform). Prove that

$$\int_{\substack{\text{closed} \\ \text{surface}}} \mathscr{B} \cdot d\mathbf{A} = 0.$$

This equation is called *Gauss's law for magnetism.* Compare with Gauss's law for electric charge (Equation 24.8).

28.30 *(4) Microcarousel, III.* An electron is accelerated through a potential difference of 1000 V and moves into a

region where there is a magnetic field of magnitude 1.35×10^{-3} T perpendicular to the velocity of the electron. Find **(a)** the radius of the electron orbit, **(b)** the rotation period of the electron, and **(c)** the angular momentum of the electron about the center of its orbit.

28.31 *(4) Microcarousel, IV.* An electron and a proton move in a plane normal to the flux lines of a uniform magnetic field. What is the ratio r_p/r_e of their orbit radii if they have **(a)** the same linear momentum? **(b)** the same angular momentum about the centers of their orbits?

28.32 *(4) Angular momentum.* A particle carrying charge q moves in a circular orbit in a magnetic field. Express the angular momentum of the particle about the center of its orbit in terms of q and the magnetic flux Φ_m that penetrates the orbit.

28.33 *(4) Helical trajectory.* A particle of mass m and charge q enters a region of uniform magnetic field $\mathscr{B}$ with arbitrary velocity $\mathbf{v}$. Call the components of $\mathbf{v}$ parallel and perpendicular to $\mathscr{B}$ $v_\parallel$ and $v_\perp$. **(a)** Show that the path of the particle is a helix whose axis lies along $\mathscr{B}$ and whose radius is $r = mv_\perp/q\mathscr{B}$. **(b)** The *pitch z* of a helix (or a screw) is defined as the distance between adjacent turns (measured along the axis). Show that the pitch of the helical path is

$$z = \frac{2\pi m v_\parallel}{q\mathscr{B}}.$$

28.34 *(4) Turn of the screw.* An electron is accelerated through a potential difference $V = 20$ kV and enters a region of uniform magnetic field 0.10 T. The angle between the electron trajectory and the flux lines is 55°. Using the result of Problem 28.33, find the radius and pitch of the helical trajectory.

28.35 *(5) Electric railroad.* The figure below shows a metal car of mass m whose light, frictionless wheels ride on rails spaced a distance d apart. A current i passes from one rail through the wheel, the axle, and the other wheel to the other rail. A uniform vertical magnetic field $\mathscr{B}$ is present. **(a)** Find the magnitude and direction of the acceleration of the car.

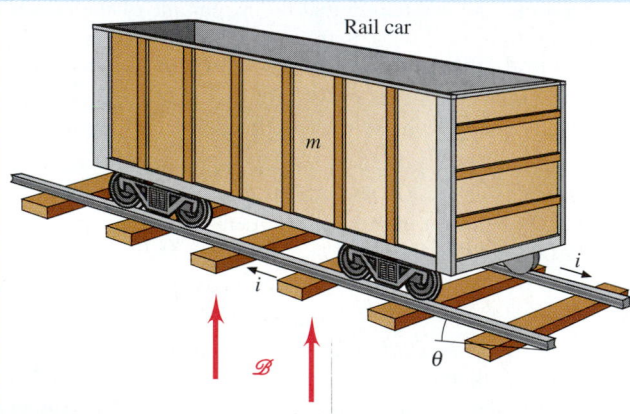

Rail car

(b) What is the incline angle of the steepest hill the rail car can climb? **(c)** Suppose that the spacing d between the rails has the standard value, 143.5 cm. Suppose also that a useful locomotive must be able to exert a force of 8×10^4 N on the train it pulls. In middle latitudes, the vertical component of the earth's magnetic field is about 5×10^{-5} T. What must the current i be? Is this a reasonable way to run a railroad?

28.36 *(5) Magnetic levitation.* A current-carrying horizontal straight wire of circular cross section is located in a uniform magnetic field $\mathscr{B}$, whose direction is perpendicular to the wire. Show that the magnetic force will support the wire if $\rho g = \mathscr{B}j$, where ρ is the mass density of the metal of which the wire is made and j is the current density. Note that this result is independent of the length and diameter of the wire.

28.37 *(6) Shifting the balance.* The balance illustrated (next column) is in balance when no current passes through the coil

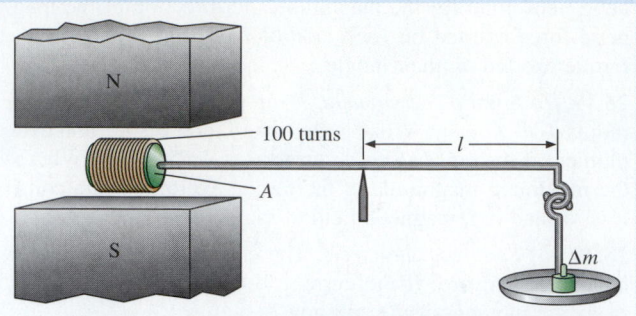

100 turns

at the end of its left arm. The coil has 100 turns, and its cross-sectional area is $A = 1.21$ cm^2. A current $i = 9.35$ mA is passed through the coil, and an extra mass $\Delta m = 63.2$ mg must be placed on the right pan to restore the balance. If the length of the right arm of the balance is $l = 25.0$ cm, what is the magnitude of the magnetic field at the location of the coil? Why doesn't the length of the left arm matter?

28.38 *(6) Currant roll?* In the figure below, a roller of radius r, made of a plastic of density ρ, rests on a plane of incline angle ϕ. A coil of N turns, carrying a current i, is embedded in the roller. The entire apparatus is located in a uniform vertical magnetic field $\mathscr{B}$. Show that, if the roller is released gently, so that it does not gain appreciable momentum as it begins to turn, it will come to a stop and not roll down the plane if the current has the value

$$i \geq \frac{\pi \rho r^2 g \sin \phi}{2N\mathscr{B}}.$$

Assume that the friction between roller and plane is large enough to prevent sliding.

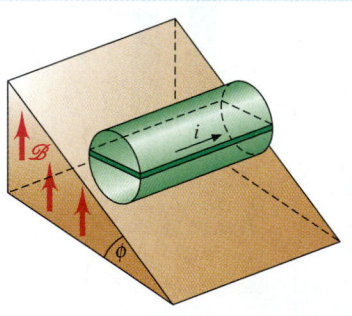

28.39 *(6) Dipole moment of a bar magnet.* A bar magnet certainly acts like a dipole in one very direct sense: When the magnet lies in a uniform magnetic field $\mathscr{B}$, there is a torque acting on the magnet that tends to align it with $\mathscr{B}$. In Problem 23.47, the torque on an *electric* dipole $\mathbf{p}$ in an *electric* field $\mathscr{E}$ turned out to be $\mathbf{T} = \mathbf{p} \times \mathscr{E}$. **(a)** What is the analogous expression for the bar magnet in the magnetic field? **(b)** You have a bar magnet, of length l, width w, and thickness t, made of an alloy whose density is ρ. You hang it from a thread, as in Figure 28.1, and set it into small oscillations about the magnetic north-south direction. You measure the period τ. Show that the dipole moment of the magnet has magnitude

$$\mu = \frac{4\pi^2 I}{\tau^2 \mathscr{B}} = \frac{\pi^2 \rho l^3 A}{3\tau^2 \mathscr{B}},$$

where I is the moment of inertia of the magnet about the thread and $A = wt$ is the cross-sectional area of the magnet. **(c)** The bar magnet of part **b** is 15 cm long, 2.5 cm wide, and 0.8 cm thick. The magnet is made of the alloy Alnico VI, whose density is $\rho = 7.4 \times 10^3$ kg/m³. The period for small oscillations is $\tau = 8.7$ s. Taking the horizontal component of the earth's magnetic field to be 4.0×10^{-5} T, calculate the magnitude μ of the magnetic dipole moment.

28.40 *(6) Measuring dipole moment.* The needle of a magnetic compass is a slender, essentially cylindrical permanent magnet of length l and mass m. The compass is placed in a known horizontal magnetic field of magnitude $\mathcal{B}$ and set into small oscillations. The oscillation period is τ. Find the magnetic moment of the needle.

28.41 *(6) Gravitational ammeter.* The square current-carrying coil shown below has 50 turns. Its sides are of length 3.5 cm, and its mass is 27 g. One side of the coil is attached to a pair of frictionless hinges. The loop lies in a uniform vertical magnetic field of magnitude 5.0×10^{-2} T. What is the current if the angle between the plane of the loop and the vertical is **(a)** 10°? **(b)** 30°? **(c)** 45°? Explain in physical terms why the coil can never be exactly horizontal.

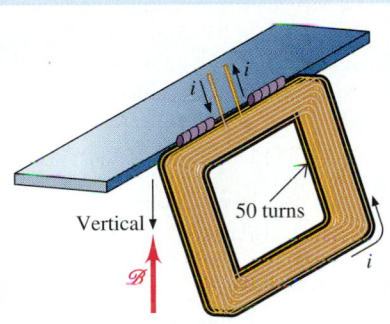

28.42 *(6) Maximizing the sensitivity.* You have a wire of length L, which you can wind into a circular coil having any integral number of turns you wish. You want to wind the coil so as to maximize the torque exerted on it when it carries current i in a uniform magnetic field $\mathcal{B}$ oriented parallel to the plane of the coil. Show that you satisfy this condition by making a single-turn loop, and express the torque in terms of L, i, and $\mathcal{B}$.

28.43 *(6) Energy of a magnetic dipole in an external magnetic field.* A current-carrying loop has magnetic dipole moment μ. When it lies in a uniform magnetic field $\mathcal{B}$, the loop experiences a torque according to Equation 28.17, $\mathbf{T} = \mu \times \mathcal{B}$. Consequently, work must be done to rotate the loop. Show that, if you assign the value $U_0 = 0$ to the potential energy of the loop when $\mu \perp \mathcal{B}$—that is, when the plane of the loop is parallel to the magnetic field—the potential energy of the loop is

$$U = -\mu \cdot \mathcal{B}.$$

28.44 *(6) Killing one bird with two stones.* A current-carrying loop has moment of inertia I. The loop is held so that the direction of its magnetic moment vector μ is antiparallel to the magnetic field $\mathcal{B}$ in which the loop is located. The loop is released and rotates freely. **(a)** What is the (scalar) angular acceleration α of the loop as a function of the angle between

μ and $\mathcal{B}$? **(b)** What is the angular velocity ω of the loop when it has rotated through half a turn so that μ is parallel to $\mathcal{B}$? **(c)** Use energy conservation and the result of Problem 28.43 to obtain the result of part **b**.

28.45 *(6) Torque and flux.* As a loop of fixed area A rotates in a uniform magnetic field, the flux Φ_m that penetrates the loop varies. Show that the magnitude of the torque exerted on a current-carrying loop can be written

$$T = -i \frac{d\Phi_m}{d\theta},$$

where θ is the angle between $\hat{\mathbf{A}}$ and $\mathcal{B}$.

28.46 *(7) Sensitive home-made galvanometer.* You wind a 600-turn coil like that of Figure 28.17 on a rectangular frame, with $c = 1.9$ cm and $d = 2.2$ cm. The frame hangs from a wire 10 cm long and 0.1 mm in radius. Unlike the string of Example 28.3, this wire has a nonnegligible torsion constant; the shear modulus of the wire material is 2.2×10^{10} Pa. A uniform, horizontal 0.200-T field is parallel to the plane of the frame. When a current i is made to flow through the coil, the frame turns through 0.75°. Find the value of i. (Hint: Refer to part **b** of Problem 15.47.)

28.47 *(7) Disk motor.* The sketch below is of a simple electric motor. A copper disk of radius r is mounted on a shaft that rests on bearings. Electrical contact is made through sliding brushes at a and b, and a current i passes through the circuit. A magnetic field $\mathcal{B}$ is directed into the page. The shaft is connected to a load, on which the motor performs work.

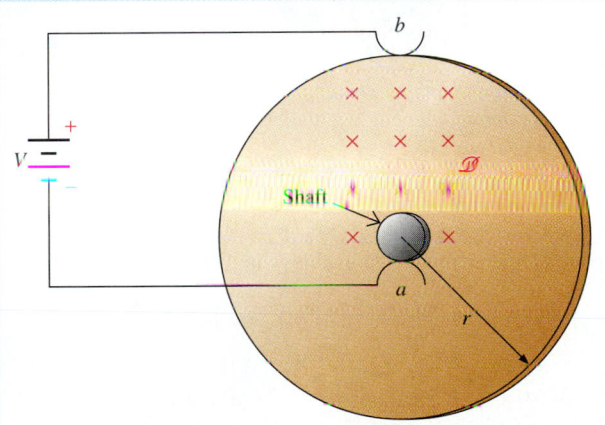

(a) Which way does the disk turn? **(b)** If the disk turns at ν rev/s, what is its power output? Express your result in terms of the disk area A. **(c)** Show that the torque exerted by the motor on its load is

$$T = \frac{1}{2\pi} iA\mathcal{B}$$

and is independent of the motor speed. (Caution: The quantity iA in this expression is not a magnetic moment because A is not the area surrounded by the current loop.) **(d)** Even though the area A of the disk shows up in the results of parts **b** and **c**, it is not necessary to place the entire disk in the magnetic field. A much smaller magnet, which imposes the field $\mathcal{B}$ only on the region of the disk between the sliding contacts, works almost as well. Explain. **(e)** Let $i = 200$ A and $\mathscr{B} = 0.20$ T. If the disk has radius 30 cm and rotates at 180 rev/min, find the output power and the output torque.

28.48 *(8) TV show.* In a television tube, the electrons are accelerated through a potential difference of 18 kV and then travel a distance 25 cm before they strike the screen, where the impact of the electrons produces a small spot of light. As you rotate the tube in the earth's magnetic field, whose magnitude is 0.5×10^{-4} T, the light spot on the screen moves slightly. If you can orient the tube in any direction you like, what is the maximum deflection of the spot you can obtain?

28.49 *(8) Electromagnetic flowmeter.* The device sketched below can be used to measure the flow speed of a conducting fluid in a pipe, provided that the electrical conductivity of the pipe is substantially smaller than that of the fluid. (This is the case, for example, for blood flowing through a blood vessel.)

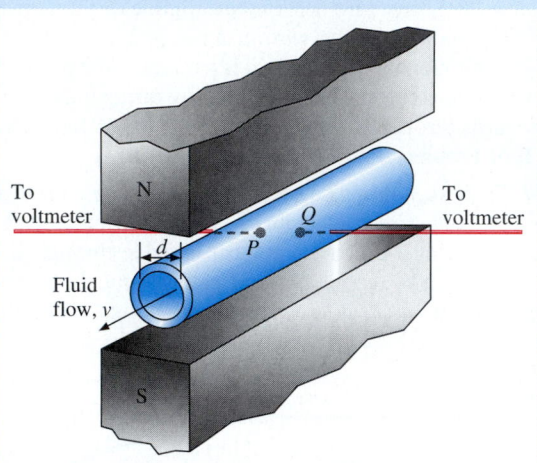

The pipe passes through a uniform magnetic field $\mathscr{B}$. Voltmeter leads are placed at P and Q, in electrical contact with the fluid; the line joining P and Q is perpendicular to the magnetic flux lines. **(a)** Express the flow speed v of the fluid in terms of $\mathscr{B}$, V, and the pipe diameter d. **(b)** If $\mathscr{B} = 7.0 \times 10^{-2}$ T, $V = 73$ μV, and $d = 2.5$ mm, find v.

28.50 *(8) Keep your sample dry!* You do a Hall-effect measurement on a sodium sample. You find that a current density $j = 2.0 \times 10^6$ A/m² and a magnetic field 1.00 T yield a Hall field $|\mathscr{E}_y| = 5.0 \times 10^{-4}$ V/m. Find the carrier concentration.

28.51 *(8) Hall mobility.* Suppose that you measure the electrical resistivity ρ and the Hall coefficient R_H of a conducting sample. Show that

$$\frac{R_H}{\rho} = \frac{q\tau}{m},$$

where q and m are the carrier charge and mass, respectively, and τ is the collision time. The ratio R_H/ρ is called the *Hall mobility* μ_H (not to be confused with the unrelated magnetic dipole moment μ); compare with the definition of mobility given in Problem 27.33.

28.52 *(8) Hall angle.* (Before working part **e** of this problem, you should work Problem 28.51.) In discussing the Hall effect, we considered the electric field component $\mathscr{E}_y$ that results from the deflection of the charge carriers toward one edge of the sample. In addition, there must be an electric field $\mathscr{E}_x$, which drives the charge through the sample in the first place. The total electric field $\mathscr{E}$ is the vector whose components are $\mathscr{E}_x$ and $\mathscr{E}_y$. The *Hall angle* α is the angle between $\mathscr{E}$ and the x direction. **(a)** Suppose that a current i is passing through the Hall sample, with the magnetic field turned off. Then imagine that the magnetic field is turned on instantaneously. Show that the initial direction of charge carrier drift is $\theta = 0$. **(b)** Why do the charge carriers not continue to move in this direction? **(c)** Show that

$$\tan \alpha = \frac{\sigma \mathscr{B}}{nq},$$

where σ is the conductivity of the sample. **(d)** Suppose that the carriers have mass m and that their collision time is τ. Express $\tan \alpha$ in terms of q, m, τ, and $\mathscr{B}$. **(e)** Express $\tan \alpha$ in terms of the Hall mobility μ_H. **(f)** Calculate the Hall angle for silver, using the values given or calculated in Example 28.4, together with a value of the conductivity calculated from Table 27.1.

GROUP C

28.53 *(4) Cyclotron resonance.* As noted in Problem 27.70, the electrons that carry charge in many solids behave as if they had an *effective mass m^** different from the mass m of a free electron. Knowledge of the value of m^* is very important to an understanding of the detailed theory of conduction in solids. *Cyclotron resonance* is a powerful technique for measuring m^*. For the time being, neglect the collisions made by the electrons; that is, assume that the mean free path is $\lambda \to \infty$. Under the influence of the magnetic field shown in the figure at right, the electrons in the block of sample material move in helical orbits (see Problem 28.33). An electric field of constant magnitude, whose direction rotates with angular frequency ω, is also imposed on the solid in the plane shown. (This can be readily done by means of microwave techniques, in a device called a *resonant cavity*.)

(a) In the experiment, the value of ω is fixed and the value of $\mathscr{B}$ is adjusted until the amount of input energy required to keep the amplitude of the electric field constant reaches a max-

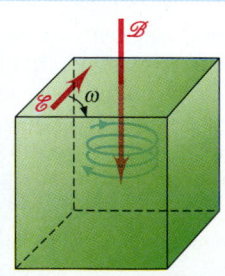

imum. Write an equation giving the condition under which electrons gain energy from the electric field. Express the effective mass m^* in terms of experimentally measured quantities. **(b)** What would happen to the electrons if λ were actually equal to infinity? **(c)** In fact, if pure samples are used and measurements are made at low temperatures and high frequencies, it is possible to attain values of λ large enough that (on the average) the electrons can make several revolutions between collisions. Describe qualitatively what happens under these circumstances. **(d)** Semiconductors (and some other materials) can contain charge carriers called *holes*, having positive charge, as well as electrons. Often, both holes and electrons are present. How does cyclotron resonance distinguish between them?

28.54 *(5) Magnetic stiffener.* A loop is made of a length l of limp wire. The loop lies on a frictionless table. A uniform magnetic field $\mathcal{B}$ is directed vertically. **(a)** What happens when a current i is passed through the loop? **(b)** Find the tension in the wire.

28.55 *(6) Magnetic pole strength.* Remembering the definition of an electric dipole, $\mathbf{p} \equiv 2q\mathbf{b}$ (Equation 23.22), write the magnetic dipole moment of a bar magnet in the analogous form

$$\mu = 2\,\Pi\mathbf{b},$$

where $2\mathbf{b}$ is the vector from the south pole to the north pole and Π is a quantity called the *magnetic pole strength.* **(a)** A bar magnet lies in a uniform magnetic field of magnitude $\mathcal{B}$. Express Π in terms of b, T, $\mathcal{B}$, and the angle θ between $\mathbf{b}$ and $\mathcal{B}$. **(b)** What are the units of Π in SI? **(c)** Calculate the pole strength of the magnet of Problem 28.39. (Assume that the poles are exactly at the ends of the magnet.) **(d)** To what electrical quantity is Π analogous? Where does the analogy between the two quantities fail? **(e)** Imagine that the bar magnet is replaced by a hollow tube of the same length l and the same cross section $A = wt$, made of a conducting material. Imagine also that a uniform current circulates around the surface of the tube. For what current i will the tube act like a magnet having the same pole strength Π as the magnet it replaces?

28.56 *(8) MHD generator.* The *magnetohydrodynamic generator* is a device that generates electric power without moving parts. It is a large-scale application of the same principle that underlies the electromagnetic flowmeter of Problem 28.49. The MHD generator operates at high temperature, making possible considerable efficiency in converting the chemical energy of fuel into electrical energy. (Remember that the thermodynamic efficiency of a conventional steam-turbine generator is determined not by the flame temperature but by the much lower steam temperature.) In the figure (next column), fuel is burned in the combustion chamber at the left. The hot, gaseous combustion products emerge at high speed v from the nozzle, which is something like the nozzle of a jet engine. The temperature is sufficiently high that a significant

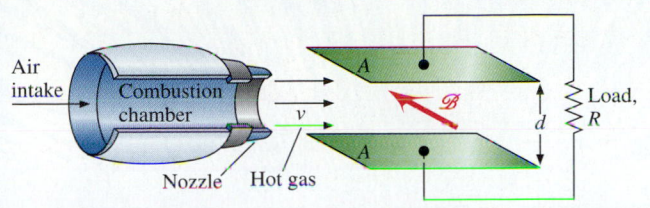

proportion of the molecules in the gas are ionized, and the electrical resistivity ρ of the gas is small. (Sometimes the gas is "seeded" with a small quantity of an easily ionizable substance such as sodium to reduce ρ still further.) The gas passes between a pair of plates of area A spaced a distance d apart. A magnetic field $\mathcal{B}$ is directed perpendicular to the gas flow and parallel to the plates. Electrical leads connect the plates to a load of resistance R. **(a)** Why does a potential difference appear between the plates? **(b)** Express the power P delivered to the load in terms of v, $\mathcal{B}$, ρ, A, d, and R. **(c)** Show that this power is maximized if $R = \rho d/A$, and that

$$P_{\text{max}} = \frac{v^2 \mathcal{B}^2 A d}{4\rho}.$$

MHD generators have been the subject of considerable engineering research, but serious technical problems still stand in the way of commercial use. The likely advent in the near future of large, cheaply operated superconducting magnets may well change the picture.

28.57 *(8) Making corrections.* In measuring the Hall effect, it is not always possible to place the leads P and Q in Figure 28.23 with sufficient accuracy that the line joining them is parallel to the y axis. This difficulty is shown below in exaggerated fashion. A common way to circumvent this difficulty is to measure the voltage between P and Q four times, reversing both current and magnetic field and measuring with all possible combinations of sense of i and direction of $\mathcal{B}$. Let us call the four measured voltages V_{++}, V_{+-}, V_{-+}, and V_{--}, where the first subscript sign refers to the sense of i and the second to the direction of $\mathcal{B}$. Express the true Hall voltage V_H in terms of these four measured quantities.

Magnetism II

Magnetic Field of an Electric Current

Left: The magnetic field patterns of the bar magnet and the current-carrying solenoid look almost exactly the same. Ampère was the first to suggest that all magnetic fields are associated with electric currents.

What immortal hand or eye
Could frame thy fearful symmetry?

—WILLIAM BLAKE, *"The Tiger"*

Introduction: Oersted's Discovery

In the early spring of 1820, the Danish physicist and chemist Hans Christian Oersted (1777–1851) was preparing demonstration apparatus for a popular lecture on electricity and magnetism when his curiosity was piqued by an accidental observation that others had probably made earlier without taking notice. When Oersted passed a current through a length of wire, he noted that a nearby compass needle changed its orientation. Evidently, a magnetic field other than the earth's field was present, and Oersted quickly surmised that the field was associated with the current in the wire. Further observation confirmed this surmise, and Oersted then proceeded to map the field. Using small magnetic needles, he found that the flux lines were circles lying in planes normal to the wire and centered on the wire, as shown in Figure 29.1. The closed form of these flux lines is consistent with what we noted in Chapter 28: no magnetic monopoles are involved.

FIGURE 29.1 A magnetic field exists in the space surrounding a current-carrying straight wire. The field can be mapped with one or more small compass needles, as shown. The sense of the magnetic flux lines is related to the sense of the electric current according to the screwdriver rule.

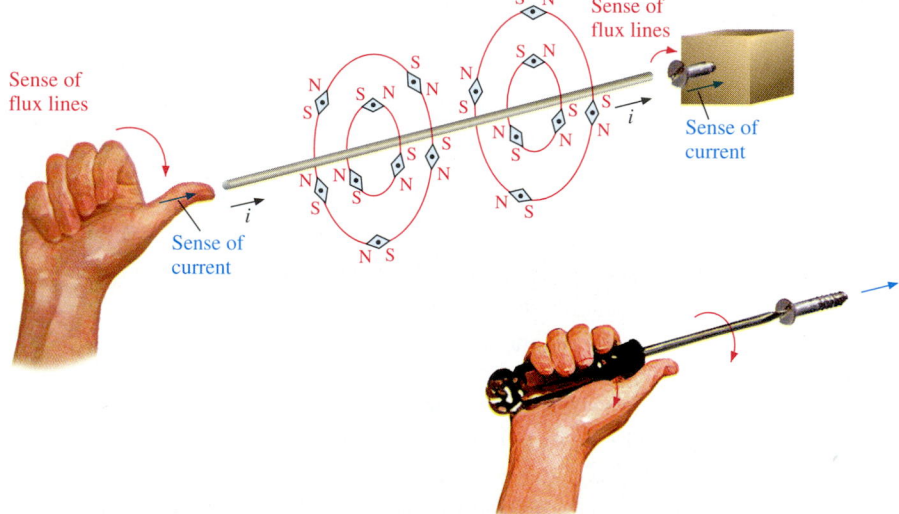

If the sense of the electric current is reversed, so is the sense of the magnetic flux lines. Two convenient ways to visualize the relation between the two senses are shown in Figure 29.1.

1. Curl the fingers of your right hand, extending the thumb. When the thumb points in the sense of the current, the fingers curl in the sense of the magnetic flux.

2. Imagine a standard right-handed wood screw whose head is turned in the sense of the magnetic flux. The screw will advance in the sense of the current.

To see that both methods express the same rule, imagine your right hand on the handle of a screwdriver, driving the screw into the wood. Your thumb points in the same direction that the screw advances. For convenience in distinguishing this rule from the right-hand rule for cross products (Section 13.4), we call this rule the **screwdriver rule**.

In what follows, we put quantitative flesh on the qualitative bones of Oersted's observation. Section 29.2 deals with the Biot-Savart law, which is the magnetic analogue of Coulomb's law for electric charge. In Sections 29.3 through 29.5, we study Ampère's law, whose relation to the Biot-Savart law is very much like the relation of Gauss's law to Coulomb's law.

Ampère's Current Balance and the Biot-Savart Law

In hindsight, we can discern a satisfying symmetry between Oersted's discovery and what we have already learned about the magnetic force:

> *A magnet exerts a force on a current-carrying wire,*
> *and*
> *a current-carrying wire exerts a force on a magnet.*

This symmetry is elegantly demonstrated in the twin apparatus invented by Faraday in 1821, shown in Figure 29.2. The system on the right side demonstrates the magnetic force considered in Chapter 28; this part of the apparatus is the subject of Query 28.12. The system on the left side shows Oersted's effect. A magnetic field arises from the current passing through the fixed wire. The magnet pole tends to follow the flux line on which it is located. Because that flux line is a closed circle centered on the wire, the pole at the free end of the magnet describes an endless circular path about the wire.

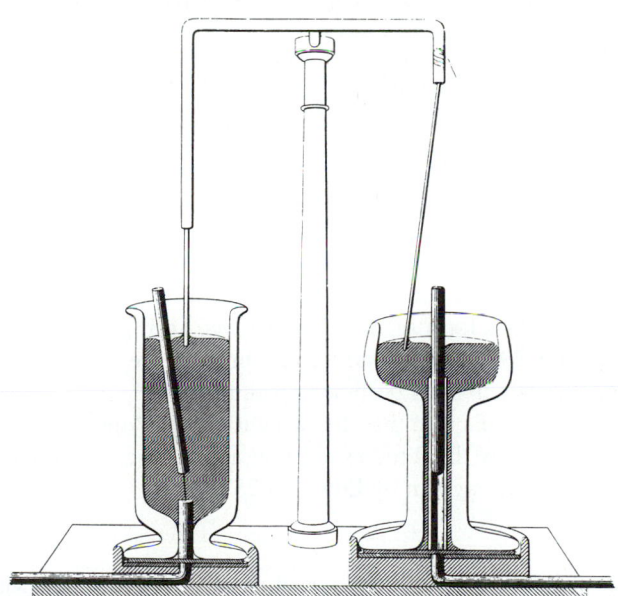

FIGURE 29.2 Faraday's first electric motors (1821). On the right, the current-carrying wire is suspended from a universal joint. The field of the fixed permanent magnet exerts a force on the wire, whose lower end rotates in a horizontal circle about the magnet pole. On the left, the magnet, which tends to float in mercury, is held down at its lower end by a universal joint. The magnetic field that arises from the current in the fixed wire exerts a force on the upper pole of the permanent magnet, which rotates in a horizontal circle about the wire. In both cases, the electric circuit consists of the magnet, the wire, the mercury pool with which both are in contact, and an external battery (not shown).

News of Oersted's discovery reached Paris about June 1820 and immediately provoked a flurry of furious activity among the highly competitive group of first-rate physicists in that city. Within a matter of weeks, the quantitative ramifications of Oersted's qualitative discovery had been worked out. The chief contributors to this work were André-Marie Ampère, aided by his friend François Arago (1786–1853), and the collaborators Jean-Baptiste Biot (1774–1862) and Félix Savart (1791–1841).

Ampère's simple but brilliant insight was this: If a current in a conductor gives rise to a magnetic field, then the conductor is behaving like a magnet. Consequently, two current-carrying wires should attract or repel one another like two magnets; the interaction between the two takes place through their magnetic fields.

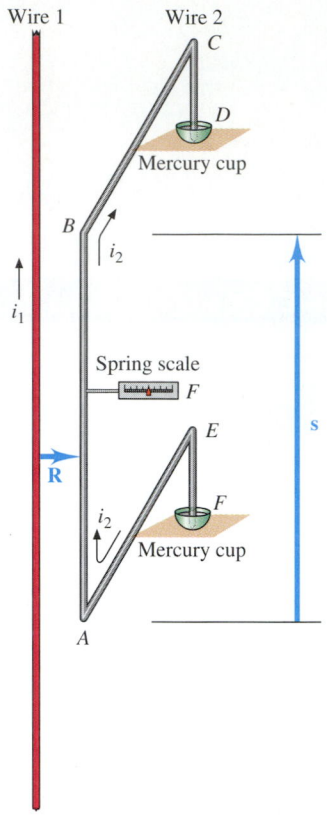

FIGURE 29.3 Ampère's current balance. The long, straight, vertical wire (1) carries current i_1, and the U-shaped wire (2) carries current i_2. (In practice, the two wires are usually connected in series so that i_1 and i_2 are equal.)

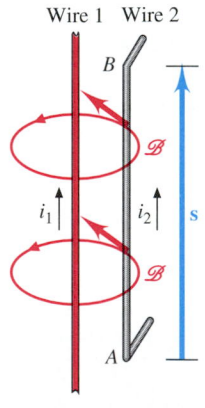

FIGURE 29.4 Part of wire 2 of Figure 29.3, and the magnetic field arising from the current in wire 1. The flux lines are circles centered on wire 1, and the magnetic field vectors are directed along the tangents to these flux lines.

To demonstrate this point quantitatively, Ampère chose the simplest possible geometry. His **current balance** is shown schematically, somewhat simplified, in Figure 29.3.* Wire 1, which is long, straight, vertically oriented, and fixed, carries current i_1, and U-shaped wire 2 carries current i_2. The vertical segment of the U, AB, has the directed length **s**. Segment AB is parallel to the fixed wire, separated from it by a distance R. The ends of wire 2 lie in two small mercury cups, which act both as low-friction bearings and as good electrical contacts. A sensitive spring scale can be used both to adjust the spacing R and to measure the force required to hold wire 2 in place when the current is turned on.

When i_1 and i_2 have the same sense, as shown in Figure 29.3, each wire is attracted toward the other. When the senses of the currents are opposite, the forces exerted by the wires on each other are repulsive. Not surprisingly, the magnitude of the force depends on the currents i_1 and i_2, the length s, and the separation distance R. Ampère found the relation

$$\mathbf{F} \propto \frac{-i_1 i_2 s}{R}\, \hat{\mathbf{R}}. \tag{29.1}$$

In this proportionality, the direction $\hat{\mathbf{R}}$ is taken from the fixed wire to the moveable wire, and the force is measured on the moveable wire. With i_1 and i_2 parallel, the product $-i_1 i_2$ is negative whether we choose the sense shown in the diagram as positive or as negative. Thus the force on wire 2 has the direction $-\hat{\mathbf{R}}$, as must be the case for an attractive force. If, on the other hand, i_1 and i_2 are antiparallel, one of the two currents must have a negative value and the product $-i_1 i_2$ is positive, as must be the case for a repulsive force on wire 2 having direction $\hat{\mathbf{R}}$.

As usual, we recast the proportionality as an equation by choosing a proportionality constant. In SI, the proportionality constant is $\mu_0/2\pi$, where we *define*

$$\mu_0 \equiv 4\pi \times 10^{-7}\ \text{N/A}^2 = 4\pi \times 10^{-7}\ \text{T·m/A}. \tag{29.2}$$

Having thus defined the proportionality constant, we can rewrite the force in the form

$$\mathbf{F} = -\frac{\mu_0}{2\pi} \frac{i_1 i_2 s}{R}\, \hat{\mathbf{R}}. \tag{29.3}$$

The quantity μ_0 is called the **permeability of free space**.[†] It is the magnetic analogue of the electrical quantity ϵ_0, the permittivity of free space.

As in the electrical case, inserting the factor 4π into the proportionality constant in Equation 29.2 makes possible the writing of other, more frequently used equations in convenient form. The first form of the SI unit given in Equation 29.2, N/A^2, arises directly from Equation 29.3. The second form, T·m/A, is the standard form. The equivalence of the two forms will soon become apparent.

Let us return to Ampère's idea that the two current-carrying wires interact like two magnets. We already know the force exerted on a current-carrying wire in a uniform magnetic field. This force is given by Equation 28.12b,

$$\mathbf{F} = i\mathbf{s} \times \mathscr{B}. \tag{29.4a}$$

Let us focus our attention on the force exerted on moveable wire 2 owing to the magnetic field that arises from the current in fixed wire 1. That is, let us consider the force

$$\mathbf{F} = i_2 \mathbf{s} \times \mathscr{B}.$$

Because wire 1 is very long, the symmetry of the situation in Figure 29.3 assures us that wire segment AB lies in a uniform magnetic field because all parts of AB are equidistant from wire 1. The direction of the magnetic field, given by the screwdriver rule of Section 29.1, is everywhere perpendicular to AB (Figure 29.4). By applying the

---

*The main complications, not shown, have to do with eliminating the effects of the earth's magnetic field. For a clear, detailed description of this and related experiments, see J. C. Maxwell, *A Treatise on Electricity and Magnetism*, 3d ed., vol. 2 (Oxford, 1891; reprint, New York: Dover Publications, 1954), pp. 158–163.

[†]Don't confuse μ_0 with the magnetic moment $\boldsymbol{\mu}$ or its magnitude μ.

right-hand rule for the cross product, you can see that the direction of $\mathbf{s} \times \mathscr{B}$ is toward wire 1; we have

$$\mathbf{s} \times \mathscr{B} = -s\mathscr{B}\hat{\mathbf{R}}.$$

Thus Equation 29.4a becomes

$$\mathbf{F} = -i_2 s \mathscr{B}\hat{\mathbf{R}}, \qquad (29.4b)$$

where $\mathscr{B}$ is the magnitude of the magnetic field due to the current i_1 in wire 1 at the location of segment AB of wire 2. We can determine the value of $\mathscr{B}$ by eliminating $\mathbf{F}$ from Equations 29.3 and 29.4b:

$$\mathscr{B} = \frac{\mu_0}{2\pi}\frac{i_1}{R} \quad \text{near a long, straight wire.} \qquad (29.5)$$

(The fact that the SI unit of μ_0 is T·m/A follows directly from this equation.)

What about the force exerted on segments BC and AE of wire 2 (Figure 29.3)? This force is not detected by the spring balance. Near their lower ends (the part shown in Figure 29.4), these segments are essentially parallel to the magnetic field, and so the cross product of Equation 29.4a is zero. Higher up, the field vector has a horizontal component because of the orientation of $\mathscr{B}$ there. But the forces on segments BC and AE are everywhere equal and opposite and add to zero. Why can we ignore the forces exerted on segments CD and EF?

EXAMPLE 29.1

Suppose the two wires in Figure 29.3 are in series, so the same current $i = 12$ A passes through both. However, the sense of the current in wire 1 is opposite that shown in Figure 29.3; see Figure 29.5. The distance between the two wires is 8.3×10^{-3} m, and the length AB is $s = 22$ cm. (a) Find the magnitude and direction of the magnetic field arising from the current in wire 1 at the location of AB. (b) Find the magnitude and direction of the force on wire 2.

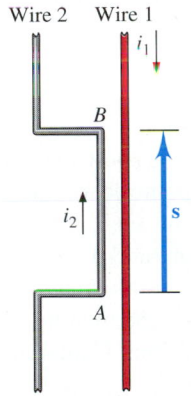

FIGURE 29.5

SOLUTION:
(a) Find $\mathscr{B}$.

The screwdriver rule shows that the direction of $\mathscr{B}$ at AB is out of the page. Using Equation 29.5, you have for the magnitude of the field

$$\mathscr{B} = \frac{4\pi \times 10^{-7} \text{ T·m/A}}{2\pi} \times \frac{12 \text{ A}}{8.3 \times 10^{-3} \text{ m}}$$

$$= 2.9 \times 10^{-4} \text{ T}.$$

This is a fairly weak field; it is an order of magnitude greater than the magnetic field of the earth.

(b) Find the force on wire 2.

You could use Equation 29.3 to evaluate the force. But, given that you already have the value of $\mathscr{B}$, it is easier to use Equation 29.4a. For the direction of $\mathbf{F}$, you use the right-hand rule to show that

$$\hat{\mathbf{s}} \times \hat{\mathscr{B}} = \hat{\mathbf{R}}.$$

That is, the force is repulsive, away from wire 1. For the magnitude of the force, you have

$$F = is\mathscr{B} = 12 \text{ A} \times 0.22 \text{ m} \times 2.9 \times 10^{-4} \text{ T}$$

$$= 7.7 \times 10^{-4} \text{ N}.$$

Definition of the Ampere and the Coulomb

In principle, the Coulomb experiment (Section 23.3) could be used to define the coulomb, the unit of electric charge. Then the ampere, the unit of electric current, could be defined as 1 A $\equiv$ 1 C/s. From a practical point of view, however, this is not a good way to proceed. Steady currents are much easier to maintain than steady static charges, and much better measurements can be made with a current balance than with a Coulomb balance. Consequently, the unit of electric current—the ampere—is defined first in terms of a current-balance measurement. In practice, a precision current balance uses a pair of coils of many turns rather than a pair of long, straight wires. But this does not change the principle. We use the magnitude part of Equation 29.3 to write the force per unit length,

$$\frac{F}{s} = \frac{\mu_0}{2\pi} \frac{i_1 i_2}{R};$$

by international convention, the **ampere** is defined to be

that constant current which, if maintained in two straight parallel conductors of infinite length, of negligible circular cross section, and placed one meter apart in vacuum, would produce between these conductors a force equal to 2×10^{-7} newton per meter of length.

The current balance has now been superseded as a measurement tool by much more precise techniques based on a special phenomenon of superconductivity called the *ac Josephson effect.*

With a definition of the ampere in hand, we can define the **coulomb** as *the quantity of charge that flows past a given point in a circuit in* 1 s, *when the circuit carries a steady current of* 1 A.

The Biot-Savart Law

The $1/R$ dependence of the force given by Equation 29.3,

$$\mathbf{F} = \frac{\mu_0}{2\pi} \frac{i_1 i_2 s}{R} \, \hat{\mathbf{R}},$$

and of the magnetic field given by Equation 29.5,

$$\mathscr{B} = \frac{\mu_0}{2\pi} \frac{i_1}{R},$$

is a dependence we have already seen in systems having cylindrical symmetry. (See, for example, Example 24.3 and the discussions immediately following that example and Example 24.4.) In the electrostatic case, the $1/R$ dependence of the electric field surrounding a long, straight, uniformly charged wire—the case of *line symmetry*—was a logical consequence of the $1/r^2$ law governing the electric field surrounding a point charge—the situation of highest symmetry. Can we find a similar relation in the present case? In searching for such a relation, we will argue from analogy with Coulomb's law to the extent that it is possible to do so. But in the end, we must insist as always that the argument lead to experimentally verifiable results.

We cannot perform an experiment analogous to the Coulomb experiment because there is nothing analogous to the approximate point charge used in the Coulomb experiment. So instead we will test our argument by seeing if it leads to Equation 29.5 for a long, straight, current-carrying wire. (Bear in mind that Equation 29.5 is analogous to Equation 24.12, $\mathscr{E} = (1/2\pi\epsilon_0)(\lambda/R)$, which gives the magnitude of the *electric* field for the analogous case of the same symmetry—in the vicinity of a long, straight, uniformly charged wire—and which can be derived from Coulomb's law.)

We start with Coulomb's law in the form

$$d\mathscr{E}(\mathbf{r}) = \frac{1}{4\pi\epsilon_0} \frac{dq}{r^2} \, \hat{\mathbf{r}}, \tag{29.6}$$

which expresses the electric field at $\mathbf{r}$ due to the infinitesimal charge dq (Figure 29.6a). Our aim is to find an analogous rule for the magnetic field, but we run into a difficulty immediately. Coulomb's law applies to a *point* charge; there is no such thing as a "point current." After all, a current must go *from* somewhere *to* somewhere else. So we do the next best thing. We consider a finite current i passing through an infinitesimal element $d\mathbf{s}$ of a conductor, as shown in Figure 29.6b. The magnitude $d\mathscr{B}(\mathbf{r})$ of the magnetic field due to this element at any point $\mathbf{r}$ must certainly depend on i. But $d\mathscr{B}(\mathbf{r})$ must also depend on the length ds of the element. (If we double the length of the element, its length is still infinitesimal. But the same current passing through twice as

much conductor should produce twice as much magnetic field.) So the magnetic analogue of the electric charge dq is the product $i\,ds$, which is called a **current element**.
In terms of the current element, we have

$$d\mathcal{B} \propto i\,ds. \tag{29.7a}$$

Next, we argue that, although the segment is not a point, its infinitesimal length makes
it look very small from any finite distance r. Thus the inverse-square law ought to
apply, as far as the *magnitude* of the field is concerned:

$$d\mathcal{B} \propto \frac{1}{r^2}. \tag{29.7b}$$

Now, as to direction. Coulomb's law is isotropic, which means that the electric flux
lines emerge from a point charge equally in all directions. But a direction perpendicular
to ds cannot be the same as a direction along ds. The simplest three-dimensional mathematical operation we know that distinguishes between perpendicular and parallel is
the cross product. And we know from Oersted's and Ampère's experiments that a
magnetic field exists in all directions perpendicular to a *finite* straight wire. So let us
argue that

$$d\mathcal{B} \propto \hat{ds} \times \hat{r}, \tag{29.7c}$$

where $\hat{ds}$ is the unit vector in the direction of ds.

Combining the three proportionalities (Equations 28.7a, b, c), we obtain

$$d\mathcal{B} \propto \frac{i\,ds}{r^2}\,\hat{ds} \times \hat{r}.$$

Although this proportionality does incorporate an inverse-square behavior just as Coulomb's law does, it also has the directional factor expressed by the cross product.
Because $ds\,\hat{ds} = d\mathbf{s}$, we rewrite the proportionality in the slightly more convenient form

$$d\mathcal{B} \propto \frac{i}{r^2}\,d\mathbf{s} \times \hat{r}. \tag{29.8}$$

We choose the proportionality constant to be $\mu_0/4\pi$. The proportionality then becomes
the equation

$$d\mathcal{B} = \frac{\mu_0}{4\pi}\frac{i}{r^2}\,d\mathbf{s} \times \hat{r}. \tag{29.9a}$$

The magnitude of $d\mathcal{B}$ is

$$d\mathcal{B} = \frac{\mu_0}{4\pi}\frac{i\,ds}{r^2}\sin\theta, \tag{29.9b}$$

where θ is the angle between $d\mathbf{s}$ and $\mathbf{r}$. We will soon verify the argument that led to
Equations 29.9a and 29.9b, which is called the **Biot-Savart law**. The choice of the
constant is made by analogy with Coulomb's law, which contains the constant $1/4\pi\epsilon_0$.

The analogy between the Biot-Savart law and Coulomb's law (Equation 29.6) is
striking when we write them side by side:

$$d\mathcal{B} = \left(\frac{\mu_0}{4\pi}\frac{i\,ds}{r^2}\right)\hat{ds} \times \hat{r} \quad \text{and} \quad d\mathscr{E} = \left(\frac{1}{4\pi\epsilon_0}\frac{dq}{r^2}\right)\hat{r}.$$

$$\text{Biot-Savart law} \qquad\qquad\qquad \text{Coulomb's law}$$

The parts of the two equations in parentheses have exactly the same mathematical form.
However, the direction of $d\mathscr{E}$ in Coulomb's law is simply the direction $\hat{r}$, whereas the
direction of $d\mathcal{B}$ in the Biot-Savart law depends on the directions of the *two* vectors $d\mathbf{s}$
and $\mathbf{r}$. In addition, the cross product $d\mathbf{s} \times \hat{r}$ contributes a factor $\sin\theta$ to the magnitude
of the magnetic field at $\mathbf{r}$.

Now, let us check our argument by comparison with Equation 29.5, which expresses
the magnetic field due to a long, straight, current-carrying wire. We wish to calculate

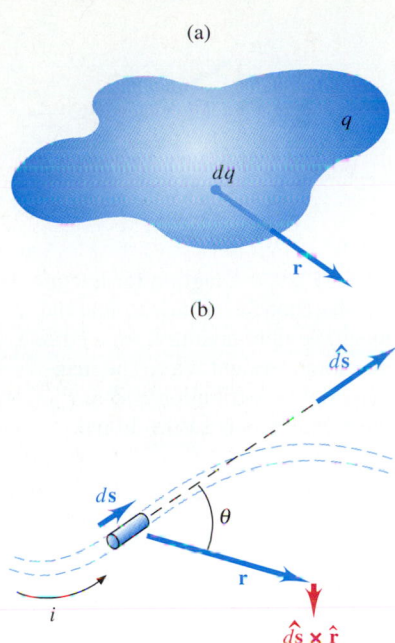

FIGURE 29.6 (*a*) The infinitesimal
charge element dq is part of a finite
charge distribution q. (The rest of
the charge distribution is shown in
light blue.) The contribution $d\mathscr{E}(\mathbf{r})$
of the element to the electric field at
the point defined by $\mathbf{r}$ depends on
three things: dq, the distance r, and
the direction $\hat{r}$. (*b*) A current i passes
through the infinitesimal element $d\mathbf{s}$
of a wire. (The rest of the wire is
shown in light blue.) The contribu-
tion $d\mathcal{B}(\mathbf{r})$ of the element to the
magnetic field at the point defined by
$\mathbf{r}$ depends on three things: the cur-
rent-length product $i\,ds$, the distance
r, and the cross product of the *two*
directions $\hat{ds}$ and $\hat{r}$. The angle θ between these two directions is shown.

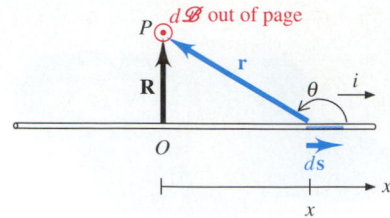

FIGURE 29.7 Diagram for integrating the Biot-Savart law to find the magnetic field arising from a current in a long, straight wire. The magnetic field contribution $d\mathscr{B}$ at P of the element $d\mathbf{s}$ is shown in red.

the magnetic field $\mathscr{B}$ at point P (in Figure 29.7), whose location with respect to the wire is given by the vector $\mathbf{R}$. We choose the x axis along the wire, with the origin at the tail of $\mathbf{R}$. We divide the wire into elements $d\mathbf{s} = dx\,\hat{\mathbf{x}}$. One of these elements is shown. By applying the right-hand rule to the cross product $d\mathbf{s} \times \hat{\mathbf{r}}$ of the Biot-Savart law, you can see that the direction of $d\mathscr{B}$ at P is out of the page. This is true for all elements $d\mathbf{s}$ along the wire. Thus the contributions of all the elements will add algebraically, and we can carry out the necessary integration using the scalar form of the Biot-Savart law (Equation 29.9b).

The elements $d\mathbf{s}$ are not equidistant from P. Let r be the distance from any particular element to P. According to Equation 29.9b, the magnitude of the contribution of that element to the field at P is

$$d\mathscr{B} = \frac{\mu_0}{4\pi} \frac{i\,dx}{r^2} \sin\theta.$$

The quantities r and θ are both variables. As we have done before, we express both in terms of the variable x:

$$r^2 = x^2 + R^2 \quad \text{and} \quad \sin\theta = \frac{R}{r} = \frac{R}{\sqrt{x^2 + R^2}}.$$

Making these substitutions in Equation 29.9b, we obtain

$$d\mathscr{B} = \frac{\mu_0 i}{4\pi} \frac{R\,dx}{(x^2 + R^2)^{3/2}}.$$

We now integrate over all elements to find the total field at P:

$$\mathscr{B} = \int d\mathscr{B} = \frac{\mu_0 i R}{4\pi} \int_{-\infty}^{\infty} \frac{dx}{(x^2 + R^2)^{3/2}}.$$

Carrying out the integration gives us

$$\mathscr{B} = \frac{\mu_0 i R}{4\pi} \left[\frac{1}{R^2} \frac{x}{\sqrt{x^2 + R^2}} \right]_{-\infty}^{\infty},$$

or

$$\mathscr{B} = \frac{\mu_0 i}{2\pi R}.$$

This is identical with Equation 29.5, which verifies the argument leading to the Biot-Savart law.

We have checked the Biot-Savart law in a highly symmetric case, but it can be applied to any current-carrying conductor, no matter what its shape. One such application is the subject of Example 29.2.

EXAMPLE 29.2

Figure 29.8 shows a circular loop of wire of radius b carrying a current i. Find the magnitude and direction of the magnetic field $\mathscr{B}$ at the point P, located on the axis of the loop at a distance x from the plane of the loop.

SOLUTION: First, note that the lead wires carry antiparallel currents. Because the leads are close together, their contributions to the magnetic field at P cancel each other. Only the loop makes a significant contribution.

You begin by choosing an element $d\mathbf{s}$ of the loop. The vector $\mathbf{r}$ from $d\mathbf{s}$ to P makes an angle α with the axis of the loop. You apply the right-hand rule to Equation 29.9a, the Biot-Savart law, to determine the direction of $d\mathscr{B}$. As you can see from Figure 29.8, the angle between the plane of the loop and $d\mathscr{B}$ is also α.

Next, you sketch the components of $d\mathscr{B}$. The component $d\mathscr{B}_x$ lies along the x axis, and the component $d\mathscr{B}_\perp$ lies in the plane perpendicular to the axis. You can use a symmetry argument to show without calculation that the contributions

FIGURE 29.8

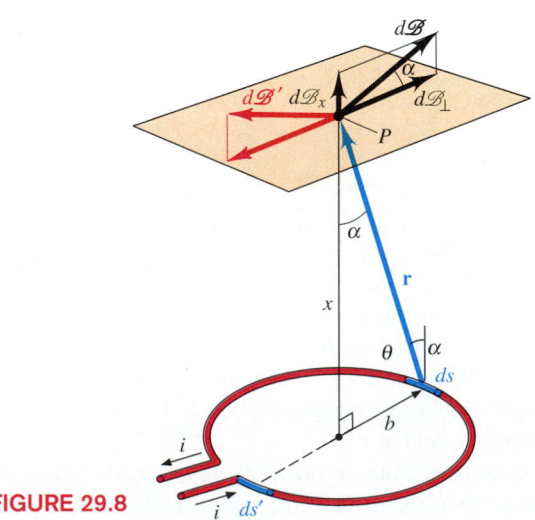

$d\mathcal{B}_\perp$ cancel. To see this, consider the element ds' that lies diametrically opposite ds. Its contribution $d\mathcal{B}'$ is shown in red, and you can easily see that the perpendicular component of this contribution is equal to and opposite $d\mathcal{B}_\perp$.

All elements ds of the loop make equal contributions $d\mathcal{B}_x$ to the axial component of the total field because all elements have the same length, carry the same current, and are the same distance from P. Each of these contributions is

$$d\mathcal{B}_x = d\mathcal{B} \sin \alpha.$$

Using the Biot-Savart law to evaluate $d\mathcal{B}$, you have

$$d\mathcal{B}_x = \left(\frac{\mu_0}{4\pi} \frac{i\,ds}{r^2} \sin \theta\right) \sin \alpha.$$

In this equation, θ is the angle between ds and $\mathbf{r}$. From the diagram, it is evident that $\theta = 90°$, and so $\sin \theta = 1$. All the other quantities in the equation are constants, and it is very simple to integrate the contributions of all the elements around the loop:

$$\mathcal{B} = \int_{\text{loop}} d\mathcal{B}_x = \frac{\mu_0 i \sin \alpha}{4\pi r^2} \int_{\text{loop}} ds.$$

But the integral of the length ds around the loop is just the circumference of the loop, $2\pi b$. So you have

$$\mathcal{B} = \frac{\mu_0 i b \sin \alpha}{2r^2}. \tag{29.10}$$

Finally, you want to express this result in terms of the distance x rather than the distance r and the angle α. You write

$$r^2 = x^2 + b^2 \quad \text{and} \quad \sin \alpha = \frac{b}{r} = \frac{b}{\sqrt{x^2 + b^2}}.$$

Substituting these values into Equation 29.10, you obtain the result

$$\mathcal{B} = \frac{\mu_0 i}{2} \frac{b^2}{(x^2 + b^2)^{3/2}}. \tag{29.11a}$$

Because the direction of $\mathcal{B}$ is the x direction, you can immediately express the magnitude and direction of the magnetic field in the vectorial form

$$\mathcal{B} = \frac{\mu_0 i}{2} \frac{b^2}{(x^2 + b^2)^{3/2}} \,\hat{\mathbf{x}}. \tag{29.11b}$$

It is interesting to compare the result of Example 29.2 with that of Example 23.5, which concerns the electric field on the axis of a loop carrying a uniformly distributed charge q.

First, consider the direction. For the charged loop, the electric field on the axis is everywhere away from the loop—in the x direction on the positive side of the loop and in the $-x$ direction on the negative side. In the plane of the loop, the electric field vanishes. This is evident from Equation 23.28, which we write here with a slight change in notation:

$$\mathscr{E} = \frac{1}{4\pi\epsilon_0} q \frac{x}{(x^2 + b^2)^{3/2}}. \tag{29.12}$$

Evidently, $\mathscr{E}$ goes to zero for $x = 0$. But, for the current-carrying loop, the direction of the magnetic field is the same everywhere along the axis, as is evident from Equation 29.11b. The magnitude of the magnetic field attains its greatest value in the plane of the loop, where $\mathcal{B} = \mu_0 i / 2b$.

To make a direct comparison between Equations 29.11a and 29.12, let us rewrite Equation 29.11a slightly. The circumference of the loop is $s = 2\pi b$. Hence we can write

$$\mathcal{B} = \frac{\mu_0}{4\pi} is \frac{b}{(x^2 + b^2)^{3/2}}. \tag{29.13}$$

The factor $\mu_0/4\pi$ is analogous to the factor $1/4\pi\epsilon_0$. The quantity is is analogous to the electric charge q in the same sense that the current element $i\,ds$ is analogous to the charge element dq. But the magnitudes $\mathcal{B}$ and $\mathscr{E}$ do not vary in the same way with distance; the fractions on the right sides of the two equations are not the same.

Consider in particular the behavior of Equations 29.11a and 29.12 at large distances, $x \gg b$. The electric field approaches the inverse-square value typical of a point charge, $\mathscr{E} = (1/4\pi\epsilon_0)(q/x^2)$. But, for the magnetic field, we have

$$\mathcal{B} \to \frac{\mu_0 i b^2}{2} \frac{1}{x^3} \quad \text{for } x \gg b.$$

Remembering that the area of the loop is $A = \pi b^2$, we can write this expression in the form

$$\mathcal{B} \to \frac{\mu_0}{2\pi} \frac{iA}{x^3} \quad \text{for } x \gg b.$$

Taking direction into consideration, we have

$$\mathcal{B} \to \frac{\mu_0}{2\pi} \frac{iA}{x^3} \hat{\mathbf{x}} = \frac{\mu_0}{2\pi} \frac{\boldsymbol{\mu}}{x^3} \quad \text{for } x \gg b, \tag{29.14}$$

where $\boldsymbol{\mu} \equiv iA\hat{\mathbf{x}}$ is the magnetic dipole moment of the loop (see Equation 28.18). The inverse-cube behavior is typical of dipoles, but bear in mind that the loop looks like a dipole only from large distances.

Ampère's Law

In electrostatics, Coulomb's law is logically equivalent to Gauss's law. Gauss's law has proved to be a powerful tool for understanding the properties of electric fields, as well as for calculating electric fields in systems exhibiting some degree of symmetry. The Biot-Savart law is the magnetic analogue of Coulomb's law. Can we fill in the missing item in the following logical structure?

Coulomb's law	$\leftrightarrow$ is analogous to $\leftrightarrow$	**the Biot-Savart law**
$\updownarrow$		$\updownarrow$
is logically equivalent to		is logically equivalent to
$\updownarrow$		$\updownarrow$
Gauss's law	$\leftrightarrow$ is analogous to $\leftrightarrow$	— ? —

The first possibility that comes to mind is a magnetic form of Gauss's law. In analogy to Equation 24.8, Gauss's law for electric charge,

$$\frac{q}{\epsilon_0} = \int_{\substack{\text{closed} \\ \text{surface}}} \mathcal{E} \cdot d\mathbf{A}, \tag{29.15}$$

we write

$$0 = \int_{\substack{\text{closed} \\ \text{surface}}} \mathcal{B} \cdot d\mathbf{A}. \tag{29.16}$$

Equation 29.16 is, indeed, **Gauss's law for magnetism**. The zero on the left side is yet another statement of the fact that magnetic monopoles do not exist; there is no magnetic quantity equivalent to the electric charge q on the left side of Equation 29.15. Although it is interesting, Equation 29.16 does not relate $\mathcal{B}$ to the electric current with which it is associated and thus is not useful for calculation.

So we must think a little more. We have a hint in Equation 29.5, which *does* relate magnetic field to electric current in the *special* case of a long, straight wire. Modifying the notation of the equation slightly to make it more general, we have

$$\mathcal{B} = \frac{\mu_0 i}{2\pi r}. \tag{29.17}$$

We multiply through by $2\pi r$ to obtain

$$u_0 i = \mathcal{B} 2\pi r. \tag{29.18}$$

Although this equation is not a form of Gauss's law, you can see from Figure 29.9 that it does have properties in common with Gauss's law for electric charge. The figure shows a magnetic flux line surrounding a long wire carrying current i. The flux line is a circle of radius r and circumference $s = 2\pi r$. According to Equation 29.17, the magnetic field has the same magnitude $\mathcal{B}$ everywhere on this flux line. The right side of Equation 29.18 is simply the product $\mathcal{B}s$. This product can in turn be written as the *line integral*

$$\mathcal{B}s = \int_{\text{circle}} \mathcal{B} \, ds,$$

where ds is an infinitesimal element of the circumference of the circle. In this special,

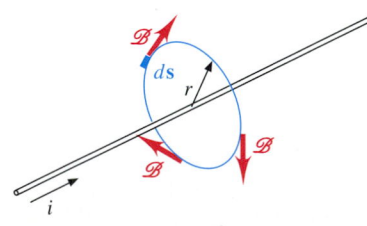

FIGURE 29.9 A long, straight wire carries current i. One of the magnetic flux lines is shown; it is a circle of radius r. The magnetic field on the flux line is everywhere tangent to the flux line and has magnitude $\mathcal{B}$. The circular path around the wire that coincides with the flux line is made up of elements $d\mathbf{s}$ and has total length $s = 2\pi r$.

highly symmetrical case, we have

$$\mu_0 i = \int_{\text{circle}} \mathscr{B}\, ds.$$

We can generalize this equation. Figure 29.10a shows an *arbitrary* closed curve surrounding the same current-carrying wire. (The view of the wire is end on.) The curve is not everywhere at the same distance from the wire. And $\mathscr{B}$ is in general not tangent to the curve. Nevertheless, we can rewrite the preceding equation in the more general form

$$\mu_0 i = \int_{\substack{\text{closed} \\ \text{curve}}} \mathscr{B} \cdot d\mathbf{s}. \qquad (29.19)$$

To show that this equation is valid, we first express the dot product in the form

$$\mathscr{B} \cdot d\mathbf{s} = \mathscr{B}\, ds \cos \theta,$$

where θ is the angle between $d\mathbf{s}$ and the tangent to the circular flux line of radius r that it crosses. As you can see from Figure 29.10b, $ds \cos \theta$ is the component of $d\mathbf{s}$ tangent to the flux line. This component subtends an angle $d\phi$ at the center of the wire. According to the definition of an angle expressed in radians, we have

$$ds \cos \theta = r\, d\phi.$$

Using this equation, together with the value of $\mathscr{B}$ given by Equation 29.5, we can write Equation 29.19 in the form

$$\mu_0 i = \int_{\substack{\text{closed} \\ \text{curve}}} \frac{\mu_0 i}{2\pi r}\, r\, d\phi.$$

The r's cancel out, leaving only constants within the integral. So the equation becomes

$$\mu_0 i = \frac{\mu_0 i}{2\pi} \int_{\substack{\text{closed} \\ \text{curve}}} d\phi.$$

Whatever the shape of the closed curve, traveling around it once involves turning through an angle 2π about the wire. So $\int d\phi = 2\pi$, and the equation becomes $\mu_0 i = \mu_0 i$. This proves the correctness of the generalization expressed by Equation 29.19,

$$\mu_0 i = \int_{\substack{\text{closed} \\ \text{curve}}} \mathscr{B} \cdot d\mathbf{s}$$

in which $\mathscr{B}$ is integrated along a closed curve of any shape.

Equation 29.19 is called **Ampère's law**. The closed curve is called an **Amperean curve**. The integral on the right side of the equation is called the **magnetic circulation** C or simply the **circulation** when there is no chance for confusion. Ampère's law can be stated quite directly in words: *The magnetic circulation around any closed curve is equal to μ_0 times the electric current threading the curve.** The simplicity of this statement makes possible a further important generalization of Ampère's law. The relation between the left and right sides of Equation 29.19 depends not on the *geometrical* relation between the current and the Amperean curve, but only on the *topological* fact of the threading of the curve by the current. Thus, distorting the long, straight wire of Figure 29.9 into some other shape makes no difference to the validity of Ampère's law.

Indeed, the key to the relation between the current and the magnetic circulation lies in the fact that the two are linked, like adjacent links of a chain. Although we have considered a long, straight wire, we know that a steady electric current requires a *closed* circuit. No matter how long we make the wire, its ends must ultimately bend back and join a source of emf somewhere. Linkage is a topological rather than a geometric relation (Figure 29.11).

---

*We speak of a current threading an Amperean curve in exactly the same sense that a length of thread passes through the eye of a needle.

(a)

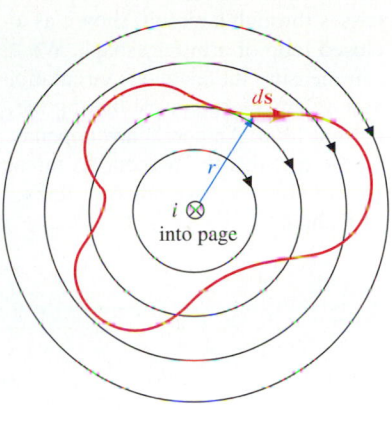

(b)

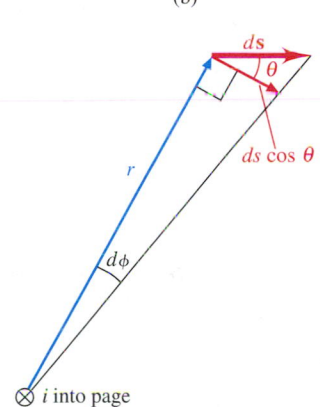

FIGURE 29.10 (*a*) An end-on view of the wire of Figure 29.9. Here we choose an arbitrary closed curve (of variable radius r) as the path of integration. Some of the circular flux lines that the path crosses are shown. (*b*) Expanded view of the particular path element $d\mathbf{s}$ shown in part *a*. The element $d\mathbf{s}$ makes an angle θ with $\mathscr{B}$, which is tangent to the local flux line and perpendicular to the radius. The tangential component of $d\mathbf{s}$ is $ds \cos \theta$; it subtends an angle $d\phi$ at the center of the wire.

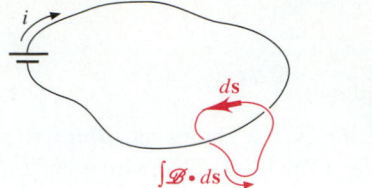

FIGURE 29.11 Electric current passes through a circuit, shown as a closed loop of arbitrary shape. We can measure the magnetic circulation around any closed curve linking the current loop. Ampère's law depends on the fact that the two curves thread one another like two adjacent links of a chain.

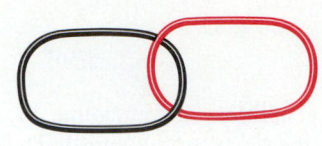

ds

$\int \boldsymbol{\mathscr{B}} \cdot d\mathbf{s}$

One final point must be settled before we can apply Ampère's law in an unambiguous way. Given an Amperean curve, which of the two possible senses shall we choose when we integrate around the curve? For consistency, we must use the screwdriver rule. *When you point your right thumb along the current, your fingers curl in the sense of the magnetic circulation.* If you choose this sense for integration, you will obtain the value $\int \boldsymbol{\mathscr{B}} \cdot d\mathbf{s}$. If you choose the opposite sense, you will obtain the value $-\int \boldsymbol{\mathscr{B}} \cdot d\mathbf{s}$, and you must insert a minus sign on the right side of Equation 29.19.

EXAMPLE 29.3

Evaluate the magnetic circulation $C = \int \boldsymbol{\mathscr{B}} \cdot d\mathbf{s}$ around the curves in the senses shown in parts *a* through *f* of Figure 29.12.

SOLUTION:

(a) Here you note that the conductor threads the Amperean curve, but the sense indicated for passage around the curve is opposite that given by the screwdriver rule for the magnetic

FIGURE 29.12

(a)

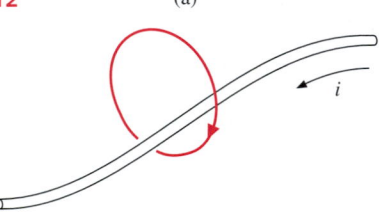

(b)

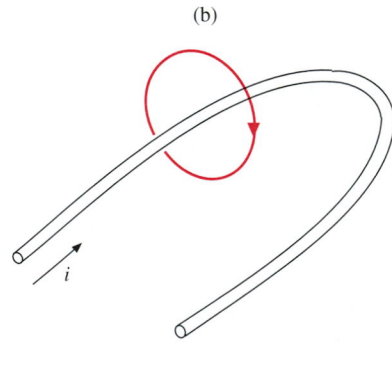

(c)

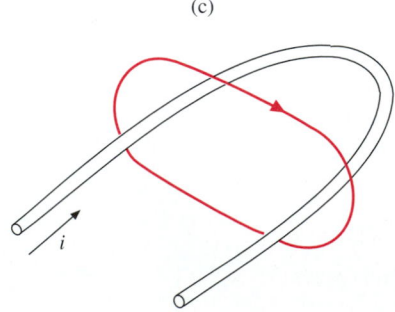

(d)

(e)

(f)

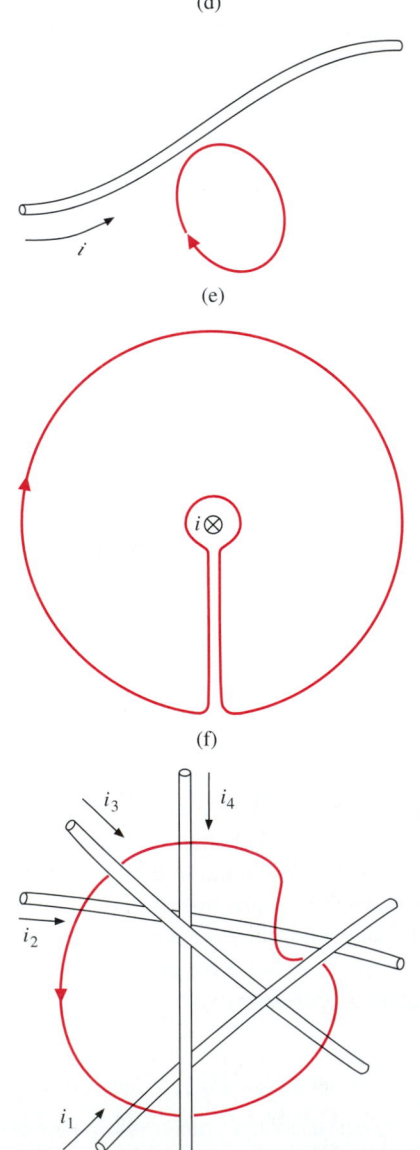

circulation. So you have

$$\mu_0 i = -C,$$

or

$$C = -\mu_0 i.$$

(b) That the conductor is bent into a "U" in the vicinity of the Amperean curve is of no consequence. The senses are consistent with the screwdriver rule, and so you have $C = \mu_0 i$.

(c) Here the conductor does not thread the Amperean curve at all; the current loop does not link with the Amperean curve. (You must assume that there are no linkages in the parts of the electric circuit not shown.) So the circulation is $C = 0$.

(d, e) In both these cases, again, there is no linkage of the current loop and the Amperean curve, and so $C = 0$.

(f) In this case, currents thread the Amperean curve in both senses. The senses of currents i_1 and i_4 and the sense indicated for the Amperean curve are in accordance with the screwdriver rule. The sense of i_3 is opposite. We can deal with this by calling i_3 a negative current. The current i_2 does not thread the Amperean curve at all. So you have

$$C = \mu_0(i_1 - i_3 + i_4).$$

EXAMPLE 29.4

Show that Ampère's law correctly describes the relation between the current i in a circular loop and the magnetic circulation around the Amperean curve shown in Figure 29.13. This curve is a very large rectangle whose upper side runs along the axis of the loop.

FIGURE 29.13

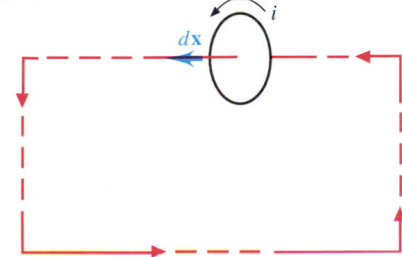

SOLUTION: The screwdriver rule yields the sense shown for the magnetic circulation around the rectangle. As you learned in Section 29.2, the loop is a magnetic dipole. At large distances, the magnetic field falls off quite rapidly, according to $1/r^3$. So, by making the rectangle large enough, you can ensure that $\mathcal{B}$ along its right, lower, and left sides is negligibly small, and thus that $\int\mathcal{B} \cdot d\mathbf{s}$ is negligible along these sides. Ampère's

law becomes

$$\mu_0 i = \int_{\substack{\text{closed} \\ \text{curve}}} \mathcal{B} \cdot d\mathbf{s} = \int_{\substack{\text{upper} \\ \text{side}}} \mathcal{B} \cdot d\mathbf{s}.$$

Equation 29.11b gives you the value of $\mathcal{B}$ along the loop axis:

$$\mathcal{B} = \frac{\mu_0 i}{2} \frac{b^2}{(x^2 + b^2)^{3/2}} \hat{\mathbf{x}}.$$

On this axis, $d\mathbf{s} = d\mathbf{x}$ and $\mathcal{B} \cdot d\mathbf{s} = \mathcal{B}\, ds$. So Ampère's law becomes

$$\mu_0 i = \frac{\mu_0 i b^2}{2} \int_{-\infty}^{\infty} \frac{dx}{(x^2 + b^2)^{3/2}}. \qquad (29.20)$$

You may as well take the limits of integration at infinity because the field at large $|x|$ is so small that it makes negligible contribution to the integral.

You can evaluate the integral directly or refer to a table of integrals. Its value is

$$\frac{x}{b^2\sqrt{x^2 + b^2}} \Bigg|_{-\infty}^{\infty} = \frac{2}{b^2}.$$

You substitute this value back into Equation 29.20 and obtain

$$\mu_0 i = \mu_0 i;$$

Ampère's law gives the correct result, even though the symmetry is quite different from the line symmetry we used in constructing the law. This gives us confidence that Ampère's law is valid for any geometry.

Applications of Ampère's Law

Although Ampère's law is quite general, its usefulness as a calculational tool is greatest in symmetrical situations. As is the case for Gauss's law, the number of possible symmetries is limited. We now consider these possibilities.

Magnetic Field inside a Current-Carrying Wire

Figure 29.14 shows a cylindrical wire of diameter b, carrying current I. We can use Ampère's law to determine the magnetic field at a point P, located inside the wire at a distance r from the wire axis. We choose as the Amperean curve a circle of radius r centered on the axis, as shown in Figure 29.14. By symmetry, the magnetic field has

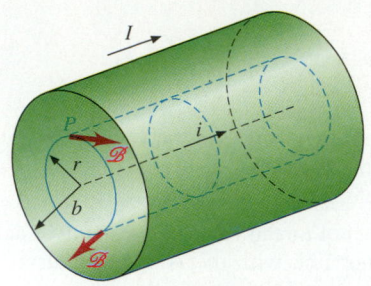

FIGURE 29.14 An Amperean curve drawn within a conductor carrying a current I with uniform current density j.

the same magnitude everywhere on the circle and is everywhere tangent to the circle, just as for an Amperean curve drawn outside the wire. Thus the magnetic circulation is

$$C = \int_{\substack{\text{Amperean} \\ \text{circle}}} \mathcal{B} \cdot d\mathbf{s} = 2\pi r \mathcal{B}.$$

We now wish to apply Ampère's law, equating the magnetic circulation to the product $\mu_0 i$. Here, however, the current i that threads the circle is not the entire current I in the wire. As you learned in Example 29.3, the part of the current passing outside the Amperean curve is of no consequence. If the current is uniformly distributed through the wire, the current density has the constant value $j = I/\pi b^2$ (see Equation 27.4a). The current i threading the Amperean curve is then the product of j with the area enclosed by the curve:

$$i = j\,\pi r^2 = I\,\frac{r^2}{b^2}.$$

We insert this value of i into Ampère's law (Equation 29.19) and obtain

$$\mu_0 I\,\frac{r^2}{b^2} = 2\pi r \mathcal{B}.$$

Solving for the magnetic field, we obtain

$$\mathcal{B} = \frac{\mu_0 I}{2\pi b^2}\,r \quad \text{for } j = \text{constant, } r < b. \tag{29.21}$$

The magnitude of the magnetic field increases linearly with distance from the axis of the wire. Compare this result with that of part **b** of Problem 24.35, which concerns the electric field inside a uniformly charged dielectric cylinder.

The Solenoid

It was Ampère who gave the name **solenoid** to the cylindrical coil of wire shown in Figure 29.15. The word comes from Greek words meaning "channel-like"; we will see that the solenoid acts like a channel for magnetic flux lines. In this, its behavior is quite similar to that of a bar magnet.

Although the winding of the solenoid is helical, we make the approximation that it consists of a stack of circular loops like the loop discussed in Example 29.2. The

FIGURE 29.15 (*a*) A solenoid. The wire is closely wound around a cylindrical core of radius b and length X; there are n turns per meter. The x axis is chosen along the axis of the solenoid. (*b*) Cross-sectional view of the solenoid.

(a)

(b)

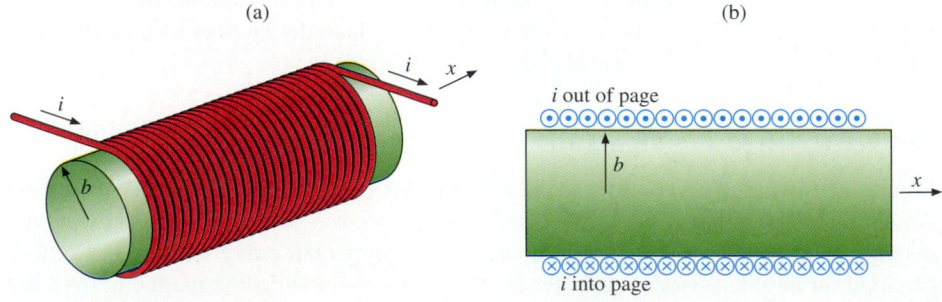

magnetic field of the solenoid is the sum of the contributions of the loops. Because the magnetic field on the axis of each loop is directed along the axis, the magnetic field on the axis of the solenoid is likewise directed along the axis.

If the solenoid is *long*—that is, if its length X is much greater than its radius b—we can show that *the magnetic field is parallel to the axis at all locations within the solenoid that are far from its ends.* The proof is based on a symmetry argument. In Figure 29.16, the magnetic field at point P is the sum of the field contributions of all the turns. Consider a pair of such turns, 1 and 2, located equidistant from P in opposite directions. At an off-axis point such as P, the field contributions $\mathscr{B}_1$ and $\mathscr{B}_2$ will not be parallel to the axis. You can see this from the schematic sketch of the flux lines that would pass through P if the electric current were run through each turn separately. The magnetic field contributions $\mathscr{B}_1$ and $\mathscr{B}_2$ lie along the tangents to these flux lines. By symmetry, they make equal and opposite angles with the x (axial) direction. The vector sum $\mathscr{B}_1 + \mathscr{B}_2$ is evidently parallel to the axis.

If P is not too close to one end of the solenoid, we can pair turns in this way for a considerable distance from P. Any remaining turns that cannot be paired are far from P. But the magnetic field of a current-carrying loop falls off rapidly with distance, and so we can neglect the contributions of these unpaired turns to the field at P.

What about the magnetic field outside the solenoid? The requirement that flux lines close on themselves means that the lines that pass through the solenoid must return through the space outside. But there is infinite space outside the solenoid, and the returning flux lines are well spread out (except at or near the ends, where most of the lines emerge from the solenoid). Consequently, the magnetic field outside the solenoid, far from its ends, is very weak. Such magnetic field as exists is essentially antiparallel to the field inside the solenoid. You can see this is indeed the case by inspecting Figure 29.17.

We are now ready to draw an Amperean curve. As with Gaussian surfaces, the idea is to draw the curve so that all segments of it are either parallel to or perpendicular to the magnetic field. The rectangle shown in Figure 29.18 satisfies this condition. Its length x along the solenoid axis is much smaller than the total length X of the solenoid, while the sides perpendicular to the axis are very long.

We must calculate the magnetic circulation around this rectangle by integrating $\mathscr{B} \cdot d\mathbf{s}$ over the paths ab, bc, cd, and da. The parts of the paths bc and da inside the solenoid are perpendicular to $\mathscr{B}$, and so $\mathscr{B} \cdot d\mathbf{s} = 0$. The parts of the same paths outside the solenoid are essentially perpendicular to $\mathscr{B}$, and $\mathscr{B}$ is in any case very small. Thus the integrals over these two sides are essentially zero. The side cd can be as far from the solenoid as we like. Although cd is parallel to the field, $\mathscr{B}$ is negligibly small, and so the integral of $\mathscr{B} \cdot d\mathbf{s}$ from c to d also is zero. This leaves the side ab, and the magnetic circulation is

$$C = \int_{\substack{\text{Amperean} \\ \text{rectangle}}} \mathscr{B} \cdot d\mathbf{s} = \int_a^b \mathscr{B} \cdot d\mathbf{s} = \mathscr{B}x.$$

What is the current threading the Amperean rectangle? Each turn of the winding carries the current i, and all turns carry current through the rectangle in the same sense. If the solenoid has n turns per meter, the number of turns threading the rectangle is nx and the total current is nxi. Using Ampère's law, we equate μ_0 times this total current to the circulation C; we have

$$\mu_0 nxi = \mathscr{B}x.$$

We solve immediately for $\mathscr{B}$, obtaining

$$\mathscr{B} = \mu_0 ni \quad \text{inside a long solenoid.} \tag{29.22}$$

The magnetic field inside the solenoid depends only on the current and on the spacing of the turns. Solenoids usually have circular cross section, as in Figure 29.15. But the derivation of Equation 29.22 does not depend on the shape of the cross section. Thus the result applies to a solenoid of any cross-sectional shape.

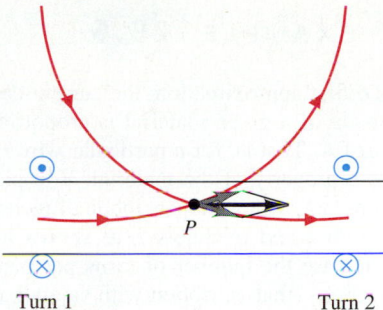

FIGURE 29.16 Diagram for the proof that the magnetic field inside a solenoid, far from either end, is everywhere parallel to the solenoid axis.

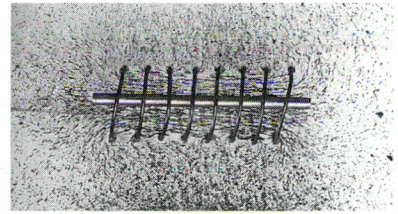

FIGURE 29.17 Photograph visualizing the magnetic flux lines of a solenoid. Note the parallel, uniformly spaced lines inside and the weakness of the pattern outside, except near the ends.

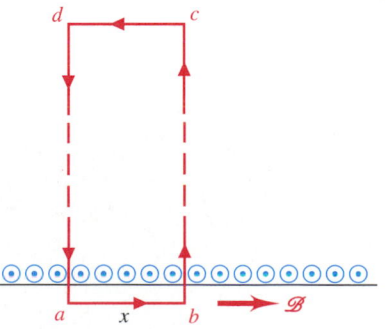

FIGURE 29.18 Amperean curve for determining the magnitude of the magnetic field inside a solenoid.

EXAMPLE 29.5

To first approximation, the current-carrying capacity of a wire made of a given material is proportional to its cross-sectional area A. That is, for a particular wire $i_{max} = j_{max}A$, where j_{max} is a property of the material. You plan to wind a solenoid of length X with a copper ribbon of rectangular cross section, with width x and thickness t, as shown in Figure 29.19. You can increase the number of turns per meter by ordering narrower ribbon—that is, ribbon with smaller x. But you are stuck with a fixed value of t. **(a)** What value of x gives you the greatest

FIGURE 29.19

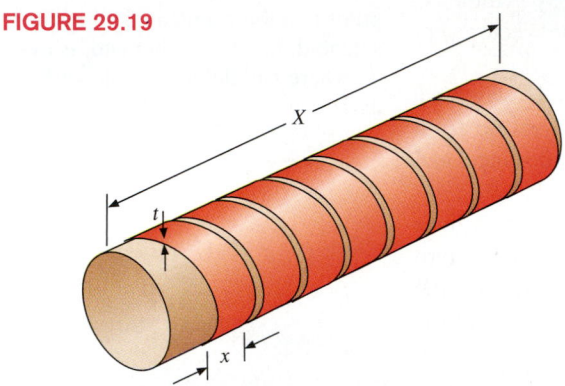

possible magnetic field? Neglect the space between turns required for insulation. **(b)** If you can wind the solenoid with l layers of copper ribbon, what is the maximum field you can obtain?

SOLUTION:

(a) Find the value of x that gives you the greatest magnetic field.

The maximum current is

$$i_{max} = j_{max}xt.$$

With a ribbon width x, you can pack $1/x$ turns into each meter of solenoid length. That is, you have

$$n = \frac{1}{x}.$$

Thus the maximum field you can obtain with one layer of winding is

$$\mathcal{B} = \mu_0 j_{max}xt\frac{1}{x} = \mu_0 j_{max}t. \qquad (29.23)$$

This value is independent of x. So it makes no difference whether you use narrow ribbon carrying a small current or wide ribbon carrying a large current. Indeed, you can wind the entire solenoid with a single turn of ribbon whose width x is equal to the length X of the solenoid. The choice of x is dictated by other considerations.

(b) What is the maximum field obtainable using l layers of ribbon?

Each of the l layers of ribbon makes an equal contribution to the total magnetic field inside the innermost layer. So you have

$$\mathcal{B} = \mu_0 l j_{max}t.$$

However, you cannot make l as large as you like. The more layers you add, the more difficult it is to remove the Joule heat generated. With very thick windings of many layers, it becomes necessary to add a cooling system. One common way of doing this is to embed copper tubing in the layers and run cold water through the tubing.

There is another way of looking at Equation 29.23. The magnetic field in the solenoid depends on the product jt, where $j = i/xt$ is the current density and i is the current through each turn of ribbon. Thus we can write the magnetic field in the form

$$\mathcal{B} = \mu_0 \frac{i}{x}.$$

In this equation, i/x is the current per unit length of the solenoid. (Remember that the current is everywhere transverse to the x axis.) If the solenoid of length X is made of a single broad turn, then i is the total current through the solenoid while i/X is still the current per unit length of the solenoid. The magnetic field is then

$$\mathcal{B} = \mu_0 \frac{i}{X}. \qquad (29.24)$$

Magnetic Field of a Current Sheet

Consider the solenoid of infinite length $(X \rightarrow \infty)$ and rectangular cross section yz shown in Figure 29.20a.* The current per unit length has the *finite* value i/x. Now, imagine that the solenoid is stretched out in the y direction until its width also is infinite, as in Figure 29.20b. This distortion does not change the field within the solenoid, which still

---

*As noted in the discussion immediately following Equation 29.22, the magnetic field inside a solenoid is independent of its cross-sectional shape.

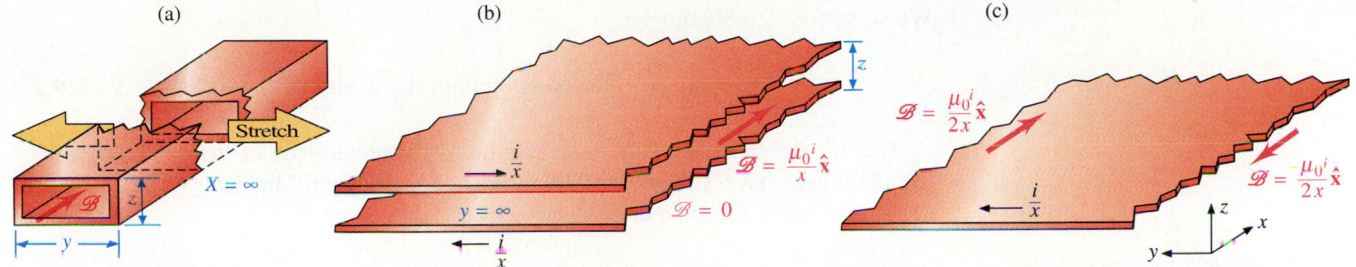

(a) (b) (c)

$$\mathscr{B} = \frac{\mu_0 i}{2x}\hat{\mathbf{x}}$$

$$\mathscr{B} = \frac{\mu_0 i}{x}\hat{\mathbf{x}}$$

$$\mathscr{B} = 0$$

$$\mathscr{B} = \frac{\mu_0 i}{2x}\hat{\mathbf{x}}$$

has the value $\mathscr{B} = \mu_0 i/x$. However, the contribution to this field of the current passing along the z direction through the very distant finite "walls" of the solenoid must be negligible; the entire field is due to the current passing in the y and $-y$ directions through the pair of infinite plates parallel to the x-y plane. Such ideal plates are often called **current sheets**. We can conclude, without calculation, that the magnetic field between a pair of infinite current sheets is

$$\mathscr{B} = \mu_0 \frac{i}{x} \hat{\mathbf{x}}. \qquad (29.25)$$

A symmetry argument, just like that for the electric field outside a pair of oppositely charged plates that make up a capacitor, should convince you that the magnetic field outside the pair of opposing current sheets is zero. This conclusion is not surprising, because the two systems share the same symmetry.

Using the result for a pair of current sheets, you can easily deduce the magnetic field due to a single sheet lying in the x-y plane and carrying current per unit length i/x in the y direction (Figure 29.20c). That field has the same magnitude, but opposite directions, on the two sides of the sheet and has the value

$$\mathscr{B} = \mu_0 \frac{i}{2x} \frac{z}{|z|} \hat{\mathbf{x}} \quad \text{for a single infinite current sheet.} \qquad (29.26)$$

The fraction $z/|z|$ provides the proper field direction: $+\hat{\mathbf{x}}$ for positive z and $-\hat{\mathbf{x}}$ for negative z. You should check this result by using the screwdriver rule. The magnitude of $\mathscr{B}$ is independent of distance from the sheet; $\mathscr{B}(z) \propto z^0$. This distance dependence is the same as that we found for the electric field of an infinite charged plate.

Equations 29.25 and 29.26 can be derived by applying Ampère's law from scratch; this alternative approach is the subject of Problems 29.64 and 29.65.

The Toroid

Figure 29.21a shows a **toroid**. You can imagine a toroid as a solenoid that has been bent into a circle so that the ends meet. We call the radius R of the circle the **major radius**. The major radius is measured to the axis, or center of symmetry, of the **cavity**. The cavity need not have a circular cross section, as does the toroid in Figure 29.21. But the cavity axis *must* be a circle; that is, R must be constant. The toroid has N turns and carries a current i.

Unlike the magnetic flux of a solenoid, the magnetic flux of a toroid needs no external return path. All of the flux lies within the cavity. By symmetry, all flux lines are circular and concentric with the cavity axis. The sense of the flux lines follows from the screwdriver rule (Figure 29.21b). Because the magnetic field at any point is tangent to the local flux line, the magnetic field is everywhere perpendicular to the major radius.

To apply Ampère's law, we choose as the Amperean curve a circle of radius r that lies within the cavity. The Amperean curve thus coincides with a flux line; by symmetry, the magnitude $\mathscr{B}$ is everywhere the same on the curve. All N turns of the winding thread through the Amperean curve. Thus the total current through the curve is Ni. Ampère's law (Equation 29.19) becomes

$$\mu_0 Ni = \int_{\text{circle}} \mathscr{B} \cdot d\mathbf{s} = \mathscr{B} 2\pi r.$$

FIGURE 29.20 (*a*) A solenoid of rectangular cross section yz and infinite length. The (transverse) current per unit length of solenoid is i/x. (*b*) The solenoid is stretched in the y direction so that it becomes infinitely wide. (*c*) Magnetic field of a single infinite sheet carrying current i/x in the y direction.

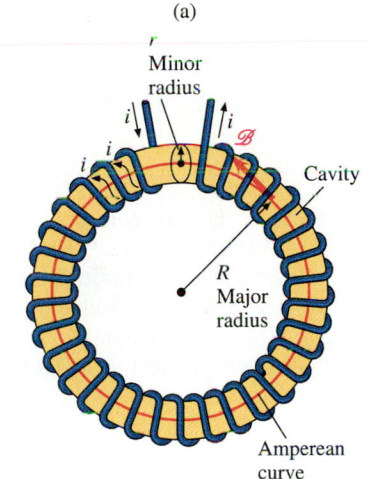

(a)

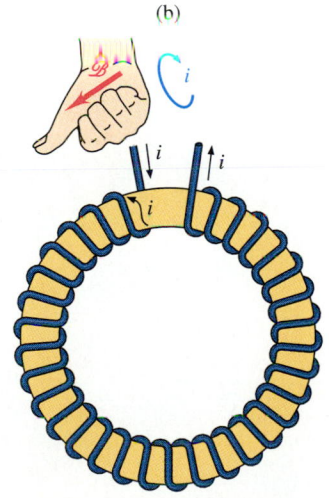

(b)

FIGURE 29.21 (*a*) A toroid of major radius R, whose cavity has circular cross section of minor radius r. The winding, which carries current i, has N turns. (*b*) The screwdriver rule used to determine the relation between the sense of the winding and the sense of the magnetic flux line.

We solve for $\mathscr{B}$, obtaining

$$\mathscr{B} = \mu_0 \frac{Ni}{2\pi r} \quad \text{for points within the cavity of the toroid.} \quad \textbf{(29.27)}$$

The $1/r$ dependence of the magnetic field is characteristic of the symmetry of the toroid. Where have you seen this dependence of magnetic field on distance before? (This question is the subject of Query 29.7.)

SECTION 29.5

Generalization of Ampère's Law*

So far, we have considered mainly currents passing through cylindrical wires. But what really counts in Ampère's law is the total current i threading the Amperean curve, no matter what form the current takes. We might wish, for example, to draw an Amperean curve in the electrolyte of a battery or in a thundercloud. In such cases, the flow of charge is not cylindrical.

We can always express the total current threading the Amperean curve in terms of current density j (Section 27.3). In Figure 29.22, we have drawn the plane surface bounded by the Amperean curve, which we call an **Amperean surface**. Because the current flow is not cylindrical, the current diverges and converges along its path, as shown. The situation is analogous to the fluid flow discussed in Section 16.2; the current lines are analogous to the fluid streamlines in Figure 16.1.

FIGURE 29.22 Electric current threads an Amperean curve $PQRSP$. All of the current passes through the plane surface of area A bounded by the curve. The current passing through any area element dA is $j\, dA$, where j is the current density.

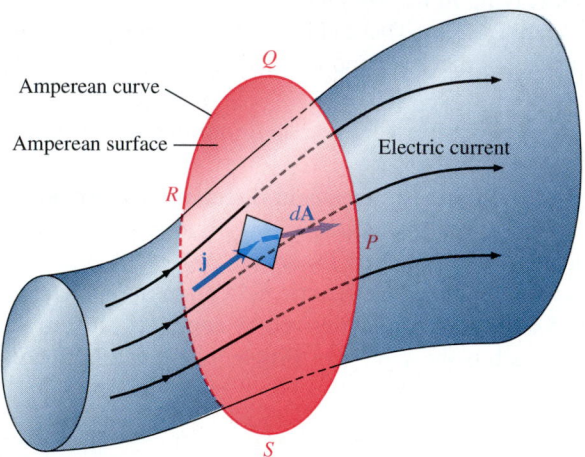

In general, the current lines are not normal to the Amperean surface, as they have been in all the situations we have considered so far. In order to calculate the total current threading the Amperean surface, we subdivide the surface into infinitesimal elements. The area and orientation of such an element are denoted by an area vector $d\mathbf{A}$. The current di passing through the element is

$$di = \mathbf{j} \cdot d\mathbf{A}. \quad \textbf{(29.28)}$$

In this equation, the current density $\mathbf{j}$ is a vector quantity. Its magnitude is the scalar current density j defined in Equation 27.3 as the current per unit area. Its direction is the direction of the current at $d\mathbf{A}$.[†] Equation 29.28 is mathematically identical with—and physically analogous to—Equation 24.6,

$$d\Phi_e = \mathscr{E} \cdot d\mathbf{A},$$

---

*This section may be omitted if the displacement current and Maxwell's equations are not to be covered.

[†]Current i can have a sense but not a direction because i describes the flow of charge through a closed circuit. But current density is a *local* quantity, and so the flow of charge has a well-defined direction at any particular point.

which relates the electric flux through an area element to the local electric field.

The total current i threading the Amperean curve $PQRSP$ is found by evaluating the integral

$$i = \int_{\substack{\text{Amperean} \\ \text{surface}}} \mathbf{j} \cdot d\mathbf{A}. \qquad (29.29)$$

To make this equation completely unambiguous, we choose one of the two opposite directions normal to the area element as the direction of $d\mathbf{A}$. In the case of Gauss's law, we adopted the convention that the positive direction was the outward normal to the (closed) Gaussian surface. But an Amperean surface is *not* closed; it is bounded by an Amperean curve. So we adopt the following convention, using the screwdriver rule:

1. Assign a sense to the bounding Amperean curve.
2. Curl the fingers of your right hand around the Amperean curve in this sense. The positive direction $d\hat{\mathbf{A}}$ is the direction in which your thumb points.

Or you can work in the opposite direction:

1a. Point your right thumb in the direction of $\mathbf{j}$, and take the direction $d\hat{\mathbf{A}}$ to be the one for which the angle between $d\hat{\mathbf{A}}$ and $\mathbf{j}$ is acute.
2a. Your fingers will curl around the Amperean surface in the positive sense.

When these rules are followed, Ampère's law yields a result whose signs are consistent. The first set of rules is more general because it assigns a direction to $d\mathbf{A}$ without reference to the direction of $\mathbf{j}$. The second set of rules, though less general, assures that both sides of the equation stating Ampère's law will have positive signs.

Using the value of i given by Equation 29.29, we write Ampère's law in the general form

$$\mu_0 \int_{\substack{\text{Amperean} \\ \text{surface}}} \mathbf{j} \cdot d\mathbf{A} = \int_{\substack{\text{Amperean} \\ \text{curve}}} \mathcal{B} \cdot d\mathbf{s}. \qquad (29.30)$$

The argument leading to Equation 29.30 began with Figure 29.22, in which the Amperean curve lies in a plane and the Amperean surface is a planar surface. But the argument does not depend on this planar geometry in any way. In particular, the same Amperean curve $PQRSP$ is shown as the boundary of a curved Amperean surface PTR in Figure 29.23. Equation 29.30 applies just as well to the new Amperean surface as to the old one. This point will be important when we make a further generalization of the concept of electric current in Chapter 34.

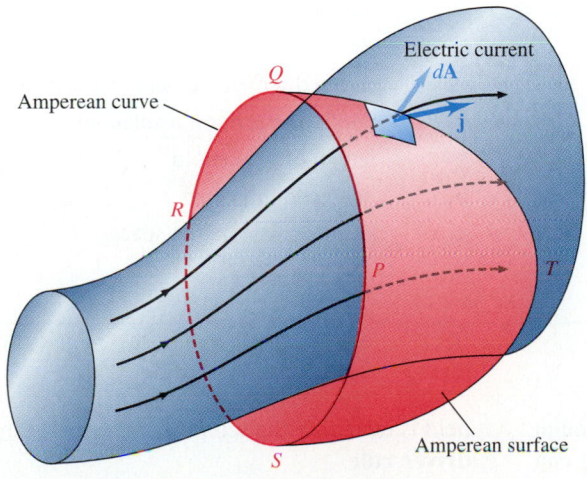

FIGURE 29.23 The same situation as that shown in Figure 29.22. This time, the Amperean surface is an arbitrary surface PTR bounded by the Amperean curve $PQRSP$.

Ampère's law and Gauss's law are *not* the same thing, as we stressed in Section 29.3. Nevertheless, the strong family resemblance between them cannot be ignored. Quite to the contrary, we used that resemblance to "cook up" Ampère's law in the first place.

In what features does the resemblance lie? At the end of Chapter 24, we reflected on the three concepts that together make Gauss's law a powerful tool: flux lines, the Gaussian surface, and symmetry. For Ampère's law, magnetic flux lines replace electric flux lines and the Amperean curve replaces the Gaussian surface. Although these differences are important, they do not hide the common conceptual ancestry of the two laws.

Ampère's law, like Gauss's law, is topological in nature. Electric flux lines penetrate the closed Gaussian surface; magnetic flux lines thread the closed Amperean curve. If we exploit the symmetry of a particular arrangement (such as the solenoid), we can shape the Amperean curve to suit our calculational purposes.

Where does the Amperean surface fit into the analogy between the two laws? Electric current lines penetrate the Amperean surface; electric flux lines penetrate the Gaussian surface. As you will see in Section 34.2, the connection between these two statements is not merely mathematical. The surprise is this: A *change* in the electric flux that penetrates a surface is just as much an electric current through the surface as a flow of charge through a surface is an electric current through the surface. Electromagnetic radiation (including light) is just one of the far-reaching consequences of this statement.

Magnetism in Matter

We have now considered two kinds of sources of magnetic field, apparently quite different from one another. The first and more familiar kind is the permanent magnet. The second, easier to analyze quantitatively, is the electric *current loop*. In view of the fact that the magnetic fields produced by the two kinds of sources have identical properties, can there be some connection between the two? Is the apparent difference between them due to our failure to perceive a fundamental property they have in common?

It was Ampère who first proposed in 1820 that *magnetism in matter is not an independent phenomenon but is due to the presence of internal electric current.* We have already noted the similarity between the magnetic field of a solenoid and that of a bar magnet of the same shape. To be more specific, a bar magnet acts as if it possesses a *surface current*—a cylindrical sheet of electric current that coincides with the surface of the bar magnet. In permanent magnets, Ampère suggested, this **Amperean surface current** is a permanent one. In induced magnets, the Amperean surface current is temporary, persisting only as long as the external inducing magnetic field is present.

The proposal that magnetism in matter is due to the presence of electric currents is called **Ampère's conjecture**. Although Ampère had no way of testing it, his conjecture remains to this day the basis of all inquiry into the magnetic properties of matter.

Industrial electromagnets exemplify both the connection between electric current and magnetic field and Ampère's conjecture. Such magnets furnish cheap, effective means both for separating magnetic from nonmagnetic scrap and for moving magnetic scrap around.

Symbols Used in Chapter 29

$\mathbf{A}, A$	area vector, area		n	number of turns per unit length of a winding
b	radius of a loop		q	electric charge
$\mathscr{B}, \mathscr{B}$	magnetic field		$d\mathbf{s}$	directed length element of a conductor
C	magnetic circulation		x	width of a turn on a solenoid
$\mathscr{E}, \mathscr{E}$	electric field		X	length of a solenoid
i, i_s	electric current, Amperean surface current		$\boldsymbol{\mu}, \mu$	magnetic moment
$\mathbf{j}, j$	current density		μ_0	permeability of free space
l	number of layers of a winding			

Summing Up

Every electric current gives rise to a magnetic field. Through this magnetic field, the conductor carrying the current can interact with a permanent magnet. The sense of the magnetic field is related to the sense of the current through the **screwdriver rule**.

Two electrical conductors exert forces on one another

through the interaction of their magnetic fields. When two long, straight, parallel wires carry currents i_1 and i_2, each wire exerts a force on the other given in SI by Equation 29.3,

$$\mathbf{F} = -\frac{\mu_0}{2\pi} \frac{i_1 i_2 s}{R} \hat{\mathbf{R}},$$

where $\mu_0 \equiv 4\pi \times 10^{-7}$ T·m/A is the **permeability of free space**. The magnitude of the magnetic field of wire 1 is given by Equation 29.5,

$$\mathscr{B} = \frac{\mu_0}{2\pi} \frac{i_1}{R}.$$

The **ampere** is defined in terms of the magnetic interaction between two current-carrying conductors. The **coulomb** is defined to be 1 A·s.

The **Biot-Savart law** (Equation 29.9a),

$$d\mathscr{B} = \frac{\mu_0}{4\pi} \frac{i}{r^2} ds \times \hat{\mathbf{r}},$$

expresses the magnetic-field contribution of a current element $i\, d\mathbf{s}$; that is, a current i passing through the directed length $d\mathbf{s}$ of a conductor. The Biot-Savart law is analogous to Coulomb's law, subject to the important limitation that the current element $i\, d\mathbf{s}$ cannot be a point, no matter how small we choose $d\mathbf{s}$. When integrated over a long, straight conductor, the Biot-Savart law yields Equation 29.5, in agreement with experiment.

Magnetic field can be related to electric current by means of **Ampère's law** (Equation 29.19),

$$\mu_0 i = \int_{\substack{\text{closed} \\ \text{curve}}} \mathscr{B} \cdot d\mathbf{s}.$$

In this equation, i is the current that threads a closed **Amperean curve**. The integral evaluated along the Amperean curve is called the **magnetic circulation** C. Ampère's law is logically equivalent to the Biot-Savart law.

Ampère's law can also be written in the more general form of Equation 29.30,

$$\mu_0 \int_{\substack{\text{Amperean} \\ \text{surface}}} \mathbf{j} \cdot d\mathbf{A} = \int_{\substack{\text{Amperean} \\ \text{curve}}} \mathscr{B} \cdot d\mathbf{s}.$$

This equation is valid for any surface bounded by the Amperean curve.

Ampère's conjecture asserts that *all* magnetic fields arise from electric currents. In magnetized matter, where no windings are present, an **Amperean surface current** i_s is present.

KEY TERMS

Section 29.2 Ampère Current Balance and the Biot-Savart Law
permeability of free space ▪ ampere, coulomb

Sections 29.3, 29.4, and 29.5 Ampère's Law, Applications, and Generalization
Gauss's law for magnetism ▪ Amperean curve, Amperean surface ▪ magnetic circulation ▪ solenoid ▪ current sheet ▪ toroid ▪ Amperean surface current

QUERIES

29.1 *(2) Tantalizing similarity, essential difference.* Compare Equation 29.14, $\mathscr{B} = (\mu_0/2\pi)(\mu/x^3)$, with the corresponding result for an electric dipole. Explain qualitatively why the similarity between the two dipoles breaks down at small distances.

29.2 *(2) Temporary magnets.* Two long, straight wires, each carrying current i, are perpendicular to one another as shown here. If the wires are free to move, what happens?

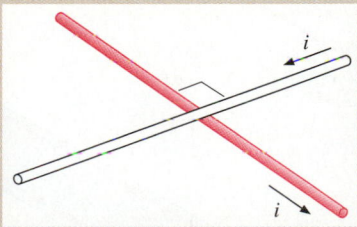

29.3 *(2) Turn of the screw.* Derive the screwdriver rule for relating the senses of electric current and magnetic flux from the right-hand rule for cross products, and thus show that the two rules are not independent.

29.4 *(3) What goes around comes around.* What is the value of the magnetic circulation around the Amperean curve shown here? Does it follow that $\mathscr{B} = 0$ everywhere on the curve? Why or why not?

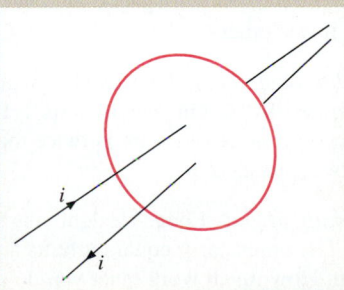

29.5 *(4) Unattainable perfection.* Figure *a* (p. 814) is a sketch of the magnetic flux lines in the gap between the poles of a pair of bar magnets whose separation is small compared with their width. Although the lines are uniform and parallel well within the gap, the lines near the edges spread outward, resulting in what is called a fringing field. To see why this must

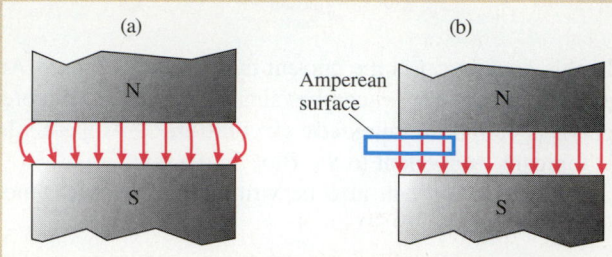

(a)   (b)

N

S

Amperean surface

N

S

be so, imagine an ideal arrangement like that in figure *b*, and show that Ampère's law is violated on the curve shown.

29.6 *(4) Relax!* In deriving Equation 29.22, which relates the magnetic field of a long solenoid to the current through its windings, we argued that $\int \mathcal{B} \cdot d\mathbf{s}$ is zero along paths *bc* and *da* in Figure 29.18. But use a symmetry argument to show that, even if the integrals were not exactly equal to zero, you would still have

$$\int_b^c \mathcal{B} \cdot d\mathbf{s} + \int_d^a \mathcal{B} \cdot d\mathbf{s} = 0.$$

29.7 *(4) Donut symmetry?* According to Equation 29.27, the magnetic field within the cavity of a toroid of *N* turns is the same as that which would arise from a current *Ni* in a conductor normal to the plane of the toroid and passing through its major center. But outside the toroid cavity, the field of the toroid is zero while the field of the normal conductor is still given by Equation 29.27 without restriction. Explain.

29.8 *(5) Circulate and penetrate.* Shown below is a closed surface of arbitrary shape. Electric currents enter and leave the region enclosed by the surface, but no charge can accumulate within the region. By drawing an arbitrary closed curve on the surface, subdivide it into two parts, 1 and 2. If the magnetic circulation around the curve is *C*, what can you say about the net currents i_1 and i_2 penetrating the two parts of the closed surface?

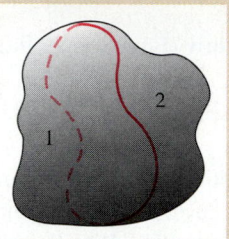

PROBLEMS

GROUP A

29.1 *(2) Wired.* A very long, straight wire carries a 5.2-A current. Find the magnitude of the magnetic field at a point 3.1 cm from the wire.

29.2 *(2) Current events, I.* At a distance of 2.4 cm from a long, straight wire, the magnetic field has magnitude 16 μT. What is the current in the wire?

29.3 *(2) Current events, II.* Two long, parallel wires carry equal currents *i*. The distance between the wires is 1.35 cm. If the force per unit length on each wire is 4.76×10^{-2} N, what is the value of *i*?

29.4 *(2) Current events, III.* For the system of Problem 29.3, what is the magnitude of the magnetic field at either wire due to the current in the other?

29.5 *(2) Double or nothing.* Two parallel wires, each 52 m long, are separated by 23 cm. Each exerts a 1.1-N force on the other. If the current in one wire is twice that in the other, what are the two currents?

29.6 *(2) Working it out.* Long, straight wires 1 and 2 lie 3.2 cm apart. The wires carry equal currents *i* = 21 A in the same direction. How much work must you do per unit length of wire to move wire 2 to a distance much greater than 3.2 cm from wire 1?

29.7 *(2) Sum difference, I.* Shown in the figure is an end-on view of two long, parallel wires that carry currents in opposite directions. If $i_1 = 22$ A and $i_2 = 34$ A, find the magnitude and direction of the magnetic field at **(a)** *A*, **(b)** *B*, and **(c)** *C*.

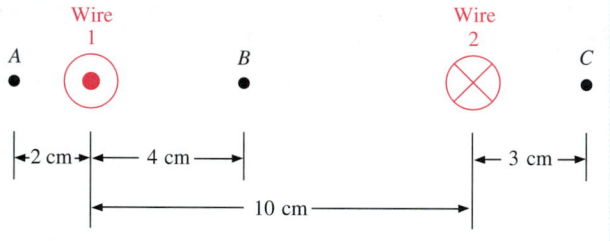

Wire 1   A • ⊙ B •   Wire 2   ⊗ C •

|←2 cm→|← 4 cm →| |← 3 cm →|
|← 10 cm →|

29.8 *(2) Sum difference, II.* The current is reversed in wire 2 of Problem 29.7, with no other changes. Find the magnitude and direction of the magnetic field at **(a)** *A*, **(b)** *B*, and **(c)** *C*.

29.9 *(2) Twins.* In the figure below, wires 2 and 3 are the two

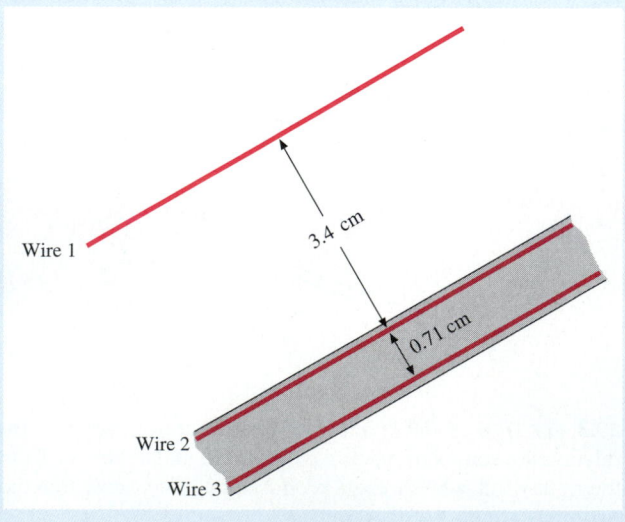

Wire 1   3.4 cm   0.71 cm   Wire 2   Wire 3

conductors in a length of "twin lead" of the kind frequently used to connect TV sets to antennas. The two wires are separated by 0.71 cm, and they carry the same 7.3-A current in opposite directions. Wire 1 is parallel to wires 2 and 3 and is 3.4 cm from wire 2; all three wires lie in a common plane. **(a)** If wire 1 carries a 5.2-A current, what is the force exerted on the twin lead per meter of its length? What is the direction of this force if the current in wire 2 is **(b)** parallel to or **(c)** antiparallel to that in wire 1?

29.10 (2) *Quadruplets?* Two strips of twin lead of the kind described in Problem 29.9 lie in a common plane; their inner conductors lie 2.8 cm apart. **(a)** What is the net force per unit length exerted on each strip if the two conductors in one strip carry 3.8-A currents in opposite directions and the two conductors in the other strip carry 6.2-A currents in opposite directions? **(b)** What is the direction of the force if the currents in the two inner conductors are antiparallel?

29.11 (2) *Going off on a tangent.* A long wire oriented in the south-north direction carries current i. A small compass needle is located directly above and 1.3 cm from the wire. What is the value of i if the compass needle points **(a)** $45°$ east of north? **(b)** $60°$ east of north? Take the horizontal component of the earth's magnetic field to be 4.2×10^{-5} T. (Note: Some of the earliest galvanometers, called *tangent galvanometers*, operated in this manner. Can you see why they are given that name?)

29.12 (2) A circular loop of radius 9.7 cm carries a 2.3-A current. Find the magnitude of the magnetic field **(a)** at the center of the loop and **(b)** on the loop axis 9.7 cm from the center of the loop.

29.13 (2) *A matter of some moment.* At the center of a circular, current-carrying loop of radius 12.3 cm, the magnetic field is 6.4×10^{-6} T. Find the magnetic moment of the loop.

29.14 (3) *Rectangular loop.* In the figure below, the long, straight wire carries a current $i_1 = 50$ A. The rectangular loop carries a current $i_2 = 15$ A in the sense shown. If $R = 1.3$ cm, $L = 1$ m, and $w = 6.7$ cm, find the magnitude and direction of the net force exerted on the loop.

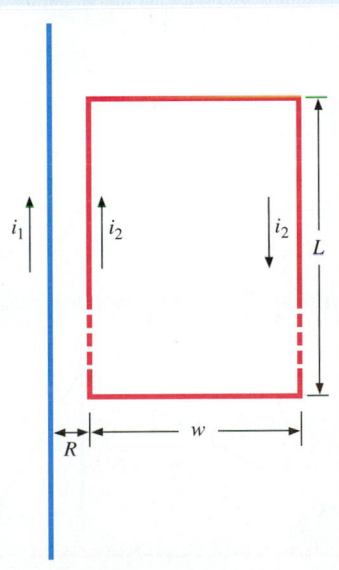

29.15 (2) *In between.* Two long, parallel wires spaced 12 cm apart carry 20-A currents. What are the magnitude and direction of the magnetic field midway between the wires if the currents have **(a)** the same sense? **(b)** opposite senses?

29.16 (2) *Cross purposes.* A long, straight wire, 1, carries a 10-A current. A second long, straight wire, 2, is perpendicular to wire 1 and carries a 25-A current in the same plane. What is $\mathscr{B}$ at a point 1.6 cm from wire 1 and 3.8 cm from wire 2?

29.17 (2) *After you, M. Biot!* . . . A segment of wire 1.00 cm long carries a current of 23.2 A in the positive x direction. **(a)** What is the contribution of this segment to the magnetic fluid $\mathscr{B}$ (magnitude and direction) at P in the figure below, 1.94 m from the wire along a line that makes an angle of $22°$ with the x axis? **(b)** Why is it permissible to use the Biot-Savart law in making this calculation, even though the length of the wire segment is finite?

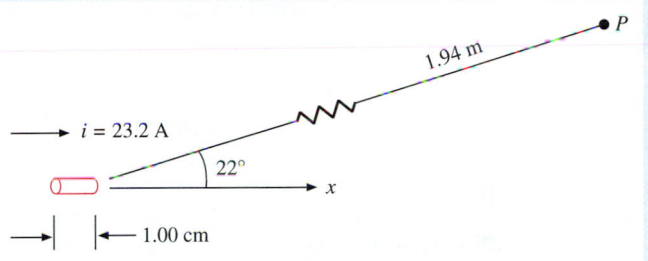

29.18 (2) *. . . No, after you, M. Savart!* A segment of wire 0.850 cm long lies on the x axis, its midpoint at the origin of an x-y coordinate system. If the wire carries a current $i = 41.3$ A, what is the contribution of the segment to the magnetic field at $x = 63.5$ cm, $y = 97.1$ cm?

29.19 (2) *No calculation needed! I.* In the figure below, points P and Q lie at the same small distance R from the current-carrying wire. (By "small" we mean that R is very much less than the length of any segment of the wire.) What is the magnitude $\mathscr{B}$ of the magnetic field **(a)** at Q? **(b)** at P? (Hint: When you don't want to calculate, think symmetry!)

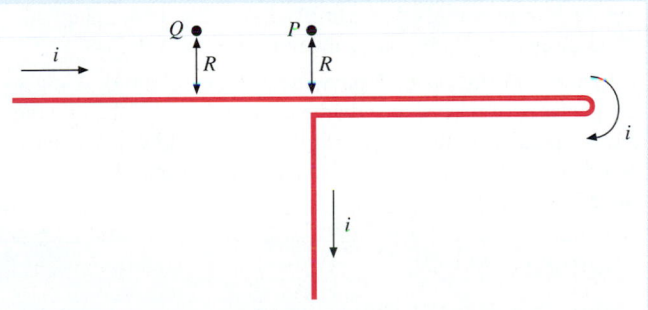

29.20 (2) *Circular reasoning, I.* A circular wire loop of radius 0.375 m carries a 132-A current. What is the magnitude of the magnetic field on the loop axis, a distance 0.410 m from the plane of the loop?

29.21 (2) *Circular reasoning, II.* A circular coil has 250 turns of fine wire, closely packed so that all the turns have nearly the same radius, 3.5 cm, and all lie nearly in the same plane. Find the magnitude of the magnetic field at the center of the coil when the current passing through it is 3.74 A.

29.22 (2) *Circular reasoning, III.* A circular coil has 175 turns of radius 5.0 cm. If the magnetic field at the center of the coil has magnitude 2.0×10^{-3} T, what is the current in the coil?

29.23 (2) *Loops within loops.* Illustrated below is a closed circuit in the form of two concentric, nearly complete circular segments joined by two straight segments lying close together. A current $i = 14.8$ A passes through the circuit in the sense shown. If the radii are $b_1 = 3.72$ cm and $b_2 = 5.63$ cm, what are the magnitude and direction of the magnetic field at the central point O?

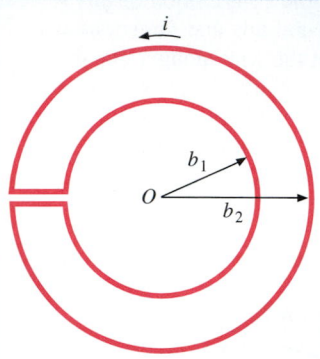

29.24 (4) *Superconducting wire.* In May 1987, a research group at IBM reported the fabrication of a superconducting material through which a current density of 10^9 A/m^2 could be made to pass without destroying the superconductivity of the sample. If you made a wire of cylindrical cross section of this material, what would be the magnitude of the magnetic field at the wire surface if the radius were **(a)** 10^{-5} m (a fine filament)? **(b)** 10^{-3} m (a wire of everyday dimensions)? (Note: This current density is about 100 times that carried in a typical light-bulb filament.)

29.25 (4) *Field work.* A long solenoid is wound with 800 turns of wire per meter. **(a)** If the current passing through the winding is 9.85 A, what is the magnitude of the magnetic field within the solenoid? **(b)** Imagine that you are looking down the axis of the solenoid. From your point of view, the sense of the current is that of a right-handed screw. Is the magnetic field directed toward you or away from you?

29.26 (4) *Bottom line.* You wind a long solenoid, using a single layer of wire whose diameter is 0.75 mm. If the spacing between adjacent turns is negligible, what field will you have inside the solenoid when you run a 1.0-A current through the wire?

29.27 (4) *Something to do when the paper towels are used up.* You build a solenoid by winding No. 14 insulated wire (diameter 0.163 cm) in a single, closely packed layer around a cardboard tube of length 28 cm and diameter 3.7 cm. The maximum safe current that the wire can carry is 20 A. What is the maximum magnetic field you can produce? Would it make any difference if you distorted the cardboard tube into a noncircular cross section?

29.28 (4) *Superconducting solenoid.* Suppose you use the material described in Problem 29.24 to make a solenoid. The solenoid winding has many layers, separated by thin insulation. But the effective thickness of the winding is 4 cm. Estimate the magnetic field that can be achieved inside this solenoid. Compare this value with the maximum magnetic field obtainable with iron-core electromagnets, about 3 T, and that obtainable with conventional steady high-current solenoids, about 20 T.

29.29 (4) *No calculation needed! II.* Use a symmetry argument to show that the magnetic field at the end of a long (but not infinite) solenoid is one-half the magnetic field inside the solenoid, far from the ends.

29.30 (4) *Doughnut, no coffee, I.* A toroid has a cavity of square cross section, with dimensions 2.0 cm × 2.0 cm, as shown below. The inner radius is 2.0 cm and the outer radius is 4.0 cm; thus the major radius is 3.0 cm. The toroid is wound with 100 turns of wire that carries a 12-A current. Find the magnetic field **(a)** at a point where r, the distance from the axis (center) of the toroid, is just a little greater than 2.0 cm, **(b)** at $r = 3.0$ cm, and **(c)** at a point where r is just a little less than 4.0 cm.

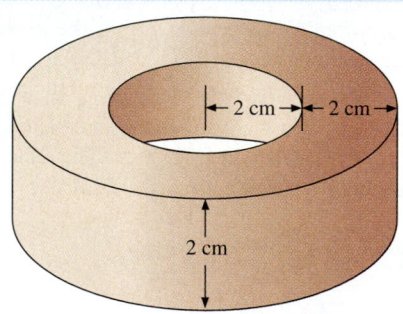

29.31 (4) *Doughnut, no coffee, II.* A toroid has 550 turns. The magnetic field within the cavity, at a point 12.5 cm from the toroid axis, is 6.48×10^{-2} T. What is the current in the winding?

GROUP B

29.32 (2) *Another way.* Beginning with the Biot-Savart law, evaluate $\mathscr{B}$ at a distance R from a long, straight wire. However, carry out the integration sketched in Figure 29.7 in terms of the variable θ instead of the variable x; that is, make the substitutions $x = -R \cot \theta$ and $r = R \csc \theta$. **(a)** Why do you need the minus sign in the substitution for x? **(b)** Show that the definite integral yields

$$\mathscr{B} = -\frac{\mu_0 i}{4\pi R} \cos \theta \Big|_0^\pi .$$

29.33 (2) *Finite wire.* The figure below shows a point P at a

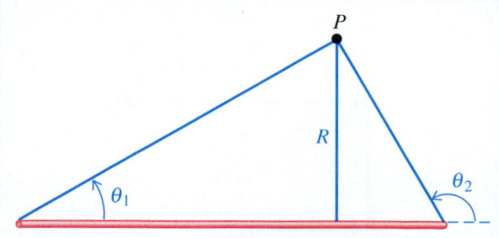

distance R from a *finite* straight wire. **(a)** Generalize the argument of Problem 29.32 to show that the magnetic field at P has magnitude

$$\mathcal{B} = \frac{\mu_0 i}{4\pi R} (\cos \theta_1 - \cos \theta_2),$$

where θ_1 and θ_2 are the angles shown in the figure. **(b)** Suppose $\theta_1 = \pi - \theta_2 \equiv \theta$. For what value of θ does $\mathcal{B}$ differ by 5% from $\mathcal{B}^*$, the magnetic field that would exist at P if the wire were infinitely long? **(c)** Let the length of the wire be L. To what ratio R/L does the result of part **b** correspond?

29.34 *(2) Knowing right from acute.* A wire bent into the form of a rectangle carries current i. The length of the diagonals is δ, and the acute angle between the diagonals is θ. Show that the magnetic field at the center of the rectangle is of magnitude

$$\mathcal{B} = \frac{4\mu_0 i}{\pi \delta \sin \theta}.$$

(Hint: See Problem 29.33.)

29.35 *(2) Polygon.* Current i flows through a wire bent into the shape of a regular polygon of n sides that is inscribed in a circle of radius b. **(a)** Using the result of Problem 29.33, show that the magnetic field at the center of the polygon has magnitude

$$\mathcal{B} = \frac{n\mu_0 i}{2\pi b} \tan \frac{\pi}{n}.$$

(b) This expression should reduce to the value of $\mathcal{B}$ at the center of a current-carrying circular loop of radius b when $n \to \infty$. Prove that this is so. (Hint: You will need to use a standard approximation.) **(c)** Compare the results for $n = 3$ (an inscribed triangle) and $n = \infty$ (the circle).

29.36 *(2) Flattened.* A wire having the form shown here carries current i. Find the magnitude and direction of the magnetic field at P.

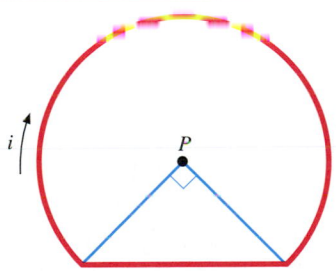

29.37 *(2) When push comes to shove.* In the illustration below, two long, parallel wires (seen end on) that are a distance

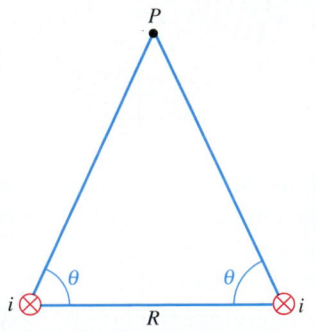

R apart carry equal currents i in the same sense. Show that the magnitude of the magnetic field at point P, which lies equidistant from the two wires at an angle θ from the plane of the wires, is

$$\mathcal{B} = \frac{\mu_0 i}{\pi R} \sin 2\theta.$$

What is the direction of $\mathcal{B}$?

29.38 *(2) Threesome.* The three long, parallel wires in the figure below carry the same current $i = 7.5$ A. The sense of the current in each wire is shown. If each wire lies 1.3 cm from both of the others, calculate the magnitude and direction of the force per unit length on each wire.

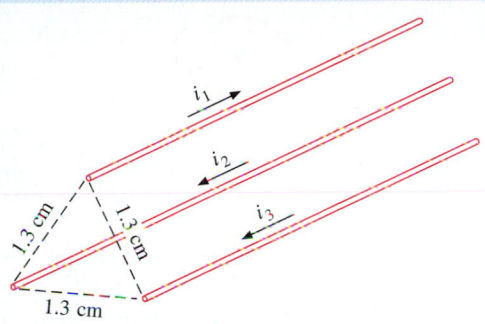

29.39 *(2) Looking for nothing, I.* Two long, parallel wires are separated by a distance R. Wire 1 carries current i_1, and wire 2 carries current i_2 in the same direction. **(a)** At what distance x from wire 1, in the plane of the wires, is the magnetic field zero? **(b)** Is there any point, out of this plane and a finite distance from wire 1, where the magnetic field is zero? Explain. **(c)** If the currents in the two wires are antiparallel, is there any point a finite distance from wire 1 where the magnetic field is zero? Explain.

29.40 *(2) Looking for nothing, II.* Two long, perpendicular wires lie along the x and y axes of a coordinate system. The first wire carries current i_x in the positive x direction, and the second wire carries current i_y in the positive y direction. Find the locus of points in the x-y plane where the magnetic field is zero.

29.41 *(2) Looking for nothing, III.* Shown below is an end-on view of three long, straight, parallel conductors spaced equal distances a apart. The outer conductors carry current i out of the page; the center conductor carries current i into the page. Where in the plane of the page is the magnetic field zero?

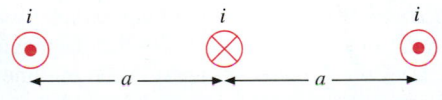

29.42 *(2) In the midst of things.* Shown below is an end-on

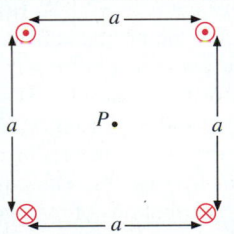

view of four long, straight, parallel conductors whose intersection with the page forms a square of side a. All four conductors carry current i, in the senses shown. Find the magnitude and direction of $\mathcal{B}$ at P, the center of the square.

29.43 *(2) Making sense of it all.* In the figure below, the conductors carry equal currents i. All straight segments are very long, and the two circular loops have equal radii. However,

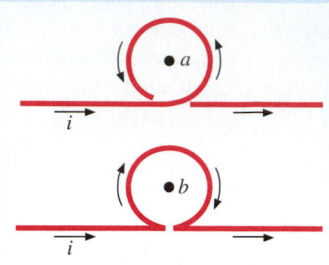

the currents around the loops have opposite senses. Show that the ratio of the magnetic fields at a and b, at the centers of the two loops, is

$$\frac{\mathcal{B}_a}{\mathcal{B}_b} = \frac{\pi + 1}{\pi - 1}.$$

29.44 *(2) Drag strip.* In the figure below, the long, straight wire carries a current i_1, and the long, thin conducting strip of width w carries a uniformly distributed current i_2. If the distance between the wire and the near edge of the strip is R, find the force per unit length exerted on the strip.

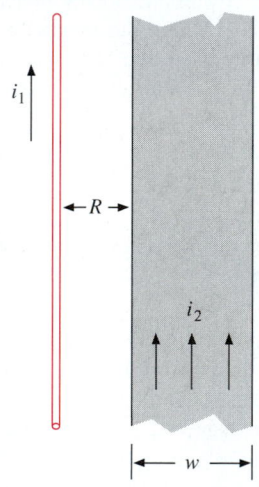

29.45 *(2) Halfway measures.* Two long, parallel wires carry equal currents in opposite senses. Show that the magnetic field in the plane of the wires is strongest at the midline between them.

29.46 *(2) Double duty, I.* A circular loop of radius $b_1 = 2.74$ cm carries a current $i_1 = 12.8$ A. **(a)** What are the magnitude and direction of the magnetic field $\mathcal{B}$ at a point on the loop axis a distance $x = 1.63$ cm from the plane of the loop? **(b)** A second loop of radius $b_2 = 6.31$ cm lies in the same plane as the first loop and is concentric with it. What current i_2 must this loop carry if it is to produce the same field at x as the first loop? **(c)** If i_2 has the value found in part b, will the magnetic field contributions of the two loops be the same for *all* values of x?

29.47 *(2) Sine galvanometer.* A variant on the tangent galvanometer (see Problem 29.11) is the *sine galvanometer*, sketched below. A magnetic compass needle lies at the center of a coil of N turns, whose plane is vertical. (The lower half of the coil is seen in top view.) With no current in the coil,

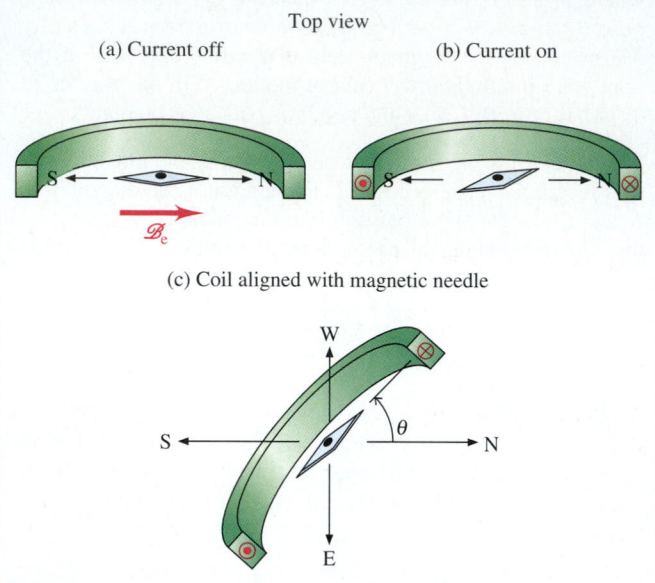

the needle aligns itself along the earth's magnetic south-north direction (figure a). Assume that, to begin with, the plane of the coil coincides with the orientation of the needle. When current i passes through the coil, the compass needle is deflected to a new orientation (figure b). The coil is now turned about a vertical axis. When this is done, the orientation of the compass needle changes further. But the coil can be turned until its plane again coincides with the orientation of the needle. Call the angle between the original and final orientations of the needle θ, and show that $i \propto \sin \theta$.

29.48 *(2) Spiral.* You wind a long wire into a flat, closely spaced spiral of N turns, as shown below. The innermost turn

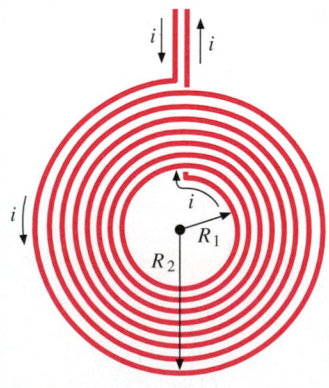

has radius R_1, and the outermost turn has radius R_2. Show that, if the wire carries current i, the magnetic field at the center is of magnitude

$$\frac{\mu_0 i N \ln \dfrac{R_2}{R_1}}{2(R_2 - R_1)}.$$

29.49 *(2) Wire sculpture, I.* In the figure below, a long, mainly straight wire contains a semicircular section of radius b. The wire carries a current i in the sense shown. Find the magnitude and direction of the magnetic field at O, the center of curvature of the semicircular section.

29.50 *(2) Wire sculpture, II.* The current-carrying wire shown below consists of two long, straight, parallel sections joined by a semicircular section of radius b. Find the magnitude and direction of the magnetic field at O, the center of curvature of the semicircular section. Compare your result with those of Problems 29.15 and 29.49.

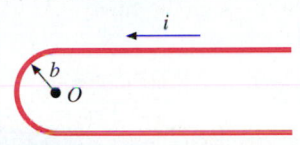

29.51 *(2) Wire sculpture, III.* The current-carrying loop shown below consists of two semicircular sections of radii b_1 and b_2, connected by straight sections. Find the magnitude and direction of $\mathscr{B}$ at O, the common center of curvature of the two semicircles. (Hint: If you have solved Problem 29.50, you won't need to do much calculating.)

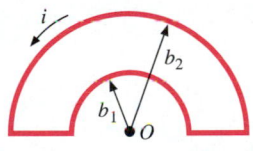

29.52 *(2) Wire sculpture, IV.* The loop of Problem 29.51 is bent by flipping the outer semicircle through 180°, as shown below. Find the magnitude and direction of $\mathscr{B}$ at O, the common center of curvature of the two semicircles. (Hint: If you have solved Problem 29.51, don't calculate from scratch!)

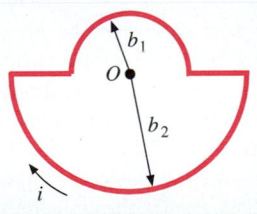

29.53 *(2) Wire sculpture, V.* For the current-carrying wire shown here, find $\mathscr{B}$ at P.

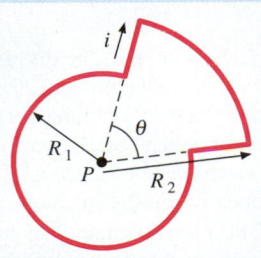

29.54 *(2) Wire sculpture, VI.* For the current-carrying wire in the figure below, show that the magnetic field at P has magnitude

$$\mathscr{B} = \frac{\mu_0 i}{4\pi}\left(\frac{3\pi}{2R} + \frac{\sqrt{2}}{d}\right).$$

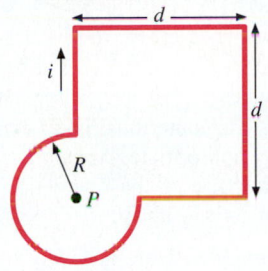

29.55 *(2) Winder's choice, I.* You have a length L of wire, with which you plan to make a ring-shaped coil of many turns. You are free to choose any coil radius b. Show that, as long as b is large compared with the thickness of the ring (so that the coil may be regarded as having a single radius), the magnetic field at the center of the coil is proportional to $1/b^2$.

29.56 *(3) Spinning up a field, I.* Electric charge q is uniformly distributed over a dielectric disk of radius R. **(a)** If the disk rotates about its axis with angular speed ω, show that there is a magnetic field at its center of magnitude

$$\mathscr{B} = \frac{\mu_0}{2\pi}\frac{\omega q}{R}.$$

What is the direction of $\mathscr{B}$? **(b)** This is not a trivial experiment; its results prove the plausible but not self-evident assumption that the magnetic field due to a moving charge distribution is identical with the magnetic field due to an electric current having the same geometry and that an electric current is thus nothing more than a flow of charge. Today we have ample experimental and theoretical evidence of the truth of the assumption. But, in the late nineteenth century, it was important precisely because it supported the developing theory. The American physicist Henry Rowland (1848–1901) conceived of the experiment in 1868 and carried it out ten years later. Rowland's disk was of diameter 21.1 cm and rotated at 61 rev/s. The disk was located between a pair of stationary disks that were grounded. With respect to these disks, the rotating disk had potential $V = 10^4$ V, and the spacing between the disks was 3.5 mm. Estimate the magnitude of the magnetic field Rowland had to detect, and see if you agree with the general opinion that Rowland was one of the greatest experimentalists of his time.

29.57 *(3) Spinning up a field, II.* The surface of a dielectric sphere of radius R carries a uniformly distributed charge q. If the sphere is rotated about a diameter with angular speed ω, find the magnetic field at the center.

29.58 *(3) Coax cable, I.* The figure (p. 820) shows a *coaxial cable*, widely used in high-frequency applications or where electrical shielding is required. The cable consists of a wire of radius r_1 surrounded by a fairly thick cylindrical insulating layer. Outside the insulation lies a second, thin cylindrical conductor of radius r_2 coaxial with the inner one. This outer conductor carries the same current as the inner one but in the

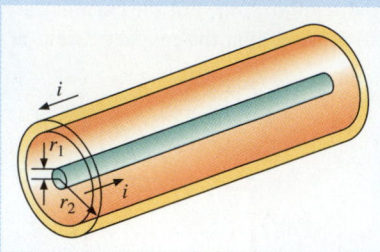

opposite direction. The outermost layer consists of more insulation. Drawing appropriate Amperean curves, find the magnitude and direction of the magnetic field at a point a distance r from the axis, with **(a)** $r_1 < r < r_2$; **(b)** $r > r_2$.

29.59 *(3) Coax cable, II.* Here is a way to use a symmetry argument to derive the result of Problem 29.58 without calculation. Imagine a toroid whose cavity has a square cross section. **(a)** How does the magnetic field inside the cavity depend on the cross-sectional shape? **(b)** How can you distort the toroid to make it resemble a very long coaxial cable? **(c)** What are the magnitude and direction of the magnetic field between the conductors of the coaxial cable?

29.60 *(3) Call it macaroni.* A copper tube has the form of a hollow cylinder with thick walls, like a piece of macaroni. It carries a current i, uniformly distributed throughout the copper. If the inner radius is b_1 and the outer radius is b_2, find the magnetic field as a function of distance r from the cylinder axis for **(a)** $r < b_1$; **(b)** $r > b_2$. **(c)** Show that, for $b_1 < r < b_2$, the magnetic field is

$$\mathscr{B} = \frac{\mu_0 i}{2\pi} \frac{1}{r} \frac{r^2 - b_1^2}{b_2^2 - b_1^2}.$$

29.61 *(3) No way.* Use Ampère's law to prove that the magnetic field within a region of space cannot be both nonuniform (that is, different in magnitude from place to place) and unidirectional (that is, having the same direction everywhere within the region). (Hint: Assume that such a region does exist, and sketch the flux-line pattern within the region. Then choose an Amperean curve and evaluate the circulation. See also Query 29.5.)

29.62 *(4) Any old Amperean curve.* Draw an Amperean curve that has an entirely arbitrary shape but is entirely located well within a long solenoid. Prove that the magnetic circulation evaluated around this curve is zero.

29.63 *(4) Winder's choice, II.* You are designing a long solenoid, to be operated with a source of emf whose output voltage has a fixed value V. You have already bought a large spool of the wire you are going to use. You are trying to decide whether to use a length L of wire, which suffices to wind the solenoid with a single layer of N closely spaced turns, or a length $2L$ to wind it with a double layer having a total of $2N$ turns. In the latter case, you can connect the two layers either in series or in parallel across the source of emf. **(a)** Compare the magnitudes of the magnetic fields obtained in the three cases. **(b)** Compare the power consumption of the three solenoids.

29.64 *(4) Magnetic field of a current sheet.* Draw an appropriate Amperean curve, and use Ampère's law to derive Equa-

tion 29.26, which gives the magnetic field of a current sheet carrying current per unit length $(i/x)\hat{\mathbf{y}}$.

29.65 *(4) Inside & outside.* Two parallel current sheets carry currents per unit length $(i/x)\hat{\mathbf{y}}$ and $(i/x)(-\hat{\mathbf{y}})$. Draw appropriate Amperean curves, and use Ampère's law to find the magnitude and direction of the magnetic field **(a)** between the sheets and **(b)** outside them.

29.66 *(4) Skinny.* For a toroid of circular cross section, we can define the ratio $\rho \equiv r/R$ of the minor to the major radius. What is the maximum allowable value of this ratio if the magnetic field within the cavity is not to vary by more than 1%?

29.67 *(4) Squaring the circle.* The N-turn toroid shown below has rectangular cross section. The ratio of the outer radius to the inner radius is $R_2/R_1 = \rho$, and the thickness of the toroid is h. Calculate the total magnetic flux Φ_m that penetrates the imaginary blue cut through the cavity.

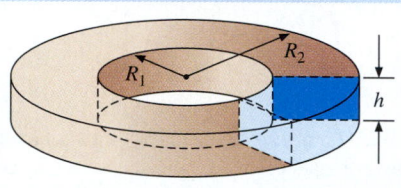

29.68 *(4) Half a donut.* A half-toroid (illustrated below) is wound with N turns that carry current i. The minor radius is r. Find the magnetic moment of the half-toroid, and show that your result is independent of the major radius.

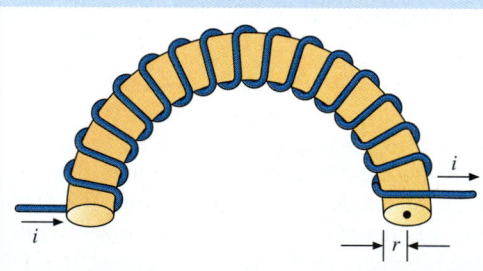

29.69 *(G) Not a good way to make geomagnetic measurements!* You have a 50-turn circular coil whose diameter is 20 cm. You orient the coil in a vertical north-south plane and place a horizontal magnetic compass needle at the coil center. You also have a 10-μF capacitor and a 100-V battery. You use an electronic switching device to charge the capacitor to 100 V and then discharge it through the coil 100 times per second. When you do this, the compass needle swings 45° from its undisturbed north-south orientation. What is the magnitude of the horizontal component of the earth's magnetic field?

29.70 *(G) Astatic galvanometer.* The disadvantage of the tangent (Problem 29.11) and sine (Problem 29.47) galvanometers is that the value of the current i obtained using them depends on the horizontal component of the earth's magnetic field, $\mathscr{B}_e$, which varies from place to place and which can be distorted by the presence of nearby magnets, masses of iron, or electric currents. The first instrument to measure i independently of $\mathscr{B}_e$ was the *astatic galvanometer*, invented in

1825 by the Italian physicist Leopoldo Nobili (1784–1835). A simple form of the device is sketched opposite. The heart of the device is a pair of well-matched magnets, each having magnetic moment μ. These two magnets are connected by a light, rigid rod so that their poles are oriented in opposite directions. The assembly is suspended from a torsion fiber of torsion constant κ. Two identical coils of N turns, whose radii b are considerably larger than the length of the magnets, surround the two magnets. The coil pair is fixed in a frame so that the coil assembly can be rotated as a unit about a vertical axis coinciding with the torsion fiber. The coils are connected in series, in such a way that the current passes through them in opposite senses. With no current in the system, the coils are rotated so that their plane coincides with the equilibrium orientation of the magnets. When a current i passes through the coils, the magnet system rotates through an angle θ. (a) Why is the operation of this galvanometer independent of the earth's magnetic field? (b) Express the current i in terms of θ, μ, κ, b, N, and any necessary constants.

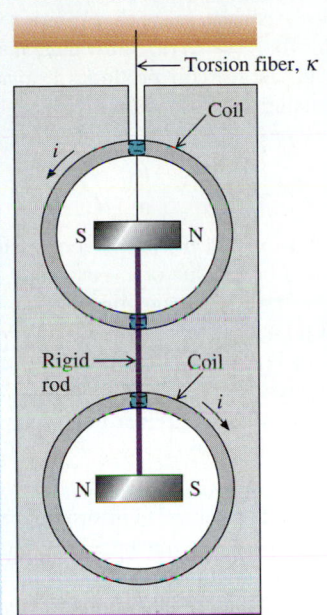

GROUP C

29.71 *(2) Square loop, I.* A stiff wire loop, having the form of a square of side $2b$, carries current i. Show that the magnitude of the magnetic field at the center of the square is

$$\mathcal{B} = \frac{\sqrt{2}\mu_0 i}{\pi} \frac{1}{b}.$$

Is this value greater or less than the corresponding value for a circular loop of radius b, $(\mu_0 i/2)(1/b)$? Can you account for the difference?

29.72 *(2) Wire sculpture, VII.* A long, current-carrying wire is bent into the form shown below. The point P lies at an arbitrary distance z from the midpoint of the middle section of the wire, whose length is $2l$. The angle subtended by the lines drawn from P to the bends in the wire is 2θ, as shown.

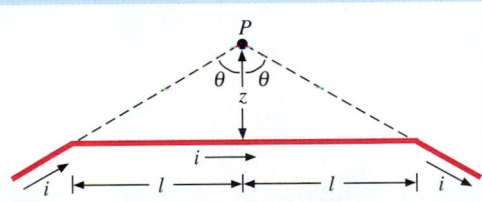

The left and right sections of the wire lie along extensions of the lines from P to the bends. Show that the magnetic field at P has magnitude

$$\mathcal{B} = \frac{\mu_0 i}{2\pi} \frac{\sin \theta}{z}$$

and that this reduces to the expression for an infinitely long straight wire when $\theta \to 90°$. Hint: Compare with Problem 23.54, remembering that $\sin \theta = l/\sqrt{l^2 + z^2}$.

29.73 *(2) Square loop, II.* A current-carrying loop is made of stiff wire formed into a square of sides $2b$. (a) Show that, at

a point on the axis of the loop a distance x from the plane of the loop, the magnetic field is directed away from the loop and has magnitude

$$\mathcal{B} = \frac{2\mu_0 i}{\pi} \frac{b^2}{(x^2 + b^2)\sqrt{x^2 + 2b^2}}.$$

(b) Show that, at large distances ($x \gg b$), the magnetic field reduces to the value given by Equation 29.14, which was derived for a *circular* loop. This is another demonstration of the assertion, made in Section 28.6, that the dipole moment of a current-carrying loop depends only on the product iA of the loop current and the enclosed area, and not on the shape of the loop.

29.74 *(2) Rectangular loop.* Find the magnitude of the magnetic field at the center of a current-carrying rectangular loop of sides L and w. To check your result, show that when $L \gg w$, the field magnitude reduces to the value at the midline between a pair of long, straight, parallel wires carrying current in opposite directions.

29.75 *(4) Double duty, II.* As in Problem 29.46, two coils of radii b_1 and b_2 lie in the same plane and are concentric. Suppose the two coils carry the same current; that is, $i_1 = i_2$. The magnetic fields produced by the two coils at x—$\mathcal{B}_1$ and $\mathcal{B}_2$—will then be different in general. (a) Show that the ratio of the two fields is

$$\frac{\mathcal{B}_2}{\mathcal{B}_1} = \left(\frac{b_2}{b_1}\right)^2 \left(\frac{x^2 + b_1^2}{x^2 + b_2^2}\right)^{3/2}.$$

(b) Show that, at large distances ($x \gg b_1, b_2$), the larger coil produces a *larger* magnetic field than the smaller one—specifically, that the ratio of the two fields approaches

$$\frac{\mathcal{B}_2}{\mathcal{B}_1} \to \left(\frac{b_2}{b_1}\right)^2.$$

Show that this is consistent with Equation 29.14. (Hint: $(1 + z)^n \simeq 1 + nz$ for $z \ll 1$.) (c) Show that, at small distances ($x \ll b_1, b_2$), the larger coil produces a *smaller* magnetic field than the smaller one—specifically, that the ratio of the two fields approaches

$$\frac{\mathscr{B}_2}{\mathscr{B}_1} \to \frac{b_1}{b_2}.$$

(Hint: See the hint for part **b**.) (d) Given the results of parts **a** and **b**, there must be a value of x—call it x^*—at which the two coils produce the *same* magnetic field; $\mathscr{B}_1 = \mathscr{B}_2$. Write an equation that relates the values x^*, b_1, and b_2. Don't try to solve it for x^*. (e) Find the value of x^* for the two loops of Problem 29.46, where $b_1 = 2.74$ cm and $b_2 = 6.31$ cm. Note that x^* is neither very much greater nor very much less than b_1 and b_2.

29.76 *(4) Helmholtz coils.* You saw in Section 29.4 how a solenoid can be used to produce a uniform magnetic field over an appreciable volume. Unfortunately, the long, narrow form of a solenoid sometimes makes its use impossible for providing a uniform field either in a confined space or where an experimental sample must lie in a uniform field but be accessible from all directions. In such cases, a pair of coils called *Helmholtz coils* can often be used. Consider the pair of identical coils shown below. Each coil of N closely spaced turns has radius b. The coils are connected in series, so each carries the same current i in the same sense. The coils have a common axis; call this the x axis. Choose $x = 0$ at the midpoint O between the coils, a distance a from each.

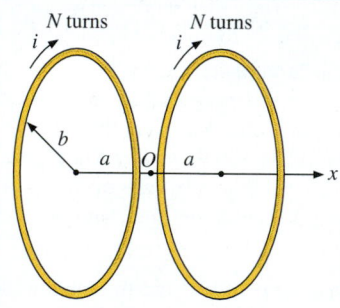

(a) Beginning with Equation 29.11a, show that the magnitude of the magnetic field at a point x on the coil axis can be written in the form

$$\mathscr{B} = \frac{\mu_0 N i b^2}{2(b^2 + a^2)^{3/2}}$$

$$\times \left[\frac{1}{\left(1 + \dfrac{x^2 + 2ax}{b^2 + a^2}\right)^{3/2}} + \frac{1}{\left(1 + \dfrac{x^2 - 2ax}{b^2 + a^2}\right)^{3/2}} \right]. \quad (1)$$

(b) Show that, if the *Helmholtz spacing* ($a = b/2$) is used, the magnetic field in the region near O ($x \ll a$) is nearly constant. To do this, expand Equation 1, using the first *three* terms of the binomial series:

$$(1 \pm z)^n = 1 + nz \pm \tfrac{1}{2}n(n - 1)z^2 \quad \text{for } z \ll 1.$$

In your result, discard all terms in powers of x greater than x^2, arguing that, if $z \ll 1$, z^4 is truly negligible. Then show

that, for the special case of the Helmholtz spacing,

$$\mathscr{B} = \frac{8\mu_0 N i}{5\sqrt{5}a}.$$

That is, $\mathscr{B}$ is independent of x up to the very small terms in x^4. (c) Why is $\mathscr{B}$ independent of terms in x^3? (Hint: Think symmetry!) Note: It is also true, but harder to show, that the magnetic field of a Helmholtz pair is nearly constant over small displacements from O in a direction perpendicular to the coil axis.

29.77 *(4) Not-so-long solenoid.* A point P lies on the axis of a solenoid of finite length, as shown below. Lines drawn from

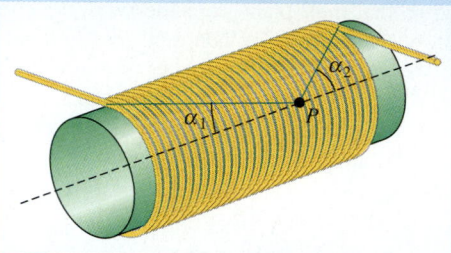

P to the ends of the solenoid make angles a_1 and a_2 with the solenoid axis. The solenoid has n turns per unit length, which carry a current i. (a) Using Equation 29.10, calculate the magnetic field $\mathscr{B}$ at P. (b) Show that this result reduces to Equation 29.22 for a very long solenoid. (c) Use the result of part **a** to find $\mathscr{B}$ on the solenoid axis, just at the end of the solenoid. (d) Compare the result of part **c** with that of Problem 29.29. Which method do you prefer?

29.78 *(4) Fan-shaped.* In the figure below, a fan-shaped plate of uniform thickness t and length Y is made of a material of relatively high resistivity. Contacts of low resistivity are soldered to its ends, of width a and b respectively. Thus the distribution of current i passing into and out of the plate is uniform. What is the magnetic field $\mathscr{B}$ at point P, just outside the plate and along its axis of symmetry, a distance y from the narrow end?

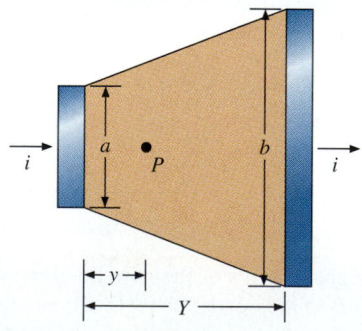

29.79 *(G) Classical electron spin.* Imagine that an electron is a little spherical ball of radius R and that its mass and charge are uniformly distributed throughout the ball. Also, assume that the electron rotates on its own axis, like the earth, with angular speed ω. (a) In imagination, divide the electron into nested cylindrical segments coaxial with the spin axis. Show that the magnitude of the electric current arising from the rotation of the charge is

$$i = \frac{e\omega}{2\pi},$$

a result independent of the radius of the electron. (Hint: Find the charge dq on each shell, and use the result to find the contribution di of that shell to the current. It is easiest to integrate in terms of the angle θ between the spin axis and the point at which the cylindrical slice terminates at the surface of the electron.) **(b)** Show that the magnetic moment of the electron is

$$\mu = \tfrac{1}{5}eR^2\omega.$$

Hints: (1) Begin with the current contribution di of a shell of radius r, and calculate the contribution $d\mu$ of this shell to the magnetic moment of the electron.

$$(2) \qquad \int_0^{\pi/2} \sin^3 \theta \cos^2 \theta \, d\theta = \tfrac{2}{15}.$$

(c) Let the spin angular momentum of the electron be $l = I\omega$, where I is the moment of inertia of the electron about its axis. Show that the *classical gyromagnetic ratio* γ_c of the electron is

$$\gamma_c \equiv \frac{\mu}{l} = \frac{e}{2m}.$$

Although it is derived on the basis of the very crude assumption that an electron is a ball, γ_c differs from the actual gyromagnetic ratio $\gamma = e/m$ by just a factor of 2! The difference is due to the essentially quantum nature of the electron.

29.80 *(G) Displacement current.* In the figure in the next column, the current i passing through the leads charges the parallel-plate capacitor whose plates are circular disks of area A (plate spacing is exaggerated for clarity). **(a)** Show that the relation between i and the electric field $\mathscr{E}$ between the plates is

$$i = \epsilon_0 A \frac{d\mathscr{E}}{dt}.$$

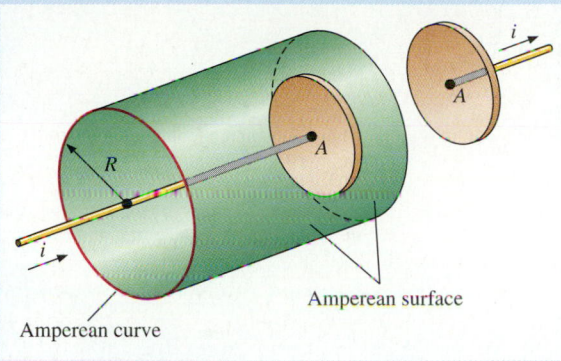

Amperean curve

Amperean surface

(b) Draw an Amperean surface shaped like a drinking glass, whose rim is an Amperean circle of radius R centered on the lead wire, as shown in the figure. How much electric charge passes through the surface per second? **(c)** Write a relation between the magnetic circulation around the Amperean circle and $d\mathscr{E}/dt$. Even though electric charge q and current $i = dq/dt$ do not appear in the quantity $\epsilon_0 A \, d\mathscr{E}/dt$, this quantity is equivalent to an electric current. Indeed, it is just as much an electric current as is dq/dt! It is called the *displacement current* and has an essential role in electromagnetic radiation. The displacement current is discussed in more detail in Chapter 34. **(d)** The current dq/dt is called the *conduction current*. If you include both conduction and displacement currents, what can you say about the total current that penetrates the closed surface consisting of the drinking-glass-shaped Amperean surface and the planar disk enclosed by the same Amperean curve? See Query 29.8.

29.81 *(G) Little solar system.* Assume that the electron of a hydrogen atom circles the proton in the same way that a planet circles the sun (except that the force is the electric force in the hydrogen atom). The ionization energy of hydrogen is 13.6 eV, and the ground-state diameter of the atom is 1.0×10^{-10} m. Using this information, calculate **(a)** the frequency of revolution of the electron and **(b)** the electric current due to this rotation.

Faraday's Law

L O O K I N G   A H E A D

Although a steady electric current gives rise to a steady magnetic flux, a steady magnetic flux does not give rise to an electric current. However, a changing magnetic flux gives rise to an electric current in a conducting loop.

The current that flows in the loop is not the result of a potential difference somewhere in the loop; rather, the changing magnetic flux gives rise to a *distributed emf*. The integral of the emf around the loop is equal to the negative of the rate of change of the magnetic flux penetrating the loop; this is Faraday's law.

The current in the loop in turn gives rise to a magnetic flux. Conservation of energy requires that this flux oppose the change in flux that gave rise to the current; this is Lenz's law.

The operation of electric generators and electric motors is based on Faraday's law. In the generator, the change in magnetic flux that gives rise to the generator emf is produced by rotating the armature coil in a magnetic field.

If Faraday's law is recast in terms of electric circulation, it can be expressed so that Faraday's law and Ampère's law constitute a symmetric pair of equations.

Left: When William E. Gladstone, then Chancellor of the Exchequer, asked Michael Faraday what use electricity was, Faraday replied, "One day, Sir, you may tax it." Perhaps even Faraday could not have foreseen that his law would make possible 300-km/h trains.

Must I ever change me, that you may remain constant?

—ANON., *"Plaint to a Fickle Mistress"* (c. 1660)

Introduction: An Experiment That Doesn't Work and One That Does

The entire discussion in Chapter 29 was based on the observation that *you can create a steady magnetic flux by setting up a steady electric current.* In view of the frequent occurrence of symmetry in physical situations, the following question probably strikes you as a worthwhile one: *Can you create a steady electric current by setting up a steady magnetic flux?*

A simple apparatus for answering this question is shown in Figure 30.1. A solenoid and a galvanometer are connected into a circuit. The galvanometer will detect any current flowing through the circuit. A bar magnet inside the solenoid exposes the circuit to considerable magnetic flux. The result of the experiment is negative: *The steady magnetic flux of the magnet results in no electric current in the circuit.*

In 1829 and 1830, two physicists working independently tried this experiment. Given the "obviousness" of the question the experiment answers, it is unlikely that they were the first to do so. But Faraday and the American Joseph Henry were perhaps more observant than their predecessors, and they noted something else that *did* happen. That something else is shown schematically in Figure 30.2. The setup is identical with that of Figure 30.1. But now we observe what happens when we *move* the magnet in and out of the coil, thus varying the magnetic flux that penetrates the coil. *As long as the magnet is moving in the vicinity of the coil, an electric current is observed in the circuit.* This effect, called **electromagnetic induction**, is the subject of the present chapter. In studying electromagnetic induction, we complete our survey of the fundamental laws of electromagnetism (though there remains much to be said about their consequences, as you will see in subsequent chapters).

In Section 30.2, we consider electromagnetic induction in its simplest context, called *motional emf.* In subsequent sections we broaden our study of the fundamental connection between changing magnetic flux and electric current, and in Section 30.5 we

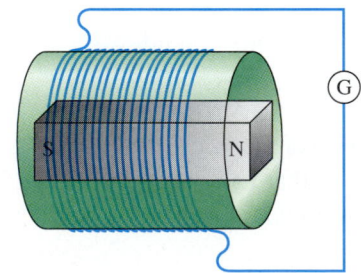

FIGURE 30.1 An experiment that has a negative result. Although the solenoid lies in the steady magnetic flux of the bar magnet, no current is detected by the galvanometer.

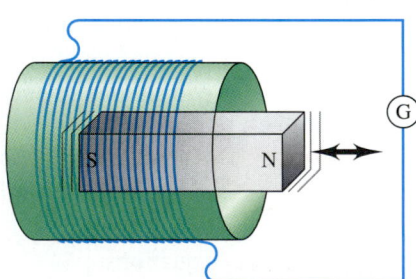

FIGURE 30.2 One form of Faraday's experiment, showing electromagnetic induction. As long as the magnet is moving in the vicinity of the solenoid, electric current passes through the circuit.

consider two of the important practical applications of electromagnetic induction—the electric generator and the electric motor. We then take a more fundamental view in Section 30.6, noting that even in the absence of a conductor, changing magnetic flux results in an electric field.

Motional emf

In Figure 30.3, a conducting wire of length Y moves with velocity $\mathbf{v}$ in a direction transverse to its length. A uniform, steady magnetic field $\mathscr{B}$ exists over the entire region through which the wire moves. The direction of $\mathscr{B}$ is perpendicular both to the length of the wire and to $\mathbf{v}$.

We already know what must happen in the wire. The free charges q in the wire experience a magnetic force $\mathbf{F}_m = q\mathbf{v} \times \mathscr{B}$ (Section 28.8). They pile up at the ends of the wire. Consequently, an electric field appears in the wire and increases until the resulting electric force $\mathbf{F}_e = q\mathscr{E}$ just balances the magnetic force. The Lorentz force (Equation 28.22) is then zero, and we have

$$\mathbf{F} = \mathbf{F}_e + \mathbf{F}_m = q(\mathscr{E} + \mathbf{v} \times \mathscr{B}) = 0,$$

or
$$\mathscr{E} = -\mathbf{v} \times \mathscr{B}. \tag{30.1}$$

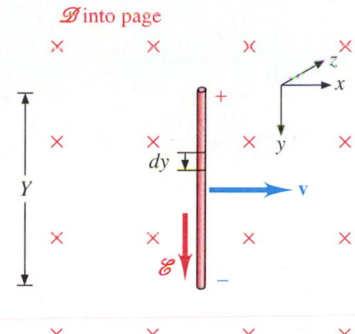

FIGURE 30.3 The motion of the wire with velocity $\mathbf{v}$ through the perpendicular magnetic field $\mathscr{B}$ results in a magnetic force that makes free charges pile up at the ends. This piling up induces an electric field $\mathscr{E}$ in the wire, and the result is a motional emf between the ends of the wire.

This electric field is shown in Figure 30.3. Its magnitude is uniform throughout the wire because $\mathbf{v}$ and $\mathscr{B}$ are everywhere the same.

Because an electric field is present, there is a voltage difference between the upper and lower ends of the wire. That is, the moving wire is a source of emf. From Figure 30.3, you can see that the direction of the electric field is the positive y direction, and the voltage difference is (Section 25.2)

$$V = -\int_0^Y \mathscr{E} \cdot d\mathbf{y} = -\mathscr{E}Y.$$

Using the value of $\mathscr{E}$ given by Equation 30.1, we obtain, for $\mathbf{v} \perp \mathscr{B}$,

$$V = v\mathscr{B}Y. \tag{30.2}$$

In Figure 30.4, we make an addition to the arrangement of Figure 30.3. The moving wire now slides without friction along the sides of a stationary U-shaped conductor.

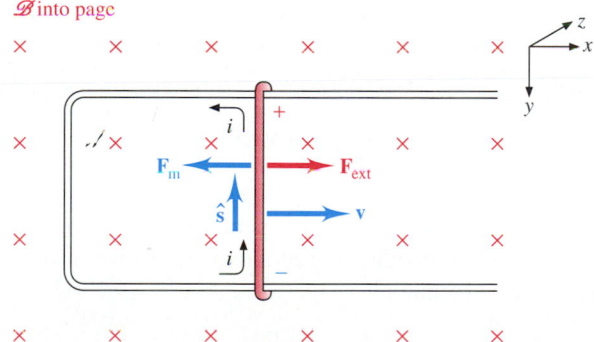

FIGURE 30.4 The motional emf drives a current i through the circuit completed by the stationary U-shaped conductor. As a result of the presence of the current, a magnetic force $\mathbf{F}_m$ tends to stop the motion of the wire. To maintain the velocity $\mathbf{v}$, an external force $\mathbf{F}_{ext} = -\mathbf{F}_m$ must be supplied.

This conductor, taken together with the moving wire, constitutes a closed circuit through which the motional emf can drive an electric current. Consequently, a current i, whose magnitude depends on the resistance of the circuit, flows in the sense shown. (The sense of the current conforms to the standard rule: from $+$ to $-$ outside the source of emf, and from $-$ to $+$ within it.)

When an emf V drives a current i through a circuit having resistance, power is dissipated. More specifically, electrical energy is transformed into heat energy as the wire warms up. According to Joule's law (Equation 27.18), this power is

$$P_{out} = -Vi. \tag{30.3}$$

The negative sign in this equation signifies that P_{out} represents a removal of electrical energy from the system. What is the source of this power? We begin to answer this question by considering a new feature introduced by the current, which was not present in the simpler arrangement of Figure 30.3. Owing to the presence of the current i, magnetic forces are exerted on the conductors that carry the current. According to Equation 28.12c, the force on a straight wire of length s has magnitude $F_m = i\mathscr{B}s \sin\theta$, where θ is the angle between $\mathbf{s}$ and $\mathscr{B}$. Let us apply this expression to the moving wire. For this wire, we have $|\mathbf{s}| = Y$ and $\sin\theta = 1$. So the magnitude of the magnetic force is

$$F_m = i\mathscr{B}Y.$$

To determine the direction of $\mathbf{F}_m$, we use the right-hand rule. We choose the direction $\hat{\mathbf{s}}$ consistent with the sense of the current in the wire, so i will be a positive quantity. We thus have $\hat{\mathbf{s}} = -\hat{\mathbf{y}}$. And the direction of $\mathscr{B}$ is $\hat{\mathbf{z}}$. It follows that

$$\hat{\mathbf{F}}_m = \hat{\mathbf{s}} \times \mathscr{B} = -\hat{\mathbf{y}} \times \hat{\mathbf{z}} = -\hat{\mathbf{x}},$$

as shown in Figure 30.4. Combining magnitude and direction, we see that the magnetic force is

$$\mathbf{F}_m = -i\mathscr{B}Y\hat{\mathbf{x}}. \tag{30.4}$$

If the wire is to continue to move with constant velocity $\mathbf{v}$, we must supply an external force

$$\mathbf{F}_{ext} = -\mathbf{F}_m = i\mathscr{B}Y\hat{\mathbf{x}}. \tag{30.5}$$

This force also is shown in Figure 30.4.

We can now determine the power $P = \mathbf{F} \cdot \mathbf{v} = F_{ext}v$ (Equation 7.28) required to keep the wire moving, so as to drive the current i through the circuit. Using Equation 30.5, we immediately have

$$P = i\mathscr{B}Yv.$$

Because $v = dx/dt$, we can write P in the form

$$P = i\mathscr{B}\frac{Y\,dx}{dt}.$$

But $(Y\,dx)/dt$ is the rate at which the area A enclosed by the electric circuit is changing. That is, we have $Y\,dx/dt = dA/dt$. So the power input to the system can be written

$$P = i\mathscr{B}\frac{dA}{dt}. \tag{30.6}$$

Because $\mathscr{B}$ is constant over the entire region, we can rewrite this expression in the form

$$P = i\frac{d(\mathscr{B}A)}{dt}.$$

The product $\mathscr{B}A$ is the magnetic flux Φ_m that penetrates the current loop at the instant when the area has the particular value A (Section 28.3). So we rewrite the power input to the system once more, obtaining

$$P = i\frac{d\Phi_m}{dt}. \tag{30.7}$$

Energy is not being stored in the system, and so the power input of Equation 30.7 must be equal to the power dissipated, given by Joule's law. Comparing Equation 30.7 with Equation 30.3, we have $-Vi = i\,d\Phi_m/dt$, or

$$V = -\frac{d\Phi_m}{dt}. \tag{30.8}$$

The motional emf that drives current through the loop is equal to the negative of the rate of change of the magnetic flux penetrating the loop. This statement is a form of

Faraday's law. The minus sign in Equation 30.8 has important physical meaning, which we consider in detail in Section 30.4. For the moment, we merely note in passing that its presence is necessary to satisfy the principle of energy conservation. For consistency with what follows, we will carry the sign along in algebraic manipulations. But in this section we will consider the physical significance of magnitudes only.

EXAMPLE **30.1**

In Figure 30.5, the width Y of the U-shaped conductor is 25 cm. The magnetic field, directed out of the page, has magnitude 1.2 T. The resistance per unit length has the same value, $\lambda = 0.50\ \Omega/m$, throughout the circuit. If you pull the sliding wire to the right at a constant speed $v = 3.3$ m/s, **(a)** at what rate does the flux penetrating the conducting loop change? **(b)** What is the potential difference V between P and Q? What is the sense of the current driven by V? **(c)** At $t = 0$, the moving wire is located at $x = 0$. What is the ammeter reading at $t = 2.5$ s? **(d)** Express the force $\mathbf{F}_{ext}$ required to maintain the constant velocity of the moving wire as a function of time. What is the value of F_{ext} at $t = 2.5$ s?

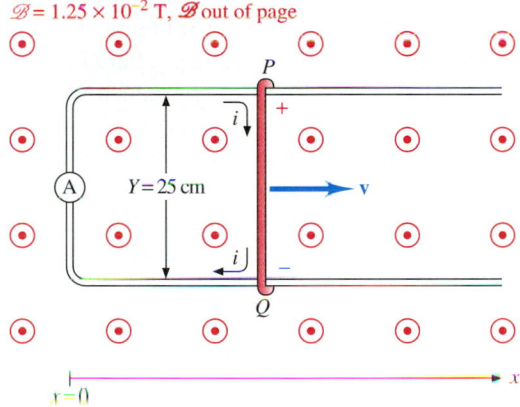

$\mathscr{B} = 1.25 \times 10^{-2}$ T, $\mathscr{B}$ out of page

$Y = 25$ cm

$x = 0$

FIGURE 30.5

SOLUTION:
(a) Evaluate $d\Phi_m/dt$ for the loop.

The flux change is due entirely to the change of area of the loop: $d\Phi_m/dt = d(\mathscr{B}A)/dt = \mathscr{B}\,dA/dt$. We have

$$\frac{dA}{dt} = Y\frac{dx}{dt} = Yv.$$

Thus the magnetic flux through the loop increases at the rate

$$\frac{d\Phi_m}{dt} = \mathscr{B}Yv = 1.2\ \text{T} \times 0.25\ \text{m} \times 3.3\ \text{m/s}$$

$$= 0.99\ \text{T·m}^2/\text{s}.$$

(b) Find the potential difference V between P and Q and the sense of the current driven by V.

Note that the direction $\hat{\mathbf{v}}$ is the same as that in Figure 30.4, but the direction of $\mathscr{B}$ is reversed. Consequently, the polarity of the moving wire, seen as a source of emf, is reversed; Q is the positive "terminal" and P is the negative "terminal." The sense of the current through the circuit is thus reversed as well,

as Figure 30.5 shows. To obtain the magnitude of V, you can use Faraday's law (Equation 30.8) to write

$$|V| = \left|\frac{d\Phi_m}{dt}\right| = 0.99\ \text{V}.$$

You can obtain the same result by inserting the numerical values given into Equation 30.2, $V = v\mathscr{B}Y$:

$$|V| = 3.3\ \text{m/s} \times 1.2\ \text{T} \times 0.25\ \text{m} = 0.99\ \text{V}.$$

(c) What is the ammeter reading at $t = 2.5$ s?

The length L of the loop varies with time as the wire moves. Given the initial condition $x = 0$ at $t = 0$, you can see from Figure 30.5 that the initial loop length is $2Y$, and the length at time t is thus

$$L = 2Y + 2vt.$$

The resistance of the wire is

$$R = \lambda L = 2\lambda(Y + vt).$$

At $t = 2.5$ s you have

$$R = 2 \times 0.50\ \Omega/m \times (0.25\ \text{m} + 3.3\ \text{m/s} \times 2.5\ \text{s})$$

$$= 8.5\ \Omega.$$

The ammeter thus reads

$$i = \frac{V}{R} = \frac{0.99\ \text{V}}{8.5\ \Omega} = 0.12\ \text{A}.$$

(d) Derive an expression for $\mathbf{F}_{ext}$ and evaluate its magnitude at $t = 2.5$ s.

There are several ways to calculate the necessary force, but the quickest is to use Joule's law. Because the circuit is ohmic (Section 27.4), you have the power output $P = Vi = V^2/R$. Writing the power input as $P = F_{ext}v$ and equating input power to output power, you obtain

$$\frac{V^2}{R} = F_{ext}v.$$

Thus the force is

$$F_{ext} = \frac{V^2}{Rv} = \frac{V^2}{2\lambda(Y + vt)v}.$$

At $t = 2.5$ s, you have

$$F_{ext} =$$

$$\frac{(0.99\ \text{V})^2}{2 \times 0.50\ \Omega/m \times (0.25\ \text{m} + 3.3\ \text{m/s} \times 2.5\ \text{s}) \times 3.3\ \text{m/s}}$$

$$= 3.5 \times 10^{-2}\ \text{N}.$$

Why does F decrease with time?

The Central Role of Magnetic Flux in Faraday's Law

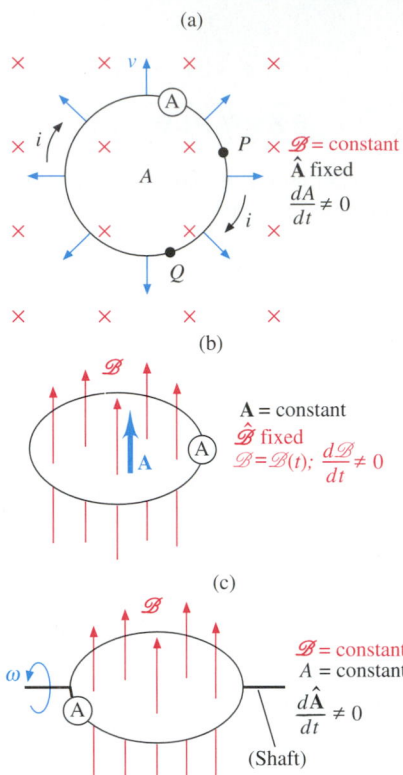

FIGURE 30.6 Three ways of changing the magnetic flux $\Phi_m = \mathscr{B} \cdot \mathbf{A}$ that penetrates a loop: (a) by changing the area A of the loop, as in Figure 30.4; (b) by changing the magnitude $\mathscr{B}$ of the magnetic field in the region of the loop; (c) by rotating the loop with respect to the magnetic field.

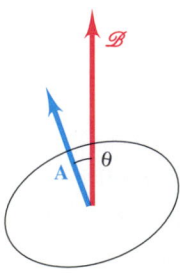

FIGURE 30.7 A loop whose area and orientation are described by the area vector $\mathbf{A}$ lies in a magnetic field $\mathscr{B}$ that is uniform over the region. The angle between $\mathscr{B}$ and $\mathbf{A}$ is θ. In general, θ may vary with time.

You might argue at this point that there is nothing *fundamentally* new in Equation 30.8, $V = -d\Phi_m/dt$, which expresses the emf generated by moving a wire through a magnetic field in terms of the rate of change of magnetic flux. The derivation leading to Equation 30.8 began with an equation for the magnetic force derived in Chapter 28. So Equation 30.8 may seem to be nothing more than a consequence of applying the magnetic force law to a special situation.

Although we have developed it in the special context of the moving-wire experiment of Figure 30.4, Equation 30.8 turns out to be valid over a much wider range of circumstances. It applies, for instance, to the quite different experiment described in Figure 30.2. Indeed, *Faraday's law is an independent law of nature based directly on experiment*, just like Ampère's law and Gauss's law. Further experiment, which we will now discuss, shows that *an emf is always induced when the magnetic flux penetrating a conducting loop is changing, no matter how that flux change is effected*. That is, the induction of motional emf in the device of Figure 30.4 is only one special case of a more general relation between changing magnetic flux and induced emf.

There are three ways to change the magnetic flux through a conducting loop and thus to induce an emf in the loop. In Figure 30.6a, the loop is made of a flexible material that is stretched uniformly, as indicated by the blue arrows. The flux penetrating the loop changes because the area is changing, just as in the device of Figure 30.4. In Figure 30.6b, the loop has fixed area, but the magnitude of the magnetic field in which it is located increases (or decreases) with time; $\mathscr{B} = \mathscr{B}(t)$. The flux penetrating the loop changes because $\mathscr{B}$ is changing. In Figure 30.6c, the loop has fixed area and the magnetic field is uniform and constant. But the flux penetrating the loop changes because the loop is made to rotate with respect to the field—that is, because $\mathscr{B} \cdot \mathbf{A}$ changes.

In all three cases, the ammeter in the loop registers a current, and so we can infer the presence of an emf, $V = iR$, that drives the current. And all three systems conform quantitatively to Faraday's law,

$$V = -\frac{d\Phi_m}{dt}.$$

To see how Faraday's law describes all these processes—or for that matter, any combination of them—let us write the magnetic flux in the form

$$\Phi_m = \mathscr{B} \cdot \mathbf{A}, \qquad \text{(30.9a)}$$

or

$$\Phi_m = \mathscr{B}A \cos \theta, \qquad \text{(30.9b)}$$

where θ is the angle between $\mathscr{B}$ and the area vector $\mathbf{A}$ of the loop, as shown in Figure 30.7. We can thus express Faraday's law in the form

$$V = -\frac{d\Phi_m}{dt} = -\frac{d}{dt}(\mathscr{B}A \cos \theta).$$

In general, any of the three quantities $\mathscr{B}$, A, and θ can vary with time. When we apply the chain rule for differentiation, the equation becomes

$$V = -\mathscr{B} \cos \theta \frac{dA}{dt} - A \cos \theta \frac{d\mathscr{B}}{dt} - \mathscr{B}A \frac{d(\cos \theta)}{dt}.$$

We carry out the differentiation in the last term on the right, and Faraday's law becomes

$$V = -\mathscr{B} \cos \theta \frac{dA}{dt} - A \cos \theta \frac{d\mathscr{B}}{dt} + \mathscr{B}A \sin \theta \frac{d\theta}{dt}. \qquad \text{(30.10)}$$

The first term on the right describes the motional emf of Figures 30.6a and 30.4, where the change of flux with time is due to the area change dA/dt. The second term describes the effect of changing field $d\mathcal{B}/dt$ shown in Figure 30.6b, and the third term describes the effect of changing orientation (rotational motion) shown in Figure 30.6c. In most applications, only one of the three flux-changing processes takes place, and thus only one of the three terms on the right side of Equation 30.10 has a nonzero value.

Distributed emf

In the moving-wire device of Figure 30.4 and Example 30.1, we thought of the moving wire as a source of emf, while the U-shaped wire was a passive element that served to complete the circuit. The moving wire had two "terminals," called P and Q in Figure 30.5, and the voltage difference between them had the value V.

In the three parts of Figure 30.6, however, no particular part of the loops has a special role in the production of emf, and there are no terminals. Nevertheless, current does flow through the loops, as evidenced by the ammeter readings. In each case, the emf that drives this current is *distributed* throughout the loop and is therefore called a **distributed emf**.

Work is done when charge carriers are driven around the loop. Indeed, if the conductor had zero resistance, the carriers would steadily gain kinetic energy as they circulated around the loop. Nevertheless, *we cannot define a potential difference between any two points on the loop*. (Indeed, suppose that there were such a potential difference between two arbitrary points on the loop, such as P and Q in Figure 30.6a. Then the senses of the currents from P to Q along the two paths connecting them would be opposite, and their magnitudes would be different. But in fact, the current has the same sense and magnitude through the entire loop.) Like all sources of emf, the loop provides the energy needed to drive the carriers by *nonelectric means*—here, by means of changing *magnetic* flux.

EXAMPLE 30.2

In Figure 30.8, a shaft-mounted coil can be made to rotate about one of its diameters. The shaft is oriented perpendicular to a uniform magnetic field $\mathcal{B}$. The coil, of radius b, has N turns. The ends of the coil are attached to **slip rings**—metal rings mounted on the shaft but insulated from it. The slip rings make sliding contact with the stationary terminals P and Q, to which a voltmeter is attached. (a) If the coil is rotated with angular speed ω, express the voltmeter reading V as a function of time. Assume that the plane of the coil is normal to the magnetic field at $t = 0$. (b) Let $N = 100$ turns, $b = 4.0$ cm, and $\mathcal{B} = 0.93$ T. If the shaft rotates at 3600 rev/min, express V as a function of time, and evaluate the maximum value attained by V.

SOLUTION:

(a) Express V as a function of t.

The change in the magnetic flux penetrating the coil is due to the rotation of the coil; $\mathcal{B}$ and the magnitude A remain constant while the direction $\hat{\mathbf{A}}$ changes as the coil rotates. Only the third term on the right side of Equation 30.10 has a nonzero value, and you can write

$$V = N\mathcal{B}A \sin \theta \frac{d\theta}{dt}, \qquad (30.11)$$

where θ is the angle between $\mathcal{B}$ and $\mathbf{A}$. The factor N accounts for the fact that the coil has more than one turn. The term $d\theta/dt$ is just the angular speed ω, and you have

$$V = \omega N\mathcal{B}A \sin \theta = \omega N\mathcal{B}A \sin \omega t. \qquad (30.12)$$

The output voltage of the coil is a sinusoidal function of time. Don't worry for the moment about the polarity of V—that is, the relative signs of terminals P and Q—in absolute terms. Instead, arbitrarily assign a positive value to V in the time interval immediately after $t = 0$. The polarity of V must change twice

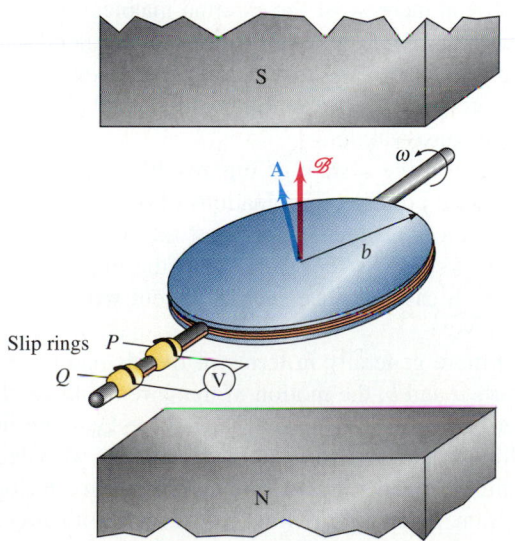

FIGURE 30.8 As the coil rotates in the magnetic field, the voltmeter monitors the induced voltage V.

per rotation of the coil, becoming negative as θ passes through 180° and positive again when θ passes through 360°. V is an *alternating voltage*, and a current driven by such an emf is an **alternating current**. The coil of Figure 30.8 is, in fact, a simplified electric generator.

(b) Express $V(t)$ numerically and evaluate V_{max}.

When inserting the numerical values into Equation 30.12, you must express ω in rad/s. The area of the coil is πb^2. You

calculate

$$\omega = 3600 \text{ rev/min} \times 2\pi \text{ rad/rev} \times \frac{1 \text{ min}}{60 \text{ s}} = 377 \text{ rad/s}.$$

The voltage V attains its maximum value when $\sin \theta = 1$. So you obtain

$$V_{max} = 377 \text{ rad/s} \times 100 \times 0.93 \text{ T} \times \pi \times (4.0 \times 10^{-2} \text{ m})^2$$
$$= 176 \text{ V}.$$

Lenz's Law and Energy Conservation

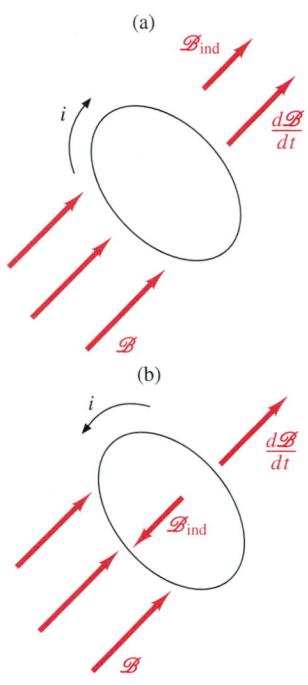

FIGURE 30.9 As the magnetic field increases, does the induced current circulate as in part *a* or as in part *b*?

H. F. E. Lenz

The German-Estonian physicist Heinrich F. E. Lenz (1804–1865), who worked mainly in St. Petersburg, first enunciated the law named after him in 1834. St. Petersburg had become a center of learning in the late eighteenth century, when Catherine the Great invited many outstanding western European thinkers to her court.

In considering the current induced by change in the magnetic flux penetrating a conducting loop, we deferred an important question, which we now address: What is the sense of the induced current?

In the interest of concreteness, let us focus on the simple geometry of Figure 30.9. A conducting loop has fixed area and orientation, expressed by the vector $\mathbf{A} = A\hat{\mathbf{A}}$. We take the loop axis as the x axis. There is a uniform magnetic field $\mathscr{B} = \mathscr{B}\hat{\mathbf{x}}$ throughout the region. Now, suppose that $\mathscr{B}$ increases slightly; that is, suppose that $d\mathscr{B}/dt > 0$. Will the sense of the current be that shown in Figure 30.9a or that shown in Figure 30.9b?

To show that Figure 30.9b is the correct picture, we demonstrate that the sense indicated in part *a* is inconsistent with the principle of energy conservation. Whenever it carries an induced current, the loop becomes a magnetic dipole. In part *a*, according to the screwdriver rule, the magnetic field of the dipole, $\mathscr{B}_{ind}$, has the same direction as the external magnetic field. The current thus leads to an increase in the flux penetrating the loop, which leads to a further increase in induced current, and so on. If this picture were correct, an infinitesimal initial fluctuation in the magnetic field would lead to an indefinite growth of both the induced current and the magnetic flux penetrating the loop. The result is a sort of perpetual-motion machine, in which the ever-increasing induced current in the loop produces Joule heating at no cost. In the correct picture (Figure 30.9b), however, the magnetic field arising from the induced current *opposes* the increase $d\mathscr{B}/dt$ of the external magnetic field. That is, *the sense of the induced current is such as to oppose the externally imposed change*. The induced current will not continue unless there is a continual increase of the external magnetic field.

The opposing sense of the induced current is one manifestation of a general statement called **Lenz's law**: *Every effect of induction acts in opposition to the cause that produces it*. As the discussion of the preceding paragraph indicates, Lenz's law is a consequence of the principle of energy conservation.

You can see Lenz's law operating in the system of Figures 30.4 and 30.5. Figure 30.10 shows this system again, with all possible combinations of directions of $\mathscr{B}$ and $\mathbf{v}$. In each case, the sense of the induced current is such that the magnetic force $\mathbf{F}_m$ tends to bring the moving wire to a stop. As in Figure 30.9, the opposite sense of induced current is impossible in each case because such a current would result in a perpetual-motion machine.

We can interpret Figure 30.10 more generally in terms of the change in magnetic flux penetrating the system. In parts *a* and *b*, the motion of the wire increases the area of the loop and thus the flux that penetrates it. The magnetic force $\mathbf{F}_m$—the induced effect—opposes the increase in flux. In parts *c* and *d*, the motion of the wire decreases the area of the loop and thus the flux that penetrates it. The magnetic force $\mathbf{F}_m$ opposes the decrease in flux. By applying Lenz's law in this way, we can infer the direction of the induced force without taking the intermediate step of determining the sense of the induced current.

832 ■ Faraday's Law

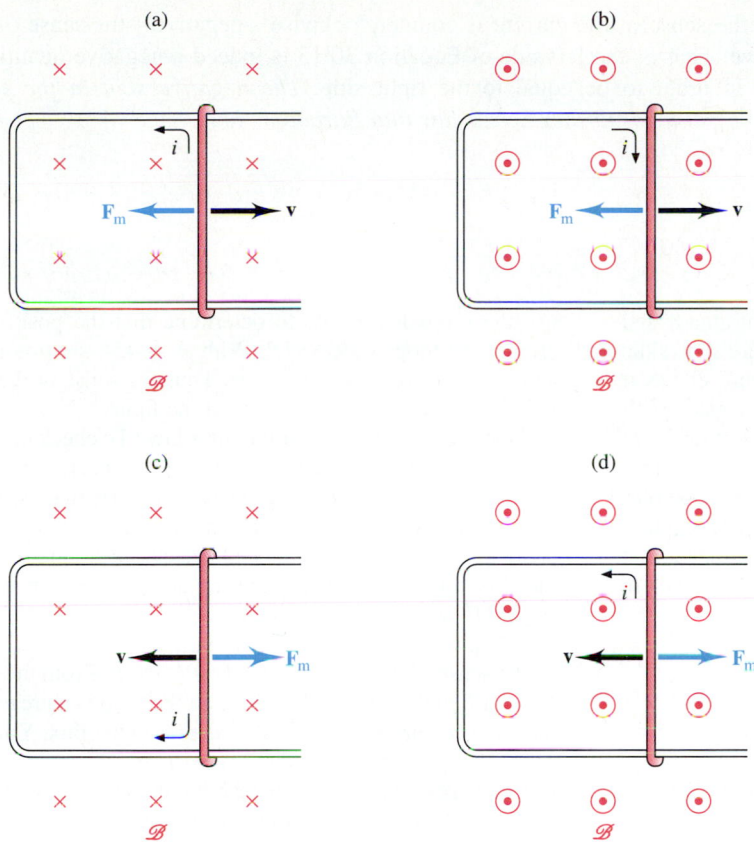

(a)

(b)

(c)

(d)

FIGURE 30.10 The device of Figure 30.4, with all possible combinations of direction of **v** and $\mathcal{B}$. (*a*) As in Figure 30.4, **v** is directed to the right and $\mathcal{B}$ is directed into the page. The sense of the induced current is counterclockwise. Consequently, $\mathbf{F_m}$ is directed to the left and tends to bring the moving wire to a stop. (*b*) As in Figure 30.5, **v** is directed to the right and $\mathcal{B}$ is directed out of the page. The sense of the induced current is clockwise. $\mathbf{F_m}$ is directed to the left and tends to bring the moving wire to a stop. (*c* and *d*) Here, **v** is to the left; in both cases, the sense of the current is such that $\mathbf{F_m}$ is directed to the right and tends to bring the moving wire to a stop.

Lenz's Law and Faraday's Law

Let us look again at Faraday's Law

$$V = -\frac{d\Phi_m}{dt},$$

with special attention to the meaning of the negative sign. Consider the system of Figure 30.11, which is the same system as that shown in Figure 30.9b. In this system, the area vector **A** of the loop remains fixed and parallel to $\mathcal{B}$, while the uniform magnetic field $\mathcal{B}$ increases with time. We can thus write Faraday's law in the special form

$$V = -\frac{d}{dt}(\mathcal{B} \cdot \mathbf{A}) = -A\frac{d\mathcal{B}}{dt}. \tag{30.13}$$

We choose the *x* axis along the external magnetic field $\mathcal{B}$, so that $\mathcal{B} = \mathcal{B}\hat{\mathbf{x}}$. Because we have assumed that $\mathcal{B}$ is increasing with time, $d\mathcal{B}/dt$ is a positive quantity. The vector magnitude $A = |\mathbf{A}|$ is positive by definition. Thus the right side of Equation 30.13 is a negative quantity.

The screwdriver rule fixes the positive sense of circulation around the loop of Figure 30.11 as the clockwise direction. To see this, imagine that the external magnetic field is produced by a large solenoid (not shown) that surrounds the loop, with the solenoid axis parallel to that of the loop. Given the direction of $\mathcal{B}$, the sense of the current through this solenoid must be clockwise.

With the clockwise sense thus fixed as positive, the counterclockwise sense of the induced current in the loop, determined in accordance with Lenz's law, is *negative*. As already noted, the induced emf that drives the current is a distributed emf; we cannot mark one point on the loop "+" and another "−," as we could if the emf were supplied by a battery located somewhere in the loop. But we can assign a sense to the distributed emf: *The sense of the distributed emf is the same as the sense of the current it drives.*

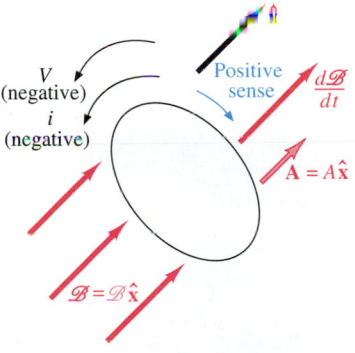

FIGURE 30.11 The system of Figure 30.9b shown again, with a view to incorporating Lenz's law into Faraday's law. The sign of $\mathcal{B}$ is chosen positive; the signs and senses of the other quantities, established in the text, are shown.

Because the sense of the current is counterclockwise—negative—the sense of V also is negative. That is, the left side of Equation 30.13 is indeed a negative quantity, as it must be in order to be equal to the right side. *The negative sign in the equation* $V = -d\Phi_m/dt$ *incorporates Lenz's law into Faraday's law.*

EXAMPLE 30.3

Figure 30.12a shows a rectangular loop of wire, of area A and resistance R, mounted on a shaft. Part b of the figure shows the loop end on (looking down the shaft from O toward O') as it appears at four instants—t_1, t_2, t_3, and t_4—as it rotates with angular velocity ω in a magnetic field $\mathscr{B}$. (The arrangement embodies the essential elements of the electric generator. This example thus serves both as an application of the principles developed in Section 30.4 and as an introduction to Section 30.5.) **(a)** At each of the four instants, is the sense of the current in the loop $PQRSTP$ or $PTSRQP$? **(b)** A small break is made in the wire between T and P. At each of the four instants, is the polarity of T positive or negative?

SOLUTION:

(a) Find the sense of the current for each of the loop positions shown in Figure 30.12b.

As in Example 30.2, Faraday's law simplifies to the form of Equation 30.12, with $N = 1$:

$$V = \omega \mathscr{B} A \sin \theta. \qquad (30.14)$$

What is the positive sense of circulation around the loop at t_1? The direction of $\mathscr{B}$ is upward. Because the loop is rotating, it doesn't matter which of the two directions normal to the loop you choose for the vector $\mathbf{A}$. You may as well choose the one shown in Figure 30.12b because it allows you to begin the sequence t_1 through t_4 with a small angle. With this choice, you

use the screwdriver rule to determine that the positive sense around the loop is $PQRSTP$. With $\theta = 45°$, $\sin \theta$ is positive, and so the sense of V is positive. Thus the sense of the current is $PQRSTP$; this sense is shown in the figure.

Does this result agree with Lenz's law? To check that it does, note that at t_1 the flux penetrating the loop is decreasing because the area vector $\mathbf{A}$ is turning away from $\mathscr{B}$; that is, the area presented by the coil to the external field is decreasing. In opposing this decrease, the induced current must tend to *increase* the flux through the loop; this requires current in the sense $PQRSTP$, as you have just seen.

At t_2, the right side of Equation 30.14 is still positive, and so the sense of the current must be $PQRSTP$. From the point of view of Lenz's law, the flux through the loop is increasing, and the induced current now tends to *decrease* the flux. You should check this point by using the screwdriver rule to show that the flux that arises from the induced current is directed along $\hat{\mathbf{A}}$ and thus opposes the increase in the external flux.

At t_3, $\sin \theta$ is negative. Thus the sense of the current is $PTSRQP$; i has reversed. From the point of view of Lenz's law, the external flux is decreasing and the sense of i tends to increase the flux through the loop.

At t_4, $\sin \theta$ is still negative, and so the sense of the current is still $PTSRQP$. Now the external flux through the loop is increasing, and the sense of i tends to decrease the flux.

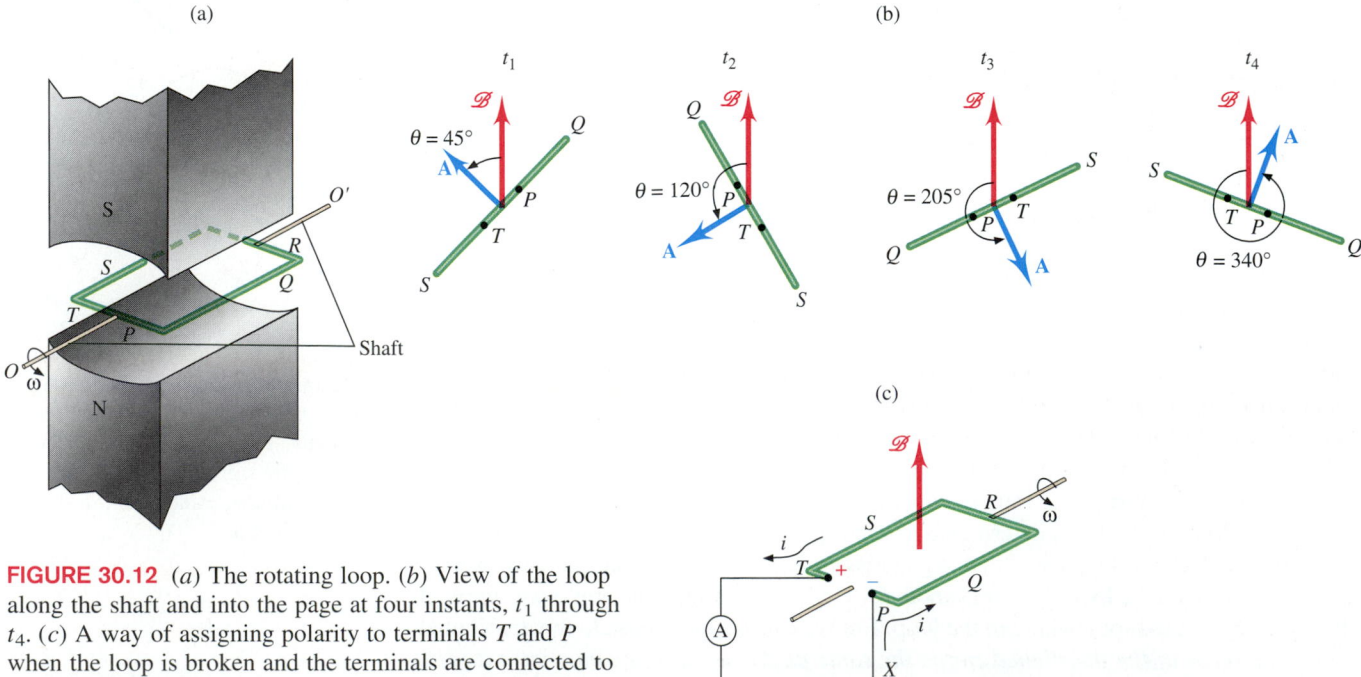

FIGURE 30.12 (*a*) The rotating loop. (*b*) View of the loop along the shaft and into the page at four instants, t_1 through t_4. (*c*) A way of assigning polarity to terminals T and P when the loop is broken and the terminals are connected to a passive external circuit.

At all four instants, Faraday's law (in the form of Equation 30.14) and Lenz's law yield the same results.

(b) A small break is made in the wire between T and P. At each of the four instants, is the polarity of T positive or negative?

When the loop is cut between T and P, current ceases to flow. We no longer need think of the emf as distributed; now the broken loop looks like a conventional source of emf, with a positive and a negative terminal. (We did the same thing for the moving wire in Figure 30.4.) Indeed, imagine that you close the loop by connecting it to a passive external circuit X, as in Figure 30.12c. At each instant, the sense of the current will be the same as that obtained in part **a**. What must be the polarities of terminals T and P to drive the current through the new circuit

in that sense? You can make the following table:

Instant	Sense Inferred from Part a	Polarity of T	Polarity of P
t_1	PQRSTXP	+	−
t_2	PQRSTXP	+	−
t_3	TSRQPXT	−	+
t_4	TSRQPXT	−	+

As always, positive charge flows from + to − outside the source of emf and from − to + inside the source (Section 27.2), and this determines the sense of the current through the entire circuit.

Electric Generators and Motors

We considered the dc electric motor briefly in Section 28.7. A complete understanding of the operation of the electric motor and its very close relative, the electric generator, requires that we reconsider these devices in the light of Faraday's law. We have already begun this reconsideration in Example 30.2.

Figure 30.13a is a sketch of the simplest type of electric generator, called a **magneto**.

(a)

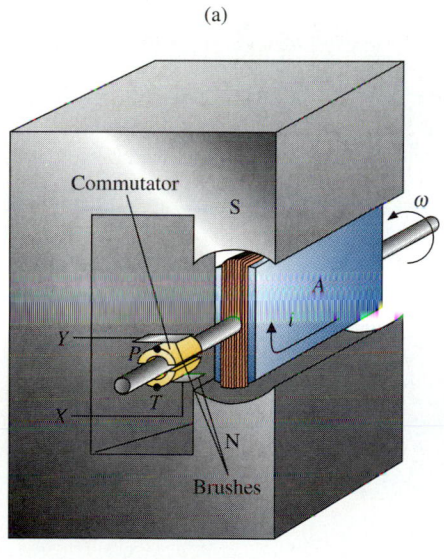

(b)

(c)

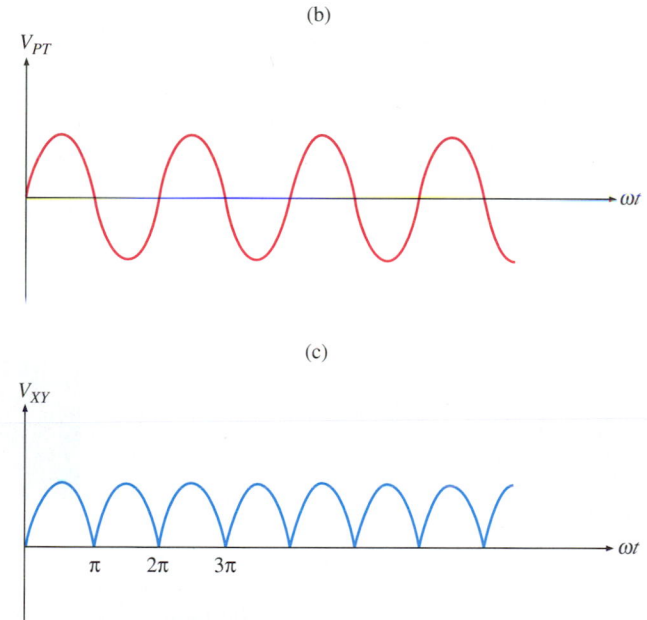

FIGURE 30.13 (*a*) A permanent-magnet dc generator, or magneto. The coil of N turns of area A is made to rotate by an external torque exerted on the shaft on which the coil is mounted. The coil ends are connected to the two halves of the split-ring commutator, identical with the one of Figure 28.19. The commutator halves make sliding contact with a pair of stationary brushes. By referring to the table at the end of Example 30.3, you can see that the polarity of the coil ends reverses when the rotor has turned 90° from the position shown. At the same instant, the connection of the brushes to the halves of the commutator reverses. As a result, the alternating emf at the coil ends (shown in part *b*) is converted into a *pulsating direct* emf (shown in part *c*) at the brushes X and Y.

A coil of cross-sectional area A and N turns is mounted on a shaft; like the corresponding component of an electric motor, this part of the generator is called the **rotor**. The coil lies in the field of a permanent magnet. In practice, the coil is wound on an iron core, which concentrates the magnetic flux; however, such a core is not central to the principles of operation.

As the rotor is turned by an external mechanical power source, the magnetic flux through the coil repeatedly increases and decreases in the fashion discussed in Examples 30.2 and 30.3. As a result, the ends of the rotor coil may be regarded as the terminals of a source of emf. The alternating emf at the coil ends P and T (which correspond to the similarly labeled terminals in Figure 30.12) is converted into an emf of constant sign (though varying voltage) at X and Y by means of the *split-ring commutator* through which the rotating coil is connected electrically to the outside world. The operation of the commutator is described in the legend for Figure 30.13; compare with Section 28.7, where the commutator is described in connection with the electric motor of Figure 28.22.

Suppose, to begin with, that there is no connection to an external circuit. Without a complete circuit there is no current, and no electrical work need be done to keep the rotor turning at constant speed. The only work done in turning the rotor is the small amount required to overcome friction.

Next, consider what happens when an external load is connected across the generator terminals, T and P. Suppose that the rotor is in the position shown in Figure 30.14, which corresponds to the position of the single-turn rotor at t_1 in Figure 30.12b. At this instant, T is positive and P is negative. Thus the sense of the current in the entire circuit must be that indicated by the arrows in Figure 30.14.

FIGURE 30.14 The magneto, rotating counterclockwise, shown at an instant when its rotor is in the same position θ as the single-turn coil of Figure 30.12b at time t_1. Because there is a current i in the rotor coil, the rotor has a magnetic dipole moment $\boldsymbol{\mu}$ whose direction makes an angle θ with the external magnetic field $\mathcal{B}$. Consequently, there is a torque that opposes the counterclockwise rotation of the rotor.

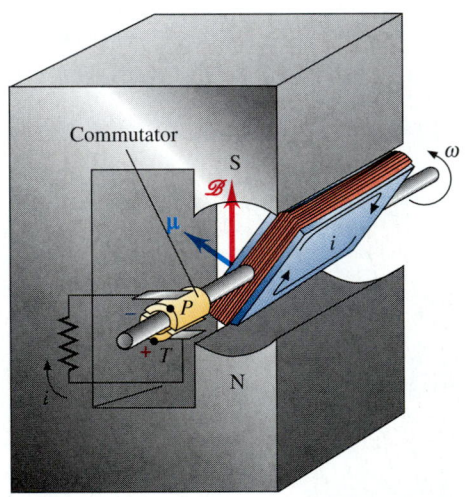

Because there is a current in the rotor coil, the coil becomes a magnetic dipole. Its magnetic moment is $\boldsymbol{\mu} = Ni\mathbf{A}$ (Equation 28.18 and Example 28.3). The screwdriver rule gives the direction of $\boldsymbol{\mu}$ as shown in Figure 30.14. The torque exerted on the dipole by the external magnetic field $\mathcal{B}$ is given by Equation 28.19, $\mathbf{T} = \boldsymbol{\mu} \times \mathcal{B}$. This torque tends to align the area vector $\mathbf{A}$ with the external field $\mathcal{B}$. That is, the current results in a clockwise torque, which *opposes* the counterclockwise rotation of the generator. Consequently, a counterclockwise torque is now required to keep the rotor turning. According to Equation 28.21, the magnitude of each of the opposing torques is

$$T = Ni\mathcal{B}A \sin\theta. \tag{30.15}$$

So the mechanical power input required is (Equation 11.42)

$$P_{\text{mech}} = T\omega = \omega Ni\mathcal{B}A \sin\theta. \tag{30.16}$$

As the current passing through the circuit is increased (say, by reducing the total resistance R of the system), the power required to turn the generator at constant speed increases as well.

What is the electrical power output of the generator? Assuming that the system is ohmic, we use Joule's law to write the power output in the form

$$P_{elec} = \frac{V^2}{R}.$$

When we substitute the value $V = \omega N \mathcal{B} A \sin \theta$ from Equation 30.12, we obtain

$$P_{elec} = \frac{(\omega N \mathcal{B} A \sin \theta)^2}{R}. \tag{30.17}$$

Not all of this power is available for the external circuit because there is always a certain amount of power dissipation due to the resistance of the rotor coil.

To show that the electrical output power is equal to the mechanical input power, we substitute $i = V/R = (\omega N \mathcal{B} A \sin \theta)/R$ into Equation 30.16. This immediately yields

$$P_{mech} = \frac{(\omega N \mathcal{B} A \sin \theta)^2}{R} = P_{elec},$$

a result that neglects frictional and similar irreversible losses. In principle, electric generators are 100% efficient. In practice, the losses are quite small; efficiencies of 98% are not unusual in large generators.

Motors Again

A dc permanent-magnet motor is nothing more than a dc magneto run in reverse. By means of an external source of emf, a current is driven through the rotor. Because it carries current, the rotor becomes a magnetic dipole and experiences a torque in the external magnetic field. Figure 30.15 shows the magneto of Figure 30.14 operated as a motor.

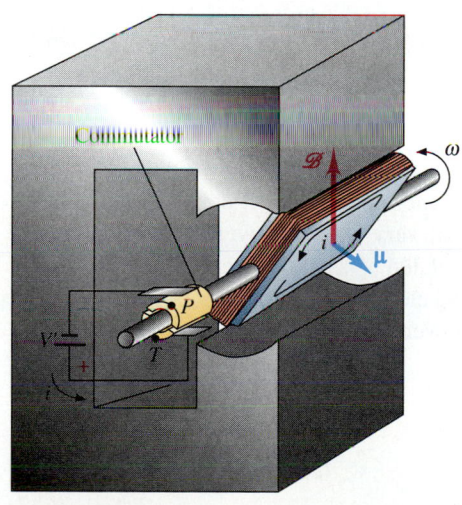

FIGURE 30.15 The device of Figure 30.14 is operated as a motor. The external source of emf V' is connected to the rotor coil through the commutator. At the instant shown, the positive pole is connected to T and the negative pole to P. The motor is now the part of the circuit external to the source of emf, and the sense of current in the rotor is from positive to negative. The direction of $\boldsymbol{\mu}$ is opposite to that of Figure 30.14, and the torque drives the rotor counterclockwise.

Suppose that the external mechanical load attached to the motor shaft is just large enough that the motor cannot turn. If the resistance of the entire circuit is R, the electrical input power is

$$P_{elec} = \frac{V'^2}{R} = i_0^2 R,$$

where $i_0 = V'/R$ is the current driven through the rotor coil by the source of emf. Equation 30.15 gives the magnitude of the *stalling torque*,

$$T_0 = N i_0 \mathcal{B} A \sin \theta = \frac{V'}{R} N \mathcal{B} A \sin \theta.$$

Because the angular speed ω is zero, the mechanical power output $P_{\text{mech}} = T_0\omega$ is zero as well.

Next, suppose that the mechanical load is reduced until the motor is rotating with angular speed ω. If the current through the rotor coil is i, the torque is $T = Ni\mathcal{B}A \sin\theta$. This torque is different from T_0 only if i is different from i_0. But why should i be different from i_0? The answer lies in Faraday's law. *Because the rotor is turning, the motor is acting as a generator.* And, because the sense of the rotation is the same as that of the generator in Figure 30.14, the motor becomes a source of induced emf with a positive pole at T and a negative pole at P. If it were not for the oppositely directed external emf, this induced emf would drive current in the sense shown in Figure 30.14, which is opposite the sense of the current in the motor. For this reason, the induced emf is called a **back emf**. The magnitude of the back emf is given by Equation 30.12,

$$V = \omega N\mathcal{B}A \sin\theta.$$

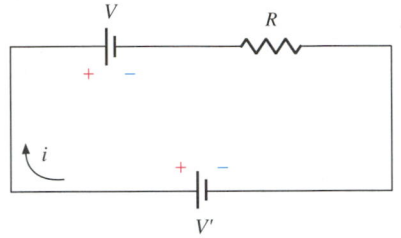

FIGURE 30.16 Reduced to ideal circuit elements, the system of Figure 30.15 consists of the external source of emf V' and the back emf V of the motor, in series with the total resistance R of the circuit.

The entire motor system, reduced to ideal circuit elements, consists of the external source of emf V', a back emf V generated by the motor, and the total resistance R of the circuit, as shown in Figure 30.16. A direct application of Kirchhoff's loop rule (Section 27.7) yields

$$V' - V - iR = 0$$

or

$$i = \frac{V' - V}{R}. \tag{30.18a}$$

Inserting $V = \omega N\mathcal{B}A \sin\theta$, we have

$$i = \frac{1}{R}(V' - \omega N\mathcal{B}A \sin\theta). \tag{30.18b}$$

Corresponding to this current is the torque delivered by the motor, given by Equation 30.15,

$$T = Ni\mathcal{B}A \sin\theta.$$

Substituting in the value of i given by Equation 30.18b, we obtain

$$T = \frac{1}{R}(V' - \omega N\mathcal{B}A \sin\theta)N\mathcal{B}A \sin\theta. \tag{30.19}$$

Although Equations 30.18b and 30.19 may look complicated, their essential message is simple: *As the motor speeds up, the back emf increases, the current through the coil decreases, and the output torque decreases.* In a dc motor, the input emf V' is constant, but the back emf is a function of θ. As a consequence, the rotor current, the torque, and the output power all vary with θ. Nevertheless, all of these quantities have mean values that can be calculated.

We now compare the electrical input power P_{elec} with the mechanical output power P_{mech}. For the input power, we use Joule's law to write

$$P_{\text{elec}} = V'i = (V + iR)i,$$

which we expand to obtain

$$P_{\text{elec}} = Vi + i^2R. \tag{30.20}$$

The term i^2R is the power dissipated as heat, owing to the resistance of the circuit. The term Vi is the product of the back emf and the current through the coil. (Both V and i are functions of ω.)

We use Equation 30.15 again to express the output power. We have $P_{\text{mech}} = T\omega$, or

$$P_{\text{mech}} = \omega Ni\mathcal{B}A \sin\theta. \tag{30.21}$$

Comparing the right side of this equation with Equation 30.12, $V = \omega N\mathcal{B}A \sin\theta$, we obtain

$$P_{\text{mech}} = Vi. \tag{30.22}$$

The output power is the product of the current and the back emf. It follows from

One of countless kitchen gadgets that use small, inexpensive electric motors, the electric coffee grinder facilitates a task that no one thought twice about doing by muscle power a generation ago.

Equations 30.20 and 30.22 that

$$P_{elec} = i^2 R + P_{mech}.$$

(30.23)

The electrical input power to the motor is equal to the sum of the Joule heating loss and the mechanical output power. With good design practices, the Joule heating loss can be made quite small compared with P_{mech}; electric motors—especially large ones—are often well over 90% efficient.

Electric Circulation and the Induced Electric Field

In Figure 30.17a, a circular wire loop of area $A = \pi b^2$ and resistance R lies in a region of uniform magnetic field $\mathscr{B}$. The magnitude of the magnetic field is changing with time, at a rate $d\mathscr{B}/dt$. (This is identical with the situation shown in Figure 30.11.) Because $\mathscr{B}$ is varying, the magnetic flux $\Phi_m = \mathscr{B}A$ through the loop changes. In accordance with Faraday's law, an electric current i is driven around the wire loop:

$$i = \frac{V}{R} = -\frac{1}{R}\frac{d\Phi_m}{dt}.$$

The quantity V is the distributed emf. How can we determine experimentally that this distributed emf is actually present? We can insert an ammeter into the loop, as in Figure 30.17b. We then infer the emf from the current reading i by using Ohm's law, $V = iR$.

The current represents a flow of free charges in the wire. Because the wire has resistance, the flow must be driven by an electric force **F**, impressed on the charges by an **induced electric field** $\mathscr{E}$. Let us establish a connection between $\mathscr{E}$ and V.

To establish this connection, we consider the work required to drive a charge q around the loop, whose circumference is $s = 2\pi b$. In this symmetrical case, the electric field is everywhere parallel to the wire. So we can write

$$W = Fs = q\mathscr{E}s.$$

The work done per unit charge is the emf V. So we have

$$V = \frac{W}{q} = \mathscr{E}s.$$

(30.24)

Using Faraday's law in the form of Equation 30.8, $V = -d\Phi_m/dt$, we can rewrite Equation 30.24 in the form

$$\mathscr{E}s = -\frac{d\Phi_m}{dt} \quad \text{for a circular loop.}$$

(30.25)

Suppose now that we remove the wire loop but change nothing else. In the absence of a closed conducting path, there is no current. But remember that the electric field is the direct *cause* of the current, not an effect of it. Consequently, removing the conductor has no effect on the field. Whether or not a conductor is present, *a changing magnetic flux induces an electric field*. This is yet another statement of Faraday's law.

We now make the new statement of Faraday's law more general. In Figure 30.18, we have drawn an arbitrary planar closed curve in the same region of space shown in Figure 30.17. Although the curve does not represent any material object, it nevertheless encloses an area. This enclosed area is penetrated by a magnetic flux Φ_m, which changes at a rate $d\Phi_m/dt$.

In imagination, we subdivide the curve into infinitesimal elements $d\mathbf{s}$; one of these elements is shown in Figure 30.18. At the location of this element, the electric field vector has some value $\mathscr{E}$, as shown. If a charge q is displaced through $d\mathbf{s}$, the work per unit charge done on it by the field is the emf $d\mathbf{V}$ across the element:

$$dV = \frac{dW}{q} = \mathscr{E}\cdot d\mathbf{s}.$$

(30.26a)

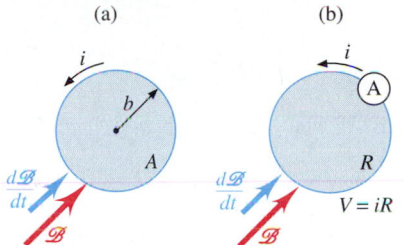

(a)   (b)

FIGURE 30.17 (a) A wire loop is located in a region of uniform magnetic field $\mathscr{B}(t)$, whose magnitude changes with time at the rate $d\mathscr{B}/dt$. Just as in Figure 30.11, there is a current i in the loop, driven by the induced emf V around the loop. (b) We confirm the existence of the induced emf by placing an ammeter of negligible resistance in the loop. The ammeter reads a current i.

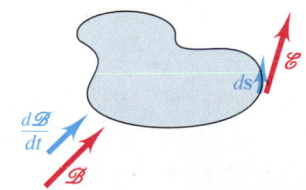

FIGURE 30.18 The wire loop has been removed from the region of Figure 30.17. However, the magnetic field is still uniform and varies at the same rate $d\mathscr{B}/dt$. An arbitrary closed curve is shown. In imagination, we subdivide this curve into infinitesimal elements $d\mathbf{s}$. At the location of a typical element $d\mathbf{s}$, there is an electric field $\mathscr{E}$.

(Compare with Equation 30.24.) To find the emf around the entire closed curve, we integrate:

$$V = \int_{\substack{\text{closed} \\ \text{curve}}} \mathscr{E} \cdot d\mathbf{s}. \tag{30.26b}$$

Let us compare Equation 30.26b with Equation 25.4, which gives the *electric potential V* at a point $\mathbf{r}$ with respect to a reference point $\mathbf{r}_0$:

$$V = -\int_{\mathbf{r}_0}^{\mathbf{r}} \mathscr{E} \cdot d\mathbf{r}. \tag{30.27}$$

Although the same symbol V is used for the emf of Equation 30.26b and the electric potential of Equation 30.27 and they are expressed in the same units (volts), there are two important differences between the quantities. First, note the difference in sign. The electric potential is a measure of the external work per unit charge done on a test charge in displacing it *against* (hence the minus sign) the electric field $\mathscr{E}$. This electric field owes its existence to the presence of some charge distribution in the region. The emf of Equation 30.26b is a measure of the work per unit charge done on a test charge *by* the induced electric field $\mathscr{E}$. This electric field owes its existence to the change over time in the magnetic flux penetrating the loop. It does *not* arise from a distribution of electric charge.

The other important difference between electric potential and induced emf has to do with the nonconservative nature of the induced electric field: If the electric potential of Equation 30.27 is integrated over a *closed curve*—that is, if $\mathbf{r} = \mathbf{r}_0$—then V must be zero, *because the Coulomb force is conservative*. But the induced emf of Equation 30.26b is *not* zero over a closed path. *The induced electric field in Equation 30.26b is nonconservative.* Indeed, the induced field cannot be conservative. By definition, a conservative force (or field) must be a function of position *only*; see Section 8.2. But the induced field is a function of time. To make this time dependence explicit, let us take Equation 30.26b through one more step. Substituting the value $-d\Phi_m/dt$ for V, we write Faraday's law in the form

$$\int_{\substack{\text{closed} \\ \text{curve}}} \mathscr{E} \cdot d\mathbf{s} = -\frac{d\Phi_m}{dt}. \tag{30.28}$$

The integral is called the **electric circulation**. Note the close analogy with the magnetic circulation that appears in Ampère's law, expressed in Equation 29.19:

$$\int_{\substack{\text{closed} \\ \text{curve}}} \mathscr{B} \cdot d\mathbf{s} = \mu_0 \frac{dq}{dt}. \tag{30.29}$$

We have rewritten Equation 29.19 slightly, using the definition $i \equiv dq/dt$. By doing this, we bring out the close symmetry of Faraday's and Ampère's laws:

Faraday's law: The electric circulation evaluated around a closed curve is equal to a negative constant (in SI, -1) times the rate of change of magnetic flux through the curve.

Ampère's law: The magnetic circulation evaluated around a closed curve is equal to a positive constant (in SI, μ_0) times the rate of flow of electric charge through the curve.

This symmetry has profound consequences for the synthesis of electromagnetism called Maxwell's equations, the main subject of Chapter 34.

EXAMPLE 30.4

Figure 30.19 shows the region between the poles of an electromagnet. The electric current through the windings of the magnets is increasing in such a way that the magnetic field between the poles increases at a rate $d\mathscr{B}/dt = 0.10$ T/s. The radius of the pole faces is $b = 8.0$ cm. Neglect the fringing field; assume that at any instant $\mathscr{B}$ has the same value at all points for which $r < b$ and is equal to zero at all points for which $r > b$. Find the magnitude and direction of the induced electric field as a function of distance r from the magnet axis **(a)** for $r < b$ and **(b)** for $r > b$. **(c)** Evaluate $\mathscr{E}$ at $r = 4.0$ cm, 8.0 cm, and 12.0 cm.

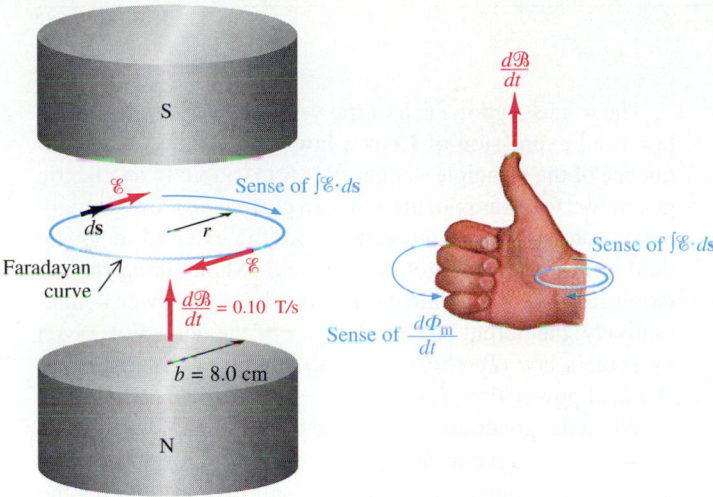

FIGURE 30.19

The minus sign in Equation 30.30a has the following meaning: If you use the screwdriver rule, pointing your thumb in the direction of $d\mathcal{B}/dt$ (because $\mathcal{B}$ is increasing, $d\mathcal{B}/dt$ has the same direction as $\mathcal{B}$), the sense of the electric circulation $\int_{\text{circle}} \mathcal{E} \cdot d\mathbf{s}$ is *opposite* that indicated by the curl of your fingers. That is, the vector $\mathcal{E}$ at any point on the Faradayan circle is tangent to the circle and is directed away from your fingertips (Figure 30.19). If there were a conducting loop present, this electric field would drive a current that tended to oppose the increase of $\mathcal{B}$, in accordance with Lenz's law.

(b) Find $\mathcal{E}(r)$ for $r > b$.

When you increase the radius of the Faradayan circle beyond the limit $r = b$, the circle encloses no additional flux. That is, at any instant all circles $r > b$ are penetrated by the same magnetic flux, whose magnitude is $\pi b^2 \mathcal{B}$. Using Equation 30.28 for a circle with $r > b$, you have

$$2\pi r \mathcal{E} = -\pi b^2 \frac{d\mathcal{B}}{dt}.$$

Solving for $\mathcal{E}$, you obtain

$$\mathcal{E} = -\frac{b^2}{2r}\frac{d\mathcal{B}}{dt} \quad \text{for } r > b. \qquad \textbf{(30.30b)}$$

This $1/r$ dependence is determined by the symmetry of the situation. (Remember that an electric current in a long, straight wire gives rise to a *magnetic* field having a $1/r$ dependence.)

(c) Evaluate $\mathcal{E}$ at $r = 4.0$ cm, 8.0 cm, and 12.0 cm.

For $r = 4.0$ cm, you must use Equation 30.30a:

$$\mathcal{E} = -\tfrac{1}{2} \times 0.040 \text{ m} \times 0.10 \text{ T/s} = -2.0 \times 10^{-3} \text{ V/m}.$$

The minus sign indicates that the sense of the circulation is the one shown in Figure 30.19. The rate of change of $\mathcal{B}$, 0.1 T/s, is about as large as you can expect for a large iron-core laboratory electromagnet. The magnitude of the resulting electric field is quite small. You will see in Chapter 34, however, that there are important circumstances under which $d\mathcal{B}/dt$ is very much larger, and the induced electric field is significant.

For $r = 8.0$ cm, you can use either Equation 30.30a or Equation 30.30b. If you use Equation 30.30a, you have

$$\mathcal{E} = -\tfrac{1}{2} \times 0.080 \text{ m} \times 0.10 \text{ T/s} = -4.0 \times 10^{-3} \text{ V/m}.$$

For $r = 12.0$ cm, you must use Equation 30.30b:

$$\mathcal{E} = -\frac{(0.080 \text{ m})^2}{2 \times 0.12 \text{ m}} \times 0.10 \text{ T/s} = -2.7 \times 10^{-3} \text{ V/m}.$$

SOLUTION:

(a) Find $\mathcal{E}(r)$ for $r < b$.

In view of the close resemblance of Faraday's law (Equation 30.28) to Ampère's law (Equation 30.27), it makes sense to proceed here very much as you did in using Ampère's law in Chapter 29. You begin by drawing a **Faradayan curve** analogous to an Amperean curve. The symmetry of the situation suggests that you draw a circle centered on the magnet axis, in the expectation that $\mathcal{E}$ will be tangent to the circumference, and have the same magnitude, at all points. You thus have for the electric circulation

$$\int_{\text{circle}} \mathcal{E} \cdot d\mathbf{s} = \mathcal{E} \int_{\text{circle}} ds = 2\pi r \mathcal{E},$$

and Equation 30.28 takes the simple form

$$2\pi r \mathcal{E} = -\frac{d(\mathcal{B}A)}{dt} = -\pi r^2 \frac{d\mathcal{B}}{dt}.$$

Now you can solve for $\mathcal{E}$:

$$\mathcal{E} = -\frac{r}{2}\frac{d\mathcal{B}}{dt} \quad \text{for } r < b. \qquad \textbf{(30.30a)}$$

It should not surprise you that this result has the same *mathematical* form as the expression for the magnetic field inside a long, straight wire that carries a uniformly distributed electric current.

According to **Faraday's law** (Equation 30.28), when the magnetic flux Φ_m penetrating a defined area changes over time, there is an electric circulation around the periphery of that area:

$$\int_{\substack{\text{closed}\\\text{curve}}} \mathcal{E} \cdot d\mathbf{s} = -\frac{d\Phi_m}{dt}.$$

This form of Faraday's law makes evident the close symmetry between Faraday's law and Ampère's law. The closed curve selected for evaluation of the electric circulation is called a **Faradayan curve**, in analogy to the Amperean curve used in Ampère's law.

Equation 30.28 implies that a changing magnetic field gives rise to an electric field. The electric circulation integral on the left side of the equation is equivalent to an **induced emf** V. Faraday's law can be expressed in terms of the induced emf, as in Equation 30.8:

$$V = -\frac{d\Phi_m}{dt}.$$

If the path of integration chosen to evaluate the electric circulation is along a closed conducting loop, the induced emf drives a current i. There are three ways to change the magnetic flux $\Phi_m = \mathcal{B} \cdot \mathbf{A}$ through such a loop:

1. by varying the area A of the loop;
2. by varying the magnitude $\mathcal{B}$ of the magnetic field; or
3. by varying the angle θ between $\mathcal{B}$ and $\mathbf{A}$—that is, by varying $\hat{\mathcal{B}} \cdot \hat{\mathbf{A}} = \cos \theta$.

Equation 30.10 expresses Faraday's law in a way that clearly separates the effects of these three processes:

$$V = -\mathcal{B} \cos \theta \frac{dA}{dt} - A \cos \theta \frac{d\mathcal{B}}{dt} + \mathcal{B}A \sin \theta \frac{d\theta}{dt}.$$

The minus sign in each of the various forms of Faraday's law is an expression of **Lenz's law**. Lenz's law is a consequence of the principle of energy conservation. In an electric generator, for example, the induced current produces a magnetic moment whose interaction with the external magnetic field opposes the rotation of the rotor. Maintaining the rotation requires the application of mechanical power. Quantitatively, the output electrical power of the generator, given by Joule's law ($P = Vi$), is always equal to the input mechanical power $P = T\omega$.

When the generator is run in reverse as a motor, the variation of the magnetic flux penetrating the rotor coil results in an induced **back emf**. The back emf always opposes the emf that drives current through the rotor. This again leads to an energy-conserving balance between input and output power, as expressed in Equation 30.23:

$$P_{\text{elec}} = i^2 R + P_{\text{mech}}.$$

KEY TERMS

Sections 30.1 and 30.2 Introduction and Motional emf
electromagnetic induction ▪ Faraday's law

Sections 30.3 and 30.4 The Central Role of Magnetic Flux in Faraday's Law; Lenz's Law and Energy Conservation
distributed emf ▪ alternating current

Section 30.5 Electric Generators and Motors
magneto ▪ rotor ▪ back emf

Section 30.6 Electric Circulation and Induced Electric Field
Faradayan curve

QUERIES

30.1 *(2) Pulling through.* The apparatus illustrated below consists of a loop of wire connected to an ammeter; the entire

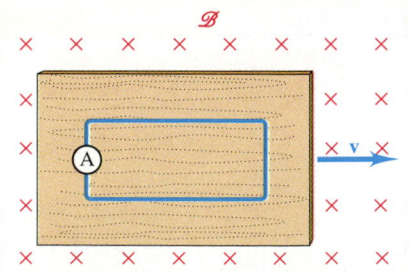

circuit is fastened to a sheet of plywood. If you pull the apparatus, without turning it, through a uniform, steady magnetic field $\mathcal{B}$ (as indicated by the arrow labeled **v**), what is the reading of the ammeter?

30.2 *(4) Topsy-turvy.* Suppose that, in the system of Figure 30.4, the sense of the current were opposite that shown. (Assume that nothing else is changed.) Show that Faraday's law would then have the form $V = +d\Phi_m/dt$, and prove that the system would violate the principle of energy conservation.

30.3 *(4) Making sense of it all.* The five closed loops shown (p. 843) are located in a uniform magnetic field directed into the page. If the magnitude of the field begins to decrease, what is the sense of the current induced in each loop?

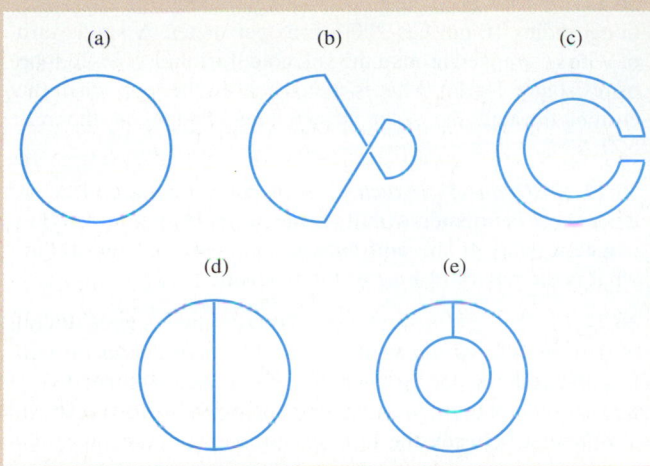

(a)　　　　(b)　　　　(c)

(d)　　　　(e)

electromagnets that are suspended a short distance above the rails. Explain in each case how Faraday's law operates to slow the train or trolley car. Electromagnetic brakes have two advantages: they operate smoothly and they do not wear out. Why are ordinary mechanical friction brakes nevertheless necessary to bring the train to a complete stop?

30.6 *(5) Which court issued the injunction?* As noted in the text, the efficiency of electric motors is often in excess of 90%, in contrast with the less than 40% typical of the most efficient heat engines. Explain. In particular, explain why the operation of an electric motor is not governed by the second law of thermodynamics.

30.7 *(5) Warming to the task.* Will an electric motor warm up more when it is operating under a small load or a heavy load?

30.4 *(4) Putting a damper on it.* Suppose you make a pendulum by attaching a copper disk to a long string. If the copper disk swings between the poles of a magnet, the pendulum comes to rest very quickly, owing to the generation of *eddy currents* in the disk. Explain. What happens to the mechanical energy of the pendulum?

30.5 *(5) And it was grand just to stand* . . . Electric locomotives and trolley cars use several forms of *electromagnetic braking*. In one form, the driving motor is disconnected from its external source of emf and is connected to a large resistor. In another form, current is driven through the coils of several

30.8 *(5) Flexible.* A gasoline-powered automobile must have a transmission—a device that makes possible a variety of gear ratios in the connection between the engine and the wheels. (If you try to start an automobile in high gear, it will probably stall.) Why is there no need for a transmission in electric automobiles? (Note: In the widely used diesel-electric locomotive, the diesel engine drives a generator that supplies electric power to motors on the wheel axles. By proper manipulation of the current through the field coils of the generator as the load on the locomotive varies, the engine can be kept running at an essentially constant speed at which its efficiency is maximum.)

PROBLEMS

GROUP A

30.1 *(2) Plane fact.* An airplane whose wingspan is 60 m flies northward at 950 km/h. If the vertical component of the earth's magnetic field is 2.0×10^{-5} T downward, what is the emf between the wingtips? Which wingtip has positive polarity?

30.2 *(2) Seining for flux, I.* A copper wire has length 85 cm. If you pull it perpendicular to its length with speed 1.8 m/s in a direction perpendicular to a uniform magnetic field of magnitude 0.76 T, what is the magnitude of the emf between the ends of the wire?

30.3 *(2) Seining for flux, II.* You now slide the wire of Problem 30.2 along a stationary U-shaped conductor, as shown in Figure 30.4. The stationary wire is much thicker than the moving wire, so the resistance of the former is negligible. **(a)** If the resistance of the moving wire is 3.5×10^{-2} Ω, what is the current through the circuit? **(b)** What force must you apply to keep the wire moving at constant speed 1.8 m/s?

30.4 *(2) Double.* The slide-wire system shown opposite lies in a uniform magnetic field $\mathcal{B}$ directed into the page. All resistances other than R_s, R_1, and R_2 are negligible. Find the current through the slide wire when it is moving with speed v.

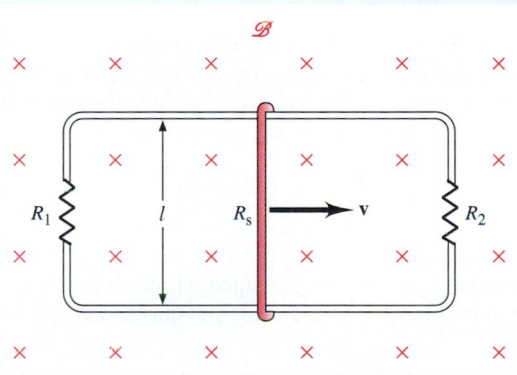

30.5 *(2) Getting into the swing, I.* A metal rod of length 1 m revolves in a horizontal plane about a vertical axis passing through one end. The magnitude of the vertical component of the earth's magnetic field is 4.8×10^{-5} T. At what angular speed, expressed in rev/min, must the rod revolve if the voltage difference between its ends is to be 1 mV?

30.6 *(2) Getting into the swing, II.* A metal rod of length l, located in a uniform magnetic field, rotates about an axis through the center of the rod and perpendicular to it. When the angular speed of the rod is ω, there is an emf V_1 between the ends. Next, the rod is remounted so that the axis passes through one of its ends. When the rod rotates with angular speed ω, the emf between the ends is now V_2. **(a)** Find V_1/V_2. **(b)** Why doesn't your result depend on the orientation of the magnetic field?

30.7 *(3) Spinning.* A circular coil of radius 9.2 cm has 550 turns. When the coil is rotated about one of its diameters at 3600 rev/min in a uniform 0.83-T magnetic field perpendicular to that diameter, what emf appears between the ends of the coil?

30.8 *(3) Zap!* A circular coil of radius 7.5 cm has 1000 turns and lies in a uniform 1-T magnetic field. The ends of the coil are placed near each other so that a spark will jump when the voltage difference between them is 3000 V. How fast must the coil be rotated about its diameter to make a spark jump?

30.9 *(3) Waning influence, I.* In the figure below, a wire of length l and resistance R slides on a U-shaped conductor of

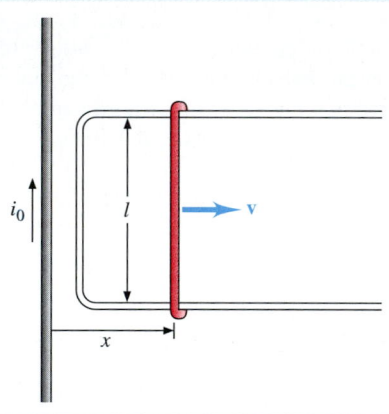

negligible resistance. The loop thus formed lies near a long wire that carries a current i_0. **(a)** Show that the current i_1 in the loop depends on the distance x between the long wire and the moving wire according to the relation

$$i_1 = \frac{\mu_0 l v i_0}{2\pi R}\frac{1}{x}.$$

(b) What is the sense of i_1? **(c)** If $i_0 = 150$ A and the values of the other quantities in the equation are the same as those given in Problems 30.2 and 30.3, find the current i_1 when $x = 0.46$ m.

30.10 *(3) Turn off, turn on.* A 200-turn coil of cross-sectional area 12 cm^2 lies in a uniform magnetic field of magnitude 0.25 T. The resistance of the coil is 42 Ω. The coil axis makes an angle of 55° with $\mathscr{B}$. The magnetic field is turned off at a uniform rate and reaches zero in 0.63 s. What is the emf between the ends of the coil?

30.11 *(3) Heated exchange.* A wire loop has resistance R. The magnetic flux penetrating the loop varies with time according to the expression $\Phi_m = at^2$, where a is a constant. How much heat energy is evolved in the loop over the time interval from $t = 0$ to $t = t_f$?

30.12 *(3) One good turn . . .* A long solenoid of cross-sectional radius 10 cm has 2000 turns per meter. A single turn of wire is wrapped around the solenoid, forming a closed loop of resistance 1 mΩ. What is the current in the loop when the current through the solenoid winding changes at the rate 300 A/s?

30.13 *(3) Deducing induction.* A 50-turn circular coil of radius 22 cm is oriented with its plane normal to a time-varying magnetic field. If the emf between the coil ends is 100 V, what is the rate of change of the magnetic field?

30.14 *(3) Doing it the easy way.* You scrounge a circular coil of diameter 40 cm and need to know how many turns the coil has. Because the coil is "potted" (the turns are embedded in a plastic compound), you cannot count directly. So you attach a voltmeter between the terminals of the coil and place the coil in a magnet whose field is 1.3 T, with its plane normal to the field. Then you turn the magnet off, and the field goes to zero in 32 ms. The voltmeter reads 230 V. Assuming to first approximation that the magnetic field decays at a constant rate, how many turns does the coil have?

30.15 *(3) Oscillating field.* A circular wire loop of radius b and resistance R lies in a plane normal to a uniform magnetic field $\mathscr{B}$. The magnetic field varies sinusoidally with time; at time t, its magnitude is $\mathscr{B}(t) = \mathscr{B}_0 \sin \omega t$, where $\mathscr{B}_0$ is a constant. What is the current in the loop at time t?

30.16 *(3) Square deal.* A copper wire of resistivity 1.72×10^{-8} $\Omega \cdot$m is formed into a square loop of sides 5 cm. The loop is oriented with its plane normal to a time-varying magnetic field $\mathscr{B} = \mathscr{B}_0 \cos \omega t$, where $\mathscr{B}_0 = 1.4 \times 10^{-2}$ T and $\omega = 300$ rad/s. Express the following quantities as a function of time and find the maximum value of each: **(a)** the flux penetrating the loop, **(b)** the induced emf in the loop, and **(c)** the current in the loop.

30.17 *(3) Bending exercise.* The wire loop illustrated below is made by taking a flat rectangular loop of sides 10 cm and 20 cm and bending the long sides at their midpoints to produce two mutually perpendicular square parts. The loop is placed in an oscillating magnetic field $\mathscr{B} = \mathscr{B}_0 \sin 2\pi v t$, with $\mathscr{B}_0 = 1.2 \times 10^{-3}$ T and $v = 60$ Hz. **(a)** Express the emf around the loop as a function of time and the angle θ. **(b)** For what angle θ does the induced emf have the largest amplitude?

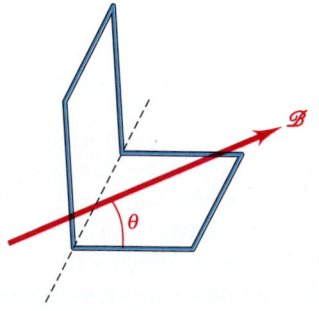

30.18 *(3) Figure eight.* The closed wire loop shown (p. 845) is a "figure eight" consisting of two squares of sides a and b. The resistance per unit length of the wire is λ. A uniform magnetic field is directed normal to the plane of the loop and

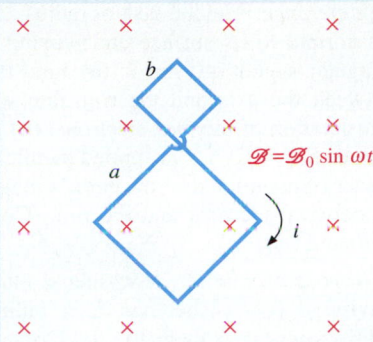

varies with time according to the function $\mathcal{B} = \mathcal{B}_0 \sin \omega t$.
(a) Is the initial sense of the current through the loop that indicated by the arrow or its opposite? (b) Show that, if you take the proper sense determined in part a to be positive, the current through the loop is

$$i = \frac{b - a}{4\lambda} \omega \mathcal{B}_0 \cos \omega t.$$

(c) What is the maximum value of i if $a = 31$ cm, $b = 12$ cm, $\lambda = 4.2 \times 10^{-2}$ $\Omega/$m, $\mathcal{B}_0 = 0.23$ T, and $\omega = 75$ rad/s?

30.19 *(3) A good turn.* A circular wire loop of radius 2.0 cm and resistance 1.0 Ω lies in a uniform magnetic field of magnitude 0.2 T. **(a)** If the loop is rotated at angular speed 40 rad/s about an axis perpendicular to the magnetic field, what is the maximum instantaneous emf around the loop? **(b)** What is the maximum instantaneous current in the loop?

30.20 *(5) Nothing ventured, nothing gained.* Show that when a dc permanent-magnet motor operates under no-load conditions, its power consumption is zero. (Neglect mechanical friction.)

30.21 *(5) Turning it out.* The 100-turn coil of a magneto has rectangular form, with length 3.0 cm and width 5.2 cm. The field of the magnet has magnitude 1.1 T. If the maximum output emf is 63 V, how fast is the rotor turning?

30.22 *(5) Self-excited generator.* Except in quite small generators, it is not practical to use a permanent magnet to provide the necessary magnetic field. Instead, an electromagnet called a *field coil* or *stator* is used. This stator is an electromagnet powered by the generator itself. In a *series-wound generator*, the stator is in series with the rotor and the load, as shown below. Assume (not too realistically for an iron-core coil) that the magnetic field in which the rotor turns is roughly proportional to the current. Suppose that the generator is driven at constant angular speed as the load resistance is decreased and

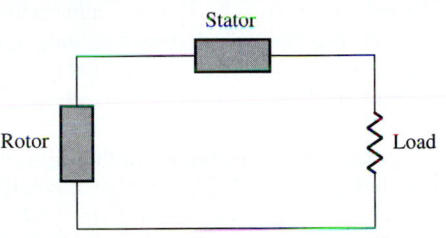

the generator is called upon to deliver increasing current. Under these circumstances, what is the relation between the current and the output voltage? Do you think such a generator is likely to have wide use?

30.23 *(6) Getting around.* A square planar loop of area 1.2×10^{-2} m^2 lies in a uniform magnetic field oriented normal to the plane of the loop. The magnitude of the field is 2.4 T. **(a)** If the field is switched off and falls to zero in 8.3 ms, what is the emf around the loop? **(b)** What is the magnitude of the electric field at a point on the loop?

GROUP B

30.24 *(2) Accelerated motion.* In the figure below, the straight wire makes sliding contact with the stationary frame whose form is the parabola $y = bx^2$. The straight wire starts from rest at the x axis and moves with constant acceleration a. A uniform magnetic field $\mathcal{B}$ is directed into the page. Show that the emf between the sliding contacts is

$$V = \mathcal{B}y \sqrt{\frac{8a}{b}}.$$

30.25 *(2) The moving wire slides,/And having slid, . . .* In the figure below, the width of the U-shaped conductor is Y. The

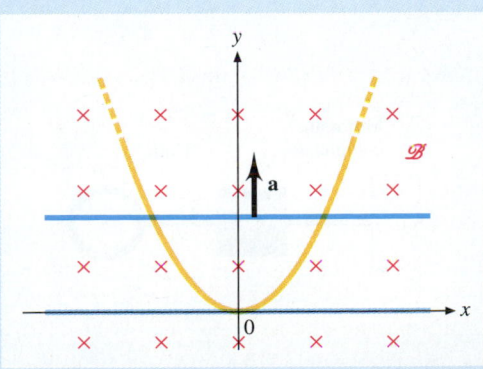

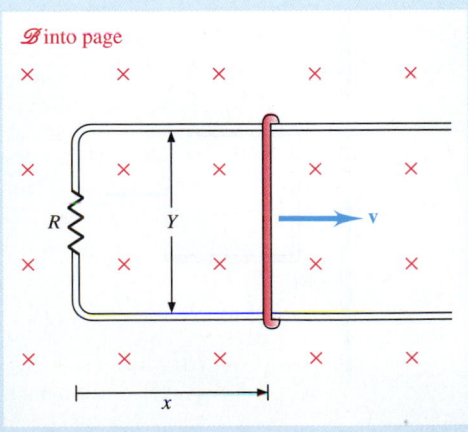

moveable wire, of mass m, leaves the left end of the U at speed v_0 and slides to the right without mechanical friction. No external force is exerted on the wire. **(a)** Assuming that the resistor R constitutes the only significant resistance in the system, show that the wire comes to rest when its distance from the left end of the U is

$$x_f = \frac{mRv_0}{\mathcal{B}^2Y^2}.$$

(b) How much heat energy is evolved in the resistor?

30.26 *(2) . . . slides on.* The sliding wire of Problem 30.25 is initially at rest at $x = 0$. At $t = 0$, a constant force F_{ext} begins to pull the sliding wire to the right. Express the speed v of the wire as a function of t.

30.27 *(2) Exponential slide?* In the figure below, the width of the U-shaped conductor is Y. The moveable wire, of mass m, leaves the left end of the U at speed v_0 and slides to the right without mechanical friction. No external force is exerted on the wire. Both the U-shaped wire and the sliding wire have resistance per unit length λ. **(a)** Express the resistance R of the loop as a function of the position x of the sliding wire with respect to the left end of the U. **(b)** Express the position x_f at which the sliding wire comes to rest in terms of λ, m, v_0, $\mathcal{B}$, and Y.

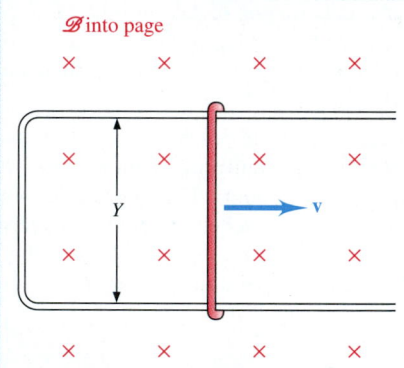

30.28 *(3) Waning influence, II.* The long wire in the figure below carries a current i_0. The square wire loop with sides l moves to the right with speed v. **(a)** Find the emf induced in the loop as a function of the distance x between the long wire and the left side of the loop. **(b)** What is the sense of the current in the loop?

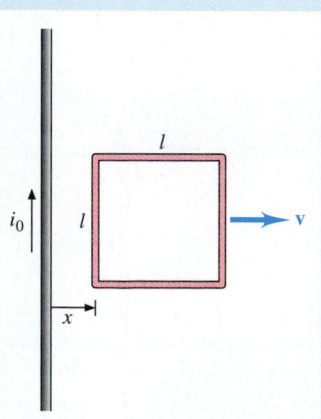

30.29 *(3) Big difference.* A metal disk of radius 25 cm rotates about an axis normal to its surface and passing through its center with angular speed 140 rad/s. **(a)** Find the potential difference between the axis and the rim, due solely to the centrifugal acceleration of the free electrons. **(b)** A magnetic field of magnitude 5.0×10^{-5} T is applied parallel to the axis. (This is the order of magnitude of the earth's magnetic field.) Find the emf between the axis and the rim. (Compare with Problems 27.70 and 28.47.)

30.30 *(3) What goes around . . .* A washer is made of metal having resistivity ρ. The washer has inner radius r_1, outer radius r_2, and thickness d. A magnetic field, oriented normal to the plane of the washer, has the time-dependent magnitude $\mathcal{B} = bt$, with b a constant. Show that the current around the washer is

$$i = (r_2{}^2 - r_1{}^2)\frac{bd}{4\rho}.$$

30.31 *(3) Flemished down.* You have a considerable length l of wire, which you wind into a flat spiral having N turns, as shown below. The outer radius of the spiral is b. The spiral lies in a uniform magnetic field directed normal to the plane of the spiral. If the field varies in time so that

$$\mathcal{B}(t) = \mathcal{B}_0 \sin \omega t,$$

what is the emf between P and Q at time t?

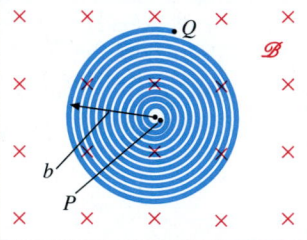

30.32 *(3) Rotating-coil magnetometer.* A rotating-coil magnetometer is shown below. An electric motor turns the small coil at constant angular speed ω. The rotating coil is connected through a split-ring commutator to a dc ammeter. The response time of the ammeter is much longer than the rotation period of the coil, so the ammeter reading i is constant. Express the magnitude $\mathcal{B}$ of the magnetic field in terms of the coil area A, the number of turns N, the total resistance R of the coil-ammeter circuit, and ω.

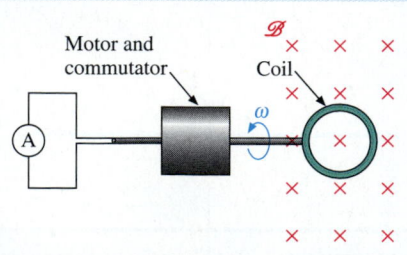

30.33 *(3) Flip-coil magnetometer.* The *flip coil*, shown below, provides a simple way to measure magnetic field $\mathcal{B}$. A small coil of N turns having cross-sectional area A is mounted on a

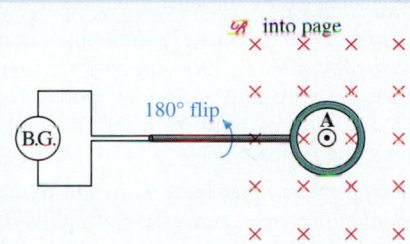

rod that serves as a probe. The coil is connected to a *ballistic galvanometer*, a device that measures the total charge flowing through it over a short time interval. The resistance of the entire circuit is R. The coil is held between the poles of a magnet and is oriented so that $\mathbf{A}$ is parallel to $\mathcal{B}$, as shown in the figure. By means of a simple trigger-and-spring mechanism, the coil can be quickly flipped through 180°, and the ballistic galvanometer reads a total charge q. **(a)** Show that the magnetic field at the location of the coil has magnitude

$$\mathcal{B} = \frac{qR}{NA}.$$

(b) Large coils of similar design are sometimes used to measure weak but uniform magnetic fields, notably the local magnetic field of the earth. A typical coil of this kind is circular in form, with diameter 30 cm, and has 1500 turns and resistance 5000 Ω. What is the approximate value of q you can expect?

30.34 *(5) Faraday disk dynamo.* In the figure below, the voltmeter is connected to the disk by means of the sliding contacts

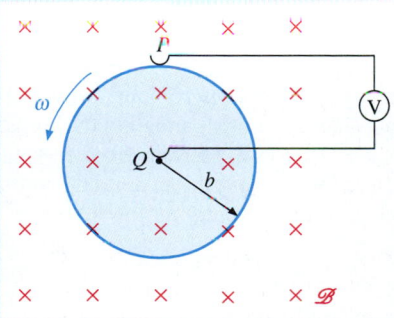

at the periphery and the axis of the disk. The disk has radius b and rotates with angular velocity ω in the sense shown. A uniform magnetic field $\mathcal{B}$ is parallel to the disk axis. **(a)** Show that there is an emf between the two sliding contacts and that the voltmeter reading is

$$V = \tfrac{1}{2}\omega b^2 \mathcal{B}.$$

(b) Which of the two terminals P and Q is positive? **(c)** Let $\mathcal{B} = 0.75$ T, $b = 12$ cm, and $\omega = 360$ rad/s, and find V.

30.35 *(5) Faraday disk motor.* The dynamo of Problem 30.34 can be converted into a motor by replacing the voltmeter with

a battery, which drives a current i through the disk. **(a)** What is the torque produced by the motor? **(b)** If the disk rotates with angular speed ω, find the power output of the motor.

30.36 *(5) Awkward motor.* The contraption illustrated below consists of a uniform metal rod of mass m and length l that

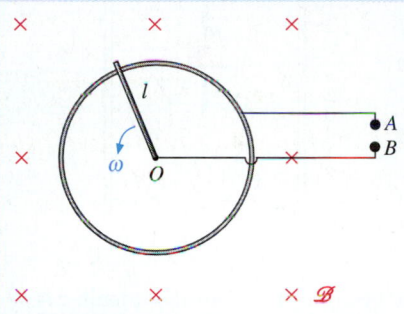

can rotate in a vertical plane about axis O, a circular wire loop with which the other end of the rod makes frictionless electrical contact, and the electrical connections shown. The entire system is located in a uniform magnetic field $\mathcal{B}$ directed into the page. At $t = 0$, the rod is oriented vertically upward. Show that, if you want to make the rod revolve about O with constant angular speed ω, you must apply a voltage between terminals A and B equal to

$$V = \frac{\pi l^2 \omega \mathcal{B}}{2} + \frac{mgR \cos \omega t}{l \mathcal{B}}.$$

30.37 *(5) Magnetic drag, I.* In the figure below, a horizontal wire of length l and resistance R slides freely on the vertical members of a U-shaped conductor of negligible resistance.

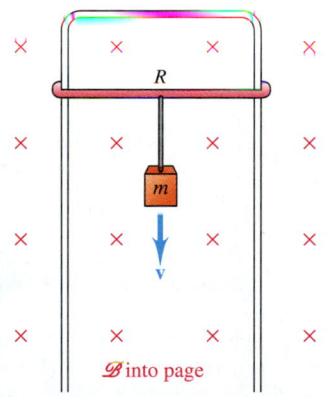

There is a uniform horizontal magnetic field of magnitude $\mathcal{B}$ over the entire region. A mass m is attached to the slide wire, whose own mass is negligible. Show that the mass descends with terminal speed

$$v = \frac{Rmg}{\mathcal{B}^2 l^2}.$$

30.38 *(5) Magnetic drag, II.* The system shown (p. 848) is similar to that of Problem 30.37. Here, however, all the conductors have negligible resistance, and a capacitor of capaci-

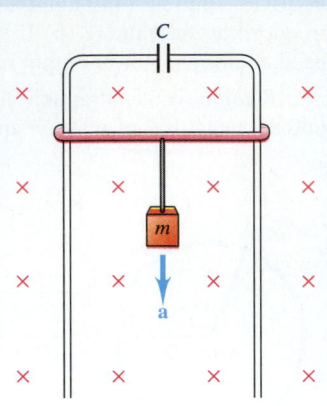

tance C has been inserted into the circuit. Show that, when the sliding wire is released, its acceleration has the constant value

$$a = \frac{1}{1 + \dfrac{l^2 \mathscr{B}^2 C}{m}} g.$$

30.39 *(6) Spinning up the charge.* In the figure below, a small dielectric body of mass m carries charge q. The body lies at the end of a rod of length b, having negligible mass. The other end of the rod is connected to a vertical shaft mounted on frictionless bearings so that the rod is perpendicular to the shaft and thus free to swing in a horizontal plane. At time $t = 0$, the rod is at rest, and a uniform magnetic field is turned on whose magnitude increases linearly with time: $\mathscr{B} = kt$, where k is a constant. The direction of the field is parallel to the shaft. **(a)** Find the angular acceleration of the body about the shaft. **(b)** What is the speed v of the body as a function of time?

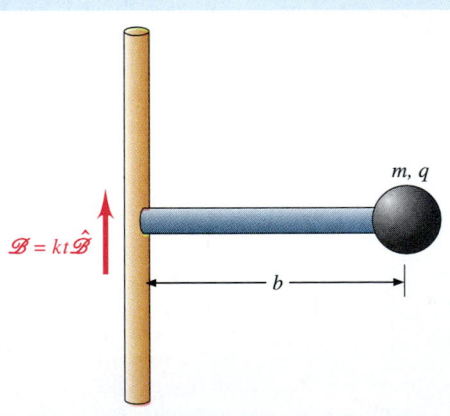

30.40 *(6) Getting a charge out of it, I.* The resistivity of a certain metal is ρ. Using a wire-drawing machine, you make a wire of uniform cross section out of a lump of the metal having volume τ. You then form the wire into a square loop, which you place in a uniform magnetic field $\mathscr{B}$ whose direction is parallel to the loop axis. Show that, when the magnetic field is turned off, the charge q passing through any imaginary

cross section through the wire is

$$q = \frac{\tau \mathscr{B}}{16\rho},$$

a result independent of the dimensions of the wire!

30.41 *(6) Getting a charge out of it, II.* Suppose that you carry out the process described in Problem 30.40, except that this time you form the wire into a circular loop. Show that the expression for q changes only by a rather small numerical factor; evaluate the factor.

30.42 *(6) Semicircular argument? I.* In the figure below, a wire is formed into an open semicircle of radius b. End O of the semicircle is attached to an axis about which the semicircle rotates with angular speed ω. The axis is normal to the plane of the semicircle. A uniform magnetic field $\mathscr{B}$, parallel to the axis, extends over the entire region. **(a)** Evaluate $\int_O^P \mathscr{E} \cdot d\mathbf{s}$ along the semicircle. **(b)** Find the magnitude and sense of the emf between O and P if $\mathscr{B} = 0.035$ T, $b = 0.83$ m, and $\omega = 120$ rad/s.

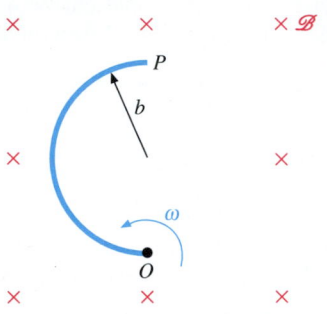

30.43 *(6) Semicircular argument? II.* The figure below represents a space that is sharply divided between a region having uniform magnetic field $\mathscr{B}$ directed into the page and a field-free region. A wire, formed into a closed semicircular loop of radius b, can rotate about an axis O normal to the page, located on the boundary between the regions. At $t = 0$, the loop, whose straight side is initially parallel to the boundary, begins to rotate counterclockwise, starting from rest with constant angular acceleration α. Find the magnitude and sense of the electric circulation induced around the loop as a function of time.

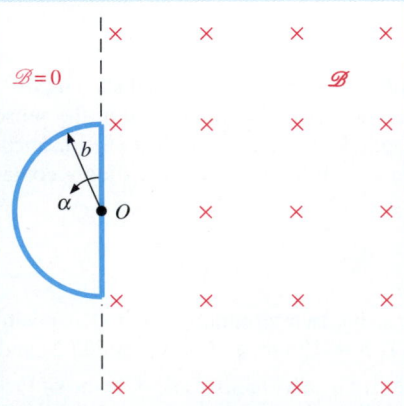

30.44 (5) *Series-wound motor.* For motors as well as for generators, use of a permanent magnet to provide the necessary magnetic field is not practical, except for quite small motors. As described in Problem 30.22, a stator coil is used. The current required by the stator, called the *exciting current*, is supplied by the same source of emf that supplies the rotor current. When the stator and rotor are connected in series, as in figure *a* below, the motor is called *series wound*. Suppose that the rotor and stator coils each have resistance R and that the external source of emf that drives the motor has a constant value V'. Assume for simplicity (and somewhat unrealistically) that no iron is present in the system, so the magnetic field of the stator is directly proportional to the current:

$$\mathscr{B} = ki; \quad k \text{ a constant.}$$

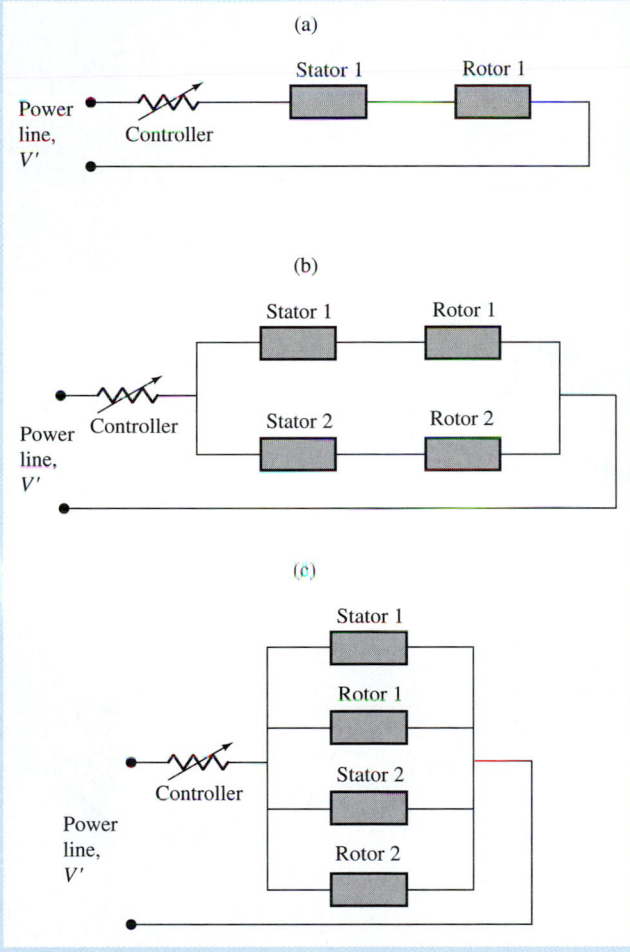

(a) Show that the instantaneous output torque of the motor is given by

$$T = \frac{cV'^2}{4R^2}\left(1 - \frac{\omega c}{2R + \omega c}\right)^2,$$

where $c \equiv kNA \sin \theta$. (b) Show that at low speeds the torque is approximately

$$T = \frac{cV'^2}{4R^2}\left(1 - \frac{\omega c}{R}\right)$$

and that therefore the mean torque decreases only slowly with increasing speed. (c) Show that at high speeds the torque is approximately

$$T = \frac{V'^2}{\omega^2 c}\left(1 - \frac{4R}{\omega c}\right)^2$$

and that therefore the mean torque decreases quite rapidly with increasing speed. (d) Explain why a lightly loaded series-wound motor tends to "run away"—that is, to turn faster and faster until it reaches a very high speed. (e) The relation between torque and current can be expressed in the proportionality $T \propto i^n$. What is the value of n? (f) If a series-wound motor is used to power a vehicle (such as a trolley car), under what circumstances is the current in the motor a maximum? (g) Why is a series-wound motor especially suitable for traction? (h) Nearly all of the older generation of trolley-car systems (those built before about 1950) ran on dc power sources because satisfactory high-power ac motors were not developed until the 1930s. Most cars had two motors, one driving each set of wheels. It was common practice to design the motors so that they could be connected in series, as in figure *b*, when the car was started from rest and then, by semiautomatic or automatic switching, reconnected in parallel, as in figure *c*, when the car achieved a moderate speed. Explain the advantage of this switching.

Inductance

If the electric current in a conducting loop is varied, the magnetic flux penetrating a nearby loop will vary, resulting in an induced emf in the second loop that is proportional to the rate of change of current in the first one.

The proportionality constant is called the mutual inductance.

Because the flux arising from the current in one turn of a current-carrying coil penetrates the other turns, a single coil exhibits self-inductance. In conformity with Lenz's law, the induced emf always opposes the change in current.

Increasing the electric current through a conductor requires work to drive the current against the induced emf. This work can be recovered when the current is decreased.

The potential energy is stored in the associated magnetic field.

Left: This magnetic resonance imaging machine uses superconducting coils to produce the requisite strong magnetic field. Such magnets store more energy per unit volume than the best batteries.

The term commonly employed to denote the electrical inertia-like effect is "self-induction," which is becoming gradually shortened to inductance.

—*Science*, July 18, 1888

SECTION 31.1 Introduction

Faraday's law tells us that electric currents are induced in electric circuits by changes in the magnetic flux that penetrates them. In Section 31.2, we study this phenomenon from a practical point of view and develop the theory of the important electric circuit element called the *inductor*.

Inductors store potential energy in their magnetic fields, in a manner analogous to the way that capacitors store potential energy in their electric fields. This magnetic energy is the subject of Section 31.3.

SECTION 31.2 Inductance

Figure 31.1 shows a modified form of the electromagnetic induction experiment of Figure 30.2. When there is a current i_1 in circuit 1, coil 1 produces a certain magnetic flux Φ_1, shown in Figure 31.1 as a family of flux lines. *Some* of this flux penetrates

FIGURE 31.1 A modified form of the electromagnetic induction experiment of Figure 30.2.

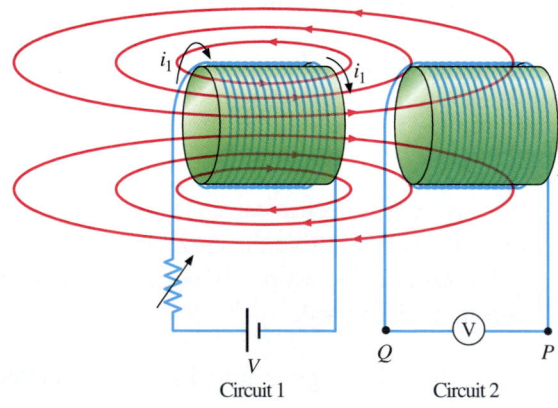

Circuit 1   Circuit 2

coil 2. We say that this portion of the flux—call it Φ_{12}—*links* the two coils. The fraction of the total flux that penetrates coil 2 depends on the size, shape, relative location, and relative orientation of the two coils. But, for any particular arrangement, we can write the proportionality

$$\Phi_{12} \propto \Phi_1 \propto i_1. \tag{31.1}$$

Mutual Inductance

To induce an emf V_2 across the terminals P and Q in circuit 2, we must (as always) vary the magnetic flux Φ_{12} that penetrates coil 2. We can readily do this by using the variable resistor in circuit 1 to vary the current i_1 through coil 1. According to Faraday's law (Equation 30.8), we have

$$V_2 = -\frac{d\Phi_{12}}{dt}. \tag{31.2}$$

Combining this result with Proportionality 31.1, we obtain

$$V_2 \propto -\frac{di_1}{dt}. \tag{31.3a}$$

As usual, we rewrite this relation as an equation by introducing a proportionality constant:

$$V_2 = -M_{12}\frac{di_1}{dt}. \tag{31.3b}$$

The proportionality constant M_{12} is called the **mutual inductance**. The SI unit of mutual inductance follows from Equation 31.3b; it is V·s/A. Because Joseph Henry was the first, in 1829, to define inductance in a clear way, this unit is called the **henry (H)**:

$$1\ \text{H} \equiv 1\ \text{V·s/A.} \tag{31.4}$$

We can readily reverse the connections to coils 1 and 2, as shown in Figure 31.2.

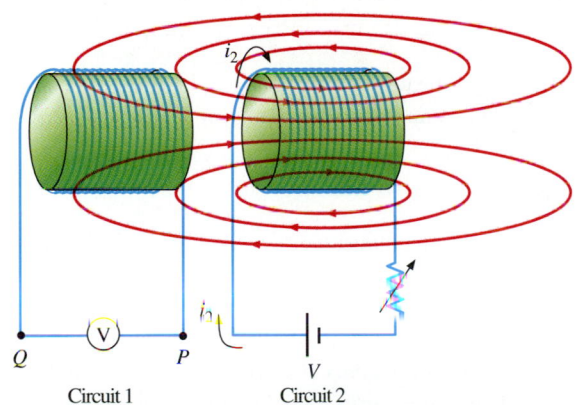

FIGURE 31.2 Switching the connections to coils 1 and 2 of Figure 31.1 so as to define the mutual inductance M_{21}.

Circuit 1   Circuit 2

Without changing the position of the coils, we arrange things so that a current i_2 in coil 2 is varied at a rate di_2/dt and induces an emf V_1 across coil 1. In complete analogy with Equation 31.3b, we define the mutual inductance M_{21}:

$$V_1 = -M_{21}\frac{di_2}{dt}.$$

We will not prove the theorem here, but it is always true that $M_{21} = M_{12}$. Problems 31.25 and 31.30 demonstrate this equality for special cases. Given that $M_{12} = M_{21}$, we drop the subscripts and write the mutual inductance as M:

$$V_2 = -M\frac{di_1}{dt} \quad \text{and} \quad V_1 = -M\frac{di_2}{dt}. \tag{31.5a,b}$$

These equations are most commonly written in the general form

$$V = -M\frac{di}{dt}, \tag{31.5c}$$

where it is understood that di/dt applies to one coil and V to the other.

EXAMPLE 31.1

In Figure 31.3, a single-turn circular loop of radius b is located inside a long solenoid having n turns per meter. The axis of the loop is parallel to the solenoid axis. **(a)** Find the mutual inductance M. **(b)** Evaluate M for $n = 1500$ turns/m and $b = 1.5$ cm.

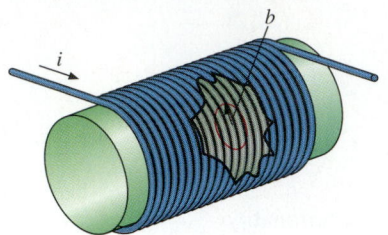

FIGURE 31.3

SOLUTION:
(a) Find M.

Suppose that the solenoid carries a current i. The magnitude of the magnetic field inside the solenoid has the uniform value given by Equation 29.22,

$$\mathscr{B} = \mu_0 n i.$$

The area of the loop is πb^2, so the flux penetrating the loop at any moment is

$$\Phi_{12} = \pi b^2 \mathscr{B} = \pi \mu_0 n b^2 i.$$

Now suppose that i varies at the rate di/dt. Differentiating both sides of the preceding equation, you obtain

$$\frac{d\Phi_{12}}{dt} = \pi \mu_0 n b^2 \frac{di}{dt}.$$

According to Faraday's law, the left side of this equation is equal to $-V$, the negative of the emf induced around the loop. So you can write

$$V = -(\pi \mu_0 n b^2) \frac{di}{dt}. \tag{31.6}$$

Compare this equation with the definition of mutual inductance given by Equation 31.5c. You can see immediately that the term in parentheses on the right side of Equation 31.6 is the mutual inductance

$$M = \pi \mu_0 n b^2. \tag{31.7}$$

(b) Evaluate M for $n = 1500$ turns/m and $b = 1.5$ cm. Inserting the numerical values, you find

$$M = \pi \times 4\pi \times 10^{-7} \text{ T·m/A} \times 1500 \text{ m}^{-1}$$
$$\times (1.5 \times 10^{-2} \text{ m})^2$$
$$= 1.3 \times 10^{-6} \text{ H}.$$

The SI unit of μ_0, the permeability of free space, is often quoted as H/m rather than the equivalent T·m/A. This equivalence is the subject of Problem 31.6.

Self-Inductance

In our discussion of mutual inductance, we regarded one coil as the source of changing magnetic flux and the other coil as the site where Faraday's law operated to induce an emf. But the phenomenon of inductance takes place even in a single coil. (Indeed, even a section of straight wire demonstrates inductive effects in a small way.) When current passes through the coil, each turn gives rise to flux lines that link with the turn itself and with some or all of the other turns. As the current changes, the flux linking each turn with the others also changes. This results in an induced emf V; according to Lenz's law, V is a back emf.

As for mutual inductance, the induced emf is proportional to di/dt:

$$V \propto -\frac{di}{dt}.$$

The proportionality constant L is called the **self-inductance**:

$$V = -L \frac{di}{dt}. \tag{31.8}$$

Like M, self-inductance L is expressed in henry.

A device having appreciable self-inductance is called an **inductor**. Inductors are important components of many circuits in which the current is not steady; the function of inductors is discussed in detail in Chapter 32. In electric circuits, an inductor is conventionally denoted by the symbol shown in Figure 31.4a. An ideal inductor possesses inductance only, and it thus has zero resistance. All real inductors (except those made of superconducting materials) possess resistance. If this resistance is significant, it is represented as shown in Figure 31.4b.

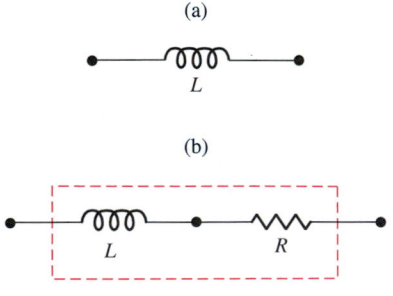

FIGURE 31.4 (*a*) Conventional symbol for an inductor of (self-)inductance L. (*b*) Representation of the inductance L and resistance R of a real inductor.

For any particular current-carrying structure, the value of L depends on the completeness of the flux linkage, which in turn depends on the geometry. The simplest case is the *toroid*. As you saw in Section 29.4, all of the flux lines in a current-carrying toroid run through the cavity. Thus *each flux line links every turn with every other turn.* Although this is not exactly true for a solenoid, it is approximately true for a long solenoid, in which nearly all the turns are far from the ends.

EXAMPLE **31.2**

(a) Derive an expression for the self-inductance of the large, skinny, ''bicycle tire'' toroid of circular cross section shown in Figure 31.5, whose major radius R is much greater than its minor radius b. (b) Find the self-inductance of a toroid having major radius $R = 15.5$ cm and minor radius $b = 1.2$ cm, wound with 1700 turns.

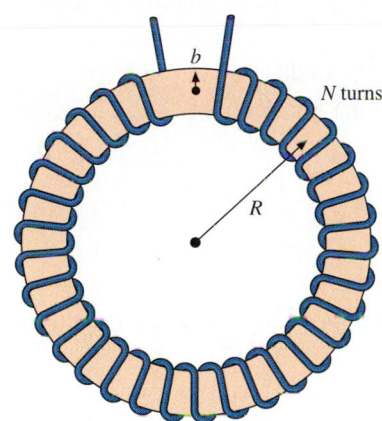

FIGURE 31.5

SOLUTION:

(a) Derive an expression for L.

According to Equation 29.27, the magnitude of the magnetic field inside the cavity of a toroid is

$$\mathcal{B} = \frac{\mu_0}{2\pi} \frac{Ni}{r},$$

where N is the number of turns and r is the distance from the toroid axis. Although $\mathcal{B}$ is a function of r, $\mathcal{B}$ does not vary

appreciably within the cavity because $b \ll R$. So to a good approximation you can take the magnetic field in the cavity to have the uniform value

$$\mathcal{B} = \frac{\mu_0}{2\pi} \frac{Ni}{R}. \tag{31.9}$$

Using this approximation, you have for the magnetic flux passing through the cavity (Section 28.3)

$$\Phi = \pi b^2 \mathcal{B} = \frac{\mu_0 Nib^2}{2R}.$$

This is the flux that links *each* turn. When i varies with time, the induced emf across the terminals is given by Faraday's law, which you can write in the form

$$V = -N \frac{d\Phi}{dt} = -\frac{\mu_0 N^2 b^2}{2R} \frac{di}{dt}.$$

The extra factor N arises from the fact that the emf induced in each turn adds to those induced in all the other turns. Comparing this value of V with that given by Equation 31.8, you obtain the self-inductance

$$L = \frac{\mu_0}{2} \frac{N^2 b^2}{R} \quad \text{for a toroid with } b \ll R. \tag{31.10}$$

(b) Evaluate L for $R = 15.5$ cm, $b = 1.2$ cm, and $N = 1700$ turns.

You find

$$L = \frac{4\pi \times 10^{-7} \text{ H/m}}{2} \times \frac{(1700)^2 \times (1.2 \times 10^{-2} \text{ m})^2}{0.155 \text{ m}}$$

$$= 1.7 \times 10^{-3} \text{ H}.$$

The result of part **a** of Example 31.2 can be readily modified to yield the self-inductance of a long solenoid. Suppose that you cut the toroid and straighten it out into a solenoid of length $l = 2\pi R$. The number of turns per unit length is then $n = N/2\pi R$. Some of the flux lines no longer link every turn to every other turn, but for a first approximation we can neglect this change in linkage. So all we need do is rewrite Equation 31.10 in terms of n (= N/l) and l:

$$L = \frac{\mu_0}{2} \frac{N^2 b^2}{R} = \pi \mu_0 n^2 l b^2. \tag{31.11}$$

You can derive the same result directly from the equation for the magnetic field of a solenoid, without reference to a toroid; this is the subject of Problem 31.30.

Energy Stored in an Inductor

As you saw in Section 26.5, work must be done in charging a capacitor, as charge is pushed onto the plates against the repulsion of the charge already present on the plates. The process of charging the capacitor is conservative, and potential energy $U = \frac{1}{2}CV^2$ is stored in the electric field of the capacitor (Equation 26.26b). Indeed, we argued that the presence of an electric field in an empty region of space always implies an energy density $u = \frac{1}{2}\epsilon_0\mathscr{E}^2$ (see Equation 26.28).

Is something similar true of current-carrying inductors and magnetic fields? It does indeed require work to change the current through an inductor. According to Equation 31.8, any attempt to increase the current i through an inductor results in the induction of a back emf $V = -L\,di/dt$. This back emf must be overcome if the current is to be increased; we must drive the current with a *forward* emf $V = L\,di/dt$. The power expended in doing this is given by Joule's law, $P = Vi$. So we have

$$P \equiv \frac{dW}{dt} = Li\,\frac{di}{dt}.$$

At a particular instant when the current is i, the work required to increase the current by an amount di is

$$dW = Li\,di.$$

Suppose that the initial current is i_i and the final current is i_f. The work required for this finite increase is

$$W = \int dW = L\int_{i_i}^{i_f} i\,di,$$

which yields

$$W = \tfrac{1}{2}L(i_f{}^2 - i_i{}^2). \tag{31.12a}$$

In the common special case where the initial current is zero, we can drop the subscripts and write

$$W = \tfrac{1}{2}Li^2. \tag{31.12b}$$

Is this work retrievable? Suppose we *reduce* the current through the inductor. The inductor responds by producing a back emf. In this case, the "back" emf is actually in the forward sense, as shown in Figure 31.6a. As the external emf driving the current through the circuit in the positive sense is reduced, the inductor supplies an emf that drives current through the circuit in the positive sense and thus does work (Figure 31.6b). The analogous situation for a capacitor is shown in Figure 31.6b.

We can thus assert that (neglecting electrical resistance) the system is conservative and can be characterized by a potential energy $U = W$. Choosing the reference value $U_0 = 0$ to correspond to $i = 0$, we write the potential energy of the system as

$$U = \tfrac{1}{2}Li^2. \tag{31.13}$$

Just as in the analogous capacitor circuit—or, indeed, any system—the potential energy is a property of the system as a whole. But it is often useful to think of potential energy as being "stored" in the system in a particular way. In the present situation, we think of the energy as being stored in the magnetic field of the inductor.

To devise a quantitative picture of energy storage in a magnetic field, we draw on the analogy with capacitors. For a parallel-plate capacitor of large area, the electric field is essentially the same everywhere in the well-defined volume between the plates and is negligible elsewhere. For inductors, the analogous case is that of the bicycle-tire toroid discussed in Example 31.2, whose magnetic field is essentially the same everywhere within the cavity and is zero elsewhere.

According to Equation 31.10, the self-inductance of the toroid is $L = \mu_0 N^2 b^2/2R$. Inserting this value into Equation 31.13, we obtain

$$U = \frac{\mu_0 N^2 i^2}{4R}\,b^2. \tag{31.14}$$

(a)

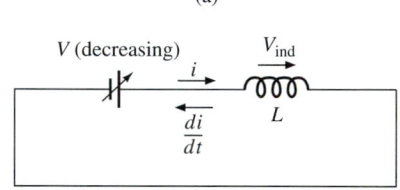

(b)

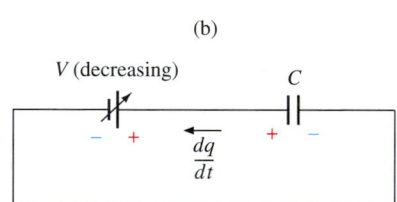

FIGURE 31.6 (*a*) A source of emf drives a current i through a circuit containing an inductor L. When the emf is reduced, the current decreases at a rate di/dt. An emf V_{ind} appears across the inductor in the forward direction, tending to maintain the current at its original value. This emf does work at an instantaneous rate $P = V_{\text{ind}}i$. (*b*) A source of emf has driven a charge $|q|$ onto each of the plates of a capacitor C. The potential across the capacitor is V. When the emf is reduced, charge is driven around the circuit at a rate $i = dq/dt$. Work is done by the capacitor at the instantaneous rate $P = V\,dq/dt$.

Let us reexpress U in terms of the magnetic field $\mathscr{B}$. To do this, we begin with Equation 31.9, $\mathscr{B} = \mu_0 Ni/2\pi R$. Combining this equation with Equation 31.14 requires some algebraic manipulation. One way to proceed is to evaluate $\mathscr{B}^2$:

$$\mathscr{B}^2 = \frac{\mu_0^2 N^2 i^2}{4\pi^2 R^2}. \qquad (31.15)$$

Then dividing both sides of Equation 31.15 by the right side gives us

$$\frac{4\pi^2 R^2 \mathscr{B}^2}{\mu_0^2 N^2 i^2} = 1.$$

Because the left side of this equation is equal to 1, we can multiply it into the right side of Equation 31.14 without changing the value of the latter. After some cancellation, we have

$$U = \frac{\mathscr{B}^2}{\mu_0} \pi^2 b^2 R. \qquad (31.16)$$

This expression does indeed give the energy stored in the inductor in terms of the magnetic field, as well as the geometric quantities b and R, which express the size of the toroid.

More interesting than the total energy U stored in the inductor is the *energy density* u, the energy per unit volume. As $\mathscr{B}$ is the same everywhere inside the toroid cavity, we simply divide U by the volume of the cavity to obtain u. The volume of the cavity is

$$\pi b^2 \times 2\pi R = 2\pi^2 b^2 R,$$

so we have

$$u = \frac{\mathscr{B}^2}{2\mu_0}. \qquad (31.17)$$

There are no contacts for making electrical connection between the electric toothbrush and its charging base. The battery in the toothbrush handle is recharged by means of the mutual inductance between a coil in the base and a corresponding coil in the handle.

Compare this with the analogous expression for electric fields, $u = \epsilon_0 \mathscr{E}^2/2$.

EXAMPLE 31.3

Figure 31.7 shows a section of a *coaxial cable*. The cable consists of two conductors in the form of thin cylindrical shells having a common axis. The inner conductor has radius r_1, and the outer conductor has radius r_2. The two conductors carry the current i in opposite directions. (This cable differs from most commercial coaxial cables, in which the inner conductor is a solid wire.) **(a)** Find U/l, the magnetic energy stored in the cable per unit length, when the cable carries a current i. **(b)** Use the result of part **a** to determine L/l, the self-inductance per unit length.

FIGURE 31.7

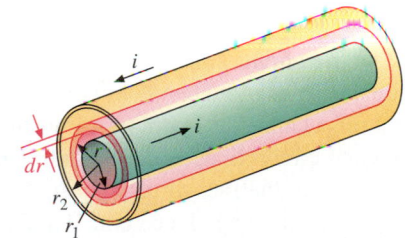

SOLUTION:
(a) Find U/l.

An Amperean circle of radius $r < r_1$ is threaded by no current. The net current threading an Amperean circle of radius $r > r_2$ is likewise zero. Thus it follows from Ampère's law that the magnetic field is zero at all points $r < r_1$ and $r > r_2$ (Problem 29.59). Between the two conductors, the field is due entirely to the current in the inner conductor and is given by Equation 29.5, with appropriate changes in notation:

$$\mathscr{B} = \frac{\mu_0}{2\pi} \frac{i}{r}.$$

Because $\mathscr{B}$ depends on r, the energy density is not uniform throughout the space between the conductors. Nevertheless, you can use Equation 31.17 to write the energy density at any point

as

$$u = \left(\frac{\mu_0}{2\pi} \frac{i}{r}\right)^2 \frac{1}{2\mu_0} = \frac{\mu_0 i^2}{8\pi^2} \frac{1}{r^2} \quad \text{for } r_1 < r < r_2.$$

Now, consider a thin cylindrical shell of radius r and thickness dr. The volume of a length l of this shell is $2\pi lr\, dr$. So the energy dU of the magnetic field contained in the shell is the product of the energy density and the volume:

$$dU = \frac{\mu_0 li^2}{4\pi} \frac{dr}{r}.$$

To find the total magnetic energy within a length l of the cable, you integrate:

$$U = \frac{\mu_0 li^2}{4\pi} \int_{r_1}^{r_2} \frac{dr}{r} = \frac{\mu_0 li^2}{4\pi} \ln \frac{r_2}{r_1}. \qquad (31.18a)$$

The energy per unit length is

$$\frac{U}{l} = \frac{\mu_0 i^2}{4\pi} \ln \frac{r_2}{r_1}.$$ **(31.18b)**

(b) Determine L/l.

Using the general relation of Equation 31.13, you can write the energy per unit length of cable as

$$\frac{U}{l} = \frac{1}{2}\frac{Li^2}{l}.$$

You combine this general equation with the specific result expressed in Equation 31.18b to obtain

$$\frac{L}{l} = \frac{\mu_0}{2\pi} \ln \frac{r_2}{r_1}$$ **(31.19)**

for the inductance per unit length.

The capacitance per unit length of a cylindrical capacitor follows directly from Equation 26.9:

$$\frac{C}{l} = \frac{2\pi\epsilon_0}{\ln \dfrac{r_2}{r_1}}.$$

Can you account for the similarities and differences between this equation and Equation 31.19? Compare the approach used in this example with that used to find the capacitance of a spherical capacitor in Example 26.5.

Symbols Used in Chapter 31

b	radius of a circle	q	electric charge
$\mathcal{B}, \mathscr{B}$	magnetic field	R	electrical resistance
$\mathcal{E}, \mathscr{E}$	electric field	V	emf, potential
i	electric current	W	work
n	number of turns per unit length	μ_0	permeability of free space
N	number of turns	Φ	magnetic flux

Summing Up

When some or all of the magnetic flux produced by a current i in an electric circuit *links* that circuit with another circuit, a change in i leads to a change in the linking magnetic flux Φ_{12}, and this change in Φ_{12} in turn leads to an induced emf in the second circuit in accordance with Faraday's law. The relation between the rate of change di/dt of the current in the first circuit and the induced emf V in the second is given by Equation 31.5c,

$$V = -M\frac{di}{dt}.$$

In this equation, the proportionality factor M is called the **mutual inductance**. Its value depends on the geometry of the complete system. The SI unit of inductance is the **henry**; according to Equation 31.4, $1\ \text{H} \equiv 1\ \text{V·s/A}$.

Because the magnetic flux arising from current in one part of a circuit can link that part of the circuit with itself and with other parts of the same circuit, the circuit possesses an internal **self-inductance** L, given by Equation 31.8,

$$V = -L\frac{di}{dt}.$$

A circuit component that possesses an appreciable self-inductance is called an **inductor**.

When the current through an inductor is changed, a back emf is induced. Consequently, the change in current requires work. The process is reversible; the energy is stored as potential energy, as expressed in Equation 31.13,

$$U = \tfrac{1}{2}Li^2.$$

It is useful to regard this energy as being stored in the magnetic field. The **energy density** u is given by Equation 31.17,

$$u = \frac{\mathscr{B}^2}{2\mu_0}.$$

KEY TERMS

Section 31.2 Inductance
mutual inductance ▪ self-inductance, inductor

QUERIES

31.1 *(2) Curious!* Compare the unit of inductance, whose dimensions can be expressed as $[\mu_0]$ [length], with the unit of capacitance.

31.2 *(2) Inducement.* The equation $V = -L \, di/dt$, which defines self-inductance, expresses a connection between i and V. Why do we need this new expression when we already have in Ohm's law a relation between i and V?

31.3 *(2) Thinking flexibly.* You wind a solenoid on an elastic core. You connect the solenoid to a battery, and current i flows through the circuit. You grab the ends of the solenoid and stretch it. As you do so, does i increase, decrease, or remain the same? Explain your answer.

PROBLEMS

GROUP A

31.1 *(2) Simon says.* Two coils have mutual inductance 0.25 H. Coil 1 is part of a circuit in which the current varies according to the relation $i_1 = 0.23t$ A. Coil 2 is part of a separate circuit whose total resistance is 150 Ω. Find the current i_2 in coil 2 at $t = 8.3$ s.

31.2 *(2) Getting the picture.* The figure below represents the current through an inductor as a function of time. At $t = 4$ s, the emf across the inductor is 2 V. **(a)** What is the inductance? **(b)** Sketch a graph of the emf across the inductor as a function of time.

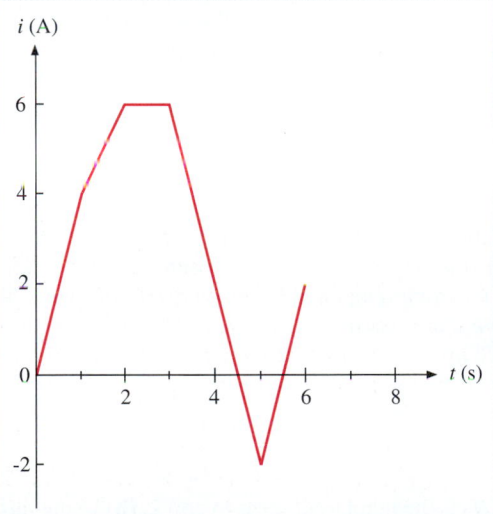

31.3 *(2) The more it changes, the more it remains the same.* A coil of self-inductance 0.50 H is part of a complete circuit. At $t = 0$, the current through the coil is 3.0 A. If the emf measured across the coil is a constant -0.3 V, at what time is the current equal to 4.3 A?

31.4 *(2) Monkey see, monkey do?* Two coils have mutual inductance 5.2 mH. The current in coil 1 is given by the expression $i = i_0 \sin 2\pi\nu t$, where $i_0 = 10.3$ A and $\nu = 55$ Hz.

(a) Express V_2, the emf induced in coil 2, as a function of time. **(b)** What is the maximum value of V_2?

31.5 *(2) Asymmetric but mutual.* A 10-turn coil of radius 7.2 cm and a single-turn coil of radius 0.85 cm lie on a common axis, their centers separated by 12 cm. Find the mutual inductance of the pair of coils. (Hint: Because it is small, the single-turn coil "sees" an essentially uniform magnetic field when a current flows through the 10-turn coil.)

31.6 *(2) Unity in diversity.* Show that the units used to express permeability—T·m/A and H/m—are equivalent.

31.7 *(2) A solenoid of any other shape.* Equation 31.11 was derived for a solenoid of circular cross section. **(a)** Reexpress the value of L in terms of the cross-sectional area A of the solenoid. **(b)** Show that the same result applies to any long solenoid of cross-sectional area A, regardless of the shape of the cross section.

31.8 *(2) Wired.* Calculate the self-inductance of a toroid with major radius 6.1 cm and minor radius 1.2 cm, wound with 450 turns.

31.9 *(2) Long count.* A closely wound toroid has major radius 14 cm and minor radius 0.85 cm. If the self-inductance is 0.54 H, how many turns are wound on the toroid?

31.10 *(2) Round and round.* A 250-turn toroid has self-inductance 0.50 H. When the winding carries a current $i = 1.5$ A, what is the magnetic flux Φ threading the cavity?

31.11 *(2) Self-induced.* Find the self-inductance of a solenoid that is 36 cm long, 4 cm in diameter, and wound with 27 turns/cm.

31.12 *(2) Inductive method.* A solenoid 25 cm long is wound with 18-gauge copper wire, whose diameter is 1.024 mm. The resistance of the wire is 0.19 Ω. What is the inductance of the solenoid? Take the resistivity of copper to be 1.724×10^{-8} Ω·m.

31.13 *(2) The current goes round and round.* A 1.3-A current flows through a circular coil of self-inductance 8.6×10^{-4} H. The magnetic flux threading the coil is 2.4×10^{-6} Wb. How many turns does the coil have?

31.14 (2) *Equivalent inductance of inductors in series.* The two inductors illustrated below have self-inductance L_1 and L_2, respectively. They are sufficiently far apart that their mu-

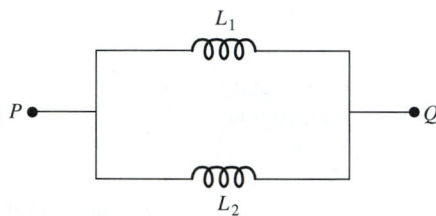

tual inductance is negligible. **(a)** Show that the total inductance L of the circuit segment between P and Q is

$$L = L_1 + L_2.$$

(b) Find L if $L_1 = 2.5$ H and $L_2 = 4.2$ H.

31.15 (2) *Equivalent inductance of inductors in parallel.* The two inductors shown below have self-inductance L_1 and L_2,

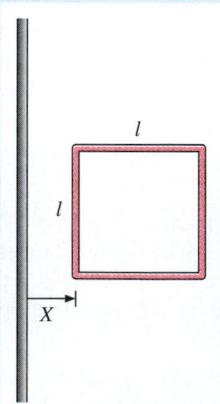

respectively. They are sufficiently far apart so that their mutual inductance is negligible. **(a)** Show that the total inductance of the circuit segment between P and Q satisfies the equation

$$\frac{1}{L} = \frac{1}{L_1} + \frac{1}{L_2}.$$

(b) Find L if $L_1 = 2.5$ H and $L_2 = 4.2$ H.

31.16 (3) *Magnifying Lenz.* A coil having self-inductance 23 mH carries a current $i = i_0 \sin 2\pi\nu t$, with $i_0 = 4.7$ A and $\nu = 94$ Hz. Express **(a)** the self-induced emf and **(b)** the energy of the magnetic field as functions of time.

31.17 (3) *Brek-ek-ek-ex . . .* The diameter of the outer conductor of a coaxial cable is 2.4 times the diameter of the inner conductor. The cable is 12 m long. Find the inductance of the cable.

31.18 (3) *Storehouse.* A solenoid of length 27 cm and radius 2.8 cm has 1000 turns. What is the magnetic field energy when the winding carries a 1.2-A current?

31.19 (3) *Current events.* A cylindrical solenoid 72 cm long with radius 2.8 cm has self-inductance 2.6×10^{-7} H. What current must flow through the solenoid if the magnetic energy density inside it is to be 1.4×10^{-3} J/m^3?

31.20 (3) *Two fields at once, I.* An electric field and a magnetic field exist in the same region of empty space. The magnitude of the magnetic field is 1.0 T. If the electric and magnetic energy densities are to be equal, what must be the magnitude of the electric field?

31.21 (3) *Independent.* **(a)** Show that the product LC of the inductance and capacitance of a coaxial cable depends only on the length of the cable and not on the radii of the conductors. **(b)** Show how, in principle, you could use measurements of the inductance and capacitance of a coaxial cable to determine the speed of light. Hint: $1/\mu_0\epsilon_0 = c^2$

31.22 (G) *Generating some interest.* A solenoid 7.2 cm long has self-inductance 1.8×10^{-4} H. The solenoid rotates at 40 rad/s in a 0.1-T magnetic field. The rotation axis is perpendicular both to the axis of the coil and to the magnetic field. Find the maximum emf that appears between the ends of the solenoid winding.

GROUP B

31.23 (2) *It's mutual.* A square loop of wire of side l lies in the same plane as a very long wire, as shown below. Find the mutual inductance of the loop and the long wire when the distance between the wire and the near side of the loop is X.

31.24 (2) *Another view of mutual inductance.* Suppose that, in the apparatus of Figure 31.1, the turns of coil 2 are closely wound (that is, all turns are in essentially the same place). When a current i_1 in circuit 1 gives rise to magnetic flux, any flux line that links *any* turn of coil 2 links *every* turn of coil 2. **(a)** Show that the mutual inductance M_{12} can be written in the alternative form

$$M_{12} = \frac{N_2\Phi_{12}}{i_1},$$

where N_2 is the number of turns in coil 2. **(b)** Using this result, show that the self-inductance of a coil having N closely wound turns is

$$L = \frac{N\Phi}{i},$$

where Φ is the flux that threads the coil when the current through the coil is i.

31.25 (2) *Nested toroids.* Toroid 1 has a cavity of square cross section. Inside the cavity lies toroid 2, whose cross section is also square (though smaller). The two toroids share a common major axis but do not necessarily share a common cavity axis;

see below. Calculate the mutual inductances M_{12} and M_{21}, and show that they are the same.

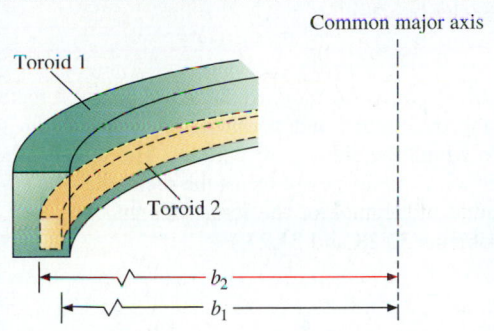

Common major axis

Toroid 1

Toroid 2

b_2

b_1

31.26 *(2) Big ring, little ring, I.* A large circular wire loop of radius r_1 is concentric and coplanar with a much smaller circular wire loop of radius r_2. **(a)** Suppose that a current i passes through the larger loop. Find the magnetic flux Φ_{12} that penetrates the smaller loop. **(b)** Calculate the mutual inductance of the two loops.

31.27 *(2) Big ring, little ring, II.* A large circular wire loop of radius r_1 carries current i. A much smaller circular wire loop of radius r_2 is concentric and coplanar with the large loop. The small loop moves along the common axis with speed v. At what distance x between the loops is the emf induced in the small loop a maximum?

31.28 *(2) Links to the faraway.* Two wire loops of radius b lie a distance d apart on a common axis. If $d \gg b$, show that

$$M = \frac{\mu_0 \pi}{2} \frac{b^4}{d^3}.$$

31.29 *(2) Superconducting ring, I.* A circular ring of radius b has self-inductance L. Suppose that the ring is made of a superconducting material. When the ring is cooled below the superconducting transition temperature T_c characteristic of the material, it loses all electrical resistance; $\rho = 0$ *exactly*, and therefore the electric field must be zero everywhere in the material. While it is in the superconducting state, the ring is moved into a region between the poles of a magnet where the magnetic field $\mathscr{B}$ is initially uniform. The ring is oriented so that the plane of the ring is normal to the field. **(a)** Using Ohm's law in the form of Equation 27.9, $\mathscr{E} = j\rho$, together with Faraday's law, show that *no magnetic flux lines must penetrate the ring as it is moved into the region between the magnet poles.* **(b)** Show that, when the ring lies between the magnet poles, it must be carrying a *persistent current*

$$i = \frac{\pi b^2 \mathscr{B}}{L}.$$

(Hint: Refer to part **b** of Problem 31.26.) **(c)** Sketch the magnetic flux lines in the vicinity of the ring.

31.30 *(2) Long solenoid.* Beginning with the expression for the magnetic field inside a solenoid that carries a current i, calculate the self-inductance of the solenoid and show that the result is the same as that given by Equation 31.11.

31.31 *(2) The long and short of it.* A solenoid of length l,

much greater than its radius, is closely wound with a single layer of wire. The length of the wire is D. **(a)** Express the self-inductance of the wire in terms of l and D only, together with any necessary constants. **(b)** How much wire will you need to make a solenoid 1 m long whose self-inductance is 1 mH?

31.32 *(2) Planning ahead.* You need an inductor of self-inductance $L = 1.0$ H. To meet this need, you plan to wind a solenoid 1 m long and 6 cm in diameter, having a single layer of winding with no space between adjacent turns. Copper wire is sold by mass. How much must you buy? (Hint: As long as the wire is thin, the result is independent of the wire diameter. Take the density of copper as $\delta = 8.9 \times 10^3$ kg/m^3.)

31.33 *(2) Weighing the windings.* A solenoid has length l much greater than its diameter. The solenoid is wound with a mass m of wire made of a metal of density δ and resistivity ρ. Show that the inductance of the solenoid is

$$L = \frac{\mu_0}{4\pi} \frac{mR}{l\rho\delta}.$$

31.34 *(2) Coaxial solenoids.* In the figure below, solenoid 1,

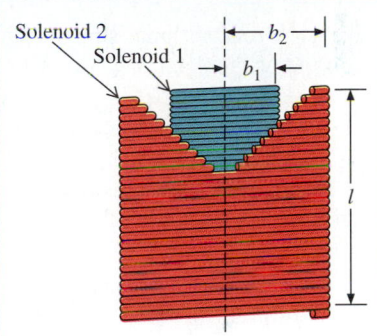

Solenoid 2

Solenoid 1

b_2

b_1

l

of radius b_1 and length l, having n_1 turns per meter, lies inside solenoid 2, of radius b_2 and (equal) length l, having n_2 turns per meter. The two solenoids are coaxial. **(a)** Suppose that solenoid 1 carries a current i_1. Show that the flux linking it with solenoid 2 is

$$\Phi_{12} = \pi\mu_0 n_1 b_1^2 i_1.$$

(b) Show that the mutual inductance M_{12} is

$$M_{12} = \pi\mu_0 n_1 n_2 l b_1^2.$$

(c) Now suppose that solenoid 2 carries a current i_2, and calculate the flux Φ_{21} linking it with solenoid 1. **(d)** Find the inductance M_{21} and show that it is equal to M_{12}, consistent with the assertion that $M_{12} = M_{21}$ always. **(e)** Consider the special case $n_1 = n_2$. Imagine that b_1 increases until it is equal to b_2, whereupon the two coils "merge together" into a single one. Elaborating on this argument, use the result of part **b** to obtain the value of the self-inductance of a single solenoid, given by Equation 31.11.

31.35 *(2) Thick toroid.* The figure (p. 862) shows a toroid of rectangular cross section, with inner radius r_1, outer radius r_2, and thickness X. The toroid is wound with N turns. Show that

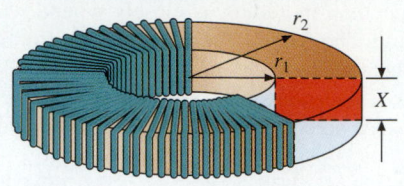

the self-inductance of the toroid is

$$L = \frac{\mu_0 N^2 X}{2\pi} \ln \frac{r_2}{r_1}.$$

31.36 *(2) Twin lead.* "Twin lead," used as the antenna input connector for FM radios and TV sets, consists of a pair of wires of radius a separated by a distance $b \gg a$ (Problem 29.9). Find the inductance per unit length of twin lead.

31.37 *(2) Sandwich.* A conducting cable consists of a pair of thin, flat metal tapes of width w, with an insulator of thickness $t \ll w$ sandwiched between them. Show that the inductance per unit length of this cable is

$$\frac{L}{l} = \mu_0 \frac{t}{w}.$$

31.38 *(2) Threaded toroid.* A long, straight wire lies along the axis of the toroid of Problem 31.35. **(a)** Show that

$$M = \frac{\mu_0 N X}{2\pi} \ln \frac{r_2}{r_1}.$$

(b) Comparing this result with the result of Problem 31.35, you can see that $L = NM$. Explain without calculation, using a topological argument.

31.39 *(2) Mutual admiration.* Suppose that the two inductors in the series circuit of Problem 31.14 are solenoids. They are brought close together along a common axis. Will the total inductance between P and Q be greater than or less than $L_1 + L_2$ if **(a)** the two solenoids are wound in the same sense? **(b)** the two solenoids are wound in opposite senses?

31.40 *(2) All kinds of linkages.* Solenoids 1 and 2 have equal length. The radius of solenoid 2 is such that it just slips inside solenoid 1. However, the difference in radius is not sufficient to make a significant difference in the self-inductances of the two; $L_1 = L_2$. What is the equivalent inductance of the two solenoids when **(a)** they are connected in series and are situated a long distance from one another? **(b)** they are connected in parallel and are situated a long distance from one another? **(c)** they are connected in series and solenoid 1 is slipped into solenoid 2 so that the senses of the windings are the same? **(d)** they are connected in parallel and solenoid 1 is slipped into solenoid 2 so that the senses of the windings are the same? **(e)** they are connected in series and solenoid 1 is slipped into solenoid 2 so that the senses of the windings are opposite? **(f)** they are connected in parallel and solenoid 1 is slipped into solenoid 2 so that the senses of the windings are opposite?

31.41 *(3) Two fields at once, II.* A ring of radius b, carrying

a uniformly distributed charge, rotates about its axis with angular speed ω. Show that, at a point on the axis a distance d from the center of the ring, the ratio of magnetic to electric field density is

$$\frac{u_m}{u_e} = \epsilon_0 \mu_0 \omega^2 b^2.$$

31.42 *(3) Rectangular loop.* In the figure below, a rectangular wire loop of length l and width w lies near a long, straight wire, to which the sides l are parallel. The distance between the long wire and the near edge of the rectangle is R. Calculate the mutual inductance of the long wire and the loop. (Hint: See Problems 30.28 and 31.23.)

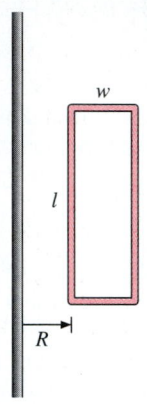

31.43 *(3) Not a very practical solenoid.* A long dielectric cylindrical shell of radius b rotates about its axis with angular speed ω. The shell carries a uniformly distributed electric charge; the charge per unit length of cylinder is λ. **(a)** Calculate the equivalent electric current per unit length around the cylinder surface. **(b)** Find the magnitude $\mathcal{B}$ of the magnetic field inside the cylinder. **(c)** What is the magnetic energy density inside the cylinder? **(d)** What is the electric energy density just outside the cylinder?

31.44 *(3) Spinning like crazy.* A circular dielectric loop of radius b carries a uniformly distributed electric charge q. **(a)** What is the magnitude $\mathcal{E}$ of the electric field at a point P on the loop axis, a distance b from the center of the loop? **(b)** The loop rotates about its axis with angular speed ω. What is the magnitude $\mathcal{B}$ of the magnetic field at the same point P? **(c)** Show that the ratio u_m/u_e of the magnetic energy density to the electric energy density at P is

$$\frac{u_m}{u_e} = \mu_0 \epsilon_0 (\omega b)^2.$$

Now, ωb is the linear speed v of the loop while (as you will see in Chapter 34) $\mu_0 \epsilon_0 = 1/c^2$, where c is the speed of light. Rewrite the energy density ratio in terms of v and c. Your result has to do with the fact that the magnetic field of moving charges is at bottom a relativistic effect; the factor you have obtained appears frequently in expressions for relativistic quantities. More on this matter in Chapter 38.

GROUP C

31.45 *(2) Two solenoids with but a single core.* You wind two separate single-layer coils of length l on a cardboard tube of radius b, as shown below. Coil 1 has n_1 turns per unit length, and coil 2 has n_2 turns per unit length. Both coils are wound like right-handed screws. **(a)** Beginning with the expressions for the self-inductances L_1 and L_2 and the expression for the mutual inductance M, derive the relation

$$M = \sqrt{L_1 L_2}.$$

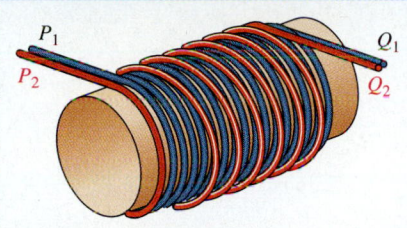

(b) What is the relation among M, L_1 and L_2 if $n_1 = n_2$? **(c)** If you connect the two coils together by joining terminals Q_1 and Q_2, what is the total inductance of the circuit segment between P_1 and P_2? **(d)** If you connect the two coils together by joining terminals Q_1 and P_2, what is the total inductance of the circuit segment between P_1 and Q_2?

31.46 *(2) Superconducting ring, II.* Consider again the superconducting ring of Problem 31.29. Show that, if you wish to flip the ring over in the magnetic field, you must do work W on it, where

$$W = \frac{\pi^3 b^4 \mathscr{B}^2}{2L}.$$

(Hint: You cannot treat the ring simply as a magnetic dipole of fixed magnitude. Indeed, the ring is a *perfect diamagnet*; that is, the dipole moment $\boldsymbol{\mu}$ of the superconducting ring tends to align itself *antiparallel* to the external magnetic field $\mathscr{B}$. This is in striking contrast with an ordinary magnetic dipole, whose dipole moment $\boldsymbol{\mu}$ tends to align itself parallel to $\mathscr{B}$. In solving this problem, you must consider how Faraday's law governs the way in which the persistent current varies with orientation of the ring, remembering that there can never be a nonzero electric field anywhere in the superconducting material.)

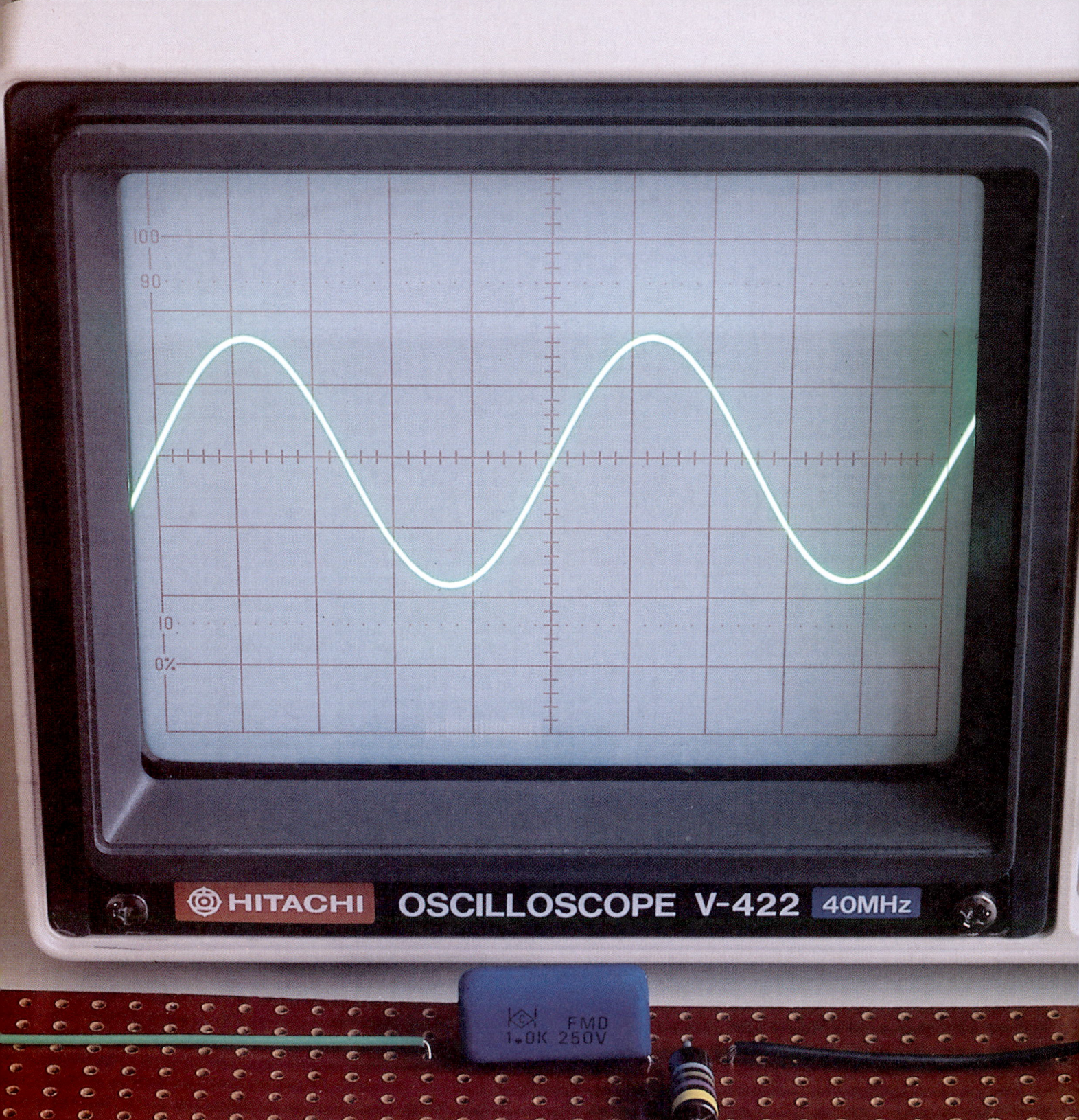

Changing Electric Currents I

RC, *RL*, and *LC* Circuits

— Resistors, capacitors, and inductors are important components of electric circuits.

— When a capacitor is charged or discharged through a resistor, the current is described by an exponential function whose time constant is *RC*.

— The current that passes through the analogous resistor-inductor circuit is described by an exponential function whose time constant is *L/R*.

— In an ideal (resistanceless) loop containing an inductor and a capacitor, the current is described by a sinusoidal function whose angular frequency is $1/\sqrt{LC}$. The system is described by the same mathematics used to describe the mass-spring oscillator.

Left: A series *LRC* circuit. The orange and green wires to the left connect the inductor (foreground), the resistor, and the capacitor (background) to a source of emf. The red and black wires to the right are connected to the oscilloscope, which displays the sinusoidal voltage across the resistor.

SECTION 32.1 Introduction

This chapter and the next one concern electric circuits in which the current varies with time. The simplest of such circuits contain a source of constant emf, a resistor, and either a capacitor or an inductor. Such circuits, called respectively *RC* and *RL circuits*, are the subject of Sections 32.2 and 32.3. In Section 32.4, we consider the *LC* circuit, which contains an inductor and a capacitor.

SECTION 32.2 The *RC* Circuit

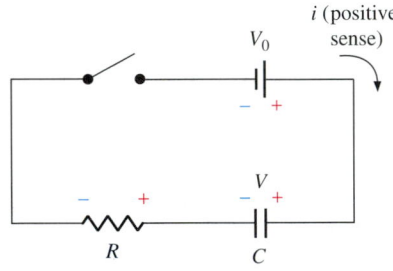

FIGURE 32.1 An *RC* circuit.

Figure 32.1 shows the simplest possible circuit containing a resistor R, a capacitor C, and a source of fixed emf V_0, whose polarity is shown. With this polarity in mind, we choose the clockwise sense around the circuit to be positive. The capacitor initially carries no charge, so the potential difference between its plates is $V_C = 0$. At time $t = 0$, the switch is closed, and there is a current i as electric charge begins to flow through the resistor and accumulates on the capacitor plates. Figure 32.1 shows the signs of the charges that accumulate on the capacitor plates, as well as the relative terminal polarities of the resistor as the charging current flows through it. As the capacitor charge increases, the potential V_C between the plates increases in a sense that opposes V_0, and thus the current decreases with time. We now develop a quantitative expression for the current as a function of time.

Consider an instant t after the switch has been closed, when charge of magnitude q has accumulated on each capacitor plate. The potential difference across the capacitor is then $V_C = q/C$ (Section 26.2), with the polarity shown in Figure 32.1. Beginning at the switch and going around the circuit in the positive sense, we use Kirchhoff's loop rule (Section 27.7) to write

$$V_0 - \frac{q}{C} - iR = 0. \tag{32.1}$$

This equation can be solved to find q as a function of time. But in studying circuits we are more often interested in finding the current i as a function of time. We use the definition $i \equiv dq/dt$ to eliminate q from Equation 32.1. To do this, we note that, if an expression has a constant value (zero included), its time derivative is equal to zero. That is, we can write

$$\frac{d}{dt}\left(V_0 - \frac{q}{C} - iR\right) = 0.$$

Because V_0 is constant, the result of carrying out the differentiation is

$$-\frac{1}{C}i - R\frac{di}{dt} = 0,$$

or

$$\frac{di}{dt} = -\frac{1}{RC}i.$$

Collecting all the terms in i, we get

$$\frac{di}{i} = -\frac{1}{RC}\,dt. \tag{32.2}$$

We have thus separated the dependent variable i from the independent variable t. We can now solve for i by integrating both sides of the equation. The integration begins at time 0. At that instant, $q = 0$ and $V_C = q/C = 0$. It follows from Equation 32.1 that the current has the initial value $i_0 = V_0/R$. The integration ends at a particular later time t, when the current has the particular value i. Thus we have

$$\int_{V_0/R}^{i} \frac{di}{i} = -\frac{1}{RC}\int_0^t dt.$$

Carrying out the integrations, we obtain

$$\ln i \bigg|_{V_0/R}^{i} = -\frac{t}{RC}.$$

Remember that $\ln a - \ln b = \ln(a/b)$. Evaluating the left side of the equation yields the result

$$\ln \frac{i}{V_0/R} = -\frac{t}{RC}. \tag{32.3a}$$

We can rewrite this equation in a form that is often more convenient. We exponentiate both sides to obtain

$$i = \frac{V_0}{R}\,e^{-t/RC}. \tag{32.3b}$$

The current i given by Equation 32.3b is plotted as a function of time in Figure 32.2. The curve is called an **exponential decay curve**.

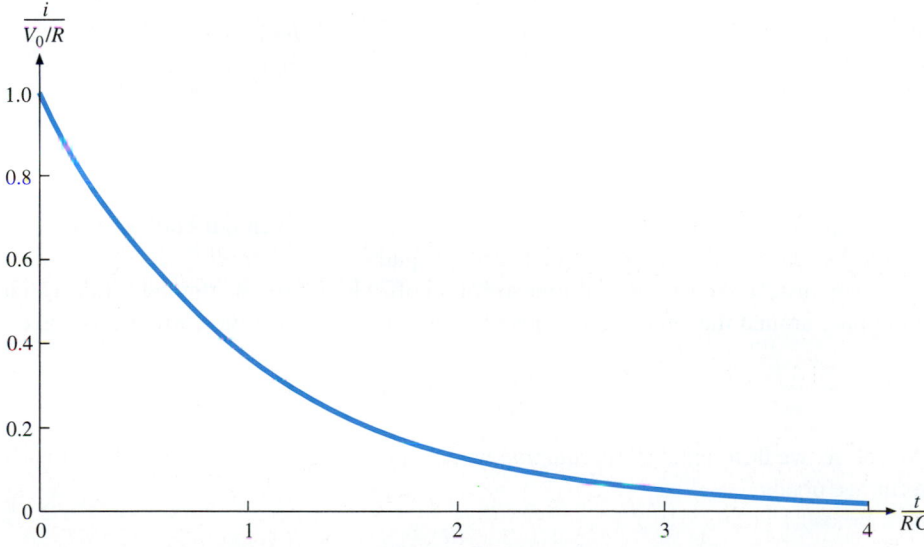

FIGURE 32.2 The dependence of current on time for the RC circuit of Figure 32.1, as the capacitor charge increases from an initially discharged state. The vertical axis gives the current as a fraction i/i_0 of its initial value $i_0 \equiv V_0/R$. The horizontal axis expresses time in multiples of the characteristic time constant $\tau_C = RC$.

During any time interval $t_f - t_i = RC$, the value of the right side of Equation 32.3b is reduced by the factor $1/e \simeq 0.37$. This time interval, which is characteristic of the circuit, is called the **time constant** τ_C:

$$\tau_C \equiv RC. \tag{32.4}$$

We can express Equations 32.3 in terms of the time constant:

$$\ln \frac{i}{V_0/R} = -\frac{t}{\tau_C} \quad \text{and} \quad i = \frac{V_0}{R}\,e^{-t/\tau_C}. \tag{32.5a,b}$$

EXAMPLE 32.1

Consider a series RC circuit like that of Figure 32.1, with $V_0 = 12.0$ V, $R = 220$ kΩ, and $C = 25.5$ μF. **(a)** What is the value of the initial current i_0? **(b)** What is the time constant τ_C of the system? **(c)** At what time $t_{1/2}$ after the capacitor begins to charge has the current decreased to one-half its initial value? **(d)** What is the charge on the capacitor at $t_{1/2}$? What is the voltage V_C across the capacitor at this time?

SOLUTION:
(a) Find i_0.

For the initial current, you have

$$i_0 = \frac{V_0}{R} = \frac{12.0 \text{ V}}{2.20 \times 10^5 \text{ }\Omega} = 5.45 \times 10^{-5} \text{ A} = 54.5 \text{ μA}.$$

(b) Evaluate τ_C.

You use Equation 32.4 to write

$$\tau_C = 2.20 \times 10^5 \text{ }\Omega \times 25.5 \times 10^{-6} \text{ F} = 5.61 \text{ s}.$$

(c) Find the time $t_{1/2}$ at which $i = \frac{1}{2}i_0$.

Because $V_0/R = i_0$, you can rewrite Equation 32.5a in the form

$$\ln \frac{i}{i_0} = -\frac{t}{\tau_C}.$$

At $t = t_{1/2}$, you have $i/i_0 = \frac{1}{2}$ by definition. So you write

$$t_{1/2} = -5.61 \text{ s} \times (-0.693) = 3.89 \text{ s}.$$

(d) Find q and V_C at time $t_{1/2}$.

One way to obtain the charge q is to begin with the definition $i \equiv dq/dt$, recast in the form $dq = i \, dt$. You integrate from $t = 0$ to $t = t_{1/2}$, using the value of i given by Equation 32.5b to obtain

$$q = \int_0^{t_{1/2}} i \, dt = i_0 \int_0^{t_{1/2}} e^{-t/\tau_C} \, dt.$$

You carry out the integration, which yields

$$q = i_0 \left[-\tau_C e^{-t/\tau_C} \right]_0^{t_{1/2}} = i_0 \tau_C (1 - e^{-t_{1/2}/\tau_C}).$$

You insert the numerical values to find the charge:

$$q = 5.45 \times 10^{-5} \text{ A} \times 5.61 \text{ s} \times (1 - e^{-(3.89 \text{ s})/(5.61 \text{ s})})$$
$$= 1.53 \times 10^{-4} \text{ C}.$$

The corresponding voltage across the capacitor is

$$V_C = \frac{q}{C} = \frac{1.53 \times 10^{-4} \text{ C}}{25.5 \times 10^{-6} \text{ F}} = 6.00 \text{ V}.$$

Can you see why the voltage across the capacitor is one-half the battery voltage just when the current is one-half the initial current? If not, refer back to this example after you have read through Example 32.2.

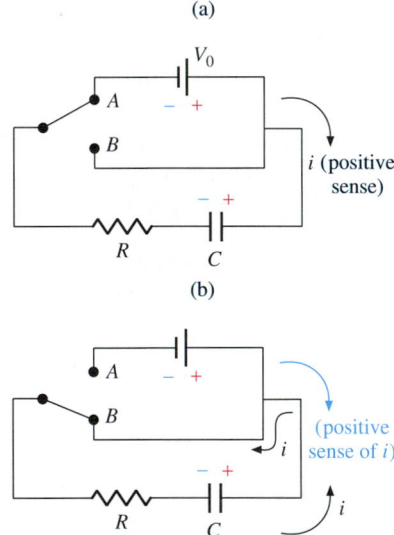

FIGURE 32.3 The RC circuit of Figure 32.1, modified to permit removal of the source of emf from the circuit. (a) With the switch in position A, the capacitor is fully charged. As in Figure 32.1, the sense of the current is taken to be positive. (b) At $t = 0$, the switch is moved to position B, and the capacitor discharges through the resistor. The sense of the current is negative.

Discharging the Capacitor

The circuit of Figure 32.3 is a slight modification of the one in Figure 32.1. With the switch in position A, the capacitor is charged exactly as before, as shown in Figure 32.3a. As in Figure 32.1, the sense of the charging current is taken to be positive. When the capacitor has been fully charged so that the potential across the plates is V_0, the switch is moved to position B, as shown in Figure 32.3b. We take the instant of switching to be $t = 0$. With the switch in position B, the source of emf is removed from the circuit, and the capacitor discharges through the resistor. Given our choice of positive sense, the sense of the discharging current is negative.

At any instant, the circuit conforms to Kirchhoff's loop rule. Beginning at the switch and going around the circuit in the positive sense (now against the current), we have

$$-\frac{q}{C} + iR = 0.$$

As before, we differentiate this equation with respect to time. Evaluating d/dt for each term, we obtain

$$-\frac{1}{C} i + R \frac{di}{dt} = 0,$$

or

$$\frac{di}{dt} = \frac{1}{RC} i.$$

As before, we rearrange to separate the variables:

$$\frac{di}{i} = \frac{1}{RC} dt. \tag{32.6}$$

Compare with Equation 32.2; the equations for charging and discharging differ only by a sign.

To integrate Equation 32.6, we must choose the proper limits. At time zero, the current is $-V_0/R$. (The sign is dictated by the negative sense of the current.) At a later particular time t, the current has the particular value i. So we have

$$\int_{-V_0/R}^{i} \frac{di}{i} = \frac{1}{RC}\int_0^t dt.$$

We evaluate the integral, which gives

$$\ln\frac{i}{-V_0/R} = \frac{t}{RC}. \tag{32.7a}$$

As with Equation 32.3a, we exponentiate both sides to obtain the alternative form

$$i = -\frac{V_0}{R}e^{-t/RC}. \tag{32.7b}$$

Except for the sense of the current, denoted by the negative value of i, the result is identical with that of Equation 32.3b. Figure 32.2 is thus a graph of the magnitude of both the charging current and the discharging current. The negative sign in Equation 32.7b makes explicit the fact that the sense of the discharging current is opposite that of the charging current of Equation 32.3b. It may seem to you that this care with sign conventions is unnecessary because in both cases the senses appear to be obvious. However, the convention will soon be very helpful in dealing with more complicated situations.

Equations 32.7a and 32.7b can also be written in terms of τ_C:

$$\ln\frac{i}{-V_0/R} = \frac{t}{\tau_C} \quad \text{and} \quad i = -\frac{V_0}{R}e^{-t/\tau_C}. \tag{32.8a,b}$$

EXAMPLE **32.2**

In both the charging process depicted in Figure 32.1 and the discharging process depicted in Figure 32.3b, the voltage V_C across the capacitor varies with time. (a) Find $V_C(t)$ for both cases. (b) Using the numerical values given in Example 32.1, find the voltage across the capacitor at $t = \tau_C$ when the capacitor is charging and when it is discharging.

SOLUTION:
(a) Find $V_C(t)$ for the charging and discharging processes.

There are several ways to carry out the necessary calculations. The easiest is to begin with Kirchhoff's loop rule and insert into it the value of $i(t)$ given by Equation 32.3b or Equation 32.8b. Then you can solve the resulting equation for $V_C(t)$.

Figure 32.4a shows the charging case, electrically identical with Figure 32.1. At any instant, Kirchhoff's loop rule gives

$$V_0 - V_C - iR = 0;$$

compare with Equation 32.1. When you substitute for i the value given by Equation 32.3a, the equation becomes

$$V_0 - V_C - V_0e^{-t/RC} = 0.$$

You can solve this immediately for V_C:

$$V_C = V_0(1 - e^{-t/RC}). \tag{32.9}$$

Figure 32.4b shows the discharging case, electrically identical with Figure 32.3b. At any instant, Kirchhoff's loop rule gives

$$-V_C - iR = 0.$$

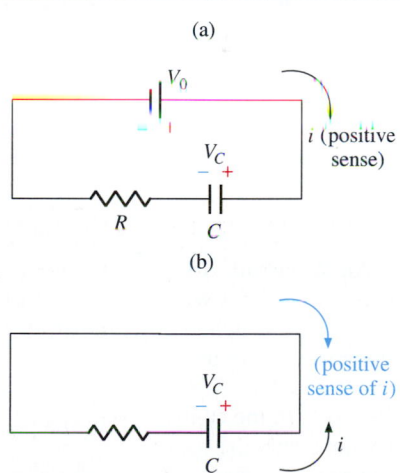

FIGURE 32.4 (a) The system of Figure 32.1, with the capacitor charging. (b) The system of Figure 32.3b, with the capacitor discharging.

(Remember that i is a negative quantity here, consistent with the minus sign in Equation 32.7b.) Inserting the value of i given by Equation 32.7b and solving for V_C, you obtain

$$V_C = V_0e^{-t/RC}. \tag{32.10}$$

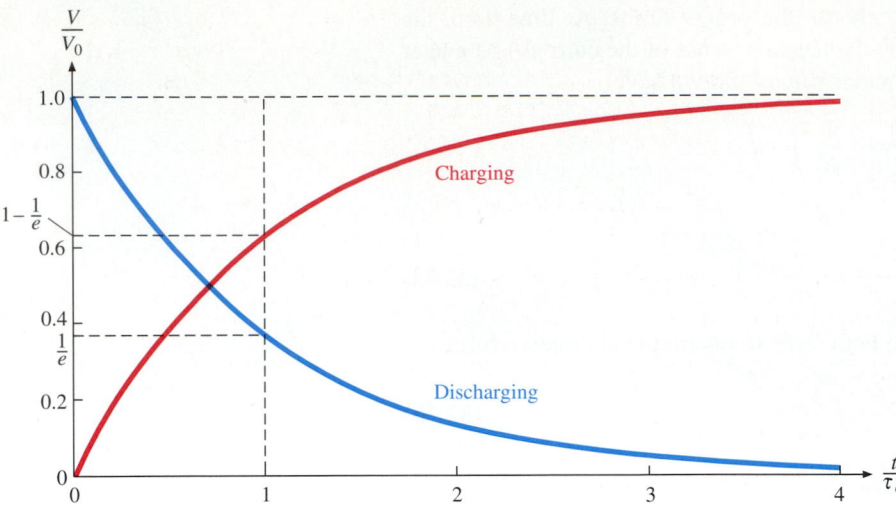

Figure 32.5 is a plot of the values of V_C given by Equations 32.9 and 32.10 as functions of time.

(b) Find V_C at $t = \tau_C$ for the charging and discharging cases.

When $t = \tau_C = RC$, the exponent in both Equation 32.9 and Equation 32.10 is -1. The only numerical value you need for solving the equations is $V_0 = 12.0$ V. For the charging capac-

itor, Equation 32.9 gives

$$V_C = V_0(1 - 1/e) = 12.0 \text{ V} \times 0.632 = 7.59 \text{ V}.$$

For the discharging capacitor, Equation 32.10 gives

$$V_C = \frac{V_0}{e} = \frac{12.0 \text{ V}}{e} = 4.41 \text{ V}.$$

The *RL* Circuit

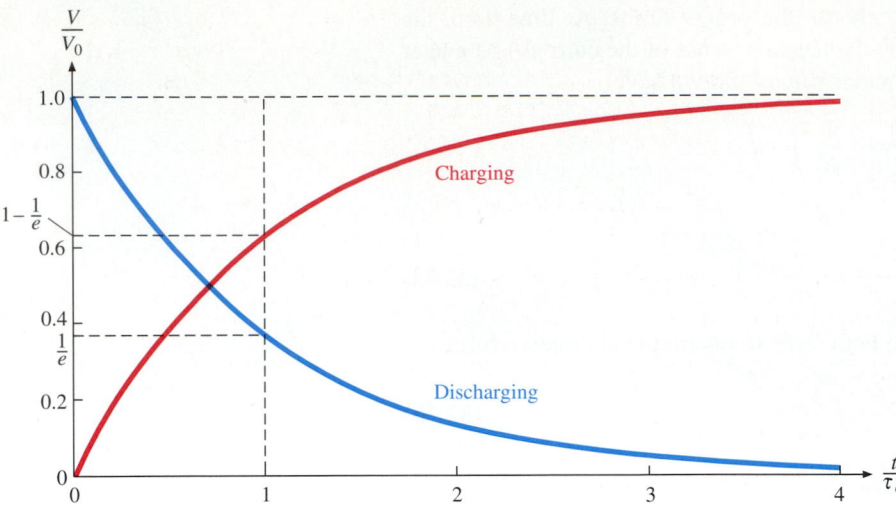

FIGURE 32.6 An *RL* circuit. A resistor R is connected in series with an inductor L. With the switch in position A, the complete circuit includes the source of emf V_0. With the switch in position B, the complete circuit contains only the resistor and the inductor.

Figure 32.6 shows a **series *RL* circuit**. In either position of the switch, the resistor and the inductor are parts of a complete circuit. With the switch in position A, a battery of fixed voltage V_0 is included in the circuit.

Suppose the switch has been in position A for a long time, so the current in the circuit has a steady value i_0. According to Equation 31.8, $V = -L\,di/dt$, the back emf produced by the inductor must be zero because di/dt is zero. The *forward* emf

$$V_L = L\frac{di}{dt} \tag{32.11}$$

required to drive charge through the inductor *against* the back emf is zero as well.* The sense of the current is clockwise, and we again take the clockwise sense to be positive. Using Kirchhoff's loop rule, we have

$$V_0 - V_L - iR = 0. \tag{32.12}$$

In the present case, this simplifies to

$$V_0 - iR = 0.$$

So the steady-state current has the simple Ohm's-law value

$$i_0 = \frac{V_0}{R}. \tag{32.13}$$

---

*It is important to bear in mind here a point made in the second paragraph of Section 31.3. When you apply Kirchhoff's loop rule to a circuit containing an inductor, you must decide whether to consider the inductor a source of emf $V = -L\,di/dt$ or a circuit element through which an external source of emf is driving a current by imposing across the inductor a voltage $V_L = L\,di/dt$. In this chapter, as in Section 31.3, we adopt the latter convention. To make this choice clear, we use V to denote the emf induced in the inductor and $V_L = -V$ to denote the voltage required to drive charge through the inductor.

Now, suppose that at time $t = 0$ the switch is suddenly moved to position B. The battery is thus excluded from the circuit. But the current continues for some time in the same sense as before, because the inductor tends to oppose change in the current in accordance with Lenz's law. The energy necessary to drive charge through the resistor now comes from the inductor, whose magnetic flux collapses as the energy is transferred to the resistor, where it is dissipated as Joule heat.

The removal of the battery from the circuit is represented mathematically by removing the term V_0 from Equation 32.12. Kirchhoff's loop rule now becomes

$$-V_L - iR = -L\frac{di}{dt} - iR = 0.$$

A little rearrangement gives us

$$\frac{di}{i} = -\frac{R}{L}\,dt. \tag{32.14}$$

This equation is *mathematically* identical with Equation 32.2,

$$\frac{di}{i} = -\frac{1}{RC}\,dt,$$

which describes the current in an RC circuit as the capacitor charges. The initial conditions also are the same; in both cases, $i_0 = V_0/R$. Thus we need not solve the equation from scratch (though we could). Instead, we write down the two forms of the solution of Equation 32.2, which are Equations 32.3a and 32.3b:

$$\ln\frac{i}{V_0/R} = -\frac{t}{RC} \quad \text{and} \quad i = \frac{V_0}{R}\,e^{-t/RC}.$$

We replace RC with L/R and immediately obtain the solutions for the inductor:

$$\ln\frac{i}{V_0/R} = -\frac{R}{L}\,t \quad \text{and} \quad i = \frac{V_0}{R}\,e^{-(R/L)t}. \tag{32.15a,b}$$

As in the charging-capacitor case, the current diminishes from its original value i_0 in exponential fashion. The **time constant** for the RL circuit is defined to be

$$\tau_L \equiv \frac{L}{R}. \tag{32.16}$$

In terms of this time constant, we can write Equations 32.15a and 32.15b in the forms

$$\ln\frac{i}{V_0/R} = -\frac{t}{\tau_L} \quad \text{and} \quad i = \frac{V_0}{R}\,e^{-t/\tau_L}. \tag{32.17a,b}$$

Equation 32.17b is plotted in blue in Figure 32.7.

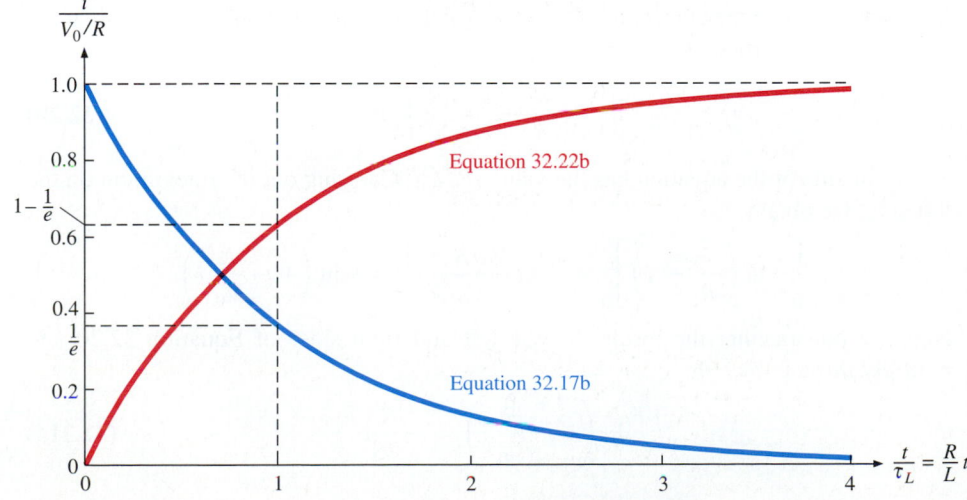

FIGURE 32.7 The blue curve is a plot of Equation 32.17b, which gives the current as a function of time for the circuit of Figure 32.6 after $t = 0$, when the switch is moved to position B. The current is expressed as a fraction of the initial current V_0/R; time is expressed as a multiple of the time constant τ_L. The red curve is a plot of Equation 32.22b, which gives the current as a function of time for the circuit of Figure 32.6 after $t = 0$, when the switch is moved to position A.

Increasing Current

Suppose now that there is no current in the circuit of Figure 32.6 when, at $t = 0$, the switch is suddenly moved to position A. The battery will begin to drive current through the circuit in the positive sense, and the current must increase from its initial zero value. Thus the value of di/dt is positive.

Consider an instant a very short time after t_0, when the current is still negligibly small. But, though small, the current is increasing—not necessarily slowly. So the inductor acts as a source of emf V whose sense opposes the battery voltage.

Under these circumstances, the iR drop through the resistor is negligibly small and Kirchhoff's loop rule (Equation 32.12) becomes $V_0 - V_L - R \cdot 0 = 0$. That is, the entire battery voltage is initially across the inductor. Again using Equation 32.11 to express V_L in terms of the inductance of the coil, we write

$$V_0 - L \left[\frac{di}{dt} \right]_{t \approx 0} = 0.$$

The initial value of di/dt is thus

$$\left[\frac{di}{dt} \right]_{t \approx 0} = \frac{V_0}{L}. \tag{32.18}$$

This value is positive, in agreement with our assertion that the current is increasing. As i increases, the iR drop across the resistor increases and the voltage across the inductor decreases. We can express this quantitatively by using Kirchhoff's loop rule again to write, for *any* time $t > 0$,

$$V_0 - L \frac{di}{dt} - iR = 0.$$

We wish to rearrange this equation into a form like that of Equation 32.14. Specifically, we wish to separate terms containing the dependent variable i from terms containing the independent variable t. We first solve for di/dt:

$$\frac{di}{dt} = \frac{V_0}{L} - i \frac{R}{L},$$

or

$$\frac{di}{dt} = \frac{R}{L} \left(\frac{V_0}{R} - i \right).$$

Then we multiply through by dt and divide through by the factor in parentheses to obtain the desired form

$$\frac{di}{V_0/R - i} = \frac{R}{L} dt. \tag{32.19}$$

To find the current i at any time $t > 0$, we integrate both sides of Equation 32.19 from time 0, when the current has the initial value 0, to any particular time t, when the current has the particular value i:

$$\int_0^i \frac{di}{V_0/R - i} = \frac{R}{L} \int_0^t dt. \tag{32.20}$$

The right side of the equation has the value $(R/L)t$. Carrying out the integration on the left side, we obtain

$$\left[-\ln \left(\frac{V_0}{R} - i \right) \right]_0^i = -\ln \frac{V_0/R - i}{V_0/R} = -\ln \left(1 - \frac{R}{V_0} i \right).$$

Now we put together the results for the left and right sides of Equation 32.20 and multiply through by -1:

$$\ln \left(1 - \frac{R}{V_0} i \right) = -\frac{R}{L} t. \tag{32.21a}$$

This form is often inconvenient because it does not give an explicit value for i. Exponentiating both sides of the equation, we have

$$1 - \frac{R}{V_0} i = e^{-(R/L)t}.$$

We now solve for i, obtaining the desired expression for i as a function of t:

$$i = \frac{V_0}{R}(1 - e^{-(R/L)t}). \qquad \text{(32.21b)}$$

Reexpressed in terms of the time constant τ_L, Equations 32.21a and 32.21b become

$$\ln\left(1 - \frac{R}{V_0} i\right) = -\frac{t}{\tau_L} \quad \text{and} \quad i = \frac{V_0}{R}(1 - e^{-t/\tau_L}). \qquad \text{(32.22a,b)}$$

Compare with Equations 32.8a and 32.8b, the corresponding expressions for the RC circuit.

The asymptotic buildup of current expressed by Equation 32.22b is plotted in red in Figure 32.7.

EXAMPLE 32.3

As the current builds up in the circuit of Figure 32.6 when the switch is in position A (in accordance with Equation 32.22b), electrical work must be performed on the inductor and on the resistor. In the inductor, the energy U stored in the magnetic field increases as i increases. In the resistor, there is Joule heating. **(a)** What is the ratio P_L/P_R of the rate at which energy is stored in the inductor to the rate at which work is done in the resistor? **(b)** Express the ratio P_L/P_R as a function of time. **(c)** If the time constant of the circuit is τ_L, what is the time t at which $P_L = P_R$?

SOLUTION:
(a) Evaluate P_L/P_R.

According to Equation 31.13, the energy stored in an inductor is $U = \frac{1}{2}Li^2$. The power required to change U is $P = dU/dt$, so at any instant you have

$$P_L = \frac{d}{dt}\left(\frac{1}{2}Li^2\right) = Li\frac{di}{dt}. \qquad \text{(32.23)}$$

For the resistor, Joule's law gives you Equation 27.19b:

$$P_R = i^2 R.$$

So the ratio is

$$\frac{P_L}{P_R} = \frac{L}{R}\frac{1}{i}\frac{di}{dt} = \tau_L \frac{1}{i}\frac{di}{dt}. \qquad \text{(32.24)}$$

(b) Express P_L/P_R as a function of time.

The value of i is given by Equation 32.22b. To find di/dt, you differentiate that equation with respect to time:

$$\frac{di}{dt} = \frac{d}{dt}\left[\frac{V_0}{R}(1 - e^{-t/\tau_L})\right] = \frac{V_0}{R\tau_L}e^{-t/\tau_L}. \qquad \text{(32.25)}$$

Inserting this value of di/dt and the value of i into Equation 32.24, you obtain

$$\frac{P_L}{P_R} = \tau_L \frac{\dfrac{V_0}{R\tau_L}e^{-t/\tau_L}}{\dfrac{V_0}{R}(1 - e^{-t/\tau_L})} = \frac{e^{-t/\tau_L}}{1 - e^{-t/\tau_L}}.$$

To simplify this equation still further, you can multiply numerator and denominator by the factor e^{t/τ_L}:

$$\frac{P_L}{P_R} = \frac{1}{e^{t/\tau_L} - 1}. \qquad \text{(32.26)}$$

When $t \simeq 0$, this ratio is quite large, reflecting the large initial value of di/dt and the small initial value of i. When t is large, the current approaches its asymptotic value and changes very slowly. As a result, P_L vanishes and the ratio goes to zero.

(c) When is $P_R = P_L$?

You set the right side of Equation 32.26 equal to 1:

$$\frac{1}{e^{t/\tau_L} - 1} = 1.$$

Simple manipulation gives you

$$e^{t/\tau_L} = 2.$$

You take the natural logarithm of both sides of this equation to obtain

$$\frac{t}{\tau_L} = \ln 2,$$

or

$$t = 0.693\tau_L.$$

Compare this result with that of part **c** of Example 32.1.

The *LC* Circuit

Figure 32.8 shows the simplest possible **LC circuit**, consisting of an ideal capacitor of capacitance C and an ideal inductor of inductance L connected to form a circuit. Suppose that, with the switch open, the capacitor initially holds a charge q_0 on the right-hand plate and a charge $-q_0$ on the left-hand plate. The potential difference between the plates is $V_0 = q_0/C$. We define this potential difference to be positive and thus specify the clockwise sense around the circuit to be the positive sense. How will current flow when the circuit is completed by closing the switch? You will see that the current oscillates back and forth through the circuit.

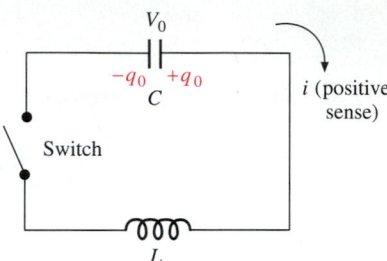

FIGURE 32.8 An *LC* circuit. With the switch open, the capacitor is initially charged to a potential difference V_0 having the polarity shown. At $t = 0$, the switch is closed and the capacitor begins to discharge, driving a current i through the inductor. The initial sense of the current is positive.

When the switch is closed at $t = 0$, a conducting path exists between the capacitor plates. The capacitor begins to discharge, and a current i begins to pass through the inductor. At first the current increases, and so di/dt is positive. In accordance with Lenz's law, the inductor produces an emf whose sense opposes the increase in current.

Because di/dt is positive, i increases with the passage of time. But the potential difference across the capacitor decreases as the charge on the plates drains off. Consequently, the rate of change di/dt decreases, falling to zero when the capacitor is fully discharged. At that instant, the current reaches its maximum value.

There is no longer a potential difference across the capacitor to drive charge through the circuit. But the inductor, in accordance with Lenz's law, produces an emf whose sense opposes a decrease in i. This emf now propels charge in the positive sense as the current decreases. That is, i is still positive, but di/dt is now negative.

Because the current persists after the capacitor has discharged, the capacitor now charges with polarity opposite its original polarity. The growing potential difference opposes the current, which falls to zero when the capacitor is fully charged. At this instant, the value of di/dt reaches a minimum (negative) value. This ends the first half of the oscillation cycle.

The second half of the cycle begins as the capacitor discharges by driving charge in the negative sense. The second half of the cycle is just the reverse of the first half. In the absence of losses in the circuit, the cycle repeats indefinitely; *the system behaves in oscillatory fashion.*

The oscillation is analogous to that of a mass-spring system. With the spring initially distended and released, the mass accelerates as the spring relaxes. When the spring is fully relaxed, the mass is in its equilibrium position. But, owing to its inertia, it moves past that position. Thus the spring is distended in the direction opposite its original distension. The half-cycle ends with the mass instantaneously at rest and the spring fully distended. The stretched spring corresponds to the charged capacitor, the mass to the inductor, and the velocity of the mass to the (signed) current.

Let us now consider the *LC* system quantitatively. As with the *RC* and *RL* circuits, our strategy is to apply Kirchhoff's loop rule. In the absence of a well-defined source of emf, however, assignment of the proper signs to the terms in the equation can be confusing. We get around this difficulty by using a trick. In Figure 32.9, we have added a battery of emf V_0 to the system. For concreteness, consider an instant when the sense of the current is positive and the relative polarities of the terminals of the circuit components are as shown. Beginning at the switch and going around the circuit in the sense

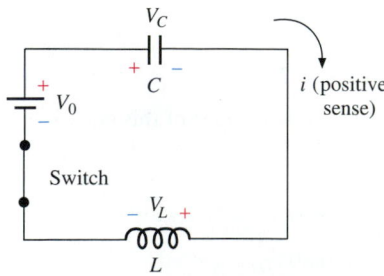

FIGURE 32.9 The *LC* circuit of Figure 32.8 with a battery of emf V_0 added.

of the current, we use Kirchhoff's loop rule to write

$$V_0 - V_C - V_L = 0.$$

Once we have written this equation, we remove the battery from the system; that is, we set $V_0 = 0$. Kirchhoff's loop rule now has the form we need in order to apply it to the circuit of Figure 32.8:

$$-V_C - V_L = 0.* \tag{32.27a}$$

We are now ready to insert the values $V_C = q/C$ and $V_L = L\,di/dt$ into Equation 32.27a:

$$-\frac{q}{C} - L\frac{di}{dt} = 0. \tag{32.27b}$$

We eliminate i from this equation by using the definition $i \equiv dq/dt$ to obtain

$$\frac{1}{C}q + L\frac{d^2q}{dt^2} = 0.$$

We now subtract the first term from both sides of the equation and divide through by L:

$$\frac{d^2q}{dt^2} = -\frac{1}{LC}q. \tag{32.28}$$

Does this equation look familiar? Compare it with the *harmonic oscillator equation* for a mass-spring system (Equation 6.32):

$$\frac{d^2x}{dt^2} = -\frac{k}{m}x.$$

The two equations describe quite different things. Equation 6.32 describes the motion of a mass m driven by the force exerted by a spring of force constant k. Equation 32.28 describes the passage of electric charge q through an inductor L driven by the potential difference across a capacitor of capacitance C. But *the two equations are mathematically identical*. That is, Equation 32.28 is just as much a harmonic-oscillator equation as Equation 6.32 is. We thus corroborate quantitatively the qualitative description with which this section began: *The passage of charge through the LC circuit is oscillatory.*

We could solve Equation 32.28 from scratch; this is the subject of Problem 30.43. But we need not do so. Instead, we write down the solutions of the mass-spring oscillator equation and make the substitutions

$$x \to q \quad \text{and} \quad \frac{k}{m} \to \frac{1}{LC}.$$

What is the angular frequency ω of the LC circuit? We begin with the mechanical oscillator, for which Equation 6.33 gives

$$\omega^2 = \frac{k}{m}.$$

We can immediately infer the angular frequency of the oscillatory motion of charge in the LC circuit. Substitution of $1/LC$ for k/m yields

$$\omega_0 = \frac{1}{\sqrt{LC}}. \tag{32.29a}$$

---

*Now you can see the point in the trick. By inserting the source of emf into the circuit, we make it unambiguous that we are treating the capacitor and the inductor as passive circuit elements, across which there is a voltage "drop." This is just what we did in studying the RC and RL circuits, where there was less possibility of confusion. But remember that whether the "drop" is in fact a decrease (as at the instant for which the signs of Figure 32.9 are valid) or an increase in the energy of a test charge pushed through the circuit depends on the signs of V_C and V_L, which vary with time.

The quantity ω_0 is called the **natural angular frequency** of the LC circuit.

As an alternative to the angular frequency, we can use the frequency ν to characterize the oscillation. From the relation $\nu = \omega/2\pi$ (Equation 6.19), we have

$$\nu_0 = \frac{1}{2\pi\sqrt{LC}}. \tag{32.29b}$$

The frequency ν_0 is called the **natural frequency** of the circuit. The same name is often used loosely for the corresponding value of ω_0; you should never use this latter terminology unless there is no possibility of confusion.

Proceeding with our solution by substitution, we use Equation 6.20 to write the position of the oscillating mass as a function of time:

$$x = A \cos(\omega_0 t + \delta),$$

where A is the amplitude and δ is a phase constant determined by the initial conditions. Making the necessary substitutions yields

$$q = q_0 \cos(\omega_0 t + \delta). \tag{32.30}$$

Just as x oscillates sinusoidally between the limits $\pm A$, the charge q on the right-hand plate of the capacitor oscillates between the limits $\pm q_0$. (At any instant, the charge on the left-hand plate is equal to and opposite that on the right-hand plate; hence the value of that charge oscillates between the same limits but with a phase difference of $180°$ between the charges on the two plates.)

We can immediately deduce the behavior of the potential difference V_C. Because $q = CV_C$ and $q_0 = CV_0$, we have from Equation 32.30

$$V_C = V_0 \cos(\omega_0 t + \delta). \tag{32.31}$$

Current in the LC Circuit

To find the current i in the LC circuit as a function of time, we again use the relation $i \equiv dq/dt$. Differentiating both sides of Equation 32.30 gives us

$$i = -\omega_0 q_0 \sin(\omega_0 t + \delta). \tag{32.32}$$

Compare this result with Equation 6.21, which gives the velocity v of the oscillating mass as

$$v = -\omega A \sin(\omega t + \delta).$$

Voltage across the Inductor

At any instant, the voltage across the inductor is the forward emf given by Equation 32.11,

$$V_L = L\frac{di}{dt}.$$

We evaluate V_L by differentiating Equation 32.32 and multiplying the result by L:

$$V_L = L[-\omega_0^2 q_0 \cos(\omega_0 t + \delta)]. \tag{32.33}$$

The constant coefficient $-L\omega_0^2 q_0$ can be simplified by using Equation 32.29a. We have $-L\omega_0^2 q_0 = -L(1/LC)(CV_0) = -V_0$. Thus V_L can be written

$$V_L = -V_0 \cos(\omega_0 t + \delta). \tag{32.34}$$

We could have obtained the same result by combining Equation 32.27a, Kirchhoff's loop rule for the circuit, with the value of V_C given by Equation 32.31.

The quantity V_L/L is analogous to the acceleration $a = F/m$ of the mass in the mass-spring system. From Equations 32.33 and 32.29a, we have

$$\frac{V_L}{L} = -\omega_0^2 q_0 \cos(\omega_0 t + \delta) = -\frac{1}{LC}q. \tag{32.35}$$

The acceleration a is given by Equation 6.22 and Equation 6.31,

$$a = -\omega^2 A \cos(\omega t + \delta) = -\frac{k}{m} x.$$

Table 32.1 is a complete list of the analogies between the LC harmonic oscillator and the mass-spring harmonic oscillator. You should study it carefully, and make sure you understand each analogy.

TABLE 32.1 Analogies between the LC Oscillator and the Mass-Spring Oscillator

LC Oscillator	Equation Reference	Mass-Spring Oscillator	Equation Reference
$\dfrac{d^2q}{dt^2} = -\dfrac{1}{LC} q$	32.28	$\dfrac{d^2x}{dt^2} = -\dfrac{k}{m} x$	6.31
$q = q_0 \cos(\omega_0 t + \delta)$	32.30	$x = A \cos(\omega t + \delta)$	6.20
$i = \dfrac{dq}{dt} = -\omega_0 q_0 \sin(\omega_0 t + \delta)$	32.32	$v = \dfrac{dx}{dt} = -\omega A \sin(\omega t + \delta)$	6.21
$\dfrac{V_L}{L} = \dfrac{d^2q}{dt^2} = -\omega_0^2 q_0 \cos(\omega t + \delta)$	32.35	$a = \dfrac{d^2x}{dt^2} = -\omega^2 A \cos(\omega t + \delta)$	6.22
$\dfrac{1}{C}$		k	
L		m	
$\omega_0 = \dfrac{1}{\sqrt{LC}}$	32.29a	$\omega = \sqrt{\dfrac{k}{m}}$	6.34a
$U_C = \dfrac{1}{2}\dfrac{1}{C} q^2$	26.26a	$U = \frac{1}{2}kx^2$	8.23
$U_L = \frac{1}{2}Li^2$	31.13	$K = \frac{1}{2}mv^2$	7.16

Phase Relations among V_C, i, and V_L

As Figure 32.10 shows, we get a surprise when we compare the current i given by Equation 32.32 with the voltages V_C and V_L given by Equations 32.31 and 32.34: The three equations are plotted in Figure 32.10. *The current is out of phase with both*

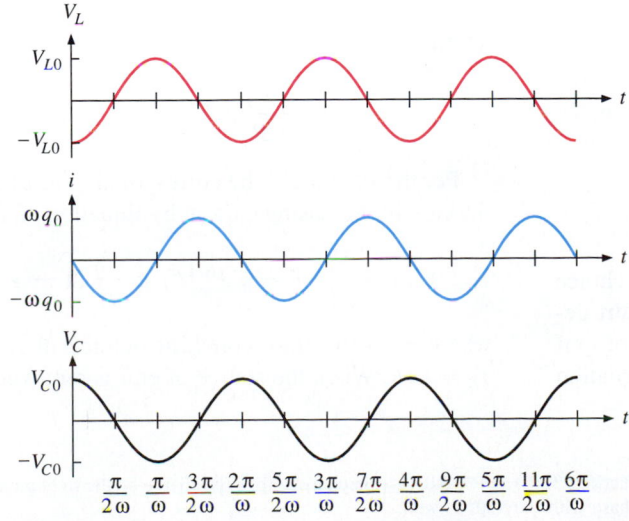

FIGURE 32.10 Plots of Equations 32.34, 32.32, and 32.31.

voltages by 90°, or one-quarter of a cycle.* The current reaches its extremal values $i = \pm \omega_0 q_0$ just when both voltages are zero; when the voltages attain their extremal values $\pm V_0$, the current is zero. Because the curve for i looks like the curve for V_L shifted to a later time, we say that the current *lags* the voltage across the inductor. Because the curve for i looks like the curve for V_C shifted to an earlier time, we say that the current *leads* the voltage across the capacitor. In contrast, the voltage V_R across a resistor and the current passing through it are always *in phase*; this is guaranteed by Ohm's law, $V_R = iR$. Suppose that a small resistor were inserted into the *LC* circuit of Figure 32.8. Which of the curves of Figure 32.10 would describe the voltage V_R across it?

EXAMPLE 32.4

In a series *LC* circuit like that of Figure 32.8, a 125-μF capacitor is initially charged to a voltage $V = 37.5$ V. The inductor has inductance 10.7 H. Find **(a)** the natural angular frequency and the natural frequency of the oscillations in the circuit; **(b)** the maximum current through the circuit.

SOLUTION:
(a) Find ω_0 and ν_0.
   For the angular frequency, you use Equation 32.29a to write

$$\omega_0 = \frac{1}{\sqrt{1.25 \times 10^{-4} \text{ F} \times 10.7 \text{ H}}} = 27.3 \text{ rad/s}.$$

The corresponding frequency is $\nu_0 = (27.3 \text{ rad/s})/2\pi = 4.35$ Hz.

(b) What is i_{max}?
   It follows from Equation 32.32 that the current achieves a maximum value $i_{max} = \omega_0 q_0$. As $q_0 = CV_0$, you write

$$i_{max} = \omega_0 CV_0$$
$$= 27.3 \text{ rad/s} \times 1.25 \times 10^{-4} \text{ F} \times 37.5 \text{ V}$$
$$= 0.128 \text{ A}.$$

Symbols Used in Chapter 32

C	capacitance		V_0	peak voltage
i	electric current		τ_C, τ_L	capacitive, inductive time constant
L	(self-)inductance		δ	phase constant determined by choice of zero of time
q	electric charge		ν	frequency
R	resistance		ω, ω_0	angular frequency, natural angular frequency
U	potential energy			

Summing Up

Kirchhoff's loop rule can be used to analyze the relation between voltages and current in series circuits even if the current is not constant. In an *RC* circuit, the current as the capacitor is charged is given by Equation 32.3b or 32.5b,

$$i = \frac{V_0}{R} e^{-t/RC} = \frac{V_0}{R} e^{-t/\tau_C}.$$

The polarity of the source of emf V_0 determines the choice of the positive sense for i, and τ_C is the **time constant** defined in Equation 32.4, $\tau_C \equiv RC$. When the source of emf is removed, the discharging current is given by Equation

32.7b or 32.8b,

$$i = -\frac{V_0}{R} e^{-t/RC} = -\frac{V_0}{R} e^{-t/\tau_C}.$$

For the *RL* circuit, the corresponding results are given for the case of increasing current by Equation 32.21b or 32.22b,

$$i = \frac{V_0}{R} (1 - e^{-(R/L)t}) = \frac{V_0}{R} (1 - e^{-t/\tau_L}),$$

where τ_L is the **time constant** defined in Equation 32.16, $\tau_L \equiv L/R$. When the source of emf is removed, the decreas-

---

*According to Equations 32.10 and 32.31, V_C is in phase with q, the charge on the right plate of the capacitor. Thus i is out of phase with q by 90° as well.

ing current is given by Equation 32.15b or 32.17b,

$$i = \frac{V_0}{R} e^{-(R/L)t} = \frac{V_0}{R} e^{-t/\tau_L}.$$

The behavior of the *LC* circuit is analogous to that of the oscillating mass-spring system. Kirchhoff's loop rule leads to Equation 32.28, a typical harmonic-oscillator equation:

$$\frac{d^2q}{dt^2} = -\frac{1}{LC} q.$$

The **natural angular frequency** of the oscillation is given by Equation 32.29a,

$$\omega_0 = \frac{1}{\sqrt{LC}}.$$

The voltage V_C across the capacitor, the current i, and the voltage V_L across the inductor are given by Equations 32.31,

32.32, and 32.34,

$$V_C = V_0 \cos(\omega_0 t + \delta), \quad i = -\omega_0 q_0 \sin(\omega_0 t + \delta),$$

and $$V_L = -V_0 \cos(\omega_0 t + \delta).$$

The current *lags* V_C by 90° and *leads* V_L by 90°. In contrast, the current through a resistor is always *in phase* with the voltage V_R across the resistor.

Queries and Problems for Chapter 32

QUERIES

32.1 *(2) It takes time.* Your friend explains the physical basis for the capacitive time constant τ_C as follows: As the capacitor discharges, the resistor restricts the passage of electrons. The greater the resistance, the fewer electrons can pass through the resistor per unit time. Consequently, it takes longer for the capacitor to discharge through a greater resistance. Increasing the capacitance increases the amount of charge that needs to flow through the resistor and likewise increases the time required. Is this explanation valid?

32.2 *(3) Of time and the circuit.* Construct an argument analogous to that of Query 32.1 for an *RL* circuit, and criticize it.

32.3 *(3) Induced variation.* In an *LR* circuit, decrease in the current results in an induced forward emf across the inductor. This extra induced emf in turn drives an extra current through the inductor, and the current is not the same in all parts of the circuit. Discuss this statement.

32.4 *(3) Wait a second!* Circuits are often protected from overload by fuses—devices containing wires of low-melting alloy that "blow" (melt and break the circuit) when their rated

current is exceeded. "Slo-blo" fuses are designed so that they withstand overload for a second or two but blow under an extended overload. Explain the advantage of using slo-blo fuses in inductive circuits.

32.5 *(4) Overshot.* An oscillating mass-spring system overshoots its equilibrium position because of the inertia of the mass and because v is not zero when $x = 0$. Using analogy, explain why the capacitor charge in an *LC* oscillator overshoots the discharged condition.

32.6 *(4) No pushing!* As Figure 32.10 shows, the current in an *LC* circuit has its extremal value just when the emf's across the inductor and the capacitor are zero. That is, nothing is driving the current at that instant. Explain how current can flow under these circumstances.

32.7 *(4) Going to extremes.* The current in an *LC* circuit can be represented by a sine curve $i(t)$. How can you conclude, based on inspection of the curve, that **(a)** V_L must be zero when the current is at one of its extremal values? **(b)** the capacitor holds maximum charge when $i = 0$?

PROBLEMS

GROUP A

32.1 *(2) So little time, I.* What is the time constant of an *RC* circuit if $C = 5.0$ pF and $R = 2.3$ MΩ?

32.2 *(2) So little time, II.* **(a)** If the time constant of an *RC* circuit is 37.5 ns, how long does it take for the current to

decrease to 25% of its initial value? **(b)** Does it make any difference whether the battery charges the capacitor through the resistor, as in Figure 32.3a, or the capacitor discharges through the resistor, as in Figure 32.3b? Explain your answer.

32.3 *(2) A little longer, I.* A 12-V battery is used to charge a 0.65-μF capacitor through a 7.2-kΩ resistor, as shown below. How long after the switch is moved to position *A* does the voltage across the capacitor reach **(a)** 3.2 V? **(b)** 8.5 V?

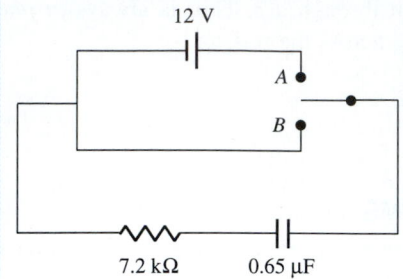

32.4 *(2) Time charges.* In the system of Problem 32.3, how long does it take for the charge on each capacitor plate to achieve the value $|q| = 2.1$ μC?

32.5 *(2) A little longer, II.* After the switch of Problem 32.3 has been in position *A* for a long time, it is moved to position *B*. How long afterward does the voltage across the resistor reach 2.1 V?

32.6 *(2) Charging time.* In Figure 32.1, let $R = 1.3$ kΩ and $V_0 = 15$ V. If the voltage across the capacitor reaches 7.2 V at $t = 2.1$ μs, find **(a)** τ_C and **(b)** C.

32.7 *(2) Measuring the leak.* Large resistors can be measured by allowing charge to leak off a known capacitor through the unknown resistor. You use a 9.0-V battery to charge a 10-μF capacitor. You disconnect the capacitor from the battery and allow the capacitor to discharge through a resistor R. If the voltage across the resistor is 6.0 V at a time 2.3 s after the discharge begins, what is the value of R?

32.8 *(2) On the skids.* A capacitor C is initially charged to a voltage V_0. **(a)** If the capacitor is discharged through a resistor R, express the voltage V_R across the resistor as a function of time. **(b)** What is the voltage V_C across the capacitor as a function of time? (Hint: Refer to Equation 32.3b.)

32.9 *(3) DC wasn't built in a second!* Beginning at $i = 0$, the current in an *RL* circuit requires 1.6 s to reach 27% of its final value. **(a)** What is the inductive time constant of the circuit? **(b)** How long does it take for the current to reach 99% of its final value?

32.10 *(3) Lack of drive?* When the current in an *RL* circuit decays, how many time constants does it take before the current falls to 10% of its initial value?

32.11 *(3) Rise and fall.* In the circuit of Figure 32.6, $V_0 = 150$ V, $R = 265$ Ω, and $L = 11.2$ H. **(a)** When the switch is moved to position *A*, how long does it take the current to reach the value $i = 0.422$ A? **(b)** After the switch has been in position *A* for some time, it is moved to position *B*. How long does it take for the current to decrease to $i' = 0.135$ A? **(c)** What is the time constant of the circuit?

32.12 *(3) A turn in time.* An inductor is in series with a 150-Ω resistor. If the time constant of the combination is 0.105 s, what is L?

32.13 *(3) Exponential relation.* A coil of resistance 12 Ω and inductance 2.5 H is connected to a 12-V battery. **(a)** At the instant when the current in the circuit is changing at 0.75 A/s, what is the value of the current? **(b)** What is the initial rate of change of the current?

32.14 *(3) Slippery slope.* The current in an *RL* circuit is decaying. At some time, its value is i. How long does the current take to decrease to $i/3$?

32.15 *(3) Winding up.* You connect a 6-V battery to a coil. The inductance of the coil is 60 μH, and its resistance is 3 Ω. Call the time when the connection is made $t = 0$. **(a)** What is the maximum value of di/dt? **(b)** At what time t does di/dt reach its maximum value? What is the value of i at this time? **(c)** When $L\,di/dt$ is equal to 2.0 V, what is i? At what time does this situation take place?

32.16 *(4) Design of the times, I.* You have a 1-mH inductor and a 1-μF capacitor. If you connect them to form an *LC* circuit, what will be the natural angular frequency ω_0?

32.17 *(4) Design of the times, II.* You want to build an *LC* circuit with natural frequency 1.00 MHz. If you plan to use a 6.83-nF capacitor, what should the inductance of the inductor be?

32.18 *(4) Not your grandfather's clock.* What is the oscillation period τ of an *LC* circuit if $L = 37.6$ mH and $C = 1.24$ μF?

32.19 *(4) Look mA, no ammeter!* You construct an *LC* circuit of negligible resistance, using a 21.5-mH inductor and a 73.2-nF capacitor. **(a)** What is the natural frequency of the system? You measure the voltage across the capacitor and find that the maximum value is 4.27 V. **(b)** What is the maximum value of the current? **(c)** How much energy is stored in the system?

32.20 *(G) Do not exceed.* A 0.45-μF capacitor is rated for a maximum 2000 V. If the capacitor is part of a circuit in which the current is $i = i_0 \sin 2\pi\nu t$, with $\nu = 250$ Hz, what is the greatest allowable value of the peak current i_0 in the circuit?

32.21 *(4) Mix and match.* You have an inductor of inductance L and two capacitors, each of capacitance C. What natural frequencies can you obtain by combining the elements in any way you choose?

32.22 *(4) Twins.* The two identical capacitors in the circuit illustrated below are initially charged until the voltage across them (between the upper plates) is 220 V. Then the switch is closed. Find **(a)** v_0 and **(b)** the maximum current that flows through the inductor.

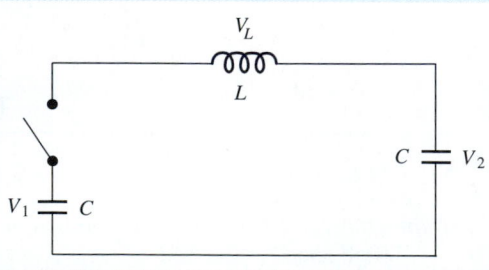

32.23 (2) *Leaking.* A capacitor C, initially charged to a voltage V_0, discharges through a resistor R. Show that the charge q remaining on one of the capacitor plates is

$$q = CV_0 e^{-t/RC}.$$

32.24 (2) *Getting a charge.* A battery of emf V_0 charges an initially uncharged capacitor C through a resistor R. Derive an expression for the charge q on the capacitor as a function of time.

32.25 (2) *Petering out.* A capacitor is discharging through a resistor. At $t = 2.17 \times 10^{-3}$ s, the voltage across the resistor is 235 V. At $t = 5.81 \times 10^{-3}$ s, the voltage is 106 V. What is the time constant of the circuit?

32.26 (2) *Electric clock, I.* In the figure below, a small neon lamp is connected across the capacitor in an RC circuit. Such lamps have the property that their resistance is very large until the voltage across the electrodes reaches a critical value V^*.

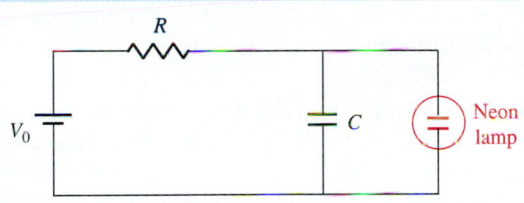

At that point, the neon gas in the lamp ionizes very quickly, and the resistance of the bulb suddenly drops to a value much less than R. The electrical energy required to ionize the atoms is reemitted as light when neon ions recombine with ionized electrons. The ionization-recombination process continues as long as charge is available to flow from one of the bulb electrodes to the other. That is, once the low-resistance condition is "triggered," it is stable until charge ceases to flow. This happens when the voltage across the lamp is nearly zero. Then the gas deionizes very quickly, and the lamp returns to its high-resistance condition. **(a)** If $V^* < V_0$, express the time interval between flashes of the lamp as a function of the ratio V^*/V_0. **(b)** Show that, if $V^* \ll V_0$, the time interval between flashes is approximately RCV^*/V_0. **(c)** A typical value of V^* is 70 V. Use reasonable values for V_0, R, and C to design a circuit that will produce one flash per second. **(d)** Suppose that $V^* \ll V_0$. Sketch a graph of the voltage across the neon lamp as a function of time.

32.27 (2) *Time constant for a time constant.* A capacitor C discharges through a resistor R. **(a)** In terms of V_0, what is the initial current i? **(b)** Suppose for the sake of argument that the current continued at the constant value determined in part **a**. Show that the time required for complete discharge of the capacitor would be τ_C.

32.28 (2) *Working capacitor, I.* A capacitor C is charged to an initial voltage V_0 and then allowed to discharge through a resistor R. **(a)** Show that the instantaneous power P dissipated in the resistor is

$$P = \frac{V_0^2}{R} e^{-2t/RC}.$$

(b) When all the charge originally stored in the capacitor has flowed through the resistor, show that the total energy dissipated in the resistor is equal to the energy $U_0 = \frac{1}{2}CV_0^2$ originally stored in the electric field of the capacitor.

32.29 (2) *Working capacitor, II.* A capacitor C is charged to an initial voltage V_0 and then allowed to discharge through a resistor R. **(a)** At what time $t_{1/2}$ after the capacitor begins to discharge has the energy stored in the capacitor diminished to one-half its original value? **(b)** Compare your result with the result of part **c** of Example 32.1, and account for the difference.

32.30 (2) *Two-timing, I.* When a capacitor is connected into an RC circuit with resistor R_1, the time constant is τ_1. When resistor R_2 is substituted for R_1, the time constant is τ_2. Express the time constant τ_s of the RC circuit, formed by connecting R_1 and R_2 in series with the capacitor, in terms of τ_1 and τ_2.

32.31 (2) *Two-timing, II.* The two resistors of Problem 32.30 are connected in parallel. The combination is then connected with the capacitor to form an RC circuit. Find the time constant τ_p.

32.32 (2) *Charge it!* The switch in the figure below is closed at $t = 0$. Using Kirchhoff's rules, show that the voltage across the capacitor is

$$V = \frac{V_0}{2}(1 - e^{-2t/RC}).$$

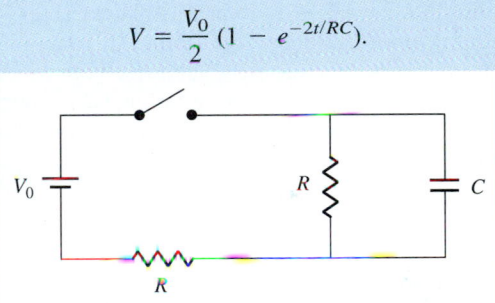

(Hint: At $t = 0$, the resistor in parallel with the capacitor is effectively short-circuited by the uncharged capacitor.)

32.33 (3) *Rising . . .* A battery of emf V_0 drives a current through an RL circuit, beginning when the circuit is completed by closing a switch at $t = 0$. Express, as functions of time, the voltages across **(a)** the resistor and **(b)** the inductor.

32.34 (3) *. . . and falling.* In the circuit of Problem 32.33, current is allowed to flow for some time. At time $t = 0$, the battery is switched out of the circuit. Express, as functions of time, the voltages across **(a)** the resistor and **(b)** the inductor.

32.35 (3) *Two-timing, III.* When an inductor is connected into an RL circuit with resistor R_1, the time constant is τ_1. When resistor R_2 is substituted for R_1, the time constant of the circuit is τ_2. Express the time constant τ_s of the RL circuit formed by connecting R_1 and R_2 in series with the inductor, in terms of τ_1 and τ_2.

32.36 (3) *Two-timing, IV.* The two resistors of Problem 32.35 are connected in parallel. The combination is then connected with the inductor to form an RL circuit. Find the time constant τ_p.

32.37 (3) *Electric clock, II.* In the figure below, a small neon lamp is connected across the resistor in an *RL* circuit. (The properties of neon lamps are described in Problem 32.26.)

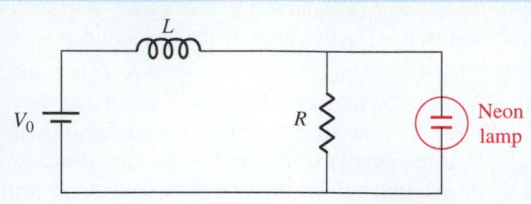

(a) If $V^* < V_0$, express the time interval between flashes of the lamp as a function of the ratio V^*/V_0. **(b)** Assuming that $V^* \ll V_0$, find the time interval between flashes. **(c)** A typical value of V^* is 10 V. Use reasonable values for V_0, R, and L to design a circuit that will produce one flash per second.

32.38 (3) *Mixed curve.* **(a)** Sketch a graph of Equation 32.26. Show that the curve is asymptotic to the P_L/P_R axis at $t = 0$ and to the t/τ_L axis at $t = \infty$. **(b)** Show that the curve approximates a hyperbola at small t and an exponential decay curve at large t.

32.39 (3) *Ad lib.* You have a mass m of metal of density δ and permission to use machinery that will form the metal into insulated wire of any diameter you wish. You make the wire and use it to make a closely wound single-layer solenoid whose length l is much greater than its diameter. **(a)** What is the inductive time constant of this solenoid? (Hint: The result is independent of the diameters you choose both for the wire and for the solenoid.) **(b)** Suppose the metal is copper and $m = 1.3$ kg. If you choose $l = 85$ cm, what is τ_L? Take the resistivity of copper as $\rho = 1.724 \times 10^{-8}$ $\Omega \cdot$m and its density as $\delta = 8.9 \times 10^3$ kg/m^3.

32.40 (3) *Grow with the flow.* The two capacitors in the circuit illustrated below are initially each charged to a 220-V potential difference. At $t = 0$, the switch is closed. Using Kirchhoff's rules, express the current through the inductor as a function of time. (Hint: See Problem 32.32.)

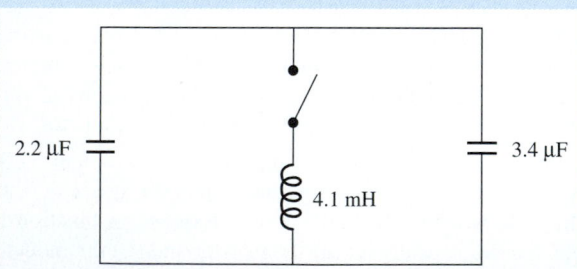

32.41 (3) *Pumping up the power, I.* In an *RL* circuit, the switch is closed at $t = 0$, and the current driven by the battery begins to build up. Express $U(t)$, the energy stored in the inductor, in terms of τ_L and U_∞, the stored energy at $t = \infty$.

32.42 (3) *Pumping up the power, II.* In an *RL* circuit, the switch is closed at $t = 0$, and the current driven by the battery begins to build up. **(a)** Show that the rate at which energy is

stored in the inductor is

$$\frac{dU}{dt} = \frac{V_0^2}{R} (1 - e^{-t/\tau_L}) e^{-t/\tau_L}.$$

(b) What is the value of dU/dt at $t = 0$? at $t = \infty$? **(c)** Show that the rate of energy storage reaches a maximum value when $t = \tau_L \ln 2$. **(d)** Express this maximum rate as a fraction of the power dissipated in the resistor at $t = \infty$. **(e)** Using a computer, make a plot of dU/dt as a function of t. Explain why dU/dt increases to a maximum value and then decreases. (Hint: As time passes, what happens to i? to di/dt?)

32.43 (4) *Do it again.* Solve Equation 32.28 step by step, using the method that leads to the solution of Equation 6.32.

32.44 (4) *Dimensional check.* Show that Equation 32.29a is dimensionally correct.

32.45 (4) *Vital difference.* **(a)** Show that, for an *LC* circuit, the maximum voltage V_{max} across the capacitor and the maximum current i_{max} through the circuit bear the relation $V_{max} = i_{max}\sqrt{L/C}$. **(b)** Show that the constant $\sqrt{L/C}$ has the dimensions of resistance. **(c)** The equation of part **a** has the mathematical form of Ohm's law. Why is the equation nevertheless *not* Ohm's law?

32.46 (4) *No change.* Two *LC* circuits have the same natural angular frequency ω_0, although neither their inductances L_1 and L_2 nor their capacitances C_1 and C_2 are necessarily the same. Show that, if the elements of the two circuits are combined to form a single series *LC* circuit, the natural angular frequency ω_0' will be equal to ω_0.

32.47 (4) *Complementary capacitors.* In the figure below, the capacitor on the left is initially charged to a potential difference V_0, and the capacitor on the right is uncharged. At $t = 0$, the switch is closed. Using Kirchhoff's loop rule, express **(a)** V_1, **(b)** V_2, and **(c)** V_L as functions of time. **(d)** What is the natural angular frequency ω of the system?

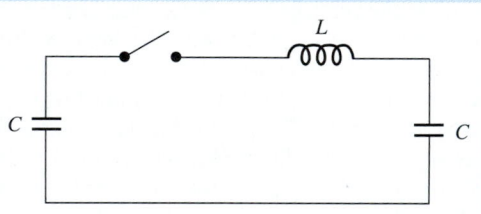

32.48 (G) *Swing time.* You connect a capacitor and a resistor into an *RC* circuit and measure the time constant τ_C. You then connect the same resistor with an inductor to form an *RL* circuit and measure the time constant τ_L. Finally, you connect the inductor and the capacitor to form an *LC* circuit and measure the period $\tau^* = 1/\nu$. Express τ^* in terms of τ_C and τ_L.

32.49 (G) *Start-up.* The inductor of an *LC* circuit lies in a magnetic field and is penetrated by flux Φ. Initially, there is no current in the circuit. At $t = 0$, the magnetic field is turned off in a time much smaller than $\sqrt{LC}$. For $t > 0$, write i as a function of time.

GROUP C

32.50 *(3) Surge.* Power transmission systems often have considerable self-inductance. We represent a dc transmission line and its load schematically by the *RL* circuit shown below.

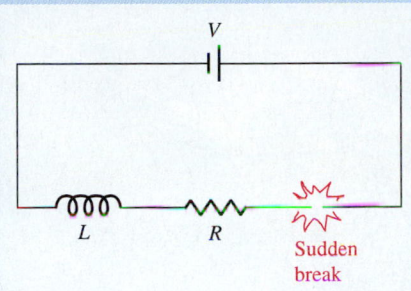

(The argument that follows also applies to ac systems.) The insulation of the system is designed for the line voltage V_0. For a power system, the current i_0 should be large, which means that $R = V_0/i_0$ is not large. Consequently, the inductive time constant $\tau_L = L/R$ can be of order 1 s or even greater. Consider what happens if there is a sudden break in the line, either by accident or because it is necessary to open a circuit breaker. When the circuit is opened, the current must decrease rapidly from its initial value i_0 over a short time δt—typically a few milliseconds. For the sake of argument, we assume that, after the break, the current is given by the expression

$$i = i_0 e^{-t/\delta t}.$$

(a) What is the voltage V_R across the resistor at a time t after the break? **(b)** What is the voltage V_L across the inductor at the same time? **(c)** If $V_0 = 440$ V, $\tau_L \approx 1$ s, and $\delta t \approx 10$ ms, what is the voltage across the inductor just after the break? **(d)** Heavy-duty circuit breakers have elaborate spark-suppression devices incorporated into them. In some, the entire switch assembly is immersed in heavy oil; in others, a spray of oil or other dielectric liquid is propelled into the gap between the contacts as they separate. Explain.

Changing Electric Currents II

The ac Series Circuit

LOOKING AHEAD

- The series alternating-current circuit, consisting of a resistor, a capacitor, an inductor, and a source of emf that varies sinusoidally with time, is of great practical importance and displays many of the features of more complicated ac circuits.

- The circuit equation has terms that contain the current, its first time derivative, and its second time derivative.

- A solution of the equation yields an expression for the current as a function of time.

- The transformer is an important device for changing ac voltages and currents; its operation depends on Faraday's law.

Left: The ease with which voltage can be transformed is a major advantage of ac power systems.

To live happily, one must be willing to accept change.

—Latin proverb

Introduction*

We continue our study of electric circuits in which the current varies with time. The *alternating-current (ac) series circuit*, which is the main concern of this chapter, is more general than the circuits we studied in Chapter 32. The ac series circuit contains a resistor, a capacitor, an inductor, and a source of emf that varies with time. The behavior of ac circuits is of fundamental importance in all branches of electrical technology. Although the series circuit is the simplest of all ac circuits, its behavior has many features common to all ac circuits; this behavior is the subject of Section 33.2.

Section 33.3 deals with *resonance*, a property of ac circuits that they share with all oscillating systems in which energy dissipated internally is replenished by a driving mechanism. Section 33.4 concerns the definition and use of mean values for voltage, current, and power in ac circuits. In Section 33.5, we consider the important ac device called the *transformer*.

The ac Series Circuit

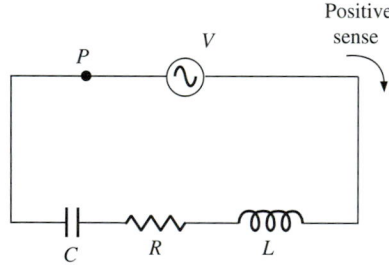

FIGURE 33.1 An ac series circuit, consisting of an ac source (represented by the standard symbol), an inductance L, a resistance R, and a capacitance C.

Figure 33.1 shows an **ac series circuit**. The circuit consists of four components in series: an inductor L, a resistor R, a capacitor C, and a source of emf that varies sinusoidally. (A simple magneto with slip rings produces an emf that is approximately sinusoidal; see Section 30.5.) Such a source of emf is called an **ac source**, and it is represented by the standard symbol shown in Figure 33.1. The emf of the ac source is given by the expression

$$V = V_0 \cos (\omega t + \delta), \tag{33.1}$$

where V_0 is the maximum emf, ω is the angular frequency, and δ is a phase constant determined by the choice of the zero of time.

We confine our study of the behavior of the circuit to the *steady state*. That is, we study the current in the circuit and the voltages across the various components only after the ac source has been running for some time. (The steady-state condition is the condition of greatest interest in electric power systems, as well as in many other systems.) In the steady state, we are free to choose the zero of time. If we choose $t = 0$ at an instant when V is zero and increasing, Equation 33.1 assumes the special form

$$V = V_0 \sin \omega t. \tag{33.2}$$

In writing Kirchhoff's loop rule, we must choose the positive sense arbitrarily because the ac source has no preferred polarity. In Figure 33.1, we have chosen the clockwise sense as positive, and we will follow this convention throughout our discussion. Beginning at point P and going around the circuit in the positive sense, we have

---

*This chapter may be omitted without loss of continuity. However, the elaboration of the phasor method in Section 33.2 is useful background for the study of single-slit diffraction in Section 36.7.

at any instant

$$V - V_L - V_R - V_C = 0. \tag{33.3a}$$

We insert the specific values of the V's into this equation to obtain

$$V_0 \sin \omega t - L\frac{di}{dt} - iR - \frac{1}{C}q = 0. \tag{33.3b}$$

What is the difference between this equation and an equation that arises from applying Kirchhoff's rules to the steady state of a dc circuit? In the dc case, each term is constant in time. In the ac case, all of the terms vary in time. Nevertheless, their *sum* has the constant value zero.

In ac circuits, we are usually interested in the current i and not the capacitor charge q. To eliminate q from Equation 33.3b, we differentiate each term with respect to time and use the relation $i = dq/dt$ to obtain

$$\omega V_0 \cos \omega t - L\frac{d^2i}{dt^2} - R\frac{di}{dt} - \frac{1}{C}i = 0. \tag{33.4}$$

What do we want to know? Given the form of V expressed in Equation 33.4, three questions arise:

1. What is the steady-state current i as a function of time?
2. How do the quantities L, R, and C affect the value of i?
3. What happens to i as the angular frequency ω is varied?

All of this important physical information is contained in Equation 33.4. It is possible, but rather complicated, to solve the equation completely. For the steady state, the solution is simpler. But even the steady-state solution involves enough mathematical steps to make it hard to follow if you don't know in advance what the various terms in the solution signify physically.

Fortunately, we can obtain the desired information without explicitly solving the equation, by means of a graphical method employing the rotating vectors called *phasors*. We considered phasors briefly in Section 6.2; here we exploit the phasor method more fully.

Phasors: The emf of an ac Source

The basic idea of the phasor is contained in Figure 33.2, which is similar to Figure 6.4 and 6.5. A vector $\mathbf{V}_0$, whose magnitude is V_0, rotates in a cartesian plane. We call the axes x and y, though they do not represent lengths. For our immediate purposes, the axes are scaled in units of V.

Suppose that, at $t = 0$, the vector $\mathbf{V}_0$ is oriented along the x axis. Suppose also that $\mathbf{V}_0$ rotates with angular velocity ω. Then, at any time t, the y component of $\mathbf{V}_0$ has the value

$$V_y = V_0 \sin \omega t.$$

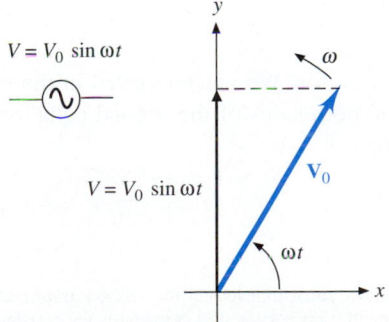

FIGURE 33.2 The emf $V = V_0 \sin \omega t$ represented as the y component of a phasor—a vector $\mathbf{V}_0$ rotating with angular velocity ω in the x-y plane.

But this quantity is identical with output emf V of the ac source, which we specified in Equation 33.2. That is, we can use this y component to represent V as a function of time:

$$V = V_y = V_0 \sin \omega t. \tag{33.5}$$

*An ac voltage $V = V_0 \sin \omega t$ can be represented as a component of a **phasor** $\mathbf{V}_0$—a vector that rotates with angular velocity ω.*

The Current in a Purely Resistive Circuit

In Figure 33.3a, the ac source is connected to a resistor R to form a circuit. Such a circuit is called **purely resistive** because neither capacitance nor inductance is present. At any instant t, Kirchhoff's loop rule gives us

$$V - iR = 0;$$

the solution of this equation for the instantaneous current i in the circuit is Ohm's law:

$$i = \frac{V}{R} = \frac{V_0}{R} \sin \omega t. \tag{33.6}$$

At any instant, this current is simply the quantity V, given by Equation 33.5, divided by the constant R. The phasor equivalent of this statement is shown in Figure 33.3b. Here the x and y axes are scaled in units of i. The phasor $\mathbf{i}_0 = \mathbf{V}_0/R$ is parallel to $\mathbf{V}_0$ and rotates with it.* The y component of $\mathbf{i}_0$ represents the instantaneous current i given by Equation 33.6.

FIGURE 33.3 (*a*) A purely resistive ac circuit. (*b*) Phasor representation of the instantaneous current i in the circuit.

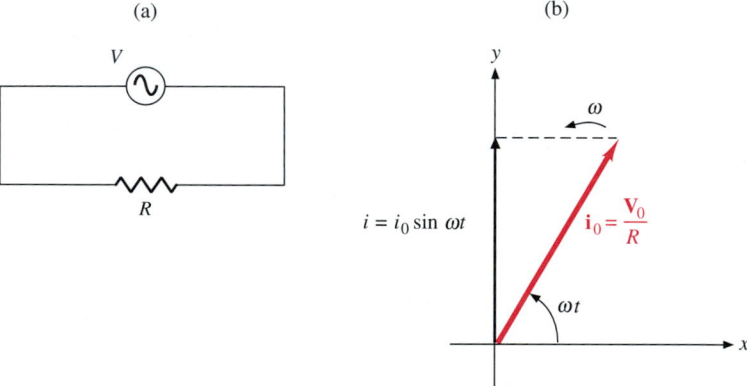

The Current in a Purely Capacitive Circuit

If purely resistive circuits were the only circuits of interest, it would not be worth the trouble to develop the phasor method. But now consider what happens in a **purely capacitive** circuit, consisting only of an ac source and a capacitor, as shown in Figure 33.4a. Here Kirchhoff's loop rule gives

$$V - \frac{q}{C} = 0 \quad \text{or} \quad q = CV,$$

where q is the instantaneous charge on the capacitor. We are interested in the instantaneous current i, not in q. So we differentiate both sides of the second equation with respect to time. We have $dq/dt = C\, dV/dt$, or

$$i = C \frac{dV}{dt}. \tag{33.7}$$

---

*You may think of the current graph, scaled in units of i, as superimposed on the voltage graph, scaled in units of V. The two scales are proportional to one another; the proportionality constant is the resistance R.

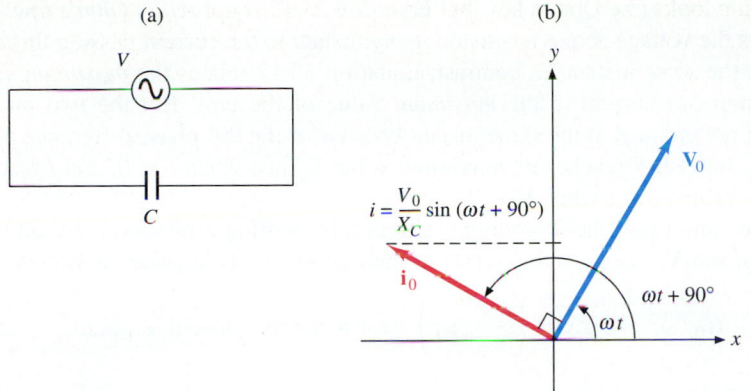

In a purely capacitive circuit, the current is proportional to the rate of change of the emf that drives it.

Now, the value of V is given by Equation 33.5, $V = V_0 \sin \omega t$. Differentiating with respect to time, we have

$$\frac{dV}{dt} = \omega V_0 \cos \omega t.$$

Inserting this value of dV/dt into Equation 33.7, we obtain for the instantaneous current

$$i = \omega C V_0 \cos \omega t. \tag{33.8}$$

In the spirit of the phasor method, we wish to represent this varying current as the *y* component of a rotating vector $\mathbf{i}_0$ whose magnitude is constant, just as we did for the purely resistive circuit. With this goal in mind, we manipulate Equation 33.8 so that it looks as much as possible like Equation 33.6. As a first step, we note that the factor $\omega C V_0$ on the right side of Equation 33.8 must have the dimension $[R]^{-1}[V]$ of an electric current. (Remember that cos ωt is dimensionless.) It follows that the dimensions of the product ωC must be $[R]^{-1}$. With this in mind, we define a quantity called the **capacitive reactance** X_C.

$$X_C \equiv \frac{1}{\omega C}. \tag{33.9}$$

In terms of the capacitive reactance, Equation 33.8 becomes

$$i = \frac{V_0}{X_C} \cos \omega t. \tag{33.10}$$

The capacitive reactance has the dimensions of resistance. Equation 33.10 has the same *mathematical* form as Ohm's law, $i = V/R$. But we must be careful; X_C *is not a resistance* because the dimensionless quantity cos ωt is not equal to 1, and Equation 33.10 is not Ohm's law.

The second step toward making Equation 33.8 look like Equation 33.6 is to rewrite Equation 33.10 in terms of a sine, rather than a cosine, so that it can be the *y* component of a vector $\mathbf{i}_0$. We apply the identity cos $\omega t = \sin (\omega t + 90°)$ to Equation 33.10 and obtain

$$i = \frac{V_0}{X_C} \sin (\omega t + 90°). \tag{33.11}$$

This is certainly the form of the *y* component of a vector oriented at an angle $\omega t + 90°$ with respect to the positive *x* direction. Because the orientation of the phasor $\mathbf{V}_0$ is given by the angle ωt, the orientation of the phasor $\mathbf{i}_0$ must be specified by the angle $\omega t + 90°$. That is, $\mathbf{i}_0$ *leads* $\mathbf{V}_0$ by 90°. This angular relation is shown in Figure 33.4b.

What is the magnitude of the phasor $\mathbf{i}_0$ whose *y* component is the instantaneous current *i*? The magnitude of $\mathbf{i}_0$ must be equal to the maximum value of *i*. So we have

$$i_0 = \frac{V_0}{X_C}. \tag{33.12}$$

This relation looks like Ohm's law, but *Equation 33.12 is not really Ohm's law*. Ohm's law relates the voltage across a resistor at any instant to the current passing through the resistor at the *same* instant. In contrast, Equation 33.12 relates the *maximum* value of the instantaneous current to the *maximum* value of the emf. But the two maximum values are not attained at the same instant because of the 90° phase difference between i_0 and V_0. Indeed, V reaches its maximum value V_0 just when $i = 0$, and i reaches its maximum value i_0 just when $V = 0$.

We can summarize the foregoing discussion by writing a relation between the two phasors, i_0 and V_0, expressed in terms of their components in polar coordinates (r, θ):

$$(i_0, \omega t + 90°) = \left(\frac{V_0}{X_C}, \omega t\right) \quad \text{for a purely capacitive circuit.} \quad \textbf{(33.13)}$$

The appearance of the term ωt in the θ component of both phasors indicates that the two phasors rotate together. Because the phase angle—the angle from V_0 to i_0—is 90°, i_0 leads V_0 by 90°.

The Current in a Purely Inductive Circuit

Figure 33.5a shows a **purely inductive** circuit, comprising only an ac source and an inductor. Here Kirchhoff's loop rule takes the form

$$V - L\frac{di}{dt} = 0. \quad \textbf{(33.14)}$$

Solving for di/dt and making the substitution $V = V_0 \sin \omega t$, we obtain

$$\frac{di}{dt} = \frac{1}{L}V = \frac{V_0}{L}\sin \omega t. \quad \textbf{(33.15)}$$

The value of i is the antiderivative

$$i = \frac{V_0}{L}\left(-\frac{1}{\omega}\cos \omega t\right),$$

or

$$i = -\frac{V_0}{\omega L}\cos \omega t. \quad \textbf{(33.16)}$$

We proceed by analogy with the purely capacitive case. We rewrite Equation 33.16 with two changes:

1. We define the **inductive reactance**

$$X_L \equiv \omega L. \quad \textbf{(33.17)}$$

2. We use the identity $-\cos \omega t = \sin(\omega t - 90°)$.

FIGURE 33.5 (*a*) A purely inductive ac circuit. (*b*) Phasor representation of the instantaneous current i in the circuit.

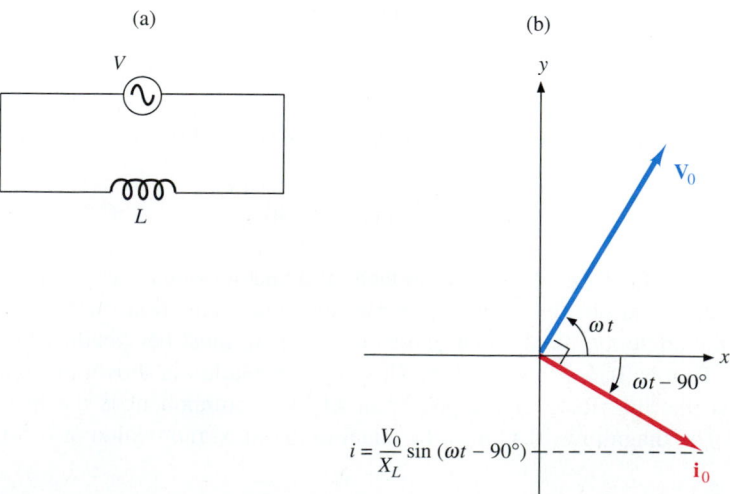

Equation 33.16 then takes the form

$$i = \frac{V_0}{X_L} \sin(\omega t - 90°).\qquad(33.18)$$

This equation has the same *mathematical* form as Equation 33.11 for the purely capacitive circuit and Equation 33.6 for the purely resistive circuit. No surprise, because that is just what we have been aiming for.

What is the phasor $\mathbf{i}_0$ of which i is the y component? The phase constant $-90°$ means that $\mathbf{i}_0$ *lags* the phasor $\mathbf{V}_0$ by $90°$. This relation is shown in Figure 33.5b.

In analogy with Equation 33.13, we can write the magnitude of the phasor $\mathbf{i}_0$ as

$$i_0 = \frac{V_0}{X_L}.\qquad(33.19)$$

Again, and for the same reasons that apply to Equation 33.12, Equation 33.19 looks like Ohm's law but is not really Ohm's law.

We can summarize the foregoing discussion by writing a relation between the two phasors, $\mathbf{i}_0$ and $\mathbf{V}_0$, expressed in polar coordinates (r, θ):

$$(i_0,\ \omega t - 90°) = \left(\frac{V_0}{X_L},\ \omega t\right) \quad \text{for a purely inductive circuit.}\qquad(33.20)$$

Here again, the two phasors rotate together. Because the phase factor—the angle from $\mathbf{V}_0$ to $\mathbf{i}_0$—is $-90°$, $\mathbf{i}_0$ lags $\mathbf{V}_0$ by $90°$.

Phasor Representation of the Series ac Circuit

We are now ready to combine the elements developed in the preceding four subsections into a complete picture of the behavior of a circuit consisting of an ac source, an inductor, a resistor, and a capacitor in series. Specifically, we are now in a position to construct a graphic solution that will give us the value of the current i through the series circuit at any instant t, in terms of ω, V_0, L, R, and C.

We begin by making a plausible assumption, which our solution will verify: *In the steady state, the current varies sinusoidally at the same frequency as the driving emf V.* That is, we can write

$$i = i_0 \sin(\omega t + \phi),\qquad(33.21)$$

where the phase constant ϕ is measured from the phasor $\mathbf{V}_0$ to the phasor $\mathbf{i}_0$. This equation is reminiscent of those for the purely resistive, inductive, and capacitive cases, where ϕ has the special values 0, $90°$, and $-90°$, respectively. Unlike the phase constant δ, discussed at the beginning of Section 33.2, ϕ is a fixed angle that depends on the characteristics of the circuit but not on the essentially arbitrary way in which we choose the instant $t = 0$. As in all series circuits, the value of i is the same everywhere in the circuit at any instant.

To evaluate i, we determine $\mathbf{i}_0$—that is, the values of the quantities i_0 and ϕ in Equation 33.21. The phasor method of doing this is best introduced in terms of a specific example.

EXAMPLE **33.1**

Figure 33.6 shows an ac series circuit in which $R = 21.0\ \Omega$, $C = 88.0\ \mu\text{F}$, and $L = 0.233\ \text{H}$. The ac source produces a sinusoidal emf whose peak value is $V_0 = 170\ \text{V}$ and whose angular frequency is $377\ \text{rad/s}$. Find **(a)** the phase constant ϕ of the current with respect to the driving emf; **(b)** the maximum voltages across the resistor, capacitor, and inductor; **(c)** the maximum current i_0 through the circuit.

SOLUTION:
(a) Find ϕ.

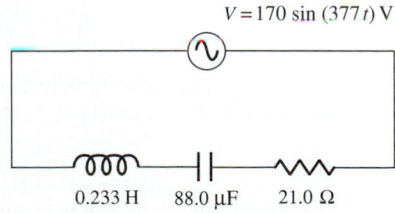

$V = 170 \sin(377t)\ \text{V}$

0.233 H   88.0 μF   21.0 Ω

FIGURE 33.6 An ac series circuit.

You begin by drawing the direction $\hat{\mathbf{i}}_0$ of the phasor $\mathbf{i}_0$ on a sheet of polar-coordinate graph paper.* This direction will serve as a reference for the other directions of interest in the construction. As with all phasors, $\hat{\mathbf{i}}_0$ varies with time. But you are interested only in the orientation of $\mathbf{i}_0$ relative to other phasors that rotate together with it. Thus you are free to choose any convenient direction, as in Figure 33.7.

Next, you construct the phasors $\mathbf{V}_{R0}$, $\mathbf{V}_{C0}$, and $\mathbf{V}_{L0}$. The magnitudes of these phasors are, respectively,

$$V_{R0} = i_0 R, \quad V_{C0} = i_0 X_C, \quad \text{and} \quad V_{L0} = i_0 X_L. \quad \text{(33.22a,b,c)}$$

You do not know i_0. But, on the voltage scale of the graph, you can set $i_0 = 1$ graph unit/Ω for the purposes of the construction. Then you have

$$V_{R0} = (1 \text{ unit}/\Omega) \times R = R \text{ units},$$

for Equation 33.22a, and similar expressions for the other two voltages.

You now insert the numerical values. For the resistor, you have

$$V_{R0} = 1 \text{ unit}/\Omega \times 21.0 \ \Omega = 21.0 \text{ units.} \quad \text{(33.23a)}$$

The voltage across a resistor is always in phase with the current, and so you draw the phasor $\mathbf{V}_{R0}$ with length 21.0 units, parallel to $\hat{\mathbf{i}}_0$, as in Figure 33.7.

For the inductor, the inductive reactance is given by Equation 33.17, $X_L = \omega L$. So you have

$$V_{L0} = 1 \text{ unit}/\Omega \times 377 \text{ rad/s} \times 0.233 \text{ H} \quad \text{(33.23b)}$$
$$= 87.8 \text{ units.}$$

For the direction of $\mathbf{V}_{L0}$, you refer to Figure 32.10 or Figure 33.5. The voltage across the inductor leads the current by 90°. Consequently, you represent $\mathbf{V}_{L0}$ by a phasor of length 87.8 graph units, drawn 90° ahead of $\hat{\mathbf{i}}_0$.

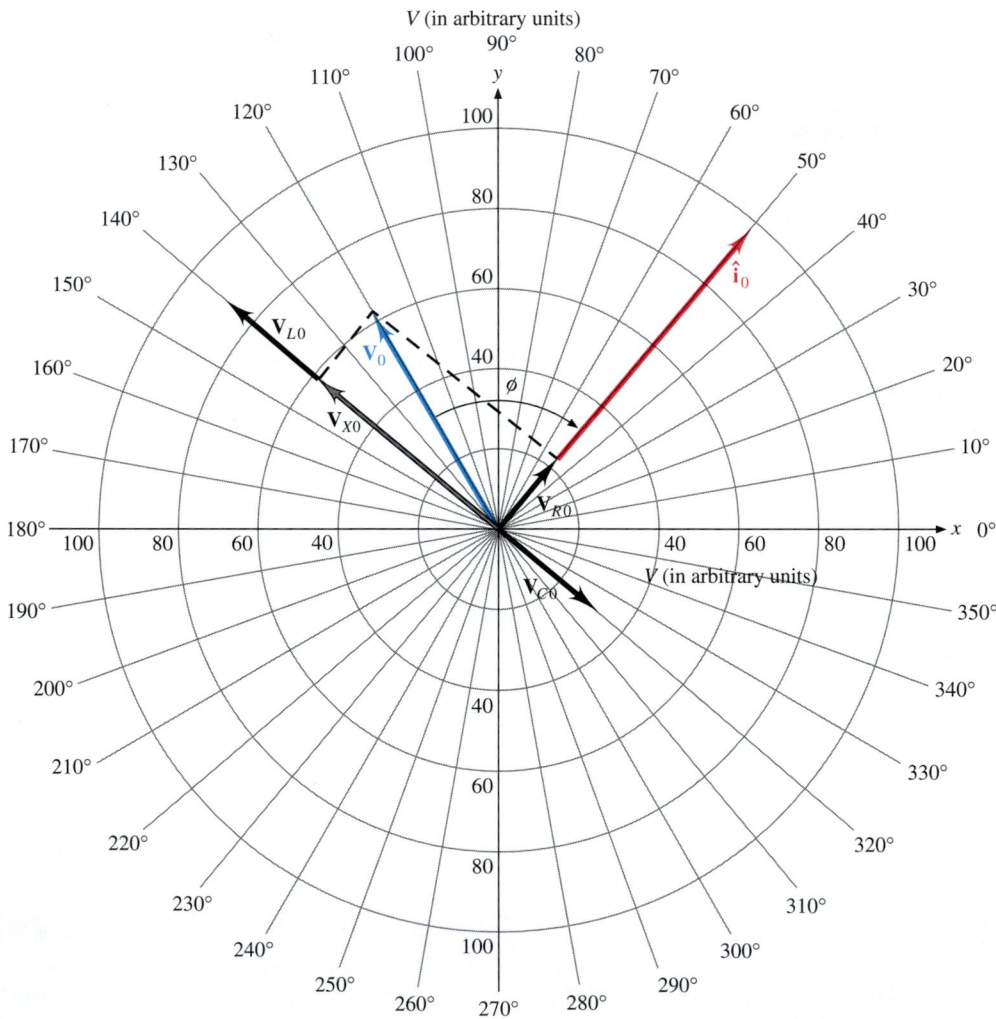

FIGURE 33.7 Phasor diagram. Because $\mathbf{i}_0$ is initially unknown, it is represented by its unit vector $\hat{\mathbf{i}}_0$, which specifies the direction of $\mathbf{i}_0$ at some arbitrarily chosen instant t.

*It is best to use polar-coordinate paper or plotting software that has polar-coordinate capability, because you can easily mark off a vector specified by its magnitude and direction, rather than the magnitude of its x and y components.

For the capacitor, the capacitive reactance is given by Equation 33.11, $X_C \equiv 1/\omega C$. So you have

$$V_{C0} = \frac{1 \text{ unit}/\Omega}{377 \text{ rad/s} \times 88.0 \times 10^{-6} \text{ F}} = 30.1 \text{ units.} \quad \textbf{(33.23c)}$$

What about the direction of $\mathbf{V}_{C0}$? As you saw in Figures 32.10 and 33.4, the voltage across the capacitor lags the current by 90°. Consequently, you represent $\mathbf{V}_{C0}$ by a phasor of length 30.1 graph units, drawn 90° behind $\hat{\mathbf{i}}_0$.

Now you are ready to construct the vector sum

$$\mathbf{V}_0 = \mathbf{V}_{R0} + \mathbf{V}_{L0} + \mathbf{V}_{C0}. \quad \textbf{(33.24)}$$

First, note that $\mathbf{V}_{C0}$ is antiparallel to $\mathbf{V}_{L0}$. You can readily construct the intermediate sum

$$\mathbf{V}_{X0} \equiv \mathbf{V}_{L0} + \mathbf{V}_{C0}. \quad \textbf{(33.25)}$$

This intermediate sum $\mathbf{V}_{X0}$ is directed along the greater of the two, $\mathbf{V}_{L0}$, and has magnitude

$$V_{X0} = V_{L0} - V_{C0} = 87.8 \text{ units} - 30.1 \text{ units} = 57.7 \text{ units.}$$

This phasor is shown in Figure 33.7.

You are now ready to find $\mathbf{V}_0$, which is the sum

$$\mathbf{V}_0 = \mathbf{V}_{R0} + \mathbf{V}_{X0}. \quad \textbf{(33.26)}$$

You can construct $\mathbf{V}_0$ by means of the parallelogram method, as shown in Figure 33.7. The result is a vector $\mathbf{V}_0$ whose magnitude and phase you can read from the graph:

$$V_0 = 61.4 \text{ units} \quad \text{and} \quad \phi = -69.5°.$$

The value of the phase constant ϕ is the solution to part **a** of this example. The negative value of ϕ indicates that the current lags the voltage. The circuit shares this property with the

purely inductive circuit. Thus the circuit is inductive on balance because, at the angular frequency $\omega = 377$ rad/s, $X_L > X_C$.

(b) Find V_{R0}, V_{C0}, and V_{L0}, the maximum voltages that appears across the resistor, the capacitor, and the inductor, respectively.

You have already calculated these voltages in arbitrary units and plotted them on the graph. Now you must translate these units into volts. To do this, you compare the value $V_0 = 170$ V, which is given, with the magnitude of V_0 in arbitrary units, which you have determined graphically.

On the graph, the magnitude V_0 is represented by 61.4 units. Thus 61.4 units must be equivalent to the given value of V_0, which is 170 V. You have the scale factor

$$\frac{170 \text{ V}}{61.4 \text{ units}} = 2.77 \text{ V/unit.}$$

Using Equations 33.23a, 33.23b, and 33.23c or reading the values from the graph, you find

$$V_{R0} = 21.0 \text{ units} \times 2.77 \text{ V/unit} = 58.1 \text{ V,}$$
$$V_{L0} = 87.8 \text{ units} \times 2.77 \text{ V/unit} = 243 \text{ V,}$$

and

$$V_{C0} = 30.1 \text{ units} \times 2.77 \text{ V/unit} = 83.3 \text{ V.}$$

The sum of these voltages is greater than V_0. Indeed, V_{L0} by itself is greater than V_0. As you will see in Problem 33.6, either V_L or V_C, or both, can be greater than V_0.

(c) Find the maximum current.

Now that you know V_R, you can use Ohm's law to find i_0:

$$i_0 = \frac{V_{R0}}{R} = \frac{58.1 \text{ V}}{21.0 \, \Omega} = 2.77 \text{ A.}$$

This result is numerically equal to the scale factor, 2.77 V/unit. Can you see why?

Let us summarize the strategy followed in applying the phasor method to the series circuit. As with all series circuits, the instantaneous current is the same everywhere in the circuit. We use this fact to construct three phasors representing the voltages across the resistor, capacitor, and inductor, as well as their relative phases, as follows:

1. We determine the relative phases (directions) of the three phasors.
   a. The voltage across the resistor is in phase with the current; that is, $\mathbf{V}_{R0}$ is parallel to $\mathbf{i}_0$.
   b. The voltage across the inductor leads the current by 90°; that is, $\mathbf{V}_{L0}$ is directed 90° ahead of $\mathbf{i}_0$.
   c. The voltage across the capacitor lags the current by 90°; that is, $\mathbf{V}_{C0}$ is directed 90° behind $\mathbf{i}_0$.

2. We determine the relative magnitudes of the three phasors.
   a. The maximum value V_{R0} of V_R is expressed in terms of the resistance R by means of Ohm's law, $V_{R0} = i_0 R$.
   b. The maximum value V_{L0} of V_L is expressed in terms of the inductive reactance $X_L \equiv \omega L$ by means of the "pseudo-Ohm's law" $V_{C0} = i_0 X_L$ (Equation 33.19).
   c. The maximum value V_{C0} of V_C is expressed in terms of the capacitive reactance $X_C \equiv 1/\omega C$ by means of the "pseudo-Ohm's law" $V_{C0} = i_0 X_C$ (Equation 33.12).

3. Having thus determined the relative magnitudes and relative directions of the three phasors, we construct their vector sum $\mathbf{V}_0$, as shown in Figure 33.8a. The phase ϕ of $\mathbf{i}_0$ relative to $\mathbf{V}_0$ is a characteristic of the circuit; we consider its physical implications later in this chapter. On the graph, the magnitude V_0 is expressed in the same arbitrary units as are V_{R0}, V_{C0}, and V_{L0}. Knowing any one of these four voltages suffices to establish an absolute voltage scale for the phasor diagram and thus to determine all four voltages

(a)

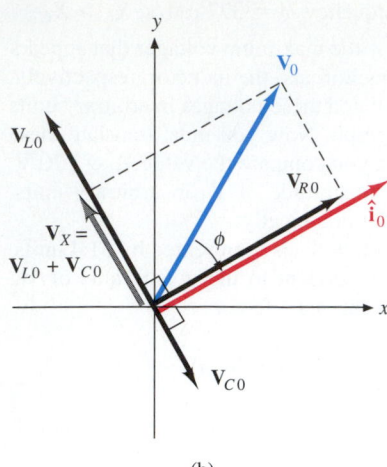

(b)

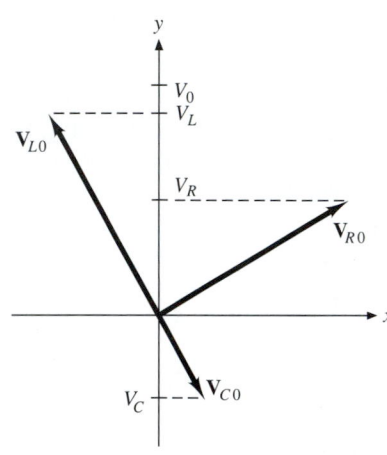

FIGURE 33.8 (a) The direction $\hat{\mathbf{i}}_0$ of the current phasor $\mathbf{i}_0$ is shown at an arbitrary instant of time. The y component of $\mathbf{i}_0$ represents the instantaneous current i in the ac series circuit of Figure 33.1, of which Figure 33.6 is a special case. We construct the vectors $\mathbf{V}_R$, $\mathbf{V}_C$, and $\mathbf{V}_L$ on an arbitrary scale, by using their known orientations relative to $\mathbf{i}_0$. The vector sum of the V's is $\mathbf{V}_0$. If $\mathbf{V}_0$ is known, an absolute voltage scale can be established and the V's and i_0 determined. The directed angle ϕ between the emf and the current it drives through the circuit is the phase constant. (b) The y components of $\mathbf{V}_R$, $\mathbf{V}_C$, and $\mathbf{V}_L$ are V_R, V_C, and V_L, the instantaneous voltages across R, C, and L. Their algebraic sum is V, the instantaneous emf of the ac source.

as well as the maximum current i_0. At any instant, the voltage across each circuit element is given by the y component of the corresponding phasor (Figure 33.8b).

Impedance

The graphical method used in Example 33.1 to construct $\mathbf{V}_0$ is laborious and limited in precision. But the method is nothing more than an application of vector addition, for which there is an algebraic equivalent of unlimited precision. We now consider the algebraic equivalent in detail.

The vector sum in question is Equation 33.24,

$$\mathbf{V}_0 = \mathbf{V}_{R0} + \mathbf{V}_{L0} + \mathbf{V}_{C0}.$$

Using Ohm's law and the "pseudo-Ohm's laws" (Equations 33.12 and 33.19), we rewrite this equation in the form

$$\mathbf{V}_0 = i_0 R \hat{\mathbf{i}}_0 + i_0 X_L(\hat{\mathbf{i}}_0 + 90°) + i_0 X_C(\hat{\mathbf{i}}_0 - 90°).$$

We can simplify this equation by using the same unit vector to express the inductive and capacitive terms. We note that $(\hat{\mathbf{i}}_0 - 90°) = -(\hat{\mathbf{i}}_0 + 90°)$. Making this substitution and dividing through by i_0, we obtain

$$\frac{\mathbf{V}_0}{i_0} = R \hat{\mathbf{i}}_0 + (X_L - X_C)(\hat{\mathbf{i}}_0 + 90°). \tag{33.27}$$

The phasor $\mathbf{V}_0/i_0$ is thus expressed as the sum of two mutually perpendicular vectors. The magnitude of one of them is the resistance R; the magnitude of the other is the difference $X_L - X_C$ of the inductive and capacitive reactances.

What are the magnitude and phase (direction) of the phasor $\mathbf{V}_0/i_0$? Let us turn first to the magnitude V_0/i_0. This magnitude is the Pythagorean sum of the two mutually perpendicular vectors on the right side of Equation 33.27:

$$\frac{V_0}{i_0} = \sqrt{R^2 + (X_L - X_C)^2}. \tag{33.28}$$

Now, look at the right side of this equation. The resistance R is a constant, and the X's are constants because they depend only on L, C, and ω. Thus the left side of the equation must be constant as well. We *define* the constant

$$Z \equiv \frac{V_0}{i_0}. \tag{33.29}$$

This quantity is called the **impedance** of the circuit at angular frequency ω. It is expressed in ohms and is the same sort of "pseudo-Ohm's law" as are Equations 33.12 and 33.19. Like those two equations, it relates a maximum voltage—the maximum emf of the ac source—to the maximum current through the circuit; the two maxima occur at different times. Although we have defined Z in terms of the special case of an ac series circuit, an impedance can be defined for any ac circuit containing a source of emf, resistance, inductance, and capacitance. Thus Equation 33.29 can be applied quite generally to ac circuits and is often called **Ohm's law for ac circuits**. In using this name, be sure to remember that it is not really Ohm's law.

By substituting Equation 33.29 into Equation 33.28, we can express the impedance entirely in terms of the resistance and the reactances:

$$Z = \sqrt{R^2 + (X_L - X_C)^2}. \tag{33.30a}$$

For purposes of calculation, it is convenient to express this result directly in terms of resistance, inductance, capacitance, and angular frequency. Substituting the values of X_L and X_C, we obtain the standard expression

$$Z = \sqrt{R^2 + \left(\omega L - \frac{1}{\omega C}\right)^2}. \tag{33.30b}$$

As this equation makes clear, Z depends not only on the circuit constants R, L, and C, but also on the driving frequency ω.

Now let us turn to the phase (direction) of the current phasor $\mathbf{i}_0$ relative to $\mathbf{V}_0/i_0$. We use the definition of the tangent of an angle to write

$$\tan \phi = -\frac{X_L - X_C}{R} \quad \text{or} \quad \phi = -\tan^{-1}\frac{X_L - X_C}{R}. \quad \text{(33.31a,b)}$$

The negative sign expresses the sense of the angle ϕ. You can see in these equations how the inductive and capacitive elements of the circuit work against one another. It is often convenient to define the difference between the inductive and capacitive reactances to be the net reactance, or simply the **reactance**:

$$X \equiv X_L - X_C. \quad \text{(33.32)}$$

In terms of the reactance, Equations 33.31a and 33.31b assume the compact form

$$\tan \phi = -\frac{X}{R} \quad \text{and} \quad \phi = -\tan^{-1}\frac{X}{R}. \quad \text{(33.33a,b)}$$

Equation 33.30a assumes the corresponding form

$$Z = \sqrt{R^2 + X^2}. \quad \text{(33.34)}$$

Can you see the graphic equivalent of Equation 33.33a in the phasor diagrams of Figures 33.7 and 33.8a?

EXAMPLE 33.2

(a) For the circuit discussed in Example 33.1, use algebraic methods to find the impedance Z and the phase constant ϕ. What are the impedance and the phase constant if the frequency is **(b)** doubled? **(c)** halved?

SOLUTION:
(a) Find Z and ϕ.

The resistance $R = 21.0 \ \Omega$ is given, and you have already effectively calculated X_L and X_C in working through Example 33.1. Because of the choice of the scale of the unknown current i_0 as 1 graph unit/Ω, the voltages calculated in graph units in Equations 33.23a, 33.23b, and 33.23c and in the equation immediately following Equation 33.25 are numerically equal to the reactances. So you have

$$R = 21.0 \ \Omega, \quad X_L = 87.8 \ \Omega,$$
$$X_C = 30.1 \ \Omega, \quad \text{and} \quad X = 57.7 \ \Omega.$$

You insert the values of R and X into Equation 33.34 to obtain

$$Z = \sqrt{(21.0 \ \Omega)^2 + (57.7 \ \Omega)^2} = 61.4 \ \Omega.$$

Using the same values, you get from Equation 33.33b

$$\phi = -\tan^{-1}\frac{57.7 \ \Omega}{21.0 \ \Omega} = -70.0°.$$

These values of Z and ϕ agree closely with the values $Z = 61.4 \ \Omega$ and $\phi = -69.5°$ obtained by the equivalent graphical method in Example 33.1.

(b) What are Z and ϕ if ω is doubled (to 754 rad/s)?

You must recalculate X_L and X_C. Using Equation 33.17, $X_L = \omega L$, you find

$$X_L = 754 \text{ rad/s} \times 0.233 \text{ H} = 176 \ \Omega;$$

doubling the frequency doubles the inductive reactance. Using Equation 33.9, $X_C \equiv 1/\omega C$, you find

$$X_C = \frac{1}{754 \text{ rad/s} \times 88.0 \times 10^{-6} \text{ F}} = 15.1 \ \Omega;$$

doubling the frequency halves the capacitive reactance.

You insert these new values of the reactances into Equation 33.30a to obtain the impedance

$$Z = \sqrt{(21.0 \ \Omega)^2 + (176 \ \Omega - 15.1 \ \Omega)^2} = 162 \ \Omega.$$

You use Equation 33.31b to find the phase constant

$$\phi = -\tan^{-1}\frac{176 \ \Omega - 15.1 \ \Omega}{21.0 \ \Omega} = -82.6°.$$

Note that there is no simple connection between these values of Z and ϕ and the values obtained for $\omega = 377$ rad/s. The circuit is inductive in both cases; increasing the frequency not only increases the net inductive reactance and thus the impedance, but also changes the phase constant in the direction of the purely inductive value $-90°$.

(c) What are Z and ϕ if ω is halved (to 189 rad/s)?

You could recalculate X_L and X_C again, but it is easier simply to halve the original inductive reactance to obtain

$$X_L = \tfrac{1}{2} \times 87.8 \ \Omega = 43.9 \ \Omega$$

and to double the original capacitive reactance to obtain

$$X_C = 2 \times 30.1 \ \Omega = 60.2 \ \Omega.$$

You now use Equation 33.30a to find

$$Z = \sqrt{(21.0 \ \Omega)^2 + (43.9 \ \Omega - 60.2 \ \Omega)^2} = 26.6 \ \Omega.$$

Equation 33.31b yields the phase constant

$$\phi = -\tan^{-1} \frac{43.9 \ \Omega - 60.2 \ \Omega}{21.0 \ \Omega} = 37.8°.$$

The reactance X is now negative; that is, the circuit is capacitive on balance. The sign of the reactance makes no difference in calculating the impedance. However, the phase constant is now positive; at low frequencies, the capacitive reactance dominates.

SECTION 33.3

Resonance

Look again at Equation 33.30b,

$$Z = \sqrt{R^2 + \left(\omega L - \frac{1}{\omega C} \right)^2}.$$

There is a frequency of special interest at which the reactance (the term in parentheses) has the value zero. This zero value occurs when

$$\omega L = \frac{1}{\omega C}.$$

The special value of ω that satisfies this condition is

$$\omega_0 = \frac{1}{\sqrt{LC}}. \tag{33.35a}$$

The angular frequency ω_0 is loosely called the **resonance frequency**. It is equal to the natural angular frequency of a pure (resistanceless) LC circuit having the same inductance and capacitance as the ac circuit we have been considering. More properly, the name "resonance frequency" applies to the corresponding quantity

$$\nu_0 = \frac{\omega_0}{2\pi} = \frac{1}{2\pi\sqrt{LC}}. \tag{33.35b}$$

Be careful that you know which of the quantities ω_0 and ν_0 is meant. In this chapter, we will use ω_0 in all discussions of resonance.

Either of Equations 33.35a and 33.35b is called the **resonance condition**; when an ac circuit satisfies the condition, we say that the circuit is **in resonance**. In resonance, the impedance attains its minimum value

$$Z = R \quad \text{for } \omega = \omega_0, \tag{33.36}$$

and the current attains its maximum value $i = V_0/R$. This is not to say that the circuit behaves as if the inductor and capacitor were not present. However, the phase constant has the value

$$\phi = -\tan^{-1} \frac{0}{R} = 0° \quad \text{for } \omega = \omega_0.$$

That is, the load "seen" by the ac source is in phase with the source emf, as if the circuit were purely resistive.

Resonance Curves and the Quality Factor*

The three curves of Figure 33.9 are called **resonance curves**. Each is a plot of the peak current i_0 in an ac series circuit as a function of angular frequency ω. The peak emf V_0 is fixed, and the inductance L and the capacitance C are the same for all three circuits. However, the resistances for the three circuits are in the ratio $1:4:16$. The absolute values of all quantities are given in the figure legend.

---

*This subsection may be omitted without loss of continuity.

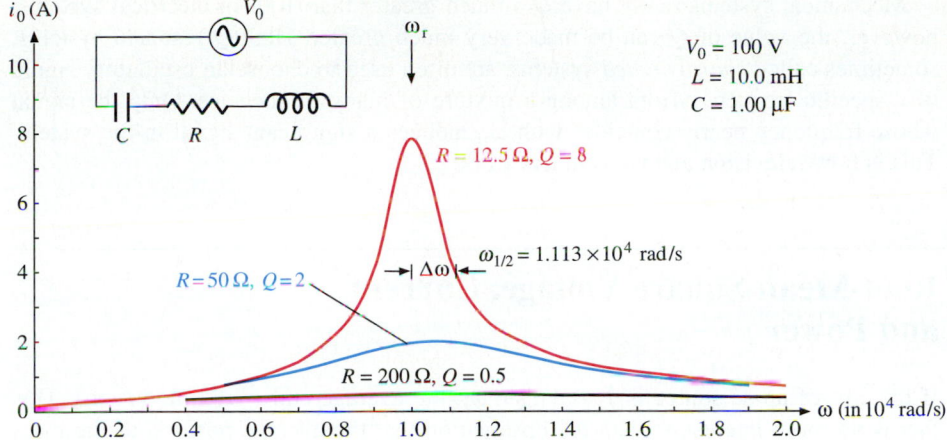

FIGURE 33.9 Resonance curves for three similar ac series circuits. The peak input emf is fixed at $V_0 =$ 100 V. For all three circuits, $L =$ 0.01 H and $C = 1$ μF. The resistances of the three circuits are 12.5 Ω (upper curve), 50 Ω (middle curve), and 200 Ω (lower curve). Each curve is also labeled with its quality factor Q. The half-width $\Delta\omega$ is shown for the upper curve.

When the resistance is small, the curve has a large maximum and falls off rapidly on both sides of the maximum; we say that the resonance is *sharp*. You can readily see why this is so. The maximum is large because at resonance, where $X = 0$, the current has the large value $i_0 = V_0/R$. Because R is small, a modest frequency difference $|\omega - \omega_0|$ results in a significant proportional increase in $|X|$ and thus in Z. Conversely, a large value of R results in a small value of i_0. Variation of ω does not strongly affect the value of Z.

The sharpness of the resonance curve is expressed quantitatively by means of the **quality factor** Q, usually called the ''Q-factor.''* The quality factor is defined to be

$$Q \equiv 2\pi \; \frac{\text{energy stored in the inductor and the capacitor at resonance}}{\text{energy dissipated in the resistor per cycle}}$$

$$= 2\pi \frac{U}{\Delta E}. \tag{33.37}$$

By means of a little calculation (see Problem 35.46), it can be shown that

$$Q = \frac{L\omega_0}{R} = \sqrt{\frac{L}{C}} \frac{1}{R}. \tag{33.38}$$

Another measure of the sharpness of a resonance curve is the loosely named ''**half-width**'' $\Delta\omega$. More properly called the *half-width at half-height*, $\Delta\omega$ is measured as shown in Figure 33.9. First, we find one of the two frequencies at which the resonance curve has one-half its maximum value. Then we measure the difference between the angular frequency $\omega_{1/2}$ at this point and the resonance value ω_0. Because the resonance curve is asymmetrical, there is some difference between the ''right side'' value of $\Delta\omega$ shown in Figure 33.9 and the value obtained by measuring from ω_0 to the left half-height point. For fairly sharp resonance curves, the discrepancy is not significant.

Resonance is not limited to electric circuits. It is a property of all oscillating systems in which energy dissipated through some kind of internal **damping** or ''friction'' is replenished by a driving source. (In the ac series circuit, the damping takes place in the resistor, and the ac source replenishes the energy.) A child's swing and a bowed violin string are typical resonating systems. If you want to swing, you must ''pump'' at a frequency close to the resonance frequency of the swing. A pumping frequency too far from ω_0 will not get you swinging very high. In a violin, the bow displaces the string sideways at irregular intervals, as the microscopic ''hooks'' on the horsehairs of the bow drag and release the string. But only those successive displacements whose frequency happens to coincide with the resonance frequency of the string produce significant vibration.

---

*Don't confuse this use of the symbol Q with its use for electric charge.

Mechanical systems rarely have Q's much greater than 100. In electrical systems, however, the value of Q can be made very much greater. High-Q resonant systems, sometimes called *sharply tuned* systems, are often used to choose an oscillatory signal of a specific frequency from among a mixture of many frequencies. Only the signal whose frequency nearly coincides with ω_0 induces a significant signal in the system. This is how television and radio tuners work.

Root-Mean-Square Voltage, Current, and Power

If the emf of an ac source is $V = V_0 \sin \omega t$, the average emf over time is zero. This fact is no more interesting than the observation that the average position of the mass in a mass-spring system is at $x = 0$. Instead, we are after a quantity that expresses the mean value of ac voltage in a useful way. A quantity commonly used is the *root-mean-square*, or rms, voltage,

$$V_{rms} = \sqrt{\langle V^2 \rangle}. \tag{33.39}$$

Compare with the use of the rms speed to characterize an ideal gas in Section 18.5.

To evaluate the quantity on the right side of Equation 33.39, we use the definition of a mean value to write $\langle V^2 \rangle$ over one oscillation period, from $t = 0$ to $t = \tau = 2\pi/\omega$:

$$\langle V^2 \rangle = \frac{V_0^2 \int_0^{2\pi/\omega} \sin^2 \omega t \, dt}{\int_0^{2\pi/\omega} dt}. \tag{33.40}$$

The value of the integral in the denominator is simply $2\pi/\omega$. The integral in the numerator has the value π/ω. So we have $\langle V^2 \rangle = V_0^2(\pi/\omega)(\omega/2\pi) = \frac{1}{2}V_0^2$, or

$$V_{rms} = \frac{1}{\sqrt{2}} V_0 \quad \text{for sinusoidal wave forms.} \tag{33.41}$$

For any other wave form, the factor in Equation 33.41 is different from $\sqrt{2}$. Root-mean-square values are almost always used in specifying power line voltages, which do have sinusoidal wave forms. Thus a standard 120-V line has a peak voltage

$$V_0 = \sqrt{2} \times 120 \text{ V} = 170 \text{ V}.$$

Voltage values are sometimes also quoted as **peak-to-peak voltages** V_{pp}. The peak-to-peak voltage is just twice the peak voltage; $V_{pp} = 2V_0$.

Except for the phase constant, the current in an ac circuit has the same mathematical form as the emf that drives it. According to Equations 33.21 and 33.29, we have

$$i = i_0 \sin (\omega t + \phi) = \frac{V_0}{Z} \sin (\omega t + \phi). \tag{33.42}$$

Consequently, the rms current i_{rms} in an ac circuit is related to the maximum current i_0 in exactly the same way that V_{rms} is related to V_0:

$$i_{rms} = \frac{1}{\sqrt{2}} i_0 \quad \text{for sinusoidal wave forms.} \tag{33.43}$$

The value of i_{rms} is always positive because i_0 is positive by definition.

Power

The instantaneous power supplied to a series circuit by an ac source is given by Joule's law, $P = Vi$ (Equation 27.18). Using Equation 33.2 to express V and Equation 33.42

to express i, we have

$$P = \frac{V_0^2}{Z} \sin \omega t \sin (\omega t + \phi). \qquad (33.44)$$

Using the trigonometric identity $\sin (x + y) = \sin x \cos y + \cos x \sin y$, we rewrite P in the form

$$P = \frac{V_0^2}{Z} \sin \omega t (\sin \omega t \cos \phi + \cos \omega t \sin \phi).$$

We expand the product on the right to obtain

$$P = \frac{V_0^2}{Z} (\cos \phi \sin^2 \omega t + \sin \phi \sin \omega t \cos \omega t).$$

To find the mean power $\langle P \rangle$, we average P over one cycle by carrying out an averaging process like that of Equation 33.40:

$$\langle P \rangle = \frac{\dfrac{V_0^2}{Z} \left(\cos \phi \displaystyle\int_0^{2\pi/\omega} \sin^2 \omega t \, dt + \sin \phi \int_0^{2\pi/\omega} \sin \omega t \cos \omega t \, dt \right)}{\displaystyle\int_0^{2\pi/\omega} dt}.$$

As in Equation 33.40, the denominator has the value $2\pi/\omega$. The first integral in the numerator is the same one that appears in Equation 33.40 and has the value π/ω. The second integral in the numerator yields the value zero. So the mean power is

$$\langle P \rangle = \frac{\omega}{2\pi} \frac{V_0^2}{Z} \cos \phi \, \frac{\pi}{\omega},$$

or

$$\langle P \rangle = \frac{1}{2} \frac{V_0^2}{Z} \cos \phi. \qquad (33.45a)$$

It is usually most convenient to write this result in terms of V_{rms}. Using Equation 33.41, we obtain

$$\langle P \rangle = \frac{V_{\mathrm{rms}}^2}{Z} \cos \phi. \qquad (33.45b)$$

Another form of this expression you can obtain by simple manipulation is

$$\langle P \rangle = V_{\mathrm{rms}} i_{\mathrm{rms}} \cos \phi. \qquad (33.45c)$$

The term $\cos \phi$ is called the **power factor**. Because the phase constant ϕ must always lie between $-90°$ (the purely capacitive case) and $90°$ (the purely inductive case), the value of the power factor always lies between 0 and 1. When $\phi = 0$, we have $\cos \phi = 1$. This is the case when the circuit is purely resistive or when the resonance condition holds and $X = 0$. When $\cos \phi = 1$, we have $Z = R$, and Equations 33.45a, 33.45b, and 33.45c assume the simple forms

$$\langle P \rangle = \frac{1}{2} \frac{V_0^2}{R}, \quad \langle P \rangle = \frac{V_{\mathrm{rms}}^2}{R}, \quad \text{and} \quad \langle P \rangle = V_{\mathrm{rms}} i_{\mathrm{rms}} \quad \text{for } \phi = 0, \cos \phi = 1.$$

$$(33.46a,b,c)$$

Equations 33.46b and 33.46c look like the corresponding expressions for dc circuits (Equations 27.19a and 27.18). But the ac equations are not the same as the dc equations because they involve the rms values of variable quantities, not constant quantities.

There is a simple relation among the quantities R, Z, and $\cos \phi$. Figure 33.10 repeats some of the information of Figure 33.7 in a modified form. In Figure 33.7, the phasors representing $\mathbf{V}_0$, $\mathbf{V}_{R0}$, and $\mathbf{V}_{X0}$ form a right triangle containing the angle ϕ. Remember that $V_0 = i_0 Z$, $V_{R0} = i_0 R$, and $V_{X0} = i_0 X$. Dividing each vector by the scalar i_0, we obtain the vector triangle shown in Figure 33.10. The sides of the triangle are R and X,

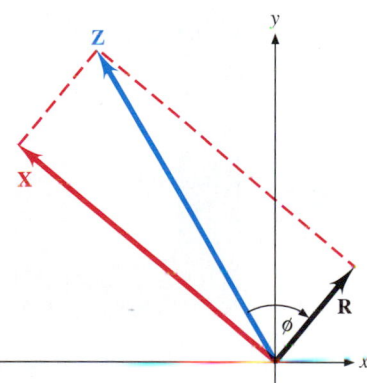

FIGURE 33.10 A vector diagram relating impedance Z, resistance R, and reactance X is drawn on the basis of the phasor triangle $\mathbf{V}_0 = \mathbf{V}_{R0} + \mathbf{V}_{X0}$ of Figure 33.7. It is evident that $\cos \phi = R/Z$.

and its hypotenuse is Z. So we have

$$\cos \phi = \frac{R}{Z} \quad \text{or} \quad Z = \frac{R}{\cos \phi}. \qquad \text{(33.47a,b)}$$

From Equation 33.29, we have $V_0 = i_0 Z$. By substituting into this expression for V_0 the value of Z given by Equation 33.47b, we obtain

$$V_0 = \frac{i_0 R}{\cos \phi}.$$

We can use Equations 33.41a and 33.41b to rewrite this expression in terms of rms values:

$$V_{\text{rms}} = \frac{i_{\text{rms}} R}{\cos \phi}.$$

Substitution of this value of V_{rms} into Equation 33.45c gives

$$\langle P \rangle = i_{\text{rms}}{}^2 R. \qquad \text{(33.48)}$$

This relation holds regardless of the value of the power factor and expresses the important fact that, *in an ac circuit, power dissipation takes place only in the resistance.* Although work is indeed required to charge a capacitor or to build up current through an inductor, the work done is conservative and averages to zero over a complete cycle.

Nevertheless, it is often undesirable to build circuits with large reactance. A case in point is the ac electric motor, whose rotor and stator coils have considerable inductance as well as resistance. With such a highly inductive load, a large current must flow back and forth from the power source to the motor during each cycle, as the inductive coils are repeatedly energized and deenergized. No useful work is done in this process, and the excess current results in wasteful joule heating of the power line. This difficulty is overcome by connecting a capacitor across the input and output leads of the ac motor. Can you explain how this expedient reduces the reactance "seen" by the ac source?

The Transformer

One of the great advantages of ac systems is that it is relatively easy to convert electric power from a high-voltage, low-current form into a low-voltage, high-current form or vice versa. To give one example, electric power is generated and (mainly) consumed at relatively low voltage—say, between 100 and a few thousand volts. At these voltages, and particularly at the lower end of the range, it is easy to insulate conductors. Also, short circuits, breaks, and other failures present relatively low risks at low voltage. On the other hand, it is important to transmit power over long distances at very high voltages. With high-voltage transmission, we minimize the current $i = P/V$ and thus the $i^2 R$ losses. (The situation may change in the not-too-distant future, with the advent of practical superconducting transmission lines having no $i^2 R$ losses at all.) Modern transmission lines are often operated at 500 kV or more.

The device that makes ac voltage conversion practical is the **transformer**, shown schematically in Figure 33.11a and represented by the standard symbol shown in Figure 33.11b. Two coils, 1 (the **primary**) and 2 (the **secondary**), having different numbers of turns N_1 and N_2, are wound in the same sense on the same iron core. This assures that essentially all the magnetic flux Φ_{m} threading either coil at any time also threads the other. Coil 1 is connected to an ac source, and coil 2 can be connected through a switch to a load, represented in Figure 33.11a by the resistance R_2.

To begin with, suppose that the switch in the secondary circuit is open while a voltage $V = V_1 \sin \omega t$ is imposed on the primary coil by the ac source. If the resistance of the primary is R_1 and its inductance is L_1, the impedance of the primary is (Equation 33.30b)

$$Z_1 = \sqrt{R_1{}^2 + (\omega L_1)^2}.$$

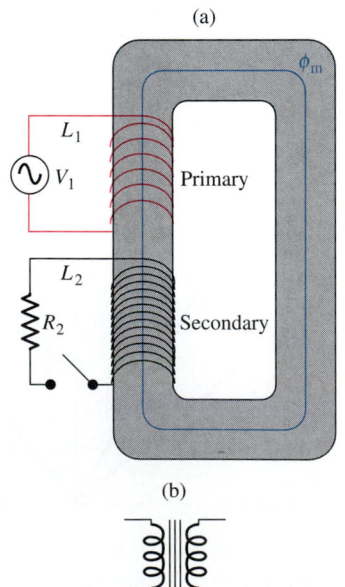

FIGURE 33.11 (*a*) A transformer, showing the primary and secondary circuits. (*b*) Standard symbol for a transformer.

By proper design, the inductive reactance can be made much larger than the resistance so that $Z_1 \simeq \omega L_1$.

According to Faraday's law, a changing current passing through the primary results in a back emf $V_b = -L_1 \, di/dt$. Thus at any instant, Kirchhoff's loop rule for the primary circuit has the form

$$V - V_b - iZ = 0,$$

or
$$V_1 \sin \omega t + L_1 \frac{di}{dt} - \omega L_1 i = 0.$$

In a purely inductive circuit, the current lags the driving voltage by 90° and can therefore be written

$$i = i_1 \sin (\omega t - 90°) = -\frac{V_1}{\omega L_1} \cos \omega t.$$

We differentiate with respect to time to find di/dt:

$$\frac{di}{dt} = \frac{V_1}{L_1} \sin \omega t.$$

Inserting this value of di/dt into Kirchhoff's loop equation (Section 27.7), we obtain

$$V_1 \sin \omega t - V_1 \sin \omega t - \omega L_1 i = 0,$$

which can be true only if $i = 0$. Thus *no current*—practically speaking, negligible current—*flows in the primary circuit when the secondary circuit is open.* The back emf is equal to and opposite the driving emf; that is, the two emf's are 180° out of phase.

Each turn of the secondary coil is threaded by the same oscillating magnetic flux that threads each turn of the primary coil. Consequently, the same emf is induced *per turn* in the secondary as in the primary. The relation between the maximum emf V_2 across the entire secondary coil and the maximum emf V_1 across the primary coil is

$$\frac{V_2}{V_1} = \frac{N_2}{N_1}. \tag{33.49}$$

However, the instantaneous emf in the secondary is 180° out of phase with the driving emf in the primary. This is because the two coils are wound in the same sense and the induced emf in the secondary is in phase with the back emf in the primary.

If $N_2 > N_1$, then $V_2 > V_1$, and the transformer is called a **step-up transformer**. If $N_2 < N_1$, then $V_2 < V_1$, and the transformer is called a **step-down transformer**.

The Transformer under Load

Now, consider what happens when the switch in the secondary circuit is closed and current passes through the load. The current is driven by the emf $V = V_2 \sin \omega t = -(N_2/N_1)V_1 \sin \omega t$. Because the secondary emf is out of phase with the primary emf, the secondary current is out of phase with the primary current. (The value of the phase difference depends on the nature of the load.) As a consequence, the oscillation of the secondary current tends to build magnetic flux in the core in the sense opposite that of the primary. Because both coils are on the same core, the net result at any instant is a reduction in the flux through the core. The effect is as if the self-inductance of the primary coil is reduced when current flows through the secondary circuit. The back emf in the primary is reduced, and the current i_1 is no longer (essentially) zero.

Indeed, we can use energy conservation and Equation 33.45c to write $\langle P_1 \rangle = \langle P_2 \rangle$, or

$$V_{1\,rms} i_{1\,rms} \cos \phi_1 = V_{2\,rms} i_{2\,rms} \cos \phi_2. \tag{33.50}$$

Using Equation 33.49, we have

$$\frac{i_{2\,rms}}{i_{1\,rms}} = \frac{N_1 \cos \phi_1}{N_2 \cos \phi_2}. \tag{33.51a}$$

Until superconducting transmission lines become practical, the world will depend on high-voltage lines carried by towers like these. The practicality of long-distance electrical energy transmission depends on the ease with which the voltage of alternating-current systems can be transformed.

It is possible to design a transformer so that under working conditions, with the design current i_2 in the secondary circuit, both power factors are approximately equal to 1. In that case, we have

$$\frac{i_{2\,\mathrm{rms}}}{i_{1\,\mathrm{rms}}} = \frac{N_1}{N_2}. \tag{33.51b}$$

Because the ac source in the primary circuit ultimately drives current through the secondary load, there must be a quantitative connection among V_1, i_1, and R_2. Using Equations 33.50 and 33.46b and assuming that the power factors are approximately 1, we obtain

$$V_{1\,\mathrm{rms}}i_{1\,\mathrm{rms}} = \frac{V_{2\,\mathrm{rms}}^2}{R_2}.$$

Using Equation 33.49 to express V_2, we have

$$\frac{V_{1\,\mathrm{rms}}}{i_{1\,\mathrm{rms}}} = \left(\frac{N_1}{N_2}\right)^2 R_2. \tag{33.52}$$

That is, the resistance in the secondary is "reflected" into the primary. More generally, there is reflection of any impedance Z_2 in the secondary. But the **reflected impedance** differs from Z_2 by the factor $(N_1/N_2)^2$. For this reason, transformers are often used to link an ac source to a load whose impedance is very different from the internal impedance of the source. This process is called **impedance matching**.

If the resistance of the primary circuit is negligible, the ratio on the left side of Equation 33.52 is the impedance Z_1 of the primary circuit. If the impedance of the secondary circuit is Z_2, we have the relation

$$Z_1 = \left(\frac{N_1}{N_2}\right)^2 Z_2. \tag{33.53}$$

By using a transformer with the proper turns ratio, we can make a load of arbitrary impedance Z_2 "appear" to the source of emf as though it had any desired value Z_1. The usefulness of doing this is the subject of Problem 33.40.

EXAMPLE 33.3

The primary coil of a transformer has 25 turns, and its secondary coil has 175 turns. The primary is connected to an ac source whose rms output is 37.5 V, and the secondary is connected to a load resistance $R_2 = 2500\ \Omega$. (a) What is the rms output emf of the secondary coil? (b) What is the rms current in the secondary circuit? (c) What is the mean power output of the ac source? (d) What is the rms current in the primary circuit? (e) What is the reflected impedance Z_1 of the load in the secondary circuit? Assume that the power factors are approximately 1.

SOLUTION:
(a) Find $V_{2\,\mathrm{rms}}$.
    You use Equation 33.49 to calculate

$$V_{2\,\mathrm{rms}} = V_{1\,\mathrm{rms}}\frac{N_2}{N_1} = 37.5\ \mathrm{V} \times \frac{175\ \mathrm{turns}}{25\ \mathrm{turns}} = 263\ \mathrm{V}.$$

(b) Find $i_{2\,\mathrm{rms}}$.
    The ac "Ohm's law" (Equation 33.29) gives you

$$i_{2\,\mathrm{rms}} = \frac{V_{2\,\mathrm{rms}}}{R_2} = \frac{263\ \mathrm{V}}{2500\ \Omega} = 105\ \mathrm{mA}.$$

(c) Find $\langle P \rangle$.
    The power output of the source must be equal to the power dissipated as heat in the resistor. So you have

$$\langle P \rangle = i_{2\,\mathrm{rms}}^2 R_2 = (0.105\ \mathrm{A})^2 \times 2500\ \Omega = 27.6\ \mathrm{W}.$$

(d) Find $i_{1\,\mathrm{rms}}$.
    There are several ways to calculate the primary rms current. A simple way is to write

$$i_{1\,\mathrm{rms}} = \frac{\langle P \rangle}{V_{1\,\mathrm{rms}}} = \frac{27.6\ \mathrm{W}}{37.5\ \mathrm{V}} = 735\ \mathrm{mA}.$$

This value is indeed $7i_{2\,\mathrm{rms}}$, as you could find on the basis of the turns ratio.

(e) Find Z_1.
    From Equation 33.53, you have

$$Z_1 = (\tfrac{1}{7})^2 \times 2500\ \Omega = 51.0\ \Omega.$$

The advantages of impedance matching are considered further in Problem 33.40.

THINKING LIKE A PHYSICIST

In this chapter and in Chapter 32, we have studied and exploited the behavior of three electrical circuit components: capacitor, resistor, and inductor. At first, the behaviors of the three seem distinct and unconnected. But there is a subtle relation among them that can really be understood *only* in mathematical terms:

$$\text{capacitor} \qquad \text{resistor} \qquad \text{inductor}$$

$$V = \frac{1}{C}q \qquad V = R\frac{dq}{dt} \qquad V = L\frac{d^2q}{dt^2}.$$

The voltages across the terminals depend, respectively, on the zeroth, first, and second time derivatives of charge. This relation is made vivid by the phasor picture, in which V_{C0}, V_{R0}, and V_{L0} are directed 90° apart (Figure 33.7).

Because the zeroth, first, and second time derivatives have a mathematical likeness to the three kinematic quantities displacement x, velocity $v = dx/dt$, and acceleration $a = d^2x/dt^2$, we are tempted to seek analogies. We have already seen one such analogy between the LC oscillator (Section 32.4) and the simple harmonic oscillator. Although we have not considered the matter in detail, the series ac circuit is analogous to such driven, damped mechanical systems as the child's swing or the bowed violin string. Even though we must always be careful not to push analogies too far, knowledge of one system is often a source of insight into the behavior of another system that is physically quite different but mathematically similar.

The mathematical relation among capacitors, resistors, and inductors also guarantees that study of the behavior of ac circuits will involve solution of second-order differential equations such as Equation 33.4. If you think back, you will see that the study of mechanics likewise involves solution of such equations, Newton's second law ($F = m\,d^2x/dt^2$) being the most important. Other fields of physics display the same mathematical characteristic. As you continue your study of physics and areas of engineering that depend on physics, you will become adept at solving second-order differential equations.

C	capacitance
ΔE	energy dissipated in a resistor per cycle
i	electric current
$\mathbf{i}_0, i_0$	current phasor, peak current
L	(self-)inductance
N	number of turns of a coil
$P, \langle P \rangle$	instantaneous power, mean power
q	electric charge
Q	quality factor
R	resistance
U	potential energy

$\mathbf{V}_0, V_0$	voltage phasor, peak voltage
X, X_L, X_C	(net) reactance, inductive reactance, capacitive reactance
Z	impedance
δ	phase constant determined by choice of zero of time
ν	frequency
$\phi, \cos\phi$	phase constant measured from $\mathbf{V}_0$ to $\mathbf{i}_0$, power factor
Φ_{m}	magnetic flux
ω, ω_0	angular frequency, resonance angular frequency

Summing Up

When applied to the ac series circuit, Kirchhoff's loop rule yields Equation 33.4,

$$\omega V_0 \cos\omega t - L\frac{d^2i}{dt^2} - R\frac{di}{dt} - \frac{1}{C}i = 0.$$

This equation can be solved to yield the current, the voltages across the components, and the phase relations among these quantities, by application of the **phasor** method. This method is summarized immediately following Example 33.1.

The solution of Equation 33.4 can be expressed algebra-ically in terms of the **impedance** Z and the **phase constant** ϕ, given respectively by Equations 33.34 and 33.33b,

$$Z = \sqrt{R^2 + X^2} \quad \text{and} \quad \phi = \tan^{-1}\frac{X}{R}.$$

In these equations, the **reactance** is defined in Equation 33.32,

$$X \equiv X_L - X_C,$$

where the **inductive reactance** $X_L \equiv \omega L$ and the **capacitive reactance** $X_C \equiv 1/\omega C$ are as defined in Equations 33.17 and 33.9.

The impedance can be used to write **"Ohm's law for ac circuits,"** given by Equation 33.29,

$$Z \equiv \frac{V_0}{i_0}.$$

This expression relates the peak voltage across the circuit to the peak current (which in general does not correspond in time to the peak voltage). The phase constant ϕ is the angle measured from the phasor $\mathbf{V}_0$ to the phasor $\mathbf{i}_0$.

When the angular frequency ω of the emf is such that $X_C = X_L$ and thus $X = 0$, the impedance is equal to the resistance and the phase constant is 0. The circuit then mimics a dc circuit in that the relation between the instantaneous driving emf V and the instantaneous current i is described by Ohm's law, $V = iR$. The **resonance condition** is given by Equation 33.35a,

$$\omega_0 = \frac{1}{\sqrt{LC}}.$$

The "sharpness" of the resonance is expressed by the **quality factor** Q given by Equation 33.38.

If the driving emf is sinusoidal, its root-mean-square value is given by Equation 33.41,

$$V_{rms} = \frac{1}{\sqrt{2}} V_0.$$

The rms value of the current is given by Equation 33.43, which has the similar form

$$i_{rms} = \frac{1}{\sqrt{2}} i_0.$$

Under the same circumstances, the mean power is given by any form of Equation 33.45,

$$\langle P \rangle = \frac{1}{2} \frac{V_0^2}{Z} \cos \phi = \frac{V_{rms}^2}{Z} \cos \phi = V_{rms} i_{rms} \cos \phi.$$

The term $\cos \phi$ is the **power factor**. It relates the resistance to the impedance, as indicated by Equation 33.47a,

$$\cos \phi = \frac{R}{Z}.$$

Regardless of the value of the power factor, power is dissipated in the circuit only in the resistance and is given by Equation 33.48,

$$\langle P \rangle = i_{rms}^2 R.$$

KEY TERMS

Section 33.2 The ac Series Circuit
ac source ▪ phasor ▪ capacitive reactance, inductive reactance, reactance ▪ impedance ▪ "Ohm's law for ac circuits"

Section 33.3 Resonance
resonance frequency, resonance condition, resonance curve ▪ quality factor ▪ damping

Section 33.4 Root-Mean-Square Voltage, Current, and Power
peak-to-peak voltage ▪ power factor

Section 33.4 The Transformer
primary coil, secondary coil ▪ step-up transformer, step-down transformer ▪ reflected impedance ▪ impedance matching

<hr/>

Queries and Problems for Chapter 33

QUERIES

33.1 *(2) A picture is worth* . . . Sketch a phasor diagram for an ideal *LC* circuit. How can you use the diagram to find **(a)** the instantaneous voltage across the capacitor? **(b)** the instantaneous voltage across the inductor? **(c)** the current in the circuit?

33.2 *(2) Add 'em up.* An ac series circuit operates at angular frequency ω. In addition to the ac source, the circuit consists of a number of components having impedance $Z_1, Z_2, Z_3, \ldots$. Can you write the impedance $Z = V_0/i_0$ of the entire circuit as $Z = Z_1 + Z_2 + Z_3 + \cdots$? Why or why not?

33.3 *(2) Angling for complements.* In this chapter, we have written ac emf's in the form $V = V_0 \sin \omega t$. Suppose instead we write $V = V_0 \cos \omega t$. What effect does this change have on the phasor picture of a series circuit?

33.4 *(3) Virtuoso.* The action of the body of a violin is analogous to that of an ac circuit. The violin receives mechanical energy from the vibrating bowed string (mainly through the

bridge) and dissipates that energy as sound into the surrounding air. The response of a good violin is smooth. That is, the intensity of the radiated sound varies only gradually as the musician plays a scale. (Cheap violins that do not satisfy this condition are said to produce "wolf notes.") Is it desirable for the violin to have large or small Q? Why is it not possible to make good violins out of sheet metal?

33.5 *(4) Adding capacity.* Answer the question with which Section 33.4 ends.

33.6 *(5) Heart of steel.* **(a)** What is the purpose of the iron core in a transformer? **(b)** For transformers designed to operate at high frequency, it is not practical to use an iron core, and *air-core* transformers are used instead. How must such a transformer be wound?

33.7 *(5) Crazy Eddy.* Currents called *eddy currents* flow in the iron core of a transformer when it is in operation. **(a)** What is the cause of eddy currents? **(b)** Why are they undesirable?

(c) In transformers, the iron core is usually *laminated*. That is, the core is made not of a solid piece but of a stack of thin sheets of iron, each of which links the two coils, bonded to- gether by thin sheets of an insulating cement. Such a lami- nated core minimizes eddy currents. Explain.

PROBLEMS

GROUP A

33.1 *(2) Is it OK to peak?* A 25-μF capacitor and a 160-Ω resistor are connected in series across a 60-Hz source of emf having peak voltage V_0. What are the peak voltages across **(a)** the resistor and **(b)** the capacitor?

33.2 *(2) Evening out.* At what angular frequency are the re- actances of a 12.5-mH inductor and a 2.75-μF capacitor equal?

33.3 *(2) Spreading out.* If the inductor and the capacitor of Problem 33.2 are part of an ac series circuit, what is the re- actance of the circuit **(a)** at $\omega = 7000$ rad/s? **(b)** at $\omega = 2000$ rad/s?

33.4 *(2) Passive resistance?* A 2750-turn solenoid of length 45 cm and radius 1.75 cm is connected across a 60-Hz ac source. If the phase constant is 64°, what is the resistance of the solenoid?

33.5 *(2) Laying it all out, I.* Make a rough phasor diagram for the circuit illustrated here. Make a rough estimate of the impedance and the phase constant.

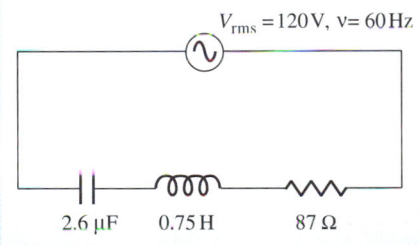

$V_{rms} = 120$ V, $\nu = 60$ Hz

2.6 μF   0.75 H   87 Ω

33.6 *(2) Laying it all out, II.* For the circuit of Problem 33.5, find **(a)** the inductive reactance; **(b)** the capacitive reactance; **(c)** the impedance; **(d)** the rms current; **(e)** the rms voltage across the resistor, inductor, and capacitor; **(f)** the phase con- stant ϕ; **(g)** the power factor; **(h)** the mean power dissipated in the resistor. **(i)** If you have done Problem 33.5, how well are your rough estimates confirmed by the calculations of this problem?

33.7 *(2) Effects of change.* What are the impedance and phase constant of the circuit shown in Problem 33.5 if **(a)** the re- sistance is doubled, with L and C unchanged? **(b)** the capac- itance is doubled, with L and R unchanged? **(c)** the inductance is doubled, with R and C unchanged?

33.8 *(2) It doesn't add up!* A series ac circuit consists of a 400-Hz source whose peak voltage is 691 V, a 125-mH in- ductor, a 375-Ω resistor, and a 1.55-μF capacitor. Find the peak voltages across **(a)** the inductor, **(b)** the resistor, and **(c)** the capacitor.

33.9 *(2) Skewing the result.* Suppose the source in Problem 33.8 has variable frequency. Other things remaining the same, find the frequency at which **(a)** the resistance of the resistor is 1% of the reactance of the inductor and **(b)** the reactance of the capacitor is 1% that of the inductor. **(c)** In each case, compare the reactance of the inductor with the total imped- ance of the series circuit.

33.10 *(2) Doing something Ohm couldn't.* A series circuit consists of a 37.5-μF capacitor, a 155-Ω resistor, and a 720-mH inductor plugged into a 120-V, 60-Hz wall socket. What is the current through the circuit?

33.11 *(2) Antiantiderivative?* Check the correctness of Equa- tion 33.16 by showing that its derivative with respect to time is Equation 33.15.

33.12 *(3) Incapacitated and choked.* An ac voltmeter, con- nected between points D and E in the circuit below, reads 50 V. Connected between E and F, the voltmeter again reads 50 V. **(a)** What will be the reading if the meter is connected between D and F? **(b)** What is the power factor of the circuit? **(c)** The capacitor is replaced with an inductor of negligible resistance. The voltmeter now reads 35 V when connected between D and E. What will the voltmeter read when con- nected between E and F? **(d)** What is the power factor of the circuit now?

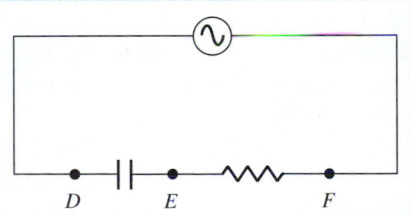

D E F

33.13 *(3) Walkperson.* The standard AM broadcast band cov- ers the frequency range from 0.550 MHz to 1.600 MHz. The tuner in a radio is essentially an LC circuit having a fixed inductor and a variable capacitor whose value is adjusted by turning the tuning knob. A typical value for the inductor is 0.3 mH. What is the capacitance range required of the variable capacitor?

33.14 *(3) Choose your partner.* You have a 25.5-mH induc- tor. If you want to build an LC circuit, what capacitance should you choose for a resonant angular frequency of **(a)** 62.8 rad/s? **(b)** 1.35×10^4 rad/s? **(c)** 8.53×10^5 rad/s?

33.15 *(3) Peak performance.* A coil of inductance $L = 340$ mH and resistance $R = 42\ \Omega$ is connected in series with a capacitor C across a 60-Hz ac source whose peak voltage is 155 V. **(a)** For what value of C is the peak voltage V_{L0} across the coil maximized? **(b)** What is this maximum value of V_{L0}? **(c)** What is the corresponding peak voltage V_{C0} across the capacitor?

33.16 *(4) Half-measure.* A series circuit contains a resistor R, an inductor L, and a source of emf whose peak voltage V_0 is fixed but whose angular frequency ω can be varied. When the source is set to deliver dc ($\omega = 0$), the power dissipated in the resistor in the steady state (that is, the power when $t \gg \tau_L$) is $\langle P \rangle$. The frequency is increased until the power is $\langle P' \rangle = \langle P \rangle / 2$. What is the new value of ω?

33.17 *(4) Going through a phase.* A coil having resistance $R = 17\ \Omega$ is connected across a 155-V, 50-Hz source. If the phase angle between current and voltage is 55° and the power consumption of the coil is 420 W, find the inductance L of the coil.

33.18 *(4) Increased capacitance, reduced capacity.* You have a 50-W, 110-V light bulb, which you want to plug into a 440-V, 60-Hz source. To keep the bulb from burning out, you plan to connect it in series with a capacitor. What value of C should you use?

33.19 *(4) Getting an angle on things.* A 120-Ω resistor and a capacitor are connected in series across a 170-V peak ac source. If the peak current is 0.52 A, what is the phase constant of the circuit?

33.20 *(5) Transformation.* A step-down transformer is illustrated below. The primary and secondary voltages are quoted in rms values. **(a)** What is the rms current i_2 in the secondary circuit? **(b)** Neglecting losses in the transformer, determine the primary current i_1. **(c)** What is the effective impedance Z_1 "seen" by the ac source?

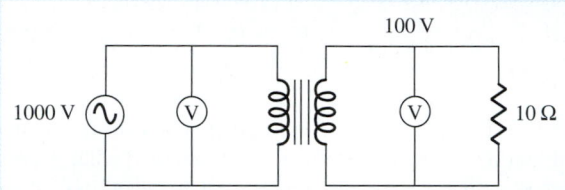

GROUP B

33.21 *(2) Dimensional check, I.* Show that the dimensions of $X_C \equiv 1/\omega C$ are the same as the dimensions of R.

33.22 *(2) Dimensional check, II.* Show that the dimensions of $X_L \equiv \omega L$ are the same as the dimensions of R.

33.23 *(2) In proportion.* In a series ac circuit, the peak voltage drop across the capacitor is twice that across the resistor, $V_{C0} = 2V_{R0}$, and the peak voltage drop across the inductor is $V_{L0} = 3V_{R0}$. Express V_{R0} in terms of the peak emf of the source.

33.24 *(2) Going to any length.* A single-layer solenoid of radius b is closely wound with insulated wire of radius a made of metal whose resistivity is ρ. There is an angular frequency $\omega_=$ at which the inductive reactance of the solenoid is equal to its resistance. **(a)** Express $\omega_=$ in terms of ρ, a, b, and necessary constants. **(b)** What is the impedance of the solenoid at this angular frequency?

33.25 *(2) Phase difference.* Two coils, having resistance as well as inductance, are connected in series across an ac line whose peak emf is $V_0 = 170$ V. The peak voltages across the coils are $V_1 = 77$ V and $V_2 = 123$ V. What is the phase difference between the phasors $\mathbf{V}_1$ and $\mathbf{V}_2$? What are the phase constants ϕ_1 and ϕ_2 of these phasors with respect to $\mathbf{V}_0$? (Hint: The vector triangle of interest is not a right triangle, but you can use the law of cosines.)

33.26 *(2) Real inductors.* You use the circuit illustrated in the next column to measure the inductance L and the resistance R of a coil. The ac ammeter indicates a current $i_{rms} = 1.20$ A. The ac source is an ordinary power line, with $V_{rms} = 120$ V and $\nu = 60$ Hz. The resistor has unknown resistance; call it R'. The rms voltage across this resistor is 60.0 V, and that across the coil is 70.0 V. **(a)** What is the resistance R' of the resistor? **(b)** What is the magnitude of the phase difference

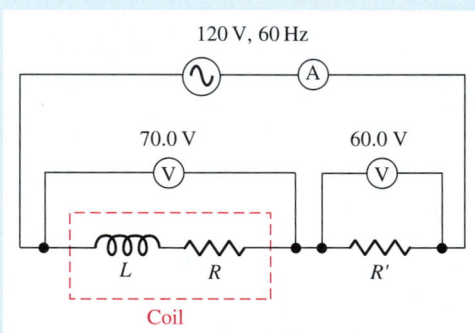

between the voltages across the resistor and the coil? **(c)** What is the inductance L of the coil? **(d)** What is the resistance R of the coil? (Hint: See the hint for Problem 33.25.)

33.27 *(2) Brightness test.* The system shown below consists of a light bulb of nominal resistance 50 Ω, an inductor of

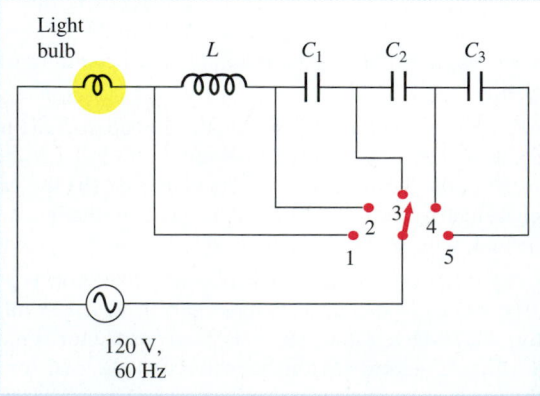

inductance 2.9 H, and three capacitors, each having capacitance 4.86 μF. The ac source is an ordinary power line, with rms voltage 120 V and frequency 60 Hz. When the switch is at position 1, the light bulb glows normally. With the switch moved to position 2, the glow is barely visible. With the switch at position 3, the glow is distinctly visible, though the bulb is substantially less bright than normal. With the switch at position 4, the bulb glows with nearly normal brightness. With the switch at position 5, the brightness is about the same as at position 3. Neglecting the variation of resistance of the bulb filament with brightness, find the rms current and the power dissipated in the bulb for each switch setting, and explain the observations.

33.28 *(2) High-pass filter.* The ac source in the figure below has a fixed peak emf V_0 but variable frequency. Show that the voltage V_{BC} across the resistor is given by

$$V_{BC} = V_0 \frac{R}{\sqrt{R^2 + 1/(\omega C)^2}}$$

and that the circuit thus terminated discriminates against the passage of low-frequency current. Make a rough sketch of V_{BC}/V_0 versus ω.

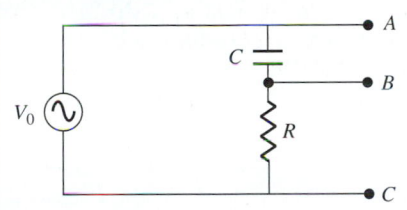

33.29 *(2) Low-pass filter.* For the circuit of Problem 33.28, find the voltage V_{AB} across the capacitor, and show that the circuit thus terminated discriminates against the passage of high-frequency current. Make a rough sketch of V_{AB}/V_0 versus ω.

33.30 *(3) Marriage of convenience.* An ac series circuit has resonant frequency ω_0. A second series circuit, whose resistance, inductance, and capacitance have values different from those of the first circuit, happens to have the same resonant frequency ω_0. Show that, if you take all six components and connect them in series to form a single circuit, the resonant frequency is still ω_0.

33.31 *(3) Geometric mean-ing.* At any frequency other than the resonance frequency ω_0, the impedance of an ac series circuit is greater than the resonance value $Z_0 = R$. Consider any frequency $\omega_1 \neq \omega_0$. Show that there is another frequency, ω_2, at which the circuit has the same impedance as at ω_1 (that is, $Z_2 = Z_1$) and that $\sqrt{\omega_1 \omega_2} = \omega_0$. (Hint: What is the relation between the reactances X_1 and X_2?)

33.32 *(3) Half-width at half-height.* Here is an analytical way to calculate the "half-width" of the resonance curve for an ac series circuit, described immediately after Equation 33.38 and shown in Figure 33.9. At resonance, $\omega = \omega_0$, the impedance is $Z_0 = R$. At the two half-height points, $Z_0 = 2R$. **(a)** Use this information to calculate the angular frequencies ω_1 and ω_2 of the two points, and show that the *half-width at*

half-height is

$$\Delta\omega = \frac{1}{2}|(\omega_1 - \omega_2)| = \frac{\sqrt{3}}{2}\frac{\omega_0}{Q}.$$

(b) Measure the "half-width" for the upper curve in Figure 33.9, and compare your result with that yielded by the equation of part **a**.

33.33 *(3) Phase versus angular frequency.* (You should use a personal computer or a programmable calculator for this problem. Otherwise, the work is tedious.) Plot the phase constant ϕ as a function of ω for an ac series circuit having $L = 10.0$ mH, $C = 1.00$ μF, and **(a)** $R = 25.0\ \Omega$, **(b)** $R = (\pi/2) \times 100\ \Omega \simeq 157.1\ \Omega$, and **(c)** $R = 250\ \Omega$. For all three plots, cover a range from $\omega = 0$ to at least $\omega = 2\omega_0$. **(d)** What is the value of Q for each circuit? **(e)** Sketch the curves for the limiting cases $Q = 0$ and $Q = \infty$.

33.34 *(3) Quality time.* In a certain series ac circuit operating at resonant frequency ω_0, the peak voltage across the capacitor is N times the peak source voltage. What is the Q-factor of the circuit?

33.35 *(4) Half-width at half-power.* (To do part **a** of this problem, you should use a personal computer or a programmable pocket calculator. Otherwise, the task is tedious. You can work parts **b** and **c** without doing part **a** first, but the insight you gain in doing part **a** will be very helpful in doing the other parts.) **(a)** For the ac series circuit whose resonance characteristics are described by the upper curve in Figure 33.9 ($L = 10.0$ mH, $C = 1.0$ μF, $R = 12.5\ \Omega$), draw a graph of the power dissipated in the resistor as a function of ω. (Don't forget the power factor!) **(b)** Let $\langle P_0 \rangle$ be the rms power at ω_0. Show that the *half-power points* satisfy the condition $Z = \sqrt{2}R$. **(c)** Define the *half-width at half-power* to be one-half the angular frequency difference between the half-power points—$\Delta\omega \equiv \frac{1}{2}|(\omega_1 - \omega_2)|$—and show that $\Delta\omega = (\omega_0)/(2Q)$.

33.36 *(4) Manipulative.* Carry out the manipulation required to obtain Equation 33.45c from Equation 33.45b.

33.37 *(4) Square root-mean-square.* The figure below shows a *square wave* of amplitude V_0. What is the value of V_{rms} for this wave? (Hint: If you do a little simple geometry in your head, you won't have to calculate anything.)

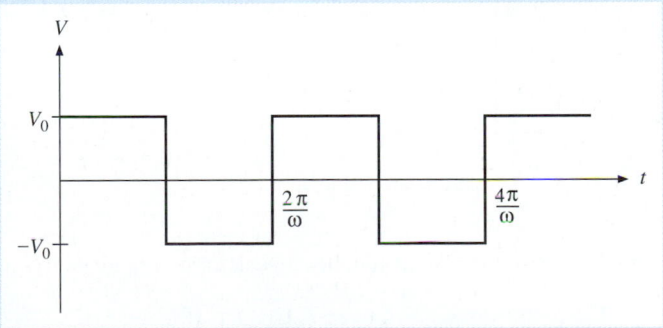

33.38 *(4) Rip saws and crosscut saws.* **(a)** Figure a on page 908 shows a *sawtooth wave* of amplitude V_0. Find V_{rms}. **(b)** Figure b shows a *triangular wave* of amplitude V_0. Find V_{rms}. (Hint: See the hint in Problem 33.37.)

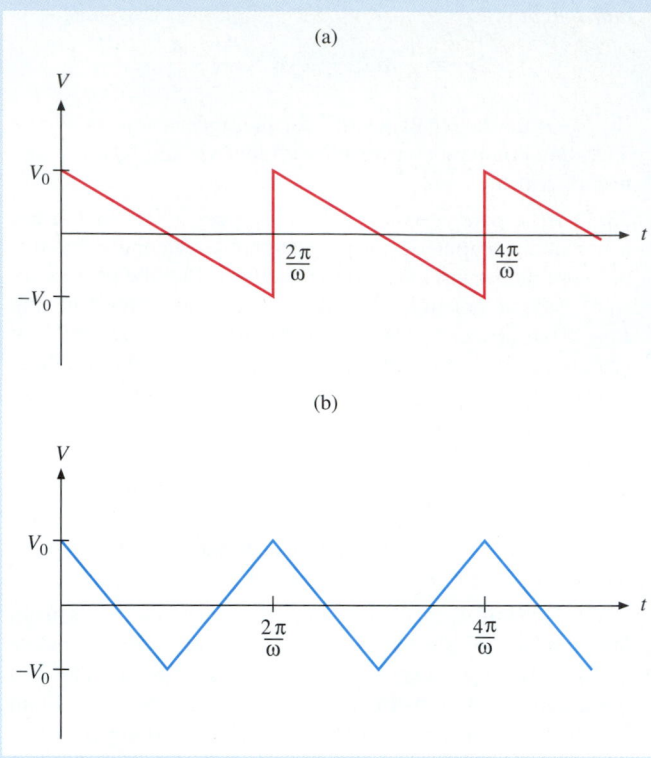

(a)

(b)

33.39 *(5) Resistance matching.* A real source of emf has internal resistance R_s. The source is connected to a load whose resistance is R. **(a)** What fraction of the total power P_s produced by the source is useful power P dissipated at the load? **(b)** For what value of R is the useful power maximized? **(c)** What is the value of P/P_s for this value of R?

33.40 *(5) Impedance matching.* A real source of emf has internal impedance Z_s. The source is connected to a load whose impedance is Z. Assume that the power factor of the circuit is $\cos \phi \simeq 1$. **(a)** What fraction of the total rms power $\langle P_s \rangle$ produced by the source is useful power $\langle P \rangle$ dissipated at the load? **(b)** For what value of Z is the useful power maximized? **(c)** The source and the load are linked indirectly through a transformer instead of directly. If the primary of the transformer has N_1 turns and the secondary has N_2 turns, for what value of the load impedance Z is the useful power maximized?

33.41 *(4) Mean energy, I.* A sinusoidal voltage is applied across a capacitor. Show that the mean energy $\langle U \rangle$ stored in the capacitor is one-half the maximum energy U_0.

33.42 *(4) Mean energy, II.* Show that the result of Problem 33.41 is also correct for an inductor.

33.43 *(4) Mean power.* A sinusoidal voltage is applied across an inductor L. **(a)** What is the instantaneous power input P to the inductor when the current is i? **(b)** Show that the mean power $\langle P \rangle$ is equal to zero. Is this result inconsistent with that of Problem 33.42? Why or why not? **(c)** Without going through the calculation in detail, show that $\langle P \rangle$ is equal to zero for a capacitor as well.

GROUP C

33.44 *(2) ac parallel circuit.* **(a)** Sketch a phasor diagram for the circuit shown below. (Hint: At any instant, the voltage across each circuit element is $V = V_0 \sin \omega t$. So, instead of beginning with $\mathbf{i}_0$ and drawing $\mathbf{V}_R$, $\mathbf{V}_L$, and $\mathbf{V}_C$, as we did for the series circuit, begin with $\mathbf{V}_0$ and draw $\mathbf{i}_R$, $\mathbf{i}_L$, and $\mathbf{i}_C$.)

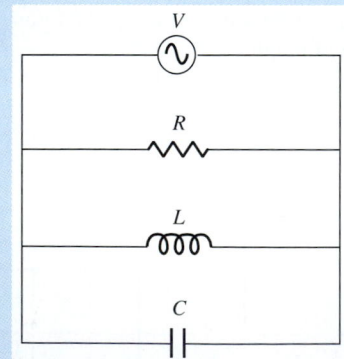

(b) Show that the source supplies a peak current i_0 given by

$$i_0 = V_0 \sqrt{\frac{1}{R^2} + \left(\frac{1}{X_C} - \frac{1}{X_L} \right)^2}.$$

(c) Show that the phase constant is

$$\phi = \tan^{-1} R\left(\frac{1}{X_C} - \frac{1}{X_L} \right).$$

33.45 *(3) Quality assurance.* Prove that Equation 33.38 is a correct expression for the quality factor Q defined in Equation 33.37.

33.46 *(G) Short circuit.* An ac series circuit is driven at an angular frequency $\omega = \omega_0$. The ac source is suddenly short circuited at an instant when $i = i_0$. If R is small (that is, if $R \ll \omega L, 1/\omega C$), it can be shown that the current is subsequently described by the expression

$$i = i_0 e^{-Rt/2L} \cos \omega_1 t,$$

where ω_1 is a little smaller than $1/\sqrt{LC}$ and $R/2L \ll \omega_1$. **(a)** Using the expression $U = \frac{1}{2}Li^2$ for the energy stored in the inductor, find the total energy stored in the circuit at time t_i and at time $t_f = t_i + 2\pi/\omega_1$. **(b)** Calculate the quality factor Q, and show that your result agrees with Equation 33.38. (Hint: For $x \ll 1$, $e^x \simeq 1 + x$.)

33.47 *(G) Energy tank.* The figure on page 909 shows an ac parallel circuit, sometimes called a *tank circuit*. The inductor and the resistor are shown as ideal, but real inductors often have significant resistance and must be represented as in the figure. The ac source produces an emf $V = V_0 \sin \omega t$. **(a)** Show that the maximum current i_0 in the circuit is

$$i_0 = V_0 \sqrt{\left(\omega C - \frac{\omega L}{R^2 + \omega^2 L^2} \right)^2 + \left(\frac{R}{R^2 + \omega^2 L^2} \right)^2}.$$

(b) Show that the current is in phase with the voltage when

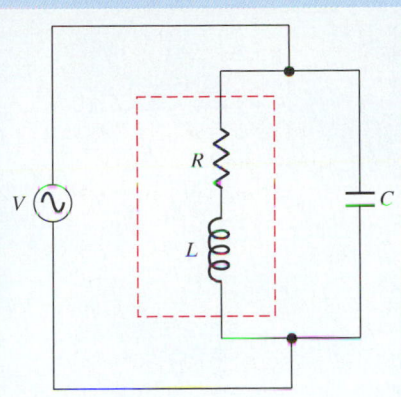

the angular frequency is

$$\omega_p = \sqrt{\frac{1}{LC} - \frac{R^2}{L^2}}.$$

The condition $\omega = \omega_p$ is sometimes called *parallel resonance*. **(c)** Is the impedance of the circuit a maximum or a minimum at parallel resonance? **(d)** Show that the impedance of the circuit at parallel resonance is

$$Z_{res} = \frac{L}{RC}.$$

(e) If the value of R is reduced, what happens to the impedance of the circuit at resonance? What happens to the Q of the circuit, defined in Equation 33.37?

33.48 *(G) Band width.* The source in an ac series circuit has fixed V_0 but variable ω. As ω is varied, the power $\langle P \rangle$ dissipated in the circuit varies. **(a)** Show that $\langle P \rangle$ attains its max-

imum value at resonance, when $\omega = \omega_0$. **(b)** At what two angular frequencies ω_- and ω_+ does the power have half its resonance value? **(c)** The *band width* of the circuit is defined to be $\Delta\omega \equiv \omega_+ - \omega_-$. Show that $\Delta\omega = \omega_0/Q$. The band width is an important property of any system whose purpose is the recording, transmission, or display of information. On the one hand, too narrow a band removes information; that is why voices sound different on the telephone than in direct conversation and why elderly people often have difficulty understanding speech even when it is amplified. On the other hand, too wide a band reduces selectivity, as when two stations are heard simultaneously on a radio.

33.49 *(5) The measurement is part of the problem.* A transformer whose secondary is a single loop of wire of resistance r is illustrated below. The points A, B, and C are uniformly spaced around the loop. An ammeter is attached to points A and C. The resistance of the ammeter and its leads is R. The ammeter and its leads can be positioned in either of the ways shown in the figure. At a certain instant, the emf induced in the secondary is V. **(a)** If the ammeter is in position D, show that it reads an instantaneous current $i_D = 3V/(9R + 2r)$. **(b)** If the ammeter is in position E, show that it reads an instantaneous current $i_E = 6V/(9R + 2r)$.

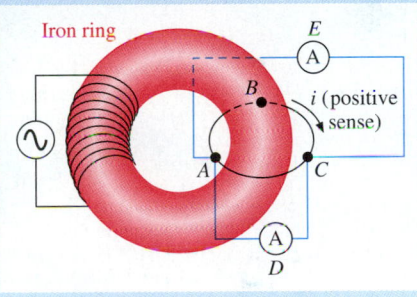

Electromagnetic Radiation

— A changing electric field is equivalent to an electric current, even though no charges move; such a current is called a displacement current.

— When displacement current is taken into account, Faraday's and Ampère's laws assume the same form in vacuum.

— Taken together with Gauss's laws for electric charge and for magnetism, Faraday's and Ampère's laws constitute Maxwell's equations, which (together with the Lorentz force equation) summarize everything we know about electromagnetism.

— Because a changing electric field gives rise to a magnetic field and a changing magnetic field gives rise to an electric field, electromagnetic waves can propagate through empty space.

— Electromagnetic waves carry energy.

— Electromagnetic waves can be generated by setting electric charge into oscillation; electromagnetic waves can be detected by means of the oscillation they induce in electric charges in a conducting antenna.

Left: Electromagnetic radiation is everywhere. This radiotelescope antenna is designed to detect natural radiation from the sky, but we make plenty ourselves for all sorts of purposes.

SECTION 34.1 Introduction*

In the preceding five chapters, you have seen how the electric and magnetic fields are linked by Ampère's law and Faraday's law. This chapter undertakes a deeper inquiry into the connection between the two fields.

In Section 34.2, we generalize Ampère's law to make it applicable to systems, such as the *RC* circuit, that contain a break in the conducting path. This generalization requires that we consider a new type of current called *displacement current*.

Generalizing to take account of the displacement current makes Ampère's law look much more like Faraday's law. This new-found symmetry between the two laws in turn makes possible a neat, beautiful presentation of everything fundamental that we know about electricity and magnetism in a set of five equations, four of which are called *Maxwell's equations*. These matters are the subject of Section 34.3. Much more than a pleasing presentation is involved. In Section 34.4, we start with Maxwell's equations and derive a pair of wave equations—the *electromagnetic wave equations*—that describe the propagation of intimately linked electric and magnetic fields through space. Out of the electromagnetic wave equations comes a truly remarkable discovery—*electromagnetic radiation*, the topic of Section 34.5. Electromagnetic radiation is manifested in light, radio waves, X rays, and many other phenomena of the first importance. Electromagnetic waves carry energy, as is discussed in Section 34.6.

It is not exaggerating to say that the discovery of electromagnetic radiation in the 1860s opened up a whole new world to physics, equal in importance to mechanics.

SECTION 34.2 Generalizing Ampère's Law Yet Again: The Displacement Current

Figure 34.1a shows a dc circuit consisting of a steady source of emf, some metal wires, and a container of dilute sulfuric acid. As indicated by the arrows, *the current i is the same everywhere in the circuit*. It does not matter that the current carriers in the wire are electrons, which move in the sense opposite that of the current, while most of the

---

*This chapter may be omitted without loss of continuity if it is taken for granted in subsequent chapters that light is a form of electromagnetic radiation.

(a)

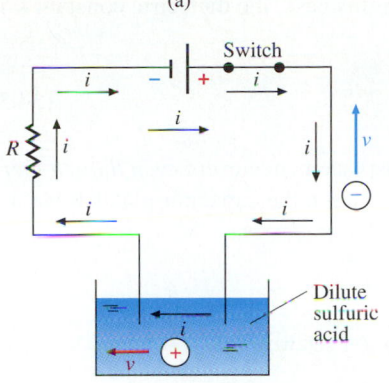

(b)

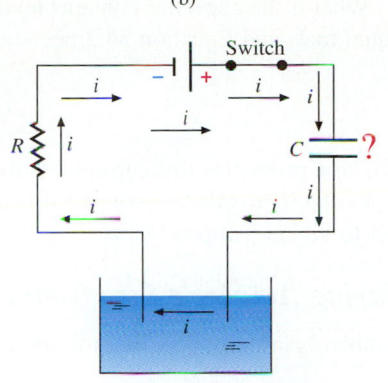

current in the acid is carried by positive ions, which move in the same sense as the current. That is, electric current is a generalization of the concept of flow of free charges.

We now generalize the electric current still further, to cases in which no free charge is present. Consider the circuit of Figure 34.1b. It is the same as the circuit of Figure 34.1a, except that a capacitor has been added. The current in such a circuit—essentially an *RC* circuit—cannot be constant in time. But once the circuit has been completed by closing the switch, an appreciable current does pass through the circuit for a finite time. Consider a single instant during this time interval. Is the current *i* the same everywhere in the circuit? There is no difficulty in answering Yes for any point outside the capacitor. But there is no conductor between the capacitor plates. Is the space between the plates anomalous? Is it possible to have current everywhere in the circuit except inside the capacitor? James Clerk Maxwell (see margin note on Maxwell in Section 18.5) first proposed a solution to this problem in 1861.

Suppose that the capacitor contains a dielectric, as shown in Figure 34.2. For concreteness, assume that the capacitor is initially discharged. When current passes through the capacitor leads, the electric field between the plates increases. As you learned in Section 26.4, this increase in the electric field between the plates results in increasing polarization of the dielectric—that is, in increasing separation of the bound positive and negative charges in the dielectric. This polarization is represented schematically in Figure 34.2 by the stretching of the "spring" that binds the positive and negative charges of the dielectric together; compare with Figure 26.11. Because positive and negative charges are in motion in opposite directions, there is a current in the dielectric, even though the moving charges are bound.

It is not difficult to derive a quantitative connection between the conduction current in the capacitor leads and the "charge-separation current" in the dielectric. Suppose that the capacitor is a parallel-plate capacitor of plate area *A* and plate separation *d* and that the dielectric has dielectric constant κ. According to Equation 26.20, the capacitance is $C = \kappa\epsilon_0 A/d$. The capacitor charge at any moment is $q = CV$, and the voltage across the capacitor can be written $V = \mathscr{E}d$. So we have $q = C\mathscr{E}d = (\kappa\epsilon_0 A/d)(\mathscr{E}d)$, or

$$q = \kappa\epsilon_0 A\mathscr{E}. \tag{34.1}$$

The magnitude of the current *i* in the conducting wires leading to and from the capacitor (and in the rest of the circuit) is dq/dt. So we differentiate both sides of Equation 34.1 with respect to time and obtain

$$i = \kappa\epsilon_0 A \frac{d\mathscr{E}}{dt}. \tag{34.2}$$

The rate of change of electric field between the capacitor plates is directly proportional to the current in the conducting part of the circuit. Indeed, *the quantity $\kappa\epsilon_0 A\, d\mathscr{E}/dt$ is an electric current*, just as much as the rate of charge flow dq/dt in the rest of the circuit is an electric current. Because it is a result of displacement of the charge in the dielectric, the current in the dielectric between the plates is called a **displacement current**. In order to make a distinction, we call the familiar current due to the flow of free charges in a conductor the **conduction current**.

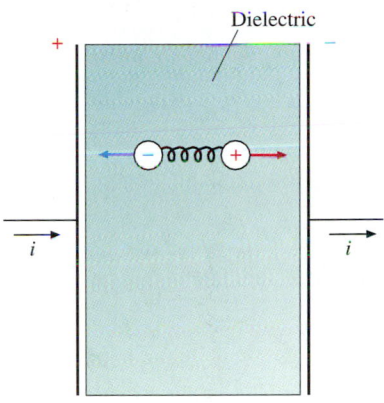

FIGURE 34.2 With current in the capacitor leads, the electric field between the plates increases. Consequently, the polarization of the dielectric increases, with effective transport of charge.

What if the capacitor contains no dielectric? In this case, the dielectric constant κ is equal to 1, and Equation 34.2 becomes

$$i = \epsilon_0 A \frac{d\mathscr{E}}{dt}. \tag{34.3}$$

We now prove that this current—still called the displacement current *even though there is no charge displaced*—must exist in the space between the capacitor plates if we are not to violate Ampère's law.

Saving Ampère's Law from Paradox

Ampère's law can be written in the form of Equation 29.30,

$$\int_{\substack{\text{Amperean} \\ \text{curve}}} \mathscr{B} \cdot d\mathbf{s} = \mu_0 \int_{\substack{\text{Amperean} \\ \text{surface}}} \mathbf{j} \cdot d\mathbf{A}. \tag{34.4}$$

In this equation, $\mathbf{j}$ is the conduction-current density over an infinitesimal area element $d\mathbf{A}$ of the Amperean surface. The integral on the left side is the magnetic circulation evaluated over the Amperean curve—the closed curve that bounds the Amperean surface.

Let us apply Equation 34.4 to the disk-shaped Amperean surface labeled S in Figure 34.3 and to the Amperean curve bounding this surface. The wire that carries the instantaneous current i from the battery to the capacitor penetrates the Amperean surface at point P. This is the only point where current threads surface S, and so the integral on the right side of Equation 34.4 has the value i. Hence the right side of the equation is equal to $\mu_0 i$.

The Amperean surface is normal to the wire and has radius r. If we locate the surface far enough from the capacitor plate, we can be sure that the magnetic field at the surface is approximately that of a long, straight, current-carrying wire. The magnetic field at any point on the Amperean curve that bounds the surface is tangent to the curve and has the constant magnitude given by Equation 29.5,

$$\mathscr{B} = \frac{\mu_0}{2\pi} \frac{i}{r}.$$

So the magnetic circulation—the integral on the right side of Equation 34.4—has the value

$$\frac{\mu_0}{2\pi} \frac{i}{r} \int_{\text{circle}} ds = \mu_0 i.$$

Not surprisingly, this value is the same as the value of the left side of Equation 34.4. We have reduced that equation to $\mu_0 i = \mu_0 i$, in concord with Ampère's law.

As was shown in the discussion accompanying Figure 29.23, Ampère's law is not restricted to planar surfaces but is valid for *any* surface and its bounding Amperean curve. The Amperean curve of Figure 34.4 is the same one shown in Figure 34.3. This curve bounds not only the disk-shaped surface S, but also the "tin cup" surface T. Surface T consists of a cylinder of radius r joined to a disk of radius r that passes through the space between the capacitor plates. As shown in Figure 34.4, the radius r is chosen greater than b, the radius of the capacitor plates.

Substituting Amperean surface T for surface S in applying Ampère's law does not affect the value of the magnetic circulation expressed by the left side of Equation 34.4; its value is still $\mu_0 i$. But the conduction current threading surface T is zero. Ampère's law (Equation 34.4) appears to be violated because it now gives us the result $\mu_0 i = 0$! But $\mu_0 i$ can be equal to zero only for the uninteresting case $t \to \infty$, when so long a time has passed that the capacitor is fully charged and there is no longer any current in the circuit.

Can we "save" Ampère's law? Specifically, can we modify the law so that it is valid not only for steady currents, but for time-dependent currents as well? Indeed we can, if we take seriously the idea that a displacement current is just as much a current as a conduction current is.

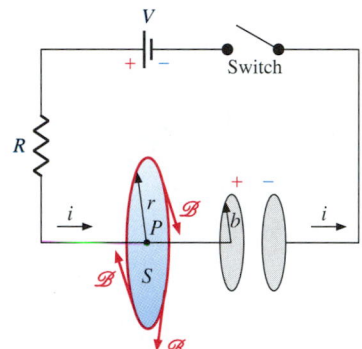

FIGURE 34.3 Ampère's law is applied to the Amperean surface S and to its bounding Amperean curve. The surface, of radius r, is normal to the straight wire that carries current from the battery to the disk-shaped capacitor plate. The space between the plates is exaggerated for clarity in Figures 34.3 through 34.6.

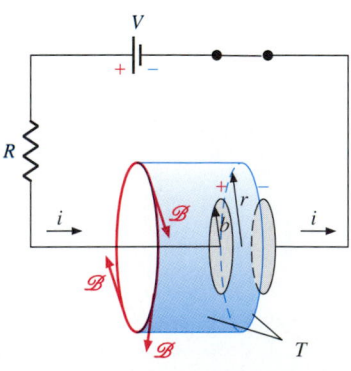

FIGURE 34.4 The Amperean curve of Figure 34.3 also bounds another surface, T. Is Ampère's law satisfied by this surface?

In order to take this idea seriously, we write the total current i threading an Amperean surface as the *sum* of the conduction current i_c and the displacement current i_d that thread it:

$$i = i_c + i_d. \tag{34.5}$$

For surface T, the conduction current is zero. But the displacement current is given by Equation 34.3. You can see that *the displacement current threading T is equal to the conduction current threading S.*

We have thus removed an anomaly we glossed over in our treatment of capacitors in Chapters 32 and 33. We have now defined a total electric current i that has the same value everywhere in the circuit, *without exception*. The space between the capacitor plates is no longer an anomalous region—a region of no current when there is current everywhere else in the circuit.

In the circuit under discussion, the current at any particular point is either entirely a displacement current or entirely a conduction current. But both kinds of current can exist simultaneously in the same region. This is the case, for example, when a capacitor has a "leaky" dielectric—that is, a dielectric whose electrical conductivity is small but not zero. In such cases, we must consider both i_d and i_c in evaluating the total current i.

Summarizing the foregoing argument, we use Equation 34.5 to rewrite Equation 34.4, Ampère's law, in the generalized form

$$\int_{\substack{\text{Amperean} \\ \text{curve}}} \mathcal{B} \cdot d\mathbf{s} = \mu_0(i_c + i_d). \tag{34.6}$$

To make this equation useful, we need to express the right side in a more explicit manner. Up to this point, we have defined the displacement current only for the ideal special case of the electric field within a parallel-plate capacitor with negligible edge effects. But we can readily generalize. In Section 27.3, we derived the special expression for conduction current that is valid when j is constant over a planar area A normal to $\hat{\mathbf{j}}$; we obtained

$$i_c = jA.$$

We then generalized to the universal expression of Equation 29.29,

$$i_c = \int_{\substack{\text{Amperean} \\ \text{surface}}} \mathbf{j} \cdot d\mathbf{A}. \tag{34.7}$$

In the same way, we generalize from Equation 34.3,

$$i_d = \epsilon_0 A \frac{d\mathcal{E}}{dt},$$

valid for $\mathcal{E}$ constant over the planar area A and normal to it, to an expression that applies to a surface of any shape over which the electric field is not uniform. We write

$$i_d = \epsilon_0 \int_{\substack{\text{Amperean} \\ \text{surface}}} \frac{\partial \mathcal{E}}{\partial t} \cdot d\mathbf{A}. \tag{34.8}$$

The reason for writing the partial derivative $\partial \mathcal{E}/\partial t$ instead of $d\mathcal{E}/dt$ (Section 21.4) is to emphasize that what is important is the change of the electric field $\mathcal{E} = \mathcal{E}(\mathbf{r}, t)$ with time and not its variation over space.

Using Equations 34.7 and 34.8, we write **Maxwell's generalization of Ampère's law**:

$$\int_{\substack{\text{Amperean} \\ \text{curve}}} \mathcal{B} \cdot d\mathbf{s} = \mu_0 \left(\int_{\substack{\text{Amperean} \\ \text{surface}}} \mathbf{j} \cdot d\mathbf{A} + \epsilon_0 \int_{\substack{\text{Amperean} \\ \text{surface}}} \frac{\partial \mathcal{E}}{\partial t} \cdot d\mathbf{A} \right). \tag{34.9}$$

We have developed this generalization in the context of an electric circuit containing a capacitor. But displacement currents are not confined to the space between the plates of a capacitor. *A displacement current exists at any location in space where an electric field varies in time, so $\partial\mathscr{E}/\partial t \neq 0$.* We will exploit this fact in our discussion of electromagnetic waves.

EXAMPLE 34.1

The disk-shaped capacitor plates in Figure 34.5 have radius $b = 10.0$ cm. There is no dielectric between the plates. Consider the Amperean surface shown in the figure, a closed-end cylinder (a tin cup) of radius $r = 6.0$ cm. At a certain instant, the current in the capacitor leads is $i = 1.00$ mA. **(a)** Find the conduction current i_c threading the Amperean surface at that instant. **(b)** Find the displacement current i_d threading the same surface. **(c)** Find the total current, and show that it is equal to the current in the capacitor leads. **(d)** Calculate the magnitude of the magnetic field at point P on the Amperean surface.

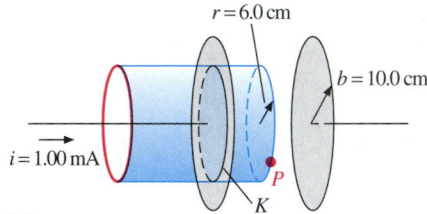

FIGURE 34.5

SOLUTION:

(a) Evaluate the conduction current i_c threading the Amperean surface.

As it flows onto the left-hand capacitor plate, charge distributes itself uniformly over the plate surface. This process requires a flow of charge through the circle K defined by the intersection of the Amperean surface with the plate; this flow is a conduction current. The charge q' that flows through the circumference K is simply the part of the total charge q that does not remain within K. Thus you can use the uniformity of the charge distribution to write

$$q' = q\,\frac{\pi b^2 - \pi r^2}{\pi b^2}.$$

The current is the time derivative of the charge. So you differentiate with respect to time to obtain

$$\frac{dq'}{dt} = i_c = i\left(1 - \frac{r^2}{b^2}\right).$$

You insert the numerical values to find

$$i_c = 1.00 \times 10^{-3}\ \text{A} \times \left(1 - \frac{(0.060\ \text{m})^2}{(0.100\ \text{m})^2}\right)$$

$$= 6.4 \times 10^{-4}\ \text{A}.$$

(b) Evaluate the displacement current i_d threading the same surface.

Because the electric field is normal to the capacitor plates, the value of $\mathscr{E}\cdot d\mathbf{A}$ is zero at any point on the curved part of the Amperean surface. Consequently, the value of $(\partial\mathscr{E}/\partial t)\cdot d\mathbf{A}$ is zero as well. To find the displacement current, you need to integrate the second term on the right side of Equation 34.9 only over the flat part of the Amperean surface. The electric

field is normal to this part of the surface, and you have

$$\frac{\partial\mathscr{E}}{\partial t}\cdot d\mathbf{A} = \frac{\partial\mathscr{E}}{\partial t}\,dA.$$

Because the flat part of the Amperean surface lies entirely between the capacitor plates, the magnitude $\mathscr{E}$ of the field is uniform. From Equation 34.1, you find

$$\mathscr{E} = \frac{q}{\epsilon_0 A} = \frac{q}{\pi\epsilon_0 b^2}.$$

The derivative of $\mathscr{E}$ with respect to time is

$$\frac{\partial\mathscr{E}}{\partial t} = \frac{1}{\pi\epsilon_0 b^2}\,i.$$

The second term on the right side of Equation 34.9 thus simplifies to

$$i_d = \epsilon_0 \int_{\substack{\text{Amperean} \\ \text{surface}}} \frac{\partial\mathscr{E}}{\partial t}\cdot d\mathbf{A} = \epsilon_0\,\frac{i}{\pi\epsilon_0 b^2}\int_{\substack{\text{Amperean} \\ \text{surface}}} dA$$

$$= \epsilon_0\,\frac{i}{\pi\epsilon_0 b^2}\,\pi r^2$$

or

$$i_d = i\,\frac{r^2}{b^2}.$$

Inserting the necessary numbers, you find

$$i_d = 1.00 \times 10^{-3}\ \text{A} \times (0.60)^2 = 3.6 \times 10^{-4}\ \text{A}.$$

(c) Find the total current threading the Amperean surface, and show that it is equal to the current in the capacitor leads.

The total current is the sum of the conduction current through the curved surface and the displacement current through the flat surface. From parts **a** and **b**, you have

$$i = i_c + i_d = 6.4 \times 10^{-4}\ \text{A} + 3.6 \times 10^{-4}\ \text{A} = 1.00\ \text{mA};$$

this is indeed equal to the current in the capacitor leads.

(d) Evaluate $\mathscr{B}$ at P.

As always in Ampère's-law calculations, you need consider only the part of the total current that passes through an Amperean curve to evaluate the magnetic field at a point such as P on the curve. The current through the flat part of the Amperean surface is entirely a displacement current. *But that does not make any difference in evaluating the magnetic circulation term on the left side of Equation 34.6 or Equation 34.9.* Using Equation 34.6, you have

$$\int_{\substack{\text{Amperean} \\ \text{curve}}} \mathscr{B}\cdot d\mathbf{s} = \mu_0 i_d.$$

By symmetry, the magnetic field on the Amperean circle of radius r is everywhere tangent to the circle and has the same magnitude everywhere on the circle. So the equation simplifies to the form

$$2\pi r\mathscr{B} = \mu_0 i_d,$$

which you solve for $\mathscr{B}$:

$$\mathscr{B} = \frac{\mu_0 i_d}{2\pi r} = \frac{\mu_0 i}{2\pi b^2} r.$$

With a minor change of notation, this is exactly the same as Equation 29.21, which gives the magnetic field at a point inside a conductor carrying a uniform *conduction* current.

Now you insert the numerical values to find

$$\mathscr{B} = \frac{4\pi \times 10^{-7} \text{ T·m/A} \times 1.00 \times 10^{-3} \text{ A}}{2 \times \pi \times (0.100 \text{ m})^2} \times 0.060 \text{ m}$$

$$= 1.20 \times 10^{-9} \text{ T}.$$

This is a quite small magnetic field.

A Changing Electric Field Gives Rise to a Magnetic Field

Part **d** of Example 34.1 gives us a new perspective on the relation between electric and magnetic fields. Up to now, we have seen Ampère's law as an expression of the relation between an electric current and the magnetic field to which it gives rise. But now we look deeper. We know that an electric current does not come into existence spontaneously but is always driven by an electric field. In the conduction-current case, the electric field exerts a force on free charges, which move.

We have usually thought of the driving electric field as fixed because a source of emf is usually present to maintain the current as energy is dissipated through electrical resistance. This continual dissipation of energy complicates and obscures the picture. It is simpler to consider an ideal *LC* circuit like that in Figure 34.6, in which charge flows back and forth indefinitely without energy loss and without the need for an external energy source. Here the continuing flow of charge—the current—clearly implies a continuing change of electric field with time. As the capacitor alternately discharges and charges, such a change must take place both between the capacitor plates and along the conductor between its ends on the plates. Thus the magnetic field observed in the vicinity of the circuit is linked with a changing electric field. The fact that the electric field makes electric charge move in a conductor is not an *essential* part of the link. That is, the fundamental causal chain is not changing electric field → moving charge → magnetic field, but changing electric field → magnetic field.

The link between the changing electric field and the magnetic field is exactly the same for a displacement current as for a conduction current. However, the link in the displacement-current case is not obscured by the distraction that arises from the fact that we usually measure currents and not electric fields.

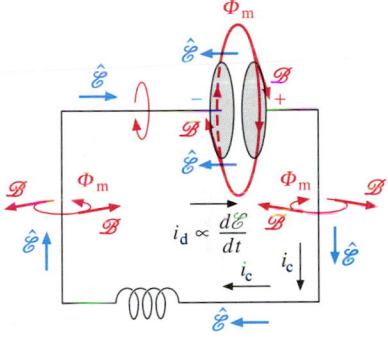

FIGURE 34.6 An ideal *LC* circuit, showing the changing electric field and the magnetic flux to which it gives rise. The magnetic field is everywhere tangent to the magnetic flux lines; the field vector $\mathscr{B}$ is shown in a few places.

Maxwell's Equations

We are now nearly ready to summarize all the fundamental laws of electromagnetism in a set of compact equations possessing considerable symmetry. One last preparatory step remains.

Equation 34.9 expresses Ampère's law in terms of the magnetic field $\mathscr{B}$ and the time derivative of the electric field $\mathscr{E}$. We can express Faraday's law in a very similar form. To do this, we begin with Faraday's law in the form of Equation 30.33,

$$\int_{\substack{\text{Faradayan} \\ \text{curve}}} \mathscr{E} \cdot d\mathbf{s} = -\frac{d\Phi_m}{dt}. \tag{34.10}$$

This equation is expressed in terms of magnetic flux Φ_m rather than magnetic field $\mathscr{B}$. But remember that the magnetic flux $d\Phi_m$ penetrating an infinitesimal surface element $d\mathbf{A}$ can be written

$$d\Phi_m = \mathscr{B} \cdot d\mathbf{A};$$

compare with Equation 30.9a. The flux penetrating a finite closed surface composed of many area elements $d\mathbf{A}$ is just the integral over the surface:

$$\Phi_m = \int_{\substack{\text{Faradayan} \\ \text{surface}}} d\Phi_m = \int_{\substack{\text{Faradayan} \\ \text{surface}}} \mathscr{B} \cdot d\mathbf{A}.$$

We want to find the time derivative of Φ_m needed to express Equation 34.10 entirely in terms of fields—that is, to express the right side of the equation in terms of $\mathscr{B}$. To do this, we exploit the fact that the order of differentiation and integration is immaterial; in general,

$$\frac{\partial}{\partial t} \int f(x, t)\, dx = \int \frac{\partial f(x, t)}{\partial t}\, dx.$$

As we have done before, we use partial differentiation to make a distinction between differentiation with respect to time and differentiation with respect to position.

The time derivative of Φ_m is thus

$$\frac{d\Phi_m}{dt} = \int_{\substack{\text{Faradayan} \\ \text{surface}}} \frac{\partial \mathscr{B}}{\partial t} \cdot d\mathbf{A}.$$

(Remember that $d\mathbf{A}$ is constant in time and need not be differentiated.)

We are now ready to write Faraday's law in terms of fields. Substituting the value of $d\Phi_m/dt$ just obtained into Equation 34.10, we obtain

$$\int_{\substack{\text{Faradayan} \\ \text{curve}}} \mathscr{E} \cdot d\mathbf{s} = -\int_{\substack{\text{Faradayan} \\ \text{surface}}} \frac{\partial \mathscr{B}}{\partial t} \cdot d\mathbf{A}. \tag{34.11}$$

Maxwell's Equations: General Statement

Everything we have learned about electricity and magnetism is based on one or more of the following five laws:

1. Gauss's law for electric charge (Equation 24.8),

$$\int_{\substack{\text{Gaussian} \\ \text{surface}}} \mathscr{E} \cdot d\mathbf{A} = \frac{q}{\epsilon_0}; \tag{34.12a}$$

2. Gauss's law for magnetism (Equation 29.16),

$$\int_{\substack{\text{Gaussian} \\ \text{surface}}} \mathscr{B} \cdot d\mathbf{A} = 0; \tag{34.12b}$$

3. Faraday's law (Equation 34.11),

$$\int_{\substack{\text{Faradayan} \\ \text{curve}}} \mathscr{E} \cdot d\mathbf{s} = -\int_{\substack{\text{Faradayan} \\ \text{surface}}} \frac{\partial \mathscr{B}}{\partial t} \cdot d\mathbf{A}; \tag{34.12c}$$

4. Maxwell's generalization of Ampère's law (Equation 34.9),

$$\int_{\substack{\text{Amperean} \\ \text{curve}}} \mathscr{B} \cdot d\mathbf{s} = \mu_0 \left(\int_{\substack{\text{Amperean} \\ \text{surface}}} \mathbf{j} \cdot d\mathbf{A} + \epsilon_0 \int_{\substack{\text{Amperean} \\ \text{surface}}} \frac{\partial \mathscr{E}}{\partial t} \cdot d\mathbf{A} \right); \tag{34.12d}$$

5. the Lorentz force (Equation 28.22),

$$\mathbf{F} = q(\mathscr{E} + \mathbf{v} \times \mathscr{B}). \tag{34.12e}$$

The first four of these equations are called **Maxwell's equations**. Each one compares either an integral over a closed surface with an integral over the volume enclosed by that surface or an integral over a closed curve with an integral over a surface area enclosed by that curve. However, there are important differences as well as similarities among the equations. In Gauss's law for magnetism, the value of the integral is zero because magnetic monopoles—that is, magnetic "charges"—are not known to exist. The right side of Faraday's law is very similar to the displacement-current term of Ampère's law. But Faraday's law does not contain a term analogous to the conduction-current term of Ampère's law for the same reason: In the absence of magnetic monopoles, there are no "magnetic conductors" containing "magnetic carriers" that can carry a magnetic current when driven by a magnetic field. Should magnetic monopoles be discovered, we would have to add new terms to the right sides of Equations 34.12b and 34.12c. Doing so would remove the major asymmetry from Maxwell's equations.

Dissatisfaction with such ''broken'' symmetries in nature has often led physicists to important discoveries and to profound new insights into the nature of the universe. You can see why physicists have not given up the search for magnetic monopoles.

Maxwell's Equations in Vacuum

In this chapter, our interest in Maxwell's equations is focused mainly on the implication they contain for the propagation of electromagnetic waves through empty space. (Drawing out this implication is the subject of the next section.) Consequently, we can restrict our consideration to the simpler form they assume in the total absence of matter. In a vacuum, q is equal to zero, and there is nothing available to carry conduction currents. Thus, **Maxwell's equations in vacuum** can be written as follows:

1.
$$\int_{\substack{\text{Gaussian} \\ \text{surface}}} \mathscr{E} \cdot d\mathbf{A} = 0, \tag{34.13a}$$

2.
$$\int_{\substack{\text{Gaussian} \\ \text{surface}}} \mathscr{B} \cdot d\mathbf{A} = 0, \tag{34.13b}$$

3.
$$\int_{\substack{\text{Faradayan} \\ \text{curve}}} \mathscr{E} \cdot d\mathbf{s} = -\int_{\substack{\text{Faradayan} \\ \text{surface}}} \frac{\partial \mathscr{B}}{\partial t} \cdot d\mathbf{A}, \tag{34.13c}$$

4.
$$\int_{\substack{\text{Amperean} \\ \text{curve}}} \mathscr{B} \cdot d\mathbf{s} = \mu_0 \epsilon_0 \int_{\substack{\text{Amperean} \\ \text{surface}}} \frac{\partial \mathscr{E}}{\partial t} \cdot d\mathbf{A}. \tag{34.13d}$$

In the absence of electric charges on which the Lorentz force can act, we omit Equation 34.12e from this list. How symmetrical these equations become when electric charges as well as magnetic monopoles are absent!

Electromagnetic Waves

In Chapter 21, we saw that, when a medium is disturbed, waves will propagate if the medium obeys Hooke's law in one form or another. The waves are described by the *wave equation*. If the waves are one-dimensional or are *plane waves* propagating with planar wave fronts in the x direction, the wave equation takes the form of Equation 21.24a,

$$\frac{\partial^2 f(x,\, t)}{\partial t^2} = v^2 \frac{\partial^2 f(x,\, t)}{\partial x^2}. \tag{34.14}$$

In this equation, $f(x, t)$ is the *wave function* that describes the wave as it varies over space and in time, and v is the *wave speed*.

Waves can be either *transverse* or *longitudinal*, as discussed in Chapter 22. In a fluid medium, which does not resist shearing forces, transverse waves cannot exist but longitudinal waves can. Sound waves in air are a familiar example of longitudinal waves.

In this section, we show that two wave equations can be deduced from Maxwell's equations in vacuum. These equations, involving an electric field and a magnetic field that both vary over space and in time, turn out to have exactly the same mathematical form as Equation 34.14. Thus they imply the existence of *electromagnetic waves*.

To simplify the discussion, we restrict our consideration to *plane waves*. That is, we assume that both the electric field and the magnetic field vary in time and also vary with position x, but do not vary with y or z. (See Section 22.2.) Under this assumption, the wave functions of interest to us must have the forms $\mathscr{E} = \mathscr{E}(x, t)$ and $\mathscr{B} = \mathscr{B}(x, t)$.

The Impossibility of Longitudinal Electromagnetic Waves

We begin by proving that electromagnetic waves cannot be longitudinal. Let us assume the contrary. This contrary assumption means that the magnitude of the x component $\mathscr{E}_x$ of $\mathscr{E}$ must vary with x. (Otherwise, the wave wouldn't be longitudinal.) The electric field may have y and z components as well. But we are not now interested in the variation of these components with x because any such variation implies the existence of transverse, not longitudinal, waves.

Figure 34.7 shows an infinitesimal cube of sides dx, dy, and dz. We know that the cube contains no electric charge. Hence we can apply Gauss's law for electric charge in the form of Equation 34.13a,

$$\int_{\substack{\text{Gaussian} \\ \text{surface}}} \mathscr{E} \cdot d\mathbf{A} = 0.$$

To evaluate the integral, we sum the quantities of electric flux $d\Phi_e$ penetrating each face of the cube. First, consider the upper and lower faces, both of area $dA = dx\,dz$. If the flux entering the cube through the lower face is to be different from the flux leaving through the upper face, the difference can be due only to a variation of $\mathscr{E}_y$, the component of $\mathscr{E}$ normal to the two faces, over the distance dy between the two faces. But we have already argued that such a variation cannot have anything to do with a longitudinal wave propagating in the x direction. Thus the net flux penetrating this pair of faces is zero. A similar argument applies to the back and front faces of area $dx\,dy$, and their contribution to the Gaussian integral is likewise zero.

This leaves the left and right faces, both of which have area $dA = dy\,dz$. The left face is located at x, and the right face at $x + dx$. The flux entering the cube through the left face is

$$d\Phi_e(x) = \mathscr{E}(x) \cdot d\mathbf{A} = -\mathscr{E}_x(x)\,dy\,dz.$$

(We use the sign convention for closed surfaces established in Chapter 24: Flux entering a Gaussian surface is taken to be negative, and flux emerging from the surface is taken to be positive.) The flux emerging from the cube through the right face is

$$d\Phi_e(x + dx) = \mathscr{E}(x + dx) \cdot d\mathbf{A} = \mathscr{E}_x(x + dx)\,dy\,dz.$$

The sum of these two terms is the net flux emerging from the infinitesimal cube as a consequence of the presence of a longitudinal wave. According to Gauss's law (Equation 34.13a), the net flux must be equal to zero in the absence of charge:

$$\mathscr{E}_x(x + dx)\,dy\,dz - \mathscr{E}_x(x)\,dy\,dz = 0.$$

But this equation can be true only if

$$\mathscr{E}_x(x + dx) = \mathscr{E}_x(x).$$

That is, Gauss's law can be satisfied only if there is *no* variation of $\mathscr{E}_x$ with x. Because longitudinal waves *are* such variations, we can conclude that *a longitudinal electric-field wave is not possible.*

An identical argument, using Equation 34.13b, applies to the longitudinal magnetic field $\mathscr{B}_x$; *a longitudinal magnetic-field wave is not possible.*

Transverse Fields

We now show that a time-independent transverse electric wave $\mathscr{E}_y(x)$ is allowed. However, Faraday's law requires that any such variation of $\mathscr{E}$ with x be linked to a *time* dependence of the magnetic field.

In general, a transverse wave propagating in the x direction can have both y and z

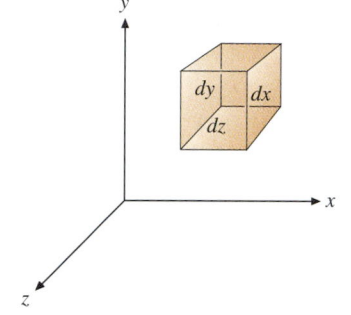

FIGURE 34.7 Construction to show that electric waves, if they exist, cannot be longitudinal.

components. For simplicity, however, we will assume that any transverse electric wave present is described by a wave function of the form $\mathscr{E} = \mathscr{E}_y(x, t)\hat{\mathbf{y}}$. That is, the wave is *polarized*. The electric vector has only a y component, which varies with x and t.

Applying Faraday's Law

Consider the infinitesimal square shown in Figure 34.8. It lies parallel to the x-y plane and has sides dx and dy (its location along the z axis is not of interest here). If there is a y-polarized electric wave, the value of $\mathscr{E}_y$ must vary along the x direction. Thus $\mathscr{E}_y(x + dx)$ is different from $\mathscr{E}_y(x)$, as shown in the figure.

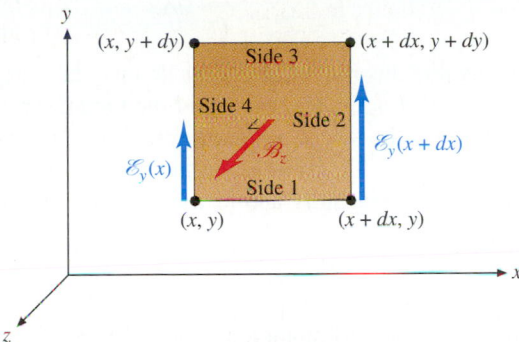

FIGURE 34.8 An infinitesimal square in a region where $\mathscr{E}_y$ varies with x, showing that Faraday's law requires the presence of a magnetic field $\mathscr{B}_z$ that varies at a rate $\partial \mathscr{B}_z / \partial t$.

Using the infinitesimal square as a Faradayan curve, we apply Faraday's law (Equation 34.13c):

$$\int_{\substack{\text{Faradayan} \\ \text{curve}}} \mathscr{E} \cdot d\mathbf{s} = -\int_{\substack{\text{Faradayan} \\ \text{surface}}} \frac{\partial \mathscr{B}}{\partial t} \cdot d\mathbf{A}.$$

The left side of the equation is the electric circulation, which we evaluate starting at (x, y) and going counterclockwise around the square. Side 1 is perpendicular to $\mathscr{E}_y$, and so the dot product $\mathscr{E} \cdot d\mathbf{s}$ is zero. The contribution of side 2 is $\mathscr{E}_y(x + dx)\, dy$. Side 3 makes a zero contribution for the same reason as side 1.* For side 4, where the path of integration is antiparallel to $\mathscr{E}$, the contribution to the circulation is $-\mathscr{E}_y(x)\, dy$. Adding the contributions of all four sides, we have the electric circulation

$$\int_{\substack{\text{Faradayan} \\ \text{curve}}} \mathscr{E} \cdot d\mathbf{s} = [\mathscr{E}_y(x + dx) - \mathscr{E}_y(x)]\, dy \tag{34.15}$$

We now reexpress the quantity on the right side of this equation in a more compact form. The term in brackets is the difference between the electric fields at two points an infinitesimal distance dx apart. This difference must be equal to the rate at which $\mathscr{E}_y$ changes per unit distance, $\partial \mathscr{E}_y / \partial x$, multiplied by the distance dx. That is, we can write the term in brackets as

$$\mathscr{E}_y(x + dx) - \mathscr{E}_y(x) = \frac{\partial \mathscr{E}_y}{\partial x}\, dx.$$

Equation 34.15 can therefore be written

$$\int_{\substack{\text{Faradayan} \\ \text{curve}}} \mathscr{E} \cdot d\mathbf{s} = \frac{\partial \mathscr{E}_y}{\partial x}\, dx\, dy. \tag{34.16}$$

The Faradayan curve around which we have just evaluated the electric circulation bounds a Faradayan surface. Our next task is to evaluate the magnetic flux penetrating this surface—that is, to evaluate the right side of Equation 34.13c,

$$-\int_{\substack{\text{Faradayan} \\ \text{surface}}} \frac{\partial \mathscr{B}}{\partial t} \cdot d\mathbf{A}.$$

---

*Even if the electric field has a component $\mathscr{E}_x$, its value must be the same along sides 1 and 3. Hence any nonzero contribution of side 1 must be canceled by the contribution of side 3.

Equation 34.16 gives the value of the left side of Equation 34.13c. If an electric wave is present, $\mathscr{E}_y$ varies with x, and the left side of Equation 34.13c is not equal to zero. Thus the integrand on the right side, $(\partial\mathscr{B}/\partial t) \cdot d\mathbf{A}$, cannot be zero either, because if it were the integral would be zero. But $(\partial\mathscr{B}/\partial t) \cdot d\mathbf{A}$ has a nonzero value only if the magnetic field has a component $\mathscr{B}_z$ (normal to the surface $dx\,dy$) that varies in time. That is, *an electric wave—an electric field that varies with x—can exist only in the presence of a perpendicular magnetic field that varies with time*. The component $\mathscr{B}_z$ is shown for a particular instant in Figure 34.8.

We can rewrite the magnetic flux integral in the simpler scalar form

$$-\int_{\substack{\text{Faradayan} \\ \text{surface}}} \frac{\partial\mathscr{B}_z}{\partial t}\, dA.$$

According to the mean value theorem of integral calculus, this integral is equal to the product of the mean value of $\partial\mathscr{B}_z/\partial t$, averaged over the Faradayan square, and the area $dx\,dy$ of the square. Because the square is infinitesimal, the necessary mean value is nearly equal to the value of $\partial\mathscr{B}_z/\partial t$ anywhere in the square, and we need not make a distinction between the variable quantity and its mean value. The integral thus yields

$$-\frac{\partial\mathscr{B}_z}{\partial t}\, dx\,dy;$$

this is the value of the right side of Equation 34.13c.

We now put the values of the left (Equation 34.16) and right sides of Equation 34.13c together to write Faraday's law for the infinitesimal square:

$$\frac{\partial\mathscr{E}_y}{\partial x} = -\frac{\partial\mathscr{B}_z}{\partial t}. \tag{34.17}$$

Note that the factor $dx\,dy$ appears on both sides of the equation, and so we have canceled it out. Equation 34.17 states in mathematical form the point made verbally in the text after Equation 34.16: An electric field whose y component $\mathscr{E}_y$ varies with x must be accompanied by a perpendicular magnetic field $\mathscr{B}_z$ that varies in time.

Applying Ampère's Law

The time-varying magnetic field $\mathscr{B}_z$ whose existence we deduced in the preceding subsection must also vary in space. This is evident from Equation 34.17, which shows that the value of $\partial\mathscr{B}_z/\partial t$ depends on the value of $\partial\mathscr{E}_y/\partial x$, which itself varies with x. We now use Ampère's law to look into the properties of $\mathscr{B}_z$, taking advantage of the close similarity between Faraday's law (Equation 34.13c) and Ampère's law (Equation 34.13d). We follow the procedure of the preceding subsection quite closely and apply Ampère's law to the infinitesimal square shown in Figure 34.9. The square lies parallel to the x-z plane (its location along the y axis is not of interest here) and has sides dx and dz. Because the existence of a y-polarized electric wave requires the existence of a z-polarized magnetic wave, the value of $\mathscr{B}_z$ must vary along the x direction. Thus $\mathscr{B}_z(x + dx)$ is different from $\mathscr{B}_z(x)$, as shown in Figure 34.9.

FIGURE 34.9 An infinitesimal square in a region where $\mathscr{B}_z$ varies with x, showing that Ampère's law requires the presence of an electric field $\mathscr{E}_y$ that varies at a rate $\partial\mathscr{E}_y/\partial t$.

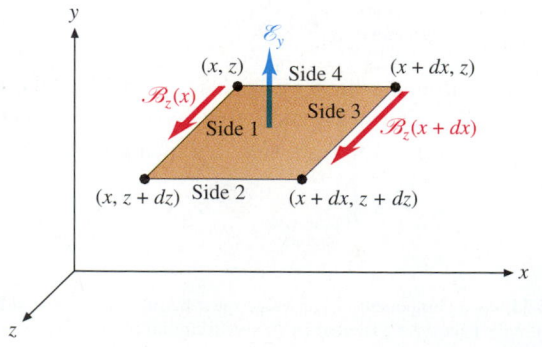

Using the infinitesimal square as an Amperean curve, we apply Ampère's law (Equation 34.13d):

$$\int_{\substack{\text{Amperean}\\\text{curve}}} \mathscr{B} \cdot d\mathbf{s} = \mu_0\epsilon_0 \int_{\substack{\text{Amperean}\\\text{surface}}} \frac{\partial \mathscr{E}}{\partial t} \cdot d\mathbf{A}.$$

The left side of the equation is the magnetic circulation, which we evaluate starting at (x, y) and traversing the sides of the square in the order 1–2–3–4. Only sides 1 and 3 make a nonzero contribution, and the value of the magnetic circulation is thus

$$\int_{\substack{\text{Amperean}\\\text{curve}}} \mathscr{B} \cdot d\mathbf{s} = -[\mathscr{B}_z(x + dx) - \mathscr{B}_z(x)]\, dz$$

$$= -\frac{\partial \mathscr{B}_z}{\partial x}\, dx\, dz. \tag{34.18}$$

Compare with Equation 34.16.

Our next task is to evaluate the electric flux penetrating the Amperean surface of Figure 34.9—that is, to evaluate the right side of Equation 34.13d. The procedure is almost exactly the same as that of the preceding subsection, and it yields

$$\mu_0\epsilon_0 \int_{\substack{\text{Amperean}\\\text{surface}}} \frac{\partial \mathscr{E}}{\partial t} \cdot d\mathbf{A} = \mu_0\epsilon_0 \int_{\substack{\text{Amperean}\\\text{surface}}} \frac{\partial \mathscr{E}_y}{\partial t}\, dA = \mu_0\epsilon_0 \frac{\partial \mathscr{E}_y}{\partial t}\, dx\, dz.$$

We now equate this expression with the right side of Equation 34.18. Again the factors $dx\, dz$ cancel, and we write Ampère's law for the infinitesimal square:

$$-\frac{\partial \mathscr{B}_z}{\partial x} = \mu_0\epsilon_0 \frac{\partial \mathscr{E}_y}{\partial t}. \tag{34.19}$$

Compare with Equation 34.17. Equation 34.19 asserts that *a magnetic field whose z component $\mathscr{B}_z$ varies with x must be accompanied by a perpendicular electric field $\mathscr{E}_y$ that varies in time*. When we take this statement together with the analogous statement given after Equation 34.17, we begin to see physically how electromagnetic waves propagate. A time variation of electric field at any location gives rise to a magnetic field that varies over space, in accordance with Equation 34.19. But such a magnetic field must also vary in time at any particular point. This time variation gives rise to an electric field that varies over space, in accordance with Equation 34.17. But such an electric field must also vary in time at any particular point. The electric and magnetic fields thus propagate as intimately connected waves, each giving rise to the other.

The Wave Equation

Equations 34.17 and 34.19,

$$\frac{\partial \mathscr{E}_y}{\partial x} = -\frac{\partial \mathscr{B}_z}{\partial t} \quad \text{and} \quad -\frac{\partial \mathscr{B}_z}{\partial x} = \mu_0\epsilon_0 \frac{\partial \mathscr{E}_y}{\partial t}, \tag{34.20a,b}$$

are a pair of simultaneous equations in $\mathscr{E}_y$ and $\mathscr{B}_z$. We now solve them to yield a wave equation in $\mathscr{E}_y$ and a wave equation in $\mathscr{B}_z$.

As always in solving simultaneous equations, we use the two equations to eliminate one variable. First, we eliminate $\mathscr{B}_z$. To do this, we begin by taking the partial derivative $\partial/\partial x$ of both sides of Equation 34.20a:

$$\frac{\partial^2 \mathscr{E}_y}{\partial x^2} = -\frac{\partial^2 \mathscr{B}_z}{\partial x\, \partial t}. \tag{34.21a}$$

Next, we take the partial derivative $\partial/\partial t$ of both sides of Equation 34.20b:

$$-\frac{\partial^2 \mathscr{B}_z}{\partial t\, \partial x} = \mu_0\epsilon_0 \frac{\partial^2 \mathscr{E}_y}{\partial t^2}. \tag{34.21b}$$

The order of partial differentiation is immaterial; that is,

$$\frac{\partial^2}{\partial x\,\partial t} = \frac{\partial^2}{\partial t\,\partial x}.$$

Thus the right side of Equation 34.21a is identical with the left side of Equation 34.21b. We combine the two equations and then divide both sides through by $\mu_0\epsilon_0$ to obtain

$$\frac{\partial^2 \mathscr{E}_y}{\partial t^2} = \frac{1}{\mu_0\epsilon_0}\frac{\partial^2 \mathscr{E}_y}{\partial x^2}.$$

(34.22)

This is indeed a wave equation; compare its form with the general form (Equation 34.14).

Before we consider Equation 34.22 further, let us derive the corresponding wave equation for $\mathscr{B}_z$. We begin again with Equations 31.20a and 31.20b, and this time we eliminate $\mathscr{E}_y$. To do this, we take the partial derivatives with respect to t and x in the order opposite that just used. Taking the partial derivative $\partial/\partial t$ of both sides of Equation 34.20a yields

$$\frac{\partial^2 \mathscr{E}_y}{\partial t\,\partial x} = -\frac{\partial^2 \mathscr{B}_z}{\partial t^2}.$$

Taking the partial derivative $\partial/\partial x$ of both sides of Equation 34.20b yields

$$-\frac{\partial^2 \mathscr{B}_z}{\partial x^2} = \mu_0\epsilon_0\frac{\partial^2 \mathscr{E}_y}{\partial x\,\partial t}.$$

We combine the two equations to obtain

$$\frac{\partial^2 \mathscr{B}_z}{\partial t^2} = \frac{1}{\mu_0\epsilon_0}\frac{\partial^2 \mathscr{B}_z}{\partial x^2}.$$

(34.23)

This magnetic wave equation is identical in form with the electric wave equation (Equation 34.22). As already noted, neither wave can exist without the other; they must propagate together as an **electromagnetic wave**, and Equations 34.22 and 34.23 are called the **electromagnetic wave equations**.

Speed of Electromagnetic Waves

We have solved Maxwell's equations to obtain the electromagnetic wave equations. Maxwell's equations have direct physical meaning. But what do the electromagnetic wave equations mean? That is, what observable physical phenomena do the electromagnetic wave equations describe? The first hint—a very strong one—emerges when we determine the speed of the electromagnetic waves.

We have already seen that all one-dimensional (or plane) waves are described by the same general equation (Equation 34.14):

$$\frac{\partial^2 f(x,\,t)}{\partial t^2} = v^2\,\frac{\partial^2 f(x,\,t)}{\partial x^2}.$$

The wave equation thus openly displays the wave speed; it is the square root of the coefficient of $\partial^2 f/\partial x^2$. For electromagnetic waves, we have $v^2 = 1/\mu_0\epsilon_0$, or

$$v = \frac{1}{\sqrt{\mu_0\epsilon_0}}.$$

(34.24)

It is important to note that this speed is *identical* for the electric and magnetic waves.

In Maxwell's time, the value of the product $\mu_0\epsilon_0$ had been determined experimentally by Weber and Kohlrausch, who used a combination of electrostatic and electromagnetic measurements.* Needless to say, Maxwell was most eager to use these ex-

---

*More precisely, Weber and Kohlrausch had measured the value of the product in terms of units then in use.

perimental results to determine the speed of the waves whose existence his equations predicted. However, he was at his summer home in the country when he worked out Equations 34.22 and 34.23 and so did not have access to scientific journals. Upon his return to Cambridge University, he wasted no time in looking up the experimental results. As the quotation at the beginning of this chapter makes evident, Maxwell was delighted to find that, within the limits of the experimental errors in the values, *the speed in vacuum of the electromagnetic waves described by Equations 34.22 and 34.23 is equal to the independently measured speed of light c.* We will soon consider further evidence that light is one of a large class of physical effects involving propagation of electric and magnetic fields in accordance with the electromagnetic wave equation. Such effects are given the generic name **electromagnetic radiation**.

SI Revisited

From the perspective of present-day knowledge, light is so clearly a typical form of electromagnetic radiation that the quantity $1/\sqrt{\mu_0 \epsilon_0}$ is *identified* with the speed of light in vacuum, c. The speed c is a fundamental constant of nature, defined to be (Section 1.2)

$$c = 299\ 792\ 458 \text{ m/s } \textit{exactly.} \tag{34.25}$$

With this defined value in mind, we use the speed of light to *define* the value of the quantity $1/\sqrt{\mu_0 \epsilon_0}$:

$$\frac{1}{\sqrt{\mu_0 \epsilon_0}} \equiv c. \tag{34.26}$$

This equation defines only the product $\mu_0 \epsilon_0$, not either quantity by itself. In SI, the value of μ_0 is fixed by definition to be *exactly*

$$\mu_0 \equiv 4\pi \times 10^{-7} \text{ T·m/A},$$

as already noted in Chapter 28. The constant ϵ_0 is then also fixed, its value being

$$\epsilon_0 = \frac{1}{\mu_0 c^2} = \frac{1}{\pi} \times 27.816\ 251\ 4 \times 10^{-12} \text{ C}^2/(\text{N·m}^2) \tag{34.27}$$

$$= 8.854\ 187\ 82 \times 10^{-12} \text{ C}^2/(\text{N·m}^2),*$$

The Relation between $\mathscr{E}_y$ and $\mathscr{B}_z$ in an Electromagnetic Wave

The electric and magnetic waves that constitute an electromagnetic wave must travel together at precisely the same speed c. We now find the relative amplitudes of the two waves and the phase relation between them. We begin by writing Equation 34.20b in the form

$$-\frac{\partial \mathscr{B}_z}{\partial x} = \frac{1}{c^2} \frac{\partial \mathscr{E}_y}{\partial t}. \tag{34.28}$$

Into this equation we insert a particular wave function $\mathscr{E}_y(x, t)$. We can then see what form the corresponding wave function $\mathscr{B}_z(x, t)$ must take. To make things easy, let us assume that the electric wave has the simplest form possible. Equation 21.17a gives the simplest form for a y-polarized transverse wave traveling in the positive x direction on a stretched string:

$$y(x, t) = A \cos\left[\frac{2\pi}{\lambda}(x - vt)\right],$$

---

*The value of ϵ_0 is exact, being defined in terms of two exact numbers. However, ϵ_0 contains π as a factor and, like π, cannot be represented exactly by a finite decimal.

where A is the amplitude of the wave, λ its wavelength, and v its speed. The corresponding form for the electric wave is

$$\mathscr{E}_y = \mathscr{E}_0 \cos\left[\frac{2\pi}{\lambda}(x - ct)\right]. \qquad (34.29)$$

In this equation, we have set $v = c$ because an electric wave in vacuum *must* have speed c.

We find the value of $\partial \mathscr{E}_y / \partial t$ that we need to substitute into Equation 34.28 by taking the partial derivative of both sides of Equation 34.29 with respect to time:

$$\frac{\partial \mathscr{E}_y}{\partial t} = \frac{2\pi c}{\lambda} \mathscr{E}_0 \sin\left[\frac{2\pi}{\lambda}(x - ct)\right].$$

When we use this value, Equation 34.28 becomes

$$-\frac{\partial \mathscr{B}_z}{\partial x} = \frac{2\pi}{\lambda c} \mathscr{E}_0 \sin\left[\frac{2\pi}{\lambda}(x - ct)\right].$$

To find $\mathscr{B}_z$, we integrate both sides of this equation with respect to x:

$$-\mathscr{B}_z = -\frac{\lambda}{2\pi}\frac{2\pi}{\lambda c}\mathscr{E}_0 \cos\left[\frac{2\pi}{\lambda}(x - ct)\right],$$

or

$$\mathscr{B}_z = \frac{1}{c}\mathscr{E}_0 \cos\left[\frac{2\pi}{\lambda}(x - ct)\right]. \qquad (34.30)$$

Compare this wave function $\mathscr{B}_z$ with the wave function we assumed for $\mathscr{E}_y$, given by Equation 34.29. You can see that

$$\boxed{\mathscr{B}_z = \frac{1}{c}\mathscr{E}_y.} \qquad (34.31)$$

We can draw three conclusions:

1. The electric and magnetic waves are *in phase* with one another.
2. The amplitude of the magnetic wave is equal to the amplitude of the electric wave divided by the speed of light.
3. The electric and magnetic vectors are mutually perpendicular, and both are perpendicular to the direction of propagation of the electromagnetic wave.

These conclusions are illustrated in Figure 34.10.

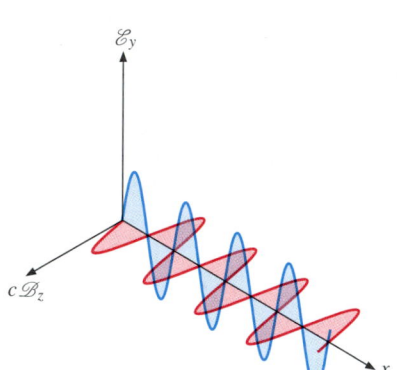

FIGURE 34.10 The electric and magnetic fields that together constitute an electromagnetic wave, at a certain instant. The quantity plotted in the x-z plane is $c\mathscr{B}_z$ rather than $\mathscr{B}_z$. This plotting of $c\mathscr{B}$ rather than $\mathscr{B}$ makes it possible to plot the two waves in the same units and on the same scale.

EXAMPLE 34.2

A sinusoidal electromagnetic plane wave travels through empty space in the x direction. The electric wave is polarized in the y direction, and the frequency is 1.5 GHz. At a certain point P in space, the oscillating electric field $\mathscr{E}_y$ attains a maximum magnitude $\mathscr{E}_0 = 63$ V/m. **(a)** What is the wavelength of the wave? **(b)** What is the maximum magnitude $\mathscr{B}_0$ of the magnetic field at the same point? What is the direction of $\mathscr{B}$? **(c)** Express the electric and magnetic fields as functions of x and t in the vicinity of P.

SOLUTION:
(a) Find λ.

As always for waves, you can use the relation $\nu\lambda = v$. Because an electromagnetic wave always propagates through empty space at speed c, you have

$$\lambda = \frac{c}{\nu} = \frac{3.0 \times 10^8 \text{ m/s}}{1.5 \times 10^9 \text{ Hz}} = 0.20 \text{ m}.$$

(b) Find $\mathscr{B}_0$.

You use Equation 34.31 to write

$$\mathscr{B}_0 = \frac{63 \text{ V/m}}{3.0 \times 10^8 \text{ m/s}} = 2.1 \times 10^{-7} \text{ T}.$$

Note how the factor c in Equation 34.31 leads to a quite small value of $\mathscr{B}_0$. By "everyday" standards, 63 V/m is a modest but appreciable electric field, whereas 2.1×10^{-7} T is a quite small magnetic field, its order of magnitude being about 1% of the earth's magnetic field. As you saw in the derivation of the electromagnetic wave equations, a y-polarized electric wave is always associated with a z-polarized magnetic wave. Hence, $\mathscr{B}$ lies in the x-z plane.

(c) Evaluate $\mathscr{E}(x, t)$ and $\mathscr{B}(x, t)$ in the vicinity of P.

Because the electric wave is sinusoidal and y-polarized, its function can be written in the form of Equation 34.29. Inserting the numerical values of $\mathcal{E}_0$, λ, and c, you obtain

$$\mathcal{E} = \mathcal{E}_y\hat{y} = 63 \text{ V/m}$$
$$\times \cos [31 \text{ m}^{-1} \times (x - 3.0 \times 10^8 t \text{ m})] \, \hat{y}.$$

The magnetic field is always in phase with the electric field.

Thus you do not need a phase constant to express $\mathcal{B}$. Using the numerical value of $\mathcal{B}_0$ you obtained in part **b**, you write

$$\mathcal{B} = \mathcal{B}_z\hat{z} = 2.1 \times 10^{-7} \text{ T}$$
$$\times \cos [31 \text{ m}^{-1} \times (x - 3.0 \times 10^8 t \text{ m})] \, \hat{z}.$$

What Is Oscillating?

The quantitative study of waves always begins with the study of transverse waves on a stretched string and proceeds with the study of various kinds of waves in solids and in fluids. We followed this path in Chapters 21 and 22. All such waves, called *mechanical waves*, are disturbances in one medium or another. It is easy to conclude, on the basis of this introductory study, that "wave" *means* "disturbance in a medium." Indeed this was the view of nearly all physicists until about 1905. From this point of view, the electric and magnetic fields that make up the electromagnetic wave are the result of some kind of disturbance in some sort of medium that is not itself directly observed. But light propagates through empty space, as you can see on any sunny day or any starry night. What medium could be present in empty space?

This question was not unique to light. Remember that Faraday, in attempting to understand how an electric charge at one location could exert a force on an electric charge at another location, postulated that the charge produced a stress, which he called the electrotonic state, in a thin, highly transparent medium that filled all space. Faraday invented the electric field as a convenient way of making his arguments quantitative, but he continued to believe that the "real" transmitter of the force was the medium. A similar medium was postulated to account for the forces exerted by magnets on one another. Explaining the propagation of light, known to be a wave phenomenon, required the presence of still another medium. As you can see from the quotation at the beginning of this chapter, Maxwell at first believed that a major contribution of his work was that a single medium, called the *luminiferous* (light-bearing) *ether*, could be used to account for all three classes of phenomena.

When the physicists of the time tried to infer the physical properties of the ether, they encountered serious difficulties. To begin with, the ether had to be a solid because electromagnetic waves are transverse waves, which cannot propagate in a fluid (Section 22.4). But solids can carry longitudinal waves as well, and no such waves were observed in free space. Moreover, the ether had to be a very stiff solid of very low density. According to Equation 22.6, the speed of transverse waves in a solid is $v = \sqrt{G/\rho}$, where G is the shear modulus and ρ the mass per unit volume. Now, the speed of light is some 10^5 times the speed of transverse waves in rigid solids, such as steel (Example 22.1). That means the ratio G/ρ for the ether must be 10^{10} as great as the same ratio for ordinary solids! Yet, in spite of its extraordinary stiffness and lightness, the planets appear to pass through the ether without any observable frictional effects. More detailed studies produced still other difficulties.

The solution to the ether dilemma is to realize that, like many dilemmas, it arises from a self-deception. Because electromagnetic waves obey a wave equation identical in form with the wave equation that describes mechanical disturbances in media, it is easy to jump to the conclusion that electromagnetic waves (and the electric and magnetic fields) are manifestations of a mechanical disturbance in a medium, the ether. The ether dilemma disappears when we realize that the jump is not logical and the conclusion is wrong. *The basic realities are the electric and magnetic fields themselves.* There is no need to postulate any mechanical medium at all. To do so is to presuppose that all physics boils down to mechanics. We have no physical justification for this presupposition. In spite of all the important connections between electromagnetism and mechanics, *electromagnetism is not mechanics*. Electromagnetism is a separate, though allied, branch of physics. It is based on a set of fundamental laws—Maxwell's equations—that are independent of, and on the same logical footing as, Newton's laws of

Reductionism Has Its Virtues, but . . .

An outstanding success of the physics of the time, in which Maxwell himself played a vital part, was the reduction of thermodynamics to mechanics, which we traced in Chapters 18 through 20. But just because the behavior of thermal systems can be ultimately explained in mechanical terms, we cannot conclude that the same is the case for electromagnetic phenomena.

mechanics. Maxwell's equations do not describe the behavior of a mechanical system; Maxwell came to this insight slowly, over a period of almost ten years, after failing in many ingenious attempts to describe the properties of the ether in a consistent way. The German physicist Heinrich Hertz summarized the new insight in his famous remark, "Maxwell's theory *is* Maxwell's equations."

Hertz's Experiments and the Electromagnetic Spectrum

The observation that Maxwell's electromagnetic waves travel at the speed of light is highly suggestive but does not prove that electromagnetic radiation and light are the same thing. More experiment is necessary, and the crucial work was performed by Heinrich Rudolf Hertz (1857–1894) in 1887–88. Hertz's experimental program comprised the following steps:

1. According to Equations 34.17 and 34.19 and the discussions immediately after them, an oscillating electric field should give rise to electromagnetic waves. Such an oscillating field can be produced by operating an *LC* circuit (Section 32.4). Today we call such a device a **transmitter**. Hertz's transmitters operated at frequencies in and just below the *microwave range*, around 10^8 Hz.

2. If an oscillating electric field generated by a transmitter indeed produces electromagnetic waves, the waves must propagate through the region surrounding the transmitter. At any point not too far away, the electric wave produces an oscillating electric field. This field should excite an oscillating electric current in a second *LC* circuit, and this current should be detectable. Today we call the second *LC* circuit a **receiver**. Hertz's receiver was a simple metal rod broken in the middle by a tiny gap; today, we call this arrangement a *dipole antenna*. The self-inductance and capacitance of the dipole antenna were such that its resonant frequency was very close to that of the transmitter. Hertz's method of detecting the current was crude. He worked in a darkened room and looked for a tiny spark that jumped across the gap when the amplitude of the electric wave was sufficient.

3. Light is known to act in a wavelike fashion. Among other things, light can be reflected from appropriate barriers, and standing waves can be set up (Sections 21.6 and 21.7). For Hertz's electromagnetic waves, sheets of metal are effective reflectors. His apparatus is shown schematically in Figure 34.11. The frequency $\nu = 1/(2\pi\sqrt{LC})$ of the transmitter is known. The wavelength λ of the standing waves can be measured by finding the nodes—the places where the amplitude of the electromagnetic wave is zero and the receiver detects no signal. From this information, the propagation speed $v = \nu\lambda$ of the electromagnetic waves can be determined and compared with the speed c of light. The details of the measurement are given in the figure legend.

4. Light exhibits interference effects other than standing waves, some of which we consider in Chapter 35. Also, light can be polarized, brought to a focus with properly curved

FIGURE 34.11 Simplified picture of Hertz's method of measuring the wavelength of electromagnetic waves. The transmitter is located at *T*, and the sheet-metal reflectors are set a few meters apart. These reflectors are nodal planes, analogous to the fixed ends of a vibrating string. The receiver *R* is moved back and forth until maximum response (the largest possible spark) is obtained. The position of the reflectors is then adjusted until the maximum response of the receiver is itself maximized. The distance between the reflectors is then one of the proper distances $n\lambda/2$ required for the production of standing waves. Intermediate nodes are located by determining the positions at which the receiver does not respond. The distance between adjacent nodes is $\lambda/2$.

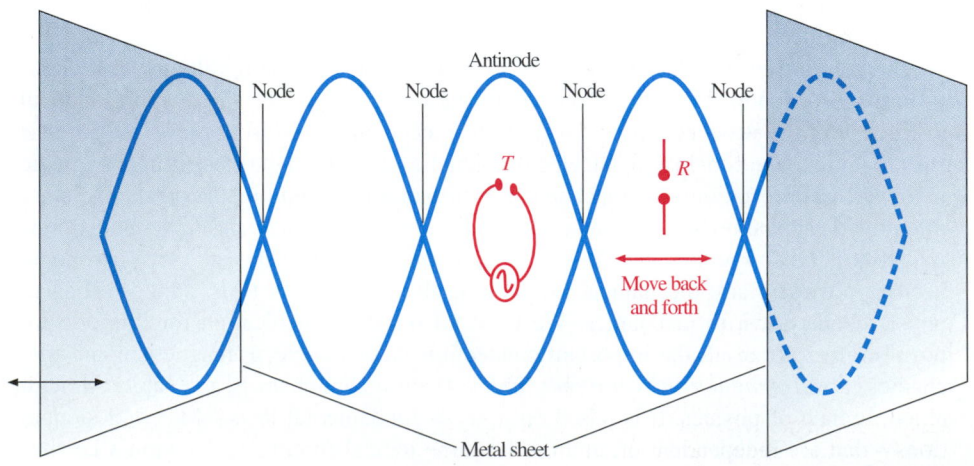

reflectors, and refracted by prisms, as discussed in Chapter 36. Hertz showed that all these effects can be produced and observed for the electromagnetic waves generated by the transmitter.

We have no reason to expect that we will see this electromagnetic radiation with our eyes. Maxwell's wave equations place no limitation on the frequency of the radiation, and eyes respond only to light within a limited frequency range, from roughly 5×10^{14} Hz to 8×10^{14} Hz. Hertz's oscillator (transmitter) did not operate in this range, so he used a receiver as a detector instead of his eyes, as we still do today. As Hertz put it in 1889,

Optics is no longer restricted to [light] waves, a small fraction of a millimeter in length; its domain is extended to waves that are measured in decimeters, meters and kilometers. And in spite of this extension, [optics] appears merely as ... a small appendage of the great domain of electricity. We see that the latter has become a mighty kingdom.*

In the absence of electronic amplifiers for detecting weak electric fields, Hertz's experiment was very difficult; success depended on his extraordinary skill as well as his deep understanding of Maxwell's theory, whose expression in modern form is due in large measure to Hertz.[†]

Today it is possible to generate, detect, and manipulate electromagnetic waves over a very large range of frequency (or wavelength). This range is called the **electromagnetic spectrum**. Working in any region of the spectrum requires the use of appropriate techniques, and highly sophisticated technologies have been developed for most regions. The regions of the spectrum have been given conventional names, shown in Figure 34.12. You are undoubtedly familiar with many of these names. The regions are not sharply defined but merge into the neighboring regions. As you can see from Figure 34.12, Maxwell's theory of electromagnetic waves can be used to account for a vast variety of observed effects.

We cannot consider the techniques for working with electromagnetic radiation in any detail here; such techniques are major concerns of electrical and electronic engineering, optics, and high-energy physics. As a general rule of thumb, however, the equipment appropriate for generating and detecting electromagnetic radiation of a given wavelength has physical dimensions comparable to the wavelength. The VHF (very high frequency) and UHF (ultrahigh frequency) waves used for television transmission, whose wavelengths are of order 1 m, are detected by antennas about 1 m long. (Look at almost any roof to check this.) X rays, whose wavelength is typically 10^{-10} m, are emitted and detected by individual atoms whose dimensions are about the same. It will not surprise you that the techniques for manipulating X rays differ in many ways from the techniques for manipulating TV signals.

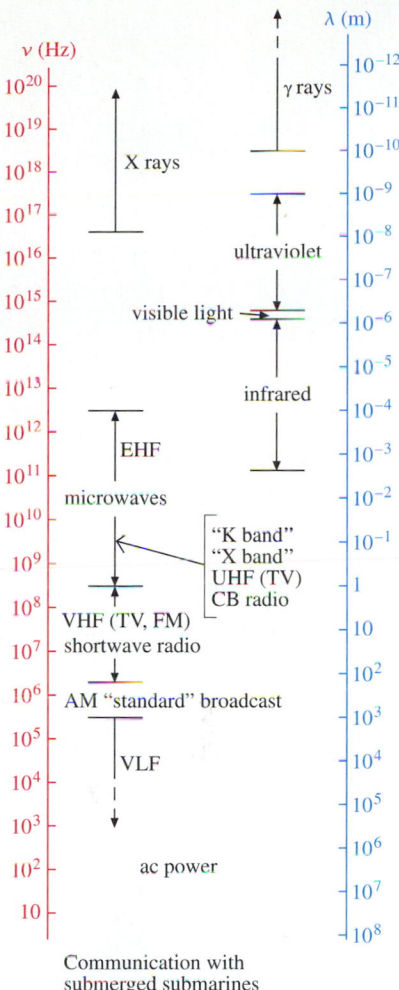

FIGURE 34.12 The electromagnetic spectrum, with the conventional names given to the various regions, or ''bands.'' Note the overlap between adjacent bands. Both the frequency scale on the left and the wavelength scale on the right are logarithmic.

Energy Carried by Electromagnetic Waves

According to Equation 26.28, there is an energy density (energy per unit volume) u_e associated with an electric field $\mathscr{E}$. In empty space, the energy density is

$$u_e = \tfrac{1}{2}\epsilon_0 \mathscr{E}^2. \tag{34.32a}$$

---

*H. Hertz, *Miscellaneous Papers* (London: Macmillan, 1986), pp. 325–326. Quoted in J. F. Mulligan, *Am. J. Phys*, **55**, 711 (1987).

[†]This combination of experimental and conceptual difficulties explains, at least in part, why more than two decades elapsed between Maxwell's theoretical work and Hertz's experimental confirmation. Remarkably, only seven more years elapsed until the Italian-British inventor Guglielmo Marconi (1874–1937) successfully demonstrated ''wireless'' transmission of messages in 1895. Marconi achieved transatlantic transmission in 1901; by then, commercial radio communication was a reality.

The energy density u_{m} associated with a magnetic field $\mathcal{B}$ is given by Equation 31.17,

$$u_{\mathrm{m}} = \frac{1}{2}\frac{\mathcal{B}^2}{\mu_0}. \tag{34.32b}$$

An electromagnetic wave possesses both electric and magnetic fields. How does the magnetic energy density compare with the electric energy density? To answer this question, we use Equation 34.31, $\mathcal{B} = \mathcal{E}/c$, and Equation 34.26, $1/\sqrt{\mu_0\epsilon_0} = c$, to rewrite Equation 34.32b as

$$u_{\mathrm{m}} = \frac{1}{2}\frac{\mathcal{E}^2}{\mu_0 c^2} = \frac{1}{2}\epsilon_0\mathcal{E}^2.$$

That is, *the energy in an electromagnetic wave is evenly divided between the electric and magnetic fields.*

The total energy density of the electromagnetic wave is just the sum

$$u = u_{\mathrm{e}} + u_{\mathrm{m}} = \epsilon_0\mathcal{E}^2 = \frac{\mathcal{B}^2}{\mu_0}. \tag{34.33}$$

It is often useful to write this result in a form that shows explicitly the presence of both electric and magnetic fields. Using the relation $\mathcal{B} = \mathcal{E}/c$ again, we have

$$u = c\epsilon_0\mathcal{E}\mathcal{B} \quad \text{or} \quad u = \frac{\mathcal{E}\mathcal{B}}{c\mu_0}. \tag{34.34a,b}$$

Energy Flux and the Poynting Vector

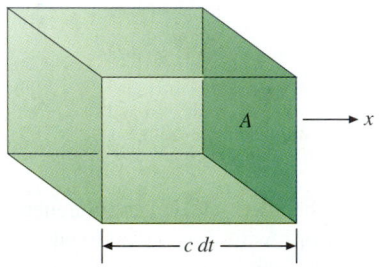

FIGURE 34.13 Construction for deriving Equation 34.35. The electromagnetic wave energy present in the volume $c\,dt\,A$ will pass through A during the next time interval dt.

As an electromagnetic wave travels from its source, it carries energy with it. In this property, it is no different from any other traveling wave (Section 21.5). As with other waves, we imagine a surface oriented normal to the direction of propagation of a plane electromagnetic traveling wave, which we assume to be the x direction, as shown in Figure 34.13. The part of the wave lying to the left of the surface, within the volume of area A and length $c\,dt$, will pass through the surface in the time interval dt. The total energy within this volume is

$$dE = u \times \text{volume} = uAc\,dt.$$

Thus the flow of energy per unit time through the area is

$$\frac{dE}{dt} = uAc.$$

We define the **energy flux** S to be the value of dE/dt *per unit area.* That is, we define

$$S \equiv \frac{1}{A}\frac{dE}{dt}.$$

You can see from this equation that the SI unit of energy flux is J/(m²·s), or W/m². For electromagnetic waves, S has the particular value

$$S = uc. \tag{34.35}$$

When we insert into this equation the value of u given by Equation 34.34b, we obtain

$$S = \frac{\mathcal{E}\mathcal{B}}{\mu_0}. \tag{34.36a}$$

So far, we have considered only the magnitude S of the energy flux, in the simple case in which the energy travels in the x direction, the electric field is in the y direction, and the magnetic field is in the z direction. But the idea of energy flux can be applied as well to more complicated situations, in which the direction of propagation is arbitrary and the waves are not necessarily plane waves. In this case, the magnitude and direction of the energy flux are given by the **Poynting vector S,*** defined as

$$\mathbf{S} \equiv \frac{1}{\mu_0} \mathscr{E} \times \mathscr{B}. \qquad (34.36b)$$

Electromagnetic radiation isn't just for communication.

For the simple geometry we have been considering, the right-hand rule $\hat{\mathbf{y}} \times \hat{\mathbf{z}} = \hat{\mathbf{x}}$ confirms that the direction of $\mathbf{S}$ is indeed the direction of propagation. Note that the direction of $\mathbf{S}$ is always positive, regardless of the instantaneous directions of $\mathscr{E}$ and $\mathscr{B}$. Because $\mathscr{E}$ and $\mathscr{B}$ are in phase, $\mathscr{B}_z$ is always positive when $\mathscr{E}_y$ is positive—and negative when $\mathscr{E}_y$ is negative. Thus their product is always positive. Because $\mathscr{E} \perp \mathscr{B}$, Equation 34.36b reduces to the scalar form of Equation 34.36a. That is, the energy flux is the magnitude of the Poynting vector.

Equation 34.36b also specifies the relative orientations of $\mathscr{E}$ and $\mathscr{B}$, by means of the right-hand rule for cross products. If you face in the direction of propagation $\hat{\mathbf{S}}$, you must turn 90° clockwise to go from $\mathscr{E}$ to $\mathscr{B}$.

Because $\mathscr{E}$ and $\mathscr{B}$ oscillate, the energy flux S and the Poynting vector $\mathbf{S}$ also are oscillatory quantities. It is often useful to express energy flux in terms of a mean value $\langle S \rangle$. This mean value is called the **intensity** I. For an electromagnetic wave of sinusoidal form

$$\mathscr{E}_y = \mathscr{E}_0 \cos(kx - \omega t) \quad \text{and} \quad \mathscr{B}_z = \frac{\mathscr{E}_0}{c} \cos(kx - \omega t),$$

we have

$$S = \frac{\mathscr{E}_0^{\,2}}{\mu_0 c} \cos^2(kx - \omega t). \qquad (34.37)$$

The mean value of $\cos^2 \theta$ over one cycle is $\frac{1}{2}$. The calculation is identical with that for the mean power in Section 33.4. So we have

$$I = \langle S \rangle = \frac{\mathscr{E}_0^{\,2}}{2\mu_0 c}. \qquad (34.38a)$$

The intensity can also be expressed in terms of the root-mean-square electric field $\mathscr{E}_{\text{rms}}$. Just as in calculating the rms value of a sinusoidal voltage in Section 33.4, you can show that $\mathscr{E}_{\text{rms}} = \mathscr{E}_0/\sqrt{2}$. It follows that

$$I = \frac{\mathscr{E}_{\text{rms}}^{\,2}}{\mu_0 c} \qquad (34.38b)$$

Just as for the Poynting vector, the SI unit of intensity is W/m^2.

EXAMPLE 34.3

Energy from the sun falls on the earth at a rate of 1353 W/m^2. This quantity is called the **solar constant**. (The solar constant represents the energy incident on the top of the atmosphere. An appreciable proportion of this energy is absorbed or reflected by the atmosphere, and so the energy flux reaching the earth's surface is considerably less. Also, note that the area is taken by definition to be normal to the sun–earth line. It follows that the energy striking a square meter of the earth's surface decreases with increasing latitude.)

Find the rms values of the electric and magnetic fields in the sunlight reaching the top of the atmosphere.

SOLUTION: Using Equation 34.38b, you can write

$$\begin{aligned}
\mathscr{E}_{\text{rms}} &= \sqrt{\mu_0 c I} \\
&= \sqrt{4\pi \times 10^{-7}\ \text{T·m/A} \times 2.998 \times 10^8\ \text{m/s} \times 1353\ \text{W/m}^2} \\
&= 714.0\ \text{V/m}.
\end{aligned}$$

You can find the value of $\mathscr{B}_{\text{rms}}$ most easily by using Equation 34.31, $\mathscr{B} = \mathscr{E}/c$:

$$\mathscr{B} = \frac{714.0\ \text{V/m}}{2.998 \times 10^8\ \text{m/s}} = 2.381 \times 10^{-6}\ \text{T}.$$

---

*All vectors point in one direction or another, but only the Poynting vector is named after the British physicist J. H. Poynting (1852–1914), who derived Equation 34.36b in 1884.

EXAMPLE 34.4

Figure 34.14 shows a section of a long, straight wire of radius b, length l, and resistance R, carrying a uniformly distributed current i. Find the magnitude and direction of the Poynting vector that describes energy flux into this wire, and use this result to evaluate the power supplied to the wire.

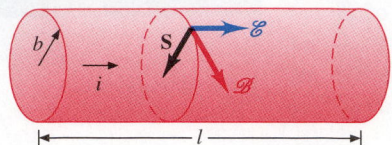

FIGURE 34.14

SOLUTION: First, consider the direction of $\mathbf{S}$. The electric field $\mathscr{E}$ drives the current, and the direction of $\mathscr{E}$ is evidently along the wire, as shown in Figure 34.14. By symmetry, $\mathscr{E}$ has the same value everywhere in the wire. The magnetic flux lines are circles centered on the wire axis. The magnetic field is tangent to these flux lines; the figure shows $\mathscr{B}$ at a particular point on the surface of the wire. According to the right-hand rule, the direction of $\mathscr{E} \times \mathscr{B}$ is radially inward, toward the axis of the wire. This inward direction is thus the direction of $\mathbf{S}$, in accordance with Equation 34.36b.

Now consider the magnitude of $\mathbf{S}$, which depends on the magnitudes $\mathscr{E}$ and $\mathscr{B}$. If the voltage drop from one end of the wire to the other is V, you have for the electric field

$$\mathscr{E} = \frac{V}{l} = \frac{iR}{l}.$$

The magnitude $\mathscr{B}$ of the magnetic field depends on the distance r from the wire axis according to Equations 29.21 and 29.5; with minor notational changes,

$$\mathscr{B} = \frac{\mu_0 i r}{2\pi b^2} \quad \text{for } r < b$$

$$\mathscr{B} = \frac{\mu_0 i}{2\pi r} \quad \text{for } r \geq b.$$

Because you are interested in the power entering the wire as a whole, you use the value of $\mathscr{B}$ appropriate to the *surface* of the wire. (Even though the source of emf does work in making charge flow from one end of the wire toward the other, the Poynting vector enters the wire through the wire's periphery. We will return to this matter shortly.) At the surface, either equation gives you $\mathscr{B} = \mu_0 i / 2\pi b$. So the Poynting vector has the magnitude

$$S = \frac{1}{\mu_0} \frac{iR}{l} \frac{\mu_0 i}{2\pi b} = \frac{i^2 R}{2\pi b l}.$$

Note that $i^2 R / 2\pi b l$ is the power dissipated by the resistance of the wire, divided by the surface area of the wire.

To find the total power flowing into the wire, you note that $\mathbf{S}$ is everywhere normal to the periphery of the wire. Because $\mathbf{S}$ is defined to be the power per unit area normal to $\hat{\mathbf{S}}$, you can write the power simply as

$$P = S\, 2\pi\, bl = i^2 R.$$

Are you surprised to find that the rate at which electrical energy enters the wire is equal to the rate at which the electrical energy is converted into heat energy?

What is the physical meaning of the radially inward energy flux in Example 34.4? We can answer this question most easily by looking at the system in a slightly different way. Let us disconnect the external source of emf that drives the current through the wire. If there is still to be an inward-directed Poynting vector, the source of energy must be elsewhere. Suppose that $\mathbf{S}$ represents electromagnetic wave energy from some distant radio transmitter.* In this case, $\mathbf{S}$ will vary as $\cos^2 \omega t$, in accordance with Equation 34.37. But at any instant, $\mathbf{S}$ has some value $\mathscr{E} \times \mathscr{B}/\mu_0$. At that instant, $\mathscr{B}$ is everywhere tangent to the surface of the wire and $\mathscr{E}$ is directed along the axis of the wire, just as in Figure 34.14. Thus the electromagnetic energy flow into the wire results in a flow of charge along the wire.

Now consider what happens one half-cycle later. The direction of $\mathbf{S}$ is still inward, but $\mathscr{E}$ and $\mathscr{B}$ have reversed. Now the electric field drives charge along the wire in the opposite direction. (Use the right-hand rule to check this.) Over time, the result is an alternating current whose frequency is the same as that of the electromagnetic wave. That is how a radio receiving antenna works. If the antenna wire is part of a resonant circuit whose resonant frequency is the same as the frequency of the incident electromagnetic wave, the amplitude of the induced current is optimized. Under normal circumstances, the current induced in the antenna is small even in this "tuned" condition, and one function of the receiver circuitry is to amplify the signal considerably.

---

*Don't worry about the transmitter geometry that would be required to produce a radially inward flow of energy.

$\mathscr{B}, \mathscr{B}$	magnetic field		P	power
c	speed of light		$\mathbf{S}, S$	Poynting vector, energy flux
E	energy		u, u_e, u_m	total, electric, magnetic energy density
$\mathscr{E}, \mathscr{E}$	electric field		v	speed
i, i_c, i_d	total current, conduction current, displacement current		κ	dielectric constant
I	intensity		ν, ω	frequency, angular frequency
$\mathbf{j}$	current density		Φ_e, Φ_m	electric, magnetic flux
q	electric charge		λ	wavelength

If Ampère's law is to be consistently applicable to all electric circuits, it is necessary to generalize the concept of current to include the **displacement current** i_d as well as the **conduction current** i_c. The displacement current is defined for a uniform field by Equation 34.3, $i_d = \epsilon_0 A\, d\mathscr{E}/dt$, and in general by Equation 34.8,

$$i_d = \epsilon_0 \int_{\substack{\text{Amperean} \\ \text{surface}}} \frac{\partial \mathscr{E}}{\partial t} \cdot d\mathbf{A}.$$

The total current i passing through any surface is then the sum given by Equation 34.5, $i = i_c + i_d$. Thus taking the displacement current into consideration, we have **Maxwell's generalization of Ampère's law**, given by Equation 34.9.

The Lorentz force, given by Equation 34.12e, $\mathbf{F} = q(\mathscr{E} + \mathbf{v} \times \mathscr{B})$, is a complete description of the force exerted by electric and magnetic fields on matter. Taken together, Maxwell's generalization of Ampère's law, Gauss's laws for electric charges and for magnetism, and Faraday's law constitute a complete description of the way in which electrically charged matter gives rise to electric and magnetic fields, and the mutual interaction of electric and magnetic fields. The five equations thus provide a complete basis for the study of classical electromagnetism. The first four are called **Maxwell's equations** (Equations 34.12a, b, c, d):

1. Gauss's law for electric charge:

$$\int_{\substack{\text{Gaussian} \\ \text{surface}}} \mathscr{E} \cdot d\mathbf{A} = \frac{q}{\epsilon_0}.$$

2. Gauss's law for magnetism:

$$\int_{\substack{\text{Gaussian} \\ \text{surface}}} \mathscr{B} \cdot d\mathbf{A} = 0.$$

3. Faraday's law:

$$\int_{\substack{\text{Faradayan} \\ \text{curve}}} \mathscr{E} \cdot d\mathbf{s} = -\int_{\substack{\text{Faradayan} \\ \text{surface}}} \frac{\partial \mathscr{B}}{\partial t} \cdot d\mathbf{A}.$$

4. Generalized Ampère's law:

$$\int_{\substack{\text{Amperean} \\ \text{curve}}} \mathscr{B} \cdot d\mathbf{s} = \mu_0 \left(\int_{\substack{\text{Amperean} \\ \text{surface}}} \mathbf{j} \cdot d\mathbf{A} + \epsilon_0 \int_{\substack{\text{Amperean} \\ \text{surface}}} \frac{\partial \mathscr{E}}{\partial t} \cdot d\mathbf{A} \right).$$

These equations possess incomplete symmetry, owing to the absence of any known magnetic monopoles analogous to electric charges. However, in a region where electric charges are absent, the two Gauss's laws assume almost identical form, as do Faraday's and Ampère's laws. **Maxwell's equations in vacuum** are given by Equations 34.13a, b, c, d:

$$\int_{\substack{\text{Gaussian} \\ \text{surface}}} \mathscr{E} \cdot d\mathbf{A} = 0;$$

$$\int_{\substack{\text{Gaussian} \\ \text{surface}}} \mathscr{B} \cdot d\mathbf{A} = 0;$$

$$\int_{\substack{\text{Faradayan} \\ \text{curve}}} \mathscr{E} \cdot d\mathbf{s} = -\int_{\substack{\text{Faradayan} \\ \text{surface}}} \frac{\partial \mathscr{B}}{\partial t} \cdot d\mathbf{A};$$

$$\int_{\substack{\text{Amperean} \\ \text{curve}}} \mathscr{B} \cdot d\mathbf{s} = \mu_0 \epsilon_0 \int_{\substack{\text{Amperean} \\ \text{surface}}} \frac{\partial \mathscr{E}}{\partial t} \cdot d\mathbf{A}.$$

From Maxwell's equations we can deduce the existence of **electromagnetic waves**, described by the twin wave equations

$$\frac{\partial^2 \mathscr{E}_y}{\partial t^2} = c^2 \frac{\partial^2 \mathscr{E}_y}{\partial x^2} \quad \text{and} \quad \frac{\partial^2 \mathscr{B}_z}{\partial t^2} = c^2 \frac{\partial^2 \mathscr{B}_z}{\partial x^2}.$$

The speed of the waves is c, the speed of light. This speed is related to the SI constants ϵ_0 and μ_0 by Equation 34.26,

$$\frac{1}{\sqrt{\mu_0 \epsilon_0}} = c.$$

The electric and magnetic fields constituting the electromagnetic wave are mutually perpendicular, and both are transverse to the direction of propagation of the wave. The two fields oscillate in phase. At any point in space, the magnitudes of the two fields are related by Equation 34.31,

$$\mathscr{B}_z = \frac{1}{c} \mathscr{E}_y.$$

The electromagnetic wave speed is the speed of light. Electromagnetic waves arising from oscillating electric and magnetic fields can be made to exhibit wavelike behavior analogous to that of light. From these two facts, we infer that light is a form of **electromagnetic radiation**. However, the **electromagnetic spectrum** includes radiation having a vast range of wavelength and frequency; only a small part of the entire spectrum is visible light. Other parts of the spectrum have a wide variety of special properties and special applications that depend on the wavelength.

The energy density of electromagnetic waves is given by any of Equations 34.33 and 33.34. The transport of energy by electromagnetic waves is described by the **Poynting vector** defined by Equation 34.36b,

$$\mathbf{S} \equiv \frac{1}{\mu_0} \boldsymbol{\mathscr{E}} \times \boldsymbol{\mathscr{B}}.$$

The magnitude S of the Poynting vector is the **energy flux** of the wave, given by Equation 34.36a. The intensity is given by Equation 34.38b,

$$I = \frac{\mathscr{E}_{\mathrm{rms}}^2}{\mu_0 c}.$$

Queries and Problems for Chapter 34

QUERIES

34.1 *(2) Subtle current.* Two parallel-plate capacitors have equal capacitance, but one contains a dielectric of dielectric constant κ and the other contains only air, for which $\kappa \simeq 1$. If the two capacitors are connected in parallel across the same source of emf, which capacitor carries the greater displacement current? If both capacitors have plates of the same area, in which one is the displacement current density greater? If both capacitors have the same spacing between their plates, in which one is the displacement current density greater?

34.2 *(4) Stringing along.* In view of the fact that ϵ_0 is defined exactly, why is its value given by an indefinitely long string of digits?

34.3 *(4) Old-fashioned ways.* Before 1983, the speed of light was defined in terms of a standard meter and a standard second, and its value was determined experimentally. The value of μ_0 was defined just as it is today. What can you say about the value of ϵ_0 used before 1983?

34.4 *(4) A long distance ago.* Although they are objects of great intrinsic brightness, quasars appear dim because they are very distant. The interest of quasars to astronomers lies largely in the fact that they represent the universe as it was when it was only one-fourth or less as old as it is today. Explain.

34.5 *(5) Is that why BBC programs are better?* In the United States and in many other countries, the TV antennas one sees on nearly every roof lie in a horizontal plane. In Britain, however, they are oriented in a vertical plane. What can you infer about the way TV signals are transmitted in the United States and in Britain?

34.6 *(5) Blue skies, smiling at me . . .* The electromagnetic waves constituting sunlight are not polarized. But, for simplicity, consider only one possible direction of the electric field vector of the sunlight. **(a)** If the sun is directly overhead,

to what set of directions is the electric field vector restricted? **(b)** The atmosphere contains plenty of electrons. What happens to an electron when polarized sunlight falls on it? **(c)** Compare the electron with the transmitting antenna implicit in Query 34.5. Why is the entire sky bright in daytime? This reradiation process is called *Rayleigh scattering*. The sky is blue because the process is much more efficient for blue light, of wavelength 400 nm, than for red light, of wavelength 700 nm.

34.7 *(6) Cross purposes.* In the discussion after Example 34.4, we did not consider the magnetic force exerted by the magnetic wave on a current carrier (assumed positive for convenience) set into motion by the electric wave. **(a)** Show that the force is always inward and that the result is a radial emf, the axis of the wire being positive with respect to the surface. **(b)** Justify the assertion that an electromagnetic wave must carry momentum as well as energy.

34.8 *(6) Getting warm.* Explain how a microwave oven cooks food. Why must the food be placed in a dielectric dish?

34.9 *(G) Chicken & egg?* The demands of an evolving technology often stimulate important developments in science. For example, Carnot's seminal contributions to modern thermodynamics, in the mid-1820s, were clearly motivated by his desire to achieve a deeper understanding of the operation of steam engines, a technology that had "caught fire" some four decades earlier. For electromagnetism, the sequence was the opposite one: Maxwell's theory (early 1860s) and Hertz's experiments (late 1880s) preceded the earliest development of radio communication (late 1890s). List five or six important nineteenth and twentieth century technologies, and determine which order was followed in each case. Can you detect a trend?

GROUP A

34.1 *(2) Displacement current, I.* A capacitor consists of two parallel disk-shaped plates, each of area 0.25 m^2. What is the displacement current through the empty space between the plates when the electric field is changing at a rate $d\mathscr{E}/dt = 8.7 \times 10^9$ V/m·s?

34.2 *(2) M. Ampère, meet Mr. Faraday!* For the situation described in Problem 34.1, what is the magnetic field $\mathscr{B}$ at a point located on an imaginary line drawn between the edges of the two plates?

34.3 *(2) Displacement current, II.* A 0.1-μF capacitor is part of a 170-V (peak), 60-Hz ac circuit. The voltage across the capacitor has the sinusoidal form $V = V_0 \sin \omega t$. **(a)** For what value of V is the displacement current a maximum? **(b)** What is the maximum displacement current?

34.4 *(2) Displacement current, III.* **(a)** Express the displacement current between the plates of a parallel-plate capacitor in terms of C and dV/dt. **(b)** Does your result agree with that obtained using Equation 33.7?

34.5 *(2) Displacement current, IV.* A 0.65-μF capacitor is part of a 9.2-kHz ac circuit. If the peak voltage across the capacitor is 780 V, what is the maximum displacement current carried by the capacitor?

34.6 *(2) Displacement current, V.* A parallel-plate capacitor is filled with material of dielectric constant κ; the capacitance is C. The capacitor is connected by wires of negligible resistance to a source whose emf is $V = V_0 \sin \omega t$. **(a)** What is the conduction current in the wires? **(b)** What is the displacement current between the capacitor plates? **(c)** What is the displacement current density between the plates?

34.7 *(2) A current with nothing flowing.* Prove that the right side of Equation 34.2 has the dimensions of electric current.

34.8 *(2) Fresh fields, I.* A point P in the space between the disk-shaped plates of an air-filled capacitor is located at a distance r from the symmetry axis of the capacitor. The electric field between the plates is changing at the rate $\partial\mathscr{E}/\partial t$. **(a)** Find the magnetic field at P. **(b)** For the capacitor of Example 34.1, what is the magnitude of $\mathscr{B}$ if the capacitor is connected to a 10-kHz ac source?

34.9 *(4) Nightline.* On television news shows, an interviewer often talks with a newsworthy person located far away, via a satellite link. The satellites are normally about 36 000 km above the surface of the earth. Estimate the time it takes for one of these persons to hear a word spoken by the other, and show that the speakers and the audience are not likely to be conscious of a delay.

34.10 *(4) Delayed action.* Chinese astronomers made detailed records of the "guest star" that appeared suddenly in 1054 C.E., became bright enough to be seen in broad daylight, and then faded from naked-eye view over about a year. We call the remains of this supernova explosion the Crab Nebula. The distance between the Crab Nebula and the earth is 6500 light years (ly), 1 ly being the distance that a light signal travels in 1 year. When did the Crab Nebula explode?

34.11 *(4) Rock 'n' roll.* An AM radio station broadcasts at 940 kHz. What is the wavelength of the electromagnetic waves it radiates?

34.12 *(4) Tuning in, I.* The standard AM radio broadcast band ranges in frequency from 550 kHz to 1600 kHz. Find the corresponding wavelengths. Is an AM radio antenna likely to satisfy the efficiency condition that its dimensions be comparable to the wavelength?

34.13 *(4) Tuning in, II.* The standard FM radio broadcast band ranges in frequency from 87.5 MHz to 107.5 MHz. Find the corresponding wavelengths. Is an FM radio antenna likely to satisfy the efficiency condition that its dimensions be comparable to the wavelength?

34.14 *(4) Chihuahuas and St. Bernards.* Referring to Figure 34.12, find the ratio of the wavelength of X rays to that of UHF waves.

34.15 *(4) Not a magnet in sight!* An electric wave has amplitude 430 V/m. Find the amplitude of the accompanying magnetic wave.

34.16 *(6) Zap!* Laser beams can be very intense indeed. When a powerful laser beam passes through air, sparks are observed if the electric field is great enough to ionize the air molecules; this is called the breakdown effect. The breakdown field for air is about 3×10^6 V/m. What is the minimum electromagnetic energy flux S required to produce the breakdown effect?

34.17 *(6) Companions.* If the electric field of a plane electromagnetic wave has maximum value 2.5×10^{-3} V/m, what is the maximum value of the magnetic field?

34.18 *(6) Manner of expression.* Express the intensity of an electromagnetic wave in terms of the rms value of the magnetic field.

34.19 *(6) What watts?* Beginning with the right side of Equation 34.38a, show that the SI unit of intensity is W/m^2.

34.20 *(6) Listening in.* A radio station radiates 50 kW of power. Assume (as is never the case in practice) that the radiation is uniform over the hemisphere centered at the base of the transmitter tower. **(a)** Find the energy flux at 10 km and 300 km from the station. **(b)** What is the amplitude of the electric field at 300 km? **(c)** The minimum electric field amplitude that can be detected by a radio receiver is called the *sensitivity* of the receiver. If a receiver has sensitivity 3.5 μV/m, what is the greatest distance at which it can "pull in" the station? Make the (often unrealistic) assumption that the wave travels from transmitter to receiver in a straight line without reflection or refraction.

34.21 *(6) Illuminating.* A 150-W light bulb radiates uniformly (well, almost uniformly) in all directions. What are the amplitudes of the electric and magnetic vectors **(a)** at 1.0 m? **(b)** at 16 m?

34.22 *(6) Alternatives.* Express the intensity I of an electromagnetic wave in terms of **(a)** the product $\mathscr{E}_{rms}\mathscr{B}_{rms}$ and **(b)** $\mathscr{B}_{rms}$ only.

34.23 *(2) Fresh fields, II.* Look again at the capacitor of Example 34.1. **(a)** Consider a point P' located at $r > b$, and show that, if the electric field between the capacitor plates is varying at the rate $\partial \mathscr{E}/\partial t$, the magnetic field at P' is

$$\mathscr{B} = \frac{\mu_0 \epsilon_0 b^2}{2r} \frac{\partial \mathscr{E}}{\partial t}.$$

(b) At what value of r does $\mathscr{B}$ attain its maximum value? Show that your result agrees with that obtained on the basis of the calculation in part **d** of Example 34.1. **(c)** Calculate the magnetic field at P' if $r = 12.0$ cm.

34.24 *(2) Leaky dielectric, I.* A parallel-plate capacitor is filled with a material of dielectric constant κ. The material is an imperfect dielectric; it has a substantial but not infinite resistivity ρ. **(a)** When an ac voltage is applied across the capacitor, what is the phase relation between the displacement current i_d and the conduction current i_c through the capacitor? **(b)** Show that the rms displacement current and the rms conduction current passing through the capacitor are equal when the frequency is $\omega = 1/(\kappa \epsilon_0 \rho)$.

34.25 *(2) Leaky dielectric, II.* The capacitor of Problem 34.24 is charged by connecting it to a battery. Then the battery is disconnected and the capacitor is allowed to discharge through its own dielectric. Find the relation between the conduction current and the displacement current as the capacitor discharges.

34.26 *(2) Spherical symmetry.* An air-filled spherical capacitor of capacitance C is part of an ac circuit. The voltage across the capacitor is $V = V_0 \cos \omega t$. Show that the displacement current through the space between the plates is the same as the displacement current through the space in a parallel-plate capacitor of equal capacitance.

34.27 *(2) Displacement current, VI.* A plane electric wave is described by the function $\mathscr{E}_y = \mathscr{E}_0 \cos (kx - \omega t)$. Find the maximum value of the displacement current density j_0.

34.28 *(2) Eddies in the channel.* An alternating current $i = i_0 \cos \omega t$ flows through a long solenoid having n turns per meter. Show that the displacement current density at a point inside the solenoid at a distance r from the solenoid axis is

$$j_d = \frac{r}{2} \frac{d^2 \mathscr{B}}{dt^2},$$

where $\mathscr{B}$ is the instantaneous magnitude of the magnetic field inside the solenoid.

34.29 *(2) Comin' atcha! I.* An electron moves in a straight path at constant speed v. **(a)** Show that the displacement current density at a point on the path at a distance r ahead of the electron is $j_d = ev/2\pi r^3$. **(b)** Evaluate j_d at a distance 0.01 mm ahead of the electron if the electron has been accelerated from rest through a potential difference of 10 V.

34.30 *(2) Comin' atcha! II.* An electric dipole of moment **p** moves with velocity **v** parallel to $\hat{\mathbf{p}}$. What is the displacement current at a point in the path of the dipole, a distance ahead of the dipole? Assume that x is much greater than the distance between the two charges of the dipole.

34.31 *(3) Nobel Prize for sure, I.* Imagine that you read in this morning's newspaper that a physicist has observed magnetic monopoles—elementary particles with a magnetic "charge" M. Rewrite Maxwell's equations (Equations 34.12a, b, c, d) to account for the existence of these monopoles.

34.32 *(3) Filling the vacuum.* How would you rewrite Equations 34.13a, b, c, d if you wanted to describe electromagnetic radiation in a region filled with material of dielectric constant κ?

34.33 *(4) Transverse only.* Prove that a magnetic wave cannot be longitudinal.

34.34 *(4) Double satisfaction.* Show that the electromagnetic wave equations (Equations 34.22 and 34.23) are satisfied by the wave functions

$$\mathscr{E}_y = \mathscr{E}_0 \cos (kx - \omega t) \quad \text{and} \quad \mathscr{B}_z = \mathscr{B}_0 \cos (kx - \omega t).$$

In these functions, $k \equiv 2\pi/\lambda$ is the wave number (Chapter 21). What condition is imposed on the ratio ω/k?

34.35 *(4) Intimate relationship.* A plane magnetic wave is described by the function $\mathscr{B}_z = \mathscr{B}_0 \cos (kx - \omega t)$. Find the electric field as a function of time at $x = 0$.

34.36 *(6) Just checking.* You are charging a parallel-plate air-gap capacitor having plate area A and plate separation D. At a certain instant, the current is i and the voltage across the capacitor is V. Assume that edge effects are negligible. **(a)** What is the total rate of energy flow, SA, into the gap? **(b)** Show that this rate is equal to the power input to the capacitor given by Joule's law, $P = Vi$.

34.37 *(6) Poynting inward.* Consider again the wire discussed in Example 34.4. This time, imagine a cylindrical surface of radius $r < b$, coaxial with the wire. Find **S** at any point in this surface, and determine the power entering the part of the wire within the surface. Show that your result is consistent with Joule's law in the form $P = i^2 R$.

34.38 *(G) Coaxial cable.* A coaxial cable consisting of two cylindrical conducting shells of radii b_1 and b_2 is shown below. The conductors are connected to the poles of a battery at

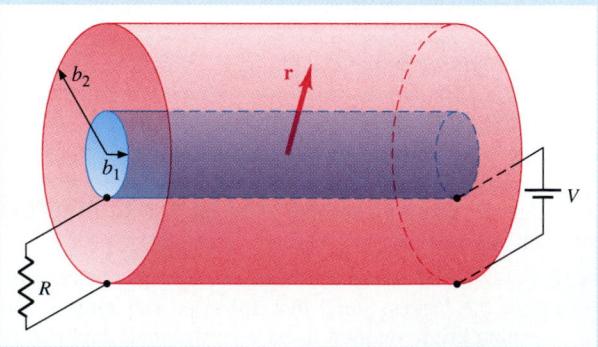

one end and to a resistor R at the other end. The resistance of the cable is negligible compared with R. **(a)** Show that the electric field at a point **r** in the space between the conductors is

$$\mathscr{E} = \frac{V}{r \ln (b_2/b_1)} \hat{\mathbf{r}}.$$

(b) Find the magnitude and direction of the magnetic field $\mathscr{B}$ at **r**. **(c)** Evaluate the Poynting vector **S**. **(d)** Integrate S over an imaginary area drawn normal to **S**, and show that the power transmitted through the coaxial cable is equal to that dissipated in the resistor.

34.39 *(6) Solenoid.* A solenoid of radius b has n turns/m. At a certain instant, the current in the solenoid is changing at a rate di/dt. **(a)** What is the electric field at a point inside the solenoid, a distance r from its axis? **(b)** Find **S** at this point.

34.40 *(6) Transmitting antenna.* A Poynting vector directed radially inward carries electromagnetic energy to a receiving antenna. In Example 34.4, energy is supplied to the same antenna wire by a source of emf. The antenna thus becomes a transmitting antenna. What is the direction of the Poynting vector representing the electromagnetic radiation? What is the direction of the Poynting vector representing the input to the antenna, from the source of emf, of the power radiated? Express the radiated energy as a "resistance" $V_{\text{rms}}/i_{\text{rms}}$ that must be added to the ordinary resistance of the antenna. Engineers call this the *radiation resistance* of the antenna. (Most antennas are designed so that the radiation resistance is much greater than the ordinary ohmic resistance.)

34.41 *(6) Hot.* **(a)** Find the total power striking the earth in the form of solar radiation. Take the solar constant to be 1353 W/m² and the earth's radius to be 6370 km. **(b)** How much energy is contained in 1 m³ of space located just above the earth's atmosphere (and not in the earth's shadow)? **(c)** The distance from the sun to the earth is 1.5×10^{11} m. What is the total power radiated by the sun? **(d)** The radius of the sun is 7.0×10^8 m. What is the energy flux S at the surface of the sun?

34.42 *(6) Solar design.* A typical single-family house consumes electric power at about 20 kWh/day if electricity is not used for water or space heating or for air conditioning. If this power is to be supplied entirely by solar cells located on the house lot, calculate the required area of the cells. Make the following assumptions for a house located at latitude 40°:

1. Mean daytime elevation of the sun (angle above the horizon): 25°

2. Mean length of day: 12 h

3. Number of clear days per year: 250

4. Efficiency of solar cells in converting light into electricity: 8%.

What area must the cells cover? Estimate the area of a typical house lot and compare your result with this area. (Note: The use of mean angles and day lengths in this problem is very crude, as are the implicit assumptions that cloudy days are evenly distributed over the year and that electrical energy can be economically stored over several months. But the calculation gives some idea of the numbers involved.)

GROUP C

34.43 *(2) Comin' atcha! III.* A dielectric ring of radius b carries uniformly distributed charge q. The ring moves along its axis with constant speed v. **(a)** At what distance z from the ring along the axis is the displacement current density j_d a maximum? **(b)** Show that at that point

$$j_d = \frac{2\sqrt{2}vq}{25\sqrt{5}\pi b^3}.$$

34.44 *(G) Nobel Prize for sure! II.* Equation 34.12e expresses the force exerted by electric and magnetic fields on a particle having charge q. What would the force on a magnetic charge M look like? (Hint: The sign is tricky. The sign you choose must be consistent with Maxwell's equations, as they are modified in Problem 34.31.)

34.45 *(G) Radiation by a current sheet.* The figure opposite shows an infinite conducting sheet lying in the y-z plane of a coordinate system. There is an alternating current in the y direction. Let the current per unit width of sheet in the z direction be J_y, where $J_y = J_0 \sin \omega t$. **(a)** Find the magnitude and direction of the magnetic field $\mathscr{B}$ at a point outside the sheet, as a function of time. (Hint: Refer to Section 29.4.) **(b)** Beginning with Equations 34.17 and 34.19, show that a magnetic wave $\mathscr{B}_z$ propagates in the x direction with speed c.

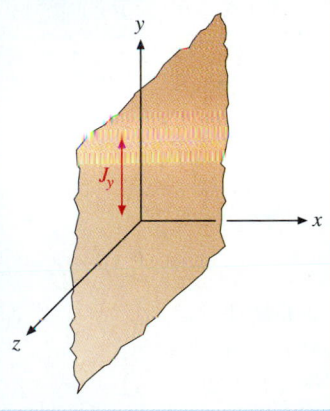

Write the function that gives the value of $\mathscr{B}_z$ for any positive x and t. **(c)** Using the same equations, show that the magnetic wave is accompanied by an electric wave $\mathscr{E}_y$ moving with the same speed with the same phase. What is the relation between the amplitudes of the magnetic and electric waves? **(d)** Show that the Poynting vector in the vicinity of the sheet is parallel to the x axis, away from the sheet, and that the intensity is $I = \frac{1}{8}\mu_0 c J_0^2$.

Part X

Optics

In the next three chapters, we study the propagation and manipulation of the special type of electromagnetic radiation called light. This study is loosely called *optics*. Some of the discussion applies only to light; other, more general considerations turn out to be valid for all kinds of electromagnetic radiation or even for all kinds of waves.

Chapters 35 and 36 deal with *physical optics*, in which light is treated explicitly as a wave phenomenon. We begin in Chapter 35 with a study of the measurement of the speed of light in vacuum and in transparent media. Because the speed of light is different in different media, light is both *reflected* at boundaries between media and *refracted* (abruptly bent) as it passes from one to another.

The wave properties of light are essential to understanding the interaction of light with structures, such as narrow slits, whose critical dimensions are very roughly comparable to the wavelength of the light. Such interactions include *interference* and *diffraction*, phenomena that underlie the op-

eration of such important de-vices as *interferometers* and *diffraction gratings*. Interference and diffraction are the main subject matter of Chapter 36.

In Chapter 37, we turn to *geometric optics*. We are often interested in the interaction of light with structures such as mirrors and lenses, whose dimensions are very large compared with the wavelength of the light. When this is the case, it is often possible to ignore the wavelike properties of light and consider only the paths it takes through a system, called *rays*. By studying the way in which the rays diverge and converge as they pass through a system of mirrors or lenses,

we can determine (or predict) the properties of the system with good accuracy. Systems of interest include the eye, telescopes, microscopes, and numerous other optical devices.

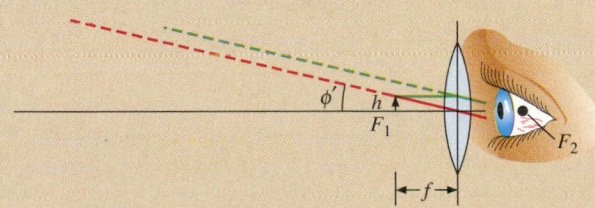

Physical Optics I
Light in Media

— The speed of light c can be measured by time-of-flight measurements.

— The speed of light v in a transparent medium is always less than c; the ratio $n = c/v$ is called the index of refraction of the medium.

— Huygens's construction is a simple way of describing the propagation of light waves.

— Reflection of light rays from a mirror is governed by the law of reflection.

— Refraction of light rays at the boundary between two transparent media is governed by Snell's law.

— The operation of light pipes, optical fibers, and reflecting prisms is based on the phenomenon of total internal reflection.

— The index of refraction of a transparent medium depends weakly on the wavelength of the light. As a consequence, white light refracted into a medium is dispersed into a spectrum of wavelengths.

— When the electric field of a light wave is restricted to just one direction, the light is polarized.

Left: There's more than one pot of gold associated with the law of refraction!

> *Light doth seize my brain.*
>
> —WILLIAM BLAKE, "Mad Song"

Introduction

In Chapter 34, we considered a few of the wavelike properties of light, our main aim being to show that light is a form of electromagnetic radiation. In this chapter and the next, we consider a number of important consequences of the wavelike behavior of light.

The propagation speed of a wave is characteristic of the wave. Maxwell made much of this point when he compared the speed of electromagnetic waves, whose existence he had predicted, with the speed of light, which had been measured by others. The speed of a wave also depends on the medium in which it travels. In Section 35.2, we consider measurements of the speed of light both in vacuum and in transparent media.

We then turn to other wavelike properties of light. In Section 35.3, we devise a simple, semiquantitative way of describing the propagation of waves, called *Huygens's construction*. We apply Huygens's construction to *reflection* in Section 35.4 and to *refraction*, the bending of light as it passes through the surface between two transparent media, in Section 35.5. Section 35.6 deals with the phenomenon of *total internal reflection*. In Section 35.7, we consider *dispersion*, and Section 35.8 concerns the *polarization* of light.

The Speed of Light

There are many methods for measuring the speed of light, but they all fall into one of two categories. The first category is *time-of-flight measurement*. The time t required for a pulse of light to travel a known distance l is measured, and the speed is determined directly by means of the relation $c = l/t$. The second category, to which most modern experiments belong, involves simultaneous measurement of the wavelength λ and the frequency ν of the light (or other electromagnetic) wave. The product $\lambda\nu = c$ is the speed of light.

Galileo was probably the first person to make a serious attempt to measure the speed of light. His time-of-flight technique, described in Query 2.2, was far too crude to measure so great a speed. Galileo could report only that the speed of light, even if it was not infinite, was very large.

The first successful quantitative evaluation of c was carried out in Paris in 1676 by the Danish astronomer Ole Rømer (1644–1710) and the Dutch physicist and astronomer Christian Huygens (1629–1695). Their method, which is the subject of Problem 35.56, also is a time-of-flight technique. It depends on measuring the variation of the observed period of revolution of a satellite of Jupiter as the Jupiter–earth distance changes over the course of the year. The method suffers from a serious shortcoming: Evaluating c requires knowledge of the sun–earth distance, a distance accurately known only since the 1960s.

The first terrestrial measurement of c was made in 1849 by the French amateur physicist Armand Hippolyte Louis Fizeau (1819–1896), who also employed a time-of-flight method. Fizeau's apparatus is shown schematically in Figure 35.1. Light from a bright source is focused into a narrow beam by a system of lenses not shown. Depending on the position of the toothed wheel, the light beam either is interrupted by a tooth or passes between two teeth.

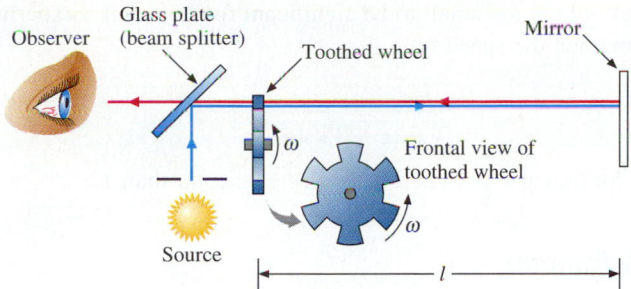

FIGURE 35.1 Schematic drawing of Fizeau's apparatus for measuring the speed of light. The wheel has 720 teeth; the wheel–mirror distance is 8.633 km.

If the wheel is at rest in a position that allows light to pass, the beam traverses the path length l to the mirror and is reflected back over nearly the same path to the observer's eye. When the wheel is made to rotate, it chops the light beam into a series of pulses. If the wheel is rotating with a large enough angular speed ω, a light pulse that has passed through the space between two of the teeth is blocked on its return from the mirror by the adjacent tooth, which has moved into place over the short time $2l/c$ required for the round-trip journey. The observer then sees no light. If the angular speed of the toothed wheel is doubled to 2ω, the light pulse passing from the source through a space on the wheel can return through the next space. Thus the observer again sees light.

A knowledge of the value of ω that results in either minimum or maximum brightness yields a value for c, as Example 35.1 shows.

EXAMPLE 35.1

Fizeau's wheel had 720 teeth, and the distance l was 8.633 km. (He located his main apparatus on one hilltop and the mirror on another to provide a clear path of this length.) As the wheel speeded up, the light seen by the observer disappeared and then reappeared, reaching maximum brightness when the wheel rotated at 25.36 rev/s. What is the corresponding value for the speed of light c?

SOLUTION: The round-trip time for the light pulse is $\Delta t = 2l/c$. Maximum brightness is achieved when the wheel turns through 1/720 of a revolution during the interval Δt. Because the wheel requires 1/25.36 s for one revolution, you have

$$\Delta t = \frac{1}{720} \frac{1}{25.36 \text{ s}} = 5.477 \times 10^{-5} \text{ s}.$$

The speed of light is then

$$c = \frac{2l}{\Delta t} = \frac{2 \times 8.633 \times 10^3 \text{ m}}{5.477 \times 10^{-5} \text{ s}} = 3.153 \times 10^8 \text{ m/s}.$$

Fizeau's value is about 4% too large. The largest experimental errors arose from the difficulty of measuring the angular speed of the wheel accurately and from the uncertainty in judging the conditions of maximum and minimum brightness.

Improvements on Fizeau's method led, over the next century, to ever more precise evaluations of c. Beginning about 1950, however, the time-of-flight method was superseded by the frequency-wavelength method. By the 1970s, the major remaining source of experimental error in the measurement of c lay in uncertainty in the length of the standard meter, in terms of which wavelength had to be expressed. In recognition of the fact that measurements of c can now be reproduced much more precisely than measurements of the meter, the speed of light is now taken as an SI base unit in place of the meter: $c \equiv 299\ 792\ 458$ m/s *exactly*, by definition (Equation 34.25). The meter is now a derived unit, defined in terms of c and and the base unit of time, the second.

Speed of Light in a Transparent Medium

In 1850, very shortly after Fizeau's work, Foucault made the first successful measurement of the speed of light in a transparent medium (water). His ingenious experiment for determining the ratio c/v_{water} is the subject of Problem 35.32. For reasons to be discussed in Section 35.3, it was important to determine whether the speed of light in transparent media was less or greater than the speed c in vacuum (or air, whose effect on the speed of light is too small to be significant for Foucault's experiment).

Foucault obtained the speed ratio

$$\frac{v_{air}}{v_{water}} \simeq \frac{c}{v_{water}} = 1.33. \tag{35.1}$$

The ratio c/v for transparent materials is *always* greater than 1.

Index of Refraction

Consider a medium in which the speed of light is v. The ratio of the speed of light in vacuum to the speed v in the medium is called the **index of refraction** n of that medium:

$$n \equiv \frac{c}{v}. \tag{35.2}$$

(The reason for the name "index of refraction" will become clear in Section 35.5.) For any transparent medium, n is a dimensionless quantity greater than 1.

Huygens's Construction

In Chapters 21 and 22, we developed a fairly elaborate mathematical apparatus for describing the propagation of waves. For many purposes, however, this propagation can be described adequately on a simpler, qualitative basis called **Huygens's construction**. Huygens's construction combines some simple notions about symmetry with the elementary concept of a wave as a disturbance that propagates through a region. (As you learned in Chapter 34, this concept of a wave is strictly applicable only to mechanical waves, where the disturbance is a mechanical motion of the elements of the medium through which the wave propagates. Electromagnetic waves can propagate through vacuum because electric and magnetic fields—not matter—do the oscillating. Fortunately, the argument that follows does not depend on *what* is oscillating.)

We begin with the concept of a *wave front*, defined in Section 22.2. In Figure 35.2, ripples propagate away from their source. Each crest forms a curve (here a circle), as does each trough. You could draw other curves (between crests and troughs) along which the height of the wave is constant. Such curves are called **wave fronts**. We say that the wave is **in phase** at all points on the same wave front.

Any normal (in two dimensions, any perpendicular) to a wave front represents a direction in which the wave propagates. Such a normal is called a **ray**. In Figure 35.3, rays are constructed for the wave-front system of Figure 35.2. The ray is an important analytical tool as well as an intuitively "obvious" concept, and we will use rays frequently.

Given any wave front (for example, one of the wave fronts in Figure 35.2) at some time $t = 0$, we can use Huygens's construction to map the wave front at any other time. That is, we can predict the way the wave propagates. The procedure is as follows:

1. Choose a set of evenly spaced points on the wave front. Figure 35.4 shows such a set of points on a linear wave front propagating in the two-dimensional space represented by the page. (Water surface waves can be satisfactorily represented in two-dimensional space.) The points lie a small distance δ apart. Each point is itself a tiny source of waves, and so we call it a **sourcelet**. The tiny waves emanating from a sourcelet are called

FIGURE 35.2 Circular wave fronts.

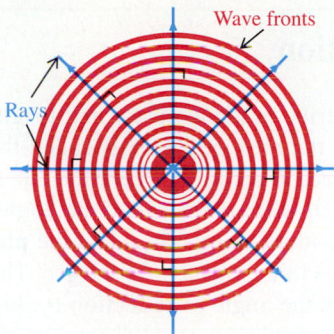

Wave fronts

Rays

FIGURE 35.3 Rays drawn for the wave-front system of Figure 35.2. In this two-dimensional system, the rays are everywhere perpendicular to the wave fronts; in a three-dimensional system, the rays are normal to the (two-dimensional) wave fronts.

wavelets.* Because the sourcelets are point sources, the wavelet fronts are independent circles until they superpose and thus merge with other wavelets.

2. In Figure 35.4, the wavelet fronts are drawn at a short time τ after being emitted by their sourcelets. The wave speed is v, and each wavelet front has traveled a distance $v\tau$ in all directions. In most directions, the wavelets have superposed with their neighbors in a complicated way, not shown. But in one particular direction—the direction perpendicular to the wave front—each wavelet has arrived at a point that no wavelets emitted from other sourcelets at $t = 0$ can have reached (because their sourcelets are farther away from that point). Because we need not consider superposition of neighboring wavelets at these points, we can easily picture the new wave front at $t = \tau$. *This new wave front is simply the curve tangent to all the wavelets.* In Figure 35.4, the original linear wave front has given rise to a new linear wave front, located a distance $v\tau$ to the right.

A planar wave front in three dimensions is analogous to a linear wave front on a two-dimensional surface. The application of Huygens's construction to plane waves is essentially the same as its application to linear waves. Indeed, we can often represent three-dimensional waves in two dimensions by considering the behavior of the waves on a cross-sectional cut through the three-dimensional space.

Figure 35.5 shows the application of Huygens's construction to a circular wave front. Evenly spaced sourcelets on the black wave front give rise to circular wavelets. At time τ, the new wave front lies a distance $v\tau$ from the original one; the symmetry of the situation dictates its circular form. The argument applies directly to spherical waves in three dimensions, as well.

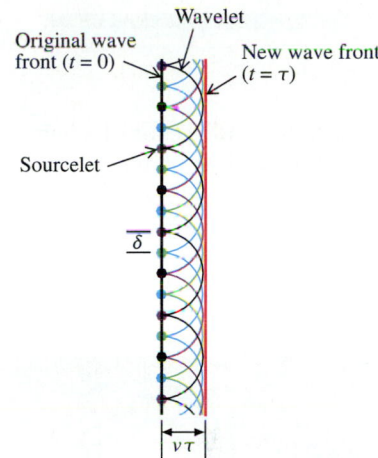

Wavelet

Original wave front ($t = 0$)

New wave front ($t = \tau$)

Sourcelet

δ

$v\tau$

FIGURE 35.4 Huygens's construction, illustrated for a linear wave front. Each sourcelet contributes a circular wavelet. These wavelets superpose in a complicated way, not shown. But there is one point on each wavelet where no superposition has taken place. The curve tangent to these points is the new wave front.

FIGURE 35.5 Application of Huygens's construction to a circular wave front.

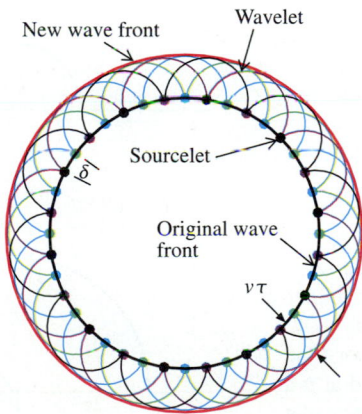

New wave front

Wavelet

Sourcelet

δ

Original wave front

$v\tau$

In the next two sections, we apply Huygens's construction to *reflection* and *refraction*.

---

*Here is a justification for considering each point on a wave front as a sourcelet for water surface waves: Suppose that the wave front is a crest. Under the influence of gravity, the crest will descend into the main body of water. As each tiny mass element on the crest descends, it displaces neighboring elements, creating a local disturbance that propagates as a wavelet.

The Law of Reflection

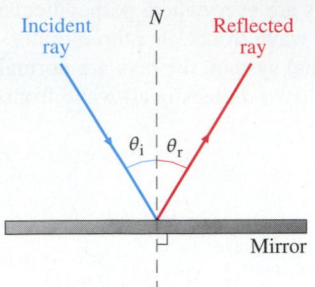

θ_i | θ_r

Mirror

FIGURE 35.6 Reflection of a light beam from a plane mirror.

In Figure 35.6, a light ray strikes a plane mirror and is reflected. The part of the light ray between the source and the mirror surface is called the **incident ray**. We erect a **normal** N to the mirror surface at the point where the incident ray strikes it. The angle between the normal and the incident ray is called the **angle of incidence** θ_i. The plane defined by the incident ray and the normal is called the **plane of incidence**.

The **reflected ray** also lies in the plane of incidence. The angle between the normal and the reflected ray is called the **angle of reflection** θ_r. Figure 35.6 illustrates the **law of reflection**: *The angle of reflection is equal to the angle of incidence*:

$$\theta_i = \theta_r. \tag{35.3}$$

You have often observed reflection of light that conforms to Equation 35.3. Such reflection is called *specular*, from the Latin word meaning "mirrorlike." You already know that a particle bouncing elastically off a smooth surface undergoes specular reflection; see Section 18.2 and Problem 10.37.

Wave trains and pulses also undergo specular reflection at smooth surfaces, as shown in the ripple-tank photograph of Figure 35.7. The reason for specular reflection of wave trains and pulses is quite different from the reason for specular reflection of particles. In Figure 35.8 and its legend, we use Huygens's construction to justify the law of reflection for waves.

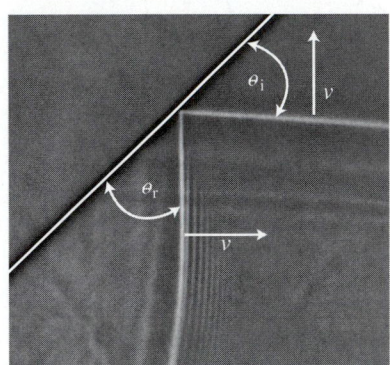

FIGURE 35.7 Ripple-tank photograph, showing a plane wave pulse undergoing specular reflection at a smooth surface, in accordance with the law of reflection.

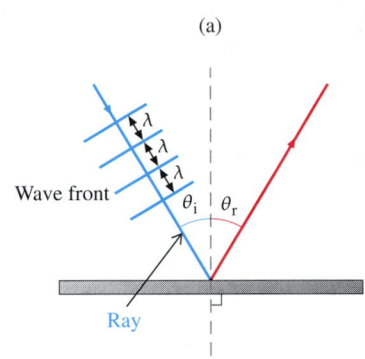

(a)

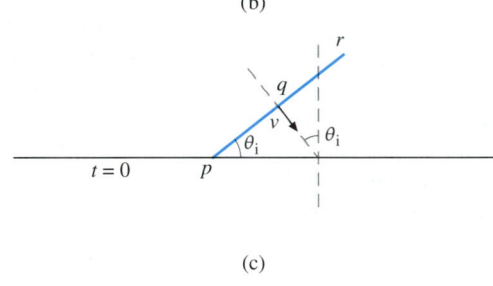

(b)

(c)

FIGURE 35.8 Justification of the law of reflection, using Huygens's construction. The plane mirror, which is oriented perpendicular to the page, is seen in cross section, as are the planar wave fronts in the incident and reflected waves. (*a*) A train of wave fronts approaches the mirror at an angle of incidence θ_i. The wave speed is v. (*b*) One of the wave fronts, pqr, is shown at $t = 0$, just as it reaches the mirror at p. (*c*) The wave front at a later time τ, when it makes contact with the mirror at q'. The wavelet produced at p at $t = 0$ now has radius $v\tau$. The line $q'p'$ is a common tangent to this wavelet and to the wavelet of very small radius just departing from q. (It is also tangent to all the wavelets produced as the wave front swept over all the points on the mirror surface between p and q'.) Thus $q'p'$ is the reflected wave front. (*d*) The wave front at $t = 2\tau$. Now the wavelet generated at p at $t = 0$ has radius $2v\tau$ and the wavelet generated at q' at $t = \tau$ has radius $v\tau$. The incident wave front has just generated, at r'', a wavelet whose radius is very small. The common tangent $r''q''p''$ is the reflected wave front. (*e*) The right triangles $pp''r''$ and prr'' are congruent; the angle of incidence θ_i and the angle of reflection θ_r must therefore be equal.

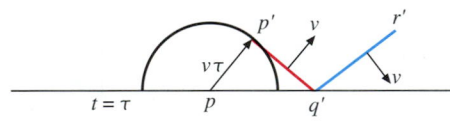

(d)

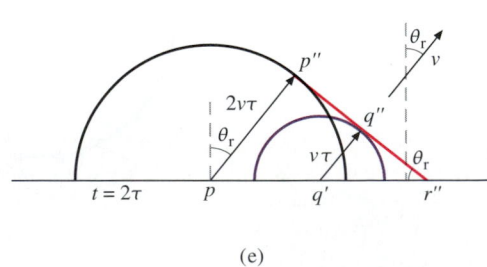

(e)

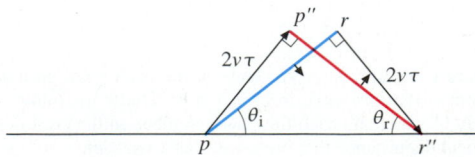

Snell's Law

When a light ray strikes a planar boundary between two transparent media, part of the light is reflected back into the first medium and part passes through the boundary into the second medium. The law of reflection holds for the reflected portion of the light, just as if the surface were a mirror. We now consider the portion of the light that passes through the boundary. The light beam bends sharply at the boundary, as shown in Figure 35.9. This bending is called **refraction**, from the Latin word *refractus*, meaning "broken." Like the reflected beam, the refracted beam lies in the plane of incidence. To describe refraction quantitatively, we compare the angle of incidence θ_1 with the **angle of refraction θ_2**. As for reflection, both angles are conventionally measured with respect to the normal to the surface.

Suppose that the speed of light is v_1 in medium 1 and v_2 in medium 2. Using Equation 35.2, we can write the corresponding indices of refraction

$$n_1 = \frac{c}{v_1} \quad \text{and} \quad n_2 = \frac{c}{v_2}. \tag{35.4a,b}$$

Experiments show that the angles of incidence and refraction bear the relation

$$n_1 \sin \theta_1 = n_2 \sin \theta_2. \tag{35.5a}$$

We now consider the special case in which medium 1 is vacuum. This case is important because it approximates very well the common situation in which light is refracted at an air-glass or air-water interface. (As you will see in Section 36.2, the index of refraction of air differs from 1 by less than 3 parts in 10^4.) For vacuum $n_1 = 1$, and we have $\sin \theta_1 = n_2 \sin \theta_2$. The only index of refraction to appear in this equation is the index of refraction of medium 2, and so it is conventional to drop the subscript and write

$$\sin \theta_1 = n \sin \theta_2. \tag{35.5b}$$

In practice, indices of refraction are not usually obtained from measurements of the speed of light. Rather, the relation expressed by Equation 35.5b is used to determine n experimentally. Either of Equations 35.5a and 35.5b is called the **law of refraction** or, more commonly, **Snell's law**, after the Dutch mathematician and physicist Willebrord Snell van Royen (1591–1626). According to Snell's law, the light beam bends toward the normal when entering a medium of larger index of refraction—called an **optically dense medium**—from a medium of smaller index of refraction—called an **optically rare medium**. You should bear in mind that these terms are always used in a relative way.

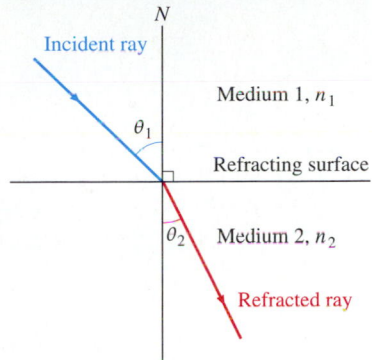

FIGURE 35.9 Diagram for Snell's law, $n_1 \sin \theta_1 = n_2 \sin \theta_2$. A ray of light passes through a planar surface from medium 1, whose index of refraction is n_1, to medium 2, whose index of refraction is n_2. The angle of incidence θ_1 and the angle of refraction θ_2 are measured with respect to the normal to the refracting surface. In the case shown, $n_2 > n_1$.

THINKING LIKE A PHYSICIST

About 1621, Snell showed that his measurements of θ_1 and θ_2 conformed to an empirical formula, expressed in a mathematical form different from but equivalent to Equation 35.56. However, his results were not published until after his death. In 1637, René Descartes derived Equations 35.5a and 35.5b on the assumption that light consists of a stream of particles. He argued that the denser medium exerts an attractive force on the particles, which therefore accelerate along the normal on approaching the boundary from the optically rare side. But Descartes's derivation implies that light travels *faster* in optically dense media than in air (Problem 35.36). Descartes's theory requires that the index of refraction of a medium be defined as $n = v/c$. This flatly contra-

dicts the requirement of the wave picture, $n = c/v$, which is derived in the next subsection. In the seventeenth century, n could be determined by applying Snell's law to measurements of refraction angles, but v was not known. The first direct experimental test of the crucial distinction between the two definitions of n was Foucault's 1850 measurement of the speed of light in water. But by that time, few physicists doubted the wave picture of the propagation of light. Science rarely proceeds along the most logical pathway!

Snell's compatriot Huygens accused Descartes of plagiarism, because Descartes had had access to Snell's unpublished work. The accusation is probably true; in any case, Snell's law is often called Descartes's law in France.

Huygens's Construction and Snell's Law

We now use Huygens's construction to justify Snell's law for a wave train. In Figure 35.10a, a planar wave front pqr strikes a refracting surface at p. We take this instant to be $t = 0$. The angle between the wave front and the surface is equal to the angle of incidence θ_1. Figure 35.10b shows the situation at a later time $t = \tau$. The wave front has reached the surface at q'. That part of the wave front still in medium 1, $q'r'$, has traveled a distance $v_1\tau$. But the left end of the wave front, now at p', has traveled through medium 2 for the entire time interval τ. It has thus traveled a distance $v_2\tau$. (In the diagram, we take $v_2 < v_1$, though the argument holds just as well for the contrary case.) The angle between the surface and the part of the wave front in medium 2 is θ_2. Figure 35.10c shows the situation at $t = 2\tau$. Now the entire wave front has entered medium 2, the right end just entering at r''. Over the entire time interval 2τ, the left end of the wave front, now at p'', has traveled a distance $2v_2\tau$, while the right end has traveled a distance $2v_1\tau$.

The initial and final situations are compared in Figure 35.10d. From triangles prr'' and $pp''r''$ we have

$$\sin \theta_1 = \frac{2v_1\tau}{l} \quad \text{and} \quad \sin \theta_2 = \frac{2v_2\tau}{l}.$$

Dividing the first equation by the second, we obtain

$$\frac{\sin \theta_1}{\sin \theta_2} = \frac{v_1}{v_2}.$$

We use Equations 35.4a and 35.4b to write

$$v_1 = \frac{c}{n_1} \quad \text{and} \quad v_2 = \frac{c}{n_2}. \tag{35.6a,b}$$

We substitute these values of v_1 and v_2 into the equation immediately preceding Equations 35.6a,b to obtain

$$\frac{\sin \theta_1}{\sin \theta_2} = \frac{n_2}{n_1},$$

or

$$n_1 \sin \theta_1 = n_2 \sin \theta_2.$$

This is Snell's law (Equation 35.5).

Snell's law is employed in a variety of instruments called **refractometers** to measure the index of refraction of transparent solids and liquids. Most solids and liquids have values of n between 1.3 and 1.7, though for some substances n is as large as 3. The index of refraction of gases is very little different from 1 and is measured by interferometric methods to be discussed in Example 35.2. The indices of refraction of selected materials are listed in Table 35.1.

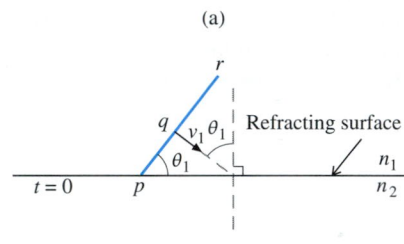

(a)

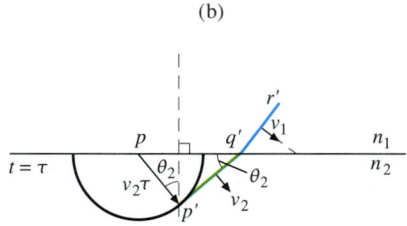

(b)

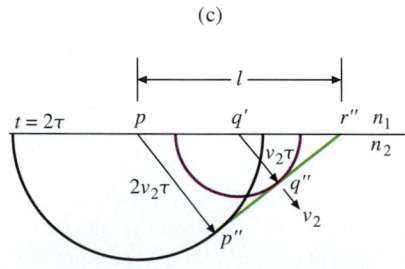

(c)

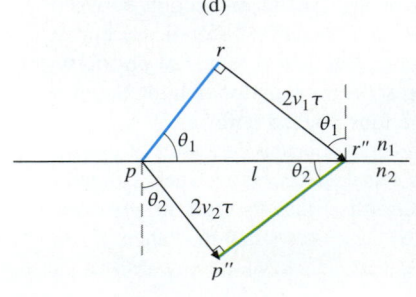

(d)

FIGURE 35.10 Illustration for deriving Snell's law from Huygens's construction. (*a*) At $t = 0$, a planar wave front pqr strikes the refracting surface at p. (*b*) At $t = \tau$, the part of the wave front still in medium 1 has proceeded a distance $v_1\tau$ in the upper medium, reaching the refracting surface at q'. The left end of the wave front has proceeded to p', a distance $v_2\tau$ through medium 2. The situation is illustrated for the case $v_2 < v_1$. (*c*) The wave front at $t = 2\tau$. The left end of the wave front has proceeded to p'', a distance $2v_2\tau$; the middle has proceeded to q'', a distance $v_2\tau$; and the right end has just entered medium 2 at r'', having traveled through medium 1 a distance $2v_1\tau$. The distance pr'' is called l. (*d*) Comparison of the wave front at $t = 0$ with the wave front at $t = 2\tau$.

TABLE 35.1 Index of Refraction of Selected Substances ($\lambda = 589$ nm, $t = 20°C$ except for gases)

Substance	Index of Refraction n
Glasses	
Fused quartz	1.458 5
Zinc crown glass	1.517 4
Light flint glass	1.575 5
Heaviest flint glass	1.890 6
Crystalline Solids	
Water ice	1.310 4
Sodium chloride	1.544 2
Aluminum oxide (ruby, sapphire)	1.769
Zirconium oxide (zircon)	2.2
Carbon (diamond)	2.417 3
Polymers	
Cellulose acetate butyrate	1.48
Polymethyl methacrylate (Lucite)	1.49
Polyvinyl chloride (PVC)	1.54
Polycarbonate (optical quality)	1.605 6
Liquids	
Water	1.333
Ethanol	1.36
Carbon disulfide	1.63
Gases ($t = 0°C, p = 1$ atm)	
Helium	1.000 036
Oxygen	1.000 271
Air (dry)	1.000 292 6
Carbon dioxide	1.000 45

EXAMPLE 35.2

A light ray enters a semicircular piece of optical glass at the center of curvature O, as shown in Figure 35.11. The angle of incidence is 45.00°. You measure the angle of refraction and find $\theta_2 = 27.49°$. **(a)** What is the index of refraction of the glass? **(b)** What is the speed of light in the glass? **(c)** Why is there no refraction at the second (circular) surface?

SOLUTION:

(a) Find n.

Using Equation 35.5b, you have

$$n = \frac{\sin \theta_1}{\sin \theta_2} = \frac{\sin 45.00°}{\sin 27.49°} = 1.532.$$

(b) Find the speed of light in the glass.

In using Equation 35.2, $v = c/n$, make sure to express c to enough significant figures so as to prevent any rounding errors. The angles are given to four significant figures, and so it certainly suffices to express c to five significant figures:

$$v = \frac{2.9979 \times 10^8 \text{ m/s}}{1.532} = 1.957 \times 10^8 \text{ m/s}.$$

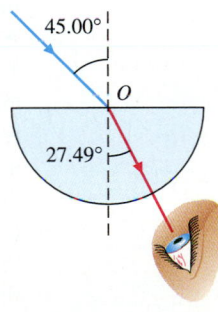

FIGURE 35.11

(c) Why is there no refraction at the circular surface?

No matter what the angles of incidence and refraction are, a light beam entering the glass at O propagates along a radius of the semicircle. Thus it always strikes the second surface at normal incidence.

Optical Path Length

For any wave of frequency ν, the wavelength λ is determined according to Equation 21.15, which can be written $\lambda = v/\nu$. As light passes from one transparent medium to another, the frequency ν does not change. This is because Huygens wavelets are excited in medium 2 by the arrival at the interface of waves with frequency ν. The waves in medium 2, constructed from these wavelets, also must have frequency ν.

What happens to the wavelength λ when the wave speed v changes at the boundary? We have $\nu = v_1/\lambda_1$ and $\nu = v_2/\lambda_2$; consequently, $v_1/\lambda_1 = v_2/\lambda_2$. We substitute into this relation the values of v_1 and v_2 given by Equation 35.6, obtaining $c/\lambda_1 n_1 = c/\lambda_2 n_2$. This simplifies to

$$\lambda_1 n_1 = \lambda_2 n_2. \tag{35.7a}$$

According to Equation 35.7a, the wavelength of light becomes shorter ($\lambda_2 < \lambda_1$) when the light enters an optically denser medium ($n_2 > n_1$). In particular, if light whose wavelength in vacuum is λ_0 passes into a transparent medium where its wavelength is λ, we have $n_1 = 1$ and $n_2 \equiv n$. In this case, Equation 35.7a assumes the simpler form

$$\lambda = \frac{\lambda_0}{n}. \tag{35.7b}$$

Because the wavelength of light in any medium is shorter than that of the same light in vacuum, a distance s in the medium "contains" more wavelengths than the same distance in vacuum. What is the (greater) distance L in vacuum that contains the *same* number of wavelengths? We have $L/\lambda_0 = s/\lambda$. Using Equation 35.7b, we obtain

$$L = ns. \tag{35.8}$$

The length L is called the **optical path length**. The concept of optical path length will be very useful when we study the superposition of light waves in Chapter 36.

SECTION 35.6

Total Internal Reflection

In Figure 35.12, six beams of light pass through a glass prism from the left. The light beams strike the glass-air surface of the prism (the right surface) at different angles of incidence. The uppermost beam strikes the surface at the smallest angle of incidence. A small portion of the light is reflected back into the glass, but most is refracted through a fairly small angle as it emerges from the prism. The next three beams strike the glass-air surface at increasing angles of incidence. In each case, a little more of the light is reflected. But most of the light emerges, the angle of refraction increasing with the angle of incidence. The last two beams are *totally* reflected; *none* of the light passes through the glass-air surface. This phenomenon is called **total internal reflection**. It takes place at all angles of incidence greater than a **critical angle** θ_c. (In Figure 35.12, the critical angle must lie somewhere between the angles of incidence of the fourth and fifth beams.)

We now derive an expression for the critical angle. This expression will make clear why total internal reflection takes place only when light strikes a boundary between two media from the optically *denser* side. Let us label the denser medium d and the rarer medium r. We write Snell's law in the form

$$\frac{n_d}{n_r} \sin \theta_d = \sin \theta_r. \tag{35.9}$$

Now, $\sin \theta_r$ can never be greater than 1. But the ratio n_d/n_r on the left side *is* greater than 1. The values of θ_d that satisfy the equation are limited to those for which $(n_d/n_r) \sin \theta_d \leq 1$. The limiting case, $(n_d/n_r) \sin \theta_d = 1$, corresponds to the critical angle $\theta_d = \theta_c$. So we have

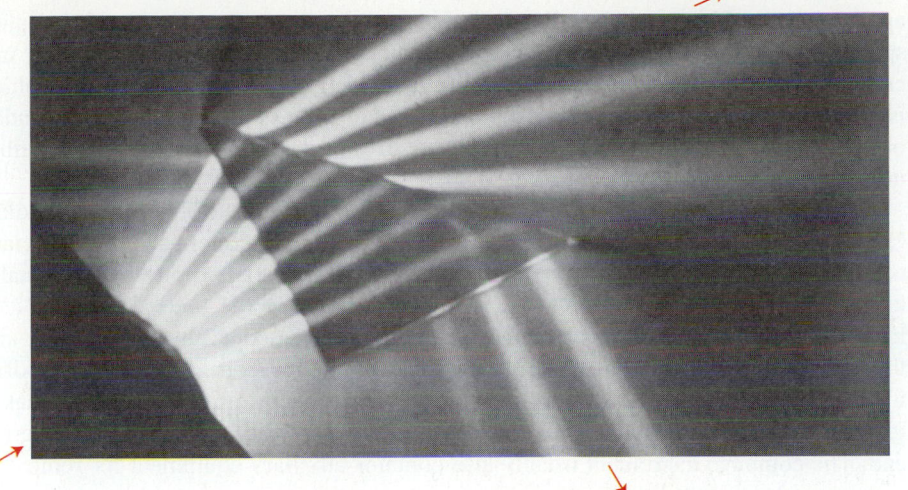

FIGURE 35.12 Six light beams strike the glass-air surface on the right at different angles of incidence. The two lowest beams, having $\theta_i > \theta_c$, are totally reflected.

(From Haber-Schaim: *PSSC Physics.* Copyright 1991 by Kendall/Hunt Publishing Company. Used with permission.)

$$\sin \theta_c = \frac{n_r}{n_d} \quad \text{or} \quad \theta_c = \sin^{-1} \frac{n_r}{n_d}. \qquad \textbf{(35.10a,b)}$$

The corresponding value of θ_r is 90°; the "emerging" light beam is tangent to the surface. For angles $\theta_d > \theta_c$, Equation 35.9 cannot be satisfied, and reflection rather than refraction takes place.

Total internal reflection is a useful way of changing the direction of light beams in optical systems because no light is lost on reflection. Even the best silvered mirrors absorb a few percent of the light incident on them. Often, this is of no consequence. But in critical applications, the loss cannot be tolerated—for example, in optical systems in which the light must undergo many reflections, as is the case in most binoculars.

EXAMPLE 35.3

Find the critical angle for a prism of glass whose index of refraction is $n_g = 1.52$ **(a)** when the prism is located in air and **(b)** when the prism is submerged in water. **(c)** Will total reflection take place under any circumstances when the prism is submerged in carbon disulfide (CS_2)?

SOLUTION:

(a) Find θ_c when the prism is in air.

You can take the index of refraction of air to be 1. Equation 35.10b gives you

$$\theta_c = \sin^{-1} \frac{1}{n_g} = \sin^{-1} \frac{1}{1.52} = 41.1°.$$

(b) Find θ_c when the prism is submerged in water.

Using the value of the index of refraction of water given in Table 35.1, you have

$$\theta_c = \sin^{-1} \frac{n_w}{n_g} = \sin^{-1} \frac{1.33}{1.52} = 61.0°.$$

The critical angle is larger when the prism is submerged in water. That is, light will be transmitted if it strikes the glass-water interface over a range of angles of incidence wider than that for the glass-air interface.

(c) What happens when the prism is submerged in CS_2?

From Table 35.1, you have $n_{CS_2} = 1.63$. Because $n_{CS_2} > n_g$, total reflection will never occur when light is incident on the glass-CS_2 interface from the glass side. However, light passing through the carbon disulfide and falling *on* the glass will be totally reflected if

$$\theta \geq \sin^{-1} \frac{1.52}{1.63}; \quad \text{that is, } \theta \geq 68.8°.$$

Light Pipes and Fiber Optics

A **light pipe** is a long, thin cylinder of transparent material. As you can see in Figure 35.13, the geometry is such that light entering at one end must always strike the wall at a large angle of incidence. The index of refraction of transparent solids is large enough that the critical angle is always exceeded (provided the light pipe is not bent through a very sharp curve). Thus total internal reflection is assured; the light is confined to the pipe by repeated reflection until it emerges at the other end.

In the simplest applications, light pipes in the form of transparent plastic rods are used to convey light to locations where it is not convenient to place a lamp; this is sometimes done in store displays and microscope lamps. But a much more important

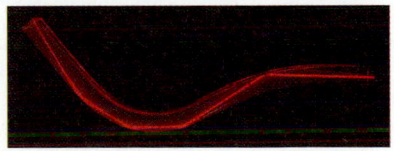

FIGURE 35.13 A light pipe. Any ray that enters one end must be internally reflected repeatedly until it emerges from the other end. One such ray is shown.

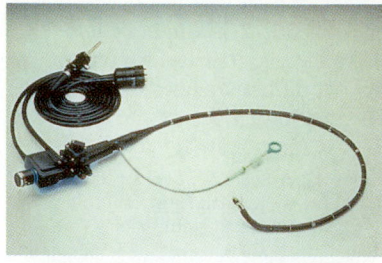

(a)

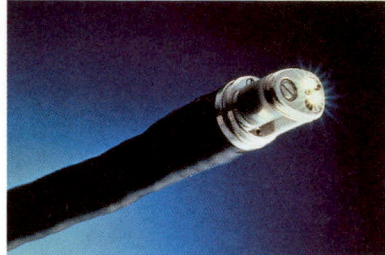

(b)

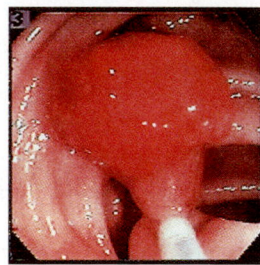

(c)

FIGURE 35.14 (*a*) A fiber-optic endoscope (a colonoscope). The instrument proper is the J-shaped cable. The tightly coiled cable at the upper left connects the endoscope to the power source and to the TV monitor. The box-shaped component contains the physician's eyepiece and various manipulator controls. (*b*) The entrant end of the colonoscope. The fiber-optic light sources are visible. (*c*) View through the colonoscope. A snare, controlled by a thin cable passing through the device, is being used to cut off a polyp at its base. Note the remarkable clarity of the image.

application is to **fiber-optic bundles**. In making the simplest type of fiber-optic bundles, special highly transparent glass is drawn out into very fine fibers, typically a few micrometers in diameter. Many such fibers are laid side by side to form a bundle, which is bound together by impregnating the bundle with a transparent polymer. The index of refraction of the polymer is less than that of the glass, so light within any fiber undergoes total internal reflection.

We now consider two important, quite different applications of fiber-optic bundles. One application is to remote imaging, as in the *flexible endoscopes* used by physicians to look inside the body in a minimally invasive manner. (An endoscope is usually known by a more specific name that signifies the part of the body for which it is intended; some examples are the colonoscope, the laparoscope, the arthroscope, and the bronchoscope.) As Figures 35.14a and b show, an endoscope consists of a flexible fiber-optic bundle together with a light source and a guidance mechanism that makes it possible for the physician to steer the entrant end of the device from the other, external end. The complete instrument usually also contains auxiliary equipment for removing tissue samples or other purposes.

The entrant end of the endoscope lies in contact with or very near some tissue structure the physician wants to see. Light is conveyed to this structure by a fiber bundle designed for the purpose, which lies parallel to the main imaging bundle. A miniature lens system forms an image of the structure on the end of the main fiber bundle. Because an image is a pattern of varying light intensity, the fibers carry light of varying intensity. The image of the structure is reproduced at the outer end of the endoscope, where it can be inspected visually, displayed on a TV monitor, or photographed. Figure 35.14c is a typical image.

A second, rapidly growing application of fiber-optic bundles is to telecommunications. The earliest telephone cables consisted of bundles of insulated copper wires, each of which carried one telephone transmission. In more modern systems, each wire carries many transmissions simultaneously. This is accomplished by modulating a high-frequency signal in one of several ways. Because the amount of information that can be carried on a conductor per unit time depends directly on the signal frequency, engineers have developed techniques for employing ever-higher frequencies. Since about 1945, microwave frequencies have been used. Since the 1980s, it has been possible to use light, rather than microwaves. Because the frequency of light is very much higher than that of microwaves, there is an improvement of many orders of magnitude in the information transmission rate. Moreover, thin glass fibers take the place of larger, more expensive microwave wave guides. Thus optical fiber cables are much lighter, more flexible, and cheaper than the cables they replace. Fiber-optic cables are essential to high-quality computer links and are widely used in this application. They will probably replace most TV cable systems and telephone lines in the near future.

Dispersion

Unlike the speed of light in vacuum, the speed of light in a transparent medium varies with the wavelength of the light. Consequently, the index of refraction n of a substance depends on the wavelength λ of the light used to measure it; $n = n(\lambda)$. This dependence of n on λ is the basis of the phenomenon called *dispersion*. The familiar and spectacular rainbow is a consequence of dispersion of sunlight by rain droplets.

No substance is transparent over the entire electromagnetic spectrum. All substances are transparent in some wavelength regions and opaque in others. The index of refraction of a transparent substance depends on wavelength only weakly, as long as the wavelength range of interest does not lie close to a value of λ at which the substance ceases to be transparent. Glass exemplifies this weak dependence of n on λ over the entire visible wavelength region—the range of electromagnetic radiation wavelengths (approximately 400 nm to 700 nm) that the human eye can see (Table 35.2).

TABLE 35.2 Conventional Subdivision of the Visible Wavelength Region into Color Ranges

Wavelength Range (nm)	Color Name
400–424	Violet
424–491	Blue
491–575	Green
575–585	Yellow
585–647	Orange
647–700	Red

A graph that displays the index of refraction of a particular sample of glass as a function of wavelength is called a **dispersion curve**. Dispersion curves are shown in Figure 35.15 for three types of glass commonly used in optical systems. As we have already noted, the proportional variation of index of refraction over the visible region is small—3% or less for each type of glass. The gradual *decrease* of n with *increasing* wavelength is typical of transparent materials and is called *normal dispersion*.

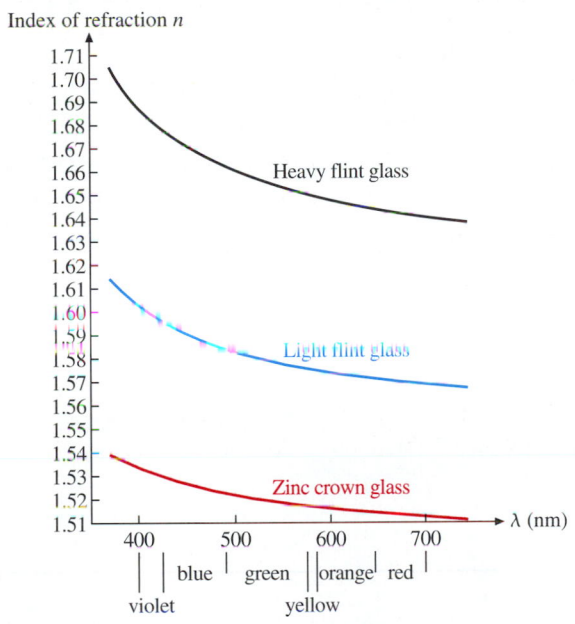

FIGURE 35.15 Dispersion curves for three optical glasses: zinc crown glass, light flint glass, and heavy flint glass. The color names assigned to approximate wavelength ranges are shown on the wavelength axis.

We now define **dispersion** in quantitative terms. Suppose a beam of light is incident on an air-glass surface, as shown in Figure 35.16. For a given angle of incidence θ_1, the angle of refraction θ_2 depends on n according to Snell's law, $\sin \theta_1 = n \sin \theta_2$. Because the index of refraction varies with wavelength, the angle of refraction depends on the wavelength of the incident light. In the figure, the light in the incident beam consists of a red component and a violet component, having wavelengths λ_r and λ_v. The corresponding indices of refraction are n_r and n_v. The two components are refracted through different angles, θ_{2r} and θ_{2v}, which satisfy the relations

$$\sin \theta_{2r} = \frac{1}{n_r} \sin \theta_1 \quad \text{and} \quad \sin \theta_{2v} = \frac{1}{n_v} \sin \theta_1.$$

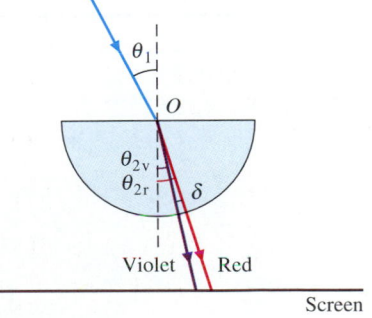

FIGURE 35.16 Dispersion of an incident beam consisting of red and violet components. (Angle δ is exaggerated for clarity.)

Section 35.7 Dispersion ■ **953**

The angle δ between the two refracted beams is called the **dispersion angle** because the two beams are *dispersed*—that is, spread apart. Because $n_v > n_r$, the angle of refraction is smaller for violet light than for red. That is, the violet component of the incident light is bent toward the normal more than the red component. The amount of spreading—the value of δ—depends on the difference between n_r and n_v and thus on the steepness of the dispersion curve.

If the incident light consists of components of many wavelengths, the light emerging from the semicircular disk of Figure 35.16 will form a fan-shaped, multicolored beam. A particular wavelength (corresponding to a particular color) emerges at every particular angle within the fan.

Light containing only a single wavelength (or, practically speaking, a very narrow range of wavelengths) is called **monochromatic light**.* However, most natural sources emit light having a mixture of wavelengths. By means of the phenomenon of dispersion, such light can be separated into its component wavelengths, often called **spectral components**. The result of the separation process can be displayed in a number of different ways. Any such display is called a **spectrum**, a word coined by Newton from the Latin meaning "something that can be inspected." Whether the spectrum takes the form of a photograph of the light emerging from the dispersing device or that of a plot of the output of a small light detector as it is moved through the fan of dispersed light, a spectrum is always a plot of $I(\lambda)$ versus λ—the intensity of the light as a function of wavelength. The process of separating light into its spectral components is called **spectrum analysis**.

EXAMPLE 35.4

A semicircular disk like that of Figure 35.16 is made of heavy flint glass. A beam consisting of violet light ($\lambda_v = 420.0$ nm) and red light ($\lambda_r = 680.0$ nm) falls on the planar surface of the disk at O, which is the center of the semicircle. If the angle of incidence is $\theta_1 = 45.00°$, find **(a)** the angle of refraction θ_v for the violet light, **(b)** the angle of refraction θ_r for the red light, and **(c)** the dispersion angle δ between the emerging violet and red light beams.

SOLUTION:
(a) Find θ_v.

You inspect Figure 35.15 to find the index of refraction of heavy flint glass corresponding to $\lambda_v = 420.0$ nm. You find that $n_v = 1.678$. You then use Snell's law in the form $\sin \theta_1 = n_v \sin \theta_v$. The solution for θ_v is

$$\theta_v = \sin^{-1}\left(\frac{1}{n_v}\sin\theta_1\right) = \sin^{-1}\left(\frac{\sin 45°}{1.678}\right) = 24.92°.$$

(b) Find θ_r.

You repeat the procedure of part **a** for $\lambda_r = 680.0$ nm. From Figure 35.15, you obtain $n_r = 1.642$. Snell's law now gives you

$$\theta_r = \sin^{-1}\left(\frac{1}{n_r}\sin\theta_1\right) = \sin^{-1}\left(\frac{\sin 45°}{1.642}\right) = 25.51°.$$

(c) Find δ.

You have

$$\delta = |\theta_r - \theta_v| = 25.51° - 24.92° = 0.59°,$$

or a little more than one-half degree.

The small angle δ calculated in Example 35.4 represents the dispersion of two wavelength components near the extremes of the visible spectrum. Two wavelength components closer together are dispersed even less. Fortunately, it is possible to build spectrometers that can discriminate between components for which the angle of dispersion is very small indeed—20 arc seconds (0.006°) is not unusual. This precision makes possible the analysis of spectral components quite close together in wavelength. In practice, semicircular disks are never used in spectrometers. Instead, dispersion is accomplished by prisms. A major advantage of the prism is that dispersion takes place twice, once at each surface. A typical *prism spectrometer* is shown in Figure 35.17.

The **dispersing prism** was the first device used for spectrum analysis, and prism spectrometers are still widely used for this purpose. Isaac Newton used prisms in his

*The word *monochromatic* is derived from Greek roots meaning "single color."

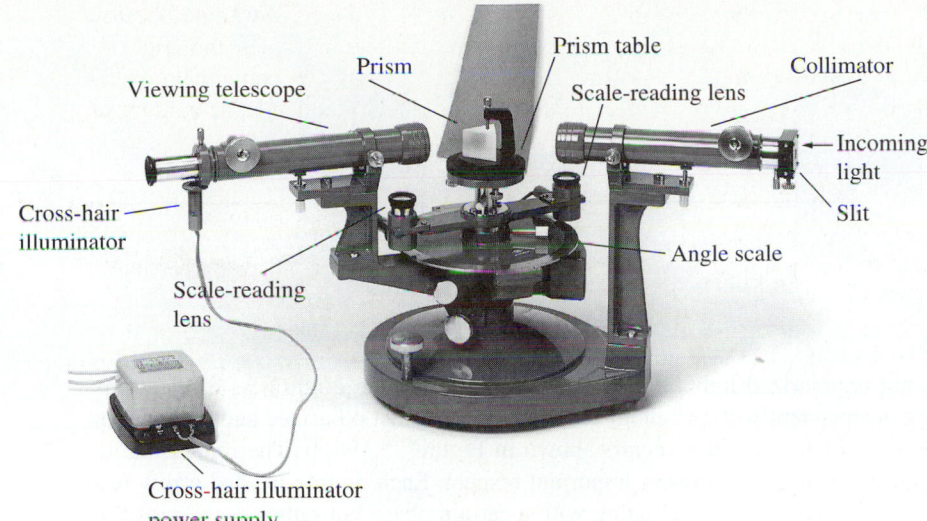

Viewing telescope • Prism • Prism table • Scale-reading lens • Collimator • Incoming light • Slit • Angle scale • Cross-hair illuminator • Scale-reading lens

Cross-hair illuminator
power supply

FIGURE 35.17 A prism spectrometer. The light source is located to the right of the telescope on the right, called a *collimator* because its lens system renders the light entering it into a parallel (collinear) beam. A narrow adjustable slit is located at the right end of the collimator. Emerging from the collimator, the light strikes the prism, where the various wavelength components are dispersed through slightly different angles. Light emerging from the prism is observed through the *viewing telescope* on the left. By turning the telescope around an axis that coincides with the prism axis, the observer aligns each spectral component in turn with cross hairs in the telescope. Because each wavelength component produces an image of the slit mounted on the collimator, each component is called a *spectral line*. For each line, the orientation of the observing telescope is determined by means of the small magnifying lenses at the bottom, which read a precise angle scale. By these means, the angular deviation of any spectral component can be measured. Careful interpolation between the deviation angles of standard-wavelength lines yields the wavelength of each line in the spectrum.

pioneering systematic studies of the dispersion of light. His *Opticks*, first published in 1704, gives a detailed account of his work. Among many other things, Newton demonstrated that what we know as **white light** is a superposition of components of all colors. Newton's demonstration is described in simplified and somewhat idealized form in Figure 35.18 and its legend.

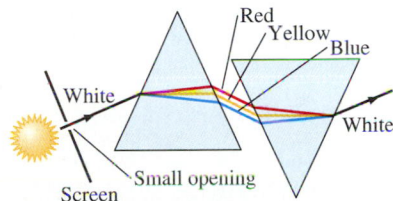

Red • Yellow • Blue • White • White • Small opening • Screen

FIGURE 35.18 White light from the sun passes through a small opening in a screen and falls on a prism. As a result of the variation of the angle of refraction with wavelength, the light is dispersed into a multicolored fan. The dispersed light then falls on an identical prism, inverted with respect to the first. The light is again refracted; here the variation of the angle of refraction results in the reconstitution of the original white light. You can convince yourself that reconstitution must follow dispersion by means of a symmetry argument. Suppose you reversed the path of light through the system. What would happen?

Polarization of Light

Like all electromagnetic radiation, light is produced when an electric charge (usually an electron) accelerates. In Figure 35.19a, an oscillating electric field is produced by a large number of electrons oscillating on the *y* axis. (This can be accomplished, in essence, in a laser.) Observers on the *x* axis see electromagnetic radiation whose electric vector is everywhere parallel (or antiparallel) to the *y* axis. We call such light **polarized**. (See Problem 21.45.)

Natural light is usually **unpolarized**. Unpolarized light consists of a superposition of waves whose electric vectors oscillate in all possible directions.* Most natural

---

*Remember from Section 34.4 that "all possible directions" means all directions lying in the plane normal to the direction of propagation.

(a)

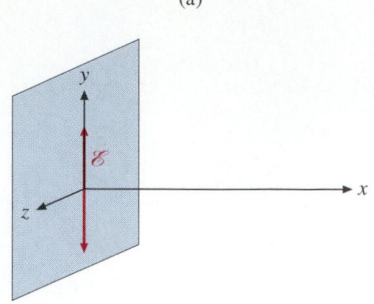

(b)

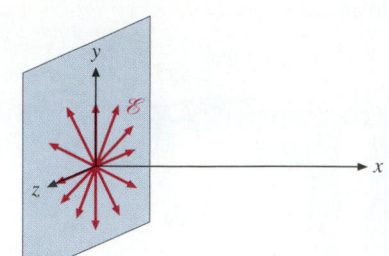

(c)

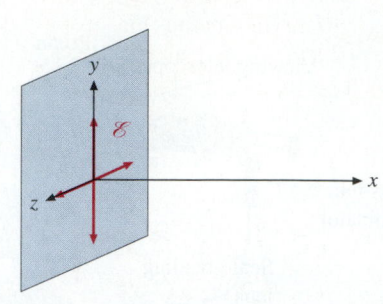

FIGURE 35.19 Schematic representation of (*a*) polarized light and (*b*) unpolarized light, propagating in the *x* direction. (*c*) Simplified representation of unpolarized light.

sources emit unpolarized light because they contain very large numbers of electrons, oscillating independently of one another. We often represent polarized and unpolarized light by the oscillating electric vectors shown in Figure 35.19a, b. The representation is loose, however, in the following important respect. Each double-headed arrow represents not a single $\mathscr{E}$ vector oscillating with a certain phase but rather a superposition of vectors, all having the same direction but oscillating with different, unrelated phases. For many purposes, we can simplify the representation of unpolarized light in part *b* to the form shown in part *c*, where all the vectors are replaced by two vectors oscillating, respectively, along the *y* axis and the *z* axis. (Here again, remember that each double-headed arrow represents a superposition of vectors having different phases.)

There are several ways to extract a beam of polarized light from a beam of unpolarized light. Any device that does this is called a **polarizer**. We describe a few such devices briefly.

Certain crystals, called *birefringent* (doubly refracting) crystals, can be made into polarizers. All birefringent crystals are anisotropic, as a consequence of which many of their properties vary with direction. One of these anisotropic properties is the speed of light waves in the crystal, which depends on the orientation of the electric vector relative to the crystal axes. As a result of this dependence, the index of refraction also depends on the orientation of $\mathscr{E}$. A single beam of unpolarized light entering the crystal is refracted into two polarized beams. These two beams emerge from the crystal at slightly different angles, as shown in Figure 35.20. The most common birefringent crystal is calcite ($CaCO_3$), which occurs naturally as Iceland spar.

(a)

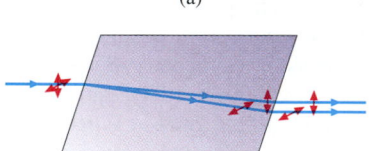

(b)

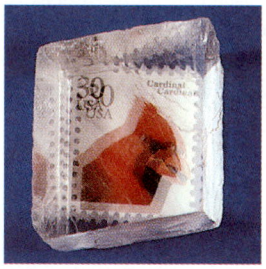

(c)

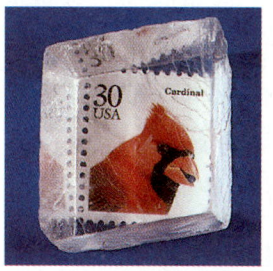

(d)

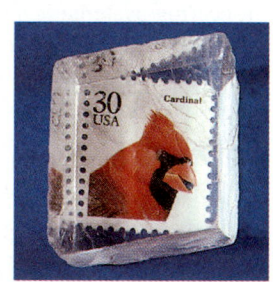

FIGURE 35.20 (*a*) Double refraction by a crystal of Iceland spar. Unpolarized light enters the crystal at a nonnormal angle of incidence at the left. The beam is refracted into two polarized beams having different angles of refraction. (The angles are exaggerated here for clarity.) (*b*) Photograph of a double image seen through an Iceland spar crystal. (*c*) Photograph of the same object and crystal through a polarizing filter that suppresses one image. (*d*) The filter is turned through 90° and suppresses the other image.

Other classes of crystals exhibit *dichroism*. A dichroic crystal is much less transparent to light whose electric vector lies in one direction relative to the axes of the crystal than to light whose electric vector lies in the perpendicular direction. The semiprecious gemstone tourmaline is a dichroic crystal. As unpolarized light passes through such a crystal, the component polarized in the favored direction is transmitted preferentially while other components are absorbed (Figure 35.21).

Although they are effective polarizers, large dichroic crystals are expensive and awkward to use. In 1932, Edwin H. Land* invented a cheap way of manufacturing

---

*American inventor (1909–1986). Some two decades later, Land revolutionized photography with his first system for "instant development," the Polaroid-Land camera.

large sheets of polarizing material called Polaroid. The organic compound iodoquinine sulfate (herapathite) is strongly dichroic. It forms tiny, needlelike crystals. These crystals, in a liquid suspension, are made to flow as a film across a sheet of plastic or glass. The crystals tend to align themselves along the direction of flow. The liquid is evaporated, leaving a sheet of Polaroid. (Electric and magnetic fields, as well as other means, also may be used to align the crystals.)

Polarization by Scattering and by Reflection

Scattering accounts for the polarization of light from the daytime sky, among other things. The mechanism is sketched in Figure 35.22. Unpolarized light from the sun travels in the x direction and strikes air located at the origin of the coordinate system. The air contains many electrons. The oscillating electric field of the sunlight sets the electrons into oscillation. The process increases the energy of the electrons, which implies that a small proportion of the incident light is absorbed. The electrons oscillate in all possible directions in the y-z plane. Because they are oscillating, the electrons emit light, thus reemitting the energy they absorbed from the incident light. Because we see this **scattered light** from all parts of the sky, the sky is bright rather than black in daytime.* In imagination, we replace the electrons oscillating in all directions in the y-z plane by two groups of electrons, one group oscillating along the y axis and the other along the z axis. Because the direction of propagation of a light wave is always perpendicular to its electric vector, electrons oscillating along the z axis do not emit

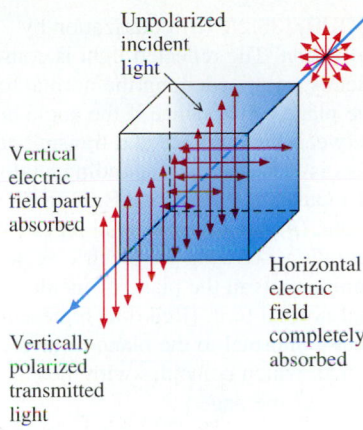

FIGURE 35.21 Selective absorption of light by a dichroic crystal.

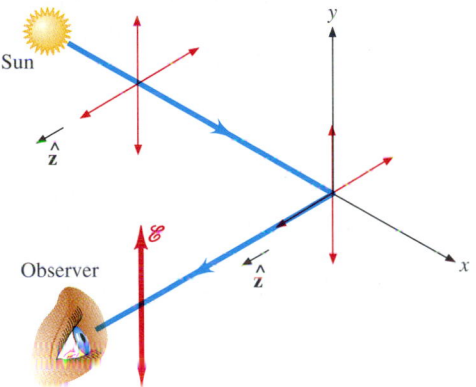

FIGURE 35.22 Polarization of scattered light. Unpolarized sunlight is incident on air located at the origin of the coordinate system. Electrons excited into oscillation by the electric vector of the incident light are represented by the two double-headed arrows at the origin. The observer on the z axis sees only light polarized with its electric vector in the y direction.

light in the z direction. However, electrons oscillating along the y axis do emit light in the z direction. Consequently, an observer located on the z axis and looking toward the origin sees only light polarized with its electric vector along the y direction. Other observers in the y-z plane likewise see polarized light. For example, an observer on the y axis receives radiation only from electrons oscillating along the z axis; this light is polarized with its electric vector along the z direction. (We will soon consider simple means of detecting the polarization of light.)

If you draw a line from yourself to the sun, the light you see coming from the sky along any direction perpendicular to that line is polarized. Consequently, there is a band in the sky, lying in the plane normal to the line, from which the light scattered toward you is polarized. Skylight from other directions is *partially polarized*; see Problem 35.55. Skylight is polarized even on cloudy days, when the sun is obscured.

In 1808, the French physicist Etienne Louis Malus (1775–1812) happened to look through a calcite crystal at the many windows of the Luxembourg Palace in Paris late in the afternoon, when the windows were ablaze with the light of the setting sun. By rotating the crystal, he found that the reflected light was partially polarized. Malus quickly showed that the degree of polarization depends on the angle of reflection. The

---

*Astronauts, above the atmosphere, see the sky as black and can see stars even when the sun is up. Even at the 10-km altitudes at which commercial aircraft fly, the sky is noticeably darker than at sea level.

Bees and Polarization

Our eyes are nearly insensitive to polarization of light, but this is not universally true of animals. Bees are notable in their use of polarized skylight for navigation. A foraging bee usually discovers a new food source by random search. The bee then returns to the hive and communicates the location of the food to other bees. Soon, many bees arrive at the source, using the information communicated at the hive. The Austrian zoologist Karl von Frisch showed that bees accomplish their remarkable navigational feat by orienting themselves with respect to the sun. But bees do not limit their food-gathering to sunny days. When the sun is obscured by clouds, they use the polarized band in the sky to sense its location.

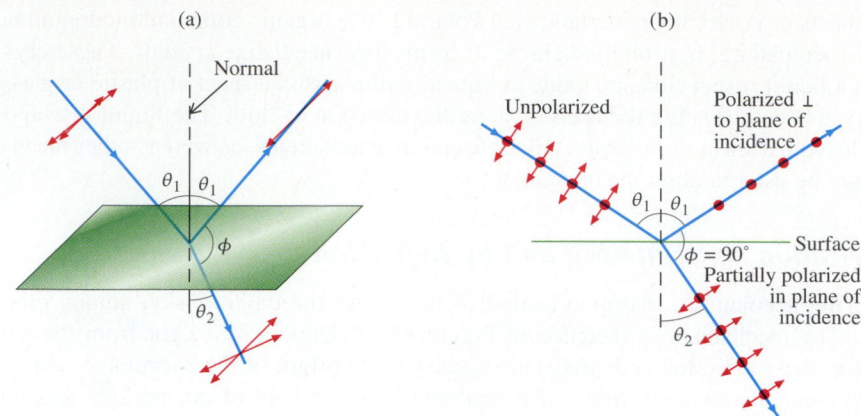

FIGURE 35.23 (*a*) Polarization by reflection. The reflected light is completely polarized along the normal to the plane of incidence if the angle ϕ between the reflected and transmitted rays is 90°. The corresponding angle of incidence θ_p is called Brewster's angle. (*b*) Two-dimensional representation, showing the electric vector components in the plane of incidence and normal to it. (Red dots represent vectors normal to the plane of incidence, which coincides with the plane of the page.)

Scottish physicist David Brewster (1781–1868) found in 1815 that light reflected from the surface of a transparent material is *completely* polarized if the angle ϕ between the reflected and refracted light rays is 90° (Figure 35.23). This condition is satisfied if the angle of incidence has the particular value $\theta_1 = \theta_p$, where

$$\theta_p = \tan^{-1}\frac{n_2}{n_1}. \tag{35.11}$$

In this equation, n_2 is the index of refraction of the medium from which the light is reflected, and n_1 is the index of refraction of the medium (usually air) through which the light propagates. The special angle θ_p is called **Brewster's angle**, and Equation 35.11 is called **Brewster's law**. Proof of Brewster's law is the subject of Problem 35.51.

Justification of Brewster's Law

The explanation for Brewster's law is akin to that for polarization by scattering. Figure 35.24 shows a ray of unpolarized light incident on an air-glass surface. The electric field of the light wave excites oscillation of electrons located in the glass, just beneath the surface. This excitation requires energy, and the incident light is absorbed. Because the electrons oscillate, they reemit the energy they absorbed from the incident light. This reemission is in the form of light that passes both deeper into the glass and back into the air. Within the glass, the reemission is along the direction of the refracted ray. The electrons must therefore be oscillating in all directions that lie in the plane normal to the refracted ray. As we have done before, we represent the oscillation of the electrons by double-headed arrows. In Figure 35.24, one of these arrows, represented by the red dot and labeled *a*, is seen head on. It is oriented normal to the plane of incidence—that is, parallel to the surface. The other arrow, labeled *b*, is drawn in the plane of incidence, perpendicular to *a*. Both *a* and *b* are perpendicular to the refracted ray.

The reflected and refracted rays must have their origins in the same oscillating electrons. The oscillation depicted by *a*, parallel to the surface, contributes to both the refracted and the reflected rays. But the oscillation depicted by *b* is parallel to the direction of the reflected ray. As we have already noted, oscillating charges do not emit light along the direction of oscillation. Light having this direction of polarization is therefore missing from the reflected ray. The reflected ray contains only light whose direction of polarization is normal to the plane of incidence.

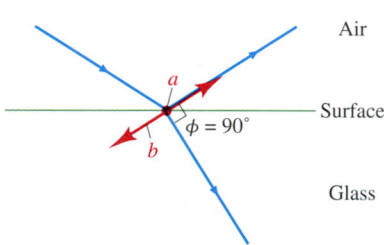

FIGURE 35.24 Justification of Brewster's law. The arrow *a*, seen head on as a red dot, represents oscillation of electrons normal to the plane of incidence. The oscillating arrow *b* lies in the plane of incidence; it is perpendicular both to the refracted ray and to *a*. Because the reflected ray is perpendicular to the refracted ray, *b* is parallel to the reflected ray.

Malus's Law

When unpolarized light falls on a polarizer, the polarizer transmits only that component of the light whose electric wave lies in a single plane (Figure 35.25). The plane defined by the electric wave is called the **plane of polarization**. The intersection of the plane of polarization with the polarizer is called the **transmission axis** of the polarizer. When the polarizer is rotated about an axis normal to its surface, the transmission axis rotates and so does the plane of polarization of the emerging light.

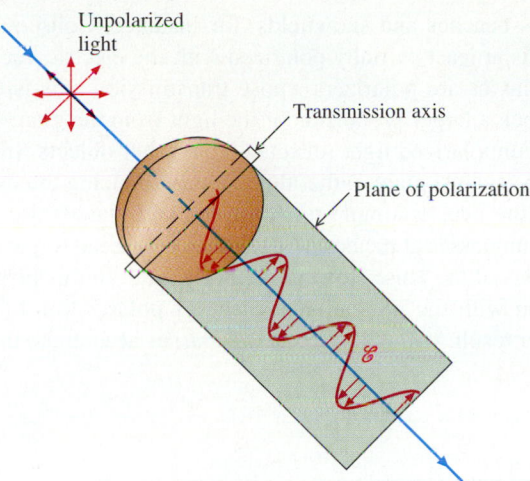

Unpolarized light

Transmission axis

Plane of polarization

$\mathscr{E}$

Suppose now that the polarized light falls on a second polarizer, conventionally called an **analyzer**. Such an arrangement is shown in Figure 35.26. The transmission axes of the polarizer and of the analyzer are shown; the angle between them is ψ.

When the transmission axes coincide, so that $\psi = 0°$, the analyzer has little effect on the light that falls on it. However, if the transmission axes are mutually perpendicular—if $\psi = 90°$—the analyzer absorbs all the light transmitted by the polarizer. A polarizer and an analyzer oriented so that $\psi = 90°$ are called **crossed polarizers**.

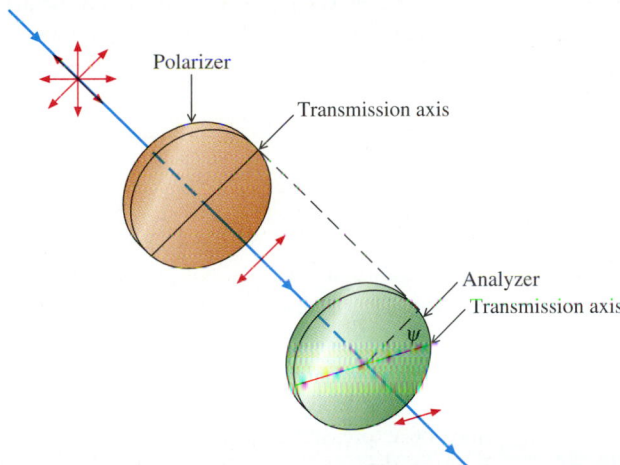

FIGURE 35.26 A beam of light passes successively through a polarizer and an analyzer. The angle between the transmission axes is ψ.

Polarizer

Transmission axis

Analyzer
Transmission axis

ψ

If ψ has some value between $0°$ and $90°$, the analyzer transmits *some* of the light that falls on it. How much? Suppose the amplitude of the electric field of the light incident on the analyzer is $|\mathscr{E}_0|$. The corresponding intensity is $I_0 \propto \mathscr{E}_0^2$ (see Equation 34.38a). We can resolve the incident electric vector into components parallel to and perpendicular to the transmission axis of the analyzer, as shown in Figure 35.27. The perpendicular component $\mathscr{E}_a$ is absorbed by the analyzer. The parallel component, which is transmitted by the analyzer, has amplitude $\mathscr{E}_t = \mathscr{E}_0 \cos \psi$. The intensity I of the transmitted light is proportional to $\mathscr{E}_t^2$. Hence we can write the ratio

$$\frac{I}{I_0} = \frac{\mathscr{E}_t^2}{\mathscr{E}_0^2} = \frac{(\mathscr{E}_0 \cos \psi)^2}{\mathscr{E}_0^2} = \cos^2 \psi.$$

That is, the intensity of the light transmitted by the analyzer is

$$I = I_0 \cos^2 \psi. \qquad \textbf{(35.12)}$$

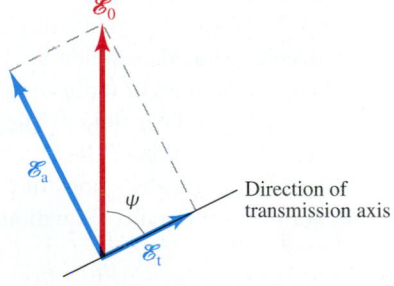

$\mathscr{E}_0$

$\mathscr{E}_a$

ψ

Direction of transmission axis

$\mathscr{E}_t$

FIGURE 35.27 Vector diagram for derivation of Malus's law.

This relation was found experimentally in 1809 by Malus and is known as **Malus's law**.

A familiar application of crossed polarizers is in Polaroid sunglasses. In high-glare situations, objects of interest are hard to see because intense, approximately uniform light is reflected from large surfaces. These glare-producing surfaces are usually more or less horizontal—beaches and snowfields, for instance. Consequently, the light reflected from them is at least partially polarized with the electric vector horizontal. The lenses of the sunglasses are polarizers whose transmission axis is oriented vertically. The lenses thus block a larger proportion of the light from the glare-producing surfaces than of the mostly unpolarized light reflected from other objects. In this way, the sunglasses discriminate against glare, rather than merely reducing the overall brightness of the light reaching the eye, as simple tinted or smoked glasses do. Next time you are wearing Polaroid sunglasses at the beach, try tilting your head to the side, so as to rotate the transmission axis of the lenses toward the horizontal. In this position, the transmission axis is aligned with the predominant plane of polarization of the light reflected from the sand. As a result, you will see that the glasses accentuate the glare, rather than reducing it.

Symbols Used in Chapter 35

c	speed of light in vacuum	θ	angle of a light ray with respect to the normal
$\mathscr{E}$	electric field	θ_p	Brewster's angle
I	intensity of light	λ	wavelength
n	index of refraction	ν	frequency
v	speed of light in a transparent medium	ψ	angle between transmission axes of polarizer and analyzer
δ	dispersion angle		

Summing Up

Light propagates through a transparent medium with speed v, which is always less than the speed c of light in vacuum. The **index of refraction** of a medium is defined by Equation 35.2,

$$n = \frac{c}{v}.$$

Huygens's construction is a useful tool for tracing the propagation of **wave fronts** of any shape in two or three dimensions. We assume that every point on the wave front is a **sourcelet**. **Wavelets** are emitted by the sourcelets at $t = 0$, and we draw circles of radius $v\tau$ to represent the wavelets as they appear at a later time τ. The common tangent to the wavelets is the position of the wave front at time τ.

We can use Huygens's construction to derive the **law of reflection** for wave trains (Equation 35.3):

$$\theta_i = \theta_r;$$

Huygens's construction also leads to the **law of refraction**, or **Snell's law** (Equation 35.5a):

$$n_1 \sin \theta_1 = n_2 \sin \theta_2.$$

If medium 1 is air, for which $n_1 \simeq 1$, we can write Snell's law in the simplified form of Equation 35.5b,

$$\sin \theta_1 = n \sin \theta_2.$$

When a light waves passes from one medium into another, the frequency of the wave remains unchanged. Because the propagation speeds are different in the two media, the wavelength must change according to Equation 35.7a,

$$\lambda_1 n_1 = \lambda_2 n_2.$$

When light propagates through a medium, any distance s may be expressed as a multiple of the wavelength of the light in the medium—that is, in terms of the number of wavelengths contained in s. The **optical path length** L is defined to be the distance in vacuum containing the same number of waves; according to Equation 35.8,

$$L = ns.$$

When light strikes the boundary between two transparent media from the optically denser side, there is refraction through the surface only for angles of incidence less than a **critical angle** θ_c, given by either of Equations 35.10a and 35.10b,

$$\sin \theta_c = \frac{n_r}{n_d} \quad \text{or} \quad \theta_c = \sin^{-1} \frac{n_r}{n_d}.$$

For angles of incidence greater than θ_c, there is **total internal reflection**.

For glass, and for many other substances that are transparent in the visible region of the spectrum, the index of refraction varies slowly with wavelength. As a consequence, light consisting of a superposition of components of different wavelengths experiences **dispersion** on refraction. The **dispersion angle** δ is always small for components of visible light; nevertheless, dispersion can be exploited to analyze light into its spectral components.

When the electric vector of a light wave is confined to a single plane, the light is said to be **polarized**. The plane of the electric wave is called the **plane of polarization**. **Unpolarized light** consists of a superposition of equal components of all possible planes of polarization. Light can be polarized by a variety of means. One means is **polarization by reflection**. According to **Brewster's law**, reflected light is completely polarized, with the electric vector normal to the plane of incidence, if the angle of incidence is that given by Equation 35.11,

$$\theta_p = \tan^{-1} \frac{n_2}{n_1}.$$

The transmission of initially unpolarized light through two polarizers (a polarizer and an analyzer) depends on the angle ψ between their **transmission axes**, according to **Malus's law** (Equation 35.12),

$$I = I_0 \cos^2 \psi.$$

QUERIES

35.1 *(2) Further reflection.* When later workers repeated Fizeau's experiment, they substituted a rotating multisided mirror for the toothed wheel. Rather than intermittently blocking the light, the mirror chops the light beam by rotating through the orientation in which the reflected light has the proper direction to pass through the apparatus. Why does this give improved results?

35.2 *(3) Breathing tube.* A long, flexible rubber tube is stretched straight under slight tension and is anchored at its ends. It lies along the surface of a container of water. As shown below, the air pressure inside the tube is varied sinu-

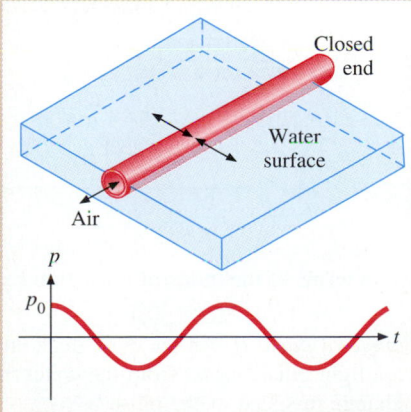

soidally; that is, $p = p_0 \cos \omega t$. Consequently, the radius of the tube varies sinusoidally as well, disturbing the surrounding water as it does so. What is the form of the wave fronts that propagate through the water?

35.3 *(4) Vanity with economy.* Make a sketch to show that you can see yourself from head to foot in a plane mirror whose height is one-half your own height. Does it make any difference where you place the mirror?

35.4 *(5) Higher than it looks.* An important part of celestial navigation is measuring the altitudes—the angles above the horizon—of certain stars. Before altitude can be used to calculate position, a correction must be made because the true altitude of a star is always greater than the observed altitude. Explain.

35.5 *(5) Getting immersed in your work.* A quick way to determine the index of refraction of a transparent solid object involves using a graduated series of liquids of different indices of refraction. Explain how to carry out the determination.

35.6 *(6) Jawohl, Herr Hauptmann!* In World War II, the German army found that their short-distance radio communications in the field were too often intercepted by their adversaries. As one solution to this problem, they used a field communication system called a *Lichtsprecher* (lightspeaker)—a wireless telephone in which the signal was carried on a light beam rather than radio waves. Because it is easy to focus light into a narrow beam, a light signal is much harder to intercept than a radio signal. The heart of the *Lichtsprecher* transmitter is shown below. The two glass prisms are mounted quite close together, so that the equilibrium spacing between them is approximately equal to the wavelength of light—say, 500 nm. The left-hand prism is connected by a system of levers to the diaphragm of the telephone mouthpiece and vibrates very slightly in response to the speaker's voice. Explain the operation of the system.

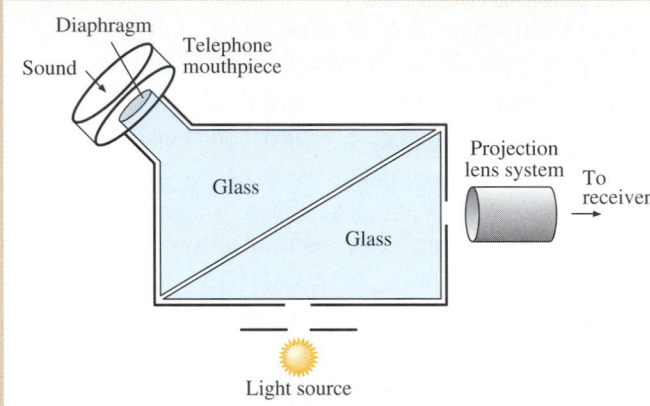

35.7 *(6) . . . even if it doesn't snap.* If bent into too sharp a curve, a flexible optical fiber will not function as a light pipe. Draw a sketch and explain.

35.8 *(7) Pretty close.* In the experiment shown in Figure 35.18, Newton found that the light dispersed by the first prism didn't *quite* merge into a beam of white light as it left the second prism, unless the two prisms were very close together. Rather, he obtained a beam whose central region was white, with a red fringe at the top and a violet fringe at the bottom. Explain why the recombination is imperfect. [Hint: Are the three rays between the prisms really parallel (as they are drawn)?]

35.9 *(8) Now you see it . . .* When Malus looked at the windows of the Luxembourg Palace through his calcite crystal, he saw two images. Describe what happened to these images as he rotated the crystal about an axis parallel to the line from the palace to his location.

35.10 *(8) Taking it all in.* When unpolarized light is incident on a surface at Brewster's angle, is the transmitted light polarized? Explain. (Hint: Compare the reflected light and the transmitted light with the incident light.)

35.11 *(8) Landing on one's feet.* When Edwin Land first developed his polarizing sheet, the major commercial application he envisioned was on automobile headlights and windshields to reduce nighttime headlight glare and to eliminate the need for motorists to drop their headlight beams for oncoming cars. Explain how this could be done. (Unfortunately, this application was never realized for a variety of practical reasons.)

35.12 *(8) Busy as a bee.* You can observe the polarization of skylight by looking at the daytime sky through Polaroid sunglasses. As you look at a part of the sky through a lens, rotate the lens and observe the effect. Do this for various parts of the sky located at different angles with respect to the sun. Try to describe an imaginary arc through the sky along which the polarization effect is greatest. Where is the arc located? (Note: It is easiest to make these observations when the sun is low, in early morning or late afternoon.)

35.13 *(8) Catching the sky.* Fill a large glass container (a fish tank works well) with slightly cloudy water. (You can obtain the necessary cloudiness by dissolving a small quantity of liquid dishwashing detergent, milk, or powdered coffee creamer in the water.) Use a slide projector or similar light source to send a beam of light through the tank the long way, and observe the light from the side of the tank using Polaroid glasses. What do you see? Explain. (Hint: The transmitted light is reddish, and the light scattered to the side is bluish.)

35.14 *(8) Stacking the deck.* A pile of glass plates spaced a short distance apart is shown below. Unpolarized light falls on the pile at Brewster's angle. Explain why the beam emerging at the bottom of the pile is almost completely polarized. What is the plane of polarization? (Hint: See Query 35.10.)

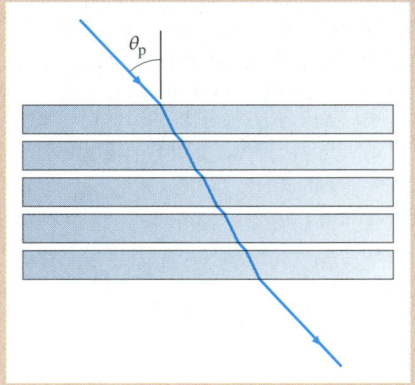

PROBLEMS

GROUP A

35.1 *(2) Faster than a speeding bullet.* Calculate the speed of light in diamond.

35.2 *(2) Faster still.* Calculate the speed of light in ethanol.

35.3 *(2) Change of pace, I.* When sunlight strikes the surface of a spacecraft window, the speed of the light undergoes a proportional change $\Delta v/c = (c - v)/c$. Express this proportional change in terms of the index of refraction n of the window glass.

35.4 *(2) Change of pace, II.* An ice cube floats in a glass of water. When a light pulse passes from the water into the ice, what is the change in speed of the pulse?

35.5 *(2) Speed zone.* Express the change Δv of the speed of light passing from medium 1 into medium 2 in terms of the indices of refraction, n_1 and n_2, and c.

35.6 *(4) A matter for reflection.* A vertical light ray strikes a plane mirror and is reflected at an angle of 45° from the vertical. By what angle is the mirror tilted from the horizontal?

35.7 *(4) Periscope.* A light ray is reflected from plane mirror 1, strikes plane mirror 2, and is reflected again. Show that, if the ray incident on mirror 1 and the ray reflected from mirror 2 are parallel, the two mirrors must be parallel.

35.8 *(4) Merit badge in optics.* A Boy Scout uses a plane hand mirror to reflect sunlight into the woods, where the light falls on a tree trunk and makes a bright patch. If he rotates the mirror through an angle θ, through what angle does the reflected beam rotate?

35.9 *(5) Turning the tables.* Using Huygens's construction as in Figure 35.10, show that Snell's law is valid for the case in which the plane wave front approaches the interface from the optically dense side.

35.10 *(5) Visiting relatives.* A plate of light flint glass is submerged in water. What is the index of refraction n_{gw} of the glass relative to the water—that is, the ratio of the speed of light in water to the speed in the glass? What is the index of refraction n_{wg} of the water relative to the glass?

35.11 *(2) Relative index of refraction.* Two transparent media, 1 and 2, have indices of refraction n_1 and n_2. Define the *relative index of refraction*, $n_{21} \equiv v_1/v_2$. Show that $n_{21} = n_2/n_1$.

35.12 *(5) Bending exercise.* A ray of light is refracted as it passes from a liquid into a submerged plate of a solid. The angle of incidence is 35.0°, and the angle of refraction is 26.2°. What is the ratio n_s/n_l of the index of refraction of the solid to that of the liquid?

35.13 *(5) Snell's construction.* Suppose a light ray is incident on a planar surface at an arbitrary angle θ_1, as shown below. Here is a construction for drawing the refracted ray.

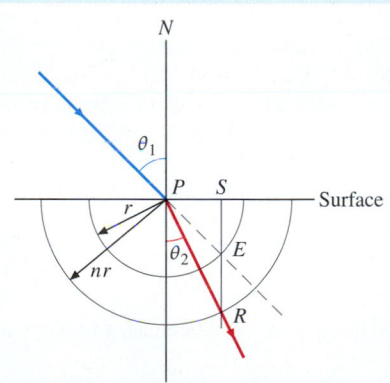

1. Construct the normal *NP* to the surface.

2. Using *P* as a center, draw two circular arcs. The first arc has any convenient radius r. The second arc has radius nr, where $n \equiv n_2/n_1$ is the index of refraction of the lower medium relative to the upper medium (Problem 35.11).

3. Drop a perpendicular *ES* from the point *E*, where the extension of the incident ray intersects the first arc, to the surface. This line, extended, intersects the second arc at *R*.

4. Draw *PR*. This is the refracted ray.

This construction is called *Snell's construction*. Explain why it works.

35.14 *(5) Disappearing act.* If a transparent solid is submerged in a liquid whose index of refraction is the same as that of the solid, the solid is nearly invisible. One way to demonstrate this invisibility is to submerge a plate of Lucite in a mixture of ethanol and carbon disulfide. To a fair approximation, the index of refraction of the mixture varies linearly with the proportion of CS_2 in the mixture. What are the proportions of ethanol and CS_2 for which the Lucite plate disappears?

35.15 *(5) Parallel . . .* In the figure below, a ray of light enters a flat plate of glass, with parallel faces, at an angle of incidence θ_1. Show that the ray emerges from the other face at an angle $\theta_4 = \theta_1$.

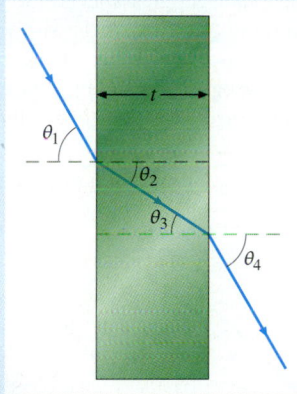

35.16 *(5) Dogleg.* A light ray strikes a flat glass plate with angle of incidence $\theta_1 = 54°$. The ray emerges from the opposite surface of the plate parallel to the incident ray but displaced from it by 1.86 cm. If the index of refraction of the glass is 1.52, how thick is the plate?

35.17 *(5) What's the angle? I.* A light ray falls at normal incidence on a prism made of glass of index of refraction 1.52. If the apex angle of the prism is 40°, what is the angle between the incident ray and the ray that emerges from the other side of the prism?

35.18 *(5) Where's the index?* A light ray falls at normal incidence on a glass prism of apex angle 17°. If the angle between the incident and emergent rays is 28°, what is the index of refraction of the glass?

35.19 *(5) Sitting duck.* A duck floats on the surface of a pond just after sunrise. The sun casts a shadow of the duck on the bottom of the pond, which is 1.75 m deep. How far is the shadow from the point on the bottom immediately below the duck?

35.20 *(5) Some things never change.* Light of wavelength $\lambda_0 = 534$ nm travels through air and strikes the surface of a plate of zinc crown glass. **(a)** What is the frequency of the light in air? **(b)** What is the frequency of the light in the glass? **(c)** What is the wavelength λ of the light in the glass?

35.21 *(5) Path-ology.* Suppose that in Problem 35.20 the light strikes the glass at normal incidence. If the glass is 12.68 cm thick, how many wavelengths does it contain?

35.22 *(5) Change of pace.* A light ray is refracted at the interface between two media. The angle of incidence is θ_1, and the angle of refraction is θ_2. Find the ratio λ_2/λ_1 of the wavelengths of the light in the two media.

35.23 *(5) Extended proceedings.* A tube 10.0 m long is filled with water. What is the optical path length for a ray of light traveling from one end of this tube to the other?

35.24 *(6) Artist's recreation.* The critical angle for total reflection of light from a turpentine-air interface is 42.13°. What is the index of refraction of turpentine?

35.25 *(6) Quick dip.* The index of refraction of a glass plate is 1.58. Find the critical angle for total reflection when the plate is **(a)** in air and **(b)** in water.

35.26 *(6) The graduate.* In search of peace and quiet, you put on your scuba gear and goggles and lie on your back at the bottom of a swimming pool. **(a)** If the surface of the pool is calm, show that all objects above the pool appear to lie within a vertical cone whose apex is located at your eye. **(b)** What is the half-apex angle α of the cone? **(c)** Find the value of α for $n = 1.33$.

35.27 *(6) Reflecting on n.* Shown below is a simple device

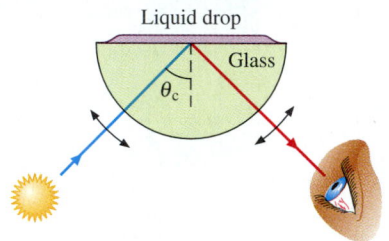

Liquid drop / Glass / θ_c

for measuring the index of refraction of liquids. A quantity of the liquid sufficient to form a large drop is placed on the flat surface of a semicircular disk of glass. By trial and error, the critical angle θ_c for total internal reflection is found. If the disk is made of heaviest flint glass and the critical angle is 48.7°, what is the index of refraction of the liquid?

35.28 *(7) Neatness counts!* The bright yellow light that comes from a flame when you sprinkle salt on it is due in large measure to the *sodium doublet*. This doublet consists of two spectral lines of wavelengths $\lambda_1 = 588.995$ nm and $\lambda_2 = 589.592$ nm. Suppose you want to use the setup of Example 35.4 to resolve (that is, separate) the light of the two wavelengths that make up the sodium doublet. What dispersion angle must you be prepared to measure? (A good prism spectrometer can readily accomplish this task.)

35.29 *(7) Splitting hairs?* For heavy flint glass, find the critical angles for total reflection for light of wavelengths **(a)** 425 nm and **(b)** 700 nm. **(c)** Suppose you aimed a narrow beam of white light at the surface of this glass from the glass side at an angle midway between the angles you calculated in parts **a** and **b**. Describe qualitatively what you would see.

35.30 *(8) Brewster booster.* Calculate Brewster's angle for light incident on **(a)** water and **(b)** zinc crown glass.

35.31 *(8) Cutting down.* Polarized light of intensity I_0 falls on an analyzer oriented to allow complete transmission of the light. Through what angles ψ must the analyzer be turned to reduce the light intensity to **(a)** $I_0/2$, **(b)** $I_0/5$, and **(c)** $I_0/10$?

GROUP B

35.32 *(2) Foucault measures c/v.* In 1850, Foucault used the apparatus shown below to determine the ratio of the speed of light in air to that in water. When mirror M^*, which can be rotated, is at rest in position 1, light travels from the source through a narrow slit to M^*, then to M_1, then back to M^*, and finally to a ground-glass scale where an image of the slit is cast at S_0. When M^* is at rest in position 2, the light reflected from M^* passes through the water-filled tube, is reflected at M_2, passes back through the tube to M^*, and again falls on S_0. **(a)** What happens when M^* is neither in position 1 nor in

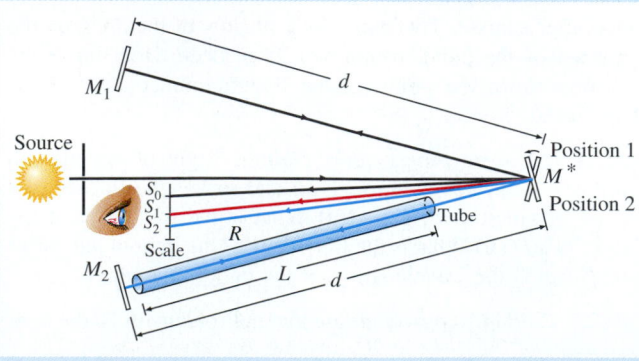

position 2? When M^* is rotating rapidly with angular speed ω, light travels alternately over both paths. However, the light reflected from M_1 falls on the scale at S_1, and the light reflected from M_2 falls at S_2. Explain qualitatively why **(b)** the light beams are displaced from S_0 and **(c)** the two displacements are not the same. **(d)** Let δ_1 be the distance between S_0 and S_1. Neglecting the difference between the speed of light in air and c, the speed in vacuum, show that

$$c = \frac{4R\omega d}{\delta_1}.$$

(e) Let δ_2 be the distance between S_0 and S_2. Show that

$$\frac{c}{v_{\text{water}}} = 1 + \left(\frac{\delta_2}{\delta_1} - 1\right)\frac{d}{L},$$

which is independent of ω and c. **(f)** Suppose the apparatus is built so that the water tube nearly fills the space between M^* and M_2. Show that the equation of part **e** reduces to a much simpler form.

35.33 *(4) Corner reflector, I.* In the figure on the next page, two plane mirrors are fixed at right angles to one another. Show that any ray of light that strikes either mirror will emerge in a direction exactly opposite its initial direction.

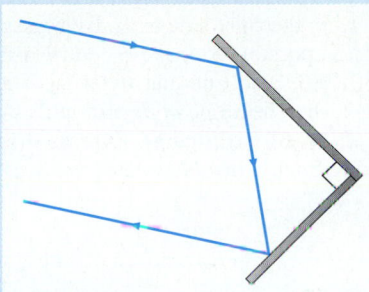

35.34 *(5) Snell's law à la Snell.* Snell originally expressed the law of refraction algebraically in the form

$$\frac{\csc \theta_2}{\csc \theta_1} = n = \text{constant}, \quad \text{for all pairs of angles } \theta_1 \text{ and } \theta_2.$$

(a) Show that this is equivalent to Equation 35.5b. **(b)** Compare Snell's form of the law of refraction with Snell's construction (Problem 35.13). Show how the equation falls naturally out of the construction, and thus see how Snell came to express the law of refraction in terms of cosecants rather than sines. In the early seventeenth century, mathematical problems were often worked out in geometric rather than algebraic form.

35.35 *(5) First approximation.* The second-century Alexandrian astronomer Claudius Ptolemy published a table of angles of incidence and corresponding angles of refraction for air-water surfaces. Although Ptolemy did not give a formula corresponding to his tabular values, they agree fairly well with the rule $\theta_{\text{air}} = 1.33\theta_{\text{water}}$. By what percentage would Ptolemy's values for θ_{water} be in error for **(a)** $\theta_{\text{air}} = 5°$? **(b)** $\theta_{\text{air}} = 10°$? **(c)** $\theta_{\text{air}} = 20°$? **(d)** $\theta_{\text{air}} = 40°$? (Ptolemy may have been aware of this pitfall; he published values only for small angles.)

35.36 *(5) Refraction of a stream of particles.* The French mathematician and philosopher René Descartes hypothesized that a beam of light consists of a stream of particles so tiny that they can pass freely through transparent materials as well as through vacuum. Taking the approach described below, he used this hypothesis to account for the law of refraction of light. In the figure below, a stream of particles moving in vacuum at velocity $\mathbf{v}_1 = (v_{1x}, v_{1y})$ approaches the planar surface of a transparent medium having index of refraction n.

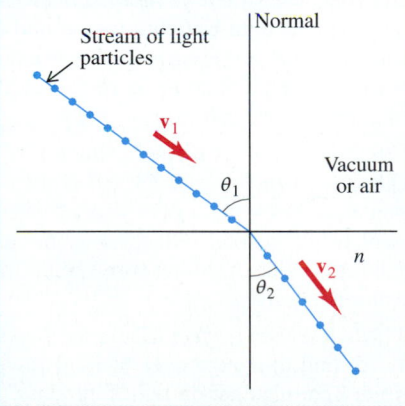

The angle of incidence is θ_1. **(a)** Particles often exert forces (for example, electric or gravitational forces) on other parti-

cles. Assume that this is true for the "light particles" and the atoms that make up the transparent medium. Use a symmetry argument to determine the direction of the force experienced by a light particle as it approaches the planar surface. **(b)** We know from observation that the light beam is refracted toward the normal; that is, $\theta_2 < \theta_1$. Is the force attractive or repulsive? **(c)** Light travels through the medium with speed $\mathbf{v}_2 = (v_{2x}, v_{2y})$. Compare v_{2x} with v_{1x}, v_{2y} with v_{1y}, and v_2 with v_1. **(d)** We know from observation that Snell's law correctly describes the process of refraction: $\sin \theta_1 = n \sin \theta_2$. Express n in terms of v_1 and v_2. **(e)** Compare the result of part **d** with that obtained in Section 35.5 on the assumption that light behaves like a wave and not like a stream of particles. Suggest a crucial experiment to distinguish between the particle and wave assumptions.

35.37 *(5) Right on!* A light ray is refracted at the surface of an object whose index of refraction is n. For a certain angle of incidence θ_1, the angle between the extension of the incident ray and the refracted ray is 45°. Express θ_1 in terms of n.

35.38 *(5) Putting a fine point on it.* It is often possible to set the index of refraction of air equal to 1 in making measurements of refraction. But suppose you are measuring angles of incidence and refraction at an air-glass surface. If the angle of incidence is about 45°, at what degree of precision of angle measurement must you take the index of refraction of air into account?

35.39 *(5) Canned light.* You have a tin can of height h and radius r, as shown below. A penny lies on the bottom at the center. From the position shown, you can barely see the far side of the bottom of the can. While you remain perfectly still,

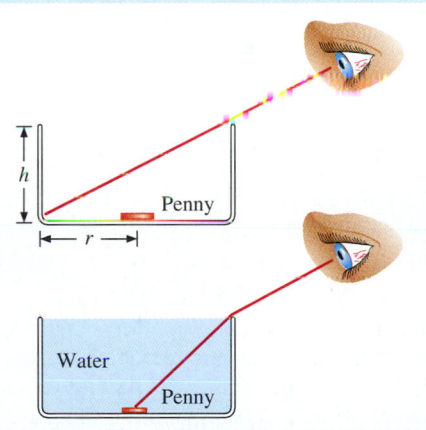

you have a friend fill the can slowly with water whose index of refraction is n. As the can is filled, you can see more and more of the bottom. When the can is full, you can just see the center of the penny. Show that the ratio of the height to the radius is

$$\frac{h}{r} = 2\sqrt{\frac{n^2 - 1}{4 - n^2}}.$$

35.40 *(5) Shadow, I.* A cube of glass whose edges are 5.00 cm is shown on the next page. Two of the vertical faces of the cube are blackened, as shown. A beam of parallel light falls on the cube in such a way that the external shadow, AB, is

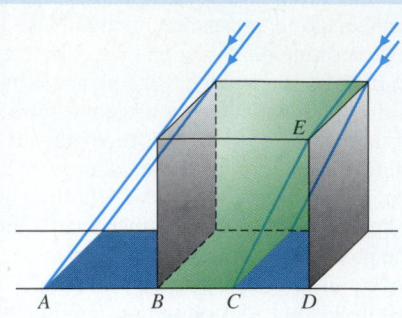

of emergence from the right side is θ_4. By adjusting the angle of incidence, it is possible to make θ_4 equal to θ_1. **(a)** Show that, if $\theta_4 = \theta_1$, it is also true that $\theta_3 = \theta_2$. Then show that $\theta_3 = \theta_2 = \alpha/2$. **(b)** The *angle of deviation* Δ is defined to be the sum $\Delta = \Delta_1 + \Delta_2$ (see figure). Express θ_4 in terms of α and Δ. **(c)** Using Snell's law, show that

$$
n = \frac{\sin\dfrac{\alpha + \Delta}{2}}{\sin\dfrac{\alpha}{2}}.
$$

This is called the *prism equation*, and it is a standard way of determining the index of refraction.

35.45 *(5) Principle of optical reversibility.* Suppose a light ray follows a certain path through an optical system as the ray is refracted or reflected. Now, suppose the path is reversed. That is, suppose a light ray begins at the exit point of the original ray and is directed opposite the original ray. An important principle is that *the second ray will follow exactly the same path as the first, in reverse.* Use this *principle of optical reversibility* to prove that **(a)** there is only one orientation of a prism relative to the incident ray that will result in *minimum deviation* of the light ray by the prism; **(b)** the prism orientation of Problem 35.44 ($\theta_1 = \theta_4$) is the proper one for minimum deviation.

35.46 *(5) Thin prism.* A prism having a small apex angle α (see Problem 35.44) is called a *thin prism*. Suppose a light ray strikes such a prism at a small angle of incidence—that is, nearly normal to the surface. Using the result of part **c** of Problem 35.44, show that the angle of deviation is

$$
\Delta = (n - 1)\alpha.
$$

Why is the assumption of near-normal incidence for the thin prism equivalent to the assumption of equal angles of incidence and emergence for the general prism of Problem 35.44?

35.47 *(6) Slick.* A layer of oil of index of refraction n_o floats on water of index of refraction n_w. A beam of light is directed upward from a source in the water. Show that the critical angle for total internal reflection is independent of n_o, even though, for angles close to θ_c, the total reflection always takes place at the oil-air surface.

35.48 *(6) Fundamental principle of light plumbing.* A right cylindrical rod has index of refraction n. If the rod is to be used as a light pipe, a light ray entering an end face at any angle of incidence must not escape until it reaches the other end face. Show that this condition is met, regardless of the length and diameter of the rod, if $n \geq \sqrt{2}$.

35.49 *(6) Critical problem.* A prism having a 120° apex angle is made of glass for which $n = 1.53$. **(a)** Is there a range of angles of incidence θ_1 for light rays striking one side of the prism that will result in total reflection at the second side? **(b)** If so, what is the value of θ_1 that corresponds to the critical angle for total reflection?

35.50 *(8) A little less than perfect.* Using the dispersion curve of Figure 35.15, find the range over which Brewster's angle varies for white light incident on light flint glass.

35.51 *(8) Proof of Brewster's law.* In Figure 35.23, let the

5.00 cm wide. The internal shadow, *CD*, is then 2.60 cm wide. Find the index of refraction of the glass.

35.41 *(5) Shadow, II.* A pole is stuck into the mud at the bottom of a pond of depth h and extends upward from the water surface an equal distance h. When the sun has reached a certain angle ϕ above the horizon, the length of the shadow cast by the pole on the bottom of the pond is equal to the length $2h$ of the pole. **(a)** Show that, under these circumstances, the angle of incidence of the sunlight on the water, $\theta_1 = 90° - \phi$, is given by the transcendental equation

$$
\tan\left(\sin^{-1}\frac{\sin\theta_1}{n}\right) + \tan\theta_1 - 1 = 0,
$$

which certainly looks horrendous. **(b)** But be of good cheer! Let $n = 1.33$, and solve this equation for θ_1 by trial and error. (A pocket calculator with a "solve" mode will do all the dirty work for you. In any case, use a calculator for the solution, and you will find that it goes quickly.)

35.42 *(5) . . . but shifted.* Let the thickness of the plate illustrated in Problem 35.15 be t. Show that, if the angle of incidence θ_1 is small, the ray emerging from the plate is shifted from its original path by an amount

$$
d = t\theta_1\,\frac{n - 1}{n}.
$$

35.43 *(5) What's the angle? II.* A glass prism ($n = 1.53$) has a 30° apex angle. A light ray is incident on one side of the prism with $\theta_1 = 12°$. Find the angle between the incident ray and the ray that emerges from the prism.

35.44 *(5) Index of refraction by the prism method.* In the figure below, a beam of monochromatic light passes through a glass prism whose apex angle is α. At the wavelength λ of the light used, the index of refraction of the glass relative to air (Problem 35.11) is $n_{ga} \equiv n_{glass}/n_{air}$. The angle of incidence of the light on the left side of the prism is θ_1; the angle

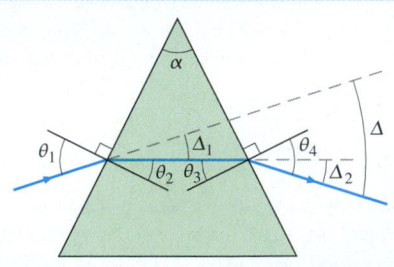

angle of incidence be Brewster's angle: $\theta_1 = \theta_p$. **(a)** Use Snell's law to relate θ_2 to θ_p. **(b)** Using the condition $\phi = 90°$, express $\sin \theta_2$ in terms of θ_p and thus arrive at Equation 35.11. (Hint: See Problem 35.37.)

35.52 *(8) Brewster's critical relations.* The critical angle for total reflection for light arriving at an interface from the optically dense side is θ_c. The Brewster angle for light arriving at the same surface from the optically rare side is θ_p. Show that

$$\sin \theta_c = \cot \theta_p.$$

35.53 *(8) Something is lost in translation.* A beam of unpolarized light falls on a polarizer. Show that the intensity of the transmitted light is one-half the intensity of the incident light.

35.54 *(8) Getting through.* A beam of unpolarized light of initial intensity I_0 passes through three polarizers in succession. The angle between the transmission axes of polarizer 1 and polarizer 2 is ψ_{12}, and the angle between the transmission axes of polarizer 2 and polarizer 3 is ψ_{23}. Find the intensity I of the light emerging from the third polarizer.

35.55 *(8) Proportion of polarization.* Light is often *partially polarized*. For example, when light is polarized by reflection at Brewster's angle from a glass plate, the transmitted light lacks some, but not all, of the incident light polarized with its electric vector normal to the plane of incidence. We can consider partially polarized light as a superposition of polarized light of intensity I_p and unpolarized light of intensity I_u, with total intensity $I_0 = I_p + I_u$. We define the *proportion of polarization P* to be

$$P = \frac{I_p}{I_0}.$$

Suppose that a beam of partially polarized light falls on an analyzer. As the analyzer is rotated, the intensity of the transmitted light varies between a maximum intensity I_{max} and a minimum intensity I_{min}. Show that

$$P = \frac{I_{max} - I_{min}}{I_{max} + I_{min}}.$$

(Hint: Use the result of Problem 35.53.)

GROUP C

35.56 *(2) Rømer's method.* In the late 1670s, the Danish astronomer Ole Rømer (1644–1710), working in Paris, observed the period of Jupiter's satellite Io. Io disappears quite suddenly as it moves behind Jupiter; this is called *occultation.* The period is measured by timing the interval between successive occultations, and Rømer had clocks that were accurate to better than 1 s/day over intervals as long as a week or two. Rømer found that the observed period of Io varied. The variation from one revolution about Jupiter to the next, $\delta \equiv |\tau_j - \tau_{j+1}|$, attains its greatest value δ_{max} when Jupiter is in quadrature—that is, when the angle θ between the sun–Jupiter direction and the sun–earth direction is a right angle (see the figure below). The orbital period of Jupiter around the sun is $T = 3.74 \times 10^8$ s. The period of the satellite Io around Jupiter is $\tau = 1.53 \times 10^5$ s, or about 42.5 hours.

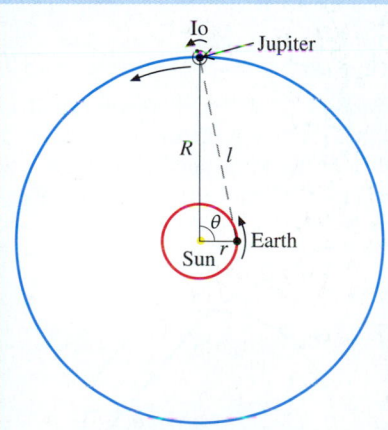

(a) Show that the Jupiter–earth distance l varies at a rate

$$\frac{dl}{dt} = \frac{2\pi}{T} \frac{Rr \sin \theta}{\sqrt{R^2 + r^2 - 2Rr \cos \theta}}.$$

Take all orbits to be circular. **(b)** Given that $R > 5r$, show that, to a good approximation, the time $2r/c$ required for a light signal to traverse a diameter of the earth's orbit is given by the relation

$$\frac{2r}{c} = \frac{T\delta_{max}}{\pi\tau}.$$

This time $2r/c$ is the key quantity reported by Rømer in 1676. **(c)** Rømer's measured value for δ, averaged over repeated observations made near quadrature, was 1.7 s. Evaluate $2r/c$. **(d)** Rømer was well aware that his data made possible an estimate of the speed of light but never published such an estimate. In 1678, Huygens used Rømer's value of $2r/c$, together with several contemporary estimates of the sun–earth distance r, to calculate c. One of these estimates yields $c \simeq 2.32 \times 10^8$ m/s. What was the value of r used? The modern value is 1.496×10^{11} m. What was the percentage error in the estimate of r? Was it the main source of the discrepancy between Huygens's value of c and the modern value? [For an excellent short account of Rømer's work and of the modern history of its misrepresentation, see A. Wróblewski, *Am. J. Phys.* **53**, 620 (1985).]

35.57 *(4) Corner reflector, II.* Three plane mirrors are arranged so that they are mutually perpendicular, as shown on the next page. For convenience, let their lines of intersection be the coordinate axes, with the origin at the point where all three mirrors come together. **(a)** A ray OP, represented by the vector $\mathbf{r}_1 = (x_1, y_1, z_1)$, is reflected from the mirror in the x-y plane. What are the components of the reflected vector $\mathbf{r}_2$? **(b)** If $\mathbf{r}_2$ is now reflected at the y-z plane, what are the components of the reflected vector $\mathbf{r}_3$? **(c)** If $\mathbf{r}_3$ is reflected at the z-x plane, what are the components of the reflected vector $\mathbf{r}_4$? **(d)** Is there any significant difference for an incident ray that is reflected from the three mirrors in some other order? **(e)** A wide variety of highway markers and reflective tapes use

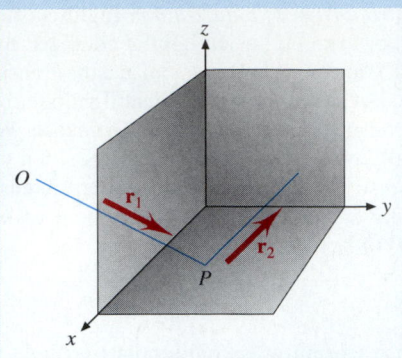

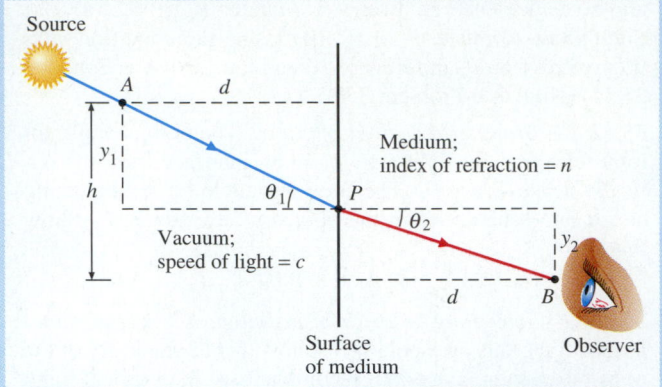

small corner reflectors in large number. Why are they useful? Several high-quality corner reflectors are located on the moon. To what uses can they be put?

35.58 *(4) Law of reflection from Fermat's principle.* The French mathematician Pierre de Fermat (1601–1665) showed that the laws of reflection and refraction could be derived from a single assumption. This assumption, now called *Fermat's principle of least time*, is as follows: Suppose a beam of light passes successively through two points, *A* and *B*. The path the light takes between *A* and *B* is the path for which the time is a minimum. Fermat's principle is "obvious" if empty space lies between *A* and *B*; the path is a straight line, which is certainly the path of minimum time.

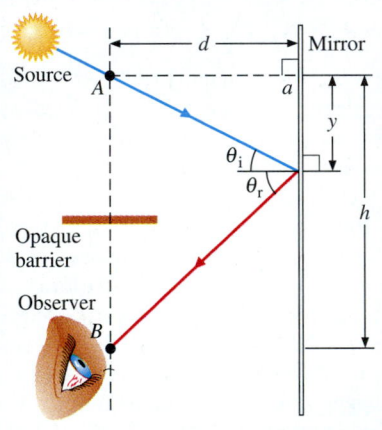

In the diagram shown here, a ray of light is emitted from a source, strikes a plane mirror at an unknown angle of incidence θ_i, reflects at an unknown angle of reflection θ_r, and is seen by an observer. In the diagram, a line is drawn parallel to the mirror at an arbitrary distance *d* from the mirror. The intersections of the line with the light ray are called *A* and *B*. The direct distance between *A* and *B* is *h*. **(a)** Suppose the point of reflection lies at a distance *y* from the point at which the perpendicular *Aa* intersects the mirror. Express the light path length *L* from *A* to *B* in terms of *h*, *d*, and *y*. **(b)** Show that the time $t = L/c$ has a minimal value for $y = h/2$ and thus that $\theta_r = \theta_i$.

35.59 *(5) Law of refraction from Fermat's principle.* In the figure in the next column, a light ray passes through point *A*, located in vacuum, and then through another point *B*, located in a medium having index of refraction *n*. The surface of the medium is planar. If we do not know Snell's law, the location of the point *P* at which the ray passes through the surface is unknown, as are the angle of incidence θ_1 and the angle of refraction θ_2. For convenience, let the perpendiculars from *A* and *B* to the surface have the same length *d*. Call the distance between the two perpendiculars *h*. **(a)** What is the time t_1 required for the light to traverse the path *AP*? **(b)** What is the time t_2 required for the light to traverse the path *PB*? **(c)** Show that the time $t_1 + t_2$ has a minimal value if $\sin \theta_1 = n \sin \theta_2$; this is Snell's law. **(d)** Refer to Problem 35.36. Suppose you did not know whether the speed of light in the medium was less than *c*, consistent with the wave theory of light, or greater than *c*, consistent with Descartes's particle theory of light. Could you use Fermat's principle to choose between the two possibilities? Why or why not?

35.60 *(6) Light cone.* A light source that emits light uniformly in all directions is located some distance beneath the surface of a clear pond with a black bottom. A fraction *F* of the total light emitted passes through the surface; the rest either is absorbed directly by the bottom or suffers total internal reflection at the surface and is then absorbed by the bottom. **(a)** Neglecting the small reflection of light at the surface, at angles where total reflection does not take place, show that

$$F = \frac{1}{2}\left(1 - \sqrt{1 - \frac{1}{n^2}}\right).$$

(b) Calculate *F* for water. (Hint: A complete solid angle comprises 4π steradians.)

35.61 *(7) Rainbow.* Everyone "knows" that rainbows are produced by the refraction of sunlight by rain droplets. But the process is a little complicated. The figure below shows a ray of monochromatic light entering a small spherical raindrop

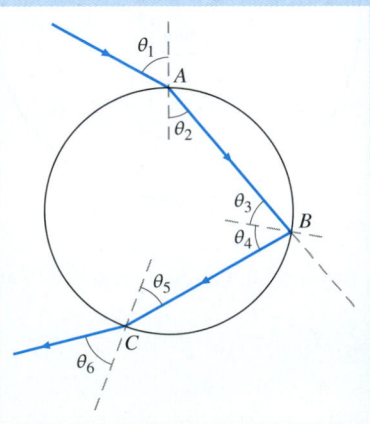

at A. The ray strikes the surface again at B, where most of the light passes out of the drop. However, we are interested in the portion of the light that is reflected internally, as happens for the upper four light beams in Figure 35.12. The reflected ray in the figure below is refracted at C, where it passes out of the drop.

(a) Show that, for such a light path, $\theta_5 = \theta_4 = \theta_3 = \theta_2$ and therefore that $\theta_6 = \theta_1$. (b) Find the total angle of deviation Δ of the light ray in terms of θ_1 and θ_2. (c) Because the sun lies in a fixed direction, all rays entering the raindrop are parallel to the incident ray shown in the figure. Consequently, the angles of incidence of such rays will differ and so will the values of Δ. In general, then, the raindrop scatters light over a wide range of directions. However, there is a particular value $\theta_1 \equiv \theta^*$, for which Δ is a minimum. Thus light rays incident on the raindrop over a range of angles in the vicinity of θ^* are deviated in nearly the same direction. An appreciable amount of light therefore leaves the raindrop in this direction, forming a bright rainbow. Set $d\Delta/d\theta_1 = 0$, and show that θ^* satisfies the relation

$$\cos^2 \theta^* = \tfrac{1}{3}(n^2 - 1).$$

(d) Find the values of θ^* and the corresponding values of Δ for red light ($n = 1.331$) and for violet light ($n = 1.342$). What is the angular width ρ of the rainbow? (To get this result to agree with observation, you must add $0.5°$ to ρ. This is because the sun is not a point source but has an angular width of $0.5°$.) (e) What is the angle between the lines from the observer to the sun and to the red part of the rainbow? (f) Is the violet part of the rainbow above or below the red part?

35.62 (7) Cauchy's formula. The ability of a prism to disperse light into its spectral components is determined by the rate at which the index of refraction of the glass of which the prism is made varies with wavelength. The dispersion of the glass (or other material) is defined to be $dn/d\lambda$. A dispersion curve whose form is like those of Figure 35.15 can be closely approximated by Cauchy's formula,

$$n = A + \frac{B}{\lambda^2}. \qquad (1)$$

(a) Using this equation, derive an expression for the dispersion $dn/d\lambda$, and show that the dispersion varies quite rapidly with wavelength over the visible range. (b) Use Equation 1 to evaluate the constants A and B for the light flint glass of Figure 35.15, and calculate the dispersion for red light ($\lambda = 700$ nm) and for violet light ($\lambda = 400$ nm).

35.63 (8) Nicol prism. Calcite crystals are often used to separate incident light into two rays called the "e" ray and the "o" ray, each of which is polarized. However, the presence of both rays is sometimes a nuisance. The Scottish physicist William Nicol (1768–1851) invented a convenient device, shown below, which prevents the transmission of one of the rays. A crystal of calcite is cut as shown and then cemented together again, using the transparent natural gum Canada balsam. When the incident light travels in the direction shown, parallel to the lower face, the two indices of refraction of the crystal for the two rays of polarized light are $n_o = 1.6583$ and $n_e = 1.4864$. (The index of refraction of Canada balsam, $n_b = 1.526$, lies between these two values; this is important in what follows.) The ends of the crystal are ground down from their natural base angle of $71.5°$ to $68°$, and the length of the crystal is such that the angle between the end face and the cemented surface is $90°$, as shown. Show that the "e" ray is transmitted through the crystal, and the "o" ray is totally reflected at the first Canada balsam surface because the angle of incidence of the "o" ray on that surface is about $8.5°$ greater than the critical angle. (Hint: See Problem 35.29.)

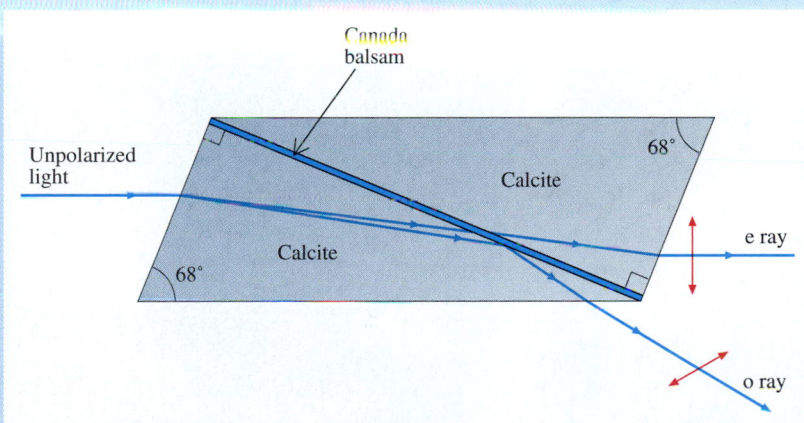

Physical Optics II

Interference and Diffraction

Left: Not an abstract painting, but the result of diffraction of white light by a soap film of varying thickness.

> *Roam on! The light we sought is shining still,*
> *Dost thou ask proof? . . .*
> *Our scholar travels yet. . . .*
>
> —Matthew Arnold, "Thyrsis"

Introduction

Because light behaves like a wave, it conforms to the *principle of superposition*; see Section 21.6. Out of this apparently simple statement arises a great wealth of optical effects to which we give the general name *diffraction effects*. In this chapter, we explore a few of the simpler and more important of these effects.

The most elementary diffraction effects are called *interference effects*. In Sections 36.2 and 36.3, we consider one-dimensional interference, in which there is superposition of light waves traveling in the same direction. One-dimensional interference is exploited in several practical devices, of which the most important is the Michelson interferometer. Sections 36.4 through 36.6 concern *two-slit* and *multiple-slit* interference—effects that arise from superposition of light waves traveling in different directions.

In Section 36.7, we study *single-slit diffraction*, an effect distinct from but closely related to multiple-slit interference. Diffraction effects set a limit on the possibility of *resolving* two objects—seeing them as separate when they are located close together. Resolution is the subject of Section 36.8.

Section 36.9 deals with a special multiple-slit interference device, the *diffraction grating*, which may well be called the workhorse of spectroscopy.

Interference in One Dimension

Many light waves can pass through the same space simultaneously. We can describe the result of superposing a number of light waves by means of the resultant electric field vector $\mathscr{E}(\mathbf{r}, t)$ at any point $\mathbf{r}$ and time t. This electric field is the sum of the electric fields of the individual waves. We begin by considering the simplest type of superposition. Two sinusoidal light waves have equal amplitude and frequency—that is, $\mathscr{E}_1 = \mathscr{E}_2 \equiv \mathscr{E}_0$, and $\nu_1 = \nu_2 \equiv \nu$. Both waves are polarized with their electric fields in the y direction, and both waves travel along the x direction. The two waves differ only in *relative phase* ϕ. Two such *component waves* can be represented by the wave functions

$$\mathscr{E}_0 \cos(kx - \omega t) \quad \text{and} \quad \mathscr{E}_0 \cos(kx - \omega t + \phi).$$

In these expressions, $k = 2\pi/\lambda$ is the *wave number*, and $\omega = 2\pi\nu$ is the *angular frequency*. (Compare with Equation 21.21 and with the discussion after Equation 33.21.) As you saw for mechanical waves in Section 21.6, the *superposed wave* $\mathscr{E}(x, t)$ is described by the sum

$$\mathscr{E}(x, t) = \mathscr{E}_0 [\cos(kx - \omega t) + \cos(kx - \omega t + \phi)]. \tag{36.1}$$

By using the trigonometric identity

$$\cos \alpha + \cos \beta = 2 \cos \frac{\alpha + \beta}{2} \cos \frac{\alpha - \beta}{2},$$

we can express the superposed wave function in the form

$$\mathscr{E}(x, t) = 2\mathscr{E}_0 \cos\frac{\phi}{2} \cos\left(kx - \omega t + \frac{\phi}{2}\right).$$

The sum of the two component wave functions is thus itself a sinusoid with the same wave number k and angular frequency ω but with amplitude

$$\mathscr{E}' = 2\mathscr{E}_0 \cos\frac{\phi}{2}. \qquad (36.2)$$

According to this equation, the amplitude $\mathscr{E}'$ of the superposed wave depends on the relative phase ϕ of the component waves. The maximum value, $\mathscr{E}' = 2\mathscr{E}_0$, is attained when $\phi = 0, 2\pi, 4\pi, \ldots$—that is, when the two component waves are in phase (the phases differ by $2n\pi$ rad), as shown in Figure 36.1a. In-phase superposition is often called **constructive interference**. The minimum value, $\mathscr{E}' = 0$, is attained when $\phi = \pi, 3\pi, 5\pi, \ldots$—that is, when the phases of the two component waves differ by $(2n + 1)\pi$ rad. We say that such waves are out of phase, as shown in Figure 36.1b.

(a)   (b)

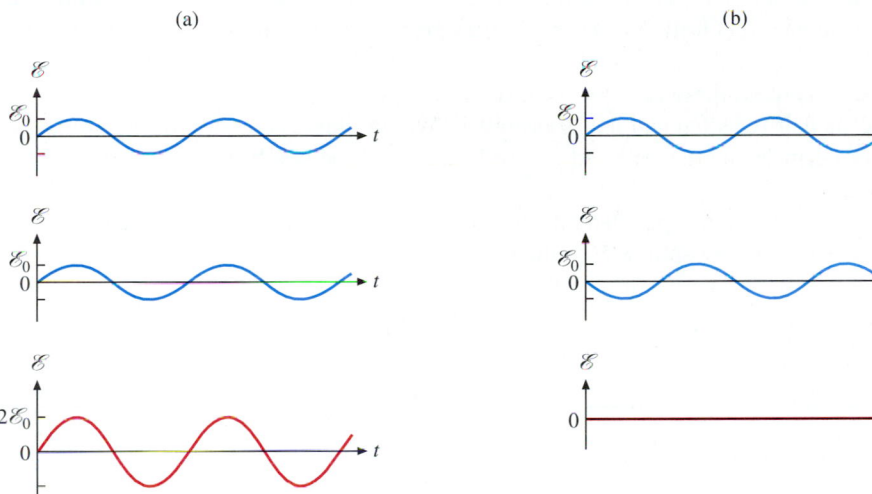

FIGURE 36.1 (a) Constructive interference. Two component one-dimensional waves of equal wavelength and amplitude are superposed in phase; $\phi = 0, 2\pi, 4\pi, \ldots$. The amplitude of the superposed wave is twice that of either component wave. (b) Destructive interference. The component waves are superposed out of phase; $\phi = \pi, 3\pi, 5\pi, \ldots$. The amplitude of the superposed wave is zero.

Out-of-phase superposition is often called **destructive interference**. For phase constants having values other than 0 or π, the amplitude $\mathscr{E}'$ has a value somewhere between 0 and $2\mathscr{E}_0$.

As you saw in Section 34.6, the *intensity* I of electromagnetic radiation is proportional to the square of the wave amplitude, $\mathscr{E}'^2$. The precise relation between I and $\mathscr{E}'^2$ is given by Equation 34.38a, but in this chapter we will be concerned with only the proportionality

$$I \propto \mathscr{E}'^2. \qquad (36.3)$$

EXAMPLE 36.1

Consider the two component light waves just discussed. Suppose the intensity of either wave separately is I_0. When the two waves interfere, the intensity of the superposed wave is I. What are the maximum and minimum possible values of the ratio I/I_0?

SOLUTION: When the superposition is out of phase, the amplitude of the resultant wave is zero, and so the ratio must be

$$\frac{I}{I_0} = \frac{\mathscr{E}'^2}{\mathscr{E}_0^2} = 0 \quad \text{for destructive interference.}$$

Even if we generalize to the case in which the component waves are not equal in intensity, the superposed intensity will be a minimum for out-of-phase superposition. We call such a minimum in intensity an **interference minimum**. When the superposition is in phase, the amplitude is $\mathscr{E}' = 2\mathscr{E}_0$. According to Equation 36.3, the intensity ratio is then

$$\frac{I}{I_0} = \frac{(2\mathscr{E}_0)^2}{\mathscr{E}_0^2} = 4 \quad \text{for constructive interference.}$$

The superposed intensity is not simply the sum of the intensities of the component waves. Rather, the intensity depends on the *square of the sum of the amplitudes.*

Whether or not the two component waves in phase have equal intensities, the superposed intensity will have its maximum value for in-phase superposition. We call such a maximum in intensity an **interference maximum**.

The Michelson Interferometer

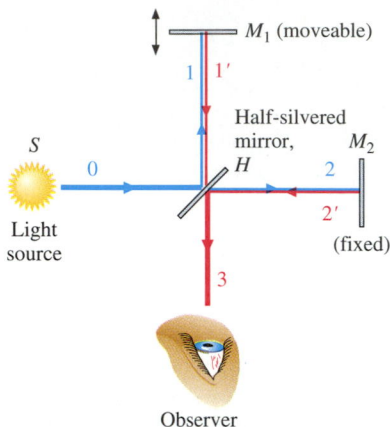

M_1 (moveable)

Half-silvered mirror, H M_2

S

Light source

(fixed)

Observer

FIGURE 36.2 The Michelson interferometer.

The very important device shown schematically in Figure 36.2 is called a **Michelson interferometer**, after its inventor. The interferometer is in a sense a variant on the Foucault apparatus (Problem 35.32), and Michelson gave full credit to this source of inspiration. But the modifications to the Foucault design are so profound as to make the interferometer a completely different instrument.

The Michelson interferometer operates as follows. The light source S is most commonly monochromatic light (Section 35.7), often a small gas laser. A beam of light from S strikes the **beam splitter** H. This beam splitter is a half-silvered mirror—a plate of glass coated with a layer of metal so thin that about half the incident light is reflected and half transmitted.* The reflected beam, 1, is directed toward mirror M_1, which can be moved back and forth by means of a very precise positioning screw. The transmitted beam, 2, is directed toward the fixed mirror, M_2. Both beams are reflected back toward the beam splitter; these reflected beams are labeled 1' and 2'. At H, each of beams 1' and 2' is half reflected and half transmitted. We are interested only in the transmitted part of beam 1' and the reflected part of beam 2'. These two beams superpose to form beam 3.

Consider the optical path lengths (Section 35.5) of the two beams 1-1' and 2-2' from the point where they split at H to the point where they recombine, again at H. If these two optical path lengths are exactly equal, the two beams interfere constructively on recombining, forming a beam 3 of high intensity. But if the path lengths differ by a

---

A. A. Michelson, Master of Measurement

The physicist Albert Abraham Michelson (1852–1931) was born in Prussian Poland but brought as a toddler to the silver-mining boom town Virginia City, Nevada. A bright, musically talented boy, Michelson was a favorite in a place where children were a rarity. He was sent to San Francisco to attend the nearest high school (which still exists as Lowell High School); it was there that he discovered his scientific interests. In 1869, he entered the U.S. Naval Academy, thus beginning a close, lifelong relation with the navy. After spending some time at sea, Michelson returned to the Academy as an instructor in physical sciences. His first research, an 1878 measurement of the speed of light, was financed by a $2000 grant from his well-to-do father-in-law. Wishing to learn more about physics than was possible at the time in the United States, Michelson studied in Germany and France during 1880–82. It was in Helmholtz's Berlin laboratory that Michelson conceived the idea of the ''interferential refractometer,'' later renamed the interferometer. A gift from Alexander Graham Bell made construction of the first model possible. In 1882, Michelson accepted a position at the Case School of Applied Science (now part of Case Western Re-serve University) and from 1881 to 1887 made a wide variety of refined measurements of quantities important in many scientific and technical fields, many of them by using the interferometer. In the late 1880s and early 1890s, the founding of German-style research universities became popular in the United States, and Michelson's growing reputation made him a prime target of faculty raids, first by Clark University in 1889 and then by the new University of Chicago in 1893. Michelson remained at Chicago until the First World War, when he returned to active duty in the navy to conduct war research. After the war, Michelson moved to the California Institute of Technology at the invitation of his former protegé, Robert Millikan, who had become its president. At Caltech, he continued in research almost until his death in 1931. An accomplished billiards player, painter, and violinist, Michelson in 1928 told the budding young writer Norman MacLean, ''Billiards is a good game. But billiards is not as good a game as painting. But painting is not as good a game as music. But then music is not as good a game as physics.'' In 1907, Michelson became the first American to win a Nobel Prize in science.

---

*Fizeau used a plain glass plate for essentially the same purpose in his apparatus for determining the speed of light (Figure 35.1). The half-silvered mirror, based on technology developed during the three intervening decades, is far more efficient.

half wavelength—a very small distance, typically 0.5 μm for visible light—beams 1' and 2' interfere destructively, and beam 3 vanishes (or, in practice, becomes very faint). When M_1 is moved slowly through a small distance, the intensity of the light seen by the observer varies in oscillatory fashion, in accordance with Equations 36.2 and 36.3.

If (as is usually the case) M_1 and M_2 are not perfectly perpendicular, the difference in optical path length may vary by several wavelengths for different parts of the two interfering light beams. As a result, the observer sees a system of bright and dark stripes called **interference fringes** (Figure 36.3). As M_1 is moved, the entire pattern moves across the field of view. A displacement of the pattern such that one bright fringe moves to the position originally occupied by its neighbor is called a **fringe shift**. One fringe shift corresponds to a change in optical path length of one wavelength λ. Because the light beam doubles back on itself, one fringe shift also corresponds to displacement of mirror M_1 through a distance $\lambda/2$.

When the interferometer is properly adjusted, the fringe shifts are very easy to see, and distances of the order of the wavelength of light are thus readily measured. To be specific, suppose that an optical path length change ΔL in the variable interferometer arm is accompanied by m fringe shifts. We then have the relation

$$\Delta L = m\lambda. \tag{36.4}$$

The corresponding displacement of the mirror is $m\lambda/2$.

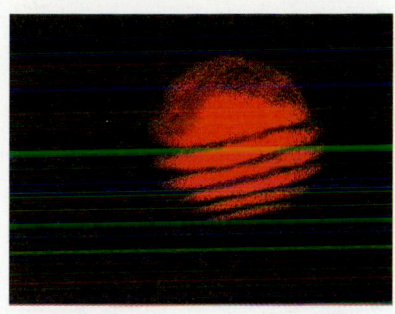

FIGURE 36.3 Typical interference fringe pattern seen in a Michelson interferometer.

EXAMPLE 36.2

You can adapt the Michelson interferometer as shown in Figure 36.4 to measure the index of refraction of a gas, which differs only slightly from 1, the vacuum value. In one arm is a gas-tight cell with glass end windows. Tubing connects the cell to a vacuum pump and to a source of the gas whose index of refraction you wish to measure.

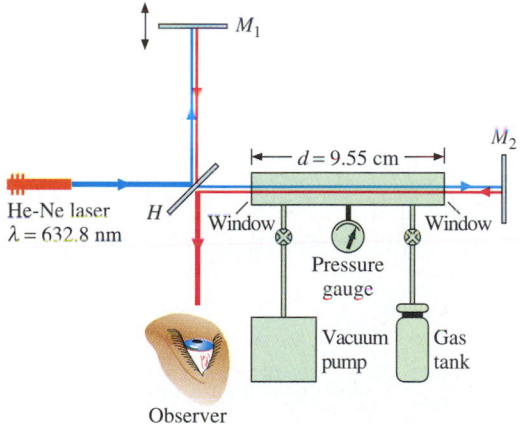

FIGURE 36.4 A Michelson interferometer adapted for measuring the index of refraction of a gas.

You begin by opening the valve between the cell and the vacuum pump, evacuating the cell, and then closing the valve. While observing the light emerging from the interferometer, you open the valve to the gas tank very slightly so that gas can flow slowly into the cell. As the gas fills the cell, you see and count fringe shifts. When the pressure gauge indicates that the pressure in the cell is equal to atmospheric pressure (or whatever final pressure you wish), you close the valve. The fringe count is the crucial part of the experiment.

You use a helium-neon laser light source because this laser is a strong source of light whose wavelength is accurately monochromatic, with λ = 632.8 nm. The inside length of the cell is d = 9.55 cm. With carbon dioxide (CO_2) gas, you count 129 fringe shifts from vacuum to atmospheric pressure. The temperature is 15°C. What is the index of refraction of CO_2 at 15°C and 1 atm pressure?

SOLUTION: The physical length of the interferometer arms does not change during the measurement. You observe fringe shifts because the *optical* path length L in the arm containing the cell changes as gas is introduced. The presence of the gaseous medium affects the speed of light and thus the index of refraction.

According to Equation 35.8, $L = ns$, the optical path length L depends on the index of refraction n of the medium and the distance s through which the light propagates. The optical path length through the gas in the cell is thus $L = 2nd$. (Remember that the light passes through the cell twice.) With the cell evacuated, the optical path length through the cell is $L_0 = 2d$. So you have

$$\Delta L = L - L_0 = 2(n - 1)d. \tag{36.5}$$

Now, you use Equation 36.4, $\Delta L = m\lambda$, to express ΔL in terms of the number of fringe shifts m. You have $m\lambda = 2(n - 1)d$, or

$$n = 1 + \frac{m\lambda}{2d}. \tag{36.6}$$

You can now insert the numerical values to obtain

$$n = 1 + \frac{129 \times 632.8 \times 10^{-9} \text{ m}}{2 \times 9.55 \times 10^{-2} \text{ m}} = 1.000\,427.$$

You can see why, under ordinary conditions of pressure and temperature, the index of refraction of gases (including air) can often be set equal to the vacuum value $n = 1$.

Thin-Film Interference

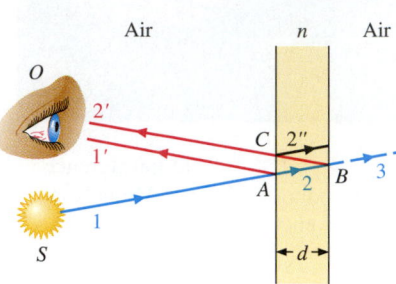

Air   n   Air

FIGURE 36.5 Thin-film interference. The discussion in the text concerns normal incidence; the angle of incidence and the angle of reflection are drawn as shown for purposes of clarity only.

You have doubtless observed the rainbow of colors displayed by an illuminated soap bubble or by a puddle of oily water. In these and many similar situations, the colors arise from **thin-film interference**.

Figure 36.5 shows a film of transparent material (such as a soap film) of thickness d and index of refraction n. The film is illuminated by light beam 1, which originates at a point source S. Let us assume for the moment that the light is monochromatic, with wavelength λ. For clarity, the observer is shown displaced from the source. But let us assume normal incidence for simplicity. That is, let us assume that the angles of incidence and reflection of the light beam are essentially zero.

A small part of beam 1 is reflected from the front surface at A as beam 1′. The remainder of the light, labeled beam 2, passes into the film. A small part of the light reaching the rear surface at B is reflected as beam 2′, while the rest passes into the air behind the film. Beam 2′ passes again through the front surface at C, where it superposes with beam 1′. (The displacement between beams 1′ and 2′ shown in Figure 36.5 vanishes for normal incidence.) We neglect the small amount of light (beam 2″) reflected a second time at C. Within this approximation, it can be shown (Problem 36.32) that the amplitudes of beams 1′ and 2′ are equal if the front and rear surfaces of the film are in contact with the same medium, as is the case in Figure 36.5.

Beams 1′ and 2′ interfere because the optical path lengths SAO and SBO are different. There are two reasons for this optical path difference. One reason is clear from the figure; optical path SBO is longer than SAO by the optical path length $2nd$.

The other reason for the optical path difference is that there is a *phase inversion* (a 180° phase change) on reflection at A but no phase change on reflection at B. The situation is analogous to that discussed for standing waves on a string in Section 21.7, and illustrated in Figure 36.6, which is the same as Figure 21.18.

We can account for the phase change on reflection from the soap film at A by adding $\lambda/2$ to the optical path length SAO. Because of this optical lengthening of SAO, the total difference between the optical path lengths SBO and SAO is

$$\Delta L = 2nd - \frac{\lambda}{2}.$$

The two beams interfere constructively if the optical path difference ΔL is equal to an integral number of wavelengths. We can express this condition mathematically in the form

$$m\lambda = \Delta L = 2nd - \frac{\lambda}{2}; \quad m = 0, 1, 2, \dots.$$

We collect terms in λ to obtain the condition

$$2nd = (m + \tfrac{1}{2})\lambda, \quad m = 0, 1, 2, \dots, \quad \text{for constructive interference.} \quad \textbf{(36.7a)}$$

The two beams interfere destructively if ΔL is equal to a half-integral number of wavelengths. We can express this condition mathematically in the form

$$\left(m - \frac{1}{2}\right)\lambda = \Delta L = 2nd - \frac{\lambda}{2}.$$

The terms $\lambda/2$ cancel, yielding the condition

$$2nd = m\lambda; \quad m = 0, 1, 2, \dots, \quad \text{for destructive interference.} \quad \textbf{(36.7b)}$$

Note that the case $m = 0$ corresponds to a film whose thickness is negligible compared with the wavelength of light (Query 36.4).

(a) (b)

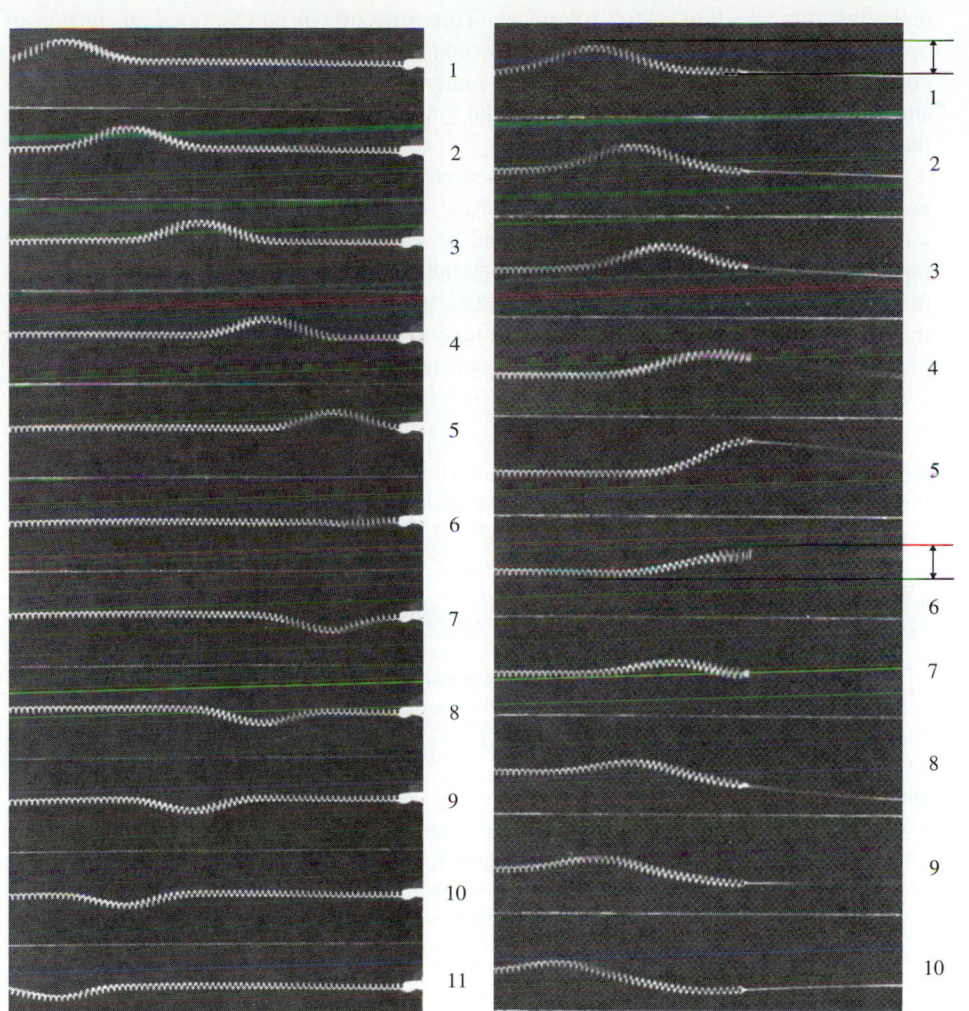

FIGURE 36.8 A wave traveling through a coil spring is reflected (a) at the
boundary between the spring and a denser medium and (b) at the boundary between the spring and a rarer medium. In part a, there is a phase
inversion on reflection; in part b, there is no phase change on reflection.

(From Haber-Schaim: *PSSC Physics*. Copyright 1991 by Kendall/Hunt Publishing Company. Used with permission.)

The conditions for both constructive and destructive interference depend on the
wavelength of the incident light. When white light, which is a mixture of many wavelengths, is incident on a film of thickness d, constructive interference occurs preferentially for those wavelengths that satisfy Equation 36.7a, and the colors corresponding
to those wavelengths are observed in the reflected beam. Soap bubbles usually have
nonuniform thickness; this accounts for the ''rainbow'' of colors seen in the light reflected from them.

Antireflection Coating

When light passes through an interface between two transparent media, a small amount
of light is reflected. For air-glass or glass-air interfaces, the resulting loss in transmitted
intensity is about 4%. In simple optical systems—an eyeglass lens, for instance—this
loss (8% for the two surfaces of the lens) is not serious. However, the loss becomes

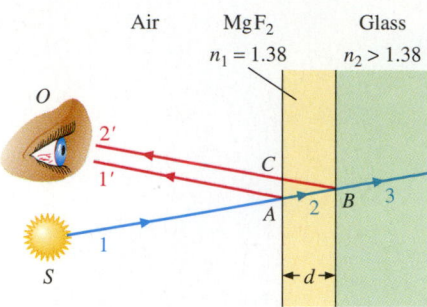

Air   MgF₂   Glass
$n_1 = 1.38$ $n_2 > 1.38$

FIGURE 36.7 A single-layer antireflection coating of thickness d. The coating material shown here is magnesium fluoride (MgF₂), whose index of refraction is $n_1 = 1.38$. The glass substrate is optically denser; typically, $n_2 = 1.52$.

unacceptable in optical systems (such as camera lenses) that contain many individual lens elements and thus many interfaces. This difficulty is ameliorated by means of **antireflection coatings**, of which we now consider the simplest type, the single-layer coating. A thin coating of a transparent material is deposited on a glass surface, as shown in Figure 36.7. The coating material must have an index of refraction less than that of glass.

The thickness of the coating is adjusted so as to produce destructive interference between beams 1′ and 2′. This requires that the difference between the optical path lengths of the two beams be equal to a half-integral number of wavelengths. Because both reflections occur at an interface with an optically denser medium—in contrast with the soap-film case—both beams experience a phase inversion on reflection. There is thus no *relative* phase inversion due to reflection. We therefore need consider only the phase difference due to the extra optical path length ABC. We have

$$2nd = (m + \tfrac{1}{2})\lambda, \quad m = 0, 1, 2, 3, \ldots, \quad \text{for destructive interference.} \quad (36.8)$$

Note that the condition for destructive interference in the absence of relative phase inversion is the same as the condition for constructive interference with relative phase inversion (Equation 36.7a).

EXAMPLE 36.3

Because the condition expressed by Equation 36.8 depends on λ, it cannot be satisfied for white light. A commonly made compromise is to adjust the coating thickness to provide an interference minimum for yellow-green light ($\lambda = 550$ nm), which lies near the middle of the visible spectrum. The yellow-green is the most intense component of sunlight and is also the component to which the eye is most sensitive. If the coating material is magnesium fluoride ($n_1 = 1.38$), find the minimum thickness for an antireflection coating.

SOLUTION: From Equation 36.8, you have

$$d = \frac{(m + \tfrac{1}{2})\lambda}{2n}.$$

For the thinnest possible coating, set $m = 0$. You then have

$$d = \frac{\lambda}{4n} = \frac{550 \text{ nm}}{4 \times 1.38} = 100 \text{ nm}.$$

This value is typical of the thicknesses used in antireflection coatings. Because the coating does not reflect light in the middle of the visible spectrum, a coated surface appears purplish owing to the preferential reflection of red and violet light. You have probably noticed that camera lenses display this purplish tinge.

Interference in Two Dimensions: Two-Slit Interference

We now extend our study of interference to cases in which the light waves that superpose are not parallel. We begin with an analysis of the simplest case, called **two-slit interference**. The experimental arrangement, devised by Young, is shown in Figure 36.8a. Figure 36.8b is a photograph of the analogous setup for ripples in water.

Light in the form of plane wave fronts from a distant single source is blocked by the opaque screen A, except at the narrow slit S_A, whose long dimension is normal to the page. In accordance with Huygens's construction, the light emerging from the slit spreads out as a series of cylindrical wave fronts, depicted in cross section as half-circles in Figure 36.8a.

The light then strikes the second opaque screen B. Two narrow slits in this screen, S_1 and S_2, are located symmetrically with respect to S_A. Each of these two slits acts as a source of cylindrical wave fronts. Because wave fronts arrive simultaneously at S_1 and S_2 from the same source, S_1 and S_2 emit light in phase. We say that the two sources are **coherent**. Screen C is illuminated by the light from the two slits. When we analyze the superposition of light waves from S_1 and S_2 at C, we need be concerned only with the phase differences due to optical path differences in the space between the slits and the screen.

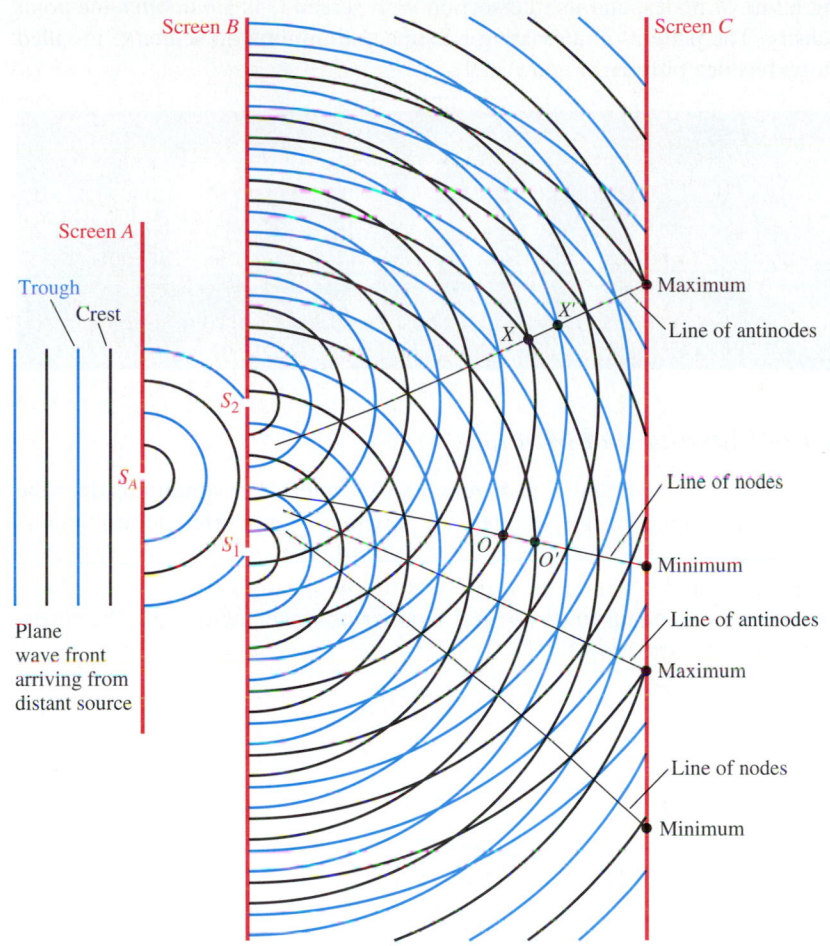

(a)

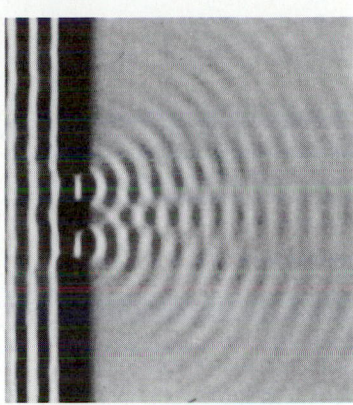

(b)

FIGURE 36.8 (*a*) Diagram of a two-slit interference experiment. All three slits are normal to the page, and the diagram is a cross-sectional view of the arrangement. Huygens's construction is used to show the propagation of waves from the slits, which all act as sources. (*b*) Ripple-tank photograph of two-slit interference for water waves.

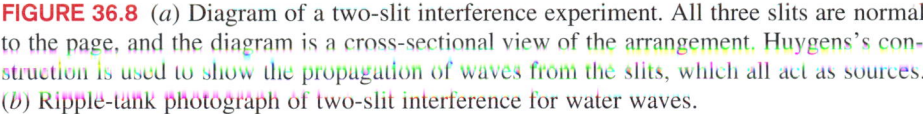

As you can see from Figure 36.8a, the two trains of cylindrical waves from slits S_1 and S_2 superpose. In the figure, the wave pattern is represented at a particular instant. The black circular arcs represent wave crests and the blue arcs represent wave troughs. Wherever two arcs of the same color cross (as they do at X and X'), the interference is constructive and the result is either a wave crest (for two black arcs) or a trough (for two blue arcs). Points such as X and X' are called **antinodal points**. Wherever two arcs of different color cross (as at O), the interference is destructive and the resultant amplitude is zero. Points such as O and O' are called **nodal points**. The interference process transforms the uniform amplitude of the cylindrical waves to the left of screen B into a pattern of localized regions. In regions like that near X, the amplitude is enhanced, at the expense of the amplitude in regions like that near O.

With the passage of time, the waves move away from the slits toward screen C. The large crest at X, for example, will move toward X'. Indeed, the path of the crest (and others preceding or following it) is the curve drawn through X, X', and neighboring similar points. Such a line is called a **line of antinodes**. The point where this line intersects screen C is a *maximum*—a point where the wave amplitude is large, and hence the intensity of the light is large.

In similar manner, the node at O moves toward O', along a **line of nodes**. Unlike points along the line of antinodes, where the wave oscillates between high crests and low troughs, points along the line of nodes are all nodal points. This is because intermediate wave fronts (not drawn in Figure 36.8a) always have equal and opposite heights when they superpose along the line of nodes. Thus the amplitude is always zero at all points along a line of nodes, and its intersection with screen C is a *minimum*—a point of zero intensity. The pattern of alternating maxima and minima on screen C is called a **two-slit interference pattern** (Figure 36.9).

FIGURE 36.9 Two-slit interference pattern.

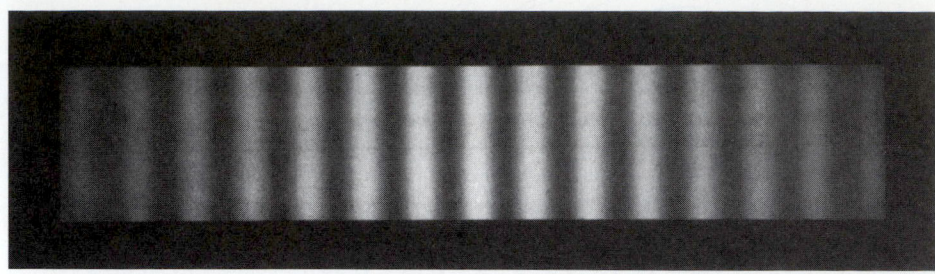

Locating the Maxima and Minima

Figure 36.10 shows screens B and C of Figure 36.8a. Under what conditions does the arbitrary point P on screen C lie at an interference maximum? an interference minimum?

Because S_1 and S_2 are coherent sources, any phase difference between the two wave trains that superpose at P is due to the difference in the optical path lengths S_1P and S_2P. If this difference is equal to an integral number of wavelengths, $m\lambda$, the superposition at P will be constructive.

FIGURE 36.10 Construction for locating two-slit interference maxima and minima on screen C of Figure 36.8a.

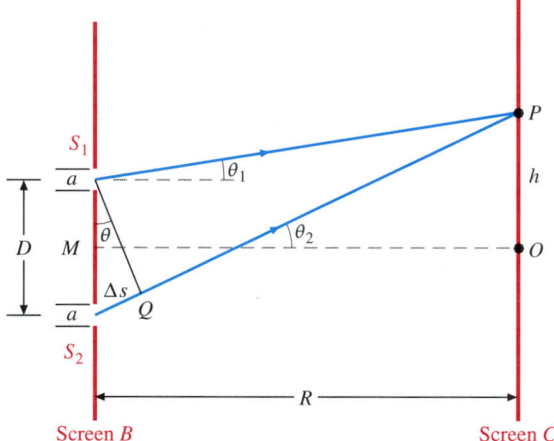

For simplicity, we limit our discussion to the commonly encountered case in which the distance R between the screens is much greater than the distance D between the slits. In addition, we assume that $D \gg \lambda$ and that the slits themselves are very narrow; that is, $D \gg a$.

Because $R \gg D$, the rays S_1P and S_2P are nearly parallel to one another (though for reasons of clarity the angle between the rays has been exaggerated). If we drop a perpendicular S_1Q from slit S_1 to the ray S_2P, the optical path difference between the two rays is the distance $\Delta s \equiv S_2Q$. Thus the condition for constructive interference at P is

$$\Delta s = m\lambda, \quad m = 0, \pm 1, \pm 2, \ldots.$$

We now use some simple trigonometry to evaluate Δs.

Since S_1P and S_2P are nearly parallel, it follows that the angles θ_1 and θ_2 in Figure 36.10 are essentially equal; let us give them the common name θ. Moreover, the angle S_2S_1Q also is θ because $S_1S_2 \perp MO$ and (approximately) $S_1Q \perp S_2P$. The diagram shows that

$$\Delta s = D \sin \theta. \tag{36.9}$$

The condition for an interference *maximum* at P is thus

$$m\lambda = D \sin \theta, \quad m = 0, \pm 1, \pm 2, \ldots, \quad \text{for two-slit maxima.} \tag{36.10a}$$

An interference minimum will be located at P if the difference in optical path lengths for the two rays is a half-integral number of wavelengths; $\Delta s = (m \pm \frac{1}{2})\lambda$. Again using the fact that $\Delta s = D \sin \theta$, we have the condition for an interference *minimum*:

$$(m \pm \tfrac{1}{2})\lambda = D \sin \theta, \quad m = 0, \pm 1, \pm 2, \ldots, \quad \text{for two-slit minima.} \quad \textbf{(36.10b)}$$

The integer m in Equations 36.10a and 36.10b is called the **order of interference**. Each value of m corresponds to a maximum and an adjacent minimum.

Interference as a Magnification Process

So far, we have specified the conditions for interference maxima and minima in terms of the angle θ. But we can readily express the conditions in terms of the distance between P and the center point O of screen C. Let us call this distance h. From Figure 36.10, you can see that $h/R = \tan \theta$. For the case $R \gg D$, we are concerned only with small angles θ. We can therefore use the approximation $\tan \theta \simeq \sin \theta$. We thus have

$$\frac{h}{R} \simeq \sin \theta. \qquad \textbf{(36.11)}$$

What is the spacing between adjacent maxima in the interference pattern? According to Equation 36.10a, the mth and $(m + 1)$th maxima must satisfy the conditions

$$m\lambda = D \sin \theta_m \quad \text{and} \quad (m + 1)\lambda = D \sin \theta_{m+1}.$$

When we use Equation 36.11, this relation becomes

$$m\lambda = \frac{D}{R} h_m \quad \text{and} \quad (m + 1)\lambda = \frac{D}{R} h_{m+1}.$$

We subtract the first of these equations from the second to obtain

$$\lambda = \frac{D}{R} (h_m - h_{m+1}).$$

The term in parentheses is the desired distance Δh between adjacent fringes. So we have

$$\Delta h = \frac{R}{D} \lambda \qquad \textbf{(36.12)}$$

The fringe spacing Δh is thus proportional to the wavelength λ. But for $R \gg D$, the proportionality constant R/D is a large number. In a sense, then, the interference pattern on screen C is a magnified picture of the sinusoidal wave incident on the pair of slits.

EXAMPLE 36.4

You use the two-slit interference apparatus shown in Figure 36.11 to measure the wavelength of the bright yellow light emitted by a sodium vapor discharge lamp. Working in the dark, you place a sheet of unexposed photographic film on surface C. Then you turn on the sodium lamp long enough to expose the film. When you develop the film, you find a pattern resembling that of Figure 36.9. Measuring the pattern, you find that 16 fringes occupy 5.66 cm. **(a)** Find the wavelength λ of the sodium light. **(b)** At what (small) angles are the maxima and minima located?

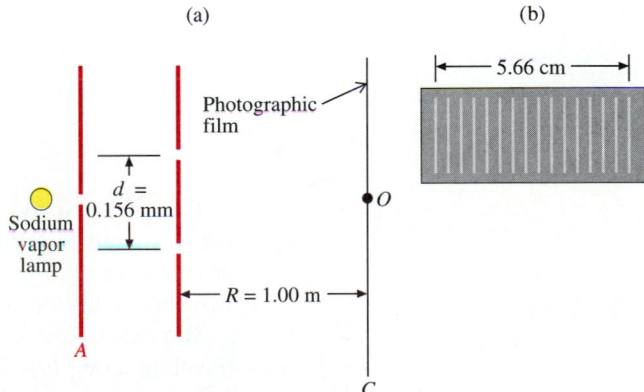

FIGURE 36.11

SOLUTION:

(a) Find λ.

According to your measurement, 5.66 cm is 15 times the distance between adjacent fringes: $15 \Delta h = 5.66$ cm, or $\Delta h = \frac{1}{15} \times 5.66 \times 10^{-2}$ m $= 3.77 \times 10^{-3}$ m. Using Equation 36.12, you have

$$\lambda = \frac{D \, \Delta h}{R} = \frac{1.56 \times 10^{-4} \text{ m} \times 3.77 \times 10^{-3} \text{ m}}{1.00 \text{ m}} = 589 \text{ nm}.$$

Thus, you can use the magnifying effect to measure the very small wavelength λ. In this case, the magnification factor is the ratio

$$\frac{R}{D} = \frac{1.00 \text{ m}}{1.56 \times 10^{-4} \text{ m}} = 6410.$$

(b) Locate the maxima and minima.

You obtain the lowest-order maximum by setting $m = 0$ in Equation 36.10a; you have $0 = D \sin \theta_0$ and thus $\theta_0 = 0$. That is, a *central maximum* lies at the midpoint O. This will be true regardless of the values of D, R, and λ. For the first-order maximum, $m = 1$, you have from Equation 36.10a

$$\sin \theta_1 = \frac{\lambda}{D} = \frac{589 \times 10^{-9} \text{ m}}{1.56 \times 10^{-4} \text{ m}} = 3.78 \times 10^{-3}.$$

For such small angles, the approximation $\sin \theta \simeq \theta$ is quite accurate, and you have $\theta_1 = 3.78 \times 10^{-3}$ rad, or about 0.216°. There will also be a maximum at $\theta_{-1} = -\theta_1$, on the other side

of the midline, corresponding to $m = -1$. Because the fringes are equally spaced, you have maxima at all small angles $\pm \theta_m = m \times 3.78 \times 10^{-3}$ rad.

You obtain the lowest-order minimum by setting $m = 0$ in Equation 36.10b. Denote the angles at which minima fall by the symbol θ'_m. Now you have

$$\theta'_0 \simeq \sin \theta'_0 = \frac{\lambda}{2D}.$$

In part **a**, you found that $\theta_1 \simeq \lambda/D$. You can thus write

$$\theta'_0 = \frac{\theta_1}{2} = 1.89 \times 10^{-3} \text{ rad},$$

or about 0.108°. The zero-order minimum lies halfway between the zero- and first-order maxima—the maxima corresponding to $m = 0$ and $m = 1$. As in the case of the maxima, the minima are symmetrically disposed about the central maximum. Higher-order minima are located at angles $\pm \theta'_m = \pm(1.89 \times 10^{-3} + m \times 3.78 \times 10^{-3})$ rad—that is, at $\pm 5.67 \times 10^{-3}$ rad, $\pm 9.45 \times 10^{-3}$ rad, $\pm 13.23 \times 10^{-3}$ rad, and so on.

Multiple-Slit Interference

The arrangement of Figure 36.12 is just like that of Figure 36.10, except that screen B now has many equally spaced slits instead of only two. In Figure 36.12, rays 1 through 5 are drawn from the first five slits to a point P on screen C. (A similar ray, j, could be drawn from the jth slit to P, as well.) Again we ask: What are the conditions for which P lies at an interference maximum?

FIGURE 36.12 Construction for locating multiple-slit interference maxima.

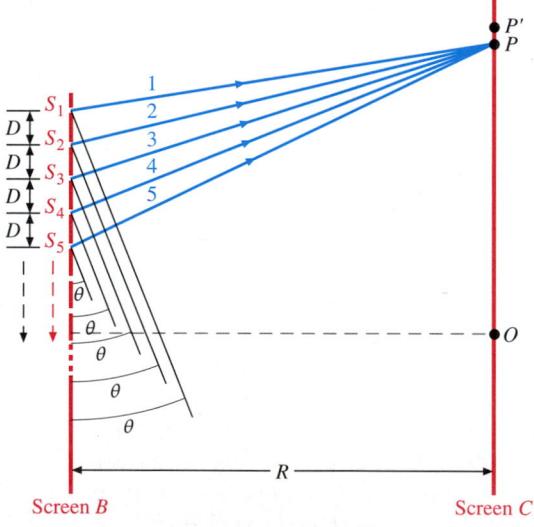

Consider first rays 1 and 2. The waves whose paths these rays describe will superpose constructively at P if the optical path difference between them is $m\lambda$. (This is the same criterion we used in the two-slit case.) An argument identical with that just used for two slits yields the condition $m\lambda = D \sin \theta$, which is simply Equation 36.10a.

But exactly the same condition applies for constructive superposition of the waves traveling along rays 2 and 3 or any other pair of adjacent rays. Indeed, if ray 2 is m wavelengths longer than ray 1, then ray 3 will be $2m$ wavelengths longer than ray 1. In general, ray j will be jm wavelengths longer than ray 1. Because jm is an integer, all the waves traveling along these rays will interfere constructively, and the condition for multiple-slit maxima is the same as that for two-slit maxima:

$$m\lambda = D \sin \theta, \quad m = 0, \pm 1, \pm 2, \ldots, \quad \text{for interference maxima.} \quad \textbf{(36.13)}$$

The maxima will be brighter than the corresponding two-slit maxima because more light can pass through many slits of a given width than through two slits of the same width. However, a much more important distinction between the multiple-slit case and the two-slit case is the way in which the intensity of the interference pattern varies between maxima. Let us consider this matter qualitatively. (A quantitative discussion is the main subject of Section 36.6.)

Assume there are 100 slits in screen B. If P lies at the first-order interference maximum, the optical path difference $S_2P - S_1P$ is equal to λ. At some point P', a little distance from P, the optical path difference $S_2P' - S_1P'$ is equal to $1.01\lambda = \lambda + 0.01\lambda$. For purposes of superposition, only the fractional part of the path difference, 0.01λ, is significant. This 1% path difference is so small that, if you were to close off all the other slits, you would find the light intensity at P' nearly the same as that at P.

The optical path difference $S_3P - S_1P$ is 2λ. But the optical path difference at P' is $S_3P' - S_1P' = 2(1.01)\lambda$. That is, the net path difference (subtracting out 2λ) is 2%, still not very large.

But consider how the waves from slits 1 and 51 superpose at P'. The optical path difference is $S_{51}P' - S_1P' = 50(1.01)\lambda$. The net path difference is now 0.5λ—a half-wavelength. Thus the two waves arrive at P' out of phase, and the superposed amplitude of the waves from slits 1 and 51 is zero. The same is true of the pair of waves arriving at P' from slits 2 and 52, 3 and 53, $\ldots$, 49 and 99, 50 and 100. Consequently, an intensity minimum lies at P', whose distance from the first-order maximum at P is only about 1% of the distance to the second-order maximum. Compare this with the two-slit case, in which the minima lie halfway between the maxima.

We say that the maxima in the multiple-slit pattern are much *sharper* than the maxima in the two-slit pattern. The sharpening effect of increasing the number of slits is made clear in the photographs of Figure 36.13, which show interference patterns for

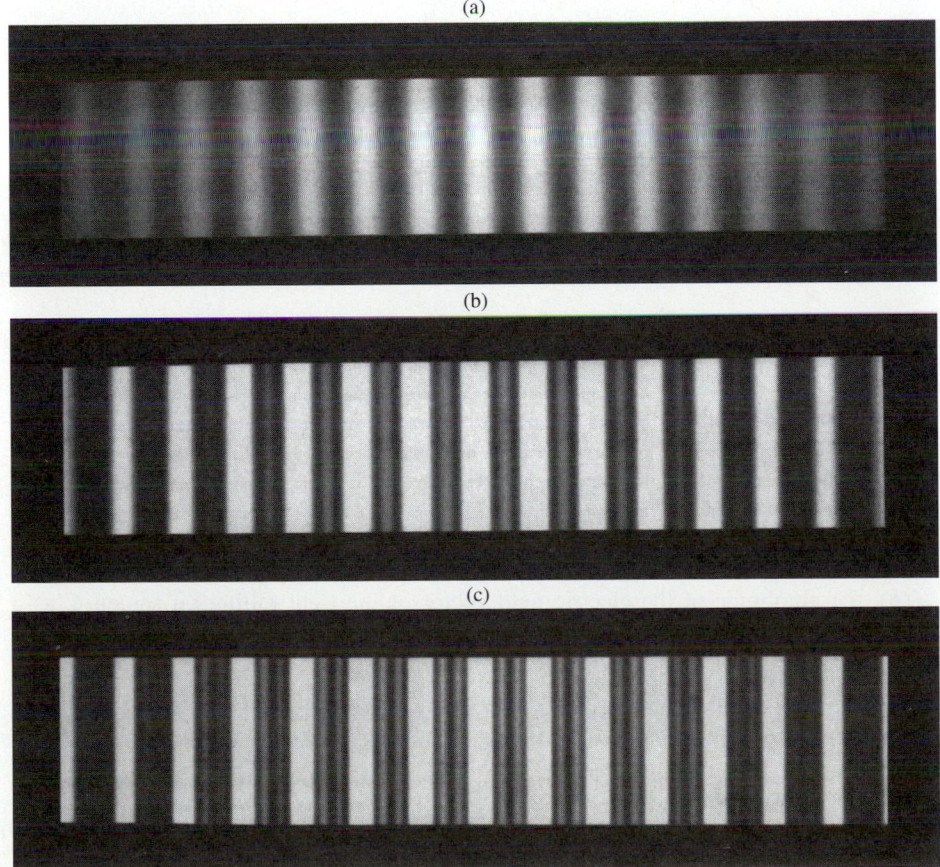

(a)

(b)

(c)

FIGURE 36.13 Photographs of (a) two-slit, (b) three-slit, and (c) four-slit interference patterns.

two, three, and four slits. Further increase of the number of slits results in further sharpening of the maxima. Indeed, it can be shown that the half-width of the **principal maxima**—the brightest ones—is inversely proportional to the number of slits N.*

In the three-slit pattern, there is one less-bright **subsidiary maximum** between adjacent principal maxima. In the four-slit pattern, the number of subsidiary maxima between adjacent principal maxima is two. In general, an N-slit pattern has $N - 2$ subsidiary maxima between adjacent principal maxima. As the number of slits increases, each additional subsidiary maximum that appears is successively fainter.

Interference: Quantitative Details

We now develop a means of describing the intensity of an interference pattern at all points on the screen. In Section 36.2, we considered the superposition of two component waves that differ only in phase. The amplitude $\mathscr{E}'$ of the superposed wave is given by Equation 36.2,

$$\mathscr{E}' = 2\mathscr{E}_0 \cos \frac{\phi}{2}.$$

Although we arrived at this result for two waves propagating along the same line, the result is equally valid for waves that differ both in phase and in direction of propagation—that is, waves that superpose as they cross one another. In contrast with the one-dimensional case, however, the value of ϕ in the two-dimensional case varies from point to point on the screen. What is the phase difference ϕ between the two light beams that arrive at the arbitrary point P on screen C in Figure 36.10?

For the reasons considered at the beginning of Section 36.4, the waves are in phase as they emerge from slits S_1 and S_2. Thus the phase difference at P depends on the optical path difference Δs between the two rays S_1P and S_2P. We know that Δs is given by Equation 36.9,

$$\Delta s = D \sin \theta.$$

According to Equation 36.11, $\sin \theta \simeq h/R$, and so we have

$$\Delta s = \frac{Dh}{R}. \tag{36.14}$$

As Δs varies from 0 to λ, ϕ varies from 0 to 2π (Figure 36.1). Thus we can write the proportion

$$\frac{\Delta s}{\lambda} = \frac{\phi}{2\pi}. \tag{36.15}$$

The phase difference is

$$\phi = 2\pi \frac{\Delta s}{\lambda}.$$

Combined with Equation 36.14, this equation yields

$$\phi = \frac{2\pi D}{\lambda R} h. \tag{36.16}$$

Finally, we insert this value of ϕ into Equation 36.2, $\mathscr{E}' = 2\mathscr{E}_0 \cos \phi/2$, to obtain $\mathscr{E}'(h)$, the amplitude of the electric field vector of the light falling on screen C as a function of distance h from the midpoint O:

$$\mathscr{E}'(h) = 2\mathscr{E}_0 \cos \frac{\pi Dh}{\lambda R}. \tag{36.17}$$

---

*The half-width w of a maximum is defined to be one-half the distance between the points at which the intensity is one-half the maximum intensity. Compare with the half-width of a resonance curve, defined in Section 33.3.

The intensity of light is proportional to the square of the amplitude of the electric field vector. If the intensity of the light falling on the screen from either slit alone is I_0, the relative intensity of the light falling on the screen varies with h according to the relation

$$\frac{I}{I_0} = 4 \cos^2 \frac{\phi}{2} = 4 \cos^2 \frac{\pi D h}{\lambda R}. \qquad (36.18)$$

Phasors*

The *phasor*—the rotating vector developed in Section 6.2 as a descriptor of harmonic motion and used in Section 33.2 for alternating currents—provides a vivid way of visualizing the superposition of light waves in an interference pattern. Consider wave 1 in Figure 36.14a. The magnitude of phasor 1 in Figure 36.14b is equal to the amplitude

(a)   (b)   (c)

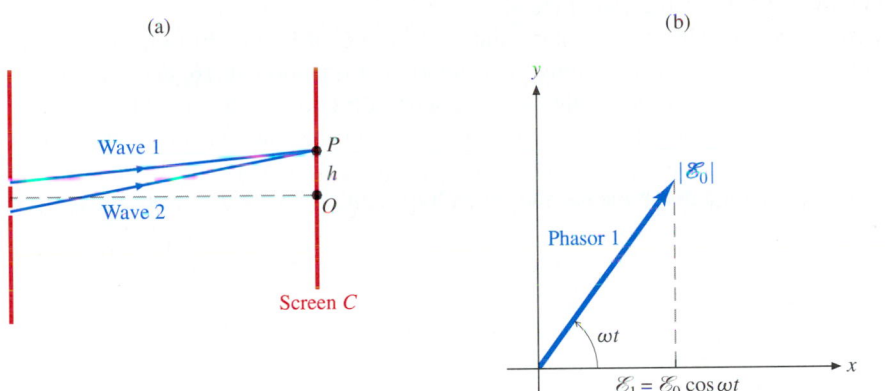

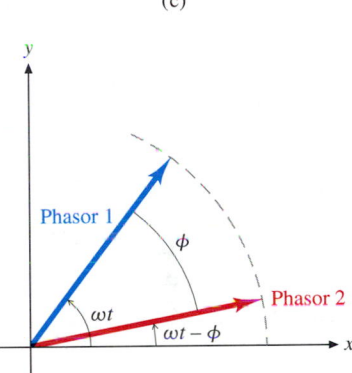

$\mathcal{E}_0$ of the electric field of wave 1. The phasor rotates counterclockwise with angular speed $\omega = 2\pi c/\lambda$, equal to the angular frequency of the light wave. Consequently, the x component of the phasor,

$$\mathcal{E}_1 = \mathcal{E}_0 \cos \omega t,$$

represents the magnitude of the electric field of wave 1 at a given point—say, at P on screen C of Figure 36.14a—as a function of time. If wave 1 were the only light wave falling on P, the intensity of the light would be proportional to the square of its amplitude: $I \propto \mathcal{E}_0^2$. However, wave 2, which passes through slit S_2, also falls at P. As already noted, this wave differs in phase from wave 1. The phase difference ϕ depends only on the position h of point P, in accordance with Equation 36.16. At P, wave 2 can be described by the expression

$$\mathcal{E}_2 = \mathcal{E}_0 \cos (\omega t - \phi) = \mathcal{E}_0 \cos \left(\omega t - \frac{2\pi D h}{\lambda R} \right).$$

That is, wave 2 at P can be described by phasor 2, shown in Figure 36.14c. Phasor 2 has the same magnitude and angular velocity ω as phasor 1 but lags behind it by the constant angle ϕ.

To find the superposed electric field $\mathcal{E}'$ at P as a function of time, we add the two phasors as we would any two vectors, by placing them head to tail, as in Figure 36.15a. The magnitude of the sum vector $\mathcal{E}'$ is equal to the amplitude $\mathcal{E}'$ of the superposed field. As the phasors rotate, the instantaneous value of the electric field varies because the value is given by the time-varying x component of $\mathcal{E}'$. But the amplitude $\mathcal{E}'$ depends only on the constant quantities $\mathcal{E}_0$ and ϕ and remains fixed at any particular point P.

Because only the relative orientation of the two phasors is significant, we conventionally orient the phasors so that the first one lies along the positive x axis, as shown

FIGURE 36.14 (*a*) Waves 1 and 2 superpose at P. (*b*) Phasor 1 represents wave 1 at P. The angular frequency of wave 1 is $\omega = 2\pi c/\lambda$, and its electric vector has amplitude $\mathcal{E}_0$. (*c*) Phasor 2 represents wave 2 at P. The phase difference between waves 1 and 2 is represented by the constant angle ϕ.

*This subsection may be omitted without loss of continuity, provided the intensity of the single-slit diffraction pattern (in Section 36.7) is not to be covered.

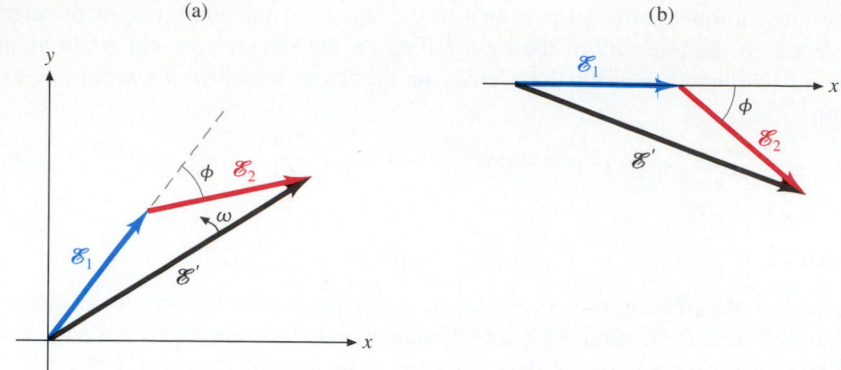

FIGURE 36.15 (*a*) Adding two phasors. The magnitude $\mathscr{E}'$ of the sum vector $\mathscr{E}'$ represents the amplitude of the superposed wave. (*b*) The conventional representation of phasors and their sum dispenses with the rotating term ωt.

in Figure 35.15b. The rotation of the phasors is neglected because it is not needed to evaluate the resultant wave amplitude.

As we move away from the center point on screen *C* of Figure 36.14a, we vary the value of phase difference ϕ according to the relation of Equation 36.16, $\phi = 2\pi Dh/R\lambda$. The effect on the superposed amplitude $\mathscr{E}'$ is seen in Figure 36.16a, in which ϕ varies through one cycle of 2π rad by steps of $\pi/3$ rad. The corresponding values of I, which is proportional to $\mathscr{E}'^2$, are plotted in Figure 36.16b. This plot is consistent with the result obtained by algebraic means and given by Equation 36.18: $I \propto \cos^2 \phi/2$.

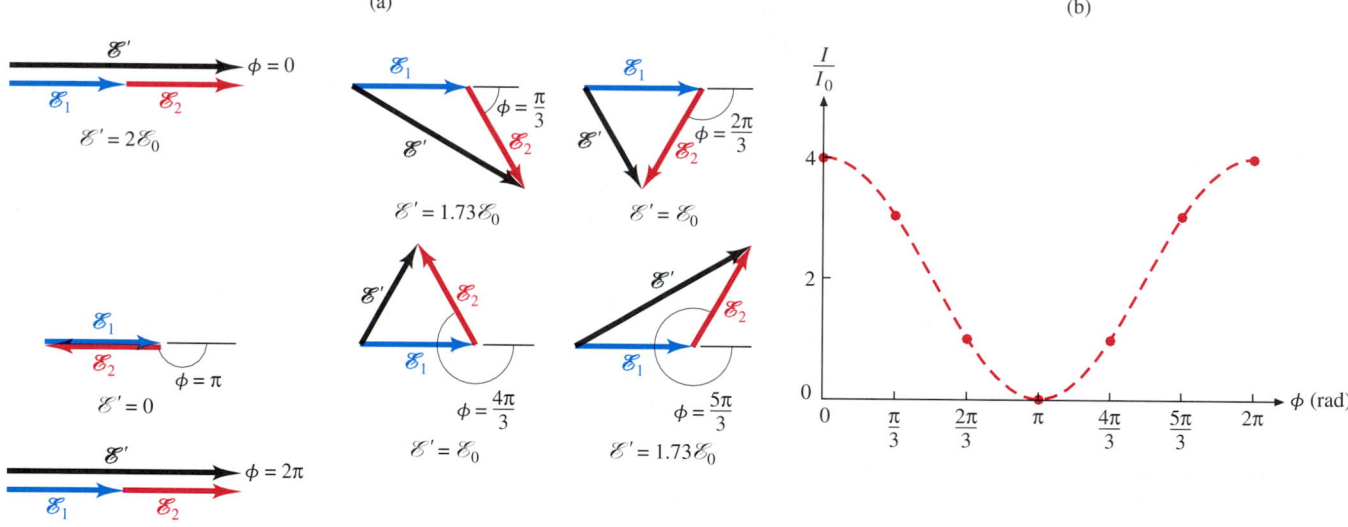

FIGURE 36.16 (*a*) Sequence of phasor diagrams showing two-slit interference. The amplitude $\mathscr{E}_0$ is set equal to 1. The phase difference increases by steps of $\pi/3$ rad through an entire cycle of 2π rad. Note that the second half of the cycle is a mirror image of the first half. (*b*) The relative intensity (square of the resultant amplitude) plotted versus ϕ. The points fall on the curve $I/I_0 = 4 \cos^2 \phi/2$.

The phasor picture becomes more useful when we apply it to multiple-slit interference. Figure 36.17 shows a cycle like that of Figure 36.16a, but for three evenly spaced slits. Three phasors now represent the three sets of waves superposing at *P*. As you can see from Figure 36.12, the optical path difference between waves 3 and 2 is the same as that between waves 2 and 1. Hence the phase differences between the corresponding phasors are equal. The new feature that appears in the three-slit interference pattern is the subsidiary maximum halfway through the cycle, at $\phi = \pi$. With more slits, more subsidiary maxima appear.

*Direct Calculation of the Multiple-Slit Interference Pattern**

Phasor diagrams for many slits are cumbersome to draw, but the phasor diagram for a small number of slits suggests a means of calculating the multiple-slit interference pattern.

---

*This subsection may be omitted without loss of continuity.

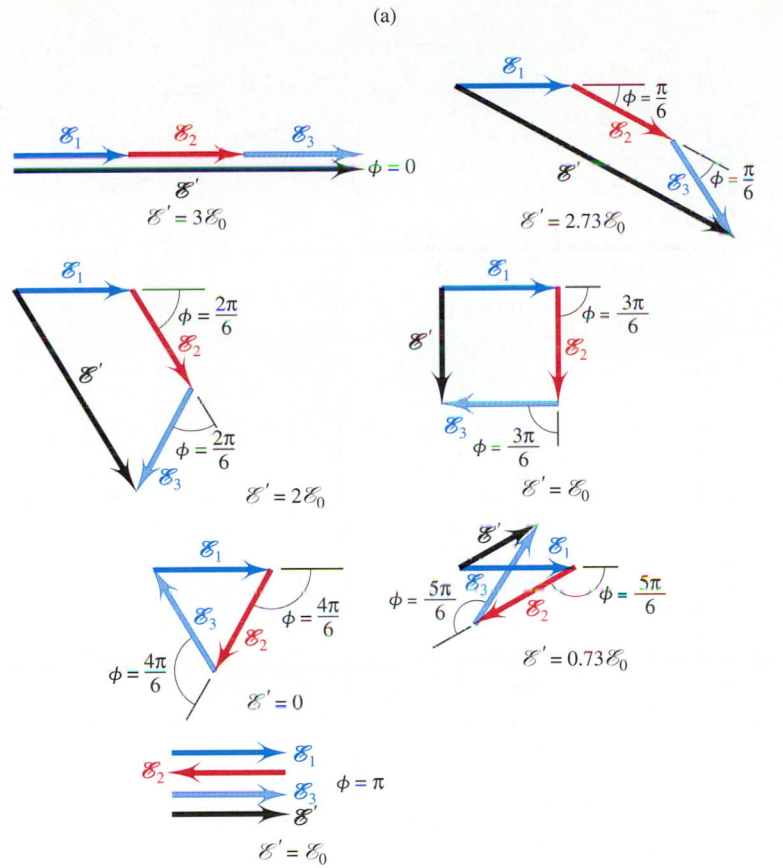

FIGURE 36.17 Sequence of phasor diagrams showing three-slit interference. The amplitude $\mathscr{E}_0$ is set equal to 1. The phase difference increases by steps of $\pi/6$ rad through one-half a cycle of 2π rad. (*b*) The relative intensity plotted versus ϕ. The points fall on a curve determined by the calculation starting with Equation 36.19a.

If there are N slits, as in Figure 36.12, the x component of the resultant phasor $\mathscr{E}'$ is

$$\mathscr{E}'_x = \mathscr{E}_0[\cos 0 + \cos \phi + \cos 2\phi + \cos 3\phi + \cdots + \cos (N - 1)\phi]. \quad \textbf{(36.19a)}$$

To see that this is so, compare this equation with Figure 36.17a, which is drawn for three slits. The y component of the resultant phasor $\mathscr{E}'$ is

$$\mathscr{E}'_y = \mathscr{E}_0[\sin 0 + \sin \phi + \sin 2\phi + \sin 3\phi + \cdots + \sin (N - 1)\phi]. \quad \textbf{(36.19b)}$$

Again, compare this equation with Figure 36.17a.

For convenience, we reexpress Equations 36.19a and 36.19b, using the summation notation:

$$\mathscr{E}'_x = \mathscr{E}_0 \sum_{j=1}^{N} \cos (j - 1)\phi \quad \text{and} \quad \mathscr{E}'_y = \mathscr{E}_0 \sum_{j=1}^{N} \sin (j - 1)\phi. \quad \textbf{(36.20a,b)}$$

We now evaluate $\mathscr{E}'^2$, using Pythagoras's theorem:

$$\mathscr{E}'^2 = \mathscr{E}'^2_x + \mathscr{E}'^2_y = \mathscr{E}_0^2 \left[\left(\sum_{j=1}^{N} \cos (j - 1)\phi \right)^2 + \left(\sum_{j=1}^{N} \sin (j - 1)\phi \right)^2 \right].$$

The intensity I at P is proportional to $\mathscr{E}'^2$. Again, we give the name I_0 to the intensity at P when only one slit is open. We can then write

$$\frac{I}{I_0} = \left(\sum_{j=1}^{N} \cos (j - 1)\phi \right)^2 + \left(\sum_{j=1}^{N} \sin (j - 1)\phi \right)^2. \quad \textbf{(36.21)}$$

To make a plot of the intensity of the interference pattern requires evaluating this equation at numerous values of ϕ spaced over the range $0° \leq \phi \leq 360°$. If done by hand, this is a formidable task. But a computer makes it easy to turn out such a plot. Figure 36.18a shows the plot obtained for five slits. Figure 36.18b is a similar plot for ten slits. The ten-slit pattern is both sharper and brighter than the five-slit pattern.

FIGURE 36.18 (*a*) Plot of intensity versus position on screen for five-slit interference, obtained using Equation 36.21. Only one-half of the pattern is shown; the pattern is symmetrical about $\phi = 0$. (*b*) Plot of intensity versus position for ten-slit interference. Note that the peak intensity is $(10/5)^2 = 4$ times as great as that in part *a*.

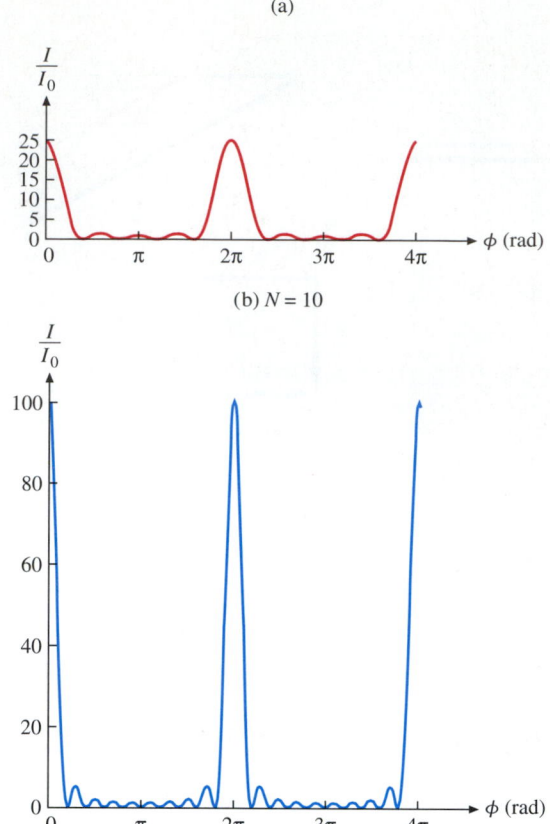

(a)

(b) $N = 10$

Diffraction

(a)

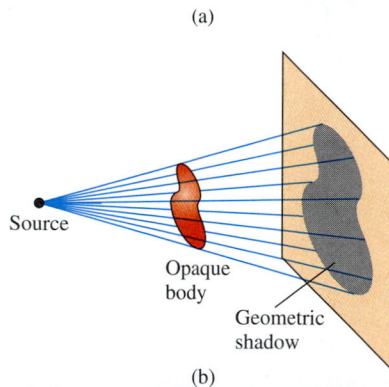

Source

Opaque body

Geometric shadow

(b)

FIGURE 36.19 (*a*) Construction of the geometric shadow of an opaque body illuminated by a point source. (*b*) Photograph of an object and its shadow.

To first approximation, objects cast *geometric shadows*. That is, an opaque object illuminated by a point source of light casts a shadow whose shape is the same as that of a geometric projection of the object from the source point. Figure 36.19 illustrates the construction of a geometric shadow. The construction works because, to first approximation, light travels in straight lines. If we do not look too closely, we can identify these straight lines with the *rays* defined in Section 35.3.

But, like all approximations, this one must not be pushed too far. It is useful only as long as the dimensions of interest are large compared with the wavelength of light. Light propagates as described by Huygens's construction (Figure 35.4), and not, strictly speaking, in straight lines. Like all waves, light bends around corners. When we look more closely at the way in which objects cast shadows, we find evidence of this typical wavelike behavior, which is called **diffraction**. All the interference patterns discussed in the preceding three sections are special, particularly simple examples of diffraction. Figure 36.20 shows more complex diffraction effects in the shadows of several opaque objects. Each illustration is a photograph of a screen on which the shadow of the object is cast by a bright point source.

Figure 36.20a shows the shadow of a fine wire. The region outside the geometric shadow of the wire is not uniformly bright but is filled with a complex pattern of bright and dark fringes of varying width and intensity. Surprisingly, the center of the geometric shadow is a bright fringe! Figure 36.20b shows the shadow of a straightedge. Here again, the region outside the geometric shadow is filled with a pattern of dark and bright fringes. The shadow of the edge is not sharp but fuzzy. Figure 36.20c shows the shadow of a saw blade. Here the pattern of light and dark is so complicated that it is difficult to distinguish the shape of the blade.

Analysis of diffraction patterns can become very complicated mathematically. But the basic principles of diffraction are all in evidence in the special case of **single-slit**

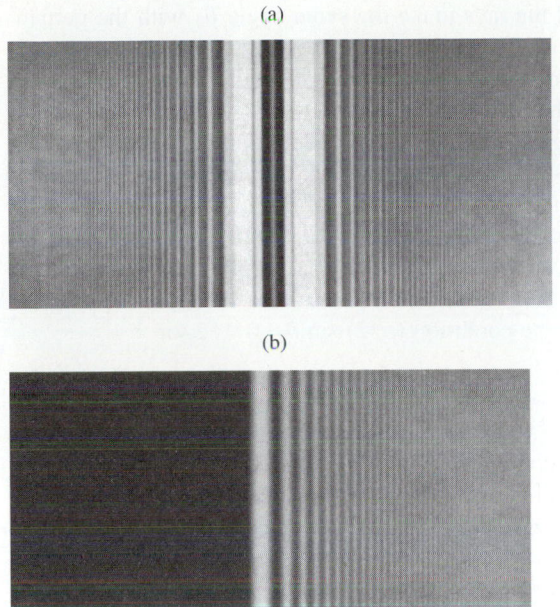

(a)

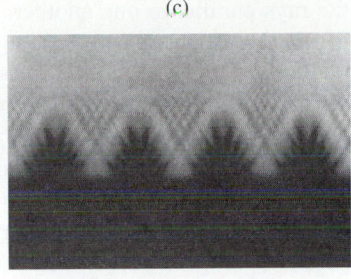

(c)

(b)

FIGURE 36.20 Diffraction effects in the shadows of (a) a fine wire, (b) a straightedge, and (c) a saw blade.

diffraction, in which light passes through a slit of small but nonnegligible width. We restrict our study to the case in which the light falls on a screen whose distance from the slit is large compared with the slit width. This case is called **Fraunhofer diffraction**.*

An apparatus for producing single-slit Fraunhofer diffraction is shown schematically in Figure 36.21. A planar wave front passes through the slit. Each point on the wave front is a sourcelet of Huygens wavelets. We choose two such points and draw a ray from each to the arbitrary point P on screen C. Because the slit has a finite width a, the optical path lengths are different for the two rays, and we must expect interference effects at P. (The path difference Δs is shown in Figure 36.21.) Other rays we might draw to P, from any of the infinite number of sourcelets on the wave front passing through the slit, have still other optical path lengths. Wavelets from all the sourcelets interfere at P. Indeed, we may regard single-slit diffraction as equivalent to an extreme case of multiple-slit interference in which the number of slits N is infinite and the spacing D between adjacent slits is zero.

Interference and Diffraction

Unfortunately, there is no uniform agreement as to the distinction to be made between interference and diffraction. We adopt the following convention: Interference refers to the effects of superposition of waves from two or more distinct sources. Diffraction refers to the effects of superposition of waves emerging from different parts of a single source of finite width. In practice, both effects are often observed simultaneously.

FIGURE 36.21 Arrangement for producing single-slit Fraunhofer diffraction.

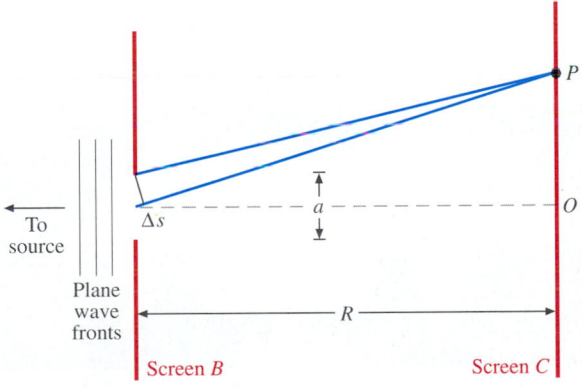

Locating the Minima

In Figure 36.22, we imagine the slit to be divided into two equal segments of width $a/2$. The evenly spaced rays all describe paths of Huygens wavelet trains to point P on the screen a distance R away. Here we take advantage of the condition $R \gg a$ to draw

---

*After the Bavarian instrument maker, astronomer, and physicist Joseph Fraunhofer (1787–1826). Fraunhofer was one of the pioneers in spectroscopy.

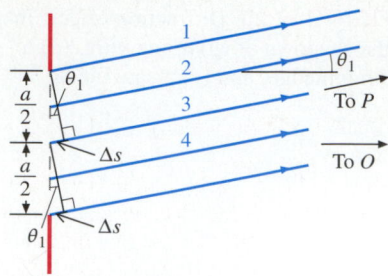

FIGURE 36.22 The wavelets whose paths are represented by the rays superpose destructively on the distant screen (not shown) to produce a diffraction minimum if $\Delta s = \lambda/2$. Here, $\Delta s = (a/2) \sin \theta_1$.

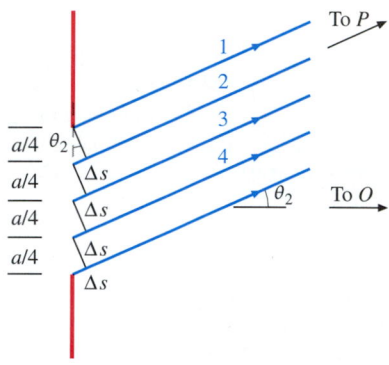

FIGURE 36.23 Another possibility for a diffraction minimum. Again, $\Delta s = \lambda/2$. But here $\Delta s = (a/4) \sin \theta_2$.

FIGURE 36.24 Photograph of a Fraunhofer diffraction pattern, showing dark and bright fringes and the central maximum.

the rays parallel to one another; all the rays make the same angle θ_1 with the perpendicular to the slit.

For what angle θ_1 will P lie at a minimum, where the superposition of all the wavelets results in zero intensity? Suppose the optical path described by ray 3 is just $\lambda/2$ longer than the path described by ray 1, so the corresponding wavelets are out of phase. Then the same will be true of rays 2 and 4. Indeed, every ray that originates in the lower half of the slit will be $\lambda/2$ longer than a corresponding ray that originates in the upper half of the slit. (Compare with the argument made in connection with the multiple-slit apparatus of Figure 36.12.) As you can see from Figure 36.22, this pairing off of rays $\lambda/2$ out of phase will take place if $\Delta s = \lambda/2$. But you can also see from the diagram that $\Delta s = (a/2)\sin \theta_1$. So we have the condition $(a/2) \sin \theta_1 = \lambda/2$, or

$$\sin \theta_1 = \frac{\lambda}{a}.$$

But this is not the only possible condition for a minimum in the diffraction pattern. Another possibility is shown in Figure 36.23, where we imagine the slit to be divided into four equal segments. If ray 2 is $\lambda/2$ longer than ray 1, then also ray 4 will be $\lambda/2$ longer than ray 3. Each pair of rays will interfere destructively. Indeed, every ray that originates in the second quarter of the slit will be $\lambda/2$ longer than a corresponding ray in the top quarter, and rays in the third and fourth quarters can be paired off in like manner. From the diagram, we see that the condition for the resulting diffraction minimum is $(a/4) \sin \theta_2 = \lambda/2$, or

$$\sin \theta_2 = 2 \sin \theta_1 = 2 \frac{\lambda}{a}.$$

We can find similar conditions for diffraction minima by dividing the slit in imagination into any even number of equal segments. We can therefore write the general condition

$$\sin \theta_m = m \frac{\lambda}{a}, \quad m = \pm1, \pm2, \pm3, \ldots, \quad \text{for diffraction minima.} \quad \textbf{(36.22)}$$

The negative values of m correspond to diffraction minima located below the midline. The bright regions between adjacent minima are called **bright fringes**; the relatively dark regions in the vicinity of minima are called **dark fringes**. The central bright fringe, between the minima corresponding to $m = 1$ and $m = -1$, is also called the **central maximum**, and the other bright fringes are called **subsidiary maxima**. Figure 36.24 is a photograph of such a Fraunhofer diffraction pattern.

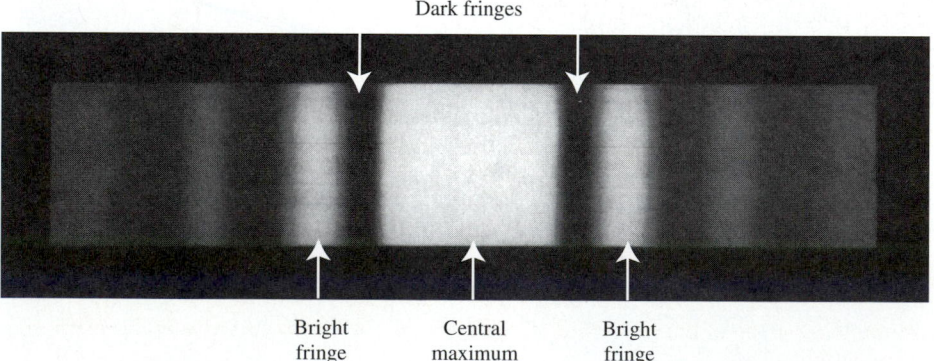

Dark fringes

Bright fringe   Central maximum   Bright fringe

EXAMPLE 36.5

Sodium light ($\lambda = 589$ nm) falls on a slit of width $a = 0.482$ mm. **(a)** Find the angles at which the first, second, and third minima occur. **(b)** What is the angular width of the central maximum? **(c)** What is the angular width of either of the first subsidiary maxima? **(d)** If the screen is located at $R = 1.50$ m from the slit, what is the distance between the central maximum and the first dark fringe?

SOLUTION:

(a) Find θ_1, θ_2, and θ_3.

Using Equation 36.22, you calculate

$$\theta_1 = \sin^{-1} \frac{589 \times 10^{-9} \text{ m}}{4.82 \times 10^{-4} \text{ m}} = 1.22 \times 10^{-3} \text{ rad, or } 0.0700°.$$

Because this angle is small, you are safe in using the approximation $\sin \theta \approx \theta$. So you have

$$\theta_2 = 2 \times 1.22 \text{ mrad} = 2.44 \text{ mrad}$$

and

$$\theta_3 = 3 \times 1.22 \text{ mrad} = 3.66 \text{ mrad}.$$

In principle, the fringes corresponding to large m are not evenly spaced because the approximation $\sin \theta \approx \theta$ is not valid. However, such fringes are almost always too faint to be visible.

(b) What is the angular width of the central maximum?

The central maximum lies between θ_1 and θ_{-1}. For the angular width, you have $\Delta\theta = \theta_1 - \theta_{-1} = 2\theta_1 = 2.44$ mrad, or $0.1400°$.

(c) What is the angular width of the first subsidiary maximum?

The first subsidiary maximum lies between θ_2 and θ_1. You have $\Delta\theta = \theta_2 - \theta_1 = 1.22$ mrad. The subsidiary maxima are half as wide as the central maximum.

(d) Find the distance between the central maximum and the first dark fringe.

The angular distance between the central maximum and the first dark fringe is θ_1. The corresponding distance h on the screen is given by $\tan \theta_1 = h/R$. But θ_1 is small, so you can use the approximation $\tan \theta \approx \theta$. You thus have

$$h = R\theta_1 = 1.50 \text{ m} \times 1.22 \times 10^{-3}$$
$$= 1.83 \times 10^{-3} \text{ m} = 1.83 \text{ mm}.$$

The fringes are actually wider than the slit that produces them. Indeed, you can see from Equation 36.22 that the fringes become wider as the slit is made narrower.

Diffraction by a Circular Aperture

Figure 36.25a shows the Fraunhofer diffraction pattern produced by a circular aperture—that is, by light passing through a circular hole in an opaque screen. The geometry of the system is sketched in Figure 36.25b. In this figure, a is the diameter of the aperture, the quantity analogous to the width a of a slit. The symbols R and θ have the same meaning as in the single-slit diffraction setup of Figure 36.21.

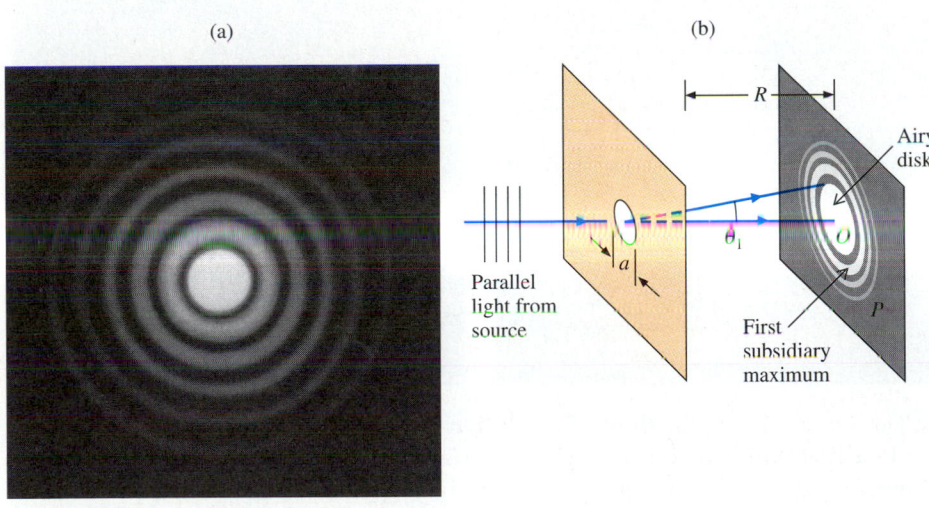

(a) (b)

FIGURE 36.25 (a) Fraunhofer diffraction pattern of a circular aperture. (b) Geometry of the system.

The main features of the circular pattern are similar to those of the single-slit diffraction pattern. There is a bright central maximum, called the **Airy disk**.* The Airy disk is surrounded by a series of subsidiary maxima whose intensity falls off rapidly. Unlike the spacing of the single-slit minima, the spacing of the circular-aperture minima is far from uniform. The mathematical analysis for finding the intensity of the circular pattern is more complicated than that for the slit, and we will not consider it here. However, we state an important result, which we will use in Section 36.8. The angle θ_1 in Figure 36.25b, which locates the first minimum (and thus measures the angular

---

*After the British astronomer and mathematician Sir George Airy (1801–1892), who first made a detailed analysis of the circular diffraction pattern in 1835.

width of the Airy disk), is given by the expression

$$\sin \theta_1 = \frac{1.22\lambda}{a} \quad \text{for a circular aperture.} \tag{36.23}$$

This result is not unlike the corresponding expression for the first minimum in a single-slit pattern: $\sin \theta_1 = \lambda/a$.

SECTION 36.8 | **Limit of Resolution**

When you see a source of light, the light must pass through *at least* one circular aperture—the pupil of your eye. As a consequence, even when you look at a point source, what you see is an Airy disk of finite angular width, surrounded by subsidiary maxima.

Usually, you are not even aware of this apparent smudging out of point sources; you have always seen things this way. But consider what happens when you wish to inspect two point sources separated by a small angle. As the angle becomes smaller, the Airy disks begin to merge (Figure 36.26). When the disks are so close that the center of each coincides with the first minimum of the other, you can just barely see that they are two separate diffraction disks produced by two separate sources. This limit on discernment occurs when the centers of the diffraction disks are just one disk radius apart. If they were any closer together, you could not *resolve* the sources—that is, you could not tell that there were two sources rather than one.

(a) (b)

FIGURE 36.26 (*a*) The angular separation between two point sources is so small that their Airy disks nearly touch. (*b*) With further decrease in the angular separation, the Airy disks merge until the center of each coincides with the first minimum of the other. This is the limit of resolution.

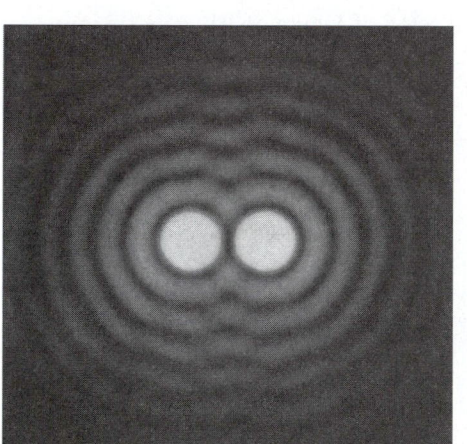
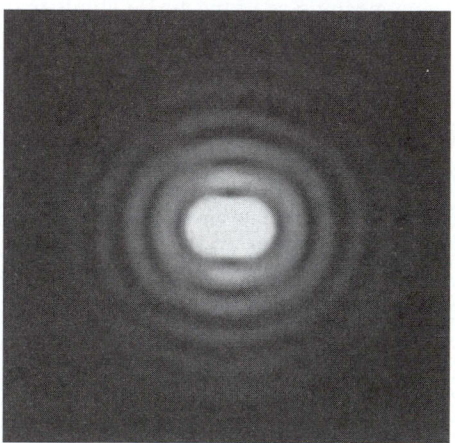

This description of the **limit of resolution** is called **Rayleigh's criterion**. We can use Equation 36.23 to express Rayleigh's criterion quantitatively. Suppose the center of one Airy disk lies at $\theta = 0$, as shown in Figure 36.27. The center of the other Airy disk lies at θ_1 (at the edge of the first disk). The angle between the centers is then $\theta_r = \theta_1$, called the **limiting angle of resolution**. The angle θ_1 is always small, and we

Compleat Physicist

The criterion for the limit of resolution is named after John W. Strutt, Lord Rayleigh (1842–1919). Rayleigh's contributions to experimental and theoretical physics cover an extraordinarily wide range of subjects, including acoustics, mechanics, hydrodynamics, and physical chemistry as well as optics. In 1899, he gave a detailed answer to the ancient question ''Why is the sky blue?''; the pertinent light-scattering process is called *Rayleigh scattering*. Rayleigh won a Nobel Prize in 1904 for his 1895 discovery of argon.

FIGURE 36.27 Limiting angle of resolution for two point sources when their light passes through a circular aperture of diameter a.

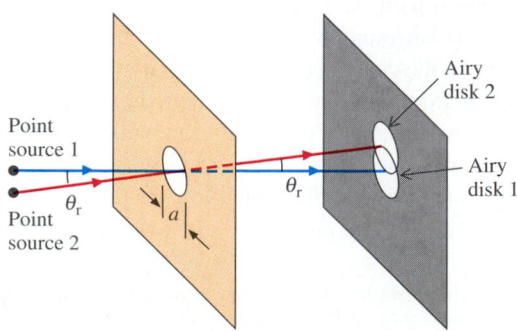

can write Equation 36.23 in the form

$$\theta_r = \frac{1.22\lambda}{a}.$$ (36.24)

In Figure 36.27, the angle between the centers of the Airy disks is equal to the angular separation of the two light sources. Hence the limiting angle of resolution applies directly to the angular separation of the sources.

Note that the limiting angle of resolution is *inversely proportional* to the aperture diameter a. This is one important advantage (though not the only one) of large-diameter telescopes.

EXAMPLE 36.6

What is the limiting angle of resolution for (a) the human eye and (b) a reflecting telescope whose diameter is 8 m? Take the diameter of the pupil of the eye to be 4 mm.

SOLUTION: In view of the fact that visible light ranges in wavelength from about 400 nm to about 700 nm, take the midpoint value $\lambda = 550$ nm to be typical. Equation 36.24 then gives you the following values.

(a) For the unaided eye:

$$\theta_r = \frac{1.22 \times 550 \times 10^{-9} \text{ m}}{4 \times 10^{-3} \text{ m}} = 1.7 \times 10^{-4} \text{ rad.}$$

This value, equivalent to 30 arc seconds, is typical; Tycho Brahe's wonderful eyes were about twice as good and could resolve two objects having an angular separation of about 15 arc seconds; see Section 14.1.

(b) For the telescope:

$$\theta_r = \frac{1.22 \times 550 \times 10^{-9} \text{ m}}{8 \text{ m}} = 8.4 \times 10^{-8} \text{ rad.}$$

This value is equivalent to 1.7×10^{-2} arc seconds—a considerable improvement over the naked eye.

The Diffraction Grating

SECTION 36.9

In Section 36.5, we studied the interference pattern produced by a system of many slits. As the number of slits increases, the principal maxima become brighter and sharper while the subsidiary maxima become relatively fainter. This characteristic of multiple-slit patterns is exploited in the **diffraction grating**, which has come to replace the dispersing prism (Section 35.7) as the basic tool of spectroscopy.

A diffraction grating is an array of very many closely spaced, highly uniform, parallel slits, made by scribing grooves, called **lines**, on a plate made of glass or some other suitable material. Gratings can have more than 10 000 lines/cm, with a total of several hundred thousand lines. Once a high-quality **master grating** is made, relatively inexpensive **replica gratings** can be manufactured by using the master as a mold for castings of plastic.

In the **transmission grating**, light passes through the spaces between the lines. In the more common **reflection grating**, the plate is covered with a thin metal reflective coating; light is reflected from the spaces between the lines. In either case, the spaces act as multiple slits. In what follows, we assume for simplicity that the grating is a transmission grating and that light is incident on it normal to the grating plane.

Under these circumstances, light of wavelength λ produces interference maxima at angles given by Equation 36.13, $m\lambda = D \sin \theta_m$. So we have

$$\sin \theta_m = m\frac{\lambda}{D}.$$ (36.25)

The distance D between grating lines is called the **grating spacing**. Now, suppose the incident light consists of a superposition of many wavelengths. In the mth order, the wavelength components λ_1 and λ_2 produce maxima at angles θ_1 and θ_2 for which

$$\sin \theta_1 = m\frac{\lambda_1}{D} \quad \text{and} \quad \sin \theta_2 = m\frac{\lambda_2}{D}.$$

The angle between the two maxima is thus

$$\Delta\theta = \left| \sin^{-1}\frac{m\lambda_1}{D} - \sin^{-1}\frac{m\lambda_2}{D} \right|. \qquad (36.26)$$

As you can see from this equation, the incident light spreads out into a series of spectra, one spectrum corresponding to each value of m. The spectrum corresponding to a particular value of m is called the **mth-order spectrum**. In some cases, orders may overlap. A grating spectrometer is shown schematically in Figure 36.28.

FIGURE 36.28 Spectra produced by a transmission grating spectrometer. The wavelength range of the incident light extends from λ_{min} (V for violet) to λ_{max} (R for red). The second- and third-order spectra overlap. For clarity, only a few orders on one side of the midline are shown.

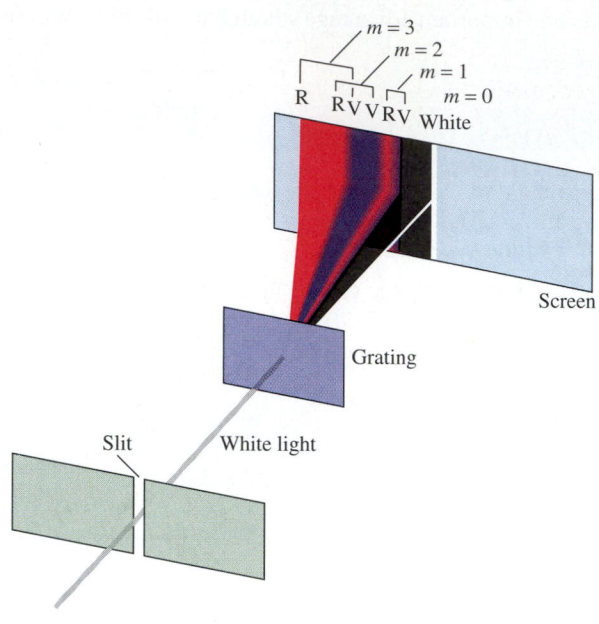

EXAMPLE 36.7

You have a grating that has $N = 49\,000$ lines. The width of the grating is 4.200 cm.

(a) Over what angle does the entire visible spectrum spread in the first order? **(b)** What is the angular separation between the sodium D lines, $\lambda_1 = 588.995$ nm and $\lambda_2 = 589.592$ nm, in the first-order spectrum?

SOLUTION:
(a) Find the angular width of the first-order visible spectrum.

First, you must calculate the grating spacing D. Call the width of the grating W. You then have

$$D = \frac{W}{N} = \frac{4.200 \times 10^{-2} \text{ m}}{49 \times 10^3} = 8.571 \times 10^{-7} \text{ m}.$$

The visible spectrum extends from the violet ($\lambda_V = 400$ nm) to the red ($\lambda_R = 700$ nm). So you use Equation 36.26, setting $m = 1$, to find

$$\Delta\theta = \left| \sin^{-1}\frac{400 \times 10^{-9} \text{ m}}{8.571 \times 10^{-7} \text{ m}} - \sin^{-1}\frac{700 \times 10^{-9} \text{ m}}{8.571 \times 10^{-7} \text{ m}} \right|$$
$$= |27.82° - 54.75°| = 26.9°.$$

(b) What is the first-order angular separation of the sodium D lines?

You have

$$\Delta\theta = \left| \sin^{-1}\frac{588.995 \times 10^{-9} \text{ m}}{8.571 \times 10^{-7} \text{ m}} - \sin^{-1}\frac{589.592 \times 10^{-9} \text{ m}}{8.571 \times 10^{-7} \text{ m}} \right|$$
$$= |43.409° - 43.463°| = 0.0550°,$$

or about 3.3 arc minutes. This is a somewhat greater dispersion angle than that produced by a typical prism; compare with Problem 35.62.

Dispersion of a Diffraction Grating*

From Example 36.7, it is evident that a diffraction grating disperses red light through a greater angle than blue light. A prism, whose dispersion curve has negative slope (Figure 35.15), does just the opposite.

---

*The remainder of the chapter may be omitted without loss of continuity.

The quantity $d\theta/d\lambda$, called the **dispersion**, is a measure of the ability of a diffraction grating (or a prism) to separate spectral components of different wavelengths. It is easy to calculate the dispersion of a diffraction grating. We write Equation 36.13 in the form $\sin\theta = m\lambda/D$ and differentiate both sides to obtain

$$\cos\theta \, d\theta = \frac{m}{D} \, d\lambda.$$

Rearrangement gives us

$$\frac{d\theta}{d\lambda} = \frac{m}{D}\frac{1}{\cos\theta}. \tag{36.27a}$$

We use the trigonometric identity $\cos\theta = \sqrt{1 - \sin^2\theta} = \sqrt{1 - m^2\lambda^2/D^2}$ to obtain

$$\frac{d\theta}{d\lambda} = \frac{m}{D}\frac{1}{\sqrt{1 - \dfrac{m^2\lambda^2}{D^2}}}.$$

If $m\lambda/D \ll 1$ (as is usually the case), we can use the approximation $(1 - \alpha)^{-1/2} \simeq 1 + \alpha/2$ to write the equation in the simplified form

$$\frac{d\theta}{d\lambda} \simeq \frac{m}{D}\left(1 + \frac{m^2\lambda^2}{2D^2}\right). \tag{36.27b}$$

According to this equation, the dispersion varies only slowly with λ. Indeed, an adequate approximation for many purposes is to write the dispersion in the form

$$\frac{d\theta}{d\lambda} \simeq \frac{m}{D}. \tag{36.27c}$$

This near constancy of the dispersion of a diffraction grating is in contrast with the dispersion of a prism, which varies significantly with λ in a complicated way (Problem 35.62). The relatively simple form of Equation 36.27b makes calculation of wavelengths from angles measured with a grating spectrometer much simpler than the corresponding calculation for a prism spectrometer.

Resolving Power of a Diffraction Grating

The usefulness of spectroscopy as a scientific tool depends in a crucial way on the possibility of distinguishing between two spectral lines whose wavelengths are very close together. The ability of a diffraction grating (or other device) to make this distinction is called the **resolving power** R of the device. Let λ_1 and λ_2 be the wavelengths of two lines that can be barely distinguished. We define

$$R \equiv \frac{\lambda}{|\lambda_2 - \lambda_1|} = \frac{\lambda}{\Delta\lambda}. \tag{36.28}$$

(We can use either of the values λ_1 and λ_2 in the numerator of this equation because they differ so little.) A device characterized by a large value of R is said to have **high resolution**. We now calculate the resolving power for a grating having N lines and grating spacing D.

Each of the two spectral components produces an N-slit interference pattern. The two patterns overlap, as shown in Figure 36.29. Under what conditions can we resolve the principal maxima of the two patterns—that is, distinguish them as separate maxima? We will be able to do so if the angle $\Delta\theta$ between the two maxima is approximately equal to the angle δ between either maximum and its adjacent diffraction minimum, as shown in Figure 36.29. If $\Delta\theta$ is substantially smaller than δ, the two principal maxima merge and are no longer resolvable. We can express this **resolution criterion** in the simple form

$$\Delta\theta = \delta. \tag{36.29}$$

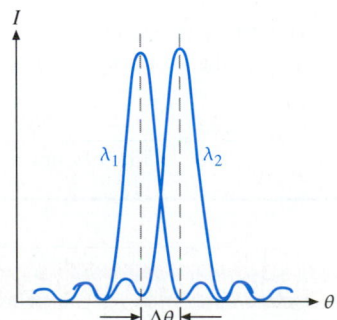

FIGURE 36.29 Overlapping multiple-slit interference patterns for two spectral components of slightly different wavelengths λ_1 and λ_2.

Our task, then, is to evaluate $\Delta\theta$ and δ and compare them.

First, we evaluate the angle $\Delta\theta = |\theta_2 - \theta_1|$ between the principal maxima of the two spectral components. We begin with Equation 36.27c, $d\theta/d\lambda = m/D$, and make the finite approximation $\Delta\theta/\Delta\lambda = m/D$. Solving for $\Delta\theta$, we have

$$\Delta\theta = \frac{m}{D}\Delta\lambda. \tag{36.30}$$

Next, we find the angle δ between either principal maximum and its adjacent minimum. The principal maxima of order m and $(m + 1)$ lie at angles given by

$$\sin\theta_m = \frac{m\lambda}{D} \quad \text{and} \quad \sin\theta_{m+1} = \frac{(m + 1)\lambda}{D}.$$

So we have

$$|\sin\theta_m - \sin\theta_{m+1}| = \frac{\lambda}{D}.$$

Although θ_m and θ_{m+1} are not necessarily small angles, the approximation $\sin\alpha \simeq \alpha$ is adequate because the resolution criterion of Equation 36.29 is itself an approximate statement. Close enough for our purpose, the angular separation between adjacent principal maxima is

$$|\theta_m - \theta_{m+1}| = \frac{\lambda}{D}.$$

Between adjacent principal maxima, there are $N - 1$ minima (Figure 36.18). For N large, we have $N - 1 \simeq N$. Thus the (small) angle δ between a principal maximum and the adjacent minimum is $1/N$ times $|\theta_m - \theta_{m+1}|$, and we have

$$\delta \simeq \sin\delta = \frac{\lambda}{ND}. \tag{36.31}$$

We are now ready to compare δ with the value of $\Delta\theta$ given by Equation 36.30. The resolution criterion (Equation 36.29) yields ($m\,\Delta\lambda/D = \lambda/ND$, or

$$R = \frac{\lambda}{\Delta\lambda} = mN. \tag{36.32}$$

The resolving power of a diffraction grating is directly proportional to the total number of lines in the grating and to the order of diffraction used. It does not depend on either the grating spacing D or the total width of the grating.

EXAMPLE 36.8

How many lines must a grating have if it is to resolve the sodium D lines in second order?

SOLUTION: You use the wavelengths given in Example 36.7 to calculate the necessary resolving power:

$$R \equiv \frac{\lambda}{\Delta\lambda} = \frac{589 \text{ nm}}{|588.995 \text{ nm} - 589.592 \text{ nm}|} = 987.$$

In second order, Equation 36.32 gives you

$$N = \frac{R}{m} = \frac{987}{2} = 493.$$

A very modest grating, having about 500 lines, suffices to resolve the sodium D lines in second order. In view of the approximate nature of the resolution criterion, you state the result in the form $N \simeq 500$.

Symbols Used in Chapter 36

a	slit width	h	distance from center of screen to a point P
d	thickness of a film or layer	I, I_0	intensity, intensity due to light from a single narrow slit
D	spacing between slits		
$\mathscr{E}$	electric field amplitude of a light wave	k	wave number

L	optical path length
m	order of interference
n	index of refraction
N	number of slits (interference) or segments (diffraction)
R	normal distance from slit(s) to screen, resolving power
$s, \Delta s$	physical path length, path difference

θ_r	limiting angle of resolution
λ	wavelength
ν	frequency
ϕ	phase difference between two waves
Φ	phase difference at P between light waves arriving from opposite edges of a slit
ω	angular frequency

The superposition of light waves leads to a wide variety of effects called **diffraction**. **Interference** is the special type of diffraction in which the superposition is of light from two or more sources having distinctly defined phase differences.

One-dimensional interference occurs between two light beams differing only in phase. If the amplitude of each beam is $\mathcal{E}_0$, the superposed amplitude can have any value between 0 and $2\mathcal{E}$, depending on the phase difference ϕ. The superposed intensity is proportional to the square of the amplitude according to Equations 36.2 and 36.3.

When light from a single source passes through two or more narrow slits spaced a distance D apart, **interference maxima** and **minima** occur at angles θ that satisfy Equations 36.10a and 36.10b:

$$m\lambda = D \sin \theta,$$
$$m = 0, \pm1, \pm2, \ldots, \quad \text{for maxima;}$$

$$(m + \tfrac{1}{2})\lambda = D \sin \theta,$$
$$m = 0, \pm1, \pm2, \ldots, \quad \text{for minima.}$$

The fringe spacing on a screen is given by Equation 36.12,

$$\Delta h = \frac{R}{D} \lambda.$$

N-slit interference patterns consist of sharp **principal maxima** between which lie $N - 2$ **subsidiary maxima**. With increasing N, the subsidiary maxima become relatively fainter as the principal maxima become sharper and brighter. The intensity distribution in the interference pattern is given by Equation 36.21.

When light passes through a slit of nonnegligible width a and falls on a screen, there is a variation in phase among the parts of the light passing through different parts of the slit. The result is a **single-slit diffraction pattern**. If the screen is far from the slit, the bright **central maximum** is surrounded by fainter **secondary maxima** whose width is one-half that of the central maximum. The **diffraction minima** are located at angles θ_m given by Equation 36.22,

$$\sin \theta_m = m \frac{\lambda}{a}, \quad m = \pm1, \pm2, \pm3, \ldots.$$

Diffraction is produced by an aperture of any shape. For a circular aperture, the width of the central maximum, or **Airy disk**, is defined as the angle θ_1 between the center of the maximum and the first minimum. This angle is given by Equation 36.23.

Light from one or more objects must always pass through at least one circular disk on its way to the retina of the eye or other detector. Two objects cannot be *resolved* if they are separated by an angle less than the **limiting angle of resolution** θ_r. This angle is related to θ_1. In the simplest case, it is given by Equation 36.24,

$$\theta_r = \frac{1.22\lambda}{a}.$$

A **diffraction grating** is often used to separate light into its spectral components. Interference maxima occur at the angles given by Equation 36.10a or by Equation 36.25,

$$\sin \theta_m = m \frac{\lambda}{D}.$$

The **dispersion** $d\theta/d\lambda$ of a grating varies only slowly with wavelength and is given approximately by Equation 36.27c,

$$\frac{d\theta}{d\lambda} \simeq \frac{m}{D}.$$

The **resolving power** R, defined by Equation 36.28, $R \equiv \lambda/\Delta\lambda$, is given for a grating by Equation 36.32,

$$R = mN.$$

KEY TERMS

Sections 36.2 and 36.3 Interference in One Dimension and Thin-Film Interference
constructive, destructive interference ▪ interference maximum, minimum ▪ Michelson interferometer ▪ beam splitter ▪ interference fringe, fringe shift ▪ antireflection coating

Section 36.4 Interference in Two Dimensions
two-slit interference ▪ coherent sources ▪ line of antinodes, line of nodes ▪ order of interference

Section 36.5 Multiple-Slit Interference

principal maximum ▪ subsidiary maximum

Sections 36.7 and 36.8 Diffraction and Limit of Resolution

Fraunhofer diffraction ▪ Airy disk ▪ Rayleigh's criterion, angle of resolution

Section 36.9 The Diffraction Grating

transmission grating, reflection grating ▪ *n*th-order spectrum

---

Queries and Problems for Chapter 36

QUERIES

36.1 *(2) On the fringe.* Make a sketch that illustrates how a slight deviation from perpendicularity of the end mirrors of a Michelson interferometer leads to the formation of a striped array of interference fringes. Explain why the width of the fringes—that is, the angular distance between adjacent fringes—becomes smaller as the misalignment angle increases.

36.2 *(2) Something extra.* In practical Michelson interferometers, an extra plain glass plate called a *compensator* is always inserted into one of the arms, as shown below. What is the purpose of this plate, and what must its thickness be?

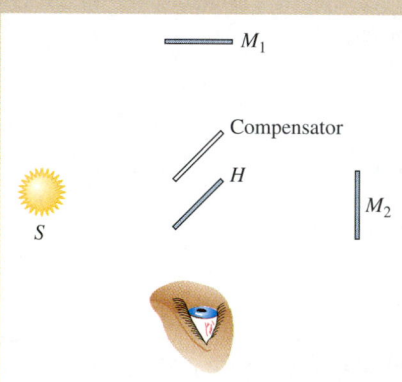

36.3 *(2) The long and short of it.* When a Michelson interferometer is used with monochromatic light, fringes can be observed even when the lengths of the two arms differ substantially. Under these conditions, one can measure a *change* in the length of an arm precisely, but not the length of the arm. With a *white* light source, however, clear fringes can be observed only when the lengths are almost exactly equal. Explain. (White light fringes are often used to adjust the arms of an interferometer to equal length. This is an important step, for example, in the process of measuring the old standard meter bar in terms of a precisely known wavelength of light, as Michelson first did in 1889.)

36.4 *(3) Walkin' in the rain.* If the oil film on a rain puddle is large enough, you often see several sequences of color. At the edges of the film, however, where the oil is very thin, the puddle looks black. Explain.

36.5 *(3) Living color.* In 1881, the French physicist Gabriel Lippmann (1845–1921) devised the first method of color pho-

tography, for which he was awarded the Nobel Prize in 1908. He used a special photographic plate prepared as shown below. A photographic emulsion is deposited on one side of a high-quality glass plate. (Photographic emulsions consist largely of silver bromide and similar salts in suspension. Exposure to light triggers a chemical process that leads, on development, to the growth of tiny grains of silver. These grains appear black.) The thickness of the emulsion is much greater than the wavelength of light. The side of the emulsion opposite the glass plate is coated with a layer of highly reflective mercury. The Lippmann plate is exposed by allowing light from the camera lens to pass through the glass into the emulsion. **(a)** A color image, consisting of light containing various mixtures of wavelengths, falls on the Lippmann plate. Make a sketch showing how standing waves are set up in the emulsion. **(b)** At what locations in the standing waves will silver grains precipitate? **(c)** After development, the plate is viewed in white light from the glass side. Assuming that development does not make the emulsion swell or shrink, explain how the image reproduces the original colors of the object photographed.

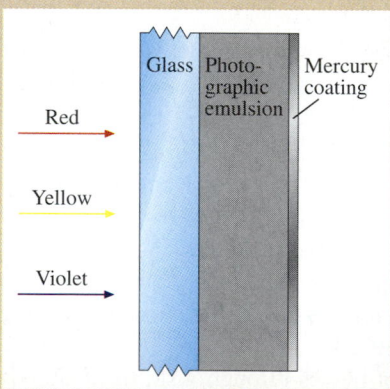

36.6 *(3) Flattering image.* The figure on the next page shows the monochromatic interference fringes obtained when an imperfectly flat piece of glass is placed in contact with an optical flat, which is a test piece of glass whose surface is flat within tolerances that are small compared with the wavelength of light. **(a)** What parts of the test piece are flattest? the least flat? **(b)** Estimate the height of the highest "peak" with respect to

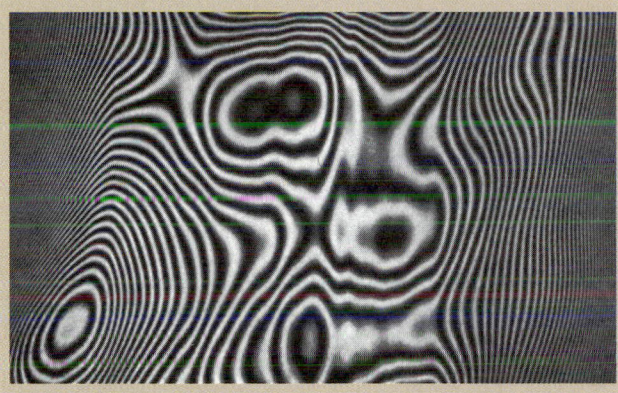

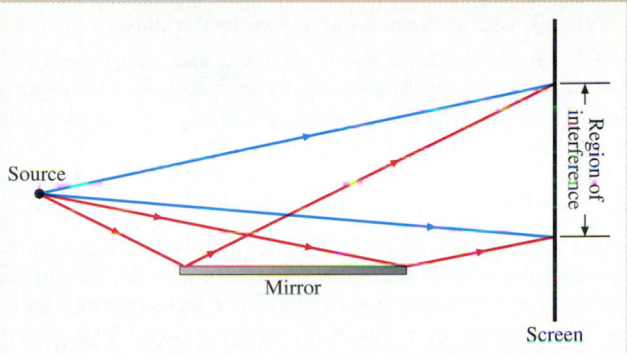

the lowest "valley." **(c)** Explain how you would make the imperfect piece into an optical flat.

36.7 *(3) Unsound practice.* Why is it not practical to use the acoustical equivalent of an antireflection coating to make soundproofing material? (There are two reasons, either of which would be sufficient by itself to render the approach impractical.)

36.8 *(4) Masking waves.* Today, everyone accepts the interpretation of two-slit interference on the basis of the wave nature of light. After Young's experiment of 1800, however, acceptance of the wave interpretation was far from universal, and many first-rate physicists remained committed to the particle theory of light. An important "convincer" for the wave theory was Fresnel's 1816 modification of Young's experiment. In this modification, Fresnel placed polarizers over the two slits. Explain what is observed when the directions of polarization of the light emerging from the two slits are mutually **(a)** parallel and **(b)** perpendicular.

36.9 *(4) LLoyd's mirror.* A modification of the Young two-slit interference experiment is shown in the next column. The mirror is aligned precisely perpendicular to the screen. The pattern on the screen is identical with the Young two-slit pattern, except that the bright and dark fringes are inter-changed. **(a)** Explain how the system produces an interference pattern. (Hint: Can you find two coherent sources?) **(b)** What can you infer about reflection from the reversal of the Young pattern?

36.10 *(5) Squaring things.* The total light energy flux that passes through a set of N identical slits is equal to NI_0; that is, the flux is proportional to N. But the intensity of the maxima is $N^2 I_0$; see Equation 36.18 and Figure 36.18, for example. How can you reconcile these conclusions?

36.11 *(7) Light in the midst of darkness.* In a single-slit diffraction pattern, do the subsidiary maxima lie exactly halfway between the adjacent minima? Why or why not?

36.12 *(7) Realm of the senses.* Two persons stand in a hallway, talking. Some distance down the hallway, an open door to the side leads to a room. The persons inside the room can hear the speakers but cannot see them. Explain.

36.13 *(7) Narrow miss.* Monochromatic light passes through a single slit and falls on a flat screen. Show that if the width of the slit is equal to the wavelength of the light, $a = \lambda$, there are no diffraction minima on the screen.

36.14 *(8) How big is big?* Why are radiotelescope dishes almost always made much larger than optical telescope mirrors?

PROBLEMS

GROUP A

36.1 *(2) Turn of the screw.* One of the end mirrors of a Michelson interferometer is mounted on a micrometer screw. As you move the mirror, you observe a shift of 425 fringes. If the distance you read on the micrometer is 0.145 mm, what is the wavelength of the light used?

36.2 *(2) Delicate measurement.* You repeat the experiment of Example 36.2, using helium in place of CO_2. You observe a shift of 10.1 fringes. **(a)** What is the index of refraction of helium at atmospheric pressure and 15°C? **(b)** What is the index of refraction of helium at atmospheric pressure and 0°C (STP)?

36.3 *(2) Going backward.* You fill the gas cell of Example 36.2 with ammonia at STP ($n = 1.000\ 38$). If you slowly pump the ammonia out of the cell, how many fringes move past the cross hairs of your observing telescope?

36.4 *(3) Playing with a bubble ring, I.* When you look directly down on a horizontal soap film illuminated by uniform white light from above, the film appears blue (take $\lambda = 450$ nm). Take the index of refraction of the soap solution to be 1.35. **(a)** What is the minimum possible thickness of the film? **(b)** What are other possible thicknesses?

36.5 *(3) Blowing bubbles.* A soap bubble ($n = 1.35$) has wall thickness 540 nm. When white light falls on a part of the bubble at normal incidence, what colors appear enhanced in the reflected light?

36.6 *(3) Yellow soap.* White light is incident on a soap film

($n = 1.35$) at 45°. If the reflected light appears yellow ($\lambda = 600$ nm), what is the minimum possible thickness of the film?

36.7 *(3) Trickle-down theory.* When a soap film is held vertically, the downward flow of liquid makes the film slightly wedge shaped. **(a)** Describe the appearance of the film when it is viewed in monochromatic light. **(b)** If 10 fringes occupy 4.2 cm when a film is illuminated with red light ($\lambda = 630$ nm), express the wedge angle in arc seconds.

36.8 *(3) Fabry-Perot interferometer.* Shown below is a sketch of the Fabry-Perot interferometer (1899). Two flat glass plates whose inner surfaces are partially silvered are aligned parallel to one another at a distance d apart. A light ray passing through the system from the left at an angle of incidence θ may pass directly through or it may be reflected one or more times before passing through. Show that constructive interference takes place at a point far to the right when the condition $2d \cos \theta = m\lambda$, $m = 0, 1, 2, 3, \ldots$, is satisfied. (Note: The mirrors of a laser constitute a Fabry-Perot interferometer.)

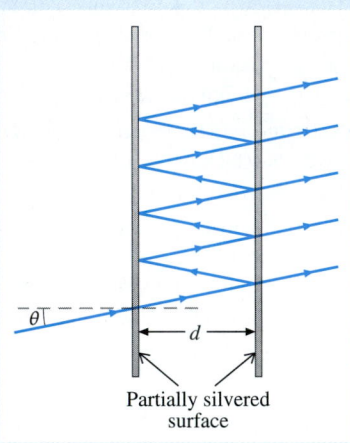

Partially silvered
surface

36.9 *(3) Stealth.* An important theme in military technology is reduction of the visibility of aircraft to enemy radars. One (partial) solution is to cover the so-called stealth aircraft with a coating of dielectric material. If the radar wavelength is 1.5 cm, what is the desired thickness of the coating, assuming there is no relative phase inversion? Assume to first approximation that the index of refraction for the coating material at radar wavelengths is slightly greater than 1.

36.10 *(4) Getting an angle on it.* The separation of the slits in a two-slit interference apparatus is 500 times the wavelength of the light that passes through them. What is the angular separation between **(a)** the first and second maxima? **(b)** the 200th and 201st maxima?

36.11 *(4) Full count.* Show that the total number of maxima produced by a two-slit apparatus cannot be greater than $(2D/\lambda) + 1$. Why is counting the number of maxima not a very precise way to determine λ?

36.12 *(4) Running interference.* Two narrow slits 0.287 mm apart are illuminated by light of wavelength 540 nm. The light falls on a screen 1.55 m away from the slits. **(a)** What is the angular separation between adjacent maxima? **(b)** At what distances from the midline do the minima fall on the screen?

36.13 *(4) Proper spacing.* You are designing a two-slit interference apparatus for use with light of wavelength 655 nm. If the slit-to-screen distance is 1.35 m and you wish the maxima to lie 1.8 mm apart, what slit spacing should you use?

36.14 *(7) Metes and bounds, I.* The slit-to-screen distance in a single-slit diffraction apparatus is 1.50 m. When light of wavelength 589 nm is used, the distance between the two first-order minima is 1.43 mm. How wide is the slit?

36.15 *(7) Metes and bounds, II.* For the apparatus of Problem 36.14, find **(a)** the distance between the second and fifth minima on the same side of the central maximum; **(b)** the distance between the second minimum on one side of the central maximum and the fifth minimum on the other side.

36.16 *(8) Long view.* When you drive at night through the West Texas plains, it is often possible to see the lights of oncoming cars at a very great distance. **(a)** At what distance can you just make out that the light comes from two separate headlights? Take the distance between the headlights to be 1.3 m and the nighttime diameter of the pupil of your eye to be 7 mm. **(b)** For modest heights h above the earth's surface, the distance d to the horizon is given by $d = \sqrt{2Rh}$, where R is the radius of the earth. Make a plausible estimate for h, and determine whether you are likely to see unresolved headlights on a road passing through a flat plain.

36.17 *(8) Look homeward, angel.* An astronaut in a low earth orbit (altitude = 300 km) is passing over her home town. Assume that the diameter of her pupils in bright sunlight is 4 mm, and determine whether she can use her unaided eye to identify **(a)** her house and **(b)** the high-school football stadium.

36.18 *(8) Spy in the sky.* Military satellites can produce photographs of the earth's surface in which objects whose dimensions are of order 1 m are clearly identifiable. **(a)** Assuming a minimum satellite altitude of 200 km, estimate the minimum diameter of the camera lens. **(b)** If the same camera is carried on a high-altitude reconnaissance airplane flying at altitude about 25 km, what is the limit of resolution? **(c)** Is the Rayleigh criterion likely to be the limiting factor in determining the detail obtainable from aircraft reconnaissance? from satellite reconnaissance (altitude about 280 km)?

36.19 *(8) Farther is better . . . up to a point.* In North American television, each picture is constructed of 525 horizontal lines. Suppose you are watching a program on a "21 inch" screen whose vertical height is 32 cm. At what minimum distance from the screen should you sit in order to see the picture as uniform? Take $\lambda = 550$ nm, and assume that the diameter of your pupils is 5 mm.

36.20 *(8) Double stars.* A large fraction of all stars are members of double-star pairs. Nevertheless, there are no stars that can be seen as doubles with the naked eye. The nearest stars are at a distance of order 10 light-years (1 ly $\approx 9.5 \times 10^{15}$ m) away from the earth. Use this information to set an order-of-magnitude upper limit on the distance between the members of a double-star pair.

36.21 *(9) Spread.* A helium-neon laser produces light of wavelength 532.8 nm. The light falls normally on a 10 250-line diffraction grating of width 2.50 cm. **(a)** What is the angular separation between the second- and third-order maxima? **(b)** What is the largest angle at which this light is diffracted?

36.22 *(9) Calibrate and measure.* When light falls normally on a diffraction grating, the first-order sodium *D* line (λ = 589 nm) lies at 17.13°. Another wavelength component is seen in the second-order spectrum at 24.20°. **(a)** Find the grating spacing. **(b)** What is the wavelength of the other component?

36.23 *(9) Discursion.* Show that the dispersion of a diffraction grating can be written in the alternative form $d\theta/d\lambda$ = $(\sin\theta)/\lambda$.

36.24 *(9) Dispersion.* **(a)** What is the second-order dispersion of the grating considered in Example 36.8 at the wavelength of the sodium *D* lines? Take *W* = 1.7 cm. **(b)** What is the angular separation between the *D* lines in second order?

36.25 *(9) Resolution.* What is the resolving power of a 12 500-line grating in second order in the green region of the spectrum ($\lambda \simeq 520$ nm)?

36.26 *(9) Disperse, resolve.* A diffraction grating has 5250 lines/cm and is 3.00 cm wide. Find **(a)** the dispersion at $\lambda \simeq 550$ nm and **(b)** the resolving power, both in first order.

36.27 *(9) Resolve, solve.* You have a diffraction grating of 15 000 lines. Suppose you are looking at the part of a first-order spectrum in the wavelength region near 500 nm. What is the minimum wavelength difference that you can resolve?

36.28 *(9) Matters of a higher order.* When light of a certain wavelength is normally incident on a diffraction grating, the first-order line appears at angle θ_1. What is the highest order in which the line can be seen?

GROUP B

36.29 *(2) Double cross.* You are observing the fringes in a Michelson interferometer, using sodium *D* light. This light consists of two spectral components, λ_1 = 588.995 nm and λ_2 = 589.592 nm. As you move mirror M_1 (Figure 36.2), you see the fringes shift, as discussed in the text. You also see a periodic fading in and out of the fringes. **(a)** Explain the fading effect. **(b)** Through what distance must you move the mirror to go from a position where the fringes fade out to the next position where they fade out?

36.30 *(2) Molecule by molecule.* For an ideal gas, the value of *n* − 1—a measure of the amount by which the gas retards the speed of light—is directly proportional to the number of gas molecules present per unit volume. **(a)** Show that *n* − 1 $\propto$ p/kT, where *p* is the pressure of the gas, *T* is its absolute temperature, and *k* is Boltzmann's constant (see Chapter 18). **(b)** Using the result of Example 36.2, calculate the index of refraction of CO_2 for *p* = 22 atm and *T* = 250°C.

36.31 *(3) Thickness gauge.* Two flat slabs of glass are separated at one edge by a thin wire, as shown below. (The diameter of the wire is exaggerated for clarity.) The system is illuminated by using light of wavelength 589 nm. **(a)** Looking down from above, do you see an interference maximum or minimum at the left edge, where the slabs are in contact? **(b)** If you see 28 interference minima, the last one located at the wire, what is the diameter of the wire?

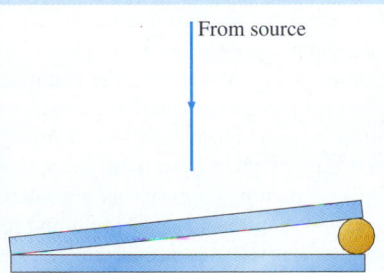

From source

36.32 *(3) Shrinking wire?* When the system of Problem 36.31 is submerged in a certain liquid, the number of interference minima is 21. Find the index of refraction of the liquid, and make an educated guess as to what the liquid is.

36.33 *(3) Newton's rings, I.* In figure *a*, a *planoconvex lens* (a piece of glass shaped with one planar surface and one spherical surface of large radius of curvature *R*) lies in contact with a flat glass plate. Both the monochromatic light source and the observer are located above the system. A set of circular fringes, called *Newton's rings*, appears as shown in figure *b*. **(a)** Show that the radii *r* of the rings are $r = \sqrt{(m + \frac{1}{2})\lambda R}$ for bright rings and $r = \sqrt{m\lambda R}$ for dark rings, with *m* = 0, 1, 2, 3, ... **(b)** Suppose the radius of curvature of the lens is *R* = 3.25 m and the diameter of the lens is 3.50 cm. How many bright rings will you see in sodium *D* light?

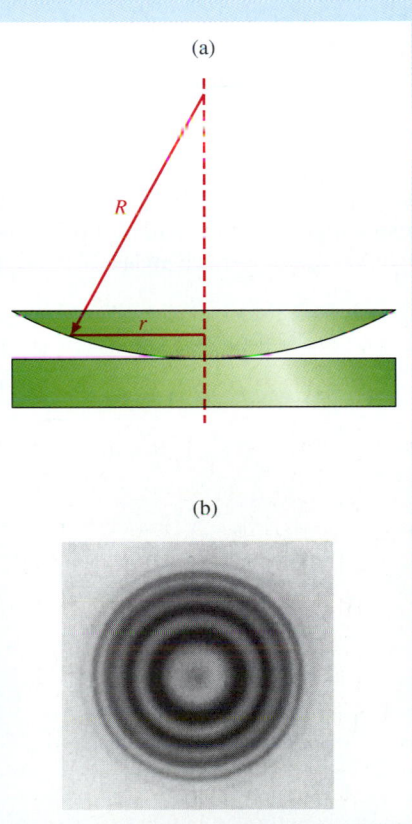

(a)

R

r

(b)

36.34 *(3) Newton's rings, II.* Suppose the system of Problem 36.33 is immersed in water. How many rings will you see?

36.35 *(3) Ringing changes on Newton.* The figure below shows a glass rod, having one cylindrical and one flat surface, in contact with a flat plate of glass. **(a)** What is the shape of the interference fringes? What are the distances from the line of contact of **(b)** the bright fringes? **(c)** the dark fringes?

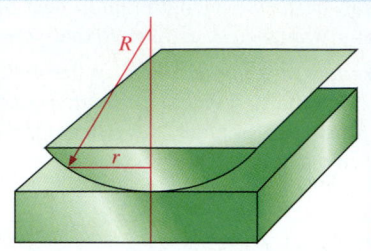

36.36 *(3) Young's variant on Newton's rings.* You assemble a Newton's-rings apparatus with a lens made of heaviest flint glass ($n = 1.89$) and a flat plate made of zinc crown glass ($n = 1.517$). You immerse the system in carbon disulfide ($n = 1.63$). Write expressions for the radii r of the bright and dark rings.

36.37 *(4) Curves.* **(a)** Show that, for any antinodal point such as X in Figure 36.8, the distances S_1X and S_2X differ by $m\lambda$, where m is an integer. **(b)** Find the corresponding relation for a nodal point such as O. **(c)** Prove that lines of antinodes and lines of nodes are hyperbolas. (Hint: A hyperbola is the locus of all points whose distances from two special points, called foci, differ by a constant.)

36.38 *(4) Dunk.* The angular separation between two adjacent two-slit interference maxima is 0.22°. If the entire apparatus is immersed in water ($n = 1.33$), what is the new angular separation between the two maxima?

36.39 *(4) Optical shim.* The thin glass wedge shown below is slid slowly in front of one of the slits of a two-slit interference apparatus illuminated by sodium light ($\lambda = 589$ nm). The length of the wedge is much greater than the width of the

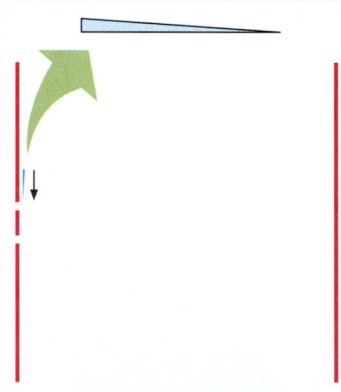

slit, so the part of the wedge in front of the slit at any moment varies only slightly in thickness. In the process, the pattern shifts so that what was originally the tenth minimum ($m = 9$) now lies at the midline. What is the thickness of the thickest part of the wedge? Take $n = 1.54$ for the glass.

36.40 *(4) Half-what? I.* For a two-slit interference pattern, show that the angular half-width of each maximum—one-half the angular separation between the points on either side of the maximum at which the intensity is one-half that of the maximum—is

$$\frac{\Delta h}{R} = \Delta\theta = \frac{\lambda}{4D}.$$

36.41 *(4) To the max.* When light of wavelength λ passes through two slits a distance D apart, what is the maximum possible number of interference fringes?

36.42 *(4) Fresnel's mirrors.* About 1815, Fresnel used a system like that diagrammed below to produce an interference pattern. The angle between the mirrors is exaggerated for clarity. **(a)** Express the distance Δh between adjacent fringes in terms of R, l, θ, and the wavelength λ of the light used. **(b)** Let $\lambda = 580$ nm, $l = 30$ cm, and $R = 1.50$ m. What should θ be in order to produce a fringe separation $\Delta h = 0.32$ mm?

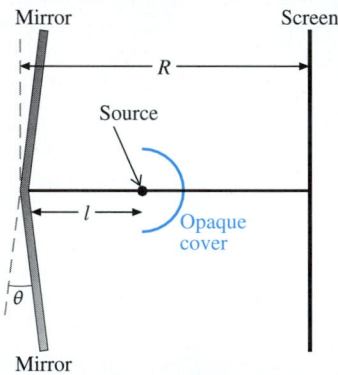

36.43 *(7) Something is missing.* A two-slit interference pattern is produced by a pair of slits whose widths are equal to one-tenth their separation. **(a)** Show that some of the maxima will be missing from the interference pattern. **(b)** What is the value of m for the first missing maximum on each side of the center?

36.44 *(8) Afternoon on the Grande Jatte.* The French neo-impressionist painter Georges Seurat (1859–1891) developed a technique called *pointillisme*. (See the example on the next page.) In this (very laborious and difficult) technique, the painting—often a very large one—is composed of many closely spaced dots of pure (unmixed) colors. From a sufficient distance, the painting appears as a continuous picture, incorporating a wide range of colors (including pastel shades). Seurat's dots are typically about 2 mm in diameter. How far must you stand from the canvas to see the picture as continuous?

36.45 *(8) Galileo and Saturn.* When Galileo turned his best telescope toward Saturn in 1610, he was able to see a peculiar bulge near the equator. But he could not resolve the bulge as a set of rings independent of the planet. (Huygens first resolved the rings in 1655.) The average distance from the earth to Saturn is about 1.5×10^{12} m. The diameter of Saturn is about 1.2×10^7 m. The inner diameter of the inner bright ring is about 1.8×10^8 m. **(a)** If Galileo's telescopes had been limited by diffraction effects, and not by imperfections, what diameter lens would have sufficed to determine that the rings are separate from the planet? **(b)** Galileo's lenses were typically of diameter 3 cm. Were his telescopes diffraction limited? Can you see Saturn's rings with a pair of ordinary binoculars of good quality?

36.46 *(8) Other worlds like ours.* The mirrors of the largest optical telescopes have diameters of order 10 m. **(a)** What is the maximum distance r at which a star could be seen by direct observation to have a solar system like our own? **(b)** No such observation has been made to date. Assume, to a crude first approximation, that the number of stars within a spherical volume of radius r, centered on the earth, is proportional to r^3. The radius of our galaxy is of order 10^4 light-years. What fraction of the stars in the galaxy have we subjected to direct

observation for planetary systems? Which of the following statements is valid?

1. Most likely, there are no other solar systems like our own in our galaxy.
2. Solar systems like our own are at best rare in our galaxy.
3. Solar systems like our own are common in our galaxy.
4. None of the above conclusions can be drawn on the basis of this crude calculation.

36.47 *(9) Grating equation.* In grating spectrometers, the incident light frequently strikes the diffraction grating not at normal incidence but at some other angle of incidence α. Show that the condition for interference maxima is $D(\sin \theta - \sin \alpha) = m\lambda$; this is often called the *grating equation.*

36.48 *(9) Bragg reflection.* Although X rays penetrate matter much more readily than does visible light, individual atoms do scatter X rays to a slight degree. In a crystalline solid, where the atoms are arrayed in an orderly fashion, there are *crystal planes* whose orientation is such that they contain large numbers of atoms. The figure (p. 1004) is an edge-on view of a series of such planes in a crystal. Although the effect is really one of diffraction rather than reflection, each crystal plane acts like a very weak mirror and reflects a tiny proportion of the

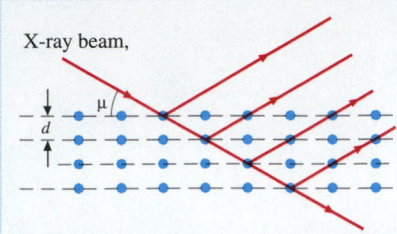

X-ray beam,

incident X-ray beam in accordance with the law of reflection. As shown in the figure, it is customary in X-ray work to use the angle θ between the crystal planes and the X-ray beam, which is the complement of the angle of incidence used in optics. Show that the crystal "reflects" X rays of wavelength λ strongly if θ satisfies the equation $m\lambda = 2d \sin \theta$, where d is the distance between adjacent planes. If λ is known, d can be determined. The equation is called *Bragg's law* after Sir William Lawrence Bragg (1890–1971). About 1913, he and his father, Sir William Henry Bragg (1862–1942), pioneered

in the exploitation of Bragg reflection to investigate crystal structures, and they shared the Nobel Prize in 1915.

36.49 (9) *Conflicting orders, I.* White light falls on a diffraction grating having 3250 lines/cm. What are the orders of the first two visible spectra that overlap? (Take the visible range to be from 400 nm to 700 nm.)

36.50 (9) *Conflicting orders, II.* Light from a gas discharge tube falls at normal incidence on a diffraction grating. Because of overlap of various orders, two lines superpose at $\theta = 41.0°$. If the wavelengths of the lines are 410.2 nm and 656.3 nm, find **(a)** the orders of the spectra of the two lines; **(b)** the grating spacing.

36.51 (9) *Resolution of the highest order.* Diffraction grating 1 has N_1 lines with grating spacing D_1. Grating 2 has N_2 lines with spacing D_2. Although $D_1 < D_2$ and $N_1 > N_2$, the products $N_1 D_1$ and $N_2 D_2$ are equal. Show that, if both gratings are used in their highest orders to examine normally incident light from the same source, they have approximately the same resolution. (Hint: See Problem 36.28.)

GROUP C

36.52 (2) *Intensity variation of interferometer fringes.* Let the amplitude of the light entering a Michelson interferometer be $2\mathscr{E}_0$. Assume that the beam splitter divides the incident light into two parts, each of amplitude $\mathscr{E}_0$. **(a)** Show that the amplitude $\mathscr{E}_y$ of the light wave reaching the observer depends on the phase difference ϕ between the two light paths according to the relation

$$\mathscr{E}_y = 2\mathscr{E}_0 \cos \frac{\phi}{2}.$$

(b) Remember that the intensity of a light beam is proportional to $\mathscr{E}^2$ (Section 34.6). The interferometer is adjusted so that an intensity maximum I_0 is observed. The mirror at the end of one arm of the interferometer is then moved away from this adjustment. Show that the observed intensity of the light depends on the distance x through which the mirror is moved according to the relation

$$I = I_0 \cos^2 \frac{2\pi x}{\lambda}.$$

36.53 (3) *Playing with a bubble ring, II.* Consider the soap film of Problem 36.4 in a more general way. Suppose that white light is normally incident on the film of thickness d. The film preferentially reflects light of wavelength λ, as shown in figure a, where the incident and reflected rays are displaced for clarity.

  (a) Show that the thickness of the film must have one of the values

$$d = \frac{(2j - 1)\lambda}{4n}, \quad j = 1, 2, 3, \ldots . \tag{1}$$

(b) When you change the angle of incidence and reflection, as shown in figure b, the preferentially reflected wavelength changes. At what angle of incidence θ_1 do you again see the wavelength λ? For a first crude analysis, neglect refraction at the first surface but continue to allow for the way the index of refraction of the film affects the optical path length through

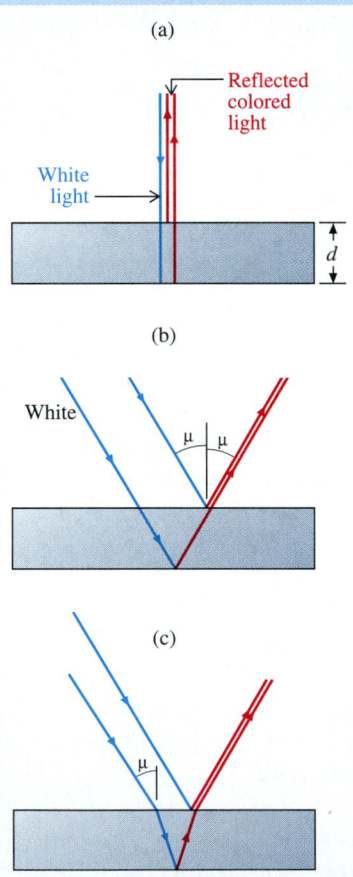

(a)

Reflected colored light

White light

d

(b)

White

μ | μ

(c)

μ

the film. Show that, to this approximation, the angle θ satisfies the equation

$$d \cos \theta_1 = \frac{(2j - 1)\lambda}{4}, \tag{2}$$

where the value of j is the same as that appearing in Equation 1; that is,

$$j = \frac{2nd}{\lambda} + \frac{1}{2}.$$

(c) In fact, neglecting refraction at the surface is not a satisfactory approximation. Show that, if you take refraction into consideration, as shown in figure c, Equation 2 is replaced by the messier expression

$$\sin \theta_1 = n \sqrt{1 - \left[\frac{4nd}{(2k-1)\lambda} \right]^2}, \qquad (3)$$

where $k = 1, 2, \ldots, j - 1$. (d) Show that Equation 3 has a solution only if the condition

$$\frac{2j-3}{2j-1} > \sqrt{1 - \frac{1}{n^2}} \qquad (4)$$

is satisfied. Let $n = 1.35$, and make trial solutions for a few values of j. By doing this, show that the wavelength visible at normal incidence cannot be seen at any other angle unless the thickness of the film corresponds to $j \geq 4$; that is, $d \geq 7\lambda/4n$. (e) What would this thickness be for blue light ($\lambda = 455$ nm)? At what angle(s) of incidence other than 0° will this wavelength be reflected preferentially?

36.54 (G) *Reflectivity.* When light strikes a surface between two transparent media, a part of the light is reflected. It can be shown that, at normal incidence, the ratio of the reflected electric field amplitude $\mathscr{E}_r$ to the incident amplitude $\mathscr{E}_0$ is

$$\frac{\mathscr{E}_r}{\mathscr{E}_0} = \left| \frac{n_2 - n_1}{n_2 + n_1} \right|, \qquad (1)$$

where n_1 and n_2 are the indices of refraction of the two media. The ratio of the intensity I_r of the reflected beam to the incident intensity I_0 is called the *reflectivity r*: $r \equiv I_r/I_0$. It follows from Equation 1 that, at normal incidence, the reflectivity at the boundary between medium 1 and medium 2 is

$$r = \left(\frac{n_2 - n_1}{n_2 + n_1} \right)^2. \qquad (2)$$

(a) Does the reflectivity depend on whether the light passes from air to glass or from glass to air? (b) Find the reflectivity for normal incidence at an air–light-flint-glass surface. Take $n_g = 1.58$. (c) Suppose you wish to design a perfect anti-reflection coating for light flint glass. What should the index of refraction n_c of the coating material be? (d) Unfortunately, there is no solid material whose value of n_c is as small as the value you found in part c. Magnesium fluoride (MgF_2) has one of the best available combinations of desirable properties: It has a relatively low index of refraction ($n_c = 1.38$) and reasonable hardness, and it can conveniently be deposited in films of controlled thickness. Show that, if you neglect multiple reflections of the light between the upper and lower surfaces of the coating, the net reflectivity of a light flint glass surface coated with an MgF_2 film of optimal thickness is only about 17% that of the uncoated glass. (Multiple reflections are in fact significant and helpful; the actual value is 8.6%.)

36.55 (4) *Straightening the curves.* In locating the maxima and minima of a two-slit interference pattern (Figure 36.9),

we assumed that the distance D between the slits was very much less than the distance R between screens B and C. If the condition $R \gg D$ does not hold, the lines S_1P and S_2P are not approximately parallel, $\theta_1 \neq \theta_2$, and the simple analysis fails. Show that the approximations made in the analysis in the text are tantamount to approximating the hyperbolic lines of nodes and antinodes of Figure 36.8a by straight lines. (Hint: See Problem 36.37.)

36.56 (7) *Babinet's principle.* Two objects are *complementary* if each is opaque where the other is transparent and transparent where the other is opaque. A simple pair of complementary objects is (1) an opaque screen with a circular hole of radius a in it and (2) an opaque circular disk of radius a. But the principle applies to pairs of objects of any shape. According to *Babinet's principle*, the two objects have identical Fraunhofer diffraction patterns. Use the principle of superposition to prove Babinet's principle.

36.57 (G) *Michelson's stellar interferometer.* The figure below shows a *stellar interferometer*, invented by Michelson in 1920. The device is used to measure the angular separation between the two members of a double star when the two are too close together to be resolved by the telescope alone. The telescope is covered with a mask, and light enters only via the pathway shown. A filter (not shown) allows light of only one wavelength λ to pass to the observer. The two outer mirrors may be regarded as the slits of a two-slit interference apparatus, and the observer sees a two-slit interference pattern for each star in the field of view. The distance D between the mirrors can be varied. Suppose that two stars of approximately equal brightness are separated by a (very small) angle θ, as seen from the earth. (a) Show that the interference pattern disappears when $\theta = \lambda/(2D)$. (b) The interferometer can be used in a similar manner to determine the size of single, large, fairly nearby stars. Show that the circular diffraction pattern of a star disappears when $\theta = 1.22\lambda/D$. (c) Betelgeuse, a red giant in the constellation Orion, was the first star thus measured, at Mount Wilson in 1920. The diffraction fringes disappeared when the outer mirrors were 3.07 m apart. Assuming $\lambda = 575$ nm, find the angular diameter of Betelgeuse. (d) The distance to Betelgeuse can be measured by other means and is known to be about 200 light-years, or 1.8×10^{18} m. Find the diameter of Betelgeuse, and show that, if our sun were a red giant, the earth would lie within its atmosphere.

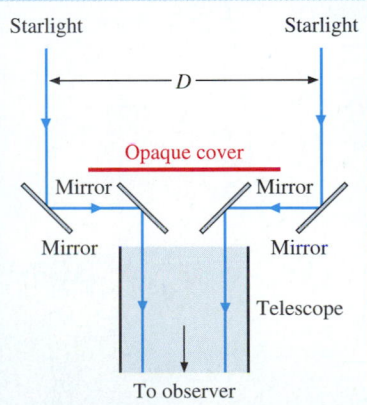

Starlight   Starlight

D

Opaque cover

Mirror   Mirror

Mirror   Mirror

Telescope

To observer

Geometric Optics

L O O K I N G   A H E A D

— Geometric optics ignores diffraction effects and concentrates on the paths of light rays through systems that contain plane or curved mirrors, lenses, or other optical components.

— Mirrors and lenses produce images of the objects from which light falls on them. The process is a one-to-one mapping; each point in the object corresponds to a point in the image.

— The mapping process is determined by the law of reflection (for mirrors) or the law of refraction (for lenses).

— The passage of paraxial rays from object to image, through a system of mirrors or lenses or both, can be analyzed geometrically or by means of a set of simple equations.

Left: The Hubble space telescope (see here in the assembly clean room) continues to revolutionize astronomy with its incomparable photographs.

More light!

—Johann Wolfgang von Goethe, reputed last words

Introduction

In *Through the Looking-Glass*, Alice (Figure 37.1) makes eloquent observations to her kitten about what she sees in the mirror:

> Now, if you'll only attend, Kitty, and not talk so much, I'll tell you all my ideas about Looking-Glass House. First, there's the room you can see through the glass—that's just the same as our drawing-room, only the things go the other way. I can see all of it when I get up on a chair—all but the bit just behind the fireplace. Oh! I do so wish I could see that bit! I want so much to know whether they have a fire in the winter: you never can tell, you know, unless our fire smokes, and then smoke comes up in

FIGURE 37.1 Alice is able to penetrate the mirror and pass from object space to image space. The illustrator, John Tenniel, has taken allowable liberties with those imaginary parts of the ''image space'' that have no counterparts in the object space; note the laughing face on the clock and the different flowers under the bell jar, neither of which could be images of objects on the other side of the mirror. But note the illustrator's monogram, seen at lower right and in mirror image at lower left of the two engravings. Compare with the signature of the engraver, visible to the left of the fireplace arch in the first engraving and to the right of it in the second.

that room too—but that may be only pretence, just to make it look as if they had a fire. Well then, the books are something like our books, only the words go the wrong way: I know that, because I've held up one of our books to the glass, and then they hold one up in the other room.

Alice's observations are of a kind familiar to all of us. Mirrors are opaque, and we see nothing of what is actually behind a mirror. Instead, we see *images* of some—but not all—of the *objects* on our side of the mirror. As Alice explains, each image strongly resembles the corresponding object, but there are important differences.

The space that appears to be behind the mirror is called an **image space**. The mirror *maps* the space on our side of the mirror, called an **object space**, into the image space. In this chapter, we see how mirrors and lenses map objects into images. By combining a little geometry with the laws of reflection and refraction, we develop a theory that makes possible reasonably accurate descriptions of the mappings. This theory is called **geometric optics**.

Using geometric optics, we will explore the properties of a number of important optical devices, including the magnifying glass, several types of telescopes, the microscope, and the human eye.

In developing geometric optics, we ignore diffraction effects. We look at objects and images in the large—that is, on such a scale that the smallest distance of interest is very large compared with the wavelength of light and the edges of shadows are sharp. On this scale, light travels in perfectly straight lines, or rays. For this reason, our development is sometimes called **ray optics**, in distinction to the **physical optics**, or **wave optics**, that was our main concern in Chapters 35 and 36.

Plane Mirrors

In Figure 37.2, a point object O emits light in all directions, and this emission is represented in the figure by a set of rays. Some of the rays strike a **plane** (flat) **mirror**; we call them **incident rays**. **Reflected rays** leave the mirror at angles determined by the law of reflection; for each ray, $\theta_{\text{reflected}} = \theta_{\text{incident}}$ (Equation 35.3).

Let us consider just those reflected rays that strike the eye of an observer. As Figure 37.3 shows, these rays all leave the object O in a diverging cone, and they arrive at the eye in a diverging cone. The eye does not "know" that the light rays have been bent en route by the mirror. The cone of rays arriving at the eye is no different from the cone of rays that would arrive at the eye from the same object placed at I with the mirror removed. We thus perceive the apex of the cone not at O but at I, situated behind the mirror. To locate I, we extend the reflected rays straight backward until they converge, as shown by the dashed lines in Figure 37.3. I is called the **image** of the point object O.

We classify all images as either *real* or *virtual*. A **real image** is one through which light rays actually pass. A **virtual image** is one through which light *appears* to pass. Point I is a virtual image because no light actually passes through it. *All images produced by plane mirrors are virtual.* A virtual image can be detected only with the aid

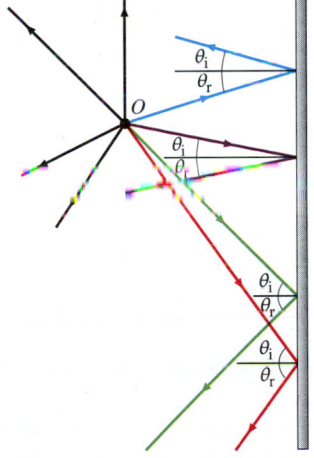

FIGURE 37.2 Some of the light rays emitted by point object O are incident on a plane mirror. The direction of each reflected ray is determined by the law of reflection, $\theta_r = \theta_i$.

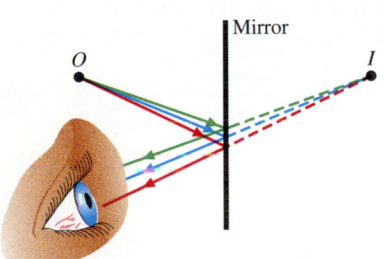

FIGURE 37.3 Construction showing that all light rays emitted by the point object O and incident on the mirror appear to come from the image point I. The image is virtual because no light rays actually pass through I.

of an optical device (the eye included). If you place a sheet of photographic film behind the mirror at I and expose it, it will record nothing. In contrast, a slide projector produces a real image on a screen. A sheet of film exposed at the screen will record the image.

All rays from O that strike the mirror contribute to the same image at I. If our purpose is to construct an image for a given object, it suffices to draw any two rays because all other rays will converge at the same point. In Figure 37.4, we construct an

FIGURE 37.4 Construction for locating the image $H'I$ of a one-dimensional object HO that lies in a plane parallel to the mirror and a distance s from it.

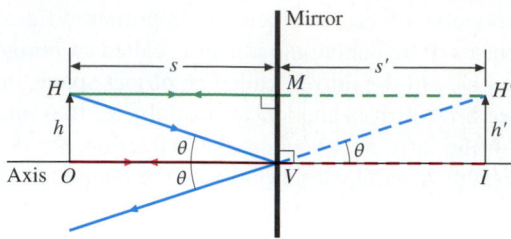

image of a one-dimensional object (represented conventionally as an arrow) lying in a plane parallel to the plane of the mirror. We choose to draw incident rays that make the construction easy—rays from the head H and the tail O of the arrow that are perpendicular to the mirror. We extend these rays through the mirror as shown. Rays HM and OV are reflected back on themselves. Next, we draw the incident ray HV and use the law of reflection to construct the reflected ray, together with its backward extension to H'.

Triangles HMV and $H'MV$ are congruent because they have equal angles and a common side. Thus the **object distance** s is equal to the **image distance** s'; *the image is as far behind the mirror as the object is in front of it.* Also, *the image height h' is equal to the object height h.* Moreover, the image is **erect**; that is, the object and image arrows have the same direction rather than opposite directions. We define the **lateral magnification** M of a mirror as the ratio

$$M \equiv \frac{\text{image height}}{\text{object height}} = \frac{h'}{h}. \tag{37.1}$$

Evidently, the lateral magnification of a plane mirror is $+1$; the positive sign denotes that the image is erect.

Any construction of the type of Figure 37.4, in which an image is located by finding the intersection of two or more rays emanating from the object, is called a **ray diagram**. Ray diagrams are widely useful in geometric optics, and we will soon apply them to systems more complicated than the plane mirror.

EXAMPLE 37.1

Your height is D. When you are standing upright, you can just see your entire image in a vertical mirror of height H, if the mirror is properly positioned. What is the value of H? Where must the upper edge of the mirror be located?

SOLUTION: As Figure 37.5 shows, you will just be able to see

your toes if the angle of incidence of the light ray from the floor to the lower edge of the mirror is equal to the angle from the bottom edge of the mirror to your eyes. Thus the height of the lower edge of the mirror above the floor must be one-half the height of your eyes above the floor. Similarly, you will just be able to see the top of your head if the angle of incidence of the

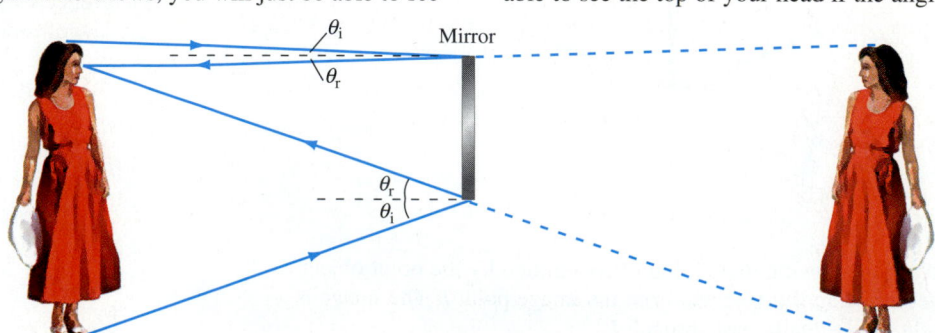

FIGURE 37.5

light ray from the top of your head to the upper edge of the mirror is equal to the angle from the upper edge of the mirror to your eyes. Thus the distance from the floor to the upper edge of the mirror must be equal to the distance from the floor to a point vertically midway between the top of your head and your eyes. So the height of the mirror is

$$H = \frac{1}{2}\left(\begin{array}{c}\text{distance from} \\ \text{floor to eyes}\end{array} + \begin{array}{c}\text{distance from eyes} \\ \text{to top of head}\end{array}\right) = \frac{D}{2}.$$

Does this result depend on your distance from the mirror?

The Mirror Image of a Three-Dimensional Object

The construction of Figure 37.4 is readily extended to three-dimensional objects and their images. We now carry out this extension so as to complete Alice's description of the properties of a plane mirror. Figure 37.6 shows a three-dimensional object, represented by a set of mutually perpendicular vectors **x**, **y**, and **z**. We construct the mirror image **x′**, **y′**, and **z′**. Now, **y** and **z** are parallel to the mirror plane. Because the lateral magnification of the mirror is +1, we conclude that **y′** = **y** and **z′** = **z**. The distance between the mirror and the *y′-z′* plane is equal to the distance between the *y-z* plane and the mirror. That is, we have $s_1' = s_1$.

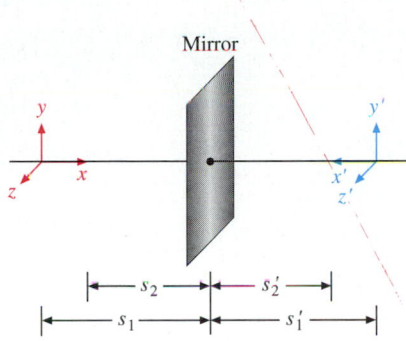

FIGURE 37.6 Construction showing that plane-mirror reflection transforms a right-handed object into a left-handed image, without change of scale.

What about the image of **x**, which lies along the normal to the mirror? The tail of **x** lies in the *y-z* plane, but its head lies closer to the mirror, at a distance s_2. Because the object and image distances must be equal for every pair of object and image points, we have $s_2' = s_2$. As you can see from Figure 37.6, **x′** = −**x**! To compare these two vectors directly, we rewrite **x′** = −**x** in the form $x'\hat{\mathbf{x}} = -x\hat{\mathbf{x}}$. This leads to the relation $x' = -x$. We define the **longitudinal magnification** L of the mirror to be the ratio

$$L \equiv \frac{x'}{x}. \qquad (37.2)$$

For a plane mirror, we have $L = -1$.

The relation between a mirror image and its object is often expressed in the familiar phrase "A mirror reverses left and right." But this statement is incorrect and confusing, as one of Lewis Carroll's young friends clearly understood. We can state the function of the mirror precisely as follows: *Plane-mirror reflection maps a right-handed object space into a left-handed image space, without change of scale.* Not surprisingly, mathematicians call this mapping **mirror inversion**. In optical practice, however, this terminology is avoided. A mirror image is called a **perverted** image (from a Latin word meaning "turned inside out"*) because the term *inverted* is restricted to the meaning "turned upside down." Although we have limited our discussion to plane mirrors, perverted images are a property of *all* mirrors; see Query 37.3 and Problem 37.77.

---

*To see the appropriateness of this terminology, think about what happens to a right glove when you turn it inside out.

Spherical Mirrors

Spherical mirrors map object space into image space in a richer variety of ways than plane mirrors do. Figure 37.7 shows just two of the many possibilities. In both cases shown, the image is *real* (otherwise it could not be cast on a screen), *inverted*, and *different in size* from the object. These are not the only possibilities; we will now develop the properties of spherical mirrors in general.

Figure 37.8 is a cross-sectional view of a **concave spherical mirror**. Concave mirrors are familiar as shaving or make-up mirrors and are the key elements in reflecting telescopes. The surface of a concave mirror is spherical, and the center of curvature *C*

(a)

(b)

FIGURE 37.7 A spherical mirror produces two real, inverted images of objects at different distances. (*a*) The image is inverted and slightly magnified. (*b*) The image is inverted and greatly demagnified (reduced).

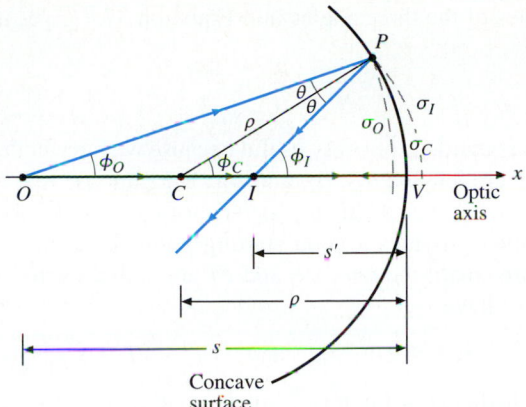

FIGURE 37.8 A concave spherical mirror of radius of curvature ρ is centered on the optic axis. The center of curvature lies at C, and a point object O is located on the axis, a distance s from the vertex V. The axial ray OV and an arbitrary ray OP are reflected by the mirror and converge at I, a distance s' from V, to form a (real) point image.

of the mirror lies *in front of* the mirror—that is, on the side from which light falls on the mirror. The radius of curvature is ρ. The line of symmetry drawn through the center of the mirror is called the **optic axis** (often shortened to **axis**). The point where the axis intersects the mirror surface is called the **vertex** V.

A point object O lies on the axis at a distance s from the vertex. As we did for the plane mirror, we draw incident rays from the object to the mirror surface. Our first choice is the ray easiest to draw—the ray emitted along the axis, toward the mirror. Its angle of incidence on the mirror surface is 90°, and it is reflected back on itself, as shown.

For our second choice, we draw the arbitrary ray OP. Because CP is the radius to P (and is thus normal to the plane that is tangent to the mirror surface at P), the angle $CPO \equiv \theta$ is the angle of incidence for OP. Using the law of reflection, we construct the equal angle $CPI = \theta$. The directed line PI is thus the reflected ray.

The two rays that left the object as OV and OP converge again at I, the image point. The distance between I and V is the image distance s'. Because light rays actually pass through the image point, the image is real.

We now derive an important equation that relates the object distance s, the image distance s', and the radius of curvature ρ of the mirror. For convenience, the angle between OP and the axis is labeled ϕ_O, the angle between CP and the axis is labeled ϕ_C, and the angle between IP and the axis is labeled ϕ_I. Now, ϕ_C is an external angle of the triangle COP and is therefore equal to the sum of the two opposite interior angles ϕ_O and θ:

$$\phi_C = \phi_O + \theta. \tag{37.3a}$$

Similarly, ϕ_I is an exterior angle of the triangle ICP, and we have

$$\phi_I = \phi_C + \theta. \tag{37.3b}$$

We eliminate θ from this pair of equations by subtracting the second one from the first. With a little rearrangement, this yields

$$\phi_O + \phi_I = 2\phi_C. \tag{37.4}$$

Next, we evaluate the angles ϕ_O, ϕ_I, and ϕ_C in terms of the distances s, s', and ρ. Let the distance from P to V, measured along the mirror surface, be σ_C. According to the definition of an angle in radians, we have

$$\phi_C = \frac{\sigma_C}{CV} = \frac{\sigma_C}{\rho}. \tag{37.5a}$$

To evaluate ϕ_O in like manner, we drop an arc from P to the axis, using O as a center; the length of this arc is σ_O. Then we drop an arc from P to the axis, using I as a center; the length of this arc is σ_I. In terms of these arc lengths, the angles are

$$\phi_O = \frac{\sigma_O}{OP} \quad \text{and} \quad \phi_I = \frac{\sigma_I}{IP}. \tag{37.5b,c}$$

Inserting these values of the three angles into Equation 37.4 gives us

$$\frac{\sigma_O}{OP} + \frac{\sigma_I}{IP} = \frac{2\sigma_C}{\rho}. \tag{37.6}$$

As it stands, this equation is not very useful because we have not explicitly evaluated OP, IP, or the three arc lengths σ_O, σ_I, and σ_C. But suppose that all three angles are small. For many optical systems, this is a satisfactory approximation. (Even when it is not, the approximation provides a good starting point for a more detailed analysis.) When ϕ_O and ϕ_I are small, the rays OP and PI are called **paraxial** (near-axis) **rays**. For paraxial rays, we have

$$OP \simeq OV = s \quad \text{and} \quad PI \simeq VI = s'. \tag{37.7a,b}$$

In addition, the dashed arcs in Figure 37.8 nearly coincide (for small ϕ) with the arc PV of the mirror; that is,

$$\sigma_O \simeq \sigma_C \quad \text{and} \quad \sigma_I \simeq \sigma_C. \tag{37.7c,d}$$

Using Equations 37.7a,b,c,d, we rewrite Equation 37.6 in the simpler form

$$\frac{\sigma_C}{s} + \frac{\sigma_C}{s'} = \frac{2\sigma_C}{\rho}.$$

Dividing through by σ_C, we obtain the **mirror equation**:

$$\frac{1}{s} + \frac{1}{s'} = \frac{2}{\rho}. \tag{37.8}$$

The fact that Equation 37.8 does *not* contain the angles ϕ is important. Although we have drawn Figure 37.8 by using only the axial ray OVI and one arbitrary paraxial ray OPI, Equation 37.8 is valid for *all* paraxial rays. Thus, all paraxial rays that emerge from O and strike the mirror will pass through the image point I.

Nonparaxial rays—rays that do not make a small angle with the axis—pass through the axis not at I but at points close to I. In this case, the image of the point source O is not a single point I but a smudgy region in the vicinity of I (Figure 37.9). This failure of all rays from a single point to converge at a single point is called **spherical aberration**. An optical system having aberration fails to map the object space into the image space perfectly. Correcting mirror and lens systems so as to minimize spherical aberration is one of the major tasks of the optical designer.

Figure 37.8 is drawn with the object O farther from the mirror than the center of curvature C is; that is, $s > \rho$. As a result, the image I lies closer to the mirror than C does; that is, $s' < \rho$. You can see from Equation 37.8 that as s increases, s' must decrease. When the object is moved farther from the mirror, the image moves closer to the mirror. In the extreme case, we have $s \rightarrow \infty$. Equation 37.8 then becomes

$$s' = \frac{\rho}{2}.$$

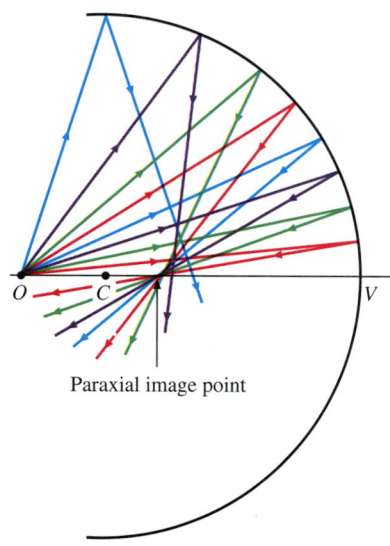

Paraxial image point

FIGURE 37.9 Spherical aberration for nonparaxial rays.

The situation is depicted in Figure 37.10, which shows the axial ray and two paraxial rays arriving at the mirror from a very distant object point. The object is so far from the mirror that the angles ϕ_O of the two paraxial rays are not appreciably different from zero. That is, the incident rays from the object are essentially parallel, even though the object is a point object. The image point is called the **focal point** or **focus** F. When $s \rightarrow \infty$, the image distance s' is assigned the special symbol f and is called the **focal length**. Using this notation, we have

$$f = \frac{\rho}{2}. \tag{37.9}$$

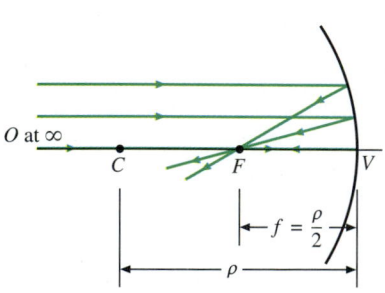

FIGURE 37.10 Essentially parallel rays from an object at infinity converge to form a point image at the focal point.

The focal length of a spherical mirror is determined entirely by its radius of curvature. Using Equation 37.9, we can rewrite the mirror equation (Equation 37.8) in the form

$$\frac{1}{s} + \frac{1}{s'} = \frac{1}{f}. \tag{37.10}$$

EXAMPLE 37.2

A concave spherical mirror has radius of curvature 50.0 cm. **(a)** Find the focal length of the mirror. **(b)** If a point object lies on the axis 65.5 cm from the vertex, where is the image? **(c)** If a point object lies on the axis 30.8 cm from the vertex, where is the image? **(d)** If a point object lies at the focus, where is the image?

SOLUTION:
(a) Find f.
   Equation 37.9 immediately gives you

$$f = \frac{50.0 \text{ cm}}{2} = 25.0 \text{ cm}.$$

(b) Find s' for $s = 65.5$ cm.
   You solve Equation 37.10 for $1/s'$, obtaining

$$\frac{1}{s'} = \frac{1}{f} - \frac{1}{s}. \tag{37.11}$$

Inserting the values of s and f, you have

$$\frac{1}{s'} = \frac{1}{25.0 \text{ cm}} - \frac{1}{65.5 \text{ cm}},$$

which yields $s' = 40.4$ cm.

As expected, the image lies between the mirror and its center of curvature.

(c) Find s' for $s = 30.8$ cm.
   In this case, the procedure of part **b** gives you

$$\frac{1}{s'} = \frac{1}{25.0 \text{ cm}} - \frac{1}{30.8 \text{ cm}}.$$

You solve for s', obtaining

$$s' = 132.8 \text{ cm}.$$

Because the object lies between the mirror and its center of curvature, the image lies beyond the center of curvature.

(d) Where is the image of an object located at f?
   When $s = f$, Equation 37.11 becomes

$$\frac{1}{s'} = \frac{1}{f} - \frac{1}{f} = 0.$$

It follows immediately that $s' = \infty$. That is, paraxial rays incident on the mirror from its focal point are reflected as parallel rays.

Virtual Images

Concave mirrors can produce virtual as well as real images. As shown in Figure 37.11, a virtual image is produced when the object distance s is less than the focal length f—that is, when the distance from the object to the vertex is less than one-half the radius of curvature. As always, the axial ray is reflected as an axial ray. But the paraxial ray OP is reflected as a *divergent* ray—a ray that moves farther and farther from the axis. To a viewer located far to the left, the rays appear to diverge from the image point I, located behind the mirror. Because light cannot pass from the object through the mirror to a point behind it, the image is virtual.

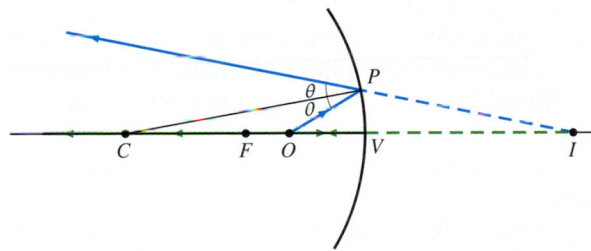

FIGURE 37.11 Formation of a virtual image by a concave mirror.

EXAMPLE 37.3

For the mirror of Example 37.2, whose focal length is 25.0 cm, find the position of the image when the object lies on the axis 20.3 cm from the vertex.

SOLUTION: Equation 37.11 gives you

$$\frac{1}{s'} = \frac{1}{f} - \frac{1}{s} = \frac{1}{25.0 \text{ cm}} - \frac{1}{20.3 \text{ cm}}.$$

Solving for s', you obtain

$$s' = -108.0 \text{ cm}.$$

The negative sign denotes a virtual image, located behind the mirror. We will expand on this and related sign conventions shortly.

Virtual Objects

Light rays from a point object always diverge. But you have seen that a concave mirror can be used to make the light converge toward a point. You will see in Section 37.5 that certain lenses can be used for the same purpose. In Figure 37.12, paraxial light rays from a point source far to the left have been made to converge by a lens somewhere to the left of the diagram. In the absence of the concave mirror, the rays would converge to form an image at O. However, the mirror is in the way and reflects the rays. They converge not at O but at I. Because light rays actually converge there, I is a real image.

FIGURE 37.12 A concave mirror produces a real image I from a virtual object O. The virtual object is itself the real image of a lens, not seen, to the left of the diagram.

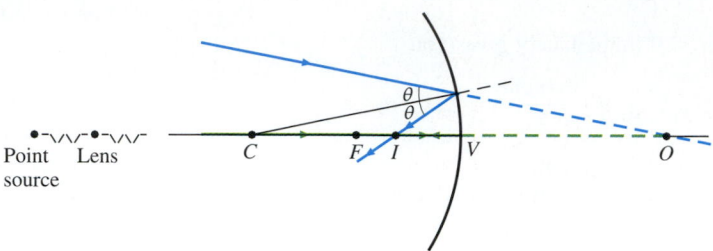

But where is the object corresponding to that image? If we consider the mirror alone, and not the optical element that makes the light converge toward O, the construction shows that the object lies at O, the point from which the light rays *appear* to diverge. Because no light passes through O, it is a **virtual object**. The point O can serve as a virtual object for the mirror only because light rays converge toward it—a situation that can exist only if another optical element is present that would produce a real image in the absence of the mirror.

Compare Figure 37.12 with Figure 37.11. You will see that they are identical, except that the object and the image have been interchanged. That is, the directions of all light rays have been reversed. As a result, the real object of Figure 37.11 is replaced by the real image of Figure 37.12, and the virtual image of Figure 37.11 is replaced by the virtual object of Figure 37.12. In optical systems, it is always possible to interchange objects with their corresponding images in this way. This interchangeability is a consequence of the **principle of optical reversibility**; see Problem 35.45.

Convex Mirrors

Spherical mirrors can be convex as well as concave. A **convex** mirror, shown in Figure 37.13, is one whose center of curvature C and focal point F lie *behind* the mirror. Convex mirrors are commonly used for security purposes in retail stores and as the right-side rear-view mirrors on automobiles. As Figure 37.13 shows, a convex mirror always produces a virtual image of a real object because diverging rays are made to diverge still more on reflection from the mirror. A real image, on the contrary, requires that actual rays (not their extensions behind the mirror) converge on an image point.

The mirror equation, $1/s + 1/s' = 1/f$, can be applied to *all* spherical mirrors, convex as well as concave. To do this, we generalize the sign convention already adopted for concave mirrors.

FIGURE 37.13 Formation of an image by a convex mirror. The image is virtual even though $|s| > |f|$.

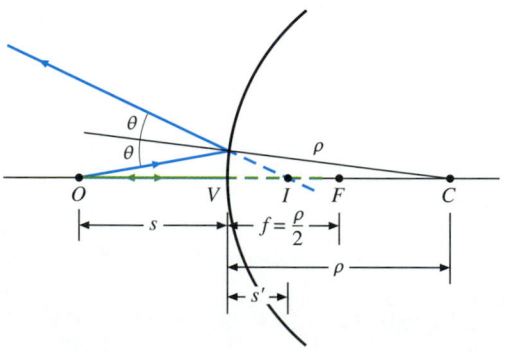

SIGN CONVENTION FOR MIRRORS

- An object O located *in front of* the mirror is *real*, and s is taken *positive*.
- An object O located *behind* the mirror is *virtual*, and s is taken *negative*.

- An image I located *in front of* the mirror is *real*, and s' is taken *positive*.
- An image I located *behind* the mirror is *virtual*, and s' is taken *negative*.

- The center of curvature C and focus F of a concave mirror are located *in front of* the mirror; ρ and f are taken *positive*.
- The center of curvature C and focus F of a convex mirror are located *behind* the mirror; ρ and f are taken *negative*.

EXAMPLE 37.4

The concave mirror of Examples 37.2 and 37.3 is replaced by a convex mirror whose radius of curvature has the same magnitude, 50.0 cm. **(a)** Find the focal length of the mirror. **(b)** If a point object lies in front of the mirror on the axis 65.5 cm from the vertex, where is the image? **(c)** If a point object lies in front of the mirror on the axis 30.8 cm from the vertex, where is the image? **(d)** If a point object lies at the focus, where is the image? **(e)** Let the object be a virtual object lying between the vertex and the focus, at a distance 20.3 cm from V. Where is the image?

SOLUTION:

(a) Find f.

According to the sign convention for mirrors, the radius of curvature is $\rho = -50.0$ cm. Equation 37.9 thus yields

$$f = \frac{-50.0 \text{ cm}}{2} = -25.0 \text{ cm}.$$

The negative sign means that the focal point lies behind the mirror, as shown in Figure 37.14a. Just as for the concave mirror, F lies midway between V and C.

(b) Find s' if $s = 65.5$ cm.

As in Example 37.2, you begin with the mirror equation in the form of Equation 37.11,

$$\frac{1}{s'} = \frac{1}{f} - \frac{1}{s}.$$

The object point O is in front of the mirror, and so $s = 65.5$ cm. Inserting the values of s and f, you have

$$\frac{1}{s'} = \frac{1}{-25.0 \text{ cm}} - \frac{1}{65.5 \text{ cm}},$$

which yields $\quad s' = -18.1$ cm.

FIGURE 37.14

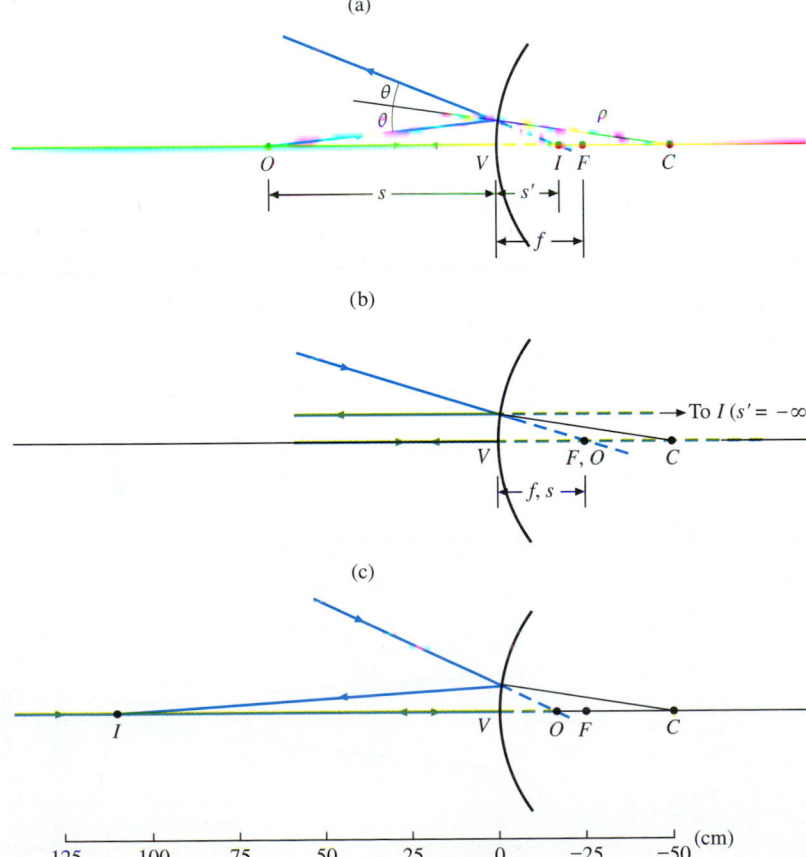

The image lies behind the mirror and is virtual. The construction of Figure 37.14a yields the same result as the calculation.

(c) Find s' if $s = 30.8$ cm.

In this case, the procedure of part **b** gives you

$$\frac{1}{s'} = \frac{1}{-25.0 \text{ cm}} - \frac{1}{30.8 \text{ cm}}.$$

You solve for s', obtaining

$$s' = -13.8 \text{ cm}.$$

Bringing the object closer to the mirror moves the image closer to the mirror as well. This contrasts with the concave-mirror case of Example 37.2.

(d) Find s' if $s = f$.

An object lying at the focus must be a virtual object because the focus is behind the mirror. When $s = f$, Equation 37.11 becomes

$$\frac{1}{s'} = \frac{1}{f} - \frac{1}{f} = 0.$$

There are two possible solutions to this equation, $s' = \infty$ and $s' = -\infty$. (In Example 37.2, we ignored the second of these solutions.) According to either solution, light rays that would have converged at F in the absence of the mirror are reflected as parallel rays. When you look into the mirror, the light appears to come from a virtual object behind the mirror, and you choose the solution $s' = -\infty$ as the physically meaningful one; see Figure 37.14b.

(e) Locate the image for a virtual object 20.3 cm from V.

For this object, you have $s = -20.3$ cm. Now Equation 37.11 gives you

$$\frac{1}{s'} = \frac{1}{-25.0 \text{ cm}} - \frac{1}{-20.3 \text{ cm}}.$$

This yields the solution

$$s' = +108.0 \text{ cm}.$$

The image lies in front of the mirror and is real (Figure 37.14c). A convex mirror *can* produce a real image, provided the (virtual) object lies between the focus and the vertex.

Lateral Magnification of Spherical Mirrors

Lateral magnification M has the same meaning for a spherical mirror as for a plane mirror; according to Equation 37.1,

$$M \equiv \frac{h'}{h}.$$

To evaluate M for a spherical mirror, we again construct a ray diagram—Figure 37.15a for a concave mirror and Figure 37.15b for a convex mirror. The following argument applies to both.

We begin with any two incident rays from the head H of the object HO and then construct the reflected rays. The intersection of the reflected rays locates the head H' of the image $H'I$. It makes sense to choose two particular incident rays that are easy to construct and make the geometrical analysis simple. In Figure 37.15, ray 1, HC, travels along a mirror radius. It therefore strikes the mirror at normal incidence and is reflected

FIGURE 37.15 Ray diagrams used to determine the lateral magnification of (*a*) a concave mirror and (*b*) a convex mirror. In part *a*, the image is real, inverted, and smaller than the object. In part *b*, the image is virtual, erect, and smaller than the object.

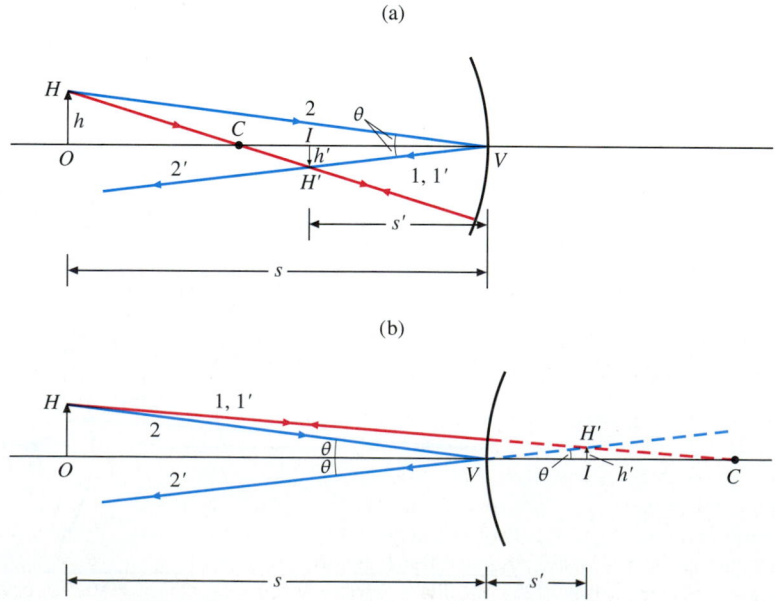

back on itself as ray 1'. (In the convex-mirror case the ray does not actually reach C.) Ray 2 strikes the mirror at the vertex V, making an angle θ with the axis. Its reflection, ray 2', therefore makes an equal angle θ on the other side of the axis. The image lies at H', where the reflected rays (or their extensions) intersect. The concave-mirror image is *real* and *inverted*; the convex-mirror image is *virtual* and *erect*.

In both cases, triangle $H'IV$ is similar to triangle HOV. We can thus write magnification M in terms of the proportionality

$$M \equiv \frac{h'}{h} = -\frac{s'}{s}. \tag{37.12}$$

In Figure 37.15a, M is negative because s and s' are both positive. The negative value of M signifies that the image is inverted. In Figure 37.15b, M is positive because s is positive and s' is negative. The positive value of M signifies that the image is erect.

More on Ray Diagrams for Mirrors

Although ray diagrams are not as accurate as the mirror equation and Equation 37.12 in predicting the location and size of the image produced by a spherical mirror, they are very useful sources of quick insight into the physical situation. A fairly rough sketch often suffices for this purpose. In making such sketches, we almost always draw two rays, choosing from four incident rays that are especially convenient. These four rays are shown in Figure 37.16 for the particular arrangement of Figure 37.15a. We have already used the first two rays in the preceding subsection.

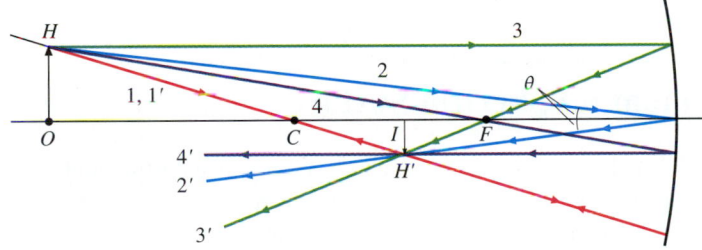

FIGURE 37.16 Four rays and their reflections, useful in constructing ray diagrams.

- 1, 1': Ray 1, passing through the center of curvature C, is reflected back on itself as ray 1', as already noted. Ray 1 is called the **radial ray** because it lies on a radius of the mirror.

- 2, 2': Ray 2, striking the vertex V at an angle θ with the axis, is reflected as ray 2' at the same angle θ on the opposite side of the axis. Ray 2 is called the **vertical ray**.

- 3, 3': Ray 3, parallel to the axis, is reflected as ray 3'. Either ray 3' or its extension passes through the focus F. Ray 3 is called the **parallel ray**.

- 4, 4': Ray 4, or its extension, passes through the focus F. It is reflected as ray 4', which is parallel to the axis. Ray 4 is called the **focal ray**.

Note that all four rays intersect at the image H'. This is not surprising; *all* paraxial rays intersect at H'.

EXAMPLE 37.5

In Figure 37.15, both mirrors produce *demagnified images*—that is, images smaller than their objects. But this is not always the case, as you will see in this example. **(a)** Draw a ray diagram, using the radial and parallel rays, to find the position and lateral magnification of an object whose distance from the vertex of a concave mirror is less than its focal length. Let $\rho = 200$ cm, $s = 53.0$ cm, and $h = 2.0$ cm. **(b)** Compare the results

of your construction with the results obtained using Equations 37.11 and 37.12.

SOLUTION:
(a) Draw a ray diagram for the system.

By comparing the values of h and s, you can see that the paraxial condition is well satisfied. However, it would be awkward to draw a ray diagram to scale because a scale convenient along the system axis (the x axis) would be too small in the perpendicular y direction. Instead, you use a scale like that shown in Figure 37.17, where the y direction is magnified 20-fold.

FIGURE 37.17

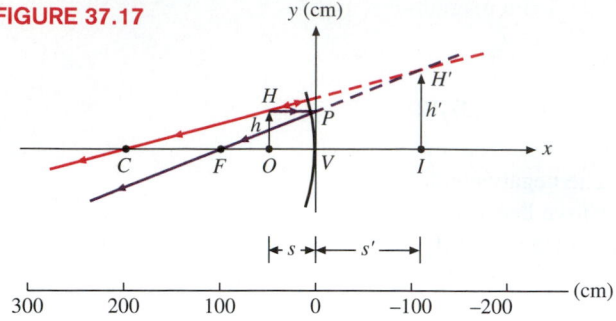

You begin by drawing the mirror and its axis, which you label as the x axis. You mark in the x and y scales and use them to locate and draw the object HO. You draw the radial ray and extend it behind the mirror. Then you draw the parallel ray HP. Note that P lies on the y axis ($x = 0$) and not on the mirror surface. This is in the spirit of the paraxial approximation. (The shape of the mirror, drawn conventionally as a circular arc, is not consistent with the expanded scale of the y axis.)

The parallel ray is reflected as a focal ray PF. You draw this ray and extend it backward. The two ray extensions intersect at H', and you drop a perpendicular to I. $H'I$ is the image of HO. The image is virtual, erect, and *magnified*—that is, larger than the object. You have doubtless exploited this effect in using the common bathroom convenience called a "magnifying mirror."

(b) Compare the results of the construction with those obtained algebraically.

The graph shows that $s' = -110$ cm. The height of the image is $h' = 4.2$ cm. For comparison, you use Equation 37.11, $1/s' = 1/f - 1/s$. You have $f = \rho/2 = 100$ cm, and thus

$$\frac{1}{s'} = \frac{1}{100 \text{ cm}} - \frac{1}{53.0 \text{ cm}}.$$

Solving, you obtain $s' = -113$ cm. The graph permits only two significant figures, and the two solutions agree within this limitation.

Equation 37.12 gives you the lateral magnification

$$M = -\frac{-113 \text{ cm}}{53 \text{ cm}} = 2.13.$$

The image size is thus $h' = Mh = 2.13 \times 2.0$ cm $= 4.26$ cm. Again, this is in reasonable agreement with the graphical solution.

SECTION 37.4 Refraction at a Spherical Surface

The geometric analysis of lenses has much in common with the analysis of mirrors. However, a lens is more complicated than a mirror in that the lens has two surfaces at which light rays are bent, rather than one. In working up to this additional complication, we begin with the hypothetical *single* spherical refracting surface shown in Figure 37.18. We grind a spherical surface of radius ρ on the end of a semi-infinite glass rod. The index of refraction of the medium in front of (to the left of) the surface is n_1 (which we can usually set equal to 1), and the index of refraction of the glass is n_2. For concreteness, we assume that $n_2 > n_1$.

FIGURE 37.18 Diagram for locating the image produced by a convex surface, of radius ρ, that is the boundary between media of refractive indices n_1 and n_2.

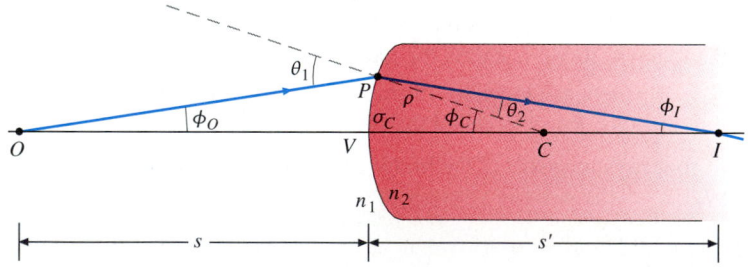

A point object O is located on the axis, a distance s from the vertex V. The axial ray OV is normally incident on the surface and therefore passes through without refraction. The arbitrary ray OP strikes the surface at an angle of incidence θ_1. The ray is refracted at the surface and continues in a direction determined by the angle of refraction θ_2. To relate θ_2 to θ_1, we use Equation 35.5a, Snell's law:

$$n_1 \sin \theta_1 = n_2 \sin \theta_2. \qquad (37.13)$$

As we did for spherical mirrors, we make the paraxial approximation that θ_1 and θ_2 are small angles. Snell's law then assumes the simpler form

$$n_1\theta_1 = n_2\theta_2. \tag{37.14}$$

Because we have assumed $n_2 > n_1$, the ray is bent toward the normal at P, which is the radius CP. The refracted ray intersects the axial ray at I. For the particular case drawn in Figure 37.18, light rays pass through I, which is thus a real image. The angles ϕ_O, ϕ_I, and ϕ_C are defined exactly as in the mirror case; compare with Figure 37.8.

In Figure 37.18, θ_1 is an external angle of triangle POC, and we thus have

$$\theta_1 = \phi_O + \phi_C. \tag{37.15a}$$

Also, ϕ_C is an external angle of triangle CIP, and thus

$$\phi_C = \phi_I + \theta_2. \tag{37.15b}$$

We eliminate θ_2 from Equations 37.14 and 37.15b. A little rearrangement gives

$$n_1\theta_1 = n_2\phi_C - n_2\phi_I.$$

Next, we multiply Equation 37.15a through by n_1 and subtract the result from the preceding equation to eliminate θ_1. A little more rearrangement yields

$$n_1\phi_O + n_2\phi_I = (n_2 - n_1)\phi_C. \tag{37.16}$$

This equation is analogous to Equation 37.4, $\phi_O + \phi_I = 2\phi_C$, which applies to spherical mirrors.

We exploit the paraxial approximation once more. Let the length of the arc PV be σ_C. We can write the ϕ's according to Equations 37.5 and 37.7:

$$\phi_C = \frac{\sigma_C}{\rho}, \quad \phi_O \simeq \frac{\sigma_C}{s}, \quad \text{and} \quad \phi_I \simeq \frac{\sigma_C}{s'}.$$

We substitute these values into Equation 37.16 and cancel the common factor σ_C to obtain the result

$$\frac{n_1}{s} + \frac{n_2}{s'} = \frac{n_2 - n_1}{\rho}. \tag{37.17}$$

This equation is analogous to Equation 37.8, the mirror equation $1/s + 1/s' = 2/\rho$. Just as in the mirror equation, the angles ϕ do not appear. That is, the analysis is valid for *all* paraxial rays, and all paraxial rays emerging from the object point O will pass through the same image point I.

We have derived Equation 37.17 for the special case of a convex surface with $n_2 > n_1$. But, just as in the mirror case, the result is universally applicable if we employ a consistent sign convention.

SIGN CONVENTION FOR REFRACTING SURFACES
- An object O located *in front of* the surface is *real*, and s is taken *positive*.
- An object O located *behind* the surface is *virtual*, and s is taken *negative*.

- An image I located *behind* the surface is *real*, and s' is taken *positive*.
- An image I located *in front of* the surface is *virtual*, and s' is taken *negative*.

- If the center of curvature C of the surface is located *behind* the surface, ρ is taken *positive*.
- If the center of curvature C of the surface is located *in front of* the surface, ρ is taken *negative*.

Although this set of conventions may seem to conflict with the one adopted for mirrors, there is a simple reason for the differences. For mirrors, real objects and images lie on the *same* side of the mirror, as do virtual objects and images. For refracting surfaces, real objects and images lie on *opposite* sides of the surface, as do virtual objects and images.

EXAMPLE 37.6

The glass sphere in Figure 37.19 has radius 25 cm. The index of refraction of the glass is 1.5. Within the sphere there is a small bubble, located 15 cm from the surface. You turn the sphere so that you are looking at the bubble nearly along a radius of the sphere. How far below the glass surface does the bubble appear to be?

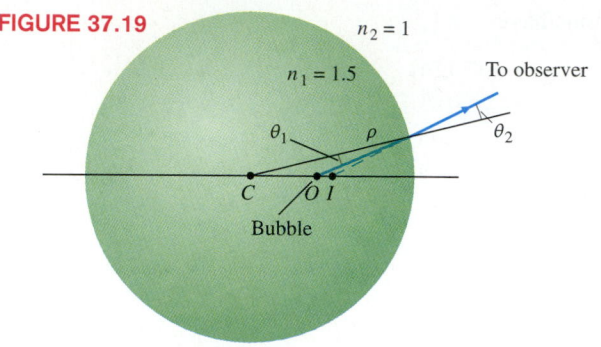

FIGURE 37.19

SOLUTION: You begin by solving Equation 37.17 for $1/s'$, the reciprocal of the image distance:

$$\frac{n_2}{s'} = \frac{n_2 - n_1}{\rho} - \frac{n_1}{s}. \qquad (37.18)$$

The light rays originate at the bubble. Thus the first medium is glass and the second is air. With this in mind, you take $n_1 = 1.5$ and $n_2 = 1$. The object is real, and it lies in front of (to the left of) the refracting surface. You use the sign convention for refracting surfaces to assign the positive value $s = 15$ cm to the object distance.

The center of curvature of the surface lies in front of the surface. According to the sign convention for refracting surfaces, the radius of curvature therefore has the negative value $\rho = -25$ cm. Equation 37.18 thus becomes

$$\frac{1}{s'} = \frac{1 - 1.5}{-25 \text{ cm}} - \frac{1.5}{15 \text{ cm}}.$$

You solve for s and obtain

$$s' = -12.5 \text{ cm}.$$

According to the sign convention for refracting surfaces, the negative sign implies that the image lies in front of the surface and is virtual. You can reach the same conclusion by drawing a ray diagram like that of Figure 37.19. As is often the case, the ray diagram gives a clear picture of the situation but does not yield highly accurate quantitative results. Aside from the limits to precision inherent in the drawing itself, the ray diagram does not really satisfy the paraxial approximation.

SECTION 37.5 Thin Lenses

We are now ready to apply the ideas developed in Section 37.4 to the **lens**, with its two refracting surfaces. Figure 37.20 shows a **thick lens**—that is, a lens in which the distance t between the two vertices V_1 and V_2 is not negligible compared with the other distances of interest. The lens in Figure 37.20 happens to have one convex surface and one concave surface, but that does not affect the following analysis.

FIGURE 37.20 Diagram showing paraxial rays passing through an arbitrary lens. For purposes of clarity, the angle between the axial and pardaxial rays is exaggerated. Also, the value chosen for the index of refraction ($n = 2$) is unrealistically large for glass.

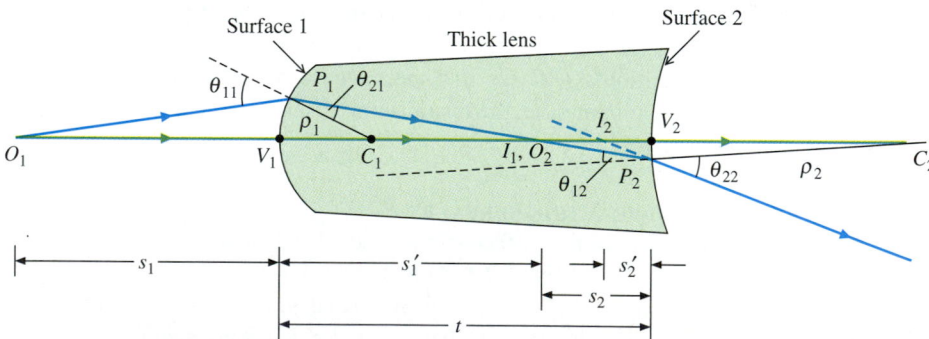

By means of a construction like that of Figure 37.18, we locate the image I_1 produced when the incident paraxial ray O_1P_1 is refracted at surface 1. The image (which happens to be real) serves as an object O_2 for surface 2. Next, we locate the image I_2 produced when the incident paraxial ray O_2P_2 is refracted at surface 2. In this case, the resulting image happens to be virtual; it is located by extending the refracted paraxial ray back from P_2 until it intersects the axial ray at I_2.

Image I_2 is thus the product of successive refractions at surfaces 1 and 2. We would like to find its position s'_2. To do this, we use Equation 37.17 twice. Let the index of refraction of the lens be n, and assume that the medium surrounding the lens is air, whose index of refraction is 1. We then have

$$\frac{n}{s_1'} = \frac{n-1}{\rho_1} - \frac{1}{s_1} \quad \text{and} \quad \frac{1}{s_2'} = \frac{1-n}{\rho_2} - \frac{n}{s_2}. \quad \textbf{(37.19a,b)}$$

As you can see from the diagram, the object distance s_2 and the image distance s_1' are connected by the thickness t of the lens:

$$s_2 = t - s_1'. \quad \textbf{(37.19c)}$$

We use this relation to eliminate s_2 from Equation 37.19b and obtain

$$\frac{1}{s_2'} = \frac{1-n}{\rho_2} - \frac{n}{t - s_1'}. \quad \textbf{(37.20)}$$

We could continue by using Equation 37.19a to eliminate s_1' from Equation 37.20. The resulting equation expresses the final image distance s_2' in terms of the object distance s_1 and the quantities ρ_1, ρ_2, and t, which are properties of the lens. In general, this equation is complicated and difficult to interpret. But there is an important special case for which the solution is simple. For a **thin lens**, t is negligible compared with all the other relevant distances. That is, the two refracting surfaces are very close together. In this case, we can make the **thin-lens approximation** $t \to 0$, and Equation 37.20 simplifies to

$$\frac{n}{s_1'} = \frac{1}{s_2'} + \frac{n-1}{\rho_2}.$$

We substitute this value of n/s_1' into Equation 37.19a and rearrange terms to obtain

$$\frac{1}{s_1} + \frac{1}{s_2'} = (n-1)\left(\frac{1}{\rho_1} - \frac{1}{\rho_2}\right).$$

The final step is simple. We are no longer interested in the intermediate image/object (I_1, O_2). Rather, we are interested in *the* image I produced by the thin lens acting as a single optical device. With this in mind, we drop the subscripts on the left side of the equation and rewrite it in the form

$$\frac{1}{s} + \frac{1}{s'} = (n-1)\left(\frac{1}{\rho_1} - \frac{1}{\rho_2}\right). \quad \textbf{(37.21)}$$

For a given object distance, the image distance is determined by the three properties of a thin lens: the radii of curvature ρ_1 and ρ_2 of the two surfaces and the index of refraction n of the lens material.

Consider what happens when the object is removed to a large distance—that is, when $s \to \infty$. As we similarly did for spherical mirrors, we call the image point the **second focal point** or the **second focus** F_2 of the lens (Figure 37.21a). The distance between the focal point and the lens is the **focal length** f. Equation 37.21 then becomes

$$\frac{1}{f} = (n-1)\left(\frac{1}{\rho_1} - \frac{1}{\rho_2}\right). \quad \textbf{(37.22)}$$

Just a few of the lenses available for one model of camera.

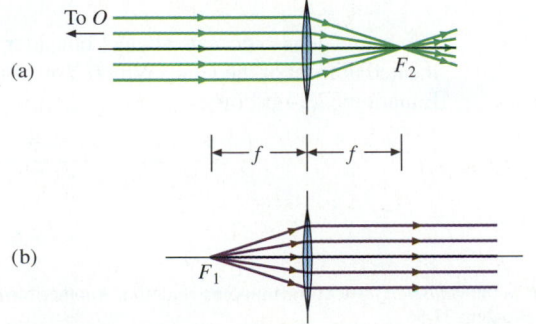

FIGURE 37.21 Definition of the focal length of a thin lens. (*a*) Parallel incident rays converge on the second focus F_2. (*b*) Rays incident on the lens from an object at the first focus F_1 are refracted to form an image at infinity. That is, the rays are parallel.

This is called the **lensmaker's equation** because it tells the lensmaker what focal length will be obtained by grinding a thin lens with surfaces of given curvature on a glass blank of known index of refraction. Note an important feature of the lensmaker's equation: It is not affected by turning the lens around so as to interchange the front and back surfaces. When you do so, you change the signs of both radii ρ. But you also interchange ρ_1 and ρ_2 with respect to the minus sign in the equation. The net result is no change at all.

Now that we have determined the value of s' when $s = \infty$, let us determine the value of s for which $s' = \infty$. That is, consider the location of the object for which the light emerges from the lens as a bundle of parallel rays, as shown in Figure 37.21b. This object location is called the **first focal point** or the **first focus** F_1. You can see from Equation 37.21 that F_1 and F_2 lie at equal distances f from the lens. Thus Equation 37.22 applies equally well to both foci.

By combining Equations 37.21 and 37.22, we obtain the frequently used relation

$$\frac{1}{s} + \frac{1}{s'} = \frac{1}{f}. \tag{37.23}$$

This equation is called the **thin-lens equation**.* In using it, we follow the sign convention for single refracting surfaces. That sign convention, taken together with the lensmaker's equation, leads to the following sign rule for the focal length f:

- A lens having *positive* focal length f has its first focus F_1 *in front of* the lens and its second focus F_2 *behind* the lens. Such a lens tends to converge incident rays and is called a **positive** or **converging lens**.

- A lens having *negative* focal length f has its first focus F_1 *behind* the lens and its second focus F_2 *in front of* the lens. Such a lens tends to diverge incident rays and is called a **negative** or **diverging lens**.

Figure 37.22 shows thin lenses of various forms. In general, a lens that is thickest at the center is a converging lens, and a lens that is thinnest at the center is a diverging lens.

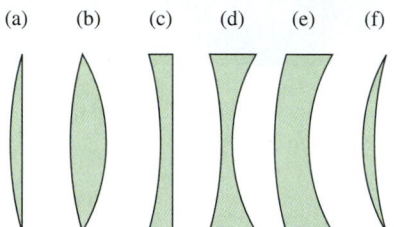

(a) (b) (c) (d) (e) (f)

FIGURE 37.22 Some common lens forms: (*a*) planoconvex, (*b*) biconvex, (*c*) planoconcave, (*d*) biconcave, (*e*) diverging convexoconcave, and (*f*) meniscus (converging convexoconcave). Simple biconvex and biconcave lenses are often (but not always) made with surfaces of equal curvature.

EXAMPLE 37.7

A biconvex lens has surfaces with equal radii of curvature 25.0 cm. The index of refraction of the glass is 1.60. **(a)** Find the focal length of the lens. **(b)** Find the image of a real point object located on the lens axis, 55.5 cm from the lens. Is the image real or virtual? **(c)** Find the image of a point object located on the lens axis, 15.0 cm from the lens. Is the image real or virtual?

SOLUTION:

(a) Find f.

Sketch the lens as in Figure 37.23. The center of curvature C_1 of the first surface lies behind the lens, and the sign convention for lenses thus requires you to set $\rho_1 = +25.0$ cm. The center C_2 of the second surface lies in front of the lens, and you

thus have $\rho_2 = -25.0$ cm. The lensmaker's equation gives you

$$\frac{1}{f} = (1.60 - 1)\left(\frac{1}{25.0 \text{ cm}} - \frac{1}{-25.0 \text{ cm}}\right).$$

Solve for f (be careful of the signs!) to obtain

$$f = 20.8 \text{ cm}.$$

The positive value of f means that the lens is a converging lens.

(b) Find the image of a real point object 55.5 cm from the lens. Is the image real or virtual?

From the thin-lens equation, you have

$$\frac{1}{s'} = \frac{1}{f} - \frac{1}{s}. \tag{37.24}$$

For a thin lens, you can always consider a real object to be located in front of the lens. (Why?) You thus set $s = +55.5$ cm. Equation 37.24 becomes

$$\frac{1}{s'} = \frac{1}{20.8 \text{ cm}} - \frac{1}{55.5 \text{ cm}}.$$

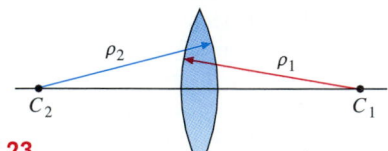

FIGURE 37.23

---

*More specifically, this is the *Gaussian form* of the thin-lens equation. Another form, called the *Newtonian form*, is the subject of Problem 37.56.

You solve for s', obtaining

$$s' = 33.4 \text{ cm}.$$

The positive value means that the image is located behind the lens and is real.

(c) Find the image of a real point object 15.0 cm from the lens. Is the image real or virtual?

Equation 37.24 now becomes

$$\frac{1}{s'} = \frac{1}{20.8 \text{ cm}} - \frac{1}{15.0 \text{ cm}},$$

which yields $s' = -53.8$ cm. This negative results denotes an image in front of the lens, which must be virtual.

Ray Diagram for a Thin Lens

The construction of a ray diagram for a thin lens is shown in Figure 37.24. The procedure is very much like that for a spherical mirror. However, there are differences in detail. A one-dimensional object HO lies at a distance s from the vertex V of the lens. In reality, a lens has two vertices, one for each surface. But in the spirit of the thin-lens approximation, the distance t between the vertices is negligible, so the two coincide at V. We take one further step in this direction; we regard the two refracting surfaces as coincident at the plane represented by the vertical line PVP', called the **principal plane**.

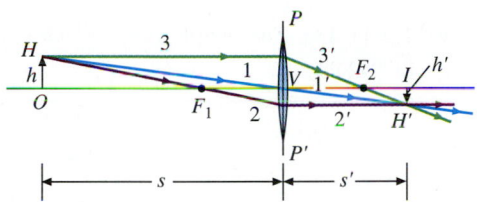

FIGURE 37.24 Ray diagram for a thin lens.

The focal points F_1 and F_2 also are drawn in, at equal distances f from the vertex.

Three rays emanating from the object arrowhead H are particularly easy to trace through the system:

- 1, 1′: Ray 1 passes through the vertex essentially undeviated because the lens is thin (Problem 37.14). Ray 1, with its extension 1′, is called the **chief ray**.

- 2, 2′: Ray 2 passes through the first focus F_1. It is refracted as ray 2′, which is parallel to the axis. Ray 2 is called the **focal ray**. Although the ray is bent twice (once at each lens surface), we regard the two refractions as coinciding at the line PVP'.

- 3, 3′: Ray 3, parallel to the axis, is refracted as ray 3′, which passes through the second focus F_2. Ray 3 is called the **parallel ray**.

All three rays converge at the image point H'. Indeed, all paraxial rays emanating from H converge at H'. Any two of the rays thus suffice to construct the image $H'I$ of the object HO. Why do we not need to repeat the construction for points on HO other than the arrowhead H?

Just as for mirrors, we define the **lateral magnification** of the lens as the ratio $M \equiv h'/h$. In Figure 37.24, the triangles HOV and $H'IV$ are similar. We thus have

$$M \equiv \frac{h'}{h} = -\frac{s'}{s}. \tag{37.25}$$

This equation is the same as Equation 37.12, which was derived for spherical mirrors. As before, the negative sign accounts for the fact that the image is inverted when both s and s' are positive, as is the case in Figure 37.24.

EXAMPLE 37.8

A concave lens has focal length $f = -20.0$ cm. **(a)** Use a ray diagram to find the image of a real object of height $h = 2.0$ cm, located a distance $s = 45.5$ cm from the lens vertex. **(b)** Check your results algebraically.

SOLUTION: Find s' and h' graphically.

You begin by choosing convenient x and y scales with the y scale exaggerated, just as in Example 37.5. Because the focal length is negative, the first focus is behind the lens, and the

FIGURE 37.25

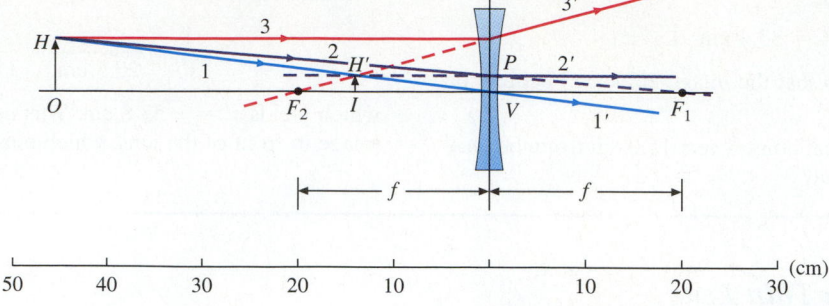

second focus is in front of the lens. You locate these foci on the diagram, as shown in Figure 37.25. Then you draw in the object *HO*.

You can draw any two of the rays shown in Figure 37.24. But, because the chief ray is the easiest to draw, it almost always makes sense to choose it as one of the two. So you draw *HV* and extend it to the right of the lens.

You need one more ray. Suppose you choose the parallel ray, 3. It intersects the lens at *P*. Its extension, ray 3′, must pass through the *second* focus F_2. In the present case, F_2 lies in front of the lens. Thus it is not ray 3′ but its backward extension (shown dashed in Figure 37.25) that passes through F_2. This is consistent with the fact that the diverging lens refracts ray 3′ away from the axis.

Ray 1 and the backward extension of ray 3′ intersect at *H*′.

This is the image of the arrowhead *H*. The image is erect.

As a check, draw ray 2, the focal ray. This is the ray that passes through the *first* focus F_1. Before it reaches F_1, however, it is refracted to form the parallel ray 2′. The backward extension of ray 2′ indeed passes through the image point *H*′. Reading from the graph, you have $s' = -14$ cm and $h' = +0.6$ cm.

(b) Check the results algebraically.

To check algebraically, use Equation 37.24, $1/s' = 1/f - 1/s$. This yields $s' = -13.9$ cm. Equation 37.25 then gives you

$$h' = Mh = -\frac{s'}{s}h = -\frac{-13.9 \text{ cm}}{45.5 \text{ cm}} \times 2.0 \text{ cm} = 0.61 \text{ cm},$$

in satisfactory agreement with the graphical results.

Series of Thin Lenses

In calculating the effect of two refracting surfaces on a light ray passing through them, we treated the surfaces in sequence. We use the same procedure in locating the image resulting from the passage of light rays from an object through two or more lenses lying on the same axis. The general solution of the problem is messy, and such systems are usually treated case by case. However, the solution is particularly simple when two thin lenses are very close to one another. If the focal lengths of the lenses are f_1 and f_2, the focal length f of the combination is given by the expression

$$\frac{1}{f} = \frac{1}{f_1} + \frac{1}{f_2}. \tag{37.26}$$

The pair of lenses can thus be treated as a single lens of focal length f. Problem 37.66 is a proof of this relation.

SECTION 37.6

Optical Instruments

The Eye

The mammalian eye is a complicated structure, but it is not difficult to understand how it produces images, at least to first approximation. The human eye is fairly typical of mammalian eyes and is shown schematically in Figure 37.26. Light enters through the transparent *cornea*, whose form is approximately spherical. The light then passes through a chamber containing clear liquid called the *aqueous humor*. At the rear of the chamber is a muscular, ring-shaped structure called the *iris*. (It is the pigmented membranous cover of the iris that determines eye color.) The circular aperture in the ring is called the *pupil*. By contracting and relaxing under control of the autonomic nervous

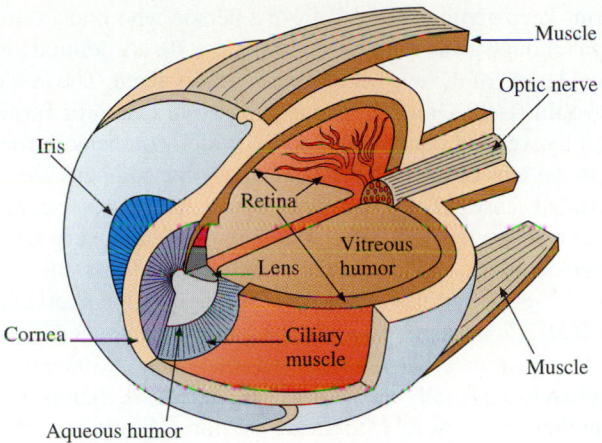

FIGURE 37.26 The human eye.

system, the iris regulates the diameter of the pupil and thus the amount of light passing farther into the eye.

Just behind the iris is the *crystalline lens*, an elastic, transparent capsule. The lens is mounted in the *ciliary muscle*. This muscle can squeeze on the lens, making it bulge and thus reducing its focal length. Behind the lens is a chamber filled with a clear, jellylike fluid called the *vitreous humor*. At the rear of the chamber is a highly complex, light-sensitive structure called the *retina*. The refracting structures of the eye—the cornea and the lens—cast an image on the retina. The retina converts light into electrical signals that are transmitted and partially processed by the *optic nerve*. The final processing and interpretation takes place in the brain.

Most of the refraction necessary for imaging takes place at the front surface of the cornea, whose index of refraction is about 1.44. The index of refraction of the lens also is about 1.44, but the humors in which it is immersed have indices of refraction close to 1.34. Thus the effective index of refraction of the lens is the relative index of refraction $1.44/1.34 = 1.08$. As you can see from the lensmaker's equation, the lens refracts only weakly, the factor $(n - 1)$ having the small value 0.08.

The function of the lens is to produce relatively small corrections in the focal length of the optical system. These corrections make it possible to produce clear images on the retina of objects at various distances.

When you look at an outdoor scene, most of the objects you see are so distant that the light from them strikes the eye as nearly parallel bundles of rays. (For practical purposes, any object more distant than 10 m or so is at infinity.) Under these circumstances, the ciliary muscle is relaxed. Consequently, the lens is relatively flat, and its focal length is maximal. We say that a normal eye is focused at infinity, as shown in Figure 37.27a. As an object approaches the eye, the image tends to move behind the retina. To compensate, the lens is squeezed by the ciliary muscle, and the focal length is reduced just enough to keep the image distance constant. We usually carry out this process, called **accommodation**, without thinking about it.

There is a limit to the accommodation process. When the object is closer to the eye than the **near point**, the eye can no longer produce a focused image. For most young persons with normal eyes, the near point lies about 25 cm from the eye. This distance increases with age as the lens loses elasticity. This condition makes its presence known in nearly everyone at some age between 35 and 50; you have probably heard the wry

(a)   (b)

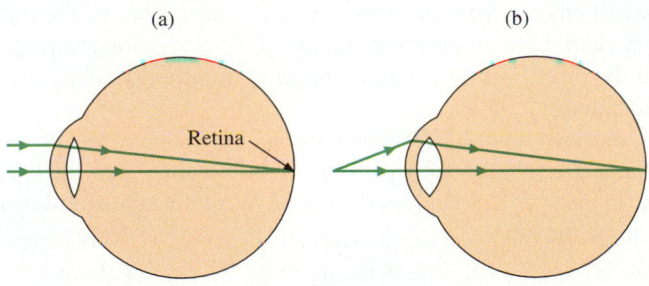

remark "My arms keep getting shorter" from a person who finds it impossible to hold reading matter far enough away to focus on it. By age 70, accommodation is completely lost, and the focal length of the eye becomes essentially fixed. The loss of lens elasticity is called **presbyopia** (from Greek roots meaning "old eye") or **farsightedness**. It is readily corrected by wearing converging lenses, which compensate for the loss of converging power in the lens of the eye. For many persons, compensation requires the use of **bifocal** or **graded** lenses. These familiar devices incorporate segments of different focal lengths into a single lens. Most persons find the results very satisfactory.

In the common condition called **myopia** or **nearsightedness**, the radius of curvature of the cornea is too small, and therefore its focal length is too short compared with the length of the eyeball. Consequently, objects at infinity are imaged in front of the retina even when the lens is in its state of maximum relaxation and maximum focal length. Myopic persons can focus readily on nearby objects; indeed, their near points are somewhat closer than those of persons with normal vision. The greatest object distance for which a myopic person can obtain a clear image is called the **far point**. Correction is obtained by wearing diverging lenses, which move the far point to infinity. The opposite condition, in which the focal length of the cornea is too large, is called **hyperopia**. Hyperopia is not nearly as common as myopia.

Another very common imperfection of vision is **astigmatism**. This condition is due to imperfect sphericity of the cornea. On account of this asphericity, a set of arrowlike objects located in the same plane normal to the axis of the eye will not be in focus simultaneously. The condition is corrected with aspherical lenses.

The Simple Magnifier

If you want to inspect a small object carefully, you bring it close to your eye. The closer the object, the greater the (small) angle ϕ that it subtends and the more detail you can see. However, it is of no use to bring the object closer than your near point, a distance d from your eye. As you can see from Figure 37.28a, the largest useful angle ϕ for an object of height h is $\phi = h/d$.

FIGURE 37.28 (a) A small object at the near point subtends an angle ϕ. (b) A converging lens of focal length f is placed just in front of the eye.

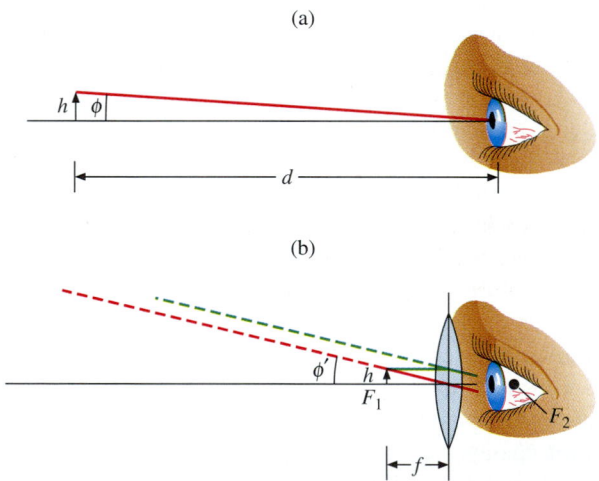

Now, place a converging lens of focal length $f < d$ immediately in front of your eye, as shown in Figure 37.28b. If you bring the object to the first focus F_1 of the lens, the light rays will emerge from the lens as parallel rays. That is, the light enters your eye just as if it came from an object at infinity. To put it more precisely, you are now looking not at the object but at its virtual image at infinity. You will have no difficulty focusing on this image.

As you can see from Figure 37.28, the angle ϕ' subtended by the image is essentially the same as the angle subtended by the object at the vertex V of the lens, because the lens is so close to the eye. This angle is $\phi' = h/f$. We define the **angular magnification** α of the lens to be the ratio

$$\alpha \equiv \frac{\phi'}{\phi} = \frac{d}{f}.$$ (37.27)

Let us use the standard value $d = 25$ cm for the normal eye. The practical lower limit on f for useful simple magnifiers is about 7 cm. We can thus obtain an angular magnification of about

$$\alpha = \frac{25 \text{ cm}}{7 \text{ cm}} \simeq 3.5.$$

This is quite adequate for many purposes. Why can we not increase α by using a shorter focal length? To do so, we must increase the curvature (reduce the radius of curvature) of the lens surfaces. As we do this, the paraxial approximation is valid for increasingly smaller bundles of rays, and the image quality degrades. In spite of this shortcoming, the Dutch linen draper Antoon van Leeuwenhoek (1632–1723) pioneered the exploration of the microscopic living world, beginning in the 1670s, with a series of tiny, superbly crafted homemade lenses of very short focal length called **simple microscopes**. No one else was able to match the quality of his lenses or the sharpness of his observations, and after his death microscopy came almost to a standstill until the development of useful compound microscopes nearly a century later.

The Compound Microscope

The **compound microscope** is so called because, in contrast with Leeuwenhoek's single-lens simple microscope, it consists of two lens systems called the **objective** and the **ocular** (or **eyepiece**). The general principle of operation can be understood in terms of the hypothetical thin-lens microscope shown in Figure 37.29.

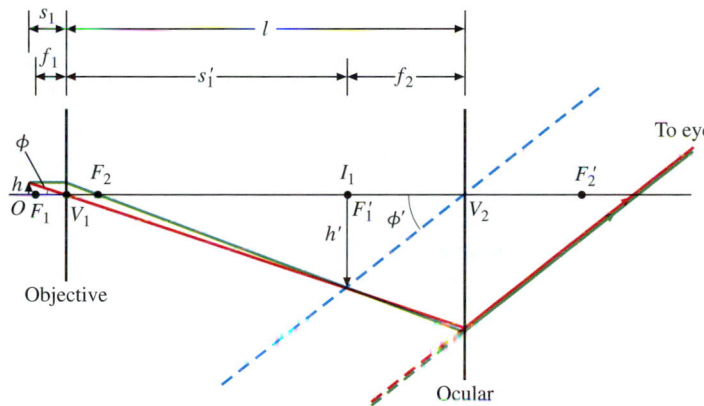

FIGURE 37.29 A thin-lens compound microscope. Each of the lenses is represented by its principal plane. The dashed vertical ray through the second vertex V_2 does not exist because it could not have come from the small objective lens. But the imaginary ray serves as a construction line; the rays that actually reach the eye are drawn parallel to it, as shown.

The objective is a converging lens of short focal length f_1, usually substantially less than 1 cm. The microscope is adjusted, by moving the entire tube, so that the distance from the (very small) object to the objective is just a little greater than f_1. This positioning of the object is the key to the way the system works. If the object were precisely at the focus, the rays emerging from the objective would be parallel, resulting in a virtual image at infinity. But with s_1 a little greater than f_1, the emerging rays converge to form a real image I_1 at a finite but large distance s_1'. The lateral magnification of the objective is given by Equation 37.25:

$$M_1 = -\frac{s_1'}{s_1} \simeq -\frac{s_1'}{f_1}.$$

Because of the large value of s_1', we can make the approximation $s_1' \simeq l$, where l is the distance between objective and ocular, called the *tube length*. We thus have

$$M \simeq -\frac{l}{f_1}.$$

The ocular, a converging lens of fairly short focal length, is located so that the image I_1 lies at its focus. That is, the ocular plays the role of a simple magnifier whose object is the magnified image of the objective. The angular magnification is given by Equation 37.27, $\alpha = d/f_2$, where d is the observer's near-point distance.

The overall magnification M of the microscope is the product of the magnifications of the two lenses:

$$M = M_1\alpha = -\frac{ld}{f_1 f_2}. \qquad \textbf{(37.28)}$$

It is not difficult to make M greater than 1000. The limit on compound microscopes is set not by magnification but by image quality and ultimately by the limit of resolution imposed by the wavelength of light, λ. Remember that ray optics fails for distances comparable to λ; when the size of the object is comparable to λ (of the order of 500 nm), the observer sees a diffraction pattern rather than an image.

The Astronomical Refracting Telescope

A simple **refracting telescope** is shown in Figure 37.30. Like the microscope, it consists of two converging lenses. Here, however, the object is located at infinity, and the incident light rays are parallel. The objective thus forms an image I_1H' at its second focus. The second focus of the objective is made to coincide with the first focus of the ocular. As in the microscope, the ocular acts as a simple magnifier.

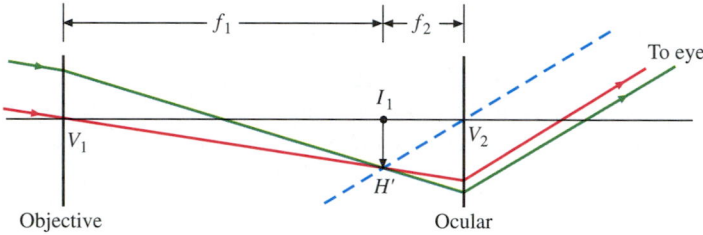

Objective   Ocular

Since both object and image lie at infinity, we consider angular magnification rather than lateral magnification. As you can see from Figure 37.30, the angle subtended by the image I_1H' at the vertex V_1 of the objective is $\phi_1 \simeq -I_1H'/f_1$. (The minus sign denotes the fact that the image is inverted—not a problem for astronomical observation but a nuisance for terrestrial uses, where other arrangements are employed.) Similarly, the angle subtended by I_1H' at the vertex V_2 of the ocular is $\phi_2 \simeq I_1H'/f_2$. The angular magnification is the ratio of these angles:

$$\alpha = \frac{\phi_2}{\phi_1} = \frac{-I_1H'/f_2}{I_1H'/f_1} = -\frac{f_1}{f_2}. \qquad \textbf{(37.29)}$$

Except for planetary observation, magnification is not a main concern of astronomers. Of greater concern are image quality, determined by the precision with which the telescope surfaces are ground; resolution, discussed in Section 36.8; and *light-gathering power*. As you can see from Figure 37.30, the telescope concentrates all of the light entering the telescope into a much narrower beam that enters the eye of the observer. This makes it possible to see objects far too faint to be detected by the unaided eye.

For both resolution and light-gathering power, objective diameter is crucial. The largest refracting telescopes, both built in the late nineteenth century, are at the Lick Observatory at Mount Hamilton (near San Jose), California, and the Yerkes Observatory in Williams Bay, Wisconsin. Both have objectives about 1 m in diameter (36 in. and 40 in.).

Reflecting Telescopes

Practical considerations severely limit the size of refracting telescopes. First, two objective lens surfaces must be polished with high precision. Second, the glass of the lens must be free of imperfections and variations in the index of refraction throughout. Third, the necessary thickness of a lens of given focal length increases with increasing diameter; this imposes severe mechanical requirements. Fourth, it is difficult to extend use of the lens beyond the wavelength range of visible light because most glasses absorb ultraviolet and infrared light.

For these and other reasons, nearly all large telescopes built in the twentieth century (and many small ones) use a large mirror in place of an objective lens. The mirror is lighter than a lens of similar diameter, can be supported at the back as well as the edges, has only one surface to grind and polish, need not be optically perfect beneath the surface, and is not limited by the transmission properties of glass.

Like an objective lens, an objective mirror produces images of objects at infinity at its focus. The ocular is not different in any significant way from those used in refracting telescopes. The only special problem is in making the image conveniently accessible for study. This is usually done in one of the three ways shown in Figure 37.31.

For many years, the largest telescope was the Mount Palomar reflector, with its 5-m (200-in.) diameter. A revolutionary new generation of telescopes is coming into service in the 1990s. These telescopes will all have effective diameters of 10 m or more. All use new technologies. The *multiple-mirror* technology, for example, uses laser and computer techniques to keep a number of relatively small mirrors aligned within a small fraction of the wavelength of light so that they behave like a single huge mirror.

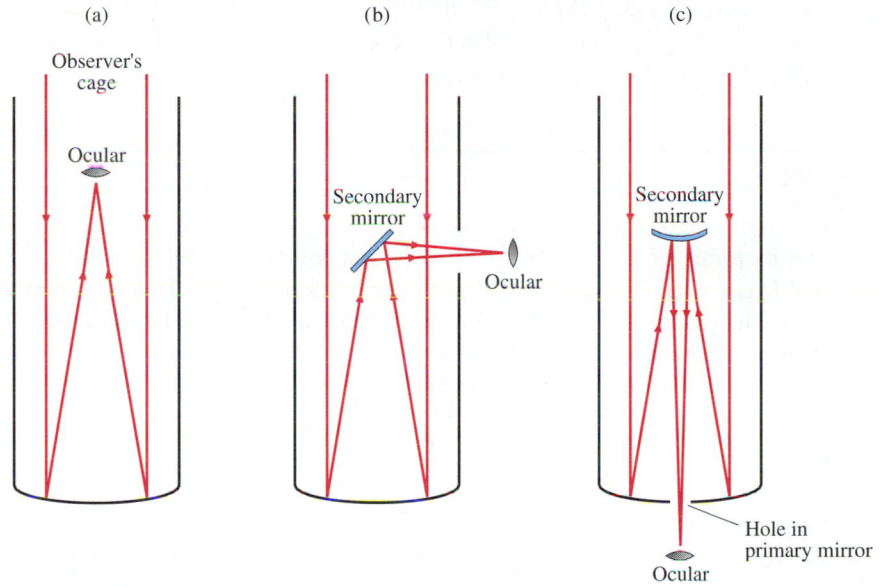

FIGURE 37.31 The major forms of the reflecting telescope. (*a*) Cassegrain configuration. The observer is located, with the ocular, in a "cage" at the focal point of the mirror. For large mirrors, the blockage of light by the cage is not serious. (*b*) Newtonian configuration. A small plane mirror, oriented at 45° to the primary mirror axis, is located a little closer to the primary mirror than the latter's focal point is. The light reflected from the primary mirror has converged here to a narrow beam, which is reflected to the side. (*c*) Gregorian configuration. The converging beam is reflected, by a slightly convex secondary mirror, through a small hole in the objective mirror, to a focus below.

THINKING LIKE A PHYSICIST

Nearly all optical instruments—binoculars, zoom lenses, microscopes, contact lenses, and countless other devices—depend on systems of lenses or mirrors or both. The instruments are the products of a sophisticated technology that is indispensable to modern life. However complex this technology may be, all of the techniques of lens design are founded on the three simple principles we have used in this chapter: the expression of light propagation in terms of rays, the law of reflection, and Snell's law. In this simplicity of foundation, geometric optics is similar to mechanics (based on Newton's laws) and electromagnetism (based on Maxwell's equations).

As is true for mechanics and electromagnetism, physicists and engineers have built a beautiful and fruitful structure on the basic principles of geometric optics. The paraxial approximation to which we have restricted our study here is not adequate for designing any but the simplest optical systems. As in mechanics and electromagnetism, powerful physical and mathematical techniques have been developed to deal with the wealth of possibilities opened up by complex systems. In geometric optics, much depends on the ability to do very complicated calculations, and the evolution of computers has made possible lens systems whose design could not have been attempted only a little earlier. Today, you can buy a camera for a few hundred dollars whose lens is far superior, in speed and sharpness of image, to the best available in 1950.

C	center of curvature		O	object point
d	near-point distance		s, s'	object distance, image distance
f	focal length		t	thickness of a lens
F	focal point		V	vertex
h, h'	height of an object, an image		α	angular magnification
H, H'	off-axis object point, image point		θ	angle of incidence or reflection for a mirror
I	image point		θ_1, θ_2	angle of incidence, angle of refraction for a lens
l	distance between two lenses		ρ	radius of curvature
L	longitudinal magnification		σ	arc length
M	lateral magnification		ϕ	angle between a ray and the system axis
n	index of refraction			

Summing Up

The central subject of **geometric optics** is the **mapping** of **object space** into **image space** by mirrors and lenses. Geometric optics is founded on a basic assumption: The propagation of light can be represented by **rays**. This assumption is valid if the wavelength of light is very small compared with any length of interest.

We follow rays through an optical system on the basis of two laws:

- the law of reflection, $\theta_{\text{reflected}} = \theta_{\text{incident}}$ (Equation 35.3);
- the law of refraction, $n_1 \sin \theta_1 = n_2 \sin \theta_2$ (Equation 37.13).

Geometric optics is much simplified if we make the **paraxial approximation**. In this approximation, we consider only rays that make small angles with the axis of the optical system. For paraxial rays, a **spherical mirror** maps object space into image space on a one-to-one basis. That is, a point object produces a point image in accordance with the **mirror equation**, which can be written in either of the forms given by Equations 37.8 and 37.10:

$$\frac{1}{s} + \frac{1}{s'} = \frac{2}{\rho} \quad \text{or} \quad \frac{1}{s} + \frac{1}{s'} = \frac{1}{f}.$$

Depending on the circumstances, the **object** may be **real** or **virtual**, and the **image** may be real or virtual, **erect** or **inverted**.

The **lateral magnification**, defined as $M \equiv h'/h$, is given for spherical mirrors (including plane mirrors, regarded as spherical mirrors for which $\rho = \infty$) by Equation 37.12,

$$M = -\frac{s'}{s}.$$

A positive value of M denotes an erect image and a negative value an inverted image. For plane mirrors, we have $M = 1$, and the **longitudinal magnification** is $L = -1$. As a consequence, the image is **perverted**; a right-handed object always results in a left-handed image.

Paraxial rays from a point image, refracted by a spherical surface, also produce a point image. This is in part a consequence of the fact that the law of refraction simplifies to Equation 37.14, $n_1\theta_1 = n_2\theta_2$, for paraxial rays. The relation between object distance and image distance is given by Equation 37.17,

$$\frac{n_1}{s} + \frac{n_2}{s'} = \frac{n_2 - n_1}{\rho}.$$

A **lens** produces a point image for paraxial rays emanating from a point object. The object and image distances are related by a set of simultaneous equations (Equations 34.19a,b,c). For the special case of a **thin lens** of index of refraction n located in air, the solution simplifies to Equation 37.23, the **thin-lens equation**,

$$\frac{1}{s} + \frac{1}{s'} = \frac{1}{f},$$

where the focal length is given by Equation 37.22, the **lens-maker's equation**:

$$\frac{1}{f} = (n - 1)\left(\frac{1}{\rho_1} - \frac{1}{\rho_2}\right).$$

These equations are valid for all thin lenses, provided the proper sign convention is observed.

When two thin lenses of focal lengths f_1 and f_2 are placed in contact, they act as a single lens of focal length f according to Equation 37.26,

$$\frac{1}{f} = \frac{1}{f_1} + \frac{1}{f_2}.$$

For an object located in a plane normal to the axis, the formation of an image by a mirror or lens is conveniently visualized by means of a **ray diagram**. The ray diagram is also convenient for analyzing optical devices containing more than one lens or mirror.

KEY TERMS

Section 37.2 Plane Mirrors

incident ray, reflected ray ▪ real image, virtual image ▪ object distance, image distance ▪ lateral magnification, longitudinal magnification ▪ ray diagram ▪ perverted image

Section 37.3 Spherical Mirrors

concave mirror, convex mirror ▪ optic axis, vertex ▪ paraxial ray ▪ spherical aberration ▪ focal point, focal length ▪ virtual object ▪ erect image, inverted image ▪ principle of optical reversibility ▪ radial ray, vertical ray,

parallel ray, focal ray

Section 37.5 Thin Lenses

thick lens, thin lens ▪ thin-lens approximation ▪ lensmaker's equation, thin-lens equation ▪ converging lens, diverging lens ▪ principal plane ▪ chief ray, focal ray, parallel ray

Section 37.6 Optical Instruments

accommodation ▪ near point, far point ▪ presbyopia, myopia, hyperopia, astigmatism ▪ angular magnification ▪ compound microscope ▪ objective, ocular ▪ refracting telescope, reflecting telescope

---

Queries and Problems for Chapter 37

QUERIES

37.1 *(1) Kitty-corner?* Alice tells her kitten that on getting up on a chair, close to the mirror, she can see the entire drawing-room in the Looking-Glass House (except for the bit just behind the fireplace). How does Alice's location in her drawing-room determine the amount of the Looking-Glass Room she can see?

37.2 *(2) Heads you win!* In constructing ray diagrams, we draw rays from the head *H* of the object only. Prove, by means of a nonmathematical, logical argument, that there is no need to draw rays from the foot *O* of the object, or from any other point, to locate the image.

37.3 *(2) The handedness of hands.* Extend the first three fingers of each of your hands in mutually perpendicular directions. Experiment with a plane mirror to see how the image of your right hand corresponds with your left hand and vice versa.

37.4 *(3) Driving to a conclusion.* In most trucks and automobiles, part or all of the right-side rear-view mirror is convex. Why is this useful? What is the reason for the standard warning, ''Objects seen in this mirror are closer than they appear''? Why does the left-side rear-view mirror usually have no convex part?

37.5 *(3) Plane or fancy.* Show that Equation 37.12 gives the correct result for the magnification of a plane mirror.

37.6 *(3) Growth policy.* Under what circumstances does a convex mirror produce an image larger than its object?

37.7 *(4) Coloring the issue.* As you learned in Section 35.7, the index of refraction of glass is smaller for blue light than for red light. Sketch a ray diagram, showing the passage of a paraxial incident ray of mixed red and blue light through a converging lens. (For simplicity, assume that $s > f$.) Describe the degradation of the image due to dispersion. This degradation is called *chromatic aberration*.

37.8 *(6) Topsy-turvy world.* When you look at an object, is the image on the retina of your eye erect or inverted? (Note: When experimental subjects are fitted with inverting spectacles, they find it very difficult to carry out the normal tasks of life at first. But within a few days they become completely adjusted and have no difficulty at all. When the spectacles are removed, the readjustment is almost instantaneous, as are subsequent readjustments when the spectacles are put on and taken off.)

37.9 *(6) But is the surgeon wearing glasses?* In the surgical procedure called *radial keratotomy*, myopia is treated—in some cases quite successfully—by cutting a number of grooves of small but carefully regulated depth into the corneal surface. The grooves are directed so that they radiate from the center of the pupil—hence the term ''radial.'' The result of the operation is to flatten the cornea slightly—that is, to increase its radius of curvature slightly. Explain how this leads to the desired result.

37.10 *(6) Nobody's perfect.* You can easily locate your near point by bringing a pencil point closer and closer to your eye until you can no longer focus on it clearly. Use a ruler or a tape measure to find the distance of your near point from your eye. If you are myopic, locate your near point both with and without glasses. Assemble a group of friends, some of whom are nearsighted and some of whom are not, and compare near-point distances.

37.11 *(6) Farsighted thinking.* If you are too young to have presbyopia, you can simulate it as follows. Find a nearsighted friend (or, if you are nearsighted yourself, a friend who is more nearsighted). Look through his or her glasses, and try to focus on objects at various distances. Locate your near point with the glasses, and compare the result with your normal near point. (For best results, try to find a friend who is not astigmatic as well as myopic. Or get your hands on a concave lens.)

37.12 *(6) Eerie glow.* Many animals (notably nocturnal animals) have a reflective membrane called the *tapetum lucidum* just behind the retina. It is owing to the tapetum lucidum that we see cats' eyes glow in the dark. Explain the utility of the structure.

37.13 *(6) All the better to see you with.* Why can't you see clearly when you swim underwater without goggles? Why can you see clearly when you wear goggles?

37.14 *(6) The hole truth.* Why is a convex secondary mirror used in a Gregorian reflecting telescope, rather than a plane mirror?

PROBLEMS

GROUP A

37.1 *(2) Emulating Alice? I.* If you walk directly toward a plane mirror with speed 1 m/s, at what speed does your image move with respect to you?

37.2 *(2) Emulating Alice? II.* If you walk with speed v toward a plane mirror at an angle of 35° to the normal, at what speed does your image move with respect to you?

37.3 *(2) Using imagination.* An object 1.0 cm high is located 45 cm from the vertex of a spherical mirror whose radius of curvature is 60 cm. Where is the image? Is it real or virtual? erect or inverted?

37.4 *(2) Narcissus.* You want to take a picture of yourself, making use of a large wall mirror. The range finder on your camera tells you that you must focus at 3.2 m. Not satisfied, you move away from the mirror, walking 3.0 m at a 50° angle to the mirror. What does your range finder say now?

37.5 *(3) Taking shape.* When an object is placed 30.0 cm in front of a certain mirror, the image appears 16.0 cm in front of the mirror. **(a)** What are the radius of curvature and focal length of the mirror? **(b)** Is the mirror concave or convex? **(c)** How big is the image? Is it erect or inverted?

37.6 *(3) Objective thinking, I.* A virtual image forms 15.0 cm behind a convex mirror whose radius of curvature is 85.5 cm. Where is the object?

37.7 *(3) Objective thinking, II.* A concave mirror of focal length 30.0 cm forms a real image 8.6 cm from the vertex. Where is the object? Is it real or virtual?

37.8 *(3) Back again.* A concave spherical mirror produces an image that coincides with its point object. How far is the object from the mirror vertex?

37.9 *(3) Reflecting on one's image, I.* The radius of curvature of a concave spherical mirror is 30 cm. An object 1.0 cm high is located 45 cm from the mirror. Find the location and height of the image. Is the image erect or inverted? real or virtual?

37.10 *(3) Reflecting on one's image, II.* The object of Problem 37.9 is now moved to a new position 10 cm from the mirror. Find the location and height of the image. Is the image erect or inverted? real or virtual?

37.11 *(3) Shrinkage.* The mirror of Problem 37.9 is to be used to produce an image one-half the size of the object. Where should the object be located?

37.12 *(3) Reflecting on one's image, III.* The radius of curvature of a convex spherical mirror is 40 cm. An object 1.7 cm high is located 32 cm from the mirror. Find the location and height of the image. Is the image erect or inverted? real or virtual?

37.13 *(4) Reversal.* You turn the sphere of Example 37.6 through 180° so that you are looking at the small bubble from the opposite side. Where does the bubble appear to be?

37.14 *(5) You can't be too rich or too thin.* **(a)** Show that, when a paraxial chief ray enters or leaves a thin lens, the angle of incidence is approximately equal to the angle between the ray and the optic axis. **(b)** Show that the displacement of the ray on passing through the thin lens is negligible and that the ray is not deviated by its passage through the lens, as the text assumes in the discussion of Figure 37.24. (Hint: See Problems 35.15 and 35.16.)

37.15 *(5) Lookalikes.* Two thin lenses of identical form are made of two kinds of glass, with $n_1 = 1.5$ and $n_2 = 1.7$. Find the ratio f_2/f_1 of the focal lengths.

37.16 *(5) Wet lensmaker.* A lens, made of glass of index of refraction n_g, is to be used immersed in a liquid of index of refraction n_l. Show that the lensmaker's equation must be modified to the form

$$\frac{1}{f} = (n_{gl} - 1)\left(\frac{1}{\rho_1} - \frac{1}{\rho_2}\right),$$

where n_{gl} is the relative index of refraction n_g/n_l.

37.17 *(5) In focus.* A real object is located 35.0 cm in front of a lens. An image forms behind the lens, at a distance of 48.5 cm. **(a)** What is the focal length of the lens? **(b)** Specify whether the lens is converging or diverging and whether the image is real or virtual, erect or inverted.

37.18 *(5) Convexoconcave.* A thin lens has a convex surface whose radius of curvature is 22.5 cm and a concave surface whose radius of curvature is 47.0 cm. The index of refraction of the glass is 1.50. **(a)** Draw a sketch of the lens. **(b)** What is the focal length of the lens? Is the lens converging or diverging? **(c)** A real point object is located 30.0 cm from the lens. Where is the image? **(d)** Is the image real or virtual? **(e)** If you reverse the lens so that the concave surface is the first surface, show that the lensmaker's equation gives the same result for the focal length.

37.19 *(5) Around here somewhere.* A real object lies 23.0 cm in front of a lens of focal length −30 cm. Where is the image? Is it erect or inverted? real or virtual?

37.20 *(5) Image maker, I.* An object of height 2.0 cm is located 15 cm from a lens of focal length +10 cm. Find the position and height of the image. Is it erect or inverted? real or virtual?

37.21 *(5) Image maker, II.* A real object 1.5 cm high is 18.0 cm from a thin lens. The image is on the opposite side of the lens, 32.2 cm from the object. **(a)** What is the focal length of the lens? **(b)** How big is the image? **(c)** Is the image erect or inverted? real or virtual?

37.22 *(5) Funny coincidence.* Is it ever possible for the object and image of a lens to coincide? Explain.

37.23 *(5) Equal and opposite.* An object and its image lie on opposite sides of a thin lens, equidistant from the lens. Express the focal length f in terms of the object distance s.

37.24 *(5) Magnificat.* Glass having index of refraction 1.5 is used to make a biconvex lens whose surfaces have radius of curvature 14 cm. An object is placed at a distance s from the lens, and an image is formed on a screen. What is the object-to-screen distance if the lateral magnification is **(a)** $M = 0.3$, **(b)** $M = 1$, and **(c)** $M = 20$?

37.25 *(5) Reading glasses.* A typical lens used in reading glasses for persons with presbyopia has a convex first surface, whose radius has magnitude 7.5 cm, and a concave second surface, whose radius has magnitude 12.0 cm. If $n = 1.52$, find the focal length of the lens. (The first surface of spectacle lenses is almost always convex for reasons of appearance. The convergent reflections from concave surfaces would make the wearer look odd.)

37.26 *(5) Hollow victory.* Repeat Example 37.7 for a biconcave lens whose surfaces have curvature of the same magnitude as those of the biconvex lens of the example.

37.27 *(5) Bis, bis.* A biconvex glass lens has surfaces of equal radius of curvature ρ. The focal length of the lens is $f = \rho$. What is the index of refraction of the glass?

37.28 *(5) Converging the divergent.* Equation 37.26 provides a convenient way of measuring the focal length f_d of a diverging lens. The lens is placed in contact with a converging lens of known focal length f_c. The focal length of the combination is then determined by measuring the distance between the combination and the real image of a distant object. Suppose that a converging lens is used for which $f_c = 20.0$ cm, and the image distance is $s' = 57.5$ cm. Find f_d.

37.29 *(6) Trade-off.* An optometrist finds that a patient has myopia, with the far point at 75 cm. (Assume both eyes are the same.) **(a)** What is the focal length of the spectacle lens required to move the far point to infinity? **(b)** When wearing the proper spectacles, the patient has a near point at 25 cm. Where is the patient's near point without spectacles?

37.30 *(6) Adjustment.* The human pupil diameter varies from 2 mm in bright light to 7 mm in dim light. **(a)** If this variation were the only mechanism for adjusting to variations in light intensity, what would be the range to which the eye could adapt? **(b)** The eye can in fact see usefully over a brightness range of about 10^{12}. Remembering that total dark adaptation takes about $\frac{1}{2}$ hour, explain the usefulness of the pupillary variation.

37.31 *(6) Visual aid.* The largest refracting telescopes have diameters of about 1 m. By what factor is the light-gathering power of the unaided eye increased? Take the diameter of the dark-adapted pupil to be 7 mm.

37.32 *(6) Moonstruck.* An astronomical refracting telescope has a tube length of 110 cm. **(a)** If the angular magnification is -10, what are the focal lengths of the objective and the ocular? **(b)** As seen by the unaided eye, the moon subtends an angle of 0.50°. What is the height of the image of the moon produced by the objective?

37.33 *(6) Choosing the right one.* The objective lens of a microscope has a focal length of 0.355 cm. The image I_1 is formed at a distance of 21.4 cm from the lens. If you wish to have an overall angular magnification of 500, what angular magnification must the ocular have?

37.34 *(6) In the small.* The focal length of a microscope objective is 0.22 cm, and the focal length of the ocular is 4.0 cm. The tube length is 18.0 cm. What is the magnification of the microscope?

GROUP B

37.35 *(2) Various -versions.* To show that the mapping accomplished by a mirror is unique, draw a right-handed x-y-z coordinate system, and then map it into an x'-y'-z' system in the following ways. Sketch **(a)** the perverted image produced by the transformation $x' = x$, $y' = y$, $z' = -z$; **(b)** the inverted image produced by the transformation $x' = -x$, $y' = -y$, $z' = z$. **(c)** Show that inversion corresponds to a rotation of the x-y-z system, and describe the rotation. **(d)** What kind of mapping results from the transformation $x' = -x$, $y' = -y$, $z' = -z$?

37.36 *(3) Twins and quintuplets.* The figure below shows two plane mirrors oriented so that the angle between them is 60°. A point source S lies at an arbitrary location between the mirrors. Show that the source has five images (not necessarily all visible to an arbitrarily located observer).

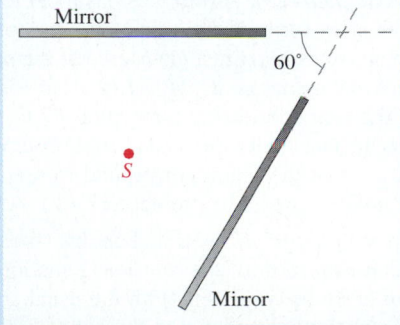

37.37 *(3) Partial realism, at least.* Show that, for a concave mirror, the object and the image cannot both be virtual.

37.38 *(3) Partial realism, at most.* Show that, for a convex mirror, the object and the image cannot both be real.

37.39 *(3) Real close.* Show that, if a concave mirror produces a virtual image, the image must be farther from the mirror than the object is.

37.40 *(3) Round and flat.* A point source lies a distance s from the vertex of a concave mirror whose radius of curvature is ρ. Opposite the convex mirror is a plane mirror, so light from the source is reflected first from the concave and then from the plane mirror. The plane mirror is moved back and forth until the reflected light forms an image that coincides with the source. **(a)** Show that the distance D between the two mirrors is

$$D = \frac{s^2}{2s - \rho}.$$

(b) If $\rho = 60$ cm and $s = 40$ cm, find D.

37.41 *(3) Changing the map, I.* You have a concave spherical mirror of radius of curvature ρ. You begin with an object far away from the mirror and bring it toward the mirror along the axis. Describe what happens to the position and lateral magnification of the image as the object moves **(a)** from $s = \infty$ to $s = \rho$, **(b)** from $s = \rho$ to $s = f$, and **(c)** from $s = f$ to $s = 0$. **(d)** When is the image real, and when is it virtual? **(e)** When does the image move toward the object, and when does it move away from the object? **(f)** Do the object and the image ever move in the same direction?

37.42 *(3) Changing the map, II.* You have a convex spherical mirror of radius of curvature ρ. You begin with an object far away from the mirror and bring it toward the mirror along the axis. Describe what happens to the position and lateral magnification of the image as the object moves.

37.43 *(3) Aberrant behavior.* Here is a simple way to prove that all light rays emitted by a point object and reflected from a spherical mirror cannot pass through the same image point. **(a)** Draw a circle. **(b)** Draw a diameter of the circle, and let it be the axis. **(c)** Pick any off-center point on the axis, and let it be the object point. **(d)** Draw a few paraxial rays and show that they are reflected to (almost) the same point. **(e)** What ray is *obviously* not reflected so as to pass through the same image point?

37.44 *(3) Range of possibilities, I.* A spherical concave mirror produces an image *I* of an object *O*. Consider each of the following combinations of object and image:

- real object, real image
- real object, virtual image
- virtual object, real image
- virtual object, virtual image

(a) Are all of these combinations possible? If not, which are not? **(b)** For each possible combination, is the image erect or inverted? **(c)** What is the range of object distances *s* for which each possible combination is observed?

37.45 *(3) Range of possibilities, II.* Solve Problem 37.44 for a spherical convex mirror.

37.46 *(3) Focal length and magnification.* A concave spherical mirror produces an inverted image of an object. The distance between object and image is *d*, and the magnification is *M*. **(a)** Show that the focal length of the mirror is

$$f = \frac{Md}{1 - M^2}.$$

(b) Let $d = 12$ cm and $M = -2.0$, and find the focal length.

37.47 *(4) Nonparaxial.* Suppose a parallel but nonparaxial ray (a ray not close to the optic axis) falls on a spherical refracting surface. Does the ray cross the optic axis closer to or farther from the surface than would a parallel paraxial ray? Explain.

37.48 *(5) Range of possibilities, III.* Solve Problem 37.44 for a thin converging lens.

37.49 *(5) Range of possibilities, IV.* Solve Problem 37.44 for a thin diverging lens.

37.50 *(5) Changing the map, III.* You have a converging thin lens, of focal length $f > 0$. You begin with an object far away from the lens and bring it toward the lens along the axis. Describe what happens to the position and lateral magnification of the image as the object moves **(a)** from $s = \infty$ to $s = f$ and **(b)** from $s = f$ to $s = 0$.

37.51 *(5) Changing the map, IV.* Answer Problem 37.50 for a diverging thin lens, of focal length $f < 0$.

37.52 *(5) Through a glass, brightly.* Equation 37.17, derived for thin lenses, bears a striking resemblance to the mirror equation (Equation 37.8). For what values of n_1 and n_2 does Equation 37.18 become identical with Equation 37.8? What

is the physical meaning of these values? (Hint: Think of *n* as a ratio of *velocities* rather than as a ratio of speeds.)

37.53 *(5) Turning into a lensmaker.* In the text, it is asserted that the shape of the surfaces in Figure 37.19 does not affect the derivation of the lensmaker's equation (Equation 37.23). To show that this is true, turn the lens of Figure 37.19 around so that the first surface is concave and the second convex. Sketch an appropriate ray diagram. Then rederive Equation 37.23, paying proper attention to sign conventions.

37.54 *(5) Look it up in the index.* Two kinds of glass, having indices of refraction n_1 and n_2, are used to make thin lenses of identical shape. What is the ratio f_1/f_2 of the focal lengths of the lenses?

37.55 *(5) The most convergence possible.* At what distance from a converging lens of focal length *f* must you place an object in order to make the object–image distance as small as possible?

37.56 *(5) Newtonian form of the thin-lens equation.* In the figure below, we define *x* to be the distance between the object point *O* and the first focus F_1 of a lens and x' to be the distance between the second focus F_2 and the image point *I*. **(a)** Show that the thin-lens equation can be written in the equivalent form

$$xx' = f^2.$$

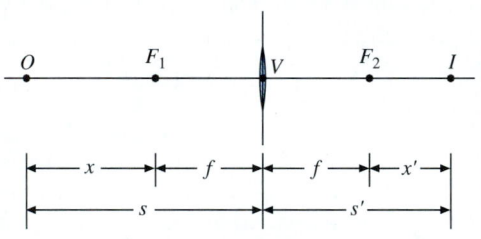

This is called the *Newtonian form of the thin-lens equation.* **(b)** Show that the lateral magnification of a thin lens can be written

$$M = -\frac{f}{x} = -\frac{x'}{f}.$$

(c) Write a set of sign conventions for *x* and x', making sure that the convention is consistent with the convention already established for *s*, s', and *f*.

37.57 *(5) Vivid picture, I.* A thin lens has focal length *f*. Using the Newtonian form of the thin-lens equation (Problem 37.56), plot x' versus *x*. [Hints: (1) Mark off the scales of the axes in units of *f*—that is, *f*, 2*f*, 3*f*, and so forth. (2) Remembering the sign convention for *x*, plot $+x$ to the left and $-x$ to the right; that is, let the *x* axis run "backward."] Indicate what parts of the graph correspond to real and virtual objects and images. (Refer to Problem 37.48.)

37.58 *(5) Vivid picture, II.* A thin lens has focal length *f*. Using the Gaussian form of the thin-lens equation, plot s' as a function of *s*. To be consistent with the usual sign convention, plot $+s$ to the left and $-s$ to the right. Indicate which parts of the graph correspond to real and virtual objects and images. (Hint: Remember the relations between s and x; s' and x'. If you begin with the graph of Problem 37.57, you can save yourself a lot of work.)

37.59 (5) *Possibilities.* (a) Show that the magnification of a thin lens can be written

$$M = \frac{f}{f - s}.$$

Assuming a real object, determine what this equation tells you about the dependence on object distance of the size and orientation of the image formed by (b) a converging lens and (c) a diverging lens. (d) Repeat parts b and c for a virtual object.

37.60 (5) *Through thick and thin.* (a) Prove that, if a lens is thicker in the center than at the edges, it is always a converging lens regardless of whether it is planoconvex, biconvex, or convexoconcave. (b) Prove the corresponding statement for a planoconcave, biconcave, or convexoconcave lens that is thinner in the center than at the edges.

37.61 (5) *Refocusing.* A lens of focal length f produces a real image on a screen at a distance s'. The screen is now moved toward the lens through a distance $\Delta s'$. (a) Through what distance Δs must the object be moved if the image is again to be in focus? (b) Let $f = 25$ cm, $s' = 5.00$ m, and $\Delta s' = 20$ cm. Find s.

37.62 (5) *Another view.* A point object lies a distance s from a converging lens. The lens forms a real image at a distance s'. Let the total object-to-image distance be l, and let the difference between the object and image distances be d. (a) Show that the focal length of the lens is

$$f = \frac{l^2 - d^2}{4l}.$$

(b) Find the smallest possible object to image distance. What are the corresponding values of s and s'?

37.63 (5) *Two ways of getting the picture.* An object of unknown height h is located 1.00 m from a screen. When a converging lens is placed at the proper point between the object and the screen, an image of height h_1' forms on the screen. If the lens is then moved 17.5 cm toward the screen, an image again forms on the screen, this time of height h_2'. (a) Find the focal length of the lens. [Hint: It is easiest to use the Newtonian form of the thin-lens equation (Problem 37.56).] (b) Show that the height of the object is $h = \sqrt{h_1' h_2'}$. (This relation is the basis for one method of finding the size of an object located at a modest but unknown distance.)

37.64 (5) *Double dip, single dip.* A thin symmetric biconvex lens is made of glass whose index of refraction is $n_g = 1.50$. Each surface has a radius of curvature 20.0 cm. (a) What is the focal length of the lens? (b) The lens is dipped into water ($n_w = 1.33$). What is the focal length of the lens? (c) The lens is sealed into the side of a tank filled with water. What is the focal length of the lens?

37.65 (5) *Not for casual dipping.* A thin lens has focal length $+10.0$ cm. It is made of glass whose index of refraction is 1.50. What is the focal length of the lens when it is immersed in carbon disulfide ($n_{CS_2} = 1.63$)?

37.66 (5) *Thin lenses in series.* (a) Sketch a ray diagram for two thin lenses, of focal lengths f_1 and f_2, located a distance l apart. Include in your sketch the object distance s_1 and the image distance s_1' for the first lens, as well as the corresponding distances s_2 and s_2' for the second lens. (b) Write the thin-lens equation for each lens, and find a relation among the quantities s_1', s_2, and l. (c) Let $l \to 0$, and show that

$$\frac{1}{s_1} + \frac{1}{s_1'} = \frac{1}{f_1} + \frac{1}{f_2}.$$

Use this result to obtain Equation 37.26.

37.67 (5) *Double duty.* In the figure below, an object of height h is located at a distance s from a converging lens, with $f < s < 2f$. A plane mirror is located at the second focus of the lens. Find the location and height of the image.

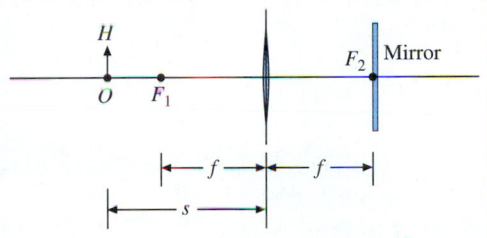

37.68 (5) *Dioptric power of a lens, I.* In assembling lens systems, it is often convenient to define the *dioptric power* (or simply the *power*) of a lens as the reciprocal of its focal length:

$$P \equiv \frac{1}{f}.$$

The unit of dioptric power is called the *diopter* (symbol Δ) in optical practice, but $1 \Delta \equiv 1$ m^{-1}. A "strong" converging lens has a large dioptric power; a strong diverging lens has a negative dioptric power of large magnitude. (a) Rewrite Equation 37.26 in terms of dioptric power. (b) An optometrist has a presbyopic patient whose near point lies at 0.50 m. What is the dioptric power of the lens that she must prescribe to bring the near point to the convenient distance of 0.25 m?

37.69 (5) *Dioptric power of a lens, II.* In older persons, the eye can be myopic and presbyopic at the same time (though myopia tends to ameliorate presbyopia). Such persons wear *bifocals*, in which two different lenses are ground into different parts of the same blank. In a certain patient, the far point lies at 0.66 m and the near point at 0.38 m. (a) What should be the dioptric powers of the far-vision (upper) and near-vision (lower) parts of the bifocal lens? Assume that the corrected near point is to be at 0.25 m. (b) The first surface of the lens has a uniform radius of curvature 0.125 m. Find the two radii of curvature of the back side of the lens. Let $n = 1.60$, a typical value for polycarbonate (plastic) lenses.

37.70 (6) *Function of the crystalline lens.* If the refracting surfaces of the eye are lumped into a single thin lens, its effective distance from the retina in a normal eye is 2.2 cm. (a) When the eye is focused on a distant object, what is the focal length of the equivalent thin lens? What is the dioptric power (Problem 37.68)? (b) When the eye is focused on an object at the near point (25 cm), what is the focal length of the equivalent thin lens? What is the dioptric power? (c) What is the focal length of a thin correcting lens, placed in contact with the equivalent thin lens of part **a**, that would convert the system into the equivalent lens of part **b**? What is the dioptric power? **(d)** What is the ratio of the power of the correcting lens to that of the equivalent thin lens of part **a**? Your result demonstrates the accommodating, or fine-focusing, function of the lens. (The power of the relaxed crystalline lens of the eye is about 5 Δ.)

37.71 *(6) Helping out the magnifier.* Equation 37.27 gives the angular magnification of a simple magnifier when it is held so that the image is at infinity. Although this is the most relaxed way to use the magnifier, you can help the lens out with the lens of your own eye. **(a)** Show that, if you place the object so that the image is at your near point instead of at infinity, the angular magnification is increased to the value

$$\alpha = \frac{d}{f} + 1.$$

(b) For a typical magnifier of focal length 7 cm, what percentage improvement do you obtain in this way?

37.72 *(6) Staying in focus.* You have a telescope whose objective has focal length 50 cm. The telescope is focused on a distant object. You refocus, moving the ocular relative to the objective so as to bring into clear focus an object whose distance is 50 m. How far must you move the ocular?

37.73 *(G) Peculiar mirror.* A biconvex thin lens is made of glass having index of refraction n. The lens has surfaces of equal curvature ρ. The lens is made into a mirror by silvering one of the surfaces. Parallel light is incident on the lens from the clear (unsilvered) side. **(a)** Show that the focal length of the lens/mirror is

$$f = \frac{\rho}{2(2n - 1)}.$$

(b) If $\rho = 35$ cm and $n = 1.5$, find the focal length.

GROUP C

37.74 *(3) Spherical aberration.* The figure below shows a ray, parallel to the axis but distant from it, falling on a concave spherical mirror of radius ρ at P. The center of curvature of the mirror is at C, and the three angles ϕ are equal. The reflected ray misses the focus F. The distances x and y are known, respectively, as the *longitudinal* and *transverse spherical aberration*. **(a)** Show that

$$x = \frac{\rho}{2}\left(\frac{1}{\cos \phi} - 1\right).$$

(b) Show that

$$y = x \tan 2\phi = \frac{\rho}{2}\left(\frac{1}{\cos \phi} - 1\right) \tan 2\phi.$$

(c) Show that spherical aberration vanishes for paraxial rays.
(d) Evaluate the spherical aberrations x and y for a mirror having $\rho = 60$ cm if the *aperture D* is equal to 50 cm.

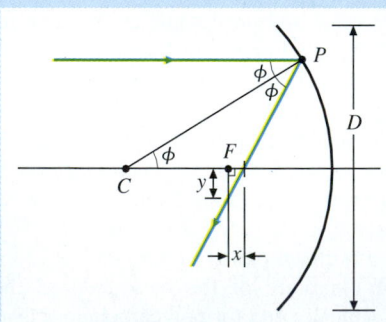

37.75 *(3) Parabolic mirror.* The spherical aberration shown in Figure 37.9 can be eliminated by using a mirror of paraboloidal rather than spherical form. The figure (next column) shows a cross section of such a mirror. Shown also is the geometrical definition of a parabola as the locus of points equidistant from a point called the focus and a line called the directrix.

   (a) Show that all rays parallel to the axis are reflected so as to pass through the focus, regardless of their distance from

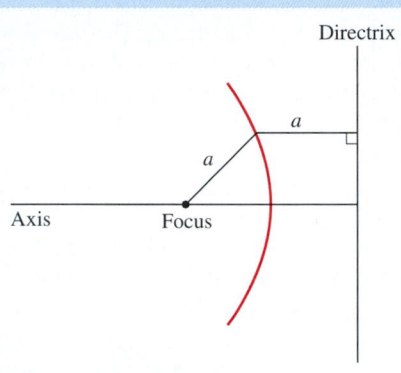

the axis. (Hint: You may find it easier to carry out the proof in terms of parallel wave fronts, using Huygens's construction.) **(b)** A team under the direction of E. F. Borra at Laval University has carried out experiments to demonstrate the practicality of building a vertically directed parabolic mirror in which the reflector is the surface of a rotating pool of mercury. Modern vibrationless air bearings and special motors are required to maintain smooth, ripple-free rotation. A 30-m-diameter mirror has been proposed, with focal length 30 m. What angular speed is required to maintain the proper surface shape? (Hint: See Problem 5.56.)

37.76 *(3) Ellipse.* According to the geometrical definition, an ellipse is the locus of all points for which the sum of the distances from two points called the foci is a constant. Using this definition, show that a mirror in the form of an ellipse, with a point object at one focus, will produce an image at the other focus.

37.77 *(6) Longitudinal magnification of a spherical mirror.* **(a)** Show that, when an object of small length Δs lies along the optic axis, a spherical mirror produces an image of length

$$\Delta s' = -\frac{1}{(s/f - 1)^2} \Delta s.$$

In this equation, the negative sign implies a reversal of orientation. Show that the image arrow and the object arrow always point in opposite directions. (Hint: Suppose that you displace a point object an infinitesimal distance ds along the optic axis. What is the corresponding displacement ds' of the image?) The quantity

$$L \equiv \frac{\Delta s'}{\Delta s} = \frac{-1}{(s/f - 1)^2}$$

is called the *longitudinal magnification*. (b) Show that this expression yields the proper value for a plane mirror. (c) Show that, unlike the lateral magnification M, L is always negative. (d) Prove that the two magnifications bear the relation $L = -M^2$. (e) Show that the mirror image of a three-dimensional object is *always* perverted, whether the mirror is concave or convex and whether the image is erect or inverted. That is, show that a "mirror image" is a mirror image for *any* mirror under *all* circumstances. (f) Because L and M are in general different for a spherical mirror, the mirror image of a three-dimensional object is usually *distorted*. That is, its dimensions are not magnified by the same factor in all directions. But show that a concave mirror produces an undistorted image of a small object when the object is placed either at its center of curvature or very close to the mirror surface.

37.78 *(5) Longitudinal magnification of a thin lens.* (a) Show that the expressions given in Problem 37.77 for the longitudinal magnification L of a spherical mirror are also valid for a thin lens. (b) Then, by carefully considering the sign convention for lenses, show that the image of a three-dimensional object produced by a lens is *not* perverted, regardless of whether the image is real or virtual, inverted or erect.

37.79 *(6) Galilean telescope, I.* Galileo's telescope [he called it a *cannocchiale* ('eye-tube')] is sketched below. As in the telescope of Figure 37.30, the second focus of the objective coincides with the first focus of the ocular. However, the ocular is a *diverging* lens. (a) Draw a ray diagram for the Galilean telescope. Is the image erect or inverted? real or virtual?

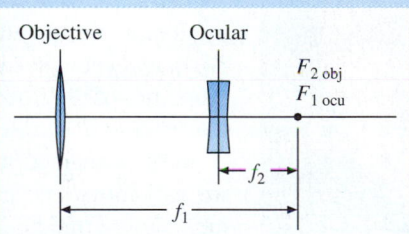

(b) Show that the angular magnification is given by Equation 37.29. (The Galilean configuration is still used in the small binoculars called opera glasses, where the compactness and lightness of the device are highly desirable and the limited magnification and field of view are not a problem.)

37.80 *(6) Galilean telescope, II.* The angular magnification of a Galilean telescope (Problem 37.79) is $\alpha = 10$. The tube length is 40 cm. What are the focal lengths of the objective and the ocular?

37.81 *(G) Achromatic doublet.* Because the index of refraction of glass is not independent of the wavelength of light, the focal length of a simple lens is not the same for all wavelengths. Consequently, the lens suffers from a colored blurring

of images called *chromatic aberration* when it is used to form an image in white light; see Query 37.7. This problem considers one method of using a pair of lenses to eliminate chromatic aberration at two chosen wavelengths λ_r and λ_b, normally chosen to be in the red and blue parts of the spectrum, respectively. (The correction is then precise for these two wavelengths and approximate for other wavelengths.) The two lenses are made of different kinds of glass—say, crown glass and flint glass—with different dispersion curves. Let the indices of refraction of the crown glass for λ_r and λ_b be n_{1r} and n_{1b}, and let the corresponding indices of the flint glass be n_{2r} and n_{2b}. (a) Using the lensmaker's equation, show that the two lenses in contact will bring the two wavelengths to the same focus if

$$(n_{1r} - n_{1b})\kappa_1 = -(n_{2r} - n_{2b})\kappa_2,$$

where κ_1 and κ_2 depend on the radii of curvature of the two lenses. (b) Now consider a wavelength λ_y that lies between λ_r and λ_b; normally, λ_y is chosen in the yellow. Let the indices of refraction of the two glasses for λ_y be n_{1y} and n_{2y}, respectively. Show that the condition of part **a** will be satisfied if the focal lengths f_{1y} and f_{2y} of the two lenses for yellow light are in the ratio

$$\frac{f_{1y}}{f_{2y}} = -\frac{\dfrac{n_{1r} - n_{1b}}{n_{1y} - 1}}{\dfrac{n_{2r} - n_{2b}}{n_{2y} - 1}}. \qquad (1)$$

The design problem thus boils down to making two lens elements with the proper ratio of focal lengths in the yellow. Why must one of the lenses be diverging? (c) You have crown glass for which $n_{1b} = 1.52262$, $n_{1y} = 1.51707$, and $n_{1r} = 1.51461$. You also have flint glass for which $n_{2b} = 1.62904$, $n_{2y} = 1.61715$, and $n_{2r} = 1.61218$. Find the desired focal length ratio. (d) If the lens is to have focal length $f = 10.0$ cm, calculate f_{1y} and f_{2y}. (e) For convenience in design, you want the two lens elements to fit snugly together so that they can be cemented into one piece. That is, the radius of curvature ρ_{12} of the second surface of the crown glass lens and the radius of curvature ρ_{21} of the first surface of the flint glass lens must be equal and opposite. Also, for simplicity, you would like to make the front surface of the doublet flat; $\rho_{11} = \infty$. Find the necessary values of the radii ρ_{12}, ρ_{21}, and ρ_{22}. [Historical notes: (1) On the basis of hasty measurements, Newton concluded erroneously that the right side of Equation 1 would always have the value 1 and that achromats were therefore impossible. That is why he turned his attention to the design of reflecting telescopes. (2) The achromatic doublet was invented in 1729 by the English amateur astronomer Chester Moor-Hall (1704–1771). Because he wanted to keep his invention secret, he ordered the crown glass elements from one lensmaker and the flint glass elements from another. He assembled the elements himself and sold them at a high profit. But he suffered a terrible piece of bad luck. The two lensmakers found themselves overworked and subcontracted the lenses out—by coincidence—to the same third lensmaker! Noting the exactly mating curvatures of the two elements, the third lensmaker became curious, put them together, and discovered the secret.]

Part *XI*

Relativity

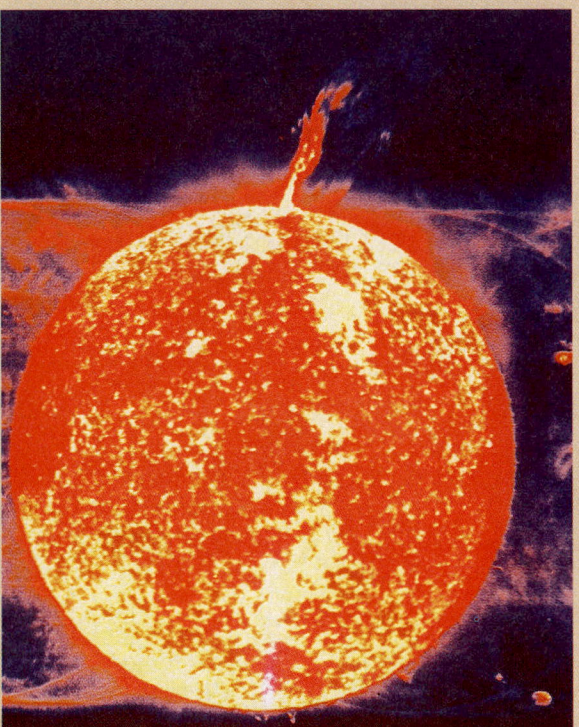

Up to this point, our study of mechanics has been limited to the classical Newtonian realm—the realm in which objects that are neither too big nor too small move at speeds much less than the speed of light. In this context, "neither too big nor too small" means that the objects are no smaller than molecules (preferably a good deal bigger) and no bigger than the sun (preferably a good deal smaller).

But the world stretches beyond the boundaries of the classical realm. We will see explicitly that, far from being universal, the writ of classical physics does not run beyond its proper boundaries. You have already seen a few examples of how applying the laws of classical physics beyond these boundaries leads to unsatisfactory agreement between theoretical predictions and experimental results. Remember just two of many possible examples. Electrons in solids do not behave like charged billiard balls (Section 27.5), and the

electrical resistance of solids reflects that fact. Electromagnetic waves cannot be consistently described in terms of the mechanical oscillation of an ether (Section 34.4), and attempts to describe the properties of the ether lead to contradictions.

In the remainder of this book, we will move decisively beyond the slow, middle-sized world where our senses operate directly and where we acquire our "common sense" into the realms of the very fast-moving, the very massive, and the very small, where classical physics fails to explain what is observed. Satisfactory explanation requires that we develop and apply one or more of several physical theories developed in the twentieth century. We will by no means abandon classical physics. Rather, we will set classical physics in the context of a much broader physics. In doing so, we will deepen our understanding of classical physics at the same time that we broaden our overall perspectives on the physical world.

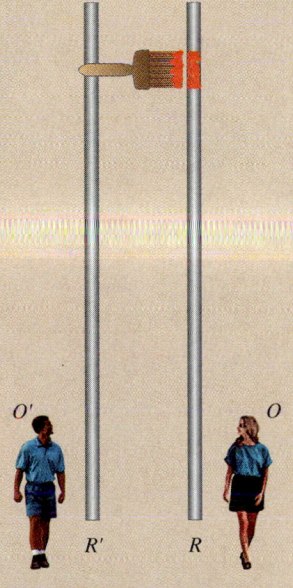

The realm where objects do not necessarily move slowly is the subject of *special relativity*, the main topic of Chapters 38 and 39. The realm of the very small is the subject of *quantum mechanics*, the topic of Part XII.

There is a price to be paid when we move away from the realm of our everyday experience. Our common sense fails because we have no familiar common-sense experience to use as a basis for new insights. We continue to rely on experimental evidence, as we have committed ourselves always to do in science. But the relevant experiments can no longer be simple extensions of familiar experience, as they often are in the classical realm. We will by no means abandon homely analogies, but we will become warier of taking them too seriously.

This price paid, we will have a reward: We will become familiar with a world far broader and far richer than that within the confines of everyday experience. We will doubt-less retain a fondness for that "homeland"; people usually do. But we will gain confidence in a wider role as citizens of a vaster physical world.

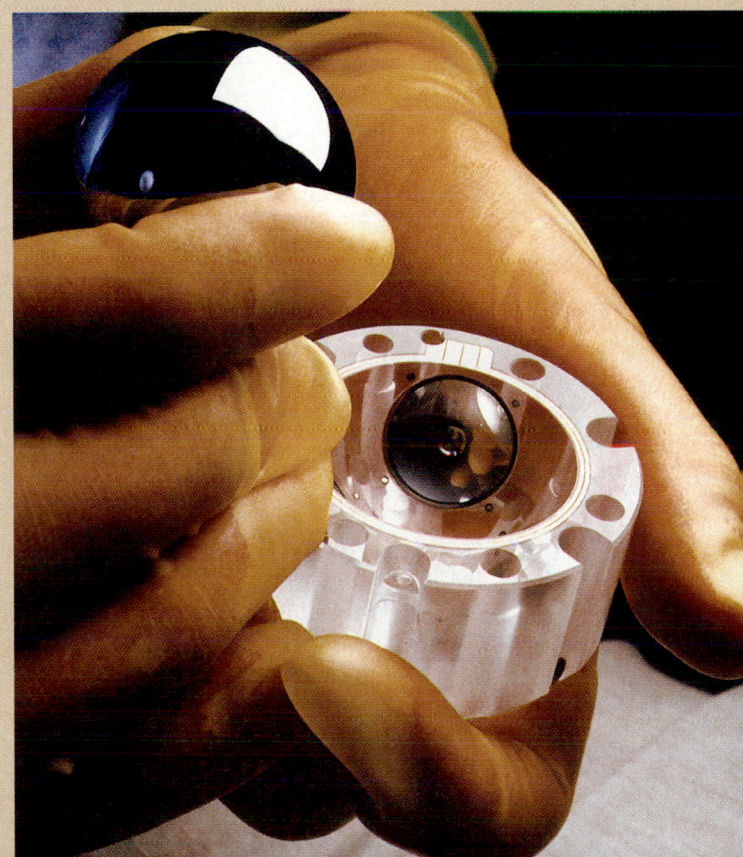

Relativistic Kinematics

———— Maxwell's equations predict that the speed of light is $c = 1/\sqrt{\mu_0 \epsilon_0}$, where the product $\mu_0 \epsilon_0$ can be measured in a laboratory. How, then, does a light wave appear to an observer who is moving with respect to the lab?

———— Careful investigation of this and related issues leads to the conclusion that the concepts of absolute time and absolute simultaneity, which are central to Newtonian mechanics, are untenable.

———— The relativistic world view reestablishes the foundations of physics on two postulates called the principles of relativity.

———— From these principles can be deduced a set of rules, called the Lorentz transformations, that enable each of two observers moving at constant velocity with respect to one another to use his own position and time measurements of an event to predict the position and time measurements for the same event made by the other observer.

Left: One of the unprecedentedly precise gyroscopes that will go into space on the Gravity Probe B satellite. The probe will provide a stringent test of general relativity theory. The quartz rotor is spherical within one wavelength of light and is coated with an equally precise thin niobium coating to render it superconducting. Part of the housing, which serves to spin the rotor up and contains suspension and orientation readout devices, is shown.

A champion fencer named Fisk
Had moves most exceedingly brisk.
So quick was his action,
The Lorentz contraction
Foreshortened his foil to a disk!

—AUTHOR UNKNOWN

Introduction

In the first twenty-two chapters of this book, we considered Newtonian (classical) mechanics and its application to a wide variety of subject areas. Some of these applications were straightforward and direct. Examples are the study of the motion of blocks sliding down inclined planes and of spinning wheels. Some applications, equally direct, required a mathematically more sophisticated approach. An example is the study of waves propagating through mechanical media. Some applications had sensational consequences. The study of celestial mechanics, for example, made it possible to understand the interactions and motions of heavenly bodies with exquisite precision centuries before it was physically possible to visit them. And some applications were in areas where it was far from obvious in advance that the science of mechanics was even relevant. An example is the study of heat and thermal interactions.

Given these far-flung triumphs—and there have been many others as well—it is small wonder that by the end of the nineteenth century scientists and others had tremendous confidence in the soundness of Newtonian mechanics. In many ways, this confidence is no less justified today. Physicists have not found, and do not expect to find, anything ''wrong'' with the way in which the subjects in this book up to this point have been treated.

Nevertheless, there is a glaring gap in the fundamental assumptions on which Newtonian mechanics is based. This gap remained hidden for a long time because it shows up directly and obviously only in the study of phenomena not accessible to experiment until after 1900. Even before 1900, indirect experimental evidence piled up gradually that, taken as a whole, suggested puzzling inconsistencies in the Newtonian world view.

As is usually true of fundamental inconsistencies, the inconsistencies underlying Newtonian mechanics crop up in several apparently independent places. One of these places lies at the boundary between mechanics and electromagnetism. It is tempting to unify electromagnetism with mechanics by ''explaining'' electromagnetic phenomena—notably electromagnetic wave propagation—in mechanical terms. But, as you saw in the final subsection of Section 34.4, the study of electromagnetism has led us beyond the bounds of mechanics. In particular, Maxwell's equations, which describe the propagation of light, do not rest on a mechanical model. Attempts to put Maxwell's equations on a mechanical basis result in a variety of paradoxes. Among them are the contradictory properties of the luminiferous ether, discussed in Section 34.4.

A number of first-rate physicists wrestled with these paradoxes for many years. It remained for Albert Einstein to cut the Gordian knot in 1905. His solution and its consequences are the subject of this chapter and the next one.

Einstein said years later that his intensive decade-long inquiry into the foundations of mechanics, leading to the development of the special theory of relativity, began when he was sixteen years old. At that time, he asked himself what a light wave, traveling

Our everyday expectations and actions are built on common-sense kinematic rules. These rules work fine as long as $v \ll c$. But remember Einstein's aphorism: ''Common sense is nothing more than a deposit of prejudices laid down by the mind before you reach eighteen.''

with speed c with respect to one observer, would look like to a second observer who was riding along with it—that is, moving with speed c with respect to the first observer. The question interested him because of the paradox it appeared to pose.

Here is one of several possible ways to look at the paradox. As you learned in Chapter 34, an electromagnetic wave is described by wave equations having the form of Equations 34.22 and 34.23:

$$\frac{\partial^2 \mathscr{E}_y}{\partial t^2} = \frac{1}{\mu_0 \epsilon_0} \frac{\partial^2 \mathscr{E}_y}{\partial x^2} \quad \text{and} \quad \frac{\partial^2 \mathscr{B}_z}{\partial t^2} = \frac{1}{\mu_0 \epsilon_0} \frac{\partial^2 \mathscr{B}_z}{\partial x^2}.$$

As with all wave equations, the constant on the right side of the equation is the square of the wave speed; according to Equation 34.24, we have

$$\frac{1}{\mu_0 \epsilon_0} = c^2.$$

But $\mu_0 \epsilon_0$ is a product of constants that can be determined by laboratory measurements made by an observer at rest with respect to the apparatus. An entirely analogous statement can be made about a stretched string: The quantity corresponding to $\mu_0 \epsilon_0$ is [tension]/[mass per unit length] $= v^2$ (Equation 21.7), and this quantity can be determined by laboratory measurements on the string—the *wave medium*. But, for the stretched string (or any other mechanical medium), v is the speed of the wave *with respect to the medium, not the observer*. An observer who is moving with respect to the medium measures a wave speed different from v.

If we insist on keeping the analogy—that is, on considering the electromagnetic wave a form of mechanical wave—we must seek the medium in which electromagnetic waves propagate. Nineteenth-century physicists called this medium the *luminiferous ether*. We have already discussed some of the difficulties posed by the hypothetical properties of such an ether. But we now consider a deeper difficulty, one that does not depend on the peculiar properties of the ether. Suppose that c is the speed of light with respect to the ether. An observer moving with speed V in the same direction as a light pulse ought to observe its speed as $c - V$. Similarly, an observer moving with speed V in the opposite direction ought to observe the speed of the pulse as $c + V$. Other directions of motion should result in other variations. But no such variations have ever been observed. In particular, the famous **Michelson-Morley experiment** of 1887 detected no effect due to the motion of an observer relative to the ether. (The details of the experiment are discussed in Problem 38.32. For a sketch of Michelson, see Section 36.2.)

Einstein's solution to this paradox is at once radical and conservative. It is radical because its fundamental postulates completely overturn many notions of what is "obvious" in the physical universe. It is also radical because it opens up to physical study horizons inconceivable in Newtonian terms. But it is conservative in a vital sense: It is compatible with Newtonian mechanics when it is applied to those areas where Newtonian mechanics is undeniably successful. That is why there is nothing "wrong" in the approach we have taken in the preceding chapters of this book.

SECTION 38.2   The Galilean Transformation

To see what is logically wrong with Newtonian mechanics, we must ask, "What is basic to Newtonian mechanics?" In large measure, the answer is the set of equations to which Einstein gave the name **Galilean transformation** (in recognition of Galileo's pioneering contribution to the study of relative motion at constant velocity, later incorporated into Newton's laws as the law of inertia).

The Galilean Position Transformation

Figure 38.1 shows observers O and O' located, respectively, at the origins of coordinate frames of the same names. The x and x' axes are collinear, and O' moves along the x axis at constant velocity $\mathbf{V} = V\hat{\mathbf{x}}$ with respect to O. (It is equally valid to say that O moves along the x' axis at constant velocity $-\mathbf{V}$ with respect to O'.) The corresponding axes y, y' and z, z' are parallel. As you saw in Section 3.1, we call such reference frames **inertial frames**. Observers O and O' make measurements of the motion of the same particle P. For convenience, the two observers synchronize their clocks, and they call the instant when the origins of their two frames coincide $t = t' = 0$.

FIGURE 38.1 The coordinate frames and clocks of observers O and O'. Both make repeated simultaneous observations on the position of P.

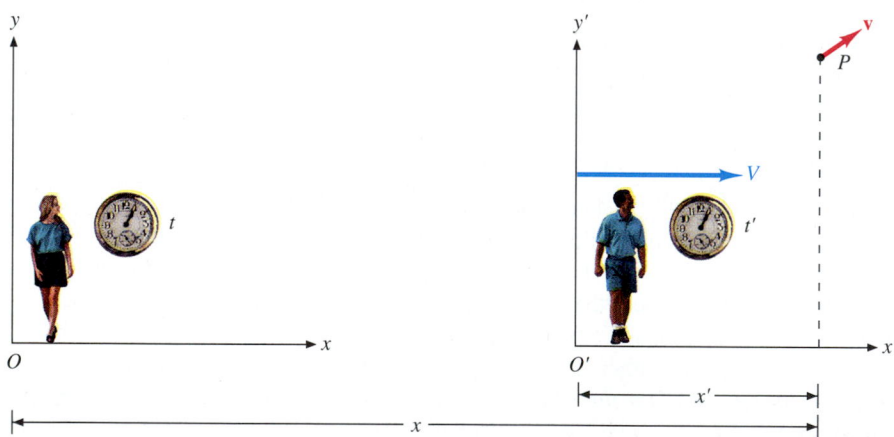

The two observers make simultaneous measurements on particle P. From the Newtonian point of view, the meaning of the word "simultaneous" is self-evident. Observers O and O' may agree in advance, for example, that each of them will measure the position of P at $t = t' = 0$ s, 0.1 s, 0.2 s, and every 0.1 second thereafter.

Each observation results, for O, in a set of position coordinates (x, y, z) and, for O', in a set (x', y', z'). From these observations, O deduces the velocity of P as seen from her coordinate frame, which she calls $\mathbf{v}$. She also deduces the acceleration of P, which she calls $\mathbf{a}$. Similarly, O' uses his data to deduce the velocity $\mathbf{v}'$ and the acceleration $\mathbf{a}'$, as seen from his coordinate frame.

It is evident from the start that the numerical values of the position (x, y, z) measured

1046 ■ Relativistic Kinematics

by O will not be the same as the values (x', y', z') measured by O'. But that is not important. What *does* count is that O and O' be able to communicate about the position of P in mutually intelligible language. That is, each must be able to check his or her own results against the other's, and the two observers must ultimately be able to agree that their data and their conclusions are mutually consistent. Or, what amounts to the same thing, O must be able to *predict*, on the basis of her own measurements (x, y, z), the values of the simultaneous measurements (x', y', z') of O'—and vice versa. When O and O' come together again after the series of measurements is complete, they can check to see if they have been successful in their predictions.

All this explanation is prelude to a result that you may consider obvious and that all physicists prior to the twentieth century considered obvious. The result,

$$x' = x - Vt \quad \text{or} \quad x = x' + Vt, \quad \text{(38.1a,b)}$$

$$y' = y \quad \text{and} \quad z' = z, \quad \text{(38.1c,d)}$$

is called the **Galilean position transformation**. It is simply the set of rules that O and O' need to predict the results of each other's measurements on the basis of their own. Because we have chosen to consider the special case in which the motion of O' relative to O is along the x axis, the time t appears only in the equations relating x and x' and not in the corresponding equations for y and z. But we have implicitly built another "obvious" assumption into Equations 38.1a and 38.1b. We should really write Equation 38.1b in the form $x = x' + Vt'$, because it is his own time measurement t' that O' uses in predicting the coordinate x measured by O. But we have taken for granted the statement

$$t' = t \quad \text{(38.1e)}$$

and have saved ourselves the trouble of writing the prime on t in Equation 38.1b.

It would never occur to anyone to "save the trouble" of omitting the prime from the quantity x' in the same equation; this would result in nonsense. It would seem that we have not thought twice about giving time a primacy over position in our logical framework. Although we are content to have transformation rules of position, we have asserted that time is *absolute*. This assertion is embodied in Equation 38.1e. Because this is one of the Galilean transformation equations, we say that time is **invariant** under the Galilean transformation. (Why do we not say the same thing about y and z?)

The Galilean Velocity Transformation

Example 38.1 suggests how you can begin with the Galilean position transformation (Equation 38.1a–e) and deduce the Galilean transformation equation for velocity.

EXAMPLE 38.1

Find the rules connecting the velocities $\mathbf{v}$ and $\mathbf{v}'$ of particle P, as measured by observers O and O' in Figure 38.1.

SOLUTION: You begin by writing the definition of velocity, $\mathbf{v}' = (v'_x, v'_y, v'_z)$, in terms of the fundamental measured quantities x', y', z', and t' (or t):

$$v'_x \equiv \frac{dx'}{dt'} = \frac{dx'}{dt}.$$

Because you want to predict the value of v'_x in terms of the measurements made by O, you use Equation 38.1a to obtain

$$v'_x = \frac{d}{dt}(x - Vt) = \frac{dx}{dt} - \frac{d}{dt}(Vt).$$

We have stipulated that $\mathbf{V}$, the velocity of O' with respect to

O, be constant, and so this equation simplifies to the form

$$v'_x = \frac{dx}{dt} - V.$$

The quantity dx/dt is simply v_x, the x component of the velocity of P as measured by O. So you have

$$v'_x = v_x - V. \quad \text{(38.2a)}$$

You can work backward in identical fashion to obtain the equivalent transformation in the opposite direction:

$$v_x = v'_x + V. \quad \text{(38.2b)}$$

These are the results we discussed informally at the end of Section 38.1.

The other two components of the velocity are found even more directly. Using Equation 38.1c, you have $v_y' \equiv dy'/dt' = dy'/dt = dy/dt$, or

$$\boxed{v_y' = v_y,} \quad \text{and likewise} \quad \boxed{v_z' = v_z.} \quad \text{(38.2c,d)}$$

Combining the results for the three components of the ve-

locity of particle P, you have

$$\mathbf{v}' = \mathbf{v} - V\hat{\mathbf{x}} \qquad (38.3)$$

for the **Galilean velocity transformation**. How would you express $\mathbf{v}$ in terms of $\mathbf{v}'$?

The Galilean velocity transformation is a precise statement of something you have probably long taken for granted—that velocities add and subtract like vectors. Your experience outside as well as within the formal study of physics supports this view. For example, suppose you are driving on a straight road at 120 km/h. How fast are you overtaking a car going 100 km/h?

Absolute Time and the Simultaneity Paradox

The problem with the Galilean transformations, as Einstein saw, was that the Newtonian view they underlie leads logically to a tangle of inconsistencies, one of which—the light-wave paradox—we mentioned in Section 38.1. As will soon become evident, the most glaring inconsistency has to do with the "obvious" assumption of Equation 38.1e, $t' = t$. But we cannot simply abandon the Galilean transformations without substituting something else. Einstein therefore proposed to substitute a set of transformation equations based on two fundamental assumptions, called the **principles of relativity**:

1. *The laws of physics are the same for all observers situated in inertial frames.*
2. *The speed of light in vacuum is the same for all observers and in all directions.*

The first principle may seem to be nothing new, but it is violated the moment we assume the presence of a luminiferous ether. To see this, let us rephrase the first principle in negative terms: No conceivable physical observation can be used to distinguish between one inertial frame and any other. But, if the ether exists, it must be at least conceivably detectable. (Otherwise there is absolutely no point in saying it exists.) If the ether is detectable, we can distinguish between a reference frame that is at rest with respect to the ether and another frame that is not. It follows logically that all inertial frames are not equivalent in every way, in contradiction to the first principle of relativity.

The restriction of the first principle of relativity to inertial frames limits us to a special case of relativity for which the mathematical treatment is fairly simple. The resulting limited theory is called the **special theory of relativity**. Removing the restriction leads to the **general theory of relativity**, originally proposed by Einstein in 1915. The general theory can be used to treat the case of two observers who accelerate with respect to one another. Although the general theory leads to profound insights into the relation between matter and space, it is mathematically and conceptually complex and we will not consider it in this book.

So confident have physicists become of the soundness of the second principle of relativity that it has been taken to underlie the definition of one of the SI base units.* As you learned in Section 35.2, the speed of light is *defined* to be $c \equiv 299\ 792\ 458$ m/s. The meter is a derived unit whose value is based on the value of c; similarly, the value of the constant ϵ_0, the permittivity of free space, is based on the value of c.

---

*You should not take this confidence in the constancy of c on faith but should reserve judgment until you have completed your study of this chapter.

The Fallacy of Absolute Simultaneity

"Everyone knows" that either two events occur at the same time or they do not. We will now show that this common-sense view is incompatible with the second principle of relativity. Consider the following thought experiment, first suggested by Einstein in a slightly different form. In Figure 38.2, observer O stands beside a railroad track. Next to her is a flashbulb. Also next to the track are two clocks. Clock 1 is located a distance L up the track from O, and clock 2 is located an equal distance down the track. On the track is a boxcar with glass sides. Its length is $2L$, and at its ends are clocks $1'$ and $2'$. In the middle of the car stands observer O'. Next to him is a flashbulb. The two observers carefully synchronize all the clocks. After nightfall, the boxcar is moved up the track and accelerated to a speed V, which it maintains thereafter throughout the experiment. (You will probably find it helpful to imagine that V is an appreciable fraction of the speed of light c, but the argument does not depend on this.)

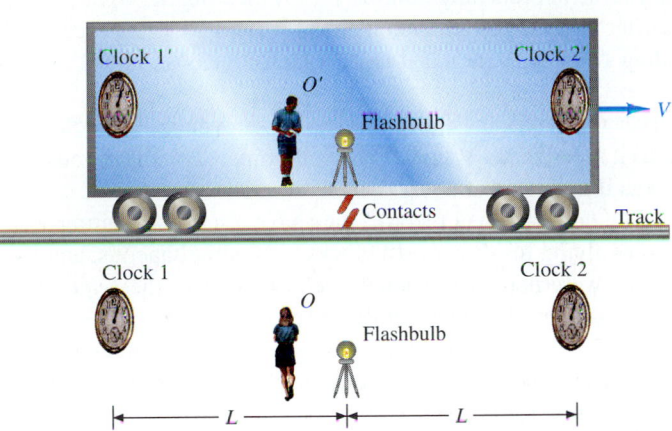

FIGURE 38.2 A thought experiment to demonstrate that simultaneity cannot be absolute.

The car rushes past observer O. Because it is dark, neither observer can see any clock. However, there is an electrical contact mounted on the underside of the boxcar just beneath O', and another one is mounted on the track next to O. Just as O' passes O, the contacts touch and set off the flashbulbs, which are adjacent to each other at that instant. By the brief flash of light, each observer can read all four clocks (Figure 38.3).

What does O see? For clocks 1 and 2 by the side of the track, the answer is simple. The light takes a time $\Delta t_{1,2} = L/c$ to reach clock faces 1 and 2. Although O will not see the light reflected from the clock faces until a later instant, it is the time $\Delta t_{1,2} = L/c$ after the flash that she will read and record for each clock. We can say that O perceives the two *events*—light reaches each of the two clock faces—as simultaneous events; that is what we commonly mean by simultaneity.

Observer O also reads the faces of clocks $1'$ and $2'$ on the boxcar, which are moving with respect to her. But, while the light pulse is traveling from the flashbulbs toward

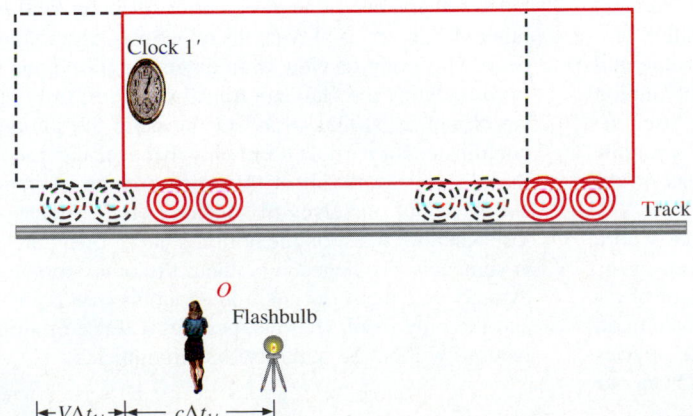

FIGURE 38.3 Illustration for Example 35.3, showing the boxcar in black at the instant when the flashbulb flashes and in red at the instant when the light pulse reaches clock $1'$ and illuminates it.

clock 1', the clock is approaching O and meets the light pulse part way. The light therefore illuminates clock face 1' at a time earlier than L/c after the flash.

Next, O turns her attention to the light flash heading toward clock 2'. She argues that this flash has a distance greater than L to travel, because clock 2' is receding from her. An argument similar to that in the preceding paragraphs shows that O reads the interval of time from the flash of the bulb until the clock face is illuminated as *longer* than $\Delta t_{1,2} = L/c$.

Observer O therefore makes the following record of the sequence of events: *First*, clock 1' was illuminated at $\Delta t_{1'}$; *then* clocks 1 and 2 were illuminated simultaneously at $\Delta t_{1,2}$; and *then* clock 2' was illuminated at $\Delta t_{2'}$.

So far so good. But now let us find what observer O' sees. From his point of view, clocks 1' and 2' are motionless and equidistant. Because, according to the second principle of relativity, the speed of light is the same for him as it is for O, he will read clocks 1' and 2' simultaneously at a time after the flash given by $\Delta t'_{1',2'} = L/c$. Because he sees clock 2 to be approaching him, he sees it meeting the light pulse part way and so reads an earlier time $\Delta t'_2$ on it than on clocks 1' and 2'. And, because O' sees clock 1 to be receding from him, he reads it later than clocks 1' and 2', the time interval being $\Delta t'_1$.

Thus observer O' makes the following record of events: *First*, clock 2 was illuminated at time $\Delta t'_2$; *then* clocks 1' and 2' were illuminated simultaneously at $\Delta t'_{1',2'}$; and *then* clock 1 was illuminated at $\Delta t'_1$.

When O and O' compare their records, they have a surprise waiting. Each one claims that the events involving his or her own clocks were simultaneous, and those involving the other's clocks were not. They cannot even agree as to the *sequence* of nonsimultaneous events. Observer O claims that the order was $1' \rightarrow (1, 2) \rightarrow 2'$, but O' claims it was $2 \rightarrow (1', 2') \rightarrow 1$. As you have just seen, this disagreement is a direct and necessary logical consequence of the second principle of relativity: The speed of light (in vacuum) is the same for all observers.

We are thus obliged to abandon the notion of absolute simultaneity and, with it, the idea of absolute time. The two are intimately connected, as Einstein pointed out in 1905:

> We have to take into account that all our judgments in which time plays a part are always judgments of simultaneous events. If, for instance, I say, "The train arrives here at seven o'clock," I mean something like this: "The pointing of the small hand of my watch to 7 and the arrival of the train are simultaneous events."

THINKING LIKE A PHYSICIST

Well, you may argue, the boxcar paradox is just a special case depending on the special way in which the thought experiment was cooked up. After all, Newton made a distinction between "absolute, true, and mathematical time," which "of itself, and from its own nature, flows equably without relation to anything external," and "relative, apparent, and common time," which "is some sensible and external . . . measure of duration by the means of motion [for example of a pendulum], which is commonly used instead of true time." Perhaps all we have here is an example of what can happen if we use common time instead of absolute time.

But, if you take this point of view, you are stuck with a terrible difficulty. The Absolute Clock—if it indeed exists at all—is forever inaccessible, at least for purposes of physical measurement. We have to make do with "common time." We had therefore better formulate the laws of physics in terms of common time and banish absolute time from our

physics books.

Can we do so satisfactorily? The answer is unequivocally Yes! For, though O and O' will have to abandon the idea that simultaneity and the order of events are identical for both of them, each one can predict perfectly well what the other sees, in terms of what his or her own observations have been. This point of view is an extension of the one we took in formulating the Galilean transformations. In Figure 38.1, we do not insist that O and O' measure the same position coordinates for particle P but only that each can predict what the other measures. It is this very breadth of vision—the ability to put ourselves in someone else's shoes—that we lose when we insist on the notion that our own point of view is somehow privileged. For then, the other person will inevitably take the same position about his own point of view, and the only result (besides perhaps a sense of smug satisfaction) will be the inability to communicate.

The Time Dilation

Now that we have disposed of the idea of absolute simultaneity, we must find the rules that two observers such as O and O' must use to transform each other's time measurements into the terms proper to their own coordinate frames. In other words, we would like to know exactly how the rate at which a clock indicates the passage of time depends on its speed V with respect to the observer who reads it.

The argument that follows is applicable to any clock at all, regardless of the details of its construction. But it is particularly simple to understand what happens if we use a **light clock**, described in Figure 38.4.

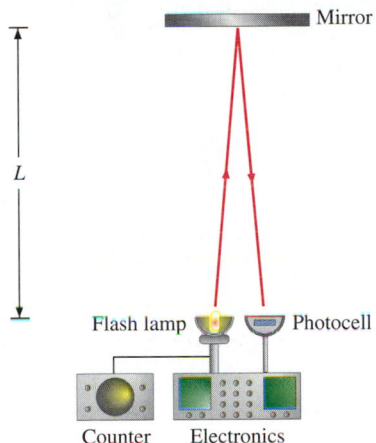

FIGURE 38.4 A light clock. When the flash lamp flashes, the time required for the light pulse to reach the photocell via the mirror is $\Delta t_0 = 2L/c$. The arriving light pulse triggers the photocell, which activates the very fast electronic circuit to flash the lamp again. Thus the lamp flashes (the clock "ticks") at intervals Δt_0. The counter tallies the flashes and thus measures the time elapsed since the clock was started.

Suppose that observer O has a light clock at rest in her reference frame. As she sees it, the time required for a light pulse to travel from the lamp to the mirror and back to the photocell depends only on the distance L and the speed of light c. Specifically, the time elapsed is $\Delta t_0 = 2L/c$. (The subscript zero is a reminder that the speed of the clock is zero with respect to O.)

Now, suppose that observer O', moving with respect to O at velocity $\mathbf{V} = V\hat{\mathbf{x}}$, has an identical clock at rest in *his* coordinate frame. He will see his clock ticking at intervals $2L/c$, just as O sees her own clock. What we want to know is *how each observer sees the other's clock.* Figure 38.5 shows how O sees the clock of O' as it passes her. Suppose that it flashes shortly before it passes her, when it is at $x = A$, and that the next flash takes place when the clock is at $x = C$. Suppose also that A and C are equidistant from O.

The first light pulse travels outward in all directions at a speed that must be c for both observers, as required by the second principle of relativity. From the point of view of O, however, the entire clock has moved by the time the reflected light hits the photocell and triggers a new flash. Because the speed of light is the same in all directions, the light must have struck the mirror when the clock was at B, directly opposite O.

Let us give the name Δt_V to the time interval between flashes, as O sees the clock of O'. (The subscript V is a reminder of the speed of the clock with respect to O.) Because of the way the clock moves with respect to O, the light from her point of view must make a round trip not of length $2L$ but of the greater length $2D$. She therefore sees the time interval between flashes to be

$$\Delta t_V = 2\frac{D}{c}.\qquad\textbf{(38.4)}$$

We use Pythagoras's theorem to find the distance D and thus to determine the value of Δt_V. The distance the clock travels between flashes is $V\,\Delta t_V$. Referring to Figure

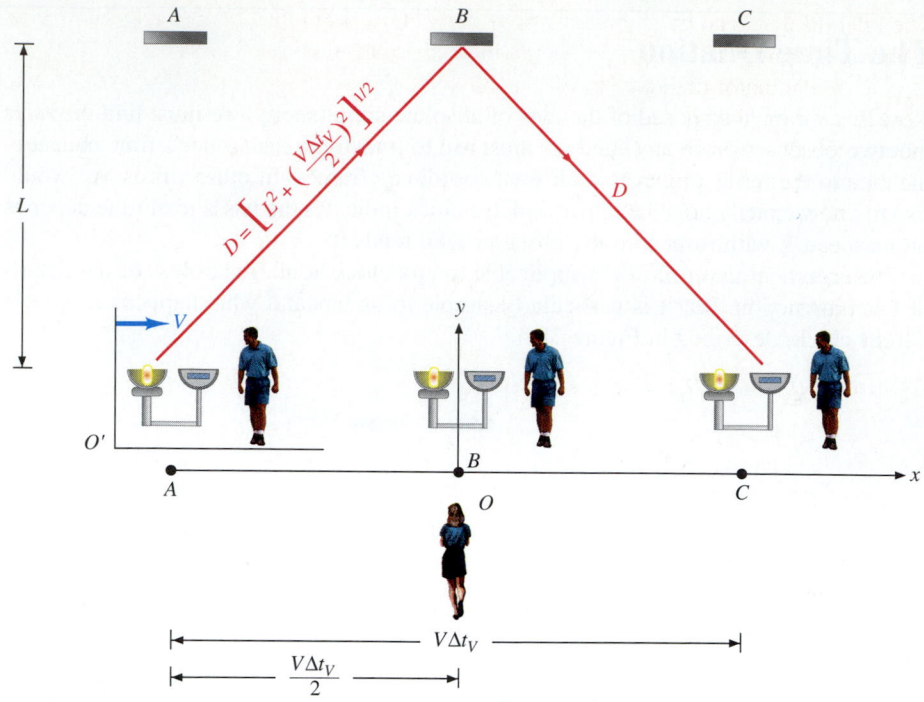

FIGURE 38.5 The light clock moves at velocity $\mathbf{V} = V\hat{\mathbf{x}}$ with respect to O. O' rides along with the clock.

38.5, you will see that the distance D is the hypotenuse of a right triangle whose sides are L and $V\,\Delta t_V/2$. Thus we have

$$D^2 = L^2 + \left(\frac{V\,\Delta t_V}{2}\right)^2. \tag{38.5}$$

We now square Equation 38.4 to obtain $\Delta t_V{}^2 = 4D^2/c^2$. Substituting the value of D^2 given by Equation 38.5 into this expression gives us

$$\Delta t_V{}^2 = \frac{4}{c^2}\left(L^2 + \frac{V^2\,\Delta t_V{}^2}{4}\right),$$

or

$$\Delta t_V{}^2 = \frac{4L^2}{c^2} + \frac{V^2\,\Delta t_V{}^2}{c^2}.$$

We bring all the terms in $\Delta t_V{}^2$ to one side of the equation and factor to obtain

$$\Delta t_V{}^2\left(1 - \frac{V^2}{c^2}\right) = \frac{4L^2}{c^2}.$$

Taking the square root of both sides of this equation, we have

$$\Delta t_V\sqrt{1 - \frac{V^2}{c^2}} = \frac{2L}{c}. \tag{38.6}$$

But the quantity $2L/c$ is just Δt_0, the time interval between flashes that O measures for her own clock. We substitute this value into Equation 38.6 and solve for Δt_V, obtaining

$$\Delta t_V = \frac{\Delta t_0}{\sqrt{1 - \dfrac{V^2}{c^2}}}. \tag{38.7}$$

The interval Δt_0, read by an observer with respect to whom the clock is at rest, is called the **proper time**.* The interval Δt_V is called the **dilated time**.

The denominator of Equation 38.7 is always equal to or less than 1, so $\Delta t_V \geq \Delta t_0$. That is, a clock always runs slower as seen by an observer with respect to whom it is moving than as seen by an observer with respect to whom it is at rest. Consequently, the relation expressed by Equation 38.7 is called the **relativistic time dilation**.

Be careful not to think that the time dilation is a result of the *clock's* being "in motion" in some absolute sense. Motion has meaning only in terms of the clock *and* the observer. Observer O' will see his clock as running at the rate Δt_0 and the clock of O as running at the slower rate Δt_V, which is the opposite of what O observes.

Applications of the Time Dilation

As always in physics, the argument leading to the time dilation must ultimately be verified by experiment. Example 38.2 shows why time dilation is not observed in everyday events—that is, why we so easily take for granted the Newtonian assumption that time is absolute. We then consider experimental evidence that supports the validity of the time dilation and exemplifies the circumstances under which it must be taken into account.

EXAMPLE 38.2

A motorist sets his wristwatch by using the phone company time service. He then drives from a point in Reno to a point in Salt Lake City, a distance of 850.000 km, at a constant speed of 110.000 km/h, or 30.5556 m/s. Finally, he checks his watch again with the phone company time service. Will the time dilation be significant? That is, will he have to reset his watch if it can be read to the nearest second? Use $c = 2.997\,92 \times 10^8$ m/s.

SOLUTION: Imagine a pair of observers, one located in Reno and the other in Salt Lake City. Each has a wristwatch synchronized with the phone service. They can respectively observe the start and finish times of the trip and thus determine a proper time Δt_0 from their point of view:

$$\Delta t_0 = \frac{850.000 \times 10^3 \text{ m}}{30.5556 \text{ m/s}} = 27\,818.1 \text{ s}.$$

On the basis of this result, they can calculate the dilated time. This is the time interval they would find directly if the first observer could read the driver's watch as he passed the starting

line and the other could read it again as he passed the finish line. According to Equation 38.7, they calculate

$$\Delta t_V = 27\,818.1 \text{ s} \times \frac{1}{\sqrt{1 - \dfrac{(30.5556 \text{ m/s})^2}{(2.997\,92 \times 10^8 \text{ m/s})^2}}}.$$

Because $[30.5556/(2.997\,92 \times 10^8)]^2 = 1.038\,85 \times 10^{-14}$, the quantity after the multiplication sign has the approximate value $(1 - 1 \times 10^{-14})^{-1/2}$. Use the approximation $(1 - x)^{-1/2} \simeq 1 + \frac{1}{2}x$, which is accurate for $x \ll 1$. This gives you the value $1 + 5 \times 10^{-15}$ for the factor that relates the dilated time to the proper time. That is, the time dilation amounts to only about 2 parts in 10^{14} of the total time. But the total time of the journey is about 3×10^4 s, and wristwatches can be read only within 1 s. Expressed as a fraction of the total time, this is about 3 parts in 10^5. Thus the time dilation effect is entirely negligible in this case. Indeed, it would have to be a billion times as great to be barely detectable by using the wristwatches.

We now consider experimental evidence, first reported by Rossi and Hall in 1941, that supports the validity of the time dilation relation (Equation 38.7).

Cosmic radiation, consisting of very energetic nuclear particles, is present everywhere in the universe and constantly bombards the earth. A cosmic ray particle, de-

---

*Note that "proper" in this sense does *not* mean "correct," because neither time is correct or incorrect. Rather the meaning "intrinsic" is intended, because it is the time read by an observer at rest in the same coordinate system as the clock.

scending toward the earth through the upper atmosphere, collides with the nucleus of an atom of air. Such collisions occur for the most part in the region between 10 and 30 km above the earth's surface. A collision results in a shower of energetic particles of various types, the most common being the type called the *pion*. The pion, however, is unstable; before it has moved very far from its point of origin, it decays into another particle called a *muon*. Subsequently the muon also decays into still other particles. As with all unstable particles, the decay of the muon is random, and there is no way of telling when a particular muon will decay. Nevertheless, the lifetime can be characterized statistically. Given a large number of muons, the *half-life* is the time it takes for just half of them to decay. This half-life can be measured by using muons produced in the laboratory, which are moving with negligible speeds with respect to the observer; its value is $\tau_0 = 1.52 \times 10^{-6}$ s.

The muons produced by pion decay at the top of the atmosphere are moving with a speed close to the speed of light c. Let us assume for the moment that the speed is $V = c$. Then we can calculate the time it takes, from the point of view of an observer on the earth, for a muon born at altitude $h = 10$ km to reach the earth's surface if it happens to be heading straight down. We have for this descent time

$$t = \frac{h}{V} \simeq \frac{h}{c} = \frac{1.0 \times 10^4 \text{ m}}{3.0 \times 10^8 \text{ m/s}} = 3.3 \times 10^{-5} \text{ s.}$$

This is equal to about 22 times the half-life τ_0. Because each half-life "sees" a reduction of the number of remaining muons by a factor 2, we would expect the number of muons reaching the earth's surface to be about 1 in 2^{22}, or about 240 in each billion of those created at $h = 10$ km (and even fewer of those created at higher altitudes).

The muon production rate is known from high-altitude measurements. If the actual attrition rate were the one we have just calculated, the detection of a muon at the earth's surface would be a rare event. In fact, cosmic-ray muons are quite commonly observed at the earth's surface. Example 38.3 explains the reason why.

EXAMPLE 38.3

Calculate the half-life τ_V of cosmic-ray muons traveling straight downward toward the earth from the point of view of an observer on the earth. From energy measurements, the average speed of cosmic-ray muons is known to be $V = 0.994c$.

SOLUTION: The muons constitute a clock, which is read by counting the fraction of the original muons that survive. But it is observed as a moving clock, and you therefore read the dilated time τ_V. So you use Equation 38.7 to write

$$\tau_V = \frac{\tau_0}{\sqrt{1 - \left(\frac{0.994c}{c}\right)^2}} = \frac{\tau_0}{\sqrt{1 - (0.994)^2}} = 9.1\tau_0.$$

In contrast with the case considered in Example 38.2, the value of V^2/c^2 here is close to 1 rather than close to 0. As a result, the dilated half-life τ_V of the muons is 9.2 times the proper half-life τ_0. Inserting the value $\tau_0 = 1.52 \times 10^{-6}$ s obtained from laboratory measurements, you have

$$\tau_V = 1.4 \times 10^{-5} \text{ s.}$$

You have seen that the descent time for the muon, as seen by the observer on the earth, is $t = 3.3 \times 10^{-5}$ s, or about $2.4\tau_V$. Consequently, between one-quarter and one-eighth of the muons produced at $h = 10$ km will survive to reach the earth's surface, and the fairly common penetration of cosmic-ray muons to the surface is explained.

The cosmic-ray experiment just described is far from being the only evidence in support of time dilation. As is often the case with such fundamental phenomena, evidence in its support arises directly or indirectly from a very wide variety of experimental results. Moreover, because we have obtained the time dilation by assuming the validity of the principles of relativity, such experimental results support those principles as well.

The Fitzgerald-Lorentz Contraction

Once we have granted that the observed rate of a clock depends on its speed with respect to the observer, we are impelled to consider that other "old faithful," the length of an object. In Newtonian terms, measuring the length of an object implies measuring the positions of its two ends simultaneously. But we must now be suspicious of any measurement process involving the uncritical use of the concept "simultaneous."

We begin by reviewing a tacit assumption we made in deriving the time dilation equation. Specifically, we assumed that the distance L between the photocell and the mirror in the light clock of Figure 38.5 was independent of the speed of the light clock with respect to the observer.

Length of an Object Perpendicular to Its Observed Velocity

We now prove that it was correct to assume L to be independent of V when, as in the light clock, the length is perpendicular to the direction of motion. Suppose that observers O and O' have rods R and R' of equal length (Figure 38.6). A paintbrush is tied to R' at a certain distance from its lower end. If O and O' bring the rods slowly together so that they are parallel with their lower ends coinciding, the brush will leave a paint mark on R at a distance from its lower end equal to the distance of the paintbrush from the lower end of R'. Now let observer O' with his rod move past O and her rod with speed V in the x direction. As R' goes by her, O must see R' either as having its length unaffected by its motion past her, or as being contracted, or as being expanded. Suppose that she sees it to be contracted. That is, suppose she observes its length to be less than what she measured it to be when it was at rest with respect to her. If this is so, the paintbrush will leave a mark on her rod R at a point below the first mark.

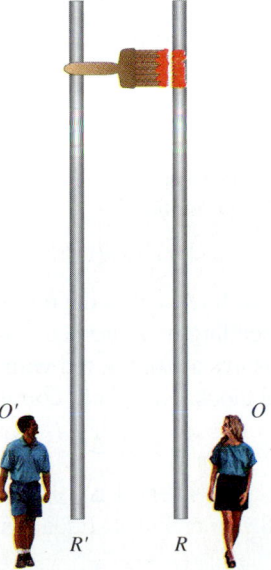

FIGURE 38.6 Thought experiment to prove that the observed length of an object perpendicular to the direction of its velocity **V** with respect to the observer is independent of V.

Observer O' must agree that the second mark is below the first mark. But from his point of view, rod R' is still at rest, and its length must therefore be unaffected. He must conclude that rod R, which he sees to be moving past him with speed V, is *expanded*. But the first principle of relativity requires that the laws of physics be the same for both observers. It is not possible that relative motion results in an observed contraction for O and an observed expansion for O'. The only remaining possibility is the conclusion that the length of an object is unaffected in a direction perpendicular to its motion relative to the observer. That is,

$$y' = y \quad \text{and} \quad z' = z \qquad \textbf{(38.8a,b)}$$

for motion in the x direction, just as for the Galilean transformation.

Length of an Object Parallel to Its Observed Velocity

When an object is observed in motion parallel to its length, the result is dramatically different from the perpendicular case. As shown in Figure 38.7, O and O' measure the length of a rod and find it to be L_0. Next, O' moves away from O, taking the rod with him. He then comes past her with speed V, the rod being parallel to his motion in the x (or x') direction. Observer O has a light clock. When the front end of the rod comes past her, she starts counting time. She continues to do so until the rear end of the rod passes her. If the passage of the rod requires a time Δt_0, she can say that its length L_V is

$$L_V = V \Delta t_0. \qquad \textbf{(38.9)}$$

(Here again, the subscripts V and 0 remind us of what is moving and what is at rest with respect to the observer.)

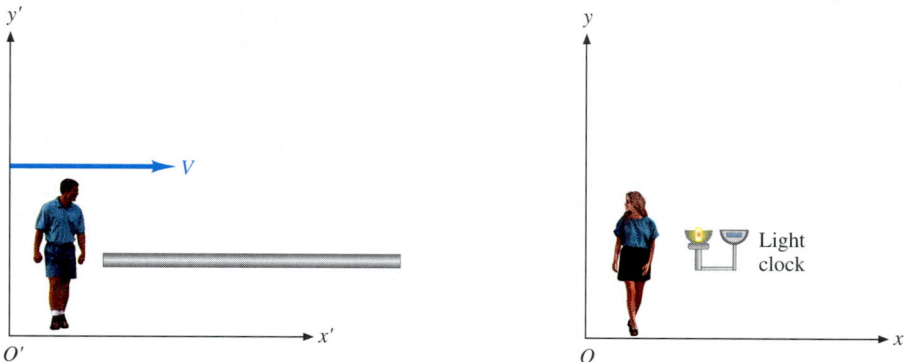

FIGURE 38.7 Thought experiment to determine the observed length of a rod as a function of its speed V with respect to the observer, when the rod is oriented parallel to the direction of motion.

Observer O', who is at rest with respect to the rod but is moving with respect to the light clock, makes the corresponding measurement. But his task is different in detail from that of O. While she measures a moving rod with a stationary clock, he measures a stationary rod with a moving clock. His result, corresponding to Equation 38.9, is

$$L_0 = V \Delta t_V. \qquad \textbf{(38.10)}$$

Observer O' is aware that the time interval Δt_V is dilated with respect to the proper time Δt_0. He uses Equation 38.7 to reexpress Equation 38.10 in the form

$$L_0 = V \frac{\Delta t_0}{\sqrt{1 - \dfrac{V^2}{c^2}}}.$$

According to Equation 38.9, the quantity $V \Delta t_0$ in this equation is L_V, the length of the rod as seen by O. Making this substitution and solving for L_V yields

$$L_V = L_0 \sqrt{1 - \frac{V^2}{c^2}}. \qquad \textbf{(38.11)}$$

The length L_0 measured by an observer with respect to whom the rod is at rest is called the **proper length**. The length L_V is called the **contracted length**; the quantity $\sqrt{1 - V^2/c^2}$ is always equal to or less than 1, so $L_V \leq L_0$. Equation 38.11 expresses the **relativistic length contraction**, also called the **Fitzgerald-Lorentz contraction** or, most commonly, the **Lorentz contraction**. Working independently in the early 1890s, the Irish physicist George Fitzgerald (1851–1901) and the Dutch physicist Hendrik A. Lorentz (1853–1928) proposed the contraction as an artificial hypothesis to account for contradictory observations, some of which are considered in Section 38.1. The details of Fitzgerald's approach are the subject of Problem 38.33.

There is a nice complementarity between the points of view of O and O' in the thought experiment leading to Equation 38.11. For O the length of the rod is contracted; for O' the time is dilated. The complementarity is explored further in Example 38.4.

EXAMPLE 38.4

Consider the muon experiment of Example 38.3 from the point of view of a hypothetical observer who moves with a group of muons as they descend through the atmosphere with speed $V = 0.994c$. From the point of view of this observer, the muons are at rest. Their half-life therefore has the proper-time value $\tau_0 = 1.52 \times 10^{-6}$ s. How can any substantial proportion of them survive to reach the earth's surface?

SOLUTION: The key to the answer lies in the fact that, from the hypothetical observer's point of view, it is the earth (including its atmosphere) that is moving with speed V. The muon's path from the point of its creation to the earth's surface is contracted from its proper length $h = 10$ km; according to Equation 38.11, its contracted length is

$$L_V = L_0 \sqrt{1 - \frac{(0.994c)^2}{c^2}}$$

$$= 1.0 \times 10^4 \text{ m} \times 0.109 = 1100 \text{ m}.$$

This is $1/9.2$ times the proper length, 10 km. Compare this result with that of Example 38.3, where you found that the dilated half-life of the muon was 9.2 times the proper half-life. The time required to traverse the distance L_V, measured by the observer's clock, is the proper time Δt_0. It is given by

$$\Delta t_0 = \frac{L_V}{V} \simeq \frac{L_V}{c}$$

$$= \frac{1100 \text{ m}}{3.0 \times 10^8 \text{ m/s}} = 3.7 \times 10^{-6} \text{ s}.$$

This time is $2.4\tau_0$. Compare with the result of Example 38.3, where the dilated time was found to be $2.4\tau_V$. Thus, no matter which coordinate frame you choose to calculate in, you will find the same proportion of the original muons reaching the earth's surface.

The Lorentz Transformation

SECTION 38.6

We are now prepared to derive from the principles of relativity the set of equations that constitute the relativistic equivalent of the Galilean transformations and thus to fulfill the task proposed at the beginning of Section 38.3.

The Position Transformation

Consider again the situation depicted in Figure 38.1. Again the observer O', fixed in the O' coordinate frame, moves past observer O at a velocity that she observes to be $\mathbf{V} = V\hat{\mathbf{x}}$. Again we choose the time $t = 0$ to be that when the two frames coincide. For the moment, let particle P be at rest in the O' frame (that is, set $\mathbf{v} = 0$), so the particle, too, moves past O at velocity $\mathbf{V}$.

At time $t = 0$, O measures the x coordinate of the position of P. She knows that this is a contracted length $x_{t=0} = x_V$, and she knows that, if O' measures the x' coordinate of P in his frame, he will obtain a proper length $x' = x_0$. The lengths measured by the two observers are related by Equation 38.11, and so we have

$$x_{t=0} = x' \sqrt{1 - \frac{V^2}{c^2}}.$$

The value of x' does not change with time because particle P is at rest with respect to O'. But the primed frame, and everything at rest with respect to it, moves away from O with speed V. At an arbitrary time t, the distance from the origin of O to the origin of O' is Vt. Thus, for O, the x coordinate of the particle P at any time is

$$x = x_{t=0} + Vt,$$

or

$$x = x'\sqrt{1 - \frac{V^2}{c^2}} + Vt.$$

We solve this equation for x' and obtain

$$x' = \frac{x - Vt}{\sqrt{1 - \dfrac{V^2}{c^2}}}. \tag{38.12a}$$

This relativistic equation corresponds to the Galilean Equation 38.1a. Note that there is really no inconsistency between the two if we restrict attention to everyday observations in which all speeds are very much smaller than the speed of light. If $V \ll c$, the value of the denominator in Equation 38.12a is approximately equal to 1 and we have $x' = x - Vt$, which is Equation 38.1a.

The transformation equation inverse to Equation 38.12a is

$$x = \frac{x' + Vt'}{\sqrt{1 - \dfrac{V^2}{c^2}}}. \tag{38.12b}$$

The derivation is the subject of Problem 38.29. Equation 38.12b reduces to Equation 38.1b for $V \ll c$.

For relative motion along the x and x' axes only, the transformations for the y and z coordinates follow directly from the argument at the beginning of Section 38.5. They are

$$y' = y \quad \text{and} \quad z' = z. \tag{38.12c,d}$$

These are identical with the corresponding Galilean transformation equations (Equations 38.1c,d).

Finally, we need the transformation equation for time. We begin by noting that Equations 38.12a and 38.12b contain t and t', respectively. We substitute the values of x' given by Equation 38.12a into Equation 38.12b. This gives

$$x = \frac{x - Vt}{1 - \dfrac{V^2}{c^2}} + \frac{Vt'}{\sqrt{1 - \dfrac{V^2}{c^2}}}.$$

We next multiply through by the denominator of the first term on the right side to obtain

$$x\left(1 - \frac{V^2}{c^2}\right) = x - Vt + Vt'\sqrt{1 - \frac{V^2}{c^2}}.$$

Collecting terms in x and dividing through by V, we have

$$-\frac{xV}{c^2} = -t + t'\sqrt{1 - \frac{V^2}{c^2}}.$$

We now solve for t', obtaining the desired result

$$t' = \frac{t - \dfrac{xV}{c^2}}{\sqrt{1 - \dfrac{V^2}{c^2}}}. \tag{38.12e}$$

Here again, we obtain $t' = t$ for $V \ll c$. This says that the concept of absolute time that underlies the Galilean transformations and all of Newtonian mechanics *is* valid if all speeds of interest are so small that we can just as well regard the speed of light to be infinite. Can you see why this is so?

The transformation equation inverse to Equation 38.12e is

$$t = \frac{t' + \dfrac{x'V}{c^2}}{\sqrt{1 - \dfrac{V^2}{c^2}}}.$$ (38.12f)

The derivation is the subject of Problem 38.30.

The Velocity Transformation

Suppose now that the particle P in Figure 38.1 is no longer at rest in the O' frame but is moving parallel to the x' axis (and the x axis as well) in such a way that observer O' measures its velocity to be v'_x. Observer O also measures the velocity of the particle and finds it to be v_x. What is the relation between v_x and v'_x? (By now you should be wary enough not to respond too quickly that $v'_x = v_x - V$.)

In the O frame, the velocity is defined to be

$$v_x \equiv \frac{dx}{dt}.$$

Similarly, in the O' frame, the velocity is defined to be

$$v'_x \equiv \frac{dx'}{dt'}.$$

Equation 38.12a gives an expression for x' in terms of x and t, and Equation 38.12e gives an expression for t' in terms of x and t. By differentiating these two equations, we will be able to express both dx' and dt' in terms of dx and dt. Having done this, we will be able to express v'_x in terms of v_x, as desired.

Beginning with Equation 38.12a, we have

$$dx' = \frac{dx - V\,dt}{\sqrt{1 - \dfrac{V^2}{c^2}}}.$$ (38.13)

And differentiating Equation 38.12e gives

$$dt' = \frac{dt - \dfrac{V}{c^2}\,dx}{\sqrt{1 - \dfrac{V^2}{c^2}}}.$$ (38.14)

We divide the first of these equations by the second to obtain

$$v'_x = \frac{dx'}{dt'} = \frac{dx - V\,dt}{dt - \dfrac{V}{c^2}\,dx}.$$

We now use the definition $dx/dt \equiv v_x$ and find

$$v'_x = \frac{v_x - V}{1 - \dfrac{Vv_x}{c^2}}.$$ (38.15a)

An argument in the same spirit leads to the transformation equations for the perpendicular velocity components. The result is

$$v_y' = \frac{\sqrt{1 - V^2/c^2}}{1 - \dfrac{Vv_x}{c^2}}\, v_y \quad \text{and} \quad v_z' = \frac{\sqrt{1 - V^2/c^2}}{1 - \dfrac{Vv_x}{c^2}}\, v_z. \tag{38.15b,c}$$

The derivation of Equations 38.15b and 38.15c is the subject of Problem 38.40. Equations 38.15a,b,c constitute the **Lorentz velocity transformation**. There is good reason for the name. In the 1890s, Lorentz wrestled with the same paradox that puzzled the younger Einstein—how could several observers moving with respect to one another apply Maxwell's equations to describe the propagation of an electromagnetic wave? Or, to put it in precise terms, how could Maxwell's equations be made *invariant* under coordinate transformations? They were certainly *not* invariant under the Galilean transformation. Lorentz showed that the transformation equations named after him would satisfy the invariance requirement but was unable to find a satisfactory physical explanation that would justify their use in preference to the Galilean transformation equations.

Equation 38.15a reduces to the familiar Galilean result $v_x' = v_x - V$ (Equation 38.2a) in the case $v_x \ll c$ or $V \ll c$. Equations 38.15b and 38.15c reduce in similar manner to Equations 38.2c and 38.2d, $v_y' = v_y$ and $v_z' = v_z$.

Equation 38.15 implies that velocities comparable to the speed of light do not simply "add," as small velocities do in the Galilean limit. This statement runs counter to everyday experience and is therefore difficult for many persons to accept at first. But everyday experience does not include observation of objects moving at speeds comparable to c!

EXAMPLE 38.5

Luke Skywalker is on an urgent interstellar mission when he suddenly sees an Empire battlecruiser headed straight for him from due galactic north at speed $0.9990c$. Before he can take evasive action, he sees a second Empire battlecruiser coming straight at him from due galactic south, also at speed $0.9990c$. He deftly executes a sharp turn and watches with understandable if not excusable satisfaction as the two less maneuverable cruisers collide with one another. Just before the collision, what is the speed of the first cruiser as seen by the unhappy captain of the second?

SOLUTION: You begin by drawing a sketch of the situation like that in Figure 38.8. Let Luke be observer O, as shown. Then cruiser 1 takes the role of particle P and has velocity $v_x = -0.9990c$, because it is moving toward Luke in the negative x direction. Observer O' must then be the captain of cruiser 2. Because V is the velocity of the primed frame seen from the unprimed frame, you have $V = 0.9990c$. You can now employ Equation 38.15:

$$v_x' = \frac{-0.9990c - 0.9990c}{1 - \dfrac{-0.9990c \times 0.9990c}{c^2}}$$

$$= \frac{-2 \times 0.9990c}{1 + (0.9990)^2} = -0.9999c.$$

FIGURE 38.8

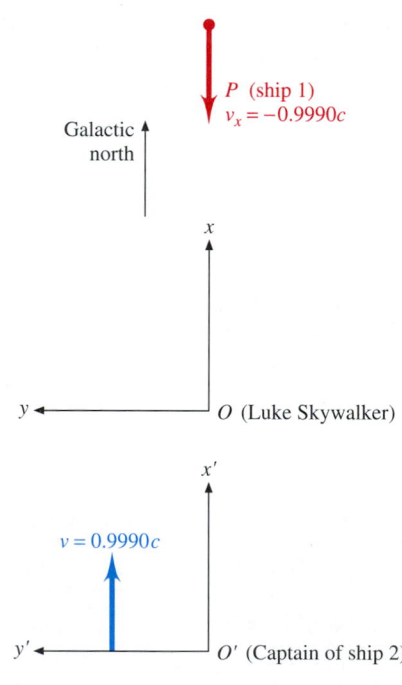

The result of the fanciful space story in Example 38.5 is instructive. It is indeed true that the captain of cruiser 2 sees cruiser 1 approaching him at a speed greater than the speed Luke observes. But the speed, though it is very close to the speed of light c, is still less than c. It is certainly not $2c$. Even if the two given speeds were $v_x = -c$ and $V = c$, the result would be $v'_x = c$, as you can quickly verify by calculation. *The speed of light constitutes an absolute upper limit on the speed of physical objects in the universe*, regardless of the observer. You will see in Chapter 39 that there are dynamical as well as kinematic reasons for this.

In hindsight, Einstein's solution to the electromagnetic wave paradox seems simple: As Lorentz and others already knew, the Galilean transformations cannot be applied to light waves. Instead of trying to "fix things up," Einstein elevated this property of light to a fundamental law of nature. According to the second principle of relativity, *all* observers see the speed of light as c. We *begin* with this principle and derive the laws of kinematics—that is, the Lorentz transformation equations—in a straightforward way, as we have done in this section and Section 38.5. One *cannot* "ride along with a light wave"!

Symbols Used in Chapter 38

c	speed of light		$\mathbf{v}, \mathbf{v}'$	velocity of an object as observed by O, O'
D	half-length of light path of light clock, as seen by an observer with respect to whom the clock is in motion		v_x, v_y, v_z	components of $\mathbf{v}$
			v'_x, v'_y, v'_z	components of $\mathbf{v}'$
			x, y, z	position coordinates measured by O
L	length of an object		x', y', z'	position coordinates measured by O'
t	time as measured by observer O, at rest in frame O		$_0$ (subscript)	labels quantities measured by observer at rest with respect to object observed
t'	time as measured by observer O', at rest in frame O'		$_V$ (subscript)	labels quantities measured by observer in motion at speed V with respect to object observed
Δt_j, $\Delta t_{j'}$	time interval indicated by clock j or j', as observed by O		τ_0, τ_V	proper half-life, dilated half-life of a muon
$\Delta t'_j$, $\Delta t'_{j'}$	time interval indicated by clock j or j', as observed by O'			
V, $\mathbf{V} = V\hat{\mathbf{x}}$	constant velocity of frame O', as observed by O			

Summing Up

When we attempt to interpret a number of crucial experiments within the framework of Newtonian mechanics, the result is a series of paradoxes. Resolution of these paradoxes begins with a careful examination of the concepts of absolute time and its twin, absolute simultaneity, on which the Galilean transformations are based.

A set of thought experiments leads to the conclusion that Equation 38.1e, $t' = t$, is untenable and that the Galilean transformations are therefore not universally valid. This conclusion in turn mandates a reformulation of the transformation equations for the kinematic quantities time, displacement, and velocity. Such a set of equations is necessary if observers who are moving with respect to one another are to be able to interpret each other's observations. The reformulation is based on two fundamental assumptions, called the **principles of relativity**, which replace the assumptions underlying Newtonian mechanics.

The resulting theoretical structure is called the **theory of relativity**. The **special theory of relativity** is limited to the case in which each observer sees the other to be in an inertial frame.

A thought experiment leads to a quantitative expression for the time dilation, Equation 38.7:

$$\Delta t_V = \frac{\Delta t_0}{\sqrt{1 - \dfrac{V^2}{c^2}}}.$$

A second thought experiment leads to the conclusion that measurements of the y and z dimensions of an object are independent of its velocity with respect to the observer. A third and final thought experiment leads to Equation 38.11, the length contraction in the x direction:

$$L_V = L_0\sqrt{1 - \frac{V^2}{c^2}}.$$

The predictions embodied in Equations 38.7 and 38.11 are amply confirmed by direct and indirect experimental evidence.

The **Lorentz transformations** fulfill the mandate for reformulation of kinematics on the basis of the principles of relativity. These equations, together with the corresponding Galilean transformation equations for purposes of comparison, are given in Table 38.1. A particularly dramatic and important implication of Equation 38.15a is that no observer can ever see a physical object moving faster than the speed of light.

TABLE 38.1 The Galilean and Lorentz Transformations

Galilean Transformations		Lorentz Transformations	
$x' = x - Vt$	(38.1a)	$x' = \dfrac{x - Vt}{\sqrt{1 - V^2/c^2}}$	(38.12a)
$y' = y$	(38.1c)	$y' = y$	(38.12c)
$z' = z$	(38.1d)	$z' = z$	(38.13d)
$t' = t$	(38.1e)	$t' = \dfrac{t - xV/c^2}{\sqrt{1 - V^2/c^2}}$	(38.12e)
$v_x' = v_x - V$	(38.2a)	$v_x' = \dfrac{v_x - V}{1 - Vv_x/c^2}$	(38.15a)
$v_y' = v_y$	(38.2c)	$v_y' = \dfrac{\sqrt{1 - V^2/c^2}}{1 - Vv_x/c^2}\, v_y$	(38.15b)
$v_z' = v_z$	(38.2d)	$v_z' = \dfrac{\sqrt{1 - V^2/c^2}}{1 - Vv_x/c^2}\, v_z$	(38.15c)

KEY TERMS

Section 38.1 Introduction
Michelson-Morley experiment

Section 38.2 The Galilean Transformation
inertial frame ▪ Galilean position transformation, velocity transformation ▪ invariant

Section 38.3 Absolute Time and the Simultaneity Paradox
principles of relativity ▪ special theory of relativity, general theory of relativity ▪ absolute time, absolute simultaneity

Sections 38.4 and 38.5 The Time Dilation and the Fitzgerald-Lorentz Contraction
light clock ▪ proper time, dilated time ▪ proper length, contracted length ▪ relativistic time dilation ▪ relativistic length contraction

Section 38.6 The Lorentz Transformation
Lorentz position and velocity transformations

Queries and Problems for Chapter 38

QUERIES

38.1 *(3) Relatedness of principles.* Show how the second principle of relativity can be derived logically from the first. Specifically, if the second principle were not true, how would this affect the first principle?

38.2 *(3) Special circumstances.* Two light clocks are located at opposite ends of a spaceship. Observer O', on the ship, sees that they flash simultaneously. Observer O, on the ground, also observes the flashes as the spaceship passes by at high speed. Are there any circumstances under which she also sees the two flashes simultaneously?

38.3 *(3) Fast Narcissus.* As a young man, Einstein imagined a runner who could run at a speed only slightly less than *c*. The runner holds a hand mirror in front of himself as he runs. Explain what the runner sees in the mirror.

38.4 *(4) Stout fellows!* A spaceship is prepared to break the record for the running-start time from Venus to Mars. A clock record is to be presented to the Guinness Book of World Records for verification. Is it better to use a clock on the spaceship or two synchronized clocks, one on Venus and the other on Mars?

38.5 *(4) Track star.* Comment on this limerick:

> There was a young lady named White,
> Who could travel much faster than light.
> She started one day
> In a relative way,
> And arrived on the previous night.

38.6 *(5) You can't get there from here!* A friend offers the following argument to show the limited utility of manned interstellar travel over large distances: A one-way trip that takes more than about 25 years of spaceship time is impractical because the crew would likely die of old age before they could return, and thus they could not tell what they saw; one might as well just send cameras and other recording devices. Be-

cause the spaceship cannot travel faster than *c*, that limits all practical manned voyages to a radius of about 25 light-years. But there are very few stars within 25 light-years. So why bother? Criticize this argument.

38.7 *(5) Einstein was wrong!* Here is a standard "disproof" of relativity offered by various persons hoping to win a Nobel Prize by showing that Einstein was wrong. A very rich man buys a very fast, very big car whose length is 12 m. His garage is only 10 m deep, but he wishes to put the car into the garage, however briefly. He installs fast-acting electric doors on both ends of the garage. While he stands next to the garage, he has his chauffeur approach the garage at such a high speed that the car is contracted to a little less than 10 m. He opens the front door, leaving the back door closed. Just as the car comes fully into the garage, he shuts the door. Thus both doors are closed at the same time, confining the car. Then he quickly opens the back door in time for the car to emerge. But the chauffeur argues that it was the garage, not the car, that was contracted. Thus, he claims, the owner could not have closed the doors, and Einstein was wrong!

Use the Lorentz transformation to reconcile what the owner sees with what the chauffeur sees, and show that there is no inconsistency.

38.8 *(G) Act of Congress?* Suppose that the speed of light were reduced to 30 m/s. Consider a number of commonplace experiences, and describe how they would be changed.

PROBLEMS

GROUP A

38.1 *(4) Speedometer check zone?* Observer *O'* has a light clock that flashes once per second. He moves along a special track at speed *V* m/s. Set up along the track are photocells that can detect and record the flashes of the clock. Checking later, *O'* finds that the distance between photocells that have detected flashes is 2*V* m. Calculate *V*.

38.2 *(4) Another point of view.* By working out the derivation leading to Equation 38.7 from the point of view of *O*, show explicitly that *O'* sees the clock of *O* running slower than his own, contrary to what *O* sees.

38.3 *(4) Longevity on the fast track.* The half-life of a stream of particles in an accelerator beam is measured and found to be 7.5 times that of the same particles observed at rest. How fast are the particles moving? Express your answer as a fraction of the speed of light and in m/s.

38.4 *(4) Real commitment.* Astronomers detect a civilization on the other side of our galaxy, a distance of 10^5 light-years away. In an attempt to make person-to-person contact, a spaceship is sent off with a crew of teenagers. If the crew are to have a reasonable chance of reaching their goal before they die of old age, approximately how fast must the ship go?

38.5 *(4) Stretching things out.* When observed at a speed of 2.90×10^8 m/s, a beam of particles has half-life 28.2 μs. What is their half-life when observed at rest?

38.6 *(4) Easy as π.* Pions have a proper half-life $\tau = 1.8 \times 10^{-8}$ s. A beam of pions passing through a set of detectors is observed to have a half-life of 3.1×10^{-8} s. **(a)** What is the speed of the pions? **(b)** What is the length of the track of a particular pion whose life happens to be just equal to the half-life?

38.7 *(4) The twin paradox, I.* Two precise, identical atomic clocks are synchronized. One of them is loaded on a commercial airplane, which flies in one direction, turns around, and returns to the starting point. The total flight time is 10 hours, and the speed of the plane is 300 m/s. **(a)** What is the difference in the readings of the two clocks? **(b)** Which clock reads the greater elapsed time? Neglect the small time during which the plane is accelerating. [Hint: To carry out the calculation, you will need to take advantage of the (very good) approximation $V \ll c$.]

38.8 *(4) The twin paradox, II.* Olivia and Oliver are twins. Olivia becomes an astronaut and, on their twenty-first birthday, bids her brother a fond farewell and makes a voyage to a nearby star. The journey takes 7 years by the clock on the spaceship. During most of the voyage, the ship travels at the constant speed 0.96*c* with respect to the earth. (Neglect the relatively brief periods of acceleration and deceleration.) After a brief period of data gathering at the star, Olivia reverses her course and returns home. When she arrives, her age is close to thirty-five. She hastens to visit her twin brother, Oliver. How old is he?

38.9 *(4) Fast, faster, fastest . . .* **(a)** Show that, for speed V close to c, you can use the approximation

$$\frac{1}{\sqrt{1 - V^2/c^2}} = \sqrt{\frac{c}{2\delta}},$$

where $\delta \equiv c - V$. What is the half-life of muons traveling past the observer at **(b)** $0.99c$, **(c)** $0.999c$, and **(d)** $0.9999c$?

38.10 *(4) . . . and still faster!* How fast is a beam of muons moving with respect to an observer who measures their half-life to be 1 s? Express your result in terms of the quantity δ defined in Problem 38.9.

38.11 *(4) It's all a plot! I.* Make a plot of $1/\sqrt{1 - V^2/c^2}$ as a function of V/c. (Hint: What is the range of interest for V/c?)

38.12 *(4) It's all a plot! II.* In doing Problem 38.11, plotting becomes awkward for values of V/c approaching 1. To get around this difficulty, repeat Problem 38.11, using semilog paper. Alternatively, plot $\log 1/\sqrt{1 - V^2/c^2}$ as a function of V/c.

38.13 *(5) It's all a plot! III.* Make a plot of $\sqrt{1 - V^2/c^2}$ as a function of V/c.

38.14 *(5) Mama's little baby loves . . .* An angry baker hurls a *baguette* of French bread 1 m long at your head. It misses narrowly. As it flies past you, you observe that its length is only 0.5 m. How fast is it going?

38.15 *(5) Deflation?* Observer O' moves past O with speed $0.62c$, holding a rod parallel to the direction of his motion. **(a)** If O sees the rod to be 1.3 m long, what is its proper length as seen by O'? **(b)** At the same time, O holds an identical rod perpendicular to the direction in which O' is moving. How long is this rod as O' sees it?

38.16 *(5) Half-measures.* If you find the length of a moving object to be one-half its proper length, how fast is it moving?

38.17 *(5) Pushing it.* You have a fast sports car whose length is 3.6 m. You drive it at full speed to impress your date, who expresses a preference for standing on the roadside and watching you go by. If you push the car to 230 km/h, how much shorter does the car appear to your date? Compare the contraction to the wavelength of blue light, $\lambda \simeq 400$ nm.

38.18 *(5) Depends on your point of view.* O and O' have identical stopwatches. O' moves past O with speed $0.9c$. Just as he passes her, both start their watches. O' waits until his watch reads 100 s and flashes a light on its face. Using a telescope, O can observe the reading of the watch at the instant of the flash. **(a)** What did the watch of O read at the instant

of the flash? **(b)** According to O, how long did the flash take to reach her? What does her watch read at the instant when she sees the flash? **(c)** According to O', how long did the flash take to reach O? What does his watch read when the flash reaches her?

38.19 *(6) Catching up.* A spaceship leaves the earth for Antares, traveling at $0.8c$ with respect to the earth. One year later, an improved model leaves for Antares, traveling at $0.9c$ with respect to the earth. **(a)** At what speed does the improved ship overtake the older one? **(b)** From the point of view of observers on the earth, how far away do the two ships meet? **(c)** At that instant, how far away do the crew members of the older ship see the earth? How long have they been en route? **(d)** How far away do the crew members of the improved ship see the earth? How long have they been en route?

38.20 *(6) Pair production.* Under proper conditions, an electron-positron pair can be created in the laboratory. (A positron has the same mass as an electron, but opposite charge.) The two particles move away from each other, each having a speed $0.95c$. What is the speed of the positron as seen by an imaginary observer located on the electron?

38.21 *(6) That fast!* **(a)** Show that, if the speed of light were infinite, all the Lorentz transformation equations would boil down to corresponding Galilean transformation equations. **(b)** Explain the implications of the conclusion of part **a** for the validity of the Galilean transformation.

38.22 *(6) Help from Hendrik.* Observer O sees event 1 occur at time t_1 at a point x. A little later, she sees event 2 occur at the same point at time t_2. The time interval between the events is $\Delta t \equiv t_2 - t_1$. Observer O' sees the same events and measures a time interval $\Delta t'$ between them. Express $\Delta t'$ in terms of Δt.

38.23 *(6) High-speed sum.* Show that if $v_x = c$ in the reference frame of O, the Lorentz velocity transformation yields c as the velocity measured by an observer O' moving at *any* velocity V.

38.24 *(G) Low-speed approximation.* **(a)** Show that, if $V \ll c$, the approximation

$$\sqrt{1 - \frac{V^2}{c^2}} \simeq 1 - \frac{1}{2}\frac{V^2}{c^2}$$

is valid. **(b)** Write a similar approximation for the factor $1/\sqrt{1 - V^2/c^2}$. **(c)** At what speed does use of the approximations of parts **a** and **b** introduce an error of 1% in the result? Express your result in m/s and in km/h, and thus show that the approximation is useful over a quite wide range of speeds V.

GROUP B

38.25 *(2) Relaxing the rules.* Equation 38.3 applies to the special case in which the coordinate frames O and O' have collinear x and x' axes and in which $\mathbf{V} = V\hat{\mathbf{x}}$. Generalize this result to any pair of coordinate systems with parallel corresponding axes and with $\mathbf{V}$ constant but having arbitrary direction.

38.26 *(5) No SLACker!* The Stanford Linear Accelerator is

3000 m long. It accelerates electrons to a speed that differs from the speed of light by only 8 parts in 10^{11}. [That is, $c - V = (8 \times 10^{-11})c$.] How long does the accelerator appear to be from the point of view of an observer riding on one of these electrons? (Hint: Your calculator will probably balk at handling the necessary numbers. Use the approximation suggested in Problem 38.9.)

38.27 *(5) Far and fast.* A spaceship on a mission to a distant star travels with constant speed V relative to the earth. When the crew members note that 1 year has passed, they flash a signal to the earth. **(a)** How long after the departure of the ship do observers on the earth receive the signal? **(b)** The observers on the earth determine that the signal was flashed when the spaceship was just 1 light-year from earth. How fast is the spaceship going?

38.28 *(6) Funny coincidence.* Observer O sees event 1 and then 3 μs later sees event 2 occurring at the same place. Observer O' sees the same two events separated in time by 4 μs. In his reference frame, what is the distance between the points at which the two events occur?

38.29 *(6) In Lorentz's footsteps, I.* Following a path similar to the one that leads to Equation 38.12a, derive Equation 38.12b.

38.30 *(6) In Lorentz's footsteps, II.* Following a path similar to the one that leads to Equation 38.12e, derive Equation 38.12f.

38.31 *(6) Speed of light in flowing water.* In 1851 and again in 1859, Fizeau measured the speed v of light in water that was flowing through a pipe with velocity V. If you approached the question classically, you might predict the result $v = c/n + V$, where n is the index of refraction of water. But Fizeau showed—and Michelson and Morley confirmed with a more precise measurement made in 1885–86—that the measured value of v was close to

$$v = \frac{c}{n} + V\left(1 - \frac{1}{n^2}\right).$$

Fresnel had predicted this value in 1818, on the basis of artificial assumptions concerning the way in which the water drags the ether along as it flows. (For this reason, the term in parentheses is called the *Fresnel drag coefficient*.) But show that the equation can be derived without special assumptions, by using the Lorentz velocity transformation.

38.32 *(G) The Michelson-Morley experiment.* This renowned 1887 experiment provides a road to the special theory of relativity, though it is one that Einstein seems not to have taken. Take the Newtonian view that light travels at a speed c with respect to the ether. Riding on the earth, you move at some speed V with respect to the ether. Now, set up a Michelson interferometer as shown in the next column. The arms have equal length L. The M_2 arm faces "upstream," in the direction in which you are moving; the M_1 arm, of equal length, is perpendicular to your direction of motion. **(a)** Show that, according to the Newtonian view, the time required for light to travel from the beam splitter H to M_2 and back is

$$t_{\parallel} = \frac{2L}{c}\left(1 + \frac{V^2}{c^2}\right).$$

(b) Show that the Newtonian speed of light "across the stream" from H to M_1 and back is $\sqrt{c^2 - V^2}$. **(c)** Find the Newtonian time required for the round trip from H to M_1 and back. **(d)** Make the assumption that $V \ll c$, and show that the time difference between the two round trips of parts **a** and **c** is $\Delta t = LV^2/c^3$. **(e)** If the wavelength of the light is λ, what is the optical path difference N between the light waves from

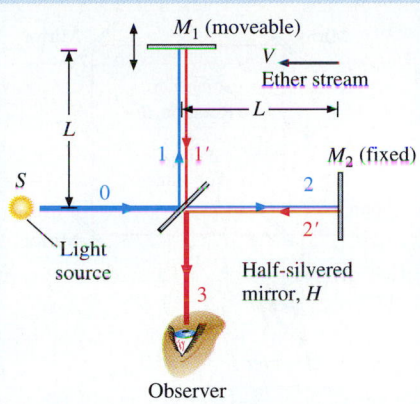

the two arms of the interferometer as they superpose along path 3? Neglect the optical path difference introduced by passage of light through the beam splitter. **(f)** You now rotate the interferometer 90° clockwise, so as to interchange the "upstream-downstream" and "across-the-stream" arms. As you do so, what fringe shift ΔN do you expect to see in the light of the foregoing analysis? **(g)** There is no way to estimate the speed of the earth with respect to the ether. For the sake of argument, use the speed of the earth relative to the sun, which is about 3.0×10^4 m/s. In the Michelson-Morley interferometer, the effective length of the arms was 11.0 m. Assuming that $\lambda = 550$ nm, show that the analysis predicts a shift of 0.4 fringe on rotating the apparatus. **(h)** In fact, Michelson and Morley found no fringe shift. They knew that they could detect a fringe shift as small as 0.1 fringe. But suppose they were very unlucky and happened to make their measurement just at an instant when the earth was at rest with respect to the ether. If they repeated the experiment six months later, what fringe shift does the classical analysis predict? **(i)** What bearing does the negative conclusion of the Michelson-Morley experiment have on the second principle of relativity? on the first principle of relativity?

38.33 *(G) Fitzgerald to the rescue!* The negative result of the Michelson-Morley experiment (see Problem 38.32) required explanation. Not surprisingly, the first attempts at explanation were carried out in a classical framework. Fitzgerald made the following bold suggestion: The interferometer arm facing into the ether stream is shortened, compared with the length it would have in the absence of an ether stream. When the interferometer is rotated through 90°, that arm lengthens as it is turned across the stream while the arm originally perpendicular to the ether stream shortens as it is turned into the stream. Suppose that the arms are adjusted to have equal lengths L before rotation. **(a)** What is the proportional length change $\Delta L/L$ of either arm needed to account for the zero fringe shift when the interferometer is rotated? Does your result look familiar? **(b)** Fitzgerald's suggestion gives a satisfactory *mathematical* explanation for the negative result of the Michelson-Morley experiment. Why is it not a satisfactory *physical* explanation?

38.34 *(G) Sagnac interferometer.* The device shown (p. 1066) is often used as part of a gyroscope or an angular accelerometer, especially in inertial guidance systems. Light travels

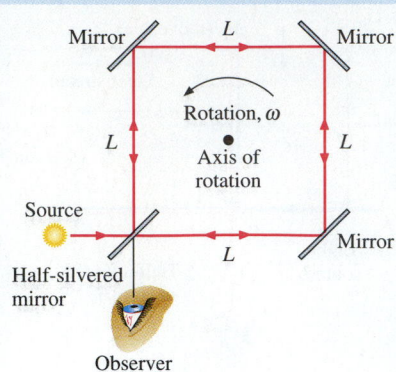

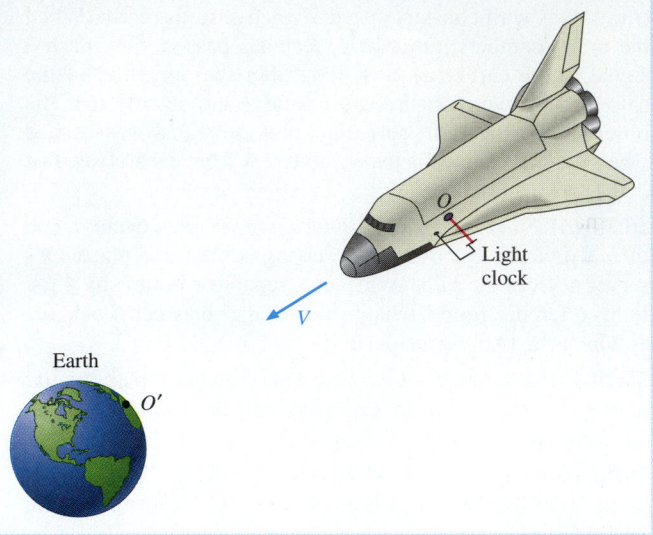

both ways around the square pathway of sides L. When the entire system is set into rotation about an axis at its center, as shown, the light path once around is greater in the direction of rotation than in the opposite direction. (You can think of the setup as a rotational version of the boxcar experiment.) As a result, the observer sees a fringe shift whose value depends on the angular speed ω. The instantaneous speed of a point on the pathway depends on its position but can be approximated by a mean speed $\langle v \rangle$. **(a)** Show that the fringe shift is given by

$$\Delta N = \frac{8L\langle v \rangle}{c\lambda} \frac{1}{1 - \langle v \rangle^2/c^2} \simeq \frac{4\omega A}{c\lambda}$$

for small angular speeds ω, where the term A is the area enclosed by the pathway. **(b)** If $L = 1$ m and visible light is used, find the approximate value of ω that will result in 1 fringe shift; $\Delta N = 1$.

This device was invented by the French physicist G. M. M. Sagnac (1869–1928) in 1913. In modern versions, the light source is often a laser incorporated directly into one or all of the arms of the closed pathway. In order to increase the sensitivity to rotation, the effective enclosed area is sometimes increased by letting the pathway be an optical fiber coiled into a many-turn coil.

38.35 *(G) Relativistic Doppler shift, I.* As you learned in Section 22.8, the wavelength of sound (or any wave traveling through a medium) is shifted if the source, the observer, or both are moving with respect to the medium. This problem concerns the relativistic Doppler shift, which affects light as it travels through empty space.

In the figure shown here, a light clock is mounted on the side of a spaceship so that its light path is perpendicular to the direction of the ship's motion. (This method of mounting makes it unnecessary to consider any Lorentz contraction of the clock itself.) The light clock ''ticks'' with frequency $\nu = 1/\Delta t$, as seen by an observer on the spaceship. The spaceship is moving toward an observer on earth at speed V. **(a)** Taking into account only the time dilation, determine what frequency would be observed on earth. **(b)** There is another effect that influences the earth observer's measurement of frequency. Because the spaceship is traveling toward him, each flash has less far to go than the preceding one. According to the earth observer, what distance does the spaceship cover between flashes? **(c)** Show that the earth observer sees light flashes

arriving with frequency

$$\nu' = \nu \sqrt{\frac{1 + \dfrac{V}{c}}{1 - \dfrac{V}{c}}}.$$

(d) Compare this with the classical result given for a moving source, a stationary medium, and a stationary observer by Equation 22.34, which here assumes the form $\nu' = \nu c/(c - V)$. Show that the two results agree if $V \ll c$. (Note: The relativistic case is simpler than the classical case, in that the only relevant velocity is that of the observer relative to the source. There is no way to make a distinction between a moving source and a moving observer, and there is no medium, moving or otherwise.) **(e)** What is the value of ν' if the spaceship is moving away from the earth at speed V? **(f)** The relativistic Doppler shift is defined as $\Delta\nu \equiv (\nu' - \nu)$. If the spaceship moves away from the earth at speed $V = 0.75c$, find the ratio $\Delta\nu/\nu$. This *red shift* is the basis for determining the rate of expansion of the universe. The most distant objects yet observed appear to be receding from us at speeds in excess of $0.9c$. Interestingly, Doppler himself suggested in 1842 that red and blue shifts might account for the observed differences among the spectra of different stars. But Doppler's suggestion soon proved far too simplistic to account for the wide variations in stellar spectra. In any case, the Doppler shifts for stars within our own galaxy are quite small.

38.36 *(G) Relativistic Doppler shift, II.* The physicist R. W. Wood (1868–1955) is said to have beaten a traffic ticket he got for passing a red light. He told the judge that as he approached the light his speed relative to the light resulted in a Doppler shift that made the red light appear green. Seeing a green light, he did not stop. The judge, knowing that Wood was a famous physicist and unwilling to expose his own total ignorance of physics, acquitted Wood. Assuming that great physicists never lie, how fast was Wood going? Take ''red'' to imply $\lambda \simeq 670$ nm and ''green'' to imply $\lambda' \simeq 525$ nm.

38.37 (G) *Dop-cop*. Police radars transmit microwave radiation at a fixed frequency. Some of the radiation is reflected off the surface of a moving car, and some of the reflected radiation is detected by a sensitive detector. The speed of the car can be deduced from the Doppler shift ratio $\Delta \nu / \nu$ of the reflected radiation (see part **f** of Problem 38.35). **(a)** What is the value of $\Delta \nu / \nu$ if an approaching car is traveling at 120 km/h? (Hint: There are *two* Doppler shifts to consider. Alternatively, imagine that the car is a mirror, and the radar receiver "sees" a mirror image of the radar transmitter approaching it. How fast is the image moving?) **(b)** The Doppler shift is measured in the detector by superposing the amplified return signal on the transmitted signal and measuring the beat frequency (see Section 22.6). Police radars are expected to be sensitive enough to record speeds within ± 1 km/h. If the transmitted signal frequency is 10 GHz, what is the corresponding change in beat frequency that must be detected?

GROUP C

38.38 (4) *The twin paradox, III*. In Problem 38.8, you found that Oliver had aged much more than Olivia. The following argument has been made to show that the two cannot have different ages: From Oliver's point of view, Olivia's time is dilated. Thus, if she returns home after 14 years of Oliver's time have passed, she will have aged more than he, which contradicts the calculation of Problem 38.8. What is wrong with this argument? (Hint: To show that the argument fails, you must find an asymmetry between the two points of view.)

38.39 (6) *Two Lorentzes in tandem*. Observer O_0 measures the speed of observer O_1 and finds it to be V_{10}. Meanwhile, observer O_1 measures the speed of observer O_2 and finds it to be V_{21}. Show that, when O_0 measures the velocity of O_2, she will find it to be

$$V_{20} = \frac{V_{21} + V_{10}}{1 + \dfrac{V_{21}V_{10}}{c^2}}.$$

38.40 (6) *In Lorentz's footsteps, III*. Derive Equation 38.15b or, what amounts to the same thing, Equation 38.15c.

38.41 (G) *Invariance of the spacetime interval*. Observer O watches a flying saucer as it passes. At time t_1, a light on the saucer flashes and she observes its position to be (x_1, y_1, z_1). At time t_2, there is another flash, and O locates the saucer at (x_2, y_2, z_2). She thus finds that, in the time interval $\Delta t \equiv t_2 - t_1$, the saucer has moved through the displacement $(\Delta x, \Delta y, \Delta z)$. Observer O' sees the same events at times t'_1 and t'_2. According to his measurements, the saucer has moved through the displacement $(\Delta x', \Delta y', \Delta z')$ in the time interval $\Delta t'$. Show that

$$c^2(\Delta t)^2 - (\Delta x)^2 - (\Delta y)^2 - (\Delta z)^2 =$$
$$c^2(\Delta t')^2 - (\Delta x')^2 - (\Delta y')^2 - (\Delta z')^2.$$

The quantity on either side of this equation is called the *spacetime interval*. It is called an *invariant* because its value is the same for all observers. What would be the corresponding invariant(s) in Newtonian kinematics?

38.42 (G) *Proper-time interval*. Observer O sees event 1 occur at (x_1, t_1) and event 2 at (x_2, t_2). Observer O' moves past O at just such a speed V that he sees both events occur at the same place; $x'_1 = x'_2$. **(a)** Show that the time interval $\Delta t'$ between the events for O' must be

$$\Delta t' = \sqrt{(\Delta t)^2 - \frac{(\Delta x)^2}{c^2}},$$

where $\Delta t = t_2 - t_1$ and $\Delta x = x_2 - x_1$. The quantity $\Delta t'$ is called the *proper-time interval*. Show that O' can make the observation just described *only* if $\Delta x < c \, \Delta t$. What is the physical meaning of this condition? **(b)** Show that, if $\Delta x > c \, \Delta t$, the time order of events depends on the frame of reference chosen. That is, event 1 occurs before, simultaneously with, or after event 2, depending on the reference frame chosen. What is the physical meaning of the condition $\Delta x > c \, \Delta t$?

38.43 (G) *Unambiguity of causal sequences*. You learned in Section 38.1 that it is impossible in general to order a sequence of events in an unambiguous way. But consider a pair of *causally related* events. Suppose, for example, that observer O sees light flash 1 and that flash 1 triggers a device that produces flash 2. The positions and relative speeds of the two flash lamps are arbitrary (but $V \le c$). Show that there is no reference frame O' in which flash 2 *precedes* flash 1. (Hint: See Problem 38.42.)

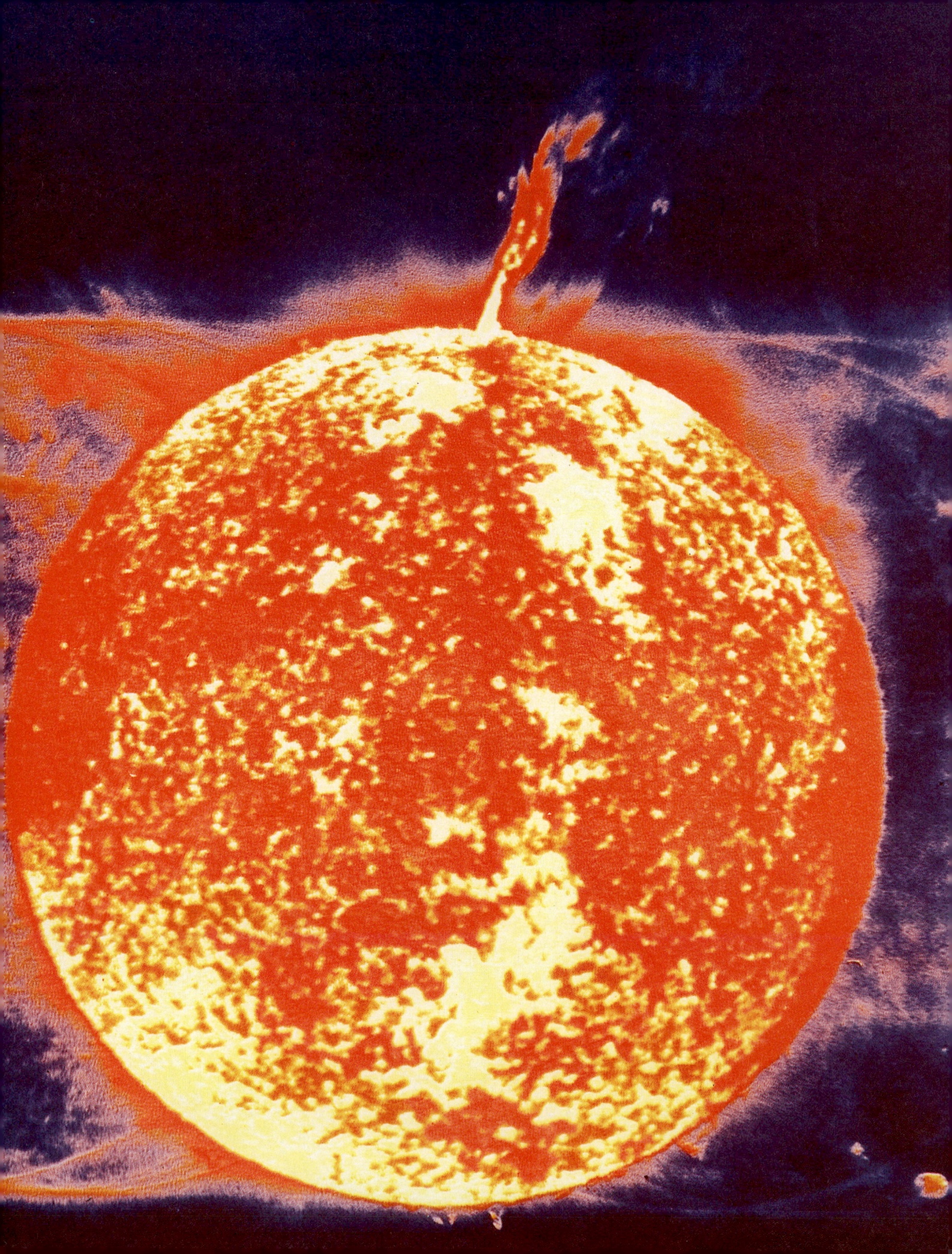

Relativistic Dynamics

The observed mass of a body depends on its speed with respect to the observer.

The rest energy, m_0c^2, may be regarded as a form of potential energy; the total energy of the body is mc^2.

Mass and energy are not distinct in this view but constitute the quantity mass-energy. The principle of mass-energy conservation is broader than, and supersedes, the separate principles of mass conservation and energy conservation.

In chemical reactions and everyday processes, the transformations of energy from one form into another do not significantly change the mass of the system. In nuclear reactions, however, the mass changes, and the transformations of mass into other forms of energy are considerable.

Left: The sun has emitted radiant energy copiously for billions of years and will continue to do so for billions more, because it is fueled by nuclear fusion processes. A little mass is equivalent to a lot of energy because c^2 is a big number!

Introduction

We now develop the theory of **relativistic mechanics**, building on the kinematics developed in Chapter 38. This is parallel to the procedure we adopted in developing the theory of classical mechanics at the beginning of the book. Not surprisingly, there will be significant differences in detail between the classical and relativistic developments. Nevertheless, we will be interested in the same dynamical quantities—notably mass, force, momentum, and energy.

Today there exists a large and varied body of experimental evidence to support the theory of relativistic mechanics. When Einstein began his work there was little or no evidence—certainly none that was clear. As in the kinematic development, he relied on thought experiments. We will follow Einstein's development fairly closely.

Section 39.2 is a study of *mass* and *momentum*. In Section 39.3, we consider *energy* and *force*. In Section 39.4, we develop the renowned concept of *relativistic mass-energy*. Section 39.5 deals with the application of these ideas to nuclear energy.

Mass and Momentum

When we want to measure the mass of a body under everyday circumstances, we almost always weigh it. That is, we depend on the *gravitational manifestation* of mass—the property that bodies exhibit when they are attracted gravitationally by other bodies (Section 14.6). But it is not possible to put a body on the pan of a scale if the body is moving past the pan at a speed V comparable to the speed of light c. Fortunately, there is another way to determine mass, which we discussed in Section 9.6. We can determine the mass m_2 of a frictionless glider by making it collide with another glider of known (or standard) mass m_1. If we know the initial and final velocities of both gliders, we have all the information needed to determine m_2. We use the principle of conservation of linear momentum in the form of Equation 9.34,

$$m_1 \, \Delta v_1 = -m_2 \, \Delta v_2. \qquad (39.1)$$

In this equation, Δv_1 and Δv_2 are the changes in velocity experienced by bodies 1 and 2. The quantities m_1 and m_2 represent the *inertial manifestation* of mass.

A particularly simple form of the experiment is sketched in Figure 39.1. We arrange for the collision to be head on and thus one-dimensional. For further simplification, we make the initial velocities equal and opposite; $v_{2i} = -v_{1i}$. Finally, we use two gliders that collide elastically.

Suppose that we find the final velocities to be

$$v_{1f} = -v_{1i} \quad \text{and} \quad v_{2f} = -v_{2i} = v_{1i}.$$

Using Equation 39.1, we write

$$m_1(-2v_{1i}) = -m_2(2v_{1i}).$$

We conclude that the two masses are equal and call their common value m_0.

$$m_1 = m_2 = m_0.$$

We now modify the experiment, placing it in a relativistic context as shown in Figure 39.2. Observer O has glider m_1, and O' has glider m_2. They have performed the experiment just described and have found that both gliders have mass m_0. Now O and O' separate, leaving their helper H with a flash lamp. They set up their coordinate systems so that the x and x' axes are collinear. Observer O positions her glider on the y axis at $y = a$. Observer O' positions his glider on the y' axis at $y' = -a$. On a signal from H, O and O' approach each other, bringing their gliders with them. Observers O and O' each see the other moving at speed V along the x (or x') axis. From the point of view of helper H, O and O' approach at equal speeds and are always equidistant from him. When all is ready, H sets off a flash lamp. Seeing the flash, O propels her glider in the $-y$ direction with an agreed-upon speed $v_\perp$ very much smaller than V. Similarly, O' propels his glider in the y' direction with the same speed $v_\perp$.

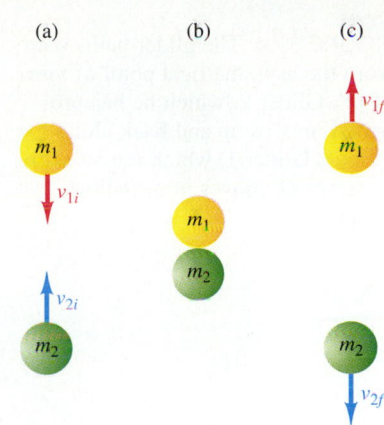

FIGURE 39.1 A particularly simple form of the Mach experiment. (a) Two bodies approach each other. Their velocities are equal and opposite and very much less than c. (b) The bodies collide. (c) The bodies move apart. Measurement shows that their directions are reversed and their speeds are unchanged. We conclude that their masses are equal.

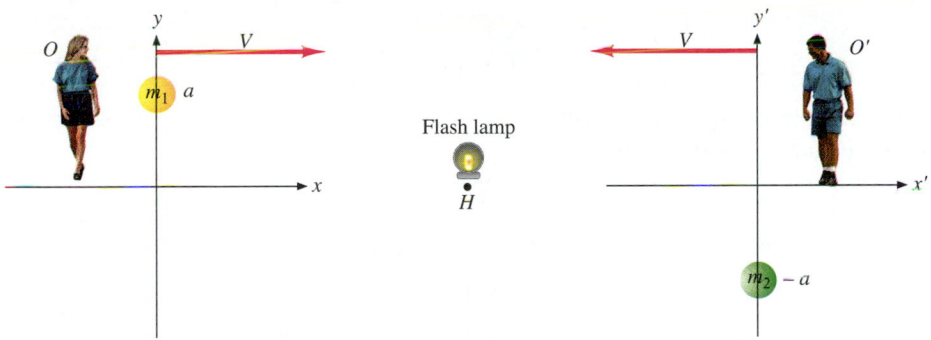

Observers O and O' pass H simultaneously. (Note that "simultaneous" is meaningful here because the positions of all three coincide.) At that instant the two gliders collide, and they subsequently rebound. Figure 39.3 shows what the experiment looks like from the point of view of H. Because we have taken care to symmetrize all the experimental conditions from his point of view, the paths of the gliders are symmetrical; $\theta_{1i} = \theta_{2i} = \theta_{1f} = \theta_{2f}$. (These angles are exaggerated for clarity; they are actually very small because $v_\perp \ll V$.) We conclude from the symmetry that the collision does not affect the velocity components of the gliders along the x (or x') axis. The components along the y (or y') axis are reversed.

FIGURE 39.2 The Mach experiment modified so that it can be performed by observers O and O', who move with relative speed V. Observer O carries m_1 with her, and O' carries m_2 with him. The situation is shown from the point of view of helper H, who is always located halfway between O and O'. At a signal from H, O and O' propel their gliders toward the origins of their own coordinate systems, each with $v_\perp$ parallel to the y and y' axes. The gliders collide just as O and O' pass H, and then they rebound.

FIGURE 39.3 The glider paths seen from the symmetrical point of view of H.

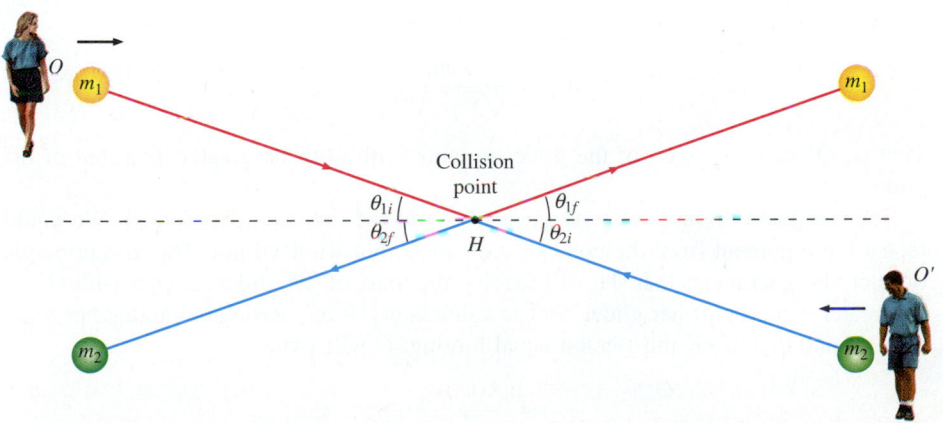

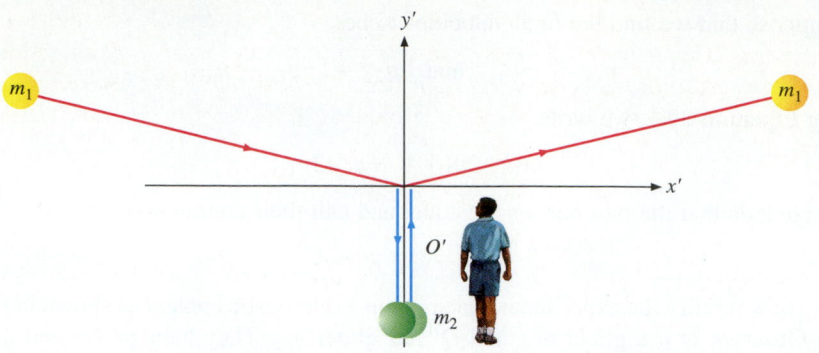

FIGURE 39.4 The glider paths seen from the asymmetrical point of view of O'. Glider 2, which he has propelled, moves up and back along the y' axis. Glider 1, which moves with observer O, moves in a shallow, symmetric V.

Figure 39.4 shows what the experiment looks like as O' sees it. From his point of view, the situation is not symmetrical at all. His own glider, m_2, travels along the y' axis with initial speed $v_\perp$ and rebounds along the y' axis with the same speed. Its velocity change in the collision is thus

$$\Delta v_{2y'} = -2v_\perp.$$

But O' sees glider m_1 traveling along the V-shaped path shown. What is its velocity component $v_{1y'}$ along the y' axis? The value of this component is *not* $v_\perp$, because O' must take into account the Lorentz velocity transformation. He must use Equation 38.15b, the second of the velocity-transformation equations, to find $v_{1y'}$. With minor changes in notation to conform to the present situation, that equation is

$$v_{1y'} = \frac{\sqrt{1 - V^2/c^2}}{1 - Vv_{1x}/c^2}\, v_{1y}.$$

Now, O' knows that O has propelled m_2 with speed $v_\perp$ along the y axis, so he sets $v_{1y} = v_\perp$ and $v_{1x} = 0$. He thus obtains

$$v_{1y'} = \sqrt{1 - V^2/c^2}\, v_\perp.$$

The rebound speed is the same, and thus the velocity change is

$$\Delta v_{1y'} = 2\sqrt{1 - V^2/c^2}\, v_\perp.$$

Now that O' has the y' components of the velocity of both gliders, he can apply the momentum-conservation principle in the form of Equation 39.1. He has $m_1\,\Delta v_{1y'} = -m_2\,\Delta v_{2y'}$, or

$$2m_1\sqrt{1 - V^2/c^2}\,v_\perp = 2m_2 v_\perp.$$

Canceling the factor $2v_\perp$ common to both sides, O' obtains

$$m_1\sqrt{1 - V^2/c^2} = m_2. \tag{39.2}$$

This equation cannot be true if $m_2 = m_1 = m_0$. But we know from wide experience with bodies moving at *low* speeds that the mass is independent of the speed of the body with respect to the observer. The speed of m_2 with respect to O' is always small. Thus it must be correct to set $m_2 = m_0$, because that is the value O and O' measured when everything was at rest. Equation 39.2 can thus be solved for m_1 to obtain

$$m_1 = \frac{m_0}{\sqrt{1 - V^2/c^2}}.$$

That is, O' sees the mass of the glider moving with O to be greater than that of his glider.

There is just one more step to our analysis of the collision experiment. We could repeat the argument from the point of view of O, but we need not. The first principle of relativity guarantees that she will see m_2, the mass of the glider moving with O', to be greater than that of her glider, whose value is $m_1 = m_0$. Recognizing that her point of view and that of O' must be on equal footing, O will write

$$m_2 = \frac{m_0}{\sqrt{1 - V^2/c^2}}.$$

We thus conclude that *the mass of a body increases with increasing speed relative to the observer*. This generalization allows us to drop the subscripts 1 and 2 and to substitute the speed v of the body with respect to an arbitrary observer for the speed V of either of two observers with respect to one another. We thus write the **relativistic mass equation**:

$$m = \frac{m_0}{\sqrt{1 - v^2/c^2}}.$$ (39.3)

In this equation, m_0 is called the **rest mass**, and m is called the **relativistic mass**.

EXAMPLE 39.1

Find the mass of a 1-kg body when it moves with respect to the observer at speed **(a)** 0.1c, **(b)** 0.8c, and **(c)** 0.9999c.

SOLUTION:
(a) For $v = 0.1c$ and $m_0 = 1$ kg, Equation 39.3 gives you

$$m = 1 \text{ kg} \times \frac{1}{\sqrt{1 - (0.1)^2}} = 1.005 \text{ kg}.$$

Even though $0.1c \, (= 3 \times 10^7 \text{ m/s})$ is a very considerable speed by everyday standards, the relativistic mass differs from the rest mass by only $\frac{1}{2}\%$.

(b) For $v = 0.8c$, you calculate

$$m = 1 \text{ kg} \times \frac{1}{\sqrt{1 - (0.8)^2}} = 1.67 \text{ kg}.$$

The relativistic mass is about 70% more than the rest mass.

(c) For $v = 0.9999c$, you have

$$m = 1 \text{ kg} \times \frac{1}{\sqrt{1 - (0.9999)^2}} = 70.7 \text{ kg}.$$

As you have seen before for quantities involving the Lorentz factor $\sqrt{1 - v^2/c^2}$, the difference between the relativistic and classical values is quite small until v becomes an appreciable fraction of the speed of light.

As its speed with respect to the observer approaches the speed of light, the relativistic mass of a body increases without limit. What would happen to such a body if you tried to accelerate it indefinitely by pushing on it with a constant force?

Momentum Conservation

Our derivation of the relativistic mass equation (Equation 39.3) is based on the assumption that momentum is conserved in all circumstances. The price we pay for holding on to this assumption is the abandonment of mass as a quantity independent of motion. That is, we have asserted that *momentum conservation is more fundamental than mass conservation*. Experiment validates this assertion. It therefore makes sense to define the **relativistic momentum p** of a body:

$$\mathbf{p} \equiv m\mathbf{v}.$$ (39.4)

In this equation, m is the relativistic mass given by Equation 39.3; that is, $m = m_0/\sqrt{1 - v^2/c^2}$.

Force and Energy

The definition of the **relativistic force** is the same as that given by Equation 9.30 for the classical force—namely,

$$\mathbf{F} = \frac{d\mathbf{p}}{dt}.$$ (39.5)

We have to be careful, however, in expressing relativistic force in terms of acceleration because the mass is not independent of the speed. When we substitute the value $\mathbf{p} = m\mathbf{v}$

given by Equation 39.4 into Equation 39.5 and use the chain rule for differentiation, we obtain

$$\mathbf{F} = \frac{d}{dt}(m\mathbf{v}) = m\frac{d\mathbf{v}}{dt} + \mathbf{v}\frac{dm}{dt}. \qquad (39.6a)$$

By definition, we have $d\mathbf{v}/dt = \mathbf{a}$, the acceleration. So we write

$$\mathbf{F} = m\mathbf{a} + \mathbf{v}\frac{dm}{dt}. \qquad (39.6b)$$

Classically, the mass of a particle does not change as it is acted on by a force; we have $m = m_0 = $ constant. Consequently, we have $dm/dt = 0$, and Equation 39.6b reduces to the familiar classical form $\mathbf{F} = m_0\mathbf{a}$. But, in general, you must take the factor dm/dt into consideration in applying either of Equations 39.6a and 39.6b.

There is one important situation, however, in which $dm/dt = 0$ for relativistic particles. Consider a charged particle moving with instantaneous velocity $\mathbf{v}$ in a uniform magnetic field. The particle is acted on by a force of constant magnitude $F = qv\mathscr{B}$, which is always directed perpendicular to $\mathbf{v}$ (Chapter 28). Thus the particle moves in a circle with constant speed v, and the relativistic mass does not change with time. Equation 39.6b then simplifies to

$$\mathbf{F} = m\mathbf{a} \quad \text{for } m = \text{constant}, \qquad (39.7)$$

where m is the relativistic mass given by Equation 39.3.

A charged particle moving at high speed in a magnetic field provides a basis for experimental verification of Equation 39.3. As early as 1901, the German physicist Walter Kaufmann (1871–1947) had repeated J. J. Thomson's e/m experiment, using β particles (the energetic electrons expelled from radioactive nuclei) in place of the less energetic electrons produced in earlier cathode-ray tube experiments (Section 28.8). Some of these β particles had speeds in excess of $0.8c$. Kaufmann found evidence that the most energetic β particles (electrons) exhibited values of e/m as small as one-third the value e/m_0 obtained for electrons at lower energies.

Before Einstein's work of 1905, several theories were proposed for interpreting Kaufmann's results, but the results were not precise enough to provide a crucial test among the theories. In 1908, Bucherer performed an experiment whose results were much more precise, at least for electrons having speeds up to about $0.7c$. The electrons were made to pass through a velocity selector (Problem 28.23) and then into a region of uniform magnetic field. The mass was then determined from the curvature of the trajectory. Kaufmann's and Bucherer's results, and those of a third investigation, are shown in Figure 39.5. Later experiments, using electrons having still greater speeds and employing more precise measuring techniques, furnish further confirmation.

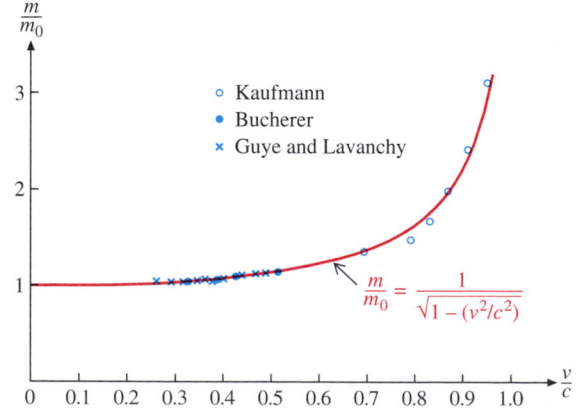

FIGURE 39.5 Electron mass as a function of speed. Speed is plotted as the ratio v/c and mass as the ratio m/m_0. The points represent the results of various experimenters; the solid curve is obtained by using Equation 39.3.

Relativistic Work

When a force acts on a particle, work is done. This is true relativistically as well as classically, and work is defined identically in both cases. In the simple case where motion is confined to the x direction and the force is parallel to the x direction, the definition has the form

$$W \equiv \int_{x_i}^{x_f} F \, dx. \tag{39.8}$$

The limits of the integral are the initial position x_i and the final position x_f of the particle.

To evaluate the work integral, we use the value of F given by Equation 39.6a,

$$F = m \frac{dv}{dt} + v \frac{dm}{dt}.$$

But we must express this value as a function of x. There is a neat way to do this simultaneously for both terms. Beginning with Equation 39.3, $m = m_0/\sqrt{1 - v^2/c^2}$, we square both sides and rearrange to obtain

$$m^2 c^2 - m^2 v^2 - m_0^2 c^2 = 0.$$

Next, we take the derivative of both sides of this equation with respect to time:

$$c^2 \frac{d(m^2)}{dt} - m^2 \frac{d(v^2)}{dt} - v^2 \frac{d(m^2)}{dt} = 0,$$

We carry through the differentiation to obtain

$$2mc^2 \frac{dm}{dt} - 2m^2 v \frac{dv}{dt} - 2mv^2 \frac{dm}{dt} = 0.$$

We divide through by $-2mv$ and transpose the first term to the right side of the equation:

$$m \frac{dv}{dt} + v \frac{dm}{dt} = c^2 \frac{1}{v} \frac{dm}{dt}.$$

We are almost there. We substitute into the right side of this equation the definition $v \equiv dx/dt$, and we have the desired result

$$F = m \frac{dv}{dt} + v \frac{dm}{dt} = c^2 \frac{dt}{dx} \frac{dm}{dt}$$

$$= c^2 \frac{dm}{dx}.$$

We can now substitute this value of F into the integrand of Equation 39.8. This gives us

$$W = \int_{x_i}^{x_f} c^2 \frac{dm}{dx} \, dx = c^2 \int_{m_i}^{m_f} dm.$$

The limits of integration m_i and m_f are the initial and final masses of the particle, corresponding to the positions x_i and x_f. Evaluating the integral, we obtain

$$W = c^2(m_f - m_i).$$

Suppose that we begin with the particle at rest at x_i. Then its initial mass is its rest mass; $m_i = m_0$. We have placed no restrictions on the final position, and hence the final mass is the relativistic mass; $m_f = m$. The work done on a particle accelerated from rest can thus be expressed in the general form

$$W = (m - m_0)c^2. \tag{39.9}$$

This equation tells us that the work done on the particle is equal to its change in mass, $m - m_0$, multiplied by c^2.

Relativistic Kinetic Energy

If no other forces act on the particle, the work W serves solely to change the kinetic energy K of the particle. That is, we can use the work-energy theorem in its simplest form, $W = K$ (Section 7.4). Equation 39.9 then gives us the **relativistic kinetic energy**

$$K = mc^2 - m_0c^2. \tag{39.10}$$

We are used to expressing the kinetic energy as a function of v, the speed of the particle with respect to the observer. To do this, we use the value of m given by Equation 39.3. Equation 39.10 then becomes

$$K = \frac{m_0c^2}{\sqrt{1 - v^2/c^2}} - m_0c^2, \tag{39.11a}$$

or

$$K = m_0c^2\left(\frac{1}{\sqrt{1 - v^2/c^2}} - 1\right). \tag{39.11b}$$

We must make sure that this relativistic equation reduces to the familiar classical equation $K = \frac{1}{2}mv^2$ when $v \ll c$. In this case, the first term in parentheses can be replaced by the approximate value $(1 + \frac{1}{2}v^2/c^2)$. Equation 39.11b then becomes

$$K = m_0c^2\left(1 + \frac{1}{2}\frac{v^2}{c^2} - 1\right) = \frac{1}{2}m_0v^2. \tag{39.12}$$

This is indeed the classical value for the kinetic energy.

Figure 39.6 is a plot of kinetic energy versus speed, according to Equation 39.11b. As in Figure 39.5, the speed is plotted as the ratio v/c. The kinetic energy is plotted as the ratio K/m_0c^2. The two crosses represent values obtained in a very direct experiment performed in 1964 by the American physicist William Bertozzi. An accelerator produced a beam of electrons of precisely controlled energy. The speed v of the electrons was measured by timing the passage of the electrons between two detectors. The electrons were then stopped by collision in a metal target. The collision transformed the kinetic energy of the electrons into thermal energy, and the temperature rise of the target was used to measure the total energy lost by the electrons. At the same time, the electric charge flowing from the target was used to measure the number of electrons in the beam. Thus the energy per electron could be calculated.

Equation 39.11b furnishes a dynamical reason why no particle having a nonzero rest mass can be accelerated to a speed c, let alone a speed greater than c. To accelerate the particle requires work. But as the speed v approaches c, the kinetic energy of the particle increases without limit, becoming mathematically infinite at $v = c$. Thus infinite work would be required to accelerate a particle, no matter how small its (nonzero) rest mass, to the speed of light. However, there is no theoretical restriction on accelerating a particle to a speed *approaching c* as closely as desired.

FIGURE 39.6 Electron energy as a function of speed. Speed is plotted as the ratio v/c and kinetic energy as the ratio K/m_0c^2. The red curve is obtained using Equation 39.11b, and the lower curve is the classical prediction. Bertozzi's experimental values are shown as crosses.

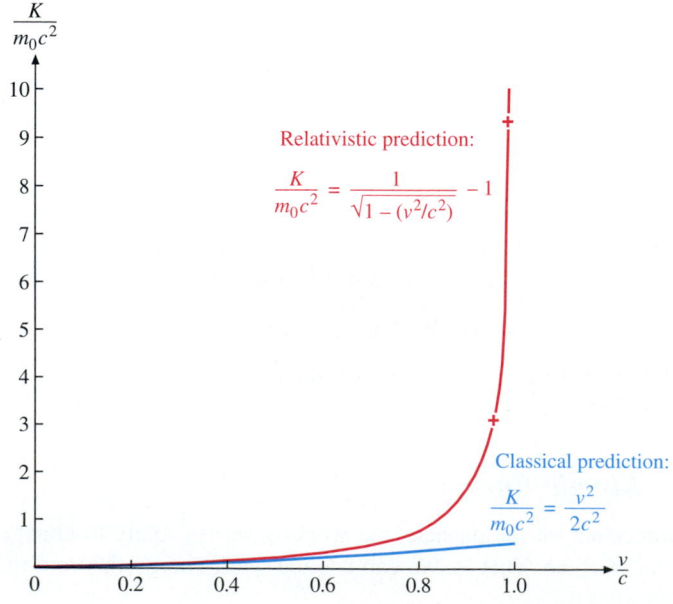

EXAMPLE **39.2**

A cyclotron (Section 28.4) is used to accelerate deuterons of rest mass $m_0 = 3.36 \times 10^{-27}$ kg. Show that the cyclotron will certainly fail to accelerate deuterons to energies in excess of 100 MeV.

SOLUTION: The key to the operation of the cyclotron is the *synchrony* of the orbits of the particles being accelerated. Regardless of the energy attained by a particular particle of charge e and mass m, its angular frequency of revolution in the magnetic field $\mathscr{B}$ is given by Equation 28.10b, $\omega_c = e\mathscr{B}/m$. On account of this synchrony, all particles arrive at the gap between the dees (Figure 28.11) just in time to acquire additional energy from the alternating electric field that is imposed across the dees.

But the isochrony condition fails if m is not constant. To see what happens to deuterons in the cyclotron at 100 MeV, suppose that a deuteron has attained that energy. Call the angular frequency of this deuteron ω_r. To allow for the possibility that relativistic effects are significant, use the relativistic mass m of the deuteron to write

$$\omega_r = \frac{e\mathscr{B}}{m} = \frac{e\mathscr{B}}{m_0}\sqrt{1 - \frac{v^2}{c^2}}.$$

Now, compare this angular frequency with the classical angular frequency ω_c, which is certainly the angular frequency of deuterons that have just begun the acceleration process. You have the ratio

$$\frac{\omega_r}{\omega_c} = \frac{e\mathscr{B}/m}{e\mathscr{B}/m_0} = \sqrt{1 - \frac{v^2}{c^2}}. \qquad \textbf{(39.13)}$$

You now need to express $\sqrt{1 - v^2/c^2}$ in terms of the ki-netic energy K of a 100-MeV deuteron. From Equation 39.11b, you have

$$\sqrt{1 - \frac{v^2}{c^2}} = \frac{1}{\dfrac{K}{m_0c^2} + 1}.$$

You insert this value into Equation 39.13 to obtain

$$\frac{\omega_r}{\omega_c} = \frac{1}{\dfrac{K}{m_0c^2} + 1}.$$

Now you can calculate numerical values. Remembering that 1 MeV = 1.6×10^{-13} J, you have

$$\frac{\omega_r}{\omega_c} = \frac{1}{\dfrac{100 \text{ MeV} \times 1.6 \times 10^{-13} \text{ J/MeV}}{3.36 \times 10^{-27} \text{ kg} \times (3 \times 10^8 \text{ m/s})^2} + 1} = 0.95.$$

That is, the 100-MeV deuterons revolve in their magnetic orbits with only 95% of the angular speed of the low-energy deuterons and thus take about 5% longer to make a circuit around the dees. Imagine that the 100-MeV deuterons start around their orbits together with the low-energy deuterons. In five revolutions, they will be out of phase by one-quarter of a cycle. That is, they will arrive at the gap between the dees just when the electric field is zero and will receive no further acceleration. When a deuteron arrives at the gap with the proper phase, each passage across the gap adds a few thousand electron-volts of energy. Thus the cyclotron will certainly not further accelerate deuterons whose energy is 100 MeV and may well fail to accelerate deuterons with considerably smaller energies.

Conservation of Mass-Energy

We now proceed to a famous result: Mass is a form of energy. Let us return to Equation 39.10, $K = mc^2 - m_0c^2$, and rearrange it in the form

$$mc^2 = K + m_0c^2. \qquad \textbf{(39.14)}$$

Each of the three terms in this equation has the dimensions of energy. Indeed, the equation looks very much like Equation 8.17, which expresses the total energy E of a system as the sum

$$E = K + U.$$

It looks as though the energy of the system has two parts: a kinetic energy K, which it possesses by virtue of its motion, and a **rest energy** m_0c^2, which it possesses when at rest. The sum of the two is called the **relativistic energy** mc^2. The relativistic energy is usually represented by the symbol E—thus Einstein's famous equation

$$E = mc^2. \qquad \textbf{(39.15)}$$

According to this equation, *the relativistic energy of a particle is the product of its relativistic mass and the square of the speed of light.*

The rest energy is usually represented by the symbol E_0; thus we have

$$E_0 = m_0c^2. \qquad \textbf{(39.16)}$$

This tells us that *mass is a form of energy.* In form, Equation 39.16 is similar to Equation 17.27, $W = JQ$, which expresses the equivalence between the frictional work W done on a body and the heat Q evolved in the process. Recall the evolution of our understanding of Equation 17.27. At first, the proportionality constant J was just an experimentally determined quantity that linked two quite different quantities, mechanical work W (expressed in joules) and heat Q (expressed in calories). But with further study we came to understand, and to become increasingly confident in, the closeness of the connection between heat and energy. We successively took the following points of view:

1. The proportionality constant connecting the heat evolved and the work done is J.

2. There is an equivalence between heat and the mechanical energy required to create it.

3. Heat is a form of energy and can be converted into and from other forms under proper conditions.

We now pass (much more quickly) through the same steps with respect to Equation 39.16:

1′. The proportionality constant between mass (expressed in kilograms) and energy (expressed in joules) is c^2.

2′. There is an equivalence between mass and energy.

3′. Mass is a form of energy and can be converted into and from other forms under proper conditions.

We have yet to make an argument for statement 3′. We begin to do so in terms of a thought experiment. In Section 39.5, we will turn to very important "real" versions of the thought experiment.

In Figure 39.7a, a short length of string holds two bodies together against the tendency of a compressed spring to force them apart. The rest masses of the two bodies are m_{01} and m_{02}. The potential energy of the compressed spring is U. In Figure 39.7b, we burn the string with a match, and the spring propels the bodies apart. The spring now has zero potential energy, but the bodies have kinetic energy K_1 and K_2. According to the principle of conservation of mechanical energy, we have

$$U = K_1 + K_2.$$

If we are to take the idea of rest energy seriously, we must add the total rest energy of the system, $m_{01}c^2 + m_{02}c^2$, to both sides of this equation. (This is certainly permissible mathematically.) Thus modified, the energy-conservation equation is

$$m_{01}c^2 + m_{02}c^2 + U = (K_1 + m_{01}c^2) + (K_2 + m_{02}c^2).$$

According to Equation 39.14, we can write this equation in the form

$$m_{01}c^2 + m_{02}c^2 + U = m_1c^2 + m_2c^2. \tag{39.17}$$

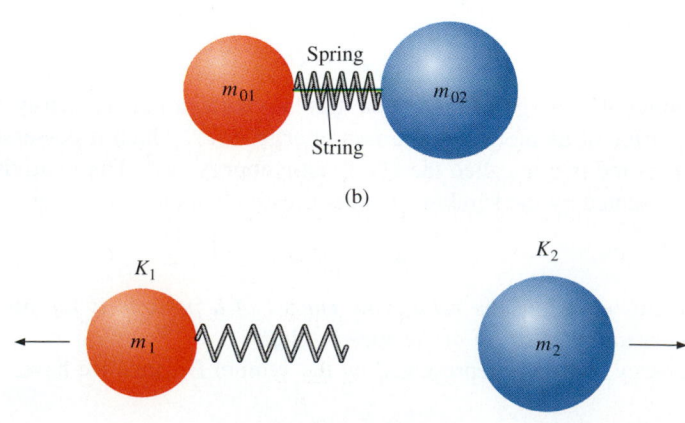

(a)

(b)

FIGURE 39.7 (*a*) Two bodies of rest mass m_{01} and m_{02} are held together by a string. Between the bodies is a compressed spring whose potential energy is U. (*b*) The string is burned and the spring expands, propelling the two bodies apart with kinetic energies K_1 and K_2.

As you can see from this equation, the energy originally in the form of potential energy U of the compressed spring has been converted into mass-energy and is now part of the total mass-energy $(m_1 + m_2)c^2$ of the system. Neither the total mechanical energy U of the system nor its total rest mass $m_{01} + m_{02}$ is conserved. *But the mass-energy is conserved.*

If we write Equation 39.17 in the form

$$\left(m_{01} + m_{02} + \frac{U}{c^2}\right)c^2 = (m_1 + m_2)c^2,$$

we have inside the parentheses on the left side a sum of three terms, each of which has the dimensions of mass. We can thus consider the term U/c^2 as a part of the rest mass of the system in Figure 39.7a. But U is the energy of the compressed spring; thus it is part of the internal energy of the system. Any change in the internal energy of a system changes its rest mass. In this case, the rest mass of the system with the spring compressed is greater than it would be if the spring were not compressed.

This thought experiment is a highly idealized model of what actually happens when an atomic nucleus breaks up into two parts. (We leave the details to Section 39.5.) But we are usually not in a position to observe the ''compressed spring'' separately, let alone evaluate its energy U. Rather, we lump the total initial rest energy of the system, given on the left side of Equation 39.17, into a single quantity M_0c^2. We then have

$$M_0c^2 = m_1c^2 + m_2c^2, \tag{39.18a}$$

or

$$M_0c^2 = m_{01}c^2 + m_{02}c^2 + K_1 + K_2. \tag{39.18b}$$

The second of these two forms makes evident the fact that the final rest energy of the system is smaller than its initial rest energy. This is because some of the initial rest energy has been converted into kinetic energy.

We can readily subject Equation 39.18b to experimental verification. We measure the rest mass M_0 on the left side directly—say, with a laboratory balance. We evaluate the two kinetic energies on the right side by measuring the speeds of the two bodies. Then we bring the two bodies to rest (removing the kinetic energy from the system) and measure their rest masses by the same means used to measure M_0. Just such experiments are considered in Section 39.5.

Let us generalize Equations 39.18. A system has an initial mass-energy

$$M_i c^2 = M_{0i}c^2 + K_i$$

and a final mass-energy

$$M_f c^2 = M_{0f}c^2 + K_f.$$

According to the **principle of conservation of mass-energy**, the two are equal:

$$M_i c^2 = M_f c^2 \qquad \text{or} \qquad M_{0i}c^2 + K_i = M_f c^2 + K_f. \tag{39.19a,b}$$

This principle supplants the separate classical principles of conservation of mass and conservation of energy.

Momentum and Energy

We now explore the relations among four important relativistic quantities: total mass-energy $E = mc^2$, rest energy m_0c^2, kinetic energy K, and momentum $p = mv$. We begin by taking the square of both sides of Equation 39.15, $E = mc^2$. This gives us

$$E^2 = m^2 c^4.$$

We express m in terms of m_0 by means of Equation 39.3, $m = m_0/\sqrt{1 - v^2/c^2}$:

$$m^2 c^4 = \frac{m_0^2 c^4}{1 - v^2/c^2}.$$

We multiply through by the denominator on the right side of this equation:

$$m^2 c^4 - m^2 v^2 c^2 = m_0^2 c^4.$$

The first term on the left is E^2, and the second term is $-(mv)^2 c^2 = -p^2 c^2$. So we have

$$E^2 = p^2 c^2 + m_0^2 c^4.$$

Taking the square root of both sides of the equation gives us the result

$$E = \sqrt{p^2 c^2 + m_0^2 c^4}. \qquad \textbf{(39.20a)}$$

It is sometimes useful to express this result in a slightly different form. We substitute mc^2 for E on the left side and divide both sides through by c^2 to obtain

$$m = \sqrt{(p/c)^2 + m_0^2}. \qquad \textbf{(39.20b)}$$

Equations 39.20a and 39.20b express the total relativistic energy E and the relativistic mass m in terms of the momentum p and the rest mass m_0.

According to Equation 39.19, the total relativistic energy is the sum of the rest energy and the kinetic energy; $E = m_0 c^2 + K$. It follows immediately that $K = E - m_0 c^2$. We use the value of E given by Equation 39.20a to obtain the relation

$$K = \sqrt{p^2 c^2 + m_0^2 c^4} - m_0 c^2. \qquad \textbf{(39.21)}$$

Equation 39.20a is a Pythagorean sum; that is, it has the form $h = \sqrt{a^2 + b^2}$. By drawing the corresponding right triangle, as shown in Figure 39.8a, you can readily remember the relation expressed by that equation, as well as the relation expressed by Equation 39.20b. Alternatively, you can divide the lengths of all three sides of this triangle by c^2 and obtain the relations in the form given by Equation 39.20b, as shown in Figure 39.8b.

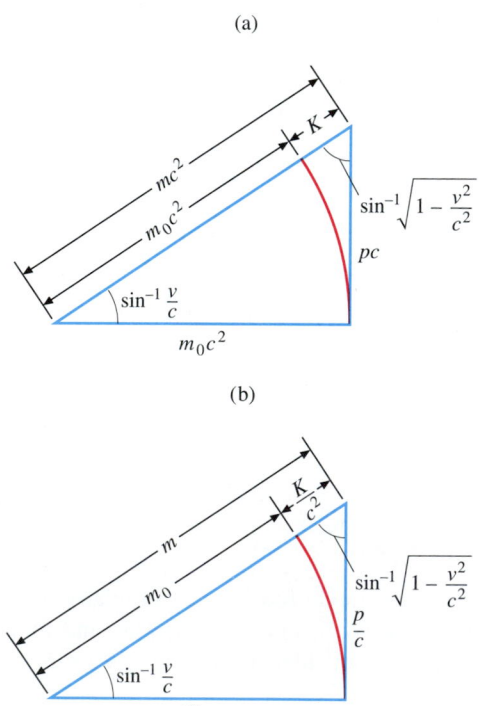

FIGURE 39.8 (a) Right triangle corresponding to the Pythagorean sum expressed in Equation 39.20a. The kinetic energy is given by the difference $K = mc^2 - m_0 c^2$. (b) Right triangle corresponding to Equation 39.20b.

EXAMPLE 39.3

An electron is accelerated through a potential difference of 500 kV. Find (a) its relativistic energy, (b) its relativistic mass, (c) its momentum, and (d) its speed.

SOLUTION:
(a) Find the relativistic energy of an electron whose kinetic energy is 500 keV,

First you must express the kinetic energy in joules:

$$K = 500 \times 10^3 \text{ eV} \times 1.60 \times 10^{-19} \text{ J/eV}$$
$$= 8.00 \times 10^{-14} \text{ J}.$$

Now you are ready to find the total relativistic energy. Using Equation 39.19, you have

$$E = m_0 c^2 + K$$
$$= 9.11 \times 10^{-31} \text{ kg} \times (3.00 \times 10^8 \text{ m/s})^2$$
$$+ 8.00 \times 10^{-14} \text{ J}$$
$$= 8.20 \times 10^{-14} \text{ J} + 8.00 \times 10^{-14} \text{ J}$$
$$= 1.62 \times 10^{-13} \text{ J}.$$

The kinetic energy of a 500-keV electron is nearly the same as its rest energy.

(b) Find the relativistic mass.

You write Equation 39.3 in the form

$$m = \frac{E}{c^2} = \frac{1.62 \times 10^{-13} \text{ J}}{(3.00 \times 10^8 \text{ m/s})^2} = 1.80 \times 10^{-30} \text{ kg}.$$

This is about twice the rest mass, just as the total energy is about twice the rest energy.

(c) Find the momentum.

It is most convenient to solve Equation 39.20b for p:

$$p = c\sqrt{m^2 - m_0^2}$$
$$= 3.00 \times 10^8 \text{ m/s}$$
$$\times \sqrt{(1.80 \times 10^{-30} \text{ kg})^2 - (9.11 \times 10^{-31} \text{ kg})^2}$$
$$= 4.66 \times 10^{-22} \text{ kg·m/s}.$$

(d) Find the speed.

There are several ways to do this, but the easiest is to refer to Figure 39.8b and note that

$$\frac{v}{c} = \sin \frac{p/c}{m} = \sin \frac{p}{mc}.$$

You thus have

$$\frac{v}{c} = \sin \frac{4.66 \times 10^{-22} \text{ kg·m/s}}{1.80 \times 10^{-30} \text{ kg} \times 3.00 \times 10^8 \text{ m/s}} = 0.759.$$

That is, the electron is traveling at about three-quarters the speed of light. You could not have used Newtonian physics to obtain the desired results. You multiply through by c to obtain

$$v = 0.759 \times 3.00 \times 10^8 \text{ m/s} = 2.28 \times 10^8 \text{ m/s}.$$

The Classical and Extreme Relativistic Limits

The argument leading to Equation 39.12 shows that the relativistic kinetic energy simplifies to the form $K = \frac{1}{2}mv^2$ in the **classical limit** $v \ll c$. The kinetic energy also assumes a simple form in the **extreme relativistic limit** $v \simeq c$.

The kinetic energy is given in general by Equation 39.21, $K = \sqrt{p^2c^2 - m_0^2c^4} - m_0c^2$. As v approaches c, the relativistic mass m increases without limit. Consequently, the momentum $p = mv$ also increases without limit. The speed-independent term $m_0^2c^4$ under the radical becomes negligibly small compared with p^2c^2, and the term m_0c^2 becomes negligibly small compared with pc. We thus have

$$K = pc \quad \text{for } pc \gg m_0c^2 \text{ or } v \simeq c. \tag{39.22}$$

Figure 39.9 shows how K approaches pc as v/c approaches 1.

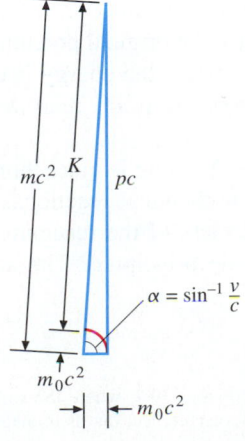

FIGURE 39.9 The triangle of Figure 39.8a in the case $v \simeq c$. As the angle α approaches $\pi/2$ rad, K approaches pc.

We have already noted that, although in principle the speed of a particle of rest mass m_0 can approach c as closely as we wish, a particle with even a very small mass cannot achieve the speed of light. This is because *any* nonzero value of m_0 will result in a relativistic mass and momentum that increase without limit as $v \to c$. However, there exist particles with *zero rest mass*. The **neutrino** has either zero rest mass or at most a very tiny rest mass.

A particle traveling at the speed of light can carry momentum and energy even if its rest mass is zero. (Indeed, from the relativistic point of view, there is no paradox in a particle's having zero rest mass, precisely because momentum has meaning in the absence of rest mass.) A particle without rest mass must always travel at *exactly* the speed of light, from the point of view of any observer. For any other value of v, the relativistic mass equation,

$$E/c^2 = m = m_0/\sqrt{1 - v^2/c^2},$$

yields $E = 0$; this contradicts the assertion that the particle carries energy. For $v = c$, the equation yields the indeterminate value $0/0$. Although this result is not informative, it is at least not contradictory. The indeterminate value of the equality $m = 0/0$ simply means that the relativistic mass m is independent of the rest mass when $m_0 = 0$.

You should not lose sleep over the concept of zero rest mass. A particle that *always* moves with speed c can never be placed on a balance pan and weighed at rest. All we really mean when we say that $m_0 = 0$ is that the relativistic mass and momentum of a neutrino depend entirely on its energy, which is purely kinetic. From Equation 39.20a, we have $p = E/c$, and from Equation 39.20b, we have $m = p/c = E/c^2$.

The significance of the speed of light is thus much broader than the fact that it is the speed at which light travels. It is, rather, the *limiting* speed for any particle of nonzero mass and the *only possible* speed for any particle of zero rest mass. Aside from neutrinos, there are a few other particles whose rest mass is zero. One is the *photon*, a zero-rest-mass particle associated with light, which will be considered in Chapter 40. The *graviton*, whose existence is postulated on the basis of gravitational theory, also has zero rest mass. But gravitons have not yet been detected.

SECTION 39.5

Chemical and Nuclear Reactions

We have already noted that the thought experiment of Figure 39.7, with two particles pushed apart by a spring, has direct counterparts in the "real" world. One such counterpart can be seen in chemical reactions. Let us consider, as a simple example, the chemical reaction in which a zinc atom and an oxygen atom combine to form a zinc oxide molecule:

$$Zn + O \to ZnO + 3.61 \text{ eV}. \tag{39.23}$$

This is a typical *exothermic* reaction—a reaction in which energy is released to the surroundings. (When the reaction takes place in a laboratory vessel, the energy ultimately becomes apparent as heat.) The original potential energy of the "spring" is thus 3.61 eV.* As in all chemical reactions, the energy is electric potential energy, and its conversion into kinetic energy arises from the rearrangement of the electrons near the outer bounds of the two atoms.

A serious question now arises: Nothing is more thoroughly established in chemistry than the conservation of mass in chemical reactions. The mass of the product (here ZnO) is always the same as the mass of the reactants (here Zn and O). How can we reconcile this fact with relativistic principles? The answer lies in the calculation of Example 39.4.

---

*Contrary to the thought experiment of Figure 39.7, where the spring pushes the bodies apart, the spring here pulls the atoms together to form a molecule. But that is immaterial for the present discussion.

EXAMPLE 39.4

Let the combined initial mass of the reactant Zn and O atoms be M. Find the proportional mass change $\Delta M/M$ of the atoms in Reaction 39.23. Assume that the zinc atom is the isotope ^{64}Zn, whose mass is $m_{Zn} = 63.9291$ u, and the oxygen atom is the isotope ^{16}O, whose mass is $m_O = 15.99491$ u.

SOLUTION: First, you need to express the mass and the energy in SI units. You have for the energy

$$U = 3.61 \text{ eV} \times 1.6 \times 10^{-19} \text{ J/eV} = 5.78 \times 10^{-19} \text{ J}.$$

The corresponding relativistic mass change is a loss, because the energy escapes from the original system. You have

$$\Delta M = \frac{U}{c^2} = \frac{5.78 \times 10^{-19} \text{ J}}{(3 \times 10^8 \text{ m/s})^2} = 6.42 \times 10^{-36} \text{ kg}.$$

For the atomic masses, you use $1 \text{ u} = 1.661 \times 10^{-27}$ kg. You have

$$M = (63.9291 + 15.994\,91) \text{ u} \times 1.661 \times 10^{-27} \text{ kg/u}$$
$$= 1.33 \times 10^{-25} \text{ kg}.$$

The proportional mass loss is

$$\frac{\Delta M}{M} = \frac{6.42 \times 10^{-36} \text{ kg}}{1.33 \times 10^{-25} \text{ kg}} = 4.84 \times 10^{-11}.$$

The ultimate sensitivity of the best modern balances is something like 1 part in 10^8. You can see why it is so often satisfactory to consider conservation of mass and conservation of energy as separate principles.

Nuclear Reactions

The situation becomes quite different from that described in Example 39.4 when we consider *nuclear reactions*. There are many types of nuclear reactions, but the most dramatic is **fission**. In this process, the nucleus of an atom—usually one of large atomic mass, comprising many protons and neutrons—splits into two pieces of more or less equal size, often together with a few small fragments. In some cases, the process is *spontaneous*; that is, it occurs without external interference. In other cases, fission is caused by a collision between the nucleus and a neutron.* This latter process, first observed in 1938, is called **neutron-induced fission**.[†] The neutron, which is electrically neutral, does not need to have great initial kinetic energy to approach the positively charged nucleus, which would strongly repel a positively charged particle such as a proton. Thus neutrons collide readily with atomic nuclei.

An atomic nucleus contains only positive charge. In view of the strong electrostatic repulsion among its components—called *nucleons*—what is the "glue" that keeps the undisturbed nucleus together? The "glue" is the *strong nuclear force*. This force is unfamiliar in everyday experience because, unlike the gravitational and electric forces, its range is very short. Indeed, its strength drops to zero at any distance larger than the radius of a typical atomic nucleus—that is, something less than 10^{-14} m. Consequently, we never experience the strong nuclear force on the macroscopic scale. At distances comparable to the nuclear radius, however, the strong nuclear force can act attractively so as to overcome the electrostatic repulsion.

The impact of a neutron disturbs the nucleus. You can picture the disturbance as distortion of an originally spherical liquid drop, as shown in Figure 39.10. If the nucleus is sufficiently distorted, the strong nuclear force can no longer hold it together. The repulsive electric force takes over and pushes the two *fission fragments* apart. (Once they are separated, each of the fission fragments constitutes an independent atomic nucleus, of atomic mass roughly half that of the parent nucleus.)

In its gross operation, the system strongly resembles the idealized system of Figure 39.7. The "string" is the strong nuclear force; the "spring" is the electric repulsion between the fission fragments; and the two bodies are the fragments themselves. The energy of the "compressed spring" is the initial electrostatic potential energy of the two positively charged fragments.

---

*A neutron is a particle having approximately the same mass as a proton but zero electric charge.

[†]The discovery was first made by the German chemist Otto Hahn (1879–1968) and his physicist associate Fritz Strassmann (b. 1902). The work was quickly followed up by many others and led to the development of the first nuclear reactor (1942) and the atomic bomb (1945).

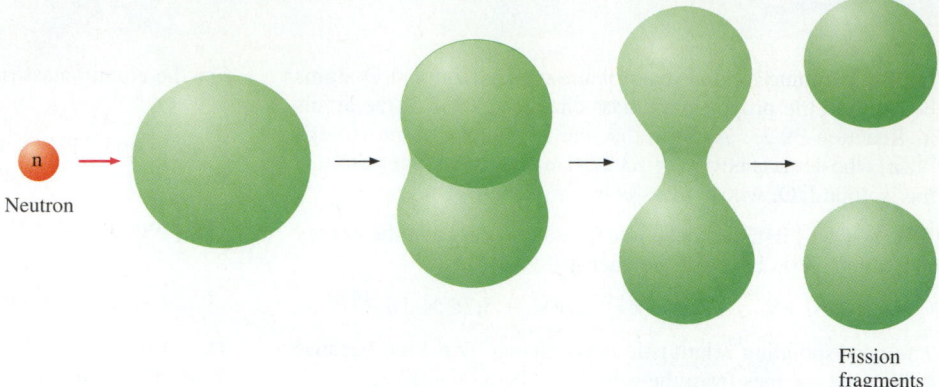

FIGURE 39.10 Sequence of drawings of a disturbed nucleus.

EXAMPLE 39.5

Make a crude calculation of the electrostatic energy released in nuclear fission. Suppose that a uranium nucleus, of charge $+92e$, splits into two fragments each having charge $+46e$. Suppose, moreover, that the electrostatic force begins to dominate when the distance between the centers of the two fragments is $r_0 \simeq 1 \times 10^{-14}$ m. **(a)** How much energy is ultimately released to the surroundings? **(b)** Let the uranium nucleus be the isotope ^{235}U, of atomic mass 235 u, and suppose that the fission is caused by collision with a neutron of atomic mass 1 u. Find the proportional mass change $\Delta M/M$.

SOLUTION:

(a) How much energy is ultimately released to the surroundings?

For a crude estimate, it is adequate to treat the fission fragments as point charges. Using Equations 25.14b and 25.5, you have

$$U = \frac{1}{4\pi\epsilon_0} \frac{q_1 q_2}{r} = \frac{1}{4\pi\epsilon_0} \frac{(46e)^2}{r}.$$

You insert the values of the constants, and the value of r_0, to obtain

$$U = 8.99 \times 10^9 \text{ N·m}^2/\text{C}^2 \times \frac{(46 \times 1.6 \times 10^{-19} \text{ C})^2}{1 \times 10^{-14} \text{ m}}$$

$$\simeq 5 \times 10^{-11} \text{ J}.$$

This value corresponds to about 300 MeV—that is, a factor of about 8×10^7 greater than the typical chemical energy given in Example 39.4, about 4 eV.

(b) Find $\Delta M/M$.

The relativistic mass of change is

$$\Delta M = \frac{5 \times 10^{-11} \text{ J}}{(3 \times 10^8 \text{ m/s})^2} \simeq 6 \times 10^{-28} \text{ kg}.$$

The original mass of the system is that of the uranium nucleus and the neutron:

$$M = (235 + 1) \text{ u} \times 1.661 \times 10^{-27} \text{ kg} \simeq 4 \times 10^{-25} \text{ kg}.$$

The ratio is of order of magnitude

$$\frac{\Delta M}{M} \simeq 1 \times 10^{-3},$$

or about one part per thousand. This is a small but far from trivial change.

As a general rule, it is not possible to refine the calculation of Example 39.5 into an accurate calculation because it is not possible—or at best very difficult—to replace the crude assumptions with accurate descriptions. But we often find it unnecessary to go into such detail. Rather, we make careful measurements of the initial and final rest masses of the system. We then use the difference in these masses to calculate the energy released.

EXAMPLE 39.6

One of the first fission reactions studied was the process

$$n + {}^{238}U \rightarrow {}^{143}Ba + {}^{93}Kr + 3 n + E,$$

which led Hahn and Strassmann to discover neutron-induced fission. Find the energy released per fission. The atomic masses are as follows:

n:	m_n	$= 1.008\ 665$ u
${}^{143}Ba$:	m_{Ba}	$= 142.9205$ u
${}^{238}U$:	m_U	$= 238.0508$ u
${}^{93}Kr$:	m_{Kr}	$= 92.931\ 13$ u

Find the total energy E released in the process.

SOLUTION: The easiest way to proceed is to find the net mass change:

$$\Delta M = m_U - 2m_n - m_{Ba} - m_{Kr}$$
$$= 238.0508\ u - 2 \times 1.008\ 665\ u$$
$$- 142.9205\ u - 92.931\ 13\ u$$
$$= 0.1818\ u.$$

Note that this change represents a loss of mass. Now, reexpress this result in kilograms:

$$\Delta M = 0.1818\ u \times 1.661 \times 10^{-27}\ kg/u$$
$$= 3.020 \times 10^{-28}\ kg.$$

In the last step, you find the energy to which this mass loss is equivalent. You have

$$E = \Delta M c^2 = 2.718 \times 10^{-11}\ J,$$

or $\qquad E = 169.9$ MeV.

THINKING LIKE A PHYSICIST

If you look back on the calculations of Examples 39.5 and 39.6, you will see that we have obtained important information concerning two complicated systems about whose details we have said very little, by substituting for the actual systems a simple model consisting of nothing more than two masses and a spring, tied together with a string. This is the model first introduced in Figure 39.7 and used as the basis for developing the principle of conservation of mass-energy. Isn't it remarkable that so simple a model can be used both to develop a profound physical principle and to interpret the experimental evidence that verifies the principle? For initial inquiry into an unfamiliar world, the simplest possible model is often the best.

Needless to say, a lot of important information about chemical and nuclear reactions depends on the structural details of molecules and nuclei and will not emerge from a spring-mass model. Nevertheless, the mileage to be gained from simple, apparently artificial models is a never-ending source of wonder to even the most experienced physicist. What other models that you have studied in this book have surprised you with their power?

The neutron-induced reaction of Example 39.6 is of the not-uncommon type in which two or more neutrons are among the fission fragments. This opens the possibility of a **chain reaction**, the process that underlies the operation of **fission reactors**. A stray neutron induces the first fission; if not too many of the product neutrons are lost through absorption or other processes, more than one neutron is available for further fission processes. The average number of neutrons produced by each reaction and surviving to induce a new reaction is called the **multiplication ratio** μ of the process. If $\mu < 1$, the process will not sustain itself, but will quickly die down. This condition is called **subcritical**. If $\mu = 1$, the fission process continues at a constant rate, and the power output is constant. This condition, called **critical**, is the condition for which nuclear reactors are adjusted. If $\mu > 1$, the rate of the process increases exponentially. This condition is called **supercritical**. The result is a runaway chain reaction, and huge amounts of energy can be released very quickly. This is what happens in "atomic bombs." The three possible processes—subcritical, critical, and supercritical—are shown schematically in Figure 39.11.

Evidently, it is of the utmost importance to control the value of μ in nuclear reactors. This is often done by using rods of materials such as cadmium, which are effective neutron absorbers. By inserting and withdrawing the rods, the reaction rate can be controlled with great precision.

Obscure Heroes

When the first nuclear reactor was allowed to reach criticality on the night of 2 December 1942, precautions had to be taken againt unforeseen events. A "suicide squad" of young (and brave) physicists and chemists was stationed on top of the reactor. Each had a glass carboy, filled with a saturated solution of a cadmium salt, and a sledgehammer. If all else failed, they were to smash the bottles so that the solution would flow into the reactor and stop the reaction. One of them, the chemist James B. Parsons, recalled much later that—at least in retrospect—the worst part of the job was the preparation. Cadmium salts are hard to dissolve, and he spent long hours rolling the heavy jugs up and down the hall, trying to get the stuff into solution.

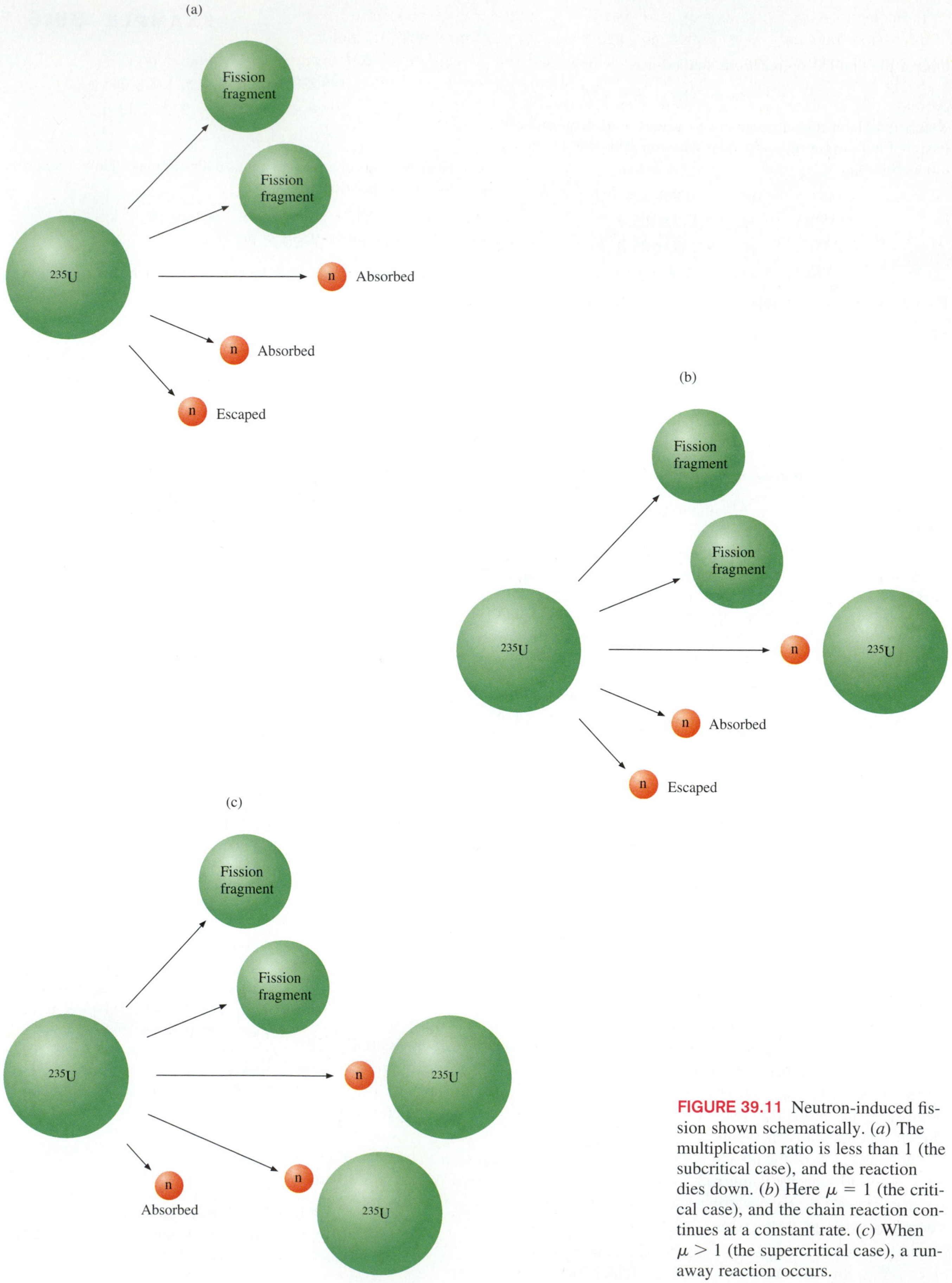

(a)

Fission
fragment

Fission
fragment

^{235}U

n Absorbed

n Absorbed

n Escaped

(b)

Fission
fragment

Fission
fragment

^{235}U

n ^{235}U

n Absorbed

n Escaped

(c)

Fission
fragment

Fission
fragment

^{235}U

n ^{235}U

n   n ^{235}U

Absorbed

FIGURE 39.11 Neutron-induced fission shown schematically. (*a*) The multiplication ratio is less than 1 (the subcritical case), and the reaction dies down. (*b*) Here $\mu = 1$ (the critical case), and the chain reaction continues at a constant rate. (*c*) When $\mu > 1$ (the supercritical case), a runaway reaction occurs.

In nuclear reactors, most of the energy is released in the form of the kinetic energy of the fission fragments. These fragments make repeated collisions with other atoms in the reactor. The kinetic energy is quickly distributed randomly among many atoms and is thus transformed into heat energy. In commercial power reactors, the heat is used to produce steam, which drives a turbine generator, very much as in a fossil-fuel reactor. The dramatic difference between nuclear and fossil-fuel reactors lies in the factor of 8×10^7 or so that distinguishes the energy of nuclear reactions from that of chemical reactions. One kilogram of nuclear fuel yields about as much energy as 800,000 tonnes of coal and reduces the carbon dioxide burden of the atmosphere by a corresponding amount. However, nuclear fuels have their own problems, mostly having to do with the radioactivity of the waste. Power production is a colossal systems-engineering problem!

$\mathbf{a}$	acceleration	$\mathbf{p}, p$	momentum
$\mathscr{B}$	magnetic field	U	potential energy
c	speed of light	$\mathbf{V}, v$	velocity, speed of a body
E	total relativistic energy	V	speed of a reference frame with respect to an observer
$\mathbf{F}$	force		
K	kinetic energy	W	work
m, m_0	relativistic mass, rest mass	μ	multiplication ratio
M	total relativistic mass of a system	ω	angular speed

Summing Up

We repeat the Mach experiment in an adaptation designed for use by two observers moving at speed V with respect to one another. Taking the principle of conservation of momentum as a starting point and using the rules of relativistic kinematics developed in Chapter 38, we find that the mass m of a body of rest mass m_0 depends on its speed v with respect to the observer. The dependence is given by Equation 39.3,

$$m = \frac{m_0}{\sqrt{1 - v^2/c^2}}.$$

When a particle of rest mass m_0 is accelerated from rest, its kinetic energy can be expressed as a function of the relativistic mass change according to Equation 39.10,

$$K = mc^2 - m_0c^2.$$

Because m is itself a function of speed v, the kinetic energy can also be expressed in terms of v according to Equation 39.11b,

$$K = m_0c^2\left(\frac{1}{\sqrt{1 - v^2/c^2}} - 1\right).$$

The total relativistic energy E of a particle can be expressed as the sum of its kinetic energy K and its rest energy m_0c^2. From this follows Equation 39.15, which expresses E

in terms of the relativistic mass according to the relation

$$E = mc^2.$$

The **principle of conservation of mass-energy** is expressed by either of Equations 39.19a and 39.19b:

$$M_ic^2 = M_fc^2 \quad \text{or} \quad M_{0i}c^2 + K_i = M_{0f}c^2 + K_f.$$

According to this principle, the removal of energy from a system must be accompanied by a change in its rest mass. For chemical reactions this change is negligibly small, but for nuclear reactions the change is significant.

Total relativistic energy E, momentum p, relativistic mass m, and rest mass m_0 are related by either of Equations 39.20a and 39.20b:

$$E = \sqrt{p^2c^2 + m_0^2c^4} \quad \text{or} \quad m = \sqrt{(p/c)^2 + m_0^2}.$$

The kinetic energy K is related to p, m, and m_0 according to Equation 39.21,

$$K = \sqrt{p^2c^2 + m_0^2c^4} - m_0c^2.$$

In the **extreme relativistic limit**, this relation simplifies to the form of Equation 39.22, $K = pc$.

KEY TERMS

Section 39.2 Mass and Momentum
rest mass, relativistic mass ▪ relativistic momentum

Section 39.3 Force and Energy
relativistic force ▪ relativistic kinetic energy

Section 39.4 Conservation of Mass-Energy
rest energy, relativistic energy ▪ classical limit, extreme relativistic limit ▪ neutrino

Section 39.5 Chemical and Nuclear Reactions
fission, neutron-induced fission ▪ chain reaction, fission reactor

---

Queries and Problems for Chapter 39

QUERIES

39.1 *(3) Shoving them along.* A machine that increases the energy of elementary particles (such as protons or electrons) to large values is conventionally called a particle *accelerator*. Why is this name not really descriptive? (Perhaps the machine should be called a *ponderator*!)

39.2 *(4) Letting go.* The spring in the system of Figure 39.7 is allowed to expand, and the two masses ultimately come to rest owing to frictional interactions with other objects. Is the total rest mass of the system increased or decreased over the entire process?

39.3 *(4) Rest-mass dependence.* Sketch a graph of relativistic kinetic energy K versus pc for **(a)** a particle with zero rest mass, **(b)** a particle with small rest mass, and **(c)** a particle with larger rest mass. (Hint: What form must curves **b** and **c** have for $v \ll c$? for $v \simeq c$?)

PROBLEMS

GROUP A

39.1 *(2) Tepid electron.* At what speed is the mass of an electron increased by 1% over its rest mass?

39.2 *(2) Warmish proton.* At what speed is the mass of a proton increased by 0.5% over its rest mass?

39.3 *(3) Hot electron, I.* What is the value of the ratio v/c for an electron having kinetic energy 100 keV?

39.4 *(3) Hot electron, II.* What is the value of the ratio v/c for an electron having kinetic energy 1 MeV?

39.5 *(3) Really hot electron.* An electron has total relativistic energy 1 J. Find **(a)** its speed, **(b)** its relativistic mass, and **(c)** its momentum. (Hint: You will need to use the approximation suggested in Problem 38.9.)

39.6 *(3) Double negative . . .* An electron has relativistic mass twice its rest mass. **(a)** What is its speed? Express your result as the ratio v/c. **(b)** What is its kinetic energy? Express your result in J and in eV.

39.7 *(3) . . . makes a positive?* A proton has relativistic mass twice its rest mass. What is its kinetic energy? Express your result in J and in eV.

39.8 *(3) Threshold.* At what speed does the relativistic kinetic energy of a body differ from its classical kinetic energy by 1%?

39.9 *(3) Dust shield.* A spaceship, traveling at high speed, strikes a cosmic dust particle of rest mass 1 μg. The collision speed is $0.9c$, and the collision is totally inelastic. How much energy is released? Compare with the explosive energy of TNT, which releases 4.2×10^6 J/kg. You will see why designing a shield for such a spaceship is more difficult than designing a cowcatcher for a locomotive!

39.10 *(4) Hidden energy.* Find the rest energy of a proton. Express your result in J and in MeV.

39.11 *(4) Positively hot.* **(a)** What is the total relativistic energy E of a proton whose speed is $0.8c$? **(b)** What is its kinetic energy K? Express your results in terms of m_0 and also in eV.

39.12 *(4) Cosmic consciousness.* The most energetic protons found in cosmic rays have energies of order 10^{20} eV. **(a)** What is the speed of such a proton? Express your result as the difference $c - v$. **(b)** What is the relativistic mass of the proton?

39.13 *(4) Gone but not forgotten.* To every elementary particle (such as an electron, a proton, or a neutron) there corresponds an *antiparticle*. When a particle and its antiparticle collide, they annihilate each other totally, producing electromagnetic radiation in the process. Suppose that a particle-antiparticle pair collide at speed $v \ll c$. What is the energy of the radiation if the particles are **(a)** an electron and its antiparticle, a positron? **(b)** a proton and an antiproton? **(c)** Why does the question make no sense for a neutrino-antineutrino pair?

39.14 *(4) Energy from the sun, I.* The sun radiates energy at a rate of about 10^{26} W. **(a)** At what rate, in kg/s, does it lose mass as a result? **(b)** The mass of the sun is 1.99×10^{30} kg. What is the fractional mass loss per year due to radiation? **(c)** If the sun lost mass in no other way, how long would it last if it continued to radiate energy at the present rate? (Note:

This must be at best an upper limit, because not all of the sun's mass can be converted into radiant energy; see Problem 39.30.)

39.15 (4) *Energy from the sun, II.* Sunlight falls on each square meter of the earth's surface normal to the sun–earth direction at a rate of 1.4 kW. (This is called the *solar constant.*) If the earth absorbed all this radiation and did not reemit it (*not* the case), what would be the resulting annual rate of increase of the earth's mass?

39.16 (5) *Decay.* A free neutron is not stable but decays into a proton (p), an electron (e), and an antineutrino ($\bar{\nu}$):

$$n = p + e + \bar{\nu} + 7.8024 \times 10^5 \text{ eV}.$$

The antineutrino has zero rest mass. The rest mass of the neutron is given in Example 39.6, and the rest mass of the electron is 5.4854×10^{-4} u. Find the rest mass of the proton.

38.17 (5) *Relativistic chemistry, I.* Of all chemical reactions, the one that releases the greatest amount of energy per unit mass of reactant is the reaction

$$H + F \rightarrow HF + 2.79 \text{ eV},$$

in which hydrogen and fluorine combine to form hydrogen fluoride. The atomic mass of the hydrogen isotope ^{1}H is 1.007 825 u; that of the fluorine isotope ^{19}F is 18.998 40 u. Find the proportional mass change $\Delta M/M$ that occurs in the reaction. (The actual overall reaction begins and ends with molecular forms: $H_2 + F_2 \rightarrow 2$ HF. But neglect the comparatively small energy difference between monatomic H and F and their molecular forms, H_2 and F_2.)

39.18 (5) *Relativistic chemistry, II.* Hydrogen and oxygen gas are combined to form 1 kmol of water. The reaction is

$$1 \text{ kmol } H_2 + \tfrac{1}{2} \text{ kmol } O_2 \rightarrow 1 \text{ kmol } H_2O + 5.75 \times 10^8 \text{ J}.$$

What is the difference in mass between the original reactants and the final product? Is the mass change a gain or a loss?

39.19 (5) *Uranium fission.* When ^{235}U undergoes fission in a nuclear reactor, the energy released depends on exactly how the nucleus splits. But on the average, the fission process yields about 200 MeV per nucleus. **(a)** What is the mass change when 1 kmol of ^{235}U undergoes fission? **(b)** The mass of 1 kmol of ^{235}U is close to 235 kg. What fraction of the initial mass is converted into other forms of energy?

GROUP B

39.20 (2) *Single variable, I.* Equation 39.4, $\mathbf{p} = m\mathbf{v}$, expresses the relativistic momentum in terms of two variables. Reexpress the magnitude p of the momentum of a body of rest mass m_0 in terms of the variable v only, together with any necessary constants.

39.21 (2) *Single variable, II.* Reexpress the magnitude p of the relativistic momentum of a body of rest mass m_0 in terms of the variable m only, together with any necessary constants.

39.22 (3) *Helping it along, I.* Through what potential difference must an electron pass if it is to acquire a speed $v = 0.95c$?

39.23 (3) *Helping it along, II.* Through what potential difference must a proton pass if it is to acquire a speed $v = 0.90c$?

39.24 (3) *Grand tour, I.* In the Fermilab accelerator, protons are subjected to a magnetic field that makes them follow an essentially circular orbit of radius 1 km. They reach a maximum energy of about 1 TeV (10^{12} eV). **(a)** What is the centripetal acceleration of such a proton? **(b)** What is its mass? **(c)** What force must the magnetic field exert on the protons? **(d)** What is the magnitude of the magnetic field?

39.25 (3) *A short life but a merry-go-round one.* The 450-MeV synchrocyclotron at the University of Chicago was the largest ever built. It remained in service for about a decade, beginning in 1953. In order to circumvent the difficulty described in Example 39.2, the deuterons to be accelerated were injected in bursts. The angular frequency of the electric potential applied to the dee system was varied as a burst of deuterons accelerated, so as to keep the accelerating potential in phase with the relativistic angular speed ω_r, as ω_r varied with increasing energy. Each cycle of modulation produced a burst of 450-MeV deuterons, and the completion of a cycle required a little more than 1 s. **(a)** Calculate the ratio ω_r/ω_c of the final to the initial angular speed of the deuterons. **(b)** Calculate ω_r and ω_c on the assumption that the magnetic field had the uniform value 1.6 T.

39.26 (3) *Constant force.* A particle of rest mass m_0 is originally located at $x = 0$, at rest with respect to observer O. Using appropriate means (for example, a uniform electric field), she exerts a constant force $\mathbf{F}$ on it, parallel to the x axis. She measures the speed v and position x of the particle at various times t. **(a)** Show that the speed of the particle depends on t according to the relation

$$v = \frac{cFt}{\sqrt{m_0{}^2c^2 + F^2t^2}}.$$

(Hint: Remember that, for a body initially at rest, impulse = momentum.) **(b)** Show that the position of the particle depends on t according to the relation

$$x = \sqrt{\left(\frac{m_0c^2}{F}\right)^2 + (ct)^2} - \frac{m_0c^2}{F}.$$

39.27 (4) *Constant in a changing world.* Observers O and O' make observations on the same particle. Show that, regardless of their speed V with respect to one another, they measure the *same* value for the quantity $E^2 - p^2c^2$. Such a quantity is said to be *Lorentz invariant.* What is the value of the quantity?

39.28 (4) *Classical simplicity.* Show that the total relativistic energy E given by Equation 39.20a simplifies in the case $v \ll c$ to the form $E = m_0c^2 + \tfrac{1}{2}mv^2$.

39.29 (4) *"Classical" self-energy of the electron.* Imagine that you could construct an electron by bringing little bits of charge together from infinity to form a sphere of charge $-e$ and radius R. **(a)** Show that the resulting electrostatic potential energy, called the *classical self-energy,* is

$$U = \frac{3}{5}\frac{e^2}{4\pi\epsilon_0 R}.$$

(b) Using the known mass of the electron, 9.1×10^{-31} kg, estimate the radius of the electron.

39.30 (5) *Energy from the sun, III.* The process by which the sun produces energy was first suggested in 1939 by the German-American physicist Hans A. Bethe (b. 1906). The process involves a sequence of *nuclear fusion* reactions, in which hydrogen nuclei combine to produce helium nuclei, with the release of energy. The overall process is

$$4\ ^1H \rightarrow\ ^4He + 2\ e + 2\ \nu.$$

In this reaction, e represents an electron and ν a neutrino. The rest mass of a hydrogen (^1H) atom is 1.007 825 u; that of a helium (^4He) atom is 4.002 60 u, that of an electron is 5.4854×10^{-4} u, and that of a neutrino is zero. Because the colliding nuclei must overcome their mutual Coulomb repulsion to combine, a high temperature—about 2×10^7 K—is required to give them adequate thermal kinetic energy. **(a)** Find the energy released in the reaction. Express your result in MeV. **(b)** The sun radiates energy at a rate of about 10^{26} W. Its mass is 1.99×10^{30} kg. If the sun were originally made of pure hydrogen and if the reaction could go on until all the hydrogen was fused into helium, how long would the sun "burn" at its current rate? For comparison, the age of the sun is about 5×10^9 years.

39.31 (5) *Senile stars and nucleosynthesis.* When all the hydrogen in a star has been converted into helium, as in Problem 39.30, the star can still use other processes to release energy, provided it is sufficiently large that the temperatures and pressures in its core are high enough. The main processes are, successively,

$$3\ ^4He \rightarrow\ ^{12}C,$$
$$^{12}C +\ ^4He \rightarrow\ ^{16}O,$$
$$2\ ^{12}C \rightarrow\ ^4He +\ ^{20}Ne \quad \text{and} \quad 2\ ^{12}C \rightarrow\ ^{24}Mg,$$
$$2\ ^{16}O \rightarrow\ ^4He +\ ^{28}Si \quad \text{and} \quad 2\ ^{16}O \rightarrow\ ^{32}S,$$

and
$$2\ ^{28}Si + 2\ e \rightarrow\ ^{56}Fe.$$

Find the energy released in each of these reactions. The atomic masses are as follows: ^4He: 4.002 50 u; ^{12}C: 12.000 00 u; ^{16}O: 15.994 91 u; ^{20}Ne: 19.992 44 u; ^{24}Mg: 23.985 04 u; ^{28}Si: 27.976 93 u; ^{32}S: 31.972 07 u; ^{56}Fe: 55.9349 u.

Each reaction begins as the "fuel" for the preceding one becomes exhausted. Exhaustion spreads from the center of the star outward, and all of the reactions can take place simultaneously in "shells" at different distances from the center. The temperatures required to sustain these reactions are successively higher, as the number of protons in the nuclei increases and the Coulomb repulsion becomes stronger. For the last reaction, a temperature of about 4×10^9 K is required. Note that the processes produce less energy per unit mass as a star goes up the ladder. Building still larger nuclei is an exothermic process. If the star is big enough, it becomes unstable when all the lighter nuclei are exhausted. The core of the star begins to collapse, the temperature rising so high that some of the fusion products dissociate back to helium—an exothermic process that removes heat from the center of the star. The star collapses violently, blowing off its outer part in a *supernova event*. The blown-off gases are later incorporated into newly forming stars. This is the source of the heavier elements in our solar system. More detailed arguments show that the sun must be at least a third-generation star, some of its matter having passed through two or more supernovae. For an excellent short account of the supernova process, see S. E. Woolsey and M. M. Phillips, *Science* **240**, 750 (6 May 1988).

39.32 (G) *Merry-go-round.* Suppose that a hydrogen atom consists of a ball-like electron in a circular orbit about a much more massive ball-like proton. When a proton captures an electron, the newly formed atom radiates 13.6 eV of electromagnetic energy. **(a)** What is the classical kinetic energy of the electron as it circles the proton? (Hint: See Section 14.5 and especially Equations 14.20 and 14.21.) **(b)** What is the total relativistic energy of the electron? **(c)** What is its relativistic kinetic energy? **(d)** By what percentage does the relativistic kinetic energy differ from the classical kinetic energy? [Note: The hydrogen atom can be treated nonrelativistically with fairly good accuracy. But the inner electrons in atoms of large atomic number (and nuclear charge) must be treated relativistically.]

GROUP C

39.33 (2) *Inelastic collision, I.* A body of rest mass m_0 is at rest with respect to an observer. Another body of equal mass, having initial speed v, strikes the first body. The collision is totally inelastic, and the two bodies stick together as they move away from the observer with speed v'. **(a)** Show that v' and v bear the relation

$$v' = \frac{\gamma v}{\gamma + 1}, \quad \text{where } \gamma \equiv \frac{1}{\sqrt{1 - v^2/c^2}}.$$

(b) Show that the final rest mass M_0' of the system is

$$M_0' = m_0\sqrt{2(\gamma + 1)}.$$

(c) Express the result of part **b** in terms of the initial kinetic energy K_0 of the system. **(d)** From the result of part **b**, you can see that $M_0' \geq 2m_0$. What is the physical reason for this?

39.34 (2) *Inelastic collision, II.* From your point of view, the relativistic mass of a particle is three times its rest mass; $m = 3m_0$. It makes a totally inelastic collision with another particle of equal rest mass. **(a)** What is the final rest mass M_0' of the system? **(b)** Just after the collision, you measure the speed v' of the system. What ratio v/c do you expect? (Hint: Use the results of Problem 39.33.)

39.35 (3) *Grand tour, II.* The now abandoned SSC (superconducting supercollider) accelerator was designed to accelerate protons to energies of about 20 TeV—about ten times the energy of the Fermilab accelerator. The radius of the ring was to be 13.5 km. What is the required value of the magnetic field?

39.36 (3) *Not always parallel.* **(a)** Using the Lorentz transformation equations (Section 38.6), show that, when a force **F** acts in an arbitrary direction on a particle moving in an arbitrary direction with velocity **v**, the acceleration **a** of the particle is not in general parallel to the force. **(b)** Show that **a** is parallel to **F** for two special cases: **F** $\parallel$ **v** and **F** $\perp$ **v**.

39.37 *(4) Photon rocket.* A galactic explorer ship makes use of a rocket engine that expels a stream of photons from the tail of the ship. **(a)** During a certain time interval, the engine expels photons having total energy E and total momentum p. Express E in terms of the ship mass m_i at the beginning of the interval and the mass m_f at the end of the interval. **(b)** Express p in terms of m_i m_f, and the speed change dv of the ship. **(c)** An observer watches the ship start from rest and achieve speed v. Using the results of parts **a** and **b**, show that, from her point of view, v bears the following relation to the initial and final rest masses:

$$v = c\,\frac{m_{0i}^2 - m_{0f}^2}{m_{0i}^2 + m_{0f}^2}.$$

(d) If the rest mass of the ship is reduced to $\frac{1}{5}$ its initial value, find v/c. **(e)** Compare the result of part **c** with the Tsiolkovskii formula for chemical rockets (Equation 9.14). Explain the much greater efficacy of photons as a propellant compared with chemical fuels.

39.38 *(G) Conservative systems.* Observer O measures the mass m_i and the velocity v_i of each of N particles moving along the x axis. The particles make many collisions with one another. Nevertheless, O finds that

$$\sum_{i=1}^{N} m_i = M_c \quad \text{and} \quad \sum_{i=1}^{N} m_i v_i = P_c,$$

where M_c and P_c are constants. (The subscript ''c'' stands for ''classical.'') The constancy of M_c expresses the classical conservation of mass. The constancy of P_c expresses the conservation of momentum. **(a)** Another observer O' moves with velocity V with respect to O. Show that, in the classical domain, where the Galilean transformation holds, O' also observes that momentum is conserved but finds a different value for P_c. **(b)** Show that, if O' uses the Lorentz velocity transformation, he does *not* obtain a constant value for P_c. **(c)** Now consider the following quantities:

$$\sum_{i=1}^{N} \frac{m_i}{\sqrt{1 - v^2/c^2}} \equiv M_r \quad \text{and} \quad \sum_{i=1}^{N} \frac{m_i v_i}{\sqrt{1 - v^2/c^2}} \equiv P_r.$$

(The subscript ''r'' stands for ''relativistic.'') Show that both O and O' will find that M_r and P_r are conserved, though the two observers will find different values for these quantities.

39.39 *(G) Lorentz transformation for momentum and energy.* Observer O finds that a particle has momentum $\mathbf{p} = (p_x, p_y, p_z)$ and total relativistic energy E. Observer O' is moving with speed V in the x direction with respect to O. Show that for O' the corresponding quantities $\mathbf{p}' = (p_x', p_y', p_z')$ and E' for the particle are

$$p_x' = \frac{p_x - VE/c^2}{\sqrt{1 - V^2/c^2}}, \tag{1}$$

$$p_y' = p_y, \quad p_z' = p_z, \tag{2,3}$$

and

$$E' = \frac{E - Vp_x}{\sqrt{1 - V^2/c^2}}. \tag{4}$$

39.40 *(G) Classical check.* Show that Equations 1 and 4 in Problem 39.39 simplify to familiar classical transformation equations in the case $V \ll c$.

39.41 *(G) Colliding beams.* An *antiproton* is a particle having the same rest mass m_0 as a proton but opposite charge. In the original 1955 Bevatron experiment used to produce antiprotons at the University of California, Berkeley, a beam of fast-moving protons (p^+) was made to collide with a metal target containing many protons essentially at rest. If the incident protons are energetic enough, new particles are created and a shower of protons and antiprotons (p^-) emerges. Because electric charge must be conserved, new matter is created in the form of proton-antiproton pairs, and the overall reaction is of the form

$$p^+ + p^+ \rightarrow 3\,p^+ + p^-.$$

(a) What is the minimum initial kinetic energy K_0 of the protons in the incident beam needed to produce the reaction? Express your result in terms of $m_0 c^2$ and also in eV. **(b)** What is the efficiency of the process? That is, what fraction of K_0 goes into the creation of new matter? What fraction of K_0 appears as kinetic energy of the four product particles? **(c)** Whenever possible, this method is supplanted in modern practice by *colliding-beam* experiments, in which the collision takes place between two beams of particles moving at equal speed in opposite directions. What is the efficiency of this process? (Hint: See Problem 39.40.)

Measuring Angles in Radians

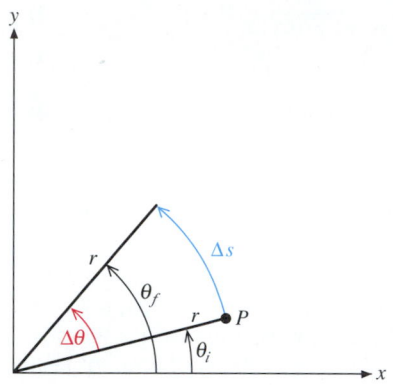

FIGURE A1.1

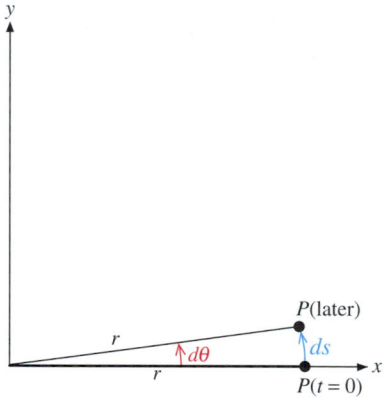

FIGURE A1.2

Angles can be expressed in a number of ways, as in degrees (°) or revolutions (rev). But it is most often convenient in mathematical and physical analysis to express them in terms of **radians** (rad), the SI unit of angle measure.

As shown in Figure A1.1, a circle of arbitrary radius r is drawn centered on the vertex of an arbitrary angle $\Delta\theta$. The angle subtends an arc whose length (measured along the arc) is Δs. The angle $\Delta\theta$, expressed in radians, is defined to be the ratio of the length Δs of the arc to the radius r:

$$\Delta\theta \equiv \frac{\Delta s}{r}. \tag{A1.1}$$

This definition is just as valid when the arc length and the angle are infinitesimal. Hence, in the case shown in Figure A1.2, the infinitesimal angle $d\theta$ is given by the ratio

$$d\theta = \frac{ds}{r}. \tag{A1.2}$$

Because the angle $\Delta\theta$ or $d\theta$ is thus defined as the ratio of two lengths, it is itself a dimensionless quantity. (The same is true of angles measured in other units, but that is not so easy to see.) As noted in Chapter 6, it follows that the unit of angular velocity, rad/s, can equally well be expressed as reciprocal seconds (s^{-1}). Similarly, as noted in Chapter 11, it follows that the unit of angular acceleration, rad/s^2, can equally well be expressed as reciprocal seconds squared (s^{-2}).

EXAMPLE

(a) Express in radians the value of the angular displacement equal to one complete revolution, or 360°. **(b)** The shaft of an electric motor rotates at 3600 rev/min. Express this angular velocity in SI units. **(c)** Find the factor for converting angular velocities from rev/min into rad/s.

SOLUTION:

(a) Express in radians the value of the angular displacement equal to one complete revolution.

When the system of Figure A1.1 experiences one complete revolution, the typical point P describes a circle whose circumference is $\Delta s = 2\pi r$. Inserting this value of Δs into Equation A1.1 gives $\Delta\theta = \Delta s/r = 2\pi r/r = 2\pi$. So one revolution (or 360°) is equal to 2π radians.

(b) Express the angular velocity 3600 rev/min in SI units.

Using the result of part **a**, you have

$$\omega = 3600\ \frac{\text{rev}}{\text{min}} \times \frac{2\pi\ \text{rad}}{1\ \text{rev}} \times \frac{1\ \text{min}}{60\ \text{s}}$$

$$= 377.0\ \text{rad/s}.$$

(c) Find the factor for converting angular velocities from rev/min into rad/s.

The principles underlying the calculation in part **b** can be expressed more generally by writing, for any angular velocity $\theta = N$ rev/min,

$$N\ \text{rev/min} = N\ \frac{\text{rev}}{\text{min}} \times \frac{2\pi\ \text{rad}}{1\ \text{rev}} \times \frac{1\ \text{min}}{60\ \text{s}}$$

$$= \frac{2\pi}{60}\ N\ \text{rad/s} = 0.1047N\ \text{rad/s}.$$

Derivation of the Work Integral

In Figure A2.1, a body moves along an arbitrary path s that begins at s_i and ends at s_f. As it moves, the body is acted on by a force whose magnitude and direction both vary in an arbitrary way. Because of this variation, it is not possible to define the work by means of Equation 7.5c, $W = \mathbf{F} \cdot \mathbf{s}$. Nevertheless, the force is doing work on the body, and we must find a way of expressing that work.

We begin by approximating the path by a large number of small, straight segments $\Delta \mathbf{s}$. As noted in Section 7.3, each segment has vectorial properties even though the path as a whole does not. This is because each segment has a definite direction $\Delta \hat{\mathbf{s}}$ as well as a length Δs:

$$\Delta \mathbf{s} = \Delta \hat{\mathbf{s}} \, \Delta s. \tag{A2.1}$$

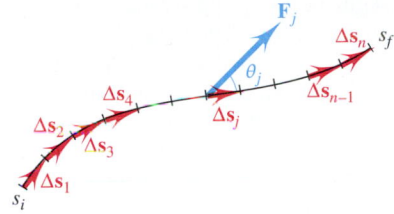

FIGURE A2.1 A body moves along the arbitrary path shown. The path may be approximated by a series of short, straight segments, of which $\Delta \mathbf{s}_j$ is a typical one. When the body lies within the segment $\Delta \mathbf{s}_j$, it is acted on by a force $\mathbf{F}_j$ which makes an angle θ_j with $\Delta \mathbf{s}_j$.

As the body moves along the path, the force $\mathbf{F}$ varies. However, we have chosen the segments $\Delta \mathbf{s}$ short enough so that $\mathbf{F}$ does not vary much over any particular segment. Thus, over each particular segment $\Delta \mathbf{s}_j$, the force does not vary much from the value $\mathbf{F}_j$ measured at, say, the middle of the segment. The small amount of work ΔW_j done on the body by the force as the body moves through the segment $\Delta \mathbf{s}_j$ is therefore given by

$$\Delta W_j \simeq \mathbf{F}_j \cdot \Delta \mathbf{s}_j. \tag{A2.2a}$$

This is just Equation 7.5c, rewritten so that it applies specifically to the present case. Because $\Delta \mathbf{s}_j = \Delta \hat{\mathbf{s}}_j \, \Delta s_j$, we have

$$\Delta W_j \simeq (\mathbf{F}_j \cdot \Delta \hat{\mathbf{s}}_j) \, \Delta s_j.$$

The quantity in parentheses in this approximation has the value $F_j \cos \theta_j$, where θ_j is the angle between the force and the direction of motion at segment j. Approximation (A2.2a) thus becomes

$$\Delta W_j \simeq F_j \cos \theta_j \, \Delta s_j. \tag{A2.2b}$$

Compare this with Equation 7.5a, $W = |\mathbf{F}| \cos \theta \, |\mathbf{s}|$.

Approximation A2.2b applies equally well to each of the many segments that make up the path. Thus the work done by the force over the entire path is closely approximated by the *sum* of the ΔW_j:

$$W \simeq \Delta W_1 + \Delta W_2 + \cdots + \Delta W_j + \cdots + \Delta W_{n-1} + \Delta W_n$$

or
$$W \simeq F_1 \cos \theta_1 \, \Delta s_1 + F_2 \cos \theta_2 \, \Delta s_2 + \cdots + F_j \cos \theta_j \, \Delta s_j + \cdots$$
$$+ F_{n-1} \cos \theta_{n-1} \, \Delta s_{n-1} + F_n \cos \theta_n \, \Delta s_n. \tag{A2.3a}$$

This approximation can be written in the condensed form

$$W \simeq \sum_{j=1}^{n} F_j \cos \theta_j \, \Delta s_j. \tag{A2.3b}$$

Approximation A2.3 in either form suggests a direct experimental way of evaluating W, at least in principle. If we measure $\mathbf{F}_j$ at each of the n midpoints of the segments Δs_j, we can evaluate each of the terms in the equation and then evaluate the sum.

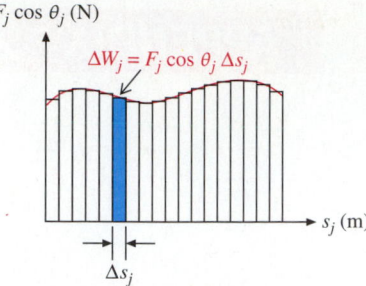

$F_j \cos \theta_j$ (N)

$\Delta W_j = F_j \cos \theta_j \Delta s_j$

s_j (m)

Δs_j

FIGURE A2.2 The work ΔW done on the body as it passes over the segment $\Delta \mathbf{s}_j$ is approximated by the area of the shaded rectangle. This area is expressed algebraically by Equation A2.2b. Over the entire path, the work W done is approximated by the sum of the areas of the rectangles, expressed algebraically by Equation A2.3b. The work W is given exactly by the area under the red curve. Evaluating Equation A2.4b is equivalent to the geometric process of measuring the area under the curve.

Figure A2.2 is a graphic representation of Approximation A2.3. According to the approximation, ΔW_j is the product of the height $F_j \cos \theta_j$ of the jth narrow rectangle in the figure with its width Δs_j. Thus ΔW_j is the *area* of the jth rectangle. The total work W, which is the sum of the ΔW_j, is approximated by the sum of the areas of these rectangles. The approximation can be improved by using a larger number of smaller segments Δs_j, but at the cost of more labor.

This method of directly measuring the areas of the small rectangles and adding them up is useful on occasion. But another method is useful much more frequently and is exact. We can often express $F \cos \theta$ as a mathematical function of the position s measured along the curve from the starting point. Suppose that the red curve in Figure A2.2 represents the graph of such a function. Because the function is continuous, we know its value for *every* value of s, at least in principle. We can therefore evaluate and sum the products $F_j \cos \theta_j \Delta s_j$, no matter how narrow we choose the segments Δs_j.

Accordingly, we divide the path into an infinite number of infinitesimal segments $d\mathbf{s}$. That is, we evaluate the sum of Approximation A2.3b in the limit where Δs approaches zero:

$$W = \lim_{\Delta s \to 0} \sum_j \Delta W_j = \lim_{\Delta s \to 0} \sum_j F_j \cos \theta_j \Delta s_j. \tag{A2.4a}$$

This is an equation rather than an approximation because the error that comes from using a fixed value of $F_j \cos \theta_j$ over each interval Δs_j vanishes as the intervals become infinitesimal. The values of j over which the sum is evaluated still range from 1 to n, just as in Approximation A2.3. But now n approaches infinity as Δs approaches zero.

When the limiting process is carried out, the right side of Equation A2.4a is written

$$W = \int_{s_i}^{s_f} F(s) \cos \theta(s)\, ds. \tag{A2.4b}$$

The quantity on the right side of the equation is called the **integral** of the function $F \cos \theta$ over the specified path s beginning at s_i and ending at s_f; the quantities s_i and s_f are called the **limits of integration**. The notations $F(s)$ and $\theta(s)$ remind us that both F and θ are variables whose values are known as functions of s. The value of the integral is exactly the area under the red curve of Figure A2.2. Equation A2.4b is identical with Equation 7.13b.

Over each infinitesimal segment $d\mathbf{s}$, an amount of work dW is done. The value of dW is implicit in Equation A2.4a; it is

$$dW = \lim_{\Delta s \to 0} \Delta W = F \cos \theta\, ds, \tag{A2.5}$$

where dW, F, and θ are all functions of s. This equation is analogous to the finite approximation A2.2b. It follows immediately that Equation A2.4b can be written in the compact form

$$W = \int_{s_i}^{s_f} dW, \tag{A2.6}$$

which is Equation 7.12. We have thus shown that the work done on a body by a varying force as the body moves over an arbitrary path can be represented by a definite integral.

Selected Mathematical Relations

Exponentials and Logarithms

$x^0 = 1; \quad x^1 = x$

$x^{-a} = \dfrac{1}{x^a}; \quad x^a x^b = x^{a+b}; \quad \dfrac{x^a}{x^b} = x^{a-b}$

$\log xy = \log x + \log y; \quad \log \dfrac{x}{y} = \log x - \log y;$

$\log x^a = a \log x$

Inverse relation between exponentiation and taking the logarithm:

$\log_a a^x = x; \quad a^{\log_a x} = x$

Change of base:

In general, $\log_b a \, \log_a x = \log_b x$

In particular, for $\log x \equiv \log_{10} x$ and $\ln x \equiv \log_e x$,

$\log x = \log e \, \ln x = 0.4343 \ln x;$

$\ln x = \ln 10 \, \log x = 2.303 \log x$

Quadratic Formula

If $ax^2 + bx + c = 0$, then $x = \dfrac{-b \pm \sqrt{b^2 - 4ac}}{2a}$

Pythagoras's Theorem and Trigonometry

For a right triangle:

$x^2 + y^2 = r^2$ or, equivalently, $\left(\dfrac{x}{r}\right)^2 + \left(\dfrac{y}{r}\right)^2 = 1$

Definitions of the trigonometric functions:

$\cos \theta = \dfrac{x}{r}$ $\qquad \sec \theta = \dfrac{1}{\cos \theta} = \dfrac{r}{x}$

$\sin \theta = \dfrac{y}{r}$ $\qquad \csc \theta = \dfrac{1}{\sin \theta} = \dfrac{r}{y}$

$\tan \theta = \dfrac{y}{x} = \dfrac{\sin \theta}{\cos \theta}$ $\qquad \cot \theta = \dfrac{1}{\tan \theta} = \dfrac{x}{y}$

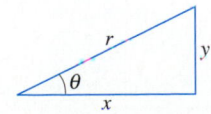

From the second form of Pythagoras's theorem and the definitions of cos θ and sin θ,

$\cos^2 \theta + \sin^2 \theta = 1$

Other trigonometric identities:

$\cos x = \sin (90° - x) = -\cos (180° - x)$

$\sin x = \cos (90° - x) = \sin (180° - x)$

$\cos (-x) = \cos x; \quad \sin (-x) = -\sin x;$
$\tan (-x) = -\tan x$

$\cos (x \pm y) = \cos x \cos y \mp \sin x \sin y$

$\sin (x \pm y) = \sin x \cos y \pm \cos x \sin y$

$\sin x \pm \sin y = 2 \sin \dfrac{x \pm y}{2} \cos \dfrac{x \mp y}{2}$

$\cos x + \cos y = 2 \cos \dfrac{x + y}{2} \cos \dfrac{x - y}{2}$

$\cos x - \cos y = -2 \sin \dfrac{x + y}{2} \sin \dfrac{x - y}{2}$

$\cos \dfrac{x}{2} = \sqrt{\dfrac{1 + \cos x}{2}}; \quad \sin \dfrac{x}{2} = \sqrt{\dfrac{1 - \cos x}{2}}$

Signs of sin x, cos x, and tan x in the four quadrants:

The quadrants in which the functions are positive are conveniently remembered by means of the mnemonic *Act Stupid, Take Chances*. All three functions are positive in the first quadrant, the *s*ine in the second quadrant, and so forth.

General Triangles

Law of sines: $\dfrac{A}{\sin \alpha} = \dfrac{B}{\sin \beta} = \dfrac{C}{\sin \gamma}$

Law of cosines:

scalar form, $C^2 = A^2 + B^2 - 2AB \cos \gamma$

vector form, $C^2 = A^2 + B^2 + 2\mathbf{A} \cdot \mathbf{B}$

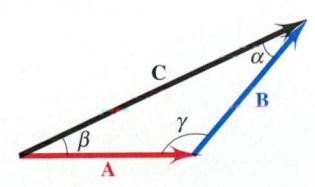

Areas

Area of a triangle of base b and altitude h: $\quad A = \frac{1}{2}bh$

Area of a rectangle or parallelogram of base b and altitude h: $\quad A = bh$

Area of a regular polygon having n sides of length l: $A = \frac{1}{4}nl^2 \cot(180°/n)$

Area of a circle of radius r: $\quad A = \pi r^2$

Area of an ellipse with major axis a and minor axis b: $A = \pi ab$

Surface area of a sphere of radius r: $\quad A = 4\pi r^2$

Volumes

Volume of a parallelepiped of base area A and altitude h: $V = Ah$

Volume of a sphere of radius r: $\quad V = \frac{4}{3}\pi r^3$

Volume of a circular cylinder of radius r and altitude h: $V = \pi r^2 h$

Volume of a cone of base radius r and altitude h: $V = \frac{1}{3}\pi r^2 h$

Selected Series Expansions

Binomial expansion:

$$(1 \pm x)^n = 1 \pm nx + \frac{n(n-1)x^2}{2!}$$
$$\pm \frac{[n(n-1)(n-2)x^3]}{3!} + \cdots$$

Any series $(x \pm y)^n$, with $y < x$, can be written

$$x^n \left(1 \pm \frac{y}{x}\right)^n,$$

and the preceding formula applied.

Taylor series:

$$f(x + h) = f(x) + hf'(x) + \frac{h^2}{2!}f''(x) + \frac{h^3}{3!}f'''(x) + \cdots$$

$$e^x = 1 + x + \frac{x^2}{2!} + \frac{x^3}{3!} + \cdots$$

$$\ln x = (x - 1) - \tfrac{1}{2}(x - 1)^2 + \tfrac{1}{3}(x - 1)^3 - \cdots$$

$$\sin x = x - \frac{x^3}{3!} + \frac{x^5}{5!} - \frac{x^7}{7!} + \cdots$$

$$\cos x = 1 - \frac{x^2}{2!} + \frac{x^4}{4!} - \frac{x^6}{6!} + \cdots$$

Equations 2.18, 2.19, 2.20, 6.4a, and 6.4b give some basic rules for differentiation; for a short table of integrals, see Table 7.1.

Selected Fundamental Physical Constants

Quantity	Symbol	Value	Units	Uncertainty (in parts per million)
Speed of light in vacuum	c	299 792 458	$m \cdot s^{-1}$	defined
Permeability of free space	μ_0	$4\pi \times 10^{-7}$	$N \cdot A^{-2}$ or $T \cdot m \cdot A^{-1}$	defined
Permittivity of free space	$\epsilon_0 \equiv \mu_0^{-1} c^{-2}$	$8.854\ 187\ 817 \ldots \times 10^{-12}$	$C^2 \cdot N^{-1} \cdot m^{-2}$	defined
Universal gravitational constant	G	$6.672\ 59 \times 10^{-11}$	$N \cdot m^2 \cdot kg^{-2}$	128
Planck's constant	h	$6.626\ 075\ 5 \times 10^{-34}$	$J \cdot s$	0.60
	$\hbar \equiv h/2\pi$	$1.054\ 572\ 66 \times 10^{-34}$	$J \cdot s$	0.60
Atomic mass unit (unified)	$u \equiv \frac{1}{12}m(^{12}C)$	$1.660\ 540\ 2 \times 10^{-27}$	kg	0.59
Avogadro's number	A	$6.022\ 136\ 7 \times 10^{26}$	$kmol^{-1}$	0.59
Boltzmann's constant	k	$1.380\ 658 \times 10^{-23}$	$J \cdot K^{-1}$	8.5
Gas constant	$R \equiv Ak$	8314.510	$J \cdot kmol^{-1} \cdot K^{-1}$	8.4
Stefan-Boltzmann constant	σ	$5.670\ 51 \times 10^{-8}$	$W \cdot m^{-2} \cdot K^{-4}$	34
Wien's constant	$b \equiv \lambda_{max}T$	$2.897\ 756 \times 10^{-3}$	$m \cdot K$	8.4
Elementary charge	e	$1.602\ 177\ 33 \times 10^{-19}$	C	0.30
Faraday's constant	F	$96\ 485.309 \times 10^3$	$C \cdot kmol^{-1}$	0.30
Electron mass	m_e	$9.109\ 389\ 7 \times 10^{-31}$	kg	0.59
		$5.485\ 799\ 03 \times 10^{-4}$	u	0.023
Electron charge/mass ratio	e/m_e	$1.758\ 819\ 62 \times 10^{11}$	$C \cdot kg^{-1}$	0.30
Proton mass	m_p	$1.672\ 623\ 1 \times 10^{-27}$	kg	0.59
		1.007 276 470	u	0.012
Neutron mass	m_n	$1.674\ 928\ 6 \times 10^{-27}$	kg	0.59
		1.008 664 904	u	0.014
Rydberg constant	$R_\infty \equiv \mu_0^2 c^3 m_e e^4 / 8h^3$	10 973 731.534	m^{-1}	0.0012
Bohr radius	$a_0 \equiv \mu_0 c e^2 / 8\pi h R_\infty$	$0.529\ 177\ 249 \times 10^{-10}$	m	0.045
Bohr magneton	$\mu_B \equiv e\hbar/2m_e$	$9.274\ 015\ 4 \times 10^{-24}$	$J \cdot T^{-1}$	0.34
Nuclear magneton	$\mu_N \equiv e\hbar/2m_p$	$5.050\ 786\ 6 \times 10^{-27}$	$J \cdot T^{-1}$	0.34
Electron magnetic moment	μ_e/μ_B	1.001 159 652 193	dimensionless	1×10^{-5}
Proton magnetic moment	μ_p/μ_N	2.792 847 386	dimensionless	0.023
Neutron magnetic moment	μ_n/μ_N	1.913 042 75	dimensionless	0.24

Abridged and slightly modified from E. R. Cohen and B. N. Taylor, *Rev. Mod. Phys.* **57**, 1121 (1987).

SI Base, Supplementary, and Derived Units

SI Base and Supplementary Units

Quantity	Unit	Symbol
Base Units		
length	meter	m
time	second	s
mass	kilogram	kg
electric current	ampere	A
temperature	kelvin	K
amount of substance	mole	mol
luminous intensity	candela	cd
Supplementary Units		
plane angle	radian	rad
solid angle	steradian	sr

Selected SI Derived Units

Quantity	Unit Name	Symbol	Equivalent
Mechanical Units			
frequency	hertz	Hz	s^{-1}
speed, velocity	meter per second	m/s	
acceleration	meter per second squared	m/s^2	
angular velocity	radian per second	rad/s	
angular acceleration	radian per second squared	rad/s^2	
force	newton	N	$kg \cdot m/s^2$
stress, pressure	pascal	Pa	N/m^2
work, energy, heat	joule	J	N·m
power	watt	W	J/s
momentum, impulse	newton-second	N·s	kg·m/s
angular momentum, angular impulse	newton-meter-second	N·m·s	$kg \cdot m^2/s$
entropy	joule per kelvin	J/K	$kg \cdot m^2/s$
Electromagnetic Units			
electric charge	coulomb	C	A·s
electric flux	newton-meter squared per coulomb, volt-meter	$N \cdot m^2/C$, V·m	
electric field	newton per coulomb, volt per meter	N/C, V/m	
electric potential, emf	volt	V	J/C
electric circulation	newton-meter per coulomb, volt	N·m/C, V	
resistance	ohm	Ω	V/A
conductance	siemens	S	A/V, Ω^{-1}
capacitance	farad	F	C/V
inductance	henry	H	Wb/A, V·s/A
magnetic flux	weber	Wb	V·s
magnetic field	tesla	T	Wb/m^2, A/m
magnetic circulation	tesla-meter, ampere	T·m, A	

Non-SI Units

Quantity	Name	Symbol	SI Equivalency
Non-SI Units in Use with SI			
time	minute	min	$1 \text{ min} = 60 \text{ s}$
	hour	h	$1 \text{ h} = 60 \text{ min} = 3600 \text{ s}$
	day	d	$1 \text{ d} = 24 \text{ h} = 86\ 400 \text{ s}$
	year	y	$1 \text{ y} = 365.24 \text{ d} = 31\ 556\ 736 \text{ s}$
plane angle	degree	°	$1° = \dfrac{\pi}{180} \text{ rad}$
	(arc) minute	'	$1' = \left(\dfrac{1}{60}\right)° = \dfrac{\pi}{10\ 800} \text{ rad}$
	(arc) second	"	$1'' = \left(\dfrac{1}{60}\right)' = \dfrac{\pi}{648\ 000} \text{ rad}$
area	hectare	ha	$1 \text{ ha} = 10^4 \text{ m}^2$
volume	liter	L	$1 \text{ L} = 10^{-3} \text{ m}^3$
mass	tonne	t	$1 \text{ t} = 10^3 \text{ kg}$
energy	electron-volt	eV	$1 \text{ eV} = e \times 1 \text{ V} = |e| \text{ J} \simeq 1.60 \times 10^{-19} \text{ J}$
	kilocalorie	kcal, Cal	$1 \text{ kcal} = 4186 \text{ J}$
Selected Metric Non-SI Units			
length	Ångstrom unit	Å	$1 \text{ Å} = 10^{-10} \text{ m} = 0.1 \text{ nm}$
	centimeter	cm	$1 \text{ cm} = 10^{-2} \text{ m}$
	astronomical unit	AU	$1 \text{ AU} = 149.5 \times 10^9 \text{ m}$
	light year	ly	$1 \text{ ly} = 9.460 \times 10^{15} \text{ m}$
	parsec	pc	$1 \text{ pc} = 3.084 \times 10^{16} \text{ m} = 3.26 \text{ ly}$
mass	gram	g	$1 \text{ g} = 10^{-3} \text{ kg}$
force	dyne	dyn	$1 \text{ dyn} \equiv 1 \text{ g·cm/s}^2 = 10^{-5} \text{ N}$
pressure	dyne per centimeter squared	dyn/cm²	$1 \text{ dyn/cm}^2 = 0.1 \text{ Pa}$
energy	erg	erg	$1 \text{ erg} \equiv 1 \text{ dyn·cm} = 10^{-7} \text{ J}$
magnetic field	gauss	G	$1 \text{ G} = 10^{-4} \text{ T}$
Selected Nonstandard Units			
length	inch	in.	$1 \text{ in.} \equiv 2.54 \text{ cm}$
	foot	ft	$1 \text{ ft} = 12 \text{ in.} \equiv 0.3048 \text{ m}$
	statute mile	mi	$1 \text{ mi} = 5280 \text{ ft} = 1.609 \text{ km}$
	nautical mile	nmi	$1 \text{ nmi} = \dfrac{2\pi}{360 \times 60} \text{ mean earth radius} = 1.852 \text{ km}$
mass	slug	slug	$1 \text{ slug} \equiv 14.59 \text{ kg}$
	pound mass	lb, lbm	$1 \text{ lbm} = 0.4536 \text{ kg}$
force	kilogram force	kgf	$1 \text{ kgf} \equiv g_{\text{std}} \text{ N} = 9.807 \text{ N}$
	poundal	pdl	$1 \text{ pdl} = 0.138 \text{ N}$
	pound force	lb, lbf	$1 \text{ lbf} \equiv 4.448 \text{ N}$
pressure	atmosphere	atm	$1 \text{ atm} \equiv 1.013 \times 10^5 \text{ Pa}$
	mm Hg (Torr)	Torr	$1 \text{ Torr} = \frac{1}{760} \text{ atm} = 133.3 \text{ Pa}$
	"kilogram-force" per square centimeter	kg/cm², kgf/cm²	$9.807 \times 10^4 \text{ Pa}$
	pound per square inch	psi, lb/in.²	$1 \text{ psi} = 6895 \text{ Pa}$
energy	foot-pound	ft·lb	$1 \text{ ft·lb} = 1.356 \text{ J}$
	British thermal unit	Btu	$1 \text{ Btu} = 1055 \text{ J}$
	kilowatt-hour	kWh	$1 \text{ kWh} = 3.6 \times 10^6 \text{ J}$
power	horsepower	hp	$1 \text{ hp} \equiv 550 \text{ ft·lb/s} = 745.7 \text{ W}$

Periodic Table of the Elements

Key

```
  1 ——— Atomic number
  H ——— Symbol
Hydrogen ——— Name
1.0079 ——— Atomic mass
```

Group IA	IIA	IIIB	IVB	VB	VIB	VIIB	VIII			IB	IIB	IIIA	IVA	VA	VIA	VIIA	Noble Gases VIIIA
1 1 H Hydrogen 1.0079																	2 He Helium 4.0026
2 3 Li Lithium 6.941	4 Be Beryllium 9.01218											5 B Boron 10.81	6 C Carbon 12.011	7 N Nitrogen 14.0067	8 O Oxygen 15.9994	9 F Fluoride 18.9984	10 Ne Neon 20.179
3 11 Na Sodium 22.98977	12 Mg Magnesium 24.305											13 Al Aluminum 26.9815	14 Si Silicon 28.0855	15 P Phosphorus 30.97376	16 S Sulfur 32.06	17 Cl Chloride 35.453	18 Ar Argon 39.948
4 19 K Potassium 39.0983	20 Ca Calcium 40.08	21 Sc Scandium 44.9559	22 Ti Titanium 47.88	23 V Vanadium 50.9415	24 Cr Chromium 51.996	25 Mn Manganese 54.9380	26 Fe Iron 55.847	27 Co Cobalt 58.9332	28 Ni Nickel 58.70	29 Cu Copper 63.546	30 Zn Zinc 65.38	31 Ga Gallium 69.72	32 Ge Germanium 72.59	33 As Arsenic 74.9216	34 Se Selenium 78.96	35 Br Bromine 79.904	36 Kr Krypton 83.80
5 37 Rb Rubidium 85.4678	38 Sr Strontium 87.62	39 Y Yttrium 88.9059	40 Zr Zirconium 91.22	41 Nb Niobium 92.9064	42 Mo Molybdenum 95.94	43 Tc Technetium 98.906	44 Ru Ruthenium 101.07	45 Rh Rhodium 102.9055	46 Pd Palladium 106.4	47 Ag Silver 107.868	48 Cd Cadmium 112.41	49 In Indium 114.82	50 Sn Tin 118.69	51 Sb Antimony 121.75	52 Te Tellurium 127.60	53 I Iodine 126.9045	54 Xe Xenon 131.30
6 55 Cs Cesium 132.9054	56 Ba Barium 137.33	57–71 *Rare earths	72 Hf Hafnium 178.49	73 Ta Tantalum 180.9479	74 W Tungsten 183.85	75 Re Rhenium 186.207	76 Os Osmium 190.2	77 Ir Iridium 192.22	78 Pt Platinum 195.09	79 Au Gold 196.9665	80 Hg Mercury 200.59	81 Tl Thallium 204.37	82 Pb Lead 207.2	83 Bi Bismuth 208.9804	84 Po Polonium (209)	85 At Astatine (210)	86 Rn Radon (222)
7 87 Fr Francium (223)	88 Ra Radium 226.0254	89–103 †Actinides	104 Rf Rutherfordium (261)	105 Ha Hahnium (262)	106 Sg Seaborgium (263)	107 Ns Neilsbohrium (262)	108 Hs Hassium (265)	109 Mt Meitnerium (266)	110 ‡ (269)	111 ‡		114					

→Stable region?

*Lanthanides 6

57 La Lanthanium 138.9055	58 Ce Cerium 140.12	59 Pr Praseodymium 140.9077	60 Nd Neodymium 144.24	61 Pm Promethium 145	62 Sm Samarium 150.4	63 Eu Europium 151.96	64 Gd Gadolinium 157.25	65 Tb Terbium 158.9254	66 Dy Dysprosium 162.50	67 Ho Holmium 164.9304	68 Er Erbium 167.26	69 Tm Thulium 168.9342	70 Yb Ytterbium 173.04	71 Lu Lutetium 174.967

†Actinides 7

89 Ac Actinium 227.0278	90 Th Thorium 232.0381	91 Pa Protactinium 231.0359	92 U Uranium 238.029	93 Np Neptunium 237.0482	94 Pu Plutonium (244)	95 Am Americium (243)	96 Cm Curium (247)	97 Bk Berkelium (247)	98 Cf Californium (251)	99 Es Einsteinium (254)	100 Fm Fermium (257)	101 Md Mendelevium (258)	102 No Nobelium (259)	103 Lr Lawrencium 262

Periods →

Note: The atomic mass value given is for naturally occurring proportions of isotopes. Values in parentheses are mass numbers for the most stable isotope.
‡Reported but not confirmed; no name proposed.

Answers to Odd-Numbered Problems

Answers are not given for problems requiring a lengthy explanation.

Chapter 1

1.1 (a) 3 m, (b) 0.01 m, (c) 3 m, (d) 2 m, (e) 0.1 m,
(f) 10 m, (g) 5×10^6 m, (h) 10^{-4} m. **1.3** 10^{-15} m.
1.5 10^{41} nuclear diameters. **1.7** $2 \times 10^{41} : 1$.
1.9 3×10^{-19} s; 0.3 as. **1.11** 1.891 km.
1.13 The number of people on earth is smaller by a factor of about 10^4. **1.15** (a) 1 terrestrial meter = 1.000 228 83 m.
(b) 229 parts per million. **1.17** 60 mi/h = 88 ft/s.
1.19 Neither is right; the figures given are exact.
1.21 (a) area = 0.7268 m^2 (carrying along an extra digit).
(b) "nearest wrong" area = 0.7284 m^2.
(c) uncertainty = 0.002. (d) uncertainty = 0.001.
(e) The uncertainty in the area is greater because it depends on the square of the diameter. The uncertainty in the diameter enters the calculation for the area twice. **1.23** (a) 0.051 m.
(b) The individual measurements each have four significant figures, whereas the difference between them has only two.
(c) No. The uncertainty in the height measurement process depends on other factors less reproducible than the length measurement—for example, how straight the girl stands, how much her hair is compressed by the ruler.
1.25 3×10^{50} atoms.

Chapter 2

2.1 $\Delta x = 3.1$ m $-$ 4.8 m $= -1.7$ m (points leftward).
2.3 $\Delta x = -2.2$ m $- (-3.9$ m$) = 1.7$ m (points leftward).
2.5 (a)

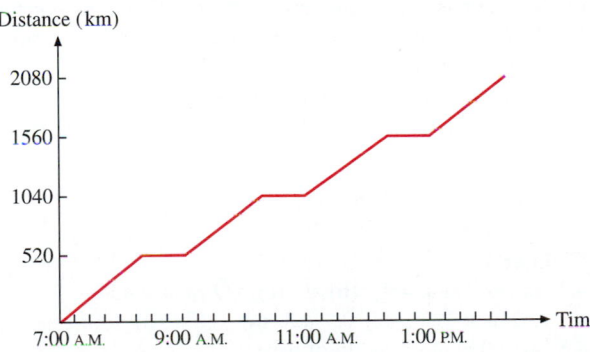

(b)

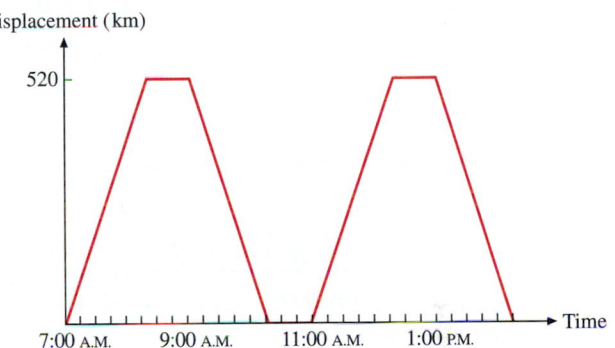

2.7 (a) For the 100-m dash, $\langle v \rangle = 10.1$ m/s. For the marathon, $\langle v \rangle = 5.45$ m/s.
(b) Acceleration to maximum speed requires a significant portion of the 100-m distance. Although most marathon runners speed up near the finish, their mean speed is determined primarily by how fast they can run over the entire distance. Thus the maximum and mean speeds are closer for the marathon.
2.9 0.016 s, which should be unnoticeable.
2.11 Mean speed is 101 km/h. The toll-taker can determine that the driver was speeding. **2.13** 66.7 km/h. **2.15** 1.90 m/s.
2.17 -1.9 m/s^2.
2.19 (a) The acceleration is constant and is -40 m/s^2.
(b) 5 m/s. (c) 0.25 s. (d) 0.32 m.
2.23 (a) Both v and a are constant. (b) Only a is constant.
(c) Neither is constant. (d) Only a is constant.
(e) Neither is constant. **2.25** (a) 14.3 s. (b) 268 m.
2.27 (a) 46 m/s. (b) 78 s. **2.29** (a) 0.049 m. (b) 22.9 m.
2.31 25.5 m. **2.33** (a) 31.9 m. (b) 6.38 s.
2.35 (a) 1.07 s, 2.09 s.
(b) 3.76 s. There is no time for which $v(t) = +21.3$ m/s.
2.37 3.68 m/s^2. **2.39** (a) $1360n$ m, where $n = 1, 2, 3, \ldots$.
(b) The answer is not unique. **2.41** (a) 12 s. (b) 333 m.

2.45 (a) $\langle v \rangle = \dfrac{n + mb}{n + m\dfrac{b}{c}}\, v_{\mathrm{w}}.$

(b) When $m = 0$, we expect $\langle v \rangle = v_{\mathrm{w}}$. When $n = 0$, we expect $\langle v \rangle = v_{\mathrm{t}} = c v_{\mathrm{w}}$. When $b = c = 1$, we expect $\langle v \rangle = v_{\mathrm{w}}$.

2.47 (a) Yes. (b) $\dfrac{c^2}{2b^2}$. **2.49** (a) 9.80 m/s^2 (downward);

(b) -1470 m/s^2 (upward); (c) zero. **2.51** (a) 7.35 m/s.
(b) Required height is 2.76 m. The student cannot juggle in his dormitory room. **2.55** 0.034 s. **2.57** $A_1/A_2 = 1.5$.
2.59 27.3 m/s. **2.61** (a) 2.19 s and 23 s.
(b) 4.8 m and 520 m.

2.63 Let the fraction of the distance of the race that Ann rides the horse be r_A, let the fraction of the distance of the race that Bob rides the horse be r_B, and let e be the ratio of the speed at which the horse can carry Ann to that at which it can carry Bob.

(a) $r_A = r_B = \dfrac{1}{2}$. **(b)** $r_A = \dfrac{1 - \dfrac{1}{c}}{1 + \dfrac{1}{b} - \dfrac{2}{c}}$, $r_B = \dfrac{\dfrac{1}{b} - \dfrac{1}{c}}{1 + \dfrac{1}{b} - \dfrac{2}{c}}$.

(c) $r_A = \dfrac{1 - \dfrac{1}{c}}{\left(1 + \dfrac{1}{b} - \dfrac{1}{c} - \dfrac{1}{e}\right)}$, $r_B = \dfrac{\dfrac{1}{b} - \dfrac{1}{c}}{\left(1 + \dfrac{1}{b} - \dfrac{1}{c} - \dfrac{1}{e}\right)}$.

2.65 (a)

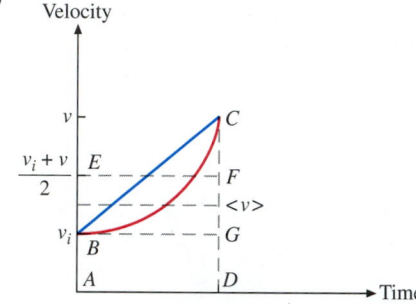

(b) The area bounded by AB, the curve, CD, and DA represents the total distance traveled.
(c) The area is less than the area bounded by rectangle $EADF$.
(d, e) The area under the curve is less than the area of the rectangle $EADF$. In other words, the distance traveled will be less than in the case in which acceleration is constant. This causes the mean velocity to be less than $(v_i + v)/2$. **2.67 (a)** 289 m; **(b)** 15.98 s. **2.69 (a)** $a/g = 0.998$.
(b) No, because a/g is so close to 1. The two bodies appear to fall with nearly the same acceleration.

Chapter 3
3.1 880 N upward. **3.3** 4.28 kg. **3.5** Yes. $F = -340$ N.
3.7 1.99 m/s. **3.9** 7300 N. **3.11** 2.0 N. **3.13** 8.00 s.
3.15 (a) 10.5 s. **(b)** 23.7 m.
3.17 (a) 49 N. **(b)** 98 000 N. **3.19 (a)** 186 N. **(b)** 1130 N.
3.21 (a) -3.09 N (upward). **(b)** -3.11 N (upward).
3.23 (a) On the earth; $F = 9.80$ N toward the mass.
(b) 1.6×10^{-24} m/s^2. **(c)** 1.7×10^{-24} m.
3.25 $\Delta x_2 = -\Delta x_1$; differentiating twice yields $a_2 = -a_1$.
3.27 3.92 m/s^2. **3.29** $g = 9.78$ m/s^2.
3.31 (a)

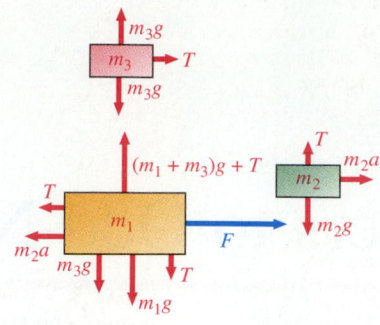

(b) $T_1 = 7200$ N, $T_2 = 4800$ N, $T_3 = 2400$ N.
3.33 $P_3 = (m_1 + m_2 + m_3)a$, $P_2 = (m_2 + m_3)a$, $P_1 = m_3a$.
3.35 (a) 3.61×10^5 N. **(b)** 0.28. **(c)** 360 m.
3.37 $a = g - A$, $a' = g$. **3.39 (a, b)** 1.76 m/s^2 upward.
3.41 735 N upward.

3.43 (a)

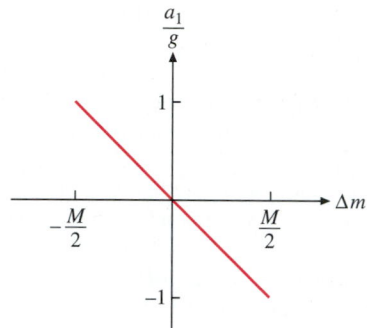

(b) $F = \dfrac{m_2}{m_3}(m_1 + m_2 + m_3)|g|$. **3.45 (a)** 2200 N.

(b) 10 900 N. **3.47 (a)** $a_1 = \dfrac{2\,\Delta m}{M}|g|$.

(b) Negative values of Δm signify that mass is taken away from m_2 and placed on m_1.

3.49 $t = t_{\text{free}}\sqrt{\dfrac{m_1 + m_2}{m_2 - m_1}}$; t is in inverse proportion to the square root of the net force $(m_2 - m_1)g$; as the propulsive force $(m_2 - m_1)g$ decreases, t increases. **3.51** $m_2 = -\dfrac{\Delta x_1}{\Delta x_2} \cdot 1$ kg.

3.53 (a) $a_1 = \dfrac{2m_2 - m_1}{4m_2 + m_1}|g|$, $a_2 = \dfrac{-4m_2 + 2m_1}{4m_2 + m_1}|g|$.

(b) $T = \dfrac{3m_1m_2}{4m_2 + m_1}|g|$.

(c) $a_1 = 0.377$ m/s^2, $a_2 = -0.754$ m/s^2, $T = 40.7$ N.
3.55 (a) $x_j' = x_j - Vt_j + $ constant. **(b)** $v_j' = v_j - V$.
(c) $a_j' = a_j$. **(d)** It depends only on the acceleration.
(e) No, there is a quadratic term in the position transformation. No; Newton's second law will be different in different frames.
3.57 (a) 25 m.
(b) 41 m/s^2 (assuming that Wonder Woman can decelerate extremely rapidly so as to match the baby's speed just before catching him).

3.59 $a_1 = \dfrac{m_1 - m_2}{m_1 + m_2}(g - A)$.

Chapter 4
4.1 (a) $\sqrt{2}$ m at 45°, **(b)** 0, **(c)** $\sqrt{2}$ m at $-45°$,
(d) $\sqrt{2}$ m at 135°, **(e)** 0. **4.3 (a)** 120°, **(b)** 120°.
4.5 (a) 60 m at 32° north of west,
(b) 94 m at 4.2° south of east. **4.7 (a)** Yes.
(b) No. The directions are different.
4.9 The components are $(-12.3$ m, 11.3 m$)$; magnitude, 16.7 m; direction, 137°. **4.11 (a)** 5.39 at $-68.2°$. **(b)** 5.39 at 111.8°.
(c) $\mathbf{D} = -\mathbf{E}$.
4.13 $(A_x, A_y) = (27.7$ km, 13.9 km$)$;
$(A_r, A_\theta) = (31.0$ km, $26.7°)$. **4.15** 252 m.

4.17 147 km/h at 61.3° north of west.
4.19 (a) 0.57°; (b) 5.7°.
(c) For $a = 0.01g$, $t = 7.8$ s; for $a = 0.1g$, $t = 2.5$ s.
4.21 (a) 390 N. (b) 290 N. **4.23** 11.5 m.
4.25 (a) $a_{max} = \mu_s g/2$. (b) 4.9 m/s². (c) $d' = d/2$.

4.27 3.2 m/s². **4.31** (a) $\mu_s = \dfrac{m_1}{m_2 + m_3}$. (b) 0.48.

4.33 $a = 6.32$ m/s² counterclockwise. **4.35** (a) 46 km.
(b) 229°. (c) 221°. **4.37** 573 km at 70° north of west.
4.39 (a) 10%, 5.7°; 20%, 11.5°; 30%, 17.5°; 40%, 23.6°.
(b) 10%, 995 m; 20%, 490 m; 30%, 318 m; 40%, 229 m.
4.41 706 N. No. The force of static friction prevents sidewise motion.
4.43 72° between 100 N and 200 N ropes; 130° between 100 N and 250 N ropes; 158° between 200 N and 250 N ropes.

4.45 (a) $\tan \alpha$. (b) $\dfrac{mg(\sin \alpha - \mu_s \cos \alpha)}{\cos \beta - \mu_s \sin \beta}$.

(c) $\dfrac{mg(\sin \alpha + \mu_s \cos \alpha)}{\cos \beta + \mu_s \sin \beta}$. (d) $\dfrac{mg(\sin \alpha + \mu_k \cos \alpha)}{\cos \beta + \mu_k \sin \beta}$.

4.47 $\mu_s mg(2n - 1)$. **4.49** (b) $T = \left(\dfrac{1 + \mu_k}{3}\right)mg$.

4.51 (a) $g[\sin \alpha - \mu_k(\sin \theta + \cos \theta) \cos \alpha]$. (b) 45°.

4.53 (a) $\beta = \tan^{-1} \mu_s$. (b) $\dfrac{\mu_s mg}{\sqrt{1 + \mu_s^2}}$. **4.55** $\dfrac{v^2}{2xg}$.

4.59 (a) $h = \left(\dfrac{3V\mu_s^2}{\pi}\right)^{1/3}$. (b) 0.88. **4.61** (a) $\beta = \tan^{-1} \mu_s$.

(b) $T = m|g|\dfrac{\mu_s \cos \alpha - \sin \alpha}{\sqrt{1 + \mu_s^2}}$.

(c) The minimizing angle is the same.

Chapter 5
5.1 20°. **5.5** 332 m. **5.7** (a) $t = 2.27$ s. (b) $x = 38.2$ m.
(c) $v = 27.9$ m/s. (d) 53.0°. **5.9** (a) $x = 84$ m, $y = 44$ m.
(b) 52 m/s at 20° above the horizontal.
5.11 (a) $\theta_1 = 1.56°$; $\theta_2 = 88.44°$. (b) 59.6 s.
5.15 (a) $Y = 2.1$ m. (b) $R = 10$ m. (c) $t = 1.3$ s.
5.17 (a) $v = 31$ m/s. (b) $a = 25$ m/s².
5.19 $\mathbf{a} = (-2.19 \times 10^{-4} \text{ m/s}^2)\hat{\mathbf{r}}$; $a_J/a_e = 3.67 \times 10^{-2}$.
5.21 (a) $\mu_s = gr/v^2$. (b) $\tau = 1.0$ s. **5.25** $\mu_s = v^2/rg$.
5.27 16 km. **5.29** $v = 71.7$ m/s.
5.33 $v_i = 14$ m/s. This is less than the speed found in Problem 5.4. **5.35** Yes; $h_{max} = 1.6$ m. **5.37** $\theta_= = 76.0°$.

5.39 $\phi(t) = \tan^{-1} \dfrac{v_i \sin \theta - \frac{1}{2}gt}{v_i \cos \theta}$. **5.41** (a) $\tau = \dfrac{2\pi r_0}{\sqrt{gr_0 \cot \theta}}$.

(b) $\tau \propto \sqrt{r}$. Thus, any decrease in r due to slipping downward will result in a tangential speed too small to hold the rider in place. **5.43** (a) $a = C/r^2$. (b) $C = gR_e^2$. (c) $a = g(R_e/r)^2$.

(d) $a = g\left[\dfrac{1}{1 + (h/R_e)}\right]^2$. (e) $\tau = 5.36 \times 10^3$ s; 0.19%.

5.45 (a) 0.34%. (b) $\tau = 1.41$ h.
(c) The calculated overall difference is 1.19%, about 40% greater than the measured difference. **5.47** $a = 2.72 \times 10^{-3}$ m/s².
5.49 (a) $v = 3.1$ m/s. (b) $\tau = 2.0$ s.
(c) The pail will be moving faster at every other point in its orbit, so τ will be less than $2\pi r/v$.

5.51 (a) $\mathbf{r}(t) = (v_i t \cos \theta)\hat{\mathbf{x}} - (v_i t \sin \theta - \frac{1}{2}gt^2)\hat{\mathbf{y}}$.
(b) The vector has magnitude $r(t) =$
$\sqrt{(v_i t \cos \theta)^2 + (v_i t \sin \theta - \frac{1}{2}gt^2)^2}$ and is tilted upward from
the positive $\hat{\mathbf{x}}$ direction by an angle $\phi = \tan^{-1} \dfrac{v_i \sin \theta - \frac{1}{2}gt}{v_i \cos \theta}$.
This is the same as the result of Problem 5.39.
(c) $\mathbf{v}(t) = (v_i \cos \theta)\hat{\mathbf{x}} + (v_i \sin \theta - gt)\hat{\mathbf{y}}$.
(d) The vector has magnitude $v(t) =$
$\sqrt{(v_i \cos \theta)^2 + (v_i \sin \theta - gt)^2}$ and is tilted upward from the
positive $\hat{\mathbf{x}}$ direction by an angle $\gamma = \tan^{-1} \left(\tan \theta - \dfrac{gt}{v_i \cos \theta}\right)$.
This agrees with the result of Problem 5.40.
5.53 (b) $\Delta t = 13.6$ s.
(c) $\theta = 42.3°$ for a championship-level throw; $\theta = 39.6°$ for a mediocre throw; both are close to the observed values.
(d) $v = 14.1$ m/s, which is not very different from the values obtained in the cruder calculations of Problems 5.33 and 5.4.
5.57 (b) $\lambda = 45°$.

Chapter 6
6.1 $x = (0.10 \text{ m}) \times \cos [(13.08 \text{ rad/s})t + \pi/3 \text{ rad}]$.
6.3 $x = (3.0 \times 10^{11} \text{ m}) \cos [(2.0 \times 10^{-7} \text{ rad/s})t]$.
6.5 $A = 0.3$ m; $\nu = 31.8$ Hz; $\tau = 0.0314$ s; $\delta = 2$ rad.
6.7 (a) $v = \beta_1 + 2\beta_2 t + 3\beta_3 t^2$; β_0 is the initial displacement of the body.
(b) $a = 2\beta_2 + 6\beta_3 t$; β_1 is the initial velocity of the body.
(c) The functions must be periodic.

6.9 (a) $v = (12.30 \text{ m/s}) \cos \left[\dfrac{2\pi t}{0.024 \text{ s}} - 4.7 \text{ rad}\right]$.

(b) $a = -(3.22 \times 10^3 \text{ m/s}^2) \sin \left[\dfrac{2\pi t}{0.024 \text{ s}} - 4.7 \text{ rad}\right]$.

(c) $A = 0.047$ m. (d) $v_{max} = 12.30$ m/s.
(e) $a_{max} = 3.22 \times 10^3$ m/s².
(f) The phase constant is significant for parts **a** and **b**.
6.11 $k = 288$ N/m. **6.13** (a) $F = 15$ N. (b) $F = 30$ N.
6.15 (b) The combined spring is stiffer than the individual springs. (c) $k_p = 565$ N/m. **6.17** $k = 2.6 \times 10^5$ N/m.
6.21 (a) $\tau = 1.02$ s; (b) $A = 0.12$ m; (c) $\delta = 0$;
(d) $v_{max} = 0.74$ m/s; (e) $a_{max} = 4.55$ m/s².
6.23 $k_1/k_2 = 1/4$. **6.25** $v = 1.95$ Hz. **6.27** $\tau = 2.84$ s.
6.29 (a) $\theta = (0.052 \text{ rad}) \cos [(3.5 \text{ rad/s})t]$.
(b) $\theta = (0.079 \text{ rad}) \cos [(3.5 \text{ rad/s})t + \pi/2 \text{ rad}]$.
(c) $\theta = (0.094 \text{ rad}) \cos [(3.5 \text{ rad/s})t + 0.98 \text{ rad}]$.
6.31 $x_2 = -29$ cm; $v_2 = -81$ cm/s.
6.33 (a) When the oscillator just reverses at the highest point, the block is first seen to leave the table. **6.37** $k' = 2k$.
6.39 (a) $y_0 = mg/k$.
(b) Yes. The release point will be located at the starting point (ending point) of the oscillation cycle. (c) $y_1 = 2mg/k$.
(d) $y = mg/k - (mg/k) \cos (\sqrt{k/m}\ t)$.

(e) $F_{max} = 2\,mg$; $F_{min} = 0$. **6.41** $\tau = 2\pi \sqrt{\dfrac{m}{k_1 + k_2}}$.

6.43 $l = \Delta$. **6.47** $\tau = 1.5$ s. **6.49** $A_{max} = 1.66 \times 10^{-2}$ m.

6.53 (a) $l = 0.32$ m. (b) $\tau = 1.14$ s.

6.55 $k_1 = \dfrac{k_p}{2}\left(1 + \sqrt{1 - 4\dfrac{k_s}{k_p}}\right)$; $k_2 = \dfrac{k_p}{2}\left(1 - \sqrt{1 - 4\dfrac{k_s}{k_p}}\right)$,

where k_1 is the stiffer of the two springs.

6.57 (a) $v = \dfrac{1}{2\pi}\sqrt{\dfrac{k}{m}}$. (b) The equilibrium lies at the center. (c) When $\Omega < \sqrt{k/m}$, the equilibrium is stable; when $\Omega = \sqrt{k/m}$, the equilibrium is neutral; when $\Omega > \sqrt{k/m}$, the equilibrium is unstable. (d) $v = \dfrac{1}{2\pi}\sqrt{\dfrac{k}{m} - \Omega^2}$.

Chapter 7

7.1 $W = 5880$ J.
7.3 (a) $F = 710.6$ N, $\theta = 6.83°$ with respect to the direction of motion. (b) $W = 3.5 \times 10^4$ J. Results should agree.
7.5 $W = 870$ J. **7.7** (a) $W = 861$ J.
(b) $s = 13.2$ m, $W = 861$ J. **7.9** $\theta = 68°$.
7.11 (a) $\hat{\mathbf{x}} \cdot \hat{\mathbf{x}} = 1$, $\hat{\mathbf{y}} \cdot \hat{\mathbf{y}} = 1$, $\hat{\mathbf{z}} \cdot \hat{\mathbf{z}} = 1$;
(b) $\hat{\mathbf{x}} \cdot \hat{\mathbf{y}} = 0$, $\hat{\mathbf{y}} \cdot \hat{\mathbf{z}} = 0$, $\hat{\mathbf{x}} \cdot \hat{\mathbf{z}} = 0$.
7.13 (a) $A = 2\sqrt{2}$, $B = \sqrt{10}$; (b) $\theta = 63°$.
7.15 $k = 246$ N/m. **7.17** $s_1/s_2 = 1.8$. **7.19** (a) $s = 3.1$ m.
(b) $W = 377$ J. (c) $W' = -267$ J. (d) $K_f = 110$ J.
7.21 $[K] = ML^2T^{-2} = [MLT^{-2}]L = [F][L] = [W]$. The SI unit of work is the joule. **7.23** $K = 1.6 \times 10^3$ J.
7.25 (a) $W = 2.25 \times 10^6$ J. (b) $d = 375$ m.
7.27 $P = 710$ W. **7.29** $P = 82.3$ kW.
7.31 $P = 49.7$ kW. **7.33** $W = \frac{1}{2}ma^2t^2$.
7.35 $W = \dfrac{1}{2}\dfrac{k_1 k_2}{k_1 + k_2}L^2$. **7.37** (a) $mg \sin\theta$.
(b) $dW = -mgl \sin\theta\, d\theta\ (d\theta < 0)$. (c) $W = mgl(1 - \cos\theta_0)$.
(d) $W = mgh$.
7.41 (a) $K(t = 0.1\text{ s}) = 4.25$ J; $K(t = 2.1\text{ s}) = 2.65 \times 10^4$ J.
(b) 13.25 kW.
(c) $P(t = 0.1\text{ s}) = -340$ W; $P(t = 2.1\text{ s}) = 26.9$ kW.
7.43 (a) 3.17 J; (b) 3.17 J; (c) 3.17 J.
7.45 $F = \dfrac{2cs}{R}\sqrt{R^2 + s^2}$. **7.47** (a) $P = mg(gt - v_i \sin\alpha)$.
(b) $\langle P \rangle = 0$.

Chapter 8

8.1 (a) $W = 200$ J. (b) $W_g = -200$ J. (c) $W_{net} = 0$.
(d) $\Delta U = 200$ J. **8.3** (a) $W = 24.0$ J. It is nonconservative.
(b) $W_s = 18.0$ J. (c) $W_m = -18.0$ J. (d) $U = 6.0$ J.
(e) $K = 18.0$ J. (f) $K' = 24.0$ J.
8.5 Mechanical advantage $= \cot\theta$. **8.7** $W_p/W_k = 2.5$.
8.9 $K = \frac{1}{2}mg^2t^2 + \frac{1}{2}mv_0^2 = 76$ J; $U = mgh - \frac{1}{2}mg^2t^2 = 86$ J;
$E = mgh + \frac{1}{2}mv_0^2 = 162$ J. **8.11** (a) $E = \frac{1}{2}mv_0^2 + mgh$.
(b) $K = \frac{1}{2}mv_0^2 + mg(h - x)$.
(c) If $x > \dfrac{v_0^2}{g} + h$, K will be negative, which is impossible.
Thus $x = H = \dfrac{v_0^2}{g} + h$ is the maximum possible height.
(e) $H = 4.59$ m. (f) $v = 8.55$ m/s.
8.13 $K = 15.9$ J (which is $[1 - (\frac{1}{2})^2]W = \frac{3}{4}W$).
8.15 (a) $K/U = \frac{1}{3}$; (b) $K/U = 1$; (c) $K/U = 3$.
8.17 (a) $K = 17$ J. (b) $l = 25.1$ m. **8.19** (a) $\mu_k = 0.19$.
(b) $W_f = -4.95$ J. **8.21** $v = 14$ m/s.
8.23 (a) $\langle P \rangle = 0.1kA^2/T$. (b) $\langle P \rangle = 1.6 \times 10^{-4}$ W.
8.27 (a) $v = \sqrt{gl/3}$. (b) $\omega = 2\sqrt{g/3l}$.
8.29 (b) $a = \dfrac{m_2 - \mu_k m_1}{m_1 + m_2}g$. **8.31** (a) $U = \frac{1}{2}mv_i^2 \sin^2\theta$.
(b) $K = \frac{1}{2}mv_i^2 \cos^2\theta$. (c) $\theta = 45°$. **8.33** $U_2/U_1 = k_1/k_2$.
8.39 (a) $h = \frac{5}{2}r$. (b) $F = mg(2 - 2\cos\theta)$.
8.41 (a) $W = \frac{1}{2}mgL$. (b) $v = \sqrt{gL}$. **8.43** $T_b - T_t = 4mg$.

8.45 (a) $\delta = 3mg/k$. (b) $\delta = 0.015$ m. **8.47** (a) $y < \frac{2}{5}l$.
(b) $\frac{2}{5}l < y < \dfrac{l}{2}$.
(c) $y > \dfrac{l}{2}$. At large displacement angles, the string will not be taut.
8.49 $x = (5.45 \times 10^{-2}\text{ m}) \cos[(2\pi/3\text{ rad/s})t - \pi/3\text{ rad}]$.
8.51 (a) $v = \sqrt{gr}$, $\Omega = \sqrt{g/r}$.
(b) $K = \frac{1}{2}mgr - mgr$; $U = mgr$; $E = \frac{3}{2}mgr$.
8.57 (a) $\Delta U_g = -mgY$. (b) $\Delta U_s = \frac{1}{2}kY^2 = mgY$.
(c) $\Delta U = \frac{1}{2}kY^2 - mgY = 0$.
(e) The system will oscillate around $Y = mg/k$ between $Y = 0$ and $Y = 2mg/k$. **8.63** (a) $T/t = 3.27$. (b) $2T/\tau = 1.29$.

Chapter 9

9.1 (a) $J = 24.5$ kg·m/s downward; (b) $p_f = 24.5$ kg·m/s;
(c) $v_f = 4.9$ m/s. **9.3** $F = 14$ N; $\theta = 135°$.
9.5 (a) $p_x = 272$ kg·m/s, $p_y = 689$ kg·m/s;
(b) $p = 740$ kg·m/s, $\theta = 68.4°$.
9.7 (a) $p = 1.8 \times 10^{29}$ kg·m/s. (b) $\Delta p = 3.6 \times 10^{29}$ kg·m/s.
(c) $\Delta p = 2.5 \times 10^{29}$ kg·m/s. **9.9** $v = 0.97$ m/s.
9.11 (a) 1.0 m/s. (b) 4.2×10^4 N. **9.13** $v = 6.1$ km/h.
9.15 $F = 35$ N. **9.17** (a) $v_f = 1.3 \times 10^3$ m/s.
(b) $F = 2.55 \times 10^5$ N. **9.19** $v_2 = 0$.
9.25 Fixing the origin at the lower left corner of the trapezoid,
$$\mathbf{R} = \left(\dfrac{2b + a}{3(b + a)}\,h,\ \dfrac{b}{2}\right).$$
9.27 The rowboat moves -1.3 m and the child moves 3.5 m with respect to the shore. **9.29** (a) $\mathbf{V} = (2.4\hat{\mathbf{x}} + 2.8\hat{\mathbf{y}})$ m/s;
(b) $\mathbf{P} = (24\hat{\mathbf{x}} + 28\hat{\mathbf{y}})$ kg·m/s.
9.31 (a) $\mathbf{V} = (3.4\hat{\mathbf{x}} - 0.31\hat{\mathbf{y}} + 1.7\hat{\mathbf{z}})$ m/s;
(b) $\mathbf{P} = (34\hat{\mathbf{x}} - 3.1\hat{\mathbf{y}} + 17\hat{\mathbf{z}})$ kg·m/s. **9.33** $m_2 = 0.39$ kg.
9.35 (a) $p_\alpha = 3.1 \times 10^{-21}$ kg·m/s; (b) $v = 8.4 \times 10^3$ m/s;
(c) $K = 1.3 \times 10^{-17}$ J. **9.37** (a) $p = 0$. (b) $\Delta t = 1.22$ s.
9.39 $0 \le t \le 2$ s: (a) $p = t^2$ kg·m/s; (b) $v = \frac{1}{3}t^2$ m/s.
$2 \le t \le 5$ s: (a) $p = 4(t - 1)$ kg·m/s; (b) $v = \frac{4}{3}(t - 1)$ m/s.
$5 \le t \le 9$ s: (a) $p = -\frac{1}{2}(t^2 - 18t + 33)$ kg·m/s;
(b) $v = -\frac{1}{6}(t^2 - 18t + 33)$ m/s.
9.41 (a) $F(t) = -m\omega^2 A \cos(\omega t)$, $J(t) = -m\omega A \sin(\omega t)$.
(b) $v(t) = -\omega A \sin(\omega t)$, $p(t) = -m\omega A \sin(\omega t)$;
(c) $J(t) = \Delta p(t)$. **9.45** $v_b = 690$ m/s.
9.47 $F = mg + \Delta p/\Delta t = \sigma(gt + \sqrt{2gh})$. When he stops pouring, the reading will decrease to $\sigma g t_f$.
9.49 (a) $\theta = \tan^{-1}\left(\dfrac{v_{ex}}{v_0}\ln\dfrac{M}{M - m}\right)$.
(b) $v' = \sqrt{v_{ex}^2\left(\ln\dfrac{M}{M - m}\right)^2 + v_0^2}$.
9.51 $\mathbf{R} = (0, 2R/\pi)$, with the origin taken at the center of curvature.
9.53 The center of mass is on the cone axis at $h/3$, measured from the base.
9.57 $v = R\sqrt{\dfrac{g}{2(h - R)}}$.
9.61 $4R/3\pi$ from the center of the semicircle, along the symmetry axis. **9.63** (a) $F = 1.25 \times 10^4$ N.
(b) It helps propel the plow. **9.65** (a) $F = \sigma v$.

(b) No, it does not violate Newton's second law, because, for constant v, $F = \dfrac{dp}{dt} = \dfrac{d(mv)}{dt} = v\dfrac{dm}{dt}$. Because $dm/dt = \sigma = $ constant, $F \propto v$.

9.67 **(a)** There is no external force acting on the system; therefore, $v = $ constant $= 0$.

(b) $x_1 = -\dfrac{m_2}{m_1 + m_2}L$, $x_2 = \dfrac{m_1}{m_1 + m_2}L$.

(c) The effective length of the spring to which m_1 attached is $\dfrac{m_2}{m_1 + m_2}L$, and that of the one to which m_2 attached is $\dfrac{m_1}{m_1 + m_2}L$. **(d)** $k_1 = \dfrac{m_1 + m_2}{m_2}k$, and $k_2 = \dfrac{m_1 + m_2}{m_1}k$.

(e) $\omega = \sqrt{k/\mu}$ for both.

(f) $\omega_1 = \sqrt{k/m_1}$, $\omega_2 = \sqrt{k/m_2}$, $\omega = \sqrt{\omega_1^2 + \omega_2^2}$.

Chapter 10

10.1 $0.5K_i$. **10.3** 7 J. **10.5** **(a)** $p_1 + p_2 = 0$;

(b) $K_2/K_1 = m_1/m_2$. **10.7** $v_2 = -(m_1/m_2)v_1$. **10.9** 13°.

10.11 $h_f = \tfrac{1}{4}h_i$. **10.13** $v_{1f} = 3.3$ m/s, $v_{2f} = 15.3$ m/s.

10.15 **(a)** $A = \dfrac{2m_1}{m_1 + m_2}\sqrt{\dfrac{m_2}{k}}\,v_{1i}$. **(b)** $A = 5.3$ cm.

10.17 $\epsilon = \sqrt{h_2/h_1} = 0.9$. The first bounce: $h_2 = 3.18$ m; the second bounce: $h_3 = 1.84$ m.

10.19 The collision is not elastic because $38° + 33° \neq 90°$.

10.21 $p_{1f} = 0.250$ kg·m/s, $p_{2f} = 0.714$ kg·m/s. The collision is inelastic. $\Delta K/K_i = 0.19$.

10.23 **(a)** $h_1 = h_i/9$; $h_2 = 16h_i/9$.

(b) The second collision will be at the same place as the first one. **(c)** $v_{1i} = \dfrac{\sqrt{2}}{3}\sqrt{gh_i}$, $v_{2i} = \dfrac{4\sqrt{2}}{3}\sqrt{gh_i}$.

(d) $h_1' = h_i$, to left side; $h_2' = 0$. **10.25** **(a)** $\Delta t_{j+1}/\Delta t_j = \epsilon$.

(c) The converging infinite series fails to describe the system accurately when the height of the bounce becomes comparable to the distance through which the ball distorts when in contact with the floor. Indeed, the bounces become smaller than the distortion distance and cease to be bounces when the ball no longer leaves the floor. **10.27** **(a)** 1.07.

(b) The collision *increases* the kinetic energy of the athlete.

(c) $\epsilon > 1$. **10.31** **(b)** $\alpha = 1$.

10.33 **(a)** $v_f = 0.83$ m/s, along Bob's original direction.

(b) $K = 1010$ J. **(c)** $\Delta E/E_i = 0.96$.

10.35 **(a)** $v_f = 1.82$ m/s at $-42.5°$, where Bob's initial velocity is at $0°$ and Alice's initial velocity is at $-135°$. **(b)** $K = 865$ J.

(c) $\Delta E/E_i = 0.82$. **10.37** $\Delta \mathbf{P} = 0$.

10.39 $x(t) = \dfrac{2m_1}{m_1 + m_2}\sqrt{\dfrac{m_2}{k}}\,v_{1i}\sin\left(\sqrt{\dfrac{k}{m_2}}\,t\right)$.

10.41 **(1)** Body 1 hits body 2 under the canopy, which Equations 10.16a and 10.16b describe.

(2) Body 1 misses body 2; that is, body 1 just passes under the canopy. Because the two pucks are identical, the experiment cannot distinguish between these two possibilities.

Chapter 11

11.1 $R = 0.17$ m. **11.3** **(a)** $\omega_e = 1.99 \times 10^{-7}$ rad/s;

(b) $\omega_m = 2.42 \times 10^{-6}$ rad/s. **11.5** **(a)** $\Delta s = 25.1$ cm.

(b) $\Delta R = 24.0$ cm.

(c) $\dfrac{\Delta s - \Delta R}{\Delta s} = 4.4\%$; the chord is shorter than the arc.

(d) $\Delta s = 0.42000$ cm; $\Delta R = 0.41998$ cm; $\dfrac{\Delta s - \Delta R}{\Delta s} = $ 0.0013%. **11.7** $t = 1.4$ s.

11.11 $\alpha = 3.6$ rad/s²; $\omega = 14.4$ rad/s. **11.13** **(a)** $I_A = 2ma^2$;

(b) $I_B = 4ma^2$; **(c)** $I_{CC'} = 2ma^2$; **(d)** $I_{DD'} = ma^2$.

11.17 **(a)** $I_A = \tfrac{1}{6}Ma^2$; **(b)** $I_B = \tfrac{5}{12}Ma^2$; **(c)** $I_C = \tfrac{2}{3}Ma^2$.

11.21 $I = 9.42 \times 10^{37} > I_{given}$. The density of the earth is not uniform; it must increase toward the center. **11.23** 7.0 m/s.

11.25 **(a)** $v = 3.6$ m/s, $\omega = 258$ rad/s;

(b) $a = 6.5$ m/s², $\alpha = 463$ rad/s²; **(c)** $T = 13.6$ N;

(d) $P(t = 2.3$ s$) = 1.2$ kW.

11.27 Low gear makes it possible to operate the engine at a speed at which its torque is significant even though the vehicle is moving slowly. As the vehicle speeds up, the engine speed exceeds that at which the torque is maximum. By shifting to a higher gear, we make it possible to operate the engine at a lower speed, thus optimizing the torque. As the vehicle continues to speed up, we shift again to successively higher gears. When the vehicle reaches constant cruising speed, little torque is required to maintain the speed, and we use a high gear to allow the engine to run slowly and quietly.

11.29 **(a)** $Mg\dfrac{L}{2}\sin\theta$. **(b)** $\alpha = \dfrac{3g}{2L}\sin\theta$.

(d) If the end of a chimney is to continue to accelerate at a rate greater than g, a downward force must be applied to it by the rest of the chimney. This force implies an upward twist on the chimney and thus a tensile force on its lower side. Therefore the chimney snaps, and the outer end falls straight down instead of continuing its circular path. **11.31** $T = mr\omega^2$.

11.33 $a_r = a_t^2 t^2/r$. **11.37** $T = 94.2$ N·m.

11.41 **(a)** $I = 0.125$ kg·m². **(b)** $I = 0.127$ kg·m². **(c)** 2%.

11.43 **(a)** $dV = 4\pi r^2\, dr$. **(b)** $dM = \dfrac{3M}{R^3}r^2\, dr$.

(c) $dI = \dfrac{2M}{R^3}r^4\, dr$. **(d)** $I = \tfrac{2}{5}Mr^2$. **11.45** **(a)** $I_0 = \tfrac{13}{24}mR^2$. **(b)** $I_P = \tfrac{23}{24}mR^2$, where m is the mass of the object shown in the figure.

11.47 The kinetic energy of the earth's rotation about its axis is 2.1×10^{29} J. The kinetic energy of its revolution about the sun is 2.7×10^{33} J. Thus $K_{rev}/K_{rot} = 1.2 \times 10^4$.

11.49 $h = 0.04$ m.

11.51 $\omega = \dfrac{x}{r}\sqrt{\dfrac{2mg}{(M + 2m)l}}$.

11.53 **(a)** $K_t = \tfrac{1}{2}Mv^2$, $K_r = 4 \times \tfrac{1}{2}mv^2$; therefore, $\dfrac{K_r}{K_t} = \dfrac{4m}{M}$.

(b) $M \simeq 1600$ kg; $m = 25$ kg. So $\dfrac{K_r}{K_t} = \dfrac{100}{1600} = 6\%$; the additional braking capacity is not significant.

11.55 **(a)** $|a| = \left|\dfrac{(m_2 - m_1)R^2}{I + (m_2 + m_1)R^2}\right|g = \left|\dfrac{m_2 - m_1}{cM + m_2 + m_1}\right|g \rightarrow$

$\left|\dfrac{m_2 - m_1}{m_2 + m_1}\right|g$ if $I \rightarrow 0$. **(b)** $|a| = \dfrac{2}{3}\left|\dfrac{m_2 - m_1}{m_2 + m_1}\right|g$.

(c) See Section 11.4. **11.57** **(a)** $\tfrac{1}{2}\kappa\theta^2$. **(b)** 1.49×10^{-7} J.

11.61 **(a)** $\Delta U(\theta) = Mg\dfrac{L}{2}(1 - \cos\theta)$.

(b) $K_r = Mg\dfrac{L}{2}(1 - \cos\theta)$.

(c) $\omega = \sqrt{\dfrac{2K}{I}} = \sqrt{\dfrac{3g}{L}(1 - \cos\theta)}$. **11.63** $I = \tfrac{3}{10}MR^2$.

11.65 (a) $M = \frac{4}{3}\rho_0 \pi R^3 - \gamma \pi R^4$. (b) $I = \frac{8}{3}\pi\left(\dfrac{R^5}{5}\rho_0 - \dfrac{R^6}{6}\gamma\right)$.

(c) $\rho_0 = 14 \times 10^3$ kg/m^3, $\gamma = 1.8 \times 10^{-3}$ kg/m^4.
(d) $\rho_R = 2.7 \times 10^3$ kg/m^3.
(e) First, the earth is not homogeneous; denser materials are concentrated toward the center. Second, under pressure, materials undergo phase changes that increase their density. Both of these effects result in a more-rapid-than-linear increase of density with depth. The goodness of the fit is due to the relatively small contribution that inner shells make to the moment of inertia.

Chapter 12
12.1 5.6 m from the heavy end (1.4 m from the light end).
12.3 At the 20-cm mark. **12.5** (a) 1.48×10^3 N;
(b) 1.1×10^3 N along the strut toward the weight.
12.7 37.8 N and -37.8 N. **12.9** 5.35 m. **12.11** 1.325 s.
12.13 1.16×10^{-4} N·m/rad. **12.15** 3.10 cm.

12.17 (a) $\tau = 2\pi\sqrt{\dfrac{R_2{}^2 + 3R_1{}^2}{2gR_1}}$. (b) $\tau = 1.06$ s.

12.19 (a) $l = z\left(1 + \dfrac{1}{12z^2}\right)$. (b) $\gamma = z\sqrt{1 + \dfrac{1}{12z^2}}$.

(d) For z large, $\tau \to 2\pi\sqrt{z/g}$.

12.21 $\nu = \dfrac{1}{2\pi}\sqrt{\dfrac{2gr}{R^2 + 2r^2}}$. **12.23** (a) 2.34 N; (b) 0.668 N.

12.25 $F_\xi = \dfrac{mg}{\cos \xi + \sin \xi \cot \eta}$; $F_\eta = \dfrac{mg}{\cos \eta + \sin \eta \cot \xi}$.

12.27 Front: vertical, $0.318mg$; horizontal, $0.111mg$. Rear: vertical, $0.182mg$; horizontal, $0.0637mg$.
12.29 Half the length of the ladder.

12.33 $\omega = -\dfrac{2\pi}{\tau}\theta_0 \sin\left(\dfrac{2\pi}{\tau}t + \delta\right)$,

$\alpha = -\dfrac{4\pi^2}{\tau^2}\theta_0 \cos\left(\dfrac{2\pi}{\tau}t + \delta\right)$. **12.35** $\nu = \dfrac{1}{\pi}\sqrt{\dfrac{k}{2M}}$.

12.37 (a) 0.211 m from the end. (b) $\tau = 1.53$ s.

12.39 (a) $\dfrac{X}{Y} = 2$. (b) $2\pi\sqrt{\dfrac{2X}{3g}}$.

12.47 (a) For $\mathbf{F}_1$ and $\mathbf{F}_2$, the Yo-Yo will roll to the left; for $\mathbf{F}_4$, it will roll to the right. For $\mathbf{F}_3$, it will not roll.

(c) $\dfrac{a_1}{a_2} = \dfrac{R_2 + R_1}{R_1}$.

(d) For $\hat{\mathbf{F}}_1$, $v = v_0\dfrac{R_2}{R_1 + R_2}$; for $\hat{\mathbf{F}}_2$, $v = v_0\dfrac{R_1}{R_2}$; for $\hat{\mathbf{F}}_4$,

$v = v_0\left(\dfrac{R_2}{R_1 - R_2}\right)$. **12.49** (a) $\frac{1}{3}mg(\mu_k \cos \theta + \sin \theta)$.

12.53 (d) $\rho = 1$: $F = mg/\sin \theta$ and $F_{\min} = mg$; $\rho = \infty$: $F = F_{\min} = 0$; $\rho < 1$: $F = \dfrac{Mg}{\sin \theta}$ and $F_{\min} = Mg$. If $\theta >$

$\sin^{-1}\left(\dfrac{1}{\rho} - 1\right)$, the result in **a** is valid. If $\theta \le \sin^{-1}\left(\dfrac{1}{\rho} - 1\right)$, the wheel cannot go over the curb.

12.55 (a) $N_f = 0.447mg$, $N_r = 0.553mg$.
(b) If you assume that both N_r and N_f are upward, you obtain $N_f = 1.09mg$ and $N_r = -0.09mg$. The negative value of N_r means that the rear wheel lifts off the pavement and the rider begins to tumble over the front wheel. Because N_f cannot be greater than mg, $a = \mu g = 5.88$ m/s^2. (c) $N_r = 0.35mg$, $N_f = 0.65mg$, $a = \mu N_r g = 2.06$ m/s^2. (d) Applying the front brake is more effective in stopping but obviously dangerous. Applying the rear brake is safe but less effective.

12.57 (b) $\dfrac{\tau_1{}^2}{4\pi^2} = \dfrac{\gamma_0{}^2 + D_1{}^2}{gD_1}$, $\dfrac{\tau_2{}^2}{4\pi^2} = \dfrac{\gamma_0{}^2 + D_2{}^2}{gD_2}$.

(d) On account of the asymmetry, $D_1 - D_2$ is not small and $\tau_1{}^2 - \tau_2{}^2$ is very small; therefore, the relatively low precision does not degrade the precision of the value of g.

12.59 $a = \dfrac{F(R_1 + R_2 \cos \theta)}{(1 + c)mR_2}$; for the case $\mathbf{F} = \mathbf{F}_3$, $a = 0$.

12.61 (a) $\theta = \cos^{-1}\dfrac{R_1}{cR_2}$.

(b) No. The condition for existence of such a value of θ is $c \ge R_1/R_2$ or $\gamma \ge \sqrt{R_1 R_2}$; that is, the radius of gyration must lie farther out than the geometric average of R_1 and R_2. If, for example, $I = \frac{1}{2}MR_2{}^2$, then the condition $R_1 \le \frac{1}{2}R_2$ must be met.
12.63 (d) $\Delta K = \frac{2}{3}(\frac{1}{4}mR^2\omega_0{}^2)$. (e) For the sphere, the speed v will be less than that of the disk; for the hoop, the speed v will be greater than that of the disk.

Chapter 13
13.1 $L = 2.9 \times 10^3$ kg·m^2/s. **13.3** (a) $\mathbf{L} = (42.6$ kg·m/s)$\hat{\mathbf{z}}$;
(b) $\mathbf{L} = (-54.5$ kg·m/s)$\hat{\mathbf{z}}$. **13.5** $L = 8 \times 10^{-3}$ kg·m^2/s.
13.7 $L = 38.2$ kg·m^2/s. **13.9** (a) $L_j/L_s = 17.4$.
(b) The sun: $K_j/K_s = 0.12$. **13.11** $G = 5.2$ kg·m^2/s.
13.13 (a) $\omega_f = 17.5$ rev/s.
(b) The centripetal force required to keep the student on the stool is proportional to ω^2. If he is even slightly unbalanced, he is in unstable equilibrium and the centripetal force needed to keep him seated will increase rapidly. At 17.5 rev/s, he will be unable to rebalance himself rapidly enough. **13.15** (a) $\mathbf{A} \times \mathbf{B} = -\hat{\mathbf{x}}$;
(b) $\mathbf{A} \times \mathbf{B} = \hat{\mathbf{x}}$. **13.17** (a) $\mathbf{A} \times \mathbf{B} = \hat{\mathbf{z}}$; (b) $\mathbf{A} \times \mathbf{B} = -\hat{\mathbf{y}}$;
(c) $\mathbf{A} \times \mathbf{B} = \hat{\mathbf{x}} - \hat{\mathbf{y}} + \hat{\mathbf{z}}$. **13.19** (a) $\mathbf{A} \times \mathbf{B} = -10\hat{\mathbf{z}}$;
(b) $\mathbf{A} \times \mathbf{B} = 12\hat{\mathbf{z}}$; (c) $\mathbf{A} \times \mathbf{B} = (6, 0, -2)$;
(d) $\mathbf{A} \times \mathbf{B} = (-4, 8, -4)$. **13.21** $\mathbf{T} = (-1, -2, -2)$ N·m.
13.23 The angle between $\mathbf{A}$ and $\mathbf{B}$ is 45°.
13.27 $\Omega = 2.14$ rad/s. **13.29** $v = 2.8 \times 10^{10}$ m/s $> c$.
13.31 (a) $W = 0$, $K_2 = K_1$.

(b) $v_r = \sqrt{\dfrac{3}{2}K_1\dfrac{r_2{}^2 - r_1{}^2}{mr_2{}^2}}$; $v_t = \sqrt{\dfrac{3}{8}K_1\dfrac{r_2{}^2 + 3r_1{}^2}{mr_2{}^2}}$.

(c) $K_2 = \dfrac{r_2{}^2 + 3r_1{}^2}{4r_2{}^2}K_1$. **13.33** (a) $\omega_f = \left(1 + \dfrac{2m}{M}\right)\omega_0$.

(b) $W = \frac{1}{2}m\left(1 + \dfrac{2m}{M}\right)R^2\omega_0{}^2$. **13.37** (a) If $\mathbf{A} \cdot \mathbf{B}$ is carried out first, the result is a scalar that cannot take part in a cross product.
13.39 (a) $\mathbf{L} = \mathbf{r} \times \mathbf{p}$.

(b) $\dfrac{d\mathbf{L}}{dt} = \dfrac{d\mathbf{r}}{dt} \times \mathbf{p} + \mathbf{r} \times \dfrac{d\mathbf{p}}{dt}$, where $\mathbf{v} = \dfrac{d\mathbf{r}}{dt} \parallel \mathbf{p}$, so $\dfrac{d\mathbf{r}}{dt} \times \mathbf{p} = 0$.

The momentum $\mathbf{p}$ of a free particle is a constant ($d\mathbf{p}/dt = 0$); therefore $d\mathbf{L}/dt = 0$ and $\mathbf{L}$ does not change with time.

13.41 (a) $L_i = mr_i^2\omega_i$, $K_i = \frac{1}{2}mr_i^2\omega_i^2$. (b) Yes.

(c) $L_f = L_i$, $\omega_f = \left(\dfrac{r_i}{r_f}\right)^2\omega_i$, $K_f = \frac{1}{2}m\dfrac{r_i^4}{r_f^2}\omega_i^2$.

(d) $F(r) = mr\omega^2$, $W = \frac{1}{2}mr_i^4\omega_i^2\left(\dfrac{1}{r_f^2} - \dfrac{1}{r_i^2}\right)$.

(e) $K_f = K_i + W = \frac{1}{2}m\dfrac{r_i^4}{r_f^2}\omega_i^2$.

13.47 (b) Among bodies of uniform radius, the solid sphere has the smallest value of c; therefore L'/L is greatest for a solid sphere.

13.49 (a) $L_s = MgR(\sin\theta)t$.

(c) With respect to this particular reference point, the force of friction produces no torque and so the total angular momentum of the sliding body and that of the rolling body are the same.

13.51 (a) $\mathbf{L'} = \mathbf{L} - \mathbf{R} \times \mathbf{p}$. (b) When $\mathbf{R} \parallel \mathbf{p}$, $\mathbf{L'} = \mathbf{L}$.

13.57 (b) $\omega = \dfrac{7v_0\cos\theta}{4R}$, the rotation is clockwise.

Chapter 14

14.3 16.7 N. **14.5** 4.77×10^7 m from the center of the moon.

14.7 (a) $F_{\text{Saturn}} = 8.5 \times 10^{-8}$ N.

(b) $F_{\text{obstetrician}} = 6.7 \times 10^{-8}$ N.

(c) $F_{\text{Saturn}} : F_{\text{obstetrician}} = 1.3:1$. No. **14.9** $h = (\sqrt{2} - 1)R$.

14.11 $M_{\text{Mars}} = 6.41 \times 10^{23}$ kg, $M_{\text{Jupiter}} = 1.90 \times 10^{27}$ kg, $M_{\text{Saturn}} = 5.74 \times 10^{26}$ kg. **14.13.** $\tau = 248$ y.

14.15 $\tau = 3.3$ y. **14.17** (a) $\tau = 2.76 \times 10^4$ s.

(b) From west to east; 0.24 martian day.

14.19 (a) Impossible, because the orbit radius for a 1-h orbit would have to be 5.09×10^6 m, which is less than the radius of the earth.

(b) Possible. The orbit radius for a 2-h orbit is 8.06×10^6 m.

14.21 $r = 4.22 \times 10^7$ m.

14.23 5.07×10^3 s, or about 85 min.

14.25 (a) $E = 9 \times 10^{-50}$ J. (b) 3×10^{-37}.

14.27 $U = -2.7 \times 10^{-5}$ J. **14.29** About $\frac{1}{2}$% more work.

14.31 (a) $M = 7.32 \times 10^{22}$ kg; (b) $a_m = 1.61$ m/s^2.

14.33 $\tau = 2\pi\sqrt{r^3/GM}$. **14.35** $v - 1.23 \times 10^4$ m/s.

14.37 $v = 1.38 \times 10^4$ m/s. **14.39** 0.49 AU.

14.41 (a) $\dfrac{da}{dr} = -\dfrac{2Gm}{(r + R)^3}$. (b) $a_c = \dfrac{Gm}{R^2}$.

(c) $a_n = \dfrac{Gm}{(R - r_e)^2}$; $a_f = \dfrac{Gm}{(R + r_e)^2}$.

(e) Both Δa_1 and Δa_2 imply high tides. Any point on the earth crosses the earth–moon line about twice a day, and one of the values Δa_1 and Δa_2 is pertinent. (g) $\Delta r = 0.52$ m.

14.45 (b) $\iota = \lambda$.

(c) The satellite can either be launched from a higher latitude or be given a northward or southward component of velocity.

(d) The orbit follows a great circle, and the angle of inclination is the highest latitude that the satellite reaches. If the satellite is launched southeastwardly from the northern hemisphere, then $\iota > \lambda$.

14.47 (a) $t = 3\pi\Delta r\sqrt{r/GM}$. (b) $t = 3.35 \times 10^6$ s = 38 d.

14.49 $\left(\dfrac{\tau_1}{\tau_2}\right)^2 = \left(\dfrac{r_1}{r_2}\right)^4$. **14.51** $r = 1.74 \times 10^6$ m.

14.53 (b) The total potential energy is the physically important quantity. Any assignment of potential energies to individual particles is valid as long as the sum is the total potential energy described in part **a**. **14.55** $v_{\text{min}} = 2330$ m/s.

14.63 There is no gravitational force inside the shell.

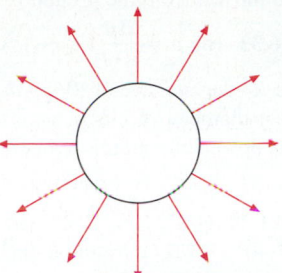

Chapter 15

15.1 (a) $\sigma = 1.9 \times 10^7$ Pa, (b) $\epsilon_p = 1.9 \times 10^{-4}$,

(c) $\Delta l = 3.1 \times 10^{-4}$ m. **15.3** $\epsilon_{\text{brass}} = 7 \times 10^{-5}$.

15.5 $\Delta w = 1.4 \times 10^{-6}$ m; $\Delta t = 1.0 \times 10^{-6}$ m.

15.7 $F = 2.2 \times 10^5$ N. **15.9** $p = 7.4 \times 10^7$ Pa.

15.11 (a) $\Delta l_1 = 1.6$ m; (b) $W_1 = W_2 = 1.8 \times 10^3$ J.

15.13 $h = 10.34$ m. **15.15** $F = 2.5 \times 10^3$ N.

15.17 (a) $F = 5.9 \times 10^{-2}$ N downward.

(b) $F = 0.15$ N upward. **15.19** $\rho = 2.70 \times 10^3$ kg/m^3.

15.21 (a) 50.4 N. (b) Yes.

15.25 (a) $\Delta l = 0.53$ cm (l_0 is the length of air column in the closed end).

(b) The cross-sectional area of the tube is uniform and cancels out. **15.27** $h = 6.7 \times 10^{-2}$ m. **15.29** (b) $\epsilon = \dfrac{\rho\omega^2}{2Y}(R^2 - r^2)$.

(e) $\omega_{\text{max}} = 3.49 \times 10^4$ rev/min $= 3.66 \times 10^3$ rad/s.

(f) $U_{\text{elastic}} = 230$ J, while $U_{\text{rot}} = 2.8 \times 10^4$ J.

15.33 $\rho = 0.816 \times 10^3$ kg/m^3. **15.35** Stable.

15.37 $h = 73.56$ cm. **15.39** 50%. **15.43** (a) $F = \dfrac{1}{2\sqrt{2}}AY\beta$.

(b) $\dfrac{F}{m} = \dfrac{1}{4}\dfrac{Y\beta}{a\rho}$. (c) $\dfrac{F'}{m'} = \dfrac{\beta G}{a\rho}$. (d) $\dfrac{F/m}{F'/m'} = \dfrac{Y}{4G} = 0.633$.

15.47 (b) $T = \dfrac{\pi G\theta}{2l}R^4$. (c) $G = \dfrac{8\pi lI}{R^4\tau^2}$. **15.49** (a) $p = \frac{1}{2}\rho gR$,

(b) $p = 1.73 \times 10^{11}$ Pa. **15.51** (a) $a(r) = \frac{4}{3}\pi G\rho_0 r - G\pi\gamma r^2$.

(c) $p_0 = 3.40 \times 10^{11}$ Pa; $\dfrac{3.62 - 3.40}{3.40} = 6.1\% < 7\%$.

(d) The matter of which the earth is made is not uniform, and its density does not increase with depth only on account of increasing pressure. Denser materials tend to concentrate toward the core, and, in addition, pressure-induced phase changes tend to increase density with depth.

Chapter 16

16.1 0.19 m/s. **16.3** 2.6 cm^2. **16.5** 1.7×10^{-4} m/s.

16.7 (a) $v_2 = v_1 = 2.4$ m/s. (b) $h = \dfrac{v_1^2}{2g} = 0.29$ m/s.

(c) They are equal. **16.9** (a) h. (b) 0.

16.11 If we neglect the buoyancy of the glycerin, $v_t = 3.26 \times 10^{-3}$ m/s. Taking the density of glycerin to be 1.26×10^3 kg/m^3, $v_t = 2.80 \times 10^{-3}$ m/s. **16.13** $\eta = 0.254$ Pa·s.

16.15 (a) $v = \sqrt{2gh_2}$;

(b) $p_1 = p_a$, $p_2 = p_a - \rho gh_2$, $p_3 = p_a - \rho g(h_1 + h_2)$,

$p_4 = p_a - \rho g(h_1 + h_2)$, $p_5 = p_a - \rho gh_2$. **16.17** $\dfrac{dV}{dt} = A\sqrt{2gh}$.

16.19 $t' = t/2$. **16.21** $W = \rho V^3/2A^2t^2$.

16.23 1.93×10^{-3} m^3/s. **16.27** 1.4 m/s. **16.29** (b) 105 W.

16.31 (e) At $h = 0$, $\theta = 45°$, R has a maximum value of $2H$. The range of a ball thrown from the ground is also maximized when $\theta = 45°$.   **16.33** (b) $v = \dfrac{\Delta p}{4\eta l}\,(R^2 - r^2)$.

(d) Because of the R^4 dependence of dV by dt, a very small variation in capillary diameter results in a substantial change in blood flow and this powerfully affects temperature regulation.

Chapter 17

17.1 (a) 10°C; (b) 55–60°C; (c) 22°C; (d) 40°C; (e) 175°C; (f) 15°C.   **17.3** $-40°$.   **17.5** $V = 48.66$ cm^3.
17.7 $\rho_{\text{summer}}/\rho_{\text{winter}} = 84\%$.   **17.9** No, 0.16%.
17.11 97.2 cm^3.   **17.13** $T = 58°C; T = -28°C$.
17.15 $d = 2.51$ cm.   **17.17** $t = 52.8°C$.   **17.19** $t = 16.6°C$.
17.21 $t = 850°C$. Because of the impossibility of thoroughly heating the end of the nail held in the pliers and the heat loss to air in the experiment.   **17.23** 70.94 kcal/kg.
17.25 0.581 kcal/kg·K.
17.27 (a) Internal friction in the turbulent water at the bottom.
(b)

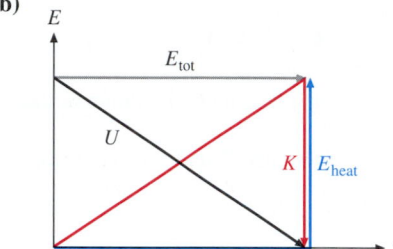

(c) $\Delta t = 0.59°C$.   **17.31** $\alpha_B = \dfrac{\Delta l}{l\,\Delta t} + \alpha_A$.

17.37 (c) 0.0432 s.   **17.39** (a) $\Delta l = 0.380$ m (no change).

17.41 (a) $\delta = \dfrac{l}{2}\sqrt{2\alpha(t - t_0)}$.   (b) $\delta = 3.1$ cm.

17.43 (a) $\gamma = \cot^{-1}\dfrac{x\alpha\,\Delta t}{y}$.

(b) $\sigma = G\beta = G\left(\dfrac{\pi}{2} - \gamma\right) = G\tan^{-1}\dfrac{x\alpha\,\Delta t}{y}$.

(c) $\Delta t = 34.6$ C°; $\sigma = 8.0 \times 10^7$ Pa.
17.45 T_{poker} depends on estimates; 500°C–800°C is plausible.
17.47 $Q = \frac{1}{4}a(T_f^4 - T_i^4)m$.   **17.49** 4 tablespoons coffee.
17.51 (a) $J = gnl/(\Delta t\, c)$.
(b) $J = 4.8 \times 10^3$ J/kcal; owing to the loss of heat, $\Delta Q = c\,\Delta t$ will be too small and J therefore too large.
17.53 (a) 5.44×10^4 W.   (b) $\Delta t = 17.3$ C°.
17.55 (a) $(h' - h)/h = 0.727\%$.
(b) α_{glass} is the relevant parameter, since the cross-sectional area is irrelevant. $h'' = h(1 + \beta_{\text{Hg}}t)(1 - \alpha_{\text{glass}}t)$. Since $\beta_{\text{glass}} = 2.6 \times 10^{-5}$ K^{-1}, $\alpha = \beta/3 \approx 0.9 \times 10^{-5}$. Even for $t = 40°C$, $\alpha_{\text{glass}}t = 3.6 \times 10^{-4}$, which is not enough to affect the reading to 1 part in 10^3.

17.59 (b) $t_1 = t_0 + \dfrac{2d}{(\alpha_{\text{Cu}} - \alpha_{\text{Fe}})l_0}\tan^{-1}\left(\dfrac{D_0}{l_0}\right)$.   (c) 0.25 mm.

Chapter 18

18.1 $\langle\epsilon\rangle = 6.21 \times 10^{-21}$ J at 300 K; $\langle\epsilon\rangle = 2.07 \times 10^{-20}$ J at 1000 K.   **18.3** $M = 1.13$ kg.
18.5 $V = 19.7$ m^3.   **18.7** $m = 7.5$ kg.   **18.9** 11.9.
18.11 $n = 0.60$ kmol.   **18.13** $\langle m\rangle = 28.8$ u.   **18.15** SO_2.

18.17 (a) $n = 1.24 \times 10^{-4}$ kmol.   (b) $V = 2.789 \times 10^{-3}$ m^3.
18.19 (a) $m = 2.59 \times 10^{-3}$ kg.   (b) $p = 1.43 \times 10^5$ Pa.
(c) $K_r = 3.35 \times 10^3$ J.
18.21 (a) $E = 5.68 \times 10^6$ J; same for O_2 gas.
(b) $E_{H_2}/E_{Ar} = 5/3$.   **18.23** $c_v = 1.58 \times 10^4$ J/kmol·K.
18.25 (a) $c_v = 2.91 \times 10^4$ J/kmol·K;
(b) $c_v' = 4.99 \times 10^4$ J/kmol·K.   **18.27** $M_{Al} = 0.13$ kg.
18.29 $p = 6.10 \times 10^4$ Pa $= 0.602$ atm.
18.31 $\Delta p = 4.56 \times 10^4$ Pa.
18.33 $v_{\text{rms}} = 1.11 \times 10^{-5}$ m/s, of order $10^{-8} \times v_{\text{rms}}(O_2) = 483$ m/s.
18.35 $v(v_{\text{rms}})/v(v_{\text{mp}}) = 3/2\sqrt{e} = 0.910$. This ratio doesn't depend on the particular gas or on the temperature.
18.37 (a) $N/V = 2.55 \times 10^{11}$ m^{-3}.
(b) $\rho = 8.47 \times 10^{-16}$ kg/m^3.   **18.39** $p = 0.42$ Pa.
18.41 $p = 6.1 \times 10^5$ Pa.
18.43 50% of the CO_2 molecules have dissociated.
18.45 $d = 3 \times 10^{-10}$ m.   **18.47** (a) $Q = \frac{5}{2}p_0V$.
(b) $Q = 5.1 \times 10^2$ J.   **18.49** $C = 4.99 \times 10^4$ J/K.
18.51 0.43%.   **18.53** (a) $h = 8.0 \times 10^3$ m.
(b) $p_h = 3.73 \times 10^4$ Pa $= p_{\text{atm}}/e$.
(c) $p_{2h} = 1.37 \times 10^4$ Pa $= p_{\text{atm}}/e^2$.   **18.57** (a) $T' = 1120$ K.
(b) $Q = 5.45 \times 10^3$ J.   (c) $p' = 4p$.
18.59 (a) $M = 1.34 \times 10^{-2}$ kg;   (b) $E = 2.53 \times 10^3$ J;
(c) $N = 2.53 \times 10^{23}$;   (d) $\langle\epsilon\rangle = 1.0 \times 10^{-20}$ J;
(e) $\langle\epsilon_t\rangle = 6.0 \times 10^{-21}$ J;   (f) $\langle v\rangle = 438$ m/s;
(g) $v_{\text{rms}} = 475$ m/s.
18.61 (e) $\tau = 8.58 \times 10^{-10}$ s; $\lambda = 3.64 \times 10^{-7}$ m.
(f) 10^{-4} Pa.
18.65 (a) The particle is sufficiently small that it participates in the random motion of the molecule with a small but nonnegligible rms speed $v_{\text{rms}} = \sqrt{3kT/m}$.   (b) 3.3×10^{-6} m/s;
(c) 3;   (d) 6.2×10^{-21} J.   **18.67** 80.6 km.

Chapter 19

19.1 $n = 2.16 \times 10^{-4}$ kmol.   **19.3** 1.7×10^6 J.
19.7 (a) $a \to c$: $|W_{\text{out}}| = 1056$ J.   (b) $c \to d$: $|W_{\text{in}}| = 616$ J.
(c) $Q_{\text{cyc}} = 440$ J.
19.9 (a)

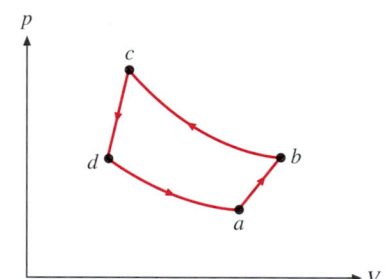

(b)

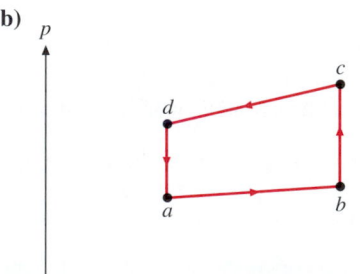

19.11 He: 2.08×10^4 J/kmol·K; O_2: 2.91×10^4 J/kmol·K.
19.13 (a) 5.66.   (b) 11.3.   **19.15** $Q_{\text{cyc}} = 410$ J.

19.17 $T_f = 886$ K.   **19.21** $\Delta S = nc_p' \ln(V_f/V_i)$.
19.23 $\Delta S = 6050$ J/K.   **19.25** (a) $T_f = 323$ K,
(b) $\Delta S_{hot} = -602$ J/K,   (c) $\Delta S_{cold} = 704$ J/K.
(d) $\Delta S_{tot} = 102$ J/K.
19.27 (a)

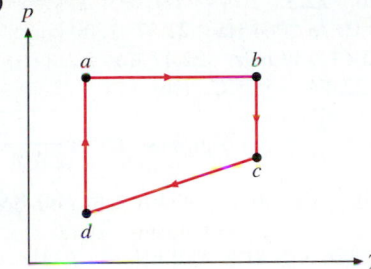

(b)

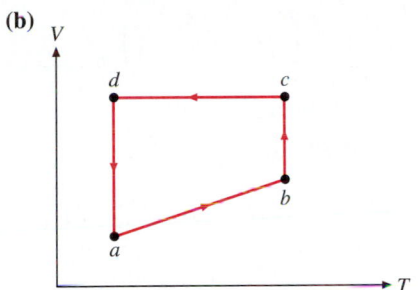

(c)

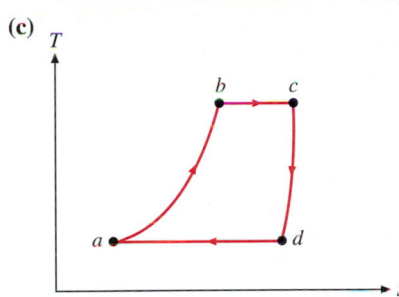

19.29 (a) $W = 169$ kJ.   (b) $Q = 2260$ kJ.   (c) $\Delta E = 2090$ kJ.
(d) $\Delta K = 0$.   (e) $\Delta P \simeq 2090$ kJ.
(f) No; it is still an isobaric process at 1 atm.
19.31 (a) No. No. Yes.   (b) $W = Q - \Delta E = 0$.   (c) Yes. The
internal energy does not change, and so the temperature should
remain the same.   (d) No. A nonideal gas has self-interactions
and potential energy. Its energy depends on pressure or volume
or both.   **19.35** (a) $W_{cyc} = 3140$ J.   (b) No.
19.37 $\Delta T = 32$ K.
19.41 (b) No. The result is true only for an ideal gas.

19.45 (a) $T_f = (T_1 + T_2)/2$.   (b) $\Delta S_1 = C \ln \dfrac{T_f}{T_1}$.

(c) $\Delta S_2 = C \ln \dfrac{T_f}{T_2}$.   (d) $\Delta S_{tot} = C \ln \dfrac{T_f^2}{T_1 T_2}$.   **19.47** $\Delta S_{cyc} = 0$.

19.49 (a) $p_b = 2.8 \times 10^5$ Pa, $p_c = 1.28 \times 10^5$ Pa, $p_d = 3.2 \times 10^5$ Pa, $V_d = 3.2 \times 10^{-3}$ m^3.
(b) $W_{a \to b} = 1280$ J, $W_{b \to c} = 560$ J, $W_{c \to d} = -940$ J, $W_{d \to a} = -560$ J.   (c) $W_{cyc} = 340$ J.
(d) $Q_{a \to b} = 1280$ J, $Q_{b \to c} = 0$, $Q_{c \to d} = -940$ J, $Q_{d \to a} = 0$.
(e) $\Delta S_{a \to b} = 3.2$ J/K, $\Delta S_{b \to c} = 0$, $\Delta S_{c \to d} = -3.2$ J/K, $\Delta S_{d \to a} = 0$.

(f)

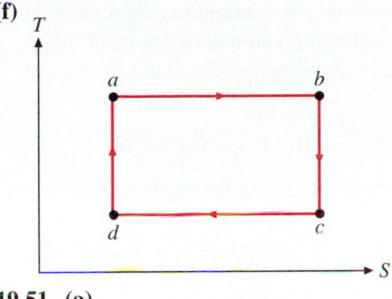

19.51 (a)

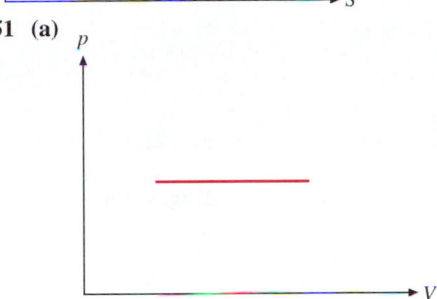

(b)

19.53 $\dfrac{(dp/dV)_{adiabatic}}{(dp/dV)_{isothermal}} = \gamma$.   **19.55** (c) $H = 2.8 \times 10^4$ m.
(e) When $h = H$, $T = 0$.
19.57 (c) $\tau_{isothermal} = 0.905\tau_0$, $\tau_{adiabatic} = 0.87\tau_0$.

Chapter 20
20.1 (a) $W_{cyc} = 29$ kJ;   (b) $\eta = 0.082$.
20.3 (a) Approximately 2000 tonnes.
(b) Efficient pumping of large amounts of water, corrosion due
to salt water, efficient heat transfer over modest temperature
gradients.   **20.5** $\eta = 0.16$.   **20.7** (a) $W_{cyc} = 720$ J.
(b) $Q_{out} = 1780$ J.   **20.9** $\eta = 0.17$.   **20.11** (a) $T_h = 444$ K.
(b) $\omega_{hp} = 2.9$.   (c) $\omega_r = 1.9$.   **20.19** (a) $\omega_r = 6.5$.
(b) $P = 131$ W.   **20.21** (a) $\omega_r = 30$.   (b) $P = 61$ W.
(c) $\omega_r = 3.5$.   (d) $\omega_r = 0.12\omega^c_r$.   **20.23** (a) $\Delta S = 156k$.
(b) $\Delta S = 2.16 \times 10^{-21}$ J/K.
(c) An ordinary thermodynamic system has many more
accessible microstates than 52!; it has $\simeq 8 \times 10^{64}$.
20.27 $\eta = 0.33$.   **20.31** (a) $\eta^c = $ Area$_{abcd}$/Area$_{abS_bS_a}$.
(b) The efficiency decreases.   (c) The efficiency decreases.
(d) A rectangular cycle.   **20.33** (a) $Q_{in} = nc_V(T_a - T_d)$.
20.35 (a) $P_5^{max} = 0.3125$; $P_4^{max} = 0.375$.
(b) $P_5^{min} = 0.03125$; $P_4^{min} = 0.0625$.
(c) The ratio of the minima decreases more rapidly than the ratio
of the maxima, resulting in a sharper distribution.
20.37 (b) $\Delta S = 0$.
(c) No. In part **a**, there are two types of gases. In part **b**, there is
only one gas, and all its molecules are indistinguishable.

(d) With indistinguishable molecules, interchanging a molecule on the left side with one on the right side does not result in a distinguishable microstate, and the microstate count implied in part **a** is too large.

20.39 **(a)** $\Delta S = \int_{T_{hi}}^{T_{lo}} C(T)\dfrac{dT}{T}$. **(b)** It approaches zero.
(c) As the temperature approaches zero, the coefficient of performance goes to zero.

20.41 The efficiency is less than or equal to $1 - \dfrac{T_{cw}}{T_{hi}}$.

20.43 **(b)** $T(h) = 366$ K.

Chapter 21

21.1 $\left[\sqrt{\dfrac{T}{\mu}}\right] = \left[\dfrac{mlt^{-2}}{ml^{-1}}\right]^{1/2} = \left[\dfrac{l}{t}\right] = [v]$. **21.3** $v_2/v_1 = 0.5$.

21.5 $v = \sqrt{\sigma/\rho}$. **21.7** $v' = \sqrt{2}v$. **21.9** 338 m/s.
21.11 $\nu = 4752$ Hz. **21.13** $\lambda = 3.15$ m.

21.15 $y = A\cos\left(\dfrac{2\pi}{\lambda}x - \dfrac{2\pi}{\tau}t\right)$. **21.17** **(a)** $y(x, t) =$
$(5.0 \times 10^{-3}$ m$)\cos(7.14 \times 10^{-3}x - 1.57t)$.
(b) $y(t) = (5.0 \times 10^{-3}$ m$)\cos(3.78 - 1.57t)$.
(c) $y(x) = (5.0 \times 10^{-3}$ m$)\cos(7.14 \times 10^{-3}x - 6.44)$.
21.19 4.3 cm. **21.21** **(a)** 1.67×10^{-3}. **(b)** 1.05×10^{-2}.
21.23 **(a)** -2.36×10^{-3} m; **(b)** 0.114 m/s; **(c)** 58.2 m/s^2.
21.25 **(a)** $dE/dx = 0.200$ J/m, $\langle\Phi\rangle = 1.50$ W.
(b) $\mu = 2.66$ kg/m. **21.27** 1.74×10^3 m.
21.29 **(a)** $Y(x, t) = 2A\cos kx\cos \omega t$; $Y(x,t) = 0$, where $x = (n + \frac{1}{2})\pi/k$.
(b) Equation 21.51 represents $y_1 - y_2$; the result in this problem expresses $y_1 + y_2$, and the phase is different. **21.31** **(a)** $n = 4$.
(b) $2A = 4.95$ cm. **21.33** $\nu = 250$ Hz.
21.35 9.85 Hz, 19.7 Hz, 29.5 Hz. **21.39** 5.2 mm.
21.45 **(a)** The amplitude is $\sqrt{2}A$, and the wave is linearly polarized in the plane midway between the x-y and x-t planes (at $45°$ with respect to either). A point on the string oscillates in this plane.
(b) The amplitude is A, and the wave is circularly polarized. A point on the string describes a circle around the equilibrium position. The wave looks like a rotating screw.
21.49 $I = 0.62$ kW/m^2. **21.55** $\nu' = 2\nu$.
21.57 **(a)** $\nu_2/\nu_1 = \sqrt{2}$. **(b)** $\epsilon = T/YA$. **(d)** 0.990.
(e) The correction factor is 0.995. Beats will be clearly heard if the strings are played together without correction.
21.61 **(b)** The shift is toward lower frequency.
21.63 **(a)** $v = \sqrt{MgL/m}$. **(b)** $\lambda = 2\pi L\sqrt{M/m}$.
(c) $\lambda = 417.7$ m.
(d) The top end of the rope is a node, and since $\lambda \gg L$, we see only a small portion of the wave near the node, where the curvature is small.
(e) As M/m decreases so does λ/L, and we can see more curvature in the string. **21.65** **(c)** $v = \sqrt{mg/\mu_0}$.
(d) $\mu = \mu_0 e^{(-\mu_0/m)z}$.

Chapter 22

22.1 **(a)** 3043 m/s; **(b)** 5092 m/s. **22.3** **(a)** 3744 m/s;
(b) 2271 m/s. **22.5** 1.7. **22.7** 1.161×10^9 Pa.
22.9 0.3945 s. **22.11** 1290 m/s. **22.13** 414 Hz.
22.15 488 m/s. **22.17** 5×10^{-11} W.
22.19 $I/I_0 = 11.5$ dB, so 14 violins sound a little more than twice as loud as one. This is not a large difference for the soloist, who is standing in front of the orchestra and is usually playing

louder, to overcome. **22.21** 24.3 m; 1.9 cm. **22.23** 7.5.
22.25 **(a)** 5.26 m. **(b)** 32.70 Hz. **22.27** 3.4 mm.
22.29 The frequency, about 8 Hz, if the pipe is open is inaudible, and the pipe is purely decorative.
22.31 35.7 Hz. No. **22.33** $\Delta\nu_1 = 1.93$ Hz; $\Delta\nu_2 = -1.93$ Hz.
22.35 From 1940 Hz to 2064 Hz. **22.37** 1200 km $\pm$ 100 km.
22.41 40.0 u. **22.43** 340 m/s. **22.47** **(a)** 5.2×10^{-5}.
(b) 3.3×10^{-4}. **22.49** $-55.5°$C.

22.51 $v = v_0\sqrt{1 - \dfrac{h}{H}} = 331.6$ m/s $\times \sqrt{1 - \dfrac{h}{2.8 \times 10^4 \text{ m}}}$.

22.53 **(a)** 2.8×10^{-5} Pa. **(b)** 3.7×10^{-12} m. **(c)** 8.98 Pa.
22.55 340 m/s.
22.57 Four: 26.2 kHz, 33.7 kHz, 41.2 kHz, 48.7 kHz.

22.59 31 m/s. **22.61** $v_f = v\dfrac{\nu}{\nu_b}\left(\sqrt{1 + \dfrac{v_b^2}{v^2}} - 1\right)$.

22.65 **(a)** $2\nu_s\dfrac{v_s}{v - v_s}$. **(c)** 1.65 m/s.

22.67 **(b)** $20\log_{10}(s^*/s)$. **22.69** **(a)** $Y = v^2L^2\rho/d^2$.

(b) 2×10^{11} Pa. **22.73** **(a)** $a\cos 2\pi\nu t + \dfrac{b}{2}\cos 4\pi\nu t + \dfrac{b}{2}$.

(b) $\Delta p(t) = a(\cos 2\pi\nu_1 t + \cos 2\pi\nu_2 t) + b[1 + \frac{1}{2}(\cos 4\pi\nu_1 t + \cos 4\pi\nu_2 t) + \cos 2\pi(\nu_1 + \nu_2)t + \cos 2\pi(\nu_1 - \nu_2)t]$.
(d) Because b/a is intensity dependent, the ear will hear a different spectrum of aural harmonies than the listener in the concert hall if the reproduced sound level differs from the original.

Chapter 23

23.1 8.99×10^7 N toward the positive charge.
23.3 24×10^{-2} N. **23.5** **(a)** $F = 2.7 \times 10^{-9}$ N.
(b) 1.6×10^{18} m/s^2. **23.7** 6.5 mm.
23.9 3.0×10^{-2} N repulsive.
23.11 **(a)** 1.4×10^{-2} N repulsive.
(b) 2.4×10^{-2} N away from the midpoint of B and C.
23.13 5×10^3 N/C at $-63°$. **23.15** 9.57×10^{-11} C.
23.17 5.6×10^4 N/C toward the middle negative charge.
23.19 **(a)** $q = 4\pi\rho r^3 g/3\mathscr{E}$. **(b)** 3 electrons. **23.21** $\theta = 28.4°$.
23.23 $\dfrac{\mathscr{E}_d}{\mathscr{E}_r} = \dfrac{2(L^2 + B^2)(\sqrt{L^2 + B^2} - L)}{B^2 L} =$

$2\left[\left(1 + \dfrac{B^2}{L^2}\right)^3 - \left(1 + \dfrac{L^2}{B^2}\right)\right]$.

23.25 **(a)** The direction is along the symmetry axis, away from the center of curvature. **23.27** 1.74×10^{15} N.
23.31 1.05×10^{-10} m.
23.33 **(a)** $x = d/(1 - \sqrt{\alpha})$. (The positive root gives the location where the field contributions are equal but in the same direction.)

(b)
$\alpha > 1$

(c)
$\alpha < 1$

23.35 $\mathscr{E} = \dfrac{1}{\sqrt{2}\pi\epsilon_0}\dfrac{q}{s^2}$, directed outward at $45°$ from the

direction from $+q$ to $-q$. **23.41** **(b)** $\mathscr{E} = -\dfrac{\lambda_0 b^2}{4\epsilon_0(z^2 + b^2)^{3/2}}\hat{x}$.

(c) For $z \gg b$, the electric field becomes $\mathscr{E} = -\dfrac{\lambda_0 b^2}{4\epsilon_0 z^3}\hat{x}$.

23.45 (a) Away from the wire for $\lambda > 0$; toward the wire for $\lambda < 0$. (b)

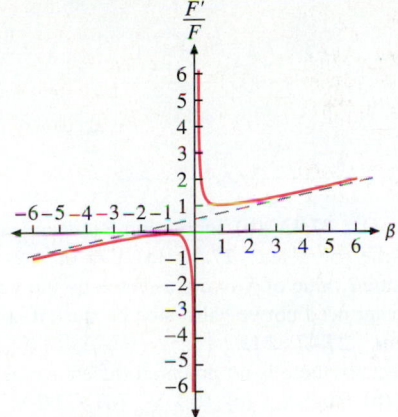

23.49 (a) $F'/F = (\beta + 1)^2/4\beta$.
(b) When $|\beta|$ is very small, the force is very small until the charge is redistributed, so F'/F is very large. The discontinuity represents the change from repulsion ($\beta > 0$) to attraction ($\beta < 0$).
(c) When $\beta = -1$, the spheres discharge each other on contact.

23.51 (a) $\mathscr{E} = \dfrac{q}{4\pi\epsilon_0} \dfrac{z^3 - (z^2 + b^2)^{3/2}}{z^2(z^2 + b^2)^{3/2}} \, \hat{\mathbf{z}}$.

(b) For $z \gg b$, $\mathscr{E} \propto \dfrac{z^3\left(1 - 1 + \dfrac{3}{2}\dfrac{b^2}{z^2}\right)}{z^5} = \dfrac{3}{2}\dfrac{b^2}{z^4}$.

23.53 $\Delta F = \dfrac{q^2}{4\pi\epsilon_0 b^2}$. **23.55** $\mathscr{E} = \dfrac{\lambda}{4\pi\epsilon_0 z}(-\hat{\mathbf{x}} + \hat{\mathbf{z}})$.

Chapter 24
24.1 N/C·m². **24.3** 55°. **24.5** $\Phi = 0$. **24.7** (a) 0,
(b) 2470 N/C, (c) 295 N/C. **24.9** (a) 704 N/C; (b) 0.
24.11 4.1×10^7 N/C. **24.17** (a) $\Phi = q/8\epsilon_0$.
(b) $\Phi = 0$ for each of the three. (c) $\Phi = q/24\epsilon_0$ for each.

24.21 $\mathscr{E} = 0$ for $r < a$; $\mathscr{E} = \dfrac{\rho}{3\epsilon_0 r^2}(r^3 - a^3)$ for $a < r < b$;

$\mathscr{E} = \dfrac{\rho}{3\epsilon_0 r^2}(b^3 - a^3)$ for $r > b$.

24.23 $\mathscr{E} = \dfrac{ar^2}{4\epsilon_0}$ for $r < R$; $\mathscr{E} = \dfrac{aR^4}{4r^2\epsilon_0}$ for $r > R$.

24.25 (a) $\mathscr{E} = \dfrac{\rho_0}{\epsilon_0}\left(\dfrac{r}{3} - \dfrac{r^2}{4R}\right)$. (b) $\mathscr{E} = \rho_0\dfrac{R^3}{12\epsilon_0 r^2}$.
(c) $\mathscr{E}(r)$ has the maximum value $\mathscr{E} = \rho_0 R/9\epsilon_0$ at $r = 2R/3$.

24.27 $t = \sqrt{\dfrac{3\pi}{4G\rho}} = 2530$ s. **24.29** (a) $\mathscr{E} = \dfrac{\rho R^2}{2\epsilon_0 r}$.

(b) $\mathscr{E} = \dfrac{\rho r}{2\epsilon_0}$. (c) $\lambda = \rho\pi R^2$. (d) They are the same.

(e)

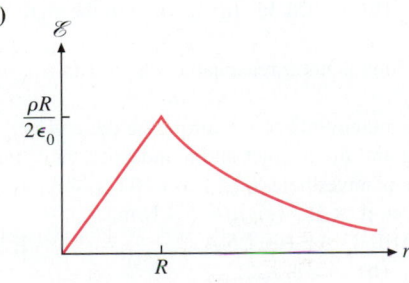

24.31 (a) $\tan\theta\sin\theta = \dfrac{q\lambda}{2\pi\epsilon_0 mgl}$.
(b) Unlike the field of a charged flat plate, the field of a charged long wire is not independent of distance. (c) 3.8×10^{-9} C.

(d) $\theta = \sqrt{\dfrac{q\lambda}{2\pi\epsilon_0 mgl}}$. (e) No. F is not proportional to θ.

24.33 (a) See Figure 26.2.
(b) $\mathscr{E} = \sigma/\epsilon_0$ between plates, 0 elsewhere.
(c) Equation 24.13: Between the plates, the contributions of the two plates add; elsewhere they cancel. Equation 24.14: All the flux emanates outward from a closed conducting surface.

24.35 (a) $\mathscr{E} = \dfrac{\rho z}{\epsilon_0}$. (c) $\nu = \dfrac{1}{2\pi}\sqrt{\dfrac{q\rho}{m\epsilon_0}}$. **24.39** $x = R/\sqrt{3}$.

Chapter 25
25.3 $\Delta V = -9.4 \times 10^{-5}$ J. **25.5** $r = 1.4 \times 10^{-13}$ m.
25.7 $W = 2.23$ J. **25.9** (a) $V = 3.0 \times 10^5$ V;
(b) $V = 2.7 \times 10^5$ V. **25.11** $v = 5.0 \times 10^5$ m/s.
25.13 13.6 eV $= 2.18 \times 10^{-18}$ J. **25.15** (a) $\sigma = q/2lw$.
(b) $\sigma = q/lw$ on the inner face, 0 on the other face. Charges move to inner faces until interiors of plates are field-free regions.

25.17 For $z \gg R$, $V = \dfrac{Q}{4\pi\epsilon_0 z}(\sqrt{R^2 + z^2} - z) \simeq$

$\dfrac{Q}{2\pi\epsilon_0 R^2}(z - z) = 0$.

25.19 (a) Graph the function $\mathscr{E}(r) = \dfrac{1}{4\pi\epsilon_0}\dfrac{q}{r^2}$ for $r > 25$ cm
($\mathscr{E} = 0$ inside the sphere), where $q = -5.37 \times 10^{-6}$ C.

(b) Graph the function $V(r) = \dfrac{1}{4\pi\epsilon_0}\dfrac{q}{r}$ for $r > 25$ cm;
for $r < 25$ cm, $V = V(0.25) = -1.93 \times 10^5$ V.

25.21 $V = -9.47$ V. **25.23** (a) $V = \dfrac{1}{2\pi\epsilon_0}\dfrac{q}{\sqrt{b^2 + z^2}}$.

(c) $\mathscr{E}(z) = \dfrac{q}{2\pi\epsilon_0}\dfrac{z}{(b^2 + z^2)^{3/2}}$.

(d) The maximum value occurs at $z = b/\sqrt{2}$;

$\mathscr{E}_{\max} = \dfrac{q}{3\sqrt{3}\pi\epsilon_0 b^2}$.

25.25 (b) In Problem 25.24, $\lim_{z\to\infty} V = \infty$ if V is defined in the same way; thus we cannot add a constant in order to define the potential at infinity as 0. **25.27** $\lambda = 3.6 \times 10^{-6}$ C/m.

25.29 For $r \le R$, $V = \dfrac{1}{8\pi\epsilon_0}\dfrac{Q}{r}\left(3 - \dfrac{r^2}{R^2}\right)$; for $r \ge R$,

$V = \dfrac{1}{4\pi\epsilon_0}\dfrac{Q}{r}$. **25.31** (b) $\mathscr{E} = \dfrac{\lambda d}{\pi\epsilon_0 r^2}$; $\phi = 2\theta$.

25.33 (a) $U = \dfrac{1}{4\pi\epsilon_0}\dfrac{q^2}{l}(\sqrt{2} - 4)$. (b) $U = -\dfrac{\sqrt{2}}{4\pi\epsilon_0}\dfrac{q^2}{l}$.

25.37 (a) $V = \dfrac{1}{2\pi\epsilon_0}\dfrac{q}{l}\ln\left[\dfrac{l}{2z}\left(1 + \sqrt{1 + \dfrac{4z^2}{l^2}}\right)\right]$.

(b) $\mathscr{E} = \dfrac{1}{2\pi\epsilon_0}\dfrac{q}{l}\left[\dfrac{1}{z} - \dfrac{4z}{l^2\left(1 + \dfrac{4z^2}{l^2} + \sqrt{1 + \dfrac{4z^2}{l^2}}\right)}\right]$.

25.39 (b) $U = -\dfrac{1}{4\pi\epsilon_0}\dfrac{q^2}{R}N\alpha$.

Chapter 26

26.1 $C = 62$ pF. **26.3** $6C$. **26.5** (a) $C = 4.40$ μF;
(b) $C = 1.02$ μF. **26.7** X to Y: $\frac{11}{5}C$; X to Z: $\frac{11}{3}C$; Y to Z: $\frac{11}{4}C$.
26.11 Distance $= 2.7 \times 10^{-3}$ cm. **26.13** (a) Nothing.

(b) Still nothing. **26.15** $C = \dfrac{\epsilon_0 A}{2d}(\kappa_1 + \kappa_2)$.

26.17 $V = 107$ V. **26.19** $\dfrac{C}{l} = \dfrac{2\pi\epsilon_0}{\ln(2b/a)}$.

26.21 (a) $C = \dfrac{2\epsilon_0 A}{3d}$. (b) $C = \dfrac{3\epsilon_0 A}{2d}$.

26.25 (a) The positive end of one electret will attract the negative end of the other and vice versa, thus causing them to align side by side in an antiparallel formation.
(b) There is no net dipole moment since the electrets align in such a way as to cancel each other out.
(c) An electric field will cause them to align in parallel formation, producing a large internal depolarization field that diminishes the total field strength. This is a large-scale version of the process that occurs in water when it is subjected to an electric field. **26.27** $C = 65$ pF. **26.31** $\mathscr{E}_{1\,max}\kappa_1 R_1 = \mathscr{E}_{2\,max}\kappa_2 r$.

26.33 (a) $U = \dfrac{q^2 x}{2\epsilon_0 A}$. (b) $F = -\dfrac{q^2}{2\epsilon_0 A}$.

26.35 (a) $C = \dfrac{4\pi\epsilon_0}{\dfrac{1}{R} - \dfrac{1}{R + \Delta R}}$.

(b) $U = \dfrac{1}{2}\dfrac{q^2}{C} = \dfrac{q^2}{8\pi\epsilon_0}\left(\dfrac{1}{R} - \dfrac{1}{R + \Delta R}\right) = \dfrac{q^2}{8\pi\epsilon_0}\left(\dfrac{\Delta R}{R(R + \Delta R)}\right)$.

(c) $\tau = 4\pi R^2\,\Delta R$; thus $\langle u \rangle = \dfrac{U}{\tau} = \dfrac{q^2}{32\pi^2\epsilon_0 R^2}\left(\dfrac{1}{R(R + \Delta R)}\right)$.

(d) $u = \lim_{\Delta R \to \infty} \langle u \rangle = \dfrac{q^2}{32\pi^2\epsilon_0 R^4}$. **26.37** (a) To the left.

(b) $F = -\dfrac{C_0 V^2 (\kappa - 1)}{2w}$. **26.39** (a) $U = \dfrac{1}{2}KQ^2$.

(b) $K = K_1 + K_2 + \cdots + K_n$. (c) $\dfrac{1}{K} = \dfrac{1}{K_1} + \dfrac{1}{K_2} + \cdots + \dfrac{1}{K_n}$.

26.41 Capacitance $= C$. **26.45** $y - y_0 = \dfrac{\epsilon_0 V^2 (\kappa - 1)}{2\rho g d^2}$.

Chapter 27

27.1 6.25×10^{18}. **27.3** 26 C. **27.5** (a) $0.94087\ \Omega$.
(b) $l = 1.0629$ m. (c) 1.00001 mm^2. **27.7** $T = 45.8$°C.
27.9 (a) $R_{RT} \approx 16.3\ \Omega$, making the very crude assumption that R varies linearly over such a large temperature range.
(b) $d = 30$ μm. **27.11** $v_d = 1.9$ mm/s.
27.13 The 240-Ω bulb is brighter and consumes 1.5 times as much power as the 360-Ω bulb.
27.15 (a) 1 quad $= 1.055 \times 10^{18}$ J. (b) 33 plants.
27.19 $639\ \Omega$. **27.21** $60\ \Omega$. **27.23** $R/3$.
27.25 (a) $V = 2.22$ V, (b) $A_2 = -6$ mA.
27.27 (a) $j = 1.01 \times 10^6$ A/m^2. (b) $v'_d/v_d = 1$.
27.29 $q = 9.64 \times 10^4$ C/mol.
27.31 (a) The aluminum wire is longer. (b) $l_{Fe}/l_{Al} = 0.31$.
27.33 (a) m^2/V·s. (b) $\mu = e\tau/m$. (c) $\sigma = ne\mu$.
(d) $j = ne\mu\mathscr{E}$. (e) $j = NAe(\mu_+ + \mu_-)\mathscr{E}$.
27.35 (a) $P = 2400$ W. (b) $l = 0.47$ m.

27.37 $\dfrac{r}{R} = \dfrac{\theta}{2\pi}\left(1 - \dfrac{\theta}{2\pi}\right)$. **27.39** (a) $R = \dfrac{V_1 R_V}{i_1 R_V - V_1}$.

(b) $R = \dfrac{V_2}{i_2} - R_A$.

27.41 (a)

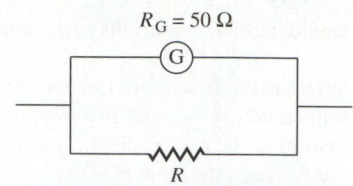

(b) $R = 0.05\ \Omega$. **27.43** (b) Increasing R_V decreases the sensitivity of the voltmeter. **27.45** (a) $V = 0$.
(c) The measured value of X is not affected by the value of V; the galvanometer need not be calibrated because it is used as a null instrument. **27.47** $2\ \Omega$.
27.49 (a) Because there is no potential difference between C and ground. (b) 0.048 A. (c) 10.1 V. (d) -7.2 V.
27.51 (a) $i_{3\,\Omega} = 3$ A to the right, $i_{4\,\Omega} = 5$ A to the left, $i_{8\,\Omega} = 2$ A to the right.
(b) $P_{3\,\Omega} = 27$ W, $P_{4\,\Omega} = 100$ W, $P_{8\,\Omega} = 32$ W.
(c) $P_{7\,V} = -21$ W, $P_{36\,V} = 180$ W (the 7-volt cell is being charged). **27.53** $q = 1.0 \times 10^{-6}$ C. **27.55** (a) $44\ \Omega$.
(b) $2\ \Omega$. (c) 2.4 A. (d) 106 V. (e) 264 W. (f) $P_s/P_t = 0.95$.
(g) $0.576\ \Omega$. (h) 4.66 A. (i) 55.9 W. (j) 0.224. (k) No.

(l) 1.3 cm; no. **27.57** (a) $P = V^2\dfrac{R}{(R + R_V)^2}$.

27.59 (b) $l_N/l_C = 62.5$. **27.61** (b) $P = \dfrac{(V_1 - V_2)^2}{R_1 + R_2}$.

27.65 (a) The voltmeter on the left.
27.67 (a) $i_{AB} = i_{AF} = i_{AG} = i_{AH} = i_\infty/4$.
(b) The total current is again i_∞; $i_{AB} = i_{CB} = i_{DB} = i_{EB} = i_\infty/4$.
(d) $i_{AB} = i_\infty/2$; $i_{AG} = i_\infty/8$; $i_{AH} = i_{AF} = 3i_\infty/16$.
Total current leaving A is i_∞. (e) $i_{AB} = V/R$. $i_\infty = 2i_{AB}$.
(f) Total current leaving A is $2i_{AB}$. $R_{eff} = R/2$.

27.71 (b) $\dfrac{R}{l} = \dfrac{1}{2\pi\sigma}\ln\left(\dfrac{r_2}{r_1}\right)$. (c) No.

Chapter 28

28.1 (a) $\Phi_m = 2.7 \times 10^{-2}$ Wb. (b) $\Phi_m = 2.4 \times 10^{-2}$ Wb.
28.3 $F = 4.6 \times 10^{-12}$ N. **28.5** (a) $v = 1.6 \times 10^8$ m/s,
(b) $K = 1.2 \times 10^{-14}$ J, (c) $V = 7.5 \times 10^4$ V. **28.7** 0.13 m.
28.9 $\tau_\alpha/\tau_p = 2$. **28.11** $D = 1.0$ m. **28.13** $i = 2.0$ A.
28.15 $i = 2.68$ A.
28.17 (a) $F_{vertical\ leg} = 12.5$ N pointing out of the page, $F_{horizontal\ leg} = 0$, $F_{hypotenuse} = 12.5$ N pointing into the page.
(b) $T = 18.8$ N·m upwards. **28.19** (a) $\mu = 2.4$ N·m/T.
(b) $T = 2.0$ N·m. **28.21** (a) $v = 3.4 \times 10^3$ m/s.
(b) $\mathscr{E}$ is perpendicular to $\mathscr{B}$; v is perpendicular to both $\mathscr{E}$ and $\mathscr{B}$.
28.23 $v = V/\mathscr{B}d$, where d is the distance between the plates.
28.25 $V_H = 7.2 \times 10^{-5}$ V. **28.27** $n = 2.47 \times 10^{28}$ m^{-3}.
28.31 (a) $r_p/r_e = 1$; (b) 1. **28.35** (a) $i\mathscr{B}d/m$ to the right.
(b) $\theta = \tan^{-1} i\mathscr{B}d/mg$.
(c) $i = 1.1 \times 10^9$ A. This is not a reasonable way to run a railroad.
28.37 $\mathscr{B} = 1.4$ T; the magnitude of the torque on the coil is determined by the coil and the magnet and is independent of the lever arm. That torque is nevertheless applied to the system as a whole. **28.39** (a) $T = \mu \times B$. (c) $\mu = 5.4$ N·m/T.
28.41 (a) 0.27 A. (b) 0.87 A. (c) 1.5 A.
28.47 (a) Clockwise. (b) $iA\mathscr{B}v$.

(d) Most of the current will flow in the region not too far from the contacts. **(e)** 34 W, 1.8 N·m. **28.49** **(a)** $v = V/\mathcal{B}d$.
(b) $v = 0.42$ m/s. **28.53** **(a)** $m^* = e\mathcal{B}/\omega$.
(b) The electrons would accelerate until they struck the edges of the block.
(c) The resonance effect is prominent because the electron can absorb energy from the field over several revolutions.
(d) The sense of rotation of the electric field for electron resonance is opposite the sense for hole resonance.

28.55 **(a)** $\Pi = \dfrac{T}{2b\mathcal{B}\sin\theta}$. **(b)** N/T. **(c)** $\Pi = 18$ N/T.

(d) Electric charge. A magnetic monopole has never been observed. **(e)** $i = 2l\pi/A$.
28.57 $V_{\mathrm{H}} = \frac{1}{4}(V_{++} + V_{--} - V_{+-} - V_{-+})$.

Chapter 29
29.1 $\mathcal{B} = 3.4 \times 10^{-5}$ T. **29.3** $i = 56.7$ A.
29.5 $i_1 = 110$ A; $i_2 = 220$ A.
29.7 **(a)** 1.6×10^{-4} T downward; **(b)** 2.2×10^{-4} T upward;
(c) 1.9×10^{-4} T downward. **29.9** **(a)** 3.8×10^{-5} N/m.
(b) Attractive. **(c)** Repulsive. **29.11** **(a)** $i = 2.7$ A;
(b) 4.7 A. **29.13** $\mu = 6.0 \times 10^{-2}$ A·m^2. **29.15** **(a)** $\mathcal{B} = 0$.
(b) $\mathcal{B} = 1.3 \times 10^{-4}$ T. **29.17** **(a)** 2.31×10^{-9} T.
(b) r is so large that the angle and the distance to the point do not vary much over the segment of wire under consideration.
29.19 **(a)** $\mu_0 i/2\pi R$; **(b)** $\mu_0 i/4\pi R$. **29.21** 1.7×10^{-2} T.
29.23 8.48×10^{-5} T into the page. **29.25** **(a)** 9.90×10^{-3} T.
(b) Away. **29.27** 1.5×10^{-2} T. No, it would not matter.
29.31 73.6 A. **29.33** **(b)** 18°. **(c)** 0.15.
29.35 **(c)** $\mathcal{B}_{\mathrm{circle}} = 0.605\mathcal{B}_{\mathrm{triangle}}$.
29.37 Parallel to R, toward the right.
29.39 **(a)** $D = i_1 R/(i_1 + i_2)$.
(b) No. At no other point are the field contributions of the two wires antiparallel so that they can cancel.
(c) $D' = i_1 R/(i_1 - i_2)$ in the plane of the wires.
29.41 In the plane perpendicular to the plane of the wires at $y = \pm a$. **29.49** $\mu_0 i/4b$ into the page.

29.51 $\dfrac{\mu_0 i}{4}\left(\dfrac{1}{b_1} - \dfrac{1}{b_2}\right)$ into the page.

29.53 $\dfrac{\mu_0 i}{2}\left(\dfrac{2\pi - \theta}{2\pi R_1} + \dfrac{\theta}{2\pi R_2}\right)$, where θ is measured in radians.

29.57 $\mathcal{B} = 3\mu_0\omega q/64R$ toward the north (right-hand rule) pole.
29.59 **(a)** Not at all.
(b) Let the length along the symmetry axis go to infinity.
(c) Circular around the inner core with magnitude $\mu_0 i/2\pi r$ for $r_1 < r < r_2$ and zero for $0 < r_1$ and $r > r_2$.
29.63 **(a)** $\mathcal{B}_{\mathrm{length\ }L} = \mathcal{B}_{\mathrm{series}} = \frac{1}{2}\mathcal{B}_{\mathrm{parallel}}$.
(b) $P_{\mathrm{length\ }L} = 2P_{\mathrm{series}} = \frac{1}{2}P_{\mathrm{parallel}}$.
29.65 **(a)** $\mathcal{B} = \mu_0 i/x$ inside; **(b)** $\mathcal{B} = 0$ outside.

29.67 $\Phi = \dfrac{\mu_0 Nih}{2\pi}\ln\rho$. **29.69** 3×10^{-5} T.

29.71 $\dfrac{\mathcal{B}_{\mathrm{square}}}{\mathcal{B}_{\mathrm{circle}}} = \dfrac{2\sqrt{2}}{\pi} < 1$, so the circular loop has a bigger field. The periphery of the square is longer, but the current is farther away from the center.

29.75 **(d)** $\left(\dfrac{b_1}{b_2}\right)^2 = \left(\dfrac{x^{*2} + b_1^2}{x^{*2} + b_2^2}\right)^{3/2}$. **(e)** 2.88 cm.

29.77 **(a)** $\mathcal{B} = \dfrac{\mu_0 ni}{2}(\cos\alpha_1 + \cos\alpha_2)$. **(c)** $\mathcal{B} = \dfrac{\mu_0 ni}{2}$.

29.81 **(a)** 7.0×10^{15} Hz. **(b)** 1.1×10^{-3} A.

Chapter 30
30.1 0.32 V. The eastward wingtip. **30.3** **(a)** 33 A. **(b)** 21 N.
30.5 4.0×10^2 rev/min. **30.7** $(4.6 \times 10^3$ V$)\sin(120\pi t)$.

30.9 **(b)** Counterclockwise. **(c)** 2.9×10^{-3} A. **30.11** $\dfrac{4a^2}{3R}t_f^3$.

30.13 13 T/s. **30.15** $i(t) = -\dfrac{\pi b^2 \mathcal{B}_0 \omega}{R}\cos\omega t$.

30.17 **(a)** $V = (4.5 \times 10^{-3}$ V$)(\sin\theta + \cos\theta)\cos(120\pi t)$.
(b) $\pi/4$. **30.19** **(a)** 0.01 V. **(b)** 0.01 A.
30.21 3.7×10^2 rad/s. **30.23** **(a)** 3.5 V. **(b)** 7.9 V/m.
30.25 **(b)** $\frac{1}{2}mv_0^2$. **30.27** **(a)** $2(x + Y)\lambda$.
(b) $Y(e^{2m\lambda v_0/\mathcal{B}^2 Y^2} - 1)$. **30.29** **(a)** 3.5×10^{-9} V.

(b) 2.2×10^{-4} V. **30.31** $\dfrac{N\pi b^2 \mathcal{B}_0 \omega}{3}\cos\omega t$.

30.33 **(b)** $q = 4.2 \times 10^{-7}$ C, assuming that $\mathcal{B} = 2.0 \times 10^{-5}$ T.
30.35 **(a)** $\frac{1}{2}ib^2\mathcal{B}$. **(b)** $\frac{1}{2}ib^2\mathcal{B}\omega$. **30.39** **(a)** $\alpha = qk/2m$.
(b) $v = qkbt/2m$. **30.41** $4/\pi$.
30.43 $\mathcal{B}b^2\alpha t/2$; counterclockwise for $0 < \theta < \pi$, clockwise for $\pi < \theta < 2\pi$.

Chapter 31
31.1 $i_2 = 3.8 \times 10^{-4}$ A. **31.3** $t = 2.2$ s.
31.5 $M = 2.7 \times 10^{-9}$ H. **31.7** **(a)** $L = \mu_0 n^2 Al$.
31.9 4.1×10^4 turns. **31.11** $L = 4.1 \times 10^{-3}$ H.
31.13 4.7×10^2 turns. **31.15** **(b)** 1.6 H.
31.17 2.1×10^{-6} H.

31.19 4.4 A. **31.23** $M = \dfrac{\mu_0 l}{2\pi}\ln\left(1 + \dfrac{l}{X}\right)$.

31.25 $\dfrac{\mu_0 h N_1 N_2}{2\pi}\ln\dfrac{b_2}{b_1}$. **31.27** $r_1/2$.

31.29 **(c)**

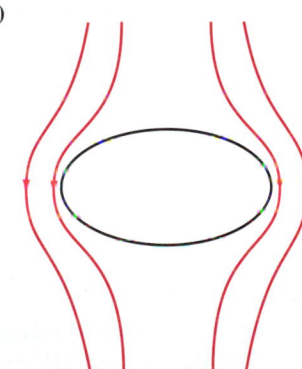

31.31 **(a)** $\mu_0 D^2/4\pi l$. **(b)** 100 m. **31.39** **(a)** greater; **(b)** less.
31.43 **(a)** $\lambda\omega/2\pi$. **(b)** $\mu_0\lambda\omega/2\pi$. **(c)** $\mu_0\omega^2\lambda^2/8\pi^2$.
(d) $\lambda^2/8\pi^2\epsilon_0 b^2$. **31.45** **(b)** $M = L_1 = L_2 = L$.
(c) $L_1 + L_2 - 2\sqrt{L_1 L_2}$. **(d)** $L_1 + L_2 + 2\sqrt{L_1 L_2}$.

Chapter 32
32.1 $\tau_C = 12$ μs. **32.3** **(a)** 1.5 ms; **(b)** 5.8 ms.
32.5 8.2 ms. **32.7** $R = 570$ kΩ. **32.9** **(a)** $\tau_L = 5.1$ s.
(b) $t = 23$ s. **32.11** **(a)** $t = 0.058$ s. **(b)** $t = 0.061$ s.
(c) $\tau_L = 0.042$ s. **32.13** **(a)** $i = 0.84$ A. **(b)** $di/dt = 4.8$ A/s.
32.15 **(a)** $di/dt = 1.0 \times 10^5$ A/s. **(b)** $t = 0$; $i = 2.0$ A.
(c) $i = 1.33$ A. **32.17** $L = 3.71$ μH.
32.19 **(a)** $\nu = 4.01$ kHz. **(b)** $i = 7.88$ mA. **(c)** $U = 0.667$ μJ.
32.21 $\nu_1 = 1/2\pi\sqrt{LC}$; $\nu_2 = 1/2\pi\sqrt{2LC}$; $\nu_3 = 1/\pi\sqrt{2LC}$.
32.25 $\tau_C = 4.6$ ms. **32.27** **(a)** $i = V_0/R$.
32.29 **(a)** $t_{1/2} = \ln\sqrt{2}RC$. **(b)** $U \propto V^2$; $i \propto V$.
32.31 $\tau_p = R_1 R_2 C/(R_1 + R_2)$.
32.33 **(a)** $V_R(t) = V_0(1 - e^{-(R/L)t})$; **(b)** $V_L(t) = V_0 e^{-(R/L)t}$.

32.35 $\dfrac{1}{\tau_s} = \dfrac{1}{\tau_1} + \dfrac{1}{\tau_2}$.   **32.37 (a)** $t = \dfrac{L}{R}\ln\left(\dfrac{V_0}{V_0 - V^*}\right)$.

(b) $t = \dfrac{LV^*}{RV_0}$.   **32.39 (a)** $\tau_L = \mu_0 m/4\pi\rho\delta l$.   **(b)** $\tau_L = 1.0$ ms.
32.41 $U_L(t) = U_\infty(1 - e^{-t/\tau_L})^2$.   **32.45 (c)** In an Ohm's law circuit the current and voltage are in phase and always strictly proportional. The current and voltage are not in phase in the LC circuit.

32.47 (a) $V_1 = \dfrac{V_0}{2}\left[\cos\left(\sqrt{\dfrac{2}{LC}}\,t\right) + 1\right]$;

(b) $V_2 = \dfrac{V_0}{2}\left[\cos\left(\sqrt{\dfrac{2}{LC}}\,t\right) - 1\right]$;   **(c)** $V_L = V_0\cos\left(\sqrt{\dfrac{2}{LC}}\,t\right)$;

(d) $\omega_0 = \sqrt{\dfrac{2}{LC}}$.   **32.49** $i(t) = \dfrac{\Phi}{L}\sin\left(\dfrac{1}{\sqrt{LC}}\,t\right)$.

Chapter 33
33.1 (a) $V_R = 0.83V_0$;   **(b)** $V_C = 0.55V_0$.
33.3 (a) $X = 35.6\ \Omega$;   **(b)** $X = 157\ \Omega$.
33.5

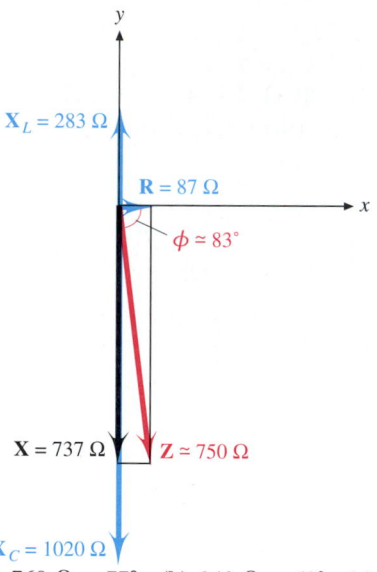

33.7 (a) $760\ \Omega$, $-77°$;   **(b)** $240\ \Omega$, $-69°$;   **(c)** $450\ \Omega$, $-79°$.
33.9 (a) $\nu = 48$ kHz;   **(b)** $\nu = 3.6$ kHz.
(c) In case **a**, $X_L = 37.5$ kΩ, $Z = 37.5$ kΩ, so $(Z - X_L)/Z =$
2.7×10^{-5}; in case **b**, $X_L = 28.4$ kΩ, $Z = 28.6$ kΩ, so
$(Z - X_L)/Z = 1.0 \times 10^{-3}$.
33.13 At the low frequency extreme, $\nu_0 = 0.55 \times 10^6$ Hz.
$C = 0.28$ nF. At the high end, $\nu_0 = 1.6 \times 10^6$ Hz. $C = 0.03$ nF.
0.03 nF $\le C \le 0.28$ nF.   **33.15 (a)** $C = 21\ \mu$F.
(b) $V_{L0} = 5.0 \times 10^2$ V.   **(c)** $V_{C0} = 4.7 \times 10^2$ V.
33.17 $L = 77$ mH.   **33.19** $\phi = 68°$; i leads V by $68.4°$.
33.23 $V_{R0} = V_0/\sqrt{2}$.   **33.25** $\phi_{12} = 66°$; $\phi_1 = 23°$; $\phi_2 = 43°$.
33.27 At position 1, $i_{rms} = 2.4$ A, $P_{rms} = 290$ W. At position
2, $i_{rms} = 0.11$ A, $P_{rms} = 0.60$ W. At this position, the bulb
glows dimly because so little power is dissipated in the bulb,
owing to the large power factor. At position 3, $i_{rms} = 0.22$ A,
$P_{rms} = 2.4$ W. Position 3 is brighter than position 2 but not
nearly as bright as position 1. At position 4, $i_{rms} = 2.4$ A,
$P_{rms} = 290$ W. At this position, the impedance is nearly the
same as at position 1 and thus the glow is the same. At position
5, $i_{rms} = 0.22$ A, $P_{rms} = 2.4$ W. At this position, the impedance
is almost exactly that at position 3.

33.29 $V_0/\sqrt{1 + \omega^2 C^2 R^2}$.

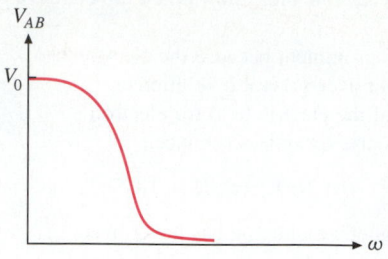

33.33 (a)

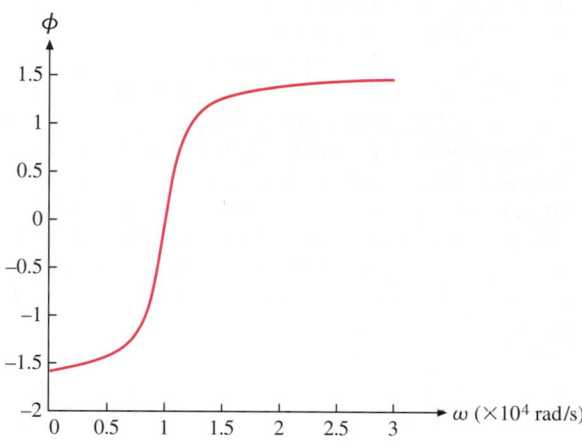

(b)

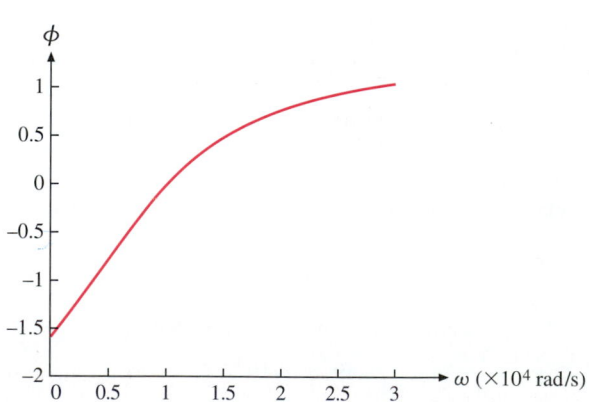

(c)

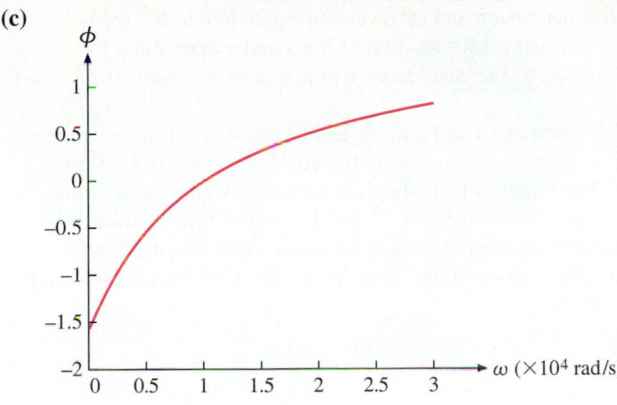

(e)

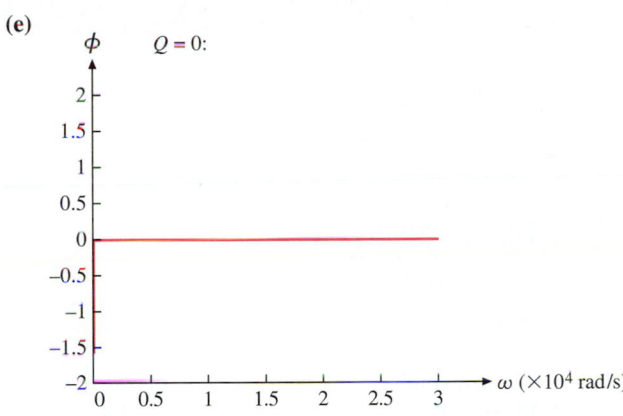

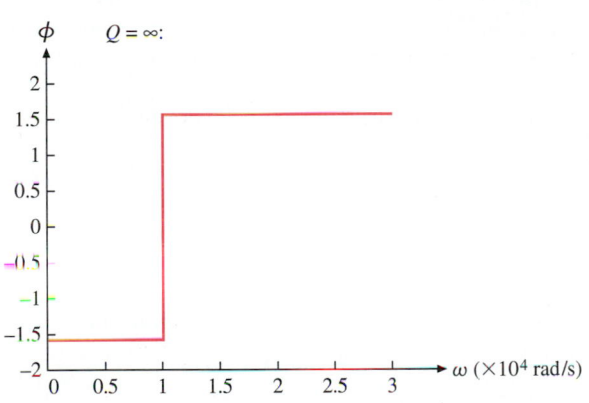

33.35

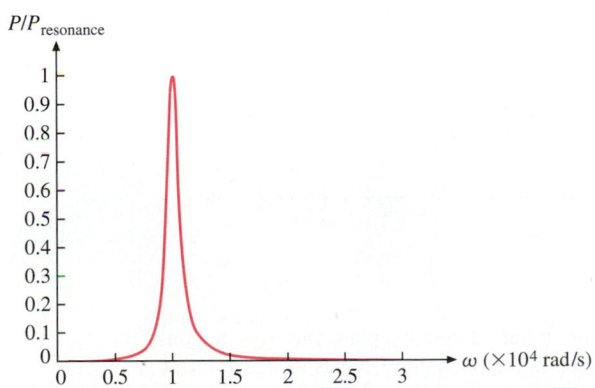

33.37 $V_{rms} = V_0$. **33.39** (a) $R/(R_s + R)$. (b) R_s. (c) $\frac{1}{2}$.
33.43 (a) $P = -V_0 i \cos \omega t$. (b) No, because of resistance.
33.47 (c) Maximum. (e) Increases, increases.

Chapter 34
34.1 $i_d = 19$ mA. **34.3** (a) $V = 0$. (b) $i_{d \, max} = 6.4$ mA.
34.5 $i_{d \, max} = 29$ A. **34.9** $t = 0.24$ s. In normal speech, the rate is about 2 words/s, so the delay is only about 40% of the duration of a single word. **34.11** $\lambda = 320$ m.
34.13 2.79 m $\leq \lambda \leq 3.43$ m. It is certainly possible to have an antenna of length comparable to the wavelengths transmitted.
34.15 $\mathscr{B}_0 = 1.4 \times 10^{-6}$ T. **34.17** $\mathscr{B}_0 = 8.3 \times 10^{-12}$ T.
34.21 (a) $\mathscr{E}_{rms} = 95$ V/m, $\mathscr{B} = 3.2 \times 10^{-7}$ T.
(b) $\mathscr{E}_{rms} = 5.9$ V/m, $\mathscr{B} = 2.0 \times 10^{-8}$ T.
34.23 (b) $r = b$. (c) $\mathscr{B} = 1.67 \times 10^{-9}$ T.

34.25 $I_D = \epsilon_0 \rho \kappa \dfrac{di}{dt}$. **34.27** $j_0 = \epsilon_0 \omega \mathscr{E}_0$. **34.29** (b) 48 A.

34.31 $\displaystyle\int_{\substack{\text{Gaussian} \\ \text{surface}}} \mathscr{E} \cdot d\mathbf{A} = q/\epsilon_0$. $\displaystyle\int_{\substack{\text{Gaussian} \\ \text{surface}}} \mathscr{B} \cdot d\mathbf{A} = \mu_0 M$.

$\displaystyle\int_{\substack{\text{Faradayan} \\ \text{curve}}} \mathscr{E} \cdot d\mathbf{s} = -\int_{\substack{\text{Faradayan} \\ \text{surface}}} \dfrac{\partial \mathscr{B}}{\partial t} \cdot d\mathbf{A} + \mu_0 \int_{\substack{\text{Faradayan} \\ \text{surface}}} \mathbf{j}_M \cdot d\mathbf{A}.$

$\displaystyle\int_{\substack{\text{Amperean} \\ \text{curve}}} \mathscr{B} \cdot d\mathbf{s} =$

$\mu_0 \left(\displaystyle\int_{\substack{\text{Amperean} \\ \text{surface}}} \mathbf{j} \cdot d\mathbf{A} + \epsilon_0 \int_{\substack{\text{Amperean} \\ \text{surface}}} \dfrac{\partial \mathscr{E}}{\partial t} \cdot d\mathbf{A} \right).$

34.35 $\mathscr{E}_y = c\mathscr{B}_0 \cos(\omega t)$. **34.37** $S = i^2 R r / 2\pi l b^2$,
$P = i^2 R(r^2/b^2)$. This result is consistent with Joule's law. If we let $r = b$, then we obtain $P = i^2 R$.

34.39 (a) $-\dfrac{\mu_0 nr}{2} \dfrac{di}{dt}$, tangent to the circle of radius r centered on

the axis. (b) $\dfrac{\mu_0 n^2 r^2}{4c} \left(\dfrac{di}{dt}\right)^2$, outward along the solenoid radius.

34.41 (a) $P = 3.4 \times 10^{17}$ W. (b) $u = 4.5 \times 10^{-6}$ J.
(c) $P = 3.8 \times 10^{26}$ W. (d) $S = 6.2 \times 10^7$ W/m².

34.43 (a) $\sqrt{\frac{3}{2}}\, b$. **34.45** (a) $\mathscr{B}_z = -\dfrac{\mu_0}{2} J_0 \sin \omega t$.

Chapter 35
35.1 $v = 1.2402 \times 10^8$ m/s. **35.3** $\Delta v / v = (n - 1)/n$.
35.5 $\Delta v = c(1/n_2 - 1/n_1)$.
35.13 $\sin \theta_1 = \overline{PS}/r$, $\sin \theta_2 = \overline{PS}/nr$, so $\sin \theta_1 = n \sin \theta_2$.
35.17 37.7°. **35.19** 1.99 m. **35.21** 3.603×10^5.
35.23 $L = 13.3$ m. **35.25** (a) $\theta_c = 39.3°$; (b) $\theta_c' = 57.5°$.
35.27 $n = 1.247$. **35.29** (a) $\theta_c = 36.6°$; (b) $\theta_c = 37.5°$.
(c) A reddish-orange beam. **35.31** (a) $\psi = 45°$; (b) $\psi = 63°$;
(c) $\psi = 72°$. **33.35** (a) 0.3%; (b) 0.5%; (c) 1.2%;
(d) 4.3%. **35.37** $\theta_1 = \tan^{-1} n$. **35.41** (b) $\theta_1 = 51.745°$.
35.43 17.3°. **35.49** (a) Yes, for $40.81° < \theta_1 < 90°$.
(b) No value of θ_1 corresponds to the critical angle on the second side. **35.51** (a) $n_1 \sin \theta_p = n_2 \sin \theta_2$.
35.57 (a) $\mathbf{r}_2 = (x_1, y_1, -z_1)$. (b) $\mathbf{r}_3 = (-x_1, y_1, -z_1)$.
(c) $\mathbf{r}_4 = (-x_1, -y_1, -z_1)$. (d) No.
(e) Corner reflectors are useful because they reflect incoming light directly back at the source. Scientists use them to find very accurate values for the distance between the earth and the moon.
35.59 (a) $t_1 = \sqrt{d^2 + y_1^2}/c$. (b) $t_2 = n\sqrt{d^2 + y_2^2}/c$.
(d) No. The refraction extremizes the light's travel time, but we do not know whether the extremum is a minimum (as it would be for waves) or a maximum (as it would be for particles).

35.61 **(b)** $\Delta = 180° - 4\theta_2 + 2\theta_1$.
(d) For red light, $\theta^* = 59.5°$ and $\Delta = 137.6°$. For violet light, $\theta^* = 58.9°$ and $\Delta = 139.2°$, $\rho = 1.6°$, and the observed angular width is $\rho + 0.5° = 2.1°$. **(e)** $42.4°$.
(f) The violet band of the rainbow is below the red band.

Chapter 36

36.1 682 nm. **36.3** $m = 115$. **36.5** Yellow and violet.
36.7 **(a)** Horizontal light and dark fringes, with possibly a broad black upper region. **(b)** 11.5 arc seconds. **36.9** $d = 3.75$ mm.
36.11 The maxima are broad and there are not a great number of them. **36.13** $D = 4.91 \times 10^{-4}$ m.
36.15 **(a)** 2.15 mm; **(b)** 5.01 mm.
36.17 **(a)** If her house is within 50 m of her neighbor's, no.
(b) Yes. **36.19** 4.5 m. **36.21** **(a)** $15.04°$. **(b)** $60.90°$.
36.25 $R = 25,000$. **36.27** $\Delta\lambda = 3 \times 10^{-2}$ nm.
36.29 **(b)** 2.90843×10^{-4} m. **36.31** **(a)** A minimum.
(b) 7.95×10^{-6} m. **36.33** **(b)** 160.
36.35 **(a)** The fringes run parallel to the axis of the cylindrical rod. In the center, there is a wide dark fringe. The fringes get narrower the farther you move away from the center line of contact. **(b)** $r = \sqrt{(m + \frac{1}{2})\lambda R}$, $m = 0, 1, 2, \ldots$.
(c) $r = \sqrt{m\lambda R}$, $m = 0, 1, 2, \ldots$.
36.37 **(b)** The distances S_1O and S_2O differ by $(m - \frac{1}{2})\lambda$, where m is an integer. **36.39** 9.82×10^{-6} m.

36.41 There are $2\dfrac{D}{\lambda} + 1$ bright fringes and $2\dfrac{D}{\lambda} - 1$ dark fringes. **36.43** **(b)** $m = 10$. **36.45** **(a)** $d = 3.1$ cm.
(b) No, they were limited by the quality of their optics. Yes.
36.49 The second and third. **36.53** **(e)** 590 nm, $71°$.
36.57 **(c)** $\theta = 2.29 \times 10^{-7}$ rad.
(d) The diameter of Betelgeuse is 4.1×10^{11} m. This implies a radius larger than 1.5×10^{11} m, the orbital radius of the earth.

Chapter 37

37.1 2 m/s. **37.3** 90 cm in front of the mirror. Real, inverted.
37.5 **(a)** $\rho = 20.9$ cm, $f = 10.4$ cm. **(b)** Concave.
(c) The image is 0.533 times the size of the object, inverted.
37.7 The object is 12.1 cm behind the mirror. It is a virtual object.
37.9 22.5 cm in front of the mirror, 0.5 cm high. Inverted, real.
37.11 The object should be a virtual object 15 cm behind the mirror.
37.13 Inside the sphere, 43.75 cm from the surface.
37.15 5/7. **37.17** **(a)** $f = 20.3$ cm.
(b) Converging, real, inverted.
37.19 13.0 cm in front of the lens. Erect, virtual.
37.21 **(a)** $f = 11.5$ cm. **(b)** 2.7 cm high. **(c)** Inverted, real.
37.23 $f = s/2$. **37.25** $f = 38.5$ cm. **37.27** $n = 1.5$.
37.29 **(a)** -75 cm. **(b)** 18.75 cm. **37.31** 2×10^4.
37.33 $\alpha = 8.43$.
37.35

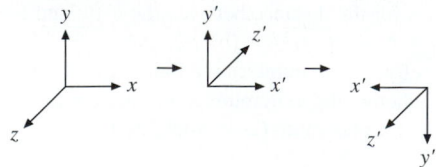

Original   (a)   (b)

(c) $180°$ rotation about the z axis.

(d) A perversion and an inversion equivalent to the reflection $z' = -z$ and a $180°$ rotation of the x and y axes about the z axis.
37.41 **(a)** s' increases from f to ρ, and M decreases from 0 to -1;
(b) s' increases from ρ to ∞, and M decreases from -1 to $-\infty$;
(c) s' increases from $-\infty$ to 0, and M decreases from ∞ to 1.
(d) The image is real when s is between ∞ and f and virtual when s is between f and 0. **(e)** The object moves toward the image when moving from ∞ to ρ and when moving from f to 0. The object moves away from the image when moving from ρ to f. **(f)** No.
37.43 **(a–d)**

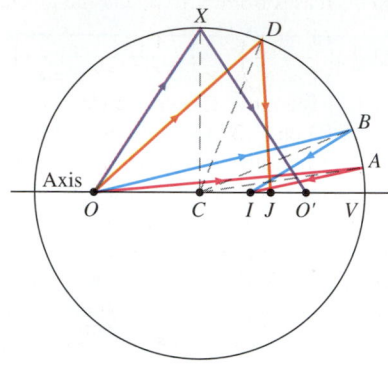

(e) The ray OXO', which is reflected at the vertical diameter of the circle.
37.45 **(a)** The first combination is not possible; the last three are possible. **(b)** The image is erect for the second and third combinations and inverted for the fourth. **(c)** For combination 2, $-|\rho|/2 < s < 0$; for combination 3, $s > 0$; for combination 4, $s < -|\rho|/2$. **37.47** Closer.
37.49 **(a)** The first combination is not possible; the last three are possible.
(b) The image is erect for the second and third combinations and inverted for the fourth.
(c) For combination 2, $-|f| < s < 0$; for combination 3, $s > 0$; for combination 4, $s < -|f|$.
37.51 As the object is moved from $s = \infty$ to $s = 0$, the image moves from $s' = -|f|$ to $s' = 0$, and the magnification increases from $M = 0$ to $M = 1$. **37.55** $s = 2f$.
37.57

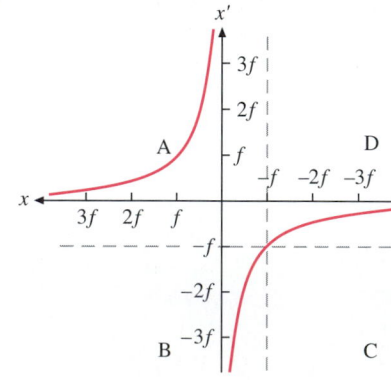

The dotted lines divide the plane into four regions. The part of the graph in region A corresponds to a real object and image. Region B corresponds to a real object and virtual image. Region D corresponds to a virtual object and real image.

37.59 **(b)** As the object moves from 0 to f, the magnification increases from 1 to ∞, and the image is erect. As the object moves from f to ∞, the magnification increases from $-\infty$ to 0, and the image is inverted.
(c) As the object moves from 0 to ∞, the magnification decreases from 1 to 0, and the image is erect.
(d) For a converging lens, as the virtual object moves from $-\infty$ to 0, the magnification increases from 0 to 1, and the image is erect. For a diverging lens, as the object moves from $-\infty$ to $-|f|$, the magnification decreases from 0 to $-\infty$, and the image is inverted. As the virtual object moves from $-|f|$ to 0, the magnification decreases from ∞ to 1, and the image is erect.
37.61 **(a)** $\Delta s = \left(\dfrac{1}{f} - \dfrac{1}{s'}\right)^{-1} - \left(\dfrac{1}{f} - \dfrac{1}{s' - \Delta s'}\right)^{-1}$, where positive Δs indicates a move toward the lens. **(b)** $s = 26.37$ cm.
37.63 **(a)** $f = 24.2$ cm. **37.65** $f = -62.7$ cm.
37.67 The image is located a distance $3f - s$ in front of the mirror. It is the same size as the object and is inverted.
37.69 **(a)** Upper: -1.52Δ; lower: 1.37Δ.
(b) Upper: $\rho_2 = 9.49$ cm; lower: $\rho_2 = 17.5$ cm.
37.71 **(b)** 28%. **37.73** **(b)** $f = 8.75$ cm.
37.75 **(b)** $\omega = 0.404$ rad/s $= 3.86$ rev/min.
37.77 **(b)** $L = -1$, as stated in Section 37.2.
37.79 **(a)**

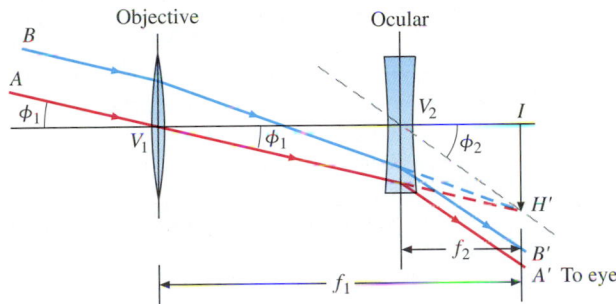

Erect and virtual. **37.81** **(c)** $f_{1y}/f_{2y} = -0.56704$.
(d) $f_{1y} = 4.33$ cm, $f_{2y} = -7.64$ cm.
(e) $\rho_{12} = -2.24$ cm, $\rho_{21} = 2.24$ cm, $\rho_{22} = 1.52$ cm.

Chapter 38
38.1 $V = \sqrt{\frac{3}{4}}c$. **38.3** $V = 0.991c = 2.97 \times 10^8$ m/s.
38.5 $\Delta t_0 = 7.22$ μs. **38.7** **(a)** $\Delta t = 1.8 \times 10^{-8}$ s.
(b) The "stationary" clock. **38.9** **(b)** $t = 1.075 \times 10^{-5}$ s.
(c) $t = 3.40 \times 10^{-5}$ s. **(d)** $t = 1.075 \times 10^{-4}$ s.
38.15 **(a)** $L_0 = 1.66$ m. **(b)** $L = 1.66$ m.
38.17 It appears 8.16×10^{-14} m shorter, or 2×10^{-7} times the wavelength of blue light.
38.19 **(a)** $0.357c$. **(b)** 7.2 ly. **(c)** 4.32 ly; 5.4 y.
(d) 3.14 ly; 3.49 y. **38.21** **(b)** For small V, $V/c \to 0$.
38.25 $v'_x = v_x - V_x$, $v'_y = v_y - V_y$, $v'_z = v_z - V_z$.
38.27 **(a)** $t = \sqrt{\dfrac{1 + \dfrac{V}{c}}{1 - \dfrac{V}{c}}}$ years. **(b)** $V = \dfrac{c}{\sqrt{2}}$.

38.33 **(a)** $\dfrac{\Delta L}{L} = 1 - \sqrt{1 - \dfrac{V^2}{c^2}}$.
(b) Fitzgerald did not explain why such a contraction should take place. It could not be due simply to ether pressure, because all materials contracted equally.

38.35 **(a)** $v_e = \dfrac{\sqrt{1 - \dfrac{V^2}{c^2}}}{\Delta t}$. **(b)** $\Delta x = \dfrac{v \, \Delta t}{\sqrt{1 - \dfrac{V^2}{c^2}}}$.

(e) $v' = v \sqrt{\dfrac{1 - \dfrac{V}{c}}{1 + \dfrac{V}{c}}}$. **(f)** $\dfrac{\Delta v}{v} = \sqrt{\dfrac{0.25}{1.75}} - 1 = -0.622$.

38.37 **(a)** $\Delta v/v = 2.22 \times 10^{-7}$. **(b)** $\Delta v = 18.5$ Hz.

Chapter 39
39.1 $v = 4.21 \times 10^7$ m/s. **39.3** $v/c = 0.548$.
39.5 **(a)** $v = c - 1.005 \times 10^{-18}$ m/s;
(b) $m = 1.11 \times 10^{-17}$ kg; **(c)** $p = 3.33 \times 10^{-9}$ kg·m/s.
39.7 $K = 1.503 \times 10^{-10}$ J $= 9.383 \times 10^8$ eV.
39.9 $E = 1.165 \times 10^8$ J; the energy released is equivalent to more than 27 kg of TNT.
39.11 **(a)** $E = 2.5 \times 10^{-10}$ J $= 1.6 \times 10^9$ eV $= 0.6 m_0 c^2$.
(b) $K = 1.0 \times 10^{-10}$ J $= 6.3 \times 10^8$ eV.
39.13 **(a)** $E = 1.02 \times 10^6$ eV; **(b)** $E = 1.88 \times 10^9$ eV.
(c) Neutrinos and antineutrinos are essentially massless and cannot collide at $v \ll c$ because they always travel at c.
39.15 $\Delta m/\Delta t = 6.3 \times 10^7$ kg/y.
39.17 $\Delta M/M = 1.498 \times 10^{-10}$.
39.19 **(a)** $\Delta M = 2.14 \times 10^{-1}$ kg. **(b)** $\Delta M/M = 9.1 \times 10^{-4}$.
39.21 $p = c\sqrt{m^2 - m_0^2}$. **39.23** $V = 1.21 \times 10^9$ V.
39.25 **(a)** $\omega_r/\omega_c = 0.808$.
(b) $\omega_r = 6.154 \times 10^7$ rad/s; $\omega_c = 7.62 \times 10^7$ rad/s.
39.27 $E^2 - p^2 c^2 = m_0^2 c^4$. **39.29** **(b)** $R = 1.687 \times 10^{-15}$ m.
39.31 $3\ ^4\text{He} \rightarrow\ ^{12}\text{C}$: $E = 1.119 \times 10^{-12}$ J. $^{12}\text{C} +\ ^4\text{He} \rightarrow\ ^{16}\text{O}$: $E = 1.133 \times 10^{-12}$ J. $2\ ^{12}\text{C} \rightarrow\ ^4\text{He} +\ ^{20}\text{Ne}$: $E = 7.551 \times 10^{-13}$. $2\ ^{12}\text{C} \rightarrow\ ^{24}\text{Mg}$: $E = 2.232 \times 10^{-12}$ J. $2\ ^{16}\text{O} \rightarrow\ ^4\text{He} +\ ^{28}\text{Si}$: $E = 1.550 \times 10^{-12}$. $2\ ^{16}\text{O} \rightarrow\ ^{32}\text{S}$: $E = 2.649 \times 10^{-12}$ J. $2\ ^{28}\text{Si} + 2\ e \rightarrow\ ^{56}\text{Fe}$: $E = 2.829 \times 10^{-12}$ J.

39.33 **(c)** $M'_0 = m_0 \sqrt{2\left(\dfrac{K}{m_0 c^2} + 2\right)}$.

(d) $M_0 > 2m_0$ because the initial kinetic energy that is classically converted to heat energy in the inelastic collision of macroscopic bodies is converted into mass in the inelastic collision of two microscopic bodies. That is why the rest mass is greater than $2m_0$. **39.35** $\mathcal{B} = 4.945$ T.
39.37 **(a)** $E = (m_i - m_f)c^2$.
(b) $|p| = E/C = (m_i - m_f)c = (m_i - m_f)v + m_i \, dv$.
(d) $v/c = 24/26 = 0.923$.
39.41 **(a)** $K_0 = 6m_0 c^2 = 5.63 \times 10^9$ eV. **(b)** $\eta = \frac{1}{3}$.
(c) 100%.

Photo and Illustration Credits

Chapter 1
Opener NASA; **1.1** National Institute of Standards and Technology Boulder Laboratories, U.S. Department of Commerce; **p. 2** Stephen Simpson/FPG International.

Chapter 2
Opener John Mutrux; **2.1** © The Harold E. Edgerton 1992 Trust, courtesy of Palm Press, Inc.; **Query 2.12a** © Gerard Vandystat, Photo Researchers; **Query 2.12b** Stanley Rowin/ THE PICTURE CUBE; **p. 33** Roger-Viollet.

Chapter 3
Opener Steven Ferry, P&F Communications; **3.2a,b** From Rutherford, F. J., Holton, G., Watson, F. G. *Project Physics.* New York: Holt, Rinehart and Winston, 1981, pp. 11, 12. Reproduced with permission of the Education Development Center, Inc.; **3.10** NASA; **3.18** From Avery, E. M. *Elements of Natural Philosophy.* New York: Sheldon & Company, 1878, p. 60, print courtesy Thomas B. Greenslade, Jr.; **p. 62** Ed Fifford/Masterfile.

Chapter 4
Opener © Neil Rabinowitz; **Problem 4.19** Courtesy of Central Scientific Company; **p. 99** Alec Pytlowany/Masterfile.

Chapter 5
Opener Steven Ferry, P&F Communications; **5.3** Ufano Diego, *Artillerie . . .* , Zutphen, 1621, detail of plate 22. Rare Books and Manuscript Division, The New York Public Library, Astor, Lenox and Tilden Foundations; **5.4** © Fundamental Photographs; **5.7** © Fundamental Photographs/Richard Megna; **5.11b** Cowboy Mike Wooldridge, Trick'n Fancy Ropin'—Member IPRA; **Problem 5.21** Paul L. Ruben; **p. 118** Michael Yelman/Photri.

Chapter 6
Opener Willard House and Clock Museum, North Grafton, Massachusetts; **6.13a,b** Author (with assistance of Thomas Douglass); **p. 145** Hall of History Foundation, Schenectady, New York; **p. 149** Patrick Ward/Stock, Boston.

Chapter 7
Opener Mark Philbrick; **p. 176** author.

Chapter 8
Opener © Fundamental Photographs/Richard Megna; **Problem 8.16** Jim Hamilton/THE PICTURE CUBE; **p. 204** Peter Christopher/Masterfile.

Chapter 9
Opener © Comstock Inc. 1995/Schaler/Contributing Library; **9.1** © The Harold E. Edgerton 1992 Trust, courtesy of Palm Press, Inc.; **9.4** Steven Ferry, P&F Communications; **9.5** From Rutherford, F. J., Holton, G., Watson, F. G. *Project Physics.* New York: Holt, Rinehart and Winston, 1981, p. 343. Reproduced with permission of the Education Development Center, Inc.; **9.8** Cover illustration from *The Little Prince* by Antoine de Saint-Exupery, copyright 1943 and renewed 1971 by Harcourt Brace & Company, reproduced by permission of the publisher, © Editions Gallimard; **9.13** © The Harold E. Edgerton 1992 Trust, courtesy of Palm Press, Inc.; **Query 9.14** Based on photo by Michael Friedlander; **p. 234** Tom Pantages.

Chapter 10
Opener © Fundamental Photographs/Richard Megna; **10.2** Francis Weston Sears, *Principles of Physics, Volume 1: Mechanics, Heat, and Sound, Second Edition*, p. 151. © 1950 by Addison-Wesley Press, Inc. Adapted by permission of Addison-Wesley Publishing Company, Inc.; **p. 259** Steven Ferry, P&F Communications.

Chapter 11
Opener Steven Ferry, P&F Communications; **p. 289** Stephen Grohe/THE PICTURE CUBE.

Chapter 12
Opener Seth Resnick/Stock, Boston; **12.8** Author (with assistance of Thomas Douglass); **12.11** Mark Philbrick; **Query 12.12** Illustration by Michael Goodman from Walker, Jearl, "The Amateur Scientist," *Scientific American*, September 1984, p. 219. Copyright © 1984 by Scientific American, Inc. All rights reserved; **Problem 12.57 (right)** Courtesy Nakamura Scientific Co., Ltd., Tokyo, Japan; **p. 313** © Superstock.

Chapter 13
Opener Steven Ferry, P&F Communications; **Query 13.3** © Comstock Inc. 1995/Russ Kinne; **p. 350** Courtesy of Central Scientific Company.

Chapter 14
Opener NASA; **14.3** From Brahe, T. *Astronomiae Instauratae Mechanica*, 1602, by permission of The British Library, 48E11; **14.9** Cavendish, H. (1798). Experiments to Determine the Density of the Earth. *Philos. Trans. Roy. Soc. London*, p. 526; **p. 374** North Wind Picture Archives.

Index

This index covers the contents of *Physics for Scientists and Engineers* and *Modern Physics for Scientists and Engineers*.